ORGANIC COMPOUNDS

Family						
Ether	**Amine**	**Aldehyde**	**Ketone**	**Carboxylic Acid**	**Ester**	**Amide**
CH_3OCH_3	CH_3NH_2	$CH_3\overset{O}{\overset{\Vert}{C}}H$	$CH_3\overset{O}{\overset{\Vert}{C}}\,CH_3$	$CH_3\overset{O}{\overset{\Vert}{C}}OH$	$CH_3\overset{O}{\overset{\Vert}{C}}OCH_3$	$CH_3\overset{O}{\overset{\Vert}{C}}NH_2$
Methoxy-methane	Methan-amine	Ethanal	Propanone	Ethanoic Acid	Methyl ethanoate	Ethanamide
Dimethyl ether	Methyl-amine	Acetal-dehyde	Acetone	Acetic acid	Methyl acetate	Acetamide
ROR	RNH_2 R_2NH R_3N	$R\overset{O}{\overset{\Vert}{C}}H$	$R\overset{O}{\overset{\Vert}{C}}R$	$R\overset{O}{\overset{\Vert}{C}}OH$	$R\overset{O}{\overset{\Vert}{C}}OR$	$R\overset{O}{\overset{\Vert}{C}}NH_2$ $R\overset{O}{\overset{\Vert}{C}}NHR$ $R\overset{O}{\overset{\Vert}{C}}NR_2$
$-\overset{\vert}{\underset{\vert}{C}}-O-\overset{\vert}{\underset{\vert}{C}}-$	$-\overset{\vert}{\underset{\vert}{C}}-\underset{\vert}{N}-$	$-\overset{O}{\overset{\Vert}{C}}-H$	$-\overset{\vert}{\underset{\vert}{C}}-\overset{O}{\overset{\Vert}{C}}-\overset{\vert}{\underset{\vert}{C}}-$	$-\overset{O}{\overset{\Vert}{C}}-OH$	$-\overset{O}{\overset{\Vert}{C}}-O-\overset{\vert}{\underset{\vert}{C}}-$	$-\overset{O}{\overset{\Vert}{C}}-\underset{\vert}{N}-$

ORGANIC CHEMISTRY

ORGANIC CHEMISTRY

Second Edition

T.W. Graham Solomons
University of South Florida

JOHN WILEY & SONS

New York Chichester Brisbane Toronto

Library of Congress Cataloging in Publication Data:

Solomons, T W Graham.
Organic chemistry.

Includes index.
1. Chemistry, Organic. I. Title.
[DNLM: 1. Biochemistry. 2. Chemistry, Organic.
QD251.2 S689o]
QD251.2.S66 1980 547 79-28839
ISBN 0-471-04213-7

Printed in the United States of America

10 9 8 7 6

For Judith

PREFACE

In writing the second edition, my foremost objective has been to make the book even more understandable to the students who will read it. I have carefully examined every section of the book and have rewritten many sections to improve clarity. In addition, I have deleted extraneous material, added new problems and examples and, to further improve the pedagogy, I have reorganized certain portions of the text. In many instances where difficult concepts are introduced I have expanded the discussion considerably. I have not, however, raised the level of the text. Without sacrificing rigor, my object has been to make the book more accessible to students.

One major change is in the treatment of nucleophilic substitution and elimination reactions. The initial treatment of this important topic has been considerably simplified and now comes much earlier. In a new chapter (Chapter 5) the classical aspects of these reactions are developed in the context of alkyl halides. The more complicated aspects (ion-pairs, neighboring group participation, etc.) are now reserved for a special topic later in the text. In the first edition these topics were combined and were not specifically treated until Chapter 17. The early introduction of S_N1, S_N2, E1, and E2 reactions makes many aspects of the chapters that follow easier to understand. Especially important in this regard are the nucleophilic substitutions and eliminations in the chapters on Alkenes, (Chapter 6), Alkynes (Chapter 9), and Alcohols, Ethers, and Phenols (Chapter 15). In the chapter on Organic Halides (Chapter 14), students will see similarities and differences in the nucleophilic substitutions of allylic and benzylic substrates on the one hand and aryl and vinylic substrates on the other.

The discussion of nucleophilic substitutions and eliminations in Chapter 5 will also allow students to see the contrasts between these important ionic reactions and the free radical reactions they studied in Chapter 4. Thus after these two chapters they will have encountered most of the fundamentals of organic reaction mechanisms.

Chapter 2 has also been the subject of a major revision. The coverage in Chapter 2 is now limited to the structural aspects of representative molecules only. This overview (which can be omitted) is intended to give students the background to explore more biological applications earlier in the course and to better prepare them for laboratory work.

Special topics have been rearranged in this edition. They are now organized as sections and follow directly after related chapters where they may be used to enhance the basic material. The special topics stand apart in the sense that they are not required for an understanding of the material in the basic chapters. Thus they should give instructors considerable flexibility in designing a course to meet the needs of their particular students. Seventeen different subjects are treated as special topics including addition polymers, condensation polymers, the photo-chemistry of vision, alkaloids, thallation reactions, mass spectroscopy, and environmental problems associated with organic halides and organometallic compounds.

This new edition, of course, retains many features of the first edition. The most important ones are described in topical form below.

An early presentation of important functional groups. Although the titles of the early chapters of the book suggest that they are concerned mainly with hydrocarbons, stereochemistry, and alkyl halides, a closer look at these chapters reveals a different picture. Much of the material in these chapters is about alcohols, aldehydes, ketones, and carboxylic acids. This diversity at an early stage results partly from the overview of functional groups given in Chapter 2 and partly from the introduction of the principles of ionic reactions in Chapter 5.

A unified development of important concepts. I have tried to develop an internal order out of the basic concepts so that as the students progress from chapter to chapter, they will be able to apply the knowledge they have gained to new situations.

An emphasis on bioorganic chemistry. There were many diversions into biochemical topics in the first edition and a number of new ones have been added. These topics have been chosen to stimulate students' interest and also to illustrate basic principles of organic chemistry and thus to reinforce their understanding of them.

A modern presentation of molecular orbital theory. Because molecular orbital theory grows in importance to organic chemists year by year, it can no longer be passed over with an outdated and imprecise treatment as is given in most books. The basic ideas are simple, and are well within the grasp of students at this level. No mathematics need be done and none is used in the treatment given here. However, by presenting some of these basic ideas, including the phase signs of orbitals, at the outset, much of what follows becomes clearer. In Chapter 1, for example, these ideas are applied to explain the bonding and antibonding orbitals of hydrogen and to make orbital hybridization clearer. In Chapter 6, they are used to account for the bonding and antibonding orbitals of alkenes. These early and simple introductions prepare the student for a better understanding of conjugated systems in Chapter 10 and of aromaticity in Chapter 11. They also make possible the presentation of the theory of electrocyclic and cycloaddition reactions as a special topic.

Extensive use of resonance theory. Although molecular orbital theory is presented in a modern way, resonance theory is not neglected. Resonance theory is introduced in Chapter 1 as an outgrowth of writing Lewis structures, and it is amplified, developed, applied, and reapplied as the student progresses through subsequent chapters. New to this edition is a special section summarizing the rules for resonance in Chapter 10.

Problems and examples. Problems are interspersed throughout the body of each chapter; others are collected in sets at the ends of chapters. The intratext problems are designed to test the students' understanding of the material up to that point. The end-of-chapter problems are graded, beginning with simple exercises and progressing to more and more difficult ones. Problems that are especially challenging have been designated as such with an asterisk. These may be made optional. Wherever possible I have tried to develop problems from material that will stimulate students' interest—especially from molecules of biological importance or from processes used in chemical industry.

Supplements. A solutions manual is available giving answers to all problems in the text. Available free to adopting departments will be transparency masters for all spectra and important figures and an expanded card file of multiple-choice test questions.

T. W. Graham Solomons

ACKNOWLEDGMENTS

I am grateful to many people for the contributions they have made during the preparation of this second edition. My colleagues at the University of South Florida, Professors Jack E. Fernandez and Robert D. Whitaker, proofread the entire book, as did Professor O. C. Dermer of Oklahoma State University. Professor George R. Wenzinger of USF prepared the index. Other USF colleagues were kind enough to offer helpful suggestions and criticisms, including Professors Stewart W. Schneller, George R. Jurch, Jr., Terence C. Owen, and Douglas Raber.

Valuable reviews came from the following people and I thank them all: Professor Newell S. Bowman, The University of Tennessee; Professor William D. Closson, State University of New York at Albany; Professor Jerry A. Hirsch, Seton Hall University; Professor John R. Holum, Augsburg College; Professor John F. Keana, University of Oregon; Professor Karl R. Kopecky, The University of Alberta; Professor Robert Levine, University of Pittsburgh; Professor Jerry March, Adelphi University; Professor Gerardo Molina, Universidad de Puerto Rico; Professor Everett Nienhouse, Ferris State College; Professor Daniel Trifan, Fairleigh Dickinson University; Professor Desmond Wheeler, University of Nebraska; Professor James K. Whitesell, The University of Texas at Austin; Professor Joseph Wolinsky, Purdue University.

I also want to express my appreciation again to the following reviewers who helped with the first edition: Paul A. Barks (North Hennepin State Junior College), Edward M. Burgess (Georgia Institute of Technology), Philip L. Hall (Virginia Polytechnic Institute and State University), John R. Holum (Augsburg College), Stanley N. Johnson (Orange Coast College), Philip W. LeQuesne (Northeastern University), Jerry March (Adelphi University), William A. Pryor (Louisiana State University), and Thomas R. Riggs (University of Michigan).

I am also grateful to Mr. Gary Carlson, the chemistry editor at Wiley, for his help and support and to Ms. Claire Egielski, the production editor, and Ms. Susan Giniger, the staff editor. I especially thank Mr. John Balbalis for the book's fine illustrations.

I am also grateful to Phil Schaeffer, Alan Goldston, and Jane Connolly for all that they have done on my behalf. And again I thank my wife, Judith Taylor Solomons, for her editing, proofreading, typing, and support. This book is affectionately dedicated to her.

T. W. Graham Solomons

CONTENTS

7 Reactions of Alkenes: Addition Reactions of the Carbon-Carbon Double Bond

8 Stereochemistry

9 Alkynes

10 Conjugated Unsaturated Systems. Visible-Ultraviolet Spectroscopy

11 Aromatic Compounds I: The Phenomenon of Aromaticity

12 Aromatic Compounds II: Reactions of Aromatic Compounds with Electrophiles

13 Physical Methods of Structure Determination — Nuclear Magnetic Resonance Spectroscopy, Infrared Spectroscopy

16 Aldehydes and Ketones

17 Carboxylic Acids and Their Derivatives: Nucleophilic Substitution at Acyl Carbon

18 Amines

21 Lipids

22 Amino Acids and Proteins

ORGANIC CHEMISTRY

1 CARBON COMPOUNDS AND CHEMICAL BONDS

1.1 INTRODUCTION

Organic chemistry is a study of *the compounds of carbon*. The compounds of carbon are the "stuff" of which all living things on this planet are made. Carbon compounds include DNA, the giant molecules that contain all the genetic information for a given species—the molecules that determine whether we are men or women, humans or frogs, whether we have blue eyes or brown, whether we have black hair or blonde, and whether in our old age we have gray hair or none. Carbon compounds make up the proteins of our muscle and skin. They make up the enzymes that catalyze the reactions that occur in our bodies. Together with oxygen in the air we breathe, carbon compounds in our diets furnish the energy that sustains life.

Considerable evidence indicates that several billion years ago most of the carbon atoms on this planet were in the form of the gas methane. This simple organic molecule, CH_4, along with water, ammonia, and hydrogen, were the main constituents of the primordial atmosphere. As lightning and highly energetic radiation passed through this atmosphere many of these simple molecules fragmented into highly reactive pieces. These recombined into more complex arrangements. Compounds called amino acids, formaldehyde, hydrogen cyanide, purines, and pyrimidines formed in this way. These and others produced in the same way, were carried by rain into the sea. The sea became richer and richer in organic molecules. It became a vast storehouse containing all of the compounds necessary for the emergence of life. Amino acids reacted with each other to form proteins. Formaldehyde became sugars, and some of these sugars, together with purines and pyrimidines, became simple molecules of DNA. At some point, and in a manner still not understood, these larger molecules collected together to form the first primitive living cells. From these first cells, through the long process of natural selection, evolved man and all the other living things present on this earth today.

Not only are we composed largely of organic compounds, not only are we derived from and nourished by them, we also live in an Age of Organic Chemistry. The clothing we wear, whether a natural substance such as wool or cotton or a synthetic such as nylon or a polyester, is made up of carbon compounds. Many of the materials that go into the houses that shelter us are organic. The gasoline that propels our automobiles, the rubber of their tires, and the plastic of their interiors are all organic. Most of the medicines that help us cure diseases and relieve suffering are organic. Organic pesticides help us eliminate many of the agents that spread diseases in both plants and animals.

Organic chemicals are also factors in some of our most serious problems. Many of the organic chemicals introduced into the environment have had consequences far beyond those originally intended. A number of insecticides, widely used for many years, have now been banned because they harm many species other than insects and they pose a danger to humans. Organic compounds called polychlorobiphenyls (PCBs) are responsible for pollution of the Hudson River that may take years and enormous amounts of money to reverse. Organic compounds used as propellants for aerosols have been banned because they threatened to destroy the ozone layer of the outer atmosphere, a layer that protects us from extremely harmful radiation. In 1976, an explosion at a factory for organic chemicals in Italy spread one of the most toxic substances known over a wide area of the surrounding countryside rendering a village uninhabitable up until now. A few organic chemicals used as food additives have now been implicated as possible cancer-causing substances.

Thus for good or bad, organic chemistry is associated with nearly every aspect of our lives. We would be wise then to understand it as best we can.

1.2 THE DEVELOPMENT OF ORGANIC CHEMISTRY AS A SCIENCE

Humans have used organic compounds and their reactions for thousands of years. Their first deliberate experience with an organic reaction probably dates from their discovery of fire. The ancient Egyptians used organic compounds (indigo and alizarin) to dye cloth. The famous "royal purple" used by the Phoenicians was also an organic substance, obtained from mollusks. The fermentation of grapes to produce ethyl alcohol and the acidic qualities of "soured wine" are both described in the Bible and were probably known earlier.

As a science, organic chemistry is less than 200 years old. Most historians of science date its origin to the early part of the nineteenth century, a time in which an erroneous belief was dispelled.

1.2A Vitalism

During the 1780s scientists began to distinguish between *organic compounds* and *inorganic compounds*. Organic compounds were defined as compounds that could be obtained from *living organisms*. Inorganic compounds were those that came from *nonliving sources*. Along with this distinction, a belief called "vitalism" grew. According to this idea, the intervention of a "vital force" was necessary for the synthesis of an organic compound. Such synthesis, chemists then held, could take place only in living organisms. It could not take place in the test tubes and flasks of a chemistry laboratory.

Between 1828 and 1850 a number of compounds that were clearly "organic" were synthesized from sources that were clearly "inorganic." The first of these syntheses was accomplished by Friedrich Wöhler in 1828. Wöhler found that the organic compound urea (a constituent of urine) could be made by heating the inorganic compound ammonium cyanate. Although "vitalism" died slowly and did

$$\underset{\text{Ammonium cyanate}}{NH_4^+NCO^-} \xrightarrow{\text{heat}} \underset{\text{Urea}}{H_2N{-}\overset{\overset{\displaystyle O}{\|}}{C}{-}NH_2}$$

not disappear completely from scientific circles until 1850, its passing made possible the flowering of the science of organic chemistry that has occurred since 1850.

1.2B Empirical and Molecular Formulas

Even while vitalism persisted* extremely important advances were made in the development of qualitative and quantitative methods for analyzing organic substances. In 1784 Antoine Lavoisier first showed that organic compounds were composed primarily of carbon, hydrogen, and oxygen. Between 1811 and 1831, *quantitative* methods for determining the composition of organic compounds were developed by Justus Liebig, J. J. Berzelius, and J. B. A. Dumas. A great confusion was dispelled in 1860 when Stanislao Cannizzaro showed that the earlier hypothesis of Amedeo Avogadro (1811) could be used to distinguish between *empirical* and *molecular formulas*. As a result, many molecules that had appeared earlier to have the same formula were seen to be composed of different numbers of atoms. For example, ethylene, cyclopentane, and cyclohexane all have the same empirical formula: CH_2. However they have molecular formulas of C_2H_4, C_5H_{10}, and C_6H_{12}, respectively. In Chapter 3 we see how empirical and molecular formulas are determined today.

1.3 THE STRUCTURAL THEORY OF ORGANIC CHEMISTRY

Between 1858 and 1861, August Kekulé, Archibald Scott Couper, and Alexander M. Butlerov, working independently, laid the basis for one of the most fundamental theories in organic chemistry: *the structural theory*. Two central ideas make up this theory:

1. The atoms of the elements in organic compounds can form a fixed number of bonds. The measure of this ability is called *valence*. Carbon is *tetravalent;* that is, carbon atoms form four bonds. Oxygen is *divalent;* oxygen atoms form two bonds. Hydrogen and the halogens are *monovalent;* their atoms form only one bond.

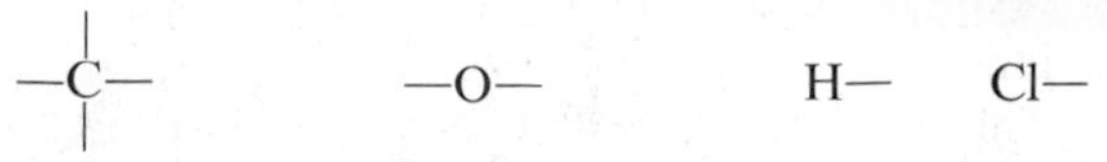

—C— (four bonds)	—O—	H— Cl—
Carbon atoms are tetravalent	Oxygen atoms are divalent	Hydrogen and halogen atoms are monovalent

2. A carbon atom can use one or more of its valences to form bonds to other carbon atoms.

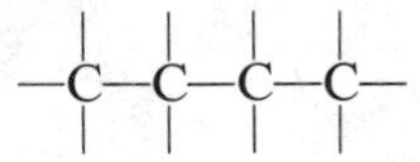

—C—C—C—C—

Carbon-carbon bonds

In his original publication Couper represented these bonds by lines much in the same way that most of the formulas in this book are drawn. In his textbook (published in 1861), Kekulé gave the science of organic chemistry its modern definition: *a study of the compounds of carbon.*

We can appreciate the importance of the structural theory if we consider now one simple example. There are two compounds that have the *same* molecular formula, C_2H_6O, but these compounds have strikingly different properties. See Table 1.1. One compound, called *dimethyl ether*, is a gas at room temperature; the

* It is a belief still held today by some groups. While there are sound arguments made against foods contaminated with pesticides, it is impossible to argue that "natural" vitamin C, for example, is healthier than the "synthetic" vitamin, since they are identical.

TABLE 1.1 Properties of Ethyl Alcohol and Dimethyl Ether

	ETHYL ALCOHOL	DIMETHYL ETHER
Boiling point	78.5°	−24.9°
Melting point	−117.3°	−138°
Reaction with sodium	Displaces H_2	No reaction

other compound, called *ethyl alcohol,* is a liquid. Dimethyl ether does not react with sodium; ethyl alcohol does, and the reaction produces hydrogen gas.

Since the molecular formula for these two compounds is the same, it gives us no basis for understanding the differences between them. The structural theory remedies this situation, however. It does so by giving us *structural formulas* for the two compounds and these structural formulas (Fig. 1.1) are different.

One glance at the structural formulas for these two compounds reveals their difference. The atoms of ethyl alcohol are connected in a way that is different from those of dimethyl ether. In ethyl alcohol there is a C—C—O linkage; in dimethyl ether the linkage is C—O—C. Ethyl alcohol has a hydrogen attached to oxygen; in dimethyl ether all of the hydrogens are attached to carbon. It is the hydrogen atom covalently bonded to oxygen in ethanol that is displaced when ethanol reacts with sodium:

$$2\,CH_3CH_2\text{—O—H} + 2\,Na \longrightarrow 2\,CH_3CH_2\text{—}\overset{-}{O}\ Na^+ + H_2$$

Hydrogen atoms that are covalently bonded to carbon are normally unreactive toward sodium. As a result, none of the hydrogens in dimethyl ether is displaced by sodium.

The hydrogen attached to oxygen also accounts for the fact that ethyl alcohol is a liquid at room temperature. As we shall see in Section 2.16, this

FIG. 1.1
Ball-and-stick models and structural formulas for ethyl alcohol and dimethyl ether.

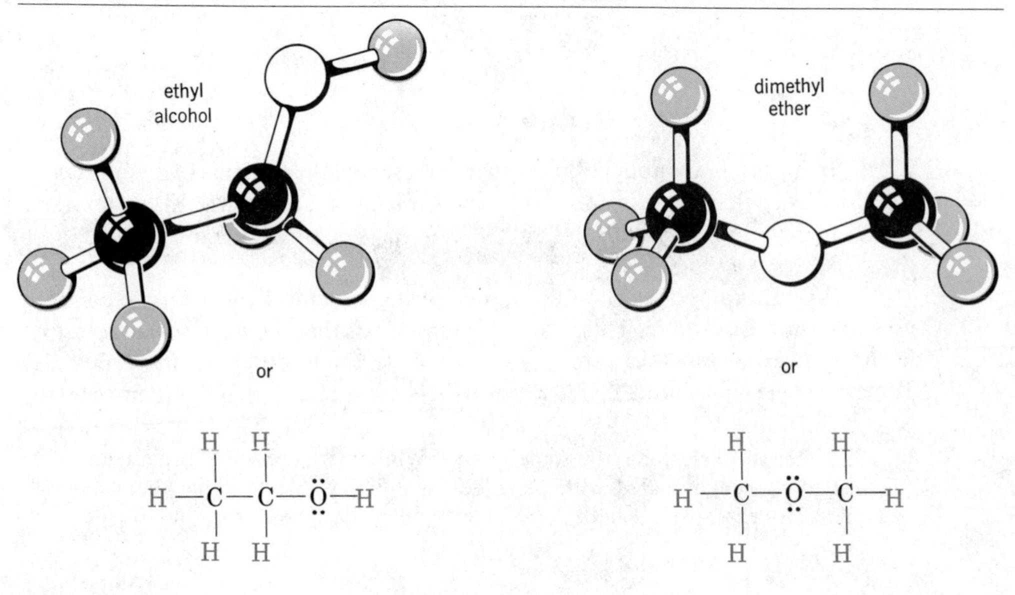

hydrogen allows molecules of ethyl alcohol to form hydrogen bonds to each other and gives ethyl alcohol a boiling point much higher than dimethyl ether.

1.4 ISOMERISM. STRUCTURAL ISOMERS

More than two million organic compounds have now been isolated in a pure state and have been characterized on the basis of their physical and chemical properties. Additional compounds are added to this list by the tens of thousands each year. A look into *Chemical Abstracts* or Beilstein's *Handbuch der Organischen Chemie,* where known organic compounds are catalogued, shows that there are dozens and sometimes hundreds of *different compounds that have the same molecular formula.* Such compounds are called *isomers.* Different compounds with the same molecular formula are said to be *isomeric,* and this phenomenon is called *isomerism.*

Ethyl alcohol and dimethyl ether are examples of *structural isomers. Structural isomers are different compounds that have the same molecular formula, but differ in the order in which their atoms are bonded together.* Structural isomers always have different physical properties (e.g., melting point, boiling point, density). The differences, however, may not always be as large as those between ethyl alcohol and dimethyl ether.

1.5 THE TETRAHEDRAL SHAPE OF METHANE

In 1874, the structural formulas originated by Kekulé, Couper, and Butlerov were expanded into three dimensions by the independent work of J. H. van't Hoff and J. A. Le Bel. Van't Hoff and Le Bel demonstrated that the four bonds of the carbon atom in methane, for example, are arranged in such a way that they would point toward the corners of a regular tetrahedron, the carbon atom being placed at its center (Fig. 1.2). The necessity for knowing the arrangement of the atoms in space, taken together with an understanding of the order in which they are connected, is central to an understanding of organic chemistry, and we will have much more to say about this later.

1.6 CHEMICAL BONDS

The first explanations of the nature of chemical bonds were advanced by W. Kössel (a German scientist) and G. N. Lewis (of the University of California, Berkeley) in 1916. They proposed two major types of chemical bonds:

1. The *ionic* or *electrovalent* bond (following the transfer of one or more electrons from one atom to another to create ions).
2. The *covalent* bond (a bond that results when atoms share electrons).

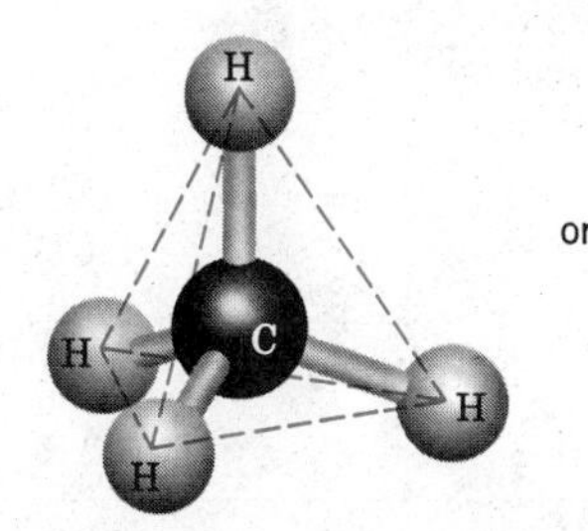

or

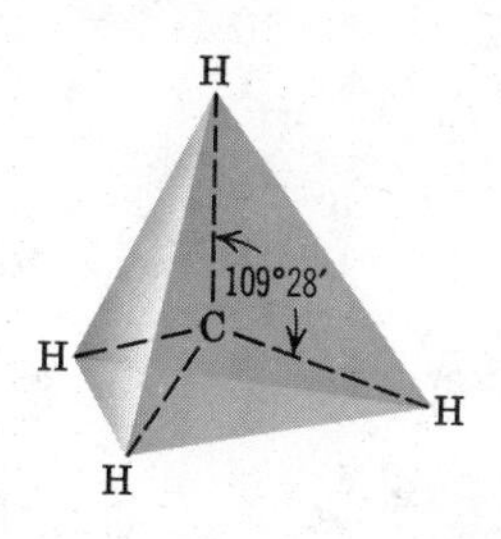

FIG. 1.2
The tetrahedral structure of methane.

The central idea in their work on bonds is that atoms without electronic configurations of noble gases generally react in a way that produces such configurations.

The concepts and explanations that arise from the original propositions of Lewis and Kössel are satisfactory for explanations of many of the problems we deal with in organic chemistry today. For this reason we will review these two types of bonds in more modern terms.

1.6A Ionic or Electrovalent Bonds

An electrovalent or ionic bond is a force of attraction between oppositely charged ions. One source of such ions is the interaction of atoms of widely differing electronegativities (Table 1.2). An example is the reaction of lithium atoms and fluorine atoms. Lithium, a typical metal, has a very low electronegativity; fluorine, a nonmetal, is the most electronegative element of all. The loss of an electron (a negatively charged species) by the lithium atom leaves a lithium cation, Li^+; the gain of an electron by the fluorine atom gives a fluoride anion, F^-. Why do these

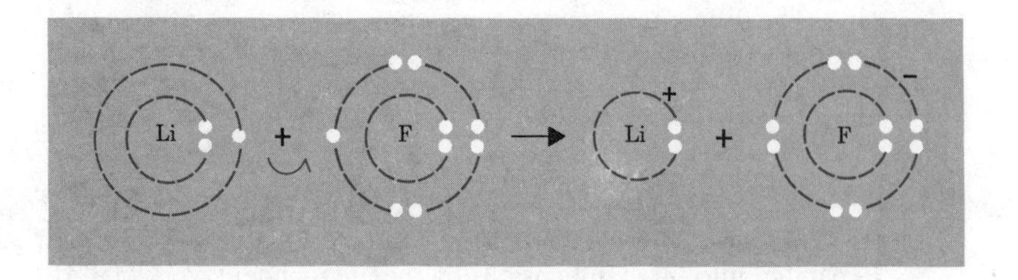

ions form? In terms of the Lewis-Kössel theory both atoms achieve the electronic structure of a noble gas by becoming ions. The lithium ion is like an atom of the noble gas helium, and the fluoride ion is like an atom of the noble gas neon (Fig. 1.3). Moreover, crystalline lithium fluoride (Fig. 1.4) forms from the individual lithium and fluoride ions. In this process negative fluoride ions become surrounded by positive lithium ions, and positive lithium ions by negative fluoride ions. In this crystalline state, the ions have substantially lower energies than the atoms from which they were formed. Lithium and fluorine are thus "stabilized" when they react to form crystalline lithium fluoride.

An ionic bond is essentially omnidirectional. It is impossible to say that any two ions (i.e., Li^+ and F^-) are bonded to each other in the sense that they are attached to each other. Crystals of ionic compounds are simply orderly arrangements of the oppositely charged ions that mutually surround each other (Fig. 1.4).

TABLE 1.2 Pauling Electronegativities of Some of the Elements

			H 2.2			
Li 1.0	Be 1.5	B 2.0	C 2.5	N 3.0	O 3.5	F 4.0
Na 0.9	Mg 1.2	Al 1.5	Si 1.8	P 2.1	S 2.5	Cl 3.0
K 0.8						Br 2.8

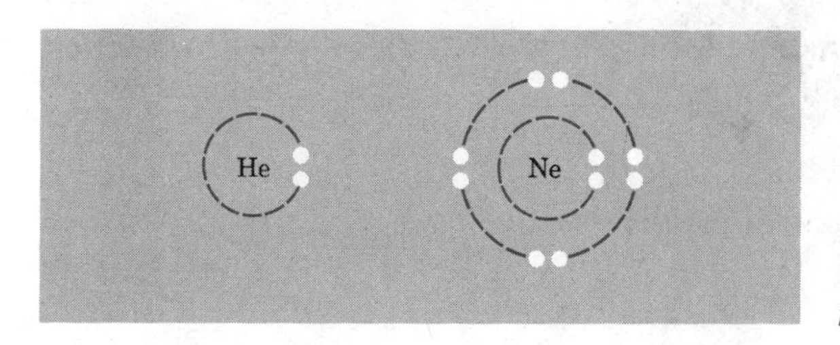

FIG. 1.3
The electronic structure of helium and neon.

Ionic substances, because of their strong internal electrostatic forces, are usually very high melting solids, often having melting points above 1000 °C. In polar solvents, such as water, the ions are solvated, and such solutions usually conduct an electric current.

Problem 1.1

Write equations and account for the formation of an ionic bond when
- (a) Sodium atoms react with chlorine atoms
- (b) Magnesium atoms react with fluorine atoms
- (c) Potassium atoms react with bromine atoms

1.6B Covalent Bonds

When two or more atoms of the same or similar electronegativities react, a complete transfer of electrons does not occur. In these instances the atoms achieve noble gas structures by *sharing electrons*. *Covalent* bonds form that hold together particles called *molecules*. Molecules may be represented by electron-dot formulas but, more conveniently, by dash formulas where each dash represents a pair of electrons shared by two atoms. Some examples are shown below. These formulas

H_2 H:H or H—H **Cl_2** $:\ddot{\underset{\cdot\cdot}{Cl}}:\ddot{\underset{\cdot\cdot}{Cl}}:$ or $:\ddot{\underset{\cdot\cdot}{Cl}}—\ddot{\underset{\cdot\cdot}{Cl}}:$

HCl $H:\ddot{\underset{\cdot\cdot}{Cl}}:$ or $H—\ddot{\underset{\cdot\cdot}{Cl}}:$ **CH_4** $\begin{array}{c}H\\ \cdot\cdot\\ H:C:H\\ \cdot\cdot\\ H\end{array}$ or $\begin{array}{c}H\\ |\\ H—C—H\\ |\\ H\end{array}$

are often called *Lewis structures* and we show only the outer level electrons.

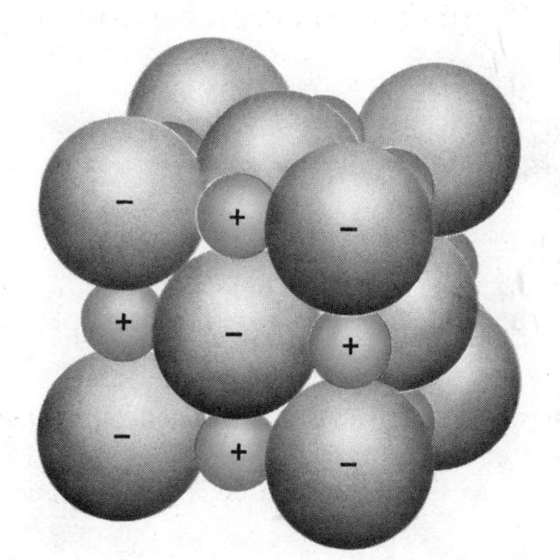

FIG. 1.4
Crystal structure of lithium fluoride, Li^+ F^-. The colored spheres represent lithium ions, Li^+. The grey spheres represent fluoride ions, F^-.

In certain cases, multiple covalent bonds are formed, for example,

N_2 :N⋮⋮N: *or* :N≡N:

and ions themselves may contain covalent bonds.

$\overset{+}{N}H_4$ H:N:H (with H above and below, electron-dot, +) or H—N⁺—H (with H above and below)

Problem 1.2

Write electron-dot and dash formulas for each of the following molecules or ions. In each case show how the atoms achieve the noble gas structure.

(a) HBr (h) PCl_3
(b) Br_2 (i) NF_3
(c) CO_2 (j) CH_3Cl
(d) CH_4 (k) H_2O
(e) H_2O_2 (l) OH^-
(f) SiH_4 (m) NH_4Cl ($NH_4^+Cl^-$)
(g) NH_3 (n) NaOH

1.7 IONS CONTAINING COVALENT BONDS: FORMAL CHARGE

Some positive or negative ions consist of atoms held together by covalent bonds. In problem 1.2, for example, you were asked to write electron-dot formulas for the ammonium ion, NH_4^+, and the hydroxide ion, OH^-.* It will be helpful at this point, as a review, to show how the charges on ions can be calculated. Let us begin with a simple example: the ammonium ion.

There are several ways to calculate the charge on the ammonium ion. The most fundamental is to calculate the arithmetic sum of all of the nuclear protons (positive charges) and of all of the extranuclear electrons (negative charges). This approach is shown below:

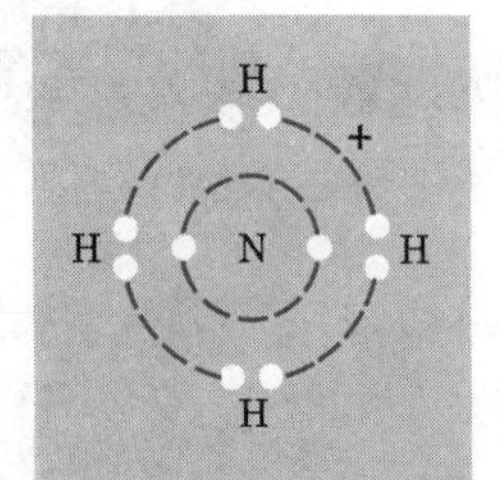

One nitrogen atom = 7 protons
Four hydrogen atoms = 4 protons
Total protons = 11

Number of outer level electrons = 8
Number of inner level electrons = 2
Total electrons = 10

Charge on the ion = $(+11) + (-10) = +1$

Another way to demonstrate that the ammonium ion has one positive

* These ions do not exist alone, of course, but are always associated with oppositely charged ions that balance the charge; the ammonium ion as ammonium *chloride*, $NH_4^+Cl^-$, for example; the hydroxide ion as *sodium* hydroxide, Na^+OH^-, and so on.

charge is to use the equation showing its formation in a chemical reaction:

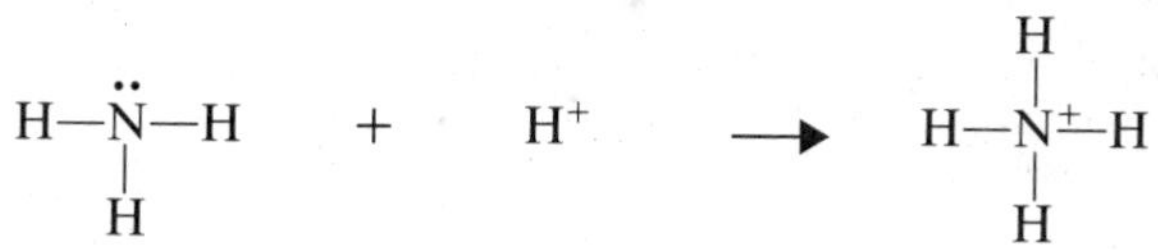

$$\begin{array}{c} \text{H—}\ddot{\text{N}}\text{—H} \\ | \\ \text{H} \end{array} \quad + \quad \text{H}^+ \quad \longrightarrow \quad \begin{array}{c} \text{H} \\ | \\ \text{H—N}^{+}\text{—H} \\ | \\ \text{H} \end{array}$$

Ammonia	+	A proton	→	Ammonium ion
(an electrically neutral molecule)	+	(one positive charge)	=	(one positive charge)
0	+	(+1)	=	+1

This approach is based on the principle that in a correctly balanced chemical equation the electrical charges must also balance.

1.7A Formal Charge

A third way of calculating the charge on a polyatomic ion is based on the concept of *formal charge*. This procedure is really a method for electron bookkeeping. In this approach, the formal charge on each atom is calculated first. Then the arithmetic sum of all of the formal charges gives the charge on the ion as a whole. The formal charge of each atom is calculated by taking the group number of that atom (from the Periodic Table) and subtracting the number of electrons associated with it using the following formula.

$$\text{Formal charge} = \text{group number} - [\tfrac{1}{2}(\text{number of shared electrons}) + (\text{number of unshared electrons})]$$

Let us consider the ammonium ion again. First we calculate the formal charge on each atom:

$$\text{Formal charge on hydrogen} = +1 - [\tfrac{1}{2}(2) + (0)] = 0$$

$$\text{Formal charge on nitrogen} = +5 - [\tfrac{1}{2}(8) + (0)] = +1$$

Then we calculate the formal charge on the ion as a whole:

$$\begin{array}{c} \text{H} \\ \text{H:}\ddot{\underset{..}{\text{N}}}\text{:H}^{+} \\ \text{H} \end{array}$$

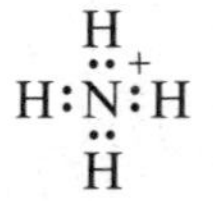

Formal charge on each hydrogen	$= 0 \times 4 = 0$
Formal charge on nitrogen	$= +1 \times 1 = +1$
Total charge on the ion	$= +1$

A moment's reflection will reveal why this formula works: the group number of an element is nothing more than the kernel charge of that element. The kernel charge is defined as the charge on that portion of the atom that includes the nuclear protons and the inner level electrons. For a nitrogen atom (Fig. 1.5) the kernel charge is +5.

Taking the sum $\tfrac{1}{2}$(number of shared electrons) + (number of unshared electrons) is nothing more than a way of apportioning the valence electrons. We divide shared electrons between the atoms that share them; we assign unshared pairs directly to the atom that possesses them. The ammonium ion has no un-

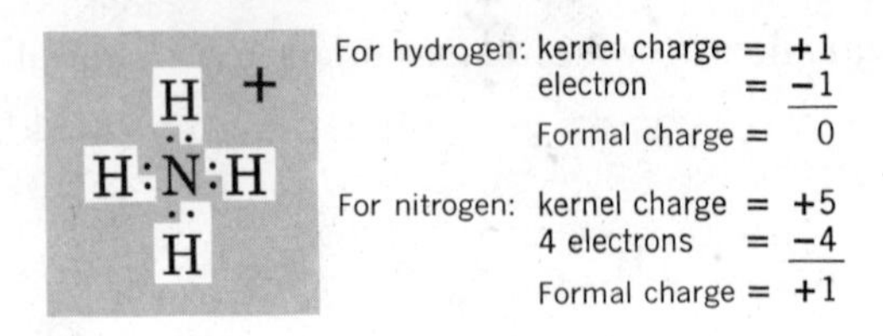

shared pairs; all of the valence electrons are divided between the sharing atoms:

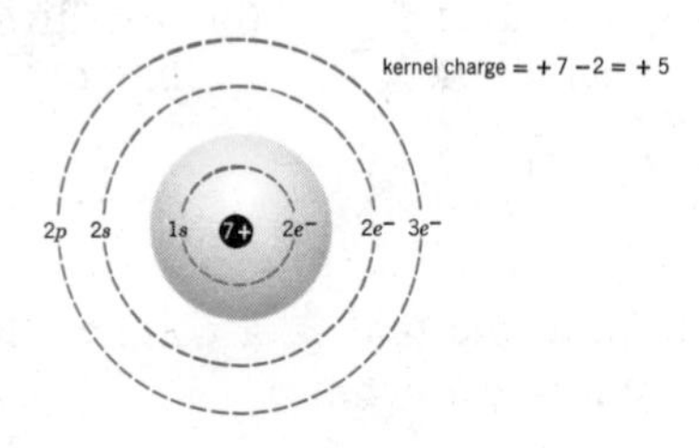

Let us consider an ion that has unshared pairs. The nitrate ion, NO_3^-, may be written in the way shown below.

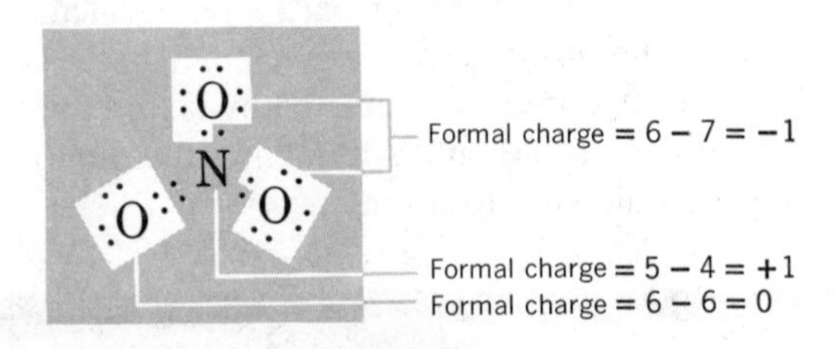

Charge on ion $= 2(-1) + 1 + 0 = -1$

The sum of the formal charges of each atom of a molecule must, of course, equal zero. Molecules by definition are neutral. Consider the following examples.

Ammonia:

H—N̈—H or H:N:H
  | H
  H

Formal charge = 5 − 5 = 0
Formal charge = 1 − 1 = 0

Charge on molecule $= 0 + 3(0) = 0$

Water:

H—Ö—H or H:O:H

Formal charge = 6 − 6 = 0
Formal charge = 1 − 1 = 0

Charge on molecule $= 0 + 2(0) = 0$

Problem 1.3

Calculate both the formal charge on each atom, and the total charge on the molecule or ion, for each of the following species.

(a) BH_4 (f) $:CH_3$ (a carbanion)
(b) OH (g) CH_3 (a carbocation)
(c) BF_4 (h) $\cdot CH_3$ (a free radical)
(d) H_3O (i) $:CH_2$ (a carbene)
(e) CO_3 (j) $:\ddot{N}H_2$

1.7B Summary of Formal Charges

With this background it should now be clear that each time an oxygen atom of the type, —Ö: appears in a molecule or ion it will have a formal charge of −1, and that each time an oxygen atom of the type =Ö: or —Ö— appears it will have a formal charge of zero. Similarly; —N— (with four bonds) will be +1, and —N̈— (with three bonds) will be zero.

It is much easier to memorize these common structures than to calculate their formal charges each time they are encountered. These common structures are summarized below.

Group	Formal charge of 0	Formal charge of +1	Formal charge of −1
3	—B—		—B⁻—
4	—C—	—C⁺—	—C̈⁻—
5	—N̈— or =N̈— or ≡N:	—N⁺— or =N⁺—	—N̈⁻— or =N̈⁻
6	—Ö— or =Ö:	—Ö⁺— or =Ö⁺—	—Ö:⁻
7	—Ẍ: (X = F, Cl, Br, or I)	—Ẍ⁺—	:Ẍ:⁻

Problem 1.4

Using the chart given above, determine the formal charge on each colored atom of the following molecules and ions. (Remember the formal charge of —Ö— = =Ö:, that of —N— = ≡N:, and so on.)

(a) CH_3—N̈(H)—H (an amine)

(d) CH_3—C≡N: (a nitrile)

(g) CH_3—C(=Ö:)—Ö: (a carboxylate ion)

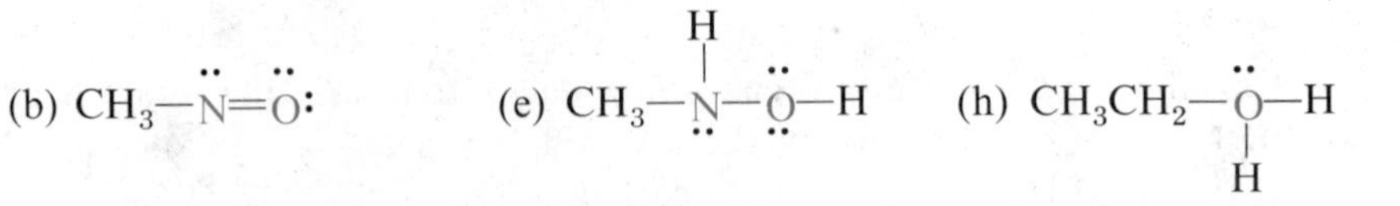

(b) $CH_3—\ddot{N}=\ddot{O}:$ (a nitroso compound)

(e) $CH_3—N(H)—\ddot{O}—H$ (a hydroxylamine)

(h) $CH_3CH_2—\ddot{O}(H)—H$ (a protonated alcohol)

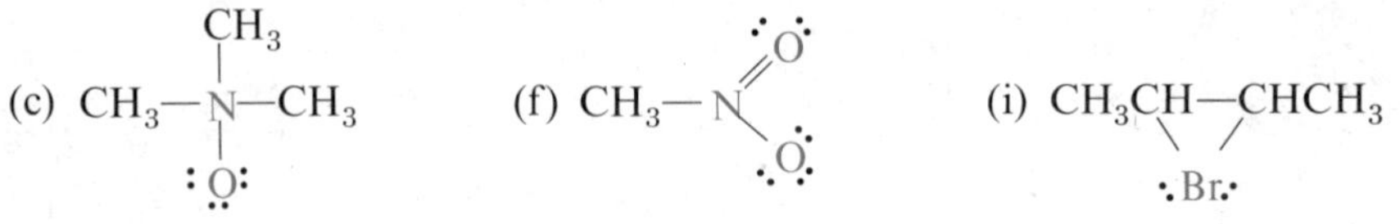

(c) $CH_3—N(CH_3)(:\ddot{O}:)—CH_3$ (an amine oxide)

(f) $CH_3—NO_2$ (a nitro compound)

(i) $CH_3CH—CHCH_3$ bridged by $\cdot\text{Br}\cdot$ (a bromonium ion)

1.8 RESONANCE

More than one *equivalent* Lewis structure can be written for many molecules and ions. Consider, for example, the carbonate ion, $CO_3^=$. We can write three *different* but *equivalent* structures, **1-3**.

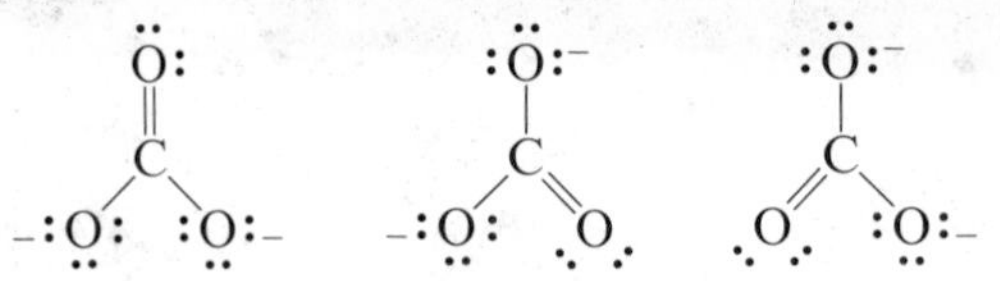

1 **2** **3**

Notice two important features of these structures. First, each atom has the noble gas configuration. Second, *and this is especially important,* we can convert one structure into any other *by changing only the positions of the electrons.* We do not need to change the relative positions of the atomic nuclei. For example, if we move the electrons in the following way we change structure **1** into structure **2:**

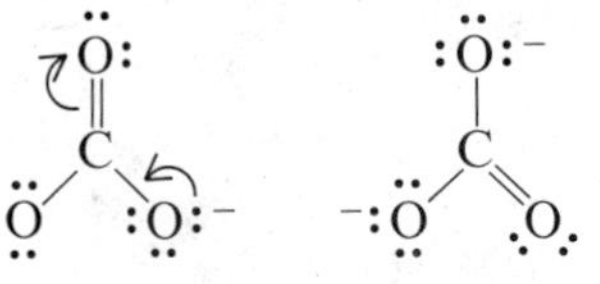

1 **2**

In a similar way we can change structure **2** into structure **3:**

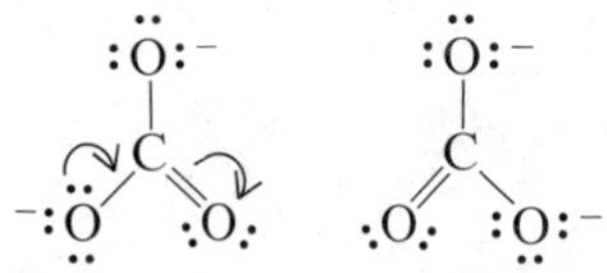

2 **3**

These structures, although not identical, *are equivalent.* None of them, however, fits important data about the carbonate ion.

X-ray studies have shown that double bonds are shorter than single bonds. The same kind of study of the carbonate ion shows, however, that all of its carbon-oxygen bonds are of equal length. One is not shorter than the others. All are equivalent. Clearly none of the three structures agrees with this evidence. In each structure, **1-3,** one carbon-oxygen bond is a double bond and the other two are single bonds. None of the structures is correct. How, then, should we represent the carbonate ion?

A situation like this can be handled by a theory called *resonance theory.* Resonance theory says that whenever a molecule or ion can be represented by two or more Lewis structures *that differ only in the positions of the electrons* two things will be true:

1. None of these structures, which we call *resonance structures* or *resonance contributors,* will be correct. None will be in accord with the physical or chemical properties of the substance.

2. The actual molecule or ion will be best represented by a *hybrid of these structures.*

Resonance structures, then, are not structures for the actual molecule or ion; they exist only in theory. As such they can never be isolated. No single contributor adequately represents the molecule or ion. In resonance theory we view the carbonate ion, which is, of course, a real entity, as having a structure that is a *hybrid* of these three *hypothetical* resonance structures.

Two often-used analogies may help make clear the relation of the contributors to the hybrid, that is, to the real structure. One might, for example, describe a *real* person as being like a combination or hybrid of Don Quixote, Sir Galahad, and Robin Hood (three fictional characters). Or one might describe a rhinoceros (a real animal) as resembling a combination (hybrid) of a unicorn and a dragon (both mythical animals). Resonance structures, therefore, are like dragons, unicorns, and Sir Galahads. While they have no real existence of their own, they are useful in helping describe real molecules.

What would a hybrid of structures **1** to **3** be like? Look at the structures and look especially at a particular carbon-oxygen bond, say, the one at the top. This carbon-oxygen bond is a double bond in one structure **(1)** and a single bond in the other two (**2** and **3**). The actual carbon-oxygen bond, since it is a hybrid, must be something in between a double bond and a single bond. Since the carbon-oxygen bond is a single bond in two of the structures and a double bond in only one it must be more like a single bond than a double bond. It must be like a one and one-third bond. We could call it a partial double bond. And, of course, what we have just said about any one carbon-oxygen bond will be equally true of the other two. Thus all of the carbon-oxygen bonds of the carbonate ion are partial double bonds, and *all are equivalent.* All of them *should be* the same length, and this is exactly what experiments tell us. They are all 1.31 Å long.

One other important point: by convention, when we draw resonance structures, we connect them by double-headed arrows to indicate clearly that they are hypothetical, not real. For the carbonate ion we write them this way:

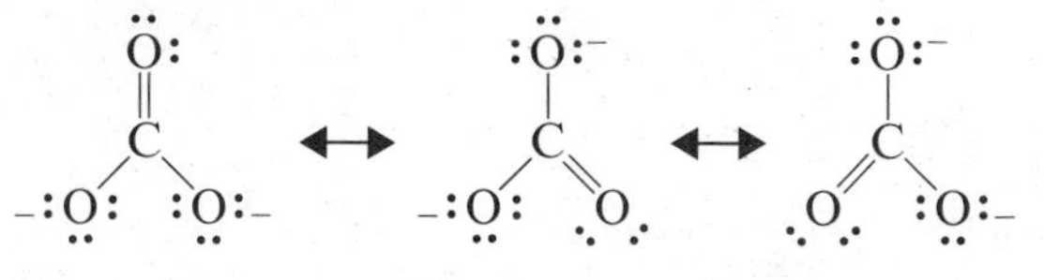

We should not let these arrows, or the word "resonance" mislead us into thinking that the carbonate ion fluctuates between one structure and another. These structures exist only on paper. The carbonate ion cannot fluctuate among them anymore than a real person can become Sir Galahad one moment, Don Quixote the next, and Robin Hood the next—no more than a rhinoceros can fluctuate between being a unicorn and a dragon.

How can we write the structure of the carbonate ion in a way that will indicate its actual structure? We may do two things: we may write all of the resonance structures as we have just done and let the reader mentally fashion the hybrid or we may write a non-Lewis structure that attempts to represent the hybrid. For the carbonate ion we might do the following:

$O^{\delta-}$
C
$^{\delta-}O$ $O_{\delta-}$

The bonds are indicated by a combination of a solid line and a dashed line. This is to indicate that the bonds are something in between a single bond and a double bond. We place a $\delta-$ (read partial minus) beside each oxygen to indicate that something less than a full negative charge resides on each oxygen. (In this instance each oxygen has two-thirds of a full negative charge.)

Problem 1.5

(a) Write two resonance structures for sulfur dioxide, SO_2. (Sulfur dioxide is a major air pollutant.)
(b) Would you expect the sulfur-oxygen bonds of SO_2 to be equivalent and thus of equal length? Explain.

1.9 POLAR COVALENT BONDS

When two atoms of different electronegativities form a covalent bond the electrons are not shared equally between them. The atom with greater electronegativity draws the electron pair closer to it, and a *polar covalent bond* results. (One definition of *electronegativity* is *the ability of an element to attract electrons that it is sharing in a covalent bond.*) An example of such a polar covalent bond is the one in hydrogen chloride. The chlorine atom, with its greater electronegativity, pulls the bonding electrons closer to it. This makes the hydrogen atom somewhat

$\delta+$ $\delta-$
H :C̤̈l:

electron deficient and gives it a *partial* positive charge ($\delta+$). The chlorine atom becomes somewhat electron rich and bears a *partial* negative charge ($\delta-$). Because the hydrogen chloride molecule has a partially positive end and a partially

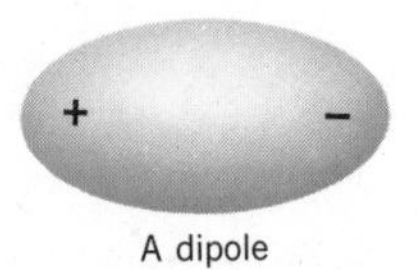

A dipole

negative end, it is a dipole, and it has a *dipole moment*. The dipole moment (μ) is a physical property that can be measured experimentally. It is expressed in Debye units (D), and it is the product of the magnitude of the charge in electrostatic units and the distance that separates them in Angstrom units times 10^{10}. (One Angstrom unit equals 10^{-8} cm.)

Dipole moment = charge (in esu) $\times$ distance (in Angstrom units)

$$\mu = e \times d \times 10^{10}$$

The dipole moment of HCl is 1.08 D.

The direction of polarity of a polar bond is sometimes symbolized by ⟶. In HCl this is expressed in the following way:

(positive end) ⟶ (negative end)
H—Cl

Dipole moments, as we will see, are very useful quantities in accounting for physical properties of compounds.

Problem 1.6

Predict the direction of the dipole (if any) in the following molecules.

(a) HBr (c) H_2
(b) ICl (d) Cl_2

1.10 SHAPES OF MOLECULES: THE VALENCE SHELL ELECTRON PAIR REPULSION (VSEPR) MODEL

Later in this chapter we discuss the shapes of molecules on the basis of theories that arise from quantum mechanics. It is possible, however, to understand the shapes of many molecules from the principle that in most compounds, electrons exist as pairs and *electrons repel one another*. This is true whether the electrons are in bonds or in unshared (nonbonding) pairs. Consequently, within the confines of the molecule, *electron pairs of the valence shell (outer level) tend to stay as far apart as possible*. Let us see how this approach—called the valence shell electron pair repulsion or VSEPR model—applies to a few simple examples.

1.10A Methane

The valence shell of methane contains four pairs of bonding electrons. Only a tetrahedral orientation will allow four pairs of electrons to have the maximum possible separation (Fig. 1.6). Any other orientation, for example, a square planar arrangement (Fig. 1.7), places the electron pairs closer together.

Thus, in the case of methane, the valence shell electron pair repulsion model accommodates what we have known since the discovery of van't Hoff and Le Bel: the molecule of methane has a tetrahedral shape.

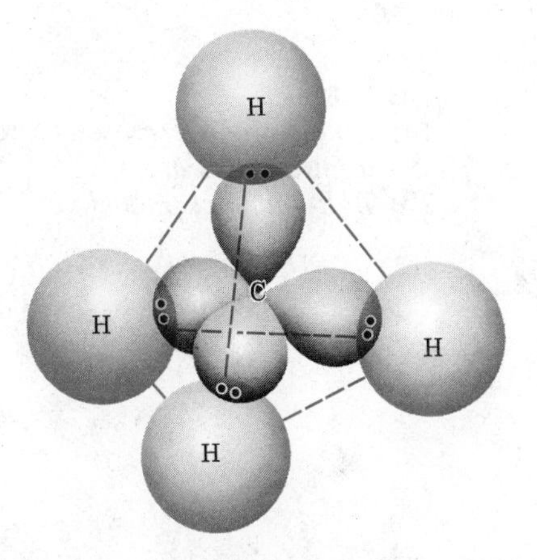

FIG. 1.6
A tetrahedral shape for methane allows the maximum separation of the bonding electron pairs.

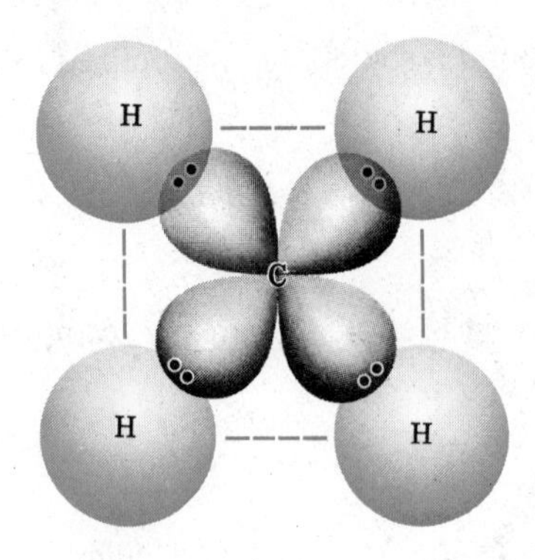

FIG. 1.7
Hypothetical square planar structure for methane. When comparing this structure to that in Fig. 1.6, remember all of the atoms in the square structure are in the same plane (of the paper), while in the tetrahedral arrangement the atoms are in three dimensions.

Problem 1.7

Part of the reasoning that led van't Hoff and Le Bel to propose a tetrahedral shape for molecules of methane was based on the number of compounds that are theoretically possible for substituted methanes, that is, for compounds in which one or more hydrogens of methane have been replaced by some other group. For example, only one compound of the type CH_2X_2 has ever been found. (a) Is this consistent with a tetrahedral shape? (b) With a square planar shape? Explain.

The bond angles for any atom that has a regular tetrahedral structure are 109.5°. One way of representing a tetrahedral atom is shown in Fig. 1.8. This representation is derived from the familiar ball-and-stick model. The lines that intersect the circle are directed out of the plane of the paper.

1.10B Water

The H—O—H bond angle in a molecule of water is 105°, an angle that is quite close to the 109.5° bond angles of methane.

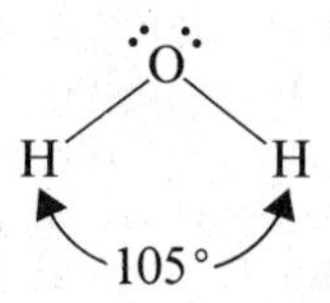

We can write a general tetrahedral structure for a molecule of water if we place the two nonbonding electron pairs at corners. Such a structure is shown in Fig. 1.9.

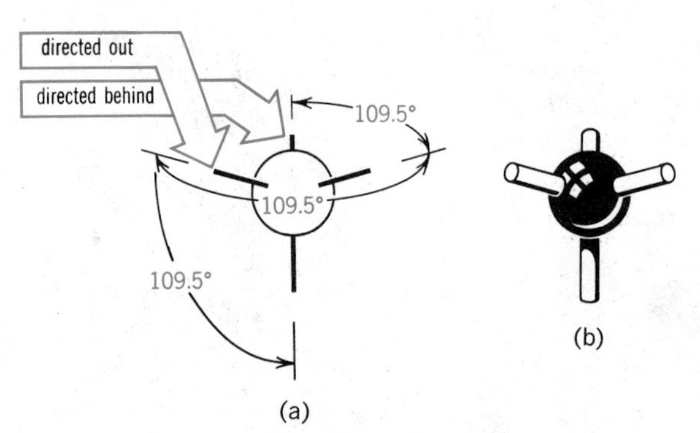

FIG. 1.8
(a) *Line-and-circle representation of a tetrahedral atom.*
(b) *Ball-and-stick model.*

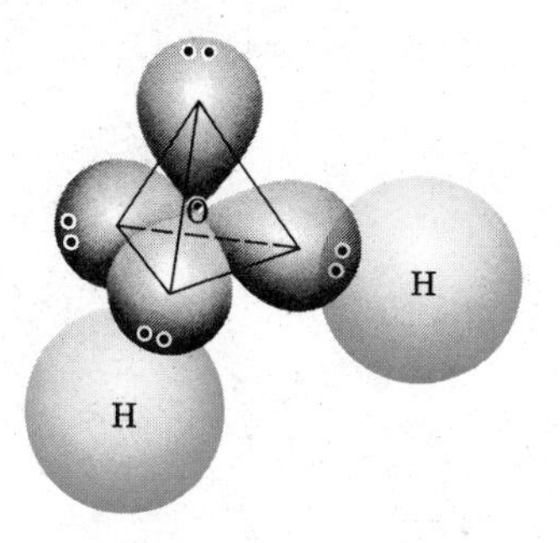

FIG. 1.9
The tetrahedral shape for a molecule of water that results when the pairs of nonbonding electrons are considered to occupy corners.

1.10C Ammonia

The bond angles in a molecule of ammonia are 107°, a value even closer to

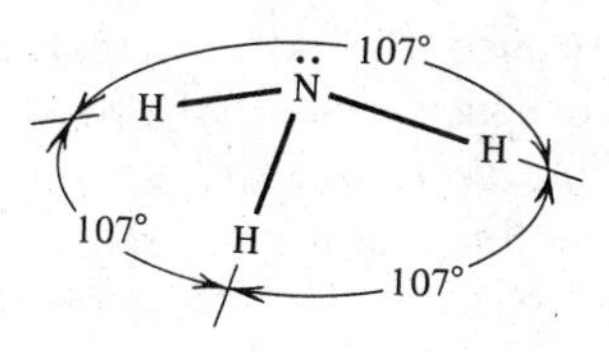

the tetrahedral angle (109.5°) than that of water. We write a general tetrahedral structure for ammonia by placing the nonbonding pair at one corner (Fig. 1.10).

1.10D Boron Trifluoride

Boron, a Group III element, has only three outer level electrons. In the compound boron trifluoride (BF_3) these three electrons are shared with three fluorine atoms. As a result, the boron atom in BF_3 has only six electrons (three bonding pairs) around it. In the boron trifluoride molecule the three fluorines lie in a plane at the corners of an equilateral triangle (Fig. 1.11). The bond angles are 120°.

This trigonal (triangular) planar structure allows for the maximum separation (120°) of the three bonding pairs of electrons.

Problem 1.8

Predict the general shapes of the following molecules and ions.

(a) SiH_4	(d) BF_4^-	(g) BH_4^-
(b) $CH_3{:}^-$	(e) CCl_4	(h) BeH_2
(c) CH_3^+	(f) $BeCl_2$	(i) BH_3

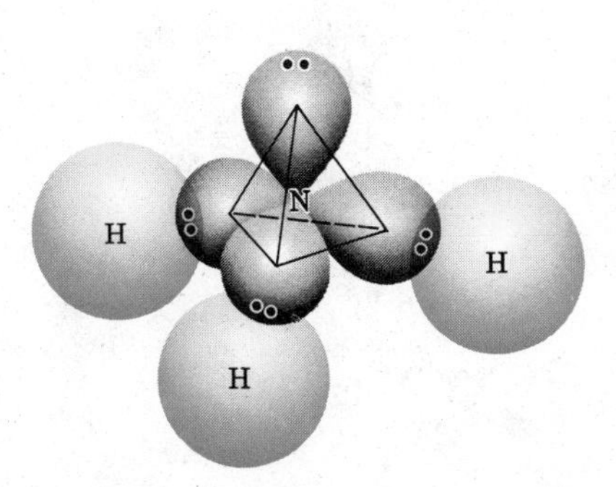

FIG. 1.10
The tetrahedral shape for an ammonia molecule that results when the nonbonding electron pair is considered to occupy one corner.

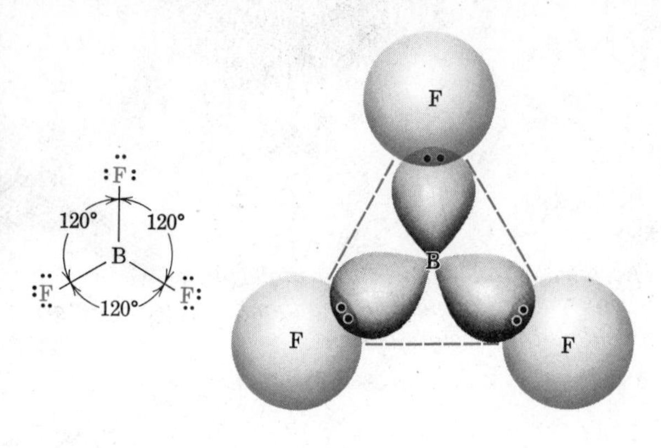

FIG. 1.11
The triangular (trigonal) shape of boron trifluoride.

Problem 1.9

The valence shell electron pair repulsion method can also be used to predict the shapes of molecules containing multiple bonds if we assume that the electrons of a multiple bond are located in the region of space between the two atoms joined by a multiple bond. That is, *we assume that the electrons of a multiple bond behave as though they were a single electron pair.* Use this method to predict the shapes and bond angles of (a) CO_2, (b) SO_2, and (c) SO_3.

1.11 POLAR AND NONPOLAR MOLECULES

In our discussion of dipole moments in Section 1.9, we restricted our attention to simple diatomic molecules. Any *diatomic* molecule in which the two atoms are *different* (and thus have different electronegativities), will, of necessity, have a dipole moment. If we examine Table 1.3, however, we find that a number of molecules (e.g., CCl_4, CO_2) consist of more than two atoms, have *polar* bonds, *but have no dipole moment.* Now that we have an understanding of the shapes of molecules we can understand how this can occur.

Consider a molecule of carbon tetrachloride (CCl_4). Because the electronegativity of chlorine is greater than that of carbon, each of the carbon-chlorine bonds in CCl_4 is polar. Each chlorine has a partial negative charge, and the carbon is considerably positive. Because a molecule of carbon tetrachloride is tetrahedral (Fig. 1.12), however, *the center of positive charge and the center of negative charge coincide, and the molecule has no net dipole moment.*

This can be illustrated in a slightly different way: if we use arrows (+→) to represent the direction of polarity of each bond we get the arrangement of bond moments shown in Fig. 1.13.

TABLE 1.3 Dipole Moments of Some Simple Molecules

FORMULA	μ, D	FORMULA	μ, D
H_2	0	CH_4	0
Cl_2	0	CH_3Cl	1.87
HF	1.91	CH_2Cl_2	1.55
HCl	1.08	$CHCl_3$	1.02
HBr	0.80	CCl_4	0
HI	0.42	NH_3	1.47
BF_3	0	NF_3	0.24
CO_2	0	H_2O	1.85

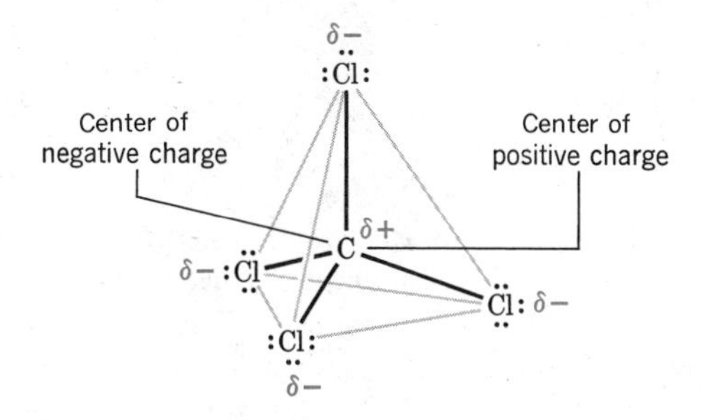

FIG. 1.12
Charge distribution in carbon tetrachloride.

Since the bond moments are vectors of equal magnitude arranged tetrahedrally, their effects cancel. Their vector sum is zero. The molecule has *no net dipole moment*.

The chloromethane molecule (CH_3Cl) has a net dipole moment of 1.87 D. Since carbon and hydrogen have electronegativities (Table 1.2) that are nearly the same, the contribution of three C—H bonds to the net dipole is negligible. The electronegativity difference beteen carbon and chlorine is large, however, and this highly polar C—Cl bond accounts for most of the dipole moment of CH_3Cl (Fig. 1.14).

Problem 1.10

A molecule of carbon dioxide (CO_2) is linear. Show how this accounts for the fact that CO_2 has no dipole moment.

Problem 1.11

(a) Sulfur dioxide (cf. Problem 1.5) has a dipole moment. (The dipole moment of SO_2 is 1.63 D) What does this fact indicate about the shape of SO_2?

Unshared pairs of electrons make large contributions to the dipole moments of water and ammonia. Because an unshared pair has no atom attached to it to partially neutralize its negative charge, an unshared electron pair contributes a large moment directed away from the central atom (Fig. 1.15). (The O—H and N—H moments are also appreciable.)

Problem 1.12

Nitrogen trifluoride ($:NF_3$) has a shape very much like that of ammonia. It has, however, a very low dipole moment ($\mu = 0.24$ D). How can you explain this?

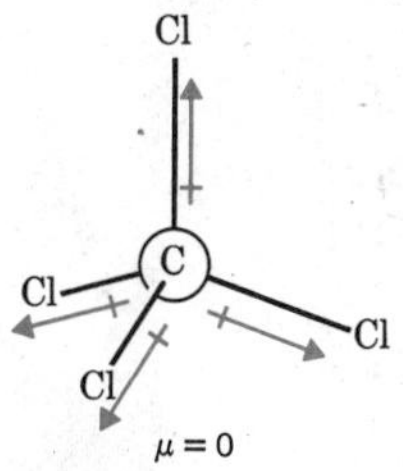

$\mu = 0$

FIG. 1.13
A tetrahedral orientation of equal bond moments causes their effects to cancel.

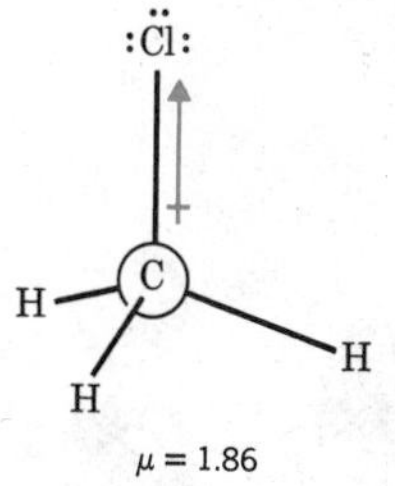

$\mu = 1.86$

FIG. 1.14
The dipole moment of chloromethane arises mainly from the highly polar carbon-chlorine bond.

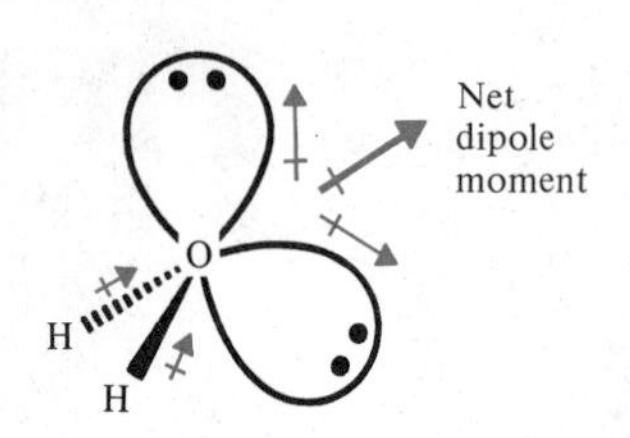

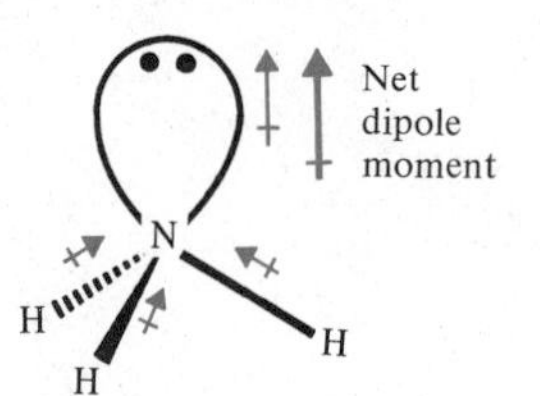

FIG. 1.15
Bond moments and the resulting dipole moment of water and ammonia.

Problem 1.13

BF_3 has no dipole moment. How can this be explained?

1.12 THE REPRESENTATION OF STRUCTURAL FORMULAS

Organic chemists use a variety of ways to write structural formulas. The representation that is chosen in a particular instance is the one that best illustrates the property being considered or the one that represents the structure most conveniently. The *dot structure* for propanol (Fig. 1.16) clearly shows the number of valence electrons and the way that they are shared. The *dash formula* is easier to write and shows us, as does the dot structure, the order in which the atoms are attached. The condensed formula is still easier to write, and when we become more familiar with it, it will impart all the information that is contained in either the dot or dash structure. In condensed formulas, the atoms that are attached to a particular carbon atom are written immediately after that atom. For example:

$$CH_3CHClCH_2CH_3 = \mathrm{H{-}\overset{H}{\underset{H}{C}}{-}\overset{H}{\underset{Cl}{C}}{-}\overset{H}{\underset{H}{C}}{-}\overset{H}{\underset{H}{C}}{-}H} = CH_3\underset{|\atop Cl}{C}HCH_2CH_3$$

and

$$CH_3CH(CH_3)CH_2CH_3 = \mathrm{H{-}\overset{H}{\underset{H}{C}}{-}\overset{H}{\underset{H-C-H \atop H}{C}}{-}\overset{H}{\underset{H}{C}}{-}\overset{H}{\underset{H}{C}}{-}H} = CH_3\underset{|\atop CH_3}{C}HCH_2CH_3$$

FIG. 1.16
Structural formulas for propanol.

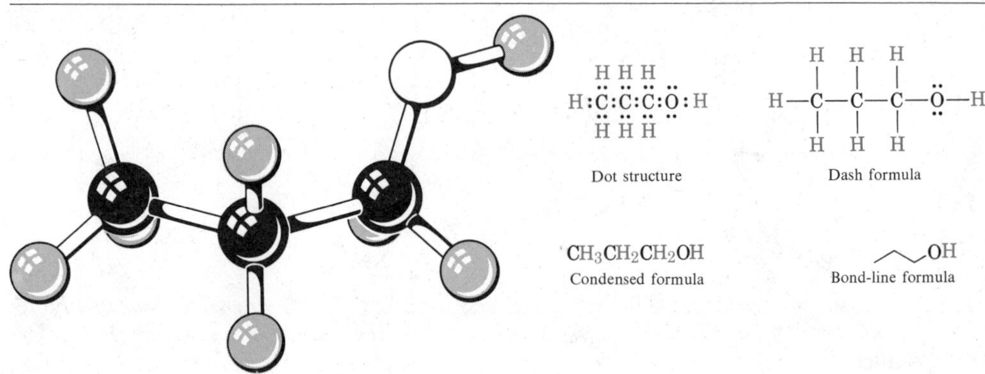

The bond-line representation is the easiest of all to write because it shows only the carbon skeleton. The number of hydrogen atoms necessary to fulfill the carbon atoms' valences are assumed to be present, but we do not write them in.* Other atoms (e.g., O, Cl, N) *are* written in:

$$CH_3CHClCH_2CH_3 = \quad \text{(bond-line, Cl)} \quad = CH_3\underset{\displaystyle Cl}{\underset{|}{C}}HCH_2CH_3$$

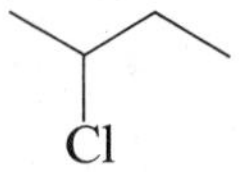

$$CH_3CH(CH_3)CH_2CH_3 = \quad \text{(bond-line)} \quad = CH_3\underset{\displaystyle CH_3}{\underset{|}{C}}HCH_2CH_3$$

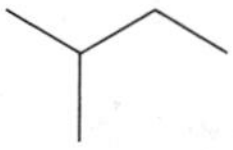

Problem 1.14

Write formulas for all of the structural isomers of C_3H_8O.

Problem 1.15

Rewrite each of the condensed structural formulas given below, as *dash formulas* and as *bond-line formulas*.

(a) $CH_3CCl_2CH_2CH_3$
(b) $CH_3CH(CH_2Cl)CH_2CH_3$
(c) $CH_3C(CH_3)_2CH_2CH_3$
(d) $CH_3CHClCHClCH_3$
(e) $CH_3CH(OH)CH_2CH_3$
(f) $CH_3CH_2CH_2CH_2OH$
(g) $CH_3\overset{\displaystyle O}{\overset{\|}{C}}CH_2CH(CH_3)_2$
(h) $CH_3CH_2CH(OH)CH(CH_3)_2$

Problem 1.16

Are any of the compounds listed in Problem 1.15 structural isomers? If so, which ones?

Problem 1.17

Write dash formulas for each of the following bond-line formulas.

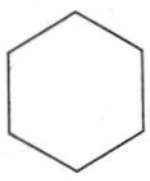

(a) 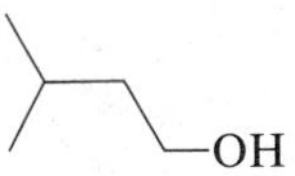 OH
(b) 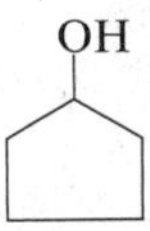 OH
(c)
(d)

1.12A Three-dimensional Formulas

None of the structures that we have described so far tells us how the atoms are arranged in space. There are, however, two frequently used formulations that do impart the three-dimensional structure of a molecule. They are the Alexander (or circle-and-line structure) and the dash-line-wedge structure (Fig. 1.17). In the

* This is comparable to classic Hebrew where the vowels are omitted. Everyone knows where they should be and a lot of space is saved. This can be important if you are carving your message into a stone tablet!

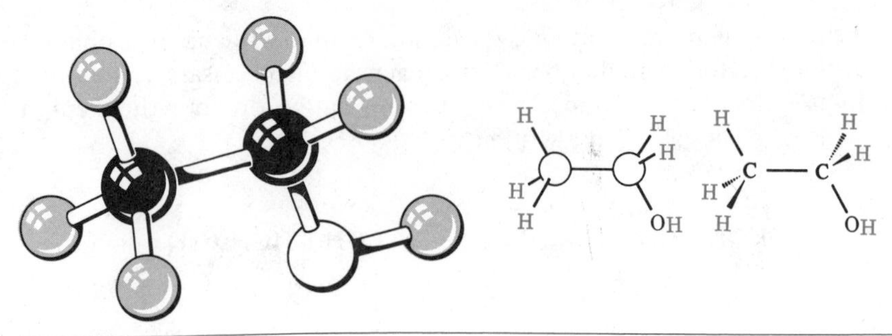

FIG. 1.17
The circle-and-line structure and the dash-line-wedge structure for ethanol.

dash-line-wedge formula, atoms that project out of the plane of the paper are connected by a wedge (—◀), those that lie behind the plane are connected with a dash (┈┈), and those atoms in the plane of the paper are connected by a line.

1.13 QUANTUM MECHANICS

In 1926 a new theory of atomic and molecular structure was advanced independently and almost simultaneously by three men: Erwin Schrödinger, Werner Heisenberg, and Paul Dirac. This theory, called *wave mechanics* by Schrödinger or *quantum mechanics* by Heisenberg, has become the basis from which we derive our modern understanding of bonding in molecules. The language of quantum mechanics is largely a mathematical language and, as such, it has a formidable appearance to those unfamiliar with differential equations, matrix algebra, and wave theory. Yet, in a sense, wave mechanics is as central to modern chemistry as the theory of natural selection is to modern biology. For this reason we have attempted to give an essentially nonmathematical account of the development of quantum mechanics as a special topic following this chapter. Some of the results of quantum mechanics are essential to our study, however, therefore we will discuss those ideas of quantum mechanics that are necessary to develop a modern understanding of bonding in covalent molecules.

The formulation of quantum mechanics that Schrödinger advanced is the form that is most often used by chemists. In Schrödinger's publication the motion of the electrons was described in terms that took into account the wave nature of the electron.* Schrödinger developed a way to convert the mathematical expression for the total energy of the system consisting of one proton and one electron—the hydrogen atom—into another expression called a *wave equation.* This equation was then solved to yield not one but a series of solutions called *wave functions.*

Wave functions are most often denoted by the Greek letter psi (ψ) and each wave function (ψ function) corresponds to a different state for the electron. Corresponding to each state and calculable from the wave equation for the state, is a particular energy.

Each state is a sublevel where one (or two) electrons can reside. The solutions to the wave equation for a hydrogen atom can also be used (with appropriate modifications) to give sublevels for the electrons of higher elements.

* The idea that the electron has the properties of a wave as well as those of a particle was proposed by Louis de Broglie in 1923.

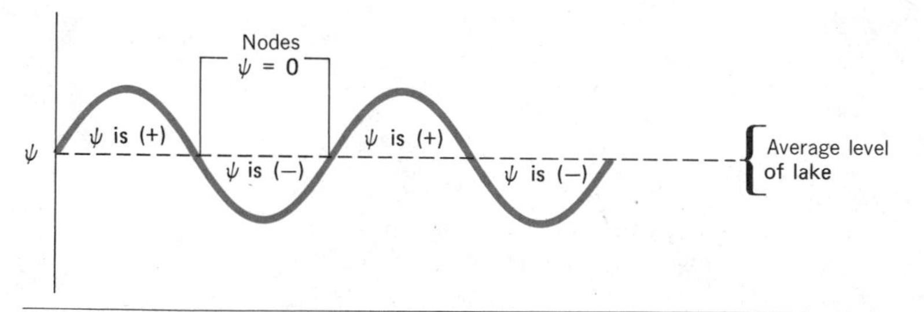

FIG. 1.18
A wave moving across a lake is viewed along a slice through the lake. For this wave the wave function, Ψ, is plus (+) in crests and minus (−) in troughs. At the average level of the lake it is zero; these places are called nodes.

A wave equation is simply a tool for calculating two important properties. These are the energy associated with the state and the relative probability of an electron residing at particular places in a sublevel (Section 1.13). When the value of a wave equation is calculated for a particular point in space relative to the nucleus, the result may be a positive number or a negative number (or zero). These signs are sometimes called *phase signs*. They are characteristic of all equations that describe waves. We do not need to go into the mathematics of waves here, but a simple analogy will help us understand the nature of these phase signs.

Imagine a wave moving across a lake. As it moves along, the wave has crests and troughs, that is, it has regions where the wave rises above the average level of the lake or falls below it (Fig. 1.18). Now, if an equation were to be written for this wave, the wave function (ψ) would be plus (+) in regions where the wave is above the average level of the lake (i.e., in crests) and it would be minus (−) in regions where the wave is below the average level (i.e., in troughs). The relative magnitude of ψ (called the amplitude) will be related to the distance the wave rises above or falls below the average level of the lake. At the places where the wave is exactly at the average level of the lake the wave function will be zero. Such a place is called a *node*.

One other characteristic of waves is their ability to reinforce each other or to interfere with one another. Imagine two waves approaching each other as they move across a lake. If the waves meet so that a crest meets a crest, that is, so that *waves of the same phase sign meet each other,* the waves *reinforce* each other, they add together, and the resulting wave is larger than either individual wave (Fig. 1.19). On the other hand, if a crest meets a trough, that is, if waves of opposite sign meet, the waves *interfere* with each other, they subtract from each other, and the resulting wave is smaller than either individual wave (Fig. 1.20). (If the two waves of opposite sign meet in precisely the right way, complete cancellation can occur.)

The wave functions that describe the motion of an electron in an atom or molecule are, of course, different from the equations that describe waves moving across lakes. And, when dealing with the electron we should be careful not to take

FIG. 1.19
When two waves of the same sign meet they reinforce *each other by combining to make a larger wave.*

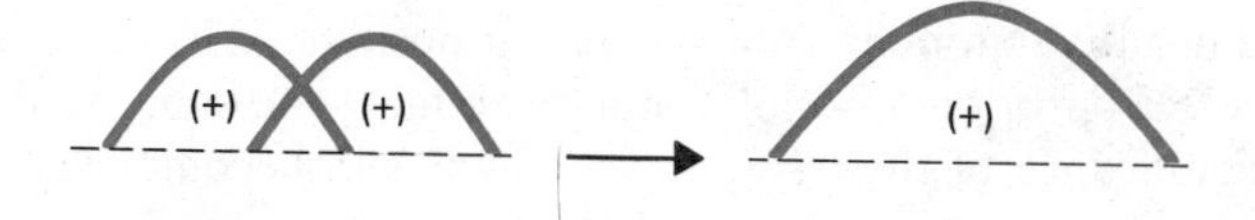

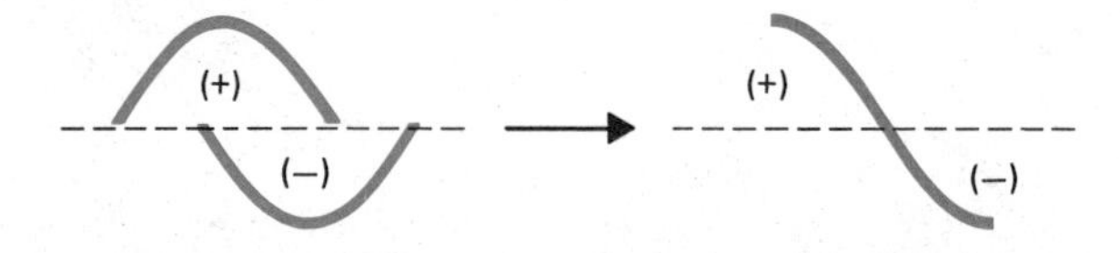

FIG. 1.20
When two waves of opposite sign meet they interfere *with each other; they combine to make a smaller wave.*

analogies like this too far. Electron wave functions, however, are like the equations that describe water waves in that they have phase signs and nodes, and *they undergo reinforcement and interference.*

1.14 ATOMIC ORBITALS

For a short time after Schrödinger's proposal in 1926, a precise physical interpretation for the electron wave function eluded early practitioners of quantum mechanics. It remained for Max Born, a few months later, to point out that the square of ψ *could* be given a precise physical meaning. According to Born, ψ^2 for a particular location (x,y,z), expresses the *probability* of finding an electron at that particular location in space. If ψ^2 is large in a unit volume of space, the probability of finding an electron in that volume is great—we say that the *electron probability density* is large. Conversely if ψ^2 for some other unit volume of space is small, the probability of finding an electron there is low. Plots of ψ^2 in three dimensions generate the shapes of the familiar *s*, *p*, and *d* atomic orbitals.

The *f* orbitals are practically never used in organic chemistry, and we will not concern ourselves with them in this book. The *d* orbitals will be discussed briefly later when we discuss compounds in which *d* orbital interactions are important. The *s* and *p* orbitals are, by far, the most important in the formation of organic molecules and, at this point, we will limit our discussion to them.

An orbital is a region of space where the probability of finding an electron is large. The shapes of *s* and *p* orbitals are shown in Fig. 1.21. There is a finite, but very small, probability of finding an electron at greater distances from the nucleus. The volumes that we typically use to illustrate an orbital are those volumes that would contain the electron 90 to 95% of the time.

Both the 1*s* and 2*s* orbitals (as are all higher *s* orbitals) are spheres. The sign of the wave function, ψ_{1s}, is positive (+) over the entire 1*s* orbital. The 2*s* orbital contains a nodal surface, that is, an area where $\psi = 0$. In the inner portion of the 2*s* orbital, ψ_{2s} is negative.

The 2*p* orbitals have the shape of two almost-touching spheres. The phase sign of the wave function, ψ_{2p}, is positive in one lobe (or sphere) and negative in the other. A nodal plane separates the two lobes of a *p* orbital, and the three *p* orbitals are arranged in space so that their axes are mutually perpendicular.

You should not associate the sign of the wave function with anything having to do with electrical charge. As we said earlier the (+) and (−) signs associated with ψ are simply the arithmetic signs of the wave function in that region of space. The (+) and (−) signs do not imply a greater or lesser probability of finding an electron either. The probability of finding an electron is ψ^2, and ψ^2 is always positive. (Squaring a negative number always makes it positive.) Thus the probability of finding the electron in the (−) lobe of a *p* orbital is the same as that of the (+) lobe. The significance of the (+) and (−) signs will become clear later

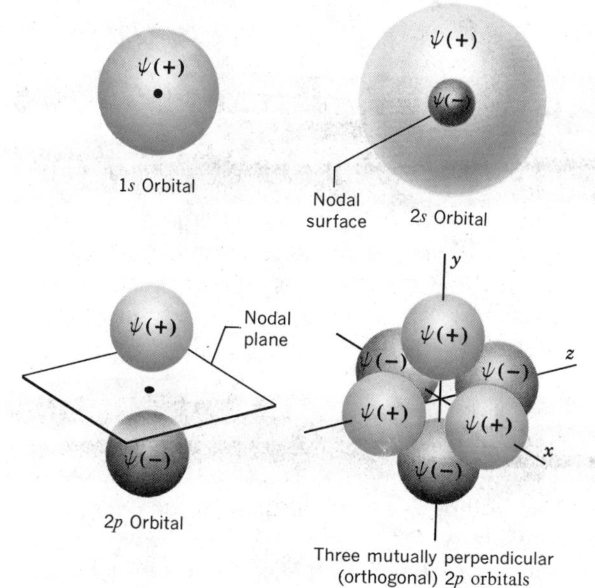

FIG. 1.21
The shapes of some s *and* p *orbitals.*

when we see how atomic orbitals combine to form molecular orbitals and when we see how covalent bonds are formed.

> There is a relationship between the number of nodes of an orbital and its energy: *the greater the number of nodes, the greater the energy*. We can see an example here; the 2*s* and 2*p* orbitals have one node each and they have greater energy than a 1*s* orbital, which has no nodes.

The relative energies of the orbitals are shown in Fig. 1.22. Electrons in 1*s* orbitals have the lowest energy because they are closest to the positive nucleus. Electrons in 2*s* orbitals are next lowest in energy. Electrons of 2*p* orbitals have equal but still higher energy. (Orbitals of equal energy are said to be *degenerate orbitals*.)

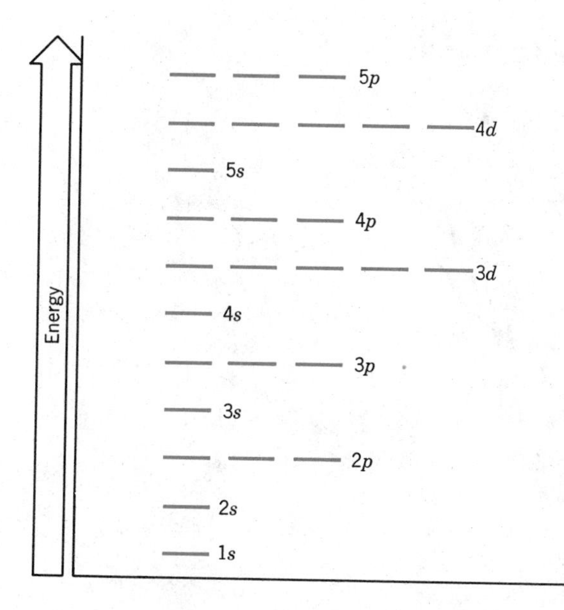

FIG. 1.22
The relative energy levels of some atomic orbitals for elements above hydrogen.

We can use Fig. 1.22 to arrive at the electronic configuration of any atom we are likely to encounter. We need only follow a few simple rules.

1. The *aufbau* principle: orbitals are filled so that those of lowest energy are filled first. (*Aufbau* is German for "building up.")
2. The Pauli exclusion principle: a maximum of two electrons may be placed in each orbital *but only when the spins of the electrons are paired.* An electron spins about its own axis. For reasons that we cannot develop here, an electron is permitted only one or another of only two possible spin orientations. We usually show these orientations by arrows, either ↑ or ↓. Thus two spin-paired electrons would be designated ↑↓ ·. Unpaired electrons, which are not permitted in the same orbital, are designated ↑↑ (or ↓↓).
3. *Hund's rule:* when we come to orbitals of equal energy (degenerate orbitals) such as the three *p* orbitals, we add one electron to each *with their spins unpaired* until each of the degenerate orbitals contains one electron. Then we begin adding a second electron to each degenerate orbital so that the spins are paired.

If we apply these rules to some of the second row elements of the Periodic Table we get the results shown in Fig. 1.23.

Before we leave the subject of atomic orbitals we should say again that the electronic configurations of the atoms given above are obtained from solutions of the Schrödinger equation calculated for the hydrogen atom. The hydrogen atom consisting of only two particles—a proton and an electron—is the only atom for which an exact solution to the Schrödinger equation has been obtained. When more complex systems are considered, approximations have to be made in order to simplify the mathematics. Even when these approximations are made, however, quantum mechanics gives surprisingly accurate correlations with results obtained experimentally for many complex molecules. In less complex molecules (consisting of only three or four interacting particles), the correlation obtained between theory and experiment is remarkable.

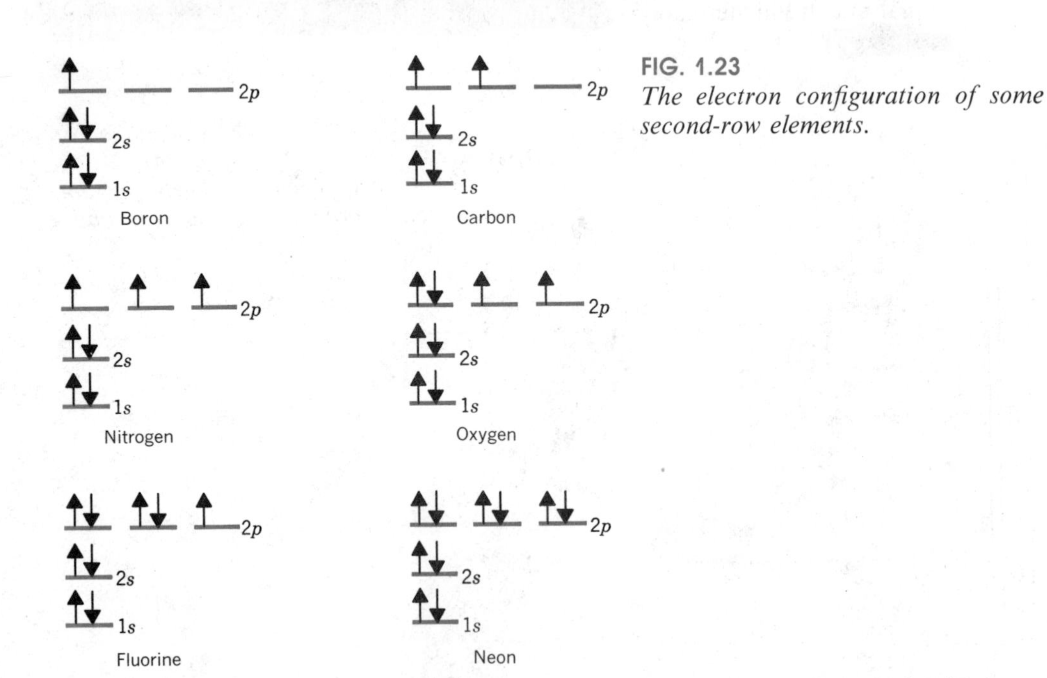

FIG. 1.23
The electron configuration of some second-row elements.

Problem 1.18

Taking into account the fact that electrons repel each other and thus prefer to stay as far apart as possible, provide an explanation for Hund's rule.

1.15 MOLECULAR ORBITALS

For the organic chemist the greatest utility of atomic orbitals is in using them to understand how atoms combine to form molecules. We will have much more to say about this subject in subsequent chapters for, as we have already said, covalent bonds are central to the study of organic chemistry. First, however, we will concern ourselves with a very simple case: the covalent bond that is formed when two hydrogen atoms combine to form a hydrogen molecule. We will see that the description of the formation of the H—H bond is the same, or at least very similar, to the description of bonds in more complex molecules.

Let us begin by examining what happens to the total energy of two hydrogen atoms with electrons of opposite spins when they are brought closer and closer together. This can best be shown with the curve shown in Fig. 1.24.

When the atoms of hydrogen are only a few Angstrom units apart (I) their total energy is simply that of two isolated hydrogen atoms. As the hydrogen atoms move closer together (II), each nucleus increasingly attracts the other's electron. This attraction more than compensates for the repulsive force between the two nuclei (or the two electrons), and the result of this attraction *is to lower the energy of the total system.* When the two nuclei are 0.74 Å (III) apart, the most stable (lowest energy) state is obtained. This distance, 0.74 Å, corresponds to the *bond length* for the hydrogen molecule. If the nuclei are moved closer together (IV) the repulsion of the two positively charged nuclei predominates, and the energy of the system rises.

Figure 1.24 shows us why hydrogen atoms do not remain in the atomic state very long when other hydrogen atoms are nearby. Hydrogen atoms will inevitably collide with each other, and as they do they will assume a state of lower energy by forming a hydrogen molecule. The extra energy originally possessed by the uncombined hydrogen atoms will be evolved as heat when the atoms combine to form molecules.

There is one serious problem with this explanation of bond formation. We have assumed that the electrons are essentially motionless and that as the nuclei

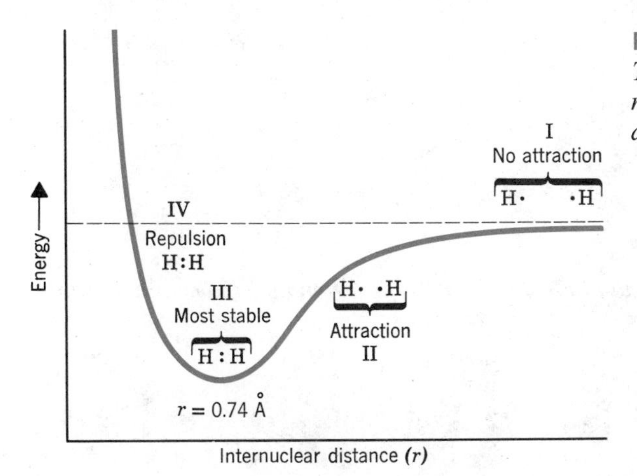

FIG. 1.24
The potential energy of the hydrogen molecule as a function of internuclear distance.

come together they will be stationary in the region between the two nuclei. Electrons do not behave that way. Electrons move about, and according to the *Heisenberg uncertainty principle,* we cannot know simultaneously the position and momentum of an electron. That is, we cannot pin the electrons down as precisely as our explanation suggests.

We avoid this problem when we use quantum mechanics, because we describe the electron in terms of wave functions (ψ) and in terms of probabilities (ψ^2) of finding it at particular places. By treating the electron this way we do not violate the uncertainty principle, because we do not talk about where the electron is precisely. We talk instead about where the *electron probability density* is large or small.

Thus a better explanation for what happens when two hydrogen atoms combine to form a hydrogen molecule is the following: as the hydrogen atoms approach each other their 1*s* orbitals (ψ_{1s}) begin to overlap. As the atoms move closer together orbital overlap increases until the *atomic orbitals* combine to become *molecular orbitals*. The molecular orbitals that are formed encompass both nuclei and, in them, the electrons can move about both nuclei. They are not restricted to the vicinity of one nucleus or the other as they were in the separate atomic orbitals. Molecular orbitals are like atomic orbitals, *they may contain a maximum of two spin-paired electrons.*

When atomic orbitals combine to form molecular orbitals *the number of molecular orbitals that result always equals the number of atomic orbitals that combine*. Thus in the formation of a hydrogen molecule the *two* atomic orbitals combine to produce *two* molecular orbitals. Two orbitals result because the mathematical properties of wave functions permit them to be combined by either *addition* or *subtraction*. That is, they can combine either *in* or *out of* phase. What are the natures of these new molecular orbitals?

One molecular orbital, called the *bonding molecular orbital* (ψ_{molec}), is formed when the atomic orbitals combine in the way shown in Fig. 1.25. Here atomic orbitals combine by *addition,* and this means that atomic *orbitals of the same phase sign overlap*. Such overlap leads to *reinforcement* of the wave function in the region between the two nuclei. Reinforcement of the wave function not only means that the value of ψ is larger between the two nuclei (recall what happens with a water wave), it means that ψ^2 is larger as well. Moreover, since ψ^2 expresses the probability of finding an electron in this region of space, we can now understand how orbital overlap of this kind leads to bonding. It does so by increasing the electron probability density in exactly the right place—in the region of space between the nuclei. When the electron density is large here, the attractive force of

FIG. 1.25
The overlapping of two hydrogen 1s atomic orbitals to form a bonding molecular orbital.

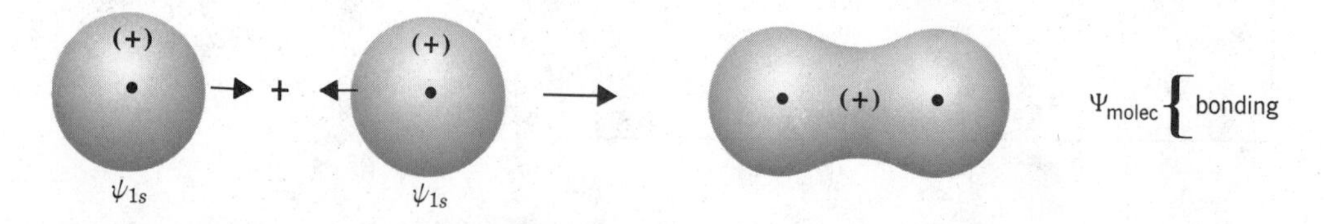

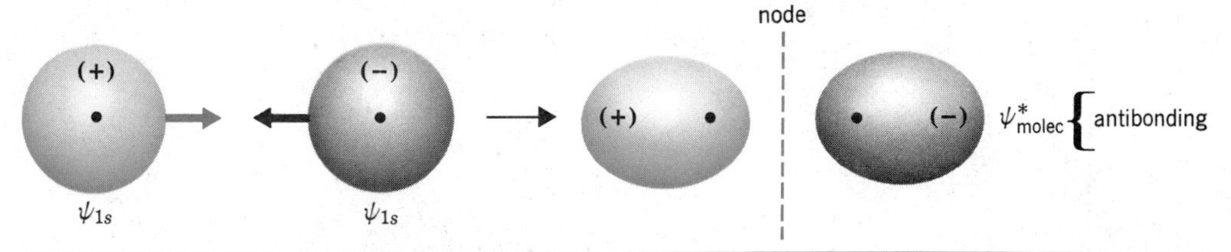

FIG. 1.26
The overlapping of two hydrogen 1s atomic orbitals to form an antibonding molecular orbital.

the nuclei for the electrons more than offsets the repulsive force acting between the two nuclei (and between the two electrons). This extra attractive force is, of course, the "glue" that holds the atoms together.

The second molecular orbital, called the *antibonding molecular orbital* (ψ^*_{molec}) is formed by subtraction in the way shown in Fig. 1.26. [Subtraction means that the phase sign of one orbital has been changed from (+) to (−).] Here, because *orbitals of opposite signs overlap* the wave functions *interfere* with each other in the region between the two nuclei and a node is produced. At the node $\psi = 0$, and on either side of the node ψ is small. This means that in the region between the nuclei ψ^2 is also small. Thus if electrons were to occupy the antibonding orbital the electrons would avoid the region between the nuclei. There would be only a small attractive force of the nuclei for the electrons. Repulsive forces (between the two nuclei and between the two electrons) would be greater than the attractive forces. Having electrons in the antibonding orbital would not tend to hold the atoms together, it would tend to make them fly apart.

What we have just described has its counterpart in a mathematical treatment called the LCAO method. LCAO stands for linear combination of atomic orbitals. In the LCAO treatment, wave functions for the atomic orbitals are combined in a linear fashion (by addition or subtraction) in order to obtain new wave functions for the molecular orbitals.

Molecular orbitals, like atomic orbitals, correspond to particular energy states for an electron. Calculations show that the relative energy of an electron in the bonding molecular orbital of the hydrogen molecule is substantially less than its energy in a ψ_{1s} atomic orbital. These calculations also show that the energy of an electron in the antibonding molecular orbital is substantially greater than its energy in a ψ_{1s} atomic orbital.

An energy diagram for the molecular orbitals of hydrogen is shown in Fig. 1.27. Notice that electrons are placed in molecular orbitals in the same way that they were in atomic orbitals. Two electrons (with their spins opposed) occupy the bonding molecular orbital, where their total energy is less than in the separate atomic orbitals. This is the *lowest electronic energy state* or *ground state* of the

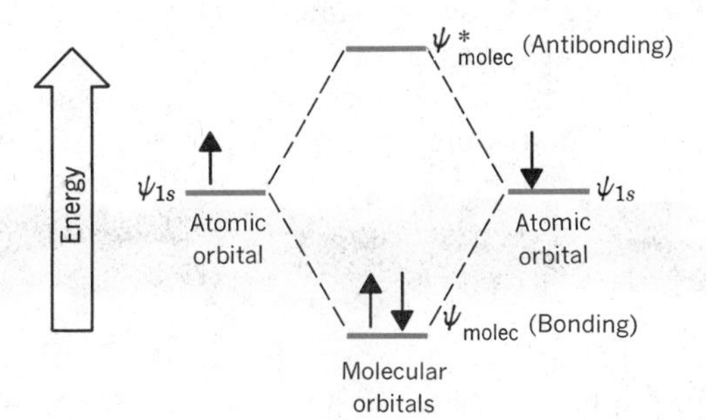

FIG. 1.27
Energy diagram for the hydrogen molecule. Combination of two atomic orbitals. ψ_{1s} gives two molecular orbitals. ψ_{molec} and ψ^*_{molec}. The energy of ψ_{molec} is lower than that of the separate atomic orbitals, and in the lowest electronic energy state of molecular hydrogen it contains both electrons.

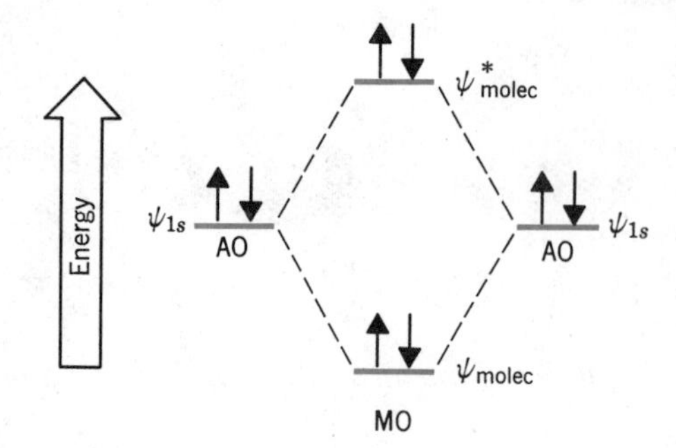

FIG. 1.28
Energy diagram for the formation of a hypothetical He_2 molecule. (AO = atomic orbital; MO = molecular orbital.)

hydrogen molecule. (An electron may occupy the antibonding orbital in what is called an *excited state* for the molecule. This state forms when the molecule in the ground state absorbs a photon of light of proper energy.)

Now that we have seen an explanation for how the reaction $H\cdot + H\cdot \longrightarrow H_2$ occurs, we can also explain why the reaction $He: + He: \longrightarrow He_2$ does not occur. We begin by assuming that the molecular orbitals for He_2 are similar to those for H_2. With electrons included, the energy diagram for the formation of He_2 is given in Fig. 1.28. In the case of He_2 the bonding orbital is fully occupied just as it is in the case of H_2; but in He_2 the antibonding orbital is also fully occupied. The two electrons in the antibonding orbital produce strong repulsive forces. Detailed calculations for He_2 show that the stability gained from the electrons in the bonding orbital is insufficient to counterbalance the instability that arises from the electrons in the antibonding orbital. As a result, He_2 is not predicted to form from separate helium atoms. The theory, thus agrees with what experiments show to be true. Helium exists as a monoatomic species, He, not as He_2.

At this point it may seem that these quantum mechanically based ideas are unnecessarily complicated and that the much simpler Lewis theory would do just as well. For certain features of simple molecules this is true. As we go on to more complicated molecules, however, we will find that a great many features of their structures and reactions require more detailed explanations than the Lewis theory permits. For many of these we will find the molecular orbital treatment to be invaluable.

1.16 ORBITAL HYBRIDIZATION AND THE STRUCTURE OF METHANE: sp^3 HYBRIDIZATION

The *s* and *p* orbitals used in the quantum mechanical description of the carbon atom, given in Section 1.13, were based on calculations for hydrogen atoms. These simple *s* and *p* orbitals do not, when taken alone, provide a satisfactory explanation for the *tetravalent-tetrahedral* carbon of methane. However, a satisfactory description of methane's structure that is based on quantum mechanics *can* be obtained through an approach called *orbital hybridization.* Orbital hybridization, in its simplest terms, is nothing more than a mathematical approach that involves the combining of individual wave functions for *s* and *p* orbitals to obtain wave functions for new orbitals. The new orbitals have, *in varying proportions,* the properties of the original orbitals taken separately. These new orbitals are called *hybrid orbitals.*

To see why the idea of orbital hybridization is necessary let us begin with two "thought" experiments. In the first of these we shall attempt to give a satis-

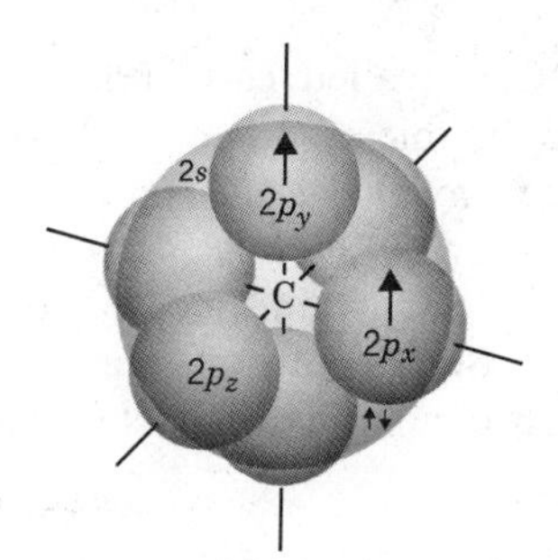

FIG. 1.29
The electronic configuration of a ground state carbon atom. The p *orbitals are designated* $2p_x$, $2p_y$, *and* $2p_z$ *to indicate their respective orientations along the* x, y, *and* z *axes. The assignment of the unpaired electrons to the* $2p_y$ *and* $2p_x$ *orbitals is arbitrary. They could also have been placed in the* $2p_x$ *and* $2p_z$ *or* $2p_y$ *and* $2p_z$ *orbitals. (To have placed them both in the same orbital would not have been correct, however, for this would have violated Hund's rule.) (Section 1.14)*

factory description of the structure of methane using four hydrogen atoms *and a carbon atom in its lowest energy state* (*or ground state*). In the second thought experiment we shall use, instead, a carbon atom *in an excited state* (i.e., a state in which one electron is promoted to a higher orbital). In both instances we shall fail to account satisfactorily for methane's structure and thus demonstrate the need for some new idea—for orbital hybridization.

We begin with the thought experiment using carbon in its ground state. According to quantum mechanical calculations the electronic configuration of the ground state of carbon should be that given below.

C $\frac{\uparrow\downarrow}{1s}$ $\frac{\uparrow\downarrow}{2s}$ $\frac{\uparrow}{2p}$ $\frac{\uparrow}{2p}$ $\frac{\quad}{2p}$

Ground state carbon

The valence electrons of carbon (those used in bonding) are those of the *outer level,* that is, the $2s$ and $2p$ electrons. If we recall the shapes of these orbitals, we see that a picture of the valence electrons of a carbon atom in its ground state might resemble that given in Fig. 1.29.

The formation of the covalent bonds of methane *from individual atoms* requires that the carbon atom overlap its orbitals containing *single electrons* with $1s$ orbitals of hydrogen atoms (which also contain a single electron). If a ground-state carbon atom were to combine with hydrogen atoms in this way, the result would be that depicted in Fig. 1.30. *Only two carbon-hydrogen bonds would be formed, and these would be at right angles to each other.*

FIG. 1.30
The Hypothetical formation of CH_2 *from a carbon atom in its ground state.*

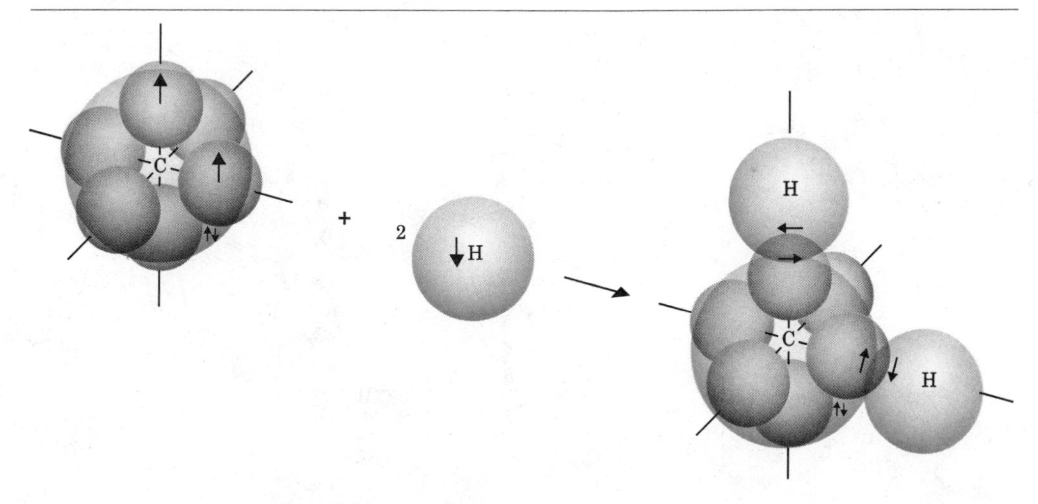

The molecule that we have described here, CH_2 (called carbene), has been detected, but only as a highly reactive substance capable of momentary existence. Clearly, then, a carbon atom in its ground state is *neither tetravalent nor tetrahedral,* and thus, it does not offer a proper model for the carbon of methane.

What, then, of carbon in an excited state? Let us imagine one electron being promoted from the $2s$ orbital to the $2p_z$ orbital. Such a carbon atom could be tetravalent, because all four orbitals would each contain a single electron.

C*	↑↓	↑	↑	↑	↑
	$1s$	$2s$	$2p_x$	$2p_y$	$2p_z$

Excited-state carbon

An excited-state carbon atom might combine with four hydrogen atoms as shown in Fig. 1.31.

The promotion of an electron from the $2s$ orbital to the $2p_z$ orbital requires energy. The amount of energy required has been determined and is equal to 96 kcal/mole. This expenditure of energy can be rationalized by arguing that the energy released when two additional covalent bonds form would more than compensate for that required to excite the electron. No doubt this is true, but it solves only one problem. The problems that cannot be solved by using an excited-state carbon as a basis for a model of methane are the problems of the carbon-hydrogen bond angles and the apparent equivalence of all four carbon-hydrogen bonds. Three of the hydrogens—those overlapping their $1s$ orbitals with the three p orbitals—would, in this model, be at angles of 90° with respect to each other; the fourth hydrogen, the one overlapping its $1s$ orbital with the $2s$ orbital of carbon, would be at some other angle, probably as far from the other bonds as the confines of the molecule would allow. Basing our model of methane on this excited state of carbon gives us a carbon that is tetravalent *but one that is not tetrahedral,* and it predicts a structure for methane in which one carbon-hydrogen bond differs from the other three.

We see then, that still another model is needed to account for the tetrahe-

FIG. 1.31
The hypothetical formation of CH_4 from an excited-state carbon.

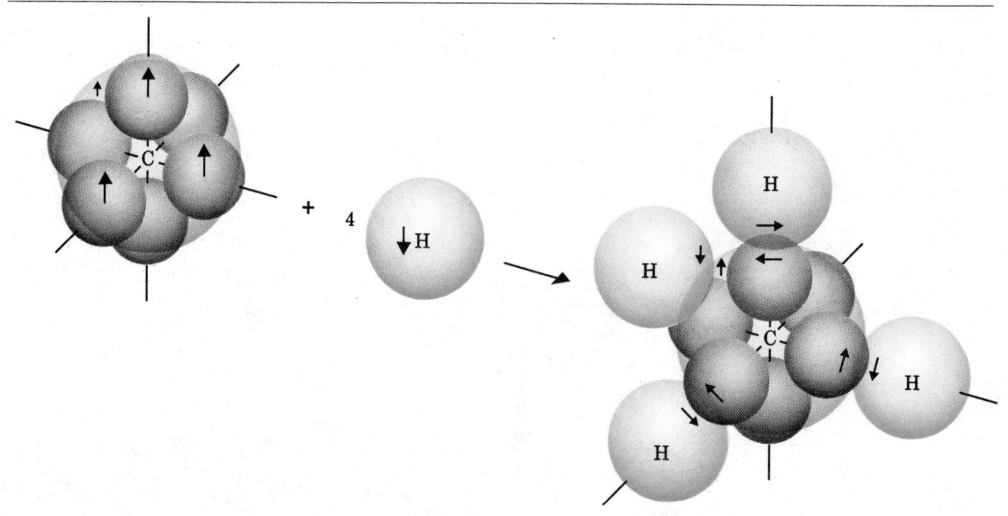

dral bond angles of methane and its four equivalent hydrogens. This model *can be provided by the theory of orbital hybridization.*

Hybrid orbitals that account for methane's structure can be obtained by combining the wave functions of the 2*s* orbital of carbon with those of the three 2*p* orbitals. The mathematical process of hybridization can be approximated by the illustration that is shown in Fig. 1.32.

In this process, four orbitals are mixed—or hybridized—and four new hybrid orbitals are obtained. The hybrid orbitals are called sp^3 orbitals to indicate that they have one part the character of an *s* orbital and three parts the character of a *p* orbital. The mathematical treatment of orbital hybridization also shows that *the four* sp^3 *orbitals should be oriented at angles of 109.5° with respect to each other.* This is, of course, what we would predict on the basis of the VSEPR method (Section 1.10) and it is precisely the spatial orientation of the four hydrogen atoms of methane.

If, in our imagination, we visualize the formation of methane from an sp^3-hybridized carbon atom and four hydrogen atoms, the process might be like that shown in Fig. 1.33. We see that an sp^3-hybridized carbon gives a *tetrahedral structure for methane, and one with four equivalent C—H bonds.*

In addition to accounting properly for the shape of methane, the theory of orbital hybridization also explains the very strong bonds that are formed between

FIG. 1.32
Hybridization of the atomic orbitals of carbon to produce sp^3*-hybrid orbitals.*

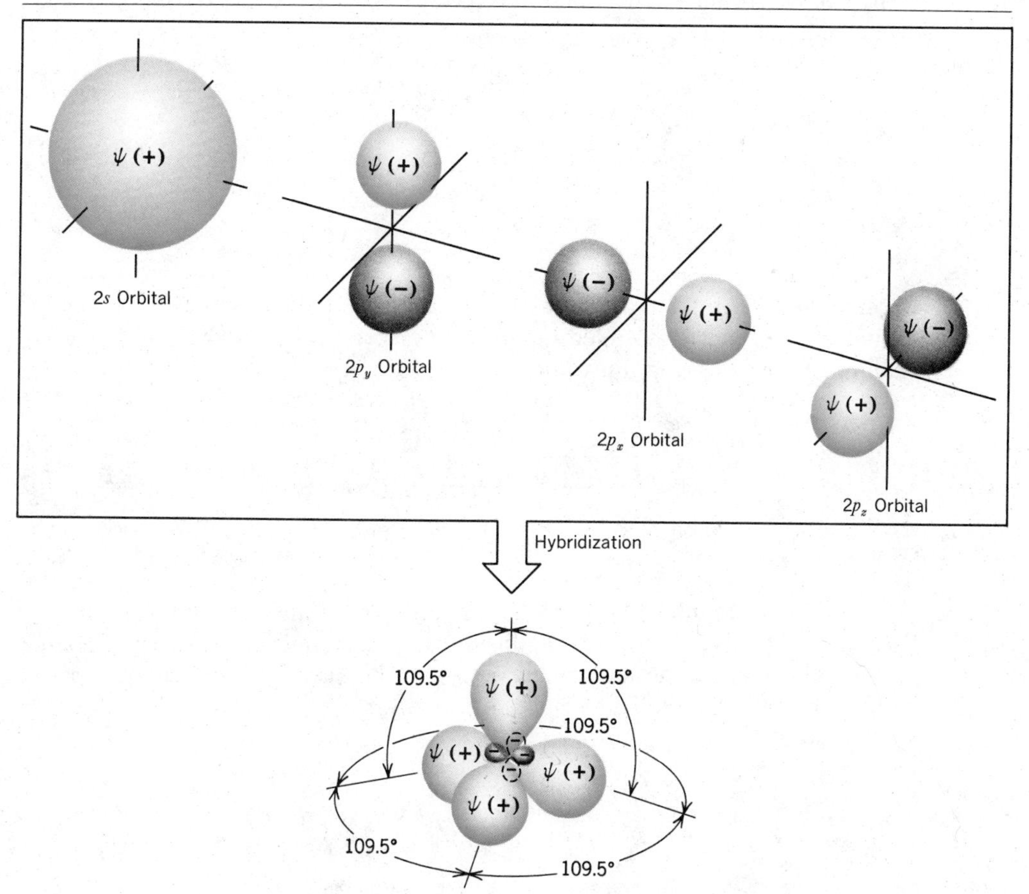

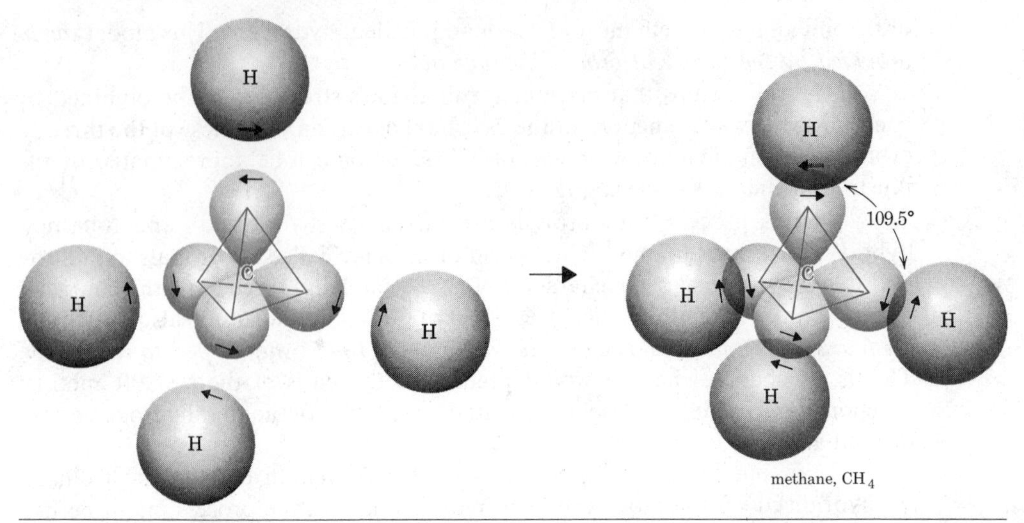

FIG. 1.33
The formation of methane from an sp^3*-hybridized carbon atom. [In orbital hybridization we combine orbitals* not *electrons. The electrons can then be replaced in the hybrid orbitals as necessary for bond formation, but always in accordance with the Pauli principle of no more than two electrons (with opposite spin) in each orbital. In this illustration we have placed one electron in each of the hybrid carbon orbitals.]*

carbon and hydrogen. To see how this is so, consider the shape of the individual sp^3 orbital shown below:

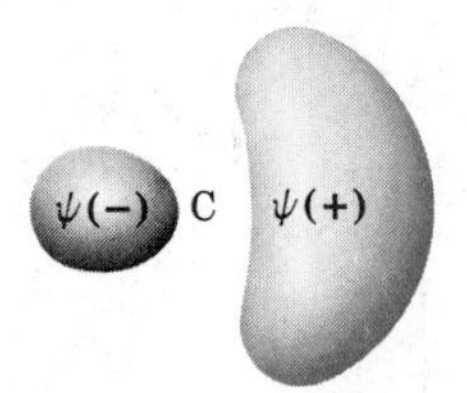

Because the sp^3 orbital has the character of a p orbital, the positive lobe of the sp^3 orbital is large and is extended quite far into space. We see why this is true by examining the following illustration.

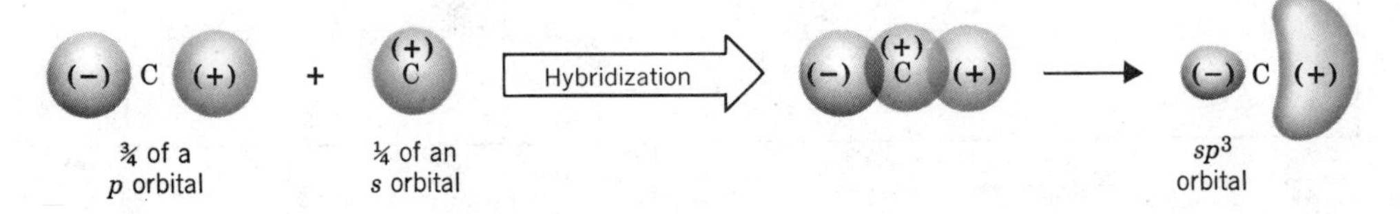

We see that on one side of the carbon atom the signs of the wave functions of both the s and p orbitals are positive. On this side the wave functions *add* to give a large *positive* lobe to the sp^3 orbital. On the other side the wave functions are of opposite signs and they *subtract*. This results in a small *negative* lobe.

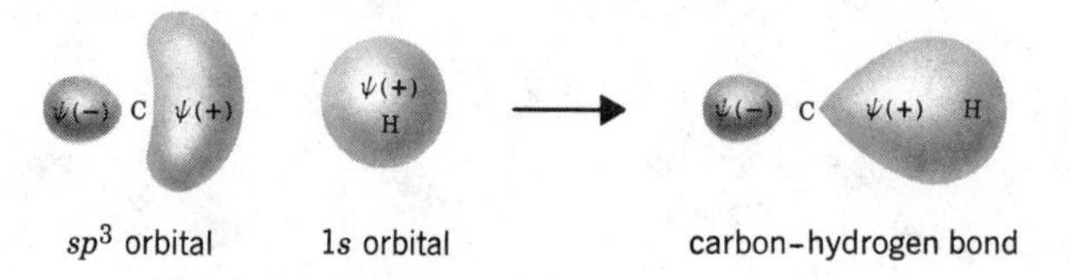

It is the positive lobe of the sp^3 orbital that overlaps with the positive 1*s* orbital of hydrogen to form a carbon-hydrogen bond.
Because the positive lobe of the sp^3 orbital is large and is extended into space, the overlap between it and the 1*s* orbital of hydrogen is also large, and the resulting carbon-hydrogen bond is quite strong.

The bond formed from the overlap of an sp^3 orbital and a 1*s* orbital is an example of a *sigma bond.* The term *sigma bond* is a general term applied to those bonds in which orbital overlap gives a bond that is *circularly symmetrical in cross section when viewed along the bond axis.*

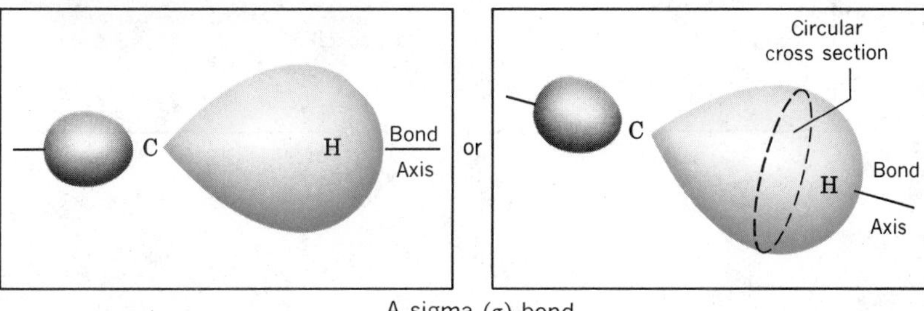

A sigma (σ) bond

1.17 ORBITAL HYBRIDIZATION AND THE STRUCTURE OF BORON TRIFLUORIDE: sp^2 HYBRIDIZATION

Boron trifluoride (Section 1.10) has a triangular (trigonal planar) shape with three equivalent boron-fluorine bonds. In its ground state boron has the following electronic configuration:

B ↑↓ 1*s* ↑↓ 2*s* ↑ 2*p* — 2*p* — 2*p*

Boron, ground state

Clearly, the *s* and *p* orbitals of the ground state will not account for the trivalent and triangular boron of BF_3.

Problem 1.19

(a) What valence would you expect a boron atom in its ground state to have?

(b) Consider an excited state of boron in which one 2*s* electron is promoted to a vacant 2*p* orbital. Show how this state of boron also fails to account for the structure of BF_3.

Once again we must resort to orbital hybridization. Here, however, we combine the 2*s* orbital with only two of the 2*p* orbitals. Mixing three orbitals as shown in Fig. 1.34 gives three equivalent hybrid orbitals and these orbitals are sp^2 orbitals. They have one part the character of an *s* orbital and two parts the character of a *p* orbital. Calculations show that these orbitals are pointed toward the corners of an equilateral triangle with angles of 120° between their axes. This, again is what we would expect on the basis of the VSEPR model. These orbitals,

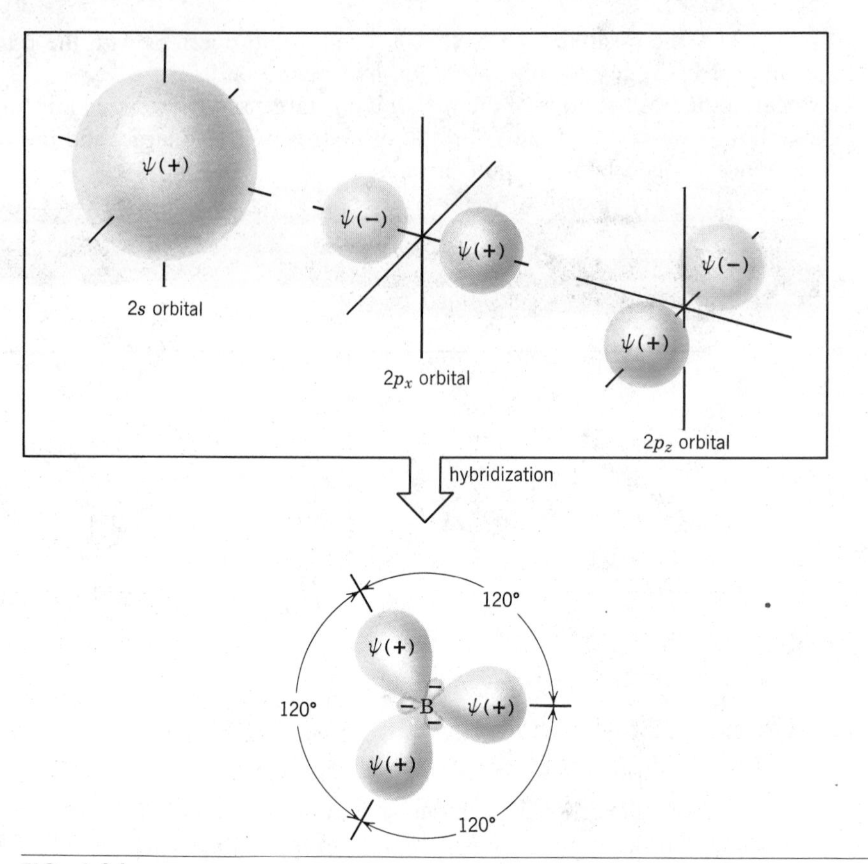

FIG. 1.34
Hybridization of one 2s *orbital and two* 2p *orbitals of boron to produce three* sp^2*-hybrid orbitals.*

then, are just what we need to account for the trivalent, triangular boron of boron trifluoride.

By placing one of the valence electrons in each of the three sp^2 orbitals and then allowing these orbitals to overlap with an orbital containing one electron from each of three fluorine atoms we obtain the structure shown in Fig. 1.35. Notice that the boron atom still has a vacant *p*-orbital, the one that we did not hybridize.

We will see in Section 6.3 that sp^2 hybridization occurs with carbon atoms that form double bonds.

FIG. 1.35
The formation of boron trifluoride from an sp^2*-hybridized boron atom and three fluorine atoms.* (For clarity, only one orbital is shown for fluorine.)

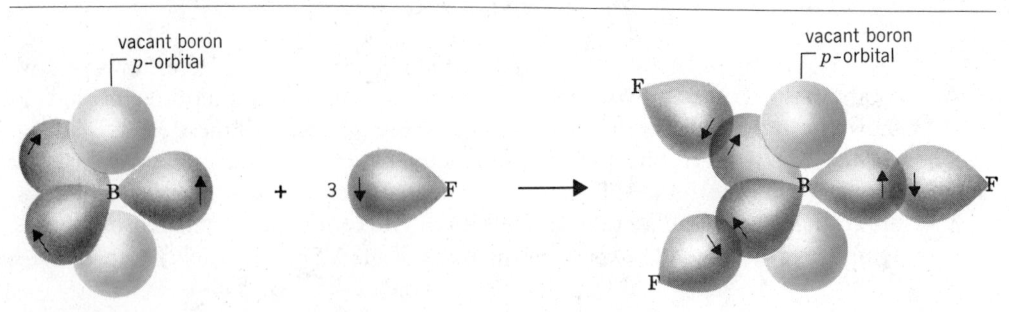

1.18 ORBITAL HYBRIDIZATION AND THE STRUCTURE OF BERYLLIUM HYDRIDE: *sp* HYBRIDIZATION

Beryllium hydride, BeH_2 is a linear molecule; the bond angle is 180°.

180°

H—Be—H

In its ground state beryllium has the following electronic configuration:

Be ↑↓ (1*s*) ↑↓ (2*s*) __ (2*p*) __ (2*p*) __ (2*p*)

In order to account for the structure of BeH_2 we again need orbital hybridization. Here, (Fig. 1.36) we hybridize one *s* orbital with one *p* orbital and obtain two *sp* orbitals. Calculations show that these *sp* orbitals are oriented at an angle of 180° again agreeing with what we would predict from the VSEPR model. The two *p* orbitals that were not mixed are vacant.

Beryllium can use these hybrid orbitals to form bonds to two hydrogen atoms in the way shown in Fig. 1.37.

Why does the beryllium atom of BeH_2 form *sp*-hybrid orbitals rather than the sp^2 or sp^3 orbitals (shown below)?

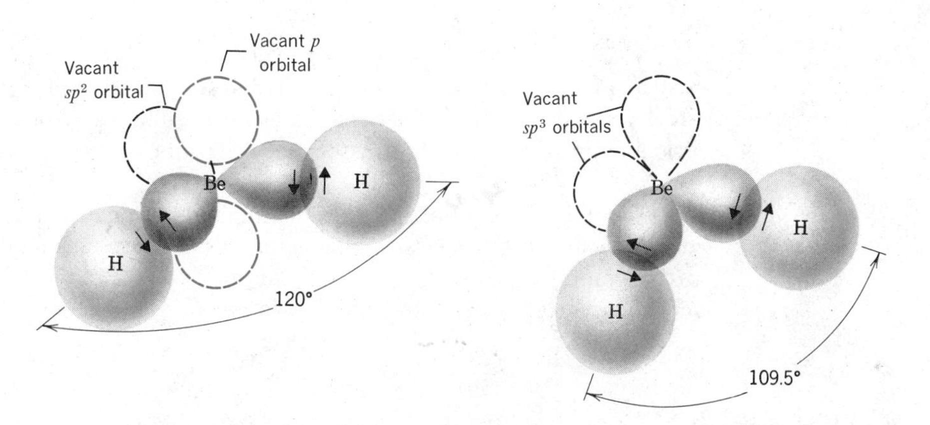

There are two reasons. Repulsions between the bonding electrons would be much greater in sp^2- or sp^3-hybridized BeH_2 because the bonds are close together. Moreover, stronger more stable bonds form when the fraction of *s*-character in an orbital increases. Thus the bonds of *sp*-hybridized BeH_2 would be strongest and most stable.

We will see in Section 9.3 that *sp* hybridization occurs with carbon atoms that form triple bonds.

1.19 A SUMMARY OF IMPORTANT CONCEPTS THAT COME FROM QUANTUM MECHANICS

1. An **atomic orbital** corresponds to a region of space about the nucleus of a single atom where there is a high probability of finding an electron. Atomic orbitals called *s*-orbitals are spherical; those called *p*-orbitals are like two almost-touching spheres. Atomic orbitals can

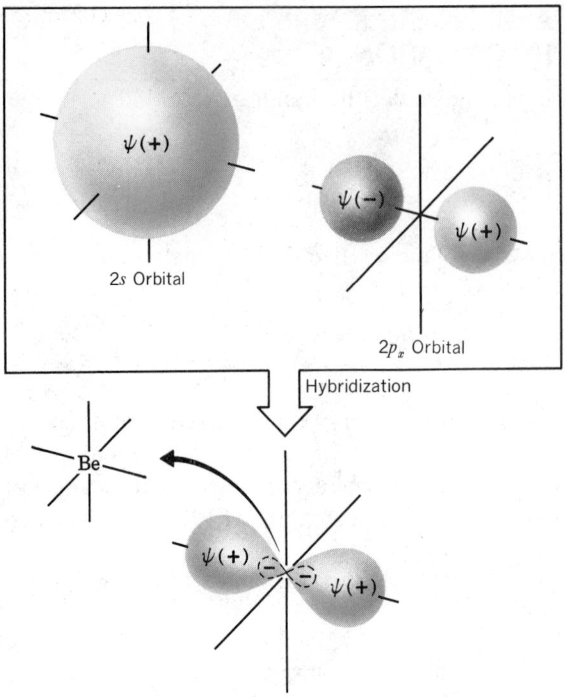

FIG. 1.36
Hybridization of one 2s orbital and one 2p orbital of beryllium to produce two sp-hybrid orbitals.

hold a maximum of two electrons when their spins are paired. Orbitals are described by a wave function, ψ, and each orbital has a characteristic energy. The phase signs associated with an orbital may be $(+)$ or $(-)$.

2. When atomic orbitals overlap, they combine to form **molecular orbitals.** Molecular orbitals correspond to regions of space encompassing two (or more) nuclei where electrons can move. Like atomic orbitals, molecular orbitals can hold up to two electrons if their spins are paired.

FIG 1.37
The formation of BeH_2 from an sp-hybridized beryllium atom and two hydrogen atoms.

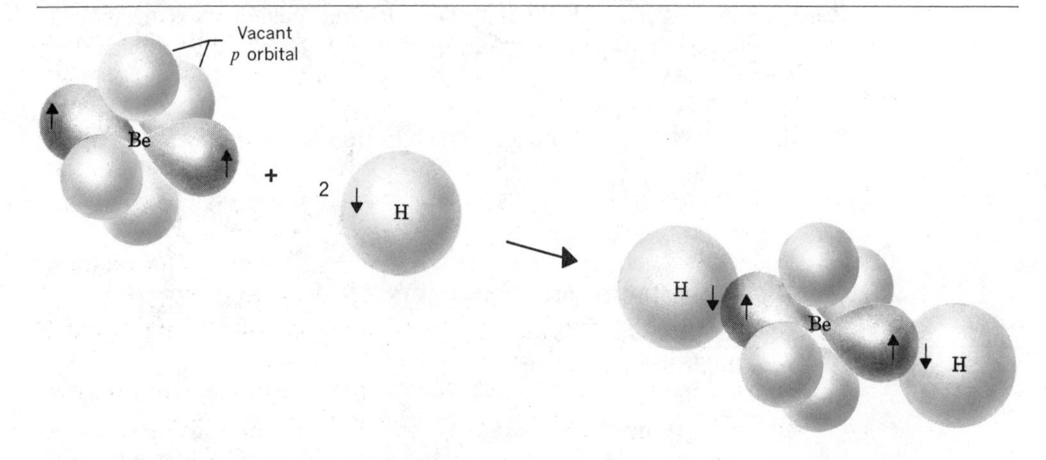

3. When atomic orbitals with the same phase sign interact they combine to form a **bonding molecular** orbital:

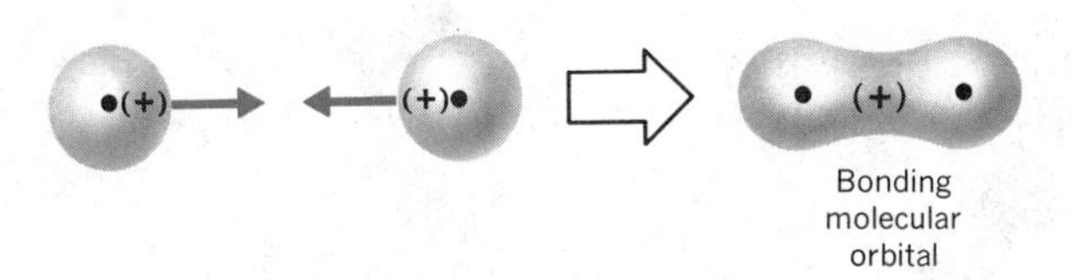

The electron probability density of a bonding molecular orbital is large in the region of space between the two nuclei where the negative electrons hold the positive nuclei together.

4. An **antibonding molecular orbital** forms when orbitals of opposite phase sign overlap.

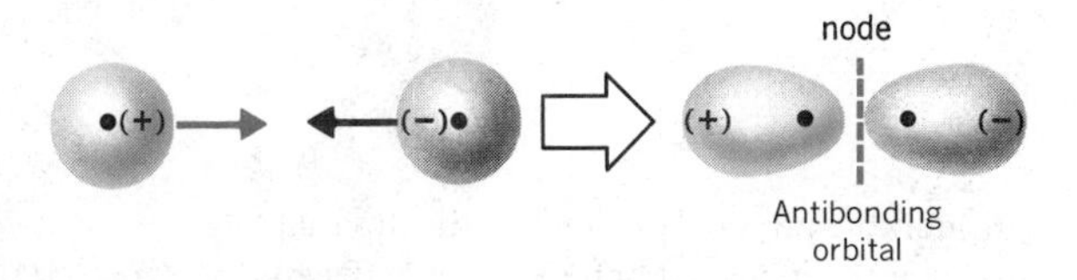

An antibonding orbital has higher energy than a bonding orbital. The electron probability density of the region between the nuclei is small and it contains a **node**—a region where $\psi = 0$. Thus, having electrons in an antibonding orbital does not help hold the nuclei together. The internuclear replusions tend to make them fly apart.

5. The **energy of electrons** in a bonding molecular orbital is less than the energy of the electrons in their separate atomic orbitals. The energy of electrons in an antibonding orbital is greater than that of electrons in their separate atomic orbitals.
6. The **number of molecular orbitals** always equals the number of atomic orbitals that combine. Combining two atomic orbitals will always yield two molecular orbitals—one bonding and one antibonding.
7. **Hybrid orbitals** form by mixing (hybridizing) the wave functions for orbitals of a different type (i.e., *s* orbitals and *p* orbitals) but from the same atom.
8. Hybridizing three *p* orbitals with one *s* orbital yields four ***sp*³ orbitals.** Atoms that are sp^3 hybridized direct the axes of their four sp^3 orbitals toward the corners of a tetrahedron. The carbon of methane is sp^3 hybridized and **tetrahedral.**
9. Hybridizing two *p* orbitals with one *s* orbital yields three ***sp*² orbitals.** Atoms that are sp^2 hybridized point the axes of three sp^2 orbitals toward the corners of an equilateral triangle. The boron of BF_3 is sp^2 hybridized and **triangular.**
10. Hybridizing one *p* orbital with one *s* orbital yields two ***sp* orbitals.** Atoms that are *sp* hybridized orient the axes of their two *sp* orbitals in opposite directions (at an angle of 180°). The beryllium atom of BeH_2 is *sp* hybridized and BeH_2 is a **linear** molecule.
11. A **sigma bond** is one in which the electron density has circular symmetry when viewed along the bond axis.

Additional Problems

1.20
Show an electron-dot formula, including any formal charge, for each of the following molecules.

(a) CH_3NCS (c) CH_3ONO_2 (e) CH_2CO (g) KNH_2
(b) CH_3CNO (d) CH_3NCO (f) CH_2N_2 (h) NaN_3

1.21
(a) Write out the electron configuration for each of the following atoms.
(b) Make a sketch of the atom showing the orbital arrangement, shape, and the disposition of the electrons in *s* and *p* orbitals.

(1) Be (3) C (5) O
(2) B (4) N

1.22
One way of writing the structure of the nitrate ion is shown below.

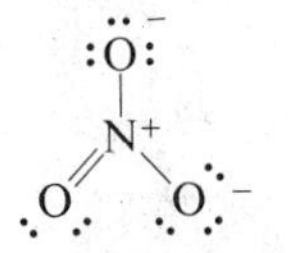

:Ö:⁻
|
N+
O O⁻

However, considerable physical evidence indicates that all three nitrogen-oxygen bonds are equivalent and that they have a bond distance between that expected for a nitrogen-oxygen single bond and a nitrogen-oxygen double bond. Explain this in terms of resonance theory.

1.23
Write two resonance structures for the nitrite ion, NO_2^-, and show the formal charge on each atom in each structure. Do these structures account for the fact that the nitrogen bonds are of equal length?

1.24
(a) Taking into account the shape of an ammonia molecule (Section 1.10), in what kind of orbital would you expect the unshared electron pair to be found? (b) What about the unshared electron pairs of a water molecule? Explain.

1.25
Multiplying dipole moments expressed in Debye units (Table 1.3) by 10^{-18} converts them to cgs units. The dipole moment of HCl, for example, is 1.08×10^{-18} esu-cm. We can assume that the distance *d* in the equation $\mu = e \times d$ is equal to the bond length; for HCl this is 1.27×10^{-8} cm. (a) What is the magnitude of the charge (in esu) on each atom in HCl? (b) Given that the electron charge is 4.8×10^{-10} esu, what fraction of an electron does the chlorine atom of the HCl have in excess?

1.26
Chloromethane, CH_3Cl, has a larger dipole moment ($\mu = 1.87$ D) than fluoromethane, CH_3F, ($\mu = 1.81$ D), even though fluorine is more electronegative than chlorine. Explain.

1.27
Cyanic acid, H—O—C≡N, and isocyanic acid, H—N=C=O, differ in the positions of their electrons but their structures do not represent resonance structures. (a) Explain. (b) Loss of a proton from cyanic acid yields the same anion as that obtained by loss of a proton from isocyanic acid. Explain.

1.28
Given that the four bonds of carbon in methane are tetrahedral, do the formulas written below represent different compounds? Explain your answer.

(a) $\mathrm{H{-}\overset{\displaystyle H}{\underset{\displaystyle Cl}{C}}{-}Cl}$ and $\mathrm{Cl{-}\overset{\displaystyle H}{\underset{\displaystyle H}{C}}{-}Cl}$

(b) $\mathrm{H{-}\overset{\displaystyle Br}{\underset{\displaystyle H}{C}}{-}Cl}$ and $\mathrm{H{-}\overset{\displaystyle Br}{\underset{\displaystyle Cl}{C}}{-}H}$

(c) $\mathrm{CH_3{-}\overset{\displaystyle CH_3}{\underset{\displaystyle H}{C}}{-}H}$ and $\mathrm{CH_3{-}\overset{\displaystyle H}{\underset{\displaystyle H}{C}}{-}CH_3}$

1.29

Although the molecule He_2 is unknown (and on the basis of our discussion in Section 1.15 we would not expect it to exist) the ion He_2^+ has been detected by spectroscopic techniques. What factor accounts for the greater stability of He_2^+?

1.30

Boron trifluoride reacts readily with ammonia to form a compound, BF_3NH_3. (a) What factors account for this reaction taking place so readily? (b) What formal charge is present on boron in the product? (c) On nitrogen? (d) What hybridization state would you expect for boron in the product? (e) For nitrogen?

1.31

Consider the substances, H_3O^+, H_2O, and OH^-; or NH_3 and NH_4^+; or H_2S and SH^- and describe the relationship between formal charge and acid strength.

1.32

Ozone, O_3, is found in the upper atmosphere where it absorbs highly energetic ultraviolet light and thus provides the surface of the earth with a protective screen (cf. Section 4.10). (a) Given that ozone molecules are not cyclic, and that all of the electrons are paired, write resonance structures for ozone. (b) Would you expect the two oxygen-oxygen bonds of ozone to be equivalent? (c) Given that ozone has a dipole moment ($\mu = 0.52$ D) what does this indicate about the shape of an ozone molecule? (d) Is this the shape you would predict on the basis of VSEPR theory? Explain.

1.33

In Problem 1.3 you wrote the Lewis structure for the methyl cation, CH_3^+. In Problem 1.8 you were able to predict its shape on the basis of VSEPR theory. Now describe the methyl cation in terms of orbital hybridization. (Pay special attention to the hybridization state of the carbon atom and be sure to include any vacant orbitals.).

A SPECIAL TOPIC

THE DEVELOPMENT OF QUANTUM MECHANICS

At the end of the last century, and early in this one, revolutionary ideas about the nature of matter and energy were advanced by Max Planck, J. J. Thomson, Ernest Rutherford, Albert Einstein, and Niels Bohr. These ideas were essential to the later development of quantum mechanics, therefore, we consider them first.

Planck's Contribution

Planck's contribution came as a consequence of his studies of the energy radiated from very hot but nonreflective objects called "black bodies." (Black only because they were not reflecting energy; actually they were red hot.) Planck found that the intensity and energy of radiation emitted from these objects at different wavelengths (or colors), could not be explained in terms of the theories that existed at that time. To explain his results, Planck made an imaginative proposal. He postulated that matter is not capable of emitting or absorbing energy (radiation) in continuous and infinitely divisible amounts but that *matter emits or absorbs energy in small bundles* (or packets) *called quanta.*

According to Planck's theory, the energy emitted by the hot objects is emitted by the very rapidly oscillating atoms on their surfaces. Moreover, the energy of these atoms is proportional to the frequency (number of oscillations per second) with which they oscillate, so that those atoms that oscillate very rapidly (with high frequency) have high energies, while those that oscillate very slowly (low frequency) have low energies. This relationship is expressed mathematically in Planck's famous equation:

$$E = h\nu$$

where E is the energy of the oscillation, ν is the frequency, and h is a proportionality constant known as Planck's constant. Most importantly, according to Planck, the energy of the oscillating atoms is *quantized.* As a result, atoms can only gain or lose energy in units (or packets) of $h\nu$. Planck's constant—called one quantum of action—has the value of 6.625×10^{-27} erg-sec and is apparently one of the most fundamental constants in the universe. It appears in all equations where the granular or quantized nature of matter and energy are expressed.

Thomson and Rutherford

Just before Planck's proposals were published in 1900, J. J. Thomson had discovered the electron as a fundamental constituent of matter. Later, in 1904, Thomson postulated a "pudding" model of the atom. According to this model, negative electrons were embedded like "plums" in a sphere of positive charge. Soon afterwards, however, Rutherford, as a result of his experiments with the scattering of alpha particles, postulated a model of the atom where the positive charge and most of the mass of the atom is intensely concentrated in a tiny nucleus. The electrons, Rutherford proposed, move in orbits about the nucleus much as the planets do around the sun.

Bohr's Model of the Atom

In 1912, Niels Bohr extended Rutherford's ideas and simultaneously introduced Planck's quantum conditions with a new model of the atom. According to Bohr, the small, massive, and positively charged nucleus is surrounded by electrons moving in circular orbits. The hydrogen atom in its lowest energy state, for example, would have its electron moving in the orbit closest to the nucleus. Moreover, according to the Bohr model, the electron can be excited to higher orbits by absorbing one or more quanta of energy. By losing energy, again in units of $h\nu$, the electron can fall back to lower orbits. The position of the electron is restricted to the discrete orbits and, according to Bohr, can not assume an energy state in between. This *quantized* state of the atom, introduced by Bohr quite arbitrarily was, at the time, a brilliant step forward. Bohr's mathematical treatment of the

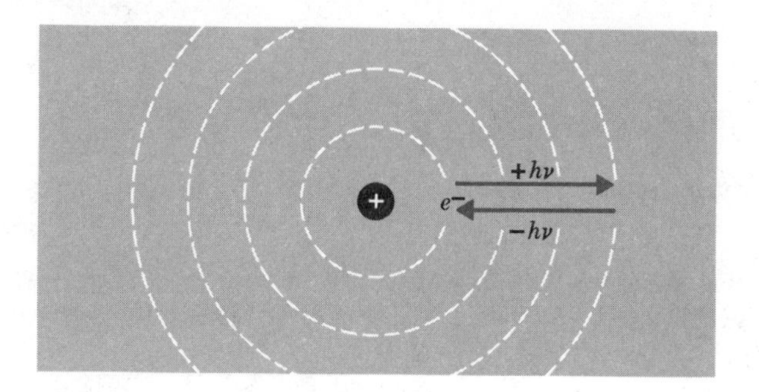

hydrogen atom, using his model as a basis, explained for the first time the spectrum of atomic hydrogen.

As time passed, however, and further experiments were done, it became clear that the Bohr model was inadequate. The Bohr treatment was successful in explaining only the spectrum of one-electron systems, and although many attempts were made to refine the Bohr model (including A. Sommerfeld's introduction of elliptical orbits), by 1920 it became clear that a new model was necessary.

Before we discuss the next steps in the development of quantum mechanics we must return to a slightly earlier time and relate an important new theory about the nature of light that had been advanced by Einstein.

Einstein and Light Photons

For almost 100 years, prior to 1905, scientists had conceived of light as a *wave phenomenon.* All of the properties of light that were known in the nineteenth century could be explained by assuming that light and other forms of electromagnetic radiation were waves occurring simultaneously in perpendicular electrical and magnetic fields. Properties of light, such as reflection and diffraction, were easily explained on that basis. Diffraction, the bending of light waves as they pass sharp edges (accounting for the fact that shadows have fuzzy edges rather than sharp ones), can be explained *only on the basis of the wave nature of light.* A new phenomenon, known as the photoelectric effect, had been discovered, however, and this new property of light, associated with light's ability to cause electrons to be emitted from the surfaces of metals, could *not* be accounted for on the basis of light's wave nature. An explanation for the photoelectric effect required that light have the nature of a particle as well. In 1905, Albert Einstein proposed that light, too, was *quantized* and that light had *both the properties of a wave and a particle.* According to this new conception we observe light's wave character when we observe diffraction; we observe its particlelike or granular nature when we observe the photoelectric effect. Einstein called these wave particles of light *photons.*

We can aid our subsequent discussion if we introduce, at this point, some of the mathematical relationships used to describe waves.

A simple wave (Fig. A.1) can be described in terms of its wavelength, λ, and its frequency, ν. The distance between two consecutive crests (or troughs) is the *wavelength;* the number of full cycles (wavelengths) that the wave completes each second, as it is propagated, is the *frequency.* Frequency has the units of cycles/sec or Hertz (Hz). Since the wave (consider a wave moving across a pond or in a long, plucked string) moves one wavelength in each cycle, the product of the wavelength, λ, and frequency, ν, gives the

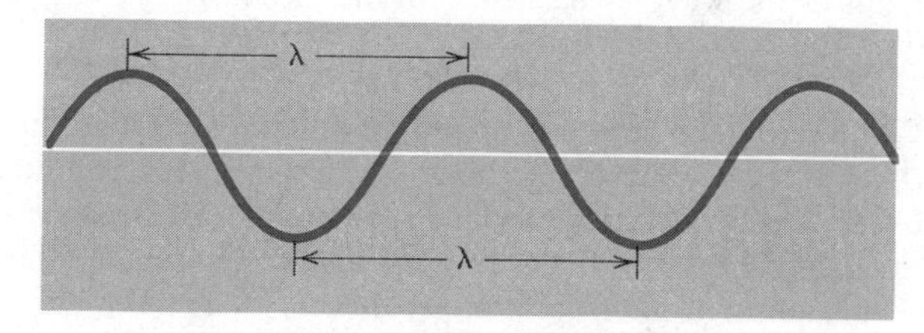

FIG. A.1
A simple wave.

wave velocity, v. Thus:

$$v = \lambda \cdot \nu$$

$$v\left(\text{wave velocity in } \frac{\text{cm}}{\text{sec}}\right) = \lambda\left(\text{wavelength in } \frac{\text{cm}}{\text{cycle}}\right) \cdot \nu\left(\text{frequency in } \frac{\text{cycle}}{\text{sec}}\right)$$

According to Einstein's theory of relativity, the velocity of a light wave in a vacuum is constant, and is equal to 3×10^{10} cm/sec. The velocity of light is usually abbreviated c, thus for a light wave:

$$c = \lambda\nu$$

Moreover, according to Einstein, light has associated with it the property of momentum because of its particlelike nature. This momentum can be expressed by the following formula:

$$\text{Momentum} = \frac{h\nu}{c} = \frac{h}{\lambda}$$

where once again, h is Planck's constant.

The energy of a photon of light, in Einstein's theory, is the product of Planck's constant and its frequency:

$$E = h\nu = \frac{hc}{\lambda}$$

Thus a photon of light of high frequency and short wavelength (blue light would be an example) will have greater energy than light of lower frequency and longer wavelength (i.e., red light).

De Broglie's Contribution

We now return to 1923, and with these simple mathematical relationships about light in mind, we can fully appreciate the bold proposal about the nature of the electron that was made by a young French graduate student, Louis de Broglie.

De Broglie was familiar with the insights gained by Planck and Einstein; he apparently asked himself the following question. If light—which until 1905 had been conceived of as purely a wave—has associated with it the properties of a particle, could the electron—explained in 1923 as being purely a particle—have associated with it *the properties of a wave?* De Broglie then proposed that the answer to this question was *yes,* and that the relationship between the wave and particle properties of the electron is expressed in the following way:

$$\text{Momentum } (mv_e) = \frac{h\nu}{v} = \frac{h}{\lambda}$$

where m = the mass of the electron
v_e = the velocity of the electron
h = Planck's constant
ν = the frequency of the wave associated with the electron
λ = the wavelength of the wave associated with the electron
and v = the velocity of the associated wave

De Broglie's hypothesis was seen, at the time, as being so wild that he almost missed getting his degree. A few years later, however, the wave nature of the electron was confirmed when Davisson and Germer observed the diffraction of a beam of electrons. De Broglie's flight of the imagination set the stage for the last step in the development of quantum mechanics and with it our modern understanding of atoms and molecules.

Schrödinger and the Quantum Mechanical Description of the Atom

This last act was performed independently and almost simultaneously by three men: Erwin Schrödinger, Werner Heisenberg, and Paul Dirac. Because Schrödinger's treatment of the atom is the most easily visualized, we will deal with it alone.

Schrödinger saw in de Broglie's hypothesis of the wave nature of the electron a way to introduce a quantized model for the atom quite naturally. (In Bohr's model, you will recall, the quantized state of the electron had been introduced arbitrarily.) Schrödinger knew that waves of a particular type, called "standing waves," were quantized, that is, in the mathematical equations that describe standing waves, a series of integers (whole numbers) inevitably appear.

To see what we mean by this, let us consider some very simple standing waves—those that move in only one dimension. Such standing waves would occur in the plucked string of a violin, or any string where the ends are fixed and cannot move. The kind of waves that occur in a string of length l are shown below:

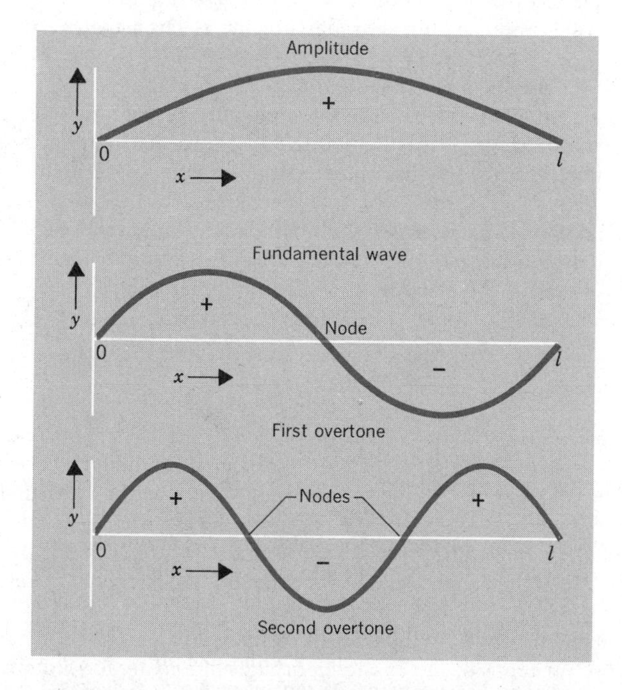

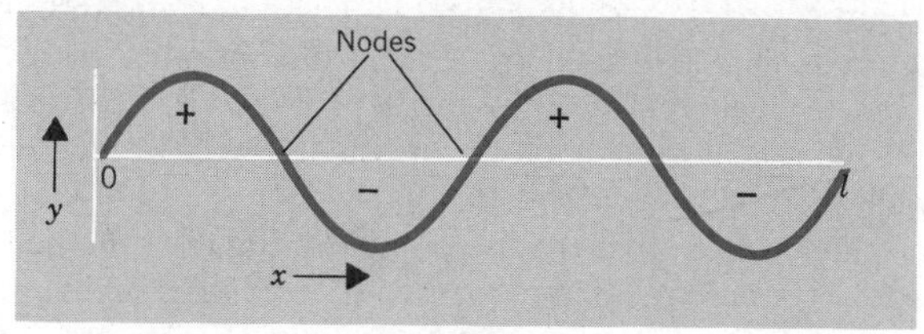

Third overtone

The height of the string along the y axis, as it moves periodically up and down, is called the *amplitude* of the wave. In the first, second, third, and higher overtones, there are parts of the string that do not move at all. These are called *nodes*. The sign of the amplitude, y, is positive (+) above the midpoint of the wave and negative (−) below it.

The mathematical equation that describes all of these waves is

$$y_n = a \sin n\pi(x/l)$$

where a is a constant and $n = 1$ for the fundamental, $n = 2$ for the first overtone, $n = 3$ for the second overtone, $n = 4$ for the third overtone, and so on. The fact that n is always a whole number and the fact that only certain vibrations occur allow us to say that the standing waves are quantized.

Standing waves occurring in two dimensions (as in the head of a vibrating drum) are described by somewhat more complicated equations, but the basic properties of such waves are the same as those in a plucked string.

In order to describe the motion of the electron in an atom in terms of the standing waves associated with it, Schrödinger had to find wave equations describing standing waves occurring in three dimensions. Fortunately for him, these equations had already been written.

One hundred years earlier, William Hamilton, a mathematical genius, had considered the problem of the standing waves that might occur on a planet uniformly flooded with water. While, at the time, these efforts by Hamilton must have appeared somewhat useless in any practical sense, it turned out that the equations that Hamilton developed for his flooded planet were precisely those that Schrödinger needed for the atom. Indeed, three-dimensional plots of the solutions obtained by Hamilton for his flooded planet resemble very closely the *s, p,* and *d* orbitals for the electron that we use so commonly today.

Schrödinger used a wave function, ψ, to describe the electron wave. Psi, ψ, is a function of the amplitude of the electron wave much as y is an amplitude function in the equation for a simple wave. It was Schrödinger's genius that led him to substitute into Hamilton's equations for the flooded planet, the de Broglie wavelength ($\lambda = h/mv_e$), for the electron.

From that point on, straightforward mathematical manipulations of the equations produced solutions for the hydrogen atom where the quantum conditions of discrete orbitals and energies for the electron appeared quite naturally.

In its most concise form, the famous Schrödinger equation can be written as follows:

$$H\psi = E\psi$$

H, called the Hamiltonian operator, tells us to do certain mathematical operations on ψ. Solutions to the equation are obtained in terms of ψ and *E,* for a particular *set of integers* called *quantum numbers. E* represents the energy of the electron. If ψ is the wave function for the 1*s* orbital, *E* is the energy of the electron in the 1*s* orbital. If ψ is the wave function for the 2*s* orbital, then the value of *E* that is obtained is the energy of the electron in the 2*s* orbital.

Wave functions—*or* ψ *functions*—can be plotted in a number of ways. One way of plotting the wave functions of the 2*s* and 2*p* orbitals of a hydrogen atom is shown in Fig. A.2. The contours given are for constant values of ψ at different distances from the nucleus.

The contours for the 2*s* orbital are a series of concentric circles centered on the nucleus. (Three-dimensional plots of the 2*s* orbital show that it is spherical.) The 2*s* orbital has a *nodal surface;* that is, it has a contour for which $\psi = 0$. Inside the nodal surface, ψ is negative; outside the nodal surface, ψ is positive.

The 2*p* orbital has two lobes that resemble two slightly squashed circles. (In three dimensions these lobes resemble slightly squashed spheres.) The lobes lie on opposite sides of the nucleus. In one lobe ψ is negative; in the other ψ is positive. A nodal plane separates the two lobes.

A 1*s* orbital (not shown) is shaped like a 2*s* orbital. In the 1*s* orbital, however, there is *no* nodal surface. The ψ function is positive throughout.

The positive and negative signs associated with ψ have nothing to do with electric charge. They are simply arithmetic signs of the wave function and in this sense they resemble the (+) and (−) signs associated with the amplitude of a simple wave (p. 45). It is the square of ψ, that is, ψ^2, that expresses the probability of finding an electron at a particular region of space (cf. Section 1.13). The quantity, ψ^2, of course, is always positive. Thus in the 2*p* orbital, for example, there is an equal probability of finding the electron in either lobe.

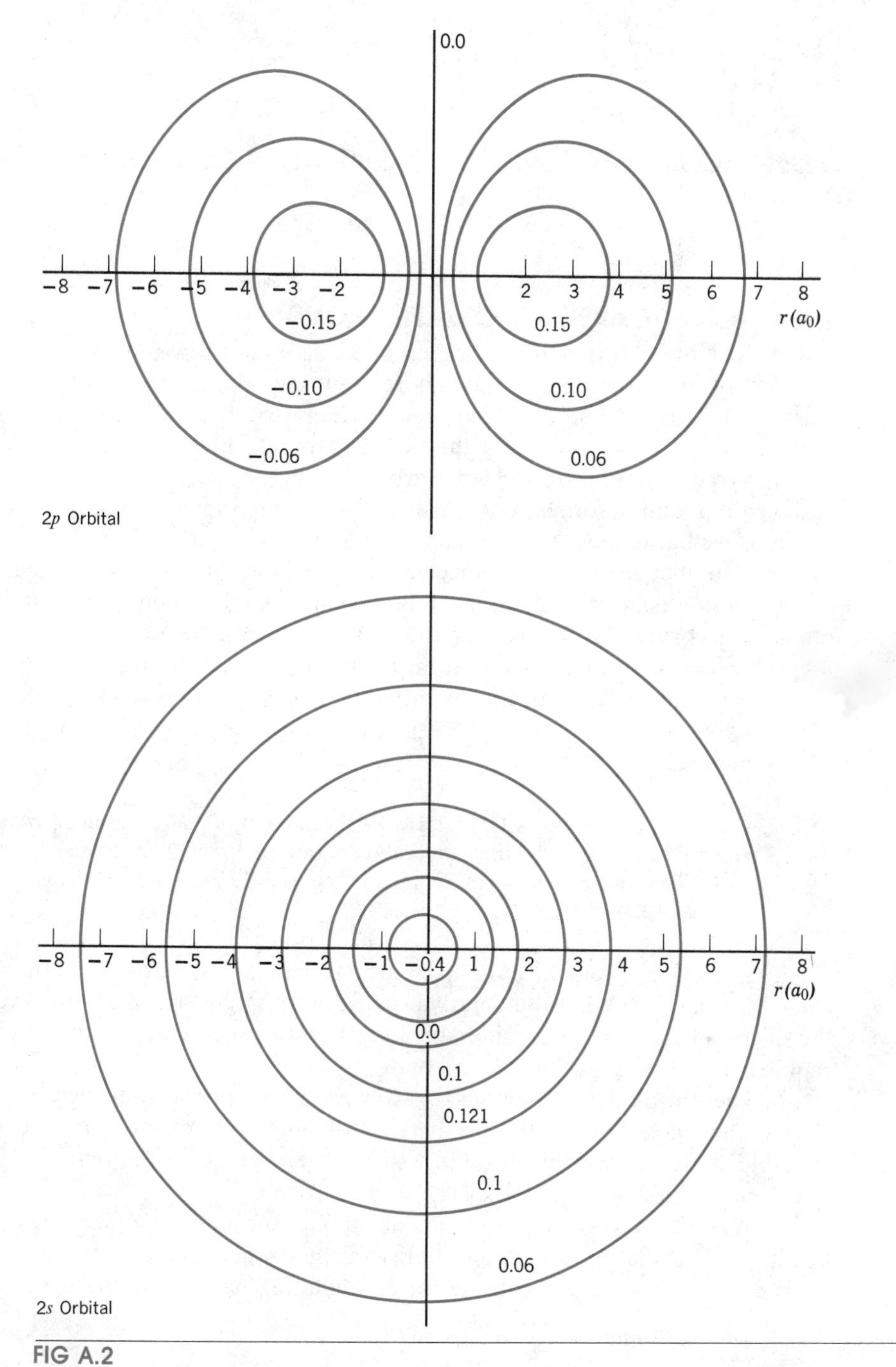

FIG A.2

Shapes of a 2s *orbital and a* 2p *orbital. The numbers on the contours give the value of* Ψ*;* r *is the distance from the nucleus in units of* a_0 *(*a_0 *is 0.529* Å). *(From K. B. Wiberg,* Physical Organic Chemistry, *Wiley, New York, 1964, pp. 29 and 30. Used with permission.)*

2 SOME REPRESENTATIVE CARBON COMPOUNDS

2.1 THE CARBON-CARBON COVALENT BOND

Carbon's ability to form strong covalent bonds to other carbon atoms is the single property of the carbon atom that—more than any other—accounts for the very existence of a field of study called organic chemistry. It is this property too that accounts in part for carbon being the element around which most of the molecules of living organisms are constructed. Carbon's ability to form strong bonds to other carbon atoms and to form strong bonds to hydrogen, oxygen, sulfur, and nitrogen atoms as well, provides the necessary versatility of structure that makes possible the vast number of different molecules required for complex living organisms.

Carbon is not unique in its ability to form covalent bonds to itself. Other atoms (e.g., oxygen, nitrogen, and silicon) have the same property. Carbon *is unique,* however, in its ability to form *very stable* bonds to other carbon atoms while at the same time being able to form strong bonds to atoms of other elements. The average amount of energy required to break a carbon-carbon bond is ~83 kcal/mole.

A kilocalorie of energy (1000 calories) is the amount of energy in the form of heat required to raise by 1 °C the temperature of one kilogram (1000 grams) of water at 15 °C. The unit of energy in SI units is the joule, J, and 1 cal = 4.184 J. (Thus 1 kcal = 4.184 kJ.)

Silicon, the element most like carbon in its atomic structure, forms silicon-silicon covalent bonds but, partly because of the larger size of the silicon atom, the silicon-silicon bond is considerably longer and weaker. Only 53 kcal/mole are required to break a silicon-silicon bond.

There also exist compounds that contain several nitrogen-nitrogen covalent bonds. Those containing as few as five linearly bonded nitrogen atoms are highly unstable, however, and this instability seems to be caused by the repulsive forces between the nonbonding electron pairs of nitrogen.

When only two oxygen atoms are linked the bond is quite weak. Once again, the explanation for the instability of the oxygen-oxygen bond lies in the large repulsive forces between the *two* nonbonding pairs on each atom.

Typical bond energy = 50 kcal/mole

$$-\ddot{N}-\ddot{N}-$$

Typical bond energy = 34 kcal/mole

$$-\ddot{\underset{\cdot\cdot}{O}}-\ddot{\underset{\cdot\cdot}{O}}-$$

2.2 METHANE AND ETHANE. REPRESENTATIVE ALKANES

Methane, CH_4, and ethane, C_2H_6, are two members of a broad family of organic compounds called *hydrocarbons.* Hydrocarbons, as the name implies, are com-

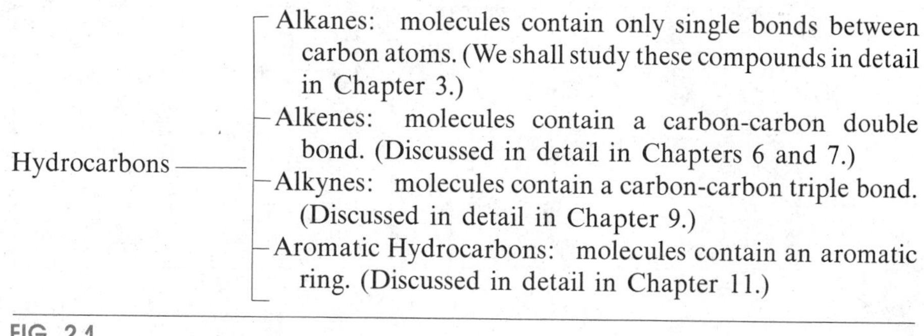

Hydrocarbons
Alkanes: molecules contain only single bonds between carbon atoms. (We shall study these compounds in detail in Chapter 3.)
Alkenes: molecules contain a carbon-carbon double bond. (Discussed in detail in Chapters 6 and 7.)
Alkynes: molecules contain a carbon-carbon triple bond. (Discussed in detail in Chapter 9.)
Aromatic Hydrocarbons: molecules contain an aromatic ring. (Discussed in detail in Chapter 11.)

FIG. 2.1
Subdivisions of the group of organic compounds called hydrocarbons.

pounds whose molecules contain only carbon and hydrogen (Fig. 2.1). Methane and ethane also belong to a subgroup of hydrocarbons known as *alkanes* whose members do not have multiple bonds between carbon atoms. Hydrocarbons whose molecules have a carbon-carbon double bond are called *alkenes,* and those with a carbon-carbon triple bond are called *alkynes.* A special group of unsaturated hydrocarbons that we shall study in Chapter 11 are called aromatic hydrocarbons.

Generally speaking, compounds whose molecules contain only single bonds are often referred to as *saturated* compounds and those whose molecules contain multiple bonds are called *unsaturated* compounds. We will see why in Section 6.4.

2.2A Sources of Methane

Methane was one major component of the early atmosphere of this planet. Methane is still found in the atmosphere of Earth, but no longer in appreciable amounts. It is, however, a major component of the atmospheres of Jupiter, Saturn, Uranus, and Neptune. Recently methane has also been detected in interstellar space—far from the earth (10^{16} km) in a celestial body that emits radio waves in the constellation Orion.

On Earth, methane is the major component of natural gas, along with ethane and other low molecular weight alkanes. The United States is currently using its large reserves of natural gas at a very high rate—one that threatens to deplete this resource in the not too distant future. Because the components of natural gas are important in industry, efforts are being made to develop coal-gasification processes to provide alternate sources. Our high rate of consumption of natural gas is caused largely by its use as a primary energy source. In 1972, for example, 34% of the primary energy consumption in the United States was derived from natural gas, whereas in Britain the amount was 12% and in Japan only 1%.

Some living organisms produce methane from carbon dioxide and hydrogen. These very primitive creatures called *methanogens* may be the Earth's oldest organisms, and they may represent a separate form of evolutionary development. Methanogens can survive only in an anaerobic (i.e., oxygen-free) environment. They have been found in ocean trenches, in mud, in sewage, and in cows' stomachs.

2.2B The Structure of Ethane

The bond angles at the carbon atoms of ethane, and of all alkanes, are also tetrahedral like those in methane. In the case of ethane (Fig. 2.2), each carbon is at one corner of the other carbon's tetrahedron; hydrogen atoms are situated at the other three corners.

The sp^3-hybridized carbon atom provides a satisfactory model for ethane

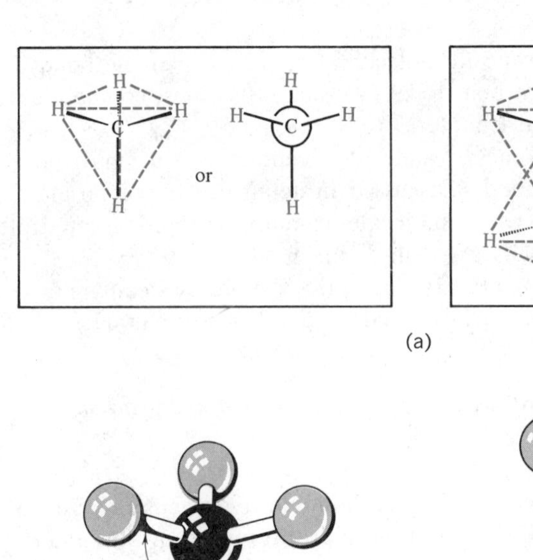

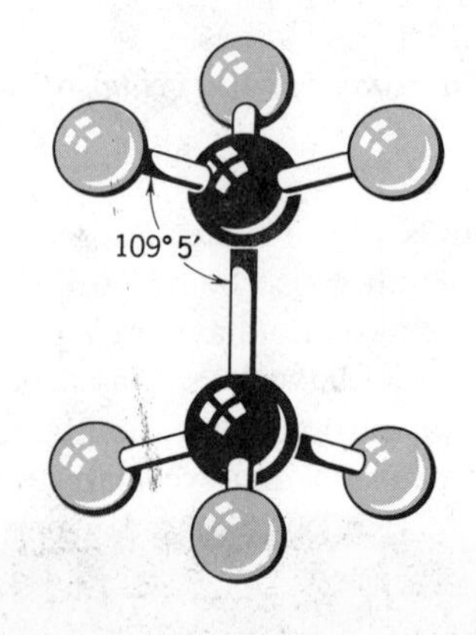

FIG. 2.2
(a) *Two ways of representing the structures of methane and ethane that show the tetrahedral arrangements of the atoms around carbon.* (b) *Ball-and-stick models of methane and ethane.*

(and for other alkanes as well). Figure 2.3 shows how we might imagine an ethane molecule being constructed from two sp^3-hybridized carbon atoms and six hydrogen atoms.

The carbon-carbon bond of ethane is a *sigma bond,* formed by two overlapping sp^3 orbitals. The carbon-hydrogen bonds are also sigma bonds. They are formed from overlapping carbon sp^3 orbitals and hydrogen s orbitals.

2.3 SIGMA BONDS AND BOND ROTATION

Groups bonded only by a sigma bond (i.e., by a single bond) can undergo rotation about that bond with respect to each other. The different arrangements of the atoms in space that result from rotations of groups about single bonds are called *conformations* of the molecule. An analysis of the energy changes that a molecule undergoes as groups rotate about single bonds is called a *conformational analysis.**

Let us consider the ethane molecule as an example. Obviously an infinite number of different conformations could result from rotations of the CH_3-groups about the carbon-carbon bond. These different conformations, however, are not all of equal stability. The conformation (Fig. 2.4) in which the hydrogen atoms attached to each carbon are perfectly staggered when viewed from one end of the

* Conformational analysis owes its modern origins largely to the work of O. Hassel of Norway and D. R. H. Barton of Britain. Hassel and Barton won the Nobel Prize in 1969, mainly for their contributions in this area. The idea that certain conformations of molecules will be favored dates from the work of van't Hoff.

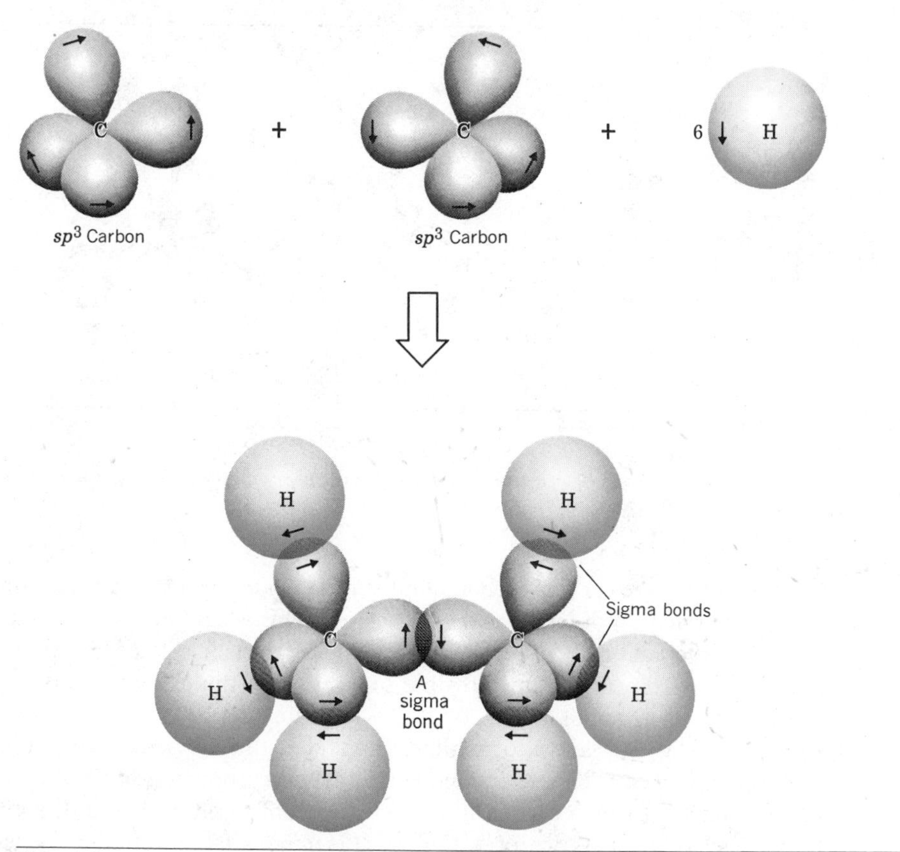

FIG. 2.3
The formation of ethane from two sp^3-hybridized carbon atoms and six hydrogen atoms. All of the bonds are sigma bonds.

FIG. 2.4
(a) *The staggered conformation of ethane.* (b) *The Newman projection formula for the staggered conformation. This formulation is obtained by viewing the molecule end-on along the carbon-carbon bond axis. Bonds of the front carbon are depicted as* ⅄, *those of the back carbon as* ⚲.

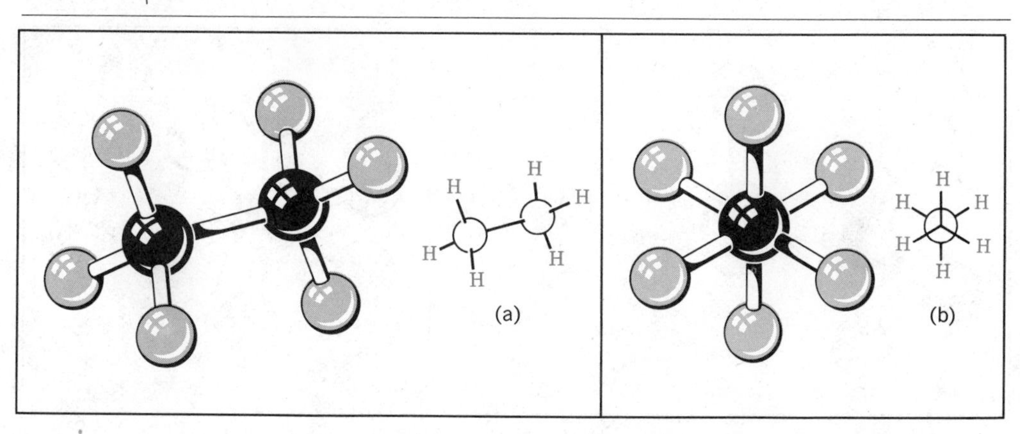

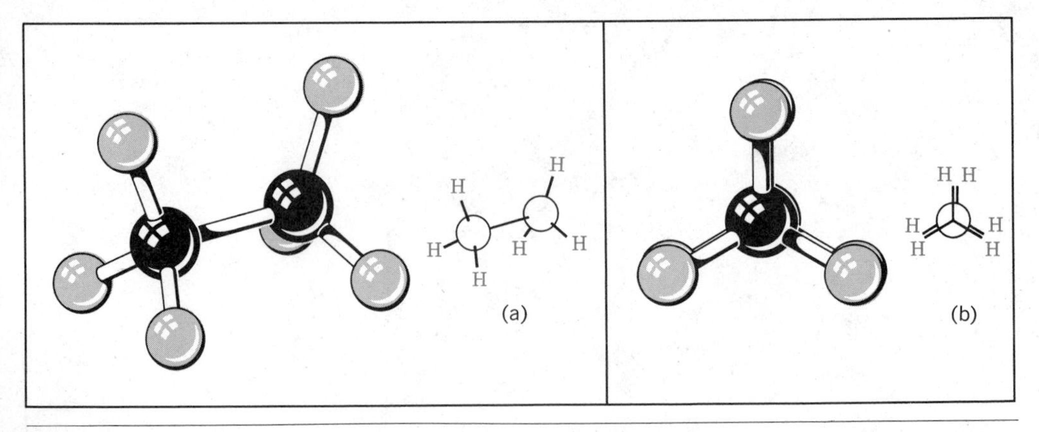

FIG. 2.5
(a) *The eclipsed conformation of ethane.* (b) *The Newman projection formula for the eclipsed conformation.*

molecule along the carbon-carbon bond axis is the *most stable* conformation (i.e., it is the conformation of *lowest potential energy*). This is easily explained in terms of repulsive interactions between bonding pairs of electrons. The staggered conformation allows the maximum possible separation of the electron pairs of the six carbon-hydrogen bonds.

The least stable conformation of ethane is the *eclipsed* conformation (Fig. 2.5). When viewed from one end along the carbon-carbon bond axis, the hydrogen atoms attached to each carbon in the eclipsed conformation are in direct opposition to each other. This conformation permits *minimum* separation of the electrons of the six carbon-hydrogen bonds. It is, therefore, of highest energy and has the least stability.

Conformations between the eclipsed and the staggered are called skew conformations (Fig. 2.6). All have stabilities between that of the staggered and that of the eclipsed conformation, that is, the relative stabilities are in the order: staggered > skew > eclipsed.

We represent this situation graphically by plotting the energy of an ethane molecule as a function of rotation about the carbon-carbon bond. The energy changes that occur are illustrated in Fig. 2.7.

FIG. 2.6
(a) *A skew conformation of ethane.* (b) *The Newman projection formula for a skew conformation.*

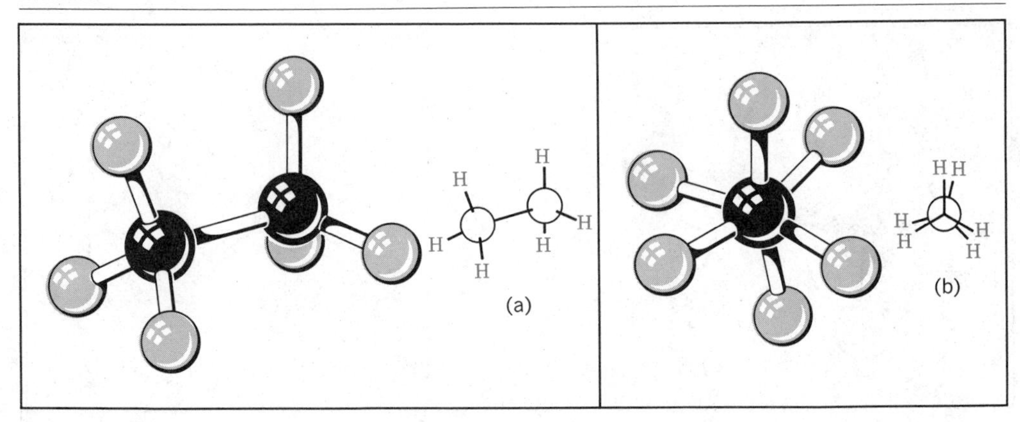

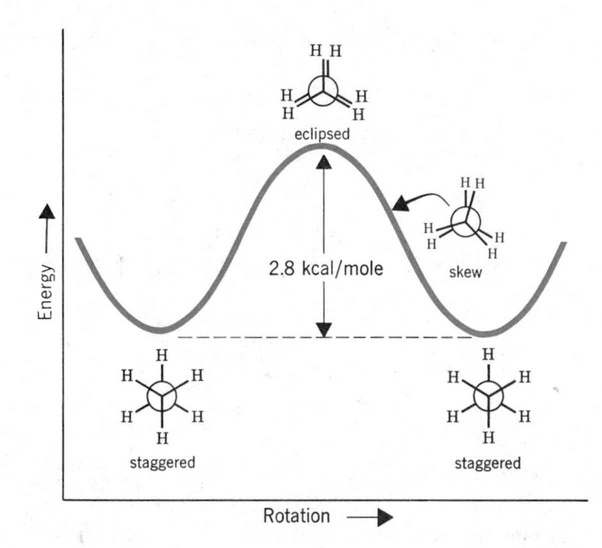

FIG. 2.7
Potential energy changes that accompany rotation of the carbon-carbon bond of ethane.

In ethane the difference in energy between the staggered and eclipsed conformations is 2.8 kcal/mole. This small barrier to rotation is called the torsional energy of the single bond. Unless the temperature is extremely low (−250°) some ethane molecules (at any given moment) will have enough energy to surmount this barrier. Most molecules will wag back and forth with their atoms in staggered or nearly staggered conformations. The more energetic ones will rotate through eclipsed conformations to other staggered conformations.

What does all this mean about ethane? We can answer this question in two different ways. If we consider a single molecule of ethane, we can say, for example, that it will spend most of its time in the lowest energy, staggered conformation or in a skew conformation very close to being staggered. Several times every second, however, it will acquire enough energy because of its thermal motion to surmount the rotational barrier and it will rotate through an eclipsed conformation. If we speak in terms of a large number of ethane molecules (a more realistic situation), we can say that at any given moment most of the molecules will be in staggered or nearly staggered conformations.

If we consider substituted ethanes such as XCH_2CH_2X, the barriers to rotation are somewhat larger but they are still far too small to allow isolation of the different conformations (see following diagram) even at temperatures considerably below room temperature. *These conformations cannot be isolated at normal temperatures.*

Problem 2.1

Two isomers have the formula $C_2H_4Cl_2$. (a) Write their structural formulas. (b) The formulas below do not represent different compounds. Explain.

$Cl{-}CH_2{-}CH_2{-}Cl$ $\quad$ $CH_2(Cl){-}CH_2(Cl)$ $\quad$ $CH_2(Cl){-}CH_2(Cl)$

Problem 2.2

(a) Write structural formulas for all isomers of C_3H_7Cl. (b) Repeat for all of the isomers of $C_3H_6Cl_2$, and (c) for $C_3H_5Cl_3$.

2.4 ENERGY CHANGES

Since we will be talking frequently about the energies of chemical systems, perhaps we should pause here for a brief review. *Energy* is defined as the capacity to do work. The two fundamental types of energy are *kinetic energy* and *potential energy*.

Kinetic energy is the energy an object has because of its motion; it equals one-half the object's mass multiplied by the square of its velocity (i.e., $\frac{1}{2}mv^2$).

Potential energy is stored energy. It exists only when an attractive or repulsive force exists between objects. Two balls attached to each other by a spring can have their potential energy increased when the spring is stretched or compressed (Fig. 2.8). If the spring is stretched, an attractive force will exist between the balls. If it is compressed, a repulsive force will exist. In either instance releasing the balls will cause the potential energy (stored energy) of the balls to be converted into kinetic energy (energy of motion).

Chemical energy—the energy stored in the bonds of the molecules of a substance—is a form of potential energy. It exists because attractive and repulsive electrical forces exist between different pieces of the molecules. Nuclei attract electrons; nuclei repel each other, and electrons repel each other. In the example of the previous section the repulsive force between the electrons of the C—H bonds gives the eclipsed conformation of ethane greater potential energy than the staggered conformation.

It is usually impractical (and often impossible) to describe the *absolute* amount of potential energy contained by a substance. Thus we usually think in terms of their *relative potential energies*. We say that one system has *more* or *less* potential energy than another. We say, for example, that the eclipsed conformation of ethane has more potential energy than the staggered conformation. Or, we say that the staggered conformation has less potential energy than a skew conformation.

FIG. 2.8
*Potential energy (P.E.) exists between objects that either attract or repel each other. When the spring is either stretched or compressed the P.E. of the two balls increases. (Adapted with permission from J.E. Brady and G.E. Humiston, General Chemistry: Principles and Structure, 1st Ed., Wiley, New York, p.*18.)

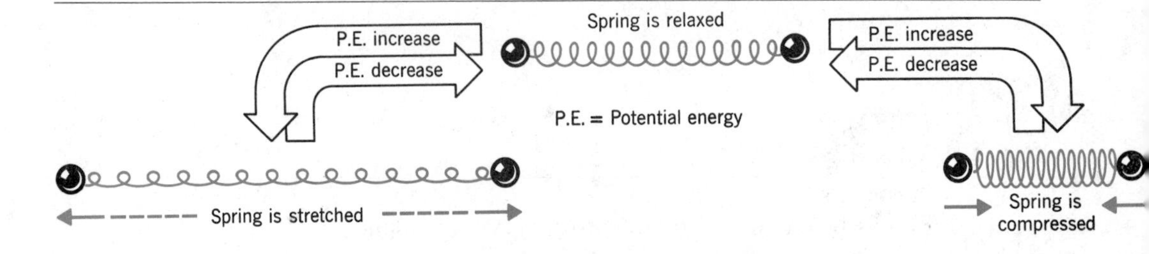

Another term chemists use frequently in this context is the term *stability*—or *relative stability*. The relative stability of a system is inversely proportional to its relative potential energy. An object that has less potential energy than another, has a *greater relative stability*. The staggered conformation of ethane, for example, has less potential energy than the eclipsed conformation and therefore it is more stable.

2.4 A Heat of Combustion

The potential energy stored in molecules can be released as heat when molecules react. (Since heat is associated with molecular motion this is essentially a change from potential energy to kinetic energy.) When methane reacts with oxygen, for example, the reaction produces a large amount of heat.

$$CH_4 + 2O_2 \longrightarrow CO_2 + 2H_2O + 192\ \text{kcal/mole}$$

The heat produced by this reaction is what makes methane such an attractive energy source.

A convenient way to represent the relative potential energies of molecules is in terms of their relative *enthalpies* or *heat contents, H*. (*Enthalpy* comes from the German word *enthalten* meaning to contain.) The difference in relative enthalpies of reactants and products in a chemical change is called the enthalpy change and is symbolized by ΔH. (The Δ (delta) in front of a quantity usually means the difference or change, in the quantity.)

For example, when methane is subjected to complete combustion in a calorimeter, a device used to determine the heat evolved or absorbed by a reaction, its *heat of combustion* can be measured. At 25° and 1 atm pressure the heat of combustion of methane is −192 kcal/mole. We write the equation this way:

$$CH_4 + 2O_2 \longrightarrow CO_2 + 2H_2O \qquad \Delta H = -192\ \text{kcal/mole}$$

By convention,* the sign of ΔH for *exothermic* reactions (those evolving heat) is negative. *Endothermic* reactions (those which absorb heat) have a positive ΔH. The heat of reaction, ΔH, measures the change in enthalpy of the atoms of the reactants as they are converted to products. For an exothermic reaction the atoms have a smaller enthalpy as products than they do as reactants. For endothermic reactions, the reverse is true. Figure 2.9 shows that a mole of methane and two

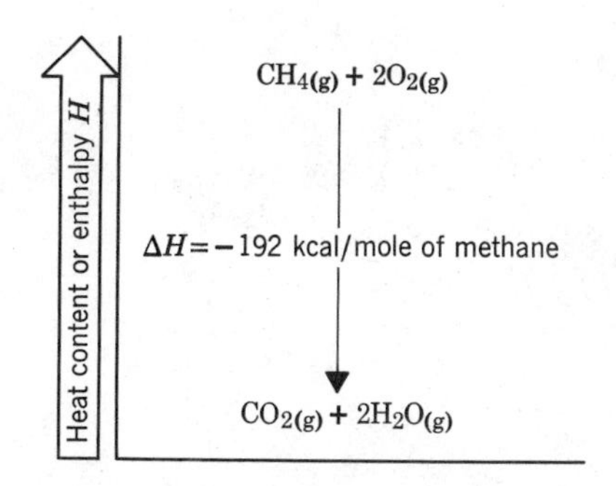

FIG. 2.9
Enthalpy (heat content) change for the complete combustion of methane.

$^*\Delta H = H_{\text{final}} - H_{\text{initial}} = H_{\text{products}} - H_{\text{reactants}}$

moles of oxygen have a much higher enthalpy than do the relatively stable products, carbon dioxide (one mole) and water (two moles). As a consequence, the enthalpy of the atoms decreases by a very large amount as reactants are converted to products.

ALKENES: COMPOUNDS CONTAINING THE CARBON-CARBON DOUBLE BOND; ETHENE AND PROPENE

The carbon atoms of all of the molecules that we have considered so far have used their four valence electrons to form four single covalent bonds to four other atoms. We find, however, that many important organic compounds exist in which carbon atoms share more than two electrons with another atom. In molecules of these compounds some bonds that are formed are multiple covalent bonds. When two carbon atoms share two pairs of electrons, for example, the result is a carbon-carbon double bond.

Hydrocarbons whose molecules contain a carbon-carbon double bond are

$$\cdot\!\!\dot{\text{C}}\!::\!\dot{\text{C}}\!\!\cdot \quad \text{or} \quad \rangle\text{C}{=}\text{C}\langle$$

called *alkenes*. Ethene, C_2H_4, and propene, C_3H_6, are both alkenes. (Ethene is also called ethylene, and propene is sometimes called propylene.)

```
H       H        H        H
 \     /          \      /
  C=C              C=C
 /     \          /      \
H       H      CH3        H
  Ethene           Propene
```

In ethene the only carbon-carbon bond is a double bond. Propene has one carbon-carbon single bond and one carbon-carbon double bond.

The spatial arrangement of the atoms of alkenes is different from that of alkanes. The six atoms of ethene are coplanar, and the arrangement of atoms around each carbon atom is triangular (Fig. 2.10). When we study alkenes in detail in Chapter 6, we will see how the structure of ethene can be explained on the basis of the same kind of orbital hybridization, sp^2, that we learned about for boron trifluoride (Sect. 1.17).

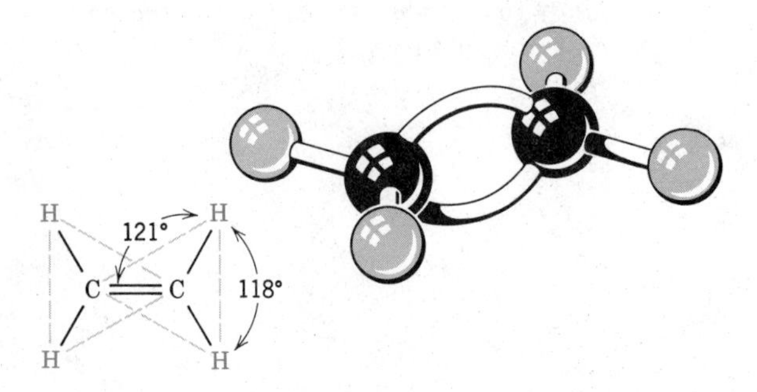

FIG. 2.10
The structure of ethene.

Problem 2.3

Explain the triangular arrangement of groups around each carbon of ethene on the basis of the Valence Shell Electron Pair (VSEPR) model. (Assume that the two electron pairs of the double bond are located in the region of space between the two carbon atoms, that is, assume that they act like a single electron pair.)

2.5A *Cis-Trans* Isomerism

Groups joined by a carbon-carbon double bond cannot rotate with respect to each other at temperatures near and considerably above room temperature. This restricted rotation of groups attached by double bonds is in marked contrast to the behavior of groups attached by single bonds; the latter rotate quite freely at these temperatures. Restricted rotation causes a new type of isomerism that we illustrate with the two dichloroethenes written below.

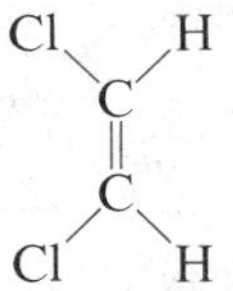

```
Cl   H          H   Cl
  \ /            \ /
   C              C
   ||             ||
   C              C
  / \            / \
Cl   H         Cl   H
```

cis-1,2-Dichloroethene 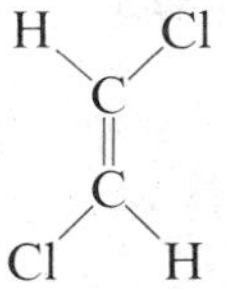 *trans*-1,2-Dichloroethene

These two compounds are isomers; they are different compounds that have the same molecular formula. We can tell that they are different compounds by trying to superpose a model of *cis*-1,2-dichloroethene on a model of *trans*-1,2-dichloroethene: we find that it cannot be done. By superpose we mean that we attempt to place one model on the other *so that all parts of each coincide.**

Cis-1,2-dichloroethene and *trans*-1,2-dichloroethene are not structural isomers because the order of attachment of atoms is the same in each. The two compounds differ only in the arrangement of their atoms in space. Isomers of this kind are classified formally as *stereoisomers,* but often they are called simply *cis-trans* isomers (*cis,* Latin: on this side; *trans,* Latin: across). (We will study stereoisomerism in detail in Chapter 8.)

The structural requirements for *cis-trans* isomerism will become clear if we consider a few additional examples. 1,1-Dichloroethene and 1,1,2-trichloroethene do not show this type of isomerism.

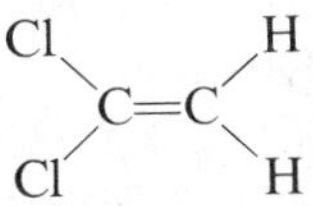

```
Cl       H          Cl       Cl
  \     /             \     /
   C=C                 C=C
  /     \             /     \
Cl       H          Cl       H
```

1,1-Dichloroethene (no *cis-trans* isomerism) 1,1,2-Trichloroethene (no *cis-trans* isomerism)

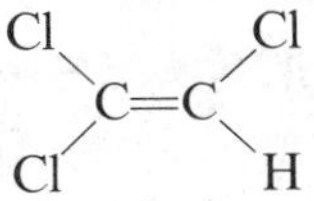

1,2-Difluoroethene and 1,2-dichloro-1,2-difluoroethene do exist as *cis-trans* isomers.

* The word *superpose,* therefore, has a more restricted meaning than the word *superimpose.* Superimpose means just to place one object on another. Models of the two dichloroethenes can be superimposed but they cannot be superposed.

F F
C=C
H H
cis-1,2-Difluoroethene

F H
C=C
H F
trans-1,2-Difluoroethene

F F
C=C
Cl Cl
cis-1,2-Dichloro-1,2-difluoroethene

F Cl
C=C
Cl F
trans-1,2-Dichloro-1,2-difluoroethene

Clearly, then, *cis-trans isomerism of this type is not possible if one carbon atom of the double bond bears two identical groups.*

Cis-trans isomers have different physical properties. They have different melting points and boiling points, and often *cis-trans* isomers differ markedly in the magnitude of their dipole moments. Table 2.1 summarizes some of the physical properties of two pairs of *cis-trans* isomers.

Problem 2.4

(a) How do you explain the fact that *trans*-1,2-dichloroethene and *trans*-1,2-dibromoethene have no dipole moments ($\mu = 0$), whereas the corresponding *cis* isomers have rather large dipole moments (for *cis*-1,2-dichloroethene, $\mu = 1.85$, and for *cis*-1,2-dibromoethene, $\mu = 1.35$)? (b) Account for the fact that *cis*-1,2-dichloroethene has a larger dipole moment than *cis*-1,2-dibromoethene?

2.5B Propene

Two of the carbon atoms of propene (Fig. 2.11) have a triangular structure; the third carbon atom is tetrahedral.

Problem 2.5

Write structural formulas for: (a) all of the compounds that could be obtained by replacing one hydrogen of propene with chlorine; (b) all of the compounds that could be obtained by replacing two hydrogens of propene with chlorine; (c) three hydrogens; (d) four hydrogens; (e) five hydrogens. (f) In each instance [(a)–(e)] designate pairs of *cis-trans* isomers.

2.5C Industrial Use of Ethene and Propene

Ethene is the single most important raw material used in the organic chemical industry. It is produced by "cracking" hydrocarbons obtained from

TABLE 2.1 Physical Properties of Cis-Trans Isomers

COMPOUND	MELTING POINT, °C	BOILING POINT, °C	DIPOLE MOMENT
cis-1,2-Dichloroethene	−80	60	1.85 D
trans-1,2-Dichloroethene	−50	48	0
cis-1,2-Dibromoethene	−53	110	1.35 D
trans-1,2-Dibromoethene	−6	108	0

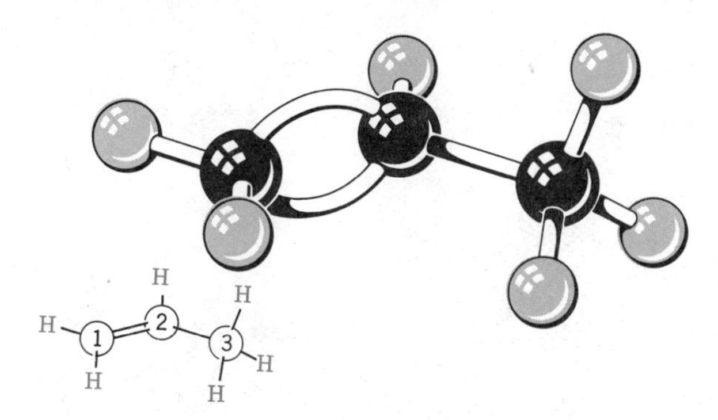

FIG. 2.11
The structure of propene. Carbon atoms 1 and 2 have the same geometry as those of ethene; carbon atom 3 is tetrahedral like that of methane.

petroleum (cf. Chapter 4) and thus it is often called a *petrochemical.* Ethene production in the United States in 1978 amounted to nearly 26 billion pounds! About half of this ethene was used to make the familiar plastic called polyethylene (cf. Special Topic C). The remainder was converted into other plastics, into a major component of antifreeze (ethylene glycol), and into a variety of other chemicals.

Propene is another important petrochemical. In 1978 nearly 14 billion pounds of propene were produced. Most was used to make plastics (e.g., polypropylene).

2.6 ALKYNES: COMPOUNDS CONTAINING THE CARBON-CARBON TRIPLE BOND; ETHYNE AND PROPYNE

Hydrocarbons in which two carbon atoms share three pairs of electrons between them and are thus bonded by a triple bond, are called alkynes. The two simplest alkynes are ethyne and propyne.

H:C⋮⋮C:H	CH_3:C⋮⋮C:H
H—C≡C—H	CH_3—C≡C—H
Ethyne	Propyne
(C_2H_2)	(C_3H_4)

Ethyne, a compound that is also called acetylene, consists of linear molecules. The H—C≡C bond angles of ethyne molecules are 180°.

H—C≡C—H

180° 180°

Problem 2.6

(a) Assuming that three bonding pairs of electrons are located generally between the two carbon atoms of ethyne, are bond angles of 180° consistent with the principle of maximum possible separation of the bonding pairs of electrons? (b) Which atoms in propyne would lie in a straight line? (c) Write out a complete three-dimensional structure for propyne.

The carbon-carbon triple bond is shorter than the carbon-carbon double bond, and the carbon-carbon double bond is shorter than the carbon-carbon single

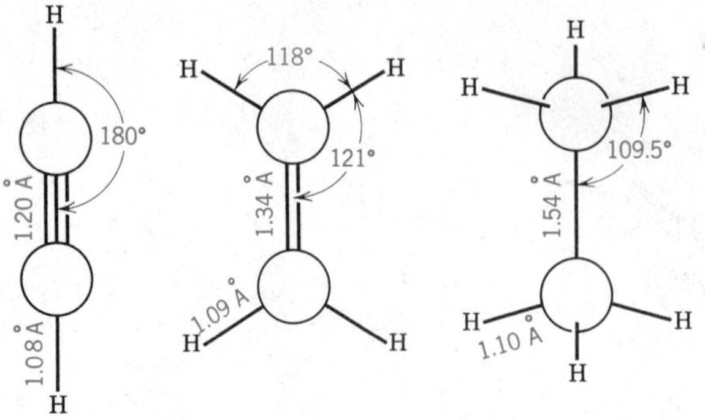

FIG. 2.12
Bond angles and bond lengths of ethyne, ethene, and ethane.

bond. The carbon-hydrogen bonds of ethyne are also shorter than those of ethene, and the carbon-hydrogen bonds of ethene are shorter than those of ethane. The differences in bond lengths and bond angles of ethyne, ethene, and ethane are summarized in Fig. 2.12.

2.7 BENZENE: A REPRESENTATIVE AROMATIC HYDROCARBON

In Chapter 11 we will study a special group of cyclic hydrocarbons known as aromatic hydrocarbons. Benzene is a typical example. Its structure is usually written in one of the ways shown below.

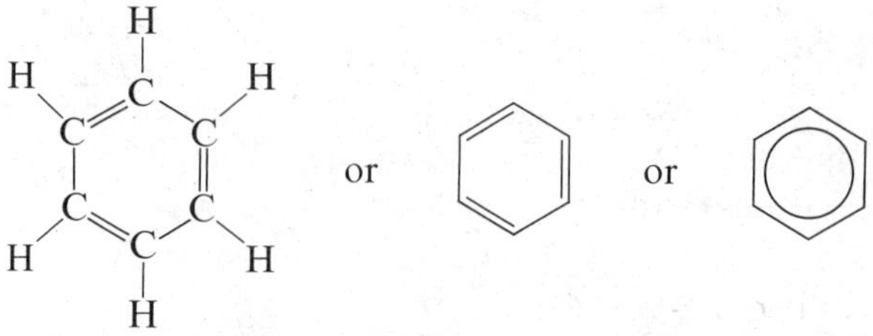

or C_6H_6

When the benzene ring is found attached to some other group of atoms in a molecule, it is called a *phenyl group* and it is represented as follows.

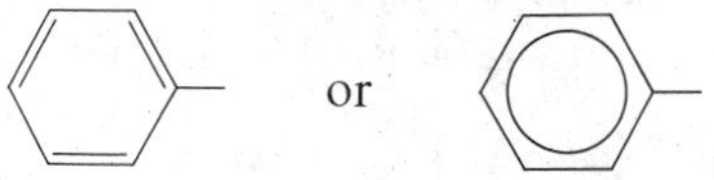

or $C_6H_5—$

Because the properties of benzene and other aromatic hydrocarbons are so special (and so interesting) we will defer any further discussions until Chapter 11 when we can take them up in detail.

2.8 FUNCTIONAL GROUPS

One great advantage of the structural theory is that it enables us to classify the vast number of organic compounds into a relatively small number of families based on their structures. (The end papers inside the front cover of this text give the most important of these families.) The molecules of compounds in a particular family are characterized by the presence of a certain arrangement of atoms called a *functional group*.

A functional group is the part of a molecule where most of its chemical reactions occur. It is the part that effectively determines the compound's chemical properties (and many of its physical properties as well). The functional group of an alkene, for example, is its carbon-carbon double bond. When we study the reactions of alkenes in greater detail in Chapter 7, we shall find that most of the chemical reactions of alkenes are the chemical reactions of the carbon-carbon double bond. One characteristic reaction of compounds containing a carbon-carbon double bond is their reaction with bromine in the following way:

$$-\overset{|}{C}=\overset{|}{C}- \;+\; Br-Br \longrightarrow -\underset{\underset{Br}{|}}{\overset{|}{C}}-\underset{\underset{Br}{|}}{\overset{|}{C}}-$$

In this reaction—called *an addition reaction*—one atom of bromine adds to each carbon of the double bond. This reaction is given by almost all alkenes. Both ethene and propene react with bromine in this way:

$$H-\overset{\overset{H}{|}}{C}=\overset{\overset{H}{|}}{C}-H \;+\; Br-Br \longrightarrow H-\underset{\underset{Br}{|}}{\overset{\overset{H}{|}}{C}}-\underset{\underset{Br}{|}}{\overset{\overset{H}{|}}{C}}-H$$

$$CH_3-\overset{\overset{H}{|}}{C}=\overset{\overset{H}{|}}{C}-H \;+\; Br-Br \longrightarrow CH_3-\underset{\underset{Br}{|}}{\overset{\overset{H}{|}}{C}}-\underset{\underset{Br}{|}}{\overset{\overset{H}{|}}{C}}-H$$

So, too, do the alkenes 1-butene and *cis*-2-butene.*

$$\underset{\text{1-Butene}}{CH_3CH_2-\overset{\overset{H}{|}}{C}=\overset{\overset{H}{|}}{C}-H} \;+\; Br_2 \longrightarrow CH_3CH_2-\underset{\underset{Br}{|}}{\overset{\overset{H}{|}}{C}}-\underset{\underset{Br}{|}}{\overset{\overset{H}{|}}{C}}-H$$

$$\underset{\textit{cis}\text{-2-Butene}}{\overset{CH_3\diagdown\quad\diagup CH_3}{\underset{H\diagup\quad\diagdown H}{C=C}}} \;+\; Br_2 \longrightarrow CH_3-\underset{\underset{Br}{|}}{\overset{\overset{H}{|}}{C}}-\underset{\underset{Br}{|}}{\overset{\overset{H}{|}}{C}}-CH_3$$

The functional group of an alkyne is its carbon-carbon triple bond. Alkanes are not usually thought of as having a functional group. Their molecules have carbon-carbon single bonds and carbon-hydrogen bonds, but these bonds are present in molecules of almost all organic molecules, and C—C and C—H bonds are, in general, much less reactive than common functional groups.

* Don't worry about where these names come from now. We will discuss the naming of organic compounds later.

2.8A Alkyl Groups and the Symbol R

Alkyl groups are the groups that we identify for purposes of naming compounds. They are made (on paper) by removing a hydrogen atom from an alkane:

$$\underset{\text{Methane}}{CH_4} \xrightarrow{-H} \underset{\text{Methyl group}}{CH_3—}$$

$$\underset{\text{Ethane}}{CH_3CH_3} \xrightarrow{-H} \underset{\text{Ethyl group}}{CH_3CH_2—} \quad (\text{or } C_2H_5—)$$

$$\underset{\text{Propane}}{CH_3CH_2CH_3} \xrightarrow[\text{end carbon}]{-H \text{ at}} \underset{\text{Propyl group}}{CH_3CH_2CH_2—}$$

$$CH_3CH_2CH_3 \xrightarrow[\text{carbon}]{-H \text{ at middle}} \underset{\text{Isopropyl group}}{CH_3\underset{|}{C}HCH_3} \quad \left(\text{or } CH_3\overset{CH_3}{\overset{|}{C}}H— \quad \text{or } (CH_3)_2CH—\right)$$

The methyl group, the ethyl group, the propyl group, and the isopropyl group are all alkyl groups. Their names and structures must be learned.

We can simplify much of our future discussion if, at this point, we introduce a symbol that is widely used in designating general structures of organic molecules: the symbol R. *R is used as a general symbol to represent any alkyl group*. R, for example, might be a methyl group, an ethyl group, a propyl group, or an isopropyl group.

$CH_3—$	Methyl	All of	
$CH_3CH_2—$	Ethyl	these	by R
$CH_3CH_2CH_2—$	Propyl	can be	
$CH_3\underset{\vert}{C}HCH_3$	Isopropyl	designated	

Using R, we can write a general formula for an alkene such as propene or 1-butene (i.e., one having only one alkyl group attached to a doubly bonded carbon). We write the formula in the following way:

$$R—CH{=}CH_2$$

We can write a general formula for any alkene like *cis*-2-butene as:

$$\begin{matrix} R & & R \\ & C{=}C & \\ H & & H \end{matrix} \quad \text{or} \quad \begin{matrix} R & & R' \\ & C{=}C & \\ H & & H \end{matrix}$$

where the primed R may be an alkyl group different from R.

We can write a general formula for any alkyne like propyne (i.e., one with only one alkyl group attached to the triple bond) as:

$$R—C{\equiv}CH$$

Problem 2.7

Using R as a symbol, how would you write the general formula for an alkane?

2.9 ALCOHOLS

Methyl alcohol (more systematically called methanol) has the structural formula CH_3OH and is the simplest member of a family of organic compounds known as *alcohols*. The characteristic functional group of this family is the hydroxyl group (OH) attached to a tetrahedral carbon atom. Other examples of alcohols are ethyl alcohol, CH_3CH_2OH (also called ethanol); propyl alcohol, $CH_3CH_2CH_2OH$ (also called 1-propanol); and isopropyl alcohol, $CH_3CHOHCH_3$ (also called 2-propanol).

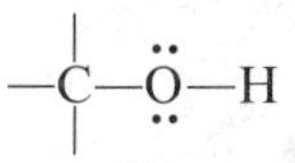

—C—Ö—H   This is the functional group of an alcohol.

Alcohols may be viewed in two ways structurally: (1) as hydroxy derivatives of alkanes, and (2) as alkyl derivatives of water. Ethyl alcohol, for example, can be seen as an ethane molecule in which one hydrogen has been replaced by a hydroxyl group, or as a water molecule in which one hydrogen has been replaced by an ethyl group. That the latter way of regarding ethyl alcohol is valid is shown by the fact that the C—O—H bond angle of ethyl alcohol is very close to the H—O—H bond angle of water.

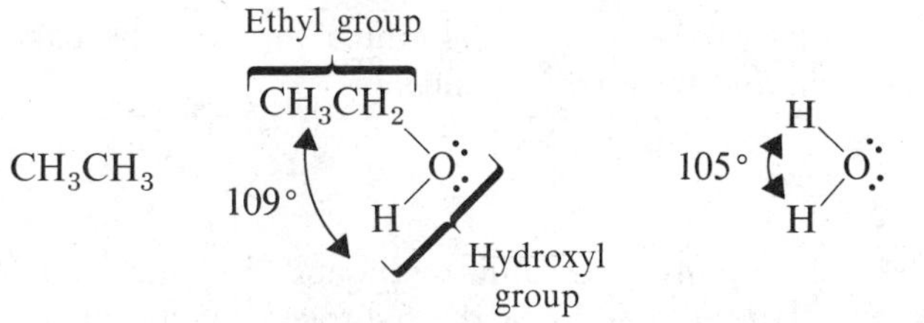

Ethyl group

CH_3CH_3 — CH_3CH_2—Ö—H (109°) — H—Ö—H (105°)

Hydroxyl group

Ethane — Ethyl alcohol (ethanol) — Water

Alcohols, generally, are classified into three groups: primary (1°), secondary (2°), or tertiary (3°) alcohols. *This classification is based on the condition of the carbon to which the hydroxyl group is directly attached.* If the carbon is itself attached to only one other carbon, the carbon is said to be a primary carbon and the alcohol is a primary alcohol.

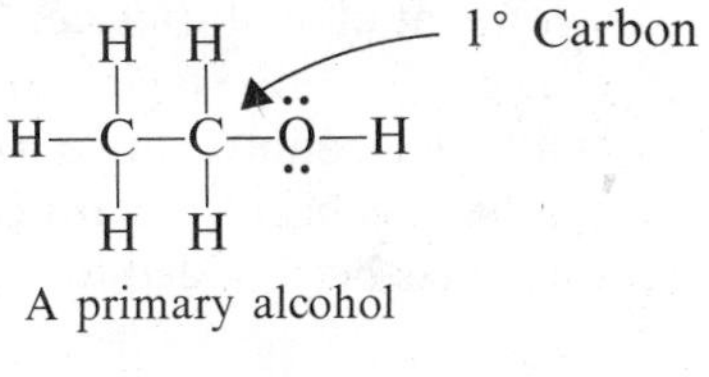

1° Carbon

H—C—C—Ö—H (with H H above and H H below)

A primary alcohol

CH_2OH

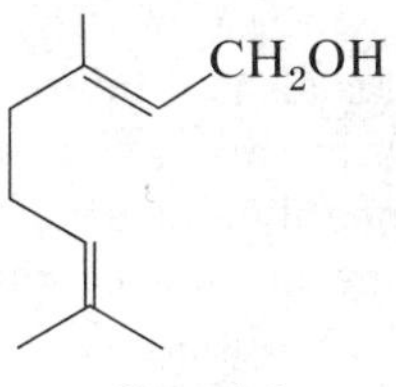

Geraniol
(A 1° alcohol with the odor of roses)

If the carbon that bears the hydroxyl group also has two other carbons attached to it, this carbon is called a secondary carbon, and the alcohol is a secondary alcohol.

```
    H  H /H        2° Carbon
    |  |▼ |
 H—C—C—C—H
    |  |  |
    H :O: H
       |
       H
```

A secondary alcohol

```
     CH3
      |
   /‾‾‾‾\
   |    |
   \____/—OH
      |
     CH
    /  \
  CH3   CH3
```

Menthol
(a 2° alcohol found
in peppermint oil)

Finally, if the carbon with the hydroxyl group is attached to three other carbons, this carbon is a tertiary carbon and the alcohol is a tertiary alcohol.

```
       H
       |
    H—C—H
    H  |  H        3° Carbon
    |  ▼  |
 H—C—C—C—H
    |  |  |
    H  |  H
      :O:
       |
       H
```

A tertiary (3°) alcohol

Problem 2.8

Using the symbol R, write a general formula for (a) a primary alcohol, (b) a secondary alcohol, and (c) a tertiary alcohol.

Problem 2.9

Alkyl halides are a family of organic compounds in which a halogen (i.e., —F, —Cl, —Br, or —I) replaces one of the hydrogens of an alkane. For example, CH_3Cl and CH_3CH_2Br are alkyl halides. Alkyl halides are classified in the same way as alcohols. (a) Using —X to represent any halogen, write the general formula for a primary alkyl halide. (b) For a secondary alkyl halide. (c) For a tertiary alkyl halide. (d) How would you write a general formula for *any* alkyl halide regardless of its classification?

Ethyl alcohol is an important industrial chemical; it is also the alcohol of alcoholic beverages. It can be produced in a variety of ways, but the oldest method for the synthesis of ethyl alcohol is by the fermentation of fruits and grains. Methyl alcohol is also an important industrial chemical and it is highly toxic. The propyl alcohols are also toxic and are used as "rubbing alcohols." In your laboratory you will probably find a bottle of ethyl alcohol that has been "denatured." This means that some substance—often methyl alcohol—has been added to it to make it unfit for consumption.

Problem 2.10

Although we will discuss the naming of organic compounds later when we consider the individual families in detail, one method for naming alkyl halides is so straightforward that it is worth describing here. We simply give the name of the alkyl group attached to the halogen and add the word *bromide, chloride,* and so forth. Write formulas for (a) propyl chloride and (b) isopropyl bromide. What are names for (c) CH_3CH_2F, (d) CH_3CHICH_3, and (e) CH_3I? (Notice that one way of naming alcohols is also like this. We just name the alkyl group attached to the hydroxyl group and add the word *alcohol,* for example, methyl alcohol, ethyl alcohol, and so forth.)

2.10 ETHERS

Ethers have the general formula R—O—R. They can be thought of as derivatives of water in which both hydrogens have been replaced by alkyl groups. The bond angle at the oxygen atom of an ether is somewhat larger than that of water.

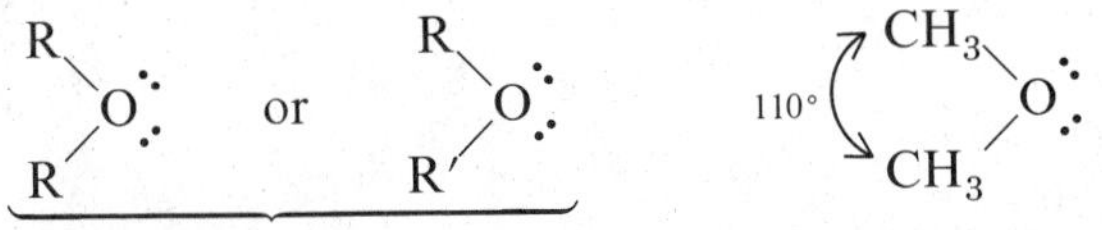

R—Ö—R or R—Ö—R′

General formula for an ether

CH_3—Ö—CH_3 (110°)

Dimethyl ether
(a typical ether)

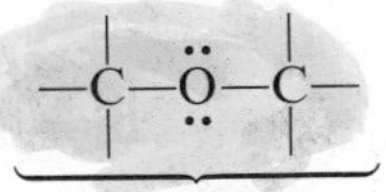

—C—Ö—C—

The functional group
of an ether

Problem 2.11

One way of naming ethers is to name the two alkyl groups attached to oxygen and add the word *ether.* If the two alkyl groups are the same, we use the prefix *di-,* for example, as in *dimethyl ether* above. Write structural formulas for (a) ethyl methyl ether, (b) dipropyl ether, (c) isopropyl methyl ether. (d) What name would you give to $CH_3CH_2OCH_2CH_2CH_3$? (e) To $(CH_3)_2CHOCH_2CH_2CH_3$?

Diethyl ether, $CH_3CH_2OCH_2CH_3$, has long been used as an anesthetic. Another ether used as an anesthetic is divinyl ether, $CH_2{=}CH{-}O{-}CH{=}CH_2$. (The vinyl group is $CH_2{=}CH{-}$.)

2.11 AMINES

Just as alcohols and ethers may be considered as organic derivatives of water, amines may be considered as organic derivatives of ammonia.

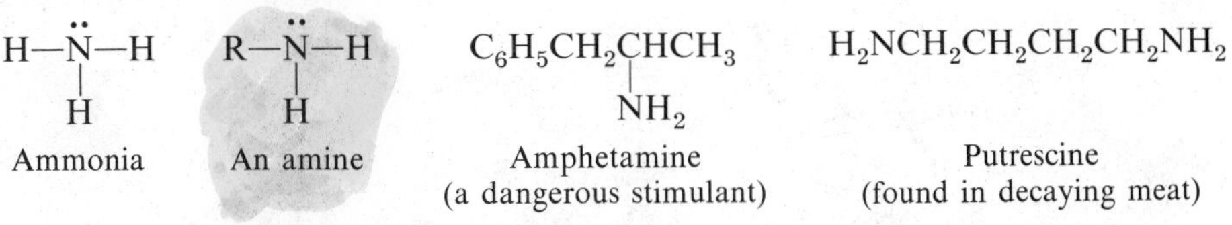

H—N̈(—H)—H	R—N̈(—H)—H	$C_6H_5CH_2CH(NH_2)CH_3$	$H_2NCH_2CH_2CH_2CH_2NH_2$
Ammonia	An amine	Amphetamine (a dangerous stimulant)	Putrescine (found in decaying meat)

Amines are classified as primary, secondary, or tertiary amines. This classification is based on *the number of organic groups that are attached to the nitrogen atom.* Notice that this is quite different from the way alcohols were

$R—\ddot{N}—H$ with H	$R—\ddot{N}—H$ with R′	$R—\ddot{N}—R''$ with R′
A primary (1°) amine	A secondary (2°) amine	A tertiary (3°) amine

classified. Isopropylamine, for example, is a primary amine even though its $—NH_2$ group is attached to a secondary carbon atom. It is a primary amine because only one organic group is attached to the nitrogen atom.

$$H_3C—CH(:NH_2)—CH_3$$

Isopropylamine
(a 1° amine)

Problem 2.12

One way of naming amines is to name the alkyl groups attached to nitrogen, using the prefixes *di-* and *tri-* if the groups are the same. Then *-amine* is added as a suffix (not as a separate word). An example is isopropylamine given above. Write formulas for (a) dimethylamine, (b) triethylamine, and (c) ethylmethylpropylamine. What are names for (d) $(CH_3)_2CHNHCH_3$, (e) $(CH_3CH_2CH_2)_2NCH_3$, and (f) $(CH_3)_2CHNH_2$?

Problem 2.13

Which amines in Problem 2.12 are (a) primary amines? (b) Secondary amines? (c) Tertiary amines?

If we consider the unshared electron pair of an amine as being at one corner, the general shape of an amine (below) is like that of ammonia; it is tetrahedral. The C—N—C bond angles of trimethylamine are 108.7°, a value very close to the H—C—H bond angles of methane. Thus, for all practical purposes, the nitrogen of an amine can be considered to be sp^3 hybridized. This means that the unshared electron pair occupies an sp^3 orbital, and thus it is considerably extended into space. This is important because, as we shall see, the unshared electron pair is involved in almost all of the reactions of amines.

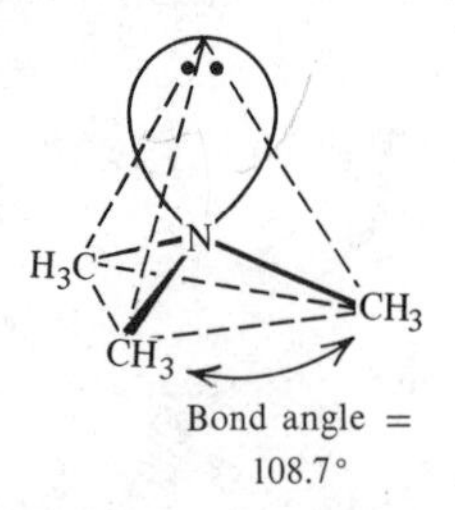

Bond angle = 108.7°

Problem 2.14

(a) What general hybridization state would you expect for the oxygen of an alcohol or an ether (cf. Section 2.9). (b) What kind of orbitals would you expect the unshared electron pairs to occupy?

2.12 ALDEHYDES AND KETONES

Aldehydes and ketones both contain the *carbonyl group*—a group in which a carbon atom forms a double bond to oxygen.

$$>C=\ddot{O}:$$

The carbonyl group

The carbonyl group in aldehydes is bonded to at least one *hydrogen atom,* and in ketones it is bonded to *two carbon atoms.* Using R, we can designate the general formula for an aldehyde as

$$R-\overset{\overset{\displaystyle \ddot{O}:}{\|}}{C}-H \qquad \text{R may also be H}$$

and the general formula for a ketone as

$$R-\overset{\overset{\displaystyle \ddot{O}:}{\|}}{C}-R \quad \text{or} \quad R-\overset{\overset{\displaystyle \ddot{O}:}{\|}}{C}-R'$$

(where R′ may be an alkyl group different from R).

Some examples of aldehydes and ketones are:

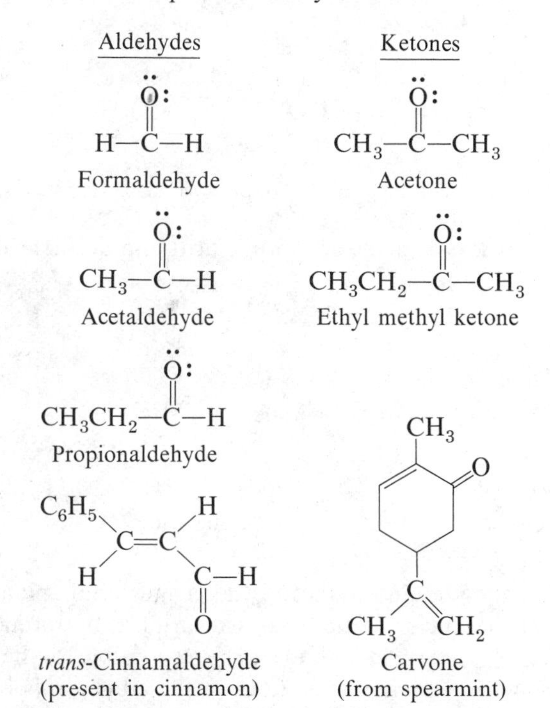

Aldehydes and ketones have a trigonal-planar arrangement of groups around the carbonyl carbon atom. In formaldehyde, for example, the bond angles are as follows:

H—C(=O:)—H with bond angles: 118° (H–C–H), 121° (H–C=O), 121° (H–C=O)

A trigonal planar arrangement of groups is just what we would expect on the basis of the VSEPR model if we consider that the two electron pairs of the double bond act as a single electron pair (cf. Problem 1.9). (We will discuss the structures of aldehydes and ketones on the basis of orbital hybridization later.)

2.13 CARBOXYLIC ACIDS, ESTERS, AND AMIDES

2.13A Carboxylic Acids

Carboxylic acids have the general formula R—C(=O)—O—H. The functional group, —C(=O)—O—H, is called the *carboxyl group* (*car*bonyl + hydro*xyl*). (Colloquially, carboxylic acids are often just called "organic acids.")

R—C(=Ö:)—Ö—H or RCOOH or RCO_2H
A carboxylic acid

—C(=Ö:)—Ö—H or —COOH or $—CO_2H$
The carboxyl group

Examples of carboxylic acids are formic acid and acetic acid.

H—C(=Ö:)—Ö—H or HCOOH or HCO_2H
Formic acid

CH_3—C(=Ö:)—Ö—H or CH_3COOH or CH_3CO_2H
Acetic acid

Formic acid is an irritating liquid produced by ants and certain plants, including nettles. (The sting of the ant or nettle is caused, in part, by formic acid being injected under the skin.) Acetic acid, the substance responsible for the sour taste of vinegar, is produced when certain bacteria act on the ethyl alcohol of wine and cause the ethyl alcohol to be oxidized.

2.13B Amides

Amides have the general formulas $RCONH_2$, RCONHR′, or RCONR′R″. Specific examples are the following:

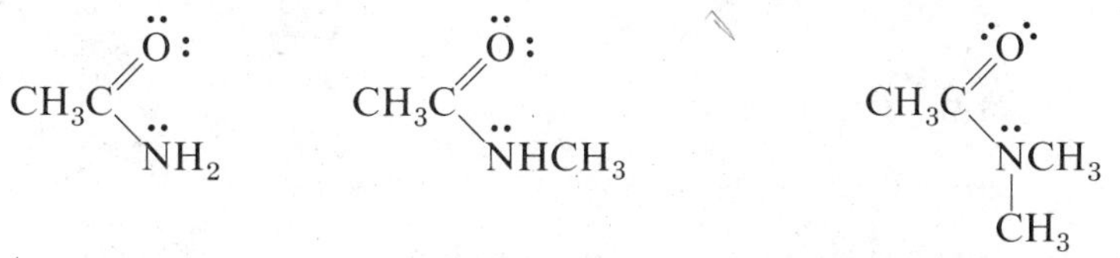

$$CH_3C(=\ddot{O}:)\ddot{N}H_2 \qquad CH_3C(=\ddot{O}:)\ddot{N}HCH_3 \qquad CH_3C(=\ddot{O}:)\ddot{N}(CH_3)CH_3$$

Acetamide — *N*-Methylacetamide (an *N*-substituted amide) — *N,N*-Dimethylacetamide (an *N,N*-disubstituted amide)

2.13C Esters

Esters have the general formula RCOOR′.

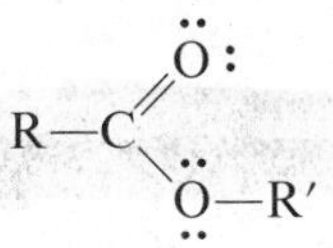

$$R{-}C(=\ddot{O}:){-}\ddot{O}{-}R' \quad \text{or} \quad RCOOR' \quad \text{or} \quad RCO_2R'$$

General formula for an ester

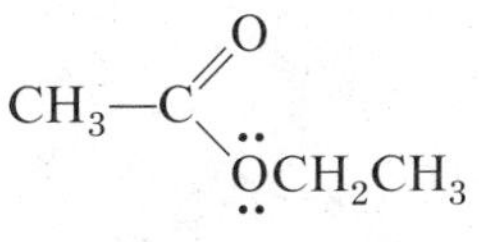

$$CH_3{-}C(=O){-}\ddot{O}CH_2CH_3 \quad \text{or} \quad CH_3COOCH_2CH_3 \quad \text{or} \quad CH_3CO_2CH_2CH_3$$

A specific ester called ethyl acetate

2.14 ACID-BASE REACTIONS: BRØNSTED-LOWRY ACIDS AND BASES

Involved at some point in the vast majority of reactions that occur with organic compounds are *acid-base reactions*. For this reason, we need to review some of the essential principles of acid-base chemistry.

2.14A Strong Acids and Bases

According to the Brønsted-Lowry theory an acid is a substance that can donate a proton, and a base is a substance that can accept a proton. Let us consider, as an example of this concept, the reaction that occurs when gaseous hydrogen chloride dissolves in water.

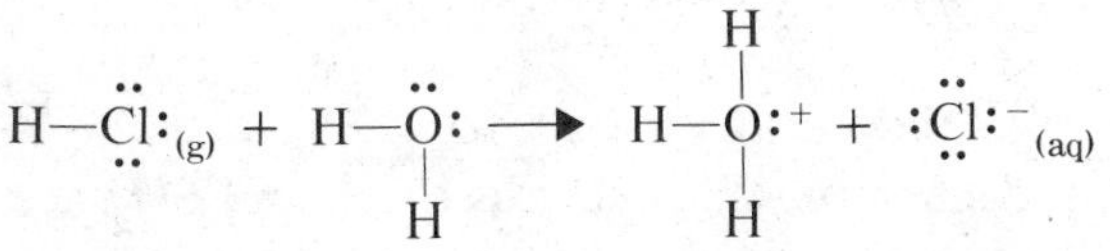

$$\underset{\text{Acid}}{H{-}\ddot{\underset{..}{Cl}}:_{(g)}} + \underset{\text{Base}}{H{-}\ddot{O}{-}H} \longrightarrow \underset{\substack{\text{Conjugate}\\\text{acid}}}{H{-}\ddot{O}(H)(H)^{+}} + \underset{\substack{\text{Conjugate}\\\text{base}}}{:\ddot{\underset{..}{Cl}}:^{-}_{(aq)}}$$

Hydrogen chloride, a very strong acid, transfers its proton to water. Water acts as a base and accepts the proton. The products that result from this reaction are the hydronium ion (H_3O^+, the conjugate acid of water) and the chloride ion (Cl^-, the conjugate base of hydrogen chloride). The reaction, for all practical purposes, goes to completion.

Other strong acids that show essentially complete proton transfer when dissolved in water are hydrogen bromide, hydrogen iodide, nitric acid, perchloric acid, and sulfuric acid.*

$$HBr + H_2O \longrightarrow H_3O^+ + Br^-$$

$$HI + H_2O \longrightarrow H_3O^+ + I^-$$

$$HNO_3 + H_2O \longrightarrow H_3O^+ + NO_3^-$$

$$HClO_4 + H_2O \longrightarrow H_3O^+ + ClO_4^-$$

$$H_2SO_4 + H_2O \longrightarrow H_3O^+ + HSO_4^-$$

$$HSO_4^- + H_2O \longrightarrow H_3O^+ + SO_4^=$$

Because sulfuric acid has two protons that it can transfer to a base, it is called a dibasic acid. The proton transfer is stepwise; the first proton transfer occurs completely and the second nearly so.

Hydronium ions, H_3O^+, and hydroxide ions, OH^-, are, for all practical purposes, the strongest acids and bases that are capable of existence in aqueous solutions. When sodium hydroxide (a crystalline compound consisting of sodium ions and hydroxide ions) dissolves in water, the result is a solution consisting of solvated sodium ions and hydroxide ions.

$$NaOH_{solid} + H_2O \longrightarrow Na^+_{(aq)} + OH^-_{(aq)}$$

When an aqueous solution of sodium hydroxide is mixed with an aqueous solution of hydrogen chloride (hydrochloric acid) the reaction that occurs is between hydroxide ions and hydronium ions. The sodium ions and chloride ions are called spectator ions because they play no part in the acid-base reaction.

$$\text{Net reaction:} \quad :\ddot{\underset{..}{O}}H^- + H\text{—}\underset{\underset{H}{|}}{\ddot{O}^+}\text{—}H \longrightarrow 2H:\ddot{\underset{..}{O}}H$$

$$\text{Spectator ions:} \quad Na^+ + \quad Cl^- \qquad\qquad Na^+ + Cl^-$$

What we have just said about hydrochloric acid and aqueous sodium hydroxide is true when aqueous solutions of all strong acids and strong bases are mixed. The net ionic reaction that occurs is simply

$$H_3O:^+ + :\ddot{\underset{..}{O}}H^- \longrightarrow 2H_2\ddot{O}:$$

2.14B Weak Acids and Bases

In contrast to the strong acids, such as HCl and H_2SO_4, acetic acid is a much weaker acid. When acetic acid dissolves in water, the reaction below does not proceed to completion.

* The extent to which an acid transfers the protons to a base like water is a measure of its "strength" as an acid. Acid strength is thus a measure of the percent of ionization and not of concentration.

$$CH_3C(=\ddot{O}:)-\ddot{O}H + H_2O \rightleftharpoons CH_3C(=\ddot{O}:)-\ddot{O}:^- + H_3O^+$$

Experiments show that in a 0.1-molar solution of acetic acid at 25 °C only 1% of the acetic acid molecules transfer their protons to water.

Because the reaction that occurs in an aqueous solution of acetic acid is an equilibrium, we can describe it with an expression for the equilibrium constant.

$$K_{eq} = \frac{[H_3O^+][CH_3COO^-]}{[CH_3COOH][H_2O]}$$

For dilute aqueous solutions the concentration of water is essentially constant (~55 moles/liter), so we can rewrite the expression for the equilibrium constant in terms of a new constant, K_a, called *the acidity constant.*

$$K_a = K_{eq}[H_2O] = \frac{[H_3O^+][CH_3COO^-]}{[CH_3COOH]}$$

At 25 °C, the acidity constant for acetic acid is 1.8×10^{-5}.

We can write similar expressions for any acid dissolved in water. Using a generalized hypothetical acid, HA, the reaction in water is

$$HA + H_2\ddot{O}: \rightleftharpoons H_3O:^+ + A^-$$

and the expression for the acidity constant is

$$K_a = \frac{[H_3O^+][A^-]}{[HA]}$$

In this standard form the molar concentrations of the products of the proton transfer reaction are written in the numerator of the expression and the molar concentration of the undissociated acid is written in the denominator. For this reason, a large value of K_a means the acid is a strong acid, and a small value of K_a means the acid is a weak acid. If the acidity constant is greater than 10 the acid will be, for all practical purposes, completely dissociated in water.

Problem 2.15

Trifluoroacetic acid (CF_3COOH) has a $K_a = 1$ at 25°. (a) What are the molar concentrations of hydronium ion and trifluoroacetate ion (CF_3COO^-) in a 0.1-M aqueous solution of trifluoroacetic acid? (b) What percentage of the trifluoroacetic acid is ionized?

Although acetic acid and other carboxylic acids containing fewer than five carbon atoms are soluble in water, many other carboxylic acids of higher molecular weight are not appreciably soluble in water. Because of their acidity, however, *water-insoluble acids dissolve in aqueous sodium hydroxide;* they do so by reacting to form *water-soluble* sodium salts.

$$\underset{\text{Water insoluble}}{C_7H_{15}COOH} + Na^+OH^- \longrightarrow \underset{\text{Water soluble}}{C_7H_{15}COO^-Na^+} + H_2O$$

Amines, like ammonia, are weak bases. The reaction of an aqueous solution of ethylamine with a strong aqueous acid, for example, is very similar to that of aqueous ammonia.

$$\underset{\text{Ammonia}}{:NH_3} + H_3O:^+ \rightleftharpoons \underset{\text{Ammonium ion}}{NH_4^+} + H_2\ddot{O}:$$

$$\underset{\text{Ethylamine}}{CH_3CH_2\ddot{N}H_2} + H_3O:^+ \rightleftharpoons \underset{\text{Ethylammonium ion}}{CH_3CH_2\overset{+}{N}H_3} + H_2\ddot{O}:$$

While ethylamine and most amines of low molecular weight are very soluble in water, higher molecular weight amines, such as hexylamine ($C_6H_{13}NH_2$), have limited water solubility. However, such water-insoluble amines dissolve readily in hydrochloric acid because the acid-base reaction produces a soluble salt.

$$\underset{\substack{\text{Hexylamine} \\ \textit{Slight water} \\ \textit{solubility}}}{C_6H_{13}\ddot{N}H_2} + HCl_{(aq)} \longrightarrow \underset{\substack{\text{Hexylammonium chloride} \\ \textit{Water-soluble} \\ \textit{salt}}}{C_6H_{13}\overset{+}{N}H_3\ Cl^-}$$

Water, itself, is a very weak acid and undergoes self-ionization even in the absence of added acids or bases.

$$H_2\ddot{O}: + H_2\ddot{O}: \rightleftharpoons H_3O:^+ + :\ddot{O}H^-$$

The self-ionization of pure water produces concentrations of hydronium and hydroxide ions equal to 10^{-7} moles/liter at 25 °C. Since the concentration of water in pure water is 55.5 moles/liter we can calculate the K_a for water.

$$K_a = \frac{[H_3O^+][OH^-]}{[H_2O]} \qquad K_a = \frac{(10^{-7})(10^{-7})}{55.5} = 1.8 \times 10^{-16}$$

In our discussion so far we have dealt only with the strengths of acids. Arising as a natural corollary to this is a principle that allows us to estimate the strengths of bases as well. Simply stated, *the conjugate base of a strong acid will be a weak base, and the conjugate base of a weak acid will be a strong base.* Moreover, *the stronger the acid, the weaker will be its conjugate base* and vice versa.

As examples of this principle consider the following:

1. The chloride ion (Cl^-), bromide ion (Br^-), iodide ion (I^-), nitrate ion (NO_3^-), and perchlorate ion (ClO_4^-) are all conjugate bases of very strong acids (K_a's > 10) and, thus, all are very weak bases. In aqueous solution these anions have virtually no affinity for protons and exist as simple solvated ions.
2. The hydroxide ion (OH^-) is the conjugate base of water; water is a weak acid ($K_a \simeq 10^{-16}$). The hydroxide ion is therefore a strong base. It is, as we have said, the strongest base that can exist in aqueous solutions.
3. The acetate ion (CH_3COO^-) is the conjugate base of the moderately weak acid, acetic acid ($K_a \simeq 10^{-5}$). As a result, the acetate ion is a moderately strong base.

2.14C Acids and Bases in Nonaqueous Solutions

Many of the organic acid-base reactions that we describe in this text occur in solutions other than aqueous solution. Many of the acids that we encounter are much weaker acids than water. The conjugate bases of these very weak acids are exceedingly powerful bases. Reactions of organic compounds, for example, are often carried out in liquid ammonia (the liquefied gas, not the aqueous solution commonly used in your general chemistry laboratory). The acidity constant, K_a, of ammonia is 10^{-34}, thus liquid ammonia is a weaker acid than water by a factor of 10^{18}. The most powerful base that can be employed in liquid ammonia is the amide ion, $:\ddot{N}H_2^-$, the conjugate base of ammonia. The amide ion, because it is the conjugate base of an extremely weak acid, is an extremely powerful base. In liquid ammonia, and in the presence of amide ion, even the hydrocarbon ethyne becomes an acid.

$$\underset{\text{Stronger acid}}{H{-}C{\equiv}C{-}H} + \underset{\text{Stronger base}}{:\ddot{N}H_2^-} \xrightarrow[NH_3]{\text{liq.}} \underset{\text{Weaker base}}{H{-}C{\equiv}C:^-} + \underset{\text{Weaker acid}}{:NH_3}$$

Cyclohexane (C_6H_{12}) is an even weaker acid ($K_a = 10^{-45}$) than ammonia. The cyclohexyl anion ($C_6H_{11}:^-$) is, thus, an even stronger base than the amide ion. If ammonia were added to a solution containing cyclohexyl anions in cyclohexane the following reaction would take place.

$$\underset{\text{Stronger base}}{C_6H_{11}:^-} + \underset{\text{Stronger acid}}{:NH_3} \longrightarrow \underset{\text{Weaker acid}}{C_6H_{12}} + \underset{\text{Weaker base}}{:\ddot{N}H_2^-}$$

The acidity constants of some of the molecules we have considered so far are listed in Table 2.2. All of the proton acids that we consider in this book have strengths in between that of cyclohexane and that of $SbF_5 \cdot FSO_3H$ (called a "superacid"). On the other hand, all of the bases that we will consider will be

TABLE 2.2 Scale of Acidities and Basicities

	ACID	APPROXIMATE K_a	CONJUGATE BASE	
INCREASING STRENGTH OF ACID (↓)	C_6H_{12}	10^{-45}	$C_6H_{11}^-$	INCREASING STRENGTH OF CONJUGATE BASE (↑)
	CH_3CH_3	10^{-42}	$CH_3CH_2^-$	
	$CH_2{=}CH_2$	10^{-36}	$CH_2{=}CH^-$	
	NH_3	10^{-34}	NH_2^-	
	$HC{\equiv}CH$	10^{-25}	$HC{\equiv}C^-$	
	CH_3CH_2OH	10^{-18}	$CH_3CH_2O^-$	
	H_2O	10^{-16}	HO^-	
	CH_3COOH	10^{-5}	CH_3COO^-	
	CF_3COOH	1	CF_3COO^-	
	HNO_3	20	NO_3^-	
	H_3O^+	50	H_2O	
	HCl	10^7	Cl^-	
	H_2SO_4	10^9	HSO_4^-	
	HI	10^{10}	I^-	
	$HClO_4$	10^{10}	ClO_4^-	
	$SbF_5 \cdot FSO_3H$	$>10^{12}$	$SbF_5 \cdot FSO_3^-$	

weaker in strength than the cyclohexyl anion. As you examine Table 2.2, however, take care not to lose sight of the vast spectrum of acidities and basicities that it represents.

2.15 LEWIS ACID-BASE THEORY

Acid-base theory was broadened immensely by G. N. Lewis in 1923. Striking at what he called "the cult of the proton," Lewis proposed that acids be defined as *electron-pair acceptors* and bases as *electron-pair donors*. In the Lewis theory, not only is the proton an acid, but many other species are as well. Aluminum chloride and boron trifluoride, for example, react with amines in much the same way a proton does.

$$H^+ + :NH_2R \longrightarrow H\text{—}\overset{+}{N}H_2\text{—}R$$

$$AlCl_3 + :NH_2R \longrightarrow Cl_3\overset{-}{Al}\text{—}\overset{+}{N}H_2\text{—}R$$

$$BF_3 + :NH_2R \longrightarrow F_3\overset{-}{B}\text{—}\overset{+}{N}H_2\text{—}R$$

In the examples above, aluminum chloride and boron trifluoride accept the electron pair of the amine just as the proton does. They do this because the central aluminum and boron atoms have only a sextet of electrons and are thus electron deficient. When they accept an electron pair, aluminum chloride and boron trifluoride are, in the Lewis definition, *acting as acids*.

Bases are much the same in both the Lewis theory and the Brønsted-Lowry theory.

The Lewis theory, by virtue of its broader definition of acids, allows acid-base theory to include all of the Lowry-Brønsted reactions and, as we will see, a great many others.

2.16 PHYSICAL PROPERTIES AND MOLECULAR STRUCTURE

So far, we have said little about one of the most obvious characteristics of organic compounds, that is, *their physical state*. Whether a particular substance is a solid, or a liquid, or a gas, would certainly be one of the first observations that we would note in any experimental work. The temperatures at which transitions occur between physical states, that is, melting points and boiling points, are also among the more easily measured physical properties. Melting points and boiling points are also useful in identifying and isolating organic compounds.

Suppose, for example, we have just carried out the synthesis of an organic compound that is known to be a liquid at room temperature and one atmosphere pressure. If we know the boiling point of our desired product, and the boiling points of other by-products and solvents that may be present in the reaction mixture, we can decide whether or not simple distillation will be a feasible method for isolating our product.

In another instance our product might be a solid. In this case, in order to isolate the substance by crystallization, we need to know its melting point and its solubility in different solvents.

The physical constants of known organic substances are easily found in handbooks and journals.* Table 2.3 lists the melting and boiling points of some of the compounds that we have discussed in this chapter.

Often in the course of research, however, the product of a synthesis is a new compound—one that has never been described before. In these instances, success in isolating the new compound depends on making reasonably accurate estimates of its melting point, boiling point, and solubilities. Estimations of these macroscopic physical properties are based on the most likely structure of the substance and on the forces that act between molecules and ions. What are these forces? How do they affect the melting point, boiling point, and solubilities of a compound?

2.16A Ion-Ion Forces

The *melting point* of a substance is the temperature at which an equilibrium exists between the well-ordered crystalline state and the more random liquid state. If the substance is an ionic compound, such as sodium acetate (Table 2.3), the forces that hold the ions together in the crystalline state are the strong electrostatic lattice forces that act between the positive and negative ions in the orderly crystalline structure. In Fig. 2.13 each sodium ion is surrounded by negatively charged acetate ions, and each acetate ion is surrounded by positive sodium ions. A large amount of thermal energy is required to break up the orderly structure of the crystal into the disorderly open structure of a liquid. As a result, the tempera-

* Two useful handbooks are: *Handbook of Chemistry,* N. A. Lange, McGraw-Hill, New York and *Handbook of Chemistry and Physics,* Chemical Rubber Co., Cleveland.

TABLE 2.3

COMPOUND	STRUCTURE	mp °C	bp °C at 1 atm
Methane	CH_4	−183	−162
Ethane	CH_3CH_3	−172	−88.2
Ethene	$CH_2{=}CH_2$	−169	−102
Ethyne	$HC{\equiv}CH$	−82	−75
Chloromethane	CH_3Cl	−97	−23.7
Chloroethane	CH_3CH_2Cl	−138.7	13.1
Ethyl alcohol	CH_3CH_2OH	−115	78.3
Acetaldehyde	CH_3CHO	−121	20
Acetic acid	CH_3COOH	16.6	118
Sodium acetate	CH_3COONa	324	Dec
Ethylamine	$CH_3CH_2NH_2$	−80	17
Diethyl ether	$(CH_3CH_2)_2O$	−116	34.6
Ethyl acetate	$CH_3COOCH_2CH_3$	−84	77

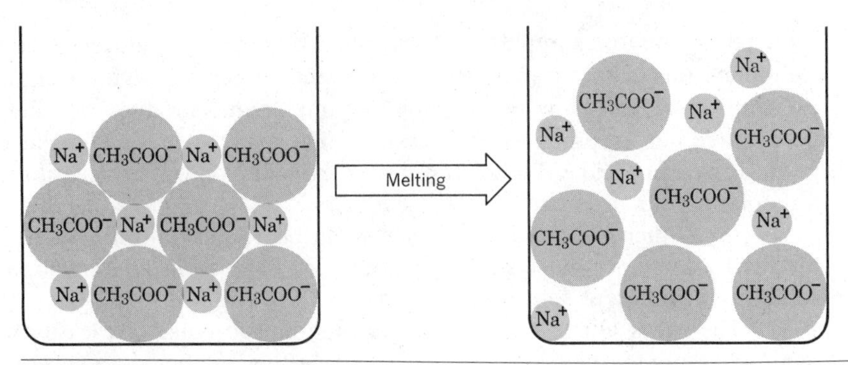

FIG. 2.13
The melting of sodium acetate.

ture at which sodium acetate melts is quite high, 324°. The *boiling points* of ionic compounds are higher still, so high that most ionic organic compounds decompose before they boil. Sodium acetate shows this behavior.

2.16B Van der Waals Forces

If we consider a substance like methane where the particles are nonpolar molecules, not oppositely charged ions, we find that the melting point and boiling point are *much* lower: −183° and −162°, respectively. Rather than ask, "Why does methane melt and boil at low temperatures?" a more appropriate question might be: "Why does methane, a nonionic, nonpolar substance, become a liquid or a solid at all?" The answer to this question can be given in terms of attractive intermolecular forces called *van der Waals forces* (or London forces).

An accurate account of the nature of van der Waals forces requires the use of quantum mechanics. We can, however, visualize the origin of these forces in the following way. The average distribution of charge in a nonpolar molecule (like methane) over a period of time is uniform. At any given instant, however, *because electrons move,* the electrons and thus the charge may not be uniformly distributed. Electrons may, in one instant, be slightly accumulated on one side of the molecule and, as a consequence, *a small temporary dipole will occur* (Fig. 2.14). This temporary dipole in one molecule can induce opposite (attractive) dipoles in surrounding molecules. It does this because the negative (or positive) charge in a portion of one molecule will distort the electron cloud of an adjacent portion of another molecule causing an opposite charge to develop there. These temporary dipoles change constantly, but the net result of their existence is to produce weak attractive forces between nonpolar molecules, and thus make possible the existence of their liquid and solid states.

An important factor that determines the magnitude of van der Waals forces

FIG. 2.14
Temporary dipoles and induced dipoles in nonpolar molecules resulting from a non-uniform distribution of electrons at a given instant.

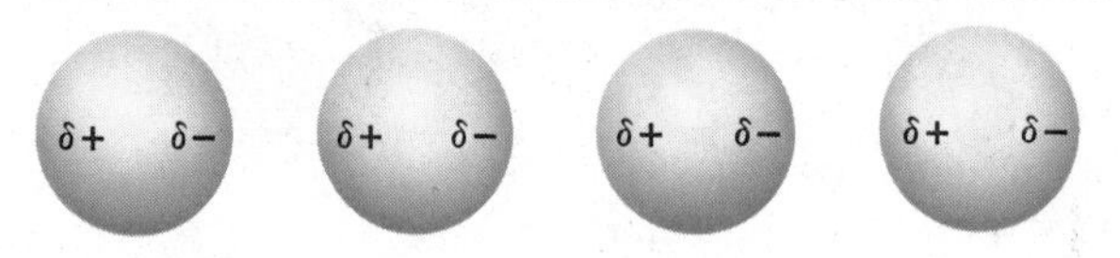

is the relative *polarizability* of the electrons of the atoms involved. By polarizability we mean *the ability of the electrons to respond to a changing electric field.* Relative polarizability depends on how loosely or tightly the electrons are held. In the halogen family, for example, polarizability increases in the order F < Cl < Br < I. Fluorine atoms show a very low polarizability because their electrons are very tightly held; they are close to the nucleus. Iodine atoms are easily polarized. Their electrons are far from the nucleus. Electrons of unshared pairs are generally more polarizable than those of bonding pairs. Thus the electrons of a halogen substituent are more polarizable than those of an alkyl group of comparable size.

The *boiling point* of a liquid is the temperature at which the vapor pressure of the liquid equals the pressure of the atmosphere above it. For this reason, the boiling points of liquids are *pressure dependent,* and boiling points are always reported as occurring at a particular pressure, at 1 atm (or at 760 Torr),* for example. A substance that boils at 150° at one atmosphere pressure will boil at a substantially lower temperature if the pressure is reduced to, for example, 0.01 Torr (a pressure easily obtained with a vacuum pump). The normal boiling point given for a liquid is its boiling point at 1 atm.

In passing from a liquid to a gaseous state the individual molecules (or ions) of the substance must separate considerably (Fig. 2.15). Because of this, we can understand why ionic organic compounds often decompose before they boil. The thermal energy required to completely separate (volatilize) the ions is so great that chemical reactions (decompositions) occur first.

Nonpolar compounds, where the intermolecular forces are very weak, usually boil at low temperatures even at 1 atm pressure. This is not always true, however, because of other factors that we have not yet mentioned: the effects of molecular weight and molecular size. Heavier molecules require greater thermal energy in order to acquire velocities sufficiently great to escape the liquid surface, and because their surface areas are usually much greater, intermolecular van der Waals attractions are also much larger. These factors explain why nonpolar ethane (bp, −88.2°) boils higher than methane (bp, −162°) at a pressure of 1 atm. It also explains why, at 1 atm, the even heavier and larger nonpolar molecule *n*-decane ($C_{10}H_{22}$) boils at +174°.

A factor (in addition to polarity) that affects the melting point of many organic compounds is the compactness and rigidity of their individual molecules. Molecules that are symmetrical and rigid fit easily into a crystal lattice and thus their compounds have higher melting points. *Tert*-butyl alcohol, for example has a much higher melting point than the other isomeric alcohols shown below.

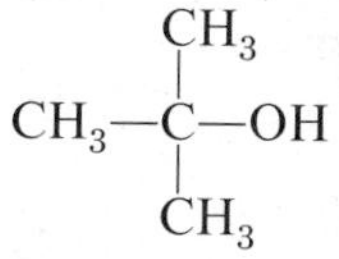

$$CH_3-\overset{\displaystyle CH_3}{\underset{\displaystyle CH_3}{\overset{|}{\underset{|}{C}}}}-OH \qquad CH_3CH_2CH_2CH_2OH \qquad CH_3\overset{\displaystyle CH_3}{\overset{|}{C}}HCH_2OH \qquad CH_3CH_2\overset{\displaystyle CH_3}{\overset{|}{C}}HOH$$

tert-Butyl alcohol, mp 25° — *n*-Butyl alcohol, mp −90° — Isobutyl alcohol, mp −108° — *sec*-Butyl alcohol, mp −114°

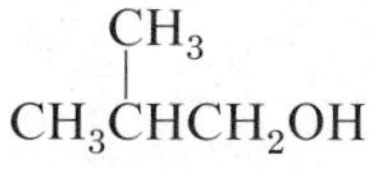

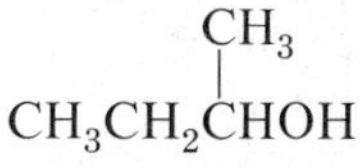

2.16C Dipole-Dipole Forces

Most organic molecules are neither ionic nor nonpolar but have *permanent dipoles* resulting from a nonuniform distribution of the bonding electrons. Chloromethane and chloroethane are examples of molecules with permanent dipoles.

* One Torr is the pressure exerted by a column of mercury 1 mm high. One atmosphere (1 atm) is equal to 760 Torr.

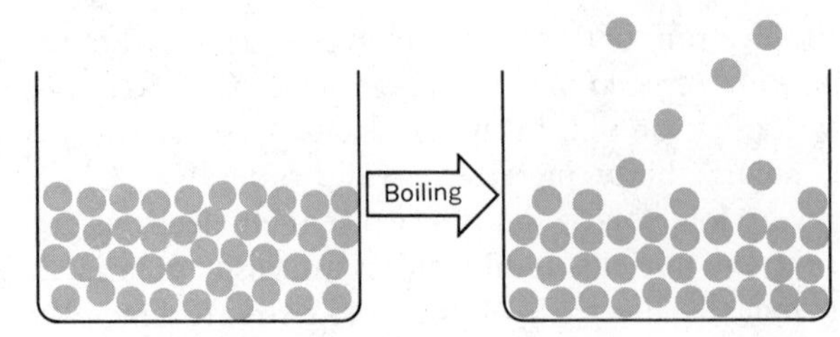

FIG. 2.15
The boiling of a liquid.

In these compounds, the attractive forces between molecules are much easier to visualize. In the liquid or solid state the molecules orient themselves so that the positive end of one molecule is directed toward, and thus attracted by, the negative end of another (Fig. 2.16). Dipole-dipole forces are usually stronger than van der Waals forces.

2.16D Hydrogen Bonds

Very strong dipole-dipole attractions occur between hydrogen atoms bonded to small, strongly electronegative elements and nonbonding electron pairs on other such electronegative elements (Fig. 2.17). This type of intermolecular force is called a *hydrogen bond.* The hydrogen bond (bond dissociation energy about 5 kcal/mole) is weaker than an ordinary covalent bond, but is much stronger than the dipole-dipole interactions that occur in chloromethane.

Hydrogen bonding accounts for the fact that ethyl alcohol has a much higher boiling point (+78.3°) than dimethyl ether (−25°) even though the two compounds have the same molecular weight. Molecules of ethyl alcohol, because they have a hydrogen covalently bonded to oxygen, can form strong hydrogen bonds to each other.

CH_2CH_3
CH_3CH_2 H etc.
etc. O: O: :O
H H CH_2CH_3

Molecules of dimethyl ether, because they lack a hydrogen attached to a strongly electronegative element, cannot form strong hydrogen bonds to each other. In dimethyl ether the intermolecular forces are weaker dipole-dipole interactions.

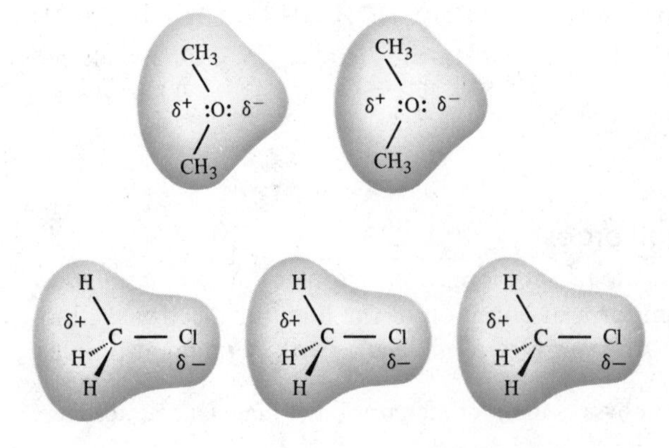

FIG. 2.16
Dipole-dipole interactions between chloromethane molecules.

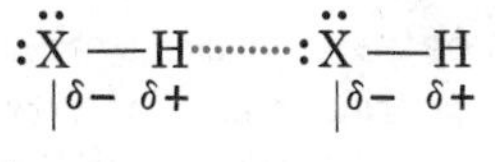

$$:\ddot{X}—H\cdots\cdots:\ddot{X}—H$$
$$\delta- \quad \delta+ \qquad \delta- \quad \delta+$$

FIG. 2.17
The hydrogen bond. X is a strongly electronegative element, usually oxygen, nitrogen, or fluorine.

Problem 2.16

Explain why trimethylamine, $(CH_3)_3N$, has a considerably lower boiling point (3°) than *n*-propylamine, $CH_3CH_2CH_2NH_2$, (49°) even though these two compounds have the same molecular weight.

2.16E Solubilities

Intermolecular forces are of primary importance in explaining the *solubilities* of substances. Dissolution of a solid in a liquid is, in many respects, like the melting of a solid. The orderly crystal structure of the solid is destroyed, and the result is the formation of the more disorderly arrangement of the molecules (or ions) in solution. In the process of dissolving, too, the molecules or ions must be separated from each other, and energy must be supplied for both changes. The energy required to overcome lattice energies and intermolecular or interionic attractions comes from the formation of new attractive forces between solute and solvent.

Consider the dissolution of an ionic substance as an example. Here both the lattice energy and interionic attractions are large. We find that water and only a few other very polar solvents are capable of dissolving ionic compounds. These solvents dissolve ionic compounds by *hydrating* or *solvating* the ions (Fig. 2.18).

Water molecules, by virtue of their great polarity, as well as their very small compact shapes, can very effectively surround the individual ions as they are freed from the crystal surface. Positive ions are surrounded by water molecules with the negative end of the water dipole pointed toward the positive ion; negative ions are solvated in exactly the opposite way. Because water is highly polar, and because water is capable of forming strong hydrogen bonds, the *dipole-ion* attractive forces

FIG. 2.18
The dissolution of an ionic solid in water showing the hydration of positive and negative ions by the very polar water molecules.

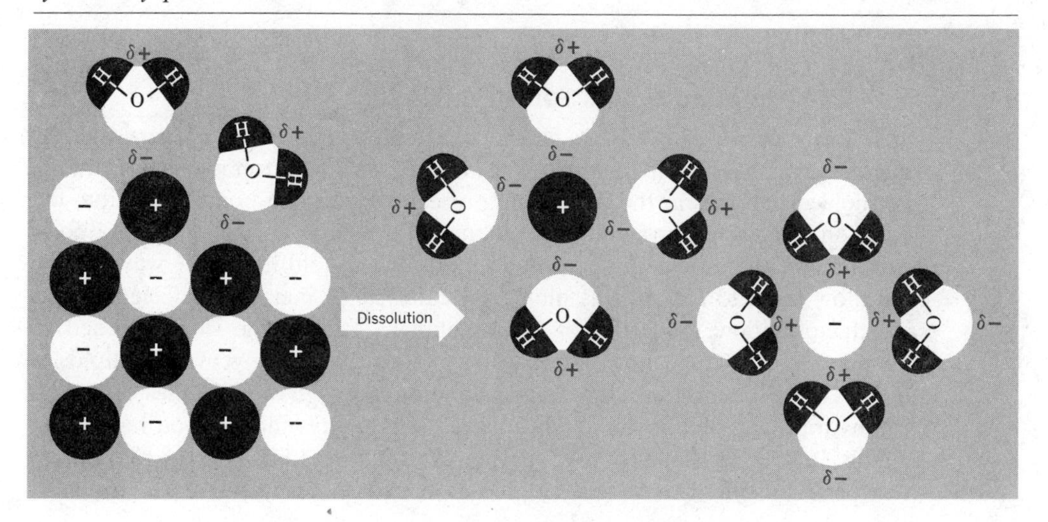

are also large. The energy supplied by the formation of these forces is great enough to overcome both the lattice energy and interionic attractions of the crystal.

A rule of thumb for predicting solubilities is that "like dissolves like." Polar and ionic compounds tend to dissolve in polar solvents. Polar liquids are generally miscible with each other in all proportions. Nonpolar solids are usually soluble in nonpolar solvents. On the other hand, nonpolar solids are insoluble in polar solvents. Nonpolar liquids are usually mutually miscible, but nonpolar liquids and polar liquids "like oil and water" do not mix.

We can understand why this is true if we understand that, when substances of similar polarities are mixed, the "new" intermolecular forces that form in the solution are very much like those that existed in the separate substances. The miscibility of nonpolar carbon tetrachloride with a nonpolar alkane would be an example. Very polar water molecules are probably capable of inducing polarities in alkane molecules that are sufficiently large to form attractive forces between them. Water and alkanes are not soluble in each other, however, because dissolution of the alkane in water requires the separation of strongly attractive water molecules from each other.

Ethanol and water, by contrast, are miscible in all proportions. In this example, both molecules are polar and the new attractive forces are as strong as those they replace and, in this instance, both compounds are capable of forming strong hydrogen bonds.

CH_3CH_2 ... H ... H ... etc.
O ... O ... O
etc. . .H ... H ... CH_2CH_3

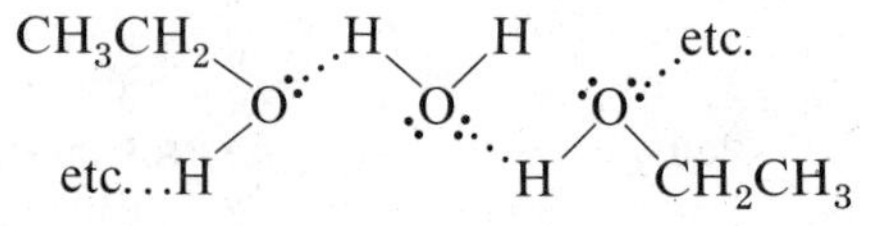

2.16F Guidelines for Water Solubility

Organic chemists usually define a compound as water soluble if at least 2 grams of the organic compound dissolve in 100 ml of water. We find that for compounds containing nitrogen or oxygen atoms—and thus capable of forming strong hydrogen bonds—the following guidelines hold. Compounds with one to three carbon atoms are water soluble, compounds with four or five carbons are borderline, and compounds with six carbons or more are insoluble.

2.16G Intermolecular Forces in Biochemistry

Later, after we have had a chance to examine in detail the properties of the molecules that make up living organisms, we will see how extremely important are intermolecular forces in the functioning of cells. Hydrogen bond formation, the hydration of polar groups, and the tendency of nonpolar groups to avoid a polar environment, all cause complex protein molecules to fold in precise ways—ways that allow them to function as biological catalysts of incredible efficiency. The same factors allow molecules of hemoglobin to assume the shape needed to transport oxygen. They allow proteins and molecules called glycosphingolipids to function as cell membranes. Hydrogen bonding alone gives molecules of certain carbohydrates a globular shape that makes them highly efficient food reserves in animals. It gives molecules of other carbohydrates a rigid linear shape that makes them perfectly suited to be structural components in plants.

Additional Problems

2.17
Classify each of the following compounds as an alkane, alkene, alkyne, alcohol, or aldehyde, and so forth.

(a) $CH_3C{\equiv}CCH_3$ (b) $CH_3CH(CH_3)CH_2C(=O)OH$ (c) cyclopentanol (ring with OH) (d) CH_2–CH_2 ring with CHCH=O

(e) $CH_3CH(CH_3)CH_2CH_2CH_2CH_2CH_2CH_2CH_2CH_2CH_2CH_2CH_2CH_2CH_2CH_3$

(a sex attractant of the female tiger moth)

(f) (obtained from peppermint oil)

2.18
Identify all of the functional groups in each of the following compounds.

(a) CH_3 ... OH

(vitamin A_1)

(b) OH ... O

(testosterone, a male sex hormone)

(c) CH_3 O ... O ... CH_3

(nepetalactone, one constituent of catnip)

(d) $[-NH-(CH_2)_6-NH\overset{O}{\overset{\|}{C}}-(CH_2)_4-\overset{O}{\overset{\|}{C}}-]_n$ (nylon)

(e)

$$\begin{array}{c} O \\ \| \\ CH \\ | \\ H-C-OH \\ | \\ HO-C-H \\ | \\ H-C-OH \\ | \\ H-C-OH \\ | \\ CH_2OH \end{array}$$

(glucose, a sugar)

(f) $CH_2{=}CH{-}O{-}CH{=}CH_2$ (an anesthetic)

(g) 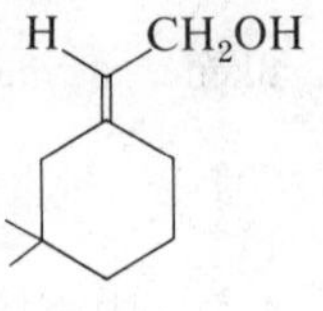 H CH$_2$OH (male boll-weevil sex attractant)

2.19

Classify the following alcohols as primary, secondary, or tertiary.

(a) $CH_3CH_2CH_2CH_2OH$

(b) $CH_3CH_2CH_2CH(OH)CH_3$

(c) $CH_3CH_2C(OH)(CH_3)CH_3$

(d) $CH_3CH(CH_3)CH(OH)CH_3$

(e) 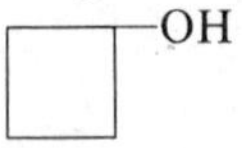 cyclobutyl–OH

(f) 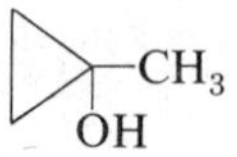 cyclopropane ring bearing CH_3 and OH on the same carbon

2.20

Classify the following amines as primary, secondary, or tertiary.

(a) $CH_3CH_2NHCH_2CH_3$

(b) $CH_3CH(NH_2)CH_2CH_3$

(c) $CH_3NCH_2CH_2CH_3$ with CH_3 on N

(d) 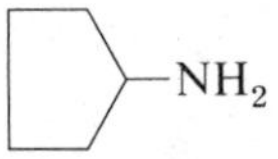 cyclopentyl–NH_2

(e) 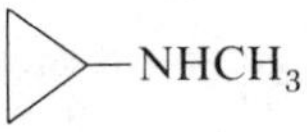 cyclopropyl–$NHCH_3$

2.21

Write net ionic equations for the following acid-base reactions.

(a) $HCl_{(aq)} + Na_2CO_{3(aq)} \longrightarrow H_2O + CO_2 + NaCl_{(aq)}$

(b) $HBr_{(aq)} + CH_3\overset{O}{\overset{\|}{C}}ONa_{(aq)} \longrightarrow CH_3\overset{O}{\overset{\|}{C}}OH + NaBr_{(aq)}$

(c) $Na_2CO_{3(aq)} + H_2O \longrightarrow NaHCO_3 + NaOH$

(d) $NaH + H_2O \longrightarrow H_2 + NaOH_{(aq)}$

(e) $CH_3Li + H_2O \longrightarrow CH_4 + LiOH$

(f) $CH_3Li + HC{\equiv}CH \longrightarrow LiC{\equiv}CH + CH_4$

(g) $HCl_{(aq)} + NH_{3(aq)} \longrightarrow NH_4Cl_{(aq)}$

(h) $NH_4Cl + NaNH_2 \xrightarrow[NH_3]{} 2NH_3 + NaCl$

(i) $CH_3CH_2ONa + H_2O \longrightarrow CH_3CH_2OH + NaOH$

2.22

Explain why almost all oxygen-containing organic compounds dissolve in concentrated sulfuric acid.

2.23

Which compound in each of the following pairs would have the higher boiling point?

(a) Ethyl alcohol, CH_3CH_2OH, or methyl ether, CH_3OCH_3

(b) Ethylene glycol, $HOCH_2CH_2OH$, or ethyl alcohol, CH_3CH_2OH

(c) Pentane, C_5H_{12}, or heptane, C_7H_{16}

(d) Acetone, $CH_3\overset{O}{\overset{\|}{C}}CH_3$ or 1-propanol, $CH_3CH_2CH_2OH$

(e) *cis*-1,2-Dichloroethene, $\overset{Cl}{\underset{H}{}}C{=}C\overset{Cl}{\underset{H}{}}$ or *trans*-1,2-dichloroethene, $\overset{Cl}{\underset{H}{}}C{=}C\overset{H}{\underset{Cl}{}}$

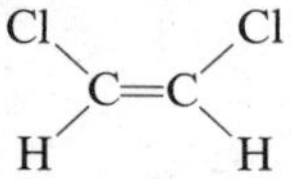

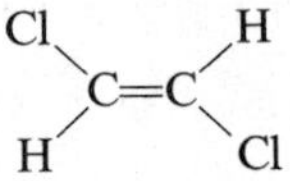

(f) Propionic acid, $CH_3CH_2\overset{\displaystyle O}{\overset{\|}{C}}OH$, or methyl acetate, $CH_3\overset{\displaystyle O}{\overset{\|}{C}}{-}OCH_3$

2.24
Write structural formulas for each of the following:
(a) An ether with the formula C_3H_8O
(b) A primary alcohol with the formula C_3H_8O
(c) A secondary alcohol with the formula C_3H_8O
(d) Two esters with the formula $C_4H_8O_2$
(e) A primary alkyl halide with the formula C_4H_9X (alkyl halides are classified in the same way as alcohols)
(f) A secondary alkyl halide with the formula C_4H_9X
(g) A tertiary alkyl halide with the formula C_4H_9X
(h) An aldehyde with the formula C_4H_8O
(i) A ketone with the formula C_4H_8O
(j) A primary amine with the formula $C_4H_{11}N$
(k) A secondary amine with the formula $C_4H_{11}N$
(l) A tertiary amine with the formula $C_4H_{11}N$
(m) An amide with the formula C_4H_9NO
(n) An *N*-substituted amide with the formula C_4H_9NO
(o) A tertiary alcohol with the formula C_4H_8O containing no multiple bonds

2.25
Write Newman projection formulas for the staggered and eclipsed forms of propane.

2.26
$(CF_3)_3N$ is a weaker base than $(CH_3)_3N$. Explain.

2.27
Compounds of the general type shown below are called lactones. What functional group do they contain?

```
   CH2—CH2
   |    |
   C    CHR
 //  \ /
O     O
```

2.28
Hydrogen fluoride has a dipole moment of 1.9 D; its boiling point is 19°. Ethyl fluoride, CH_3CH_2F, has an almost identical dipole moment and has a larger molecular weight, yet its boiling point is −32°. Explain.

***2.29**
What factors might account for the fact that tetrafluoromethane, CF_4, has a much lower boiling point (bp = −129°) than hexane, C_6H_{14}, (bp = 68°) even though both compounds are nonpolar and both have approximately the same molecular weight?

***2.30**
The boiling points of nonpolar molecules increase with increasing size and increasing numbers of electrons. The boiling points of the halogens and of the noble gases, for example, show the following orders: $I_2 > Br_2 > Cl_2 > F_2$, and Xe > Kr > Ar > Ne > He. Explain.

* An asterisk beside a problem indicates that it is somewhat more challenging. Your instructor may tell you that these problems are optional.

3 ALKANES AND CYCLOALKANES

3.1 INTRODUCTION

We noted earlier that the family of organic compounds called hydrocarbons can be divided into several groups based on the type of bond that exists between the individual carbon atoms. Those hydrocarbons in which all of the carbon-carbon bonds are single bonds are called *alkanes;* those hydrocarbons that contain a carbon-carbon double bond are called *alkenes;* and those with a carbon-carbon triple bond are called *alkynes.*

Cycloalkanes are alkanes in which the carbon atoms are arranged in a ring. Alkanes have the general formula C_nH_{2n+2}; cycloalkanes have two fewer hydrogen atoms and thus have the general formula C_nH_{2n}. Several examples of alkanes and cycloalkanes are given below.

Alkanes	Cycloalkanes
C_nH_{2n+2}	C_nH_{2n}
$CH_3CH_2CH_3$ Propane (C_3H_8)	CH_2 ring: CH_2—CH_2—CH_2 Cyclopropane (C_3H_6)
$CH_3CH_2CH_2CH_3$ Butane (C_4H_{10})	CH_2—CH_2 CH_2—CH_2 Cyclobutane (C_4H_8)
$CH_3CH_2CH_2CH_2CH_3$ Pentane (C_5H_{12})	CH_2 CH_2 CH_2 CH_2—CH_2 Cyclopentane (C_5H_{10})
$CH_3CH_2CH_2CH_2CH_2CH_3$ Hexane (C_6H_{14})	CH_2 CH_2 CH_2 CH_2 CH_2 CH_2 Cyclohexane (C_6H_{12})

Alkanes and cycloalkanes are so similar that many of their basic properties can be considered side by side. Some differences remain, however, and certain structural features arise from the rings of cycloalkanes that are more conveniently studied separately. We will point out the chemical and physical similarities of alkanes and cycloalkanes as we go along.

3.2 SHAPES OF ALKANES

A general tetrahedral orientation of groups—and thus sp^3 hybridization—is the rule for the carbon atoms of all alkanes and cycloalkanes. Using line-and-circle formulas, we can represent the shapes of alkanes in the following way.

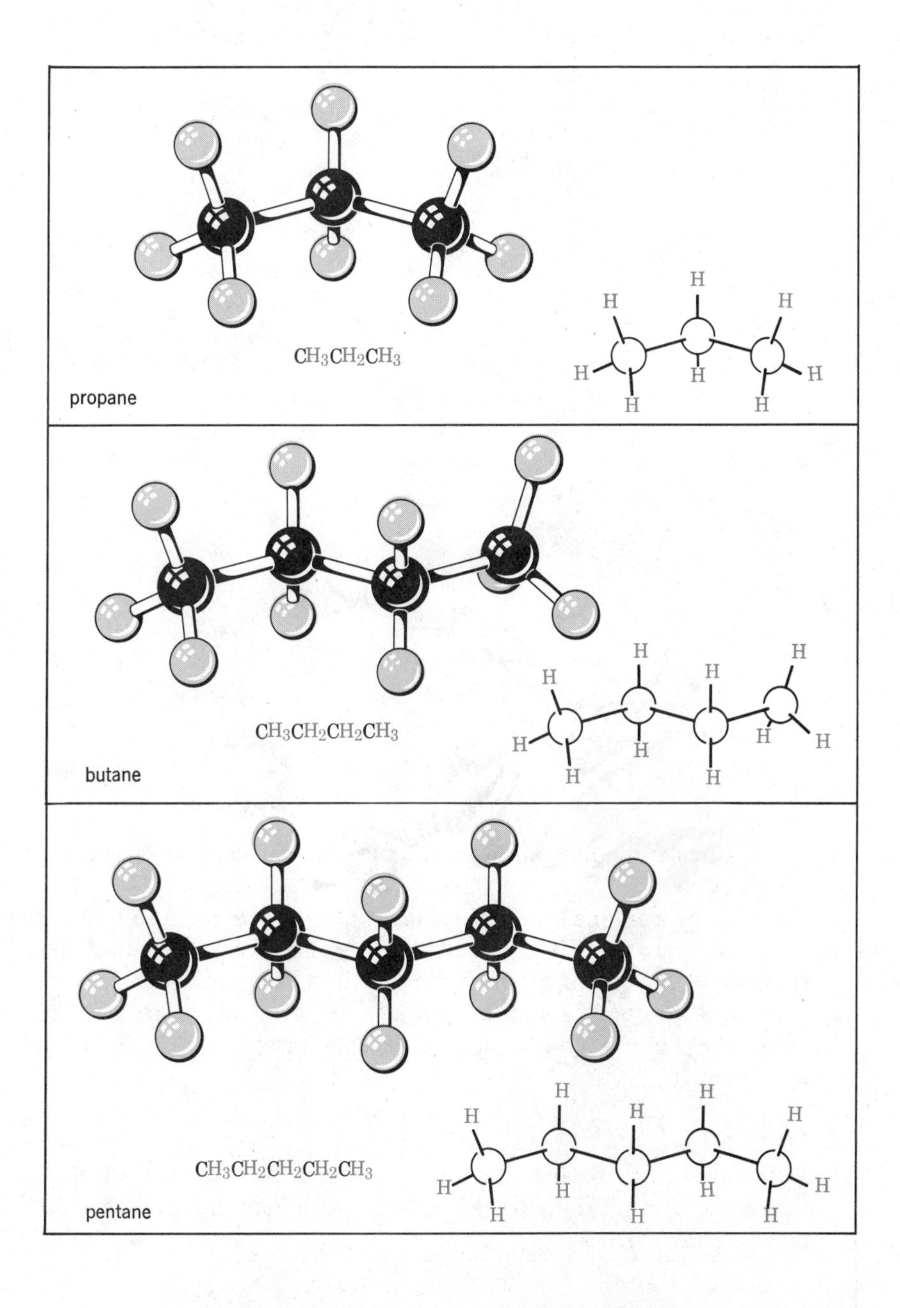

Butane and pentane are examples of alkanes that are sometimes called "straight-chain" alkanes. One glance at their three-dimensional structures shows that because of the tetrahedral carbon atoms their chains are zigzagged and not at all straight. Indeed, the structures that we have depicted above are the straightest possible arrangements of the chains, for rotations about the carbon-carbon single bonds create arrangements that are even less straight. The better description is *unbranched*. This means that each carbon of the chain is bonded to no more than two other carbon atoms. Unbranched alkanes are also often called "normal" alkanes or *n*-alkanes.

Isobutane and isopentane are examples of branched-chain alkanes.

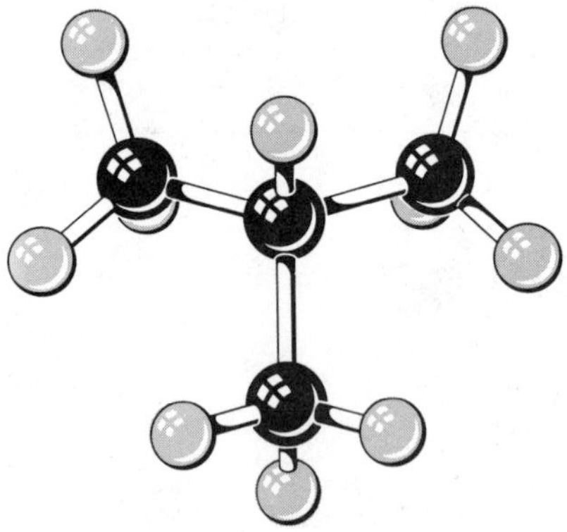

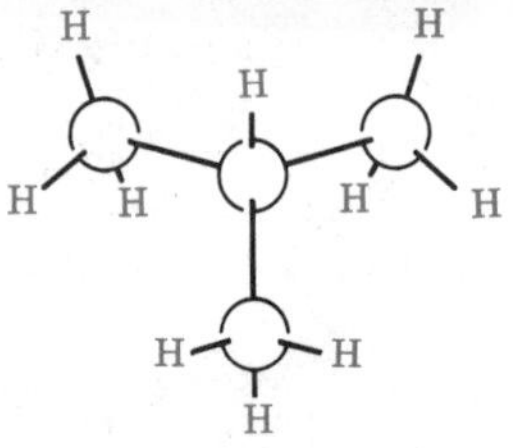

$CH_3—CH—CH_3$ (with CH_3 on the central carbon)

isobutane

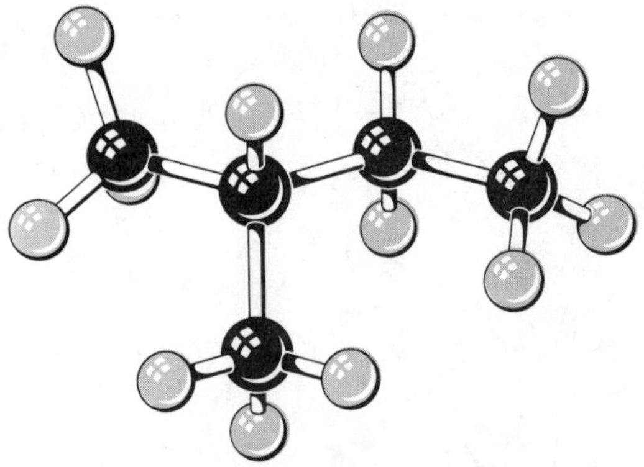

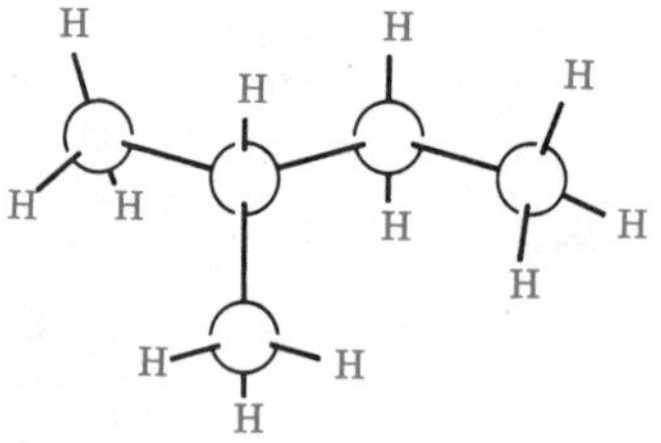

$CH_3—CH—CH_2—CH_3$ (with CH_3 on the second carbon)

isopentane

In each of the compounds above one carbon is attached to three other carbon atoms.

In neopentane (p. 87) the central carbon is bonded to four carbon atoms.

Butane and isobutane have the same molecular formula: C_4H_{10}. The two compounds have their atoms connected in a different order and are, therefore, *structural isomers*. Pentane, isopentane, and neopentane are also structural isomers. They, too, have the same molecular formula (C_5H_{12}) but have different structures.

Problem 3.1

Write line-and-circle and condensed structural formulas for all of the structural isomers of C_6H_{14}. Compare your answers with the condensed structural formulas given in Table 3.1.

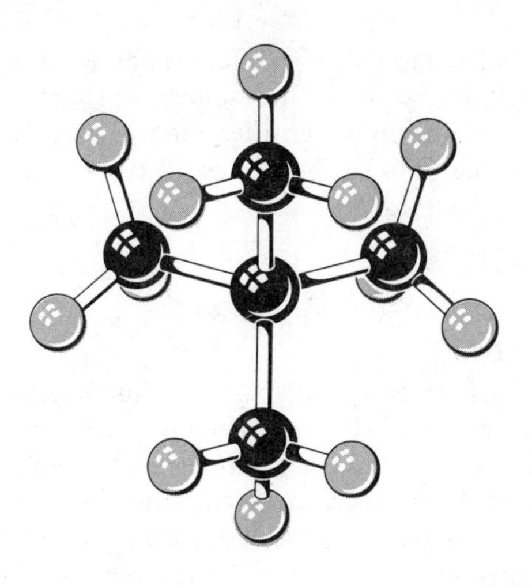

$$CH_3-\underset{CH_3}{\overset{CH_3}{C}}-CH_3$$

neopentane

TABLE 3.1 Physical Constants of the Butane, Pentane, and Hexane Isomers

MOLECULAR FORMULA	STRUCTURAL FORMULA	mp °C	bp °C	DENSITY[a]	INDEX OF REFRACTION[b] n_D 20 °C
C_4H_{10}	$CH_3CH_2CH_2CH_3$	−138.3	−0.5	0.6012_4^0	1.3543
C_4H_{10}	$CH_3\underset{CH_3}{CH}CH_3$	−159	−12	0.603_4^0	—
C_5H_{12}	$CH_3CH_2CH_2CH_2CH_3$	−129.72	36	0.6262_4^{20}	1.3579
C_5H_{12}	$CH_3\underset{CH_3}{CH}CH_2CH_3$	−160	27.9	0.6197_4^{20}	1.3537
C_5H_{12}	$CH_3-\underset{CH_3}{\overset{CH_3}{C}}-CH_3$	−20	9.45	0.61350_4^{20}	1.3476
C_6H_{14}	$CH_3CH_2CH_2CH_2CH_2CH_3$	−95	68	0.65937_4^{20}	1.3748
C_6H_{14}	$CH_3\underset{CH_3}{CH}CH_2CH_2CH_3$	−153.67	60.3	0.6532_4^{20}	1.3714
C_6H_{14}	$CH_3CH_2\underset{CH_3}{CH}CH_2CH_3$	−118	63.265	0.6643_4^{20}	1.3765
C_6H_{14}	$CH_3\underset{CH_3}{CH}-\underset{CH_3}{CH}CH_3$	−128.8	58	0.6616_4^{20}	1.3750
C_6H_{14}	$CH_3-\underset{CH_3}{\overset{CH_3}{C}}-CH_2CH_3$	−98	49.7	0.6492_4^{20}	1.3688

[a] Density is relative to water. The superscript indicates the temperature of the alkane and the subscript refers to the temperature of water to which the density is referred.

[b] The index of refraction is a measure of the ability of the alkane to bend (refract) light rays. The values reported are for light of the *D* line of the sodium spectrum (n_D).

Structural isomers, as stated earlier, have different physical properties. The differences may not always be large, but structural isomers will always be found to have different melting points, boiling points, densities, indexes of refraction, and so forth. Table 3.1 gives some of the physical properties of the C_4H_{10}, C_5H_{12}, and C_6H_{14} isomers.

3.3 NOMENCLATURE OF ALKANES

The development of a formal system for naming organic compounds did not come about until near the end of the nineteenth century. Prior to that time many organic compounds had already been discovered. The names given these compounds sometimes reflected a source of the compound. Acetic acid, for example, can be obtained from vinegar; it got its name from the Latin word for vinegar, *acetum*. Formic acid can be obtained from ants; it got its name from the Latin word for ants, *formicae*. Ethanol (or ethyl alcohol) was at one time called grain alcohol because it was obtained by the fermentation of grains. Methanol (or methyl alcohol) was called wood alcohol because it was obtained by the destructive distillation of wood. The names of other compounds were based on their method of preparation. Ethylene chloride (CH_2ClCH_2Cl), for example, got its name because it was made by allowing ethylene to react with chlorine.

These older names for organic compounds are now called "common" or "trivial" names. Many of these names are still widely used by chemists, biochemists, and the companies that sell chemicals. For this reason it is still necessary to learn the common names for some of the common compounds. We will point out these common names as we go along, and we will use them occasionally. Most of the time, however, the names that we will use will be those called IUPAC names.

The formal system of nomenclature used today was one proposed by the International Union of Pure and Applied Chemistry (IUPAC). This system was first developed in 1892 and has been revised at irregular intervals to keep it up to date. Underlying the IUPAC system of nomenclature for organic compounds is a fundamental principle: *each different compound should have a different name*. Thus, through a systematic set of rules, the IUPAC system provides different names for the more than two million known organic compounds, and names can be devised for any one of millions of other compounds yet to be synthesized. In addition, the IUPAC system is simple enough to allow any chemist familiar with the rules (or with the rules at hand) to write the name for any compound that might be encountered. In the same way, one is also able to derive the structure of a given compound from its IUPAC name.

The IUPAC system for naming alkanes is quite simple, and the principles involved are used in naming compounds in other families as well. For these reasons we begin our study of the IUPAC system with the rules for naming alkanes.

The names for several of the unbranched alkanes (*n*-alkanes) are listed in Table 3.2. The ending for all of the names of alkanes is *-ane*. The prefixes of the names of most of the alkanes (above C_4) are of Greek and Latin origin. Learning the prefixes is like learning to count in organic chemistry. Thus, one, two, three, four, five, becomes meth-, eth-, prop-, but-, pent-. If you take the time now to memorize only the first 10 prefixes, you will, through application of the rules that follow, have learned the names for nearly 100,000 alkanes. In addition, you will have formed the basis for the names of perhaps 100,000 alkenes, 100,000 alkynes, 100,000 alcohols, and 100,000 aldehydes. (The specific rules for alkenes, alkynes,

TABLE 3.2 The Unbranched Alkanes (or *n*-Alkanes)

NAME	NO. OF CARBONS	STRUCTURE	NAME	NO. OF CARBONS	STRUCTURE
Methane	1	CH_4	Heptadecane	17	$CH_3(CH_2)_{15}CH_3$
Ethane	2	CH_3CH_3	Octadecane	18	$CH_3(CH_2)_{16}CH_3$
Propane	3	$CH_3CH_2CH_3$	Nonadecane	19	$CH_3(CH_2)_{17}CH_3$
Butane	4	$CH_3(CH_2)_2CH_3$	Eicosane	20	$CH_3(CH_2)_{18}CH_3$
Pentane	5	$CH_3(CH_2)_3CH_3$	Heneicosane	21	$CH_3(CH_2)_{19}CH_3$
Hexane	6	$CH_3(CH_2)_4CH_3$	Docosane	22	$CH_3(CH_2)_{20}CH_3$
Heptane	7	$CH_3(CH_2)_5CH_3$	Tricosane	23	$CH_3(CH_2)_{21}CH_3$
Octane	8	$CH_3(CH_2)_6CH_3$	Triacontane	30	$CH_3(CH_2)_{28}CH_3$
Nonane	9	$CH_3(CH_2)_7CH_3$	Hentriacontane	31	$CH_3(CH_2)_{29}CH_3$
Decane	10	$CH_3(CH_2)_8CH_3$	Tetracontane	40	$CH_3(CH_2)_{38}CH_3$
Undecane	11	$CH_3(CH_2)_9CH_3$	Pentacontane	50	$CH_3(CH_2)_{48}CH_3$
Dodecane	12	$CH_3(CH_2)_{10}CH_3$	Hexacontane	60	$CH_3(CH_2)_{58}CH_3$
Tridecane	13	$CH_3(CH_2)_{11}CH_3$	Heptacontane	70	$CH_3(CH_2)_{68}CH_3$
Tetradecane	14	$CH_3(CH_2)_{12}CH_3$	Octacontane	80	$CH_3(CH_2)_{78}CH_3$
Pentadecane	15	$CH_3(CH_2)_{13}CH_3$	Nonacontane	90	$CH_3(CH_2)_{88}CH_3$
Hexadecane	16	$CH_3(CH_2)_{14}CH_3$	Hectane	100	$CH_3(CH_2)_{98}CH_3$

alcohols and aldehydes will be given in subsequent chapters.) One has few opportunities to gain such a large body of knowledge from such a small investment of time.

3.3A Nomenclature of Alkyl Groups

Groups derived from alkanes by removal of one hydrogen have names that end in *yl,* and these groups are called *alkyl groups.* When the alkane is *unbranched,* and the hydrogen that is removed is a *terminal* hydrogen, the names are straightforward:

Alkane		Alkyl Group	Abbreviation
CH_4 Methane	becomes	$CH_3—$ Methyl	Me
CH_3CH_3 *Ethane*	becomes	$CH_3CH_2—$ Ethyl	Et
$CH_3CH_2CH_3$ Propane	becomes	$CH_3CH_2CH_2—$ Propyl	Pr
$CH_3CH_2CH_2CH_3$ Butane	becomes	$CH_3CH_2CH_2CH_2—$ Butyl	Bu

For alkanes with more than two carbons, however, more than one derived group is possible. Two groups can be derived from propane; the propyl group (above) is derived by removal of a terminal hydrogen, and the isopropyl group (below) is derived by removal of an inner hydrogen.

$$\begin{array}{c} CH_3 \\ | \\ CH_3CH— \end{array} \quad \text{or} \quad \begin{array}{c} CH_3—CH—CH_3 \\ | \end{array}$$

Isopropyl group

There are four alkyl groups that contain four carbon atoms each:

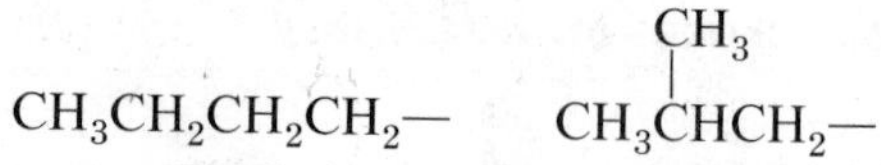

$CH_3CH_2CH_2CH_2—$
Butyl
(or *n*-butyl)

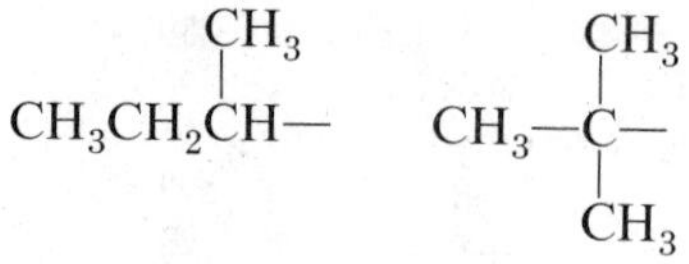

$CH_3\overset{CH_3}{\overset{|}{C}}HCH_2—$
Isobutyl

$CH_3CH_2\overset{CH_3}{\overset{|}{C}}H—$
sec-Butyl
(or *s*-butyl)

$CH_3—\overset{CH_3}{\overset{|}{\underset{CH_3}{\underset{|}{C}}}}—$
tert-Butyl
(or *t*-butyl)

All of these groups should be memorized so well that you are able to recognize them when they are written backward or upside down. As an aid in learning these groups it may be helpful to point out the following characteristics.

1. The base name of any group relates to the total number of carbon atoms in the group. (The propyl and isopropyl groups have *three* carbons and all of the "butyl" groups have *four* carbons.)

2. The isopropyl and isobutyl groups both have an iso structure.* They have a carbon branch, (i.e., a methyl group, $CH_3—$) at the second carbon of the chain.

$CH_3—\overset{CH_3}{\overset{|}{C}}H—$ $\quad$ $CH_3\overset{CH_3}{\overset{|}{C}}H—CH_2—$

Iso structures

3. The incomplete valence of the *sec*-butyl group is directed away from a *secondary* carbon.

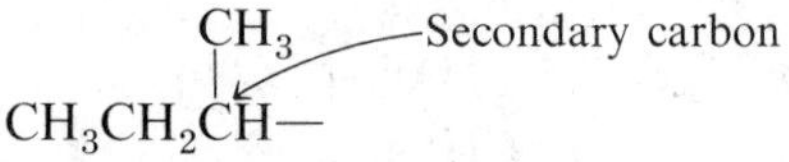

$CH_3CH_2\overset{CH_3}{\overset{|}{C}}H—$ ← Secondary carbon
sec-Butyl

4. The incomplete valence of the *tert*-butyl group is directed away from a *tertiary* carbon.

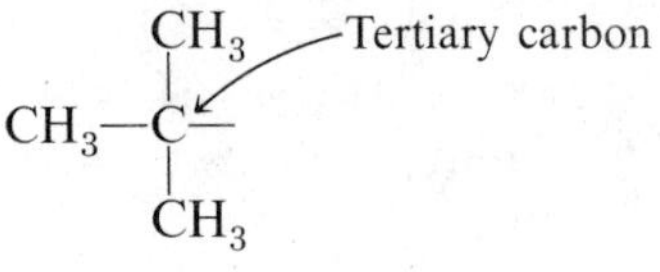

$CH_3—\overset{CH_3}{\overset{|}{\underset{CH_3}{\underset{|}{C}}}}—$ ← Tertiary carbon
tert-Butyl

* So too, do the isopentyl and isohexyl groups, $CH_3\overset{CH_3}{\overset{|}{C}}HCH_2CH_2—$ and $CH_3\overset{CH_3}{\overset{|}{C}}HCH_2CH_2CH_2—$, respectively. These names are also provided in the IUPAC system, and the IUPAC system allows the names isobutane (for $CH_3\overset{CH_3}{\overset{|}{C}}HCH_3$), isopentane (for $CH_3\overset{CH_3}{\overset{|}{C}}HCH_2CH_3$), and isohexane (for $CH_3\overset{CH_3}{\overset{|}{C}}HCH_2CH_2CH_3$). The iso system stops after isohexyl (for groups) and isohexane (for alkanes) however.

3.3B Nomenclature of Branched-Chain Alkanes

Branched-chain alkanes are named according to the following rules:

1. *Locate the longest continuous chain of carbon atoms; this chain determines the base name for the alkane.*

   We designate the compound listed below, for example, as a *hexane* because the longest continuous chain contains six carbon atoms.

$$\underset{\displaystyle CH_3}{CH_3CH_2CH_2CH_2\overset{|}{C}HCH_3}$$

   The longest continuous chain may not always be obvious from the way the formula is written. Notice, for example, that the alkane written below is designated as a *heptane* because the longest chain contains seven carbon atoms.

$$CH_3CH_2CH_2CH_2CH(CH_2CH_3)—CH_3$$

2. *Number the longest chain beginning with the end of the chain nearest the branching.*

   Applying this rule, we number the two alkanes that we illustrated above in the following way.

$$\overset{6}{C}H_3\overset{5}{C}H_2\overset{4}{C}H_2\overset{3}{C}H_2\overset{2}{C}H(CH_3)\overset{1}{C}H_3 \qquad \overset{7}{C}H_3\overset{6}{C}H_2\overset{5}{C}H_2\overset{4}{C}H_2\overset{3}{C}H(\overset{2}{C}H_2\overset{1}{C}H_3)CH_3$$

3. *Use the numbers obtained by application of rule 2 to designate the location of the substituent group.* The base name is placed last, and the substituent group, preceded by the number designating its location on the chain, is placed first. Numbers are separated from words by a dash. Our two examples are 2-methylhexane and 3-methylheptane, respectively.

$$\overset{6}{C}H_3\overset{5}{C}H_2\overset{4}{C}H_2\overset{3}{C}H_2\overset{2}{C}H(CH_3)\overset{1}{C}H_3 \qquad \overset{7}{C}H_3\overset{6}{C}H_2\overset{5}{C}H_2\overset{4}{C}H_2\overset{3}{C}H(\overset{2}{C}H_2\overset{1}{C}H_3)CH_3$$

2-Methylhexane | 3-Methylheptane

4. *When two or more substituents are present, give each substituent a number corresponding to its location on the longest chain.* For example, we designate the compound below as 4-ethyl-2-methylhexane.

$$\begin{array}{llll} CH_3CH & -CH_2- & CHCH_2CH_3 \\ \;\;\;| & & \;| \\ \;\;CH_3 & & CH_2 \\ & & \;| \\ & & CH_3 \end{array}$$

4-Ethyl-2-methylhexane

The groups should be listed *alphabetically* (i.e., ethyl before methyl). In deciding on alphabetical order disregard multiplying prefixes such as *di* and *tri* and disregard structure-defining prefixes that are separated from the name by a hyphen such as *sec*- and *tert*-. Thus ethyl precedes dimethyl, and *tert*-butyl precedes ethyl, but ethyl precedes isobutyl.

5. *When two or more substituents are present on the same carbon, use that number twice.*

$$\begin{array}{ccc} & CH_3 & \\ & | & \\ CH_3CH_2- & C & -CH_2CH_2CH_3 \\ & | & \\ & CH_2 & \\ & | & \\ & CH_3 & \end{array}$$

3-Ethyl-3-methylhexane

6. *When two or more substituents are identical indicate this by the use of the prefixes di-, tri-, tetra-*, and so on. Commas are used to separate numbers from each other.

$$\begin{array}{l} CH_3CH-CHCH_3 \\ \;\;\;\;\;|\;\;\;\;\;\;\;| \\ \;\;\;CH_3\;CH_3 \end{array} \qquad \begin{array}{l} \;\;\;\;\;\;\;\;\;\;\;\;\;\;\;CH_3 \\ \;\;\;\;\;\;\;\;\;\;\;\;\;\;\;\;| \\ CH_3CHCHCHCH_3 \\ \;\;\;\;\;\;\;|\;\;\;\;\;\;\;\;\;| \\ \;\;\;\;CH_3\;\;\;CH_3 \end{array}$$

2,3-Dimethylbutane 2,3,4-Trimethylpentane

Application of these six rules allows us to name most of the alkanes that we will encounter. Two other rules, however, may be required occasionally.

7. *When two chains of equal length compete for selection as the base chain, choose the chain with the greater number of substituents.*

$$\begin{array}{cccccc} {}^{7}\;\;{}^{6} & {}^{5} & {}^{4} & {}^{3} & {}^{2} & {}^{1} \\ CH_3CH_2- & CH- & CH- & CH- & CH- & CH_3 \\ & | & | & | & | & \\ & CH_3 & CH_2 & CH_3 & CH_3 & \\ & & | & & & \\ & & CH_2 & & & \\ & & | & & & \\ & & CH_3 & & & \end{array}$$

2,3,5-Trimethyl-4-propylheptane
(four substituents)
not 4-*sec*-Butyl-2,3-dimethylheptane
(three substituents)

8. *When branching first occurs at an equal distance from either end of the longest chain, choose the name that gives the lower number at the first point of difference.*

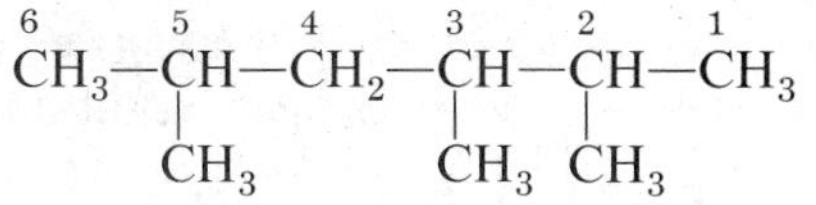

$$\overset{6}{CH_3}-\underset{\displaystyle CH_3}{\overset{5}{C}H}-\overset{4}{CH_2}-\underset{\displaystyle CH_3}{\overset{3}{C}H}-\underset{\displaystyle CH_3}{\overset{2}{C}H}-\overset{1}{CH_3}$$

2,3,5-Trimethylhexane

not 2,4,5-Trimethylhexane

Problem 3.2

(a) Give correct IUPAC names for all the C_6H_{14} isomers in Table 3.1. (b) Write structural formulas for the nine isomers of C_7H_{16} and give IUPAC names for each. (Hint: You may find it helpful to name each compound as you write its structure. This will help you to decide whether or not two structures are really different. If their IUPAC names are different then so are the structures.)

Alkanes bearing halogen substituents are named, in the IUPAC system, as *haloalkanes;* for example,

CH_3CH_2Cl — Chloroethane

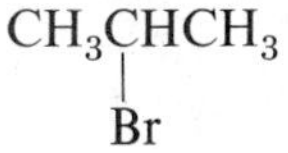

$CH_3\underset{\displaystyle Br}{C}HCH_3$ — 2-Bromopropane

Common names for many simple haloalkanes are still widely used, however. In this common nomenclature system haloalkanes are named as alkyl halides.

CH_3CH_2Cl — Ethyl chloride

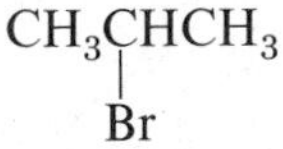

$CH_3\underset{\displaystyle Br}{C}HCH_3$ — Isopropyl bromide

3.4 NOMENCLATURE OF CYCLOALKANES

3.4A Monocyclic Compounds

Cycloalkanes with only one ring are named by attaching the prefix *cyclo* to the names of the alkanes possessing the same number of carbon atoms. For example,

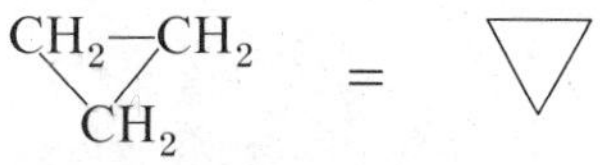

CH_2—CH_2 (ring closed through CH_2) = ▽

Cyclopropane

CH_2—CH_2, CH_2, CH_2, CH_2 (five-membered ring) = ⬠

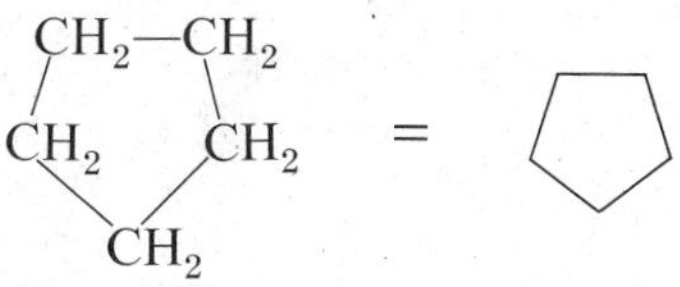

Cyclopentane

Naming substituted cycloalkanes is straightforward: we name them as

alkylcycloalkanes. When more than one substituent is present we number the ring *beginning with one substituent* in the way that gives the next substituent the lower number possible.

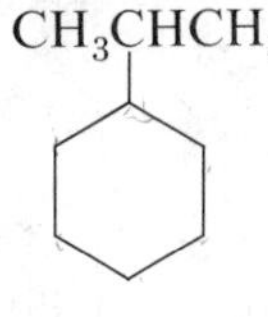

CH_3CHCH_3 CH_3 CH_2CH_3

Isopropylcyclohexane

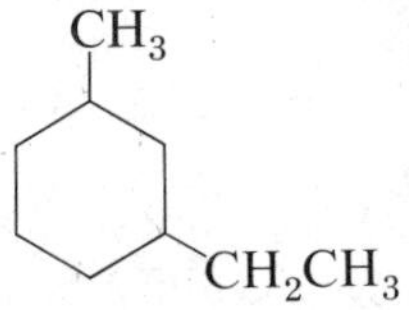

1-Ethyl-3-Methylcyclohexane
(not 1-ethyl-5-methylcyclohexane)

If the alkane chain contains a greater number of carbon atoms than the ring, then we designate the ring as the substituent.

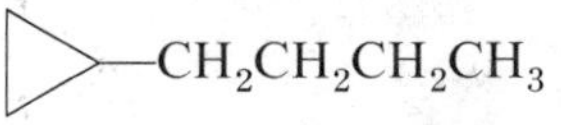

$—CH_2CH_2CH_2CH_3$

1-Cyclopropylbutane
(not butylcyclopropane)

Problem 3.3

Give names for the following substituted cycloalkanes.

(a) 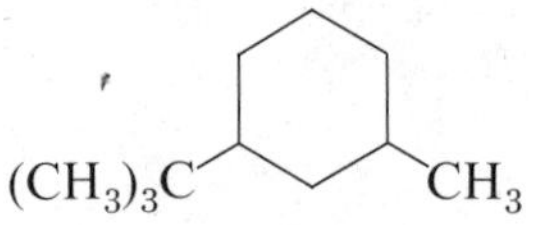 $(CH_3)_3C$ CH_3 (b) 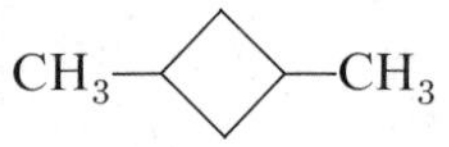 $CH_3—$ $—CH_3$

(c) 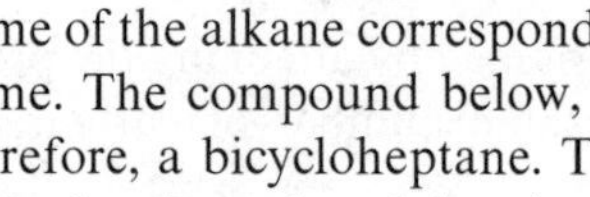 $CH_3(CH_2)_5CH_2—$

3.4B Bicyclic Compounds

We name compounds containing two rings as *bicycloalkanes* and we use the name of the alkane corresponding to the total number of carbon atoms as the base name. The compound below, for example, contains seven carbon atoms and is, therefore, a bicycloheptane. The carbon atoms common to both rings are called bridgeheads, and each bond, or chain of atoms connecting the bridgehead atoms, is called a bridge.

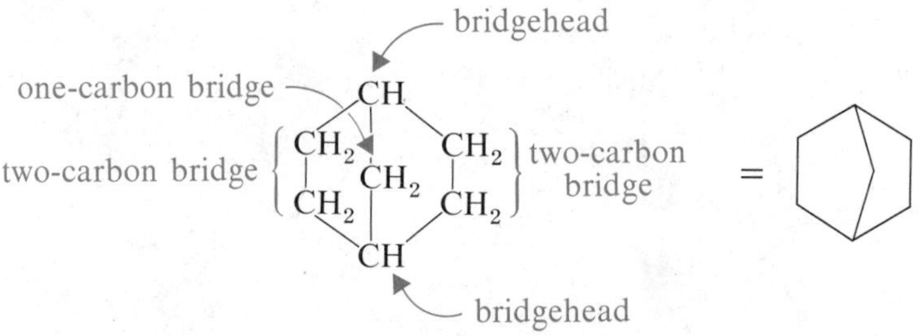

bridgehead
one-carbon bridge
two-carbon bridge
CH
CH_2 CH_2 CH_2 CH_2 CH_2
CH
two-carbon bridge
=
bridgehead

A bicycloheptane

Then we interpose in the name an expression in brackets that denotes the number of carbon atoms in each bridge (in descending order). For example:

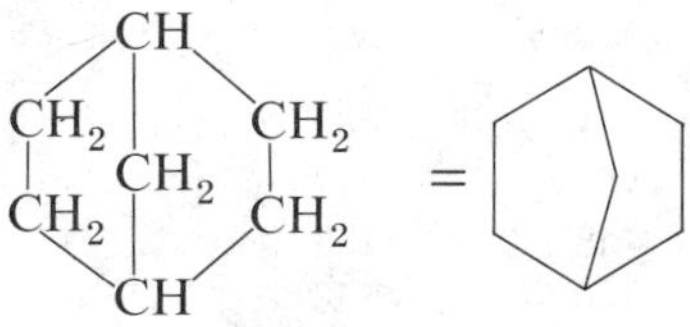

Bicyclo [2.2.1] heptane

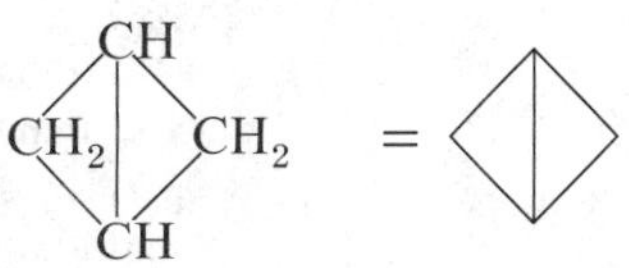

Bicyclo [1.1.0] butane

Problem 3.4

Give names for each of the following bicyclic alkanes.

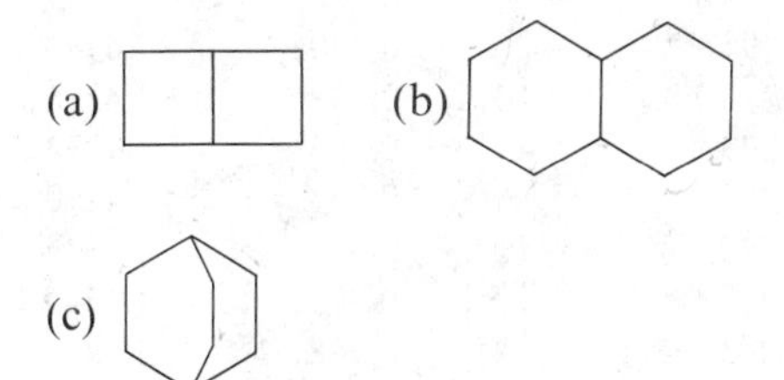

(d) Write the structure of a bicyclic compound that is an isomer of bicyclo [2.2.1]-heptane and give its name.

3.5 PHYSICAL PROPERTIES OF ALKANES AND CYCLOALKANES

If we examine the unbranched alkanes (*n*-alkanes) in Table 3.2 we notice that each alkane differs from the preceding one by one —CH_2— group. Butane, for example, is $CH_3(CH_2)_2CH_3$ and pentane is $CH_3(CH_2)_3CH_3$. A series of compounds like this, where each member differs from the next member by a constant unit, is called a *homologous* series. Members of a homologous series are called *homologs.*

At room temperature (25 °C) and one atmosphere pressure the first four members of the homologous series of normal alkanes (Table 3.3) are gases; the C_5 to C_{17} *n*-alkanes (pentane to heptadecane) are liquids; and the *n*-alkanes with 18 and more carbon atoms are solids.

Boiling Points. The boiling points of the *n*-alkanes show a regular increase with increasing molecular weight (Fig. 3.1). Branching of the alkane chain, however, dramatically lowers the boiling point. (Examples of the effect of chain branching can be seen in Table 3.1.) Cycloalkanes, however, have higher boiling points than unbranched alkanes with the same number of carbon atoms (Table 3.4).

Melting Points. The unbranched alkanes do not show the same smooth increase in melting points with increasing molecular weight (black line Fig. 3.2) that they show in their boiling points. There is an alternation as one progresses from an unbranched alkane with an even number of carbon atoms to the next one with an odd number of carbon atoms. For example, propane (mp −188°) melts lower than ethane (mp −172°) and even lower than methane (mp −183°). Butane (mp −135°) melts 53° higher than propane and only 5° lower than pentane (mp −130°). If, however, the even- and odd-numbered alkanes are plotted on *separate*

TABLE 3.3 Physical Constants of *n*-Alkanes

NUMBER OF CARBON ATOMS	NAME	bp °C (1 atm.)	mp °C	DENSITY d_4^{20}	REFRACTIVE INDEX n_D^{20}
1	Methane	−161.5	−183		1.3543
2	Ethane	−88.6	−172		
3	Propane	−42.1	−188		
4	Butane	−0.5	−138		
5	Pentane	36.1	−130	0.626	1.3579
6	Hexane	68.7	−95	0.659	1.3748
7	Heptane	98.4	−91	0.684	1.3876
8	Octane	125.7	−57	0.703	1.3974
9	Nonane	150.8	−54	0.718	1.4054
10	Decane	174.1	−30	0.730	1.4119
11	Undecane	195.9	−26	0.740	1.4176
12	Dodecane	216.3	−10	0.749	1.4216
13	Tridecane	243	−5.5	0.756	1.4233
14	Tetradecane	253.5	6	0.763	1.4290
15	Pentadecane	270.5	10	0.769	1.4315
16	Hexadecane	287	18	0.773	1.4345
17	Heptadecane	303	22	0.778	1.4369
18	Octadecane	305–307	28	0.777	1.4349
19	Nonadecane	330	32	0.777	1.4409
20	Eicosane	343	36.8	0.789	1.4425

curves (white and colored lines in Fig. 3.2), there *is* a smooth increase in melting point with increasing molecular weight.

X-ray-diffraction studies have revealed the reason for this apparent anomaly. Alkane chains with an even number of carbon atoms pack more closely in the

FIG. 3.1
Boiling points of n-alkanes (in color) and cycloalkanes (in white).

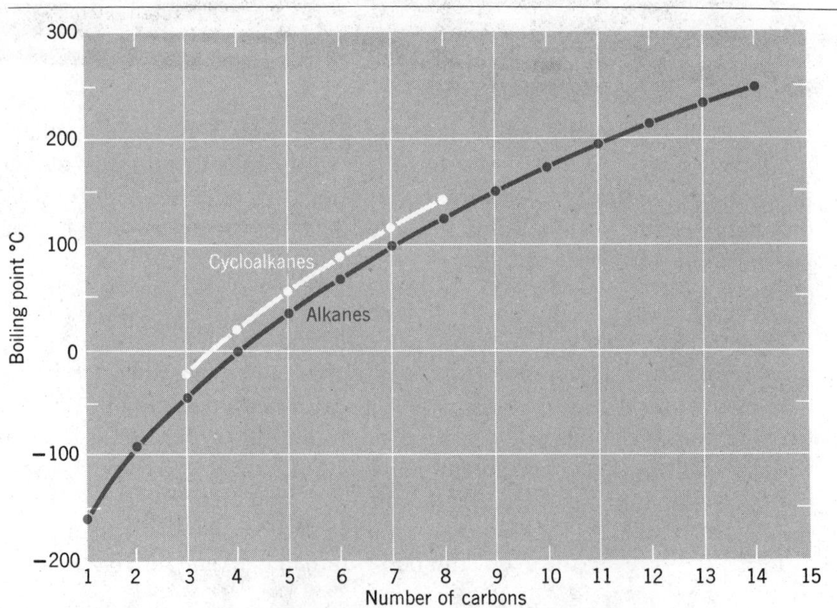

TABLE 3.4 Physical Constants of Cycloalkanes

NUMBER OF CARBON ATOMS	NAME	bp °C (1 atm)	mp °C	DENSITY d_4^{20}	REFRACTIVE INDEX n_D^{20}
3	Cyclopropane	−33	−126.6		
4	Cyclobutane	13	−50		1.4260
5	Cyclopentane	49	−94	0.751	1.4064
6	Cyclohexane	81	6.5	0.779	1.4266
7	Cycloheptane	118.5	−12	0.811	1.4449
8	Cyclooctane	149	13.5	0.834	

crystalline state. As a result, attractive forces between individual chains are greater and melting points are higher.

The effect of chain branching on the melting points of alkanes is more difficult to predict. Generally, however, branching that produces highly symmetrical structures results in abnormally high melting points. The compound 2,2,3,3-tetramethylbutane, for example, melts at 100.7°. Its boiling point is only six degrees higher, 106.3°.

$$
\begin{array}{c}
\quad CH_3\;\; CH_3 \\
\quad | \qquad | \\
CH_3{-}C{-}{-}C{-}CH_3 \\
\quad | \qquad | \\
\quad CH_3\;\; CH_3
\end{array}
$$

2,2,3,3-Tetramethylbutane

Cycloalkanes, also have much higher melting points than their open-chain counterparts (Table 3.4). Because of their greater symmetry they pack more tightly into a crystal lattice.

Density. As a class, the alkanes and cycloalkanes are the least dense of all groups of organic compounds. All alkanes and cycloalkanes have densities con-

FIG. 3.2
Melting points of n-alkanes.

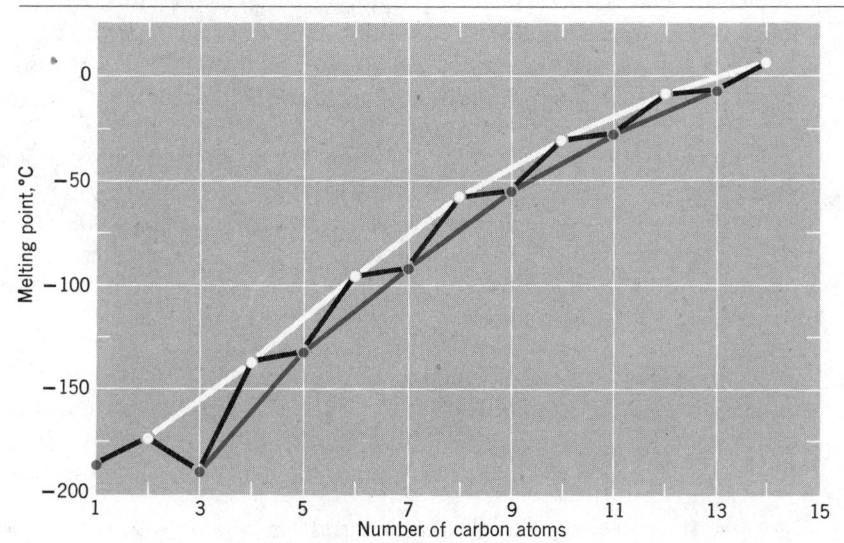

siderably less than 1.00 g/ml (the density of water at 4 °C). As a result, petroleum (a mixture of hydrocarbons rich in alkanes) floats on water.

Solubility. Alkanes and cycloalkanes are almost totally insoluble in water because of their very low polarity and their inability to form hydrogen bonds. Liquid alkanes and cycloalkanes are miscible with each other, and they generally dissolve in solvents of low polarity. Good solvents for them are benzene, carbon tetrachloride, chloroform, and other hydrocarbons.

3.6 CONFORMATIONAL ANALYSIS OF BUTANE

The study of the energy changes that occur in a molecule when groups rotate about single bonds is called *conformational analysis*. We saw the results of such a study for ethane in Chapter 2. Ethane has a slight barrier (2.8 kcal/mole) to free rotation about the carbon-carbon single bond. This barrier causes the energy of the ethane molecule to rise to a maximum when rotation brings the hydrogen atoms into an eclipsed conformation. This barrier to free rotation, which can be accounted for by quantum mechanics but for which no simple physical picture can be presently given, is called the *torsional* strain of the molecule.

If we consider rotation about the C-2—C-3 bond of butane, torsional strain plays a part, too. There are, however, additional factors. To see what these are, we should look at the important conformations of butane I–VI.

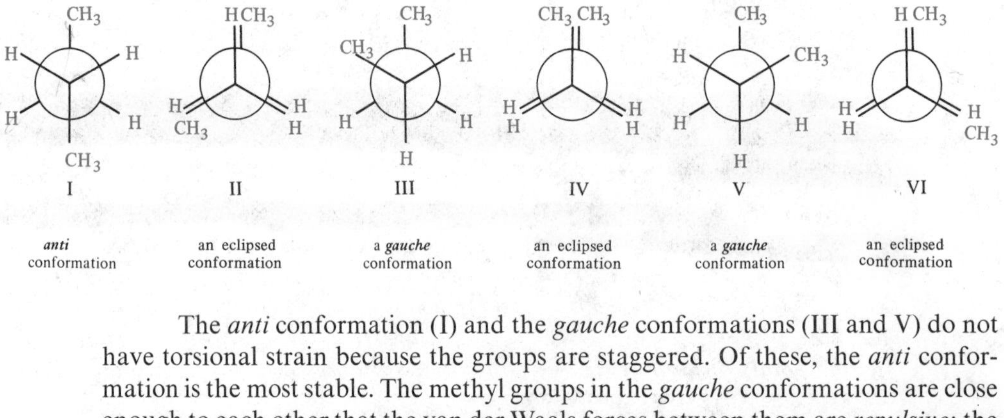

I — *anti* conformation

II — an eclipsed conformation

III — a *gauche* conformation

IV — an eclipsed conformation

V — a *gauche* conformation

VI — an eclipsed conformation

The *anti* conformation (I) and the *gauche* conformations (III and V) do not have torsional strain because the groups are staggered. Of these, the *anti* conformation is the most stable. The methyl groups in the *gauche* conformations are close enough to each other that the van der Waals forces between them are *repulsive;* the electron clouds of the two groups are so close that they repel each other. This repulsion causes the *gauche* conformations to have approximately 0.8 kcal/mole more energy than the *anti* conformation.

The eclipsed conformations (II, IV, and VI) represent energy maxima in the potential-energy diagram (Fig. 3.3). Eclipsed conformations II and VI not only have torsional strain, they have van der Waals repulsions arising from the eclipsed methyls and hydrogens. Eclipsed conformation IV has the greatest energy of all because, in addition to torsional strain, there is the large van der Waals repulsive force between the eclipsed methyl groups.

While the barriers to rotation in a butane molecule are larger than those of ethane (Section 2.3), they are still far too small to permit isolation of the *gauche* and *anti* conformations at normal temperatures. Only at extremely low temperatures would the molecules have insufficient energies to surmount these barriers.

We saw earlier that van der Waals forces can be *attractive*. Here, however, we find that they can also be *repulsive*. Whether or not van der Waals interactions

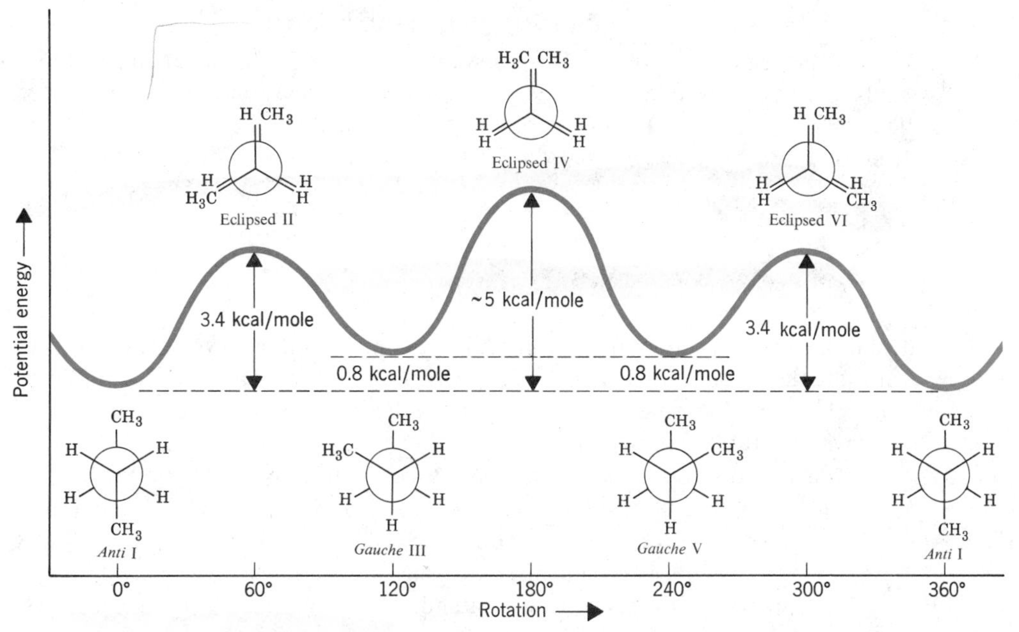

FIG. 3.3
Energy changes that arise from rotation of the C-2—C-3 bond of butane.

lead to attraction or repulsion depends on the distance that separates the two groups. As two nonpolar groups are brought closer and closer together, the first effect is one in which a momentarily unsymmetrical distribution of electrons in one group induces an opposite polarity in the other. The opposite charges induced in those portions of the two groups that are in closest proximity lead to attraction between them. This attraction increases to a maximum as the internuclear distance of the two groups decreases. The internuclear distance at which the attractive force is at a maximum is equal to the sum of what are called the *van der Waals radii* of the two groups. The van der Waals radius of a group is, in effect, a measure of its size. If the two groups are brought still closer—closer than the sum of their van der Waals radii—the interaction between them becomes repulsive: their electron clouds begin to penetrate each other, and strong electron-electron interactions begin to occur.

Problem 3.5

Sketch a curve similar to that in Fig. 3.3 showing the energy changes that occur when the groups rotate about one carbon-carbon bond of propane.

3.7 THE RELATIVE STABILITIES OF CYCLOALKANES: RING STRAIN

Cycloalkanes do not all have the same relative stability. Data from heats of combustion (Section 3.7A) show that cyclohexane is the most stable cycloalkane and cyclopropane and cyclobutane are much less stable. The relative instability of cyclopropane and cyclobutane is a direct consequence of their cyclic structures and for this reason their molecules are said to possess *ring strain*. To see how this can be demonstrated experimentally, we need to examine the relative heats of combustion of cycloalkanes.

3.7A Heats of Combustion of Cycloalkanes

The heat of combustion of a compound (Section 2.4) is the quantity of heat evolved when one mole of a substance is oxidized to carbon dioxide and water. The cycloalkanes constitute a homologous series; each member of the series differs from the one immediately preceding it by the constant amount of one $-CH_2-$ group. Thus, the general equation for combustion of a cycloalkane can be formulated as follows:

$$(CH_2)_n + \tfrac{3}{2}n\, O_2 \longrightarrow nCO_2 + nH_2O + \text{heat}$$

From the total amount of heat evolved in the combustion of a cycloalkane, we can calculate the amount of heat evolved *per CH_2 group*. On this basis, the stabilities of the cycloalkanes become directly comparable. The results of such an investigation are given in Table 3.5.

Several observations emerge from a consideration of these results.

1. Cyclohexane has the lowest heat of combustion per CH_2 group (157.4 kcal/mole). This amount does not differ from that of *n*-alkanes which, having no ring, can have no ring strain. Thus we can assume that cyclohexane has no ring strain and that it can serve as our standard for comparison with other cycloalkanes. We can calculate ring strain for the other cycloalkanes (Table 3.6) by multiplying 157.4 kcal/mole by n and then subtracting the result from the heat of combustion of the cycloalkane.
2. The combustion of cyclopropane evolves the greatest amount of heat per CH_2 group. Therefore, molecules of cyclopropane must have the greatest ring strain (27.6 kcal/mole, cf. Table 3.6). Since cyclopropane molecules evolve the greatest amount of heat energy per CH_2 group on combustion they must contain the greatest amount of potential energy per CH_2 group. Thus what we call ring strain is a form of potential energy that the cyclic molecule contains. The more ring strain a molecule possesses, the more potential energy it has and the less stable it is compared to its ring homologs.
3. The combustion of cyclobutane evolves the second largest amount of heat per CH_2 group and thus has the second largest amount of ring strain (26.3 kcal/mole).
4. While other cycloalkanes possess ring strain to varying degrees, the

TABLE 3.5 Heats of Combustion of Cycloalkanes

CYCLOALKANE $(CH_2)_n$	n	HEAT OF COMBUSTION kcal/mole	HEAT OF COMBUSTION/CH_2 GROUP kcal/mole
Cyclopropane	3	499.8	166.6
Cyclobutane	4	655.9	164.0
Cyclopentane	5	793.5	158.7
Cyclohexane	6	944.5	157.4
Cycloheptane	7	1108.2	158.3
Cyclooctane	8	1269.2	158.6
Cyclononane	9	1429.5	158.8
Cyclodecane	10	1586.0	158.6
Cyclopentadecane	15	2362.5	157.5
n-Alkane			157.4

TABLE 3.6 Ring Strain of Cycloalkanes

CYCLOALKANE	RING STRAIN kcal/mole
Cyclopropane	27.6
Cyclobutane	26.3
Cyclopentane	6.5
Cyclohexane	0
Cycloheptane	6.4
Cyclooctane	10.0
Cyclononane	12.9
Cyclodecane	12.0
Cyclopentadecane	1.5

relative amounts are not large. Cyclopentane and cycloheptane have about the same modest amount of ring strain. Rings of 8, 9, and 10 members have slightly larger amounts of ring strain and then the amount falls off. A 15-membered ring has only a very slight amount of ring strain.

Keep one other factor in mind. The large ring strain of cyclopropane (27.6 kcal/mole) is concentrated in just three carbon-carbon bonds (~9 kcal/bond). By contrast, cyclodecane has its strain (12.0 kcal/mole) spread over a much larger ring (~1.2 kcal/C-C bond). One might expect, therefore, some unusual chemistry from cyclopropane but expect cyclodecane to behave like a normal alkane. We will see later that this is true.

3.8 THE ORIGIN OF RING STRAIN IN CYCLOPROPANE AND CYCLOBUTANE: ANGLE STRAIN AND TORSIONAL STRAIN

The carbon atoms of alkanes are sp^3 hybridized. The normal tetrahedral bond angle of an sp^3-hybridized atom is 109.5°. In cyclopropane (a molecule with the shape of a regular triangle) the internal angles must be 60° and therefore they must depart from this ideal value by a very large amount—by 49.5°.

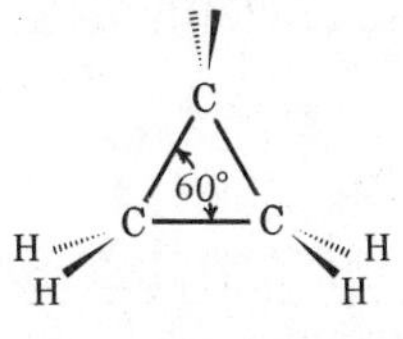

H H
C
60°
C C
H H
H H

This compression of the internal bond angle causes what chemists call *angle strain*. Angle strain exists in a cyclopropane ring because the sp^3 orbitals of carbon cannot overlap as effectively (Fig. 3.4) as they do in alkanes (where perfect end-on overlap is possible). The bonds of cyclopropane are often described as being "bent." Because orbital overlap is less effective, the carbon-carbon bonds of cyclopropane are weaker and the molecule has greater potential energy.

While angle strain accounts for most of the ring strain in cyclopropane it does not account for it all. Because the ring is (of necessity) planar, the hydrogen atoms of the ring are all *eclipsed* (Fig. 3.5), and the molecule has torsional strain as well.

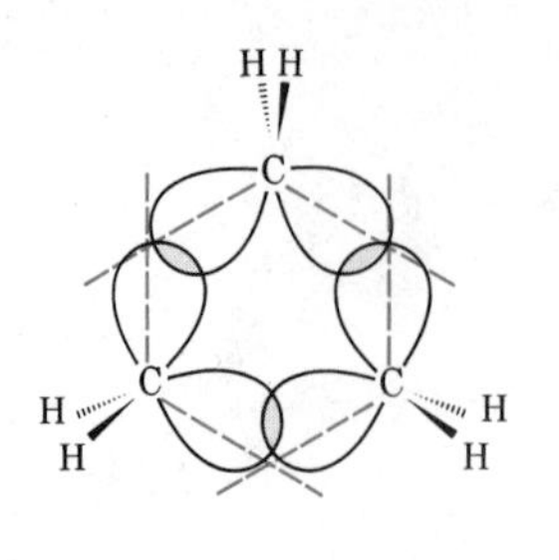

FIG. 3.4
Orbital overlap in the carbon-carbon bonds of cyclopropane cannot occur perfectly end-on. This leads to weaker "bent" bonds and to angle strain.

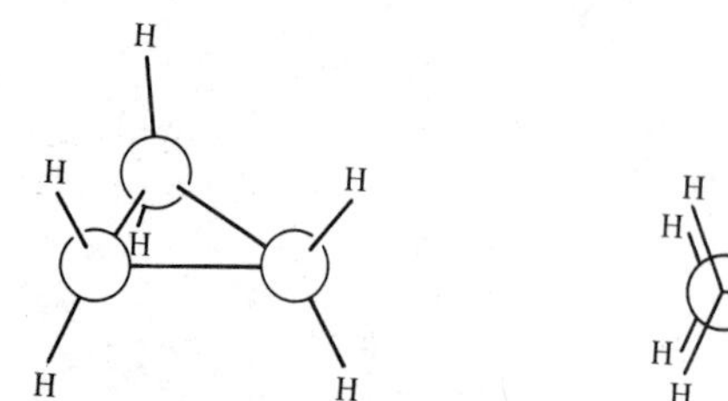

FIG. 3.5
(a) *A line-and-circle drawing of the cyclopropane ring.*
(b) *A Newman projection formula as viewed along one carbon-carbon bond shows the eclipsed hydrogens. (Viewing along either of the other two bonds would show the same picture.)*

Cyclobutane also has considerable angle strain. The internal angles are 88°— a departure of more than 21° from the normal tetrahedral bond angle. The cyclobutane ring is not planar but is slightly "folded" (Fig. 3.6). If the cyclobutane ring were planar the angle strain would be somewhat less (the internal angles would be 90° instead of 88°), but torsional strain would be considerably larger because all eight hydrogens would be eclipsed. By folding, or bending, slightly the cyclobutane ring relieves more of its torsional strain than it costs in the slight increase in its angle strain.

3.9 CYCLOPENTANE

The internal angles of a regular pentagon are 108°, a value very close to the normal tetrahedral bond angles of 109.5°. Thus if cyclopentane molecules were planar, they would have very little angle strain. Planarity, however, would introduce considerable torsional strain because all 10 hydrogens would be eclipsed. Thus, like cyclobutane, cyclopentane assumes a slightly bent conformation in which one or two of the atoms of the ring are out of the plane of the others (Fig. 3.7). This relieves torsional strain. Slight twisting of carbon-carbon bonds causes the out-of-plane atoms to move into plane and causes others to move out. Thus, the molecule is flexible and shifts from one conformation to another constantly. With little torsional strain and angle strain, cyclopentane is almost as stable as cyclohexane.

3.10 CONFORMATIONS OF CYCLOHEXANE

There is considerable evidence that the most stable conformation of the cyclohexane ring is the "chair" conformation illustrated in Fig. 3.8.* In this nonplanar structure the carbon-carbon bond angles are all 109.5° and are thus free of angle strain. The chair conformation is free of torsional strain as well. When viewed along the carbon-carbon bonds of any side (viewing the structure from an end, Fig. 3.9) the atoms are seen to be perfectly staggered. Moreover, the hydrogen atoms on carbons at opposite corners of the cyclohexane ring are maximally separated. The separation between two such hydrogens is illustrated in Fig. 3.10.

* An understanding of this and subsequent discussions of conformational analysis can be aided immeasurably through the use of appropriate molecular models.

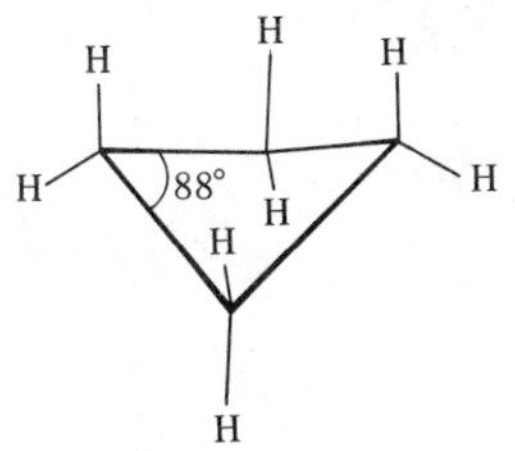

FIG. 3.6
The "folded" or "bent" conformation of cyclobutane

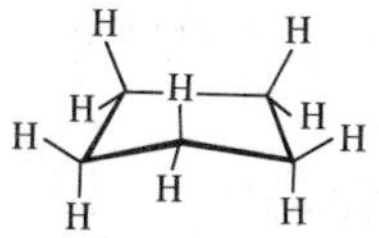

FIG. 3.7
The "bent" or "envelope" form of cyclopentane. In this structure the front carbon atom is bent upwards. In actuality, the molecule is flexible and shifts conformations constantly.

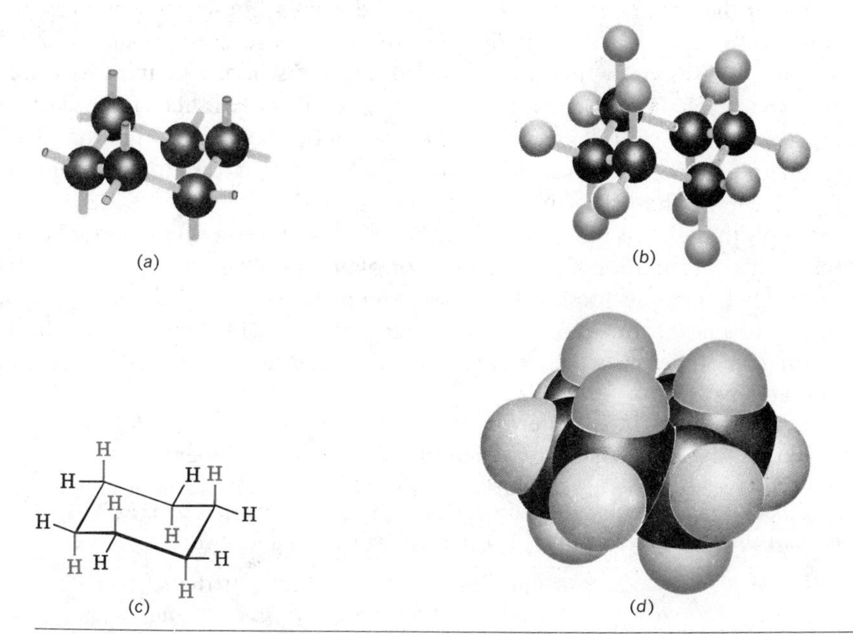

FIG. 3.8
Representations of the chair conformation of cyclohexane: (a) *carbon skeleton only;* (b) *carbon and hydrogen atoms;* (c) *line drawing;* (d) *space-filling model of cyclohexane.*

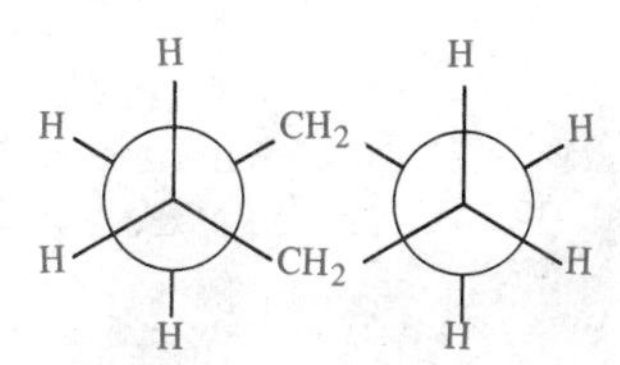

FIG. 3.9
A Newman projection of the chair conformation of cyclohexane. (Comparison with an actual molecular model will make this formulation clearer, and will show that similar staggered arrangements are seen when other carbon-carbon bonds are chosen for sighting.)

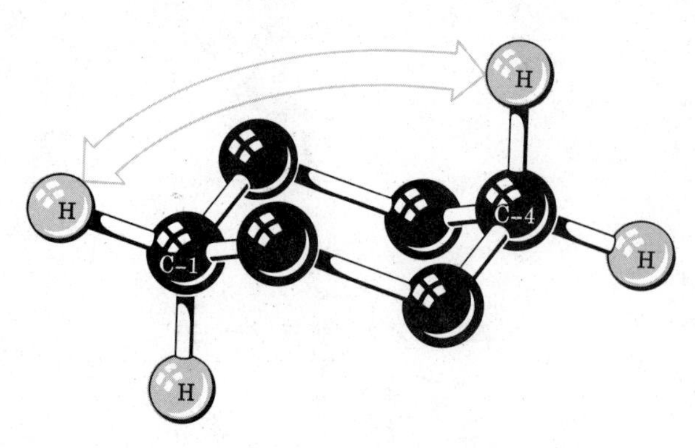

FIG. 3.10
Illustration of large separation between hydrogens at opposite corners of the ring (designated as C-1 and C-4) in the chair conformation of cyclohexane.

Another possible conformation of cyclohexane that is also free of angle strain is the "boat" conformation (Fig. 3.11).

The boat conformation, however, is not free of torsional strain. When a model of the boat conformation is viewed down carbon-carbon bond axes along either side (Fig. 3.12) the hydrogen atoms at those carbons are found to be eclipsed. Additionally, two of the hydrogen atoms on C—1 and C—4 are close enough to each other to cause van der Waals repulsion (Fig. 3.13). This latter effect has been called the "flagpole" interaction of the boat conformation. Torsional strain and flagpole interactions cause the boat conformation to have considerably higher energy than the chair conformation.

Although it is more stable, the chair conformation is much more rigid than the boat conformation. The boat conformation appears to be quite flexible—in fact, a better name for the boat conformation is the *flexible conformation.* By flexing to a new form—the twist conformation (Fig. 3.14)—the boat conformation can relieve some of its torsional strain and, at the same time, reduce the flagpole interactions.

FIG. 3.11
The boat-conformation of cyclohexane: (a) *carbon skeleton only;* (b) *carbon and hydrogen atoms;* (c) *line drawing;* (d) *space-filling model.*

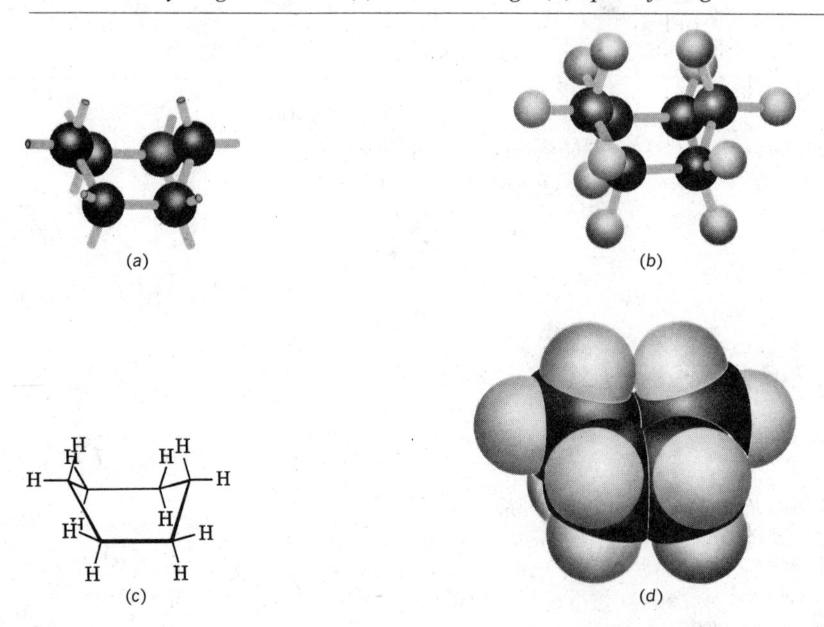

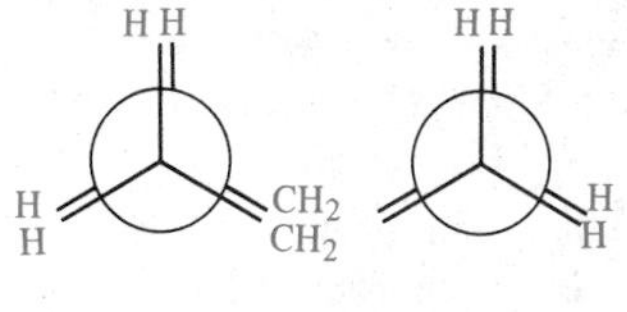

FIG. 3.12
Illustration of eclipsed hydrogens of the boat conformation of cyclohexane.

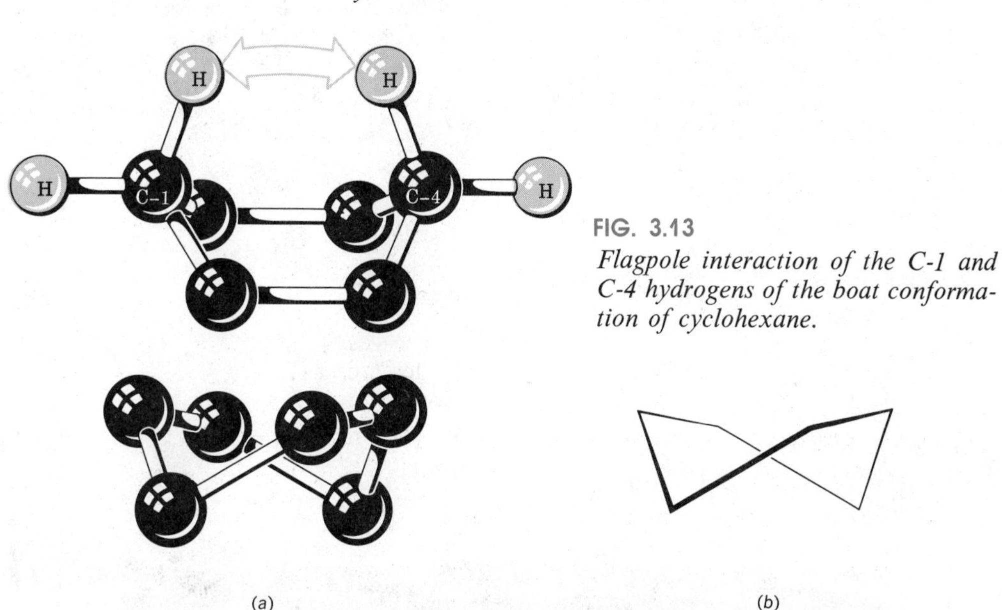

FIG. 3.13
Flagpole interaction of the C-1 and C-4 hydrogens of the boat conformation of cyclohexane.

FIG. 3.14
(a) *Carbon skeleton and* (b) *line drawing of the twist conformation of cyclohexane.*

This flexing causes the twist conformation to have lower energy than the boat conformation. *The stability gained by flexing is insufficient, however, to cause the twist conformation of cyclohexane to be more stable than the chair conformation.* The chair conformation is estimated to be lower in energy than the twist conformation by approximately 5 kcal/mole.

The energy barriers between the chair, boat, and twist conformations of cyclohexane are low enough to make their separation impossible at room temperature. At room temperature, the thermal energies of the molecules are great enough to cause approximately one million interconversions to occur each second and, *because of its greater stability, more than 99% of the molecules are estimated to be in a chair conformation at any given moment.*

3.11 CONFORMATIONS OF HIGHER CYCLOALKANES

Cycloheptane, cyclooctane, and cyclononane and other higher cycloalkanes also exist in nonplanar conformations. The small instabilities of these higher cycloalkanes (Table 3.6) appear to be caused primarily by torsional strain and van der Waals repulsions between hydrogens across rings. The nonplanar conformations of these rings, however, are essentially free of angle strain. Although they are not known with certainty, the most stable conformations of cycloheptane and cyclooctane appear to be those shown in Fig. 3.15.

X-ray crystallographic studies of cyclodecane reveal that the most stable conformation has carbon-carbon bond angles of 117°. This indicates some angle strain. The wide bond angles apparently allow the molecule to expand and thereby minimize unfavorable repulsions between hydrogens across the ring.

There is very little free space in the center of a cycloalkane unless the ring is quite large. Calculations indicate that cyclooctadecane, for example, is the smallest ring through which a $—CH_2CH_2CH_2—$ chain can be threaded. Molecules have been synthesized, however, which have large rings threaded on chains and which have large rings that are interlocked like links in a chain. These latter molecules are called *catenanes*.

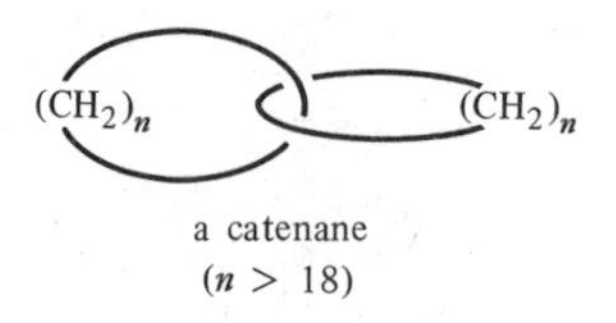

a catenane
($n > 18$)

3.12 SUBSTITUTED CYCLOHEXANES: AXIAL AND EQUATORIAL HYDROGENS

The six-membered ring is the most common ring found among nature's organic molecules. For this reason, we will give it special attention. We have already seen that the chair conformation of cyclohexane is the most stable one and that it is the predominant conformation of the molecules in a sample of cyclohexane. With this fact in mind, we are in a position to undertake a limited analysis of the conformations of substituted cyclohexanes.

If we look carefully at the chair conformation of cyclohexane (Fig. 3.16) we can see that there are only two different kinds of hydrogens. One hydrogen attached to each of the six carbons lies in a plane generally defined by the ring of carbon atoms. These hydrogens, by analogy with the equator of the earth, are called *equatorial* hydrogens. Six other hydrogens, one on each carbon, are oriented in a direction that is generally perpendicular to the average plane of the ring. These hydrogens, again by analogy with the earth, are called *axial* hydrogens. There are three axial hydrogens on each face of the cyclohexane ring and their orientation (up or down) alternates from one carbon to the next.

The question one might next ask is what is the most stable conformation of a cyclohexane derivative *in which one hydrogen has been replaced by a substituent?* That is, what is the most stable conformation of a *monosubstituted* cyclohexane? We can answer this question by considering methylcyclohexane as an example.

Methylcyclohexane has two possible chair conformations (Fig. 3.17), and these are interconvertible (via a flexible conformation) through partial rotations about the single bonds of the ring. In one conformation (Fig. 3.17*a*) the methyl group occupies an *axial* position, and in the other (Fig. 3.17*b*) the methyl group occupies an *equatorial* position. Studies indicate that the conformation with the methyl group equatorial is more stable than the conformation with the methyl group axial by about 1.6 kcal/mole. Thus, in the equilibrium mixture, the conformation with the methyl group in the equatorial position is the predominant one. Calculations show that it constitutes about 93% of the equilibrium mixture.

The greater stability of methylcyclohexane with an equatorial methyl group can be understood through an inspection of the two forms as they are shown in Fig. 3.18.

Studies done with scale models of the two conformations show that when the methyl group is axial, it is so close to the two axial hydrogens on the same side

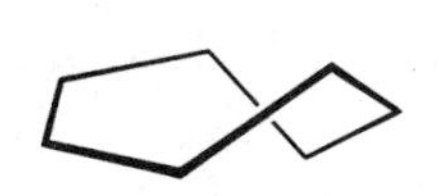

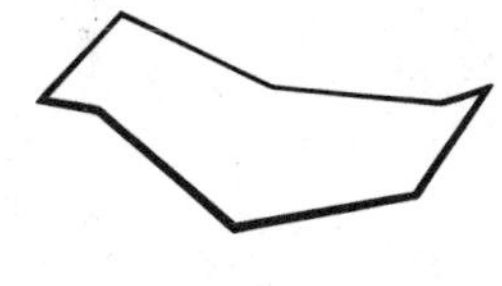

FIG. 3.15
Most stable conformations of cycloheptane and cyclooctane.

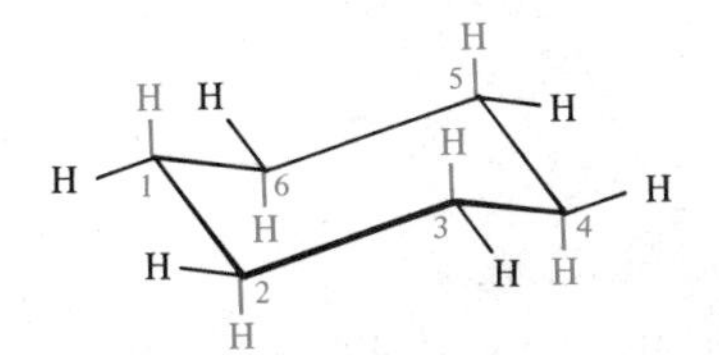

FIG. 3.16
The chair conformation of cyclohexane. The axial hydrogens are shown in color.

of the molecule that the van der Waals forces between them are repulsive. Similar studies also reveal that *any group has considerably more room when it occupies an equatorial position.*

In cyclohexane derivatives with larger substituents, this effect is even more pronounced. The conformation of *tert*-butylcyclohexane with the *tert*-butyl group equatorial is estimated to be more than 5 kcal/mole more stable than the axial form. This large energy difference between the two conformations means that, at room temperature, virtually 100% of the molecules of *tert*-butylcyclohexane have the *tert*-butyl group in the equatorial position.

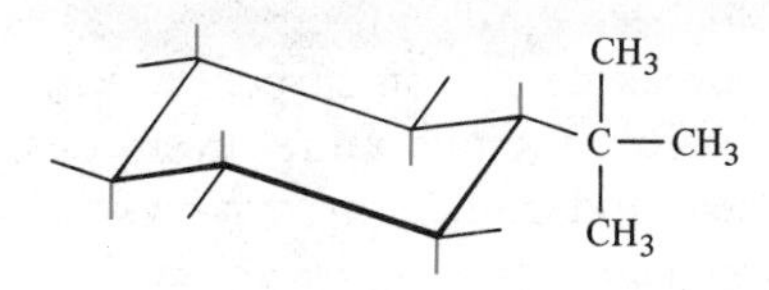

Equatorial *tert*-butylcyclohexane

FIG. 3.17
The conformations of methylcyclohexane with the methyl group axial (a) *and equatorial* (b).

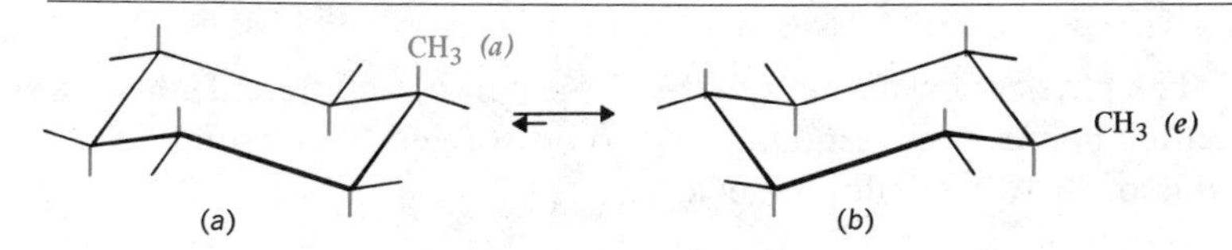

FIG. 3.18
(a) *1,3-Diaxial interactions between the two axial hydrogens and the axial methyl group in the axial conformation of methylcyclohexane.* (b) *Less crowding occurs in the equatorial conformation.*

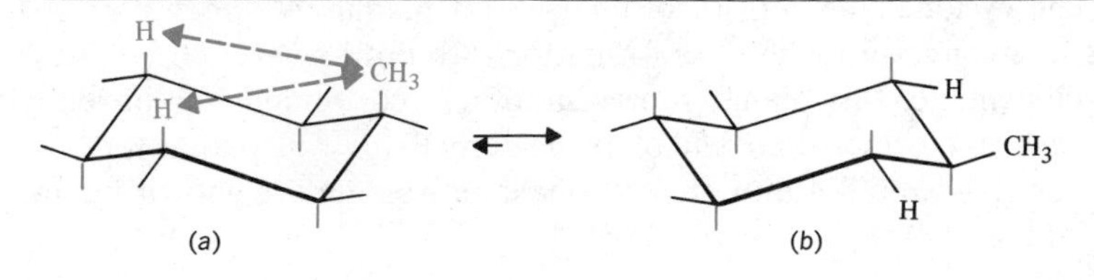

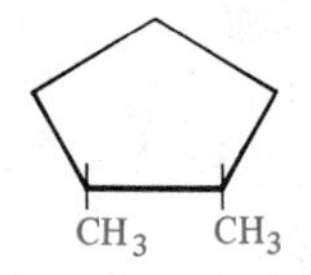

cis-1, 2-Dimethylcyclopentane
bp 130°

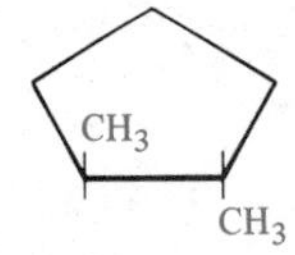

trans-1, 2-Dimethylcyclopentane
bp 123.7°

FIG. 3.19
Cis- *and* trans-*1,2-dimethylcyclopentanes.*

3.13 DISUBSTITUTED CYCLOALKANES: CIS-TRANS ISOMERISM

The presence of two substituents on the ring of a molecule of any cycloalkane allows for the possibility of *cis-trans* isomerism. We can see this most easily if we begin by examining cyclopentane derivatives because the cyclopentane ring is essentially planar. (At any given moment the ring of cyclopentane is, of course, slightly bent, but we know that the various bent conformations are rapidly interconverted. Over a period time, the average conformation of the cyclopentane ring is planar.) Since the planar representation is much more convenient for an initial presentation of *cis-trans* isomerism in cycloalkanes we will use it here.

Let us consider 1,2-dimethylcyclopentane as an example. We can write the structures shown in Fig. 3.19. In the first structure the methyl groups are on the same side of the ring, that is, they are *cis*. In the second structure the methyl groups are on opposite sides of the ring; they are *trans*.

The *cis*- and *trans*- 1,2-dimethylcyclopentanes are stereoisomers: they differ from each other only in the arrangement of the atoms in space. The two forms cannot be interconverted without breaking carbon-carbon bonds. This requires a great deal of energy, and does not occur at temperatures even considerably above room temperature. As a result, the *cis* and *trans* forms can be separated, placed in separate bottles, and kept indefinitely.

1,3-Dimethylcyclopentanes show *cis-trans* isomerism as well:

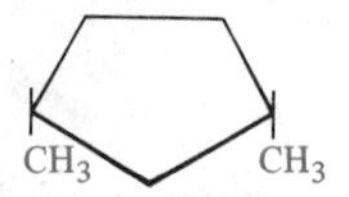

cis-1, 3-Dimethylcyclopentane

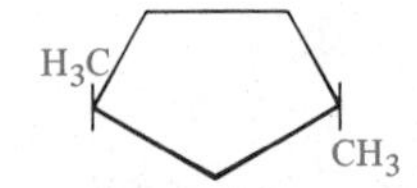

trans-1, 3-Dimethylcyclopentane

The physical properties of *cis-trans* isomers are different; they have different melting points, boiling points, and so on. Table 3.7 lists some of the physical constants of the dimethylcyclohexanes.

Problem 3.6

Write structures for the *cis* and *trans* isomers of: (a) 1,2-dimethylcyclopropane; (b) 1,2-dibromocyclobutane.

The cyclohexane ring is, of course, not planar. A "time average" of the various interconverting chair conformations would, however, be planar and, as with cyclopentane, this planar representation is convenient for introducing the topic of *cis-trans* isomerism of cyclohexane derivatives. The planar representations of the 1,2-, 1,3-, and 1,4-dimethylcyclohexane isomers are shown below.

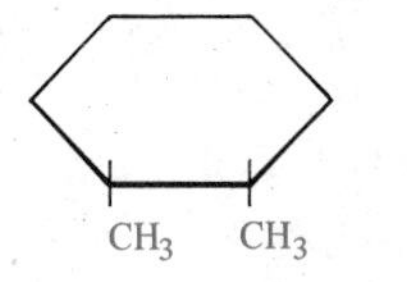

cis-1, 2-Dimethylcyclohexane

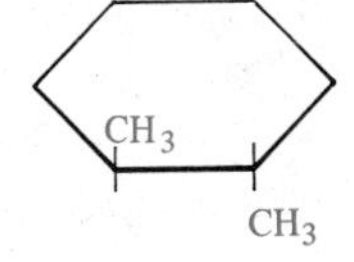

trans-1, 2-Dimethylcyclohexane

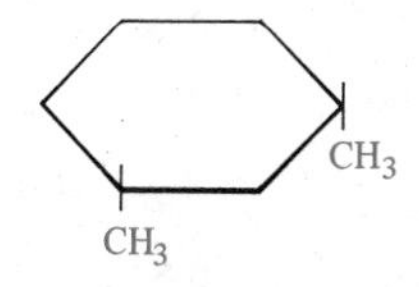

cis-1, 3-Dimethylcyclohexane

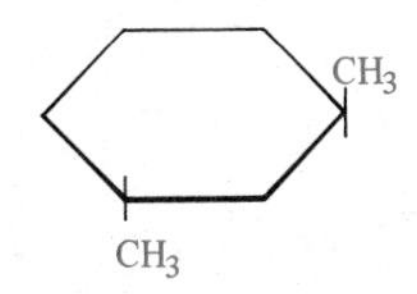

trans-1, 3-Dimethylcyclohexane

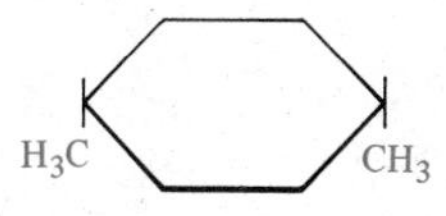

cis-1, 4-Dimethylcyclohexane

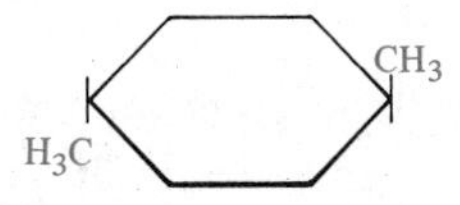

trans-1, 4-Dimethylcyclohexane

If we consider the *actual* conformations of these isomers, the structures are somewhat more complex. Beginning with *trans*-1,4-dimethylcyclohexane, since it is easiest to visualize, we find there are two possible chair conformations (Fig. 3.20). In one conformation both methyl groups are axial; in the other both are equatorial. The diequatorial conformation is, as we would expect it to be, the more stable

TABLE 3.7 Physical Constants of *cis*- and *trans*-Cyclohexane Derivatives

SUBSTITUENTS	ISOMER	mp °C	bp °C[a]
1,2-Dimethyl	*cis*	−50.1	130.04^{760}
1,2-Dimethyl	*trans*	−89.4	123.7^{760}
1,3-Dimethyl	*cis*	−85	124.9^{760}
1,3-Dimethyl	*trans*	−79.4	123.5^{760}
1,2-Dichloro	*cis*	−6	93.5^{22}
1,2-Dichloro	*trans*	−7	74^{16}

[a] The pressures (in units of Torr) at which the boiling points were measured are given as superscripts.

FIG. 3.20
The two chair conformations of trans-*1,4-dimethylcyclohexane. (Note: all other C—H bonds have been omitted for clarity.)*

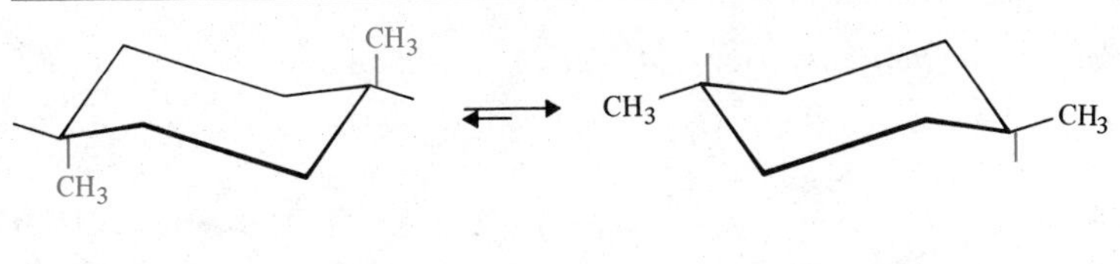

conformation, and it represents the structure of at least 99% of the molecules at equilibrium.

That the diaxial form of *trans*-1,4-dimethylcyclohexane is a *trans* isomer is easy to see; the two methyl groups are clearly on opposite sides of the ring. The *trans* relationship of the methyl groups in the diequatorial form is not as obvious, however. The *trans* relationship of the methyls becomes more apparent if we imagine ourselves "flattening" the molecule by turning one end up and the other down.

CH_3 CH_3 → CH_3 CH_3

A second way to recognize a *trans*-disubstituted cyclohexane is to notice that one group is attached by the *upper* bond (of the two to its carbon) and one by the *lower* bond.

cis-1,4-Dimethylcyclohexane also exists in two conformations. The *cis* relationship of the methyl groups, however, precludes the possibility of a structure with both groups in an equatorial position. There are, thus, two equivalent conformations (Fig. 3.21) of *cis*-1,4-dimethylcyclohexane.

Problem 3.7

(a) Write structural formulas for the two chair conformations of *cis*-1-*tert*-butyl-4-methylcyclohexane. (b) Are these two conformations equivalent? (c) If not, which would be more stable? (d) Which would be the preferred conformation at equilibrium?

trans-1,3-Dimethylcyclohexane is like the *cis*-1,4-compound in that no conformation is possible with both methyl groups in the favored equatorial position. The two conformations below are of equal energy and are equally populated at equilibrium.

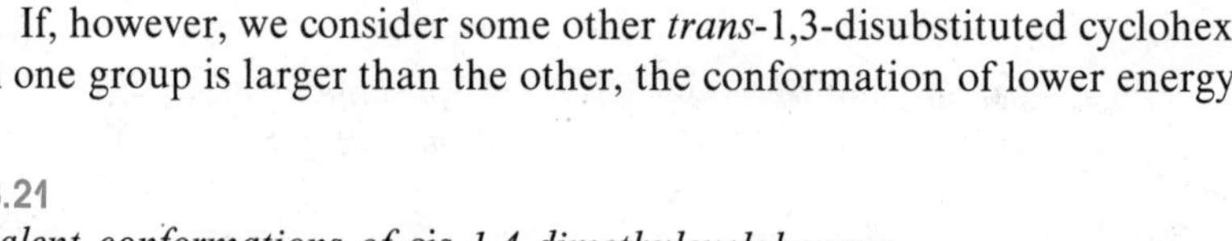

(e) CH_3 CH_3 (a) ⇆ CH_3 (a) CH_3 (e)

trans-1, 3-Dimethylcyclohexane

If, however, we consider some other *trans*-1,3-disubstituted cyclohexane in which one group is larger than the other, the conformation of lower energy is the

FIG. 3.21
Equivalent conformations of cis-*1,4-dimethylcyclohexane.*

CH_3 CH_3 ⇆ CH_3 CH_3

Equatorial-axial

Axial–equatorial

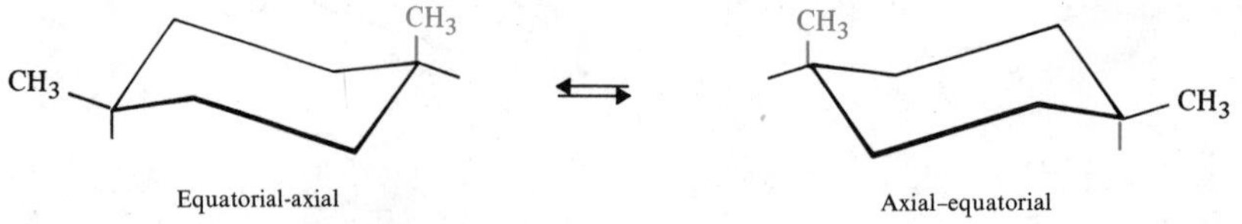

one having the larger group in the equatorial position. For example, the more stable conformation of *trans*-1-*tert*-butyl-3-methylcyclohexane, shown below, has the large *tert*-butyl group occupying the equatorial position.

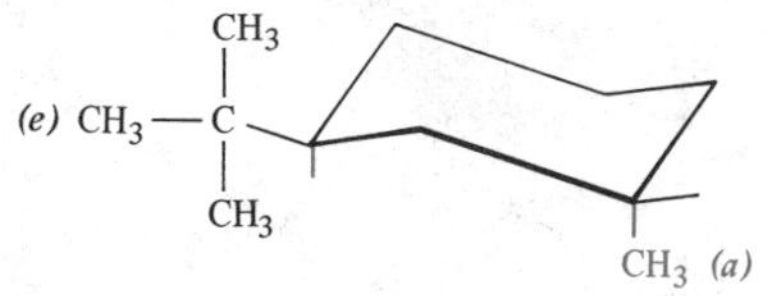

Problem 3.8

(a) Write chair conformations for *cis*- and *trans*-1,2-dimethylcyclohexane. (b) For which isomer (*cis* or *trans*) are the two conformations equivalent? (c) For the isomer where the two conformations are not equivalent, which conformation is more stable? (d) Which conformation would be more highly populated at equilibrium? (Check your answer with Table 3.8.)

The different conformations of the dimethylcyclohexanes are summarized in Table 3.8. The more stable conformation, where one exists, is set in heavy type.

3.14 BICYCLIC AND POLYCYCLIC ALKANES

Many of the molecules that we encounter in our study of organic chemistry contain more than one ring. One of the most important bicyclic systems is bicyclo[4.4.0]-decane, a compound that is usually called by its common name, *decalin*.

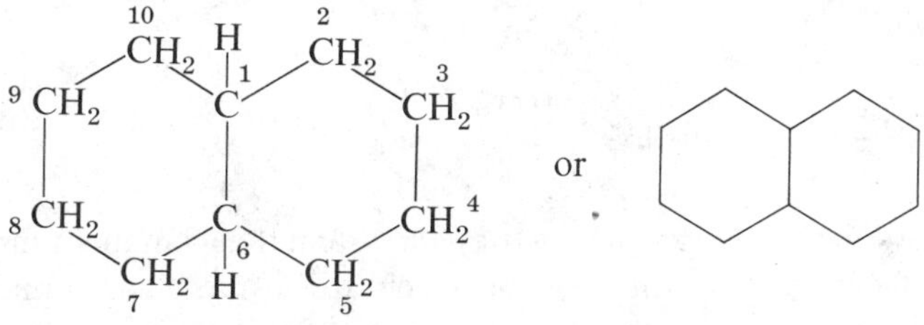

Decalin (or bicyclo [4.4.0] decane)
(carbons 1 and 6 are bridgehead carbons)

Decalin shows *cis-trans* isomerism:

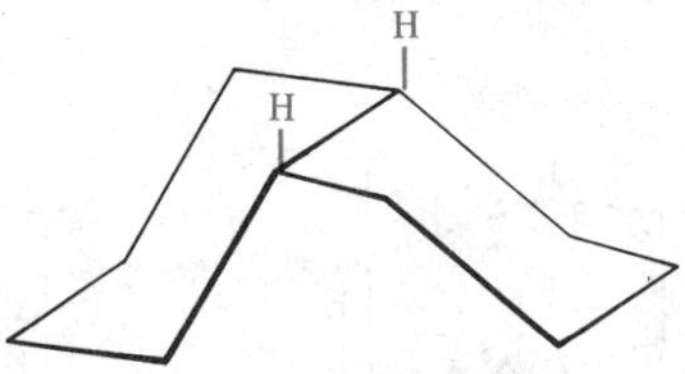

cis-Decalin

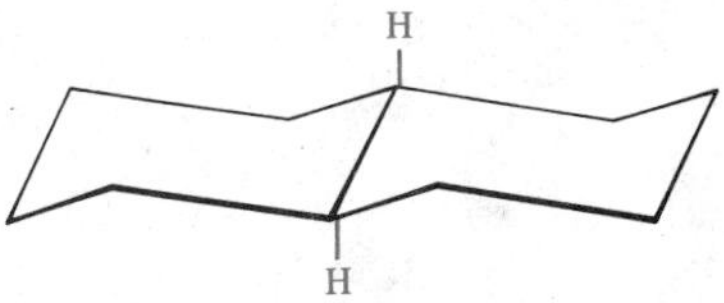

trans-Decalin

TABLE 3.8 Conformations of Dimethylcyclohexanes

COMPOUND	*cis* ISOMER			*trans* ISOMER		
1,2-Dimethyl-	*a,e*	or	*e,a*	***e,e***	or	*a,a*
1,3-Dimethyl-	***e,e***	or	*a,a*	*a,e*	or	*e,a*
1,4-Dimethyl-	*a,e*	or	*e,a*	***e,e***	or	*a,a*

In *cis*-decalin the two hydrogens attached to the bridgehead atoms lie on the same side of the ring; in *trans*-decalin they are on opposite sides. We often indicate this by writing their structures in the following way.

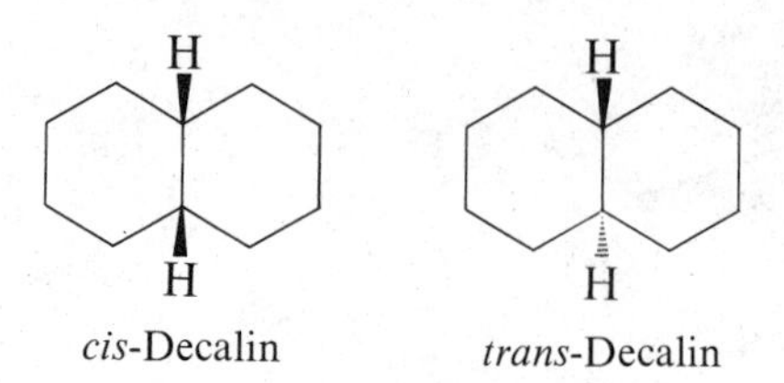

cis-Decalin *trans*-Decalin

Simple rotations of groups about carbon-carbon bonds do not interconvert *cis*- and *trans*-decalins. In this respect they resemble the isomeric *cis*- and *trans*-disubstituted cyclohexanes. (We can, in fact, regard them as being *cis*- or *trans*-1,2-disubstituted cyclohexanes in which the 1,2-substituents are the two ends of a four carbon bridge, that is, $—CH_2CH_2CH_2CH_2—$.)

The *cis*- and *trans*-decalins can be separated. *cis*-Decalin boils at 195° (at 760 Torr) and *trans*-decalin boils at 185.5° (at 760 Torr).

Norbornane (below) is another bicyclic system. In norbornane the cyclohexane ring is held in a boat form by a methylene bridge ($—CH_2—$) between the 1- and 4- carbons.

Norbornane
(or bicyclo [2.2.1] heptane)

Adamantane (below) is a tricyclic system that contains a three-dimensional array of cyclohexane rings, all of which are in the chair form. Extending the structure of adamantane in three dimensions gives the structure of diamond. The great hardness of diamond results from the fact that the entire diamond crystal is actually one very large molecule—a molecule that is held together by millions of strong covalent bonds.

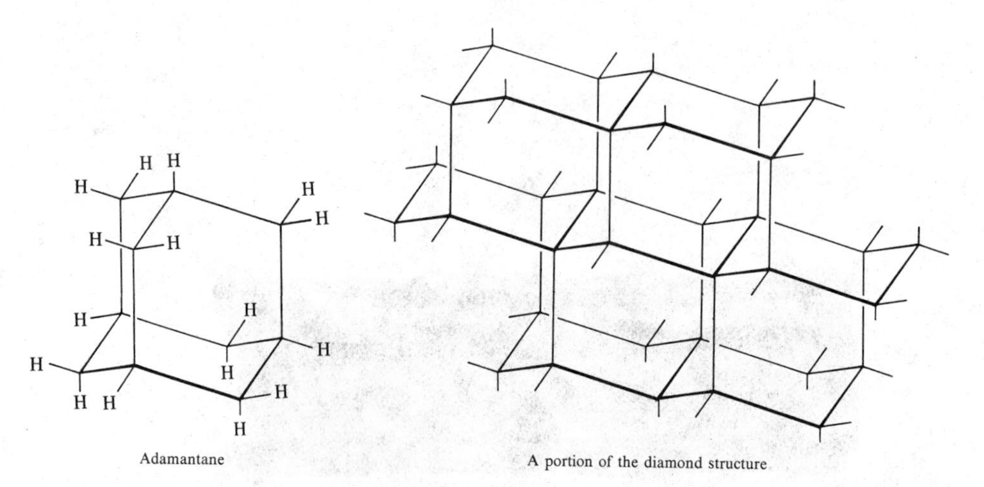

Adamantane

A portion of the diamond structure

A Newman projection as viewed down any carbon-carbon bond in the diamond structure shows an exactly staggered conformation.

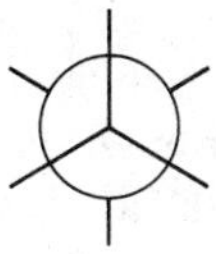

Another form of diamond called "wurtzite diamond" has been synthesized at laboratories of the General Electric Company (by using explosive shock waves). It has also been found in traces around a large meteorite crater in the Southwest. Wurtzite diamond has a slightly different structure. It has a network of carbon atoms based on the following structure:

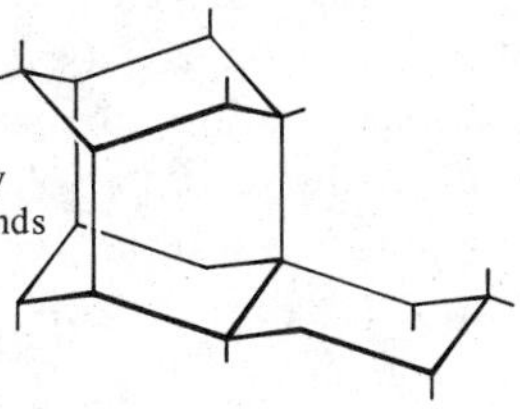

Cyclohexane rings joined by three axial bonds

There are many eclipsing interactions in this structure and this probably explains why wurtzite diamond is very much rarer than ordinary diamond.

The tetracyclic ring system of perhydrocyclopentanophenanthrene (below) is incorporated into a broad group of biologically potent compounds known as steroids. Many steroids function as hormones: they regulate numerous biochemical processes including sexual development, fertility, and growth. Cholesterol, a steroid that is found in nearly all body tissues, apparently serves as a precursor for many steroidal hormones. Accumulation of cholesterol in the arteries of the heart, however, has been strongly implicated in the development of heart disease.

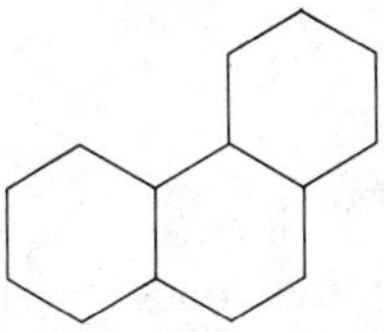

Perhydrophenanthrene*

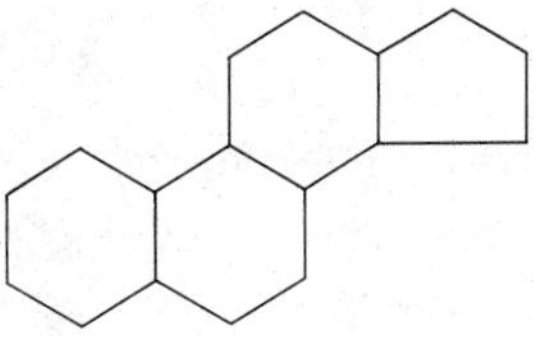

Perhydrocyclopentanophenanthrene

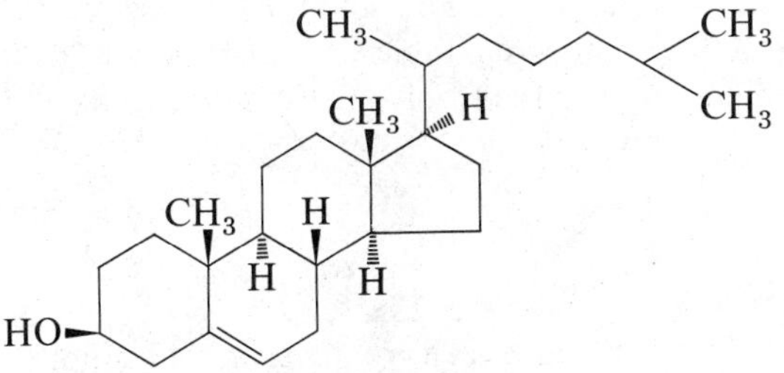

Cholesterol

* *Perhydro* as a prefix means that the ring contains no multiple bonds.

3.15 SOURCES OF ALKANES: PETROLEUM

The primary source of alkanes is petroleum. Petroleum is a complex mixture of organic compounds, most of which are hydrocarbons. It also contains small amounts of oxygen, nitrogen, and sulfur-containing compounds as well as water.

The primary use of alkanes is as a source of energy through combustion. Today more than 50% of the world's energy requirement is met from petroleum. Most of the fuels that we commonly use—natural gas, gasoline, kerosene, and diesel fuel, for example—are composed of mixtures of alkanes, and most are obtained from petroleum. In a sense the energy of these alkanes is *stored sunlight*. Solar energy was absorbed by tiny microorganisms in the seas of the earth some 300 to 400 million years ago. While alive, these organisms used the sun's energy to make organic molecules. When they died they settled and were covered by layers of sediments. Gradually, under great heat and pressure generated as the sediments became rock, their organic compounds were converted to petroleum. Where the rock formations were porous, the petroleum collected into great underground pools. These pools, located in various parts of the earth, supply today's petroleum industry. Nonporous formations became oil shales.

Because of the way it was formed, petroleum is sometimes called a "fossil fuel." Other fossil fuels are coal, natural gas (cf. Section 2.2A), and the organic matter in oil shales. All of these are the remnants of prehistoric organisms as acted upon by millions of years of geology.

As we now all know, the problem with petroleum and other fossil fuels is that *their supplies are finite*. Estimates vary, and much depends on how much we use, but most estimates agree that the world's supply of petroleum will run out sometime between 2020 and 2075. It is this hard fact and the fact that our societies are so largely organized around machines that use petroleum-based fuels, that constitutes the core of the energy crisis.

The solution to the energy crisis lies in developing alternate sources of energy and *in conservation*. Science can solve the technical aspects, but much will depend on the social, political, and economic decisions that are taken. It is almost certainly true that we will not, in the next 20 to 30 years, have the abundant and cheap sources of energy that we have had in the recent past and we will all have to adjust our way of living accordingly.

3.15A Petroleum Refining

The first step in refining petroleum is distillation; the object here is to separate the petroleum into fractions based on the volatility of its components. Complete separation into fractions containing individual compounds is economically impractical and virtually impossible technically. More than 500 different compounds are contained in the petroleum distilates boiling below 200° and many have almost the same boiling points. Thus the fractions taken contain mixtures of alkanes of similar boiling points (cf. Table 3.9). Mixtures of alkanes, fortunately, are perfectly suitable for uses as fuels, solvents, and lubricants, the primary uses of petroleum.

3.15B Cracking

The demand for gasoline is much greater than that supplied by the gasoline fraction of petroleum. Important processes in the petroleum industry, therefore, are concerned with converting hydrocarbons from other fractions into gasoline. When a mixture of alkanes from the gas oil (C_{12} and higher) fraction are heated at

TABLE 3.9 Typical Fractions Obtained by Distillation of Petroleum

BOILING RANGE OF FRACTION C°	NUMBER OF CARBON ATOMS PER MOLECULE	USE
Below 20°	C_1–C_4	Natural gas, bottled gas, petrochemicals
20–60°	C_5–C_6	Petroleum ether, solvents
60–100°	C_6–C_7	Ligroin, solvents
40–200°	C_5–C_{10}	Gasoline
175–325°	C_{12}–C_{18}	Kerosene and jet fuel
250–400°	C_{12} and higher	Gas oil, fuel oil, and diesel oil
Nonvolatile liquids	C_{20} and higher	Refined mineral oil, lubricating oil, grease
Nonvolatile solids	C_{20} and higher	Paraffin wax, asphalt, and tar

Adapted with permission from John R. Holum, *Organic Chemistry: A Brief Course,* Wiley, New York, 1975.

very high temperatures (~500°) in the presence of a variety of catalysts, the molecules break apart and rearrange to smaller, more highly branched alkanes containing 5 to 10 carbons (see Table 3.9). This process is called *catalytic cracking.* Cracking can also be done in the absence of a catalyst—called *thermal cracking*—but in this process the products tend to have unbranched chains and alkanes with unbranched chains have a very low "octane rating" (cf. Section 7.6). Gasoline components can also be made from smaller molecules (alkenes) by a process called alkylation. We will see how this is done later.

3.16 SYNTHESIS OF ALKANES AND CYCLOALKANES

Mixtures of alkanes as they are obtained from petroleum are reasonably suitable as fuels. However, in our laboratory work we often have the need for a pure sample of a particular alkane. For these purposes, the chemical preparation—or synthesis—of that particular alkane is often the most reliable way of obtaining it. The preparative method that we choose should be one that will lead to the desired product alone or, at least, to products that can be easily and effectively separated.

Several such methods are available, and three are outlined here. In subsequent chapters we will encounter others.

1. *Hydrogenation of alkenes.* Alkenes add hydrogen in the presence of metal catalysts to produce alkanes. Examples are:

$$CH_3CH{=}CHCH_3 + H_2 \xrightarrow[C_2H_5OH\ (25°,\ 50\ atm)]{Ni} CH_3\underset{H}{\underset{|}{C}}H{-}\underset{H}{\underset{|}{C}}HCH_3$$

2-Butene → Butane

$$CH_3{-}\overset{CH_3}{\overset{|}{C}}{=}CH_2 + H_2 \xrightarrow[C_2H_5OH\ (25°,\ 50\ atm)]{Ni} CH_3\underset{H}{\underset{|}{\overset{CH_3}{\overset{|}{C}}}}{-}\underset{H}{\underset{|}{C}}H_2$$

2-Methylpropene → Isobutane

Cyclohexene $+ H_2 \xrightarrow[C_2H_5OH\ (25°,\ 1\ atm)]{Pt}$ Cyclohexane

These reactions are very easy to carry out in the laboratory. We will see several applications of hydrogenation in Chapter 7.

2. *Reduction of alkyl halides.* Most alkyl halides react with zinc and aqueous acid to produce an alkane. In this reaction the more electronegative halogen is replaced by the less electronegative hydrogen. For this reason the reaction is one of reduction.

$$2CH_3CH_2\underset{\displaystyle Br}{\underset{|}{C}}HCH_3 \xrightarrow[Zn]{H^+} 2CH_3CH_2\underset{\displaystyle H}{\underset{|}{C}}HCH_3 + ZnBr_2$$

2-Bromobutane → Butane

$$2CH_3\overset{\displaystyle CH_3}{\overset{|}{C}}HCH_2CH_2—Br \xrightarrow[Zn]{H^+} 2CH_3\overset{\displaystyle CH_3}{\overset{|}{C}}HCH_2CH_2—H + ZnBr_2$$

Isopentyl bromide → Isopentane

Cyclopropane, but not higher cycloalkanes, can be prepared by a similar reaction—one that results in the formation of a new carbon-carbon bond.

$$I—CH_2CH_2CH_2—I \xrightarrow[Alcohol]{Zn} \text{cyclo-}(CH_2)_3 + ZnI_2$$

1,3-Diiodopropane → (70%)

3. *The Corey-House alkane synthesis.* A highly versatile method for the synthesis of alkanes from alkyl halides has been developed independently by Professors E. J. Corey (Harvard University) and Herbert O. House (Georgia Institute of Technology). The overall result of the synthesis is:

$$R—X + R'—X \xrightarrow[\text{steps }(-2X)]{\text{Several}} R—R'$$

In the last step a lithium dialkylcuprate, R_2CuLi, is treated with an alkyl halide, R′—X; the product of the reaction is the alkane, R—R′.

$$R_2CuLi + R'—X \longrightarrow R—R' + RCu + LiX$$

Lithium dialkylcuprates are prepared in the following way.

$$R—X \xrightarrow[Ether]{2Li} 2RLi \xrightarrow{CuI} R_2CuLi$$

Alkyllithium; Lithium dialkylcuprate

An alkyl halide (R—X) reacts with lithium metal to produce an alkyllithium (R—Li). The alkyllithium is then treated with cuprous iodide to produce the lithium dialkylcuprate (R_2CuLi).

The overall synthesis, summarized on the next page, gives excellent yields when the alkyl halide (R′—X) is a primary halide and when the alkyl groups of the dialkylcuprate are also primary.

The overall scheme for the Corey-House synthesis:

$$2RLi \xrightarrow{CuI} R_2CuLi \xrightarrow{R'X} R{-}R' + RCu + LiX$$

An alkyllithium — A lithium dialkylcuprate

$2R{-}X \xrightarrow[\text{ether}]{4Li}$ 2RLi

$2R{-}X$: 1° Alkyl halide; $R'{-}X$: 1° Alkyl halide

These are the organic starting materials. The R— and R′— groups need not be different.

Two specific examples are outlined below.

$$(CH_3)_2CuLi + CH_3CH_2CH_2CH_2CH_2{-}I \xrightarrow[\text{3.5 hr, 25°}]{\text{Diethyl ether}} CH_3{-}CH_2CH_2CH_2CH_2CH_3 \ (98\%)$$

$$(CH_3CH_2CH_2CH_2)_2CuLi + CH_3CH_2CH_2CH_2CH_2CH_2CH_2{-}Cl \xrightarrow[\text{5 days, 0°}]{\text{Diethyl ether}}$$

$$CH_3CH_2CH_2CH_2{-}CH_2CH_2CH_2CH_2CH_2CH_2CH_3 \ (75\%)$$

The Corey-House synthesis is one of the most useful methods for forming a carbon-carbon bond between two different groups. We shall see other applications of it in Chapters 14 and 16.

Problem 3.9 * Know how to do

Outline a synthesis of each of the following alkanes from appropriate alkyl halides using the Corey-House method.

(a) Propane
(b) Butane
(c) 2-Methylbutane
(d) 2,7-Dimethyloctane

Problem 3.10

Outline methods showing how hexane could be prepared starting with:

(a) A propyl bromide
(b) A butyl bromide
(c) A pentyl bromide
(d) A hexyl bromide
(e) A hexene

3.17 EMPIRICAL AND MOLECULAR FORMULAS

In Section 1.2, we discussed briefly the pioneering work of Berzelius, Dumas, Liebig, and Cannizzaro in devising methods for determining the formulas of organic compounds. Although the experimental procedures for these analyses have been refined, the basic methods for determining the elemental composition of an organic compound today are not substantially different from those used in the nineteenth century. A carefully weighed quantity of the compound to be analyzed is oxidized completely to carbon dioxide and water. The weights of carbon dioxide and water are carefully measured and used to find the percentages of carbon and hydrogen in the compound. The percentage of nitrogen is usually determined by measuring the volume of nitrogen (N_2) produced in a separate procedure.

Special techniques for determining the percentage composition of other elements typically found in organic compounds have also been developed, but the direct determination of the percentage of oxygen is difficult. However, if the percentage composition of all the other elements is known, then the percentage of oxygen can be determined by difference. The following examples will illustrate how these calculations can be carried out.

Example A

A new organic compound is found to have the following elemental analysis.

Carbon	67.95%
Hydrogen	5.69
Nitrogen	26.20
Total:	99.84%

Since the total of these percentages is very close to 100% (within experimental error) we can assume that no other element is present. For the purpose of our calculation it is convenient to assume that we have a 100-g sample. If we did, it would contain the following:

67.95 g of carbon
5.69 g of hydrogen
26.20 g of nitrogen

In other words, we use percents *by weight* to give us the ratios *by weight* of the elements in the substance. To write a formula for the substance, however, we need *ratios by moles.*

We now divide each of these weight-ratio numbers by the atomic weight of the particular element and obtain the number of moles of each element, respectively, in 100 g of the compound. This operation gives us the ratios *by moles* of the elements in the substance:

$$\text{C:} \quad \frac{67.95 \text{ g}}{12.01 \text{ g/mole}} = 5.66 \text{ moles}$$

$$\text{H:} \quad \frac{5.69 \text{ g}}{1.008 \text{ g/mole}} = 5.64 \text{ moles}$$

$$\text{N:} \quad \frac{26.20 \text{ g}}{14.01 \text{ g/mole}} = 1.87 \text{ moles}$$

One possible formula for the compound, therefore, is $C_{5.66}H_{5.64}N_{1.87}$.

By convention, however, we use *whole* numbers in formulas. Therefore, we convert these fractional numbers of moles to whole numbers by dividing each by 1.87, the smallest number.

$$\text{C:}\quad \frac{5.66}{1.87} = 3.03 \text{ which is } \sim 3$$

$$\text{H:}\quad \frac{5.64}{1.87} = 3.02 \text{ which is } \sim 3$$

$$\text{N:}\quad \frac{1.87}{1.87} = 1.00$$

Thus within experimental error, the ratios by moles are 3C to 3H to 1N, and C_3H_3N is the *empirical formula.* By empirical formula we mean the formula in which the subscripts are the smallest integers that give the ratio of atoms in the compound. In contrast, a *molecular* formula discloses the complete composition of one molecule. The molecular formula of this particular compound could be C_3H_3N or some whole-number multiple of C_3H_3N; that is, $C_6H_6N_2$, $C_9H_9N_3$, $C_{12}H_{12}N_4$, and so on. If, in a separate determination, we find that the molecular weight of the compound is 108 ± 3, we can be certain that the *molecular formula* of the compound is $C_6H_6N_2$.

Formula	Molecular Weight
C_3H_3N	53.06
$C_6H_6N_2$	106.13 (which is within the range 108 ± 3)
$C_9H_9N_3$	159.19
$C_{12}H_{12}N_4$	212.26

The most accurate method for determining molecular weights is by mass spectroscopy; this method (which can also be used to determine molecular formulas and structures) is described in Special Topic H. A variety of other methods based on freezing point depression, boiling point elevation, osmotic pressure, and vapor density can also be used to determine molecular weights.

Example B

Histidine, an amino acid isolated from protein, has the following elemental analysis:

Carbon	46.38%
Hydrogen	5.90
Nitrogen	27.01
Total:	79.29
Difference	20.71 (assumed to be oxygen)
	100.00%

Since no elements, other than oxygen, are found to be present in histidine the difference is assumed to be oxygen. Again, we assume a 100-g sample and divide the weight of each element by its gram-atomic weight. This gives us the ratio of moles (A).

(A) (B) (C)

$$\text{C:}\quad \frac{46.38}{12.01} = 3.86 \quad \frac{3.86}{1.29} = 2.99 \times 2 = 5.98 \sim 6 \text{ carbon atoms}$$

$$\text{H:}\quad \frac{5.90}{1.008} = 5.85 \quad \frac{5.85}{1.29} = 4.53 \times 2 = 9.06 \sim 9 \text{ hydrogen atoms}$$

$$\text{N:}\quad \frac{27.01}{14.01} = 1.94 \qquad \frac{1.94}{1.29} = 1.50 \times 2 = 3.00 = 3 \text{ nitrogen atoms}$$

$$\text{O:}\quad \frac{20.71}{16.00} = 1.29 \qquad \frac{1.29}{1.29} = 1.00 \times 2 = 2.00 = 2 \text{ oxygen atoms}$$

Dividing each of the moles (A) by the smallest does not give a set of numbers (B) that is close to a set of whole numbers. Multiplying each of the numbers in column (B) by 2 does, however, as seen in column (C). The empirical formula of histidine is, therefore, $C_6H_9N_3O_2$.

In a separate determination the molecular weight of histidine was found to be 158 ± 5. The empirical formula weight of $C_6H_9N_3O_2$ (155.15) is within this range; thus the molecular formula for histidine is the same as the empirical formula.

Problem 3.11

What is the empirical formula of each of the following compounds?

(a) Hydrazine, N_2H_4
(b) Benzene, C_6H_6
(c) Dioxane, $C_4H_8O_2$
(d) Nicotine, $C_{10}H_{14}N_2$
(e) Cyclodecane, $C_{10}H_{20}$
(f) Acetylene, C_2H_2

Problem 3.12

The empirical formulas and molecular weights of several compounds are given below. In each case calculate the molecular formula for the compound.

Empirical Formula	Molecular Weight
(a) CH_2O	179 ± 5
(b) CHN	80 ± 5
(c) CCl_2	410 ± 10

Problem 3.13

The widely used antibiotic, penicillin G, gave the following elemental analysis: C, 57.45%; H, 5.40%; N, 8.45%; S, 9.61%. The molecular weight of penicillin G is 330 ± 10. Assume that no other elements except oxygen are present and calculate the empirical and molecular formulas for penicillin G.

3.18 SOME IMPORTANT TERMS AND CONCEPTS

Alkanes are hydrocarbons with the general formula C_nH_{2n+2}. Molecules of alkanes have no rings (i.e., they are *acyclic*) and they have only single bonds between carbon atoms. Their carbon atoms are sp^3 hybridized.

Cycloalkanes are hydrocarbons with the general formula C_nH_{2n} whose molecules have their carbon atoms arranged into a ring. They have only single bonds between carbon atoms and their carbon atoms are sp^3 hybridized.

Conformational analysis. A study of the energy changes that occur in a molecule when groups rotate about single bonds.

Torsional strain. A small barrier (2.8 kcal/mole) to free rotation about the carbon-carbon single bond of ethane.

Van der Waals forces. Weak forces that act between nonpolar molecules or between parts of the same molecule. Bringing two groups together first results in an *attractive* van der Waals force between them because a temporary unsymmetrical distribution of electrons in one group induces an opposite polarity in the other. When the groups are brought closer than their *van der Waals radii,* the force between them becomes repulsive because their electron clouds begin to interpenetrate each other. The methyl groups of the *gauche* form of butane, for example, are close enough for the van der Waals forces to be repulsive.

Ring strain. A strain that gives certain cycloalkanes greater potential energy than others. The principal sources of ring strain are *torsional strain* and *angle strain.*

Angle strain is strain introduced into a molecule because some factor (e.g., ring size) causes the bond angles of its atoms to deviate from the normal bond angle. The normal bond angles of an sp^3 carbon are 109.5° but in cyclopropane, for example, one pair of bonds at each carbon is constrained to a much smaller angle. This introduces considerable angle strain into the molecule causing molecules of cyclopropane to have greater potential energy per CH_2 group than cycloalkanes with less (or no) angle strain.

***Cis-trans* isomerism.** A type of isomerism possible with certain alkenes (Section 2.5A) and with disubstituted cycloalkanes. *Cis-trans* isomers of 1,2-dimethylcyclopropane are shown below.

CH_3 CH_3 — *cis*

CH_3 CH_3 — *trans*

These two isomers can be separated and they have different physical properties. The two forms cannot be interconverted without breaking carbon-carbon bonds.

Conformations of molecules of cyclohexane. The most stable conformation is a chair conformation. Twist conformations and boat conformations have greater potential energy. These conformations (chair, boat, and twist) can be interconverted by rotations of single bonds. In a sample of cyclohexane more than 99% of the molecules are in a chair conformation at any given moment. A group attached to a carbon of a chair conformation of a molecule of cyclohexane can assume either of two positions: *axial* or *equatorial,* and these are interconverted when the ring flips from one chair conformation to another. A group has more room when it is equatorial; thus most of the molecules of substituted cyclohexanes at any given moment will assume that chair conformation that has the largest group (or groups) equatorial.

Additional Problems

3.14

Write structural formulas for each of the following compounds:

(a) 2,3-Dichloropentane
(b) *tert*-Butyl iodide
(c) 3-Ethylpentane
(d) 2,3,4-Trimethyldecane
(e) 4-Isopropylnonane
(f) 1,1-Dimethylcyclopropane
(g) *cis*-1,2-Dimethylcyclobutane
(h) *trans*-1,3-Dimethylcyclobutane
(i) Isopropylcyclohexane (more stable conformation)
(j) *trans*-1-Isopropyl-3-methylcyclohexane (more stable conformation)

(k) Isohexyl chloride
(l) 2,2,4,4-Tetramethyloctane
(m) Neopentyl chloride
(n) Isopentane

3.15

Name each of the following compounds by the IUPAC system.

(a) $CH_3CH(C_2H_5)CH(CH_3)CH_2CH_3$

(b) $CH_3CH(CH_3)CH_2CH_3$

(c) 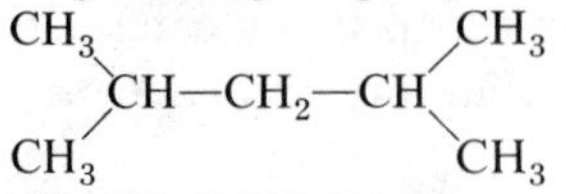
$$\begin{array}{ccccc} CH_3 & & & & CH_3 \\ & \diagdown & & \diagup & \\ & CH{-}CH_2{-}CH & & & \\ & \diagup & & \diagdown & \\ CH_3 & & & & CH_3 \end{array}$$

(d) $CH_3CH_2CH(C_2H_5)CH_3$

(e) 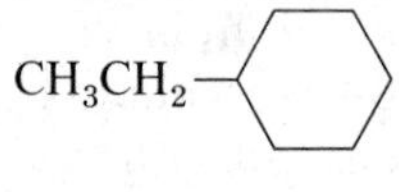 CH_3CH_2-(cyclohexane ring)

(f)

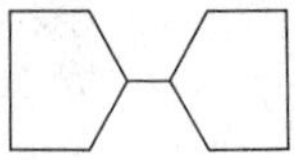

(g) 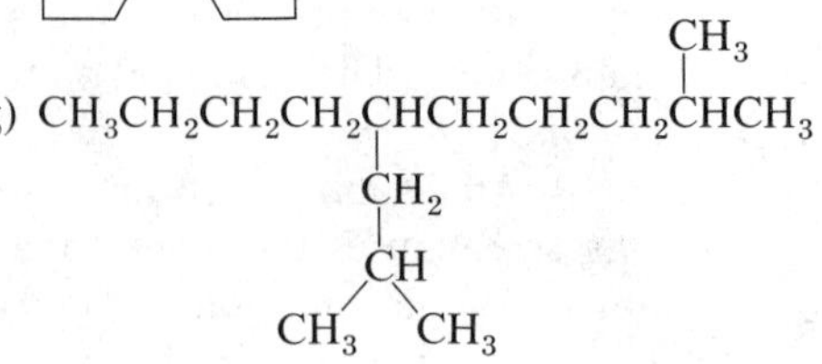
$$\begin{array}{c} \qquad\qquad\qquad\qquad\qquad\qquad\qquad\qquad\quad CH_3 \\ \qquad\qquad\qquad\qquad\qquad\qquad\qquad\qquad\quad | \\ CH_3CH_2CH_2CH_2\underset{\displaystyle\underset{\displaystyle CH_3\diagup\ \diagdown CH_3}{\overset{|}{\underset{|}{CH_2}}\atop CH}}{CH}CH_2CH_2CH_2CHCH_3 \end{array}$$

3.16

Give both common and IUPAC names for the isomers of

(a) C_3H_7Cl
(b) C_4H_9Br.

3.17

Write the structure and give the IUPAC name of an alkane or cycloalkane with the formula:

(a) C_5H_{12} that has only primary hydrogens (i.e. hydrogens attached to primary carbons)
(b) C_5H_{12} that has only one tertiary hydrogen
(c) C_5H_{12} that has only primary and secondary hydrogens
(d) C_5H_{10} that has only secondary hydrogens
(e) C_6H_{14} that has only primary and tertiary hydrogens

3.18

An alternative system of nomenclature that is occasionally used for naming alkanes with a high degree of symmetry is based on naming them as alkyl-substituted methanes. In this system, 3-ethylpentane would be called triethylmethane:

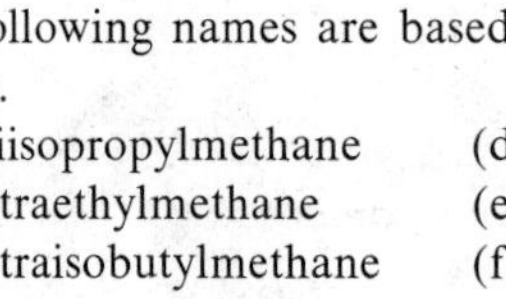

$$\begin{array}{c} CH_3 \\ | \\ CH_2 \\ | \\ CH_3CH_2{-}C{-}CH_2CH_3 \\ | \\ H \end{array}$$

3-Ethylpentane
(or triethylmethane)

The following names are based on this system. Give their structures and their IUPAC names.

(a) Triisopropylmethane
(b) Tetraethylmethane
(c) Tetraisobutylmethane
(d) Tri-*sec*-butylmethane
(e) Di-*tert*-butylmethane
(f) Tetra-*n*-butylmethane

3.19

Three different alkenes will yield 2-methylbutane when they add hydrogen in the presence of a metal catalyst. Give their structures and write equations for the reactions involved.

3.20

An alkane with the formula C_6H_{14} can be synthesized by treating (in separate reactions) five different alkyl chlorides ($C_6H_{13}Cl$) with zinc and aqueous acid. Give the structure of the alkane and the structures of the alkyl chlorides.

3.21

The carbon-carbon bond angles of isobutane are ~111.5°. These angles are larger than those expected from purely tetrahedral carbon (i.e., 109.5°). Explain.

3.22

Sketch approximate potential-energy diagrams for rotations about
(a) The C-2—C-3 bond of 2,3-dimethylbutane
(b) The C-2—C-3 bond of 2,2,3,3-tetramethylbutane

3.23

Without referring to tables, decide which member of each of the following pairs has the higher boiling point.
(a) Hexane or isohexane
(b) Hexane or pentane
(c) Pentane or neopentane
(d) Ethane or chloroethane

3.24

Cis-1,2-dimethylcyclopropane has a larger heat of combustion than *trans*-1,2-dimethylcyclopropane. (a) Which compound is more stable? (b) Give a reason that would explain your answer to part (a).

3.25

Write structural formulas for (a) the two chair conformations of *cis*-1-isopropyl-3-methylcyclohexane, (b) the two chair conformations of *trans*-1-isopropyl-3-methylcyclohexane. (c) Designate which conformation in part (a) and (b) is more stable.

3.26

Which member of each of the following pairs of compounds is more stable?
(a) *cis*- or *trans*-1,2-Dimethylcyclohexane
(b) *cis*- or *trans*-1,3-Dimethylcyclohexane
(c) *cis*- or *trans*-1,4-Dimethylcyclohexane

3.27

Professor Norman L. Allinger of the University of Georgia has obtained evidence indicating that while *cis*-1,3-di-*tert*-butylcyclohexane exists predominantly in a chair conformation, *trans*-1,3-di-*tert*-butylcyclohexane adopts a twist conformation. Explain.

3.28

The important sugar glucose exists in the cyclic form shown below:

CH_2OH
CH
CHOH O
CHOH CHOH
CH
OH

The six-membered ring of glucose has the chair conformation. All of the hydroxyl groups and the —CH_2OH group are equatorial. Write a structure for glucose.

3.29

Calculate the percentage composition of each of the following compounds.
(a) $C_6H_{12}O_6$

(b) $CH_3CH_2NO_2$
(c) $CH_3CH_2CBr_3$

3.30
An organometallic compound called *ferrocene* contains 30.02% iron. What is the minimum molecular weight of ferrocene?

3.31
A gaseous compound gave the following analysis: C, 40.04%; H, 6.69%. At standard temperature and pressure, 1.00 g of the gas occupied a volume of 746 ml. What is the molecular formula of the compound?

3.32
A gaseous hydrocarbon has a density of 1.251 g/liter at standard temperature and pressure. When subjected to complete combustion, a 1.000-liter sample of the hydrocarbon gave 3.926 g of carbon dioxide and 1.608 g of water. What is the molecular formula for the hydrocarbon?

3.33
Nicotinamide, a vitamin that prevents the occurrence of pellagra, gave the following analysis: C, 59.10%; H, 4.92%; N, 22.91%. The molecular weight of nicotinamide was shown in a separate determination to be 120 ± 5. What is the molecular formula for nicotinamide?

3.34
The antibiotic chloramphenicol gave the following analysis: C, 40.88%; H, 3.74%; Cl, 21.95%; N, 8.67%. The molecular weight was found to be 300 ± 30. What is the molecular formula for chloramphenicol?

*** 3.35**
When 1,2-dimethylcyclohexene (below) is allowed to react with hydrogen in the presence of a platinum catalyst, the product of the reaction is a cycloalkane that has a melting point of $-50°$ and a boiling point of $130°$ (at 760 Torr). (a) What is the structure of the product of this reaction? (b) Consult an appropriate table and tell which stereoisomer it is. (c) What does this experiment suggest about the mode of addition of hydrogen to the double bond?

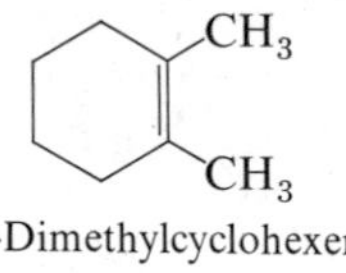

CH_3
CH_3
1,2-Dimethylcyclohexene

*** 3.36**
When cyclohexene is dissolved in an appropriate solvent and allowed to react with chlorine, the product of the reaction, $C_6H_{10}Cl_2$, has a melting point of $-7°$ and a boiling point (at 16 Torr) of $74°$. (a) Which stereoisomer is this? (b) What does this experiment suggest about the mode of addition of chlorine to the double bond?

*** 3.37**
Compounds with rings containing atoms other than carbon are called *heterocyclic compounds*. Molecules of the heterocyclic compound given below have been shown to exist predominantly in a chair conformation with the hydroxyl group axial. (a) Write this conformation. (b) What factor will account for the molecule preferring to have the hydroxyl group axial rather than equatorial?

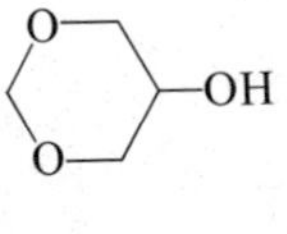

O
—OH
O

*** 3.38**

(Use models in solving this problem) In contrast to most cyclohexane compounds *trans*-decalin is *conformationally rigid,* that is, its rings do not exhibit the usual chair ⇄ chair "flipping." Thus, it is possible for a substituent of *trans*-decalin to be unequivocally defined as being axial or equatorial and we will see how this is useful when we study steroids in Chapter 23. (a) Consider what would happen to the carbons of one ring were the other to be "flipped" into another chair conformation and explain why *trans*-decalin is conformationally rigid. (b) Would you expect *cis*-decalin to show the same rigidity?

*** 3.39**

The following data have been advanced to support the assertion that *trans*-1,2-dibromocyclohexane exists to a considerable extent in a diaxial conformation. (1) *cis*-1,2-Dibromocyclohexane has a dipole moment equal to 3.09 D, a value that is quite close to that of *cis*-3-bromo-*trans*-4-bromo-1-*tert*-butylcyclohexane, which is assumed to exist primarily in the conformation shown below.

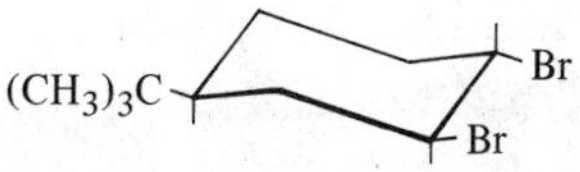

$(CH_3)_3C$ — Br — Br

μ = 3.28 D

cis-3-bromo-***trans***-4-bromo-1-***tert***-butylcyclohexane

(2) *trans*-1,2-Dibromocyclohexane has a much lower dipole moment, μ = 2.11 D. (a) Assume that C—H and C—C bonds make negligible contributions to the dipole moments of the compounds given, and explain how the assertion is justified. (b) We saw earlier that *trans*-1,2-dimethylcyclohexane exists almost exclusively in a diequatorial conformation. What factor might account for the different behavior of *trans*-1,2-dibromocyclohexane?

4 CHEMICAL REACTIVITY I: REACTIONS OF ALKANES AND CYCLOALKANES

4.1 INTRODUCTION: HOMOLYSIS AND HETEROLYSIS OF COVALENT BONDS

In this chapter and the next we will begin to look at some of the important kinds of reactions that organic compounds undergo. As we examine these reactions we will not only want to know what the products are, we will also be interested in *how the reaction takes place*. We will be interested in what chemists call the *mechanism of the reaction, the events that take place at the molecular level as reactants become products*. If the reaction takes place in more than one step, then what are these steps, and what kinds of *intermediates* intervene between reactants and products?

We would also like to know about the energy changes that accompany reactions, and about how they affect *the rates* of chemical reactions. If the mechanism involves the formation of an intermediate, we would like to know about this intermediate's *relative stability*. Quite often one mixture of reactants can become *different* mixtures of products according to the temperature used. This means that two or more reactions *compete* with each other. Which products form and in what relative amounts depends, therefore, on this competition. Will the intermediate formed in one reaction, for example, be more energetic (and thus less stable) than the intermediate formed in another competing reaction? If so, we will find that this can be an important factor in determining how rapidly the competing reactions take place. The major product will often be the result of the most rapid of the competing reactions.

4.1A Homolysis and Heterolysis of Covalent Bonds

Reactions of organic compounds almost inevitably involve the making and breaking of covalent bonds. If we consider a hypothetical molecule A:B, its covalent bond may break in three possible ways:

$$A{:}B \xrightarrow{(1)} A\cdot + B\cdot \qquad \textbf{Homolysis}$$

$$A{:}B \xrightarrow{(2)} A{:}^- + B^+ \qquad \textbf{Heterolysis}$$

$$A{:}B \xrightarrow{(3)} A^+ + {:}B^- \qquad \textbf{Heterolysis}$$

In (1) above the bond breaks so that A and B each retain one of the electrons of the bond, and cleavage leads to the neutral fragments A· and B·. This type of bond breaking is called *homolysis* (Gr: *homo-*, the same, + *lysis*, loosening or cleavage); the bond is said to have broken *homolytically*. The neutral fragments A· and B· are called free *radicals*, or often simply radicals. Radicals always contain unpaired electrons.

In (2) and (3) bond cleavage leads to charged fragments or *ions* ($A{:}^-$ and

B^+ or A^+ and $^-$:B). This kind of bond cleavage is called *heterolysis* (Gr: *hetero,* different, + *lysis*); the bond is said to have broken *heterolytically*.

4.2 REACTIVE INTERMEDIATES IN ORGANIC CHEMISTRY

Organic reactions that take place in more than one step involve the formation of an *intermediate*—one that results from either homolysis or heterolysis of a bond. Homolysis of a bond to carbon leads to an intermediate known as a carbon *radical* (or free radical).

$$-\overset{|}{\underset{|}{C}}:Z \xrightarrow{\text{Homolysis}} -\overset{|}{\underset{|}{C}}\cdot \quad + Z\cdot$$

A carbon radical
(or *free* radical)

Heterolysis of a bond to carbon can lead either to a carbon cation or carbon anion.

$$-\overset{|}{\underset{|}{C}}:Z \xrightarrow{\text{Heterolysis}} -\overset{|}{\underset{|}{C}}^{+} \quad + :Z^-$$

Carbocation
(or *carbonium ion*)

$$-\overset{|}{\underset{|}{C}}:^- \quad + Z^+$$

Carbanion

Carbon cations are called either *carbocations* or *carbonium ions*. The term *carbocation* has a clear and distinct meaning. The older term *carbonium ion* has taken on a different meaning in some of the chemical literature.* Because of this, we shall always refer to trivalent, positively charged species such as $-\overset{|}{\underset{|}{C}}^{+}$ as carbocations.

Carbon anions are called *carbanions*.

Carbon radicals and carbocations are electron-deficient species. A carbon radical has seven electrons in its valence shell; a carbocation has only six. As a consequence, both species are *electrophiles. In their reactions they seek the extra electron or electrons that will give them a stable octet.*

Carbanions are usually strong *bases* and strong *nucleophiles. Nucleophiles are Lewis bases—they are electron-pair donors. Carbanions, therefore seek either a proton* or some other *positively charged center to neutralize their negative charge.*

Carbon radicals, carbocations, and carbanions are usually highly reactive species. In most instances they exist only as short-lived intermediates in an organic reaction. Under certain conditions however, these species may exist long enough for chemists to study them using special techniques. We will have more to say

* For example, the highly reactive species, CH_5^+, is called a carbonium ion by some chemists. CH_5^+ is clearly a different kind of chemical entity than $-\overset{|}{\underset{|}{C}}^{+}$, so the term carbonium ion has become ambiguous.

about how this is done in later chapters. A few carbon radicals, carbocations, and carbanions are stable enough to be isolated. This only happens, however, when special groups are attached to the central carbon that allow the charge or the odd electron to be stabilized.

4.3 BOND DISSOCIATION ENERGIES

When atoms combine to form molecules, energy is released as covalent bonds form. The molecules of the products have lower enthalpy than the separate atoms. When hydrogen atoms combine to form hydrogen molecules, for example, the reaction is *exothermic;* it evolves 104 kcal of heat for every mole of hydrogen that is produced. Similarly, when chlorine atoms combine to form chlorine molecules the reaction evolves 58 kcal/mole of chlorine produced.

$$\left.\begin{array}{ll} \text{H}\cdot + \text{H}\cdot \longrightarrow \text{H—H} & \Delta H = -104 \text{ kcal/mole} \\ \text{Cl}\cdot + \text{Cl}\cdot \longrightarrow \text{Cl—Cl} & \Delta H = -58 \text{ kcal/mole} \end{array}\right\} \begin{array}{l} \textit{Bond formation} \\ \textit{is an exothermic process} \end{array}$$

To break covalent bonds, energy must be supplied. Reactions in which only bond breaking occurs are always endothermic. The energy required to break the covalent bonds of hydrogen or chlorine homolytically is exactly equal to that evolved when the separate atoms combine to form molecules. In the bond cleavage reaction, however, ΔH is positive.

$$\text{H—H} \longrightarrow \text{H}\cdot + \text{H}\cdot \qquad \Delta H = +104 \text{ kcal/mole}$$

$$\text{Cl—Cl} \longrightarrow \text{Cl}\cdot + \text{Cl}\cdot \qquad \Delta H = +58 \text{ kcal/mole}$$

The energies required to break covalent bonds homolytically have been determined experimentally for many types of covalent bonds. These energies are called *bond dissociation energies,* and they are usually abbreviated by the letter $DH°$. The bond dissociation energies of hydrogen and chlorine, for example, might be written in the following way.

$$\underset{(DH° = 104 \text{ kcal/mole})}{\text{H—H}} \qquad \underset{(DH° = 58 \text{ kcal/mole})}{\text{Cl—Cl}}$$

The bond dissociation energies of a variety of covalent bonds are listed in Table 4.1.

4.3A Bond Dissociation Energies and Heats of Reaction

Bond dissociation energies have, as we will see, a variety of uses. They can be used, for example, to calculate the enthalpy change ($\Delta H°$) for a reaction.* To make such a calculation (below) we must remember only that for bond breaking $\Delta H°$ is positive and for bond formation $\Delta H°$ is negative. Let us consider, for example, the reaction of hydrogen and chlorine to produce two moles of hydrogen chloride. From Table 4.1 we get the following values of $DH°$.

* $\Delta H°$ is the enthalpy change when the components are in standard states and at the same temperature, usually 25°.

$$\underset{(DH^\circ = 104)}{H{-}H} + \underset{(DH^\circ = 58)}{Cl{-}Cl} \longrightarrow \underset{(DH^\circ = 103) \times 2}{2H{-}Cl}$$

+162 kcal/mole is required for bond cleavage — −206 kcal/mole is evolved in bond formation

Overall, the reaction is exothermic: $\Delta H^\circ = (-206$ kcal/mole + 162 kcal/mole) = -44 kcal/mole.

For the purpose of our calculation, we have assumed a particular pathway, that amounts to:

$$H{-}H \longrightarrow 2H\cdot$$

and $$Cl{-}Cl \longrightarrow 2Cl\cdot$$

then $$2H\cdot + 2Cl\cdot \longrightarrow 2H{-}Cl$$

This is not the way the reaction actually occurs. Nonetheless, the heat of reaction, ΔH°, is a thermodynamic quantity that is dependent *only* on the initial and final

TABLE 4.1 Single-Bond Dissociation Energies DH° at 25°

$A{:}B \longrightarrow A\cdot + B\cdot$

	KCAL/MOLE		KCAL/MOLE
$H{-}H$	104	$(CH_3)_2CH{-}H$	94.5
$D{-}D$	106	$(CH_3)_2CH{-}F$	105
$F{-}F$	38	$(CH_3)_2CH{-}Cl$	81
$Cl{-}Cl$	58	$(CH_3)_2CH{-}Br$	68
$Br{-}Br$	46	$(CH_3)_2CH{-}I$	53
$I{-}I$	36	$(CH_3)_2CH{-}OH$	92
$H{-}F$	136	$(CH_3)_2CH{-}OCH_3$	80.5
$H{-}Cl$	103	$(CH_3)_2CHCH_2{-}H$	98
$H{-}Br$	87.5	$(CH_3)_3C{-}H$	91
$H{-}I$	71	$(CH_3)_3C{-}Cl$	78.5
$CH_3{-}H$	104	$(CH_3)_3C{-}Br$	63
$CH_3{-}F$	108	$(CH_3)_3C{-}I$	49.5
$CH_3{-}Cl$	83.5	$(CH_3)_3C{-}OH$	90.5
$CH_3{-}Br$	70	$(CH_3)_3C{-}OCH_3$	78
$CH_3{-}I$	56	$C_6H_5CH_2{-}H$	85
$CH_3{-}OH$	91.5	$CH_2{=}CHCH_2{-}H$	85
$CH_3{-}OCH_3$	80	$CH_2{=}CH{-}H$	103
$CH_3CH_2{-}H$	98	$C_6H_5{-}H$	103
$CH_3CH_2{-}F$	106	$HC{\equiv}C{-}H$	125
$CH_3CH_2{-}Cl$	81.5	$CH_3{-}CH_3$	88
$CH_3CH_2{-}Br$	69	$CH_3CH_2{-}CH_3$	85
$CH_3CH_2{-}I$	53.5	$CH_3CH_2CH_2{-}CH_3$	85
$CH_3CH_2{-}OH$	91.5	$CH_3CH_2{-}CH_2CH_3$	82
$CH_3CH_2{-}OCH_3$	80	$(CH_3)_2CH{-}CH_3$	84
		$(CH_3)_3C{-}CH_3$	80
$CH_3CH_2CH_2{-}H$	98	$HO{-}H$	119
$CH_3CH_2CH_2{-}F$	106	$HOO{-}H$	90
$CH_3CH_2CH_2{-}Cl$	81.5	$HO{-}OH$	51
$CH_3CH_2CH_2{-}Br$	69	$CH_3CH_2O{-}OCH_3$	44
$CH_3CH_2CH_2{-}I$	53.5		
$CH_3CH_2CH_2{-}OH$	91.5		
$CH_3CH_2CH_2{-}OCH_3$	80		

states of the reacting molecules. $\Delta H°$ is independent of the path followed and, for this reason, our calculation is valid.

Problem 4.1

Calculate the heat of reaction, $\Delta H°$, for the following reactions.

(a) $H_2 + Br_2 \longrightarrow 2HBr$
(b) $CH_3CH_3 + F_2 \longrightarrow CH_3CH_2F + HF$
(c) $CH_3CH_3 + I_2 \longrightarrow CH_3CH_2I + HI$
(d) $CH_4 + Cl_2 \longrightarrow CH_3Cl + HCl$
(e) $(CH_3)_3CH + Cl_2 \longrightarrow (CH_3)_3CCl + HCl$
(f) $(CH_3)_3CH + Br_2 \longrightarrow (CH_3)_3CBr + HBr$
(g) $CH_3CH_2CH_3 \longrightarrow CH_3CH_2\cdot + CH_3\cdot$
(h) $2CH_3CH_2\cdot \longrightarrow CH_3CH_2CH_2CH_3$

4.3B Bond Dissociation Energies and the Relative Stabilities of Free Radicals

Bond dissociation energies also provide us with a convenient way to estimate the relative stabilities of free radicals. If we examine the data given in Table 4.1 we find the following values of $DH°$ for the primary and secondary C—H bonds of propane:

$CH_3CH_2CH_2—H$ $DH° = 98$ kcal/mole

$(CH_3)_2CH—H$ $DH° = 94.5$ kcal/mole

This means that for the reaction in which the designated C—H bonds are broken homolytically, the values of $\Delta H°$ are those given below.

$$CH_3CH_2CH_2—H \longrightarrow CH_3CH_2CH_2\cdot + H\cdot \qquad \Delta H° = +98 \text{ kcal/mole}$$

Propyl radical (a 1° radical)

$$CH_3\underset{\displaystyle H}{\underset{|}{C}}HCH_3 \longrightarrow CH_3\underset{\bullet}{C}HCH_3 + H\cdot \qquad \Delta H° = +94.5 \text{ kcal/mole}$$

Isopropyl radical (a 2° radical)

These reactions resemble each other in two respects: they both begin with the same alkane (propane) and they both produce an alkyl radical and a hydrogen atom. They differ, however, in the amount of energy required and in the type of carbon radical being produced.* These two differences must be related to each other.

More energy must be supplied to produce a primary carbon radical (the propyl radical) from propane than is required to produce a secondary carbon radical (the isopropyl radical) from the same compound. This must mean that the primary radical has absorbed more energy and thus has greater *potential energy*. Since the relative stability of a chemical species is inversely related to its potential energy, the secondary radical must be the *more stable* radical (Fig. 4.1a). In fact, the secondary isopropyl radical is more stable than the primary propyl radical by 3.5 kcal/mole.

We can use the data in Table 4.1 to make a similar comparison of the

* Carbon radicals are classified as being 1°, 2°, or 3° on the basis of the carbon that has the unpaired electron.

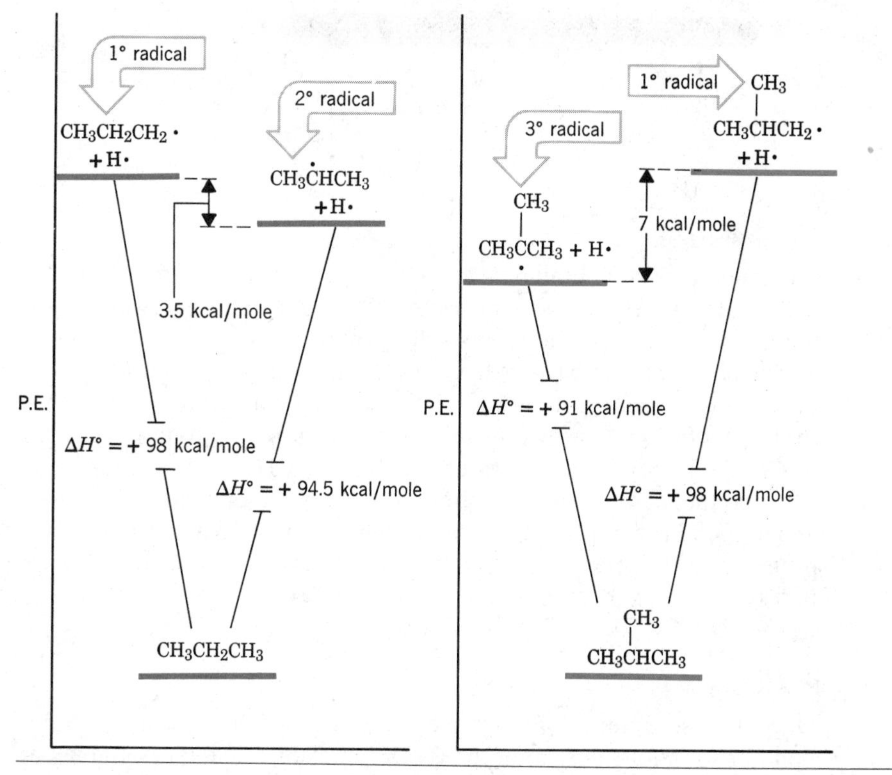

FIG. 4-1
(a) A comparison of the potential energies of the propyl radical (+ H·) and the isopropyl radical (+ H·) relative to propane. The isopropyl radical—a 2° radical—is more stable than the 1° radical by 3.5 kcal/mole. (b) A comparison of the potential energies of the tert-*butyl radical (+ H·) and the isobutyl radical (+ H·) relative to isobutane. The 3° radical is more stable than the 1° radical by 7 kcal/mole.*

tert-butyl radical (a 3° radical) and the isobutyl radical (a 1° radical) relative to isobutane.

$$(CH_3)_3C\text{—}H \longrightarrow (CH_3)_3C\cdot + H\cdot \qquad \Delta H° = +91 \text{ kcal/mole}$$

tert-Butyl radical (a 3° radical)

$$CH_3CH(CH_3)CH_2\text{—}H \longrightarrow CH_3CH(CH_3)CH_2\cdot + H\cdot \qquad \Delta H° = +98 \text{ kcal/mole}$$

Isobutyl radical (a 1° radical)

Here we find (Fig. 4.1b) that the difference in stability of the two radicals is even larger. The tertiary radical is more stable than the primary radical by 7 kcal/mole.

The kind of pattern that we find in these examples is found with alkyl radicals generally; overall their relative stabilities are:

Tertiary > Secondary > Primary > Methyl

```
   C          C           H          H
   |          |           |          |
C—C·  >   C—C·   >   C—C·  >   H—C·
   |          |           |          |
   C          H           H          H
```

Problem 4.2

(a) Sketch diagrams similar to those in Fig. 4.1 showing the potential energy of $(CH_3)_2CH\cdot + H\cdot$ relative to propane and the potential energy of $CH_3CH_2\cdot + H\cdot$ relative to ethane. Align the two diagrams so that the potential energy of the alkane is the same in each. What does this indicate about the stability of an ethyl radical and an isopropyl radical relative to the alkane from which each is derived? (b) Repeat this process by drawing potential energy diagrams showing the energy of $CH_3CH_2\cdot + H\cdot$ relative to ethane and of $CH_3\cdot + H\cdot$ relative to methane. What do these graphs indicate about the relative stabilities of an ethyl radical and a methyl radical? (c) Make similar sketches that compare an ethyl radical with a propyl radical. (d) Account for the similarity of the potential energy diagrams in part (c).

Problem 4.3

One can also estimate the relative stabilities of free radicals by comparing the bond dissociation energies of the C—X bonds of haloalkanes. Show how this can be done with CH_3—Cl, CH_3CH_2—Cl, $(CH_3)_2CH$—Cl, and $(CH_3)_3C$—Cl.

4.4 CHEMICAL REACTIONS OF ALKANES

Alkanes, as a class, are characterized by a general inertness to many chemical reagents. Carbon-carbon and carbon-hydrogen bonds are quite strong; they do not break unless alkanes are heated to very high temperatures. Because carbon and hydrogen have nearly the same electronegativity, the carbon-hydrogen bonds of alkanes are only slightly polarized. As a consequence, they are generally unaffected by most bases. Molecules of alkanes have no unshared electrons to offer sites for attack by acids. This low reactivity of alkanes toward many reagents accounts for the fact that alkanes were originally called *paraffins* (Latin: *parum affinis,* low affinity).

The term paraffin, however, is probably not an appropriate one. We all know that alkanes react vigorously with oxygen when an appropriate mixture is ignited. This combustion occurs in the cylinders of automobiles, and in oil furnaces, for example. When heated, alkanes also react with chlorine, and they react explosively with fluorine.

Most of the reactions of alkanes are characterized by an attack on the alkane by a reagent with unpaired electrons. Oxygen, the reagent responsible for combustion of alkanes, has unpaired electrons. (Its structure does not conform to the Lewis rules.)

$$\cdot\ddot{\underset{\cdot\cdot}{O}}:\ddot{\underset{\cdot\cdot}{O}}\cdot$$

The reactions of methane with halogens, reactions that we will soon consider, involve the attack of a halogen atom on the methane molecule.

$$X\cdot + H{:}CH_3 \longrightarrow X{:}H + \cdot CH_3$$

Exceptions to what we have just related are the reactions of alkanes with very powerful acids, called superacids. Superacids react with alkanes by donating protons to their carbon-carbon and carbon-hydrogen sigma bonds. We will describe these reactions later in this chapter.

4.5 THE REACTIONS OF ALKANES WITH HALOGENS: SUBSTITUTION REACTIONS

Methane, ethane, and other alkanes react with the first three members of the halogen family: fluorine, chlorine, and bromine. Alkanes do not react appreciably with iodine. With methane the reaction produces a mixture of halomethanes and a hydrogen halide.

$$\underset{\text{Methane}}{CH_4} + \underset{\text{Halogen}}{X_2} \longrightarrow$$

$$\underset{\text{Halomethane}}{CH_3X} + \underset{\text{Dihalomethane}}{CH_2X_2} + \underset{\text{Trihalomethane}}{CHX_3} + \underset{\text{Tetrahalomethane}}{CX_4} + \underset{\text{Hydrogen halide}}{H\text{—}X}$$

X = F, Cl, or Br

The reaction of an alkane with a halogen is called *halogenation*. The general reaction to produce a monohaloalkane can be written as follows.

$$R\text{—}H + X_2 \longrightarrow R\text{—}X + HX$$

In these reactions a halogen atom replaces one or more of the hydrogen atoms of the alkane. Reactions of this type, *where one group replaces another,* are called *substitution reactions.*

$$-\overset{|}{\underset{|}{C}}\text{—}H + Cl_2 \longrightarrow -\overset{|}{\underset{|}{C}}\text{—}Cl + H\text{—}Cl \qquad \textbf{A substitution reaction}$$

Substitution reactions are not limited to the replacement of hydrogen. When chloromethane, for example, is treated with sodium hydroxide, a substitution reaction takes place in which an —OH group replaces the —Cl. (We will study this reaction in detail in the next chapter.)

$$\underset{\text{Chloromethane}}{H{-}\overset{\overset{H}{|}}{\underset{\underset{H}{|}}{C}}{-}Cl} + NaOH \xrightarrow[\text{Heat}]{\text{Solvent}} \underset{\text{Methyl alcohol}}{H{-}\overset{\overset{H}{|}}{\underset{\underset{H}{|}}{C}}{-}OH} + NaCl$$

A substitution reaction

When bromomethane reacts with potassium iodide, —I replaces —Br; this, too, is a substitution reaction.

$$\underset{\text{Bromomethane}}{H{-}\overset{\overset{H}{|}}{\underset{\underset{H}{|}}{C}}{-}Br} + KI \xrightarrow{\text{Solvent}} H{-}\overset{\overset{H}{|}}{\underset{\underset{H}{|}}{C}}{-}I + KBr$$

A substitution reaction

Substitution reactions are among the most common and most useful reactions that we will encounter in our study of organic chemistry. We will see many other examples in succeeding chapters.

4.5A Halogenation Reactions

One characteristic of alkane halogenations is that multiple substitution reactions almost always occur. As we saw at the beginning of this section, the halogenation of methane produces a mixture of monohalomethane, dihalomethane, trihalomethane, and tetrahalomethane.

This happens because all hydrogens attached to carbon are capable of reacting with fluorine, chlorine, or bromine.

Let us consider the reaction that takes place between chlorine and methane as an example. If we bring together a mixture of methane and chlorine (both substances are gases at room temperature) and then either heat the mixture or irradiate it with light, a reaction begins to occur vigorously. At the outset, the only compounds that are present in the mixture are chlorine and methane, and the only reaction that can take place is one that produces chloromethane and hydrogen chloride.

$$H{-}\overset{\overset{H}{|}}{\underset{\underset{H}{|}}{C}}{-}H + Cl_2 \longrightarrow H{-}\overset{\overset{H}{|}}{\underset{\underset{H}{|}}{C}}{-}Cl + H{-}Cl$$

As the reaction progresses, however, the concentration of chloromethane in the mixture increases, and a second substitution reaction begins to occur. Chloromethane reacts with chlorine to produce dichloromethane.

$$H{-}\overset{\overset{H}{|}}{\underset{\underset{H}{|}}{C}}{-}Cl + Cl_2 \longrightarrow H{-}\overset{\overset{Cl}{|}}{\underset{\underset{H}{|}}{C}}{-}Cl + H{-}Cl$$

Dichloromethane can then produce trichloromethane,

$$\text{H—}\overset{\text{Cl}}{\underset{\text{H}}{\overset{|}{\underset{|}{\text{C}}}}}\text{—Cl} + Cl_2 \longrightarrow \text{H—}\overset{\text{Cl}}{\underset{\text{Cl}}{\overset{|}{\underset{|}{\text{C}}}}}\text{—Cl} + \text{H—Cl}$$

and trichloromethane, as it accumulates in the mixture, can react with chlorine to produce tetrachloromethane.

$$\text{H—}\overset{\text{Cl}}{\underset{\text{Cl}}{\overset{|}{\underset{|}{\text{C}}}}}\text{—Cl} + Cl_2 \longrightarrow \text{Cl—}\overset{\text{Cl}}{\underset{\text{Cl}}{\overset{|}{\underset{|}{\text{C}}}}}\text{—Cl} + \text{H—Cl}$$

Each time a substitution of —Cl for —H takes place a molecule of H—Cl is produced.

When ethane and chlorine react similar substitution reactions occur. Ultimately all six hydrogen atoms of ethane may be replaced. We notice below that the second substitution reaction of ethane results in the formation of two different molecules: 1,1-dichloroethane and 1,2-dichloroethane. These two molecules have the same molecular formula, $C_2H_4Cl_2$, but they have *different structures*. They are *structural isomers*.

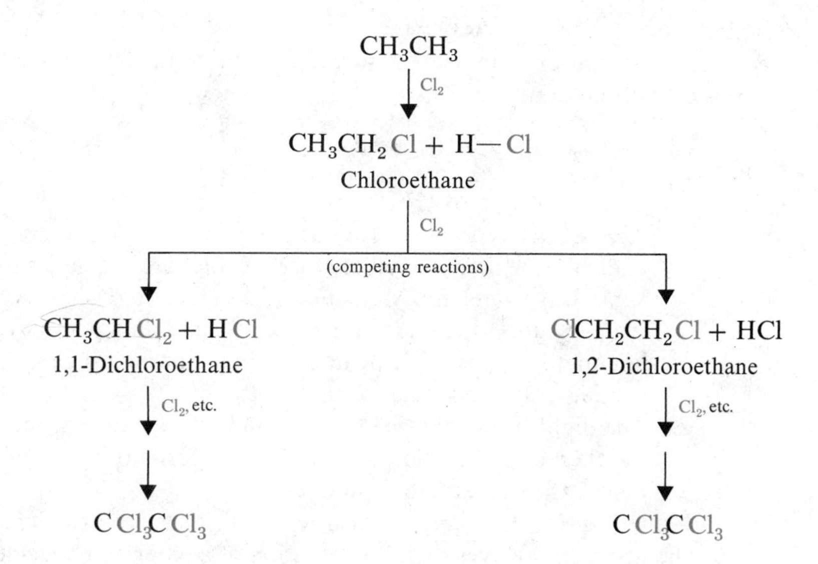

Problem 4.4

(a) How many trichloroethanes would be produced when 1,1-dichloroethane reacts with chlorine? (b) How many trichloroethanes would be produced when 1,2-dichloroethane reacts with chlorine? (c) Write structural formulas for these trichloroethanes. (d) Are they structural isomers? (e) How many tetrachloroethanes might be produced in the next chlorination step? (f) Write structural formulas for these tetrachloroethanes. (g) How many pentachloroethanes are theoretically possible?

Problem 4.5

When propane reacts with chlorine the reaction yields (in addition to more highly chlorinated compounds) *two compounds with the molecular formula* C_3H_7Cl. One compound has a boiling point of 46.6°, the other a boiling point of 36.5°. (a) Write possible structural formulas for these compounds. (b) Can you tell which one is which? (c) When the compound with a boiling point of 46.6° is allowed to react with chlorine the reaction yields three compounds with the formula $C_3H_6Cl_2$ (again in addition to more highly chlorinated compounds). When the compound with boiling point 36.5° reacts with chlorine only two isomers with the formula $C_3H_6Cl_2$ are produced. Show how this information allows you to assign structural formulas to the two C_3H_7Cl compounds. (d) Write all of the reactions involved.

Problem 4.6

Three isomeric alkanes **A, B,** and **C** have the formula C_5H_{12}. **A** reacts with chlorine to yield only one compound with the formula $C_5H_{11}Cl$. **B** reacts with chlorine to yield three isomers with the formula $C_5H_{11}Cl$ and **C** reacts to yield four isomers with the formula $C_5H_{11}Cl$. Assign structural formulas to **A, B,** and **C,** and show the isomers of $C_5H_{11}Cl$ formed when each reacts with chlorine. (Other more highly chlorinated compounds are also formed but these are not important for this problem.)

4.6 CHLORINATION OF METHANE: THE REACTION MECHANISM

Several important experimental observations can be made about the reaction of methane with chlorine:

$$CH_4 + Cl_2 \longrightarrow CH_3Cl + HCl$$
$$(+\ CH_2Cl_2,\ CHCl_3,\ \text{and}\ CCl_4)$$

1. *The reaction is promoted by heat or light.* At room temperature methane and chlorine do not react at a perceptible rate as long as the mixture is kept away from light. Methane and chlorine do react, however, at room temperature if the reaction mixture is irradiated with ultraviolet light, and methane and chlorine do react in the dark, if the mixture is heated to temperatures greater than 100°C.
2. The light-promoted reaction is highly efficient. A relatively small number of light photons permits the formation of relatively large amounts of chlorinated product.

A mechanism that is most consistent with these observations has several steps. The first step involves the fragmentation of a chlorine molecule, by heat or light, into two chlorine atoms.

$$\text{Step 1:}\quad :\underset{\cdot\cdot}{\overset{\cdot\cdot}{Cl}}:\underset{\cdot\cdot}{\overset{\cdot\cdot}{Cl}}: \xrightarrow[\text{or light}]{\text{Heat}} 2\,:\underset{\cdot\cdot}{\overset{\cdot\cdot}{Cl}}\cdot$$

Chlorine is known, from other evidence, to undergo such reactions. It can be shown, moreover, that the frequency of light that promotes the chlorination of methane is a frequency that is absorbed by chlorine molecules and not by methane molecules.

The second and third steps of the mechanism are as follows:*

* These conventions are used in illustrating reaction mechanisms in this text.
1. Arrows ⌒ or ⌒ always show the direction of movement of electrons.
2. Single-barbed arrows ⌒ show the attack (or movement) of an unpaired electron.
3. Double-barbed arrows, ⌒ show the attack (or movement) of an electron pair.

$$\text{Step 2:}\quad :\ddot{\text{Cl}}\cdot + \text{H}:\text{CH}_3 \longrightarrow \text{H}:\ddot{\text{Cl}}: + \cdot\text{CH}_3$$

$$\text{Step 3:}\quad \text{H}_3\text{C}\cdot + :\ddot{\text{Cl}}:\ddot{\text{Cl}}: \longrightarrow \text{H}_3\text{C}:\ddot{\text{Cl}}: + \cdot\ddot{\text{Cl}}:$$

Step 2 is the abstraction of a hydrogen atom from the methane molecule by a chlorine atom. This step results in the formation of a molecule of hydrogen chloride and a methyl radical.

In step 3 the highly reactive methyl radical reacts with a chlorine molecule by abstracting a chlorine atom. This results in the formation of a molecule of chloromethane (one of the ultimate products of the reaction) and a *chlorine atom.* This latter product is particularly significant, for the chlorine atom formed in step 3 can attack another methane molecule and cause a repetition of step 2. Then,

$$:\ddot{\text{Cl}}\cdot + \text{H}:\text{CH}_3 \longrightarrow \text{H}:\ddot{\text{Cl}}: + \cdot\text{CH}_3 \qquad \text{Step 2 is repeated}$$

$$\text{H}_3\text{C}\cdot + :\ddot{\text{Cl}}:\ddot{\text{Cl}}: \longrightarrow \text{H}_3\text{C}:\ddot{\text{Cl}}: + \cdot\ddot{\text{Cl}}: \qquad \text{Step 3 is repeated}$$

Step 2 is repeated again, then step 3, and so forth. With each repetition of step 3 a molecule of chloromethane is produced. This type of sequential, stepwise mechanism, in which each step generates the reactive intermediate that causes the next step to occur, is called a *chain reaction.*

The chain nature of the reaction is what accounts for the observation that the light-promoted reaction is highly efficient. The presence of a relatively few atoms of chlorine at any given moment is all that is needed to cause the formation of many thousands of molecules of chloromethane.

What causes the chains to terminate? Why does one photon of light not promote the chlorination of all of the methane molecules present? We know that this does not happen because we find that at low temperatures continuous irradiation is required or the reaction slows and stops. The answer to these questions is the existence of *chain-terminating steps:* steps that occur infrequently, but occur often enough to use up one or both of the reactive intermediates. The continuous replacement of intermediates used up by chain-terminating steps requires continuous irradiation. Plausible chain-terminating steps are

$$\text{H}_3\text{C}\cdot + \cdot\ddot{\text{Cl}}: \longrightarrow \text{H}_3\text{C}:\ddot{\text{Cl}}:$$

$$\text{H}_3\text{C}\cdot + \cdot\text{CH}_3 \longrightarrow \text{H}_3\text{C}:\text{CH}_3$$

and

$$:\ddot{\underset{..}{Cl}}\cdot + \cdot\ddot{\underset{..}{Cl}}: \longrightarrow :\ddot{\underset{..}{Cl}}:\ddot{\underset{..}{Cl}}:$$

This last step probably occurs least frequently. The two chlorine atoms are highly energetic; as a result, the simple diatomic chlorine molecule that is formed has to dissipate its excess energy rapidly by colliding with some other molecule or the walls of the container. Otherwise it simply flies apart again. By contrast, chloromethane and ethane, formed in the other two chain-terminating steps, can dissipate their excess energy through vibrations of their C—H bonds.

Problem 4.7

(a) Write mechanisms showing how CH_2Cl_2, $CHCl_3$, and CCl_4 might be formed in the reaction mixture when methane is chlorinated. (b) The use of a large excess of methane minimizes the formation of CH_2Cl_2, $CHCl_3$, and CCl_4. The use of a large excess of chlorine maximizes the formation of these more highly chlorinated compounds. Explain these observations.

4.7 THE CHLORINATION OF METHANE: ENERGY CHANGES

We saw earlier that we can calculate the overall heat of reaction from bond dissociation energies. We can also calculate the heat of reaction for each individual step of a mechanism.

Chain-initiating step

$$\underset{(DH^\circ = 58)}{Cl{-}Cl} \longrightarrow 2Cl\cdot \qquad \Delta H^\circ = +58 \text{ kcal/mole}$$

Chain-propagating steps

$$\underset{(DH^\circ = 104)}{Cl\cdot + CH_3{-}H} \longrightarrow \underset{(DH^\circ = 103)}{CH_3\cdot + H{-}Cl} \qquad \Delta H^\circ = +1 \text{ kcal/mole}$$

$$\underset{(DH^\circ = 58)}{CH_3\cdot + Cl{-}Cl} \longrightarrow \underset{(DH^\circ = 83.5)}{CH_3{-}Cl} + Cl\cdot \qquad \Delta H^\circ = -25.5 \text{ kcal/mole}$$

Chain-terminating steps

$$CH_3\cdot + Cl\cdot \longrightarrow \underset{(DH^\circ = 83.5)}{CH_3{-}Cl} \qquad \Delta H^\circ = -83.5 \text{ kcal/mole}$$

$$CH_3\cdot + \cdot CH_3 \longrightarrow \underset{(DH^\circ = 88)}{CH_3{-}CH_3} \qquad \Delta H^\circ = -88 \text{ kcal/mole}$$

$$Cl\cdot + Cl\cdot \longrightarrow \underset{(DH^\circ = 58)}{Cl{-}Cl} \qquad \Delta H^\circ = -58 \text{ kcal/mole}$$

In the chain-initiating step only one bond is broken—the bond between two chlorine atoms—and no bonds are formed. The heat of reaction for this step is simply the bond dissociation energy for a chlorine molecule and it is highly endothermic.

In the chain-terminating steps bonds are formed, but no bonds are broken. As a result, all of the chain-terminating steps are highly exothermic.

Each of the chain-propagating steps, on the other hand, requires the breaking of one bond and the formation of another. The value of ΔH° for each of these steps is the difference between the bond dissociation energy of the bond that is broken and the bond dissociation energy for the bond that is formed. The first chain-propagating step is slightly endothermic ($\Delta H^\circ = +1$ kcal/mole), but the second is exothermic by a large amount ($\Delta H^\circ = -25.5$ kcal/mole).

Problem 4.8

Assuming the same mechanism occurs, calculate $\Delta H°$ for the chain-initiating, chain-propagating, and chain-terminating steps involved in the bromination of methane.

The addition of the chain-propagating steps yields the overall equation for the chlorination of methane:

$$\begin{array}{lll} \text{Cl}\cdot + CH_3\text{—H} \longrightarrow CH_3\cdot + \text{H—Cl} & & \Delta H° = +1 \text{ kcal/mole} \\ CH_3\cdot + \text{Cl—Cl} \longrightarrow CH_3\text{—Cl} + \text{Cl}\cdot & & \Delta H° = -25.5 \text{ kcal/mole} \\ \hline CH_3\text{—H} + \text{Cl—Cl} \longrightarrow CH_3\text{—Cl} + \text{H—Cl} & & \Delta H° = -24.5 \text{ kcal/mole} \end{array}$$

and the addition of the values of $\Delta H°$ for the individual chain-propagating steps yields the overall value of $\Delta H°$ for the reaction.

Problem 4.9

Why would it be incorrect to include the chain-initiating and chain-terminating steps in the calculation of the overall value of $\Delta H°$ given above?

We will return to the individual steps of the reaction of methane with halogens soon. Before doing so, however, we will need to introduce some of the basic ideas of the collision theory of reaction rates.

4.8 REACTION RATES: COLLISION THEORY

Reaction rates are important in guiding our predictions about the eventual outcome of chemical reactions. Many reactions that have favorable energy changes occur so slowly as to be imperceptible. In other instances several reaction pathways may compete with each other, and the actual distribution of products of such a reaction may be governed not by a position of equilibrium—or by the magnitude of $\Delta H°$—*but by the rate of the reaction that occurs most rapidly.*

In order to understand the factors that affect the rate of a chemical reaction, let us consider the hypothetical example:

$$A + B \longrightarrow C$$

Let us assume that the reaction occurs in the gas phase, where interactions of the reacting substances with solvent molecules are not involved, and let us further assume that A and B react to produce C by colliding with each other.

The rate of the reaction $A + B \longrightarrow C$ can be determined experimentally by measuring the rates at which A or B disappear from the reaction mixture, or by measuring the rate at which C is formed in the mixture. We can make either of these measurements in the laboratory by simply withdrawing small samples of the mixture at measured intervals of time and analyzing the samples for the concentrations of A, B, or C.

After determining the rate, we can analyze our data to find certain relationships. In this instance, we would find that the overall rate of the reaction is proportional to the concentrations of A *and* B present in the mixture at any given moment, that is,

$$\text{Rate} \propto [A]$$
$$\text{Rate} \propto [B]$$
$$\text{therefore,} \quad \text{Rate} \propto [A][B]$$

This proportionality can be expressed as an equation by the introduction of a proportionality constant, k, called the rate constant:

$$\text{Rate} = k[A][B]$$

When the concentrations of A and B are 1.0 mole/liter then,

$$\text{Rate} = k$$

4.8A Collision Theory of Reaction Rates

Much of what we observe about the rates of reactions is explained by the collision theory of reaction rates. As its name suggests, the basic premise of this theory is that for a chemical reaction to take place the reacting particles must collide. *Not all collisions between molecules are effective, however.* By "effective" we mean that a collision produces products. The rate at which molecules collide is called the *collision frequency*. For a gas at room temperature and one atmosphere pressure, the collision frequency is about 10^{27} collisions per milliliter per second. This is an enormous number—so large, in fact, that if all collisions did lead to reaction then all chemical reactions would take place exceedingly rapidly.

Collisions between particles may not lead to a product for two reasons: (1) the colliding particles may not have enough energy and (2) they may not have the correct orientation with respect to each other when they collide. The total collision frequency, usually abbreviated by the letter **Z,** is *the number of collisions between* A *and* B *that occur in each unit volume of the reaction mixture in each second.* To determine the frequency of successful collisions we must multiply the total collision frequency by two other factors: by a factor, **P,** *that measures the fraction of collisions in which the orientation of the colliding molecules with respect to each other allows a reaction to take place;* and a factor, **f,** *that measures the fraction of collisions in which the collision energy is greater than a certain minimum amount, called the energy of activation.* Thus the rate of reaction can be expressed in the following way.

$$\text{Reaction rate} = \mathbf{Z} \times \mathbf{P} \times \mathbf{f}$$

4.8B The Collision Frequency, Z

Many experiments show that the rates of chemical reactions are directly proportional to *the collision frequency*—the greater the collision frequency, the faster the reaction. Two major factors that determine the magnitude of the collision frequency are *concentration and temperature.*

The more concentrated the reacting molecules are (or for a gas-phase reaction, the greater is the pressure), the greater will be the number of collisions occurring in each unit volume of the mixture each second. A simple analogy will help make this clear. Consider, for example, a room in which blindfolded people walk about randomly. It is easy to see that if the room is crowded (a high concentration of people), people will bump into each other more often (have a higher collision frequency) than they will if only a few are present.

The higher the temperature, the faster the molecules move and, as a result, the greater are the number of collisions in a unit volume per unit of time. In our analogy, this would correspond to having our blindfolded people run rather than walk. Clearly in this situation, more collisions will occur in each unit of time, at any given population (concentration) of the room.

The relation between the collison frequency **Z**, and the concentrations of the reactants can be expressed mathematically as follows,

$$\mathbf{Z} = \mathbf{C} \times [A] \times [B]$$

where **C** is a constant that depends only on the speeds (and sizes) of the molecules involved. Overall then:

$$\text{Reaction rate} = \mathbf{C} \times \mathbf{P} \times \mathbf{f} \times [A] \times [B] = k[A][B]$$

We will have more to say about the effect of concentration on reaction rate in the next chapter.

4.8C The Orientation or Probability Factor, P

A great deal of experimental evidence indicates that collisions between molecules must occur in a particular way in order to be effective. While the probability factor for a particular reaction is difficult to estimate, it is easy to see why it should exist. Consider, for example, the reaction of a chlorine atom with a molecule of hydrogen bromide to produce a bromine atom and a molecule of hydrogen chloride: $Cl\cdot + H{-}Br \longrightarrow H{-}Cl + Br\cdot$.

There are, naturally, an infinite number of ways that collisions between chlorine atoms and molecules of hydrogen bromide could occur. Some of these possibilities are shown in Fig. 4.2. Since the reaction that we are considering is one in which the chlorine atom abstracts a hydrogen atom from hydrogen bromide, it is reasonable to assume that in order to be effective, the chlorine atom must collide with the hydrogen end of the hydrogen bromide molecule. Collisions in which the chlorine atom collides with the bromine end of the hydrogen bromide molecule are not likely to be effective.

4.8D The Energy Factor, f, and the Energy of Activation

Not all collisions produce a chemical reaction even when the colliding particles have the proper orientation. A collision will lead to a product only when the colliding particles bring to the collision a certain minimum amount of energy called *the energy of activation* (and abbreviated E_{act}). We know that molecules possess energy called kinetic energy* because of their motion through space.

In a properly aligned collision this kinetic energy, if it is large enough, can be used to provide the energy of activation. If the sum of the kinetic energies of the colliding particles is too small, no reaction occurs.

How do we explain this requirement for a minimum amount of energy for a collision to be effective? Let us answer by considering a specific example: the reaction of a fluorine atom with a methane molecule.

$$\underset{}{F\cdot} + \underset{(DH^\circ = 104)}{CH_3{-}H} \longrightarrow \underset{(DH^\circ = 136)}{H{-}F} + \cdot CH_3 \qquad \begin{array}{l} \Delta H^\circ = -32 \text{ kcal/mole} \\ E_{act} = +1.2 \text{ kcal/mole} \end{array}$$

This reaction, the first chain-propagating step in the reaction of fluorine with

* The kinetic energy of a molecule is proportional to its mass and to the square of its velocity ($KE = \frac{1}{2}mv^2$); and the temperature of a gas is a measure of the average kinetic energy of the molecules present in the gas.

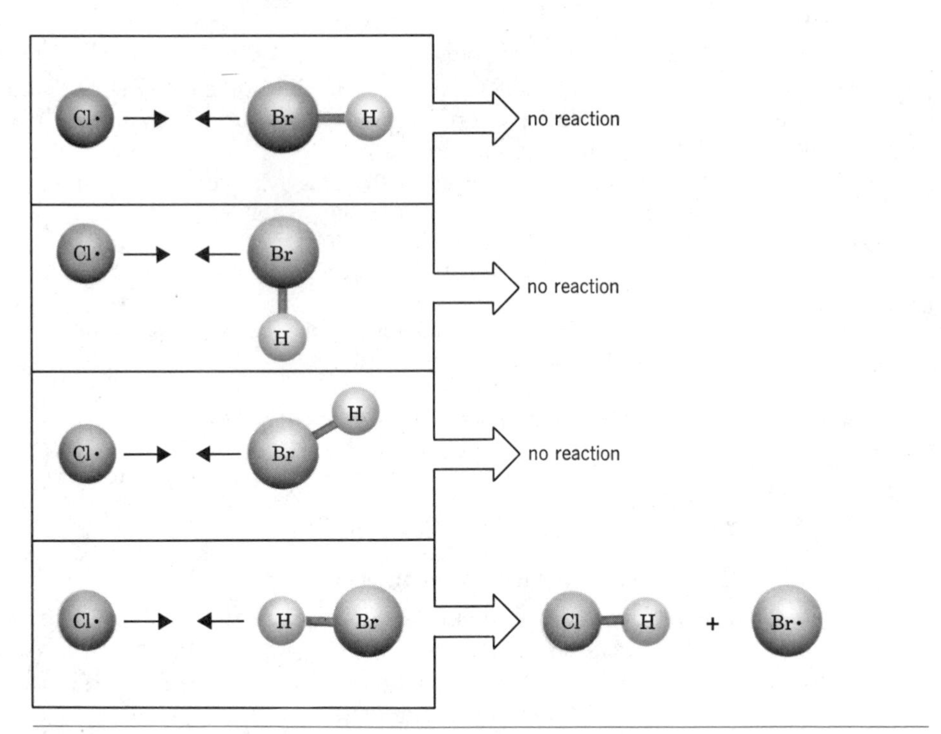

FIG. 4.2
Some of the possible orientations of collisions between chlorine atoms and molecules of hydrogen bromide.

methane, is a highly exothermic one. The energy released in forming the H—F bond is considerably greater than the energy required to break the C—H bond of methane. If all of this energy released in forming the H—F bond were available when needed to break the C—H bond, then breaking the C—H bond should be no problem. Apparently, however, this is not the case. We find, experimentally, that 1.2 kcal of energy per mole—*the energy of activation*—has to be supplied from the kinetic energy of the colliding particles for products to form. The reason for this is that the C—H bond of methane must be largely broken before the H—F bond is completely formed. In general, bond formation appears to lag behind bond rupture and it is this lag that explains the need for an energy of activation.

We can illustrate this situation graphically by plotting the potential energy of the reacting particles as they are transformed from reactants to products during the course of a collision against what is called the *reaction coordinate*. Such an illustration is given in Fig. 4.3.

The reaction coordinate is a measure of the *progress of the reaction* during the collision. For most reactions a point along the reaction coordinate represents a collection of distances of all of the particles of the reactants at a given moment during the reaction. We may think of it as representing the changes in geometry that the colliding particles undergo as atoms are forced apart in bond breaking and as atoms are drawn together in bond formation.

In this illustration (Fig. 4.3), we can see that the energy of activation is an energy barrier between reactants and products. It is, in a sense, an energy hill that the reacting species must traverse in order to become products. The height of this barrier (in kcal/mole) above the level of reactants is the energy of activation.

The top of the energy hill, called the *transition state,* corresponds to a particular arrangement of the atoms of the reacting species as they are converted

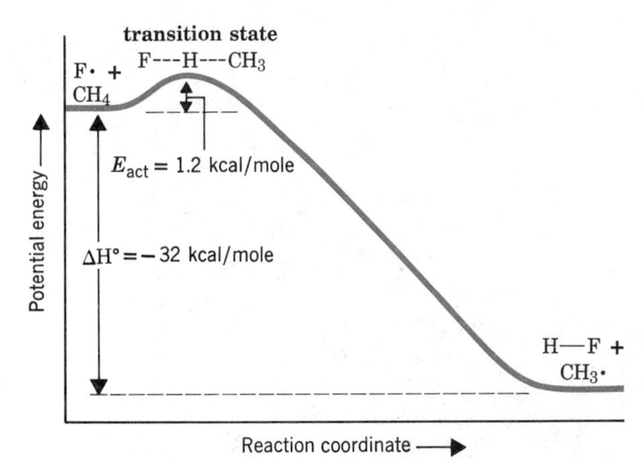

FIG. 4.3
The energy changes that occur in the reaction of a fluorine atom and a methane molecule.

from reactants to products. It represents a configuration in which bonds are partially broken and partially formed. In the present example, the bond between the hydrogen atom and the carbon of the methyl group is partially broken; the bond between the fluorine and the hydrogen atom is partially formed. Although the transition state has an extremely brief existence ($\sim 10^{-12}$ sec), its structure and relative potential energy are, as we will see, highly important aspects of a chemical reaction.

The difference in potential energy between the reactants and the transition state is the energy of activation, E_{act}. The difference in potential energy between the reactants and products is the heat of reaction, $\Delta H°$. For the reaction $F\cdot + CH_4 \longrightarrow H{-}F + CH_3\cdot$, the energy level of the products is 32 kcal/mole lower than that of the reactants. In terms of our analogy, we can say that the reactants on one energy plateau must traverse an energy hill (the energy of activation) in order to arrive at the lower energy plateau of products.

When a three-dimensional plot of potential energy versus the reaction coordinate is made, the transition state is found to resemble a mountain pass or *col* (Fig. 4.4) rather than the top of an energy hill as we have shown in our two-dimensional plot above. That is, the reactants and products appear to be separated by an energy barrier resembling a mountain range. While an infinite number of possible routes lead from reactants to products, the transition state lies at the top of the route that requires the lowest (energy) climb. Whether or not the pass is a wide or narrow one depends on the orientation or probability factor. A wide pass means that there is a relatively large probability that collisions will occur with an orientation that allows a reaction to take place. A narrow pass means just the opposite.

How does one know what the energy of activation for a reaction will be? Could we, for example, have predicted from bond-dissociation energies that the energy of activation for the reaction, $F\cdot + CH_4 \longrightarrow HF + CH_3\cdot$, would be precisely 1.2 kcal/mole? The answer is *no*. The energy of activation must be determined from other experimental data. It cannot be directly measured—it is calculated. Certain principles can be established, however, that enable one to arrive at estimates of energies of activation:

1. Any reaction in which *bonds are broken* will have an energy of activation greater than zero.
2. Activation energies of *endothermic reactions that involve both bond formation and bond rupture will be greater than the heat of reaction, $\Delta H°$.* Two examples illustrate this principle: the first chain-propagating step in the chlorination of methane and the corresponding step in the

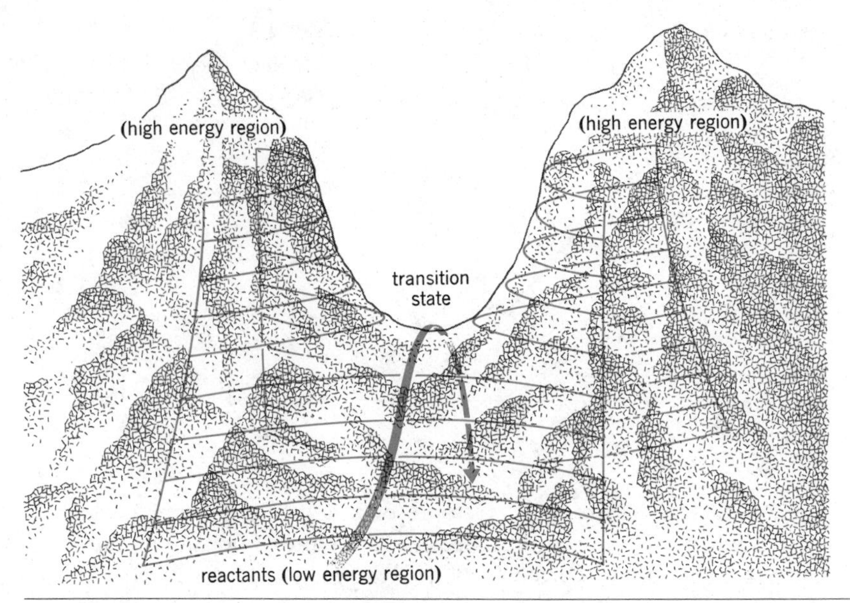

FIG. 4.4
Mountain pass or col *analogy for the transition state. (Adapted with permission from J. E. Lefler and E. Grunwald,* Rates and Equilibria of Organic Reactions, *Wiley, New York, 1963, page 6.)*

bromination of methane:

$$\begin{array}{llllll} Cl\cdot + & CH_3{-}H & \longrightarrow & H{-}Cl & + CH_3\cdot & \Delta H^\circ = +1 \text{ kcal/mole} \\ & (DH^\circ = 104) & & (DH^\circ = 103) & & E_{act} = +3.8 \text{ kcal/mole} \\ Br\cdot + & CH_3{-}H & \longrightarrow & H{-}Br & + CH_3\cdot & \Delta H^\circ = +16.5 \text{ kcal/mole} \\ & (DH^\circ = 104) & & (DH^\circ = 87.5) & & E_{act} = +18.6 \text{ kcal/mole} \end{array}$$

In both of these reactions the energy released in bond formation is less than that required for bond rupture; both reactions are, therefore, endothermic. We can easily see why the energy of activation for each reaction is greater than the heat of reaction by looking at the potential energy diagrams in Fig. 4.5. In each case the path from reactants to products is from a lower energy plateau to a higher one. In each case the intervening energy hill (or *col*) is higher still, and since the energy of activation is the vertical (energy) distance between the plateau of reactants and the top of this hill the energy of activation exceeds the heat of reaction.

3. *The energy of activation of a gas-phase reaction where bonds are broken homolytically but no bonds are formed is equal to* ΔH°.* An example of this type of reaction is the chain-initiating step in the fluorination of methane: the dissociation of fluorine molecules into fluorine atoms.

$$\begin{array}{lll} F{-}F \longrightarrow 2F\cdot & \Delta H^\circ = +38 \text{ kcal/mole} \\ (DH^\circ = 38) & E_{act} = +38 \text{ kcal/mole} \end{array}$$

The potential energy diagram for this reaction is shown in Fig. 4.6.

* This rule only applies to radical reactions taking place in the gas phase. It does not apply to reactions taking place in solution, especially where ions are involved.

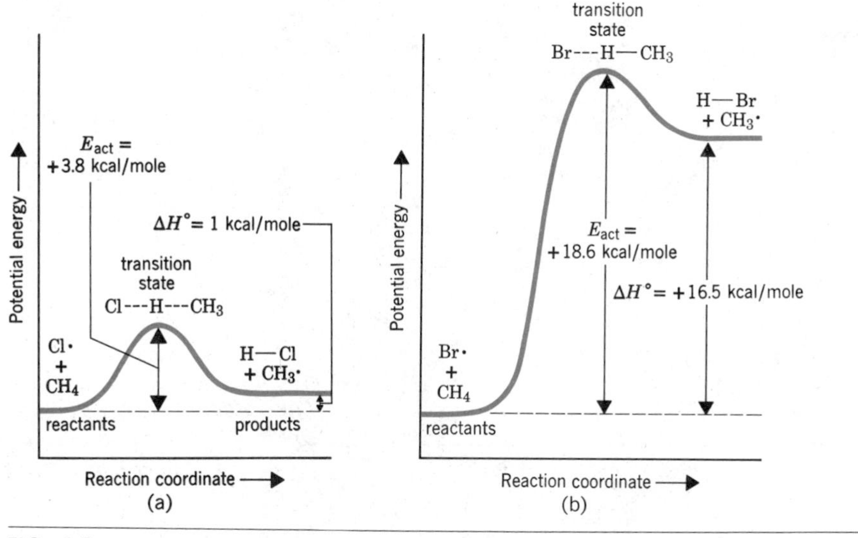

FIG. 4.5
Potential energy diagrams for (a) *the reaction of a chlorine atom with methane and* (b) *the reaction of a bromine atom with methane.*

4. The energy of activation for a *gas-phase* reaction in which *radicals combine to form molecules is usually zero.** In reactions of this type the problem of nonsimultaneous bond formation and bond rupture does not exist; only one process occurs: that of bond formation. All of the chain-terminating steps in the chlorination of methane fall into this category. An example is the combination of two methyl radicals to form a molecule of ethane.

$$2CH_3\cdot \longrightarrow \underset{(88)}{CH_3—CH_3} \quad \Delta H^\circ = -88 \text{ kcal/mole} \quad E_{act} = 0$$

* This rule only applies to radical reactions taking place in the gas phase. It does not apply to reactions taking place in solution, especially where ions are involved.

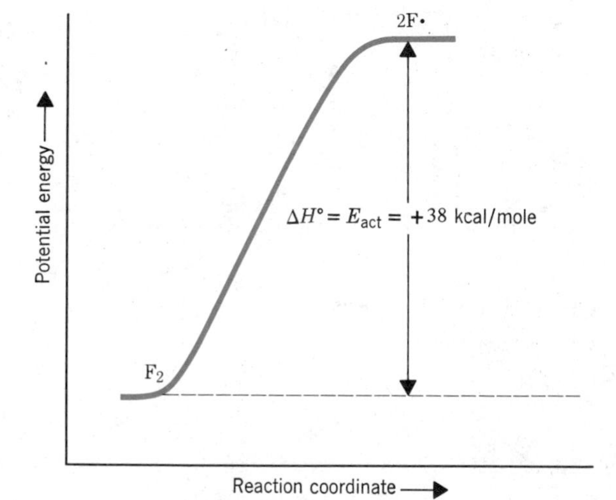

FIG. 4.6
The potential energy diagram for the dissociation of a fluorine molecule into fluorine atoms.

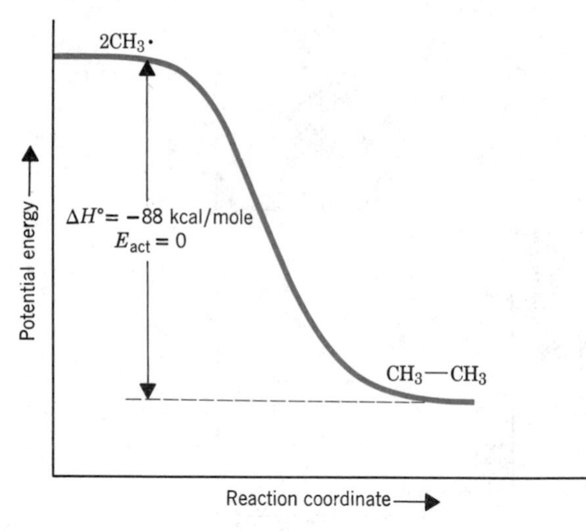

FIG. 4.7
The potential energy diagram for the combination of two methyl radicals to form a molecule of ethane.

Figure 4.7 illustrates the potential energy changes that occur in this reaction.

In Section 4.10 we will see how we can estimate energies of activation by taking advantage of the fact that the transition states of reactions resemble the reactants and products.

Problem 4.10

When gaseous ethane is heated to a very high temperature free-radical reactions take place that produce (among other products) methane and butane. This type of change is called thermal cracking (cf. Section 3.15). Among the reactions that take place when ethane undergoes thermal cracking are the following.

(1) $CH_3CH_3 \longrightarrow 2CH_3\cdot$
(2) $CH_3\cdot + CH_3CH_3 \longrightarrow CH_4 + CH_3CH_2\cdot$
(3) $2CH_3CH_2\cdot \longrightarrow CH_3CH_2CH_2CH_3$
(4) $CH_3CH_2\cdot \longrightarrow CH_2{=}CH_2 + H\cdot$
(5) $CH_3\cdot + H\cdot \longrightarrow CH_4$

(a) For which reaction(s) would you expect E_{act} to equal zero?
(b) For which would you expect E_{act} to be greater than zero?
(c) For which would you expect E_{act} to equal $\Delta H°$?

Problem 4.11

Sketch potential energy diagrams for the following reactions. Label the heat of reaction ($\Delta H°$) and the energy of activation (E_{act}) in each case.

(a) $CH_3\cdot + HCl \longrightarrow CH_3{-}H + Cl\cdot$
(b) $CH_3\cdot + HBr \longrightarrow CH_3{-}H + Br\cdot$
(c) $CH_3{-}CH_3 \longrightarrow 2CH_3\cdot$
(d) $Br{-}Br \longrightarrow 2Br\cdot$
(e) $2Cl\cdot \longrightarrow Cl{-}Cl$

4.8E Reaction Rates: an Explanation of the Effect of Temperature Changes

With an understanding of the origin and nature of the energy of activation behind us, we can now provide an explanation for the remarkable increase in the rate of most reactions that accompanies a modest increase in temperature. For most reactions *increasing the temperature only 10% will cause the reaction rate to double.*

In a given sample of a gas at a given temperature the molecules move with widely varying velocities. If we could actually see them, we would observe that, at a given instant, some molecules would be moving quite slowly and others very rapidly. Most of the molecules would have velocities at or near the average velocity of all of the molecules present. Since the kinetic energy of individual molecules is proportional to the square of their velocities ($KE = \frac{1}{2}mv^2$), the statistical distribution of the values of the kinetic energies of the molecules should parallel the distribution of their velocities. This distribution is given in the curve shown in Fig. 4.8. Note that while a few of the molecules have very low kinetic energies and a few have very high kinetic energies, most of the molecules have kinetic energies near the average.

The energies brought to collisions between molecules in the gas phase show the same general distribution. Occasionally molecules having very low kinetic energies collide and these collisions have, as a result, very low energy. Occasionally, too, molecules possessing very high kinetic energies collide and these collisions have very high energy. Most of the collisions, however, are between molecules with kinetic energies at or near the average and have collision energies at or near the average.

Let us designate a particular collision energy (E_{act}) as that required, as *a minimum,* to bring about a reaction between the colliding molecules. Then the number of collisions having sufficient energy to cause a reaction is proportional to the area under that portion of the curve that represents collision energies greater than or equal to E_{act}. If, at the same time, we also examine the distribution of collision energies at two different temperatures, T_1 and T_2, where T_2 is a higher temperature than T_1, we get the result shown in Fig. 4.9.

Because of the shapes of the curves, and because the curve for the higher temperature is shifted to higher kinetic energies, the number of collisions with energies great enough to cause a reaction, that is, with energies greater than E_{act}, *is very much larger at the higher temperature.* The area under curve T_1 with energy greater than E_{act} is quite small, but that under curve T_2 with energy greater than E_{act} is quite large. It is this difference that accounts, primarily, for the very

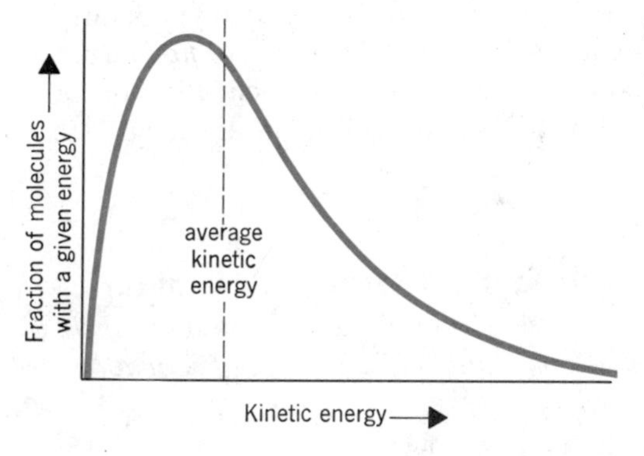

FIG. 4.8
A plot showing the distribution of kinetic energies among molecules in the gas phase.

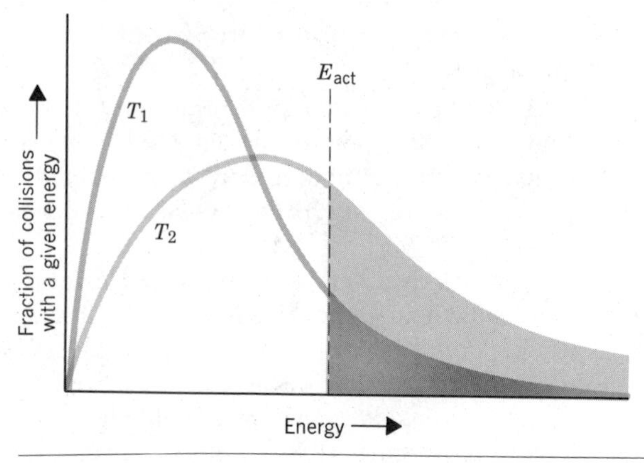

FIG. 4.9
The distribution of collision energies at two different temperatures, T_1 *and* T_2 ($T_2 > T_1$). *The number of collisions with energies greater than the activation energy is indicated by the appropriately shaded area under each curve.*

dramatic increase in reaction rate that can occur with a very modest increase in temperature.

What we have just illustrated graphically can also be explained mathematically on the basis of an equation developed by the Swedish chemist Svante Arrhenius*—an equation that underlies the collision theory. Arrhenius discovered, experimentally, how the rates of many chemical reactions depend on the absolute temperature and how this relationship may be expressed mathematically. If the concentrations of the reactants are unity (i.e., 1.0 mole per liter) the Arrhenius equation can be written in a way that corresponds directly to the expression for reaction rates we gave on page 140.

$$\text{Rate} = \mathbf{Z} \times \mathbf{P} \times 10^{-E_{act}/2.3RT}$$

Z is the collision frequency.
P is the orientation or probability factor.
$10^{-E_{act}/2.3RT}$ is the energy factor, ***f***. It equals the fraction of collisions with energy greater than E_{act}.

The exponential nature of the energy factor is what accounts for the remarkable increase in the rate of a reaction that usually accompanies an increase in temperature. This point can be illustrated in the following way. If we assume that a particular reaction has an energy of activation of 10 kcal/mole, we can show through a simple calculation that the small change from room temperature, 300 K, to the temperature of a steambath, 373 K, should cause a *20-fold increase* in the number of collisions with the energy greater than the activation energy. Although increasing the temperature also increases the collision frequency, **Z**, this effect is much less important.

4.8F Reaction Rates: The Effect of the Energy of Activation

The energy factor also accounts for another important observation: *there is a relation between the reaction rate and the magnitude of the energy of activation for*

* Arrhenius received the Nobel prize for chemistry in 1903.

different reactions of the same type occurring at the same temperature. The nature of this relationship can be illustrated by considering, as examples, three different reactions—all having very different energies of activation—but all occurring at the same temperature. The reactions that we have chosen are the first of the two chain-propagating steps that occur when methane is fluorinated, chlorinated, and brominated.

$$F\cdot + CH_3\text{—}H \longrightarrow H\text{—}F + CH_3\cdot \quad E_{act} = 1.2 \text{ kcal/mole}$$
$$Cl\cdot + CH_3\text{—}H \longrightarrow H\text{—}Cl + CH_3\cdot \quad E_{act} = 3.8 \text{ kcal/mole}$$
$$Br\cdot + CH_3\text{—}H \longrightarrow H\text{—}Br + CH_3\cdot \quad E_{act} = 18.6 \text{ kcal/mole}$$

Substituting these values for E_{act} into the expression $10^{-E_{act}/2.3RT}$ gives the following results. (The temperature that we have chosen for our calculation is 300 °C or 573 K.) E_{act} is part of an *exponent* and we now see what a dramatic difference that makes.

REACTION	E_{act}	FRACTION OF COLLISIONS WITH ENERGY GREATER THAN E_{act}
$F\cdot + CH_4 \longrightarrow HF + CH_3\cdot$	1.2 kcal/mole	0.35
$Cl\cdot + CH_4 \longrightarrow HCl + CH_3\cdot$	3.8 kcal/mole	0.035
$Br\cdot + CH_4 \longrightarrow HBr + CH_3\cdot$	18.6 kcal/mole	0.00000008

The magnitude of these results can be more fully appreciated if we express them in the following way: out of every 100 million collisions, 35 million will be sufficiently energetic to make products in the fluorination reaction, 3.5 million will be sufficiently energetic in the chlorination reaction, but only 8 collisions will be sufficiently energetic in the bromination reaction! Thus relatively small differences in E_{act} may mean very large differences in rates. Figure 4.10 may help us see this more clearly.

We see from the shape of the curve that when E_{act} is large the area under the curve representing the fraction of collisions with sufficient energy to allow a reaction is quite small. On the other hand, when the energy of activation is small, the area under the curve is quite large. The first situation might represent the reaction $Br\cdot + CH_4 \longrightarrow HBr + CH_3\cdot$—a reaction that at room temperature occurs very slowly. The latter might represent the reaction $F\cdot + CH_4 \longrightarrow HF + CH_3\cdot$—a reaction that under the same conditions, occurs very rapidly.

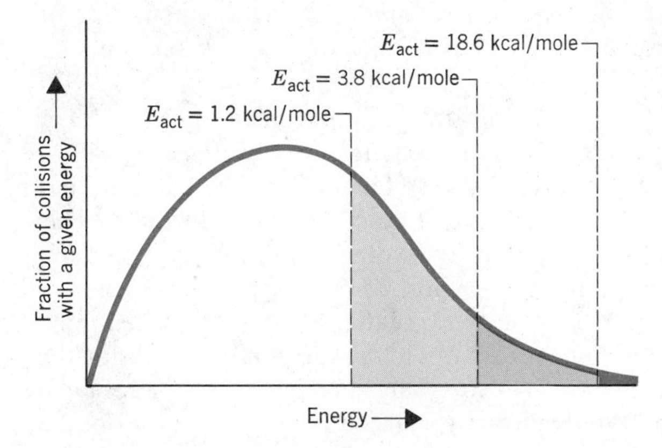

FIG. 4.10
The fraction of collisions with energies greater than the energy of activation at three different activation energies.

It thus becomes clear that in planning experiments, or in explaining their results, we need to give careful consideration to the relative sizes of the energies of activation.

4.9 THERMODYNAMICS AND KINETICS OF THE REACTIONS OF METHANE WITH HALOGENS

The *reactivity* of one substance toward another is measured by the *rate* at which the two substances react. A reagent that reacts very rapidly with a particular substance is said to be highly reactive toward that substance. One that reacts slowly or not at all under the same experimental conditions (e.g., concentration, pressure, and temperature) is said to have a low relative reactivity or to be unreactive. The reactions of the halogens (fluorine, chlorine, bromine, and iodine) with methane show a wide spread of relative reactivities. Fluorine is most reactive—so reactive, in fact, that without special precautions mixtures of fluorine and methane explode. Chlorine is next most reactive. However, the chlorination of methane is easily controlled by the judicious control of heat and light. Bromine is much less reactive toward methane than chlorine, and iodine is so unreactive that for all practical purposes we can say that no reaction takes place at all.

If the mechanisms for fluorination, bromination, and iodination of methane are the same as for its chlorination, we can explain the wide variation in reactivity of the halogens by a careful examination of $\Delta H°$ and E_{act} for each step.

Fluorination	$\Delta H°$(kcal/mole)	E_{act}(kcal/mole)
Chain initiating:		
$F_2 \longrightarrow 2F\cdot$	+38	+38
Chain propagating:		
$F\cdot + CH_4 \longrightarrow HF + CH_3\cdot$	−32	+1.2
$CH_3\cdot + F_2 \longrightarrow CH_3F + F\cdot$	−70	small
Overall $\Delta H°$ =	−102	

The chain-initiating step in fluorination is highly endothermic and thus has a high energy of activation.

If we did not know otherwise, we might carelessly conclude from the energy of activation of the chain-initiating step alone that fluorine would be quite unreactive toward methane. (If we then proceeded to try the reaction, as a result of this careless assessment, the results would be literally disastrous.) We know, however, that the chain-initiating step occurs only infrequently relative to the chain-propagating steps. One initiating step is able to produce thousands of fluorination reactions. As a result, the high activation energy for this step is not an impediment to the reaction.

Chain-propagating steps, by contrast, cannot afford to have high energies of activation. If they do, the highly reactive intermediates are consumed by chain-terminating steps before the chains progress very far. Both of the chain-propagating steps in fluorination have very small energies of activation. This allows a relatively large fraction of energetically favorable collisions even at room temperature. Moreover, the overall heat of reaction, $\Delta H°$, is very large. This means that as the reaction occurs, a large quantity of heat is evolved. This heat may accumulate in the mixture faster than it dissipates to the surroundings, causing the temperature to rise and with it a rapid increase in the frequency of additional chain-initiating steps that would generate additional chains. These two factors, the low energy of activation for the chain-propagating steps and the large overall heat of reaction, account for the high reactivity of fluorine toward methane.*

Chlorination	$\Delta H°$(kcal/mole)	E_{act}(kcal/mole)
Chain initiating:		
$Cl_2 \longrightarrow 2Cl\cdot$	+58	+58
Chain propagating:		
$Cl\cdot + CH_4 \longrightarrow HCl + CH_3\cdot$	+ 1	+3.8
$CH_3\cdot + Cl_2 \longrightarrow CH_3Cl + Cl\cdot$	−25.5	small
Overall $\Delta H°$ =	−24.5	

The higher energy of activation of the first chain-propagating step in chlorination of methane (+3.8 kcal/mole), versus the lower energy of activation (+1.2 kcal/mole) in fluorination, partly explains the lower reactivity of chlorine. The greater energy required to break the chlorine-chlorine bond in the initiating step (+58 kcal/mole for Cl_2 versus +38 kcal/mole for F_2) has some effect, too. However, the much greater overall heat of reaction in fluorination probably plays the greatest role in accounting for the much greater reactivity of fluorine.

Bromination	$\Delta H°$(kcal/mole)	E_{act}(kcal/mole)
Chain initiating:		
$Br_2 \longrightarrow 2Br\cdot$	+46	+46
Chain propagating:		
$Br\cdot + CH_4 \longrightarrow HBr + CH_3\cdot$	+16.5	+18.6
$CH_3\cdot + Br_2 \longrightarrow CH_3Br + Br\cdot$	−24	small
Overall $\Delta H°$ =	− 7.5	

In contrast to chlorination, the first chain-propagating step in bromination has a very high energy of activation (E_{act} = 18.6 kcal/mole). This means that only a very tiny fraction of all of the collisions between bromine atoms and methane molecules will be energetically effective even at a temperature of 300 °C. Bromine, as a result, is much less reactive toward methane than chlorine even though the net reaction is slightly exothermic.

Iodination	$\Delta H°$(kcal/mole)	E_{act}(kcal/mole)
Chain initiating:		
$I_2 \longrightarrow 2I\cdot$	+36	+36
Chain propagating:		
$I\cdot + CH_4 \longrightarrow HI + CH_3\cdot$	+31	+33.5
$CH_3\cdot + I_2 \longrightarrow CH_3I + I\cdot$	−20	small
Overall $\Delta H°$ =	+11	

The thermodynamic quantities for iodination of methane make it clear that the chain-initiating step is not responsible for the observed order of reactivities: $F_2 > Cl_2 > Br_2 > I_2$. The iodine-iodine bond is even weaker than the fluorine-fluorine bond. On this basis alone, one would predict that iodine would be the most reactive of the halogens. This clearly is not the case. Once again, it is the first

* Fluorination reactions can be controlled. This is usually accomplished by diluting both the hydrocarbon and the fluorine with an inert gas like helium before bringing them together. The reaction is also carried out in a reactor packed with copper shot. The copper, by absorbing the heat produced, moderates the reaction.

chain-propagating step that correlates with the experimentally determined order of reactivities. The energy of activation of this step in the iodine reaction (33.5 kcal/mole) is so large that only two collisions out of every 10^{12} have sufficient energy to produce reactions at 300 °C. As a result, iodination is not a feasible reaction experimentally.

Before we leave this topic one further point needs to be made. We have given explanations of the relative reactivities of the halogens toward methane that have been based on energy considerations alone. This has been possible *only because the reactions are quite similar and thus have similar orientation factors*. Had the reactions been of different types, this kind of analysis would not have been proper and might have given incorrect explanations.

4.10 HALOGENATION OF HIGHER ALKANES

Higher alkanes react with halogens by the same kind of chain mechanisms as those that we have just seen.

Chlorination of most alkanes whose molecules contain more than three carbons gives a mixture of isomeric monochloro products as well as more highly halogenated compounds. An example is the light-promoted chlorination of isobutane.

$$CH_3\underset{|}{\overset{CH_3}{\overset{|}{C}}}HCH_3 \xrightarrow[\text{light}]{Cl_2} CH_3\overset{CH_3}{\overset{|}{C}}HCH_2Cl + CH_3\underset{\underset{Cl}{|}}{\overset{CH_3}{\overset{|}{C}}}CH_3 + \text{Polychlorinated products} + HCl$$

Isobutane — Isobutyl chloride (48%) — *tert*-Butyl chloride (29%) — (23%)

The ratios of products that we obtain from chlorination reactions of higher alkanes are not identical with what we would expect if all the hydrogens of the alkane were equally reactive. We find that there is a correlation between reactivity of different hydrogens and the type of hydrogen (1°, 2°, or 3°) being replaced. The tertiary hydrogens of an alkane are most reactive, secondary hydrogens are next most reactive, and primary hydrogens are the least reactive.

Problem 4.12

If we examine just the monochloro products of the reaction of isobutane with chlorine given above, we find that isobutyl chloride represents 62.5% of the monochlorinated product, while *tert*-butyl chloride represents 37.5%. Explain how this demonstrates that the tertiary hydrogen is more reactive. (Hint: Consider what percentages of the butyl chlorides would be obtained if the nine primary hydrogens and the single tertiary hydrogen were all equally reactive.)

We can account for the relative reactivities of the primary, secondary, and tertiary hydrogens in a chlorination reaction on the basis of bond dissociation energies we saw earlier (Table 4.1). Of the three types, breaking a tertiary C—H bond requires the least energy, and breaking a primary C—H bond requires the most. Since the step in which the C—H bond is broken (i.e., the hydrogen abstraction step) determines the location or orientation of the chlorination, we would expect the E_{act} for abstracting a tertiary hydrogen to be least and for abstracting a primary hydrogen to be greatest. Thus tertiary hydrogens should be most reactive, secondary hydrogens should be next most reactive, and primary hydrogens should be the least reactive.

The differences in the rates with which primary, secondary, and tertiary hydrogens are replaced by chlorine are not large, however. Chlorine, as a result, does not discriminate between the different types of hydrogens in a way that makes chlorination of higher alkanes a generally useful laboratory procedure. We will find later that there are much better laboratory methods for introducing chlorine into a molecule. (Alkane chlorinations do find use in some industrial processes, especially in those instances where mixtures of alkyl chlorides can be used.)

Problem 4.13

Chlorination reactions of certain higher alkanes can be used for laboratory preparations. Examples are the preparation of neopentyl chloride from neopentane and cyclopentyl chloride from cyclopentane. What structural feature of these molecules makes this possible?

Problem 4.14

The hydrogen abstraction steps for most alkane chlorinations are exothermic. Show that this is true by calculating $\Delta H°$ for the reaction by which Cl· abstracts

(a) a primary hydrogen of ethane.
(b) a secondary hydrogen of propane.
(c) a primary hydrogen of propane.

4.10A Reactivity and Selectivity

Bromine is less reactive toward alkanes in general than chlorine, but bromine is more *selective* in the site of attack when it does react. Bromine shows a much greater ability to discriminate among the different types of hydrogens. The reaction of isobutane and bromine, for example, gives almost exclusive replacement of the tertiary hydrogen.

$$CH_3-\underset{\underset{H}{|}}{\overset{\overset{CH_3}{|}}{C}}-CH_3 \xrightarrow[\text{Light, }127°]{Br_2} \underset{>99\%}{CH_3-\underset{\underset{Br}{|}}{\overset{\overset{CH_3}{|}}{C}}-CH_3} + \underset{\text{Trace}}{CH_3-\underset{\underset{H}{|}}{\overset{\overset{CH_3}{|}}{C}}-CH_2Br}$$

The greater selectivity of bromine can be explained in terms of transition state theory and bromine's greater selectivity is directly related to its lower reactivity. According to a postulate made by Professor G. S. Hammond (of the California Institute of Technology) *the structure of the transition state of an endothermic step of a reaction resembles the products of that step more than it does the reactants. For an exothermic step the structure of the transition state is more like the reactants than the products.*

This principle can be better understood through consideration of the potential energy versus reaction coordinate diagrams given in Fig. 4.11.

Highly exothermic and highly endothermic steps are shown in Fig. 4.11 because they illustrate Hammond's postulate most dramatically.* In the highly

* Hammond's postulate is quite general, however, and applies to reactions where values of $\Delta H°$ are not so exaggerated.

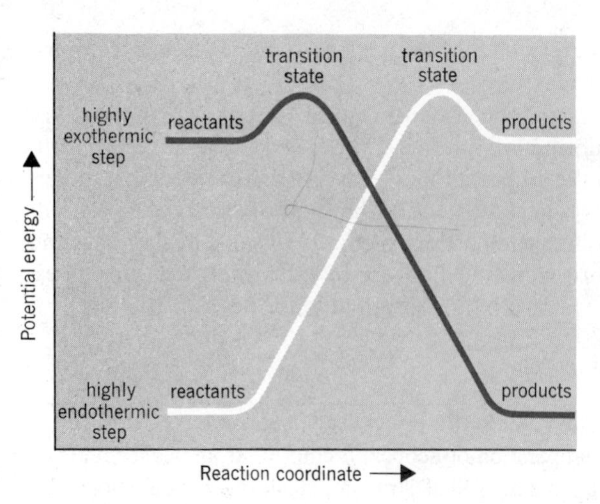

FIG. 4.11
Energy diagrams for highly exothermic and highly endothermic steps of reactions (From William A. Pryor, Introduction to Free Radical Chemistry *© 1966. Reprinted by permission of Prentice-Hall, Inc., Englewood Cliffs, N.J., p. 53.)*

exothermic step the energy levels of the reactants and the transition state are close to each other. The transition state also lies close to the reactants along the *reaction coordinate.* This means that in the highly exothermic step, bond-breaking has not proceeded very far when the transition state is reached. In the exothermic reactions of isobutane molecules with chlorine atoms, for example, relatively little carbon-hydrogen bond-breaking has developed in the transition states. The transition states for the two hydrogen abstraction steps might resemble those shown below.

$$CH_3CH(CH_3)CH_3 + Cl\cdot \longrightarrow CH_3CH(CH_3)\overset{\delta\cdot}{C}H_2{-}{-}H{-}{-}{-}{-}{-}\overset{\delta\cdot}{Cl} \longrightarrow CH_3CH(CH_3)CH_2\cdot + HCl$$

Reactantlike transition state; 1° Radical

$$\downarrow Cl_2$$

$$CH_3CH(CH_3)CH_2Cl + Cl\cdot$$

Isobutyl chloride

$$(CH_3)_3C{-}H + Cl\cdot \longrightarrow (CH_3)_2\overset{\delta\cdot}{C}(CH_3){-}{-}H{-}{-}{-}{-}{-}\overset{\delta\cdot}{Cl} \longrightarrow (CH_3)_3C\cdot \;+ HCl$$

Reactantlike transition state; 3° Radical

$$\downarrow Cl_2$$

$$(CH_3)_3C{-}Cl + Cl\cdot$$

tert-Butyl chloride

Since the transition states in both cases are reactantlike in both structure and energy, they show little resemblance to the products of the hydrogen abstraction step, a 1° radical and a 3° radical. And, since the reactants in both cases are the same, the exact type of C—H bond being broken (primary or tertiary) has a relatively small influence on the relative rates of the reactions. The two reactions proceed with similar (but not identical) rates because their respective activation energies are quite similar (Fig. 4.12).

The transition states of highly endothermic steps, on the other hand, lie close to products on the potential energy coordinate and *along the reaction coordinate.* In highly endothermic steps the bond has broken to a considerable extent by the time that the transition state is reached. The two hydrogen abstraction steps in the reaction of isobutane with bromine are both highly endothermic. In these reactions considerable carbon-hydrogen bond breaking has occurred when the transition state is reached. These transition states might be depicted in the following way.

$$CH_3CH(CH_3)CH_3 + Br\cdot \longrightarrow \underset{\text{Productlike transition state}}{CH_3CH(CH_3)\overset{\delta\cdot}{C}H_2\text{-----}H\text{--}\overset{\delta\cdot}{Br}} \longrightarrow \underset{1^\circ\ \text{Radical}}{CH_3CH(CH_3)CH_2\cdot} + HBr$$

$$\underset{1^\circ\ \text{Radical}}{CH_3CH(CH_3)CH_2\cdot} \xrightarrow{Br_2} \underset{\text{Isobutyl bromide}}{CH_3CH(CH_3)CH_2Br} + Br\cdot$$

$$CH_3C(CH_3)_2\text{—}H + Br\cdot \longrightarrow \underset{\text{Productlike transition state}}{CH_3\overset{\delta\cdot}{C}(CH_3)_2\text{-----}H\text{--}\overset{\delta\cdot}{Br}} \longrightarrow \underset{3^\circ\ \text{Radical}}{CH_3C(CH_3)_2\cdot} + HBr$$

$$\underset{3^\circ\ \text{Radical}}{CH_3C(CH_3)_2\cdot} \xrightarrow{Br_2} \underset{\textit{tert}\text{-Butyl bromide}}{CH_3C(CH_3)_2\text{—}Br} + Br\cdot$$

Since the transition states for both steps in bromination are productlike in structure and energy, and since the products of each hydrogen abstraction step are, in fact, quite different (a 1° radical versus a 3° radical), the type of C—H bond being broken will have a marked influence on the relative rates of the reactions. In fact, they proceed with very different rates. Abstraction of the 3° hydrogen takes place much faster. Bromine, as a result, discriminates more effectively between the primary and tertiary hydrogens. A comparison of potential energy diagrams for the abstraction of the primary and tertiary hydrogens by bromine is given in Fig. 4.13.

Problem 4.15

Fluorine is far less selective than bromine and is even less selective than chlorine. The products that one obtains from alkane fluorinations are, in fact, almost those

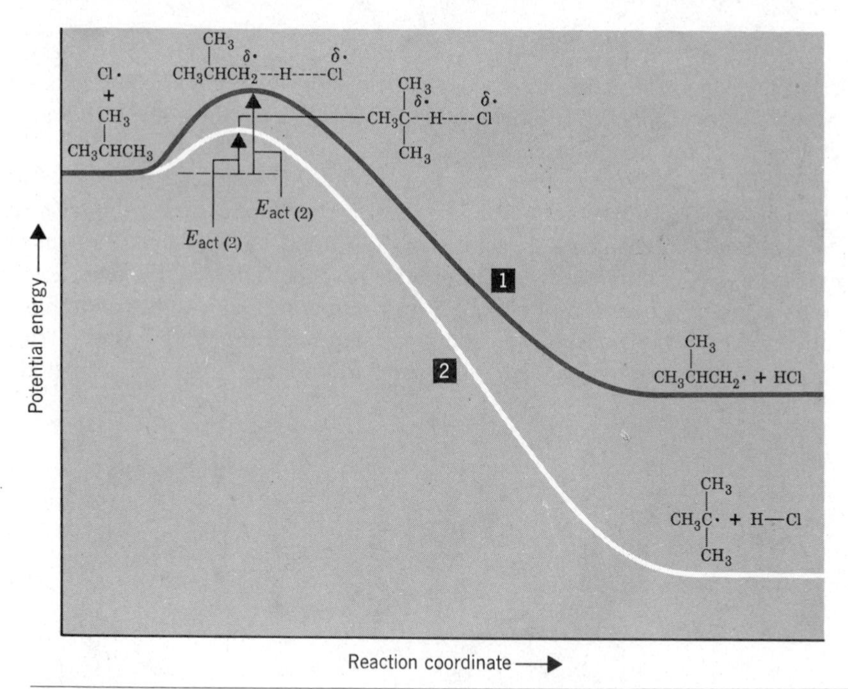

FIG. 4.12
Potential energy diagrams for the two hydrogen abstraction steps in the reaction of isobutane with Cl·. Both steps are exothermic and both transition states resemble the reactants. The activation energies are similar; but, because 3° C—H bonds are broken more easily than 1° C—H bonds, reaction (2) has a lower activation energy and proceeds at a somewhat faster rate.

that one would expect if the different types of hydrogen were equally reactive. Explain.

4.11 OTHER IMPORTANT FREE RADICAL CHAIN REACTIONS

Free-radical chain mechanisms are important in understanding many other organic reactions. We will see other examples in later chapters but let us examine two here: the combustion of alkanes and some reactions of chlorofluoromethanes that have threatened the protective layer of ozone in the stratosphere.

4.11A Combustion of Alkanes

When alkanes react with oxygen (e.g., in oil furnaces and in internal combustion engines) a complex series of reactions takes place ultimately converting the alkane to carbon dioxide and water (Section 2.4A). Although our understanding of the detailed mechanism of combustion is incomplete, we do know that the important reactions occur by free-radical chain mechanisms with chain-propagating steps like those shown below.

$$R\cdot + O_2 \longrightarrow R{-}OO\cdot$$

$$R{-}OO\cdot + R{-}H \longrightarrow R{-}OOH + R\cdot$$

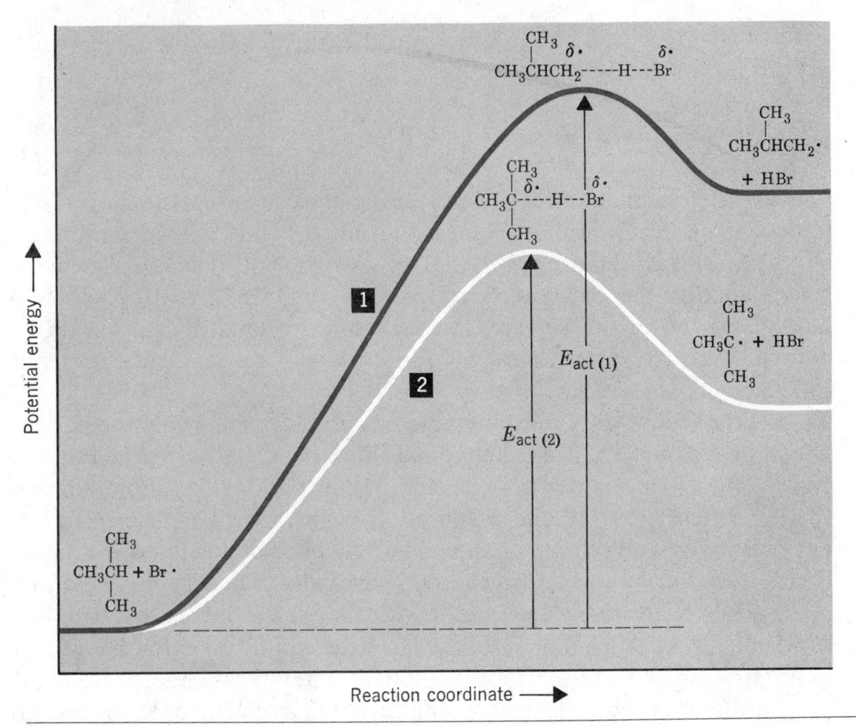

FIG. 4.13
Potential energy diagrams for the two hydrogen abstraction steps in the reaction of isobutane with Br·. Both steps are highly endothermic and, in both, the transition states resemble the products. Since the products—a 3° radical and a 1° radical—have quite different potential energies (stabilities) the transition states for the two steps are also quite different. The transition state for reaction (1) resembles a 1° radical. It occurs at a much higher potential energy than the transition state for reaction (2) because the state for reaction (2) resembles a much more stable 3° radical. The activation energy for reaction (2) is much lower than that for reaction (1). Reaction (2), consequently, proceeds at a much faster rate. The ultimate product that arises from reaction (2) is tert-*butyl bromide, and this is the predominant product of the reaction.*

One product of the second step is R—OOH, called an alkyl hydroperoxide. The oxygen-oxygen bond of an alkyl hydroperoxide is quite weak, and it can break and produce other free radicals that can initiate other chains:

$$RO—OH \longrightarrow RO\cdot + HO\cdot$$

4.11B Freons and Ozone Depletion

In the stratosphere at altitudes of about 25 km, very high-energy (very short wavelength) ultraviolet light converts oxygen (O_2) into ozone (O_3). The reactions that take place are:

1. $O_2 + h\nu \longrightarrow O + O$
2. $O + O_2 + M \longrightarrow O_3 + M + heat$

where M is some other particle that can absorb some of the energy released in the second step.

The ozone produced in step 2 can also interact with high-energy ultraviolet light in the following way.

3. $O_3 + h\nu \longrightarrow O_2 + O + \text{heat}$

The oxygen atom formed in step 3 can cause a repetition of step 2 and so forth. The net result of these steps is to convert highly energetic ultraviolet light into heat. This is important because the existence of this cycle shields the earth from radiation that is destructive to living organisms. This shield makes life possible on the earth's surface. Even a relatively small increase in high-energy ultraviolet radiation at the earth's surface would cause a large increase in the incidence of skin cancers.

Production of chlorofluoromethanes (and chlorofluoroethanes) called *Freons* began in 1930. These compounds have been used as refrigerants, solvents, and as propellants in aerosol cans. Typical Freons are trichlorofluoromethane, $CFCl_3$ (called Freon-11), and dichlorodifluoromethane, CF_2Cl_2 (called Freon-12).

By 1974 world Freon production was about 2 billion pounds annually. Most Freon, even that used in refrigeration, eventually makes its way into the atmosphere where it diffuses unchanged into the stratosphere. In June 1974 F. S. Rowland and M. J. Mollina published an article indicating, for the first time, that in the stratosphere Freon is able to initiate free radical chain reactions that can upset the natural ozone balance. Using Freon-12 as an example, the reactions that take place are:

Chain-initiating step (1) $CF_2Cl_2 + h\nu \longrightarrow CF_2Cl\cdot + Cl\cdot$

Chain-propagating steps
(2) $Cl\cdot + O_3 \longrightarrow ClO\cdot + O_2$
(3) $ClO\cdot + O \longrightarrow O_2 + Cl\cdot$

In the chain-initiating step, ultraviolet light causes homolytic cleavage of one C—Cl bond of the Freon. The chlorine atom, thus produced, is the real villain; it can set off a chain reaction that destroys thousands of molecules of ozone before it diffuses out of the stratosphere or reacts with some other substance.

In 1975 a study by the National Academy of Science supported the predictions of Rowland and Mollina and since January 1978 the use of Freons in aerosol cans in the United States has been banned.

4.12 REACTIONS OF ALKANES WITH SUPERACIDS

Alkanes, as we noted earlier, are normally unreactive towards acids. Alkanes can be boiled for days with concentrated hydrochloric acid, for example, and no detectable reaction will occur. However, Professor George Olah (of Case Western Reserve University) has discovered that when alkanes are treated with very powerful acids, called superacids, sigma bonds break *heterolytically,* and the reactions give carbocations.*

* Olah and his co-workers have isolated and studied the properties of a number of carbocations. In doing so, they have contributed greatly to our understanding of them.

$$-\overset{|}{\underset{|}{C}}{:}H \quad + \quad H^+ \longrightarrow -\overset{|}{\underset{|}{C}}{}^+ \quad + H{:}H$$

Heterolytic cleavage of sigma bond — From a superacid — Carbocation

Superacids are mixtures of very powerful Lewis acids and very powerful proton donors. One of the strongest superacids is a mixture of antimony pentafluoride, SbF_5, (the Lewis acid) and fluorosulfonic acid, FSO_3H (the proton donor).

The reaction between an alkane and a superacid is an example of *protolysis* (cleavage by a proton). The process begins with the donation of a proton to a sigma bond of the alkane and the formation of an intermediate with a pentacoordinate carbon. In the reaction of isobutane with SbF_5—FSO_3H (below), three of the bonds to the central carbon of the intermediate are ordinary carbon-carbon sigma bonds; the two hydrogens, however, are attached through a three-center bond, that is, *a bond in which three nuclei are held together by only two electrons.*

$$CH_3-\overset{CH_3}{\underset{CH_3}{C}}-H + FSO_3H\cdot SbF_5 \longrightarrow \left[CH_3-\overset{CH_3}{\underset{CH_3}{C}}\langle{}^{H}_{H}\right]^+ \longrightarrow CH_3-\overset{CH_3}{\underset{CH_3}{C^+}} + H_2$$

Pentacoordinate carbon compound containing a three-center bond — *tert*-Butyl cation

The intermediate then loses a molecule of hydrogen through cleavage of the three-center bond: this gives a *tert*-butyl cation.

Other three-center bonds are known. Diborane, B_2H_6, contains two three-

$$\begin{matrix} H & & H & & H \\ & B & & B & \\ H & & H & & H \end{matrix}$$

Diborane

center bonds. When we examine the reactions of carbocations in subsequent chapters we will see other examples of intermediates containing three-center bonds.

Carbocations, like free radicals, are highly reactive species. In most chemical environments, they have only transient existence, but in superacids, carbocations have lifetimes long enough to allow them to be analyzed spectroscopically (Chapter 13).

4.13 THE STRUCTURE OF CARBOCATIONS AND FREE RADICALS: sp^2 HYBRIDIZATION

4.13A Carbocations

Considerable experimental evidence indicates that the structure of carbocations is *trigonal planar* like that of BF_3 (Section 1.16). Just as the trigonal-planar structure of BF_3 can be accounted for on the basis of sp^2 hybridization so, too, can the trigonal-planar structure of carbocations.

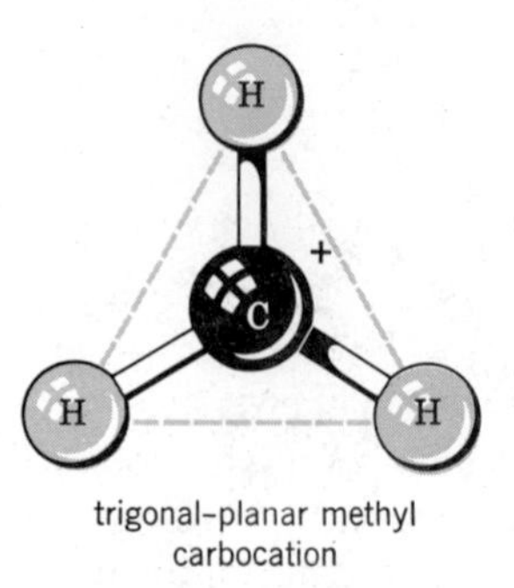

trigonal-planar methyl carbocation

Mathematically, the model for sp^2-hybridized carbon is obtained by mixing, or hybridizing, one *s* orbital with two *p* orbitals. In the illustration below, we hybridize the *s* orbital with the p_x and p_z orbitals.

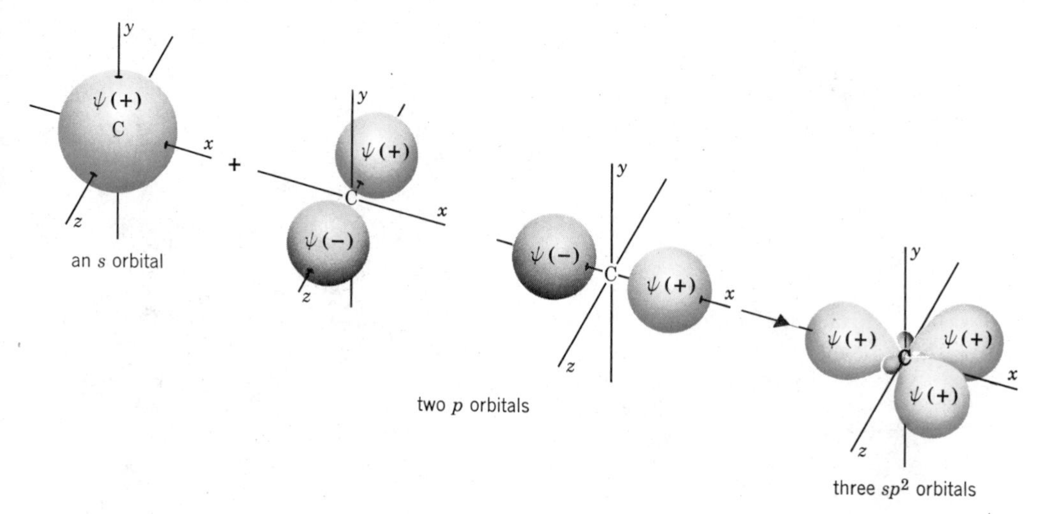

The three sp^2 orbitals that result are directed toward the corners of a regular triangle (with angles of 120° between them). The carbon *p* orbital (the p_y orbital) that is not hybridized with the *s* orbital is perpendicular to the plane of the triangle formed by the hybrid sp^2 orbitals.

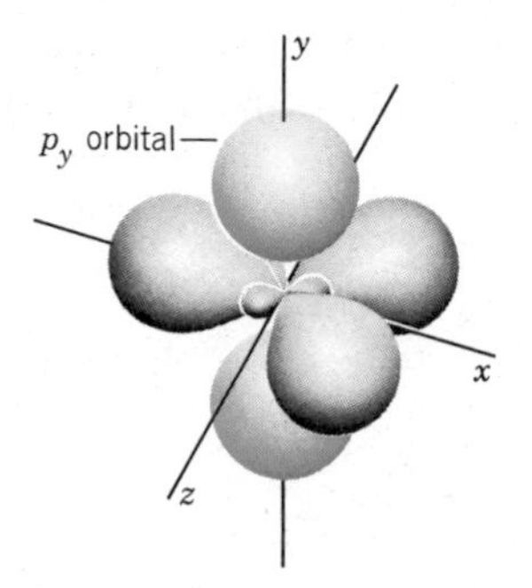

The central carbon in a carbocation is electron deficient; it has only six electrons in its outside energy level. In our model (Fig. 4.14) these six electrons are used to form sigma covalent bonds to hydrogens (or to alkyl groups). The *p* orbital contains no electrons.

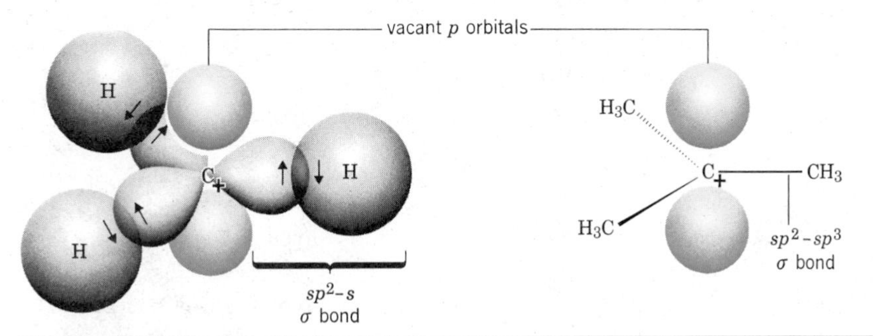

FIG. 4.14
(a) *Orbital structure of the methyl cation. The bonds to hydrogen are sigma bonds* (σ) *formed by overlap of the carbon's* sp^2 *orbitals with the 1*s *orbital of hydrogen. The* p *orbital is vacant.* (b) *A line-dash-wedge representation of the* tert-*butyl cation. The bonds between carbon atoms are formed by overlap of* sp^3 *orbitals of the methyl groups.*

4.13B Free Radicals

Experimental evidence indicates that the structure of most alkyl radicals is also trigonal planar at the carbon having the unpaired electron. This structure, too, can be accommodated by an sp^2-hybridized central carbon. In an alkyl radical, however, the *p* orbital is not vacant, but contains the unpaired electron (Fig. 4.15).

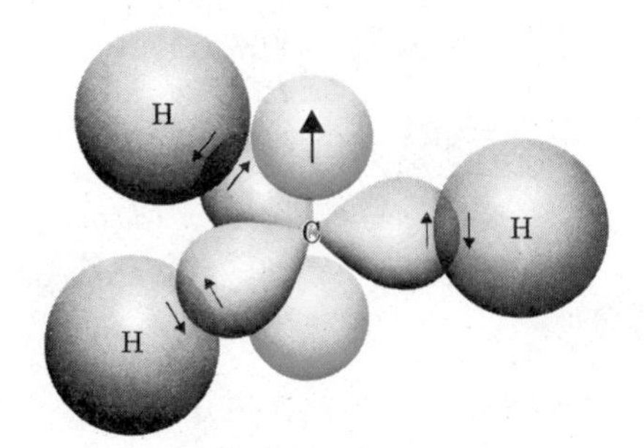

FIG. 4.15
sp^2-*Hybridized methyl radical showing the odd electron in one lobe of the vacant* p *orbital. It could be pictured in the other lobe.*

Some free radicals, such as the trifluoromethyl radical, appear to be sp^3 hybridized (Fig. 4.16). Experiments also indicate sp^3-hybridized radicals undergo rapid inversion about the central carbon.

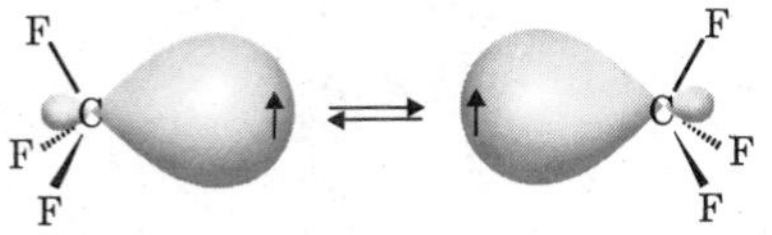

FIG. 4.16
Inverting trifluoromethyl radical.

4.14 THE STABILITY OF CARBOCATIONS

A large body of experimental evidence indicates that the relative stabilities of carbocations parallel those of free radicals. Tertiary carbocations are the most stable, and the methyl cation is the least stable. The overall order of stability is

$$R_3C^+ > R_2HC^+ > RH_2C^+ > H_3C^+$$

Most stable ... **Least stable**

3° > 2° > 1° > methyl

This order of stability of carbocations can be explained on the basis of a law of physics that states that *a charged system is stabilized when the charge is dispersed or delocalized.* Alkyl groups, when compared to hydrogen atoms, are *electron releasing*. This means that alkyl groups have a tendency to shift electron density (negative charge) toward a positive charge. Through electron release, *alkyl groups* attached to the positive carbon of a carbocation *delocalize* the positive charge. In doing so, the attached alkyl groups assume part of the positive charge themselves and thus *stabilize* the carbocation. We can see how this occurs by inspecting Fig. 4.17.

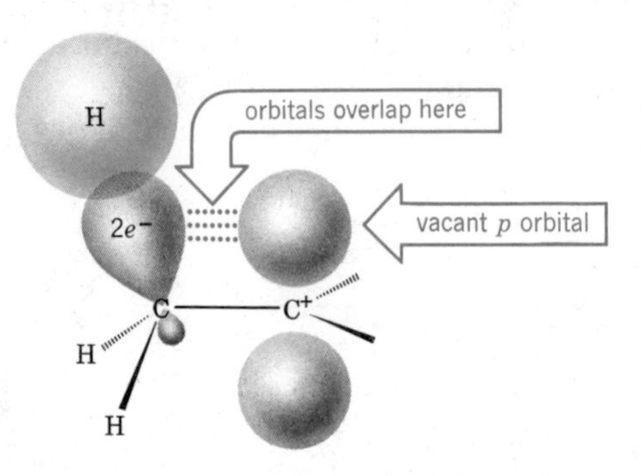

FIG. 4.17

How a methyl group helps stabilize the positive charge of a carbocation. Electron density from one of the carbon-hydrogen sigma bonds of the methyl group flows into the vacant p *orbital of the carbocation because the orbitals can partly overlap. Shifting electron density in this way makes the* sp^2*-hybridized carbon of the carbocation somewhat less positive and the hydrogens of the methyl group assumes some of the positive charge. Delocalization (dispersal) of the charge in this way leads to greater stability.*

In the *tert*-butyl cation three electron-releasing methyl groups surround the central carbon atom and assist in delocalizing the positive charge (Fig. 4.18). In the isopropyl cation there are only two attached methyl groups that can serve to delocalize the charge. In the ethyl cation there is only one attached methyl group, and in the methyl cation there is none at all. As a result, *the delocalization of charge and the order of stability of the carbocations parallel the number of attached methyl groups.*

$$(\overset{\delta+}{CH_3})_3C^{\delta+} \text{ is more stable than } (\overset{\delta+}{CH_3})_2CH^{\delta+} \text{ is more stable than } \overset{\delta+}{CH_3}CH_2^{\delta+} \text{ is more stable than } H_3C^+$$

tert-Butyl cation (3°)	Isopropyl cation (2°)	Ethyl cation (1°)	Methyl cation
Most stable			**Least stable**

FIG. 4.18

The relative stability of carbocations is 3° > 2° > 1° > methyl.

4.15 REACTIONS OF CYCLOALKANES: RING-OPENING REACTIONS OF CYCLOPROPANE

The reactions of cycloalkanes whose molecules have rings of five carbon atoms or larger are similar to those of alkanes. Cyclopentane and cyclohexane, for example,

$$\text{Cyclopentane} \xrightarrow[\text{Heat, light}]{Cl_2} \text{Chlorocyclopentane}$$

$$\text{Cyclohexane} \xrightarrow[\text{Heat, light}]{Br_2} \text{Bromocyclohexane}$$

undergo halogenation by substitution. Cyclopropane, however, undergoes ring opening when treated with bromine (and a Lewis acid) in carbon tetrachloride.

$$\text{cyclo-}(CH_2)_3 \xrightarrow[AlBr_3,\ CCl_4]{Br_2} BrCH_2CH_2CH_2Br + CH_3CHBrCH_2Br + CH_3CH_2CHBr_2$$

Ring opening also occurs when cyclopropane is treated with hydroiodic acid or aqueous sulfuric acid.

$$\text{cyclo-}(CH_2)_3 \xrightarrow{H{-}I} CH_3CH_2CH_2I$$

$$\text{cyclo-}(CH_2)_3 \xrightarrow[H_2O]{H_2SO_4} CH_3CH_2CH_2OH$$

Although we will not consider their mechanisms in detail, the reactions of cyclopropane with HI and H_2SO_4 both involve heterolytic cleavage of a carbon-carbon sigma bond of the cyclopropane ring through protolysis. The ring opening of cyclopropane by bromine (given above) apparently proceeds through a similar mechanism in which the agent responsible for bond cleavage is either the Lewis acid or a positive bromine ion.

An explanation for the ring-opening reactions of cyclopropane can be found in the great strain of the three-membered ring. As a result of this strain, the carbon-carbon bonds of cyclopropane are considerably weakened. An indication of the strain present can be obtained by considering the value of $\Delta H°$ for opening the cyclopropane ring by homolytic cleavage of one of the carbon-carbon bonds:

$$\text{cyclo-}(CH_2)_3 \longrightarrow \cdot CH_2CH_2CH_2 \cdot \qquad \Delta H° = 55 \text{ kcal/mole}$$

By contrast, the bond dissociation energy for the carbon-carbon bonds of most alkanes is ~82 kcal/mole.

The cyclobutane ring is strained as well, but not nearly as much as the cyclopropane ring. Cyclobutane does not undergo a ring-opening reaction when it is treated with bromine in carbon tetrachloride.

4.16 SOME IMPORTANT TERMS AND CONCEPTS

A **reaction mechanism** is a description of how a chemical reaction takes place. If the mechanism is multistep, it includes the steps involved and the *intermediates* that form.

Cleavage of covalent bonds can take place by *homolysis* (i.e., $A{-}B \longrightarrow A\cdot + B\cdot$) or *heterolysis* ($A{-}B \longrightarrow A^+ + {:}B^-$, or $A{-}B \longrightarrow A{:}^- + B^+$).

Carbon radicals (also called **free radicals**) are formed by homolysis of a bond to a carbon atom. Carbon radicals have unpaired electrons and show the following order of stabilities:

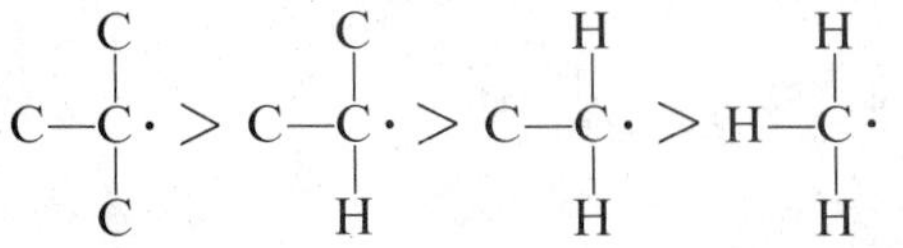

$$\begin{array}{ccccccc} \begin{array}{c} C \\ | \\ C{-}C\cdot \\ | \\ C \end{array} & > & \begin{array}{c} C \\ | \\ C{-}C\cdot \\ | \\ H \end{array} & > & \begin{array}{c} H \\ | \\ C{-}C\cdot \\ | \\ H \end{array} & > & \begin{array}{c} H \\ | \\ H{-}C\cdot \\ | \\ H \end{array} \\ 3^\circ & & 2^\circ & & 1^\circ & & \text{Methyl} \end{array}$$

Carbocations (also called **carbonium ions**) are formed by heterolysis of a bond to carbon as follows, ${-}C{:}Y \longrightarrow {-}C^+ + {:}Y^-$. Carbocations show the following order of stabilities:

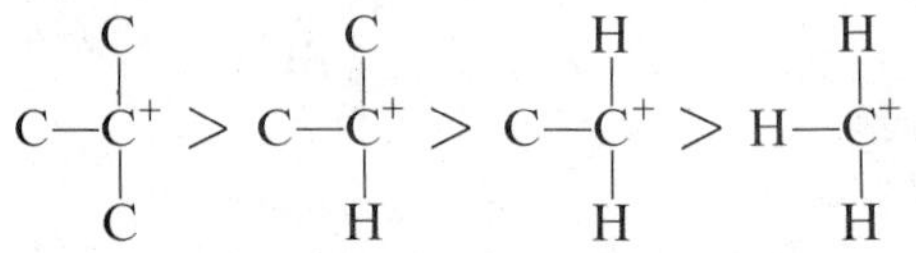

$$\begin{array}{ccccccc} \begin{array}{c} C \\ | \\ C{-}C^+ \\ | \\ C \end{array} & > & \begin{array}{c} C \\ | \\ C{-}C^+ \\ | \\ H \end{array} & > & \begin{array}{c} H \\ | \\ C{-}C^+ \\ | \\ H \end{array} & > & \begin{array}{c} H \\ | \\ H{-}C^+ \\ | \\ H \end{array} \\ 3^\circ & & 2^\circ & & 1^\circ & & \text{Methyl} \end{array}$$

Carbanions are formed by heterolysis of a bond to carbon as follows,

$$\begin{array}{c} | \\ {-}C{:}Y \\ | \end{array} \longrightarrow \begin{array}{c} | \\ {-}C{:}^- \\ | \end{array} + Y^+$$

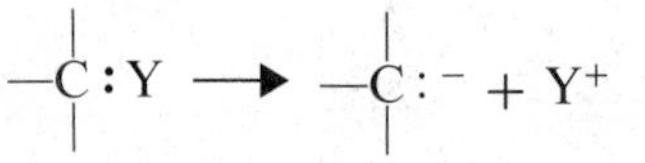

Bond dissociation energy (abbreviated DH°) is the amount of energy required for homolysis of a covalent bond.

Heat of reaction is the enthalpy change (ΔH°) for a chemical reaction and is equal to $H^\circ_{\text{products}} - H^\circ_{\text{reactants}}$. For an exothermic reaction ΔH° is negative; for an endothermic reaction ΔH° is positive.

A **substitution reaction** is one in which one group replaces another.

Halogenations of alkanes are substitution reactions in which a halogen replaces one (or more) of the alkane's hydrogens.

$$RH + X_2 \longrightarrow RX + HX$$

The reactions occur by a free radical mechanism.

$$\begin{array}{lrcl} \textbf{1.} & X_2 & \longrightarrow & 2X\cdot \\ \textbf{2.} & RH + X\cdot & \longrightarrow & R\cdot + HX \\ \textbf{3.} & R\cdot + X_2 & \longrightarrow & RX + X\cdot \end{array}$$

Chain reactions are reactions whose mechanisms involve a series of steps with each step producing a reactive intermediate that causes the next step to occur. The halogenation of an alkane (above) is a chain reaction.

Reaction rate is the rate at which reactants are converted to products in a chemical reaction. The rate of a reaction can be determined experimentally by measuring the rate at which reactants disappear from the mixture or the rate products form in the mixture.

Energy of activation is the minimum amount of energy (on a molar basis) that must be provided for a reaction to take place. It is the potential energy difference between the reactants and the transition state.

Hammond's postulate holds that the structure of the transition state of an endothermic step of a reaction resembles the products of that step more than it does the reactants. Conversely, the structure of the transition state of an exothermic step is more like the reactants than the products.

Additional Problems

4.16

Define and give an example of each of the following.

(a) Homolysis of a covalent bond
(b) Heterolysis of a covalent bond
(c) Bond dissociation energy
(d) A free radical
(e) A carbocation
(f) A carbanion

4.17

Arrange the following free radicals in order of decreasing stability.

$$CH_3\overset{CH_3}{\overset{|}{C}}HCH_2CH_2\cdot \qquad CH_3\overset{CH_3}{\overset{|}{C}}H\underset{\bullet}{C}HCH_3 \qquad CH_3\underset{\bullet}{\overset{CH_3}{\overset{|}{C}}}CH_2CH_3 \qquad \cdot CH_2\overset{CH_3}{\overset{|}{C}}HCH_2CH_3$$

4.18

Arrange the following carbocations in order of decreasing stability:

(a) $CH_3\overset{CH_3}{\overset{|}{C}}HCH_2CH_2^+ \qquad CH_3\overset{CH_3}{\overset{|}{C}}H\underset{+}{C}HCH_3 \qquad CH_3\underset{+}{\overset{CH_3}{\overset{|}{C}}}CH_2CH_3 \qquad ^+CH_2\overset{CH_3}{\overset{|}{C}}HCH_2CH_3$

(b) CH_3, + $\qquad$ CH_3, + $\qquad$ CH_2^+ $\qquad$ CH_3, +

4.19

An alkane with the molecular formula C_8H_{18} yields only a single product with the formula $C_8H_{17}Cl$ when it is subjected to free radical chlorination. Propose a structural formula for the alkane.

4.20

How many isomers with the formula $C_6H_{11}Cl$ would you expect to be formed when methylcyclopentane is chlorinated? Write their formulas.

4.21

How many isomers with the formula $C_6H_{10}Cl_2$ would you expect to be formed when cyclohexane is chlorinated (with excess chlorine)? Write their formulas.

4.22

(a) Bromination of 2-methylbutane gives a predominance of one product. What is it? (b) Chlorination of 2-methylbutane gives (in addition to polychloro compounds) a mixture of monochloro products in the following proportions: 1-chloro-2-methylbutane (34%), 2-chloro-2-methylbutane (22%), 2-chloro-3-methylbutane (28%), and 1-chloro-3-methylbutane (16%). Show how each of these products is formed. (c) Account for the different results obtained from the bromination and chlorination reactions.

4.23

Give an example and sketch a potential energy diagram for:

(a) A reaction in which a covalent bond is broken but no bond is formed.

(b) A reaction in which a covalent bond is formed but no bond is broken
(c) A highly exothermic reaction
(d) A highly endothermic reaction

4.24
The reaction $CH_3CH_3 + Cl_2 \xrightarrow[\text{light}]{\text{heat or}} CH_3CH_2Cl + HCl$ proceeds by a mechanism similar to that for the chlorination of methane. (a) Write equations for the chain-initiating, chain-propagating, and chain-terminating steps. (b) Calculate values of $\Delta H°$ for each of the chain-propagating steps. (c) Ethane undergoes free-radical chlorination more readily than methane. This can be demonstrated by allowing a mixture of equimolar amounts of ethane and methane to react with chlorine; the reaction produces more chloroethane than chloromethane. What factor accounts for this?

4.25
A possible reaction of ethane with chlorine is

$$Cl_2 + CH_3CH_3 \longrightarrow 2CH_3{-}Cl$$

This reaction could conceivably occur by the following chain reaction:

Initiating (1) $Cl{-}Cl \longrightarrow 2Cl\cdot$

Propagating $\begin{cases} (2)\ Cl\cdot + CH_3{-}CH_3 \longrightarrow CH_3Cl + CH_3\cdot \\ (3)\ CH_3\cdot + Cl{-}Cl \longrightarrow CH_3Cl + Cl\cdot \end{cases}$

then (2), (3), (2), (3), etc.

(a) Do you think this overall reaction would compete effectively with the one you gave as answer to problem 4.24? (Base your answer on the $\Delta H°$ and E_{act} for each step.)
(b) Speculate about the possibility of a similar reaction of ethane with fluorine, that is,

Initiating (1) $F{-}F \longrightarrow 2F\cdot$

Propagating $\begin{cases} (2)\ F\cdot + CH_3{-}CH_3 \longrightarrow CH_3{-}F + CH_3\cdot \\ (3)\ CH_3\cdot + F{-}F \longrightarrow CH_3{-}F + F\cdot \end{cases}$

then (2), (3), (2), (3), etc.

4.26
Free-radical fluorination of methane occurs in the absence of light. A mechanism that has been proposed for the dark reaction is

$$CH_4 + F_2 \xrightarrow{\text{slow}} CH_3\cdot + HF + F\cdot$$

$$CH_3\cdot + F\cdot \xrightarrow{\text{fast}} CH_3F$$

(a) Basing your answer on bond dissociation energies, assess the likelihood of the reaction occurring by this mechanism.
(b) What is the likelihood of a similar mechanism occurring when a mixture of methane and chlorine is heated in the dark?

4.27
We saw in Sect. 4.12 that treating isobutane with $FSO_3H{-}SbF_5$ gives hydrogen gas and a solution containing *tert*-butyl cations. A side reaction also takes place in the mixture; this reaction produces methane and isopropyl cations. $CH_3C^+HCH_3$. Propose a mechanism that accounts for the side reaction.

4.28
The carbon-carbon bonds of cyclopropane are sometimes described as being "bent" because overlap of the carbon sp^3 orbitals occurs as shown on the next page. While bent bonds probably relieve some angle strain they also make the carbon-carbon sigma bonds

more susceptible to attack by an electrophile such as a proton. (a) Show how protolysis of cyclopropane by HI could lead to an *n*-propyl cation, $CH_3CH_2CH_2^+$ and (b) show how a propyl cation could stabilize itself by reacting with an iodide ion to form *n*-propyl iodide.

bent bonds of cyclopropane

4.29

Use bond dissociation energies in Table 4.1 to account for the following: (a) Thermal cracking of a C—H bond of methane requires a higher temperature (~1200°) than does a similar breaking of a C—H bond of ethane (500–600°). (b) When ethane undergoes homolysis at high temperatures, the C—C bond breaks more readily than the C—H bonds. (c) When *n*-butane "cracks" the reaction $CH_3CH_2CH_2CH_3 \longrightarrow 2CH_3CH_2\cdot$, occurs more readily than the reaction $CH_3CH_2CH_2CH_3 \longrightarrow CH_3CH_2CH_2\cdot + CH_3\cdot$.

4.30

When propane is heated to a very high temperature it undergoes thermal cracking through homolysis of C—C and C—H bonds. The major products of the reaction are methane and ethene. A chain mechanism has been proposed for this reaction.
(a) Which of the following reactions is most likely to be the major chain-initiating step? Explain your answer by estimating activation energies for each reaction.

$$CH_3CH_2CH_3 \longrightarrow CH_3CH_2\cdot + CH_3\cdot$$
$$CH_3CH_2CH_3 \longrightarrow CH_3CH_2CH_2\cdot + H\cdot$$
$$CH_3CH_2CH_3 \longrightarrow CH_3\dot{C}HCH_3 + H\cdot$$

Possible chain-propagating steps are:

$$(1)\ CH_3\cdot + CH_3CH_2CH_3 \longrightarrow CH_4 + \cdot CH_2CH_2CH_3$$
$$(2)\ \cdot CH_2{-}CH_2{:}CH_3 \longrightarrow CH_2{=}CH_2 + \cdot CH_3$$

(b) Both reactions have reasonably low activation energies (low enough to occur at very high temperatures). Show that this is likely for step 1 by calculating $\Delta H°$ for step 1.
(c) An alternative to step 1 is:

$$CH_3\cdot + CH_3CH_2CH_3 \longrightarrow CH_4 + CH_3\dot{C}HCH_3$$

Comment on the likelihood of this reaction occurring in terms of energy and probability factors.

4.31

Carbocations are powerful Lewis acids. (a) What factor accounts for this? (b) Write the Lewis acid-base reactions that take place when an ethyl cation, $CH_3CH_2^+$, reacts with a chloride ion. (c) With a hydrogen sulfate ion. (d) With a molecule of water. (e) Carbocations often undergo reactions in which they are transformed into alkenes. What is involved here? (f) Illustrate your answer to (e) by showing the reaction in which an ethyl cation is converted to ethene.

***4.32**

Compounds with an oxygen-oxygen single bond—called peroxides—are often used to initiate free radical chain reactions such as alkane halogenations. (a) Examine the bond energies in Table 4.1 and give reasons that will explain why peroxides are especially effective as free-radical initiators. (b) Illustrate your answer by outlining how di-*tert*-butyl peroxide, $(CH_3)_3CO{-}OC(CH_3)_3$, might initiate an alkane halogenation.

*4.33

The reactions below show comparisons between two sets of similar reactions. In each set we compare reactions in which a hydrogen atom is abstracted from methane and from ethane. In the first set **(A)** the abstracting agent is a methyl radical; in the second set **(B)** it is a bromine atom. (a) Sketch potential energy diagrams for each set of reactions that take Hammond's postulate into account. Take care to locate each transition state properly not only along the energy axis but along the reaction coordinate as well. For convenience in making comparisons, you should align the curves so that the potential energies of the reactants are the same. (b) For which reaction will bond-breaking have occurred to the *least* extent when the transition state is reached? (c) To the *greatest* extent? (d) To what approximate extent will bond breaking have occurred in reaction **A**(1)? (e) For which set of reactions will the transition states more resemble products? (f) Notice that the difference in ΔH° for the two sets of reactions is the same (6 kcal/mole). Why is this so? (g) The difference in E_{act} for the first set of reactions is relatively small (2.8 kcal/mole). For the second set of reactions, however, the difference in E_{act} is large (5.0 kcal/mole); it is nearly as large as the difference in ΔH°. Explain.

	Reaction	ΔH° (KCAL/MOLE)	E_{act} (KCAL/MOLE)
(A)	(1) $CH_3\cdot + H{-}CH_3 \longrightarrow CH_3{-}H + \cdot CH_3$	0	14.5
	(2) $CH_3\cdot + H{-}CH_2CH_3 \longrightarrow CH_3{-}H + \cdot CH_2CH_3$	−6.0	11.7
	difference	6.0	2.8
(B)	(1) $Br\cdot + H{-}CH_3 \longrightarrow Br{-}H + \cdot CH_3$	16.5	18.6
	(2) $Br\cdot + H{-}CH_2CH_3 \longrightarrow Br{-}H + \cdot CH_2CH_3$	10.5	13.6
	difference	6.0	5.0

B SPECIAL TOPIC

ELEMENTARY THERMODYNAMICS: $\Delta H°$, $\Delta S°$, AND $\Delta G°$

With a chemical reaction we are usually concerned with two features: *the extent* to which it takes place and *the rate* of the reaction. By the extent of a reaction we mean how much of the reactants will be converted to products when equilibrium is established between them. By the rate of reaction we mean how rapidly will the reactants be converted to products. In simpler terms then, with chemical reactions we are concerned with "how far" and "how fast."

In Chapter 4 we devoted a great deal of attention to the question of how fast reactions occur. We saw that most reactions have an energy barrier between reactants and products and that the most important factor determining how rapidly a reaction takes place is the height of that barrier, the energy of activation, E_{act}.

When we consider the question of how far a reaction goes, one important factor is the relative energy (stability) of the reactants and products. We learned that the spontaneous tendency for reactions to proceed downhill (in energy terms) favors the formation of those products that are downhill. The energy change with which we have been mainly concerned is the enthalpy change, $\Delta H°$. The change in enthalpy is a function largely of the potential energy in chemical bonds. Reactions in which strong bonds are formed at the expense of weak bonds are exothermic. Those in which weak bonds are formed from strong bonds are endothermic. Consider the reaction of methane with chlorine as an example:

$$\underset{(DH° = 104)}{CH_3\text{—}H} + \underset{(DH° = 58)}{Cl\text{—}Cl} \longrightarrow \underset{(DH° = 83.5)}{CH_3\text{—}Cl} + \underset{(DH° = 103)}{H\text{—}Cl}$$

$$\Delta H° = (104 + 58) - (83.5 + 103) = -24.5 \text{ kcal/mole}$$

Because the bonds of CH_3Cl and HCl are collectively stronger than those of CH_4 and Cl_2, the reaction is exothermic. In this case the reaction is highly exothermic; $\Delta H° = -24.5$ kcal/mole.

Many studies of a large number of chemical reactions have shown that reactions that have a negative enthalpy change as large as this almost always go to completion at equilibrium. A *small* negative value for $\Delta H°$, however, is not a guarantee that a reaction will produce substantial proportions of products at equilibrium. To account for this we need to consider two other thermodynamic quantities; $\Delta G°$, the *free-energy change,* and $\Delta S°$, *the entropy change.* To see how these thermodynamic quantities determine the position of equilibrium of a chemical reaction let us consider the general reaction written below:

$$A + B \rightleftarrows C$$

An expression can be written for the equilibrium constant, K_{eq},

$$K_{eq} = \frac{[C]}{[A][B]}$$

where $[A]$ and $[B]$ are the equilibrium concentrations of the reactants A and B in moles/liter, and $[C]$ is the equilibrium concentration of the product C in moles/liter.

By convention, the equilibrium concentration of the products of a reaction are written in the numerator of this expression and those of the reactants are written in the denominator. This means that a large value of K_{eq} is associated with a reaction that goes essentially to completion, and a small value of K_{eq} is associated with a reaction that produces only small amounts of products at equilibrium.

Problem B.1

Consider the simple reaction:

$$X \rightleftarrows Y$$

Assume an initial concentration of X equal to 1.0 mole/liter, and calculate the equilibrium concentration of Y that is formed if the equilibrium constant is (a) 10, (b) 1, (c) 10^{-3}. In each case, what percentage of the reactant is converted to product?

The equilibrium constant for a reaction is directly related to the thermodynamic quantity called the free-energy change for the reaction. The free-energy change is symbolized as $\Delta G°$ and the relationship between $\Delta G°$ and the equilibrium constant, K_{eq}, is:

$$\log K_{eq} = \frac{\Delta G°}{-2.3RT}$$

where R is the gas constant (1.986 cal/degree-mole) and T is the absolute temperature.

It is easy to show with this equation that *the more negative $\Delta G°$ is, the larger will be the value of the equilibrium constant, K_{eq}. And the larger the value for K_{eq}, of course, the more the reaction will favor the formation of products.* Table B.1 gives the results of several calculations for the simple equilibrium X $\rightleftarrows$ Y.

What Table B.1 shows us is that a relatively modest negative $\Delta G°$ (e.g., -4.1 kcal/mole) means that the product will be the major component (99.9%) present at equilibrium. For a reaction X $\rightleftarrows$ Y, it means for all practical purposes that the reaction goes to completion.

Problem B.2

Table B.1 is especially useful when applied to equilibria involving two forms of a cyclic molecule such as those we studied in Chapter 3. (a) The free energy change $\Delta G°$, for the boat $\rightleftarrows$ chair form of cyclohexane is -5 to -6 kcal/mole. What does this mean about the relative population of the two forms at equilibrium? (b) For the axial $\rightleftarrows$ equatorial form of ethylcyclohexane $\Delta G° \simeq -1.8$ kcal/mole; what percentage of the axial form is present at equilibrium?

The free-energy change ($\Delta G°$) and the enthalpy change ($\Delta H°$) are related to one another through another equation that involves a third thermodynamic quantity, the entropy change ($\Delta S°$), and the absolute temperature, T. This important equation is:

$$\Delta G° = \Delta H° - T\Delta S°$$

The entropy of any system is a measure of the relative order or randomness of that system. The more random (less ordered) a system is, the greater is its entropy. Thus, a *positive* entropy change ($+\Delta S°$) is always associated with a change from a more ordered arrangement to a more random one. A negative entropy change ($-\Delta S°$) always accompanies the reverse process.

TABLE B.1 Relation between $\Delta G°$ and K_{eq} at 25° for Reactions of the Type X $\rightleftarrows$ Y

$\Delta G°$, KCAL/MOLE	−0.41	−0.65	−0.82	−0.95	−1.4	−1.8	−2.7	−4.1	−5.5
K_{eq}	2	3	4	5	10	20	100	1000	10,000
Percentage of Y at equilibrium	67	75	80	83	91	95	99	99.9	99.99

Adapted from E. Eliel, Stereochemistry of Carbon Compounds, Wiley, N. Y., (1962), p. 207.

For a chemical system the relative order (or randomness) of the molecules can be related to the number of *degrees of freedom* available to the molecules and their constituent atoms. Degrees of freedom are associated with ways in which *movement* or *changes in relative position* can occur. Molecules have three sorts of degrees of freedom: translational degrees of freedom associated with movements of the whole molecule through space, rotational degrees of freedom associated with the tumbling motions of the molecule, and vibrational degrees of freedom associated with the stretching and bending motion of atoms about the bonds that connect them.

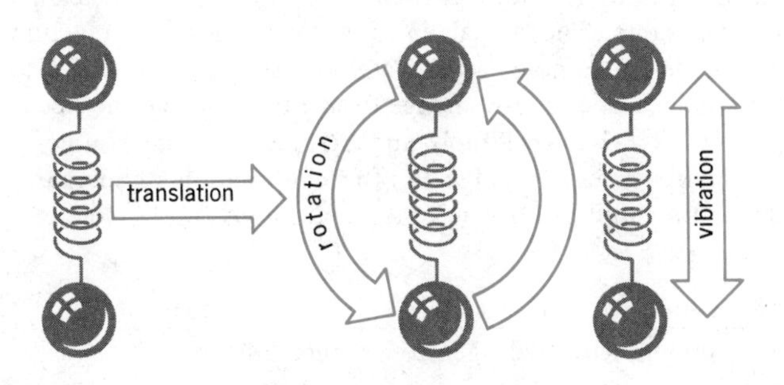

If the atoms of the products of a reaction have more degrees of freedom available than they did as reactants, the entropy change for the reaction will be positive. If, on the other hand, the atoms of the products are more constrained (have fewer degrees of freedom) than they were as reactants, a negative $\Delta S°$ will result.

With these ideas and relationships in mind, we are now in a position to understand how these three thermodynamic quantities, $\Delta H°$, $\Delta G°$, and $\Delta S°$, account for the course of a chemical reaction. For reactions occurring near room temperature*, if $\Delta H°$ has a large negative value ($\Delta H° > -15$ kcal/mole), a small negative entropy change (associated with the atoms being more constrained in the products) will not generally result in a positive free-energy change ($+\Delta G°$). For these reactions $\Delta G°$ will be negative and the formation of products at equilibrium will be favored:

$$\Delta G° = \Delta H° - [T \times \Delta S°]$$
$$\Delta G° = \text{(large negative quantity)} - [\text{(positive quantity)} \times \text{(small negative quantity)}]$$
$$\Delta G° = \text{(large negative quantity)} - \text{(smaller negative quantity)}$$
$$\Delta G° = \text{negative quantity}$$

We can now see how, in some reactions that are endothermic (have a $+\Delta H$), the formation of products at equilibrium will still be favored. These reactions have an accompanying positive entropy change large enough to offset the unfavorable enthalpy change. This can occur when the atoms of the products are much less constrained than those of the reactants. In such cases the large positive entropy term $(T) \times (+\Delta S°)$ more than compensates for the unfavorable $+\Delta H°$.

$$\Delta G° = \Delta H° - [T \times \Delta S°]$$
$$\Delta G° = \text{(small positive quantity)} - [\text{(positive quantity)} \times \text{(large positive quantity)}]$$
$$\Delta G° = \text{(small positive quantity)} - \text{(larger positive quantity)}$$
$$\Delta G° = \text{negative quantity}$$

For many reactions the entropy change is so small that the term $T\Delta S°$ is almost zero and $\Delta H° \simeq \Delta G°$. This happens when the reaction is one in which the number of degrees of freedom available to molecules of reactants and to molecules of products is

* At very high temperatures the entropy term ($T\Delta S°$) in the equation $\Delta G° = \Delta H° - T\Delta S°$, becomes much more important because T becomes very large. Most of the reactions that we will consider, however, are carried out between 250 K and 450 K. Under these conditions $\Delta H°$ is the important term if it is more negative than -15 kcal/mole.

essentially the same. Consider the reaction of methane and chlorine again:

$$CH_4 + Cl_2 \longrightarrow CH_3Cl + HCl$$

Here, two moles of product molecules are formed from the same number of moles of reactant molecules. Thus the number of translational degrees of freedom available to products and reactants will be approximately the same. Moreover, CH_3Cl is a tetrahedral molecule like CH_4 and HCl is a diatomic molecule like Cl_2. This means that vibrational and rotational degrees of freedom available to products and reactants should also be approximately the same. The actual entropy change for this reaction is quite small, $\Delta S° = +0.67$ cal deg K^{-1} $mole^{-1}$. Thus at room temperature (298 K) the $T\Delta S$ term is only +0.2 kcal/mole. The enthalpy change for the reaction and the free energy change are almost equal, $\Delta H° = -24.5$ kcal/mole and $\Delta G° = -24.7$ kcal/mole.

Since the free energy change for the chlorination of methane is such a large negative number the equilibrium constant is enormous, $K_{eq} \simeq 10^{18}$.

Problem B.3

The reaction of ethyne with hydrogen to produce ethene,

$$H—C{\equiv}C—H_{(g)} + H_{2(g)} \longrightarrow CH_2{=}CH_{2(g)}$$

has the following values of $\Delta H°$ and $\Delta S°$ at 27 °C (300 K): $\Delta H° = -41.7$ kcal/mole, $\Delta S° = -26.6$ cal deg K^{-1} $mole^{-1}$. (a) Calculate the value of $\Delta G°$ for this reaction. (Remember: 1000 cal = 1.0 kcal.) (b) Would you expect the position of equilibrium to favor the formation of ethene? (c) Does the entropy change for this reaction, taken alone, favor the formation of ethene? (d) How can you account for the negative entropy change?

The reaction in Problem B.3 takes place in the gas phase and is therefore, relatively simple. For reactions taking place in solution the situation is much more complex because an analysis must also take into account enthalpy and entropy changes of the solvent.

Remember: a negative free-energy change for a reaction only ensures that a substantial proportion of products will be formed *when the reaction reaches equilibrium.* The overall free-energy change for a reaction tells us nothing, however, about how long that particular reaction will take to reach equilibrium. The reaction of hydrogen and oxygen to produce water, for example,

$$H_2 + O_2 \longrightarrow 2H_2O$$

has associated with it a very large negative free-energy change. In the absence of a catalyst (or applied heat), however, mixtures of hydrogen and oxygen react so slowly that the formation of water is essentially imperceptible.

For many other reactions (Fig. B.1) pathways to several different products may have favorable free-energy changes. In such cases, the products that we actually obtain in greatest proportions are the ones that come from pathways with lowest energies of

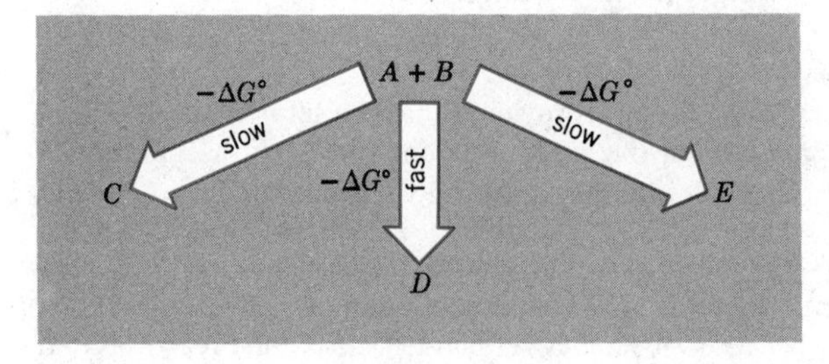

FIG. B.1
The hypothetical reaction of A *and* B *to form different products,* C, D, *and* E. *All of the reactions have favorable free-energy changes.* D *is the product that is actually obtained, however, because it is produced by the reaction that proceeds most rapidly.* E *and* C *are not isolated in appreciable amounts because the reactions by which they are formed are slow.*

activation. They will form the most rapidly. We can see from this that a knowledge of the factors that determine reaction rates is of considerable importance as well.

An example of the kind of reaction illustrated in Figure B.1 is the bromination of 2-methylbutane. The major product (by far) is 2-bromo-2-methylbutane because the energy of activation for the hydrogen abstraction step in its formation is lowest.

$$\mathrm{CH_3\overset{\overset{CH_3}{|}}{C}HCH_2CH_3 \xrightarrow[h\nu]{Br_2} \underset{(93\%)}{CH_3\underset{\underset{Br}{|}}{\overset{\overset{CH_3}{|}}{C}}CH_2CH_3} + \underset{(6\%)}{CH_3\overset{\overset{CH_3}{|}}{C}H\underset{\underset{Br}{|}}{C}HCH_3} + \underset{(1\%)}{CH_3\overset{\overset{CH_3}{|}}{C}HCH_2CH_2Br}}$$

5 CHEMICAL REACTIVITY II: AN INTRODUCTION TO NUCLEOPHILIC SUBSTITUTION AND ELIMINATION REACTIONS OF ALKYL HALIDES

5.1 ALKYL HALIDES IN ORGANIC SYNTHESIS

Alkyl halides are extremely useful compounds in *organic synthesis*. When, for one reason or another, we find ourselves in need of an organic compound that is not available in the stockroom, or perhaps even of one that has never been made before, the task of organic synthesis starts. We will have the job of making the compound we need from other compounds that are available. Very often, as we do this, we will find ourselves turning to an alkyl halide as one of our starting compounds. There are several reasons for this.

Alkyl halides are readily available; they can be synthesized from other compounds in many specific ways. The halogen substituents of alkyl halides are easily replaced by other groups in *substitution reactions* and halogen substituents offer us the opportunity to introduce multiple bonds between carbon atoms through *elimination reactions.*

We will discuss how alkyl halides are made in detail later (Chapter 14). Here we shall examine their highly useful substitution and elimination reactions.

5.2 NUCLEOPHILIC SUBSTITUTION REACTIONS

There are many reactions of the general type shown below.

$$\underset{\text{Nucleophile}}{Nu{:}^-} + \underset{\text{Alkyl halide}}{R{-}\ddot{\underset{..}{X}}{:}} \longrightarrow \underset{\text{Product}}{R{-}Nu} + \underset{\text{Halide ion}}{{:}\ddot{\underset{..}{X}}{:}^-}$$

Following are some examples.

$$H\ddot{\underset{..}{O}}{:}^- + CH_3{-}\ddot{\underset{..}{Cl}}{:} \longrightarrow CH_3{-}\ddot{\underset{..}{O}}H + {:}\ddot{\underset{..}{Cl}}{:}^-$$

$$CH_3\ddot{\underset{..}{O}}{:}^- + CH_3CH_2{-}\ddot{\underset{..}{Br}}{:} \longrightarrow CH_3CH_2{-}\ddot{\underset{..}{O}}CH_3 + {:}\ddot{\underset{..}{Br}}{:}^-$$

$${:}\ddot{\underset{..}{I}}{:}^- + CH_3CH_2CH_2{-}\ddot{\underset{..}{Cl}}{:} \longrightarrow CH_3CH_2CH_2{-}\ddot{\underset{..}{I}}{:} + {:}\ddot{\underset{..}{Cl}}{:}^-$$

In this type of reaction a *nucleophile, a species with an unshared electron pair,* reacts with an alkyl halide by replacing the halogen substituent. A *substitution reaction* takes place and the halogen substituent leaves as a halide ion. Because the substitution reaction is initiated by a nucleophile it is called a *nucleophilic substitution reaction.*

Later in this chapter when we examine the mechanisms of nucleophilic substitution reactions (and elimination reactions) we will find them quite unlike the free-radical reactions of alkanes. The reactions that we will study here are *ionic reactions*. They involve *heterolytic* cleavage of covalent bonds and the reaction of and formation of *ions,* not radicals. They take place in solution (not in the gas phase) and the ability of the solvent to stabilize ions is an important aspect of these reactions.

5.2A Nucleophiles

The word nucleophile comes from *nucleus* (the positive part of an atom) plus *phile* from the Greek word *philein* meaning to love. Thus a nucleophile is a reagent that "loves" or "seeks" a positive center in an organic molecule. In an alkyl halide the positive center is the carbon to which the halogen is attached. This carbon bears a partial positive charge because the electronegative halogen pulls the electrons of the carbon-halogen bond in its direction.

$$\overset{\delta+}{-\mathrm{C}-} \longrightarrow \overset{\delta-}{\mathrm{:\ddot{X}:}}$$

This is the positive center that the nucleophile seeks

The electronegative halogen polarizes the C—X bond

Any molecule or negative ion that has an unshared pair of electrons is a potential nucleophile. The hydroxide ion, for example, is a nucleophile and it can react with an alkyl halide to form an alcohol.

$$\mathrm{HO:^-} + \mathrm{R{-}X} \longrightarrow \mathrm{R{-}OH} + \mathrm{:X^-}{}^*$$

Specific example:

$$\mathrm{HO:^-} + \mathrm{CH_3CH_2Br} \longrightarrow \mathrm{CH_3CH_2OH} + \mathrm{:Br^-}$$

Although nucleophiles are often negatively charged ions, they do not have to be. Neutral molecules with unshared electron pairs act as nucleophiles as well. Water molecules can act as nucleophiles, for example, and water can react with an alkyl halide to yield an alcohol.

$$\underset{\text{Nucleophile}}{\mathrm{H{-}\underset{|}{\overset{}{O}}:}\atop \mathrm{H}} + \underset{\text{Alkyl halide}}{\mathrm{R{-}X}} \longrightarrow \mathrm{R{-}\underset{\substack{|\\ \mathrm{H}}}{O^+}{-}H} + \mathrm{:X^-}$$

$$\Big\updownarrow \mathrm{H_2O}$$

$$\mathrm{R{-}OH} + \mathrm{H_3O^+} + \mathrm{:X^-}$$

Specific example:

$$\mathrm{H{-}\underset{\substack{|\\ \mathrm{H}}}{O}:} + \mathrm{(CH_3)_3C{-}Cl} \longrightarrow \mathrm{(CH_3)_3C{-}\underset{\substack{|\\ \mathrm{H}}}{O^+}{-}H} + \mathrm{:Cl^-}$$

$$\Big\updownarrow \mathrm{H_2O}$$

$$\mathrm{(CH_3)_3C{-}OH} + \mathrm{H_3O^+} + \mathrm{:Cl^-}$$

In this reaction the first product is an alkyloxonium ion, $\mathrm{R{-}\underset{\substack{|\\ \mathrm{H}}}{O^+}{-}H}$, which

* A hydroxide ion actually has three unshared electron pairs, not just one, as we have shown in this reaction. A halide ion actually has four. They are $\mathrm{H\ddot{\underset{..}{O}}:^-}$ and $\mathrm{:\ddot{\underset{..}{X}}:^-}$, respectively. But, since only one electron pair is important in this reaction, it is convenient to write the reaction the way we have. In the future we will often neglect to write in all of the unshared electron pairs. When we do this, it does not mean that they are not there; it just means that we have left them out because they were not important in the reaction being illustrated.

then loses a proton to a water molecule to form an alcohol. Alkyloxonium ions are like hydronium ions, $H—\overset{+}{\underset{|}{O}}—H$ (with H below O), and just as hydronium ions can donate protons to bases, so too can alkyloxonium ions. Just as hydronium ions are "protonated" water molecules, alkyloxonium ions of the type $R—\overset{+}{\underset{|}{O}}—H$ (with H below O) are often called "protonated alcohols." Both hydronium ions and alkyloxonium ions are strong Brønsted acids.

Problem 5.1

Write an electron dot structure for each of the following molecules and ions that shows that each is a potential nucleophile.

(a) Ethyl alcohol, C_2H_5OH
(b) Ethoxide ion, $C_2H_5O^-$
(c) Ammonia, NH_3
(d) Methylamine, CH_3NH_2
(e) Cyanide ion, CN^-
(f) Acetic acid, CH_3COOH
(g) Acetate ion, CH_3COO^-
(h) Formic acid, $HCOOH$
(i) Formate ion, $HCOO^-$
(j) Ethanethiol, C_2H_5SH
(k) Ethanethiolate ion, $C_2H_5S^-$
(l) Azide ion, N_3^-

5.2B Leaving Groups

Alkyl halides are not the only substances whose molecules undergo nucleophilic substitution reactions. We will see later that other compounds can also react in the same way. To be reactive—that is, to be able to act as the *substrate* in a nucleophilic substitution reaction—a molecule must have a good *leaving group*. In alkyl halides the leaving group is the halogen substituent—it leaves as a halide ion. To be a good leaving group the substituent must be able to leave as a relatively stable, weakly basic molecule or ion. Because halide ions are relatively stable and are very weak bases, they are good leaving groups. Other groups can function as good leaving groups as well. We can write more general equations for nucleophilic substitution reactions using L to represent a leaving group.

$$Nu:^- + R—L \longrightarrow R—Nu + :L^-$$

or

$$Nu: + R—L \longrightarrow R—Nu^+ + :L^-$$

Specific examples:

$$HO:^- + CH_3—Cl \longrightarrow CH_3—OH + :Cl^-$$

$$H_3N: + CH_3—Br \longrightarrow CH_3—NH_3^+ + :Br^-$$

Later we will also see reactions where the substrate bears a positive charge and a reaction like the following take place.

$$Nu: + R—L^+ \longrightarrow R—Nu^+ + :L$$

Specific example:

$$CH_3{-}\underset{\substack{|\\H}}{\ddot{O}}{:} + CH_3{-}\underset{\substack{|\\H}}{O^+}{-}H \longrightarrow CH_3{-}\underset{\substack{|\\H}}{O^+}{-}CH_3 + {:}\underset{\substack{|\\H}}{\ddot{O}}{-}H$$

5.3 MOLECULARITY OF NUCLEOPHILIC SUBSTITUTION REACTIONS: REACTION KINETICS

Nucleophilic substitution reactions will be more understandable and useful if we know something about their mechanisms. How does the nucleophile replace the leaving group? Does the reaction take place in one step, or is more than one step involved? If more than one step is involved, what kind of intermediates are formed? Which steps are fast and which are slow?

5.3A Multistep Reactions and the Rate Limiting Step

If a reaction takes place in a series of steps, one step will usually be intrinsically slower than all the others and the rate of the overall reaction will be essentially the same as the rate of this slow step. This slow step, consequently, is called the *rate-limiting step* or the *rate-determining* step.

Consider a multistep reaction such as the following. (We will see a specific example in Section 5.6.)

1. Reactant $\xrightarrow{\text{slow}}$ Intermediate-1

2. Intermediate-1 $\xrightarrow{\text{fast}}$ Intermediate-2

3. Intermediate-2 $\xrightarrow{\text{fast}}$ Product

Here, the first step is the slow step. The rate of the overall reaction will be essentially the same as the rate of this first step; it will be *limited* (or *determined*) by the rate at which intermediate-1 forms from the reactant.

When we say that the first step is the slow step, we mean that the rate constant for step **1** is very much smaller than the rate constant for step **2** or for step **3:**

Step **1.** Rate = $\mathbf{k_1}$ [Reactant]
Step **2.** Rate = $\mathbf{k_2}$ [Intermediate-1]
Step **3.** Rate = $\mathbf{k_3}$ [Intermediate-2]

$$\mathbf{k_1} \ll \mathbf{k_2} \text{ or } \mathbf{k_3}$$

When we say that steps **2** and **3** are *fast,* we mean that because their rate constants are larger, they could (in theory) take place rapidly if the concentrations of the two intermediates ever became high. In actuality, the concentrations of the intermediates are always very small because of the slowness of step **1,** and steps **2** and **3** actually occur at the same rate as step **1.**

An analogy may help clarify this. Imagine an hour glass modified in the way shown in Figure 5.1. The opening between the top chamber and the one just below is considerably smaller than the other two. The overall rate at which sand falls from the top to the bottom of the hour glass is limited by the rate at which sand passes through this small orifice. This step, in the passage of sand, is analogous to the rate-limiting step of the multistep reaction.

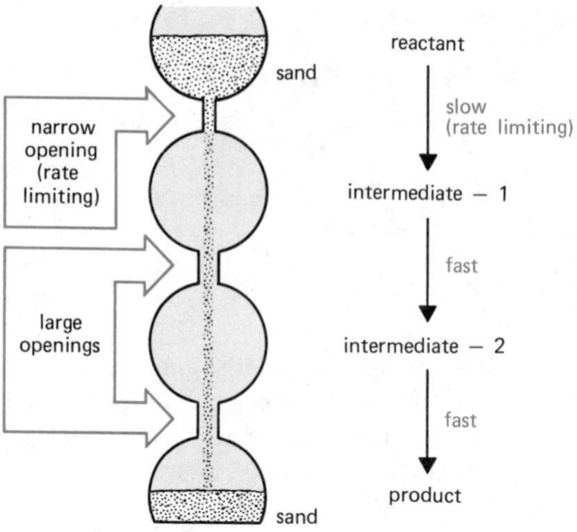

narrow opening (rate limiting)
large openings
sand
sand
reactant
slow (rate limiting)
intermediate − 1
fast
intermediate − 2
fast
product

FIG. 5.1
A modified hour glass that serves as an analogy for a multistep reaction. The overall rate is limited by the rate of the slow step.

5.3B The Molecularity of a Reaction

Of primary importance in formulating a mechanism for a reaction is knowledge of the *molecularity of the reaction*. The molecularity of a reaction is *the number of molecules or ions that participate in the transition state of the rate-limiting step of its mechanism*. If only one molecule or ion is involved in the transition state of the slow step, the reaction is said to be *unimolecular*. If two species are involved the reaction is *bimolecular*, and if three are involved the reaction is *termolecular*.

5.3C Reaction Kinetics

The most common way to determine the molecularity of a chemical reaction is through a study of reaction *kinetics*. That is, we measure the way in which *the rate of the reaction varies as we vary the concentration of each reactant*.

Let us assume that we are studying a reaction of the type:

$$\mathrm{Nu{:}^- + R{-}X \longrightarrow Nu{-}R + X{:}^-}$$

Let us assume, further, that we carry out the same reaction several times *at the same temperature*, but with *different initial concentrations of Nu:⁻ and R—X*. In each case we will measure the rate at which the reaction takes place by measuring the rate *soon after each reaction begins*.* We can do this in several ways. We can withdraw small samples from the reaction mixture at specific intervals of time and determine the concentrations of reactants or products. The rate of the reaction will be equal to the rate at which $\mathrm{Nu{:}^-}$ or R—X disappear from the reaction mixture or to the rate at which $\mathrm{X{:}^-}$ and Nu—R are formed in the reaction mixture. Just which quantity we determine is often a matter of experimental convenience. We can, for example, measure the rate of disappearance of halide ions by precipitating them with silver nitrate and weighing the precipitate. We can measure the rate of disappearance of hydroxide ions by titrating the samples with acid.

Now let us suppose that by using one of these methods we measure the rates of several reactions of a specific $\mathrm{Nu{:}^-}$ and a specific R—X with different initial

* We want to obtain the initial reaction rate because as the reaction progresses, the concentrations of the reactants will decrease.

concentrations of Nu:⁻ and R—X. According to the collision theory, the different rates that we obtain must arise from different collision frequencies. When the reacting molecules are more concentrated, they are more crowded, and thus they collide with each other more frequently. The different rates cannot arise from different energy factors because we carried out all of the reactions at the same temperature. They cannot arise from different orientation factors because the reacting species are the same in each reaction and the fraction of collisions having the proper orientation will be unchanged.

If we find that the reaction rate is directly proportional to the concentration of Nu:⁻ and to the concentration of R—X, it is reasonable to assume that Nu:⁻ and R—X collide with each other in order to bring about a reaction and *that both Nu:⁻ and R—X are involved in the transition state of the rate-limiting step*. We can assume, therefore, that the reaction is *bimolecular*.

We can describe this bimolecular reaction with equations. Since the rate is directly proportional to the concentration of Nu:⁻ and to the concentration of R—X, we can write the following proportionality:

$$\text{Rate} \propto [\text{Nu:}^-]\,[\text{R—X}]$$

We can make this proportionality into an equation—called the *reaction rate equation*—by including a constant, k (called the *rate constant*).

$$\text{Rate} = k[\text{Nu:}^-]\,[\text{R—X}]$$

A reaction rate equation like this is said to be *first order* with respect to both [Nu:⁻] and [R—X]. That is, the reaction rate is proportional to [Nu:⁻] raised to the first power and to [R—X] raised to the first power. However, since the reaction rate is proportional to the first powers of two concentrations, that is, [Nu:⁻] and [R—X], the reaction rate is said to be *second order* overall.* The kinds of numbers that we might get from a kinetic study of this reaction are shown in Table 5.1.

TABLE 5.1 Hypothetical Second-Order Reaction

$$\text{Nu:}^- + \text{R—X} \longrightarrow \text{R—Nu} + \text{X}^-$$

EXPT. NO.	INITIAL CONCENTRATION		INITIAL RATE OF FORMATION OF
	Nu:⁻	R—X	R—Nu OR X⁻
(1)	1 mole/liter	1 mole/liter	0.01 mole/liter sec⁻¹
(2)	2 mole/liter	1 mole/liter	0.02 mole/liter sec⁻¹
(3)	1 mole/liter	2 mole/liter	0.02 mole/liter sec⁻¹
(4)	2 mole/liter	2 mole/liter	0.04 mole/liter sec⁻¹

* In general the overall order of a reaction is equal to the sum of the exponents a and b in the rate equation.

$$\text{Rate} = k[\text{A}]^a[\text{B}]^b$$

If in some other reaction, for example, we found that the

$$\text{Rate} = k[\text{A}]^2[\text{B}]$$

then we would say that the reaction rate is second order with respect to [A], first order with respect to [B], and third order overall.

5.4 KINETICS OF NUCLEOPHILIC SUBSTITUTION REACTIONS

Let us now examine two actual nucleophilic substitution reactions: the reaction of methyl chloride (1) and of *tert*-butyl chloride (2) with hydroxide ion. We will find that these two similar reactions have quite different kinetics and different mechanisms:

1. $CH_3{-}Cl + OH^- \xrightarrow[H_2O]{} CH_3{-}OH + Cl^-$
2. $(CH_3)_3C{-}Cl + OH^- \xrightarrow[H_2O]{} (CH_3)_3C{-}OH + Cl^-$

5.4A The Reaction of Methyl Chloride: An S_N2 Reaction

When methyl chloride reacts with aqueous sodium hydroxide, experiments have shown that the rate depends on the concentrations of both methyl chloride and hydroxide ion. When one *doubles* the concentration of methyl chloride while keeping the concentration of the hydroxide ion constant, the rate of the reaction doubles. When one *doubles* the concentration of hydroxide ion while keeping the concentration of methyl chloride constant, the rate *doubles*. And, finally, when one simultaneously *doubles both concentrations* (hydroxide ion and methyl chloride) the rate increases four times. (In all of these experiments the temperature is the same.) Thus the rate equation for the reaction must be first order with respect to methyl chloride, first order with respect to hydroxide ion, and second order overall.

$$CH_3{-}Cl + OH^- \xrightarrow[H_2O]{} CH_3{-}OH + Cl^-$$

$$\text{Rate} \propto [CH_3Cl]\ [OH^-]$$

$$\text{Rate} = k[CH_3Cl]\ [OH^-]$$

We can conclude, therefore, that the transition state for the rate-limiting step involves both a methyl chloride molecule and a hydroxide ion, and that the reaction is *bimolecular*. We call this kind of reaction an S_N2 reaction, meaning Substitution, Nucleophilic, bimolecular.

5.4B The Reaction of *Tert*-Butyl Chloride: An S_N1 Reaction

When *tert*-butyl chloride reacts with aqueous sodium hydroxide the kinetic results are quite different. The rate of formation of *tert*-butyl alcohol is dependent on the concentration of *tert*-butyl chloride, but it is *independent of the concentration of hydroxide ion*. Doubling the *tert*-butyl chloride concentration *doubles* the rate of the reaction. But changing the hydroxide ion concentration (within limits) has no appreciable effect. *Tert*-butyl chloride reacts by substitution at virtually the same rate in pure water (where the hydroxide ion is 10^{-7} M) as it does in 0.05 M aqueous sodium hydroxide (where the hydroxide ion concentration is 500,000 times larger). (We will see in Section 5.6 that the important nucleophile in this reaction is a molecule of water.)

Thus the rate equation for this substitution reaction is first order with respect to *tert*-butyl chloride and *first-order overall.*

$$(CH_3)_3C{-}Cl + OH^- \xrightarrow[H_2O]{} (CH_3)_3C{-}OH + Cl^-$$

$$\text{Rate} \propto [(CH_3)_3CCl]$$

$$\text{Rate} = k[(CH_3)_3CCl]$$

We can conclude, therefore, that hydroxide ions do not participate in the transition state of the rate-limiting step, and that only molecules of *tert*-butyl chloride are involved. This reaction is *unimolecular*. We call this type of reaction an S_N1 reaction (Substitution, Nucleophilic, unimolecular).

How can we account for these results in terms of reaction mechanisms? Let us begin with the S_N2 reaction.

5.5 A MECHANISM FOR THE S_N2 REACTION

One possible mechanism for the S_N2 reaction—one that is based on a mechanism proposed in the 1930s by Sir Christopher Ingold* of the University College, London—is outlined below.

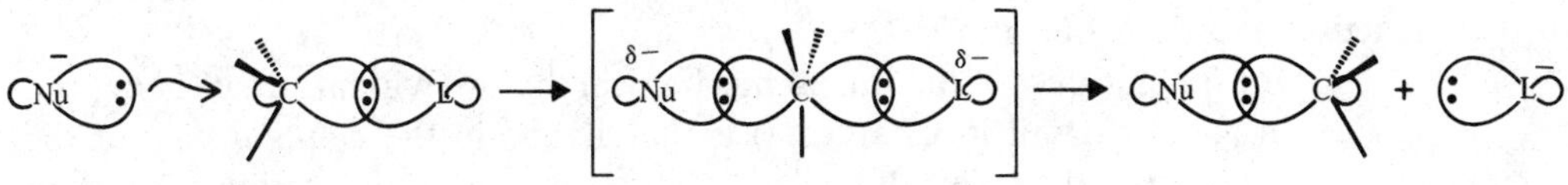

$$\overset{-}{Nu}: + C{-}L \longrightarrow \left[\overset{\delta-}{Nu} \cdots C \cdots \overset{\delta-}{L} \right] \longrightarrow Nu{-}C + :\overset{-}{L}$$

transition state

According to this mechanism the nucleophile approaches the carbon bearing the leaving group from the *back side*. It attacks from the side directly opposite the leaving group. The orbital that contains the electron pair of the nucleophile begins to overlap with the small back lobe of an orbital of the carbon bearing the leaving group. As the reaction progresses the lobe between the nucleophile and carbon atom grows and the lobe between the carbon atom and the leaving group shrinks. As this happens the leaving group is pushed away. The formation of the bond between the nucleophile and the carbon provides most of the energy necessary to break the bond between the carbon atom and the leaving group. We can represent this mechanism with methyl chloride and hydroxide ion in the following way.

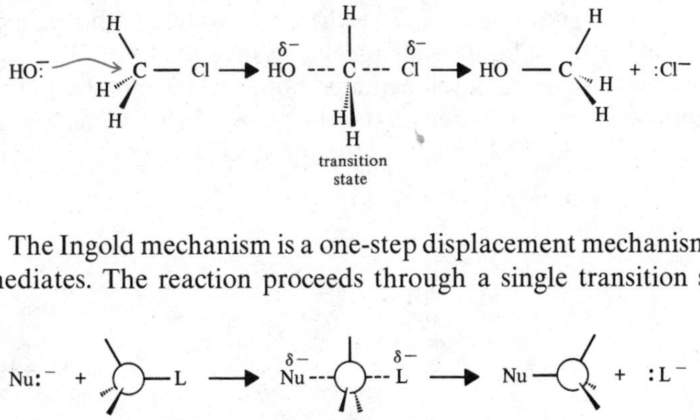

$$HO^{-} + H_3C{-}Cl \longrightarrow \overset{\delta-}{HO} \cdots CH_3 \cdots \overset{\delta-}{Cl} \longrightarrow HO{-}CH_3 + :Cl^{-}$$

transition state

The Ingold mechanism is a one-step displacement mechanism. There are no intermediates. The reaction proceeds through a single transition state:

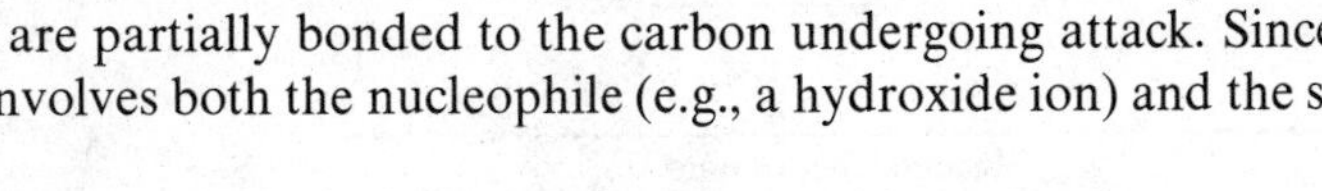

$$Nu:^{-} + C{-}L \longrightarrow \overset{\delta-}{Nu} \cdots C \cdots \overset{\delta-}{L} \longrightarrow Nu{-}C + :L^{-}$$

transition state
(not an intermediate)

The transition state is one in which both the nucleophile and the leaving group are partially bonded to the carbon undergoing attack. Since this transition state involves both the nucleophile (e.g., a hydroxide ion) and the substrate (e.g., a

* Ingold and his co-workers were the pioneers in this field. Their work provided the foundation on which our understanding of nucleophilic substitution and elimination is built.

molecule of methyl chloride), this mechanism accounts for the second-order reaction kinetics that we observe.

A potential energy diagram for the reaction of methyl chloride with hydroxide ion is shown in Figure 5.2. The energy of activation is about 26 kcal/mole. This is a relatively high energy of activation when compared with those of the free-radical alkane halogenations that we studied in Chapter 4. Remember, however, that alkane halogenations are *chain reactions;* their individual steps must occur very rapidly because otherwise the chains will be terminated. A nonchain reaction like that of methyl chloride with hydroxide can afford to have a much higher energy of activation. One as high as 26 kcal/mole does not prevent the reaction from taking place at a reasonable rate. The reaction of methyl chloride with hydroxide ion will be essentially complete in a matter of several hours if the reaction is carried out at 50°.

One other aspect of the Ingold mechanism that we will introduce here (and discuss in greater detail in Chapter 8) is the change in the *configuration* of the carbon atom that is the object of nucleophilic attack. The configuration of an atom *is the particular arrangement of groups around that atom in space.* We notice in the Ingold mechanisms that as the displacement takes place the configuration of the carbon atom under attack *inverts*—it is turned inside out in much the same way that an umbrella is turned inside out, or inverts, when caught in a strong wind.

$$HO:^- + H_3C{-}Cl \longrightarrow \overset{\delta-}{HO}\,{-}{-}{-}\,CH_3\,{-}{-}{-}\,\overset{\delta-}{Cl} \longrightarrow HO{-}CH_3 + :Cl^-$$

an inversion of configuration

With a molecule like methyl chloride, however, there is no way to prove that attack by the nucleophile inverts the configuration of the carbon atom because one form of methyl chloride is identical to its inverted form. With a cyclic molecule like *cis*-1-chloro-3-methylcyclopentane, however, we can observe the results of a *configuration inversion.* When *cis*-1-chloro-3-methylcyclopentane reacts with hydroxide ion in an S_N2 reaction the product is *trans*-3-methylcyclopentanol.

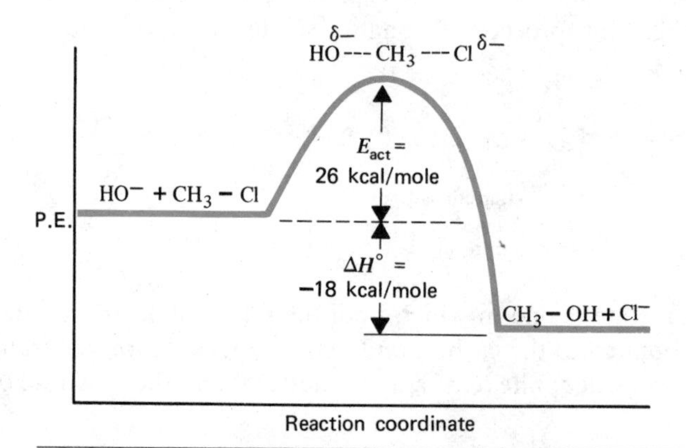

FIG. 5.2

A potential energy diagram for the reaction of methyl chloride with hydroxide ion.

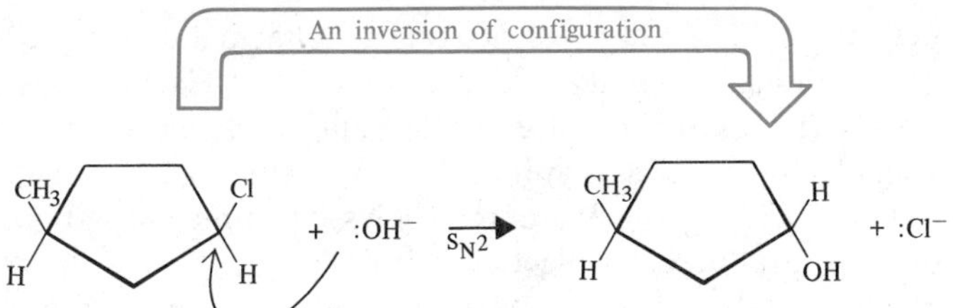

An inversion of configuration

$$\text{cis-1-chloro-3-methylcyclopentane} + :OH^- \xrightarrow{S_N2} \text{trans-3-methylcyclopentanol} + :Cl^-$$

cis–1–chloro–3–methylcyclopentane

trans–3–methylcyclopentanol

Presumably, the transition state for this reaction is like that shown below.

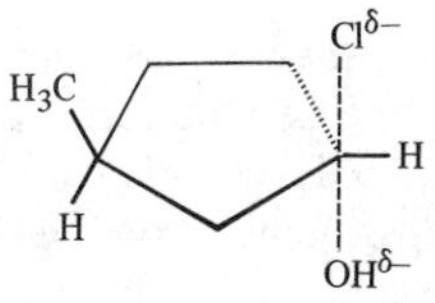

H_3C, H, $Cl^{\delta-}$, H, $OH^{\delta-}$

Problem 5.2

What product would result from the reaction just given, if attack by the hydroxide ion had occurred from the same side as the leaving group—that is, what product would have been formed if no inversion of configuration had taken place?

5.6 A MECHANISM FOR THE S_N1 REACTION

The mechanism for the reaction of *tert*-butyl chloride (Section 5.4) apparently involves three steps. Two distinct *intermediates* are formed. The first step is the slow step—it is the rate-limiting step. In it a molecule of *tert*-butyl chloride ionizes and becomes a *tert*-butyl cation and a chloride ion. Carbocation formation in general takes place slowly.

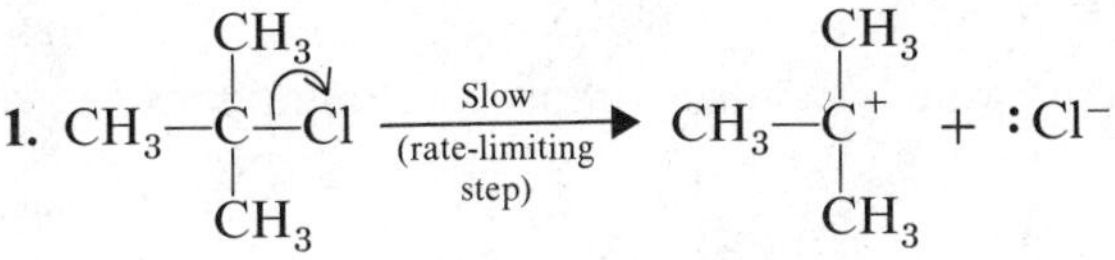

$$\text{1. } (CH_3)_3C{-}Cl \xrightarrow[\text{(rate-limiting step)}]{\text{Slow}} (CH_3)_3C^+ + :Cl^-$$

The next two steps are the following:

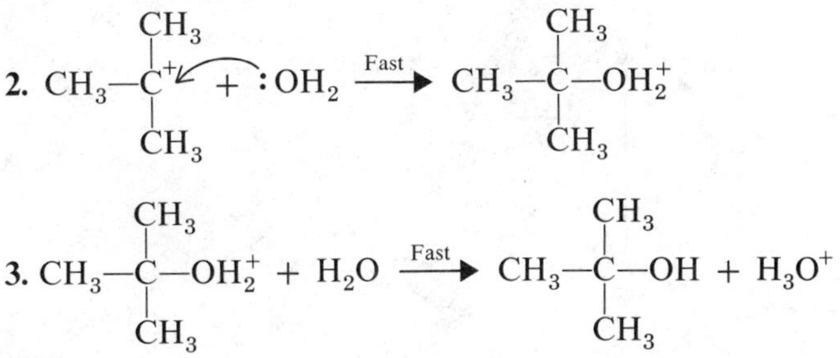

$$\text{2. } (CH_3)_3C^+ + :OH_2 \xrightarrow{\text{Fast}} (CH_3)_3C{-}OH_2^+$$

$$\text{3. } (CH_3)_3C{-}OH_2^+ + H_2O \xrightarrow{\text{Fast}} (CH_3)_3C{-}OH + H_3O^+$$

In the second step the intermediate *tert*-butyl cation reacts rapidly with water to produce a *tert*-butyloxonium ion (another intermediate) which, in the third step, rapidly transfers a proton to a molecule of water producing *tert*-butyl alcohol.

The first step requires heterolytic cleavage of the carbon-chlorine bond. Since no other bonds are formed in this step, it should be highly endothermic and it should have a high energy of activation. That it takes place at all is largely because of the ionizing ability of the solvent, water. Calculations indicate that in the gas phase (i.e., in the absence of a solvent), the energy of activation is nearly 160 kcal/mole! In aqueous solution, however, the energy of activation is much lower—about 20 kcal/mole. Water molecules surround and stabilize the cation and anion that are produced (cf. Section 2.16E).

Even though the *tert*-butyl cation produced in step 1 is stabilized by solvation, it is still a highly reactive species. Almost immediately after it is formed, it reacts with one of the surrounding water molecules to form the *tert*-butyloxonium ion, $(CH_3)_3COH_2^+$. (It may also occasionally react with a hydroxide ion, but water molecules are far more plentiful.)

> The *tert*-butyl cation is so reactive that its presence cannot be detected in aqueous solution—it reacts with water too rapidly. In the very powerful Lewis acid, SbF_5, however, the environment is different. Not only can the *tert*-butyl cation be detected, it can be kept indefinitely. In an experiment carried out by Olah and his co-workers (p. 158), the *tert*-butyl cation was generated by dissolving *tert*-butyl fluoride in SbF_5. The SbF_6^- anion is apparently a very weak nucleophile and the carbocation simply does not react further—there is nothing for it to react with.
>
> $$(CH_3)_3C\text{—}F + SbF_5 \longrightarrow (CH_3)_3C^+ + SbF_6^-$$
>
> We have more to say about this in Chapter 13.

A potential energy diagram for the S_N1 reaction of *tert*-butyl chloride and water is given in Fig. 5.3.

The important transition state for the S_N1 reaction is the transition state of the rate-limiting step [TS(1)]. In it the carbon-chlorine bond of *tert*-butyl chloride is largely broken and ions are beginning to develop:

$$CH_3\text{—}\overset{\overset{\displaystyle CH_3}{|}}{\underset{\underset{\displaystyle CH_3}{|}}{C^{\delta+}}}\text{-----}Cl^{\delta-}$$

The solvent (water) stabilizes these developing ions by solvation.

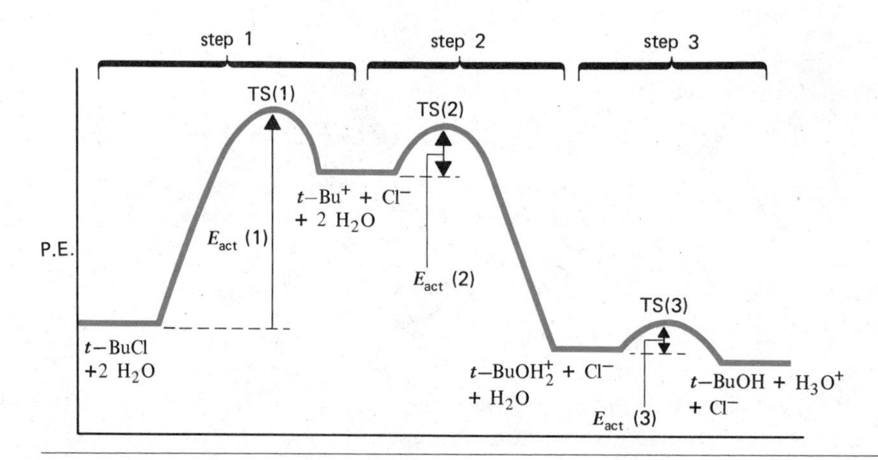

FIG. 5.3
A potential energy diagram for the S_N1 reaction of $(CH_3)_3C$—Cl (t-Bu—Cl). The energy of activation of the first step, $E_{act}(1)$, is much larger than $E_{act}(2)$ or $E_{act}(3)$.

Problem 5.3

Keeping in mind that carbocations have a trigonal planar structure, (a) write a structure for the carbocation intermediate and (b) write structures for the alcohol (or alcohols) you would expect from the following reaction.

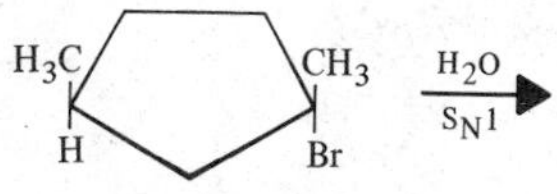

H_3C CH_3 H Br $\xrightarrow[S_N1]{H_2O}$

5.6A Solvolysis

The S_N1 reaction of *tert*-butyl chloride with water is an example of *solvolysis*. A solvolysis is a nucleophilic substitution in which *the nucleophile is a molecule of the solvent* (solvent + lysis: cleavage by the solvent). Since the solvent in this instance is water we could also call the reaction a *hydrolysis*. If the reaction had taken place in ethanol we would call the reaction an *ethanolysis*.

Problem 5.4

(a) What product would be obtained from the ethanolysis of *tert*-butyl chloride? (b) Outline the steps of this S_N1 reaction.

5.7 FACTORS AFFECTING THE RATES OF S_N1 AND S_N2 REACTIONS

Now that we have an understanding of the mechanisms of S_N2 and S_N1 reactions, our next task is to explain why methyl chloride reacts by an S_N2 mechanism and *tert*-butyl chloride by an S_N1 mechanism. We would also like to be able to predict which pathway—S_N1 or S_N2—would be followed by the reaction of any alkyl halide with any nucleophile under varying conditions.

The answer to this kind of question is connected directly to *relative rates of reaction*. If a given alkyl halide and nucleophile react rapidly by an S_N2 mechanism but slowly by an S_N1 mechanism under a given set of conditions, then an S_N2 pathway will be followed by most of the molecules. On the other hand another alkyl halide and another nucleophile may react very slowly (or not at all) by an S_N2 pathway. If they react rapidly by an S_N1 mechanism, then the reactants will follow an S_N1 pathway.

Experiments have shown that a number of factors affect the relative rates of unimolecular and bimolecular nucleophilic substitution reactions. The most important factors are:

1. The structure of the carbon bearing the leaving group.
2. The concentration and reactivity of the nucleophile (for bimolecular reactions only).
3. The nature of the leaving group.
4. The ionizing ability of the solvent.

5.7A The Effect of the Nature of the Carbon Bearing the Leaving Group

S_N2 Reactions. Simple alkyl halides show the following general order of reactivity in S_N2 reactions:

Methyl > primary > secondary > (tertiary)

TABLE 5.2 Relative Rates of Reactions of Alkyl Halides in S_N2 Reactions

SUBSTITUENT	COMPOUND	RELATIVE RATE
Methyl	$CH_3—X$	30
1°	$CH_3CH_2—X$	1
2°	$(CH_3)_2CHX$	0.02
Neopentyl	$(CH_3)_3CCH_2X$	0.00001
3°	$(CH_3)_3CX$	~0

Table 5.2 gives the relative rates of typical S_N2 reactions. Neopentyl halides, even though they are primary halides, are very unreactive.

$$CH_3—\overset{\displaystyle CH_3}{\underset{\displaystyle CH_3}{\overset{|}{\underset{|}{C}}}}—CH_2—X$$

A neopentyl halide

The important factor behind this order of reactivity is a *steric effect*. A steric effect is an effect on relative rates caused by the space-filling properties of those parts of a molecule attached at or near the reacting site. One kind of steric effect—the kind that is important here—is called *steric hindrance*. By this we mean that the spatial arrangement of the atoms or groups at or near the reacting site of a molecule hinders or retards a reaction.

For particles (molecules and ions) to react, their reactive centers must be able to come within bonding distance of each other. Although most molecules are reasonably flexible, very large and bulky groups can often hinder the formation of the required transition state. In some cases they can prevent its formation altogether.

An S_N2 reaction requires an approach by the nucleophile to a distance within bonding range of the carbon bearing the leaving group. Because of this, bulky substituents on *or near* that carbon have a dramatic inhibiting effect (Fig. 5.4). Of the simple alkyl halides, methyl halides react most rapidly in S_N2 reactions because only three small hydrogens interfere with the approaching nucleophile. Neopentyl and tertiary halides are the least reactive because bulky

FIG. 5.4
Steric effects in the S_N2 reaction.

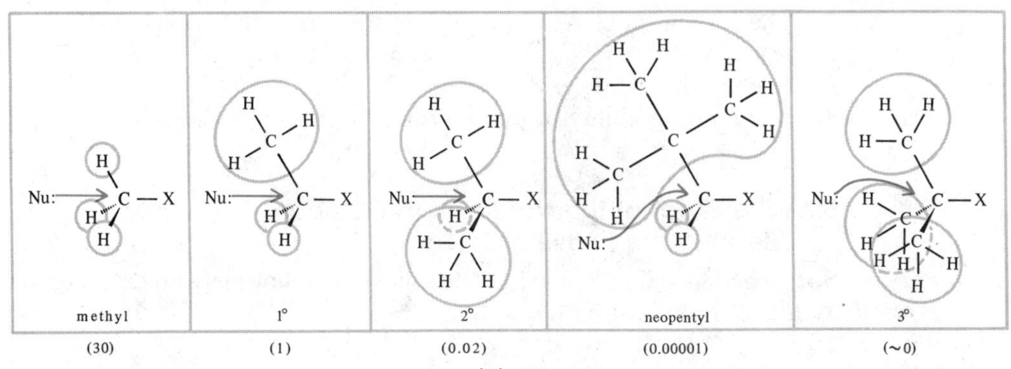

groups present a strong hindrance to the approaching nucleophile. (Tertiary substrates, for all practical purposes, do not react by an S_N2 mechanism.)

S_N1 Reactions. *The primary factor that determines the reactivity of organic substrates in an S_N1 reaction is an electrical effect* (*or electronic effect*). By this we mean that *some molecular feature stabilizes an electrical charge* on a transition state or intermediate. In S_N1 reactions the charge that must be stabilized is the positive charge that develops on the carbocation produced in the rate-limiting step.

Except for those reactions that take place in strong acids (which we will study later), the only organic compounds that undergo reaction by an S_N1 path at a reasonable rate are *those that are capable of forming relatively stable carbocations.* Of the simple alkyl halides that we have studied so far, this means (for all practical purposes) that only tertiary halides react by an S_N1 mechanism. (Later we will see that organic halides, called allylic halides and benzylic halides, can also react by an S_N1 mechanism because they can form relatively stable carbocations, cf., Section 14.2.)

Tertiary carbocations are stabilized because three alkyl groups release electrons to the positive carbon and thereby disperse its charge (see Section 4.13).

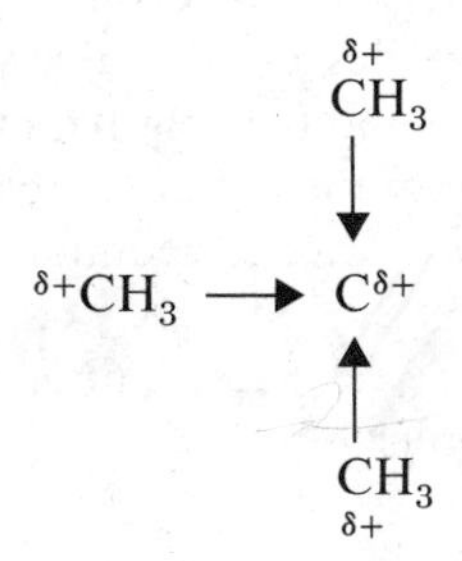

3° Carbocation
Stabilized because three electron-releasing alkyl groups disperse its charge

Formation of a relatively stable carbocation is important in an S_N1 reaction because it means that the energy of activation for the slow step (*i.e.*, $R{-}X \longrightarrow R^+ + X^-$) will be low enough for the overall reaction to take place at a reasonable rate. This step is endothermic (see Fig. 5.3 again) and according to Hammond's postulate (Section 4.10A) the transition state should bear a strong resemblance to the product. Since the product (actually an intermediate) is a carbocation, any factor that stabilizes it—such as dispersal of the positive charge by electron-releasing groups—will also stabilize the transition state in which the charge is developing. Stabilization of the transition state lowers its potential energy and thus lowers the energy of activation.

5.7B The Effect of the Concentration and Strength of the Nucleophile

Since the nucleophile does not participate in the rate-limiting step of an S_N1 reaction, the rates of S_N1 reactions are unaffected by either the concentration or the identity of the nucleophile. The rates of S_N2 reactions, however, depend on *both* the concentration *and* the identity of the attacking nucleophile. We saw in Section 5.4 how increasing the concentration of the nucleophile increases the rate of an S_N2 reaction. We can now examine how the rate of an S_N2 reaction depends on the identity of the nucleophile.

We describe nucleophiles as being *strong* or *weak*. When we do this we are really describing their relative reactivities in S_N2 reactions. A strong nucleophile is one that reacts rapidly with a given substrate. A weak nucleophile is one that reacts slowly with the same substrate under the same reaction conditions.

The relative strengths of nucleophiles can be correlated with several structural features:

1. A negatively charged nucleophile is always a stronger nucleophile than its conjugate acid. Thus OH^- is a stronger nucleophile than H_2O and RO^- is stronger than ROH.
2. In a group of nucleophiles in which the nucleophilic atom is the same, nucleophilicities parallel basicities. Oxygen compounds, for example, show the following order of reactivity:

   $$RO^- > HO^- \gg RCOO^- \gg ROH > H_2O$$

   This is also their order of basicity. An alkoxide ion (RO^-) is a slightly stronger base than a hydroxide ion, a hydroxide ion is a much stronger base than a carboxylate ion ($RCOO^-$) and so on.
3. The relative strengths of nucleophiles do not always parallel their basicities, however. When we examine the relative nucleophilicity of compounds within the same group of the periodic table we find that *in solvents such as alcohols and water* the nucleophile with the larger nucleophilic atom is stronger. Thiols (R—SH) are stronger nucleophiles than alcohols (ROH); RS^- ions are stronger than RO^- ions; and the halide ions show the following order:

   $$I^- > Br^- > Cl^- > F^-$$

This effect is related to the strength of the interactions between the nucleophile and its surrounding layer of solvent molecules. A small ion, because its charge is more concentrated, is more strongly solvated than a larger one. For a nucleophile to react it must shed some of its solvent molecules because it must closely approach the carbon bearing the leaving group. A large ion can shed some of its solvent molecules more easily and thus it will be more nucleophilic.*

Problem 5.5

Which would you expect to be the stronger nucleophile: (a) the amide ion (NH_2^-) or ammonia? (b) RS^- or RSH? (c) PH_3 or NH_3?

5.7C The Nature of the Leaving Group

The best leaving groups are those that become the most stable molecules or ions after they depart. This means, in general, that the best leaving groups are the ions or molecules that are the weakest bases. Of the halogens, an iodide ion is the best leaving group and a fluoride ion is the poorest:

$$I^- > Br^- > Cl^- \gg F^-$$

* We will see in Special Topic L that in solvents that do not solvate anions effectively (called aprotic solvents), the relative nucleophilicities of halide ions is just the reverse of that given here.

The order is the opposite of the basicity:

$$F^- \gg Cl^- > Br^- > I^-$$

Other weak bases that are good leaving groups are alkane sulfonate ions and alkyl sulfate ions:

$$^{-}O-\overset{\overset{O}{\|}}{\underset{\underset{O}{\|}}{S}}-R \qquad ^{-}O-\overset{\overset{O}{\|}}{\underset{\underset{O}{\|}}{S}}-O-R$$

An alkane sulfonate ion An alkyl sulfate ion

S_N1 reactions of alkyl halides are promoted by the addition of a silver salt such as silver nitrate. The silver ion acts as a Lewis acid and coordinates with the electron pair of the halogen. This weakens the carbon-halogen bond. It also assists departure of the leaving group because the leaving group becomes *a molecule of silver halide*—an even weaker base than a halide ion. If an alkyl halide is treated with silver nitrate in ethanol the following reactions take place:

$$R-\ddot{\underset{..}{X}}: \xrightleftharpoons{Ag^+} R-\overset{..+}{\underset{..}{X}}-Ag \longrightarrow R^+ + AgX\downarrow$$

then

$$R^+ + CH_3CH_2\ddot{O}H \longrightarrow R-\underset{\underset{H}{|}}{\overset{+}{O}}CH_2CH_3 \xrightarrow{-H^+} \underset{\text{An ether}}{ROCH_2CH_3}$$

Silver nitrate in ethanol is often used as a test for alkyl halides (Section 9.17). The formation of a silver halide precipitate indicates a positive test. As we might expect, tertiary halides react most rapidly and methyl halides are the slowest to react. Tertiary halides give an immediate precipitate at room temperature, secondary halides react within a few minutes, and methyl halides or primary halides require heating.

Strongly basic ions rarely act as leaving groups. The hydroxide ion, for example, is a strong base and thus reactions like the following do not take place.

$$X:^- \; R-OH \not\longrightarrow R-X + OH^-$$

(this reaction does not take place because the leaving group is a strongly basic hydroxide ion)

However, we will see in Section 14.6 that when an alcohol is dissolved in a strong acid it can react with a halide ion. Because the acid protonates the —OH group of the alcohol, the leaving group no longer needs to be a hydroxide ion, it is now a molecule of water—a much weaker base than a hydroxide ion.

$$X: \rightarrow R-\underset{\underset{H}{|}}{\overset{+}{O}}H \longrightarrow R-X + H_2O$$

(this reaction takes place because the leaving group is a weak base)

Problem 5.6

The reaction of methyl chloride with aqueous sodium hydroxide to produce methyl alcohol and chloride ion is essentially irreversible—the reaction proceeds virtually to completion. What factors account for this?

5.7D The Ionizing Ability of the Solvent

Polar solvents are able to solvate and thereby stabilize cations and anions. They do this by surrounding the ions as we showed in Fig. 2.18 (p. 79). Solvents with —OH groups can also solvate and stabilize some anions by forming hydrogen bonds to them. Water molecules, for example, can solvate halide ions in the following way.

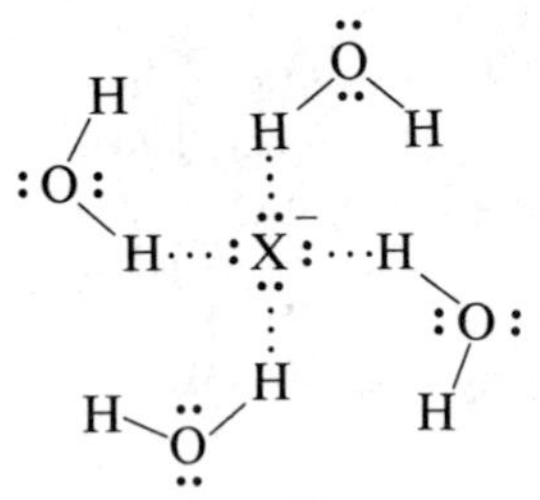

Because of its polarity, water molecules also stabilize cations effectively.

A solvent that stabilizes ions effectively will greatly increase the rate of ionization of an alkyl halide in any S_N1 reaction. It does this because solvation stabilizes the transition state leading to the intermediate carbocation and halide ion more than it does the reactants, thus the energy of activation is lower. The transition state for this endothermic step is one in which separated charges are developing and thus it resembles the ions that are ultimately produced.

$$(CH_3)_3C{-}Cl \longrightarrow (CH_3)_3\overset{\delta+}{C}\text{----}\overset{\delta-}{Cl} \longrightarrow (CH_3)_3C^+ + Cl^-$$

Reactant — Transition state — Products

Separated charges are developing

The use of a polar solvent also increases the rate of an S_N2 reaction of alkyl halides when the nucleophile is a neutral molecule.

$$Nu{:} + R{-}X \longrightarrow \overset{\delta+}{Nu}\text{---}R\text{---}\overset{\delta-}{X} \longrightarrow Nu^{+}{-}R + X^-$$

Reactants — Transition state — Products

Separated charges are developing

Here, too, solvation stabilizes the transition state with its developing electrical charges more than it does the electrically neutral reactants and this lowers the energy of activation.

On the other hand, the use of a polar solvent decreases the rate of an S_N2 reaction of an alkyl halide when the nucleophile is negatively charged. In this case the electrical charge is dispersed in the transition state.

$$Nu{:}^- + R{-}X \longrightarrow \overset{\delta-}{Nu}\text{---}R\text{---}\overset{\delta-}{X} \longrightarrow Nu{-}R + X^-$$

Reactants — Transition state — Products

Charge is concentrated — *Charge is dispersed*

TABLE 5.3 Dielectric Constants of Common Solvents

	SOLVENT	FORMULA	DIELECTRIC CONSTANT
Increasing solvent polarity (↑)	Water	H_2O	80
	Formic acid	$HCOH$ (with C=O)	59
	Methanol	CH_3OH	33
	Ethanol	CH_3CH_2OH	24
	Acetic acid	CH_3COH (with C=O)	6

A rough indication of a solvent's polarity is a quantity called the *dielectric constant*. The dielectric constant is a measure of the solvent's ability to insulate opposite charges from each other. The greater the dielectric constant, the greater is the solvent's ability to solvate both ions and developing ions. Table 5.3 gives the dielectric constants of some common solvents.

Water is the most effective solvent for promoting ionization, but most organic compounds do not dissolve appreciably in water. They usually dissolve, however, in alcohols, and often mixed solvents are used. Methanol-water and ethanol-water are common mixed solvents for nucleophilic substitution reactions. (Other highly polar solvents are discussed in Special Topic L.)

Problem 5.7

Each of the following reactions can be carried out in a mixed water-ethanol solvent. State whether you would expect the reaction rate to increase or decrease as the percentage of water in the mixed solvent is increased.

(a) $NH_3 + CH_3CH_2I \xrightarrow[C_2H_5OH]{H_2O} CH_3CH_2NH_3^+ + I^-$

(b) $CN^- + CH_3Br \xrightarrow[C_2H_5OH]{H_2O} CH_3CN + Br^-$

(c) $(CH_3)_3CBr + OH^- \xrightarrow[C_2H_5OH]{H_2O} (CH_3)_3COH + (CH_3)_3COC_2H_5 + Br^-$

5.7E Summary. S_N1 versus S_N2

Reactions of alkyl halides by an S_N1 mechanism are favored by the use of substrates that can form relatively stable carbocations, by the use of weak nucleophiles, and by the use of highly ionizing solvents. S_N1 mechanisms, therefore, are important in solvolysis reactions of tertiary halides, especially when the solvent is highly polar. In a solvolysis the nucleophile is weak because it is a neutral molecule (of the solvent) rather than an anion.

If we want to favor the reaction of an alkyl halide by an S_N2 mechanism we should use a relatively unhindered alkyl halide, a strong nucleophile, and a high concentration of the nucleophile. For substrates, the order of reactivity in S_N2 reactions is:

$$\mathrm{CH_3{-}X > C{-}CH_2{-}X > C{-}\underset{}{\overset{C}{\overset{|}{C}}}H{-}X}$$
$$\text{methyl} > 1° > 2°$$

Tertiary halides do not react by an S_N2 mechanism.

The effect of the leaving group is the same both in S_N1 and S_N2 reactions: alkyl iodides react fastest; fluorides react slowest. (Alkyl fluorides react so slowly they are seldom used in nucleophilic substitution reactions.)

$$\mathrm{R{-}I > R{-}Br > R{-}Cl}\} \; S_N1 \text{ or } S_N2$$

5.8 THE UNREACTIVITY OF VINYLIC HALIDES

Compounds that have a halogen attached to one carbon of a double bond are called *vinylic halides*.

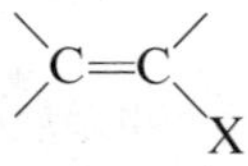

$$\mathrm{\rangle C{=}C\langle^{X}}$$

A vinylic halide

This name comes from the common name for a haloethene. Chloroethene, for example, is often called vinyl chloride.

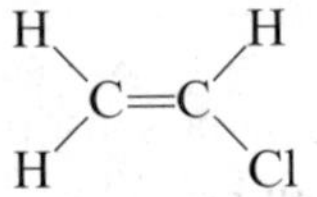

$$\mathrm{H_2C{=}CHCl}$$

Chloroethene
or
Vinyl chloride

The group, $CH_2{=}CH{-}$ is called the *vinyl* group, and a position at a carbon-carbon double bond is called a *vinylic* site.

Vinylic halides are generally unreactive in S_N1 or S_N2 reactions. Vinylic cations are highly unstable. This explains their unreactivity in S_N1 reactions. The carbon-halogen bond of a vinylic halide is stronger than that of an alkyl halide (we will see why later) and the electrons of the π bond repel an approaching nucleophile. These factors explain the unreactivity of a vinylic halide in an S_N2 reaction.

5.9 ELIMINATION REACTIONS OF ALKYL HALIDES

In an elimination reaction the pieces of some molecule, YZ, are removed (eliminated) from adjacent atoms of the reactant. This leads to the introduction of a multiple bond:

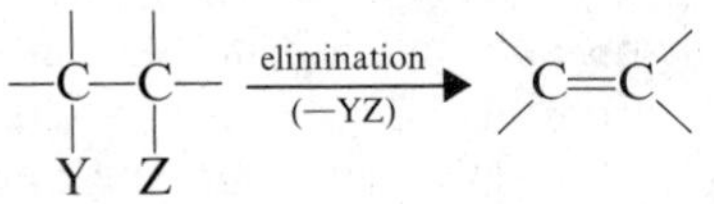

$$\mathrm{-\underset{Y}{\overset{|}{\underset{|}{C}}}-\underset{Z}{\overset{|}{\underset{|}{C}}}-} \xrightarrow[(-YZ)]{\text{elimination}} \mathrm{\rangle C{=}C\langle}$$

5.9A Dehydrohalogenation

A widely used method for synthesizing alkenes is the elimination of HX from adjacent atoms of an alkyl halide. Heating the alkyl halide with a strong base causes the reaction to take place. Two examples are shown below.

$$CH_3CH(Br)CH_3 \xrightarrow[C_2H_5OH,\ 55°]{C_2H_5ONa} CH_2{=}CH{-}CH_3 + NaBr + C_2H_5OH$$

(79%)

$$CH_3{-}C(CH_3)_2{-}Br \xrightarrow[C_2H_5OH,\ 25°]{C_2H_5ONa} CH_3{-}C(CH_3){=}CH_2 + NaBr + C_2H_5OH$$

(91%)

Reactions like these are not limited to the elimination of hydrogen bromide. Chloroalkanes also undergo the elimination of hydrogen chloride, iodoalkanes undergo the elimination of hydrogen iodide and, in all cases, alkenes are produced. When the elements of a hydrogen halide are eliminated from a haloalkane in this way, the reaction is often called *dehydrohalogenation*.

$$-\overset{H}{\underset{|}{\overset{|}{C}}}{}^{\beta}-\overset{|}{\underset{X}{C}}{}^{\alpha}- + :B^- \longrightarrow \;\rangle C{=}C\langle\; + H{:}B + :X^-$$

(a base)

Dehydrohalogenation

In these eliminations, as in S_N1 and S_N2 reactions, there is a leaving group and an attacking particle (the base) that possesses an electron pair.

Chemists often call the carbon that bears the halogen (above) the *alpha* (α) *carbon* and any carbon adjacent to it a *beta* (β) *carbon*. A hydrogen attached to the β carbon is called a β *hydrogen*. Since the hydrogen that is eliminated in dehydrohalogenation is from the β carbon these reactions are often called β *eliminations*. They are also often referred to as 1,2 eliminations.

We will have more to say about dehydrohalogenation in the next chapter, but we can examine some basic features here.

5.9B Bases Used in Dehydrohalogenation

A variety of strong bases have been used for dehydrohalogenations. Potassium hydroxide dissolved in ethyl alcohol is a reagent sometimes used, but the sodium salts of alcohols often offer distinct advantages.

The sodium salt of an alcohol (a sodium alkoxide) can be prepared by treating an alcohol with sodium metal:

$$\underset{\text{Alcohol}}{2R{-}\ddot{\underset{..}{O}}H} + 2Na \longrightarrow \underset{\substack{\text{Sodium}\\\text{alkoxide}}}{2R{-}\ddot{\underset{..}{O}}{:}^-Na^+} + H_2$$

This reaction involves the displacement of hydrogen from the alcohol and is, thus, an *oxidation-reduction reaction*. Sodium, an alkali metal, is a very powerful reducing agent and always displaces hydrogen atoms that are bonded to oxygen atoms. The vigorous (at times explosive) reaction of sodium with water is of the same type.

$$2H{-}\ddot{\underset{..}{O}}H + 2Na \longrightarrow 2H\ddot{\underset{..}{O}}{:}^{-}Na^{+} + H_2$$

Sodium hydroxide

Sodium alkoxides can also be prepared by allowing an alcohol to react with sodium hydride, NaH. The hydride ion, $H{:}^{-}$ is a very strong base.

$$R{-}\ddot{\underset{..}{O}}H + Na^{+}{:}H^{-} \longrightarrow R{-}\ddot{\underset{..}{O}}{:}^{-}\ Na^{+} + H_2$$

Sodium (and potassium) alkoxides are usually prepared by using an excess of the alcohol, and the excess alcohol becomes the solvent for the reaction. Sodium ethoxide is frequently employed in this way.

$$2CH_3CH_2OH + 2Na \longrightarrow 2CH_3CH_2O^{-}Na^{+} + H_2$$

Ethyl alcohol (excess) — Sodium ethoxide

Potassium *tert*-butoxide is another highly effective dehydrohalogenating reagent.

$$2CH_3\underset{CH_3}{\overset{CH_3}{C}}{-}\ddot{\underset{..}{O}}H + 2K \longrightarrow 2CH_3\underset{CH_3}{\overset{CH_3}{C}}{-}\ddot{\underset{..}{O}}{:}^{-}K^{+} + H_2$$

tert-Butyl alcohol (excess) — Potassium *tert*-butoxide

5.9C Mechanisms of Dehydrohalogenations

Elimination reactions occur by a variety of mechanisms. With alkyl halides, two mechanisms are especially important because they are closely related to the S_N2 and S_N1 reactions that we have just studied. One mechanism is a bimolecular mechanism called the E2 reaction; the other is a unimolecular mechanism called the E1 reaction.

5.10 THE E2 REACTION

When isopropyl bromide is heated with sodium ethoxide in ethanol to form propene, the reaction rate depends on the concentration of isopropyl bromide and on the concentration of ethoxide ion. The rate equation is first order in each reactant and second order overall.

$$\text{Rate} \propto [CH_3CHBrCH_3][C_2H_5O^{-}]$$
$$\text{Rate} = k[CH_3CHBrCH_3][C_2H_5O^{-}]$$

From this we infer that the transition state for the rate-limiting step must involve both the alkyl halide and the alkoxide ion. The reaction must be bimolecular. Considerable experimental evidence indicates that the reaction takes place in the following way.

$$C_2H_5O{:}^{-} + H{-}\underset{H}{\overset{H}{C^{\beta}}}{-}\underset{Br}{\overset{H}{C^{\alpha}}}{-}CH_3 \longrightarrow \underset{H}{\overset{H}{\;}}C{=}C\underset{CH_3}{\overset{H}{\;}} + C_2H_5OH + {:}Br^{-}$$

The ethoxide ion, using its electron pair, acts as a base and begins to remove one of the β hydrogens by forming a covalent bond to it. At more or less the same time the electron pair that had joined the β hydrogen to its carbon moves in to become the second bond of the double bond, and the bromine begins to depart with its electron pair (as a bromide ion). The transition state (below) is one in which partial bonds exist between the ethoxide ion and the β hydrogen, between the β hydrogen and the β carbon, and between the α carbon and the bromine. The carbon-carbon bond has also begun to develop some double-bond character.

$$C_2H_5\overset{\delta-}{O}\text{---}H \quad -\overset{|}{\underset{|}{C}}\text{=-=}\overset{|}{\underset{\vdots}{C}}- \quad Br^{\delta-}$$

Transition state for an E2 reaction

5.11 THE E1 REACTION

Eliminations may take a different pathway from that given in the previous section. When *tert*-butyl chloride is treated with 80% aqueous ethanol at 25°, for example, one obtains *substitution products* in 83% yield and an elimination product (2-methylpropene) in 17% yield.

$$(CH_3)_3C\text{—}Cl \xrightarrow[25^\circ]{80\%\ C_2H_5OH,\ 20\%\ H_2O} \begin{cases} \xrightarrow{S_N1} (CH_3)_3C\text{—}OH + (CH_3)_3C\text{—}OCH_2CH_3 \quad (83\%) \\ \xrightarrow{E1} CH_2\text{=}C(CH_3)_2 \quad (17\%) \end{cases}$$

tert-Butyl ethyl ether

2-Methylpropene

The initial step for both reactions is the formation of a *tert*-butyl cation. This is also the rate-limiting step for both reactions; thus both reactions are unimolecular.

$$CH_3\text{—}\underset{CH_3}{\overset{CH_3}{C}}\text{—}\ddot{\underset{..}{Cl}}: \xrightarrow{\text{Slow}} CH_3\underset{CH_3}{\overset{CH_3}{C^+}} + :\ddot{\underset{..}{Cl}}:^-$$

(solvated) (solvated)

Whether substitution or elimination takes place depends on the next step (the fast step). If a solvent molecule reacts as a nucleophile at the positive carbon of the *tert*-butyl cation the product is *tert*-butyl ethyl ether or *tert*-butyl alcohol and the reaction is S_N1.

$$\mathrm{CH_3{-}\underset{CH_3}{\overset{CH_3}{C^+}}\ \ Sol{-}\ddot{O}H \xrightarrow{Fast} CH_3\underset{CH_3}{\overset{CH_3}{C}}{-}\overset{+}{O}(Sol)(H)\!: \rightleftharpoons CH_3\underset{CH_3}{\overset{CH_3}{C}}{-}O{-}Sol + H^+} \quad S_N1 \text{ reaction}$$

($Sol = C_2H_5—$ or $H—$)

If, however, a solvent molecule acts as a base and abstracts one of the β hydrogens as a proton, the product is 2-methylpropene and the reaction is E1. E1 reactions always accompany S_N1 reactions.

$$\mathrm{Sol{-}\underset{H}{\ddot{O}}\!:\ H{-}CH_2{-}\underset{CH_3}{\overset{CH_3}{C^+}} \xrightarrow{Fast} Sol{-}\underset{H}{\ddot{O}^+}{-}H + CH_2{=}C(CH_3)_2} \quad \text{E1 reaction}$$

2-Methylpropene

5.12 SUBSTITUTION VERSUS ELIMINATION

Since the reactive part of a nucleophile or a base is an unshared electron pair, all nucleophiles are potential bases and all bases are potential nucleophiles. It should not be surprising, then, that nucleophilic substitution reactions and elimination reactions often compete with each other.

5.12A S_N1 versus E1

Since the E1 reaction and the S_N1 reaction proceed through the formation of a common intermediate, the two types respond in similar ways to factors affecting reactivities. E1 reactions are favored with substrates that can form stable carbocations; they are also favored by the use of weak nucleophiles (bases) and they are generally favored by the use of polar solvents.

It is usually difficult to influence the relative partition between S_N1 and E1 products because the energy of activation for either reaction of the carbocation (loss of a proton or combination with a molecule of the solvent) is very small.

In most unimolecular reactions the S_N1 reaction is favored over the E1 reaction. Increasing the temperature of the reaction, however, favors reaction by the E1 mechanism at the expense of the S_N1 mechanism. If the elimination product is desired however, it is more convenient to add a strong base and force an E2 reaction to take place instead.

5.12B S_N2 versus E2

Since eliminations occur best by an E2 path when carried out with a high concentration of a strong base (and thus a high concentration of a strong nucleophile), substitution reactions by an S_N2 path often compete with the elimination reaction. When the nucleophile (base) attacks a β hydrogen, elimination occurs. When the nucleophile attacks the carbon bearing the leaving group substitution results.

$$\mathrm{Nu\!:^- + H{-}\overset{|}{C}{-}\underset{|}{C}{-}X} \begin{cases} \xrightarrow[\text{E2}]{\text{(a) Elimination}} \mathrm{>C{=}C< + Nu{-}H + :X^-} \\ \xrightarrow[S_N2]{\text{(b) Substitution}} \mathrm{H{-}\overset{|}{C}{-}\underset{|}{C}{-}Nu + X:^-} \end{cases}$$

When the substrate is a primary halide and the base is ethoxide ion, substitution is highly favored.

$$C_2H_5O^-Na^+ + CH_3CH_2Br \xrightarrow[55^\circ\ (-NaBr)]{C_2H_5OH} \underset{\substack{(90\%) \\ S_N2}}{CH_3CH_2OCH_2CH_3} + \underset{\substack{(10\%) \\ E2}}{CH_2{=}CH_2}$$

With secondary halides, however, the elimination reaction is favored.

$$C_2H_5O^-Na^+ + CH_3CHBrCH_3 \xrightarrow[55^\circ\ (-NaBr)]{C_2H_5OH} \underset{\substack{(21\%) \\ S_N2}}{CH_3CH(OC_2H_5)CH_3} + \underset{\substack{(79\%) \\ E2}}{CH_2{=}CHCH_3}$$

And with tertiary halides an S_N2 reaction cannot take place and thus the elimination reaction is highly favored, especially when the reaction is carried out at higher temperatures. Any substitution that occurs probably takes place through an S_N1 mechanism.

$$C_2H_5O^-Na^+ + CH_3CBr(CH_3)CH_3 \xrightarrow[25^\circ\ (-NaBr)]{C_2H_5OH} \underset{\substack{(9\%) \\ S_N1}}{CH_3C(CH_3)(OC_2H_5)CH_3} + \underset{\substack{(91\%) \\ E2\ +\ E1}}{CH_2{=}C(CH_3)CH_3}$$

$$C_2H_5O^-Na^+ + CH_3CBr(CH_3)CH_3 \xrightarrow[55^\circ\ (-NaBr)]{C_2H_5OH} \underset{\substack{(100\%) \\ E2}}{CH_2{=}C(CH_3)CH_3}$$

Increasing the reaction temperature is only one way of favorably influencing an elimination reaction of an alkyl halide. Another way is to use a strong sterically hindered base such as the *tert*-butoxide ion. The bulky methyl groups of the *tert*-butoxide ion appear to inhibit its reacting by substitution, so elimination reactions take precedence. We can see an example of this effect in the two reactions shown below. The relatively unhindered methoxide ion reacts with *n*-octadecyl bromide primarily by *substitution;* the bulky *tert*-butoxide ion gives mainly *elimination.*

$$CH_3O^- + CH_3(CH_2)_{15}CH_2CH_2{-}Br \xrightarrow[\text{reflux}]{CH_3OH}$$

$$\underset{\substack{(1\%) \\ E2}}{CH_3(CH_2)_{15}CH{=}CH_2} + \underset{\substack{(99\%) \\ S_N2}}{CH_3(CH_2)_{15}CH_2CH_2OCH_3}$$

$$\mathrm{CH_3{-}\underset{CH_3}{\overset{CH_3}{C}}{-}O^- + CH_3(CH_2)_{15}CH_2CH_2{-}Br \xrightarrow[40^\circ]{(CH_3)_3COH}}$$

$$\underset{\substack{(85\%)\\ \text{E2}}}{\mathrm{CH_3(CH_2)_{15}CH{=}CH_2}} + \underset{\substack{(15\%)\\ \text{S}_\text{N}2}}{\mathrm{CH_3(CH_2)_{15}CH_2CH_2{-}O{-}\underset{CH_3}{\overset{CH_3}{C}}{-}CH_3}}$$

One other factor that affects the relative rates of E2 and S_N2 reactions is the relative basicity and polarizability of the base/nucleophile. Using a strong, slightly polarizable base such as amide ion (NH_2^-) or alkoxide ion (especially a hindered one) tends to increase the likelihood of elimination (E2). Using a weakly basic ion such as acetate ion, CH_3COO^-, or a weakly basic and highly polarizable one such as RS^- increases the likelihood of substitution (S_N2). Acetate ion, for example, reacts with isopropyl bromide almost exclusively by the S_N2 path:

$$\mathrm{CH_3\overset{O}{\overset{\|}{C}}{-}O^- + CH_3\overset{CH_3}{CH}{-}Br \longrightarrow \underset{\substack{(\sim 100\%)\\ S_N2}}{CH_3\overset{O}{\overset{\|}{C}}{-}O{-}\overset{CH_3}{C}HCH_3} + Br^-}$$

The more strongly basic ethoxide ion (p. 197) reacts with the same compound mainly by an E2 mechanism.

5.12C Tertiary Halides. S_N1 versus E2

S_N2 reactions do not take place with tertiary halides. Suppose, for example, that we are given the task of preparing *tert*-butyl ethyl ether from *tert*-butyl bromide. If we were to attempt an S_N2 reaction using sodium ethoxide in ethanol, an elimination reaction would take place instead:

$$\mathrm{C_2H_5O^-Na^+ + CH_3{-}\underset{CH_3}{\overset{CH_3}{C}}{-}Br \xrightarrow[\substack{25^\circ\\(-NaBr)}]{C_2H_5OH} \underset{\substack{91\%\\(E2\,+\,E1)}}{CH_2{=}\overset{CH_3}{C}CH_3} + \underset{\substack{9\%\\(S_N1\ product)}}{C_2H_5O{-}\underset{CH_3}{\overset{CH_3}{C}}{-}CH_3}}$$

At higher temperatures our yield of *tert*-butyl ethyl ether would be even lower (cf., p. 197).

Our best choice is an S_N1 reaction. To maximize the amount of substitution, we should keep the temperature low and use no added strong base. In other words, we should carry out an ethanolysis:

$$\mathrm{CH_3{-}\underset{CH_3}{\overset{CH_3}{C}}{-}Br \xrightarrow[25^\circ]{C_2H_5OH} \underset{\substack{19\%\\(E1\ product)}}{CH_2{=}\overset{CH_3}{C}{-}CH_3} + \underset{\substack{81\%\\(S_N1\ product)}}{C_2H_5O{-}\underset{CH_3}{\overset{CH_3}{C}}{-}CH_3}}$$

In this instance the substitution product is obtained in relatively good yield. In general, however, substitution reactions of tertiary halides are not very useful as synthetic methods. Such halides undergo eliminations much too easily.

5.13 A BIOLOGICAL NUCLEOPHILIC SUBSTITUTION REACTION: BIOLOGICAL METHYLATION

The cells of living organisms synthesize many of the compounds they need from other, smaller molecules. Often these biosyntheses resemble the syntheses organic chemists carry out in their laboratories. Let us examine one example now.

A number of reactions take place in the cells of plants and animals that involve the transfer of a methyl group from an amino acid called methionine to some other compound. That this transfer takes place can be demonstrated experimentally by feeding a plant or animal methionine containing a radioactive carbon atom (^{14}C) in its methyl group. Later, other compounds containing the "labeled" methyl group can be isolated from the organism. Some of the compounds that get their methyl groups from methionine are listed below. The radioactively labeled carbon is shown in color.

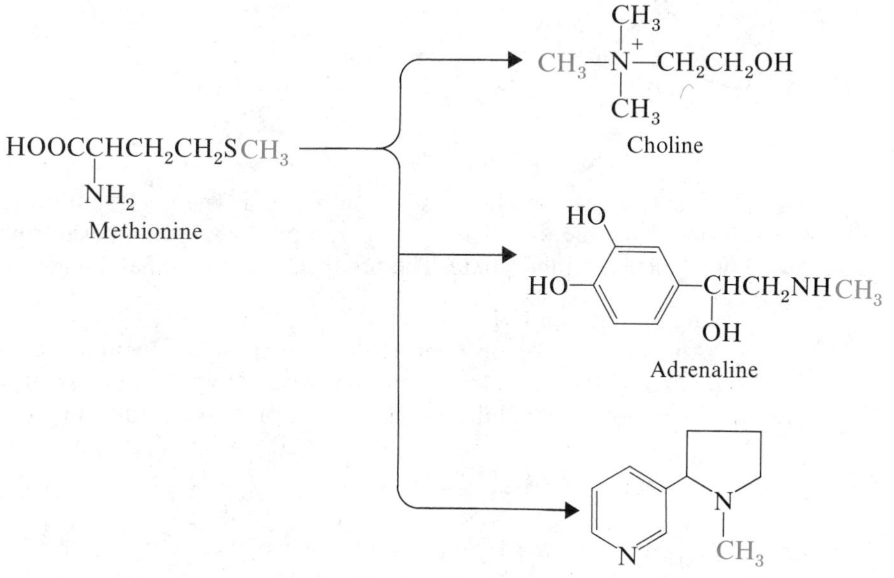

$$HOOC\underset{\displaystyle NH_2}{\underset{|}{C}}HCH_2CH_2SCH_3$$

Methionine

$$\longrightarrow CH_3-\overset{\displaystyle CH_3}{\overset{|}{\underset{\displaystyle CH_3}{\underset{|}{N^+}}}}-CH_2CH_2OH$$

Choline

$$\longrightarrow (HO)_2C_6H_3-\underset{\displaystyle OH}{\underset{|}{C}}HCH_2NHCH_3$$

Adrenaline

$$\longrightarrow \text{(3-pyridyl)–pyrrolidine ring with } N-CH_3$$

Nicotine

Choline is important in the transmission of nerve impulses, adrenaline causes blood pressure to increase, and nicotine is the compound contained in tobacco that makes smoking tobacco addictive. (In larger doses nicotine is poisonous.)

The transfer of the methyl group from methionine to these other compounds does not take place directly. The actual methylating agent is not methionine; it is *S*-adenosylmethionine, a compound that results when methionine reacts with adenosine triphosphate (ATP):

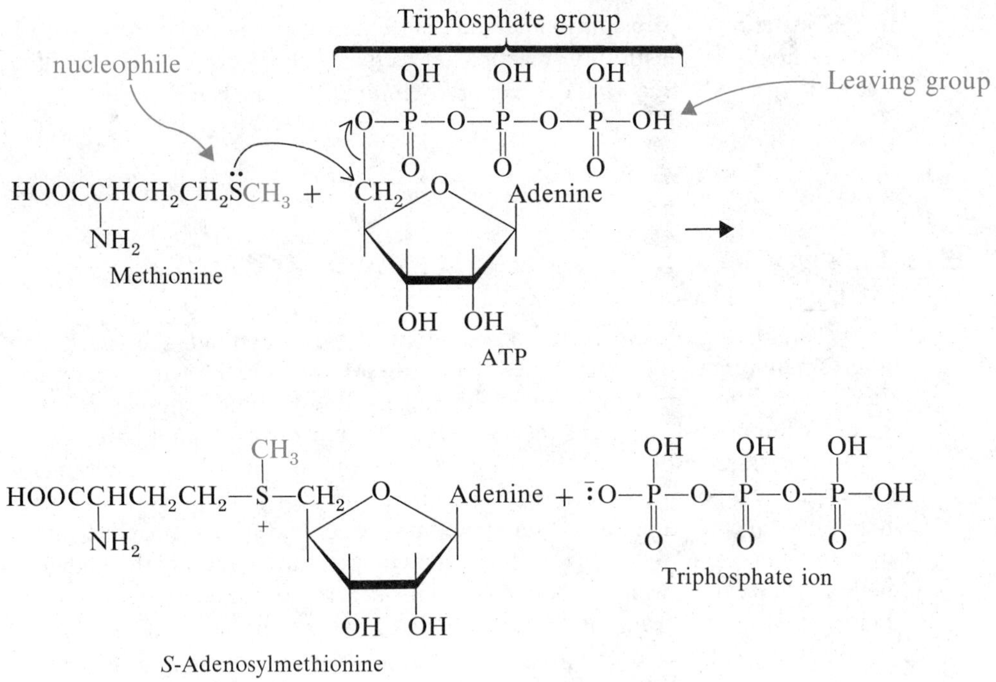

Triphosphate group

nucleophile

Leaving group

Methionine

ATP

S-Adenosylmethionine

Triphosphate ion

Adenine =

This reaction is a nucleophilic substitution reaction. The nucleophilic atom is the sulfur atom of methionine. The leaving group is the weakly basic triphosphate group of adenosine triphosphate. The product, *S*-adenosylmethionine, contains a methylsulfonium group, $CH_3{-}\overset{|}{S}^{+}{-}$.

S-Adenosylmethionine then acts as the substrate for other nucleophilic substitution reactions. In the biosynthesis of choline, for example, it transfers its methyl group to a nucleophilic nitrogen atom of *N,N*-dimethylethanolamine:

N,N-Dimethylethanolamine

Choline

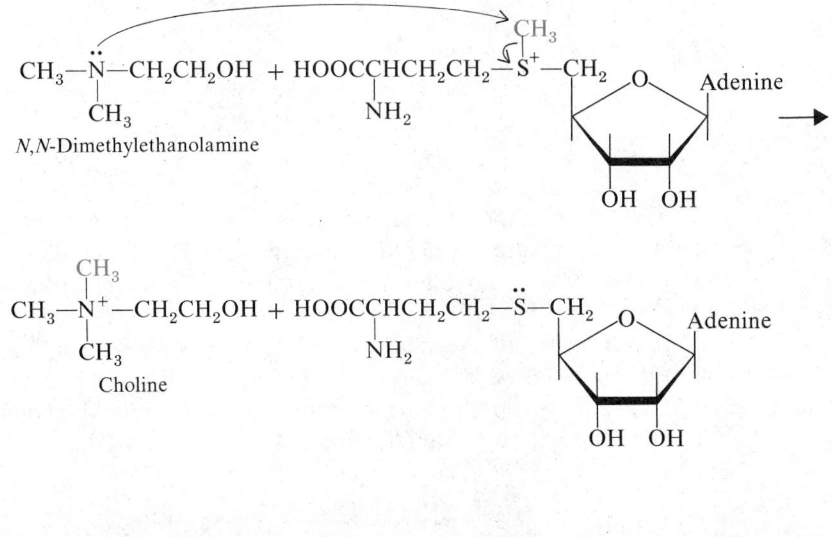

These reactions appear complicated only because the structures of the nucleophiles and substrates are complex. Yet conceptually they are simple and they illustrate many of the principles we have encountered in this chapter. In them we see how nature makes use of the high nucleophilicity of sulfur atoms. We also see how a weakly basic group (e.g., the triphosphate group of ATP) functions as a leaving group. In the reaction of *N,N*-dimethylethanolamine we see that the more basic $(CH_3)_2\ddot{N}$ group acts as the nucleophile rather than the less basic $-\ddot{O}H$ group. And, when a nucleophile attacks *S*-adenosylmethionine we see that the attack takes place at the less hindered CH_3- group rather than at one of the more hindered $-CH_2-$ groups.

Problem 5.8

(a) What is the leaving group when *N,N*-dimethylethanolamine reacts with *S*-adenosylmethionine? (b) What would the leaving group have to be if methionine itself were to react with *N,N*-dimethylethanolamine? (c) Of what special significance is this difference?

5.14 SOME IMPORTANT TERMS AND CONCEPTS

Nucleophile. A molecule or negative ion that has an unshared pair of electrons. In a chemical reaction a nucleophile attacks a positive center of some other molecule or positive ion.

Nucleophilic substitution reaction (abbreviated as S_N reaction). A substitution reaction brought about when a nucleophile reacts with a *substrate* that bears a *leaving group*.

S_N2 reaction. A nucleophilic substitution reaction for which the rate-limiting step is *bimolecular* (i.e., the transition state involves two species). The reaction of methyl chloride with hydroxide ion is an S_N2 reaction. According to the Ingold mechanism it takes place in a *single step* as follows.

$$HO:^- + CH_3{-}Cl \longrightarrow \underset{\text{Transition state}}{\overset{\delta^-}{HO}{-}{-}{-}CH_3{-}{-}{-}\overset{\delta^-}{Cl}} \longrightarrow HO{-}CH_3 + :Cl^-$$

The order of reactivity of alkyl halides in S_N2 reaction is:

$$\underset{\text{Methyl}}{CH_3{-}X} > \underset{1°}{RCH_2X} > \underset{2°}{R_2CHX}$$

S_N1 reaction. A nucleophilic substitution reaction for which the rate-limiting step is *unimolecular*. The hydrolysis of *tert*-butyl chloride is an S_N1 reaction that takes place in three steps as follows. The rate-limiting step is step (1).

1. $(CH_3)_3CCl \xrightarrow{\text{slow}} (CH_3)_3C^+ + Cl^-$
2. $(CH_3)_3C^+ + H_2O: \xrightarrow{\text{fast}} (CH_3)_3COH_2^+$
3. $(CH_3)_3COH_2^+ + H_2O \xrightarrow{\text{fast}} (CH_3)_3COH + H_3O^+$

S_N1 reactions are important with tertiary halides and with other substrates that can form relatively stable carbocations.

Solvolysis. A nucleophilic substitution reaction in which the nucleophile is a molecule of the solvent.

Steric effect. An effect on relative reaction rates caused by the space-filling properties of those parts of a molecule attached at or near the reacting site. *Steric hindrance* is an important steric effect in S_N2 reactions. It explains why methyl halides are most reactive and tertiary halides are least reactive.

Electrical effect (or **electronic effect**). An effect on relative reaction rates when some molecular feature stabilizes the electrical charge on a transition state or an intermediate. The electron-releasing ability of methyl groups stabilizes the transition state and the carbocation formed by ionization of a 3° alkyl halide in the slow step of an S_N1 reaction, for example.

Elimination reaction. A reaction in which the pieces of some molecule are eliminated from adjacent atoms of the reactant leading to the introduction of a multiple bond. Dehydrohalogenation is an elimination reaction in which HX is eliminated from an alkyl halide, leading to the formation of an alkene.

$$-\overset{H}{\underset{|}{\overset{|}{C}}}-\overset{|}{\underset{X}{C}}- + :B^- \longrightarrow \;\rangle C{=}C\langle\; + H{:}B + :X^-$$

E1 reaction. A unimolecular elimination. The first step of an E1 reaction, formation of a carbocation, is the same as that of an S_N1 reaction, consequently E1 and S_N1 reactions compete with each other. E1 reactions are important when tertiary halides are subjected to solvolysis in polar solvents especially at higher temperatures. The steps in the E1 reaction of *tert*-butyl chloride are:

$$\text{1. } CH_3-\overset{CH_3}{\underset{CH_3}{C}}-Cl \xrightarrow{\text{Slow}} CH_3-\overset{CH_3}{\underset{CH_3}{C^+}} + :Cl^-$$

$$\text{2. } Sol-\ddot{O}H + H-CH_2-\overset{CH_3}{\underset{CH_3}{C^+}} \longrightarrow CH_2{=}C(CH_3)_2 + Sol-\overset{+}{\ddot{O}}H_2$$

E2 reaction. A bimolecular elimination that often competes with S_N2 reactions. E2 reactions are favored by the use of a high concentration of a strong, bulky, and slightly polarizable base. The order of reactivity of alkyl halides toward E2 reactions is: $3° \gg 2° > 1°$. The mechanism of the E2 reaction involves a single step:

$$B{:}^- + -\overset{H}{\underset{|}{\overset{|}{C}}}-\overset{|}{\underset{X}{C}}- \longrightarrow B-H + \;\rangle C{=}C\langle\; + :X^-$$

Additional Problems

5.9

Show how you might use a nucleophilic substitution reaction of propyl bromide to synthesize each of the following compounds. (You may use any other compounds that are necessary.)

(a) $CH_3CH_2CH_2OH$

(b) $CH_3CH_2CH_2I$

(c) $CH_3CH_2OCH_2CH_2CH_3$
(d) $CH_3CH_2CH_2{-}S{-}CH_3$

(e) $CH_3\overset{\overset{\displaystyle O}{\|}}{C}OCH_2CH_2CH_3$
(f) $CH_3CH_2CH_2N_3$

(g) $CH_3{-}\underset{\underset{\displaystyle CH_3}{|}}{\overset{\overset{\displaystyle CH_3}{|}}{N^+}}{-}CH_2CH_2CH_3\ Br^-$

(h) $CH_3CH_2CH_2CN$
(i) $CH_3CH_2CH_2SH$

5.10

Which alkyl halide would you expect to react more rapidly by an S_N2 mechanism? Explain your answer.

(a) $CH_3CH_2CH_2CH_2Br$ or $CH_3CH_2\underset{\underset{\displaystyle Br}{|}}{C}HCH_3$

(b) $CH_3CH_2\underset{\underset{\displaystyle Br}{|}}{C}HCH_3$ or $CH_3\underset{\underset{\displaystyle Br}{|}}{\overset{\overset{\displaystyle CH_3}{|}}{C}}CH_3$

(c) $CH_3CH_2CH_2{-}Cl$ or $CH_3CH_2CH_2{-}Br$

(d) $CH_3\underset{\underset{\displaystyle CH_3}{|}}{C}HCH_2CH_2Br$ or $CH_3CH_2\underset{\underset{\displaystyle CH_3}{|}}{C}HCH_2Br$

(e) CH_3CH_2Cl or $CH_2{=}CHCl$

5.11

Which S_N2 reaction of each pair would you expect to take place more rapidly? Explain your answer.

(a) $CH_3CH_2CH_2Br + CH_3OH \longrightarrow CH_3CH_2CH_2OCH_3 + HBr$
or
$CH_3CH_2CH_2Br + CH_3O^- \longrightarrow CH_3CH_2CH_2OCH_3 + Br^-$

(b) $CH_3CH_2I + OH^- \longrightarrow CH_3CH_2OH + I^-$
or
$CH_3CH_2I + SH^- \longrightarrow CH_3CH_2SH + I^-$

(c) $CH_3Br + CH_3OH \longrightarrow CH_3OCH_3 + HBr$
or
$CH_3Br + CH_3SH \longrightarrow CH_3SCH_3 + HBr$

(d) $CH_3CH_2I + CH_3S^-(1.0\ M) \longrightarrow CH_3CH_2SCH_3 + I^-$
or
$CH_3CH_2I + CH_3S^-(2.0\ M) \longrightarrow CH_3CH_2SCH_3 + I^-$

5.12

Which S_N1 reaction would you expect to take place more rapidly? Explain your answer.

(a) $(CH_3)_3CI + CH_3OH \longrightarrow (CH_3)_3COCH_3 + HI$
or
$(CH_3)_3CCl + CH_3OH \longrightarrow (CH_3)_3COCH_3 + HCl$

(b) $(CH_3)_3CBr + H_2O \longrightarrow (CH_3)_3COH + HBr$
or
$(CH_3)_3CBr + CH_3OH \longrightarrow (CH_3)_3COCH_3 + HBr$

(c) $(CH_3)_3CCl + CH_3O^-(0.01\ M) \xrightarrow[CH_3OH]{} (CH_3)_3COCH_3 + Cl^-$
or
$(CH_3)_3CCl + CH_3O^-(0.001\ M) \xrightarrow[CH_3OH]{} (CH_3)_3COCH_3 + Cl^-$

(d) $(CH_3)_3CCl + H_2O \longrightarrow (CH_3)_3COH + HCl$
or
$(CH_3)_2C{=}CHCl + H_2O \longrightarrow (CH_3)_2C{=}CHOH + HCl$

5.13
Listed below are several hypothetical nucleophilic substitution reactions. None is synthetically useful because the product indicated is *not* formed at an appreciable rate. In each case account for the failure of the reaction to take place.
(a) $HO^- + CH_3CH_3 \not\rightarrow CH_3CH_2OH + H{:}^-$
(b) $HO^- + CH_3CH_2CH_3 \not\rightarrow CH_3CH_2OH + CH_3{:}^-$
(c) cyclopentane (CH_2 ring) $+ H_2O \not\rightarrow CH_3CH_2CH_2CH_2CH_2OH$
(d) $CN^- + (CH_3)_3CBr \not\rightarrow (CH_3)_3C{-}CN + Br^-$
(e) $CH_3CH{=}CHBr + CH_3S^- \not\rightarrow CH_3CH{=}CHSCH_3 + Br^-$
(f) $Cl^- + CH_3OCH_3 \not\rightarrow CH_3Cl + CH_3O^-$
(g) $NH_3 + CH_3CH_2\overset{+}{O}H_2 \not\rightarrow CH_3CH_2NH_3^+ + H_2O$
(h) $CH_3{:}^- + CH_3CH_2OH \not\rightarrow CH_3CH_2CH_3 + OH^-$

5.14
You are given the task of preparing propene by dehydrohalogenating one of the propyl halides (i.e., $CH_3CH_2CH_2Br$ or $CH_3CHBrCH_3$). Which halide would you choose to give the alkene in maximum yield? Why?

5.15
Which product (or products) would you expect to obtain from each of the following reactions? In each case give the mechanism (S_N1, S_N2, E1, or E2) by which each product is formed and predict the relative amount of each (i.e., would the product be the only product, the major product, a minor product, etc?).

(a) $CH_3CH_2CH_2CH_2Br + CH_3O^- \xrightarrow[CH_3OH]{50°}$

(b) $CH_3CH_2CH_2CH_2Br + (CH_3)_3CO^- \xrightarrow[(CH_3)_3COH]{50°}$

(c) $(CH_3)_3CO^- + CH_3I \xrightarrow[(CH_3)_3COH]{50°}$

(d) $(CH_3)_3CI + CH_3O^- \xrightarrow[CH_3OH]{50°}$

(e) cyclobutane ring bearing CH_3, H on one carbon and I, H on the 1,3-carbon $+ CH_3O^- \xrightarrow[CH_3OH]{50°}$

(f) cyclopentane ring bearing CH_3, H on one carbon and I, CH_3 on another $\xrightarrow[CH_3OH]{25°}$

(g) $CH_3CH_2CHCH_2CH_3$ (Br on C-3) $+ C_2H_5O^- \xrightarrow[C_2H_5OH]{50°}$

(h) $(CH_3)_3CO^- + CH_3CHCH_3$ (Br on C-2) $\xrightarrow[(CH_3)_3COH]{50°}$

5.16
Many S_N2 reactions of alkyl chlorides and alkyl bromides are catalyzed by the addition of sodium or potassium iodide. For example, the hydrolysis of methyl bromide takes place much faster in the presence of sodium iodide. Explain.

5.17
When *tert*-butyl chloride undergoes hydrolysis (p. 180) in aqueous sodium hydroxide the rate of formation of *tert*-butyl alcohol does not increase appreciably as the hydroxide ion

concentration is increased. Increasing hydroxide ion concentration, however, causes a marked increase in the rate of disappearance of *tert*-butyl chloride. Explain.

5.18
(a) Consider the general problem of converting a tertiary alkyl halide to an alkene, for example, the conversion of *tert*-butyl chloride to 2-methylpropene. What experimental conditions would you choose to insure that elimination is favored over substitution? (b) Consider the opposite problem, that of carrying out a substitution reaction on a tertiary alkyl halide. Use as your example the conversion of *tert*-butyl chloride to ethyl *tert*-butyl ether. What experimental conditions would you employ to insure the highest possible yield of the ether?

5.19
Bridged cyclic compounds like those shown below are extremely *unreactive* in S_N2 reactions.

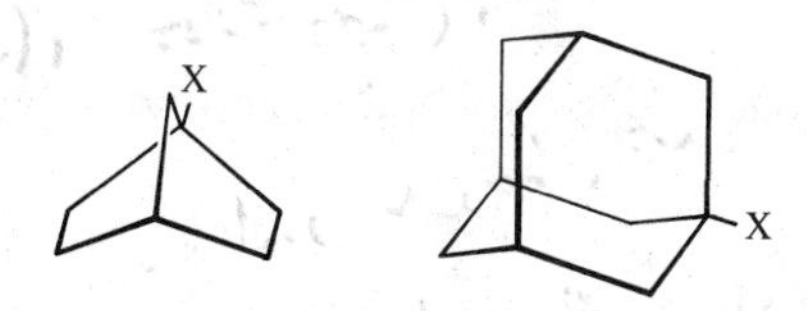

(a) Give a reason that will explain this. (b) How can you explain the fact that compounds of this type are also less reactive in S_N1 reactions than similar noncyclic compounds? (Consider the fact that carbocations are generally sp^2 hybridized.)

5.20
When CH_3Br reacts with CN^- the major product is CH_3CN but some CH_3NC is formed as well. Write the Lewis structure for both products and explain.

*** 5.21**
The relative rates of ethanolysis of several primary alkyl halides are as follows: CH_3CH_2Br, 1.0; $CH_3CH_2CH_2Br$, 0.28; $(CH_3)_2CHCH_2Br$, 0.030; $(CH_3)_3CCH_2Br$, 0.0000042. (a) Are these reactions S_N1 or S_N2? (b) What factor will account for these relative reactivities?

*** 5.22**
In contrast to S_N2 reactions, S_N1 reactions show relatively little nucleophile selectivity. That is, when more than one nucleophile is present in the reaction medium, S_N1 reactions show only a slight tendency to discriminate between weak nucleophiles and strong nucleophiles, whereas S_N2 reactions show a marked tendency to discriminate. (a) Provide an explanation for this behavior. (b) Show how your answer accounts for the fact that $CH_3CH_2CH_2CH_2Cl$ reacts with 0.01 M NaCN in ethanol to yield primarily $CH_3CH_2CH_2CH_2CN$ whereas under the same conditions, $(CH_3)_3CCl$ reacts to give primarily $(CH_3)_3COCH_2CH_3$.

*** 5.23**
When *tert*-butyl bromide undergoes S_N1 hydrolysis, adding a "common ion" (i.e., NaBr) to the aqueous solution has no effect on the rate. On the other hand when $(C_6H_5)_2CHBr$ undergoes S_N1 hydrolysis, adding NaBr retards the reaction. Given that the $(C_6H_5)_2CH^+$ cation is known to be much more stable than the $(CH_3)_3C^+$ cation (and we will see why in Section 12.10B), provide an explanation for the different behavior of the two compounds.

*** 5.24**
When the alkyl bromides (below) were subjected to hydrolysis in a mixture of ethanol and water (80% C_2H_5OH/20% H_2O) at 55°, the rates of the reaction showed the following order:

$$(CH_3)_3CBr > CH_3Br > CH_3CH_2Br > (CH_3)_2CHBr$$

Provide an explanation for this order of reactivity.

6 ALKENES: STRUCTURE AND SYNTHESIS

6.1 INTRODUCTION

Alkenes are hydrocarbons whose molecules contain the carbon-carbon double bond. An old name for this family of compounds that is still often used is the name *olefins*. Ethylene, the simplest olefin (alkene), was called olefiant gas (Latin: *oleum*, oil + *facere*, to make) because gaseous ethylene (C_2H_4) reacts with chlorine to form $C_2H_4Cl_2$, a liquid (oil).

The double bond is the *functional group* of alkenes and it determines their chemical properties.

6.2 NOMENCLATURE OF ALKENES AND CYCLOALKENES

Many older names for alkenes are still in common use. Propene is often called propylene, and 2-methylpropene frequently bears the name isobutylene.

$CH_2{=}CH_2$
IUPAC: Ethene or ethylene*

$CH_3CH{=}CH_2$
IUPAC: Propene
Common: Propylene

$CH_3{-}C(CH_3){=}CH_2$
IUPAC: 2-Methylpropene
Common: Isobutylene

The IUPAC rules for naming alkenes are similar in many respects to those for naming alkanes:

1. We determine the base name by selecting the longest chain *that contains the double bond* and we change the ending of the name of the alkane of identical length from *ane* to *ene*. Thus, if this longest chain contains five carbon atoms, the base name for the alkene is *pentene;* if it contains six carbons, the base name is hexene, and so on
2. We number the chain so as to include both carbons of the double bond, and *we begin numbering at the end of the chain nearer the double bond.* We designate the location of the double bond by using the number of the first atom of the double bond as a prefix:

$\overset{1}{C}H_2{=}\overset{2}{C}H\overset{3}{C}H_2\overset{4}{C}H_3$
1-Butene
(not 3-butene)

$CH_3CH{=}CHCH_2CH_2CH_3$
2-Hexene
(not 4-hexene)

3. We indicate the locations of the substituent groups by the number of the carbon atom to which they are attached.

$\underset{1}{C}H_3{-}\underset{2}{C}(CH_3){=}\underset{3}{C}H\underset{4}{C}H_3$
2-Methyl-2-butene
(not 3-methyl-2-butene)

$\underset{1}{C}H_3\underset{2}{C}(CH_3){=}\underset{3}{C}H\underset{4}{C}H_2\underset{5}{C}H(CH_3)\underset{6}{C}H_3$
2,5-Dimethyl-2-hexene

* The IUPAC system also retains the name ethylene when no substituents are present.

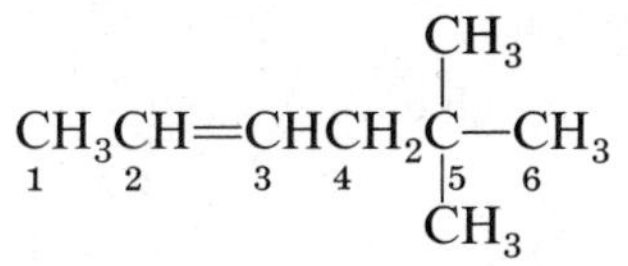

$$\underset{1\quad 2\quad\quad 3\quad 4\quad\quad 5\quad 6}{CH_3CH{=}CHCH_2\overset{CH_3}{\underset{CH_3}{C}}{-}CH_3}$$

5,5-Dimethyl-2-hexene

4. We number substituted cycloalkenes in the way that gives the carbon atoms of the double bond the 1- and 2- positions and that also gives the substituent groups the lowest numbers at the first point of difference. With substituted cycloalkenes it is not necessary to specify the position of the double bond since it will always be on C—1. The two examples listed below illustrate the application of these rules.

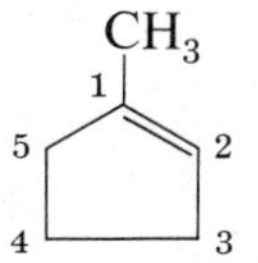

1-Methylcyclopentene
(not 2-methylcyclopentene)

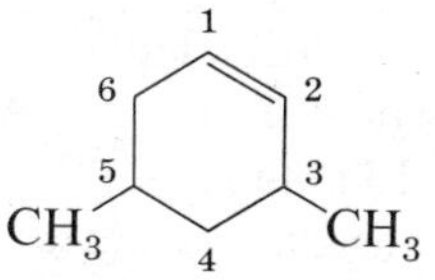

3,5-Dimethylcyclohexene
(not 4,6-dimethylcyclohexene)

5. Two frequently encountered alkenyl groups are often named as though they were substituents. These groups are the *vinyl group* and the *allyl group*.

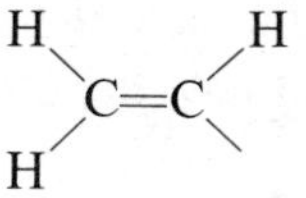

$H_2C{=}CH{-}$ or $CH_2{=}CH{-}$

The vinyl group

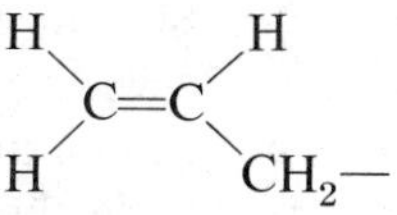

$H_2C{=}CH{-}CH_2{-}$ or $CH_2{=}CHCH_2{-}$

The allyl group

The following examples illustrate how these names are employed.

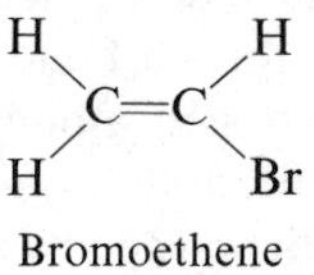

$H_2C{=}CHBr$

Bromoethene
or
Vinyl bromide
(common)

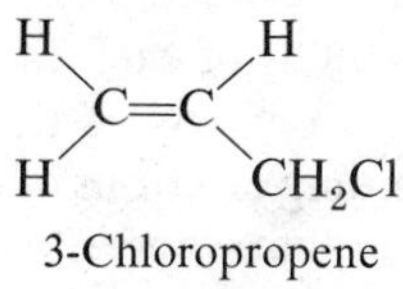

$H_2C{=}CHCH_2Cl$

3-Chloropropene
or
Allyl chloride
(common)

6. We designate the geometry of the double bond with the prefixes *cis*- and *trans*-. If two similar groups (usually hydrogens) are on the same side of the double bond it is *cis;* if they are on opposite sides, it is *trans.*

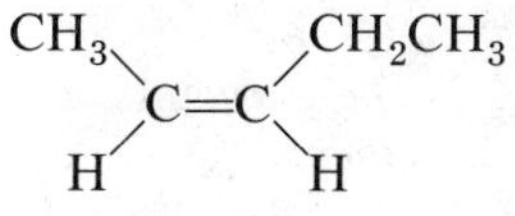

$CH_3(H)C{=}C(H)CH_2CH_3$ (CH_3 and CH_2CH_3 on the same side)

cis-2-Pentene

$CH_3(H)C{=}C(H)CH_2CH_3$ (CH_3 and CH_2CH_3 on opposite sides)

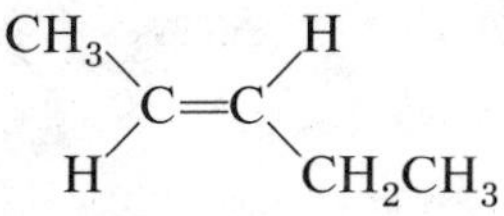

trans-2-Pentene

In Section 8.21 we will see another method for designating the geometry of the double bond.

Problem 6.1

Give IUPAC names for the following alkenes:

(a) $(CH_3)_2C{=}C(H)CH_3$ — CH_3 and CH_3 on one carbon; H and CH_3 on the other

(b) $CH_3CH_2CH_2CH{=}CHCH_2CH_2CH_3$ — $CH_3CH_2CH_2$ and $CH_2CH_2CH_3$ on the same side, H and H on the same side

(c) $(CH_3)_2C{=}CHBr$ — CH_3 and CH_3 on one carbon; H and Br on the other

(d) Cyclohexene ring bearing a CH_3 substituent

Problem 6.2

Write structural formulas for

(a) *cis*-3-Hexene

(b) *trans*-2-Pentene

(c) 3-Ethylcyclohexene

(d) Vinylcyclohexane

(e) 4,4-Dimethyl-1-hexene

6.3 ORBITAL HYBRIDIZATION AND THE STRUCTURE OF ALKENES

We can account for the structure of the carbon-carbon double bond in terms of orbital hybridization. As with free radicals and carbocations, the basis for our model is *sp^2-hybridized carbon* atoms.

Consider, for example, the simplest alkene, ethene. Molecules of ethene are known to be planar as shown in Fig. 6.1. The H—C—C bond angles of ethene are known to be $\sim 121°$, and the H—C—H bond angles are $\sim 118°$.

In our model for ethene (Fig. 6.2) we see that two sp^2-hybridized carbon atoms form a sigma (σ) bond between them by the overlap of one sp^2 orbital from each. The remaining sp^2 orbitals of each carbon form σ bonds to four hydrogens through overlap with the 1*s* orbitals of the hydrogen atoms. These five bonds account for 10 of the 12 bonding electrons of ethene, and they are called the *σ-bond framework*. The bond angles that we would predict on the basis of sp^2-hybridized carbon atoms (120° all around) are quite close to the bond angles that are actually found (Fig. 6.2).

The remaining two bonding electrons in our model are located in the *p* orbitals of each carbon. We can better visualize how these *p* orbitals interact with each other if we imagine the σ-bond framework, shown in Fig. 6.2, as being rotated perpendicular to the plane of this page. (It is also helpful, in this regard, to replace

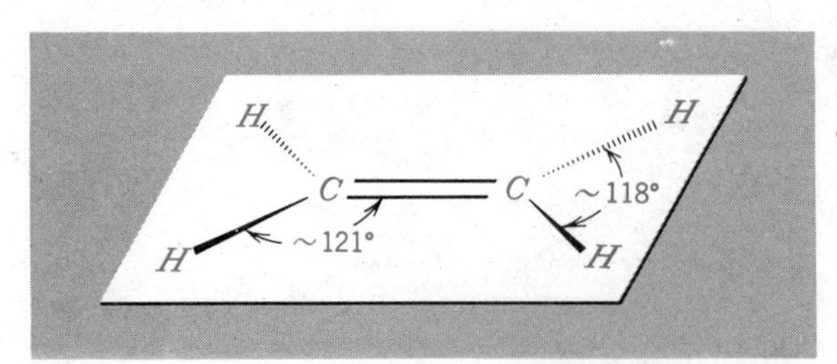

FIG. 6.1
The structure and bond angles of ethene.

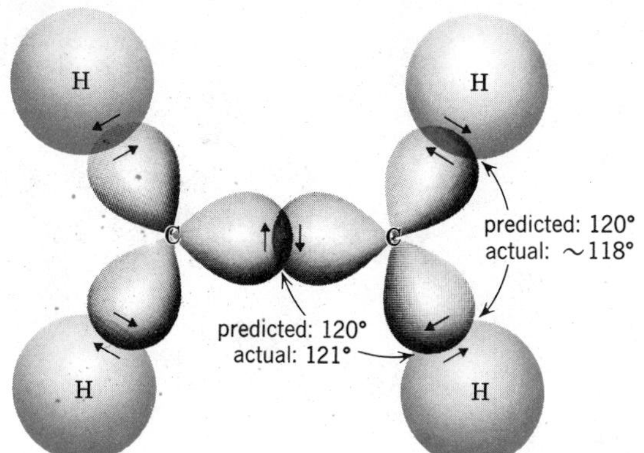

FIG. 6.2
The sigma-bond framework for ethene based on sp^2*-hybridized carbon atoms.*

the σ bonds by lines.) This new perspective is shown in Fig. 6.3. We see that the parallel *p* orbitals *overlap above and below the plane of the* σ *framework*. This sideways overlap of the *p* orbitals results in a new type of covalent bond, known as a *pi* (π) *bond*. Note the difference in shape of the bonding molecular orbital of a π bond as contrasted to that of a σ bond. A σ bond has cylindrical symmetry about a line connecting the two bonded nuclei. A π bond has a shape similar to that of a *p* orbital. In each case it is the shape of the bonding molecular orbital that is related to the name.

According to molecular orbital theory both bonding and antibonding π molecular orbitals are formed when *p* orbitals interact in this way to form a π bond. The bonding π orbital (Fig. 6.4) results when *p*-orbital lobes of like sign overlap; the antibonding π orbital is formed when *p*-orbital lobes of opposite signs overlap.

The bonding π orbital is the lower-energy orbital and contains both pi electrons (with opposite spins) in the ground state of the molecule. The region of greatest probability of finding the electrons in the bonding π orbital is a region generally situated between the two carbon atoms above and below the plane of the σ-bond framework. The antibonding π orbital is of higher energy, and it is not occupied by electrons when the molecule is in the ground state. It can become occupied, however, if the molecule absorbs light of the right frequency, and an electron is promoted from the lower energy level to the higher one. The antibonding π orbital has a nodal plane between the two carbon atoms.

To summarize: In our model based on orbital hybridization, the carbon-carbon double bond consists of two different kinds of bonds, *a* σ *bond and a* π *bond*. The σ bond is formed by two overlapping sp^2 orbitals and is symmetrical about an axis linking the two carbon atoms. The π bond is formed by a sideways overlap of two *p* orbitals; it has a shape similar to that of a *p* orbital. In the ground state the electrons of the π bond are located between the two carbon atoms but generally above and below the plane of the σ-bond framework.

Electrons of the π bond have greater energy than electrons of the σ bond. The relative energies of the σ and π orbitals (with the electrons in the ground state) are shown on the next page.

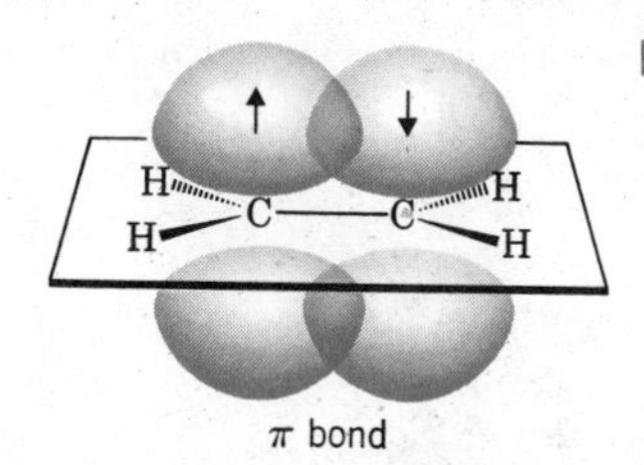

FIG. 6.3
The overlapping p *orbitals of ethene.*

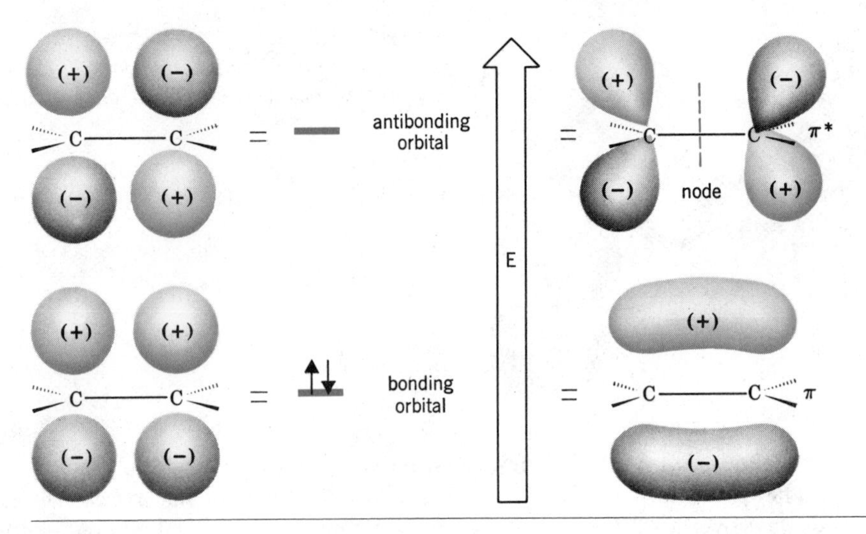

FIG. 6.4
The interaction of p *orbitals to form the bonding and antibonding molecular orbitals of a* π *bond.*

E
σ* orbital
π* orbital
antibonding
π orbital
σ orbital
bonding

6.3A Bond Lengths of Alkenes

Our model accounts for the fact that the carbon-carbon double bond is shorter than the carbon-carbon single bond (Fig. 6.5). With greater electron density between the two carbon nuclei in the double bond, the nuclei are drawn closer to each other.

The carbon-hydrogen bonds of ethene are slightly shorter than those of ethane. The sp^2 orbitals of ethene contain more s character than the sp^3 orbitals of ethane, and carbon-hydrogen bonds formed by sp^2 orbitals are generally shorter (Fig. 6.6).

6.3B Restricted Rotation and the Double Bond

The σ-π model for the carbon-carbon double bond also accounts for a property of the double bond that we first encountered in Chapter 2: *the large*

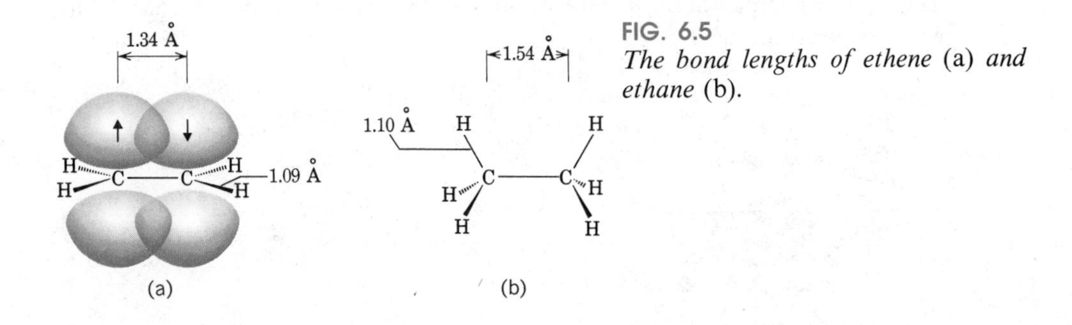

FIG. 6.5
The bond lengths of ethene (a) *and ethane* (b).

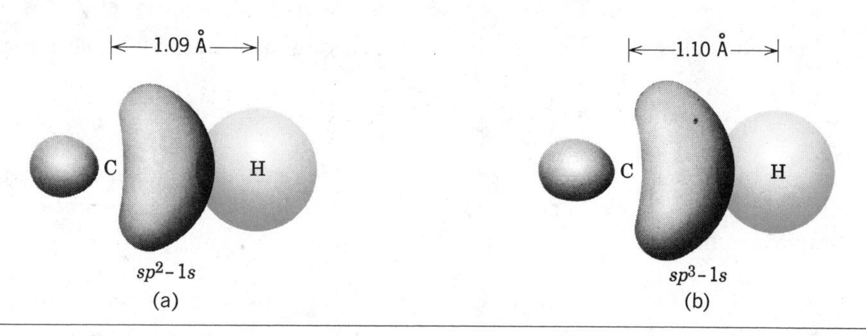

FIG. 6.6
Carbon-hydrogen bonds formed by sp^2 *(a) and* sp^3 *(b) carbon atoms. Because the* sp^2 *orbital has more s character the electrons are (on average) closer to the nucleus. This makes the C—H bond slightly shorter.*

barrier to free rotation associated with groups joined by a double bond. Maximum overlap between the *p* orbitals of a π bond occurs when the axes of the *p* orbitals are exactly parallel. Rotating one carbon of the double bond 90° (Fig. 6.7) breaks the π bond, for then the axes of the *p* orbitals are perpendicular and there is no net overlap between them. Estimates based on thermochemical calculations indicate that the strength of the π bond is 63 kcal/mole. This, then, is the barrier to rotation of the double bond. It is markedly higher than the rotational barrier of groups joined by carbon-carbon single bonds (3–6 kcal/mole). While groups joined by single bonds rotate freely at room temperature, those joined by double bonds do not. Whereas the separation of *anti-* and *gauche* isomers of butane (Section 3.6) is *impossible* at room temperature, the separation of *cis-* and *trans-*2-butene is not only possible but, once separated, they can be stored indefinitely at the normal laboratory temperature. *Cis-trans* isomers *can* be interconverted, but only if they are heated (to temperatures usually greater than 300°). Under these conditions the molecules acquire sufficient thermal energy to surmount the rotational barrier. *Cis-trans* isomers can also be interconverted by exposure to light of a wavelength absorbed by the double bond or by chemical means.

Problem 6.3

Can you explain how the light-activated interconversion of *cis-trans* isomers might occur?

The σ bond of ethene has been estimated to have a bond dissociation energy of 109 kcal/mole. Taken together, the σ bond and π bond give a total bond dissociation energy for the double bond of ~172 kcal/mole.

$$\begin{aligned} \sigma\text{-bond strength} &\simeq 109 \text{ kcal/mole} \\ \pi\text{-bond strength} &\simeq \underline{\ \ 63 \text{ kcal/mole}} \\ \text{Total} &\simeq 172 \text{ kcal/mole} \end{aligned}$$

FIG. 6.7
Rotation of a carbon atom of a double bond through an angle of 90° results in the breaking of the π bond.

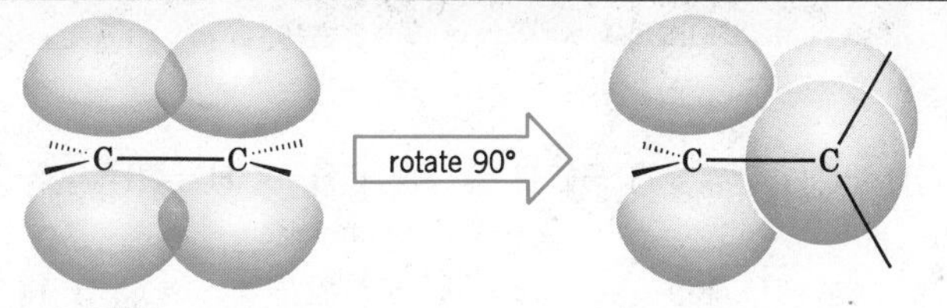

This is the energy needed for a homolytic scission of the ethene molecule, that is,

$$CH_2::CH_2 \longrightarrow 2CH_2: \qquad DH^\circ = 172 \text{ kcal/mole}$$

In the next chapter, when we discuss the reactions of alkenes, we will find that the relatively weaker π bond (~63 kcal/mole) *and its exposed electrons* are important in explaining the reactivity of alkenes.

6.4 HYDROGENATION OF ALKENES

Alkenes react with hydrogen in the presence of a variety of finely divided metal catalysts (cf. Section 3.16). The reaction that takes place is an *addition reaction;* one atom of hydrogen *adds* to each carbon of the double bond. The catalysts that are most commonly used are finely divided nickel, palladium, and platinum. Without a catalyst the reaction does not take place. (We will see how the catalyst functions in Section 6.14.)

$$CH_2{=}CH_2 + H_2 \xrightarrow[\text{or Pt} \atop 25^\circ]{\text{Ni, Pd}} CH_3{-}CH_3$$

$$CH_3CH{=}CH_2 + H_2 \xrightarrow[\text{or Pt} \atop 25^\circ]{\text{Ni, Pd}} CH_3CH_2{-}CH_3$$

The product that results from the addition of hydrogen to an alkene is an alkane. Alkanes have only single bonds and contain the maximum number of hydrogen atoms that a hydrocarbon can possess. For this reason, alkanes are said to be *saturated compounds.* Alkenes, because they contain a double bond and possess fewer than the maximum number of hydrogens, are capable of adding hydrogen and are said to be *unsaturated.* The process of adding hydrogen to an alkene is sometimes described as being one of *saturation.* Most often, however, the term used to describe the addition of hydrogen, is *hydrogenation.*

6.5 MOLECULAR FORMULAS OF HYDROCARBONS: THE INDEX OF HYDROGEN DEFICIENCY

Alkenes whose molecules contain only one double bond have the general formula C_nH_{2n}. They are isomeric with cycloalkanes. For example, 1-hexene and cyclohexane have the same molecular formula, C_6H_{12}:

$CH_2{=}CHCH_2CH_2CH_2CH_3$

1-Hexene
C_6H_{12}

Cyclohexane
C_6H_{12}

Cyclohexane and 1-hexene are structural isomers.

Alkenes with two double bonds (alkadienes) have the general formula C_nH_{2n-2}; alkenes with three double bonds (alkatrienes) have the general formula C_nH_{2n-4}, and so forth.

$CH_2{=}CH{-}CH{=}CH_2$
1,3-Butadiene
C_4H_6

$CH_2{=}CH{-}CH{=}CH{-}CH{=}CH_2$
1,3,5-Hexatriene
C_6H_8

A chemist working with an unknown hydrocarbon can obtain considerable information about its structure from its molecular formula and its *index of hydrogen deficiency*. The index of hydrogen deficiency is defined as the number of *pairs* of hydrogen atoms that must be subtracted from the molecular formula of the corresponding alkane to give the molecular formula of the compound under consideration.

For example, both cyclohexane and 1-hexene have an index of hydrogen deficiency equal to one (meaning one *pair* of hydrogen atoms). The corresponding alkane (i.e., the alkane with the same number of carbon atoms) is hexane.

C_6H_{14} = formula of corresponding alkane (hexane)
C_6H_{12} = formula of compound (1-hexene or cyclohexane)
H_2 = difference = 1 pair of hydrogens
Index of hydrogen deficiency = 1

The index of hydrogen deficiency of 1,3-butadiene equals 2; the index of hydrogen deficiency of 1,3,5-hexatriene equals 3. (Do the calculations.)

Because the index of hydrogen deficiency *equals the number of double bonds plus the number of rings* it tells us something about the structural possibilities for the molecular formula and it rules out many. To find out how the index of hydrogen deficiency breaks down into the number of double bonds and the number of rings is easily done experimentally. Molecules with double bonds add hydrogen in the presence of a catalyst at room temperature, whereas molecules with rings alone do not. Hydrogenation therefore allows us to distinguish between double bonds and rings. For example, 1-hexene reacts with one mole of hydrogen to yield hexane; under the same conditions cyclohexane does not react.

$$CH_2{=}CH(CH_2)_3CH_3 + H_2 \xrightarrow[25^\circ]{Pt} CH_3(CH_2)_4CH_3$$

$$\text{(cyclohexane)} + H_2 \xrightarrow[25^\circ]{Pt} \text{No reaction}$$

Or consider another example. Cyclohexene and 1,3-hexadiene have the same molecular formula, C_6H_{10}. Both compounds react with hydrogen in the presence of a catalyst, but cyclohexene, because it has a ring and only one double bond, reacts with only one mole. 1,3-hexadiene adds two moles.

$$\text{Cyclohexene} + H_2 \xrightarrow[25^\circ]{Pt} \text{(cyclohexane)}$$

Cyclohexene

$$CH_2{=}CHCH{=}CHCH_2CH_3 + 2H_2 \xrightarrow[25^\circ]{Pt} CH_3(CH_2)_4CH_3$$

Problem 6.4

(a) What is the index of hydrogen deficiency of 2-hexene? (b) Of methylcyclopentane? (c) Does the index of hydrogen deficiency reveal anything about the location of the double bond in the chain? (d) About the size of the ring? (e) Alkynes also react with hydrogen to yield alkanes. What is the index of hydrogen deficiency of an alkyne? (f) In general terms, what structural possibilities exist for a compound with the molecular formula $C_{10}H_{16}$?

Problem 6.5

Zingiberene, a fragrant compound isolated from ginger, has the molecular formula $C_{15}H_{24}$ and is known not to contain any triple bonds. (a) What is the index of hydrogen deficiency of zingiberene? (b) When zingiberene is subjected to catalytic hydrogenation using an excess of hydrogen, one mole of zingiberene absorbs three moles of hydrogen and produces a compound with the formula $C_{15}H_{30}$. How many double bonds does a molecule of zingiberene have? (c) How many rings?

6.6 HEATS OF HYDROGENATION: THE STABILITY OF ALKENES

Hydrogenation also provides us with a way to measure the relative stabilities of certain alkenes. The reaction of an alkene with hydrogen is an exothermic reaction; the amount of heat evolved is called *the heat of hydrogenation.* Most alkenes have heats of hydrogenation of approximately −30 kcal/mole. Individual alkenes, however, have heats of hydrogenation that differ from this average value by more than 2 kcal/mole.

$$>C{=}C< + H{-}H \xrightarrow{Pt} -\underset{H}{\underset{|}{C}}-\underset{H}{\underset{|}{C}}- \qquad \Delta H^\circ \simeq -30 \text{ kcal/mole}$$

These differences allow us to measure the relative stabilities of alkenes *when hydrogenation converts them to the same product.*

Consider, as examples, the three butene isomers shown below:

$$\underset{\text{1-Butene}}{CH_3CH_2CH{=}CH_2} + H_2 \xrightarrow{Pt} \underset{\text{Butane}}{CH_3CH_2CH_2CH_3} \qquad \Delta H^\circ = -30.3 \text{ kcal/mole}$$

$$\underset{\textit{cis}\text{-2-Butene}}{(CH_3)(H)C{=}C(CH_3)(H)} + H_2 \xrightarrow{Pt} \underset{\text{Butane}}{CH_3CH_2CH_2CH_3} \qquad \Delta H^\circ = -28.6 \text{ kcal/mole}$$

$$\underset{\textit{trans}\text{-2-Butene}}{(CH_3)(H)C{=}C(H)(CH_3)} + H_2 \xrightarrow{Pt} \underset{\text{Butane}}{CH_3CH_2CH_2CH_3} \qquad \Delta H^\circ = -27.6 \text{ kcal/mole}$$

In each reaction the product (butane) is the same. In each case, too, one of the reactants (hydrogen) is the same. Different amounts of *heat* are evolved in each reaction, however, and these differences must be related to different relative stabilities (different heat contents) of the individual butenes. 1-Butene evolves the greatest amount of heat when hydrogenated, and *trans*-2-butene evolves the least. Therefore 1-butene must have the greatest potential energy and be the least stable isomer. *Trans*-2-butene must have the lowest potential energy and be the most stable isomer. The potential energy (and stability) of *cis*-2-butene falls in between. The order of stabilities of the butenes is easier to see if we examine the potential energy diagram in Fig. 6.8.

The greater stability of the *trans*-2-butene when compared to *cis*-2-butene

FIG. 6.8
A potential energy diagram for the three butene isomers. The order of stability is trans-*2-butene* > cis-*2-butene* > *1-butene.*

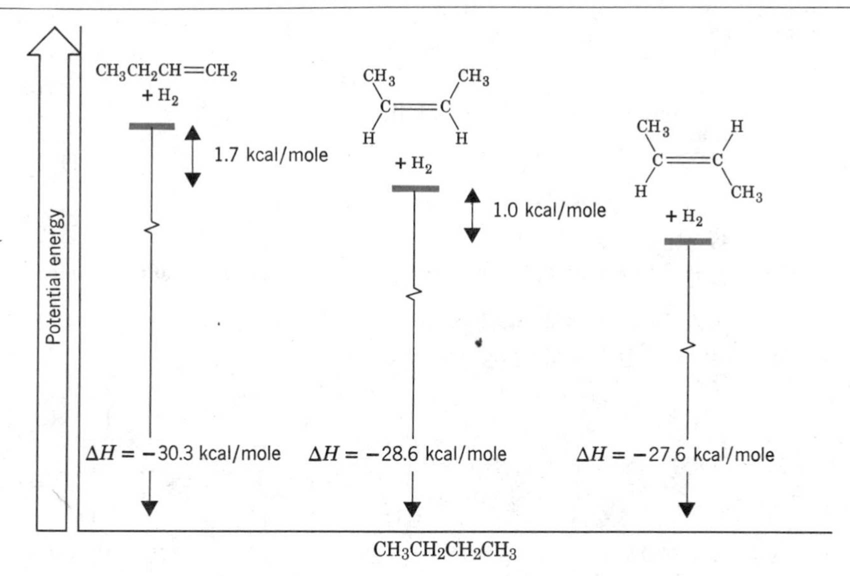

illustrates a general pattern found in *cis-trans* alkene pairs. The 2-pentenes, for example, show the same stability relationship: *trans* isomer > *cis* isomer.

$$\underset{cis\text{-2-Pentene}}{CH_3CH_2(H)C{=}C(H)CH_3} + H_2 \xrightarrow{Pt} \underset{\text{Pentane}}{CH_3CH_2CH_2CH_2CH_3} \qquad \Delta H^\circ = -28.6 \text{ kcal/mole}$$

$$\underset{trans\text{-2-Pentene}}{CH_3CH_2(H)C{=}C(H)CH_3} + H_2 \xrightarrow{Pt} \underset{\text{Pentane}}{CH_3CH_2CH_2CH_2CH_3} \qquad \Delta H^\circ = -27.6 \text{ kcal/mole}$$

The greater potential energy of *cis* isomers can be attributed to strain caused by the crowding of two alkyl groups on the same side of the double bond (Fig. 6.9)—a strain that is relieved when the molecule becomes saturated and therefore more flexible.

Studies of numerous alkenes also reveal a pattern of stabilities that is related to the number of alkyl groups attached to the carbon atoms of the double bond. *The greater the number of attached alkyl groups, (i.e., the more highly*

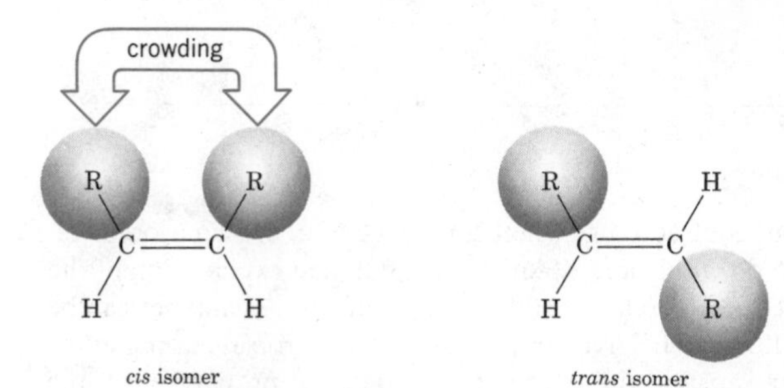

FIG. 6.9
Cis-*and* trans-*alkene isomers. The less stable* cis *isomer has greater strain.*

substituted the double bond) the greater is the alkene's stability. We can represent this order of stabilities in general terms, as follows:*

Relative stabilities of alkenes

$R_2C{=}CR_2$ > $R_2C{=}CRH$ >
Tetrasubstituted Trisubstituted

$RHC{=}CHR$ (trans) > $R_2C{=}CH_2$ ≃ $RHC{=}CHR$ (cis) > $RHC{=}CH_2$ > $H_2C{=}CH_2$

← Disubstituted → Monosubstituted Unsubstituted

A portion of this overall pattern can be seen by inspecting the heats of hydrogenation of the following pentenes.

$(CH_3)_2C{=}CHCH_3$ $\Delta H° = -26.9$ kcal/mole

$CH_3CH_2(CH_3)C{=}CH_2$ $\Delta H° = -28.5$ kcal/mole

$(CH_3)_2CHCH{=}CH_2$ $\Delta H° = -30.3$ kcal/mole

$R_2C{=}CHR$ > $R_2C{=}CH_2$ > $RHC{=}CH_2$

Trisubstituted Disubstituted Monosubstituted

Direct comparisons can be made for these pentenes because hydrogenation of each yields the same alkane, isopentane.

$CH_3CH(CH_3){-}CH_2CH_3$

Isopentane

Problem 6.6

Another C_4H_8 isomer, 2-methylpropene, cannot be compared directly with the butene isomers that we have already discussed (*cis*-2-butene, *trans*-2-butene, and 1-butene) because 2-methylpropene yields a different product on hydrogenation. (a) What is this product? (b) What products would all four C_4H_8 isomers yield on complete combustion? (c) Could heats of combustion be used to determine the relative stabilities of these four C_4H_8 isomers? (d) What are two other C_4H_8 isomers? (e) Could their relative stabilities be determined in the same way?

Problem 6.7

Select the more stable alkene of each pair

(a) 1-Heptene or *cis*-2-heptene

* This order of stabilities may seem contradictory when compared with explanation given for the relative stabilities of *cis* and *trans* isomers. Although a detailed explanation of the trend given here is beyond our scope, the relative stabilities of substituted alkenes can be rationalized. Part of the explanation can be given in terms of the electron-releasing effect of alkyl groups; an effect that satisfies the electron-withdrawing properties of the sp^2-hybridized carbons of the double bond.

(b) *cis*-2-Heptene or *trans*-2-heptene
(c) *trans*-2-Heptene or 2-methyl-2-hexene
(d) 2-Methyl-2-hexene or 2,3-dimethyl-2-pentene

Problem 6.8

Reconsider the pairs of alkenes given in Problem 6.7. For which pairs could you use heats of hydrogenation to determine their relative stabilities? For which pairs would you be required to use heats of combustion (as in problem 6.6)?

6.7 CYCLOALKENES

The rings of cycloalkenes containing six carbon atoms or fewer exist only in the *cis* form (Fig. 6.10). The introduction of a *trans* double bond into rings this small would, if it were possible, introduce greater strain than the bonds of the ring atoms could accommodate. If it existed, *trans*-cyclohexene might resemble the structure shown in Fig. 6.11.

trans-Cycloheptene has been observed with instruments called spectrometers, but it is a substance with a very short lifetime and has not been isolated.

trans-Cyclooctene (Fig. 6.12) has been isolated, however. Here the ring is large enough to accommodate the geometry required by a *trans* double bond.

6.8 SYNTHESIS OF ALKENES THROUGH ELIMINATION REACTIONS

Eliminations are the most widely used reactions for synthesizing alkenes. In this chapter we will examine four methods: dehydrohalogenation of alkyl halides (—HX), dehydration of alcohols (—H_2O), debromination of dibromoalkanes (—Br_2), and dehydrogenation of alkanes (—H_2).

FIG. 6.10
Cis-*cycloalkenes.*

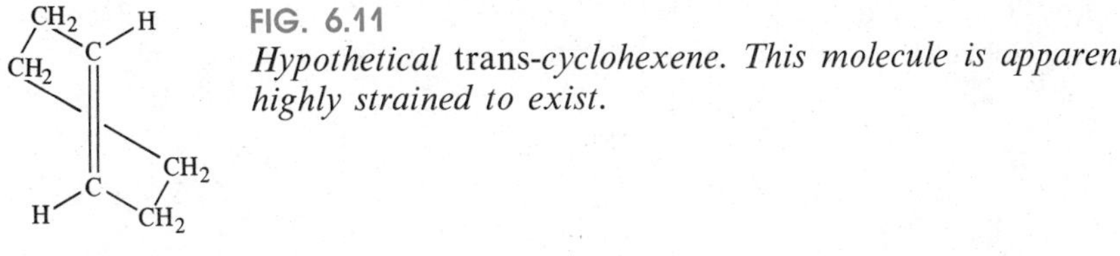

Cyclopropene · Cyclobutene · Cyclopentene · Cyclohexene

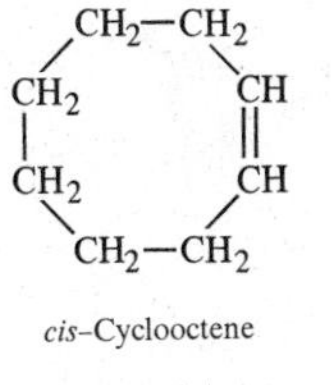

FIG. 6.11
Hypothetical trans-*cyclohexene. This molecule is apparently too highly strained to exist.*

cis-Cyclooctene · *trans*-Cyclooctene

FIG. 6.12
Cis *and* trans *forms of cyclooctene.*

6.9 DEHYDROHALOGENATION: THE QUESTION OF THE ORIENTATION OF THE DOUBLE BOND

When we attempt to synthesize an alkene by dehydrohalogenation we almost always attempt to create conditions for an E2 reaction:

$$\text{B:} \rightarrow \text{H}-\overset{|}{\underset{|}{\text{C}}}\beta-\overset{|}{\underset{|}{\text{C}}}\alpha-\text{X} \xrightarrow{\text{E2}} \text{>C=C<} + \text{B:H} + \text{:X}^-$$

The reason for this choice is that dehydrohalogenation by an E1 mechanism is too unreliable. Too many competing events are possible, one being rearrangement of the carbon skeleton (Section 6.12). In order to bring about an E2 reaction we use a secondary or tertiary alkyl halides if possible. (If we must begin with a primary halide we use a bulky base.) To try to avoid E1 conditions we use a high concentration of a strong, relatively nonpolarizable base such as an alkoxide ion, and we use a relatively nonpolar solvent such as an alcohol. To favor elimination generally, we use a relatively high temperature. The typical reagents for dehydrohalogenation are sodium ethoxide in ethanol and potassium *tert*-butoxide in *tert*-butyl alcohol. Potassium hydroxide in ethanol is also used sometimes; in this reagent the reactive bases probably include the ethoxide ion formed by the following equilibrium.

$$OH^- + C_2H_5OH \rightleftharpoons H_2O + C_2H_5O^-$$

In our earlier examples of dehydrohalogenations (Sections 5.9–5.12) only a single elimination product was possible. For example:

$$CH_3\underset{\text{Br}}{\underset{|}{C}}HCH_3 \xrightarrow[\substack{C_2H_5OH \\ 55°}]{C_2H_5O^-Na^+} \underset{(79\%)}{CH_2{=}CHCH_3}$$

$$CH_3\overset{CH_3}{\overset{|}{\underset{\text{Br}}{\underset{|}{C}}}}CH_3 \xrightarrow[\substack{C_2H_5OH \\ 55°}]{C_2H_5O^-Na^+} \underset{(100\%)}{CH_2{=}\overset{CH_3}{\overset{|}{C}}{-}CH_3}$$

$$CH_3(CH_2)_{15}CH_2CH_2Br \xrightarrow[\substack{(CH_3)_3COH \\ 40°}]{(CH_3)_3CO^-K^+} CH_3(CH_2)_{15}CH{=}CH_2$$

Dehydrohalogenation of most alkyl halides, however, yields more than one product. For example, dehydrohalogenation of 2-bromo-2-methylbutane (below) can yield two products: 2-methyl-2-butene or 2-methyl-1-butene.

$$\text{B:} \xrightarrow{(a)} CH_3\overset{H}{\overset{|}{C}}H-\overset{\overset{H}{|}}{\overset{CH_2}{\overset{|}{\underset{CH_3}{\underset{|}{C}}}}}-Br \begin{cases} \xrightarrow{(a)} CH_3CH{=}C(CH_3)CH_3 \quad \text{2-Methyl-2-butene} \\ \xrightarrow{(b)} CH_3CH_2C({=}CH_2)CH_3 \quad \text{2-Methyl-1-butene} \end{cases}$$

2-Bromo-2-methylbutane

2-Methyl-2-butene

2-Methyl-1-butene

Which product predominates depends on the base we use, particularly on its steric bulk near the site of the negative charge. If we use a relatively unhindered base such as ethoxide ion, the major product will be *the more stable alkene*. The more stable alkene, as we know from Section 6.6, has the more highly substituted double bond.

$$CH_3CH_2O^- + CH_3CH_2{-}\underset{CH_3}{\overset{CH_3}{C}}{-}Br \xrightarrow[CH_3CH_2OH]{70^\circ} CH_3CH{=}C(CH_3)_2 + CH_3CH_2C(=CH_2)CH_3$$

(69%) 2-Methyl-2-butene (more stable)

(31%) 2-Methyl-1-butene (less stable)

2-Methyl-2-butene is a trisubstituted alkene (three methyl groups are attached to carbons of the double bond), whereas 2-methyl-1-butene is only disubstituted. 2-Methyl-2-butene is the major product.

The reason for this behavior appears to be related to double-bond character that develops in the transition state (cf. Section 5.10) for each reaction:

$$C_2H_5\overset{\delta-}{O}{\cdots}H{-}C{\cdots}C{-}Br^{\delta-}$$

Transition state for an E2 reaction

The carbon-carbon bond has some of the character of a double bond

The transition state for the reaction leading to 2-methyl-2-butene resembles the product of the reaction: a trisubstituted alkene. The transition state for the reaction leading to 2-methyl-1-butene resembles its product: a disubstituted alkene. Since the transition state leading to 2-methyl-2-butene resembles a more stable alkene, this transition state is more stable. Since this transition state is more stable (occurs at lower potential energy), the energy of activation for this reaction is lower and 2-methyl-2-butene is formed faster. This explains why 2-methyl-2-butene is the major product.

Whenever an elimination occurs to give the more highly substituted alkene chemists say that the elimination follows the *Zaitzev rule* to honor the nineteenth-century-Russian chemist A. N. Zaitzev (1841–1910) who formulated it. (Zaitzev's name is also transliterated as Saytseff or Saytzev.)

Carrying out dehydrohalogenations with a sterically hindered base such as potassium *tert*-butoxide favors the formation of *the less-substituted alkene:*

$$CH_3{-}\underset{CH_3}{\overset{CH_3}{C}}{-}O^- + CH_3CH_2{-}\underset{CH_3}{\overset{CH_3}{C}}{-}Br \xrightarrow[(CH_3)_3COH]{75^\circ} CH_3CH{=}C(CH_3)_2 + CH_3CH_2C(=CH_2)CH_3$$

(27.5%) 2-Methyl-2-butene (more substituted)

(72.5%) 2-Methyl-1-butene (less substituted)

This behavior seems to be related to the steric bulk of the base. The large *tert*-butoxide ion appears to have difficulty removing one of the internal (2°) hydrogens. It removes one of the more exposed (1°) hydrogens of the methyl group instead.

Whenever the less substituted alkene forms in an elimination, chemists say that the elimination follows the *Hofmann rule* to honor the nineteenth-century German chemist A. W. Hofmann (1818–1895). Hofmann originally developed his rule to account for the results of a special elimination reaction now called *the Hofmann elimination* (Section 18.13).

Problem 6.9

Dehydrohalogenation of 2-bromobutane with ethoxide ion yields a mixture of 2-butene and 1-butene. (a) Which butene would you expect to predominate? (b) The 2-butene formed in the reaction is a mixture of *cis*-2-butene and *trans*-2-butene. Which 2-butene would you expect to predominate?

6.10 SYNTHESIS OF ALKENES BY DEHYDRATION OF ALCOHOLS

Heating most alcohols with a strong acid causes them to lose a molecule of water (to *dehydrate*) and form an alkene:

$$-\underset{H}{\overset{|}{\underset{|}{C}}}-\underset{OH}{\overset{|}{\underset{|}{C}}}- \xrightarrow[\text{heat}\ (-H_2O)]{H^+} \gt C=C \lt$$

The most commonly used acids are Bronsted acids—proton donors such as sulfuric acid and phosphoric acid, or Lewis acids such as alumina (Al_2O_3).

Dehydration reactions of alcohols show several important characteristics:

1. The experimental conditions—temperature and acid concentration—that are required to bring about dehydration are closely related to the structure of individual alcohols. Alcohols in which the hydroxyl group is attached to a primary carbon (primary alcohols) are the most difficult to dehydrate. Dehydration of ethyl alcohol, for example, requires concentrated sulfuric acid and a temperature of 180°.

$$H-\underset{H}{\overset{H}{\underset{|}{\overset{|}{C}}}}-\underset{O-H}{\overset{H}{\underset{|}{\overset{|}{C}}}}-H \xrightarrow[H_2SO_4\ 180°]{\text{conc.}} H-\overset{H}{\overset{|}{C}}=\overset{H}{\overset{|}{C}}-H + H_2O$$

Ethyl alcohol
(a 1° alcohol)

Dehydration of propyl alcohol, also a primary alcohol, requires concentrated acid at a similarly high temperature.

$$\underset{\text{Propyl alcohol}}{CH_3CH_2CH_2OH} \xrightarrow[170°]{\text{conc. } H_2SO_4} \underset{\text{Propene}}{CH_3CH=CH_2} + H_2O$$

Secondary alcohols usually dehydrate under milder conditions. Cyclohexanol, for example, dehydrates in 85% phosphoric acid at 165 to 170°.

$$\text{Cyclohexanol} \xrightarrow[165\text{–}170°]{85\%\ H_3PO_4} \text{Cyclohexene (80\%)} + H_2O$$

Tertiary alcohols are usually so easily dehydrated that extremely mild conditions can be used. *tert*-Butyl alcohol, for example, dehydrates in 20% aqueous sulfuric acid at a temperature of 85°.

$$\underset{\substack{tert\text{-Butyl}\\ \text{alcohol}}}{CH_3-\overset{\overset{\displaystyle CH_3}{|}}{\underset{\underset{\displaystyle CH_3}{|}}{C}}-OH} \xrightarrow[85°]{20\%\ H_2SO_4} \underset{\substack{\text{2-Methylpropene}\\ (84\%)}}{CH_3-\overset{\overset{\displaystyle CH_3}{|}}{\underset{\underset{\displaystyle CH_2}{\|}}{C}}} + H_2O$$

Thus, overall, the relative ease with which alcohols undergo dehydration is in the following order:

$$\text{Ease of dehydration:}\quad \underset{3°\ \text{Alcohol}}{R-\overset{\overset{\displaystyle R}{|}}{\underset{\underset{\displaystyle R}{|}}{C}}-OH} > \underset{2°\ \text{Alcohol}}{R-\overset{\overset{\displaystyle R}{|}}{\underset{\underset{\displaystyle H}{|}}{C}}-OH} > \underset{1°\ \text{Alcohol}}{R-\overset{\overset{\displaystyle H}{|}}{\underset{\underset{\displaystyle H}{|}}{C}}-OH}$$

2. Certain alcohols dehydrate to give more than one product. When 1-butanol dehydrates, for example, three products form: *trans*-2-butene, *cis*-2-butene, and 1-butene.

$$CH_3CH_2CH_2CH_2OH \xrightarrow[170°]{\text{conc. } H_2SO_4} \underset{\substack{trans\text{-2-Butene}\\ \text{(major product)}}}{\overset{CH_3}{\underset{H}{}}\!\!>C=C<\!\!\overset{H}{\underset{CH_3}{}}} + \underset{\substack{cis\text{-2-Butene}\\ \text{(minor product)}}}{\overset{CH_3}{\underset{H}{}}\!\!>C=C<\!\!\overset{CH_3}{\underset{H}{}}} + \underset{\substack{\text{1-Butene}\\ \text{(minor product)}}}{\overset{CH_3CH_2}{\underset{H}{}}\!\!>C=C<\!\!\overset{H}{\underset{H}{}}} + H_2O$$

The dehydration of 2-methylcyclohexanol gives both 1-methylcyclohexene and 3-methylcyclohexene.

$$\underset{\substack{\text{2-Methyl-}\\ \text{cyclohexanol}}}{\text{(cyclohexane ring with }CH_3\text{ and OH)}} \xrightarrow{\substack{85\%\\ H_3PO_4}} \underset{\substack{\text{1-Methyl-}\\ \text{cyclohexene}\\ \text{(major product)}}}{\text{(ring, }CH_3\text{)}} + \underset{\substack{\text{3-Methyl-}\\ \text{cyclohexene}\\ \text{(minor product)}}}{\text{(ring, }CH_3\text{)}} + H_2O$$

3. Some primary and secondary alcohols also undergo *rearrangements of their carbon skeleton* during dehydration. Such a rearrangement occurs in the dehydration of 3,3-dimethyl-2-butanol.

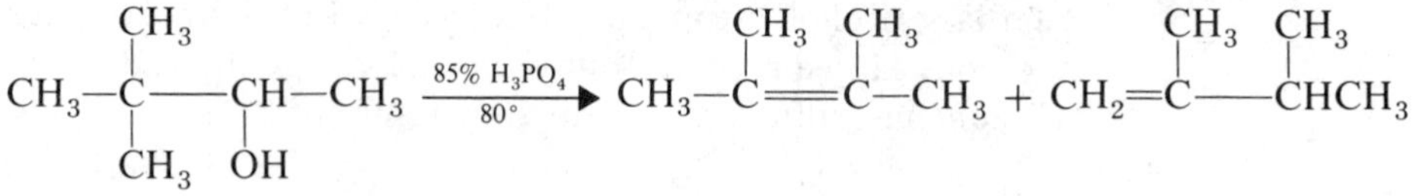

$$CH_3{-}\overset{\overset{CH_3}{|}}{\underset{\underset{CH_3}{|}}{C}}{-}\underset{\underset{OH}{|}}{CH}{-}CH_3 \xrightarrow[80^\circ]{85\%\ H_3PO_4} CH_3{-}\overset{\overset{CH_3}{|}}{C}{=}\overset{\overset{CH_3}{|}}{C}{-}CH_3 + CH_2{=}\overset{\overset{CH_3}{|}}{C}{-}\overset{\overset{CH_3}{|}}{C}HCH_3$$

3,3-Dimethyl-2-butanol — 2,3-Dimethyl-2-butene (80%) — 2,3-Dimethyl-1-butene (20%)

Notice that the carbon skeleton of the reactant is

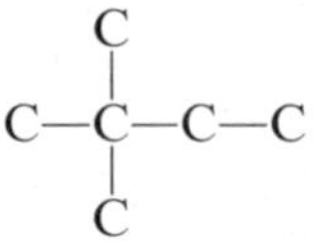

$$C{-}\overset{\overset{C}{|}}{\underset{\underset{C}{|}}{C}}{-}C{-}C$$

while that of the products is

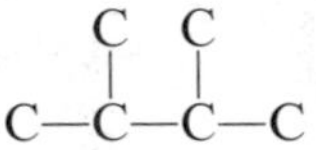

$$C{-}\overset{\overset{C}{|}}{C}{-}\overset{\overset{C}{|}}{C}{-}C$$

Explanations for all of these observations can be given on the basis of a stepwise mechanism originally proposed by F. Whitmore (of Pennsylvania State University). If we examine the mechanism carefully we will see that it is *an E1 reaction in which the substrate is a protonated alcohol* (*or an alkyloxonium ion,* see Section 5.6). We use the dehydration of *tert*-butyl alcohol as an example.

Step 1
$$CH_3{-}\overset{\overset{CH_3}{|}}{\underset{\underset{CH_3}{|}}{C}}{-}\ddot{\underset{\cdot\cdot}{O}}{-}H + H{-}\overset{\overset{H}{|}}{\underset{\underset{H}{|}}{\ddot{O}}}{:}^{+} \rightleftharpoons CH_3{-}\overset{\overset{CH_3}{|}}{\underset{\underset{CH_3}{|}}{C}}{-}\overset{\overset{H}{|}}{\underset{\cdot\cdot}{O}}{}^{+}{-}H + H{:}\ddot{\underset{\cdot\cdot}{O}}{-}H$$

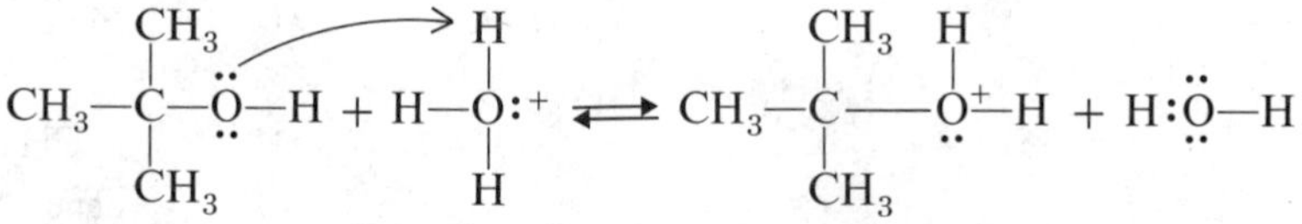

Protonated alcohol or alkyloxonium ion

In this step, an acid-base reaction, a proton is transferred from the acid to one of the unshared electron pairs of the alcohol. In dilute sulfuric acid the acid is a hydronium ion; in concentrated sulfuric acid the proton donor is sulfuric acid itself.

The presence of the positive charge on the oxygen of the protonated alcohol weakens all bonds from oxygen including the carbon-oxygen bond, and in step 2 the carbon-oxygen bond breaks. The leaving group is a molecule of water:

Step 2
$$CH_3{-}\overset{\overset{CH_3}{|}}{\underset{\underset{CH_3}{|}}{C}}{-}\overset{\overset{H}{|}}{\underset{\cdot\cdot}{O}}{}^{+}{-}H \rightleftharpoons CH_3{-}\overset{\overset{CH_3}{|}}{\underset{\underset{CH_3}{|}}{C}}{}^{+} + {:}\overset{\overset{H}{|}}{\underset{\cdot\cdot}{O}}{-}H$$

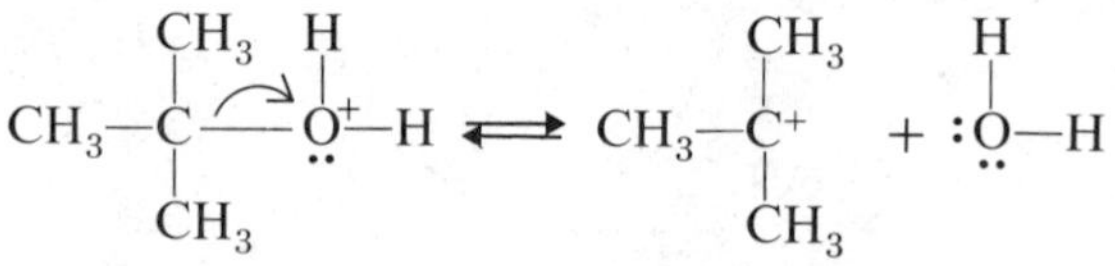

A carbocation

The carbon-oxygen bond breaks *heterolytically*. The bonding electrons depart with the water molecule and leave behind a carbocation. The carbocation is, of course, highly reactive because the central carbon atom has only six electrons in its valence level, not eight.

Finally, in step 3, the carbocation stabilizes itself by transferring a proton to a molecule of water. The result is the formation of a hydronium ion and an alkene.

$$\text{Step 3}\quad CH_3-\underset{CH_3}{\overset{H-\overset{H}{\overset{|}{C}}-H}{\overset{|}{C^+}}} + :\ddot{O}(H)-H \rightleftharpoons CH_3-\underset{CH_3}{\overset{CH_2}{\overset{\|}{C}}} + H-\overset{H}{\overset{|}{\ddot{O}^+}}-H$$

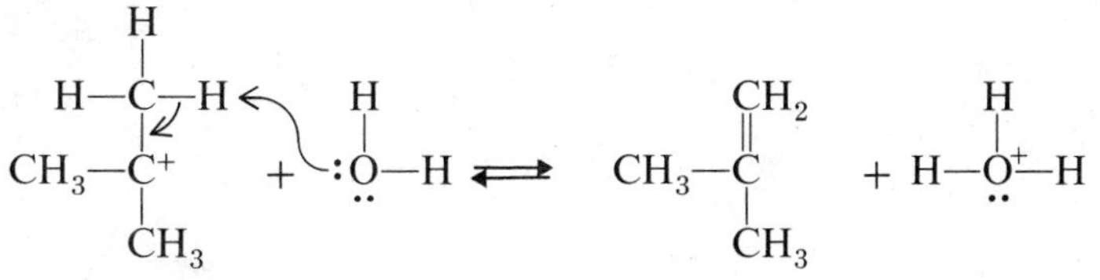

2-Methylpropene

In step 3, also an acid-base reaction, any one of the nine protons available at the three methyl groups can be transferred to a molecule of water. The electron pair that bonded the hydrogen to the carbon in the carbocation becomes the second bond of the double bond of the alkene.

Problem 6.10

(a) What would the leaving group have to be for the alcohol itself (rather than the protonated alcohol) to undergo dehydration? (b) How does this explain the requirement for an acid catalyst in alcohol dehydrations?

By itself, the Whitmore mechanism does not explain the observed order of reactivity of alcohols: tertiary > secondary > primary. Taken alone, it does not explain the formation of more than one product in the dehydration of certain alcohols nor the occurrence of a rearranged carbon skeleton in the dehydration of others. But, when coupled with what is known about *the stability of carbocations,* the Whitmore mechanism *does* eventually account for all of these observations.

6.11 CARBOCATION STABILITY AND THE TRANSITION STATE

We saw in Section 4.13 that the order of stability of carbocations is tertiary > secondary > primary > methyl:

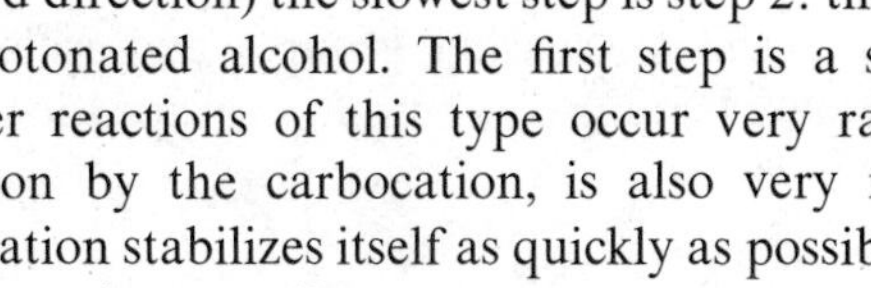

$$\underset{3^\circ}{R-\overset{R}{\overset{|}{\underset{R}{\underset{|}{C^+}}}}} > \underset{2^\circ}{R-\overset{H}{\overset{|}{\underset{R}{\underset{|}{C^+}}}}} > \underset{1^\circ}{R-\overset{H}{\overset{|}{\underset{H}{\underset{|}{C^+}}}}} > \underset{\text{Methyl}}{H-\overset{H}{\overset{|}{\underset{H}{\underset{|}{C^+}}}}}$$

In the dehydration of alcohols, (i.e., following the equations below in the forward direction) the slowest step is step 2: the formation of the carbocation from the protonated alcohol. The first step is a simple acid-base reaction. Proton-transfer reactions of this type occur very rapidly. The third step, the loss of a proton by the carbocation, is also very rapid because the highly reactive carbocation stabilizes itself as quickly as possible by transferring a proton to water and becoming an alkene.

General Mechanism for the Acid-Catalyzed Dehydration of an Alcohol

$$\text{Step 1}\quad -\underset{H}{\underset{|}{\overset{|}{C}}}-\underset{|}{\overset{|}{C}}-\ddot{O}H + H_3O:^+ \rightleftharpoons -\underset{H}{\underset{|}{\overset{|}{C}}}-\underset{|}{\overset{|}{C}}-\overset{H}{\overset{|}{\ddot{O}^+}}-H + H_2\ddot{O}: \qquad \text{fast}$$

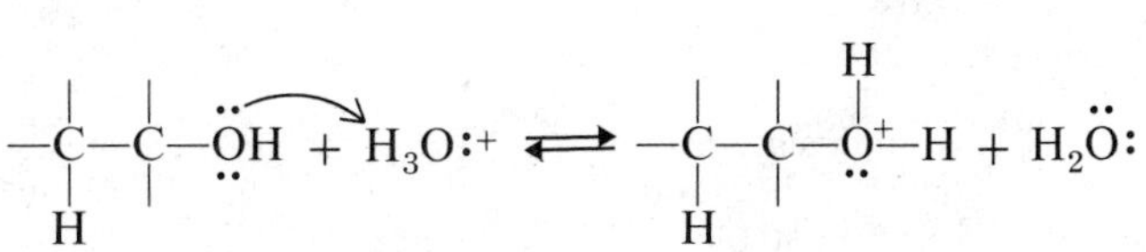

$$\text{Step 2}\quad -\overset{|}{\underset{\mathrm{H}}{\overset{|}{\mathrm{C}}}}-\overset{|}{\underset{|}{\mathrm{C}}}-\overset{\mathrm{H}}{\overset{|}{\ddot{\mathrm{O}}}}{}^{+}-\mathrm{H} \rightleftharpoons -\underset{\mathrm{H}}{\overset{|}{\mathrm{C}}}-\overset{|}{\underset{+}{\mathrm{C}}}- + \mathrm{H_2\ddot{O}}: \qquad \text{slow (rate-limiting)}$$

$$\text{Step 3}\quad -\underset{\mathrm{H}}{\overset{|}{\mathrm{C}}}-\overset{|}{\underset{+}{\mathrm{C}}}- + \mathrm{H_2\ddot{O}}: \rightleftharpoons -\overset{|}{\mathrm{C}}=\overset{|}{\mathrm{C}}- + \mathrm{H_3O}:^{+} \qquad \text{fast}$$

Because step 2 is, then, the rate-limiting step, step 2 is the step that determines the reactivity of alcohols toward dehydration. With this in mind, we can now understand why tertiary alcohols are the most easily dehydrated. The formation of a tertiary carbocation is easiest because the energy of activation for step 2 of a reaction leading to a tertiary carbocation is lowest.

The reactions by which carbocations are formed from protonated alcohols are all highly *endothermic*. According to Hammond's postulate (Section 4.10A), there should be a strong resemblance between the transition state and the product in each case. Of the three, *the transition state that leads to the tertiary carbocation is lowest in potential energy because it resembles the most stable product*. By contrast, the transition state that leads to the primary carbocation occurs at highest potential energy because it resembles the least stable product. In each instance, moreover, the transition state is stabilized by the same factor that stabilizes the carbocation itself: *by delocalization of the charge*. We can understand this if we examine the process by which the transition state is formed.

The oxygen atom of the protonated alcohol bears a full positive charge. As the transition state develops this oxygen begins to separate from the carbon to which it is attached. The carbon, because it is losing the electrons that bonded it to the oxygen, begins to develop a partial positive charge. This developing positive charge *is most effectively delocalized in the transition state leading to a tertiary carbocation because of the presence of three electron-releasing alkyl groups*. The positive charge is less effectively delocalized in the transition state leading to a

$$-\overset{|}{\underset{|}{\mathrm{C}}}-\overset{\mathrm{H}}{\overset{|}{\ddot{\mathrm{O}}}}{}^{+}-\mathrm{H} \rightleftharpoons -\overset{|}{\underset{|}{\mathrm{C}}}{}^{\delta+}\text{- - -}\overset{\mathrm{H}}{\overset{|}{\ddot{\mathrm{O}}}}{}^{\delta+}-\mathrm{H} \rightleftharpoons -\overset{|}{\underset{|}{\mathrm{C}}}{}^{+} + :\overset{\mathrm{H}}{\overset{|}{\ddot{\mathrm{O}}}}-\mathrm{H}$$

Protonated alcohol — Transition state — Carbocation

secondary carbocation (*two* electron-releasing groups) and is least effectively delocalized in the transition state leading to a primary carbocation (*one* electron-releasing group).

$$\overset{\delta+}{\mathrm{R}}\!\downarrow,\ {}^{\delta+}\mathrm{R}\!\rightarrow\!\mathrm{C}^{\delta+}\text{- - - - - -}\overset{\mathrm{H}}{\overset{|}{\underset{\delta+}{\ddot{\mathrm{O}}}}}-\mathrm{H},\ \underset{\delta+}{\mathrm{R}}\!\uparrow$$

Transition state leading to 3° carbocation (most stable)

$$\overset{\delta+}{\mathrm{R}}\!\downarrow,\ \overset{\delta+}{\mathrm{R}}\!\rightarrow\!\underset{\mathrm{H}}{\overset{|}{\mathrm{C}}}{}^{\delta+}\text{- - -}\overset{\mathrm{H}}{\overset{|}{\underset{\delta+}{\ddot{\mathrm{O}}}}}-\mathrm{H}$$

Transition state leading to 2° carbocation

$$\overset{\delta+}{\mathrm{R}}\!\rightarrow\!\overset{\mathrm{H}}{\underset{\mathrm{H}}{\overset{|}{\underset{|}{\mathrm{C}}}}}{}^{\delta+}\text{- - -}\overset{\mathrm{H}}{\overset{|}{\underset{\delta+}{\ddot{\mathrm{O}}}}}-\mathrm{H}$$

Transition state leading to 1° carbocation (least stable)

6.12 CARBOCATION STABILITY AND THE OCCURRENCE OF MOLECULAR REARRANGEMENTS

With an understanding of carbocation stability and its effect on transition states behind us, we now proceed to explain the rearrangements of carbon skeletons that occur in some alcohol dehydrations. For example, let us consider again the rearrangement that occurs when 3,3-dimethyl-2-butanol is dehydrated.

$$CH_3-C(CH_3)_2-CH(OH)-CH_3 \xrightarrow[\text{heat}]{85\%\ H_3PO_4} CH_3-C(CH_3){=}C(CH_3)-CH_3 + CH_2{=}C(CH_3)-CH(CH_3)CH_3$$

3,3-Dimethyl-2-butanol → 2,3-Dimethyl-2-butene (major product) + 2,3-Dimethyl-1-butene (minor product)

The first step of this dehydration is the formation of the protonated alcohol in the usual way:

$$(1)\quad CH_3-C(CH_3)_2-CH(\ddot{O}-H)-CH_3 + H-\ddot{O}^{+}H_2 \rightleftharpoons CH_3-C(CH_3)_2-CH(\overset{+}{O}H_2)CH_3 + H_2\ddot{O}:$$

Protonated alcohol

In the second step the protonated alcohol loses water and a secondary carbocation forms:

$$(2)\quad CH_3-C(CH_3)_2-CH(\overset{+}{O}H_2)-CH_3 \rightleftharpoons CH_3-C(CH_3)_2-\overset{+}{C}HCH_3 + H_2\ddot{O}:$$

A 2° carbocation

Now the rearrangement occurs. Before anything else can happen to it, *a less stable, secondary carbocation rearranges to a more stable tertiary carbocation.*

$$(3)\quad CH_3-C(CH_3)_2-\overset{+}{C}HCH_3 \longrightarrow CH_3-C(CH_3)\overset{+}{\cdots}(CH_3)\cdots CHCH_3 \longrightarrow CH_3-\overset{+}{C}(CH_3)-CH(CH_3)-CH_3$$

2° Carbocation (less stable) → Transition state → 3° Carbocation (more stable)

The rearrangement occurs through the migration of an alkyl group (methyl) from the carbon adjacent to the one with the positive charge. The methyl group migrates *with its pair of electrons,* that is, as a methanide, $^{-}:CH_3$, ion. After the migration is complete, the carbon that the methanide ion left has become a carbocation, and the positive charge on the carbon to which it migrated has been neutralized.

The transition state for the methyl migration has a *three-center bond.* In the transition state the shifting methyl is partly bonded to both carbon atoms by the pair of electrons with which it migrates. It never leaves the carbon skeleton.

The final step of the reaction is the loss of proton from the new carbocation and the formation of an alkene. This step, however, can occur in two ways.

$$(a)\ H-CH_2-\overset{+}{C}(CH_3)-C(H)(CH_3)-CH_3 \xrightarrow{(a)} CH_2{=}C(CH_3)-CHCH_3(CH_3) \text{ (minor product) Less stable alkene}$$

$$\xrightarrow{(b)} CH_3-C(CH_3){=}C(CH_3)-CH_3 \text{ (major product) More stable alkene}$$

The more favored route is dictated by the type of the alkene being formed. Path (*b*) leads to the highly stable tetrasubstituted alkene and this is the path followed by most of the carbocations. Path (*a*), on the other hand, leads to a less stable, disubstituted alkene and produces the minor product of the reaction. *The formation of the more stable alkene is the general rule in the acid-catalyzed dehydration reactions of alcohols.*

When we examine the rearrangement that occurs in the dehydration of *n*-butyl alcohol we find that the mechanism is similar to that for 3,3-dimethyl-2-butanol. In this instance, however, the migrating group is a hydride ion, $H:^-$—*a group that also migrates with a pair of electrons.*

$$(1)\ CH_3CH_2CH_2CH_2OH + H^+ \rightleftharpoons CH_3CH_2CH_2CH_2\overset{+}{O}H_2$$

$$(2)\ CH_3CH_2CH_2CH_2\overset{+}{O}H_2 \rightleftharpoons CH_3CH_2CH_2CH_2^{\ +} + H_2O$$

A 1° carbocation

$$(3)\ CH_3CH_2CH(H)-\overset{+}{C}H_2 \longrightarrow CH_3CH_2CH\overset{+}{\cdots}H\cdots CH_2 \longrightarrow CH_3CH_2\overset{+}{C}H-CH_2(H)$$

1° Carbocation — Transition state — 2° Carbocation

$$(4)\ CH_3CH(H)-\overset{+}{C}H-CH_2(H) \xrightarrow[H_2\ddot{O}:]{(a)} CH_3CH{=}CHCH_3 + CH_3CH{=}CHCH_3 + H_3O^+$$

trans (major product) *Most stable alkene*; *cis* (minor product) *Less stable alkene*

$$\xrightarrow{(b)} CH_3CH_2CH{=}CH_2 + H_3O^+$$

(minor product) *Least stable alkene*

In the final step, three different alkenes are possible products: *trans*-2-butene, *cis*-2-butene, and 1-butene. Of these, *trans*-2-butene is the most stable and it is, therefore, the major product.

Studies of thousands of reactions involving carbocations show that rearrangements like those just described are general phenomena. They occur almost inevitably when the migration of an alkyl group or hydride ion can lead to a more stable carbocation.

Problem 6.11

On page 221 we illustrated the dehydration of 2-methylcyclohexanol. The products of this reaction were 1-methylcyclohexene (major product) and 3-methylcyclohexene (minor product).

(a) Is a rearrangement involved here?
(b) If not, how can you account for the formation of two products?
(c) How do you account for the fact that 1-methylcyclohexene is the major product?

Problem 6.12

Heating neopentyl iodide, $(CH_3)_3CCH_2I$, in formic acid (a solvent of very high ionizing ability) slowly leads to the formation of 2-methyl-2-butene. Propose a mechanism for this reaction.

6.13 THE FORMATION OF ALKENES BY DEBROMINATION OF VICINAL DIBROMIDES

Vicinal (or *vic*) dihalides are dihalo compounds in which the halogens are situated on adjacent carbon atoms. The name *geminal* (or *gem*) dihalide is used for those dihalides where both halogens are attached to the same carbon atom.

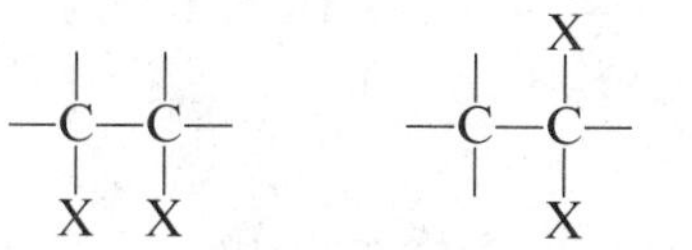

$$-\underset{X}{\overset{|}{\underset{|}{C}}}-\underset{X}{\overset{|}{\underset{|}{C}}}- \qquad -\overset{|}{\underset{|}{C}}-\underset{X}{\overset{X}{\underset{|}{\overset{|}{C}}}}-$$

A *vic*-dihalide A *gem*-dihalide

Vic-dibromides undergo the loss of a molecule of bromine (debromination) when they are treated with a solution of sodium iodide in acetone or with a mixture of zinc dust in acetic acid (or ethanol).

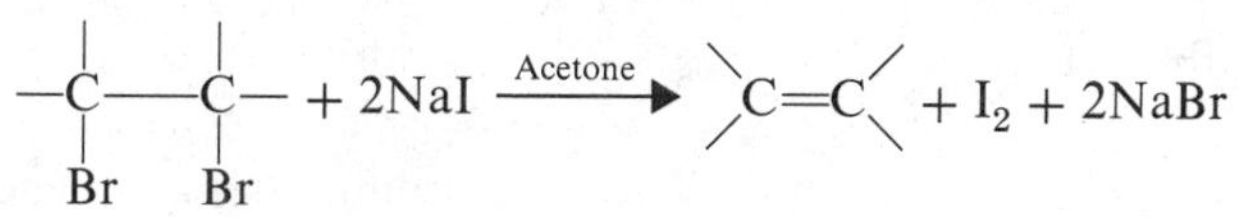

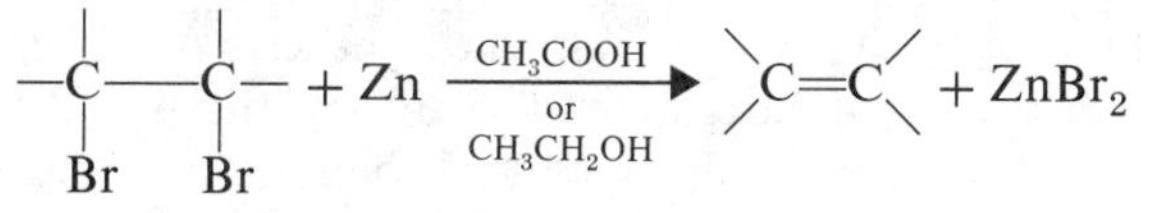

$$-\underset{Br}{\overset{|}{\underset{|}{C}}}-\underset{Br}{\overset{|}{\underset{|}{C}}}- + 2NaI \xrightarrow{\text{Acetone}} \rangle C{=}C\langle + I_2 + 2NaBr$$

$$-\underset{Br}{\overset{|}{\underset{|}{C}}}-\underset{Br}{\overset{|}{\underset{|}{C}}}- + Zn \xrightarrow[\text{or } CH_3CH_2OH]{CH_3COOH} \rangle C{=}C\langle + ZnBr_2$$

Debromination by sodium iodide takes place by an E2 mechanism similar to that for dehydrohalogenation.

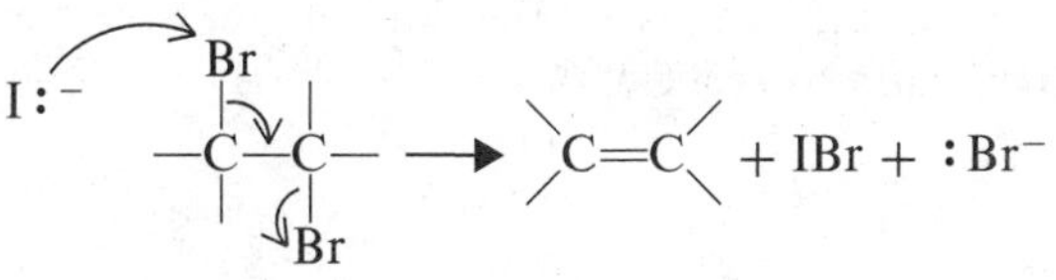

$$I:^- \quad -\overset{Br}{\underset{|}{\overset{|}{C}}}-\overset{|}{\underset{Br}{\underset{|}{C}}}- \longrightarrow \rangle C{=}C\langle + IBr + :Br^-$$

then $I^- + IBr \longrightarrow I_2 + Br^-$

Debromination by zinc takes place on the surface of the metal and the mechanism is uncertain. Other electropositive metals (Na, Ca, Mg, for example) also cause debromination of *vic*-dibromides.

Vic-dibromides are usually prepared by the addition of bromine to an alkene (cf., Sections 2.8 and 7.9). Consequently, dehalogenation of a *vic*-dibromide is of little use as a general preparative reaction. Bromination followed by debromination is useful, however, in the purification of alkenes and in "protecting" the double bond.

As an example, of the first use let us consider the problem of separating an alkene from an alkane of a similar molecular weight—for example, the separation of 2-methyl-2-butene from 2-methylbutane. Since the two compounds have very similar boiling points, separation by simple distillation is difficult. However, if we treat the mixture of the two compounds with bromine at room temperature and in the absence of light, a reaction takes place that produces a *vic*-dibromide from the alkene but no new product results from the alkane.

$$CH_3C(CH_3){=}CHCH_3 \xrightarrow[\text{25°, Addition}]{Br_2} CH_3C(CH_3)(Br){-}CH(Br)CH_3$$

Bp 38.6° → Bp 72°

$$CH_3CH(CH_3)CH_2CH_3 \xrightarrow[\text{25°, No reaction}]{Br_2} CH_3CH(CH_3)CH_2CH_3$$

Bp 27.9° → Bp 27.9°

Difficult to separate by distillation (starting materials); Easy to separate by distillation (products)

These two compounds, 2-methylbutane and the *vic*-dibromide, can be separated easily by distillation, then pure 2-methyl-2-butene can be obtained from the *vic*-dibromide by treating it with sodium iodide or zinc.

$$CH_3C(CH_3)(Br){-}CH(Br)CH_3 + Zn \xrightarrow{CH_3COOH} CH_3C(CH_3){=}CHCH_3 + ZnBr_2$$

We can also use the bromination-debromination sequence to protect a double bond while we carry out some other reaction on another part of the molecule. We find an example of this in the synthesis of a steroid (Fig. 6.13).

The overall reaction that is desired in this synthesis is the oxidation of a tertiary hydrogen to a tertiary alcohol by chromium trioxide. Double bonds, unfortunately, are not stable toward this reagent. Direct oxidation (dotted arrow) would cause oxidation of both the double bond and the tertiary hydrogen. However, conversion of the double bond to a *vic*-dibromide "protects" the double bond by transforming it temporarily into a saturated grouping, and the desired oxidation can be carried out. Afterward, the double bond can be regenerated by allowing the *vic*-dibromide to react with zinc.

6.14 HYDROGENATION AND DEHYDROGENATION: THE FUNCTION OF THE CATALYST

We saw in Section 6.4 that alkenes can be converted to alkanes by hydrogenation. The reverse of a hydrogenation reaction is called *dehydrogenation*. Hydrogenation

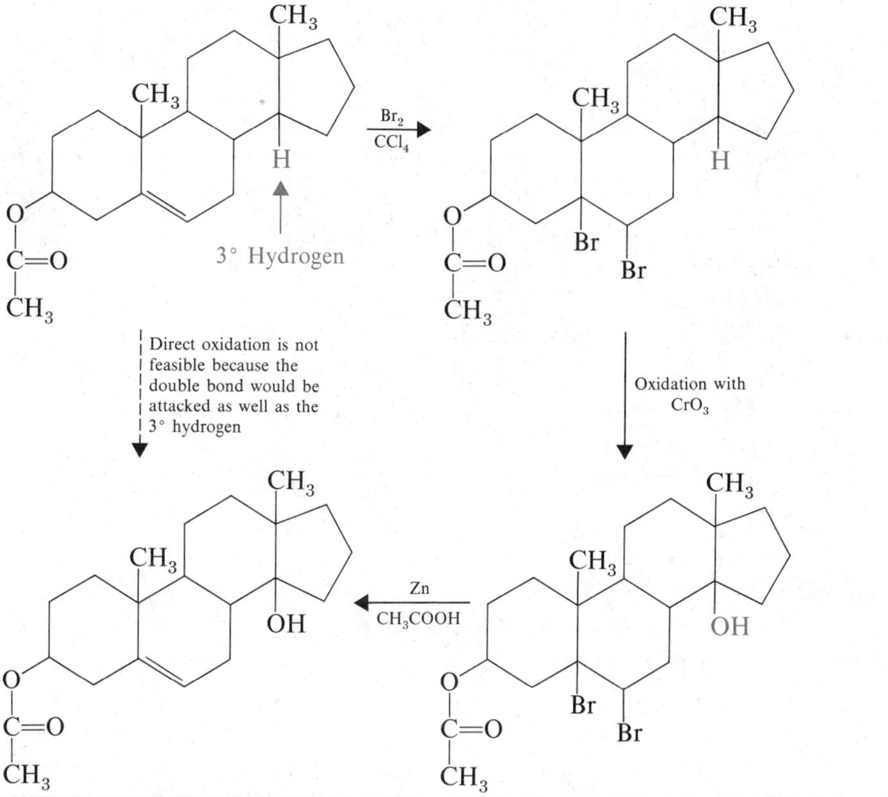

FIG. 6.13
Steps in the synthesis of a steroid in which bromination is used to protect a double bond. Debromination removes the "protecting" bromine atoms and regenerates the double bond.

of an alkene is an exothermic reaction ($\Delta H° \cong -30$ kcal/mole). Dehydrogenation of the same alkene is *endothermic* and by the same amount.

$$\text{R—CH=CH—R} + H_2 \underset{\text{Dehydrogenation}}{\overset{\text{Hydrogenation}}{\rightleftharpoons}} \text{R—CH}_2\text{—CH}_2\text{—R} + \text{heat}$$

Both hydrogenation and dehydrogenation reactions usually have high energies of activation. The reaction of an alkene with molecular hydrogen does not take place at room temperature in the absence of a catalyst, but often *does* take place at room temperature when a metal catalyst is added. The catalyst provides a new pathway for the reaction with a *lower energy of activation* (Fig. 6.14).

The most commonly used catalysts for hydrogenation are finely divided platinum, nickel, palladium, rhodium and ruthenium. (These same catalysts are often used in carrying out dehydrogenation.) In a hydrogenation reaction the metal catalyst apparently serves to adsorb hydrogen molecules on its surface. This adsorption of hydrogen is essentially a chemical reaction; unpaired electrons on the surface of the metal *pair* with the electrons of hydrogen (Fig. 6.15*a*) and bind the hydrogen to the surface. The collision of an alkene with the surface bearing adsorbed hydrogen causes adsorption of the alkene as well (Fig. 6.15*b*). A stepwise transfer of hydrogen atoms takes place, and this produces an alkane before the organic molecule leaves the catalyst surface (Fig. 6.15*c* and *d*). As a conse-

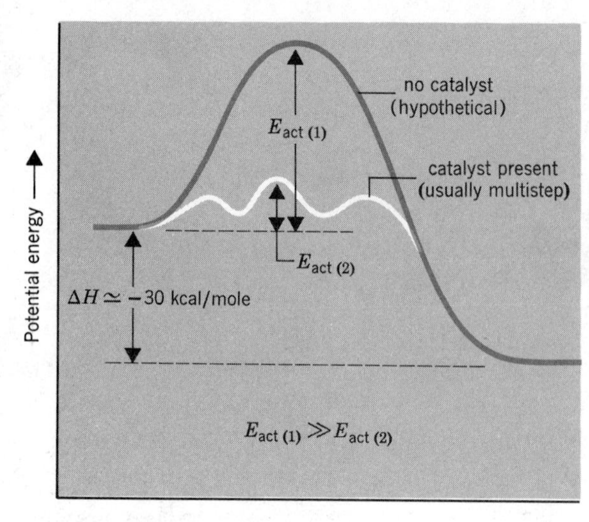

FIG. 6.14
Potential energy diagram for the hydrogenation of an alkene in the presence of a catalyst and the hypothetical reaction in the absence of a catalyst. The energy of activation for the uncatalyzed reaction [$E_{act(1)}$] *is very much larger than the largest energy of activation for the catalyzed reaction* [$E_{act(2)}$].

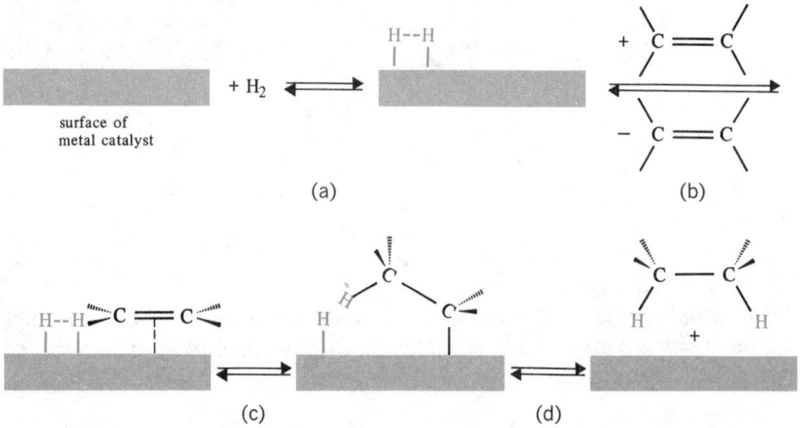

FIG. 6.15
The mechanism for the hydrogenation of an alkene (forward reaction) and the dehydrogenation of an alkane (reverse reaction) as catalyzed by finely divided platinum metal.

quence, the addition of hydrogen to an alkene is usually a *syn addition; that is, both hydrogens usually add from the same side of the molecule.*

The mechanism for a dehydrogenation reaction is simply the reverse of that of a hydrogenation reaction. The overall direction of this reversible reaction can often be determined by the experimental conditions we employ. We usually carry out hydrogenation with a large excess of hydrogen and at lower temperatures. We bring about dehydrogenation by using higher temperatures because the reaction is endothermic. In dehydrogenations we usually sweep the reaction mixture with an inert gas (nitrogen) to remove the hydrogen as it is formed. This minimizes the concentration of hydrogen in the reaction flask and on the catalyst surface.

Reversible hydrogenation-dehydrogenation reactions occur in living cells as well as in the apparatus of an organic laboratory. An example of a biochemical hydrogenation-dehydrogenation equilibrium is the one catalyzed by the enzyme, *succinate dehydrogenase* (Fig. 6.16).

Enzymes are very large complex protein molecules that catalyze biochemical reactions. The enzyme *succinate dehydrogenase* has a molecular weight of 175,000. *Succinate dehydrogenase* also contains four atoms of iron, and it appears

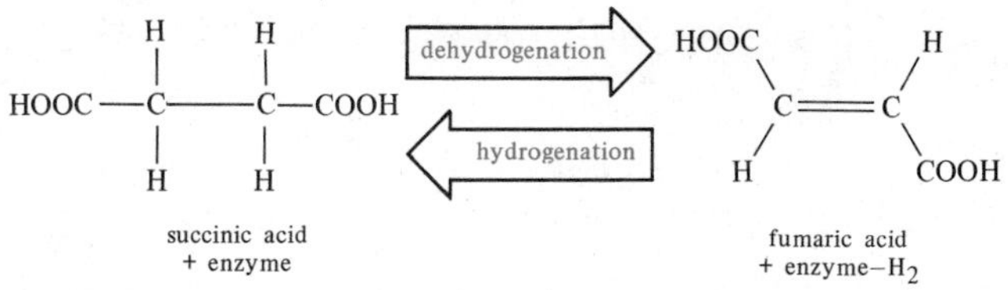

$$\mathrm{HOOC{-}\underset{H}{\overset{H}{C}}{-}\underset{H}{\overset{H}{C}}{-}COOH} \underset{\text{hydrogenation}}{\overset{\text{dehydrogenation}}{\rightleftharpoons}} \mathrm{\overset{HOOC}{\underset{H}{}}C{=}C\overset{H}{\underset{COOH}{}}}$$

succinic acid + enzyme → fumaric acid + enzyme–H_2

FIG. 6.16
The succinic dehydrogenase *equilibrium. The enzyme serves as a hydrogen acceptor in dehydrogenation and a reduced form of the enzyme (enzyme—H_2) is produced. In hydrogenation hydrogens are transferred to fumaric acid producing succinic acid and the oxidized enzyme.*

likely that there is an $Fe^{++} \rightleftharpoons Fe^{+++}$ oxidation-state change as the enzyme is oxidized and reduced. The cell's reliance on enzymatically bound metals shows a superficial resemblance to the chemist's reliance on catalytic metals in laboratory hydrogenations and dehydrogenations. The mechanisms of the reactions are quite different.

In theory, dehydrogenation of succinic acid could produce either the *trans* isomer, fumaric acid, or the *cis* isomer, maleic acid.

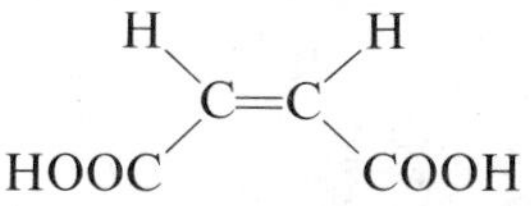

$$\mathrm{\overset{H}{\underset{HOOC}{}}C{=}C\overset{H}{\underset{COOH}{}}}$$

Maleic acid

The enzymatically catalyzed dehydrogenation produces the single isomer, fumaric acid. Moreover, the enzymatically catalyzed hydrogenation utilizes only fumaric acid not maleic acid, to produce succinic acid. This type of selectivity is a general rule in biochemical reactions.

6.15 SUMMARY OF METHODS FOR THE PREPARATION OF ALKENES

In this chapter we described four general methods for the preparation of alkenes. All are based on eliminations.

1. Dehydrohalogenation of alkyl halides (Section 6.9)

General: $$-\underset{H}{\overset{|}{C}}-\underset{X}{\overset{|}{C}}- \xrightarrow[\text{Heat}\ (-HX)]{\text{Base}} \rangle C{=}C\langle$$

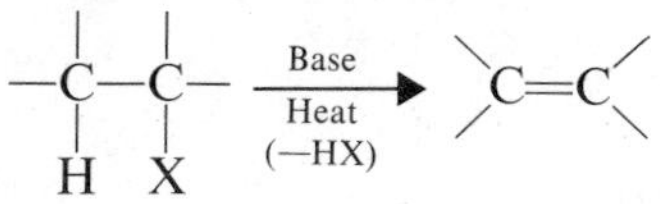

Specific examples:

$$\mathrm{CH_3CH_2\underset{Br}{\underset{|}{C}}HCH_3} \xrightarrow[C_2H_5OH]{C_2H_5ONa} \underset{(cis\ \text{and}\ trans\ 81\%)}{\mathrm{CH_3CH{=}CHCH_3}} + \underset{(19\%)}{\mathrm{CH_3CH_2CH{=}CH_2}}$$

$$CH_3CH_2\underset{\displaystyle Br}{\underset{|}{C}}HCH_3 \xrightarrow[70°]{(CH_3)_3COK} CH_3CH{=}CHCH_3 + CH_3CH_2CH{=}CH_2$$

(*cis* and *trans*, 47%) Disubstituted alkenes; (53%) Monosubstituted alkene

2. Dehydration of alcohols (Sections 6.10–6.12)

General: $$-\underset{H}{\overset{|}{\underset{|}{C}}}-\underset{OH}{\overset{|}{\underset{|}{C}}}- \xrightarrow[Heat]{Acid} \;>C{=}C<\; + H_2O$$

Specific examples: $$CH_3CH_2OH \xrightarrow[180°]{conc.\ H_2SO_4} CH_2{=}CH_2 + H_2O$$

$$CH_3\underset{CH_3}{\overset{CH_3}{\underset{|}{\overset{|}{C}}}}-OH \xrightarrow[85°]{20\%\ H_2SO_4} CH_3\overset{CH_3}{\overset{|}{C}}{=}CH_2 + H_2O$$

(83%)

3. Dehalogenation of *vic*-dibromides (Section 6.13)

General: $$-\underset{Br}{\overset{|}{\underset{|}{C}}}-\underset{Br}{\overset{|}{\underset{|}{C}}}- \xrightarrow[CH_3COOH]{Zn} \;>C{=}C<\; + ZnBr_2$$

Specific example:

[5,6-dibromo steroid ketone (CH₃, CH₃, C₈H₁₇, O, Br, Br)] $\xrightarrow[\text{Diethyl ether, reflux}]{Zn/CH_3COOH}$ [Δ5-steroid ketone (CH₃, CH₃, C₈H₁₇, O)] (72%)

4. Dehydrogenation of alkanes (Section 6.14)

General: $$-\underset{H}{\overset{|}{\underset{|}{C}}}-\underset{H}{\overset{|}{\underset{|}{C}}}- \xrightarrow[Heat]{Catalyst} \;>C{=}C<\; + H_2$$

Specific example: $$CH_3CH_2CH_2CH_3 \xrightarrow[500°]{Cr_2O_3 \cdot Al_2O_3} CH_3CH{=}CHCH_3$$

(*cis* and *trans*)

+ other butenes + H_2

In subsequent chapters we will see a number of related methods for alkene synthesis.

Additional Problems

6.13

Each of the following names is incorrect. Tell how and give the correct name.

(a) *cis*-3-Pentene
(b) 1,1,2,2-Tetramethylethene
(c) 2-Methylcycloheptene
(d) 1-Methyl-1-heptene
(e) 3-Methyl-2-butene
(f) 4,5-Dichlorocyclopentene

6.14

Write structural formulas for each of the following:

(a) 1-Methylcyclobutene
(b) 3-Methylcyclopentene
(c) 2,3-Dimethyl-2-pentene
(d) *trans*-2-Hexene
(e) *cis*-3-Heptene
(f) 3,3,3-Trichloropropene
(g) Isobutylene
(h) Propylene
(i) 4-Cyclopentyl-1-pentene
(j) Cyclopropylethene

6.15

Write structural formulas and give IUPAC names for all alkene isomers of (a) C_5H_{10} and (b) C_6H_{12}. (c) What other isomers are possible for C_5H_{10} and C_6H_{12}? Write their structures.

6.16

Give the IUPAC names for each of the following:

(a) cyclohexene ring bearing CH_3 on a double-bond carbon and CH_3 on the ring

(b) $CH_2{=}CCH_2CH_3$ with $CH_2CH_2CH_3$ on the second carbon

(c) CH_3CH_2 and $CH_3CH_2CH_2$ attached to $C{=}CH_2$

(d) cyclopentene ring bearing CH_3 and CH_2CH_3 on the double-bond carbons

6.17

Write structural formulas for the products of the following reactions. If more than one product is possible tell which one would be the major product.

(a) $CH_3C(CH_3)(Cl)CH_2CH_2CH_3 \xrightarrow[C_2H_5OH]{C_2H_5ONa}$

(b) $CH_3C(CH_3)(Cl)CH_2CH_2CH_3 \xrightarrow[(CH_3)_3COH]{(CH_3)_3COK}$

(c) $CH_3C(CH_3)(OH)CH_2CH_2CH_3 \xrightarrow[\text{Heat}]{H^+}$

(d) 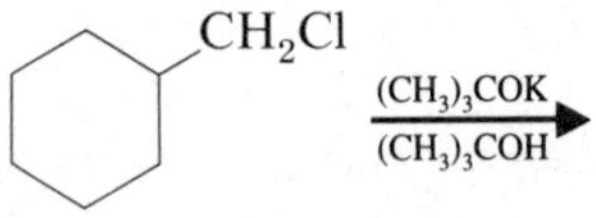 CH_2Cl $\xrightarrow[(CH_3)_3COH]{(CH_3)_3COK}$

(e) Product of (d) $\xrightarrow{Br_2}$

(f) Product of (e) $\xrightarrow[\text{Acetone}]{\text{NaI}}$

(g) Product of (f) $\xrightarrow[\text{Pt}]{H_2}$

(h) 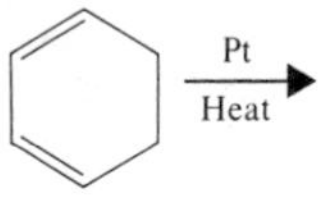 $\xrightarrow[\text{Heat}]{\text{Pt}}$

6.18

Cyclohexane and 1-hexene have the same molecular formula. Suggest a simple chemical reaction that will distinguish one from the other.

6.19

For which of the following compounds is *cis-trans* isomerism possible? Where *cis-trans* isomerism is possible, write structural formulas for the isomeric compounds.

(a) 1-Butene
(b) 2-Methylpropene
(c) 2-Heptene
(d) 2-Methyl-2-heptene
(e) 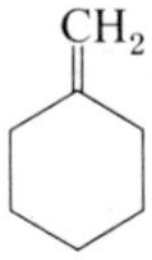 CH_2
(f) 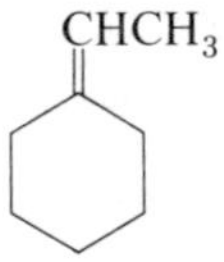 $CHCH_3$
(g) $CHCH_3$ CH_3

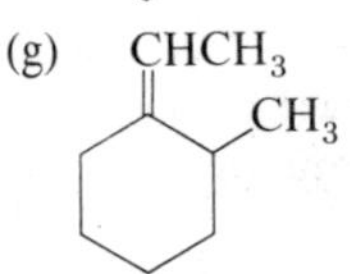

(h) 1-Chloro-1-butene
(i) 1,1-Dichloro-1-butene
(j) 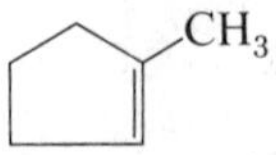 CH_3
(k) 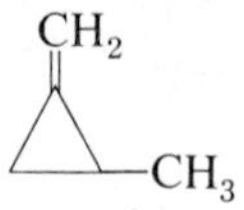 CH_2 CH_3
(l) 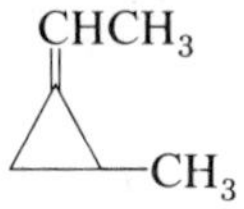 $CHCH_3$ CH_3

6.20

(a) Arrange the following alkenes in order of their relative stabilities:

trans-3-hexene; 1-hexene; 2-methyl-2-pentene; *cis*-2-hexene; 2,3-dimethyl-2-butene

(b) For which of the alkenes listed in part (a) could comparative heats of hydrogenation be used to measure their relative stabilities?

6.21

Which compound would you expect to have the larger heat of hydrogenation (in kcal/mole): *cis*-cyclooctene or *trans*-cyclooctene? Explain.

6.22

When *cis*-2-butene is heated to a temperature greater than 300°, a mixture of two isomeric 2-butenes results. (a) What chemical change takes place? (b) Which butene isomer would you expect to predominate in the mixture when equilibrium is established between them?

6.23

Write the structural formulas for the alkenes that could be formed when each of the following alkyl halides is subjected to dehydrohalogenation by the action of ethoxide ion in ethanol. When more than one product results, designate the major product. (Neglect *cis-trans* isomerism in this problem.)

(a) $CH_3CHBrCH(CH_3)CH_3$
(b) $CH_3CH_2CH(CH_3)CH_2Br$
(c) $CH_3CH_2CHBrCH_2CH_3$
(d) $CH_3CHBrCH_2CH_2CH_3$

(e) cyclohexyl—CH_2Br　　(f) 1-bromo-1-methylcyclohexane (Br and CH_3 on the same ring carbon)

6.24

Repeat Problem 6.23 giving the results expected when dehydrohalogenation is carried out with potassium *tert*-butoxide in *tert*-butyl alcohol.

6.25

Give structural formulas for the alcohols that, on dehydration, would yield each of the following alkenes as major products.

(a) 2-Methylpropene　　(c) Cyclopentene
(b) 2,3-Dimethyl-2-butene　　(d) Cyclohexene

6.26

Arrange the following alcohols in order of their reactivity toward acid-catalyzed dehydration and explain your reasoning.

$CH_3CH(CH_3)CH_2CH_2OH$　　$CH_3C(OH)(CH_3)CH_2CH_3$　　$CH_3CH(CH_3)CH(OH)CH_3$

6.27

(a) When 1,2-dimethylcyclopentene reacts with hydrogen in the presence of finely divided platinum, only one of the isomers of 1,2-dimethylcyclopentane forms. Which is it? (Assume a mechanism similar to that in Fig. 6.15) (b) What predominant product would you expect from a similar hydrogenation of 1,2-dimethylcyclohexene? Write a conformational formula for this product. (c) Which isomer would you expect to obtain from the reaction of cyclohexene and deuterium (D_2) in the presence of finely divided platinum?

6.28

Write step-by-step mechanisms that account for each product of the following reactions and explain the relative proportions of the isomers obtained in each instance.

(a) $CH_3—C(CH_3)_2—CH_2OH \xrightarrow[\text{heat}]{H^+}$ $(CH_3)_2C{=}CHCH_3$ (major product) + $H_2C{=}C(CH_2CH_3)(CH_3)$ (minor product)

(b) 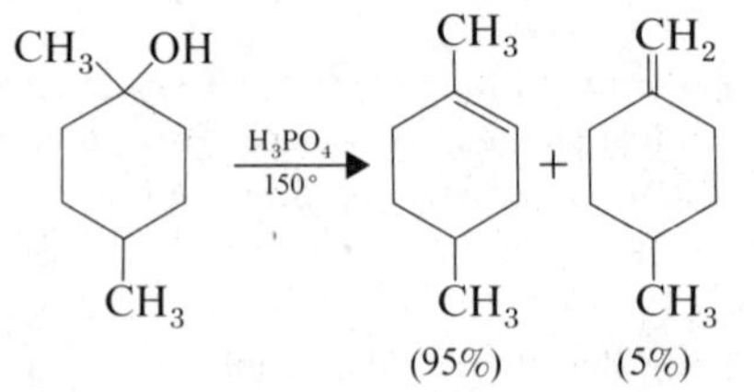 1,4-dimethylcyclohexanol $\xrightarrow[150°]{H_3PO_4}$ 1,4-dimethylcyclohexene (95%) + 4-methyl-1-methylenecyclohexane (5%)

(c) 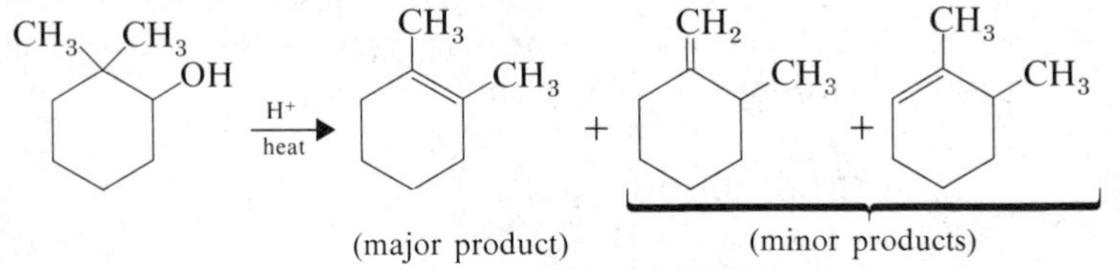 1,2,2-trimethylcyclohexanol $\xrightarrow[heat]{H^+}$ 1,2-dimethylcyclohexene (major product) + methylene product + cyclohexene product (minor products)

6.29

An alkyl halide with the formula $C_7H_{15}Br$ yields a single alkene (not a mixture of *cis-trans* isomers) when dehydrohalogenation is carried out with either sodium ethoxide in ethanol or potassium *tert*-butoxide in *tert*-butyl alcohol. What is the structure of the alkyl halide?

6.30

Cholesterol (below) is an important steroid found in nearly all body tissues; it is also the major component of gallstones. Impure cholesterol can be obtained from gallstones by extracting them with an organic solvent. The crude cholesterol thus obtained can be purified by (a) treatment with Br_2 in $CHCl_3$, (b) careful crystallization of the product, and (c) treatment of the latter with zinc in ethyl alcohol. What reactions are involved in this procedure?

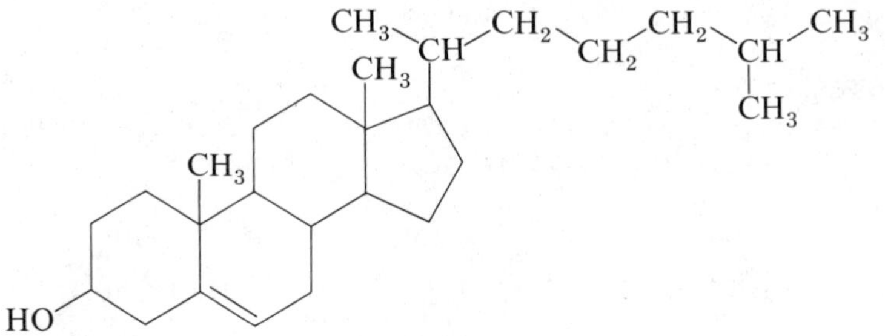

Cholesterol

6.31

Caryophyllene, a compound found in oil of cloves, has the molecular formula $C_{15}H_{24}$. Reaction of caryophyllene with an excess of hydrogen in the presence of a platinum catalyst produces a compound with the formula $C_{15}H_{28}$. How many (a) double bonds, and (b) rings does a molecule of caryophyllene have?

6.32

Squalene, an important intermediate in the biosynthesis of steroids, has the molecular formula $C_{30}H_{50}$. (a) What is the index of hydrogen deficiency of squalene? (b) Squalene undergoes catalytic hydrogenation to yield a compound with the molecular formula $C_{30}H_{62}$. How many double bonds does a molecule of squalene have? (c) How many rings?

6.33

Reconsider the interconversion of *cis*-2-butene and *trans*-2-butene given in Problem 6.22. (a) What is the value of $\Delta H°$ for the reaction, *cis*-2-butene $\longrightarrow$ *trans*-2-butene? (b) What minimum value of E_{act} would you expect for this reaction? (c) Sketch a potential energy diagram for the reaction and label $\Delta H°$ and E_{act}.

***6.34**

The sp^2-hybridized carbons of alkenes are often said to be "more electronegative" than the sp^3-hybridized carbons of alkanes. (a) Explain this statement. (b) Can you account for the fact that ethane ($K_a = 10^{-42}$) is a weaker acid than ethene ($K_a = 10^{-36}$)?

***6.35**

A spectacular example of the kind of rearrangement that can occur in the acid-catalyzed dehydration of an alcohol is given below. When 3β-friedelanol is treated with acid, a carbocation forms, and then a total of *seven* methyl and hydrogen migrations occur before the loss of a proton produces 13(18)-oleanene. Each migration is a 1,2-shift, that is, a methanide ion ($CH_3{:}^-$) or hydride ion ($H{:}^-$) migrates from one atom to an adjacent atom. Show how each of these migrations occur.

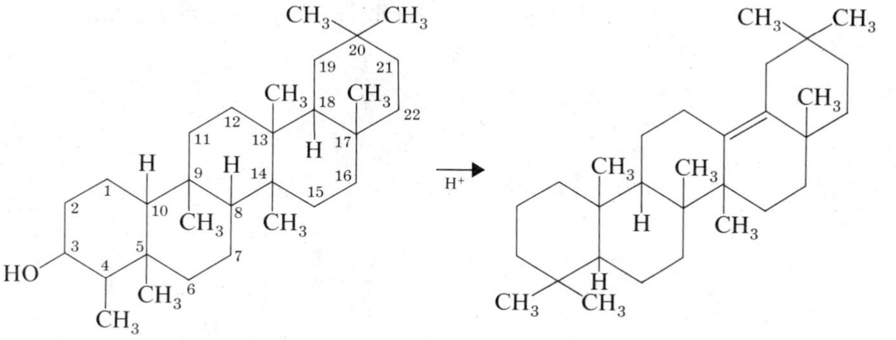

3β-Friedelanol $\xrightarrow{H^+}$ 13(18)-Oleanene

7 REACTIONS OF ALKENES: ADDITION REACTIONS OF THE CARBON-CARBON DOUBLE BOND

7.1 INTRODUCTION

The most commonly encountered reaction of compounds containing a carbon-carbon double bond is addition. Addition may involve a symmetrical reagent:

$$>C=C< + Z-Z \longrightarrow -\underset{Z}{\overset{|}{C}}-\underset{Z}{\overset{|}{C}}-$$

or an unsymmetrical reagent:

$$>C=C< + Z-Y \longrightarrow -\underset{Z}{\overset{|}{C}}-\underset{Y}{\overset{|}{C}}-$$

Examples of addition of symmetrical reagents are the additions of hydrogen or bromine:

$$\underset{\text{Alkene}}{>C=C<} + H_2 \xrightarrow{Pt} \underset{\text{Alkane}}{-\underset{H}{\overset{|}{C}}-\underset{H}{\overset{|}{C}}-}$$

$$\underset{\text{Alkene}}{>C=C<} + Br_2 \xrightarrow[CCl_4]{} \underset{vic\text{-Dibromide}}{-\underset{Br}{\overset{|}{C}}-\underset{Br}{\overset{|}{C}}-}$$

Examples of additions of unsymmetrical reagents are the additions of hydrogen halides or the acid-catalyzed addition of water:

$$\underset{\text{Alkene}}{>C=C<} + HX \longrightarrow \underset{\text{Alkyl halide}}{-\underset{H}{\overset{|}{C}}-\underset{X}{\overset{|}{C}}-}$$

$$\underset{\text{Alkene}}{>C=C<} + H-OH \xrightarrow{H^+} \underset{\text{Alcohol}}{-\underset{H}{\overset{|}{C}}-\underset{OH}{\overset{|}{C}}-}$$

The large numbers of reactions of alkenes that fall into the category of addition reactions can be accounted for on the basis of two characteristics of the carbon-carbon double bond:

1. An addition reaction results in the conversion of one π bond and one σ bond into two σ bonds. The result of this change is usually energetically

$$\rangle C{=}C\langle + X{-}Y \longrightarrow -\overset{|}{\underset{X}{C}}-\overset{|}{\underset{Y}{C}}-$$

π bond, σ bond (Bonds broken) $\longrightarrow$ 2 σ bonds (Bonds formed)

favorable. The heat evolved in making 2 σ bonds exceeds that needed to break one σ and one π bond, and addition reactions are usually exothermic.

2. The electrons of the π bond are exposed because the π orbital has considerable p character. Thus, a π bond is particularly susceptible to electron-seeking reagents. Such electron-seeking reagents are said to be *electrophilic* (electron loving) and are called *electrophiles*. Electrophiles include positive reagents like the proton, H^+, or neutral reagents such as bromine (because it can be polarized), and the Lewis acids BF_3 and $AlCl_3$. Metal ions that contain vacant orbitals—the silver ion, Ag^+, the mercuric ion, Hg^{++}, and the platinum ion, Pt^{++}, for example—also act as electrophiles.

Halogens and metal ions often react with the exposed π electrons to form an intermediate called *a π complex:*

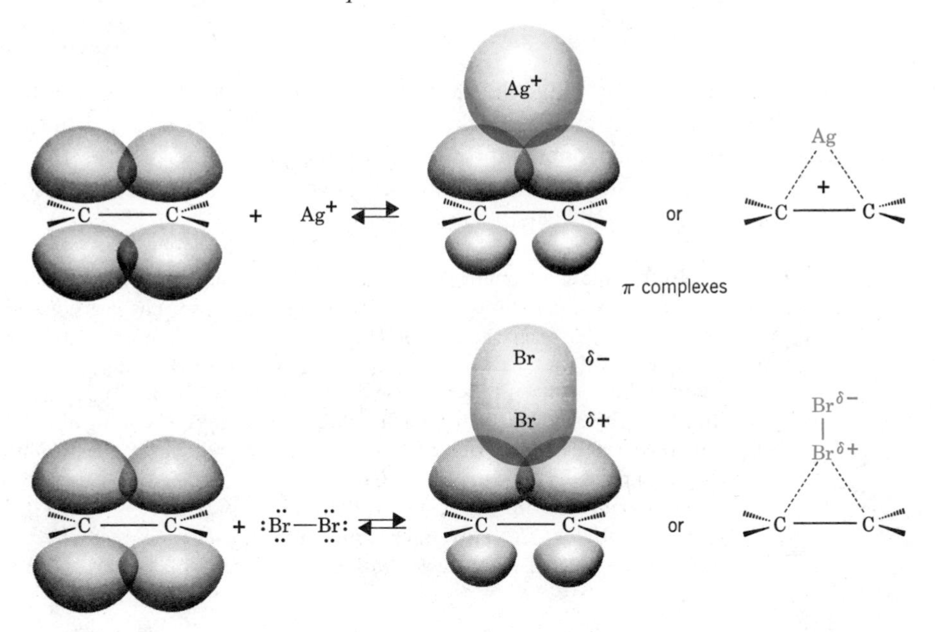

Strong acids (proton donors) react with alkenes to yield *carbocations*. In this reaction the proton acts as an electrophile and attacks the π bond. It uses the π electrons to form a σ bond to one carbon of the alkene. We represent this reaction in the following way:

$$H^+ \; \rangle C{=}C\langle \longrightarrow H-\overset{|}{\underset{|}{C}}-\overset{|}{\underset{|}{C}}^+$$

Carbocation

The carbocation, thus formed, then goes on to react further. For example, if the proton donor is HCl, the carbocation can combine with a chloride ion to form an alkyl halide:

$$\mathrm{H{-}\overset{|}{\underset{|}{C}}{-}\overset{|}{\underset{|}{C}}^{+} + :Cl^{-} \longrightarrow H{-}\overset{|}{\underset{|}{C}}{-}\overset{|}{\underset{|}{C}}{-}Cl}$$

Alkyl halide

Electrophiles are really Lewis acids: they are molecules or ions that can accept an electron pair. Nucleophiles are molecules or ions that can furnish an electron pair (i.e., Lewis bases). Any reaction of an electrophile also involves a nucleophile. In the protonation of an alkene the electrophile is the proton donated by an acid; the nucleophile is the alkene.

$$\underset{\text{Electrophile}}{\mathrm{H^{+}}} + \underset{\text{nucleophile}}{\mathrm{>C{=}C<}} \longrightarrow \mathrm{{-}\overset{H}{\underset{|}{C}}{-}\overset{+}{\underset{|}{C}}{-}}$$

In the next step, the reaction of the carbocation with a chloride ion, the carbocation is the electrophile and the chloride ion is the nucleophile.

$$\underset{\text{Electrophile}}{\mathrm{{-}\overset{H}{\underset{|}{C}}{-}\overset{+}{\underset{|}{C}}{-}}} + \underset{\text{Nucleophile}}{\mathrm{:Cl^{-}}} \rightleftharpoons \mathrm{{-}\overset{H}{\underset{|}{C}}{-}\overset{Cl}{\underset{|}{C}}{-}}$$

Either reaction could be described in two ways: the first step could be described as an electrophilic attack by a proton on the alkene, or as a nucleophilic attack by the alkene on the proton. Chemists generally consider the organic molecule as the substance being attacked, and would usually describe this reaction as an *electrophilic attack by a proton on the alkene.*

Since electrophile-nucleophile reactions are Lewis acid-base reactions, why do chemists use two different sets of words for them? To answer this question, one must first recall that the terms acidty and basicity are ways of talking about relative equilibrium constants. A strong acid, in this sense, is one that in aqueous solution produces a high concentration of hydronium ions and its conjugate base at equilibrium. Electrophilicity and nucleophilicity are ways of talking about relative rates of reactions. These terms refer to the rates at which the electrophiles and nucleophiles react. Ethoxide ion, $CH_3CH_2O{:}^-$, for example, is a stronger base than the ethyl mercaptide ion, $CH_3CH_2S{:}^-$. (The K_a for C_2H_5OH is $\sim 10^{-18}$; that for C_2H_5SH is $\sim 10^{-12}$). The ethyl mercaptide ion, however, is in many cases the stronger nucleophile; it reacts faster with electrophiles—particularly with electrophilic carbon atoms.

7.2 THE ADDITION OF HYDROGEN HALIDES TO ALKENES: MARKOVNIKOV'S RULE

Hydrogen halides (HF, HCl, HBr, and HI) add readily to the double bond of alkenes:

$$\mathrm{>C{=}C< + HX \longrightarrow {-}\overset{|}{\underset{H}{C}}{-}\overset{|}{\underset{X}{C}}{-}}$$

Two examples are shown below.

$$CH_3CH{=}CHCH_3 + H{-}Cl \longrightarrow CH_3CH_2\underset{\displaystyle Cl}{\underset{|}{C}}HCH_3$$

2-Butene (*cis* or *trans*) → 2-Chlorobutane

$$\text{Cyclohexene} + HBr \longrightarrow \text{Cyclohexyl bromide}$$

Cyclohexene → Cyclohexyl bromide

In carrying out these reactions, the hydrogen halide may be dissolved in acetic acid and mixed with the alkene, or gaseous hydrogen halide may be bubbled directly into the alkene with the alkene, itself, being used as the solvent.

The addition of HX to an unsymmetrical alkene could conceivably occur in two ways. In practice, however, one product usually predominates. The addition of HCl to propene, for example, could conceivably lead to either 1-chloropropane or 2-chloropropane. The actual product, however, is 2-chloropropane.

$$CH_2{=}CHCH_3 + HCl \longrightarrow CH_3\underset{\displaystyle Cl}{\underset{|}{C}}HCH_3 \quad (\text{not } ClCH_2CH_2CH_3)$$

2-Chloropropane (not 1-Chloropropane)

When 2-methylpropene reacts with HI the product is 2-iodo-2-methylpropane, not 1-iodo-2-methylpropane.

$$(CH_3)_2C{=}CH_2 + HI \longrightarrow CH_3{-}\overset{\displaystyle CH_3}{\overset{|}{\underset{\displaystyle I}{\underset{|}{C}}}}{-}CH_3 \quad \left(\text{not } CH_3{-}\overset{\displaystyle CH_3}{\overset{|}{C}}H{-}CH_2{-}I\right)$$

2-Methylpropene (isobutylene) → 2-Iodo-2-methylpropane (*tert*-butyl iodide)

Consideration of many examples like this led the Russian chemist Vladimir Markovnikov (in 1869) to formulate what is now known as *Markovnikov's rule.* As originally stated, this rule said that in the addition of HX to an alkene, *the hydrogen adds to the carbon of the double bond with the greater number of hydrogens.* Using propene as an illustration, this original statement of the rule takes the form shown below:

$$\underset{\uparrow\,H}{CH_2}{=}\underset{\uparrow\,Cl}{CH}CH_3 \longrightarrow CH_3\underset{\displaystyle Cl}{\underset{|}{C}}HCH_3$$

Carbon with the greater number of hydrogens (arrow to CH_2)

Markovnikov addition product

Reactions that follow Markovnikov's rule are said to undergo *Markovnikov addition.*

Zaitzev (Section 6.9), like Markovnikov, originally formulated his rule in terms of the number of hydrogens on each carbon atom. Zaitzev stated that in elimination reactions a hydrogen is preferentially lost from the adjacent carbon that is poorer in hydrogen. However, Professor Louis Fieser of Harvard University pointed out, several years ago, that both rules were anticipated by St. Matthew: "unto him that hath even more shall be added, but from him that hath not shall be taken away even that which he hath" (Matt. XXV, 29).

7.2A Theoretical Explanation of Markovnikov's Rule

The mechanism for addition of a hydrogen halide to an alkene involves the two steps shown below:

$$\textbf{1.}\quad \gt C{=}C\lt \; + \; H{-}X \xrightarrow{\text{Slow}} {-}\overset{+}{C}{-}\overset{H}{\underset{|}{C}}{-} \; + \; :X^{-}$$

$$\textbf{2.}\quad {-}\overset{+}{C}{-}\overset{H}{\underset{|}{C}}{-} \; + \; :X^{-} \xrightarrow{\text{Fast}} {-}\underset{X}{C}{-}\overset{H}{\underset{|}{C}}{-}$$

The important step—because it is the *rate-limiting step*—is step 1. In step 1 the alkene accepts a proton from the hydrogen halide and forms a carbocation. This step (Fig. 7.1) is highly endothermic and has a high energy of activation. Consequently, it takes place slowly. In step 2 the highly reactive carbocation stabilizes itself by combining with a halide ion. This exothermic step has a very low energy of activation and takes place very rapidly.

If the alkene that undergoes addition of a hydrogen halide is an unsymmetrical alkene such as propene, then step 1 could conceivably lead to two different carbocations:

$$CH_3CH{=}CH_2 + H^+ \longrightarrow CH_3CH_2CH_2^+ \qquad \text{1° Carbocation (less stable)}$$

$$CH_3CH{=}CH_2 + H^+ \longrightarrow CH_3\overset{+}{C}HCH_3 \qquad \text{2° Carbocation (more stable)}$$

These two carbocations are not of equal stability, however. The secondary carbocation is *more stable* and it is the greater stability of the secondary carboca-

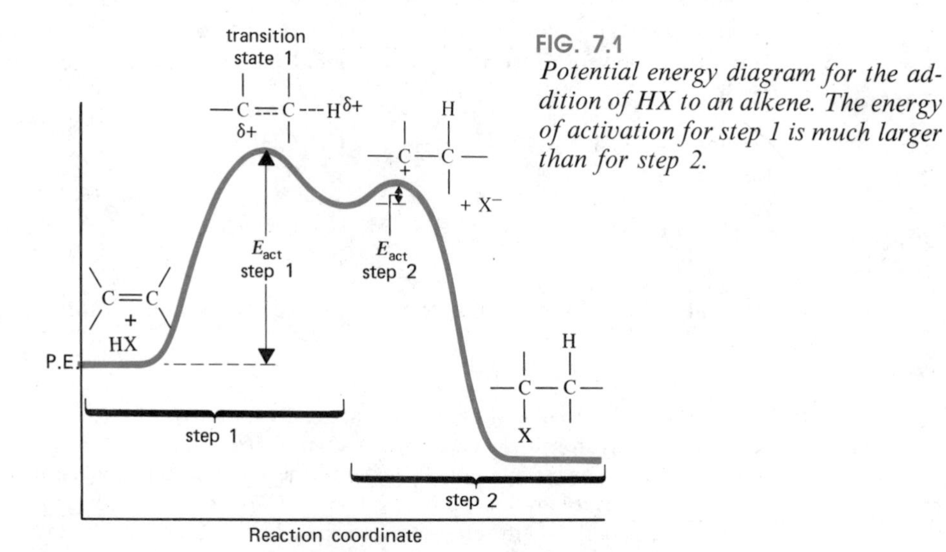

FIG. 7.1
Potential energy diagram for the addition of HX to an alkene. The energy of activation for step 1 is much larger than for step 2.

tion that accounts for the correct prediction of the overall addition by Markovnikov's rule. In the addition of HCl to propene, for example, the reaction takes the following course:

$$CH_3CH{=}CH_2 \xrightarrow[\text{Slow}]{H^+} \begin{cases} \not\rightarrow CH_3CH_2\overset{+}{C}H_2 \ (1°) \xrightarrow{Cl^-} CH_3CH_2CH_2Cl \ \text{(1-Chloropropane)} & \text{(not formed)} \\ \rightarrow CH_3\overset{+}{C}HCH_3 \ (2°) \xrightarrow[\text{Fast}]{Cl^-} CH_3CH(Cl)CH_3 \ \text{(2-Chloropropane)} & \text{(actual product)} \end{cases}$$

|—Step 1——————————|——Step 2———|

The ultimate product of the reaction is 2-chloropropane because the more stable 2° carbocation is formed preferentially in the first step.

The more stable carbocation is formed preferentially because it is formed faster. We can understand why this is true if we examine the potential energy diagrams in Fig. 7.2.

The reaction (Fig. 7.2) leading to the 2° carbocation (and ultimately to 2-chloropropane) has the lower energy of activation. That is reasonable because its transition state resembles the more stable carbocation. The reaction leading to the 1° carbocation (and ultimately to 1-chloropropane) has a higher energy of activa-

FIG. 7.2
Potential energy diagram for the addition of HCl to propene. E_{act} *(2°) is less than* E_{act} *(1°).*

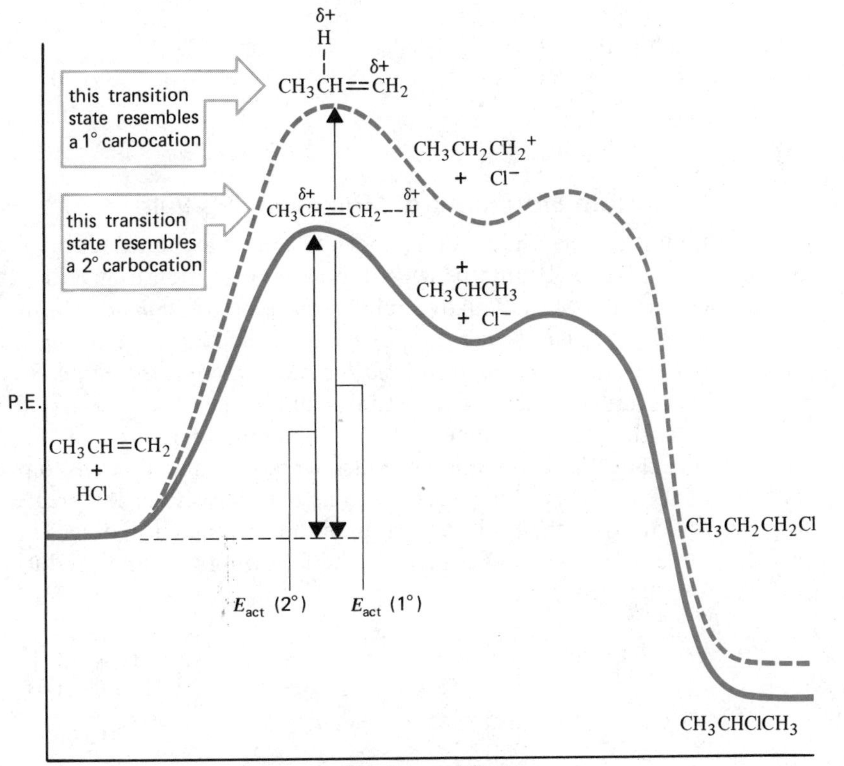

tion because its transition state resembles a less stable 1° carbocation. This second reaction is much slower and does not compete with the first reaction.

The reaction of HI with 2-methylpropene produces only 2-iodo-2-methylpropane, and for the same reason. Here, in the first step (i.e., the attachment of the proton) the choice is even more pronounced—between a 3° carbocation and a 1° carbocation.

Thus, 1-iodo-2-methylpropane is *not* obtained as a product of the reaction because its formation would require the formation of a primary carbocation. Such a reaction would have a much higher energy of activation than that leading to a tertiary carbocation.

$$(CH_3)_2C{=}CH_2 \quad \text{2-Methylpropene} \xrightarrow{HI}$$

$$CH_3\text{—}\overset{+}{C}(CH_3)\text{—}CH_3 \quad \text{3° Carbocation} \xrightarrow{:\ddot{I}:^-} CH_3\text{—}C(CH_3)(I)\text{—}CH_3$$

2-Iodo-2-methylpropane
Actual product

$$CH_3\text{—}CH(CH_3)\text{—}CH_2^+ \quad \text{1° Carbocation} \xrightarrow{:\ddot{I}:^-} CH_3\text{—}CH(CH_3)\text{—}CH_2\text{—}I$$

1-Iodo-2-methylpropane
Not formed

7.2B Modern Statement of Markovnikov's Rule

With this understanding of the mechanism for the ionic addition of hydrogen halides to alkenes behind us, we are now in a position to give the following modern statement of Markovnikov's rule: *in the ionic addition of an unsymmetrical reagent to a double bond, the positive portion of the adding reagent attaches itself to a carbon of the double bond so as to yield the more stable carbocation.* Since this is the step that occurs first (prior to the addition of the negative portion of the adding reagent), it is the step that determines the overall orientation of the reaction.

Notice that this formulation of Markovnikov's rule allows us to predict the outcome of the addition of a reagent such as ICl. Because of the greater electronegativity of chlorine, the positive portion of this molecule is iodine. The addition of ICl to 2-methylpropene takes place in the following way and produces 1-iodo-2-chloro-2-methylpropane.

$$(CH_3)_2C{=}CH_2 + \overset{\delta+}{:\ddot{I}}\text{—}\overset{\delta-}{\ddot{Cl}:} \longrightarrow (CH_3)_2\overset{+}{C}\text{—}CH_2\text{—}\ddot{I}: + :\ddot{Cl}:^- \longrightarrow CH_3\text{—}C(CH_3)(Cl)\text{—}CH_2\text{—}I$$

2-Methylpropene → 2-chloro-1-Iodo-2-methylpropane

Problem 7.1

Give the structure and name of the product that would be obtained from Markovnikov addition of ICl to propene.

Problem 7.2

Outline mechanisms for the ionic additions (a) of hydrogen iodide to 1-butene, (b) of IBr to 2-methyl-2-butene, (c) of hydrogen chloride to 1-methylcyclohexene.

Problem 7.3

The addition of hydrogen chloride to 3,3-dimethyl-1-butene (below) yields two products: 3-chloro-2,2-dimethylbutane and 2-chloro-2,3-dimethylbutane. Write mechanisms that account for the formation of each product.

$$CH_3-\underset{CH_3}{\overset{CH_3}{\underset{|}{\overset{|}{C}}}}-CH=CH_2 + HCl \longrightarrow CH_3-\underset{CH_3}{\overset{CH_3}{\underset{|}{\overset{|}{C}}}}-\underset{Cl}{\underset{|}{CH}}-CH_3$$

3,3-Dimethyl-1-butene → 3-Chloro-2,2-dimethylbutane

$+$

$$CH_3-\underset{Cl}{\overset{CH_3}{\underset{|}{\overset{|}{C}}}}-\underset{CH_3}{\underset{|}{CH}}-CH_3$$

2-Chloro-2,3-dimethylbutane

7.3 REGIOSELECTIVE REACTIONS

Chemists describe reactions like the Markovnikov additions of hydrogen halides to alkenes as being *regioselective*. *Regio* comes from the Latin word *regere* meaning to rule. When a reaction that can potentially yield two or more structural isomers actually produces only one (or a predominance of one) the reaction is said to be *regioselective*. The addition of HX to an unsymmetrical alkene such as propene could conceivably yield two structural isomers, for example. However, as we have seen, the reaction yields only one, and therefore it is regioselective.

7.4 THE ADDITION OF WATER TO ALKENES: ACID-CATALYZED HYDRATION

The acid-catalyzed addition of water to the double bond of an alkene (hydration of an alkene) is a convenient method for the preparation of secondary and tertiary alcohols. The acids most commonly used to catalyze the hydration of alkenes are sulfuric acid and phosphoric acid. These reactions, too, are usually regioselective and the addition of water to the double bond follows Markovnikov's rule. In general the reaction takes the form shown below.

$$>C=C< + HOH \xrightarrow{H^+} -\underset{H}{\underset{|}{\overset{|}{C}}}-\underset{OH}{\underset{|}{\overset{|}{C}}}-$$

Two specific examples are:

$$CH_3-\underset{}{\overset{CH_3}{\overset{|}{C}}}=CH_2 + HOH \xrightarrow[25°]{H^+} CH_3-\underset{OH}{\overset{CH_3}{C}}-CH_3$$

2-Methylpropene — *tert*-Butyl alcohol

$$\text{Methylenecyclopentane} + HOH \xrightarrow[25°]{H^+} \text{1-Methylcyclopentanol}$$

Methylenecyclopentane — 1-Methylcyclopentanol

Because the reactions follow Markovnikov's rule, acid-catalyzed hydrations of alkenes do not yield primary alcohols except in the case of the hydration of ethene.

$$CH_2=CH_2 + HOH \xrightarrow[240°]{H^+} CH_3CH_2OH$$

The mechanism for the hydration of an alkene is simply the reverse of the mechanism for the dehydration of an alcohol. We can illustrate this by giving the mechanism for the *hydration* of isobutylene and by comparing it with the mechanism for the *dehydration* of *tert*-butyl alcohol given in Section 6.10.

Step 1
$$CH_3-\underset{CH_3}{\overset{CH_2}{\overset{\|}{C}}} + H-\underset{+}{\overset{H}{\ddot{O}}}-H \underset{}{\overset{\text{Slow}}{\rightleftharpoons}} CH_3-\underset{CH_3}{\overset{CH_3}{C^+}} + :\overset{H}{\ddot{O}}-H$$

Step 2
$$CH_3-\underset{CH_3}{\overset{CH_3}{C^+}} + :\overset{H}{\ddot{O}}-H \overset{\text{Fast}}{\rightleftharpoons} CH_3-\underset{CH_3}{\overset{CH_3}{C}}-\underset{+}{\overset{H}{\ddot{O}}}-H$$

Step 3
$$CH_3-\underset{CH_3}{\overset{CH_3}{C}}-\overset{H}{\underset{..}{O^+}}-H + :\overset{H}{\ddot{O}}-H \overset{\text{Fast}}{\rightleftharpoons} CH_3-\underset{CH_3}{\overset{CH_3}{C}}-\ddot{O}-H + H-\underset{+}{\overset{H}{\ddot{O}}}-H$$

The relation between the mechanisms for hydration of an alkene and dehydration of an alcohol illustrates the *principle of microscopic reversibility*. According to this principle, a reaction and its reverse proceed through *the same path but in opposite directions when both occur under the same conditions*.

The rate-limiting step in the *hydration* mechanism is step 1: the formation of the carbocation. It is this step, too, that accounts for the Markovnikov addition of water to the double bond. The reaction produces *tert*-butyl alcohol because step 1 leads to the formation of the more stable *tert*-butyl cation rather than the much less stable isobutyl cation:

$$CH_3-\underset{CH_3}{\overset{CH_2}{\overset{\|}{C}}} + H-\underset{+}{\overset{H}{\ddot{O}}}-H \underset{\text{slow}}{\overset{\text{Very}}{\rightleftharpoons}} CH_3\underset{CH_3}{\overset{CH_2^+}{CH}} + :\overset{H}{\ddot{O}}-H$$

For all practical purposes this reaction does not take place because it produces a 1° carbocation

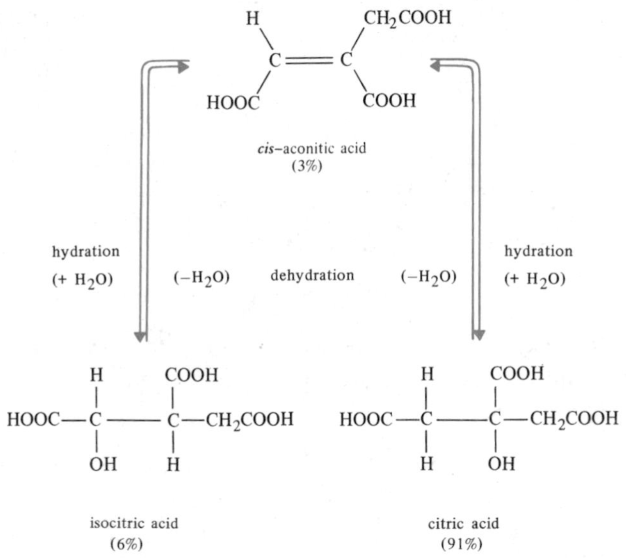

$HOOC(H)C{=}C(COOH)(CH_2COOH)$

cis-aconitic acid
(3%)

hydration (+ H_2O) (−H_2O) dehydration (−H_2O) hydration (+ H_2O)

$HOOC{-}CH(OH){-}CH(COOH){-}CH_2COOH$

isocitric acid
(6%)

$HOOC{-}CH_2{-}C(OH)(COOH){-}CH_2COOH$

citric acid
(91%)

FIG. 7.3
The aconitase *equilibrium.*

Problem 7.4

(a) Show all steps in the acid-catalyzed hydration of propene. (b) Account for the fact that the product of the reaction is isopropyl alcohol (in accordance with Markovnikov's rule) and not propyl alcohol, that is,

$$H_2O + CH_3{-}CH{=}CH_2 \xrightarrow{H^+} \underset{\text{Isopropyl alcohol}}{CH_3\underset{\displaystyle OH}{\underset{|}{C}H}CH_3} \qquad (\text{not } \underset{\text{Propyl alcohol}}{CH_3CH_2CH_2OH})$$

A spectacular biochemical example of both hydration of a double bond and dehydration of an alcohol occurs in the reactions catalyzed by the enzyme *aconitase*. Aconitase is an important enzyme in the biochemical pathway called the tricarboxylic acid or Krebs cycle—a biological pathway by which much of the energy content of carbohydrates is "extracted" and made available for various energy-requiring processes in living organisms. The reactions that are catalyzed by aconitase are shown in Fig. 7.3.

In the aconitase equilibrium, four reactions are catalyzed by a single enzyme. Two involve dehydration, and two involve hydration. The composition of the equilibrium mixture at 25° and pH 7.4 is indicated by the percentages given in parentheses in Fig. 7.3.

7.5 ADDITION OF SULFURIC ACID TO ALKENES

When alkenes are treated with cold concentrated sulfuric acid they react by addition to form alkyl hydrogen sulfates. In the first step of this reaction the alkene accepts a proton from sulfuric acid to form a carbocation; in the second step the carbocation reacts with a hydrogen sulfate ion to form an alkyl hydrogen sulfate:

$$\underset{\text{Alkene}}{\rangle C{=}C\langle} + \underset{\substack{\text{Sulfuric}\\ \text{acid}}}{HO{-}\overset{\overset{O}{\|}}{\underset{\underset{O}{\|}}{S}}{-}OH} \longrightarrow \underset{\text{Carbocation}}{-\overset{+}{C}{-}\underset{H}{\overset{|}{C}}-} + \underset{\substack{\text{Hydrogen}\\ \text{sulfate}\\ \text{ion}}}{{}^{-}O{-}\overset{\overset{O}{\|}}{\underset{\underset{O}{\|}}{S}}{-}OH} \longrightarrow \underset{\substack{\text{Alkyl hydrogen}\\ \text{sulfate}}}{-\overset{\overset{OH}{|}\;O{=}S{=}O\;\overset{|}{O}}{\overset{|}{C}}{-}\underset{H}{\overset{|}{C}}-}$$

The addition of sulfuric acid is also regioselective, and it follows Markovnikov's rule. Propene, for example, reacts to yield isopropyl hydrogen sulfate rather than propyl hydrogen sulfate.

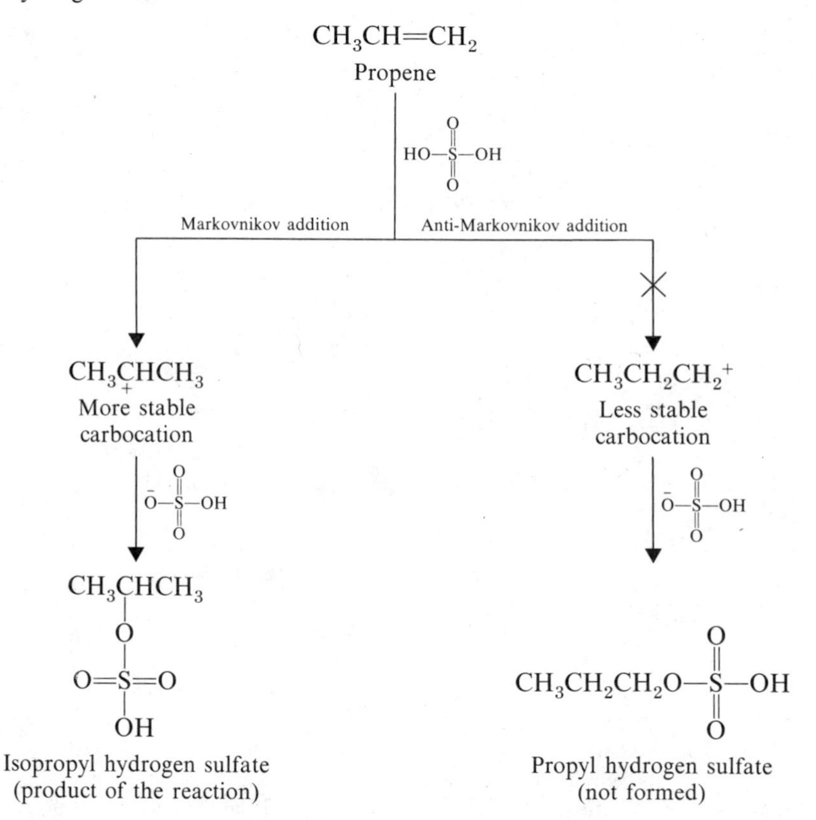

Alkyl hydrogen sulfates can be hydrolyzed to alcohols. Thus, the overall result of the addition of sulfuric acid to an alkene followed by hydrolysis is the same as that obtained by acid-catalyzed hydration: Markovnikov addition of H— and —OH.

$$CH_3CH{=}CH_2 \xrightarrow[H_2SO_4]{\text{cold}} \underset{\underset{OSO_3H}{|}}{CH_3CHCH_3} \xrightarrow{H_2O,\ \text{heat}} \underset{\underset{OH}{|}}{CH_3CHCH_3} + H_2SO_4$$

7.6 DIMERIZATION OF ALKENES: ALKYLATION OF ALKENES BY CARBOCATIONS

Heating isobutylene (2-methylpropene) in 60% sulfuric acid at 70° causes the formation of two main products—two isomeric compounds called "diisobutylenes." The diisobutylenes are examples of dimers (di- = two + Gr., *meros* = part); isobutylene is said to have been dimerized by the reaction. (The dashed lines through the structures divide them into their original units.)

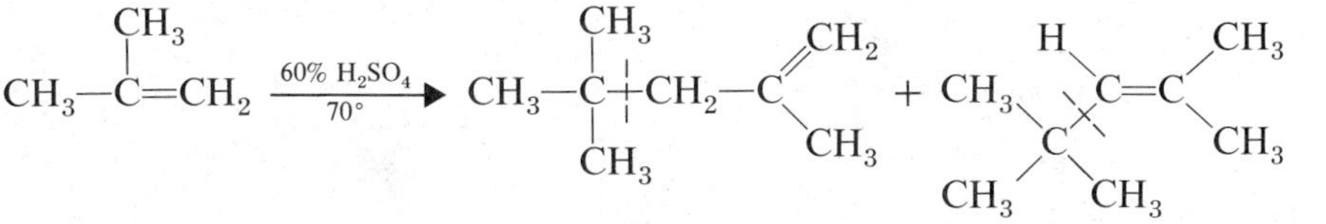

$$\mathrm{CH_3{-}\underset{}{\overset{CH_3}{\overset{|}{C}}}{=}CH_2 \xrightarrow[70^\circ]{60\%\ H_2SO_4} CH_3{-}\underset{CH_3}{\overset{CH_3}{C}}{-}CH_2{-}C({=}CH_2)CH_3 + (CH_3)_3C{-}CH{=}C(CH_3)_2}$$

80% (first product); 20% (second product)

Diisobutylenes

In this reaction the less highly substituted alkene is the major product because it is apparently the more stable. In molecules of the more highly substituted alkene considerable internal repulsion exists between the large *tert*-butyl group and a *cis* methyl group. This is an exception to the general rule concerning alkene stability (Section 6.6).

The dimerization of isobutylene is important in the petroleum industry because hydrogenation of the mixture of dimers produces the single product called "isooctane."*

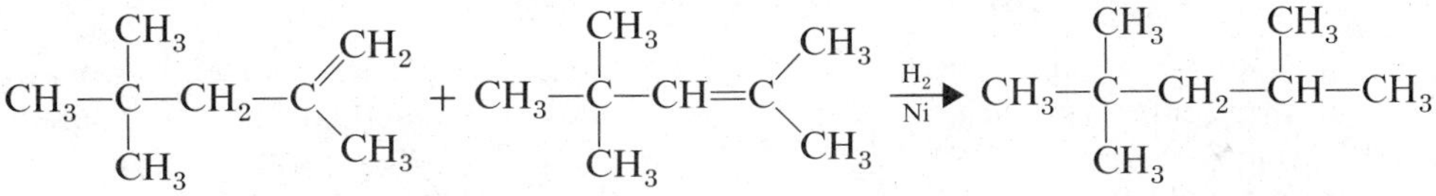

$$\mathrm{CH_3{-}\underset{CH_3}{\overset{CH_3}{C}}{-}CH_2{-}C({=}CH_2)CH_3 + CH_3{-}\underset{CH_3}{\overset{CH_3}{C}}{-}CH{=}C(CH_3)_2 \xrightarrow[Ni]{H_2} CH_3{-}\underset{CH_3}{\overset{CH_3}{C}}{-}CH_2{-}\overset{CH_3}{CH}{-}CH_3}$$

"Isooctane"
(2,2,4-trimethylpentane)

Isooctane burns very smoothly (without knocking) in internal combustion engines and is used as one of the standards by which the octane rating of gasolines is established. According to this scale isooctane has an octane rating of 100. Heptane, a compound that produces much knocking when it is burned in an internal combustion engine, is given an octane rating of zero. Mixtures of isooctane and heptane are used as standards for octane ratings in between zero and 100. A gasoline, for example, that has the same characteristics in an engine as a mixture of 90% isooctane-10% heptane would be rated as 90-octane gasoline.

The mechanism by which isobutylene dimerizes (below), is straightforward. The second step shows us another property of carbocations: *their ability to react with alkenes.*

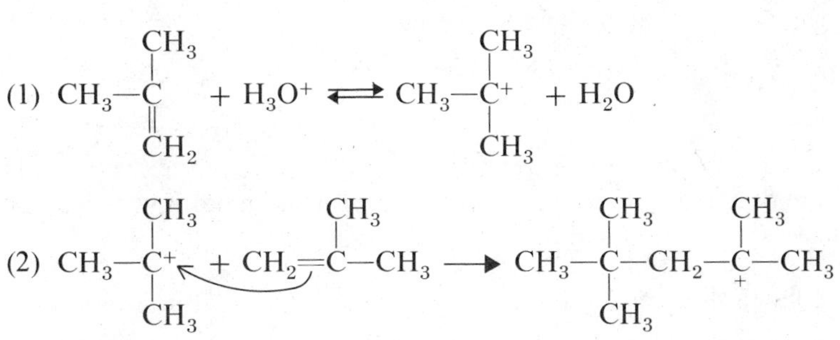

$$(1)\quad \mathrm{CH_3{-}\underset{\underset{CH_2}{\|}}{\overset{CH_3}{C}} + H_3O^+ \rightleftharpoons CH_3{-}\underset{CH_3}{\overset{CH_3}{C^+}} + H_2O}$$

$$(2)\quad \mathrm{CH_3{-}\underset{CH_3}{\overset{CH_3}{C^+}} + CH_2{=}\overset{CH_3}{C}{-}CH_3 \longrightarrow CH_3{-}\underset{CH_3}{\overset{CH_3}{C}}{-}CH_2{-}\underset{+}{\overset{CH_3}{C}}{-}CH_3}$$

* Although calling this compound isooctane is incorrect, the name has been used for many years in the petroleum industry. The correct IUPAC name is 2,2,4-trimethylpentane.

$$(3)\ CH_3-\underset{CH_3}{\overset{CH_3}{C}}-\underset{H\,(b)}{CH}-\overset{+}{C}(CH_3)-CH_2(H\,(a)) \xrightarrow{H_2O} \begin{cases} (a)\ CH_3-\underset{CH_3}{\overset{CH_3}{C}}-CH_2-C(=CH_2)CH_3 & \text{(major product)} \\ (b)\ CH_3-\underset{CH_3}{\overset{CH_3}{C}}-CH=C(CH_3)CH_3 & \text{(minor product)} \end{cases}$$

In step 1 the acid donates a proton to isobutylene to give a *tert*-butyl cation. We have seen similar reactions several times before. In step 2 the carbocation—acting as an electrophile—attacks another molecule of isobutylene. Isobutylene acts as a nucleophile; its π electrons form a σ bond to the *tert*-butyl cation. The larger carbocation that is formed as a result may lose a proton in two different ways (step 3) to form the mixture of diisobutylenes.

7.7 ALCOHOLS FROM ALKENES THROUGH OXYMERCURATION-DEMERCURATION (SOLVOMERCURATION-DEMERCURATION)

A highly useful procedure for synthesizing alcohols from alkenes, one that complements acid-catalyzed hydration, is a two-step method called oxymercuration-demercuration.

Alkenes react with mercuric acetate in a mixture of tetrahydrofuran* and water to produce (hydroxyalkyl)mercury compounds. These (hydroxyalkyl)mercury compounds can be reduced to alcohols with sodium borohydride:

$$-\overset{|}{C}=\overset{|}{C}- + H_2O + Hg\left(\overset{O}{\overset{\|}{O}CCH_3}\right)_2 \xrightarrow[\text{Oxymercuration}]{\text{THF}} -\underset{HO}{\overset{|}{C}}-\underset{Hg-\overset{O}{\overset{\|}{O}CCH_3}}{\overset{|}{C}}- + CH_3\overset{O}{\overset{\|}{C}}OH$$

$$-\underset{HO}{\overset{|}{C}}-\underset{Hg-\overset{O}{\overset{\|}{O}CCH_3}}{\overset{|}{C}}- + OH^- + NaBH_4 \xrightarrow{\text{Demercuration}} -\underset{HO}{\overset{|}{C}}-\underset{H}{\overset{|}{C}}- + Hg + CH_3\overset{O}{\overset{\|}{C}}O^-$$

In the first step, *oxymercuration,* water and mercuric acetate add to the double bond; in the second step, *demercuration,* sodium borohydride reduces the mercuriacetate group and replaces it with hydrogen.

* Tetrahydrofuran, or THF, is a cyclic ether with the structure shown below.

$$\begin{matrix} CH_2-CH_2 \\ | \quad\quad | \\ CH_2 \quad CH_2 \\ \ddot{O} \end{matrix}$$

Both steps can be carried out in the same vessel, and both reactions take place very rapidly at room temperature or below. The first step—oxymercuration—usually goes to completion within a period of 20 seconds to 10 minutes. The second step—demercuration—normally requires less than an hour. The overall reaction gives alcohols in very high yields, usually greater than 90%.

Oxymercuration-demercuration is also highly regioselective. The net orientation of the addition of the elements of water is in accordance with Markovnikov's rule.

$$n\text{-}C_3H_7CH{=}CH_2 \xrightarrow[\text{THF-}H_2O\ (15\ \text{sec})]{Hg(OAc)_2^*} n\text{-}C_3H_7\underset{OH}{CH}{-}\underset{HgOAc}{CH_2} \xrightarrow[OH^-\ (1\ \text{hr})]{NaBH_4} n\text{-}C_3H_7\underset{OH}{CH}CH_3 + Hg$$

1-Pentene → 2-Pentanol (93%)

$$CH_3CH_2\overset{CH_3}{C}{=}CH_2 \xrightarrow[\text{THF-}H_2O\ (10\ \text{sec})]{Hg(OAc)_2} CH_3CH_2\overset{CH_3}{\underset{OH}{C}}{-}\underset{HgOAc}{CH_2} \xrightarrow[OH^-\ (5\ \text{min})]{NaBH_4} CH_3CH_2\overset{CH_3}{\underset{OH}{C}}CH_3 + Hg$$

2-Methyl-1-butene → 2-Methyl-2-butanol (90%)

1-Methylcyclopentene $\xrightarrow[\text{THF-}H_2O\ (20\ \text{sec.})]{Hg(OAc)_2}$ 1-methyl-1-hydroxy-2-(HgOAc)cyclopentane (CH_3, OH, HgOAc, H) $\xrightarrow[OH^-\ (6\ \text{min.})]{NaBH_4}$ 1-Methylcyclopentanol (CH_3, OH) + Hg

Rearrangements of the carbon skeleton seldom occur in oxymercuration-demercuration. A striking example that illustrates this is the oxymercuration-demercuration of 3,3-dimethyl-1-butene shown below.

$$CH_3\overset{CH_3}{\underset{CH_3}{C}}{-}CH{=}CH_2 \xrightarrow[(2)\ NaBH_4,\ OH^-]{(1)\ Hg(OAc)_2/\text{THF-}H_2O} CH_3\overset{CH_3}{\underset{CH_3}{C}}{-}\underset{OH}{CH}CH_3$$

3,3-Dimethyl-1-butene → 3,3-Dimethyl-2-butanol (94%)

Analysis of the mixture of products by gas-liquid chromatography failed to reveal the presence of any 2,3-dimethyl-2-butanol. The acid-catalyzed hydration of 3,3-dimethyl-1-butene, by contrast, gives 2,3-dimethyl-2-butanol as the major product.

* $Hg(OAc)_2 = Hg(O\overset{O}{\overset{\|}{C}}CH_3)_2$

$$CH_3-C(CH_3)_2-CH=CH_2 \xrightarrow[H_2O]{H_2SO_4} CH_3-C(OH)(CH_3)-CH(CH_3)-CH_3$$

3,3-Dimethyl-1-butene → 2,3-Dimethyl-2-butanol (major product)

A mechanism that accounts for the orientation of addition in the oxymercuration stage, and one that also explains the general lack of accompanying rearrangements, has been proposed by Professor H. C. Brown of Purdue University. According to this mechanism, the first step of the oxymercuration reaction is an electrophilic attack by the mercury species, $\overset{+}{Hg}OAc$, at the less-substituted carbon of the double bond (i.e., at that carbon that bears the greater number of hydrogens). We can illustrate this step using 1-pentene in the following example.

$$CH_3(CH_2)_2CH=CH_2 + \overset{+}{Hg}OAc \longrightarrow CH_3(CH_2)_2\overset{+}{C}H-CH_2(HgOAc)$$

1-Pentene → Mercury-substituted carbocation

The mercury-substituted carbocation produced in this way then reacts very rapidly with water to produce a (hydroxyalkyl)mercury compound.

$$CH_3(CH_2)_2\overset{+}{C}H-CH_2(HgOAc) + :\ddot{O}H_2 \xrightarrow{-H^+} CH_3(CH_2)_2CH(\ddot{O}H)-CH_2(HgOAc)$$

Mercury-substituted carbocation → (Hydroxyalkyl)mercury compound

Calculations indicate that mercury-substituted carbocations like those formed in this reaction retain much of the positive charge on the mercury moiety:

$$CH_3(CH_2)_2\overset{\delta+}{C}H \cdots CH_2 \cdots \overset{\delta+}{Hg}OAc$$

Only a small portion of the positive charge resides on carbon. The charge on carbon is large enough to account for the observed Markovnikov addition, but it is too small to allow the usual rapid carbon-skeleton rearrangements that take place with fully developed carbocations.

The mechanism for the demercuration stage is not well understood. Free radicals are thought to be involved and rearrangements occasionally occur.

7.7A Solvomercuration-Demercuration Reactions

When mercuration-demercuration is carried out in solvents other than H_2O-THF, it can be used to synthesize ethers. H. C. Brown, who developed these general procedures, proposed that these reactions (including oxymercuration) be called *solvomercuration-demercuration.*

In oxymercuration, water acts as a nucleophile and attacks the mercury-substituted carbocation; after demercuration the product is an *alcohol.* When the reaction is carried out in an alcohol solvent, however, the alcohol acts as a nucleophile and attacks the mercury-substituted carbocation; after demercuration the product is an *ether:*

$$\text{C=C (Alkene)} + \overset{+}{\text{Hg}}\text{OAc} \longrightarrow -\overset{+}{\text{C}}-\underset{\text{HgOAc}}{\text{C}}- \xrightarrow[-\text{H}^+]{\text{R}-\ddot{\text{O}}-\text{H} \text{ alcohol}} -\overset{\text{OR}}{\text{C}}-\underset{\text{HgOAc}}{\text{C}}- \text{ (Alkoxyalkyl)mercury compound} \xrightarrow{\text{NaBH}_4/\text{OH}^-} -\overset{\text{OR}}{\text{C}}-\underset{\text{H}}{\text{C}}- \text{ Ether}$$

The following examples illustrate this synthesis of ethers.

$$\underset{\text{2-Methyl-1-butene}}{CH_3CH_2\overset{CH_3}{\overset{|}{C}}{=}CH_2} \xrightarrow[\text{(2) NaBH}_4\text{, OH}^-]{\text{(1) Hg(OAc)}_2/\text{CH}_3\text{OH}} \underset{\text{2-Methoxy-2-methylbutane (100\%)}}{CH_3CH_2\overset{CH_3}{\underset{OCH_3}{C}}{-}CH_3}$$

$$\underset{\text{1-Hexene}}{CH_3(CH_2)_3CH{=}CH_2} \xrightarrow[\text{(2) NaBH}_4\text{, OH}^-]{\text{(1) Hg(OAc)}_2/\text{C}_2\text{H}_5\text{OH}} \underset{\text{2-Ethoxyhexane (98\%)}}{CH_3(CH_2)_3\underset{OC_2H_5}{CH}CH_3}$$

When tertiary alcohols are used in this synthesis, best results are obtained by replacing mercuric acetate by mercuric trifluoroacetate, $Hg(OOCCF_3)_2$. This is a good method for synthesizing ethers with tertiary groups.

$$\underset{\text{1-Hexene}}{CH_3(CH_2)_3CH{=}CH_2} \xrightarrow[\text{(2) NaBH}_4\text{, OH}^-]{\text{(1) Hg(OOCCF}_3)_2/(\text{CH}_3)_3\text{COH}} \underset{\text{2-}\textit{tert}\text{-Butoxyhexane (100\%)}}{CH_3(CH_2)_3\underset{O-C(CH_3)_3}{CH}CH_3}$$

Problem 7.5

Outline all steps in the mechanism for the synthesis of 2-ethoxyhexane from 1-hexene by solvomercuration-demercuration.

Problem 7.6

Show how solvomercuration (including oxymercuration)-demercuration syntheses could be used to prepare the following alcohols and ethers from the alkenes given.

(a) 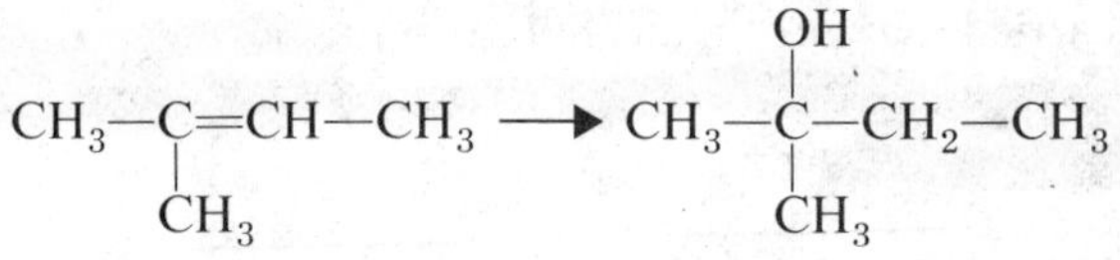 $CH_3-\underset{CH_3}{\underset{|}{C}}=CH-CH_3 \longrightarrow CH_3-\underset{CH_3}{\underset{|}{\overset{OH}{\overset{|}{C}}}}-CH_2-CH_3$

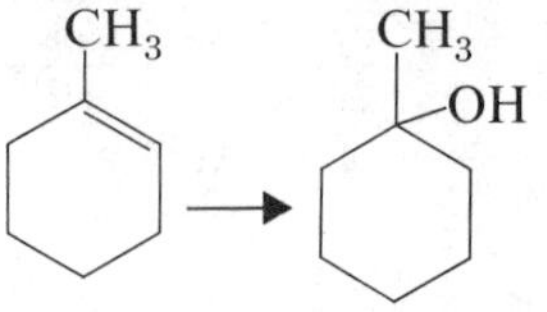

(b) 1-methylcyclohexene (CH_3) $\longrightarrow$ 1-methylcyclohexanol (CH_3, OH)

(c) cyclohexene $\longrightarrow$ methoxycyclohexane (OCH_3)

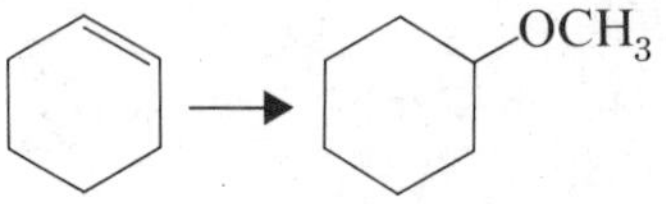

(d) 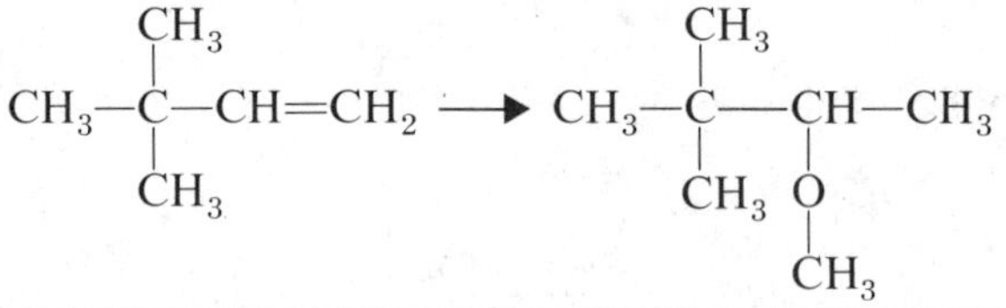 $CH_3-\underset{CH_3}{\underset{|}{\overset{CH_3}{\overset{|}{C}}}}-CH=CH_2 \longrightarrow CH_3-\underset{CH_3}{\underset{|}{\overset{CH_3}{\overset{|}{C}}}}-\underset{O-CH_3}{\underset{|}{CH}}-CH_3$

7.8 HYDROBORATION-OXIDATION

Addition of the elements of water to a double bond can also be achieved in the laboratory through the use of the boron hydride $(BH_3)_2$, called diborane. The addition of water is indirect and two reactions are involved. The first is the addition of boron hydride to the double bond (hydroboration); the second is the oxidation and hydrolysis of the organoboron intermediate to an alcohol and boric acid. We can illustrate these steps with the hydroboration-oxidation of propene.*

$$3CH_3CH=CH_2 + \tfrac{1}{2}(BH_3)_2 \longrightarrow (CH_3CH_2CH_2)_3B$$

$$\xrightarrow{H_2O_2,\ OH^-} \underset{n\text{-Propyl alcohol}}{3CH_3CH_2CH_2OH} + \underset{\text{Boric acid}}{H_3BO_3}$$

Both diborane and alkylboranes ignite spontaneously when exposed to air.

To avoid this danger, diborane is usually generated in the presence of the alkene (*in situ*) by the addition of boron trifluoride in ether to a mixture of sodium borohydride and the alkene. Boron trifluoride and sodium borohydride react to produce diborane according to the following equation:

$$3Na^+BH_4^- + 4BF_3 \longrightarrow 2(BH_3)_2 + 3Na^+BF_4^-$$

Boron trifluoride and sodium borohydride are much more easily handled. Boron trifluoride is usually obtained in diethyl ether solution, where it forms the stable

* Hydroboration reactions were discovered by H. C. Brown (p. 252). They have proved to be of great synthetic versatility. We will see other examples in subsequent chapters.

complex $(C_2H_5)_2O:BF_3$. Hydroboration reactions are usually carried out in ethers; either in diethyl ether, $(C_2H_5)_2O$, itself or in some higher-molecular-weight ether such as "diglyme," $(CH_3OCH_2CH_2)_2O$, *di*ethylene *gly*col di*me*thyl ether.

The alkylboranes produced by hydroboration usually are not isolated. They are oxidized and hydrolyzed to alcohols in the same reaction vessel by the addition of hydrogen peroxide in an aqueous base.

$$R_3B \xrightarrow[\text{NaOH, 25°}]{H_2O_2} 3R{-}OH + H_3BO_3$$
Oxidation

Hydroboration reactions are regioselective and the net result of hydroboration-oxidation is an apparent *anti-Markovnikov addition of water.* As a consequence, *hydroboration-oxidation gives us a method for the preparation of alcohols that cannot normally be obtained through the acid-catalyzed hydration of alkenes or by oxymercuration-demercuration.* For example, acid-catalyzed hydration (or oxymercuration-demercuration) of 1-hexene yields 2-hexanol:

$$CH_3CH_2CH_2CH_2CH{=}CH_2 \xrightarrow{H_3O^+,\ H_2O} CH_3CH_2CH_2CH_2\underset{\displaystyle OH}{\underset{|}{C}}HCH_3$$

1-Hexene — 2-Hexanol

Hydroboration-oxidation, by contrast, yields 1-hexanol:

$$CH_3CH_2CH_2CH_2CH{=}CH_2 \xrightarrow[(2)\ H_2O_2,\ OH^-]{(1)\ (BH_3)_2} CH_3CH_2CH_2CH_2CH_2CH_2OH$$

1-Hexene — 1-Hexanol (90%)

Other examples of hydroboration-oxidation are the following:

$$CH_3{-}\overset{\displaystyle CH_3}{\underset{\displaystyle CH_3}{C}}{-}CH{=}CH_2 \xrightarrow[(2)\ H_2O_2,\ OH^-]{(1)\ (BH_3)_2} CH_3{-}\overset{\displaystyle CH_3}{\underset{\displaystyle CH_3}{C}}{-}CH_2CH_2OH$$

3,3-Dimethyl-1-butene — 3,3-Dimethyl-1-butanol (67%)

$$CH_3{-}\overset{\displaystyle CH_3}{C}{=}CHCH_3 \xrightarrow[(2)\ H_2O_2,\ OH^-]{(1)\ (BH_3)_2} CH_3{-}\overset{\displaystyle CH_3}{C}H{-}\underset{\displaystyle OH}{C}HCH_3$$

2-Methyl-2-butene — 3-Methyl-2-butanol (59%)

$$\text{1-Methylcyclopentene} \xrightarrow[(2)\ H_2O_2,\ OH^-]{(1)\ (BH_3)_2} \textit{trans}\text{-2-Methylcyclopentanol}$$

1-Methylcyclopentene — *trans*-2-Methylcyclopentanol (86%)

7.8A Mechanism of Hydroboration

Hydroboration, the first step of the hydroboration-oxidation synthesis, occurs in a stepwise manner. Diborane $(BH_3)_2$, acts as if it were in a monomeric

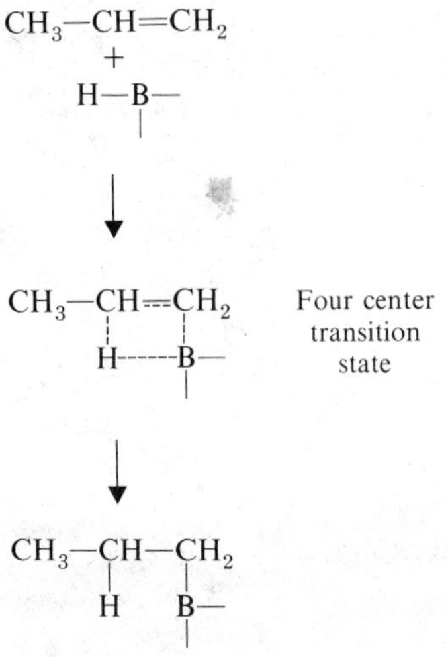

$$CH_3-CH=CH_2 + H-B\!\!<$$

↓

$$CH_3-CH\text{---}CH_2 \;\; (H\text{------}B) \quad \text{Four center transition state}$$

↓

$$CH_3-CH(H)-CH_2(B\!\!<)$$

FIG. 7.4
The one-step mechanism for the addition of a boron hydride to propene.

form, as if it were borane, BH_3. Borane adds successively to the double bonds of three molecules of the alkene until all of its hydrogens are replaced by alkyl groups.

$$CH_3CH=CH_2 + H-BH_2 \longrightarrow CH_3CH_2CH_2-BH_2 \xrightarrow{CH_3CH=CH_2} (CH_3CH_2CH_2)_2BH \xrightarrow{CH_3CH=CH_2} (CH_3CH_2CH_2)_3B$$

Tripropylboron

In each addition step *the boron atom becomes attached to the less substituted carbon of the double bond* and a hydrogen is transferred from the boron to the other carbon of the double bond.

Experimental evidence indicates that the mechanism of addition of borane to the double bond occurs *in a single step* through a four-center transition state. The observed attachment of boron to the less substituted carbon of the double bond seems to result primarily from *steric factors*—the bulky boron-containing group can approach the less-substituted carbon more easily. An addition occurring through a four-center transition state is shown in Fig. 7.4 using propene as an example.

7.8B Syn and Anti Additions

An addition that places the parts of the adding reagent on the same side (or face) of the reactant is called a *syn* addition.

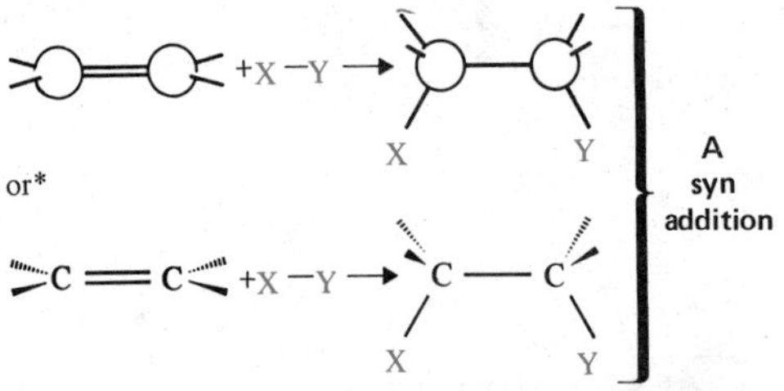

$$>C=C< + X-Y \longrightarrow \underset{X}{C}-\underset{Y}{C}$$

or*

$$>C=C< + X-Y \longrightarrow \underset{X}{C}-\underset{Y}{C}$$

A syn addition

* Notice in these formulas a dashed wedge (''''') means the bond is directed behind the plane of the page; a solid wedge (◄) means the bond is directed out of the page; an ordinary bond (—) means the bond lies in the plane of the page.

The opposite of a syn addition is an *anti* addition. An anti addition places the parts of the adding reagent on opposite faces of the reactant.

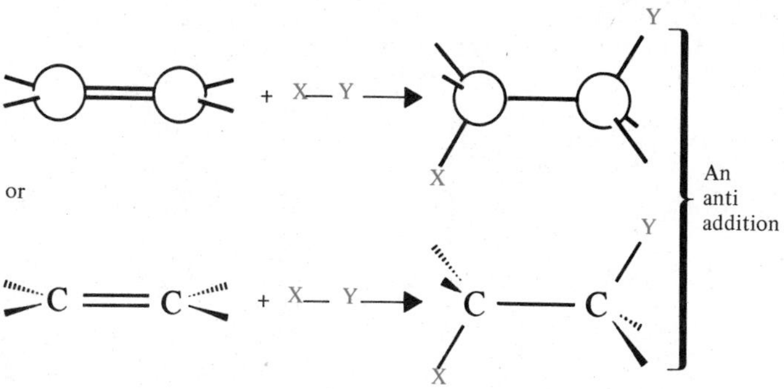

7.8C Syn Addition of Boron Hydrides

The four-center transition state for the addition of boron hydrides to a double bond requires the addition of both boron and hydrogen to the same face of the molecule. The addition of boron hydrides is then, of necessity, a syn addition. With 1-methylcyclopentene, for example, both boron and hydrogen add to the same face of the ring (Equation 1, Fig. 7.5).

7.8D Oxidation of Alkylboranes

The mechanism for the oxidation in the hydroboration-oxidation synthesis of alcohols is beyond the scope of our present discussion. But, one important point can be made. *The hydroxyl group replaces the boron atom where it stands in the organoboron compound.* The net result of the two steps (hydroboration and oxidation) is the syn addition of —H and —OH, that is, of the pieces of a water molecule. We can see this if we examine the hydroboration-oxidation of 1-methylcyclopentene (Fig. 7.5).

FIG. 7.5

The hydroboration-oxidation of 1-methylcyclopentene. The first reaction is a syn *addition of boron hydride. (In this illustration we have shown the boron and hydrogen both entering from the bottom side of 1-methylcyclopentene. We could also have shown the reaction with both groups entering from the top.) In the second reaction the boron atom is replaced by a hydroxyl group on the same side of the molecule. The product is a* trans *compound* (trans-*2-methylcyclopentanol) and the overall result is the syn addition of —H and —OH.*

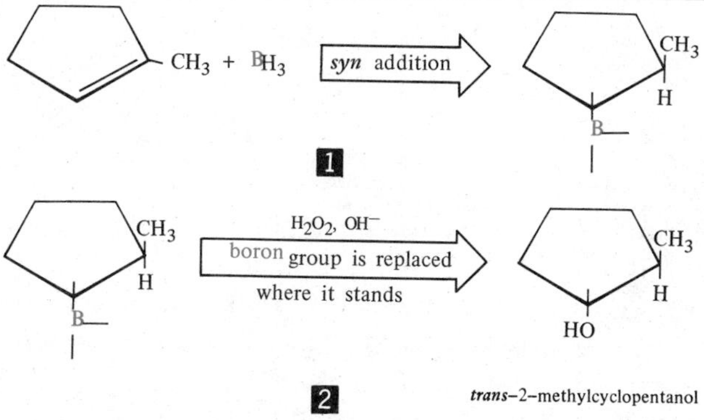

Problem 7.7

Show how you might employ hydroboration-oxidation reactions to carry out the following syntheses.

(a) 1-Butene $\longrightarrow$ $CH_3CH_2CH_2CH_2OH$

(b) 2-Methyl-2-butene $\longrightarrow$ $CH_3CH(CH_3)CH(OH)CH_3$

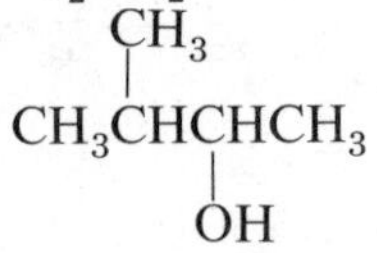

(c) 1-Methylcyclohexene $\longrightarrow$ (cyclohexane chair with CH_3, H, H, OH) $\rightleftarrows$ (chair with H, CH_3, HO, H)

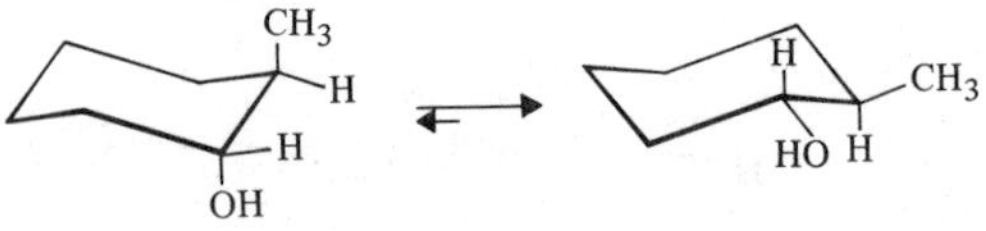

Problem 7.8

Alkylboranes react with acetic acid in the following way:

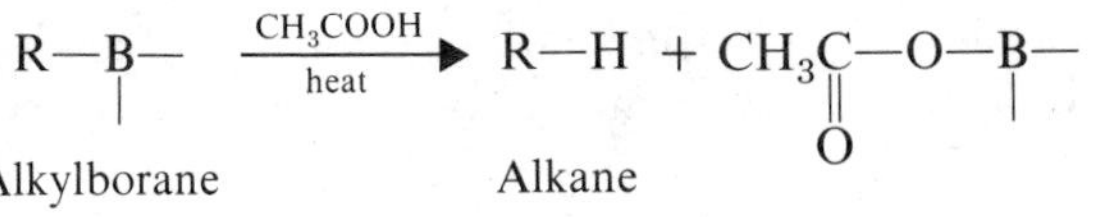

$$\underset{\text{Alkylborane}}{R{-}\overset{|}{B}{-}} \xrightarrow[\text{heat}]{CH_3COOH} \underset{\text{Alkane}}{R{-}H} + CH_3\overset{O}{\overset{\|}{C}}{-}O{-}\overset{|}{B}{-}$$

In this reaction hydrogen replaces boron *where it stands* in the alkylborane. With this fact in mind, and assuming you have deuterioacetic acid (CH_3COOD) available, can you suggest syntheses of the following deuterium-labeled compounds?

(a) 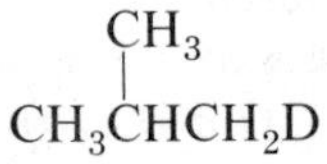 $CH_3CH(CH_3)CH_2D$

(b) 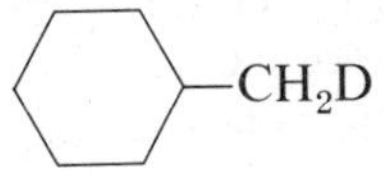 Cyclohexyl$-CH_2D$

(c) 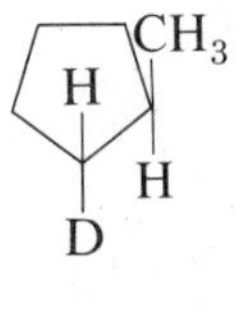 Cyclopentane with CH_3, H, H, D

(d) Assume you also have available $(BD_3)_2$. Can you suggest a synthesis of the following?

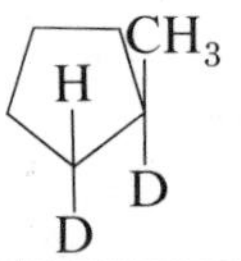

Cyclopentane with CH_3, H, D, D

7.9 ADDITION OF HALOGENS TO ALKENES

In the absence of light, *alkanes* do not react appreciably with chlorine or bromine at room temperature. If we add an alkane to a solution of bromine in carbon tetrachloride, the red-brown color of the bromine will persist in the solution as long as we keep the mixture away from sunlight and as long as the solution is not heated.

$$\underset{\substack{\text{Alkane}\\\text{(colorless)}}}{R{-}H} + \underset{\substack{\text{Bromine}\\\text{(red-brown)}}}{Br_2} \xrightarrow[\text{in the dark, } CCl_4]{\text{Room temperature}} \text{no appreciable reaction}$$

On the other hand, if we expose the reactants to sunlight, the bromine color will fade rapidly. If we now place a small piece of moist blue litmus in the region above the liquid, the litmus will turn red because of the hydrogen bromide that evolves as the alkane and bromine react. (Hydrogen bromide is not very soluble in carbon tetrachloride.)

$$\underset{\substack{\text{Alkane}\\\text{(colorless)}}}{R{-}H} + \underset{\substack{\text{Bromine}\\\text{(red-brown)}}}{Br_2} \xrightarrow[\text{sunlight, } CCl_4]{\text{Room temperature}} \underset{\substack{\text{Alkyl halide}\\\text{(colorless)}}}{R{-}Br} + \underset{\substack{\text{(detected by}\\\text{moist blue litmus)}}}{HBr}$$

The behavior of *alkenes* toward bromine in carbon tetrachloride contrasts markedly with that of alkanes. Alkenes react rapidly with bromine at room temperature and in the *absence of light*. If we add bromine to an alkene the red-brown color of the bromine disappears almost instantly as long as the alkene is present in excess. If we test the atmosphere above the solution with moist blue litmus, we will find that no hydrogen bromide is present. The reaction is one of addition.

$$\underset{\substack{\text{An alkene}\\\text{(colorless)}}}{>C{=}C<} + Br_2 \xrightarrow[\text{in the dark, } CCl_4]{\text{Room temperature}} \underset{\substack{vic\text{-Dibromide}\\\text{(colorless)}}}{-\underset{Br}{\overset{|}{C}}-\underset{Br}{\overset{|}{C}}-}$$

Because of these differences, bromine in carbon tetrachloride is a useful reagent for distinguishing between alkanes and alkenes.

The addition reaction between alkenes and chlorine or bromine is a general one. The products are vicinal dihalides.

$$CH_3CH{=}CHCH_3 + Cl_2 \xrightarrow[-9^\circ]{O_2^*} CH_3\underset{Cl}{CH}\underset{Cl}{CH}CH_3 \quad (100\%)$$

$$CH_3CH_2CH{=}CH_2 + Cl_2 \xrightarrow[-9^\circ]{O_2} CH_3CH_2\underset{Cl}{CH}{-}\underset{Cl}{CH_2} \quad (97\%)$$

$$\text{Cyclohexene} + Br_2 \xrightarrow[CCl_4/C_2H_5OH]{-5^\circ} \underset{trans\text{-1,2-Dibromocyclohexane}}{\text{(Br, H, H, Br)}} \quad (95\%)$$

trans-1,2-Dibromocyclohexane

7.9A Mechanism of Halogen Addition

One mechanism that has been proposed for halogen addition is an ionic mechanism.† In the first step the exposed electrons of the π bond of the alkene

* Oxygen acts as a free-radical inhibitor.

† There is evidence that in the absence of oxygen some reactions between alkenes and chlorine proceed through a free-radical mechanism. We will not discuss this mechanism here, however.

attack the halogen to form a π complex. In a bromine addition, for example, the π complex probably forms in the following way:

Step 1 C=C + :Br—Br: ⇌ C=C ↓ :Br:$^{\delta+}$ | :Br:$^{\delta-}$ ⇌ —C—C— :Br:$^{\delta+}$ | :Br:$^{\delta-}$

π complex

As the π electrons of the alkene approach the bromine molecule, the electrons of the bromine-bromine bond drift in the direction of the bromine atom more distant from the approaching alkene. The bromine molecule becomes *polarized* as a result. The more distant bromine develops a partial negative charge; the nearer bromine becomes partially positive.

Polarization weakens the bromine-bromine bond, and in the next step *it breaks heterolytically*. A bromide ion departs, and a *bromonium ion* forms. In the

Step 2 C—C :Br:$^{\delta+}$ | :Br:$^{\delta-}$ ⟶ C—C Br^{+} + :Br:$^{-}$

π complex Bromonium ion Bromide ion

bromonium ion a bromine atom is bonded to two carbon atoms by *two pairs of electrons*. This is possible because of the unshared electron pairs on bromine.

C—C Br^{+} or C—C Br^{+}

Bromonium ion

In the third step, one of the bromide ions produced in step 2 attacks one of the carbon atoms of the bromonium ion. The nucleophilic attack results in the formation of a *vic*-dibromide by opening the three-membered ring.

Step 3 :Br:$^{-}$ C—C Br^{+} ⟶ —C(:Br:)—C(:Br:)—

vic-Dibromide

This ring-opening (above) is an S_N2 reaction. The bromide ion, acting as a nucleophile, uses a pair of electrons to form a bond to one carbon atom of the bromonium ion while the positive bromine of the bromonium ion acts as a leaving group.

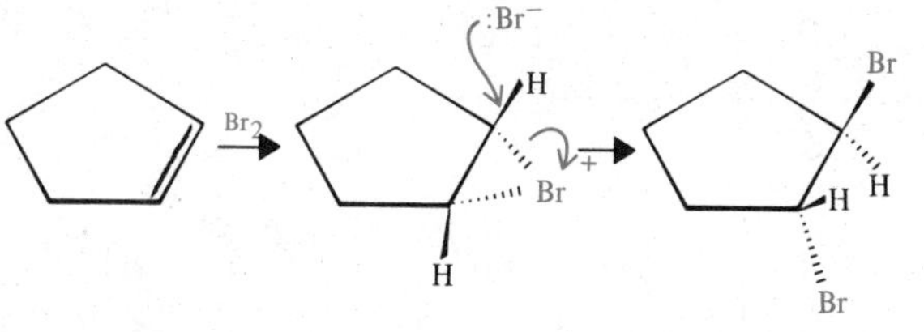

trans–1, 2 Dibromo–cyclopentane

FIG. 7.6
Anti addition of bromine to cyclopentene. (The formation of the π complex is not shown.) In this reaction the bromide ion attacks a carbon atom of the bromonium ion in an S_N2 reaction causing an inversion of configuration.

7.10 ANTI ADDITION OF HALOGENS

The addition of bromine to cyclopentene provides additional evidence for bromonium ion intermediates in bromine additions. When cyclopentene reacts with bromine in carbon tetrachloride, *anti addition* occurs, and the product of the reaction is *trans*-1,2-dibromocyclopentane. The anti addition of bromine to cyclopentene can be accounted for by the mechanism given in Fig. 7.6—one in which a bromide ion attacks a carbon of the ring from the side opposite that of the bromonium ion. This nucleophilic attack by the bromide causes the configuration of the carbon being attacked *to invert* (Section 5.5). Inversion of configuration at one carbon of the ring leads to the formation of *trans*-1,2-dibromocyclopentane.

Anti addition also occurs when chlorine adds to simple alkenes by an ionic mechanism. However, not all halogen additions to double bonds are anti additions, nor do all additions proceed through a halonium ion intermediate. We will see, in Chapter 12, that syn additions occur frequently when phenyl groups (C_6H_5 groups) are attached to one or both carbons of the double bond. In these instances it appears that carbocations and carbocationlike intermediates are formed instead of halonium ions, because the carbocations are quite stable.

7.11 HALOHYDRIN FORMATION

If the halogenation of an alkene is carried out in aqueous solution (rather than in carbon tetrachloride) the major product of the overall reaction is not a *vic*-dihalide, but rather it is a *halo alcohol* called a *halohydrin*. Molecules of the solvent become reactants, too.

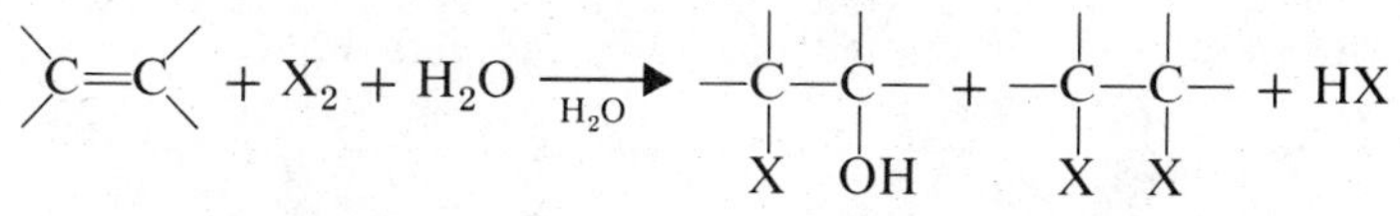

$$>C=C< + X_2 + H_2O \xrightarrow{H_2O} -\underset{X}{\overset{|}{C}}-\underset{OH}{\overset{|}{C}}- + -\underset{X}{\overset{|}{C}}-\underset{X}{\overset{|}{C}}- + HX$$

X = Cl or Br — Halohydrin — *vic*-Dihalide

Halohydrin formation can be explained by the following mechanism.

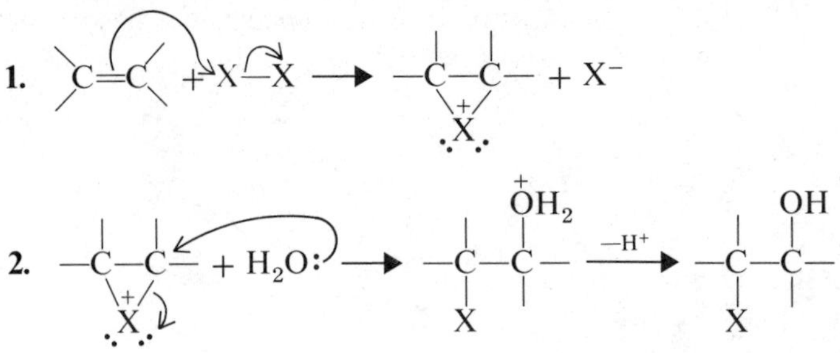

1. $>C{=}C< + X{-}X \longrightarrow -C{-}C- \text{ (bridged by } X^+) + X^-$

2. $-C{-}C- \text{ (bridged by } X^+) + H_2O: \longrightarrow -C(X){-}C(\overset{+}{O}H_2)- \xrightarrow{-H^+} -C(X){-}C(OH)-$

The first step is the same as that for halogen addition. In the second step, however, the two mechanisms differ. In halohydrin formation, water acts as the nucleophile and attacks one carbon of the halonium ion. The three-membered ring opens, and a protonated halohydrin is produced. Loss of a proton then leads to the formation of the halohydrin itself.

Water, because of its unshared electron pairs, acts as a nucleophile in this and in many other reactions. In this instance water molecules far outnumber halide ions because water is the solvent for the reactants. This accounts for the halohydrin being the major product.

Problem 7.9

Outline a mechanism that accounts for the formation of *trans*-2-chlorocyclopentanol from cyclopentene and chlorine in aqueous solution.

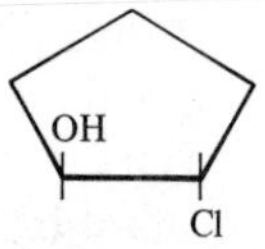

OH, Cl

trans-2-chlorocyclopentanol

If the alkene is unsymmetrical the halogen ends up on the carbon with the greater number of hydrogens. Bonding in the intermediate bromonium ion is *unsymmetrical.* The more highly substituted carbon bears the greater positive charge because it resembles the more stable carbocation. Consequently water attacks this carbon preferentially even though it is more hindered.

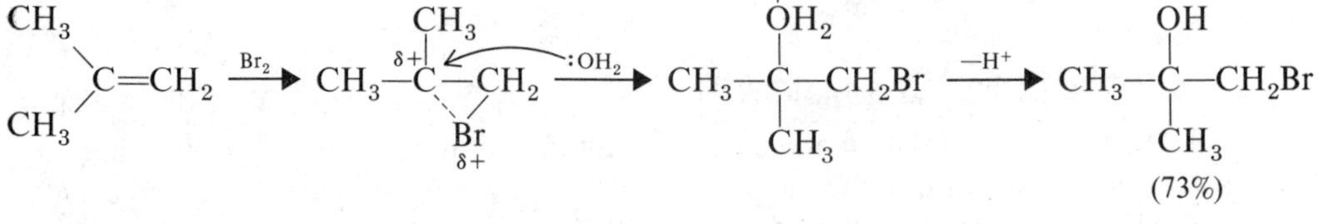

$$(CH_3)_2C{=}CH_2 \xrightarrow{Br_2} CH_3{-}\overset{\delta+}{C}(CH_3){-}CH_2 \text{ (bridged by } Br^{\delta+}) \xrightarrow{:OH_2} CH_3{-}C(\overset{+}{O}H_2)(CH_3){-}CH_2Br \xrightarrow{-H^+} CH_3{-}C(OH)(CH_3){-}CH_2Br$$

(73%)

7.12 EPOXIDES: EPOXIDATION OF ALKENES

Epoxides are cyclic ethers with three-membered rings:

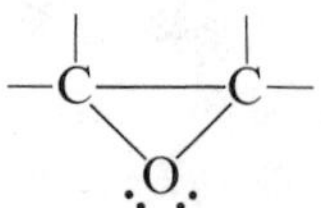

$$-C{-}C- \text{ (bridged by } \ddot{O})$$

An epoxide

In IUPAC nomenclature epoxides are called *oxiranes*. More commonly, however, they are called *alkene oxides*.

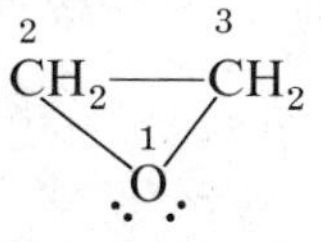

$$\overset{2}{C}H_2—\overset{3}{C}H_2 \text{ (ring through } \overset{1}{O}\text{)}$$

IUPAC name: oxirane
Common name: ethylene oxide

The common names for epoxides originate from the most widely used method for synthesizing them: *epoxidation*, the reaction of an alkene and an organic *peroxy acid* (sometimes called simply a *peracid*).

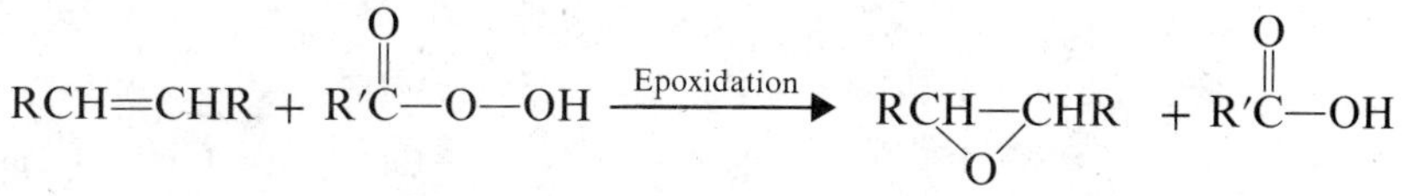

$$RCH{=}CHR + R'\overset{O}{\overset{\|}{C}}—O—OH \xrightarrow{\text{Epoxidation}} RCH\overset{O}{—}CHR + R'\overset{O}{\overset{\|}{C}}—OH$$

An alkene | A peroxy acid | An alkene oxide (or oxirane)

In this reaction the peroxy acid transfers an oxygen atom to the alkene. The following mechanism has been proposed.

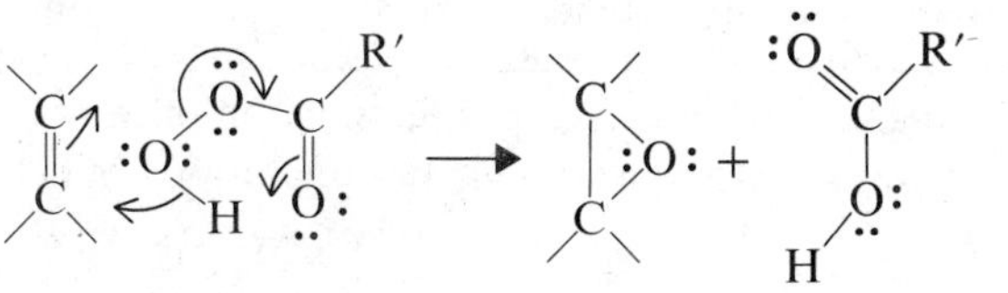

The addition of oxygen to the double bond in an epoxidation reaction is, of necessity, a *syn* addition. With cyclopentene, for example, epoxidation gives the following result.

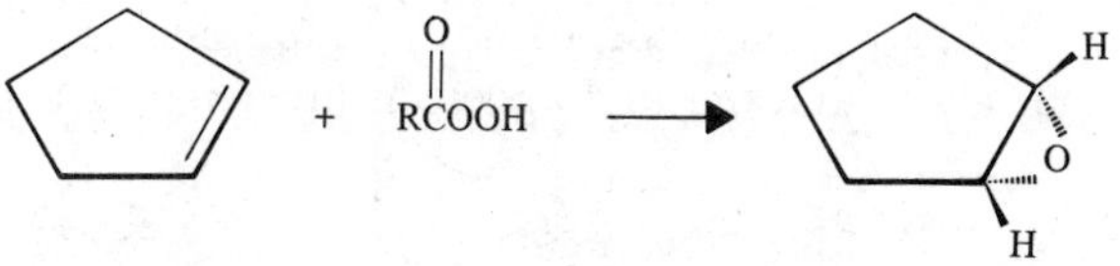

$$\text{Cyclopentene} + R\overset{O}{\overset{\|}{C}}OOH \longrightarrow \text{Cyclopentene oxide}$$

Cyclopentene | Cyclopentene oxide

The peroxy acids most commonly used are peroxyformic acid ($H\overset{O}{\overset{\|}{C}}OOH$), peroxyacetic acid ($CH_3\overset{O}{\overset{\|}{C}}OOH$), and peroxybenzoic acid ($C_6H_5\overset{O}{\overset{\|}{C}}OOH$). Cyclohexene, for example, reacts with peroxybenzoic acid to give cyclohexene oxide in a quantitative yield.

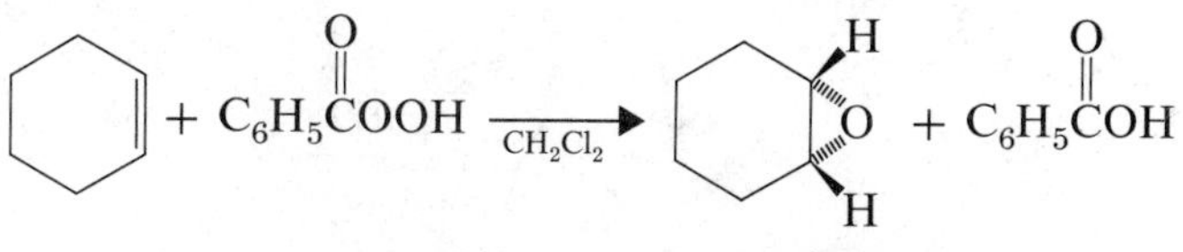

$$\text{Cyclohexene} + C_6H_5\overset{O}{\overset{\|}{C}}OOH \xrightarrow{CH_2Cl_2} \text{Cyclohexene oxide} + C_6H_5\overset{O}{\overset{\|}{C}}OH$$

Peroxybenzoic acid | Cyclohexene oxide (100%)

7.12A Acid-Catalyzed Hydrolysis of Epoxides

Although most ethers react with few reagents, the strained three-membered ring of epoxides makes them highly susceptible to ring-opening reactions. Ring opening takes place through cleavage of one of the carbon-oxygen bonds. It can be initiated by either electrophiles or nucleophiles, or catalyzed by either acids or bases. The acid-catalyzed hydrolysis of an epoxide, for example, is a useful procedure for preparing vicinal-dihydroxy compounds called glycols.

$$CH_2{-}CH_2\ (\text{epoxide, } \ddot{O}) + H_3O{:}^+ \rightleftharpoons CH_2{-}CH_2\ (\ddot{O}^+{-}H) + H_2\ddot{O}{:} \rightleftharpoons$$

$$\underset{:\ddot{O}H}{CH_2}{-}\overset{:\overset{+}{O}H_2}{CH_2} \xrightleftharpoons{-H^+} \underset{:\ddot{O}H}{CH_2}{-}\overset{:\ddot{O}H}{CH_2} \text{ or } \underset{:\ddot{O}H}{CH_2}{-}\underset{:\ddot{O}H}{CH_2}$$

Ethylene glycol

Problem 7.10

The hydrolysis of an epoxide (above) takes place much more rapidly in the presence of aqueous acid than in a neutral solution (pH = 7). (a) What general type of reaction is this, and what is the specific function of the acid? (b) Epoxides also undergo ring opening rapidly in strongly basic solutions. What factor accounts for this?

7.12B Anti Hydroxylation of Alkenes

Acid-catalyzed hydrolysis of cyclopentene oxide yields a *trans*-dialcohol, *trans*-1,2-cyclopentanediol. (Dialcohols are called *diols* or *glycols*). Water acting as a nucleophile attacks the protonated epoxide from the side opposite the epoxide group. The carbon being attacked undergoes an inversion of configuration:

$$\text{cyclopentene oxide} + H_3\overset{+}{\ddot{O}}{:} \rightleftharpoons \text{protonated epoxide } ({:}\overset{+}{O}H) \xrightarrow{H_2O:} \text{(}OH_2^+\text{, H, OH, H)} \xrightarrow{-H^+} \text{(OH, H, OH, H) or (:}\ddot{O}H\text{, H/H, :}\ddot{O}H\text{)}$$

trans – 1, 2 – Cyclopentanediol

Epoxidation followed by acid-catalyzed hydrolysis gives us, therefore, a method for *hydroxylating* a double bond (i.e., a method for adding a hydroxyl group to each carbon). This technique, moreover, amounts to a net antihydroxylation, and its mechanism parallels closely the mechanism for the bromination of an alkene given earlier (Section 7.9A).

7.13 OXIDATION OF ALKENES

Alkenes undergo a number of other reactions in which the carbon-carbon double bond is oxidized. Potassium permanganate or osmium tetroxide, for example, can also be used to oxidize alkenes to glycols.

$$CH_2{=}CH_2 + KMnO_4 \xrightarrow[OH^-]{Cold} \underset{OH}{CH_2}-\underset{OH}{CH_2}$$

Ethylene → Ethylene glycol

$$CH_3CH{=}CH_2 \xrightarrow[(2)\ Na_2SO_3]{(1)\ OsO_4^*} CH_3\underset{OH}{CH}-\underset{OH}{CH_2}$$

Propylene → Propylene glycol

7.13A Syn Hydroxylation of Alkenes

The mechanisms for the formation of glycols by permanganate ion and osmium tetroxide oxidations first involve the formation of cyclic intermediates. Then in the second step cleavage of the oxygen-metal bond takes place (at the dashed lines below).

$$>C{=}C< + MnO_4^- \longrightarrow \text{cyclic manganate intermediate} \xrightarrow[H_2O]{OH^-} -\underset{OH}{C}-\underset{OH}{C}- + MnO_2$$

$$>C{=}C< + OsO_4 \longrightarrow \text{An osmate ester} \xrightarrow{Na_2SO_3} -\underset{OH}{C}-\underset{OH}{C}- + Os$$

An osmate ester

* Writing reagents above and below the arrow like this, $\xrightarrow[(2)\ Na_2SO_3]{(1)\ OsO_4}$, means that two steps are involved. In the first we treat the alkene with OsO_4. Then, after that reaction is complete, we add Na_2SO_3.

The course of these reactions is *syn hydroxylation*. This can be seen, readily, when cyclopentene reacts with cold dilute potassium permanganate (in base) or with osmium tetroxide (followed by treatment with Na_2SO_3). The product in either case is *cis*-1,2-cyclopentanediol.

+ MnO_4^- —cold→ [cyclic manganate ester: O, O, Mn, O, O^-]$^-$ —H_2O, OH^-→ OH OH

cis-1,2-Cyclopentanediol

+ OsO_4 —25°→ [cyclic osmate ester: O, O, Os, O, O] —Na_2SO_3→ OH OH

cis-1,2-Cyclopentanediol

Of the two reagents used for syn hydroxylation, osmium tetroxide gives the higher yields. Unfortunately, however, osmium tetroxide is highly toxic and very expensive. Potassium permanganate is a very powerful oxidizing agent and is easily capable of causing further oxidation of the glycol. Limiting the reaction to hydroxylation alone is often difficult, but is usually attempted by using cold, dilute, and basic solutions of potassium permanganate.

7.13B Oxidative Cleavage of Alkenes

Alkenes are oxidatively cleaved to salts of carboxylic acids, by hot permanganate solutions. We can illustrate this reaction with the oxidative cleavage of either *cis*- or *trans*-2-butene to two moles of acetate ion.

$$CH_3CH{=}CHCH_3 \xrightarrow[\text{heat}]{KMnO_4,\ OH^-} 2CH_3C(=O)O^- \xrightarrow{H^+} 2CH_3C(=O)OH$$

(*cis* or *trans*) Acetate ion Acetic acid

Acidification of the mixture, after the oxidation is complete, produces two moles of acetic acid for each mole of 2-butene.

The terminal CH_2 group of a 1-alkene is completely oxidized to carbon dioxide and water by hot permanganate. A disubstituted carbon of a double bond becomes the $\gt C{=}O$ group of a ketone.

$$CH_3CH_2C(CH_3){=}CH_2 \xrightarrow[\text{(2) } H^+]{\text{(1) } KMnO_4,\ OH^-,\ \text{heat}} CH_3CH_2C(CH_3){=}O + CO_2 + H_2O$$

The oxidative cleavage of alkenes has frequently been used to prove the location of the double bond in an alkene chain or ring. We can see how this might be done with the following example:

Example

An unknown alkene with the formula C_8H_{16} was found, on oxidation with hot permanganate, to yield a five-carbon carboxylic acid (pentanoic acid) and a three-carbon carboxylic acid (propanoic acid).

$$C_8H_{16} \xrightarrow[(2)\ H^+]{(1)\ KMnO_4,\ H_2O,\ OH^-,\ heat} CH_3CH_2CH_2CH_2\overset{O}{\overset{\|}{C}}{-}OH + CH_3CH_2\overset{O}{\overset{\|}{C}}{-}OH$$

Pentanoic acid — Propanoic acid

We can see, then, from what we know about oxidative cleavage by permanganate, that the original alkene must have been either *cis-* or *trans*-3-octene:

$$CH_3CH_2CH{=}CHCH_2CH_2CH_2CH_3$$

3-Octene
(*cis* or *trans*)

7.13C Ozonization (Ozonolysis) of Alkenes

A more widely used method for locating the double bond of an alkene involves the use of ozone, O_3. Ozone reacts vigorously with alkenes to form unstable compounds called *initial ozonides.*

Initial ozonide — Ozonide

The initial ozonide that is formed rearranges spontaneously to produce a compound known simply as an *ozonide.* This rearrangement is thought to occur through dissociation of the initial ozonide into reactive fragments that recombine to yield the ozonide.

Initial ozonide

Ozonide

Ozonides, themselves, are very unstable compounds and often explode violently. Because of this property they are not usually isolated, but are reduced directly by treatment with zinc and water. The reduction produces carbonyl compounds (either aldehydes or ketones) that can be safely isolated and identified.

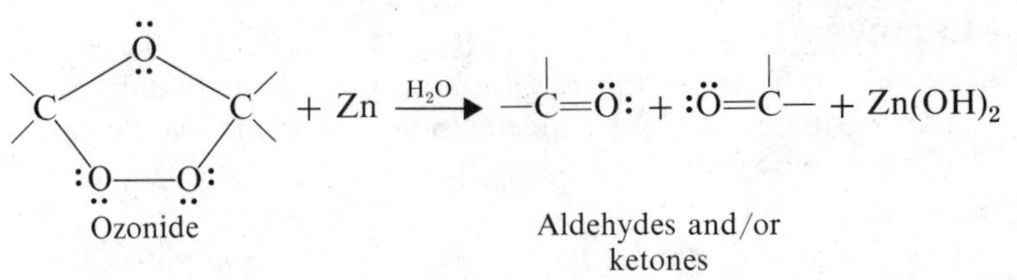

$$\text{Ozonide} + Zn \xrightarrow{H_2O} -\overset{|}{C}=\ddot{O}: + :\ddot{O}=\overset{|}{C}- + Zn(OH)_2$$

Ozonide — Aldehydes and/or ketones

The identities of the aldehydes or ketones disclose the location of the double bond in the original alkene. The examples listed below will illustrate the kinds of products that result from ozonization and subsequent treatment with zinc and water.

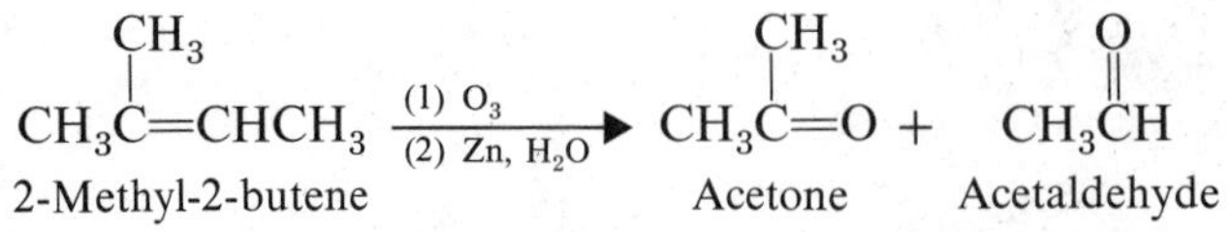

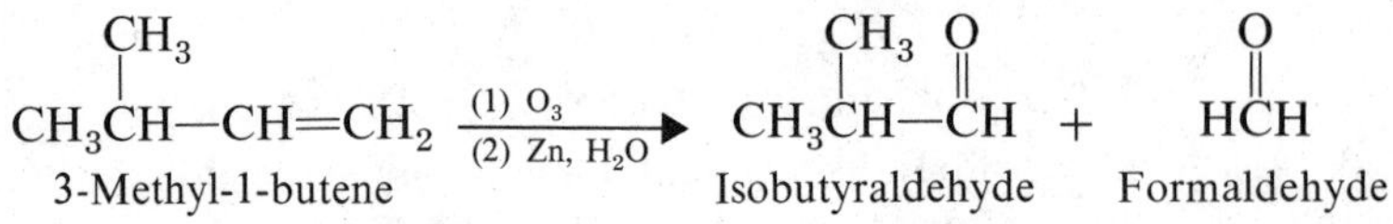

$$\underset{\text{2-Methyl-2-butene}}{CH_3\overset{CH_3}{\overset{|}{C}}=CHCH_3} \xrightarrow[(2)\ Zn,\ H_2O]{(1)\ O_3} \underset{\text{Acetone}}{CH_3\overset{CH_3}{\overset{|}{C}}=O} + \underset{\text{Acetaldehyde}}{CH_3\overset{O}{\overset{\|}{C}}H}$$

$$\underset{\text{3-Methyl-1-butene}}{CH_3\overset{CH_3}{\overset{|}{C}}H-CH=CH_2} \xrightarrow[(2)\ Zn,\ H_2O]{(1)\ O_3} \underset{\text{Isobutyraldehyde}}{CH_3\overset{CH_3}{\overset{|}{C}}H-\overset{O}{\overset{\|}{C}}H} + \underset{\text{Formaldehyde}}{H\overset{O}{\overset{\|}{C}}H}$$

Problem 7.11

Write the general structures of the alkenes that would produce the following products when treated with ozone and then with zinc and water.

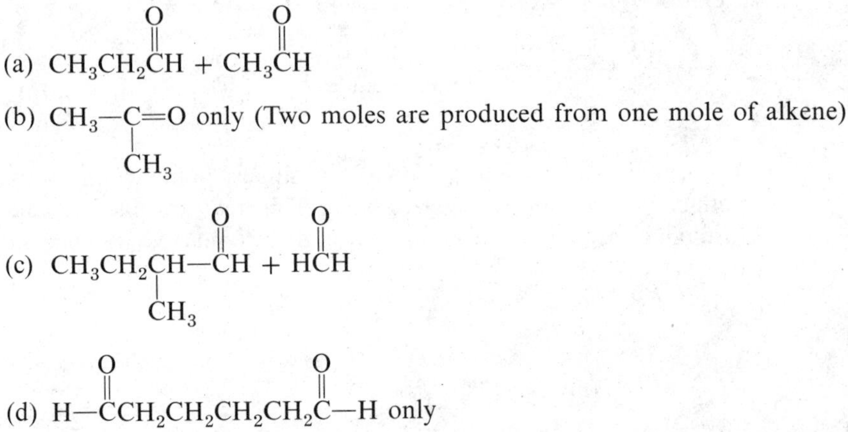

(a) $CH_3CH_2\overset{O}{\overset{\|}{C}}H + CH_3\overset{O}{\overset{\|}{C}}H$

(b) $CH_3-\underset{CH_3}{\underset{|}{C}}=O$ only (Two moles are produced from one mole of alkene)

(c) $CH_3CH_2\underset{CH_3}{\underset{|}{C}H}-\overset{O}{\overset{\|}{C}}H + H\overset{O}{\overset{\|}{C}}H$

(d) $H-\overset{O}{\overset{\|}{C}}CH_2CH_2CH_2CH_2\overset{O}{\overset{\|}{C}}-H$ only

7.14 FREE-RADICAL ADDITION TO ALKENES: THE ANTI-MARKOVNIKOV ADDITION OF HYDROGEN BROMIDE

Before 1933, the orientation of the addition of hydrogen bromide to alkenes was the subject of much confusion. At times addition occurred in accordance with Markovnikov's rule; at other times it occurred in just the opposite manner. Many instances were reported where, under what seemed to be the same experimental conditions, Markovnikov additions were obtained in one laboratory and anti-

Markovnikov additions in another. At times even the same chemist would obtain different results using the same conditions but on different occasions.

The mystery was solved in 1933 by the research of M. S. Kharasch and F. R. Mayo. The culprit turned out to be organic peroxides present in the alkenes—peroxides that were formed by the action of atmospheric oxygen on the alkenes. Kharasch and Mayo found that when alkenes that contained peroxides reacted with hydrogen bromide, anti-Markovnikov addition of hydrogen bromide occurred.

$$R\text{—}\ddot{\underset{..}{O}}\text{—}\ddot{\underset{..}{O}}\text{—}R$$

An organic peroxide

Under these conditions, for example, propene yields 1-bromopropane. In the absence of peroxides, or in the presence of compounds that would "trap" free radicals, normal Markovnikov addition occurs.

$$CH_3CH{=}CH_2 + HBr \xrightarrow{ROOR} CH_3CH_2CH_2Br \qquad \text{Anti-Markovnikov addition}$$

$$CH_3CH{=}CH_2 + HBr \xrightarrow[\text{peroxides}]{\text{No}} CH_3\underset{\underset{Br}{|}}{C}HCH_3 \qquad \text{Markovnikov addition}$$

2-Bromopropane

Neither hydrogen chloride nor hydrogen iodide give anti-Markovnikov addition even when peroxides are present.

According to Kharasch and Mayo, the mechanism for anti-Markovnikov addition of hydrogen bromide is a *free-radical chain reaction* initiated by peroxides.

Chain-initiating steps

$$(1)\ R\text{—}\ddot{\underset{..}{O}}{:}\ddot{\underset{..}{O}}\text{—}R \xrightarrow{\text{Heat}} 2R\text{—}\ddot{\underset{..}{O}}\cdot$$

$$(2)\ R\text{—}\ddot{\underset{..}{O}}\cdot + H{:}\ddot{\underset{..}{Br}}{:} \longrightarrow R\text{—}\ddot{\underset{..}{O}}{:}H + {:}\ddot{\underset{..}{Br}}\cdot$$

Chain-propagating steps

$$(3)\ {:}\ddot{\underset{..}{Br}}\cdot + CH_2\text{—}CHCH_3 \longrightarrow {:}\ddot{\underset{..}{Br}}{:}CH_2\dot{C}HCH_3$$

2° Free radical

$$(4)\ {:}\ddot{\underset{..}{Br}}\text{—}CH_2\dot{C}HCH_3 + H{:}\ddot{\underset{..}{Br}}{:} \longrightarrow BrCH_2\underset{\underset{H}{..}}{C}HCH_3 + \cdot\ddot{\underset{..}{Br}}{:}$$

1-Bromopropane

then (3), (4), (3), (4), and so on.

Step 1 is the simple homolytic cleavage of the peroxide molecule to produce two peroxy free radicals. The oxygen-oxygen bond of peroxides is weak and such reactions are known to occur readily.

$$R\text{—}O{:}O\text{—}R \rightarrow 2R\text{—}O\cdot \quad \Delta H^\circ \cong +35 \text{ kcal/mole}$$

A question arises about step 2 of the mechanism, however. Why does not the following reaction occur instead?

$$R\text{—}\ddot{\underset{..}{O}}\cdot + H{:}\ddot{\underset{..}{Br}}{:} \longrightarrow R\text{—}\ddot{\underset{..}{O}}{:}\ddot{\underset{..}{Br}}{:} + H\cdot \quad \Delta H^\circ \text{ is relatively high}$$

The answer can be found in the thermodynamic quantities associated with the two possibilities. Step 2 of the mechanism, abstraction of a hydrogen atom by the peroxy radical, is exothermic, and has a low energy of activation. The alternative

$$\mathrm{R{-}\ddot{\underset{\cdot\cdot}{O}}\cdot + H{:}\ddot{\underset{\cdot\cdot}{Br}}{:} \longrightarrow R{-}\ddot{\underset{\cdot\cdot}{O}}{:}H + {:}\ddot{\underset{\cdot\cdot}{Br}}\cdot} \qquad \Delta H^\circ \cong -23 \text{ kcal/mole}, \; E_{act} \text{ is low}$$

reaction, the abstraction of a bromine atom by the peroxy radical, is highly endothermic and, consequently, has a high energy of activation.

$$\mathrm{R{-}\ddot{\underset{\cdot\cdot}{O}}\cdot + H{:}\ddot{\underset{\cdot\cdot}{Br}}{:} \longrightarrow R{-}\ddot{\underset{\cdot\cdot}{O}}{-}\ddot{\underset{\cdot\cdot}{Br}}{:} + H\cdot} \qquad \Delta H^\circ \cong +39 \text{ kcal/mole}, \; E_{act} > +39 \text{ kcal/mole}$$

Step 3 of the mechanism determines the final orientation of bromine in the product. It occurs, as it does, because a *more stable secondary radical* is produced. Had the bromine attacked propene at the central carbon atom, a less stable, primary radical would have been the result,

$$\mathrm{Br\cdot + CH_2{=}CHCH_3 \not\longrightarrow \cdot CH_2\underset{\underset{Br}{|}}{C}HCH_3}$$

1° Free radical

and this reaction would have had a higher energy of activation.

Step 4 of the mechanism is simply the abstraction of a hydrogen atom from hydrogen bromide by the radical produced in step 3. This hydrogen abstraction produces a bromine atom that can bring about step 3 again, then step 4 occurs again—a chain reaction.

Problem 7.12

You may have noticed that step 1 of the mechanism is almost as endothermic as the reaction that we disallowed for step 2, that is,

$$\mathrm{R{-}\ddot{\underset{\cdot\cdot}{O}}{:}\ddot{\underset{\cdot\cdot}{O}}{-}R \longrightarrow 2R\ddot{\underset{\cdot\cdot}{O}}\cdot} \qquad \Delta H^\circ \cong +35 \text{ kcal/mole}$$

$$\mathrm{R{-}\ddot{\underset{\cdot\cdot}{O}}\cdot + HBr \not\longrightarrow R{-}\ddot{\underset{\cdot\cdot}{O}}{-}Br + H\cdot} \qquad \Delta H^\circ \cong +39 \text{ kcal/mole}$$

Yet, to invoke the first reaction as a chain-initiating step, and to exclude the second from any important role, is perfectly reasonable. How can you explain this?

Many molecules, other than hydrogen bromide, add to alkenes under the influence of a peroxide catalyst. The following reactions are examples.

$$\mathrm{CH_3CH_2CH_2CH{=}CH_2 + HCCl_3 \xrightarrow{\text{Peroxides}} CH_3CH_2CH_2CH_2CH_2{-}CCl_3}$$

1,1,1-Trichlorohexane

$$\mathrm{CH_3\overset{\overset{CH_3}{|}}{C}{=}CH_2 + CH_3CH_2SH \xrightarrow{\text{Peroxides}} CH_3{-}\overset{\overset{CH_3}{|}}{C}H{-}CH_2{-}S{-}CH_2CH_3}$$

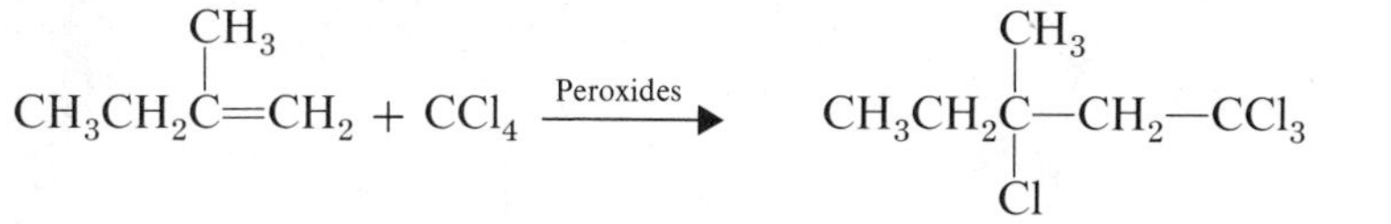

$$\mathrm{CH_3CH_2\overset{\overset{\large CH_3}{|}}{C}{=}CH_2 + CCl_4 \xrightarrow{\text{Peroxides}} CH_3CH_2\underset{\underset{\large Cl}{|}}{\overset{\overset{\large CH_3}{|}}{C}}{-}CH_2{-}CCl_3}$$

1,1,1,3-Tetrachloro-3-methylpentane

Problem 7.13

Write free-radical, chain-reaction mechanisms that account for the products formed in each of the reactions listed above.

Free radicals can also cause alkenes to add to each other to form large molecules called addition polymers. These reactions are described in Special topic C.

7.15 A BIOLOGICAL TOUR DE FORCE

In Section 7.6 we saw how carbocations can react with an alkene to form a new carbocation:

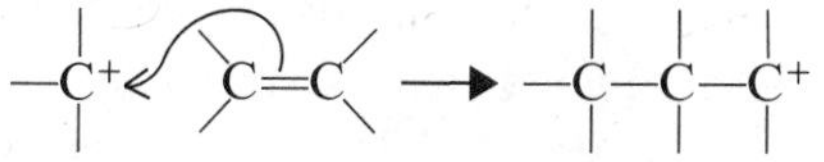

$$-\overset{|}{\underset{|}{C}}{}^{+} \;\; \mathrm{C{=}C} \longrightarrow -\overset{|}{\underset{|}{C}}-\overset{|}{\underset{|}{C}}-\overset{|}{\underset{|}{C}}{}^{+}$$

This type of reaction, *the alkylation of an alkene by a carbocation,* has a number of counterparts in biochemistry. An extraordinary example is the conversion of squalene 2,3-epoxide into lanosterol.

The first step, ring opening of the three-membered epoxide ring, produces a tertiary carbocation. The second step—a series of alkylation reactions occurring in concert—is initiated by the original carbocation. This series of electron shifts brings about the *closure of four rings.* Reactions involving hydride and methanide shifts occur next and these ultimately produce lanosterol. All of these steps are catalyzed by a single enzyme—*squalene oxide cyclase* (cf. Section N.2).

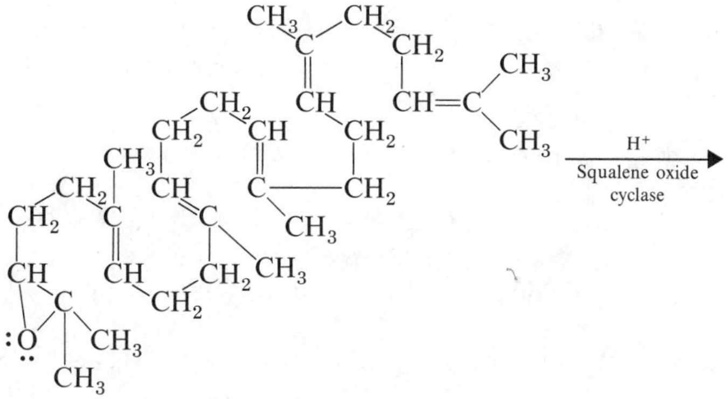

$\xrightarrow[\text{Squalene oxide cyclase}]{H^+}$

Squalene 2,3-epoxide

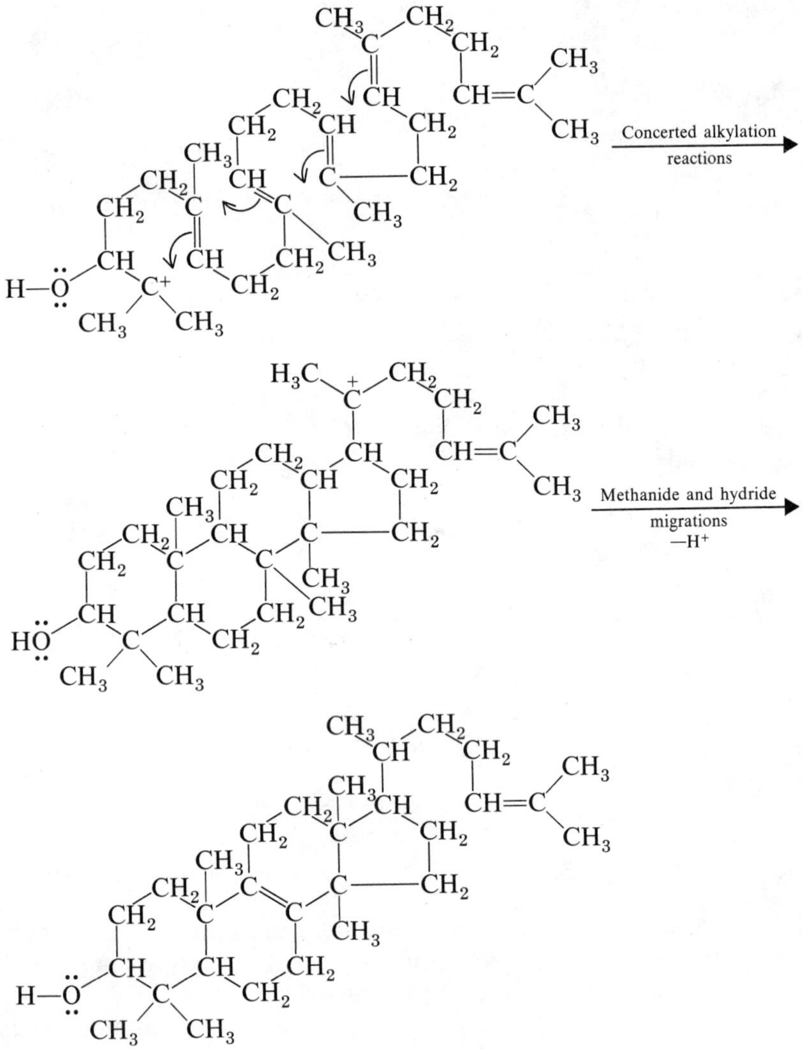

Concerted alkylation reactions

Methanide and hydride migrations
$-H^+$

Lanosterol

7.16 SUMMARY OF ADDITION REACTIONS OF ALKENES

1. Addition of hydrogen (Sections 6.4, 6.6, and 6.14)

General: $\rangle C{=}C\langle \; + \; H{-}H \xrightarrow{\text{Cat.}} -\underset{H}{\overset{|}{C}}-\underset{H}{\overset{|}{C}}-$

Specific examples: $CH_3CH_2CH{=}CH_2 + H_2 \xrightarrow[25°]{Pt} CH_3CH_2CH_2CH_3$ (100%)

$CH_3CH{=}CHCH_3 + H_2 \xrightarrow[25°]{Pt} CH_3CH_2CH_2CH_3$ (100%)
cis or *trans*

2. Addition of hydrogen halides (Sections 7.2 and 7.14)

General: $R{-}CH{=}CH_2 + HX \longrightarrow R{-}\underset{X}{\underset{|}{C}H}{-}CH_3$ Markovnikov addition

$$R{-}CH{=}CH_2 + HBr \xrightarrow{ROOR} R{-}CH_2{-}CH_2Br$$ Anti-Markovnikov addition

Specific examples: $$CH_3{-}\underset{\underset{CH_3}{|}}{C}{=}CH_2 + HCl \xrightarrow{-80°} CH_3{-}\overset{\overset{Cl}{|}}{\underset{\underset{CH_3}{|}}{C}}{-}CH_3 \quad (94\%)$$

$$CH_3CH_2CH_2CH_2CH{=}CH_2 + HBr \xrightarrow{Peroxides} CH_3CH_2CH_2CH_2CH_2CH_2Br \quad (83\%)$$

3. Addition of water (Section 7.4)

General: $$\rangle C{=}C\langle + H{-}OH \xrightarrow{H^+} {-}\underset{\underset{H}{|}}{\overset{|}{C}}{-}\underset{\underset{OH}{|}}{\overset{|}{C}}{-}$$

Specific examples: $$CH_2{=}CH_2 \xrightarrow[(2)\ H_2O]{(1)\ 98\%\ H_2SO_4} CH_3CH_2OH$$

$$(CH_3)_2C{=}CH_2 \xrightarrow[25°]{10\%\ H_2SO_4} CH_3{-}\overset{\overset{CH_3}{|}}{\underset{\underset{OH}{|}}{C}}{-}CH_3$$

4. Addition of sulfuric acid (Section 7.5)

General: $$R{-}CH{=}CH_2 + HOSO_2OH \longrightarrow R\underset{\underset{OSO_2OH}{|}}{C}HCH_3$$

Specific example: $$CH_3CH{=}CH_2 + HOSO_2OH \longrightarrow CH_3\underset{\underset{OSO_2OH}{|}}{C}HCH_3$$

5. Addition of other alkenes (Sections 7.6 and 7.15)

General: $$R{-}CH{=}CH_2 \xrightarrow{H^+} R{-}\overset{\overset{CH_3}{|}}{C}H^+ \xrightarrow{CH_2{=}CHR} R{-}\overset{\overset{CH_3}{|}}{C}H{-}CH_2{-}\overset{\overset{R}{|}}{C}H^+$$

$\downarrow$

Mixture of alkenes or a polymer

Specific example: $$CH_3\overset{\overset{CH_3}{|}}{C}{=}CH_2 \xrightarrow[80°]{60\%\ H_2SO_4} CH_3{-}\overset{\overset{CH_3}{|}}{\underset{\underset{CH_3}{|}}{C}}{-}CH_2{-}C({=}CH_2)CH_3 + (CH_3)_3C(H)C{=}C(CH_3)_2$$

(80%) (20%)

6. Addition of boron hydrides (Section 7.8)

General: $$R{-}CH{=}CH_2 + \tfrac{1}{2}(BH_3)_2 \longrightarrow (RCH_2CH_2)_3B$$

Specific examples: $CH_3CH_2CH_2CH{=}CH_2 \xrightarrow[\text{(an ether solvent)}]{(BH_3)_2} (CH_3CH_2CH_2CH_2CH_2)_3B$ (94%)

1-Methylcyclopentene + $(BH_3)_2$ $\xrightarrow{\text{syn addition}}$ (trans-2-methylcyclopentyl)borane (CH₃ and H cis; H and B added syn)

Reactions of alkylboranes:

$CH_3(CH_2)_7CH{=}CH_2 \xrightarrow[\text{Diglyme, }25°]{(BH_3)_2} [CH_3(CH_2)_7CH_2CH_2]_3B \xrightarrow[H_2O_2]{NaOH} CH_3(CH_2)_7CH_2CH_2OH$
(93%)
Alcohol

$CH_3(CH_2)_3CH{=}CH_2 \xrightarrow[\text{Diglyme, }25°]{(BH_3)_2} [CH_3(CH_2)_3CH_2CH_2]_3B \xrightarrow{CH_3\overset{O}{\overset{\|}{C}}OH} CH_3(CH_2)_3CH_2CH_3$
(91%)
Alkane

7. Addition of halogens (Sections 7.9 and 7.10)

General: $>C{=}C< + X_2 \xrightarrow{CCl_4} -\underset{X}{\overset{|}{C}}-\underset{X}{\overset{|}{C}}-$

Specific example: Cyclohexene $\xrightarrow[CCl_4/C_2H_5OH,\ -5°]{Br_2}$ trans-1,2-Dibromocyclohexane (95%)

$CH_3CH{=}CHCH_3 \xrightarrow[CH_3COOH,\ 25°]{Cl_2} CH_3\underset{Cl}{CH}\underset{Cl}{CH}CH_3$ (100%)

8. Halohydrin formation (Section 7.11)

General: $>C{=}C< \xrightarrow[H_2O]{X_2} -\underset{X}{\overset{|}{C}}-\underset{OH}{\overset{|}{C}}$

Specific example:

$(CH_3)_2C{=}CH_2 + Br_2 + H_2O \longrightarrow (CH_3)_2\underset{OH}{C}-\underset{Br}{CH_2}$
(73%)
A bromohydrin

9. Anti hydroxylation: epoxidation (Section 7.12)

General: $\text{C=C} + \text{RCOOH} \longrightarrow -\text{C}-\text{C}-\text{(O)} \xrightarrow[H_2O]{H^+} -\underset{OH}{\text{C}}-\overset{OH}{\text{C}}-$

$\text{1,2-dimethylcyclohexene } (CH_3, CH_3) \xrightarrow[CHCl_3]{C_6H_5COOH}$ epoxide $(CH_3, O, CH_3) \xrightarrow[H_2O]{H_2SO_4}$ diol (OH, CH_3, OH, CH_3)

(60%)

10. Syn hydroxylation (Section 7.13)

General: $\text{C=C} \xrightarrow[\text{(1) } OsO_4 \text{ (2) } Na_2SO_3]{\text{Cold, dilute } KMnO_4 \text{ or}} -\underset{OH}{\text{C}}-\underset{OH}{\text{C}}-$

Specific example:

$\text{norbornene} \xrightarrow[H_2O/(CH_3)_3COH,\ 0°]{KMnO_4,\ OH^-}$ cis-diol (OH, OH, H, H)

(40%)

11. Ozonolysis (Section 7.13)

General: $\text{C=C} \xrightarrow{O_3} -\text{C}(O)(O{-}O)\text{C}- \xrightarrow[H_2O]{Zn} -\text{C=O} + \text{O=C}-$

Specific example: $(CH_3CH_2)_2C{=}C(CH_3)H \xrightarrow[\text{(2) Zn, } H_2O]{\text{(1) } O_3} (CH_3CH_2)_2C{=}O + O{=}C(CH_3)H$

Additional Problems

7.14 $CH_2{=}CHCH_2CH_2CH_3$

Write structural formulas for the products that form when 1-pentene reacts with each of the following reagents.

(a) HCl
(b) Br_2 in CCl_4, room temperature
(c) H_3O^+, H_2O, heat
(d) Cold concentrated H_2SO_4
(e) Cold concentrated H_2SO_4, then H_2O and heat
(f) HBr
(g) $Hg(OAc)_2$ in THF-H_2O, then $NaBH_4$, OH^-
(h) $NaBH_4 + BF_3$, then H_2O_2, OH^-
(i) $Hg(OAc)_2$ in CH_3OH, then $NaBH_4$, OH^-
(j) D_2, Pt
(k) Br_2 in CCl_4, then KI in acetone
(l) $KMnO_4$, OH^-, cold

(m) OsO_4, then Na_2SO_3
(n) $KMnO_4$, OH^-, heat, then H^+
(o) O_3, then Zn, H_2O
(p) $NaBH_4 + BF_3$, then CH_3COOH
(q) HBr, peroxides

7.15
Repeat Problem 7.14 using cyclopentene instead of 1-pentene.

7.16
Write structural formulas for the products that would form in the following reactions:

(a) 1-Butene + $Br_2 \xrightarrow[CCl_4]{}$

(b) Cyclohexene + dilute $KMnO_4 \xrightarrow[Cold]{OH^-}$

(c) Cyclohexene + $OsO_4 \xrightarrow[(2)\ Na_2SO_3]{}$

(d) Cyclohexene + $C_6H_5\overset{O}{\overset{\|}{C}}OOH \longrightarrow \xrightarrow[H_2O]{H^+}$

(e) Ethene + conc. $H_2SO_4 \xrightarrow[Cold]{}$

(f) Product of (e) + $H_2O \longrightarrow$

(g) 2-Methylpropene + $(BH_3)_2 \longrightarrow$

(h) Product of (g) + $NaOH/H_2O_2 \longrightarrow$

(i) *sec*-Butyl alcohol + hot conc. $H_2SO_4 \longrightarrow$

(j) Bromocyclopentane + KOH $\xrightarrow[Heat]{Alcohol}$

(k) 1,2-Dibromopentane + Zn $\longrightarrow$

(l) 2,3-Dimethyl-1-butene + HBr $\xrightarrow[Heat]{(free\text{-}radical\ inhibitor)}$

(m) 1-Hexene + HCl $\xrightarrow[Heat]{}$

(n) 1-Hexene + HBr $\xrightarrow[Peroxides]{}$

(o) *cis*-3-Hexene + $O_3 \longrightarrow$

(p) Product of (o) + Zn $\xrightarrow[H_2O]{}$

(q) *cis*-3-Hexene + $KMnO_4 \xrightarrow[Heat]{OH^-}$

(r) 1-Methylcyclopentene + $O_3 \longrightarrow$

(s) Product of (r) + Zn $\xrightarrow[H_2O]{}$

(t) 1-Methylcyclopentene + $KMnO_4 \xrightarrow[(2)\ H^+]{(1)\ OH^-/heat}$

(u) 1-Methylcyclopentene + $H_2O \xrightarrow[H^+]{}$

(v) 1-Methylcyclopentene + $H_2 \xrightarrow[Ni]{}$

(w) 2-Methyl-2-pentene + HI $\xrightarrow{Heat}$

(x) 1,2-Dimethylcyclopentene + HCl $\longrightarrow$

(y) 1-Methylcyclopentene + $Cl_2 \xrightarrow[H_2O]{}$

7.17
(a) Write all steps of a mechanism for the acid-catalyzed dimerization of propene.
(b) What would be the major product?

7.18
When either *cis*- or *trans*-2-butene is treated with hydrogen chloride in ethyl alcohol, one of the products of the reaction is

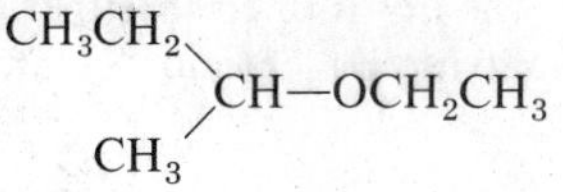

$$\begin{matrix} CH_3CH_2 \diagdown \\ \quad CH{-}OCH_2CH_3 \\ CH_3 \diagup \end{matrix}$$

Write a mechanism that accounts for the formation of this product.

7.19
When alkenes add HX the relative rates of reaction are:

$$R_2C{=}CH_2 > RCH{=}CH_2 > CH_2{=}CH_2$$

What factor accounts for this?

7.20
Write three-dimensional formulas for the products of the following reactions. In each case, designate the location of deuterium atoms.

(a) 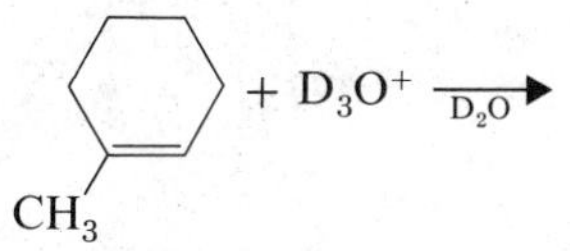 1-methylcyclohexene (ring with CH_3 on the double-bond carbon) $+ D_3O^+ \xrightarrow{D_2O}$

(b) 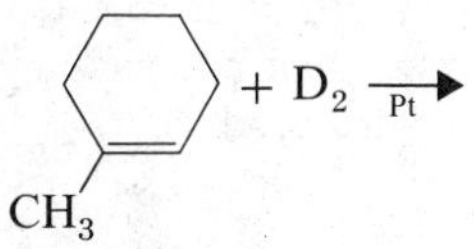 1-methylcyclohexene (ring with CH_3 on the double-bond carbon) $+ D_2 \xrightarrow{Pt}$

(c) 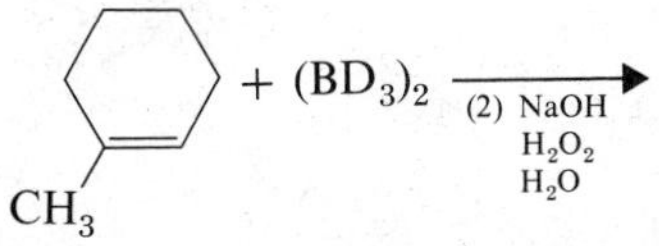 1-methylcyclohexene (ring with CH_3 on the double-bond carbon) $+ (BD_3)_2 \xrightarrow[\text{(2) NaOH, } H_2O_2\text{, } H_2O]{}$

7.21
Write the products and show how many moles of each would be formed when squalene is subjected to ozonolysis and the ozonide is subsequently treated with zinc and water.

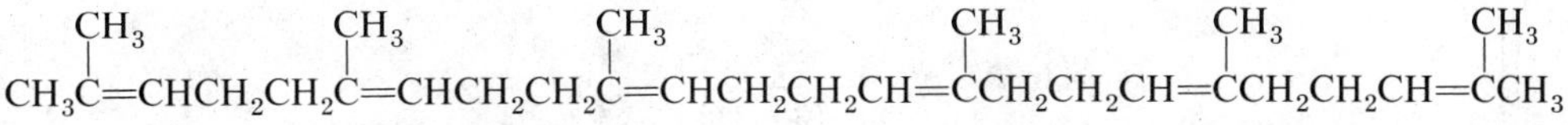

$$CH_3\overset{CH_3}{\overset{|}{C}}{=}CHCH_2CH_2\overset{CH_3}{\overset{|}{C}}{=}CHCH_2CH_2\overset{CH_3}{\overset{|}{C}}{=}CHCH_2CH_2CH{=}\overset{CH_3}{\overset{|}{C}}CH_2CH_2CH{=}\overset{CH_3}{\overset{|}{C}}CH_2CH_2CH{=}\overset{CH_3}{\overset{|}{C}}CH_3$$

Squalene

7.22
Arrange the following alkenes in order of their reactivity toward acid-catalyzed hydration and explain your reasoning.

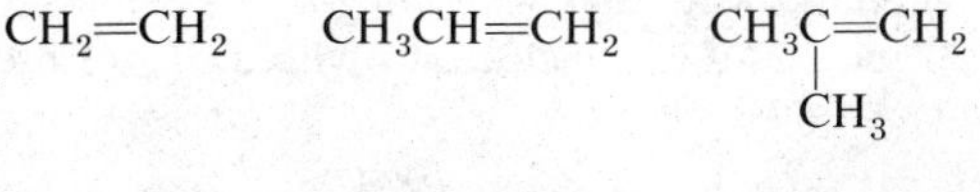

$$CH_2{=}CH_2 \qquad CH_3CH{=}CH_2 \qquad CH_3\underset{CH_3}{\underset{|}{C}}{=}CH_2$$

7.23

(a) When treated with strong acid at 25°, either *cis*-2-butene or *trans*-2-butene is converted to a mixture of *trans*-2-butene, *cis*-2-butene, and 1-butene. *Trans*-2-butene predominates in the mixture. (The mixture contains 74% *trans*-2-butene, 23% *cis*-2-butene, and 3% 1-butene.) Write mechanisms for the reactions that occur, and account for the relative amounts of the alkene isomers that are formed.

(b) When treated with strong acid, 1-butene is converted to the same mixture of alkenes referred to in part (a). How can you explain this?

(c) Can you also explain why 2-methylpropene is *not* formed in either of the reactions referred to in parts (a) or (b) even through 2-methylpropene is more stable than 1-butene?

7.24

Write a mechanism that explains the course of the following reaction.

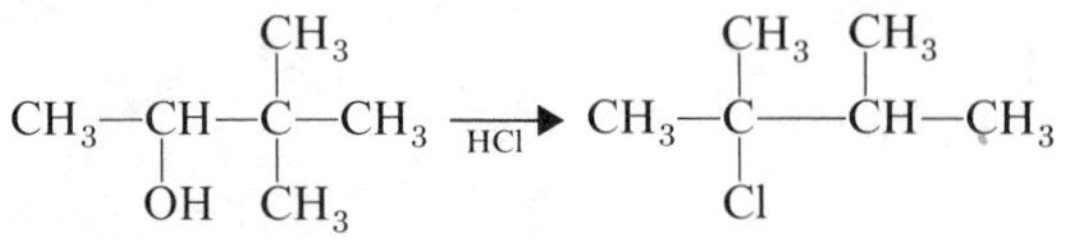

$$CH_3-\underset{OH}{CH}-\overset{CH_3}{\underset{CH_3}{C}}-CH_3 \xrightarrow{HCl} CH_3-\overset{CH_3}{\underset{Cl}{C}}-\overset{CH_3}{CH}-CH_3$$

7.25

A cycloalkene reacts with hydrogen and a catalyst to yield methylcyclohexane. On vigorous oxidation with potassium permanganate the cycloalkene yields only

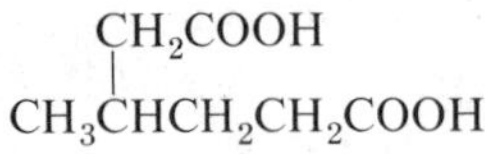

$$CH_3\overset{CH_2COOH}{CH}CH_2CH_2COOH$$

What is the structure of the cycloalkene?

7.26

Outline all steps in a laboratory synthesis of each of the following compounds. You should begin with the organic compound indicated and you may use any needed solvents or inorganic compounds. These syntheses may require more than one step and should be designed to give reasonably good yields of reasonably pure products.

(a) Propene from propane
(b) 2-Bromopropane from propane
(c) 1-Bromopropane from propane
(d) 2-Methylpropene from 2-methylpropane
(e) *tert*-Butyl alcohol from 2-methylpropane
(f) 1,2-Dichlorobutane from 1-chlorobutane
(g) Ethylene bromohydrin from ethyl bromide
(h) 2,5-Dimethylhexane from 2-methylpropene
(i) 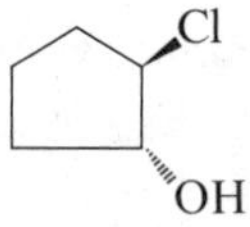 *trans*-2-chlorocyclopentanol (cyclopentane ring with Cl and OH, *trans*) from cyclopentane

trans

(j) 2-Bromobutane from 1-bromobutane
(k) 3,4-Dimethylhexane from 1-chlorobutane
(l) 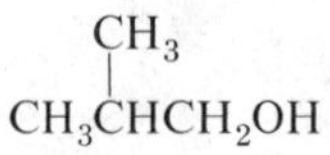 $CH_3\overset{CH_3}{CH}CH_2OH$ from $CH_3\overset{CH_3}{\underset{CH_3}{C}}-OH$

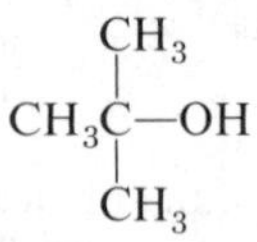

7.27

(a) How many grams of bromine will react with 7.0 g of 1-pentene? (b) What is the molecular weight of the alkene 2.24 g of which reacts with 3.20 g of bromine?

7.28

Write a mechanism that accounts for the following cyclization reaction:

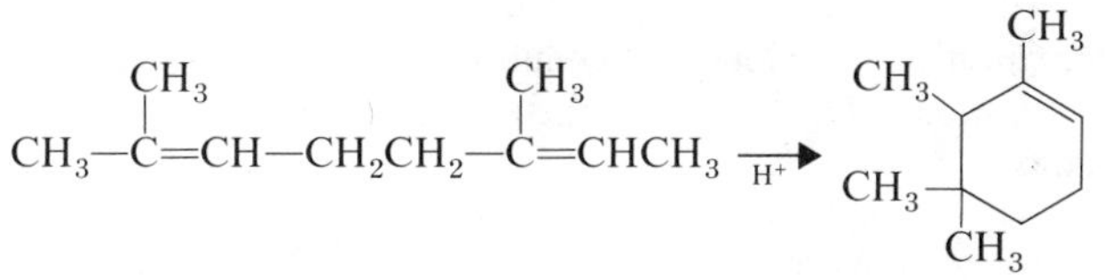

$$CH_3-\underset{}{\overset{CH_3}{\overset{|}{C}}}=CH-CH_2CH_2-\overset{CH_3}{\overset{|}{C}}=CHCH_3 \xrightarrow{H^+}$$ (1,2,6,6-tetramethylcyclohexene: ring bearing CH_3 on the double-bond carbon and two CH_3 groups on the adjacent carbon, plus CH_3)

7.29

2-Fluoropentane cannot be synthesized in good yield from an alkane by direct fluorination because the reaction is highly nonselective and tends to give multifluorination. Suggest a specific synthesis of 2-fluoropentane from an alkene.

7.30

Halohydrins can be synthesized by treating epoxides with HX. (a) Show how you would use this method to synthesize 2-chlorocyclopentanol (p. 262) from cyclopentene. (b) Would you expect the product to be *cis*-2-chlorocyclopentanol or *trans*-2-chlorocyclopentanol, that is, would you expect a net *syn* addition or a net *anti* addition of —Cl and OH? Explain.

7.31

Pheromones are substances secreted by animals (especially insects) that produce a specific behavioral reaction in other members of the same species. Pheromones are effective at very low concentrations and include sex attractants, warning substances, and "aggregation" compounds. After many years of research, the sex attractant of the gypsy moth has been identified and synthesized in the laboratory. This sex pheromone is unusual in that it appears to be equally attractive to male and female gypsy moths. (It promises to be useful in their control even though this may seem somewhat unfair.) The final step in the synthesis of the pheromone involves treatment of *cis*-2-methyl-7-octadecene with a peroxy acid. What is the structure of the gypsy moth sex pheromone?

7.32

The green peach aphid is repelled by its own defensive pheromone. (It is also repelled by other squashed aphids.) This alarm pheromone has been isolated and has been shown to have the molecular formula $C_{15}H_{24}$. On catalytic hydrogenation it absorbs four moles of hydrogen and yields 2,6,10-trimethyldodecane, that is,

$$CH_3\overset{CH_3}{\overset{|}{C}}HCH_2CH_2CH_2\overset{CH_3}{\overset{|}{C}}HCH_2CH_2CH_2\overset{CH_3}{\overset{|}{C}}HCH_2CH_3$$

When subjected to ozonolysis followed by treatment with zinc and water one mole of the alarm pheromone produces:

Two moles of formaldehyde, $H-\overset{O}{\overset{\|}{C}}-H$

One mole of acetone, $CH_3\overset{O}{\overset{\|}{C}}CH_3$

One mole of $CH_3\overset{O}{\overset{\|}{C}}CH_2CH_2\overset{O}{\overset{\|}{C}}H$

One mole of $H\overset{O}{\overset{\|}{C}}CH_2CH_2\overset{O}{\overset{\|}{C}}-\overset{O}{\overset{\|}{C}}H$

Neglecting *cis-trans* isomerism, propose a structure for the green peach aphid alarm pheromone.

7.33

Although formation of the more highly substituted alkene is the general rule in acid-catalyzed dehydrations, the following reaction yields mainly the less highly substituted alkene. Explain.

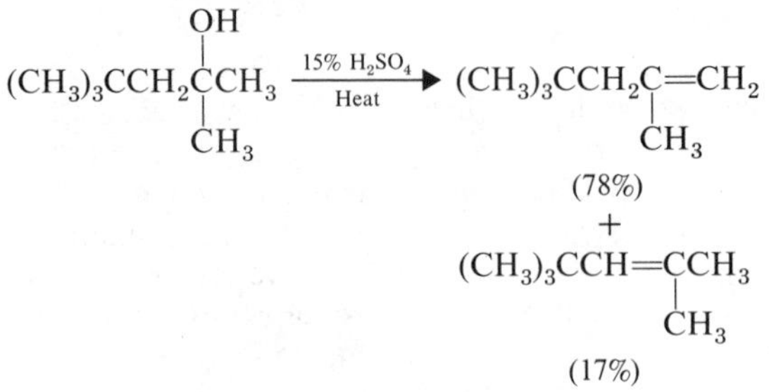

$$(CH_3)_3CCH_2C(OH)(CH_3)CH_3 \xrightarrow[\text{Heat}]{15\%\ H_2SO_4} (CH_3)_3CCH_2C(CH_3){=}CH_2\ (78\%) + (CH_3)_3CCH{=}C(CH_3)CH_3\ (17\%)$$

7.34

(a) What product would you expect to form when isobutyl bromide is heated with sodium

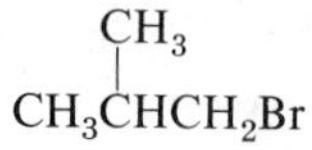

$CH_3CH(CH_3)CH_2Br$

Isobutyl bromide

ethoxide in ethanol? (b) Can you suggest a method for the conversion of isobutyl bromide into *tert*-butyl bromide?

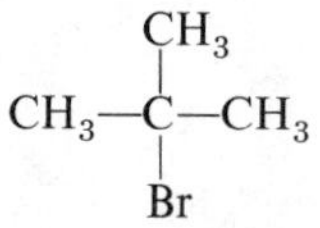

$CH_3{-}C(CH_3)(Br){-}CH_3$

tert-Butyl bromide

7.35

When cyclopentene is allowed to react with bromine in an aqueous solution of sodium chloride, the products of the reaction are *trans*-1,2-dibromocyclopentane, the *trans*-bromohydrin of cyclopentene, *and trans-1-bromo-2-chlorocyclopentane.* Write a mechanism that explains the formation of this last product.

7.36

Write a mechanism that accounts for the following reaction.

(1) $Hg(OAc)_2$, H_2O

(2) $NaBH_4$, OH^-

HO

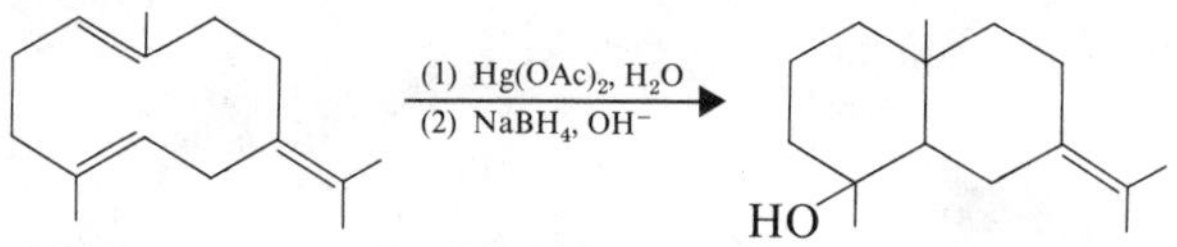

7.37

When an alkene, $RCH{=}CH_2$, and hydrogen sulfide are irradiated with light (of wavelength that can be absorbed by H_2S), a chain reaction takes place producing a thiol, RCH_2CH_2SH. (a) Outline a possible mechanism for this reaction. (b) A side-product of the reaction is a thioether, $(RCH_2CH_2)_2S$. Suggest how it is formed.

7.38

The structures of the diisobutylene dimers (Section 7.6) were determined by F. C. Whitmore and his students on the basis of the products formed when each dimer was subjected to ozonolysis. Show how this might have been done.

7.39

In an industrial process, propene is heated with phosphoric acid at 205° under a pressure of 1000 atm. The major products of the reaction are two isomers with the molecular formula $C_{12}H_{24}$. Propose structures for the isomers and write a mechanism that explains their formation. (Note: at one time these isomers were used extensively in the synthesis of a nonbiodegradable detergent. This synthesis is given in Problem 12.43.)

***7.40**

When hydrogen chloride adds to the double bond of 3,3,3-trifluoropropene, $CF_3CH{=}CH_2$, the product is $CF_3CH_2CH_2Cl$ rather than $CF_3CHClCH_3$. Although the orientation of addition in this reaction is not the one that would have been predicted on the basis of the original version of Markovnikov's rule (p. 241), it is consistent with the modern version. Explain.

***7.41**

Listed below are estimated values of $\Delta H°$ for the chain-propagating steps in the addition of HX to ethene. Use these data to account for the observation (Section 7.14) that only HBr adds to an alkene by a free-radical mechanism.

$$X\cdot + CH_2{=}CH_2 \longrightarrow XCH_2CH_2\cdot$$

$X = F$, $\Delta H° = -46$ kcal/mole
$X = Cl$, $\Delta H° = -21$ kcal/mole
$X = Br$, $\Delta H° = -9$ kcal/mole
$X = I$, $\Delta H° = +7$ kcal/mole

$$XCH_2CH_2\cdot + HX \longrightarrow X{-}CH_2CH_3 + X\cdot$$

$X = F$, $\Delta H° = +38$ kcal/mole
$X = Cl$, $\Delta H° = +5$ kcal/mole
$X = Br$, $\Delta H° = -11$ kcal/mole
$X = I$, $\Delta H° = -27$ kcal/mole

8 STEREOCHEMISTRY

8.1 INTRODUCTION

In 1877, Hermann Kolbe, one of the most eminent organic chemists of the time, wrote the following:

"Not long ago, I expressed the view that the lack of general education and of thorough training in chemistry was one of the causes of the deterioration of chemical research in Germany. . . . Will anyone to whom my worries seem exaggerated please read, if he can, a recent memoir by a Herr van't Hoff on 'The Arrangements of Atoms in Space,' a document crammed to the hilt with the outpourings of a childish fantasy . . . This Dr. J. H. van't Hoff, employed by the Veterinary College at Utrecht, has, so it seems, no taste for accurate chemical research. He finds it more convenient to mount his Pegasus (evidently taken from the stables of the Veterinary College) and to announce how, on his bold flight to Mount Parnassus, he saw the atoms arranged in space."

Kolbe, nearing the end of his career, was reacting to a publication of a 22-year-old Dutch scientist. This publication had appeared two years earlier in September 1874, and in it, van't Hoff had argued that the spatial arrangement of four groups around a central carbon atom is tetrahedral. A young French scientist, J. A. Le Bel, had independently advanced the same idea in a publication in November 1874. Within 10 years after Kolbe's comments, however, abundant evidence had accumulated to substantiate the "childish fantasy" of van't Hoff, and in 1901 he was named the first recipient of the Nobel Prize for chemistry.

Together, the publications of van't Hoff and Le Bel marked the beginning of a field of study that is concerned with the structures of molecules in three-dimensions: *stereochemistry.*

Until now in our study of organic chemistry, we have been concerned primarily with the order in which the atoms of molecules are attached to each other. We will find, however, that an understanding of the properties of many organic compounds requires that we also concern ourselves with the arrangement of their atoms in space.

8.2 ISOMERISM: STRUCTURAL ISOMERS AND STEREOISOMERS

Isomers are different compounds that have the same molecular formula. In our study of carbon compounds, thus far, most of our attention has been directed toward those isomers that we have called structural isomers.

*Structural isomers are isomers that differ because their atoms are joined in a different order.** Several examples of structural isomers are the following.

*Another term that is used for what we have called structural isomers is the term *constitutional isomers.* Both terms are defined in the same way. The term constitutional isomers may eventually become widely adopted, but at present most texts and monographs use the term structural isomers. For this reason we have retained the latter terminology in this book. It would be wise, however, to learn both terms.

Molecular Formula	Structural Isomers	
C_4H_{10}	$CH_3CH_2CH_2CH_3$ and Butane	$CH_3CH(CH_3)CH_3$ Isobutane
C_3H_7Cl	$CH_3CH_2CH_2Cl$ and 1-Chloropropane	$CH_3CH(Cl)CH_3$ 2-Chloropropane
C_2H_6O	CH_3CH_2OH and Ethanol	CH_3OCH_3 Dimethyl ether

Structural isomers are sometimes classified into subcategories. Butane and isobutane, for example, are sometimes called *chain isomers,* 1-chloropropane and 2-chloropropane are called *position isomers,* and ethanol and dimethyl ether are sometimes called *functional group isomers.* The origin of these subclassifications is self-evident: Butane and isobutane differ in the structure of their carbon chains, 1-chloropropane and 2-chloropropane differ in the position of attachment of the chlorine atom, and ethanol and dimethyl ether differ in their functional groups.

Stereoisomers are not structural isomers—they have their constituent atoms attached in the same order. *Stereoisomers differ only in arrangement of their atoms in space.* The *cis* and *trans* isomers of alkenes are stereoisomers; we can see that this is true if we examine the *cis-* and *trans*-2-butenes shown below.

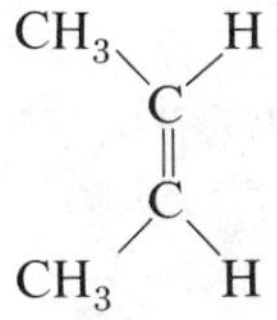

$$\begin{array}{ccc} CH_3 & & H \\ & \diagdown\;\diagup & \\ & C & \\ & \| & \\ & C & \\ & \diagup\;\diagdown & \\ CH_3 & & H \end{array} \qquad \begin{array}{ccc} CH_3 & & H \\ & \diagdown\;\diagup & \\ & C & \\ & \| & \\ & C & \\ & \diagup\;\diagdown & \\ H & & CH_3 \end{array}$$

cis-2-Butene 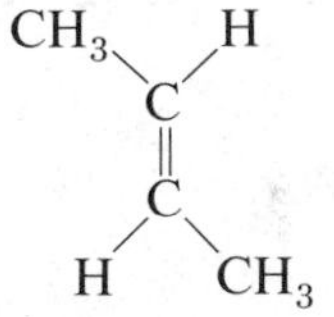 *trans*-2-Butene

cis-2-Butene and *trans*-2-butene are isomers because both compounds have the same molecular formula, C_4H_8. They are *not* structural isomers, because the order of attachment of the atoms in both compounds is the same. Both compounds have a continuous chain of four carbon atoms, both compounds have two central atoms joined by a double bond, and both compounds have one methyl group and one hydrogen atom attached to the two central atoms. *cis-* and *trans*-2-Butene are isomers that differ only in the arrangement of their atoms in space. In *cis*-2-butene the methyl groups are on the same side of the molecule, and in *trans*-2-butene the methyl groups are on opposite sides. Thus, *cis-* and *trans*-2-butene are stereoisomers.

Stereoisomers can be subdivided into two general categories: *enantiomers* and *diastereomers.* Enantiomers are stereoisomers whose molecules *are* mirror reflections of each other. Diastereomers are stereoisomers whose molecules *are not* mirror reflections of each other.

Molecules of *cis-* and *trans*-2-butene *are not* mirror reflections of each other. If one holds a model of *cis*-2-butene up to a mirror, the model that one sees in the mirror is not *trans*-2-butene. But, *cis-* and *trans*-2-butene *are* stereoisomers and, since they are not related to each other as an object and its mirror reflection, they are diastereomers.

8.3 ENANTIOMERS AND CHIRAL MOLECULES

Enantiomers occur only with those compounds whose molecules are *chiral.* A chiral molecule is defined as *one that is not superposable on its mirror reflection.*

The word chiral comes from the Greek word *cheir,* meaning "hand." Chiral objects (including molecules) are said to possess "handedness." The term chiral is used to describe molecules of enantiomers because they are related to each other in the same way that a left hand is related to a right hand. When you view your left hand in a mirror, the mirror reflection of your left hand is a right hand (Fig. 8.1). Your left and right hands, moreover, are not *superposable* (Fig. 8.2). (This fact becomes obvious when one attempts to put a "left-handed" glove on a right hand or vice versa.)

Many familiar objects are chiral and the chirality of some of these objects is clear because we normally speak of them as having "handedness." We speak, for example, of nuts and bolts as having right- or left-handed threads or of a propeller as having a right- or left-handed pitch. The chirality of many other objects is not obvious in this sense, but becomes obvious when we apply the test of nonsuperposability of the object and its mirror reflection.

Objects (and molecules) that *are* superposable on their mirror images are *achiral.* Most socks, for example, are achiral whereas gloves are chiral.

Problem 8.1

Classify the following objects as to whether they are chiral or achiral.

(a) Screw
(b) Plain spoon
(c) Fork
(d) Cup
(e) Foot
(f) Ear
(g) Shoe
(h) Spiral staircase

The chirality of molecules can be demonstrated with relatively simple compounds. Consider, for example, 2-butanol.

$$\underset{\displaystyle OH}{CH_3\overset{}{C}HCH_2CH_3}$$

2-Butanol

FIG. 8.1
The mirror reflection of a left hand is a right hand.

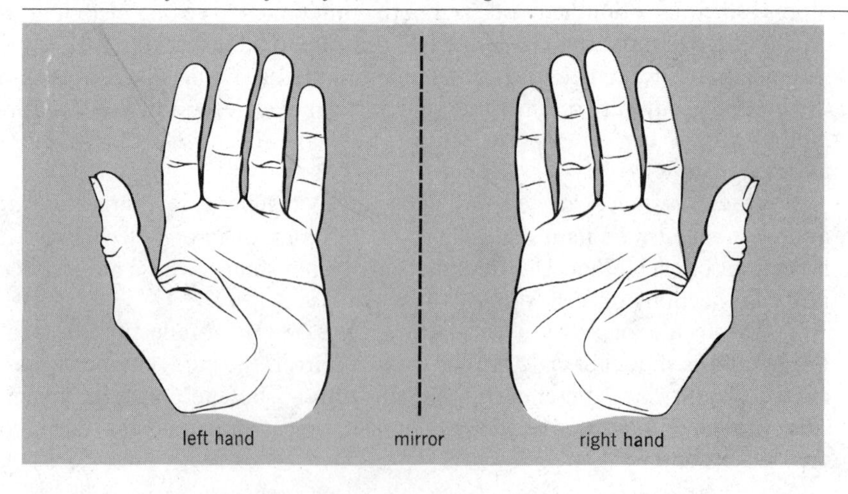

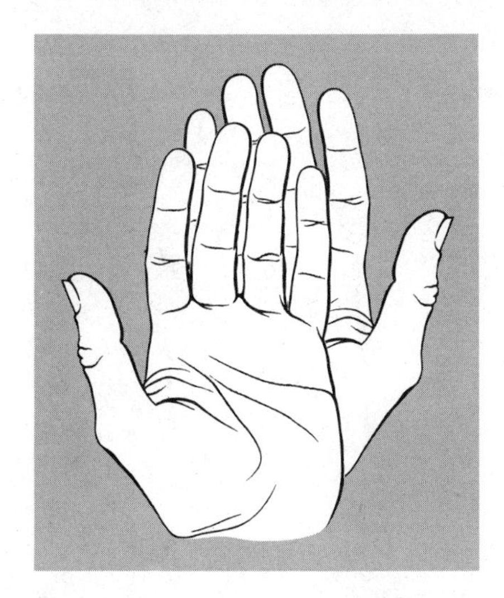

FIG. 8.2
Left and right hands are not superposable.

Until now, we have presented the formula just written as though it represented only one compound; for we have not mentioned that molecules of 2-butanol are chiral. Because they are, there are actually two different 2-butanols and these two 2-butanols are enantiomers. We can understand this if we examine the drawings and models in Fig. 8.3.

If model I is held before a mirror, model II is seen in the mirror and vice versa. Models I and II are not superposable on each other, therefore they represent different, but isomeric, molecules. Because models I and II are nonsuperposable mirror reflections of each other, the molecules that they represent are enantiomers.

Problem 8.2

(a) If models are available, construct the 2-butanols represented in Fig. 8.3 and demonstrate for yourself that they are not mutually superposable. (b) Make similar models of 2-propanol, $CH_3CHOHCH_3$. Are they superposable? (c) Is 2-propanol chiral? (d) Would you expect to find enantiomeric forms of 2-propanol?

A pair of enantiomers will be possible for all molecules that contain *a single* chiral carbon. *A chiral carbon is a carbon atom that has four different groups attached to it.** In 2-butanol (Fig. 8.4) the chiral carbon is carbon-2. The four different groups that are attached to 2-butanol are a hydroxyl group, a hydrogen, a methyl group, and an ethyl group.

Figure 8.5 demonstrates the validity of the generalization that enantiomeric compounds are possible whenever a molecule contains a single chiral carbon.*

Problem 8.3

Demonstrate the validity of what we have represented in Fig. 8.5 by constructing models. Arrange four different colored atoms at each corner of a tetrahedral carbon atom. Demonstrate for yourself that III and IV are related as an object and

* Any tetrahedral atom with four different groups attached to it is called a *chiral* atom or chiral center. An older term for a chiral atom is an *asymmetric atom.*

* This generalization is not necessarily true of molecules that contain more than one chiral carbon. It is also not necessary for a molecule to have a chiral carbon in order to exist in enantiomeric forms. All that is necessary is that the molecule as a whole be chiral. We shall see examples later.

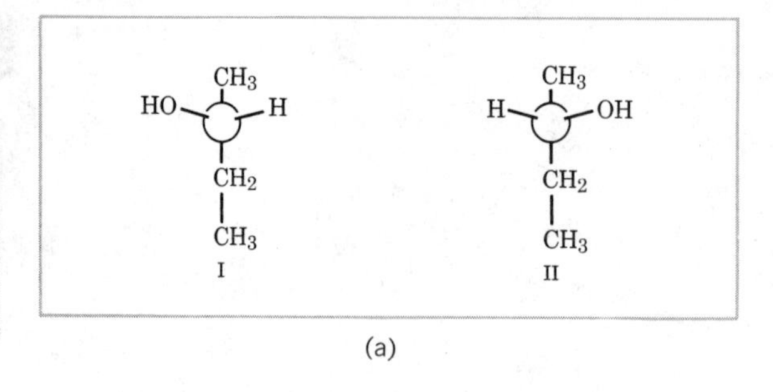

(a)

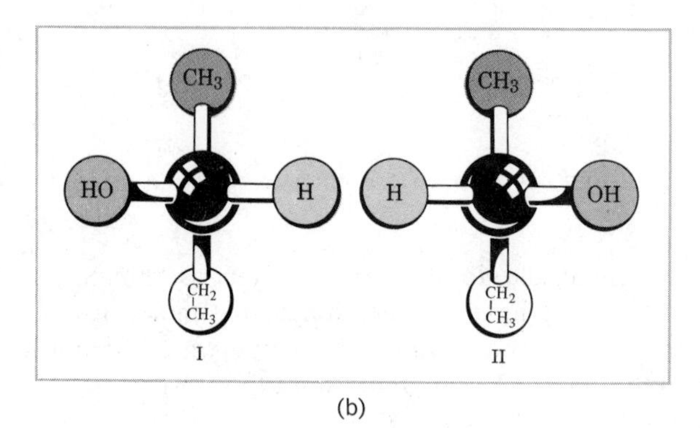

(b)

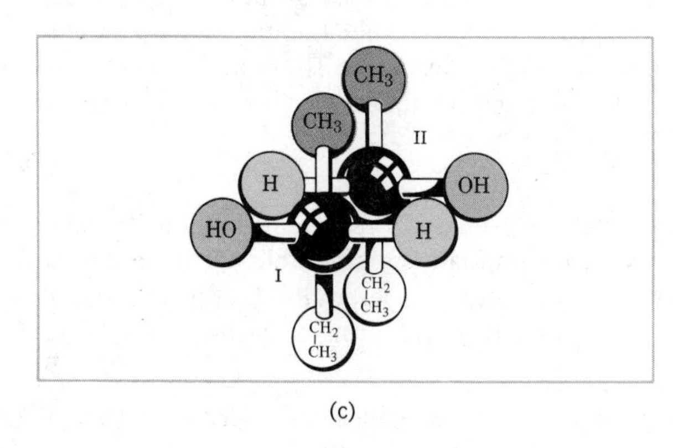

(c)

FIG. 8.3
(a) *Three-dimensional drawings of the 2-butanol enantiomers I and II.* (b) *Models of the 2-butanol enantiomers.* (c) *An unsuccessful attempt to superpose models of I and II.*

its mirror reflection *and that they are not superposable* (i.e., that III and IV are chiral molecules and are enantiomers). (a) Replace one atom on each model so that each model has two atoms of the same color arranged around the central carbon. Are the molecules that these models represent mirror reflections of each other? (b) Are they superposable? (c) Are they chiral? (d) Are they enantiomers?

If two or more of the groups that are attached to a tetrahedral atom *are the same,* the molecule is superposable on its mirror image and is achiral. An example

(hydrogen)

$$\text{(methyl)}\quad \overset{1}{CH_3}-\overset{\overset{H}{|}\,2\,*}{C}-\overset{3}{CH_2}\overset{4}{CH_3}\quad \text{(ethyl)}$$
$$\underset{\text{(hydroxyl)}}{OH}$$

FIG. 8.4
The chiral carbon of 2-butanol. (By convention chiral carbons are often designated with an asterisk.)

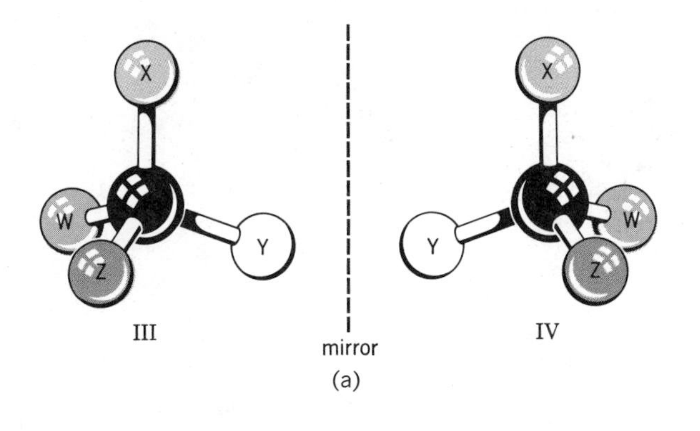

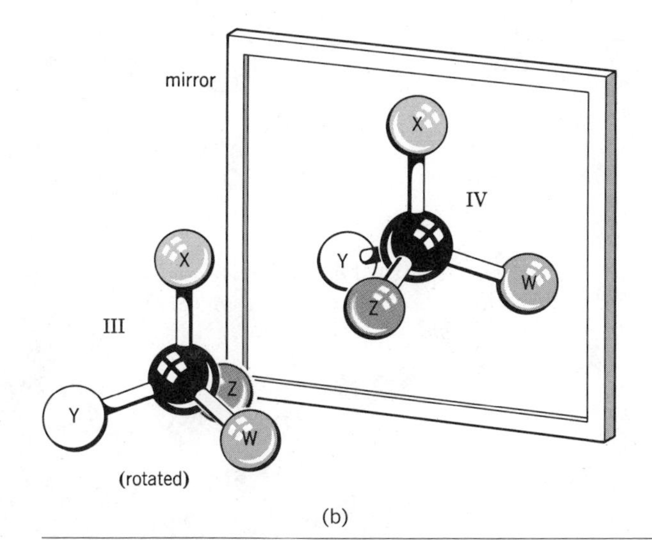

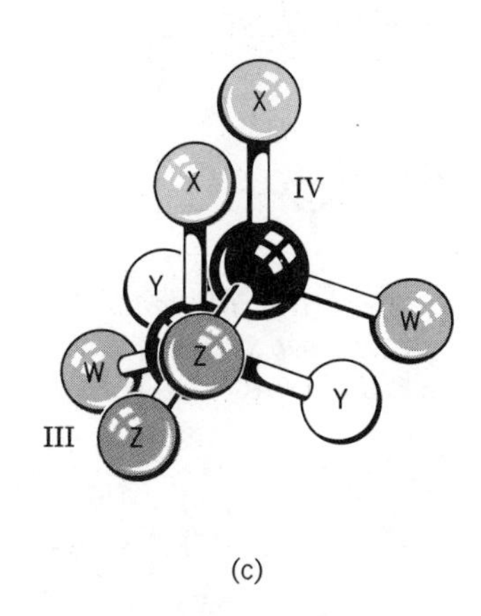

FIG. 8.5
A demonstration of chirality of a generalized molecule containing one chiral carbon. (a) *The four different groups around the carbon atom in III and IV are arbitrary.* (b) *III is rotated and placed in front of a mirror. III and IV are found to be related as an object and its mirror reflection.* (c) *III and IV are not superposable; therefore, the molecules that they represent are chiral and are enantiomers.*

of a molecule of this type is 2-propanol, since two identical methyl groups are attached to the central atom. If we write three-dimensional formulas for 2-propanol we find (Fig. 8.6) that one structure can be superposed on its mirror reflection.

Thus, we would not predict the existence of enantiomeric forms of 2-propanol, and experimentally only one form of 2-propanol has ever been found.

Problem 8.4

Some of the molecules listed below have chiral carbons; some do not. Write three-dimensional formulas for the enantiomers of those molecules that do have chiral carbons.

(a) 1-Chloropropane
(b) Bromochloroiodomethane
(c) 1-Chloro-2-methylpropane
(d) 2-Chloro-2-methylpropane

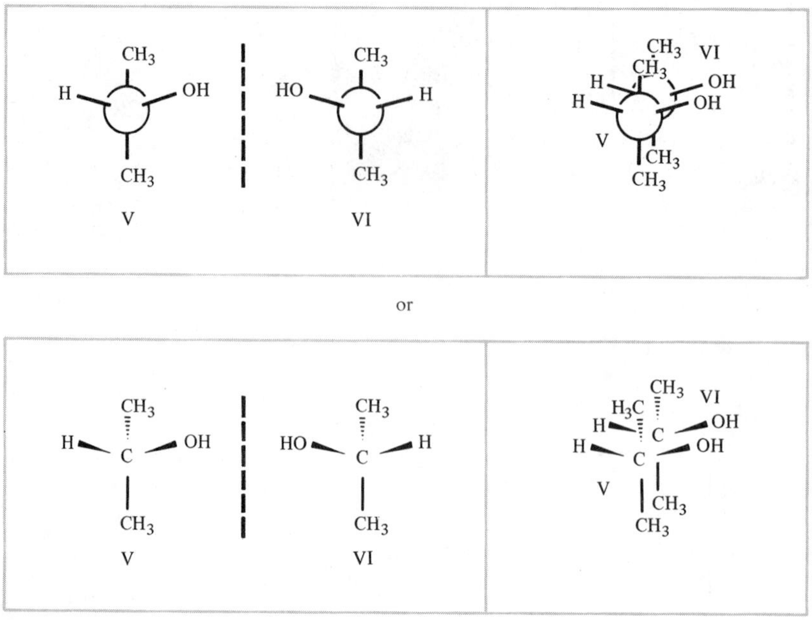

FIG. 8.6
(a) *2-Propanol (V) and its mirror reflection (VI).* (b) *When one is rotated, the two structures* are *superposable and thus do not represent enantiomers. They represent two molecules of the same compound.*

(e) 2-Bromobutane
(f) 1-Chloropentane
(g) 2-Chloropentane
(h) 3-Chloropentane

It was reasoning based on many observations such as those we have just presented that led van't Hoff to the conclusion that the spatial orientation of groups around carbon atoms is tetrahedral when carbon is bonded to four other atoms. The following information was available to van't Hoff.

1. Only one compound with the general formula CH_3X is ever found.
2. Only one compound with the formula CH_2X_2 or CH_2XY is ever found.
3. Two enantiomeric compounds with the formula CHXYZ are found.

Problem 8.5

(a) Prove to yourself the correctness of van't Hoff's reasoning by writing tetrahedral representations for carbon compounds of the three types given above. (b) How many isomers would be possible in each instance if the carbon atom were at the center of a square? (c) At the center of a rectangle? (d) At one corner of a regular pyramid?

8.4 SYMMETRY ELEMENTS: PLANES OF SYMMETRY

The ultimate way to test for molecular chirality is to construct models of the molecule and its mirror reflection and then determine whether they are superposable. If the two models are superposable, the molecule that they represent is achiral. If the models are not superposable, then the molecules that they represent are chiral. We can apply this test with actual models, as we have just described, or we can apply it by drawing three-dimensional structures and attempting to superpose them in our minds.

There are other aids, however, that will assist us in recognizing chiral molecules. We have mentioned one already: the presence of a *single* chiral atom. The other aids are based on the presence or absence in the molecule of certain symmetry elements. A molecule *will not be chiral,* for example, if it possesses (1) a plane of symmetry, (2) a center of symmetry, or (3) any n-fold (n = even number) alternating axis of symmetry. The latter two symmetry elements are beyond the scope of our discussion at this point, but an ability to recognize planes of symmetry will serve us very well. An ability to recognize planes of symmetry will, in fact, enable us to make a decision about the existence or nonexistence of chirality in most of the molecules that we will encounter.

A plane of symmetry is defined as an imaginary plane that bisects a molecule in such a way that the two halves of the molecule are mirror reflections of each other. For example, 2-chloropropane has a plane of symmetry (Fig. 8.7*a*), while 2-chlorobutane does not (Fig. 8.7*b*).

Problem 8.6

Which of the objects listed in Problem 8.1 possess a plane of symmetry?

Problem 8.7

Write three-dimensional formulas and designate a plane of symmetry for all of the achiral molecules in Problem 8.4. (In order to be able to designate a plane of symmetry you may have to write the molecule in an appropriate conformation. This is permissible with all of these molecules because they have only single bonds and groups joined by single bonds are capable of essentially free rotation at room temperature.)

8.5 NOMENCLATURE OF ENANTIOMERS: THE *R-S* SYSTEM

The two enantiomers of 2-butanol are shown below.

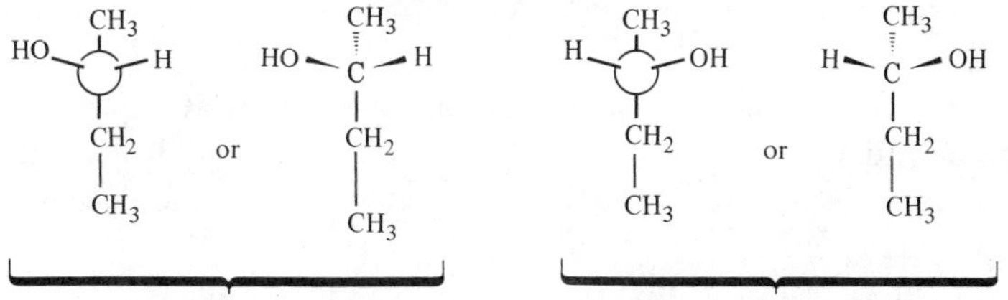

CH_3 / HO—C—H / CH_2 / CH_3 or CH_3 / HO—C—H / CH_2 / CH_3

I

CH_3 / H—C—OH / CH_2 / CH_3 or CH_3 / H—C—OH / CH_2 / CH_3

II

If we name these two enantiomers using the IUPAC system of nomenclature both enantiomers will have the same name—2-butanol (Section 15.1A). This is undesirable because *each compound must have its own name.* Moreover, the name that is given a compound should allow a chemist who is familiar with the rules of nomenclature to write the structure of the compound from its name alone. Given the name 2-butanol a chemist could write either structure I or structure II.

Three chemists, Professors R. S. Cahn (England), C. K. Ingold (England),

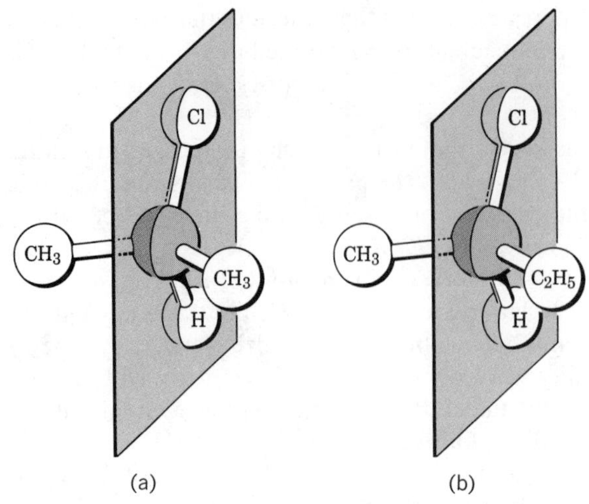

FIG. 8.7
(a) *2-Chloropropane has a plane of symmetry and is achiral.* (b) *2-Chlorobutane does not possess a plane of symmetry and is chiral.*

and V. Prelog (Switzerland), devised a system of nomenclature that, when added to the IUPAC system, solves both of these problems. This system, called the *R-S* system, or the Cahn-Ingold-Prelog system, is now widely used.

According to the system one enantiomer of 2-butanol should be designated *R*-2-butanol and the other enantiomer should be designated *S*-2-butanol. (*R* and *S* are from the Latin words *rectus* and *sinister* meaning right and left, respectively.)

R and *S* designations are assigned as follows.

1. Each group attached to the chiral carbon is assigned a *preference* or *priority a, b, c,* or *d*. Preference is first assigned on the basis of the atomic number of the atom that is directly attached to the chiral carbon. The group with the lowest atomic number is given the lowest preference, *d*; the group with next highest atomic number is given the next higher preference, *c*; and so on.

We can illustrate the application of this rule with the 2-butanol enantiomer, I.

CH_3 *(b* or *c)*
(a) HO— —H *(d)*
CH_2 *(b* or *c)*
|
CH_3

I

Oxygen has the highest atomic number of the four atoms attached to the chiral carbon and is assigned the highest preference, *a*. Hydrogen has lowest atomic number and is assigned the lowest preference, *d*. A preference cannot be assigned for the methyl group and the ethyl group because the atom that is directly attached to the chiral carbon is a carbon atom in both groups.

2. When a preference cannot be assigned on the basis of the atomic number of the atoms that are directly attached to the chiral carbon then the next sets of atoms in the unassigned groups are examined. This process is continued until a decision can be made. *We assign a preference at the first point of difference.*

When we examine the methyl group of enantiomer I, we find that the next set of atoms consists of three hydrogens (H, H, H). In the ethyl group of I the next set of atoms consists of one carbon and two hydrogens (C, H, H). Carbon has a higher atomic number than hydrogen so we assign the ethyl group the higher preference, *b*, and the methyl group the lower preference, *c*.

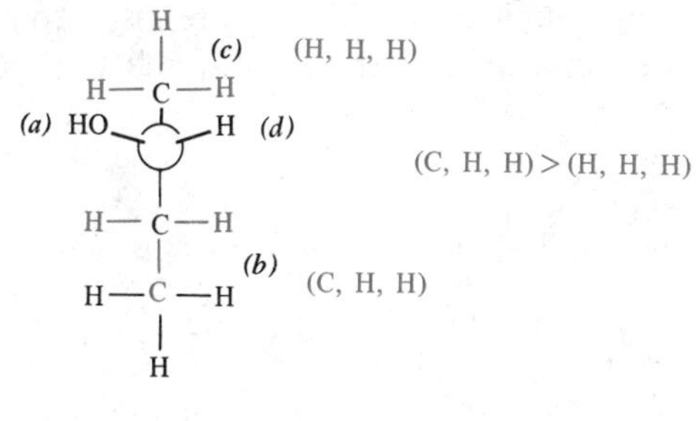

(*c*) (H, H, H)

(*a*) HO, H (*d*)

(C, H, H) > (H, H, H)

(*b*) (C, H, H)

I

3. We now rotate the formula (or model) so that the group with lowest preference (*d*) is directed away from us.

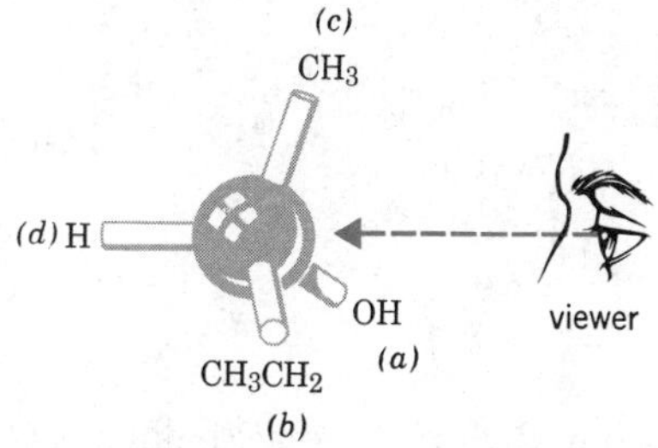

(*c*) CH_3

(*d*) H

OH (*a*)

viewer

CH_3CH_2 (*b*)

I

Then we trace a path from *a* to *b* to *c*. If, as we do this, the direction of our finger (or pencil) is *clockwise* the enantiomer is designated *R*. If the direction is *counter-clockwise*, the enantiomer is designated *S*. On this basis the 2-butanol enantiomer I is *R*-2-butanol.

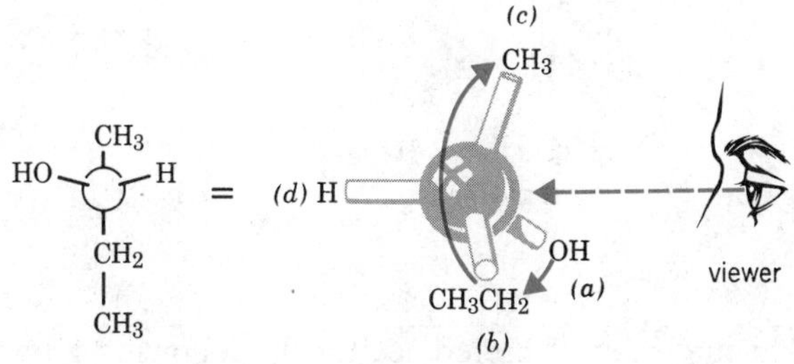

CH_3, HO, H, CH_2, CH_3 = (*d*) H

(*c*) CH_3

OH (*a*)

CH_3CH_2 (*b*)

viewer

arrows are clockwise

Problem 8.8

Apply the procedure, just given, to the 2-butanol enantiomer II and show that it is *S*-2-butanol.

Problem 8.9

Give *R* and *S* designations for each pair of enantiomers given as answers to Problem 8.4.

The first three rules of the Cahn-Ingold-Prelog system allow us to make an *R* or *S* designation for most compounds containing single bonds. For compounds containing multiple bonds one other rule is necessary.

4. Groups containing double or triple bonds are assigned preferences as if both atoms were duplicated or triplicated, that is,

$$-\overset{|}{C}{=}Y \text{ as if it were } -\underset{(Y)}{\overset{|}{C}}-\underset{(C)}{Y} \quad \text{and} \quad -C{\equiv}Y \text{ as if it were } -\underset{(Y)}{\overset{(Y)}{C}}-\underset{(C)}{\overset{(C)}{Y}}$$

where the atoms in parentheses are duplicate or triplicate representations of the atoms at the other end of the double bond.

Thus, the vinyl group, $-CH{=}CH_2$, is of higher preference than the isopropyl group, $-CH(CH_3)_2$.

$$-CH{=}CH_2 \text{ is treated as though it were } -\underset{(C)}{\overset{H}{C}}-\underset{(C)}{\overset{H}{C}}-H \text{ which has higher preference than } -\overset{H}{\underset{H-\underset{H}{C}-H}{C}}-\underset{H}{\overset{H}{C}}-H$$

because at the third set of atoms out, the vinyl group (below) is C, H, H, whereas the isopropyl group along either branch is H, H, H. (At the first and second set of atoms both groups are the same: C, then C, C, H.)

$$-\underset{(C)}{\overset{H}{C}}-\underset{(C)}{\overset{H}{C}}-H > -\overset{H}{\underset{H-\underset{H}{C}-H}{C}}-\underset{H}{\overset{H}{C}}-H$$

C, H, H	>	H, H, H
Vinyl group		Isopropyl group

Problem 8.10

Demonstrate that the order of preference for the following groups is as follows:

$$-C_6H_5 \text{ (phenyl)} > -C{\equiv}CH > -C(CH_3)_3$$

Problem 8.11

An important compound in stereochemistry and biology is the compound glyceraldehyde (below). Write three-dimensional formulas for the glyceraldehyde enantiomers and give each its proper *R-S* designation.

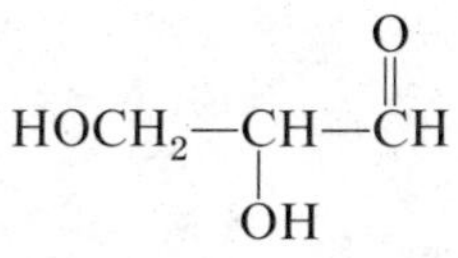

Glyceraldehyde

Problem 8.12

Assign R and S designations to each of the following compounds.

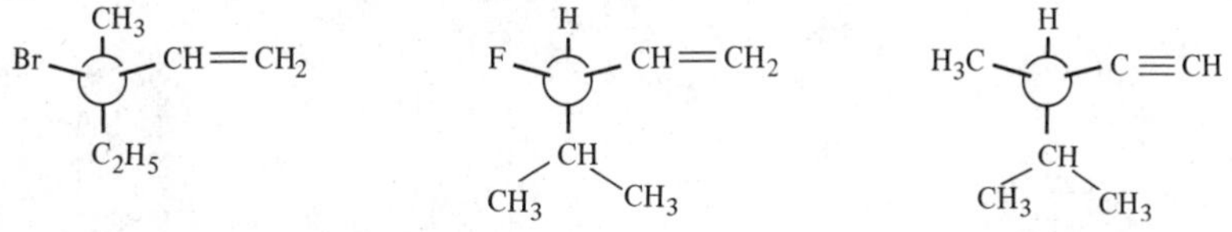

(a) (b) (c)

8.6 PROPERTIES OF ENANTIOMERS: OPTICAL ACTIVITY

The molecules of enantiomers are not superposable one on the other, and on this basis alone, we have concluded that enantiomers are different compounds. How are they different? Do enantiomers resemble structural isomers and diastereomers in having different melting and boiling points? The answer is *no*. Enantiomers have *identical* melting and boiling points. Do enantiomers have different indexes of refraction, different solubilities in common solvents, different infrared spectra, and different rates of reaction with ordinary reagents? The answer to each of these questions is also no.

We can see examples if we examine Table 8.1 where some of the physical properties of the 2-butanol enantiomers are listed.

Enantiomers show different behavior only when they interact with other chiral substances. Enantiomers show different solubilities in solvents that consist of a single enantiomer or an excess of a single enantiomer. Enantiomers also show different rates of reaction toward other chiral molecules—that is, toward reagents that consist of a single enantiomer or an excess of a single enantiomer.

One easily observable way in which enantiomers differ is in *their behavior toward plane-polarized light*. Plane-polarized light has chiral properties. When a beam of plane-polarized light passes through an enantiomer, the plane of polarization *rotates*. Moreover, separate enantiomers rotate the plane of plane-polarized light equal amounts *but in opposite directions*. Because of their effect on plane-polarized light, separate enantiomers are said to be *optically active compounds*.

In order to understand this behavior of enantiomers we need to understand the nature of plane-polarized light. We also need to understand how an instrument called a *polarimeter* operates.

TABLE 8.1 Physical Properties of *R*- and *S*-2-Butanol

PHYSICAL PROPERTY	*R*-2-BUTANOL	*S*-2-BUTANOL
Boiling point (1 atm)	99.5°	99.5°
Density (20°/4°)	0.808	0.808
Index of refraction (20°)	1.397	1.397

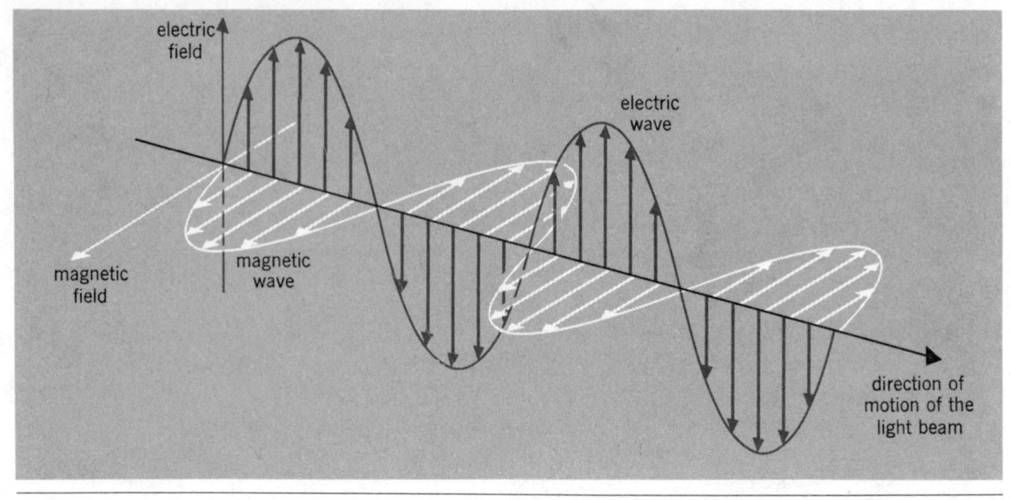

FIG. 8.8
The oscillating electric and magnetic fields of a beam of ordinary light.

8.6A Plane-Polarized Light

Light is an electromagnetic phenomenon. A beam of light consists of two mutually perpendicular oscillating fields: an oscillating electric field and an oscillating magnetic field. The planes in which the electrical and magnetic oscillations occur are also perpendicular to the direction of propagation of the beam of light (Fig. 8.8).

If we were to view a beam of ordinary light from one end, and if we could actually see the planes in which the electrical oscillations were occurring, we would find that oscillations of the electric field were occurring in all possible planes perpendicular to the direction of propagation (Fig. 8.9). (The same would be true of the magnetic field.)

When ordinary light is passed through a polarizer, the polarizer interacts with the electrical field so that the electrical field of the light that emerges from the polarizer (and the magnetic field perpendicular to it) is oscillating only in one plane. Such light is called plane-polarized light (Fig. 8.10).

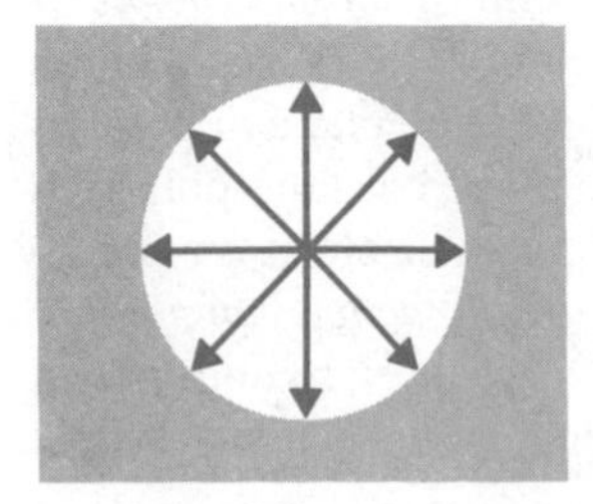

FIG. 8.9
Oscillation of the electrical field of ordinary light occurs in all possible planes perpendicular to the direction of propagation.

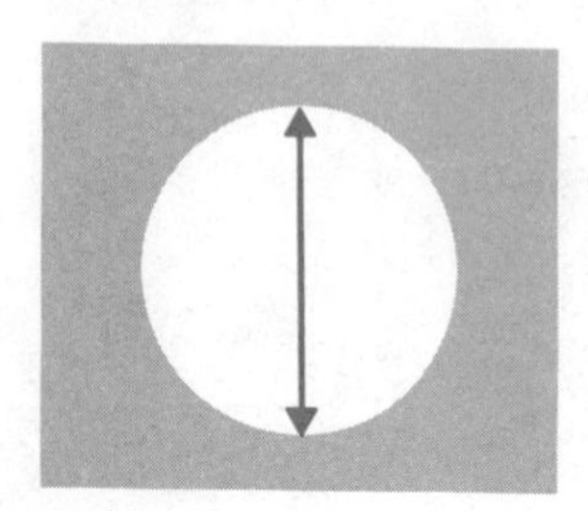

FIG. 8.10
The plane of oscillation of the electrical field of plane-polarized light. In this example the plane of polarization is vertical.

The lenses of Polaroid sunglasses have this effect. You can demonstrate for yourself that this is true with two pairs of Polaroid sunglasses. If two lenses are placed one on top of the other so that the axes of polarization coincide, then light passes through both normally. Then if one lens is rotated 90° with respect to the other, no light passes through.

8.6B The Polarimeter

The device that is used for measuring the effect of plane-polarized light on optically active compounds is a polarimeter. A sketch of a polarimeter is shown in Fig. 8.11. The principal working parts of a polarimeter are (1) a light source (usually a sodium lamp), (2) a polarizer, (3) a tube for holding the optically active substance (or solution) in the light beam, (4) an analyzer, and (5) a scale for measuring the number of degrees that the plane of polarized light has been rotated.

FIG. 8.11
The principal working parts of a polarimeter and the measurement of optical rotation. (From John R. Holum, Organic Chemistry: A Brief Course, *Wiley, New York, 1975, p. 316.)*

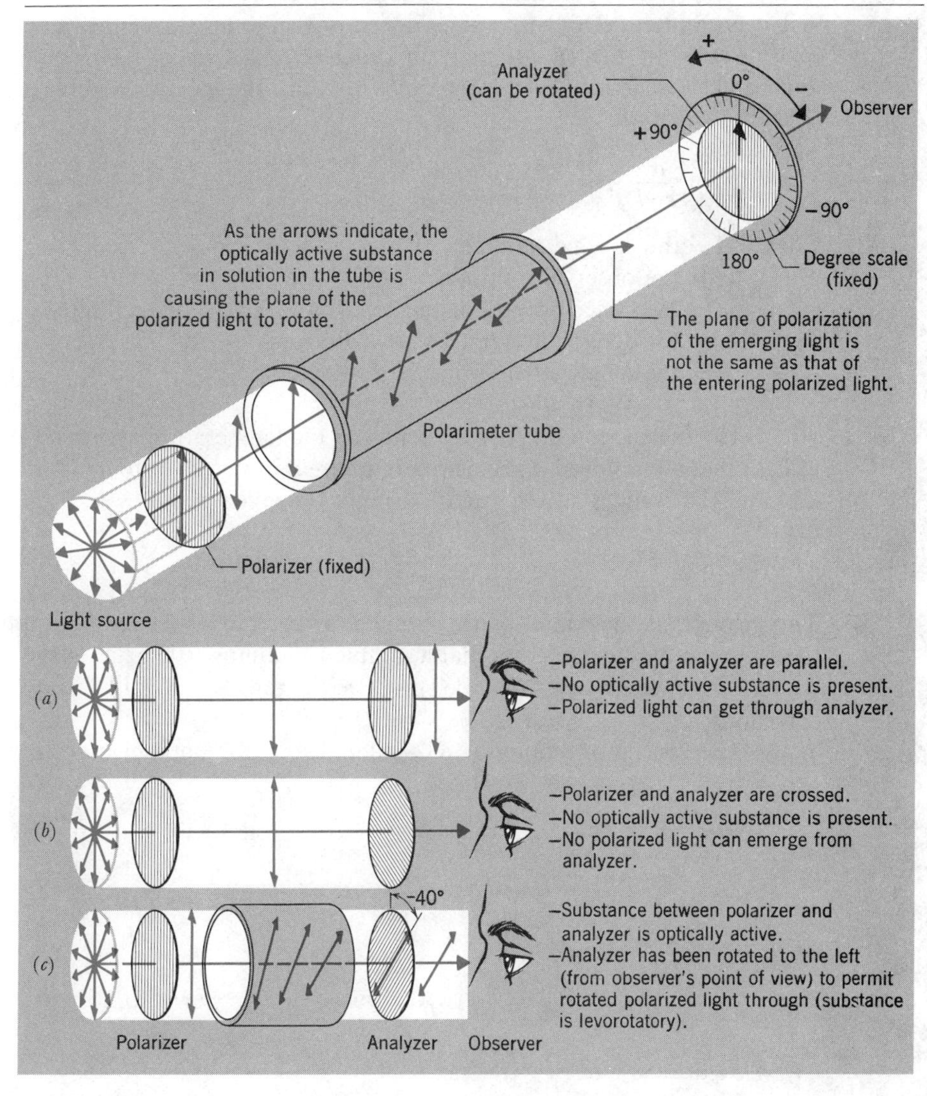

The analyzer of a polarimeter (Fig. 8.11) is nothing more than another polarizer. If the tube of the polarimeter is empty, or if an optically *inactive* substance is present, the axes of the plane-polarized light and the analyzer will be exactly parallel when the instrument reads 0°, and the observer will detect the maximum amount of light passing through. If, by contrast, the tube contains an optically active substance, a solution of one enantiomer, for example, the plane of polarization of the light will be rotated as it passes through the tube. In order to detect the maximum brightness of light the observer will have to rotate the axis of the analyzer in either a clockwise or counterclockwise direction. If the analyzer is rotated in a clockwise direction, the rotation, α, (measured in degrees) is said to be positive (+). If the rotation is counterclockwise, the rotation is said to be negative (−). A substance that rotates plane-polarized light in the clockwise direction is also said to be *dextrorotatory,* and one that rotates plane-polarized light in a counterclockwise direction is said to be *levorotatory.*

8.6C Specific Rotation

The number of degrees that the plane of polarization is rotated as it passes through a solution of an enantiomer depends on the number of chiral molecules that it encounters. This, of course, depends on the length of the tube and the concentration of the enantiomer. In order to place measured rotations on a standard basis, chemists calculate a quantity called the specific rotation, $[\alpha]$, by the following equation:

$$[\alpha] = \frac{\alpha}{c \cdot l}$$

where $[\alpha]$ = the specific rotation
α = the observed rotation
c = the concentration of the solution in grams per milliliter (or density in g/ml for neat liquids)
l = the length of the tube in decimeters (1 dm = 10 cm)

The specific rotation also depends on the temperature and the wavelength of light that is employed. Specific rotations are reported so as to incorporate these quantities as well. A specific rotation might be given as follows:

$$[\alpha]_D^{25} + 3.12°$$

This means that, the D line* of a sodium lamp was used for the light, a temperature of 25 °C was maintained, and that a sample containing 1.00 g/ml of the optically active substance, in a 1-dm tube, produced a rotation of 3.12° in a clockwise direction.

The specific rotations of *R*-2-butanol and *S*-2-butanol are given below.

***R*-2-butanol**
$[\alpha]_D^{25}$ −13.52°

***S*-2-butanol**
$[\alpha]_D^{25}$ +13.52°

* Wavelength = 5896 Å.

The direction of rotation of plane-polarized light is often incorporated into the names of optically active compounds. The two sets of enantiomers below show how this is done.

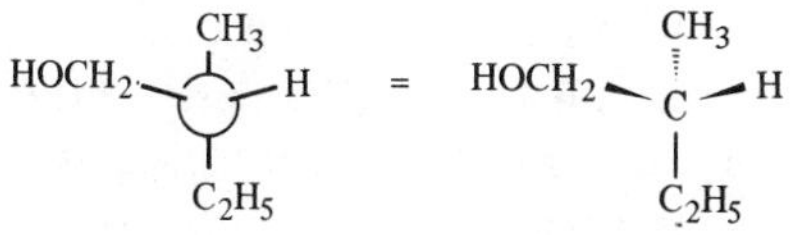

R-(+)-2-methyl-1-butanol
$[\alpha]_D^{25}$ +5.756°

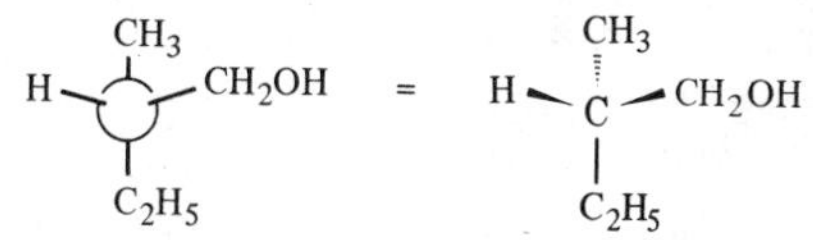

S-(−)-2-methyl-1-butanol
$[\alpha]_D^{25}$ −5.756°

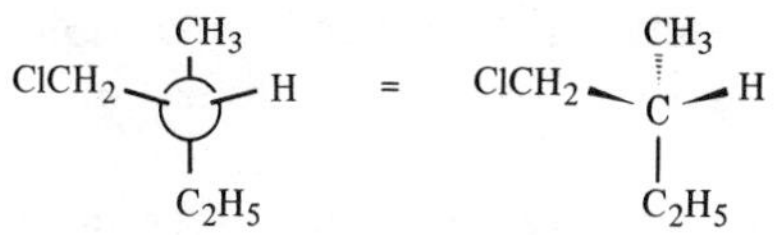

R-(−)-1-chloro-2-methylbutane
$[\alpha]_D^{25}$ −1.64°

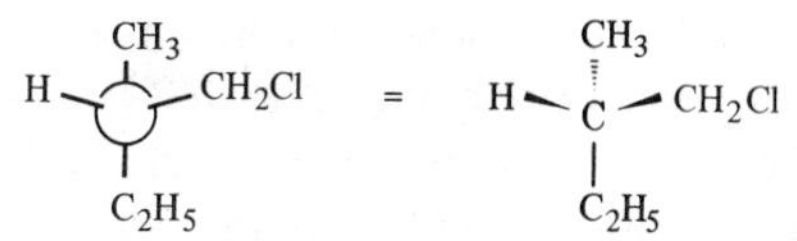

S-(+)-1-chloro-2-methylbutane
$[\alpha]_D^{25}$ +1.64°

The compounds above also illustrate an important principle: *no obvious correlation exists between the configurations of enantiomers and the direction in which they rotate plane-polarized light.*

R-(+)-2-Methyl-1-butanol and *R*-(−)-1-chloro-2-methylbutane have the same *configuration,* that is, they have the same general arrangement of their atoms in space.

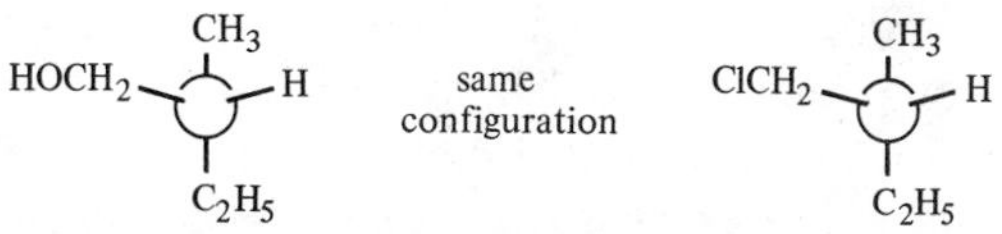

R-(+)-2-methyl-1-butanol — same configuration — **R-(−)-1-chloro-2-methylbutane**

They have, however, an opposite effect on the direction of rotation of the plane of plane-polarized light.

These same compounds also illustrate a second important principle: *no necessary correlation exists between the R and S designation and the direction of rotation of plane-polarized light. R*-2-Methyl-1-butanol is dextrorotatory, (+), and *R*-1-chloro-2-methylbutane is levorotatory, (−).

A method based on the measurement of optical rotation measured at many different wavelengths, called optical rotatory dispersion, has been used to correlate configurations of chiral molecules. A discussion of the technique of optical rotatory dispersion, however, is beyond the scope of this text.

8.7 THE ORIGIN OF OPTICAL ACTIVITY

It is not possible to give a complete, condensed account of the origin of the optical activity observed for separate enantiomers. An insight into the source of this phenomenon can be obtained, however, by comparing what occurs when a beam of plane-polarized light passes through a solution of *achiral* molecules, with what occurs when a beam of plane-polarized light passes through a solution of *chiral* molecules.

Almost all *individual* molecules, whether chiral or achiral, are theoretically capable of producing a slight rotation of the plane of plane-polarized light. The direction and magnitude of the rotation produced by an individual molecule depends, in part, on its orientation at the precise moment that it encounters the beam. In a solution, of course, billions of molecules are in the path of the light beam and at any given moment these molecules will be present in all possible orientations. If the beam of plane-polarized light passes through a solution of the achiral compound 2-propanol, for example, it should encounter at least two molecules in the exact orientations shown below in Fig. 8.12. The effect of the first encounter might be to produce a very slight rotation of the plane of polarization to the right. Before the beam emerges from the solution, however, it should encounter at least one molecule of 2-propanol that is in exactly the mirror reflection orientation of the first. The effect of this second encounter will be to produce an equal and opposite rotation of the plane: a rotation that exactly cancels the first rotation. The beam, therefore, emerges with no net rotation.

What we have just described for the two encounters shown in Fig. 8.12 can be said of all possible encounters of the beam with molecules of 2-propanol. Because so many molecules are present, it is statistically certain that *for each encounter with a particular orientation there will be an encounter with a molecule that is in a mirror-reflection orientation.* The result of all of these encounters will be such that all of the rotations produced by individual molecules will be canceled and 2-propanol will be found to be *optically inactive.*

What, then, is the situation when a beam of plane-polarized light passes through a solution of one enantiomer of a chiral compound? We can answer this question by considering what might occur when plane-polarized light passes through a solution of pure *R*-2-butanol. Figure 8.13 illustrates one possible encounter of a beam of plane-polarized light with a molecule of *R*-2-butanol.

When a beam of plane-polarized light passes through a solution of *R*-2-butanol, *no molecule is present that can ever be exactly oriented as a mirror-reflection of any given orientation of an R-2-butanol molecule.* The only molecules that could do this would be molecules of *S*-2-butanol, and they are not present. Exact cancellation of the rotations produced by all of the encounters of the beam with random orientations of *R*-2-butanol does not happen and, as a result, a net rotation of the plane of polarization is observed. *R*-2-Butanol is found to be *optically active.*

FIG. 8.12
A beam of plane-polarized light encountering a molecule of 2-propanol (*an achiral molecule*) *in orientation* (a) *and then a second molecule in the mirror-reflection orientation* (b). *The beam emerges from these two encounters with no net rotation of its plane of polarization.*

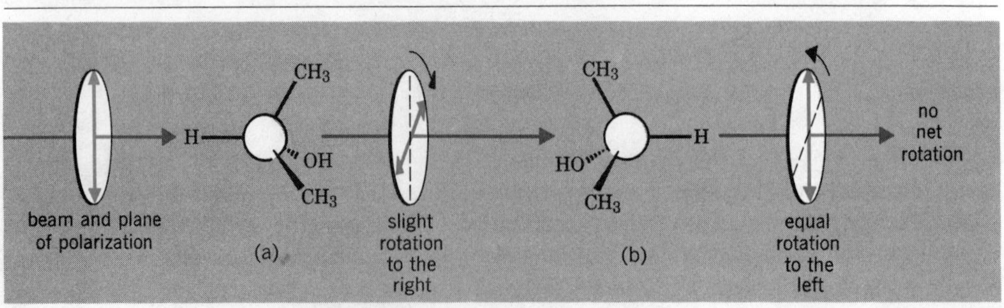

beam and plane of polarization
CH_3 H OH CH_3 (a)
slight rotation to the right
CH_3 HO H CH_3 (b)
equal rotation to the left
no net rotation

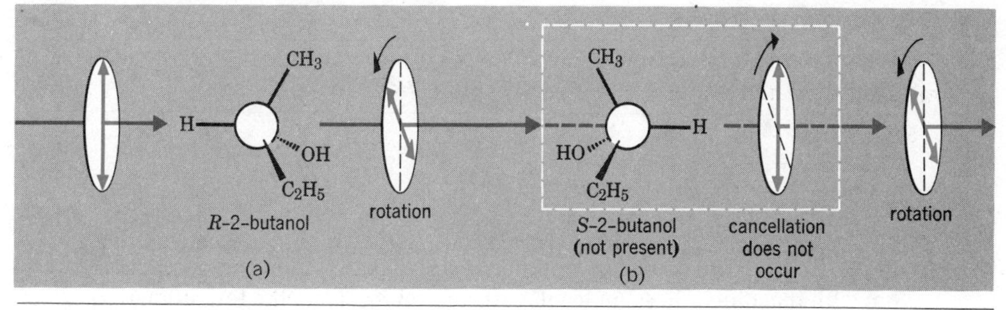

FIG. 8.13
(a) *A beam of plane-polarized light encounters a molecule of* R-2-butanol *(a chiral molecule) in a particular orientation. This encounter produces a slight rotation of the plane of polarization.* (b) *Exact cancellation of this rotation requires that a second molecule be oriented as an exact mirror-reflection. This cancellation does not occur because the only molecule that could ever be oriented as an exact mirror reflection the first encounter is a molecule of* S-2-*butanol which is not present. As a result, a net rotation of the plane of polarization occurs.*

8.7A Racemic Modifications

The net rotation of the plane of polarization that we observe for a solution consisting of molecules of *R*-2-butanol alone would not be observed if we passed the beam through a solution that contained equimolar amounts of *R*-2-butanol and *S*-2-butanol. In the latter instance, molecules of *S*-2-butanol would be present in a quantity equal to those of *R*-2-butanol and for every possible orientation of one enantiomer, a molecule of the other enantiomer would be in a mirror-reflection orientation. Exact cancellations of all rotations would occur, and the solution of the equimolar mixture of enantiomers would be *optically inactive.*

Equimolar mixtures of two enantiomers are called *racemic modifications* or *racemates.* Racemic modifications show no rotation of plane-polarized light; as such, they are often designed as being (±). A racemic modification of *R*-(−)-2-butanol and *S*-(+)-2-butanol might be indicated as

(±)-2-butanol

or as (±) $CH_3CH_2CHOHCH_3$

8.7B Optical Purity

A sample of an optically active substance that consists of a single enantiomer is said to be *optically pure.* An optically pure sample of *S*-(+)-2-butanol shows a specific rotation of +13.52° ($[\alpha]_D^{25}$ + 13.52°). On the other hand, a sample of *S*-(+)-2-butanol that contains less than an equimolar amount of *R*-(−)-2-butanol will show a specific rotation that is less than +13.52° but greater than 0°. Such a sample is said to have an *optical purity* less than 100%.

Let us suppose, for example, that the sample showed a specific rotation of +6.76°. We would then say that the optical purity of the *S*-(+)-2-butanol is 50%.*

$$\text{Optical purity} = \frac{\text{observed specific rotation}}{\text{specific rotation of the pure enantiomer}} \times 100$$

$$\text{Optical purity} = \frac{+6.76°}{+13.52°} \times 100 = 50\%$$

* The term optical purity is applied to a single enantiomer or to mixtures of enantiomers only. It should not be applied to mixtures in which some other compound is present.

Problem 8.13

What relative molar proportions of S-(+)-2-butanol and R-(−)-2-butanol would give a specific rotation, $[\alpha]_D^{25}$, equal to +6.76°?

8.8 THE SYNTHESIS OF ENANTIOMERS

Many times in the course of working in the organic laboratory a reaction carried out with reactants whose molecules are achiral results in the formation of products whose molecules are chiral. The usual outcome of such a reaction is the formation of an optically inactive racemic modification. The reason: the chiral molecules of the product are obtained as a 50:50 mixture of enantiomers.

An example is the synthesis of 2-butanol by the nickel-catalyzed hydrogenation of 2-butanone.

$$\underset{\substack{\text{2-Butanone}\\ \text{(achiral}\\ \text{molecules)}}}{CH_3CH_2\underset{\underset{O}{\|}}{C}CH_3} + \underset{\substack{\text{Hydrogen}\\ \text{(achiral}\\ \text{molecules)}}}{H{-}H} \xrightarrow{Ni} \underset{\substack{(\pm)\text{ 2-Butanol}\\ \text{(chiral molecules}\\ \text{but 50:50 mixture } R \text{ and } S)}}{(\pm)\ CH_3CH_2\overset{*}{\underset{\underset{OH}{|}}{C}}HCH_3}$$

Molecules of neither reactant (2-butanone nor hydrogen) are chiral. The molecules of the product (2-butanol) are chiral. The product, however, is obtained as a racemic modification because the two enantiomers, R-(−)-2-butanol and S-(+)-2-butanol, are obtained in equal amounts.*

Figure 8.14 shows why a racemic modification of 2-butanol is obtained. Hydrogen, adsorbed on the surface of the nickel catalyst, adds with equal facility at either face of 2-butanone. Reaction at one face produces one enantiomer;

* This is not the result if reactions like this are carried out in the presence of a chiral influence such as an optically active solvent or, as we will see later, an enzyme. The nickel catalyst used in this reaction does not exert a chiral influence.

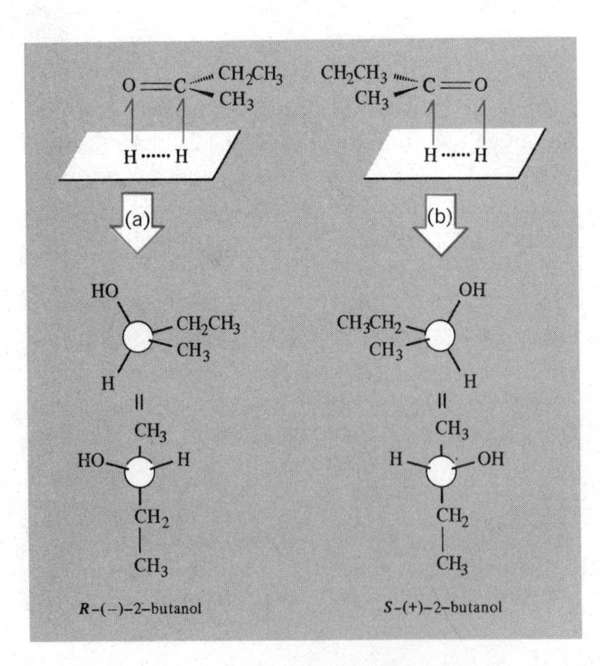

FIG. 8.14
The reaction of 2-butanone with hydrogen in the presence of a nickel catalyst. The reaction rate by path (a) *is equal to that by path* (b). R-(−)-2-*butanol and* S-(+)-2-*butanol are produced in equal amounts,* as a racemic modification.

reaction at the other face produces the other enantiomer, and the two reactions occur at the same rate.

Another example, of the same type, is the catalytic hydrogenation of pyruvic acid to produce a racemic modification of lactic acid.

$$\underset{\text{Pyruvic acid (achiral molecules)}}{CH_3\overset{\displaystyle O}{\overset{\|}{C}}COOH} + \underset{\text{Hydrogen (achiral molecules)}}{H_2} \xrightarrow{Ni} \underset{\text{(±) Lactic acid (chiral molecules but 50:50 mixture of } R \text{ and } S \text{ forms)}}{(\pm)\ CH_3\underset{OH}{\underset{|}{C}}HCOOH}$$

Pyruvic acid (achiral molecules) — Hydrogen (achiral molecules) — (±) Lactic acid (chiral molecules but 50 : 50 mixture of *R* and *S* forms)

Problem 8.14

What products would you expect from each of the following reactions? What relative amounts of each product would be formed?

(a) $(CH_3CH_2CH_2)(CH_3CH_2)C{=}CH_2 \xrightarrow[Pt]{H_2}$ (drawn as $CH_3CH_2CH_2$ and CH_3CH_2 on one carbon, H and H on the other)

(b) $CH_3CH_2CH{=}CH_2 \xrightarrow[H_2O]{H^+}$

(c) $CH_3CH{=}CHCH_3 \xrightarrow[(2)\ OH^-,\ H_2O_2,\ heat]{(1)\ (BH_3)_2}$

8.8A Stereoselective Reactions

The stereochemical course of reactions that occur in living cells are often dramatically different from those that are carried out in the glassware of the laboratory. The laboratory hydrogenation of pyruvic acid in the presence of a nickel catalyst, as we have just seen, produces a racemic modification of lactic acid. In muscle cells, however, the conversion of pyruvic acid to lactic acid is catalyzed by an enzyme called *lactic acid dehydrogenase*.* The enzyme-catalyzed reaction results in the formation of a single enantiomer, *S*-(+)-lactic acid:

$$CH_3\overset{\displaystyle O}{\overset{\|}{C}}COOH \xrightarrow[\text{dehydrogenase, NADH}^\dagger]{\text{lactic acid}} \text{HO}-\overset{COOH}{\underset{CH_3}{C}}-\text{H}$$

pyruvic acid — *S*-(+)-lactic acid, $[\alpha] = +3.82$

The enzyme-catalyzed reaction is said to be *stereoselective*. A stereoselective reaction is *a reaction that yields exclusively* (*or predominately*) *only one of a set of stereoisomers*. The stereoselectivity of the enzyme-catalyzed reaction is a result of

* The name of the enzyme is derived from the fact that this enzyme also catalyzes the reverse reaction, that is, the dehydrogenation of *S*-(+)-lactic acid to produce pyruvic acid.

† The reducing agent in this reaction is a compound, NADH, called the reduced form of nicotinamide adenine dinucleotide, NAD^+, (Sec. 11.10). NADH is a chemical part of the enzyme.

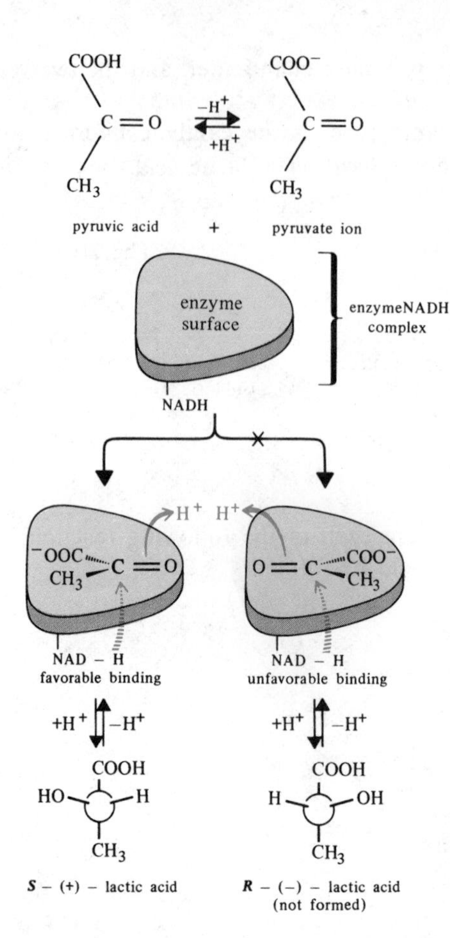

FIG. 8.15
A mechanism that accounts for stereoselectivity in the enzyme-catalyzed reduction of pyruvic acid to S-(+)-*lactic acid. Pyruvate ion can bind itself to the surface of the enzyme-NADH complex in only one favorable way. As a result, NADH can transfer hydrogen (H:⁻) to only one surface of the pyruvate ion. This leads to the formation only of* S-(+)-*lactic acid.* R-(−)-*lactic acid is not produced because its formation would require an unfavorable binding between the NADH-enzyme complex.*

a special relation between the reacting molecule and the enzyme surface as shown in Fig. 8.15.

8.9 MOLECULES WITH MORE THAN ONE CHIRAL CARBON

Thus far all of the chiral molecules that we have considered have contained only one chiral carbon. Many organic molecules, especially those important in biology, contain more than one chiral carbon. Cholesterol (p. 956), for example, contains eight chiral carbons. (Can you locate them?) We can begin, however, with simpler molecules. Let us consider the structure shown below—a structure that has two chiral carbons and that is characteristic of molecules that are called *carbohydrates* or *sugars* (Chap. 19).

```
  O
  ‖
  CH
  |
 *CHOH
  |
 *CHOH
  |
  CH2OH
```

A simple carbohydrate or sugar

There is a useful rule that helps us to know how many stereoisomers to expect from structures like this one. *The total number of stereoisomers will not exceed 2^n where* n *is equal to the number of chiral carbons.* For the structural formula given above we should not expect more than four stereoisomers ($2^2 = 4$).

Our next task is to write three-dimensional formulas for the stereoisomers of our simple sugar. We begin by writing a three-dimensional formula for one stereoisomer and then by writing the formula for *its* mirror reflection. This is shown below.

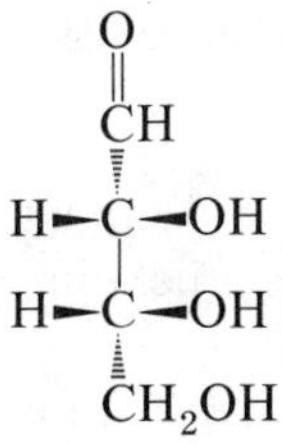

```
     O            O
     ‖            ‖
     CH           CH
     ⋮            ⋮
  H—C—OH      HO—C—H
     |            |
  H—C—OH      HO—C—H
     ⋮            ⋮
    CH2OH        CH2OH
     1            2
```

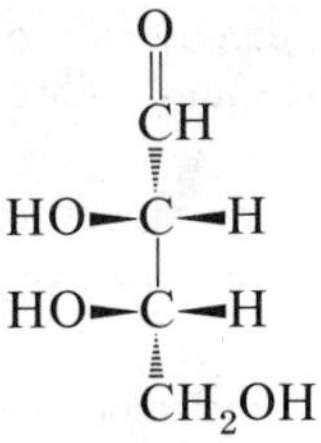

It is helpful to follow certain conventions when we write these three-dimensional formulas. For example, we usually write our structures in eclipsed conformations. When we do this we do not mean to imply that eclipsed conformations are the most stable ones—they most certainly are not. We write eclipsed conformations because, as we shall see later, they make it easy for us to recognize planes of symmetry when they are present. We also write the longest carbon chain in a generally vertical orientation on the page; this makes the structures that we write directly comparable. As we do these things, however, *we must remember that molecules can rotate in their entirety* and that *at normal temperatures groups attached by single bonds can also rotate.* If rotations of the structure itself, or rotations of groups joined by single bonds make one structure superposable with another, then *the structures do not represent different compounds;* instead, they represent different orientations or different conformations of two molecules of the same compound.

Since structures **1** and **2** are not superposable, they represent different compounds. Since structures **1** and **2** differ *only* in the arrangement of their atoms in space, they represent stereoisomers. Structures **1** and **2** are also mirror reflections of each other, thus **1** and **2** represent enantiomers.

We will see in Chapter 19 that structures **1** and **2** are enantiomers of a sugar called *erythrose*. Compound **1** has the common name D-erythrose and compound **2** has the common name L-erythrose.

Structures **1** and **2** are not the only possible structures, however. We find that we can write a structure **3** (below) that is different from either **1** or **2**, and we can write a structure **4** that is a nonsuperposable mirror reflection of structure **3**.

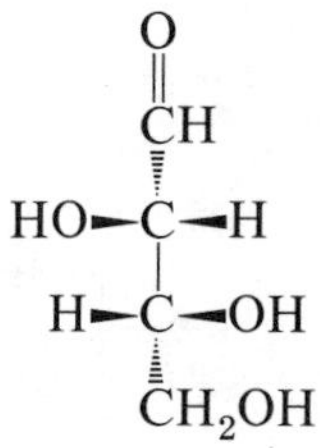

```
     O            O
     ‖            ‖
     CH           CH
     ⋮            ⋮
 HO—C—H        H—C—OH
     |            |
  H—C—OH      HO—C—H
     ⋮            ⋮
    CH2OH        CH2OH
     3            4
```

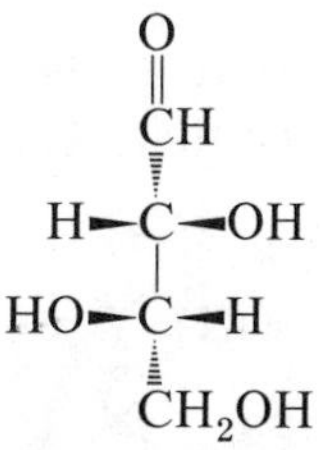

Structures **3** and **4** correspond to another pair of enantiomers. Structures **1**

to **4** are all different, so there are a total of four stereoisomers of this simple sugar. At this point you should convince yourself that there are no other stereoisomers by writing other structural formulas. You will find that rotation of the single bonds (or of the entire structure) of any other arrangement of the atoms will cause the structure to become superposable with one of the structures that we have written here. Better yet, using different-colored balls make molecular models as you work this out.

> Structures **3** and **4** represent the enantiomers of another simple sugar called *threose*. Sugar chemists call **3** D-threose and **4** L-threose.

The compounds represented by structures **1** to **4** are all optically active compounds. Any one of them, if placed separately in a polarimeter, would show optical activity.

The compounds represented by structures **1** and **2** are enantiomers. The compounds represented by structures **3** and **4** are also enantiomers. But, what is the isomeric relation between the compounds represented by **1** and **3**?

We can answer this question by observing that **1** and **3** are stereoisomers and that they are *not* mirror reflections of each other. They are, therefore, *diastereomers*. Diastereomers have different physical properties—different melting points, boiling points, different solubilities, and so forth. In this respect these diastereomers are just like diastereomeric alkenes such as *cis*- and *trans*-2-butene.

Problem 8.15

(a) What is the stereoisomeric relation between compounds **2** and **3**? (b) Between **1** and **4**? (c) Between **2** and **4**? (d) Make a table showing all of the stereoisomeric relations between compounds **1** to **4**. (e) Would compounds **1** and **2** have the same boiling point? (f) Would compounds **1** and **3**?

8.9A Meso Compounds

A structure with two chiral carbons will not always give rise to four stereoisomers. Sometimes there are only *three*. This happens because some molecules with chiral centers are, overall, *achiral*.

To understand this, let us write stereochemical formulas for the structure shown below.

```
 CH3
  |
*CHOH
  |
*CHOH
  |
 CH3
```

2,3-Butanediol

We begin in the same way as we did before. We write the formula for one stereoisomer and for its mirror reflection.

```
    CH3            CH3
     |              |
HO — C — H      H — C — OH
     |              |
 H — C — OH    HO — C — H
     |              |
    CH3            CH3
     A              B
```

Structures **A** and **B** are nonsuperposable and represent a pair of enantiomers.

When we write structure **C** (below) and its mirror reflection **D,** however, the situation is different. *The two structures are superposable.* This means that **C** and **D** do not represent a pair of enantiomers. Formulas **C** and **D** represent two different orientations of the same compound.

$$\begin{array}{c} CH_3 \\ H-C-OH \\ H-C-OH \\ CH_3 \\ \mathbf{C} \end{array} \qquad \begin{array}{c} CH_3 \\ HO-C-H \\ HO-C-H \\ CH_3 \\ \mathbf{D} \end{array}$$

This structure when turned in the plane end for end can be superposed on **C**

The molecule represented by structure **C** (or **D**) is not chiral even though it contains chiral carbons. *Achiral molecules that contain chiral centers* are called *meso compounds. Meso* compounds are optically inactive.

The ultimate test for molecular chirality is to construct a model (or write the structure) of the molecule and then test whether or not the model (or structure) is superposable on its mirror reflection. If it is, the molecule is achiral: if it *is not,* the molecule is chiral.

We have already carried out this test with structure **C** and found that it is achiral. We can also demonstrate that **C** is achiral in another way. Figure 8.16 shows that structure **C** *has a plane of symmetry.*

Problem 8.16

Which of the following would be optically active?

(a) **A** alone
(b) **B** alone
(c) **C** alone
(d) An equimolar mixture of **A** and **B**

Problem 8.17

Shown below are formulas for compounds **A, B,** or **C** written in noneclipsed conformations. In each instance tell which compound (**A, B,** or **C**) each formula represents.

(1) Upper carbon: HO, CH_3, H; lower carbon: H, H_3C, OH

(2) Upper carbon: HO, CH_3, H; lower carbon: H_3C, H, OH

(3) Upper carbon: H_3C, H, OH; lower carbon: HO, H, CH_3

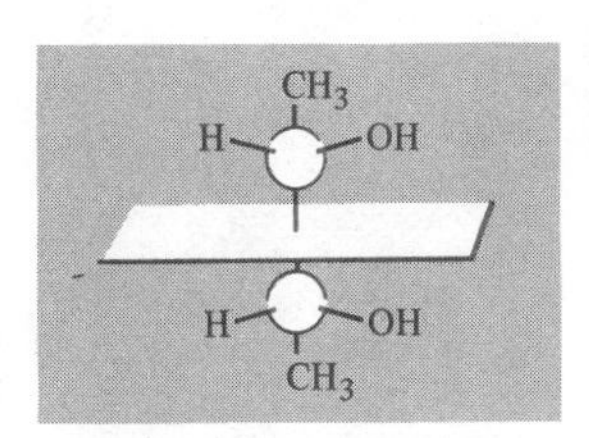

FIG. 8.16
The plane of symmetry of meso-*2,3-butanediol. This plane divides the molecule into two halves that are mirror reflections of each other.*

Problem 8.18

Write three-dimensional formulas for all of the stereoisomers of each of the following compounds.

(a) $CH_3CHBrCHBrCH_3$
(b) $CH_3CHBrCHClCH_3$
(c) $CH_3CHBrCHBrCH_2Br$
(d) $CH_2BrCHBrCHBrCH_2Br$
(e) $CH_3CHClCHClCHClCH_3$
(f) In answers to parts (a) to (e) label pairs of enantiomers and *meso* compounds.

Problem 8.19

Tartaric acid, HOOCCH(OH)CH(OH)COOH, was an important compound in the history of stereochemistry. Two naturally occurring forms of tartaric acid are optically inactive. One form has a melting point of 206°; the other a melting point of 140°. The inactive tartaric acid with a melting point of 206° can be separated into two optically active forms of tartaric acid with the same melting point (170°). One optically active tartaric acid has $[\alpha]_D^{25°}$ +12°, the other $[\alpha]_D^{25°}$ −12°. All attempts to separate the other inactive tartaric acid (melting point 140°) into optically active compounds fail. (a) Write the three-dimensional structure of the tartaric acid with melting point 140°. (b) What are possible structures for the optically active tartaric acids with melting points of 170°? (c) Can you be sure which tartaric acid in (b) has a positive rotation and which has a negative rotation? (d) What is the nature of the form of tartaric acid with melting point of 206°?

8.9B Stereoisomerism of Cyclic Compounds

Cyclic compounds also exist in stereoisomeric forms. For example, 1,2-cyclopentanediol has two chiral carbons and exists in three stereoisomeric forms, **5**, **6**, and **7**.

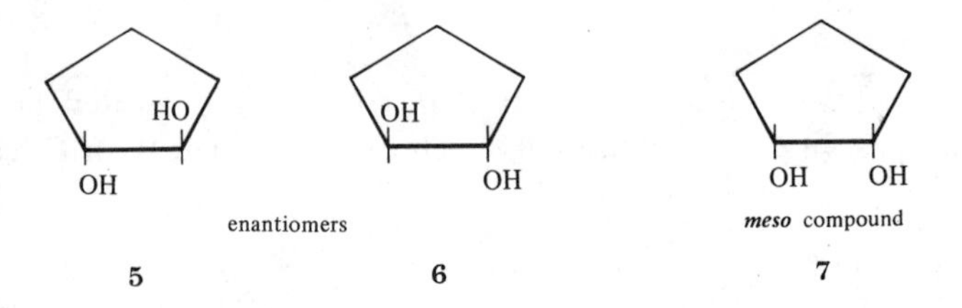

The *trans* compound exists as a pair of enantiomers **5** and **6.** *cis*-1,2-Cyclopentanediol is a *meso* compound. It has a plane of symmetry that is perpendicular to the plane of the ring.

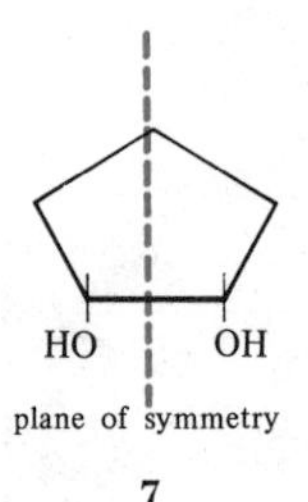

Problem 8.20

(a) Is the *trans*-1,2-cyclopentanediol, **5**, superposable on its mirror image (i.e., compound **6**)? (b) Is the *cis*-1,2-cyclopentanediol, **7**, superposable on its mirror image? (c) Is the *cis*-1,2-cyclopentanediol a chiral molecule? (d) Would *cis*-1,2-cyclopentanediol show optical activity? (e) What is the stereoisomeric relation between **5** and **7**? (f) Between **6** and **7**?

In Chapter 7 we saw that *cis*-1,2-cyclopentanediol could be synthesized by the syn hydroxylation of cyclopentene.

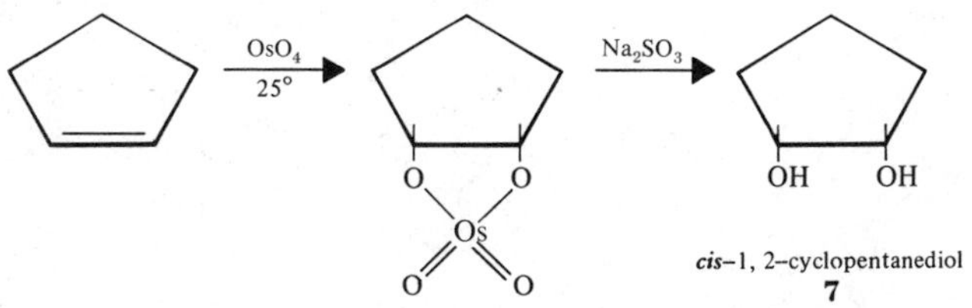

OsO_4, 25° → (cyclic osmate ester) → Na_2SO_3 →

cis-1, 2-cyclopentanediol
7

A single product is obtained from this reaction because the product is a *meso* compound. (Addition of OsO_4 from either above or below the plane of the ring produces the same compound.)

Cyclopentene undergoes anti hydroxylation when we use peroxybenzoic acid ($C_6H_5\overset{O}{\overset{\|}{C}}OOH$), and subsequently treat the product with aqueous acid. The product that is obtained from the overall reaction is a racemic modification of the *trans*-1,2-cyclopentanediols.

$$\text{Cyclopentene} \xrightarrow[\text{(2) } H^+,\ H_2O]{\text{(1) } C_6H_5\overset{O}{\overset{\|}{C}}OOH,\ CHCl_3,\ 25°} (\pm)\text{-}trans\text{-1,2-cyclopentanediol}$$

We can see why a racemic modification is obtained if we examine the two steps in detail

Cyclopentene reacts with peroxybenzoic acid and forms an epoxide that is a *meso* compound.

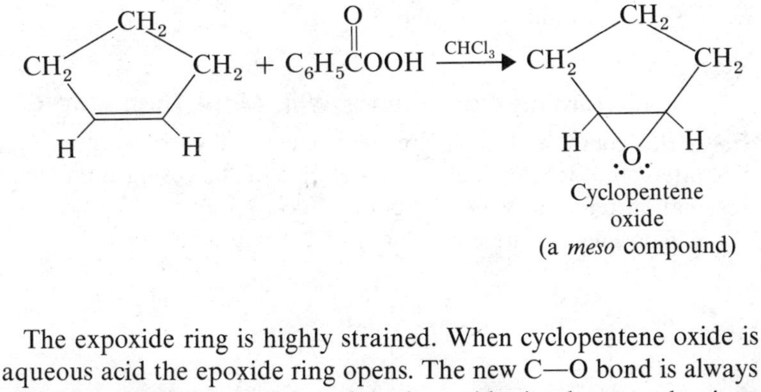

$$\text{Cyclopentene} + C_6H_5\overset{O}{\overset{\|}{C}}OOH \xrightarrow{CHCl_3} \text{Cyclopentene oxide}$$

Cyclopentene
oxide
(a *meso* compound)

The expoxide ring is highly strained. When cyclopentene oxide is treated with aqueous acid the epoxide ring opens. The new C—O bond is always formed from the side opposite that of the original epoxide ring because the ring-opening reaction (below) is an S_N2 reaction with water acting as the nucleophile and the protonated epoxide acting as the substrate.

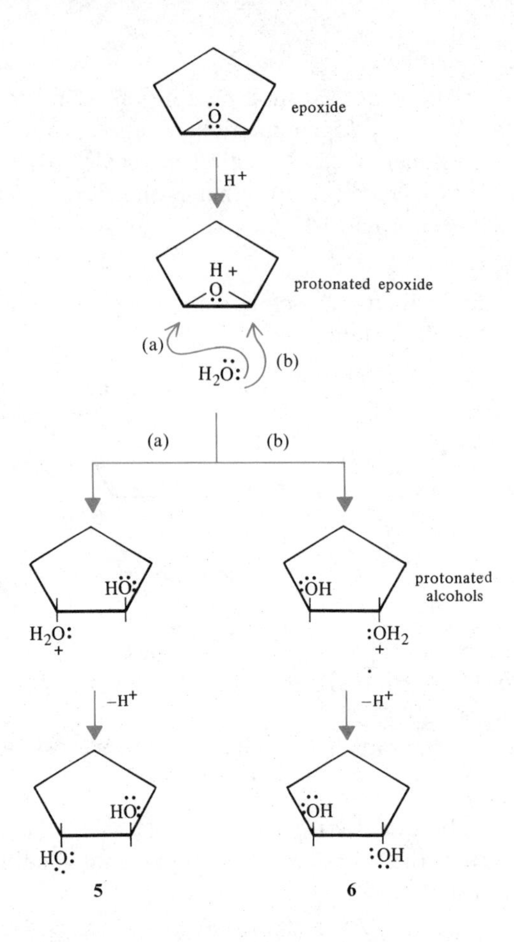

The rates of ring opening by paths (a) and (b) are equal, and the enantiomers **5** and **6** are obtained in equal amounts. Since the *meso* epoxide is achiral, it yields an optically inactive racemic modification of the *trans*-1,2-cyclopentanediols when it reacts with the achiral reagent, water (see Section 8.8).

Problem 8.21

In Section 7.10 we gave a mechanism for the addition of bromine to cyclopentene. (a) What products would you expect from this reaction? (b) In what form would they be obtained? (c) If a syn addition of bromine to cyclopentene were to occur, what product would be obtained?

8.9C Naming Compounds With More Than One Chiral Carbon

If a compound has more than one chiral carbon we analyze each center separately and decide whether it is *R* or *S*. Then, using numbers, we tell which designation refers to which carbon.

Consider the stereoisomer **A** of 2,3-butanediol.

$^1CH_3$; HO–C2–H; H–C3–OH; $^4CH_3$ or $^1CH_3$; HO–$_2$C–H; H–$_3$C–OH; $^4CH_3$

A

(2, 3 – butanediol)

When this formula is rotated so that the group of lowest preference attached to carbon-2 is directed away from the viewer it resembles the following.

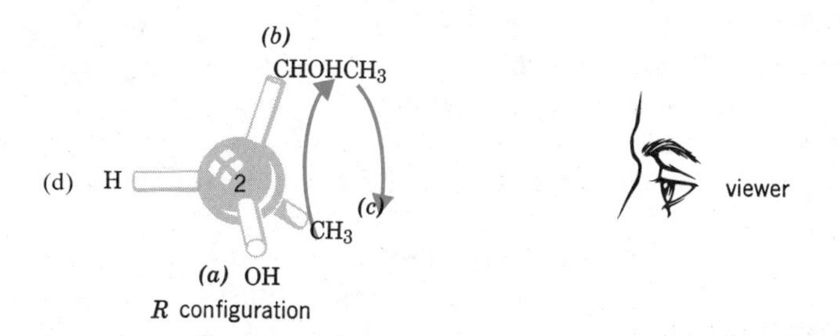

The order of progression from the group of highest preference to that of next highest preference (from —OH, to —$CHOHCH_3$, to —CH_3) is clockwise. So carbon-2 has the ***R*** configuration.

When we repeat this procedure with carbon-3 we find that carbon-3 also has the ***R*** configuration. Compound **A,** therefore, is (2*R*, 3*R*)-2,3-butanediol.

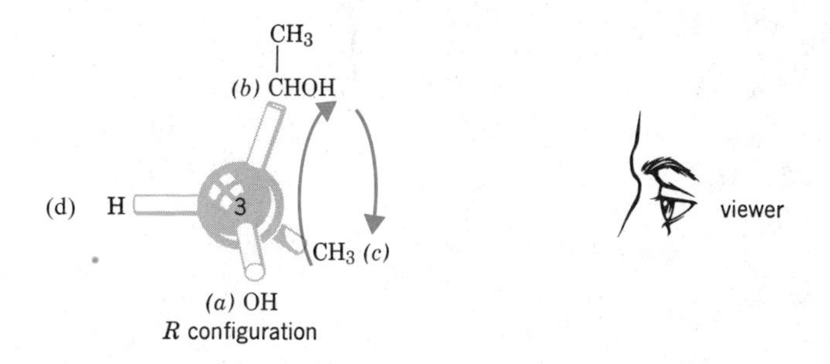

Problem 8.22

Give names that include *R* and *S* designations for compounds **B** and **C** pages 304 and 305.

Problem 8.23

Give names that include *R* and *S* designations for your answers to Problem 8.18 (a) and (b).

8.10 STEREOSPECIFIC REACTIONS

Reactions in which stereoisomerically different reactants give stereoisomerically different products are called *stereospecific reactions.*

The distinction between the terms stereoselective and stereospecific is a subtle one. A stereoselective reaction is one that yields a predominance of one stereoisomer *regardless* of the stereochemistry of the reactant. On p. 301 we described the enzymatic reduction of pyruvic acid as being stereoselective. This is true because molecules of pyruvic acid are achiral and do not exist in different stereoisomeric forms. When they associate with the chiral molecules of the enzyme they produce a predominance of one diastereomeric form of the pyruvic acid-enzyme complex. This complex then goes on to yield a single enantiomeric form of lactic acid, that is, *S*-(+)-lactic acid.

On the other hand, when we say that a reaction is *stereospecific* we mean that *a particular stereoisomeric form of the starting material* reacts in such a way that it gives a *specific stereoisomeric form of the product*. It does this because the reaction mechanism requires that the configurations of the atoms involved change in a characteristic way.

We should notice that the terms stereoselective and stereospecific are not mutually exclusive. All reactions that are stereospecific are, of necessity, stereoselective but not all stereoselective reactions are stereospecific. The enzymatic reduction of pyruvic acid is stereoselective, but it is not stereospecific. It is not stereospecific because the reactant cannot exist in stereoisomeric forms. Other reactions may be stereoselective but not be stereospecific, because two reactants of different stereochemistry may yield one predominant stereoisomeric product.

The reaction of alkenes and peroxy acids takes place in a stereospecific way: *cis*-2-butene, for example, yields only *cis*-2,3-dimethyloxirane, and *trans*-2-butene yields only the *trans*-2,3-dimethyloxiranes.

(*a*) $CH_3CH{=}CHCH_3$ + $RCOOH$ (with C=O) ⟶ CH_3, CH_3, H, H on oxirane ring (C-2, C-3, O-1)

cis-2-Butene

cis-2,3-Dimethyloxirane
(a *meso* compound)

(*b*) $CH_3CH{=}CHCH_3$ + $RCOOH$ (with C=O) ⟶ oxirane + oxirane

trans-2-Butene

Enantiomeric *trans*-2,3-dimethyloxiranes

The reactants *cis*-2-butene and *trans*-2-butene are stereoisomers; they are *diastereomers*. The product of reaction (*a*), *cis*-2,3-dimethyloxirane, is a *meso* compound, and it is a stereoisomer of either of the products of reaction (*b*) (the enantiomeric *trans*-2,3-dimethyloxiranes). Thus, by definition, both reactions are stereospecific. One stereoisomeric form of the reactant (e.g., *cis*-2-butene) gives one product (the *meso* compound) while the other stereoisomeric form of the reactant (*trans*-2-butene) gives a stereoisomerically different product (the enantiomers).

We can better understand the results of these two reactions if we examine their mechanisms. *cis*-2-Butene (Fig. 8.17) reacts with the peroxy acid through a syn addition. A syn addition of oxygen to either face of *cis*-2-butene produces the same compound: the *meso* compound, *cis*-2,3-dimethyloxirane.

On the other hand, when *trans*-2-butene undergoes a syn addition of oxygen at one face it produces one enantiomeric form of *trans*-2,3-dimethyloxirane; when it undergoes syn addition at the other face it produces the other enantiomeric form of *trans*-2,3-dimethyloxirane. These reactions are shown in Fig. 8.18.

The reactions that lead to the enantiomeric *trans*-2,3-dimethyloxiranes, moreover, have identical energies of activation. As a result, they occur at identical rates and produce the enantiomers in equal amounts, as a racemic modification.

When either *cis*-2,3-dimethyloxirane or the racemic modification of *trans*-2,3-dimethyloxirane is subjected to acid-catalyzed hydrolysis, stereospecific reactions also take place.

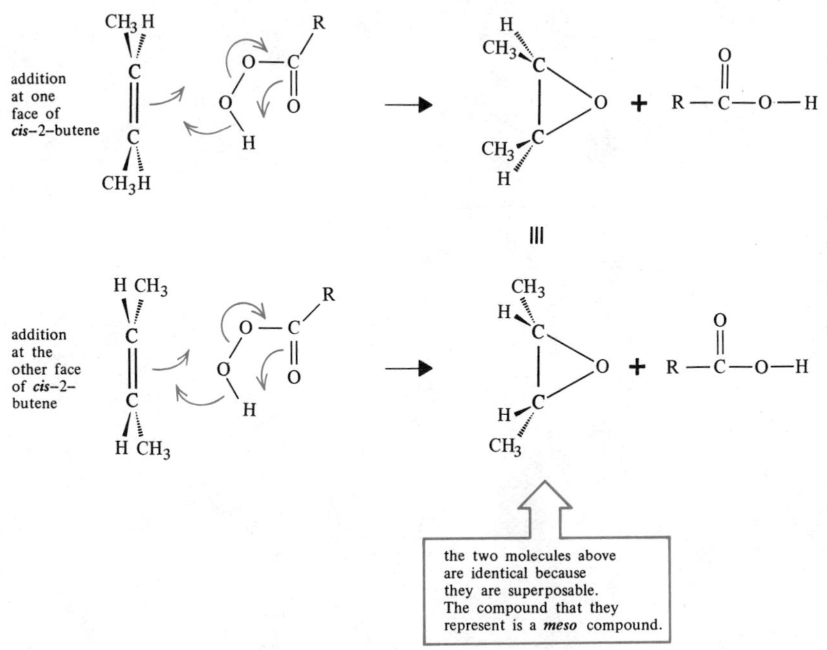

addition at one face of *cis*–2–butene

addition at the other face of *cis*–2–butene

the two molecules above are identical because they are superposable. The compound that they represent is a ***meso*** compound.

FIG. 8.17
Epoxidation of cis-*2-butene yields* cis-*2,3-dimethyloxirane (a* meso *compound).*

FIG. 8.18
Epoxidation of trans-*2-butene yields enantiomeric* trans-*2,3-dimethyloxiranes.*

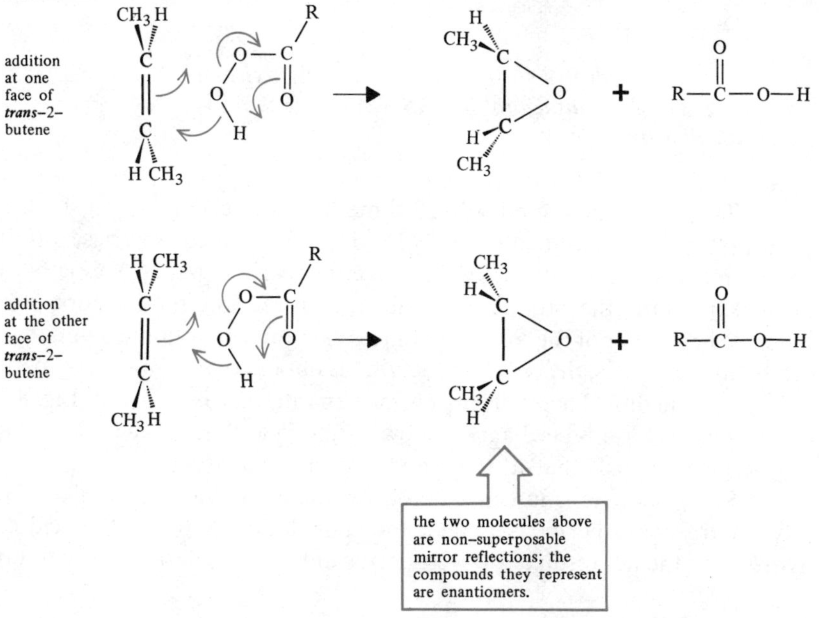

addition at one face of ***trans***–2–butene

addition at the other face of ***trans***–2–butene

the two molecules above are non–superposable mirror reflections; the compounds they represent are enantiomers.

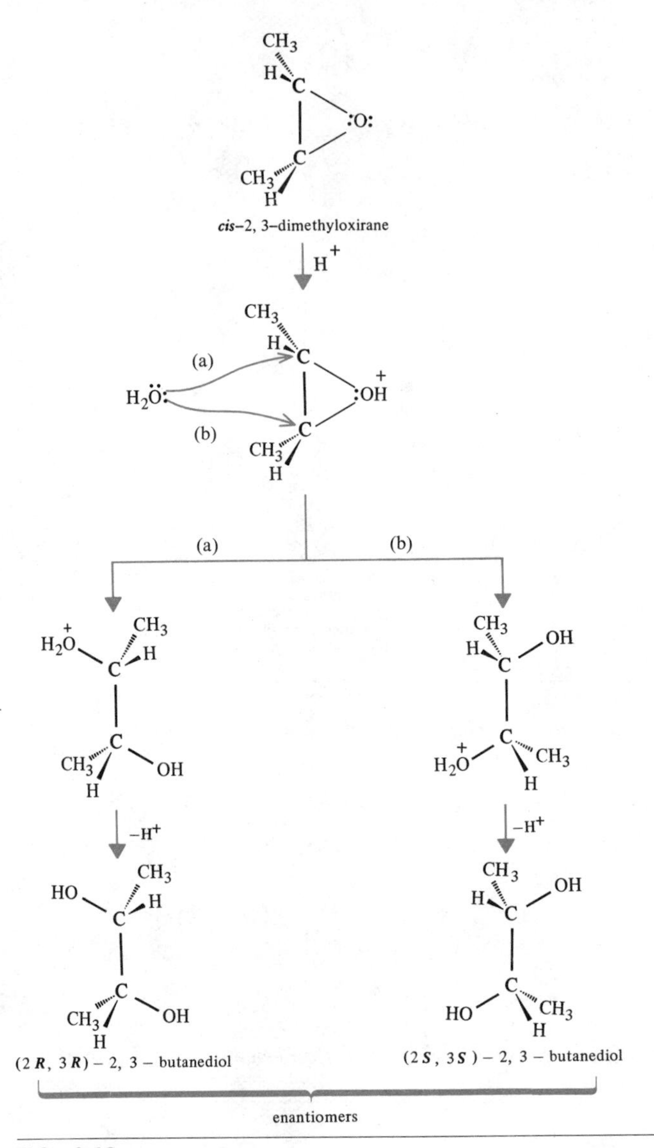

FIG. 8.19
Acid-catalyzed hydrolysis of cis-*2-3-dimethyloxirane yields* (2R, 3R)-*2,3-butanediol by path* (a) *and* (2S, 3S)-*2,3-butanediol by path* (b).

The *meso* compound, *cis*-2,3-dimethyloxirane (Fig. 8.19), hydrolyzes to yield (2*R*, 3*R*)-2,3-butanediol and (2*S*, 3*S*)-2,3-butanediol. These products are enantiomers. Since the attack by water at either carbon [path (*a*) or path (*b*)] occurs at the same rate, the product is obtained as a racemic modification.

When either of the *trans*-2,3-dimethyloxirane enantiomers undergoes acid-catalyzed hydrolysis, the only product that is obtained is the *meso* compound, (2*R*, 3*S*)-2,3-butanediol. The hydrolysis of one enantiomer is shown in Fig. 8.20. (You might construct a similar diagram showing the hydrolysis of the other enantiomer to convince yourself that it, too, yields the same product.)

Since both steps in this method for the conversion of an alkene to a diol (glycol) are stereospecific (i.e., both the epoxidation step and the acid-catalyzed hydrolysis), the net result is a stereospecific anti hydroxylation of the double bond (Fig. 8.21).

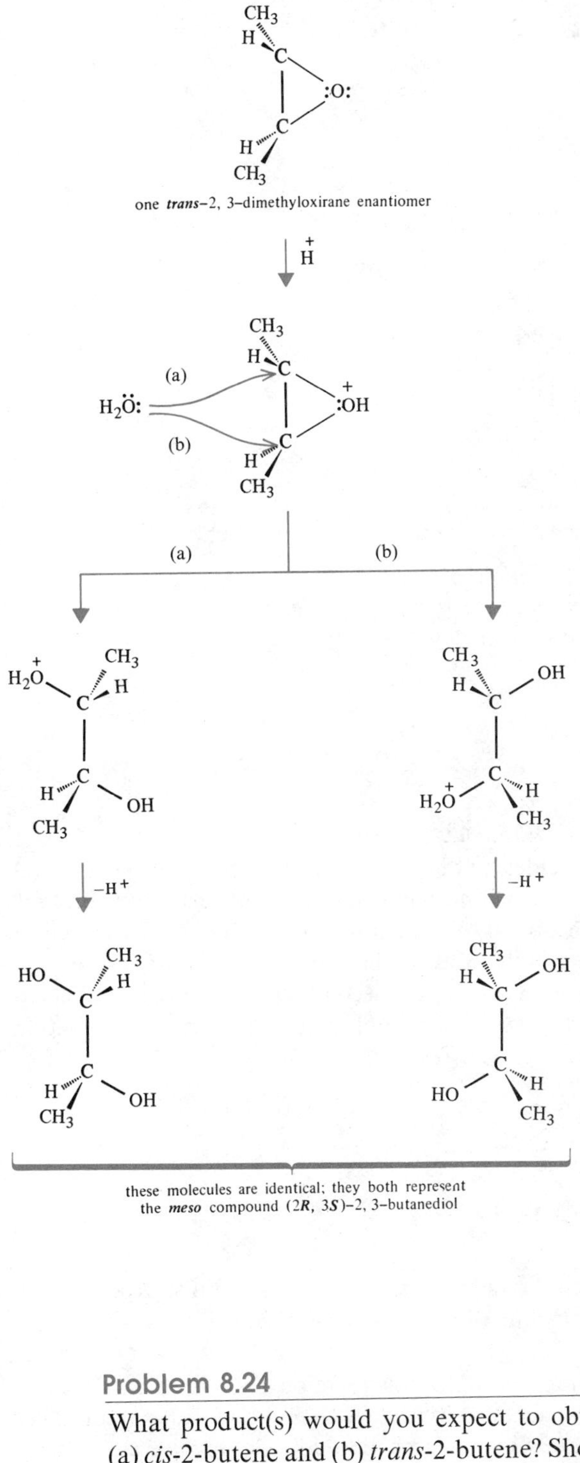

one *trans*-2, 3-dimethyloxirane enantiomer

these molecules are identical; they both represent the *meso* compound (2*R*, 3*S*)-2, 3-butanediol

FIG. 8.20
The acid-catalyzed hydrolysis of one trans-*2,3-dimethyloxirane enantiomer produces the* meso *compound, (2*R*, 3*S*)-2,3-butanediol, by path* (a) *or* (b). *Hydrolysis of the other enantiomer (or the racemic modification) would yield the same product. (You should convince yourself that the two structures given for the products above do represent the same compound.)*

Problem 8.24

What product(s) would you expect to obtain from the addition of bromine to (a) *cis*-2-butene and (b) *trans*-2-butene? Show all steps in each addition reaction in ways that clearly illustrate their stereochemistry.

Problem 8.25

What products would you expect to obtain from the syn hydroxylation of (a) *cis*-2-butene and (b) *trans*-2-butene using either $KMnO_4$ or OsO_4 and Na_2SO_3?

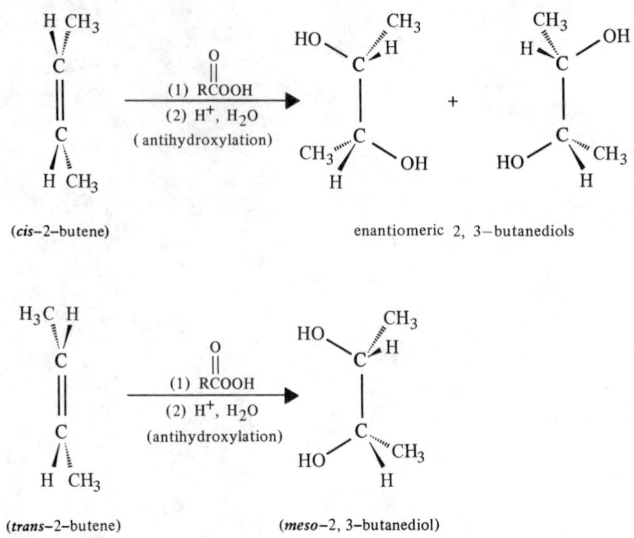

(1) RCOOH
(2) H^+, H_2O
(antihydroxylation)

(*cis*-2-butene) → enantiomeric 2, 3-butanediols

(1) RCOOH
(2) H^+, H_2O
(antihydroxylation)

(*trans*-2-butene) → (*meso*-2, 3-butanediol)

FIG. 8.21
The overall result of epoxidation followed by acid-catalyzed hydrolysis is a stereospecific anti hydroxylation of the double bond. Cis-*2-butene yields the enantiomeric 2,3-butanediols;* trans-*2-butene yields the* meso *compound.*

8.11 REACTIONS OF CHIRAL MOLECULES

The reactions that chiral molecules undergo can be divided into two general categories: (1) those reactions in which no bonds to the chiral carbon are broken, and (2) those reactions in which a bond to the chiral carbon *is* broken. Reactions of the first type (Section 8.12) are simpler and can be used *to relate configurations* of chiral molecules (Section 8.13). Reactions of the second type are more complicated but as we will see in Section 8.14 if we understand the mechanism of the reaction we can often predict the stereochemical result.

8.12 REACTIONS IN WHICH NO BONDS TO THE CHIRAL CARBON ARE BROKEN

Reactions in which no bonds to the chiral carbon are broken must proceed with *retention of configuration*. We can illustrate this by the reaction of *R*-(−)-2-butanol with acetic anhydride shown in Fig. 8.22.

The reaction of anhydrides with alcohols (Section 17.7) is an easy way to synthesize esters. The reaction of acetic anhydride with *R*-(−)-2-butanol produces the ester, *R*-(−)-*sec*-butyl acetate. Since the reaction does not involve cleavage of any of the bonds to the chiral carbon, we can be certain that the general spatial arrangement of the groups in the product will be the same as that of the reactant. The configuration of the reactant is said to be *retained* in the product; the reaction is said to proceed with *retention of configuration.*

In the example that we just cited, the optically active reactant and product both rotate plane-polarized light in the same direction. This is not always the case as the following example illustrates.

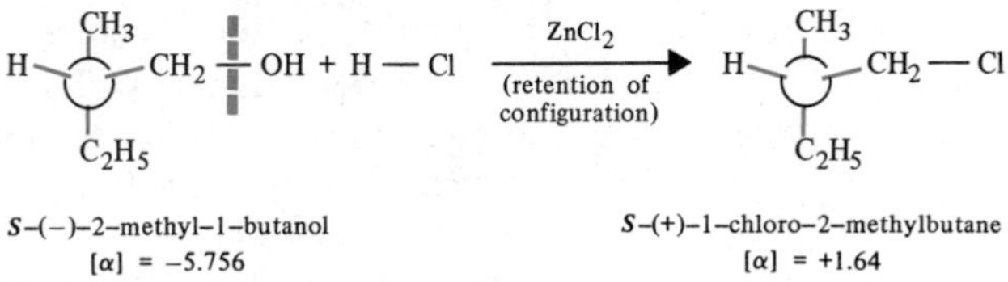

$$\text{S-(−)-2-methyl-1-butanol} + H—Cl \xrightarrow[\text{(retention of configuration)}]{ZnCl_2} \text{S-(+)-1-chloro-2-methylbutane}$$

S-(−)-2-methyl-1-butanol [α] = −5.756

S-(+)-1-chloro-2-methylbutane [α] = +1.64

And in some instances the *R* and *S* designation may change even though the reaction proceeds with retention of configuration.

$$\text{R-1-bromo-2-butanol} + Zn \xrightarrow[\text{(retention of configuration)}]{H^+} \text{S-2-butanol} + ZnBr_2$$

R-1-bromo-2-butanol

S-2-butanol

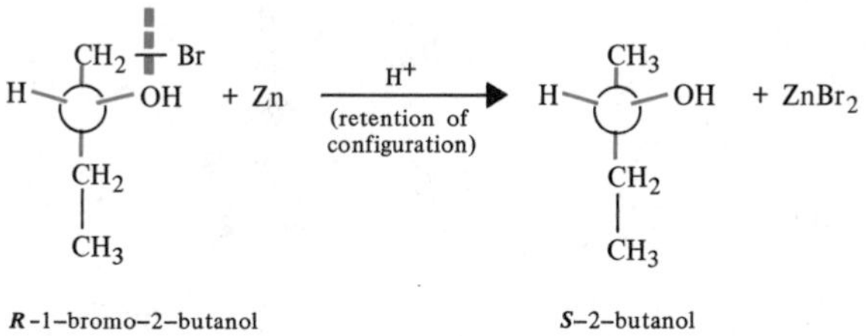

In this example the *R-S* designation changes because the $—CH_2Br$ group of the reactant ($—CH_2Br$ has a higher preference than $—CH_2CH_3$) changes to a $—CH_3$ group in the product ($—CH_3$ has a lower preference than $—CH_2CH_3$).

8.13 RELATIVE AND ABSOLUTE CONFIGURATIONS

Reactions in which no bonds to the chiral carbon are broken are useful in relating configurations of chiral molecules. That is, they allow us to demonstrate that certain compounds have the same *relative configuration.* In each of the three examples that we have just cited, the products of the reactions have the same *relative configurations* as the reactants. (They also, we now know, have the same *absolute configuration.*)

FIG. 8.22
The reaction of R-(−)-2-*butanol with acetic anhydride. The bonds that are broken are shown with dashed lines. A bond between one of the carbonyl groups and the oxygen of acetic anhydride is broken. The only bond that is broken in* R-(−)-2-*butanol is the one between the hydrogen and oxygen of the hydroxyl group.* No bonds to the chiral carbon (*shown in color*) are broken.

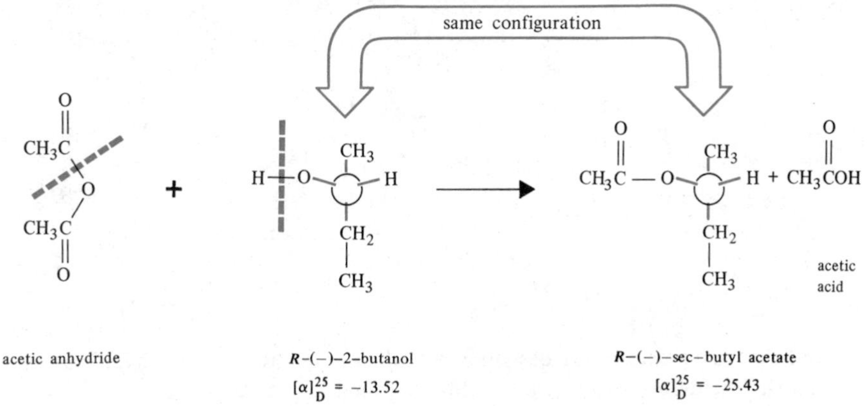

same configuration

$$(CH_3CO)_2O + \text{R-(−)-2-butanol} \longrightarrow CH_3C(=O)—O—CH(CH_3)(CH_2CH_3) + CH_3COOH$$

acetic anhydride

R-(−)-2-butanol $[\alpha]_D^{25} = -13.52$

R-(−)-sec-butyl acetate $[\alpha]_D^{25} = -25.43$

acetic acid

Before 1951, only relative configurations of chiral molecules were known. No one, prior to that time, had been able to demonstrate with certainty what the actual spatial arrangement of groups was in any chiral molecule. To say this another way, no one had been able to determine the *absolute configuration* of an optically active compound.

Configurations of chiral molecules were related to each other *through reactions of known stereochemistry*. Attempts were also made to relate all configurations back to a single compound that had been chosen arbitrarily to be the standard. This standard compound was glyceraldehyde.

$$\begin{array}{c} O \\ \| \\ CH \\ | \\ {}^{*}CHOH \\ | \\ CH_2OH \end{array}$$

Glyceraldehyde

Glyceraldehyde molecules have one chiral carbon; therefore, glyceraldehyde exists as a pair of enantiomers.

$$\begin{array}{c} O \\ \| \\ C-H \\ H \text{—}\bigcirc\text{—} OH \\ | \\ CH_2OH \end{array} \qquad \text{and} \qquad \begin{array}{c} O \\ \| \\ C-H \\ HO \text{—}\bigcirc\text{—} H \\ | \\ CH_2OH \end{array}$$

R-glyceraldehyde* *S*-glyceraldehyde

One glyceraldehyde enantiomer is dextrorotatory (+) and the other, of course, is levorotatory (−). Before 1951, no one could be sure, however, which configuration belonged to which enantiomer. Chemists decided arbitrarily to assign the *R* configuration to the (+)-enantiomer. Then configurations of other molecules were related to one glyceraldehyde enantiomer or the other through reactions of known stereochemistry.

For example, the configuration of (−)-lactic acid can be related to (+)-glyceraldehyde through the following sequence of reactions.

$$\underset{(+)\text{-Glyceraldehyde}}{\begin{array}{c} O \\ \| \\ C-H \\ | \\ H-C-OH \\ | \\ CH_2OH \end{array}} \xrightarrow[\text{(oxidation)}]{HgO} \underset{(-)\text{-Glyceric acid}}{\begin{array}{c} O \\ \| \\ C-OH \\ | \\ H-C-OH \\ | \\ CH_2OH \end{array}} \xleftarrow{HNO_2} \underset{(+)\text{-Isoserine}}{\begin{array}{c} O \\ \| \\ C-OH \\ | \\ H-C-OH \\ | \\ CH_2-NH_2 \end{array}} \xrightarrow[HBr]{HNO_2}$$

$$\underset{(-)\text{-3-Bromo-2-hydroxy-propanoic acid}}{\begin{array}{c} O \\ \| \\ C-OH \\ | \\ H-C-OH \\ | \\ CH_2-Br \end{array}} \xrightarrow{Zn,\ H^+} \underset{(-)\text{-Lactic acid}}{\begin{array}{c} O \\ \| \\ C-OH \\ | \\ H-C-OH \\ | \\ CH_3 \end{array}}$$

* In the older system for designating configurations R-glyceraldehyde was called D-glyceraldehyde and S-glyceraldehyde was called L-glyceraldehyde.

The stereochemistry of all of these reactions is known. Because bonds to the chiral carbon are not broken in any of them, they all proceed with retention of configuration. If the assumption is made that the configuration of (+)-glyceraldehyde is

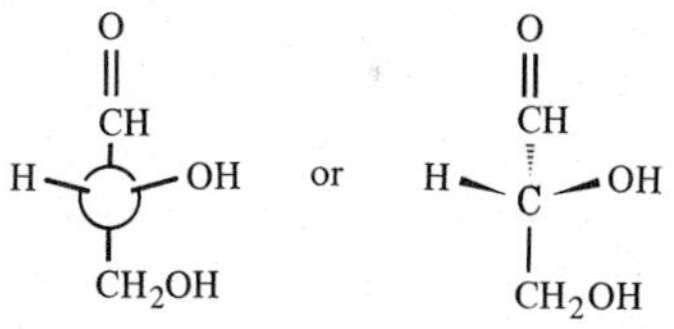

R-(+)-glyceraldehyde

then the configuration of (−)-lactic acid is

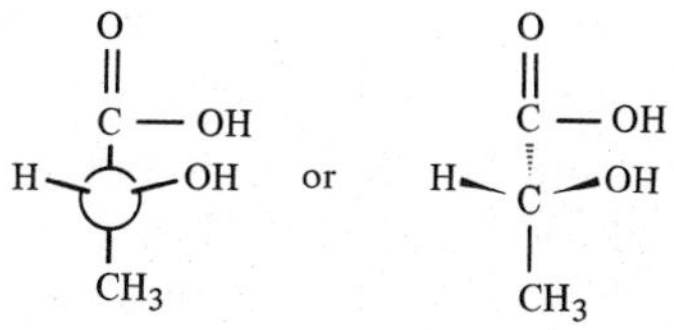

R-(−)-lactic acid

Problem 8.26

(a) Write three-dimensional structures for the relative configurations of (−) glyceric acid and (−)-3-bromo-2-hydroxypropanoic acid. (b) What is the *R*-*S* designation of (−) glyceric acid? (c) Of (+)-Isoserine? (d) Of (−)-3-bromo-2-hydroxypropanoic acid? (e) Of (−)-lactic acid?

The configuration of (−)-glyceraldehyde was also related through reactions of known stereochemistry to (+)-tartaric acid.

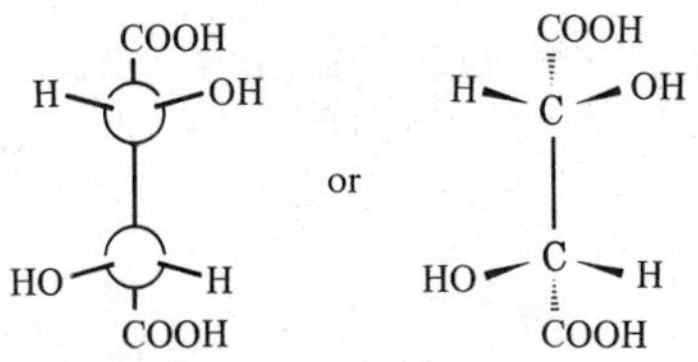

(+)-tartaric acid

In 1951, J. M. Bijvoet, the director of the van't Hoff Laboratory of the University of Utrecht in Holland, using a special technique of X-ray diffraction, was able to show conclusively that (+)-tartaric acid had the absolute configuration given above. This meant that the original guess about the configurations of (+)- and (−)-glyceraldehyde was also correct. It also meant that the configurations of all of the compounds that had been related to one glyceraldehyde enantiomer or the other were now absolute configurations.

Problem 8.27

Lactic acid with a specific rotation of +3.82° can be converted to the ester, methyl lactate, $CH_3CHOHCO_2CH_3$. The methyl lactate that is produced has a specific rotation of −8.25°. What is the absolute configuration of (−) methyl lactate?

8.14 REACTIONS IN WHICH A BOND TO A CHIRAL CARBON IS BROKEN

Reactions that involve the breaking of a bond to a chiral carbon can have three possible results:

1. Inversion of configuration.
2. Racemization.
3. Retention of configuration.

The important factor that determines which result we obtain from a given reaction is the *reaction mechanism*. To see how this is true, let us examine reactions of each type.

8.15 REACTIONS THAT INVOLVE INVERSION OF CONFIGURATION

We saw in Section 5.5 how S_N2 reactions proceed with inversion of configuration of the carbon that undergoes nucleophilic attack. The examples that we studied there involved cyclic compounds where inversion of configuration at one carbon caused the formation of a *cis-* product from a *trans-* reactant or vice versa. Now that we have a greater knowledge of stereochemistry we can examine reactions where nucleophilic attack takes place at a chiral carbon. Here, too, we find that *S_N2 reactions always lead to inversion of configuration.*

A compound that contains one chiral carbon and, therefore, exists as a pair of enantiomers is 2-bromooctane. These enantiomers have been obtained separately and are known to have the configurations and rotations shown below.

C_6H_{13} / H — Br / CH_3 $\qquad$ C_6H_{13} / Br — H / CH_3

R-(−)-2-bromooctane
[α] − 34.6°

S-(+)-2-bromooctane
[α] +34.6°

The alcohol 2-octanol is also chiral. The configurations and rotations of the 2-octanol enantiomers have also been determined:

C_6H_{13} / H — OH / CH_3 $\qquad$ C_6H_{13} / HO — H / CH_3

R-(−)-2-octanol
[α] −9.9°

S-(+)-2-octanol
[α] +9.9°

When *R*-(—)-2-bromooctane reacts with sodium hydroxide, the product that is obtained from the reaction is (*S*)-(+)-2-octanol. The reaction (below) is S_N2 and it takes place with *complete inversion of configuration.*

C_6H_{13} / H — Br / CH_3 + NaOH $\xrightarrow[S_N2]{}$ C_6H_{13} / HO — H / CH_3

R-(−)-2-bromooctane
[α] −34.6°
optical purity = 100%

S-(+)-2-octanol
[α] +9.9°
optical purity = 100%

Problem 8.28

Reactions that involve breaking bonds to chiral carbons can also be used to relate configurations of molecules when the mechanism of the reaction *and its stereochemistry* are known. Illustrate how this is true by explaining how you would use the configuration and rotations of the 2-butanol enantiomers given on page 296 to discover the configurations of the enantiomers of 2-bromobutane.

8.16 REACTIONS THAT INVOLVE RACEMIZATION

A reaction that transforms an optically active compound whose molecules contain a single chiral carbon into a racemic modification is said to proceed with *racemization*. If the original compound loses all of its optical activity in the course of the reaction, chemists describe the reaction as having taken place with *complete* racemization. If the original compound loses only part of its optical activity, as would be the case if an enantiomer were only partially converted to a racemic modification, then chemists describe this as proceeding with *partial* racemization.

Racemization will take place whenever either of events occur:

1. The reaction causes the chiral molecules to be converted to an achiral intermediate, or,
2. The reaction results in the inversion of configuration of some of the molecules.

Examples of the first type are S_N1 reactions in which the leaving group departs from a chiral carbon. These reactions almost always result in extensive and often complete racemization. For example, heating optically active *S*-3-bromo-3-methylhexane with aqueous acetone results in the formation of a racemic modification of 3-methyl-3-hexanol.

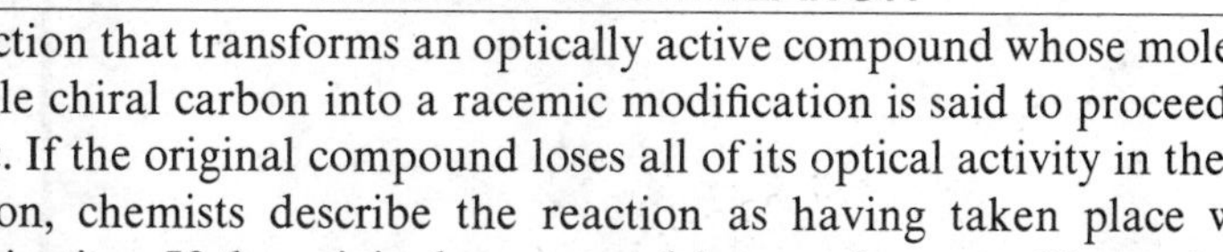

$$(CH_3CH_2CH_2)(H_3C)(CH_3CH_2)C{-}Br \xrightarrow[\text{acetone}]{H_2O} (CH_3CH_2CH_2)(H_3C)(CH_3CH_2)C{-}OH + HO{-}C(CH_2CH_2CH_3)(CH_3)(CH_2CH_3)$$

S-3-bromo-3-methylhexane (optically active) → *S*-3-methyl-3-hexanol + *R*-3-methyl-3-hexanol (optically inactive, a racemic modification)

The reason: the S_N1 reaction proceeds through the formation of an intermediate carbocation (Fig. 8.23) and the carbocation, because of its trigonal-planar configuration, *is achiral*. It reacts with water with equal ease from either side to form the enantiomers of 3-methyl-3-hexanol in equal amounts.

An example of racemization by the inversion of the configuration of *some* of the molecules is the racemization of *S*-2-bromobutane that takes place when it is heated with bromide ions in acetone:

$$S\text{-2-Bromobutane} \xrightarrow[\text{acetone}]{Br^-} (\pm)\text{-2-Bromobutane}$$

Here the reaction is S_N2. When a bromide ion displaces a bromide ion from *S*-2-bromobutane it produces a molecule of *R*-2-bromobutane:

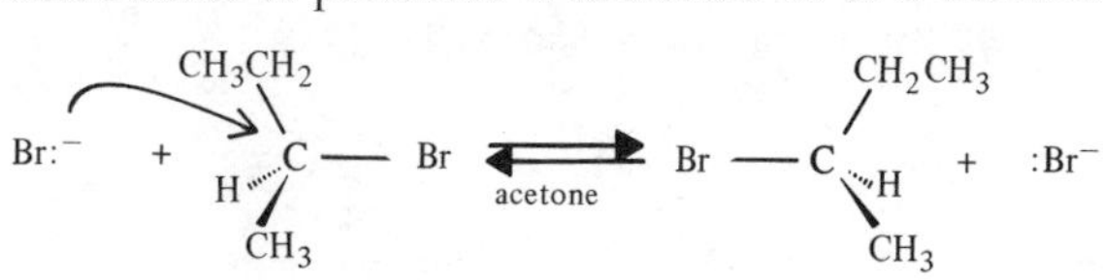

$$Br:^- + (CH_3CH_2)(H)(CH_3)C{-}Br \underset{\text{acetone}}{\rightleftharpoons} Br{-}C(CH_2CH_3)(H)(CH_3) + :Br^-$$

S-2-bromobutane ⇌ *R*-2-bromobutane

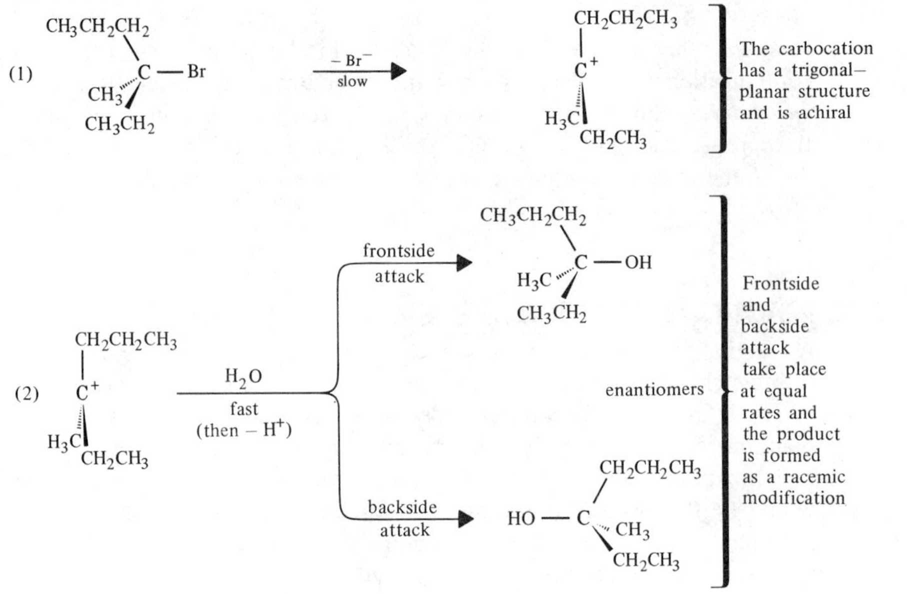

(1) $CH_3CH_2CH_2$, CH_3, CH_3CH_2 C—Br $\xrightarrow[\text{slow}]{-Br^-}$ C^+ ($CH_2CH_2CH_3$, H_3C, CH_2CH_3)

The carbocation has a trigonal-planar structure and is achiral

(2) C^+ ($CH_2CH_2CH_3$, H_3C, CH_2CH_3) $\xrightarrow[\text{fast (then } -H^+)]{H_2O}$

frontside attack → $CH_3CH_2CH_2$, H_3C, CH_3CH_2 C—OH

backside attack → HO—C ($CH_2CH_2CH_3$, CH_3, CH_2CH_3)

enantiomers

Frontside and backside attack take place at equal rates and the product is formed as a racemic modification

FIG. 8.23
The S_N1 reaction of 3-bromo-3-methylhexane proceeds with racemization because the intermediate carbocation is achiral.

Of course, the *R*-2-bromobutane can also react with bromide ion to form the *S* compound again. But, initially the concentration of *R*-2-bromobutane will be very low and, consequently the rate of formation of *S*-2-bromobutane (from *R*-2-bromobutane) will be much lower than the rate of formation of the *R* enantiomer (from *S*-2-bromobutane). Eventually the two enantiomers will be present in equal amounts—as a racemic modification—and the two reactions will take place at equal rates.

Problem 8.29

When *R*-2-butanol (p. 296) is heated with an aqueous solution of a strong acid the solution gradually loses its optical activity. If heating is continued long enough the solution is found to contain (±)-2-butanol. Provide a mechanistic explanation for what happens.

8.17 REACTIONS THAT TAKE PLACE WITH RETENTION OF CONFIGURATION

When a bond to the chiral carbon breaks in a reaction, the usual result is inversion of configuration or racemization. Retention of configuration is rare, but it can take place. For example, when a chiral alcohol is treated with $SOCl_2$ (thionyl chloride) the product of the reaction is a chiral alkyl chloride with the same configuration as the alcohol. The other products are SO_2 and HCl.

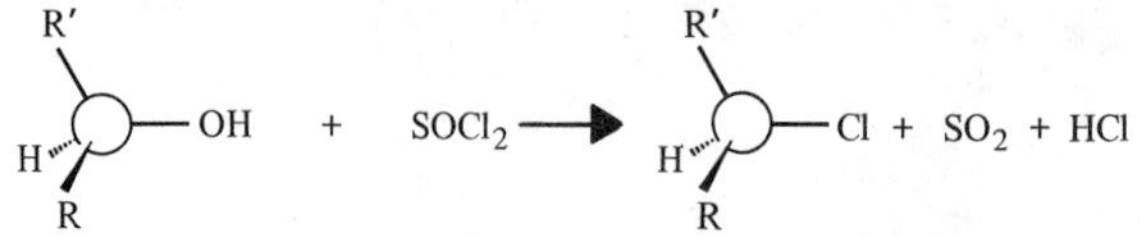

$$R'R(H)C\text{—OH} + SOCl_2 \longrightarrow R'R(H)C\text{—Cl} + SO_2 + HCl$$

The mechanism for the reaction has three steps. In the first, the alcohol reacts with $SOCl_2$ to form an alkyl chlorosulfite; no bond to the chiral carbon breaks:

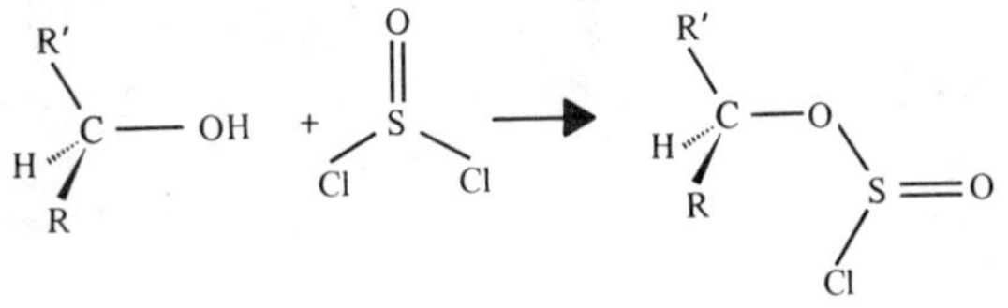

alkyl chlorosulfite

In the second step, the alkyl chlorosulfite loses sulfur dioxide and forms an *intimate ion pair:*

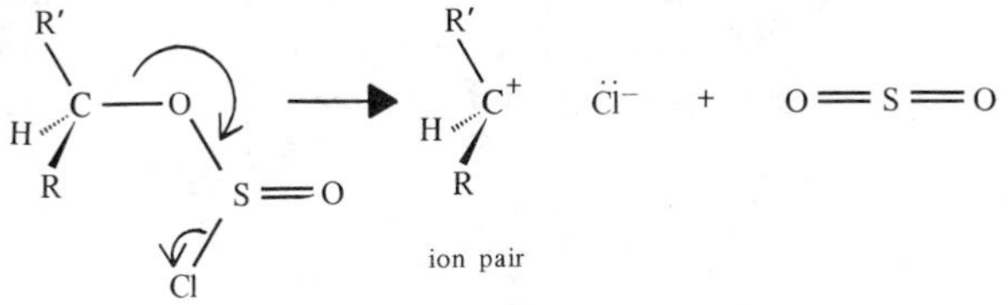

ion pair

An intimate ion pair is a short-lived intermediate in which a carbocation and an anion are in close proximity and are not separated by molecules that might solvate them. Before the carbocation of the intimate ion pair has time to invert its configuration it combines with the chloride ion to form the alkyl chloride:

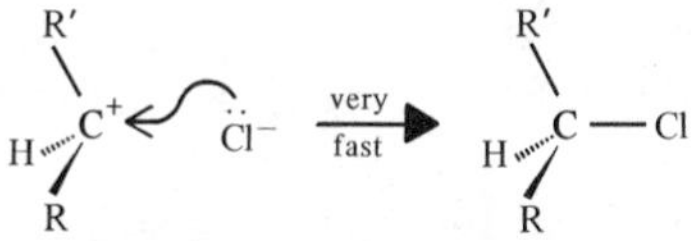

very fast

Through all these steps, the configuration of the chiral carbon is retained.

8.18 SEPARATION OF ENANTIOMERS: RESOLUTION

So far we have left unanswered an important question about optically active compounds and racemic modifications: How are enantiomers separated? Enantiomers have identical solubilities in ordinary solvents, and they have identical boiling points. Consequently, the conventional methods for separating organic compounds such as crystallization and distillation fail when applied to a racemic modification.

It was, in fact, Louis Pasteur's separation of a racemic modification of a salt of tartaric acid in 1848 that led to the discovery of the phenomenon called enantiomerism and to the later founding of the field of stereochemistry.

Tartaric acid is one of the by-products of wine making. Pasteur had obtained a sample of racemic tartaric acid* from the owner of a chemical plant. In the course of his investigation Pasteur began examining the crystal structure of the sodium ammonium salt of racemic tartaric acid. He noticed that two types of

* At the time racemic tartaric acid was simply called racemic acid (L. *racemis,* a bunch of grapes).

crystals were present. One was identical with crystals of the sodium ammonium salt of (+)-tartaric acid that had been discovered earlier and had been shown to be dextrorotatory. Crystals of the other type were *non*superposable mirror reflections of the first kind. The two types of crystals were actually chiral. Using tweezers, Pasteur separated the two kinds of crystals under a microscope, dissolved them in water, and placed the solutions in a polarimeter. The solution of crystals of the first type was dextrorotatory, and the crystals themselves proved to be identical with the sodium ammonium salt of (+)-tartaric acid that was already known. The solution of crystals of the second type was levorotatory: they rotated plane-polarized light in the opposite direction and by an equal amount. The crystals of the second type were the sodium ammonium salt of (−)-tartaric acid. The chirality of the crystals themselves disappeared, of course, as the crystals dissolved into their solutions *but the optical activity* remained. Pasteur reasoned, therefore, that the molecules themselves must be chiral.

Pasteur's discovery of enantiomerism and his demonstration that the optical activity of the two forms of tartaric acid was a property of the molecules themselves led, in 1874, to the proposal of the tetrahedral structure of carbon by van't Hoff and Le Bel.

Problem 8.30

On page 317 we gave the absolute configuration of (+)-tartaric acid. (a) What is the absolute configuration of (−)-tartaric acid? (b) What other isomer of tartaric acid is possible? (c) Would it be optically active?

Unfortunately, few organic compounds give chiral crystals as do the (+)- and (−)-tartaric acid salts. Few organic compounds crystallize into separate crystals (containing separate enantiomers) that are visibly chiral like the crystals of the sodium ammonium salt of tartaric acid. Pasteur's method, therefore, is not one that is generally applicable.

The most useful procedure for separating enantiomers is based on allowing a racemic modification to react with a single enantiomer of some other compound. This changes *a racemic modification into a mixture of diastereomers;* and *diastereomers, because they have different melting and boiling points, can be separated by conventional means.* The separation of the enantiomers of a racemic modification is called *resolution.*

We can illustrate this procedure in general terms by showing how a racemic modification of an organic acid might be resolved (separated) into its enantiomers using the single enantiomer of an amine (Fig. 8.24) as a resolving agent.

In this procedure the single enantiomer of an amine [$(-)R\text{NH}_2$] is added to the racemic modification of the acid. The salts that form—$(+)R\text{COO}^-(-)R\text{NH}_3^+$ and $(-)R\text{COO}^-(-)R\text{NH}_3^+$—are not enantiomers: they are diastereomers. (The chiral carbon of the acid portion of the salts are enantiomerically related to each other, but the chiral carbons of the amine portion are not.) The diastereomers have different solubilities and can be separated by careful crystallization. The separated salts are then acidified with hydrochloric acid and the enantiomeric acids precipitate from the separate solutions. The amines remain in solution as hydrochloride salts.

The single enantiomers that are employed as resolving agents are readily available from natural sources. Since most of the chiral organic molecules that occur in living organisms are synthesized by enzymatically catalyzed reactions, most of them occur as single enantiomers. Naturally occurring optically active amines such as (−)-quinine, (−)-strychnine, and (−)-brucine are often employed

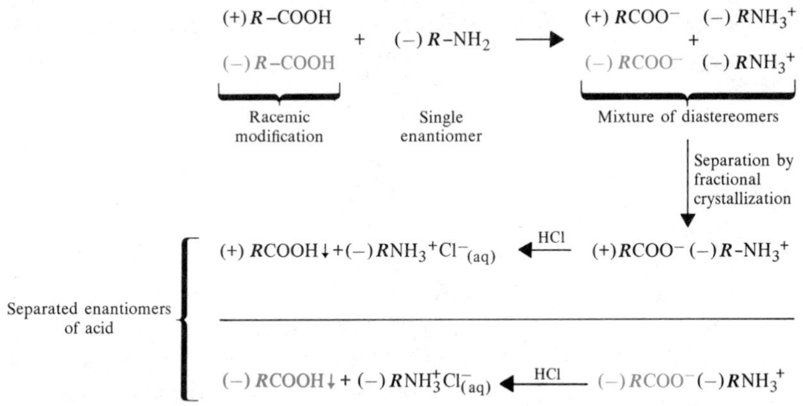

$$\underbrace{\begin{matrix}(+)R\text{-COOH}\\(-)R\text{-COOH}\end{matrix}}_{\text{Racemic modification}} + \underbrace{(-)R\text{-}NH_2}_{\text{Single enantiomer}} \longrightarrow \underbrace{\begin{matrix}(+)RCOO^- \ (-)RNH_3^+\\ + \\ (-)RCOO^- \ (-)RNH_3^+\end{matrix}}_{\text{Mixture of diastereomers}}$$

Separation by fractional crystallization

Separated enantiomers of acid:

$$(+)RCOOH\downarrow + (-)RNH_3^+Cl^-_{(aq)} \xleftarrow{HCl} (+)RCOO^- (-)R\text{-}NH_3^+$$

$$(-)RCOOH\downarrow + (-)RNH_3^+Cl^-_{(aq)} \xleftarrow{HCl} (-)RCOO^- (-)RNH_3^+$$

FIG. 8.24
The resolution of the racemic modification of an organic acid.

as resolving agents for racemic acids. Acids such as (+)- or (−)-tartaric acid are often used for resolving racemic bases.

8.19 COMPOUNDS WITH CHIRAL CENTERS OTHER THAN CARBON

Any tetrahedral atom with four different groups attached to it is a chiral atom. Listed below are examples of compounds whose molecules contain chiral atoms other than carbon.

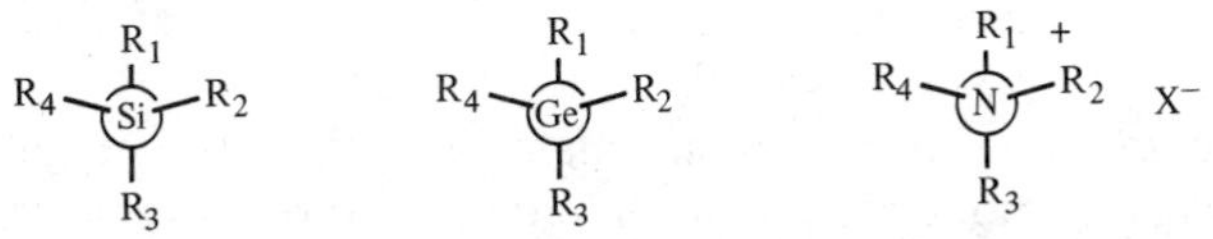

R_1, R_2, R_3, R_4 around Si; R_1, R_2, R_3, R_4 around Ge; R_1, R_2, R_3, R_4 around N^+ X^-

Silicon and germanium are in the same group of the Periodic Table as carbon. They form tetrahedral compounds as carbon does. When four different groups are situated around the central atom in silicon and germanium compounds, the molecules are chiral and the enantiomers can be separated.

Nitrogen forms four covalent bonds in ammonium compounds. Numerous examples of chiral ammonium compounds have been resolved; listed below are two.

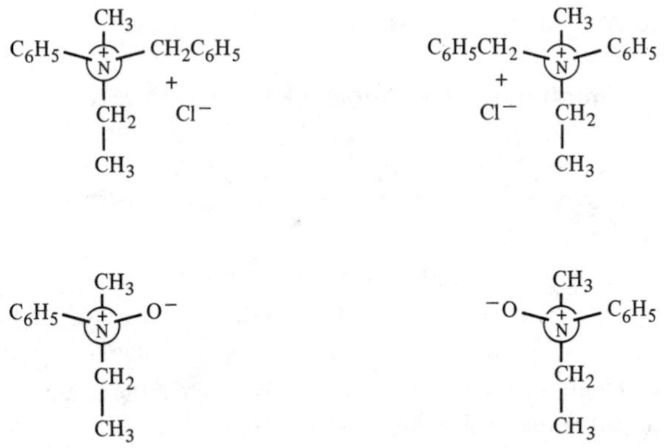

CH_3, C_6H_5, $CH_2C_6H_5$, CH_2CH_3 around N^+ Cl^- — and its mirror image: CH_3, $C_6H_5CH_2$, C_6H_5, CH_2CH_3 around N^+ Cl^-

CH_3, C_6H_5, O^-, CH_2CH_3 around N^+ — and its mirror image: CH_3, ^-O, C_6H_5, CH_2CH_3 around N^+

The nitrogen atom of a tertiary amine is also tetrahedral if one considers the unshared pair as occupying one corner of the tetrahedron.

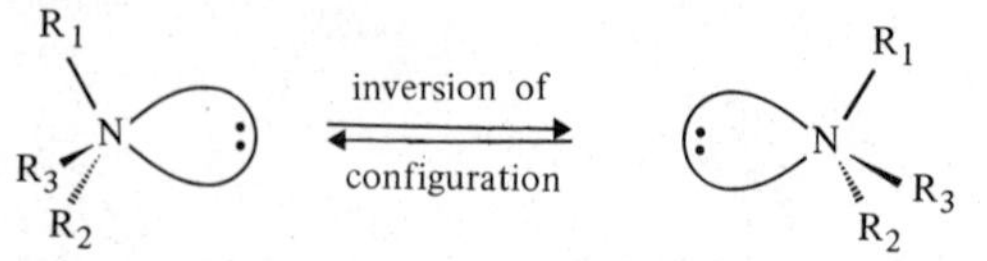

inversion of configuration

The nitrogen atoms of most tertiary amines invert their configurations so rapidly at room temperature that they cannot be resolved into separate enantiomers even when three different groups are attached at the other corners.*

Phosphorus analogs of amines (called phosphines) invert much more slowly and many have been resolved.

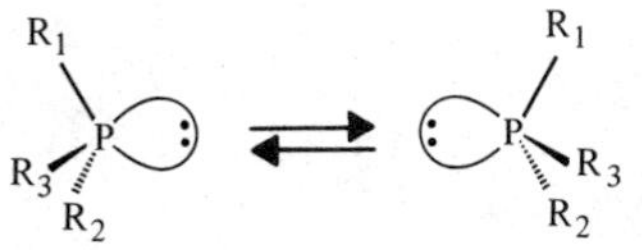

E_{act} = 25–30 kcal/mole

Sulfur compounds with three different attached groups have been resolved. The sulfonium compounds shown below are examples.

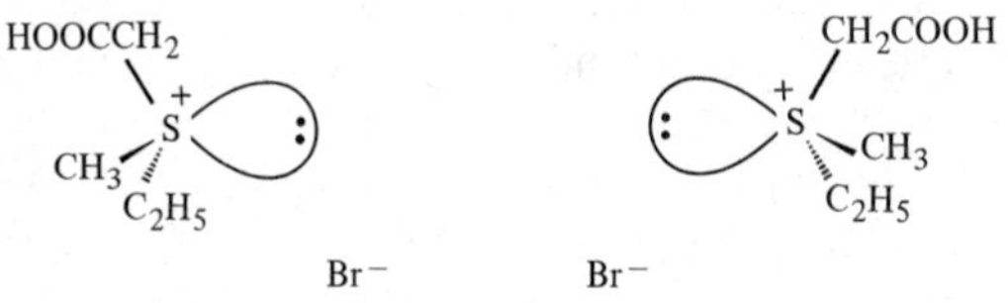

Br^- Br^-

Molecules containing sulfur atoms of this type retain their configuration at room temperature and do not undergo the characteristic rapid inversion associated with tertiary amines.

8.20 CHIRAL MOLECULES THAT DO NOT POSSESS A CHIRAL ATOM

A molecule is chiral if it is not superposable on its mirror reflection. The presence of a chiral atom is only one focus that will confer chirality on a molecule. Most of the molecules that we will encounter do have chiral atoms (usually chiral carbons). Many chiral molecules are known, however, that do not. Two examples of chiral molecules that do not have chiral atoms are 1,3-dichloroallene and *trans*-cyclooctene.

Allenes are compounds whose molecules contain the double bond sequence shown below.

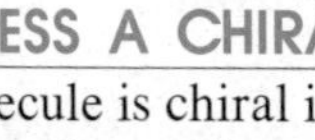

$$>C=C=C<$$

* Tertiary amines with chiral centers have been resolved when the tertiary nitrogen is a part of a ring that prevents inversion of its configuration.

(The double bonds of an allene are said to be *cumulated*.) The planes of the π bonds of allenes are perpendicular to each other.

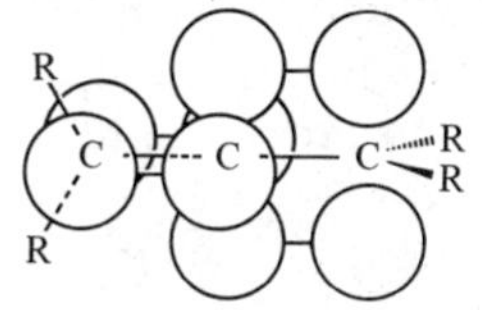

This geometry of the π bonds causes the groups attached to the end carbons to lie in perpendicular planes and, because of this, allenes with different substituents on the end carbons are chiral (Fig. 8.25).

Trans-cyclooctene (Fig. 8.26) also exists in enantiomeric forms because of the geometry of the *trans*-cycloalkene ring.

Problem 8.31

What is the stereoisomeric relation between *cis*-cyclooctene and either of the *trans*-cyclooctene enantiomers?

8.21 THE *E-Z* SYSTEM FOR DESIGNATING ALKENE DIASTEREOMERS

The terms *cis* and *trans,* when used to designate the stereochemistry of alkene diastereomers, are unambiguous only when applied to disubstituted alkenes. If the alkene is trisubstituted or tetrasubstituted the terms *cis* and *trans* are either ambiguous or do not apply at all. Consider the alkenes shown below as examples.

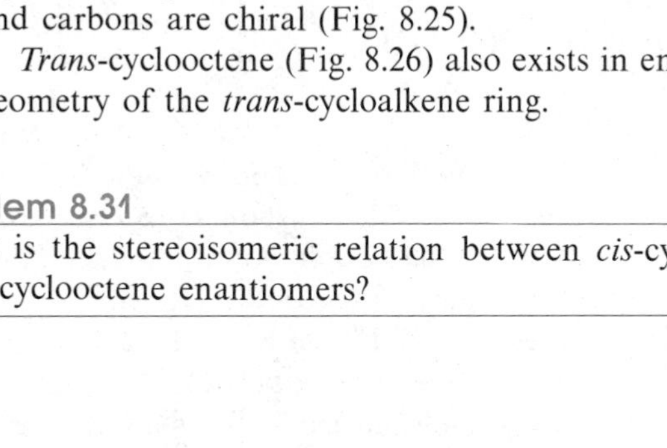

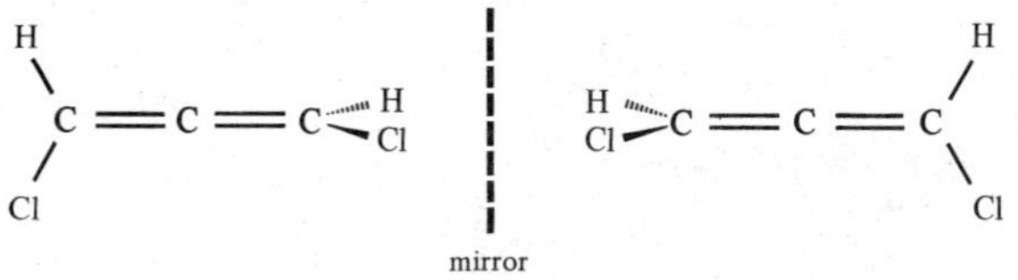

FIG. 8.25
Enantiomeric forms of 1,3-dichloroallene. These two molecules are non-superposable mirror reflections of each other and are, therefore, chiral. They do not possess a chiral atom, however.

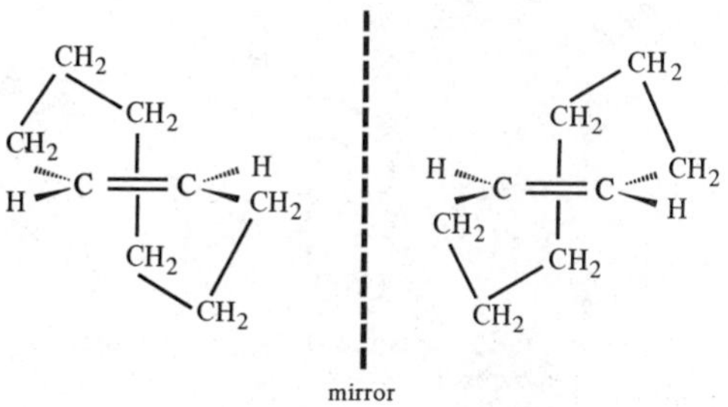

FIG. 8.26
The trans-*cyclooctene enantiomers.*

It is impossible to decide whether A is *cis* or *trans* since no two groups are the same. The same is true of compound B.

A newer system is based on the preferences of groups in the Cahn-Ingold-Prelog convention. This system, called the *E-Z* system, applies to alkene diastereomers of all types. In the *E-Z* system, we examine the two groups attached to one carbon of the double bond and arrange them in order of preference. Then we repeat that operation at the other carbon.

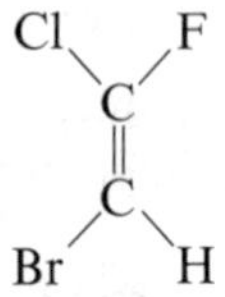

Cl F
C
‖
C
Br H
Z

Cl > F
Br > H

F Cl
C
‖
C
Br H 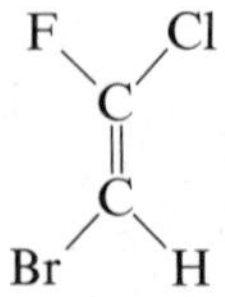
E

We then take the group of higher preference on one carbon and compare it with the group of higher preference on the other carbon. If the two groups of higher preference are on the same side of the double bond the alkene is designated *Z* (from the German word *zusammen,* meaning together). If the two groups of higher preference are on opposite sides of the double bond the alkene is designated *E* (from the German word *entgegen,* meaning opposite).

Most compounds that we would normally designate *cis* are designated *Z*. Similarly most compounds that we would designate *trans* are designated *E*.

$CH_3 > H$ 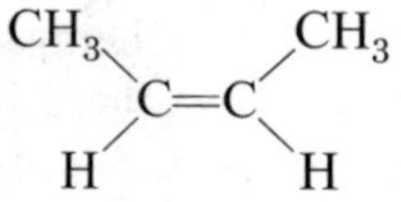 CH_3 CH_3 C=C H H $CH_3 > H$

(*Z*)-2-Butene
(*cis*-2-butene)

$CH_3 > H$ 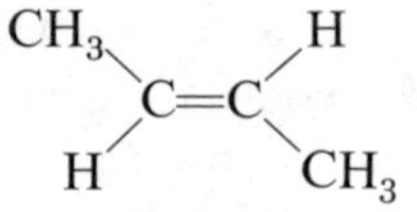 CH_3 H C=C H CH_3 $CH_3 > H$

(*E*)-2-Butene
(*trans*-2-butene)

$CH_3 > H$ CH_3 $CH_2CH_2CH_3$ C=C H H $C_3H_7 > H$

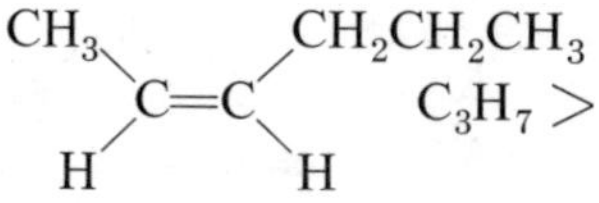

(*Z*)-2-Hexene
(*cis*-2-hexene)

$CH_3 > H$ CH_3 H C=C H $CH_2CH_2CH_3$ $C_3H_7 > H$

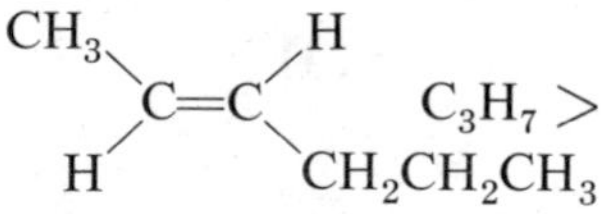

(*E*)-2-Hexene
(*trans*-2-hexene)

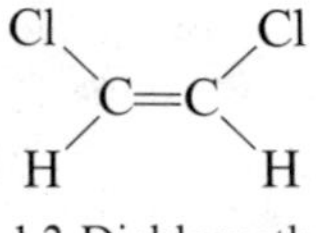

Cl Cl C=C H H

(*Z*)-1,2-Dichloroethene
(*cis*-1,2-dichloroethene)

Cl > H
Cl > H

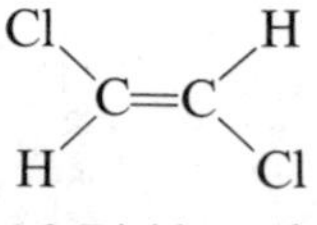

Cl H C=C H Cl

(*E*)-1,2-Dichloroethene
(*trans*-1,2-dichloroethene)

There are exceptions, however.

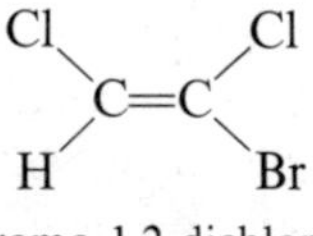

Cl Cl C=C H Br

(*E*)-1-Bromo-1,2-dichloroethene
(*cis*-1-bromo-1,2-dichloroethene)

Cl > H
Br > Cl

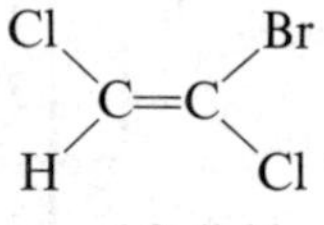

Cl Br C=C H Cl

(*Z*)-1-Bromo-1,2-dichloroethene
(*trans*-1-bromo-1,2-dichloroethene)

Problem 8.32

Give the *E-Z* designation and name of each of the following.

(a) $Cl(Br)C{=}CH(CH_2CH_3)$ — Cl and Br on the left carbon; H and CH_2CH_3 on the right carbon

(b) $I(Cl)C{=}C(Br)CH_3$ — I and Cl on the left carbon; Br and CH_3 on the right carbon

(c) $CH_3(H)C{=}C(CH_2CH_3)CH(CH_3)_2$ — CH_3 and H on the left carbon; CH_2CH_3 and $CH(CH_3)_2$ on the right carbon

(d) $Cl(F)C{=}C(CH_3)CH_2CH_3$ — Cl and F on the left carbon; CH_3 and CH_2CH_3 on the right carbon

8.22 SOME IMPORTANT TERMS AND CONCEPTS

Stereochemistry. Chemical studies that take into account the three-dimensional aspects of molecules.

Isomers are different compounds that have the same molecular formula. All isomers fall into either of two groups: *structural* isomers or *stereoisomers.*

Structural isomers (also called **constitutional isomers**) are isomers that have their atoms joined in a different order.

Stereoisomers have their atoms joined in the same order but differ in the way their atoms are arranged in space. Stereoisomers can be subdivided into two categories: *enantiomers* and *diastereomers.*

Enantiomers are stereoisomers that are related as an object and its mirror reflection. Enantiomers only occur with compounds whose molecules are chiral, that is, with molecules that are *not* superposable on their mirror reflections. Separate enantiomers rotate the plane of polarized light and are said to be *optically active.* They have equal but opposite specific rotations.

Diastereomers are stereoisomers that are not enantiomers, that is, they are stereoisomers that are not related as an object and its mirror reflection.

Chirality is equivalent to "handedness." A chiral molecule is one that is not superposable on its mirror reflection. An *achiral* molecule is one that can be superposed on its mirror reflection. A *chiral carbon* (also called an *asymmetric carbon* or a *chiral center*) is any carbon atom that has four different groups attached to it. A pair of enantiomers will be possible for all molecules that contain *a single* chiral carbon. For molecules with more than one chiral carbon, the number of stereoisomers will not exceed 2^n where n is the number of chiral carbons.

Plane of symmetry. An imaginary plane that bisects a molecule in such a way that the two halves of the molecule are mirror reflections of each other. Any molecule that has a plane of symmetry will be achiral.

Configuration. The particular arrangement of atoms (or groups) in space that is characteristic of a given stereoisomer. The configuration at each chiral carbon can be designated as *R* or *S* using the rules given in Section 8.5.

Racemic modification or racemate. An equimolar mixture of enantiomers.

Meso compound. An optically inactive compound whose molecules are achiral even through they contain chiral centers.

Stereoselective reaction. One that yields exclusively (or predominantly) one of a set of stereoisomers.

Stereospecific reaction. One in which stereoisomerically different reactants give stereoisomerically different products. That is, a reaction in which a given stereoisomeric form of the reactant is converted to a specific stereoisomeric form of product.

Resolution. The separation of the enantiomers of a racemic modification.

Subdivision of Isomers:

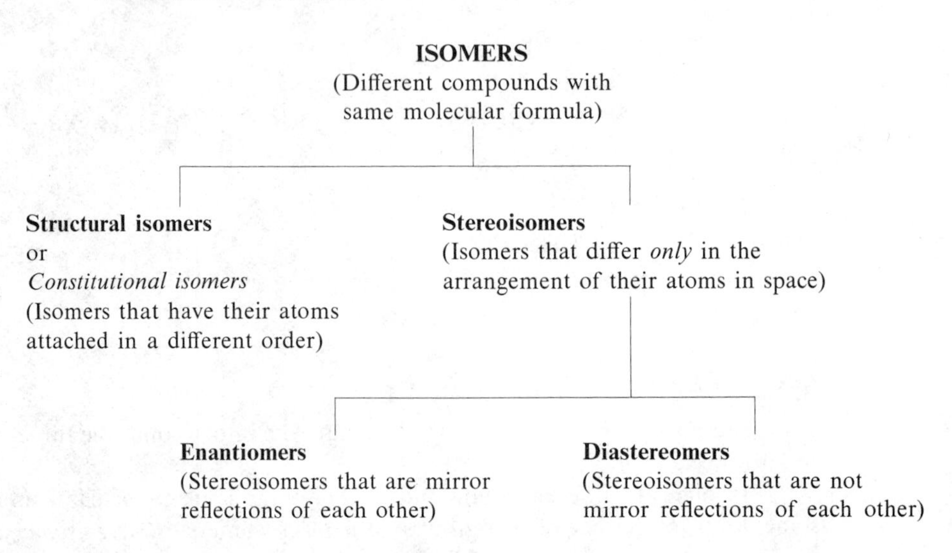

Additional Problems

8.33

Give definitions of each of the following terms and examples that illustrate their meaning.

(a) Isomers
(b) Structural isomers
(c) Stereoisomers
(d) Diastereomers
(e) Enantiomers
(f) *Meso* compound
(g) Racemic modification
(h) Plane of symmetry
(i) Chiral atom
(j) Chiral molecule
(k) Achiral molecule
(l) Optical activity
(m) Dextrorotatory
(n) Retention of configuration

8.34

Consider the following pairs of structures. Identify the relation between them by describing them as representing enantiomers, diastereomers, structural isomers, or two molecules of the same compound.

(a) 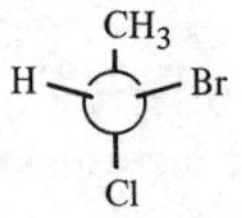 CH_3, H, Br, Cl and CH_3, H, Cl, Br

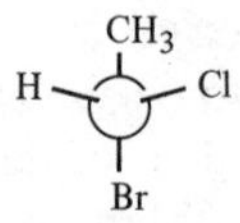

(b) 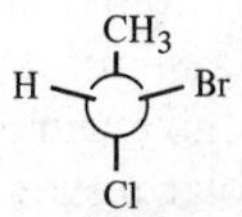 CH_3, H, Br, Cl and CH_3, Cl, H, Br

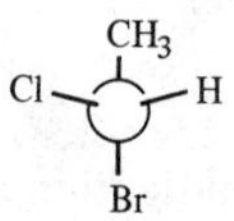

(c) 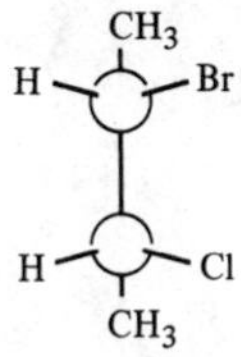 CH_3, H, Br / H, Cl, CH_3 and CH_3, H, Cl / H, Br, CH_3

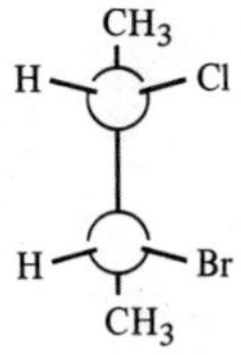

(d) 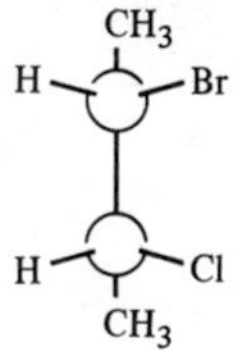 CH_3, H, Br / H, Cl, CH_3 and 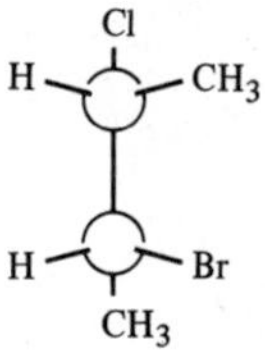 Cl, H, CH_3 / H, Br, CH_3

(e) 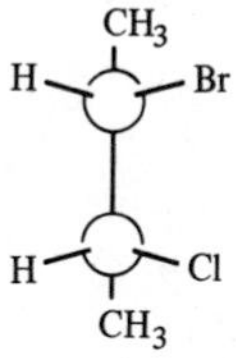 CH_3, H, Br / H, Cl, CH_3 and 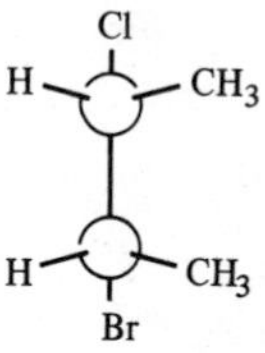 Cl, H, CH_3 / H, CH_3, Br

(f) 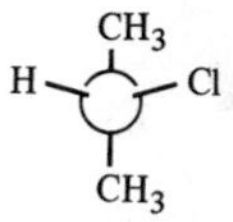 CH_3, H, Cl, CH_3 and CH_3, H, H, CH_2Cl

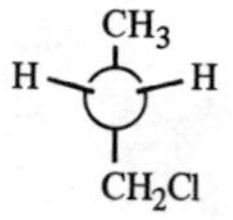

(g) 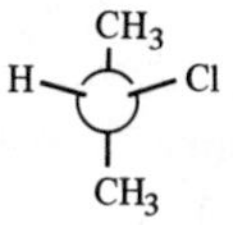 CH_3, H, Cl, CH_3 and CH_3, Cl, H, CH_3

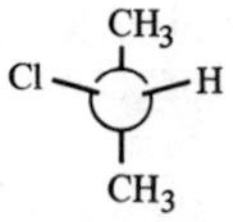

(h) 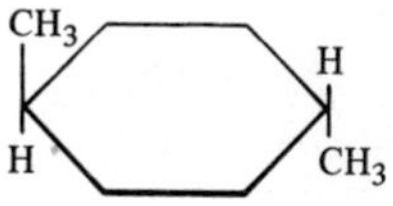 CH_3, H, H, CH_3 and H, CH_3, H, CH_3

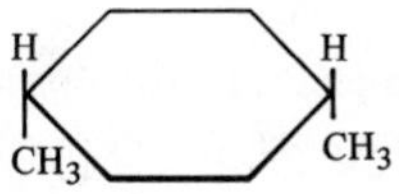

(i)  CH_3, CH_3 and CH_3, CH_3

(j)  CH_3, CH_3 and CH_3, CH_3

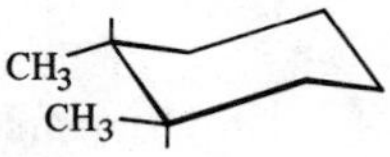

(k) 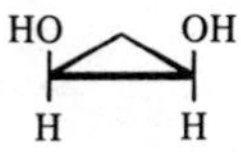 HO, OH, H, H and 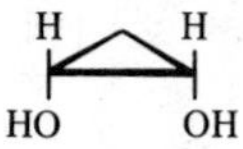 H, H, HO, OH

(l) 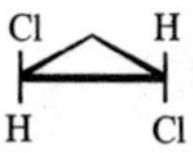 Cl, H, H, Cl and 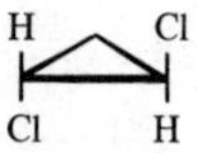 H, Cl, Cl, H

(m) 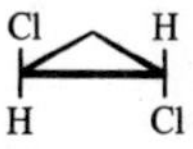 Cl, H, H, Cl and 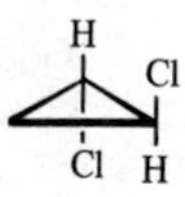 H, Cl, Cl, H

(n) 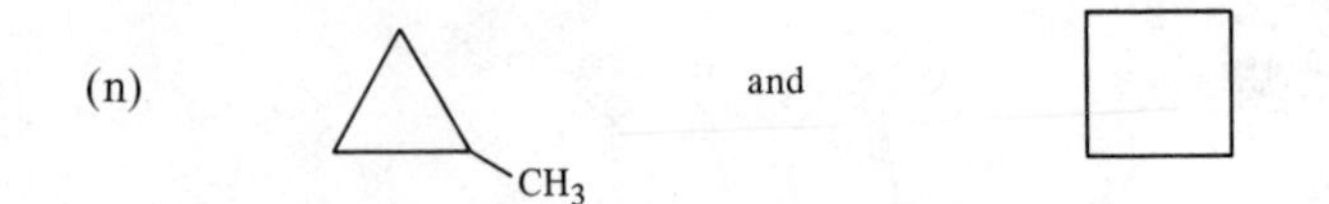

(o) CH_3, CH_3 C=C H, Cl and CH_3, CH_3 C=C Cl, H

(p) CH_3, Cl C=C H, CH_3 and CH_3, Cl C=C CH_3, H

(q) H, CH_3 C=C=C H, CH_3 and H, CH_3 C=C=C H, CH_3

8.35

There are four dimethylcyclopropane isomers. (a) Write three-dimensional formulas for them. (b) Which dimethylcyclopropane isomers would, if taken separately, show optical activity? (c) Which dimethylcyclopropane isomer is a *meso* compound? (d) If a mixture consisting of one mole of each of the four dimethylcyclopropane isomers were subjected to fractional distillation, how many fractions would be obtained? (e) How many of these fractions would show optical activity?

8.36

Write stereochemical formulas for all of the products that you would expect from each of the following reactions.

(a) CH_3, H C=C CH_2CH_3, H $\xrightarrow[\text{(2) } Na_2SO_3]{\text{(1) } OsO_4}$

(b) CH_3, H C=C CH_2CH_3, H $\xrightarrow[\text{(2) } H^+,\ H_2O]{\text{(1) } C_6H_5COOOH}$

(c) CH_3, H C=C H, CH_2CH_3 $\xrightarrow[\text{(2) } Na_2SO_3]{\text{(1) } OsO_4}$

(d) CH_3, H C=C H, CH_2CH_3 $\xrightarrow[\text{(2) } H^+,\ H_2O]{\text{(1) } C_6H_5COOOH}$

(e) CH_3, H C=C H, CH_2CH_3 $\xrightarrow{Br_2,\ CCl_4}$

(f) CH_3, H C=C CH_2CH_3, H $\xrightarrow{Br_2,\ CCl_4}$

8.37

Give *R-S* designations for each different compound given as an answer to Problem 8.36.

8.38

(±)-Lactic acid reacts with (*S*)-(−)-2-methyl-1-butanol (p. 315) to produce a mixture of esters. (a) Write three-dimensional formulas for these esters. (b) What is the stereochemical relation between the esters? (c) Could the esters be separated by fractional distillation? (d) Could this reaction be used as the basis for resolving (±) lactic acid? (e) If so, how would you go about obtaining the lactic acid enantiomers?

8.39

Consider each of the reactions given below and decide whether you would expect it to take place with retention of configuration, inversion of configuration, or racemization. In each instance give the reason for your decision.

$$\text{(a)}\ (-)CH_3\underset{\underset{\displaystyle OH}{|}}{C}HCHO \xrightarrow[H_2O]{Br_2} CH_3\underset{\underset{\displaystyle OH}{|}}{C}HCOOH$$

$$\text{(b)}\ (+)CH_3\underset{\underset{\displaystyle OH}{|}}{C}H\overset{\overset{\displaystyle O}{\|}}{C}{-}Cl \xrightarrow{NH_3} CH_3\underset{\underset{\displaystyle OH}{|}}{C}H\overset{\overset{\displaystyle O}{\|}}{C}{-}NH_2$$

$$\text{(c)}\ (-)n\text{-}C_6H_{13}\underset{\underset{\displaystyle Br}{|}}{C}HCH_3 + I^- \longrightarrow n\text{-}C_6H_{13}\underset{\underset{\displaystyle I}{|}}{C}HCH_3 + Br^-$$

$$\text{(d)}\ (-)n\text{-}C_6H_{13}\underset{\underset{\displaystyle OH}{|}}{C}HCH_3 \xrightarrow{H_3O^+} n\text{-}C_6H_{13}\underset{\underset{\displaystyle OH}{|}}{C}HCH_3$$

$$\text{(e)}\ (+)CH_3CH_2\overset{\overset{\displaystyle CH_3}{|}}{\underset{\underset{\displaystyle I}{|}}{C}}CH_2CH_2CH_3 + CH_3OH \longrightarrow CH_3CH_2\overset{\overset{\displaystyle CH_3}{|}}{\underset{\underset{\displaystyle OCH_3}{|}}{C}}CH_2CH_2CH_3 + HI$$

$$\text{(f)}\ (+)CH_3CH_2CH_2CHDBr + :CN^- \longrightarrow CH_3CH_2CH_2CHDCN + Br^-$$

$$\text{(g)}\ (-)CH_3CH_2\overset{\overset{\displaystyle CH_3}{|}}{C}HCH_2Cl \xrightarrow[S_N2]{OH^-} CH_3CH_2\overset{\overset{\displaystyle CH_3}{|}}{C}HCH_2OH$$

8.40

(a) Which stereoisomer would you expect to obtain from the following reaction?

$$\text{1,2-Dimethylcyclopentene} \xrightarrow[Pt]{H_2}$$

(b) Would the product be optically active?

8.41

When 1-methylcyclopentene is subjected to hydroboration-oxidation the product is obtained as a racemic modification. (a) Write three-dimensional structures for the enantiomers that are formed. (b) Write a mechanism that shows how they are formed.

8.42

Ricinoleic acid, a compound that can be isolated from castor oil, has the structure $CH_3(CH_2)_5CHOHCH_2CH{=}CH(CH_2)_7COOH$. (a) How many stereoisomers of this structure are possible? (b) Write these structures.

8.43

There are two dicarboxylic acids with the general formula $HOOCCH{=}CHCOOH$. One dicarboxylic acid is called maleic acid; the other is called fumaric acid. In 1880, Kekulé found that on treatment with cold dilute $KMnO_4$, maleic acid yields *meso*-tartaric acid and that fumaric yields (±)-tartaric acid. Show how this information allows one to write stereochemical formulas for maleic acid and fumaric acid.

8.44
Use your answers to the preceding problem to predict the stereochemical outcome of the addition of bromine to maleic acid and to fumaric acid. (a) Which dicarboxylic acid would add bromine to yield a *meso*-compound? (b) Which would yield a racemic modification?

8.45
An optically active compound **A** (assume that it is dextrorotatory) has the molecular formula $C_7H_{11}Br$. **A** reacts with hydrogen bromide, in the absence of peroxides, to yield isomeric products, **B** and **C,** with molecular formula $C_7H_{12}Br_2$. **B** is optically active; **C** is not. Treating **B** with one mole of potassium *tert*-butoxide yields **(+)A.** Treating **C** with one mole of potassium *tert*-butoxide yields **(±)A.** Treating **A** with potassium *tert*-butoxide yields **D** (C_7H_{10}). Subjecting one mole of **D** to ozonolysis followed by treatment with zinc and water yields two moles of formaldehyde and one mole of 1,3-cyclopentanedione.

O O

1,3-Cyclopentanedione

Propose stereochemical formulas for **A, B, C,** and **D** and outline the reactions involved in these transformations.

8.46
Propose structures for compounds **E-H.**

(a) Compound **E** has the molecular formula C_5H_8 and is optically active. On catalytic hydrogenation **E** yields **F. F** has the molecular formula C_5H_{10}, is optically inactive, and cannot be resolved.

(b) Compound **G** has the molecular formula C_6H_{10} and is optically active. **G** contains no triple bonds. On catalytic hydrogenation **G** yields **H. H** has the molecular formula C_6H_{14}, is optically inactive, and cannot be resolved.

8.47
Compounds **I** and **J** both have the molecular formula C_7H_{14}. **I** and **J** are both optically active and both rotate plane-polarized light in the same direction. On catalytic hydrogenation **I** and **J** yield the same compound **K,** C_7H_{16}. **K** is optically active. Propose possible structures for **I, J,** and **K.**

8.48
Compounds **L** and **M** have the molecular formula C_7H_{14}. **L** and **M** are optically inactive, are nonresolvable, and are diastereomers of each other. Catalytic hydrogenation of either **L** or **M** yields **N. N** is optically inactive but can be resolved. Propose possible structures for **L, M,** and **N.**

8.49
One stereoisomeric form of 1,3-di-*sec*-butylcyclohexane is optically inactive. Write a conformational structure for this isomer.

***8.50**
When the 3-bromo-2-butanol with the stereochemical structure **A** is treated with concentrated HBr it yields *meso*-2,3-dibromobutane; a similar reaction of the 3-bromo-2-butanol **B** yields (±)-2,3-dibromobutane. This classic experiment performed in 1939 by S. Winstein

and H. J. Lucas was the starting point for a series of investigations of what are called *neighboring group effects* (see also Sec. L.5). Propose mechanisms that will account for the stereochemistry of these reactions.

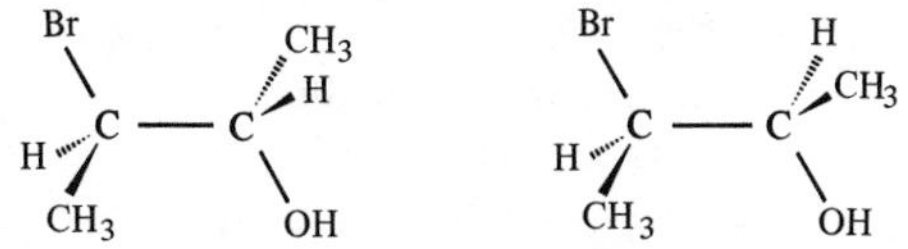

Br, H, CH_3 on C — C with CH_3, H, OH

A

Br, H, CH_3 on C — C with H, CH_3, OH

B

*** 8.51**

When 2-iodooctane reacts with radioactive iodide ions ($^{*}I^{-}$) the starting materials and the products are chemically equivalent.

$$n\text{-}C_6H_{13}\underset{\displaystyle I}{\underset{|}{C}}HCH_3 + {}^{*}I^{-} \longrightarrow n\text{-}C_6H_{13}\underset{\displaystyle {}^{*}I}{\underset{|}{C}}HCH_3 + I^{-}$$

When the reaction is carried out with optically active 2-iodooctane, the rate of racemization is found to be exactly *twice* the rate of incorporation of radioactive iodine. This experiment is considered to provide convincing proof that an S_N2 reaction takes place with inversion of configuration. (a) How is this true? (b) What would you expect the relative rates of racemization and incorporation of radioactive iodine to be if the reaction took place through the formation of an achiral intermediate such as a carbocation?

SPECIAL TOPIC

POLYMERIZATION OF ALKENES: ADDITION POLYMERS

The names *Orlon, Plexiglas, Lucite, polyethylene,* and *Teflon* are now familiar names to most of us. These "plastics" or polymers are used in the construction of many objects around us—from the clothing we wear, to portions of the houses we live in. Yet, all of these compounds were unknown 50 years ago. The development of the processes by which synthetic polymers are made, more than any other single factor, has been responsible for the remarkable growth of chemical industry in this century.

At the same time, some scientists are now expressing concern about the reliance we have placed on these "man-made" materials. Because they are the products of laboratory and industrial processes rather than processes that occur in nature, nature often has no way of disposing of many of them. Although progress has been made in the development of "biodegradable plastics" in recent years, many materials are still used that are not biodegradable. Although all of these objects are combustible, incineration is not always a feasible method of disposal because of attendant air pollution.

Not all polymers are synthetic. Many naturally occurring compounds are polymers as well. Silk and wool are polymers that we call proteins. The starches of our diet are polymers and so is the cellulose of cotton and wood.

Polymers are compounds that consist of very large molecules made up of many repeating subunits. The molecular subunits that are used to synthesize polymers are called *monomers,* and the reactions by which monomers are joined together are called polymerization reactions.

Propylene, for example, can be polymerized to form *polypropylene.* This polymerization occurs by an addition reaction and, as a consequence, polymers such as polypropylene are called *addition polymers.*

$$\underset{\text{Propylene}}{CH_2{=}\underset{\displaystyle CH_3}{\underset{|}{C}}H} \xrightarrow{\text{Polymerization}} \underbrace{-CH_2\underset{\displaystyle CH_3}{\underset{|}{C}}H\left(CH_2\underset{\displaystyle CH_3}{\underset{|}{C}}H\right)_n CH_2-\underset{\displaystyle CH_3}{\underset{|}{C}}H-}_{\text{Polypropylene}}$$

As we might expect, alkenes are convenient starting materials for the preparation of addition polymers. The addition reactions occur through free-radical, cationic, or anionic mechanisms depending on how they are initiated. The following examples illustrate these mechanisms.

Free-Radical Polymerization

$$W\cdot + \gt C{=}C\lt \longrightarrow W{:}\overset{|}{\underset{|}{C}}-\overset{|}{\underset{|}{C}}\cdot \xrightarrow{\gt C=C\lt} W-\overset{|}{\underset{|}{C}}-\overset{|}{\underset{|}{C}}-\overset{|}{\underset{|}{C}}-\overset{|}{\underset{|}{C}}\cdot \xrightarrow{\gt C=C\lt} \text{etc.}$$

Cationic Polymerization

$$Y^+ + \gt C{=}C\lt \longrightarrow Y-\overset{|}{\underset{|}{C}}-\overset{|}{\underset{|}{C}}^+ \xrightarrow{\gt C=C\lt} Y-\overset{|}{\underset{|}{C}}-\overset{|}{\underset{|}{C}}-\overset{|}{\underset{|}{C}}-\overset{|}{\underset{|}{C}}^+ \xrightarrow{\gt C=C\lt} \text{etc.}$$

Anionic Polymerization

$$Z{:}^- + \gt C{=}C\lt \longrightarrow Z-\overset{|}{\underset{|}{C}}-\overset{|}{\underset{|}{C}}{:}^- \xrightarrow{\gt C=C\lt} Z-\overset{|}{\underset{|}{C}}-\overset{|}{\underset{|}{C}}-\overset{|}{\underset{|}{C}}-\overset{|}{\underset{|}{C}}{:}^- \xrightarrow{\gt C=C\lt} \text{etc.}$$

All of these reactions are chain reactions. Ethylene, for example, polymerizes by a free-radical mechanism when it is heated at a pressure of 1000 atm with a small amount of an organic peroxide. The peroxide dissociates to produce free radicals which in turn initiate chains.

Chain Initiation

$$R{-}\overset{O}{\overset{\|}{C}}{-}O{:}O{-}\overset{O}{\overset{\|}{C}}{-}R \longrightarrow 2R{:}\overset{O}{\overset{\|}{C}}{-}O\cdot \longrightarrow 2CO_2 + 2R\cdot$$

$$R\cdot + CH_2{=}CH_2 \longrightarrow R{:}CH_2{-}CH_2\cdot$$

Chain Propagation

$$R{-}CH_2CH_2\cdot + nCH_2{=}CH_2 \longrightarrow R{-}(CH_2CH_2)_n{-}CH_2CH_2\cdot$$

Chains propagate by adding successive ethylene units, until their growth is stopped by combination or disproportionation.

Chain Termination

$$2R{-}(CH_2CH_2)_nCH_2CH_2\cdot \xrightarrow{\text{Combination}} \left[R{-}(CH_2CH_2)_nCH_2CH_2{-}\right]_2$$

$$2R{-}(CH_2CH_2)_nCH_2CH_2\cdot \xrightarrow{\text{Disproportionation}} R{-}(CH_2CH_2)_nCH{=}CH_2 + R{-}(CH_2CH_2)_nCH_2CH_3$$

The free radical at the end of the growing polymer chain can also abstract a hydrogen atom from itself by what is called "back biting." This leads to chain branching:

$$R{-}CH_2\overset{H}{C}H{-}(CH_2CH_2)_n{-}CH_2\dot{C}H_2 \longrightarrow RCH_2\dot{C}H(CH_2CH_2)_nCH_2CH_2{-}H$$

$$\xrightarrow{CH_2=CH_2} RCH_2CH(CH_2CH_2)_nCH_2CH_3 \text{ (with branch } {-}CH_2{-}\dot{C}H_2\text{)} \xrightarrow{\text{etc.}}$$

The polyethylene produced by free radical polymerization is not generally useful unless it has a molecular weight of nearly 1,000,000. Very high molecular weight polyethylene can be obtained by using a low concentration of the initiator. This initiates the growth of only a few chains and ensures that each chain will have a large excess of the monomer available. More initiator may be added as chains terminate during the polymerization and, in this way, new chains are begun.

Polyethylene has been produced commercially since 1943. It is used in manufacturing flexible bottles, films, sheets, and insulation for electric wires. Polyethylene produced by free radical polymerization has a softening point of about 110°.

Free-radical polymerization of chloroethene (vinyl chloride) produces a polymer called poly(vinyl chloride).

$$\underset{\text{Vinyl chloride}}{CH_2{=}\underset{\underset{Cl}{|}}{CH}} \longrightarrow \underset{\text{Poly(vinyl chloride)}}{\left(CH_2{-}\underset{\underset{Cl}{|}}{CH} \right)_n}$$

This reaction produces a polymer that has a molecular weight of about 1,500,000 and that is a hard, brittle, and rigid material. In this form it is often used to make pipes, rods, and phonograph records. Poly(vinyl chloride) can be softened by mixing it with esters (called plasticizers). The softer material is used for making "vinyl leather," plastic raincoats, shower curtains, and garden hoses.

Exposure to vinyl chloride has been linked to the development of a rare cancer of the liver called angiocarcinoma. This link was first noted in 1974 and 1975 among workers in vinyl chloride factories. Since that time, standards have been set to limit workers' exposure to less than one part per million average over an eight-hour day. The Food and Drug Administration has banned the use of poly(vinyl chloride) in packages for food. [There is evidence that poly(vinyl chloride) contains traces of vinyl chloride.]

Acrylonitrile ($CH_2{=}CHCN$) polymerizes to form polyacrylonitrile or *Orlon*. The initiator for the polymerization is a mixture of ferrous sulfate and hydrogen peroxide. These two compounds react to produce hydroxyl radicals, $\cdot OH$, which act as chain initiators.

$$\underset{\text{Acrylonitrile}}{CH_2{=}\underset{\underset{CN}{|}}{CH}} \xrightarrow[H{-}O{-}O{-}H]{FeSO_4} \underset{\substack{\text{Polyacrylonitrile} \\ \textit{Orlon}}}{\left(CH_2{-}\underset{\underset{CN}{|}}{CH} \right)_n}$$

Polyacrylonitrile decomposes before it melts, thus melt spinning cannot be used for the production of fibers. Polyacrylonitrile, however, is soluble in dimethylformamide, and these solutions can be used to spin fibers. Fibers produced in this way are used in making carpets and clothing.

Teflon is made by polymerizing tetrafluoroethylene in aqueous suspension.

$$nCF_2{=}CF_2 \xrightarrow[\substack{H_2O_2 \\ H_2O}]{Fe^{++}} {+}CF_2{-}CF_2{+}_n$$

The reaction is highly exothermic and water helps dissipate the heat that is produced. Teflon has a melting point (327°) that is unusually high for an addition polymer. It is also highly resistant to chemical attack and has a low coefficient of friction. Because of these properties Teflon is used in greaseless bearings, in liners for pots and pans, and in many special situations that require a substance that is highly resistant to corrosive chemicals.

Vinyl alcohol is an unstable compound that rearranges spontaneously to acetaldehyde. Consequently, the water-soluble polymer, poly(vinyl alcohol) can not be made

$$\underset{\text{Vinyl alcohol}}{CH_2{=}\underset{\underset{OH}{|}}{CH}} \rightleftharpoons \underset{\text{Acetaldehyde}}{CH_3{-}\underset{\underset{O}{\|}}{CH}}$$

directly. It can be made, however, by an indirect method that begins with the polymerization of vinyl acetate to poly(vinyl acetate). This is then hydrolyzed to poly(vinyl alcohol). Hydrolysis is rarely carried to completion, however, because the presence of a few ester

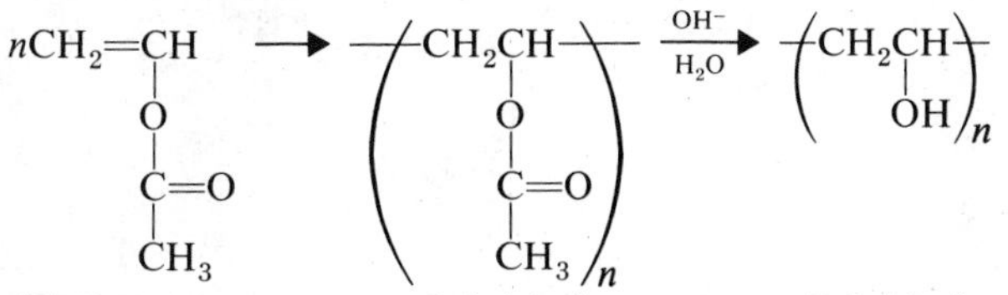

$$n\,CH_2{=}CH{-}O{-}C({=}O){-}CH_3 \longrightarrow \left(CH_2CH(O{-}C({=}O){-}CH_3)\right)_n \xrightarrow[H_2O]{OH^-} \left(CH_2CH(OH)\right)_n$$

Vinyl acetate — Poly(vinyl acetate) — Poly(vinyl alcohol)

groups helps confer water solubility on the product. The ester groups apparently help keep the polymer chains apart and this permits hydration of the hydroxyl groups. Poly(vinyl alcohol) in which 10% of the ester groups remain dissolves readily in water. Poly(vinyl alcohol) is used to manufacture water-soluble films and adhesives. Poly(vinyl acetate) is used as an emulsion in water-base paints.

A polymer with excellent optical properties can be made by the free-radical

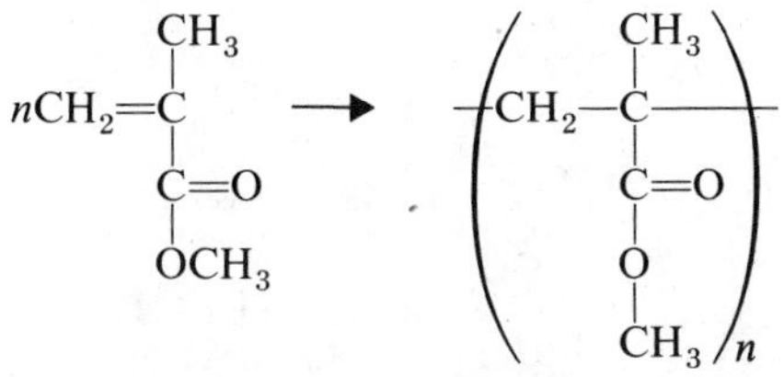

$$n\,CH_2{=}C(CH_3){-}C({=}O){-}OCH_3 \longrightarrow \left(CH_2{-}C(CH_3)(C({=}O){-}O{-}CH_3)\right)_n$$

Poly(methyl methacrylate)

polymerization of methyl methacrylate. Poly(methyl methacrylate) is marketed under the names *Lucite, Plexiglas,* and *Perspex.*

A mixture of vinyl chloride and vinylidene chloride polymerizes to form what is known as a *copolymer.* The familiar *Saran Wrap* used in food packaging is made by polymerizing a mixture in which the vinylidene chloride predominates.

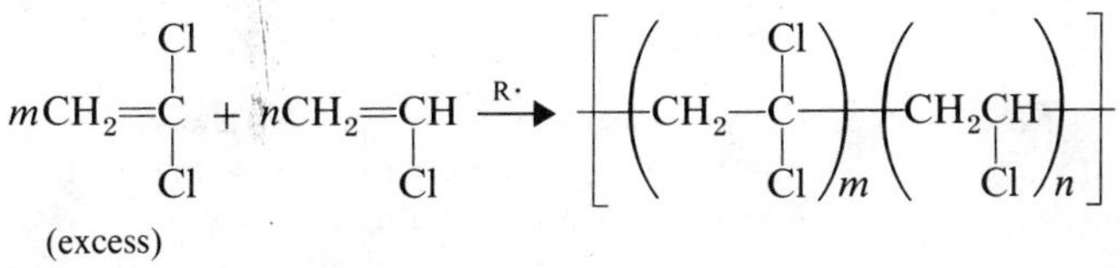

$$m\,CH_2{=}CCl_2 + n\,CH_2{=}CHCl \xrightarrow{R\cdot} \left[\left(CH_2{-}CCl_2\right)_m\left(CH_2CHCl\right)_n\right]$$

(excess) Vinylidene chloride — Vinyl chloride — Saran Wrap

The subunits do not necessarily alternate regularly along the polymer chain.

Problem C.1

Can you suggest an explanation that accounts for the fact that the radical polymerization of propylene occurs in a head-to-tail fashion

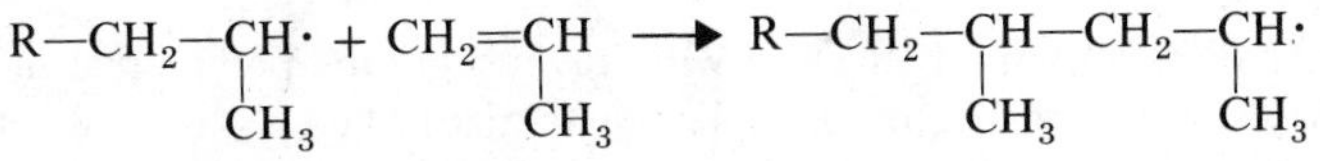

$$R{-}CH_2{-}\underset{\text{"head"}}{CH(CH_3)\cdot} + \underset{\text{"tail"}}{CH_2{=}CH(CH_3)} \longrightarrow R{-}CH_2{-}CH(CH_3){-}CH_2{-}CH(CH_3)\cdot$$

rather than the head-to-head manner, shown below?

$$\underset{\text{"head"}}{\mathrm{R{-}CH_2{-}\underset{\displaystyle CH_3}{CH}\cdot}} + \underset{\text{"head"}}{\mathrm{\underset{\displaystyle CH_3}{CH}{=}CH_2}} \longrightarrow \mathrm{R{-}CH_2{-}\underset{\displaystyle CH_3}{CH}{-}\underset{\displaystyle CH_3}{CH}{-}CH_2\cdot}$$

Problem C.2

Outline general methods for the synthesis of each of the following polymers by free radical polymerization. Assume that the appropriate monomers are available.

(a) Poly(vinyl fluoride) (Tedlar), $\mathrm{+CH_2CHF+_{\mathit{n}}}$

(b) Poly(chlorotrifluoroethylene) (Kel—F), $\mathrm{+CF_2{-}CFCl+_{\mathit{n}}}$

(c) *Viton,* a copolymer of hexafluoropropene, $CF_2{=}CFCF_3$, and vinylidene fluoride, $CH_2{=}CF_2$

Alkenes also polymerize when they are treated with strong acids. The growing chains in acid-catalyzed polymerizations are *cations* rather than free radicals. The cationic polymerization of isobutylene is illustrated below.

$$(1)\quad \mathrm{H_2O + BF_3 \rightleftharpoons H^+ + BF_3(OH)^-}$$

$$(2)\quad \mathrm{H^+ + CH_2{=}\overset{\displaystyle CH_3}{\underset{\displaystyle CH_3}{C}} \longrightarrow CH_3{-}\overset{\displaystyle CH_3}{\underset{\displaystyle CH_3}{C^+}}}$$

$$(3)\quad \mathrm{CH_3{-}\overset{\displaystyle CH_3}{\underset{\displaystyle CH_3}{C^+}} + CH_2{=}\overset{\displaystyle CH_3}{\underset{\displaystyle CH_3}{C}} \longrightarrow CH_3{-}\overset{\displaystyle CH_3}{\underset{\displaystyle CH_3}{C}}{-}CH_2{-}\overset{\displaystyle CH_3}{\underset{\displaystyle CH_3}{C^+}}}$$

$$(4)\quad \mathrm{CH_3{-}\overset{\displaystyle CH_3}{\underset{\displaystyle CH_3}{C}}{-}CH_2{-}\overset{\displaystyle CH_3}{\underset{\displaystyle CH_3}{C^+}} \xrightarrow{CH_2=C(CH_3)_2} CH_3\overset{\displaystyle CH_3}{\underset{\displaystyle CH_3}{C}}{-}CH_2{-}\overset{\displaystyle CH_3}{\underset{\displaystyle CH_3}{C}}{-}CH_2{-}\overset{\displaystyle CH_3}{\underset{\displaystyle CH_3}{C^+}} \xrightarrow{etc.}}$$

The catalysts used for cationic polymerizations are usually Lewis acids that contain a small amount of water. The polymerization of isobutylene illustrates how the catalyst (BF_3 and H_2O) functions to produce growing cationic chains.

Problem C.3

How can you account for the fact that isobutylene polymerizes in the way we indicated above, rather than in the manner shown below?

$$\mathrm{H^+ + \overset{\displaystyle CH_3}{\underset{\displaystyle CH_3}{C}}{=}CH_2 \longrightarrow \overset{\displaystyle CH_3}{\underset{\displaystyle CH_3}{CH}}{-}CH_2^+ \xrightarrow{(CH_3)_2C=CH_2} \overset{\displaystyle CH_3}{\underset{\displaystyle CH_3}{CH}}{-}CH_2{-}\overset{\displaystyle CH_3}{\underset{\displaystyle CH_3}{C}}{-}CH_2^+ \xrightarrow{etc.}}$$

Alkenes containing electron-withdrawing groups polymerize in the presence of strong bases. Acrylonitrile, for example, polymerizes when it is treated with sodium amide ($NaNH_2$) in liquid ammonia. The growing chains in this polymerization are anions.

$$H_2\ddot{N}:^- + CH_2{=}\underset{\displaystyle CN}{\underset{|}{CH}} \xrightarrow{NH_3} H_2N{-}CH_2{-}\underset{\displaystyle CN}{\underset{|}{CH}}:^-$$

$$H_2N{-}CH_2{-}\underset{\displaystyle CN}{\underset{|}{CH}}:^- \xrightarrow{CH_2{=}CHCN} H_2N{-}CH_2\underset{\displaystyle CN}{\underset{|}{CH}}{-}CH_2{-}\underset{\displaystyle CN}{\underset{|}{CH}}:^- \xrightarrow{etc.}$$

Anionic polymerization of acrylonitrile is less important in commercial production than the free-radical process we illustrated earlier.

Problem C.4

If alkene monomers used in anionic polymerization are extremely pure, the chains continue growing until all of the monomer is consumed. Even then, however, most of the chains ends are still anions. These chains are said to be "living" chains. The "living" chains can be terminated—"killed"—by the addition of water. (a) How does water terminate the chain? (b) Speculate about what happens when one adds first ethylene oxide and then water to a "living" anionic polymer.

$$\left(CH_2\underset{\displaystyle R}{\underset{|}{CH}}\right)_n CH_2\underset{\displaystyle R}{\underset{|}{CH}}:^- \xrightarrow{\text{Excess } CH_2{-}CH_2 \text{ (O)}} ? \xrightarrow{H_2O} ?$$

Hint: Ethylene oxide, $\underset{O}{CH_2{-}CH_2}$, can also be polymerized by anions. The reaction involves ring opening of the highly strained three-membered ring.

$$CH_3{-}\ddot{\underset{..}{O}}:^- + \underset{O}{CH_2{-}CH_2} \longrightarrow CH_3OCH_2CH_2\ddot{\underset{..}{O}}:^- \xrightarrow[etc.]{\underset{O}{CH_2CH_2}} \underset{\text{A polyether}}{CH_3O(CH_2CH_2O)_n}$$

Stereochemistry of Addition Polymers

Head-to-tail polymerization of propylene produces a polymer in which every other atom is chiral. Many of the physical properties of the polypropylene produced in this way depend on the stereochemistry of these chiral centers.

$$CH_2{=}\underset{\displaystyle CH_3}{\underset{|}{CH}} \xrightarrow[\text{(head-to-tail)}]{\text{Polymerization}} {-}CH_2\underset{\displaystyle CH_3}{\underset{|}{\overset{*}{C}H}}CH_2\underset{\displaystyle CH_3}{\underset{|}{\overset{*}{C}H}}CH_2\underset{\displaystyle CH_3}{\underset{|}{\overset{*}{C}H}}CH_2\underset{\displaystyle CH_3}{\underset{|}{\overset{*}{C}H}}{-}$$

There are three general arrangements of the methyl groups and hydrogens along the chain. These arrangements are described as being *atactic, syndiotactic,* and *isotactic.*

If the stereochemistry at the chiral centers is random (Fig. C.1) the polymer is said to be atactic (*a,* without + Gr: *taktikos,* order).

In atactic polypropylene the methyl groups are randomly disposed on either side of the stretched carbon chain. [If we were to arbitrarily designate one end of the chain as having higher preference than the other we could give *R-S* designations (Section 8.5) to the chiral carbons. In atactic polypropylene the *R-S* designations would also be random, that is, *R, R, S, R, S, S,* and so on.

Polypropylene produced by free-radical polymerization at high pressures is atactic. Because the polymer is atactic, it is noncrystalline, has a low softening point, and has poor mechanical properties.

A second possible arrangement of the groups along the carbon chain is that of *syndiotactic* polypropylene. In syndiotactic polypropylene the methyl groups alternate regularly from one side of the stretched chain to the other (Fig. C.2). If we were to

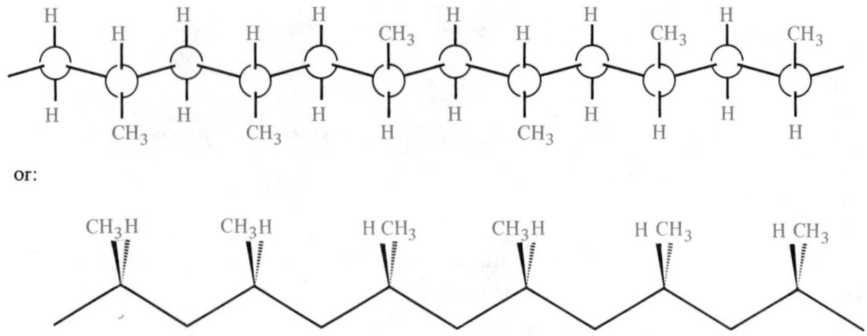

FIG. C.1
Atactic polypropylene. (In this illustration a "stretched" carbon chain is used for clarity.)

arbitrarily designate one end of the chain of syndiotactic polypropylene as having higher preference, the configuration of the chiral carbons would alternate, *R*, *S*, *R*, *S*, *R*, *S*, *R*, *S*, and so on.

The third possible arrangement of chiral centers is the *isotactic* arrangement shown in Fig. C.3. In the isotactic arrangement all of the methyl groups are on the same side of the stretched chain. The configurations of the chiral centers are either all *R* or all *S* depending on which end of the chain is assigned higher preference.

The names isotactic and syndiotactic come from the Greek term *taktikos* (order) plus *iso* (same) and *syndyo* (two together).

Before 1953, isotactic and syndiotactic addition polymers were unknown. In that year, however, a German chemist, Karl Ziegler, and an Italian chemist, Giulio Natta, announced independently the discovery of catalysts that permit stereochemical control of polymerization reactions.* The Ziegler-Natta catalysts, as they are now called, are prepared from transition metal halides and a reducing agent. The catalysts most commonly used are prepared from titanium tetrachloride ($TiCl_4$) and a trialkyl aluminum (R_3Al).

Ziegler-Natta catalysts are generally employed as suspended solids, and polymerization probably occurs at metal atoms on the surfaces of the particles. The mechanism for the polymerization is an ionic mechanism, but its details are not fully understood. There is evidence that polymerization occurs through an insertion of the alkene monomer between the metal and the growing polymer chain.

Both syndiotactic and isotactic polypropylene have been made using Ziegler-Natta catalysts. The polymerizations occur at much lower pressures and the polymers that are produced are much higher melting than atactic polypropylene. Isotactic polypropylene, for example, melts at 175°. Isotactic and syndiotactic polymers are also much more crystalline than atactic polymers. The regular arrangement of groups along the chains allows them to fit together better in a crystal structure.

* Ziegler and Natta were awarded the Nobel Prize for their discoveries in 1963.

FIG. C.2
Syndiotactic polypropylene.

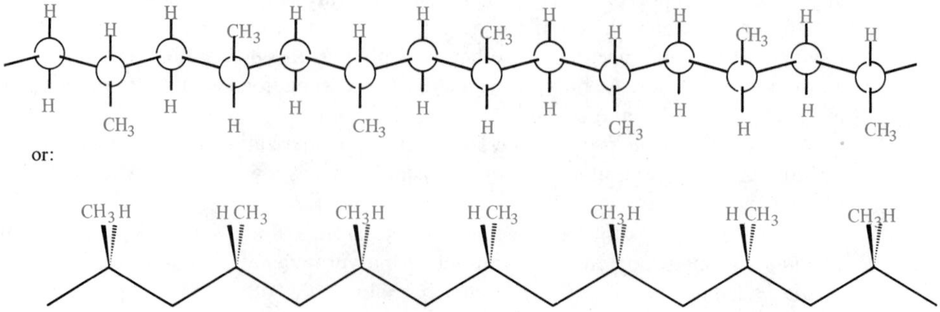

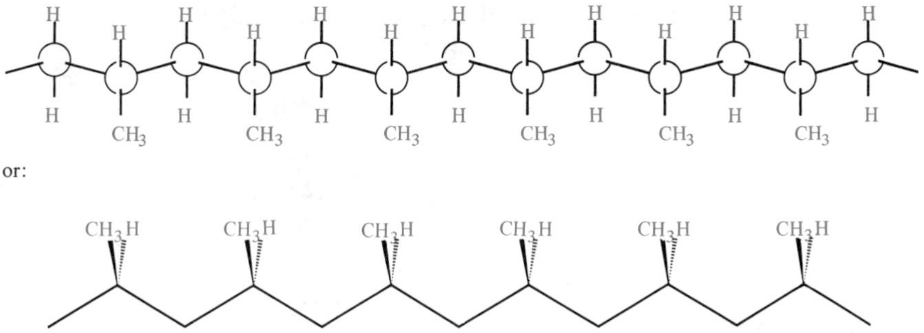

or:

FIG. C.3
Isotactic polypropylene.

Atactic, syndiotactic, and isotactic forms of poly(methyl methacrylate) (p. 337) are also known. The atactic form is a noncrystalline glass. The crystalline syndiotactic and isotactic forms melt at 160° and 200°, respectively.

Problem C.5

Write structural formulas for portions of the chain of:

(a) Atactic poly(methyl methacrylate)
(b) Syndiotactic poly(methyl methacrylate)
(c) Isotactic poly(methyl methacrylate)

SPECIAL TOPIC

DIVALENT CARBON COMPOUNDS: CARBENES

In recent years considerable research has been devoted to investigating the structures and reactions of a group of compounds in which carbon forms only *two bonds*. These neutral divalent carbon compounds are called *carbenes*.

Most carbenes are highly unstable compounds that are capable of only fleeting existence. Soon after carbenes are formed they usually react with another molecule. The reactions of carbenes are especially interesting because, in many instances, the reactions show a remarkable degree of stereospecificity. The reactions of carbenes are also of great synthetic use in the preparation of compounds that have three-membered rings.

Structure of Methylene

The simplest carbene is the compound called methylene, CH_2. Methylene can be prepared by the decomposition of diazomethane,* CH_2N_2. This decomposition can be

*Diazomethane is a resonance hybrid of the three structures, I, II, and III, shown below.

$$\underset{\text{I}}{:\overset{-}{C}H_2—\overset{+}{N}\equiv N:} \longleftrightarrow \underset{\text{II}}{CH_2=\overset{+}{N}=\overset{-}{\ddot{N}}:} \longleftrightarrow \underset{\text{III}}{\overset{..-}{C}H_2—\overset{..}{N}=\overset{+}{N}:}$$

We have chosen resonance structure I to illustrate the decomposition of diazomethane because with I it is readily apparent that heterolytic cleavage of the carbon-nitrogen bond results in the formation of methylene and molecular nitrogen.

$$:\overset{-}{C}H_2 \,\vdots\, :\overset{+}{N}\equiv N: \longrightarrow :CH_2 + :N\equiv N:$$

accomplished by heating diazomethane (thermolysis) or by irradiating it with light of a wavelength that it can absorb (photolysis).

$$:\bar{C}H_2-\overset{+}{N}{\equiv}N: \xrightarrow[\text{or light}]{\text{Heat}} :CH_2 + :N{\equiv}N:$$

Diazomethane Methylene Nitrogen

The structure of methylene is more complicated than its simple formula, CH_2, suggests. There are actually three different forms of methylene. These forms **1, 2,** and **3,** differ in the arrangement of their nonbonding electrons.

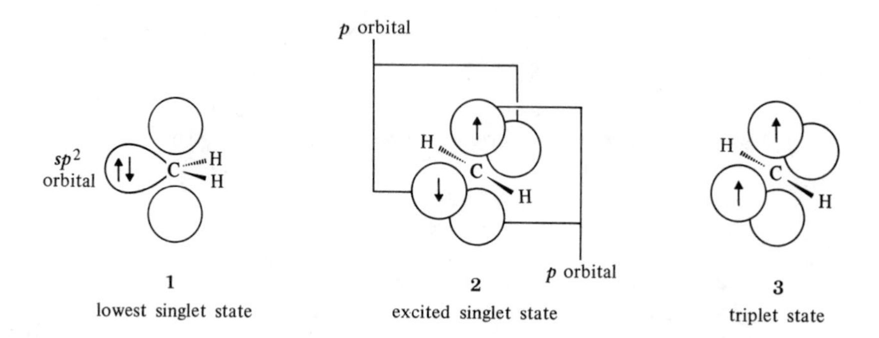

The two states, **1** and **2,** that have spin-paired electrons are called *singlet states.* State **3,** in which the electron spins are *not* paired, is called a *triplet state.*† In the lower singlet state **1,** the *spin-paired* nonbonding electrons occupy an sp^2 orbital of an essentially sp^2-hybridized carbon. (The bond angle at the carbon is close to 120°.) The p orbital of **1** is vacant.

The carbon atoms of the excited singlet state **2** and the triplet state **3** are sp hybridized and both are linear molecules. The electrons of the excited singlet state occupy separate p orbitals *but their spins are paired.* The electrons of the triplet state also occupy separate p orbitals but their spins are *unpaired.* The triplet state of methylene resembles a diradical and, as we will see, it reacts like one.

The relative energies of the three states of methylene are shown in Fig. D.1. The

†The names singlet and triplet are derived from spectroscopic designations for species of this general type.

FIG. D.1
The relative energies of the three states of methylene.

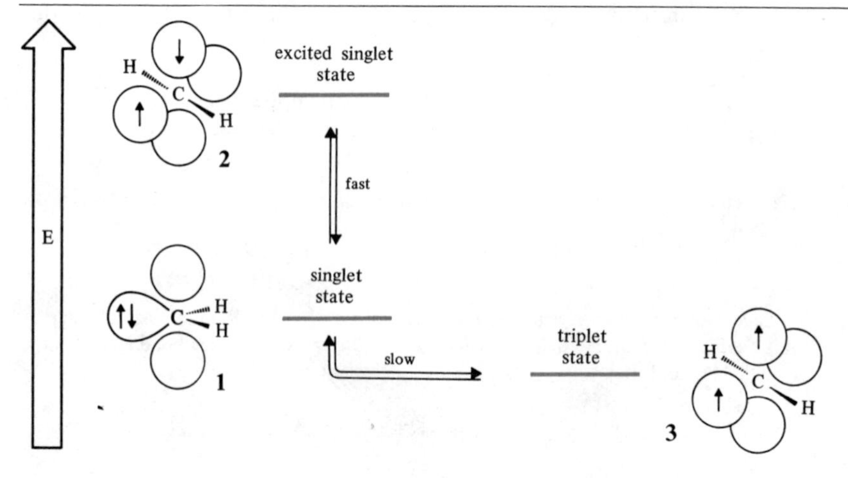

photolysis of diazomethane produces singlet methylene initially, even though triplet methylene has slightly lower energy. Transitions between singlet methylene and triplet methylene, however, do not occur rapidly. The transition from singlet methylene to triplet methylene requires that the singlet methylene lose energy. It can only do this by colliding with some other molecule or the walls of the container, to which it transfers the excess energy.

Reactions of Methylene

Singlet and triplet methylene both react with alkenes by adding to the double bond to form cyclopropanes. Additions of singlet and triplet methylene, however, give different stereochemical results.

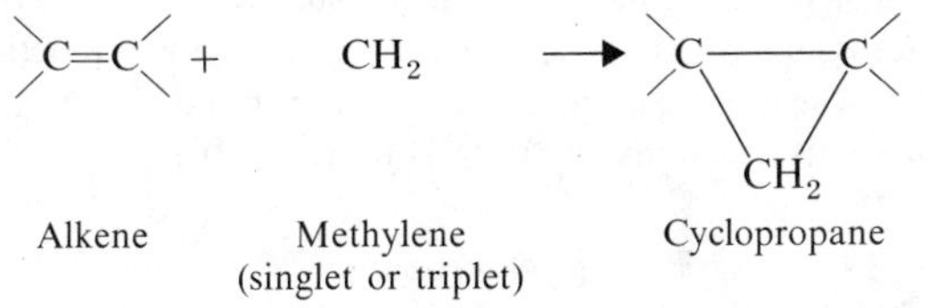

$$>C=C< \; + \; CH_2 \longrightarrow \text{cyclopropane (}>C\text{—}C<\text{ bridged by } CH_2\text{)}$$

Alkene — Methylene (singlet or triplet) — Cyclopropane

Singlet methylene reacts *stereospecifically*. It reacts with *cis*-2-butene, for example, to give *cis*-1,2-dimethylcyclopropane.

$$CH_3(H)C=C(H)CH_3 \; + \; \uparrow\downarrow CH_2 \longrightarrow$$ *cis*-1,2-dimethylcyclopropane

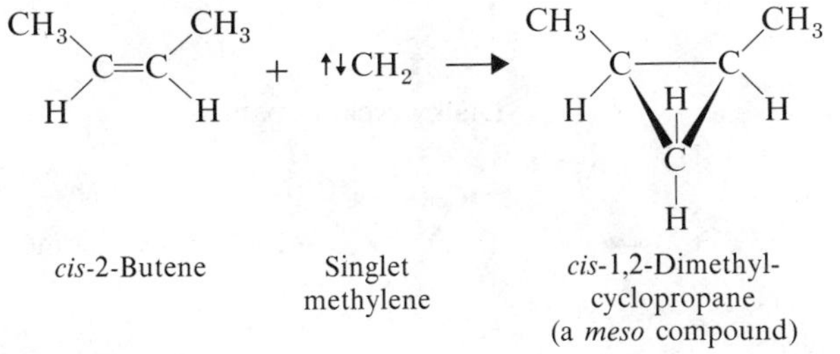

cis-2-Butene — Singlet methylene — *cis*-1,2-Dimethyl-cyclopropane (a *meso* compound)

Singlet methylene reacts with *trans*-2-butene to give a racemic modification of the *trans*-1,2-dimethylcyclopropane enantiomers.*

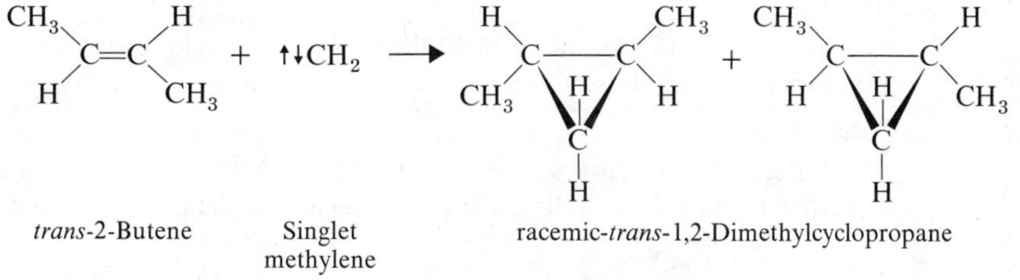

$$CH_3(H)C=C(CH_3)H \; + \; \uparrow\downarrow CH_2 \longrightarrow$$ (both *trans* enantiomers) $+$

trans-2-Butene — Singlet methylene — racemic-*trans*-1,2-Dimethylcyclopropane

When triplet methylene reacts with either *cis*- or *trans*-2-butene the reactions are *nonstereospecific*. The reaction of either *cis*-2-butene or *trans*-2-butene with triplet methylene gives a mixture of the *cis*- and *trans*-1,2-dimethylcyclopropanes.

*These reactions are stereospecific (rather than stereoselective) because in them, stereoisomerically different reactants (*cis*- and *trans*-2-butene) give stereoisomerically different products (*cis*-1,2-dimethylcyclopropane and the *trans*-1,2-dimethylcyclopropanes, respectively).

$CH_3CH{=}CHCH_3$ + $\uparrow\overset{\uparrow}{C}H_2$ ⟶

cis- or *trans*-
2-Butene

Triplet
methylene

CH3 CH3 H CH3 CH3 H
C—C + C—C + C—C
H H CH3 H H CH3
CH2 CH2 CH2

Mixture of *cis*- or *trans*-1,2-dimethylcyclopropanes

As we might expect, different stereochemical results are obtained from the reactions of singlet and triplet methylene because the mechanisms of the two reactions are different. Singlet methylene adds to the double bond in a single step; bonds form between the carbon of singlet methylene and the carbons of the alkene at the same time. As a result, the

cis-Alkene ⟶ *cis*-Diakylcyclopropane

R R R R
C===C C—C
H H ⟶ H H
CH2 CH2
Singlet methylene

trans-Alkene ⟶ *trans*-Dialkylcyclopropane

R H R H
C===C C—C
H R ⟶ H R + enantiomer
CH2 CH2
Singlet
methylene

stereochemistry of the reactant is preserved in the product: *cis*-alkenes yield *cis*-dialkylcyclopropanes; *trans*-alkenes yield *trans*-dialkylcyclopropanes.

Problem D.1

How can you account for the fact that the addition of singlet methylene to *trans*-2-butene produces equal amounts of the two enantiomeric *trans*-1,2-dimethylcyclopropanes?

Because its electrons are not paired, triplet methylene reacts in a two-step mechanism. Triplet methylene (a diradical itself) reacts with the alkene to form an intermediate

cis-Alkene

R R R R H R
C—C ⟶ C—C —Rotation⟶ C—C
H H H H R H
·CH2 ·CH2 ·CH2
Triplet
methylene
A B
↓ ↓

R R H R
C—C C—C
H H R H
CH2 CH2

cis-1,2-Dialkylcyclopropane

trans-1,2-Dialkylcyclopropane

diradical in conformation **A.** The intermediate diradical is also in a triplet state (with its electrons unpaired) and therefore it has a lifetime long enough to allow rotation of groups joined by single bonds. Rotations lead to a mixture of conformations **A** and **B**, and ring closures of **A** and **B** give diastereomerically different cyclopropanes.

When we plan experiments we must take into account the stereochemical differences between the reactions of singlet and triplet methylene. But how do we know which form of methylene will be the predominant one in a reaction? The answer is that the predominant form will depend on the reaction conditions that we use.

In the liquid phase the reaction of singlet methylene will predominate because it forms first in the decomposition of diazomethane and because *it reacts with the alkene before it can undergo transition to triplet methylene.* In the liquid state the singlet methylene is in close proximity to many alkene molecules. The reaction of singlet methylene and an alkene, therefore, occurs rapidly. The transition of singlet to triplet methylene, by contrast, is slow and in the liquid state does not occur to an appreciable extent. We find that reactions of methylene in the liquid state are largely stereospecific.

When we carry out methylene additions in the gas phase, particularly when we dilute the mixture by the addition of an inert gas, the molecules will be widely separated. In gas-phase reactions singlet methylene will have an opportunity to undergo transition to the lower-energy triplet state through collisions with molecules of the inert gas. Thus, gas-phase reactions give primarily those products expected from triplet methylene; the reactions are largely nonstereospecific.

Problem D.2

What products would you expect from the following reactions?

(a) *cis*-2-Pentene + $CH_2N_2 \xrightarrow[\text{(liquid state)}]{\text{Light}}$

(b) *trans*-2-Pentene + $CH_2N_2 \xrightarrow[\text{(liquid state)}]{\text{Light}}$

(c) *cis*-2-Pentene + $CH_2N_2 \xrightarrow[\text{He (gas phase)}]{\text{Light}}$

(d) *trans*-2-Pentene + $CH_2N_2 \xrightarrow[\text{He (gas phase)}]{\text{Light}}$

Other Reactions of Methylene

Singlet methylene not only reacts with double bonds to form three-membered rings, it also reacts with carbon-hydrogen bonds by *insertion.* This reaction is particularly

$$\geq C{-}H + {\uparrow\downarrow}CH_2 \xrightarrow{\text{Insertion}} \geq C{-}{-}{-}{-}H \text{ (bridged by } CH_2\text{)} \longrightarrow \geq C{-}CH_3$$

(Singlet methylene)

important if the reactant does not contain a double bond. Butane, for example, reacts with singlet methylene to give a mixture of pentane and isopentane. Insertion reactions of

$$CH_3CH_2CH_2CH_3 \xrightarrow[\text{light}]{CH_2N_2} CH_3CH_2CH_2CH_2CH_3 + CH_3CH_2\underset{\displaystyle CH_3}{\underset{|}{C}}HCH_3$$

methylene are highly exothermic. As a consequence, singlet methylene is relatively unselective. It does not discriminate between primary, secondary, and tertiary hydrogens to an appreciable extent.

When triplet methylene reacts with alkanes, it reacts primarily by abstracting hydrogens.

$$CH_3CH_2CH_3 + \uparrow\overset{\uparrow}{C}H_2 \longrightarrow CH_3CH_2CH_2\cdot \text{ or } CH_3\underset{\bullet}{C}HCH_3 + CH_3\cdot$$

Triplet methylene

Problem D.3

What eventual products would you expect from the reaction of propane with triplet methylene given above?

Reactions of Other Carbenes

Dihalocarbenes are also frequently employed in the synthesis of cyclopropane derivatives from alkenes. The singlet state of dihalocarbenes appears to be the more stable and most reactions of dihalocarbenes are stereospecific.

Dichlorocarbene can be synthesized by the α *elimination* of the elements of hydrogen chloride from chloroform. This reaction resembles the β elimination reactions by

$$R{-}\ddot{\underset{\cdot\cdot}{O}}{:}^{-}K^{+} + H{:}CCl_3 \rightleftharpoons R{-}\ddot{\underset{\cdot\cdot}{O}}{:}H + {}^{-}{:}CCl_3 + K^{+}$$

$${}^{-}{:}CCl_3 \xrightarrow{\text{Slow}} {:}CCl_2 + {:}\ddot{\underset{\cdot\cdot}{Cl}}{:}^{-}$$

Dichlorocarbene

which alkenes are synthesized from alkyl halides (Section 5.9). Compounds *with a β hydrogen* react by β elimination preferentially. Compounds with no β hydrogen (such as chloroform) react by α elimination.

A variety of cyclopropane derivatives have been prepared by generating dichlorocarbene in the presence of alkenes. Cyclohexene, for example, reacts with

$$\rangle C{=}C\langle + {:}CCl_2 \longrightarrow \text{(cyclopropane: } \rangle C{-}C\langle \text{ bridged by } CCl_2)$$

dichlorocarbene generated by treating chloroform with potassium *tert*-butoxide to give a bicyclic product.

$$\text{cyclohexene} \xrightarrow[CHCl_3]{KOC(CH_3)_3} \text{7,7-dichlorobicyclo[4.1.0]heptane (Cl, Cl)} \quad (59\%)$$

Carbenoids: The Simmons-Smith Cyclopropane Synthesis

A useful cyclopropane synthesis has been developed by H. E. Simmons and R. D. Smith of the du Pont Company. In this synthesis diiodomethane and a zinc-copper couple are stirred together with an alkene. The diiodomethane and zinc react to produce a carbenelike species called a *carbenoid.*

$$CH_2I_2 + Zn(Cu) \longrightarrow ICH_2ZnI$$

A carbenoid

The carbenoid then brings about the stereospecific addition of a CH_2 group directly to the double bond.

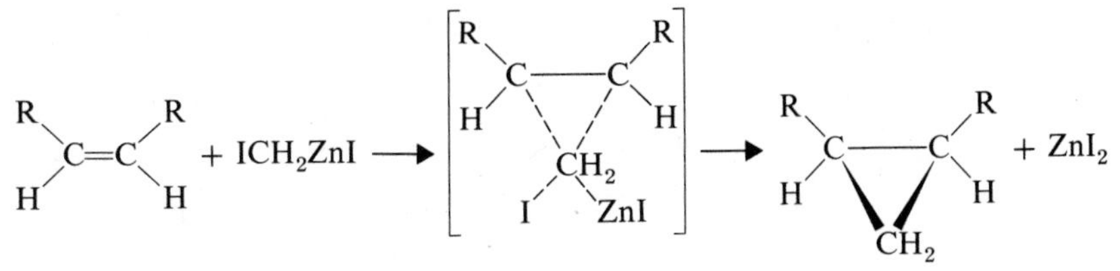

$$\text{RHC=CHR} + ICH_2ZnI \longrightarrow [\text{transition state: } CH_2 \text{ bridging the two C atoms, with I and ZnI}] \longrightarrow \text{cyclopropane (R, H on each C; } CH_2\text{)} + ZnI_2$$

This synthesis has been used widely. One example is the synthesis of methyl dihydrosterculate from methyl oleate.

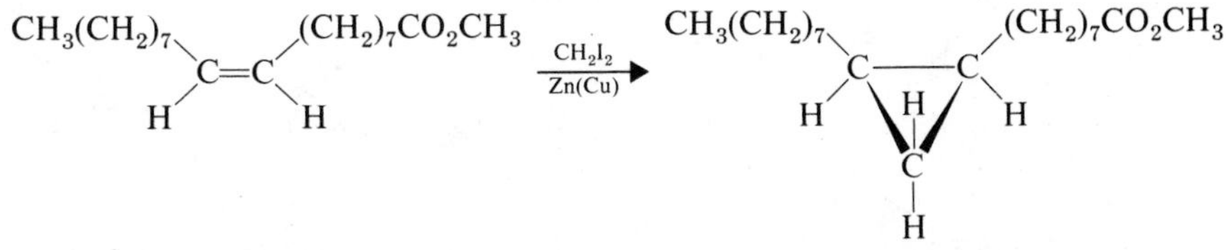

$$CH_3(CH_2)_7CH{=}CH(CH_2)_7CO_2CH_3 \xrightarrow[Zn(Cu)]{CH_2I_2} CH_3(CH_2)_7\text{-}\underset{\text{cyclopropane, }CH_2}{CH\text{—}CH}\text{-}(CH_2)_7CO_2CH_3$$

Methyl oleate — Methyl dihydrosterculate

Methyl dihydrosterculate is related to sterculic acid, an interesting compound that has been isolated from the kernel oil of the tropical tree *Sterculia foetida*. Sterculic acid was the first naturally occurring compound found to have the highly strained cyclopropene ring.

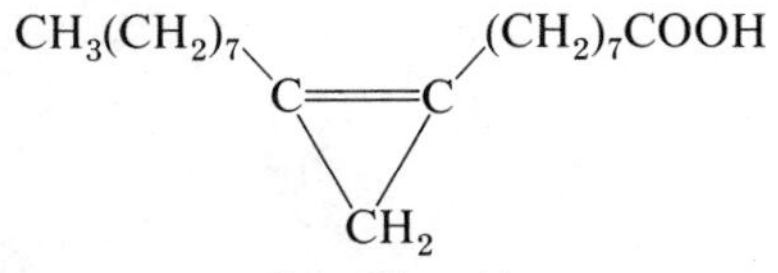

$$CH_3(CH_2)_7\text{-}C{=}C\text{-}(CH_2)_7COOH \text{ (ring closed by } CH_2\text{)}$$

Sterculic acid

Sterculic acid, itself, has been synthesized using the Simmons-Smith method.

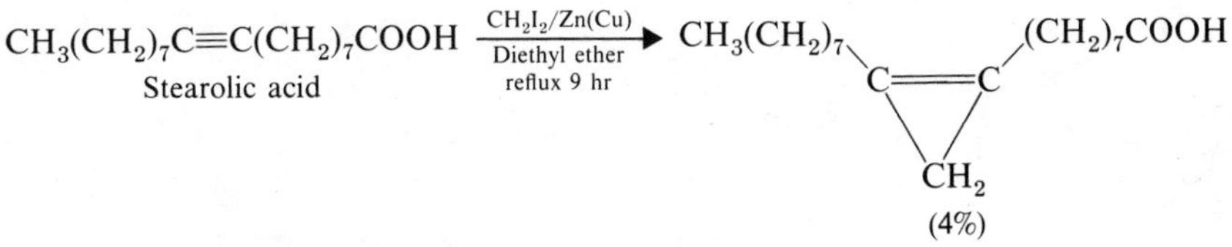

$$CH_3(CH_2)_7C{\equiv}C(CH_2)_7COOH \xrightarrow[\text{Diethyl ether, reflux 9 hr}]{CH_2I_2/Zn(Cu)} CH_3(CH_2)_7\text{-}C{=}C\text{-}(CH_2)_7COOH \text{ (ring closed by } CH_2\text{)}$$

Stearolic acid — (4%)

This reaction illustrates the addition of a carbenoid to a carbon-carbon triple bond. (In Section 9.14 we will see how stearolic acid is prepared.)

The zinc-copper couple used in the Simmons-Smith synthesis can also be prepared *in situ* (in the reaction mixture) as the following example illustrates.

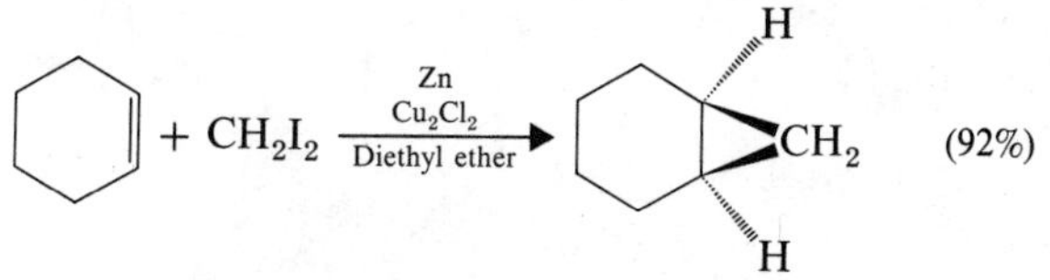

$$\text{Cyclohexene} + CH_2I_2 \xrightarrow[\text{Diethyl ether}]{Zn,\ Cu_2Cl_2} \text{bicyclo[4.1.0]heptane (H, H; } CH_2\text{)} \quad (92\%)$$

Problem D.4

How might the following compounds be synthesized?

(a) [cis-1,1-dichloro-2,3-dimethylcyclopropane: CH_3, H on each ring carbon; CCl_2]

(b) [bicyclo[3.1.0]hexane: H, CH_2, H]

(c) [6-tert-butyl-6-chlorobicyclo[3.1.0]hexane: $C(CH_3)_3$, Cl]

(d) [7,7-dibromobicyclo[4.1.0]heptane: H, CBr_2, H]

Problem D.5

(a) Provide a plausible explanation of the fact that the triplet state of methylene is of lower energy than the singlet state. (b) Explain why the singlet state of CCl_2 is lower in energy than the triplet state.

9 ALKYNES

9.1 INTRODUCTION

Hydrocarbons whose molecules contain the carbon-carbon triple bond are called alkynes. The common name for this family is *acetylenes,* after the first member, HC≡CH. Alkynes or acetylenes are also unsaturated compounds. They have the general formula C_nH_{2n-2} and therefore have an index of hydrogen deficiency (Section 6.5) equal to 2.

	Alkane	Cycloalkane	Alkene	Alkyne
General formula	C_nH_{2n+2}	C_nH_{2n}	C_nH_{2n}	C_nH_{2n-2}
Representative compound	$CH_3(CH_2)_3CH_3$ Pentane C_5H_{12}	(pentagon) Cyclopentane C_5H_{10}	$CH_3(CH_2)_2CH{=}CH_2$ 1-Pentene C_5H_{10}	$CH_3(CH_2)_2C{\equiv}CH$ 1-Pentyne C_5H_8
Index of hydrogen deficiency	0	1	1	2
	Saturated compounds		Unsaturated compounds	

9.2 NOMENCLATURE OF ALKYNES

9.2A IUPAC Nomenclature

Alkynes are named in much the same way as alkenes. Unbranched alkynes, for example, are named by replacing the *-ane* of the name of the corresponding alkane with the ending *-yne.* The chain is numbered in order to give the carbon atoms of the triple bond the lowest possible numbers. The lower number of the two carbons of the triple bond is used to designate the location of the triple bond. The IUPAC names of several unbranched alkynes are shown below.

HC≡CH	Ethyne or acetylene*
$CH_3C{\equiv}CH$	Propyne
$CH_3CH_2C{\equiv}CH$	1-Butyne
$CH_3C{\equiv}CCH_3$	2-Butyne

* The name acetylene is retained by the IUPAC system for the compound HC≡CH and is used much more frequently than the name ethyne.

$CH_3CH_2CH_2C{\equiv}CH$ 1-Pentyne
$CH_3CH_2C{\equiv}CCH_3$ 2-Pentyne

The locations of substituent groups of branched alkynes and substituted alkynes are also indicated with numbers.

$\overset{3}{Cl{-}CH_2}\overset{2}{C}{\equiv}\overset{1}{CH}$
3-Chloropropyne

$\overset{4}{CH_3}\overset{3}{C}{\equiv}\overset{2}{C}\overset{1}{CH_2}Cl$
1-Chloro-2-butyne

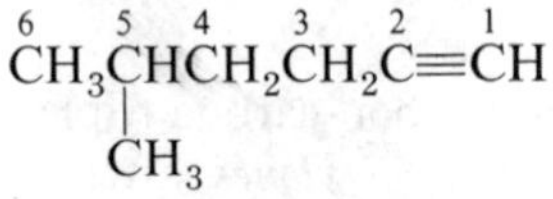

$\overset{6}{CH_3}\overset{5}{CH}(CH_3)\overset{4}{CH_2}\overset{3}{CH_2}\overset{2}{C}{\equiv}\overset{1}{CH}$
5-Methyl-1-hexyne

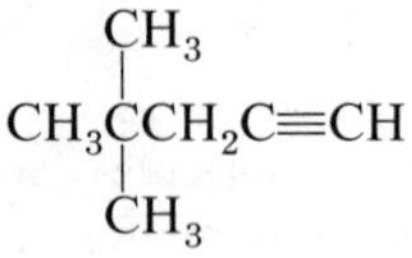

$CH_3C(CH_3)_2CH_2C{\equiv}CH$
4,4-Dimethyl-1-pentyne

Problem 9.1

Give the IUPAC names of all of the alkyne isomers of (a) C_4H_6, (b) C_5H_8, (c) C_6H_{10}.

9.2B Common Names

In an older system of nomenclature, alkynes are named as substituted acetylenes. Propyne, for example, a monosubstituted acetylene, is sometimes called methylacetylene. The disubstituted acetylene, 2-butyne, is called dimethylacetylene. The following examples illustrate this older system of nomenclature.

$CH_3C{\equiv}CH$
Propyne
(methylacetylene)

$CH_3CH_2C{\equiv}CH$
1-Butyne
(ethylacetylene)

$CH_3C{\equiv}CCH_3$
2-Butyne
(dimethylacetylene)

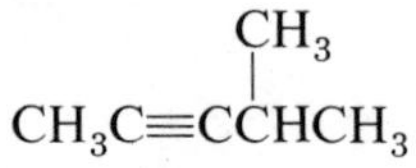

$CH_3C{\equiv}CCH(CH_3)CH_3$
4-Methyl-2-pentyne
(methylisopropylacetylene)

Monosubstituted acetylenes or 1-alkynes are called terminal acetylenes, and the hydrogen attached to the carbon of the triple bond is called the acetylenic hydrogen.

$R{-}C{\equiv}C{-}H$ ← Acetylenic hydrogen
A terminal acetylene

The anion obtained when the acetylenic hydrogen is removed is known as *an alkynide ion* or an acetylide ion.

$R{-}C{\equiv}C{:}^-$
An alkynide ion
(an acetylide ion)

9.3 *sp*-HYBRIDIZED CARBON: THE STRUCTURE OF ACETYLENE

The molecule of acetylene is linear. The carbon-carbon triple bond (1.20 Å) is considerably shorter than the carbon-carbon double bond of ethene (1.34 Å). The carbon-hydrogen bonds of acetylene (1.08 Å) are also shorter than those of ethene (1.09 Å).

$$\underset{\longleftarrow 180° \longrightarrow}{H\overset{1.08\ Å}{—}C\overset{1.20\ Å}{\equiv}C—H}$$

We can account for the structure of acetylene on the basis of orbital hybridization in much the same way that we did for ethane and ethene. In our model for ethane (Section 2.2B) we saw that the carbon orbitals are sp^3 hybridized, and in our model for ethene (Section 6.3) we saw that they are sp^2 hybridized. In our model for acetylene we will see that the carbon atoms are *sp hybridized.*

The mathematical process for obtaining the *sp*-hybrid orbitals of acetylene can be visualized in the way shown in Fig. 9.1. The 2*s* orbital and one 2*p* orbital of carbon are hybridized to form two *sp* orbitals. (In our illustration we have used the $2p_x$ orbital for this purpose but we could have used either of the other two.) The remaining two 2*p* orbitals (in this illustration the $2p_z$ and $2p_y$ orbitals) are not hybridized. When the centers of the orbitals are superposed, the *sp*-hybrid orbitals are found to have their large positive lobes oriented at an angle of 180° with respect to each other. The 2*p* orbitals that were not hybridized are perpendicular to the axis that passes through the center of the two *sp* orbitals.

We envision our model for acetylene being formed in the following way. Two carbon atoms overlap *sp* orbitals to form a sigma bond between them (this is one bond of the triple bond). The remaining two *sp* orbitals at each carbon overlap with *s* orbitals from hydrogen atoms to produce sigma C—H bonds. (In the illustration below we have omitted the small negative lobes of the *sp* orbitals.)

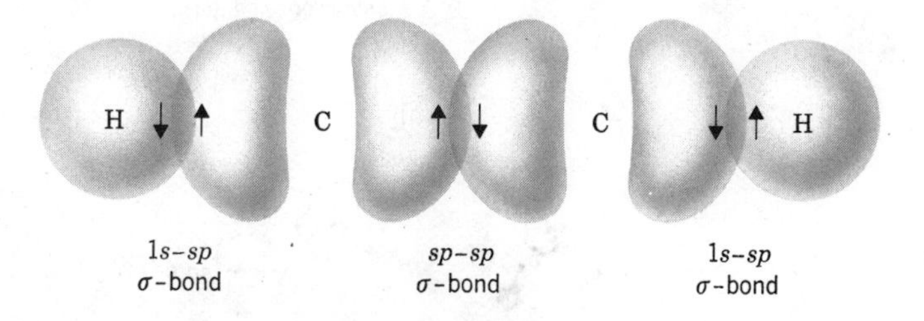

The two *p* orbitals on each carbon also overlap side to side to form two π bonds. If we replace the σ bonds of the illustration above with lines, it is easier to see how the *p* orbitals overlap.

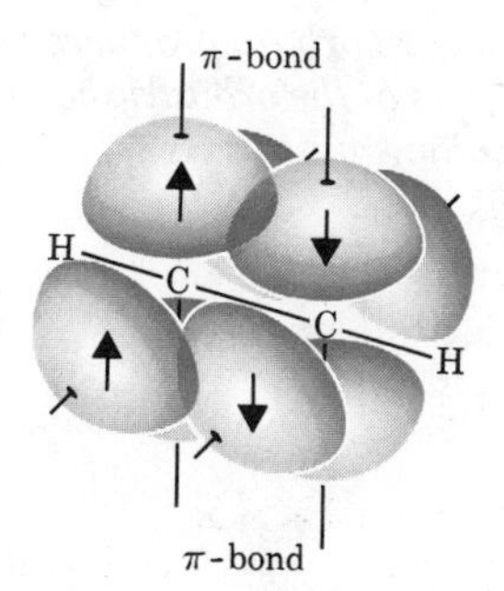

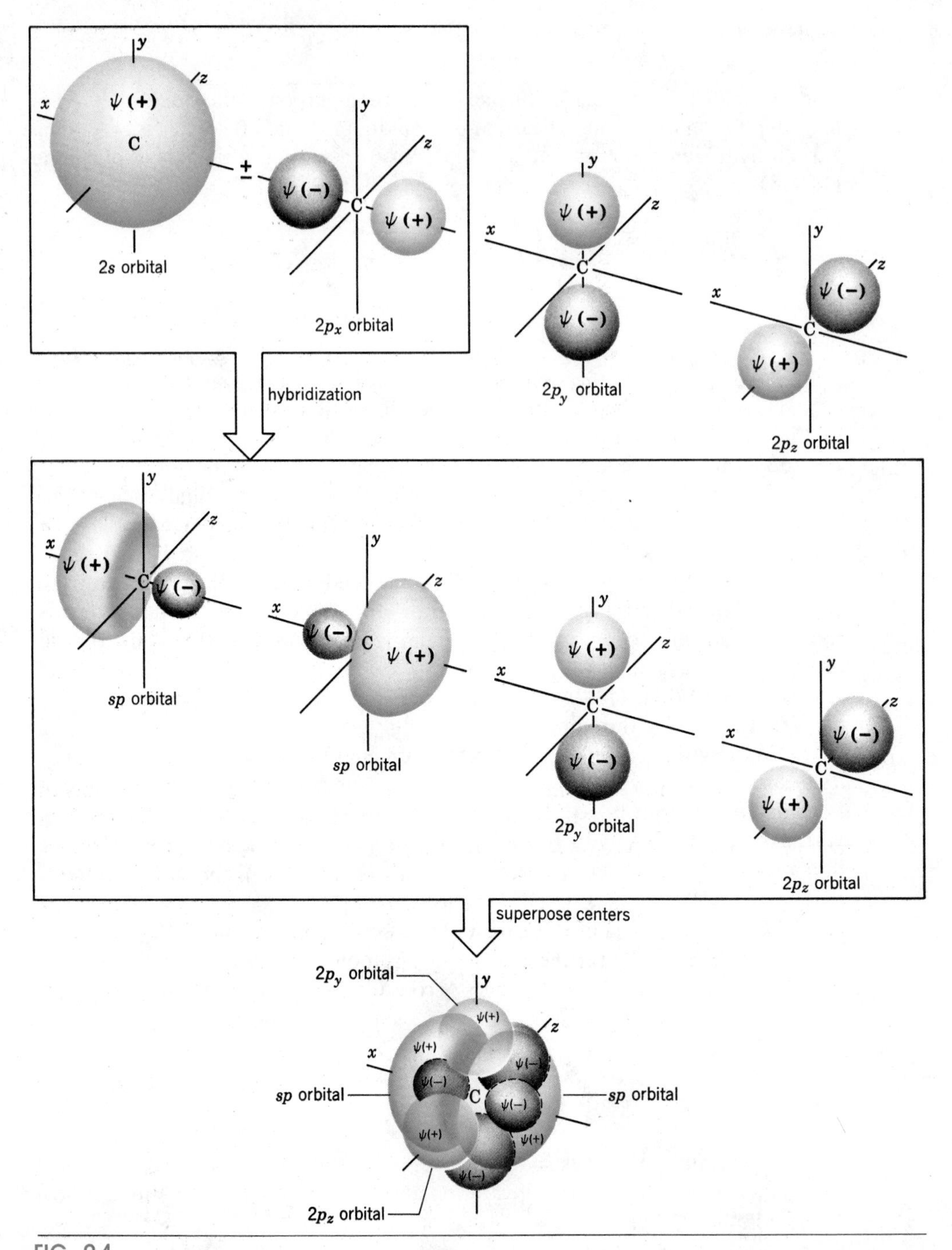

FIG. 9.1
An sp *hybridized carbon atom. Mixing a* 2s *orbital with a* 2p *orbital produces two* sp *orbitals. Two* 2p *orbitals are not hybridized. Superposing the centers of the orbitals shows that the large* (+) *lobes of* sp *orbitals point in opposite directions (along the* x *axis). The* 2p *orbitals are perpendicular to each other and lie along the* y *and* z *axes.*

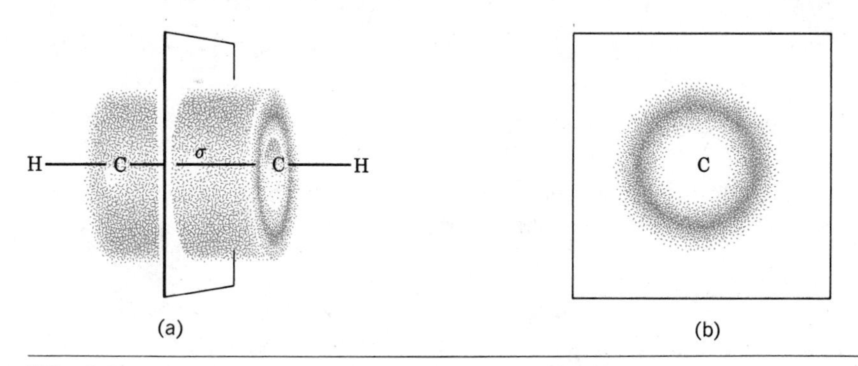

FIG. 9.2
(a) *The shape of the π-electron cloud in acetylene.* (b) *A cross section. The electrons of the carbon-carbon σ bond are located in the "hollow" part of the cylinder where the π-electron density is low. The σ and π bonds, however, share some space in common.*

Thus we see that the carbon-carbon triple bond consists of two π bonds and one σ bond.

The *p* orbitals giving rise to the π bonds overlap both above and below the carbon atoms and in front of and behind them as well. Because the *p* orbitals overlap in this way, the π-electron density of the triple bond is actually cylindrical in cross section (Fig. 9.2).

9.3A Bond Lengths of Acetylene

In addition to accounting for the linear structure of acetylene, the model based on *sp*-hybridized carbons also helps us to understand why the carbon-carbon and carbon-hydrogen bonds of acetylene are shorter than those of ethene. Orbitals that are *sp*-hybridized contain more *s* character than orbitals that are sp^2 hybridized and *sp* orbitals, generally, form shorter bonds than sp^2 orbitals. This effect, along with the greater electron density of the triple bond, drawing the nuclei closer to each other, accounts for the fact that the carbon-carbon triple bond is shorter. The greater *s* character of *sp* orbitals accounts for the shorter carbon-hydrogen bonds of acetylene as well.

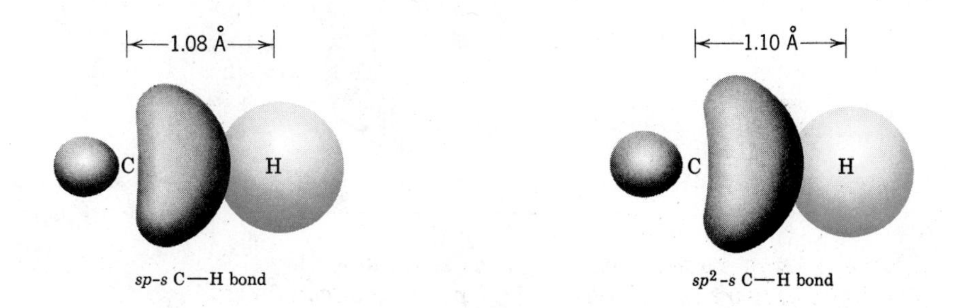

9.4 THE ACIDITY OF ACETYLENE AND TERMINAL ALKYNES

The C—H bonds of acetylene are *more difficult* to break *homolytically* than are the C—H bonds of ethene or ethane

$$\equiv C{:}H \xrightarrow[\text{cleavage}]{\text{Homolytic}} \equiv C\cdot \;+ H\cdot$$

but they are *easier* to break in the following *heterolytic* way.

$$\equiv C{:}H \xrightarrow[\text{cleavage}]{\text{Heterolytic}} \equiv C{:}^- + H^+$$

As a result of this ease of heterolytic cleavage, the hydrogens of acetylene are considerably more acidic than those of ethene or ethane.

H—C≡C—H	$H_2C{=}CH_2$	$H_3C{-}CH_3$
$K_a \simeq 10^{-25}$	$K_a \simeq 10^{-36}$	$K_a \simeq 10^{-42}$

We can account for this order of acidities on the basis of the hybridization state of carbon in each compound. Electrons of 2*s* orbitals, because they are spherical, tend on thé average to be much closer to the nucleus than electrons of 2*p* orbitals. With hybrid orbitals, therefore, *having more s character means that the electrons will, on average, be closer to the nucleus.* The *sp* orbitals used to make the sigma bonds of acetylene have 50% *s* character (because they arise from one *s* orbital and one *p* orbital), those of sp^2 orbitals have 33.3% *s* character, while those of sp^3 orbitals have only 25% *s* character. This means, in effect, that the *sp* orbitals of acetylene act as if they were the most electronegative when compared to sp^2 and sp^3 orbitals. (Remember that electronegativity measures an atom's ability to hold bonding electrons close to it.) The order of electronegativity of carbon in each hybridization state is: $sp > sp^2 > sp^3$.

Now we can see how the order of relative acidities of acetylene, ethene, and ethane parallels the effective electronegativity of carbon:

Relative acidity: $HC{\equiv}CH > CH_2{=}CH_2 > CH_3CH_3$

Being most electronegative, the carbon of acetylene is best able to accommodate the electron pair in the anion left after the proton is lost.

$$HC{\equiv}CH + B{:}^- \rightleftarrows B{-}H + HC{\equiv}C{:}^-$$

This anion is the most stable because the *sp* carbon is most electronegative. Thus, acetylene is most acidic.

$$CH_2{=}CH_2 + B{:}^- \rightleftarrows B{-}H + CH_2{=}CH{:}^-$$

$$CH_3CH_3 + B{:}^- \rightleftarrows B{-}H + CH_3CH_2{:}^-$$

This anion is the least stable because the sp^3 carbon is the least electronegative. Thus, ethane is the least acidic.

Notice that the explanation given here is the same as that given to account for the relative acidities of HF, H_2O, and NH_3.

Relative acidity: $HF > H_2O > NH_3$

Since the atoms being compared are in the same horizontal row of the periodic table the acidity of these hydrides parallels the electronegativity of their atoms:

$$F > O > N$$

Because fluorine is most electronegative, fluoride ions are the most stable (more stable than OH^- or NH_2^-) and thus HF is the strongest acid.

The order of basicities of the anions is opposite their relative stability. The ethanide ion is least stable and thus it is the most basic.

Relative basicity: $CH_3CH_2:^- > CH_2{=}CH:^- > HC{\equiv}C:^-$

What we have said about acetylene and acetylide ions is true of any terminal alkyne, $RC{\equiv}CH$, and any alkynide ion, $RC{\equiv}C:^-$. If we include other hydrogen compounds of the first-row elements of the periodic table we can write the following orders of relative acidities and basicities.

Relative acidity: $H{-}\ddot{\underset{..}{O}}H;\ H{-}\ddot{\underset{..}{O}}R > H{-}C{\equiv}CR > H{-}\ddot{N}H_2 > H{-}CH{=}CH_2 > H{-}CH_2CH_3$

Relative basicity: $^-:\ddot{\underset{..}{O}}H;\ ^-:\ddot{\underset{..}{O}}R < ^-:C{\equiv}CR < ^-:\ddot{N}H_2 < ^-:CH{=}CH_2 < ^-:CH_2CH_3$

The arguments that we have just made apply only to acid-base reactions that take place in solution. In the gas phase, acidities and basicities are very much different. For example, in the gas phase the hydroxide ion is a stronger base than the acetylide ion. The explanation for this shows us again the important roles solvents play in reactions that involve ions (cf. Section 5.7). In solution, smaller ions (e.g., hydroxide ions) are more effectively solvated than larger ones (e.g., acetylide ions). Because they are more effectively solvated, smaller ions are more stable and are therefore less basic. In the gas phase, large ions are stabilized by polarization of their bonding electrons and the bigger a group is the more polarizable it will be. Thus in the gas phase larger ions are less basic.

Problem 9.2

Predict the products of the following acid-base reactions. If the equilibrium would not result in the formation of appreciable amounts of products you should so indicate. In each case label the stronger acid, the stronger base, the weaker acid, and the weaker base.

(a) $H{-}C{\equiv}C{-}H + NaNH_2$
(b) $CH_2{=}CH_2 + NaNH_2$
(c) $CH_3CH_3 + NaNH_2$
(d) $H{-}C{\equiv}C:^-Na^+ + CH_3CH_2OH$
(e) $H{-}C{\equiv}C:^-Na^+ + H_2O$

9.5 ACETYLENE AS AN INDUSTRIAL CHEMICAL

Acetylene, an important industrial chemical, is used as a starting material for the preparation of many useful compounds. Acetylene can be prepared from relatively inexpensive starting materials. In one process coke and lime are heated in an electric furnace to produce calcium carbide. This is allowed to react with water and acetylene is generated.

$$\underset{\text{Coke}}{3C} + \underset{\text{Lime}}{CaO} \xrightarrow{2500^\circ} \underset{\text{Calcium carbide}}{CaC_2} + CO$$

$$CaC_2 + 2H_2O \xrightarrow[\text{temperature}]{\text{Room}} H{-}C{\equiv}C{-}H + Ca(OH)_2$$

In another process methane (from natural gas) is heated briefly to a very high temperature; acetylene is one of the products.

$$2CH_4 \xrightarrow[0.1-0.01 \text{ sec}]{1500^\circ} H{-}C{\equiv}C{-}H + 3H_2$$

Acetylene is a colorless gas at room temperature. It is thermodynamically unstable; and when shocked, acetylene explodes to form elemental carbon and hydrogen.

$$C_2H_2 \longrightarrow 2C + H_2 \qquad \Delta H^\circ = -56 \text{ kcal/mole}$$

The explosive decomposition of acetylene is more likely if the acetylene is at a high pressure, and liquid acetylene (bp -84°) has to be handled with extreme care. Special techniques have been developed, however, that allow for the relatively safe manipulation of acetylene even in large quantities and at pressures up to 30 atm. When large-diameter pipes are used for transporting acetylene, these pipes are filled with smaller pipes to minimize the free volume. Acetylene is also often diluted with an unreactive gas such as water vapor or nitrogen.

Acetylene is used extensively in welding because it burns with an extremely hot flame (up to 2800°). When we examine the heats of combustion of ethane, ethene, and acetylene below, we might expect that any one of these gases could be used for welding. We might even expect that ethane and ethene would produce hotter flames than acetylene.

$$CH_3CH_3 + 3\tfrac{1}{2}O_2 \longrightarrow 2CO_2 + 3H_2O \qquad \Delta H^\circ = -373 \text{ kcal/mole}$$

$$CH_2{=}CH_2 + 3O_2 \longrightarrow 2CO_2 + 2H_2O \qquad \Delta H^\circ = -337 \text{ kcal/mole}$$

$$CH{\equiv}CH + 2\tfrac{1}{2}O_2 \longrightarrow 2CO_2 + H_2O \qquad \Delta H^\circ = -317 \text{ kcal/mole}$$

Ethane and ethene produce more heat on a molar basis but they also require more oxygen, and heat is consumed in bringing the oxygen up to the flame temperature. Because more water is produced in the combustion of ethane and ethene as well, the number of product molecules, among which the heat must be distributed, is also greater. Because of these factors, of the three hydrocarbons, acetylene produces the highest flame temperature.

For safe handling of acetylene for use in welding, it is dissolved in acetone under a pressure of 300 psi in cylinders filled with a porous but inert substance.

9.6 REACTIONS OF ALKYNES

Many of the reactions of alkynes are characterized by addition reactions of the triple bond. Depending on the conditions used, the additions can occur once or twice.

$$-C{\equiv}C- + Y{-}Z \longrightarrow \overset{}{\underset{Y}{>}}C{=}C\underset{Z}{<} \xrightarrow{Y-Z} -\underset{Y}{\overset{Y}{C}}-\underset{Z}{\overset{Z}{C}}- \text{ or } -\underset{Y}{\overset{Z}{C}}-\underset{Z}{\overset{Y}{C}}-$$

We will see several examples of reactions of this general type in Sections 9.7–9.10.

Other reactions of alkynes result from the acidity of acetylenic hydrogen. In the presence of strong bases the following reaction takes place.

$$-C{\equiv}C{-}H + :B^- \longrightarrow -C{\equiv}C:^- + H{:}B$$

Strong base — Alkynide ion

When we study these reactions in Section 9.11 we will find that alkynide ions are important reagents for the synthesis of other alkynes.

9.7 ADDITION OF HYDROGEN

Depending on the conditions and the catalyst employed, one or two moles of hydrogen will add to a carbon-carbon triple bond. When a platinum catalyst is used, the alkyne generally reacts with two moles of hydrogen to give an alkane.

$$CH_3C{\equiv}CCH_3 \xrightarrow[H_2]{Pt} [CH_3CH{=}CHCH_3] \xrightarrow[H_2]{Pt} CH_3CH_2CH_2CH_3$$

However, reduction of an alkyne to an alkene can be accomplished through the use of special catalysts or reagents. Moreover, these special methods allow the preparation of either *cis-* or *trans*-alkenes from disubstituted acetylenes.

A catalyst that permits hydrogenation of an alkyne to an alkene is the nickel boride compound called P-2 catalyst. This catalyst can be prepared by the reduction of nickel acetate with sodium borohydride.

$$Ni(O\overset{O}{\overset{\|}{C}}CH_3)_2 \xrightarrow[C_2H_5OH]{NaBH_4} Ni_2B \qquad \text{(P-2)}$$

Hydrogenation of alkynes in the presence of P-2 catalyst causes *syn addition of hydrogen* to take place and the alkene that is formed from an alkyne with an internal triple bond has the *cis* configuration. The hydrogenation of 3-hexyne (below) illustrates this method.

$$CH_3CH_2C{\equiv}CCH_2CH_3 \xrightarrow[(syn\ addition)]{H_2/Ni_2B\ (P\text{-}2)} \underset{H}{\overset{CH_3CH_2}{}}\!\!C{=}C\!\!\underset{H}{\overset{CH_2CH_3}{}} \qquad (97\%)$$

3-Hexyne — *cis*-3-Hexene

Other specially conditioned catalysts can be used to prepare *cis*-alkenes from disubstituted acetylenes. Metallic palladium deposited on calcium carbonate can be used in this way after it has been conditioned with lead acetate and

$$R{-}C{\equiv}C{-}R \xrightarrow[H_2,\ quinoline\ (syn\ addition)]{Pd/CaCO_3\ (Lindlar's\ catalyst)} \underset{H}{\overset{R}{}}\!\!C{=}C\!\!\underset{H}{\overset{R}{}}$$

quinoline (Section 18.2). This special catalyst is known as Lindlar's catalyst. Palladium deposited on barium sulfate gives similar results.

$$CH_3O_2C{-}(CH_2)_3{-}C{\equiv}C{-}(CH_2)_3{-}CO_2CH_3 \xrightarrow[\text{H}_2\text{, quinoline (}syn\text{ addition)}]{5\%\ \text{Pd/BaSO}_4} \underset{\text{H}\qquad\qquad\text{H}}{CH_3O_2C{-}(CH_2)_3{-}C{=}C{-}(CH_2)_3{-}CO_2CH_3} \quad (97\%)$$

Diborane can also be used to prepare *cis*-alkenes from appropriate alkynes. Diborane adds once to the triple bond of 3-hexyne to form the vinylic borane shown below.

$$3C_2H_5C{\equiv}CC_2H_5 + \tfrac{1}{2}(BH_3)_2 \xrightarrow{0°} [(C_2H_5)(H)C{=}C(C_2H_5)]_3B$$

A vinylic borane

The vinylic borane can be converted to *cis*-3-hexene by treating the solution with acetic acid (cf. Problem 7.8).

$$[(C_2H_5)(H)C{=}C(C_2H_5)]_3B \xrightarrow[0°]{CH_3COOH} (C_2H_5)(H)C{=}C(H)(C_2H_5) \quad (90\%\ \text{overall})$$

cis-3-Hexene

Diborane cannot be used to reduce terminal alkynes to alkenes because diborane adds twice to the triple bond of terminal alkynes (p. 362).

trans-Alkenes are obtained when alkynes are reduced with lithium metal in ammonia or ethylamine at low temperatures.

$$\underset{\text{4-Octyne}}{CH_3(CH_2)_2{-}C{\equiv}C{-}(CH_2)_2CH_3} \xrightarrow[C_2H_5NH_2,\ -78°]{Li} \underset{trans\text{-4-Octene}}{CH_3(CH_2)_2(H)C{=}C(H)(CH_2)_2CH_3} \quad (52\%)$$

9.8 ADDITION OF HALOGENS

Alkynes show the same kind of reactions toward chlorine and bromine that alkenes do: *they react by addition*. However, with alkynes the addition may occur twice.

$$-C{\equiv}C- \xrightarrow[CCl_4]{Br_2} \underset{\text{Dibromoalkene}}{>C(Br){=}C(Br)<} \xrightarrow[CCl_4]{Br_2} \underset{\text{Tetrabromoalkane}}{-CBr_2{-}CBr_2-}$$

$$-C{\equiv}C- \xrightarrow[CCl_4]{Cl_2} \underset{\text{Dichloroalkene}}{>C(Cl){=}C(Cl)<} \xrightarrow[CCl_4]{Cl_2} \underset{\text{Tetrachloroalkane}}{-CCl_2{-}CCl_2-}$$

The dihaloalkenes that result from the first addition are usually less reactive toward further addition than the alkyne itself. Thus, it is usually possible to prepare a dihaloalkene by simply adding one mole of the halogen.

$$CH_3CH_2CH_2CH_2C{\equiv}CCH_2OH \xrightarrow[CCl_4,\ 0^\circ]{Br_2\ (one\ mole)} CH_3CH_2CH_2CH_2CBr{=}CBrCH_2OH\ (80\%)$$

Most additions of chlorine and bromine to alkynes are anti additions and yield *trans*-dihaloalkenes. Addition of bromine to acetylenedicarboxylic acid, for example, gives the *trans* isomer in 70% yield.

$$HOOC{-}C{\equiv}C{-}COOH \xrightarrow{Br_2} (HOOC)(Br)C{=}C(Br)(COOH)\ (70\%)$$

Acetylenedicarboxylic acid

9.9 ADDITION OF HYDROGEN HALIDES

Alkynes react with hydrogen chloride and hydrogen bromide to form haloalkenes or geminal dihalides depending, once again, on whether one or two moles of the hydrogen halide are used. Both additions follow Markovnikov's rule: the hydrogen

$$-C{\equiv}C- \xrightarrow{HCl} (H)C{=}C(Cl) \xrightarrow{HCl} -CH_2{-}CCl_2-$$

Chloroalkene — *gem*-Dichloride

$$-C{\equiv}C- \xrightarrow[Radical\ inhibitor]{HBr} (H)C{=}C(Br) \xrightarrow[Radical\ inhibitor]{HBr} -CH_2{-}CBr_2-$$

Bromoalkene — *gem*-Dibromide

of the hydrogen halide becomes attached to the carbon that has the greater number of hydrogens. Propyne, for example, reacts with one mole of hydrogen chloride to yield 2-chloropropene and with two moles to yield 2,2-dichloropropane.

$$CH_3C{\equiv}CH \xrightarrow{HCl} CH_3CCl{=}CH_2 \xrightarrow{HCl} CH_3CCl_2{-}CH_3$$

2-Chloropropene — 2,2-Dichloropropane

The initial addition of a hydrogen halide to an alkyne usually occurs in an anti manner. This is especially likely if an ionic halide corresponding to the halogen of the hydrogen halide is present in the mixture.

$$CH_3CH_2C{\equiv}CCH_2CH_3 + HCl \xrightarrow[CH_3COOH,\ 25^\circ]{Cl^-} (CH_3CH_2)(H)C{=}C(Cl)(CH_2CH_3)\ (97\%)$$

Anti-Markovnikov addition of hydrogen bromide to alkynes occurs when peroxides are present in the reaction mixture.

$$CH_3CH_2CH_2CH_2C{\equiv}CH \xrightarrow[\text{peroxides}]{\text{HBr}} CH_3CH_2CH_2CH_2CH{=}CHBr \quad (74\%)$$

9.10 ADDITION OF WATER

Alkynes add water readily when the reaction is catalyzed by strong acids and mercuric (Hg^{++}) ions. Aqueous solutions of sulfuric acid and mercuric sulfate are often used for this purpose. The vinylic alcohol that is initially produced is usually unstable, and it rearranges rapidly to an aldehyde or a ketone. The rearrangement

$$-C{\equiv}C- + H{-}OH \xrightarrow[H_2SO_4]{HgSO_4} \left[-CH{=}\underset{\underset{OH}{|}}{C}- \right] \longrightarrow -\underset{\underset{H}{|}}{\overset{\overset{H}{|}}{C}}-\underset{\underset{O}{\|}}{C}-$$

A vinylic alcohol → Aldehyde or ketone

involves the loss of a proton from the hydroxyl group, the addition of a proton to the adjacent carbon, and the relocation of the double bond. This kind of rearrangement, known as a *tautomerization,* is acid catalyzed and occurs in the following way.

$$H^+ + -\underset{\underset{H}{|}}{C}{=}\underset{\underset{\ddot{O}-H}{|}}{C}- \longrightarrow -\underset{\underset{H}{|}}{\overset{\overset{H}{|}}{C}}-\underset{\underset{\ddot{O}-H}{\|+}}{C}- \longrightarrow -\underset{\underset{H}{|}}{\overset{\overset{H}{|}}{C}}-\underset{\underset{\ddot{O}:}{\|}}{C}- + H^+$$

Vinylic alcohol → Aldehyde or Ketone

The vinylic alcohol accepts a proton at one carbon atom of the double bond to yield a cationic intermediate that then loses a proton from the oxygen atom to produce an aldehyde or ketone.

Vinylic alcohols are often called *enols* (*-en,* the ending for alkenes, plus *-ol,* the ending for alcohols). The product of the rearrangement is often a ketone and these rearrangements are known as *keto-enol tautomerizations.*

$$-\overset{|}{C}{=}\underset{\underset{:\ddot{O}-H}{|}}{C}- \overset{H^+}{\rightleftharpoons} -\underset{\underset{H}{|}}{\overset{|}{C}}-\underset{\underset{\ddot{O}:}{\|}}{C}-$$

Enol form ⇌ Keto form

We examine this phenomenon in greater detail in Section 16.11.

When acetylene itself undergoes addition of water the product is an aldehyde.

$$H{-}C{\equiv}C{-}H + H_2O \xrightarrow[H_2SO_4]{HgSO_4} \left[\begin{matrix} H & & H \\ & C{=}C & \\ H & & OH \end{matrix} \right] \longrightarrow H{-}\underset{H}{\overset{H}{C}}{-}\underset{O}{\overset{\|}{C}}{-}H$$

Acetaldehyde

This method has been important in the commercial production of acetaldehyde.

The addition of water to alkynes also follows Markovnikov's rule—the hydrogen becomes attached to the carbon with the greater number of hydrogens. Therefore, when higher alkynes are hydrated, ketones, rather than aldehydes, are the products.

$$R{-}C{\equiv}C{-}H \xrightarrow[H_2O,\ H^+]{Hg^{++}} \left[R\underset{OH}{C}{=}CH_2 \right] \longrightarrow R{-}\underset{O}{\overset{}{C}}{-}CH_3$$

A ketone

Several examples of this ketone synthesis are listed below.

$$CH_3C{\equiv}CH \xrightarrow[H_2O,\ H^+]{Hg^{++}} \left[CH_3{-}\underset{OH}{C}{=}CH_2 \right] \longrightarrow CH_3{-}\underset{O}{\overset{}{C}}{-}CH_3$$

Acetone

$$CH_3CH_2CH_2CH_2C{\equiv}CH \xrightarrow[H_2SO_4,\ H_2O]{HgSO_4} CH_3CH_2CH_2CH_2\underset{O}{C}CH_3$$

(80%)

$$\text{1-ethynylcyclohexanol } (C_6H_{10}(OH)C{\equiv}CH) \xrightarrow[H_2SO_4,\ H_2O]{HgO} \text{1-acetylcyclohexanol } (C_6H_{10}(OH)\overset{O}{\overset{\|}{C}}CH_3)$$

(65–67%)

9.10A Hydroboration–Oxidation

Boron hydrides can also be used to synthesize ketones from alkynes. The overall method is very similar to the hydroboration-oxidation of alkenes that we saw in Chapter 7. Diborane adds only once to alkynes that have an internal triple bond (p. 358).

$$C_2H_5{-}C{\equiv}C{-}C_2H_5 \xrightarrow[0°]{(BH_3)_2} \left(\begin{matrix} C_2H_5 & & C_2H_5 \\ & C{=}C & \\ H & & \end{matrix} \right)_3 B$$

A vinylic borane

The vinylic borane formed in the reaction above can be treated with hydrogen peroxide and aqueous sodium hydroxide to produce an enol. The enol rearranges spontaneously to a ketone.

$$\underset{\text{H}}{\overset{\text{C}_2\text{H}_5}{}}\!\!\text{C}{=}\text{C}\!\!\underset{\!\!\!\!)_3\text{B}}{\overset{\text{C}_2\text{H}_5}{}} \xrightarrow[\text{H}_2\text{O}_2]{\text{OH}^-} \left[\underset{\text{H}}{\overset{\text{C}_2\text{H}_5}{}}\!\!\text{C}{=}\text{C}\!\!\underset{\text{OH}}{\overset{\text{C}_2\text{H}_5}{}}\right] \longrightarrow C_2H_5CH_2\underset{\underset{O}{\|}}{C}C_2H_5$$

(62% overall)

Terminal alkynes generally react with one mole of diborane to produce *gem*-dibora derivatives:

$$R{-}C{\equiv}C{-}H + (BH_3)_2 \longrightarrow R{-}CH_2\underset{\underset{|}{B-}}{\overset{\overset{|}{B-}}{C}}H$$

gem-Dibora compound

However, by using a highly hindered borane such as disiamylborane, terminal

$$\left(CH_3\overset{CH_3}{\overset{|}{C}}H{-}\overset{CH_3}{\overset{|}{C}}H\right)_2BH$$

Disiamylborane
(Sia_2BH)

alkynes can be converted to vinyl boranes. The bulky alkyl groups of disiamylborane hinder its reaction with terminal acetylenes and the addition occurs only once.

$$n\text{-}C_4H_9C{\equiv}CH + Sia_2BH \longrightarrow n\text{-}C_4H_9CH{=}CH{-}BSia_2 \quad (100\%)$$

Oxidation of the product with alkaline hydrogen peroxide produces an aldehyde, probably by way of an enol.

$$n\text{-}C_4H_9CH{=}CH{-}Sia_2B \xrightarrow[OH^-/H_2O]{H_2O_2} n\text{-}C_4H_9CH_2\overset{\overset{O}{\|}}{C}H$$

Hexanal

Overall, this procedure provides an anti-Markovnikov addition of H— and —OH to an alkyne, and thus it is a convenient synthesis of aldehydes.

Problem 9.3

What other organic product would you expect to obtain from the oxidation of $n\text{-}C_4H_9CH{=}CH{-}BSia_2$ with hydrogen peroxide in base?

Problem 9.4

Show how each of the following transformations could be carried out. [Hint: In parts (c) and (d) assume that you have D_2, $(BD_3)_2$, and CH_3COOD available.]

(a) $CH_3CH_2C{\equiv}CH \longrightarrow CH_3CH_2\overset{O}{\overset{\|}{C}}CH_3$

(b) $C_5H_9{-}C{\equiv}CH \longrightarrow C_5H_9{-}CH_2\overset{O}{\overset{\|}{C}}H$ (cyclopentyl)

(c) $CH_3C{\equiv}CCH_3 \longrightarrow$ (CH$_3$)(D)C=C(CH$_3$)(D), with the two CH$_3$ groups cis

(d) $CH_3CH_2C{\equiv}CH \longrightarrow$ (CH$_3$CH$_2$)(H)C=C(H)(D), with CH$_3$CH$_2$ and D trans

9.11 REPLACEMENT OF THE ACETYLENIC HYDROGEN OF TERMINAL ALKYNES

Sodium acetylide and other sodium alkynides can be prepared by treating terminal alkynes with sodium amide in liquid ammonia.

$$H{-}C{\equiv}C{-}H + Na^{+}\,{}^{-}{:}\ddot{N}H_2 \longrightarrow H{-}C{\equiv}C{:}^{-}Na^{+} + \ddot{N}H_3$$

$$CH_3C{\equiv}C{-}H + Na^{+}\,{}^{-}{:}\ddot{N}H_2 \longrightarrow CH_3C{\equiv}C{:}^{-}Na^{+} + \ddot{N}H_3$$

$$CH_3CH_2C{\equiv}CH + Na^{+}\,{}^{-}{:}\ddot{N}H_2 \longrightarrow CH_3CH_2C{\equiv}C{:}^{-}Na^{+} + \ddot{N}H_3$$

These are acid-base reactions. The amide ion, by virtue of its being the anion of the very weak acid, ammonia ($K_a \simeq 10^{-34}$), is able to remove the acetylenic protons of terminal alkynes ($K_a \simeq 10^{-25}$). These reactions, for all practical purposes, go to completion.

Sodium alkynides are useful intermediates for the synthesis of other alkynes. This can be accomplished by treating the sodium alkynide with a primary alkyl halide.

$$\underset{\text{Sodium alkynide}}{(H)R{-}C{\equiv}C{:}^{-}Na^{+}} + \underset{\text{Primary alkyl halide}}{R'CH_2Br} \longrightarrow \underset{\text{Mono- or disubstituted acetylene}}{(H)R{-}C{\equiv}CCH_2R'} + NaBr$$

The following examples illustrate this synthesis of alkynes.

$$HC{\equiv}C{:}^{-}Na^{+} + CH_3{-}Br \xrightarrow[\text{5 hr}]{\text{liq. } NH_3} \underset{\text{Propyne (84\%)}}{H{-}C{\equiv}C{-}CH_3}$$

$$H{-}C{\equiv}C{:}^{-}Na^{+} + CH_3CH_2CH_2Br \xrightarrow[\text{5 hr}]{\text{liq. } NH_3} \underset{\text{1-Pentyne (85\%)}}{HC{\equiv}CCH_2CH_2CH_3}$$

$$H{-}C{\equiv}C{:}^{-}Na^{+} + CH_3\overset{\overset{\large CH_3}{|}}{C}HCH_2CH_2Br \xrightarrow[6\ hr]{liq.\ NH_3} H{-}C{\equiv}CCH_2CH_2\overset{\overset{\large CH_3}{|}}{C}HCH_3$$

5-Methyl-1-hexyne
(68%)

$$CH_3CH_2C{\equiv}C{:}^{-}Na^{+} + CH_3CH_2Br \xrightarrow[6\ hr]{liq.\ NH_3} CH_3CH_2C{\equiv}CCH_2CH_3$$

3-Hexyne
(75%)

In all of these examples the alkynide ion acts as a nucleophile and displaces a halide ion from the primary alkyl halide. The result is *an* S_N2 *reaction* (Sect. 5.5).

$$(H)RC{\equiv}C{:}^{-}\ Na^{+} + R'(H)(H)C{-}\ddot{B}r{:} \xrightarrow[\text{substitution } S_N2]{\text{Nucleophilic}} (H)RC{\equiv}C{-}CH_2R' + {:}\ddot{B}r{:}^{-}\ Na^{+}$$

Sodium alkynide — 1° Alkyl halide

The unshared electron pair of the alkynide ion attacks the carbon that bears the halogen and forms a bond to it. The halogen departs as a halide ion.

When secondary or tertiary halides are used, the alkynide ion acts as a base rather than as a nucleophile, and the major result is an E2 *elimination* (Sect. 5.10). The products of the elimination are an alkene and the alkyne from which the sodium alkynide was originally formed.

$$(H)RC{\equiv}C{:}^{-} + H{-}C(R')(H){-}C(H)(R''){-}Br \xrightarrow{E2} (H)RC{\equiv}CH + R'CH{=}CHR'' + Br^{-}$$

2° alkyl halide

As we have seen, elimination reactions nearly always compete with nucleophilic substitution reactions because all nucleophiles are also potential bases. Primary alkyl halides and methyl halides are more susceptible to nucleophilic substitution because attack at the carbon bearing the halogen is relatively unhindered.

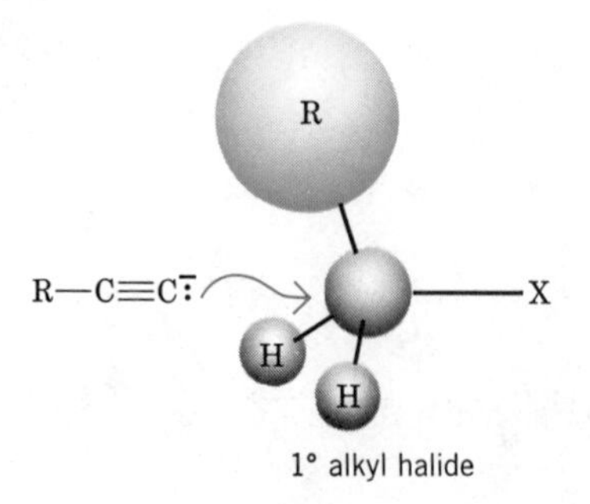

1° alkyl halide

Alkyl groups are large in comparison to hydrogens, and in secondary and tertiary alkyl halides the additional alkyl groups hinder attack at the carbon that bears the halogen. With secondary and tertiary halides the alkynide ion attacks a β hydrogen instead, and this brings about an elimination reaction.

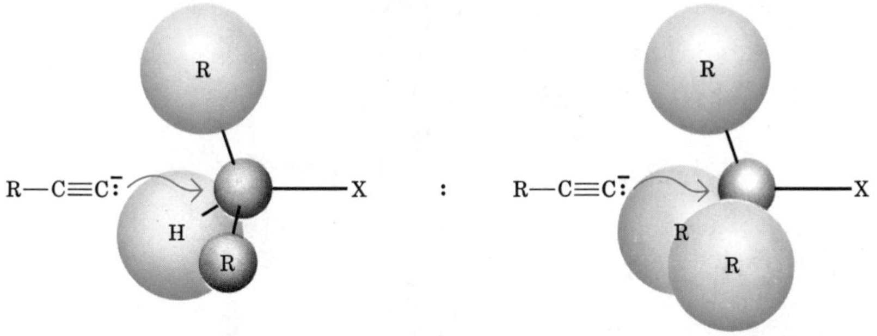

2° alkyl halide
(attack is hindered)

3° alkyl halide
(attack is greatly hindered)

Problem 9.5

In addition to sodium amide and liquid ammonia, assume that you have the following four compounds available and want to carry out a synthesis of 2,2-

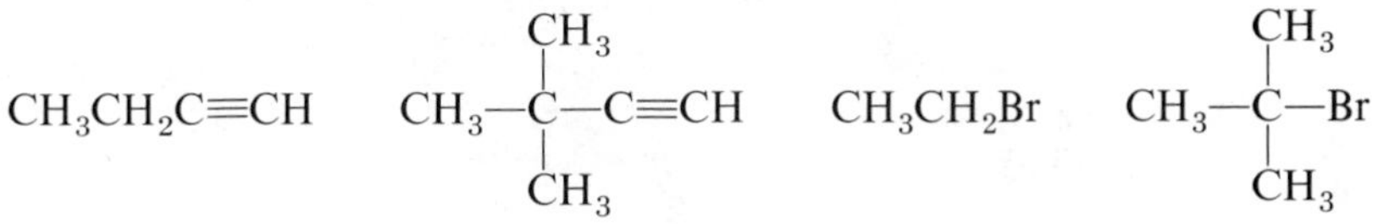

$$CH_3CH_2C{\equiv}CH \qquad CH_3{-}\underset{CH_3}{\overset{CH_3}{C}}{-}C{\equiv}CH \qquad CH_3CH_2Br \qquad CH_3{-}\underset{CH_3}{\overset{CH_3}{C}}{-}Br$$

dimethyl-3-hexyne. Which synthetic route would you choose?

9.12 OTHER METAL ACETYLIDES

Acetylene and terminal alkynes also form metal derivatives with silver and copper(I) ions.

$$RC{\equiv}CH + Cu(NH_3)_2^+ + OH^- \xrightarrow{H_2O} R{-}C{\equiv}CCu + H_2O + 2NH_3$$

$$RC{\equiv}CH + Ag(NH_3)_2^+ + OH^- \xrightarrow{H_2O} RC{\equiv}CAg + H_2O + 2NH_3$$

Silver and copper alkynides differ from sodium alkynides in several ways. The metal-carbon bond in silver and copper alkynides is largely covalent. As a result, silver and copper alkynides are poor bases and poor nucleophiles. Silver and copper alkynides can be prepared in water, whereas sodium alkynides react vigorously with water.

$$R{-}C{\equiv}CNa + H_2O \longrightarrow R{-}C{\equiv}CH + NaOH$$

Silver and copper alkynides are also quite insoluble in water and precipitate when they are prepared. This is the basis for an old and still convenient test for terminal alkynes as well as a method for separating terminal alkynes from alkynes that have an internal triple bond.

$$R—C{\equiv}C—R + Ag(NH_3)_2^+OH^- \longrightarrow \text{no precipitate}$$
$$R—C{\equiv}CH + Ag(NH_3)_2^+OH^- \longrightarrow RC{\equiv}CAg\downarrow$$

Once a separation has been carried out the terminal alkyne can be regenerated by treating it with sodium cyanide (or with a strong acid).

$$R—C{\equiv}CAg + 2CN^- + H_2O \longrightarrow R—C{\equiv}CH + Ag(CN)_2^- + OH^-$$

Silver and copper alkynides must be handled cautiously; when dry they are likely to explode.

9.13 COUPLING REACTIONS OF TERMINAL ALKYNES

Terminal alkynes undergo an interesting coupling (or dimerization) reaction when they are treated with copper salts and amines. One method for dimerizing terminal alkynes uses copper(I) chloride, ammonia, and oxygen. The general reaction can be illustrated as follows.

$$R—C{\equiv}C—H + CuCl \xrightarrow[O_2,\ CH_3OH]{NH_3} \underset{\text{A diyne}}{R—C{\equiv}C—C{\equiv}C—R}$$

Propyne has been dimerized in this way to 2,4-hexadiyne.

$$CH_3C{\equiv}CH + CuCl \xrightarrow[O_2,\ CH_3OH]{NH_3} \underset{(74\%)}{CH_3—C{\equiv}C—C{\equiv}C—CH_3}$$

The mechanism of the reaction is not understood completely, but it is thought that copper(I) ions form a π complex with the alkyne and that, at some point in the coupling reaction, oxygen causes a $Cu^+ \longrightarrow Cu^{++}$ oxidation.

$$HC{\equiv}C(CH_2)_5C{\equiv}CH \xrightarrow[\text{pyridine}]{\text{Cupric acetate}}$$

Cyclic dimer: ring of two $(CH_2)_5$ units joined by two $C{\equiv}C—C{\equiv}C$ units (10%)

+

Cyclic trimer: ring of three $(CH_2)_5$ units joined by three $C{\equiv}C—C{\equiv}C$ units, e.g. $(CH_2)_5—C{\equiv}C—C{\equiv}C—(CH_2)_5$ (13%)

+

cyclic tetramer (11%)

+

cyclic pentamer (4%)

Professor F. Sondheimer (of University College, London) has made extensive use of a modification of this procedure—one that utilizes copper(II) in pyridine (Sect. 11.9) in the absence of oxygen. This method is particularly useful because terminal diacetylenes are dimerized, trimerized, and so on, to cyclic compounds. This has proved to be one of the most valuable methods for the synthesis of compounds whose molecules have large carbon rings.

9.14 SYNTHESIS OF ALKYNES

9.14A From Other Alkynes

Alkynes can be prepared from other terminal alkynes through the nucleophilic substitution reaction that we saw in Section 9.11.

$$R{-}C{\equiv}CH \xrightarrow[\text{liq. } NH_3]{NaNH_2} R{-}C{\equiv}C{:}^{-}Na^{+} \xrightarrow[\text{Primary halide}]{R'{-}X} R{-}C{\equiv}C{-}R'$$

9.14B By Elimination Reactions

Alkynes can also be synthesized from alkenes. In this method an alkene is first treated with bromine to form a *vic*-dibromo compound, and the *vic*-dibromide

$$RCH{=}CHR + Br_2 \longrightarrow R{-}\overset{H}{\underset{Br}{C}}{-}\overset{H}{\underset{Br}{C}}{-}R$$

vic-Dibromide

is dehydrohalogenated through its reaction with a strong base. The dehydrohalogenation occurs in two steps. The first step yields a bromoalkene.

$$R{-}\overset{H}{\underset{Br}{C}}{-}\overset{H}{\underset{Br}{C}}{-}R \quad {:}B^{-} \longrightarrow R{-}\overset{H}{C}{=}\underset{Br}{C}{-}R + H{:}B + Br^{-}$$

vic-Dibromide — Strong base — Bromoalkene

The second step is more difficult; it yields an alkyne.

$$B{:}^{-} \quad R{-}\overset{H}{C}{=}\underset{Br}{C}{-}R \longrightarrow R{-}C{\equiv}C{-}R + H{:}B + Br^{-}$$

Alkyne

Depending on the conditions, these two dehydrohalogenations may be carried out as separate reactions or they may be carried out consecutively in a single mixture. Potassium hydroxide in ethanol usually brings about a single dehydrohalogenation unless the reaction is carried out at a very high temperature.

The stronger base, sodium amide, is capable of effecting both dehydrohalogenations in a single reaction mixture. (Two moles of sodium amide per mole of the dihalide must be used.) Dehydrohalogenations with sodium amide are usually carried out in liquid ammonia or in an inert medium such as mineral oil.

The following examples illustrate both methods.

1. $$CH_3CH_2CH{=}CH_2 \xrightarrow[CCl_4]{Br_2} CH_3CH_2\underset{\displaystyle Br}{CH}CH_2Br \xrightarrow[\text{Ethanol reflux}]{KOH} \left[\begin{array}{c} CH_3CH_2CH{=}CHBr \\ + \\ CH_3CH_2\underset{\displaystyle Br}{C}{=}CH_2 \end{array}\right]$$

$$\xrightarrow[\text{Mineral oil, }110\text{–}160^\circ]{NaNH_2} [CH_3CH_2C{\equiv}CH] \xrightarrow{NaNH_2} CH_3CH_2C{\equiv}C{:}^-Na^+$$

$$\xrightarrow{H^+} CH_3CH_2C{\equiv}CH$$

2. $$C_6H_5CH{=}CH_2 \xrightarrow[CCl_4]{Br_2} C_6H_5CHBrCH_2Br$$

$$\xrightarrow[\text{liq. }NH_3]{NaNH_2} C_6H_5C{\equiv}C{:}^-Na^+ \xrightarrow{H^+} C_6H_5C{\equiv}CH$$

3. $$\underset{\text{Oleic acid}}{CH_3(CH_2)_7CH{=}CH(CH_2)_7COOH} \xrightarrow[\text{ether}]{Br_2} CH_3(CH_2)_7\underset{\displaystyle Br}{CH}{-}\underset{\displaystyle Br}{CH}(CH_2)_7COOH$$

$$\xrightarrow[(2)\ H^+]{(1)\ NaNH_2,\ \text{liq. }NH_3} \underset{\text{Stearolic acid}}{CH_3(CH_2)_7C{\equiv}C(CH_2)_7COOH}$$

Stearolic acid (above) has been used to synthesize sterculic acid, a naturally occurring compound that contains a cyclopropene ring (Special Topic D).

Ketones can be converted to *gem*-dichlorides through their reaction with phosphorus pentachloride and these can also be used to synthesize alkynes.

$$\underset{\text{Methyl cyclohexyl ketone}}{C_6H_{11}{-}\overset{\displaystyle O}{\overset{\|}{C}}{-}CH_3} \xrightarrow[0^\circ]{PCl_5} \underset{\substack{(70\text{–}80\%)\\ \text{A } gem\text{-dichloride}}}{C_6H_{11}{-}CCl_2{-}CH_3} + POCl_3$$

$$\xrightarrow[\text{Mineral oil, heat}]{NaNH_2} \underset{\substack{\text{Cyclohexylacetylene}\\ (46\%)}}{C_6H_{11}{-}C{\equiv}CH}$$

Problem 9.6

(a) Suggest a method for converting cyclohexylacetylene into methyl cyclohexyl ketone. (b) Suggest a method for carrying out the following transformation:

$$C_6H_{11}-\overset{O}{\overset{\|}{C}}CH_3 \longrightarrow C_6H_{11}-CH_2\overset{O}{\overset{\|}{C}}H$$

9.15 PHYSICAL PROPERTIES OF ALKYNES

Alkynes have boiling points and melting points that are similar to those of corresponding alkenes. The lower-molecular-weight alkynes (Table 9.1) are gases at room temperature. As we might expect, alkynes dissolve in other nonpolar solvents (or in solvents of low polarity) such as carbon tetrachloride, ether, or liquid alkanes. Alkynes are only very slightly soluble in water although they are more soluble than alkenes and alkanes. The densities of alkynes are lower than that of water.

Problem 9.7

The dipole moment of 1-butyne (0.80D) is substantially larger than that of 1-butene (0.30D). Given the hint that these dipole moments arise primarily from the polarity of one carbon-carbon single bond in each molecule, provide an explanation for the larger dipole moment of 1-butyne.

TABLE 9.1 Physical Constants of Alkynes

NAME	FORMULA	mp °C	bp, °C (760 mmHg unless otherwise noted)	SPECIFIC GRAVITY d_4^{20}
Acetylene	$HC{\equiv}CH$	−81.8	−83.6	—
Propyne	$CH_3C{\equiv}CH$	−101.51	−23.2	—
1-Butyne	$CH_3CH_2C{\equiv}CH$	−122.5	8.1	—
2-Butyne	$CH_3C{\equiv}CCH_3$	−32.3	27	0.691
1-Pentyne	$CH_3(CH_2)_2C{\equiv}CH$	−90	39.3	0.695
2-Pentyne	$CH_3CH_2C{\equiv}CCH_3$	−101	55.5	0.714
1-Hexyne	$CH_3(CH_2)_3C{\equiv}CH$	−132	71	0.715
2-Hexyne	$CH_3(CH_2)_2C{\equiv}CCH_3$	−88	84	0.730
3-Hexyne	$CH_3CH_2C{\equiv}CCH_2CH_3$	−51	81.8	0.724
1-Heptyne	$CH_3(CH_2)_4C{\equiv}CH$	−81	100	0.734
2-Heptyne	$CH_3(CH_2)_3C{\equiv}CCH_3$	—	112	0.748
3-Heptyne	$CH_3(CH_2)_2C{\equiv}CCH_2CH_3$	—	105	0.752
1-Octyne	$CH_3(CH_2)_5C{\equiv}CH$	−80	125.2	0.746
2-Octyne	$CH_3(CH_2)_4C{\equiv}CCH_3$	−60.2	137.2	0.759
3-Octyne	$CH_3(CH_2)_3C{\equiv}CCH_2CH_3$	−105	133	0.752
4-Octyne	$CH_3(CH_2)_2C{\equiv}C(CH_2)_2CH_3$	−102	131	0.751
1-Nonyne	$CH_3(CH_2)_6C{\equiv}CH$	−65	160.7	0.760
2-Nonyne	$CH_3(CH_2)_5C{\equiv}CCH_3$	—	155^{747}	0.769
3-Nonyne	$CH_3(CH_2)_4C{\equiv}CCH_2CH_3$	—	154^{745}	0.762
4-Nonyne	$CH_3(CH_2)_3C{\equiv}C(CH_2)_2CH_3$	—	152^{752}	0.757
1-Decyne	$CH_3(CH_2)_7C{\equiv}CH$	−44	174	0.765
3-Decyne	$CH_3(CH_2)_5C{\equiv}CCH_2CH_3$	—	175	0.765

9.16 OXIDATIVE CLEAVAGE OF ALKYNES

Treating alkynes with ozone or with basic potassium permanganate leads to cleavage of the carbon-carbon triple bond. The products are carboxylic acids.

$$R{-}C{\equiv}C{-}R' \xrightarrow[(2)\ H_2O]{(1)\ O_3} RCOOH + R'COOH$$

or

$$R{-}C{\equiv}C{-}R' \xrightarrow[(2)\ H^+]{(1)\ KMnO_4,\ OH^-} RCOOH + R'COOH$$

Ozone has been used to locate the position of the triple bond in an alkyne. Stearolic acid, for example, reacts with ozone to yield an ozonide that decomposes to a nine-carbon carboxylic acid and a nine-carbon dicarboxylic acid.

$$\underset{\text{Stearolic acid}}{CH_3(CH_2)_7C{\equiv}C(CH_2)_7COOH} \xrightarrow[(2)\ H_2O]{(1)\ O_3,\ CCl_4} CH_3(CH_2)_7COOH + HOOC(CH_2)_7COOH$$

Problem 9.8

Give the name and structure of each of the following alkynes used in the following reactions.

(a) $C_7H_{12} \xrightarrow[(2)\ H_2O]{(1)\ O_3} CH_3\underset{\displaystyle CH_3}{\underset{|}{C}}HCOOH + CH_3CH_2COOH$

(b) $C_8H_{12} \xrightarrow[(2)\ H_2O]{(1)\ O_3} HOOC{-}(CH_2)_6{-}COOH$ (only)

(c) $C_7H_{12} \xrightarrow[Pt]{2H_2} CH_3CH_2CH_2CH_2CH_2CH_2CH_3$

$C_7H_{12} \xrightarrow[OH^-]{Ag(NH_3)_2^+} C_7H_{11}Ag\downarrow$

9.17 CHEMICAL ANALYSIS OF ALKANES, ALKENES, ALKYNES, ALKYL HALIDES, AND ALCOHOLS

Very often in the course of laboratory work we need to decide what functional groups are present in a compound that we have isolated. We may have isolated a compound from a synthesis, for example, and the presence of a particular functional group may tell us whether our synthesis has succeeded or failed. Or, we may have isolated a compound from some natural material. Before we subject it to elaborate procedures for structure determination it is often desirable to know something about the kind of compound we have.

Spectroscopic methods are available that will do all of these things for us and we will study these procedures in Chapters 10 and 13. Spectrometers are expensive instruments, however, and spectroscopic procedures are sometimes time consuming. It is helpful, therefore, to have simpler and more rapid means to identify a particular functional group.

Very often this can be done by a simple chemical test. Such a test will often consist of a single reagent that, when mixed with the compound in question, will indicate the presence of a particular functional group. Not all reactions of a functional group serve as chemical tests, however. To be useful the reaction ought to proceed with a clear signal; a color change, the evolution of a gas, or the appearance of a precipitate.

Let us consider an example. Suppose we have just finished carrying out what we hope was an elimination reaction of an alkyl halide to produce an alkene. We have isolated a colorless liquid by distillation of the reaction mixture. Both alkenes and alkyl halides are colorless when pure. The question now is, "Is this colorless liquid our product (an alkene), or is it our starting material (an alkyl halide)?" To decide we might remember that alkenes undergo rapid addition of bromine, and that if the alkene is present in excess, the red-brown color of the bromine will disappear within seconds.

$$\underset{\text{Colorless}}{R{-}CH{=}CHR'} + \underset{\text{Red-brown}}{Br_2} \xrightarrow{CCl_4} \underset{\text{Colorless}}{R{-}CHBrCHBr{-}R'}$$

Alkyl halides, on the other hand, do not react with bromine in carbon tetrachloride at room temperature and in the absence of steady irradiation by a strong light source.

$$\underset{\text{Colorless}}{R{-}Br} + \underset{\text{Red-brown}}{Br_2} \xrightarrow{CCl_4} \text{no reaction}$$

(Red-brown color persists as long as exposure to a strong light source is avoided.)

So by simply adding a drop of bromine in carbon tetrachloride to a small amount of our compound, we will be able to make a decision. If the red-brown color of bromine disappears within a few seconds we can be reasonably certain that our elimination reaction has succeeded. If the color persists, then the elimination reaction did not occur. Now we may decide that we need to use a stronger base or a higher temperature or a different solvent to bring about the reaction.

The test that we have just described will serve us very well if the elimination that we were attempting was one in which an alkyl halide was being converted to an alkene. But if the elimination were one in which we were attempting to convert a *vic*-dihalide to an alkyne (below), testing with bromine in carbon tetrachloride would not be definitive.

$$\underset{vic\text{-Dibromide}}{R{-}CHBrCHBr{-}R} \xrightarrow{\text{Base}} \underset{\text{Vinylic bromide}}{R{-}CH{=}CHBr{-}R} \xrightarrow{\text{Base}} \underset{\text{Alkyne}}{R{-}C{\equiv}C{-}R}$$

or

$$\underset{vic\text{-Dibromide}}{R{-}CHBrCH_2Br} \xrightarrow{\text{Base}} \underset{\text{Vinylic bromide}}{R{-}CHBr{=}CH_2} \xrightarrow{\text{Base}} \underset{\text{Terminal alkyne}}{R{-}C{\equiv}CH}$$

In either of the reactions above, the material that we isolate might be the starting *vic*-dihalide, it might be the intermediate vinylic bromide, or it might be an alkyne. Decolorization of bromine in carbon tetrachloride by the liquid that we isolate would only tell us that our product is not our starting material but it would not tell us whether it was the vinylic halide or the alkyne. Both compounds react with bromine, and both would decolorize bromine in carbon tetrachloride.

$$\underset{\text{Vinylic bromide}}{-\overset{|}{C}=\underset{\underset{\text{Br}}{|}}{C}-} + Br_2 \xrightarrow{CCl_4} -\underset{\underset{\text{Br}}{|}}{\overset{|}{C}}-\underset{\underset{\text{Br}}{|}}{\overset{\overset{\text{Br}}{|}}{C}}-$$

$$\underset{\text{Alkyne}}{-C{\equiv}C-} + 2Br_2 \xrightarrow{CCl_4} -\underset{\underset{\text{Br}}{|}}{\overset{\overset{\text{Br}}{|}}{C}}-\underset{\underset{\text{Br}}{|}}{\overset{\overset{\text{Br}}{|}}{C}}-$$

An additional test is required. If the reaction was one that would yield a terminal alkyne, then we could use ammoniacal silver nitrate.

$$RC{\equiv}CH + Ag(NH_3)_2^+\,OH^- \xrightarrow[H_2O]{} RC{\equiv}CAg\downarrow$$

$$RCBr{=}CH_2 + Ag(NH_3)_2^+\,OH^- \xrightarrow[H_2O]{} \text{no reaction (and no precipitate)}$$

The formation of a precipitate would tell us that our liquid was a terminal acetylene. If no precipitate appeared and if the liquid had also given a positive test with bromine in carbon tetrachloride, we could be reasonably sure that we had isolated the vinylic bromide.

If, however, the reaction was one that would yield a disubstituted acetylene, testing with ammoniacal silver nitrate would not allow us to make a decision.

$$R-C{\equiv}C-R + Ag(NH_3)_2^+\,OH^- \longrightarrow \text{no precipitate}$$

$$R-CH{=}CBrR + Ag(NH_3)_2^+\,OH^- \longrightarrow \text{no precipitate}$$

Neither compound would give a precipitate.

In this instance we would need still another method. If our product gave a positive test with bromine in carbon tetrachloride, then we would be reasonably safe in assuming that it was either the vinylic bromide or the disubstituted acetylene. We could then distinguish between these two possibilities by adding a drop of the compound to a small amount of molten sodium. This reaction would reduce the bromine of the vinylic bromide to sodium bromide.

$$R-CH{=}CBrR + Na \xrightarrow[\text{Fusion}]{} Na^+Br^- + \text{other products}$$

Acidification of the resulting mixture with nitric acid, followed by treatment with aqueous silver nitrate, would give a precipitate of silver bromide.

$$Na^+Br^- + Ag^+NO_3^- \xrightarrow{H_2O} AgBr\downarrow + Na^+NO_3^-$$

The disubstituted acetylene, since it does not contain a halogen, would not give a positive test.

9.17A Chemical Tests

A number of reagents that are used as tests for some of the functional groups that we have studied so far are summarized below. We are restricting our attention at this point to alkanes, alkenes, alkynes, alkyl halides, and alcohols.

9.17B Concentrated Sulfuric Acid

Alkenes, alkynes, and alcohols are protonated and thus dissolve when they are added to cold concentrated sulfuric acid.

$$\text{>C=C<} + H_2SO_4 \longrightarrow \underbrace{-\overset{|}{\underset{\underset{H}{|}}{C}}-\overset{|}{\underset{+}{C}}-}_{\text{Soluble}} + HSO_4^-$$

$$-C\equiv C- + H_2SO_4 \longrightarrow \underbrace{-\underset{\underset{H^+}{\downarrow}}{C}\equiv C- \text{ or } -\underset{\underset{H}{|}}{C}=\underset{+}{C}-}_{\text{Soluble}} + HSO_4^-$$

$$-\overset{|}{\underset{|}{C}}-\ddot{\underset{\cdot\cdot}{O}}-H + H_2SO_4 \longrightarrow \underbrace{-\overset{|}{\underset{|}{C}}-\overset{\overset{H}{|}}{\underset{+}{\ddot{O}}}-H \text{ or } -\overset{|}{\underset{|}{C}}{}^+}_{\text{Soluble}} + HSO_4^-$$

Alkanes and alkyl halides are insoluble in cold concentrated sulfuric acid.

9.17C Bromine in Carbon Tetrachloride

Alkenes and alkynes both add bromine at room temperature and in the absence of light. Alkanes, alkyl halides, and alcohols do not react with bromine unless the reaction mixture is heated or exposed to strong irradiation. Thus, rapid decolorization of bromine in carbon tetrachloride at room temperature and in the absence of strong irradiation by light indicates the presence of a carbon-carbon double bond or a carbon-carbon triple bond.

9.17D Cold Dilute Potassium Permanganate

Alkenes and alkynes are oxidized by cold dilute solutions of potassium permanganate. If the alkene or alkyne is present in excess, the deep-purple color of the permanganate solution disappears and is replaced by the brown color of precipitated manganese dioxide.

$$\text{>C=C<} + \underset{\text{Purple}}{MnO_4^-} \xrightarrow[H_2O]{25^\circ} -\overset{|}{\underset{\underset{OH}{|}}{C}}-\overset{|}{\underset{\underset{OH}{|}}{C}}- + \underset{\text{Brown}}{MnO_2} + \text{other oxidation products}$$

$$-C\equiv C- + \underset{\text{Purple}}{MnO_4^-} \xrightarrow[H_2O]{25^\circ} -\underset{\underset{O}{\|}}{C}-\underset{\underset{O}{\|}}{C}- + \underset{\text{Brown}}{MnO_2} + \text{other oxidation products}$$

Alkanes, alkyl halides, and pure alcohols do not react with cold dilute potassium permanganate. When these compounds are tested the purple color is not discharged and a precipitate of manganese dioxide does not appear. (Impure alcohols often contain aldehydes and aldehydes give a positive test with cold dilute potassium permanganate.)

Cold dilute potassium permanganate is often called Baeyer's reagent.

9.17E Alcoholic Silver Nitrate

Alkyl and allylic halides (Section 10.2) react with silver ion to form a precipitate of silver halide. Ethyl alcohol is a convenient solvent because it dissolves silver nitrate and the alkyl halide. It does not dissolve the silver halide.

$$\underset{\text{Alkyl halide}}{R\text{—}X} + AgNO_3 \xrightarrow[\text{Alcohol}]{} AgX\downarrow + R^+ \longrightarrow \text{other products}$$

$$\underset{\text{An allylic halide}}{R\text{—}CH\text{=}CHCH_2X} + AgNO_3 \xrightarrow[\text{Alcohol}]{} AgX\downarrow + RCH\text{=}CH\overset{+}{C}H_2 \longrightarrow \text{other products}$$

Vinylic halides (Section 5.8) and phenyl halides (Chapter 11) do not give a silver halide precipitate when treated with silver nitrate in alcohol.

$$\underset{\text{A vinylic halide}}{R\text{—}CH\text{=}CHX} + AgNO_3 \xrightarrow[\text{Alcohol}]{} \text{no reaction}$$

$$\underset{\text{Phenyl halide}}{C_6H_5X} + AgNO_3 \xrightarrow[\text{Alcohol}]{} \text{no reaction}$$

9.17F Silver Nitrate in Ammonia

Silver nitrate reacts with ammonia to give a solution containing $Ag(NH_3)_2OH$. This reacts with terminal alkynes to form a precipitate of the silver alkynide.

$$R\text{—}C\equiv CH + Ag(NH_3)_2^+\,OH^- \longrightarrow R\text{—}C\equiv CAg\downarrow + HOH + 2NH_3$$

Nonterminal alkynes do not give a precipitate. *Silver alkynides can be distinguished from silver halides on the basis of their solubility in nitric acid; silver alkynides dissolve whereas silver halides do not.*

9.17G Sodium Fusion

The halogens of all organic halides including vinylic halides and phenyl halides are reduced to halide ions when these compounds are fused with sodium metal. Afterwards, the solutions can be acidified with nitric acid and treated with aqueous silver nitrate. A precipitate of silver halide indicates the presence of halogen in the original compound.

$$\begin{array}{c} \text{R—X} \\ \text{or} \\ \text{R—CH=CHX} \\ \text{Vinylic halide} \\ \text{or} \\ C_6H_5X \\ \text{Phenyl halide} \end{array} + \text{Na} \xrightarrow[\text{Fuse}]{} \text{Na}^+\text{X}^- + \text{other products} \xrightarrow{(1)\ HNO_3,\ (2)\ AgNO_3} \underset{\text{Silver halide}}{\text{AgX}\downarrow}$$

9.17H Chromic Oxide

Primary and secondary alcohols are oxidized to aldehydes or ketones by a solution of chromic oxide, CrO_3, in aqueous sulfuric acid. Chromic oxide dissolves in aqueous sulfuric acid to give a solution containing $HCrO_4^-$ ions. A positive test is indicated when the clear orange solution containing $HCrO_4^-$ ions becomes opaque and takes on a greenish cast within two seconds. Tertiary alcohols do not give this test within the specified time because they are more difficult to oxidize.

$$\begin{array}{c} \text{R—CH}_2\text{OH} \\ \text{or} \\ \text{R—CHOH} \\ | \\ \text{R} \end{array} + \underset{\substack{\text{Clear} \\ \text{orange} \\ \text{solution}}}{HCrO_4^-} \longrightarrow \begin{array}{c} \text{O} \\ \| \\ \text{RCH} \\ \text{or} \\ \text{R—C=O} \\ | \\ \text{R} \end{array} + \underset{\text{(green)}}{Cr^{3+}} + H_2O$$

9.17I Hydrogenation

This test is more time consuming than the others that we have mentioned. It also requires more elaborate equipment. None of the tests that we have described thus far will distinguish between an alkene and a nonterminal alkyne. If the approximate formula weight of the compound is known, however, quantitative hydrogenation will distinguish between an alkene and a nonterminal alkyne because the alkyne will ultimately absorb twice as much hydrogen per mole.

$$\text{R—CH=CH—R}' + H_2 \xrightarrow{\text{Pt}} \text{R—CH}_2\text{CH}_2\text{—R}'$$

$$\text{R—C}\equiv\text{C—R}' + 2H_2 \xrightarrow{\text{Pt}} \text{R—CH}_2\text{CH}_2\text{—R}'$$

This test is usually carried out by hydrogenating a known weight of the substance in an apparatus that permits an accurate measurement of the volume of hydrogen consumed. From this volume of hydrogen, and the number of moles of the unknown compound used, a simple calculation will tell whether it is an alkene or an alkyne.

Quantitative hydrogenation, of course, will not distinguish between an alkyne and an alkadiene; both compounds absorb two moles of hydrogen.

$$\text{R—CH=CH—CH=CH—R}' + 2H_2 \xrightarrow{\text{Pt}} \text{RCH}_2\text{CH}_2\text{CH}_2\text{CH}_2\text{R}'$$

$$\text{R—C}\equiv\text{C—R}' + 2H_2 \xrightarrow{\text{Pt}} \text{RCH}_2\text{CH}_2\text{R}'$$

TABLE 9.2 Simple Chemical Tests

	CONC. H_2SO_4	Br_2/CCl_4	DILUTE $KMnO_4$	$AgNO_3$/ C_2H_5OH	$Ag(NH_3)_2OH$	(1) SODIUM (2) DIL. HNO_3 (3) $AgNO_3$	CrO_3 in H_2SO_4
Alkane	Insol.	—	—	—	—	—	—
>C=C<	Sol.	*Rapid Br_2 addition:* red-brown ↓ colorless	*Oxidation:* purple soln. ↓ brown ppt.	—	—	—	—
RC≡CH	Sol.	*Rapid Br_2 addition:* red-brown ↓ colorless	*Oxidation:* purple soln. ↓ brown ppt.	*	Precipitation of R—C≡CAg Ppt. dissolves in dil. HNO_3	—	—
RC≡CR	Sol.	*Rapid Br_2 addition:* red-brown ↓ colorless	*Oxidation:* purple soln. ↓ brown ppt.	—	—	—	—
R—X	Insol.	—	—	Precipitation of AgX	—	Precipitation of AgX	—
—C=C—C—X	Sol.	*Rapid Br_2 addition:* red-brown ↓ colorless	*Oxidation:* purple soln. ↓ brown ppt.	Precipitation of AgX	—	Precipitation of AgX	—
—C=C—X	Sol.	*Rapid Br_2 addition:* red-brown ↓ colorless	*Oxidation:* purple soln. ↓ brown ppt.	—	—	Precipitation of AgX	—
H—C(H)—OH (—CH_2—OH)	Sol.	—	—	—	—	—	*Oxidation:* orange soln. ↓ green opaque
C—C(C)(H)—OH	Sol.	—	—	—	—	—	*Oxidation:* orange soln. ↓ green opaque
C—C(C)(C)—OH	Sol.	—	—	—	—	—	—

*Some terminal alkynes may give a precipitate (R—C≡CAg) when treated with $AgNO_3$ in alcohol. Silver alkynides, however, can be distinguished from silver halides on the basis of their solubility in dilute HNO_3 (see p. 374).

Table 9.2 summarizes, in tabular form, all of the simple chemical tests that we have presented. Hydrogenation is not included.

9.18 SUMMARY OF THE CHEMISTRY OF ALKYNES

9.18A Synthesis of Alkynes

1. From other alkynes (Sections 9.11 and 9.14)
General:

$$\text{R—C}\equiv\text{C—H} \xrightarrow[\text{Liq. NH}_3]{\text{NaNH}_2} \text{R—C}\equiv\text{C:}^-\ \text{Na}^+ \xrightarrow[\text{Primary halide}]{\text{R}'\text{—X}} \text{R—C}\equiv\text{C—R}' + \text{NaX}$$

Specific example:

$$CH_3CH_2C{\equiv}CH \xrightarrow[\text{Liq. } NH_3]{NaNH_2} CH_3CH_2C{\equiv}C{:}^- \, Na^+ \xrightarrow{CH_3CH_2Br} CH_3CH_2C{\equiv}CCH_2CH_3 \; (75\%)$$

2. Through elimination reactions (Section 9.14)
   General:

$$R{-}\underset{X}{\overset{H}{C}}{-}\underset{X}{\overset{H}{C}}{-}R' \xrightarrow[(-2HX)]{\text{Base}} R{-}C{\equiv}C{-}R'$$

vic-Dihalide

$$R{-}\underset{H}{\overset{H}{C}}{-}\underset{X}{\overset{X}{C}}{-}R' \xrightarrow[(-2HX)]{\text{Base}} R{-}C{\equiv}C{-}R'$$

gem-Dihalide

Specific example:

$$CH_3(CH_2)_7\underset{Br}{CH}{-}\underset{Br}{CH_2} + 2NaNH_2 \xrightarrow[\text{Heat}]{} CH_3(CH_2)_7C{\equiv}CH + 2NaBr \; (54\%)$$

9.18B Reactions of Alkynes

1. Addition of hydrogen (Section 9.7)
   General:

$$R{-}C{\equiv}C{-}H \xrightarrow[\text{Cat.}]{H_2} R{-}CH{=}CH_2 \xrightarrow[\text{Cat.}]{H_2} R{-}CH_2CH_3$$

$$R{-}C{\equiv}C{-}R' \xrightarrow[\text{syn addition}]{\text{Special catalyst}} \text{(R)(H)C=C(R')(H)} \quad \textit{cis}\text{-Alkene}$$

$$R{-}C{\equiv}C{-}R' \xrightarrow[\text{anti addition}]{\text{Li, } NH_3} \text{(R)(H)C=C(H)(R')} \quad \textit{trans}\text{-Alkene}$$

Specific examples:

$$CH_3{-}\underset{CH_3}{\overset{CH_3}{C}}{-}CH_2C{\equiv}CH \xrightarrow[\text{Ni}]{H_2} \left[CH_3{-}\underset{CH_3}{\overset{CH_3}{C}}{-}CH_2{-}CH{=}CH_2\right] \xrightarrow[\text{Ni}]{H_2} CH_3{-}\underset{CH_3}{\overset{CH_3}{C}}{-}CH_2CH_2CH_3$$

$$\text{2-(C}{\equiv}C(CH_2)_5CH_3\text{)cyclopentanol} \xrightarrow[\text{Pd/CaCO}_3\text{, Quinoline}]{H_2} \text{cis-2-(CH=CH(CH}_2)_5CH_3\text{)cyclopentanol} \; (86\%)$$

$$CH_3(CH_2)_2C{\equiv}C(CH_2)_2CH_3 \xrightarrow[\text{liq. } NH_3]{Li} \underset{(52\%)}{(CH_3CH_2CH_2)HC{=}CH(CH_2CH_2CH_3)}$$

2. Addition of halogens (Section 9.8)
General:

$$R{-}C{\equiv}C{-}R' + X_2 \longrightarrow (R)(X)C{=}C(X)(R') \xrightarrow{X_2} R{-}CX_2{-}CX_2{-}R'$$

Specific example:

$$C_6H_5{-}C{\equiv}C{-}CH_3 + Br_2 \xrightarrow[\substack{CH_3COOH \\ 25°}]{LiBr} \underset{(98\%)}{(C_6H_5)(Br)C{=}C(Br)(CH_3)}$$

3. Addition of hydrogen halides (Section 9.9)
General:

$$R{-}C{\equiv}C{-}H + HX \longrightarrow (R)(X)C{=}C(H)(H) \xrightarrow{HX} R{-}CX_2{-}CH_3$$

Specific example:

$$CH_3{-}C{\equiv}CH + HCl \longrightarrow (CH_3)(Cl)C{=}C(H)(H) \xrightarrow{HCl} CH_3{-}CCl_2{-}CH_3$$

4. Addition of water (Section 9.10)
General:

$$R{-}C{\equiv}C{-}R \xrightarrow[\substack{H_2O \\ H_2SO_4}]{Hg^{++}} \left[R{-}C(OH){=}C(H){-}R\right] \longrightarrow R{-}C(=O){-}CH_2R$$

Specific examples:

$$H{-}C{\equiv}C{-}H + H_2O \xrightarrow[\substack{18\%\ H_2SO_4 \\ 90°}]{Hg^{++}} CH_3{-}C(=O){-}H$$

$$\text{steroid (HO, CH}_3\text{, CH}_3\text{, OH, C}{\equiv}\text{CH)} \xrightarrow[\substack{H_2O \\ C_2H_5OH}]{Hg^{++}} \underset{(95\%)}{\text{steroid (HO, CH}_3\text{, CH}_3\text{, OH, C(=O)}{-}\text{CH}_3\text{)}}$$

5. Addition of boron hydrides (Sections 9.7 and 9.10)
General:

$$R{-}C{\equiv}C{-}R + (BH_3)_2 \longrightarrow \left(\begin{matrix} R \\ H \end{matrix}\right)C{=}C\left(\begin{matrix} R \\ \end{matrix}\right)_3 B$$

$$\xrightarrow[H_2O_2]{^-OH} \left[\begin{matrix} R \\ H \end{matrix} C{=}C \begin{matrix} R \\ OH \end{matrix}\right] \longrightarrow R{-}CH_2\underset{\underset{O}{\|}}{C}{-}R$$

$$\xrightarrow{CH_3COOH} \begin{matrix} R \\ H \end{matrix} C{=}C \begin{matrix} R \\ H \end{matrix}$$

Specific example:

$$C_3H_7{-}C{\equiv}C{-}C_3H_7 + (BH_3)_2 \xrightarrow{0^\circ} \begin{matrix} C_3H_7 \\ H \end{matrix} C{=}C \begin{matrix} C_3H_7 \\ B{-} \end{matrix}$$

$$\xrightarrow{OH^-,\ H_2O_2} C_3H_7\underset{\underset{O}{\|}}{C}CH_2C_3H_7$$

$$\xrightarrow[CH_3COOH]{0^\circ} \begin{matrix} C_3H_7 \\ H \end{matrix} C{=}C \begin{matrix} C_3H_7 \\ H \end{matrix}$$

6. Replacement of the acetylenic hydrogen (Sections 9.11 and 9.12)
Specific examples:

$$H{-}C{\equiv}C{-}H + NaNH_2 \xrightarrow[NH_3]{liq.} H{-}C{\equiv}C{:}^-Na^+ + NH_3$$

$$CH_3(CH_2)_6C{\equiv}CH + Ag(NH_3)_2{}^+OH^- \xrightarrow[H_2O]{} CH_3(CH_2)_6C{\equiv}C{:}^-Ag^+ + 2NH_3 + H_2O$$

7. Oxidative Coupling (Section 9.13)

$$2CH_3\overset{\overset{CH_3}{|}}{\underset{\underset{OH}{|}}{C}}{-}C{\equiv}CH + CuCl \xrightarrow[O_2,\ CH_3OH]{Pyridine} CH_3{-}\overset{\overset{CH_3}{|}}{\underset{\underset{OH}{|}}{C}}{-}C{\equiv}C{-}C{\equiv}C{-}\overset{\overset{CH_3}{|}}{\underset{\underset{OH}{|}}{C}}{-}CH_3$$

(80%)

8. Oxidative cleavage of alkynes (Section 9.16)
General:

$$R{-}C{\equiv}C{-}R' \xrightarrow[(2)\ H^+]{(1)\ KMnO_4,\ OH^-} R{-}COOH + R'{-}COOH$$

$$R{-}C{\equiv}C{-}R' \xrightarrow[(2)\ H_2O]{(1)\ O_3,\ CCl_4} R{-}COOH + R'{-}COOH$$

Additional Problems

9.9

Give IUPAC names for each of the following alkynes.

(a) $CH_3CHC{\equiv}CH$ with CH_3 on the second carbon
(b) $(CH_3)_3CC{\equiv}CCH_2CH_3$
(c) $CH_3CH_2C{\equiv}C(CH_2)_4CH_3$
(d) $CH_3CH_2C{\equiv}CCH_2CH_3$
(e) di-*tert*-Butylacetylene
(f) Diisopropylacetylene
(g) Methylpropylacetylene
(h) Methyl-*sec*-butylacetylene
(i) Diisobutylacetylene
(j) *n*-Hexylacetylene

9.10

Which of the compounds in Problem 9.9 would give a positive test with $Ag(NH_3)_2^+ OH^-$?

9.11

(a) Which of the compounds in Problem 9.9 would yield *n*-hexane when reduced with two moles of hydrogen and a platinum catalyst?
(b) Which would yield *n*-heptane?

9.12

Starting only with coke, lime, water, methane, and any other needed inorganic reagents show how you might synthesize each of the following compounds. (You need not show the synthesis of the same compound twice.)

(a) Acetylene
(b) Ethene
(c) Propyne
(d) Propene
(e) Acetone
(f) 2-Butyne
(g) 1-Butyne
(h) $CH_3CH_2CCH_3$ with $=O$ on the third carbon
(i) Bromoethane
(j) Methylethylacetylene
(k) Ethanol
(l) 1,2-Dibromoethane
(m) 2,2-Dichloropropane
(n) *cis*-2-Butene
(o) *trans*-2-Butene
(p) 1-Butene
(q) 2-Bromobutane
(r) 1-Bromobutane
(s) *n*-Butyl alcohol
(t) *sec*-Butyl alcohol
(u) *meso*-2,3-Dibromobutane
(v) Racemic-2,3-dibromobutane
(w) (*Z*)-2-Chloro-2-butene
(x) 1-Bromopropane
(y) 3-Hexyne
(z) 1-Bromo-1-butene

9.13

Give the structure of the products that you would expect from the reaction of 1-pentyne with:

(a) One mole Br_2
(b) One mole HCl
(c) Two moles HCl
(d) One mole HBr and peroxides
(e) H_2O, H^+, Hg^{++}
(f) H_2, Pd-$CaCO_3$, quinoline
(g) $(Sia)_2BH$ then CH_3COOH
(h) $(Sia)_2BH$ then OH^-, H_2O_2
(i) $NaNH_2$ in liq. NH_3
(j) $NaNH_2$ in liq. NH_3, then CH_3I
(k) CuCl, NH_3, O_2, CH_3OH
(l) $Ag(NH_3)_2OH$
(m) $Cu(NH_3)_2OH$

9.14

Give the structure of the products you would expect from the reaction (if any) of 3-hexyne with:

(a) One mole HCl
(b) Two moles HCl
(c) One mole Br_2
(d) Two moles Br_2
(e) Ni_2B(P-2), H_2
(f) $Pd/BaSO_4$, quinoline, H_2
(g) Li/NH_3
(h) H_2O, H^+, Hg^{++}
(i) $Ag(NH_3)_2OH$
(j) $Cu(NH_3)_2OH$
(k) $(BH_3)_2$ then CH_3COOH
(l) $(BH_3)_2$ then OH^-, H_2O_2
(m) CuCl, NH_3, O_2, CH_3OH
(n) Two moles H_2, Pt
(o) $KMnO_4$, OH^-, then H^+
(p) O_3, H_2O
(q) $NaNH_2$, liq. NH_3

9.15

Show how each of the following compounds might be transformed into 1-pentyne.
(a) 1-Pentene
(b) 1-Chloropentane
(c) 1-Chloro-1-pentene
(d) 1,1-Dichloropentane
(e) 1-Bromopropane and acetylene

9.16

Describe with equations a simple test that would distinguish between each of the following pairs of compounds. (In each case tell what you would see.)
(a) Propane and propyne
(b) Propene and propyne
(c) 1-Bromopropene and 2-bromopropane
(d) 2-Bromo-2-butene and 1-butyne
(e) 2-Bromo-2-butene and 2-butyne
(f) 2-Butyne and *n*-butyl alcohol
(g) 2-Butyne and 2-bromobutane
(h) $CH_3C{\equiv}CCH_2OH$ and $CH_3CH_2CH_2CH_2OH$
(i) $CH_3CH{=}CHCH_2OH$ and $CH_3CH_2CH_2CH_2OH$

9.17

Three compounds **A, B,** and **C** all have the formula C_5H_8. All three compounds rapidly decolorize bromine in carbon tetrachloride, all three give a positive test with Baeyer's reagent, and all three are soluble in cold concentrated sulfuric acid. Compound **A** gives a precipitate when treated with ammoniacal silver nitrate but compounds **B** and **C** do not. Compounds **A** and **B** both yield *n*-pentane (C_5H_{12}) when they are treated with excess hydrogen in the presence of a platinum catalyst. Under these same conditions, compound **C** absorbs only one mole of hydrogen and gives a product with the formula C_5H_{10}.
(a) Suggest a possible structure for **A, B,** and **C.**
(b) Are other structures possible for **B** and **C?**
(c) Oxidative cleavage of **B** with hot, basic $KMnO_4$ gives after acidification acetic acid and CH_3CH_2COOH. What is the structure of **B?**
(d) Oxidative cleavage of **C** with ozone gives $H\underset{\underset{\displaystyle O}{\|}}{C}CH_2CH_2CH_2\underset{\underset{\displaystyle O}{\|}}{C}H$. What is the structure of **C?**

9.18

Starting with 3-methyl-1-butyne and any inorganic reagents show how the following compounds could be synthesized.

(a) $CH_3CH(CH_3)C(Cl){=}CH_2$ (d) $CH_3CH(CH_3)CCl_2CH_2Cl$

(b) $CH_3CH(CH_3)CH_2CH_2Br$ (e) $CH_3CH(CH_3)CClBrCH_3$

(c) $CH_3CH(CH_3)CH(Cl)CH_3$ (f) $CH_3CH(CH_3)COOH$

9.19

In addition to 1-pentyne and ordinary inorganic reagents, assume that you also have available the following deuterium compounds: D_2, DCl, CH_3COOD, and $(Sia)_2BD$. Show how you might prepare the following deuterium-labeled compounds.

(a) $CH_3CH_2CH_2(D)C{=}C(H)D$ ($CH_3CH_2CH_2$ and H trans; D and D cis... as drawn: $CH_3CH_2CH_2$ upper left, D lower left, H upper right, D lower right)

(b) $CH_3CH_2CH_2(H)C{=}C(H)D$ ($CH_3CH_2CH_2$ upper left, H lower left, H upper right, D lower right)

(c) $CH_3CH_2CH_2(Cl)C{=}C(D)H$ ($CH_3CH_2CH_2$ upper left, Cl lower left, D upper right, H lower right)

(d) $CH_3CH_2CH_2CH(D)CHO$

9.20

A number of industrially important chloroethenes and chloroethanes can be prepared from acetylene in ways that allow the synthesis of specific isomers. Examples are the compounds listed below. Show how these syntheses might be carried out.

(a) *trans*-1,2-Dichloroethene
(b) 1,2-Dichloroethane
(c) 1,1,2,2-Tetrachloroethane
(d) 1,1,2-Trichloroethene
(e) 1,1,2-Trichloroethane
(f) 1,1,1,2,2-Pentachloroethane
(g) 1,1,2,2-Tetrachloroethene

9.21

Disiamylborane (p. 362) is prepared by allowing diborane to react with an alkene. (a) Give the structure of this alkene and show how it reacts with diborane. (b) What factor favors the formation of disiamylborane, Sia_2BH, in this reaction by preventing the formation of Sia_3B?

9.22

A naturally occurring antibiotic called mycomycin has the structure shown below. Mycomycin is optically active. Explain this by writing structures for the enantiomeric forms of mycomycin.

$$HC{\equiv}C{-}C{\equiv}C{-}CH{=}C{=}CH{-}(CH{=}CH)_2CH_2COOH$$

Mycomycin

9.23

An optically active compound **D** has the molecular formula C_6H_{10}. The compound gives a precipitate when treated with a solution containing $Ag(NH_3)_2OH$. On catalytic hydrogenation **D** yields **E**, C_6H_{14}. **E** is optically inactive and cannot be resolved. Propose structures for **D** and **E**.

***9.24**

Alkenes are more reactive than alkynes toward addition of electrophilic reagents (i.e., Br_2, Cl_2, HCl, etc.). Yet when alkynes are treated with these same electrophilic reagents it is easy to stop the addition at the "alkene stage." This appears to be a paradox and yet it is not. Explain.

***9.25**

A number of compounds containing carbon-carbon triple bonds have been found in nature. One such compound obtained from the seeds of an African tree is called *erythrogenic acid. Erythro* comes from a Greek word meaning red, and when exposed to light erythrogenic acid forms a bright red polymer. Erythrogenic acid is a monocarboxylic acid (i.e., each molecule has one —COOH group) and has the molecular formula $C_{18}H_{26}O_2$. When treated with excess hydrogen in the presence of a platinum catalyst, erythrogenic acid absorbs five moles of hydrogen and yields a compound with the structural formula $CH_3(CH_2)_{16}COOH$. (a) What important clues does this experiment give us about the structure of erythrogenic acid? (b) At this point many possible structures can be written for erythrogenic acid. Write at least four. (c) Erythrogenic acid reacts with ozone and when the ozonide is subsequently treated with water, four products can be isolated: $H-\overset{\overset{\displaystyle O}{\|}}{C}-H$, $OHC(CH_2)_4COOH$, OHC—COOH, and $OHC(CH_2)_7COOH$. This experiment narrows the structural possibilities for erythrogenic acid down to two formulas. What are they? (d) When an alcoholic solution of $CH_2{=}CH(CH_2)_4C{\equiv}CH$ and $HC{\equiv}C(CH_2)_7COOH$ is treated with cuprous chloride, ammonia, and oxygen, three products can be isolated from the reaction mixture; one is non-acidic, one is a monocarboxylic acid, and one is a dicarboxylic acid. This monocarboxylic acid is erythrogenic acid. What is the structure of erythrogenic acid?

10 CONJUGATED UNSATURATED SYSTEMS. VISIBLE-ULTRAVIOLET SPECTROSCOPY

10.1 INTRODUCTION

In our study of the reactions of alkenes in Chapter 7 we saw how important the π bond is in understanding the chemistry of unsaturated compounds. In this chapter we will study a special group of unsaturated compounds and again we will find that the π bond is the important part of the molecule. Here we will examine *species that have a p orbital on an atom adjacent to a double bond.* The *p* orbital may be one that contains a single electron as in the allyl radical, $CH_2{=}CHCH_2\cdot$ (Section 10.2); it may be a vacant *p* orbital as in the allyl cation, $CH_2{=}CHCH_2^+$ (Section 10.4); or it may be the *p* orbital of another double bond as in 1,3-butadiene, $CH_2{=}CH{-}CH{=}CH_2$ (Section 10.6). We will see that having a *p* orbital on an atom adjacent to a double bond allows the formation of an extended π bond—one that encompasses more than two nuclei.

Systems that have a *p* orbital on an atom adjacent to a double bond—molecules with extended π bonds—are called *conjugated unsaturated systems.* This general phenomenon is called *conjugation.* As we will see, conjugation gives these systems special properties. We will find, for example, that conjugated radicals, ions, or molecules, are more stable than nonconjugated ones. We will demonstrate this with the allyl radical, the allyl cation, and 1,3-butadiene. Conjugation also allows molecules to undergo unusual reactions, and we will study these, too, including an important reaction for forming rings called the Diels-Alder reaction (Section 10.8). Finally we will study how molecules absorb light (Section 10.9) and here again we will find that conjugation has an important effect.

10.2 ALLYLIC SUBSTITUTION AND THE ALLYL RADICAL

When propene reacts with bromine or chlorine at low temperatures the reaction that takes place is the usual addition of halogen to the double bond.

$$CH_2{=}CH{-}CH_3 + X_2 \xrightarrow[\text{CCl}_4 \text{ (addition reaction)}]{\text{Low temperature}} \underset{\substack{| \\ X}}{CH_2}{-}\underset{\substack{| \\ X}}{CH}{-}CH_3$$

However, when propene reacts with chlorine or bromine at very high temperatures or under conditions in which the concentration of the halogen is very small, the reaction that occurs is a *substitution.* These two examples illustrate how we can often change the course of an organic reaction simply by changing the conditions. (They also illustrate the need for specifying the conditions of a reaction carefully when we report experimental results.)

$$\underset{\text{Propene}}{CH_2{=}CH{-}CH_3} + X_2 \xrightarrow[\text{low conc. of } X_2 \text{ (substitution reaction)}]{\text{High temperature or}} CH_2{=}CH{-}CH_2X + HX$$

In this substitution a halogen replaces one of the hydrogens of the methyl group of propene. These hydrogens are called the *allylic hydrogens* and the substitution reaction is known as an *allylic substitution.**

$$\mathrm{H_2C{=}CH{-}CH_3} \quad \text{(the three H of } \mathrm{CH_3}\text{: Allylic hydrogens)}$$

Propene undergoes allylic chlorination when propene and chlorine react in the gas phase at 400°.

$$\underset{}{\mathrm{CH_2{=}CH{-}CH_3 + Cl_2}} \xrightarrow[\text{Gas phase}]{400^\circ} \underset{\text{Allyl chloride}}{\mathrm{CH_2{=}CH{-}CH_2Cl}} + \mathrm{HCl}$$

Propene undergoes allylic bromination when it is treated with *N*-bromosuccinimide in carbon tetrachloride at room temperature. The reaction is initiated by light or peroxides.

$$\mathrm{CH_2{=}CH{-}CH_3} + \underset{\substack{N\text{-Bromosuccinimide}\\ \text{(NBS)}}}{\mathrm{(CH_2C{=}O)_2NBr}} \xrightarrow[\substack{\text{(or}\\ \text{peroxides)}\\ \mathrm{CCl_4}}]{h\nu} \underset{\text{Allyl bromide}}{\mathrm{CH_2{=}CH{-}CH_2Br}} + \underset{\text{Succinimide}}{\mathrm{(CH_2C{=}O)_2NH}}$$

N-Bromosuccinimide is insoluble in CCl_4 and it provides a constant but very low concentration of bromine in the reaction mixture. It does this by reacting with the hydrogen bromide formed in the substitution reaction. Each molecule of HBr that is formed is replaced by one molecule of Br_2.

$$\mathrm{(CH_2C{=}O)_2N{-}Br + H{-}Br \longrightarrow (CH_2C{=}O)_2N{-}H + Br_2}$$

* These are general terms as well. The hydrogens of any saturated carbon adjacent to a double bond, that is,

$$\mathrm{>C{=}C<} \text{ with } \mathrm{{-}\overset{|}{\underset{H}{C}}{-}}$$

are called *allylic* hydrogens and any reaction in which an allylic hydrogen is replaced is called an *allylic substitution.*

Under these conditions, that is, in a nonpolar solvent and with a very low concentration of bromine, very little bromine adds to the double bond; it reacts by substitution and replaces an allylic hydrogen instead.

The mechanism for these substitution reactions is much the same as the chain mechanism for alkane halogenations that we saw in Chapter 4. In the chain-initiating step, the halogen (chlorine or bromine) dissociates into halogen atoms.

Chain-initiating step

$$:\ddot{X}:\ddot{X}: \xrightarrow{h\nu} 2:\ddot{X}\cdot$$

In the first chain-propagating step the halogen atom abstracts one of the allylic hydrogens.

First chain-propagating step

$$H_2C{=}CH{-}CH_2{:}H + \cdot\ddot{X}: \longrightarrow H_2C{=}CH{-}\dot{C}H_2 + H{:}\ddot{X}:$$

Allyl radical

The radical that is produced in this step is called an *allyl radical.*

In the second chain-propagating step the allyl radical reacts with a molecule of the halogen.

Second chain-propagating step

$$H_2C{=}CH{-}CH_2\cdot + :\ddot{X}:\ddot{X}: \longrightarrow H_2C{=}CH{-}CH_2{:}X + :\ddot{X}\cdot$$

Allyl halide

This step results in the formation of a molecule of allyl halide and a halogen atom. The halogen atom then brings about a repetition of the first chain-propagating step. The chain reaction continues until the usual chain-terminating steps consume the radicals.

The allylic substitution reactions of propene will seem somewhat less surprising if we examine the bond dissociation energy of an allylic carbon-hydrogen bond and compare it with the bond dissociation energies of other carbon hydrogen bonds (cf. Table 4.1.)

$$\underset{\text{Propene}}{CH_2{=}CHCH_2{-}H} \longrightarrow \underset{\text{Allyl radical}}{CH_2{=}CHCH_2\cdot} + H\cdot \qquad DH^\circ = 85\ \text{kcal/mole}$$

$$\underset{\text{Isobutane}}{(CH_3)_3C{-}H} \longrightarrow \underset{3^\circ\ \text{Radical}}{(CH_3)_3C\cdot} + H\cdot \qquad DH^\circ = 91\ \text{kcal/mole}$$

$$\underset{\text{Propane}}{(CH_3)_2CH{-}H} \longrightarrow \underset{2^\circ\ \text{Radical}}{(CH_3)_2CH\cdot} + H\cdot \qquad DH^\circ = 94.5\ \text{kcal/mole}$$

$$\underset{\text{Propane}}{CH_3CH_2CH_2{-}H} \longrightarrow \underset{1^\circ\ \text{Radical}}{CH_3CH_2CH_2\cdot} + H\cdot \qquad DH^\circ = 98\ \text{kcal/mole}$$

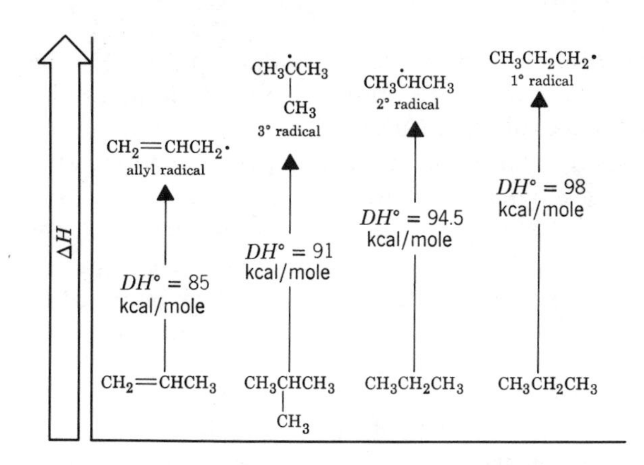

FIG. 10.1
The relative stability of the allyl radical compared to 1°, 2°, and 3° radicals. (The stabilities of the radicals are relative to the hydrocarbon from which each was formed and the overall order of stability is: allyl > 3° > 2° > 1°.

We see that an allylic carbon-hydrogen bond of propene is broken with greater ease than even the tertiary carbon-hydrogen bond of isobutane and with far greater ease than either a secondary carbon-hydrogen bond or a primary carbon-hydrogen bond of propane.

The ease with which an allylic carbon-hydrogen bond is broken means that relative to primary, secondary, and tertiary free radicals the allyl radical is the *most stable* (Fig. 10.1).

10.3 THE STABILITY OF THE ALLYL RADICAL

An explanation of the stability of the allyl radical can be approached in two ways: in terms of molecular orbital theory and in terms of resonance theory. The molecular orbital approach is easier to visualize so let us begin with it.

10.3A Molecular Orbital Description of the Allyl Radical

As an allylic hydrogen is abstracted from propene, the sp^3-hybridized carbon of the methyl group changes its hybridization state to sp^2 (cf. Section 4.13B).

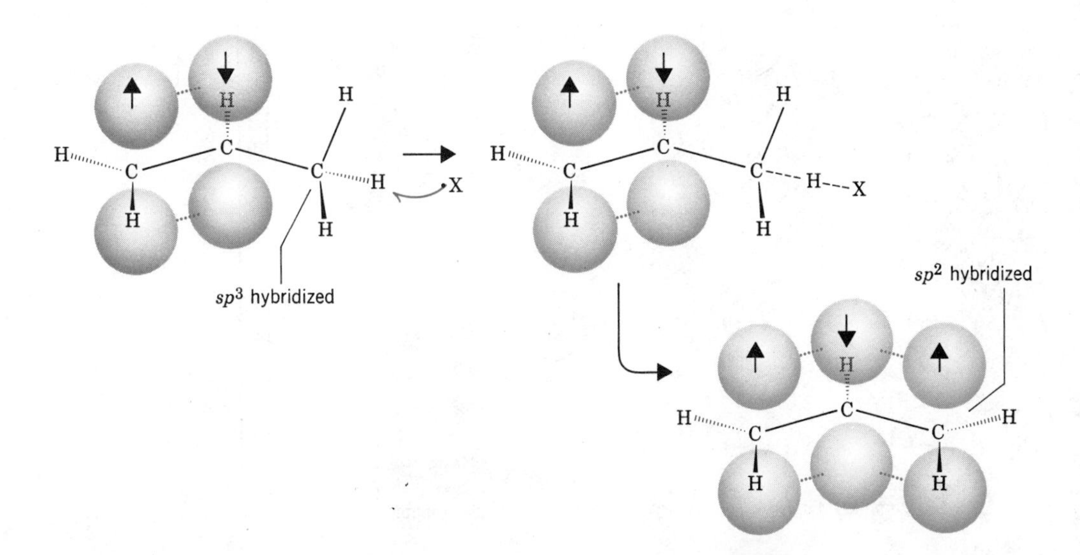

The *p* orbital of this new sp^2-hybridized carbon overlaps with the *p* orbital of the central carbon. Thus, in the allyl radical three *p* orbitals overlap to form a set of π-molecular orbitals that encompass all three carbons. The new *p* orbital of the allyl radical is said to be *conjugated* with those of the double bond and the allyl radical is said to be a *conjugated unsaturated system.*

The unpaired electron of the allyl radical and the two electrons of the π bond are *delocalized* over all three carbons. This delocalization of the unpaired electron accounts for the greater stability of the allyl radical when compared to primary, secondary, and tertiary radicals. Although some delocalization occurs in primary, secondary, and tertiary radicals, delocalization is not as effective because it occurs through σ bonds.

The diagram in Figure 10.2 illustrates how the three *p* orbitals of the allyl radical combine to form three π molecular orbitals. (Remember: the number of molecular orbitals that result always equals the number of atomic orbitals that combine, cf. Section 1.14). The bonding π molecular orbital is of lowest energy; it

FIG. 10.2
The combination of atomic orbitals to form π molecular orbitals in the allyl radical. The bonding π molecular orbital is formed by the combination of three p *orbitals with lobes of the same sign overlapping above and below the plane of the atoms. The nonbonding π molecular orbital is formed by a combination of* p *orbitals from only two carbon atoms: carbons* **1** *and* **3.** *This nonbonding π molecular orbital has a node at carbon* **2.** *The antibonding π molecular orbital has two nodes: between carbons* **1** *and* **2,** *and between carbons* **2** *and* **3.**

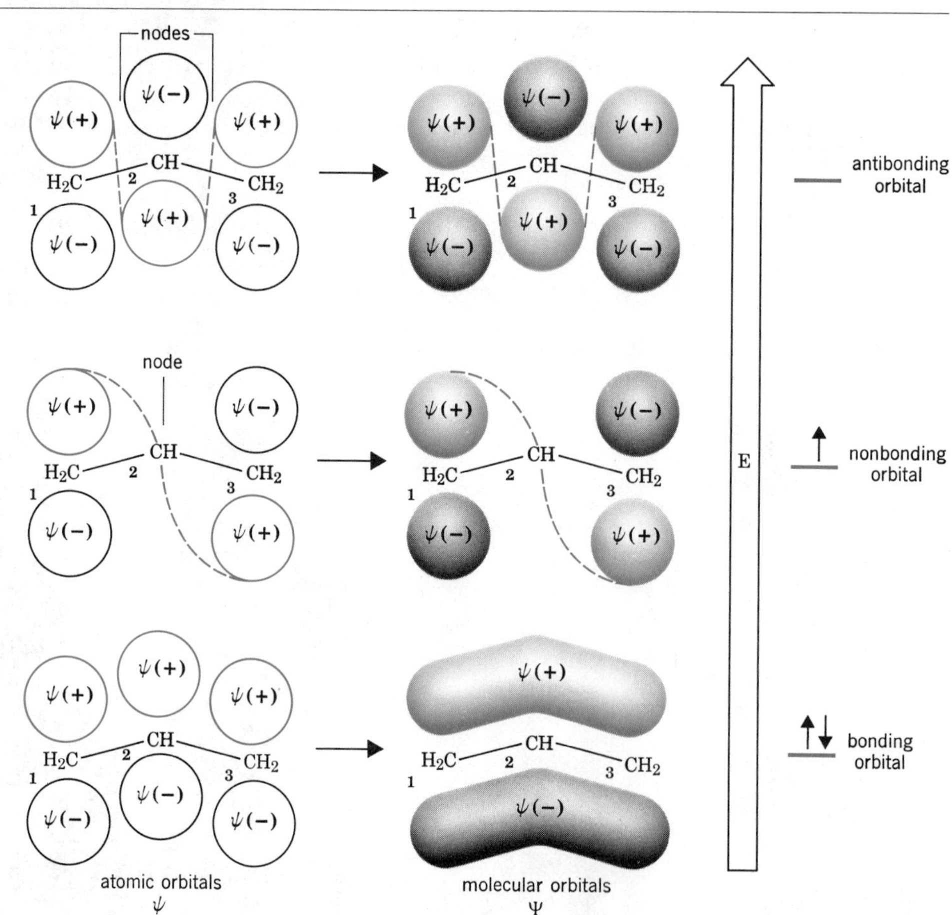

encompasses all three carbons and is occupied by two spin-paired electrons. This bonding π orbital is the result of having p orbitals with lobes of the same sign overlap between adjacent carbon atoms. This type of overlap, as we recall, increases the π electron density in the regions between the atoms where it is needed for bonding. The nonbonding π orbital is occupied by one unpaired electron and it has a node at the central carbon. This node means that the unpaired electron is located in the vicinity of carbon **1** and **3** only. The antibonding π molecular orbital results when orbital lobes of opposite sign overlap between adjacent carbon atoms: such overlap means that in the antibonding π orbital there is a node between each pair of carbon atoms. This antibonding orbital of the allyl radical is of highest energy and is empty in the ground state of the radical.

We can illustrate the picture of the allyl radical given by molecular orbital theory in simpler terms with the structure shown below.

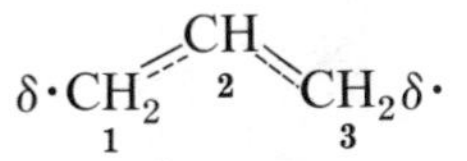

$$\delta\cdot CH_2 \overset{CH}{\underset{2}{\cdots}} CH_2\delta\cdot$$
1 3

We indicate with dotted lines that both carbon-carbon bonds are partial double bonds. This accommodates one of the things that molecular-orbital theory tells us: *that there is a π bond encompassing all three atoms.* We also place the symbol $\delta\cdot$ beside carbons **1** and **3**. This denotes a second thing molecular-orbital theory tells us: *that the unpaired electron spends its time in the vicinity of carbons **1** and **3**.* Finally, implicit in the molecular orbital picture of the allyl radical is this: the two ends of the allyl radical are *equivalent.* This aspect of the molecular-orbital description is also implicit in the formula given above.

10.3B Resonance Description of the Allyl Radical

Earlier in this section we wrote the stucture of the allyl radical as **A**.

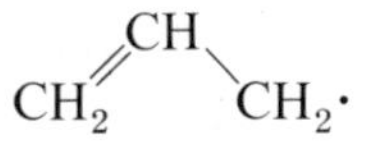

$$CH_2{=}CH{-}CH_2\cdot$$
A

However, we might just as well have written the equivalent structure, **B**.

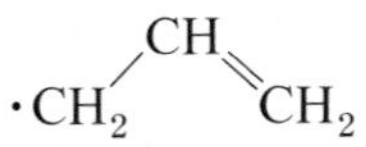

$$\cdot CH_2{-}CH{=}CH_2$$
B

In writing structure **B** we do not mean to imply that we have simply taken structure **A** and turned it over. What we have done is moved the electrons in the following way.

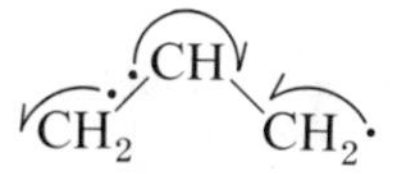

$$CH_2{=}CH{-}CH_2\cdot$$

We have not moved the atomic nuclei themselves.

Resonance theory (Section 1.8) tells us that whenever we can write two structures for a chemical entity *that differ only in the positions of the electrons,* the entity cannot be represented by either structure alone but is a *hybrid* of both. We can represent the hybrid in two ways: we can write both structures **A** and **B**, and connect them with a double-headed arrow, a special sign in resonance theory, that indicates they are resonance structures.

$$\underset{1}{CH_2}{=}\underset{2}{CH}{-}\underset{3}{CH_2}\cdot \longleftrightarrow \cdot\underset{1}{CH_2}{-}\underset{2}{CH}{=}\underset{3}{CH_2}$$

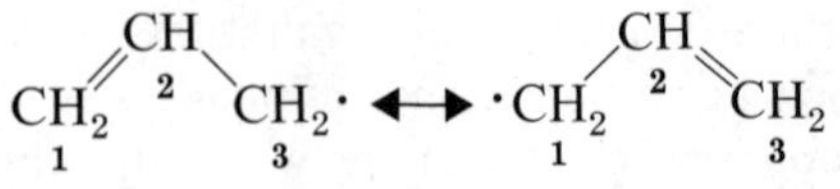

A **B**

Or, we can write a single structure, **C**, that blends the features of both resonance structures.

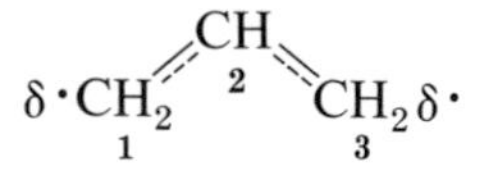

$$\delta\cdot\underset{1}{CH_2}\overset{\cdots}{-}\underset{2}{CH}\overset{\cdots}{-}\underset{3}{CH_2}\delta\cdot$$

C

Structure **C** describes the carbon-carbon bonds of the allyl radical as partial double bonds. This indicates one feature of both **A** and **B.** In **A** the bond between carbon **1** and carbon **2** is a double bond and the bond between carbon **2** and carbon **3** is a single bond. In structure **B** just the reverse is true.

The resonance structures **A** and **B** also tell us that the unpaired electron is associated only with carbons **1** and **3.** We indicate this in structure **C** by placing a $\delta\cdot$ beside carbons **1** and **3.*** Because resonance structures **A** and **B** are equivalent *carbons **1** and **3** are also equivalent.*

Another rule in resonance theory is that *whenever equivalent resonance structures* can be written for a chemical species, *the chemical species is much more stable than either resonance structure* (*when taken alone*) *would indicate.* If we were to examine either **A** or **B** alone we might decide that they resemble primary radicals. Thus, we might estimate the stability of the allyl radical as approximately that of a primary radical. In doing so, we would greatly underestimate the stability of the allyl radical. Resonance theory tells us, however, that since **A** and **B** are *equivalent resonance structures,* the allyl radical should be much more stable than either, that is, much more stable than a primary radical. This correlates with what experiments have shown to be true: the allyl radical is even more stable than a tertiary radical.

Problem 10.1

(a) What product(s) would you expect to obtain if propene labeled with ^{14}C at carbon **1** were subjected to allylic chlorination or bromination? (b) Explain your answer.

$$^{14}CH_2{=}CHCH_3 + X_2 \xrightarrow[\text{low conc. of } X_2]{\text{High temperature or}} ?$$

* A resonance structure such as the one shown below would indicate that an unpaired electron is associated with carbon **2.** This structure is not a proper resonance structure because resonance theory dictates that *all resonance structures must have the same number of unpaired electrons* (cf. Section 10.10).

$$\cdot CH_2{-}\dot{C}H{-}CH_2\cdot$$
(an incorrect resonance structure)

(c) If more than one product would be obtained what relative proportions would you expect?

10.4 THE ALLYL CATION

Although we cannot go into the experimental evidence here, the allyl cation, $CH_2{=}CHCH_2^+$, is an unusually stable carbocation. It is even more stable than a secondary carbocation and is almost as stable as a tertiary carbocation. In general terms, the relative order of stabilities of carbocations is that given below.

Relative order of carbocation stability

$$\underset{3^\circ}{\mathrm{C{-}\overset{C}{\underset{C}{C^+}}}} > \underset{\text{Allyl}}{-\mathrm{C}{=}\mathrm{C}{-}\overset{+}{\mathrm{C}}\mathrm{H}_2} > \underset{2^\circ}{\mathrm{C{-}\overset{C}{\underset{H}{C^+}}}} > \underset{1^\circ}{\mathrm{C{-}\overset{H}{\underset{H}{C^+}}}}$$

As we might expect, the unusual stability of the allyl cation can also be accounted for in terms of molecular orbital or resonance theory.

The molecular orbital description of the allyl cation is shown in Fig. 10.3.

The bonding π molecular orbital of the allyl cation, like that of the allyl radical (Fig. 10.2), contains two spin-paired electrons. The nonbonding π molecular orbital of the allyl cation, however, is empty. Since an allyl cation is what we

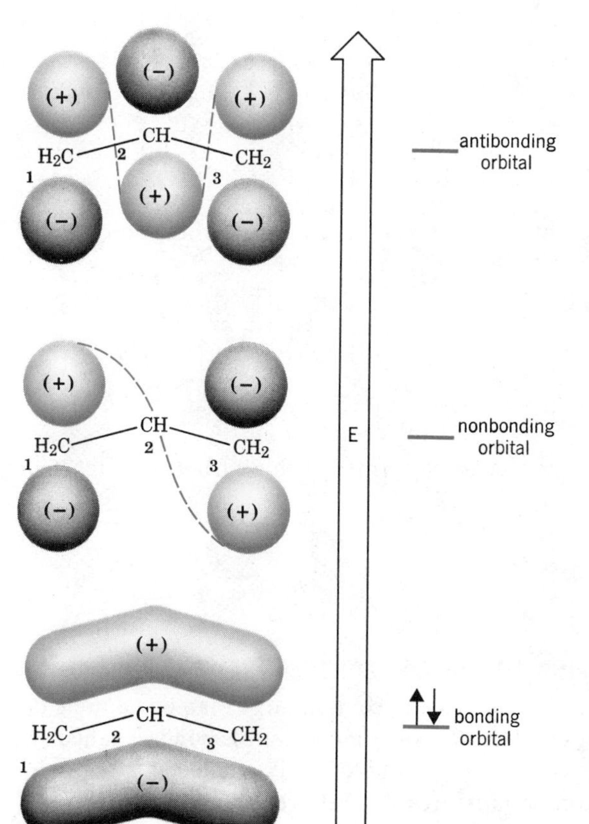

FIG. 10.3
The π molecular orbitals of the allyl cation. The allyl cation, like the allyl radical (Fig. 10.2), is a conjugated unsaturated system.

would get if we removed an electron from an allyl radical, we can say, in effect, that we remove the electron from the nonbonding molecular orbital.

$$CH_2{=}CHCH_2\cdot \xrightarrow{-e^-} CH_2{=}CHCH_2^+$$

Removal of an electron from a nonbonding orbital (cf. Fig. 10.2) is known to require less energy than removal of an electron from a bonding orbital. In addition, the positive charge that forms on the allyl cation is *effectively delocalized* between carbons **1** and **3.** Thus, in molecular orbital theory these two factors, the ease of removal of a nonbonding electron and the delocalization of charge, account for the unusual stability of the allyl cation.

Resonance theory depicts the allyl cation as a hybrid of structures **D** and **E** below.

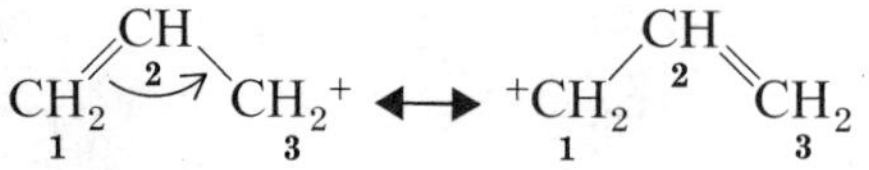

$$\underset{\mathbf{D}}{\overset{1}{CH_2}{=}\overset{2}{CH}{-}\overset{3}{CH_2}{}^+} \longleftrightarrow \underset{\mathbf{E}}{{}^+\overset{1}{CH_2}{-}\overset{2}{CH}{=}\overset{3}{CH_2}}$$

Because **D** and **E** are *equivalent* resonance structures, resonance theory predicts that the allyl cation should be unusually stable. Since the positive charge is located on carbon **3** in **D** and on carbon **1** in **E,** resonance theory also tells us that the positive charge should be delocalized over both carbons. Carbon **2** carries none of the positive charge. The hybrid structure **F** (below) includes charge and bond features of both **D** and **E.**

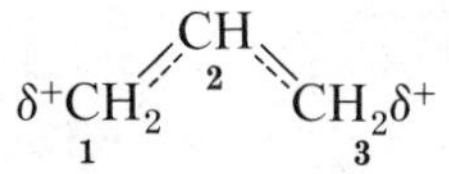

$$\underset{\mathbf{F}}{\delta^+\overset{1}{CH_2}\cdots\overset{2}{CH}\cdots\overset{3}{CH_2}\delta^+}$$

Problem 10.2

(a) Write structures corresponding to D, E, and F for the carbocation shown below.

$$CH_3{-}\overset{+}{C}H{-}CH{=}CH_2$$

(b) This carbocation appears to be even more stable than a tertiary carbocation; how can you explain this?

(c) What product(s) would you expect to be formed if this carbocation reacted with a chloride ion?

10.5 ALKADIENES AND POLYUNSATURATED HYDROCARBONS

Many hydrocarbons are known whose molecules contain more than one double or triple bond. A hydrocarbon whose molecules contain two double bonds is called an *alkadiene;* one whose molecules contain three double bonds is called an *alkatriene,* and so on. Colloquially, these compounds, are often referred to simply as "dienes" or "trienes." A hydrocarbon with two triple bonds is called an *alkadiyne,* and a hydrocarbon with a double and triple bond is called an *alkenyne.*

The following examples of polyunsaturated hydrocarbons illustrate how specific compounds are named.

$CH_2{=}C{=}CH_2$
1,2-Propadiene
(Allene)

$\overset{1}{C}H_2{=}\overset{2}{C}H{-}\overset{3}{C}H{=}\overset{4}{C}H_2$
1,3-Butadiene

cis-1,3-Pentadiene

trans, trans-2,4-Hexadiene

cis, trans-2,4-Hexadiene

$HC{\equiv}C{-}CH_2CH{=}CH_2$
1-Penten-4-yne*

trans, trans, trans-2,4,6-Octatriene

1,3-Cyclohexadiene

1,4-Cyclohexadiene

The multiple bonds of polyunsaturated compounds are classified as being *cumulated, conjugated,* or *isolated.* The double bonds of allene (1,2-propadiene) are said to be cumulated because one carbon (the central carbon) participates in two double bonds. Hydrocarbons whose molecules have cumulated double bonds are called *cumulenes.* The name *allene* is also applied generally to molecules with two cumulated double bonds.

$CH_2{=}C{=}CH_2$
Allene

$>C{=}C{=}C<$
A cumulated diene

An example of a conjugated diene is 1,3-butadiene. In conjugated polyenes the double and single bonds *alternate* along the chain.

$CH_2{=}CH{-}CH{=}CH_2$
1,3-Butadiene

A conjugated diene

* Where there is a choice, the double bond is given the lower number.

Trans, trans, trans-2,4,6-octatriene is an example of a conjugated alkatriene.

If one or more saturated carbons intervene between the double bonds of an alkadiene, the double bonds are said to be *isolated*. An example of an isolated diene is 1,4-pentadiene.

$$CH_2{=}CH{-}CH_2{-}CH{=}CH_2$$

1,4-Pentadiene

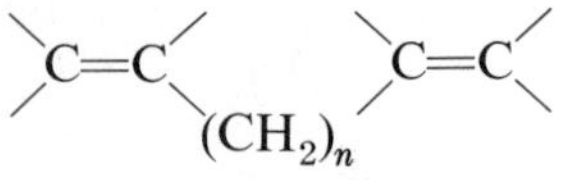

$$\gt C{=}C\lt\!(CH_2)_n\!\gt C{=}C\lt$$

An isolated diene ($n \neq 0$)

Problem 10.3

(a) Which other compounds on page 393 are conjugated dienes? (b) Which other compound is an isolated diene? (c) Which compound is an isolated enyne?

In Chapter 8 we saw that appropriately substituted cumulated dienes (allenes) give rise to chiral molecules even though the molecules do not have chiral carbon atoms. Cumulated dienes have had some commercial importance and cumulated double bonds are occasionally found in naturally occurring molecules (cf. Problem 9.22). In general, cumulated dienes are less stable than isolated dienes.

The double bonds of isolated dienes behave just as their name suggests—as isolated "enes." They undergo all of the reactions of alkenes; and, except for the fact that they are capable of reacting twice, their behavior is not unusual. Conjugated dienes are far more interesting because we find that their double bonds interact with each other. This interaction leads to unexpected properties and reactions. We will, therefore, consider the chemistry of conjugated dienes in detail.

10.6 1,3-BUTADIENE: ELECTRON DELOCALIZATION

10.6A Bond Lengths of 1,3-Butadiene

The carbon-carbon bond lengths of 1,3-butadiene have been determined and are shown below.

$$\overset{1}{C}H_2 \underset{1.34\,Å}{=\!=} \overset{2}{C}H \underset{1.47\,Å}{-\!-} \overset{3}{C}H \underset{1.34\,Å}{=\!=} \overset{4}{C}H_2$$

The C-1—C-2 bond and the C-3—C-4 bond are (within experimental error) the same length as the carbon-carbon double bond of ethene. The central bond of 1,3-butadiene (1.47 Å), however, is considerably shorter than the single bond of ethane (1.54 Å).

This should not be surprising. All of the carbons of 1,3-butadiene are sp^2 hybridized and, as a result, the central bond of butadiene results from overlapping sp^2 orbitals. The carbon-carbon bond of ethane, by contrast, results from overlapping sp^3 orbitals. And, as we know, a sigma bond that is sp^3-sp^3 is *longer*. There is, in fact, a steady decrease in bond length of carbon-carbon single bonds as the hybridization state of the bonded atoms changes from sp^3 to sp (Table 10.1).

TABLE 10.1 Carbon-Carbon Single Bond Lengths and Hybridization State

COMPOUND	HYBRIDIZATION STATE	BOND LENGTH, Å
$H_3C—CH_3$	sp^3-sp^3	1.54
$CH_2{=}CH—CH_3$	sp^2-sp^3	1.50
$CH_2{=}CH—CH{=}CH_2$	sp^2-sp^2	1.47
$HC{\equiv}C—CH_3$	sp-sp^3	1.46
$HC{\equiv}C—CH{=}CH_2$	sp-sp^2	1.43
$HC{\equiv}C—C{\equiv}CH$	sp-sp	1.37

10.6B Conformations of 1,3-Butadiene

There are two possible planar conformations of 1,3-butadiene: the *s-cis* and the *s-trans* conformation.

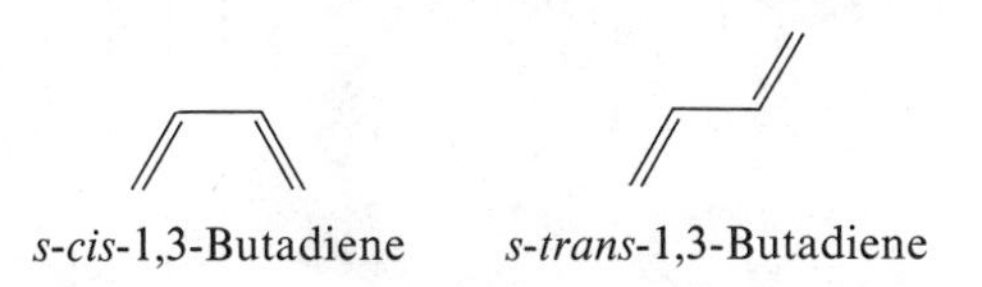

These are not true *cis* and *trans* forms since the *s-cis* and *s-trans* conformations of 1,3-butadiene can be interconverted through rotation about the single bond (hence the prefix *s*). The *s-trans* conformation is the predominant one at room temperature.

10.6C Molecular Orbitals of 1,3-Butadiene

The central carbons of 1,3-butadiene (Fig. 10.4) are close enough for overlap to occur between the *p* orbitals of carbon-2 and carbon-3. This overlap is not as great as that between the orbitals of C-1 and C-2 (or those of C-3 and C-4). The C-2, C-3 orbital overlap, however, gives the central bond partial double bond character and allows the four π electrons of 1,3-butadiene to be delocalized over all four atoms.

Figure 10.5 shows how the four *p* orbitals of 1,3-butadiene combine to form a set of four π molecular orbitals.

Two of the π molecular orbitals of 1,3-butadiene are bonding molecular orbitals. In the ground state these orbitals hold the four π electrons with two spin-paired electrons in each. The other two π molecular orbitals are antibonding molecular orbitals. In the ground state these orbitals are unoccupied. An electron can be excited from the highest occupied molecular orbital to the lowest empty molecular orbital when 1,3-butadiene absorbs light with a wavelength of 217 nanometers (2170 Å). (We will study the absorption of light by unsaturated molecules in Section 10.11).

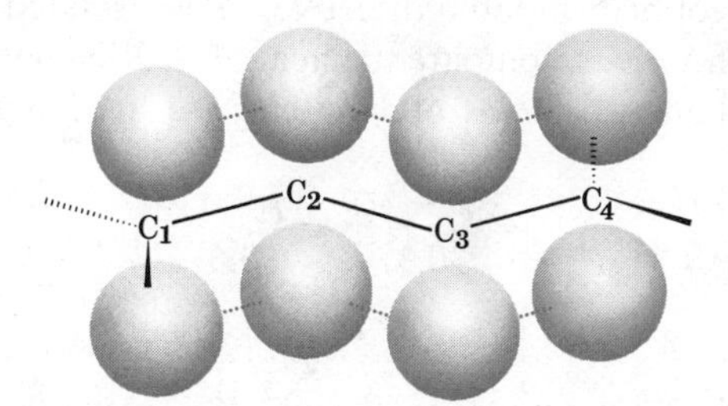

FIG. 10.4
The p *orbitals of 1,3-butadiene.*

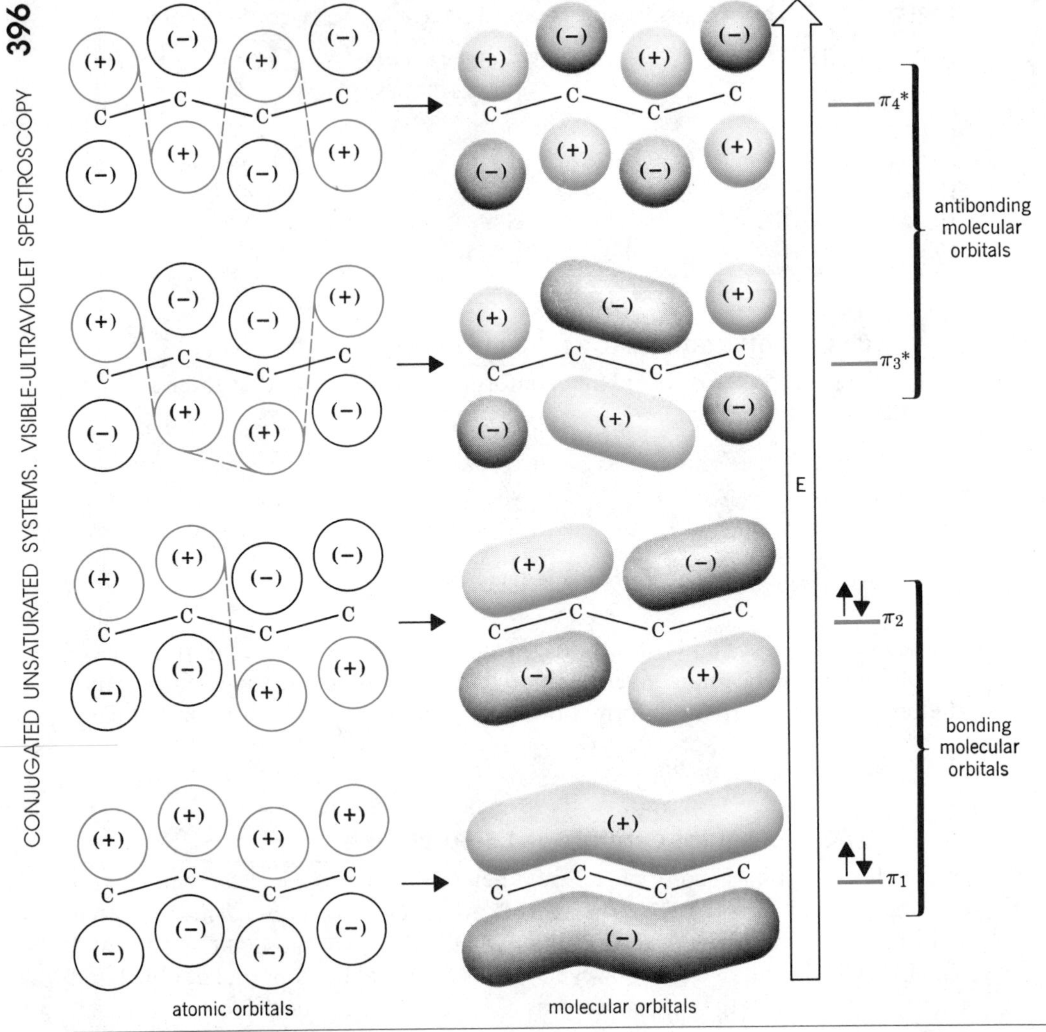

FIG. 10.5
The π molecular orbitals of 1,3-butadiene.

The delocalized bonding that we have just described for 1,3-butadiene is characteristic of all conjugated polyenes.

10.7 THE STABILITY OF CONJUGATED DIENES

Conjugated alkadienes are thermodynamically more stable than corresponding isolated alkadienes. Two examples of this extra stability of conjugated dienes can be seen if we analyze the heats of hydrogenation given in Table 10.2.

In itself, 1,3-butadiene cannot be compared directly with an isolated diene of the same chain length. We can, however, compare the heat of hydrogenation of 1,3-butadiene to that obtained when two moles of 1-butene are hydrogenated.

$$\Delta H°\text{, kcal/mole}$$

$$2CH_2{=}CHCH_2CH_3 + 2H_2 \longrightarrow 2CH_3CH_2CH_2CH_3\ (2 \times -30.3) = -60.6$$
1-Butene

$$CH_2{=}CHCH{=}CH_2 + 2H_2 \longrightarrow CH_3CH_2CH_2CH_3 \qquad -57.1$$

Difference 3.5 kcal/mole

TABLE 10.2 Heats of Hydrogenation of Alkenes and Alkadienes

COMPOUND	MOLES H_2	$\Delta H°$, KCAL/MOLE
1-Butene	1	−30.3
1-Pentene	1	−30.1
trans-2-Pentene	1	−27.6
1,3-Butadiene	2	−57.1
trans-1,3-Pentadiene	2	−54.1
1,4-Pentadiene	2	−60.8
1,5-Hexadiene	2	−60.5

Since 1-butene has the same kind of monosubstituted double bond as either of those in 1,3-butadiene, we might expect that hydrogenation of 1,3-butadiene would liberate the same amount of heat (−60.6 kcal/mole) as two moles of 1-butene. We find, however, that 1,3-butadiene liberates only 57.1 kcal/mole, 3.5 kcal/mole *less* than expected. We conclude, therefore, that conjugation imparts some extra stability to the conjugated system (Fig. 10.6).

With the pentadienes we can make a direct comparison between conjugated and isolated dienes of the same chain length:

$CH_2{=}CH{-}CH_2{-}CH{=}CH_2$ — $\Delta H° = -60.8$ kcal/mole
1,4-Pentadiene

$CH_2{=}CH(H)C{=}C(H)CH_3$ — $\Delta H° = -54.1$ kcal/mole
trans-1,3-Pentadiene

Difference = 6.7 kcal/mole

This direct comparison is not really appropriate, however. Both double bonds of 1,4-pentadiene are monosubstituted double bonds, while one of the double bonds of *trans*-1,3-pentadiene is a *trans*-disubstituted double bond. Hydrocarbons with disubstituted double bonds are known to be more stable (Section 6.6).

A better assessment of the stabilization that conjugation provides can be made by comparing the heat of hydrogenation of *trans*-1,3-pentadiene to the sum of the heats of hydrogenation of 1-pentene and *trans*-2-pentene. This way we will be comparing double bonds of comparable types.

$CH_2{=}CHCH_2CH_2CH_3$ — $\Delta H° = -30.1$ kcal/mole
1-Pentene

$CH_3CH_2(H)C{=}C(H)CH_3$ — $\Delta H° = -27.6$ kcal/mole
trans-2-Pentene

Sum = −57.7 kcal/mole

$CH_2{=}CH(H)C{=}C(H)CH_3$ — $\Delta H° = -54.1$ kcal/mole
trans-1,3-Pentadiene

Difference = 3.6 kcal/mole

We see from these calculations that conjugation affords *trans*-1,3-

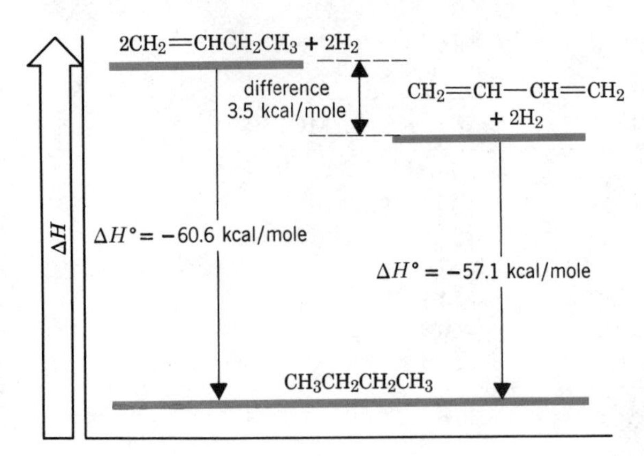

FIG. 10.6
Heats of hydrogenation of two moles of 1-butene and one mole of butadiene.

pentadiene an extra stability of 3.6 kcal/mole, a value that is very close to the one we obtained for 1,3-butadiene (3.5 kcal/mole).

When calculations like these are carried out for other conjugated dienes, similar results are obtained; conjugated dienes are found to be more stable than isolated dienes. The question, then, is this: What is the source of the extra stability associated with conjugated dienes? This question is still being debated. Some chemists argue that the extra stability of conjugated dienes arises solely from the stronger sp^2-sp^2 bond that they contain. Others argue that the extra stability arises from the additional delocalization of the π-electrons that occurs in conjugated dienes. Both arguments may be correct; the additional stability may be derived from both sources.

10.8 ELECTROPHILIC ATTACK AND CONJUGATED DIENES: 1,4 ADDITION

Not only are conjugated dienes somewhat more stable than nonconjugated dienes, they also display unusual behavior when they react with electrophilic reagents. For example, 1,3-butadiene reacts with one mole of hydrogen chloride to produce two products: 3-chloro-1-butene and 1-chloro-2-butene.

$$CH_2{=}CH{-}CH{=}CH_2 \xrightarrow[25^\circ]{HCl} CH_2{=}CH{-}\underset{\displaystyle Cl}{\underset{|}{C}H}{-}CH_3 + ClCH_2{-}CH{=}CH{-}CH_3$$

1,3-Butadiene — 3-Chloro-1-butene (78%) — 1-Chloro-2-butene (22%)

If only the first product (3-chloro-1-butene) were formed, we would not be particularly surprised. We would conclude that hydrogen chloride had added to one double bond of 1,3-butadiene in the usual way.

$$\overset{1}{C}H_2{=}\overset{2}{C}H{-}\overset{3}{C}H{=}\overset{4}{C}H_2 + H{-}Cl \xrightarrow{\text{1,2-Addition}} \underset{\displaystyle H}{\underset{|}{C}H_2}{-}\underset{\displaystyle Cl}{\underset{|}{C}H}{-}CH{=}CH_2$$

3-Chloro-1-butene

It is the second product, 1-chloro-2-butene, that is unusual. Its double bond is between the central atoms and the elements of hydrogen chloride have added to carbons 1 and 4.

$$\overset{1}{C}H_2{=}\overset{2}{C}H{-}\overset{3}{C}H{=}\overset{4}{C}H_2 + H{-}Cl \xrightarrow{\text{1,4-Addition}} \underset{H}{C}H_2{-}CH{=}CH{-}\underset{Cl}{C}H_2$$

1-Chloro-2-butene

This unusual behavior of 1,3-butadiene can be attributed directly to the stability and the delocalized nature of an allylic cation. In order to see this, consider a mechanism for the addition of hydrogen chloride.

Step 1 $H^+ + CH_2{=}CH{-}CH{=}CH_2 \longrightarrow CH_3{-}\overset{+}{C}H{-}CH{=}CH_2 \longleftrightarrow CH_3{-}CH{=}CH{-}\overset{+}{C}H_2$

An allylic cation equivalent to

$$CH_3{-}\overset{\delta+}{C}H{\cdots}CH{\cdots}\overset{\delta+}{C}H_2$$

Step 2 $CH_3\overset{\delta+}{C}H{\cdots}CH{\cdots}\overset{\delta+}{C}H_2 + :\ddot{Cl}:^-$ →

$\longrightarrow CH_3\underset{Cl}{C}H{-}CH{=}CH_2$ 1,2 Addition

$\longrightarrow CH_3CH{=}CHCH_2Cl$ 1,4 Addition

In step (1) a proton adds to one of the terminal carbons of 1,3-butadiene to form a resonance-stabilized allylic cation. Addition to one of the inner carbons would have produced a much less stable primary cation, one that could not be stabilized by resonance.

$$CH_2{=}CH{-}CH{=}CH_2 + H^+ \not\longrightarrow {}^+CH_2{-}CH_2{-}CH{=}CH_2$$

A 1° carbocation

In step (2) a chloride ion forms a bond to one of the carbons of the allylic cation that bears a partial positive charge. Reaction at one carbon results in the 1,2-addition product; reaction at the other gives the 1,4-addition product.

Problem 10.4

(a) What products would you expect to obtain if hydrogen chloride were allowed to react with a 2,4-hexadiene, $CH_3CH{=}CHCH{=}CHCH_3$? (b) With 1,3-pentadiene, $CH_2{=}CHCH{=}CHCH_3$? (neglect *cis-trans* isomerism.)

Butadiene shows 1,4-addition reactions with electrophilic reagents other than hydrogen chloride. Two examples are shown below, the addition of hydrogen bromide and the addition of bromine.

$$CH_2{=}CHCH{=}CH_2 + HBr \xrightarrow{40°} \underset{(20\%)}{CH_3CHBrCH{=}CH_2} + \underset{(80\%)}{CH_3CH{=}CHCH_2Br}$$

$$CH_2{=}CHCH{=}CH_2 + Br_2 \xrightarrow[\text{Hexane}]{-15°} \underset{(54\%)}{CH_2BrCHBrCH{=}CH_2} + \underset{(46\%)}{CH_2BrCH{=}CHCH_2Br}$$

Reactions of this type are quite general with other conjugated dienes. Conjugated trienes often show 1,6 addition. An example is the 1,6 addition of bromine to

1,3,5-cyclooctatriene:

$$\text{cyclooctatriene} \xrightarrow[CHCl_3]{Br_2} \text{dibromocyclooctadiene}$$

(> 68%)

10.8A Rate Control Versus Equilibrium Control of a Chemical Reaction

The addition of hydrogen bromide to 1,3-butadiene is interesting in another respect. The relative amounts of 1,2- and 1,4-addition product that we obtain are dependent on the temperature at which we carry out the reaction.

When 1,3-butadiene and hydrogen bromide react at a low temperature ($-80°$), the major reaction is 1,2 addition; we obtain about 80% of the 1,2 product and only about 20% of the 1,4 product. At a higher temperature (40°) the result is reversed. The major reaction is 1,4 addition; we obtain about 80% of the 1,4 product and only about 20% of the 1,2 product.

When the mixture formed at the lower temperature is brought to the higher temperature, moreover, the relative amounts of the two products change. This new reaction mixture eventually contains the same proportion of products given by the reaction carried out at the higher temperature.

It can also be shown that at the higher temperature and in the presence of hydrogen bromide, the 1,2-addition product rearranges to the 1,4-product and that an equilibrium exists between them.

$$CH_2{=}CHCH{=}CH_2 + HBr \xrightarrow{-80°} \underset{(80\%)}{CH_3\underset{|}{\underset{Br}{C}}HCH{=}CH_2} + \underset{(20\%)}{CH_3CH{=}CHCH_2Br}$$

$$\downarrow 40°$$

$$CH_2{=}CHCH{=}CH_2 + HBr \xrightarrow{40°} \underset{(20\%)}{CH_3\underset{|}{\underset{Br}{C}}HCH{=}CH_2} + \underset{(80\%)}{CH_3CH{=}CHCH_2Br}$$

$$\underset{\text{1,2-Addition product}}{CH_3\underset{|}{\underset{Br}{C}}HCH{=}CH_2} \underset{HBr}{\overset{80°}{\rightleftharpoons}} \underset{\text{1,4-Addition product}}{CH_3CH{=}CHCH_2Br}$$

Since this equilibrium favors the 1,4-addition product, *it must be more stable.*

The reactions of hydrogen bromide and 1,3-butadiene serve as a striking illustration of the way that the outcome of a chemical reaction can be determined, in one instance, by relative rates of competing reactions and, in another, by the relative stabilities of the final products. At the lower temperature, the relative amounts of the products of the addition are determined by the relative rates at which the two additions occur: 1,2 addition occurs faster so the 1,2-addition

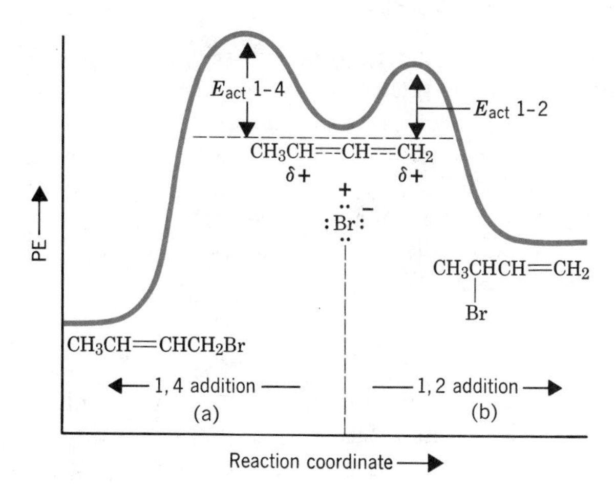

FIG. 10.7
Potential energy versus *reaction coordinate diagram for the reactions of an allyl cation with a bromide ion. One reaction pathway* (a) *leads to the 1,4-addition product and the other* (b) *leads to the 1,2-addition product.*

product is the major product. At the higher temperature, the relative amounts of the products are determined by the position of an equilibrium: the 1,4-addition product is the more stable so it is the major product.

This behavior of 1,3-butadiene and hydrogen bromide can be more fully understood if we examine the diagram shown in Fig. 10.7.

The step that determines the overall outcome of the reaction is the step in which the hybrid allylic cation combines with a bromide ion, that is,

$$CH_2{=}CH{-}CH{=}CH_2 \xrightarrow{H^+}$$

$$CH_3{-}\overset{\delta^+}{C}H{=}{=}{=}CH{=}{=}{=}\overset{\delta^+}{C}H_2 \begin{cases} \xrightarrow{Br^-} CH_3{-}\underset{\displaystyle Br}{\underset{|}{C}H}{-}CH{=}CH_2 & \text{1,2 Product} \\ \xrightarrow{Br^-} CH_3{-}CH{=}CH{-}CH_2Br & \text{1,4 Product} \end{cases}$$

This step determines the orientation of the reaction

We see in Fig. 10.7 that, for this step, the energy of activation leading to the 1,2-addition product is less than the energy of activation leading to the 1,4-addition product, even though the 1,4 product is more stable. At low temperatures, a larger fraction of collisions between the intermediate ions will have enough energy to cross the lower barrier (leading to the 1,2 product), and only a very small fraction of collisions will have enough energy to cross the higher barrier (leading to the 1,4 product). In either case (and this is the *key point*) whichever barrier is crossed, product formation is *irreversible* because there is not enough energy available to lift either product out of its deep potential energy valley. Since 1,2 addition occurs faster, the 1,2 product predominates and the reaction is said to be under *rate control* or *kinetic control.*

At higher temperatures, the intermediate ions have sufficient energy to cross both barriers with relative ease. More importantly, however, *both reactions are reversible.* Sufficient energy is also available to take the products back over their energy barriers to the intermediate level of allylic cations and bromide ions. The 1,2 product is still formed faster, but being less stable than the 1,4 product, it also reverts to the allylic cation faster. Under these conditions, that is, at higher temperatures, the relative proportions of the products *do not reflect* the relative heights of the energy barriers leading from allylic cation to products. Instead, *they reflect the relative stabilities of the products themselves.* Since the 1,4 product is more stable, it is formed at the expense of the 1,2 product because the overall change from 1,2 product to 1,4 product is energetically favored. Such a reaction is said to be under *equilibrium control* or *thermodynamic control.*

Before we leave this subject one final point should be made. This example clearly demonstrates that predictions of relative reaction rates made on the basis of product stabilities alone can be wrong. This is not always the case, however. For many reactions in which a common intermediate leads to two or more products, the most stable product is formed fastest.

Problem 10.5

(a) Can you suggest a possible explanation for the fact that the 1,2-addition reaction of 1,3-butadiene and hydrogen bromide occurs faster than 1,4-addition? (Hint: Consider the relative contributions that the two forms $CH_3\overset{+}{C}HCH{=}CH_2$ and $CH_3CH{=}CH\overset{+}{C}H_2$ make to the resonance hybrid of the allylic cation.) (b) How can you account for the fact that the 1,4-addition product is more stable?

10.9 SUMMARY OF RULES FOR RESONANCE

We have used resonance theory extensively in earlier sections of this chapter because we have been describing radicals and ions with delocalized electrons (and charges) in π bonds. Resonance theory is especially useful with systems like this, and we will use it again and again in the chapters that follow. We had an introduction to resonance theory in Section 1.8 and it should be helpful now to summarize the rules for writing resonance structures and for estimating the relative contribution a given structure will make to the overall hybrid.

1. *Resonance structures exist only on paper.* They are like Don Quixotes, or unicorns (Section 1.8); they have no real existence of their own. Resonance structures are useful because they allow us to describe molecules, radicals, and ions for which a single Lewis structure is inadequate. We write two or more Lewis structures, calling them resonance structures or resonance contributors. We connect these structures by double-headed arrows (⟷), and we say that the real molecule, radical, or ion is like a hybrid of all of them.
2. *In writing resonance structures we are only allowed to move electrons.* The positions of the nuclei must remain the same in all of the structures. Structure **G** below is not a resonance structure for the allyl cation, for example, because in order to form the three-membered ring we would have to move carbon 1 and 3 of the allyl cation closer together and this is not permitted.

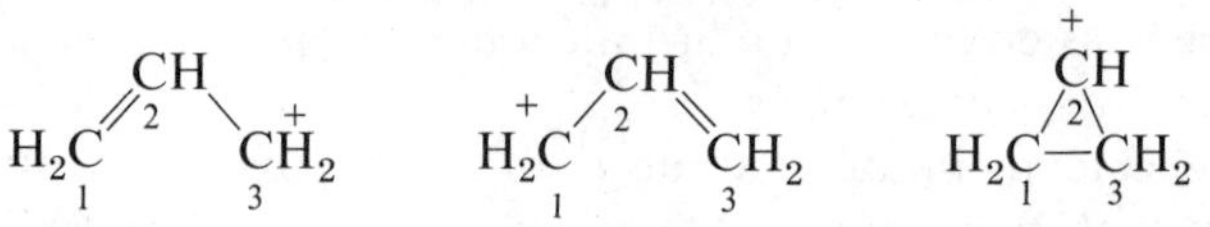

$$\underset{\mathbf{D}}{H_2\overset{1}{C}{=}\overset{2}{C}H{-}\overset{+}{\overset{3}{C}}H_2} \qquad \underset{\mathbf{E}}{\overset{+}{H_2\overset{1}{C}}{-}\overset{2}{C}H{=}\overset{3}{C}H_2} \qquad \underset{\mathbf{G}}{\text{cyclic } H_2\overset{1}{C}{-}\overset{3}{C}H_2 \text{ with } \overset{+}{\overset{2}{C}H}}$$

This is not a proper resonance structure for the allyl cation because carbon atoms 1 and 3 have been moved.

3. *All of the structures must be proper Lewis structures.* We should not write structures in which carbon has five bonds, for example.

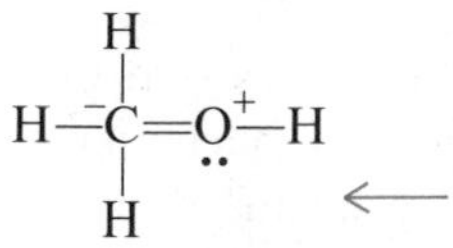

$$H-\overset{\overset{H}{|}}{\underset{\underset{H}{|}}{\bar{C}}}=\ddot{O}^{+}-H$$

This is not a proper resonance structure because carbon has five bonds. Elements of the first major row of the periodic table cannot have more than eight electrons in their valence shell.

4. *All resonance structures must have the same number of unpaired electrons.* The structure below is not a resonance structure for the allyl radical because it contains three unpaired electrons and the allyl radical contains only one.

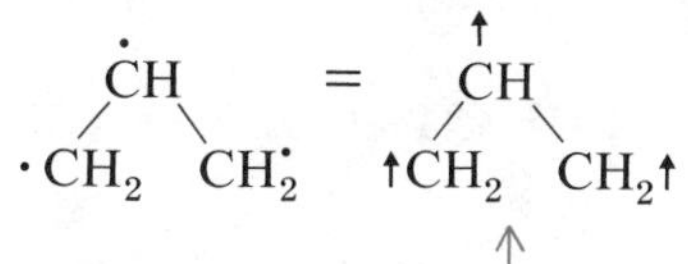

$$\cdot CH_2-\dot{C}H-CH_2\cdot \quad = \quad \uparrow CH_2-\overset{\uparrow}{C}H-CH_2\uparrow$$

This is not a proper resonance structure for the allyl radical because it does not contain the same number of unpaired electrons as $CH_2{=}CHCH_2\cdot$.

5. *The energy of the actual molecule is lower than the energy that might be estimated for any contributing structure.* The actual allyl cation, for example, is more stable than either resonance structure **D** or **E** (p. 402) taken separately would indicate. **D** and **E** resemble primary carbocations and yet the allyl radical is more stable (has lower energy) than a secondary carbocation. Chemists often call this kind of stabilization *resonance stabilization.*
6. *Equivalent resonance structures make equal contributions to the hybrid, and systems described by them have a large resonance stabilization.* Structures **D** and **E** make equal contributions to the allyl cation because they are equivalent. They also make a large stabilizing contribution and account for the allyl cation being more stable than a 1° or 2° cation. The same can be said about the contributions made by **A** and **B** (p. 390) to the allyl radical. Several other rules will help in estimating the relative contributions made to the hybrid by non-equivalent resonance structures.
7. Structures that are not equivalent do not make equal contributions. Generally speaking, *the more stable a structure is* (*when taken by itself*) *the greater is its contribution to the hybrid.* For example, the cation at the top of the next page is a hybrid of structures **H** and **I.** Structure **H** makes a greater contribution than **I** because structure **H** resembles a tertiary carbocation while structure **I** resembles a primary cation and tertiary cations are more stable.

$$\overset{1}{CH_3}-\overset{\overset{CH_3}{|}}{\underset{\delta+}{\overset{2}{C}}}\cdots\overset{3}{CH}\cdots\underset{\delta+}{\overset{4}{CH_2}} = \underset{\mathbf{H}}{CH_3-\overset{\overset{CH_3}{|}}{\overset{+}{C}}-CH=CH_2} \longleftrightarrow \underset{\mathbf{I}}{CH_3-\overset{\overset{CH_3}{|}}{C}=CH-\overset{+}{C}H_2}$$

That **H** makes a larger contribution means that the partial positive charge on carbon 2 of the hybrid will be larger than the partial positive charge on carbon 4. It also means that the bond between carbons 3 and 4 will be more like a double bond than the bond between carbons 2 and 3.

8. *The more covalent bonds a structure has, the more stable it is.* This is exactly what we would expect because we know that forming a covalent bond lowers the energy of atoms. This means that of the structures for 1,3-butadiene below, **J** is by far the most stable and makes by far the largest contribution because it contains one more bond. (It is also most stable for the reason given under rule 10 below.)

$$\underset{\mathbf{J}}{CH_2=CH-CH=CH_2} \longleftrightarrow \underset{\mathbf{K}}{\overset{+}{C}H_2-CH=CH-\ddot{C}H_2^-} \longleftrightarrow \underset{\mathbf{L}}{\overset{-}{\ddot{C}}H_2-CH=CH-\overset{+}{C}H_2}$$

This structure is the most stable because it contains more covalent bonds.

9. *Structures in which all of the atoms have a complete valence shell of electrons* (i.e., the noble gas structure) *are especially stable and make large contributions to the hybrid.* Again this is what we would expect from what we know about bonding. This means, for example, that **N** makes a larger stabilizing contribution to the cation below than **M** because all of its atoms have a complete valence shell. (Notice too, that **N** has more covalent bonds than **M,** cf. Rule 8).

$$\underset{\mathbf{M}}{\overset{+}{C}H_2-\ddot{\underset{..}{O}}-CH_3} \longleftrightarrow \underset{\mathbf{N}}{CH_2=\overset{+}{\underset{..}{O}}-CH_3}$$

Here this carbon has only six electrons.

Here the carbon has eight electrons.

10. *Charge separation decreases stability.* Separating opposite charges requires energy. Therefore, structures in which opposite charges are separated have greater energy (lower stability) than those that have no charge separation. This means that of the two structures for vinyl chloride below, structure **O** makes a larger contribution because it does not have separated charges. (This does not mean that structure **P** does not contribute to the hybrid, it just means that the contribution made by **P** is smaller.)

$$\underset{\mathbf{O}}{CH_2=CH-\ddot{\underset{..}{Cl}}:} \longleftrightarrow \underset{\mathbf{P}}{:\overset{-}{C}H_2-CH=\overset{+}{\underset{..}{Cl}}:}$$

11. *A structure in which a negative charge is on an electronegative atom is more stable than one in which the negative charge is carried by a less electronegative atom.* The two structures **Q** and **R** below are resonance contributors to the anion formed when acetaldehyde loses a proton. **R** makes a greater contribution to the hybrid because the more electronegative oxygen atom carries the negative charge.

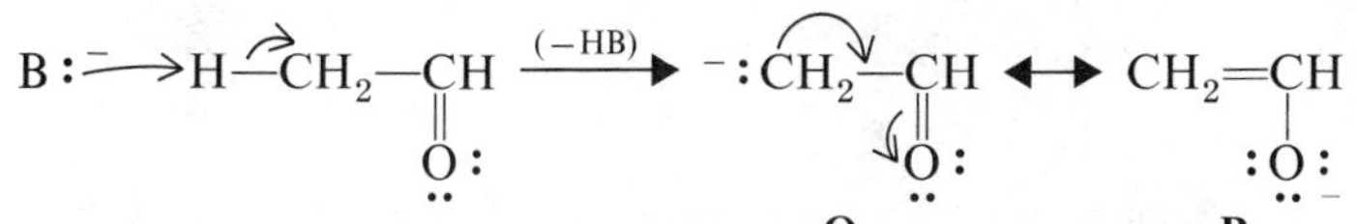

$$B:^- \rightarrow H-CH_2-\underset{\underset{\ddot{O}:}{\|}}{CH} \xrightarrow{(-HB)} {}^-:CH_2-\underset{\underset{\ddot{O}:}{\|}}{CH} \longleftrightarrow CH_2=\underset{\underset{:\ddot{O}:^-}{|}}{CH}$$

Q R

Problem 10.6

Give all of the important resonance structures for each of the following.

(a) 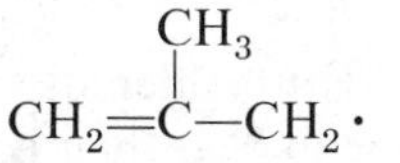 $CH_2=\overset{\overset{CH_3}{|}}{C}-CH_2\cdot$

(b) $CH_2=CH-\underset{+}{CH}-CH=CH_2$

(c) 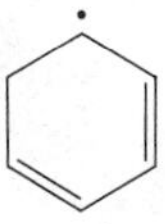 cyclohexadienyl radical (ring with two double bonds, radical dot on the sp^3-adjacent carbon)

(d)  cyclohexadienyl cation (ring with two double bonds, + on the carbon at top)

(e) $CH_3CH=CH-CH=\overset{+}{\ddot{O}}H$

Problem 10.7

From each set of resonance structures below designate the one that would contribute most to the hybrid and explain your choice.

(a) 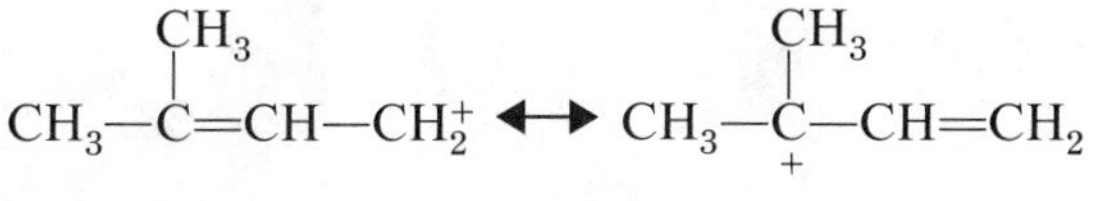 $CH_3-\overset{\overset{CH_3}{|}}{C}=CH-CH_2^+ \longleftrightarrow CH_3-\underset{+}{\overset{\overset{CH_3}{|}}{C}}-CH=CH_2$

(b) 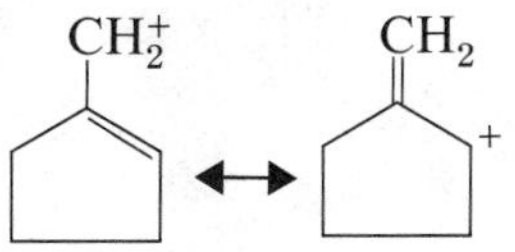 cyclopentene ring bearing CH_2^+ $\longleftrightarrow$ cyclopentane ring bearing $=CH_2$ with + on ring carbon

(c) $\overset{+}{C}H_2-\ddot{N}(CH_3)_2 \longleftrightarrow CH_2=\overset{+}{N}(CH_3)_2$

(d) 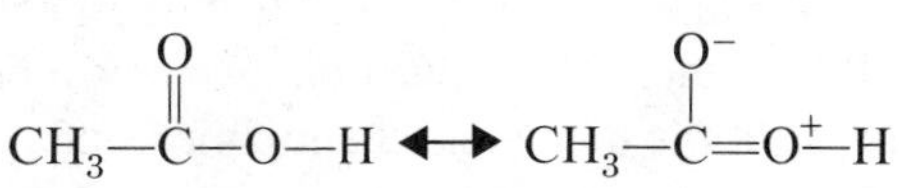 $CH_3-\overset{\overset{O}{\|}}{C}-O-H \longleftrightarrow CH_3-\overset{\overset{O^-}{|}}{C}=O^+-H$

(e) $\dot{C}H_2CH=CHCH=CH_2 \longleftrightarrow CH_2=CH\dot{C}HCH=CH_2 \longleftrightarrow$

$CH_2=CHCH=CH\dot{C}H_2$

Problem 10.8

The keto and enol forms below differ in the positions for their electrons but they are not resonance structures. Explain.

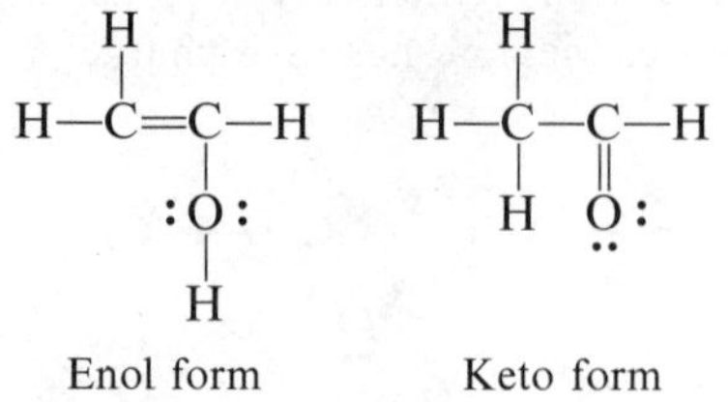

```
     H              H
     |              |
 H—C=C—H        H—C—C—H
       |            |  ||
      :O:           H  O:
       |               ··
       H
  Enol form       Keto form
```

10.10 THE DIELS-ALDER REACTION: A 1,4-CYCLOADDITION REACTION OF DIENES

In 1928 two German chemists, Otto Diels and Kurt Alder, discovered a 1,4-cycloaddition reaction of dienes that has since come to bear their names. The reaction proved to be one of such great versatility and synthetic utility that Diels and Alder were awarded the Nobel Prize for chemistry in 1950.

An example of the Diels-Alder reaction is the reaction that takes place when 1,3-butadiene and maleic anhydride are heated together at 100°. The product is obtained in quantitative yield.

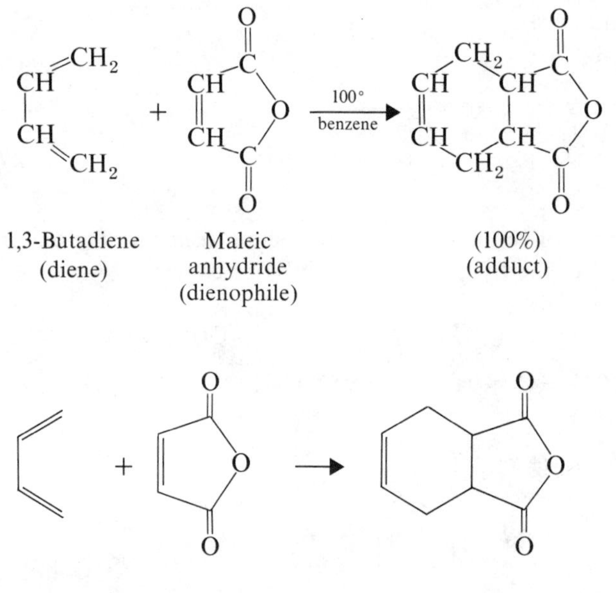

1,3-Butadiene (diene) + Maleic anhydride (dienophile) $\xrightarrow[\text{benzene}]{100°}$ (100%) (adduct)

In general terms, the reaction is one between a conjugated *diene* (a 4π-electron system) and compound containing a double bond (a 2π-electron system) called a *dienophile* (diene + Gr: *philein,* to love). The product of a Diels-Alder reaction is often called an *adduct*. In the Diels-Alder reaction, two new σ bonds are formed at the expense of two π bonds of the diene and dienophile. Since σ bonds are usually stronger than π bonds, formation of the adduct is usually favored energetically, *but most Diels-Alder reactions are reversible.*

The simplest example of a Diels-Alder reaction is the one that takes place between 1,3-butadiene and ethylene. This is also one of the poorest examples

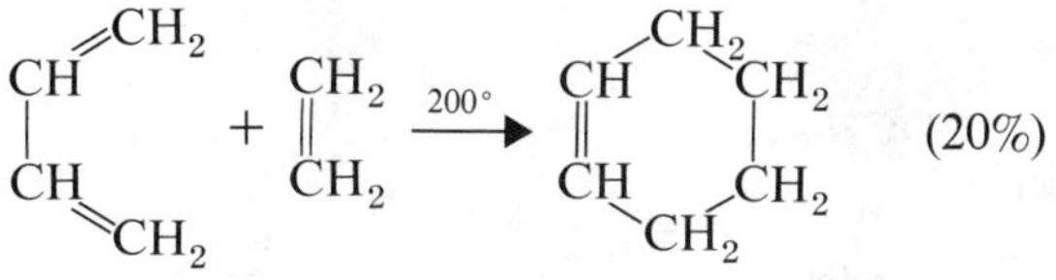

$$CH_2{=}CH{-}CH{=}CH_2 + CH_2{=}CH_2 \xrightarrow{200^\circ} \text{cyclohexene} \quad (20\%)$$

because the product, cyclohexene, is obtained in only 20% yield.

Alder originally stated that the Diels-Alder reaction is favored by the presence of electron-withdrawing groups in the dienophile and by electron-releasing groups in the diene. Maleic anhydride, a very potent dienophile, has two carbonyl groups on carbons adjacent to the double bond. Carbonyl groups are electron-withdrawing because of the electronegativity of their oxygen atoms and because resonance structures such as those shown below contribute to the hybrid.

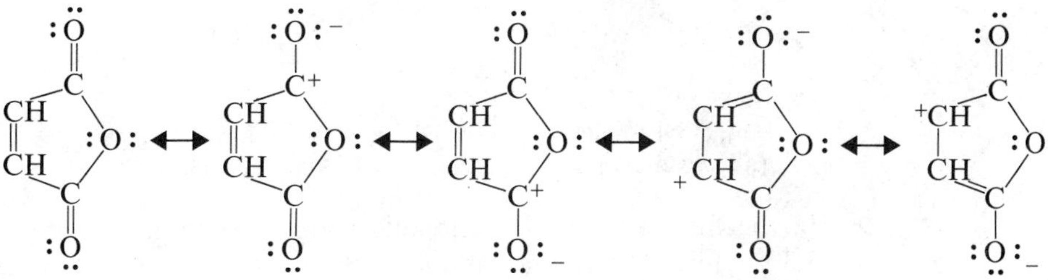

[Resonance structures of maleic anhydride: five structures connected by ↔, showing C^+/$:\ddot{O}:^-$ and ^+CH charge-separated forms]

The comparative yields of the two examples that we have given (1,3-butadiene + maleic anhydride and 1,3-butadiene + ethylene) illustrate the help that electron-withdrawing groups in the dienophile gives the Diels-Alder reaction.

The helpful effect of electron-releasing groups in the diene can also be demonstrated; 2,3-dimethyl-1,3-butadiene, for example, is nearly five times as reactive in Diels-Alder reactions as is 1,3-butadiene. When 2,3-dimethyl-1,3-butadiene reacts with acrolein at only 30° the adduct is obtained in quantitative yield.

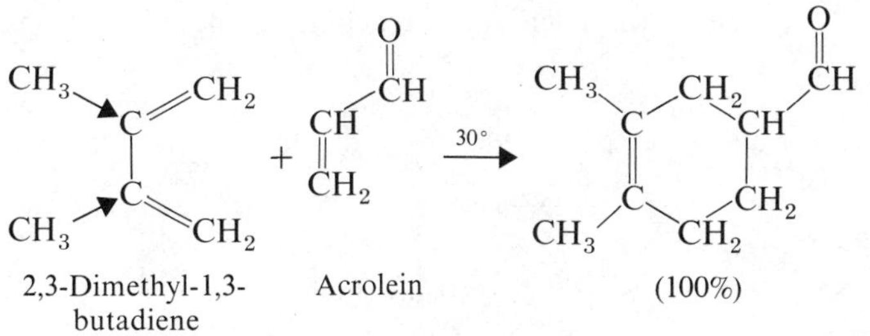

$$\underset{\text{2,3-Dimethyl-1,3-butadiene}}{CH_2{=}C(CH_3){-}C(CH_3){=}CH_2} + \underset{\text{Acrolein}}{CH_2{=}CH{-}CHO} \xrightarrow{30^\circ} \underset{(100\%)}{\text{3,4-dimethylcyclohex-3-ene-1-carbaldehyde}}$$

Recent research has shown that the locations of electron-withdrawing and electron-releasing groups in the dienophile and diene can be reversed without reducing yield of the adducts. Dienes with electron-withdrawing groups have been found to react readily with dienophiles containing electron-releasing groups. Additional facts about the reaction are these.

The Diels-Alder reaction is highly stereospecific:

1. The reaction is a syn addition and the configuration of the dienophile is *retained* in the products. Two examples that illustrate this aspect of the reaction are shown below.

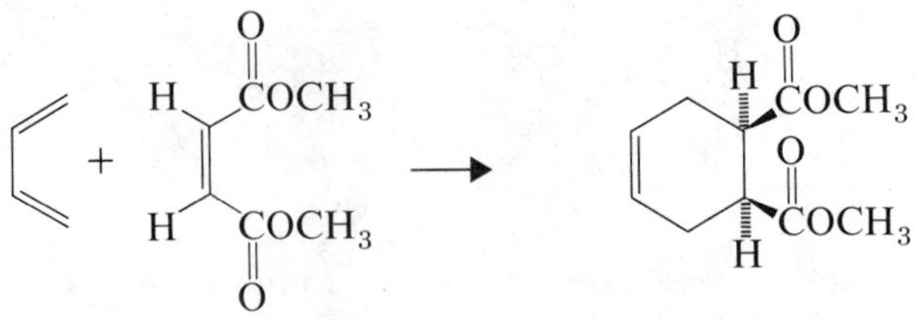

Dimethyl maleate (a *cis*-dienophile) → Dimethyl *cis*-4-cyclohexene-1,2-dicarboxylate

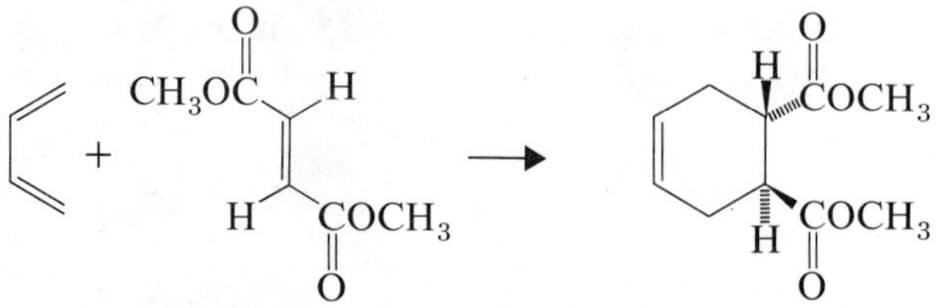

Dimethyl fumarate (a *trans* dienophile) → Dimethyl *trans*-4-cyclohexene-1,2-dicarboxylate

In the first example, a dienophile with *cis* ester groups reacts with 1,3-butadiene to give an adduct with *cis* ester groups. In the second example just the reverse is true. A *trans* dienophile gives a *trans* adduct.

2. The diene, of necessity, must react in the *s-cis* conformation rather than the *s-trans*.

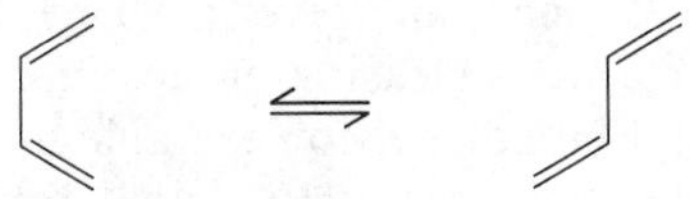

s-cis Conformation ⇌ *s-trans* Conformation

Reaction in the *s-trans* conformation would, if it occurred, produce a six-membered ring with a highly strained *trans* double bond. This course of the Diels-Alder reaction has never been observed.

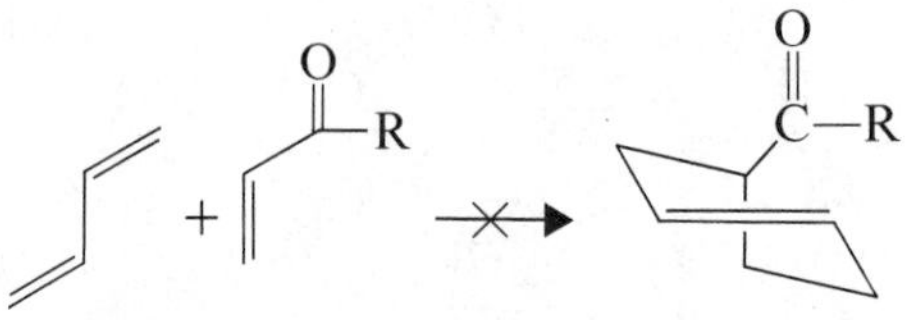

Highly strained

Cyclic dienes in which the double bonds are held in the *s-cis* configuration are usually highly reactive in the Diels-Alder reaction. Cyclopentadiene, for example, reacts with maleic anhydride at room temperature to give the adduct shown below in quantitative yield.

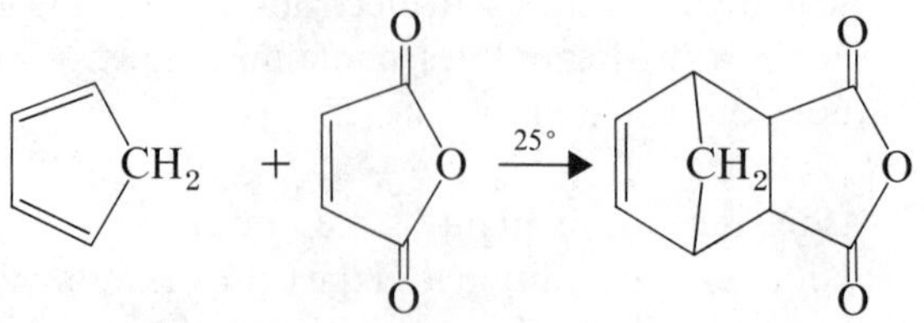

Cyclopentadiene + Maleic anhydride $\xrightarrow{25°}$ (100%)

Cyclopentadiene is so reactive that on standing at room temperature it slowly undergoes a Diels-Alder reaction with itself.

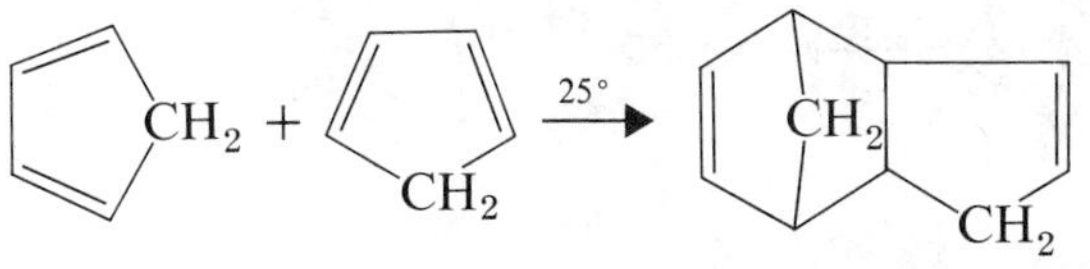

CH_2 + CH_2 $\xrightarrow{25°}$ CH_2 CH_2

"Dicyclopentadiene"

The reaction is reversible, however. When "dicyclopentadiene" is distilled, it dissociates into two moles of cyclopentadiene.

The reactions of cyclopentadiene illustrate a third stereochemical characteristic of the Diels-Alder reaction.

3. The Diels-Alder reaction occurs primarily in an *endo* rather than an *exo* fashion when the reaction is kinetically controlled (cf. Problem 10.27). *Endo* and *exo* are terms used to designate the stereochemistry of bridged bicyclic rings like norbornane below: The point of reference is the one-carbon bridge. A group that is on the same side of the six-membered ring as the one carbon bridge is said to be *exo;* if it is on the opposite side, it is *endo.**

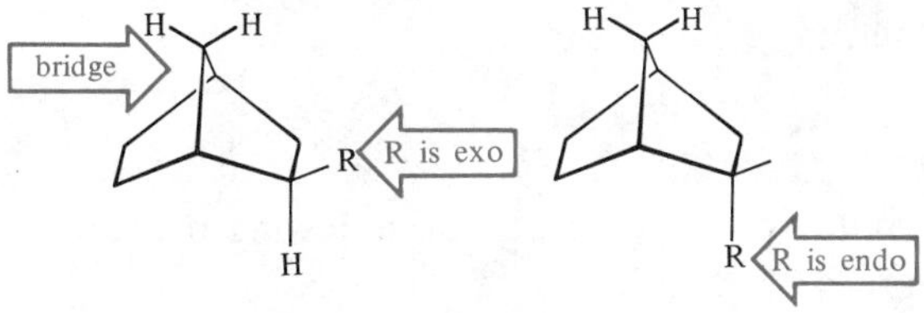

H H bridge R R is exo H   H H R R is endo

In the Diels-Alder reaction of cyclopentadiene with maleic anhydride the major product is the one in which the anhydride linkage, $-\underset{\underset{O}{\|}}{C}-O-\underset{\underset{O}{\|}}{C}-$ has assumed the *endo* configuration. See the illustration at the top of the next page. This favored *endo* stereochemistry seems to arise from favorable interactions between the π electrons of the developing double bond in the diene and the π electrons of unsaturated groups of the dienophile. In the example above, the π electrons of the $-\underset{\underset{O}{\|}}{C}-O-\underset{\underset{O}{\|}}{C}-$ linkage of the anhydride interact with the π electrons of the developing double bond in cyclopentadiene.

* In general, the *exo* substituent is always on the same side as the *shorter* bridge of a bicyclic structure, for example:

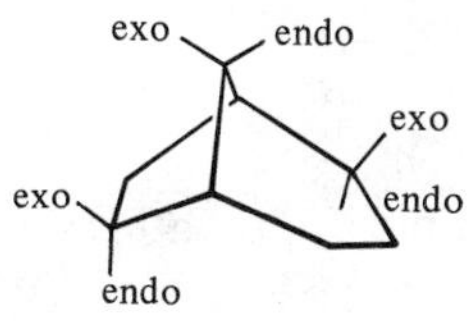

exo endo exo endo exo endo

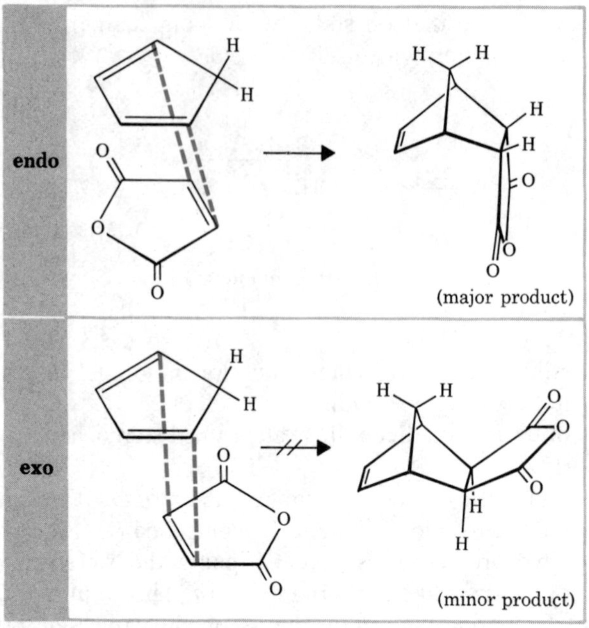

endo

(major product)

exo

(minor product)

Problem 10.9

The dimerization of cyclopentadiene also occurs in an *endo* way. (a) Show how this happens. (b) Which π electrons interact? (c) What is the three-dimensional structure of the product?

Problem 10.10

What products would you expect from the following reactions?

(a) 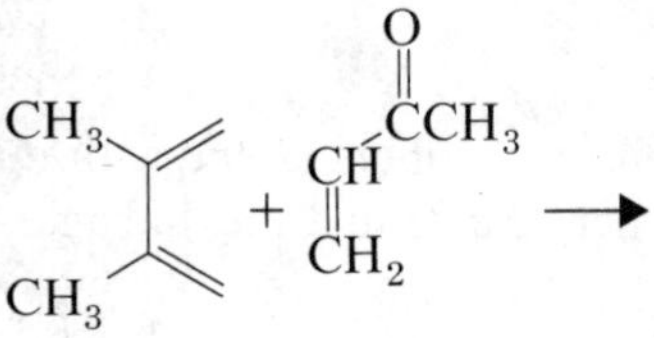 2,3-dimethyl-1,3-butadiene + $CH_2{=}CHCCH_3$ (methyl vinyl ketone) →

(b) 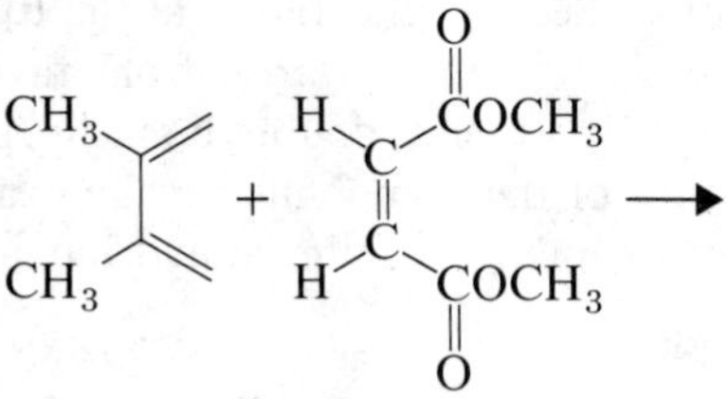 2,3-dimethyl-1,3-butadiene + $H(COCH_3)C{=}C(H)COCH_3$ →

(c) 1,3-cyclohexadiene + maleic anhydride →

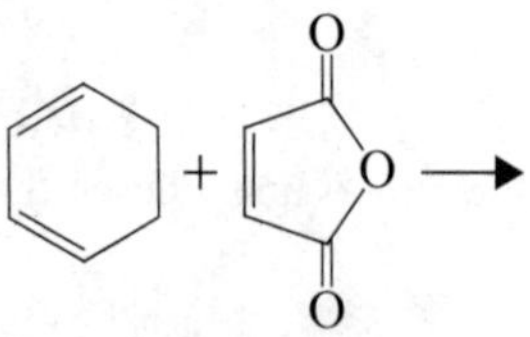

Problem 10.11

Which diene and dienophile would you employ to synthesize the following compound?

CO_2CH_3
H
H
CO_2CH_3

Problem 10.12

Diels-Alder reactions also take place with triple-bonded (acetylenic) dienophiles. Which diene and which dienophile would you use to prepare:

CO_2CH_3
CO_2CH_3

10.11 VISIBLE AND ULTRAVIOLET SPECTROSCOPY

10.11A The Electromagnetic Spectrum

The names of most forms of electromagnetic energy have become familiar terms. The *X rays* used in medicine, the *light* that we see, the *ultraviolet* rays that produce sunburns, and the *radio* and *radar* waves used in communication are all different forms of the same phenomenon: electromagnetic radiation.

According to quantum mechanics, electromagnetic radiation has a dual and seemingly contradictory nature. Electromagnetic radiation has the properties of both a wave and a particle (cf. Special Topic A). Electromagnetic radiation can be described as a wave occurring simultaneously in electrical and magnetic fields. It can also be described as if it consisted of particles called quanta or photons. Different experiments disclose these two different aspects of electromagnetic radiation. They are not seen together in the same experiment.

A wave is usually described in terms of its *wavelength* (λ) or its frequency (ν). A simple wave is shown in Fig. 10.8. The distance between consecutive crests (or troughs) is the wavelength. The number of full cycles of the wave that pass a given point each second, as the wave moves through space, is called the *frequency*.

An analogy may make the term frequency more understandable. Imagine that you are anchored offshore in a small boat. Since your boat is anchored, your position on the surface of the water is fixed. As waves pass by, the boat rises to one crest and falls into another trough. Now suppose that you decide to count the number of times the boat rises from one crest to the next each minute. In doing this you will be measuring the frequency of the waves in cycles per minute.

The frequencies of most electromagnetic waves are much higher than those

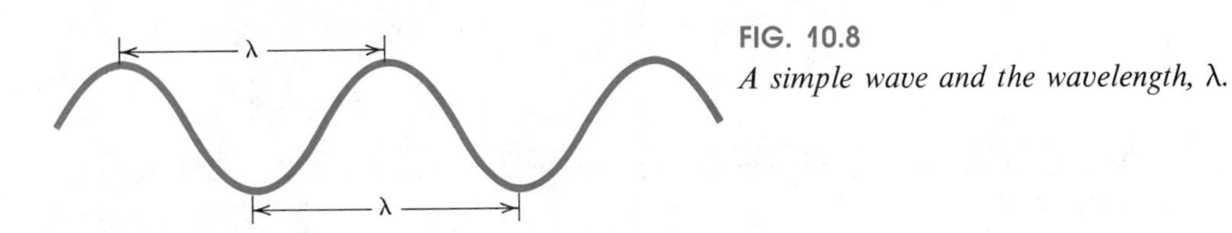

FIG. 10.8
A simple wave and the wavelength, λ.

of waves on an ocean or lake. As a result, the frequencies of electromagnetic waves are usually reported in cycles per second or Hertz.* The wavelengths of electromagnetic radiation are expressed in either meters (m), millimeters, (1 mm = 10^{-3} m), micrometers (1 μm = 10^{-6} m), or nanometers (1 nm = 10^{-9} m). [An older term for micrometer is *micron* (abbreviated μ) and an older term for nanometer is *millimicron*.]

The energy of a quantum of electromagnetic energy is directly related to its frequency.

$$E = h\nu$$

where h = Planck's constant, 6.625×10^{-27} erg-sec,

and

ν = the frequency in cycles/sec.

This means that the higher the frequency of radiation the greater is its energy. X rays, for example, are much more energetic than rays of visible light. The frequencies of X rays are of the order of 10^{19} cycles/sec, while those of visible light are of the order of 10^{15} cycles/sec.

Since $\nu = c/\lambda$, the energy of electromagnetic radiation is inversely proportional to its wavelength,

$$E = \frac{hc}{\lambda} \qquad (c = \text{the velocity of light})$$

Thus, per quantum electromagnetic radiation of long wavelength has low energy, while that of short wavelength has high energy. X rays have wavelengths of the order of 0.1 nm, while visible light has wavelengths between 400 nm and 750 nm.*

It may be helpful to point out, too, that for visible light, wavelengths (and, thus, frequencies) are related to what we perceive as colors. The light that we call red light has a wavelength of approximately 750 nm. The light we call violet light has a wavelength of approximately 400 nm. All of the other colors of the visible spectrum (the rainbow) lie in between these wavelengths.

The different regions of the electromagnetic spectrum are shown in Fig. 10.9. Nearly every portion of the electromagnetic spectrum from the region of X rays to those of microwave and radio wave has been used in elucidating structures of atoms and molecules. In Chapter 13 we discuss the use that can be made of the infrared and radio regions when we take up infrared spectroscopy and nuclear magnetic resonance spectroscopy. At this point we direct our attention to electromagnetic radiation in the near ultraviolet and visible regions and see how it interacts with conjugated polyenes.

* The term Hertz (after the German physicist H. R. Hertz), abbreviated Hz, is now often used in place of *cycles per second*. Frequency of electromagnetic radiation is also sometimes expressed in *wavenumbers*, that is, the number of waves per centimeter.

* A convenient formula that relates wavelength (in nm) to the energy of electromagnetic radiation is the following.

$$E\,(\text{in kcal/mole}) = \frac{28{,}600}{\text{wavelength in nanometers}}$$

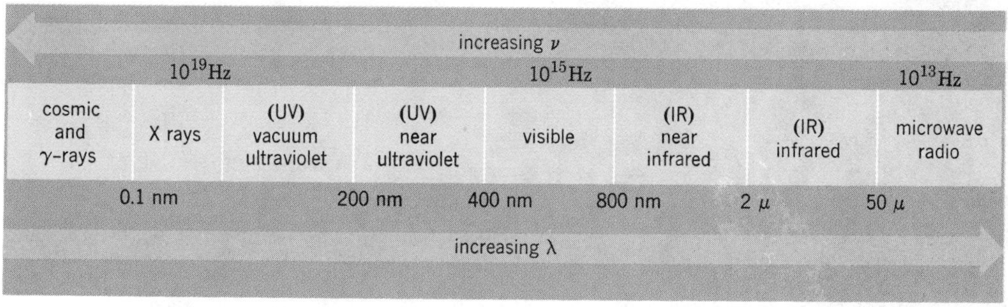

increasing ν

10^{19} Hz — 10^{15} Hz — 10^{13} Hz

cosmic and γ-rays | X rays | (UV) vacuum ultraviolet | (UV) near ultraviolet | visible | (IR) near infrared | (IR) infrared | microwave radio

0.1 nm — 200 nm — 400 nm — 800 nm — 2 μ — 50 μ

increasing λ

FIG. 10.9
The electromagnetic spectrum.

10.11B Visible and Ultraviolet Spectra

When electromagnetic radiation in the ultraviolet and visible regions passes through a compound containing multiple bonds, a portion of the radiation is usually absorbed by the compound. Just how much of the radiation is absorbed depends on the wavelength of the radiation and the structure of the compound. The absorption of radiation is caused by the subtraction of energy from the radiation beam when electrons in orbitals of lower energy are excited into orbitals of higher energy.

Instruments, called visible-ultraviolet spectrometers, are used to measure the amount of light absorbed at each wavelength of the visible and ultraviolet region. In these instruments a beam of light is split; half the beam (the sample beam) is directed through a transparent cell containing a solution of the compound being analyzed, and half (the reference beam) is directed through an identical cell that does not contain the compound but contains the solvent. Solvents are chosen to be transparent in the region being analyzed. The instrument is designed so that it can make a comparison of the intensities of the two beams at each wavelength of the region. If the compound absorbs light at a particular wavelength, the intensity of the sample beam (I_S) will be less than that of the reference beam (I_R). The instrument indicates this by producing a graph—a plot of the wavelength of the entire region *versus* the absorbance (A) of light at each wavelength. [The absorbance at a particular wavelength is defined by the equation: $A_\lambda = \log (I_R/I_S)$.] Such a graph is called an *absorption spectrum.*

A typical ultraviolet absorption spectrum, that of 2,5-dimethyl-2,4-hexadiene, is shown in Fig. 10.10. It shows a broad absorption band in the region between 210 and 260 nm. The absorption is at a maximum at 241.5 nm. It is this wavelength that is usually reported in the chemical literature.

In addition to reporting the wavelength of maximum absorption (λ_{max}), chemists often report another quantity that indicates the strength of the absorption, called the molar absorptivity, ϵ.*

The molar absorptivity is simply the proportionality constant that relates the observed absorbance (A) at a particular wavelength, λ, to the molar concentration, C, of the sample and the length, l, (in centimeters) of the path of the light beam through the sample cell.

$$A = \epsilon \times C \times l \quad \text{or} \quad \epsilon = \frac{A}{C \times l}$$

* The molar absorptivity, ϵ, is often referred to as the molar extinction coefficient.

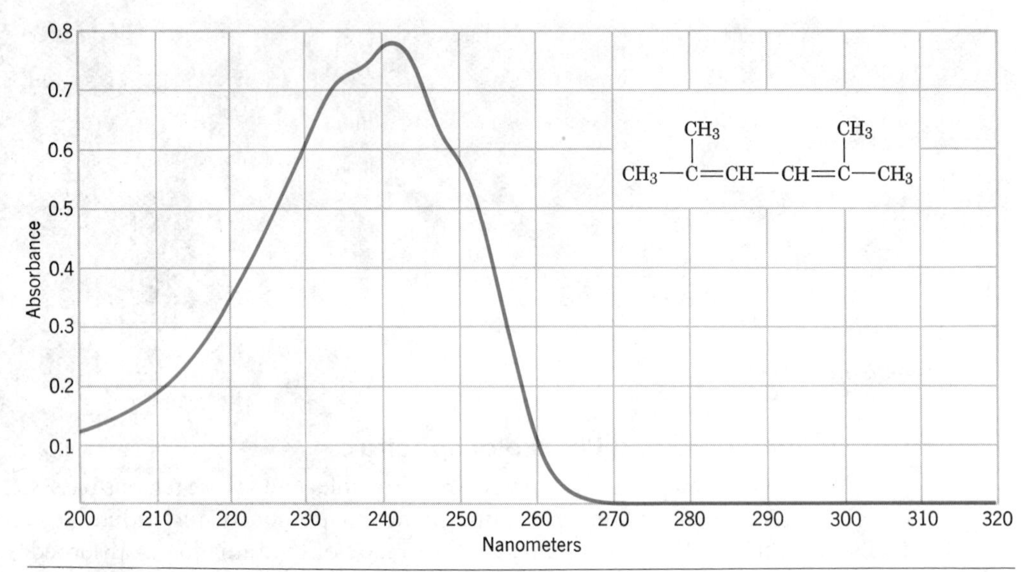

FIG. 10.10
The ultraviolet absorption spectrum of 2,5-dimethyl-2,4-hexadiene in methanol. (Spectrum courtesy of Sadtler Research Laboratories, Philadelphia, Pa.)

For 2,5-dimethyl-2,4-hexadiene dissolved in methanol the molar absorptivity at the wavelength of maximum absorbance (241.5 nm) is 13,000 $M^{-1}\ cm^{-1}$. In the chemical literature this would be reported as

$$\text{2,5-dimethyl-2,4-hexadiene, } \lambda_{max}^{methanol}\ 241.5\text{ nm } (\epsilon = 13{,}100)$$

As we noted earlier, when compounds absorb light in the ultraviolet and visible regions, electrons are excited from lower electronic energy levels to higher ones. For this reason, visible and ultraviolet spectra are often called *electronic spectra.* The absorption spectrum of 2,5-dimethyl-2,4-hexadiene is a typical electronic spectrum because the absorption band (or peak) is very broad. Most absorption bands in the visible and ultraviolet region are broad because each electronic energy level has, associated with it, vibrational and rotational levels. Thus, electron transitions may occur between any of several vibrational and rotational states of one electronic level to any of several vibrational and rotational states of a higher level.

Alkenes and isolated dienes usually absorb light of a wavelength shorter than 200 nm. Ethene, for example, gives an absorption maximum at 171 nm; 1,4-pentadiene gives an absorption maximum at 178 nm. These absorptions occur at wavelengths that are out of the range of operation of most visible-ultraviolet spectrometers because they occur where air also absorbs. Special vacuum techniques must be employed in measuring them.

Compounds whose molecules contain *conjugated* multiple bonds absorb light at wavelengths longer than 200 nm. For example, 1,3-butadiene absorbs at 217 nm. This longer-wavelength absorption by conjugated dienes is a direct consequence of conjugation.

We can understand how conjugation of multiple bonds brings about longer-wavelength absorption of light if we examine Fig. 10.11.

When a molecule absorbs light at its longest wavelength, an electron is excited from its highest occupied molecular orbital (HOMO) to the lowest unoccupied molecular orbital (LUMO). For alkenes and alkadienes the highest occu-

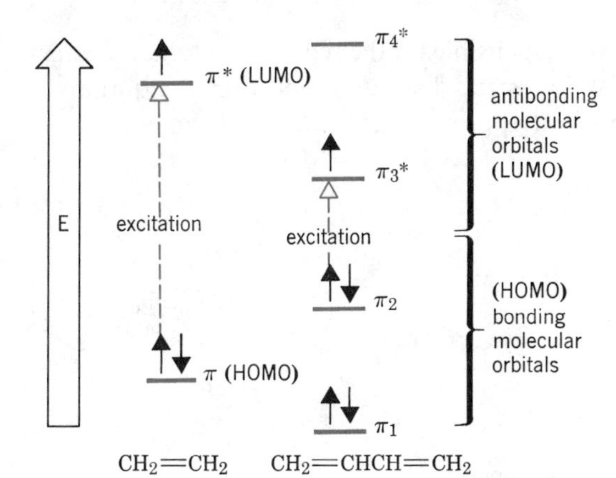

FIG. 10.11
The relative energies of the π molecular orbitals of ethene and 1,3-butadiene.

pied molecular orbital is a bonding π orbital and the lowest unoccupied molecular orbital is an antibonding π^* orbital. The wavelength of the absorption maximum is determined by the difference in energy between these two levels. The energy gap between the highest occupied molecular orbital and lowest unoccupied molecular orbital of ethene is greater than that between the corresponding orbitals of 1,3-butadiene. Thus, the $\pi \longrightarrow \pi^*$ electron excitation of ethene requires absorption of light of greater energy (shorter wavelength) than the corresponding $\pi_2 \longrightarrow \pi_3^*$ excitation in 1,3-butadiene. The energy difference between the highest occupied molecular orbitals and the lowest unoccupied molecular orbitals of the two compounds is reflected in their absorption spectra. Ethene has its λ_{max} at 171 nm; 1,3-butadiene has a λ_{max} at 217 nm.

The narrower gap between the highest occupied molecular orbital and the lowest unoccupied molecular orbital in 1,3-butadiene results from the conjugation of the double bonds. Molecular orbital calculations indicate that a much larger gap should occur in isolated alkadienes. This is borne out experimentally. Isolated alkadienes give absorption spectra similar to those of alkenes. Their λ_{max} are at shorter wavelengths, usually below 200 nm. As we mentioned, 1,4-pentadiene has its λ_{max} at 178 nm.

Conjugated alkatrienes absorb at longer wavelengths than conjugated alkadienes, and this too can be accounted for in molecular orbital calculations. The energy gap between the highest occupied molecular orbital and the lowest unoccupied molecular orbital of an alkatriene is even smaller than that of an alkadiene. In fact, there is a general rule that states that *the greater the number of conjugated multiple bonds a compound contains, the longer will be the wavelength at which the compound absorbs light.*

Polyenes with eight or more conjugated double bonds absorb light in the visible region of the spectrum. For example, β-carotene, a precursor of Vitamin A, and a compound that imparts its orange color to carrots, has 11 conjugated double bonds; β-carotene has an absorption maximum at 497 nm, well into the visible region.

β-Carotene

Lycopene, a compound partly responsible for the red color of tomatoes, also has 11 conjugated double bonds. Lycopene has an absorption maximum at 505 nm, and it absorbs there intensely. (Approximately 0.02 g of lycopene can be isolated from 1 kg of fresh, ripe tomatoes.)

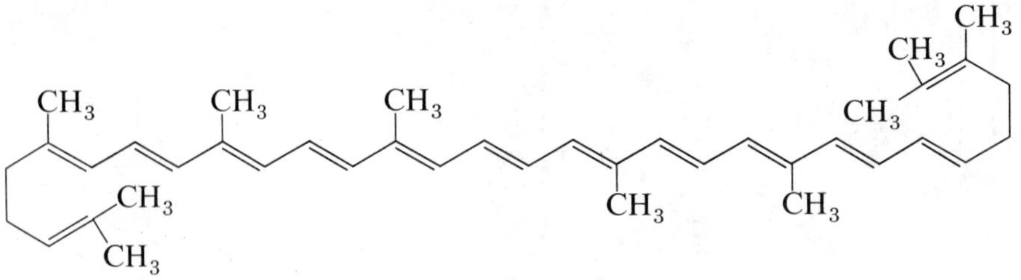

Lycopene

Table 10.3 gives the values of λ_{max} for a number of unsaturated compounds.

TABLE 10.3 Long Wavelength Absorption Maxima of Unsaturated Hydrocarbons

COMPOUND	STRUCTURE	λ_{max} (nm)	ϵ_{max}
Ethene	$CH_2{=}CH_2$	171	15,530
trans-2-Hexene	$CH_3CH_2(H)C{=}C(H)CH_2CH_3$	184	10,000
Cyclohexene		182	7,600
1-Octene	$CH_3(CH_2)_5CH{=}CH_2$	177	12,600
1-Octyne	$CH_3(CH_2)_5C{\equiv}CH$	185	2,000
1,3-Butadiene	$CH_2{=}CHCH{=}CH_2$	217	21,000
cis-1,3-Pentadiene	$CH_3(H)C{=}C(H)CH{=}CH_2$	223	22,600
trans-1,3-Pentadiene	$CH_3(H)C{=}C(H)CH{=}CH_2$	223.5	23,000
1-Buten-3-yne	$CH_2{=}CHC{\equiv}CH$	228	7,800
1,4-Pentadiene	$CH_2{=}CHCH_2CH{=}CH_2$	178	
1,3-Cyclopentadiene		239	3,400
1,3-Cyclohexadiene		256	8,000
trans-1,3,5-Hexatriene	$CH_2{=}CH(H)C{=}C(H)CH{=}CH_2$	274	50,000

Compounds with carbon-oxygen double bonds also absorb light in the ultraviolet region. Acetone, for example, has a broad absorption peak at 280 nm that corresponds to the excitation of an electron from one of the unshared pairs (a nonbonding or "*n*" electron) excited to the π^* orbital of the carbon-oxygen double bond:

$$(CH_3)_2C{=}\ddot{O}: \xrightarrow{n \rightarrow \pi^*} (CH_3)_2\dot{C}{-}\dot{\ddot{O}}\cdot$$

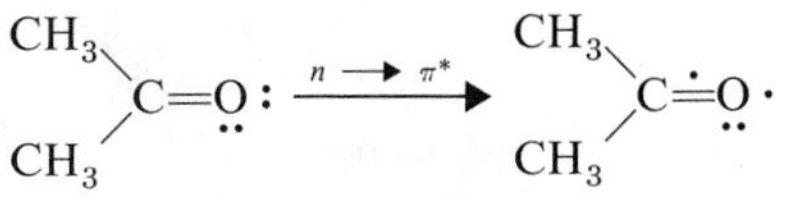

Acetone
$\lambda_{max} = 280$
$\epsilon_{max} = 15$

Compounds in which the carbon-oxygen double bond is conjugated with a carbon-carbon double bond have absorption maxima corresponding to $n \longrightarrow \pi^*$ excitations and $\pi \longrightarrow \pi^*$ excitations. The $n \longrightarrow \pi^*$ absorption maximum occurs at longer wavelengths but is much weaker.

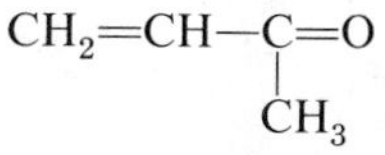

$$CH_2{=}CH{-}C(CH_3){=}O$$

$n \longrightarrow \pi^*$ $\lambda_{max} = 324$ nm, $\epsilon_{max} = 24$
$\pi \longrightarrow \pi^*$ $\lambda_{max} = 219$ nm, $\epsilon_{max} = 3{,}600$

Additional Problems

10.13

Outline a synthesis of 1,3-butadiene starting from

(a) 1,4-dibromobutane
(b) $HOCH_2(CH_2)_2CH_2OH$
(c) $CH_2{=}CHCH_2CH_2OH$
(d) $CH_2{=}CHCH_2CH_2Cl$
(e) $CH_2{=}CHCHClCH_3$
(f) $CH_2{=}CHCHOHCH_3$
(g) $HC{\equiv}CCH{=}CH_2$

10.14

What product would you expect from the following reaction?

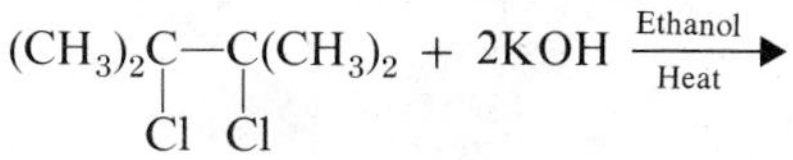

$$(CH_3)_2CCl{-}CCl(CH_3)_2 + 2KOH \xrightarrow[\text{Heat}]{\text{Ethanol}}$$

10.15

What products would you expect from the reaction of 1,3-butadiene and each of the following reagents. (If no reaction would occur you should indicate that as well.)

(a) One mole Cl_2
(b) Two moles Cl_2
(c) Two moles Br_2
(d) Two moles H_2, Ni
(e) $Ag(NH_3)_2^+ OH^-$
(f) One mole Cl_2 in H_2O
(g) Hot $KMnO_4$
(h) H^+, H_2O

10.16

Show how you might carry out each of the following transformations. (In some transformations several steps may be necessary.)

(a) 1-Butene $\longrightarrow$ 1,3-butadiene
(b) 1-Pentene $\longrightarrow$ 1,3-pentadiene
(c) $CH_3CH_2CH_2CH_2OH \longrightarrow CH_2BrCH{=}CHCH_2Br$
(d) $CH_3CH{=}CHCH_3 \longrightarrow CH_3CH{=}CHCH_2Br$

(e) 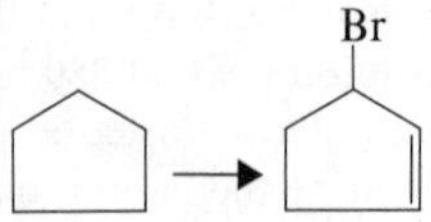 Br

(f) 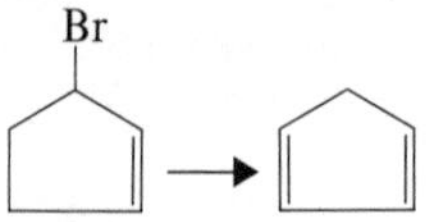 Br

10.17

Conjugated dienes react with free radicals by both 1,2 and 1,4 addition. Account for this fact by using the peroxide-catalyzed addition of one mole of HBr to 1,3-butadiene as an illustration.

10.18

Outline a simple chemical test that would distinguish between the members of each of the following pairs of compounds.

(a) 1,3-Butadiene and 1-butyne
(b) 1,3-Butadiene and *n*-butane
(c) 1,3-Butadiene and $CH_2{=}CHCH_2CH_2OH$
(d) 1,3-Butadiene and $CH_2{=}CHCH_2CH_2Br$
(e) $CH_2BrCH{=}CHCH_2Br$ and $CH_3CBr{=}CBrCH_3$

10.19

(a) The hydrogen atoms of carbon-3 of 1,4-pentadiene are unusually susceptible to abstraction by free radicals. How can you account for this? (b) Can you also provide an explanation for the fact that the protons of carbon-3 of 1,4-pentadiene are more acidic than the methyl hydrogens of propene?

10.20

When 2-methyl-1,3-butadiene (isoprene) undergoes a 1,4-addition of hydrogen chloride, the major product that is formed is 1-chloro-3-methyl-2-butene. Little or no 1-chloro-2-methyl-2-butene is formed. How can you explain this?

10.21

Which diene and dienophile would you employ in a synthesis of each of the following?

(a) 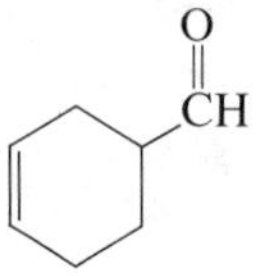 O ‖ CH

(d) 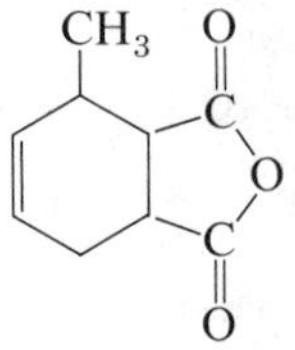 CH_3 O ‖ C O C ‖ O

(b) CH_3 CH_3 O ‖ $COCH_3$ $COCH_3$ ‖ O

(e) CN

(c) CH_3 CH_3 O ‖ $COCH_3$ $COCH_3$ ‖ O

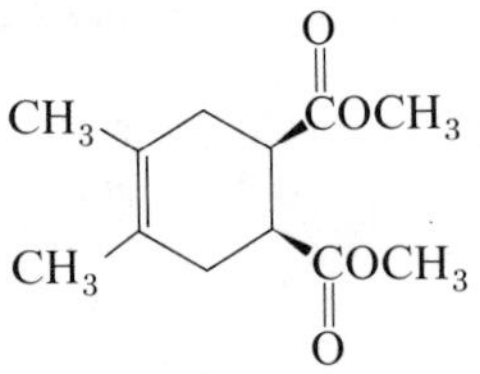

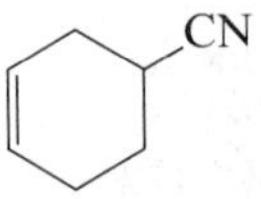

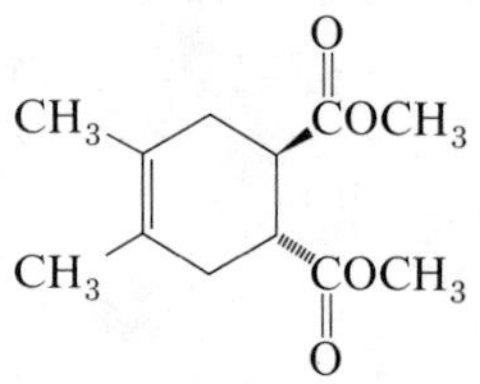

10.22
Account for the fact that neither of the following compounds undergoes a Diels-Alder reaction with maleic anhydride.

HC≡C—C≡CH or (cyclohexene ring)=CH_2

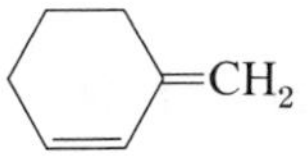

10.23
Acetylenic compounds may be used as dienophiles in the Diels-Alder reaction (cf. Problem 10.12). Write structures for the adducts that you expect from the reaction of 1,3-butadiene with:

(a) $CH_3O\overset{O}{\overset{\|}{C}}C\equiv C\overset{O}{\overset{\|}{C}}OCH_3$ (dimethyl acetylenedicarboxylate)
(b) $CF_3C\equiv CCF_3$ (hexafluoro-2-butyne)

10.24
Two compounds, **A** and **B**, have the same molecular formula C_6H_8. Both **A** and **B** decolorize bromine in carbon tetrachloride and both give positive tests with cold dilute potassium permanganate. Both **A** and **B** react with two moles of hydrogen in the presence of platinum to yield cyclohexane. **A** shows an absorption maximum at 256 nm, while **B** shows no absorption maximum beyond 200 nm. What are the structures of **A** and **B?**

10.25
Cyclopentadiene undergoes a dimerization reaction at room temperature that results in the formation of "dicyclopentadiene" (p. 409). Can you suggest a simple method for following the rate of this reaction?

10.26
Three compounds, **D, E,** and **F,** have the same molecular formula C_5H_6. In the presence of a platinum catalyst, all three compounds absorb three moles of hydrogen and yield *n*-pentane. Compounds **E** and **F** give a precipitate when treated with ammoniacal silver nitrate; compound **D** gives no reaction. Compounds **D** and **E** show an absorption maximum near 230 nm. Compound **F** shows no absorption maximum beyond 200 nm. Propose structures for **D, E,** and **F.**

10.27
When furan and maleimide (below) undergo a Diels-Alder reaction at 25°, the major product is the *endo* adduct **G.** When the reaction is carried out at 90°, however, the major product is the *exo* isomer **H.** The *endo* adduct isomerizes to the *exo* adduct when it is heated to 90°. Propose an explanation that will account for these results.

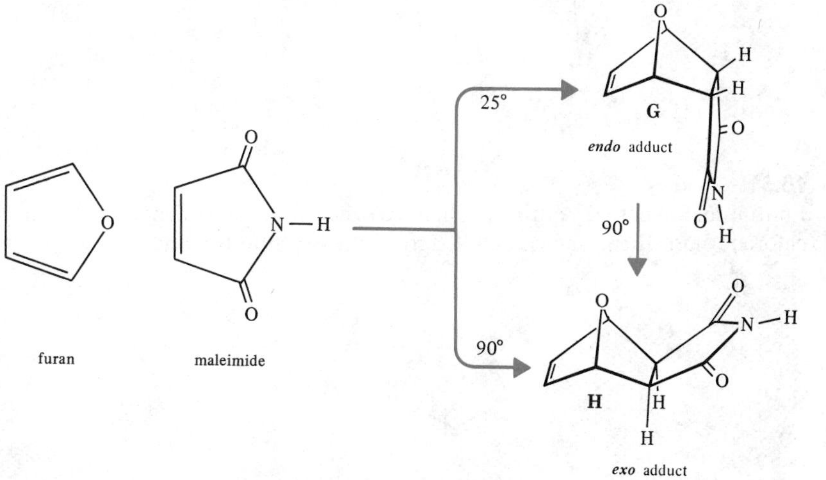

furan + maleimide —25°→ **G** *endo* adduct; —90°→ *exo* adduct (**H**); **G** —90°→ *exo* adduct

10.28

Two controversial "hard" insecticides are Aldrin and Dieldrin (see following diagram). (The Environmental Protection Agency has recommended discontinuance of use of these insecticides because of possible harmful side effects and because they are not biodegradable.) The commercial synthesis of Aldrin begins with hexachlorocyclopentadiene and norbornadiene. Dieldrin is synthesized from Aldrin. Show how these syntheses might be carried out.

aldrin

dieldrin

hexachlorocyclopentadiene

norbornadiene

10.29

(a) Norbornadiene for the Aldrin synthesis (Problem 10.28) can be prepared from cyclopentadiene and acetylene. Show the reaction involved. (b) It can also be prepared by allowing cyclopentadiene to react with vinyl chloride and then treating the product with base. Outline this synthesis.

10.30

Two other hard insecticides are Chlordan and Heptachlor. Their commercial syntheses begin with cyclopentadiene and hexachlorocyclopentadiene. Show how these syntheses might be carried out.

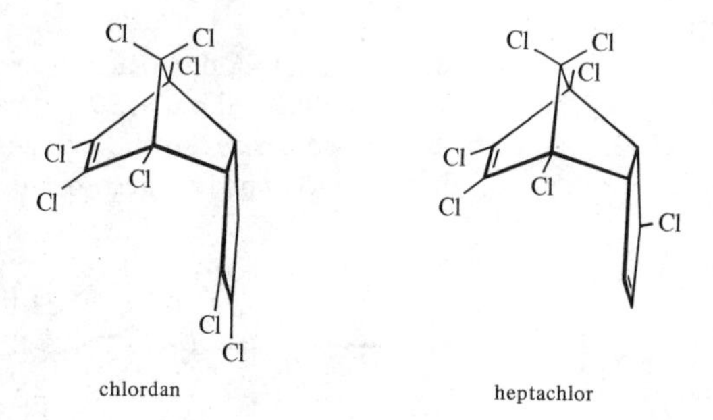

chlordan

heptachlor

10.31

Endrin, an isomer of Aldrin, is obtained when cyclopentadiene reacts with the hexachloronorbornadiene shown below. Propose a structure for Endrin.

Endrin

10.32

When $CH_3CH{=}CHCH_2OH$ is treated with concentrated HCl, two products are produced, $CH_3CH{=}CHCH_2Cl$ and $CH_3CHClCH{=}CH_2$. Outline a mechanism that will explain this.

10.33

When a solution of 1,3-butadiene in CH_3OH is treated with chlorine, the products are $ClCH_2CH{=}CHCH_2OCH_3$ (30%) and $ClCH_2CH(OCH_3)CH{=}CH_2$ (70%). Write a mechanism that accounts for their formation.

10.34

Dehydrohalogenation of *vic*-dihalides (with the elimination of two moles of HX) normally leads to an alkyne rather than to a conjugated diene. However, when 1,2-dibromocyclohexane is dehydrohalogenated, 1,3-cyclohexadiene is produced in good yield. What factor accounts for this?

10.35

When 1-pentene reacts with *N*-bromosuccinimide, two products with the formula C_5H_9Br are obtained. What are these products and how are they formed?

10.36

Treating either 1-chloro-3-methyl-2-butene or 3-chloro-3-methyl-1-butene with Ag_2O in water gives (in addition to AgCl) the same mixture of alcohols: $(CH_3)_2C{=}CHCH_2OH$ (15%) and $(CH_3)_2C(OH)CH{=}CH_2$ (85%). (a) Write a mechanism that accounts for the formation of these products. (b) What might explain the relative proportions of two alkenes that are formed?

10.37

The heat of hydrogenation of allene is 71.3 kcal/mole while that of propyne is 69.3 kcal/mole. (a) Which compound is more stable? (b) Treating allene with a strong base causes it to isomerize to propyne. Explain.

*** 10.38**

Mixing furan (Problem 10.27) with maleic anhydride in ether yields a crystalline solid with a melting point of 125°. When melting of this compound takes place, however, one can notice that the melt evolves a gas. If the melt is allowed to resolidify, one finds that it no longer melts at 125° but instead it melts at 56°. Consult an appropriate chemistry handbook and provide an explanation for what is taking place.

*** 10.39**

Propose explanations for the following facts: (a) The reactivity of *trans*-1,3-pentadiene in Diels–Alder reactions is much like that of 1,3-butadiene. On the other hand, *cis*-1,3-pentadiene reacts much more slowly. (b) 2-*tert*-Butyl-1,3-butadiene undergoes a Diels-Alder reaction much more rapidly than 1,3-butadiene, while *cis*-5,5-dimethyl-1,3-hexadiene (*cis*-1-*tert*-butyl-1,3-butadiene) does not react at all.

*** 10.40**

Treating hexachloropropene with aluminum chloride leads to the formation of a stable organic salt. Propose a reasonable structure for this salt and account for its formation.

E SPECIAL TOPIC

NATURALLY OCCURRING ALKENES

People have isolated organic compounds from plants since antiquity. By gently heating or by steam distilling certain plant materials, one can obtain mixtures of odoriferous compounds known as "essential oils." These compounds have had a variety of uses, particularly in early medicine and in the making of perfumes.

As the science of organic chemistry developed, chemists separated the various components of these mixtures, determined their molecular formulas and then later their structural formulas. Even today these natural products offer challenging problems for chemists interested in structure determination and synthesis. Research in this area has also given us important information about the ways the plants, themselves, synthesize these compounds.

Compounds known generally as *terpenes* and *terpenoids* are the most important constituents of essential oils. Most terpenes have carbon skeletons of 10, 15, 20, or 30 atoms and are classified in the following way.

NUMBER OF CARBONS	CLASS
10	Monoterpenes
15	Sesquiterpenes
20	Diterpenes
30	Triterpenes

One can view terpenes as being built up from two or more five-carbon units known as *isoprene units*. Isoprene is 2-methyl-1,3-butadiene. Isoprene and the isoprene unit can be represented in various ways.

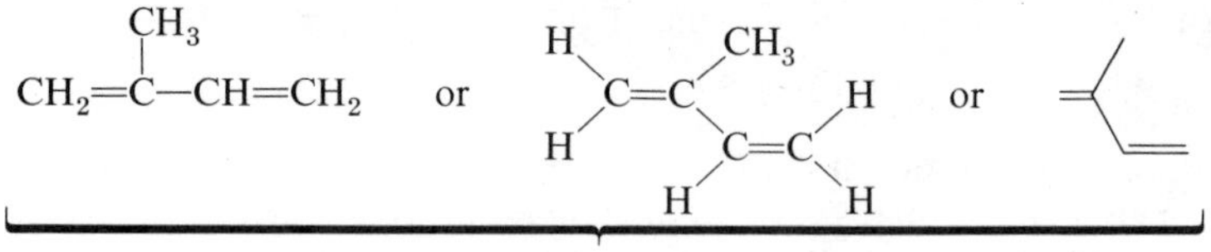

$$CH_2{=}C(CH_3){-}CH{=}CH_2$$ or $(H)_2C{=}C(CH_3){-}CH{=}CH_2$ or [skeletal formula]

Isoprene

C—C(C)—C—C or [skeletal formula]

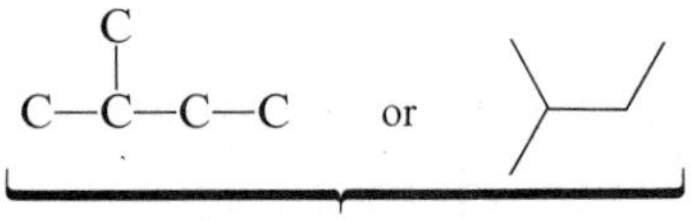

An isoprene unit

We now know that plants do not synthesize terpenes from isoprene. However, recognition of the isoprene unit as a component of the structure of terpenes has been a great aid in elucidating their structures. We can see how, if we examine the structures shown below.

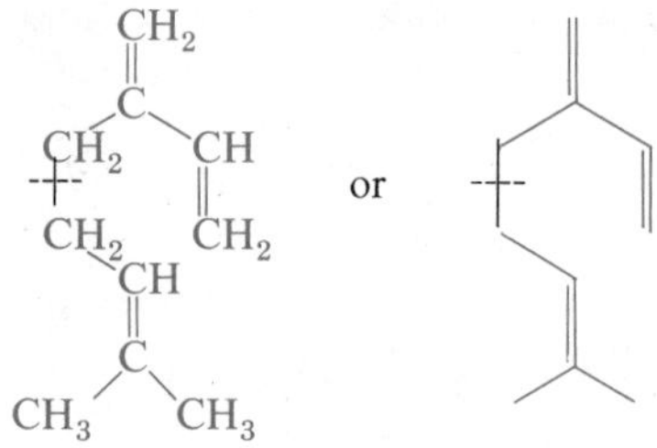

$$CH_2{=}C(CH{=}CH_2){-}CH_2{-}CH_2{-}CH{=}C(CH_3){-}CH_3$$ or [skeletal formula]

Myrcene
(isolated from bay oil)

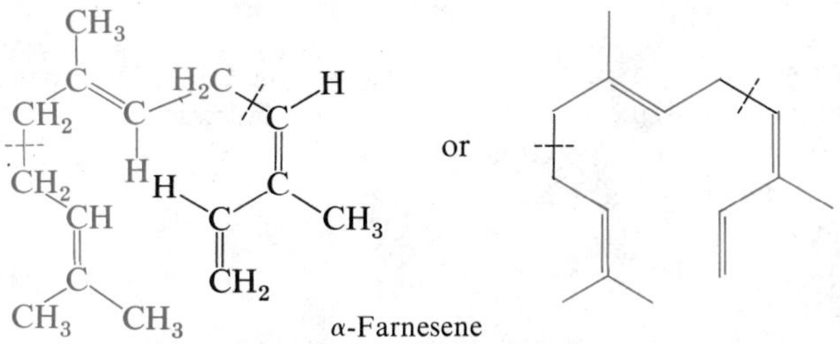

α-Farnesene
(from oil of citronella)

Using dashed lines to separate isoprene units we can see that the monoterpene, myrcene, has two isoprene units; and that the sesquiterpene, α-farnesene, has three. In both compounds the isoprene units are linked head to tail.

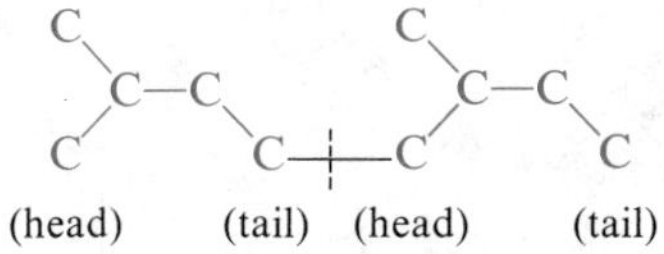

(head) (tail) (head) (tail)

Many terpenes also have isoprene units linked in rings, and others (terpenoids) contain oxygen.

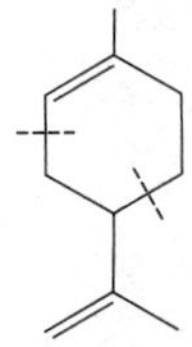

Limonene
(from oil of lemon or orange)

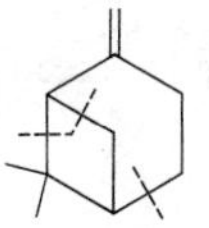

β-Pinene
(from oil of turpentine)

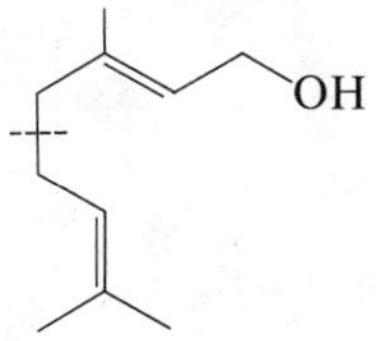

Geraniol
(from roses and other flowers)

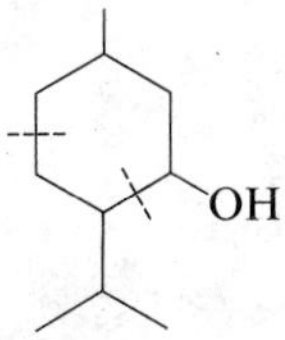

Menthol
(from peppermint)

Problem E.1

(a) Show the isoprene units in each of the following terpenes. (b) Classify each as a monoterpene, sesquiterpene, diterpene, and so on.

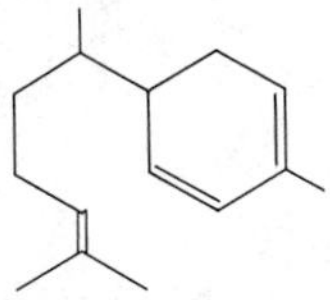

Zingiberene
(from oil of ginger)

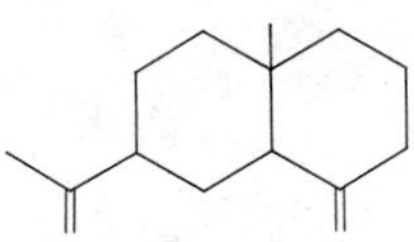

β-Selinene
(from oil of celery)

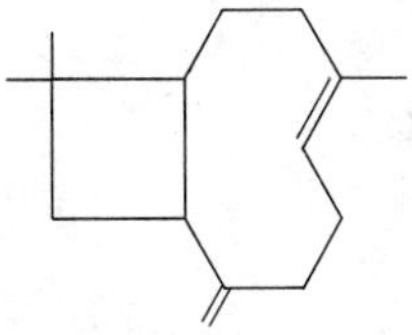

Caryophyllene
(from oil of cloves)

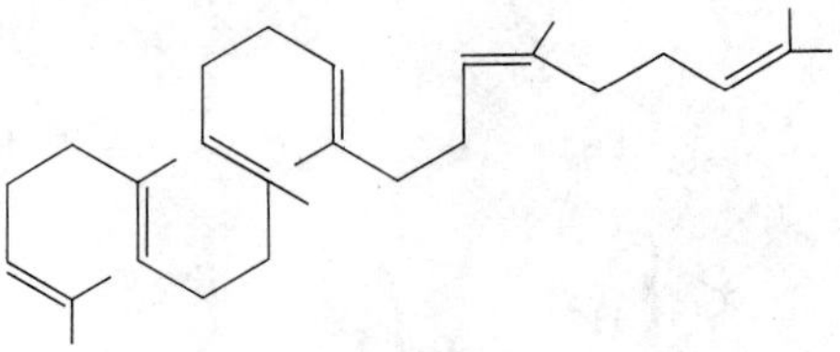

Squalene
(from shark liver oil)

Problem E.2

What products would you expect to obtain if each of the following terpenes were subjected to ozonization and subsequent treatment with zinc and water?

(a) Myrcene
(b) Limonene
(c) α-Farnesene
(d) Geraniol
(e) Squalene

Problem E.3

Give structural formulas for the products that you would expect from the following reactions.

(a) β-Pinene + hot $KMnO_4 \longrightarrow$

(b) Zingiberene + $H_2 \xrightarrow{Pt}$

(c) Caryophyllene + HCl $\longrightarrow$

(d) β-Selinene + $2(BH_3)_2 \xrightarrow[(2)\ H_2O_2,\ OH^-]{}$

Problem E.4

What simple chemical test could you use to distinguish between geraniol and menthol?

The carotenes are tetraterpenes. They can be thought of as two diterpenes linked in tail-to-tail fashion.

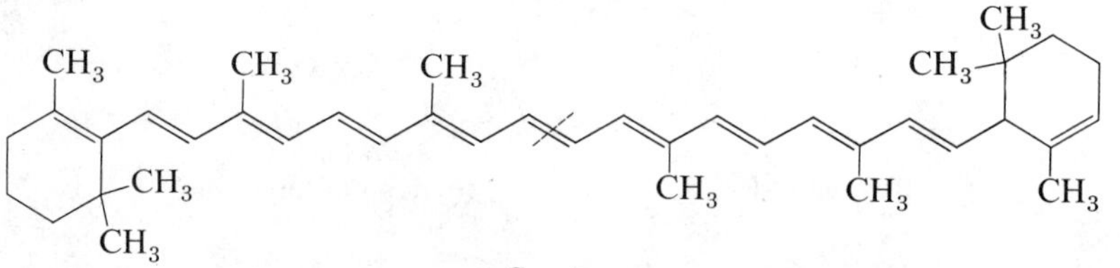

α-Carotene

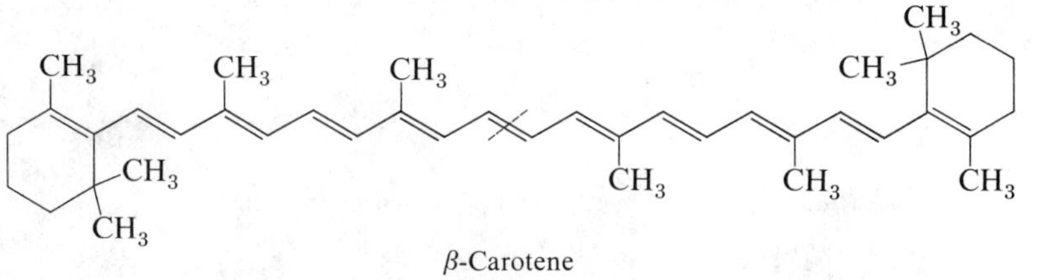

β-Carotene

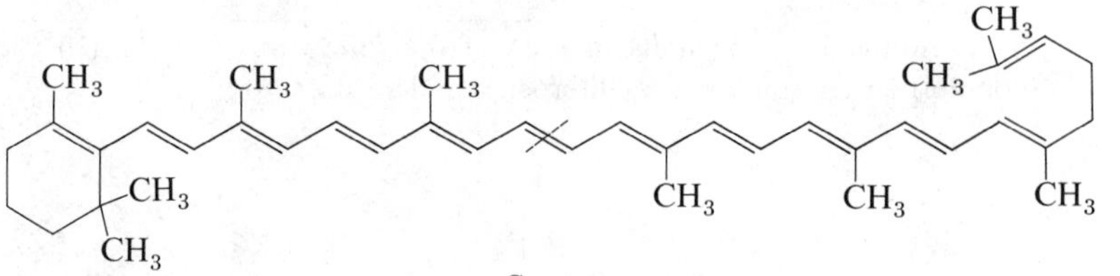

γ-Carotene

The carotenes are present in almost all green plants. All three carotenes serve as precursors for vitamin A, for they all can be converted to vitamin A by enzymes in the liver.

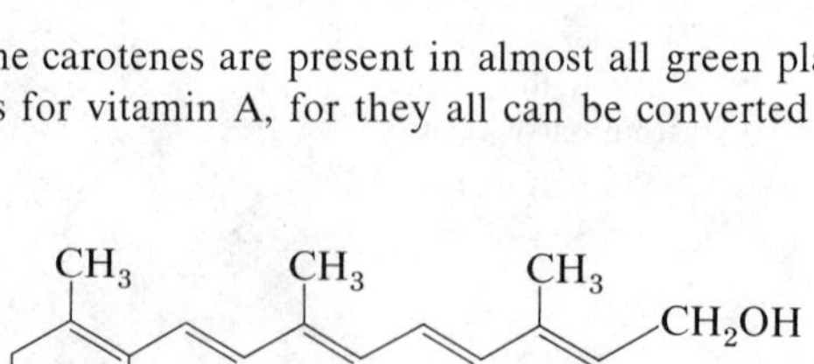

Vitamin A

In this conversion, one molecule of β-carotene yields two of vitamin A; α- and γ-carotene give only one. Because β-carotene became available before vitamin A, β-carotene was adopted as the standard for vitamin A content in foods. One international unit (I.U.) of vitamin A is equivalent to the vitamin A activity of 0.6 μg of crystalline β-carotene. Vitamin A is important not only in vision but in many other ways as well. For example, young animals whose diets are deficient in vitamin A fail to grow.

Natural Rubber

Natural rubber can be viewed as a 1,4-addition polymer of isoprene. In fact, pyrolysis degrades natural rubber to isoprene. Pyrolysis (Greek: *pyros,* a fire + *lysis*) is heating something in the absence of air until it decomposes. The isoprene units of natural rubber are all linked in a head-to-tail fashion and all of the double bonds are *cis.*

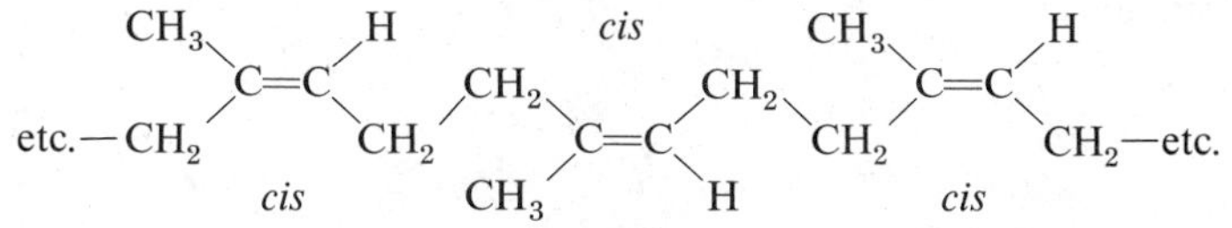

$$\text{etc.}-CH_2(CH_3)C{=}CH-CH_2-CH_2(CH_3)C{=}CH-CH_2-CH_2(CH_3)C{=}CH-CH_2-\text{etc.} \quad (\textit{cis}, \textit{cis}, \textit{cis})$$

Natural rubber
(*cis*-1,4-polyisoprene)

Using Ziegler-Natta catalysts (Special Topic C), it is now possible to polymerize isoprene and obtain a synthetic product that is identical with the rubber obtained from natural sources.

Pure natural rubber is soft and tacky. To be useful, natural rubber has to be "vulcanized." In vulcanization, natural rubber is heated with sulfur. A reaction takes place that produces cross-links between the *cis*-polyisoprene chains and makes the rubber much harder. Sulfur reacts both at the double bonds and at allylic hydrogens.

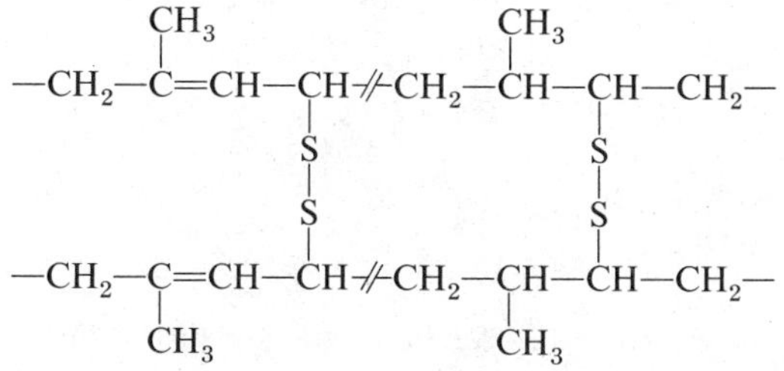

$$\begin{array}{c}
-CH_2-\underset{}{\overset{CH_3}{\overset{|}{C}}}{=}CH-\underset{\underset{S}{|}}{CH}\not\!-CH_2-\overset{CH_3}{\overset{|}{CH}}-\underset{\underset{S}{|}}{CH}-CH_2- \\
\quad\quad\quad\quad | \quad\quad\quad\quad\quad\quad | \\
-CH_2-\underset{\underset{CH_3}{|}}{C}{=}CH-\overset{\overset{S}{|}}{CH}\not\!-CH_2-\underset{\underset{CH_3}{|}}{CH}-\overset{\overset{S}{|}}{CH}-CH_2-
\end{array}$$

Vulcanized rubber

Biosynthesis of Isoprenoids

One of the most interesting areas of research today is the discovery of the pathways—the specific reaction sequences—by which plants and animals synthesize the compounds that they need. We will have more to say about this in later chapters, but we can show now, in outline form, how some of those compounds composed of isoprene units are constructed in living organisms.

The basic building block for the synthesis of terpenes and terpenoids is 3-methyl-

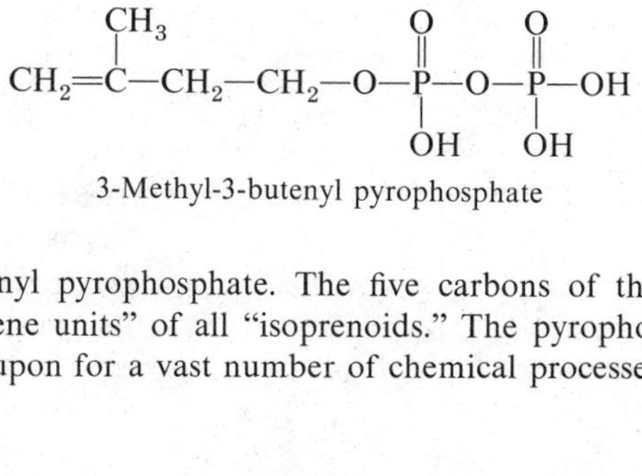

$$CH_2{=}\overset{CH_3}{\overset{|}{C}}-CH_2-CH_2-O-\overset{O}{\overset{\|}{\underset{OH}{\underset{|}{P}}}}-O-\overset{O}{\overset{\|}{\underset{OH}{\underset{|}{P}}}}-OH$$

3-Methyl-3-butenyl pyrophosphate

3-butenyl pyrophosphate. The five carbons of this compound are the source of the "isoprene units" of all "isoprenoids." The pyrophosphate group is a group that nature relies upon for a vast number of chemical processes. In the reactions that we will show below the pyrophosphate group functions as a natural "leaving group" in enzymatic reactions that resemble the alkylation reactions of alkenes which we studied in Section 7.6.

3-Methyl-3-butenyl pyrophosphate is isomerized by an enzyme to 3-methyl-2-butenyl pyrophosphate. The isomerization establishes an equilibrium that makes both compounds available to the cell.

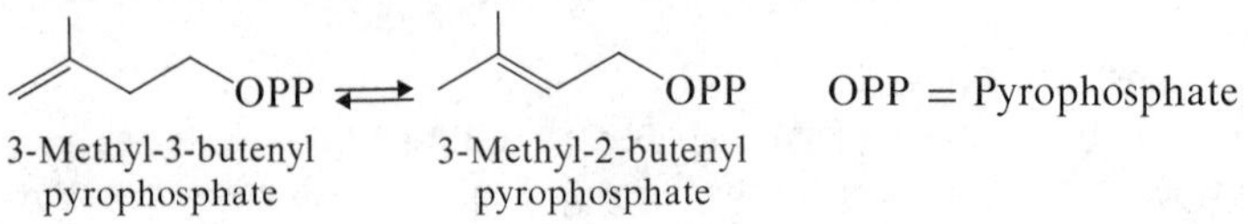

OPP ⇌ OPP   OPP = Pyrophosphate

3-Methyl-3-butenyl pyrophosphate   3-Methyl-2-butenyl pyrophosphate

These two 5-carbon compounds condense with each other in another enzymatic reaction to yield the 10-carbon compound, geranyl pyrophosphate. The first step involves the formation of an allylic cation.

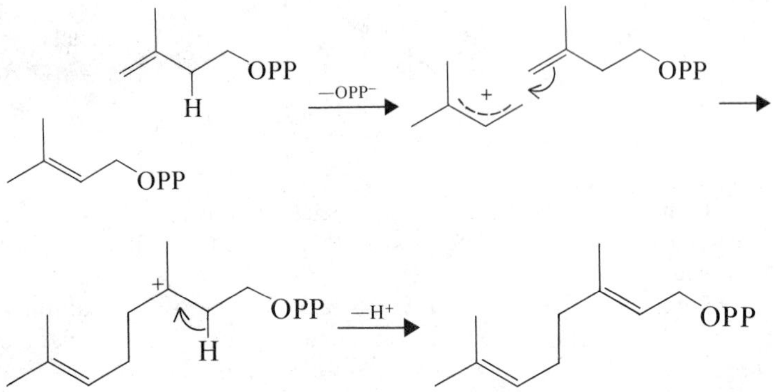

OPP H OPP —OPP⁻→ + OPP →

+ OPP H —H⁺→ OPP

Geranyl pyrophosphate

Geranyl pyrophosphate is the precursor of the monoterpenes; hydrolysis of geranyl pyrophosphate, for example, yields geraniol.

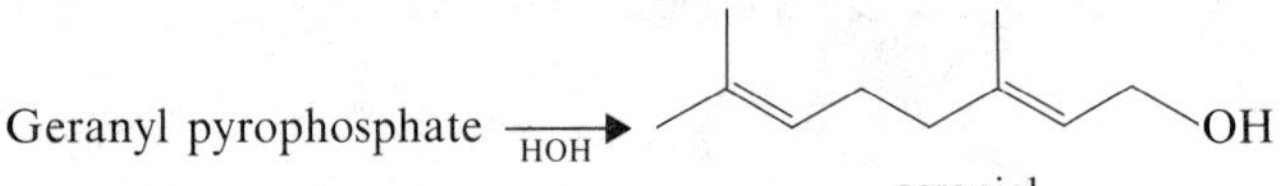

Geranyl pyrophosphate —HOH→ OH

geraniol

Geranyl pyrophosphate can also condense with 3-methyl-3-butenyl pyrophosphate to form the 15-carbon precursor for sesquiterpenes, farnesyl pyrophosphate.

OPP + H H OPP

Geranyl pyrophosphate

↓ −OPP⁻

OPP

Farnesyl pyrophosphate

OH

Farnesol

Other sesquiterpenes

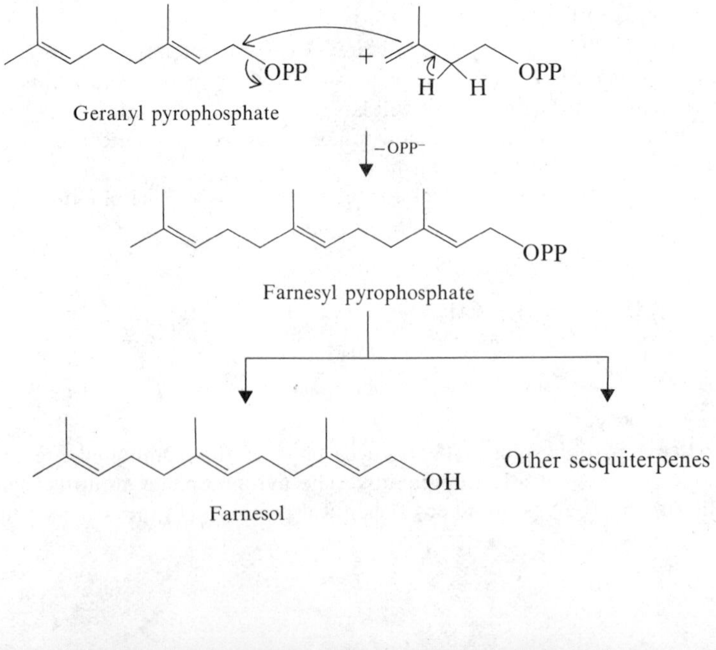

Farnesol has been isolated from ambrette seed oil. It has the odor of lily of the valley. Farnesol also functions as a hormone in insects and initiates the change from caterpillar to pupa to moth.

Similar condensation reactions yield the precursors for all of the other terpenes (Fig. E.1). In addition, a tail-to-tail reductive coupling of two molecules of farnesyl pyrophosphate produces squalene, the precursor for another important group of isoprenoids known as *steroids* (cf. Chapter 21 and Special Topic N).

Problem E.5

(a) Show how reductive tail-to-tail coupling of two molecules of farnesyl pyrophosphate would yield squalene.

Squalene

(b) Show how an oxidative tail-to-tail coupling of two of the C_{20}-pyrophosphates would yield a precursor for the carotenes.

Problem E.6

When farnesol is treated with sulfuric acid it is converted to bisabolene. Outline a possible mechanism for this reaction.

Farnesol $\xrightarrow{H_2SO_4}$

Bisabolene

FIG. E.1
The biosynthetic paths for terpenes and steroids

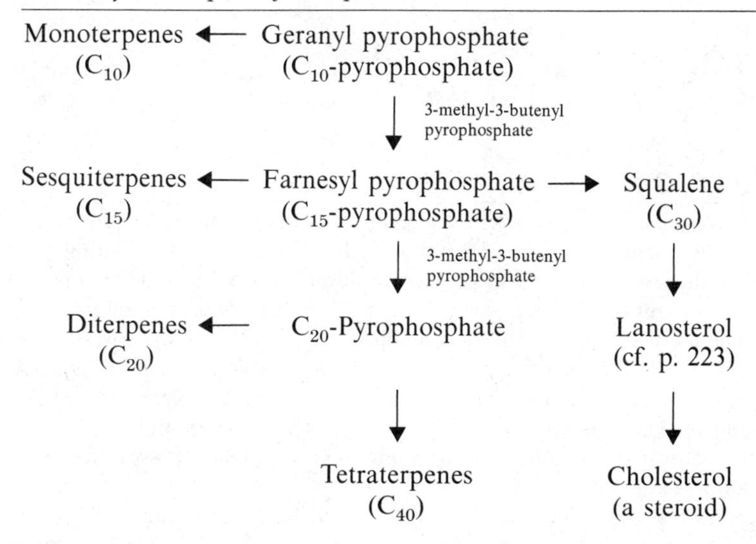

Problem E.7

When limonene (p. 423) is heated strongly, it yields two moles of isoprene. What kind of reaction is involved here?

* Problem E.8

α-Phellandrene and β-phellandrene are isomeric compounds that are minor constituents of spearmint oil; they have the molecular formula $C_{10}H_{16}$. Each compound has an ultraviolet absorption maximum in the 230–270 nm range. On catalytic hydrogenation each compound yields 1-isopropyl-4-methylcyclohexane. On vigorous oxidation with potassium permanganate, α-phellandrene yields $CH_3\overset{O}{\overset{\|}{C}}COOH$ and $CH_3\underset{CH_3}{\underset{|}{C}}HCH(COOH)CH_2COOH$. A similar oxidation of β-phellandrene yields $CH_3\underset{CH_3}{\underset{|}{C}}HCH(COOH)CH_2CH_2\overset{O}{\overset{\|}{C}}COOH$ as the only isolable product. Propose structures for α- and β-phellandrene.

* Problem E.9

The double bonds of vitamin A all have a *trans*-configuration. An isomer of vitamin A, called neovitamin A, that can be obtained from shark-liver oil has a 13-*cis*-double bond. Both compounds react with maleic anhydride at the 11,13-diene unit; however, neovitamin A reacts much more slowly. (The difference in their reactivities is so great that the Diels–Alder reaction can be used as an assay for neovitamin A in fish liver oils once the total concentration of vitamin A isomers is known.) Propose an explanation for the difference in reactivity of vitamin A and neovitamin A.

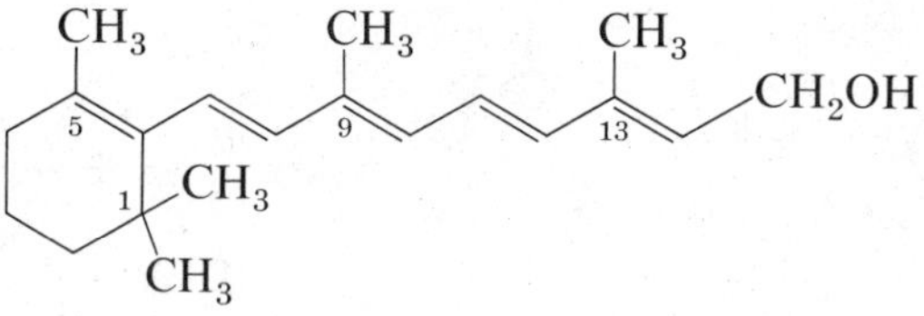

CH_3 CH_3 CH_3 CH_2OH 5 9 13 1 CH_3 CH_3

Vitamin A

F SPECIAL TOPIC

THE PHOTOCHEMISTRY OF VISION

The chemical changes that occur when light impinges on the retina of the eye involve several of the phenomena that we have studied in earlier chapters. Central to an understanding of the visual process at the molecular level are two phenomena in particular: the absorption of light by conjugated polyenes and the interconversion of *cis-trans* isomers.

The retina of the human eye contains two types of receptor cells. Because of their shapes, these cells have been named *rods* and *cones*. Rods are located primarily at the periphery of the retina and are responsible for vision in dim light. Rods, however, are color-blind and "see" only in shades of gray. Cones are found mainly in the center of the retina and are responsible for vision in bright light. Cones also possess the pigments that are responsible for color vision.

Some animals do not possess both rods and cones. The retinas of pigeons contain only cones. Thus, while pigeons have color vision, they see only in the bright light of day.

The retinas of owls, on the other hand, have only rods; owls see very well in dim light, but are color blind.

> A recent discovery about the vision of cats is of interest here. Considerable confusion had existed as to whether or not cats are color blind (as one might expect for a nocturnal predator). Experiments have shown, however, that cats have enough cones to distinguish colors easily if the object is large enough (about the size of a credit card). The report of these experiments concluded with the statement, that as far as cats are concerned, "apples are red, but cherries are gray."

The chemical changes that occur in rods are much better understood than those in cones. For this reason we will concern ourselves here with rod vision alone.

When light strikes rod cells, it is absorbed by a compound called rhodopsin. This initiates a series of chemical events that ultimately results in the transmission of a nerve impulse to the brain.

Our understanding of the chemical nature of rhodopsin and the conformational changes that occur when rhodopsin absorbs light have resulted largely from the research of Professor George Wald and his co-workers at Harvard University. Wald's research began in 1933 when he was a graduate student in Berlin; work with rhodopsin, however, began much earlier in other laboratories.

Rhodopsin was discovered in 1877 by the German physiologist Franz Boll. Boll noticed that the initial red-purple color of a pigment in the retina of frogs was "bleached" by the action of light. The bleaching process led first to a yellow retina and then to a colorless one. A year later, another German scientist, Willy Kuhne isolated the red-purple pigment and named it, because of its color, *sehpurpur* or "visual purple." The name visual purple is still commonly used for rhodopsin.

In 1952, Wald and one of his students, Ruth Hubbard, showed that the chromophore (light-absorbing group) of rhodopsin is the polyunsaturated aldehyde, 11-*cis*-retinal.

Rhodopsin is formed in the retina by a reaction between 11-*cis*-retinal and a protein called opsin (Fig. F.1). The reaction is between the aldehyde group of 11-*cis*-retinal and an amino group on the surface of the protein and involves the loss of a molecule of

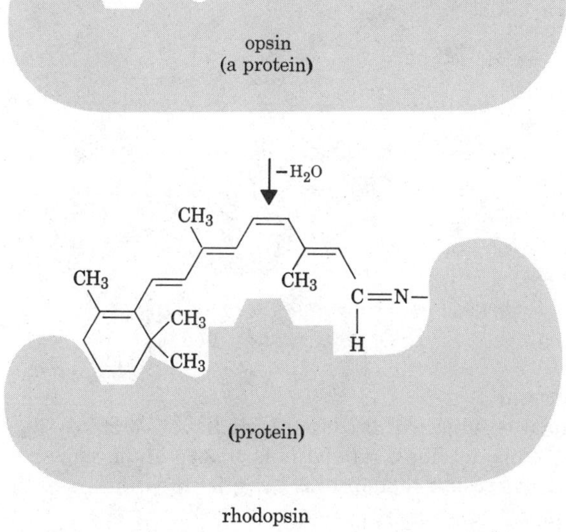

FIG. F.1
*The formation of rhodopsin from 11-*cis*-retinal and opsin.*

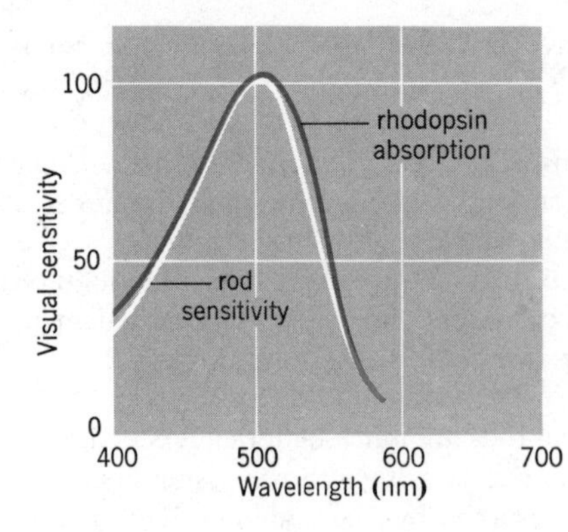

FIG. F.2
A comparison of the visible absorption spectrum of rhodopsin and the sensitivity curve for rod vision. [Adapted from S. Hecht, S. Shlaer, and M. H. Pirenne, J. Gen. Chem. Physiol., 25, *819 (1942).]*

water. Other secondary interactions involving —SH groups of the protein probably also hold the *cis*-retinal in place. The site on the surface of the protein is one on which *cis*-retinal fits precisely.

The conjugated polyunsaturated chain of 11-*cis*-retinal gives rhodopsin the ability to absorb light over a broad region of the visible spectrum. Figure F.2 shows the absorption curve of rhodopsin in the visible region and compares it with the sensitivity curve for human rod vision. The fact that these two curves coincide provides strong evidence that rhodopsin is the light-sensitive material in rod vision.

When rhodopsin absorbs a photon of light the visual process begins. Two very important phenomena accompany the absorption of the photon: a nerve impulse is generated, and 11-*cis*-retinal of rhodopsin is ultimately isomerized to the all-*trans* form of metarhodopsin II. The all-*trans* configuration of retinal does not fit the site on the surface of the protein and, because it does not, hydrolysis of the —CH=N— group occurs. Hydrolysis produces opsin and all-*trans*-retinal. These steps are illustrated in Fig. F.3.

Rhodopsin has an absorption maximum at 498 nm. This gives rhodopsin its red-purple color. Together, all-*trans*-retinal and opsin have an absorbance maximum at 387 nm and thus, are yellow. The light-initiated transformation of rhodopsin to all-*trans*-retinal and opsin corresponds to the initial bleaching that Boll observed in the retinas of frogs. Further bleaching to a colorless form occurs when all-*trans*-retinal is reduced enzymatically to all-*trans*-vitamin A. This reduction converts the aldehyde group of retinal to the primary alcohol of vitamin A.

all-*trans*-retinal

(H) ↓ enzyme

all-*trans*-vitamin A

Regeneration of Rhodopsin

If the retina of a live animal is subjected to constant irradiation, a steady-state concentration of rhodopsin in the retina develops: rhodopsin is created at the same rate that it is destroyed. This is important because rhodopsin is essential to vision.

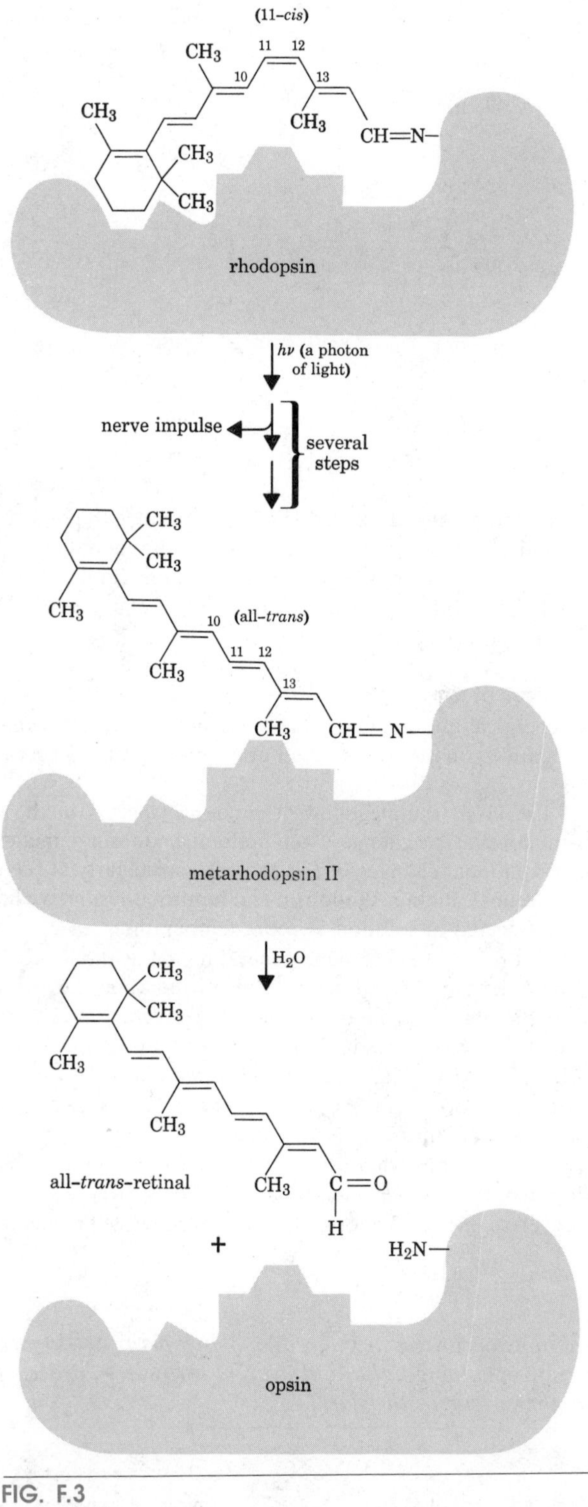

(11-*cis*)

rhodopsin

$h\nu$ (a photon of light)

nerve impulse

several steps

(all-*trans*)

metarhodopsin II

H_2O

all-*trans*-retinal

+

H_2N—

opsin

FIG. F.3

*The important chemical steps of the visual process. Absorption of a photon of light by the 11-*cis*-retinal portion of rhodopsin generates a nerve impulse and sets off an isomerization that leads, through a series of steps, to metarhodopsin II. Then hydrolysis of metarhodopsin II produces all-*trans*-retinal and opsin.*

TABLE F.1

COMPOUND	λ_{MAX}
Rhodopsin	498 nm
↓ $h\nu$	
Prelumirhodopsin	534 nm
↓	
Lumirhodopsin	500 nm
↓	
Metarhodopsin I	478 nm
↓	
Metarhodopsin II	380 nm

If an animal is placed in the dark for about 25 minutes a process of "dark adaptation" takes place and retinal rhodopsin reaches its maximum value.

Rhodopsin regeneration occurs in two important ways: one (in light) occurs in the retina itself; the other (in the dark) involves enzymes of the liver. Let us begin by describing how rhodopsin regeneration occurs in light.

Photoregeneration of Rhodopsin

Several different chemical intermediates that occur between rhodopsin and metarhodopsin II have been identified on the basis of their absorption spectra. These are shown in Table F.1.

Isomerization of the 11-*cis* double bond of rhodopsin occurs in the first step, rhodopsin ⟶ prelumirhodopsin. The changes from prelumirhodopsin to metarhodopsin II probably involve conformational changes of the protein, particularly at the site of its attachment to the all-*trans*-retinal. Prelumirhodopsin and lumirhodopsin have lifetimes so short that they cannot be detected in solution at room temperature.

Metarhodopsin I absorbs light in the same general region as rhodopsin itself, and there is evidence that when the retina is exposed to strong irradiation (i.e., when a large number of photons fall on the rod cells) metarhodopsin I is converted back to rhodopsin. This regeneration of rhodopsin from metarhodopsin I is a photochemical process, too. Metarhodopsin I absorbs a photon of light and its all-*trans* double bond structure is reconverted to the 11-*cis* form of rhodopsin. We can summarize the reactions that occur in the retina in the following way (Figure F.4).

The bleaching sequence begins when rhodopsin absorbs a photon of light and is converted to metarhodopsin I. If, however, metarhodopsin I absorbs a second photon of equal energy it can be converted back to rhodopsin before it is converted to metarhodopsin

FIG. F.4
A summary of the reactions occurring in the retina. Wavy arrows, ⇝, are used to represent reactions involving light; and straight arrows, →, are used to represent reactions that occur in the dark.

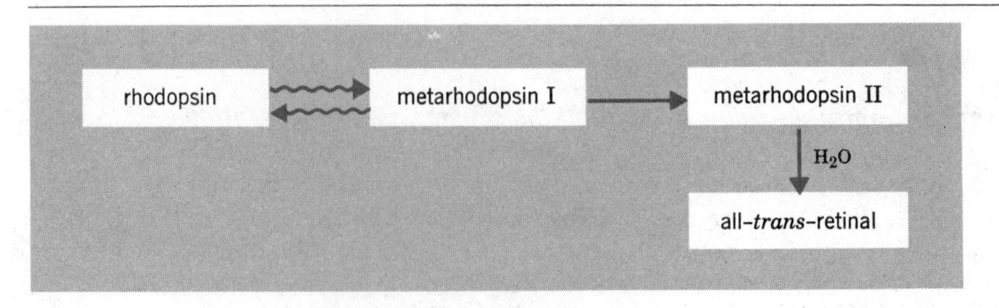

II. This is not likely to occur in dim light when the number of photons striking the retina is low. However, in bright light photoreversal bcomes important because the number of photons striking the retina is large, and the probability that the same molecule will absorb a second photon before it is further transformed is also large. Under these conditions the amount of rhodopsin being bleached reaches a limiting value.

The all-*trans*-retinal that is formed when metarhodopsin II is hydrolyzed can be isomerized to 11-*cis*-retinal by an enzyme in the retina. This enzymatic reaction also requires light but it requires light of shorter wavelength than that for the photoregeneration of rhodopsin from metarhodopsin I. Rhodopsin is resynthesized in the retina when the 11-*cis*-retinal recombines with opsin.

Regeneration of Rhodopsin in the Dark

The all-*trans* retinal that is not isomerized to 11-*cis*-retinal in the retina is reduced to all-*trans*-vitamin A by an enzyme, retinal reductase. The all-*trans*-vitamin A is then transported to the liver and the next step in the dark synthesis of rhodopsin occurs there.

Enzymes in the liver convert all-*trans*-vitamin A into 11-*cis*-vitamin A.

CH_3 CH_3 CH_3 CH_3 11 CH_2OH 10 12 CH_3

all-*trans*-vitamin A

↓

CH_3 CH_3 CH_3 11 12 10 CH_3 CH_3 CH_2OH

11-*cis*-vitamin A

Then, 11-*cis*-vitamin A is returned to the eye where it is reoxidized to 11-*cis*-retinal and used to synthesize rhodopsin.

The entire visual cycle is summarized in Fig. F.5.

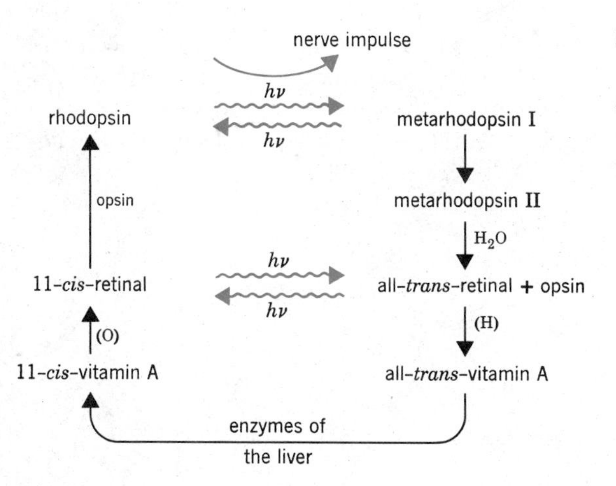

FIG. F.5
The visual cycle. (All of the photoreactions require light of a wavelength that can be absorbed by reacting molecules.)

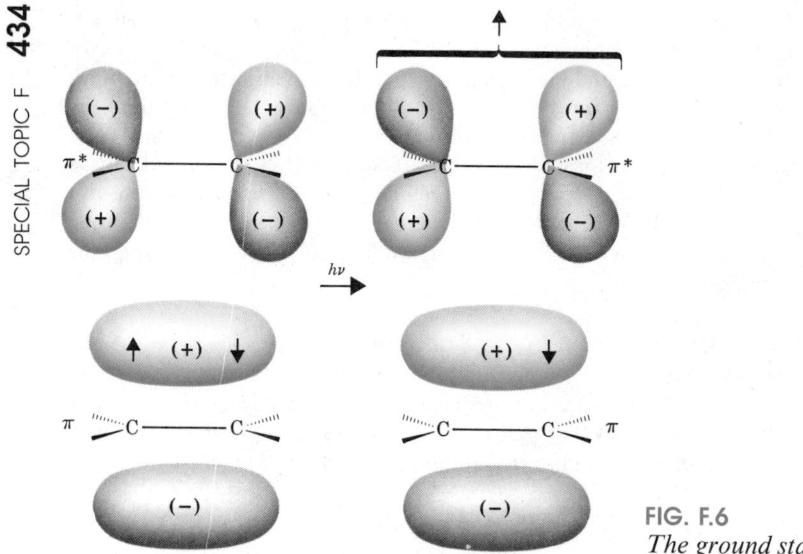

FIG. F.6
The ground state and an excited state of an alkene.

Photochemical Isomerization of Alkenes

The photochemical isomerizations of *cis* and *trans* alkenes that play such an important part in the visual process require extra comment. The barrier to rotation of groups joined by a carbon-carbon double bond is quite large. The energy of activation for the interconversion of 11-*cis* retinal and all-*trans*-retinal, for example, has been estimated to be approximately 25 kcal/mole. Yet, when rhodopsin absorbs light of the proper wavelength, the reaction occurs very rapidly.

We can understand how light absorption brings about the rapid interconversion of *cis-trans* isomers if we examine the molecular orbitals given in Fig. F.6.

In the ground state of an alkene, both electrons are in the bonding molecular orbital, π. In this bonding orbital the electrons are located in a region of space above and below but generally between the two carbon atoms. In the excited state, however, one electron is promoted to the antibonding orbital, π^*. The antibonding orbital has four lobes that are directed generally away from the area between the two carbon atoms. Thus in the excited state the carbon-carbon bond is much more like that of a single bond and has a very low barrier to rotation, as a result.

This does not mean, however, that in photoisomerization we have avoided the energy barrier. The energy for the isomerization is provided by the photon, and it is supplied just where it is needed: at the carbon-carbon double bond.

The Sensitivity of the Eye

The human eye is truly an amazing instrument. Although each rod contains at least 10 billion molecules of rhodopsin, it has been shown that the absorption of as few as *five* photons of light by five molecules of rhodopsin is detectable if the rhodopsin molecules are in different rods. Each rod apparently possesses an extraordinary mechanism for amplification of the nerve impulse generated by absorption of the photon.

11 AROMATIC COMPOUNDS I: THE PHENOMENON OF AROMATICITY

11.1 INTRODUCTION

In 1825, Michael Faraday, a scientist most often remembered for his work with electricity, isolated a new substance from a gas used at that time for illumination. This compound, now called *benzene,* was an example of a new class of organic substances called *aromatic compounds.*

In 1834, benzene was shown to have the empirical formula CH. Later its molecular formula was shown to be C_6H_6. This in itself was a surprising discovery. Benzene has *only* as many hydrogen atoms as it has carbon atoms! Most compounds that were then known had a far greater proportion of hydrogen atoms, usually twice as many. Benzene, with a formula of C_6H_6, or C_nH_{2n-6}, must consist of *highly unsaturated* molecules. It has an index of hydrogen deficiency equal to four. Thus even if it contains a ring, it must also contain multiple bonds. However, as we will see, benzene and other aromatic compounds are characterized by a tendency to undergo the substitution reactions characteristic of saturated compounds, rather than the addition reactions characteristic of unsaturated compounds.

During the latter part of the nineteenth century the Kekulé-Couper-Butlerov theory of valence was systematically applied to all known organic compounds. One result of this effort was the placing of organic compounds in either of two broad categories; compounds were classified as being either *aliphatic* or *aromatic*. To be classified as aliphatic meant then that the chemical behavior of a compound was "fatlike." (Now it means that the compound reacts like an alkane, an alkene, an alkyne, or one of their cyclic counterparts.) To be classified as aromatic meant then that the compound had a low hydrogen-to-carbon ratio and that it was "fragrant." Most of the early aromatic compounds were obtained from balsams, resins, or essential oils. Included among these were benzaldehyde (from oil of bitter almonds), benzoic acid and benzyl alcohol (from gum benzoin), and toluene (from tolu balsam).

Kekulé was the first to recognize that these early aromatic compounds all contain a six-carbon unit and that they retain this six-carbon unit through most chemical transformations and degradations. Benzene was eventually recognized as being the parent compound of this new series.

Since this new group of compounds proved to be distinctive in ways that are far more important than their odors, the term "aromatic" began to take on a purely chemical connotation. We will see in this chapter that the meaning of the term aromatic has evolved as chemists have learned more about the reactions and properties of aromatic compounds.

11.2 REACTIONS OF BENZENE

The benzene molecule is highly unsaturated, and because of this we might expect that it would react accordingly. We might expect, for example, that benzene would decolorize bromine in carbon tetrachloride by *adding* bromine; that it would decolorize aqueous potassium permanganate by being *oxidized;* that it would *add* hydrogen easily in the presence of a catalyst; and that it would *add* water in the presence of acids.

Benzene does none of these. When benzene is treated with bromine in carbon tetrachloride in the dark or with aqueous potassium permanganate or with dilute acids, no reaction occurs. Benzene does add hydrogen in the presence of finely divided nickel, but only at high temperatures and under high pressures.

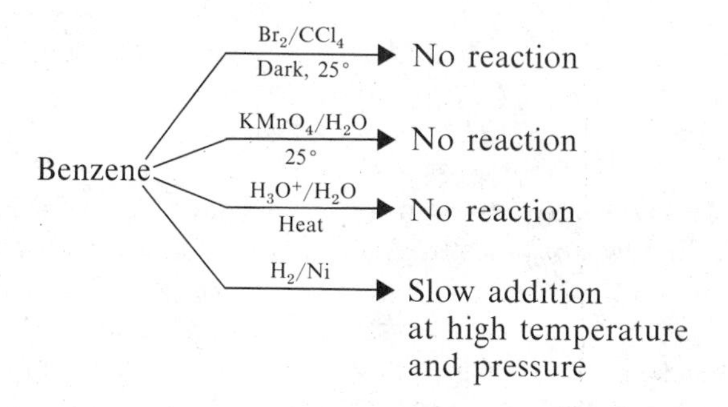

Benzene *does* react with bromine but only in the presence of a Lewis-acid catalyst such as ferric bromide. Most surprisingly, however, it reacts not by addition but by *substitution.*

Substitution: $C_6H_6 + Br_2 \xrightarrow{FeBr_3} C_6H_5Br + HBr$ Observed

Addition: $C_6H_6 + Br_2 \not\rightarrow C_6H_6Br_2 + C_6H_6Br_4 + C_6H_6Br_6$ Not observed

Benzene undergoes similar substitution reactions with chlorine in the presence of a ferric salt, with nitric acid in the presence of sulfuric acid, with SO_3 in the presence of sulfuric acid, and with alkyl halides in the presence of aluminum chloride.

$$C_6H_6 + Cl_2 \xrightarrow{FeCl_3} \underset{\text{Chlorobenzene}}{C_6H_5Cl} + HCl$$ Chlorination

$$C_6H_6 + HONO_2 \xrightarrow{H_2SO_4} \underset{\text{Nitrobenzene}}{C_6H_5NO_2} + H_2O$$ Nitration

$$C_6H_6 + SO_3 \xrightarrow{H_2SO_4} \underset{\text{Benzenesulfonic acid}}{C_6H_5SO_3H}$$ Sulfonation

$$C_6H_6 + CH_3Cl \xrightarrow{AlCl_3} \underset{\text{Toluene}}{C_6H_5CH_3} + HCl$$ Alkylation

When benzene reacts with bromine *only one monobromobenzene* is formed. That is, only one compound with the formula C_6H_5Br is found among the products. Similarly, when benzene is chlorinated *only one monochlorobenzene* results, and when benzene is nitrated *only one mononitrobenzene* is formed. The same kind of behavior is true of the sulfonation and alkylation reactions of benzene.

Two possible explanations can be given for these observations. The first is that only one of the six hydrogens in benzene is reactive toward these reagents. The second is that all six hydrogens in benzene are equivalent , and replacing any one of them with a substituent results in the same product. As we will see, the second explanation is correct.

Problem 11.1

Listed below are several compounds that have the formula C_6H_6. (a) For which of these compounds, if any, would a substitution of bromine for hydrogen yield only one *mono*bromo product? (b) Which of these compounds would you expect to react with bromine by substitution alone?

$$\underset{(a)}{H-C\equiv C-CH_2CH_2-C\equiv C-H} \qquad \underset{(d)}{H-C\equiv C-CH_2-C\equiv C-CH_3}$$

$$\underset{(b)}{CH_2=CH-CH=CH-C\equiv CH} \qquad \underset{(e)}{CH_2=CH-C\equiv C-CH=CH_2}$$

$$\underset{(c)}{CH_3CH_2-C\equiv C-C\equiv C-H}$$

11.3 THE KEKULÉ STRUCTURE FOR BENZENE

In 1865, August Kekulé proposed a structure for benzene. Kekulé suggested that the carbon atoms of benzene are in a ring, that they are bonded to each other by alternating single and double bonds, and that one hydrogen atom is attached to each carbon. This structure satisfied the requirements that carbon atoms form four bonds, and that all of the hydrogen atoms of benzene are equivalent.

H H C C H H C C H H C C or

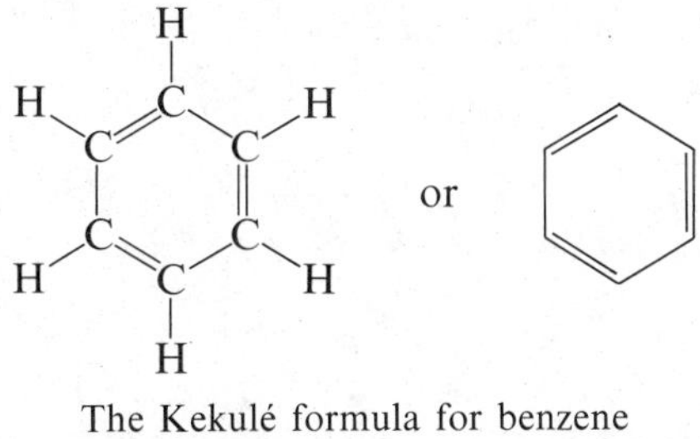

The Kekulé formula for benzene

A problem soon arose with the Kekulé structure, however. The Kekulé structure predicts that there should be two different 1,2-dibromobenzenes. In one of these hypothetical compounds (below), the carbons that bear the bromines are separated by a single bond, and in the other they are separated by a double bond. *Only one 1,2-dibromobenzene, however, has ever been found.*

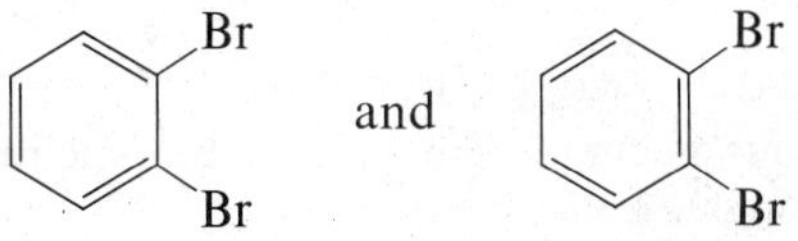

Br Br and Br Br

In order to accommodate this objection, Kekulé proposed that the two forms of benzene (and of benzene derivatives) are in a state of equilibrium, and that this equilibrium is so rapidly established that it prevents isolation of the separate compounds.

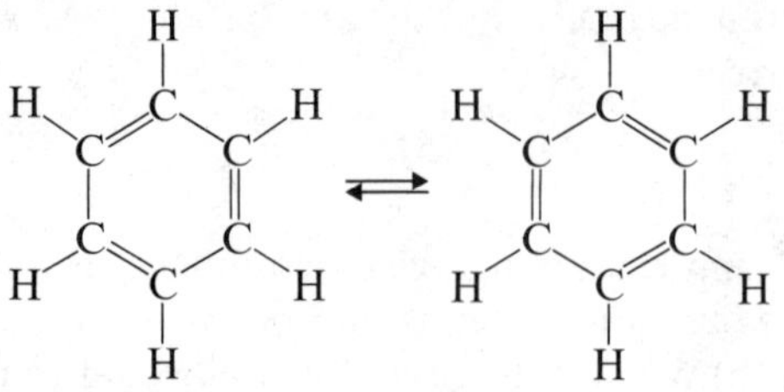

Thus, the two 1,2-dibromobenzenes would also be rapidly equilibrated, and this would explain why chemists had not been able to isolate the two forms.

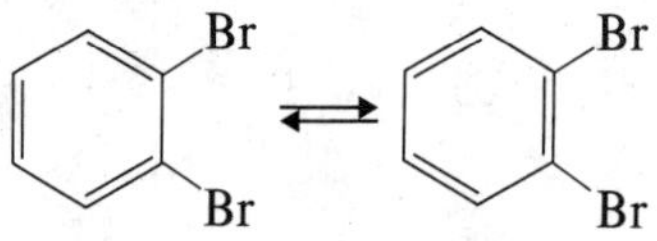

We now know that this proposal was wrong and that *no such equilibrium exists*. Nonetheless, the Kekulé formulation of benzene's structure was an important step forward and, for very practical reasons, it is still used today. We understand its meaning differently, however.

With the Kekulé formula we are able to reformulate the substitution reactions of benzene in a clearer way.

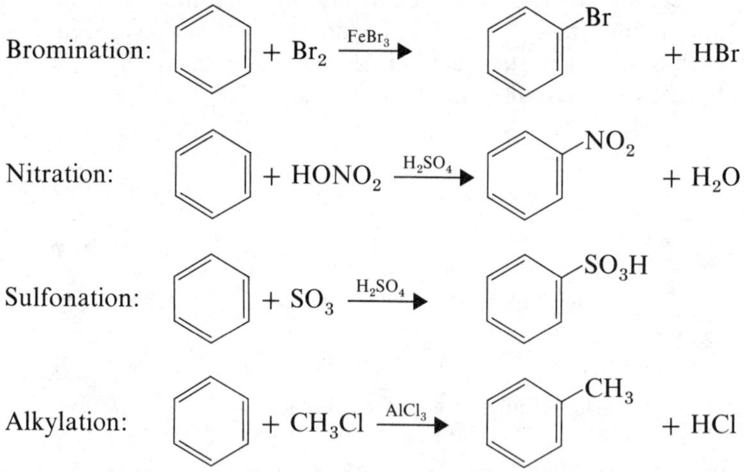

Bromination: $C_6H_6 + Br_2 \xrightarrow{FeBr_3} C_6H_5Br + HBr$

Nitration: $C_6H_6 + HONO_2 \xrightarrow{H_2SO_4} C_6H_5NO_2 + H_2O$

Sulfonation: $C_6H_6 + SO_3 \xrightarrow{H_2SO_4} C_6H_5SO_3H$

Alkylation: $C_6H_6 + CH_3Cl \xrightarrow{AlCl_3} C_6H_5CH_3 + HCl$

In each of these substitution reactions the benzene ring with its alternating single and double bonds is retained in the product.* This would not have been true if benzene had reacted by addition. Had an addition reaction occurred, the ring of alternating single and double bonds would no longer exist in the products. Addition of bromine, for example, would have resulted in the following reaction.

* As we will see in Section 11.5, our modern formula for the benzene ring does not have "alternating single bonds and double bonds."

(by 1,4-addition)

The tendency of benzene to react by substitution rather than addition gave rise to another concept of aromaticity. Since benzene reacts by substitution, and thereby retains its ring of alternating single and double bonds, aromatic compounds were seen as being compounds whose molecules retain their cyclic, unsaturated structures in most of their reactions. For a compound to be called aromatic meant, experimentally, that it gave substitution reactions rather than addition reactions even though it was highly unsaturated.

Before 1900, chemists assumed that the ring of alternating single and double bonds was the structural feature that gave rise to the aromatic properties. Since benzene and benzene derivatives (i.e., compounds with six-membered rings) were the only aromatic compounds known, chemists naturally sought other examples. The compound cyclooctatetraene seemed to be a likely candidate.

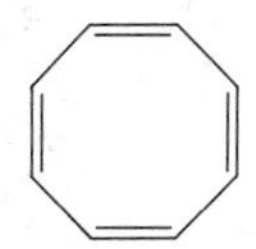

Cyclooctatetraene

In 1911, through a brilliant synthesis that we will discuss later, Richard Willstätter succeeded in synthesizing cyclooctatetraene. Willstätter found, however, that it is not at all like benzene. Cyclooctatetraene reacts with bromine by addition, it adds hydrogen readily, it decolorizes solutions of potassium permanganate, and thus it is clearly *not aromatic*. While these findings must have been a keen disappointment to Willstätter, they were very significant for what they did not prove. Chemists, as a result, had to look deeper to discover the origin of benzene's aromaticity.

11.4 THE STABILITY OF BENZENE

We have seen that benzene shows unusual behavior by undergoing substitution reactions when, on the basis of its Kekulé structure, we should expect it to undergo addition. Benzene is unusual in another sense: it is *more stable* than the Kekulé structure suggests. To see how, consider the following thermochemical results.

Cyclohexene, a six-membered ring containing one double bond, can be hydrogenated easily to cyclohexane. When the $\Delta H°$ for this reaction is measured it is found to be -28.6 kcal/mole, very much like that of any alkene.

Cyclohexene $+ H_2 \xrightarrow{Pt}$ Cyclohexane $\quad \Delta H° = -28.6$ kcal/mole

We would expect that hydrogenation of 1,3-cyclohexadiene would liberate roughly twice as much heat and thus have a $\Delta H°$ equal to about -57.2 kcal/mole. When this experiment is done, the result is a $\Delta H° = -55.4$ kcal/mole. This result is quite close to what we calculated, and the difference can be explained by taking into account the fact that compounds containing conjugated double bonds are usually somewhat more stable than those that contain isolated double bonds.

1,3-Cyclohexadiene $+ 2H_2 \xrightarrow{Pt}$ Cyclohexane

Calculated
$\Delta H° = (2 \times -28.6) = -57.2$ kcal/mole

Observed
$\Delta H° = -55.4$

If we extend this kind of thinking, and if benzene is simply 1,3,5-cyclohexatriene, we would predict that benzene would liberate approximately 85.8 kcal/mole (3×-28.6) when it is hydrogenated. When the experiment is actually done the result is surprisingly different. The reaction is exothermic, but only by 49.8 kcal/mole.

Benzene $+ 3H_2 \xrightarrow{Pt}$ Cyclohexane

Calculated	$\Delta H° = (3 \times -28.6)$	$= -85.8$ kcal/mole
Observed	$\Delta H°$	$= -49.8$
Difference		$=$ 36.0 kcal/mole

When these results are represented in Fig. 11.1 it becomes clear that benzene is much more stable than we calculated it to be. Indeed, it is more stable

FIG. 11.1
Relative stabilities of cyclohexene, 1,3-cyclohexadiene, 1,3,5-cyclohexatriene (hypothetically), and benzene.

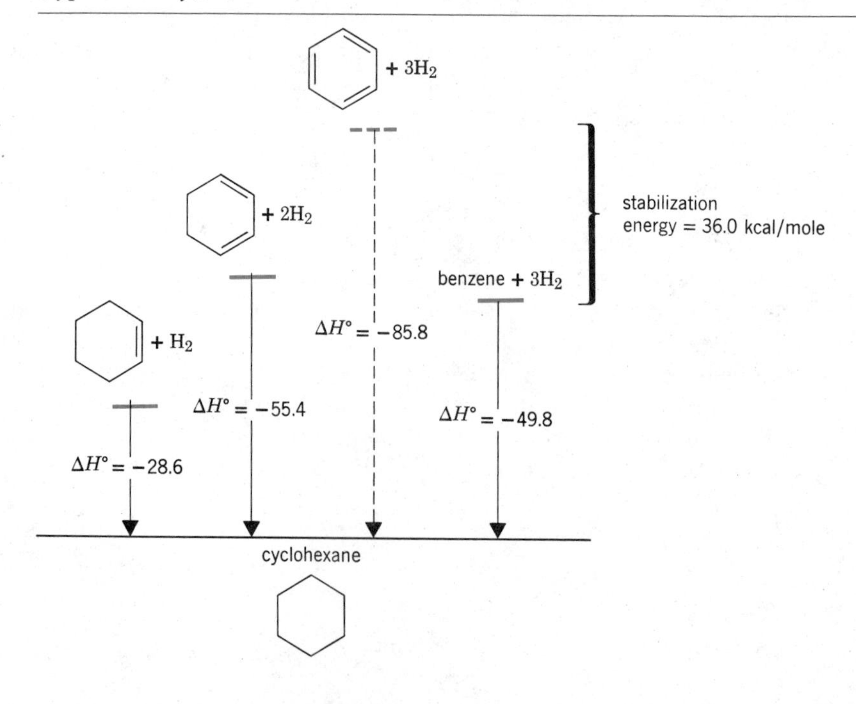

than the hypothetical 1,3,5-cyclohexatriene by 36 kcal/mole. This difference between the amount of heat actually released and that calculated on the basis of the Kekulé structure is now called the *stabilization energy* of the compound. Other terms that are used to describe this energy difference are *resonance energy* or *delocalization energy.*

11.5 MODERN THEORIES OF THE STRUCTURE OF BENZENE

It was not until the development of quantum mechanics in the 1920s that the unusual behavior and stability of benzene began to be understood. Quantum mechanics, as we have seen, produced two ways of viewing bonds in molecules: resonance theory and molecular orbital theory. We now look at both of these as they apply to benzene.

11.5A The Resonance Explanation of the Structure of Benzene

A basic postulate of resonance theory (Sections 1.8 and 10.10) is that whenever two or more structures can be written for a molecule *differing only in the positions of the electrons,* none of the structures will be in complete accord with the compound's chemical and physical properties. If we recognize this, we can now understand the true nature of the two Kekulé structures (I and II) for benzene. The two Kekulé structures differ only in the positions of the electrons. Structures I and II, then, do not represent two separate molecules in equilibrium as Kekulé had proposed. Instead, they are the closest we can get to a structure for benzene within the limitations of its molecular formula, the classical rules of valence, and the fact that the six hydrogens are chemically equivalent. That they are false structures is not the fault of nature; it is the fault of earlier theories of molecular structure. Resonance theory, fortunately, does not stop with telling us when to expect this kind of trouble; it also gives us a way out. Resonance theory tells us to use structures I and II as resonance contributors to a picture of the real molecule of benzene. As such I and II should be connected with a double-headed arrow and not with two separate ones (because we must reserve the symbol of two separate arrows for chemical equilibria). Resonance contributors, we emphasize again, are not in equilibrium. They are not structures of real molecules. They are the closest we can get if we are bound by simple rules of valence, and they are very useful in helping us visualize the actual molecule as a hybrid.

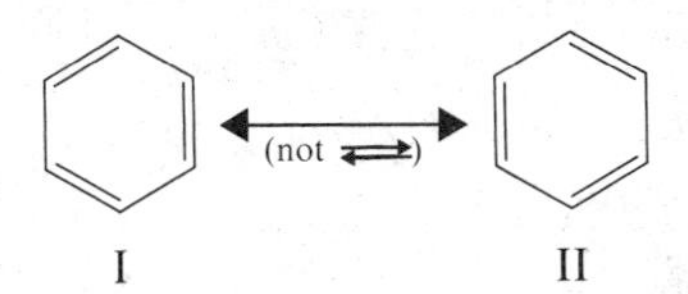

Look at the structures carefully. All of the single bonds in structure I are double bonds in structure II. If we blend I and II, that is, if we fashion a hybrid of them, then the carbon-carbon bonds in benzene are neither single bonds nor double bonds. Rather, they have a bond order between that of a single bond and that of a double bond. This is exactly what we find experimentally. Spectroscopic measurements show that molecules of benzene are planar and that all of its carbon-carbon bonds are of equal length. Moreover, the carbon-carbon bond lengths in benzene (Fig. 11.2) are 1.39 Å, a value in between that for a carbon-carbon single bond between sp^2-hybridized atoms (1.47 Å) (cf. Table 10.1) and that for a carbon-carbon double bond (1.33 Å).

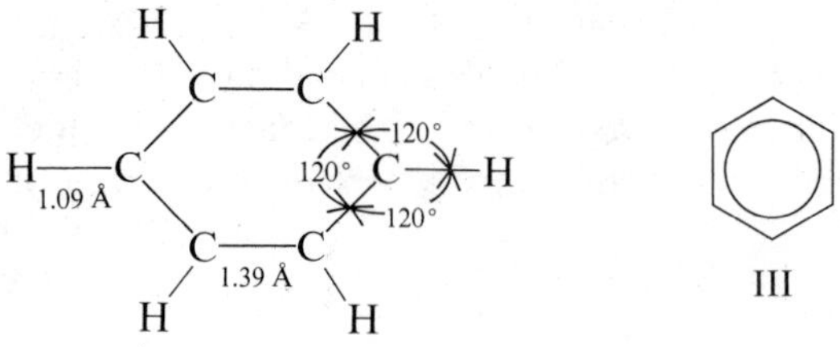

H—C 1.09 Å; C—C 1.39 Å; 120°, 120°, 120°

III

FIG. 11.2
Bond lengths and angles in benzene.

The hybrid structure is represented by inscribing a circle in the hexagon, and it is this new formula, III, that is most often used for benzene today. There are times, however, when an accounting of the electrons must be made, and for these purposes we may use one or the other of the Kekulé structures. We do this simply because the electron count in a Kekulé structure is obvious, while the number of electrons represented by a circle or portion of a circle is ambiguous.

Problem 11.2

If benzene were 1,3,5-cyclohexatriene the carbon-carbon bonds would be alternately long and short as indicated in the structures below. However, to consider the structures below as resonance contributors (or to connect them by a double-headed arrow) violates a basic principle of resonance theory. Explain.

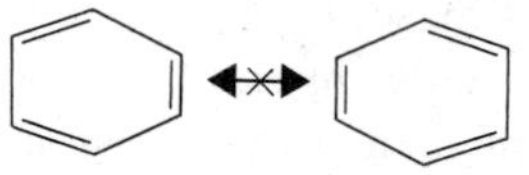

Resonance theory (Section 10.9) also tells us that whenever equivalent resonance structures can be drawn for a molecule the molecule (or hybrid) is much more stable than any of the resonance structures could be individually if they could exist. In this way resonance theory accounts for the much greater stability of benzene when compared to the hypothetical 1,3,5-cyclohexatriene. For this reason the stabilization energy is sometimes called the *resonance energy.*

11.5B The Molecular Orbital Explanation of the Structure of Benzene

The fact that the bond angles of the carbon atoms in the benzene ring are 120° strongly suggests that the carbon atoms are sp^2 *hybridized.* If we accept this and construct a planar six-membered ring from sp^2 carbon atoms as shown in Fig. 11.3, another picture of benzene begins to emerge.

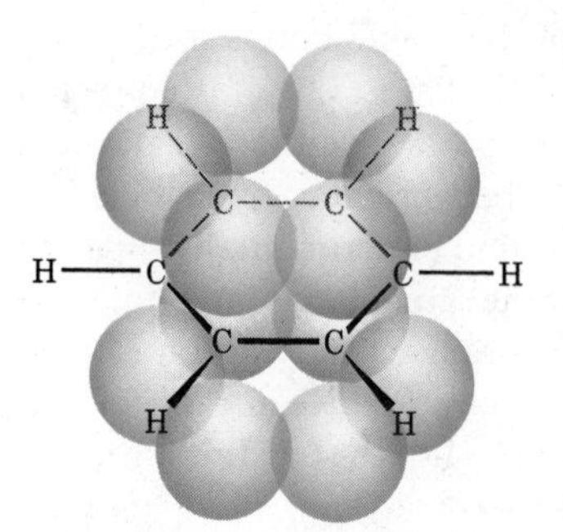

FIG. 11.3
Overlapping p *orbitals in benzene.*

Since the carbon-carbon bond lengths are all 1.39 Å, the *p* orbitals are close enough to overlap effectively. The *p* orbitals overlap equally all around the ring.

According to molecular orbital theory, the six overlapping *p* orbitals combine to form a set of six π molecular orbitals. Molecular orbital theory also allows us to calculate the relative energies of the π-molecular orbitals. These calculations are beyond the scope of our discussion but the energy levels are shown in Fig. 11.4.

Molecular orbitals, as we have seen, can accommodate two electrons if their spins are opposed. Thus, the electronic structure of the ground state of benzene is obtained by adding the six electrons to the π molecular orbitals starting with those of lowest energy as shown in Fig. 11.4. Notice that in benzene, all of the bonding orbitals are filled, all of the electrons have their spins paired, and there are no electrons in antibonding orbitals. Benzene is, thus, said to have a *closed bonding shell* of delocalized π electrons. This closed bonding shell accounts, in part, for the stability of benzene.

The shapes of the π molecular orbitals of benzene are given in Fig. 11.5.

Further insight into the unusual stability of benzene can be obtained by inspecting the lowest energy molecular orbital, ψ_1 (Fig. 11.5). In quantum mechanics electrons are described in terms of their *wave nature* and the energy of an electron is inversely proportional to its wavelength (Special Topic A). That is, *the longer the wavelength* of the electron; *the lower is its energy*. Molecular orbital ψ_1 encompasses the entire ring of carbon atoms and has no nodes between carbon atoms; the π electrons that occupy ψ_1 can move about this entire distance. Thus, these electrons can have a wavelength much longer than electrons in the molecular orbital of an isolated double bond (where the motion of the electrons is restricted to the vicinity of only two atoms). As a result, the electrons in the lowest π molecular orbital of benzene have unusually low energies, and this factor also helps us account for the exceptional stability of benzene.

11.6 HUCKEL'S RULE

In 1931, the German physicist E. Hückel carried out a series of mathematical calculations based on the kind of theory that we have just described. Hückel concerned himself with the general situation of *coplanar monocyclic* rings in which

FIG. 11.4
The energy levels of the π molecular orbitals of benzene. Energy levels below the dotted line are bonding while those above the dotted line are antibonding.

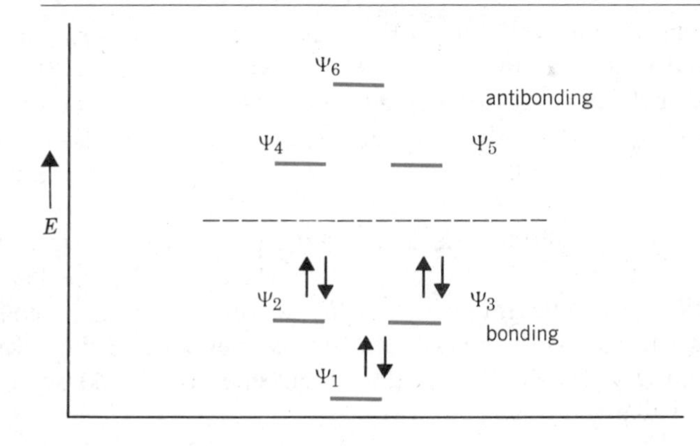

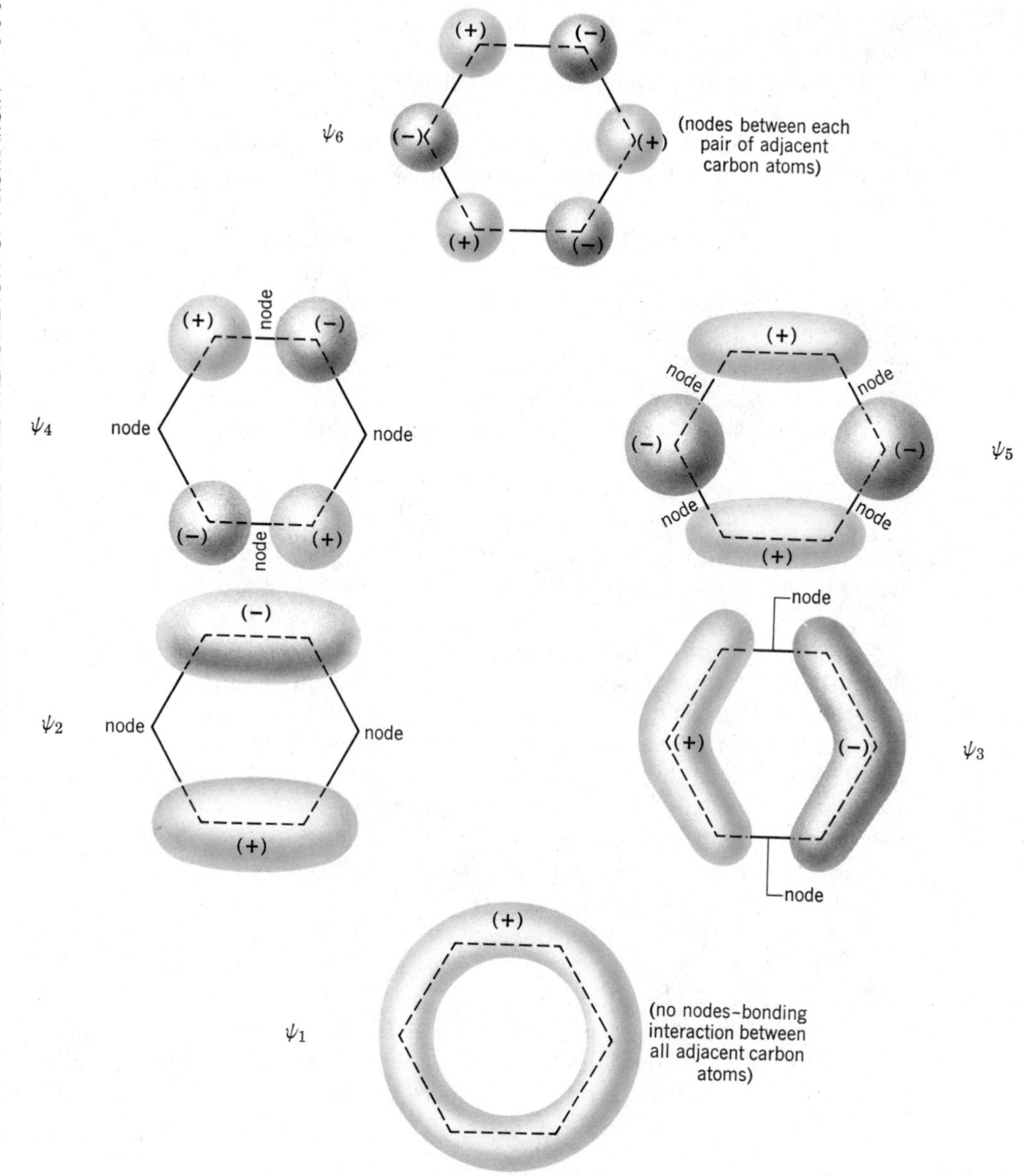

FIG. 11.5
Shapes of the π molecular orbitals of benzene as viewed from above.

each atom of the ring has an available p orbital as in benzene. His calculations indicate that coplanar rings containing $(4n + 2)$ π electrons, where $n = 0, 1, 2, 3$ . . . etc., have closed shells of delocalized electrons like benzene, and should have substantial stabilization (or delocalization) energies. In other words, *coplanar monocyclic rings with 2, 6, 10, 14, 18, and 22 delocalized π electrons should be aromatic.*

The fact that cyclooctatetraene lacks aromatic properties correlates with Hückel's rule. Cyclooctatetraene has a total of 8 π electrons. On the basis of Hückel's calculations, monocyclic rings with 8 π electrons are not predicted to have a closed shell of π electrons or substantial stabilization energies and they are not expected to be aromatic. A molecule of cyclooctatetraene, we now know, is not planar, but is shaped like a tub (Fig. 11.6).

FIG. 11.6
The actual (tub) shape of cyclooctatetraene.

The bonds of cyclooctatetraene are alternately long and short; electron diffraction crystallographic studies indicate that they are 1.50 Å and 1.35 Å, respectively. These values are those of single bonds and double bonds, and are not at all like the hybrid bonds of benzene. Cyclooctatetraene is then, clearly, a simple cyclic polyene.

11.6A The Annulenes

The name annulene has been proposed as a general name for monocyclic compounds that can be represented by structures having alternating single and double bonds. The ring size of an annulene is indicated by a number in brackets. Thus, benzene is [6] annulene and cyclooctatetraene is [8] annulene.* Hückel's rule

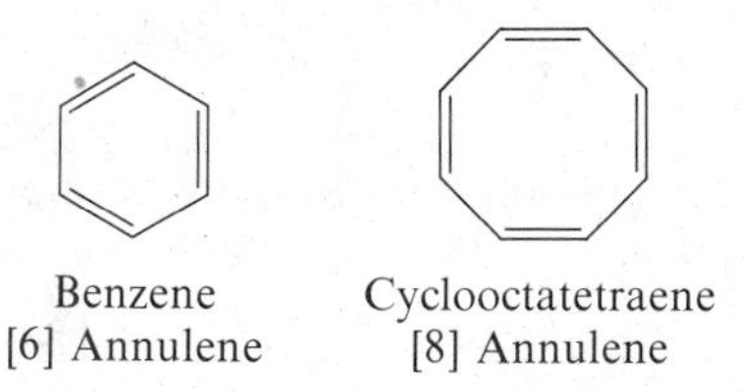

predicts that annulenes will be aromatic, provided their molecules have $(4n + 2)\ \pi$ electrons and have a planar carbon skeleton.

Before 1960, the only annulenes that were available to test Hückel's predictions were benzene and cyclooctatetraene. During the 1960s, and largely as a result of research by Professor Franz Sondheimer (now at University College, London), annulenes up to [24] annulene were synthesized, and the predictions of Hückel's rule were verified in every instance.†

Sondheimer's syntheses of large-ring annulenes involve reactions that we have seen before. Consider, as an example, his synthesis of [18] annulene.

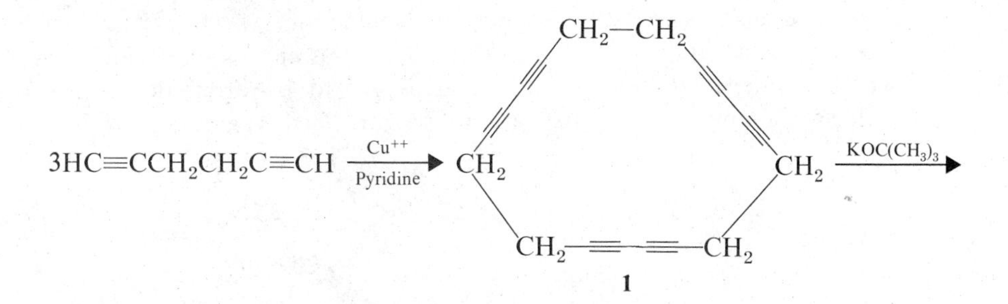

* These names are seldom used for benzene and cyclooctatetraene. They are often used, however, for conjugated rings of 10 or more carbon atoms.

† A [30] annulene has also been synthesized. However, calculations by Professor M. J. S. Dewar of the University of Texas indicate that aromatic stabilization of monocyclic rings with $(4n + 2)\ \pi$ electrons gradually decreases as the rings get larger. In other words, Hückel's rule has an upper limit lying between [22] annulene and [26] annulene. Thus, even though this [30] annulene obeys Hückel's rule ($4n + 2$, where $n = 7$), it is above this limit. The [30] annulene is not aromatic.

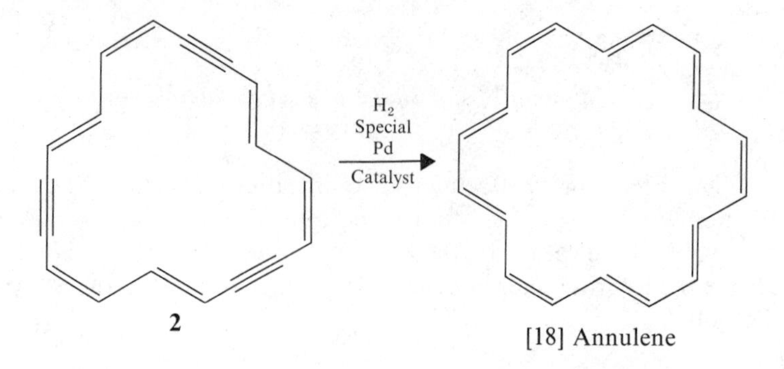

In the first step, oxidative coupling of 1,5-hexadiyne with cupric ion in pyridine (Section 9.13) gave the 18-membered cyclic "trimer" **1** in ~6% yield. When **1** was treated with potassium *tert*-butoxide, the more stable system, tridehydro- [18] annulene* **2** was produced. In the last step, hydrogenation of the triple bonds of **2** using a specially prepared palladium catalyst (Section 9.7) gave [18] annulene.

Examples of [14], [16], [20], [22], and [24] annulenes have been synthesized by similar methods. Of these, *as Hückel's rule predicts,* the [14], [18], and [22] annulenes ($4n + 2$ when $n = 3, 4, 5$ respectively) have been found to be aromatic. The [16] annulene and the [24] annulene are not aromatic. They are $4n$ compounds, not $4n + 2$ compounds.

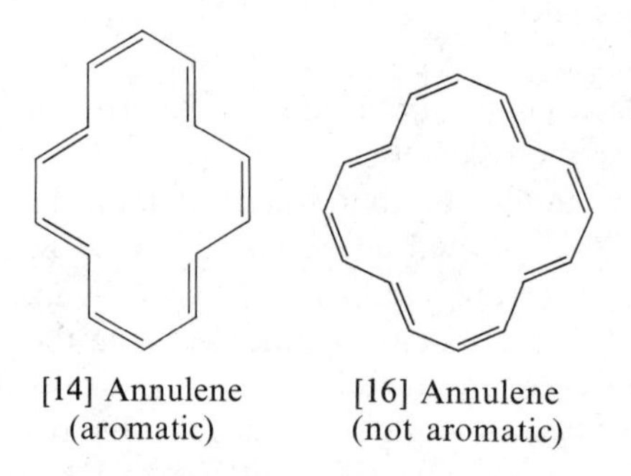

Examples of [10] and [12] annulenes have been synthesized by other methods and none are aromatic. We would not expect [12] annulenes to be aromatic since they have 12 π electrons and, thus, do not obey Hückel's rule. The [10] annulenes (below) would be expected to be aromatic on the basis of electron count, but their rings are not planar. The [10] annulene, **4**, has two *trans*

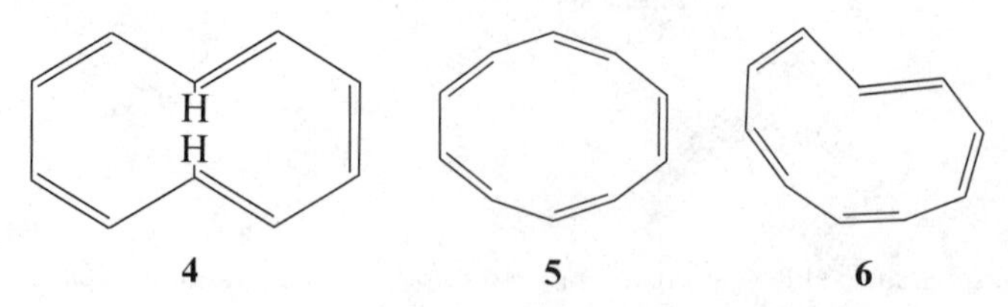

[10] Annulenes
(none are aromatic because none are planar)

double bonds. The carbons of its ring are prevented from becoming coplanar

* The prefix *tridehydro* indicates that the annulene has three triple bonds.

because the two hydrogens in the center of the ring interfere with each other. The [10] annulene with all *cis* double bonds, **5**, and the [10] annulene with one *trans* double bond, **6**, are not planar because of ring strain.

After many unsuccessful attempts [4] annulene (or cyclobutadiene) has also been synthesized. As we would expect, it is a highly reactive compound and *it is not aromatic.*

Cyclobutadiene
or [4] annulene
(not aromatic)

11.6B Aromatic Ions

In addition to the neutral molecules that we have discussed so far, there are a number of monocyclic species that bear either a positive or negative charge. Some of these ions show unexpected stabilities that suggest that they, too, are aromatic. Hückel's rule is helpful in accounting for the properties of these ions as well. We will consider two examples: the cyclopentadienyl anion and the cycloheptatrienyl cation.

Cyclopentadiene is not aromatic; however, it is unusually acidic for a hydrocarbon. (The K_a for cyclopentadiene is 10^{-15} and, by contrast, the K_a for cycloheptatriene is 10^{-36}.) Because of its acidity, cyclopentadiene can be converted to its anion by treatment with moderately strong bases. The cyclopentadienyl anion, moreover, is unusually stable and nuclear magnetic resonance (nmr) spectroscopy (Chapter 13) shows that all five hydrogens in the cyclopentadienyl anion are equivalent.

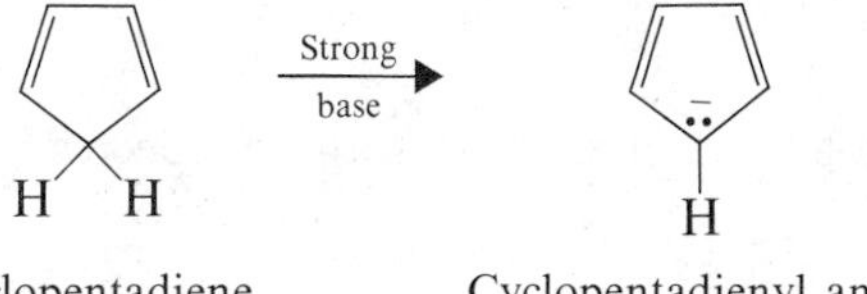

Cyclopentadiene Cyclopentadienyl anion

The orbital structure of cyclopentadiene (Fig. 11.7*a*) shows why cyclopentadiene, itself, is not aromatic. Not only does it not have the proper number of π electrons, but the π electrons cannot be delocalized about the entire ring because of the intervening sp^3-hybridized $—CH_2—$ group with no available *p* orbital.

On the other hand, if the $—CH_2—$ carbon atom becomes sp^2 hybridized after it loses a proton (Fig. 11.7*b*), the two electrons left behind can occupy the new *p* orbital that is produced. Moreover, this new *p* orbital can overlap with the *p* orbitals on either side of it and give rise to a ring with *six* delocalized π electrons. Since the electrons are delocalized, all of the hydrogens are equivalent and this agrees with what nmr spectroscopy tells us.

Six is, of course, a Hückel number ($4n + 2$, where $n = 1$), and the cyclopentadienyl anion is, in fact, an *aromatic anion.* The unusual acidity of cyclopentadiene is a result of the unusual stability of its anion.

Problem 11.3

(a) Write all of the resonance structures for the cyclopentadienyl anion. (b) Can the equivalence of the five hydrogen atoms be explained by resonance?

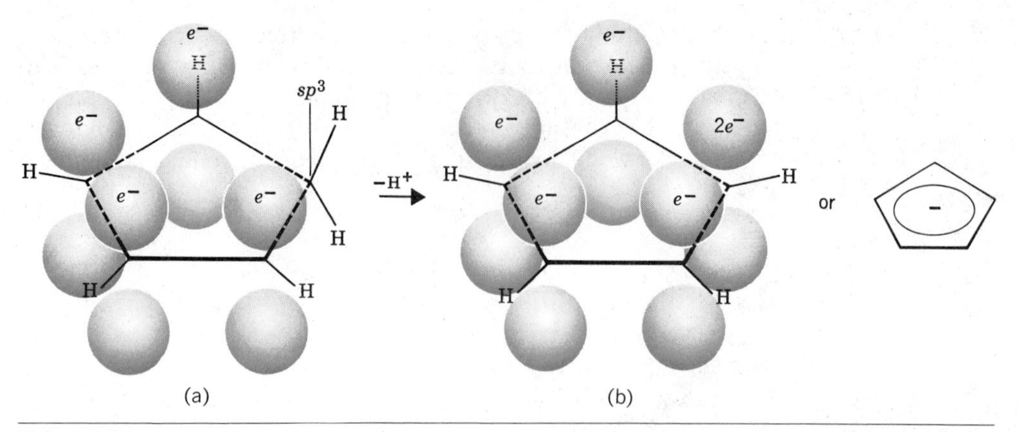

FIG. 11.7
(a) p *Orbitals of cyclopentadiene and* (b) *the* p *orbitals of the cyclopentadienyl anion.*

Cycloheptatriene below has six π electrons. However, the six π electrons of cycloheptatriene cannot be fully delocalized because of the presence of the $—CH_2—$ group—a group that does not have an available *p* orbital (Fig. 11.8).

When cycloheptatriene is treated with a reagent that can abstract a hydride ion, it is converted to the cycloheptatrienyl (or tropylium) cation. The loss of a

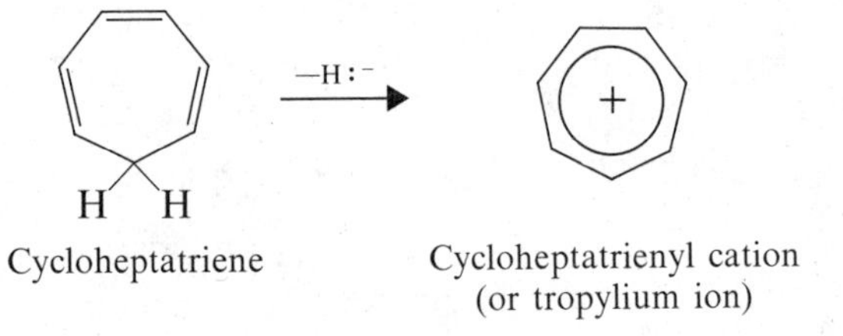

hydride ion from cycloheptatriene occurs with unexpected ease, and the cyclo-

FIG. 11.8
p *Orbitals of cycloheptatriene* (a) *and the cycloheptatrienyl* (*tropylium*) *cation* (b).

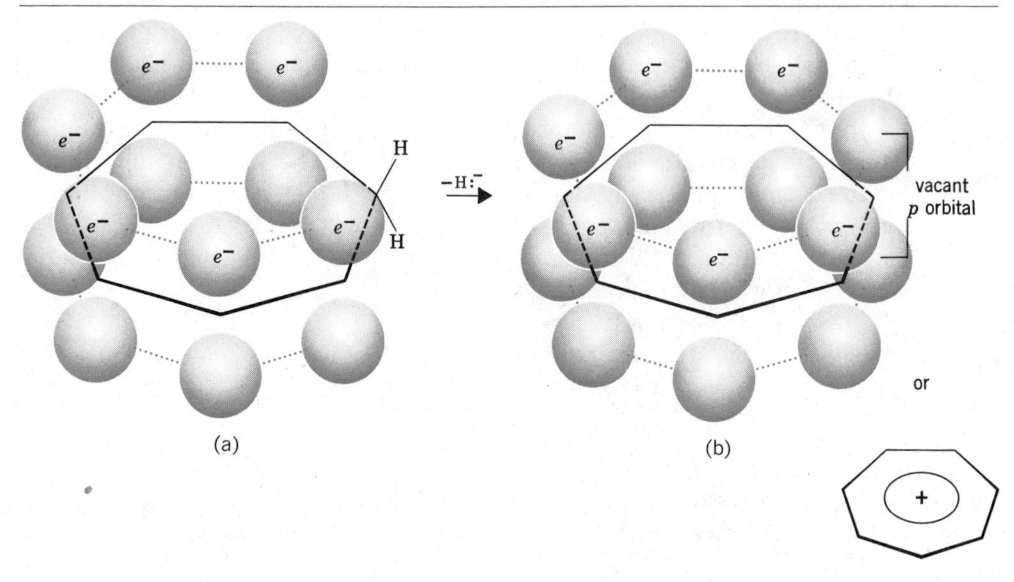

heptatrienyl cation is found to be unusually stable. The nmr spectrum of the cycloheptatrienyl cation indicates that all seven hydrogens are equivalent. If we look closely at Fig. 11.8 we see how we can account for these observations.

As a hydride ion is removed from the —CH_2— group of cycloheptatriene, a vacant *p* orbital is created, and the carbon atom becomes sp^2 hybridized. The cation that results has seven overlapping *p* orbitals containing *six* delocalized π electrons. The cycloheptatrienyl cation is, therefore, an aromatic cation and all of its hydrogens should be equivalent; again this is exactly what we find experimentally.

Problem 11.4

The conversion of cycloheptatriene to the cycloheptatrienyl cation can be accomplished by treating cycloheptatriene with triphenylcarbonium perchlorate [$(C_6H_5)_3C^+ \ ClO_4^-$]. (a) Write the structure of triphenylcarbonium perchlorate and show how it abstracts a hydride ion from cycloheptatriene. (b) What other product is formed in this reaction? (c) What anion is associated with the cycloheptatrienyl cation that is produced?

Problem 11.5

Tropylium bromide, C_7H_7Br, is insoluble in nonpolar solvents but dissolves readily in water. When an aqueous solution of tropylium bromide is treated with silver nitrate a precipitate of AgBr forms immediately. The melting point of tropylium bromide is above 200°, quite high for an organic compound. How can you account for these facts?

11.6C Aromatic, Antiaromatic, and Nonaromatic Compounds

When we say that the cyclopentadienyl anion is aromatic, we do not mean that we can brominate, nitrate, or sulfonate it the way we can benzene. If we were, for example, to place the cyclopentadienyl anion in a mixture of nitric and sulfuric acids in an attempt to nitrate it, the anion would immediately accept a proton and become cyclopentadiene. From that point on, any reactions that took place would not be reactions of the cyclopentadienyl anion but would be reactions of cyclopentadiene.

Many of the higher annulenes that we have classified as aromatic are too highly reactive to nitrate, sulfonate, and so on.

What do we mean, then, when we say that these compounds are aromatic? We mean that their π electrons are *delocalized* over the entire ring and that they are *stabilized* by the π-electron delocalization.

One of the best ways to determine whether or not the π electrons of a cyclic system are delocalized is through the use of nuclear magnetic resonance spectroscopy. It provides direct physical evidence of whether or not the π electrons are delocalized. We will have more to say about how this is done in Chapter 13.

But, what do we mean by saying that a compound is stabilized by π-electron delocalization? We have an idea of what this means from our comparison of the heat of hydrogenation of benzene and that calculated for the hypothetical 1,3,5-cyclohexatriene. We saw that benzene—in which the π electrons are delocalized—is much more stable than 1,3,5-cyclohexatriene (a model in which the π electrons are not delocalized). We call the energy difference between them the stabilization energy (delocalization energy) or resonance energy.

In order to make similar comparisons for other aromatic compounds we need to choose proper models. But what should these models be?

One proposal is that we should compare the π-electron energy of the cyclic system with that of the corresponding open-chain compound. This approach is particularly useful because it furnishes us with models not only for annulenes but for aromatic cations and anions as well. (Corrections need to be made, of course, when the cyclic system is strained.)

When we use this approach we take as our model a linear chain of sp^2-hybridized atoms that carries the same number of π electrons as our cyclic compound. Then we imagine ourselves removing two hydrogens from the end of this chain and joining the ends to form a ring. If the ring has *lower* π-electron energy than the open chain then the ring is *aromatic*. If the ring and chain have *the same* π-electron energy then the ring is *nonaromatic*. If the ring has *greater* π-electron energy than the open chain then the ring is *antiaromatic*.

The actual calculations and experiments used in determining π-electron energies are beyond our scope, but we can study four examples that illustrate how this approach has been used.

Cyclobutadiene. For cyclobutadiene we consider the change in π-electron energy for the following transformation.

H H H H — π Electron Energy increases → + H_2

1,3-Butadiene
4 π electrons

Cyclobutadiene
(4 π electrons, **antiaromatic**)

Calculations indicate and experiments appear to confirm that the π-electron energy of cyclobutadiene is higher than that of its open-chain counterpart. Thus cyclobutadiene is classified as being antiaromatic.

Benzene. Here our comparison is based on the following transformation.

H H H H — π Electron Energy decreases → + H_2

1,3,5-Hexatriene
6 π electrons

Benzene
(6 π electrons, **aromatic**)

Calculations indicate and experiments confirm that benzene has a much lower π-electron energy than 1,3,5-hexatriene. Benzene is classified as being aromatic on the basis of this comparison as well.

Cyclopentadienyl Anion. Here we use a linear anion:

H H HC:$^-$ H H — π Electron Energy decreases → + H_2

6 π electrons

Cyclopentadienyl anion
(6 π electrons, **aromatic**)

Both calculations and experiments confirm that the cyclic anion has a lower π-electron energy than its open-chain counterpart. Therefore the cyclopentadienyl anion is classified as being aromatic.

[30] Annulene. Calculations indicate that when the number of π electrons is greater than 22–26, the open-chain conjugated polyene and its cyclic counterpart have the same energy. Thus, we would expect [30] annulene to be *nonaromatic*. Experiments indicate that this is true.

Problem 11.6

(a) What open-chain compound would you use for comparison in assessing the π-electron energy of the cycloheptatrienyl cation? (b) Both theory and experiments indicate that the cycloheptatrienyl cation has a lower π-electron energy than its open-chain counterpart. What conclusion does this justify?

Problem 11.7

(a) The cyclopentadienyl cation is apparently *antiaromatic*. Show what this means through a comparison of π-electron energies of the cyclic and open-chain compounds. (b) Make a similar comparison for the cyclopropenyl cation (below)—a

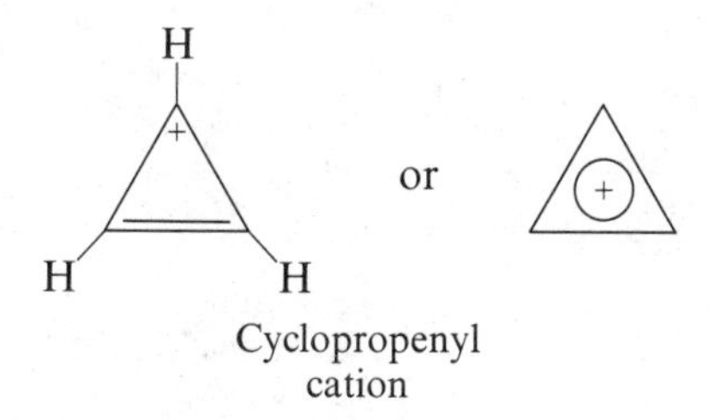

Cyclopropenyl
cation

system that is known from experiments to be *aromatic*. (c) What does Hückel's rule predict for the cyclopropenyl cation? (d) Calculations and experiments indicate that the cyclopropenyl *anion* is *antiaromatic*. What does this mean?

11.7 OTHER AROMATIC COMPOUNDS

11.7A Benzenoid Aromatic Compounds

In addition to those that we have seen so far, there are many other examples of aromatic compounds. Representatives of one broad class of aromatic compounds, called *polycyclic benzenoid aromatic hydrocarbons,* are illustrated in Fig. 11.9.

All of these consist of molecules having two or more benzene rings *fused* together. A close look at one, naphthalene, will illustrate what we mean by this.

According to resonance theory, a molecule of naphthalene can be considered to be a hybrid of three Kekulé structures. One of these Kekulé structures is shown in Fig. 11.10. There are two carbon atoms in naphthalene (C-9 and C-10) that are common to both rings. These two atoms are said to be at the points of *ring fusion.* They direct all of their bonds toward other carbon atoms and do not bear hydrogen atoms.

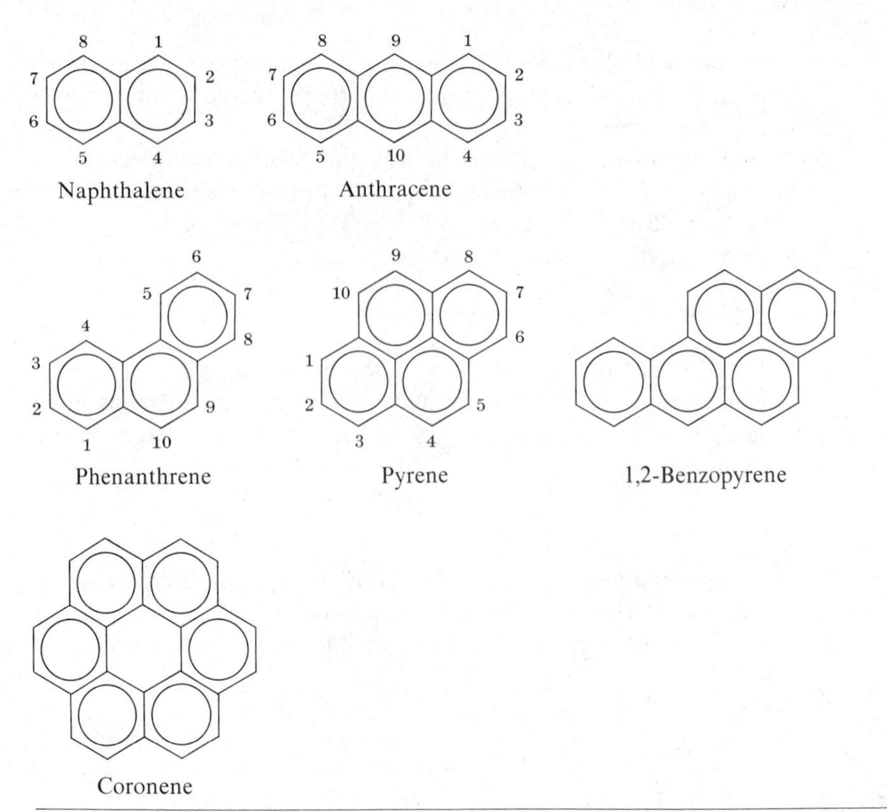

FIG. 11.9
Benzenoid aromatic hydrocarbons.

Problem 11.8

(a) Write the three resonance structures for naphthalene. (b) The C-1—C-2 bond of naphthalene is shorter than the C-2—C-3 bond. Do the resonance structures you have written account for this? Explain.

Molecular orbital calculations for naphthalene begin with the model shown in Fig. 11.11. The *p* orbitals overlap around the periphery of both rings and across the points of ring fusion.

When molecular-orbital calculations are carried out for naphthalene using the model shown in Fig. 11.11, the results of the calculations correlate well with our experimental knowledge of naphthalene. The calculations indicate that delocalization of the 10 π electrons over the two rings produces a structure with considerably lower energy than that calculated for any individual Kekulé struc-

FIG. 11.10
One Kekulé structure for naphthalene.

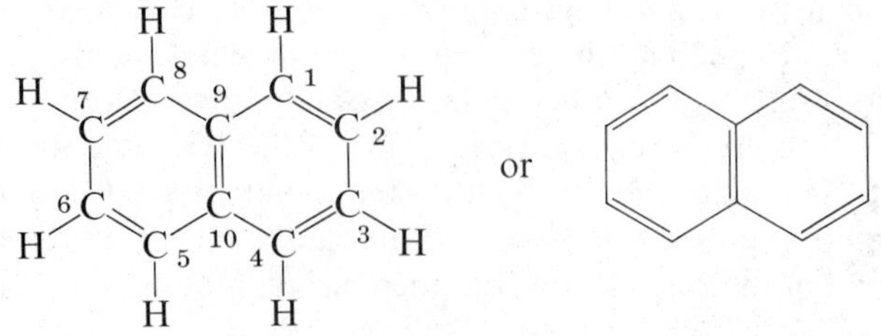

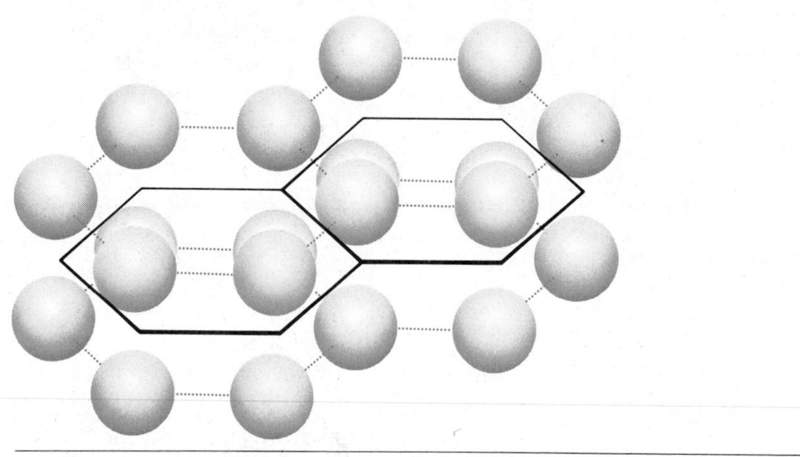

FIG. 11.11
The p *orbitals of naphthalene.*

ture. Naphthalene, consequently, has a substantial stabilization energy. Based on what we know about benzene, moreover, naphthalene's tendency to react by substitution rather than addition, and to show other properties associated with aromatic compounds, is understandable.

Anthracene and phenanthrene are isomers. In anthracene the three rings are fused in a linear way, and in phenanthrene they are fused so as to produce an angular molecule. Both of these molecules also show large stabilization energies and chemical properties typical of aromatic compounds.

Pyrene and coronene are also aromatic. Pyrene itself has been known for a long time; a pyrene derivative, however, has been the object of research that shows another interesting application of Hückel's rule.

In order to understand this particular research we need to pay special attention to the Kekulé structure for pyrene (Fig. 11.12). The total number of π electrons in pyrene is 16 (8 double bonds = 16 π electrons). Sixteen is a non-Hückel number, but Hückel's rule is intended to be applied only to monocyclic compounds and pyrene is clearly tetracyclic. If we disregard the internal double bond of pyrene, however, and look only at the periphery, we see that the periphery resembles a monocyclic ring with 14 π electrons. The periphery is, in fact, very much like that of the [14] annulene shown below. Fourteen *is* a Hückel number ($4n + 2$, where $n = 3$) and one might then predict that the periphery of pyrene would be aromatic by itself, in the absence of the internal double bond.

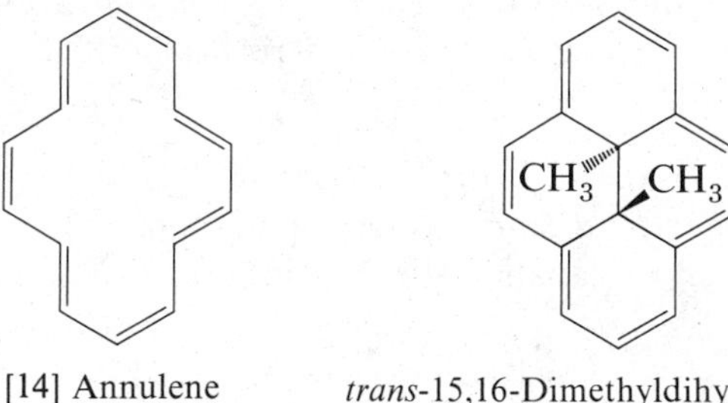

[14] Annulene *trans*-15,16-Dimethyldihydropyrene

This prediction was confirmed when Professor V. Boekelheide (University of Oregon) synthesized the very unusual *trans*-15,16-dimethyldihydropyrene and showed that it is aromatic.

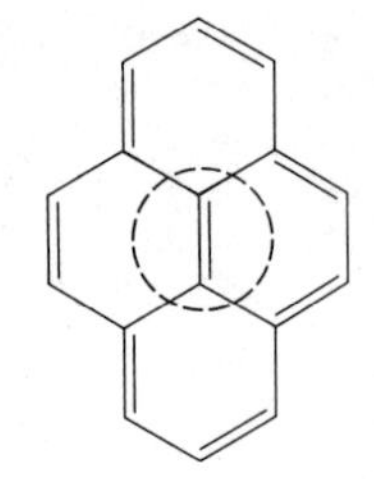

FIG. 11.12
One Kekulé structure for pyrene. The internal double bond is enclosed in a dotted circle for emphasis.

11.7B Nonbenzenoid Aromatic Compounds

Naphthalene, phenanthrene, and anthracene are examples of *benzenoid* aromatic compounds. On the other hand, the cyclopentadienyl anion, the cycloheptatrienyl cation, *trans*-15,16-dimethyldihydropyrene, and the aromatic annulenes (except for [6] annulene) are classified as *nonbenzenoid* aromatic compounds.

Another example of a *nonbenzenoid* aromatic hydrocarbon is the compound azulene. Azulene has a stabilization energy of 49 kcal/mole.

Azulene

This deep-blue hydrocarbon (its name is derived from the word azure) is an isomer of naphthalene. It has the same number of π electrons as naphthalene and, for this reason, azulene is also said to be *isoelectronic* with naphthalene. In addition to its deep-blue color (naphthalene by contrast is colorless), azulene differs from naphthalene in another respect that seems, at first, to be peculiar. Azulene is found to have a substantial dipole moment. The dipole moment of azulene is 1.0 D, whereas the dipole moment of naphthalene is zero.

That azulene has a dipole moment at all indicates that charge separation exists in the molecule. If we recognize this and begin writing resonance structures for azulene that involve charge separation, we find that we can write a number of structures like the one shown in Fig. 11.13.

When we inspect this resonance structure we see that the five-membered ring is very much like the *aromatic* cyclopentadienyl anion and that the seven-membered ring resembles the *aromatic* cycloheptatrienyl cation. If resonance structures of this type contribute to the overall hybrid for azulene, then we not only understand why azulene has a dipole moment, but we also have some insight into why it is aromatic. Such speculation is strengthened by the results of studies done with substituted azulenes that show quite conclusively that the five-membered ring is negatively charged and the seven-membered ring is positive.

Problem 11.9

Diphenylcyclopropenone (I) has a much larger dipole moment than benzophenone (II). Can you think of an explanation that would account for this?

C_6H_5–C=C–C_6H_5 (ring closed by C=Ö:) $\qquad$ C_6H_5–C(=Ö:)–C_6H_5

I $\qquad\qquad$ II

FIG. 11.13
One resonance structure for azulene that has separated charges.

11.8 NOMENCLATURE OF BENZENE DERIVATIVES

Two systems are used in naming monosubstituted benzenes. In compounds of one type, *benzene* is the base name and the substituent is simply indicated by a prefix. We have, for example,

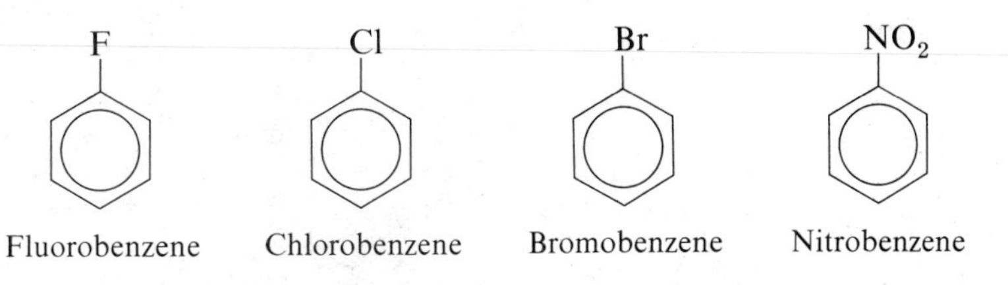

Fluorobenzene Chlorobenzene Bromobenzene Nitrobenzene

For compounds of the other type, the substituent and the benzene ring taken together may form a new base name. Methylbenzene is usually called *toluene,* hydroxybenzene is almost always called *phenol,* and aminobenzene is almost always called *aniline*. These and other examples are indicated below.

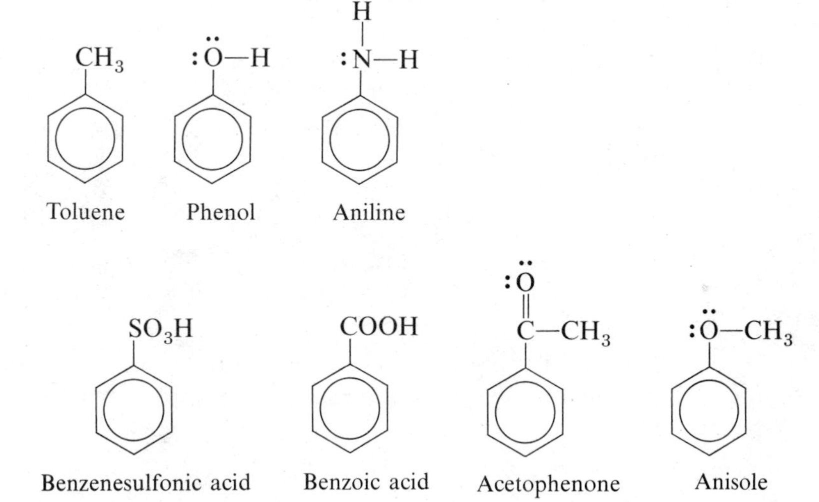

Toluene Phenol Aniline

Benzenesulfonic acid Benzoic acid Acetophenone Anisole

When two substituents are present, their relative positions are indicated by the use of numbers or by the prefixes: *ortho, meta,* and *para* (abbreviated *o*-, *m*-, and *p*-). For the dibromobenzenes we have

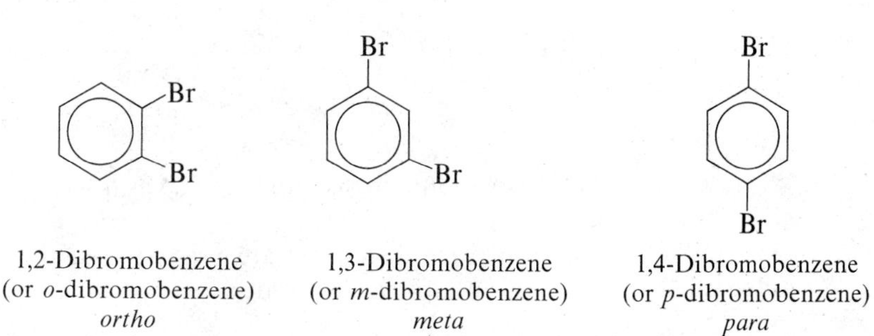

1,2-Dibromobenzene (or *o*-dibromobenzene) *ortho*

1,3-Dibromobenzene (or *m*-dibromobenzene) *meta*

1,4-Dibromobenzene (or *p*-dibromobenzene) *para*

For the chlorotoluenes we have

2-Chlorotoluene (or *o*-chlorotoluene) | 3-Chlorotoluene (or *m*-chlorotoluene) | 4-Chlorotoluene (or *p*-chlorotoluene)

and for the nitrobenzoic acids:

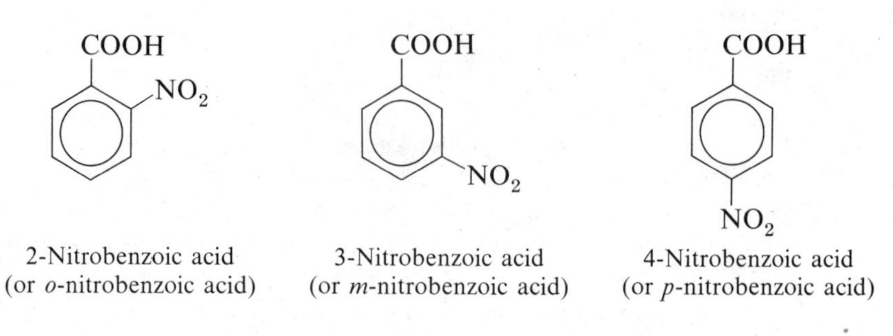

2-Nitrobenzoic acid (or *o*-nitrobenzoic acid) | 3-Nitrobenzoic acid (or *m*-nitrobenzoic acid) | 4-Nitrobenzoic acid (or *p*-nitrobenzoic acid)

The dimethylbenzenes are called *xylenes*.

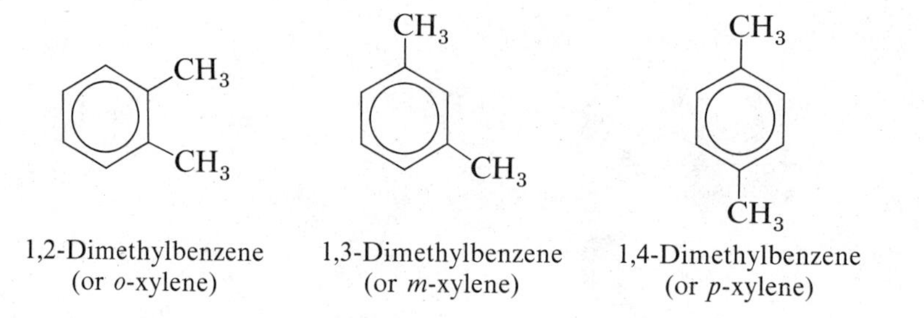

1,2-Dimethylbenzene (or *o*-xylene) | 1,3-Dimethylbenzene (or *m*-xylene) | 1,4-Dimethylbenzene (or *p*-xylene)

If more than two groups are present on the benzene ring their positions must be indicated by the use of *numbers*. As examples, consider the two compounds below.

1,2,3-Trichlorobenzene

1,2,4-Tribromobenzene (not 1,3,4-tribromobenzene)

We notice, too, that the benzene ring is numbered so as to give *the lowest possible numbers to the substituents*.

When more than two substituents are present, and the substituents are different, they are listed in alphabetical order:

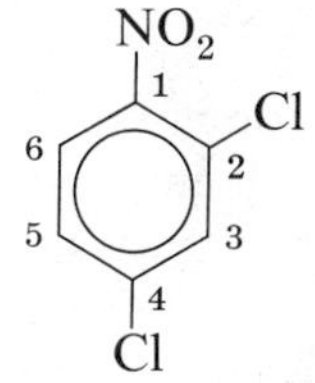

2,4-Dichloro-1-nitrobenzene

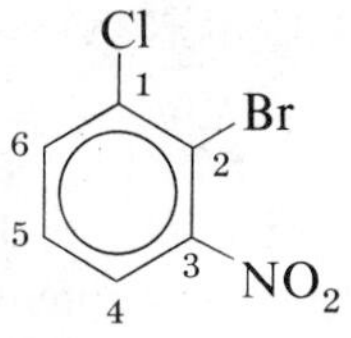

2-Bromo-1-chloro-3-nitrobenzene

When a substituent is one that when taken together with the benzene ring gives a new base name, that substituent is assumed to be in position 1 and the new base name is used:

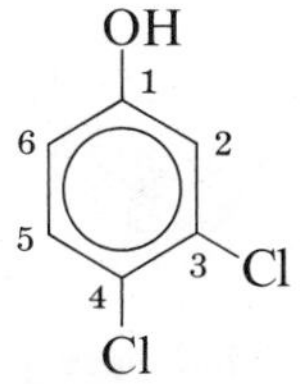

3,4-Dichlorophenol

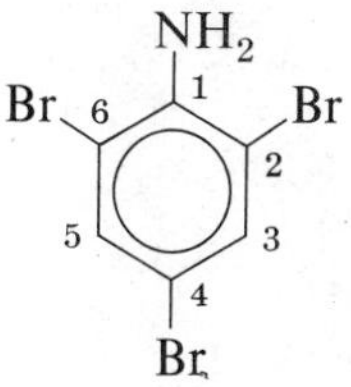

2,4,6-Tribromoaniline

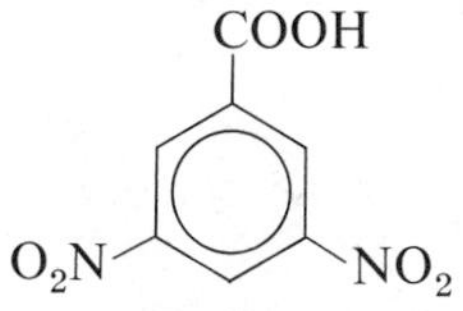

3,5-Dinitrobenzoic acid

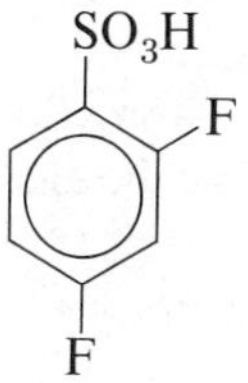

2,4-Difluorobenzenesulfonic acid

When the benzene ring is attached to a chain of six or more carbons it is named as a substituent, and it is called a *phenyl group*. For example,

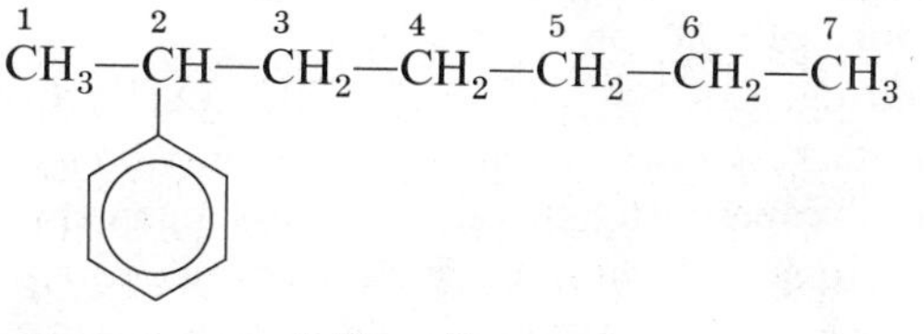

$$\overset{1}{C}H_3-\overset{2}{C}H(C_6H_5)-\overset{3}{C}H_2-\overset{4}{C}H_2-\overset{5}{C}H_2-\overset{6}{C}H_2-\overset{7}{C}H_3$$

2-Phenylheptane
(not 2-benzylheptane)

The name *benzyl* is reserved for the group derived from toluene by the removal of one hydrogen of the methyl group.

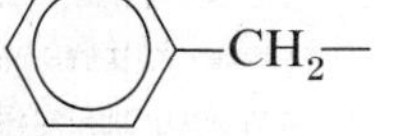

$C_6H_5-CH_2-$

The benzyl group
(or the phenylmethyl group)

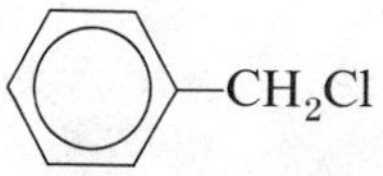

$C_6H_5-CH_2Cl$

Benzyl chloride
(or phenylmethyl chloride)

11.9 HETEROCYCLIC AROMATIC COMPOUNDS

All of the aromatic molecules that we have discussed so far have had rings composed solely of carbon atoms. However, in molecules of many aromatic compounds an element other than carbon is present in the ring. These compounds are called *heterocyclic* aromatic compounds. Heterocyclic molecules are quite commonly encountered in nature. For this reason, and because the structures of some of these molecules are closely related to the compounds that we have discussed earlier, we will now describe a few of the most important examples.

Heterocyclic compounds containing nitrogen, oxygen, and sulfur are by far the most common. Four important examples are given below in their Kekulé forms.

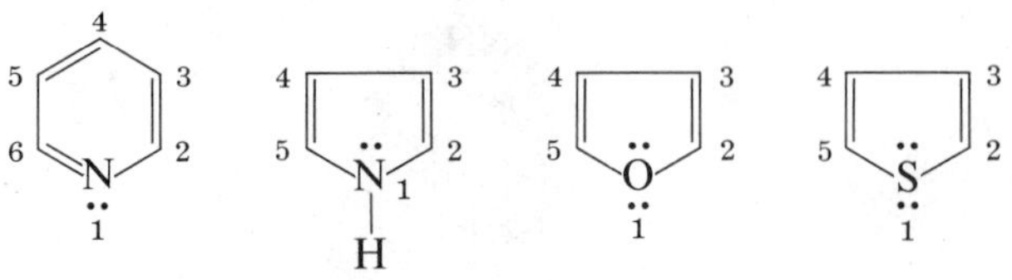

Pyridine | Pyrrole | Furan | Thiophene

If we examine these structures we will see that pyridine is structurally related to benzene, and that pyrrole, furan, and thiophene are related to the cyclopentadienyl anion.

The nitrogen atoms in molecules of both pyridine and pyrrole are sp^2 hybridized. In pyridine the electrons of the sp^2 nitrogen can be assigned in the way shown in Fig. 11.14.

In the pyridine ring the two sp^2 orbitals of nitrogen, each with only one electron, overlap with sp^2 orbitals of the 2- and 6-carbon atoms to form σ bonds. The remaining sp^2 orbital of nitrogen (with two electrons) does not form a bond. Its axis is in the same plane as the ring but directed away from it (Fig. 11.15). The p orbital of nitrogen (containing one electron) is perpendicular to the plane of the ring and this orbital overlaps with the p orbitals of the 2- and 6-carbon atoms to form a π system of six π electrons. The π system of pyridine is isoelectronic with the π system of benzene. Pyridine, consequently, shows aromatic properties. It has a stabilization energy, for example, of 27 kcal/mole.

In pyrrole the nitrogen atom is sp^2 hybridized but the electrons are arranged differently. The sp^2 orbitals each contain one electron and *the p orbital contains two electrons* (Fig. 11.16). The reason for this different arrangement of the electrons becomes clear when we look at the orbital structure of the entire pyrrole ring (Fig. 11.17). Since only four π electrons are contributed by the carbon atoms in pyrrole, the nitrogen atom must contribute two electrons to give an aromatic sextet. Two of the sp^2 orbitals of the pyrrole nitrogen form sigma bonds to the 2- and 5-carbon atoms; the remaining sp^2 orbital overlaps with the 1s orbital of a hydrogen atom to give the N—H sigma bond.

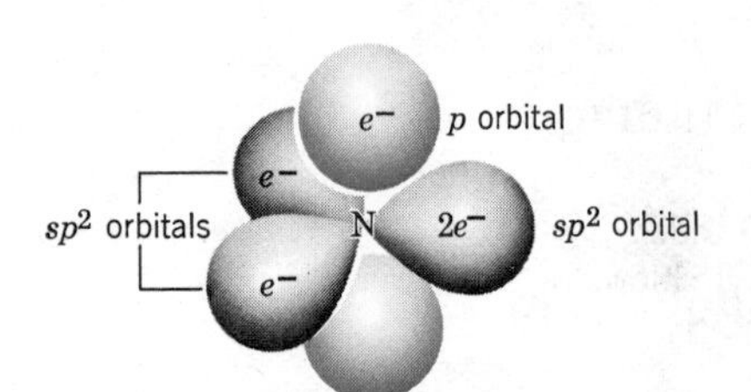

FIG. 11.14
The arrangement of electrons in an sp² *nitrogen atom of the pyridine type.*

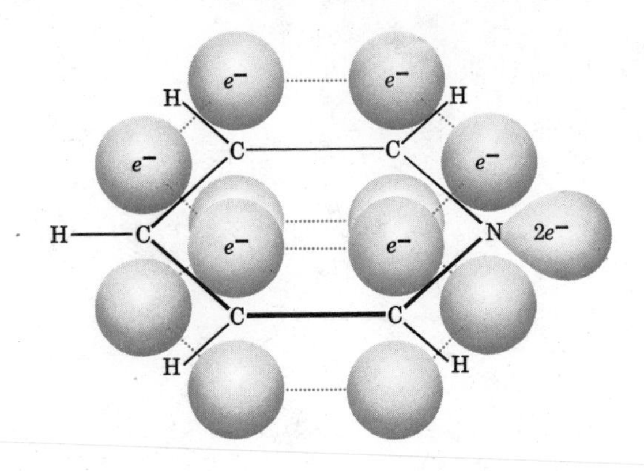

FIG. 11.15
Orbital structure of pyridine.

Furan and thiophene are structurally quite similar to pyrrole. The oxygen atom in furan and the sulfur atom in thiophene are sp^2 hybridized. In both compounds the p orbital of the heteroatom donates two electrons to the π system. The oxygen and sulfur atoms of furan and thiophene carry an unshared pair of electrons in an sp^2 orbital (Fig. 11.18).

11.10 AROMATIC COMPOUNDS IN BIOCHEMISTRY

Compounds with aromatic rings occupy numerous and important positions in reactions that occur in living systems. It would be impossible to describe them all in this chapter. We will, however, point out a few examples now and we will see others later.

Two amino acids necessary for protein synthesis contain the benzene ring:

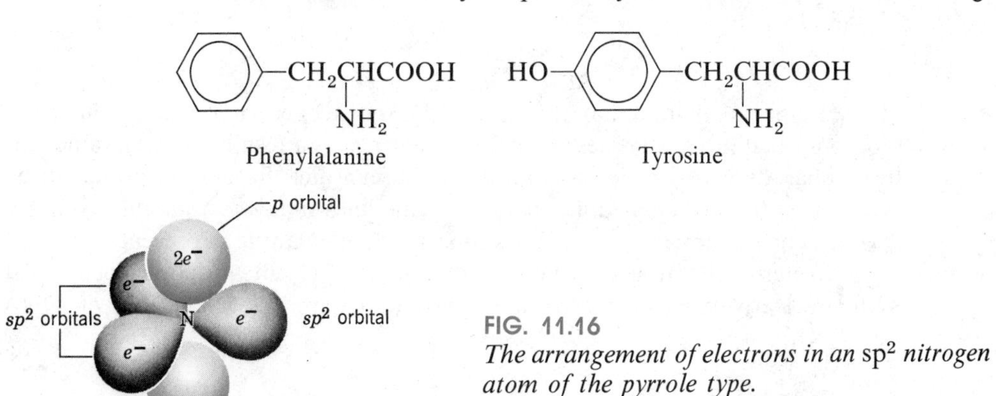

FIG. 11.16
The arrangement of electrons in an sp^2 *nitrogen atom of the pyrrole type.*

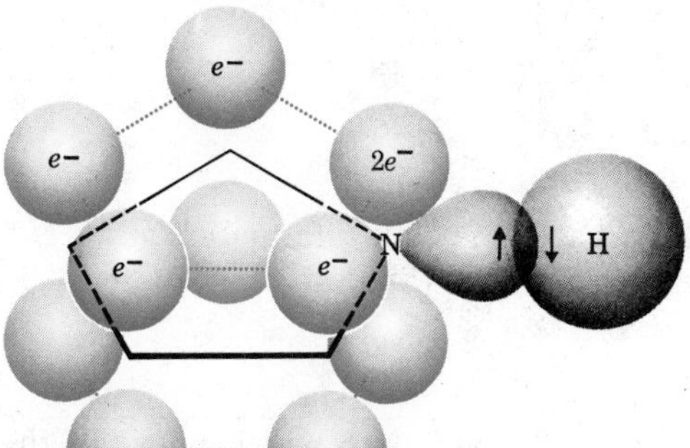

FIG. 11.17
The orbital structure of pyrrole. (Compare with the orbital structure of the cyclopentadienyl anion in Fig. 11.7.)

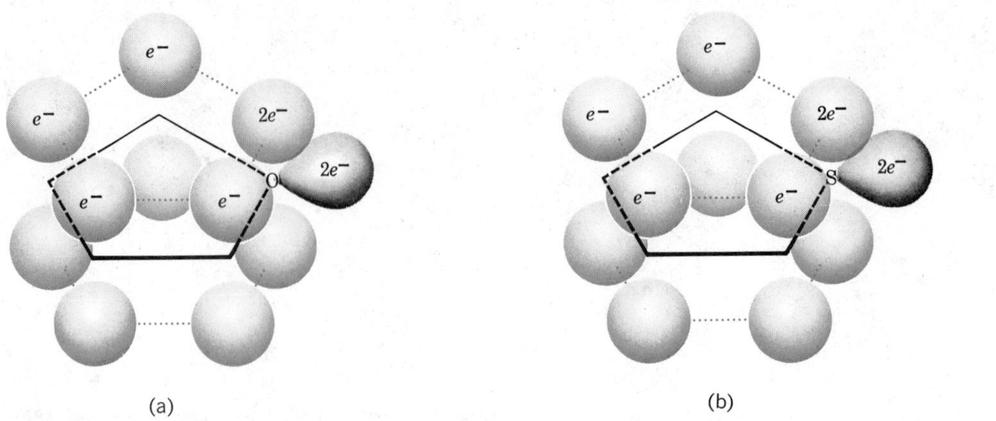

FIG. 11.18
The orbital structures of furan (a) *and thiophene* (b).

A third aromatic amino acid, tryptophan, contains a benzene ring fused to a pyrrole ring. (This aromatic ring system is called an indole system, cf. Special Topic O.)

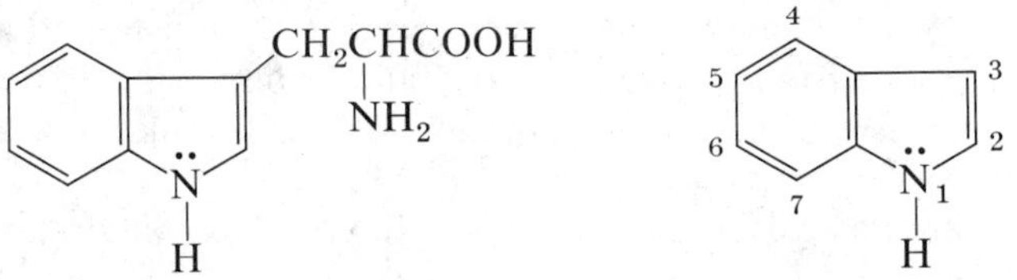

It appears that humans, because of the course of evolution, do not have the biochemical ability to synthesize the benzene ring. As a result, phenylalanine and tryptophan derivatives are essential in the human diet. Because tyrosine can be synthesized from phenylalanine by an enzyme known as *phenylalanine hydroxylase,* it is not essential in the diet as long as phenylalanine is present.

Heterocyclic aromatic compounds are also present in many biochemical systems. Derivatives of purine and pyrimidine are essential parts of DNA and RNA.

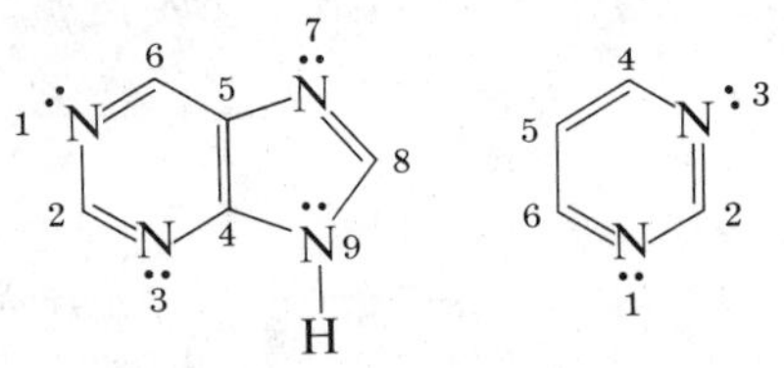

DNA is the molecule responsible for the storage of genetic information and RNA is prominently involved in the synthesis of enzymes (Special Topic P).

Problem 11.10

Classify each nitrogen atom in the purine molecule as to whether it is of the pyridine type or of the pyrrole type.

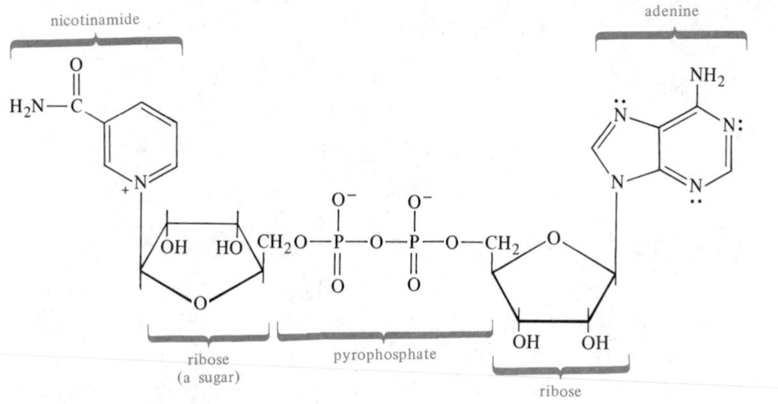

nicotinamide, adenine, ribose (a sugar), pyrophosphate, ribose

FIG. 11.19
Nicotinamide adenine dinucleotide (NAD^+).

Both a pyridine derivative (nicotinamide) and a purine derivative (adenine) are present in one of the most important coenzymes in biological oxidations. This molecule, *nicotinamide adenine dinucleotide* (NAD^+), is shown in Fig. 11.19.

NAD^+, together with another compound in the liver (an apoenzyme), is capable of oxidizing alcohols to aldehydes. (We saw an example of this in the oxidation of vitamin A to retinal in Special Topic F). While the overall change is quite complex, a look at one aspect of it will illustrate a *biological use* of the extra stability (stabilization or delocalization energy) associated with an aromatic ring.

A simplified version of the oxidation of an alcohol to an aldehyde is illustrated below:

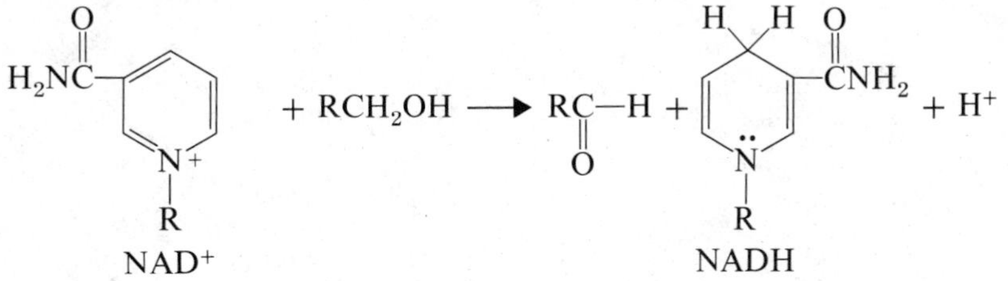

$$NAD^+ + RCH_2OH \longrightarrow RCHO + NADH + H^+$$

NAD+ NADH

The *aromatic* pyridine ring (actually a *pyridinium* ring, because it is positively charged) in NAD^+ is converted to a *nonaromatic* ring in NADH. The extra stability of the pyridine ring is lost in this change; and, as a result, the potential energy of NADH is greater than that of NAD^+. The conversion of the alcohol to the aldehyde, however, occurs with a decrease in potential energy. Because these reactions are coupled in biological systems (Fig. 11.20), a portion of the potential energy contained in the alcohol becomes chemically contained in NADH. This stored energy in NADH is used to bring about other biochemical reactions that require energy and that are necessary to life.

Although many aromatic compounds are essential to life, others are hazardous. Many are quite toxic and several polycyclic benzenoid compounds are *carcinogenic* (i.e., cancer causing). Two examples are benzo[*a*]pyrene and 7-methylbenz[*a*]anthracene.

CH_3

Benzo[*a*]pyrene　　　7-Methylbenz[*a*]anthracene

The hydrocarbon benzo[*a*]pyrene has been found in cigarette smoke and in the exhaust from automobiles. It is also formed in the incomplete combustion of any fossil fuel. It is found on charcoal-broiled steaks and exudes from asphalt streets on a hot summer day. Benzo[*a*]pyrene is so carcinogenic that one can induce skin cancers in mice with almost total certainty by simply shaving an area of the body of the mouse and applying a coating of benzo[*a*]pyrene.

11.11 VISIBLE AND ULTRAVIOLET ABSORPTION SPECTRA OF AROMATIC COMPOUNDS

Benzene and alkylbenzenes have two absorption bands in the ultraviolet region. One band lies near 200 nm and is, by far, the more intense. The second band, called the *benzenoid band*, is found at longer wavelengths, near 260 mm, and is quite weak.

Both of these bands are shifted to longer wavelengths when groups are attached to the ring in a way that causes the conjugated system to be extended. We can see this effect if we compare the absorption maxima of benzene and styrene.

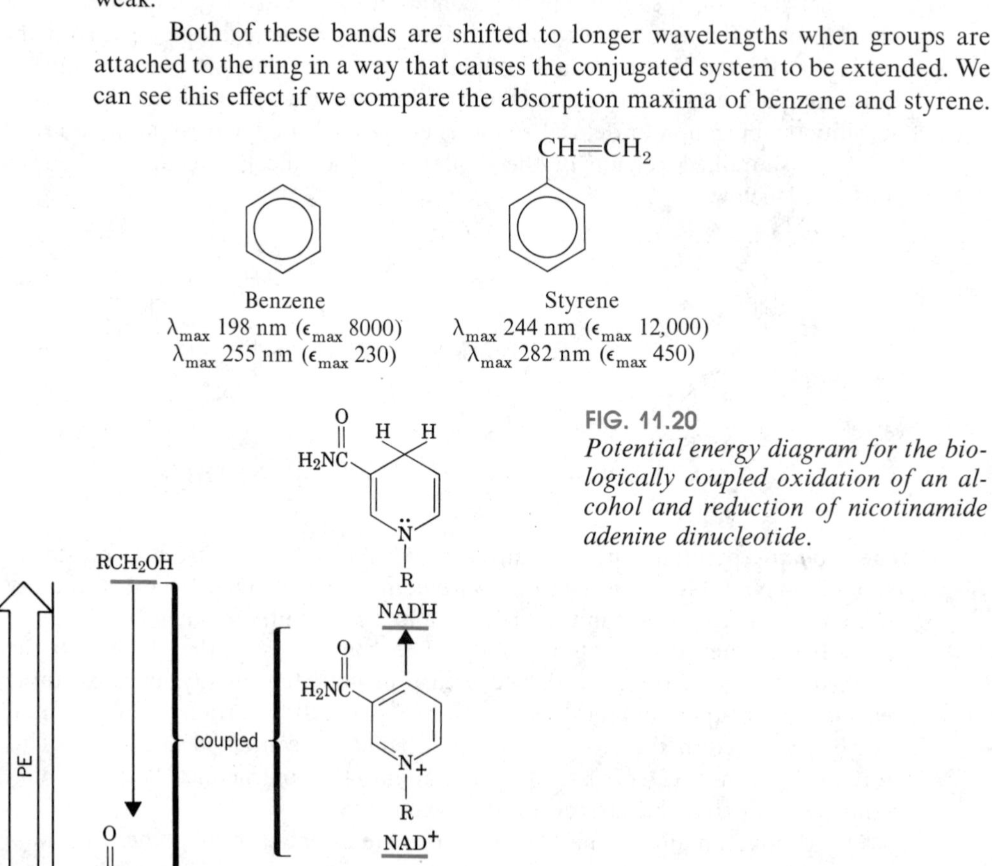

FIG. 11.20
Potential energy diagram for the biologically coupled oxidation of an alcohol and reduction of nicotinamide adenine dinucleotide.

The benzenoid band of polycyclic compounds is shifted to longer wavelengths in a dramatic way when benzene rings are fused in a linear fashion.

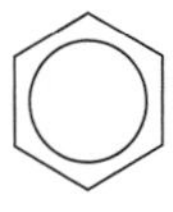

Benzene
λ_{max} 255 nm
ϵ_{max} 230

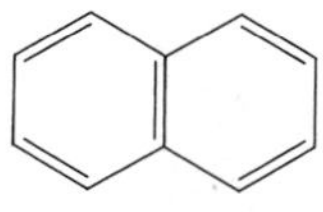

Naphthalene
λ_{max} 314 nm
ϵ_{max} 316

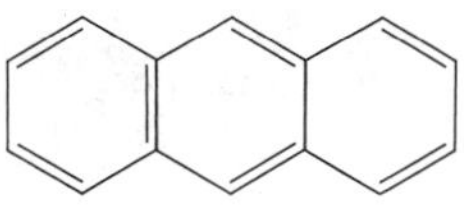

Anthracene
λ_{max} 380 nm
ϵ_{max} 7900

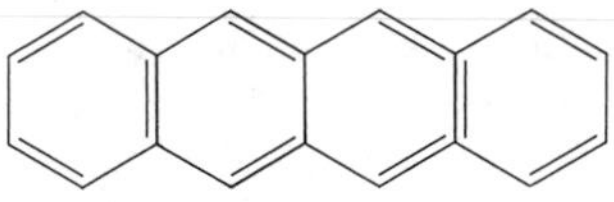

Naphthacene
λ_{max} 480 nm
ϵ_{max} 11,000

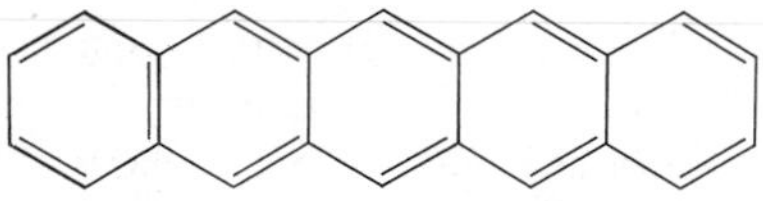

Pentacene
λ_{max} 580 nm
ϵ_{max} 12,600

Benzene, naphthalene, and anthracene are colorless because their absorption maxima lie in the ultraviolet region. Naphthacene and pentacene, however, are colored because their absorption maxima lie in the visible region. Naphthacene is orange-yellow and pentacene is purple.

Additional Problems

11.11

Draw structural formulas for the following:

(a) 4-Nitrobenzenesulfonic acid
(b) *o*-Chlorotoluene
(c) *m*-Dichlorobenzene
(d) *p*-Dinitrobenzene
(e) 4-Bromoanisole
(f) *m*-Nitrobenzoic acid
(g) *p*-Iodophenol
(h) 2-Chloroacetophenone
(i) 2-Bromonaphthalene
(j) 9-Chloroanthracene
(k) 3-Nitrophenanthrene
(l) 4-Nitropyridine
(m) 2-Methylpyrrole
(n) 2,4-Dichloro-1-nitrobenzene
(o) *p*-Nitrobenzyl bromide
(p) *o*-Chloroaniline
(q) 2,5-Dibromo-3-nitrobenzoic acid
(r) 1,3,5-Trimethylbenzene (mesitylene)
(s) *p*-Hydroxybenzoic acid
(t) Vinylbenzene (styrene)
(u) Benzo[*a*]pyrene
(v) 2-Phenylcyclohexanol
(w) 2,4,6-Trinitrotoluene (TNT)
(x) A [12] annulene
(y) A [14] annulene
(z) An [18] annulene

11.12
Write structural formulas and give names for all of the following:
(a) Trichlorobenzenes
(b) Dibromonitrobenzenes
(c) Dichlorotoluenes
(d) Monochloronaphthalenes
(e) Nitropyridines
(f) Methylfurans
(g) Chlorodinitrobenzenes
(h) Chloroxylenes
(i) Cresols (hydroxytoluenes)

11.13
(a) Write the five principal resonance structures for phenanthrene.
(b) On the basis of these can you speculate about the length of the 9,10 bond?
(c) About its double-bond character?
(d) Phenanthrene, in contrast to most aromatic molecules, tends to *add* one mole of bromine to form a molecule with the formula $C_{14}H_{10}Br_2$. How can you account for this?

11.14
Methoxydimethylcyclopropene (III) has been found to react with fluoroboric acid to yield methanol and a compound with the formula $C_6H_9^+BF_4^-$ (IV).

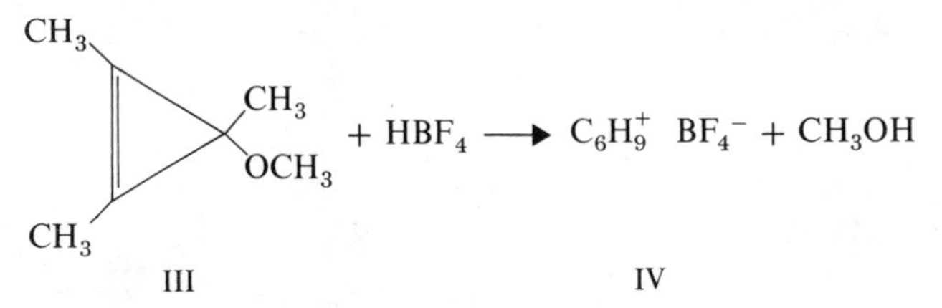

(a) What is the structure of $C_6H_9^+BF_4^-$?
(b) How can you account for its formation?

11.15
Diphenylcyclopropenone (cf. Problem 11.9) reacts with hydrogen bromide to form a stable crystalline hydrobromide. What is its structure?

11.16
Cyclooctatetraene has been shown by Prof. Thomas Katz of Columbia University to react with 2 moles of potassium to yield an unusually stable compound with the formula $2K^+ C_8H_8^=$ (V). The nmr spectrum of V indicates that all of its hydrogens are equivalent.
(a) What is the structure of V?
(b) How can you account for the formation of V?

$$2K + \text{(Cyclooctatetraene)} \longrightarrow 2K^+ \; C_8H_8^{2-}$$

Cyclooctatetraene V

11.17
The bicyclic molecule, VI, reacts with 2 moles of lithium to yield lithium chloride and a compound with the formula $Li^+ C_9H_9^-$ (VII).

$$\text{VI (H, Cl)} + 2Li \longrightarrow LiCl + Li^+ \; C_9H_9^-$$

VI VII

(a) What is VII?
(b) Account for its formation.

11.18

Using the data in Section 11.4 and Fig. 11.1 calculate $\Delta H°$ for the following reactions.

(a) 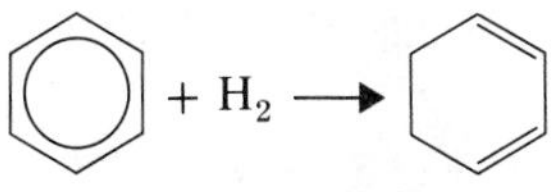 [benzene] $+ H_2 \longrightarrow$ [1,3-cyclohexadiene]

(b) 2 [1,3-cyclohexadiene] $\longrightarrow$ [benzene] + [cyclohexene]

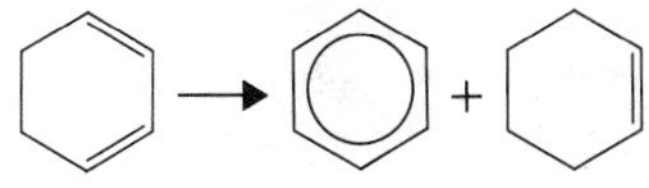

(c) On the basis of what you have found in (a) and (b), what is the likelihood of reaction (a) occurring successfully?

11.19

Consider whether each of the following molecules or ions would or would not be aromatic. Explain you answer in each instance.

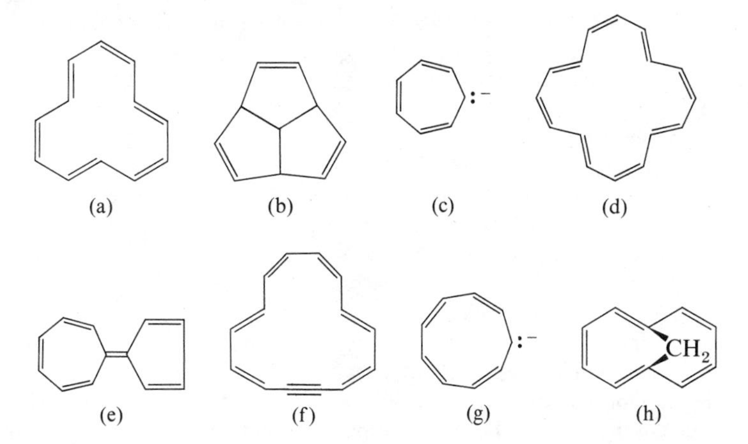

*** 11.20**

Cycloheptatrienone, **I**, is very stable. Cyclopentadienone, **II**, by contrast is quite unstable and rapidly undergoes a Diels-Alder reaction with itself. (a) Propose an explanation for the different stabilities of these two compounds. (b) Write the structure of the Diels-Alder adduct of cyclopentadienone.

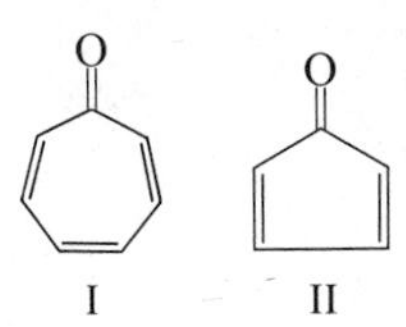

*** 11.21**

The final product **C** of the following synthesis is an aromatic compound. What is its structure and what are **A** and **B?**

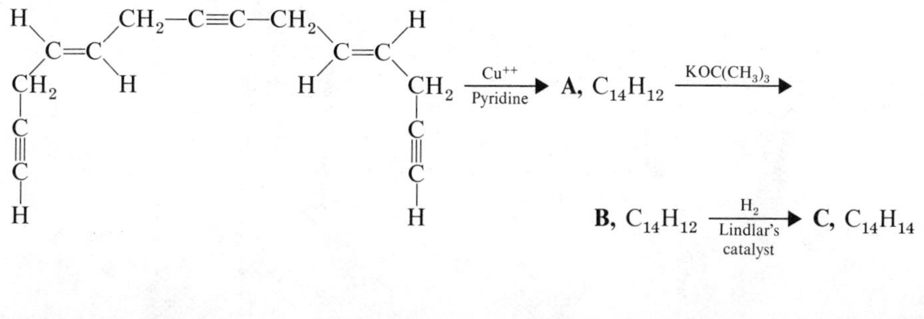

$$HC{\equiv}C{-}CH_2{-}CH{=}CH{-}CH_2{-}C{\equiv}C{-}CH_2{-}CH{=}CH{-}CH_2{-}C{\equiv}CH \xrightarrow[\text{Pyridine}]{Cu^{++}} \textbf{A}, C_{14}H_{12} \xrightarrow{KOC(CH_3)_3}$$

$$\textbf{B}, C_{14}H_{12} \xrightarrow[\text{Lindlar's catalyst}]{H_2} \textbf{C}, C_{14}H_{14}$$

*** 11.22**

The relative energies of the π molecular orbitals of conjugated monocyclic systems can be derived in a relatively simple way. One inscribes a regular polygon (corresponding to the ring system) in a circle so that *one corner of the polygon is at the bottom.* The points where the corners of the polygon touch the circle correspond to the energy levels of the π molecular orbitals of the system. With benzene, for example, this method (below) furnishes the energy levels given in Fig. 11.4.

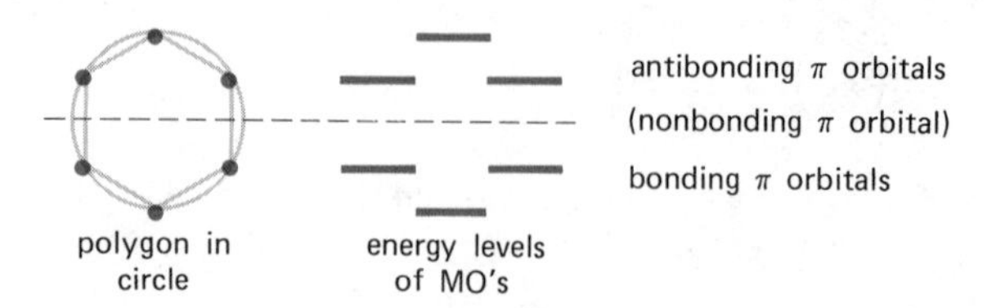

A horizontal line halfway up the circle divides the bonding orbitals from the antibonding orbitals. (If an orbital falls on this line, it is a nonbonding orbital.) Use this method to derive the relative energies of the π-molecular orbitals of the following molecules, radicals, and ions. In each case you should also give the π electron distribution of the lowest energy state.

(a) The cyclopropenyl cation (cf. Problem 11.7)
(b) The cyclopropenyl radical
(c) The cyclopropenyl anion
(d) Cyclobutadiene
(e) The cyclobutadienyl dication
(f) The cyclopentadienyl cation
(g) The cyclopentadienyl anion
(h) The cycloheptatrienyl cation
(i) The cycloheptatrienyl anion
(j) Cyclooctatetraene
(k) The cyclooctatetraenyl dication
(l) The cyclooctatetraenyl dianion
(m) Which of the systems above have unpaired electrons? (n) Which have closed bonding shells of π electrons? (o) Which systems would you expect to be aromatic. (p) Does your answer to (o) correspond to the predictions of Hückel's rule?

12 AROMATIC COMPOUNDS II. REACTIONS OF AROMATIC COMPOUNDS WITH ELECTROPHILES

12.1 ELECTROPHILIC AROMATIC SUBSTITUTION REACTIONS

Aromatic hydrocarbons are known generally as *arenes*. An *aryl group* is one derived from an arene by removal of a hydrogen and its symbol is Ar—. Thus arenes are designated ArH just as alkanes are designated RH.

The most characteristic reactions of benzenoid arenes are the substitution reactions that occur when they react with electrophilic reagents. These reactions are of the general type shown below.

$$ArH + E^+ \longrightarrow Ar{-}E + H^+$$

or

$$C_6H_6 + E^+ \longrightarrow C_6H_5E + H^+$$

Electrophilic aromatic substitution reactions allow us to introduce a wide variety of groups into aromatic rings and because of this, they give us synthetic access to a vast number of aromatic compounds that would not be available otherwise

In this chapter we will see how we can use electrophilic aromatic substitution reactions to introduce:

1. A halo substituent, —X (Section 12.3).
2. A nitro substituent, $—NO_2$ (Section 12.4).
3. A sulfonic acid group, $—SO_3H$ (Section 12.5).
4. An alkyl group, —R (Section 12.6).
5. An acyl group, —COR (Section 12.7).

All of these reactions involve an attack on the benzene ring by an electron-deficient species—by an electrophile.

12.2 A GENERAL MECHANISM FOR ELECTROPHILIC AROMATIC SUBSTITUTION: ARENIUM IONS

Benzene is susceptible to electrophilic attack primarily because of its exposed π electrons. In this respect benzene resembles an alkene, for in the reaction of an alkene with an electrophile the site of attack is the exposed π bond.

We saw in Chapter 11, however, that benzene differs from an alkene in a very significant way. Benzene's closed shell of six π electrons gives it a special stability. So, while benzene is susceptible to electrophilic attack, benzene undergoes *substitution reactions* rather than *addition reactions.* Substitution reactions allow the aromatic sextet of π electrons of benzene to be regenerated after attack by the electrophile has occurred. We can see how this happens if we examine a general mechanism for electrophilic aromatic substitution.

A considerable body of experimental evidence indicates that electrophiles attack the π system of benzene to form a *delocalized carbocation* known as an *arenium ion* (or sometimes as a σ *complex*).

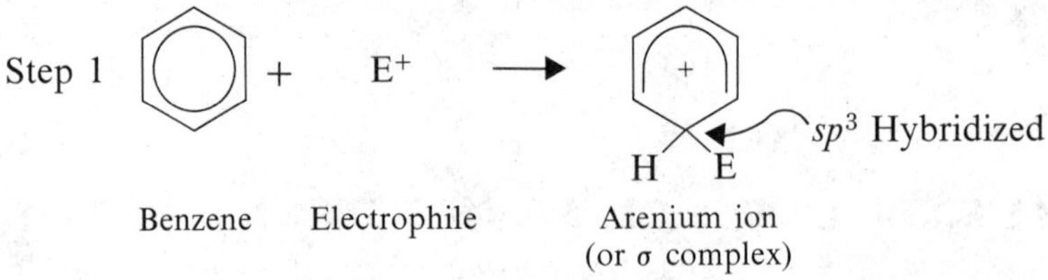

Step 1 Benzene + E^+ (Electrophile) → Arenium ion (or σ complex) [H, E; sp^3 Hybridized]

In step 1 the electrophile takes two electrons of the π system to form a σ bond between it and one carbon of the benzene ring. This interrupts the cyclic system of π electrons, because in the formation of the arenium ion one carbon becomes sp^3 hybridized and thus, no longer has an available p orbital. The four remaining π electrons of the arenium ion are delocalized over the five remaining sp^2 carbons.

In step 2 the arenium ion loses a proton from the carbon that bears the electrophile. The two electrons that bonded this proton to carbon become a part of the π system. The carbon that bears the electrophile becomes sp^2 hybridized again and a benzene derivative with six fully delocalized π electrons is formed.

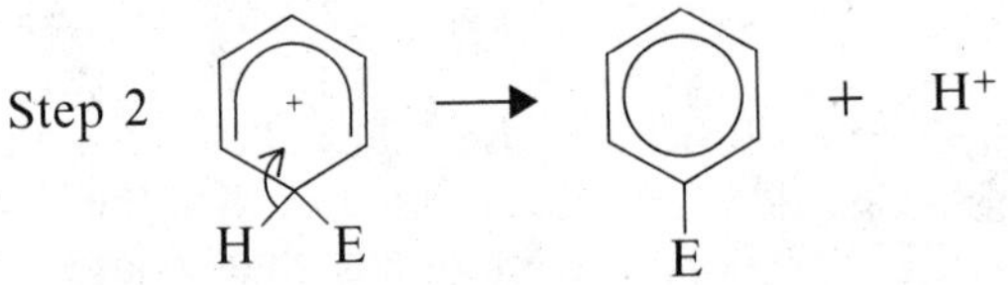

Step 2 (arenium ion, H E) → (benzene ring, E) + H^+

We can represent both steps of this mechanism with one of the Kekulé formulas for benzene. When we do this it becomes much easier to account for the π electrons. The first step can be shown in the following way:

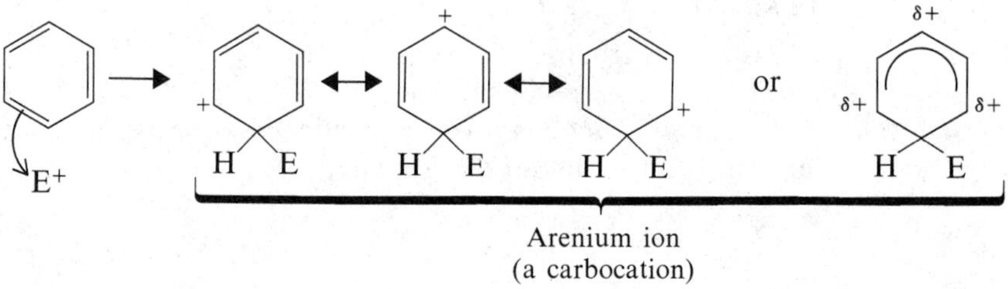

E^+ → H E ↔ H E ↔ H E or ($\delta+$, $\delta+$, $\delta+$) H E

Arenium ion
(a carbocation)

By using a Kekulé structure we can also see that the arenium ion is a hybrid of three *allylic-type* resonance structures each of which has a positive charge *on an ortho or para carbon of the ring.*

We can represent the second step, the loss of a proton, with any one of the resonance structures for the arenium ion.

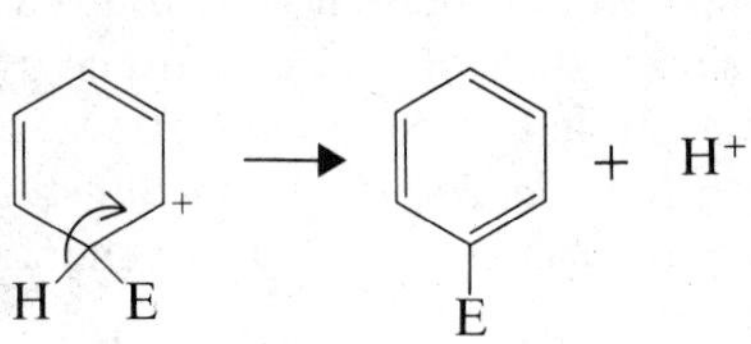

H E (+) → E + H^+

When we do this we can see clearly that two electrons from the carbon-hydrogen bond serve to regenerate the ring of alternating single and double bonds.

Problem 12.1

Show the second step of this mechanism using each of the other two resonance structures for the arenium ion.

There is firm experimental evidence that the arenium ion is a true *intermediate* in electrophilic substitution reactions. It is not a transition state. This means that in a potential-energy diagram (Fig. 12.1) the arenium ion lies in an energy valley between two transition states.

The energy of activation for the reaction leading from benzene and the electrophile, E^+, to the arenium ion [$E_{act(1)}$] has been shown to be much greater than the energy of activation [$E_{act(2)}$] leading from the arenium ion to the final product. This is consistent with what we would expect. The reaction leading from benzene and an electrophile to the arenium ion is highly endothermic, because the benzene ring loses its stabilization energy. The reaction leading from the arenium ion to the substituted benzene, by contrast, is highly exothermic because in it the benzene ring regains it stabilization energy.

Of the two steps shown below, step 1—the formation of the arenium ion—is the rate-limiting step in electrophilic aromatic substitution.

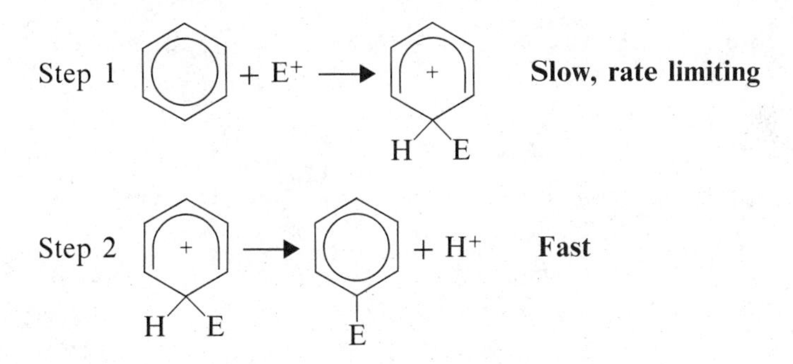

Step 2, the loss of a proton, occurs rapidly relative to step 1 and has no effect on the overall rate of reaction.

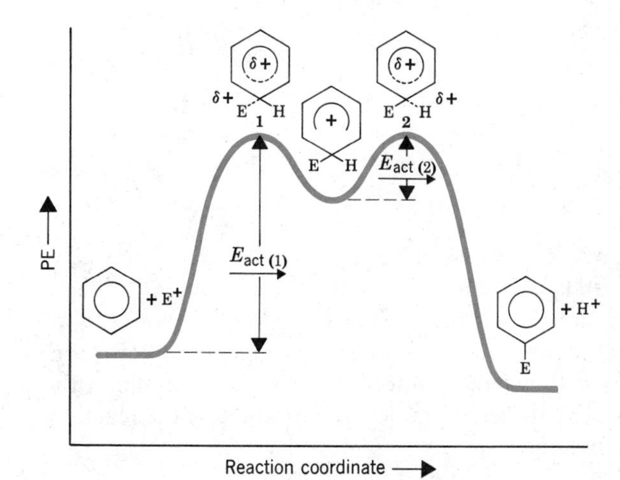

FIG. 12.1
*The potential energy diagram for an electrophilic aromatic substitution reaction. The arenium ion is a true intermediate lying between transition states **1** and **2**. In transition state **1** the bond between the electrophile and one carbon of the benzene ring is only partially formed. In transition state **2** the bond between the same benzene carbon and its hydrogen is partially broken.*

12.3 HALOGENATION OF BENZENE

Benzene does not react with bromine or chlorine unless a Lewis acid is present in the mixture. (As a consequence, benzene does not decolorize a solution of bromine in carbon tetrachloride.) When Lewis acids are present, however, benzene reacts readily with bromine or chlorine, and the reactions give bromobenzene and chlorobenzene in good yields.

$$C_6H_6 + Cl_2 \xrightarrow[25°]{FeCl_3} C_6H_5Cl + HCl$$

Chlorobenzene (90%)

$$C_6H_6 + Br_2 \xrightarrow[heat]{FeBr_3} C_6H_5Br + HBr$$

Bromobenzene (75%)

The Lewis acids most commonly used to effect chlorination and bromination reactions are $FeCl_3$, $FeBr_3$, and $AlCl_3$.

The mechanism for aromatic bromination is shown below.

Step 1 $$:\ddot{Br}-\ddot{Br}: + FeBr_3 \rightleftharpoons :\ddot{Br}-\overset{+}{\ddot{Br}}:\overset{-}{Fe}Br_3$$

Step 2 $$C_6H_6 + :\ddot{Br}-\overset{+}{\ddot{Br}}-\overset{-}{Fe}Br_3 \xrightarrow{Slow}$$

$$[C_6H_6Br]^+ \text{(resonance structures)} \longleftrightarrow \text{etc.} + \overset{-}{Fe}Br_4$$

Arenium ion

Step 3 $$[C_6H_6Br]^+ \; :\ddot{Br}-\overset{-}{Fe}Br_3 \xrightarrow{Fast} C_6H_5Br + H-\ddot{Br}: + FeBr_3$$

The function of the Lewis acid can be seen in steps 1 and 2. In step 1 the Lewis acid forms a complex with bromine in a Lewis acid-base reaction. In step 2 this complex acts as an electrophile and transfers a positive bromine ion, Br^+, to the benzene ring. The result is the formation of an arenium ion and an $FeBr_4^-$ ion.

In step 3 the arenium ion transfers a proton to $FeBr_4^-$. This results in the formation of bromobenzene and hydrogen bromide—the products of the reaction. At the same time this step regenerates the catalyst—$FeBr_3$.

One might ask, why aromatic brominations require the presence of a Lewis acid, whereas alkene brominations do not? The answer lies in the relative strengths with which the π electrons are held. The π electrons of alkenes are rather loosely held: so loosely, in fact, they can polarize the bromine-bromine bond and cause it, ultimately, to break.

$$\mathrm{C{=}C} + \overset{\delta+}{:\ddot{Br}}-\overset{\delta-}{\ddot{Br}:} \longrightarrow \mathrm{C{-}C}(\ddot{Br}:^{+}) + :\ddot{Br}:^{-}$$

An alkene

The π electrons of benzene, by contrast, are held more tightly. The π electrons of benzene are not able to polarize the bromine-bromine bond to an extent that will cause it to break. In the bromination of benzene, the Lewis acid assists by weakening the bromine-bromine bond through the formation of a complex.

We will see aromatic compounds later (e.g., anisole, aniline and phenol) in which the π electrons are more loosely held. These compounds react with bromine directly, without the intervention of a Lewis acid.

The mechanism of the chlorination of benzene in the presence of ferric chloride is analogous to the one for bromination. Ferric chloride serves the same purpose in aromatic chlorinations as ferric bromide does in aromatic brominations. It assists in the generation and transfer of a positive halogen ion.

It should be mentioned here that in *alkene* chlorinations Lewis acids are often added to the reaction mixture. The chlorine-chlorine bond is stronger than the bromine-bromine bond, and in alkene chlorinations Lewis acids ensure the transfer of Cl^{+} ions.

Problem 12.2

Very often in aromatic chlorinations and brominations, iron metal is added to the reaction mixture instead of ferric bromide or ferric chloride. Can you suggest what happens in these reactions?

Fluorine reacts so rapidly with benzene that aromatic fluorination requires special conditions and special types of apparatus. Even then, it is difficult to limit the reaction to monofluorination. Fluorobenzene can be made, however, by an indirect method that we will see in Chapter 18.

Iodine, on the other hand, is so unreactive that special techniques have to be used to effect direct iodination. One way is by adding silver perchlorate to the mixture.

$$C_6H_6 + I_2 + AgClO_4 \xrightarrow{25°} C_6H_5I\ (80\%) + AgI\downarrow + HClO_4$$

The mechanism of this reaction is not understood.

12.4 NITRATION OF BENZENE

Benzene reacts slowly with hot concentrated nitric acid to yield nitrobenzene. The reaction is much faster if it is carried out by heating benzene with a mixture of concentrated nitric acid and concentrated sulfuric acid.

$$C_6H_6 + HNO_3 + H_2SO_4 \xrightarrow{50\text{–}55°} C_6H_5NO_2 + H_3O^+ + HSO_4^-$$

(85%)

Concentrated sulfuric acid increases the rate of the reaction by increasing the concentration of the electrophile—the nitronium ion, NO_2^+.

Step 1 $H\text{—}\ddot{\underset{..}{O}}\text{—}NO_2 + HOSO_3H \rightleftharpoons H\text{—}\overset{+}{\ddot{O}}(H)\text{—}NO_2 + HSO_4^-$

(H_2SO_4)

Step 2 $H\text{—}\overset{+}{\ddot{O}}(H)\text{—}NO_2 + H_2SO_4 \rightleftharpoons \overset{+}{N}O_2 + H_3O^+ + HSO_4^-$

Nitronium ion

In the first step (above) nitric acid acts as a base and accepts a proton from the stronger acid, sulfuric acid. In the second step protonated nitric acid dissociates and produces a nitronium ion.

The nitronium ion reacts with benzene by attacking the π cloud and forming an arenium ion.

Step 3 $C_6H_6 + \overset{+}{N}O_2 \xrightarrow{\text{Slow}}$ [arenium ion resonance structures ↔ ↔]

Arenium ion

The arenium ion then transfers a proton to some base in the mixture such as HSO_4^- and becomes nitrobenzene.

Step 4 $\text{arenium ion} + HSO_4^- \xrightarrow{\text{Fast}} C_6H_5NO_2 + H_2SO_4$

The nitronium ion has been observed spectroscopically in mixtures of nitric and sulfuric acids. Further evidence that the nitronium ion is the electrophile in aromatic nitrations comes from research in which stable nitronium salts such as $NO_2^+ClO_4^-$ and $NO_2^+PF_6^-$ have been prepared and used to carry out nitration reactions.

Problem 12.3

Write equations that show how nitronium ions might be formed in nitration reactions in which concentrated nitric acid is used by itself.

12.5 SULFONATION OF BENZENE

Benzene reacts with fuming sulfuric acid at room temperature to produce benzenesulfonic acid. Fuming sulfuric acid is sulfuric acid that contains added sulfur trioxide, SO_3. Sulfonation also takes place in concentrated sulfuric acid alone, but more slowly.

$$C_6H_6 + SO_3 \xrightarrow[\text{conc. } H_2SO_4]{25^\circ} C_6H_5SO_3H$$

Sulfur trioxide

Benzenesulfonic acid (56%)

In either reaction the electrophile appears to be sulfur trioxide. In concentrated sulfuric acid, sulfur trioxide is produced in the following equilibrium in which H_2SO_4 acts as both an acid and a base.

Step 1 $2H_2SO_4 \rightleftharpoons SO_3 + H_3O^+ + HSO_4^-$

When sulfur trioxide reacts with benzene the following steps occur.

Step 2 $C_6H_6 + SO_3 \xrightleftharpoons{\text{Slow}} C_6H_6SO_3$ (arenium ion, three resonance structures with H and SO_3^-)

Arenium ion

Step 3 $C_6H_6SO_3 + HSO_4^- \rightleftharpoons C_6H_5SO_3^- + H_2SO_4$ *Fast*

Step 4 $C_6H_5SO_3^- + H_3O^+ \rightleftharpoons C_6H_5SO_3H + H_2O$ *Fast*

All of the steps are equilibria, including step 1 in which sulfur trioxide is formed from sulfuric acid. This means that the overall reaction is an equilibrium as well. In concentrated sulfuric acid, the overall equilibrium is the sum of steps 1 to 4.

$$C_6H_6 + H_2SO_4 \rightleftharpoons C_6H_5SO_3H + H_2O$$

In fuming sulfuric acid, step 1 is unimportant because the dissolved sulfur trioxide reacts directly.

Since all of the steps are equilibria, the position of equilibrium can be influenced by the conditions we employ. If we want to sulfonate benzene we use concentrated sulfuric acid—or better yet—fuming sulfuric acid. Under these conditions the position of equilibrium lies appreciably to the right and we obtain benzenesulfonic acid in good yield.

On the other hand, we may want to remove a sulfonic acid group from a benzene ring. To do this we employ dilute sulfuric acid and usually pass steam through the mixture. Under these conditions—with a high concentration of water—the equilibrium lies appreciably to the left and desulfonation occurs. The equilbrium is shifted even further to the left with volatile aromatic compounds because the aromatic compound distills with the steam.

We will see later that sulfonation and desulfonation reactions are often used in synthetic work. We may, for example, introduce a sulfonic acid group into a benzene ring to influence the course of some further reaction. Later, we may remove the sulfonic acid group by desulfonation.

Problem 12.4

In most desulfonation reactions the electrophile is a proton. Other electrophiles may be used, however. (a) Show all steps of the desulfonation reaction that would occur when benzenesulfonic acid is desulfonated with deuterium sulfate, D_2SO_4, dissolved in D_2O. (b) When benzenesulfonic acid reacts with bromine in the presence of ferric bromide, bromobenzene is obtained from the reaction mixture. What is the electrophile in this reaction? Show all steps in the mechanism for this desulfonation.

12.6 FRIEDEL-CRAFTS ALKYLATION

In 1877 a French chemist, Charles Friedel, and his American collaborator, James M. Crafts, discovered new methods for the preparation of alkylbenzenes (ArR) and acylbenzenes (ArCOR). These reactions are now called the Friedel-Crafts alkylation and acylation reactions. We will study the Friedel-Crafts alkylation reaction here and take up the Friedel-Crafts acylation reaction in Section 12.7.

A general equation for a Friedel-Crafts alkylation reaction is the following:

$$C_6H_6 + R{-}X \xrightarrow{AlCl_3} C_6H_5{-}R + HX$$

The mechanism for the reaction (below—using isopropyl chloride as R—X) starts with the formation of a carbocation (step 1). The carbocation then acts as an electrophile (step 2) and attacks the benzene ring forming an arenium ion. The arenium ion (step 3) then loses a proton generating isopropylbenzene.

Step 1

$$(CH_3)_2CH{-}Cl + AlCl_3 \rightleftharpoons (CH_3)_2CH^+ + AlCl_4^-$$

Isopropyl chloride → Carbocation

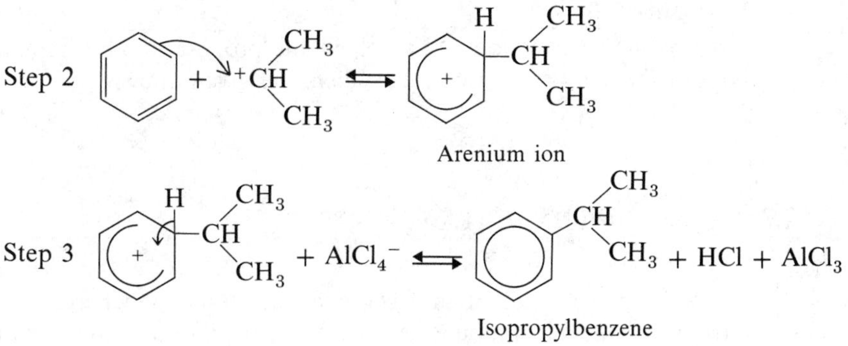

Step 2 (benzene + $^{+}CH(CH_3)_2$ ⇌ arenium ion, H and $CH(CH_3)_2$ on the same ring carbon)

Arenium ion

Step 3 (arenium ion + $AlCl_4^-$ ⇌ $C_6H_5CH(CH_3)_2$ + $HCl + AlCl_3$)

Isopropylbenzene

When R—X is a primary halide a full carbocation probably does not form. Rather the aluminum chloride forms a complex with the alkyl halide and this complex acts as the electrophile. The complex is one in which the carbon-halogen is nearly broken—and one in which the carbon has a considerable positive charge:

$$\overset{\delta+}{R}CH_2\text{---}Cl:\overset{\delta-}{Al}Cl_3$$

Even though this complex is not a full carbocation, it acts as if it were and it transfers a positive alkyl group to the aromatic ring. As we will see in Section 12.12, these complexes are so carbocationlike that they also undergo typical carbocation rearrangements.

Friedel-Crafts alkylations are not restricted to the use of alkyl halides and aluminum chloride. Many other pairs of reagents that form carbocations (or carbocationlike species) may be used as well. These possibilities include the use of a mixture of an alkene and an acid.

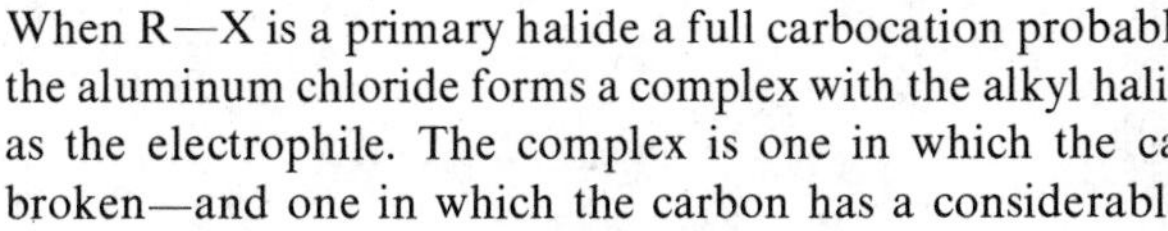

$$C_6H_6 + CH_3CH{=}CH_2 \xrightarrow[HF]{0^\circ} C_6H_5CH(CH_3)_2$$

Propene — Isopropylbenzene (84%)

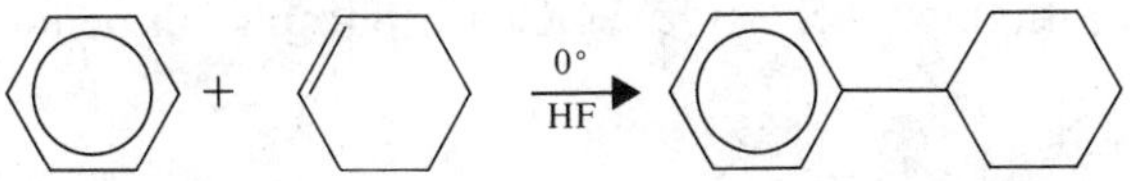

Benzene + Cyclohexene $\xrightarrow[HF]{0^\circ}$ Cyclohexylbenzene (62%)

A mixture of an alcohol and an acid may also be used

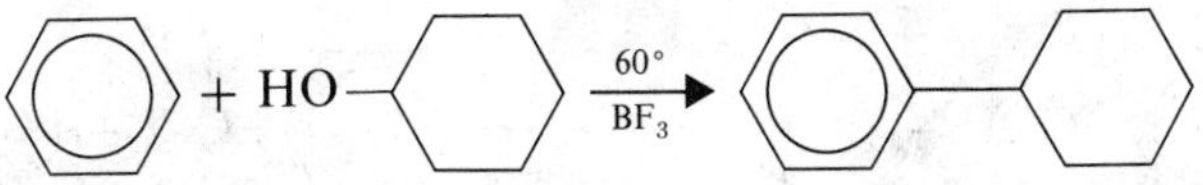

Benzene + HO–Cyclohexane $\xrightarrow[BF_3]{60^\circ}$ Cyclohexylbenzene (56%)

Cyclohexanol — Cyclohexylbenzene (56%)

There are several important limitations of the Friedel-Crafts alkylation reaction. These are discussed in Section 12.12B.

Problem 12.5

Assume that carbocations are involved and propose step-by-step mechanisms for both of the syntheses of cyclohexylbenzene given above.

12.7 FRIEDEL-CRAFTS ACYLATION

The $RC(=O)—$ group is called an *acyl group,* and a reaction whereby an acyl group is introduced into a compound is called an *acylation* reaction. Two common acyl groups are the acetyl group and the benzoyl group.

$$CH_3C(=O)— \qquad C_6H_5—C(=O)—$$

Acetyl group (ethanoyl group) — Benzoyl group

The Friedel-Crafts acylation reaction is an effective means of introducing an acyl group into an aromatic ring. The reaction is often carried out by treating the aromatic compound with an acyl halide. Unless the aromatic compound is one that is highly reactive, the reaction requires the addition of at least one equivalent of a Lewis acid (such as $AlCl_3$) as well. The product of the reaction is an aryl ketone.

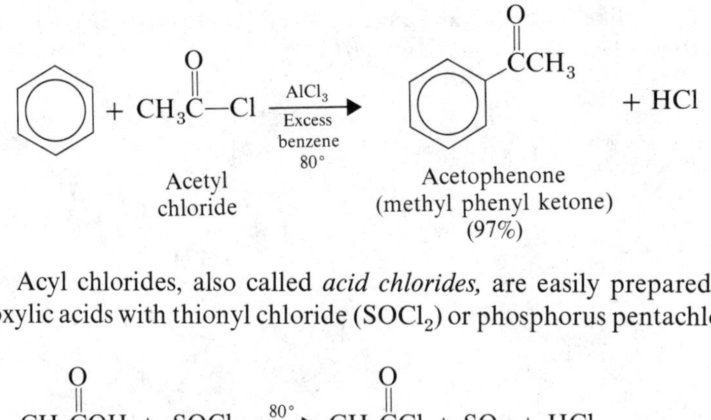

$$C_6H_6 + CH_3C(=O)—Cl \xrightarrow[\text{80°}]{AlCl_3,\ \text{Excess benzene}} C_6H_5C(=O)CH_3 + HCl$$

Acetyl chloride → Acetophenone (methyl phenyl ketone) (97%)

Acyl chlorides, also called *acid chlorides,* are easily prepared by treating carboxylic acids with thionyl chloride ($SOCl_2$) or phosphorus pentachloride (PCl_5).

$$CH_3C(=O)OH + SOCl_2 \xrightarrow{80°} CH_3C(=O)Cl + SO_2 + HCl$$

Acetic acid + Thionyl chloride → Acetyl chloride (80–90%)

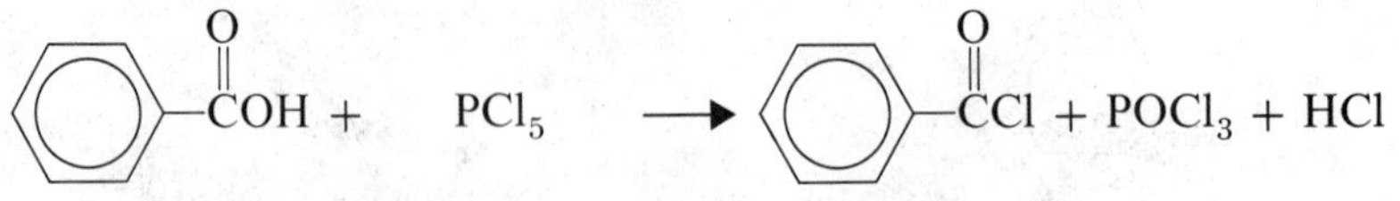

$$C_6H_5—C(=O)OH + PCl_5 \longrightarrow C_6H_5—C(=O)Cl + POCl_3 + HCl$$

Benzoic acid + Phosphorus pentachloride → Benzoyl chloride (90%)

Friedel-Crafts acylations can also be carried out using carboxylic acid anhydrides. For example:

$$C_6H_6 + (CH_3CO)_2O \xrightarrow[\text{Excess benzene, } 80^\circ]{AlCl_3} C_6H_5COCH_3 + CH_3COOH$$

Acetic anhydride (a carboxylic acid anhydride) — Acetophenone (82–85%)

In most Friedel-Crafts acylations the electrophile appears to be an *acylium ion* formed from an acyl halide in the following way.

$$R{-}C({=}O){-}Cl + AlCl_3 \rightleftharpoons R{-}C({=}O){-}\overset{+}{Cl}{:}\overset{-}{AlCl_3}$$

$$R{-}C({=}O){-}\overset{+}{Cl}{:}\overset{-}{AlCl_3} \rightleftharpoons \underbrace{R{-}\overset{+}{C}{=}O{:} \longleftrightarrow R{-}C{\equiv}\overset{+}{O}{:}}_{\text{An acylium ion (a resonance hybrid)}} + AlCl_4^-$$

An acylium ion
(a resonance hybrid)

Problem 12.6

Show how an acylium ion could be formed from an acid anhydride.

The remaining steps in the Friedel-Crafts acylation of benzene are the following.

$$C_6H_6 + R{-}C{\equiv}\overset{+}{O}{:} \longrightarrow C_6H_6^+(H)(COR)$$

Arenium ion

$$C_6H_6^+(H)(COR) + AlCl_4^- \longrightarrow C_6H_5COR + HCl + AlCl_3$$

$$C_6H_5C(R){=}\ddot{O}{:} + AlCl_3 \rightleftharpoons C_6H_5C(CH_3){=}\overset{+}{O}{:}\overset{-}{AlCl_3}$$

In the last step (above) aluminum chloride (a Lewis acid) forms a complex with the ketone (a Lewis base). After the reaction is over, treating the complex with water liberates the ketone.

$$(R)(C_6H_5)C{=}\overset{+}{O}{:}\overset{-}{AlCl_3} + H_2O \longrightarrow (R)(C_6H_5)C{=}\ddot{O}{:} + AlCl_2(OH) + HCl$$

Several important synthetic applications of the Friedel-Crafts acylation reaction are given in Section 12.12C.

12.8 EFFECT OF SUBSTITUENTS: REACTIVITY AND ORIENTATION

When substituted benzenes undergo electrophilic attack, groups already on the ring affect both the rate of the reaction and the site of attack. We say, therefore, that substituent groups affect both *reactivity* and *orientation* in electrophilic aromatic substitutions.

We can divide substituent groups into two classes according to their influence on the reactivity of the ring. Those that cause the ring to be more reactive than benzene itself we call *activating groups*. Those that cause the ring to be less reactive than benzene we call *deactivating groups*.

We also find that we can divide substituent groups into two classes according to the way they influence the orientation of attack by the incoming electrophile. Substituents in one class tend to bring about electrophilic substitution primarily at the positions *ortho* and *para* to themselves. We call these groups *ortho-para directors* because they tend to *direct* the incoming group into the ortho and para positions. Substituents in the second category tend to direct the incoming electrophile to the *meta* position. We call these groups *meta directors*.

Several examples will illustrate more clearly what we mean by these terms.

12.8A Activating Groups—Ortho-Para Directors

The methyl is an *activating* group and an ortho-para director. Toluene reacts considerably faster than benzene in all electrophilic substitutions.

More reactive toward electrophilic substitution | Less reactive toward electrophilic substitution

We observe the greater reactivity of toluene in several ways. We find, for example, that with toluene, milder conditions—lower temperatures and lower concentrations of the electrophile—can be used in electrophilic substitutions than with benzene. We also find that under the same conditions, toluene reacts faster than benzene. In nitration, for example, toluene reacts 25 times as fast as benzene.

We find, moreover, that when toluene undergoes electrophilic substitution, most of the substitution takes place at its ortho and para positions. When we nitrate toluene using nitric and sulfuric acid we get mononitrotoluenes in the following relative proportions.

$$C_6H_5CH_3 \xrightarrow[H_2SO_4]{HNO_3} o\text{-}CH_3C_6H_4NO_2 + p\text{-}CH_3C_6H_4NO_2 + m\text{-}CH_3C_6H_4NO_2$$

Ortho	*Para*	*Meta*
o-Nitrotoluene (59%)	*p*-Nitrotoluene (37%)	*m*-Nitrotoluene (4%)

Of the mononitrotoluene obtained from the reaction, 96% (59% + 37%) has the nitro group in an ortho or para position. Only 4% has the nitro group in a meta position.

Problem 12.7

What percentage of each nitrotoluene would you expect if substitution were to take place on a purely *statistical* basis?

Predominant substitution at the ortho and para positions of toluene is not restricted to nitration reactions as the following examples show.

CH_3 (toluene) $\xrightarrow[FeBr_3]{Br_2}$ (40%) *o*-Bromotoluene + (60%) *p*-Bromotoluene + (trace) *m*-Bromotoluene

CH_3 (toluene) $\xrightarrow[H_2SO_4,\ 0°]{SO_3}$ (43%) *o*-Toluenesulfonic acid + (53%) *p*-Toluenesulfonic acid + (4%) *m*-Toluenesulfonic acid

Problem 12.8

When toluene is sulfonated at 100° the following proportions of monosubstituted products are obtained: *o*-toluenesulfonic acid (13%), *m*-toluenesulfonic acid (8%), *p*-toluenesulfonic acid (79%). (a) Keeping in mind that sulfonations are reversible, how can you account for the different relative percentages formed at the lower and higher temperature? (b) Which toluenesulfonic acid appears to be more stable, *o*-toluenesulfonic acid or *p*-toluenesulfonic acid?

All alkyl groups are activating groups, and they are all also ortho-para directors. Nitration of *tert*-butylbenzene, for example, takes place at a rate that is roughly 16 times as fast as for benzene, and gives the following percentages of mononitro products.

$C(CH_3)_3$ (tert-butylbenzene) $\xrightarrow[H_2SO_4]{HNO_3}$ (12%) Ortho + (80%) Para + (8%) Meta

If we compare the nitration reaction of *tert*-butylbenzene with the nitration reaction of toluene (p. 478), we can see an effect on the relative amounts of ortho and para product produced by the steric bulk of the *tert*-butyl group. As we would expect, this effect is most pronounced in ortho substitution. *Tert*-butylbenzene gives much less ortho-nitro product (12% versus 59%) and gives much more para-nitro product (80% versus 37%) than toluene. Steric effects are, as we might expect, always greater at positions ortho to the group on the ring.

The methoxyl group, CH_3O—, and the acetamido group, CH_3CONH—, are strong activating groups and both are ortho-para directors. Bromination of anisole proceeds readily in the absence of a catalyst.

OCH_3 OCH_3 Br OCH_3 OCH_3

$\xrightarrow[CH_3COOH]{Br_2}$ + +

Br Br

Anisole (4%) *o*-Bromo-anisole (96%) *p*-Bromo-anisole (trace) *m*-Bromo-anisole

Nitration of acetanilide takes place at zero degrees.

$NHCOCH_3$ $NHCOCH_3$ NO_2 $NHCOCH_3$ $NHCOCH_3$

$\xrightarrow[H_2SO_4]{HNO_3,\ 0°}$ + +

NO_2 NO_2

Acetanilide (19%) *o*-Nitro-acetanilide (79%) *p*-Nitro-acetanilide (2%) *m*-Nitro-acetanilide

The hydroxyl group and the amino group are very powerful activating groups and are also powerful ortho-para directors. Phenol and aniline react with bromine in water (no catalyst is required) to produce products in which both of the ortho positions and the para position are substituted. These tribromo products are obtained in nearly quantitative yield.

OH OH

$\xrightarrow[H_2O]{Br_2}$ Br Br ($\sim$100%)

Br 2,4,6-Tribromophenol

NH_2 NH_2

$\xrightarrow[H_2O]{Br_2}$ Br Br ($\sim$100%)

Br 2,4,6-Tribromoaniline

12.8B Deactivating Groups—Meta Directors

The nitro group is a very strong *deactivating* group. Nitrobenzene undergoes nitration at a rate that is 10^4 times slower than benzene. The nitro group is a meta director. When nitrobenzene is nitrated with nitric and sulfuric acid, 93% of the substitution occurs at the meta position.

NO_2 — $\xrightarrow[H_2SO_4]{HNO_3}$ — NO_2, NO_2 (6%) + NO_2, NO_2 (1%) + NO_2, NO_2 (93%)

The carboxyl group (—COOH) and the trifluoromethyl group (—CF_3) are also strong deactivating groups; they are also meta directors.

COOH + HNO_3 $\xrightarrow{H_2SO_4}$ COOH, NO_2 (19%) + COOH, NO_2 (1%) + COOH, NO_2 (80%)

Benzoic acid

CF_3 $\xrightarrow[H_2SO_4]{HNO_3}$ CF_3, NO_2 (~100%)

12.8C Halo Substituents: Deactivating Ortho-Para Directors

The chloro and bromo groups are moderate deactivating groups. Chlorobenzene and bromobenzene undergo nitration at rates that are, respectively, 33 and 30 times slower than for benzene. The chloro and bromo groups are ortho-para directors, however. The relative percentages of monosubstituted products that are obtained when chlorobenzene is chlorinated, brominated, nitrated, and sulfonated are shown in Table 12.1.

TABLE 12.1 Electrophilic Substitutions of Chlorobenzene

REACTION	ORTHO PRODUCT (percent)	PARA PRODUCT (percent)	TOTAL ORTHO AND PARA (percent)	META PRODUCT (percent)
Chlorination	39	55	94	6
Bromination	11	87	98	2
Nitration	30	70	100	—
Sulfonation	—	100	100	—

The corresponding results from electrophilic substitutions of bromobenzene are given in Table 12.2.

TABLE 12.2 Electrophilic Substitutions of Bromobenzene

REACTION	ORTHO PRODUCT (percent)	PARA PRODUCT (percent)	TOTAL ORTHO AND PARA (percent)	META PRODUCT (percent)
Chlorination	45	53	98	2
Bromination	13	85	98	2
Nitration	38	62	100	—
Sulfonation	—	100	100	—

12.8D Classification of Substituents

Studies like the ones that we have presented in this section have been done for a number of other substituted benzenes. The effects of these substituents on reactivity and orientation are included in Table 12.3.

Problem 12.9

What would be the major monochloro product (or products) formed when each of the following compounds reacts with chlorine in the presence of ferric chloride?

(a) Ethylbenzene, $C_6H_5CH_2CH_3$
(b) (Trifluoromethyl)benzene, $C_6H_5CF_3$
(c) Phenyltrimethylammonium chloride, $C_6H_5\overset{+}{N}(CH_3)_3\ Cl^-$
(d) Methyl benzoate, $C_6H_5COOCH_3$

TABLE 12.3 Effect of Substituents on Electrophilic Aromatic Substitution

ORTHO-PARA DIRECTORS	META DIRECTORS
Strongly Activating	*Moderately Deactivating*
$-\ddot{N}H_2$, $-\ddot{N}HR$, $-\ddot{N}R_2$	$-C{\equiv}N$
$-\ddot{O}H$	$-SO_3H$
Moderately Activating	$-COOH$, $-COOR$
$-\ddot{N}HCOCH_3$, $-\ddot{N}HCOR$	$-CHO$ $-COR$
$-\ddot{O}CH_3$, $-\ddot{O}R$	*Strongly Deactivating*
Weakly Activating	$-NO_2$
$-CH_3$, $-C_2H_5$, $-R$	$-NR_3^+$
$-C_6H_5$	$-CF_3$, $-CCl_3$
Weakly Deactivating	
$-\ddot{F}:$, $-\ddot{Cl}:$, $-\ddot{Br}:$, $-\ddot{I}:$	

12.9 THEORY OF ELECTROPHILIC AROMATIC SUBSTITUTION

12.9A Reactivity: The Effect of Electron-Releasing and Electron-Withdrawing Groups

We have now seen that certain groups *activate* the benzene ring toward electrophilic substitution, while other groups *deactivate* the ring. When we say that a group activates the ring, what we mean, of course, is that the group increases the relative rate of the reaction. We mean that an aromatic compound with an activating group reacts faster in electrophilic substitutions than benzene. When we say that a group deactivates the ring, we mean that an aromatic compound with a deactivating group reacts slower than benzene.

We have also seen that we can account for relative reaction rates by examining the transition state for the rate-limiting steps. We know that any factor that increases the energy of the transition state relative to that of the reactants decreases the relative rate of the reaction. It does this because it increases the energy of activation of the reaction. In the same way, any factor that decreases the energy of the transition state relative to that of the reactants, lowers the energy of activation, and increases the relative rate of the reaction.

The rate-limiting step in electrophilic substitutions of substituted benzenes is the step that results in the formation of the arenium ion. We can write the formula for a substituted benzene in a generalized way if we use the letter **S** to represent any ring substituent including hydrogen. (If **S** is hydrogen the compound is benzene itself.)

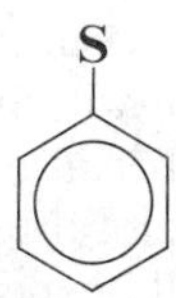

We can write the structure for the arenium ion in the way shown below. By this formula we mean that **S** can be in any position—ortho, meta, or para—relative to the electrophile, E.

S
+
E H

Using these conventions, then, we are able to write the rate-determining step for electrophilic aromatic substitution in the following general way.

S + E^+ ⟶ [S δ+ δ+ E H] ⟶ S + E H

Transition state

Arenium ion

When we examine this step for a large number of reactions we find that the relative rates of the reactions depend on whether **S** *withdraws* or *releases* electrons. If **S** is an electron-releasing group (relative to hydrogen) the reaction occurs faster than the corresponding reaction of benzene. If **S** is an electron-withdrawing group, the reaction is slower than that of benzene.

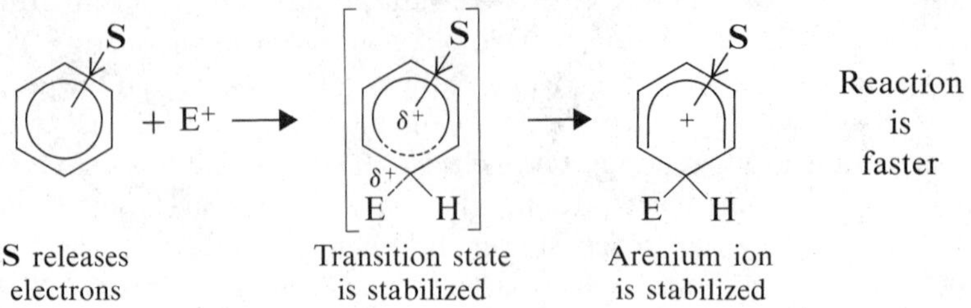

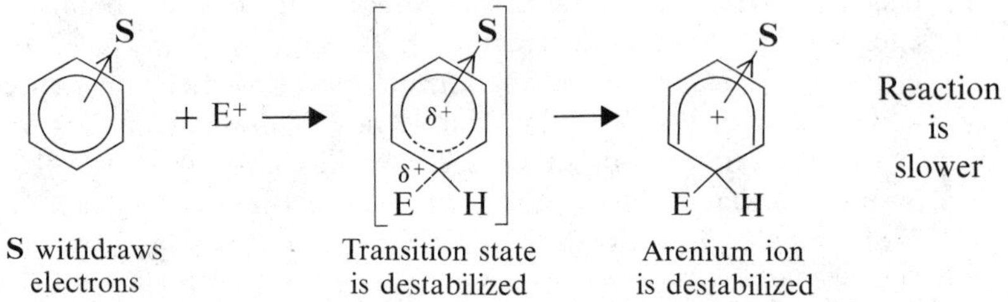

It appears then, that the substituent, **S**, must affect the stability of the transition state relative to that of the reactants. Electron-releasing groups apparently make the transition state more stable, while electron-withdrawing groups make it less stable. That this is so is entirely reasonable, because the transition state resembles the arenium ion, and the arenium ion is a delocalized *carbocation*.

Since the arenium ion is positively charged, we would expect an electron-releasing group to stabilize it *and the transition state leading to the arenium ion,* for the transition state is a developing delocalized carbocation. We can make the same kind of arguments about the effect of electron-withdrawing groups. An electron-withdrawing group should make the arenium ion *less* stable and in a corresponding way it should make the transition state leading to the arenium ion less stable.

Figure 12.2 shows how the electron-withdrawing abilities of substituents affect the relative energies of activation of electrophilic aromatic substitution reactions.

12.9B Inductive and Resonance Effects: Theory of Orientation

We can account for the electron-withdrawing and electron-releasing properties of groups on the basis of two factors: *an inductive effect and a resonance effect*. We will also see that these two factors determine orientation in aromatic substitution reactions.

The inductive effect of a group of atoms is the change in the electron density at a nearby atom caused by differences in electronegativity. The inductive effect can be "electron releasing" or "electron withdrawing," meaning that the electron density at the nearby site is either increased or decreased, respectively. Inductive effects are important in accounting for a great many phenomena in organic chemistry. In Chapter 17 we will see how inductive influences affect the

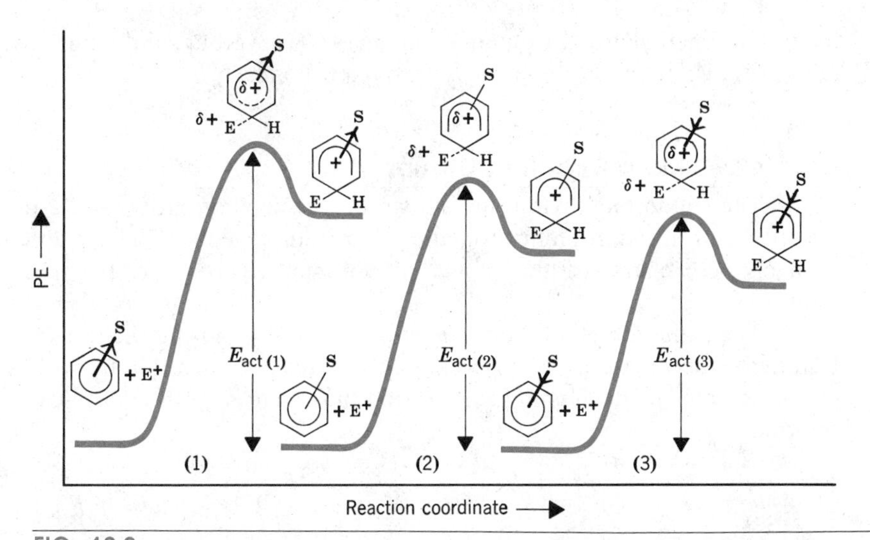

FIG. 12.2
*Energy profiles for the formation of the arenium ion in three electrophilic aromatic substitution reactions. In (1), **S** is an electron-withdrawing group. In (2) **S** = H. In (3) **S** is an electron-releasing group.* $E_{act(1)} > E_{act(2)} > E_{act(3)}$.

strengths of acids. Another example of an inductive effect—one that we will soon be able to relate to orientation in electrophilic aromatic substitution reactions—occurs in the addition of hydrogen chloride to (trifluoromethyl)ethylene (3,3,3-trifluoropropene). Hydrogen chloride adds to 3,3,3-trifluoropropene more slowly than to propene and the addition takes place in the manner shown below.

$$CF_3CH{=}CH_2 \xrightarrow{HCl} CF_3CH_2CH_2Cl$$

At first inspection this addition appears to occur in an anti-Markovnikov fashion (i.e., the hydrogen adds to the carbon with *fewer* hydrogens). If we examine the two possible carbocations that could be formed by addition of a proton to trifluoromethylethylene we will see that the addition is consistent with a modern statement of Markovnikov's rule. *The reaction proceeds through the formation of the more stable carbocation.*

$$CF_3{\leftarrow}CH{=}CH_2 \xrightarrow{H^+} \begin{cases} \xrightarrow[\times]{(1)} CF_3{\leftarrow}\overset{+}{C}H{-}CH_3 & \text{(not formed)} \\ \xrightarrow{(2)} CF_3{\leftarrow}CH_2{-}\overset{+}{C}H_2 & \text{(more stable)} \end{cases}$$

The trifluoromethyl group, because of the three highly electronegative fluorines, is strongly electron withdrawing. The carbocation that would be formed by path 1 would be highly unstable because even though it is a secondary carbocation, the positive charge is on the carbon adjacent to the highly electronegative trifluoromethyl group. Electron withdrawal from a positively charged site destabilizes the system. Consequently, this carbocation does not form, and the reaction follows path 2. The carbocation formed by path 2 is more stable even though it is a primary carbocation, because the positive charge is separated from the trifluoromethyl group by an intervening —CH_2— group, and inductive effects are not transmitted very effectively through σ bonds.

12.9C Meta-Directing Groups

The trifluoromethyl group is a strong deactivating group and a powerful meta director in electrophilic aromatic substitution reactions. We can account for both of these characteristics of the trifluoromethyl group on the basis of its inductive effect.

The trifluoromethyl group affects reactivity by causing the transition state leading to the arenium ion to be highly unstable. It does this by withdrawing electrons from the developing carbocation thus increasing the positive charge in the ring.

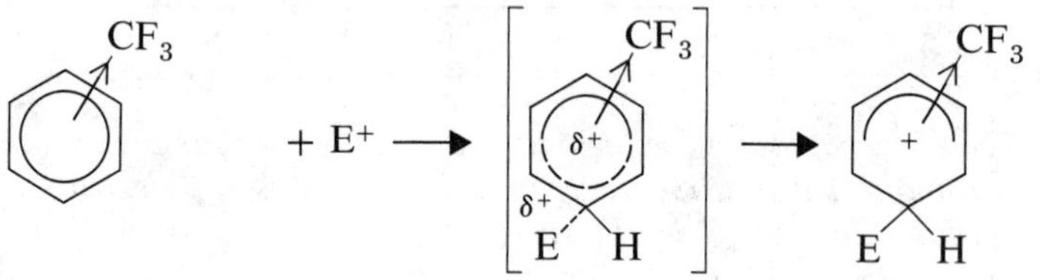

CF_3 + E^+ → [CF_3, $\delta+$, $\delta+$, E, H] → CF_3, +, E, H

(Trifluoromethyl)benzene Transition state Arenium ion

We can understand how the trifluoromethyl group affects *orientation* in electrophilic aromatic substitution if we examine the resonance structures for the arenium ion that would be formed when an electrophile attacks the ortho, meta, and para positions of (trifluoromethyl)benzene.

Ortho attack:

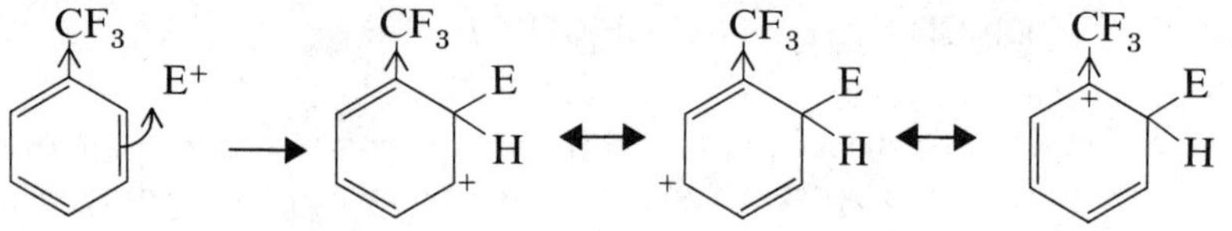

CF_3, E^+ → CF_3, E, H, + ↔ CF_3, E, H, + ↔ CF_3, +, E, H

Highly unstable

Meta attack:

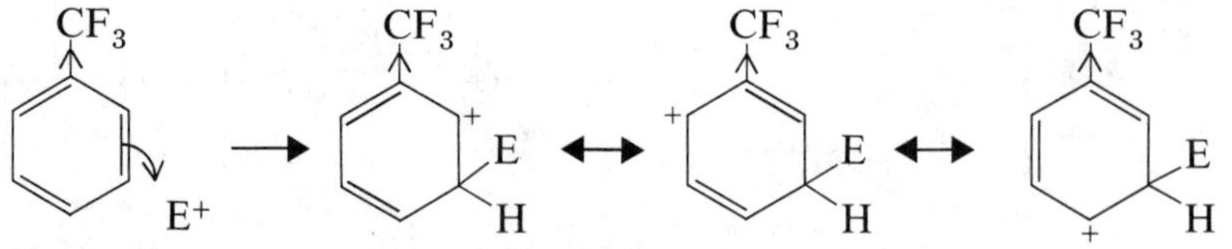

CF_3, E^+ → CF_3, +, E, H ↔ CF_3, +, E, H ↔ CF_3, E, H, +

Para attack:

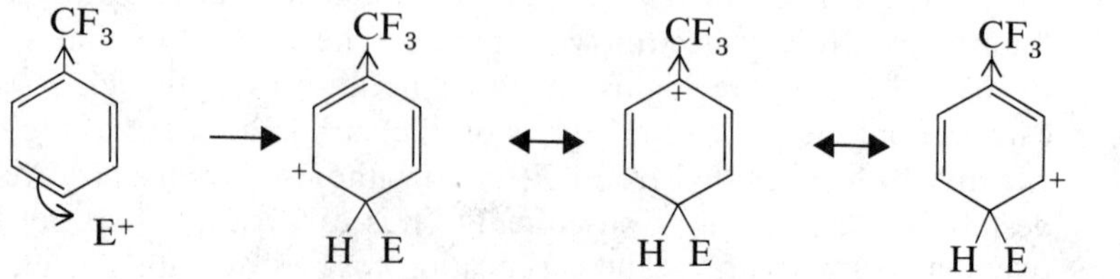

CF_3, E^+ → CF_3, +, H E ↔ CF_3, +, H E ↔ CF_3, +, H E

Highly unstable

We see in the resonance structures for the arenium ion arising from ortho and para attack, that *one contributing structure is highly unstable relative to all the others because the positive charge is located on the ring carbon that bears the electron-withdrawing group*. We see *no* such highly unstable resonance structure in the arenium ion arising from *meta* attack. This means that the arenium ion formed by meta attack should be the most stable of the three. By the usual reasoning we would also expect the transition state leading to the meta arenium ion to be the most stable and, therefore, that meta attack would be favored. This is exactly what we find experimentally. The trifluoromethyl group is a powerful meta director.

$$C_6H_5CF_3 + HNO_3 \xrightarrow{H_2SO_4} m\text{-}O_2NC_6H_4CF_3$$

(Trifluoromethyl)benzene (benzotrifluoride) — (100%)

Problem 12.10

(a) What product would you expect from the addition of hydrogen chloride to vinyltrimethylammonium chloride, $CH_2{=}CH\overset{+}{N}(CH_3)_3\ Cl^-$? (b) Would you expect this addition to occur slower or faster than the addition of hydrogen chloride to propene? (c) How do you account for the fact that the trimethylammonium group, $(CH_3)_3\overset{+}{N}—$, is a strong deactivating group in electrophilic aromatic substitution? (d) Write resonance structures that account for the fact that the trimethylammonium group is also a meta-directing group.

Problem 12.11

(a) When the following trimethylammonium compounds were subjected to nitration they gave the following percentages of meta product: **A,** 100%; **B,** 88%; **C,** 19%; **D,** 5%.
How can you account for these results?

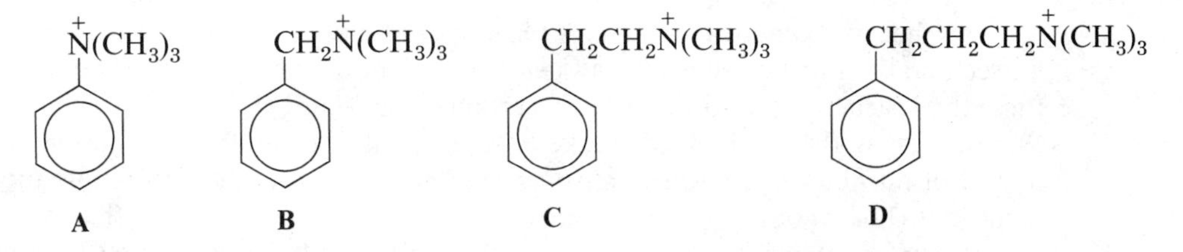

(b) How do you account for the percentages of meta product obtained from nitration of the following compounds?

CCl_3	$CHCl_2$	CH_2Cl	CH_3	
Benzotri-chloride	Benzal dichloride	Benzyl chloride	Toluene	
(64%)	(34%)	(14%)	(4%)	% meta product

The nitro group is another powerful electron-withdrawing group. The nitro group withdraws electrons through an inductive effect *and a resonance effect*. These two effects combine to cause a large deactivation of the ring.

The inductive effect of the nitro group arises from three electronegative atoms—a nitrogen and two oxygens. The electronegativities of these atoms combine to make the nitro group as a whole strongly electronegative. The inductive effect of the nitro group increases the energy of the transition state leading to the arenium ion by withdrawing electrons from it. This accounts, in part, for the deactivating effect of the nitro group.

In order to see how the nitro group asserts its resonance effect and how this affects the energy of activation in electrophilic aromatic substitution, we need to examine some of the resonance structures for nitrobenzene, itself, and then some of the resonance structures for the arenium ion.

Resonance structures for nitrobenzene

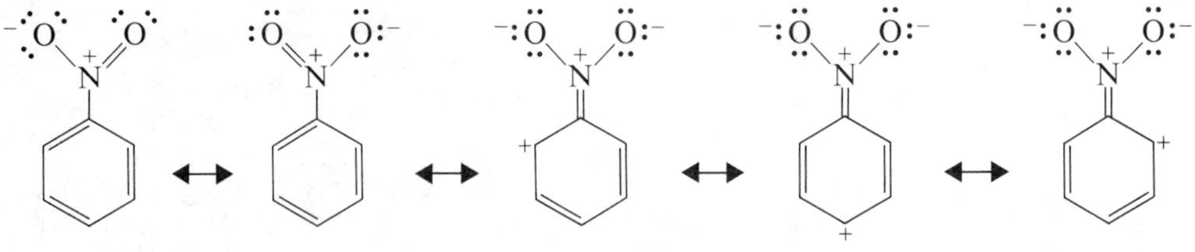

Resonance structures for the meta-substituted arenium ion

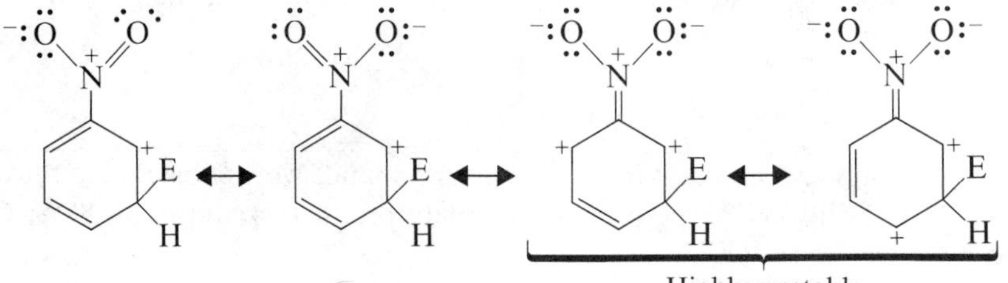

Highly unstable

When we examine the last three resonance structures for nitrobenzene, itself, we see that the nitro group withdraws electrons from the benzene ring. In each of these three structures an ortho or para carbon of the ring bears a positive charge. We know that these structures make a substantial contribution to the overall hybrid of nitrobenzene because nitrobenzene has an unusually high dipole moment (3.95 D). Another important effect of these contributors is one that bears directly on nitrobenzene's reactivity in electrophilic substitution: contributions made by these structures stabilize nitrobenzene. That is, they reduce its potential energy relative to that of benzene.

When we examine the last two structures for the arenium ion we see that they correspond to the last three contributing structures for nitrobenzene in the sense that in them electrons are withdrawn from the ring. These structures make only a *small contribution* to the hybrid for the arenium ion, however, because *they involve electron withdrawal from a ring that is already positively charged.* As a consequence, resonance stabilization *is also of little importance in the transition state leading to the arenium ion.*

Thus, the resonance effect and the inductive effect of the nitro group combine to make the energy of activation for an electrophilic attack on nitroben-

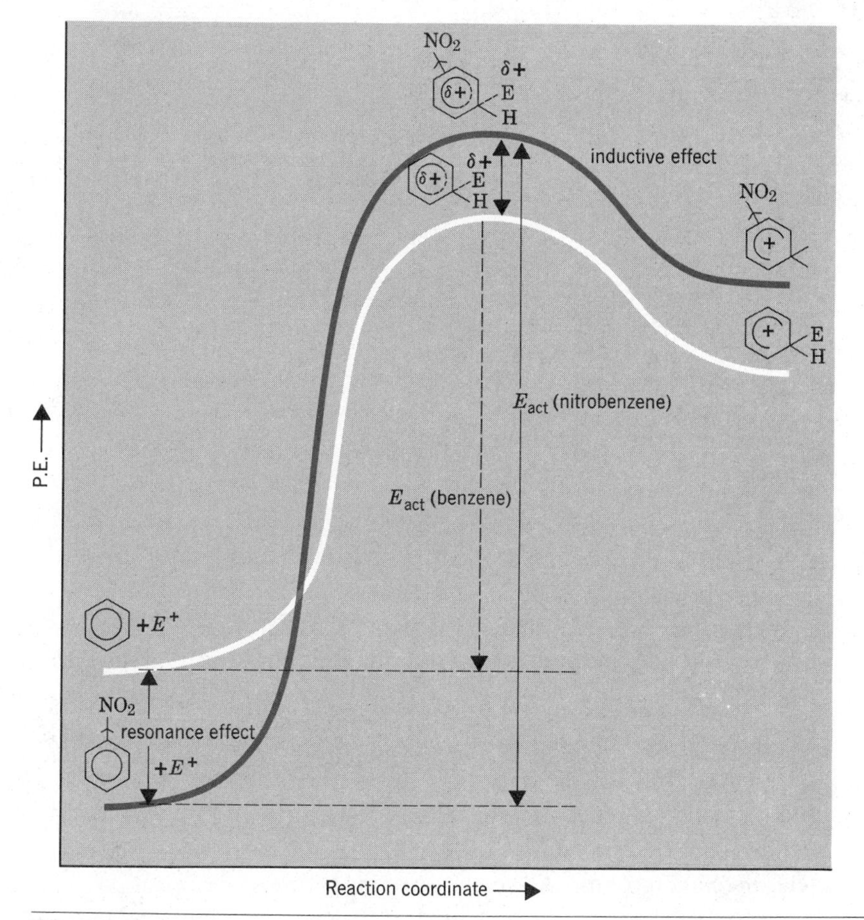

FIG. 12.3
Resonance and inductive effects combine to make the energy of activation for an electrophilic attack on nitrobenzene substantially greater than that for an electrophilic attack on benzene. (The two curves are superimposed only to make comparison of the relative effects easier.)

zene much greater than that for benzene (Fig. 12.3). The resonance effect lowers the potential energy of the reactants; the inductive effect raises the potential energy of the transition state. Thus, we can understand why the electrophilic substitution of nitrobenzene occurs at a rate that is approximately 10,000 times slower than that of benzene.

We can account for the meta-directing property of the nitro group in much the same way as we did for (trifluoromethyl) benzene. Since the nitro group is an electron-withdrawing group we can write resonance structures (below) for the arenium ions formed by ortho, meta, and para attack. When we do, we find that

Ortho attack:

NO_2 E^+ → NO_2 E H ⟷ NO_2 E H ⟷ NO_2 E H

Highly unstable

Meta attack:

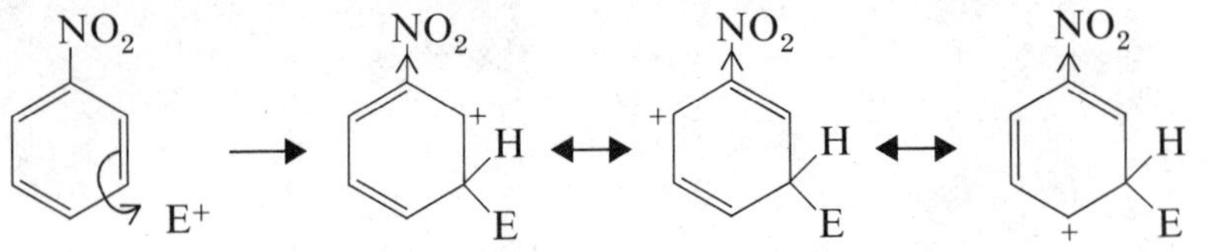

NO_2 NO_2 NO_2 NO_2

E+ H E H E H E

Para attack:

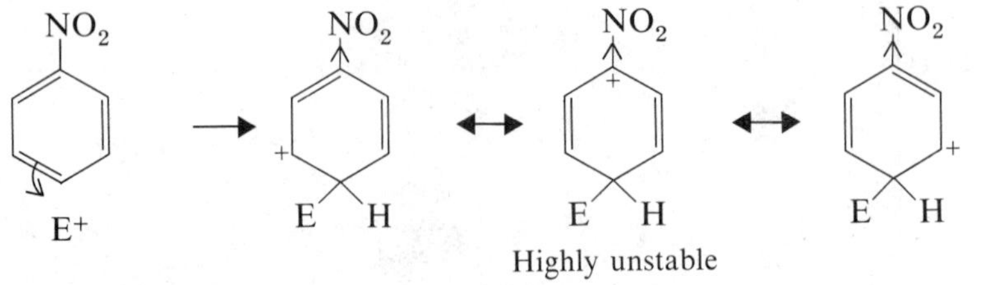

NO_2 NO_2 NO_2 NO_2

E+ E H E H E H

Highly unstable

the meta-substituted arenium ion is most stable of the three because no highly unstable structure contributes to its hybrid. Thus, the transition state leading to the meta-substituted arenium ion occurs at a lower potential energy than those for ortho and para attack, and meta substitution is favored. Meta substitution is favored, however, only in the sense that it is the least unfavorable of three unfavorable pathways.

The resonance explanation that we have given for the directive influence of the nitro and trifluoromethyl groups can be given for the directive influence of all meta-directing groups. All meta-directing groups are electron withdrawing; therefore, all destabilize the transition states leading to ortho and para-substituted arenium ions to a greater extent than they destabilize the transition state leading to meta-substituted arenium ions.

Resonance effects similar to those we saw in nitrobenzene also play a part in determining the reactivity of other aromatic compounds. Greater resonance stabilization of the reactant relative to that of the transition state leading to the arenium ion will be found for all those compounds containing meta-directing groups of the following type.

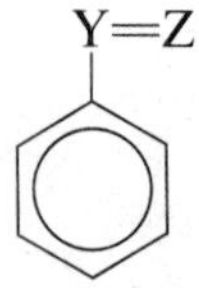

Y=Z

Z is an electronegative element

where Y=Z could be:

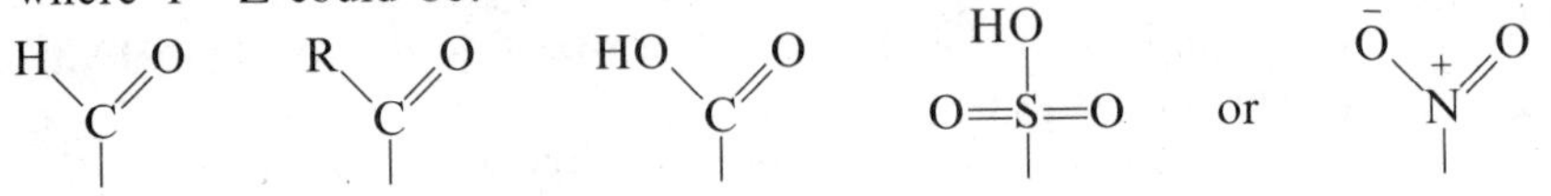

H–C=O, R–C=O, HO–C=O, HO–S(=O)=O or $^{-}$O–N^{+}=O

All of these groups make stabilizing resonance contributions to the hybrid of the reactant like those that follow:

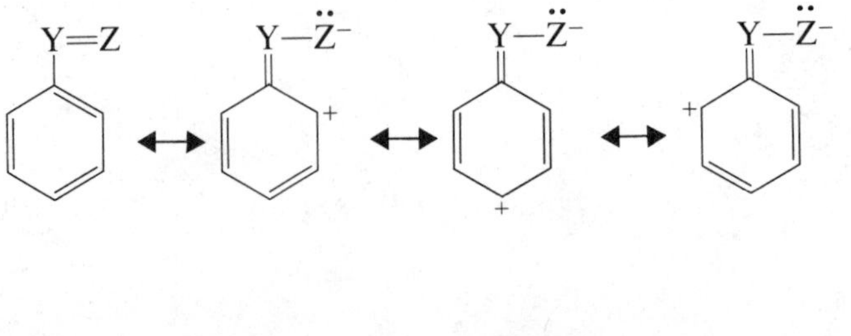

Y=Z ↔ Y–Z:$^{-}$ ↔ Y–Z:$^{-}$ ↔ Y–Z:$^{-}$

Since resonance structures of this type require the withdrawal of electrons from the ring, *similar stabilizing structures do not make significant contributions to the developing delocalized carbocation of the transition state.* This results in a large energy of activation for the electrophilic substitution reaction.

Problem 12.12

All of the following groups are of the general type shown above: —C(=O)—OH, —C(=O)—R, —C(=O)—H, —C≡N:. Write resonance structures that show how benzene compounds containing each of these groups would be stabilized by resonance.

12.9D Ortho-Para Directing Groups

Except for the alkyl and phenyl substituents, all of the ortho-para directing groups in Table 12.3 are of the general type shown below.

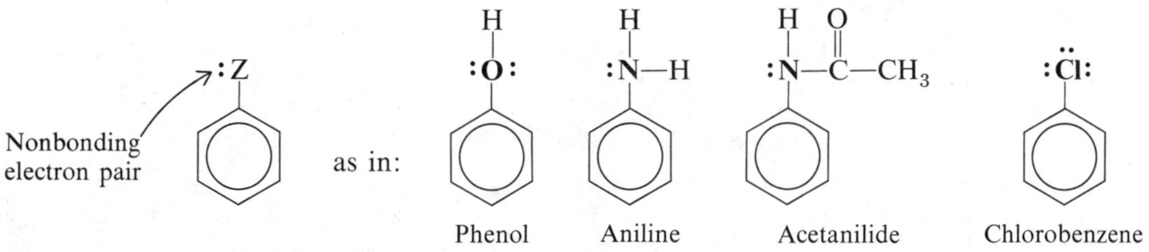

Nonbonding electron pair :Z (on benzene ring) as in: Phenol (:ÖH), Aniline (:NH—H), Acetanilide (:N(H)—C(=O)—CH_3), Chlorobenzene (:Cl:)

All of these ortho-para directors have at least one pair of nonbonding electrons on the atom adjacent to the benzene ring.

This structural feature—an unshared electron pair on the atom adjacent to the ring—determines the orientation and influences reactivity in electrophilic substitution reactions.

The *directive effect* of these groups with an unshared pair is predominantly caused by a resonance effect. This resonance effect, moreover, operates primarily in the arenium ion and, consequently, in the transition state leading to it.

Except for the halogens, which represent a special case, the primary effect on reactivity of these groups is also caused by a resonance effect. And, again, this effect operates primarily in the transition state leading to the arenium ion.

In order to understand these resonance effects let us begin by recalling the effect of the amino group on electrophilic aromatic substitution reactions. The amino group is not only a powerful activating group, it is also a powerful ortho-para director. We saw earlier (p. 480) that aniline reacts with bromine at room temperature and in the absence of a catalyst to yield a product in which both ortho positions and the para position are substituted.

The inductive effect of the amino group makes it slightly electron withdrawing. Nitrogen, as we know, is more electronegative than carbon. The difference between the electronegativities of nitrogen and carbon in aniline is not large, however, because the carbon of the benzene ring is sp^2 hybridized and thus is somewhat more electronegative than it would be if it were sp^3 hybridized.

The resonance effect of the amino group is far more important than its inductive effect in electrophilic aromatic substitution, and this resonance effect makes the amino group electron releasing. We can understand this effect if we write the resonance structures for the arenium ions that would arise from ortho, meta, and para attack on aniline.

Ortho attack:

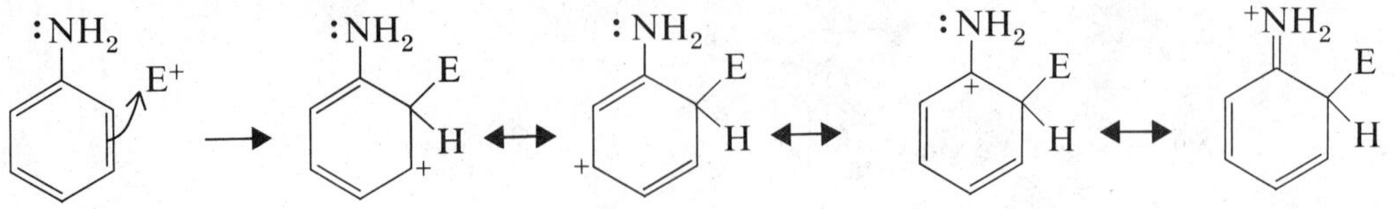

Relatively stable contributor

Meta attack:

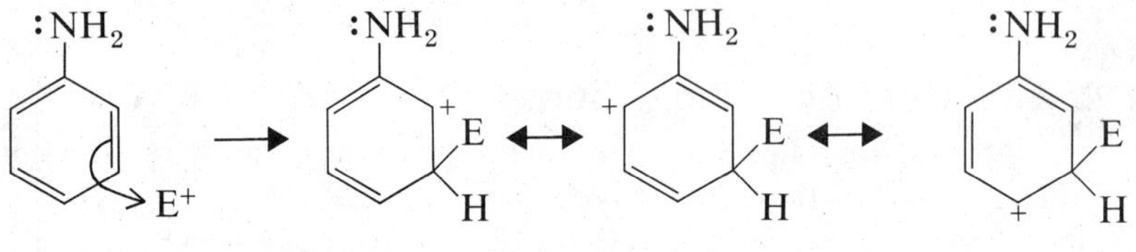

Para attack:

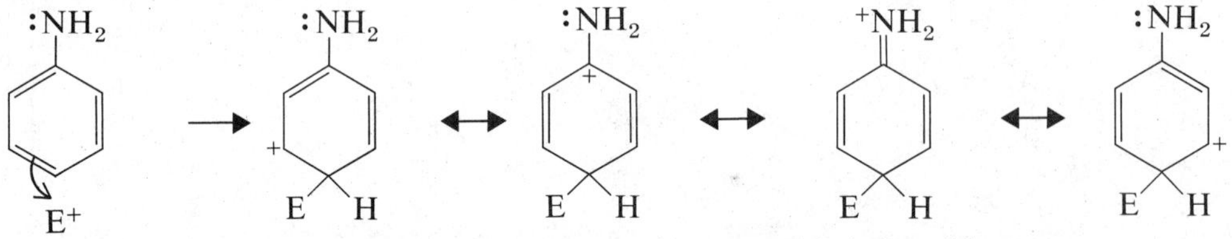

Relatively stable contributor

We see that four reasonable resonance structures can be written for the arenium ions resulting from ortho and para attack, whereas only three can be written for the arenium ion that results from meta attack. This, in itself, suggests that the ortho and para arenium ions should be more stable. Of greater importance, however, are the relatively stable structures that contribute to the hybrid for the ortho and para arenium ions. In these structures, nonbonding pairs of electrons from nitrogen form an extra bond to the carbon of the ring. This extra bond—and the fact that every atom in each of these structures has a complete outer octet of electrons—makes these structures the most stable of all of the contributors. Because these structures are unusually stable, they make a large—*and stabilizing*—contribution to the hybrid. This means, of course, that the ortho and para arenium ions themselves are considerably more stable than the arenium ion that results from the meta attack. The transition states leading to the ortho and para arenium ions occur at unusually low potential energies. As a result, electrophiles react at the ortho and para positions very rapidly.

Problem 12.13

(a) Write resonance structures for the arenium ions that would result from electrophilic attack on the ortho, meta, and para positions of phenol. (b) Can you account for the fact that phenol is highly susceptible to electrophilic attack? (c) Can you account for the fact that the hydroxyl group is an ortho and para director? (d) Would you expect the phenoxide ion, $C_6H_5—O^-$, to be more or less reactive than phenol in electrophilic substitution? (e) Explain.

Problem 12.14

(a) Ignore resonance structures involving electrons of the ring and write *one* other resonance structure for acetanilide. (Your structure will contain + and − charges.)

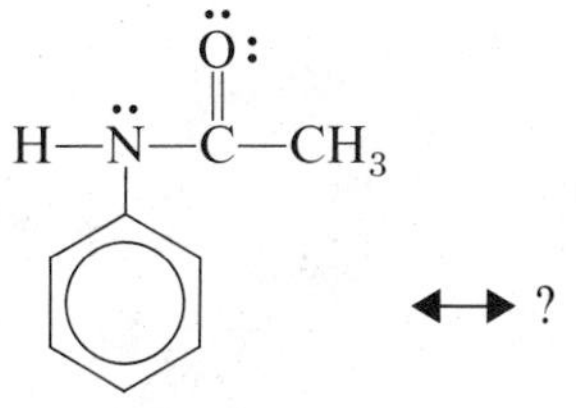

$$\text{H—}\ddot{\text{N}}\text{—C(=}\ddot{\text{O}}\text{:)—CH}_3 \;(\text{N bonded to }C_6H_5) \longleftrightarrow ?$$

Acetanilide

(b) Acetanilide is less reactive toward electrophilic substitution than aniline. How can you explain this on the basis of the resonance structure you have just written? (c) Acetanilide, however, is much more reactive than benzene and the acetamido group, $CH_3CONH—$, is an ortho-para director. Can you account for these facts in terms of resonance structures that involve the ring? (d) Would you expect phenyl acetate to be *more* or *less* reactive than phenol? Explain.

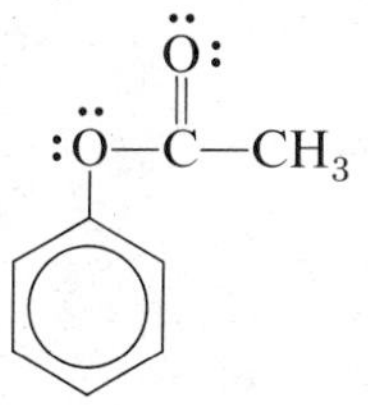

$$:\ddot{\text{O}}\text{—C(=}\ddot{\text{O}}\text{:)—CH}_3 \;(\text{O bonded to }C_6H_5)$$

Phenyl acetate

(e) What kind of directional influence would you expect the acetoxy group, $CH_3C(=\ddot{O}:)—\ddot{O}—$, to show? (f) Would you expect phenyl acetate to be *more* or *less* reactive in electrophilic substitution than benzene? Explain.

The directive and reactivity effects of halo substituents seem to be contradictory. The halo groups are the only ortho-para directors that are also deactivating groups. All other deactivating groups (Table 12.3) are meta directors. We can account for the seemingly contradictory behavior of halo substituents if we assume that their inductive effect influences reactivity and their resonance effect governs orientation.

Let us apply these assumptions specifically to chlorobenzene. The chloro group is highly electronegative. Thus, we would expect a chloro group to withdraw electrons from the benzene ring and thereby deactivate it.

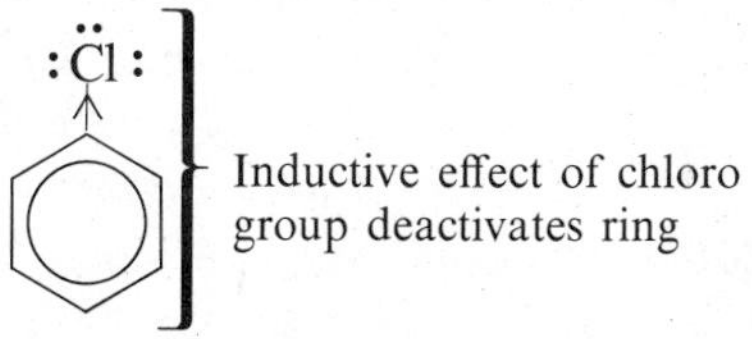

$:\ddot{\text{Cl}}:$ (↑ from benzene ring) — Inductive effect of chloro group deactivates ring

On the other hand, when electrophilic attack does take place, the chloro group stabilizes the arenium ions resulting from ortho and para attack. The chloro group does this in the same way as amino groups and hydroxyl groups do—*by donating an unshared pair of electrons*. These electrons give rise to relatively stable

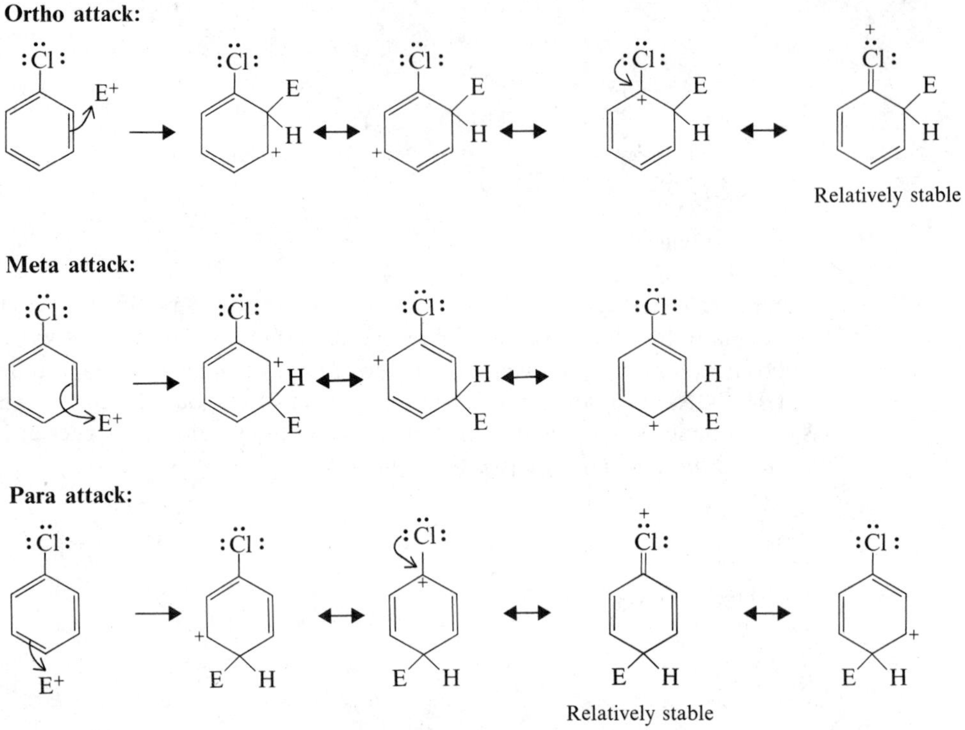

Ortho attack:

:Cl: E+ → :Cl: E H + ↔ :Cl: E H + ↔ :Cl: + E H ↔ +Cl: E H

Relatively stable

Meta attack:

:Cl: E+ → :Cl: + H E ↔ :Cl: + H E ↔ :Cl: H + E

Para attack:

:Cl: E+ → :Cl: + E H ↔ :Cl: + E H ↔ +Cl: E H ↔ :Cl: + E H

Relatively stable

resonance structures contributing to the hybrids for the ortho and para arenium ions.

What we have said about chlorobenzene is, of course, true of bromobenzene.

We can summarize the inductive and resonance effects of halo substituents in the following way. Through their inductive effect halo groups make the ring more positive than that of benzene. This causes the energy of activation for any electrophilic aromatic substitution reaction to be greater than that for benzene, and, therefore, halo groups are deactivating. Through their resonance effect, however, halo substituents cause the energies of activation leading to ortho and para substitution to be lower than the energy of activation leading to meta substitution. This makes halo substituents ortho-para directors.

Problem 12.15

The trifluoromethyl group and the chloro group both withdraw electrons inductively, and both (trifluoromethyl) ethylene and chloroethene add hydrogen chloride more slowly than ethene. The mode of addition of hydrogen chloride to chloroethene, however, is opposite that of 3,3,3-trifluoropropene. How can you account for this?

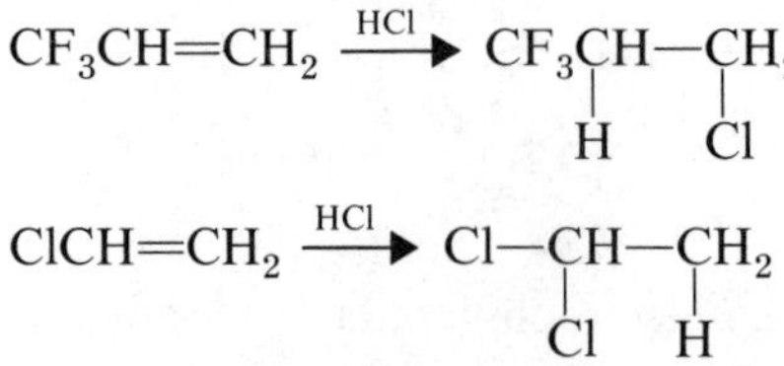

$$CF_3CH{=}CH_2 \xrightarrow{HCl} CF_3\underset{\displaystyle H}{\underset{|}{C}}H{-}\underset{\displaystyle Cl}{\underset{|}{C}}H_2$$

$$ClCH{=}CH_2 \xrightarrow{HCl} Cl{-}\underset{\displaystyle Cl}{\underset{|}{C}}H{-}\underset{\displaystyle H}{\underset{|}{C}}H_2$$

You may have noticed an apparent contradiction between the rationale offered for the unusual effects of the halogens and that offered earlier for amino or hydroxyl groups. That is, oxygen is *more* electronegative than chlorine or bromine (and especially iodine). Yet, the hydroxyl group is an activating group while these halogens are deactivating groups. An explanation for this can be obtained if we consider the relative stabilizing contributions made to the transition state leading to the arenium ion by resonance structures involving a group —Z̈ (—Z̈ = —N̈H$_2$, —Ö—H, —F̈:, —C̈l:, —B̈r:, —Ï:) that is directly attached to the benzene ring in which Z donates an electron pair. If —Z̈ is —ÖH or —N̈H$_2$, these resonance structures arise because of the overlap of a 2*p* orbital of carbon with that of oxygen or nitrogen. Such overlap is favorable because the atoms are almost the same size. With chlorine, however, donation of an electron pair to the benzene ring requires overlap of a carbon 2*p* orbital with a chlorine 3*p* orbital. Such overlap is less effective; the chlorine atom is much larger and its 3*p* orbital is much further from its nucleus. With bromine and iodine overlap is even less effective. Justification for this explanation can be found in the observation that fluorobenzene (Z = —F̈:) is the most reactive halobenzene in spite of the high electronegativity of fluorine and that —F̈: is the most powerful ortho-para director of the halogens. With fluorine, donation of an electron pair arises from overlap of a 2*p* orbital of fluorine with a 2*p* orbital of carbon (as with —N̈H$_2$ and —ÖH). This overlap is effective because the atoms =C— and—F̈: are of the same relative size.

12.9E Reactivity and Orientation of Alkylbenzenes

Alkyl groups can be much better electron-releasing groups than hydrogen. Because of this they can activate a benzene ring toward electrophilic substitution by stabilizing the transition state lending to the arenium ion:

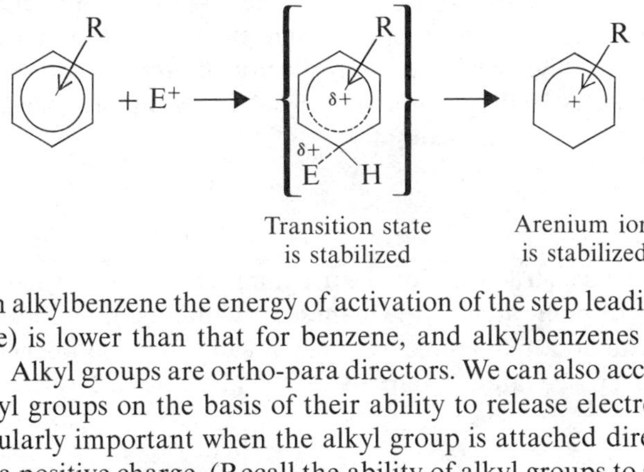

Transition state is stabilized

Arenium ion is stabilized

For an alkylbenzene the energy of activation of the step leading to the arenium ion (above) is lower than that for benzene, and alkylbenzenes react faster.

Alkyl groups are ortho-para directors. We can also account for this property of alkyl groups on the basis of their ability to release electrons—an effect that is particularly important when the alkyl group is attached directly to a carbon that bears a positive charge. (Recall the ability of alkyl groups to stabilize carbocations that we discussed in Section 4.14.)

If, for example, we write resonance structures for the arenium ions formed when toluene undergoes electrophilic substitution we get the following result:

Ortho attack:

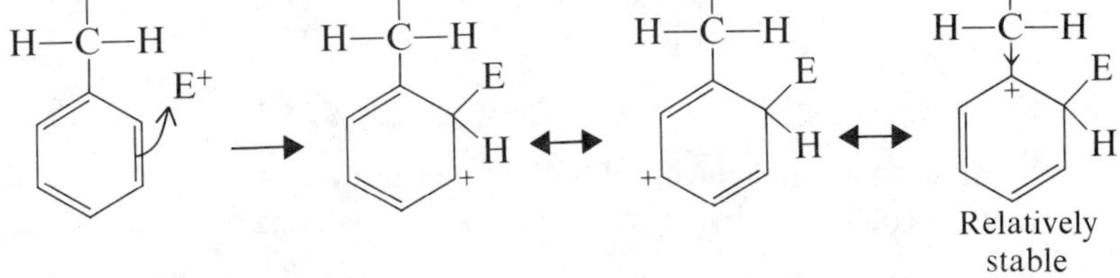

Relatively stable

Meta attack:

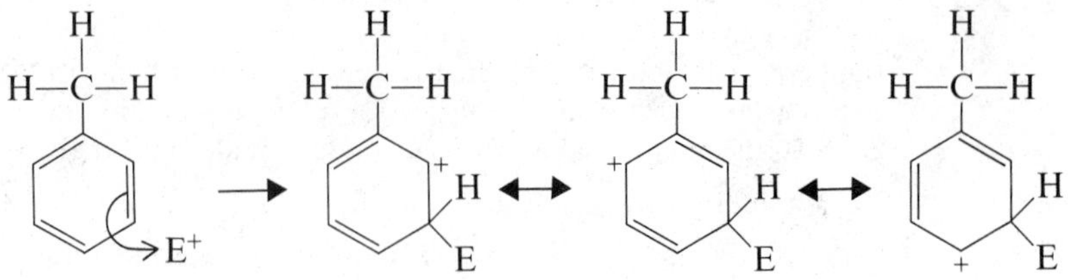

Para attack:

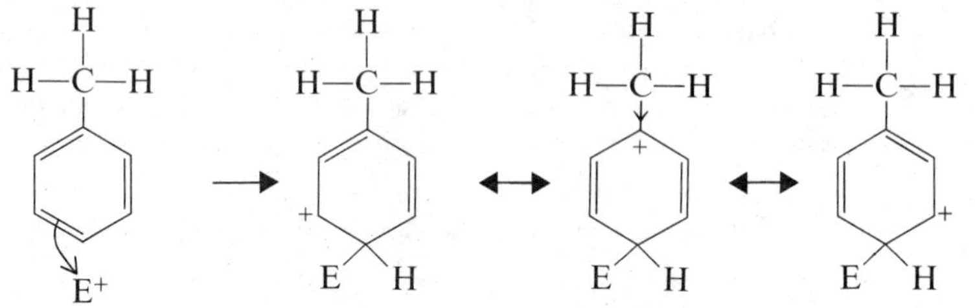

Relatively stable

In ortho attack and para attack we find that we can write resonance structures in which the methyl group is directly attached to a positively charged carbon of the ring. These structures are more *stable* relative to any of the others because in them the stabilizing influence of the methyl group (by electron release) is most effective. These structures, therefore, make a large (stabilizing) contribution to the overall hybrid for ortho and para-substituted arenium ions. No such relatively stable structure contributes to the hybrid for the meta-substituted arenium ion and as a result, it is less stable than the ortho or para-substituted arenium ion. Since the ortho and para-substituted arenium ions are more stable, the transition states leading to them occur at lower energy and ortho and para substitution takes place most rapidly.

Problem 12.16

(a) Write resonance structures for the arenium ions formed when ethylbenzene undergoes electrophilic attack. (b) Do these structures account for the fact that the ethyl group is an ortho-para director? (c) How can you account for the fact that the ethyl group is an activating group?

Problem 12.17

Resonance structures can also be used to account for the fact that phenyl group is an ortho-para director and that it is an activating group. Show how this is possible.

12.10 REACTIONS OF THE SIDE CHAIN OF ALKYLBENZENES

Hydrocarbons that consist of both aliphatic and aromatic groups are also known as *arenes*. Toluene, ethylbenzene, and isopropylbenzene are *alkylbenzenes*.

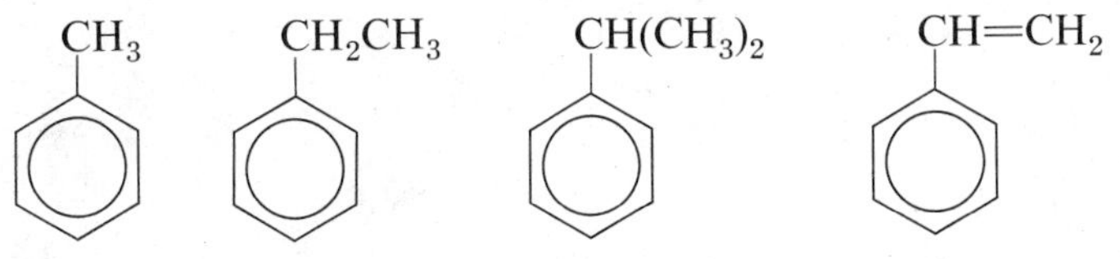

CH_3 CH_2CH_3 $CH(CH_3)_2$ $CH{=}CH_2$

Toluene Ethylbenzene Isopropylbenzene Styrene (phenylethene or vinylbenzene)

Styrene is an example of an *alkenylbenzene*. The aliphatic portion of these compounds is commonly called the *side chain*.

Styrene is one of the most important industrial chemicals—more than six billion pounds are produced each year. The starting material for the commercial synthesis of styrene is ethylbenzene, produced by Friedel-Crafts alkylation of benzene:

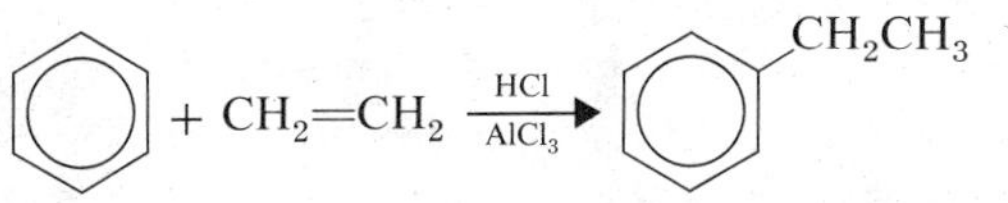

$$C_6H_6 + CH_2{=}CH_2 \xrightarrow[AlCl_3]{HCl} C_6H_5CH_2CH_3$$

Ethylbenzene

Ethylbenzene is then dehydrogenated in the presence of a catalyst (zinc or chromium oxide) to produce styrene.

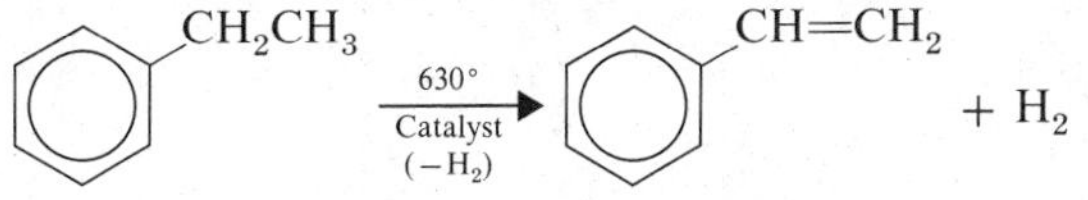

$$C_6H_5CH_2CH_3 \xrightarrow[\text{Catalyst}\ (-H_2)]{630^\circ} C_6H_5CH{=}CH_2 + H_2$$

Styrene (90–92% yield)

Most styrene is polymerized to the familiar plastic, polystyrene (cf. Special Topic C).

$$C_6H_5CH{=}CH_2 \xrightarrow{\text{Catalyst}} -CH_2CH(C_6H_5)-(CH_2CH(C_6H_5))_n-CH_2CH(C_6H_5)-$$

Polystyrene

12.10A Halogenation of the Side Chain

We have seen that bromine and chlorine replace hydrogens of the ring of toluene when the reaction takes place in the presence of a Lewis acid. In ring halogenations the electrophiles are *positive* chlorine or bromine ions or they are Lewis-acid complexes that have positive halogens. These positive electrophiles attack the π electrons of the benzene ring and aromatic substitution takes place.

Chlorine and bromine can also be made to replace hydrogens of the methyl group of toluene. Side-chain halogenation takes place when the reaction is carried out *in the absence of Lewis acids* and under conditions that favor the formation of free radicals. When toluene reacts with *N*-bromosuccinimide in the presence of peroxides, for example, the major product is benzyl bromide. (*N*-Bromosuccinimide furnishes a low concentration of Br_2 (cf. Section 10.2.)

$$C_6H_5CH_3 + \text{NBS} \xrightarrow{\text{Peroxides}} C_6H_5CH_2Br + \text{succinimide}$$

Benzyl bromide
(α-bromotoluene)
(64%)

Side-chain chlorination of toluene also takes place in the gas phase at 400 to 600° or in the presence of ultraviolet light. When an excess of chlorine is used multiple chlorinations of the side chain occur.

$$C_6H_5CH_3 \xrightarrow[\text{heat or } h\nu]{Cl_2} C_6H_5CH_2Cl \xrightarrow[\text{heat or } h\nu]{Cl_2} C_6H_5CHCl_2 \xrightarrow[\text{heat or } h\nu]{Cl_2} C_6H_5CCl_3$$

Benzyl chloride — (Dichloromethyl)-benzene — (Trichloromethyl)-benzene

These halogenations take place through free-radical mechanisms like those we saw for alkanes in Section 4.10. The halogens dissociate to produce halogen atoms and then the halogen atoms initiate chains by abstracting hydrogens of the methyl group.

Chain-initiating step

$$(1)\ X_2 \xrightarrow[\text{or } h\nu]{\text{Peroxides, heat}} 2X\cdot$$

Chain-propagating steps

$$(2)\ C_6H_5CH_3 + X\cdot \longrightarrow C_6H_5CH_2\cdot + HX$$

Benzyl radical

$$(3)\ C_6H_5CH_2\cdot + X_2 \longrightarrow C_6H_5CH_2X + X\cdot$$

Benzyl radical — Benzyl halide

Abstraction of a hydrogen from the methyl group of toluene produces *a benzyl radical.* The benzyl radical then reacts with a halogen molecule to produce a benzyl halide and a halogen atom. The halogen atom then brings about a repetition of step 2, then step 3 occurs again, and so on.

The name benzyl radical is used not only as a specific name for the radical produced in the reaction above, but also as a general name (benzylic radical) for all radicals that have an unpaired electron on the side chain carbon that is directly attached to the benzene ring. The hydrogens of the carbon directly attached to the benzene ring are called benzylic hydrogens.

$C_6H_5\dot{C}H_2$ — The benzyl radical

$C_6H_5\dot{C}<$ — A benzylic radical

C_6H_5—C—H: Benzylic carbon; A benzylic hydrogen

Benzylic halogenations are similar to allylic halogenations in that they involve the formation of *unusually stable free radicals*. Benzyl and allyl radicals are even more stable than tertiary radicals, as data for bond dissociation energies show:

$$C_6H_5CH_2{-}H \longrightarrow \underset{\text{Benzyl radical}}{C_6H_5CH_2\cdot} + H\cdot \qquad \Delta H^\circ = 85 \text{ kcal/mole}$$

$$CH_2{=}CHCH_2{-}H \longrightarrow \underset{\text{Allyl radical}}{CH_2{=}CHCH_2\cdot} + H\cdot \qquad \Delta H^\circ = 85 \text{ kcal/mole}$$

$$(CH_3)_3C{-}H \longrightarrow \underset{3^\circ\text{ Radical}}{(CH_3)_3C\cdot} + H\cdot \qquad \Delta H^\circ = 91 \text{ kcal/mole}$$

We can account for the unusual stability of the benzyl radical in much the same way that we did for the allyl radical: through resonance theory and molecular orbital theory.

We can understand how resonance theory accounts for the stability of the benzyl radical by examining the following structures.

[Resonance structures of the benzyl radical: $\cdot CH_2$–ring $\leftrightarrow$ CH_2= ring (radical at ortho) $\leftrightarrow$ CH_2= ring (radical at para) $\leftrightarrow$ CH_2= ring (radical at other ortho)]

We see that four reasonable resonance structures contribute to the hybrid for the benzyl radical. Since all are reasonable, they all make stabilizing contributions to the hybrid. They also tell us that the unpaired electron is *delocalized* over the benzylic carbon and the ortho and para carbons of the ring.

[Hybrid structure: $\delta\cdot CH_2$ on ring with $\delta\cdot$ at both ortho positions and para position]

In molecular orbital theory the benzyl radical is treated by combining the wave functions for the *p* orbitals of the ring with the *p* orbital of the benzylic carbon.

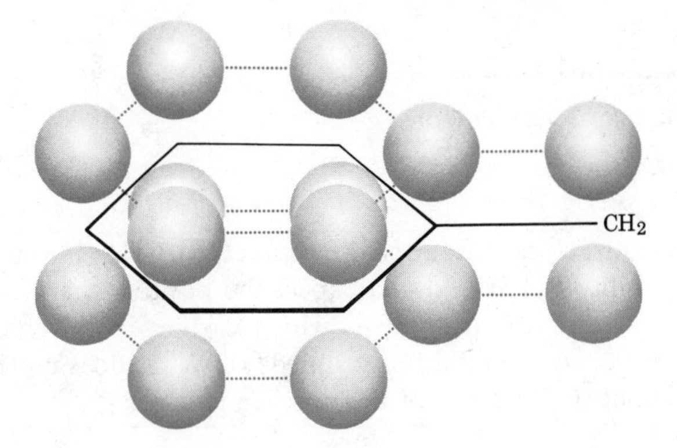

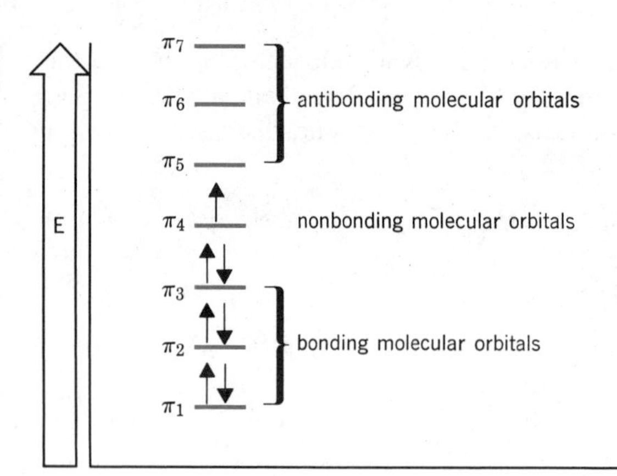

FIG. 12.4
The energy levels of the π-molecular orbitals of the benzyl radical.

When this is done, calculations give energy levels for each of the π-molecular orbitals. These are shown in Fig. 12.4.

The molecular orbital that contains the unpaired electron is π_4. Orbital π_4 is nonbonding and the calculations show that it involves only the *p* orbitals of the benzylic carbon and the ortho and para carbons of the ring. It has nodes located at the other benzene carbons (Fig. 12.5).

Thus, molecular orbital theory tells us the same thing that resonance theory does—that the unpaired electron of the benzyl radical is delocalized over the benzylic carbon and the ortho and para carbons of the ring.

Problem 12.18

When propylbenzene reacts with chlorine in the presence of ultraviolet radiation the major product is 1-chloro-1-phenylpropane. Both 2-chloro-1-phenylpropane and 1-chloro-3-phenylpropane are minor products. Write the structure of the radical leading to each product and account for the fact that 1-chloro-1-phenylpropane is the major product.

12.10B Benzyl and Benzylic Cations

Benzyl and benzylic cations are unusually stable carbocations, more stable even than tertiary cations.

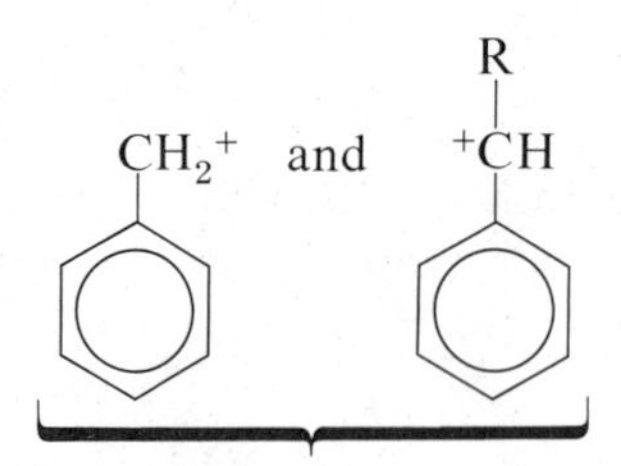

More stable than 3° carbocations

Problem 12.19

(a) Write resonance structures for the *benzyl cation* $C_6H_5CH_2^+$. (b) Can you account for its unusual stability? (c) Would you expect the positive charge of the benzyl cation to be delocalized? (d) If so, over which carbons? (e) Which molecular orbitals of the benzyl cation would be occupied? (f) What does molecular orbital theory tell you about the location of the positive charge?

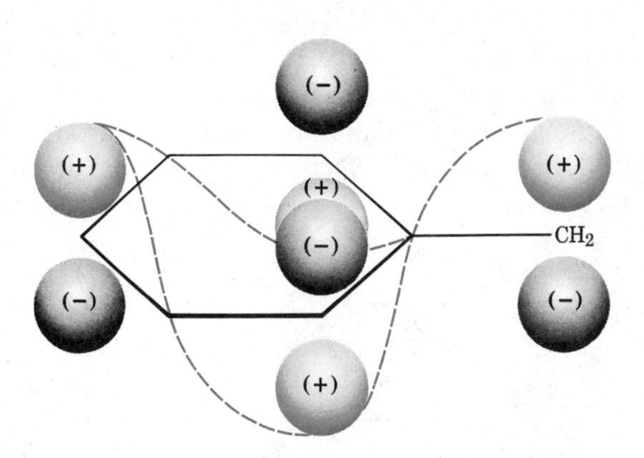

FIG. 12.5
The p *orbitals of the nonbonding orbital* (π_4) *of the benzyl radical.*

12.10C Oxidation of the Side Chain

Strong oxidizing agents oxidize toluene to benzoic acid. The oxidation can be carried out by heating toluene with dilute nitric acid in a sealed tube, by the action of hot chromic acid, or by the action of hot alkaline potassium permanganate. The last method gives benzoic acid in almost quantitative yield.

$$C_6H_5-CH_3 \xrightarrow[(2)\ H_3O^+]{(1)\ KMnO_4,\ OH^-,\ heat} C_6H_5-\overset{O}{\overset{\|}{C}}OH$$

Benzoic acid
(~100%)

Other alkylbenzenes can be oxidized in the same way. Hot alkaline potassium permanganate oxidizes *o*-chlorotoluene to *o*-chlorobenzoic acid.

$$o\text{-}ClC_6H_4CH_3 \xrightarrow[(2)\ H_3O^+]{(1)\ KMnO_4,\ OH^-,\ heat} o\text{-}ClC_6H_4COOH$$

o-Chlorotoluene → *o*-Chlorobenzoic acid
(77%)

An important characteristic of side-chain oxidations is that oxidation takes place at the benzylic carbon and, consequently, alkylbenzenes with alkyl groups longer than methyl are degraded to benzoic acids.

$$C_6H_5CH_2CH_2CH_2R \xrightarrow[heat\ \ (2)\ H_3O^+]{(1)\ KMnO_4,\ OH^-} C_6H_5-\overset{O}{\overset{\|}{C}}-OH$$

An alkylbenzene → Benzoic acid

Side-chain oxidations are similar to benzylic halogenations, because in the first step the oxidizing agent abstracts a benzylic hydrogen. Once oxidation is begun at the benzylic carbon, it continues at that site. Ultimately, the oxidizing agent oxidizes the benzylic carbon to a carboxyl group and, in the process, it scissors the remaining carbons of the side chain. (*tert*-Butylbenzene is resistant to side-chain oxidation. Why?)

Benzene derivatives with acyl side chains can also be oxidized to benzoic acids with strong oxidizing agents. Solutions of bromine in sodium hydroxide oxidize acetyl groups to carboxylic acids but do not affect alkyl groups that are present on the ring (cf. Section 16.16).

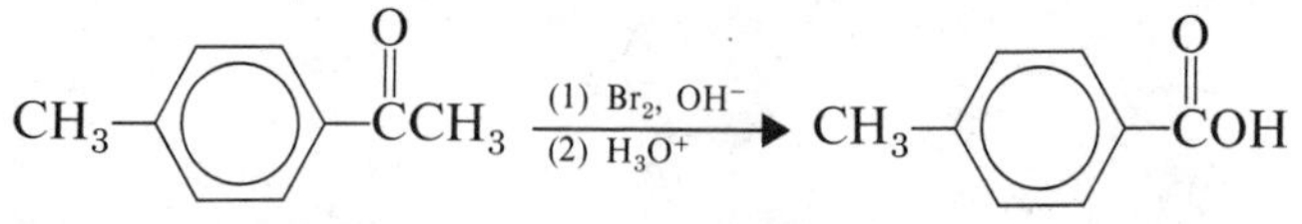

$$CH_3-C_6H_4-\overset{O}{\overset{\|}{C}}CH_3 \xrightarrow[(2)\ H_3O^+]{(1)\ Br_2,\ OH^-} CH_3-C_6H_4-\overset{O}{\overset{\|}{C}}OH$$

p-Toluic acid

12.11 ALKENYLBENZENES

12.11A Stability of Conjugated Alkenylbenzenes

Alkenylbenzenes that have their double bond conjugated with the benzene ring are more stable than those that do not.

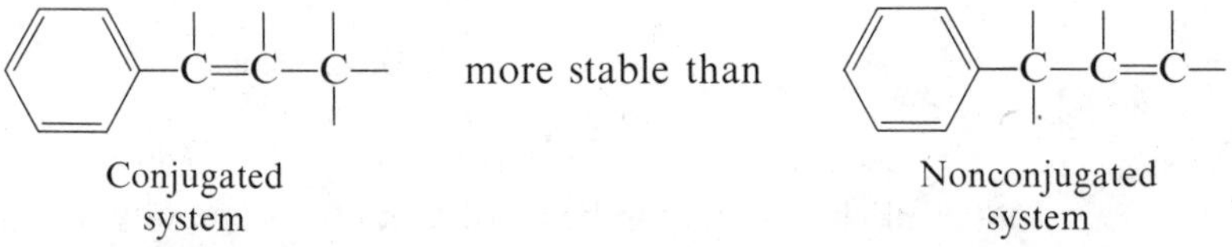

$$C_6H_5-\overset{|}{C}=\overset{|}{C}-\underset{|}{\overset{|}{C}}- \quad \text{more stable than} \quad C_6H_5-\underset{|}{\overset{|}{C}}-\overset{|}{C}=\overset{|}{C}-$$

Conjugated system — Nonconjugated system

Part of the evidence for this comes from acid-catalyzed alcohol dehydrations which are known to yield the more stable alkene (Section 6.12). For example, dehydration of the alcohol shown below yields exclusively the conjugated system.

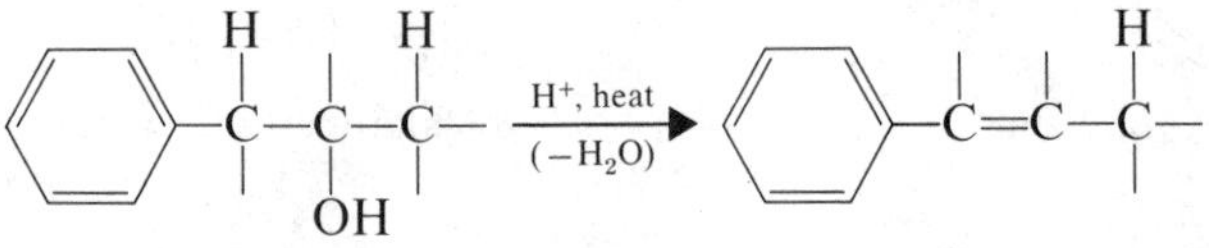

$$C_6H_5-\underset{|}{\overset{H}{\overset{|}{C}}}-\underset{OH}{\underset{|}{\overset{|}{C}}}-\underset{|}{\overset{H}{\overset{|}{C}}}- \xrightarrow[(-H_2O)]{H^+,\ heat} C_6H_5-\overset{|}{C}=\overset{|}{C}-\underset{|}{\overset{H}{\overset{|}{C}}}-$$

Because conjugation always lowers the energy of an unsaturated system by allowing the π-electrons to be delocalized this behavior is just what we would expect.

12.11B Additions to the Double Bond of Alkenylbenzenes

The phenyl group of *conjugated* alkenylbenzenes affects the reactivity, orientation, and *stereochemistry* of reactions that take place at the double bond.

For example, bromine adds to styrene three times as fast as it does to ethene.

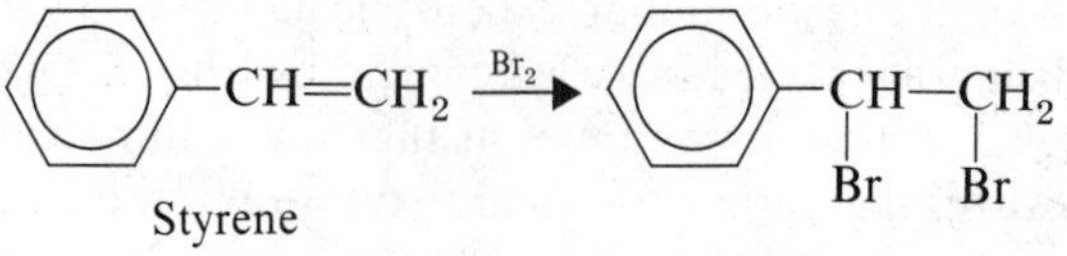

$$C_6H_5-CH=CH_2 \xrightarrow{Br_2} C_6H_5-\underset{Br}{\underset{|}{C}H}-\underset{Br}{\underset{|}{C}H_2}$$

Styrene

In the presence of peroxides, hydrogen bromide adds to the double bond of 1-phenylpropene to give 1-phenyl-2-bromopropane as the major product.

$$C_6H_5\text{-}CH{=}CHCH_3 \xrightarrow[\text{Peroxides}]{\text{HBr}} C_6H_5\text{-}CH_2\underset{\displaystyle Br}{\underset{|}{C}}HCH_3$$

1-Phenylpropene — 2-Bromo-1-phenylpropane

In the absence of peroxides, HBr adds in just the opposite way.

$$C_6H_5\text{-}CH{=}CHCH_3 \xrightarrow[\text{(no peroxides)}]{\text{HBr}} C_6H_5\text{-}\underset{\displaystyle Br}{\underset{|}{C}}HCH_2CH_3$$

1-Phenylpropene — 1-Phenyl-1-bromopropane

Reactivity and orientation in all of these reactions are consequences of the unusual stabilities of benzylic cations (cf. Problem 12.19) and benzylic radicals.

The addition of bromine to styrene appears to occur through a bromonium ion, but one that has considerable carbocationic character.

$$C_6H_5\text{-}CH{=}CH_2 \xrightarrow{Br_2} C_6H_5\text{-}\overset{\delta+}{C}H\text{—}CH_2 \; {:}Br^- \;\; (\text{bridged by } Br^{\delta+})$$

$$\downarrow$$

$$C_6H_5\text{-}\overset{\displaystyle Br}{\overset{|}{C}}H\text{—}CH_2Br$$

Because the intermediate resembles a relatively stable benzylic cation, bromination of styrene proceeds faster than the corresponding reaction of ethene.

The addition of hydrogen bromide to 1-phenylpropene proceeds through a benzylic radical in the presence of peroxides, and through a benzylic cation in their absence [cf. **1** below and **2** on the next page].

1. Hydrogen bromide addition in the presence of peroxides.

Chain-initiating steps

$$R\text{—}O\text{—}O\text{—}R \longrightarrow 2R\text{—}O\cdot$$

$$RO\cdot + H\text{—}Br \longrightarrow R\text{—}O\text{—}H + Br\cdot$$

Chain-propagating steps

$$Br\cdot + C_6H_5CH{=}CHCH_3 \longrightarrow C_6H_5\dot{C}H\text{—}\underset{\displaystyle Br}{\underset{|}{C}}HCH_3$$

A benzylic radical

$$C_6H_5\dot{C}H\underset{\displaystyle Br}{\underset{|}{C}}HCH_3 + H\text{—}Br \longrightarrow C_6H_5CH_2\underset{\displaystyle Br}{\underset{|}{C}}HCH_3 + Br\cdot$$

1-Phenyl-2-bromopropane

The mechanism for the addition of hydrogen bromide to 1-phenylpropene in the presence of peroxides is a chain mechanism analogous to the one we discussed when we described anti-Markovnikov addition in Section 7.14. The step that determines the orientation of the reaction is the first chain-propagating step. Bromine attacks the second carbon of the chain because by doing so the reaction produces a more stable benzylic radical. Had the bromine atom attacked the double bond in the opposite way a less stable secondary radical would have been formed.

$$C_6H_5CH{=}CHCH_3 + Br\cdot \not\longrightarrow C_6H_5\underset{\substack{|\\ Br}}{C}H{-}\dot{C}HCH_3$$

A secondary radical

2. Hydrogen bromide addition in the absence of peroxides.

$$C_6H_5CH{=}CHCH_3 + HBr \longrightarrow C_6H_5\overset{+}{C}HCH_2CH_3 + Br^-$$

A benzylic cation

$$\downarrow$$

$$C_6H_5\underset{\substack{|\\ Br}}{C}HCH_2CH_3$$

In the absence of peroxides hydrogen bromide adds through an ionic mechanism. The step that determines the orientation in the ionic mechanism is the first, where the proton attacks the double bond to give the more stable benzylic cation. Had the proton attacked the double bond in the opposite way a less stable secondary cation would have been formed.

$$C_6H_5CH{=}CHCH_3 + HBr \not\longrightarrow C_6H_5\underset{\substack{|\\ H}}{C}H{-}\overset{+}{C}HCH_3$$

A secondary cation

12.11C Syn Additions

Studies of the peroxide-free addition of DBr rather than HBr, show that the addition is primarily (88%) *a syn addition* (Fig. 12.6). This means that in the addition, the benzylic cation and the bromide ion are not completely independent of each other during the reaction. They remain in close proximity as an *ion pair* and the bromide ion adds from the same side of the double bond.

Problem 12.20

(a) What products would be obtained from the anti addition of DBr to *cis*-1-phenylpropene? (b) Would knowing the stereochemistry of the products of the addition of DBr to *cis*-1-phenylpropene allow you to decide whether or not the addition occurred in a syn or anti fashion?

Problem 12.21

(a) What products would be obtained from the syn addition of DBr to *trans*-1-phenylpropene? (b) From the anti addition?

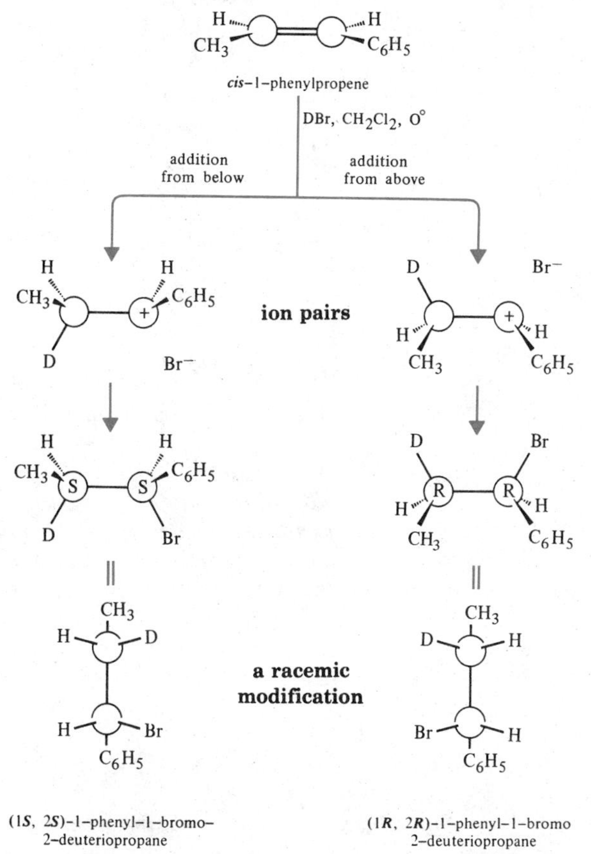

cis-1-phenylpropene

$DBr, CH_2Cl_2, 0°$

addition from below

addition from above

ion pairs

a racemic modification

(1*S*, 2*S*)-1-phenyl-1-bromo-2-deuteriopropane

(1*R*, 2*R*)-1-phenyl-1-bromo 2-deuteriopropane

FIG. 12.6
Syn *addition of DBr to cis-1-phenylpropene. The product of* syn *addition is a racemic modification of* (*1*S, 2S) (*1*R, 2R)-*1-bromo-1-phenyl-2-deuteriopropane.*

Syn additions also predominate when chlorine and fluorine add to 1-phenylpropene (Table 12.4). When bromine adds, however, the addition is predominantly *anti.*

These and other experimental results allow us to unify all the mechanisms for electrophilic additions to alkenes. This is done in Fig. 12.7 with the addition of reagents X-Y to *cis*-alkenes of the general type RCH=CHR′.

The overall stereochemical course of electrophilic additions to alkenes appears to depend on the relative stabilities of the *onium ions* (Fig. 12.7) and *carbocations* that can be formed. And these stabilities depend on *the nature of the adding reagent* and *the structure of the alkene* undergoing addition.

TABLE 12.4 Addition Reactions to ***cis*** and ***trans*** 1-Phenylpropenes

ADDING REAGENT	CONDITIONS	ALKENE	PERCENT SYN ADDITION	PERCENT ANTI ADDITION
DBr	CH_2Cl_2, 0°	1-Phenylpropene	88	12
F_2	CCl_3F, −126°	*cis*-1-Phenylpropene	78	22
F_2	CCl_3F, −126°	*trans*-1-Phenylpropene	73	27
Cl_2	CH_2Cl_2, 0°	*cis*-1-Phenylpropene	75	25
Cl_2	CH_2Cl_2, 0°	*trans*-1-Phenylpropene	67	33
Br_2	CCl_4, 2-5°	*cis*-1-Phenylpropene	17	83
Br_2	CCl_4, 2-5°	*trans*-1-Phenylpropene	12	88

From an article by W. R. Dolbier, Jr., *J. Chem. Education, 46,* 342 (1969). The experimental results were obtained by M. J. S. Dewar, R. C. Fahey, C. Schubert, H. J. Schneider, and R. F. Merrit.

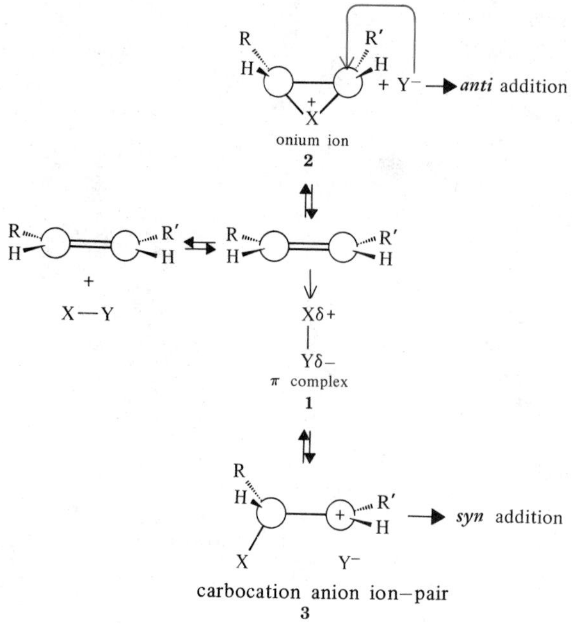

+ Y⁻ → *anti* addition

onium ion
2

X—Y

Xδ+
|
Yδ−
π complex
1

→ *syn* addition

carbocation anion ion–pair
3

FIG. 12.7
*A unified scheme that accounts for the stereochemistry of ionic additions to alkenes. If the onium ion **2** is more stable than the carbocation **3** then the reaction gives mainly the anti addition product. If the carbocation (formed as an ion pair with Y) is more stable than the onium ion then the reaction gives mainly the syn addition product.*

A number of studies indicate that the relative stabilities of *onium ions* depend on the nature of X.

Relative stability: X = Br > X = Cl > X = F > X = H or D

When R and R′ are aliphatic groups, the onium ion appears to be more stable than the carbocation even when X = H or D. Thus, in most additions to aliphatic alkenes (Section 7.9) the initially formed π complex **1** collapses to the onium ion **2** and this ultimately gives anti addition.

Anti addition is also the result when R or R′ = C_6H_5 and when X = Br. Here the onium ion **2** is still more stable than the carbocation **3**.

However, when R or R′ = C_6H_5 and X = Cl, F, H, or D, the carbocation **3** is formed because it is more stable than the onium ion. Since **3** is an ion pair in which the charged groups do not separate, it collapses to give syn addition.

12.12 SYNTHETIC APPLICATIONS

The substitution reactions of aromatic rings and the reactions of the side chains of alkyl and alkenylbenzenes, when taken together offer us a powerful set of reactions for organic synthesis. By using these reactions skillfully, we will be able to synthesize a large number of benzene derivatives.

Part of the skill in planning a synthesis is in deciding the order in which reactions should be carried out. Let us suppose, for example, that we want to synthesize *o*-bromonitrobenzene. We can see very quickly that we should introduce the bromine into the ring first because it is an ortho-para director.

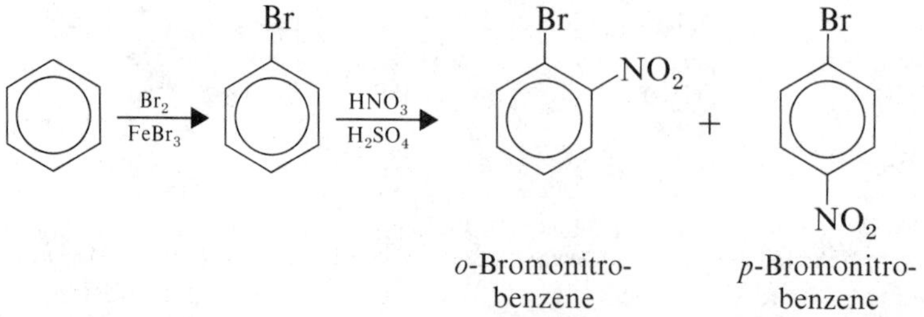

Br_2 / $FeBr_3$ → bromobenzene; HNO_3 / H_2SO_4 →

o-Bromonitrobenzene + *p*-Bromonitrobenzene

The ortho and para compounds that we get as products can be separated by fractional distillation. However, had we introduced the nitro group first, we would have obtained *m*-nitrobromobenzene as the major product.

Other examples in which choosing the proper order for the reactions are important are the syntheses of the ortho-, meta-, and para-nitrobenzoic acids. We can synthesize the ortho- and para-nitrobenzoic acids from toluene by nitrating it, separating the ortho- and para-nitrotoluenes, and then oxidizing the methyl groups to carboxyl groups.

CH_3 (toluene) $\xrightarrow[H_2SO_4]{HNO_3}$ *p*-Nitrotoluene (separate) + *o*-Nitrotoluene

p-Nitrotoluene $\xrightarrow[(2)\ H_3O^+]{(1)\ KMnO_4,\ OH^-,\ heat}$ *p*-Nitrobenzoic acid

o-Nitrotoluene $\xrightarrow[(2)\ H_3O^+]{(1)\ KMnO_4,\ OH^-,\ heat}$ *o*-Nitrobenzoic acid

We can synthesize *m*-nitrobenzoic acid by reversing the order of the reactions.

CH_3 (toluene) $\xrightarrow[(2)\ H_3O^+]{(1)\ KMnO_4,\ OH^-,\ heat}$ Benzoic acid $\xrightarrow[H_2SO_4,\ heat]{HNO_3}$ *m*-Nitrobenzoic acid

Problem 12.22

Suppose you needed to synthesize 1-(*p*-chlorophenyl)propene from propylbenzene.

$CH_2CH_2CH_3$ (propylbenzene) - - - - - → $CH{=}CHCH_3$ / Cl (1-(*p*-chlorophenyl)propene)

You could introduce the double bond into the side chain through a benzylic halogenation and subsequent dehydrohalogenation. You could introduce the chlorine into the benzene ring through a Lewis-acid catalyzed chlorination. Which reaction would you carry out first? Why?

Very powerful activating groups like the amino group and the hydroxyl group cause the benzene ring to be so reactive that undesirable reactions take place. Some reagents used for electrophilic substitution reactions, such as nitric acid, are also strong *oxidizing agents*. (Both electrophiles and oxidizing agents seek electrons.) Thus, amino groups and hydroxyl groups not only activate the ring toward electrophilic substitution, they also activate it toward oxidation. Nitration of aniline, for example, results in considerable destruction of the benzene ring because it is oxidized by the nitric acid. Direct nitration of aniline, consequently, is not a satisfactory method for the preparation of *o*- and *p*-nitroaniline.

Treating aniline with acetyl chloride, CH_3COCl, or acetic anhydride $(CH_3CO)_2O$ converts aniline (below) to acetanilide. The amino group is converted to an acetamido group, $—NHCOCH_3$, a group that is only moderately activating and one that does not make the ring highly susceptible to oxidation. With acetanilide direct nitration becomes possible.

NH_2 (aniline) $\xrightarrow[\text{base}]{CH_3COCl}$ $NHCOCH_3$ (Acetanilide) $\xrightarrow[H_2SO_4]{HNO_3}$ *p*-Nitroacetanilide (90%) + *o*-Nitroacetanilide (trace)

p-Nitroacetanilide $\xrightarrow{H^+,\ H_2O}$ *p*-Nitroaniline + CH_3COOH

Nitration of acetanilide gives *p*-nitroacetanilide in excellent yield with only a trace of the ortho isomer. Acidic hydrolysis of *p*-nitroacetanilide removes the acetyl group and gives *p*-nitroaniline, also in good yield.

Suppose, however, we need *o*-nitroaniline. The synthesis that we outlined above would obviously not be a satisfactory method, for only a trace of *o*-nitroacetanilide is obtained in the nitration reaction. (The acetamido group is purely a para director in many reactions. Bromination of acetanilide, for example, gives *p*-bromoacetanilide almost exclusively.)

We can synthesize *o*-nitroacetanilide, however, through the reaction shown below.

Acetanilide $\xrightarrow[H_2SO_4]{\text{conc.}}$ $NHCOCH_3$ / SO_3H (para) $\xrightarrow{HNO_3}$ $NHCOCH_3$ / NO_2 (ortho) / SO_3H (para) $\xrightarrow[H_2SO_4,\ \text{heat}]{H_2O}$ NH_2 / NO_2 (ortho) (56%)

Here we see how a sulfonic acid group can be used as a "blocking group." We can remove the sulfonic acid group by desulfonation at a later stage. In this example, the reagent used for desulfonation (dilute H_2SO_4) also conveniently removes the acetyl group that we employed to "protect" the benzene ring from oxidation by nitric acid.

12.12A Orientation in Disubstituted Benzenes

The problem of orientation is somewhat more complicated when two substituents are present on a benzene ring. We find, however, that in many instances we can make very good estimates of the outcome of the reaction by relatively simple analyses.

If two groups are located so that their directive effects reinforce each other then predicting the outcome is easy. Consider the examples shown below. In each case the entering substituent is directed by both groups into the position indicated by the arrows.

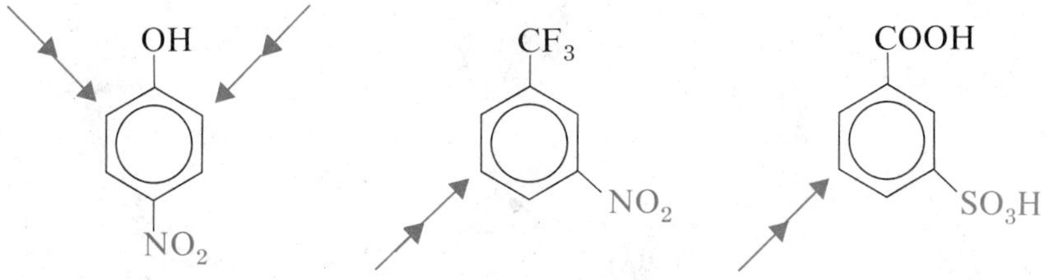

OH, NO_2 — **1** CF_3, NO_2 — **2**   COOH, SO_3H — **3**

When the directive effect of two groups oppose each other, *the more powerful activating group* (Table 12.3) *generally determines the outcome of the reaction.* Let us consider, as an example, the orientation of electrophilic substitution of *p*-methylacetanilide.

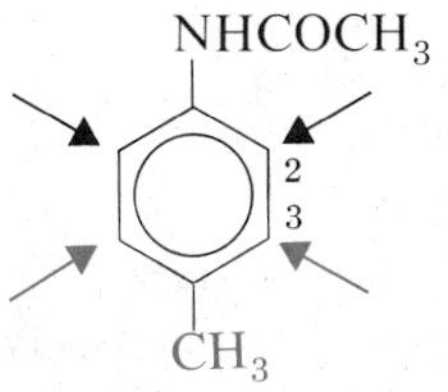

$NHCOCH_3$, CH_3

p-Methylacetanilide

The acetamido group is a much stronger activating group than the methyl group. The following examples show that the acetamido group determines the outcome of the reaction. Substitution occurs primarily at position 2.

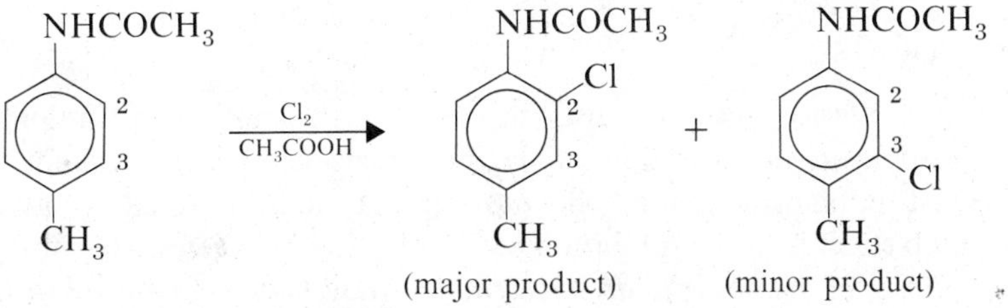

$$\text{CH}_3\text{C}_6\text{H}_4\text{NHCOCH}_3 \xrightarrow[\text{CH}_3\text{COOH}]{\text{Cl}_2} \text{2-Cl product (major product)} + \text{3-Cl product (minor product)}$$

$NHCOCH_3$ (2, 3), CH_3 $\xrightarrow[CH_3COOH]{Br_2}$ $NHCOCH_3$, Br (2, 3), CH_3 (major product) + $NHCOCH_3$ (2, 3), Br, CH_3 (minor product)

Other examples that illustrate the greater directive influence of the more powerful activating group are show below.

CH_3, OCH_3 $\xrightarrow[CH_3COOH]{Br_2}$ CH_3, Br, OCH_3 (major product)

Cl, COOH $\xrightarrow[H_2SO_4]{HNO_3}$ Cl, COOH, NO_2 (major product)

OH, CH_3 $\xrightarrow{HNO_3}$ OH, NO_2, CH_3 (major product)

When two opposing groups have approximately the same activating effect the results are not nearly so clear-cut. The following reaction is a typical example.

CH_3, Cl $\xrightarrow[H_2SO_4]{HNO_3}$ CH_3, Cl, NO_2 (19%) + CH_3, Cl, NO_2 (17%) + CH_3, Cl, O_2N (43%) + O_2N, CH_3, Cl (21%)

Steric effects are also important in aromatic substitutions. We saw one example in the nitration of *tert*-butylbenzene (p. 479), where we found that substitution ortho to the bulky *tert*-butyl group is inhibited. Another example of a steric effect is in the inhibition of substitution between meta substituents. *Substitution does not occur to an appreciable extent between meta substituents if another position is open.* A good example of this effect can be seen in the nitration of *m*-bromochlorobenzene.

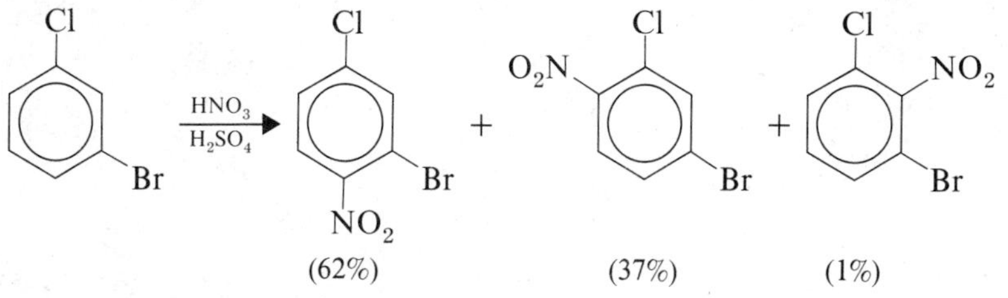

Cl, Br $\xrightarrow[H_2SO_4]{HNO_3}$ Cl, Br, NO_2 + O_2N, Cl, Br + Cl, NO_2, Br

(62%) (37%) (1%)

Only 1% of the mononitro product has the nitro group between the bromine and chlorine.

Problem 12.23

Predict the major product (or products) that would be obtained when each of the following compounds is nitrated.

(a) 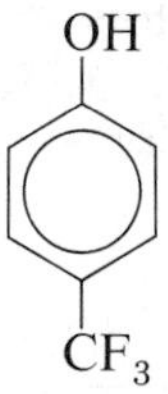 OH, CF_3 (b) 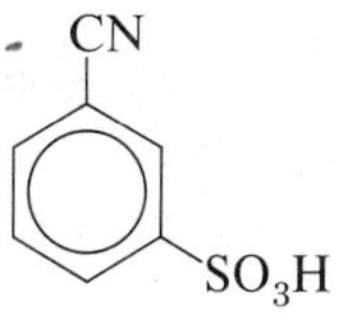 CN, SO_3H (c) 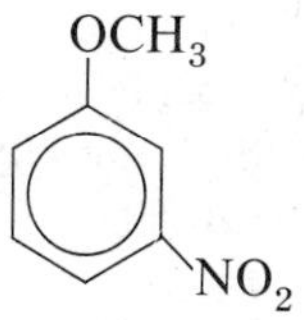 OCH_3, NO_2

(d) 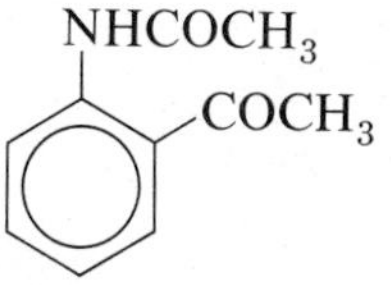 $NHCOCH_3$, $COCH_3$ (e) 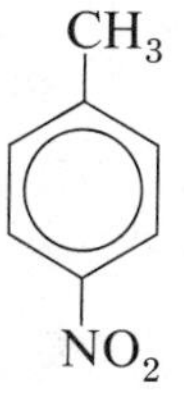 CH_3, NO_2 (f) 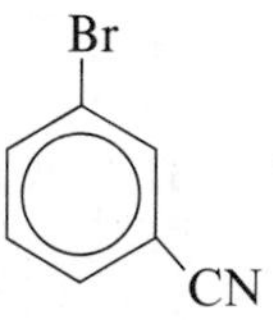 Br, CN

(g) 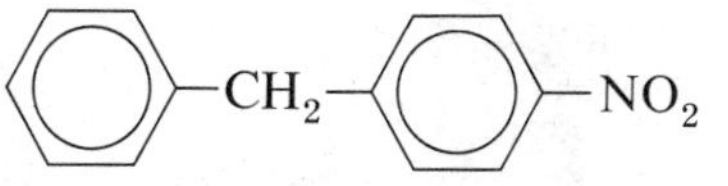 C_6H_5—CH_2—C_6H_4—NO_2

12.12B Limitations of Friedel-Crafts Alkylation Reactions

Several restrictions limit the usefulness of Friedel-Crafts alkylations.

1. Aryl and vinylic halides cannot be used as the halide component because they do not form carbocations readily (cf. Sec. 14.3).

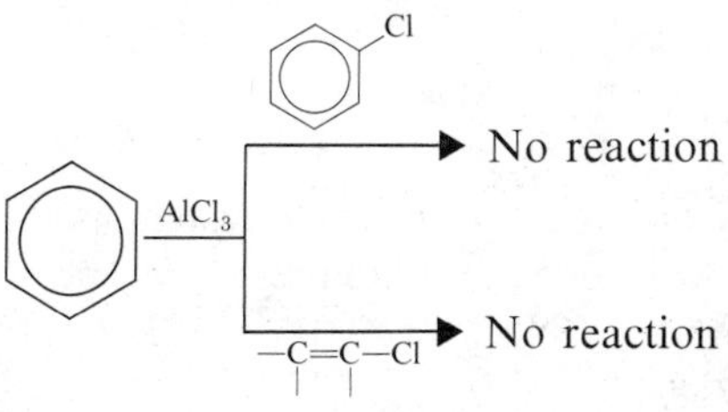

$AlCl_3$; Cl (chlorobenzene) → No reaction

—C=C—Cl → No reaction

2. Polyalkylations often occur. Alkyl groups are electron-releasing groups and once one is introduced into the benzene ring it activates the ring toward further substitution (cf. Section 12.8).

$$C_6H_6 + (CH_3)_2CHOH \xrightarrow[60^\circ]{BF_3} C_6H_5CH(CH_3)_2 + p\text{-}C_6H_4[CH(CH_3)_2]_2$$

Isopropylbenzene (24%) — *p*-Diisopropylbenzene (14%)

3. When the carbocation formed from an alkyl halide, alkene, or alcohol can rearrange to a more stable carbocation, it usually does so and the major product obtained from the reaction is usually the one from the more stable carbocation.

When benzene is alkylated with butyl bromide, for example, some of the developing *n*-butyl cations rearrange by a hydride shift—some developing 1° carbocations (below) become more stable 2° carbocations. Then benzene reacts with both kinds of carbocations to form both *n*-butylbenzene and *sec*-butyl benzene:

$$CH_3CH_2CH_2CH_2Br \xrightarrow{AlCl_3} CH_3CH_2\underset{H}{C}H\overset{\delta+}{C}H_2\text{---}Br\overset{\delta-}{A}lCl_3 \xrightarrow{(-AlCl_3Br^-)} CH_3CH_2\overset{+}{C}HCH_3$$

$$\xrightarrow[(-H^+)]{C_6H_6} C_6H_5CH(CH_3)CH_2CH_3 + C_6H_5CH_2CH_2CH_2CH_3$$

sec-Butylbenzene (64–68%) — *n*-Butylbenzene (32–36%)

4. Migrations of alkyl groups on the ring often take place in Friedel-Crafts reactions.

$$\text{1,2,4-}C_6H_3(CH_3)_3 \xrightarrow[HCl]{AlCl_3} \text{1,3,5-}C_6H_3(CH_3)_3 + AlCl_3 + HCl$$

1,2,4-Trimethylbenzene — Mesitylene

5. Friedel-Crafts reactions do not occur when powerful electron-withdrawing groups (Section 12.8) are present on the aromatic ring or when the ring bears an $—NH_2$, $—NHR$, or $—NR_2$ group.

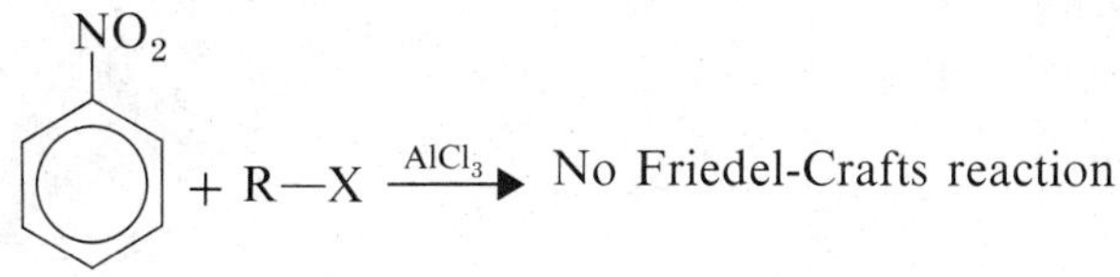

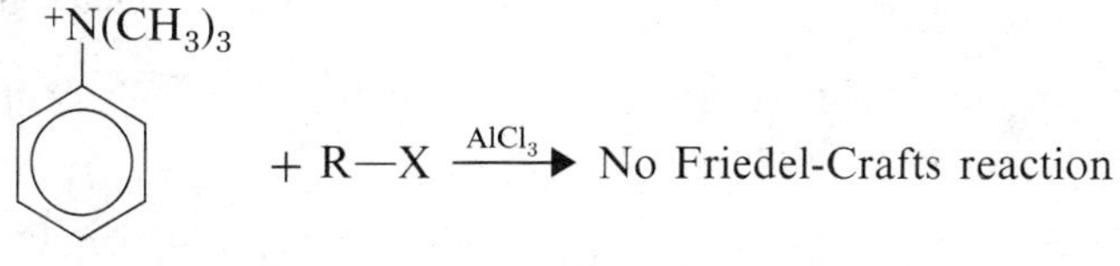

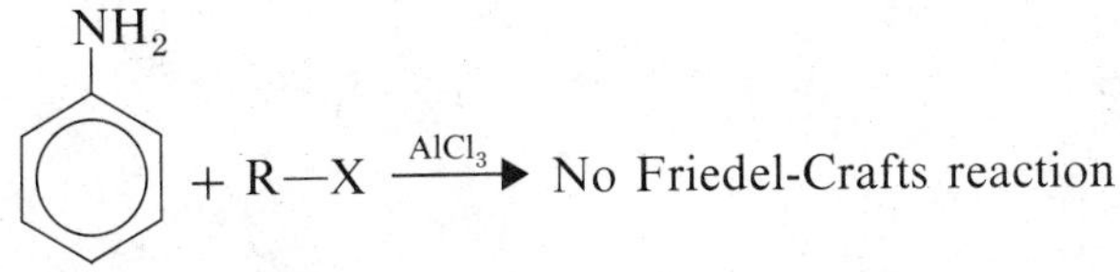

NO_2

+ R—X $\xrightarrow{AlCl_3}$ No Friedel-Crafts reaction

$^+N(CH_3)_3$

+ R—X $\xrightarrow{AlCl_3}$ No Friedel-Crafts reaction

NH_2

+ R—X $\xrightarrow{AlCl_3}$ No Friedel-Crafts reaction

Any substituent more electron withdrawing (or deactivating) than a halogen (cf. Table 12.3) usually makes an aromatic ring too electron deficient to undergo electrophilic substitution by a carbocation or a carbocation-like electrophile. The amino groups, $—NH_2$, —NHR, and $—NR_2$, are changed into powerful electron-withdrawing groups by the Lewis acids used to catalyze Friedel-Crafts reactions. For example:

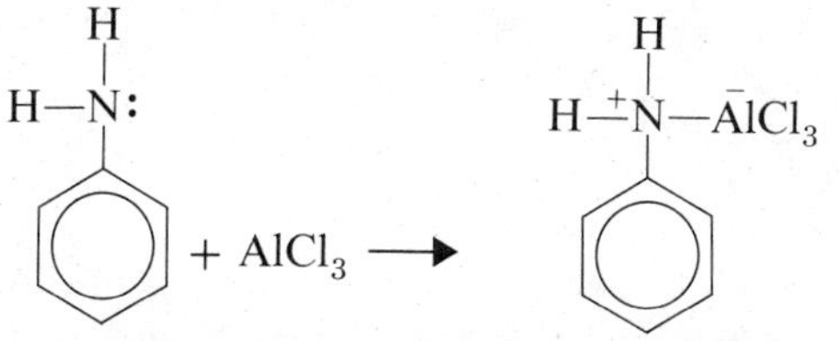

$C_6H_5NH_2 + AlCl_3 \longrightarrow C_6H_5\overset{+}{N}H_2—\overset{-}{Al}Cl_3$

Does not undergo
a Friedel-Crafts reaction

Problem 12.24

When benzene reacts with *n*-propyl alcohol in the presence of boron trifluoride, both *n*-propylbenzene and isopropyl benzene are obtained as products. Write a mechanism that accounts for this.

12.12C Synthetic Applications of Friedel-Crafts Acylations

Rearrangements of the carbon chain do not occur in Friedel-Crafts acylations. The acylium ion, because it is stabilized by resonance, is more stable than most other carbocations. Thus, there is no driving force for a rearrangement. Because rearrangements do not occur, Friedel-Crafts acylations often give us much better synthetic routes to *n*-alkylbenzenes than do Friedel-Crafts alkylations.

As an example, let us consider the problem of synthesizing *n*-propylbenzene. If we attempt this synthesis through a Friedel-Crafts alkylation, a rearrangement occurs and the major product is isopropylbenzene (see also Problem 12.24).

$$C_6H_6 + CH_3CH_2CH_2Br \xrightarrow{AlCl_3} C_6H_5CH(CH_3)_2 + C_6H_5CH_2CH_2CH_3 + HBr$$

Isopropylbenzene (major product) — *n*-Propylbenzene (minor product)

By contrast, the Friedel-Crafts acylation of benzene with propionyl chloride produces a ketone with an unrearranged carbon chain in excellent yield.

$$C_6H_6 + CH_3CH_2\overset{O}{\overset{\|}{C}}Cl \xrightarrow{AlCl_3} C_6H_5\overset{O}{\overset{\|}{C}}{-}CH_2CH_3 + HCl$$

Propionyl chloride — Ethyl phenyl ketone (90%)

This ketone can then be reduced to *n*-propylbenzene by several methods. One method—called the Clemmensen reduction—consists of refluxing the ketone with amalgamated zinc and hydrochloric acid.

$$C_6H_5\overset{O}{\overset{\|}{C}}CH_2CH_3 \xrightarrow[\text{HCl, reflux}]{\text{Zn(Hg)}} C_6H_5CH_2CH_2CH_3$$

Ethyl phenyl ketone — *n*-Propylbenzene (80%)

Friedel-Crafts acylations can be used in a variety of other ways. Diaryl ketones are formed when the acid chloride of an aromatic acid is used as one component. The reaction of benzene with benzoyl chloride is an example.

$$C_6H_6 + Cl{-}\overset{O}{\overset{\|}{C}}{-}C_6H_5 \xrightarrow{AlCl_3} C_6H_5{-}\overset{O}{\overset{\|}{C}}{-}C_6H_5$$

Benzoyl chloride — Diphenyl ketone (benzophenone) (82%)

When cyclic anhydrides are used as one component, the Friedel-Crafts acylation provides a means of adding a new ring to an aromatic compound. One illustration is shown below.

$$C_6H_6 + \text{succinic anhydride} \xrightarrow[(88\%)]{AlCl_3} C_6H_5\overset{O}{\overset{\|}{C}}CH_2CH_2\overset{O}{\overset{\|}{C}}OH \xrightarrow[\text{reflux (83–90\%)}]{\text{Zn(Hg), HCl}}$$

(excess) — Succinic anhydride — 3-Benzoylpropanoic acid

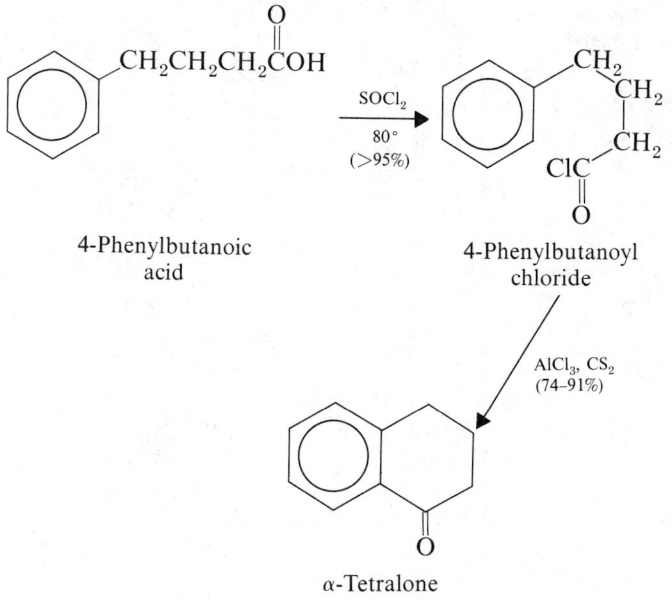

$C_6H_5CH_2CH_2CH_2\overset{O}{\overset{\|}{C}}OH \xrightarrow[80^\circ\ (>95\%)]{SOCl_2} C_6H_5CH_2CH_2CH_2\overset{O}{\overset{\|}{C}}Cl$

4-Phenylbutanoic acid

4-Phenylbutanoyl chloride

$\xrightarrow[(74\text{–}91\%)]{AlCl_3,\ CS_2}$

α-Tetralone

Problem 12.25

Starting with benzene and the appropriate acid chloride or anhydride, outline a synthesis of each of the following.

(a) n-Hexylbenzene
(b) Isobutylbenzene
(c) Anthrone

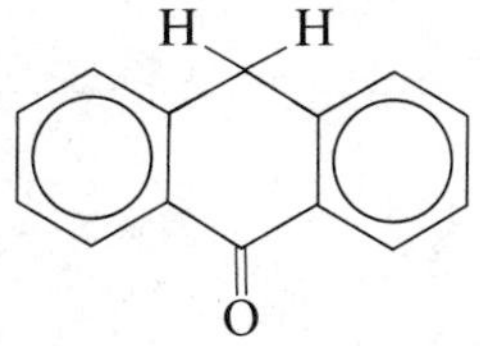

Problem 12.26

Polyacylations are not a problem in Friedel-Crafts acylations. What factor (or factors) explain(s) this?

Additional Problems

12.27

Outline ring bromination, nitration, and sulfonation reactions of the following compounds. In each case give the structure of the major reaction product or products. Also indicate whether the reaction would occur faster or slower than the corresponding reaction of benzene.

(a) Anisole, $C_6H_5OCH_3$
(b) (Difluoromethyl)benzene, $C_6H_5CHF_2$
(c) Ethylbenzene
(d) Nitrobenzene
(e) Chlorobenzene
(f) Benzenesulfonic acid
(g) Ethyl benzoate, $C_6H_5\overset{O}{\overset{\|}{C}}OC_2H_5$
(h) Phenoxybenzene, $C_6H_5OC_6H_5$
(i) Biphenyl, $C_6H_5-C_6H_5$

(j) *tert*-Butylbenzene
(k) Fluorobenzene
(l) Ethyl phenyl ketone, $C_6H_5\overset{O}{\overset{\|}{C}}C_2H_5$
(m) Benzonitrile (cyanobenzene), C_6H_5CN
(n) Phenylacetate, $C_6H_5O\overset{O}{\overset{\|}{C}}CH_3$
(o) Benzamide, $C_6H_5\overset{O}{\overset{\|}{C}}NH_2$
(p) Iodobenzene

12.28
Predict the major products of the following reactions.
(a) Sulfonation of *p*-methylacetophenone
(b) Nitration of *m*-dichlorobenzene
(c) Nitration of 1,3-dimethoxybenzene
(d) Monobromination of *p*-$CH_3CONHC_6H_4NH_2$
(e) Nitration of *p*-$HO_3SC_6H_4OH$
(f) Nitration of $C_6H_5-CH_2-C_6H_4-COOH$
(g) Chlorination of $C_6H_5CCl_3$

12.29
Give the structures of the major products of the following reactions:
(a) Styrene + HCl $\longrightarrow$
(b) 2-Bromo-1-phenylpropane + $C_2H_5ONa \longrightarrow$
(c) $C_6H_5CH_2CHOHCH_2CH_3 \xrightarrow{H^+, \text{ heat}}$
(d) Product of (c) + HBr $\xrightarrow{\text{Peroxides}}$
(e) Product of (c) + $H_2O \xrightarrow[\text{Heat}]{H^+}$
(f) Product of (c) + H_2(1mole) $\xrightarrow[25°]{Pt}$
(g) Product of (f) $\xrightarrow[(2)\ H_3O^+]{(1)\ KMnO_4,\ OH^-,\ \text{heat}}$

12.30
Starting with benzene, toluene, or aniline, show how you might synthesize each of the following compounds.
(a) *m*-Chlorobenzoic acid
(b) *p*-Bromoaniline
(c) *o*-Bromoaniline
(d) 2-Bromo-4-nitroaniline
(e) 4-Bromo-2-nitroaniline
(f) *o*-Acetyltoluene
(g) *o*-Toluic acid, *o*-$CH_3C_6H_4COOH$
(h) *p*-Iodobenzenesulfonic acid
(i) 2-Bromo-4-nitrotoluene
(j) *p*-Bromobenzoic acid
(k) *n*-Butylbenzene
(l) 1-(*p*-Bromophenyl)-1-butene
(m) 3-Nitro-1-(trichloromethyl)benzene
(n) 2,4,6-Trinitrotoluene (TNT)

12.31
(a) Which ring of benzanilide would you expect to undergo electrophilic substitution more readily? (b) Write resonance structures that explain your choice.

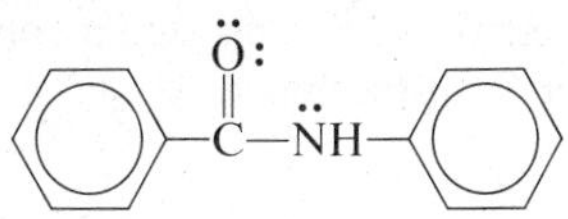

Benzanilide

12.32
What products would you expect from the nitration of phenyl benzoate?

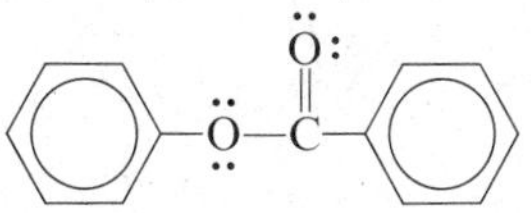

Phenyl benzoate

12.33
We have seen that benzene undergoes ring substitution when it reacts with chlorine in the presence of a Lewis acid. However, benzene can be made to undergo *addition* of chlorine by irradiating a mixture of benzene and chlorine with ultraviolet light. The addition reaction produces a mixture of 1,2,3,4,5,6-hexachlorocyclohexanes. One of these hexachlorocyclohexanes is *Lindane,* a very effective (but potentially hazardous) insecticide. The chloro groups of Lindane at carbons 1, 2, and 3 are equatorial, those at 4, 5, and 6 are axial. (a) Write the structure of Lindane. (b) Would you expect Lindane to exist in enantiomeric forms? (c) If not, why not? (d) One 1,2,3,4,5,6-hexachlorocyclohexane isomer does exist in enantiomeric forms. Write its structure.

12.34
Naphthalene undergoes electrophilic attack at the 1 position much more rapidly than it does at the 2 position.

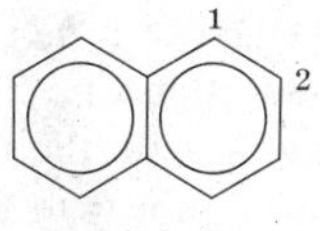

Naphthalene

The greater reactivity at the 1 position can be accounted for by writing resonance structures for the ring that undergoes electrophilic attack. Show how this is possible.

12.35
Naphthalene can be synthesized from benzene through the sequence of reactions shown below. Write the structures of each intermediate.

$$\text{Benzene} + \text{succinic anhydride} \xrightarrow{AlCl_3} \underset{(C_{10}H_{10}O_3)}{A} \xrightarrow[HCl]{Zn(Hg)} \underset{(C_{10}H_{12}O_2)}{B} \xrightarrow{SOCl_2}$$

$$\underset{(C_{10}H_{11}ClO)}{C} \xrightarrow{AlCl_3} \underset{(C_{10}H_{10}O)}{D} \xrightarrow{H_2,\ Pt} \underset{(C_{10}H_{12}O)}{E} \xrightarrow[heat]{H^+} \underset{(C_{10}H_{10})}{F} \xrightarrow[heat]{Pt} \text{naphthalene} + H_2$$

12.36
Anthracene and many other polycyclic aromatic compounds have been synthesized by a cyclization reaction known as the *Bradsher reaction* or *aromatic cyclodehyration.* This method, developed by Professor C. K. Bradsher of Duke University, can be illustrated by the conversion of an *o*-benzylphenyl ketone to a substituted anthracene.

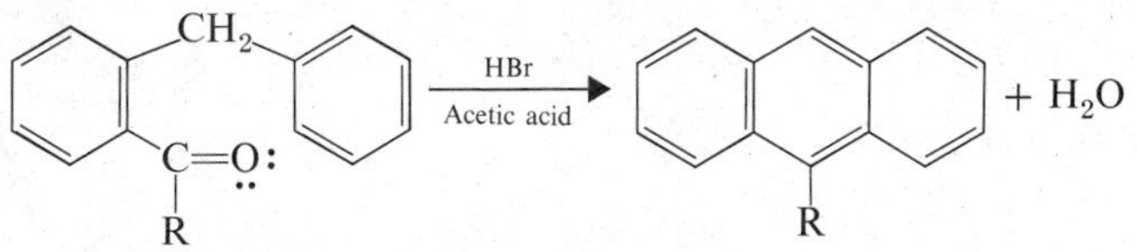

$$\xrightarrow[\text{Acetic acid}]{HBr} \quad + H_2O$$

An *o*-benzylphenyl ketone

Substituted anthracene

An arenium ion is an intermediate in this reaction and the last step involves the dehydration of an alcohol. Propose a mechanism for the Bradsher reaction.

12.37
Propose structures for compounds **G-I.**

$$\text{1,3-}C_6H_4(OH)_2 \xrightarrow[60\text{-}65°]{\text{Conc. } H_2SO_4} \underset{(C_6H_6S_2O_8)}{G} \xrightarrow[\text{Conc. } H_2SO_4]{\text{Conc. } HNO_3} \underset{(C_6H_5NS_2O_{10})}{H} \xrightarrow[\text{Heat}]{H_3O^+/H_2O} \underset{(C_6H_5NO_4)}{I}$$

12.38
2,6-Dichlorophenol has been isolated from the females of two species of ticks (*A. Americanum* and *A. maculatum*) where it apparently serves as a sex attractant. Each female tick yields about 5 nanograms of 2,6-dichlorophenol. Assume that you need larger quantities than this, and outline a synthesis of 2,6-dichlorophenol from phenol. (Hint: when phenol is sulfonated at 100°, the product is chiefly *p*-hydroxybenzenesulfonic acid.)

*** 12.39**
Treating propene with acetyl chloride in the presence of aluminum chloride gives two isomeric compounds **A** and **B** with the molecular formula C_5H_8O. Carrying the reaction out at low temperatures permits the isolation of a compound **C** with the molecular formula C_5H_9ClO. Propose structures for **A, B,** and **C** and write mechanisms that explain their formation.

*** 12.40**
The addition of a hydrogen halide (hydrogen bromide or hydrogen chloride) to 1-phenyl-1,3-butadiene produces (only) 1-phenyl-3-halo-1-butene. (a) Write a mechanism that accounts for the formation of this product. (b) Is this 1,4-addition or 1,2-addition to the butadiene system? (c) Is the product of the reaction consistent with the formation of the most stable intermediate carbocation? (d) Does the reaction appear to be under kinetic control or equilibrium control? Explain.

*** 12.41**
Treating *cis*-stilbene (*cis*-$C_6H_5CH{=}CHC_6H_5$) with chlorine in 1,2-dichloroethane yields an optically inactive addition product. All attempts to resolve the addition product into separate enantiomers fail. What does this suggest about (a) the stereochemistry of the addition product and (b) the mode of addition? (c) What result would you expect from a similar reaction starting with *trans*-stilbene?

*** 12.42**
Write mechanisms that account for the products of the following reactions:

(a) 9-fluorenylmethanol (CH_2OH on the 9-position of fluorene) $\xrightarrow[(-H_2O)]{H^+}$ phenanthrene

(b) $2\ CH_3{-}\underset{C_6H_5}{\underset{|}{C}}{=}CH_2 \xrightarrow{H^+}$ indane bearing CH_3 and C_6H_5 on C-1 and CH_3, CH_3 on C-3

*** 12.43**
At one time extensive use was made of detergents manufactured from the propene

tetramers whose synthesis we saw in Problem 7.39. In the industrial process, heating benzene and a mixture of the propene tetramers with aluminum chloride at 35–45° gave a mixture of isomers with the molecular formula $C_{18}H_{30}$. This mixture was then heated with sulfuric acid and the products of this reaction (isomers with the molecular formula $C_{18}H_{30}SO_3$) were treated with aqueous sodium hydroxide to give the detergent (also a mixture of isomers). Propose structures for the detergent isomers and write the reactions involved in their formation. (Note: use of these detergents was discontinued a number of years ago because they proved not to be biodegradable.)

*** 12.44**

The compound phenylbenzene, $C_6H_5—C_6H_5$, is called *biphenyl* and the rings are numbered in the manner shown below.

3 2 2′ 3′
4 4′
5 6 6′ 5′

Use models to answer the following questions about substituted biphenyls. (a) When certain large groups occupy three or four of the *ortho* positions (i.e., 2,6,2′ and 6′) the substituted biphenyl may exist in enantiomeric forms. An example of a biphenyl that exists in enantiomeric forms is the compound in which the following substituents are present: 2-NO_2, 6-COOH, 2′-NO_2, 6′-COOH. What factors account for this? (b) Would you expect a biphenyl with 2-Br, 6-COOH, 2′-COOH, 6′-H to exist in enantiomeric forms? (c) The biphenyl with 2-NO_2, 6-NO_2, 2′-COOH, 6′-Br cannot be resolved into enantiomeric forms. Explain.

*** 12.45**

Give structures (including stereochemistry where appropriate) for compounds **A** to **I**.

(a) Benzene + $CH_3CH_2\overset{O}{\overset{\|}{C}}Cl \longrightarrow$ **A** $\xrightarrow[0°]{PCl_5}$

B ($C_9H_{10}Cl_2$) $\xrightarrow[\text{Mineral oil, Heat}]{2NaNH_2}$ **C** (C_9H_8) $\xrightarrow{H_2, Ni_2B(P2)}$ **D** (C_9H_{10})

(b) **C** $\xrightarrow{\text{Li, liq. } NH_3}$ **E** (C_9H_{10})

(c) **D** $\xrightarrow[2-5°]{Br_2, CCl_4}$ **F** + enantiomer (major products)

(d) **E** $\xrightarrow[2-5°]{Br_2, CCl_4}$ **G** + enantiomer (major products)

(e) **D** $\xrightarrow[0°]{Cl_2, CH_2Cl_2}$ **H** + enantiomer (major products)

(f) **E** $\xrightarrow[0°]{Cl_2, CH_2Cl_2}$ **I** + enantiomer (major products)

*** 12.46**

Friedel-Crafts acylation of azulene gives mainly one isomer:

Azulene $\xrightarrow[AlCl_3]{CH_3\overset{O}{\overset{\|}{C}}—Cl}$ $C_{10}H_7—\overset{O}{\overset{\|}{C}}CH_3$ (mainly one isomer)

One ring of azulene is attacked by $CH_3C{\equiv}O^+$ preferentially because an especially stable arenium ion forms. (a) What is the structure of this arenium ion and (b) why is it especially stable. (c) What is the structure of the acetylazulene that forms as the major product?

SPECIAL TOPIC

THALLATION REACTIONS

Experimental work done by Professors Edward C. Taylor (of Princeton University) and Alexander McKillop (of the University of East Anglia) has shown that compounds of the metal *thallium* can be of great use in organic synthesis. We will see other examples of organo*metallic* chemistry in Chapter 14, but we can begin our study of organometallic chemistry here by examining the reaction of thallium(III) trifluoroacetate with benzene compounds. As we do this, we will see that the use of thallium(III) trifluoroacetate gives a method for introducing halogens (and other groups) into a benzene ring in highly specific ways.

Arenes react with thallium(III) trifluoroacetate* by electrophilic aromatic substitution. The reaction is usually carried out in trifluoroacetic acid and takes place in the manner shown below.

$$\underset{}{ArH} + \underset{\text{Thallium(III)-trifluoro-acetate}}{Tl(OOCCF_3)_3} \xrightarrow{CF_3COOH} \underset{\text{Arylthallium bis-trifluoroacetate}}{ArTl(OOCCF_3)_2} + CF_3COOH$$

This reaction usually gives the aryl thallium bis-trifluoroacetate in very high yields.

Because the carbon-thallium bond of arylthallium bis-trifluoroacetates is very weak (25–30 kcal/mole), it is easily cleaved either homolytically or heterolytically. Cleavage of the carbon-thallium bond allows us to replace the thallium bis-trifluoroacetate group by a number of other substituents.

Replacement by Iodine

Arylthallium bis-trifluoroacetates react with potassium iodide to give aryl iodides.

$$ArTl(OOCCF_3)_2 + KI \longrightarrow \underset{\text{Aryl iodide}}{ArI}$$

Replacement by a Cyano Group

The thallium bis-trifluoroacetate group can be replaced by a cyano group in two ways: through the reaction of arylthallium bis-trifluoroacetates with cuprous cyanide in dimethylformamide (DMF) or through their reaction with aqueous potassium cyanide under the influence of ultraviolet light.

$$ArTl(OOCCF_3)_2 \xrightarrow{CuCN} ArCN$$

$$ArTl(OOCCF_3)_2 \xrightarrow[h\nu]{KCN} ArCN$$

Aryl cyanide

Replacement by a Hydroxyl Group

The thallium bis-trifluoroacetate group can be replaced by a hydroxyl group through the treatment of arylthallium bis-trifluoroacetates with lead tetraacetate and triphenylphosphine followed by treatment with sodium hydroxide. This gives us a method for the synthesis of phenols.

$$ArTl(OOCCF_3)_2 \xrightarrow[(2)\ OH^-]{(1)\ Pb(OAc)_4/(C_6H_5)_3P} \underset{\text{A phenol}}{ArOH}$$

* Thallium compounds are extremely toxic and must be handled with great care.

The most remarkable aspect of aromatic thallations is that they are highly *regioselective,* as we will now see.

Para Substitution

Thallation takes place almost exclusively at the para position of benzene derivatives with a single alkyl group, halo group, or methoxyl group. The regioselectivity of these thallation reactions is attributed to the steric bulk of the electrophile. Thallium(III) trifluoroacetate, because of its size, reacts faster at the uncrowded para position.

$$C_6H_5Z + Tl(OOCCF_3)_3 \xrightarrow{CF_3COOH} p\text{-}Z\text{-}C_6H_4\text{-}Tl(OOCCF_3)_2$$

Para thallation

Z— = R—, X—, or CH_3O—

The examples listed below show how thallation reactions have been used to synthesize para-disubstituted benzene derivatives in a regioselective way.

$$C_6H_5C(CH_3)_3 \xrightarrow[(2)\ KI/H_2O]{(1)\ Tl(OOCCF_3)_3/CF_3COOH} p\text{-}(CH_3)_3C\text{-}C_6H_4\text{-}I \quad (93\%)$$

$$C_6H_5C(CH_3)_3 \xrightarrow[(2)\ KCN/H_2O,\ h\nu]{(1)\ Tl(OOCCF_3)_3/CF_3COOH} p\text{-}(CH_3)_3C\text{-}C_6H_4\text{-}CN \quad (80\%)$$

$$C_6H_5Cl \xrightarrow[(2)\ Pb(OOCCH_3)_4/(C_6H_5)_3P;\ (3)\ OH^-]{(1)\ Tl(OOCCF_3)_3/CF_3COOH} p\text{-}Cl\text{-}C_6H_4\text{-}OH \quad (56\%)$$

(only isomer)

Para-bromophenyl compounds can be prepared by using thallium(III) trifluoroacetate as the Lewis acid in bromination reactions. Although ring thallation does not take place in these reactions, para orientation is again a result of the steric bulk of the electrophile. The electrophile is a bromine-thallium(III) trifluoroacetate complex, and it reacts fastest at the more-open para position.

$$C_6H_5OCH_3 \xrightarrow[Tl(OOCCF_3)_3]{Br_2} p\text{-}CH_3O\text{-}C_6H_4\text{-}Br \quad (91\%)$$

Para-acetylanisole can be prepared in a similar way by using thallium(III) trifluoroacetate and acetyl chloride.

OCH_3 (anisole) $\xrightarrow[Tl(OOCCF_3)_3]{CH_3CCl (CH_3C(=O)Cl)}$ p-$CH_3OC_6H_4C(=O)CH_3$ (80%)

Ortho Substitution

Ortho thallation takes place when benzoic acid or methyl benzoate are treated with thallium(III) trifluoroacetate. Ortho thallation also occurs with benzyl alcohol, benzyl methyl ether, 2-phenylethanol, 2-phenylethyl methyl ether, phenylacetic acid, and methyl phenylacetate.

$C_6H_5Y \xrightarrow[CF_3COOH, \text{ Room temperature}]{Tl(OOCCF_3)_3} o\text{-}Y\text{-}C_6H_4Tl(OOCCF_3)_2$ Ortho thallation

Y =
- $-COOH$ (benzoic acid)
- $-COOCH_3$ (methyl benzoate)
- $-CH_2OH$ (benzyl alcohol)
- $-CH_2OCH_3$ (benzyl methyl ether)
- $-CH_2CH_2OH$ (2-phenylethanol)
- $-CH_2CH_2OCH_3$ (2-phenylethyl methyl ether)
- $-CH_2COOH$ (phenylacetic acid)
- $-CH_2COOCH_3$ (methyl phenylacetate)

Ortho substitution is the preferred course in all of these reactions because each compound contains an oxygen that can form a complex with thallium(III) trifluoroacetate and deliver it to the ortho position.

$C_6H_5C(=O)OR \xrightarrow[Tl(OOCCF_3)_3, \text{ Room temperature}]{CF_3COOH}$ [$C_6H_5C(OR)=O\cdots Tl(OOCCF_3)_3$] $\rightarrow$ $o\text{-}(ROC(=O))C_6H_4Tl(OOCCF_3)_2 + CF_3COOH$

R = H
R = CH_3

$C_6H_5CH_2CH_2OH \xrightarrow[Tl(OOCCF_3)_3, \text{ Room temperature}]{CF_3COOH}$ $C_6H_5CH_2CH_2OH\cdots Tl(OOCCF_3)_3$ $\rightarrow$ $o\text{-}(HOCH_2CH_2)C_6H_4Tl(OOCCF_3)_2 + CF_3COOH$

We see that in each case the intermediate complex resembles a stable five- or six-membered ring.

Problem G.1

Write the structure of complexes that would account for ortho thallation of the following compounds:

(a) Benzyl alcohol
(b) Benzyl methyl ether
(c) 2-Phenylethyl methyl ether
(d) Phenylacetic acid
(e) Methyl phenylacetate

Problem G.2

The ester, 2-phenylethyl acetate, undergoes para thallation when it is treated with thallium(III) trifluoroacetate in trifluoroacetic acid at room temperature.

$$C_6H_5CH_2CH_2OC(=O)CH_3 \xrightarrow[Tl(OOCCF_3)_3]{CF_3COOH} p\text{-}(CF_3COO)_2Tl\text{-}C_6H_4CH_2CH_2OC(=O)CH_3$$

2-Phenylethyl acetate

Can you suggest a reason that will account for the orientation of this reaction? (Hint: Complex formation probably occurs with the carbonyl oxygen of the ester group.)

The two examples listed below illustrate how thallations can be used for the regioselective synthesis of ortho-disubstituted benzenes.

$$C_6H_5COOH \xrightarrow[(2)\ KI/H_2O]{(1)\ Tl(OOCCF_3)_3} o\text{-}I\text{-}C_6H_4COOH \quad (79\%)$$

$$C_6H_5CH_2OCH_3 \xrightarrow[(2)\ KCN/H_2O,\ h\nu]{(1)\ Tl(OOCCF_3)_3/CF_3COOH} o\text{-}NC\text{-}C_6H_4CH_2OCH_3 \quad (55\%)$$

Meta Substitution

Thallation is reversible. In the absence of complex formation, thallation occurs at the para position because reaction at the para position takes place faster. Meta-thallated benzenes appear to be more stable, however. So at higher temperatures meta thallation is the preferred course of the reaction. An example that illustrates the effect of a temperature change on the course of a thallation reaction is the following.

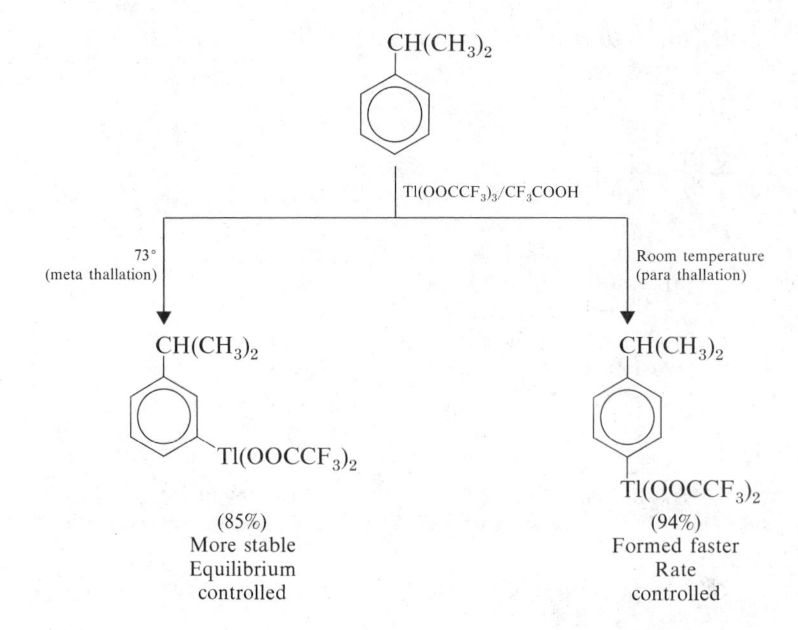

This example also furnishes us with another illustration of rate versus equilibrium control of an organic reaction (cf. Section 10.8).

Ortho, Meta, and Para Substitution of the Same Compound

The impressive versatility and control offered by thallations can be illustrated by one other example—one that is spectacular. Thallation reactions allow 2-phenylethanol to be *selectively converted* to ortho-, meta-, and para-iodo isomers.

$C_6H_5CH_2CH_2OH \longrightarrow C_6H_5CH_2CH_2OC(=O)CH_3$

(1) $Tl(OOCCF_3)_3$, 25° (2) KI, H_2O → CH_2CH_2OH (ortho-I) **A** (83%)

(1) $Tl(OOCCF_3)_3$, 73° (2) KI/H_2O → CH_2CH_2OH (meta-I) **B** (54%)

(1) $Tl(OOCCF_3)_3$ (2) KI, H_2O (3) H_3O^+, heat → CH_2CH_2OH (para-I) **C** (84%)

The ortho isomer results from reaction **A** because the oxygen of the alcohol substituent delivers the thallium to that position.

The meta isomer results from reaction **B** because the reaction is carried out at a higher temperature and the meta-thallated compound is *more stable*.

The para isomer is obtained from the reaction sequence **C** because the alcohol group is first converted to an acetate ester. Complex formation of thallium with the carbonyl oxygen of the acetate group places the thallium at too great a distance to be delivered to the ortho position (cf. Problem G.2). Instead, thallium substitutes at the para position where it *reacts faster*.

Problem G.3

Using thallium(III) trifluoroacetate and any other necessary starting materials, show how the following compounds could be synthesized in a regioselective way.

(a) 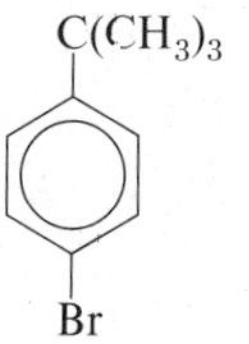 $C(CH_3)_3$, Br

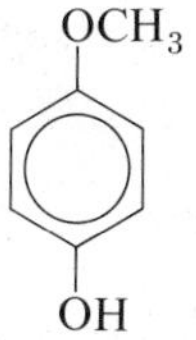

(c) OCH_3, OH

(b) 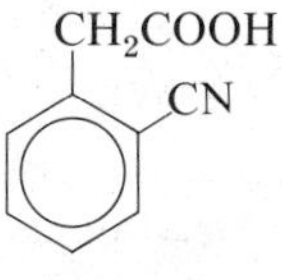 CH_2COOH, CN

(d) 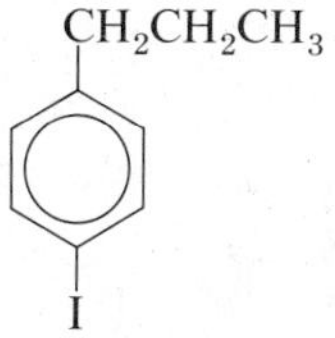 $CH_2CH_2CH_3$, I

(e) 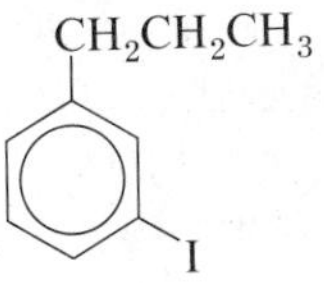 $CH_2CH_2CH_3$, I

(g) 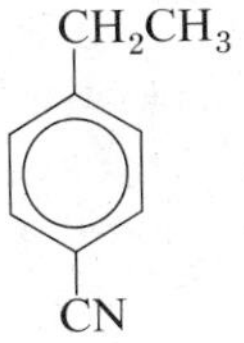 CH_2CH_3, CN

(f) 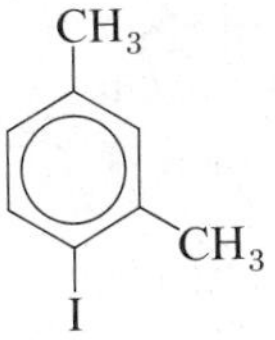 CH_3, CH_3, I

Problem G.4

Show how you might synthesize each of the following compounds from toluene.

(a) *p*-Bromotoluene

(b) *p*-Iodotoluene

(c) 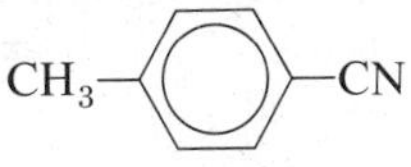 CH_3—⟨ ⟩—CN

(d) 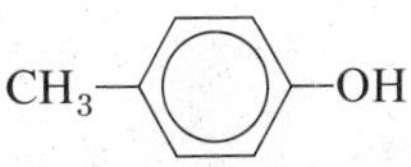 CH_3—⟨ ⟩—OH

(e) 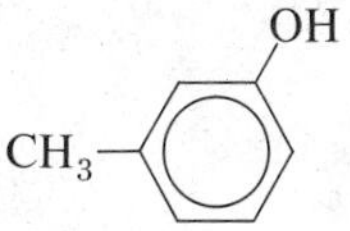 CH_3—⟨ ⟩, OH

(f) 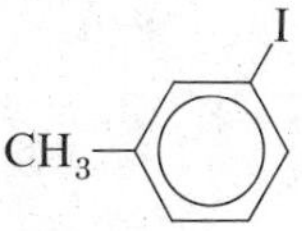 CH_3—⟨ ⟩, I

13 PHYSICAL METHODS OF STRUCTURE DETERMINATION · NUCLEAR MAGNETIC RESONANCE SPECTROSCOPY · INFRARED SPECTROSCOPY

13.1 NUCLEAR MAGNETIC RESONANCE SPECTROSCOPY

The hydrogen nucleus, or proton, has magnetic properties. When one places a compound containing hydrogen in a very strong magnetic field and simultaneously irradiates it with electromagnetic energy, the hydrogen nuclei of the compound may absorb energy through a process known as *magnetic resonance.** This absorption of energy, like all processes that occur on the atomic and molecular scale, is *quantized.* Absorption of energy does not occur until the strength of the magnetic field and the frequency of electromagnetic radiation are at specific values.

Instruments, known as nuclear magnetic resonance (nmr) spectrometers (Fig. 13.1), allow chemists to measure the absorption of energy by hydrogen nuclei. These instruments use very powerful magnets and irradiate the sample with electromagnetic radiation in the radio frequency region.

Nuclear magnetic resonance spectrometers are usually designed so that they irradiate the compound with electromagnetic energy of a constant frequency while the magnetic field strength is varied. When the magnetic field reaches the correct strength, the nuclei absorb energy and resonance occurs. This cause a tiny electrical current to flow in an antenna coil surrounding the sample. The instrument then amplifies this current and displays it as a signal (a peak or series of peaks) on a strip of calibrated chart paper.

If hydrogen nuclei were stripped of their electrons and isolated from other nuclei, all hydrogen nuclei (protons) would absorb energy at the same magnetic

* Magnetic resonance is an entirely different phenomenon from the resonance theory that we have discussed in earlier chapters.

FIG. 13.1
Essential parts of a nuclear magnetic resonance spectrometer.

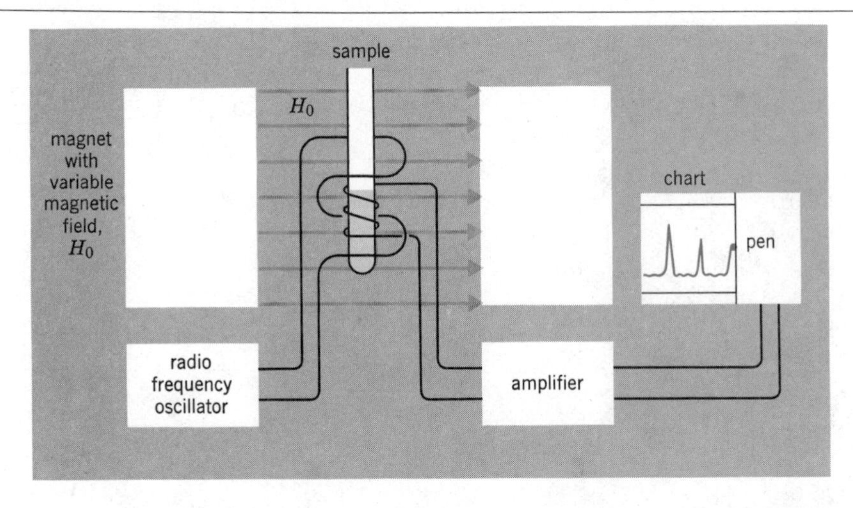

field strength for a given frequency of electromagnetic radiation. If this were the case, nuclear magnetic resonance spectrometers would only be very expensive instruments for hydrogen analysis.

Fortunately, the nuclei of hydrogen atoms of compounds of interest to the organic chemist are not stripped of their electrons, and they are not isolated from each other. Some hydrogen nuclei are in regions of greater electron density than others. Because of this, the protons of these compounds absorb energy at *slightly different* magnetic field strengths. The actual field strength at which absorption occurs is highly dependent on the magnetic environment of each proton. This magnetic environment depends on two factors: magnetic fields generated by circulating electrons and magnetic fields that result from other nearby protons (or other magnetic nuclei).

We will discuss the theory of nuclear magnetic resonance in more detail later. Before we do, however, it will be helpful to examine the proton magnetic resonance spectra of some simple compounds.

Figure 13.2 shows the proton magnetic resonance spectra of *p*-xylene and *tert*-butyl acetate.

Magnetic field strength is measured along the bottom of the spectra on a delta (δ) scale in units of parts per million (ppm) and along the top in Hertz (cycles per second). We will have more to say about these units later; for the moment, we need only to point out that the externally applied magnetic field strength increases from left to right. A signal that occurs at $\delta = 7$ ppm occurs at a lower external magnetic field strength than one that occurs at $\delta = 2$ ppm. Signals on the left of the spectrum are also said to occur *downfield* and those on the right are said to be *upfield.*

Both spectra in Fig. 13.2 show a small signal at $\delta = 0$ ppm. This arises from a compound that has been added to the sample to allow calibration of the instrument.

The first feature we want to notice is the relation between the number of signals in each spectrum and the number of different types of hydrogens in each compound.

Both *p*-xylene and *tert*-butyl acetate have only *two* different types of hydrogens, and each compound gives only *two* signals in its nuclear magnetic resonance spectrum.

The two different types of hydrogens of *p*-xylene are the hydrogens of the methyl groups and the hydrogens of the benzene ring. The six methyl hydrogens of *p*-xylene are all *equivalent* and they are in a different environment from the four hydrogens of the ring. The six methyl hydrogens give rise to the signal that occurs at $\delta = 2.30$ ppm. The four hydrogens of the benzene ring are also equivalent; they give rise to the signal at $\delta = 7.05$ ppm.

The two different kinds of hydrogens of *tert*-butyl acetate are the three equivalent hydrogens of the methyl group (*b*) and the nine equivalent hydrogens of the *tert*-butyl group (*a*). The *tert*-butyl hydrogens absorb at $\delta = 1.45$ ppm and the methyl hydrogens absorb at $\delta = 1.97$ ppm.

The next features of these two spectra that we want to examine are the relative magnitudes of the peaks (or signals), for these are often helpful in assigning peaks to particular groups of hydrogens. What is important here is not necessarily the height of each peak, but *the area underneath it*. These areas, when accurately measured (the spectrometers do this automatically), are in the same ratio as the number of hydrogens causing each signal. We can see however, without measuring, that the area under the signal for the methyl groups of *p*-xylene (6H) is larger than that for the benzene hydrogens (4H). When these areas are measured accurately they are found to be in ratio of 3 : 2 or 6 : 4. When measurements are

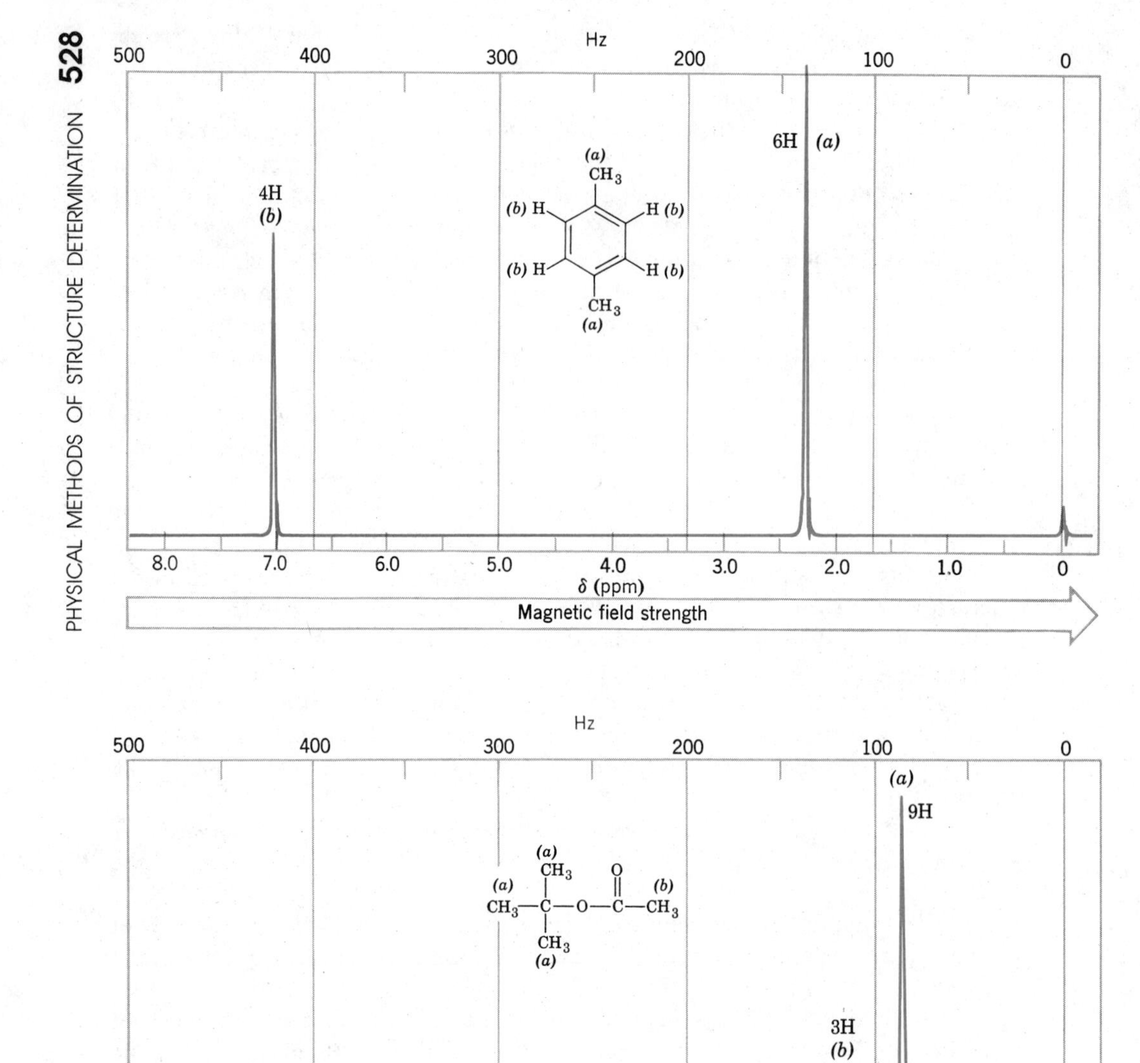

FIG. 13.2
The proton nmr spectra of p-*xylene and* tert-*butyl acetate. (Spectra courtesy of Varian Associates, Palo Alto, Calif.)*

made of the areas under the signals from *tert*-butyl acetate, they are found to be in a ratio of 3 : 1 or 9 : 3 for the *tert*-butyl and methyl hydrogens, respectively.

A third feature of proton nmr spectra that provides us with information about the structure of a compound can be illustrated if we examine the spectrum for 1,1,2-trichloroethane (Fig. 13.3).

In Fig. 13.3 we have an example of signal splitting. Signal splitting is a phenomenon that arises from magnetic influences of hydrogens on atoms adjacent

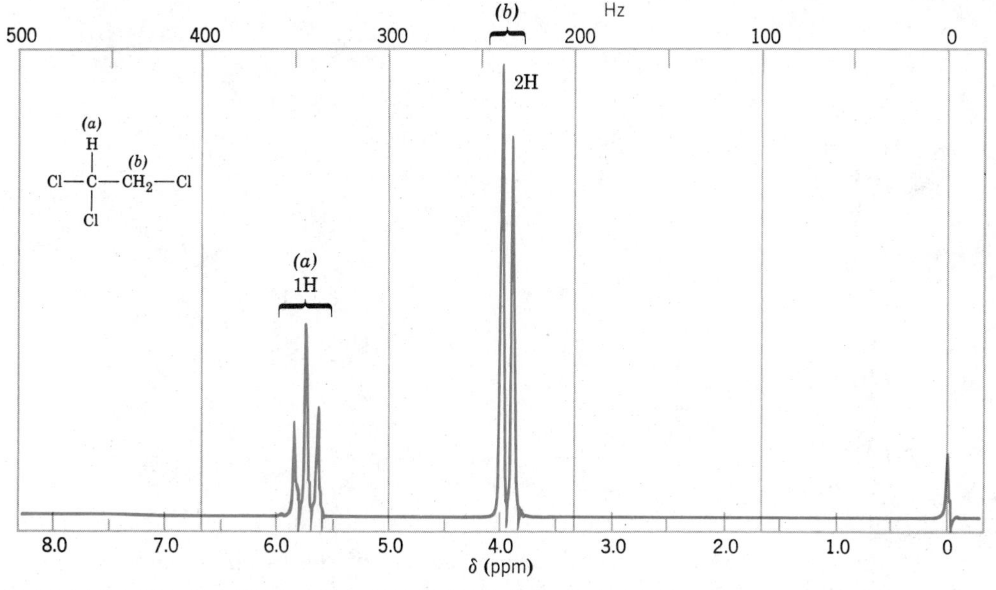

FIG. 13.3
The proton nmr spectrum of 1,1,2-trichloroethane. (Spectrum courtesy of Varian Associates, Palo Alto, Calif.)

to those bearing the hydrogens causing the general signal. The signal (*b*) from the two equivalent hydrogens of the $—CH_2Cl$ group is split into two peaks (a doublet) by the magnetic influence of the hydrogen of the $—CHCl_2$ group. Conversely, the signal (*a*) from the hydrogen of the $—CHCl_2$ group is split into three peaks (a triplet) by the magnetic influences of the two equivalent hydrogens of the $—CH_2Cl$ group.

At this point signal splitting may seem like an unnecessary complication. As we gain experience in interpreting proton nmr spectra we will find that because signal splitting occurs in a predictable way, it often provides us with important information about the structure of the compound.

Now that we have had an introduction to the important features of proton nmr spectra we are in a position to consider them in greater detail.

13.2 NUCLEAR SPIN: THE ORIGIN OF THE SIGNAL

We are already familiar with the concept of electron spin and with the fact that the spins of electrons confer on them the spin quantum states of $+\frac{1}{2}$ or $-\frac{1}{2}$. Electron spin is the basis for the Pauli exclusion principle (Section 1.13); it allows us to understand how two electrons with paired spins may occupy the same atomic or molecular orbital.

The nuclei of certain isotopes also spin and therefore these nuclei possess spin quantum numbers, I. The nucleus of ordinary hydrogen, ^1H (i.e., a proton), is like the electron; its spin quantum number I is $\frac{1}{2}$ and it can assume either of two spin states: $+\frac{1}{2}$ or $-\frac{1}{2}$. These correspond to the magnetic moments allowed for $I = \frac{1}{2}, m = +\frac{1}{2}$ or $-\frac{1}{2}$. Other nuclei with spin quantum numbers $I = \frac{1}{2}$ are ^{13}C, ^{19}F, and ^{31}P. Some nuclei, such as ^{12}C, ^{16}O, and ^{32}S, have no spin ($I = 0$); other nuclei have spin quantum numbers greater than $\frac{1}{2}$. In our treatment here we will be primarily concerned with the spectra that arise from protons.

Since the proton is electrically charged, the spinning proton generates a tiny

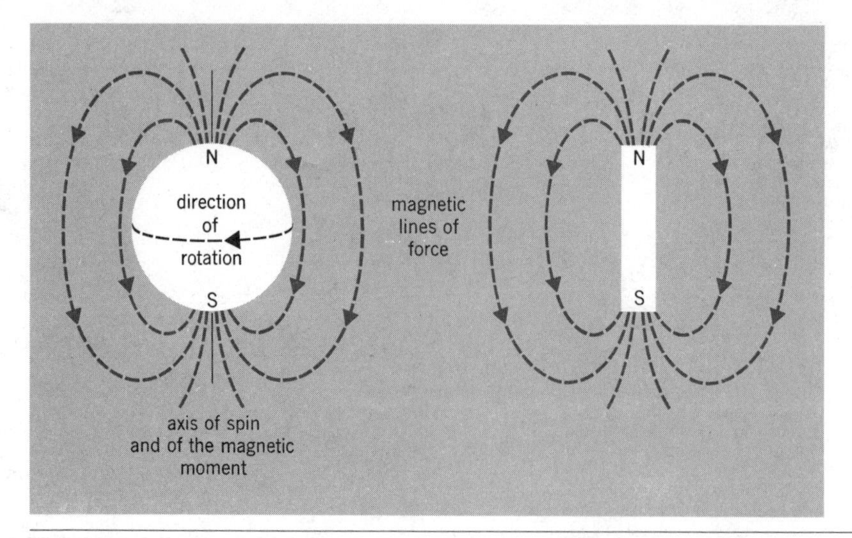

FIG. 13.4
The magnetic field associated with a spinning proton (a). *The spinning proton resembles a tiny bar magnet* (b).

magnetic moment—one that coincides with the axis of spin (Fig. 13.4). This tiny magnetic moment confers on the spinning proton the properties of a tiny bar magnet.

When a compound containing hydrogen (and thus protons) is placed in an external magnetic field, the protons may assume one of two possible orientations with respect to the external magnetic field. The magnetic moment of the proton may be aligned "with" the external field or "against" it (Fig. 13.5).

As we might expect, the two alignments of the proton in an external field are not of equal energy. When the proton is aligned with the magnetic field, its energy is lower than when it is aligned against the magnetic field (Fig. 13.6).

Energy is required to "flip" the proton from its lower-energy state (with the field) to its higher-energy state (against the field). In a nuclear magnetic resonance spectrometer this energy is supplied by electromagnetic radiation in the radio frequency region. The energy required is proportional to the strength of the magnetic field. One can show by relatively simple calculations, that in a magnetic field of approximately 14,100 gauss, electromagnetic radiation of 60×10^6 cycles per second (60 MHz) supplies the correct amount of energy for protons.*

* The relationship between the frequency of the radiation, v, and the strength of the magnetic field, H_0, is,

$$v = \frac{\mu H_0}{2\pi}$$

where μ is the magnetogyric ratio. For a proton, $\mu = 26{,}753$.

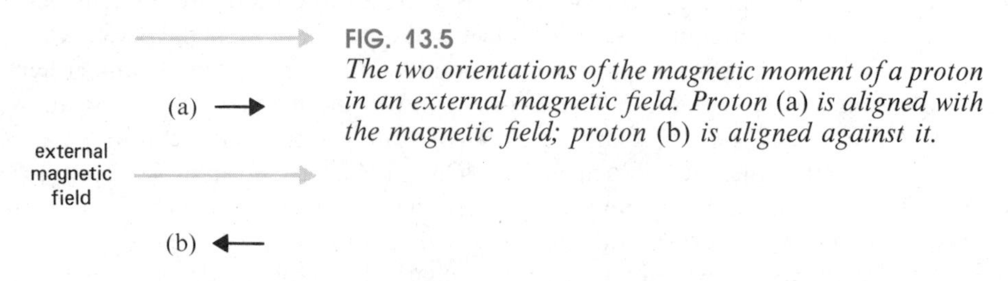

FIG. 13.5
The two orientations of the magnetic moment of a proton in an external magnetic field. Proton (a) *is aligned with the magnetic field; proton* (b) *is aligned against it.*

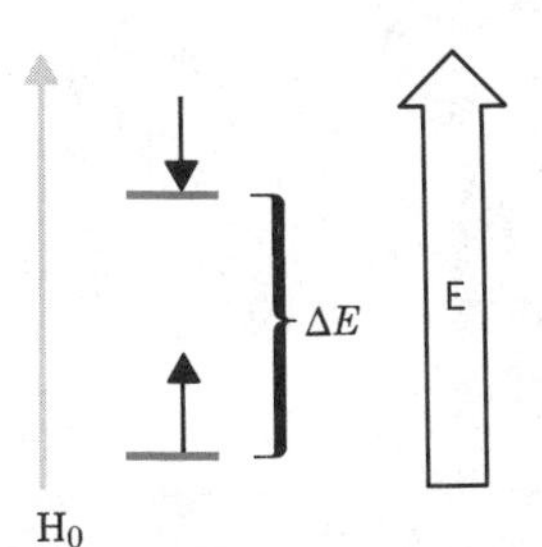

FIG. 13.6
The two energy states of a proton in an external magnetic field. The difference between these two energy states is ΔE.

As we mentioned earlier, nmr spectrometers are designed so that the radio frequency is kept constant (at 60 MHz, for example) and the magnetic field is varied. When the magnetic field is tuned to precisely the right strength the protons of a particular set in the molecule flip from one state to the other. In doing so they absorb radio frequency energy. This flipping of the protons generates a small electric current in a coil of wire surrounding the sample. After being amplified, the current is displayed as a signal in the spectrum.

13.3 SHIELDING AND DESHIELDING OF PROTONS

All protons do not absorb energy at the same external magnetic field strength. The three spectra that we examined earlier demonstrate this for us. The aromatic protons of *p*-xylene absorb at lower field strength ($\delta = 7.05$ ppm); the various alkyl protons of *p*-xylene, *tert*-butyl acetate, and 1,1,2-trichloroethane all absorb at higher magnetic field strengths.

The general position of a signal in a nuclear magnetic resonance spectrum—that is, the strength of the magnetic field required to bring about absorption of energy—can be related to electron densities and electron circulations in the compounds. Under the influence of an external magnetic field the electrons move in certain preferred paths. Because they do, and because electrons are charged particles, they generate tiny magnetic fields.

We can see how this happens if we consider the electrons around the proton in a σ bond of a C—H group. In doing so, we will oversimplify the situation by assuming that σ electrons move in generally circular paths. The magnetic field generated by these σ electrons is shown in Fig. 13.7.

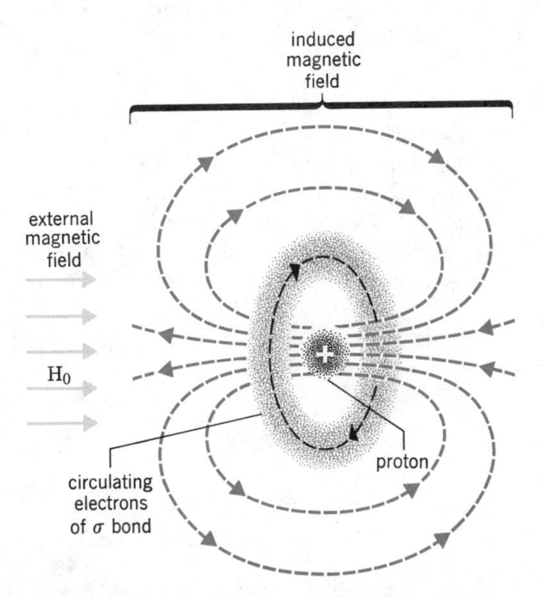

FIG. 13.7
The circulations of the electrons of a C—H bond under the influence of an external magnetic field. The electron circulations generate a small magnetic field (an induced field) *that* shields *the proton from the external field.*

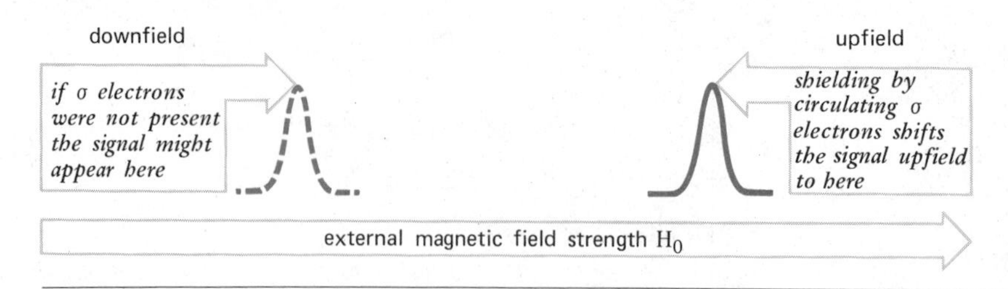

FIG. 13.8
Shielding by σ electrons causes proton nmr absorption to be shifted to higher external magnetic field strengths.

The small magnetic field generated by the electrons is called *an induced field. At the proton, the induced magnetic field opposes the external magnetic field.* This means that the actual magnetic field sensed by the proton is slightly less than the external field. The electrons are said *to shield* the proton.

A proton shielded by electrons will not, of course, absorb at the same external field strength as a proton that has no electrons. A shielded proton will absorb *at higher external field strengths;* the external field must be made larger by the spectrometer in order to compensate for the small induced field (Fig. 13.8).

The extent to which a proton is shielded by the circulations of σ electrons depends on the relative electron density around the proton. This electron density depends largely on the presence or absence of electronegative groups. Electronegative groups withdraw electron density from the C—H bond, particularly if they are attached to the same carbon. We can see an example of this effect in the spectrum of 1,1,2-trichloroethane (Fig. 13.3). The proton of carbon-1 absorbs at a lower magnetic field strength ($\delta = 5.77$ ppm) than the protons of carbon-2 ($\delta = 3.95$ ppm). Carbon-1 bears two highly electronegative chloro groups whereas carbon-2 bears only one. The protons of carbon-2, consequently, are more effectively shielded because the σ-electron density around them is greater.

The circulations of delocalized π electrons generate magnetic fields that can either *shield or deshield* nearby protons. Whether shielding or deshielding occurs, depends on the location of the proton in the *induced* field. The aromatic protons of benzene derivatives (Fig. 13.9) are *deshielded* because their locations are such that the induced magnetic field reinforces the applied magnetic field.

Because of this deshielding effect the absorption of energy by phenyl protons occurs downfield at relatively low magnetic field strength. The protons of benzene, itself, absorb at $\delta = 7.27$ ppm. The aromatic protons of *p*-xylene (Fig. 13.2) absorb at $\delta = 7.05$ ppm.

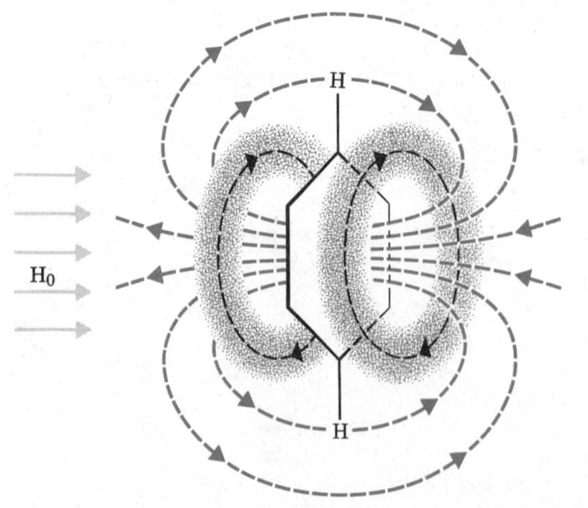

FIG. 13.9
The induced magnetic field of the π electrons of benzene deshields the benzene protons. This happens because at the location of the protons the induced field is in the same direction as the applied field.

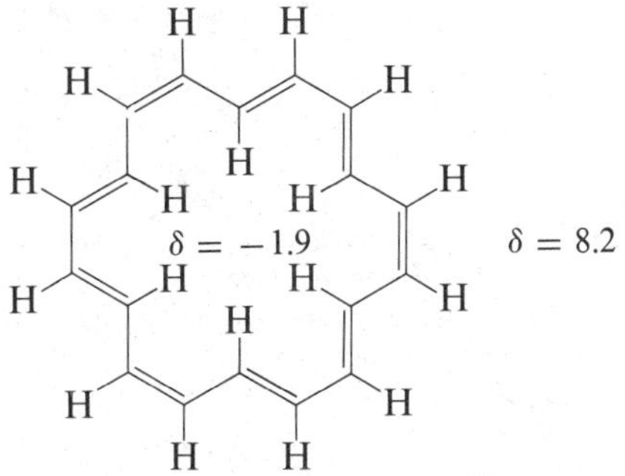

$\delta = -1.9$ $\delta = 8.2$

FIG. 13.10
18-Annulene. The internal protons are highly shielded and absorb at $\delta = -1.9$. The external protons are highly deshielded and absorb at $\delta = 8.2$.

The deshielding of external aromatic protons that results from the circulating π electrons is one of the best pieces of physical evidence that we have for π electron delocalization in aromatic rings. In fact, low field strength proton absorption is often used as a criterion for aromaticity in newly synthesized conjugated cyclic compounds.

Not all aromatic protons absorb at low magnetic field strengths, however. Large-ring aromatic compounds have been synthesized that have hydrogens *in the center of the ring* (in the π electron cavity). The protons of these internal hydrogens absorb at unusually high magnetic field strengths because they are highly shielded by the opposite induced field in the center of the ring (cf. Fig. 13.9). These internal protons often absorb at field strengths greater than that used for the reference point, $\delta = 0$. The internal protons of 18-annulene (Fig. 13.10) absorb at $\delta = -1.9$ ppm.

Problem 13.1

The methyl protons of 15,16-dimethyldihydropyrene (Section 11.7) absorb at very high magnetic field strengths, $\delta = -4.2$. Can you account for this?

Pi-electron circulations also *shield* the protons of acetylene causing them to absorb at higher magnetic field strengths than we might otherwise expect. If we were to consider *only* the relative electronegativities of carbon in its three hybridization states we might expect the following order of protons attached to each type of carbon:

$$\text{(low field strength) } sp < sp^2 < sp^3 \text{ (high field strength)}$$

In fact, acetylenic protons absorb between $\delta = 2.0$ and $\delta = 3.0$ and the order is:

$$\text{(low field strength) } sp^2 < sp < sp^3 \text{ (high field strength)}$$

This upfield shift of the absorption of acetylenic protons is a result of shielding produced by the circulating π electrons of the triple bond. The origin of this shielding is illustrated in Fig. 13.11.

13.4 THE CHEMICAL SHIFT

We see now that shielding and deshielding effects cause the absorptions of protons to be shifted from the position that a bare proton would absorb (i.e., a proton stripped of its electrons). Since these shifts result from the circulations of electrons in *chemical* bonds, they are called *chemical shifts*.

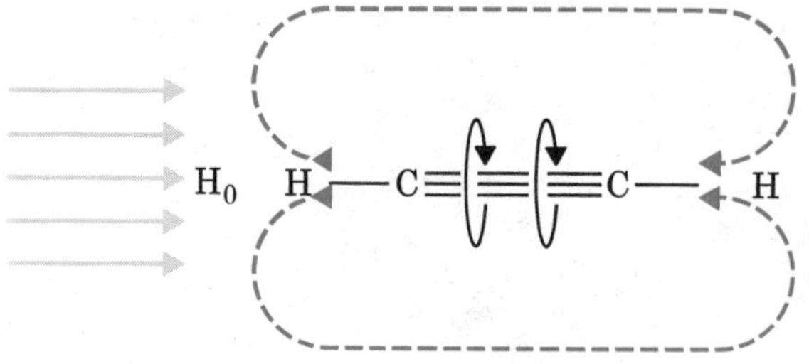

H_0 H—C≡C—H

FIG. 13.11
The shielding of acetylenic protons by π-electron circulations. Shielding causes acetylenic protons to absorb further upfield than vinyl protons.

Chemical shifts are measured with reference to the absorption of protons of reference compounds. A reference is used because it is impractical to measure the actual value of the magnetic field at which absorptions occur. The reference compound most often used is tetramethylsilane (TMS). A small amount of tetramethylsilane is usually added to the sample whose spectrum is being measured, and the signal from the 12 equivalent protons of tetramethylsilane is used to establish the zero point on the delta scale.

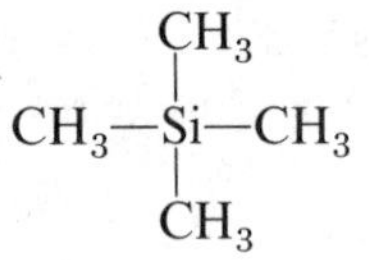

$$\begin{array}{c} CH_3 \\ | \\ CH_3—Si—CH_3 \\ | \\ CH_3 \end{array}$$

Tetramethylsilane

Tetramethylsilane was chosen as a reference compound for several reasons. It has 12 hydrogens and, therefore, a very small amount of tetramethylsilane gives relatively large signal. Because the hydrogens are all equivalent they give a *single signal.* Since silicon is less electronegative than carbon the protons of tetramethylsilane are in regions of high electron density. They are, as a result, highly shielded and the signal from tetramethylsilane occurs in a region of the spectrum where few other hydrogens absorb. Thus, their signal seldom interferes with the signals from other hydrogens. Tetramethylsilane, like an alkane, is relatively inert. Finally, it is volatile; its boiling point is 27 °C. After the spectrum has been determined the tetramethylsilane can be removed easily by evaporation.

Chemical shifts are measured in Hertz (cycles per second), as if the frequency of the electromagnetic radiation were being varied. In actuality it is the magnetic field that is changed. But, since the values of frequency and the strength of the magnetic field are mathematically related, frequency units (Hertz) are appropriate ones.

The chemical shift of a proton, when expressed in Hertz, will be proportional to the strength of the external magnetic field. Since spectrometers with different magnetic field strengths are commonly used, it is desirable to express chemical shifts in a form that is independent of the strength of the external field. This can be done easily by dividing the chemical shift by frequency of the spectrometer, with both numerator and denominator of the fraction expressed in frequency units (Hertz). Since chemical shifts are always very small (typically less than 500 Hz) compared with the total field strength (commonly the equivalent of 30, 60, and 100 *million* Hz), it is convenient to express these fractions in units of *parts per million* (ppm). This is the origin of the delta scale for the expression of chemical shifts relative to TMS.

$$\delta = \frac{(\text{observed shift from TMS in Hertz}) \times 10^6}{(\text{operating frequency of the instrument in Hertz})}$$

TABLE 13.1 Typical Proton Chemical Shifts

TYPE OF PROTON	CHEMICAL SHIFT, PPM
1° Alkyl, RCH_3	0.8–1.0
2° Alkyl, RCH_2R	1.2–1.4
3° Alkyl, R_3CH	1.4–1.7
Allylic, $R_2C{=}C(R){-}CH_3$	1.6–1.9
Benzylic, $ArCH_3$	2.2–2.5
Alkyl chloride, RCH_2Cl	3.6–3.8
Alkyl bromide, RCH_2Br	3.4–3.6
Alkyl iodide, RCH_2I	3.1–3.3
Ether, $ROCH_2R$	3.3–3.9
Alcohol, $HOCH_2R$	3.3–4.0
Ketone, $RC({=}O)CH_3$	2.1–2.6
Aldehyde, $RC({=}O)H$	9.5–9.6
Vinylic, $R_2C{=}CH_2$	4.6–5.0
Vinylic, $R_2C{=}CH(R)$	5.2–5.7
Aromatic, ArH	6.0–9.5
Acetylenic, $RC{\equiv}CH$	2.5–3.1
Alcohol hydroxyl, ROH	0.5–6.0[a]
Carboxylic, $RC({=}O)OH$	10–13[a]
Phenolic, ArOH	4.5–7.7[a]
Amino $R{-}NH_2$	1.0–5.0[a]

[a]The chemical shifts of these groups vary in different solvents and with temperature and concentration.

Table 13.1 gives the values of proton chemical shifts for some common hydrogen-containing groups.

13.5 CHEMICAL SHIFT EQUIVALENT AND NONEQUIVALENT PROTONS

Two or more protons that are in identical magnetic environments have the same chemical shift and, therefore, give only one proton nmr signal. How do we know when protons are in the same magnetic environment? For most compounds, protons that are in the same magnetic environment are also equivalent in chemical reactions. This is, *chemically equivalent* protons are *chemical shift equivalent* in proton nmr spectra.

We saw, for example, that the six methyl protons of *p*-xylene give a single proton nmr signal. We probably recognize, intuitively, that these six hydrogens are chemically equivalent. We can demonstrate their equivalence, however, by replacing each hydrogen in turn with some other group. If in making these substitutions we get the same compound from each replacement then the protons are chemically equivalent and are chemical shift equivalent. The replacements can be replacements that occur in an actual chemical reaction or they can be purely imaginary. For the methyl hydrogens of *p*-xylene we can think of an actual chemical reaction that demonstrates their equivalence, *benzylic bromination.*

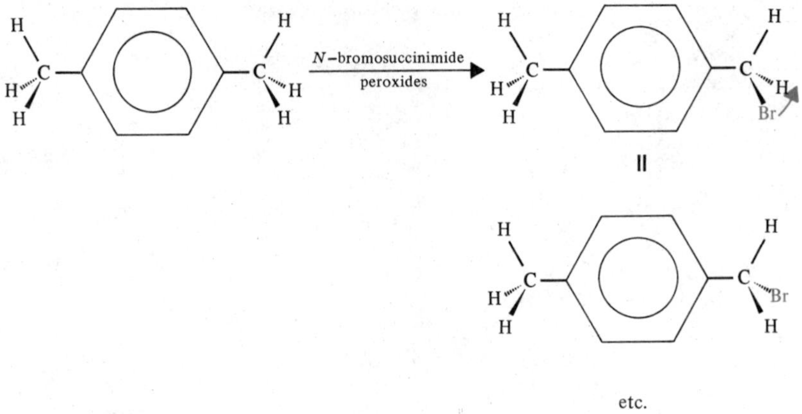

N-bromosuccinimide
peroxides

etc.

Benzylic bromination produces the same monobromo product regardless of which of the six hydrogens is replaced.

We can also think of a chemical reaction that demonstrates the equivalence of the four aromatic hydrogens of *p*-xylene, *ring bromination*. Once again, we get the same compound regardless of which of the four hydrogens is replaced.

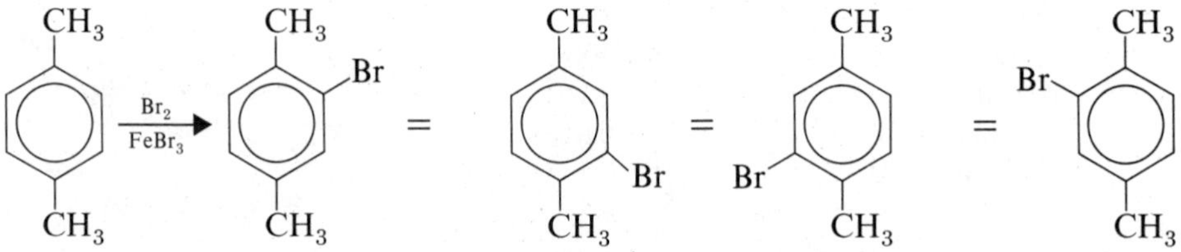

Br_2
$FeBr_3$

With *tert*-butyl acetate we can demonstrate that the three hydrogens of the methyl group are equivalent by replacing each with an imaginary group **W**.

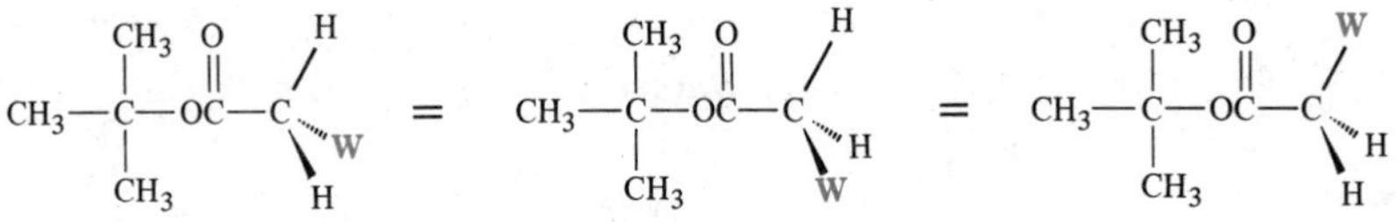

We can demonstrate the equivalence of the nine *tert*-butyl hydrogens by the same process.

Problem 13.2

How many different sets of equivalent protons do each of the following compounds have? How many signals would each compound give in its proton nmr spectrum?

(a) CH_3CH_3 (b) $CH_3CH_2CH_3$

(c) CH_3OCH_3

(d) 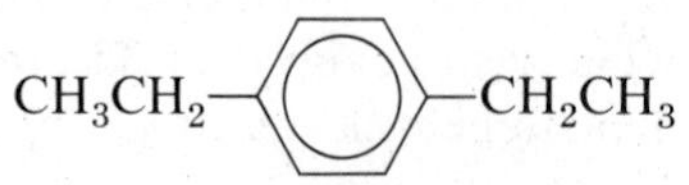 $CH_3CH_2-C_6H_4-CH_2CH_3$

(e) $CH_3C(=O)-OCH_3$

(f) $CH_3C(=O)-OCH(CH_3)_2$

13.5A Enantiotopic and Diastereotopic Hydrogens

If replacement of each of two hydrogens by the same group yields compounds that are enantiomers, the two hydrogens are said to be *enantiotopic. The protons of enantiotopic hydrogens have the same chemical shift and give only one proton nmr signal.**

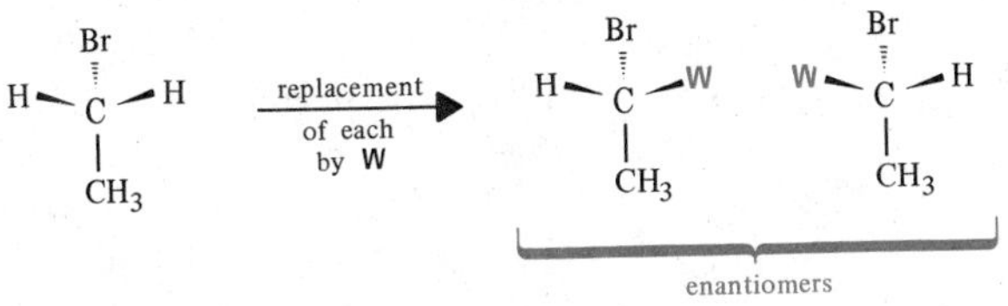

Br, H, C, H, CH_3 —replacement of each by **W**→ Br, H, C, W, CH_3 ; Br, W, C, H, CH_3

enantiomers

The two hydrogens of the $—CH_2Br$ group of ethyl bromide are enantiotopic. Ethyl bromide, then, gives two signals in its pmr spectrum. The three equivalent protons of the $CH_3—$ group give one signal; the two enantiotopic protons of the $—CH_2Br$ group give the other signal. [The proton nmr spectrum of ethyl bromide as we will see, actually consists of seven peaks. This is a result of signal splitting. One signal (from the $CH_3—$ group) is split into three peaks; the other signal (from the $—CH_2Br$ group) is split into four peaks.]

If replacement of each of two hydrogens by a group, **W,** gives compounds that are diastereomers, the two hydrogens are said to be *diastereotopic*. Except for accidental coincidence, *diastereotopic protons do not have the same chemical shift and give rise to different proton nmr signals.*

The two protons of the $=CH_2$ group of chloroethene are diastereotopic.

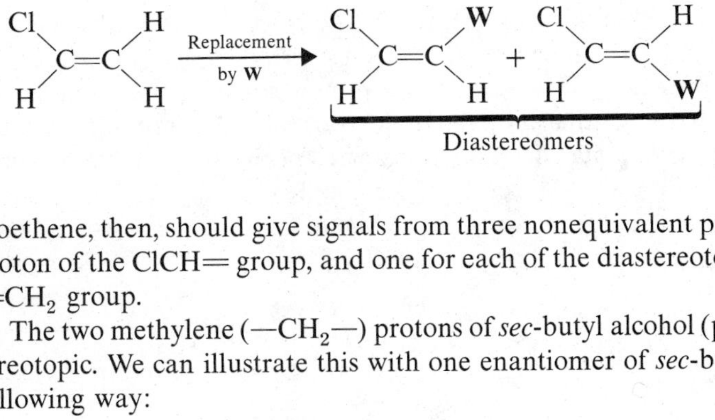

Cl, H, C=C, H, H —Replacement by **W**→ Cl, **W**, C=C, H, H + Cl, H, C=C, H, **W**

Diastereomers

Chloroethene, then, should give signals from three nonequivalent protons; one for the proton of the ClCH= group, and one for each of the diastereotopic protons of the $=CH_2$ group.

The two methylene $(—CH_2—)$ protons of *sec*-butyl alcohol (p. 538) are also diastereotopic. We can illustrate this with one enantiomer of *sec*-butyl alcohol in the following way:

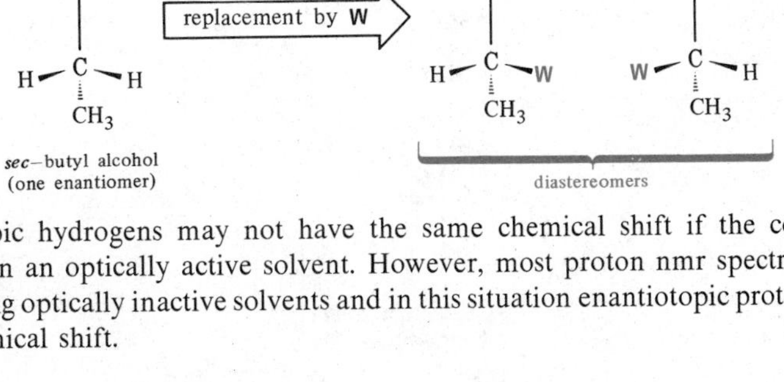

CH_3, H, C, OH, H, C, H, CH_3 —replacement by **W**→ CH_3, H, C, OH, H, C, W, CH_3 ; CH_3, H, C, OH, W, C, H, CH_3

sec-butyl alcohol
(one enantiomer)

diastereomers

* Enantiotopic hydrogens may not have the same chemical shift if the compound is dissolved in an optically active solvent. However, most proton nmr spectra are determined using optically inactive solvents and in this situation enantiotopic protons have the same chemical shift.

Problem 13.3

(a) Show that replacing each of the two methylene protons of the other *sec*-butyl alcohol enantiomer by **W** also leads to a pair of diastereomers. (b) How many chemically different kinds of protons are there in *sec*-butyl alcohol? (c) How many proton nmr signals would you expect to find in the spectrum of *sec*-butyl alcohol?

Problem 13.4

How many proton nmr signals would you expect from each of the following compounds?

(a) $CH_3CH_2CH_2CH_3$
(b) CH_3CH_2OH
(c) $CH_3CH{=}CH_2$
(d) *trans*-2-Butene
(e) 1,2-Dibromopropane
(f) 1,1-Dimethylcyclopropane
(g) *trans*-1,2-Dimethylcyclopropane
(h) *cis*-1,2-Dimethylcyclopropane
(i) 1-Pentene

13.6 SIGNAL SPLITTING: SPIN-SPIN COUPLING

Signal splitting is caused by magnetic fields of protons on nearby atoms. We have seen an example of signal splitting in the spectrum of 1,1,2-trichloroethane (Fig. 13.3). The signal from the two equivalent protons of the $—CH_2Cl$ group of 1,1,2-trichloroethane is split into two peaks by the single proton of the $CHCl_2—$ group. The signal from the proton of the $CHCl_2—$ group is split into three peaks by the two protons of the $—CH_2Cl$ group. This is further illustrated in Fig. 13.12.

Signal splitting arises from a phenomenon known as *spin-spin coupling*. This spin-spin coupling occurs because protons tend to align themselves in particular ways in external magnetic fields. Since protons have magnetic moments themselves, the magnetic moment of one proton can affect the magnetic environment of a proton on an adjacent atom. (The spins of the protons are said to be coupled.) Spin-spin coupling effects are transferred primarily through the bonding electrons and are not usually observed if the coupled protons are separated by more than three σ bonds. Thus, we observe signal splitting from the protons of *adjacent*

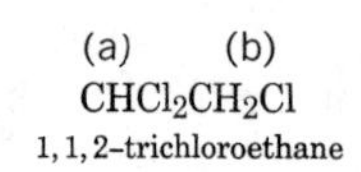

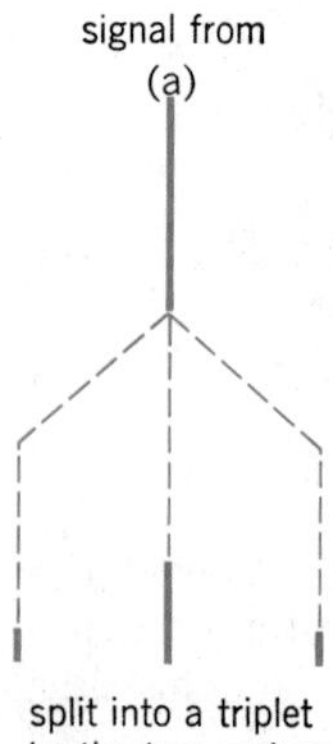

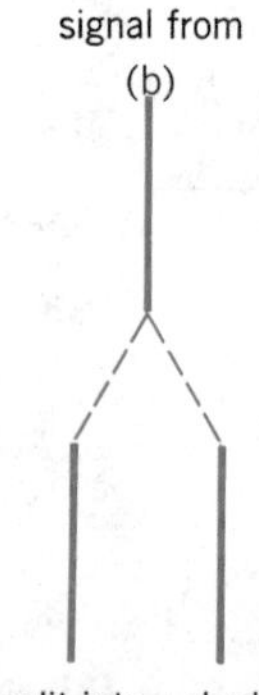

FIG. 13.12
Signal splitting in 1,1,2-trichloroethane.

σ-bonded atoms as in 1,1,2-trichloroethane (Fig. 13.3). However, we do not observe splitting of either signal of *tert*-butyl acetate (Fig. 13.2) because the protons labeled (b) are separated from those labeled (a) by more than three bonds.

$$\overset{(a)}{CH_3}-\underset{\underset{CH_3\,(a)}{|}}{\overset{\overset{(a)\,CH_3}{|}}{C}}-O-\overset{\overset{O}{\|}}{C}-\overset{(b)}{CH_3}$$

tert-Butyl acetate
(no signal splitting)

Spin-spin splitting is not observed for protons that are *chemically equivalent or enantiotopic*. That is, spin-spin splittings do not occur between protons that have *exactly the same chemical shift*. Thus, we would not expect, and do not find, signal splitting in the signal from the six equivalent hydrogens of ethane.

CH_3CH_3 (no signal splitting)

Nor do we find signal splitting occurring from enantiotopic protons of methoxyacetonitrile (Fig. 13.13).

We do, however, observe signal splitting from *protons that are not chemical shift equivalent* and are not too far apart. How can we account for this?

We have seen that protons can be aligned in only two ways in an external magnetic field: with the field or against it. Therefore, the magnetic moment of a proton on an adjacent atom may affect the magnetic field at the proton whose signal we are observing in only two ways. This causes the appearance of a smaller peak somewhat upfield (from where the signal might have occurred) and another peak somewhat downfield.

FIG. 13.13
The proton nmr spectrum of methoxyacetonitrile. No signal splitting occurs with the enantiotopic protons (b). (*Spectrum courtesy of Varian Associates, Palo Alto, Calif.*)

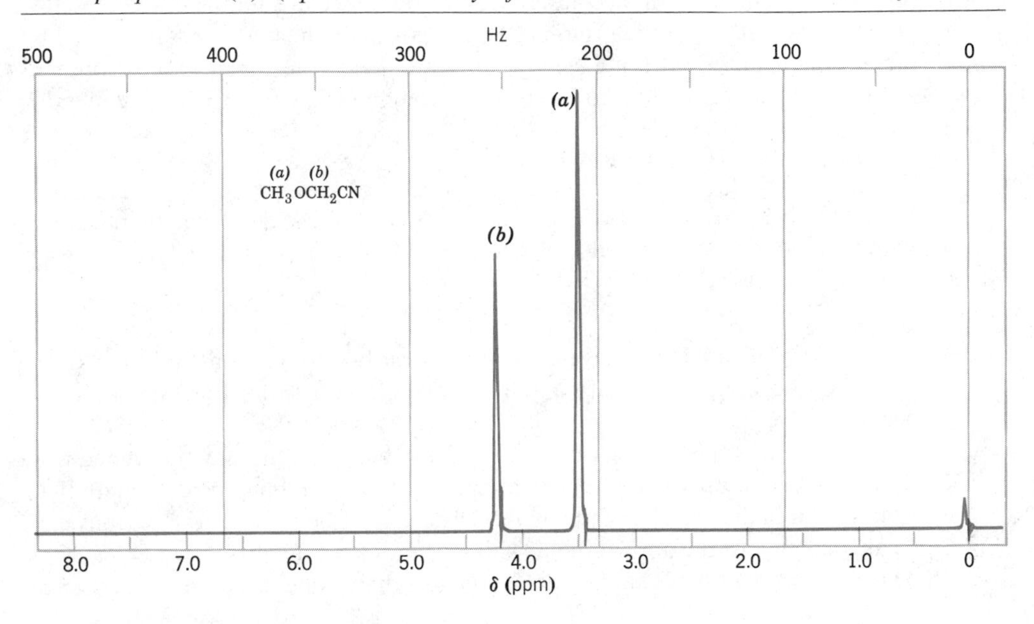

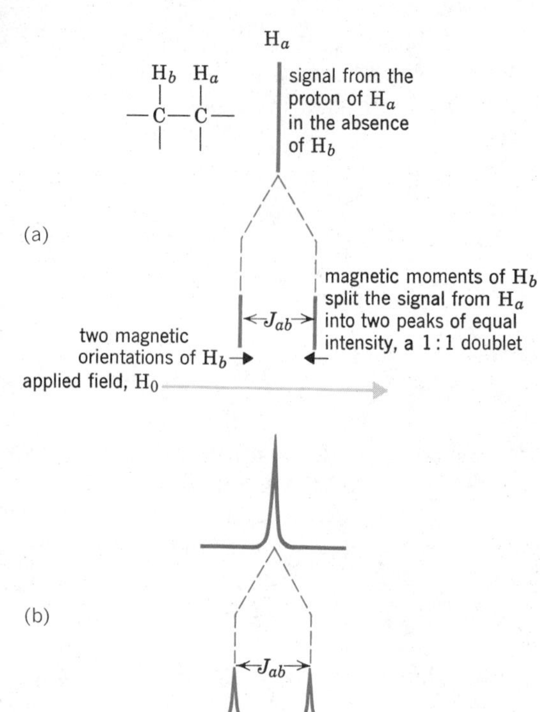

FIG. 13.14
Signal splitting arising from spin-spin coupling with one nonequivalent proton of a neighboring hydrogen. A theoretical spectrum is shown in (a) *and the actual appearance of the spectrum in* (b). *The distance between the centers of the peaks of the doublet is called the* coupling constant, J_{ab}. J_{ab} *is measured in Hertz (cycles per second). The magnitudes of coupling constants are* not *dependent upon the magnitude of the applied field and their values, in Hertz, will be the same regardless of the operating frequency of the spectrometer used.*

Figure 13.14 shows how two possible orientations of the proton of a neighboring hydrogen, H_b, splits the signal of the proton H_a. (H_b and H_a are not equivalent.)

The separation of these peaks in frequency units is called the *coupling constant* and is abbreviated J_{ab}. Coupling constants are generally reported in Hertz (cycles per second). Because they are caused entirely by internal forces, the magnitudes of coupling constants *are not* dependent on the magnitude of the applied field. Coupling constants measured on an instrument operating at 60 MHz will be the same as those measured on an instrument operating at 100 MHz.

When we determine proton nmr spectra we are, of course, observing effects produced by billions of molecules. Since the difference in energy between the two possible orientations of the proton of H_b is very small, the two orientations will be present in roughly (but not exactly) equal numbers. The signal that we observe from H_a is, therefore, split into two peaks of roughly equal intensity, *a 1:1 doublet.*

Problem 13.5

Sketch the proton nmr spectrum of $CHBr_2CHCl_2$. Which signal would you expect to occur at lower magnetic field strength; that of the proton of the $CHBr_2$— group or of the —$CHCl_2$ group? Why?

Two equivalent protons on an adjacent carbon (or carbons) split the signal from an absorbing proton into a 1:2:1 *triplet*. Figure 13.15 illustrates how this occurs.

In compounds of either type (Fig. 13.15), both protons may be aligned with the applied field. This causes a peak to appear at lower applied field strength than would occur in the absence of the two hydrogens H_b. Conversely, both protons may be aligned against the applied field. This orientation of the protons of H_b causes a peak to appear at higher applied field strengths than would occur in their absence.

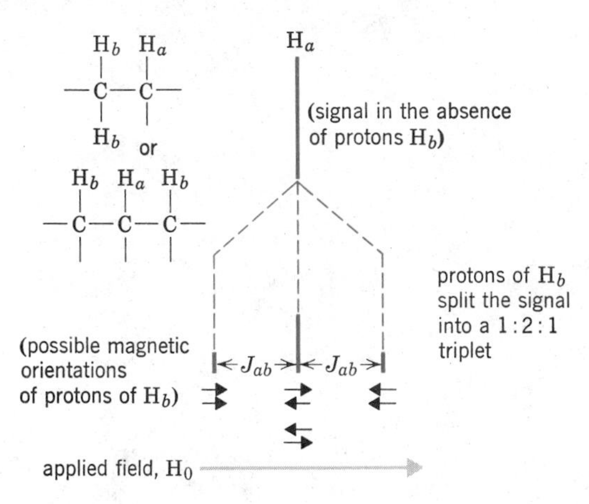

FIG. 13.15
Two equivalent protons (H_b) on an adjacent carbon split the signal from H_a into a 1:2:1 triplet.

Finally there are two ways in which the two protons may be aligned in which one opposes the applied field and one reinforces it. These arrangements do not displace the signal. Since the probability of this last arrangement is twice that of either of the other two, the center peak of the triplet is twice as intense.

The proton of the $—CHCl_2$ group of 1,1,2-trichloroethane is an example of a proton of the type having two equivalent protons on an adjacent carbon. The signal from the $—CHCl_2$ group (Fig. 13.3) appears as a 1:2:1 triplet and, as we would expect, the protons of the $—CH_2Cl$ group of 1,1,2-trichloroethane are split into a 1:1 doublet by the proton of the $—CHCl_2$ group.

The spectrum of 1,1,2,3,3-pentachloropropane (Fig. 13.16) is similar to that of 1,1,2-trichloroethane in that it also consists of a 1:2:1 triplet and a 1:1 doublet. The hydrogens H_b of 1,1,2,3,3-pentachloropropane are equivalent even though they are on separate carbons.

FIG. 13.16
The proton nmr spectrum of 1,1,2,3,3-pentachloropropane. (Spectrum courtesy of Varian Associates, Palo Alto, Calif.)

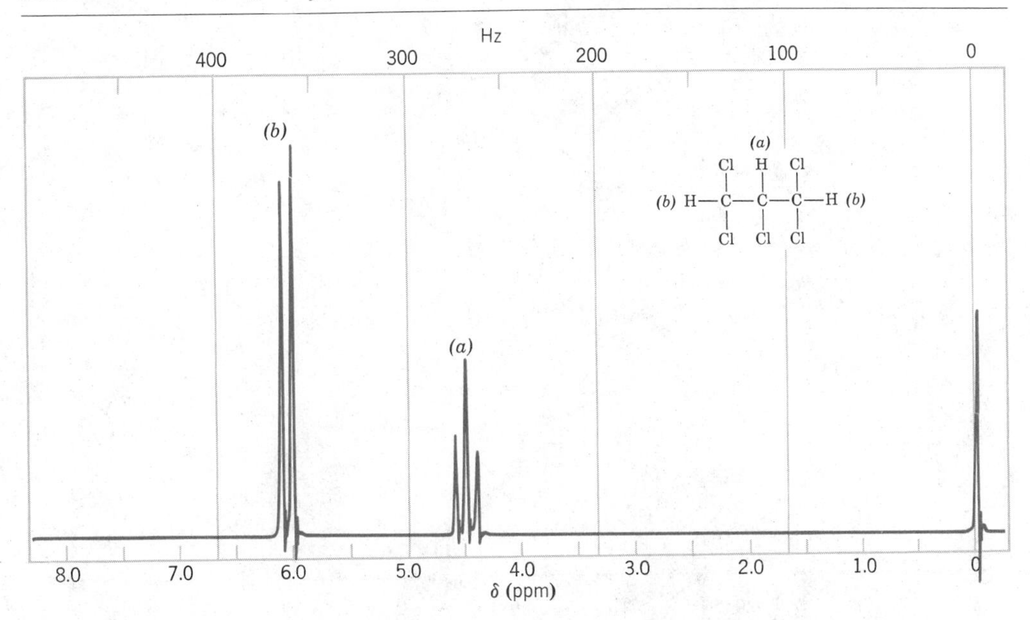

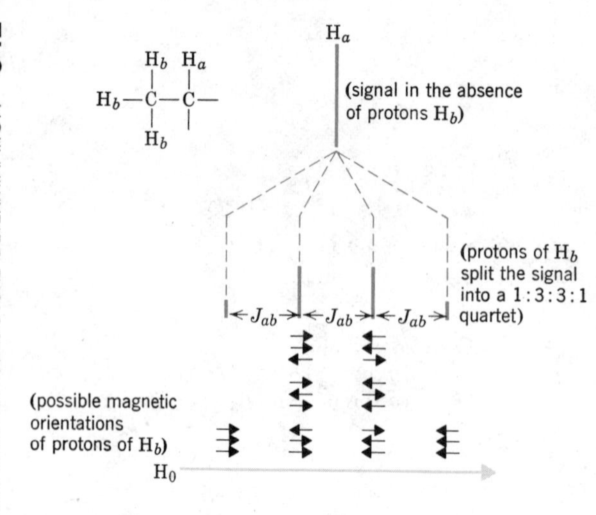

FIG. 13.17
Three equivalent protons (H_b) on an adjacent carbon split the signal from H_a into a 1:3:3:1 quartet.

Problem 13.6

The relative positions of the doublet and triplet of 1,1,2-trichloroethane (Fig. 13.3) and 1,1,2,3,3-pentachloropropane (Fig. 13.16) are reversed. Explain this.

Three equivalent protons (H_b) on a neighboring carbon split the signal from the H_a into a 1:3:3:1 quartet. This is shown in Fig. 13.17.

The signal from two equivalent protons of the $—CH_2Br$ group of ethyl bromide (Fig. 13.18) appears as a 1:3:3:1 quartet because of this type of signal splitting. The three equivalent protons of the $CH_3—$ group are split into a 1:2:1 triplet by the two protons of the $—CH_2Br$ group.

The kind of analysis that we have just given can be extended to compounds with even larger numbers of equivalent protons on adjacent atoms. These analyses show that *if there are n equivalent protons on adjacent atoms these will split a signal*

FIG. 13.18
The proton nmr spectrum of ethyl bromide.

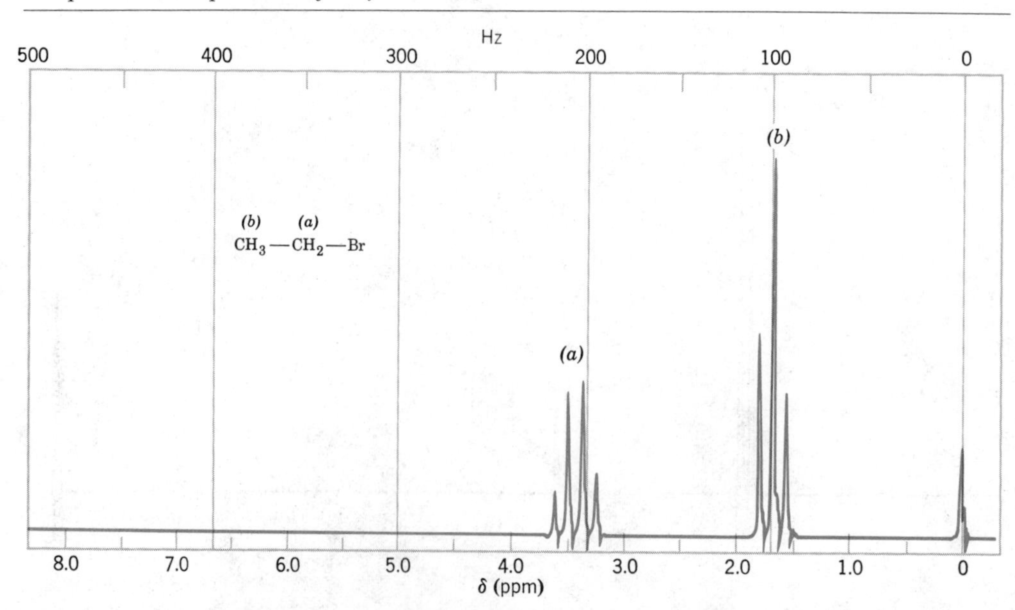

into n + 1 peaks. (We may not always see all of these peaks in actual spectra, however, because some of them may be very small.)

Problem 13.7

What kind of proton nmr spectrum would you expect the following compound to give?

$(Cl_2CH)_3CH$

Sketch the spectrum showing the splitting patterns and relative position of each signal.

Problem 13.8

Propose structures for each of the compounds shown in Fig. 13.19, and account for the splitting pattern of each signal.

The splitting patterns shown in Fig. 13.19 are fairly easy to recognize because in each compound there are only two sets of nonequivalent hydrogens. One feature present in all spectra, however, will help us recognize splitting patterns in more complicated spectra: the *reciprocity of coupling constants.*

The separation of the peaks in Hertz gives us the value of the coupling constants. Therefore, if we look for doublets, triplets, quartets, and so on, that have *the same coupling constants* the chances are good that these multiplets are related to each other because they arise from reciprocal spin-spin couplings.

The two sets of protons of an ethyl group, for example, appear as a triplet and a quartet as long as the ethyl group is attached to an atom that does not bear any hydrogens. The spacings of the peaks of the triplet and the quartet of an ethyl group will be the same because the coupling constants, J_{ab}, are the same (Fig. 13.20).

Proton nmr spectra have other features, however, that are not at all helpful when we try to determine the structures of a compound.

1. Signals may overlap. This happens when the chemical shifts of the signals are very nearly the same. In the spectrum of ethyl chloroacetate (Fig. 13.21) we see that the singlet of the $—CH_2Cl$ group falls directly on top of one of the outermost peaks of the ethyl quartet.

We see a more dramatic example of signal overlapping in the spectrum of *n*-octane (Fig. 13.22). The methyl protons absorb at $\delta = 0.88$, but all of the three different sets of methylene ($—CH_2—$) protons have approximately the same chemical shift, $\delta = 1.27 \pm 0.03$.

2. Spin-spin couplings between the protons of nonadjacent atoms may occur. This long-range coupling happens frequently when π-bonded atoms intervene between the atoms bearing the coupled protons. We can see an example in the expanded signals of the spectrum of methyl formate (Fig. 13.23). Even though the proton of the formyl group, $—\overset{\overset{\large O}{\|}}{C}—H$, is separated by two atoms from the proton of the methyl group, the signal from the proton of the $—\overset{\overset{\large O}{\|}}{C}H$ group is split into a quartet and the signal from the protons of the methyl group is split into a doublet. The coupling constants are small, however, and in the "unexpanded" spectrum both peaks simply look like "ragged" singlets.

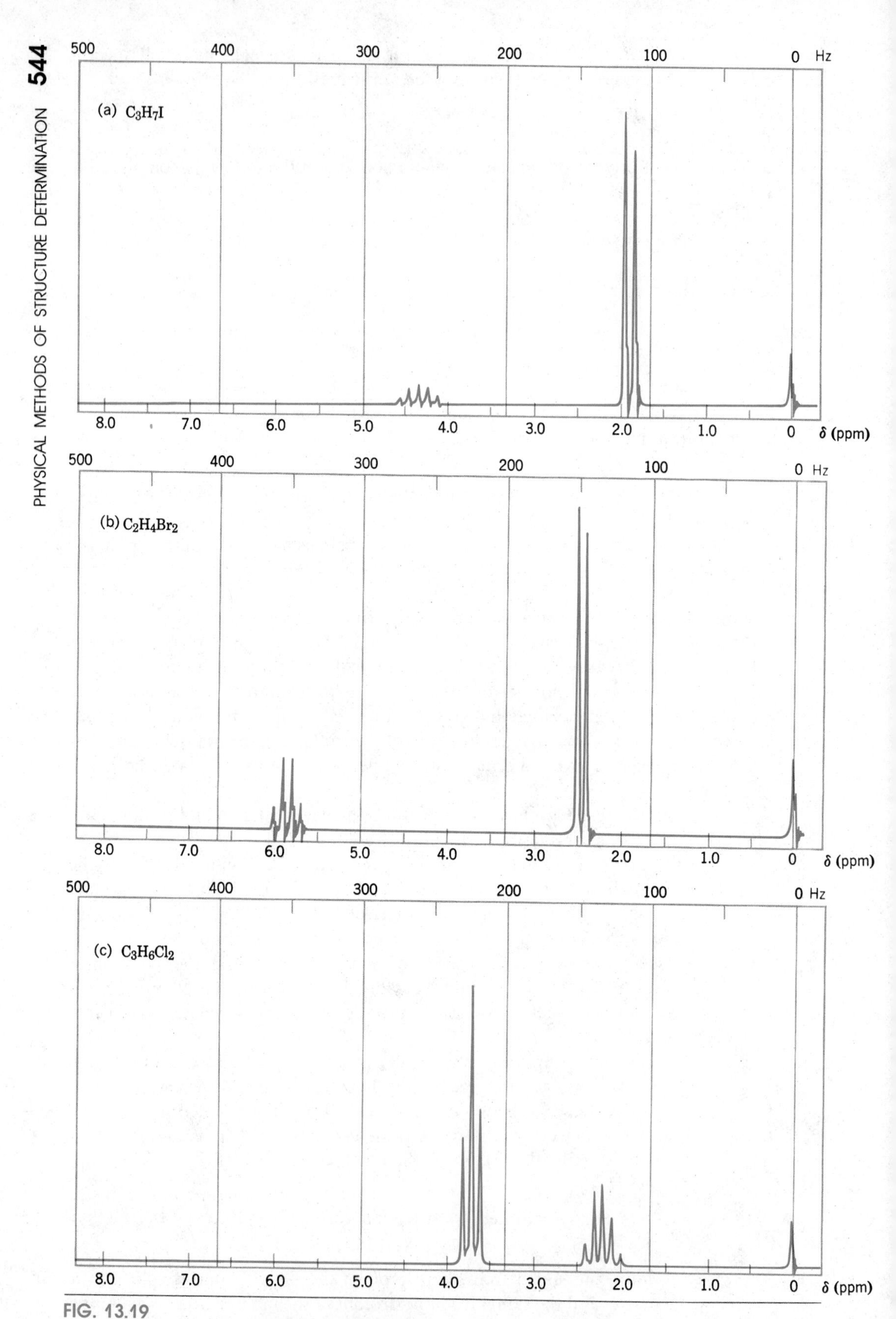

FIG. 13.19
Proton nmr spectra for Problem 13.8. (Spectra courtesy of Varian Associates, Palo Alto, Calif.)

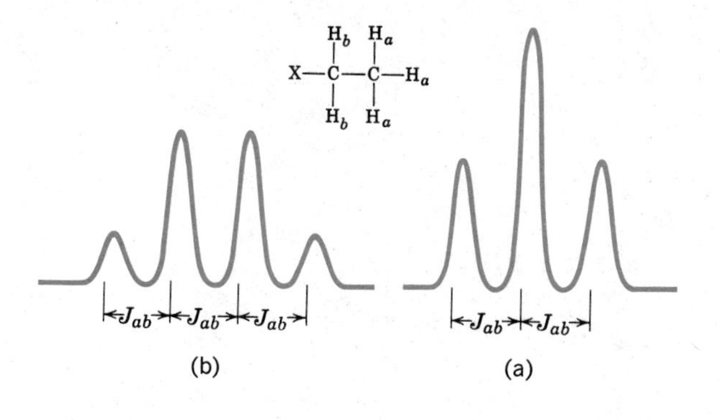

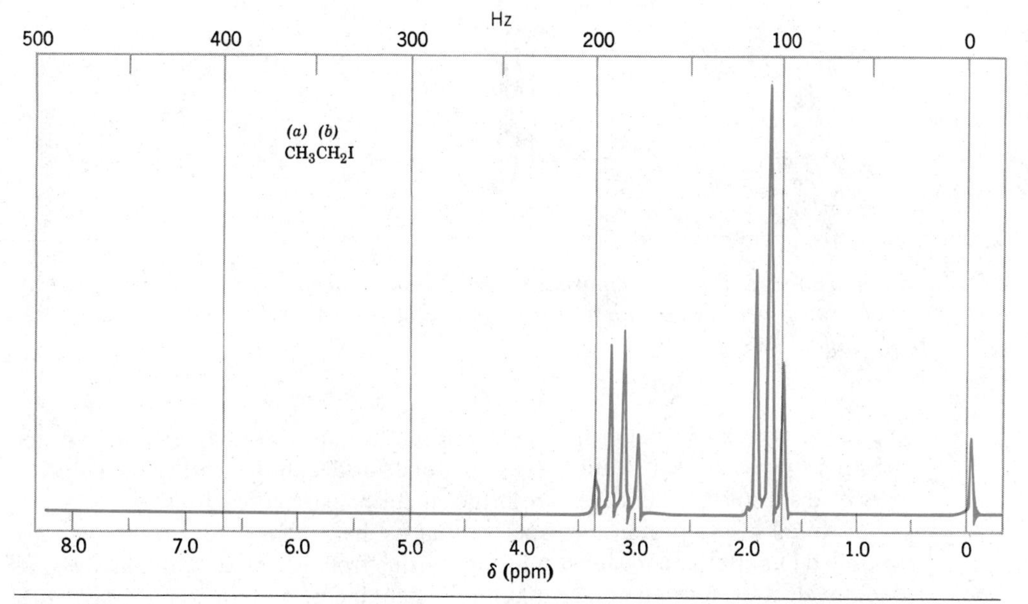

FIG. 13.20
(Top) A theoretical splitting pattern for an ethyl group. (Bottom) The spectrum of ethyl iodide. (Spectrum courtesy of Varian Associates, Palo Alto, Calif.)

3. The splitting patterns of aromatic groups are difficult to analyze. A monosubstituted benzene ring (a phenyl group) has three different kinds of protons.

X
H_a H_a
H_b H_b
H_c

The chemical shifts of these protons may be so similar that the phenyl group will give a signal that resembles a singlet. This happens, for example, in the spectrum of toluene (Fig. 13.24). Or the chemical shifts may be different and because of long-range couplings, the phenyl group appears as a very complicated multiplet. The signal from the aromatic protons of anisole (Fig. 13.25) gives us an example of this.

Disubstituted benzenes show a range of complicated splitting patterns. In many instances these patterns can be analyzed using techniques that are beyond

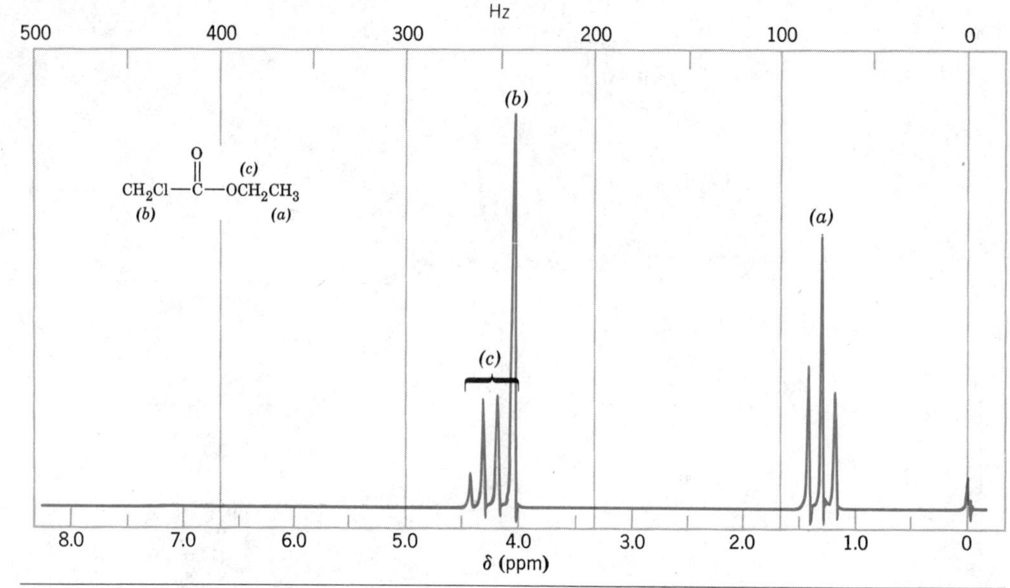

FIG. 13.21
The proton nmr spectrum of ethyl chloroacetate. The singlet from the protons of (b) *falls on one of the outermost peaks of the quartet from* (c). (*Spectrum courtesy of Varian Associates, Palo Alto, Calif.*)

the scope of our discussion here. We will see later in this chapter that infrared spectroscopy gives us a relatively easy method for deciding whether the substituents of disubstituted benzenes are ortho, meta or para to each other.

In all of the proton nmr spectra that we have considered so far, we have restricted our attention to signal splittings arising from interactions of only two sets of equivalent protons on adjacent atoms. What kind of patterns should we expect

FIG. 13.22
The proton nmr spectrum of n-octane. (*Spectrum courtesy of Varian Associates, Palo Alto, Calif.*)

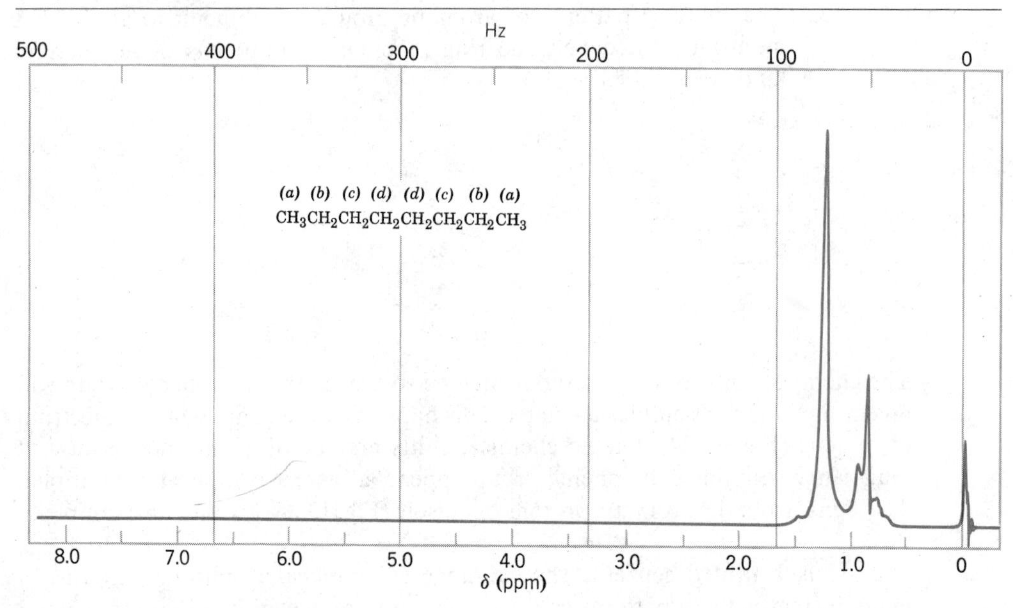

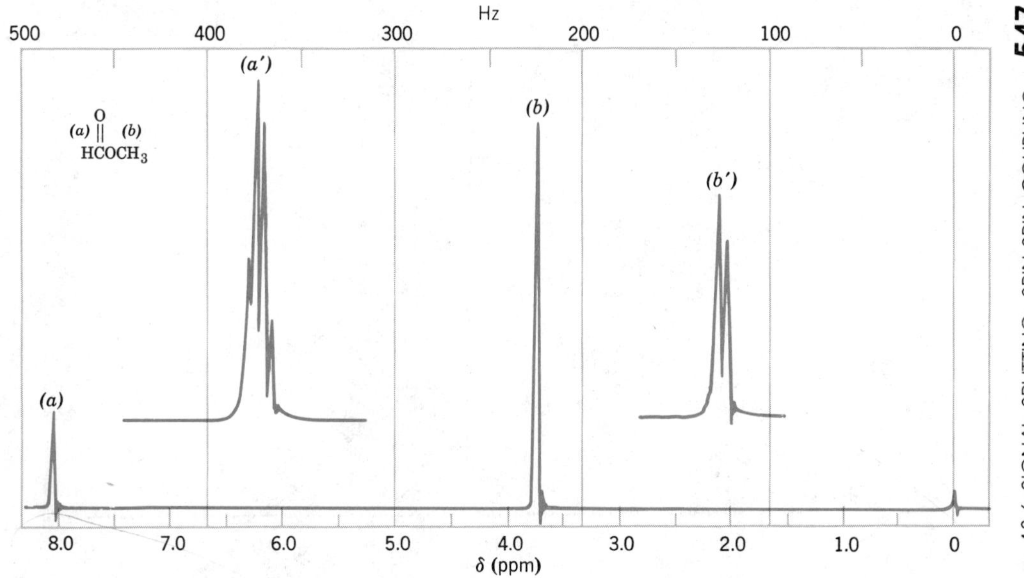

FIG. 13.23
The proton nmr spectrum of methyl formate. The signals (a′) *and* (b′) *are expanded versions of* (a) *and* (b). *The expanded signals are offset on the δ scale for clarity and their relative areas are not related to the number of protons giving rise to each signal. (Spectrum courtesy of Varian Associates, Palo Alto, Calif.)*

FIG. 13.24
The proton nmr spectrum of toluene. (Spectrum courtesy of Varian Associates, Palo Alto, Calif.)

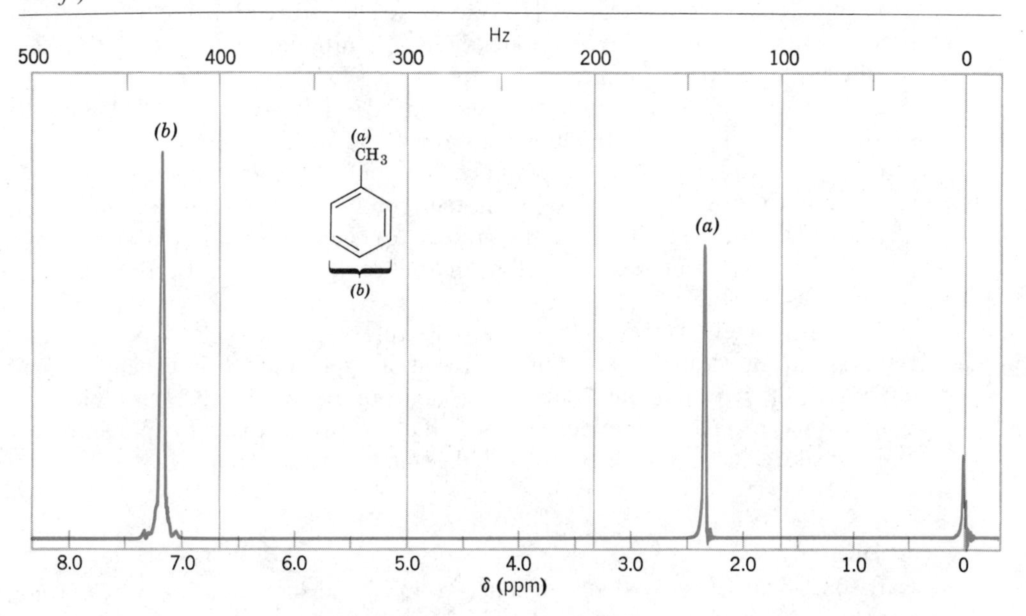

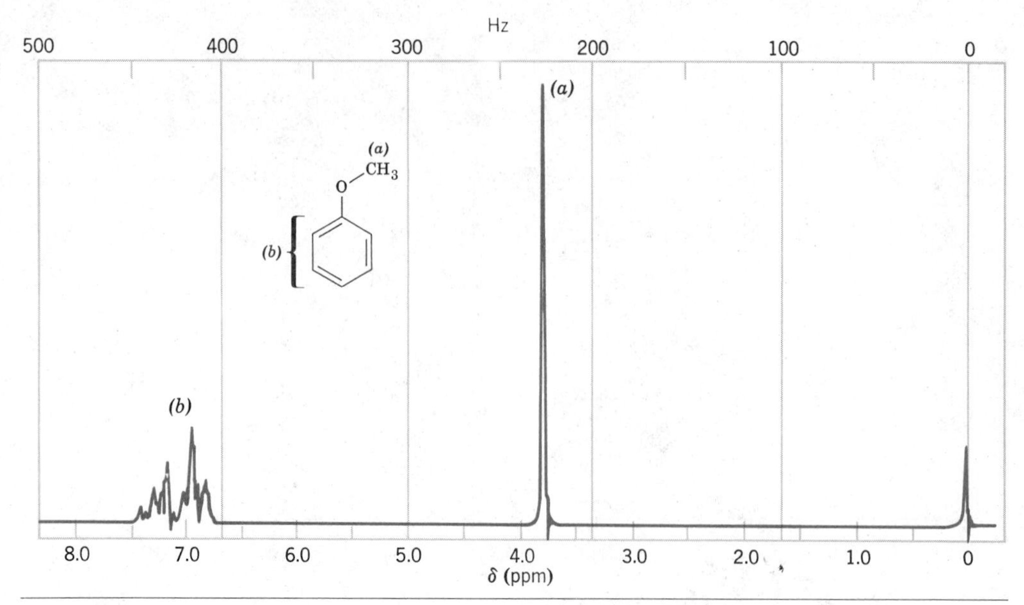

FIG. 13.25
The proton nmr spectrum of anisole (methoxybenzene). (Spectrum courtesy of Varian Associates, Palo Alto, Calif.)

from compounds in which more than two sets of equivalent protons are interacting? We cannot answer this question completely but we can give an example that illustrates the kind of analysis that is involved. Let us consider a 1-substituted propane.

$$\overset{(a)}{CH_3}-\overset{(b)}{CH_2}-\overset{(c)}{CH_2}-Z$$

Here, there are three sets of equivalent protons. We have no problem in deciding what kind of signal splitting to expect from the protons of the CH_3— group or the —CH_2Z group. The methyl group is spin-spin coupled only to the two protons of the central —CH_2— group. Therefore, the methyl group should appear as a triplet. The protons of the —CH_2Z group are similarly coupled only to the two protons of the central —CH_2— group. Thus, the protons of the —CH_2Z group should also appear as a triplet.

But what about the protons of the central —CH_2— group (*b*)? They are spin-spin coupled with the three protons at (*a*) and with two protons at (*c*). The protons at (*a*) and (*c*), moreover, are not equivalent. If the coupling constants J_{ab} and J_{bc} have quite different values, then, the protons at (*b*) could give as many as 12 peaks. That is, the signal from the protons (*b*) could be split into a quartet by the three protons (*a*) and each line of the quartet could be split into a triplet by the two protons (*c*) (Fig. 13.26).

It is unlikely, however, that we would observe as many as 12 peaks in an actual spectrum because the coupling constants are such that peaks usually fall on top of peaks. The proton nmr spectrum of 1-nitropropane (Fig. 13.27) is typical of 1-substituted propyl compounds. We see that the (*b*) protons are split into six major peaks, each of which shows a slight sign of further splitting.

Problem 13.9

Carry out an analysis like that shown in Fig. 13.26 and show how many peaks the signal from (*b*) would be split into if $J_{ab} = 2J_{bc}$ and if $J_{ab} = J_{bc}$. (Hint: in both cases

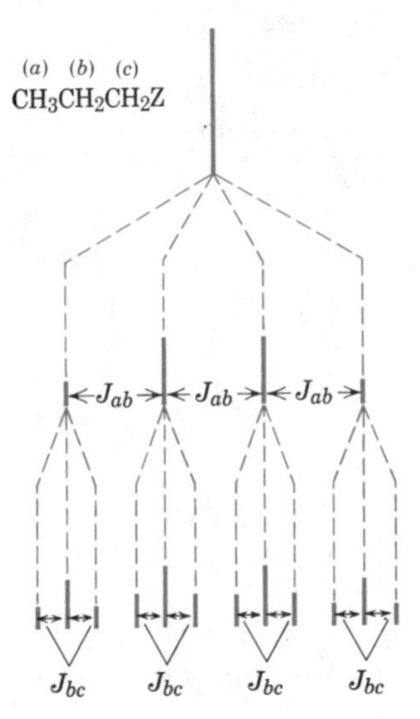

FIG. 13.26
The splitting pattern that would occur for the (b) *protons of* $CH_3CH_2CH_2Z$ *if* J_{ab} *is much larger than* J_{bc}. *Here* $J_{ab} = 4J_{bc}$.

peaks will fall on top of peaks so that the total numbers of peaks in the signal is less than 12.)

Problem 13.10

Assign structures to each of the compounds in Fig. 13.28.

The presentation we have given here applies only to what are called *first-order spectra*. In first-order spectra, the distance in Hertz, Δv, that separates the coupled signals is very much larger than the coupling constant, J. That is, $\Delta v \gg J$. In *second-order spectra* (which we have not discussed) Δv approaches J in magnitude and the situation becomes much more complex. The number of peaks increases and the intensities are not those that might be expected from first-order considerations.

FIG. 13.27
The proton nmr spectrum of 1-nitropropane. (*Spectrum courtesy of Varian Associates, Palo Alto, Calif.*)

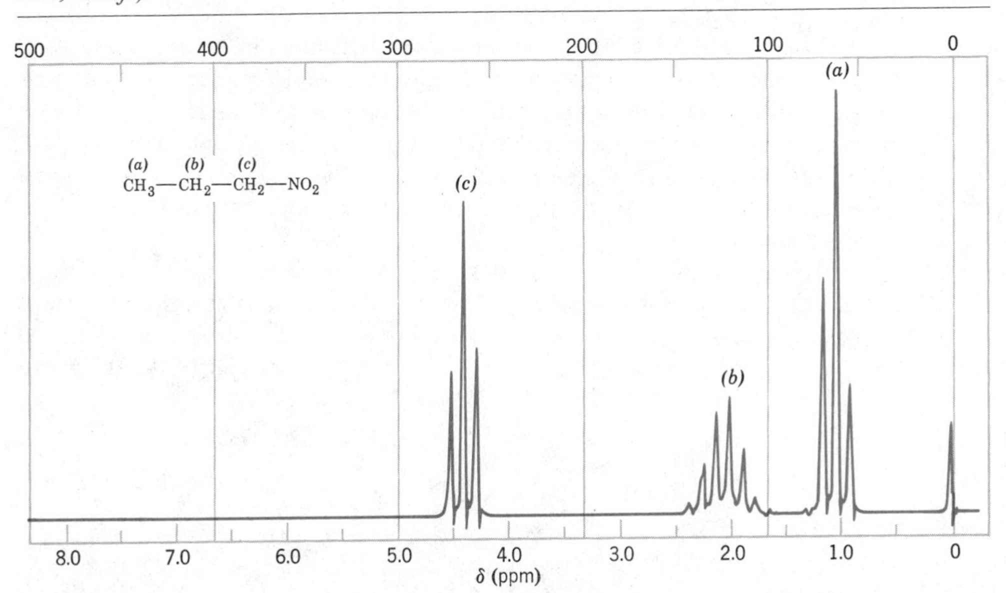

13.7 PROTON NUCLEAR MAGNETIC RESONANCE SPECTRA OF COMPOUNDS CONTAINING FLUORINE AND DEUTERIUM

The fluorine (^{19}F) nucleus has spin quantum numbers of $+\frac{1}{2}$ and $-\frac{1}{2}$. In this respect ^{19}F nuclei resemble protons and fluorine magnetic spectra can be observed. When measured at the same radiofrequency, the signals from ^{19}F absorptions occur at considerably different magnetic field strengths than those of protons, so we do not see peaks due to ^{19}F absorption in proton magnetic resonance spectra. We do, however, see splitting of proton signals caused by spin-spin couplings between protons and ^{19}F nuclei. The signal from the two protons of 1,2-dichloro-1,1-difluoroethane, for example, is split into a triplet by the fluorine atoms on the adjacent carbon (Fig. 13.29).

The nucleus of a deuterium atom (a deuteron) has a much smaller magnetic moment than a proton and signals from deuteron absorption do not occur in proton magnetic resonance spectra.

Spin-spin couplings between deuterons and protons are small but the presence of deuterium on an adjacent atom can cause splitting of the proton signal.

Problem 13.11

Sketch the pmr spectra that you would expect from each of the following compounds.

(a) $CH_3CF_2CH_3$ (c) CH_3CFCl_2
(b) CH_3CF_2Cl (d) CH_3CF_3

13.8 PROTON NUCLEAR MAGNETIC RESONANCE SPECTRA AND RATE PROCESSES

Professor J. D. Roberts (of the California Institute of Technology), a pioneer in the application of NMR spectroscopy to problems of organic chemistry, has compared the nuclear magnetic resonance spectrometer to a camera with a relatively slow shutter speed. Just as a camera with a slow shutter speed blurs photographs of objects that are moving rapidly, the nuclear magnetic resonance spectrometer blurs its picture of molecular processes that are occurring rapidly.

What are some of the rapid processes that occur in organic molecules?

At temperatures near room temperature, groups connected by carbon-carbon single bonds rotate very rapidly. Because of this, when we determine spectra of compounds with single bonds, the spectra that we obtain often reflect the individual hydrogens in their average environment. That is, in an environment that is an average of all the environments that the protons have as a result of the group rotations.

To see an example of this effect, let us consider the spectrum of ethyl iodide again. The most stable conformation of ethyl iodide is the one in which the groups are perfectly staggered. In this staggered conformation one hydrogen of the methyl group (in color below) is in a different environment from that of the other two

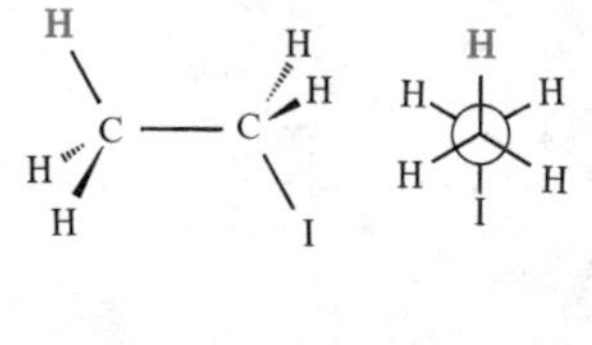

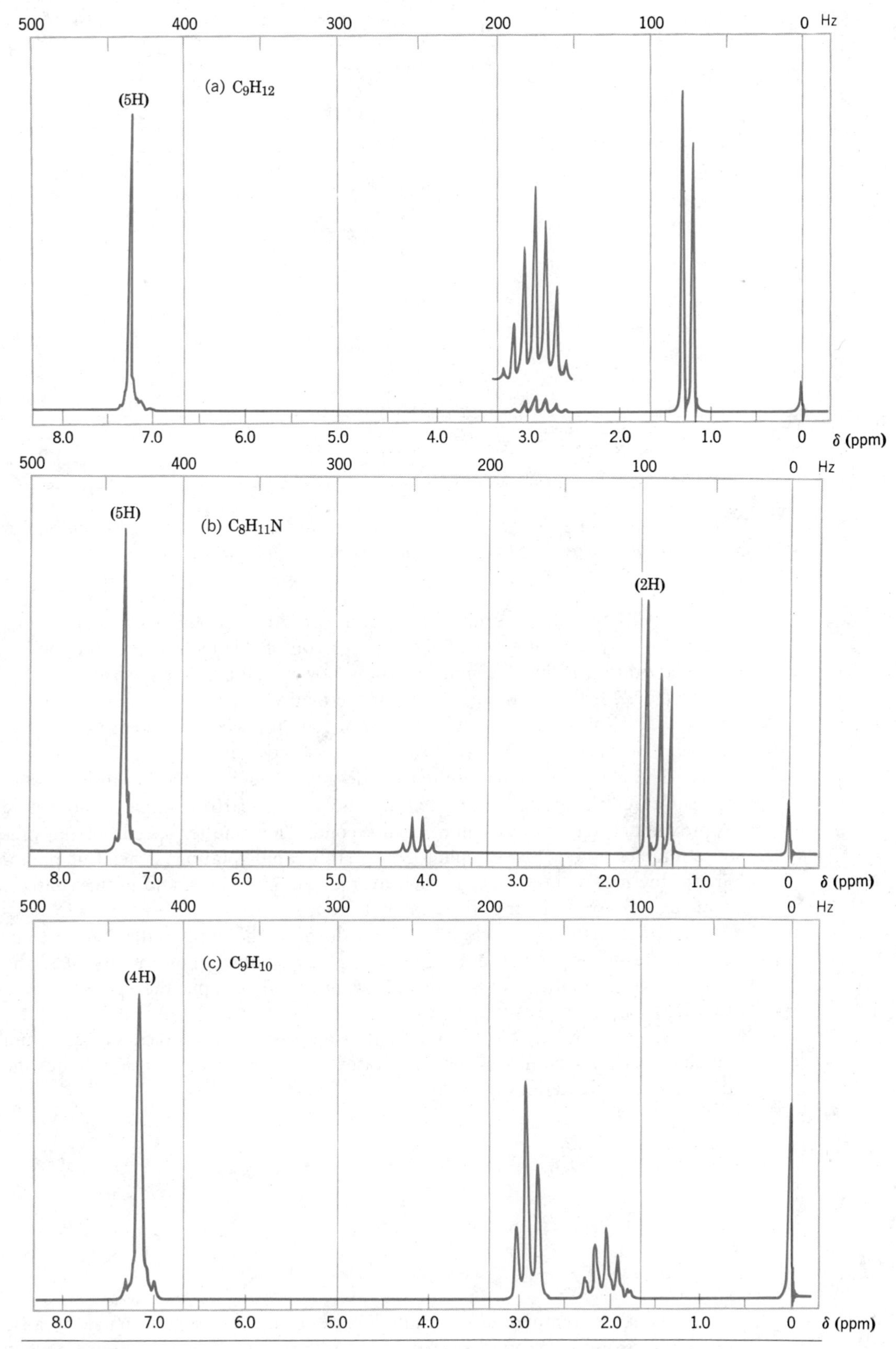

FIG. 13.28
Proton nmr spectra for Problem 13.10. (Spectra courtesy of Varian Associates, Palo Alto, Calif.)

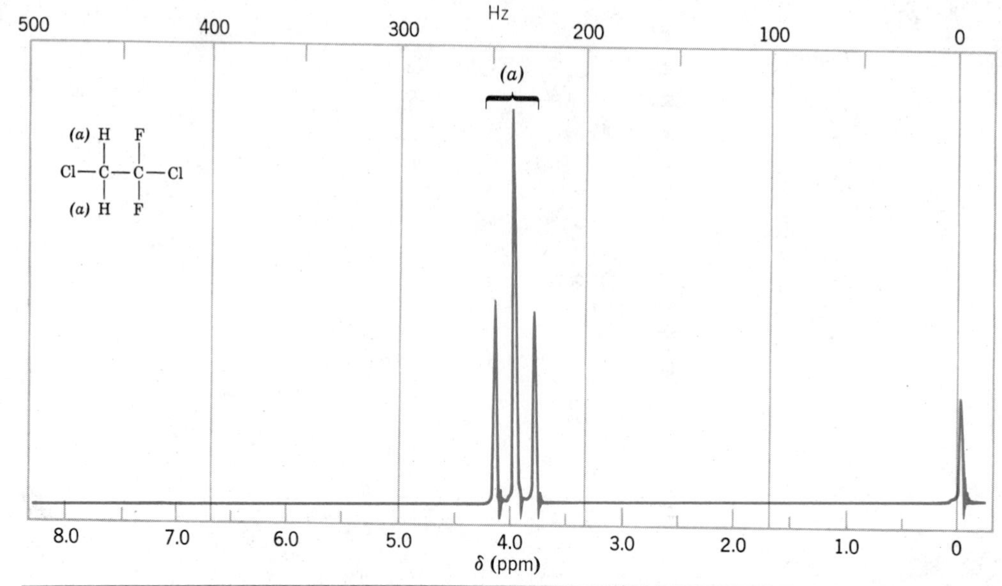

FIG. 13.29
The proton nmr spectrum of 1,2-dichloro-1,1-difluoroethane. The J_{HF} *coupling constant is ~12 Hz. (Spectrum courtesy of Varian Associates, Palo Alto, Calif.)*

methyl hydrogens. If the nmr spectrometer were to detect this particular conformation of ethyl iodide it would show the proton of this hydrogen of the methyl group at *a different chemical shift*. We know, however, that in the spectrum of ethyl iodide (Fig. 13.20), the three protons of the methyl group give *a single signal* (a signal that is split into a triplet by spin-spin coupling with the two protons of the adjacent carbon).

The methyl protons of ethyl iodide give a single signal because at room temperature the groups connected by the carbon-carbon single bond rotate approximately one million times each second. The "shutter speed" of the nmr spectrometer is too slow to "photograph" this rapid rotation; instead, it photographs the methyl hydrogens in their average environments, and in this sense, it gives us a blurred picture of the methyl group.

Rotations about single bonds slow down as the temperature of the compound is lowered. Sometimes, this slowing of rotations allows us to "see" the different conformations of a molecule when we determine the spectrum at a sufficiently low temperature.

An example of this, and one that also shows the usefulness of deuterium labeling, can be seen in the low temperature proton nmr spectra of cyclohexane and of undecadeuteriocyclohexane.

H D
D_2 D_2
D_2 D_2
D_2

undecadeuteriocyclohexane

At room temperature, ordinary cyclohexane gives one signal because interconversions between the various chair forms occur very rapidly. At low temperatures, however, ordinary cyclohexane gives a very complex proton nmr spectrum. At low temperatures interconversions are slow; the axial and equatorial protons have different chemical shifts; and complex spin-spin couplings occur.

At $-100°$, however, undecadeuteriocyclohexane gives only two signals of equal intensity. These signals correspond to the axial and equatorial hydrogens of the two chair conformations shown below.

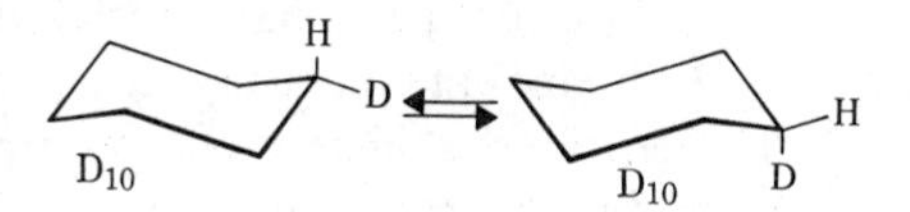

Interconversions between these conformations occur at this low temperature, but they happen slowly enough for the nmr spectrometer to detect the individual conformations.

Problem 13.12

What kind of proton nmr spectrum would you expect to obtain from undecadeuteriocyclohexane at room temperature?

Another example of a rapidly occurring process can be seen in proton nmr spectra of ethanol. The proton nmr spectrum of ordinary ethanol shows the hydroxyl proton as a singlet and the protons of the $—CH_2—$ group as a quartet. (Fig. 13.30). In ordinary ethanol we observe *no signal splitting arising from coupling between the hydroxyl proton and the protons of the $—CH_2—$ group even though they are on adjacent atoms.*

If we examine a proton nmr spectrum of *very pure* ethanol, however, we find that the signal from the hydroxyl proton is split into a triplet, and that the signal from the protons of $—CH_2—$ group is split into a multiplet of eight peaks. Clearly, in very pure ethanol the spin of the proton of the hydroxyl group couples with the spins of the protons of the $—CH_2—$ groups.

FIG. 13.30
The proton nmr spectrum of ordinary ethanol. (Spectrum courtesy of Varian Associates, Palo Alto, Calif.)

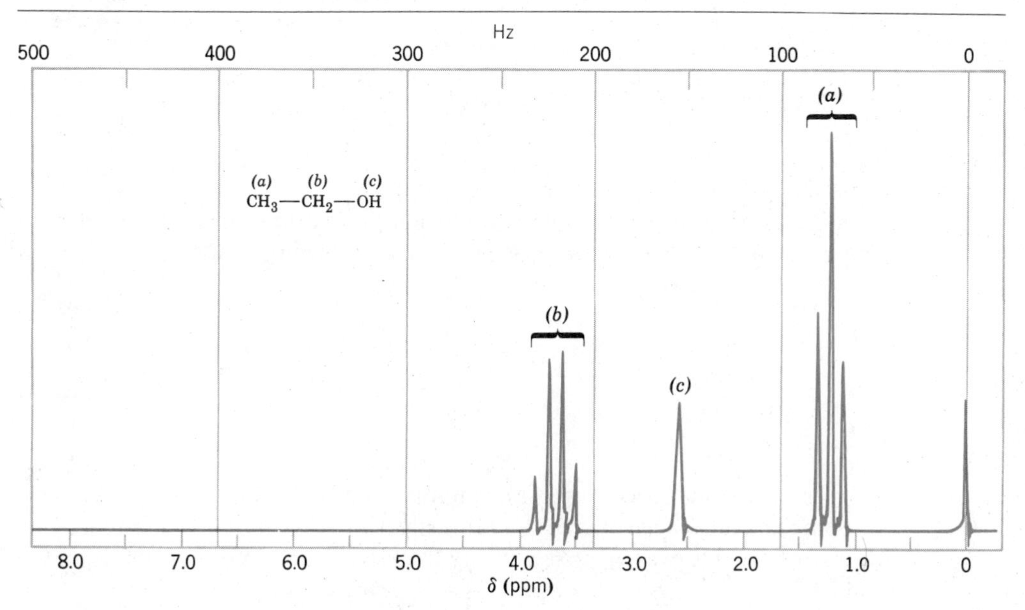

Whether or not coupling occurs between the hydroxyl protons and the methylene protons depends on the length of time the proton spends on a particular ethanol molecule. Protons attached to electronegative atoms with lone pairs such as oxygen can undergo rapid *chemical exchange*. That is they can be transferred rapidly from one molecule to another. The rate of chemical exchange in very pure ethanol is slow and, as a consequence, we see the signal splitting of and by the hydroxyl proton in the spectrum. In ordinary ethanol, acidic and basic impurities catalyze the rate of chemical exchange; the exchange occurs so rapidly that the hydroxyl proton gives an unsplit signal and the methylene protons are split only by coupling with the protons of the methyl group. We say, then, that rapid exchange causes *spin decoupling*.

Spin decoupling is often found in the proton nmr spectra of alcohols, amines, and carboxylic acids.

13.9 PROTON NMR SPECTRA OF CARBOCATIONS

Olah (p. 158) has developed methods for preparing carbocations under conditions where they are stable enough to be studied by nmr spectroscopy. Olah has found, for example, that in liquid sulfur dioxide, alkyl fluorides react with antimony pentafluoride to yield solutions of carbocations.

$$\mathrm{R{-}F} + \mathrm{SbF_5} \xrightarrow[\text{liq. SO}_2]{} \mathrm{R^+\ SbF_6^-}$$

Antimony pentafluoride is a powerful Lewis acid.

When the proton nmr spectrum of *tert*-butyl fluoride is measured in liquid sulfur dioxide the nine protons appear as a doublet centered at δ 1.3. (Why do the protons of *tert*-butyl fluoride appear as a doublet?) When antimony pentafluoride is added to the solution, the doublet at δ 1.35 is replaced by a singlet at δ 4.35. Both the change in the splitting pattern of the methyl protons and the downfield shift are consistent with the formation of a *tert*-butyl cation. (Why?)

$$\underset{\substack{\delta\,1.35\\ \text{(doublet)}}}{\mathrm{(CH_3)_3C{-}F}} + \mathrm{SbF_5} \xrightarrow[\text{SO}_2]{\text{liq.}} \underset{\substack{\delta\,4.35\\ \text{(singlet)}}}{\mathrm{(CH_3)_3C^+}} \quad \mathrm{SbF_6^-}$$

When a solution of isopropyl fluoride in liquid sulfur dioxide is treated with antimony pentafluoride an even more remarkable downfield shift occurs.

$$\underset{\substack{\delta\,1.23 \quad \delta\,4.64\\ \text{(quartet) (multiplet)}}}{\mathrm{(CH_3)_2CHF}} + \mathrm{SbF_5} \longrightarrow \underset{\substack{\delta\,5.03 \quad \delta\,13.50\\ \text{(doublet) (septet)}}}{\mathrm{(CH_3)_2CH^+}} \quad \mathrm{SbF_6^-}$$

The proton nmr spectrum of the arenium ion that results when mesitylene is protonated has also been observed. In this experiment a solution of mesitylene in liquid sulfur dioxide was treated with hydrogen fluoride and antimony pentafluoride.

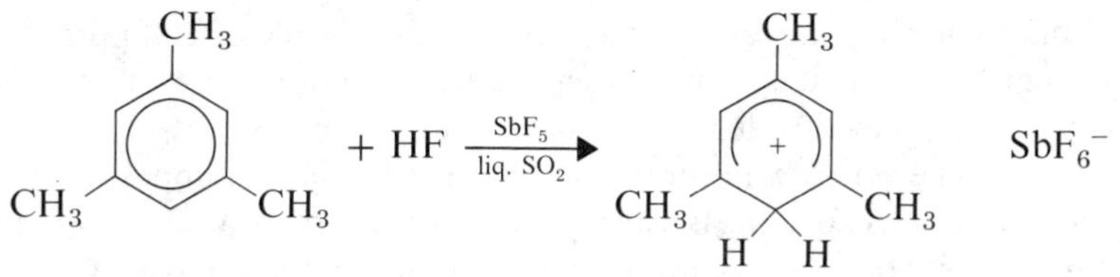

$$\text{mesitylene (1,3,5-trimethylbenzene)} + HF \xrightarrow[\text{liq. } SO_2]{SbF_5} \text{arenium ion (CH}_3\text{, CH}_3\text{, CH}_3\text{; H H)} \quad SbF_6^-$$

(a) Singlet, δ 2.35 (9H)
(b) Singlet, δ 6.70 (3H)

(a) Singlet, δ 2.8 (6H)
(b) Singlet, δ 2.9 (3H)
(c) Singlet, δ 4.6 (2H)
(d) Singlet, δ 7.7 (2H)

Problem 13.13

Make assignments for each of the proton nmr signals of mesitylene and the arenium ion given above.

Problem 13.14

When 1,1-diphenyl-2-methyl-2-propanol in liquid sulfur dioxide was treated with the "superacid" FSO_3H—SbF_5 the solution showed the proton nmr absorptions given below.

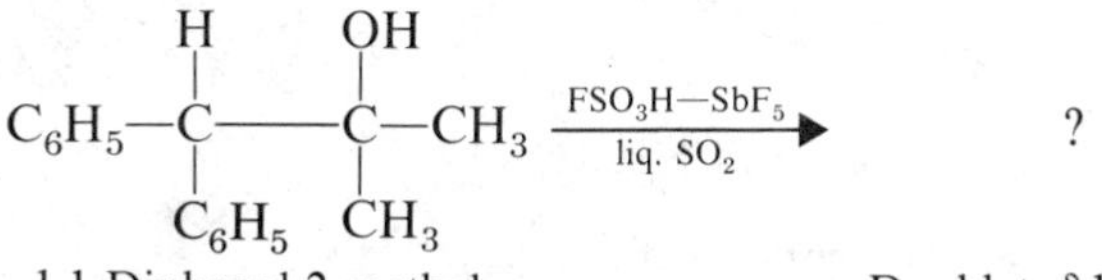

$$C_6H_5-\underset{C_6H_5}{\overset{H}{C}}-\underset{CH_3}{\overset{OH}{C}}-CH_3 \xrightarrow[\text{liq. } SO_2]{FSO_3H-SbF_5} ?$$

1,1-Diphenyl-2-methyl-2-propanol

Doublet, δ 1.48 (6H)
Multiplet, δ 4.45 (1H)
Multiplet, δ 8.0 (10H)

What carbocation is formed in this reaction?

13.10 CARBON-13 NMR SPECTROSCOPY

The most abundant isotope of the element carbon is carbon-12 (^{12}C) (natural abundance ~ 99%). Nuclei of carbon-12 have no net magnetic spin and therefore they cannot give nmr signals. This is not true, however, of nuclei of the much less abundant isotope of carbon, ^{13}C (natural abundance ~ 1%). Carbon-13 nuclei have a net magnetic spin and can give nmr signals. The nuclei of ^{13}C are like the nuclei of ^{1}H in that they can assume spin states of $+\frac{1}{2}$ or $-\frac{1}{2}$.

The low natural abundance of ^{13}C means that highly sensitive spectrometers employing special techniques must be used to measure ^{13}C spectra. These spectrometers have now become widely available and ^{13}C spectroscopy is rapidly becoming another powerful method for determining the structures of organic molecules.

With ^{1}H spectroscopy (proton nmr) we obtain indirect information about the carbon skeleton of an organic molecule because *most* (but not all) of the carbon atoms have at least one attached hydrogen. In ^{13}C spectroscopy, we observe the carbon skeleton directly and, therefore, we see peaks arising from all of the carbon atoms, whether they bear hydrogens or not.

One great advantage of ^{13}C spectroscopy is the wide range of chemical shifts over which ^{13}C nuclei absorb. In ^{13}C spectroscopy signals from organic compounds are spread over a chemical shift range of 200 ppm, compared with a range less than 20 ppm in proton spectra. Carbon-13 spectra are generally simpler because signals are less likely to overlap.

The very low natural abundance of ^{13}C has an important effect that further simplifies ^{13}C spectra. Because of its low natural abundance, there is a very low probability that two adjacent carbon atoms will both have ^{13}C nuclei. This means that in ^{13}C spectra, we do not observe spin-spin couplings between the carbon nuclei.

Electronic techniques are also available to allow *decoupling* of spin-spin interactions between ^{13}C nuclei and ^{1}H nuclei. Thus, it is possible to obtain ^{13}C spectra in which all carbon resonances appear as singlets. Spectra obtained in this mode of operation of the spectrometer are called *proton decoupled* spectra.

Carbon-13 spectrometers can also be operated in another mode, one that allows one-bond couplings between ^{13}C and ^{1}H nuclei to occur. This mode of operation is called *proton off-resonance decoupling*. It produces spectra in which —CH_2— groups appear as triplets, ⋟CH groups as doublets, and carbons with no attached hydrogens as singlets.

An excellent illustration of the application of ^{13}C spectroscopy is shown in spectra of 4-(*N,N*-diethylamino)benzaldehyde (Fig. 13.31).

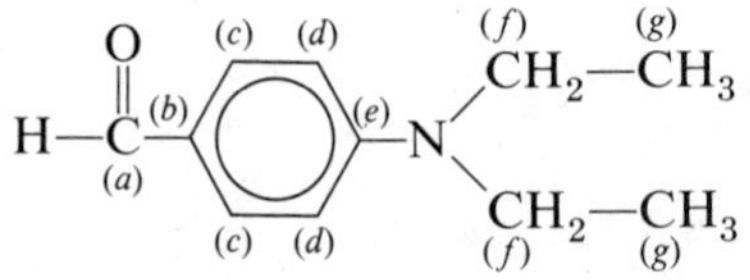

O (c) (d) (f) (g)
CH_2—CH_3
H—C (b) (e) N
(a)
CH_2—CH_3
(c) (d) (f) (g)

4-(*N,N*-Diethylamino)benzaldehyde

The bottom spectrum in Fig. 13.31 is the *proton decoupled* spectrum in which all the signals from 4-(*N,N*-diethylamino)benzaldehyde appear as singlets. The triplet centered at $\delta 79$ (ppm) is caused by the solvent, $CDCl_3$. (The ^{13}C signal of $CDCl_3$ is split into a triplet by coupling with the deuterium atom, spin quantum number = 1, spin states $+1, 0, -1$). The signal at $\delta 0$ arises from $(CH_3)_4Si$.

The top spectrum in Fig. 13.31 is the *proton off-resonance decoupled* spectrum. It shows us immediately which signals belong to the ^{13}C nuclei of the ethyl groups. The triplet at $\delta 47$ is caused by the equivalent —CH_2— groups and the quartet at $\delta 13$ arises from the equivalent —CH_3 groups.

The two singlets in the top spectrum at $\delta 126$ and $\delta 154$ correspond to the carbons of the benzene ring that do not bear hydrogens, (*b*) and (*e*). The greater electronegativity of nitrogen (when compared to carbon) causes the signal from (*e*) to be further downfield (at $\delta 154$). The doublet at $\delta 193$ arises from the carbon of the aldehyde group. Its chemical shift is the most downfield of all the peaks because of the great electronegativity of its attached oxygen and because of resonance contribution of the second structure shown below. Both factors cause the electron density at this carbon to be very low and, therefore, the carbon is not well shielded.

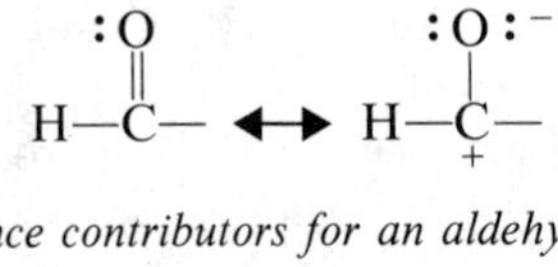

:Ö: :Ö:⁻
H—C— ⟷ H—C⁺—

Resonance contributors for an aldehyde group

This leaves the signals at $\delta 112$ and $\delta 135$ and the two sets of carbons of the benzene ring labeled (*c*) and (*d*) to be accounted for. Both signals appear as doublets in the proton off-resonance decoupled spectrum because both types of

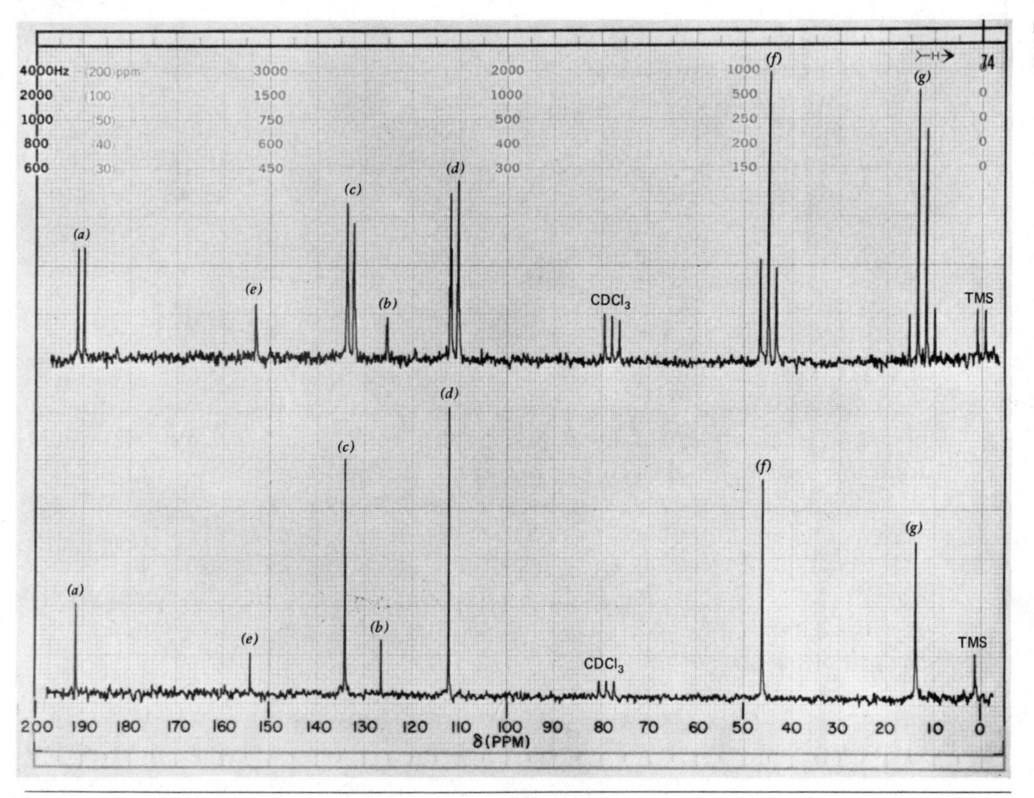

FIG. 13.31
The ^{13}C *nmr spectrum of 4-(N,N-diethylamino)benzaldehyde,* $HC(=O)$–C_6H_4–$N(CH_2CH_3)_2$. *(Spectrum courtesy of Philip L. Fuchs, Purdue University.)*

carbon have one attached hydrogen. But which signal belongs to which set of carbons? Here we find another interesting application of resonance theory.

If we write resonance structures **A** to **D** involving the unshared electron pair of the amino group, we see that contributions made by **B** and **D** increase the

A ⟷ **B** ⟷ **C** ⟷ **D**

electron density at the set of carbons labeled (*d*). On the other hand, writing structures **E** to **H** involving the aldehyde group shows us that contributions made

E ⟷ **F** ⟷ **G** ⟷ **H**

by **F** and **H** decrease the electron density at the set of carbons labeled (*c*). (Other resonance structures are possible but are not pertinent to the argument here.)

Increasing the electron density at a carbon should increase its shielding and should shift its signal upfield. Therefore, we assign the signal at $\delta 112$ to the set of carbons labeled (d). Conversely, decreasing the electron density at a carbon should shift its signal downfield, so we assign the signal at $\delta 135$ to the set labeled (*c*).

13.11 INFRARED SPECTROSCOPY

We saw in Section 10.11 that many organic compounds absorb light in the visible and ultraviolet regions of the electromagnetic spectrum. We also saw that when compounds absorb light of the visible and ultraviolet regions, electrons are excited from lower-energy molecular orbitals to higher ones.

Organic compounds also absorb electromagnetic energy in the infrared region of the spectrum. Infrared radiation does not have sufficient energy to cause the excitation of electrons but it does cause atoms and groups of organic compounds to vibrate about the covalent bonds that connect them. The vibrations are *quantized* and as they occur, the compounds absorb infrared energy in particular regions of the spectrum.

Infrared spectrometers operate in a manner similar to that of visible-ultraviolet spectrometers. A beam of infrared radiation is passed through the sample and this beam is constantly compared with a reference beam as the frequency of the incident radiation is varied. The spectrometer plots the results as a graph showing absorption versus frequency or wavelength.

The location of an infrared absorption band (or peak) can be specified in *frequency units* by its wavenumber $\bar{\nu}$, measured in reciprocal centimeters (cm^{-1}), or by its *wavelength,* λ, measured in micrometers (μm; old name micron, μ). The wavenumber is the number of cycles of the wave in each centimeter along the light beam, and the wavelength is the length of the wave, crest to crest.

$$\bar{\nu} = \frac{1}{\lambda} \text{ (with } \lambda \text{ in cm)} \qquad \text{or} \qquad \bar{\nu} = \frac{10{,}000}{\lambda} \text{ (with } \lambda \text{ in } \mu\text{m)}$$

In their vibrations covalent bonds behave as if they were tiny springs connecting the atoms. When the atoms vibrate they can do so only at certain frequencies, as if the bonds were "tuned." Because of this, covalently bonded atoms have only particular vibrational energy levels. The excitation of a molecule from one vibrational energy level to another occurs only when the compound absorbs infrared radiation of a particular energy, meaning a particular wavelength or frequency (since $\Delta E = h\nu$).

Molecules can vibrate in a variety of ways. Two atoms joined by a covalent bond can undergo a stretching vibration where the atoms move back and forth as if joined by a spring.

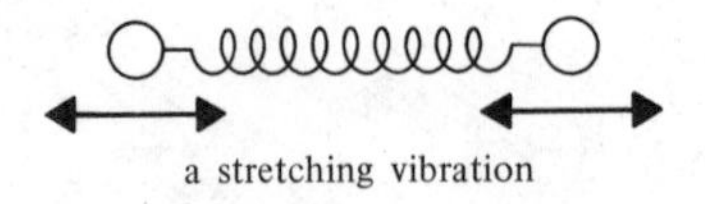

a stretching vibration

Three atoms can also undergo a variety of stretching and bending vibrations:

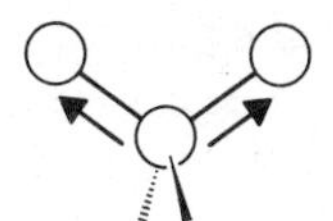
symmetric stretching

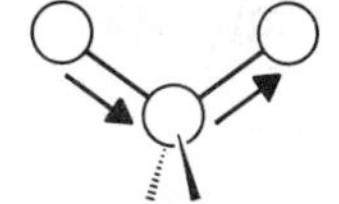
asymmetric stretching

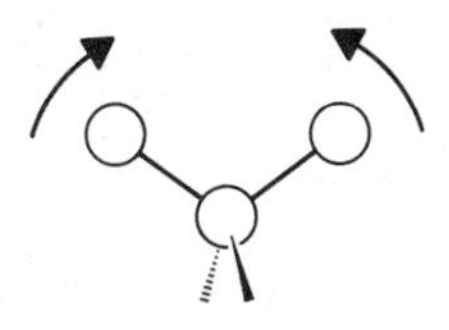
an in–plane bending vibration

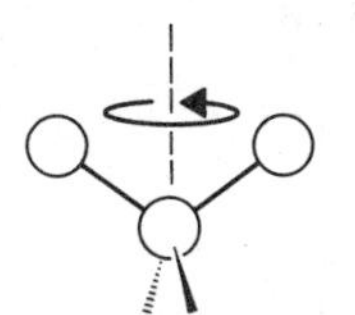
an out–of–plane bending vibration

The *frequency* of a given stretching vibration and thus *its location in an infrared spectrum* can be related to two factors. These are *the masses of the bonded atoms*—light atoms vibrate at higher frequencies than heavier ones—*and the relative stiffness of the bond.* Triple bonds are stiffer (and vibrate at higher frequencies) than double bonds and double bonds are stiffer (and vibrate at higher frequencies) than single bonds. We can see some of these effects in Table 13.2. Notice that stretching frequencies of groups involving hydrogen (a light atom) such as C—H, N—H, and O—H all occur at relatively high frequencies:

Bond		Frequency Range, cm^{-1}
alkyl	C—H	2853–2962
alcohol	O—H	3590–3650
amine	N—H	3300–3500

Notice too, that triple bonds vibrate at higher frequencies than double bonds:

Bond	Frequency Range cm^{-1}
C≡C	2100–2260
C≡N	2220–2260
C=C	1620–1680
C=O	1690–1750

The infrared spectra of even relatively simple compounds contain many absorption peaks. It can be shown that a nonlinear molecule of n atoms has $3n - 6$ possible *fundamental* vibrational modes that can be responsible for the absorption of infrared radiation. This means that, theoretically, methane has 9 possible fundamental absorption peaks and benzene has 30.

Not all molecular vibrations result in the absorption of energy, however. In order for a vibration to occur with the absorption of infrared energy, *the dipole moment of the molecule must change as the vibration occurs.* Thus when the four hydrogens of methane vibrate symmetrically, methane does not absorb infrared energy. Symmetrical vibrations of the carbon-carbon double and triple bonds of ethene and ethyne do not result in the absorption of infrared radiation, either.

Vibrational absorption may occur outside the region measured by a particular infrared spectrophotometer and vibrational absorptions may occur so closely together that peaks fall on top of peaks. These factors, together with the absence of absorptions because of vibrations that have no dipole moment change, cause infrared spectra to contain fewer peaks than the formula $3n - 6$ would predict.

However, other factors bring about even more absorption peaks. Overtones

TABLE 13.2 Characteristic Infrared Absorptions of Functional Groups

GROUP	FREQUENCY RANGE cm^{-1}	INTENSITY[a]
A. **Alkyl**		
C—H (Stretching)	2853–2962	(m–s)
Isopropyl, $—CH(CH_3)_2$	1380–1385	(s)
	and 1365–1370	(s)
tert-Butyl, $—C(CH_3)_3$	1385–1395	(m)
	and ~1365	(s)
B. **Alkenyl**		
C—H (stretching)	3010–3095	(m)
C=C (stretching)	1620–1680	(v)
$R—CH=CH_2$ (out-of-plane C—H bendings)	985–1000	(s)
	and 905–920	(s)
$R_2C=CH_2$ (out-of-plane C—H bendings)	880–900	(s)
cis-RCH=CHR (out-of-plane C—H bendings)	675–730	(s)
trans-RCH=CHR (out-of-plane C—H bendings)	960–975	(s)
C. **Alkynyl**		
≡C—H (stretching)	~3300	(s)
C≡C (stretching)	2100–2260	(v)
D. **Aromatic**		
Ar—H (stretching)	~3030	(v)
Aromatic substitution type (C—H out-of-plane bendings)		
Monosubstituted	690–710	(very s)
	and 730–770	(very s)
o-Disubstituted	735–770	(s)
m-Disubstituted	680–725	(s)
	and 750–810	(very s)
p-Disubstituted	790–840	(very s)
E. **Alcohols, Phenols, Carboxylic Acids**		
OH (alcohols, phenols, dilute solns)	3590–3650	(sharp, v)
OH (alcohols, phenols, hydrogen bonded)	3200–3550	(broad, s)
OH (carboxylic acids, hydrogen bonded)	2500–3000	(broad, v)
F. **Aldehydes, Ketones, Esters, and Carboxylic Acids**		
C=O stretch	1630–1780	(s)
aldehydes	1690–1740	(s)
ketones	1650–1730	(s)
esters	1735–1750	(s)
carboxylic acids	1710–1780	(s)
amides	1630–1690	(s)
G. **Amines**		
N—H	3300–3500	(m)
H. **Nitriles**		
C≡N	2220–2260	(m)

[a] Abbreviations: s = strong, m = medium, w = weak, v = variable, ~ = approximately.

(harmonics) of fundamental absorption bands may be seen in infrared spectra even though these overtones occur with greatly reduced intensity. Bands called combination bands and difference bands also appear in infrared spectra.

Because infrared spectra contain so many peaks, the possibility that two compounds will have the same infrared spectrum is exceedingly small. It is because

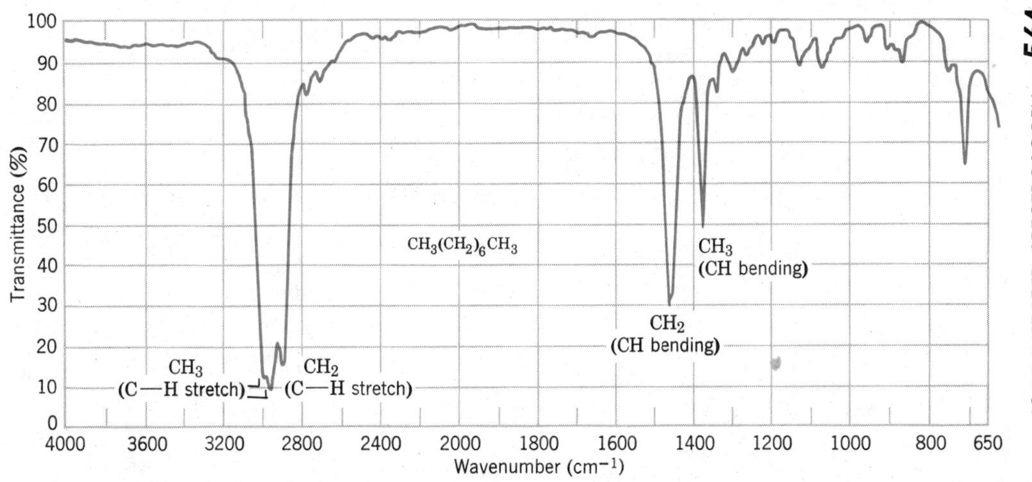

FIG. 13.32
The infrared spectrum of n-octane. *(Notice that, in infrared spectra, the peaks are "upside down." This is simply a result of the way infrared spectrophotometers operate.)*

of this that an infrared spectrum has been called the "fingerprint" of a molecule. Thus, with organic compounds, if two pure samples give different infrared spectra, one can be certain that they are different compounds. If they give the same infrared spectrum then they are the same compound.

In the hands of one skilled in their interpretation, infrared spectra contain a wealth of information about the structures of compounds. We show some of the information that can be gathered from the spectra of *n*-octane and toluene in Figs.13.32 and 13.33. We have neither the time nor the space here to develop the skill that would lead to complete interpretations of infrared spectra, but we can learn how to recognize the presence of absorption peaks in the infrared spectrum that result from vibrations of characteristic functional groups in the compound. By doing only this, however, we will be able to use the information we gather from infrared spectra in a powerful way, particularly when we couple it with the information we gather from nuclear magnetic resonance spectra.

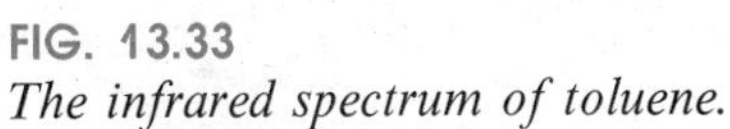
FIG. 13.33
The infrared spectrum of toluene.

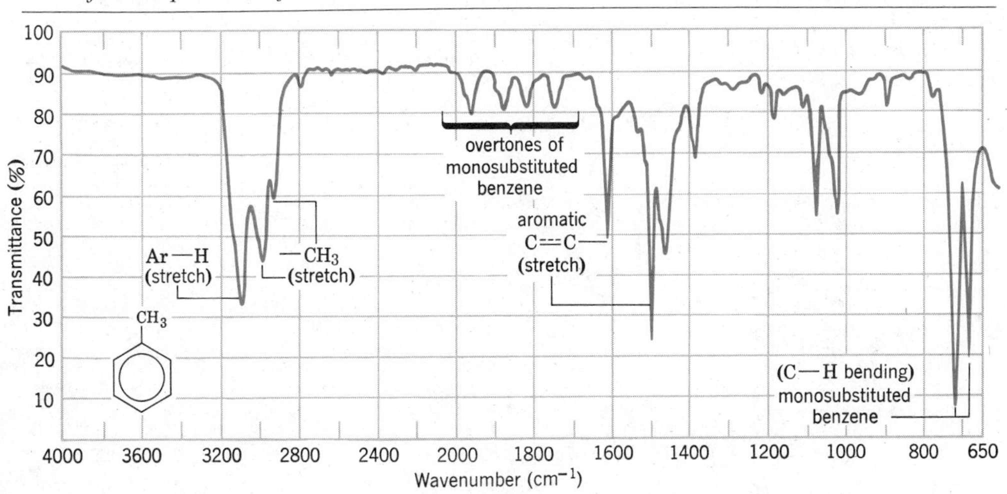

Let us now see how we can apply the data given in Table 13.2 to the interpretation of infrared spectra.

13.11A Hydrocarbons

All hydrocarbons give absorption peaks in the 2800 to 3300 cm^{-1} region that are associated with carbon-hydrogen stretching vibrations. We can use these peaks in interpreting infrared spectra because the exact location of the peak depends on the strength (and stiffness) of the C—H bond which in turn depend on the hybridization state of the carbon that bears the hydrogen. We have already seen that C—H bonds involving *sp*-hybridized carbon are strongest and those involving sp^3-hybridized carbon are weakest. The order of bond strength is:

$$sp > sp^2 > sp^3$$

This too is the order of the bond stiffness.

The carbon-hydrogen stretching peaks of hydrogens attached to *sp*-hybridized carbons occur at highest frequencies, ~3300 cm^{-1}. Thus, ≡C—H groups of terminal alkynes give peaks in this region. We can see the absorption of the acetylenic hydrogen of 1-hexyne at 3320 cm^{-1} in Fig. 13.34.

The carbon-hydrogen stretching peaks of hydrogens attached to sp^2-hybridized carbons occur in the 3000–3100 cm^{-1} region. Thus, alkenyl hydrogens and the C—H groups of aromatic rings give absorption peaks in this region. We can see the alkenyl C—H absorption peak at 3080 cm^{-1} in the spectrum of 1-hexene (Fig. 13.35) and we can see the C—H absorption of the aromatic hydrogens at 3090 cm^{-1} in the spectrum of toluene (Fig. 13.33).

The carbon-hydrogen stretching bands of hydrogens attached to sp^3-hybridized carbons occur at lowest frequencies, in the 2800–3000 cm^{-1} region. We can see methyl and methylene absorption peaks in the spectra of *n*-octane (Fig. 13.32), toluene (Fig. 13.33), 1-hexyne (Fig. 13.34), and 1-hexene (Fig. 13.35).

Hydrocarbons also give absorption peaks in their infrared spectra that result from carbon-carbon bond stretchings. Carbon-carbon single bonds normally give rise to very weak peaks that are usually of little use in assigning structures. More useful peaks arise from multiple carbon-carbon bonds, however. Carbon-carbon double bonds give absorption peaks in the 1620–1680 cm^{-1} region and carbon-carbon triple bonds give absorption peaks between 2100 cm^{-1} and

FIG. 13.34
The infrared spectrum of 1-hexyne. (Spectrum courtesy of Sadtler Research Laboratories, Inc., Philadelphia, Pa.)

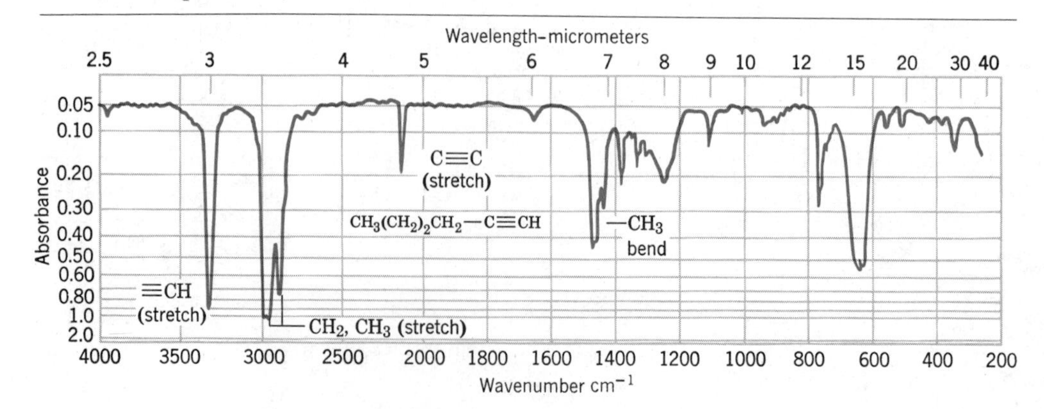

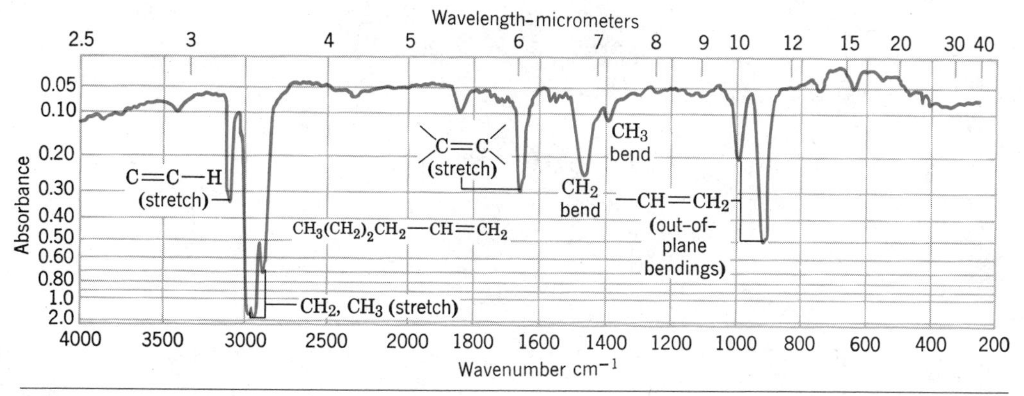

FIG. 13.35
The infrared spectrum of 1-hexene. (*Spectrum courtesy of Sadtler Research Laboratories, Inc., Philadelphia, Pa.*)

2260 cm^{-1}. These absorptions are not usually strong ones and they will not be present at all if the double or triple bond is symmetrically substituted. (No dipole moment change will be associated with the vibration.) The stretchings of the carbon-carbon bonds of benzene rings usually give a set of characteristic sharp peaks in the 1450–1600 cm^{-1} region.

Absorptions arising from carbon-hydrogen bending vibrations of alkenes occur in the 600–1000 cm^{-1} region. The exact location of these peaks can often be used to determine the *nature of the double bond and its configuration.*

Monosubstituted alkenes give two strong peaks in the 905–920 cm^{-1} and the 985–1000 cm^{-1} regions. Disubstituted alkenes of the type $R_2C{=}CH_2$ give a strong peak in the 880–900 cm^{-1} range. *Cis*-alkenes give an absorption peak in the 675–730 cm^{-1} region and *trans*-alkenes give a peak between 960 cm^{-1} and 975 cm^{-1}. These ranges for the carbon-hydrogen bending vibrations can be used with fair reliability for alkenes that do not have an electron-releasing or electron-withdrawing substituent (other than an alkyl group) on one of the carbons of the double bond. When electron-releasing or electron-withdrawing substituents are present on a double-bond carbon, the bending absorption peaks may be shifted out of the regions we have given.

13.11B Substituted Benzenes

If we compare the proton nmr spectra of the three xylenes (Fig. 13.2 and Fig. 13.36) we find that they are very similar. Using proton nmr spectra alone, we would have little difficulty in deciding whether an unknown compound was a xylene or not. We would, however, have considerable difficulty in deciding just which xylene it was, *o*-xylene, *m*-xylene, or *p*-xylene.

Infrared spectroscopy provides us with a way around this difficulty. Ortho-, meta-, and para-substituted benzenes give absorption peaks in the 680–840 cm^{-1} region that characterize their substitution patterns. Ortho-substituted benzenes show a strong absorption peak arising from bending motions of the aromatic hydrogens between 735 and 770 cm^{-1}. Meta-substituted benzenes show two peaks; one strong peak between 680 and 725 cm^{-1} and one very strong peak between 750 and 810 cm^{-1}. Para-substituted benzenes give a single very strong absorption between 790 and 840 cm^{-1}.

These characteristic peaks are designated in the infrared spectra of ortho-, *meta*-, and *para*-xylene (Figs. 13.37 to 39).

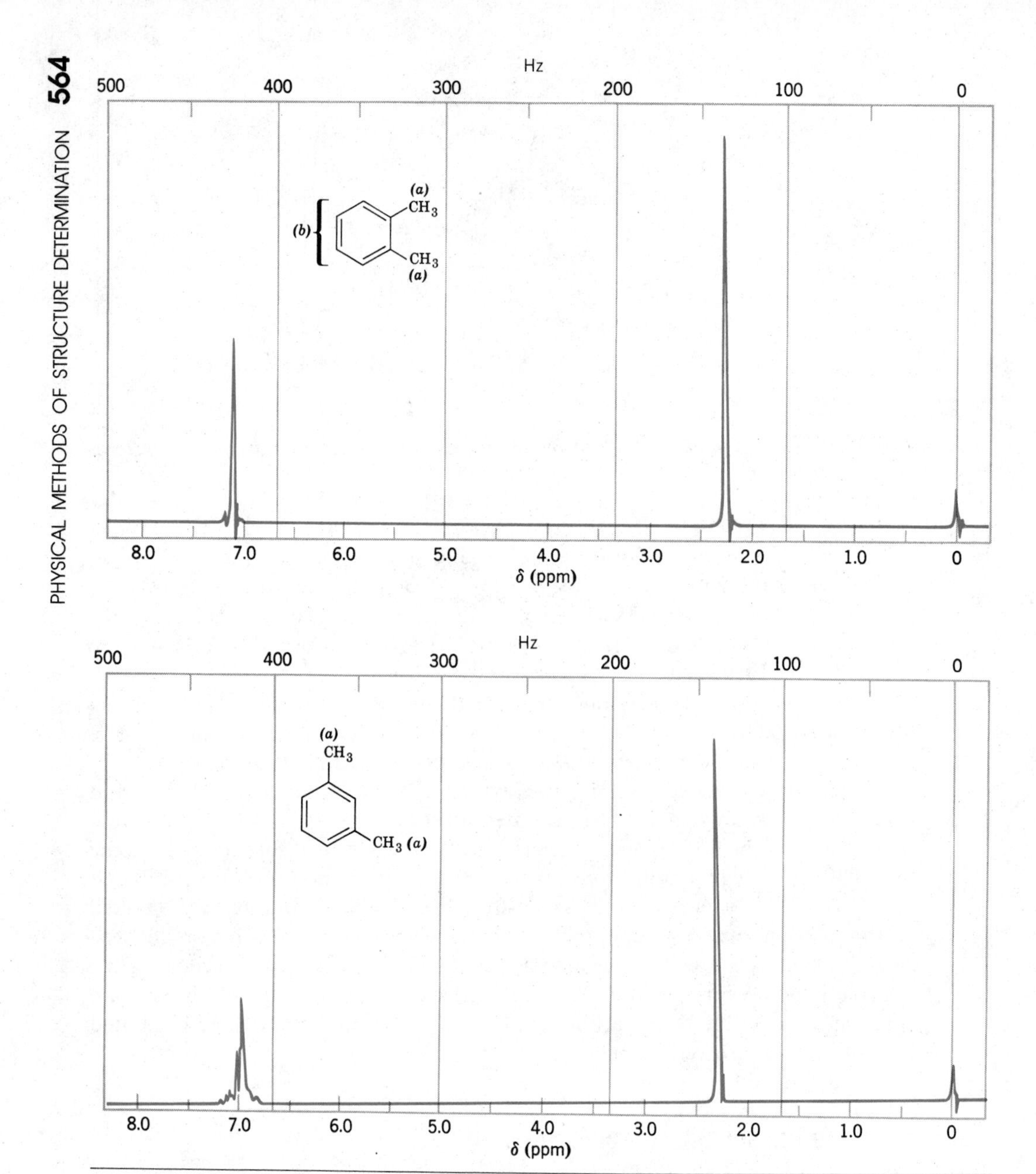

FIG. 13.36
The proton nmr spectra of ortho- *and* meta-*xylene.* (*Spectra courtesy of Varian Associates, Palo Alto, Calif.*)

Monosubstituted benzenes give two very strong peaks, between 690 and 710 cm^{-1} and between 730 and 770 cm^{-1}.

Problem 13.15

Using Table 13.2, make assignments to as many other peaks as you can in the spectra of *o*-, *m*-, and *p*-xylene.

Problem 13.16

Four benzenoid compounds, all with the formula C_7H_7Br, gave the following infrared peaks in the 680–840 cm^{-1} region.

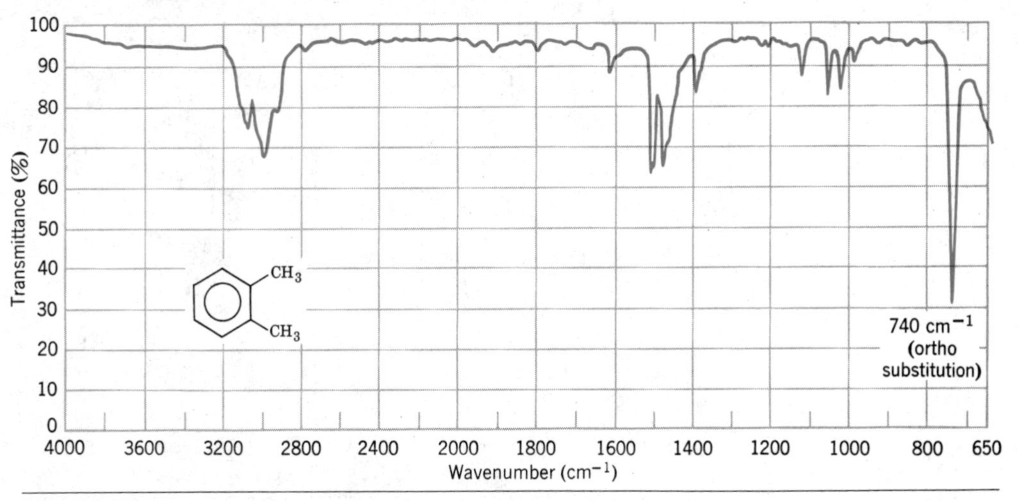

FIG. 13.37
The infrared spectrum of o-*xylene.*

A, 740 cm^{-1} (s) **C,** 680 cm^{-1} (s) and 760 cm^{-1} (very s)
B, 795 cm^{-1} (very s) **D,** 693 cm^{-1} (very s) and 765 cm^{-1} (very s)

Propose structures for **A, B, C,** and **D.**

13.11C Other Functional Groups

Infrared spectroscopy gives us an invaluable method for recognizing quickly and simply the presence of certain functional groups in a molecule. One important functional group that gives a prominent absorption peak in infrared spectra is the **carbonyl group,** $>C{=}O$. This group is present in aldehydes, ketones, esters, carboxylic acids, amides, and so forth. The carbon-oxygen double bond stretching frequency of all these groups gives a strong peak between 1630 and 1780 cm^{-1}. The exact location of the peak depends on whether it arises from an aldehyde, ketone, ester, and so forth. These locations are the following and we will

FIG. 13.38
The infrared spectrum of m-*xylene.*

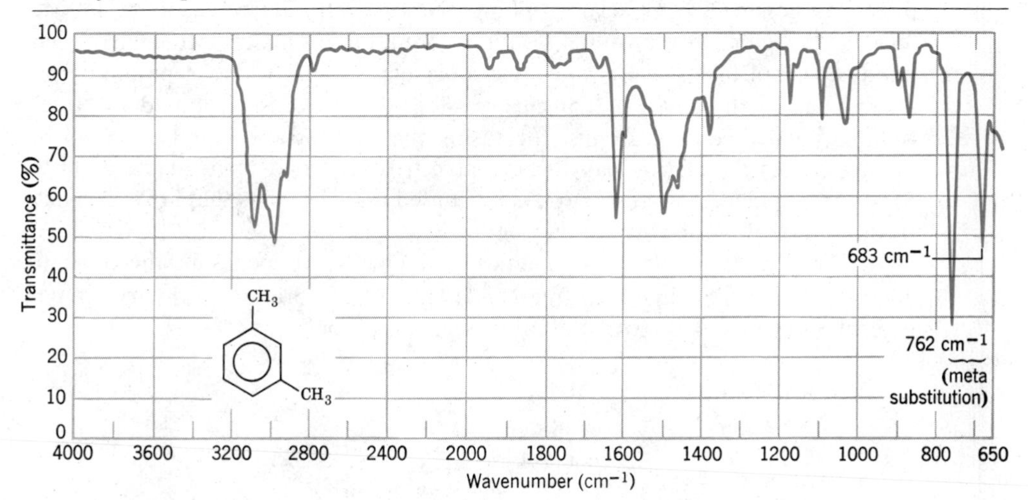

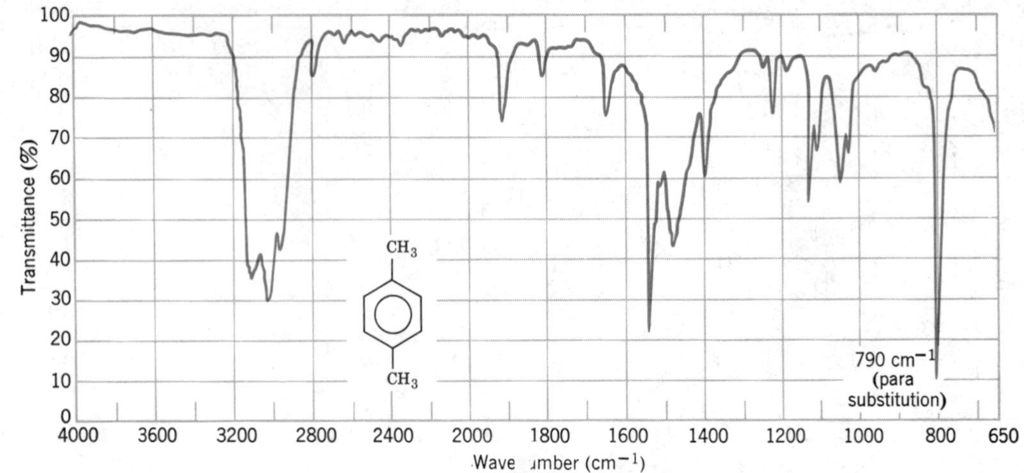

FIG. 13.39
The infrared spectrum of p-*xylene.*

have more to say about carbonyl absorption peaks when we discuss these compounds in later chapters.

$R{-}\overset{O}{\overset{\|}{C}}{-}H$	$R{-}\overset{O}{\overset{\|}{C}}{-}R$	$R{-}\overset{O}{\overset{\|}{C}}{-}OR$
Aldehyde	Ketone	Ester
1690–1740 cm^{-1}	1650–1730 cm^{-1}	1735–1750 cm^{-1}

$R{-}\overset{O}{\overset{\|}{C}}{-}OH$	$R{-}\overset{O}{\overset{\|}{C}}{-}NH_2$
1710–1780 cm^{-1}	1630–1690 cm^{-1}
Carboxylic acid	Amide

The **hydroxyl groups** of alcohols and phenols are also easy to recognize in infrared spectra by their O—H stretching absorptions. These bonds also give us direct evidence for hydrogen bonding. If an alcohol or phenol is present as a very dilute solution in CCl_4, O—H absorption occurs as a very sharp peak in the 3590 to 3650 cm^{-1} region. In very dilute solution, formation of intermolecular hydrogen bonds does not take place because the molecules are too widely separated. The sharp peak in the 3590 to 3650 cm^{-1} region, therefore, is attributed to "free" (unassociated) hydroxyl groups. Increasing the concentration of the alcohol or phenol causes the sharp peak to be replaced by a broad band in the 3200 to 3550 cm^{-1} region. This absorption is attributed to OH groups that are associated through intermolecular hydrogen bonding.

Very dilute solutions of **amines** also give sharp peaks in the 3300 to 3500 cm^{-1} region arising from free N—H stretching vibrations. Primary amines give two sharp peaks; secondary amines give only one.

RNH_2	R_2NH
1° Amine	2° Amine
Two peaks in 3300–3500 cm^{-1} region	One peak in 3300–3500 cm^{-1} region

Hydrogen bonding causes these peaks to broaden. The NH groups of **amides** also give similar absorption peaks.

Additional Problems

13.17

Listed below are proton nmr absorption peaks for several compounds. Propose a structure that is consistent with each set of data. (In some cases characteristic infrared absorptions are given as well.)

(a) $C_4H_{10}O$	proton nmr spectrum	
	singlet, $\delta 1.28$ (9H)	
	singlet, $\delta 1.35$ (1H)	
(b) C_3H_7Br	proton nmr spectrum	
	doublet, $\delta 1.71$ (6H)	
	septet, $\delta 4.32$ (1H)	
(c) C_4H_8O	proton nmr spectrum	IR spectrum
	triplet, $\delta 1.05$ (3H)	1720 cm^{-1}
	singlet, $\delta 2.13$ (3H)	
	quartet, $\delta 2.47$ (2H)	
(d) C_7H_8O	proton nmr spectrum	IR spectrum
	singlet, $\delta 2.43$ (1H)	broad peak in
	singlet, $\delta 4.58$ (2H)	3200–3550 cm^{-1}
	multiplet, $\delta 7.28$ (5H)	region
(e) C_4H_9Cl	proton nmr spectrum	
	doublet, $\delta 1.04$ (6H)	
	multiplet, $\delta 1.95$ (1H)	
	doublet, $\delta 3.35$ (2H)	
(f) $C_{15}H_{14}O$	proton nmr spectrum	IR spectrum
	singlet, $\delta 2.20$ (3H)	strong peak
	singlet, $\delta 5.08$ (1H)	near 1720 cm^{-1}
	multiplet, $\delta 7.25$ (10H)	
(g) $C_4H_7BrO_2$	proton nmr spectrum	IR spectrum
	triplet, $\delta 1.08$ (3H)	broad peak in
	multiplet, $\delta 2.07$ (2H)	2500–3000 cm^{-1}
	triplet, $\delta 4.23$ (1H)	region and a peak
	singlet, $\delta 10.97$ (1H)	at 1715 cm^{-1}
(h) C_8H_{10}	proton nmr spectrum	
	triplet, $\delta 1.25$ (3H)	
	quartet, $\delta 2.68$ (2H)	
	multiplet, $\delta 7.23$ (5H)	
(i) $C_4H_8O_3$	proton nmr spectrum	IR spectrum
	triplet, $\delta 1.27$ (3H)	broad peak in
	quartet, $\delta 3.66$ (2H)	2500–3000 cm^{-1}
	singlet, $\delta 4.13$ (2H)	region and a peak
	singlet, $\delta 10.95$ (1H)	at 1715 cm^{-1}
(j) $C_3H_7NO_2$	proton nmr spectrum	
	doublet, $\delta 1.55$ (6H)	
	septet, $\delta 4.67$ (1H)	
(k) $C_4H_{10}O_2$	proton nmr spectrum	
	singlet, $\delta 3.25$ (6H)	
	singlet, $\delta 3.45$ (4H)	
(l) $C_5H_{10}O$	proton nmr spectrum	IR spectrum
	doublet, $\delta 1.10$ (6H)	strong peak
	singlet, $\delta 2.10$ (3H)	near 1720 cm^{-1}
	septet, $\delta 2.50$ (1H)	
(m) C_8H_9Br	proton nmr spectrum	
	doublet, $\delta 2.0$ (3H)	
	quartet, $\delta 5.15$ (1H)	
	multiplet, $\delta 7.35$ (5H)	

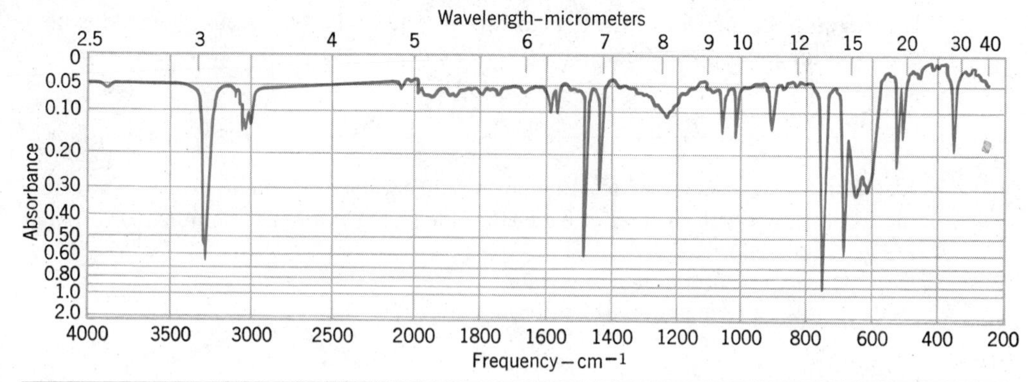

FIG. 13.40
The infrared spectrum of compound E. (Spectrum courtesy of Sadtler Research Laboratories Inc., Philadelphia, Pa.)

13.18
The infrared spectrum of compound **E**, C_8H_6, is shown in Fig. 13.40. **E** decolorizes bromine in carbon tetrachloride and gives a precipitate when treated with ammoniacal silver nitrate. What is the structure of **E**?

13.19
The proton nmr spectrum of cyclooctatetraene consists of a single line located at $\delta 5.78$. What does this suggest about electron delocalization in cyclooctatetraene?

13.20
The [14] annulene shown below obeys Huckel's rule. Its proton nmr spectrum shows signals

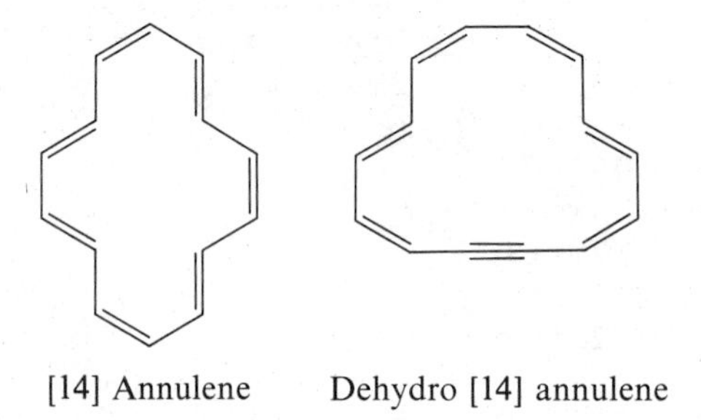

[14] Annulene Dehydro [14] annulene

at $\delta 7.78$. (10H) and $\delta -0.61$ (4H). Dehydro [14] annulene gives proton nmr signals at $\delta 8.0$ (10H) and $\delta 0.0$ (2H). How can you account for the relative intensities of the signals given by the two compounds?

13.21
Give a structure for compound **F** that is consistent with the proton nmr and IR spectra in Fig. 13.41.

13.22
(a) When either *n*-butyl fluoride or *sec*-butyl fluoride is treated with excess SbF_5, their solutions give identical proton nmr spectra. These spectra, moreover, are the same as that obtained when *tert*-butyl fluoride is treated with SbF_5. Explain these results. (b) Treating the seven isomeric fluoropentanes with excess SbF_5 furnishes solutions that give identical proton nmr spectra. What species is formed in these reactions? Sketch the spectrum that you would expect to obtain.

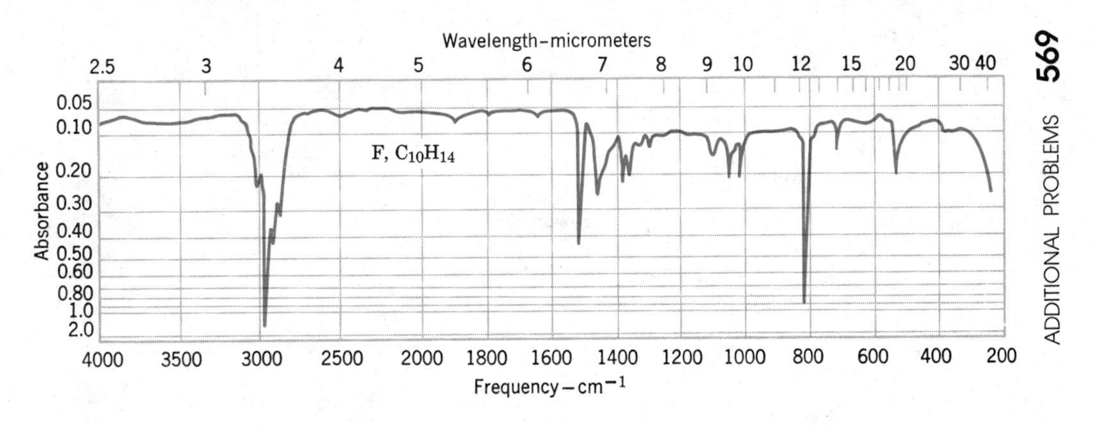

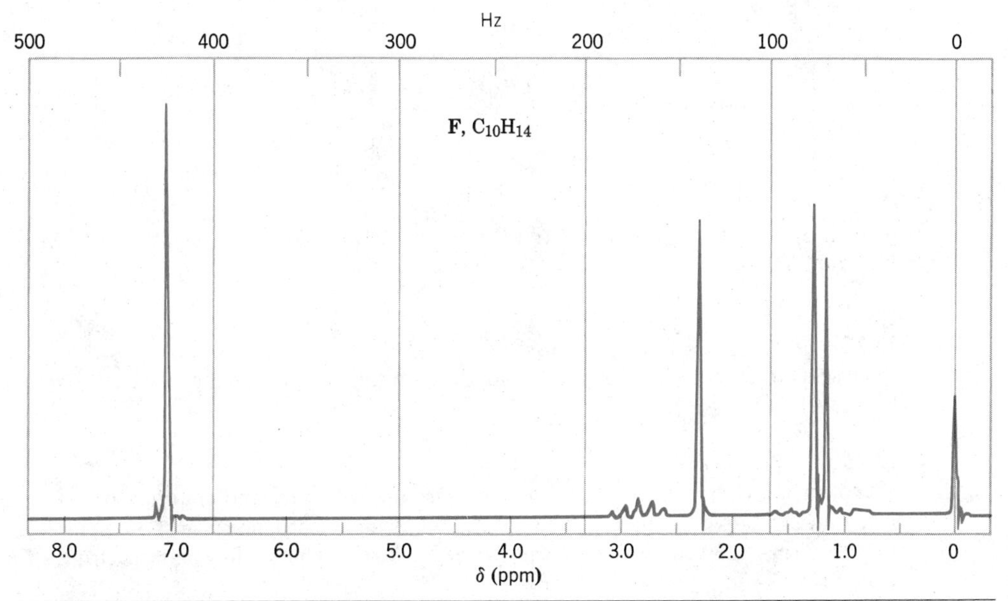

FIG. 13.41
*The proton nmr and IR spectra of compound **F**, Problem 13.21. (Proton nmr spectrum adapted from Varian Associates, Palo Alto, Calif. IR spectrum adapted from Sadtler Research Laboratories, Inc., Philadelphia, Pa.)*

13.23

(a) How many peaks would you expect to find in the proton nmr spectrum of caffeine?

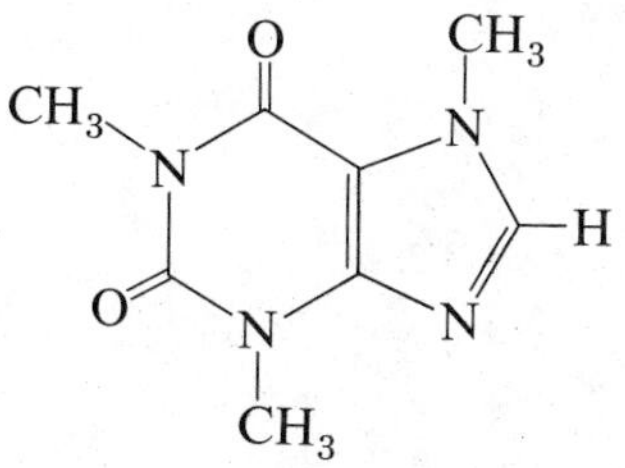

Caffeine

(b) What characteristic peaks would you expect to find in the infrared spectrum of caffeine?

13.24

Propose structures for the compounds **G** and **H** whose proton nmr spectra are shown in Fig. 13.42 and Fig. 13.43.

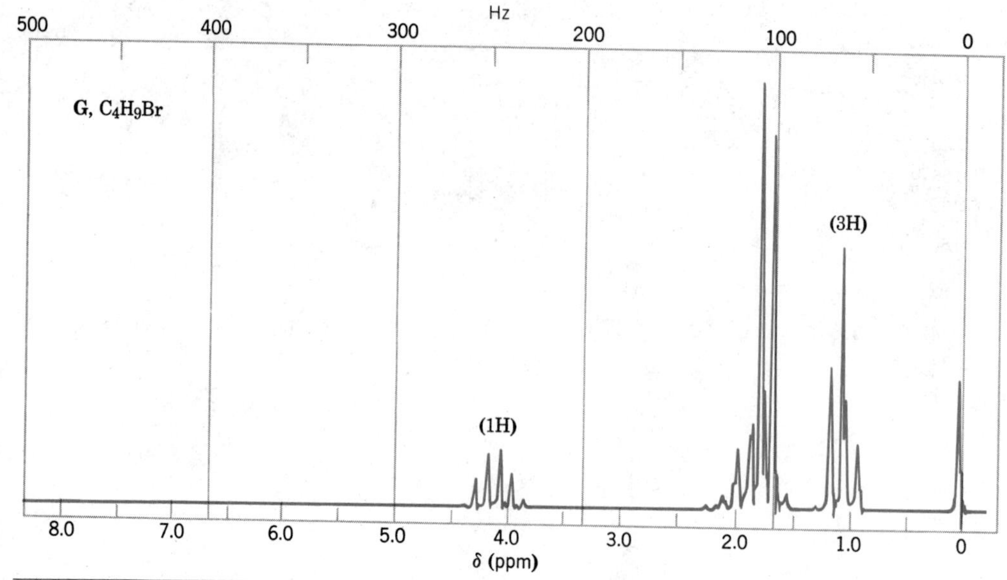

FIG. 13.42
*The proton nmr spectrum of compound **G** (Problem 13.24). (Spectrum courtesy of Varian Associates, Palo Alto, Calif.)*

13.25

Propose a structure for compound **I** whose proton nmr and infrared spectra are given in Fig. 13.44 and Fig. 13.45.

13.26

A two-carbon compound, **J**, contains only carbon, hydrogen, and chlorine. Its infrared spectrum is relatively simple and shows the following absorbance peaks: 3125 cm^{-1}(m), 1625 cm^{-1}(m), 1280 cm^{-1}(m), 820 cm^{-1}(s), 695 cm^{-1}(s). The proton nmr spectrum of **J**

FIG. 13.43
*The proton nmr spectrum of compound **H** (Problem 13.24). (Spectrum courtesy of Varian Associates, Palo Alto, Calif.)*

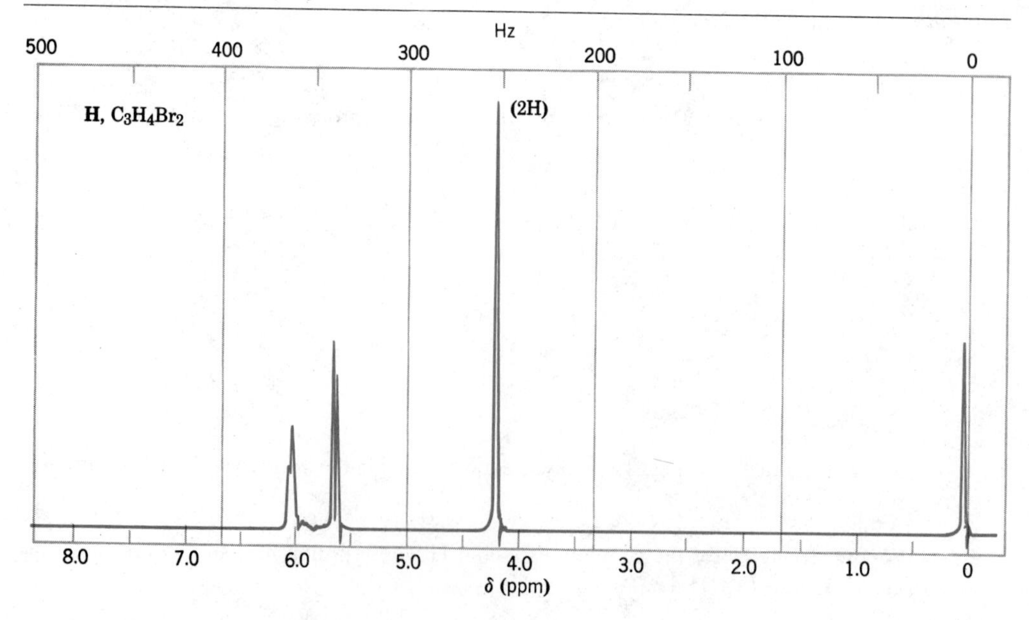

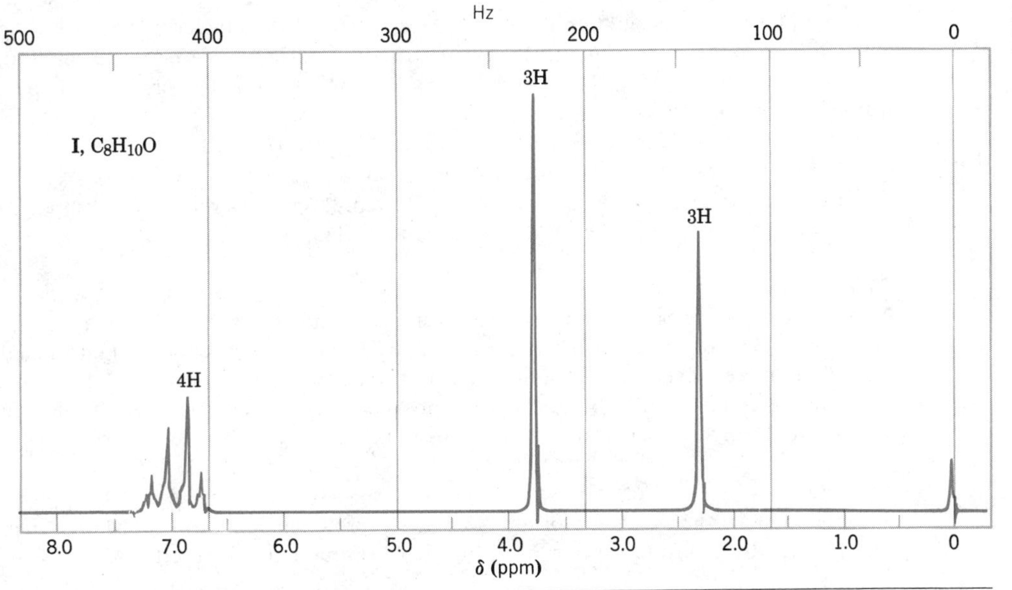

FIG. 13.44
*The proton nmr spectrum of compound **I** (Problem 13.25). (Spectrum courtesy of Varian Associates, Palo Alto, Calif.)*

consists of a singlet at δ6.3. Using Table 13.2 make as many infrared assignments as you can and propose a structure for compound **J**.

13.27

When dissolved in $CDCl_3$, a compound, **K**, with the molecular formula $C_4H_8O_2$ gives a proton nmr spectrum that consists of a doublet at δ1.35, a singlet at δ2.15, a broad singlet at δ3.75 (1H) and a quartet at δ4.25 (1H). When dissolved in D_2O, the compound gives a similar proton nmr spectrum with the exception that the signal at δ3.75 has disappeared. The infrared spectrum of the compound shows a strong absorption peak near 1720 cm^{-1}. (a) Propose a structure for compound **K** and (b) explain why the nmr signal at δ3.75 disappears when D_2O is used as the solvent.

13.28

A compound, **L**, with the molecular formula C_9H_{10} decolorizes bromine in carbon

FIG. 13.45
*The infrared spectrum of compound **I** (Problem 13.25). (Spectrum courtesy of Sadtler Research Laboratories, Inc., Philadelphia, Pa.)*

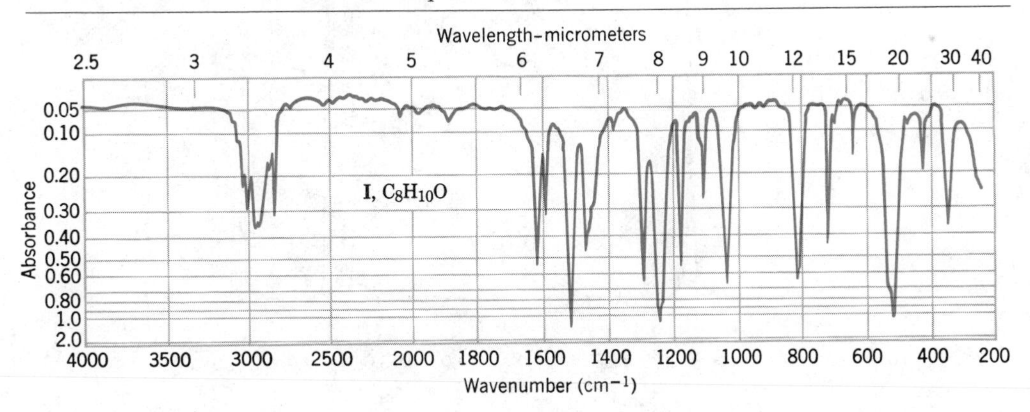

tetrachloride and gives an infrared absorption spectrum that includes the following absorption peaks: 3035 cm^{-1}(m), 3020 cm^{-1}(m), 2925 cm^{-1}(m), 2853 cm^{-1}(w), 1640 cm^{-1}(m), 990 cm^{-1}(s), 915 cm^{-1}(s), 740 cm^{-1}(s), 695 cm^{-1}(s). The proton nmr spectrum of **L** consists of:

Doublet δ3.1(2H) Multiplet δ5.1 Multiplet δ7.1(5H)
Multiplet δ4.8 Multiplet δ5.8

The ultraviolet spectrum shows a maximum at 255 nm. Propose a structure for compound **L** and make assignments for each of the infrared peaks.

13.29

Assume that in a certain proton nmr spectrum, you find two peaks of roughly equal intensity. You are not certain whether these two peaks are *singlets* arising from uncoupled protons at different chemical shifts, or whether they are two peaks of a *doublet* that arises from protons coupling with a single adjacent proton. What simple experiment would you perform to distinguish between these two possibilities?

13.30

Compound **M** has the molecular formula C_9H_{12}. The proton nmr spectrum of **M** is given in Fig. 13.46 and the infrared spectrum in Fig. 13.47. Propose a structure for **M**.

13.31

A compound, **N**, with the molecular formula $C_9H_{10}O$ gives a positive test with cold dilute aqueous potassium permanganate. The proton nmr spectrum of **N** is shown in Fig. 13.48 and the infrared spectrum of **N** is shown in Fig. 13.49. Propose a structure for **N**.

***13.32**

When 2,3-dibromo-2,3-dimethylbutane is treated with SbF_5 in liquid SO_2 at $-60°$, the proton nmr spectrum does not show the two signals that would be expected of a carbocation like $CH_3\underset{CH_3}{\underset{|}{C}}Br-\underset{CH_3}{\underset{|}{\overset{+}{C}}}CH_3$. Instead only one signal (at δ2.9) is observed. What cation formed in this reaction and of what special significance is this experiment?

FIG. 13.46

*The proton nmr spectrum of compound **M**, Problem 13.30. (Spectrum courtesy of Aldrich Chemical Co., Milwaukee, Wisconsin.)*

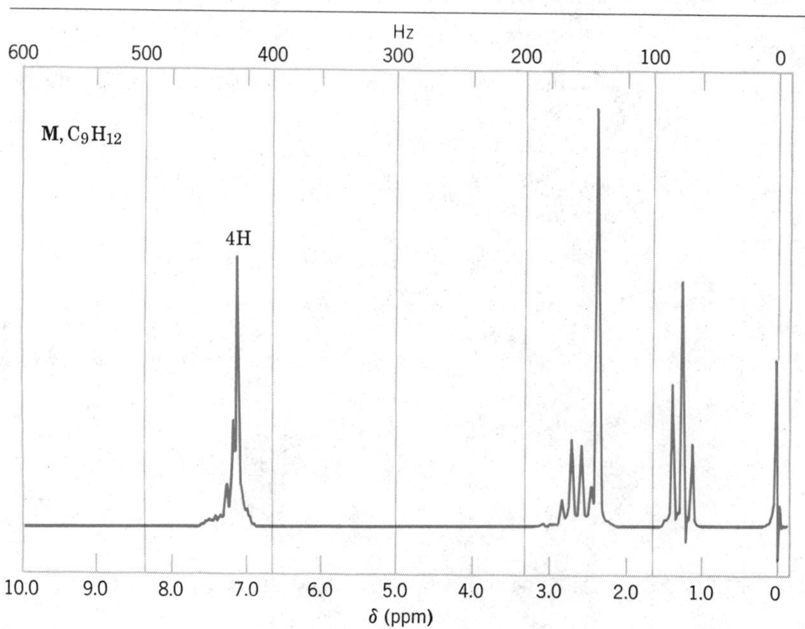

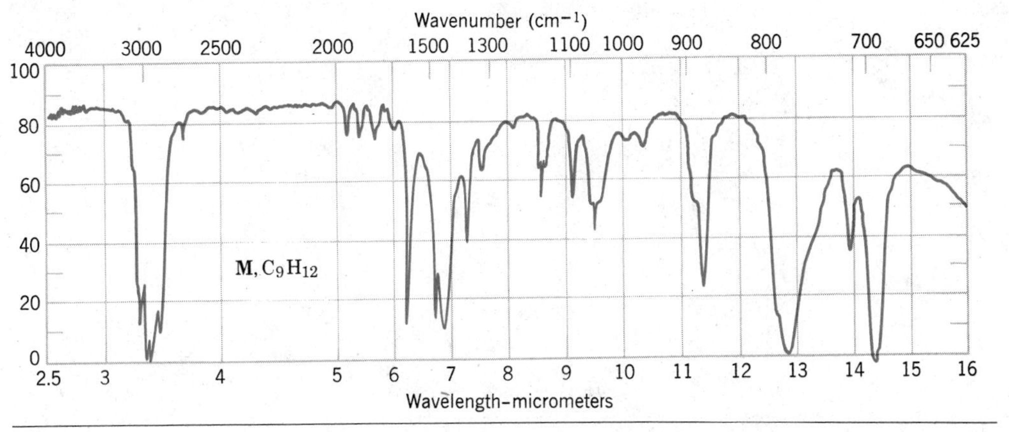

FIG. 13.47
*The infrared spectrum of compound **M**, Problem 13.30. (Spectrum courtesy of Aldrich Chemical Co., Milwaukee, Wis.)*

*** 13.33**

1,2,3,4-Tetramethyl-3,4-dichlorocyclobutene, I, gives a proton nmr signal at $\delta 1.15$ corresponding to the protons labeled **A** and a signal at $\delta 1.26$ corresponding to the protons labeled **B**.

A { CH_3, CH_3 } — cyclobutene ring — { CH_3, Cl, Cl, CH_3 } B

I

FIG. 13.48
*The proton nmr spectrum of compound **N**, Problem 13.31. (Spectrum courtesy of Aldrich Chemical Co., Milwaukee, Wis.)*

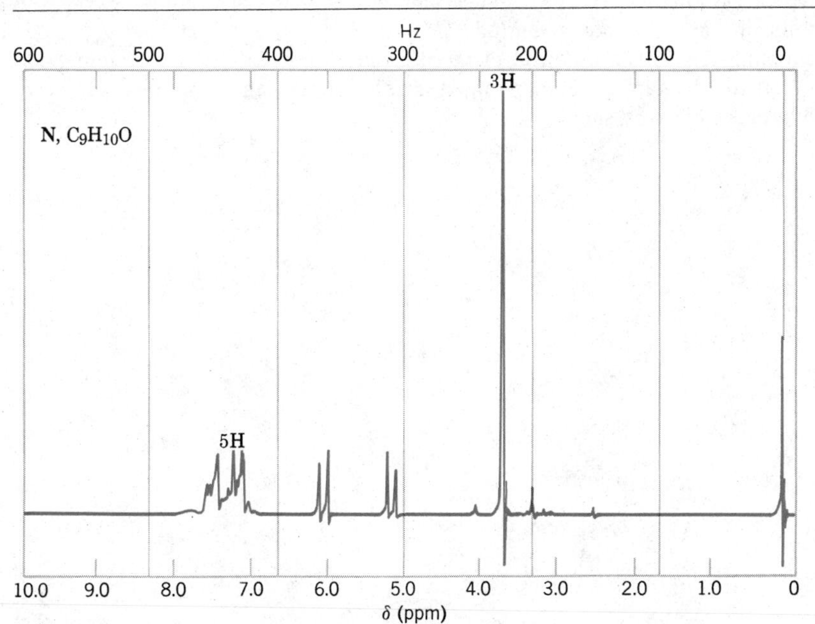

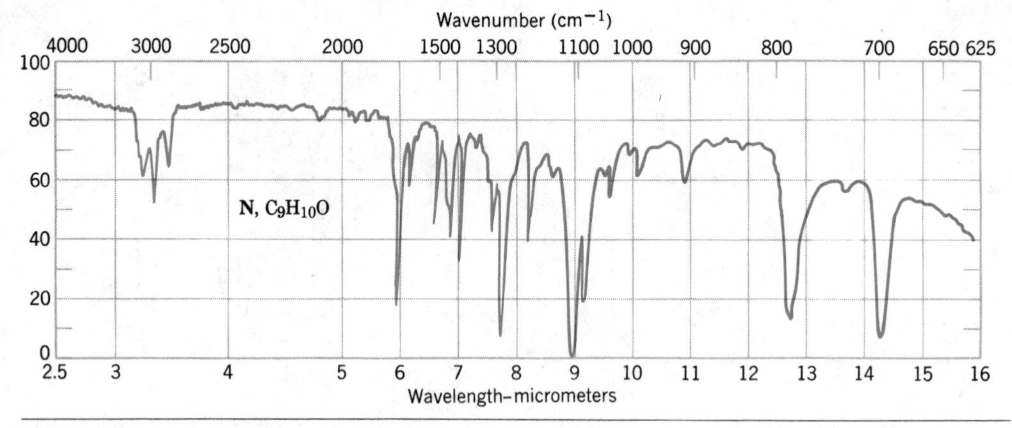

FIG. 13.49
The infrared spectrum of compound **N**, *Problem 13.31. (Spectrum courtesy of Aldrich Chemical Co., Milwaukee, Wis.)*

When I is added to SbF_5—SO_2 at $-78°$ a pale-yellow solution is formed whose proton nmr spectrum shows the following singlets: $\delta 2.05$ (3H), $\delta 2.20$ (3H), $\delta 2.65$ (6H). After several minutes, these peaks begin to be replaced by a sharp singlet at $\delta 3.68$. Recall that SbF_5 is a powerful Lewis acid and explain what is taking place.

*** 13.34**

Given the following information predict the appearance of the proton nmr spectrum given by the vinyl hydrogens of *p*-chlorostyrene.

(*c*) H H (*b*)
C=C
H (*a*)
Cl

Deshielding by the induced magnetic field of the ring is greatest at proton *c* ($\delta 6.7$) and is least at proton *b* ($\delta 5.3$). The chemical shift of *a* is $\sim \delta 5.7$. The coupling constants have the following approximate magnitudes: $J_{ac} \simeq 18$ Hz, $J_{bc} \simeq 11$ Hz, and $J_{ab} \simeq 2$ Hz. (These coupling constants are typical of those given by vinylic systems: coupling constants for *trans* hydrogens are larger than those for *cis* hydrogens, and coupling constants for geminal hydrogens are very small.)

H SPECIAL TOPIC

MASS SPECTROSCOPY

H.1 THE MASS SPECTROMETER

In a mass spectrometer (Fig. H.1) molecules in the gaseous state under low pressure are bombarded with a beam of high-energy electrons. The energy of the beam of electrons is usually 70 eV (electron volts) and one of the things this bombardment can do is dislodge one of the electrons of the molecule and produce a positively charged ion called *the molecular ion.*

$$\underset{\textit{Molecule}}{\mathrm{M}} + \underset{\textit{High-energy electron}}{e^-} \longrightarrow \underset{\textit{Molecular ion}}{\mathrm{M}^{\ddagger}} + 2e^-$$

The molecular ion is not only a cation, but because it contains an odd number of electrons, it also is a free radical. Thus it belongs to a general group of ions called *radical cations.* If, for example, the molecule under bombardment is a molecule of ammonia the following reaction will take place.

$$\mathrm{H:\ddot{\underset{\cdot\cdot}{N}}:H} + e^- \longrightarrow \left[\mathrm{H:\dot{\underset{\cdot\cdot}{N}}:H}\right]^+ + 2e^-$$
$$\quad\ \ \mathrm{H} \qquad\qquad\qquad\qquad\quad \mathrm{H}$$

Molecular ion, M$^+$.
(a radical cation)

An electron beam with an energy of 70 electron volts ($\sim$1600 kcal/mole) not only dislodges electrons from molecules, producing molecular ions, it also imparts to the molecular ions considerable surplus energy. Not all molecular ions will have the same amount of surplus energy, but for most, the surplus will be far in excess of that required to break covalent bonds (50–100 kcal/mole). Thus, soon after they are formed, most molecular ions literally fly apart—they undergo *fragmentation.* Fragmentation can take place in a variety of ways depending on the nature of the particular molecular ion and as we will see later, the way a molecular ion fragments can give us highly useful information about the structure of a complex molecule. Even with a relatively simple molecule like ammonia, however, fragmentation can produce several new cations. The molecular ion can eject a hydrogen atom, for example, and produce the cation, $NH_2^{\ddagger}$.

$$\mathrm{H:\overset{\cdot\,+}{\underset{\cdot\cdot}{N}}:H} \longrightarrow \mathrm{H:\underset{\cdot\cdot}{N}:^+} + \mathrm{H\cdot}$$
$$\quad \mathrm{H} \qquad\qquad\quad \mathrm{H}$$

This $NH_2^{\ddagger}$ cation can then lose a hydrogen atom to produce $NH^{\ddagger}$, which can lead, in turn, to N^+.

$$\mathrm{H:\underset{\cdot\cdot}{N}:^+} \longrightarrow \mathrm{H:\underset{\cdot}{N}:^+} + \mathrm{H\cdot}$$
$$\quad \mathrm{H}$$

$$\mathrm{H:\underset{\cdot}{N}:^+} \longrightarrow \mathrm{:N:^+} + \mathrm{H\cdot}$$

The mass spectrometer then *sorts* these cations on the basis of their mass/charge or m/e ratio. Since for all practical purposes the charge on all of the ions is +1, this amounts

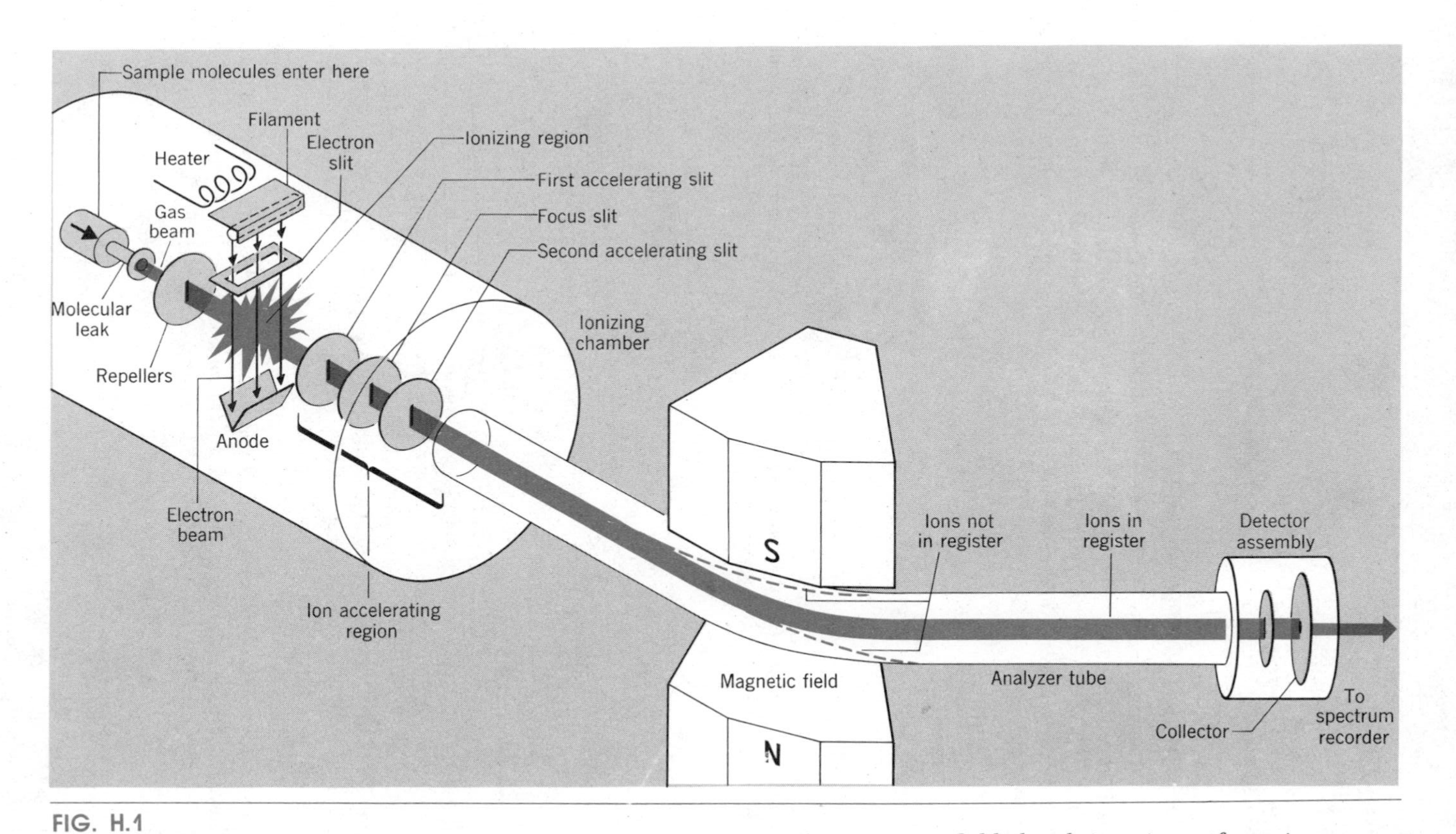

FIG. H.1
Mass spectrometer. Schematic diagram of CEG model 21-103. The magnetic field that brings ions of varying mass-to-charge ratios into register is perpendicular to the page. (From John R. Holum, Organic Chemistry: A Brief Course, *Wiley, New York, 1975. Used with permission.)*

to sorting them on the basis of their mass. The conventional mass spectrometer does this by accelerating the ions through a series of slits and then it sends the ion beam into a curved tube (see Fig. H.1 again). This curved tube passes through a variable magnetic field and the magnetic field exerts an influence on the moving ions. Depending on its strength at a given moment, the magnetic field will cause ions with a particular m/e ratio to follow a curved path that exactly matches the curvature of the tube. These ions are said to be "in register." Because they are in register, these ions pass through another slit and impinge on an ion collector where the intensity of the ion beam is measured electronically. The intensity of the beam is simply a measure of the relative abundance of the ions with a particular m/e ratio. Some mass spectrometers are so sensitive that they can detect the arrival of a *single ion*.

The actual sorting of ions takes place in the magnetic field and this sorting takes place because laws of physics govern the paths followed by charged particles when they move through magnetic fields. Generally speaking, a magnetic field such as this will cause ions moving through it to move in a path that represents part of a circle. The radius of curvature of this circular path (r, cm) is related to the m/e ratio of the ions, to the strength of the magnetic field (**H**, gauss) and to the accelerating voltage (V, volts). The equation that describes the relationship is,

$$\frac{m}{e} = \frac{4.82 \times 10^{-5} r^2 \mathbf{H}^2}{V}$$

We can see that if r and V are made to be constant, then,

$$\frac{m}{e} \propto \mathbf{H}^2$$

This equation tells us that if we keep the accelerating voltage constant and progressively increase the magnetic field, ions whose m/e ratios are progressively larger will travel in a circular path of radius r that exactly matches that of the curved tube. Hence by steadily increasing **H**, ions with progressively increasing m/e ratios will be brought into register and thus will be detected at the ion collector. Since, as we said earlier, the charge on nearly all of the ions is unity, this means that *ions of progressively increasing mass arrive at the collector and are detected.*

What we have described is called "magnetic focusing" (or "magnetic scanning") and all of this is done automatically by the mass spectrometer. The spectrometer displays the results by plotting a series of peaks of varying intensity in which each peak corresponds to ions of a particular m/e ratio. This display (Fig. H.2) is one form of a *mass spectrum*.

Ion sorting can also be done with "electrical focusing." In this technique, the magnetic field is held constant and the accelerating voltage is varied. Both methods, of course, accomplish the same thing, and some high-resolution mass spectrometers employ both techniques.

To summarize: a mass spectrometer bombards organic molecules with a beam of high-energy electrons causing them to ionize and fragment. It then separates the resulting

FIG. H.2
A portion of the mass spectrum of n-*octane.*

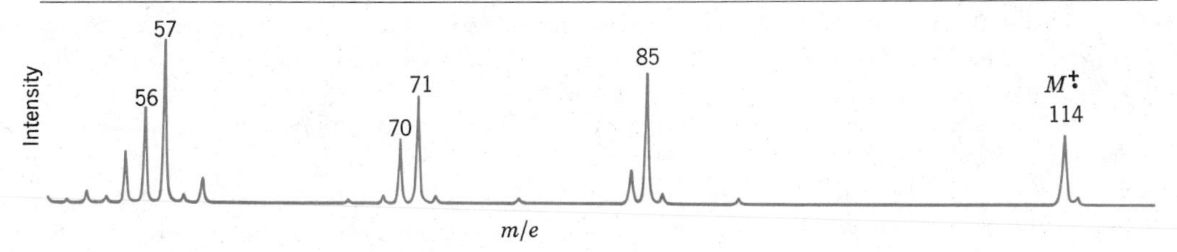

mixture of ions on the basis of their mass/charge ratio and records the relative abundance of each ionic fragment. It displays this result as a plot of ion abundance versus m/e.

H.2 THE MASS SPECTRUM

Mass spectra are usually published as bar graphs or in tabular form, as illustrated in Figure H.3 for the mass spectrum of ammonia. In either presentation, the most intense peak—called the *base peak*—is arbitrarily assigned an intensity of 100%. The intensities of all other peaks are given proportionate values, as percentages of the base peak.

The masses of the ions given in a mass spectrum are those that we would calculate for the ion by assigning the constituent atoms *masses rounded off to the nearest whole number*. For the commonly encountered atoms the nearest whole-number masses are:

$H = 1$
$C = 12$
$N = 14$
$O = 16$
$F = 19$

In the mass spectrum of ammonia we see peaks at $m/e = 14, 15, 16,$ and 17. These correspond to the molecular ion and to the fragments we saw earlier.

$$\underset{m/e\ =}{NH_3} \xrightarrow{-e^-} \underset{\substack{17 \\ (molecular \\ ion)}}{[NH_3]^{\dot{+}}} \xrightarrow{-H\cdot} \underset{16}{[NH_2]^+} \xrightarrow{-H\cdot} \underset{15}{[NH]^{\dot{+}}} \xrightarrow{-H\cdot} \underset{14}{[N]^+}$$

By convention we express,

$H:\dot{\underset{\cdot\cdot}{N}}:H^+$ (with H below N) as $[NH_3]^{\dot{+}}$

$H:\underset{\cdot\cdot}{N}:^+$ (with H below N) as $[NH_2]^+$

$H:\underset{\cdot}{N}:^+$ as $[NH]^{\dot{+}}$

and $:N:^+$ as $[N]^+$

In the case of ammonia, the base peak is the peak arising from the molecular ion. This is not always the case, however; in many of the spectra that we will see later the base peak (the most intense peak) will be at an m/e value different from that of the molecular ion. This happens because in many instances the molecular ion fragments so rapidly that some other ion at a smaller m/e value becomes the most intense peak. In a few cases the molecular ion peak is extremely small, and sometimes it is absent altogether.

One other feature in the spectrum of ammonia requires explanation: the small peak that occurs at m/e 18. In the bar graph we have labeled this peak $M^{\dot{+}} + 1$ to indicate that it is one mass unit greater than the molecular ion. The $M^{\dot{+}} + 1$ peak appears in the spectrum because most elements (e.g., nitrogen and hydrogen) have more than one naturally occurring isotope (Table H.1). Although most of the NH_3 molecules in a sample of ammonia are composed of $^{14}N^1H_3$, a small but detectable fraction of molecules will be composed of $^{15}N^1H_3$. (A very tiny fraction of molecules will also be composed of $^{14}N^1H_2{}^2H$.) These molecules—$^{15}N^1H_3$ or $^{14}N^1H_2{}^2H$—will produce molecular ions at m/e 18, that is at $M^{\dot{+}} + 1$.

The spectrum of ammonia begins to show us with a simple example how the masses (or m/e's) of individual ions can give us information about the composition of the ions and

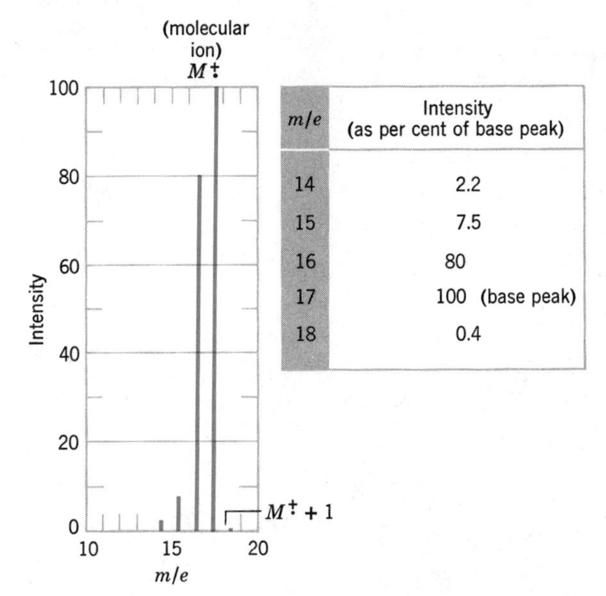

m/e	Intensity (as per cent of base peak)
14	2.2
15	7.5
16	80
17	100 (base peak)
18	0.4

FIG. H.3
The mass spectrum of NH_3 presented as a bar graph and in tabular form.

how this information can allow us to arrive at possible structures for a compound. Problems H.1 to H.3 will allow us further practice with this technique.

Problem H.1

Propose a structure for the compound whose mass spectrum is given in Fig. H.4 and make reasonable assignments for each peak.

Problem H.2

Propose a structure for the compound whose mass spectrum is given in Fig. H.5 and make reasonable assignments for each peak.

TABLE H.1 Principal Stable Isotopes of Common Elements[a]

ELEMENT	MOST COMMON ISOTOPE		NATURAL ABUNDANCE OF OTHER ISOTOPES (BASED ON 100 ATOMS OF MOST COMMON ISOTOPE)			
Carbon	^{12}C	100	^{13}C	1.08		
Hydrogen	^{1}H	100	^{2}H	0.016		
Nitrogen	^{14}N	100	^{15}N	0.38		
Oxygen	^{16}O	100	^{17}O	0.04	^{18}O	0.20
Fluorine	^{19}F	100				
Sulfur	^{32}S	100	^{33}S	0.78	^{34}S	4.40
Chlorine	^{35}Cl	100	^{37}Cl	32.5		
Bromine	^{79}Br	100	^{81}Br	98.0		
Iodine	^{127}I	100				

[a] Data obtained from R. M. Silverstein, G. C. Bassler, and T. C. Morrill, *Spectrometric Identification of Organic Compounds,* 3rd Ed., Wiley, N.Y. 1974, p. 13.

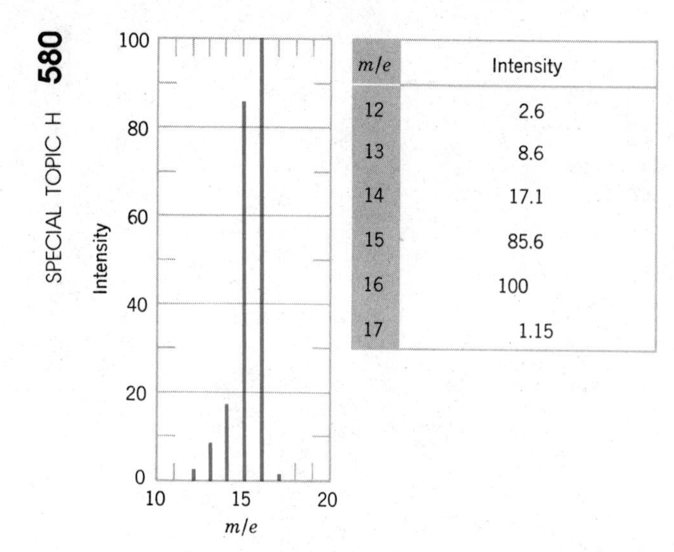

m/e	Intensity
12	2.6
13	8.6
14	17.1
15	85.6
16	100
17	1.15

FIG. H.4
Mass spectrum for Problem H.1.

Problem H.3

The compound whose mass spectrum is given in Fig. H.6 contains three elements, one of which is fluorine. Propose a structure for the compound and make reasonable assignments for each peak.

H.3 DETERMINATION OF MOLECULAR FORMULAS AND MOLECULAR WEIGHTS

H.3A The Molecular Ion and Isotopic Peaks

Look at Table H.1 for a moment. Notice that most of the common elements found in organic compounds have naturally occurring *heavier* isotopes. For three of the elements—carbon, hydrogen, and nitrogen—the principal heavier isotope is one mass unit greater than the most common isotope. The presence of these elements in a compound will give rise to a small isotopic peak one unit greater than the molecular ion—at $M^{\ddagger} + 1$. For four of the elements—oxygen, sulfur, chlorine, and bromine—the principal heavier isotope

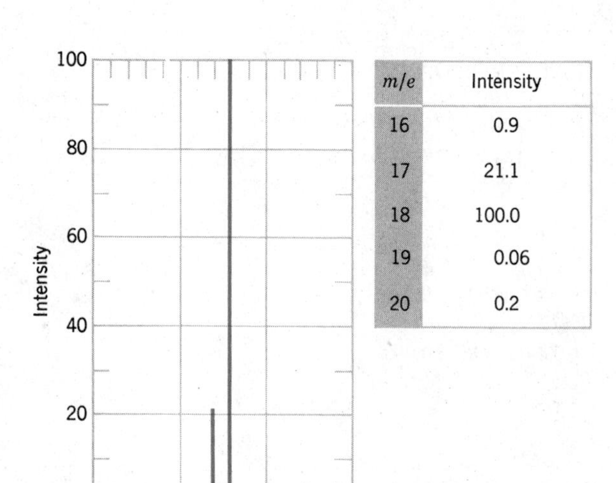

m/e	Intensity
16	0.9
17	21.1
18	100.0
19	0.06
20	0.2

FIG. H.5
Mass spectrum for Problem H.2.

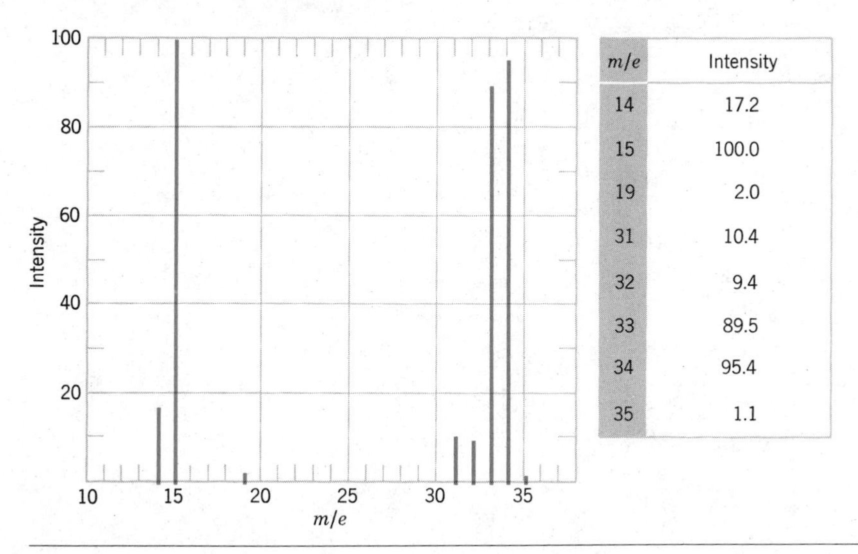

m/e	Intensity
14	17.2
15	100.0
19	2.0
31	10.4
32	9.4
33	89.5
34	95.4
35	1.1

FIG. H.6
Mass spectrum for Problem H.3.

is two mass units greater than the most common isotope. The presence of these elements in a compound give rise to an isotopic peak at $M^{+\cdot} + 2$.

$M^{+\cdot}$ + 1 Elements: C, H, N
$M^{+\cdot}$ + 2 Elements: O, S, Br, Cl

Isotopic peaks give us one method for determining molecular formulas. To understand how this can be done, let us begin by noticing that the isotope abundances in Table H.1 are based on 100 atoms of the normal isotope. Now let us suppose, as an example, that we have 100 molecules of methane, CH_4. On the average there will be 1.08 molecules that contain ^{13}C and 4×0.016 molecules that contain ^{2}H. Altogether then, these heavier isotopes should contribute an $M^{+\cdot} + 1$ peak whose intensity is ~1.14% of the intensity of the molecular ion.

$$1.08 + 4(0.016) \simeq 1.14\%$$

This correlates well with the observed intensity of the $M^{+} + 1$ peak in the actual spectrum of methane given in Fig. H.4.

For molecules with a modest number of atoms we can determine molecular formulas in the following way. If the $M^{+\cdot}$ peak is not the base peak the first thing we do with the mass spectrum of an unknown compound is to recalculate the intensities of the $M^{+\cdot} + 1$ and $M^{+\cdot} + 2$ to express them as percents of the intensity of the $M^{+\cdot}$ peak. Consider, for example, the mass spectrum given in Fig. H.7. The $M^{+\cdot}$ peak at $m/e = 72$ is not the base peak. Therefore, we need to recalculate the intensities of the peaks in our spectrum at m/e 72, 73, and 74 as percents of the peak at m/e 72. We do this by dividing each intensity by the intensity of the $M^{+\cdot}$ peak (which is 73%) and multiply by 100. These results are shown below and in the second column of Fig. H.7.

m/e		Intensity % of $M^{+\cdot}$
72	$73.0/73 \times 100 =$	100
73	$3.3/73 \times 100 =$	4.5
74	$0.3/73 \times 100 =$	0.3

Then we use the following guides to determine the molecular formula.

m/e	Intensity (as per cent of base peak)		m/e	Intensity (as per cent of $M^{\ddagger}$)
27	59.0		72 $M^{\ddagger}$	100.0
28	15.0		73 $M^{\ddagger}+1$	4.5
29	54.0		74 $M^{\ddagger}+2$	0.3
39	23.0			
41	60.0		Recalculated to base on $M^{\ddagger}$	
42	12.0			
43	79.0			
44	100.0 (base)			
72	73.0	$M^{\ddagger}$		
73	3.3			
74	0.2			

FIG. H.7
Mass spectrum of an unknown compound.

1. Is $M^{\ddagger}$ odd or even? According to the nitrogen rule, if it is even, then the compound must contain an even number of nitrogen atoms (zero is an even number).

For our unknown $M^{\ddagger}$ is even. The compound must have an even number of nitrogens.

2. The relative abundance of the $M^{\ddagger}+1$ peak indicates the number of carbon atoms.

Number of C atoms = Relative Abundance of $(M^{\ddagger}+1)/1.1$

For our unknown (Fig. H.7), Number of C atoms $= \frac{4.5}{1.1} = 4$

(This formula works because ^{13}C is the most important contributor to the $M^{\ddagger}$ peak and the approximate natural abundance of ^{13}C is 1.1%.)

3. The relative abundance of the $M^{\ddagger}+2$ peak indicates the presence (or absence) of S, (4.4%); Cl, (33%); or Br (100%) (see Table H.1).

For our unknown, $M^{\ddagger}+2 = 0.2\%$; thus we can assume that S, Cl, and Br are absent.

4. The molecular formula can now be established by determining the number of hydrogen atoms and adding the appropriate number of oxygens, if necessary.

For our unknown the $M^{\ddagger}$ peak at m/e 72 gives us the molecular weight. It also tells us (since it is even) that nitrogen is absent because C_4N_2 has a molecular weight (76) greater than that of our compound.

For a molecule composed of C and H only:

$H = 72 - (4 \times 12) = 24$, but C_4H_{24} is impossible.

For a molecule composed of C, H, and one O:

$H = 72 - (4 \times 12) - 16 = 8$ and thus our unknown has the molecular formula C_4H_8O.

Problem H.4

(a) Write structural formulas for at least 14 stable compounds that have the formula C_4H_8O. (b) The infrared spectrum of the unknown compound shows a strong peak near 1730 cm^{-1}. Which structures now remain as possible formulas for the compound? (We continue with this compound in Problem H.14.)

Problem H.5

Determine the molecular formula of the following compound. (The complete mass spectrum of this compound is given in Fig. H.18, cf. Problem H.19.)

m/e	INTENSITY (AS PERCENT OF BASE PEAK)
86 $M^{+\cdot}$	10.00
87	0.56
88	0.04

Problem H.6

(a) What approximate intensities would you expect for the $M^{+\cdot}$ and $M^{+\cdot} + 2$ peaks of CH_3Cl? (b) For the $M^{+\cdot}$ and $M^{+\cdot} + 2$ peaks of CH_3Br? (c) An organic compound gives an $M^{+\cdot}$ peak at m/e 122 and a peak of nearly equal intensity at m/e 124. What is a likely molecular formula for the compound?

Problem H.7

Use the mass spectral data given in Fig. H.8 to determine the molecular formula for the compound.

m/e	INTENSITY (AS PERCENT OF BASE PEAK)
14	8.0
15	38.6
18	16.3
28	39.7
29	23.4
42	46.6
43	10.7
44	100.0 (base)
73	86.1 $M^{+\cdot}$
74	3.2
75	0.2

FIG. H.8
Mass Spectrum for Problem H.7

Problem H.8

(a) Determine the molecular formula of the compound whose mass spectrum is given below. (b) The proton nmr spectrum of this compound consists only of a large doublet and a small septet. What is the structure of the compound?

m/e	INTENSITY (AS PERCENT OF BASE PEAK)
27	34
39	11
41	22
43	100 (base)
63	26
65	8
78	24 $M^{+\cdot}$
79	0.8
80	8

TABLE H.2 Relative Intensities of $M^{\dot{+}} + 1$ and $M^{\dot{+}} + 2$ Peaks for Various Combinations of C, H, N, and O for Mass 72 and 73

		PERCENT OF $M^{\dot{+}}$ INTENSITY				PERCENT OF $M^{\dot{+}}$ INTENSITY	
$M^{\dot{+}}$	FORMULAS	$M^{\dot{+}} + 1$	$M^{\dot{+}} + 2$	$M^{\dot{+}}$	FORMULAS	$M^{\dot{+}} + 1$	$M^{\dot{+}} + 2$
72	CH_2N_3O	2.30	0.22	73	CHN_2O_2	1.94	0.41
	CH_4N_4	2.67	0.03		CH_3N_3O	2.31	0.22
	$C_2H_2NO_2$	2.65	0.42		CH_5N_4	2.69	0.03
	$C_2H_4N_2O$	3.03	0.23		C_2HO_3	2.30	0.62
	$C_2H_6N_3$	3.40	0.04		$C_2H_3NO_2$	2.67	0.42
	$C_3H_4O_2$	3.38	0.44		$C_2H_5N_2O$	3.04	0.23
	C_3H_6NO	3.76	0.25		$C_2H_7N_3$	3.42	0.04
	$C_3H_8N_2$	4.13	0.07		$C_3H_5O_2$	3.40	0.44
	C_4H_8O	4.49	0.28		C_3H_7NO	3.77	0.25
	$C_4H_{10}N$	4.86	0.09		$C_3H_9N_2$	4.15	0.07
	C_5H_{12}	5.60	0.13		C_4H_9O	4.51	0.28
					$C_4H_{11}N$	4.88	0.10
					C_6H	6.50	0.18

Data from J. H. Beynon, *Mass Spectrometry and Its Application to Organic Chemistry,* Elsevier, Amsterdam, 1960.

As the number of atoms in a molecule increases, calculations like this become more and more complex and time consuming. Fortunately, however, these calculations can be done readily with computers, and tables are now available that give relative values for the $M^{\dot{+}} + 1$ and $M^{\dot{+}} + 2$ peaks for all combinations of common elements with molecular formulas up to mass 500. Part of the data obtained from one of these tables is given in Table H.2. Use Table H.2 to check the results of our example (Fig. H.7) and your answer to Problem H.8.

H.3B High-Resolution Mass Spectroscopy

All of the spectra that we have described so far were determined on what are called "low-resolution" mass spectrometers. These spectrometers, as we noted earlier, measure m/e values to the nearest whole-number mass unit. Most laboratories are equipped with this type of mass spectrometer.

Many laboratories, however, are equipped with the more expensive "high-resolution" mass spectrometers. These spectrometers can measure m/e values to three or four decimal places and thus they provide an extremely accurate method for determining molecular weights. And, because molecular weights can be measured so accurately, these spectrometers also allow us to determine molecular formulas.

The determination of a molecular formula by an accurate measurement of a molecular weight is possible because the actual masses of atomic particles (nuclides) are not integers (see Table H.3). Consider, as examples, the three molecules, O_2, N_2H_4, and CH_3OH. The actual atomic masses of the molecules are all different.

$$^{16}O_2 = 2(15.9949) = 31.9898$$
$$N_2H_4 = 2(14.0031) + 4(1.00783) = 32.0375$$
$$CH_4O = 12.0000 + 4(1.00783) + 15.9949 = 32.0262$$

TABLE H.3 Exact Masses of Nuclides

ISOTOPE	MASS
^{1}H	1.00783
^{2}H	2.01410
^{12}C	12.00000 (std)
^{13}C	13.00336
^{14}N	14.0031
^{15}N	15.0001
^{16}O	15.9949
^{17}O	16.9991
^{18}O	17.9992
^{19}F	18.9984
^{32}S	31.9721
^{33}S	32.9715
^{34}S	33.9679
^{35}Cl	34.9689
^{37}Cl	36.9659
^{79}Br	78.9183
^{81}Br	80.9163
^{127}I	126.9045

High-resolution mass spectrometers are available that are capable of measuring mass with an accuracy of 1 part in 40,000. Thus, such a spectrometer can easily distinguish between these three molecules and tell us the molecular formula.

H.4 FRAGMENTATION

In most instances the molecular ion is a highly energetic species and in the case of a complex molecule a great many things can happen to it. The molecular ion can break apart in a variety of ways and the fragments that are produced can then undergo further fragmentation and so on. In a certain sense mass spectroscopy is a "brute force" technique. Striking an organic molecule with 70 eV electrons is a little like firing a howitzer at a house made of matchsticks. That fragmentation takes place in any sort of predictable way is truly remarkable—and yet it does. Many of the same factors that govern ordinary chemical reactions seem to apply to fragmentation processes and many of the principles that we have learned about the relative stabilities of carbocations, free radicals, and molecules will help us to make some sense out of what takes place. And, as we learn something about what kind of fragmentations to expect, we will be much better able to use mass spectra as aids in determining the structures of organic molecules.

We cannot, of course, in the limited space that we have here, look at these processes in great detail, but we can examine some of the more important ones.

As we begin, keep two important principles in mind. (1) The reactions that take place in a mass spectrometer are usually *unimolecular*—that is, they involve only a *single* molecular fragment. This is true because the pressure in a mass spectrometer is kept so low ($\sim 10^{-6}$ torr) that reactions requiring bimolecular collisions usually do not occur. (2) The relative ion abundances, as measured by ion intensities, are extremely important. We will see that the appearance of certain prominent peaks in the spectrum gives us important information about the structures of the fragments produced and about their original locations in the molecule.

H.4A Fragmentation by Cleavage of a Single Bond

One important type of fragmentation is the simple cleavage of a single bond. With a radical cation this cleavage can take place in at least two ways; each way produces a *cation* and a *free radical.* With the molecular ion obtained from propane, for example, the two possibilities are:

$$CH_3CH_2\overset{+}{\cdot}CH_3 \longrightarrow \underset{m/e\ 29}{CH_3CH_2^+} + \cdot CH_3$$

$$CH_3CH_2\overset{+}{\cdot}CH_3 \longrightarrow CH_3CH_2\cdot + \underset{m/e\ 15}{{}^+CH_3}$$

These two modes of cleavage do not take place at equal rates, however. While the relative abundance of ions produced by such a cleavage is influenced both by the stability of the carbocation and by the stability of the free radical, *the carbocation's stability is more important.** In the spectrum of propane the peak at m/e 29 ($CH_3CH_2^+$) is the most intense peak; the peak at m/e 15 (CH_3^+) has an intensity of only 5.6%. This reflects the greater stability of $CH_3CH_2^+$ when compared to CH_3^+.

H.4B Fragmentation Equations

Before we go further, we need to examine some of the conventions that are used in writing equations for fragmentation reactions. In the two equations for cleavage of the single bond of propane that we have just written, we have localized the odd electron and the charge on one of the carbon-carbon sigma bonds of the molecular ion. When we write structures this way, the choice of just where to localize the odd electron and the charge is sometimes arbitrary. When possible, however, we write the structure showing the molecular ion that would result from the removal of one of the most loosely held electrons of the original molecule. Just which electrons these are can usually be estimated from ionization potentials (Table H.4). [The ionization potential of a molecule is the amount of energy (in eV) required to remove an electron from the molecule.] As we might expect, ionizaion potentials indicate that the nonbonding electrons of nitrogen and oxygen and the pi electrons of alkenes and aromatic molecules are held more loosely than the electrons of carbon-carbon and carbon-hydrogen sigma bonds. Thus the convention of localizing the odd electron and charge is especially applicable when the molecule contains an oxygen, nitrogen, double bond, or aromatic ring. If the molecule contains only carbon-carbon and carbon-hydrogen sigma bonds, and if it contains a great many of these, then the choice of where to localize the odd electron and the charge is so arbitrary as to be impractical. In these instances we usually resort to another convention: we write the formula for the radical cation in brackets and place the odd electron and charge outside. Using this convention we would write the two fragmentation reactions of propane in the following way.

$$[CH_3CH_2CH_3]^{+\cdot} \longrightarrow CH_3CH_2^+ + \cdot CH_3$$

$$[CH_3CH_2CH_3]^{+\cdot} \longrightarrow CH_3CH_2\cdot + CH_3^+$$

* This can be demonstrated through thermochemical calculations that we cannot go into here. The interested student is referred to F. W. McLafferty, *Interpretation of Mass Spectra,* 2nd Ed., W. A. Benjamin, Reading, Mass., 1973, p. 41 and pp. 210–211.

TABLE H.4 Ionization Potentials of Selected Molecules

COMPOUND	IONIZATION POTENTIAL (ELECTRON VOLTS)
$CH_3(CH_2)_3NH_2$	8.7
C_6H_6	9.2
C_2H_4	10.5
CH_3OH	10.8
C_2H_6	11.5
CH_4	12.7

Problem H.9

The most intense peak in the mass spectrum of 2,2-dimethylbutane occurs at m/e 57. (a) What carbocation does this peak represent? (b) Using the convention that we have just described, write an equation that shows how this carbocation arises from the molecular ion.

Figure H.9 shows us the kind of fragmentation a longer-chain alkane can undergo. The example here is n-hexane and we see a reasonably intense molecular ion at m/e 86 accompanied by a small $M^{+\cdot} + 1$ peak. There is also a smaller peak at m/e 71 ($M^{+\cdot} - 15$) corresponding to the loss of $\cdot CH_3$ and the base peak is at m/e 57 ($M^{+\cdot} - 29$) corresponding to the loss of $\cdot CH_2CH_3$. The other prominent peaks are at m/e 43 ($M^{+\cdot} - 43$) and m/e 29 ($M^{+\cdot} - 57$) corresponding to the loss of $\cdot CH_2CH_2CH_3$ and $\cdot CH_2CH_2CH_2CH_3$, respectively. The important fragmentations are just the ones we would expect:

$$[CH_3CH_2CH_2CH_2CH_2CH_3]^{+\cdot} \longrightarrow \begin{cases} \underset{m/e\ 71}{CH_3CH_2CH_2CH_2CH_2^+} + \cdot CH_3 \\ \underset{m/e\ 57}{CH_3CH_2CH_2CH_2^+} + \cdot CH_2CH_3 \\ \underset{m/e\ 43}{CH_3CH_2CH_2^+} + \cdot CH_2CH_2CH_3 \\ \underset{m/e\ 29}{CH_3CH_2^+} + \cdot CH_2CH_2CH_2CH_3 \end{cases}$$

Chain branching increases the likelihood of cleavage at a branch point because a more stable carbocation can result. When we compare the mass spectrum of 2-methylbutane (Fig. H.10) with the spectrum of n-hexane we see a much more intense peak at $M^{+\cdot} - 15$. Loss of a methyl radical from 2-methylbutane can give a secondary carbocation:

$$\underset{\substack{m/e\ 72 \\ M^{+\cdot}}}{\left[\begin{matrix} CH_3 \\ | \\ CH_3CHCH_2CH_3 \end{matrix}\right]^{+\cdot}} \longrightarrow \underset{\substack{m/e\ 57 \\ M^{+\cdot} - 15}}{CH_3\overset{+}{C}HCH_2CH_3} + \cdot CH_3$$

FIG. H.9
Mass spectrum of n-*hexane.*

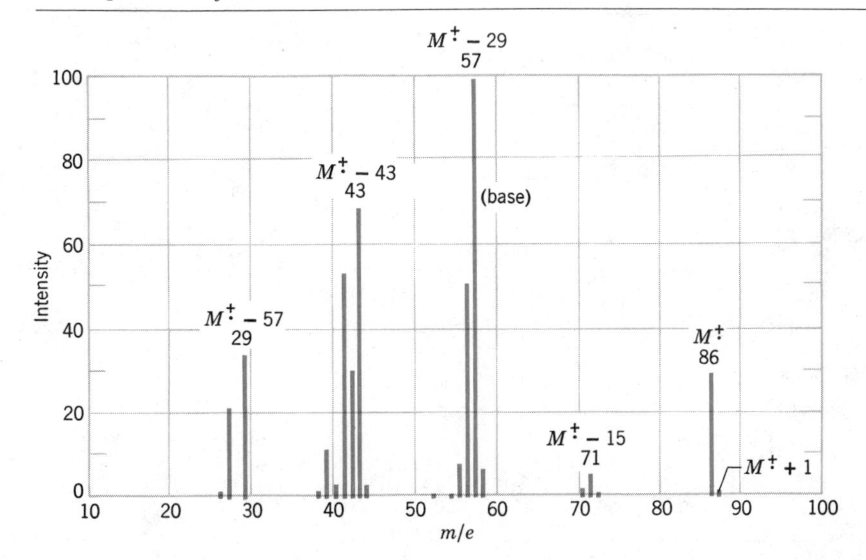

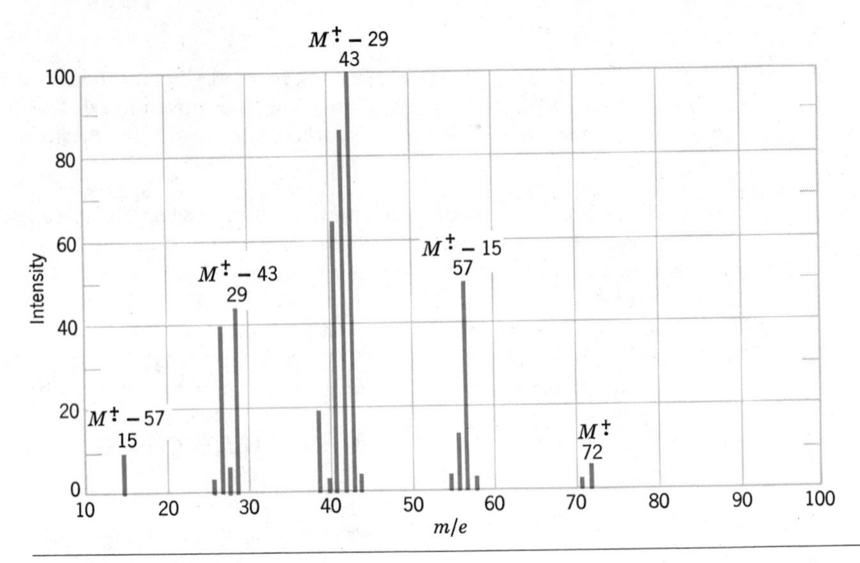

FIG. H.10
The mass spectrum of 2-methylbutane.

whereas with *n*-hexane, loss of a methyl radical can yield only a primary carbocation.

With 2,2-dimethylpropane (Fig. H.11), this effect is even more dramatic. Loss of a methyl radical by the molecular ion produces a *tertiary* carbocation and this reaction takes place so readily that virtually none of the molecular ions survive long enough to be detected.

$$\left[CH_3{-}\overset{\overset{\displaystyle CH_3}{|}}{\underset{\underset{\displaystyle CH_3}{|}}{C}}{-}CH_3 \right]^{+\cdot} \longrightarrow CH_3{-}\overset{\overset{\displaystyle CH_3}{|}}{\underset{\underset{\displaystyle CH_3}{|}}{C^+}} + \cdot CH_3$$

m/e 72 — $M^{+\cdot}$; m/e 57 — $M^{+\cdot} - 15$

FIG. H.11
Mass spectrum of 2,2-dimethylpropane.

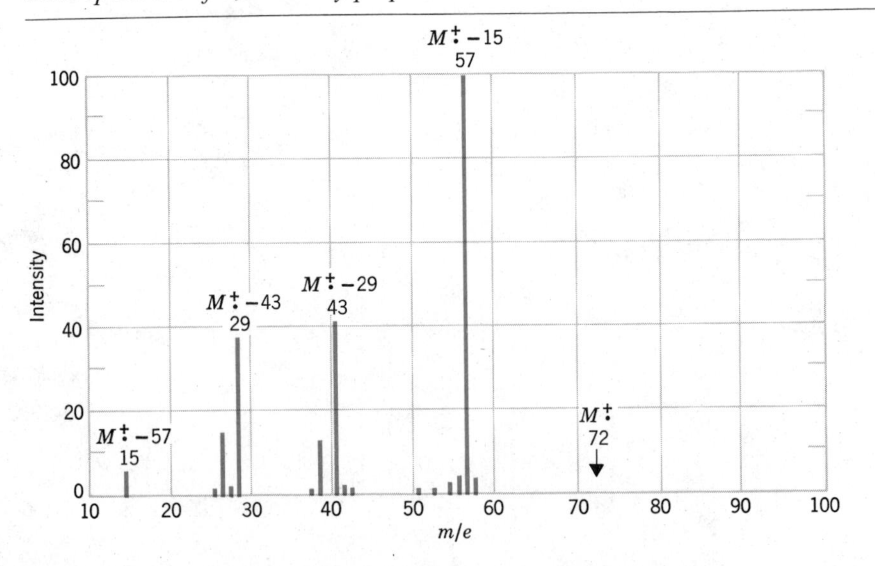

Problem H.10

In contrast to 2-methylbutane and 2,2-dimethylpropane, the mass spectrum of 3-methylpentane (not given) has a peak of very low intensity at $M^{+\cdot} - 15$. It has a peak of very high intensity at $M^{+\cdot} - 29$, however. Explain.

Carbocations stabilized by resonance are usually also prominent in mass spectra. Several ways that resonance-stabilized carbocations can be produced are outlined below.

1. Alkenes frequently undergo fragmentations that yield allylic cations.

$$CH_2\overset{+\cdot}{-}CH-CH_2\!:\!R \longrightarrow \overset{+}{C}H_2-CH=CH_2 + \cdot R$$

m/e 41

2. Carbon-carbon bonds next to an atom with an unshared electron pair usually break readily because the resulting carbocation is resonance stabilized.

$$R-\overset{+\cdot}{Z}-CH_2\!:\!CH_3 \longrightarrow R-\overset{+}{Z}=CH_2 + \cdot CH_3$$

$$\updownarrow$$

$$R-\ddot{Z}-\overset{+}{C}H_2$$

Z = N, O, or S; R may also be H.

3. Carbon-carbon bonds next to the carbonyl group of an aldehyde or ketone break readily because resonance-stabilized ions called acylium ions are produced.

$$\begin{matrix}R\\ \quad C=\overset{\cdot +}{O}: \\ R'\end{matrix} \longrightarrow R'-C\equiv\overset{+}{O}: + R\cdot$$

$$\updownarrow$$

$$R'-\overset{+}{C}=\ddot{O}:$$

or Acylium ion

$$\begin{matrix}R\\ \quad C=\overset{\cdot +}{O}: \\ R'\end{matrix} \longrightarrow R-C\equiv\overset{+}{O}: + R'\cdot$$

$$\updownarrow$$

$$R-\overset{+}{C}=\ddot{O}:$$

Acylium ion

4. Alkyl-substituted benzenes undergo loss of a hydrogen atom or methyl group to yield the relatively stable tropylium ion (cf. Section 11.6). This fragmentation gives a prominent peak (sometimes the base peak) at *m/e* 91.

$$C_6H_5CH_3^{+\cdot} \xrightarrow{-H\cdot} C_7H_7^{+}$$

m/e 91

$$CH_3C_6H_4CH_3^{+\cdot} \xrightarrow{-CH_3\cdot} C_7H_7^{+}$$

m/e 91

5. Substituted benzenes also lose their substituent and yield a phenyl cation at m/e 77.

$$\text{C}_6\text{H}_5\text{—Y}^{+\cdot} \xrightarrow{-\text{Y}\cdot} \text{C}_6\text{H}_5^{+}$$

m/e 77

Y = halogen, $-NO_2$, $—\overset{\text{O}}{\overset{\|}{\text{C}}}\text{R}$, —R, etc.

Problem H.11

The mass spectrum of 4-methyl-1-hexene (not given) shows intense peaks at m/e 57 and m/e 41. What fragmentation reactions account for these peaks?

Problem H.12

Explain the following observations that can be made about the mass spectra of alcohols: (a) The molecular ion peak of a primary or secondary alcohol is very small; with a tertiary alcohol it is usually undetectable. (b) Primary alcohols show a prominent peak at m/e 31. (c) Secondary alcohols usually give prominent peaks at m/e 45, 59, 73 and so forth. (d) Tertiary alcohols have prominent peaks at m/e 59, 73, 87 and so forth.

Problem H.13

The mass spectra of isopropyl butyl ether and propyl butyl ether are given in Figs. H.12 and H.13. (a) Which spectrum represents which ether? (b) Explain your choice.

H.4C Fragmentation by Cleavage of Two Bonds

Many peaks in mass spectra can be explained by fragmentation reactions that involve the breaking of two covalent bonds. When a radical-cation undergoes this type of fragmentation the products are *a new radical-cation* and *a neutral molecule*. Some important examples are the following.

FIG. H.12
Mass spectrum for Problem H.13.

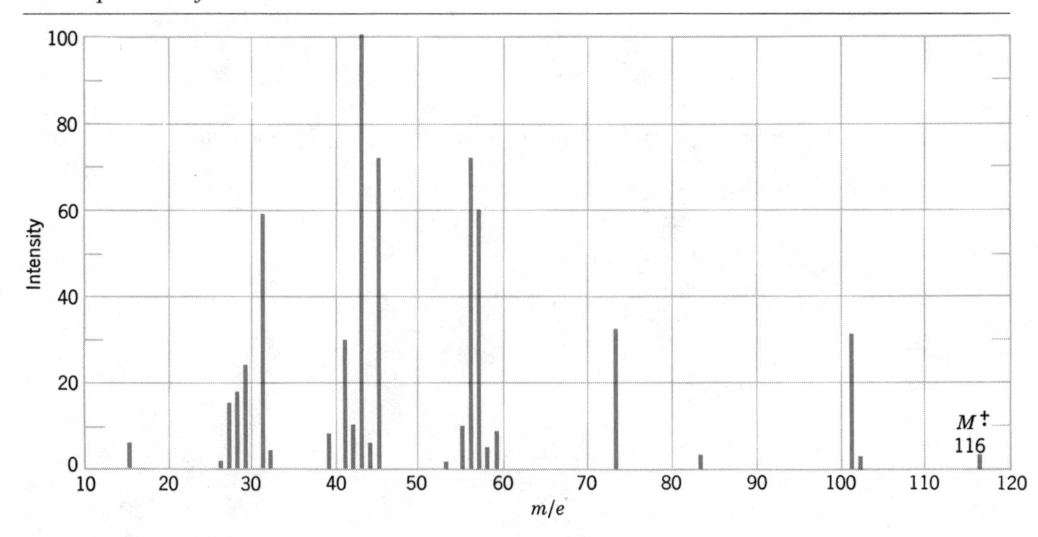

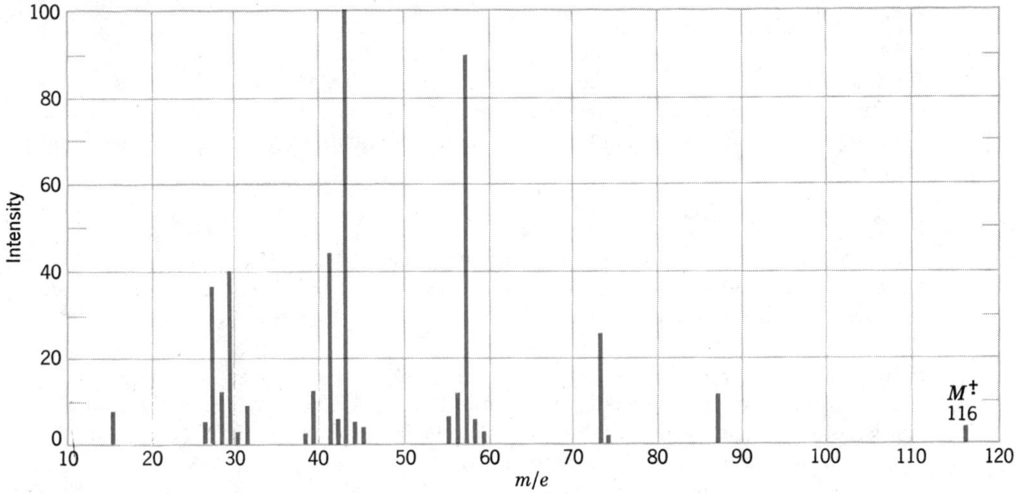

FIG. H.13
Mass spectrum for Problem H.13.

1. Alcohols frequently show a prominent peak at $M^{+\cdot} - 18$. This corresponds to the loss of a molecule of water.

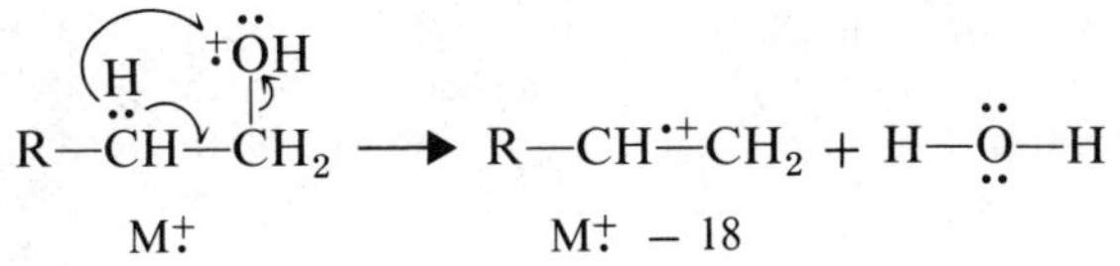

$$R-\ddot{C}H-CH_2(\overset{+}{\ddot{O}}H)(H) \longrightarrow R-CH\overset{\cdot +}{-}CH_2 + H-\ddot{\underset{\cdot\cdot}{O}}-H$$

$M^{+\cdot}$ → $M^{+\cdot} - 18$

or

$$[R-CH_2-CH_2-OH]^{+\cdot} \longrightarrow [R-CH=CH_2]^{+\cdot} + H_2O$$

$M^{+\cdot}$ → $M^{+\cdot} - 18$

2. Cycloalkenes can undergo a retro-Diels-Alder reaction that produces an alkene and an alkadienyl radical cation.

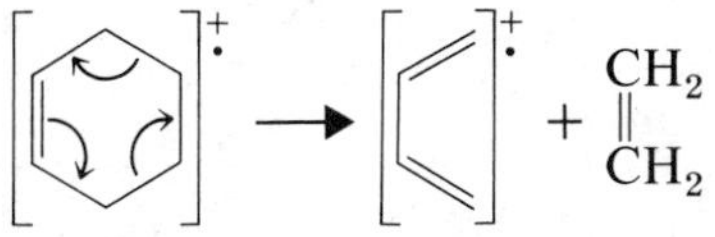

$$[\text{cyclohexene}]^{+\cdot} \longrightarrow [\text{butadiene}]^{+\cdot} + CH_2{=}CH_2$$

3. Carbonyl compounds with a hydrogen on their γ-carbon undergo a fragmentation called the *McLafferty rearrangement*.

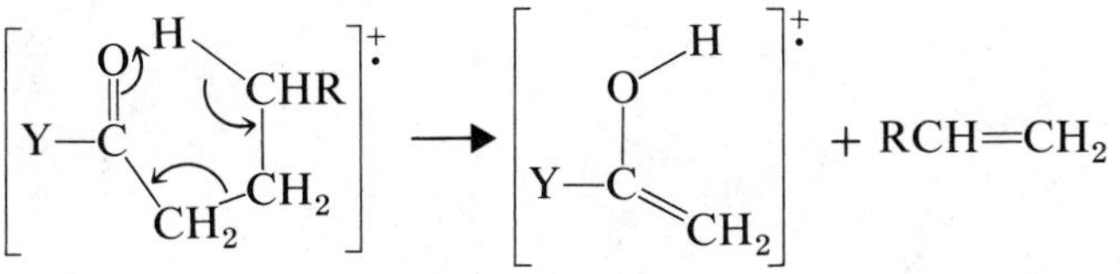

$$[Y-C(=O)-CH_2-CH_2-CHR-H]^{+\cdot} \longrightarrow [Y-C(-O-H)=CH_2]^{+\cdot} + RCH=CH_2$$

Y may be R, H, OR, OH, etc.

In addition to these reactions, we frequently find peaks in mass spectra that result from the elimination of other small stable neutral molecules, for example, H_2, NH_3, CO, HCN, H_2S, alcohols, and alkenes.

Additional Problems

H.14

Reconsider Problem H.4 and the spectrum given in Fig. H.7. Important clues to the structure of this compound are the peaks at m/e 44 (the base peak) and m/e 29. Propose a structure for the compound and write fragmentation equations showing how these peaks arise.

H.15

The homologous series of primary amines, $CH_3(CH_2)_nNH_2$, from CH_3NH_2 to $CH_3(CH_2)_{13}NH_2$ all have their base (largest) peak at m/e 30. What ion does this peak represent and how is it formed?

H.16

The mass spectrum of compound **A** is given in Fig. H.14. The proton nmr spectrum of **A** consists of two singlets with area ratios of 9 : 2. The larger singlet is at $\delta 1.2$, the smaller one at $\delta 1.3$. Propose a structure for compound **A.**

H.17

The mass spectrum of compound **B** is given in Fig. H.15. The infrared spectrum of **B** shows a broad peak between 3200–3550 cm^{-1}. The proton nmr spectrum of **B** shows the following peaks: a triplet at $\delta 0.9$, a singlet at $\delta 1.1$, and a quartet at $\delta 1.6$. The area ratios of these peaks is 3 : 7 : 2 respectively. Propose a structure for **B.**

H.18

The mass spectrum of compound **C** is given in Fig. H.16. Compound **C** is an isomer of **B** and the infrared spectrum of **C** also shows a broad peak in the 3200–3550 region. The proton nmr spectrum of **C** is given in Fig. H.17. Propose a structure for **C.**

H.19

The mass spectrum of compound **D** is given in Fig. H.18. (**D** is also the subject of problem H.5) **D** shows a strong infrared peak at 1710 cm^{-1}. The proton nmr spectrum of **D** is given in Fig. H.19. Propose a structure for **D.**

H.20

Propose a structure for compound **E** whose mass spectrum is given in Fig. H.20.

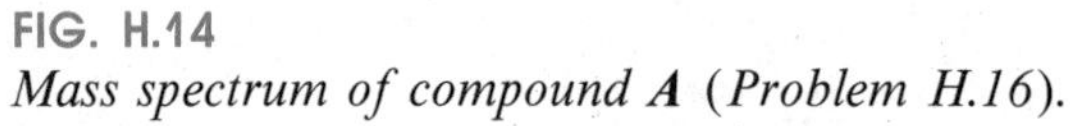

FIG. H.14
*Mass spectrum of compound **A** (Problem H.16).*

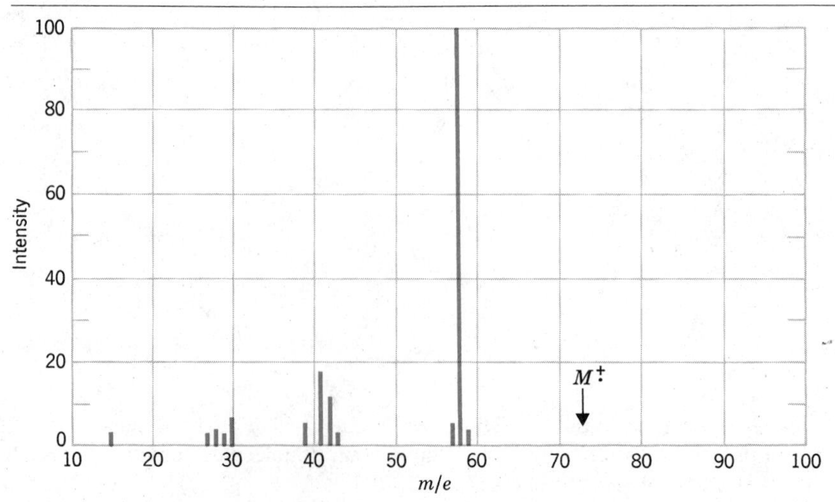

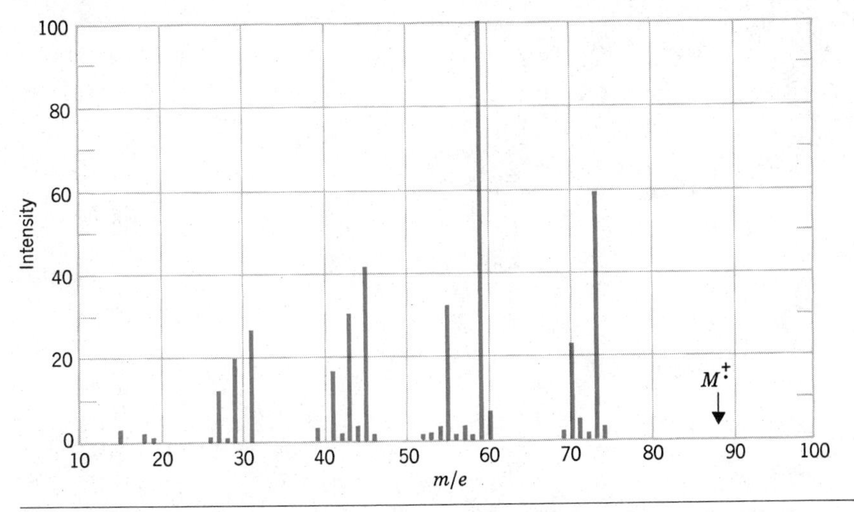

FIG. H.15
*Mass spectrum of compound **B** (Problem H.17).*

H.21

Propose a structure for compound **F** whose mass spectrum is given in Fig. H.21 and whose proton nmr spectrum is given in Fig. H.22.

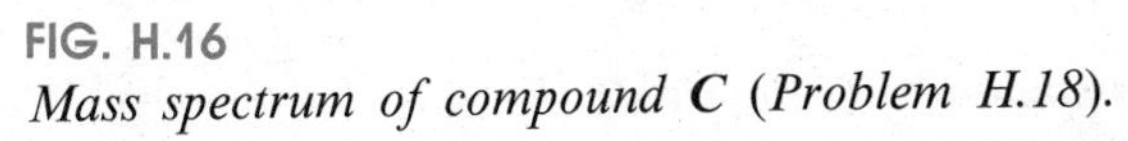

FIG. H.16
*Mass spectrum of compound **C** (Problem H.18).*

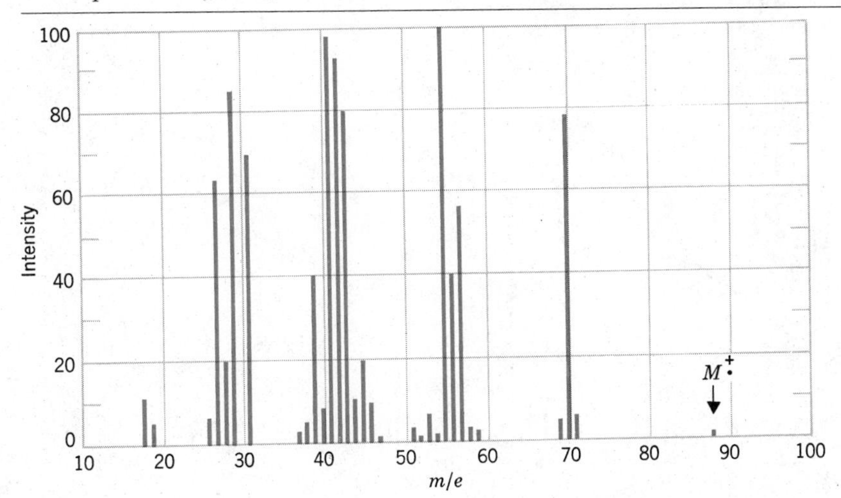

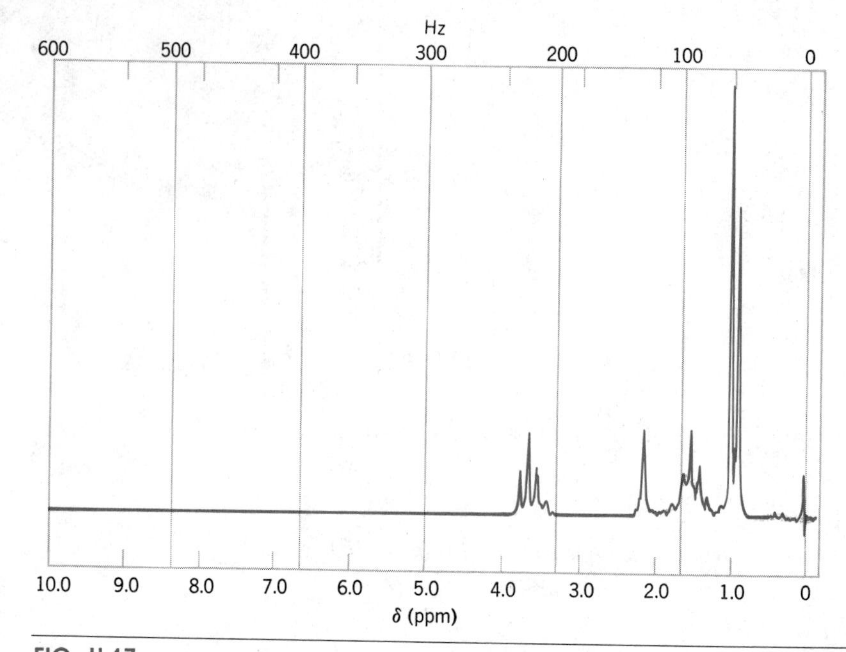

FIG. H.17
The proton nmr spectrum of compound **C** *(Problem H.18). (Courtesy of Aldrich Chemical Co., Milwaukee, Wis.)*

FIG. H.18
The mass spectrum of compound **D** *(Problem H.19).*

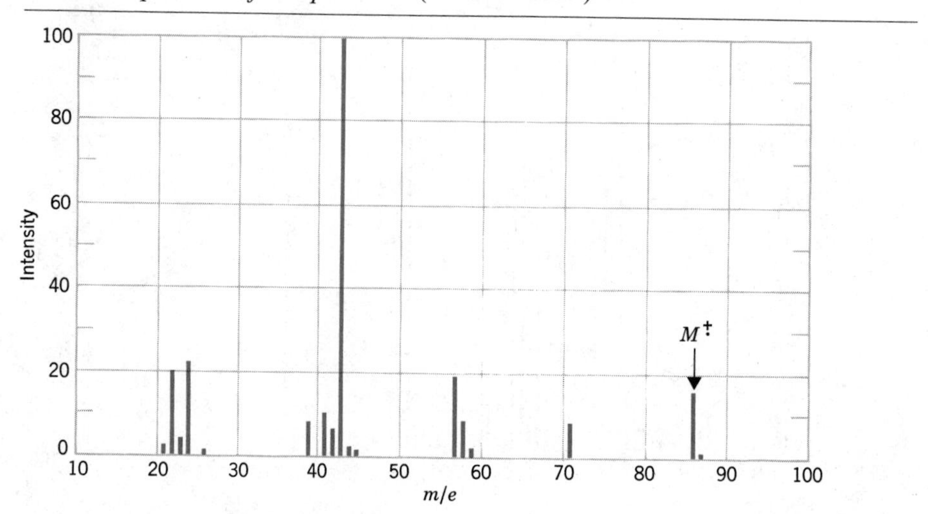

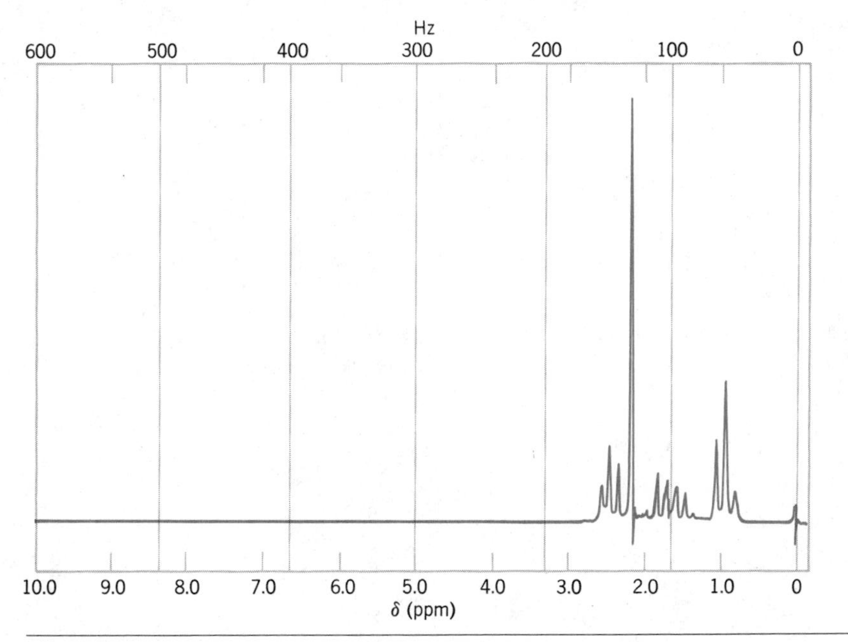

FIG. H.19
The proton nmr spectrum of compound **D** *(Problem H.19). (Courtesy of Aldrich Chemical Co., Milwaukee, Wis.)*

FIG. H.20
The mass spectrum of compound **E** *(Problem H.20).*

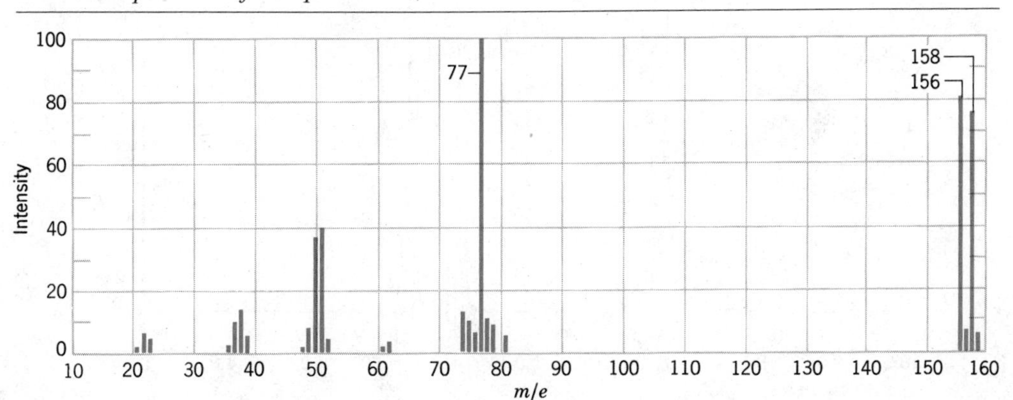

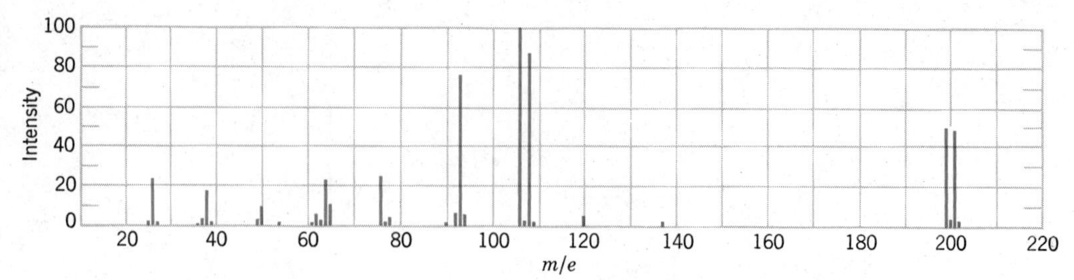

FIG. H.21
*The mass spectrum of compound **F** (Problem H.21). (Adapted with permission from E. Stenhagen, S. Abrahamsson, and F. W. McLafferty,* Registry of Mass Spectra Data, *Vol. II, Wiley, New York, 1974, p. 992.)*

FIG. H.22
*The proton nmr spectrum of compound **F** (Problem H.21). (Courtesy of Aldrich Chemical Co., Milwaukee, Wis.)*

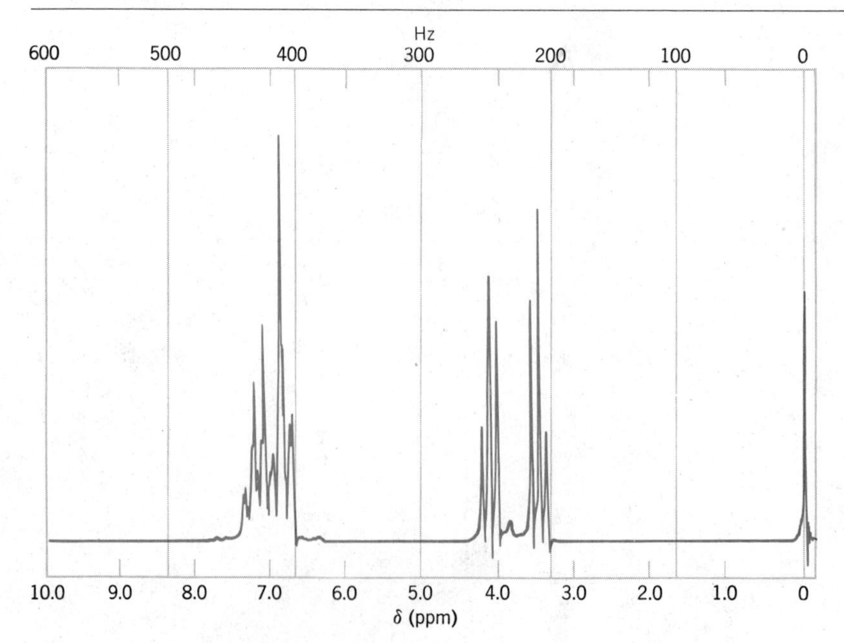

14 ORGANIC HALIDES AND ORGANOMETALLIC COMPOUNDS

14.1 INTRODUCTION

We have already encountered organic halogen compounds at many points in this text and we will see them again at many others. The reason for this is simple. Halogen substituents, particularly chlorine and bromine, can be introduced into organic compounds by a variety of methods and they can usually be easily replaced by other groups. Organic halogen compounds, therefore, are used frequently in synthetic procedures.

Halogen substituents on saturated carbon atoms can be replaced in nucleophilic substitution reactions. We saw our first examples of this type of reaction in Chapter 5.

$$\text{Nu:}^- + \text{R—X} \longrightarrow \text{R—Nu} + \text{X}^-$$

Specific examples:

$$\text{OH}^- + \text{CH}_3\text{Cl} \xrightarrow{S_N2} \text{CH}_3\text{OH} + \text{Cl}^-$$

$$2\text{H}_2\text{O} + (\text{CH}_3)_3\text{CCl} \xrightarrow{S_N1} (\text{CH}_3)_3\text{COH} + \text{H}_3\text{O}^+ + \text{Cl}^-$$

In Chapter 9 we saw how nucleophilic substitution reactions (S_N2) of primary alkyl halides could be used to synthesize alkynes:

$$\text{R—C}\equiv\text{C:}^-\text{Na}^+ + \text{R}'\text{—CH}_2\text{—X} \longrightarrow \text{R—C}\equiv\text{C—CH}_2\text{—R}' + \text{NaX}$$

Halogen substituents also give us the possibility of carrying out elimination reactions and thus provide a method for introducing carbon-carbon multiple bonds. We saw examples of this synthetic application in the preparations of alkenes and alkynes (Sections 5.9–5.12, 6.9, and 9.14).

$$\text{R—CH}_2\text{—}\underset{\text{X}}{\underset{|}{\text{CH}}}\text{—R}' + \text{NaOR}'' \longrightarrow \text{R—CH=CH—R}' + \text{R}''\text{OH} + \text{NaX}$$

$$\text{R—}\underset{\text{X}}{\underset{|}{\text{CH}}}\text{—}\underset{\text{X}}{\underset{|}{\text{CH}}}\text{—R}' + 2\text{NaNH}_2 \longrightarrow \text{R—C}\equiv\text{C—R}' + 2\text{NaX} + 2\text{NH}_3$$

Halogen substituents also allow us to prepare compounds in which organic groups are bonded to metals. *Organometallic compounds,* as we will see later in this chapter, are of considerable synthetic importance because the metal portion of many organometallic compounds can be replaced by a variety of other groups.

Organometallic compounds also give us valuable synthetic methods for creating new carbon-carbon single bonds.

Organometallic compounds are almost always prepared from organic halides, and thus it is convenient to describe organometallic compounds in this chapter as well. As we do this, we will focus our attention particularly on the organic compounds of magnesium and lithium.

Problem 14.1

Starting with an appropriate alkyl halide and any other necessary reagents show how you might synthesize each of the following.

(a) *S*-2-butanol, that is, *S*-$CH_3CH(OH)CH_2CH_3$

(b) CH_3–(cyclohexane ring)–OH

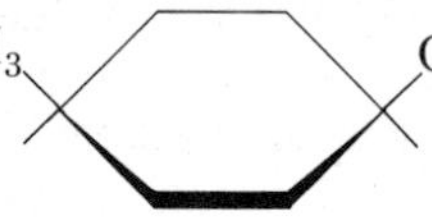

(c) $C_6H_5CH{=}CHCH_3$
(d) $C_6H_5C{\equiv}CH$
(e) $C_6H_5CH_2C{\equiv}CH$
(f) $CH_3CH_2CH_2CH{=}CH_2$

14.2 ALLYLIC AND BENZYLIC HALIDES IN NUCLEOPHILIC SUBSTITUTION REACTIONS

Allylic and benzylic halides can be classified in the same way that we have classified other organic halides:

$-\overset{\vert}{C}{=}\overset{\vert}{C}-CH_2X$	$-\overset{\vert}{C}{=}\overset{\vert}{C}-\overset{R}{\overset{\vert}{C}}HX$	$-\overset{\vert}{C}{=}\overset{\vert}{C}-\underset{R'}{\underset{\vert}{\overset{R}{\overset{\vert}{C}}}}X$
1° Allylic	2° Allylic	3° Allylic

$ArCH_2X$	$Ar\overset{R}{\overset{\vert}{C}}HX$	$Ar\underset{R'}{\underset{\vert}{\overset{R}{\overset{\vert}{C}}}}X$
1° Benzylic	2° Benzylic	3° Benzylic

All of these compounds undergo nucleophilic substitution reactions. As with other tertiary halides (Section 5.7A), the steric hindrance associated with having three bulky groups on the carbon bearing the halogen prevents tertiary allylic and tertiary benzylic halides from reacting by an S_N2 mechanism. They react with nucleophiles only by an S_N1 mechanism.

Primary and secondary allylic and benzylic halides can react either by an S_N2 mechanism or by an S_N1 mechanism in ordinary nonacidic solvents. We would expect these halides to react by an S_N2 mechanism because they are structurally similar to primary and secondary alkyl halides. (Having only one or two groups attached to the carbon bearing the halogen does not prevent S_N2 attack). But primary and secondary allylic and benzylic halides can also react by an S_N1 mechanism because they can form relatively stable carbocations and in this regard they differ from primary and secondary alkyl halides.*

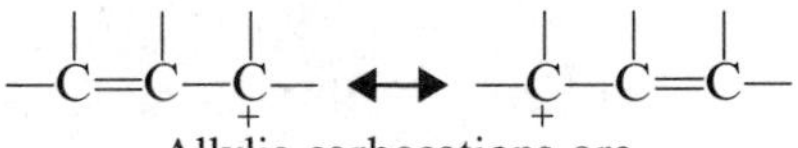

$$-\overset{|}{C}=\overset{|}{C}-\underset{+}{\overset{|}{C}}- \longleftrightarrow -\underset{+}{\overset{|}{C}}-\overset{|}{C}=\overset{|}{C}-$$

Allylic carbocations are
stabilized by resonance

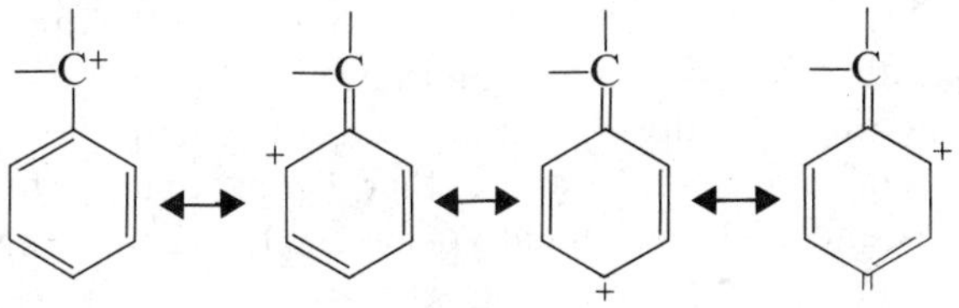

Benzylic carbocations are
stabilized by resonance

Overall we can summarize the effect of structure on reactivity of alkyl, allylic, and benzylic halides in the following way.

These halides give only S_N2 reactions	These halides give only S_N1 reactions
CH_3-X $R-CH_2-X$ $R-\underset{R'}{\underset{\mid}{C}}H-X$	$R'-\overset{R}{\overset{\mid}{\underset{R''}{\underset{\mid}{C}}}}-X$
These halides may give either S_N1 or S_N2 reactions	
$Ar-CH_2-X$ $Ar-\underset{R}{\underset{\mid}{C}}H-X$	$Ar-\overset{R}{\overset{\mid}{\underset{R'}{\underset{\mid}{C}}}}-X$
$CH_2=CHCH_2-X$ $CH_2=CH\underset{R}{\underset{\mid}{C}}H-X$	$CH_2=CH-\overset{R}{\overset{\mid}{\underset{R'}{\underset{\mid}{C}}}}-X$

Problem 14.2

Account for the following observations: (a) When 1-chloro-2-butene is allowed to react with a relatively concentrated solution of sodium ethoxide in ethyl alcohol the reaction rate depends on the concentration of the allylic halide and on the concentration of ethoxide ion. The product of the reaction is almost exclusively $CH_3CH=CHCH_2OCH_2CH_3$. (b) When 1-chloro-2-butene is allowed to react with very dilute solutions of sodium ethoxide in ethyl alcohol (or with ethyl

* There is some dispute as to whether 2° alkyl halides can react by an S_N1 mechanism in ordinary nonacidic solvents such as mixtures of water and alcohol or acetone. But it is clear that reaction by an S_N2 mechanism is, for all practical purposes, the most important pathway.

alcohol alone) the reaction rate is independent of the concentration of ethoxide ion; it depends only on the concentration of the allylic halide. Under these conditions the reaction produces a mixture of $CH_3CH{=}CHCH_2OCH_2CH_3$ and $\underset{\displaystyle OCH_2CH_3}{CH_3\underset{|}{C}HCH{=}CH_2}$. (c) In the presence of traces of water pure 1-chloro-2-butene is slowly converted to a mixture of 1-chloro-2-butene and 3-chloro-1-butene.

Problem 14.3

1-Chloro-3-methyl-2-butene undergoes hydrolysis in a mixture of water and dioxane at a rate that is more than a thousand times that of 1-chloro-2-butene. (a) What factor accounts for the difference in reactivity? (b) What products would you expect to obtain? (Dioxane is a cyclic ether (below) that is miscible with water in all proportions and is a convenient co-solvent for conducting reactions like these.)

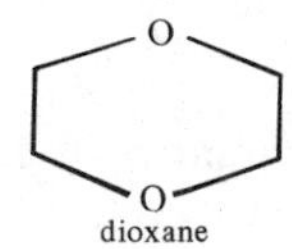

dioxane

Problem 14.4

Primary halides of the type $ROCH_2X$ apparently undergo S_N1 type reactions, whereas most primary halides do not. Can you propose a resonance explanation that will account for this?

Problem 14.5

Each of the following chlorides undergo solvolysis in ethanol at the relative rates given in parentheses. How can you explain these results?

$C_6H_5CH_2Cl$	$C_6H_5CH(Cl)CH_3$	$(C_6H_5)_2CHCl$	$(C_6H_5)_3CCl$
(0.08)	(1)	(300)	(3×10^6)

14.3 ARYL HALIDES AND NUCLEOPHILIC AROMATIC SUBSTITUTION

Most aryl halides are like vinylic halides (Section 5.8) in that they are relatively unreactive toward nucleophilic substitution. Chlorobenzene, for example, can be

$$C_6H_5Cl + NaOH \xrightarrow[\text{reflux}]{H_2O} \text{No substitution}$$

boiled with sodium hydroxide for days without producing a detectable amount of phenol (or sodium phenoxide). Similarly, when vinyl chloride is heated with sodium hydroxide no substitution occurs:

$$CH_2{=}CHCl + NaOH \xrightarrow[\text{reflux}]{H_2O} \text{No substitution}$$

Aryl halides and vinylic halides do not give a positive test (a silver halide precipitate) when treated with alcoholic silver nitrate (Section 9.17E).

We can understand this lack of reactivity on the basis of several factors. The benzene ring of an aryl halide prevents backside attack in an S_N2 reaction:

Nu:-------(benzene ring)—X ⟶✕ No reaction

Phenyl cations are very unstable, thus S_N1 reactions do not occur. The carbon-halogen bonds of aryl (and vinylic halides) are stronger than those of alkyl, allylic, and benzylic halides. Stronger carbon-halogen bonds mean that bond breaking by either an S_N1 or S_N2 mechanism will require more energy.

Two effects make the carbon-halogen bonds of aryl and vinylic halides stronger. (1) The carbon of either type of halide is sp^2 hybridized and thus the electrons of the carbon orbital are closer to the nucleus than those of an sp^3-hybridized carbon. (2) Resonance of the type shown below strengthens the carbon-halogen bond by giving it *double-bond character.*

$$RCH{=}CH{-}\ddot{\underset{..}{X}}: \longleftrightarrow R{-}\overset{-}{\underset{..}{C}}H{-}CH{=}\ddot{X}:^{+}$$

:X: (aryl halide) ⟷ :X⁺ (ortho carbanion) ⟷ :X⁺ (para carbanion) ⟷ :X⁺ (ortho carbanion)

14.3A Nucleophilic Aromatic Substitution by an Addition-Elimination Mechanism

Nucleophilic substitution reactions of aryl halides *do* occur readily when an electronic factor makes the aryl carbon susceptible to nucleophilic attack. Nucleophilic substitution can occur *when strong electron-withdrawing groups are ortho or para to the halogen.* Examples are reactions of the compounds shown below.

o-chloronitrobenzene (Cl, NO_2) $+ OH^- \xrightarrow[130°]{\text{aq. } NaHCO_3} \xrightarrow{H^+}$ o-nitrophenol (OH, NO_2)

2,4-dinitrochlorobenzene (Cl, NO_2, NO_2) $+ OH^- \xrightarrow[100°]{\text{aq. } NaHCO_3} \xrightarrow{H^+}$ 2,4-dinitrophenol (OH, NO_2, NO_2)

2,4,6-trinitrochlorobenzene (O_2N, Cl, NO_2, NO_2) $+ OH^- \xrightarrow[35°]{\text{aq. } NaHCO_3} \xrightarrow{H^+}$ 2,4,6-trinitrophenol (O_2N, OH, NO_2, NO_2)

We also see in these examples that the temperature required to bring about the reaction is related to the number of ortho or para nitro groups. Of the three compounds, *o*-nitrochlorobenzene requires the highest temperature (*p*-nitrochlorobenzene reacts at 130° as well) and 2,4,6 trinitrochlorobenzene requires the lowest temperature.

A *meta*-nitro group does not produce a similar activating effect. For example, *m*-nitrochlorobenzene gives no corresponding reaction.

The mechanism that operates in these reactions is an *addition-elimination* mechanism involving the formation of a delocalized *carbanion.* In the first step (below) addition of a hydroxide ion to *p*-nitrochlorobenzene, for example, produces the delocalized carbanion; then elimination of a chloride ion yields the substitution product as the aromaticity of the ring is recovered.

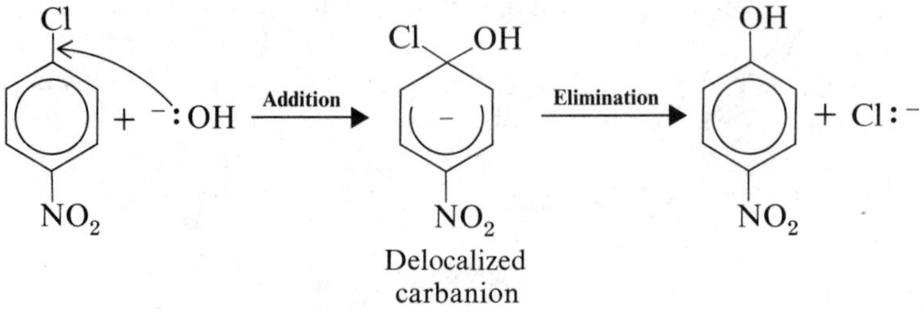

Cl, NO_2 + $^{-}$:OH —Addition→ Cl, OH, NO_2 Delocalized carbanion —Elimination→ OH, NO_2 + Cl:$^{-}$

The delocalized carbanion is stabilized by *electron-withdrawing groups* in the positions *ortho* and *para* to the halogen. If we examine the following resonance structures, we can see how.

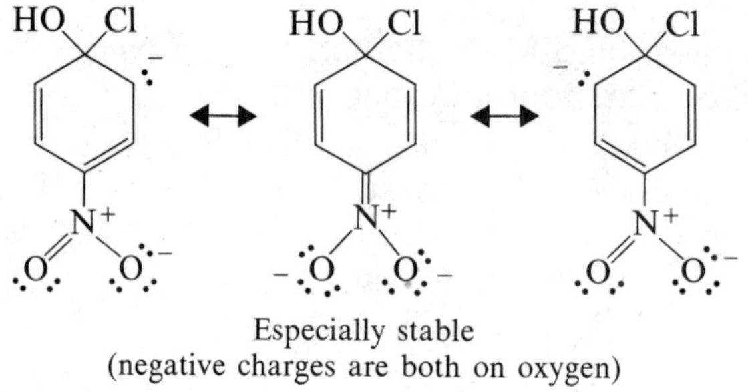

HO Cl, N+, O, O^{-} ↔ HO Cl, N+, $^{-}$O, O^{-} ↔ HO Cl, N+, O, O^{-}

Especially stable
(negative charges are both on oxygen)

Problem 14.6

What products would you expect from each of the following nucleophilic substitution reactions?

(a) *p*-Nitrochlorobenzene + $CH_3ONa \xrightarrow{100°}$

(b) *o*-Nitrochlorobenzene + $CH_3NH_2 \xrightarrow[160°]{C_2H_5OH}$

(c) 2,4-Dinitrochlorobenzene + $C_6H_5NH_2 \xrightarrow{95°}$

Vinylic halides also undergo nucleophilic substitution when they are activated by a strong electron-withdrawing group. An example is the following.

$$C_6H_5\overset{Cl}{\overset{|}{C}}{=}CH{-}C{\equiv}N: + C_2H_5O^- \longrightarrow C_6H_5{-}\overset{Cl}{\underset{OC_2H_5}{C}}{-}\ddot{C}H^-{-}C{\equiv}N:$$

$$\updownarrow$$

$$C_6H_5{-}\underset{OC_2H_5}{C}{=}CH{-}C{\equiv}N: + Cl^- \longleftarrow C_6H_5{-}\overset{Cl}{\underset{OC_2H_5}{C}}{-}CH{=}C{=}\ddot{N}:^-$$

14.3B Nucleophilic Aromatic Substitution Through an Elimination-Addition Mechanism: Benzyne

Although aryl halides such as chlorobenzene and bromobenzene do not react with most nucleophiles under ordinary circumstances, they do react under highly forcing conditions. Chlorobenzene can be converted to phenol by heating it with aqueous sodium hydroxide in a pressurized reactor at 340°.

$$C_6H_5Cl + NaOH \xrightarrow[H_2O]{340^\circ} C_6H_5ONa \xrightarrow{H^+} C_6H_5OH$$

phenol

Bromobenzene reacts with the very powerful base, NH_2^-, in liquid ammonia:

$$C_6H_5Br + \overset{+}{K}\ :\ddot{N}H_2^- \xrightarrow[NH_3]{-33^\circ} C_6H_5\ddot{N}H_2 + KBr$$

aniline

These reactions take place through an *elimination-addition mechanism* that involves the formation of an interesting intermediate called *benzyne*. We can illustrate this mechanism with the reaction of bromobenzene and amide ion.

In the first step, p. 604, the amide ion initiates an elimination by abstracting an ortho proton. This elimination produces the highly unstable, and thus highly reactive, *benzyne*. Benzyne then reacts with any available nucleophile (in this case, amide ion) by a two-step addition reaction to produce aniline.

The nature of benzyne, itself, will become clearer if we examine the orbital diagram below.

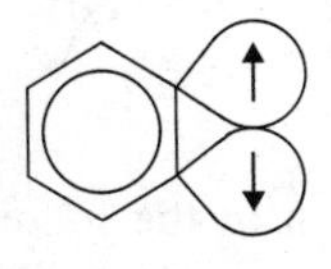

Benzyne

Elimination

Br, H, $:\bar{N}H_2$ $\xrightarrow{(-NH_3)}$ Br $\xrightarrow{(-Br^-)}$ Benzyne (or dehydrobenzene)

Addition: $:\ddot{N}H_2^-$ → $\ddot{N}H_2$ (carbanion) → $:NH_3$ → $\ddot{N}H_2$ + $:\ddot{N}H_2^-$

The extra bond in benzyne results from the overlap of sp^2 orbitals on adjacent carbons of the ring. The axes of these sp^2 orbitals lie in the same plane as that of the ring, and consequently they do not overlap with the π orbitals of the aromatic system. They do not appreciably disturb the aromatic system and they do not make an appreciable resonance contribution to it. The extra bond is weak. Even though the ring hexagon is probably somewhat distorted in order to bring the sp^2 orbitals closer together, overlap between them is not large. Benzyne, as a result, is highly unstable and highly reactive.

What, then, is some of the evidence for the existence of benzyne and for an elimination-addition mechanism in some nucleophilic aromatic substitutions?

The first piece of clear-cut evidence was an experiment done by Professor J. D. Roberts in 1953—one that marked the beginning of benzyne chemistry. Roberts showed that when ^{14}C-labeled bromobenzene is treated with amide ion in liquid ammonia, the aniline that is produced has the label equally divided between the 1 and 2 positions. This result is consistent with the elimination-addition mechanism below but is, of course, not at all consistent with a direct displacement or with an addition-elimination mechanism (Why?).

Br $\xrightarrow{K^+NH_2^-}$ benzyne $\xrightarrow[NH_3]{NH_2^-}$ *NH_2 (50%) + *…NH_2 (50%)

Elimination **Addition**

An even more striking illustration can be seen in the following reaction. When the ortho derivative **1** is treated with sodium amide, the only organic product obtained is *meta*-(trifluoromethyl)aniline.

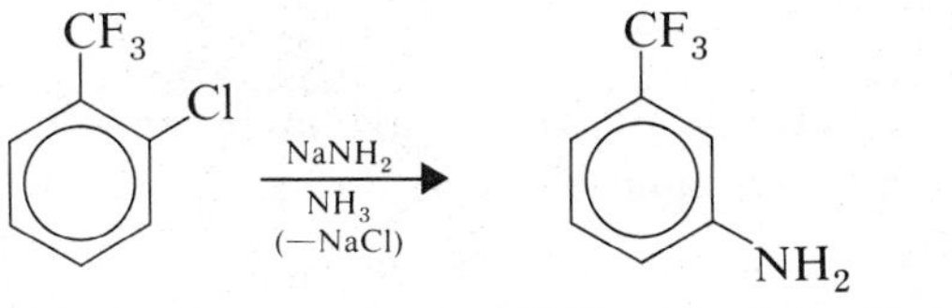

CF_3, Cl; $\xrightarrow[NH_3\ (-NaCl)]{NaNH_2}$ CF_3, NH_2

1 *m*-Trifluoromethylaniline

This result can also be explained by an elimination-addition mechanism. The first step produces the benzyne **2**:

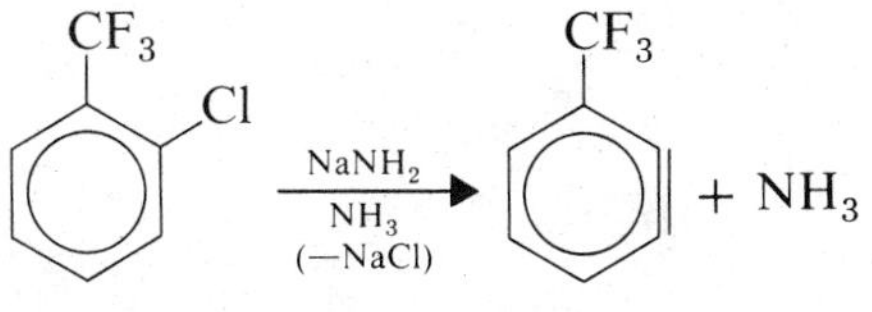

CF_3, Cl; $\xrightarrow[NH_3\ (-NaCl)]{NaNH_2}$ CF_3 + NH_3

1 2

This benzyne then adds an amide ion in the way that produces the more stable carbanion **3** rather than the less stable carbanion **4**.

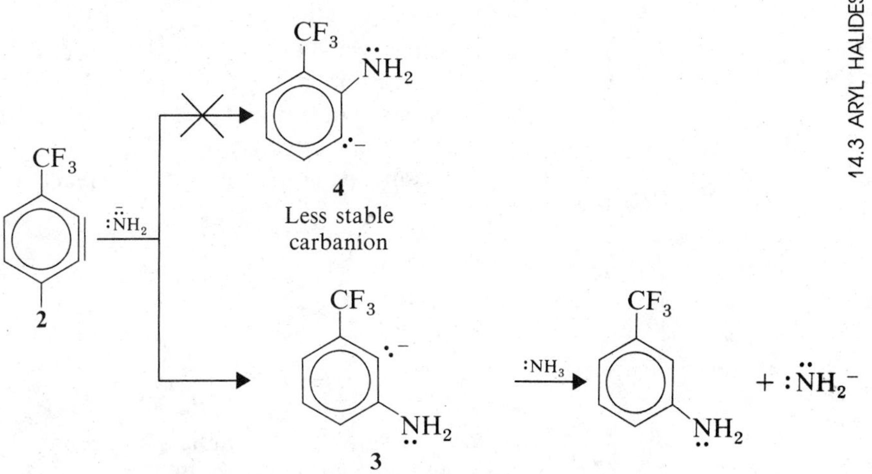

CF_3 2 $:\bar{N}H_2$

CF_3, $\ddot{N}H_2$

4
Less stable
carbanion

CF_3, $\ddot{N}H_2$

3
More stable carbanion
(negative charge is ortho
to the electronegative
trifluoromethyl group)

$\xrightarrow{:NH_3}$ CF_3, $\ddot{N}H_2$ $+ :\ddot{N}H_2^-$

Carbanion **3** then accepts a proton from ammonia to form *m*-(trifluoromethyl)-aniline.

Carbanion **3** is more stable than **4** because the carbon bearing the negative charge is closer to the highly electronegative trifluoromethyl group. The trifluoromethyl group stabilizes the negative charge through its inductive effect. (Resonance effects are not important here because the sp^2 orbital that contains the electron pair does not overlap with the π orbitals of the aromatic system).

Benzyne intermediates have been "trapped" through the use of Diels-Alder reactions. When benzyne is generated in the presence of the diene *furan*, the product is a Diels-Alder adduct.

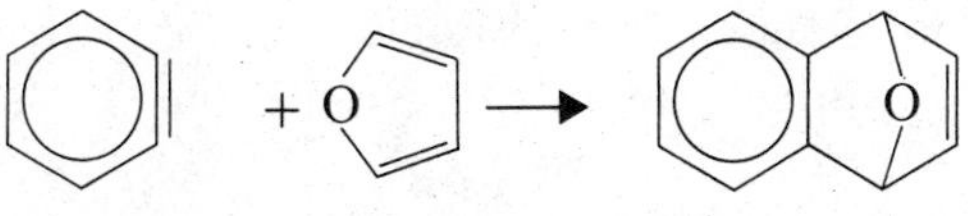

+ O → O

Benzyne
(generated by
an elimination
reaction)

Furan

Diels-Alder adduct

Problem 14.7

When *p*-chlorotoluene is heated with aqueous sodium hydroxide at 340°, *p*-cresol and *m*-cresol are obtained in equal amounts. Write a mechanism that would account for this result.

Problem 14.8

When 2-bromo-3-methylanisole is treated with amide ion in liquid ammonia no substitution takes place. This has been interpreted as providing evidence for the elimination-addition mechanism. Explain.

14.4 PREPARATION OF ORGANIC HALIDES

We have already seen a variety of methods for synthesizing alkyl, aryl, allylic, vinylic, and benzylic halides. Before we go on to new procedures, let us review these.

1. **Alkyl Halides From Alkanes** (discussed in Section 4.10)
We saw in Chapter 4 that alkanes and cycloalkanes undergo free radical substitution reactions when they react with halogens. Although alkane chlorinations find considerable use in industrial preparations they are less important in laboratory work because chlorine reacts with most alkanes in relatively nonspecific ways. If all of the hydrogens of an alkane or cycloalkane are equivalent then the reaction is more useful. For example:

$$\text{Cyclopentane (large excess)} + Cl_2 \xrightarrow[h\nu]{\text{Heat}} \text{Chlorocyclopentane (high yield)}$$

Bromine is more selective in free radical halogenations than chlorine (Section 4.11) and we can make use of this:

$$(CH_3)_3C{-}H + Br_2 \xrightarrow[\text{or } h\nu]{\text{Heat}} (CH_3)_3C{-}Br \quad (99\%)$$

2. **Alkyl and Allylic Halides From Alkenes**
Alkyl halides can be prepared from alkenes by additions of hydrogen halides (Section 7.2).

$$R{-}CH{=}CH{-}R + HX \longrightarrow R{-}CH_2\underset{\displaystyle X}{\underset{|}{C}}H{-}R$$

Cyclohexyl fluoride, for example, can be prepared through the addition of hydrogen fluoride to cyclohexene.

$$\text{Cyclohexene} + HF \longrightarrow \text{Cyclohexyl fluoride} \quad (60\%)$$

Hydrogen bromide adds to alkenes in a Markovniknov fashion in the absence of peroxides and in an anti-Markovniknov way in their presence (Section 7.14).

$$CH_3CH_2CH_2CH{=}CH_2 + HBr \xrightarrow[CH_3COOH]{} \underset{\displaystyle Br}{CH_3CH_2CH_2CHCH_3} \quad (84\%)$$ **Markovnikov Addition**

$$CH_3CH_2CH_2CH{=}CH_2 + HBr \xrightarrow[\text{peroxides}]{} CH_3CH_2CH_2CH_2CH_2Br \quad (81\%)$$ **AntiMarkovnikov Addition**

Allylic halides can be prepared by allowing alkenes to react with chlorine or bromine at high temperatures or under conditions in which the concentration of halogen is low (Section 10.2).

$$>C{=}C<\text{—}\underset{H}{C}\text{—} + Cl_2 \xrightarrow{400^\circ} >C{=}C<\text{—}\underset{Cl}{C}\text{—} + HCl$$

An allylic chloride

or

$$>C{=}C<\text{—}\underset{H}{C}\text{—} + N\text{—}Br \xrightarrow[\text{(or peroxides) } CCl_4]{h\nu} >C{=}C<\text{—}\underset{Br}{C}\text{—} + N\text{—}H$$

N-Bromo-succinimide

An allylic bromide

Benzylic halides can be prepared in the same way (Sect. 12.10)

$$C_6H_5\text{—}\underset{H}{C}\text{—} + Cl_2 \xrightarrow{h\nu} C_6H_5\text{—}\underset{Cl}{C}\text{—}$$

A benzylic chloride

$$C_6H_5\text{—}\underset{H}{C}\text{—} + N\text{—}Br \xrightarrow[\text{(or peroxides) } CCl_4]{h\nu} C_6H_5\text{—}\underset{Br}{C}\text{—} + N\text{—}H$$

3. Vinylic Halides from Alkynes

Vinylic halides can be prepared by addition of one mole of hydrogen halide to an alkyne (Section 9.9).

$$\text{—}C{\equiv}C\text{—} + HX \longrightarrow \overset{H}{>}C{=}C\underset{X}{<}$$

Vinylic halide

4. Dihalides from Alkenes and Alkynes

Alkenes add bromine or chlorine to form *vic*-dihalides (Section 7.9).

$$\text{>C=C<} + X_2 \longrightarrow \text{—}\underset{X}{\underset{|}{\overset{|}{C}}}\text{—}\underset{X}{\underset{|}{\overset{|}{C}}}\text{—}$$

A *vic*-dihalide

Alkynes add two moles of hydrogen halide to yield a *gem*-dihalide (Section 9.8),

$$\text{—C≡C—} + 2HX \longrightarrow \text{—}\underset{H}{\underset{|}{\overset{H}{\overset{|}{C}}}}\text{—}\underset{X}{\underset{|}{\overset{X}{\overset{|}{C}}}}\text{—}$$

A *gem*-dihalide

Gem-dichlorides can be prepared by allowing a ketone to react with PCl_5 (Section 9.14B).

$$R\text{—}\underset{O}{\underset{\|}{C}}\text{—}R + PCl_5 \longrightarrow R\text{—}\underset{Cl}{\underset{|}{\overset{Cl}{\overset{|}{C}}}}\text{—}R + POCl_3$$

5. Aryl Halides from Arenes

Aryl bromides and chlorides can be prepared by treating arenes with bromine or chlorine in the presence of FeX_3 (Section 12.3).

$$ArH + X_2 \xrightarrow{FeX_3} ArX + HX$$

Problem 14.9

Starting with an appropriate alkane, alkene, alkyne, or arene show at least one method for preparing each of the following organic halides.

(a) 1-Bromopentane (from an alkene)
(b) 2-Bromopentane (from an alkene)
(c) 2-Bromo-2-methylbutane (from an alkane)
(d) 2-Chloro-1-butene
(e) 3-Bromocyclohexene (from cyclohexene)
(f) 2,2-Dibromobutane
(g) 2,3-Dibromobutane
(h) 1-Bromo-1-phenylheptane
(i) 1-(*p*-Bromophenyl)heptane
(j) Benzyl chloride
(k) Cyclopentyl fluoride
(l) Cyclohexyl iodide

14.5 ALKYL HALIDES FROM ALCOHOLS

Alcohols react with a variety of reagents to yield alkyl halides. The most commonly used reagents are hydrogen halides (HCl, HBr, or HI), phosphorus tribromide (PBr_3), and thionyl chloride ($SOCl_2$). Examples of the use of these reagents are shown below.

$$CH_3-\underset{CH_3}{\overset{CH_3}{C}}-OH + HCl_{(concd.)} \xrightarrow{25°} CH_3-\underset{CH_3}{\overset{CH_3}{C}}-Cl + H_2O$$

94%

$$CH_3CH_2CH_2CH_2OH + HBr_{(concd.)} \xrightarrow[Reflux]{H_2SO_4} CH_3CH_2CH_2CH_2Br$$

95%

$$3(CH_3)_2CHCH_2OH + PBr_3 \xrightarrow[4\ hr]{-10°\ to\ 0°} 3(CH_3)_2CHCH_2Br + H_3PO_3$$

(55–60%)

$$m\text{-}CH_3OC_6H_4CH_2OH + SOCl_2 \xrightarrow{Pyridine} m\text{-}CH_3OC_6H_4CH_2Cl + SO_2 + HCl$$

(91%)

14.6 ALKYL HALIDES FROM THE REACTION OF ALCOHOLS WITH HYDROGEN HALIDES

When alcohols react with a hydrogen halide a substitution takes place producing an alkyl halide and water:

$$R+OH + HX \longrightarrow R-X + H_2O$$

The order of reactivity of the hydrogen halides is HI > HBr > HCl (HF is generally unreactive), and the order of reactivity of alcohols is benzylic and allylic > 3° > 2° > 1°.

The reaction is *acid catalyzed;* it does not proceed at an appreciable rate unless a strong acid is present. The strong acid may be the hydrogen halide itself, or it may be concentrated sulfuric acid that has been added to the mixture. Primary and secondary alcohols are often converted to alkyl iodides and bromides by allowing them to react with a mixture of a sodium halide and concentrated sulfuric acid. This not only generates the hydrogen halide in the mixture (*in situ*); it also provides the acid catalyst.

$$ROH + NaX \xrightarrow{H_2SO_4} RX + NaHSO_4 + H_2O$$

14.6A Mechanisms of the Reaction of Alcohols with HX

Secondary, tertiary, allylic, and benzylic alcohols appear to react by a mechanism (p. 610) that involves the formation of a carbocation—one that is recognizable *as an S_N1-type reaction with the protonated alcohol acting as the substrate.* We illustrate this mechanism with the reaction of *tert*-butyl alcohol and hydrochloric acid.

The first two steps are the same as the mechanism for the dehydration of an alcohol (Section 6.10). The alcohol accepts a proton and then the protonated alcohol dissociates to form a carbocation and water.

$$\text{Step 1}\quad CH_3-\underset{CH_3}{\overset{CH_3}{C}}-\ddot{O}-H + H-\underset{H}{\overset{+}{\ddot{O}}}-H \overset{\text{fast}}{\rightleftharpoons} CH_3\underset{CH_3}{\overset{CH_3}{C}}-\overset{H}{\underset{..}{O^+}}-H + :\underset{H}{\ddot{O}}-H$$

$$\text{Step 2}\quad CH_3-\underset{CH_3}{\overset{CH_3}{C}}-\overset{H}{O^+}-H \overset{\text{slow}}{\rightleftharpoons} CH_3-\underset{CH_3}{\overset{CH_3}{C^+}} + :\overset{H}{\ddot{O}}-H$$

In step 3 the mechanisms for the dehydration of an alcohol and the formation of an alkyl halide differ. In dehydration reactions the carbocation loses a proton in an E1-type reaction to form an alkene. In the formation of an alkyl halide, the carbocation reacts with a nucleophile (a halide ion) in an S_N1-type reaction.

$$\text{Step 3}\quad CH_3-\underset{CH_3}{\overset{CH_3}{C^+}} + :\ddot{Cl}:^- \overset{\text{fast}}{\rightleftharpoons} CH_3-\underset{CH_3}{\overset{CH_3}{C}}-\ddot{Cl}:$$

How can we account for the different course of these two reactions?

When we dehydrate alcohols we usually carry out the reaction in concentrated sulfuric acid. The only nucleophiles present in this reaction mixture are water and hydrogen sulfate (HSO_4^-) ions. Both are poor nucleophiles and both are usually present in low concentrations. Under these conditions, the highly reactive carbocation ion stabilizes itself by losing a proton and becoming an alkene.The net result is *an E1* reaction.

In the reverse reaction, that is, the hydration of an alkene (Section 7.4) the carbocation ion *does* react with a nucleophile. It reacts with water. Alkene hydrations are carried out in dilute sulfuric acid where the water concentration is high. In some instances, too, carbocations may react with HSO_4^- ions or with sulfuric acid, itself. When they do they form alkyl hydrogen sulfates, $R-OSO_2OH$.

When we convert an alcohol to an alkyl halide, we carry out the reaction in the presence of acid and *in the presence of halide ions*. Halide ions are good nucleophiles (much stronger nucleophiles than water), and since they are present in high concentration, most of the carbocations stabilize themselves by accepting the electron pair of a halide ion and the overall result is an S_N1 reaction.

These two reactions, dehydration and the formation of an alkyl halide, also furnish us another example of the competition between nucleophilic substitution and elimination (cf. Section 5.12). Very often, in conversions of alcohols to alkyl halides, we find that the reaction is accompanied by the formation of some alkene (i.e., by elimination). Thus, not all of the carbocations react with nucleophiles; some stabilize themselves by losing protons.

Not all acid-catalyzed conversions of alcohols to alkyl halides proceed through the formation of carbocations. Primary alcohols and methyl alcohol apparently react through a mechanism that we will recognize as *an S_N2 type.*

$$:\ddot{X}:^- + -\overset{|}{\underset{|}{C}}-\overset{H}{\underset{..}{O^+}}-H \longrightarrow :\ddot{X}-\overset{|}{\underset{|}{C}}- + :\overset{H}{\ddot{O}}-H$$

(protonated 1° or methyl alcohol) (a good leaving group)

In these reactions the function of the acid is to produce a protonated alcohol. The carbon-oxygen bond of the protonated alcohol is weakened but not completely broken. However, because the bond is weakened, the halide ion is able to displace a molecule of water (a good leaving group) from carbon and form an alkyl halide.

Although halide ions (particularly iodide and bromide ions) are strong nucleophiles they are not strong enough to carry out substitution reactions with alcohols themselves. That is, reactions of the following type do not occur to any appreciable extent.

$$:\ddot{\underset{\cdot\cdot}{Br}}:^- + -\overset{|}{\underset{|}{C}}-\ddot{\underset{\cdot\cdot}{O}}H \not\longrightarrow :\ddot{\underset{\cdot\cdot}{Br}}-\overset{|}{\underset{|}{C}}- + :\ddot{\underset{\cdot\cdot}{O}}H^-$$

They do not occur because the leaving group would have to be a strongly basic hydroxide ion.

The reverse reaction, that is, the reaction of an alkyl halide with hydroxide ion, does occur and is a method for the synthesis of alcohols. We saw this reaction in Chapter 5.

We can see now why the reactions of alcohols with hydrogen halides are acid catalyzed. With allylic, benzylic, tertiary, and secondary alcohols the function of the acid is to help produce a carbocation. With methyl alcohol and primary alcohols the function of the acid is to produce a substrate in which the leaving group is a weakly basic water molecule rather than a strongly basic hydroxide ion.

As we might expect, many reactions of alcohols with hydrogen halides, particularly those in which carbocations are formed, are accompanied by rearrangements.

Because the chloride ion is a weaker nucleophile, hydrogen chloride does not react with primary alcohols unless zinc chloride or some similar Lewis acid is added to the reaction mixture as well. Zinc chloride, a good Lewis acid, acts as an acid catalyst in this reaction. It forms a complex with the alcohol through association with an unshared pair of electrons on the oxygen atom. This weakens the carbon-oxygen bond and attack by the chloride ion can cause it to break to give the alkyl halide.

$$R-\underset{\underset{H}{|}}{\ddot{O}}: + ZnCl_2 \rightleftharpoons R-\underset{\underset{H}{|}}{\overset{\cdot\cdot +}{O}}-\overset{-}{Zn}Cl_2$$

$$:\ddot{\underset{\cdot\cdot}{Cl}}:^- + R-\underset{\underset{H}{|}}{\overset{\cdot\cdot +}{O}}-\overset{-}{Zn}Cl_2 \longrightarrow :\ddot{\underset{\cdot\cdot}{Cl}}-R + [Zn(OH)Cl_2]^-$$

$$[Zn(OH)Cl_2]^- + H^+ \rightleftharpoons ZnCl_2 + H_2O$$

A solution of zinc chloride in concentrated hydrochloric acid is often used as the basis for a simple test (called the Lucas test) to differentiate between primary, secondary, and tertiary alcohols with fewer than eight carbons. When treated with a solution of zinc chloride in concentrated HCl (Lucas reagent) tertiary alcohols react immediately at room temperature to give an insoluble layer of the alkyl chloride. Secondary alcohols take several minutes to react, and primary alcohols do not react with Lucas reagent unless they are heated. The test is used *only* when other evidence shows that an unknown is an alcohol of fewer than eight carbons. Alcohols composed of more than eight carbons are generally too insoluble in the reagent.

Problem 14.10

Treating 3-methyl-2-butanol (below) with HBr yields 2-bromo-2-methylbutane as the sole product. Outline a mechanism for the reaction.

$$\underset{\text{3-Methyl-2-butanol}}{CH_3CH(CH_3)CH(OH)CH_3} \xrightarrow{HBr} \underset{\text{2-Bromo-2-methylbutane}}{CH_3CBr(CH_3)CH_2CH_3}$$

Problem 14.11

Phenols do not react with HX to yield aryl halides. What factors might account for this?

14.7 ALKYL HALIDES FROM THE REACTIONS OF ALCOHOLS WITH PBr_3 or $SOCl_2$

Primary and secondary alcohols react with phosphorus tribromide to yield alkyl halides.

$$\underset{(1° \text{ or } 2°)}{3R{-}OH} + PBr_3 \longrightarrow 3R{-}Br + H_3PO_3$$

Unlike the reaction of an alcohol with HBr, the reaction of an alcohol with PBr_3 does not involve the formation of a carbocation and usually does not occur with rearrangement of the carbon skeleton (especially if the temperature is kept below 0°). For this reason phosphorus tribromide is often preferred as a reagent for the transformation of an alcohol to the corresponding alkyl bromide.

The mechanism for the reaction involves the initial formation of a protonated alkyl dibromophosphite (below) by a nucleophilic displacement on phosphorus; the alcohol acts as the nucleophile:

$$RCH_2\ddot{O}H + Br{-}PBr{-}Br \longrightarrow \underset{\text{Protonated alkyl dibromophosphite}}{R{-}CH_2\overset{+}{O}(H){-}PBr_2} + Br:^-$$

Then a bromide ion acts as a nucleophile and displaces $HOPBr_2$.

$$Br:^- + RCH_2{-}\overset{+}{O}(H)PBr_2 \longrightarrow RCH_2Br + HOPBr_2$$

A good leaving group

The $HOPBr_2$ can react with more alcohol so the net result is the conversion of three moles of alcohol to alkyl bromide by one mole of phosphorus tribromide.

Thionyl chloride ($SOCl_2$) converts primary and secondary alcohols to alkyl chlorides (usually without rearrangement).

$$\underset{(1° \text{ or } 2°)}{R{-}OH} + SOCl_2 \xrightarrow{\text{Reflux}} R{-}Cl + SO_2 + HCl$$

Often a tertiary amine is added to the mixture to promote the reaction by reacting with the HCl.

$$R_3N\!: + HCl \longrightarrow R_3NH^+ + Cl^-$$

The reaction mechanism involves initial formation of the alkyl chlorosulfite:

$$RCH_2\ddot{O}H + Cl{-}\underset{\underset{O}{\|}}{S}{-}Cl \longrightarrow RCH_2{-}O{-}\underset{\underset{O}{\|}}{S}{-}Cl + HCl$$

Alkyl chlorosulfite

Then a chloride ion (from $R_3N + HCl \longrightarrow R_3NH^+ + Cl^-$) can bring about an S_N2 displacement:

$$Cl\!:^- + RCH_2{-}O{-}\underset{\underset{O}{\|}}{S}{-}Cl \longrightarrow RCH_2Cl + {}^-O{-}\underset{\underset{O}{\|}}{S}{-}Cl \longrightarrow RCH_2Cl + SO_2 + Cl^-$$

If no tertiary amine is present in the mixture the alkyl chlorosulfite from a secondary alcohol can form an ion-pair by losing SO_2, which can then collapse to an alkyl chloride:

$$R{-}\overset{\overset{R'}{|}}{C}H{-}O{-}S(=O){-}Cl \xrightarrow{-SO_2} R{-}\overset{\overset{R'}{|}}{C}H^+ \;\; \ddot{C}l^- \longrightarrow R{-}\overset{\overset{R'}{|}}{C}H{-}Cl$$

Ion pair

As we saw in Section 8.17, if the secondary alcohol is chiral this reaction can take place with retention of configuration.

Problem 14.12

(a) Would you expect retention of configuration in the reaction of a chiral secondary alcohol with $SOCl_2$ in the presence of a tertiary amine? Explain. (b) What stereochemical result would you expect?

14.8 PHYSICAL PROPERTIES OF ORGANIC HALIDES

Alkyl and aryl halides have very low solubilities in water, but as we might expect, they are miscible with each other and with other relatively nonpolar solvents. Methylene chloride (CH_2Cl_2), chloroform ($CHCl_3$), and carbon tetrachloride are often used as solvents for nonpolar and moderately polar organic compounds. Many chlorohydrocarbons including $CHCl_3$ and CCl_4 have a cumulative toxicity, however, and should, therefore, be used only in fume hoods and with great care.

Methyl iodide (bp 42°) is the only monohalomethane that is a liquid at room temperature and atmospheric pressure. Ethyl bromide (bp 38°) and ethyl iodide (bp 72°) are both liquids but ethyl chloride (bp 13°) is a gas. The propyl chlorides, bromides, and iodides are all liquids. In general alkyl chlorides, bromides, and iodides tend to have boiling points near those of alkanes of similar molecular weights.

Polyfluoroalkanes, however, tend to have unusually low boiling points. Hexafluoroethane boils at −79°, even though its molecular weight (138) is near that of decane (bp 174°).

Table 14.1 lists some of the physical properties of organic halides.

14.9 ORGANOMETALLIC COMPOUNDS

Compounds that contain carbon-metal bonds are called *organometallic compounds*. The natures of the carbon-metal bonds vary widely, ranging from bonds that are essentially ionic to those that are primarily covalent. While the structure of the organic portion of the organometallic compound has some effect on the nature of the carbon-metal bond, the identity of the metal itself is of far greater importance. Carbon-sodium and carbon-potassium bonds are largely ionic in character; carbon-lead, carbon-tin, carbon-thallium(III), and carbon-mercury bonds are essentially covalent. Carbon-lithium and carbon-magnesium bonds fall in between these extremes.

$—\overset{\mid}{\underset{\mid}{C}}:^{-}\ \overset{+}{M}$	$—\overset{\mid}{\underset{\mid}{C}}\overset{\delta-}{:}\ \overset{\delta+}{M}$	$—\overset{\mid}{\underset{\mid}{C}}—M$
Primarily ionic ($M = Na^+$ or K^+)	(M = Mg or Li)	Primarily covalent (M = Pb, Sn, Hg, Tl)

As we might expect, the reactivity of organometallic compounds increases

TABLE 14.1 Organic Halides

NAME OF GROUP	FLUORIDE bp °C	FLUORIDE SPECIFIC GRAVITY	CHLORIDE bp °C	CHLORIDE SPECIFIC GRAVITY	BROMIDE bp °C	BROMIDE SPECIFIC GRAVITY	IODIDE bp °C	IODIDE SPECIFIC GRAVITY
Methyl	−78.4	0.84^{-60}	−23.8	0.92^{20}	3.6	1.73^{0}_{0}	42.5	2.28^{20}_{4}
Ethyl	−37.7		13.1	0.91^{15}_{4}	38.4	1.46^{20}_{4}	72	1.95^{20}_{20}
n-Propyl	−2.5	0.78^{-3}	46.6	0.89^{20}_{4}	70.8	1.35^{20}_{4}	102	1.74^{20}_{4}
Isopropyl	−9.4	0.72^{20}_{4}	34	0.86^{20}_{4}	59.4	1.31^{20}_{4}	89.4	1.70^{20}_{4}
n-Butyl	32	0.78^{20}_{4}	78.4	0.89^{20}_{4}	101	1.27^{20}_{4}	130	1.61^{20}_{4}
sec-Butyl			68	0.87^{20}_{4}	91.2	1.26^{20}_{4}	120	1.60^{20}_{4}
Isobutyl			69	0.87^{20}_{4}	91	1.26^{20}_{4}	119	1.60^{20}_{4}
tert-Butyl	12	0.75^{12}_{4}	51	0.84^{20}_{4}	73.3	1.22^{20}_{4}	100 dec.	1.57^{0}_{0}
n-Pentyl	62	0.79^{20}_{4}	108.2	0.88^{20}_{4}	129.6	1.22^{20}_{4}	155^{740}	1.52^{20}_{4}
Neopentyl			84.4	0.87^{20}_{4}	105	1.20^{20}_{4}	127 dec.	1.53^{13}
Vinyl	−51	0.68^{26}_{4}	13.9		16	1.52^{14}_{4}	56	2.04^{20}
Allyl	−3		45	0.94^{20}_{4}	70	1.40^{20}_{4}	102–3	1.84^{22}_{4}
Phenyl (halobenzene)	85	1.02^{20}_{4}	132	1.10^{20}_{4}	155	1.52^{20}_{4}	189	1.82^{20}_{4}
Benzyl	140	1.02^{25}_{4}	179	1.10^{25}_{4}	201	1.44^{22}_{0}	93^{10}	1.73^{25}

with the percent ionic character of the carbon metal bond. Alkylsodium and alkylpotassium compounds are highly reactive and are among the most powerful of bases. They react explosively with water and burst into flame when exposed to air. Organomercury and lead compounds are much less reactive; they are often volatile and are stable in air. They are all poisonous. They are generally soluble in nonpolar solvents. Tetraethyllead, for example, is used as an "antiknock" compound in gasoline.

Organometallic compounds of lithium and magnesium are of great importance in organic synthesis. They are relatively stable in ether solutions but their carbon-metal bonds have considerable ionic character. Because of this, the carbon atom of an organolithium or organomagnesium compound that is bonded to the metal is a strong base and a powerful nucleophile. We will soon see reactions that illustrate both of these properties.

14.10 PREPARATION OF ORGANOLITHIUM AND ORGANOMAGNESIUM COMPOUNDS

14.10A Organolithium Compounds

Organolithium and organomagnesium compounds are usually prepared by the reaction of an organic halide with lithium or magnesium metal. These reactions are usually carried out in ether solvents, and since organolithium and organomagnesium compounds are strong bases, care must be taken to exclude moisture. (Why?) Exclusion of oxygen is also important. The ethers most commonly used as solvents are diethyl ether and tetrahydrofuran. (Tetrahydrofuran is a cyclic ether.)

$CH_3CH_2\ddot{O}CH_2CH_3$ — Diethyl ether (Et_2O)

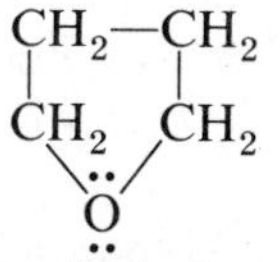

CH_2-CH_2, CH_2 CH_2 bonded through $\ddot{O}$ (ring) — Tetrahydrofuran (THF)

For example, *n*-butyl bromide reacts with lithium metal in diethyl ether to give a solution of *n*-butyllithium.

$$\underset{n\text{-Butyl bromide}}{CH_3CH_2CH_2CH_2Br} + 2Li \xrightarrow[Et_2O]{-10°} \underset{\substack{(80\text{–}90\%)\\ n\text{-Butyllithium}}}{CH_3CH_2CH_2CH_2Li} + LiBr$$

Other organolithium compounds, such as methyllithium, ethyllithium, and phenyllithium, can be prepared in the same general way.

$$\underset{(\text{or Ar—X})}{R—X} + 2Li \xrightarrow{\text{ether}} \underset{(\text{or ArLi})}{RLi} + LiX$$

The order of reactivity of halides is RI > RBr > RCl. (Alkyl and aryl fluorides are seldom used in the preparation of organolithium compounds.)

Vinyllithium can be prepared by treating vinyl chloride with a lithium-sodium mixture.

$$CH_2{=}CHCl + \underset{(2\%\ Na)}{2Li} \xrightarrow{Et_2O} \underset{(60\%)}{CH_2{=}CHLi} + LiCl$$

Most organolithium compounds slowly attack ethers by bringing about an elimination reaction.

$$\overset{\delta-}{R}{:}\overset{\delta+}{Li} + H{-}CH_2{-}CH_2{-}OCH_2CH_3 \longrightarrow RH + CH_2{=}CH_2 + \overset{+}{Li}\,\overset{-}{O}CH_2CH_3$$

For this reason, ether solutions of organolithium reagents are not usually stored but are used directly in some other reaction. Organolithium compounds are much more stable in hydrocarbon solvents. Several alkyl- and aryllithium reagents are commercially available in hexane, paraffin wax, or mineral oil.

14.10B Grignard Reagents

Organomagnesium halides were discovered by the French chemist Victor Grignard in 1900. Grignard received the Nobel Prize for his discovery in 1912 and organomagnesium halides are now called Grignard reagents in his honor. Grignard reagents probably have a greater use in organic synthesis than any other single analogous organic reagent.

Grignard reagents are usually prepared by the reaction of an organic halide and magnesium metal (turnings) in an ether solvent.

$$\left.\begin{array}{l} R{-}X + Mg \xrightarrow{Ether} RMgX \\ Ar{-}X + Mg \xrightarrow{Ether} ArMgX \end{array}\right\} \text{Grignard reagents}$$

The order of reactivity of halides with magnesium is also RI > RBr > RCl. Very few organomagnesium fluorides have been successfully prepared. Aryl Grignard reagents are more easily prepared from aryl bromides and aryl iodides. Aryl chlorides react sluggishly.

Grignard reagents are seldom isolated but are used for further reactions in ether solution. The ether solutions can be analyzed for the content of the Grignard reagent, however, and the yields of Grignard reagent are almost always very high (85–95%). Several examples are shown below.

$$CH_3I + Mg \xrightarrow[35°]{Et_2O} \underset{\text{Methylmagnesium iodide}}{CH_3MgI} \quad (95\%)$$

$$C_2H_5Br + Mg \xrightarrow[35°]{Et_2O} \underset{\text{Ethylmagnesium bromide}}{C_2H_5MgBr} \quad (93\%)$$

$$C_6H_5Br + Mg \xrightarrow[35°]{Et_2O} \underset{\text{Phenylmagnesium bromide}}{C_6H_5MgBr} \quad (95\%)$$

$$CH_2{=}CHBr + Mg \xrightarrow[66°]{THF} \underset{\text{Vinylmagnesium bromide}}{CH_2{=}CHMgBr} \quad (90\%)$$

The actual structures of Grignard reagents are more complex than the general formula RMgX indicates. Experiments done with radioactive magnesium have established that, for most Grignard reagents, there is an equilibrium between an alkylmagnesium halide and a dialkylmagnesium.

$$\underset{\text{Alkylmagnesium halide}}{2RMgX} \rightleftharpoons \underset{\text{Dialkylmagnesium}}{R_2Mg} + MgX_2$$

For convenience in this text, however, we will write the formula for the Grignard reagent as though it were simply RMgX. As we do this, we should always remember that the formula RMgX is an oversimplification.

Grignard reagents also form complexes with their ether solvents like the one shown below.

```
R     R
 \ .. /
   O
   ..
R—Mg—X
   ..
   O
 / .. \
R     R
```

Complex formation with molecules of ether is an important factor in the formation and stability of Grignard reagents. Organomagnesium compounds can be prepared in nonethereal solvents but the preparations are more difficult.

The mechanism by which Grignard reagents are formed is still not fully understood. The most likely path, however, appears to be the two-step radical mechanism shown below.

$$R{-}X + {:}Mg \longrightarrow R\cdot + \cdot MgX$$
$$R\cdot + \cdot MgX \longrightarrow RMgX$$

14.11 REACTIONS OF ORGANOLITHIUM AND ORGANOMAGNESIUM COMPOUNDS

14.11A Reactions with Compounds Containing Acidic Hydrogens

Grignard reagents and organolithium compounds are very strong bases. They react with any compound that has a hydrogen more acidic than the hydrogens of the hydrocarbon from which the Grignard reagent or organolithium is derived. We can understand how these reactions occur if we represent the Grignard reagent and organolithium compounds in the following ways.

$$\overset{\delta-}{R}{:}\overset{\delta+}{MgX} \quad \text{and} \quad \overset{\delta-}{R}{:}\overset{\delta+}{Li}$$

When we do this, we can see that the reactions of Grignard reagents with water and alcohols (below) are nothing more than acid-base reactions; they lead to the formation of the weaker conjugate acid and weaker conjugate base. The Grignard reagent behaves as if it contained the anion of an alkane, *as if it contained a carbanion.*

$$\overset{\delta-}{R}\!:\!\overset{\delta+}{Mg}X + H\!:\!\ddot{O}H \longrightarrow R\!:\!H + H\ddot{O}\!:^{-} + Mg^{2+} + X^{-}$$

Grignard reagent (stronger base) | Water (stronger acid) | Alkane (weaker acid) | Hydroxide ion (weaker base)

$$\overset{\delta-}{R}\!:\!\overset{\delta+}{Mg}X + H\!:\!\ddot{O}R \longrightarrow R\!:\!H + R\ddot{O}\!:^{-} + Mg^{2+} + X^{-}$$

Grignard reagent (stronger base) | Alcohol (stronger acid) | Alkane (weaker acid) | Alkoxide ion (weaker base)

Problem 14.13

Write similar equations for the reactions that take place when *n*-butyllithium is treated with (a) water, (b) ethanol. Designate the stronger and weaker acids and the stronger and weaker bases.

Problem 14.14

Assuming you have *tert*-butyl bromide, magnesium, dry ether, and deuterium oxide (D_2O) available, show how you might synthesize the following deuterium-labeled alkane.

$$CH_3\overset{\displaystyle CH_3}{\underset{\displaystyle D}{C}}CH_3$$

Grignard reagents and organolithium compounds abstract protons that are much less acidic than those of water and alcohols. They react with the terminal hydrogens of 1-alkynes, for example, and this is a useful method for the preparation of alkynylmagnesium halides and alkynyllithiums. These reactions are also acid-base reactions.

$$RC{\equiv}CH + \overset{\delta-}{R'}\!:\!\overset{\delta+}{Mg}X \longrightarrow RC{\equiv}\overset{\delta-}{C}\!:\!\overset{\delta+}{Mg}X + R'\!:\!H$$

Terminal alkyne (stronger acid) | Grignard reagent (stronger base) | Alkynylmagnesium halide (weaker base) | Alkane (weaker acid)

$$R{-}C{\equiv}CH + \overset{\delta-}{R'}\!:\!\overset{\delta+}{Li} \longrightarrow R{-}C{\equiv}\overset{\delta-}{C}\!:\!\overset{\delta+}{Li} + R'\!:\!H$$

Terminal alkyne (stronger acid) | Alkyllithium (stronger base) | Alkynyllithium (weaker base) | Alkane (weaker acid)

The fact that these reactions go to completion is not surprising when we recall that the acidity constants of alkanes are of the order of $10^{-42} - 10^{-45}$, while those of terminal alkynes are 10^{-25} (Table 2.2).

Grignard reagents are not only strong bases, they are also *powerful nucleo-*

philes. We will see in the next section, when we study these reactions in detail, that reactions in which Grignard reagents act as nucleophiles are by far the most important. At this point, let us consider general examples that illustrate the ability of a Grignard reagent to act as a nucleophile by attacking saturated and unsaturated carbons.

14.11B Reactions of Grignard Reagents with Ethylene Oxide

Grignard reagents carry out nucleophilic attack at a saturated carbon when they react with ethylene oxide. These reactions take the general form shown below and they give us a convenient synthesis of primary alcohols.

The nucleophilic alkyl group of the Grignard reagent attacks the partially positive carbon of the ethylene oxide ring. The highly strained ring opens, and the reaction leads to the salt of a primary alcohol. Subsequent acidification yields the alcohol.

$$\overset{\delta-}{R}{:}\overset{\delta+}{Mg}X + \underbrace{\overset{\delta+}{CH_2}-\overset{\delta+}{CH_2}}_{\underset{\delta-}{\ddot{O}}} \longrightarrow R-CH_2CH_2-\ddot{O}{:}^{-}\overset{2+}{Mg}\,X^{-} \xrightarrow{H^+} R-CH_2CH_2\ddot{O}H$$

Ethylene oxide

A primary alcohol

14.11C Reactions of Grignard Reagents with Carbonyl Compounds

From a synthetic point of view, the most important reactions of Grignard reagents and organolithium compounds are those in which these reagents act as nucleophiles and attack an unsaturated carbon—*especially the carbon of a carbonyl group*.

The carbon atom of a carbonyl group is positively charged because of the inductive effect of the oxygen and because of the resonance contribution of the second structure shown below.

$$>C=\ddot{O}{:} \longleftrightarrow >\overset{+}{C}-\ddot{O}{:}^{-}$$

Resonance structures of the carbonyl group

Compounds that contain carbonyl groups are, therefore, highly susceptible to nucleophilic attack. Grignard reagents react with carbonyl compounds (aldehydes and ketones) in the following way:

$$\overset{\delta-}{R}{:}\overset{\delta+}{Mg}X + >C=\ddot{O}{:} \longrightarrow R-\overset{|}{\underset{|}{C}}-\ddot{O}{:}^{-}\,Mg^{2+}X^{-}$$

This reaction is a nucleophilic addition to the carbon-oxygen double bond. The nucleophilic carbon of the Grignard reagent uses its electron pair to form a bond to the carbonyl carbon. The carbonyl carbon can accept this electron pair because one pair of electrons of the carbon-oxygen double bond can shift out to the oxygen.

The product formed when a Grignard reagent adds to a carbonyl group is an alkoxide ion $\mathrm{R{-}\overset{|}{\underset{|}{C}}{-}\ddot{\underset{..}{O}}:^-}$ that is associated with $Mg^{2+}X^-$. When water or dilute acid is added to the reaction mixture after the Grignard addition is over, an acid-base reaction takes place to produce an alcohol.

$$\mathrm{R{-}\overset{|}{\underset{|}{C}}{-}\ddot{\underset{..}{O}}:\,MgX + H{-}\overset{+}{\ddot{O}}{-}H \longrightarrow R{-}\overset{|}{\underset{|}{C}}{-}\ddot{\underset{..}{O}}{-}H + MgX_2 + H_2\ddot{O}:}$$

$$\mathrm{\quad\quad\quad\quad\quad\quad\quad\quad\quad\quad\quad\underset{X^-}{\overset{|}{H}}}$$

Magnesium halide alkoxide; X^-; Alcohol

Grignard reagents react in a similar way with the carbon of carbon dioxide. Acidification of the salts that are formed in these reactions gives us a synthesis of *carboxylic acids*. (We will study this reaction in Section 17.3).

$$\mathrm{\overset{\delta-}{R}\!:\overset{\delta+}{Mg}X + \overset{\ddot{O}:}{\underset{\ddot{O}:}{\overset{\|}{\underset{\|}{C}}}} \longrightarrow R{-}\overset{\ddot{O}:}{\overset{\|}{C}}{-}\ddot{\underset{..}{O}}:\,MgX \xrightarrow{HX} R\overset{O}{\overset{\|}{C}}OH + MgX_2}$$

14.12 ALCOHOLS FROM GRIGNARD REAGENTS

Grignard additions to carbonyl compounds are especially useful because they can be used to prepare primary, secondary, or tertiary alcohols.

1. A Grignard reagent reacts with formaldehyde, for example, to give a primary alcohol.

$$\mathrm{\overset{\delta-}{R}\!:\,\overset{\delta+}{Mg}X + \underset{\text{Formaldehyde}}{\overset{H}{\underset{H}{>}}C{=}\ddot{\underset{..}{O}}:} \longrightarrow R{-}\overset{H}{\underset{H}{\overset{|}{\underset{|}{C}}}}{-}\ddot{\underset{..}{O}}:\,MgX \xrightarrow{H_3O^+} \underset{1^\circ\ \text{Alcohol}}{R{-}\overset{H}{\underset{H}{\overset{|}{\underset{|}{C}}}}{-}\ddot{\underset{..}{O}}H}}$$

2. Grignard reagents react with higher aldehydes to give secondary alcohols.

$$\mathrm{\overset{\delta-}{R}\!:\,\overset{\delta+}{Mg}X + \underset{\text{Higher aldehyde}}{\overset{R'}{\underset{H}{>}}C{=}\ddot{\underset{..}{O}}:} \longrightarrow R{-}\overset{R'}{\underset{H}{\overset{|}{\underset{|}{C}}}}{-}\ddot{\underset{..}{O}}:\,MgX \xrightarrow[H_3O^+]{} \underset{2^\circ\ \text{Alcohol}}{R{-}\overset{R'}{\underset{H}{\overset{|}{\underset{|}{C}}}}{-}\ddot{\underset{..}{O}}H}}$$

3. And, Grignard reagents react with ketones to give tertiary alcohols.

$$\mathrm{\overset{\delta-}{R}\!:\,\overset{\delta+}{Mg}X + \underset{\text{Ketone}}{\overset{R'}{\underset{R''}{>}}C{=}\ddot{\underset{..}{O}}:} \longrightarrow R{-}\overset{R'}{\underset{R''}{\overset{|}{\underset{|}{C}}}}{-}\ddot{\underset{..}{O}}:\,MgX \xrightarrow[H_3O^+]{} \underset{3^\circ\ \text{Alcohol}}{R{-}\overset{R'}{\underset{R''}{\overset{|}{\underset{|}{C}}}}{-}\ddot{\underset{..}{O}}H}}$$

Specific examples of these reactions are shown below.

$$C_6H_5MgBr + \underset{\text{Formaldehyde}}{H_2C{=}O} \xrightarrow{\text{Ether}} C_6H_5CH_2OMgBr \xrightarrow{H_3O^+} C_6H_5CH_2OH$$

Phenylmagnesium bromide; Formaldehyde; Benzyl alcohol (90%)

$$CH_3CH_2MgBr + CH_3(H)C{=}O \xrightarrow{\text{Ether}} CH_3CH_2C(CH_3)(H){-}OMgBr \xrightarrow{H_3O^+} CH_3CH_2CH(OH)CH_3$$

Ethylmagnesium bromide; Acetaldehyde; 2-Butanol (80%)

$$CH_3CH_2CH_2CH_2MgBr + (CH_3)_2C{=}O \xrightarrow{\text{Ether}} CH_3CH_2CH_2CH_2C(CH_3)_2{-}OMgBr \xrightarrow{H_3O^+} CH_3CH_2CH_2CH_2C(CH_3)(OH){-}CH_3$$

n-Butylmagnesium bromide; Acetone; 2-Methyl-2-hexanol (92%)

4. A Grignard reagent also adds to the carbonyl group of an ester. The initial product is unstable and it loses magnesium alkoxide to form a ketone. Ketones are more reactive toward Grignard reagents than esters. Therefore as soon as a molecule of the ketone is formed in the mixture, it reacts with a second molecule of the Grignard reagent. After hydrolysis, the product is a tertiary alcohol with two identical alkyl groups, groups that correspond to the alkyl portion of the Grignard reagent.

$$R{:}MgX + R'(R''\ddot{O})C{=}\ddot{O}{:} \longrightarrow \left[R{-}C(R')(\ddot{O}{-}R''){-}\ddot{O}{-}MgX\right] \xrightarrow[\text{Spontaneously}]{-R''OMgX}$$

Ester; Initial product (unstable)

$$\left[R'(R)C{=}\ddot{O}{:}\right] \xrightarrow{RMgX} R{-}C(R')(R){-}\ddot{O}MgX \xrightarrow{H_3O^+} R{-}C(R')(R){-}OH$$

A ketone; Salt of an alcohol (not isolated); 3° Alcohol

A specific example of this reaction is shown below.

$$CH_3CH_2MgBr + \underset{C_2H_5O}{\overset{CH_3}{>}}C{=}O \longrightarrow \left[CH_3CH_2{-}\underset{OC_2H_5}{\overset{CH_3}{\underset{|}{\overset{|}{C}}}}{-}OMgBr \right] \xrightarrow{-C_2H_5OMgBr}$$

Ethylmagnesium bromide — Ethyl acetate

$$\left[\underset{CH_3CH_2}{\overset{CH_3}{>}}C{=}O \right] \xrightarrow{CH_3CH_2MgBr} CH_3CH_2\underset{OMgBr}{\overset{CH_3}{\underset{|}{\overset{|}{C}}}}{-}CH_2CH_3 \xrightarrow{H_3O^+} CH_3CH_2\underset{OH}{\overset{CH_3}{\underset{|}{\overset{|}{C}}}}CH_2CH_3$$

3-Methyl-3-pentanol (67%)

Grignard reagents also react with ethylene oxide to form alcohols by opening the three-membered ring (Section 14.11B). After hydrolysis, the product is a primary alcohol with two carbons more than the organic group of the Grignard reagent.

$$CH_3CH_2CH_2{:}\ MgBr + \overset{\delta+}{C}H_2\overset{\delta+}{C}H_2\ (\text{ring } \overset{\delta-}{O}) \longrightarrow CH_3CH_2CH_2CH_2CH_2OMgBr \xrightarrow{H_3O^+} CH_3CH_2CH_2CH_2CH_2OH$$

Propylmagnesium bromide — Ethylene oxide — 1-Pentanol (76%)

14.12A Planning a Grignard Synthesis

By using the Grignard synthesis skillfully we can synthesize almost any alcohol we wish. In planning a Grignard synthesis we must simply choose the correct Grignard reagent and the correct aldehyde, ketone, ester, or epoxide. We do this by examining the alcohol we wish to prepare and by paying special attention to the groups attached to the carbon bearing the —OH group. Many times there may be more than one way of carrying out the synthesis. In these cases our final choice will probably be dictated by the availability of starting compounds. Let us consider an example.

Example

Suppose we want to prepare 3-phenyl-3-pentanol. We examine its structure and we see that the groups attached to the carbon bearing the —OH are a *phenyl*

$$\boxed{CH_3CH_2}{-}\underset{OH}{\overset{\boxed{C_6H_5}}{\underset{|}{\overset{|}{C}}}}{-}\boxed{CH_2CH_3}$$

3-Phenyl-3-pentanol

group and *two ethyl groups*. This means that we can synthesize this compound in three different ways.

1. We can use a ketone with two ethyl groups (3-pentanone) and allow it to react with phenylmagnesium bromide:

$$C_6H_5MgBr + CH_3CH_2\underset{\underset{O}{\|}}{C}CH_2CH_3 \xrightarrow{(2)H_3O^+} CH_3CH_2-\underset{OH}{\overset{C_6H_5}{C}}-CH_2CH_3$$

Phenylmagnesium bromide — 3-Pentanone — 3-Phenyl-3-pentanol

2. We can use a ketone containing an ethyl group and a phenyl group (ethyl phenyl ketone) and allow it to react with ethylmagnesium bromide:

$$CH_3CH_2MgBr + (C_6H_5)(CH_3CH_2)C{=}O \xrightarrow{2(H_3O^+)} CH_3CH_2-\underset{OH}{\overset{C_6H_5}{C}}-CH_2CH_3$$

Ethylmagnesium bromide — Ethyl phenyl ketone — 3-Phenyl-3-pentanol

3. Or, we can use an ester of benzoic acid and allow it to react with two moles of ethylmagnesium bromide:

$$2CH_3CH_2MgBr + C_6H_5\overset{\overset{O}{\|}}{C}OCH_3 \xrightarrow{(2)H_3O^+} CH_3CH_2-\underset{OH}{\overset{C_6H_5}{C}}-CH_2CH_3$$

Ethylmagnesium bromide — Methyl benzoate — 3-Phenyl-3-pentanol

All of these methods will be likely to give us our desired compound in yields greater than 80%.

Problem 14.15

Show how Grignard reactions could be used to synthesize each of the following compounds.

(a) *tert*-Butyl alcohol (two ways)

(b) $CH_3CH_2CH_2CHOHCH_3$ (two ways)

(c) $C_6H_5\underset{OH}{\overset{CH_3}{C}}CH_2CH_3$ (three ways)

(d) $CH_3CH_2CH_2CH_2CH_2CH_2OH$ (two ways)

14.12B Restrictions on the Use of Grignard Reagents

While the Grignard synthesis is one of the most versatile of all general synthetic procedures, it is not without its limitations. Most of these limitations arise from the very feature of the Grignard reagent that makes it so useful, its *extraordinary reactivity as a nucleophile and a base.*

The Grignard reagent is a very powerful base; it effectively contains a carbanion. Thus, it is not possible to prepare a Grignard reagent from an organic group that contains an *acidic hydrogen;* and by an acidic hydrogen we mean any hydrogen more acidic than the hydrogens of an alkane or alkene. We cannot, for example, prepare a Grignard reagent from a compound containing an —OH group, an —NH— group, an —SH group, a —COOH group, or an $—SO_3H$ group. If we were to attempt to prepare a Grignard reagent from an organic halide containing any of these groups, the formation of the Grignard reagent would simply fail to take place. (Even if a Grignard reagent were to form, it would immediately react with the acidic group.)

Grignard reactions are so sensitive to acidic compounds that when we prepare a Grignard reagent we must take special care to exclude moisture from our apparatus, and we must use anhydrous ether as our solvent.

Even acetylenic hydrogens are acidic enough to react with Grignard reagents. This is a limitation that we can use, however. We can make acetylenic Grignard reagents by allowing terminal alkynes to react with alkyl Grignard reagents (cf. Section 14.8). We can then use these acetylenic Grignard reagents to carry out other syntheses. For example:

$$C_6H_5C{\equiv}CH + C_2H_5MgBr \longrightarrow C_6H_5C{\equiv}CMgBr + C_2H_6\uparrow$$

$$C_6H_5C{\equiv}CMgBr + C_2H_5\overset{\overset{\displaystyle O}{\|}}{C}H \xrightarrow[(2)H^+]{} C_6H_5C{\equiv}C{-}\underset{\underset{\displaystyle OH}{|}}{C}HC_2H_5$$

(52%)

When we plan Grignard syntheses we must also take care not to plan a reaction in which a Grignard reagent is treated with an aldehyde, ketone, or ester that contains an acidic group (other than when we deliberately let it react with a terminal alkyne). If we were to do this, the Grignard reagent would simply react as a base with the acidic hydrogen rather than react at the carbonyl or epoxide carbon as a nucleophile. If we were to treat 4-hydroxy-2-butanone with methylmagnesium bromide, for example, the following reaction would take place first,

$$CH_3MgBr + HOCH_2CH_2\underset{\underset{\displaystyle O}{\|}}{C}CH_3 \longrightarrow CH_4\uparrow + BrMgOCH_2CH_2\underset{\underset{\displaystyle O}{\|}}{C}CH_3$$

4-Hydroxy-2-butanone

rather than

$$CH_3MgBr + HOCH_2CH_2\underset{\underset{\displaystyle O}{\|}}{C}CH_3 \not\longrightarrow HOCH_2CH_2\overset{\overset{\displaystyle CH_3}{|}}{\underset{\underset{\displaystyle OMgBr}{|}}{C}}{-}CH_3$$

Since Grignard reagents are powerful nucleophiles we cannot prepare a Grignard reagent from any organic halide that contains a carbonyl, epoxy, or cyano (—CN) group. If we were to attempt to carry out this kind of reaction, any Grignard reagent that formed would only react with the unreacted starting material.

14.12C The Use of Lithium Reagents

Organolithium reagents, RLi, react with carbonyl compounds in the same way as Grignard reagents and thus provide an alternative method for preparing alcohols.

$$\overset{\delta-}{R}:\overset{\delta+}{Li} + \;\rangle C{=}\ddot{O}: \longrightarrow R{-}\overset{|}{\underset{|}{C}}{-}\ddot{\underset{\cdot\cdot}{O}}:Li \xrightarrow{H_3O^+} R{-}\overset{|}{\underset{|}{C}}{-}OH$$

Organolithium reagent — Aldehyde or Ketone — Lithium alkoxide — Alcohol

Organolithium reagents have the advantage of being somewhat more reactive than Grignard reagents.

14.13 REACTION OF GRIGNARD AND ORGANOLITHIUM REAGENTS WITH LESS ELECTROPOSITIVE METAL HALIDES

Grignard reagents and alkyllithium reagents react with a number of halides of less electropositive metals to produce new organometallic compounds. Since Grignard reagents and alkyllithiums are easily prepared, these reactions furnish us with useful syntheses of alkyl derivatives of mercury, zinc, cadmium, copper, silicon, and phosphorus, for example. In general terms the reactions using Grignard reagents take the form shown below.

$$nRMgX + MX_n \longrightarrow R_nM + nMgX_2$$

(M is less electropositive than Mg)

Several specific examples are shown below.

$$2CH_3MgCl + HgCl_2 \longrightarrow (CH_3)_2Hg + 2MgCl_2$$

Dimethylmercury

$$2CH_3CH_2MgBr + ZnCl_2 \longrightarrow (CH_3CH_2)_2Zn + 2MgClBr$$

(100%) Diethylzinc

$$2C_6H_5MgBr + CdCl_2 \longrightarrow (C_6H_5)_2Cd + 2MgClBr$$

(>83%) Diphenylcadmium

$$3CH_3CH_2CH_2CH_2MgBr + PCl_3 \longrightarrow (CH_3CH_2CH_2CH_2)_3P + 3MgClBr$$

(57%) Tributylphosphine

$$4CH_3CH_2MgCl + SiCl_4 \longrightarrow (CH_3CH_2)_4Si + 4MgCl_2$$

(80%) Tetraethylsilane

We will see in Chapter 16 that organocadmium compounds are useful reagents for preparing ketones.

Alkyllithiums are used in the preparation of lithium dialkylcuprates for the Corey-House synthesis of alkanes (cf. Section 3.16) and alkenes.

$$2CH_3Li + CuI \xrightarrow[Et_2O]{0°} (CH_3)_2CuLi$$

Lithium dimethylcuprate

Coupling reactions of lithium dimethylcuprate with an alkyl and an allylic halide are shown below,

$$\text{Iodocyclohexane} + (CH_3)_2CuLi \xrightarrow[0°,\ Et_2O]{10\text{ hr}} \text{Methylcyclohexane} + CH_3Cu + LiI$$

(75%)
Methylcyclohexane

$$\text{3-Bromocyclohexene} + (CH_3)_2CuLi \xrightarrow[Et_2O]{0°} \text{3-Methylcyclohexene} + CH_3Cu + LiBr$$

(75%)
3-Methylcyclohexene

Vinylic halides also couple with lithium dialkylcuprates. They do so in a stereospecific way as the following example illustrates.

$$(n\text{-}C_8H_{17})HC{=}CHI + n\text{-}(C_4H_9)_2CuLi \xrightarrow[Et_2O]{-95°} (n\text{-}C_8H_{17})HC{=}CH(C_4H_9)$$

E-1-Iodo-1-decene

(74%)
E-5-Tetradecene

Aryl Grignard reagents undergo coupling reactions in very high yield when they are treated with thallium(I) bromide. Two examples are shown below.

$$2\ C_6H_5\text{—}MgBr \xrightarrow{TlBr} C_6H_5\text{—}C_6H_5 \quad (92\%)$$

Phenylmagnesium bromide — Biphenyl

$$C_{10}H_7\text{—}MgBr \xrightarrow{TlBr} C_{10}H_7\text{—}C_{10}H_7$$

2-Naphthylmagnesium bromide — (84%) 2,2′-Binaphthyl

14.14 METALLOCENES: ORGANOMETALLIC SANDWICH COMPOUNDS

Cyclopentadiene reacts with phenylmagnesium bromide to give the Grignard reagent of cyclopentadiene. This reaction is not unusual, for it is simply another acid-base reaction like those we saw earlier. The methylene hydrogens of cyclo-

$$\text{Cyclopentadiene} + C_6H_5MgBr \xrightarrow{\text{Ether}} \text{Cyclopentadienylmagnesium bromide} + C_6H_6$$

Cyclopentadiene + Phenylmagnesium bromide → Cyclopentadienylmagnesium bromide + Benzene

pentadiene are much more acidic than the hydrogens of benzene and, therefore, the reaction goes to completion. (The methylene hydrogens of cyclopentadiene are acidic relative to ordinary methylene hydrogens because the cyclopentadienyl anion is aromatic, cf. Section 11.7.)

When the Grignard reagent of cyclopentadiene is treated with ferrous chloride a reaction takes place that produces a highly interesting product, called *ferrocene*.

$$2\,C_5H_5^{-}\,\overset{2+}{Mg}\overset{-}{Br} + FeCl_2 \longrightarrow (C_5H_5)_2Fe + 2MgBrCl$$

Ferrocene
(71% overall yield from cyclopentadiene)

Ferrocene is an orange solid with a melting point of 174°. It is a highly stable compound; ferrocene can be sublimed at 100° and is not damaged when heated to 400°.

Many studies, including X-ray analysis, show that ferrocene is a compound in which the iron (II) ion is located betweeen two cyclopentadienide rings.

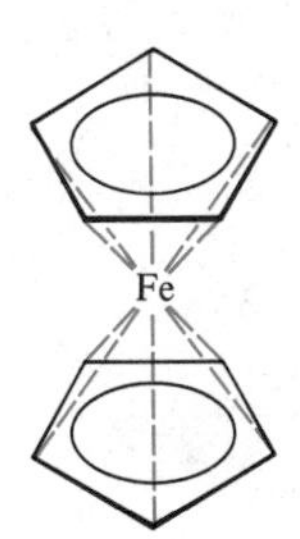

The carbon-carbon bond distances are all 1.40 Å and the carbon-iron bond distances are all 2.04 Å. Because of their structures, molecules such as ferrocene have been called "sandwich" compounds.

The carbon-iron bonding in ferrocene results from overlap between the inner lobes of the p orbitals of the cyclopentadienyl anions and $3d$ orbitals of the iron atom. Studies have shown, moreover, that this bonding is such that the rings of ferrocene are capable of essentially free rotation about an axis that passes through the iron atom and that is perpendicular to the rings.

Ferrocene is an *aromatic compound.* It undergoes a number of electrophilic aromatic substitution, including sulfonation and Friedel-Crafts acylation.

The discovery of ferrocene (in 1951) was followed by the preparation of a number of similar aromatic compounds. These compounds, as a class, are called *metallocenes.** Metallocenes with five-, six-, seven-, and even eight-membered

* Professors Ernst O. Fischer (of the Technical University, Munich) and Geoffrey Wilkinson (of Imperial College, London) received the Nobel Prize in 1973 for their pioneering work (performed independently) on the chemistry of organometallic sandwich compounds—or metallocenes.

rings have been synthesized from metals as diverse as manganese, cobalt, nickel, chromium, and uranium.

"Half-sandwich" compounds have been prepared through the use of metal carbonyls. Several are shown below.

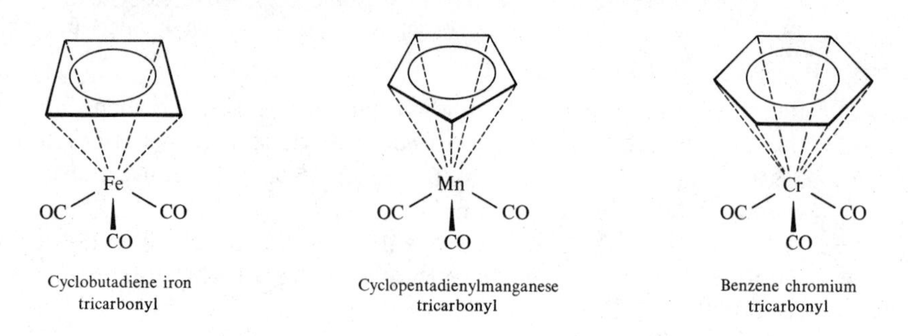

Cyclobutadiene iron tricarbonyl

Cyclopentadienylmanganese tricarbonyl

Benzene chromium tricarbonyl

Although cyclobutadiene itself is *not* stable, the tricarbonylcyclobutadieneiron is.

Problem 14.16

When ferrocene is acetylated twice in sequential fashion, one acetyl group becomes attached to each ring (rather than both acetyl groups to the same ring). How can you account for this?

Additional Problems

14.17

Show how you might prepare 2-bromobutane from
(a) 2-Butanol, $CH_3CH_2CHOHCH_3$
(b) 1-Butanol, $CH_3CH_2CH_2CH_2OH$
(c) 1-Butene
(d) 1-Butyne

14.18

Show how you might prepare 1-bromobutane from each of the compounds listed in Problem 14.17.

14.19

Show how you might carry out the following transformations.
(a) Cyclohexanol $\longrightarrow$ chlorocyclohexane
(b) Cyclohexene $\longrightarrow$ chlorocyclohexane
(c) 1-Methylcyclohexene $\longrightarrow$ 1-bromo-1-methylcyclohexane
(d) 1-Methylcyclohexene $\longrightarrow$ 2-bromo-1-methylcyclohexane
(e) 1-Bromo-1-methylcyclohexane $\longrightarrow$ 1-deuterio-1-methylcyclohexane

14.20

What product would be formed from the reaction of isobutyl bromide, $(CH_3)_2CHCH_2Br$, with each of the following reagents?
(a) OH^-, H_2O
(b) CN^-, alcohol
(c) $(CH_3)_3CO^-$, $(CH_3)_3COH$
(d) CH_3O^-, CH_3OH
(e) Li, ether, then $CH_3\overset{\overset{\displaystyle O}{\|}}{C}CH_3$, then H_3O^+

(f) Mg, ether, then $CH_3\overset{\overset{O}{\|}}{C}H$, then H_3O^+

(g) Mg, ether, then $CH_3\overset{\overset{O}{\|}}{C}OCH_3$, then H_3O^+

(h) Mg, ether, then $\overset{O}{\overbrace{CH_2\text{—}CH_2}}$, then H_3O^+

(i) Mg, ether, then $H\text{—}\overset{\overset{O}{\|}}{C}\text{—}H$, then H_3O^+

(j) Li, ether, then CH_3OH

(k) Li, ether, then $CH_3C{\equiv}CH$

14.21

What products would you expect from the reaction of ethylmagnesium bromide, CH_3CH_2MgBr, with each of the following reagents?

(a) H_2O

(b) D_2O

(c) $C_6H_5\overset{\overset{O}{\|}}{C}H$, then H_3O^+

(d) $C_6H_5\overset{\overset{O}{\|}}{C}C_6H_5$, then H_3O^+

(e) $C_6H_5\overset{\overset{O}{\|}}{C}OCH_3$, then H_3O^+

(f) $C_6H_5\overset{\overset{O}{\|}}{C}CH_3$, then H_3O^+

(g) $CH_3CH_2C{\equiv}CH$, then $CH_3\overset{\overset{O}{\|}}{C}H$, then H_3O^+

(h) Cyclopentadiene

(i) $HgCl_2$

(j) $CdCl_2$

(k) PCl_3

14.22

What products would you expect from the reaction of propyllithium, $CH_3CH_2CH_2Li$, with each of the following reagents?

(a) $(CH_3)_2CH\overset{\overset{O}{\|}}{C}H$, then H_3O^+

(b) $(CH_3)_2CH\overset{\overset{O}{\|}}{C}CH_3$, then H_3O^+

(c) 1-Pentyne, then $CH_3\overset{\overset{O}{\|}}{C}CH_3$, then H_3O^+

(d) Ethyl alcohol

(e) CuI, then $CH_2{=}CHCH_2Br$

(f) CuI, then cyclopentyl bromide

(g) CuI, then *Z*-1-iodopropene

(h) $(CH_3)_2CuLi$

(i) CH_3COOD

(j) $SiCl_4$

(k) $ZnCl_2$

14.23

Show how you might prepare each of the following alcohols through a Grignard synthesis. (Assume that you have available any necessary organic halides, aldehydes, ketones, esters, and epoxides as well as any necessary inorganic reagents.)

(a) $CH_3CH_2C(CH_3)_2OH$ (three ways)

(b) $C_6H_5C(OH)(CH_2CH_3)CH_2CH_3$ (three ways)

(c) 1-phenylcyclohexanol ($C_6H_{10}(OH)C_6H_5$)

(d) cyclopentyl–CH_2CH_2OH

(e) cyclobutyl–$CH(OH)CH_3$ (two ways)

14.24

Outline all steps in a synthesis that would transform isopropyl alcohol, $CH_3CHOHCH_3$, into each of the following:

(a) $(CH_3)_2CHCHOHCH_3$

(b) $(CH_3)_2CHCH_2OH$

(c) $(CH_3)_2CHCH_2CH_2Cl$

(d) $(CH_3)_2CHCHOHCH(CH_3)_2$

(e) CH_3CHDCH_3

(f) cyclohexyl–$CH(CH_3)_2$

14.25

Although $(C_6H_5)_2CHCl$ is a secondary halide, it undergoes S_N1 solvolysis in 80% ethanol at a rate that is 1300 times that for the tertiary halide, $(CH_3)_3CCl$. Explain.

14.26

When Grignard reagents are prepared from allyl halides, that is,

$$RCH{=}CHCH_2X + Mg \xrightarrow{\text{Ether}} RCH{=}CHCH_2MgX$$

unavoidable by-products of the reactions are compounds with the formula $RCH{=}CHCH_2CH_2CH{=}CHR$. By-product formation is especially prevalent when concentrated solutions are used. Explain.

14.27

(a) How can you account for the following relative S_N1 reaction rates?

p-$CH_3OC_6H_4CH_2X$	$C_6H_5CH_2X$	p-$O_2NC_6H_4CH_2X$
(5×10^3)	(1)	(5×10^{-3})

(b) How can you explain the fact that a *m*-methoxybenzyl halide undergoes S_N1 reaction at about the same rate as an unsubstituted benzyl halide?

14.28

How might you carry out the following transformations?

(a) Bromocyclopentane ⟶ methylcyclopentane
(b) 3-Bromocyclopentene ⟶ 3-methylcyclopentene
(c) Allyl bromide ⟶ 1-pentene
(d) (*E*)-2-Iodo-2-butene ⟶ (*E*)-3-methyl-2-heptene

14.29

What products would you expect from the following reactions?

(a) Phenyllithium + acetic acid ⟶
(b) Phenyllithium + methyl alcohol ⟶
(c) Methylmagnesium bromide + ammonia ⟶
(d) (Four moles) methylmagnesium bromide + $SiCl_4$ ⟶
(e) (Three moles) phenylmagnesium bromide + PCl_3 ⟶
(f) (Two moles) ethylmagnesium bromide + $CdCl_2$ ⟶
(g) Phenylmagnesium bromide + H—C(=O)—H $\xrightarrow[(2) H_3O^+]{}$

14.30

Propose simple chemical tests that could be used to distinguish between the following pairs of compounds.

(a) Allyl bromide and *n*-propyl bromide
(b) Benzyl bromide and *p*-bromotoluene
(c) Vinyl chloride and benzyl chloride
(d) Phenyllithium and diphenylmercury
(e) Bromobenzene and bromocyclohexane

14.31

Compounds **A** and **B** are isomers with the molecular formula $C_4H_6Cl_2$. Both compounds decolorize bromine in carbon tetrachloride. The proton nmr spectrum of **A** shows only two signals: a singlet at $\delta 4.25$ and a singlet at $\delta 5.35$; the signal area ratio is 2:1. The proton nmr spectrum of **B** shows a singlet at $\delta 2.2$, a doublet at $\delta 4.15$, and a triplet at $\delta 5.7$; the signal area ratios are 3:2:1. Propose structures for **A** and **B.**

14.32

Propose a structure for compound **C,** $C_4H_8Cl_2$, whose proton nmr spectrum is given in Fig. 14.1.

14.33

The proton nmr spectrum of compound **D,** C_4H_7Cl, is given in Fig. 14.2. Compound **D** decolorizes bromine in carbon tetrachloride and gives a precipitate when treated with alcoholic $AgNO_3$. Propose a structure for **D.**

14.34

Compound **E,** C_4H_6, reacts with one mole of chlorine to yield a mixture of isomeric compounds **F** and **G** with molecular formula $C_4H_6Cl_2$. Compounds **F** and **G** react with chlorine to yield the same compound **H,** $C_4H_6Cl_4$. The proton nmr spectrum of compound **H** is shown in Fig. 14.3. Propose structures for **E, F, G,** and **H.**

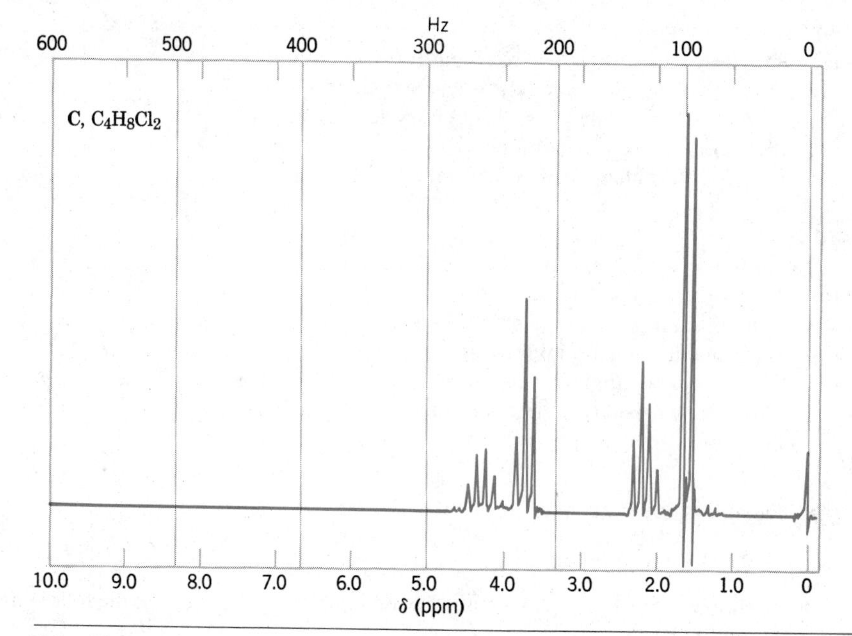

FIG. 14.1
The proton nmr spectrum of compound **C,** *Problem 14.32. (Spectrum courtesy of Aldrich Chemical Co., Milwaukee, Wis.)*

FIG. 14.2
The proton nmr spectrum of compound **D,** *Problem 14.33. (Spectrum courtesy of Aldrich Chemical Co., Milwaukee, Wis.)*

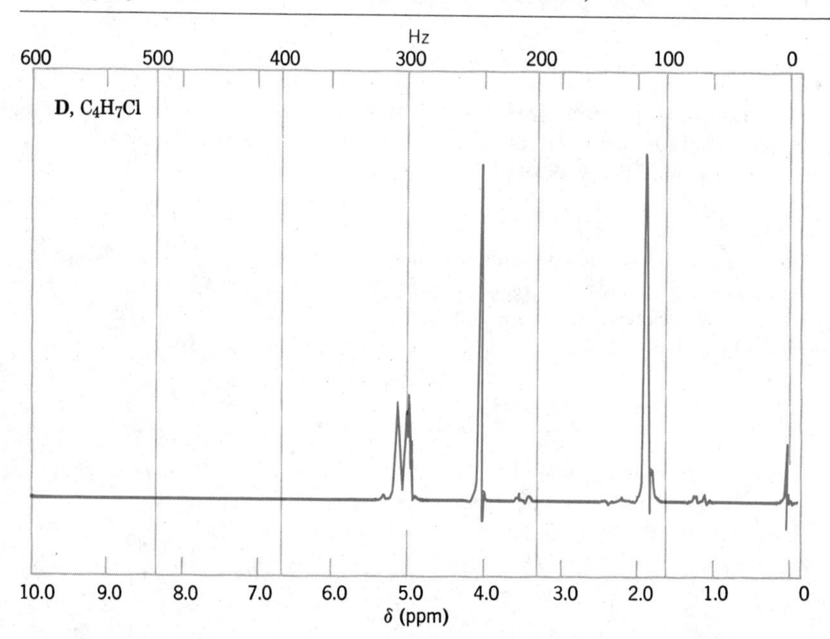

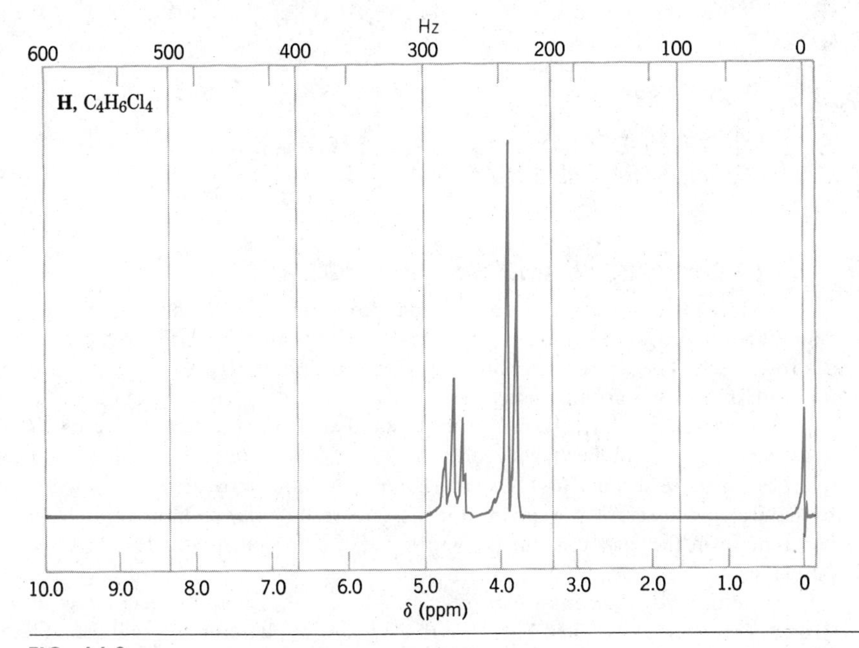

FIG. 14.3
*The proton nmr spectrum of compound **H**, Problem 14.34. (Spectrum courtesy of Aldrich Chemical Co., Milwaukee, Wis.)*

SPECIAL TOPIC

ORGANIC HALIDES AND ORGANOMETALLIC COMPOUNDS IN THE ENVIRONMENT

I.1 ORGANIC HALIDES AS INSECTICIDES

Since the discovery of the insecticidal properties of DDT in 1942, vast quantities of chlorinated hydrocarbons have been sprayed over the surface of the earth in an effort to destroy insects. These efforts initially met with incredible success in ridding large areas of the earth of disease-carrying insects, particularly those of typhus and malaria. As time has passed, however, we have begun to understand that this prodigious use of chlorinated hydrocarbons has not been without harmful—indeed tragic—side effects. Chlorinated hydrocarbons are usually highly stable compounds and they are only slowly destroyed by natural processes in the environment. As a result, many chloroorganic insecticides will remain in the environment for years. These persistent pesticides are called "hard" pesticides.

Chlorohydrocarbons are also fat soluble and they tend to accumulate in the fatty tissues of most animals. The food chain that runs from plankton to small fish to larger fish to birds and to larger animals including man tends to magnify the concentrations of chloroorganic compounds at each step.

The chlorohydrocarbon DDT is prepared from inexpensive starting materials, chlorobenzene and trichloroacetaldehyde. The reaction is catalyzed by acid.

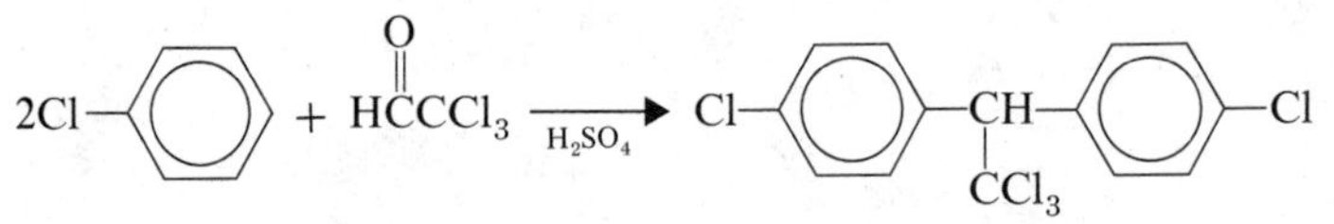

$$2\,Cl{-}C_6H_5 + HC(=O)CCl_3 \xrightarrow{H_2SO_4} Cl{-}C_6H_4{-}CH(CCl_3){-}C_6H_4{-}Cl$$

DDT
1,1,1-Trichloro-2,2-bis(*p*-chlorophenyl)-ethane

In nature the principal decomposition product of DDT is DDE.

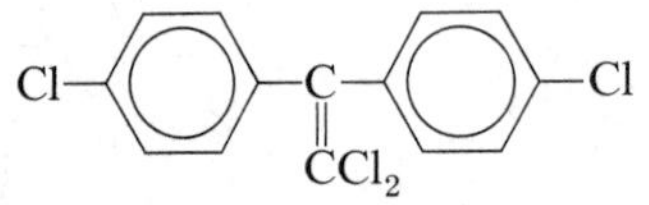

$$Cl{-}C_6H_4{-}C(=CCl_2){-}C_6H_4{-}Cl$$

DDE
1,1-Dichloro-2,2-bis(*p*-chlorophenyl)ethene

Estimates indicate that nearly one billion pounds of DDE are now spread throughout the world ecosystem. One pronounced environmental effect of DDE has been in its action on egg-shell formation of many birds. DDE inhibits the enzyme *carbonic anhydrase* that controls the calcium supply for shell formation. As a consequence, the shells are often very fragile and do not survive until hatching occurs. During the late 1940s the populations of eagles, falcons, and hawks dropped dramatically. There can be little doubt that DDE was primarily responsible.

DDE also accumulates in the fatty tissues of man. Although man appears to have a short-range tolerance to moderate DDE levels, the long-range effects are far from certain.

Other hard insecticides are Aldrin, Dieldrin, and Chlordan. Aldrin can be manufactured through the Diels-Alder reaction of hexachlorocyclopentadiene and norbornadiene.

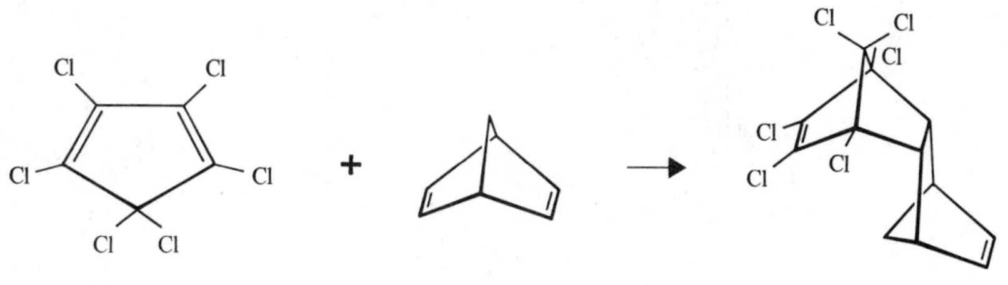

hexachloro-cyclopentadiene + norbornadiene → Aldrin

Chlordan can be made by adding chlorine to the unsubstituted double bond of the Diels-Alder adduct obtained from hexachlorocyclopentadiene and cyclopentadiene. Dieldrin can be made by converting an Aldrin double bond to an epoxide. (This reaction also takes place in nature.)

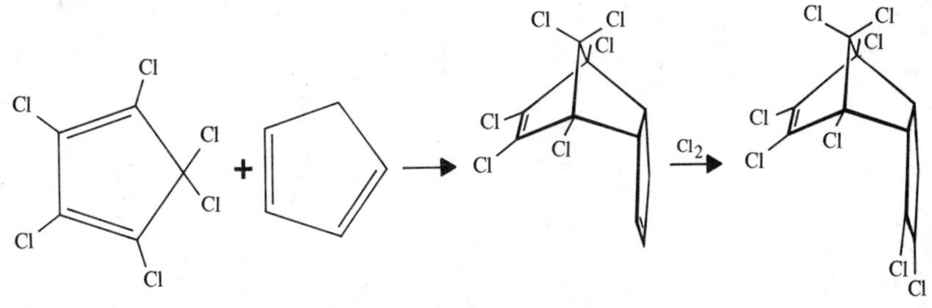

$\xrightarrow{Cl_2}$ chlordan

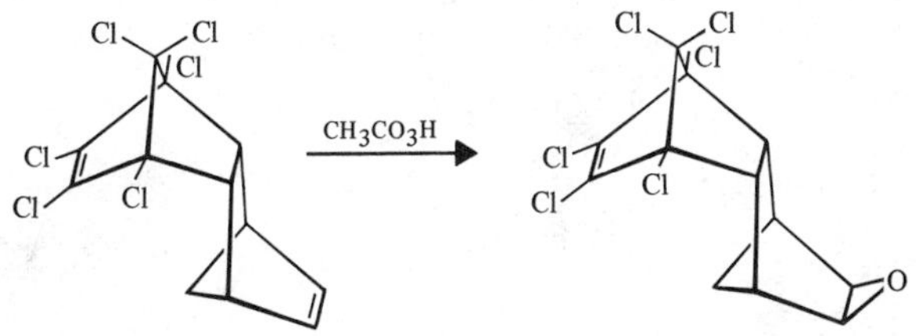

Aldrin $\xrightarrow{CH_3CO_3H}$ Dieldrin

During the 1970s the Environmental Protection Agency (EPA) banned the use of DDT, Aldrin, Dieldrin, and Chlordan because of known or suspected hazards to human life. All of the compounds are suspected of causing cancers.

Problem I.1

The mechanism for the formation of DDT from chlorobenzene and trichloroacetaldehyde in sulfuric acid involves two electrophilic aromatic substitution reactions. In the first electrophilic substitution reaction the electrophile is protonated trichloroacetaldehyde. In the second the electrophile is a carbocation. Propose a mechanism for the formation of DDT.

Problem I.2

What kind of reaction is involved in the conversion of DDT to DDE?

Mirex, Kepone, and lindane (p. 636) are also hard insecticides whose use has been banned.

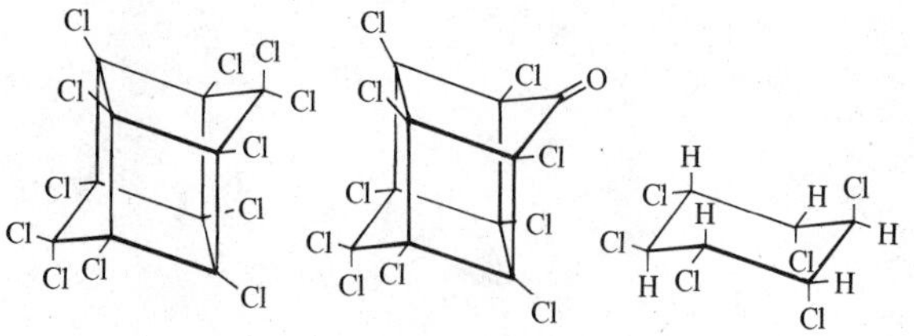

mirex kepone lindane

I.2 ORGANIC HALIDES AS HERBICIDES

Other chlorinated organic compounds have been used extensively as herbicides. Two examples are 2,4-D and 2,4,5-T shown below.

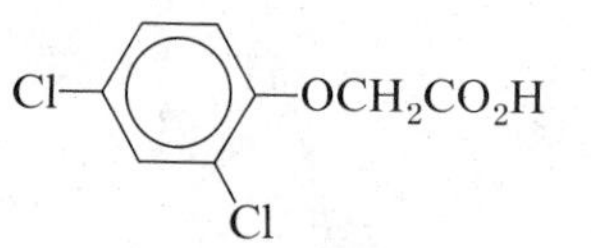

2,4-D
2,4-Dichlorophenoxy-
acetic acid

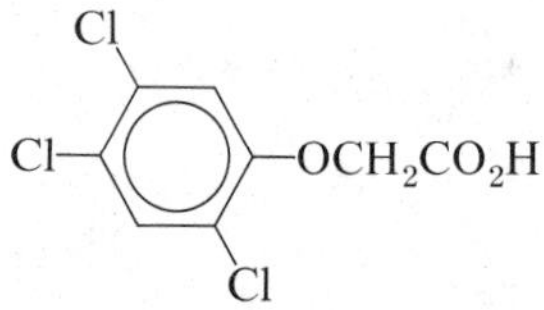

2,4,5-T
2,4,5-Trichlorophenoxy-
acetic acid

Enormous quantities of these two compounds were used as defoliants in the jungles of Indochina during the Vietnam war. Some samples of 2,4,5-T have been shown to be a teratogen (a fetus-deforming agent). This teratogenic effect was the result of an impurity present in commercial 2,4,5-T, the compound 2,3,7,8-tetrachlorodibenzodioxin. 2,3,7,8-Tetrachlorodibenzodioxin is also highly toxic; it is more toxic, for example, than cyanide ion, strychnine, and the nerve gases.

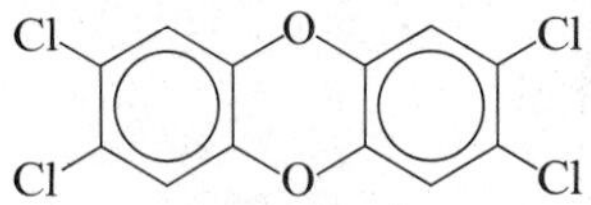

2,3,7,8-Tetrachlorodibenzodioxin
(Also called TCDD)

This dioxin is also highly stable; it persists in the environment and because of its fat solubility can be passed up the food chain. In sublethal amounts it can cause a disfiguring skin disease called chloracne.

In July 1976 an explosion at a chemical plant in Seveso, Italy, caused the release of between 22 and 132 pounds of this dioxin into the atmosphere. The plant was engaged in the manufacture of 2,4,5-trichlorophenol (used in making 2,4,5-T) using the following method:

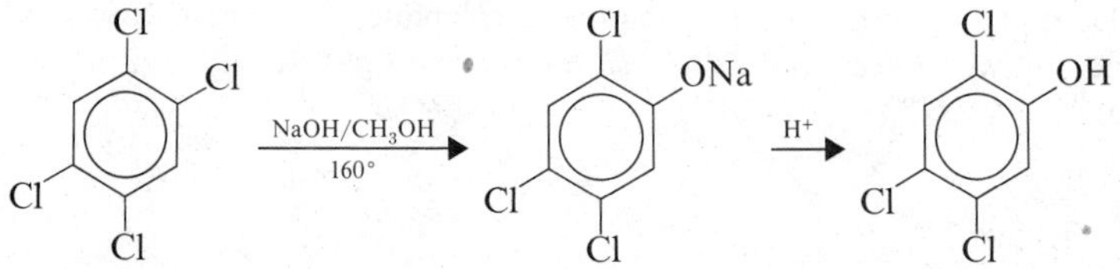

1,2,4,5-Tetra-
chlorobenzene $\xrightarrow[160°]{NaOH/CH_3OH}$ Sodium 2,4,5-
trichlorophenoxide $\xrightarrow{H^+}$ 2,4,5-Trichloro-
phenol

The temperature of the first reaction must be very carefully controlled, if it is not, this dioxin forms in the reaction mixture:

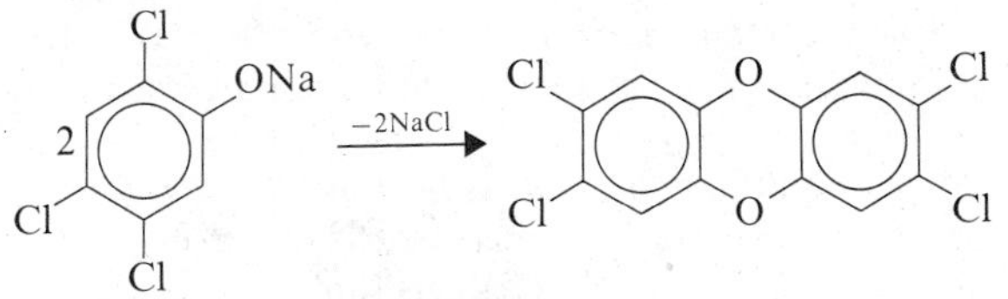

Apparently at the Italian factory the temperature got out of control causing the pressure to build up. Eventually a valve opened, releasing a cloud of trichlorophenols and the dioxin into the atmosphere. Many wild and domestic animals were killed and many people, especially children, were afflicted with severe skin rashes.

Problem I.3

(a) Assume that the ortho and para chlorine atoms provide enough activation by electron withdrawal for nucleophilic substitution to occur by an addition-elimination pathway and outline a possible mechanism for the conversion of 1,2,4,5-tetrachlorobenzene to sodium 2,4,5-trichlorophenoxide. (b) Do the same for the conversion of 2,4,5-trichlorophenoxide to the dioxin of Section I.2.

Problem I.4

2,4,5-T is made by allowing sodium 2,4,5-trichlorophenoxide to react with sodium chloroacetate, $ClCH_2COONa$. (This produces the sodium salt of 2,4,5-T, which, on acidification, gives 2,4,5-T itself.) What kind of mechanism accounts for the reaction of sodium 2,4,5-trichlorophenoxide with $ClCH_2COONa$? Write the equation.

I.3 Germicides

2,4,5-Trichlorophenol is also used in the manufacture of hexachlorophene, a germicide once widely used in soaps, shampoos, deodorants, mouthwashes, aftershave lotions, and other over-the-counter products.

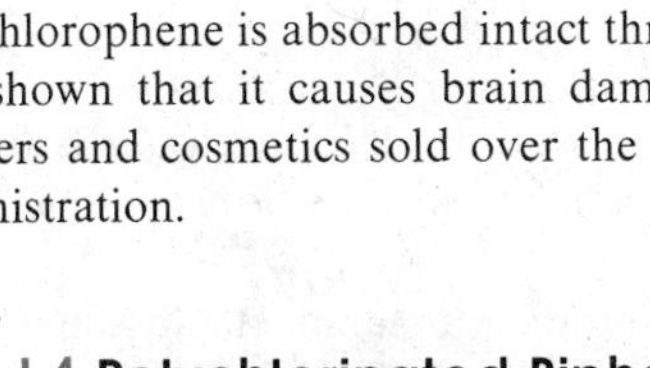

Hexachlorophene

Hexachlorophene is absorbed intact through the skin and tests with experimental animals have shown that it causes brain damage. Since 1972, the use of hexachlorophene in cleansers and cosmetics sold over the counter has been banned by the Food and Drug Administration.

I.4 Polychlorinated Biphenyls (PCBs)

Mixtures of polychlorinated biphenyls have been produced and used commercially since 1929. In these mixtures, biphenyls with chlorine atoms at any of the numbered positions (below) may be present. In all, there are 210 possible compounds. A typical

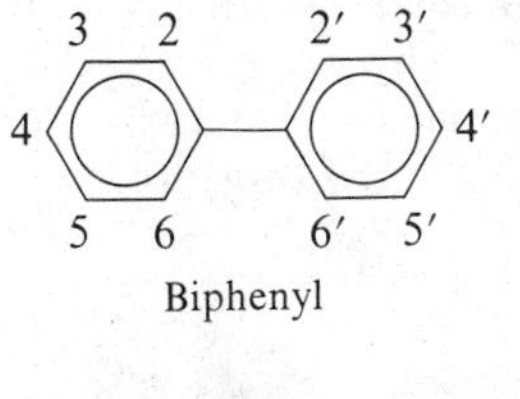

Biphenyl

commercial mixture may contain as many as 50 different PCBs. Mixtures are usually classified on the basis of their chlorine content, and most industrial mixtures contain from 40 to 60 percent chlorine.

PCBs have had a multitude of uses: as heat exchange agents in transformers; in capacitors, thermostats, and hydraulic systems; as plasticizers in polystyrene coffee cups, frozen food bags, bread wrappers, and plastic liners for baby bottles. They have been used in printing inks, in carbonless carbon paper, and as waxes for making molds for metal castings. Between 1929 and 1972 about 500,000 metric tons of PCBs were manufactured.

Although they were never intended for release into the environment, PCBs have become, perhaps more than any other chemical, the most widespread pollutant. They have been found in rain water, in many species of fish, birds, and others animals (including polar bears) all over the globe, and in human tissue.

PCBs are highly persistent and being fat soluble they tend to accumulate in the food chain. Fish that feed in PCB-contaminated waters, for example, have PCB levels 1000 to 100,000 times the level of the surrounding water and this amount is further magnified in birds that feed on the fish. The toxicity of PCBs depends on the composition of the individual mixture. The largest incident of human poisoning by PCBs occurred in Japan in 1968 when about 1000 people ingested a cooking oil accidentally contaminated with PCBs.

As late as 1975 industrial concerns were legally discharging PCBs into the Hudson river. In 1977 the Environmental Protection Agency banned the direct discharge into waterways and since 1979 their manufacture, processing, and distribution have been prohibited.

I.5 Polybromobiphenyls (PBBs)

Polybromobiphenyls are bromine analogs of PCBs that have been used as flame retardants. In 1973, in Michigan, a mistake at a chemical company led to PBBs being mixed into animal feeds that were sold to farmers. Before the mistake was recognized, PBBs had affected thousands of dairy cattle, hogs, chickens, and sheep, necessitating their destruction.

I.6 Organometallic Compounds

With few exceptions, organometallic compounds are toxic. This toxicity varies greatly depending on the nature of the organometallic compound and the identity of the metal. Organic compounds of arsenic, antimony, lead, thallium, and mercury are toxic because the metal ions, themselves, are toxic. Certain organic derivatives of silicon are toxic even though silicon and most of its inorganic compounds are nontoxic.

Early in this century the recognition of the biocidal effects of organoarsenic compounds led Paul Ehrlich to his pioneering work in chemotherapy. Ehrlich sought compounds (which he called "magic bullets") that would show greater toxicity toward disease-causing microorganisms than they would toward their hosts. Ehrlich's research led to the development of Salvarsan and Neosalvarsan, two organoarsenic compounds that were used successfully in the treatment of diseases caused by spirochetes (e.g., syphilis) and trypanosomes (e.g., sleeping sickness). Salvarsan and Neosalvarsan are no longer used in the treatment of these diseases; they have been displaced by safer and more effective antibiotics. Ehrlich's research, however, initiated the field of chemotherapy (cf. Section 18.11).

Many microorganisms actually synthesize organometallic compounds, and this discovery has an alarming ecological aspect. Mercury metal is toxic, but mercury metal is also unreactive. In the past untold tons of mercury metal present in industrial wastes have been disposed of by simply dumping such wastes into lakes and streams. Since mercury is toxic, many bacteria protect themselves from its effect by converting mercury metal to methylmercury ions (CH_3Hg^+) and to gaseous dimethylmercury ($(CH_3)_2Hg$). These organic mercury compounds are passed up the food chain (with modification) through fish to humans where methylmercury ions act as a deadly nerve poison. Between 1953 and 1964, 116 people in Minamata, Japan, were poisoned by eating fish containing methylmercury compounds. Arsenic is also methylated by organisms to the poisonous dimethylarsine, $(CH_3)_2AsH$.

Ironically, chlorinated hydrocarbons appear to inhibit the biological reactions that bring about mercury methylation. Lakes polluted with organochlorine pesticides show significantly lower mercury methylation. While this particular interaction of two pollutants may, in a certain sense, be beneficial, it is also instructive of the complexity of the environmental problems that we may face.

Tetraethyllead and other alkyllead compounds have been used as antiknock agents in gasoline since 1923. In the years since then, more than one trillion pounds of lead have been introduced into the atmosphere. In the northern hemisphere, gasoline burning alone has spread about 10 mg of lead on each square meter of the earth's surface. In highly industrialized areas the amount of lead per square meter is probably several hundred times higher. Because of the well-known toxicity of lead, these facts must also be of great concern.

15 ALCOHOLS, PHENOLS, AND ETHERS

15.1 STRUCTURE AND NOMENCLATURE

Alcohols are compounds whose molecules have a hydroxyl group attached to a *saturated* carbon.* The saturated carbon may be a carbon of a simple alkyl group:

CH_3OH	CH_3CH_2OH	$CH_3CH(OH)CH_3$
Methanol (methyl alcohol)	Ethanol (ethyl alcohol)	2-Propanol (isopropyl alcohol)

The carbon may be a saturated carbon of an alkenyl or alkynyl group.

$CH_2{=}CHCH_2OH$	$H{-}C{\equiv}CCH_2OH$
2-Propenol (allyl alcohol)	2-Propynol (propargyl alcohol)

Or, the carbon may be a saturated carbon that is attached to an aromatic ring.

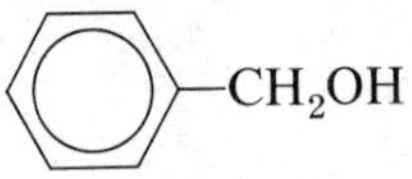

$C_6H_5{-}CH_2OH$

Benzyl alcohol

Compounds that have a hydroxyl group attached directly to an aromatic ring are called *phenols*.

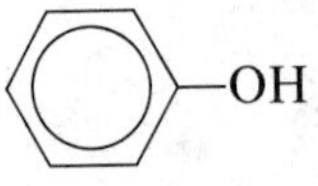

$C_6H_5{-}OH$	$CH_3{-}C_6H_4{-}OH$
Phenol	*p*-Cresol (a phenol)

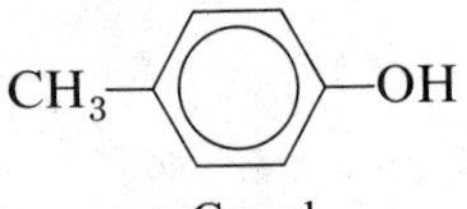

Compounds that have a hydroxyl group attached to a polycyclic benzenoid ring are also phenols, but they are called naphthols and phenanthrols, for example.

* Compounds in which a hydroxyl group is attached to an unsaturated carbon of a double bond (i.e., C=C) are called enols, cf. Section 16.11.

1-Naphthol
(α-naphthol)

2-Naphthol
(β-naphthol)

9-Phenanthrol

Ethers differ from alcohols and phenols in that the oxygen atom of an ether is bonded to two carbon atoms. The hydrocarbon groups may be alkyl, alkenyl, vinyl, alkynyl, or aryl. Several examples are shown below.

$CH_3CH_2{-}O{-}CH_2CH_3$
Diethyl ether

$CH_2{=}CHCH_2{-}O{-}CH_3$
Methyl allyl ether

$CH_2{=}CH{-}O{-}CH{=}CH_2$
Divinyl ether

$C_6H_5{-}OCH_3$
Anisole

15.1A Nomenclature of Alcohols

Simple alcohols are often called by their *common* names. We have seen many examples already. In addition to *methyl alcohol, ethyl alcohol, isopropyl alcohol,* and *allyl alcohol,* for example, there are several others including the following:

$CH_3CH_2CH_2OH$
n-Propyl alcohol

$CH_3CH_2CH_2CH_2OH$
n-Butyl alcohol

$CH_3CH_2CH(OH)CH_3$
sec-Butyl alcohol

$CH_3CH(CH_3)CH_2OH$
Isobutyl alcohol

$(CH_3)_3C{-}OH$
tert-Butyl alcohol

$(CH_3)_3CCH_2OH$
Neopentyl alcohol

$C_6H_5{-}CH{=}CHCH_2OH$
Cinnamyl alcohol

$(C_6H_5)_3C{-}OH$
Triphenylcarbinol

The IUPAC rules for naming alcohols are as follows:

1. Select the longest continuous carbon chain *to which the hydroxyl is directly attached.* Change the name of the alkane corresponding to this chain by dropping the final *e* and adding *ol.* This gives the base name of the alcohol.
2. Number the longest continuous carbon chain so as to give the carbon bearing the hydroxyl group the lower number. Indicate the position of the hydroxyl group by using this number; indicate the positions of other substituents (or multiple bonds) by using the numbers corresponding to their position along the carbon chain.

The examples below show how these rules are applied.

$\overset{3}{C}H_3\overset{2}{C}H_2\overset{1}{C}H_2OH$

1-Propanol

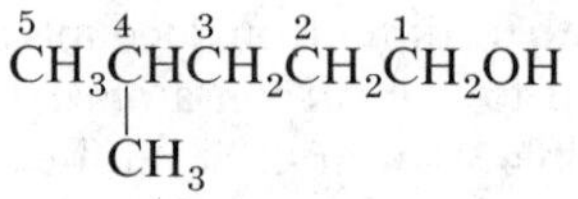

$\overset{5}{C}H_3\overset{4}{C}H(CH_3)\overset{3}{C}H_2\overset{2}{C}H_2\overset{1}{C}H_2OH$

4-Methyl-1-pentanol
(Not 2-methyl-5-pentanol)

$Cl\overset{3}{C}H_2\overset{2}{C}H_2\overset{1}{C}H_2OH$

3-Chloro-1-propanol

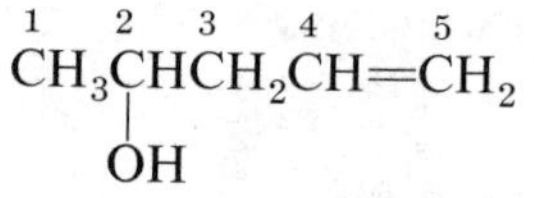

$\overset{1}{C}H_3\overset{2}{C}H(OH)\overset{3}{C}H_2\overset{4}{C}H{=}\overset{5}{C}H_2$

4-Penten-2-ol
(Not 1-penten-4-ol)

CH_3CHOH (attached to benzene ring)

1-Phenylethanol

$C_6H_5{-}CH_2CH_2OH$

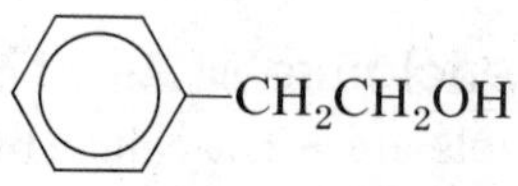

2-Phenylethanol

Alcohols containing two hydroxyl groups are commonly called glycols. In the IUPAC system they are named as *diols.*

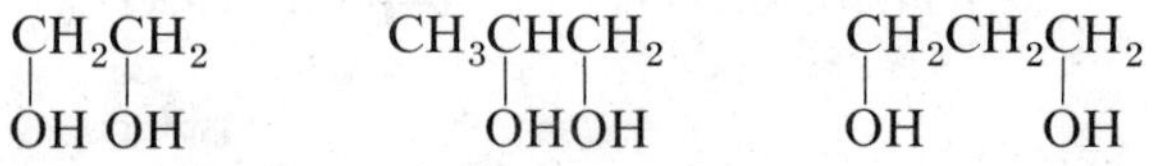

	$CH_2(OH)CH_2(OH)$	$CH_3CH(OH)CH_2(OH)$	$CH_2(OH)CH_2CH_2(OH)$
Common:	Ethylene glycol	Propylene glycol	Trimethylene glycol
IUPAC:	1,2-Ethanediol	1,2-Propanediol	1,3-Propanediol

15.1B Nomenclature of Phenols

We studied the nomenclature of some of the phenols in Chapter 11. In many compounds *phenol* is the base name:

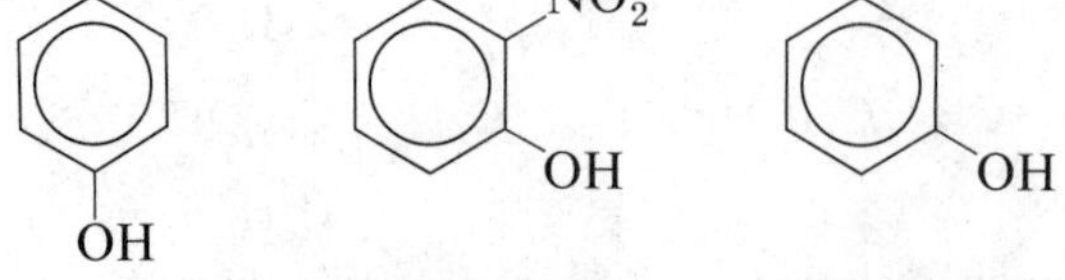

Cl, OH on benzene ring (para)	NO_2, OH on benzene ring (ortho)	Br, OH on benzene ring (meta)
4-Chlorophenol (*p*-Chlorophenol)	2-Nitrophenol (*o*-Nitrophenol)	3-Bromophenol (*m*-Bromophenol)

The methylphenols are often called *cresols:*

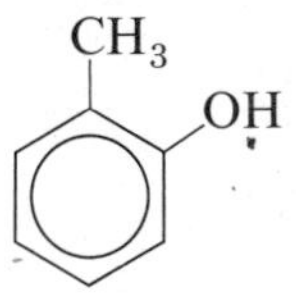

2-Methylphenol (*o*-cresol) 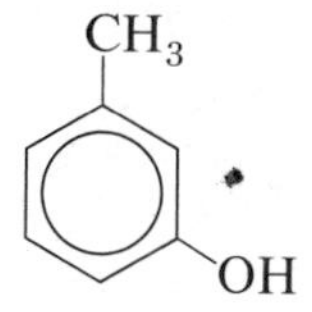 3-Methylphenol (*m*-cresol) 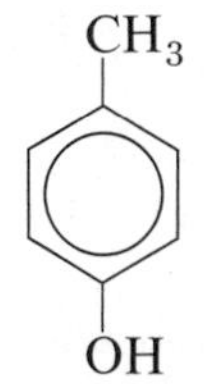 4-Methylphenol (*p*-cresol)

The benzenediols have special names:

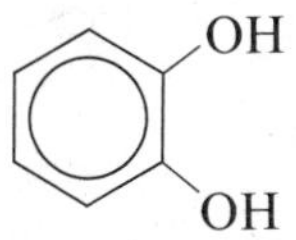

1,2-Benzenediol (catechol) 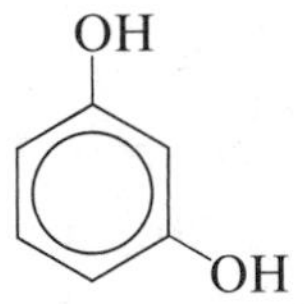 1,3-Benzenediol (resorcinol) 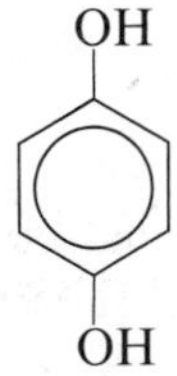 1,4-Benzenediol (hydroquinone)

15.1C Nomenclature of Ethers

Ethers are frequently given common names and to do so is quite easy. One simply names both groups that are attached to the oxygen:

$CH_3OCH_2CH_3$
Methyl ethyl ether

$CH_3CH_2OCH_2CH_3$
Diethyl ether

$C_6H_5OC(CH_3)_2-CH_3$
Phenyl *tert*-butyl ether

$CH_2{=}CHOCH{=}CH_2$
Divinyl ether

IUPAC names are seldom used for simple ethers. IUPAC names are used for complicated ethers, however, and for compounds with more than one ether linkage. In the IUPAC system ethers are named as alkoxyalkanes, alkoxyalkenes, and alkoxyarenes.

$CH_3CH(OCH_3)CH_2CH_2CH_3$
2-Methoxypentane

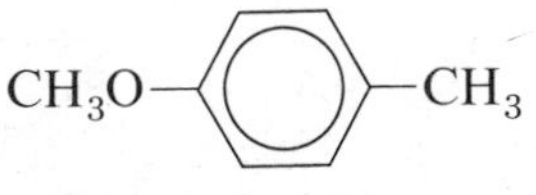

$CH_3O-C_6H_4-CH_3$
1-Methoxy-4-methylbenzene

$CH_3OCH_2CH_2OCH_3$
1,2-Dimethoxyethane

$CH_3OCH{=}CHCH_3$
1-Methoxy-1-propene

Two cyclic ethers that are frequently used as solvents have the common names tetrahydrofuran (THF) and dioxane.

Tetrahydrofuran (THF) Dioxane

Problem 15.1

Give appropriate names for all of the alcohols and ethers with the formulas (a) C_3H_6O, (b) C_4H_8O, and (c) $C_5H_{12}O$.

15.2 PHYSICAL PROPERTIES OF ALCOHOLS, PHENOLS, AND ETHERS

When we compare the boiling points of alcohols and phenols with the boiling points of hydrocarbons of roughly the same molecular weight, we notice (Table 15.1) that alcohols and phenols boil at much higher temperatures.

The boiling point of methanol is more than 150° higher than that of ethane even though the two compounds have approximately the same molecular weight. Ethanol boils 123° higher than propane and phenol boils almost 70° higher than toluene.

On the other hand, when we compare the boiling points of ethers with those of hydrocarbons of roughly the same molecular weight, the differences are not at all large (Table 15.2). Diethyl ether and pentane, for example, have almost the same boiling point.

We can account for these contrasts and similarities on the basis of a single phenomenon: the presence or absence of strong *hydrogen bonds.*

In alcohols and phenols a hydrogen atom is attached to an oxygen atom.

TABLE 15.1 Comparison of Boiling Points of Alcohols, Phenols, and Hydrocarbons of Approximately the Same Molecular Weight

COMPOUND	MOL WT	bp °C (1 atm)
CH_3OH	32.04	65
CH_3CH_3	30.07	−88.6
CH_3CH_2OH	46.07	78.5
$CH_3CH_2CH_3$	44.09	−44.5
C_6H_5–OH	94.11	182
C_6H_5–CH_3	92.13	110.6

TABLE 15.2 A Comparison of Boiling Points of Ethers and Hydrocarbons of Approximately the Same Molecular Weight

COMPOUND	MOL WT	bp °C (1 atm)
CH_3OCH_3	46.07	−25
$CH_3CH_2CH_3$	44.09	−44.5
$CH_3CH_2OCH_2CH_3$	74.12	34.6
$CH_3CH_2CH_2CH_2CH_3$	72.15	36
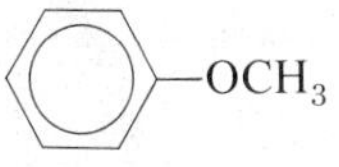 C_6H_5—OCH_3	108.13	158.3
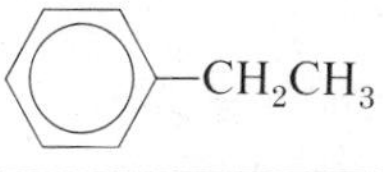 C_6H_5—CH_2CH_3	106.17	136

Because oxygen is highly electronegative, molecules of alcohols and phenols are capable of forming strong hydrogen bonds to each other. Hydrogen bonding holds the molecules together (causes them to be associated). Alcohols and phenols, as a result, have boiling points that correspond to those of compounds with much greater molecular weight because these hydrogen bonds must be broken in the process of volatilization.

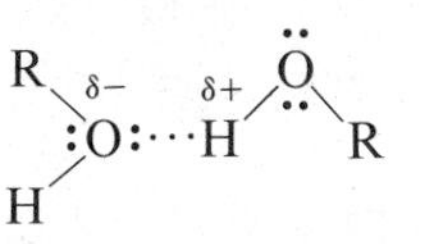

R(H)$\overset{\delta-}{\ddot{O}}$:···$\overset{\delta+}{H}$—$\ddot{O}$—R

Hydrogen bonding in an alcohol

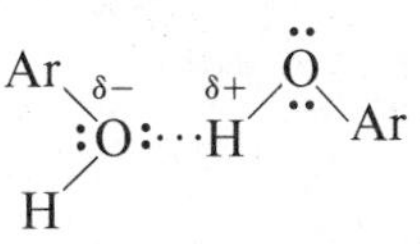

Ar(H)$\overset{\delta-}{\ddot{O}}$:···$\overset{\delta+}{H}$—$\ddot{O}$—Ar

Hydrogen bonding in a phenol

Molecules of ethers, by contrast, do not have a hydrogen attached to a strongly electronegative element and in this respect ethers are like hydrocarbons. Ethers are not able to associate with each other through hydrogen bonding. Although ethers are more polar than hydrocarbons, their boiling points are comparable.

Ethers, however, *are* able to form hydrogen bonds with compounds such as water. Ethers, therefore, have solubilities in water that are similar to those of alcohols of the same molecular weight, and that are very different from those of hydrocarbons.

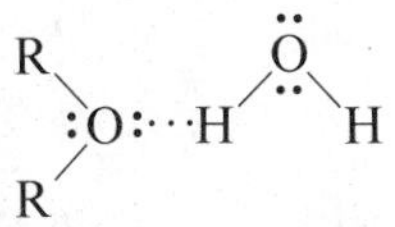

R(R):O:···H—$\ddot{O}$—H

Hydrogen bonding between an ether and water

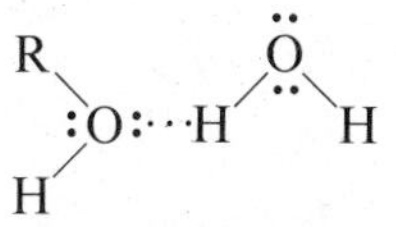

R(H):O:···H—$\ddot{O}$—H

Hydrogen bonding between an alcohol and water

Diethyl ether and 1-butanol, for example, have the same solubility in water, approximately 8.0 g/100 ml at room temperature. Pentane, by contrast, is virtually insoluble in water.

Methyl alcohol, ethyl alcohol, both propyl alcohols, and *tert*-butyl alcohol are completely miscible with water (Table 15.3). The remaining butyl alcohols have solubilities in water between 7.9 and 12.5 g/100 ml. The solubility of alcohols in water gradually decreases as the hydrocarbon portion of the molecule lengthens; long-chain alcohols are more "alkanelike" and are, therefore, less like water.

Phenols also show appreciable solubilities in water and apparently form even stronger hydrogen bonds to water molecules than alcohols. Phenol, itself, has a solubility in water of 9.3 g/100 ml at 20°, whereas the solubility of cyclohexanol is only 3.6 g/100 ml.

One of the reasons we are interested in solubilities in water is that we need to learn the structural features that help or hinder a substance's solubility in this important solvent. This solvent is the medium, for example, of the cells of all living organisms.

The physical properties of a number of alcohols, phenols, and ethers are presented in Tables 15.3, 15.4, and 15.5.

Problem 15.2

How can you account for the fact that the boiling point of ethylene glycol is much higher than that of either *n*-propyl alcohol or isopropyl alcohol even though all three compounds have roughly the same molecular weight?

TABLE 15.3 Physical Properties of Alcohols

COMPOUND	NAME	mp °C	bp °C (1 atm)	SPECIFIC GRAVITY	WATER SOLUBILITY g/100 g H_2O
Monohydroxy Alcohols					
CH_3OH	Methyl alcohol	−97	64.7	0.792	∞
CH_3CH_2OH	Ethyl alcohol	−114	78.3	0.789	∞
$CH_3CH_2CH_2OH$	*n*-Propyl alcohol	−126	97.2	0.804	∞
$CH_3CH(OH)CH_3$	Isopropyl alcohol	−88	82.3	0.786	∞
$CH_3CH_2CH_2CH_2OH$	*n*-Butyl alcohol	−90	117.7	0.810	7.9
$CH_3CH(CH_3)CH_2OH$	Isobutyl alcohol	−108	108.0	0.802	10.0
$CH_3CH_2CH(OH)CH_3$	*sec*-Butyl alcohol	−114	99.5	0.808	12.5
$(CH_3)_3COH$	*tert*-Butyl alcohol	25	82.5	0.789	∞
$CH_3(CH_2)_3CH_2OH$	*n*-Pentyl alcohol	−78.5	138.0	0.817	2.4
$CH_3(CH_2)_4CH_2OH$	*n*-Hexyl alcohol	−52	156.5	0.819	0.6
$CH_3(CH_2)_5CH_2OH$	*n*-Heptyl alcohol	−34	176	0.822	0.2
$CH_3(CH_2)_6CH_2OH$	*n*-Octyl alcohol	−15	195	0.825	0.05
$CH_3(CH_2)_7CH_2OH$	*n*-Nonyl alcohol		212	0.827	—
$CH_3(CH_2)_8CH_2OH$	*n*-Decyl alcohol	6	228	0.829	—
$CH_2{=}CHCH_2OH$	Allyl alcohol	−129	97	0.855	∞
$(CH_2)_4CHOH$ (ring)	Cyclopentyl alcohol		140	0.949	—
$(CH_2)_5CHOH$ (ring)	Cyclohexyl alcohol	24	161.5	0.962	3.6
$C_6H_5CH_2OH$	Benzyl alcohol	−15	205	1.046	4
Diols and Triols					
CH_2OHCH_2OH	Ethylene glycol	−16	197	1.113	∞
$CH_3CHOHCH_2OH$	Propylene glycol		187	1.040	∞
$CH_2OHCH_2CH_2OH$	Trimethylene glycol		215	1.060	∞
$CH_2OHCHOHCH_2OH$	Glycerol	18	290	1.261	∞

TABLE 15.4 Physical Properties of Ethers

NAME	FORMULA	mp °C	bp °C	DENSITY 20°
Dimethyl ether	CH_3OCH_3	−140	−24.9	0.661
Methyl ethyl ether	$CH_3OCH_2CH_3$		7.9	0.697
Diethyl ether	$CH_3CH_2OCH_2CH_3$	−116	34.6	0.714
Di-*n*-propyl ether	$(CH_3CH_2CH_2)_2O$	−122	90.5	0.736
Diisopropyl ether	$(CH_3)_2CHOCH(CH_3)_2$	−86	68	0.735
Di-*n*-butyl ether	$(CH_3CH_2CH_2CH_2)_2O$	−95	141	0.769
1,2-Dimethoxyethane	$CH_3OCH_2CH_2OCH_3$	−58	83	0.863
Tetrahydrofuran	$(CH_2)_4O$ (cyclic)	−108	65.4	0.888
1,4-Dioxane	O(CH$_2$CH$_2$)$_2$O (cyclic: CH_2CH_2 above and below, O at each end)	11	101	1.033
Anisole	C_6H_5–OCH_3 (benzene ring–OCH_3)	−37.3	158.3	0.994

TABLE 15.5 Physical Properties of Phenols

NAME	FORMULA	mp °C	bp °C	WATER SOLUBILITY g/100 g H_2O
Phenol	C_6H_5OH	43	181	9.3
o-Cresol	*o*-$CH_3C_6H_4OH$	30	191	2.5
m-Cresol	*m*-$CH_3C_6H_4OH$	11	201	2.6
p-Cresol	*p*-$CH_3C_6H_4OH$	35.5	201	2.3
o-Chlorophenol	*o*-ClC_6H_4OH	8	176	2.8
m-Chlorophenol	*m*-ClC_6H_4OH	33	214	2.6
p-Chlorophenol	*p*-ClC_6H_4OH	43	220	2.7
o-Nitrophenol	*o*-$O_2NC_6H_4OH$	45	217	0.2
m-Nitrophenol	*m*-$O_2NC_6H_4OH$	96		1.4
p-Nitrophenol	*p*-$O_2NC_6H_4OH$	114		1.7
2,4-Dinitrophenol	O_2N–(benzene ring)–OH, with NO_2 ortho to OH	113		0.6
2,4,6-Trinitrophenol (picric acid)	O_2N–(benzene ring)–OH, with NO_2 at both positions ortho to OH	122		1.4

15.3 SPECTROSCOPIC PROPERTIES OF ALCOHOLS, PHENOLS, AND ETHERS

15.3A Nuclear Magnetic Resonance Spectra

The signals for protons of —CH—O— groups of alcohols and ethers are shifted downfield by the electronegative influence of the oxygen. Typical chemical shifts for these protons are $\delta 3.4$ to 4.0. The position of the —OH proton of alcohols and phenols is dependent on the extent of hydrogen bonding and this depends on the concentration, solvent, and temperature. Hydrogen bonding usually causes these protons to be deshielded. Therefore, increasing the concentration of the alcohol in the solvent used for the nmr study causes the —OH absorption to shift to lower field strength. For alcohols chemical shifts of —OH protons are $\delta 1$ to 6; for phenols $\delta 4$ to 12. Rapid exchange of —OH protons can cause spin-spin decoupling (cf., Section 13.8). The ring protons of phenols occur at low magnetic field strengths ($\delta 6$–8), typical of aromatic protons.

15.3B Infrared Spectra

We discussed the effect of hydrogen bonding on the O—H stretching frequencies in Section 13.11. The characteristic O—H stretching frequency is 3200 to 3650 cm^{-1}. There is also an O—H bending frequency in the 1260 to 1410 cm^{-1} range. The C—O bond of alcohols, phenols, and ethers has a strong stretching absorption between 1050 and 1200 cm^{-1}.

15.3C Visible-Ultraviolet Spectra

Saturated alcohols and ethers have no absorption higher than 190 nm, and are transparent, therefore, in the region covered by most visible-ultraviolet spectrometers. Saturated alcohols and ethers are often used as solvents for visible-ultraviolet spectra. Phenol shows a typical benzenoid absorption band (Section 11.11) at 270 nm (ϵ_{max} 1450). The absorption bands of nitrophenols are at longer wavelengths; 2,4,6-trinitrophenol, for example, has an absorption maximum at 380 nm that tails past 400 nm, into the visible region, imparting a yellow color to this compound.

15.4 IMPORTANT ALCOHOLS AND ETHERS

15.4A Methyl Alcohol

At one time, most methyl alcohol was produced by the destructive distillation of wood (i.e., heating wood to a high temperature in the absence of air). It was because of this method of preparation that methyl alcohol came to be called "wood alcohol." Today, most methyl alcohol is prepared by the catalytic hydrogenation of carbon monoxide. This reaction takes place under high pressure and at a temperature of 300° to 400°.

$$CO + H_2 \xrightarrow[\substack{200\text{–}300\ \text{atm.} \\ ZnO\text{—}Cr_2O_3}]{300\text{–}400^\circ} CH_3OH$$

Methyl alcohol is highly toxic. Ingestion of even small quantities of methyl alcohol can cause blindness; large quantities cause death. Methyl alcohol poisoning can also occur by inhalation of the vapors or by prolonged exposure of the skin.

15.4B Ethyl Alcohol

Ethyl alcohol can be made by the fermentation of sugars and it is the alcohol of alcoholic beverages. The synthesis of ethyl alcohol in the form of wine by the fermentation of the sugars of fruit juices was probably man's first accomplishment in the field of organic synthesis. Sugars from a wide variety of sources can be used in the preparation of alcoholic beverages. Often, these sugars are from grains and it is this that accounts for ethyl alcohol having the synonym "grain alcohol."

Fermentation is usually carried out by adding yeast to a mixture of sugars and water. Yeast contains enzymes that promote a long series of reactions that ultimately convert a simple sugar ($C_6H_{12}O_6$) to ethyl alcohol and carbon dioxide.

$$C_6H_{12}O_6 \xrightarrow{\text{Yeast}} \underset{(\sim 95\%\ \text{yield})}{2CH_3CH_2OH} + 2CO_2$$

Fermentation alone does not produce beverages with an alcohol content greater than 12 to 15% because the yeast is destroyed at high concentrations. To produce beverages of higher alcohol content the aqueous solution must be distilled. Brandy, whiskey, and vodka are produced in this way. The "proof" of an alcoholic beverage is simply twice the percentage of alcohol (by volume). One hundred proof whiskey is 50% alcohol. The flavors of the various distilled liquors result from other organic compounds that distill with the alcohol and water.

Distillation of a solution of ethyl alcohol and water will not yield ethyl alcohol more concentrated than 95%. A mixture of 95% ethyl alcohol and 5% water boils at a lower temperature (78.15°) than either pure ethyl alcohol (bp 78.3°) or pure water (bp 100°). Such a mixture is an example of an *azeotrope.** Pure ethyl alcohol can be prepared by adding benzene to the mixture of 95% ethyl alcohol and water and then distilling this solution. Benzene forms a different azeotrope with ethyl alcohol and water that is 7.5% water. This azeotrope boils at 64.9° and allows removal of the water (along with some ethyl alcohol). Eventually pure ethyl alcohol distills over. Pure ethyl alcohol is called *absolute alcohol.*

Ethyl alcohol is quite cheap, but when it is used for beverages it is highly taxed. (The tax is greater than $20 per gallon in most states.) Federal law requires that some ethyl alcohol used for scientific and industrial purposes be adulterated or "denatured" to make it undrinkable. A variety of denaturants are used including methyl alcohol.

Ethyl alcohol is an important industrial chemical. Most ethyl alcohol for industrial purposes is produced by the acid-catalyzed hydration of ethene.

$$CH_2{=}CH_2 + H_2O \xrightarrow[\text{acid}]{} CH_3CH_2OH$$

Ethyl alcohol is a *hypnotic* (sleep producer). It depresses activity in the

* Azeotropes can also have boiling points that are higher than that of either of the pure components.

upper brain even though it gives the illusion of being a stimulant. Ethyl alcohol is also toxic but it is much less toxic than methyl alcohol. In rats the lethal dose of ethyl alcohol is 13.7 g per kilogram of body weight. Abuse of the use of ethyl alcohol is a major drug problem in most countries.

15.4C Isopropyl Alcohol

This alcohol is more toxic than ethyl alcohol. Isopropyl alcohol is commonly used as rubbing alcohol.

15.4D Ethylene Glycol

Ethylene glycol has a low molecular weight, a high boiling point, and is miscible with water. These properties make ethylene glycol an ideal automobile antifreeze. Much ethylene glycol is sold for this purpose under a variety of trade names.

15.4E Diethyl Ether

Diethyl ether is a very low-boiling, highly flammable compound. Care should always be taken when diethyl ether is used in the laboratory because open flames or sparks from light switches can cause explosive combustion of mixtures of diethyl ether and air.

Most ethers react slowly with oxygen in the air to form organic peroxides. These peroxides, which may accumulate in ethers that have been left standing for long periods in contact with air, are dangerously explosive and may detonate without warning when ether solutions are distilled to near dryness. Since ethers are used frequently in extractions, one should take care to test for and decompose any peroxides present in the ether before a distillation is carried out. (Consult a laboratory manual for instructions.)

Diethyl ether was first employed as a surgical anesthetic by C. W. Long of Jefferson, Georgia, in 1842. Long's use of diethyl ether was not published, but shortly thereafter, diethyl ether was introduced into surgical use at the Massachusetts General Hospital in Boston by J. C. Warren.

The idea of chemical anesthesia dates from 1799 with Sir Humphrey Davy's discovery of the anesthetic properties of nitrous oxide (laughing gas). During the nineteenth century chloroform was also used as an anesthetic; Queen Victoria, for example, gave birth to her children under chloroform anesthesia. Chloroform is not used as an anesthetic any longer because its use causes liver damage. (It did not, however, appear to shorten Queen Victoria's life appreciably.)

Two popular modern anesthetics are divinyl ether (vinethene) and cyclopropane. Cyclopropane acts very rapidly producing unconsciousness within seconds. Like diethyl ether, divinyl ether and cyclopropane are highly flammable and produce explosive mixtures in air. Thus, special skills are required in the application of these compounds.

15.5 PREPARATION OF ALCOHOLS

We have already studied three methods for the synthesis of alcohols from alkenes: hydration, oxymercuration-demercuration, and hydroboration-oxidation. We have also seen how alcohols can be prepared from Grignard reagents.

15.5A Hydration of Alkenes (discussed in Section 7.4)

Alkenes add water in the presence of an acid catalyst. The addition follows Markovnikov's rule; thus, except for the hydration of ethylene, the reaction produces secondary and tertiary alcohols. The reaction is reversible and the mechanism for the hydration of an alkene is simply the reverse of that for the dehydration of an alcohol (Section 6.10).

$$\underset{\text{Alkene}}{>C{=}C<} \underset{-H^+}{\overset{+H^+}{\rightleftharpoons}} -\underset{H}{\overset{|}{\underset{|}{C}}}-\overset{|}{\underset{+}{C}}- \underset{-H_2O}{\overset{+H_2O}{\rightleftharpoons}} -\underset{H}{\overset{|}{\underset{|}{C}}}-\underset{{}^+OH_2}{\overset{|}{\underset{|}{C}}}- \underset{+H^+}{\overset{-H^+}{\rightleftharpoons}} \underset{\text{Alcohol}}{-\underset{H}{\overset{|}{\underset{|}{C}}}-\underset{OH}{\overset{|}{\underset{|}{C}}}-}$$

Problem 15.3

Show how you would prepare each of the following alcohols by acid-catalyzed hydration of the appropriate alkene.

(a) *tert*-Butyl alcohol
(b) 2-Hexanol
(c) Cyclopentanol
(d) 1-Methylcyclohexanol

Problem 15.4

When 3,3-dimethyl-1-butene is subjected to acid-catalyzed hydration the major product is 2,3-dimethyl-2-butanol. How can you explain this?

15.5B Oxymercuration-Demercuration (discussed in Section 7.7)

Addition of water to an alkene can be achieved indirectly through oxymercuration-demercuration. In the first step of this synthesis an alkene is allowed to react with mercuric acetate in a water-tetrahydrofuran solution. This step produces a hydroxyalkylmercury compound (below) in which mercury is bonded to the less-substituted carbon. In the second step, sodium borohydride reduction of this intermediate results in replacement of the mercury substituent by hydrogen. The net result of the two steps is Markovnikov addition of —H and —OH. Both steps can be carried out rapidly and the yields obtained are usually high. An example is the following synthesis of 2-pentanol.

$$CH_3CH_2CH_2CH{=}CH_2 \xrightarrow[\text{(15 sec.)}]{Hg(OAc)_2,\ THF{-}H_2O} \underset{\substack{(93\%) \\ \text{Hydroxyalkylmercury compound}}}{CH_3CH_2CH_2\underset{OH}{\underset{|}{C}}H{-}\underset{HgOAc}{\underset{|}{C}}H_2} \xrightarrow[\text{(1 hr)}]{NaBH_4,\ OH^-} \underset{\substack{\text{2-pentanol} \\ (93\%)}}{CH_3CH_2CH_2\underset{OH}{\underset{|}{C}}HCH_3}$$

Problem 15.5

Show how you might employ the oxymercuration-demercuration synthesis to prepare each of the following alcohols from an appropriate alkene.

(a) 2-Hexanol
(b) 1-Methylcyclohexanol
(c) 2,4,4-Trimethyl-2-pentanol
(d) 2-Phenyl-2-propanol

15.5C Hydroboration-Oxidation (discussed in Section 7.8)

Hydroboration-oxidation gives us a method for adding water to a double bond in a manner opposite to that of acid-catalyzed hydration or oxymercuration-demercuration. The reaction takes place in two steps: in the first step boron hydride adds to the double bond with the boron atom becoming attached to the less-substituted carbon; in the second step oxidation with hydrogen peroxide in aqueous base replaces the boron group with a hydroxyl group. The result is an apparent anti-Markovnikov addition of —H and —OH.

$$3R{-}CH{=}CH_2 + \tfrac{1}{2}(BH_3)_2 \longrightarrow (R{-}CH_2CH_2)_3B$$

$$(R{-}CH_2CH_2)_3B \xrightarrow[OH^-,\ H_2O]{H_2O_2} 3R{-}CH_2CH_2OH + B(OH)_3$$

The overall stereochemistry of hydroboration-oxidation is a *syn* addition of —H and —OH.

Problem 15.6

Show how you would utilize the hydroboration-oxidation procedure to prepare each of the following alcohols.

(a) 3,3-Dimethyl-1-butanol
(b) 1-Hexanol
(c) 2-Phenylethanol
(d) *trans*-2-Methylcyclohexanol

15.5D Alcohols from Grignard Reagents and Organolithium Compounds

We saw in the last chapter (Sections 14.11 and 14.12) that Grignard reagents and organolithium compounds can be used to synthesize alcohols through their reactions with carbonyl compounds and with ethylene oxide.

$$\underset{\text{(or RLi)}}{RMgX} + \underset{\text{Aldehyde or Ketone}}{{>}C{=}O} \longrightarrow R{-}\overset{|}{\underset{|}{C}}{-}OMgX \;\left(\text{or } R{-}\overset{|}{\underset{|}{C}}{-}OLi\right) \xrightarrow{H^+} \underset{\text{Alcohol}}{R{-}\overset{|}{\underset{|}{C}}{-}OH}$$

$$RMgX + \underset{\text{Ester}}{{-}\overset{O}{\overset{\|}{C}}{-}OR'} \longrightarrow R{-}\overset{R}{\overset{|}{\underset{|}{C}}}{-}OMgX \xrightarrow{H^+} R{-}\overset{R}{\overset{|}{\underset{|}{C}}}{-}OH$$

$$RMgX + \underset{\diagdown O \diagup}{CH_2{-}CH_2} \longrightarrow RCH_2CH_2OMgX \xrightarrow{H^+} RCH_2CH_2OH$$

In Section 15.6 we will see that alcohols can also be prepared by reduction of various carbonyl compounds.

15.6 OXIDATION-REDUCTION REACTIONS IN ORGANIC CHEMISTRY

Reduction of an organic molecule usually corresponds to *increasing its hydrogen content* or to *decreasing its oxygen content*. For example, converting a carboxylic acid to an aldehyde is a reduction because the oxygen content is decreased.

Oxygen content decreases

$$\underset{\text{Carboxylic acid}}{\mathrm{R{-}\overset{\overset{O}{\|}}{C}{-}OH}} \xrightarrow[\text{reduction}]{(H)} \underset{\text{Aldehyde}}{\mathrm{R{-}\overset{\overset{O}{\|}}{C}{-}H}}$$

Converting an aldehyde to an alcohol is also a reduction.

Hydrogen content increases

$$\mathrm{R{-}\overset{\overset{O}{\|}}{C}{-}H} \xrightarrow[\text{reduction}]{(H)} \mathrm{RCH_2OH}$$

Converting an alcohol to an alkane is also a reduction.

Oxygen content decreases

$$\mathrm{RCH_2OH} \xrightarrow[\text{reduction}]{(H)} \mathrm{RCH_3}$$

In these examples we have used the symbol (H) to indicate that a reduction of the organic compound has taken place. We do this when we want to write a general equation without specifying what the reducing agent is.

The opposite of reduction is oxidation. Thus, *increasing the oxygen content* of an organic molecule or *decreasing its hydrogen content* is an *oxidation* of the organic molecule. The reverse of each reaction that we have given is an oxidation of the organic molecule and we can summarize these oxidation-reduction reactions as follows below. We use the symbol (O) to indicate in a general way that the organic molecule has been oxidized.

$$\underset{\substack{\text{Lowest}\\ \text{oxidation}\\ \text{state}}}{\mathrm{RCH_3}} \underset{(H)}{\overset{(O)}{\rightleftharpoons}} \mathrm{RCH_2OH} \underset{(H)}{\overset{(O)}{\rightleftharpoons}} \mathrm{R\overset{\overset{O}{\|}}{C}H} \underset{(H)}{\overset{(O)}{\rightleftharpoons}} \underset{\substack{\text{Highest}\\ \text{oxidation}\\ \text{state}}}{\mathrm{R\overset{\overset{O}{\|}}{C}OH}}$$

Of course, when an organic compound is reduced, something else—*the reducing agent*—must be oxidized. And when an organic compound is oxidized, something else—*the oxidizing agent*—is reduced. These oxidizing and reducing agents are often inorganic compounds and in the next two sections we will see what some of them are.

Problem 15.7

One method for assigning an oxidation state to a carbon atom of an organic compound is to base that assignment on the groups attached to the carbon; a bond to hydrogen (or anything less electronegative than carbon) makes it -1, a bond to oxygen, nitrogen, or halogen (or to anything more electronegative than carbon) makes it +1, and a bond to another carbon 0. Thus the carbon of methane is assigned an oxidation state of −4, and that of carbon dioxide +4. (a) Use this method to assign oxidation states to the carbons of methyl alcohol (CH_3OH), formic acid ($HCOH$ with C=O), and formaldehyde (HCH with C=O). (b) Arrange the compounds methane, carbon dioxide, methyl alcohol, formic acid, and formaldehyde in order of increasing oxidation state. (c) What change in oxidation state accompanies the reaction, methyl alcohol ⟶ formaldehyde? (d) Is this an oxidation or a reduction? (e) When H_2CrO_4 acts as an oxidizing agent in this reaction the chromium of H_2CrO_4 becomes Cr^{+3}. What change in oxidation state does chromium undergo? (f) How many moles of H_2CrO_4 will be required to oxidize one mole of methyl alcohol to formaldehyde?

Problem 15.8

(a) Use the method described in the preceding problem to assign oxidation states to each carbon of ethyl alcohol and to each carbon of acetaldehyde. (b) What do these numbers reveal about the site of oxidation when ethyl alcohol is oxidized to acetaldehyde? (c) Repeat this procedure for the oxidation of acetaldehyde to acetic acid. (d) When silver oxide oxidizes acetaldehyde to acetic acid, the silver oxide is reduced to metallic silver. How many moles of Ag_2O will be required to oxidize one mole of acetaldehyde?

Problem 15.9

(a) Although we have described the hydrogenation of an alkene as an addition reaction, organic chemists often refer to it as a "reduction." Refer to the method described in Problem 15.7 and explain. (b) Make similar comments about the reversible reaction:

$$CH_3-\overset{\overset{\large O}{\|}}{C}-H + H_2 \underset{}{\overset{Ni}{\rightleftharpoons}} CH_3CH_2OH$$

15.7 ALCOHOLS BY REDUCTION OF CARBONYL COMPOUNDS

Primary and secondary alcohols can be synthesized by the reduction of a variety of compounds that contain the carbonyl ($\rangle C{=}O$) group. Several general examples are shown below.

$$\underset{\text{Carboxylic acid}}{R-\overset{\overset{\large O}{\|}}{C}-OH} \xrightarrow{(H)} \underset{1^\circ\ \text{Alcohol}}{R-CH_2OH}$$

$$\underset{\text{Ester}}{R-\overset{\overset{\large O}{\|}}{C}-OR'} \xrightarrow{(H)} \underset{1^\circ\ \text{Alcohol}}{R-CH_2OH}\ (+\ R'OH)$$

$$\underset{\text{Aldehyde}}{R-\overset{\overset{\large O}{\|}}{C}-H} \xrightarrow{(H)} \underset{1^\circ\ \text{Alcohol}}{R-CH_2OH}$$

$$R{-}\overset{\overset{O}{\|}}{C}{-}R' \xrightarrow{(H)} R{-}\underset{\underset{OH}{|}}{C}H{-}R'$$

Ketone 2° Alcohol

Reductions of carboxylic acids are the most difficult, and prior to 1946 direct reduction of acids was not possible. However, the discovery in 1946 of the powerful reducing agent *lithium aluminum hydride* gave organic chemists a method for reducing acids to primary alcohols in excellent yields. Two examples are the lithium aluminum hydride reductions of acetic acid and 2,2-dimethylpropanoic acid.

$$4RCO_2H + 3LiAlH_4 \xrightarrow{Et_2O} [(RCH_2O)_4Al]Li + 4H_2 + 2LiAlO_2$$

Lithium aluminum hydride

$$\xrightarrow{H_2O} 4RCH_2OH + Al(OH)_3 + LiOH$$

$$CH_3\overset{\overset{O}{\|}}{C}{-}OH \xrightarrow[(2)\ H_2O]{(1)\ LiAlH_4/Et_2O} CH_3CH_2OH$$

Acetic acid → Ethyl alcohol (100%)

$$CH_3{-}\overset{\overset{CH_3}{|}}{\underset{\underset{CH_3}{|}}{C}}{-}COOH \xrightarrow[(2)\ H_2O]{(1)\ LiAlH_4/Et_2O} CH_3\overset{\overset{CH_3}{|}}{\underset{\underset{CH_3}{|}}{C}}{-}CH_2OH$$

2,2-Dimethylpropanoic acid → Neopentyl alcohol (92%)

Esters can be reduced in several ways: through the use of sodium in alcohol (called the Bouveault-Blanc reduction), by high-pressure hydrogenation (a reaction preferred for industrial process and often referred to as "hydrogenolysis"), or through the use of lithium aluminum hydride.

$$R{-}\overset{\overset{O}{\|}}{C}OC_2H_5 \xrightarrow[C_2H_5OH]{Na} R{-}CH_2OH + C_2H_5OH$$

$$R{-}\overset{\overset{O}{\|}}{C}OR' + H_2 \xrightarrow[175°,\ 5000\ psi]{CuO\cdot CuCr_2O_4} RCH_2OH + R'OH$$

$$R{-}\overset{\overset{O}{\|}}{C}OR' + LiAlH_4 \xrightarrow[(2)H_2O]{Et_2O} RCH_2OH + R'OH$$

The last method is the one most commonly used now in small-scale laboratory synthesis.

Aldehydes and ketones can also be reduced to alcohols by hydrogen and a metal catalyst, by sodium in alcohol, and by lithium aluminum hydride. The reducing agent most often used, however, is sodium borohydride, $NaBH_4$.

$$4RCH(=O) + NaBH_4 + 3H_2O \longrightarrow 4RCH_2OH + NaOB(OH)_2$$

$$\underset{\text{Butanal}}{CH_3CH_2CH_2CH(=O)} \xrightarrow[H_2O]{NaBH_4} \underset{\substack{\text{1-Butanol} \\ (85\%)}}{CH_3CH_2CH_2CH_2OH}$$

$$\underset{\text{2-Butanone}}{CH_3CH_2C(=O)CH_3} \xrightarrow[H_2O]{NaBH_4} \underset{\substack{\text{2-Butanol} \\ (87\%)}}{CH_3CH_2CH(OH)CH_3}$$

The key step in the reduction of a carbonyl compound by either lithium aluminum hydride or sodium borohydride is the transfer of a *hydride ion* from the metal to the carbonyl carbon. In this transfer the hydride ion acts as a *nucleophile.* Since compounds of trivalent-boron and -aluminum are Lewis acids they react as electrophiles at the carbonyl oxygen and facilitate the hydride transfer. The mechanism for the reduction of a ketone by sodium borohydride is illustrated below.

$$\underset{\delta+\ \ \delta-}{R_2C{=}\ddot{O}{:}} \quad H_4B^- \ Na^+ \longrightarrow R{-}CH(R){-}\ddot{O}{:}\bar{B}H_3Na^+$$

This step is then repeated until all hydrogens attached to boron have been transferred. The boron complex decomposes in water to form the secondary alcohol.

$$(R_2CHO)_4B^-\ Na^+ + 3H_2O \longrightarrow 4RCH(OH)R + NaH_2BO_3$$

Sodium borohydride is a milder reducing agent than lithium aluminum hydride. Lithium aluminum hydride will reduce acids, esters, aldehydes, and ketones; but sodium borohydride will reduce only aldehydes and ketones. Lithium aluminum hydride *reacts violently with water* and therefore reductions with lithium aluminum hydride must be carried out in anhydrous solutions, usually in anhydrous ether. (Water is added cautiously after the reaction is over to decompose the aluminum complex.)* Sodium borohydride reductions, by contrast, can be carried out in water or alcohol solutions.

Problem 15.10

Which reducing agent ($NaBH_4$ or $LiAlH_4$) would you use to carry out each of the following transformations?

* Unless special precautions are taken, lithium aluminum hydride reductions can be very dangerous. You should consult an appropriate laboratory manual before attempting such a reduction and the reaction should be carried out on a small scale.

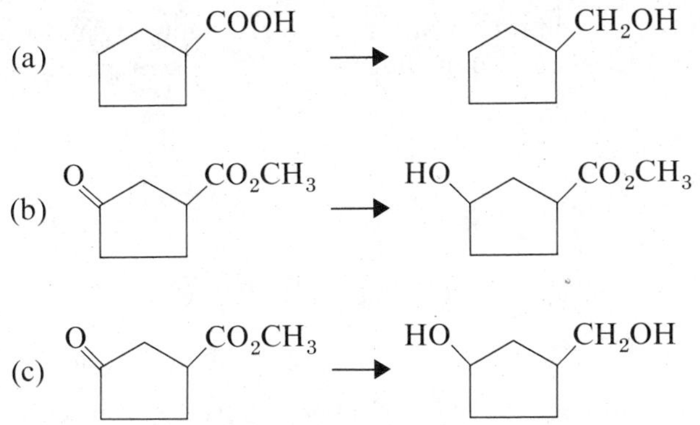

(a) cyclopentane–COOH → cyclopentane–CH_2OH

(b) O= cyclopentane–CO_2CH_3 → HO–cyclopentane–CO_2CH_3

(c) O= cyclopentane–CO_2CH_3 → HO–cyclopentane–CH_2OH

15.8 OXIDATION OF ALCOHOLS

15.8A Oxidation of 1° Alcohols

Primary alcohols can be oxidized to aldehydes and carboxylic acids.

$$\underset{\text{1° Alcohol}}{R{-}CH_2OH} \xrightarrow{(O)} \underset{\text{Aldehyde}}{R{-}\overset{\overset{\displaystyle O}{\|}}{C}{-}H} \xrightarrow{(O)} \underset{\text{Carboxylic acid}}{R{-}\overset{\overset{\displaystyle O}{\|}}{C}{-}OH}$$

The oxidation of aldehydes to carboxylic acids usually takes place with milder oxidizing agents than those required to oxidize primary alcohols to aldehydes; thus it is difficult to stop the oxidation at the aldehyde stage. One way of avoiding this problem, is to remove the aldehyde as soon as it is formed. This can often be done because aldehydes have lower boiling points than alcohols (why?), and therefore aldehydes can be distilled from the reaction mixture as they are formed. An example is the synthesis of butanal from 1-butanol using a mixture of $K_2Cr_2O_7$ and sulfuric acid:

$$\underset{\substack{\text{1-Butanol}\\ \text{bp 117.5°}}}{CH_3CH_2CH_2CH_2OH} \xrightarrow[H_2SO_4]{K_2Cr_2O_7} \underset{\substack{\text{Butanal}\\ \text{bp 75.7°}}}{CH_3CH_2CH_2\overset{\overset{\displaystyle O}{\|}}{C}H} \quad \text{(50\% yield)}$$

This procedure does not give good yields with aldehydes that boil above 100°, however.

An industrial process for preparing low-molecular-weight aldehydes is *dehydrogenation* of a primary alcohol:

$$CH_3CH_2OH \xrightarrow[300°]{Cu} CH_3\overset{\overset{\displaystyle O}{\|}}{C}H + H_2$$

(Notice that dehydrogenation of an organic compound corresponds to oxidation whereas hydrogenation (cf. Problem 15.9) corresponds to reduction.)

In most laboratory preparations we must rely on special oxidizing agents to prepare aldehydes from primary alcohols. An excellent reagent for this purpose is the complex formed when CrO_3 is added carefully to excess pyridine. (Carelessness may cause a fire.)

$$CrO_3 + C_5H_5N\!: \longrightarrow CrO_3 \cdot 2C_5H_5N$$

Pyridine (C_5H_5N) — Chromic oxide-pyridine complex

The complex of chromic oxide and pyridine will oxidize a primary alcohol to an aldehyde and stop at that stage.

$$(C_2H_5)_2C(CH_3){-}CH_2OH + CrO_3 \cdot 2C_5H_5N \longrightarrow (C_2H_5)_2C(CH_3){-}\overset{O}{\overset{\|}{C}}H$$

2-Methyl-2-ethyl-1-butanol — 2-Methyl-2-ethylbutanal

The chromic oxide-pyridine complex does not attack double bonds, as the following example shows.

$$CH_2{=}C(CH_3)(CH_2)_2CH{=}C(CH_3)(CH_2)_3CH_2OH + CrO_3 \cdot 2C_5H_5N \xrightarrow[25^\circ]{CH_2Cl_2}$$

$$CH_2{=}C(CH_3)(CH_2)_2CH{=}C(CH_3)(CH_2)_3\overset{O}{\overset{\|}{C}}H$$

(92%)

Primary alcohols can be oxidized to **carboxylic acids** by potassium permanganate. The reaction is usually carried out in basic aqueous solution where MnO_2 precipitates as the oxidation takes place. After the oxidation is complete, filtration allows removal of the MnO_2 and acidification of the filtrate gives the carboxylic acid.

$$R{-}CH_2OH + KMnO_4 \xrightarrow[H_2O,\ heat]{OH^-} RCOO^-K^+ + MnO_2$$

$$RCOO^-K^+ \xrightarrow{H^+} RCOOH$$

15.8B Oxidation of 2° Alcohols

Secondary alcohols can be oxidized to ketones. The reaction usually stops at the ketone stage because further oxidation requires the breaking of a carbon-carbon bond.

$$R{-}\underset{}{\overset{OH}{\overset{|}{C}}}H{-}R' \xrightarrow{(O)} R{-}\overset{O}{\overset{\|}{C}}{-}R'$$

2° Alcohol — Ketone

A variety of oxidizing agents have been used to oxidize secondary alcohols to ketones. The most commonly used reagent is chromic acid, H_2CrO_4. Chromic acid is usually prepared by adding chromic oxide (CrO_3) or sodium dichromate ($Na_2Cr_2O_7$) to sulfuric acid. Oxidations of secondary alcohols are generally carried out in acetone or acetic acid solutions. The balanced equation is shown below.

$$3\ R_2CHOH + 2H_2CrO_4 + 6H^+ \longrightarrow 3\ R_2C{=}O + 2Cr^{+++} + 8H_2O$$

As chromic acid oxidizes the alcohol to the ketone, chromium is reduced from the +6 oxidation state (H_2CrO_4) to the +3 oxidation state (Cr^{+++}).* Chromic acid oxidations of secondary alcohols generally give ketones in excellent yields. Two specific examples are the oxidation of cyclooctanol to cyclooctanone and the oxidation of (−)-menthol to (−)-menthone.

Cyclooctanol $\xrightarrow[\text{Acetone, } 35^\circ]{H_2CrO_4}$ Cyclooctanone (92–96%)

(−)-Menthol $\xrightarrow[55^\circ]{K_2Cr_2O_7,\ \text{dil. } H_2SO_4}$ (−)-Menthone (85%)

The use of chromic oxide in aqueous acetone is usually called the Jones oxidation (or oxidation by the Jones reagent). This procedure rarely affects double bonds present in the molecule.

15.8C Oxidation of 3° Alcohols

Tertiary alcohols can be oxidized but only under very forcing conditions. Tertiary alcohols are not easily oxidized because their oxidation requires the breaking of a carbon-carbon bond. Oxidations of tertiary alcohols, when they do occur, are of little synthetic utility.

$$R'{-}C(R)(R''){-}OH$$

3° Alcohol — *Difficult to oxidize*

* It is the color change that accompanies this change in oxidation state that allows chromic acid to be used as a test for primary and secondary alcohols.

The relative ease of oxidation of primary and secondary alcohols compared with the difficulty in oxidizing tertiary alcohols forms the basis for the chromic acid test we gave earlier (Section 9.17).

15.9 REACTIONS OF ALCOHOLS INVOLVING O—H BOND BREAKING

We can classify the reactions of alcohols into two general groups: (1) those reactions that take place with cleavage of the O—H bond, and (2) those that take place with cleavage of the C—O bond.

—C—O—H (O—H bond cleavage) —C—O—H (C—O bond cleavage)

O—H bond cleavage C—O bond cleavage

We begin our study of the reactions of alcohols with those in which the O—H bond is broken.

15.9A Alcohols as Acids

Alcohols are weak acids. For reactions taking place in solution, the acidity constants of most alcohols are of the order of 10^{-18}. This means that alcohols are slightly weaker acids than water ($K_a \sim 10^{-16}$), but they are much stronger acids than terminal alkynes ($K_a \sim 10^{-25}$) or ammonia ($K_a \sim 10^{-34}$). Alcohols are, of course, very much stronger acids than alkanes ($K_a \sim 10^{-42}$).

The conjugate base of an alcohol is an *alkoxide ion*. Because an alcohol is a weaker acid than water the alkoxide ion is a stronger base than the hydroxide ion. This means that if we add an alcohol to aqueous solution of sodium hydroxide the position of equilibrium of the acid-base reaction (shown below) will favor hydroxide ions and alcohol molecules. Very little alkoxide ion will be present in the solution.

$$\underset{\substack{\text{Hydroxide ion}\\\text{(weaker base)}}}{:\ddot{O}H^-} + \underset{\substack{\text{Alcohol}\\\text{(weaker acid)}}}{R\ddot{O}H} \rightleftarrows \underset{\substack{\text{Alkoxide ion}\\\text{(stronger base)}}}{R\ddot{O}:^-} + \underset{\substack{\text{Water}\\\text{(stronger acid)}}}{H\ddot{O}H}$$

On the other hand, since ammonia is a much weaker acid than an alcohol, the amide ion is a much stronger base than the alkoxide ion. If we were to add an alcohol to a solution of sodium amide in liquid ammonia, the position of equilibrium would favor the formation of alkoxide ions and ammonia (from the amide ion).

$$\underset{\substack{\text{Amide ion}\\\text{(stronger base)}}}{:\ddot{N}H_2^-} + \underset{\substack{\text{Alcohol}\\\text{(stronger acid)}}}{R\ddot{O}H} \rightleftarrows \underset{\substack{\text{Alkoxide ion}\\\text{(weaker base)}}}{R\ddot{O}:^-} + \underset{\substack{\text{Ammonia}\\\text{(weaker acid)}}}{:NH_3}$$

Problem 15.11

Write equations for the acid-base reactions that would occur if ethanol were added to ether solutions of each of the following compounds. In each equation label the stronger acid, the stronger base, and so on.

(a) $CH_3C{\equiv}CNa$
(b) $CH_3CH_2CH_2CH_2Li$
(c) CH_3CH_2MgBr

Sodium and potassium alkoxides are often used as bases in organic syntheses (Section 5.9). We use alkoxides when we carry out reactions that require stronger bases than hydroxide ion, but do not require exceptionally powerful bases such as the amide ion or the anion of an alkane. We also use alkoxide ions when (for reasons of solubility) we need to carry out a reaction in an alcohol solvent rather than in water.

15.9B Formation of Inorganic Esters

We have already seen examples of reactions in which alcohols react with acyl chlorides or acid anhydrides to form esters of carboxylic acids. We will study the properties of carboxylic acid esters in detail in Chapter 17.

$$\underset{\text{Acyl chloride}}{R'\overset{O}{\overset{\|}{C}}{-}Cl} + \underset{\text{Alcohol}}{H{-}OR} \xrightarrow[(-HCl)]{OH^-} \underset{\text{An ester of a carboxylic acid}}{R'\overset{O}{\overset{\|}{C}}{-}OR}$$

$$\underset{\text{Carboxylic acid anhydride}}{R'\overset{O}{\overset{\|}{C}}{-}O\overset{O}{\overset{\|}{C}}R'} + \underset{\text{Alcohol}}{H{-}OR} \xrightarrow[(-R'\underset{\underset{O}{\|}}{C}OH)]{} \underset{\text{An ester of a carboxylic acid}}{R'\overset{O}{\overset{\|}{C}}{-}OR}$$

Sulfonates. Alcohols also form esters when they react with certain derivatives of inorganic acids. Among the more important esters of this type are *sulfonates*. Ethyl alcohol, for example, reacts with methanesulfonyl chloride to form *ethyl methanesulfonate* and with *p*-toluenesulfonyl chloride to form *ethyl p-toluenesulfonate*. These reactions also involve cleavage of the O—H bond of the alcohol:

$$\underset{\text{Methanesulfonyl chloride}}{CH_3\overset{O}{\overset{\|}{\underset{\underset{O}{\|}}{S}}}{-}Cl} + \underset{\text{Ethyl alcohol}}{H{-}OCH_2CH_3} \xrightarrow[(-HCl)]{OH^-} \underset{\text{Ethyl methanesulfonate (ethyl mesylate)}}{CH_3\overset{O}{\overset{\|}{\underset{\underset{O}{\|}}{S}}}{-}OCH_2CH_3}$$

$$\underset{p\text{-Toluenesulfonyl chloride}}{CH_3{-}C_6H_4{-}\overset{O}{\overset{\|}{\underset{\underset{O}{\|}}{S}}}{-}Cl} + \underset{\text{Ethyl alcohol}}{H{-}OCH_2CH_3} \xrightarrow[(-HCl)]{OH^-} \underset{\text{Ethyl }p\text{-toluenesulfonate (ethyl tosylate)}}{CH_3{-}C_6H_4{-}\overset{O}{\overset{\|}{\underset{\underset{O}{\|}}{S}}}{-}OCH_2CH_3}$$

Sulfonyl chlorides are usually prepared by treating sulfonic acids with phosphorus pentachloride.

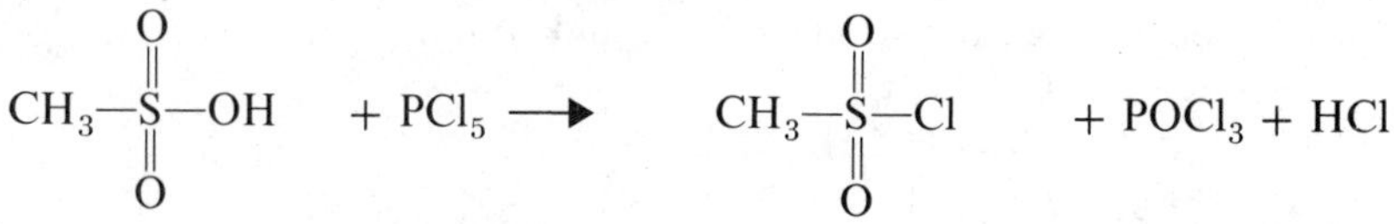

$$CH_3{-}\underset{\underset{O}{\|}}{\overset{\overset{O}{\|}}{S}}{-}OH + PCl_5 \longrightarrow CH_3{-}\underset{\underset{O}{\|}}{\overset{\overset{O}{\|}}{S}}{-}Cl + POCl_3 + HCl$$

Methanesulfonic acid — Methanesulfonyl chloride (mesyl chloride)

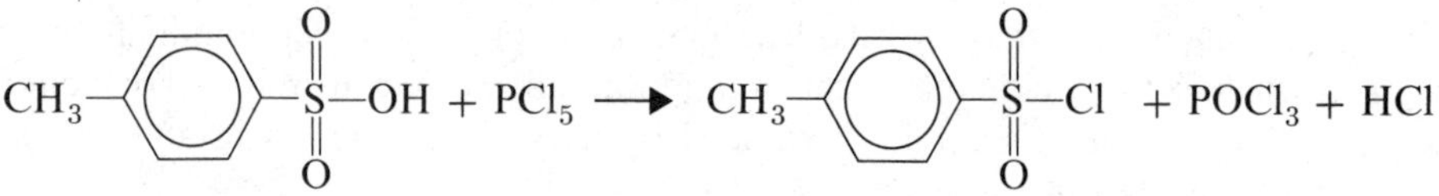

$$CH_3{-}C_6H_4{-}\underset{\underset{O}{\|}}{\overset{\overset{O}{\|}}{S}}{-}OH + PCl_5 \longrightarrow CH_3{-}C_6H_4{-}\underset{\underset{O}{\|}}{\overset{\overset{O}{\|}}{S}}{-}Cl + POCl_3 + HCl$$

p-Toluenesulfonic acid — *p*-Toluenesulfonyl chloride (tosyl chloride)

Methanesulfonyl chloride and *p*-toluenesulfonyl chloride are used so often that organic chemists have shortened their rather long names to "mesyl chloride" and "tosyl chloride," respectively. The methanesulfonyl group is often called a "mesyl" group and the *p*-toluenesulfonyl group is called a "tosyl" group. Methanesulfonates are known as "mesylates" and *p*-toluenesulfonates are known as "tosylates."

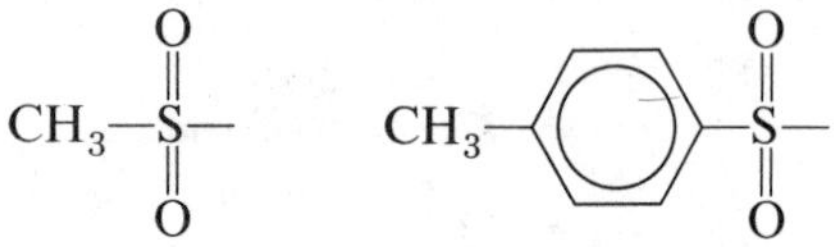

$$CH_3{-}\underset{\underset{O}{\|}}{\overset{\overset{O}{\|}}{S}}{-} \qquad CH_3{-}C_6H_4{-}\underset{\underset{O}{\|}}{\overset{\overset{O}{\|}}{S}}{-}$$

or Ms— the mesyl group — or Ts— the tosyl group

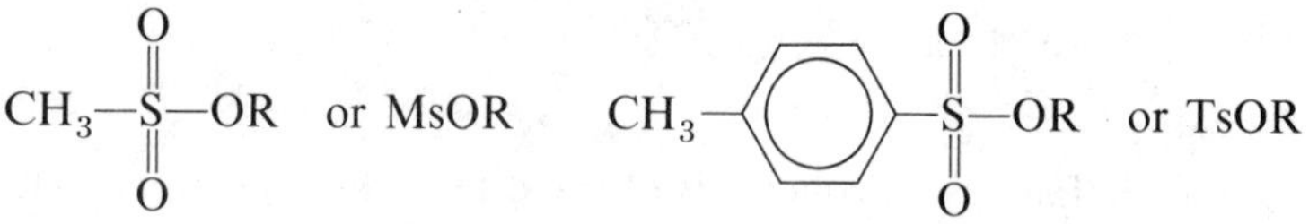

$$CH_3{-}\underset{\underset{O}{\|}}{\overset{\overset{O}{\|}}{S}}{-}OR \text{ or MsOR} \qquad CH_3{-}C_6H_4{-}\underset{\underset{O}{\|}}{\overset{\overset{O}{\|}}{S}}{-}OR \text{ or TsOR}$$

An alkyl mesylate — An alkyl tosylate

Problem 15.12

Starting with benzene or toluene and any necessary alcohols or inorganic reagents, show how you would prepare each of the following sulfonates.

(a) Methyl benzenesulfonate
(b) Isopropyl *p*-toluenesulfonate (isopropyl tosylate)
(c) *tert*-Butyl *p*-bromobenzenesulfonate (*tert*-butyl brosylate)

Alkyl sulfonates are frequently used as substrates for nucleophilic substitution reactions because sulfonate ions are excellent leaving groups.

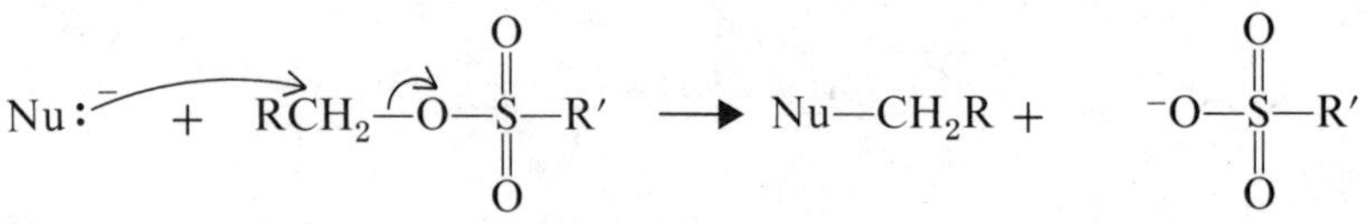

$$Nu:^- + RCH_2{-}O{-}\underset{\underset{O}{\|}}{\overset{\overset{O}{\|}}{S}}{-}R' \longrightarrow Nu{-}CH_2R + {}^-O{-}\underset{\underset{O}{\|}}{\overset{\overset{O}{\|}}{S}}{-}R'$$

Alkyl sulfonate (tosylate, mesylate, etc.) — Sufonate ion (weak base— a good leaving group)

Thus sulfonates give us an indirect method for carrying out nucleophilic substitution reactions on alcohols. We first convert the alcohol to a sulfonate and then we allow the sulfonate to react with a nucleophile. When the carbon bearing the —OH is chiral, the first step—sulfonate formation—proceeds with *retention of configuration* because no bonds to the chiral carbon are broken. Only the O—H bond breaks. The second step—if the reaction is S_N2—proceeds with *inversion of configuration.*

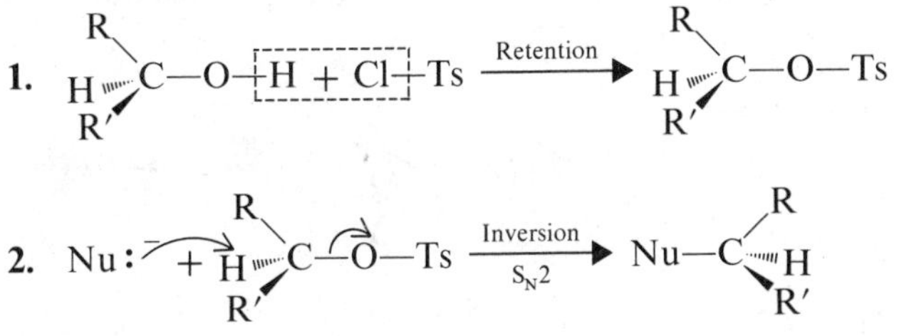

1. (R)(R′)(H)C—O—H + Cl—Ts $\xrightarrow{\text{Retention}}$ (R)(R′)(H)C—O—Ts

2. Nu: + (R)(R′)(H)C—O—Ts $\xrightarrow[S_N2]{\text{Inversion}}$ Nu—C(R)(H)(R′)

Problem 15.13

Show the configurations of products formed when (a) (R)-2-butanol is converted to a tosylate, and (b) when this tosylate reacts with hydroxide ion by an S_N2 reaction. (c) Converting *cis*-4-methylcyclohexanol to a tosylate and then allowing the tosylate to react with LiCl (in an appropriate solvent) yields *trans*-1-chloro-4-methylcyclohexane. Outline the stereochemistry of these steps.

Alkyl sulfates (Section 7.5) differ from alkyl sulfonates. Sulfates are esters of alcohols and sulfuric acid and sulfonates are esters of alcohols and a sulfonic acid, RSO_2OH.

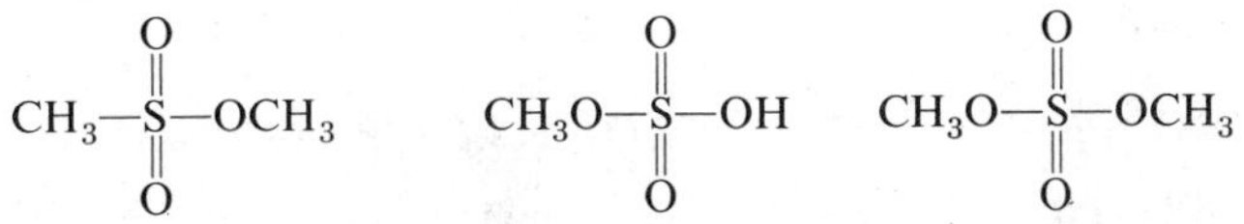

$CH_3—S(=O)_2—OCH_3$	$CH_3O—S(=O)_2—OH$	$CH_3O—S(=O)_2—OCH_3$
Methyl methanesulfonate	Methyl hydrogen sulfate	Dimethyl sulfate

Alkyl sulfonates contain one carbon-sulfur bond and one carbon-oxygen-sulfur linkage, whereas alkyl sulfates and dialky sulfates contain only carbon-oxygen-sulfur linkages. Alkyl and dialkyl sulfates are made by treating alkenes or alcohols with sulfuric acid. Dimethyl sulfate, for example, can be obtained by distilling a mixture of methyl alcohol and sulfuric acid at reduced pressure. Organic sulfates are also frequently used in nucleophilic substitution reactions because sulfate ions are weak bases and therefore good *leaving groups.*

15.9C Alkyl Phosphates

Alcohols react with phosphoric acid to yield alkyl phosphates:

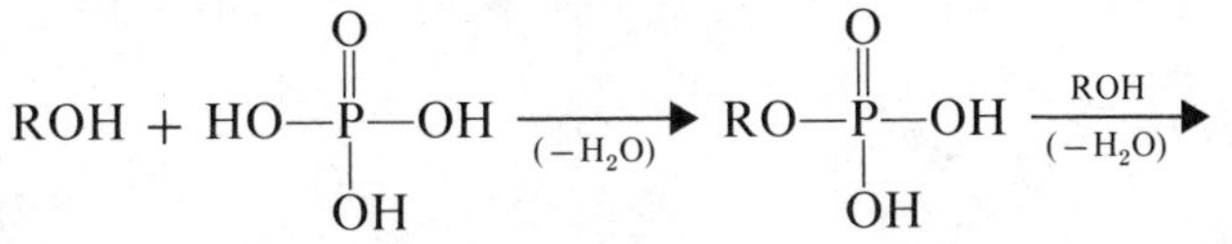

$$ROH + HO—P(=O)(OH)—OH \xrightarrow[(-H_2O)]{} RO—P(=O)(OH)—OH \xrightarrow[(-H_2O)]{ROH}$$

Phosphoric acid — Alkyl phosphate

$$\mathrm{RO{-}\overset{O}{\overset{\|}{P}}{-}OH\ (OR)} \xrightarrow[(-H_2O)]{ROH} \mathrm{RO{-}\overset{O}{\overset{\|}{P}}{-}OR\ (OR)}$$

Dialkyl hydrogen phosphate → Trialkyl phosphate

When phosphoric acid is heated it forms phosphoric *anhydrides* called diphosphoric acid and triphosphoric acid.

$$\mathrm{2HO{-}\overset{O}{\overset{\|}{P}}(OH){-}OH} \xrightarrow{-H_2O} \mathrm{HO{-}\overset{O}{\overset{\|}{P}}(OH){-}O{-}\overset{O}{\overset{\|}{P}}(OH){-}OH}$$

Anhydride linkage

Diphosphoric acid
(pyrophosphoric acid)

$$\mathrm{3HO{-}\overset{O}{\overset{\|}{P}}(OH){-}OH} \xrightarrow[(-2H_2O)]{} \mathrm{HO{-}\overset{O}{\overset{\|}{P}}(OH){-}O{-}\overset{O}{\overset{\|}{P}}(OH){-}O{-}\overset{O}{\overset{\|}{P}}(OH){-}OH}$$

Anhydride linkages

Triphosphoric acid

These phosphoric acid anhydrides can also react with alcohols to form esters such as those shown below.

$$\mathrm{RO{-}\overset{O}{\overset{\|}{P}}(OH){-}O{-}\overset{O}{\overset{\|}{P}}(OH){-}OH} \qquad \mathrm{RO{-}\overset{O}{\overset{\|}{P}}(OH){-}O{-}\overset{O}{\overset{\|}{P}}(OH){-}O{-}\overset{O}{\overset{\|}{P}}(OH){-}OH}$$

An alkyl trihydrogen diphosphate — An alkyl tetrahydrogen triphosphate

Esters of phosphoric acids are extremely important in biochemical reactions. Especially important are triphosphate esters. Although hydrolysis of the ester group or of one of the anhydride linkages of an alkyl triphosphate is exothermic, these reactions occur very slowly in aqueous solutions.

$$\mathrm{RO{-}\overset{O}{\overset{\|}{P}}(OH){-}O{-}\overset{O}{\overset{\|}{P}}(OH){-}O{-}\overset{O}{\overset{\|}{P}}(OH){-}OH} \xrightarrow[Slow]{H_2O}$$

Ester linkage (RO—P); *Anhydride linkages* (P—O—P)

$$\rightarrow \mathrm{ROH + HO{-}\overset{O}{\overset{\|}{P}}(OH){-}O{-}\overset{O}{\overset{\|}{P}}(OH){-}O{-}\overset{O}{\overset{\|}{P}}(OH){-}OH}$$

$$\rightarrow \mathrm{RO{-}\overset{O}{\overset{\|}{P}}(OH){-}OH + HO{-}\overset{O}{\overset{\|}{P}}(OH){-}O{-}\overset{O}{\overset{\|}{P}}(OH){-}OH}$$

$$\rightarrow \mathrm{RO{-}\overset{O}{\overset{\|}{P}}(OH){-}O{-}\overset{O}{\overset{\|}{P}}(OH){-}OH + HO{-}\overset{O}{\overset{\|}{P}}(OH){-}OH}$$

Thus alkyl triphosphates are relatively stable compounds in the aqueous medium of a living cell.

Enzymes, on the other hand, are able to catalyze reactions of these triphosphates in which the energy made available when their anhydride linkages break helps the cell make other chemical bonds. We have more to say about this in Chapter 19 when we discuss the important triphosphate called adenosine triphosphate (or ATP).

15.10 REACTIONS OF ALCOHOLS INVOLVING C—O BOND BREAKING

We have already seen two reactions that involve cleavage of the C—O bond of an alcohol.

15.10A Dehydration of Alcohols (discussed in Sections 6.10–6.12)

When alcohols are heated with strong acids they undergo elimination of water (dehydration) and form alkenes. The mechanism (discussed earlier) is given below.

$$\underset{\mathrm{H}\ \ \mathrm{OH}}{-\mathrm{C}-\mathrm{C}-} \underset{-\mathrm{H}}{\overset{+\mathrm{H}^+}{\rightleftharpoons}} \underset{\mathrm{H}\ \ {}^+\mathrm{OH}_2}{-\mathrm{C}-\mathrm{C}-} \underset{+\mathrm{H}_2\mathrm{O}}{\overset{-\mathrm{H}_2\mathrm{O}}{\rightleftharpoons}} \underset{\mathrm{H}}{-\mathrm{C}-\overset{+}{\mathrm{C}}-} \underset{+\mathrm{H}^+}{\overset{-\mathrm{H}^+}{\rightleftharpoons}} \mathrm{>C{=}C<}$$

Two specific examples of alcohol dehydrations are the following:

$$\underset{\text{2-Methyl-2-butanol}}{CH_3CH_2C(CH_3)(OH)CH_3} \xrightarrow[90\text{–}95^\circ]{46\%\ H_2SO_4} \underset{\substack{\text{2-Methyl-2-butene}\\(84\%)}}{CH_3CH{=}C(CH_3){-}CH_3} + H_2O$$

$$\underset{\text{2-Pentanol}}{CH_3CH_2CH_2CH(OH)CH_3} \xrightarrow[90\text{–}95^\circ]{62\%\ H_2SO_4} \underset{\substack{\text{2-Pentene}\\(80\%)}}{CH_3CH_2CH{=}CHCH_3} + H_2O$$

As these examples demonstrate, alcohol dehydrations generally produce the more highly substituted and, thus, *the more stable* alkene. The dehydration of 2-methyl-2-butanol produces, mainly, 2-methyl-2-butene rather than 2-methyl-1-butene, and the dehydration of 2-pentanol yields, mainly, 2-pentene rather than 1-pentene.

Problem 15.14

(a) Which 2-pentene isomer would you expect to obtain in the greater amount from the dehydration of 2-pentanol, *cis*-2-pentene or *trans*-2-pentene? (b) Dehydration of 1-phenyl-2-propanol yields mainly *trans* (and some *cis*)-1-phenylpropene. The reaction produces practically no 3-phenylpropene. Explain.

Problem 15.15

When 3,3-dimethyl-2-butanol is treated with 85% phosphoric acid the following products are obtained.

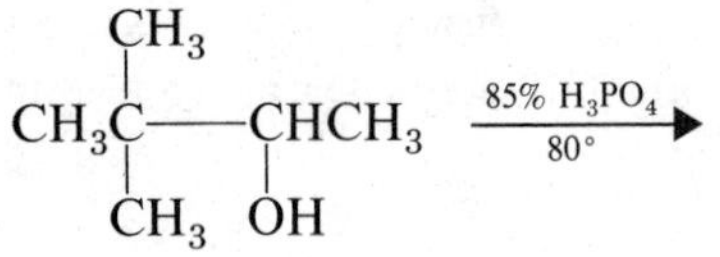

$$\underset{\text{3,3-Dimethyl-2-butanol}}{CH_3\overset{CH_3}{\underset{CH_3}{C}}—\underset{OH}{C}HCH_3} \xrightarrow[80°]{85\%\ H_3PO_4}$$

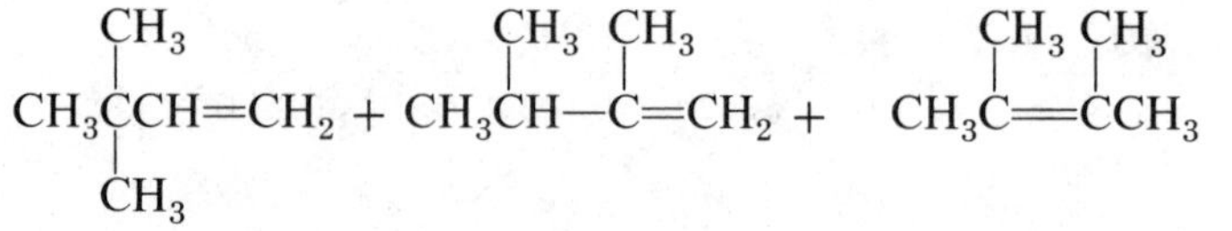

$$\underset{\substack{\text{3,3-Dimethyl-1-}\\\text{butene}\\(0.4\%)}}{CH_3\overset{CH_3}{\underset{CH_3}{C}}CH{=}CH_2} + \underset{\substack{\text{2,3-Dimethyl-1-}\\\text{butene}\\(20\%)}}{CH_3\overset{CH_3}{C}H—\overset{CH_3}{C}{=}CH_2} + \underset{\substack{\text{2,3-Dimethyl-2-butene}\\(80\%)}}{CH_3\overset{CH_3}{C}{=}\overset{CH_3}{C}CH_3}$$

(a) Write a mechanism that accounts for the formation of each product. (b) Why is 2,3-dimethyl-2-butene the major product?

15.10B Reactions of Alcohols with HX, PBr_3, and $SOCl_2$

We saw in Sections 14.5 to 14.7 that alcohols react with HX, PBr_3, and $SOCl_2$ to form alkyl halides. These reactions also involve cleavage of the carbon-oxygen bond of the alcohol.

$$R—OH + HX \longrightarrow R—X + H_2O$$

$$3R—OH + PBr_3 \longrightarrow R—Br + H_3PO_3$$

$$R—OH + SOCl_2 \longrightarrow R—Cl + SO_2 + HCl$$

15.11 POLYHYDROXYL ALCOHOLS

Compounds containing two hydroxyl groups are called "glycols" or diols. The simplest possible diol is the unstable compound methylene glycol or methanediol.

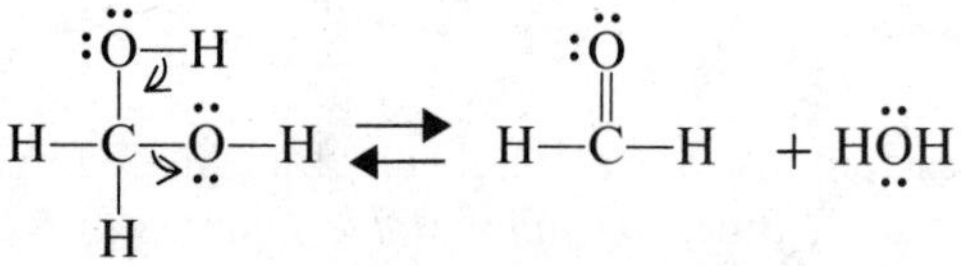

$$\underset{\text{Methanediol}}{H—\overset{:\ddot{O}—H}{\underset{H}{C}}—\ddot{O}—H} \rightleftharpoons \underset{\text{Formaldehyde}}{H—\overset{:\ddot{O}}{\overset{\|}{C}}—H} + H\ddot{O}H$$

Methanediol is a *gem*-diol (a diol that has both hydroxyl groups attached to the same carbon). Most *gem*-diols are unstable except in an aqueous solution. When the water is removed, the *gem*-diol dehydrates and forms an aldehyde or ketone. When methanediol dehydrates it produces formaldehyde.

Gem-diols with strong electron-withdrawing groups can usually be isolated. One example is chloral hydrate (2,2,2-trichloroethanediol). Chloral hydrate is occasionally used as a sleep-inducing drug.

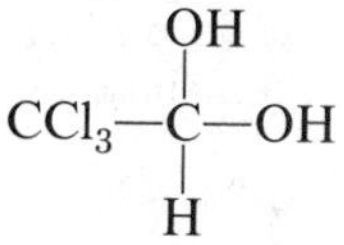

$$CCl_3\text{—}C(OH)(H)\text{—}OH$$

Chloral hydrate
(mp 51.7°)

Vic-diols are much more stable than *gem*-diols. *Vic*-diols can be prepared by the hydroxylation of alkenes (cf. Section 7.12 and Section 7.13).

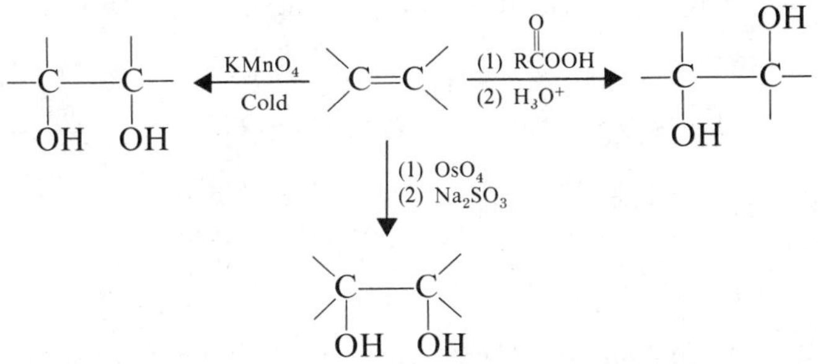

$$\text{—}C(OH)\text{—}C(OH)\text{—} \xleftarrow[\text{Cold}]{KMnO_4} >C{=}C< \xrightarrow[(2)\ H_3O^+]{(1)\ RCOOH} \text{—}C(OH)\text{—}C(OH)\text{—}$$

$$>C{=}C< \xrightarrow[(2)\ Na_2SO_3]{(1)\ OsO_4} >C(OH)\text{—}C(OH)<$$

An important triol is the compound glycerol (1,2,3-propanetriol).

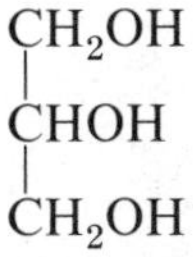

$$CH_2OH\text{—}CHOH\text{—}CH_2OH$$

Glycerol

Glycerol esters are, as we will see, very important compounds in biochemistry. Glycerol, itself, is a viscous hygroscopic liquid that in moderate amounts is nontoxic. It is often used as a moistening agent in food, tobacco, and cosmetics.

The ester formed when glycerol reacts with nitric acid is the well-known explosive nitroglycerin.

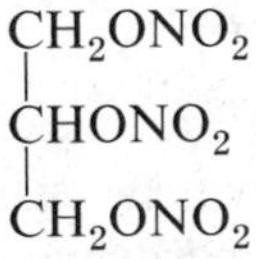

$$CH_2ONO_2\text{—}CHONO_2\text{—}CH_2ONO_2$$

Glyceryl trinitrate or "nitroglycerin"

Nitroglycerin is very sensitive to shock but becomes much more stable and, therefore, safer when it is absorbed by sawdust or diatomaceous earth. In this form nitroglycerin is called "dynamite." Dynamite was invented by the Swedish industrial chemist Alfred Nobel.

In 1895 Nobel established a trust fund for the purpose of awarding annual prizes for exceptional contributions to the fields of chemistry, physics, medicine, and literature and to the cause of world peace. Prizes have been awarded since 1900. The first recipient of the Nobel Prize for chemistry was J. H. van't Hoff (p. 282). Marie Sklodowska Curie, a Polish chemist who worked in France, won two Nobel Prizes for science; one for her work in physics (1903) and one for her work in chemistry (1911). Professor Linus Pauling of the California Institute of Technology is the only person to have won two Nobel Prizes in distinctly separate areas. Pauling won the Nobel Prize for his contributions to chemistry in 1954, and for his contributions toward world peace in 1962.

15.12 SYNTHESIS OF ETHERS

We have already studied two methods for synthesizing ethers.

1. **By Solvomercuration-Demercuration.** Ethers can be prepared by solvomercuration-demercuration (Section 7.7) using an alcohol as the solvent in the first (solvomercuration) stage.

$$\rangle C{=}C\langle + Hg(OAc)_2 \xrightarrow{ROH} -\underset{OR}{\overset{|}{\underset{|}{C}}}-\underset{HgOAc}{\overset{|}{\underset{|}{C}}}- \xrightarrow[OH^-]{NaBH_4} -\underset{OR}{\overset{|}{\underset{|}{C}}}-\underset{H}{\overset{|}{\underset{|}{C}}}-$$

Alkoxyalkylmercury compound — ether

2. **By Epoxidation.** Cyclic ethers with three-membered rings (epoxides) can be synthesized by treating an alkene with a peroxy acid (Section 7.12).

$$\rangle C{=}C\langle + R\overset{O}{\overset{\|}{C}}-O-O-H \xrightarrow[CH_2Cl_2]{} -\overset{|}{C}\underset{\ddot{O}}{\diagdown\!\diagup}\overset{|}{C}-$$

Epoxide

We can now examine two new methods for preparing ethers.

3. **By Intermolecular Dehydration of Alcohols.** Alcohols can dehydrate to form an alkene. We studied this in Sections 6.10 and 15.10. Primary alcohols can also dehydrate to form an ether:

$$R-O\boxed{H + HO-}R \xrightarrow[(-H_2O)]{H^+} R-O-R$$

Dehydration to an ether usually takes place at a lower temperature than dehydration to the alkene and dehydration to the ether can be aided by distilling the ether as it is formed. Diethyl ether is made commercially by dehydration of ethyl alcohol. Diethyl ether is the predominant product at 140°; ethene is the major product at 180°:

$$CH_3CH_2OH \xrightarrow[180^\circ]{H_2SO_4} CH_2{=}CH_2 \text{ (Ethene)}$$

$$CH_3CH_2OH \xrightarrow[140^\circ]{H_2SO_4} CH_3CH_2OCH_2CH_3 \text{ (Diethyl ether)}$$

The mechanism for the formation of the ether is probably an S_N2 mechanism with one molecule of the alcohol acting as the nucleophile and with another protonated molecule of the alcohol acting as the substrate.

$$CH_3CH_2\ddot{O}H + CH_3CH_2—\overset{+}{O}H_2 \rightleftharpoons CH_3CH_2—\underset{H}{\overset{+}{O}}—CH_2CH_3 + H_2O$$

$$\rightleftharpoons CH_3CH_2OCH_2CH_3 + H_3O^+$$

This method of preparing ethers is of limited usefulness, however. Attempts to synthesize ethers with 2° alkyl groups by intermolecular dehydration of 2° alcohols are usually unsuccessful, because alkenes form too easily. Attempts to make ethers with 3° alkyl groups lead exclusively to the alkene. And, finally, this method cannot be used to prepare unsymmetrical ethers from primary alcohols because the reaction leads to a mixture of products.

$$\underbrace{ROH + R'OH}_{1° \text{ alcohols}} \underset{H_2SO_4}{\rightleftharpoons} \begin{matrix} ROR \\ + \\ ROR' + H_2O \\ + \\ R'OR' \end{matrix}$$

Problem 15.16

An exception to what we have just said has to do with syntheses of unsymmetrical ethers in which one alkyl group is a *tert*-butyl group and the other group is either primary or secondary. These syntheses can be accomplished in two ways (a) by adding *tert*-butyl alcohol to a mixture of the primary or secondary alcohol and an acid at room temperature or (b) by adding isobutylene, $CH_2{=}C(CH_3)_2$, to a mixture of the primary or secondary alcohol and an acid. Give likely mechanisms for these reactions and explain why they are successful.

In addition to the method given in problem 15.16, unsymmetrical ethers can also be prepared from alkenes using solvomercuration-demercuration (Section 7.7). Another important route to unsymmetrical ethers is a nucleophilic substitution reaction known as the Williamson synthesis.

4. **The Williamson Synthesis of Ethers.** This synthesis consists of an S_N2 reaction of a sodium alkoxide with an alkyl halide, alkyl sulfonate, or alkyl sulfate:

$$R—O^- \ Na^+ + R'—L \longrightarrow R—O—R' + Na^+ \ L^-$$

$$(L = Br, I, OSO_2R'', \text{ or } OSO_2OR'')$$

A specific example of the Williamson synthesis is shown below.

$$\underset{n\text{-Propyl alcohol}}{CH_3CH_2CH_2OH} + Na \longrightarrow \underset{\text{Sodium } n\text{-propoxide}}{CH_3CH_2CH_2\ddot{O}:^- \ Na^+} + \tfrac{1}{2}H_2$$

$$\downarrow CH_3CH_2I$$

$$\underset{\substack{\text{Ethyl } n\text{-propyl ether} \\ (70\%)}}{CH_3CH_2OCH_2CH_2CH_3} + Na^+ \ I^-$$

The usual limitations of S_N2 reactions apply here. Best results are obtained when the alkyl halide, sulfonate, or sulfate is primary (or methyl). If the substrate is tertiary, elimination is the exclusive result. Substitution is also favored over elimination at lower temperatures.

Problem 15.17

(a) Outline two methods for preparing isopropyl methyl ether by a Williamson synthesis. (b) One method gives a much better yield of the ether than the other. Explain which is the better method and why.

Problem 15.18

The two syntheses of 1-phenyl-2-ethoxypropane shown below give products with opposite optical rotations.

$$C_6H_5CH_2CH(OH)CH_3 \; (\alpha = +33.0^\circ) \xrightarrow{K} \text{Potassium salt} \xrightarrow{C_2H_5Br} C_6H_5CH_2CH(OC_2H_5)CH_3 \; (\alpha = +23.5^\circ)$$

$$C_6H_5CH_2CH(OH)CH_3 \xrightarrow{TsCl/base} C_6H_5CH_2CH(OTs)CH_3 \xrightarrow[K_2CO_3]{C_2H_5OH} C_6H_5CH_2CH(OC_2H_5)CH_3 \; (\alpha = -19.9^\circ)$$

How can you explain this?

Problem 15.19

Write a mechanism that explains the formation of tetrahydrofuran from the reaction of 4-chloro-1-butanol and aqueous sodium hydroxide.

Problem 15.20

Epoxides can be synthesized by treating halohydrins with aqueous base. For example, treating $ClCH_2CH_2OH$ with aqueous sodium hydroxide yields ethylene oxide. (a) Propose a mechanism for this reaction. (b) *Trans*-2-chlorocyclohexanol reacts readily with sodium hydroxide to yield cyclohexene oxide. *Cis*-2-chlorocyclohexanol does not undergo this reaction, however. How can you account for this?

15.13 REACTIONS OF ETHERS

15.13A Dialkyl Ethers

Dialkyl ethers react with very few reagents. The only reactive sites that molecules of a dialkyl ether present to another reactive substance are the C—H bonds of the alkyl groups and the $-\ddot{\underset{\cdot\cdot}{O}}-$ group of the ether linkage.

Ethers are like alkanes in that they undergo free-radical substitution reactions but these are of little synthetic importance.

The oxygen of the ether linkage makes ethers basic. Ethers can react with proton donors or Lewis acids to form *oxonium salts.*

$$CH_3CH_2\ddot{\underset{\cdot\cdot}{O}}CH_2CH_3 + HBr \rightleftharpoons CH_3CH_2-\overset{+}{\underset{H}{\underset{|}{\ddot{O}}}}-CH_2CH_3\ Br^-$$

An oxonium salt

$$CH_3CH_2\ddot{\underset{\cdot\cdot}{O}}CH_2CH_3 + BF_3 \rightleftharpoons CH_3CH_2-\overset{+}{\underset{\overset{-}{BF_3}}{\underset{|}{\ddot{O}}}}-CH_2CH_3$$

(an oxonium salt)
Boron trifluoride etherate

The oxonium salt formed when diethyl ether reacts with boron trifluoride is stable enough to allow its distillation at low pressures (bp 48° at 10 Torr).

Heating dialkyl ethers with very strong acids (HI, HBr, and H_2SO_4) causes them to undergo reactions in which the carbon-oxygen bond breaks. Diethyl ether, for example, reacts with hot concentrated hydrobromic acid to give two moles of ethyl bromide.

$$CH_3CH_2OCH_2CH_3 + 2HBr \longrightarrow 2CH_3CH_2Br + H_2O$$

Cleavage of an ether

The mechanism for this reaction begins with formation of an oxonium ion. Then a S_N2 reaction with a bromide ion acting as the nucleophile produces ethyl alcohol and ethyl bromide.

$$CH_3CH_2\ddot{\underset{\cdot\cdot}{O}}CH_2CH_3 + H\ddot{\underset{\cdot\cdot}{Br}}: \rightleftharpoons CH_3CH_2\overset{+}{\underset{H}{\underset{|}{\ddot{O}}}}-CH_2CH_3 + :\ddot{\underset{\cdot\cdot}{Br}}:^- \longrightarrow$$

$$CH_3CH_2\underset{H}{\underset{|}{\ddot{O}}}: + CH_3CH_2Br$$

Ethyl alcohol　　Ethyl bromide

In the next step the ethyl alcohol (formed above) reacts with HBr to form a second mole of ethyl bromide.

$$CH_3CH_2\ddot{\underset{\cdot\cdot}{O}}H + H\ddot{\underset{\cdot\cdot}{Br}}: \rightleftharpoons :\ddot{\underset{\cdot\cdot}{Br}}:^- + CH_3CH_2-\underset{H}{\underset{|}{\ddot{O}^+}}-H \longrightarrow$$

$$CH_3CH_2-\ddot{\underset{\cdot\cdot}{Br}}: + :\underset{H}{\underset{|}{\ddot{O}}}-H$$

Problem 15.21

Account for the fact that dialkyl ethers are cleaved readily when they are heated with strong acids but are not cleaved in weak acids or in neutral media.

15.13B Epoxides

The highly strained three-membered ring in molecules of epoxides makes them much more reactive toward nucleophilic substitution than other ethers. We have seen two examples: acid-catalyzed hydration of epoxides (Sections 7.12 and 8.10) is a method for preparing *vic*-diols (glycols) and the reaction of ethylene oxide with Grignard reagents (Section 14.10) is a method for preparing primary alcohols with a carbon chain two carbons longer than the Grignard reagent.

Acid-catalysis assists epoxide ring-opening by providing a better leaving group (an alcohol) at the carbon undergoing nucleophilic attack. This is especially important if the nucleophile is a weak nucleophile such as water or an alcohol:

$$-\overset{|}{C}-\overset{|}{C}-\ (\text{ring } \ddot{O}) \underset{-H^+}{\overset{+H^+}{\rightleftharpoons}} -\overset{|}{C}-\overset{|}{C}-\ (\text{ring } \overset{+}{O}-H) \xrightarrow{H-\ddot{O}-H}$$

$$\longrightarrow HO-\overset{|}{\underset{|}{C}}-\overset{|}{\underset{|}{C}}-\overset{+}{\underset{H}{O}}-H \xrightarrow{-H^+} HO-\overset{|}{\underset{|}{C}}-\overset{|}{\underset{|}{C}}-OH$$

In the absence of an acid catalyst the leaving group must be a strongly basic alkoxide ion. Such reactions do not occur with other ethers, but they are possible with epoxides (because of ring strain), provided that the attacking nucleophile is strong.

$$Nu:^- + -\overset{|}{C}-\overset{|}{C}-\ (\text{ring } O) \longrightarrow Nu-\overset{|}{\underset{|}{C}}-\overset{|}{\underset{|}{C}}-O^-$$

A strong nucleophile — An alkoxide ion

$$\xrightarrow{+H^+} Nu-\overset{|}{\underset{|}{C}}-\overset{|}{\underset{|}{C}}-OH$$

Problem 15.22

Propose structures for each of the following products.

(a) Ethylene oxide $\xrightarrow[CH_3OH]{H^+}$ $C_3H_8O_2$ (an industrial solvent called Methyl Cellosolve)

(b) Ethylene oxide $\xrightarrow[CH_3CH_2OH]{H^+}$ $C_4H_{10}O_2$ (Ethyl Cellosolve)

(c) Ethylene oxide $\xrightarrow[H_2O]{KI}$ C_2H_5IO

(d) Ethylene oxide $\xrightarrow{NH_3}$ C_2H_7NO

(e) Ethylene oxide $\xrightarrow[CH_3OH]{CH_3ONa}$ $C_3H_8O_2$

Problem 15.23

Treating propene oxide with sodium ethoxide in ethanol gives primarily 1-ethoxy-2-propanol and very little 2-ethoxy-1-propanol. What factor accounts for this?

Problem 15.24

When sodium ethoxide reacts with epichlorohydrin, labeled with ^{14}C as shown by the asterisk in **I** below, the major product is an epoxide bearing the label as in **II**. Provide an explanation.

$$\underset{\text{Epichlorohydrin}}{\underset{\mathbf{I}}{Cl{-}CH_2{-}\overset{}{CH}{-}\overset{*}{C}H_2 \ (\text{O bridge})}} \xrightarrow{NaOC_2H_5} \underset{\mathbf{II}}{C_2H_5O\overset{*}{C}H_2{-}CH{-}CH_2 \ (\text{O bridge})}$$

15.14 SYNTHESIS OF PHENOLS

Phenol is a highly important industrial chemical; it serves as the raw material for a large number of commercial products ranging from aspirin to a variety of plastics. Worldwide production of phenol is more than three million tons per year! Several methods are used to synthesize phenol commercially.

1. **Hydrolysis of Chlorobenzene (Dow Process).** In this process chlorobenzene is heated at 350° (under high pressure) with aqueous sodium hydroxide. This produces sodium phenoxide which on acidification yields phenol. The mechanism for the reaction probably involves the formation of benzyne (Section 14.3).

$$C_6H_5Cl + 2NaOH \xrightarrow[\text{(High pressure)}]{350^\circ} C_6H_5ONa + NaCl + H_2O$$

$$C_6H_5ONa \xrightarrow{HCl} C_6H_5OH + NaCl$$

2. **Alkali Fusion of Sodium Benzenesulfonate.** This, the first commercial process for synthesizing phenol, was developed in Germany in 1890. Sodium benzenesulfonate is melted (fused) with sodium hydroxide (at 350°) to produce sodium phenoxide. Acidification then yields phenol.

$$\underset{\text{Sodium benzene-sulfonate}}{C_6H_5SO_3Na} + 2NaOH \xrightarrow{350^\circ} C_6H_5ONa + Na_2SO_3 + H_2O$$

This procedure can also be used in the laboratory and works quite well for the preparation of *p*-cresol as the following example shows. However, the conditions required to bring about the reaction are so vigorous that its use in the preparation of many phenols is limited.

$$p\text{-}CH_3C_6H_4SO_3Na \xrightarrow[300\text{–}330^\circ]{NaOH(72\%)\text{-}KOH(28\%)} p\text{-}CH_3C_6H_4O^-Na^+ \xrightarrow{H_3O^+} p\text{-}CH_3C_6H_4OH$$

Sodium *p*-toluenesulfonate

p-Cresol
(63–70% overall)

3. **From Cumene Hydroperoxide.** This process illustrates industrial chemistry at its best. Overall it is a method for converting two relatively inexpensive organic compounds—benzene and propene—into two more expensive ones—phenol and acetone. The only other substance consumed in the process is oxygen from air. Most of the worldwide production of phenol is now based on this method.

The synthesis begins with the Friedel-Crafts alkylation of benzene with propene to produce cumene (isopropylbenzene).

$$\textbf{1.}\quad C_6H_6 + CH_2{=}CHCH_3 \xrightarrow[H_3PO_4,\ \text{pressure}]{250^\circ} C_6H_5CH(CH_3)_2$$

Cumene

Then cumene is oxidized to cumene hydroperoxide:

$$\textbf{2.}\quad C_6H_5{-}CH(CH_3)_2 + O_2 \xrightarrow{95^\circ\text{–}135^\circ} C_6H_5{-}C(CH_3)_2{-}O{-}O{-}H$$

Cumene hydroperoxide

Finally, when treated with 10% sulfuric acid, cumene hydroperoxide undergoes a hydrolytic rearrangement that yields phenol and acetone:

$$\textbf{3.}\quad C_6H_5{-}C(CH_3)_2{-}O{-}OH \xrightarrow[50\text{–}90^\circ]{H^+,\ H_2O} C_6H_5OH + (CH_3)_2C{=}O$$

Phenol Acetone

The mechanisms of each of these steps require some comment. The first is a familiar one. The isopropyl carbocation generated by the reaction of propene with the acid (H_3PO_4) alkylates benzene in a typical electrophilic aromatic substitution:

$$CH_2{=}CHCH_3 \xrightarrow{H^+} CH_3\overset{+}{C}HCH_3 \xrightarrow{C_6H_6} C_6H_5^{+}(H)\text{—}CH(CH_3)_2 \xrightarrow{-H^+} C_6H_5\text{—}CH(CH_3)_2$$

The second reaction is a free radical chain reaction. A radical initiator abstracts the benzylic hydrogen of cumene producing a 3° benzylic free radical. Then a chain reaction with oxygen produces cumene hydroperoxide:

Chain-initiating step

$$C_6H_5\text{—}C(CH_3)_2\text{—}H + R\cdot \longrightarrow C_6H_5\text{—}C(CH_3)_2\cdot + R\text{—}H$$

Chain-propagating steps

$$C_6H_5\text{—}C(CH_3)_2\cdot + O_2 \longrightarrow C_6H_5\text{—}C(CH_3)_2\text{—}O\text{—}O\cdot$$

$$C_6H_5\text{—}C(CH_3)_2\text{—}O\text{—}O\cdot + H\text{—}C(CH_3)_2\text{—}C_6H_5 \longrightarrow C_6H_5\text{—}C(CH_3)_2\text{—}O\text{—}O\text{—}H + C_6H_5\text{—}C(CH_3)_2\cdot$$

The third reaction—the hydrolytic rearrangement—resembles the carbocation rearrangements that we have studied before. In this instance, however, the rearrangement involves *a cationic oxygen atom.* All of the steps of the mechanism are shown below.

$$C_6H_5\text{—}C(CH_3)_2\text{—}\ddot{\underset{..}{O}}\text{—}\ddot{O}H + H^+ \longrightarrow C_6H_5\text{—}C(CH_3)_2\text{—}\ddot{\underset{..}{O}}\text{—}\overset{+}{\ddot{O}}H_2 \xrightarrow{-H_2O}$$

$$C_6H_5\text{—}C(CH_3)_2\text{—}\ddot{\underset{..}{O}}^{+} \xrightarrow[\text{migration to oxygen}]{\text{Phenyl anion}} {}^{+}C(CH_3)_2\text{—}\ddot{\underset{..}{O}}\text{—}C_6H_5$$

$$\xrightarrow{H_2O} H{-}\ddot{O}^{+}{-}C(CH_3)_2{-}\ddot{O}{-}C_6H_5 \rightleftarrows H{-}\ddot{O}{-}C(CH_3)_2{-}\ddot{O}^{+}(H){-}C_6H_5$$

$$\xrightarrow{-H^+} \ddot{O}{=}C(CH_3)_2 + H\ddot{O}C_6H_5$$

Acetone Phenol

The second and third steps of the mechanism may actually take place at the same time, i.e., the loss of H_2O and the migration of C_6H_5— may be concerted.

4. **Other Methods.** In Chapter 18 we will discuss a useful laboratory procedure for synthesizing phenols from arylamines. And if you studied Special Topic G, you saw there that phenols can be prepared from arylthallium compounds.

15.15 REACTIONS OF PHENOLS

15.15A Phenols as Acids

Although phenols are structurally similar to alcohols, they are much stronger acids. The acidity constants of most alcohols are of the order of 10^{-18}. However, as we see in Table 15.6 the acidity constants of phenols are of the order of 10^{-11} or greater.

We can explain the greater acidity of phenols relative to alcohols through the application of resonance theory. Before we do this, however, it will be helpful if we review acid-base theory in general terms.

TABLE 15.6 The Acidity Constants of Phenols

NAME	ACIDITY CONSTANT K_a (in H_2O at 25°)
Phenol	1.3×10^{-10}
o-Cresol	6.3×10^{-11}
m-Cresol	9.8×10^{-11}
p-Cresol	6.7×10^{-11}
o-Chlorophenol	7.7×10^{-9}
m-Chlorophenol	1.6×10^{-9}
p-Chlorophenol	6.3×10^{-10}
o-Nitrophenol	6.8×10^{-8}
m-Nitrophenol	5.3×10^{-9}
p-Nitrophenol	7×10^{-8}
2,4-Dinitrophenol	1.1×10^{-4}
2,4,6-Trinitrophenol (picric acid)	4.2×10^{-1}
1-Naphthol	4.9×10^{-10}
2-Naphthol	2.8×10^{-10}

Acid-base reactions are under *equilibrium control.* That is, they are reversible reactions, and the relative concentrations of conjugate acid and base formed are governed by the position of an equilibrium. We have seen other reactions under equilibrium control; for example, 1,4-additions to conjugated dienes (Section 10.8) and aromatic sulfonations (Section 12.5). We have also seen that we can account for reactions that are under equilibrium control by comparing the relative energies of the products and the reactants, rather than the relative magnitudes of energies of activation. The reason for this is simple; in reactions under equilibrium control both the products and the reactants have sufficient energy to surmount the energy barrier between them.

Thus, with acid-base reactions, we can assess the effects of variations in structure on K_a by estimating the energy difference between the acid and its conjugate base. Structural factors that stabilize the conjugate base, A:$^-$, more than they stabilize the acid, HA, cause the value of K_a to be larger. (HA will be, therefore, a stronger acid). Conversely, structural factors that stabilize the acid more than they do the conjugate base cause the value of K_a to be smaller. (In this case HA will be a weaker acid.)

Resonance effects are often an important factor in acid-base reactions, and resonance effects usually stabilize the conjugate base more than they stabilize the acid. We can see a vivid example of this if we return now to one of our original tasks and account for the greater acidity of a phenol relative to that of an alcohol.

Let us compare two *superficially* similar compounds, cyclohexanol and phenol.

Cyclohexanol $K_a \simeq 10^{-18}$ Phenol $K_a = 1.3 \times 10^{-10}$

Although phenol is a weak acid when compared with a carboxylic acid such as acetic acid ($K_a = 10^{-5}$), phenol is a much stronger acid than cyclohexanol (by a factor of almost 10^8).

We can write resonance structures for both phenol and the phenoxide ion such as those shown below. (No analogous resonance structures are possible for cyclohexanol, of course.)

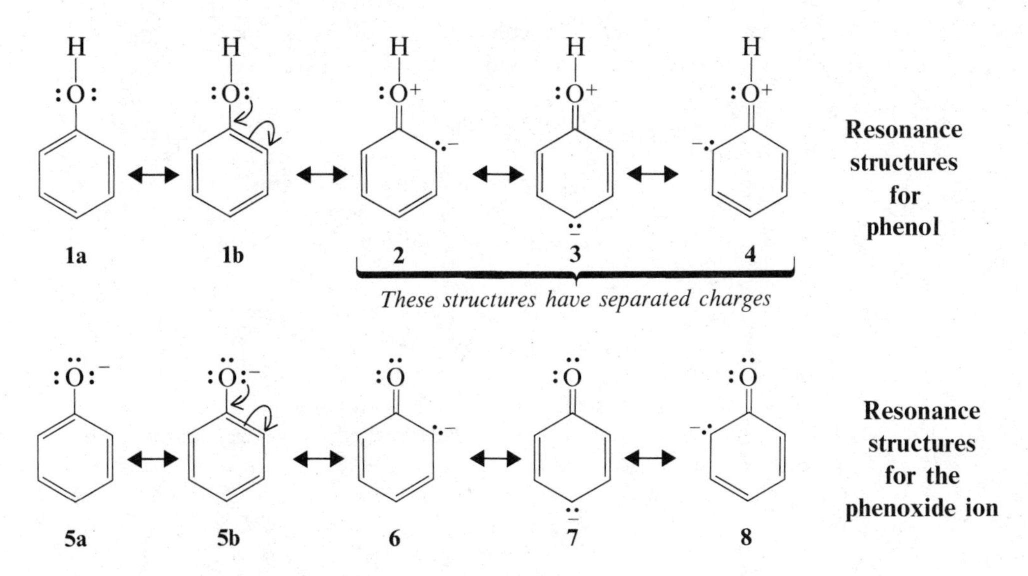

With the exception of the Kekulé structures (**1a** and **1b**), we see that the resonance structures for phenol (**2–4**) require separation of opposite charges while the corresponding resonance structures (**6–8**) for the phenoxide ion do not. *Energy is required to separate opposite charges,* and therefore we can conclude that this type of resonance makes a *smaller* stabilizing contribution to phenol than it does to the phenoxide ion. That is, *resonance stabilizes the conjugate base* (the phenoxide ion) *more than it does the acid* (phenol).

Resonance stabilization of the phenoxide ion is particularly important because the negative charge is *delocalized* over the benzene ring. It is not localized on the oxygen atom as it would be in the anion of cyclohexanol:

OH

$+ H_2\ddot{O}: \rightleftharpoons$

δ− O δ− δ− δ−

$+ H_3O^+$

Phenol
(moderate resonance stabilization

Phenoxide ion
(large resonance stabilization—
charge is delocalized)

Anion is stabilized more than the acid— K_a is larger

OH

$+ H_2\ddot{O}: \rightleftharpoons$

$:\ddot{O}:^-$

$+ H_3O^+$

Cyclohexanol
(no resonance stabilization)

Cyclohexoxide ion
(no resonance stabilization—
charge is localized)

Neither acid nor anion is stabilized— K_a is smaller

Since resonance stabilizes the phenoxide ion more than it does phenol itself, ionization of phenol is a less endothermic reaction than the corresponding ionization of cyclohexanol (Fig. 15.1). The equilibrium between phenol and its conjugate base will, as a consequence, favor the formation of the conjugate base and the hydronium ion to a greater extent than the corresponding equilibrium involving cyclohexanol (where neither the alcohol nor the alkoxide ion is resonance stabilized). Both compounds are weak acids, but, of the two, phenol is the stronger.

In the analysis given here, we have made our comparisons solely on the basis of enthalpy changes ($\Delta H°$'s) for the two reactions. Such an analysis contains a simplification. In actuality, the equilibrium constant for a reaction is directly related to the standard free-energy change for the reaction, $\Delta G°$ (cf. Special Topic B). The equation that relates these two quantities is

$$\log K_{eq} = -\frac{\Delta G°}{2.303\, RT}$$

where R is the gas constant and T is the absolute temperature.

Since $\Delta G°$ is related both to the enthalpy change $\Delta H°$ and to the entropy change $\Delta S°$, that is,

$$\Delta G° = \Delta H° - T\Delta S°$$

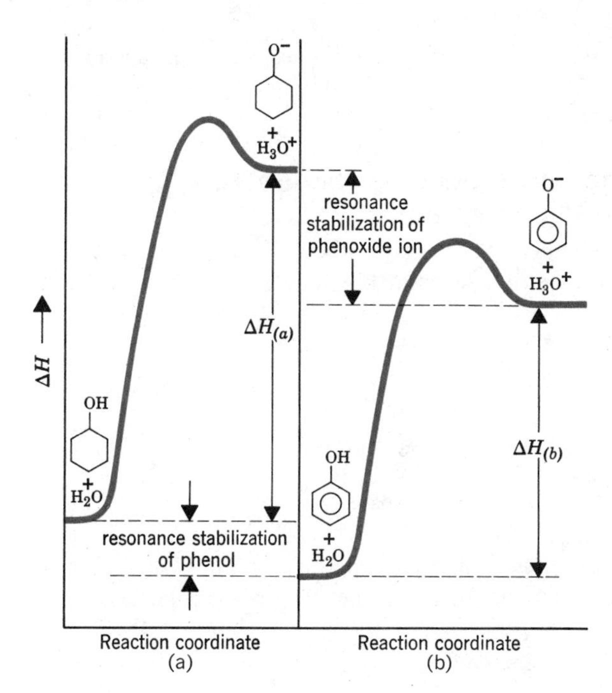

FIG. 15.1
Potential energy diagrams for the ionization of cyclohexanol (a) *and phenol* (b). *Resonance stabilizes the phenoxide ion to a greater extent than it does phenol, itself; thus, the ionization of phenol is a less endothermic reaction than the ionization of cyclohexanol, that is,* $\Delta H_{(a)} > \Delta H_{(b)}$. *Phenol, therefore, is the stronger acid.* (*The curves are aligned for comparison only.*)

entropy effects can be important. In the two reactions considered here, the entropy changes have been of the same order of magnitude and have, therefore, canceled out. In other reactions, however, this may not be the case and, in fact, entropy changes may be the predominant influence.

Problem 15.25

The carbon-oxygen bond of phenol is much stronger than that of an alcohol. Phenol, for example, is not converted to bromobenzene when it is refluxed with concentrated hydrobromic acid. Similar treatment of cyclohexanol, however, does give bromocyclohexane.

$$C_6H_5\text{—OH} + HBr \xrightarrow{\text{Reflux}} \text{No reaction}$$

$$C_6H_{11}\text{—OH} + HBr \xrightarrow{\text{Reflux}} C_6H_{11}\text{—Br} + H_2O$$

Although resonance structures **2** to **4** do not make as great a contribution to the phenol hybrid as the corresponding structures (**6**–**8**) do to the phenoxide ion, structures **2** to **4** do help us understand why the carbon-oxygen bond of phenols is very strong. Explain.

Problem 15.26

If we examine Table 15.6 we see that phenols having electron-withdrawing groups (Cl— or O_2N—) attached to the benzene ring are more acidic than phenol itself. On the other hand, those phenols bearing electron-releasing groups (e.g. CH_3—) are less acidic than phenol. Account for this trend on the basis of resonance and inductive effects. [Notice that 2,4,6-trinitrophenol (called *picric acid*) is exceptionally acidic ($K_a = 4.2 \times 10^{-1}$)—over 20,000 times as acidic as acetic acid ($K_a = 1.8 \times 10^{-5}$).]

15.15B Distinguishing and Separating Phenols from Alcohols and Carboxylic Acids

Because phenols are more acidic than water, the following reaction goes to completion and produces water-soluble sodium phenoxide.

$$C_6H_5\text{—}OH + NaOH \xrightarrow{H_2O} C_6H_5\text{—}O^-Na^+ + H_2O$$

(slight soluble) Stronger acid $K_a \simeq 10^{-12}$ | Stronger base | (soluble) Weaker base | Weaker acid $K_a \simeq 10^{-16}$

The corresponding reaction of cyclohexanol with aqueous sodium hydroxide does not occur to any appreciable extent because cyclohexanol is a weaker acid than water.

$$C_6H_{11}\text{—}OH + NaOH \underset{H_2O}{\rightleftarrows} C_6H_{11}\text{—}O^-Na^+ + H_2O$$

(very slightly soluble) Weaker acid $K_a \simeq 10^{-18}$ | Weaker base | (soluble) Stronger base | Stronger acid $K_a \simeq 10^{-16}$

The fact that phenols dissolve in aqueous sodium hydroxide whereas most alcohols with six carbons or more do not, gives us a convenient means for distinguishing and separating phenols from most alcohols. (Alcohols with five carbons or fewer are quite soluble in water—some are infinitely so—and thus they dissolve in aqueous sodium hydroxide even though they are not converted to sodium alkoxides in appreciable amounts.)

Most phenols, however, are not soluble in aqueous sodium bicarbonate ($NaHCO_3$), but carboxylic acids are soluble. Thus, aqueous $NaHCO_3$ provides a method for distinguishing and separating most phenols from carboxylic acids.

Problem 15.27

The apparent acidity constant for the first ionization of carbonic acid ($H_2CO_3 + H_2O \rightleftarrows HCO_3^- + H_3O^+$) is 4.3×10^{-7}. Which of the following compounds would you expect to dissolve in aqueous sodium bicarbonate (aq. $NaHCO_3$)? Explain your answers.

(a) Phenol
(b) *p*-Cresol
(c) *o*-Chlorophenol
(d) 2,4-Dinitrophenol
(e) 2,4,6-Trinitrophenol
(f) Benzoic acid ($K_a = 6.4 \times 10^{-5}$)

15.15C Other Reactions of the O—H Group of Phenols.

Phenols react with carboxylic acid anhydrides and acid chlorides to form esters. These reactions are quite similar to those of alcohols (Section 15.9).

$$C_6H_5\text{-}OH \xrightarrow[\text{Base}]{(RCO)_2O} C_6H_5\text{-}O\text{-}\overset{O}{\overset{\|}{C}}R$$

$$C_6H_5\text{-}OH \xrightarrow[\text{Base}]{RCOCl} C_6H_5\text{-}O\text{-}\overset{O}{\overset{\|}{C}}R$$

Phenols can be converted to ethers through the Williamson synthesis. Since phenols are more acidic than alcohols they can be converted to sodium phenoxides through the use of sodium hydroxide (rather than metallic sodium, the reagent used to convert alcohols to alkoxide ions). A specific example is the synthesis of anisole from phenol.

$$C_6H_5OH + NaOH \xrightarrow{H_2O} C_6H_5O^-Na^+ \xrightarrow{CH_3OSO_2OCH_3} C_6H_5OCH_3 + NaOSO_2OCH_3$$

(72–75% overall)

When alkyl aryl ethers react with strong acids such as HI and HBr, the reaction produces an alkyl halide and a phenol. The phenol does not react further to produce an aryl halide because the carbon-oxygen bond is very strong (cf. problem 15.24).

$$CH_3\text{-}C_6H_4\text{-}OCH_3 + HBr \xrightarrow{H_2O} CH_3\text{-}C_6H_4\text{-}OH + CH_3Br$$

p-Methylanisole → *p*-Cresol + Methyl bromide

p-Cresol $\xrightarrow{HBr}$ No reaction

15.15D Reactions of the Benzene Ring of Phenols

Bromination. The hydroxyl group is a powerful activating group—and an ortho-para director—in electrophilic substitutions. Phenol itself reacts with bromine in aqueous solution to yield 2,4,6-tribromophenol in nearly quantitative yield.

$$C_6H_5OH + 3Br_2 \xrightarrow{H_2O} 2,4,6\text{-}Br_3C_6H_2OH + 3HBr$$

2,4,6-Tribromophenol
(~100%)

Monobromination of phenol can be achieved by carrying out the reaction in carbon disulfide at a low temperature. The major product is the para isomer.

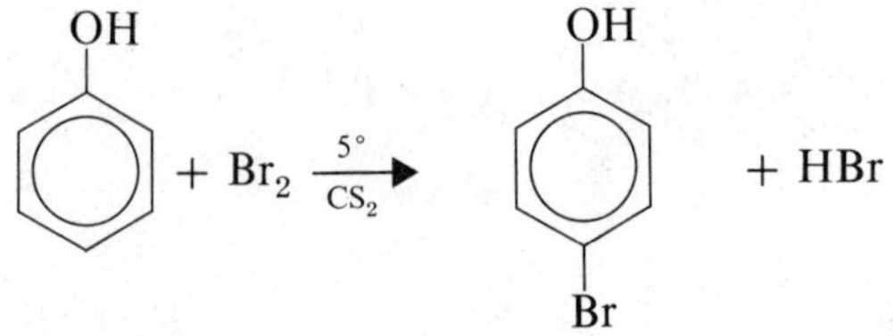

$$C_6H_5OH + Br_2 \xrightarrow[CS_2]{5^\circ} p\text{-}BrC_6H_4OH + HBr$$

p-Bromophenol
(80–84%)

Nitration. Phenol reacts with dilute nitric acid to yield a mixture of *o*- and *p*-nitrophenol. Although the yield is relatively low (because of oxidation of the ring) the ortho and para isomers can be separated by steam distillation. *o*-Nitrophenol is the more volatile isomer because its hydrogen bonding (below) is intramolecular (its effective molecular weight is lower). Thus, *o*-nitrophenol passes over with the steam, and *p*-nitrophenol remains in the distillation flask.

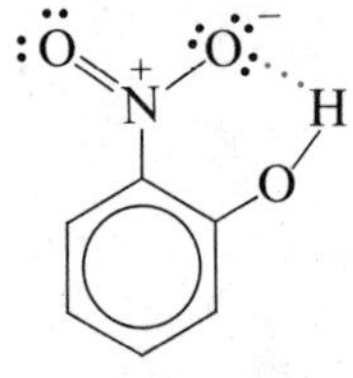

o-Nitrophenol
(more volatile because of
intramolecular hydrogen bonding)

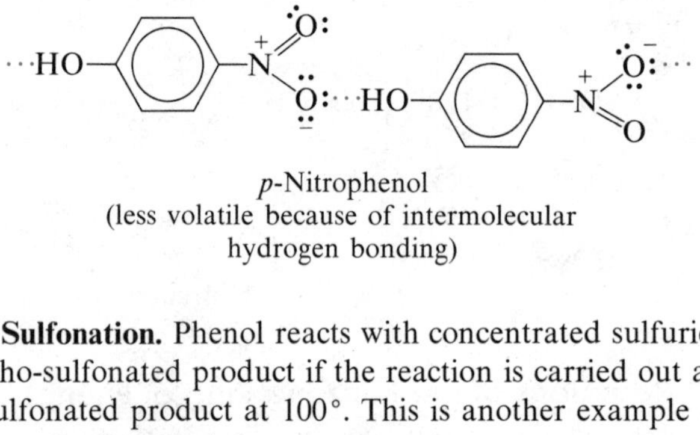

p-Nitrophenol
(less volatile because of intermolecular
hydrogen bonding)

Sulfonation. Phenol reacts with concentrated sulfuric acid to yield mainly the ortho-sulfonated product if the reaction is carried out at 25° and mainly the para-sulfonated product at 100°. This is another example of equilibrium versus rate control of a reaction.

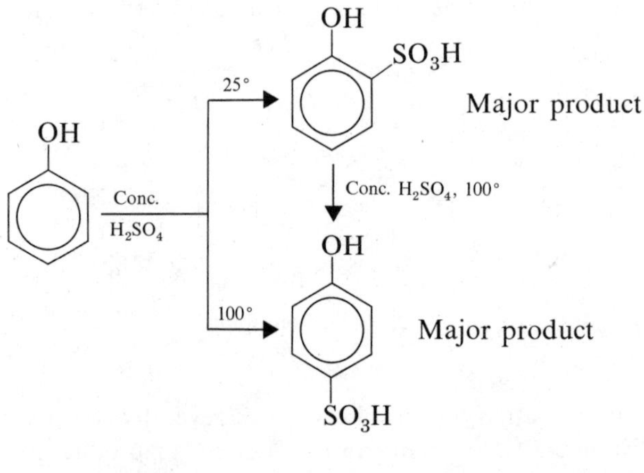

Conc. H_2SO_4; 25°: *o*-$HOC_6H_4SO_3H$ — Major product

100°: *p*-$HOC_6H_4SO_3H$ — Major product

Conc. H_2SO_4, 100° (ortho product → para product)

Problem 15.28

(a) Which sulfonic acid (above) is more stable? (b) For which sulfonation (ortho or para) is the energy of activation lower?

Kolbe Reaction. The phenoxide ion is even more susceptible to electrophilic aromatic substitution than phenol itself. (Why?) Use is made of the high reactivity of the phenoxide ring in a reaction called the *Kolbe reaction*. In the Kolbe reaction carbon dioxide acts as the electrophile.

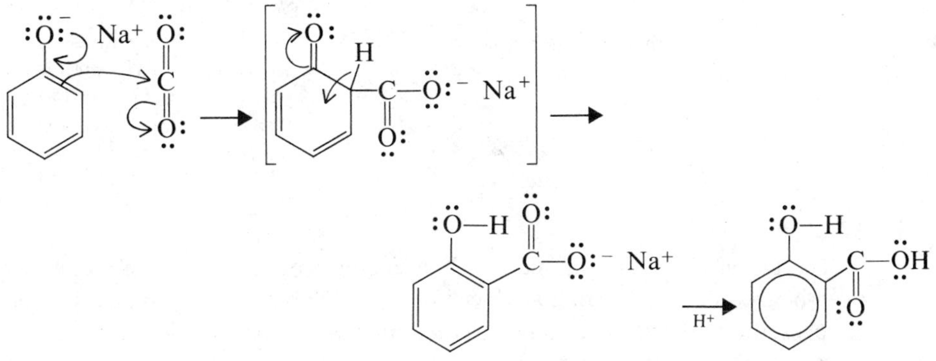

:Ö:⁻ Na⁺ Ö=C=Ö: → [O=C₆H₅(H)–C(=O)–Ö:⁻ Na⁺] → :Ö–H / C(=O)–Ö:⁻ Na⁺ $\xrightarrow{H^+}$:Ö–H / C(=O)–ÖH

Salicylic acid

The reaction is usually carried out by allowing sodium phenoxide to absorb carbon dioxide (sodium phenyl carbonate is formed) and then heating the product to 125° under a pressure of several atmospheres of carbon dioxide. Subsequent acidification of the mixture produces *salicylic acid.*

Reaction of salicylic acid with acetic anhydride yields the widely used pain reliever—*aspirin.*

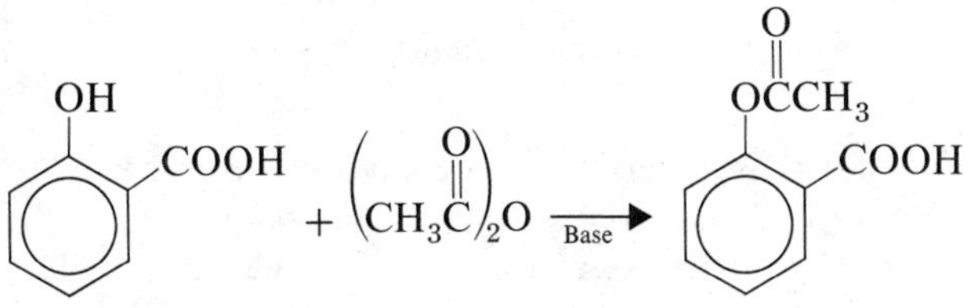

OH / COOH + $(CH_3\overset{O}{\overset{\|}{C}})_2O$ $\xrightarrow{\text{Base}}$ $O\overset{O}{\overset{\|}{C}}CH_3$ / COOH

Salicylic acid — Acetylsalicylic acid (aspirin)

15.16 NATURALLY OCCURRING PHENOLS

A number of phenols occur in nature. Eugenol, for example, is found in cloves and vanillin is a component of the vanilla flavoring obtained from vanilla beans. The urushiols are blistering agents (vesicants) found in poison ivy and poison oak.

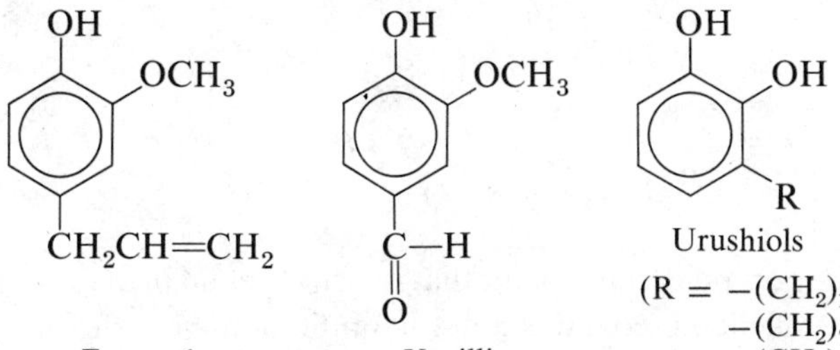

OH, OCH_3, $CH_2CH{=}CH_2$ — Eugenol

OH, OCH_3, $\overset{O}{\overset{\|}{C}}{-}H$ — Vanillin

OH, OH, R — Urushiols

($R = -(CH_2)_{14}CH_3$,
$-(CH_2)_7CH{=}CH(CH_2)_5CH_3$,
or $-(CH_2)_7CH{=}CHCH_2CH{=}CH(CH_2)_2CH_3$)

Tetrahydrocannabinol is one of the active components of marijuana. The tetracyclines are antibiotics.

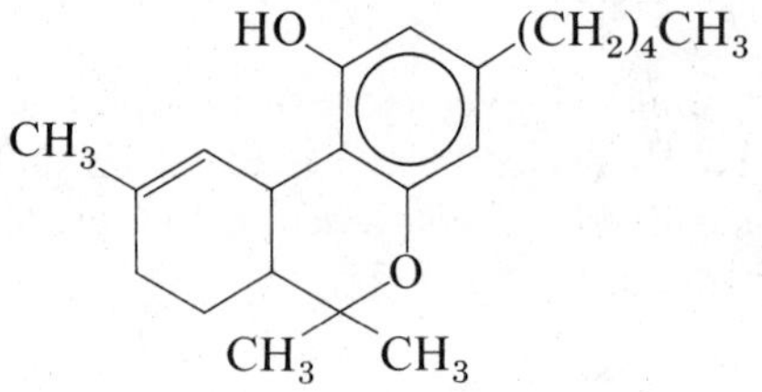

HO $(CH_2)_4CH_3$ CH_3 O CH_3 CH_3

Tetrahydrocannabinol

Y CH_3 OH Z $N(CH_3)_2$ OH C—NH_2 OH O OH OH O O

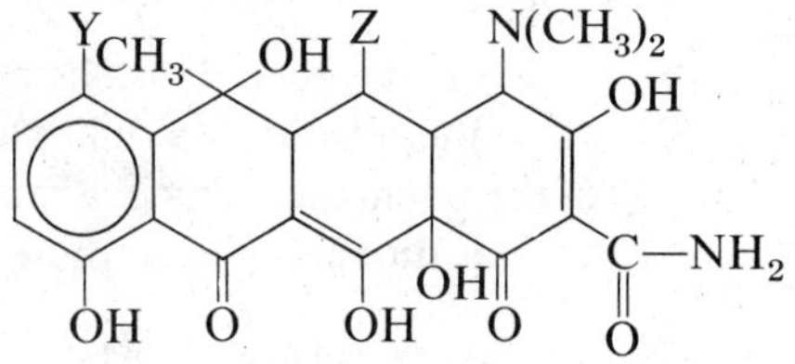

Tetracyclines
(Y = Cl, Z = H; aureomycin)
(Y = H, Z = OH; terramycin)

15.17 QUINONES

Oxidation of hydroquinone (1,4-benzenediol) produces a compound known as *p*-benzoquinone. The oxidation can be brought about by mild oxidizing agents and overall the oxidation amounts to the removal of a pair of electrons ($2e^-$) and two protons from hydroquinone. (Another way of visualizing the oxidation is as the loss of a hydrogen molecule, H : H, making it a dehydrogenation.)

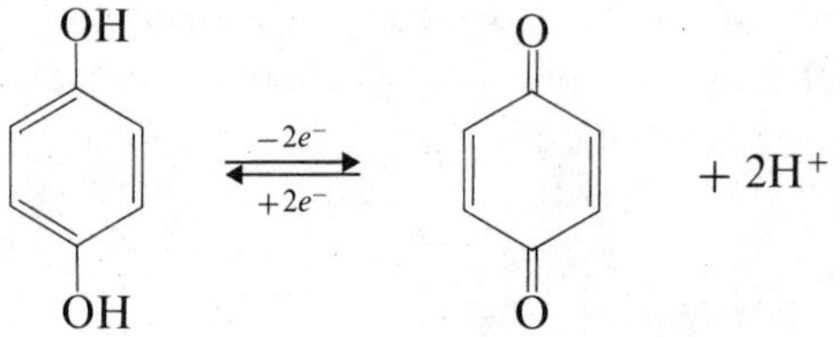

$$\text{Hydroquinone} \underset{+2e^-}{\overset{-2e^-}{\rightleftharpoons}} p\text{-Benzoquinone} + 2H^+$$

Hydroquinone *p*-Benzoquinone

The reaction (above) is reversible; *p*-benzoquinone is easily reduced by mild reducing agents to hydroquinone.

Nature makes much use of this type of reversible oxidation-reduction to transport a pair of electrons from one substance to another in enzyme-catalyzed reactions. Important compounds in this respect are the compounds called *ubiquinones* (from *ubiquitous* + quinone—these quinones are found everywhere in biological systems). Ubiquinones are also called coenzyme Q.

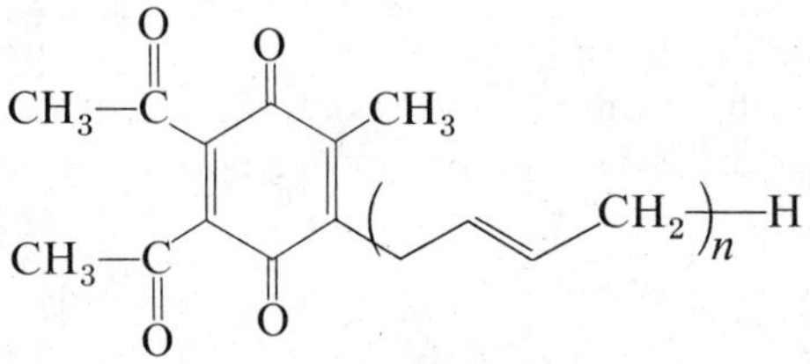

CH_3—C(=O) O CH_3 $(CH_2)_n$—H CH_3—C(=O) O

Ubiquinones (n = 1.2, ... 10)
(coenzyme Q)

Vitamin K_1, the important dietary factor that is instrumental in maintaining the coagulant properties of blood, contains a 1,4-naphthoquinone structure.

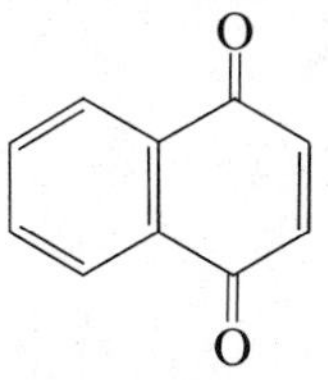

1,4-Naphthoquinone

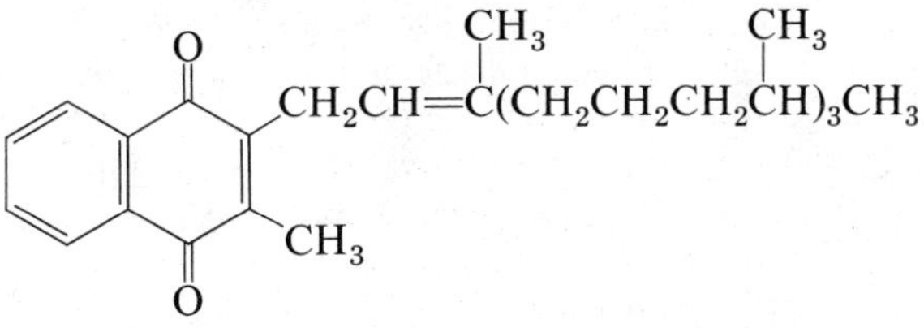

$CH_2CH{=}C(CH_3)(CH_2CH_2CH_2CH(CH_3))_3CH_3$, CH_3

Vitamin K_1

Problem 15.29

p-Benzoquinone and 1,4-naphthoquinone act as dienophiles in Diels-Alder reactions. Give the structures of the products of the following reaction:

(a) *p*-Benzoquinone + butadiene
(b) 1,4-Naphthoquinone + butadiene
(c) *p*-Benzoquinone + 1,3-cyclopentadiene

Problem 15.30

Outline a possible synthesis of the following compound.

OH

OH

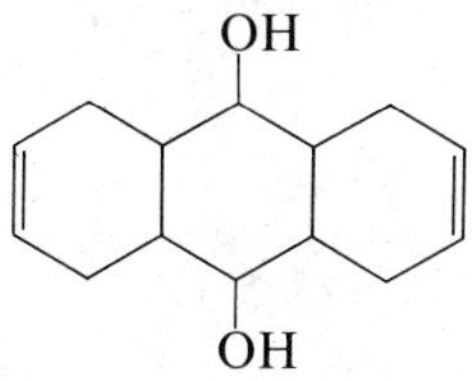

Additional Problems

15.31

Show how each of the following transformations could be carried out.

(a) Styrene ⟶ 1-phenylethanol (two ways)
(b) Styrene ⟶ 2-phenylethanol
(c) Styrene ⟶ 1-methoxy-1-phenylethane
(d) 1-Phenylethanol ⟶ 1-phenylethyl ethyl ether
(e) Phenylacetic acid ($C_6H_5CH_2COOH$) ⟶ 2-phenylethanol
(f) Phenyl methyl ketone ($C_6H_5COCH_3$) ⟶ 1-phenylethanol
(g) Toluene ⟶ 2-phenylethanol
(h) Benzene ⟶ 2-phenylethanol
(i) Methyl phenylacetate ($C_6H_5CH_2CO_2CH_3$) ⟶ 2-phenylethanol

15.32

Show how 1-butanol could be transformed into each of the following compounds. (You may use any necessary inorganic reagents and you need not show the synthesis of a particular compound more than once.)

(a) 1-Butene
(b) 2-Butanol
(c) 2-Butanone ($CH_3COCH_2CH_3$)
(d) 1-Bromobutane
(e) 2-Bromobutane
(f) 1-Pentanol
(g) 1-Hexene
(h) 3-Methyl-3-heptanol
(i) 1-Butanal ($CH_3CH_2CH_2CHO$)

(j) 4-Octanol
(k) 3-Methyl-4-heptanol
(l) Pentanoic acid ($CH_3CH_2CH_2CH_2COOH$)
(m) *n*-Butyl *sec*-butyl ether (two ways)
(n) Di-*n*-butyl ether (two ways)
(o) *n*-Butyllithium
(p) *n*-Octane

15.33
Give structures and names for the compounds that would be formed when 1-propanol is treated with each of the following reagents.
(a) Sodium metal
(b) Sodium metal then 1-bromobutane
(c) Methanesulfonyl chloride
(d) *p*-Toluenesulfonyl chloride
(e) Acetyl chloride
(f) Hot basic $KMnO_4$
(g) Phosphorus trichloride
(h) Thionyl chloride
(i) Sulfuric acid at 140°
(j) Refluxing concentrated hydrobromic acid
(k) 2-Methylpropene and mercuric acetate, followed by treatment with $NaBH_4$ and OH^-
(l) Benzene and BF_3

15.34
Give structures and names for the compounds that would be formed when 2-propanol is treated with each of the reagents given in Problem 15.32.

15.35
What products would be obtained from each of the following acid-base reactions?
(a) Sodium ethoxide in ethanol + phenol $\longrightarrow$
(b) Phenylmagnesium bromide in ether + ethanol $\longrightarrow$
(c) Phenol + aqueous sodium hydroxide $\longrightarrow$
(d) Sodium phenoxide + aqueous hydrochloric acid $\longrightarrow$
(e) Sodium ethoxide in ethanol + H_2O $\longrightarrow$

15.36
What compounds would you expect to be formed when each of the following ethers is refluxed with excess concentrated hydrobromic acid?
(a) Methyl ethyl ether
(b) Phenyl ethyl ether
(c) Tetrahydrofuran
(d) Dioxane (i.e., six-membered ring with O at positions 1 and 4)

15.37
Complete the following equations.

(a) $CH_3CH{-}CH_2$ (epoxide, O bridging) + H_3O^+ $\xrightarrow{H_2O}$

(b) CH_3CH_2OH + CH_2CH_2 (epoxide, O bridging) $\xrightarrow{OH^-}$

(c) C_6H_5OH + CH_2CH_2 (epoxide, O bridging) $\xrightarrow{OH^-}$

(d) 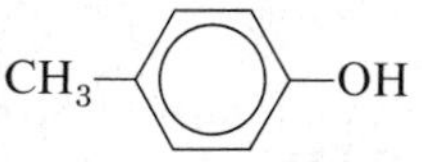 + *p*-toluenesulfonyl chloride $\xrightarrow[OH^-]{}$

(e) Cyclohexanol + acetic anhydride $\longrightarrow$

(f) Phenol + 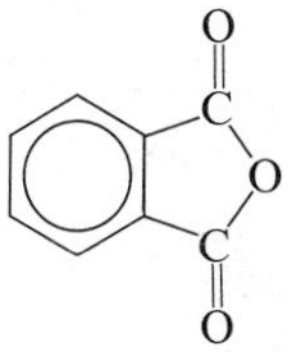

(g) *p*-Cresol + Br_2 $\xrightarrow{H_2O}$

(h) Benzyl alcohol + $KMnO_4$ $\xrightarrow[heat]{OH^-}$

(i) Benzyl alcohol + $CrO_3 \cdot 2C_5H_5N$ $\xrightarrow[25°]{CH_2Cl_2}$

(j) Benzyl alcohol + NaH $\longrightarrow$

(k) Product of (j) + $CH_3OSO_2OCH_3$ $\longrightarrow$

(l) *cis*-3-Hexene + $C_6H_5\overset{O}{\overset{\|}{C}}OOH$ $\longrightarrow$

(m) Product of (l) + H_3O^+/H_2O $\longrightarrow$

15.38

Describe a simple chemical test that could be used to distinguish between each of the following pairs of compounds.

(a) *p*-Cresol and benzyl alcohol
(b) Cyclohexanol and cyclohexane
(c) Cyclohexanol and cyclohexene
(d) Allyl propyl ether and dipropyl ether
(e) Anisole and *p*-cresol
(f) Picric acid and 2,4,6-trimethylphenol

15.39

Write a mechanism that accounts for the following reaction.

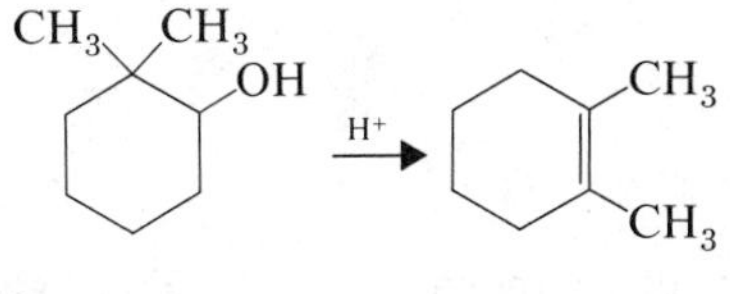

15.40

Thymol (below) can be obtained from thyme oil. Thymol is an effective disinfectant and is used in many antiseptic preparations. (a) Suggest a synthesis of thymol from *m*-cresol and propylene. (b) Suggest a method for transforming thymol into menthol (p. 659).

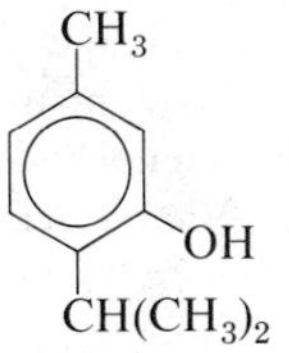

Thymol

15.41
Carvacrol is another naturally occurring phenol and it is an isomer of thymol (Problem 15.40). Carvacrol can be synthesized from *p*-cymene (*p*-isopropyltoluene) by ring sulfonation and alkali fusion of the sulfonic acid. Explain why this synthesis yields mainly carvacrol and very little thymol.

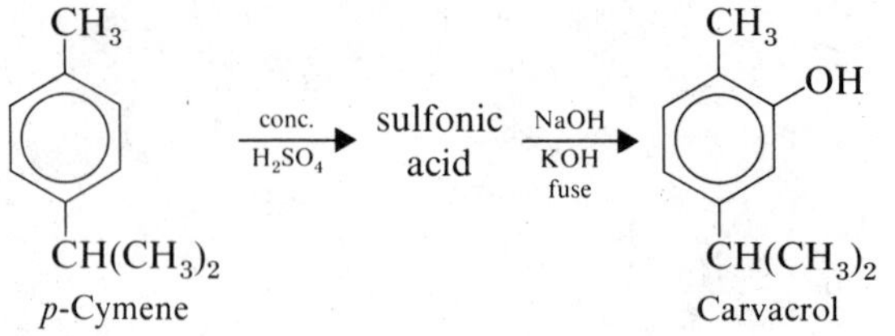

CH_3 / $CH(CH_3)_2$
p-Cymene

$\xrightarrow[H_2SO_4]{\text{conc.}}$ sulfonic acid $\xrightarrow[\text{KOH fuse}]{\text{NaOH}}$

CH_3 / OH / $CH(CH_3)_2$
Carvacrol

15.42
A widely used synthetic antiseptic is 4-*hexylresorcinol.* Suggest a synthesis of 4-hexylresorcinol from resorcinol and hexanoic acid.

15.43
Anethole (below) is the chief component of anise oil. Suggest a synthesis of anethole from anisole and propanoic acid.

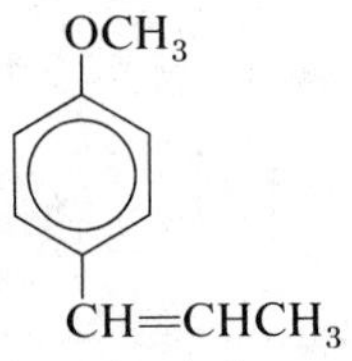

OCH_3 / $CH{=}CHCH_3$
Anethole

15.44
A compound **X,** $C_{10}H_{14}O$, dissolves in aqueous sodium hydroxide but is insoluble in aqueous sodium bicarbonate. Compound **X** reacts with bromine in water to yield a dibromo derivative, $C_{10}H_{12}Br_2O$. The 3000 to 4000 cm^{-1} region of the infrared spectrum of **X** shows a broad peak centered at 3250 cm^{-1}; the 680–840 cm^{-1} region shows a strong peak at 830 cm^{-1}. The proton nmr spectrum of **X** gives the following:

Singlet	δ1.3 (9H)
Singlet	δ4.9 (1H)
Multiplet	δ7.0 (4H)

What is the structure of **X?**

15.45
The infrared and proton nmr spectra of compound **Y,** $C_9H_{12}O$, are given in Fig. 15.2. Propose a structure for **Y.**

15.46
The widely used antioxidant and food preservative called **BHA** (**B**utylated **H**ydroxy **A**nisole) is actually a mixture of 2-*tert*-butyl-4-methoxyphenol and 3-*tert*-butyl-4-methoxyphenol. **BHA** is synthesized from *p*-methoxyphenol and 2-methylpropene. (a) Suggest how this is done. (b) Another widely used antioxidant is **BHT** (**B**utylated **H**ydroxy **T**oluene). **BHT** is actually 2,6-di-*tert*-butyl-4-methylphenol and the raw materials used in its production are *p*-cresol and 2-methylpropene. What reaction is used here?

15.47
The herbicide **2,4-D** (cf. Special Topic I) can be synthesized from phenol and chloroacetic acid. Outline the steps involved.

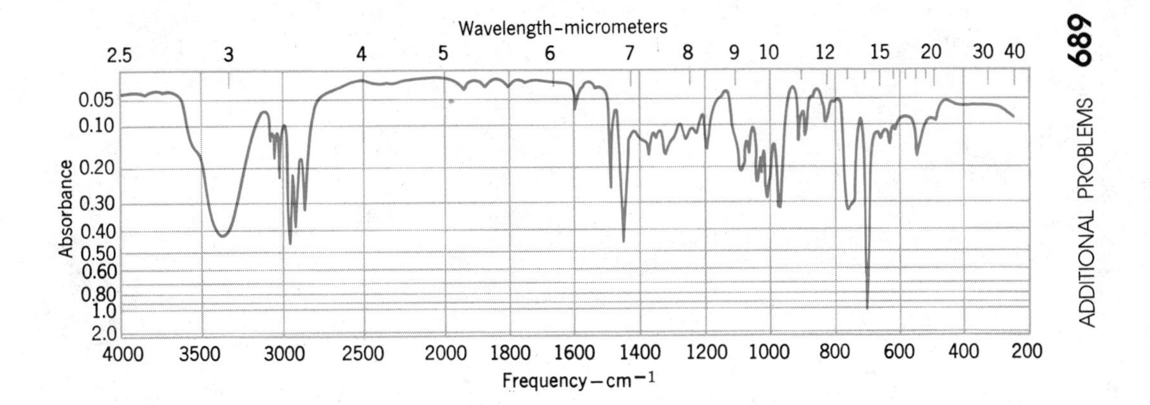

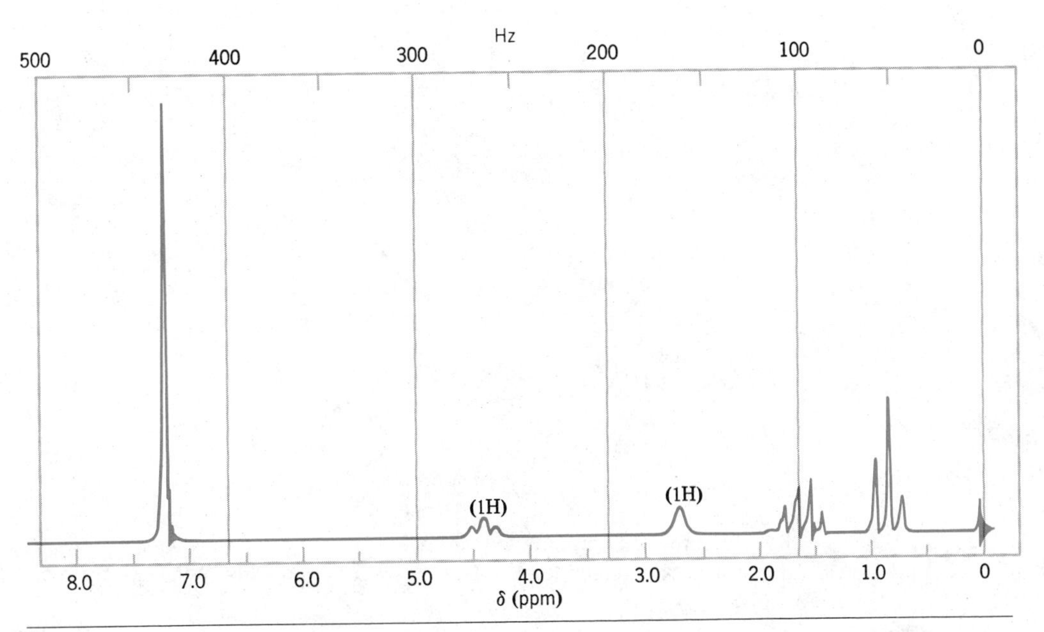

FIG. 15.2
The infrared and proton nmr spectra of compound Y, Problem 15.45. (Infrared spectrum courtesy of Sadtler Research Laboratories, Philadelphia, Pa. The proton nmr spectrum courtesy of Varian Associates, Palo Alto, Calif.)

Cl—C_6H_3(Cl)—OCH_2COOH $ClCH_2COOH$

2,4-D
(2,4-Dichlorophenoxyacetic acid)

Chloroacetic acid

*** 15.48**
Compound **Z**, $C_5H_{10}O$, decolorizes bromine in carbon tetrachloride. The infrared spectrum of **Z** shows a broad peak in the 3200 to 3600 cm^{-1} region. The proton nmr spectrum of **Z** is given in Fig. 15.3. Propose a structure for **Z**.

*** 15.49**
A Grignard reagent that is a key intermediate in an industrial synthesis of Vitamin A (Section 16.13D) can be prepared in the following way:

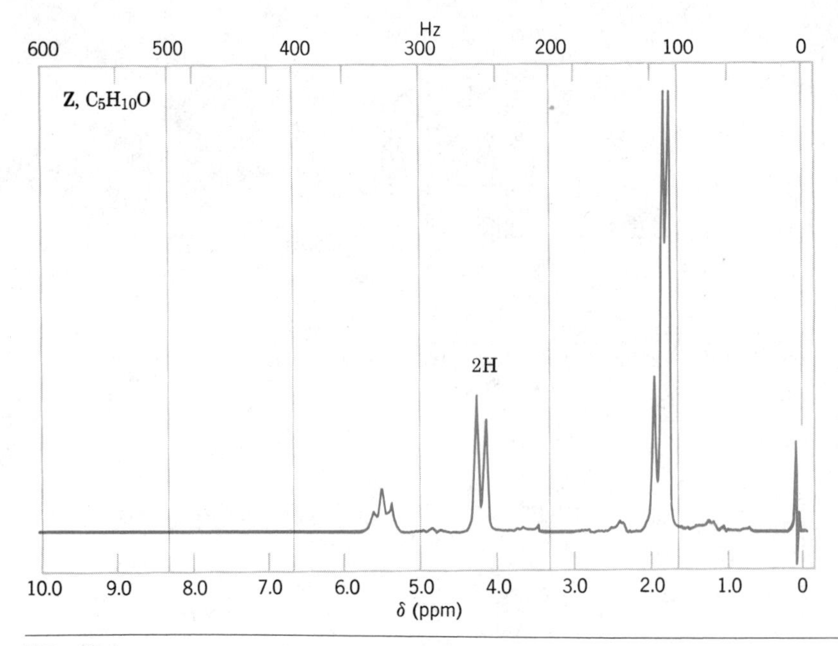

FIG. 15.3
The proton nmr spectrum of compound ***Z****, Problem 15.48. (Spectrum courtesy of Aldrich Chemical Co., Milwaukee, Wis.)*

$$HC{\equiv}CLi + CH_2{=}CH\overset{O}{\overset{\|}{C}}CH_3 \xrightarrow[(2)\ NH_4^+]{} \mathbf{A}\ (C_6H_8O) \xrightarrow{H^+}$$

$$\mathbf{B},\ HOCH_2CH{=}\overset{CH_3}{\overset{|}{C}}{-}C{\equiv}CH \xrightarrow{2C_2H_5MgBr} \mathbf{C}\ (C_6H_6Mg_2Br_2O)$$

(a) What are the structures of compounds **A** and **C**?
(b) The acid-catalyzed rearrangement of **A** to **B** takes place very readily. What two factors account for this?

*** 15.50**

The remaining steps in the industrial synthesis of Vitamin A (as in acetate) are as follows: The Grignard reagent **C** from problem 15.49 is allowed to react with the aldehyde shown below. (See Problems 16.25 and 20.36 for its synthesis.)

$$\text{[cyclohexenyl structure]}{-}CH_2CH{=}C(CH_3){-}\underset{O}{\underset{\|}{C}H}$$

After acidification, the product obtained from this step is a diol **D**. Selective hydrogenation of the triple bond of **D** using Lindlar's catalyst yields **E** ($C_{20}H_{32}O_2$). Treating **E** with one mole of acetic anyhydride yields a monoacetate, **F**, and dehydration of **F** yields Vitamin A acetate. What are the structures of **D** to **F**?

*** 15.51**

Heating acetone with an excess of phenol in the presence of hydrogen chloride is the basis for an industrial process used in the manufacture of a compound called "bisphenol A." (Bisphenol A is used in the manufacture of epoxy resins and a polymer called "Lexan," cf. Problem K.5.) Bisphenol A has the molecular formula $C_{15}H_{16}O_2$ and the reactions involved in its formation are similar to those involved in the synthesis of DDT (Problem I.1). Write out these reactions and give the structure of bisphenol A.

J SPECIAL TOPIC

THIOLS, THIOETHERS, AND THIOPHENOLS

Sulfur is directly below oxygen in Group VI of the Periodic Table and, as we might expect, there are sulfur counterparts of the oxygen compounds that we studied in Chapter 15.

Some important examples of organosulfur compounds are:

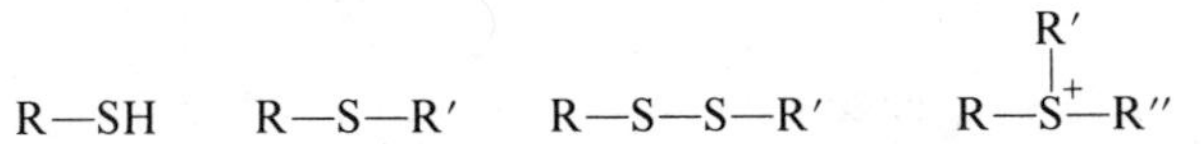

R—SH	R—S—R′	R—S—S—R′	$R—\overset{R'}{\overset{|}{S^+}}—R''$
Thiols	Thioethers	Disulfides	Trialkylsulfonium ions

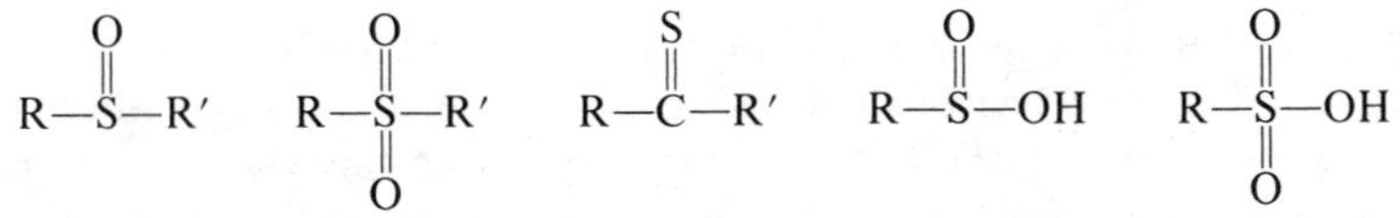

$R—\overset{O}{\overset{\|}{S}}—R'$	$R—\overset{O}{\overset{\|}{\underset{O}{\underset{\|}{S}}}}—R'$	$R—\overset{S}{\overset{\|}{C}}—R'$	$R—\overset{O}{\overset{\|}{S}}—OH$	$R—\overset{O}{\overset{\|}{\underset{O}{\underset{\|}{S}}}}—OH$
Sulfoxides	Sulfones	Thioketones	Sulfinic acids	Sulfonic acids

The sulfur counterpart of an alcohol is called a *thiol* or a *mercaptan*. The name mercaptan comes from the Latin, *mercurium captans,* meaning "capturing mercury." Mercaptans react with mercuric ions and the ions of other heavy metals to form precipitates. The compound, $\underset{SH}{\underset{|}{CH_2}}\underset{SH}{\underset{|}{CH}}CH_2OH$, known as British Anti-Lewisite (BAL), was developed as an antidote for poisonous arsenic compounds used as war gases. BAL is also an effective antidote for mercury poisoning.

Several simple thiols are shown below.

CH_3CH_2SH	$CH_3CH_2CH_2SH$
Ethanethiol (ethyl mercaptan)	Propanethiol (propyl mercaptan)

$CH_3\overset{CH_3}{\overset{|}{C}}HCH_2CH_2SH$	$CH_2{=}CHCH_2SH$
3-Methyl-1-butanethiol (isopentyl mercaptan)	2-Propenethiol (allyl mercaptan)

Compounds of sulfur, in general, and the low-molecular-weight thiols, in particular, are noted for their disagreeable odors. Anyone who has passed anywhere near a general chemistry laboratory when hydrogen sulfide (H_2S) was being used has noticed the strong odor of that substance—the odor of rotten eggs. Another sulfur compound, 3-methyl-1-butanethiol, is one unpleasant constituent of the liquid that skunks use as a defensive weapon. Propanethiol evolves from freshly chopped onions, and allyl mercaptan is one of the compounds responsible for the odor and flavor of garlic.

Aside from their odors, analogous sulfur and oxygen compounds show other chemical differences. These arise largely from the following features of sulfur compounds.

1. The sulfur atom is larger and more polarizable than the oxygen atom. As a result, sulfur compounds are more powerful nucleophiles and compounds containing —SH groups are stronger acids than their oxygen analogs. The ethyl mercaptide ion ($CH_3CH_2\ddot{\underset{..}{S}}:^-$), for example, is a much stronger nucleophile

when it reacts at carbon atoms than is the ethoxide ion ($CH_3CH_2\ddot{O}:^-$). On the other hand since ethanol is a weaker acid than ethanethiol, the ethoxide ion is the stronger of the two conjugate bases.

2. The bond dissociation energy (~80 kcal/mole) of the S—H bond of thiols is much less than that (~100 kcal/mole) of the O—H bond of alcohols. The weakness of the S—H bond allows thiols to undergo an oxidative coupling reaction when they react with mild oxidizing agents; the product is a disulfide:

$$\underset{\text{A thiol}}{2RS{-}H} + H_2O_2 \longrightarrow \underset{\text{A disulfide}}{RS{-}SR} + 2H_2O$$

Alcohols do not undergo an analogous reaction. When alcohols are treated with oxidizing agents, oxidation takes place at the weaker C—H bond (~85 kcal/mole) rather than at the strong O—H bond.

3. Because sulfur atoms are easily polarized they can stabilize a negative charge on an adjacent atom. This means that hydrogens on carbons that are adjacent to an alkylthio group are more acidic than those adjacent to an alkyloxy group. Thioanisole, for example, reacts with *n*-butyllithium in the following way.

$$\underset{\text{Thioanisole}}{C_6H_5{-}SCH_3} + C_4H_9{:}^-\ Li^+ \longrightarrow C_6H_5{-}SCH_2{:}^-\ Li^+ + C_4H_{10}$$

Anisole ($CH_3OC_6H_5$) does not undergo an analogous reaction.

The $>$S=O group of sulfoxides and the positive sulfur of sulfonium ions are even more effective in delocalizing negative charge on an adjacent atom:

$$\underset{\text{Dimethyl sulfoxide}}{CH_3\overset{\overset{:\ddot{O}}{\|}}{\ddot{S}}CH_3} \xrightarrow[(-H^+)]{\overset{+}{Na}\ :\overset{-}{H}} \left[CH_3\overset{\overset{:\ddot{O}}{\|}}{\ddot{S}}{-}\overset{\ddot{-}}{C}H_2 \longleftrightarrow CH_3\overset{\overset{:\ddot{O}:^-}{|}}{\ddot{S}}{=}CH_2\right] + H_2$$

$$\underset{\text{Trimethylsulfonium bromide}}{CH_3\overset{\overset{CH_3}{|}}{\underset{+}{\ddot{S}}}CH_3\ Br^-} \xrightarrow[(-HBr)]{\text{base}} \underset{\text{An ylide*}}{CH_3\overset{\overset{CH_3}{|}}{\underset{+}{\ddot{S}}}{-}\overset{\ddot{-}}{C}H_2}$$

The anions formed in the reactions given above are of synthetic use. They can be used to synthesize epoxides, for example (cf. Section 16.10B).

J.1 PREPARATION OF THIOLS

Alkyl bromides and iodides react with potassium hydrogen sulfide to form thiols. (Potassium hydrogen sulfide can be generated by passing gaseous H_2S into an alcoholic solution of potassium hydroxide.)

$$R{-}Br + KOH + \underset{\text{(excess)}}{H_2S} \xrightarrow[\text{heat}]{C_2H_5OH} RSH + KBr + H_2O$$

The thiol that forms is sufficiently acidic to form a thiolate ion in the presence of potassium hydroxide. Thus, if excess H_2S is not employed in the reaction, the major product of the

* An ylide (pronounced ill/id) is a neutral molecule that can be represented as a resonance hybrid, one structure of which has negative carbon directly attached to a positive heteroatom.

reaction will be a thioether. The thioether results from the following reactions:

$$R{-}SH + KOH \longrightarrow R{-}\ddot{\underset{..}{S}}{:}^- \; K^+$$

$$R{-}\ddot{\underset{..}{S}}{:}^- \; K^+ + R{-}\ddot{\underset{..}{Br}}{:} \longrightarrow \underset{\text{Thioether}}{R{-}\ddot{\underset{..}{S}}{-}R} + KBr$$

Alkyl halides also react with thiourea to form (stable) *S*-alkylisothiouronium salts. These can be used to prepare thiols.

$$\underset{\text{Thiourea}}{(H_2\ddot{N})_2C{=}\ddot{S}{:}} + CH_3CH_2{-}\ddot{\underset{..}{Br}}{:} \xrightarrow{C_2H_5OH} \underset{\substack{S\text{-Ethylisothiouronium}\\ \text{bromide}\\ (95\%)}}{(H_2\ddot{N})_2C{=}\overset{+}{\underset{..}{S}}{-}CH_2CH_3 \; Br^-}$$

$$\xrightarrow{OH^-/H_2O \text{ then } H^+} \underset{\text{Urea}}{(H_2N)_2C{=}O} + \underset{\substack{\text{Ethanethiol}\\ (90\%)}}{CH_3CH_2SH}$$

J.2 PHYSICAL PROPERTIES OF THIOLS

Thiols form very weak hydrogen bonds; their hydrogen bonds are not nearly as strong as those of alcohols. Because of this, low-molecular-weight thiols have lower boiling points than corresponding alcohols. Ethanethiol, for example, boils more than 40° lower than ethanol (37° versus 78°). The relative weakness of hydrogen bonds between molecules of thiols is also evident when we compare the boiling points of ethanethiol and its isomer dimethyl sulfide:

$$\underset{\text{bp } 37°}{CH_3CH_2SH} \qquad \underset{\text{bp } 38°}{CH_3SCH_3}$$

Physical properties of several thiols are given in Table J.1.

J.3 THIOLS AND DISULFIDES IN BIOCHEMISTRY

Thiols and disulfides are important compounds in living cells, and in many biochemical oxidation-reduction reactions they are interconverted.

$$2RSH \underset{[H]}{\overset{[O]}{\rightleftharpoons}} R{-}S{-}S{-}R$$

TABLE J.1 Physical Properties of Thiols

COMPOUND	STRUCTURE	mp C°	bp C°
Methanethiol	CH_3SH	−123	6
Ethanethiol	CH_3CH_2SH	−144	37
1-Propanethiol	$CH_3CH_2CH_2SH$	−113	67
2-Propanethiol	$(CH_3)_2CHSH$	−131	58
1-Butanethiol	$CH_3(CH_2)_2CH_2SH$	−116	98

Lipoic acid, for example, an important cofactor in biological oxidations, undergoes this oxidation-reduction reaction.

$$\text{Lipoic acid (cyclic S—S)-(CH}_2)_4\text{COOH} \underset{[O]}{\overset{[H]}{\rightleftharpoons}} \text{Dihydrolipoic acid (HS, HS)-(CH}_2)_4\text{COOH}$$

Lipoic acid — Dihydrolipoic acid

The amino acids *cysteine* and *cystine* are interconverted in a similar way.

$$2\text{HOOCCH(NH}_2)\text{CH}_2\text{SH} \underset{[H]}{\overset{[O]}{\rightleftharpoons}} \text{HOOCCH(NH}_2)\text{CH}_2\text{S—SCH}_2\text{CH(NH}_2)\text{COOH}$$

Cysteine — Cystine

As we will see later, the disulfide linkages of cystine units are important in determining the overall shapes of protein molecules.

Problem J.1

Give structures for the products of the following reactions:
(a) Benzyl bromide + thiourea ⟶
(b) Product of (a) + OH^-/H_2O ⟶
(c) Product of (b) + H_2O_2 ⟶
(d) Product of (c) + NaOH ⟶
(e) Product of (d) + benzyl bromide ⟶

Problem J.2

Allyl disulfide, $CH_2{=}CHCH_2S{-}SCH_2CH{=}CH_2$, is another important component of oil of garlic. Suggest a synthesis of allyl disulfide starting with allyl bromide.

Problem J.3

Starting with allyl alcohol, outline a synthesis of British Anti-Lewisite, $CH_2SHCHSHCH_2OH$

Problem J.4

A synthesis of lipoic acid (above) is outlined below. Supply the missing reagents and intermediates.

$$\text{Cl—C(=O)(CH}_2)_4\text{CO}_2\text{C}_2\text{H}_5 \xrightarrow[\text{AlCl}_3]{\text{CH}_2=\text{CH}_2} \text{(a) C}_{10}\text{H}_{17}\text{ClO}_3 \xrightarrow{\text{NaBH}_4}$$

$$\text{ClCH}_2\text{CH}_2\text{CH(OH)(CH}_2)_4\text{CO}_2\text{C}_2\text{H}_5 \xrightarrow{\text{(b)}} \text{ClCH}_2\text{CH}_2\text{CH(Cl)(CH}_2)_4\text{CO}_2\text{C}_2\text{H}_5$$

$$\xrightarrow[\text{(d)}]{\text{(c)}} \text{C}_6\text{H}_5\text{CH}_2\text{SCH}_2\text{CH}_2\text{CH(SCH}_2\text{C}_6\text{H}_5)(\text{CH}_2)_4\text{CO}_2\text{H} \xrightarrow[\text{(2) H}^+]{\text{(1) Na, NH}_3}$$

$$\text{(e) C}_8\text{H}_{16}\text{S}_2\text{O}_2 \xrightarrow{\text{O}_2} \text{lipoic acid}$$

Problem J.5

One chemical-warfare agent used in World War I is a powerful vesicant called "mustard gas." (The name comes from its mustardlike odor; mustard gas, however, is not a gas but a high-boiling liquid that was dispersed as a mist of tiny droplets.) Mustard gas can be synthesized from ethylene oxide in the manner shown below. Outline the reactions involved.

$$2\underset{\diagdown O \diagup}{CH_2{-}CH_2} + H_2S \longrightarrow C_4H_{10}SO_2 \xrightarrow[ZnCl_2]{HCl} \underset{\text{"mustard gas"}}{C_4H_8SCl_2}$$

16 ALDEHYDES AND KETONES

16.1 INTRODUCTION

The carbonyl group, $\rangle C{=}O$, is one of the most important functional groups in organic chemistry.

Although we have had some experience with carbonyl compounds in earlier chapters, we will now consider their chemistry in detail. In this chapter we will study aldehydes—compounds in which the carbonyl group is bonded to carbon and hydrogen—and ketones—compounds in which the carbonyl group is bonded to two carbons. In Chapter 17 we will study compounds in which the carbonyl group is bonded to an oxygen (acids, esters, anhydrides), to a nitrogen (amides), and to halogens (acyl halides).

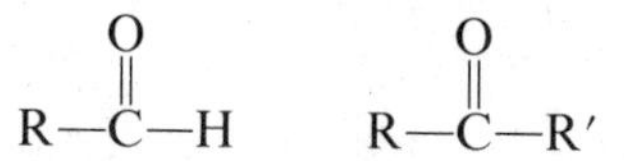

$$\underset{\text{An aldehyde}}{R{-}\overset{O}{\overset{\|}{C}}{-}H} \qquad \underset{\text{A ketone}}{R{-}\overset{O}{\overset{\|}{C}}{-}R'}$$

16.2 NOMENCLATURE OF ALDEHYDES AND KETONES

In the IUPAC system we name aliphatic aldehydes by replacing the final *e* of the name of the corresponding alkane with *al*. Since the aldehyde group must be at the end of the chain of carbons, there is no need to designate its position. When other substituents are present, however, we assume that the carbonyl group occupies position -1. (Many aldehydes also have common names; these are given below in parentheses.)

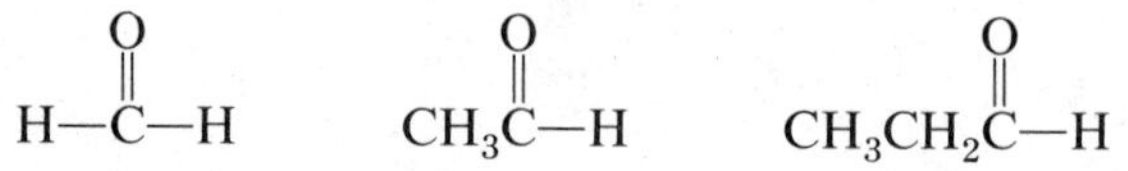

$$H{-}\overset{O}{\overset{\|}{C}}{-}H \qquad CH_3\overset{O}{\overset{\|}{C}}{-}H \qquad CH_3CH_2\overset{O}{\overset{\|}{C}}{-}H$$

Methanal (formaldehyde) — Ethanal (acetaldehyde) — Propanal (propionaldehyde)

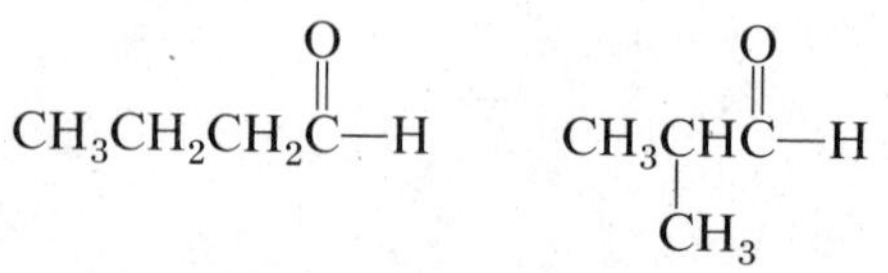

$$CH_3CH_2CH_2\overset{O}{\overset{\|}{C}}{-}H \qquad CH_3\underset{CH_3}{\underset{|}{C}}H\overset{O}{\overset{\|}{C}}{-}H$$

Butanal (butyraldehyde) — 2-Methylpropanal (isobutyraldehyde)

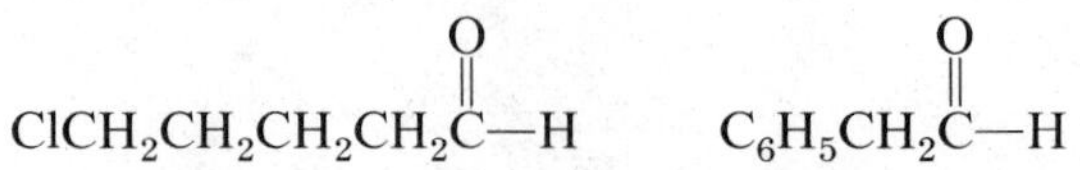

$$ClCH_2CH_2CH_2CH_2\overset{O}{\overset{\|}{C}}{-}H \qquad C_6H_5CH_2\overset{O}{\overset{\|}{C}}{-}H$$

5-Chloropentanal — Phenylethanal (phenylacetaldehyde)

When the carbonyl group is attached to an aromatic ring we name the compound as a benzaldehyde, tolualdehyde, naphthaldehyde, and so on.

Benzaldehyde

p-Chloro-benzaldehyde

o-Hydroxybenzaldehyde (salicylaldehyde)

o-Tolualdehyde

2-Naphthaldehyde

We name aliphatic ketones by replacing the final *e* of the name of the corresponding alkane with *one*. We then number the chain in the way that gives the carbonyl carbon the lower possible number and we use this number to designate its position. (No number is necessary for propanone and butanone. Why?)

CH_3CCH_3 (C=O)
Propanone (acetone)

$CH_3CH_2CCH_3$ (C=O)
Butanone (methyl ethyl ketone)

$CH_3CCH_2CH_2CH_3$ (C=O)
2-Pentanone (methyl propyl ketone)

Common names for ketones (in parentheses above and below) are obtained simply by separately naming the two groups attached to the carbonyl group and adding the word *ketone* as a separate word.

Some aromatic ketones have special names.

Acetophenone (methyl phenyl ketone)

Benzophenone (diphenyl ketone)

When we find it necessary to name the —CH(=O) group as a substituent, we call it the *formyl group*. It is often written —CHO. When RC(=O)— groups are named as substituents, they are called *acyl groups*.

o-Formylbenzoic acid

p-Acetylbenzenesulfonic acid

Problem 16.1

(a) Give IUPAC names for the seven isomeric aldehydes and ketones with the formula $C_5H_{10}O$. (b) How many aldehydes and ketones contain a benzene ring and have the formula C_8H_8O? (c) What are their names and structures?

16.3 PHYSICAL PROPERTIES

The carbonyl group is a polar group; therefore aldehydes and ketones have higher boiling points than hydrocarbons of the same molecular weight. However, since aldehydes and ketones cannot have strong hydrogen bonds between their molecules, they have lower boiling points than corresponding alcohols.

$CH_3CH_2CH_2CH_3$	$CH_3CH_2CH{=}O$	$CH_3C({=}O)CH_3$	$CH_3CH_2CH_2OH$
Butane	Propanal	Acetone	Propanol
bp −0.5°	49°	56.1°	97.2°

Problem 16.2

Which compound in each pair listed below has the higher boiling point? (Answer this problem without consulting tables.)

(a) Pentanal or 1-pentanol
(b) 2-Pentanone or 2-pentanol
(c) *n*-Pentane or pentanal
(d) Acetophenone or 2-phenylethanol
(e) Benzaldehyde or benzyl alcohol

The carbonyl oxygen allows molecules of aldehydes and ketones to form strong hydrogen bonds to molecules of water. As a result, low-molecular-weight aldehydes and ketones show appreciable solubilities in water. Acetone and acetaldehyde are soluble in water in all proportions.

Table 16.1 lists the physical properties of a number of common aldehydes and ketones.

TABLE 16.1 Physical Properties of Aldehydes and Ketones

FORMULA	NAME	mp °C	bp °C	SOLUBILITY IN WATER
HCHO	Formaldehyde	−92	−21	Very sol.
CH_3CHO	Acetaldehyde	−125	21	∞
CH_3CH_2CHO	Propanal	−81	49	Very sol.
$CH_3(CH_2)_2CHO$	Butanal	−99	76	Sol.
$CH_3(CH_2)_3CHO$	Pentanal	−91.5	102	Sl. sol.
$CH_3(CH_2)_4CHO$	Hexanal	−56	128	Sl. sol.
C_6H_5CHO	Benzaldehyde	−57	178	Sl. sol.
$C_6H_5CH_2CHO$	Phenylacetaldehyde	33	193	Sl. sol.
CH_3COCH_3	Acetone	−95	56.1	∞
$CH_3COCH_2CH_3$	Butanone	−86	79.6	Very sol.
$CH_3COCH_2CH_2CH_3$	2-Pentanone	−78	102	Sol.
$CH_3CH_2COCH_2CH_3$	3-Pentanone	−42	102	Sol.
$C_6H_5COCH_3$	Acetophenone	21	202	Insol.
$C_6H_5COC_6H_5$	Benzophenone	48	306	Insol.

A number of aromatic aldehydes obtained from natural sources have very pleasant fragrances. Some of these are the following.

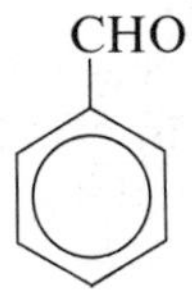

Benzaldehyde
(from bitter almonds)

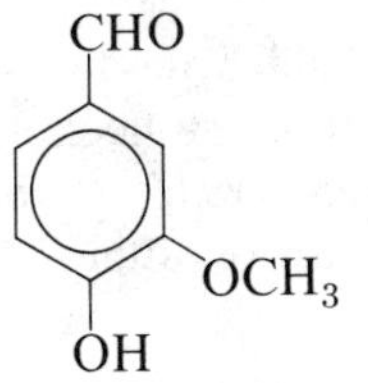

Vanillin
(from vanilla bean)

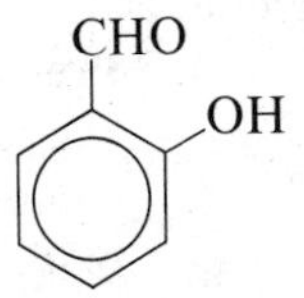

Salicylaldehyde
(from meadow sweet)

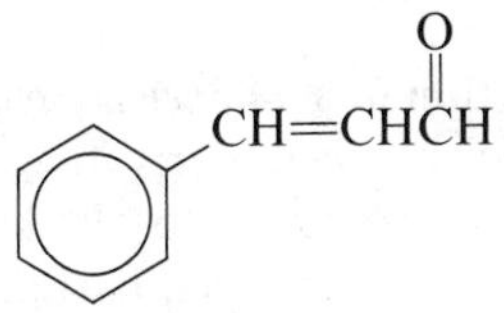

Cinnamaldehyde
(from cinnamon)

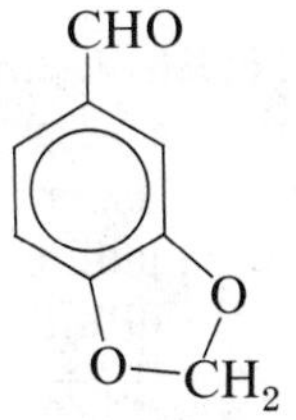

Piperonal
(made from safrole,
odor of heliotrope)

16.3A Spectroscopic Properties of Aldehydes and Ketones

Carbonyl groups of aldehydes and ketones give rise to very strong C=O stretching bands in the 1665–1780 cm^{-1} region of the *infrared spectrum*. The exact location of the peak (Table 16.2) depends on the structure of the aldehyde or ketone.

The CHO group of aldehydes also gives two weak bands in the 2700 to 2775 and 2820–2900 cm^{-1} region of the infrared spectrum.

The aldehydic proton gives a peak far downfield ($\delta = 9$–10) in proton nmr *spectra*.

The carbonyl groups of saturated aldehydes and ketones give a weak absorption band in the *ultraviolet region* between 270 and 300 nm. This band is shifted to longer wavelengths (300–350 nm) when the carbonyl group is conjugated with a double bond.

TABLE 16.2 Carbonyl Stretching Bands of Aldehydes and Ketones

C=O STRETCHING FREQUENCIES			
R—CHO	1720–1740 cm^{-1}	RCOR	1705–1720 cm^{-1}
Ar—CHO	1695–1715 cm^{-1}	ArCOR	1680–1700 cm^{-1}
—C=C—CHO	1680–1690 cm^{-1}	—C=C—COR	1665–1680 cm^{-1}
		Cyclohexanone	1715 cm^{-1}
		Cyclopentanone	1751 cm^{-1}

16.4 PREPARATION OF ALDEHYDES

Aldehydes can be prepared by methods that involve oxidation or reduction. However, since aldehydes are easily oxidized and reduced, we must use special reagents or techniques.

1. **Preparation of aldehydes by oxidation of primary alcohols** (Discussed in Section 15.8). Primary alcohols can be oxidized to aldehydes. A reagent commonly used for this purpose is a complex formed when chromic oxide reacts with pyridine.

$$\underset{\text{1-Heptanol}}{C_6H_{13}CH_2OH} + \underset{\text{Chromic oxide-pyridine}}{CrO_3 \cdot 2C_5H_5N} \xrightarrow{CH_2Cl_2} \underset{\substack{\text{Heptanal} \\ (93\%)}}{C_6H_{13}CHO}$$

2. **Preparation of aldehydes by reduction of acid derivatives.** A number of special methods can be used to convert acid derivatives to aldehydes. Acyl chlorides, for example, can be reduced to aldehydes by using hydrogen and a palladium catalyst that has been partially deactivated by treatment with sulfur. This technique, called the *Rosenmund reduction,* usually gives aldehydes in excellent yields.

$$R{-}\overset{\overset{O}{\|}}{C}{-}OH \xrightarrow{SOCl_2} R{-}\overset{\overset{O}{\|}}{C}{-}Cl \xrightarrow[Pd(S)]{H_2} \underset{(80\text{–}90\%)}{R{-}\overset{\overset{O}{\|}}{C}{-}H}$$

Acid chlorides can also be reduced to aldehydes by treating them with lithium tri-*tert*-butoxyaluminum hydride, $LiAlH[OC(CH_3)_3]_3$.

$$CH_3(OCH_3)C_6H_3{-}\overset{\overset{O}{\|}}{C}{-}Cl \xrightarrow{LiAlH[OC(CH_3)_3]_3} CH_3(OCH_3)C_6H_3{-}\overset{\overset{O}{\|}}{C}{-}H$$

3. **Preparation of aldehydes from terminal alkynes.** (Discussed in Section 9.10). Aldehydes can also be prepared by hydroboration of terminal alkynes using a hindered borane (e.g., Sia_2BH) followed by oxidation with hydrogen peroxide, H_2O_2, in basic solution.

$$R{-}C{\equiv}CH + Sia_2BH \longrightarrow RCH{=}CH{-}BSia_2 \xrightarrow[OH^-/H_2O]{H_2O_2} RCH_2\overset{\overset{O}{\|}}{C}H$$

16.5 PREPARATION OF KETONES

We have seen three methods for the preparation of ketones in earlier chapters.

1. **Preparation of ketones by Friedel-Crafts acylations** (discussed in Section 12.7).

$$\text{ArH} + \text{R—}\overset{\overset{\displaystyle \text{O}}{\|}}{\text{C}}\text{—Cl} \xrightarrow{\text{AlCl}_3} \text{Ar—}\overset{\overset{\displaystyle \text{O}}{\|}}{\text{C}}\text{—R}$$

An aryl alkyl ketone

or

$$\text{ArH} + \text{Ar—}\overset{\overset{\displaystyle \text{O}}{\|}}{\text{C}}\text{—Cl} \xrightarrow{\text{AlCl}_3} \text{Ar—}\overset{\overset{\displaystyle \text{O}}{\|}}{\text{C}}\text{—Ar}$$

A diaryl ketone

2. **Preparation of ketones by oxidation of secondary alcohols** (discussed in Section 15.8).

$$\text{R—}\overset{\overset{\displaystyle \text{OH}}{|}}{\text{C}}\text{H—R}' \xrightarrow{\text{(O)}} \text{R—}\overset{\overset{\displaystyle \text{O}}{\|}}{\text{C}}\text{—R}'$$

3. **Preparation of Ketones from Alkynes** (discussed in Section 9.10). Alkynes can be converted to ketones by hydration (with H_3O^+, Hg^{++}, and H_2O) or by hydroboration (with $(BH_3)_2$) followed by oxidation.

$$\text{R—C}\equiv\text{C—R} \xrightarrow[\text{H}_2\text{O}]{\text{H}_3\text{O}^+,\ \text{Hg}^{++}} \left[\begin{matrix} \text{R} & & \text{H} \\ & \text{C=C} & \\ \text{HO} & & \text{R} \end{matrix}\right] \longrightarrow \text{R—}\underset{\underset{\displaystyle \text{O}}{\|}}{\text{C}}\text{—CH}_2\text{R}$$

$$\text{R—C}\equiv\text{C—R} \xrightarrow[\text{(2) H}_2\text{O}_2,\ \text{OH}^-]{\text{(1) (BH}_3)_2} \left[\begin{matrix} \text{R} & & \text{R} \\ & \text{C=C} & \\ \text{HO} & & \text{H} \end{matrix}\right] \longrightarrow \text{R—}\underset{\underset{\displaystyle \text{O}}{\|}}{\text{C}}\text{—CH}_2\text{R}$$

Two other methods for the preparation of ketones are based on the use of organometallic compounds.

4. **Preparation of ketones by the reaction of organocadmium compounds with acid chlorides.** Grignard reagents are too reactive toward ketones to be used to prepare ketones from esters or acid chlorides. We saw in Section 14.11 that Grignard reagents react with esters to yield tertiary alcohols. Grignard reagents also react with acid chlorides in the same general way for the same reason.

$$\text{R—}\overset{\overset{\displaystyle \text{O}}{\|}}{\text{C}}\text{—Cl} + 2\text{R}'\text{MgX} \xrightarrow[\text{(2)HOH}]{} \text{R—}\underset{\underset{\displaystyle \text{R}'}{|}}{\overset{\overset{\displaystyle \text{OH}}{|}}{\text{C}}}\text{—R}' + \text{MgXCl} + \text{MgXOH}$$

However, if a Grignard reagent is first converted to a dialkylcadmium by treating it with anhydrous cadmium chloride, subsequent addition of an acid chloride gives a ketone a good yield.

$$2RMgX + CdCl_2 \longrightarrow \underset{\substack{\text{Dialkylcadmium} \\ + \\ 2MgXCl}}{R_2Cd} \xrightarrow{2R'-\overset{O}{\overset{\|}{C}}-Cl} 2R-\overset{O}{\overset{\|}{C}}-R'$$

This synthesis succeeds because cadmium reagents are too unreactive toward ketones to attack the product, but are reactive enough toward acid chlorides. An example of this method is shown below.

$$2(CH_3)_2CHMgBr \xrightarrow[\text{ether}]{CdCl_2} [(CH_3)_2CH]_2Cd \xrightarrow{2CH_3CH_2\overset{O}{\overset{\|}{C}}Cl} \underset{(60\%)}{2(CH_3)_2CH\overset{O}{\overset{\|}{C}}CH_2CH_3}$$

5. **Preparation of ketones from lithium dialkylcuprates.** When an ether solution of a lithium dialkylcuprate is treated with an acid chloride at $-78°$, the product is a ketone. This ketone synthesis is a variation of the Corey-House alkane synthesis (Sections 3.16 and 14.12)

$$\underset{\substack{\text{Lithium} \\ \text{dialkylcuprate}}}{R_2CuLi} + R'-\overset{O}{\overset{\|}{C}}-Cl \longrightarrow R'-\overset{O}{\overset{\|}{C}}-R$$

A specific example is the preparation of methyl cyclohexyl ketone.

$$C_6H_{11}-\overset{O}{\overset{\|}{C}}-Cl + (CH_3)_2CuLi \xrightarrow[\text{ether}]{-78°} \underset{(81\%)}{C_6H_{11}-\overset{O}{\overset{\|}{C}}-CH_3}$$

Problem 16.3

Which reagents would you use to carry out each of the following reactions?

(a) Benzene $\longrightarrow$ bromobenzene $\longrightarrow$ phenylmagnesium bromide $\longrightarrow$ benzyl alcohol $\longrightarrow$ benzaldehyde

(b) Toluene $\longrightarrow$ benzoic acid $\longrightarrow$ benzoyl chloride $\xrightarrow[\text{(two ways)}]{}$ benzaldehyde

(c) Ethyl bromide $\longrightarrow$ 1-butyne $\longrightarrow$ butanal

(d) 2-Butyne $\longrightarrow$ 2-butanone

(e) 1-Phenylethanol $\longrightarrow$ acetophenone

(f) Benzene $\longrightarrow$ acetophenone

(g) Benzoyl chloride $\xrightarrow[\text{(two ways)}]{}$ acetophenone

16.6 GENERAL CONSIDERATION OF THE REACTIONS OF CARBONYL COMPOUNDS

Before we continue our detailed study of carbonyl compounds it will be helpful if we examine the structures of carbonyl compounds in a general way. As we do this we will find that certain structural features of carbonyl compounds underlie—*and thus unify*—most of the reactions that we will discuss in this chapter and the next.

16.6A Structure of the Carbonyl Group

The carbonyl carbon is sp^2 hybridized; thus it, and the three atoms attached to it, lie in the same plane. The bond angles between the three attached atoms are what we would expect of a trigonal coplanar structure: they are approximately 120°.

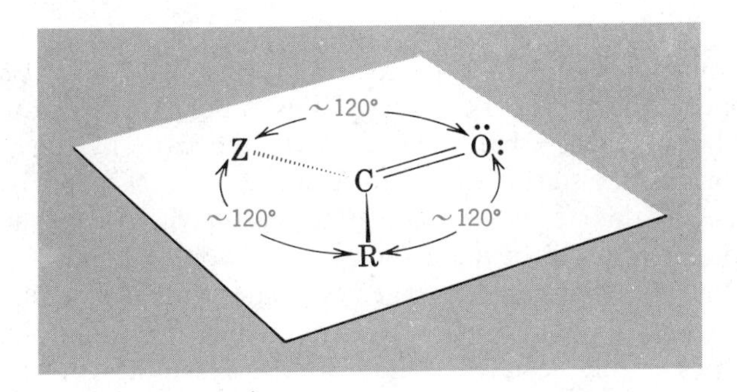

The carbonyl carbon bears a substantial partial *positive* charge; the carbonyl oxygen bears a substantial partial *negative* charge.

$$\delta+ \;\rangle C{=}\ddot{O}: \;\delta-$$

This charge distribution arises from two effects: (1) the inductive effect of the electronegative oxygen and (2) the resonance contribution of the second structure shown below.

$$\rangle C{=}\ddot{O}: \longleftrightarrow \rangle \overset{+}{C}{-}\ddot{\underset{\cdot\cdot}{O}}:^- \quad \text{or} \quad \rangle \overset{\delta+}{C}{=\!=}\overset{\delta-}{O}$$

Hybrid

Resonance structures for the carbonyl group

16.6B Nucleophilic Addition to the Carbon-Oxygen Double Bond

One highly characteristic reaction of aldehydes and ketones is a *nucleophilic addition* to the carbon-oxygen double bond. Aldehydes and ketones are especially susceptible to nucleophilic addition because of the structural features that we have just mentioned. The trigonal coplanar arrangement of groups around the carbonyl carbon means that the carbonyl carbon is relatively open to attack from above or below. The positive charge on the carbonyl carbon means that it is especially susceptible to attack by a nucleophile. The negative charge on the carbonyl oxygen means that nucleophilic addition is susceptible to acid catalysis. We can visualize nucleophilic addition to the carbon-oxygen double bond occurring in either of two general ways:

1. When the reagent is a particularly strong nucleophile, addition will usually take place in the following way.

$$\mathrm{Nu:^-} + \underset{\substack{\text{Trigonal}\\ \text{coplanar}}}{\overset{\delta+}{>C}=\overset{\delta-}{\ddot{O}}:} \rightleftarrows \underset{\text{Tetrahedral}}{\mathrm{Nu}-C-\ddot{O}:^-} \overset{H^+}{\rightleftarrows} \underset{\text{Tetrahedral}}{\mathrm{Nu}-C-O-H}$$

In this type of addition the nucleophile uses its electron pair to form a bond to the carbonyl carbon. As this happens an electron pair of the carbon-oxygen π bond shifts out to the carbonyl oxygen and the hybridization state of the carbon changes from sp^2 to sp^3. The important aspect of this step is the ability of the carbonyl oxygen to accommodate the electron pair of the carbon-oxygen double bond.

In the second step the oxygen associates with an electrophile (usually a proton). This happens because the oxygen is now much more basic; it carries a full negative charge, and it resembles the oxygen of an alkoxide ion. (In some reactions the oxygen of the carbonyl group actually becomes the oxygen of an alkoxide ion.)

2. A second general mechanism that operates in nucleophilic additions to carbon-oxygen double bonds is an acid-catalyzed mechanism:

$$\mathrm{R'R}C=\ddot{O}: + \underset{\text{(or a Lewis acid)}}{H^+} \rightleftarrows \mathrm{R'R}C=\overset{+}{\ddot{O}}H \longleftrightarrow \mathrm{R'R}\overset{+}{C}-\ddot{O}H$$

$$\mathrm{R'R}C=\overset{+}{\ddot{O}}-H + \mathrm{Nu:^-} \rightleftarrows \mathrm{Nu}-C(\mathrm{R'})(\mathrm{R})-\ddot{O}-H$$

This mechanism operates when carbonyl compounds are treated with reagents that are *strong acids* but *weak nucleophiles*. In the first step the acid attacks an electron pair of the carbonyl oxygen: the resulting protonated carbonyl compound is highly reactive toward nucleophilic attack at the carbonyl carbon (in the second step) because of the contribution made by the second resonance structure below:

$$\mathrm{R_2}C=\overset{+}{\ddot{O}}-H \longleftrightarrow \mathrm{R_2}\overset{+}{C}-\ddot{O}-H$$

16.6C Reversibility of Nucleophilic Additions to the Carbon-Oxygen Double Bond

Many nucleophilic additions to carbon-oxygen double bonds are reversible; the overall results of these reactions will depend, therefore, on the position of an equilibrium. This behavior contrasts markedly with most electrophilic additions to carbon-carbon double bonds and with nucleophilic substitutions at saturated carbons. The latter reactions are essentially irreversible, and overall results are a function of relative reaction rates.

16.6D Subsequent Reactions of Addition Products

Nucleophilic addition to a carbon-oxygen double bond may lead to a product that is stable under the reaction conditions that we employ. If this is the case we are then able to isolate products with the general structure shown below:

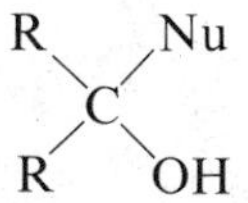

$$\begin{array}{c} R \quad\quad Nu \\ \diagdown \;\; \diagup \\ C \\ \diagup \;\; \diagdown \\ R \quad\quad OH \end{array}$$

In other reactions the product formed initially may be unstable and may spontaneously undergo additional reactions. Even if the initial addition product is stable, however, we may deliberately bring about a subsequent reaction by changing the reaction conditions. When we begin our study of specific reactions we will see that one common subsequent reaction is an *elimination reaction.*

Problem 16.4

The reaction of an aldehyde or ketone with a Grignard reagent (Section 14.11C) is a nucleophilic addition to the carbon-oxygen double bond. (a) What is the nucleophile? (b) The magnesium portion of the Grignard reagent plays an important part in this reaction. What is its function? (c) What product is formed initially? (d) What product forms when water is added?

Problem 16.5

The reactions of aldehydes and ketones with $LiAlH_4$ and $NaBH_4$ (Section 15.7) are nucleophilic additions to the carbonyl group. What is the nucleophile in these reactions?

16.7 THE ADDITION OF ALCOHOLS: ACETALS AND KETALS

Dissolving an aldehyde in an alcohol establishes an equilibrium between these substances and a product called a *hemiacetal.* The hemiacetal is formed by a nucleophilic addition of the alcohol to the carbonyl group:

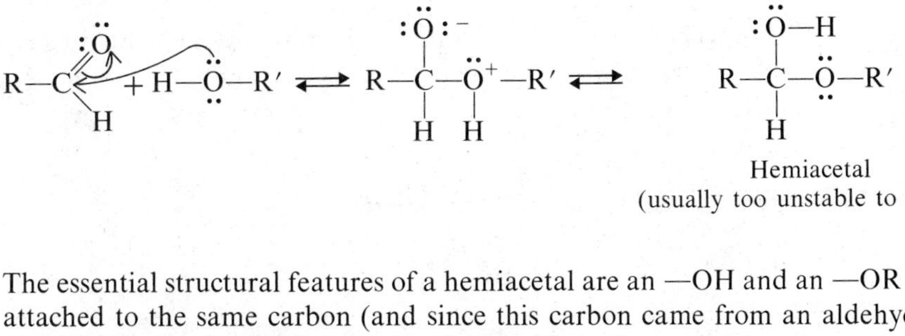

$$R{-}\overset{\ddot{O}:}{\overset{\|}{C}}{-}H + H{-}\ddot{O}{-}R' \rightleftharpoons R{-}\overset{:\ddot{O}:^-}{\underset{H}{\overset{|}{\underset{|}{C}}}}{-}\underset{H}{\underset{|}{\ddot{O}^+}}{-}R' \rightleftharpoons R{-}\overset{:\ddot{O}{-}H}{\underset{H}{\overset{|}{\underset{|}{C}}}}{-}\ddot{O}{-}R'$$

Hemiacetal
(usually too unstable to isolate)

The essential structural features of a hemiacetal are an —OH and an —OR group attached to the same carbon (and since this carbon came from an aldehyde the carbon also has one hydrogen attached to it).

Most open-chain hemiacetals are not sufficiently stable to allow their isolation. Cyclic hemiacetals with five- or six-membered rings, however, are usually much more stable:

$$HOCH_2CH_2CH_2CH_2CH{=}O \rightleftharpoons \text{cyclic hemiacetal } (CH_2CH_2CH_2CH_2\text{–}O\text{–}CH(OH) \text{ ring})$$

Most simple sugars (Chapter 19) exist, primarily, in a cyclic hemiacetal form.

$$\underset{\substack{\text{A hexose}\\ \text{(a sugar)}}}{{}^{1}CHO\text{–}{}^{2}CHOH\text{–}{}^{3}CHOH\text{–}{}^{4}CHOH\text{–}{}^{5}CH(OH)\text{–}{}^{6}CH_2OH} \rightleftharpoons \underset{\substack{\text{Cyclic hemiacetal form}\\ \text{of a hexose}\\ \text{(stable enough to isolate)}}}{\text{ring: } {}^{5}CH(^{6}CH_2OH)\text{–}O\text{–}{}^{1}CH(OH)\text{–}{}^{2}CHOH\text{–}{}^{3}CHOH\text{–}{}^{4}CHOH}$$

Hemiacetal group

Ketones undergo similar reactions when they are dissolved in an alcohol. The products (also unstable in open-chain compounds) are called *hemiketals.*

$$R_2C{=}O + R'\text{—}OH \rightleftharpoons \underset{\text{A hemiketal}}{R\text{—}C(OH)(R)\text{—}OR'}$$

If we take an alcohol solution of an aldehyde and pass into it a small amount of gaseous HCl, a second reaction takes place. The hemiacetal reacts with a second mole of the alcohol and produces an *acetal.* An acetal has two —OR groups attached to the same CH group.

$$\underset{\text{Hemiacetal}}{R\text{—}CH(OH)\text{—}OR'} \xrightarrow[R'\text{—}OH]{HCl_{(g)}} \underset{\text{An acetal}}{R\text{—}CH(OR')\text{—}OR'} + H_2O$$

The mechanism for acetal formation involves an acid-catalyzed elimination of water followed by a second *addition* of the alcohol.

$$\underset{\text{Hemiacetal}}{R(H)C(OH)(OR')} \underset{-H^+}{\overset{+H^+}{\rightleftharpoons}} R(H)C(\overset{+}{O}H_2)(OR') \underset{+H_2O}{\overset{-H_2O}{\rightleftharpoons}} R(H)C{=}\overset{+}{O}\text{—}R'$$

Acid-catalyzed elimination of water

$$\underset{H}{\overset{R}{>}}C{=}\overset{+}{O}{-}R' \underset{-R'OH}{\overset{+R'\ddot{O}-H}{\rightleftharpoons}} \underset{H}{\overset{R}{>}}C\genfrac{}{}{0pt}{}{\overset{H}{\overset{|+}{O}}R'}{OR'} \underset{+H^+}{\overset{-H^+}{\rightleftharpoons}} \underset{H}{\overset{R}{>}}C\genfrac{}{}{0pt}{}{O{-}R'}{O{-}R'}$$

Acetal

Addition of a second mole of the alcohol

Problem 16.6

(a) Write another resonance structure for the intermediate, $RCH{=}O^+R$. (b) Which structure would you expect to make a larger contribution to the hybrid? (c) Why?

All steps in the formation of an acetal from an aldehyde are reversible. If we dissolve an aldehyde in a large excess of an anhydrous alcohol and add a small amount of an anhydrous acid (i.e., gaseous HCl or conc. H_2SO_4), the equilibrium will strongly favor the formation of an acetal. After the equilibrium is established we can isolate the acetal by neutralizing the acid and evaporating the excess alcohol.

If we then place the acetal in water and add a small amount of acid, all of the steps reverse. Under these conditions (an excess of water) the equilibrium favors the formation of the aldehyde. The acetal undergoes *hydrolysis*.

$$\underset{\text{Acetal}}{\underset{H}{\overset{R}{>}}C\genfrac{}{}{0pt}{}{OR'}{OR'}} + H_2O \underset{\text{(several steps)}}{\overset{H^+}{\rightleftharpoons}} \underset{\text{Aldehyde}}{R{-}\overset{O}{\overset{\|}{C}}{-}H} + 2R'OH$$

Ketal formation is not favored when ketones are treated with simple alcohols and gaseous HCl. Cyclic ketal formation *is* favored, however, when a ketone is treated with an excess of a 1,2-diol and a trace of acid.

$$\underset{R}{\overset{R}{>}}C{=}O + \underset{\text{(excess)}}{\genfrac{}{}{0pt}{}{HOCH_2}{HOCH_2}} \overset{H^+}{\rightleftharpoons} \underset{\text{Cyclic ketal}}{\underset{R}{\overset{R}{>}}C\genfrac{}{}{0pt}{}{O{-}CH_2}{O{-}CH_2}} + H_2O$$

This reaction, too, can be reversed by treating the ketal with aqueous acid.

$$\underset{R}{\overset{R}{>}}C\genfrac{}{}{0pt}{}{O{-}CH_2}{O{-}CH_2} + H_2O \overset{H^+}{\rightleftharpoons} \underset{R}{\overset{R}{>}}C{=}O + \genfrac{}{}{0pt}{}{CH_2OH}{CH_2OH}$$

Problem 16.7

Outline all steps in the mechanism for the formation of a cyclic ketal from acetone and ethylene glycol in the presence of gaseous HCl.

16.7A Acetals and Cyclic Ketals as Protecting Groups

Although acetals and cyclic ketals are hydrolyzed to aldehydes and ketones in aqueous acid, *they are stable in aqueous base.*

$$RHC(OR')_2 + H_2O \xrightarrow{OH^-} \text{No reaction}$$

$$R_2C(OCH_2)_2 + H_2O \xrightarrow{OH^-} \text{No reaction}$$

Because of this property, acetals and ketals give us a convenient method for protecting aldehyde and ketone groups from undesired reactions in basic solutions. (Acetals and ketals are really *gem*-diethers and, like ethers, they are relatively unreactive.) We can convert an aldehyde or ketone to an acetal or cyclic ketal, carry out a reaction on some other part of the molecule, and then hydrolyze the acetal or ketal with aqueous acid.

As an example, let us consider the problem of converting

3-oxocyclopentyl–COC_2H_5 (ester, C=O) **A** to 3-oxocyclopentyl–CH_2OH **B**

Keto groups are more easily reduced than ester groups. Any reducing agent (i.e., $LiAlH_4$ or H_2/Ni) that will reduce the ester group of **A** will reduce the keto group as well. But, if we "protect" the keto group by converting it to a cyclic ketal, we can reduce the ester group without affecting the cyclic ketal. After we finish the ester reduction we can hydrolyze the cyclic ketal and obtain our desired product, **B.**

$$\mathbf{A} \xrightarrow[CH_2OHCH_2OH]{H^+} \text{cyclic ketal ester } (COC_2H_5) \xrightarrow[(C_2H_5)_2O \;\; (2)\ H_2O]{(1)\ LiAlH_4} \text{cyclic ketal–}CH_2OH \xrightarrow[H_2O]{H^+} \text{3-oxocyclopentyl–}CH_2OH$$

Problem 16.8

(a) Show how you might use a cyclic ketal in carrying out the following transformation.

3-oxocyclopentyl–$CO_2C_2H_5$ (**A**) $\longrightarrow$ 3-oxocyclopentyl–$C(OH)(CH_3)CH_3$ (**C**)

(b) Why would a direct addition of methylmagnesium bromide to **A** fail to give **C?**

16.7B Example—Synthesis of Vitamin C

Cyclic ketal formation can also be used to protect hydroxyl groups from undesired reactions. The synthesis of vitamin C from (−)-sorbose provides an especially interesting application of this technique and, at the same time, it illustrates a number of other familiar reactions.

The starting compound, (−)-sorbose, is a pentahydroxy ketone. (It is also a simple sugar that is almost as sweet as ordinary table sugar.) In solution (−)-sorbose exists as an equilibrium mixture of an open-chain form, **1a,** and a cyclic form, **1b.**

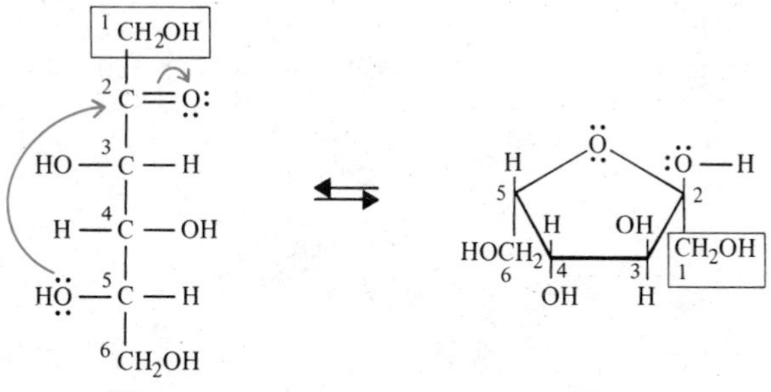

1a
(−)−Sorbose
(open chain form)

1b
(−)−Sorbose
(cyclic hemiketal form)

The synthesis of vitamin C requires that the $—CH_2OH$ group at position 1 of (−)-sorbose be oxidized to a —COOH group. However, direct treatment of (−)-sorbose with $KMnO_4$ (the oxidizing agent) not only causes oxidation of the alcohol group at position 1, it causes oxidation of the other four alcohol groups as well; carbon-carbon bonds break and the whole molecule falls apart.

However, the cyclic form of (−)-sorbose reacts with two moles of acetone (in the presence of a trace of acid) to form a *tricyclic* ketal, **2.** The hydroxyl groups of the cyclic hemiketal form of (−)-sorbose are properly positioned for this reaction to take place. (There are actually three cyclic ketal linkages in **2.** Can you find them?)

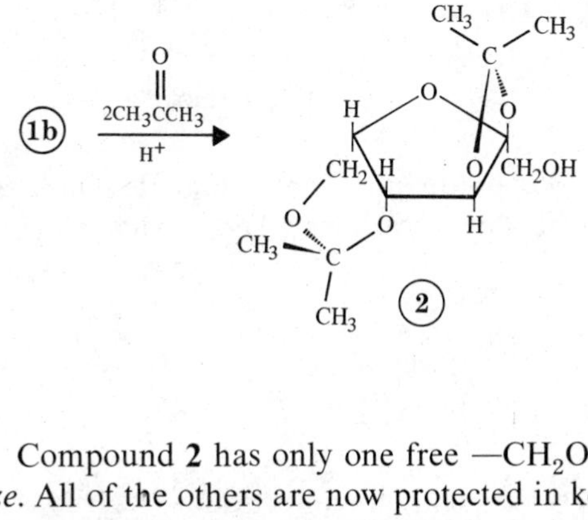

$$\mathbf{1b} \xrightarrow[H^+]{2CH_3COCH_3} \mathbf{2}$$

Compound **2** has only one free $—CH_2OH$ group: *the one that we need to oxidize.* All of the others are now protected in ketal linkages. Oxidation of **2** with potassium permanganate gives **3;** and hydrolysis of **3** with aqueous acid gives **4.**

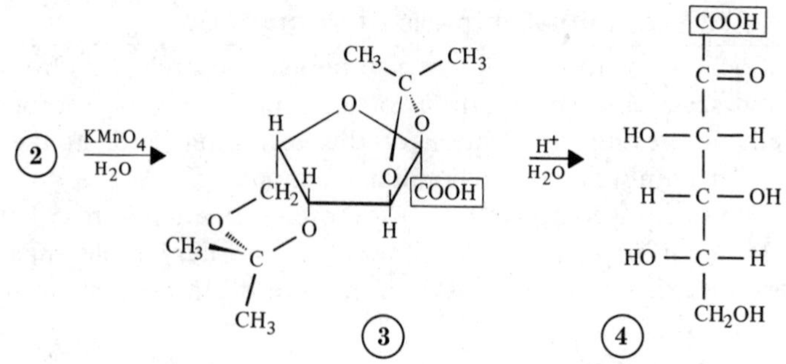

$$\textcircled{2} \xrightarrow[H_2O]{KMnO_4} \textcircled{3} \xrightarrow[H_2O]{H^+} \textcircled{4}$$

The remaining steps are the following. The carboxyl group of **4** reacts with methyl alcohol in the presence of acid to produce a methyl ester, **5.** Compound **5** enolizes when it is treated with base to give **6** and when **6** is heated it cyclizes to vitamin C.

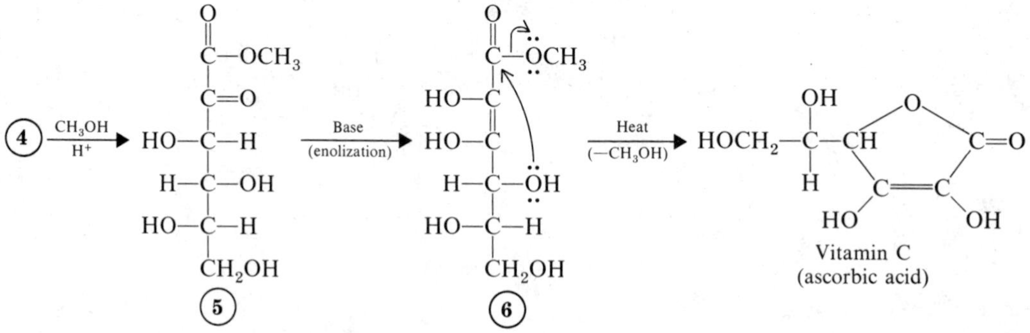

$$\textcircled{4} \xrightarrow[H^+]{CH_3OH} \textcircled{5} \xrightarrow[\text{(enolization)}]{\text{Base}} \textcircled{6} \xrightarrow[(-CH_3OH)]{\text{Heat}} \text{Vitamin C (ascorbic acid)}$$

Vitamin C
(ascorbic acid)

16.7C Thioacetals and Thioketals

Aldehydes and ketones react with thiols to form *thioacetals* and *thioketals.*

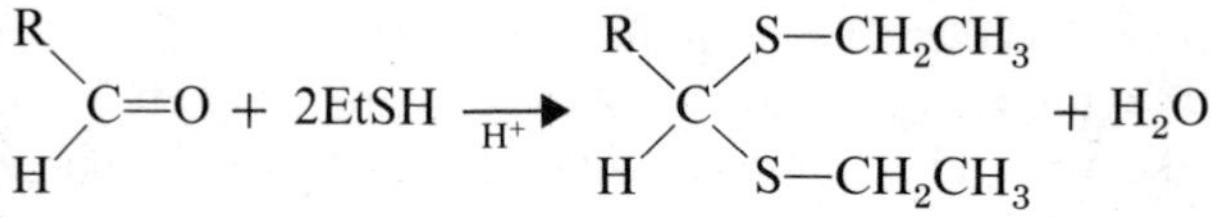

$$RCHO + 2EtSH \xrightarrow{H^+} RCH(S{-}CH_2CH_3)_2 + H_2O$$

Thioacetal

$$R_2C{=}O + HSCH_2CH_2SH \xrightarrow{BF_3} R_2C(SCH_2CH_2S) + H_2O$$

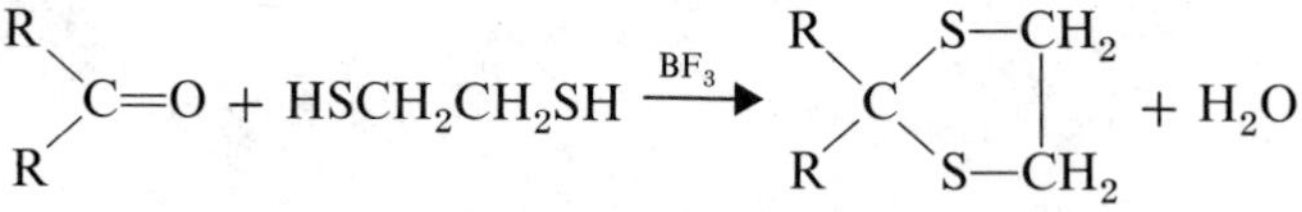

Cyclic thioketal

Thioacetals and thioketals are important in orgnaic synthesis because they react with Raney nickel to yield hydrocarbons.* These reactions, (i.e., thioacetal or

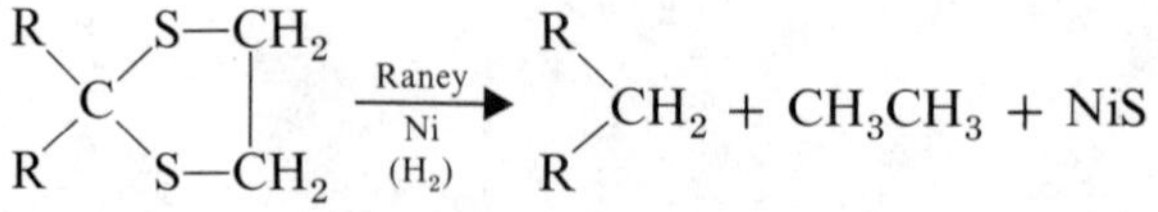

$$R_2C(SCH_2CH_2S) \xrightarrow[\text{Ni } (H_2)]{\text{Raney}} R_2CH_2 + CH_3CH_3 + NiS$$

thioketal formation and subsequent "desulfurization") give us an additional

* Raney nickel is a special nickel catalyst that contains adsorbed hydrogen.

method for converting carbonyl groups of aldehydes and ketones to $—CH_2—$ groups The other method we have studied is the Clemmensen reduction (Section 12.12C). In the next section (16.8C), we will see how this can also be accomplished with the Wolff-Kishner reduction.

Problem 16.9

Show how you might use thioacetal or thioketal formation and Raney nickel desulfurization to convert: (a) cyclohexanone to cyclohexane; (b) $CH_3\overset{\overset{O}{\|}}{C}CH_2CH_2CO_2C_2H_5$ to $CH_3CH_2CH_2CH_2CO_2C_2H_5$.

16.8 THE ADDITION OF DERIVATIVES OF AMMONIA

Aldehydes and ketones react with a number of derivatives of ammonia in the general way shown below.

Addition

$$—\ddot{N}(H)—H + \diagdown C{=}\ddot{O}: \rightleftharpoons —\overset{H}{\underset{H}{N^+}}—\overset{|}{\underset{|}{C}}—\ddot{O}:^- \rightleftharpoons —\overset{H}{\ddot{N}}—\overset{|}{\underset{|}{C}}—\ddot{O}—H \xrightarrow{-H_2O} —\ddot{N}{=}C\diagup\diagdown$$

Elimination

Table 16.3 lists several important examples of this general reaction.

16.8A 2,4-Dinitrophenylhydrazones, Semicarbazones, and Oximes

The products of the reactions of aldehydes and ketones with 2,4-dinitrophenylhydrazine, semicarbazide, and hydroxylamine are often used to identify unknown aldehydes and ketones. These compounds, that is, 2,4-dinitrophenylhydrazones, semicarbazones, and oximes, are usuually relatively insoluble solids that have sharp characteristic melting points. Table 16.4 gives representative examples from the very extensive tables of these derivatives.

16.8B Imines and Enamines

Aldehydes and ketones react with primary amines to form *imines*. (Imines are also called Schiff bases.)

$$\underset{\text{Acetaldehyde}}{CH_3\overset{H}{\overset{|}{C}}{=}O} + \underset{\text{Methylamine}}{H_2\ddot{N}—CH_3} \xrightarrow[Na_2SO_4\ (-H_2O)]{\text{ether}} \underset{\substack{(40\%)\\ \text{Acetaldimine}\\ \text{(an imine)}}}{CH_3\overset{H}{\overset{|}{C}}{=}\ddot{N}CH_3}$$

Imines are important in many biochemical reactions because many enzymes use an $—NH_2$ group of an amino acid to react with an aldehyde or ketone to

TABLE 16.3 Reaction of Aldehydes and Ketones with Derivatives of Ammonia

1. Reactions with Hydroxylamine

General Reaction

$$\rangle C{=}O + H_2N{-}OH \longrightarrow \rangle C{=}N{-}OH + H_2O$$

Aldehyde or ketone — Hydroxylamine — An oxime

Specific Example

$$(CH_3)(H)C{=}O + H_2NOH \xrightarrow[\text{base}]{\text{Weak}} (CH_3)(H)C{=}NOH$$

Acetaldehyde — Acetaldoxime

2. Reactions with Hydrazine, Phenylhydrazine, and 2,4-Dinitrophenylhydrazine

General Reactions

Aldehyde or ketone

$$\rangle C{=}O + H_2NNH_2 \longrightarrow \rangle C{=}NNH_2 + H_2O$$

Hydrazine — A hydrazone

$$\rangle C{=}O + H_2NNHC_6H_5 \longrightarrow \rangle C{=}NNHC_6H_5 + H_2O$$

Phenylhydrazine — A phenylhydrazone

$$\rangle C{=}O + H_2NNH{-}C_6H_3(NO_2)_2 \longrightarrow \rangle C{=}NNH{-}C_6H_3(NO_2)_2 + H_2O$$

2,4-Dinitrophenylhydrazine — A 2,4 dinitrophenylhydrazone

Specific Examples

$$(C_6H_5)(CH_3CH_2)C{=}O + H_2NNH_2 \xrightarrow{\text{Heat}} (C_6H_5)(CH_3CH_2)C{=}NNH_2 + H_2O$$

Propiophenone — Propiophenone hydrazone

$$(C_6H_5)(CH_3)C{=}O + H_2NNHC_6H_5 \xrightarrow[CH_3COOH]{H_3O^+} (C_6H_5)(CH_3)C{=}NNHC_6H_5 + H_2O$$

Acetophenone — Acetophenone phenylhydrazone

$$(C_6H_5)(H)C{=}O + H_2NNH{-}C_6H_3(NO_2)_2 \xrightarrow[C_2H_5OH,\ H_2O]{HCl} (C_6H_5)(H)C{=}NNH{-}C_6H_3(NO_2)_2 + H_2O$$

Benzaldehyde — Benzaldehyde 2,4-dinitrophenylhydrazone

TABLE 16.3 Reaction of Aldehydes and Ketones wit Derivatives of Ammonia (Continued)

3. Reactions with Semicarbazide

General Reaction

$$\text{>C=O} + H_2NNH\overset{O}{\overset{\|}{C}}NH_2 \longrightarrow \text{>C=}NNH\overset{O}{\overset{\|}{C}}NH_2 + H_2O$$

Aldehyde or ketone — Semicarbazide — A semicarbazone

Specific Example

$$\text{Cyclohexanone}{=}O + H_2NNH\overset{O}{\overset{\|}{C}}NH_2 \xrightarrow[C_2H_5OH]{CH_3COO^-K^+} \text{C}_6\text{H}_{10}{=}NNH\overset{O}{\overset{\|}{C}}NH_2 + H_2O$$

Cyclohexanone semicarbazone

form an imine linkage. We saw an example of this in the reaction of rhodopsin with all-*trans*-retinal (Special Topic F).

Imines are also formed as intermediates in a useful synthesis of amines (Section 18.5).

Aldehydes and ketones react with secondary amines to form *enamines* (*ene* + *amine*):

$$CH_3\underset{H}{\overset{CH_3}{C}}HC{=}O + H{-}\ddot{N}(CH_2)_5 \xrightarrow[0^\circ\ (-H_2O)]{K_2CO_3} CH_3\overset{CH_3}{C}{=}CH{-}\ddot{N}(CH_2)_5$$

Piperidine — (75%) An enamine

Enamines are useful synthetic intermediates. We will discuss their chemistry in Section 20.10.

16.8C Hydrazones: The Wolff-Kishner Reduction

Hydrazones are occasionally used to identify aldehydes and ketones. But, unlike 2,4-dinitrophenylhydrazones, the melting points of simple hydrazones are

TABLE 16.4 Derivatives of Aldehydes and Ketones

ALDEHYDE OR KETONE	mp OF 2,4-DINITRO-PHENYLHYDRAZONE	mp OF SEMICARBAZONE	mp OF OXIME
Acetaldehyde	168.5°	162°	46.5°
Acetone	128	187 dec.	61
Benzaldehyde	237	214	35
o-Tolualdehyde	195	208	49
m-Tolualdehyde	211	213	60
p-Tolualdehyde	239	221	79
Phenylacetaldehyde	121	156	103

often low. Hydrazones, however, do form the basis for a useful method to reduce carbonyl groups of aldehydes and ketones to $-CH_2-$ groups, called the *Wolff-Kishner reduction:*

$$\gt C{=}\ddot{O}: + H_2N{-}NH_2 \xrightarrow[\text{heat}]{\text{Base}} [\gt C{=}N{-}NH_2] + H_2O \longrightarrow \gt CH_2 + N_2$$

Aldehyde or ketone; Hydrazone (not isolated)

Specific Example

$$C_6H_5{-}\overset{O}{\overset{\|}{C}}CH_2CH_3 + H_2NNH_2 \xrightarrow[\text{Triethylene glycol, } 200^\circ]{\text{NaOH}} C_6H_5{-}CH_2CH_2CH_3 \quad (82\%)$$

The Wolff-Kishner reduction complements the Clemmensen reduction (Section 12.12C) and the reduction of thioacetals, (Section 16.7C) because all three reactions convert $\gt C{=}O$ groups into $-CH_2-$ groups. The Clemmensen reduction takes place in strongly acidic media and can be used for those compounds that are sensitive to base. The Wolff-Kishner reduction takes place in strongly basic solutions and can be used for those compounds that are sensitive to acid. The reduction of thioacetals takes place in neutral solution and can be used for compounds that are sensitive to both acids and bases.

The mechanism of the Wolff-Kishner reduction is outlined below. The first step is the formation of the hydrazone. Then the strong base brings about an isomerization of the hydrazone to a derivative with the structure $\gt CH{-}N{=}NH$. This derivative then undergoes the base-catalyzed elimination of a molecule of nitrogen. The loss of the especially stable molecule of nitrogen provides the driving force for the reaction.

1. $$-\overset{|}{C}{=}O + H_2N{-}NH_2 \rightleftharpoons -\overset{|}{C}{=}N{-}NH_2 + H_2O$$

2. $$-\overset{|}{C}{=}N{-}NH_2 \underset{H_2O}{\overset{OH^-}{\rightleftharpoons}} \left[-\overset{|}{C}{=}N{-}\ddot{N}H^- \longleftrightarrow -\overset{..}{\underset{|}{C}}{}^-{-}N{=}NH\right] \underset{OH^-}{\overset{H_2O}{\rightleftharpoons}}$$

$$-\overset{H}{\underset{|}{\overset{|}{C}}}{-}N{=}N{-}H \underset{H_2O}{\overset{OH^-}{\rightleftharpoons}} -\overset{H}{\underset{|}{\overset{|}{C}}}{-}N{=}\ddot{N}:^- \xrightarrow{-N_2}$$

$$-\overset{H}{\underset{|}{\overset{|}{C}}}:^- \xrightarrow{H_2O} -\overset{H}{\underset{|}{\overset{|}{C}}}{-}H$$

16.9A Hydrogen Cyanide Addition

Hydrogen cyanide adds to the carbonyl groups of aldehydes and most ketones to form compounds called *cyanohydrins*. Ketones in which the carbonyl group is highly hindered do not undergo this reaction.

$$\text{RCH(=O)} + \text{HCN} \rightleftharpoons \text{R(H)C(OH)(CN)}$$

$$\text{R—C(=O)—R}' + \text{HCN} \rightleftharpoons \text{R(R}'\text{)C(OH)(CN)}$$

Cyanohydrins

Hydrogen cyanide is a very weak acid but the cyanide ion is a strong nucleophile. Therefore, cyanohydrin formation is initiated by a nucleophilic attack by the cyanide ion on the carbonyl carbon.

$$\text{:}\ddot{\text{O}}\text{=C} + \text{:}\bar{\text{C}}\equiv\text{N:} \rightleftharpoons \text{—C(—}\ddot{\text{O}}\text{:}^-\text{)—C}\equiv\text{N:} \underset{-\text{H}^+}{\overset{+\text{H}^+}{\rightleftharpoons}} \text{—C(—}\ddot{\text{O}}\text{—H)—C}\equiv\text{N:}$$

Liquid hydrogen cyanide can be used for this reaction, but since HCN is very toxic and volatile, it is safer to generate it in the reaction mixture. This can be done by mixing the aldehyde or ketone with aqueous sodium cyanide and then slowly adding sulfuric acid slowly to the mixture. *Even with this procedure, however, great care must be taken and the reaction must be carried out in a very efficient fume hood.*

Cyanohydrins are useful intermediates in organic synthesis. Depending on the conditions used, acidic hydrolysis converts cyanohydrins to α-hydroxy acids or to α,β-unsaturated acids:

$$CH_3CH_2\text{—C(=O)—}CH_3 \xrightarrow{\text{HCN}} CH_3CH_2\text{—C(OH)(}CH_3\text{)—CN} \xrightarrow[\text{heat}]{\text{HCl, } H_2O} CH_3CH_2\text{—C(HO)(}CH_3\text{)—C(=O)OH}$$

α-Hydroxy acid

$$CH_3CH_2\text{—C(OH)(}CH_3\text{)—CN} \xrightarrow[\text{heat}]{\text{conc. } H_2SO_4} CH_3CH\text{=C(}CH_3\text{)—C(=O)OH}$$

α,β-Unsaturated acid

Treating acetone cyanohydrin with acid and methanol converts it to methyl methacrylate. Methyl methacrylate is the starting material for the synthesis of the polymer known as Plexiglas or Lucite (Special Topic C).

$$\underset{\text{Acetone}}{CH_3-\overset{O}{\overset{\|}{C}}-CH_3} \xrightarrow{HCN} \underset{\substack{\text{Acetone cyanohydrin} \\ (78\%)}}{CH_3-\underset{CH_3}{\overset{OH}{C}}-CN} \xrightarrow[CH_3OH]{H_2SO_4} \underset{\substack{\text{Methyl methacrylate} \\ (90\%)}}{CH_2=\underset{CH_3}{C}-\overset{O}{\overset{\|}{C}}OCH_3}$$

Reducing a cyanohydrin with lithium aluminum hydride gives a β-amino alcohol:

$$\text{Cyclohexanone} \xrightarrow{HCN} \text{1-hydroxycyclohexane-CN (OH, CN)} \xrightarrow{LiAlH_4} \text{1-hydroxycyclohexane-}CH_2NH_2 \text{ (OH, } CH_2NH_2)$$

Problem 16.10

(a) Show how you might prepare lactic acid ($CH_3CHOHCOOH$) from acetaldehyde through a cyanohydrin intermediate. (b) What stereoisomeric form of lactic acid would you expect to obtain?

16.9B Sodium Bisulfite Addition

Sodium bisulfite ($NaHSO_3$) adds to a carbonyl group in a way that is very similar to the addition of HCN.

$$\overset{O}{\overset{\|}{C}} \quad \underset{Na^+}{:\overset{-}{S}O_3H} \rightleftharpoons -\overset{:\ddot{O}:^- \; Na^+}{\underset{|}{\overset{|}{C}}}-SO_3H \rightleftharpoons \underset{\substack{\text{Bisulfite addition} \\ \text{product}}}{-\overset{:\ddot{O}-H}{\underset{|}{\overset{|}{C}}}-SO_3^- \; Na^+}$$

This reaction takes place with aldehydes and some ketones. Most aldehydes react with one equivalent of sodium bisulfite to give the addition product in 70 to 90% yield. Under the same conditions, methyl ketones give yields varying from 12 to 56%. Most higher ketones do not give bisulfite addition products in appreciable amounts because the addition is very sensitive to steric hindrance. However, since the reaction involves an equilibrium, yields from aldehydes and methyl ketones can be improved by using an excess of sodium bisulfite.

Because bisulfite addition compounds are crystalline salts, a bisulfite addition reaction is often used in separating aldehydes and methyl ketones from other substances. Since bisulfite addition is reversible, the aldehyde or methyl ketone can be regenerated, after a separation has been made, by adding either an acid or a base.

16.10A The Wittig Reaction

Aldehydes and ketones react with phosphorus ylides to yield *alkenes* and triphenylphosphine oxide. [An ylide (pronounced ill/id) is a neutral resonance hybrid having one contributor with a negative carbon adjacent to a positive heteroatom.]

$$\underset{\text{Aldehyde or ketone}}{RR'C{=}O} + \underset{\text{Phosphorus ylide}}{R''R'''C{=}P(C_6H_5)_3 \longleftrightarrow R''R'''\ddot{C}^{-}{-}\overset{+}{P}(C_6H_5)_3} \longrightarrow \underset{\text{Alkene}}{RR'C{=}CR''R'''} + \underset{\text{Triphenylphosphine oxide}}{O{=}P(C_6H_5)_3}$$

This reaction, known as the *Wittig reaction**, has proved to be a valuable method for synthesizing alkenes. The Wittig reaction is applicable to a wide variety of compounds and it gives a great advantage over most other alkene syntheses in that no ambiguity exists as to the location of the double bond in the product.

Phosphorus ylides are easily prepared from triphenylphosphine and alkyl halides. Their preparation involves two reactions:

General Reaction

1. $$\underset{\text{Triphenylphosphine}}{(C_6H_5)_3P\!:} + R''R'''CH{-}X \longrightarrow \underset{\text{Alkyltriphenylphosphonium halide}}{(C_6H_5)_3\overset{+}{P}{-}CHR''R'''\ X^-}$$

2. $$(C_6H_5)_3\overset{+}{P}{-}\overset{-}{C}R''R'''{-}H \ {:}B \longrightarrow \underset{\text{Phosphorus ylide}}{(C_6H_5)_3P{=}CR''R'''} + H{:}B$$

Specific Example

1. $$(C_6H_5)_3P\!: + CH_3Br \xrightarrow{C_6H_6} \underset{\text{Methyltriphenylphosphonium bromide}}{\underset{(89\%)}{(C_6H_5)_3\overset{+}{P}{-}CH_3\ Br^-}}$$

2. $$\underset{Br^-}{(C_6H_5)_3\overset{+}{P}{-}CH_3} + \underset{Na^+}{\overset{-}{:}CH_2SOCH_3} \xrightarrow{CH_3SOCH_3} (C_6H_5)_3P{=}CH_2 + CH_3SOCH_3 + NaBr$$

* Discovered in 1954 by Professor Georg Wittig, then at the University of Tübingen. Wittig was a cowinner of the Nobel prize for chemistry in 1979.

The first reaction is a nucleophilic substitution reaction. Triphenylphosphine acts as a nucleophile and displaces a halide ion from the alkyl halide to give an alkyltriphenylphosphonium salt. The second reaction is an acid-base reaction. A strong base (usually an alkyllithium, or $Na^+\bar{C}H_2SOCH_3$) removes a proton from the carbon that is attached to phosphorus to give the ylide.

Phosphorus ylides can be represented as a hybrid of the two resonance structures shown below. The contribution made to the hybrid by the second

$$(C_6H_5)_3P{=}C(R'')(R''') \longleftrightarrow (C_6H_5)_3\overset{+}{P}{-}\overset{-}{C}{:}(R'')(R''')$$

structure explains the reaction of the ylide with an aldehyde or ketone because the ylide acts as a nucleophile—in effect as a carbanion—and attacks the carbonyl carbon. This step yields an intermediate with separated charges called a *betaine*. The betaine then loses triphenylphosphine oxide (often spontaneously) to give the alkene.

General Mechanism

$$RR'C{=}\ddot{O}{:} + {:}\overset{-}{C}R''R'''{-}\overset{+}{P}(C_6H_5)_3 \rightleftharpoons R'{-}CR({:}\ddot{O}{:}^-){-}CR''({\overset{+}{P}}(C_6H_5)_3){-}R'''$$

$$\updownarrow \quad R''R'''C{=}P(C_6H_5)_3$$

A betaine

$$RR'C{=}CR''R''' + {:}\ddot{O}{=}P(C_6H_5)_3 \longleftarrow R'{-}CR({:}\ddot{O}){-}CR''(P(C_6H_5)_3){-}R'''$$

Specific Example

$$C_6H_{10}{=}\ddot{O}{:} + {:}\overset{-}{C}H_2{-}\overset{+}{P}(C_6H_5)_3 \rightleftharpoons C_6H_{10}({:}\ddot{O}{:}^-){-}CH_2{-}\overset{+}{P}(C_6H_5)_3$$

$$\updownarrow \quad CH_2{=}P(C_6H_5)_3$$

$$C_6H_{10}{=}CH_2 + O{=}P(C_6H_5)_3 \longleftarrow C_6H_{10}({:}\ddot{O}){-}CH_2{-}P(C_6H_5)_3$$

Methylenecyclohexane
(86% from cyclohexanone
and methyltriphenylphosphonium
bromide)

The elimination of triphenylphosphine oxide from the betaine may occur in two separate steps as we have shown above, or both steps may occur simultaneously.

While Wittig syntheses may appear to be complicated, in actual practice they are easy to carry out. Most of the steps can be carried out in the same reaction vessel and the entire synthesis can be accomplished in a matter of hours.

The overall result of a Wittig synthesis is:

$$\begin{matrix}R \\ R'\end{matrix}\!\!>C{=}O + \begin{matrix}X \\ H\end{matrix}\!\!>C<\!\!\begin{matrix}R'' \\ R'''\end{matrix} \xrightarrow[\text{steps}]{\text{Several}} \begin{matrix}R \\ R'\end{matrix}\!\!>C{=}C<\!\!\begin{matrix}R'' \\ R'''\end{matrix}$$

Planning a Wittig synthesis begins with recognizing in the desired alkene what can be the aldehyde or ketone component and what can be the halide component. Any or all of the R groups may be hydrogen. The halide must be 1°, 2°, or methyl.

The Wittig reaction is most often employed to introduce a $=CH_2$ group into a compound using $CH_2{=}P(C_6H_5)_3$ as the ylide. However, the Wittig reaction can be used in other ways as the following examples show. Methyl and primary halides give higher yields of alkyltriphenylphosphonium salts (the first step) than secondary halides (why?). The last step of a Wittig synthesis can lead to a mixture of *cis* and *trans* isomers.

$$C_6H_5\overset{\overset{O}{\|}}{C}H + (\text{cyclopentylidene}){=}P(C_6H_5)_3 \longrightarrow C_6H_5CH{=}(\text{cyclopentylidene})$$

(65%)

$$CH_3\overset{\overset{O}{\|}}{C}H + CH_3(CH_2)_4CH{=}P(C_6H_5)_3 \longrightarrow CH_3CH{=}CH(CH_2)_4CH_3 + (C_6H_5)_3P{=}O$$

(69%)
(*cis* and *trans*)

Problem 16.11

Starting with triphenylphosphine and the appropriate alkyl halide, show how you would prepare each of the ylides just illustrated.

Problem 16.12

In addition to triphenylphosphine assume that you have available as starting materials any necessary aldehydes, ketones, and organic halides. Show how you might synthesize each of the following alkenes using the Wittig reaction.

(a) $C_6H_5CH{=}C(CH_3){-}CH_3$

(b) $C_6H_5C(CH_3){=}CH_2$

(c) $C_6H_5C(CH_3){=}CHCH_3$

(d) $(CH_3)_2C{=}CH_2$

(e) methylenecyclopentane ($CH_2{=}$cyclopentane)

(f) $CH_3CH_2CH{=}C(CH_3)CH_2CH_3$

(g) $C_6H_5CH{=}CHCH{=}CH_2$

(h) $C_6H_5CH{=}CHC_6H_5$

Problem 16.13

Triphenylphosphine can be used to convert epoxides to alkenes, for example,

$$C_6H_5CH\overset{O}{—}CHCH_3 + (C_6H_5)_3P: \longrightarrow C_6H_5CH{=}CHCH_3 + (C_6H_5)_3PO$$

Propose a likely mechanism for this reaction.

Problem 16.14

The Wittig reaction can be used in the synthesis of aldehydes, for example,

$$CH_3O{-}C_6H_4{-}\overset{O}{\overset{\|}{C}}CH_3 + CH_3OCH{=}P(C_6H_5)_3 \longrightarrow CH_3O{-}C_6H_4{-}\overset{CH_3}{\overset{|}{C}}{=}CHOCH_3 \quad (60\%)$$

$$\xrightarrow{H_3O^+/H_2O} CH_3O{-}C_6H_4{-}\overset{CH_3}{\overset{|}{C}H}{-}\underset{\underset{O}{\|}}{CH} \quad (85\%)$$

(a) How would you prepare $CH_3OCH{=}P(C_6H_5)_3$? (b) Why does the second reaction (above) yield an aldehyde? (c) How would you use this method to prepare cyclohexyl–CHO from cyclohexanone?

16.10B The Addition of Sulfur Ylides

Sulfur ylides also react as nucleophiles at the carbonyl carbon of aldehydes and ketones. The betaine that forms usually decomposes to an *epoxide* rather than to an alkene.

$$(CH_3)_2\ddot{S}: + CH_3I \longrightarrow (CH_3)_2\overset{+}{S}{-}CH_3 \; I^- \xrightarrow[CH_3SOCH_3,\ 0°]{Na^+\bar{C}H_2SOCH_3} [(CH_3)_2\overset{+}{S}{-}\overset{-}{\ddot{C}}H_2 \longleftrightarrow (CH_3)_2S{=}CH_2]$$

Trimethylsulfonium iodide

Resonance-stabilized sulfur ylide

$$C_6H_5{-}\overset{:\ddot{O}}{\overset{\|}{C}H} + :\overset{-}{C}H_2\overset{+}{S}(CH_3)_2 \longrightarrow C_6H_5{-}\overset{:\ddot{O}:^-}{\overset{|}{C}H}{-}CH_2{-}\overset{+}{S}(CH_3)_2 \longrightarrow C_6H_5{-}\overset{O}{CH{-}CH_2} + :\ddot{S}(CH_3)_2 \quad (75\%)$$

Benzaldehyde

Sulfur ylides are also discussed in Special Topic J.

Problem 16.15

Show how you might use a sulfur ylide to prepare

(a) 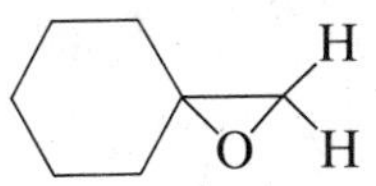 cyclohexane spiro-epoxide (O, H, H)

(b) 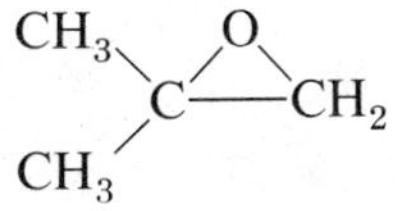 $(CH_3)_2C$—CH_2 epoxide (O)

16.11 REACTIONS INVOLVING THE α-POSITION OF ALDEHYDES AND KETONES: KETO-ENOL TAUTOMERISM

16.11A The Acidity of the α-Hydrogen of Carbonyl Compounds

In addition to a propensity toward nucleophilic attack at the carbonyl carbon a second characteristic of carbonyl compounds is an unusual acidity of hydrogens on carbons adjacent to the carbonyl group. (These hydrogens are usually called the α-hydrogens, and the carbon to which they are attached is called the α-carbon.)

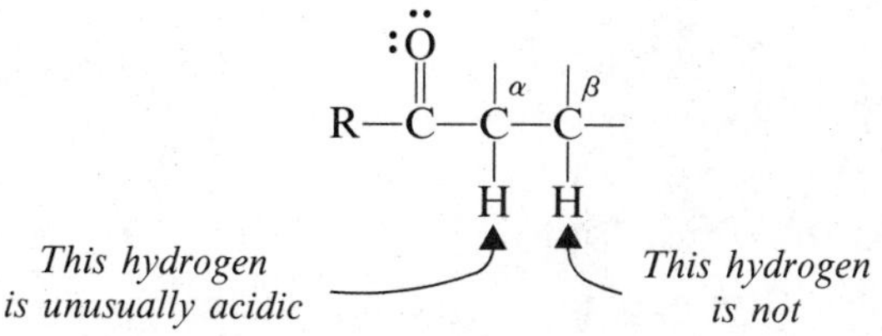

$$\mathrm{R{-}\overset{\displaystyle :\ddot{O}}{\overset{\|}{C}}{-}\underset{H}{\underset{|}{\overset{|\,\alpha}{C}}}{-}\underset{H}{\underset{|}{\overset{|\,\beta}{C}}}{-}}$$

This hydrogen is unusually acidic (α-H) *This hydrogen is not* (β-H)

When we say that the α-hydrogens are acidic, *we mean that they are unusually acidic for hydrogens attached to carbon.* The acidity constants (K_a's) for the α-hydrogens of most simple aldehydes or ketones are of the order of 10^{-19}–10^{-20}. This means that they are more acidic than hydrogens of acetylene ($K_a = 10^{-25}$) and are far more acidic than the hydrogens of ethene ($K_a = 10^{-36}$) or of ethane ($K_a = 10^{-42}$).

The reason for the unusual acidity of the α-hydrogens of carbonyl compounds is straightforward: when a carbonyl compound loses an α-proton the anion that is produced is *stabilized by resonance.*

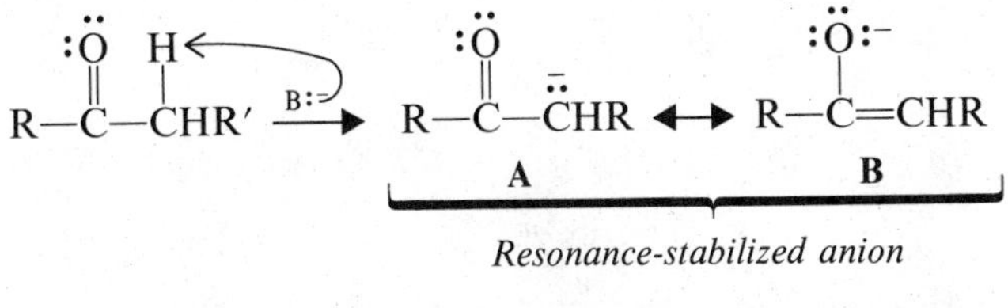

$$\mathrm{R{-}\overset{\displaystyle :\ddot{O}}{\overset{\|}{C}}{-}\overset{H}{\overset{|}{C}}HR'} \xrightarrow{B:^-} \underbrace{\underset{\mathbf{A}}{\mathrm{R{-}\overset{\displaystyle :\ddot{O}}{\overset{\|}{C}}{-}\ddot{\bar{C}}HR}} \longleftrightarrow \underset{\mathbf{B}}{\mathrm{R{-}\overset{\displaystyle :\ddot{O}:^-}{\overset{|}{C}}{=}CHR}}}_{\textit{Resonance-stabilized anion}}$$

Resonance-stabilized anion

We see (above) that two resonance structures, **A** and **B,** can be written for the anion. In structure **A** the negative charge is on carbon and in structure **B** the negative charge is on oxygen. Both structures contribute to the hybrid. **A** is favored by the greater strength of its carbon-oxygen π bond relative to the weaker carbon-carbon π bond of **B**. However, we would expect structure **B** to make a substantial contribution to the hybrid because oxygen, being highly electronegative, is better able to accommodate the negative charge. We can depict the hybrid in the following way.

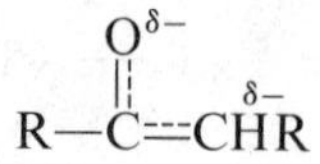

$$\mathrm{R{-}\overset{\overset{\displaystyle O^{\delta-}}{\|}}{C}{=\!=}\overset{\delta-}{C}HR}$$

When this resonance-stabilized anion accepts a proton, it can do so in either of two ways: it can accept the proton at carbon to form the original carbonyl compound (in the *keto* form), or it may accept the proton at oxygen to form an *enol.*

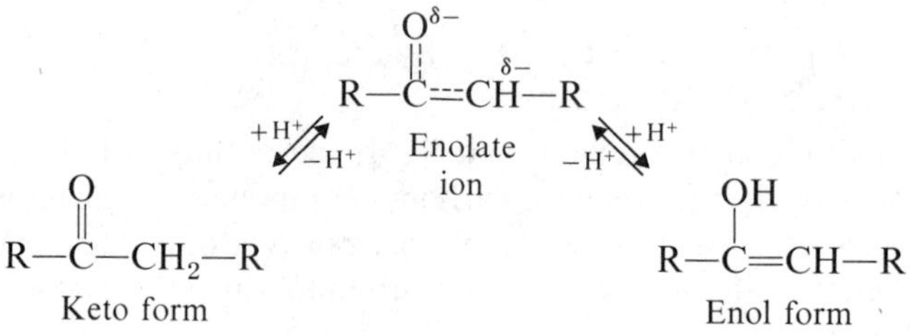

$$\mathrm{R{-}\overset{\overset{\displaystyle O}{\|}}{C}{-}CH_2{-}R}\ \underset{-H^+}{\overset{+H^+}{\rightleftharpoons}}\ \underset{\text{Enolate ion}}{\mathrm{R{-}\overset{\overset{\displaystyle O^{\delta-}}{\|}}{C}{=\!=}\overset{\delta-}{C}H{-}R}}\ \underset{-H^+}{\overset{+H^+}{\rightleftharpoons}}\ \mathrm{R{-}\overset{\overset{\displaystyle OH}{|}}{C}{=}CH{-}R}$$

Keto form — Enol form

Both of these reactions are reversible. Because of its relation to the enol, the resonance stabilized anion is often called an *enolate ion.*

16.11B Keto and Enol Tautomers

The keto and enol forms of carbonyl compounds are structural isomers, but of a special type. Because they are easily interconverted in the presence of traces of acids and bases, chemists use a special term to describe this type of structural isomerism. Interconvertible keto and enol forms are said to be *tautomers,* and their interconversion is called *tautomerization.*

Under most circumstances, we encounter keto-enol tautomers in a state of equilibrium. (The surfaces of ordinary laboratory glassware are able to catalyze the interconversion and establish the equilibrium.) For simple monocarbonyl compounds such as acetone and acetaldehyde, the amount of the enol form present at equilibrium is *very small.* In acetone it is much less than 1%; in acetaldehyde the enol concentration is too small to be detected. The greater stability of the keto forms of monocarbonyl compounds (below) can be related to the greater strength of the carbon-oxygen π bond compared to the carbon-carbon π bond (~87 kcal/mole versus ~60 kcal/mole).

	Keto form		Enol form
Acetaldehyde:	$\mathrm{CH_3\overset{\overset{\displaystyle O}{\|}}{C}H}$ (~100%)	$\rightleftharpoons$	$\mathrm{CH_2{=}\overset{\overset{\displaystyle OH}{|}}{C}H}$ (extremely small)
Acetone:	$\mathrm{CH_3\overset{\overset{\displaystyle O}{\|}}{C}CH_3}$ (>99%)	$\rightleftharpoons$	$\mathrm{CH_2{=}\overset{\overset{\displaystyle OH}{|}}{C}CH_3}$ (1.5×10^{-4})

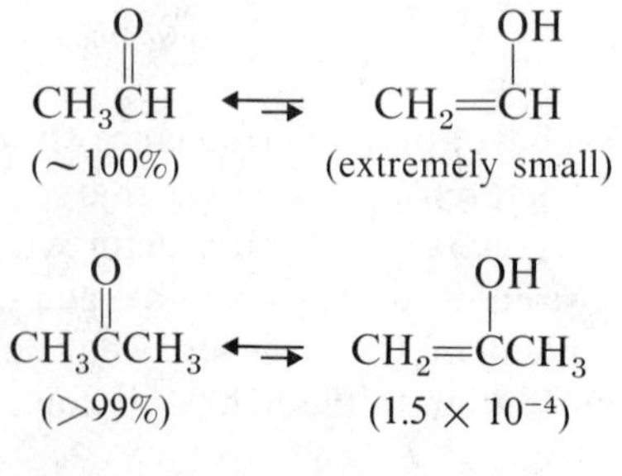

Cyclohexanone: (98.8%) ⇌ (1.2%)

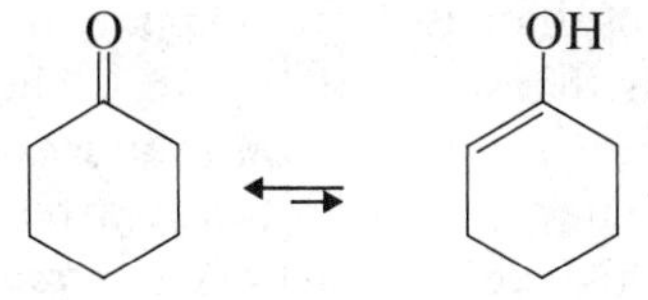

In compounds whose molecules have two carbonyl groups separated by one saturated carbon (called β-dicarbonyl compounds), the amount of enol present at equilibrium is far higher. For example, 2,4-pentanedione exists in the enol form to an extent of 76%.

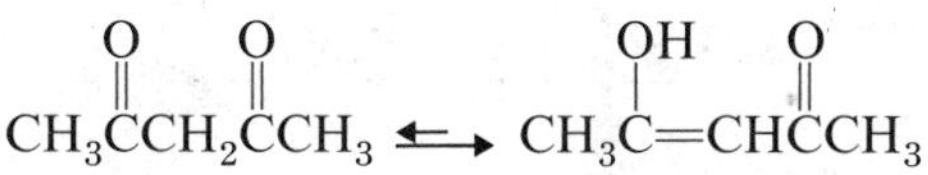

$$CH_3\overset{O}{\overset{\|}{C}}CH_2\overset{O}{\overset{\|}{C}}CH_3 \rightleftharpoons CH_3\overset{OH}{\overset{|}{C}}{=}CH\overset{O}{\overset{\|}{C}}CH_3$$

(24%) 2,4-Pentanedione ⇌ (76%) Enol form

The greater stability of the enol form of β-dicarbonyl compounds can be attributed to stability gained through internal hydrogen bonding in a cyclic form and to resonance:

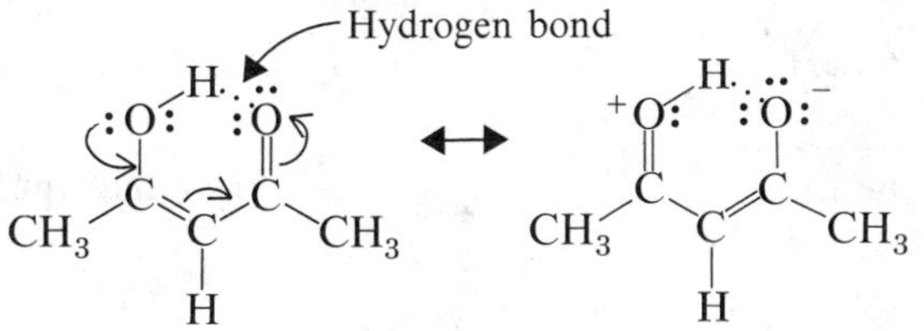

Hydrogen bond

CH_3, H, CH_3 ⟷ CH_3, H, CH_3

16.11C Enols in Biochemistry

An important reaction in the synthesis and degradation of sugars is one in which a three-carbon *ketone* and a three-carbon *aldehyde* are interconverted.

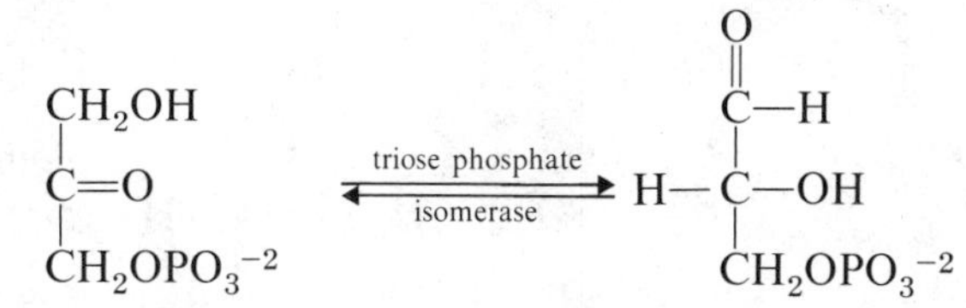

$$\begin{array}{c} CH_2OH \\ | \\ C{=}O \\ | \\ CH_2OPO_3^{-2} \end{array} \xrightleftharpoons[\text{isomerase}]{\text{triose phosphate}} \begin{array}{c} O \\ \| \\ C{-}H \\ | \\ H{-}C{-}OH \\ | \\ CH_2OPO_3^{-2} \end{array}$$

Dihydroxyacetone phosphate ⇌ (*R*)-Glyceraldehyde 3-phosphate

This remarkable reaction is catalyzed by an enzyme called *triose phosphate isomerase*. Although all the details of the enzymatic reaction are not fully understood, there is substantial evidence that the interconversion proceeds through the enol intermediate (actually an *enediol*) shown below:

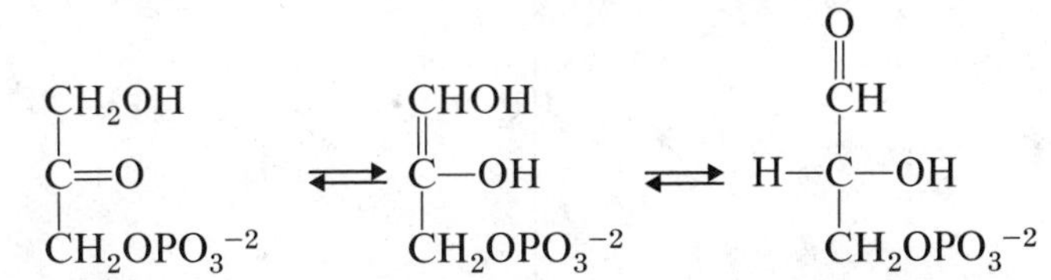

$$\begin{array}{c} CH_2OH \\ | \\ C{=}O \\ | \\ CH_2OPO_3^{-2} \end{array} \rightleftharpoons \begin{array}{c} CHOH \\ \| \\ C{-}OH \\ | \\ CH_2OPO_3^{-2} \end{array} \rightleftharpoons \begin{array}{c} O \\ \| \\ CH \\ | \\ H{-}C{-}OH \\ | \\ CH_2OPO_3^{-2} \end{array}$$

Dihydroxyacetone phosphate ⇌ "Enediol" 3-phosphate ⇌ (*R*)-Glyceraldehyde 3-phosphate

Triose phosphate isomerase apparently binds the enediol 3-phosphate in a manner similar to that shown in Fig. 16.1. Then, if the enediol 3-phosphate accepts a proton at carbon-1, the reaction produces dihydroxyacetone phosphate. If the enediol 3-phosphate accepts a proton at carbon-2, the product is (*R*)-glyceraldehyde 3-phosphate. (This overall reaction is *stereoselective*. Achiral dihydroxyacetone phosphate interconverts only with the *R*-enantiomer of glyceraldehyde 3-phosphate.)

We will see one biochemical fate of dihydroxyacetone phosphate and (*R*)-glyceraldehyde 3-phosphate in Section 16.13E.

16.12 REACTIONS VIA ENOLS AND ENOLATE IONS

16.12A Racemization

When a solution of (+)-*sec*-butyl phenyl ketone (below) in aqueous ethanol is treated with either acids or bases the solution gradually loses its optical activity. After a time, isolation of the ketone shows that it is a racemate.

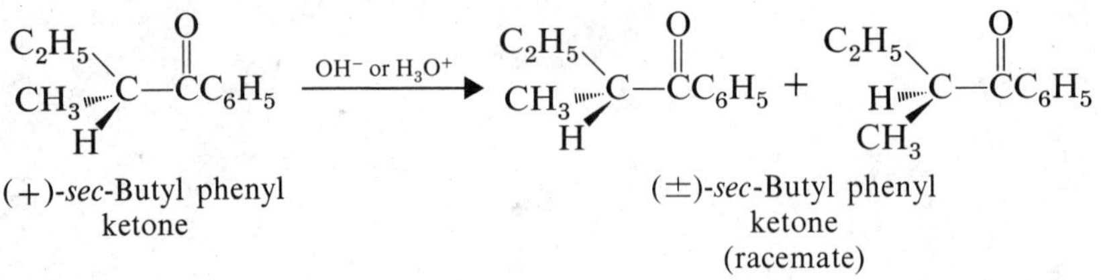

$$C_2H_5(CH_3)(H)C{-}C({=}O)C_6H_5 \xrightarrow{OH^- \text{ or } H_3O^+} C_2H_5(CH_3)(H)C{-}C({=}O)C_6H_5 + C_2H_5(H)(CH_3)C{-}C({=}O)C_6H_5$$

(+)-*sec*-Butyl phenyl ketone → (±)-*sec*-Butyl phenyl ketone (racemate)

FIG. 16.1
A proposed mechanism for triose phosphate isomerase interconversion of dihydroxyacetone phosphate and R*-glyceraldehyde 3-phosphate. (Colored arrows show the electron flow when protonation occurs at carbon-1.)*

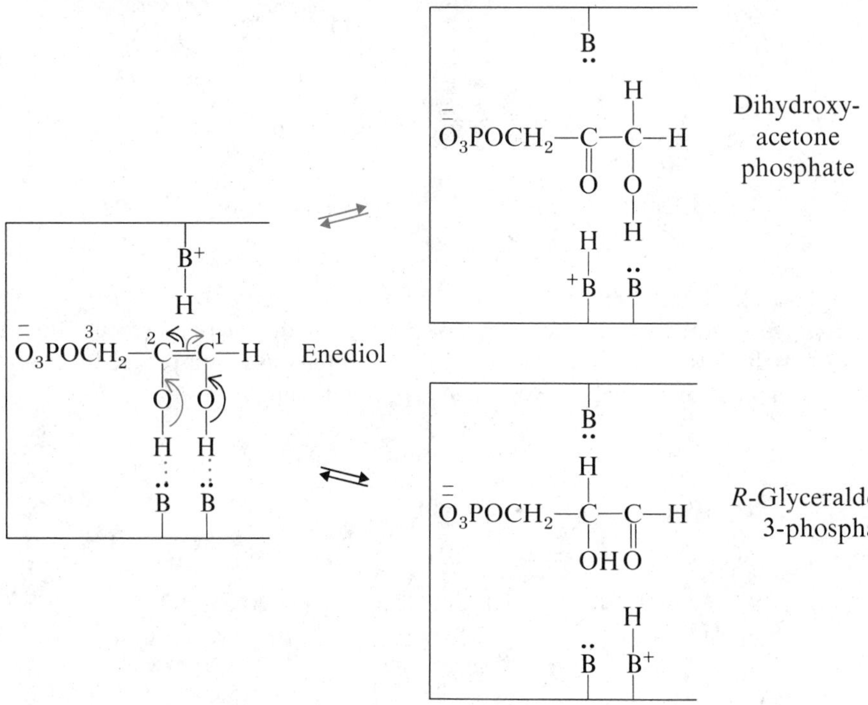

Racemization takes place in the presence of acids or bases because the ketone slowly but reversibly changes to its enol *and the enol is achiral.* When the enol reverts to the keto form it produces equal amounts of the two enantiomers.

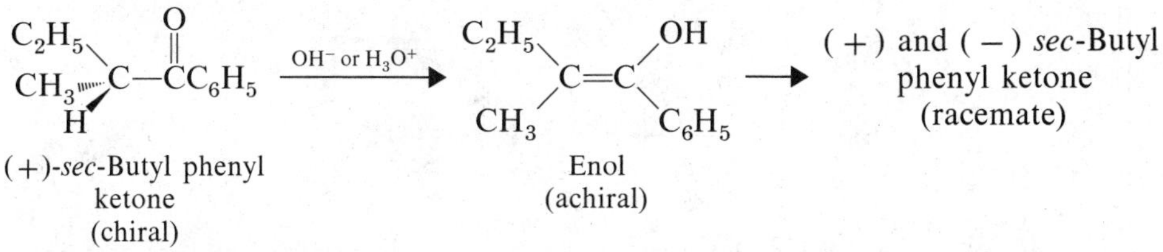

C_2H_5, CH_3, H–C–C(=O)C_6H_5 —(OH^- or H_3O^+)→ (C_2H_5)(CH_3)C=C(OH)(C_6H_5) → (+) and (−) *sec*-Butyl phenyl ketone (racemate)

(+)-*sec*-Butyl phenyl ketone (chiral) — Enol (achiral)

Base catalyzes the formation of an enol through the intermediate formation of an enolate ion:

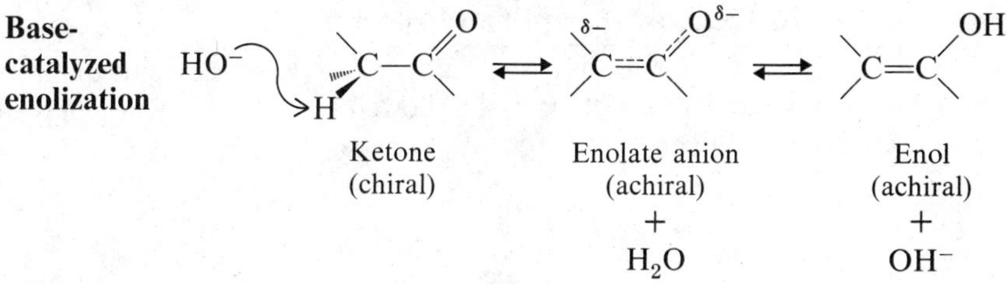

Base-catalyzed enolization

HO^- + H–C–C=O ⇌ C=C–$O^{\delta-}$ ⇌ C=C–OH

Ketone (chiral) — Enolate anion (achiral) + H_2O — Enol (achiral) + OH^-

Acid can catalyze enolization in the following way:

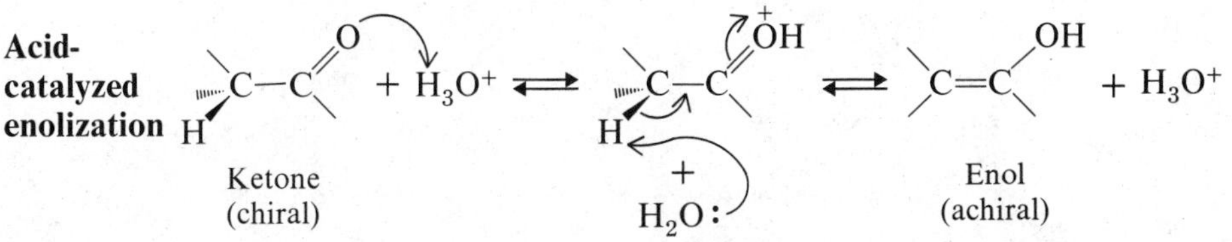

Acid-catalyzed enolization

H–C–C=O + H_3O^+ ⇌ H–C–C=$\overset{+}{O}$H + H_2O: ⇌ C=C–OH + H_3O^+

Ketone (chiral) — Enol (achiral)

Problem 16.16

Would you expect optically active ketones such as those shown below to undergo acid- or base-catalyzed racemization? Explain your answer.

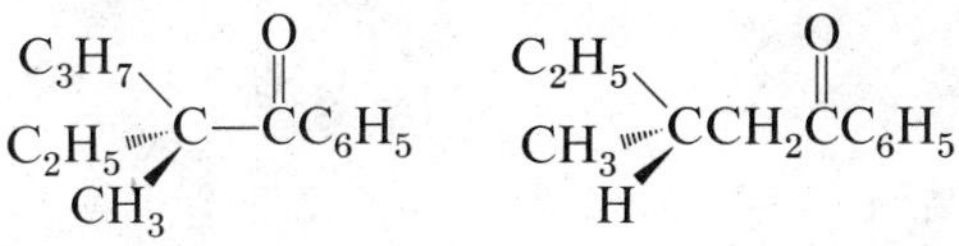

(C_3H_7)(C_2H_5)(CH_3)C–C(=O)C_6H_5 (C_2H_5)(CH_3)(H)CCH_2C(=O)C_6H_5

Problem 16.17

When *sec*-butyl phenyl ketone is treated with either OD^- or D_3O^+ in the presence of D_2O, the ketone undergoes hydrogen-deuterium exchange and produces:

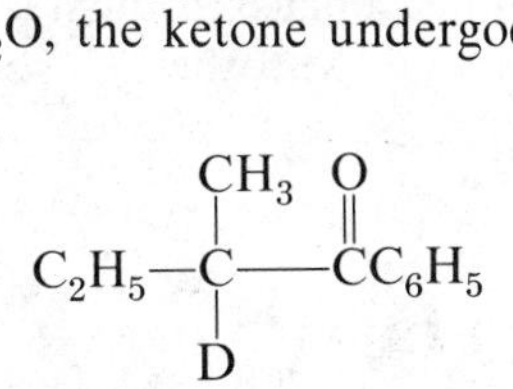

C_2H_5–C(CH_3)(D)–C(=O)C_6H_5

Write mechanisms that account for this.

16.12B Halogenation of Ketones

Ketones that have an α-hydrogen react readily with halogens by substitution. The rates of these halogenation reactions *increase when acids or bases are added and substitution takes place almost exclusively at the α-carbon:*

$$-\overset{\text{H}}{\underset{|}{\overset{|}{\text{C}}}}-\overset{\text{O}}{\overset{\|}{\text{C}}}- + X_2 \xrightarrow[\text{or base}]{\text{Acid}} -\overset{\text{X}}{\underset{|}{\overset{|}{\text{C}}}}-\overset{\text{O}}{\overset{\|}{\text{C}}}- + HX$$

This behavior of ketones can be accounted for in terms of two related properties that we have already encountered: the acidity of the α-hydrogens of ketones and the tendency of ketones to form enols.

16.12C Base-Promoted Halogenation

In the presence of bases, halogenation takes place through the slow formation of an enolate ion or an enol, followed by a rapid reaction of the enolate ion or enol with halogen.

1. B:⁻ + H–C–C(=O) ⇌ (Slow) B:H + Enolate ion [C⩵C–O$^{\delta-}$, $\delta-$ on C] ⇌ (Fast) Enol [C=C–OH]

2a. X–X + ⁻:C–C(=O) → (Fast) X–C–C(=O) + X⁻

↕

Enolate ion [C=C–O:⁻]

or

2b. X–X + C=C–O–H (Enol) ⇌ (Fast) X–C–C(=O) + X⁻

16.12D Acid-Catalyzed Halogenation

In the presence of acids, halogenation takes place through slow formation of an enol followed by rapid reaction of the enol with the halogen.

1. H–C–C(=O) + H:B ⇌ (Fast) H–C–C(=O⁺H) [:B⁻ removes H] ⇌ (Slow) C=C–OH (Enol) + H:B

$$\text{2. } X\!-\!X + \;\rangle C\!=\!C\langle^{O-H} \;\underset{}{\overset{\text{Fast}}{\rightleftharpoons}}\; -\overset{X}{\overset{|}{\underset{|}{C}}}-C\overset{O}{\underset{\diagdown}{\diagup\!\!\diagup}} + HX$$

Part of the evidence that supports these mechanisms comes from studies of reaction kinetics. Both base-promoted and acid-catalyzed halogenations of ketones *show initial rates that are independent of the halogen concentration.* The mechanisms that we have written are in accord with this observation: in both instances the slow step of the mechanism occurs prior to the intervention of the halogen.

Problem 16.18

Why do we say that the first reaction is "base promoted" rather than "base catalyzed?"

Problem 16.19

Additional evidence for the halogenation mechanisms that we presented above comes from the following facts:

(a) Optically active phenyl *sec*-butyl ketone undergoes acid-catalyzed racemization at exactly the same rate that it undergoes acid-catalyzed halogenation.

(b) Phenyl *sec*-butyl ketone undergoes acid-catalyzed iodination at the same rate that it undergoes acid-catalyzed bromination.

(c) Phenyl *sec*-butyl ketone undergoes base-catalyzed hydrogen-deuterium exchange at the same rate that it undergoes base-promoted halogenation.

Explain how each of these observations support the mechanisms that we have presented.

16.13 THE ALDOL CONDENSATION: THE ADDITION OF ENOLATE IONS TO ALDEHYDES AND KETONES

When acetaldehyde reacts with dilute sodium hydroxide at room temperature (or below) dimerization takes place and 3-hydroxybutanal forms. Since 3-hydroxy-

$$2CH_3\overset{O}{\overset{\|}{C}}H \xrightarrow[5^\circ]{10\%\ NaOH/H_2O} CH_3\overset{OH}{\overset{|}{C}}HCH_2\overset{O}{\overset{\|}{C}}H$$

(50%)
3-Hydroxybutanal
"aldol"

butanal is both an *ald*ehyde and an alcoh*ol* it has been given the common name "aldol", and reactions of this general type have come to be known as *aldol additions* (or *aldol condensations*).

The mechanism for the aldol addition illustrates two important characteristics of carbonyl compounds: the acidity of their α-hydrogens and the tendency of their carbonyl groups to undergo nucleophilic addition.

In the first step, the base (hydroxide ion) abstracts a proton from the α-carbon of one molecule of acetaldehyde to give a resonance-stabilized enolate ion:

$$\text{Step 1}\quad H\ddot{O}:^- + H{-}CH_2\overset{\overset{\ddot{O}:}{\|}}{C}H \rightleftharpoons HOH + \left[:\bar{C}H_2{-}\overset{\overset{\ddot{O}:}{\|}}{C}H \longleftrightarrow CH_2{=}\overset{\overset{:\ddot{O}:^-}{|}}{C}H\right]$$

Enolate ion

In the second step, the enolate ion acts as a nucleophile (actually as a carbanion) and attacks the carbonyl carbon of a second molecule of acetaldehyde. This step gives an alkoxide ion.

$$\text{Step 2}\quad CH_3\overset{\overset{:\ddot{O}}{\|}}{C}H + :\bar{C}H_2{-}\overset{\overset{\ddot{O}:}{\|}}{C}H \rightleftharpoons CH_3\overset{\overset{:\ddot{O}:^-}{|}}{C}HCH_2\overset{\overset{O}{\|}}{C}H$$

An alkoxide ion

$$\updownarrow$$

$$CH_2{=}\overset{\overset{:\ddot{O}:^-}{|}}{C}H$$

In the third step, the alkoxide ion abstracts a proton from water to form aldol. This step takes place because the alkoxide ion is a stronger base than a hydroxide ion.

$$\text{Step 3}\quad \underset{\text{Stronger base}}{CH_3\overset{\overset{:\ddot{O}:^-}{|}}{C}HCH_2\overset{\overset{\ddot{O}:}{\|}}{C}H} + H\ddot{O}H \rightleftharpoons CH_3\overset{\overset{:\ddot{O}H}{|}}{C}HCH_2\overset{\overset{\ddot{O}:}{\|}}{C}H + \underset{\text{Weaker base}}{:\ddot{\bar{O}}H}$$

16.13A Dehydration of Addition Product

If the basic mixture containing the aldol is heated, dehydration takes place and crotonaldehyde (2-butenal) is formed. Dehydration occurs readily because of the acidity of the remaining α-hydrogens even though the leaving group is a hydroxide ion.

$$H\ddot{O}:^- + CH_3\overset{\overset{OH}{|}}{C}H{-}\underset{\underset{H}{|}}{C}H{-}\overset{\overset{O}{\|}}{C}H \xrightarrow{-H_2O} CH_3\overset{\overset{OH}{|}}{C}H{-}\ddot{C}H^-{-}\overset{\overset{O}{\|}}{C}H \xrightarrow{-OH^-} \underset{\text{Crotonaldehyde (2-butenal)}}{CH_3CH{=}CH\overset{\overset{O}{\|}}{C}H}$$

In some aldol reactions, dehydration occurs so readily that we cannot isolate the product in the aldol form; we obtain the derived *enal* instead. An aldol *condensation* occurs rather than an aldol *addition*. A condensation reaction is one which molecules are joined through the intermolecular elimination of a small molecule such as water or an alcohol.

$$2RCH_2\overset{\overset{O}{\|}}{C}H \xrightarrow{\text{base}} \underset{\text{Not isolated}}{\left[RCH_2\overset{\overset{OH}{|}}{C}H\underset{\underset{R}{|}}{C}H\overset{\overset{O}{\|}}{C}H\right]}\ \textit{Addition product} \xrightarrow{-H_2O}$$

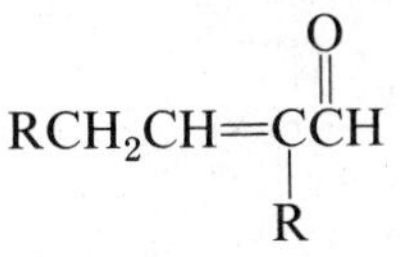

$$RCH_2CH{=}\overset{\displaystyle O \atop \|}{\underset{\underset{\displaystyle R}{|}}{C}}CH \quad \textit{Condensation product}$$

An enal

16.13B Synthetic Applications

The aldol addition or condensation is a general reaction of aldehydes that possess an α-hydrogen. Propanal, for example, reacts with aqueous sodium hydroxide to give 3-hydroxy-2-methylpentanal.

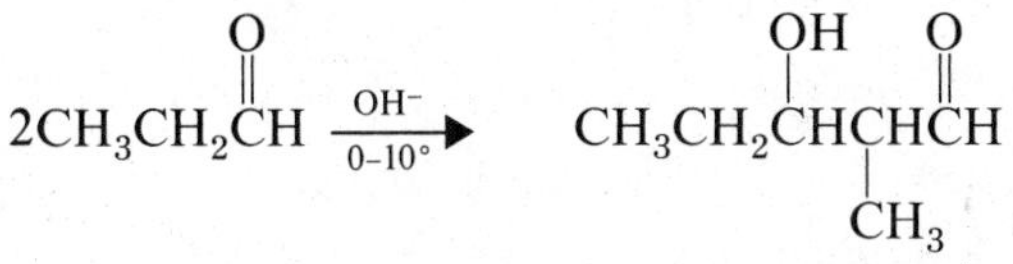

$$2CH_3CH_2\overset{\overset{\displaystyle O}{\|}}{C}H \xrightarrow[0\text{–}10^\circ]{OH^-} CH_3CH_2\overset{\overset{\displaystyle OH}{|}}{C}H\underset{\underset{\displaystyle CH_3}{|}}{C}H\overset{\overset{\displaystyle O}{\|}}{C}H$$

3-Hydroxy-2-methylpentanal
(55–60%)

Problem 16.20

(a) Show all steps in the aldol addition that occurs when propanal ($CH_3CH_2\overset{\overset{O}{\|}}{C}H$) is treated with base. (b) How can you account for the fact that the product of the aldol addition is $CH_3CH_2\overset{\overset{OH}{|}}{C}H\underset{\underset{CH_3}{|}}{C}H\overset{\overset{O}{\|}}{C}H$ and not $CH_3CH_2\overset{\overset{OH}{|}}{C}HCH_2CH_2\overset{\overset{O}{\|}}{C}H$? (c) What product would be formed if the reaction mixture were heated?

The aldol addition is important in organic synthesis because it gives us a method for linking two smaller molecules by introducing a carbon-carbon bond between them. Because aldol products contain two functional groups, —OH and —CHO, we can use them to carry out a number of subsequent reactions. Examples are shown at the top of page 730.

Problem 16.21

Show how each of the following products could be synthesized from butanal.

(a) 3-Hydroxy-2-ethylhexanal
(b) 2-Ethyl-2-hexen-1-ol
(c) 2-Ethyl-1-hexanol
(d) 2-Ethyl-1,3-hexanediol (the insect repellent "6–12")

Problem 16.22

Most simple ketones do not undergo aldol additions because the equilibrium is unfavorable. Acetone can be induced to react if the reaction is carried out in a special apparatus that allows the addition product to be removed from contact with the base as soon as it forms. (a) Write a mechanism for an aldol-type (ketol) addition of acetone in base. (b) What compound would you obtain when the ketol product is dehydrated?

$$RCH_2CH{=}O \xrightarrow{OH^-,\ H_2O} \underset{\text{The aldol}}{RCH_2CH(OH)CH(R)CH{=}O} \xrightarrow{NaBH_4} \underset{\text{A 1,3-diol}}{RCH_2CH(OH)CH(R)CH_2OH}$$

$$\text{The aldol} \xrightarrow{H^+,\ -H_2O} \underset{\text{An }\alpha,\beta\text{-unsaturated aldehyde}}{RCH_2CH{=}C(R)CH{=}O} \xrightarrow{NaBH_4} \underset{\text{An allylic alcohol}}{RCH_2CH{=}C(R)CH_2OH}$$

$$\text{An }\alpha,\beta\text{-unsaturated aldehyde} \xrightarrow{H_2,\ Ni} \underset{\text{A saturated alcohol}}{RCH_2CH_2CH(R)CH_2OH}$$

16.13C Crossed Aldol Additions

"Mixed" or "crossed" aldol additions are of little synthetic importance if both reactants have α-hydrogens because these reactions give a complex mixture of products. If, for example, we were to carry out a crossed aldol addition using acetaldehyde and propanal we might expect to obtain at least four products.

$$CH_3CH{=}O + CH_3CH_2CH{=}O \xrightarrow{\text{Base}} \underset{\substack{\text{3-Hydroxybutanal}\\ \text{(from two moles of acetaldehyde)}}}{CH_3CH(OH)CH_2CH{=}O} + \underset{\substack{\text{2-Methyl-3-hydroxypentanal}\\ \text{(from two moles of propanal)}}}{CH_3CH_2CH(OH)CH(CH_3)CH{=}O}$$

$$+ \underset{\text{2-Methyl-3-hydroxybutanal}}{CH_3CH(OH)CH(CH_3)CH{=}O} \text{ and } \underset{\text{3-Hydroxypentanal}}{CH_3CH_2CH(OH)CH_2CH{=}O}$$

(from one mole of acetaldehyde and one mole of propanal)

Problem 16.23

We have already seen how the first two products (above) are formed. Write mechanisms that illustrate the formation of 2-methyl-3-hydroxybutanal and 3-hydroxypentanal.

Crossed aldol reactions *are* practical, however, when one reactant does not have an α-hydrogen and, thus, cannot undergo self-condensation. We can avoid other side reactions by placing this component in base and by slowly adding the reactant with an α-hydrogen to the mixture. Under these conditions the concentration of the reactant with an α-hydrogen will always be low and most of it will be present as an enolate ion. The main reaction that will take place is one between this enolate ion and the component that has no α-hydrogen. The examples listed in Table 16.5 illustrate this technique.

As the examples in Table 16.5 also show, the crossed aldol reaction is often accompanied by dehydration. Whether or not dehydration occurs can, at times, be determined by our choice of reaction conditions, but dehydration is especially likely when it leads to an extended conjugated system.

Problem 16.24

When excess formaldehyde in basic solution is treated with acetaldehyde, the following reaction takes place.

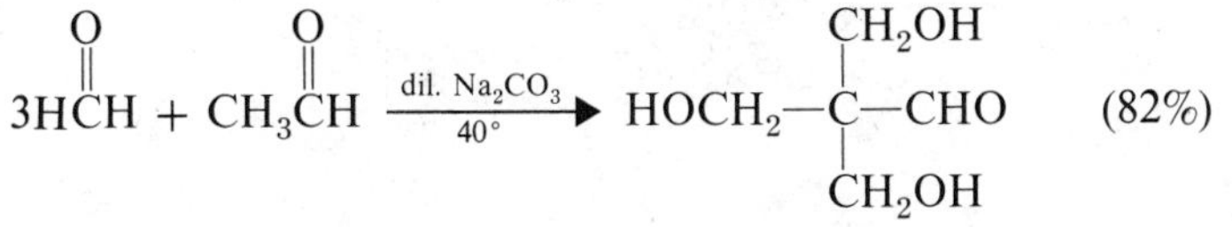

$$3\,\mathrm{H\overset{O}{\overset{\|}{C}}H} + \mathrm{CH_3\overset{O}{\overset{\|}{C}}H} \xrightarrow[40°]{\text{dil. }Na_2CO_3} \mathrm{HOCH_2{-}\underset{CH_2OH}{\underset{|}{\overset{CH_2OH}{\overset{|}{C}}}}{-}CHO} \quad (82\%)$$

Write a mechanism that accounts for the formation of the product.

TABLE 16.5 Crossed Aldol Condensations

THIS REACTANT WITH NO α-HYDROGEN IS PLACED IN BASE		THIS REACTANT WITH AN α-HYDROGEN IS ADDED SLOWLY		PRODUCT
$C_6H_5\overset{O}{\overset{\|}{C}}H$ Benzaldehyde	+	$CH_3CH_2\overset{O}{\overset{\|}{C}}H$ Propionaldehyde	$\xrightarrow[10°]{OH^-}$	$C_6H_5CH{=}\overset{CH_3}{\overset{|}{C}}{-}\overset{O}{\overset{\|}{C}}H$ (68%) 2-Methyl-3-phenyl-2-propenal (α-methylcinnamaldehyde)
$C_6H_5\overset{O}{\overset{\|}{C}}H$ Benzaldehyde	+	$C_6H_5CH_2\overset{O}{\overset{\|}{C}}H$ Phenylacetaldehyde	$\xrightarrow[20°]{OH^-}$	$C_6H_5CH{=}\underset{C_6H_5}{\underset{|}{C}}\overset{O}{\overset{\|}{C}}H$ 2,3-Diphenyl-2-propenal
$H\overset{O}{\overset{\|}{C}}H$ Formaldehyde	+	$CH_3\overset{CH_3}{\overset{|}{C}H}{-}\overset{O}{\overset{\|}{C}}H$ 2-Methylpropanal	$\xrightarrow[40°]{\text{dil. }Na_2CO_3}$	$CH_3{-}\underset{CH_2OH}{\underset{|}{\overset{CH_3}{\overset{|}{C}}}}{-}\overset{O}{\overset{\|}{C}}H$ 3-Hydroxy-2,2-dimethylpropanal (>64%)

16.13D Claisen-Schmidt Reactions

Ketones can be used as one component in crossed aldol reactions that are called *Claisen-Schmidt* reactions. Two examples are shown below.

$$C_6H_5\overset{O}{\overset{\|}{C}}H + CH_3\overset{O}{\overset{\|}{C}}CH_3 \xrightarrow[100°]{OH^-} C_6H_5CH{=}CH\overset{O}{\overset{\|}{C}}CH_3$$

4-Phenyl-3-buten-2-one
(benzalacetone)
(70%)

$$C_6H_5\overset{O}{\overset{\|}{C}}H + CH_3\overset{O}{\overset{\|}{C}}C_6H_5 \xrightarrow[20°]{OH^-} C_6H_5CH{=}CH\overset{O}{\overset{\|}{C}}C_6H_5$$

1,3-Diphenyl-2-propenone
(benzalacetophenone)
(85%)

An important step in a commerical synthesis of Vitamin A makes use of a Claisen-Schmidt reaction between geranial and acetone:

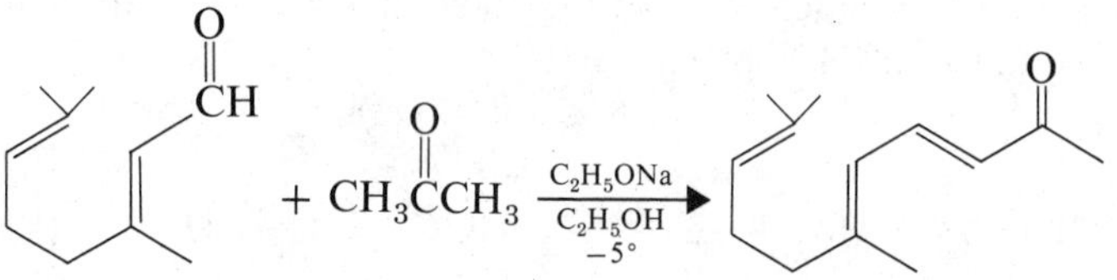

$$\text{Geranial} + CH_3\overset{O}{\overset{\|}{C}}CH_3 \xrightarrow[C_2H_5OH,\ -5°]{C_2H_5ONa} \text{Pseudoionone}$$

Geranial

Pseudoionone
(49%)

Geranial is a naturally occurring aldehyde that can be obtained from lemongrass oil. (This vitamin A synthesis is also the subject of Problems 15.49, 15.50, 16.25, and 20.36.)

Problem 16.25

When pseudoionone is treated with BF_3 in acetic acid, ring closure takes place and α- and β-ionone are produced. This is the next step in the Vitamin A synthesis.

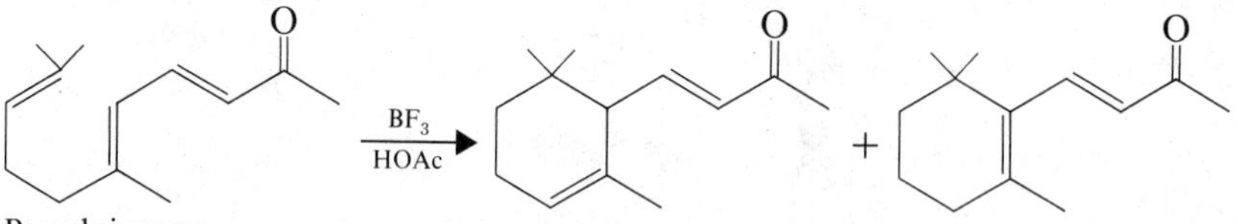

$$\text{Pseudoionone} \xrightarrow[HOAc]{BF_3} \alpha\text{-Ionone} + \beta\text{-Ionone}$$

Pseudoionone

α-Ionone

β-Ionone

(a) Write mechanisms that explain the formation of α- and β-ionone. (b) β-Ionone is the major product. How can you explain this? (c) Which ionone would you expect to absorb at longer wavelengths in the visible-ultraviolet region? Why?

16.13E Aldol Additions in Biochemistry

A number of aldol-type additions take place in the living cells of plants and animals. One *in vivo* reaction that is an important step in the synthesis of glucose involves a crossed aldol addition between (*R*)-glyceraldehyde 3-phosphate and dihydroxyacetone phosphate (cf. Sect. 16.11C) to produce D-fructose-1,6-diphosphate.* The enzyme that catalyzes this reaction has the appropriate name, *aldolase*. The reaction is completely stereospecific; it produces only D-fructose-1,6-diphosphate, although three other stereoisomers are theoretically possible.

$$\underset{\text{(R)-Glyceraldehyde 3-phosphate}}{HC(=O)-C(H)(OH)-CH_2OPO_3H^-} + \underset{\text{Dihydroxyacetone phosphate}}{CH_2OPO_3H^- -C(=O)-CH_2OH} \underset{}{\overset{\text{aldolase}}{\rightleftharpoons}} \underset{\text{D-Fructose 1,6-diphosphate}}{CH_2OPO_3H^- -C(=O)-C(HO)(H)-C(H)(OH)-C(H)(OH)-CH_2OPO_3H^-}$$

The aldolase reaction, like other aldol additions, is reversible. Thus, the cell can use the reaction to synthesize glucose (via fructose 1,6-diphosphate) or it can use the reaction to degrade glucose (via glyceraldehyde 3-phosphate and dihydroxyacetone phosphate) to carbon dioxide and water. Since glucose has far greater chemical potential energy than carbon dioxide and water, glucose *synthesis* gives the cell a method for storing energy. Glucose *degradation* gives the cell a source of energy.

16.14 THE CANNIZZARO REACTION

When aldehydes *that have no α-hydrogen* are placed in concentrated alkali they undergo a reaction known as the Cannizzaro reaction. The products of the Cannizzaro reaction are an alcohol and the salt of a carboxylic acid. Three examples of Cannizzaro reactions are the following.

$$2C_6H_5-\overset{O}{\overset{\|}{C}}-H \xrightarrow{60\%\ KOH} C_6H_5CH_2OH + C_6H_5\overset{O}{\overset{\|}{C}}-O^-$$

(80–90% yield)

$$2R-\underset{R}{\overset{R}{\overset{|}{\underset{|}{C}}}}-\overset{O}{\overset{\|}{C}}H \xrightarrow{Ca(OH)_2} R-\underset{R}{\overset{R}{\overset{|}{\underset{|}{C}}}}-CH_2OH + R-\underset{R}{\overset{R}{\overset{|}{\underset{|}{C}}}}-\overset{O}{\overset{\|}{C}}-O^-$$

$$2H\overset{O}{\overset{\|}{C}}H \xrightarrow{50\%\ NaOH} CH_3OH + H\overset{O}{\overset{\|}{C}}-O^-$$

The Cannizzaro reaction is an *oxidation-reduction* in which the aldehyde

* We will discuss the meaning of the designation D in Chapter 19.

acts both as an oxidizing agent and as a reducing agent. That is, one mole of the aldehyde reduces another to an alcohol; in this process the first aldehyde is oxidized to a carboxylate ion.

A likely mechanism for the Cannizzaro reaction is the following. ($G = Ar$, H, or R_3C)

Step 1

$$G{-}\overset{\overset{\displaystyle :\ddot{O}}{\|}}{C}{-}H + :\ddot{O}H^- \rightleftharpoons G{-}\overset{\overset{\displaystyle :\ddot{O}:^-}{|}}{\underset{\underset{\displaystyle H{-}\ddot{O}:}{|}}{C}}{-}H$$

Step 2

$$G{-}\overset{\overset{\displaystyle :\ddot{O}:^-}{|}}{\underset{\underset{\displaystyle H{-}\ddot{O}:}{|}}{C}}{-}H + G{-}\overset{\overset{\displaystyle :\ddot{O}}{\|}}{C}{-}H \longrightarrow G{-}C\overset{\displaystyle \ddot{O}:}{\underset{\displaystyle :\ddot{O}{-}H}{}} + G{-}\overset{\overset{\displaystyle :\ddot{O}:^-}{|}}{\underset{\underset{\displaystyle H}{|}}{C}}{-}H$$

reducing agent — hydride transfer → oxidizing agent

fast

$$GCOO^- + GCH_2\ddot{O}H$$

In step 1, a hydroxide ion attacks the carbonyl carbon of one aldehyde molecule and in step 2, oxidation-reduction takes place through a transfer of a hydride ion. (In step 2, the Cannizzaro reaction resembles $LiAlH_4$ and $NaBH_4$ reductions of aldehydes and ketones cf. Sect. 15.7).

Cannizzaro reactions themselves are of little synthetic use since most aldehydes are "expensive" reducing agents. However, crossed Cannizzaro reactions, using inexpensive formaldehyde as one component (and in excess), are practical. Formaldehyde has the advantage, too, that it will preferentially accept the hydroxide ion in step 1 (why?) and thus act as the reducing agent in step 2. The reaction shown below illustrates this method.

$$m\text{-}CH_3OC_6H_4\overset{\overset{\displaystyle O}{\|}}{C}H + H\overset{\overset{\displaystyle O}{\|}}{C}H \xrightarrow[\text{(excess) } H_2O/CH_3OH]{30\%\ NaOH} m\text{-}CH_3OC_6H_4CH_2OH + HCOO^-$$

(85–90%)

Problem 16.26

When acetaldehyde is treated with four equivalents of formaldehyde in aqueous calcium hydroxide (at 15–45°) the product of the reaction is pentaerythritol, $HOCH_2C(CH_2OH)_3$ (71% yield). Using your answer to Problem 16.24, show how pentaerythritol forms.

16.15 α,β-UNSATURATED ALDEHYDES AND KETONES

When α,β-unsaturated aldehydes and ketones react with nucleophilic reagents they may do so in two ways: they may react by a *simple addition,* that is, one in which the nucleophile adds across the double bond of the carbonyl group; or they may react by a *conjugate addition,* one that resembles the 1,4-addition reactions of conjugated dienes (Section 10.8).

$$-\overset{|}{C}{=}\underset{|}{C}-\overset{\ddot{O}:}{\overset{\|}{C}}- + \mathrm{Nu}:^- \longrightarrow \begin{cases} \xrightarrow{+H^+} -\underset{|}{C}{=}\underset{|}{C}-\overset{:\ddot{O}H}{\underset{Nu}{C}}- & \text{Simple addition} \\[2ex] \xrightarrow{+H^+} \underset{\text{Enol form}}{-\overset{|}{\underset{Nu}{C}}-\underset{|}{C}{=}\overset{:\ddot{O}H}{C}-} \rightleftarrows \underset{\text{Keto form}}{-\overset{|}{\underset{Nu}{C}}-\overset{H}{\underset{|}{C}}-\overset{\ddot{O}:}{\overset{\|}{C}}-} & \text{Conjugate addition} \end{cases}$$

In many instances both modes of addition occur in the same mixture. As examples, let us consider the Grignard reactions shown below.

$$CH_3CH{=}CH\overset{O}{\overset{\|}{C}}CH_3 + CH_3MgBr \xrightarrow[(2)\ H^+]{} \underset{(72\%)}{CH_3CH{=}CH\overset{OH}{\underset{CH_3}{C}}CH_3} \quad \text{Simple addition product}$$

$$+$$

$$\underset{(20\%)}{CH_3\underset{CH_3}{C}HCH_2\overset{O}{\overset{\|}{C}}CH_3} \quad \text{Conjugate addition product (in keto form)}$$

$$CH_3CH{=}CH\overset{O}{\overset{\|}{C}}CH_3 + C_6H_5MgBr \xrightarrow[(2)\ H^+]{} \underset{(55\%)}{CH_3CH{=}CH\overset{:\ddot{O}H}{\underset{C_6H_5}{C}}CH_3} \quad \text{Simple addition product}$$

$$+$$

$$\underset{(20\%)}{CH_3\underset{C_6H_5}{C}HCH_2\overset{O}{\overset{\|}{C}}CH_3} \quad \text{Conjugate addition product (in keto form)}$$

In each of these examples we see that simple addition is favored, but that conjugate addition accounts for a substantial amount of the product.

If we examine the resonance structures that contribute to the overall hybrid for an α,β-unsaturated aldehyde or ketone (below) we will be in a better position to understand these reactions.

$$\underset{\mathbf{A}}{-\overset{|}{C}{=}\overset{|}{C}-\overset{\ddot{O}:}{\overset{\|}{C}}-} \longleftrightarrow \underset{\mathbf{B}}{-\overset{|}{C}{=}\overset{|}{C}-\overset{:\ddot{O}:^-}{\underset{+}{C}}-} \longleftrightarrow \underset{\mathbf{C}}{-\overset{|}{\underset{+}{C}}-\overset{|}{C}{=}\overset{:\ddot{O}:^-}{C}}$$

Although structures **B** and **C** involve separated charges they make a significant contribution to the hybrid because, in each, the negative charge is carried by electronegative oxygen. Structures **B** and **C** not only indicate that the oxygen of the hybrid should bear a partial negative charge; but they also indicate that *both the carbonyl carbon and the β-carbon should bear a partial positive charge.* They indicate that we should represent the hybrid in the following way:

$$\overset{\delta+}{-\text{C}}{=}{=}\text{C}{=}{=}\overset{\delta+}{\text{C}}(=\overset{\delta-}{\text{O}})-$$

This structure tells us that we should expect an electrophilic reagent to attack the carbonyl oxygen and a nucleophilic reagent to attack either the carbonyl carbon or the β-carbon.

This is exactly what happens in the Grignard reactions that we saw earlier. The electrophilic magnesium attacks the carbonyl oxygen; the nucleophilic carbon of the Grignard reagent attacks either the carbonyl carbon or the β-carbon.

Simple Addition

$$\overset{\delta+}{-\text{C}}{=}\text{C}-\overset{\delta+}{\text{C}}(=\overset{\delta-}{\ddot{\text{O}}}\text{:}\cdots\overset{\delta+}{\text{Mg}}\text{X})\ \overset{\delta-}{\text{R}} \longrightarrow -\text{C}{=}\text{C}-\text{C}(\text{:}\ddot{\text{O}}-\text{MgX})-\text{R} \xrightarrow{\text{H}^+} -\text{C}{=}\text{C}-\text{C}(\text{OH})-\text{R}$$

Conjugate Addition

$$\overset{\delta+}{\text{C}}(=\overset{\delta-}{\ddot{\text{O}}}\text{:}\cdots\overset{\delta+}{\text{Mg}}(\text{X})\overset{\delta-}{\text{R}})-\text{C}{=}\overset{\delta+}{\text{C}}- \longrightarrow \text{C}(-\ddot{\text{O}}-\text{MgX}){=}\text{C}-\text{C}(\text{R}) \xrightarrow{\text{H}^+}$$

$$\text{C}(-\ddot{\text{O}}-\text{H}){=}\text{C}-\text{C}(\text{R}) \rightleftarrows \text{C}(=\ddot{\text{O}}\text{:})-\text{C}(\text{H})-\text{C}(\text{R}) = -\underset{\text{R}}{\text{C}}-\underset{\text{H}}{\text{C}}-\overset{\text{O}}{\overset{\|}{\text{C}}}-$$

Addition of Other Nucleophiles

Grignard reagents are not the only nucleophilic reagents that add in a conjugate manner to α,β-unsaturated aldehydes and ketones. Almost every nucleophilic reagent that adds at the carbonyl carbon of a simple aldehyde or ketone is capable of adding at the β-carbon of an α,β-unsaturated carbonyl compound. In many instances conjugate addition is the major reaction path:

$$C_6H_5CH{=}CH\overset{O}{\overset{\|}{C}}C_6H_5 + CN^- \xrightarrow[CH_3COOH]{C_2H_5OH} C_6H_5\underset{CN}{\underset{|}{C}}H-CH_2\overset{O}{\overset{\|}{C}}C_6H_5$$

(95%)

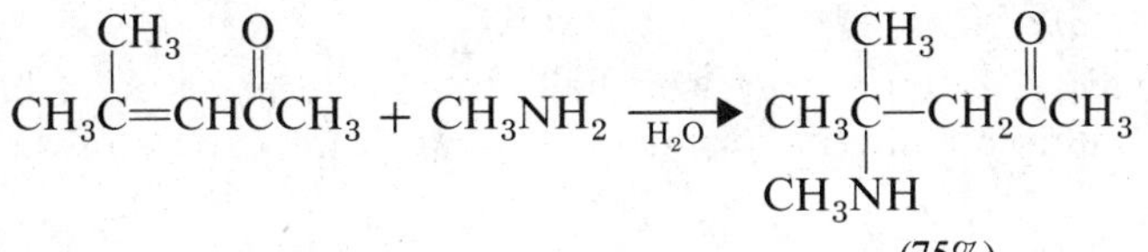

$$CH_3C(CH_3){=}CHC(O)CH_3 + CH_3NH_2 \xrightarrow{H_2O} CH_3C(CH_3)(CH_3NH){-}CH_2C(O)CH_3$$

(75%)

The sequence shown below illustrates how a conjugate aldol addition followed by a simple aldol condensation may be used to build one ring on to another. This procedure is known as the *Robinson annellation* ("ring forming") reaction.

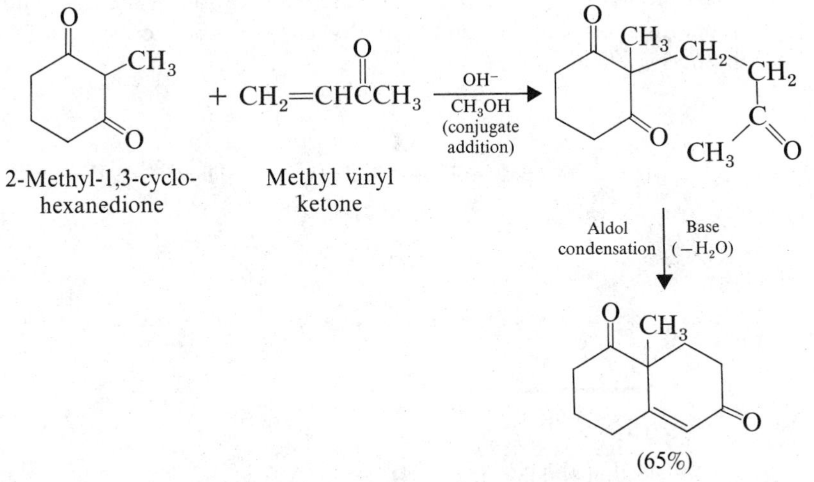

2-Methyl-1,3-cyclohexanedione + $CH_2{=}CHCCH_3$ (Methyl vinyl ketone) $\xrightarrow[\text{(conjugate addition)}]{OH^-,\ CH_3OH}$ 2-methyl-2-($CH_2CH_2C(O)CH_3$)-1,3-cyclohexanedione

Aldol condensation, Base $(-H_2O)$ ↓

(65%)

Problem 16.27

(a) Propose step-by-step mechanisms for both steps of the Robinson annellation sequence shown above. (b) Would you expect 2-methyl-1,3-cyclohexanedione to be more or less acidic than cyclohexanone? Explain your answer.

Michael Additions

Conjugate nucleophilic additions of enolate ions to α,β-unsaturated carbonyl compounds are known generally as *Michael additions*. We will see other examples of Michael additions in later chapters.

Problem 16.28

What product would you expect to obtain from the base-catalyzed Michael reaction (a) of benzalacetophenone (p. 732) and acetophenone? (b) of benzalacetophenone and cyclopentadiene? Show all steps in each mechanism.

Problem 16.29

When acrolein reacts with hydrazine, the product is a dihydropyrazole:

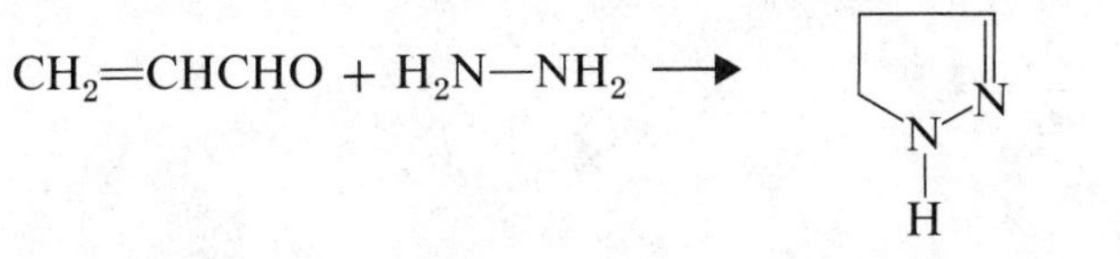

$$CH_2{=}CHCHO + H_2N{-}NH_2 \longrightarrow$$ (dihydropyrazole ring, N–H)

Acrolein — Hydrazine — A dihydropyrazole

Suggest a mechanism that explains this reaction.

16.16 THE HALOFORM REACTION

When methyl ketones react with halogens in the presence of base (cf. Section 16.12), multiple halogenations tend to occur at the carbon of the methyl group.

$$R{-}\overset{O}{\overset{\|}{C}}{-}CH_3 + 3X_2 \xrightarrow{\text{base}} R{-}\overset{O}{\overset{\|}{C}}{-}CX_3 + 3X^-$$

Multiple halogenations occur because substitution of the first halogen makes the remaining α-hydrogens on the methyl carbon more acidic. It does this through its inductive effect.

$$R{-}\overset{O}{\overset{\|}{C}}{-}CH_3 + :\bar{B} \rightleftharpoons R{-}\overset{O}{\overset{\|}{C}}{-}\ddot{C}H_2^- \xrightarrow{X{-}X} R{-}\overset{O}{\overset{\|}{C}}{-}CH_2X + X^-$$

$$R{-}\overset{O}{\overset{\|}{C}}{-}CH_2X + :\bar{B} \rightleftharpoons R{-}\overset{O}{\overset{\|}{C}}{-}\ddot{C}HX^- \xrightarrow{X{-}X} R{-}\overset{O}{\overset{\|}{C}}{-}CHX_2 \xrightarrow{:\bar{B}\text{ then }X_2} R{-}\overset{O}{\overset{\|}{C}}{-}CX_3$$

Acidity is increased by electron-withdrawing halogen

When methyl ketones react with halogens in aqueous sodium hydroxide (i.e., in *hypohalite solutions*) an additional reaction takes place. Hydroxide ion attacks the carbonyl carbon of the trihalo ketone and causes a cleavage of the carbon-carbon bond between the carbonyl group and the trihalomethyl group. This ultimately produces a carboxylate ion and a *haloform* (i.e., either $CHCl_3$,

$$R{-}\overset{:\ddot{O}}{\overset{\|}{C}}{-}CX_3 + {}^{-}:\ddot{O}H \rightleftharpoons R{-}\overset{:\ddot{O}:^-}{\underset{:\ddot{O}H}{C}}{-}CX_3$$

$$R{-}\overset{:\ddot{O}:^-}{\underset{:\ddot{O}H}{C}}{-}CX_3 \rightleftharpoons R{-}\overset{:\ddot{O}}{\overset{\|}{C}}{-}\ddot{O}{-}H + {}^{-}:CX_3$$

$$\rightleftharpoons R{-}\overset{:\ddot{O}}{\overset{\|}{C}}{-}\ddot{O}:^- + H{-}CX_3$$

Carboxylate ion   Haloform

$CHBr_3$ or CHI_3). The initial step is a nucleophilic attack by hydroxide ion on the carbonyl carbon. In the next step carbon-carbon bond cleavage occurs and the haloform anion, $:CX_3^-$, departs. This step can occur because the haloform anion is unusually stable; its negative charge is dispersed by the three electronegative halogens. Finally, a proton transfer takes place between the carboxylic acid and the haloform anion.

The haloform reaction is of synthetic utility as a means of converting methyl ketones to carboxylic acids. When the haloform reaction is used in synthesis, chlorine and bromine are most commonly used as the halogen component. Chloroform ($CHCl_3$) and bromoform ($CHBr_3$) are both liquids and are easily separated from the acid.

16.16A The Iodoform Reaction

The haloform reaction using iodine and aqueous sodium hydroxide is called the *iodoform reaction.* The iodoform reaction is frequently used in structure determinations because it allows identification of the two groups shown below:

$$-\underset{\underset{O}{\|}}{C}-CH_3 \quad \text{and} \quad -\underset{\underset{OH}{|}}{CH}-CH_3$$

Compounds containing either of these groups react with iodine in sodium hydroxide to give bright yellow precipitates of *iodoform* (CHI_3, mp 119°). Compounds containing the $-CHOHCH_3$ group give a positive iodoform test because they are first oxidized to methyl ketones:

$$-\underset{\underset{OH}{|}}{C}HCH_3 + I_2 + 2OH^- \longrightarrow -\underset{\underset{O}{\|}}{C}CH_3 + 2I^- + 2H_2O$$

Methyl ketones then react with iodine and hydroxide ion to produce iodoform:

$$-\underset{\underset{O}{\|}}{C}-CH_3 + 3I_2 + 3OH^- \longrightarrow -\underset{\underset{O}{\|}}{C}-CI_3 + 3I^- + 3H_2O$$

$$-\underset{\underset{O}{\|}}{C}-CI_3 + OH^- \longrightarrow -\underset{\underset{O}{\|}}{C}-O^- + \underset{\text{Yellow precipitate}}{CHI_3\downarrow}$$

The group to which the $-COCH_3$ or $-CHOHCH_3$ function is attached can be aryl, alkyl, or hydrogen. Thus, even ethyl alcohol and acetaldehyde give positive iodoform tests.

Problem 16.30

Which of the following compounds would give a positive iodoform test?

(a) Acetone
(b) Acetophenone
(c) Pentanal
(f) 1-Phenylethanol
(g) 2-Phenylethanol
(h) 2-Butanol

(d) 2-Pentanone (i) Methyl 2-naphthyl ketone
(e) 3-Pentanone (j) 3-Pentanol

16.17 CHEMICAL TESTS FOR ALDEHYDES AND KETONES

In addition to the iodoform reaction (Section 16.16), a variety of other chemical methods can be used in identifying aldehydes and ketones.

Aldehydes and ketones can be differentiated from noncarbonyl compounds through their reactions with derivatives of ammonia (Section 16.8). Semicarbazide, 2,4-dinitrophenylhydrazine, and hydroxylamine react with aldehydes and ketones to form precipitates. Semicarbazones and oximes are usually colorless, while 2,4-dinitrophenylhydrazones are usually orange. The melting point of these derivatives can also be used in identifying specific aldehydes and ketones.

The ease with which aldehydes undergo oxidation provides a useful test that differentiates aldehydes from most ketones.

16.17A Tollens' Test (Silver Mirror Test)

Mixing aqueous silver nitrate with aqueous ammonia produces a solution known as Tollens' reagent. The reagent contains the silver diammine ion, $Ag(NH_3)_2^+$. Although this ion is a very weak oxidizing agent it will oxidize aldehydes to carboxylate ions. As it does this, silver is reduced from the +1 oxidation state (of $Ag(NH_3)_2^+$) to metallic silver. If the rate of reaction is slow and the walls of the vessel are clean, metallic silver deposits on the walls of the test tube as a mirror; if not, it deposits as gray to black precipitate. Tollens' reagent gives a negative result with all ketones except α-hydroxy ketones.

$$\underset{\text{Aldehyde}}{R{-}\overset{\overset{\displaystyle O}{\|}}{C}{-}H} + Ag(NH_3)_2^+ \xrightarrow{\text{aq. } NH_3} R{-}\overset{\overset{\displaystyle O}{\|}}{C}{-}\bar{O} + \underset{\substack{\text{Silver}\\ \text{mirror}}}{Ag\downarrow}$$

$$\underset{\alpha\text{-Hydroxy ketone}}{R{-}\overset{\overset{\displaystyle O}{\|}}{C}{-}\overset{\overset{\displaystyle OH}{|}}{C}HR'} + Ag(NH_3)_2^+ \xrightarrow{\text{aq. } NH_3} R{-}\overset{\overset{\displaystyle O}{\|}}{C}{-}\overset{\overset{\displaystyle O}{\|}}{C}R' + \underset{\substack{\text{Silver}\\ \text{mirror}}}{Ag\downarrow}$$

$$\underset{\text{Ketone}}{R{-}\overset{\overset{\displaystyle O}{\|}}{C}{-}R'} + Ag(NH_3)_2^+ \xrightarrow{\text{aq. } NH_3} \text{No reaction}$$

16.18 SUMMARY OF THE ADDITION REACTIONS OF ALDEHYDES AND KETONES

Table 16.6 summarizes the nucleophilic addition reactions of aldehydes and ketones that occur at the carbonyl carbon.

Not only do α,β-unsaturated aldehydes and ketones undergo all of these reactions, they also undergo nucleophilic addition at their β-carbons (Section 16.15).

General Reaction

$$Nu:^- + -\overset{|}{C}=\overset{|}{C}-\overset{O}{\overset{\|}{C}}- \rightleftharpoons -\underset{Nu}{\overset{|}{C}}-\overset{|}{C}=\overset{O^-}{\overset{|}{C}}- \xrightleftharpoons{H^+} -\underset{Nu}{\overset{|}{C}}-\underset{H}{\overset{|}{C}}-\overset{O}{\overset{\|}{C}}-$$

TABLE 16.6 Nucleophilic Addition Reactions of Aldehydes and Ketones

1. Addition of Organometallic Compounds

General Reaction

$$\overset{\delta-}{R}:\overset{\delta+}{M} + \rangle C=\ddot{O}: \longrightarrow R-\overset{|}{\underset{|}{C}}-O^-M^+ \xrightarrow{H^+} R-\overset{|}{\underset{|}{C}}-O-H$$

Specific Example using a Grignard Reagent (Section 14.10)

$$CH_3CH_2MgBr + CH_3\overset{O}{\overset{\|}{C}}-H \xrightarrow{(2)H^+} CH_3CH_2\overset{OH}{\overset{|}{C}}HCH_3$$

(67%)

2. Addition of Hydride Ion

General Reaction

$$\overset{-}{H}: + \rangle C=O \longrightarrow H-\overset{|}{\underset{|}{C}}-O^- \xrightarrow{H^+} H-\overset{|}{\underset{|}{C}}-OH$$

Specific Examples Using Metal Hydrides (Section 15.7)

$$\text{cyclobutanone} + LiAlH_4 \xrightarrow{(2)H^+} \text{cyclobutanol (–OH)}$$

(90%)

$$CH_3\overset{O}{\overset{\|}{C}}CH_2CH_2CH_3 + NaBH_4 \xrightarrow[OH^-]{CH_3OH} CH_3\overset{OH}{\overset{|}{C}}HCH_2CH_2CH_3$$

(100%)

The Cannizzaro Reaction (Section 16.14)

$$\text{(3,4-methylenedioxyphenyl)}-\overset{O}{\overset{\|}{C}}-H \xrightarrow[H_2O/CH_3OH]{HCHO,\ 30\%\ NaOH} \text{(3,4-methylenedioxyphenyl)}-CH_2OH + HCOO^-$$

(85–90%)

3. Addition of Hydrogen Cyanide and Sodium Bisulfite (Section 16.9)

General Reaction

$$N\equiv C:^- + \rangle C=O \rightleftharpoons N\equiv C-\overset{|}{\underset{|}{C}}-O^- \xrightleftharpoons{H^+} N\equiv C-\overset{|}{\underset{|}{C}}-OH$$

Specific Example

$$(CH_3)_2C=O \xrightarrow[H^+]{NaCN} (CH_3)_2C(OH)CN$$

Acetone cyanohydrin (77–78%)

TABLE 16.6 Nucleophilic Addition Reactions of Aldehydes and Ketones (Continued)

General Reaction

$$NaHSO_3 + \text{>C=O} \rightleftharpoons \text{>C(SO}_3^-\ Na^+\text{)(OH)}$$

Specific Example

$$CH_3CH{=}O + NaHSO_3 \rightleftharpoons CH_3CH(SO_3^-\ Na^+)(OH)$$

(88%)

4. The Aldol Addition. The Addition of Enolate Ions (Section 16.13)
General Reaction

$$-\overset{O^-}{\overset{|}{C}}{=}\overset{}{C}- + \text{>C=O} \rightleftharpoons -\overset{O}{\overset{\|}{C}}-\overset{|}{\underset{|}{C}}-\overset{|}{\underset{|}{C}}-O^- \overset{H^+}{\rightleftharpoons} -\overset{O}{\overset{\|}{C}}-\overset{|}{\underset{|}{C}}-\overset{|}{\underset{|}{C}}-OH$$

Specific Example

$$H\overset{O}{\overset{\|}{C}}CH_3 + OH^- \rightleftharpoons H-\overset{O^-}{\overset{|}{C}}{=}CH_2 \xrightarrow[\longleftarrow]{CH_3\overset{O}{\overset{\|}{C}}H,\ H_2O} H\overset{O}{\overset{\|}{C}}-CH_2\overset{OH}{\overset{|}{\underset{H}{\underset{|}{C}}}}CH_3$$

5. Addition of Ylides (Section 16.10)
The Wittig Reaction

$$Ar_3P{=}\overset{}{\underset{|}{C}}- + \text{>C=O} \rightleftharpoons -\overset{|}{\underset{Ar_3\overset{+}{P}}{\underset{|}{C}}}-\overset{|}{\underset{O^-}{\underset{|}{C}}}- \xrightarrow{-Ar_3PO} \text{>C=C<}$$

The Addition of Sulfur Ylides

$$R_2S{=}CH_2 + \text{>C=O} \longrightarrow -\overset{|}{\underset{R_2\overset{+}{S}}{\underset{|}{C}}}-\overset{|}{\underset{O^-}{\underset{|}{C}}}- \xrightarrow{-R_2S} -\overset{|}{C}-\overset{|}{C}-\ (\text{bridged by O})$$

6. Addition of Alcohols (Section 16.7)
General Reaction

$$R-\ddot{\underset{..}{O}}-H + \text{>C=O} \rightleftharpoons R-O-\overset{|}{\underset{|}{C}}-OH \xrightarrow[H^+]{ROH} R-O-\overset{|}{\underset{|}{C}}-O-R$$

Hemiacetal or hemiketal — Acetal or ketal

Specific Example

$$C_2H_5OH + CH_3\overset{O}{\overset{\|}{C}}H \rightleftharpoons C_2H_5O-\overset{CH_3}{\overset{|}{\underset{H}{\underset{|}{C}}}}-OH \xrightarrow[H^+]{C_2H_5OH} C_2H_5O\overset{CH_3}{\overset{|}{\underset{H}{\underset{|}{C}}}}-OC_2H_5$$

TABLE 16.6 Nucleophilic Addition Reactions of Aldehydes and Ketones (Continued)

7. **Addition of Derivatives of Ammonia** (Section 16.8)
General Reaction

$$-\ddot{N}(H)-H + \rangle C=O \rightleftharpoons -\ddot{N}(H)-\overset{|}{\underset{|}{C}}-OH \xrightarrow{-H_2O} -N=C\langle$$

Specific Examples

$$CH_3\overset{O}{\overset{\|}{C}}H + NH_2OH \longrightarrow CH_3CH{=}NOH$$

Acetaldoxime

$$C_6H_5\overset{O}{\overset{\|}{C}}H + H_2NNHC_6H_5 \longrightarrow C_6H_5CH{=}NNHC_6H_5$$

Benzaldehyde phenylhydrazone

Additional Problems

16.31

Give structural formulas for each of the following compounds.

(a) Formaldehyde
(b) Acetaldehyde
(c) Phenylacetaldehyde
(d) Acetone
(e) Methyl ethyl ketone
(f) Acetophenone
(g) Benzalacetone
(h) Benzalacetophenone
(i) Benzophenone
(j) Salicylaldehyde
(k) Piperonal
(l) Vanillin

16.32

Write structural formulas for the products formed when propanal reacts with each of the following reagents.

(a) $NaBH_4$ in aqueous NaOH
(b) C_6H_5MgBr then H_2O
(c) $LiAlH_4$ then H_2O
(d) OH^-, H_2O
(e) OH^-, H_2O then heat
(f) H_2 and Pt
(g) $HOCH_2CH_2OH$ and H^+
(h) $CH_3CH{=}P(C_6H_5)_3$
(i) Br_2 in acetic acid
(j) $Ag(NH_3)_2^+$
(k) Hydroxylamine
(l) Semicarbazide
(m) Phenylhydrazine
(n) Cold dilute $KMnO_4$
(o) $HSCH_2CH_2SH$, H^+
(p) $HSCH_2CH_2SH$, H^+, then Raney nickel

16.33

Give structural formulas for the products formed (if any) from the reaction of acetone with each reagent in Problem 16.32.

16.34

What products would be obtained from each of the following reactions of *p*-tolualdehyde?

(a) *p*-Tolualdehyde + acetaldehyde $\xrightarrow{OH^-}$

(b) *p*-Tolualdehyde $\xrightarrow[NaOH]{conc.}$

(c) *p*-Tolualdehyde + formaldehyde $\xrightarrow[\text{NaOH}]{\text{conc.}}$

(d) *p*-Tolualdehyde + cold dilute $KMnO_4$ $\xrightarrow[\text{(2) H}^+]{}$

(e) *p*-Tolualdehyde + $KMnO_4$ $\xrightarrow[\text{(2) H}^+]{\text{(1) heat}}$

(f) *p*-Tolualdehyde + $CH_2{=}P(C_6H_5)_3$ $\longrightarrow$

16.35

What products would be obtained from each of the following reactions of acetophenone?

(a) Acetophenone + HNO_3 $\xrightarrow[H_2SO_4]{}$

(b) Acetophenone + Cl_2 (excess) $\xrightarrow[OH^-]{}$

(c) Acetophenone + $CH_2{=}P(C_6H_5)_3$ $\longrightarrow$

(d) Acetophenone + $NaBH_4$ $\xrightarrow[OH^-]{H_2O}$

(e) Acetophenone + C_6H_5MgBr $\xrightarrow[\text{(2) H}_2\text{O}]{}$

16.36

(a) Give three methods for synthesizing phenyl *n*-propyl ketone from benzene and any other needed reagents. (b) Give three methods for transforming phenyl *n*-propyl ketone into *n*-butylbenzene.

16.37

Review the mechanism given for the Cannizzaro reaction on p. 734. (a) What products would you expect to obtain when benzaldehyde is treated with concentrated NaOD in D_2O? (b) What products would you expect to obtain when benzaldehyde is treated with NaOH in H_2O? (c) Would these experiments help verify the proposed mechanism? How?

16.38

Outline simple chemical tests that would distinguish between each of the following:

(a) Benzaldehyde and benzyl alcohol
(b) Hexanal and 2-hexanone
(c) 2-Hexanone and 3-hexanone
(d) 2-Hexanol and 2-hexanone
(e) 2-Hexanol and 3-hexanol
(f) 3-Hexanone and 3-hexanol
(g) Benzalacetophenone and benzophenone
(h) 1-Phenylethanol and 2-phenylethanol
(i) Pentanal and diethyl ether

(j) $CH_3\overset{O}{\overset{\|}{C}}CH_2\overset{O}{\overset{\|}{C}}CH_3$ and $CH_3\overset{OH}{\overset{|}{C}}{=}CH\overset{O}{\overset{\|}{C}}CH_3$

(k) 2-hydroxytetrahydrofuran (ring: CH_2–CH_2–CH_2–O–C(H)(OH)) and 2-methoxytetrahydrofuran (ring: CH_2–CH_2–CH_2–O–C(H)(OCH_3))

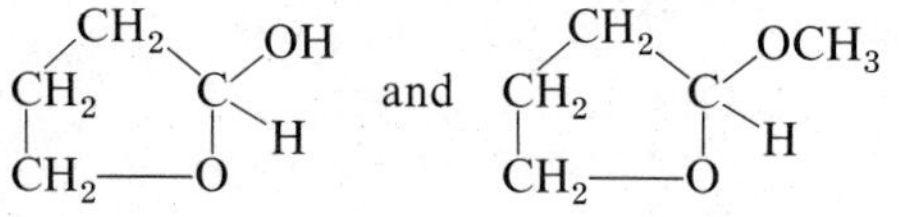

16.39

(a) Infrared spectroscopy gives an easy method for deciding whether the product obtained from the addition of a Grignard reagent to an α,β-unsaturated ketone is the simple addition product or the conjugate addition product. Explain. (What peak or peaks would you look for?) (b) How might you follow the rate of the following reaction using ultraviolet spectroscopy?

$$(CH_3)_2C{=}CH\overset{O}{\overset{\|}{C}}CH_3 + CH_3NH_2 \xrightarrow[H_2O]{} (CH_3)_2\underset{CH_3NH}{\underset{|}{C}}CH_2\overset{O}{\overset{\|}{C}}CH_3$$

16.40
The hydrogens of the γ-carbon of crotonaldehyde are appreciably acidic ($K_a \simeq 10^{-20}$).
(a) Write resonance structures that will explain this fact.

$$\overset{\gamma}{C}H_3\overset{\beta}{C}H{=}\overset{\alpha}{C}HCHO$$
Crotonaldehyde

(b) Write a mechanism that accounts for the following reaction.

$$C_6H_5CH{=}CHCHO + CH_3CH{=}CHCHO \xrightarrow[C_2H_5OH]{\text{base}} C_6H_5(CH{=}CH)_3CHO \quad (87\%)$$

16.41
What reagents would you use to bring about each step of the following syntheses?

(a)
(b)
(c) HIO$_4$ (Section 19.4D)
(d)

16.42
The α-hydrogens of nitroalkanes (i.e., R—CH_2NO_2) are appreciably acidic ($K_a = \sim 10^{-10}$). (a) Write resonance structures for the anion of nitromethane that will account for the acidity of nitromethane. (b) Assuming that you have available, aldehydes, ketones, and nitroalkanes, show how you might synthesize each of the following.
(a) $HOCH_2CH_2NO_2$
(b) $C_6H_5CH{=}CHNO_2$
(c) $C_6H_5CH{=}C(CH_3)NO_2$

16.43

A useful synthesis of sesquiterpene ketones, called *cyperones*, was accomplished through a modification of the Robinson annellation procedure shown below.

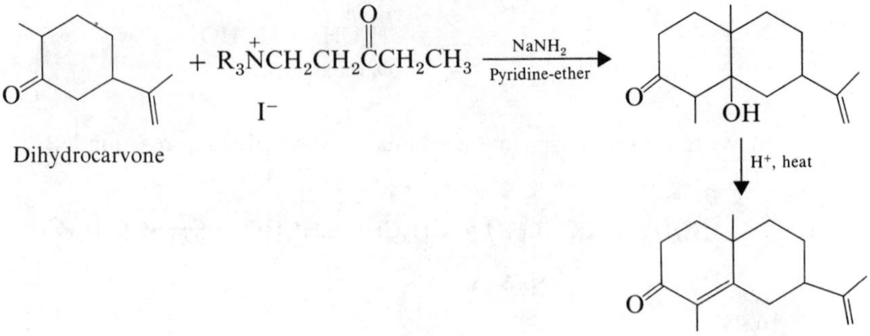

$$\text{Dihydrocarvone} + R_3\overset{+}{N}CH_2CH_2\overset{O}{\overset{\|}{C}}CH_2CH_3\ I^- \xrightarrow[\text{Pyridine-ether}]{NaNH_2}$$

Dihydrocarvone

$\downarrow$ H$^+$, heat

OH

A cyperone

Write a mechanism that accounts for each step of this synthesis.

16.44

Dimethyl sulfoxide reacts with methyl iodide to give trimethylsulfoxonium iodide. Trimethylsulfoxonium iodide forms an ylide when it is treated with sodium hydride.

$$CH_3\ddot{S}(=O)CH_3 + CH_3{-}I \longrightarrow CH_3\overset{+}{S}(CH_3)(=O)CH_3\ I^- \xrightarrow{NaH} CH_2{=}S(CH_3)(=O)CH_3 + NaI + H_2$$

Trimethylsulfoxonium iodide

Ylide

Professor E. J. Corey of Harvard University has shown that this ylide reacts with ketones to give epoxides, that is,

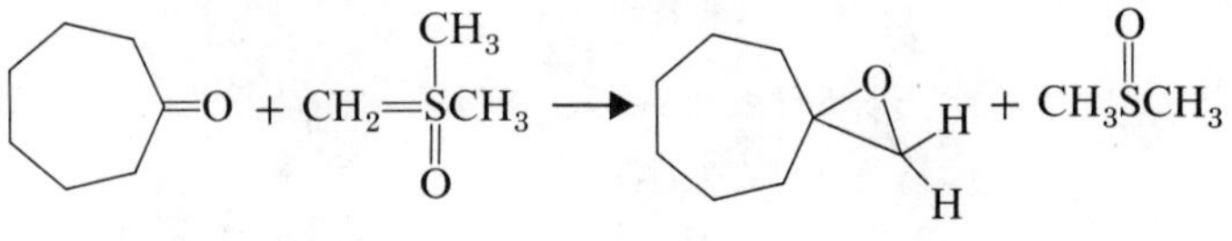

$$\text{cycloheptanone}{=}O + CH_2{=}S(CH_3)(=O)CH_3 \longrightarrow \text{epoxide} + CH_3\overset{O}{\overset{\|}{S}}CH_3$$

(71%)

Propose a mechanism for this epoxide synthesis.

16.45

(a) A compound **U** ($C_9H_{10}O$) gives a negative iodoform test. The infrared spectrum of **U** shows a strong absorption peak at 1690 cm^{-1}. The proton nmr spectrum of **U** gives the following:

Triplet	δ1.2 (3H)
Quartet	δ3.0 (2H)
Multiplet	δ7.7 (5H)

What is the structure of **U**?

(b) A compound **V** is an isomer of **U**. Compound **V** gives a positive iodoform test; its infrared spectrum shows a strong peak at 1705 cm^{-1}. The proton nmr spectrum of **V** gives the following:

Singlet	δ2.0 (3H)
Singlet	δ3.5 (2H)
Multiplet	δ7.1 (5H)

What is the structure of **V**?

16.46
Compounds **W** and **X** are isomers; they have the molecular formula C_9H_8O. The infrared spectra of both compounds show a strong absorption band near 1715 cm^{-1}. Oxidation of either compound with hot, basic potassium permanganate followed by acidification yields phthalic acid. The proton nmr spectrum of **W** shows a multiplet at δ7.3 and a singlet at δ3.4. The proton nmr spectrum of **X** shows a multiplet at δ7.5, a triplet at δ3.1, and a triplet at δ2.5. Propose structures for **W** and **X.**

16.47
Compounds **Y** and **Z** are isomers with the molecular formula $C_{10}H_{12}O$. The infrared spectra of both compounds show a strong absorption band near 1710 cm^{-1}. The proton nmr spectra of **Y** and **Z** are given in Fig. 16.2 and Fig. 16.3. Propose structures for Y and Z.

16.48
Compound **A** has the molecular formula $C_6H_{12}O_3$ and shows a strong infrared absorption peak at 1710 cm^{-1}. When treated with iodine in aqueous sodium hydroxide **A** gives a yellow precipitate. When **A** is treated with Tollens' reagent no reaction occurs; however if **A** is treated first with water containing a drop of sulfuric acid and then treated with Tollens' reagent, a silver mirror forms in the test tube. Compound **A** shows the following proton nmr spectrum.

Singlet δ2.1
Doublet δ2.6
Singlet δ3.2 (6H)
Triplet δ4.7

Write a structure for **A**.

*** 16.49**
3-Hydroxybenzaldehyde undergoes a Cannizzaro reaction readily; 2-hydroxybenzaldehyde and 4-hydroxybenzaldehyde, however, fail to react. Explain.

*** 16.50**
When semicarbazide, $H_2NNHCONH_2$, reacts with a ketone (or an aldehyde) to form a

FIG. 16.2
The proton nmr spectrum of compound Y, Problem 16.47. (Courtesy Aldrich Chemical Co., Milwaukee, Wis.)

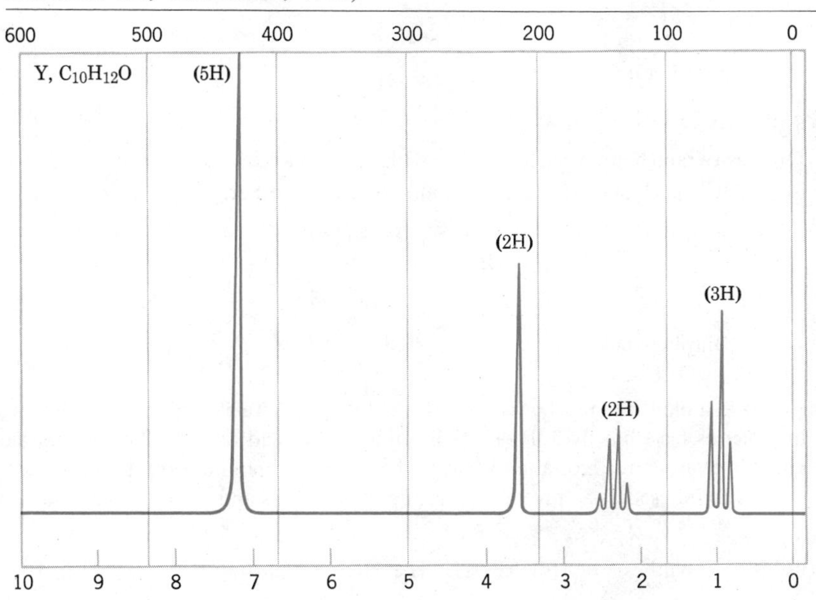

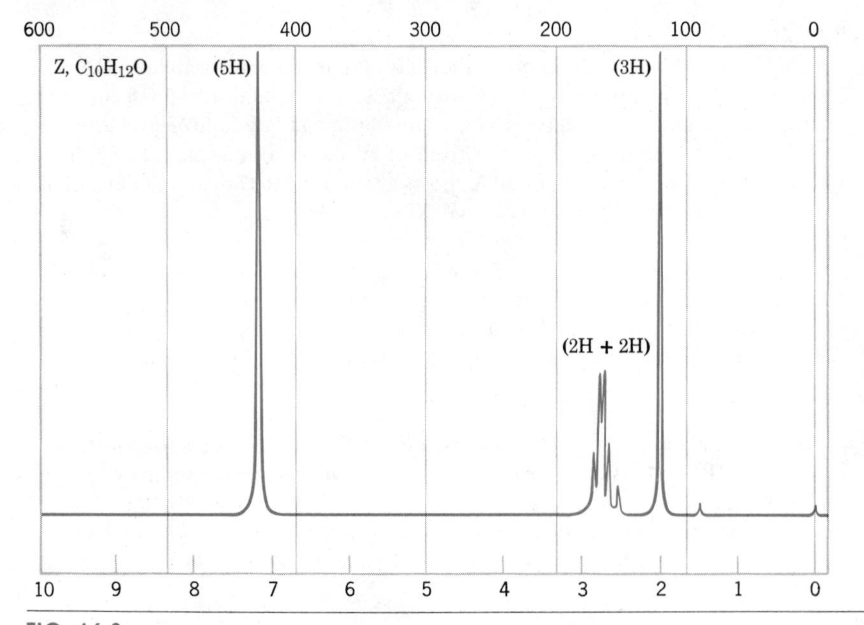

FIG. 16.3
The proton nmr spectrum of compound Z, Problem 16.47. (Courtesy Aldrich Chemical Co., Milwaukee, Wis.)

semicarbazone (Section 16.8) only one nitrogen of semicarbazide acts as a nucleophile and attacks the carbonyl carbon of the ketone. The product of the reaction, consequently, is $R_2C{=}NNHCONH_2$ rather than $R_2C{=}NCONHNH_2$. What factor accounts for the fact that two nitrogens of semicarbazide are relatively nonnucleophilic?

*** 16.51**

Treating a solution of *cis*-1-decalone (below) with base causes an isomerization to take place. When the system reaches equilibrium the solution is found to contain about 95% *trans*-1-decalone and about 5% *cis*-1-decalone. Explain.

H O

H

cis-1-Decalone

*** 16.52**

Dihydropyran (below) reacts readily with an alcohol in the presence of a trace of anhydrous HCl or H_2SO_4, to form a tetrahydropyranyl ether.

$$\text{Dihydropyran} + ROH \xrightarrow{H^+} \text{Tetrahydropyranyl ether (O, OR)}$$

Dihydropyran   Tetrahydropyranyl ether

(a) Write a plausible mechanism for this reaction. (b) Tetrahydropyranyl ethers are stable in aqueous base but hydrolyze rapidly in aqueous acid to yield the original alcohol and another compound. Explain. (What is the other compound?) (c) The tetrahydropyranyl group can be used as a "protecting" group for alcohols and phenols. Show how you might use it in a synthesis of $HOCH_2CH_2CH_2CH_2C(CH_3)_2OH$ starting with $HOCH_2CH_2CH_2CH_2Cl$.

17 CARBOXYLIC ACIDS AND THEIR DERIVATIVES: NUCLEOPHILIC SUBSTITUTION AT ACYL CARBON

17.1 INTRODUCTION

The carboxyl group, $-\overset{O}{\overset{\|}{C}}OH$, is one of the most widely occurring functional groups in chemistry and biochemistry. Not only are carboxylic acids themselves important, but the carboxyl group is the parent group of a large family of related compounds (Table 17.1).

All of these carboxylic acid derivatives contain the acyl group:

$$R-C{\overset{\displaystyle O}{\diagup\!\!\diagup}}\atop{\diagdown}$$

An acyl group

As a result, they are often called *acyl compounds.*

TABLE 17.1 Carboxylic Acid Derivatives

STRUCTURE	NAME
$R-\overset{O}{\overset{\|}{C}}-Cl$	Acyl (or acid) chloride
$R-\overset{O}{\overset{\|}{C}}-O-\overset{O}{\overset{\|}{C}}-R$	Acid anhydride
$R-\overset{O}{\overset{\|}{C}}-O-R'$	Ester
$R-\overset{O}{\overset{\|}{C}}-NH_2$	Amides
$R-\overset{O}{\overset{\|}{C}}-NHR'$	Amides
$R-\overset{O}{\overset{\|}{C}}-NR_2$	Amides
$R-\overset{O}{\overset{\|}{C}}-SR'$	Thiol ester

17.2A Carboxylic Acids

IUPAC names for carboxylic acids are obtained by dropping the final *e* of the name of the alkane corresponding to the longest chain in the acid and by adding *oic acid.* The carboxyl carbon is assigned number 1. The examples listed below illustrate how this is done.

$$\underset{\substack{\text{Methanoic}\\\text{acid}\\\text{(formic acid)}}}{\text{H}\overset{\text{O}}{\overset{\|}{\text{C}}}\text{OH}} \qquad \underset{\substack{\text{Ethanoic}\\\text{acid}\\\text{(acetic acid)}}}{\text{CH}_3\overset{\text{O}}{\overset{\|}{\text{C}}}\text{OH}} \qquad \underset{\substack{\text{Propanoic}\\\text{acid}\\\text{(propionic acid)}}}{\text{CH}_3\text{CH}_2\overset{\text{O}}{\overset{\|}{\text{C}}}\text{OH}} \qquad \underset{\substack{\text{Butanoic}\\\text{acid}\\\text{(butyric acid)}}}{\text{CH}_3\text{CH}_2\text{CH}_2\overset{\text{O}}{\overset{\|}{\text{C}}}\text{OH}}$$

$$\underset{\text{4-Methylhexanoic acid}}{\overset{6}{\text{C}}\text{H}_3\overset{5}{\text{C}}\text{H}_2\underset{\substack{|\\\text{CH}_3}}{\overset{4}{\text{C}}}\text{H}\overset{3}{\text{C}}\text{H}_2\overset{2}{\text{C}}\text{H}_2\overset{\text{O}}{\overset{\|}{\text{C}}}\text{OH}} \qquad \underset{\text{4-Hexenoic acid}}{\overset{6}{\text{C}}\text{H}_3\overset{5}{\text{C}}\text{H}{=}\overset{4}{\text{C}}\text{H}\overset{3}{\text{C}}\text{H}_2\overset{2}{\text{C}}\text{H}_2\overset{\text{O}}{\overset{\|}{\text{C}}}\text{OH}}$$

Many carboxylic acids have common names that are derived from Latin or Greek words that indicate one of their natural sources (Table 17.2). Methanoic

TABLE 17.2 Carboxylic Acids

STRUCTURE	IUPAC NAME	COMMON NAME	mp °C	bp °C	WATER SOL. (g/100 g H_2O) 25°	K_a (at 25°)
HCOOH	Methanoic acid	Formic acid	8	100.5	∞	1.77×10^{-4}
CH_3COOH	Ethanoic acid	Acetic acid	16.6	118	∞	1.76×10^{-5}
CH_3CH_2COOH	Propanoic acid	Propionic acid	−21	141	∞	1.34×10^{-5}
$CH_3(CH_2)_2COOH$	Butanoic acid	Butyric acid	− 6	164	∞	1.54×10^{-5}
$CH_3(CH_2)_3COOH$	Pentanoic acid	Valeric acid	−34	187	4.97	1.52×10^{-5}
$CH_3(CH_2)_4COOH$	Hexanoic acid	Caproic acid	− 3	205	1.08	1.31×10^{-5}
$CH_3(CH_2)_6COOH$	Octanoic acid	Caprylic acid	16	239	0.07	1.28×10^{-5}
$CH_3(CH_2)_8COOH$	Decanoic acid	Capric acid	31	269	0.015	1.43×10^{-5}
$CH_3(CH_2)_{10}COOH$	Dodecanoic acid	Lauric acid	44	179^{18}	0.006	
$CH_3(CH_2)_{12}COOH$	Tetradecanoic acid	Myristic acid	54	200^{20}	0.002	
$CH_3(CH_2)_{14}COOH$	Hexadecanoic acid	Palmitic acid	63	219^{17}	0.0007	
$CH_3(CH_2)_{16}COOH$	Octadecanoic acid	Stearic acid	70	235^{20}	0.0003	
$CH_2ClCOOH$	Chloroacetic acid		63	189	Very sol.	1.40×10^{-3}
$CHCl_2COOH$	Dichloroacetic acid		10.8	192	Very sol.	3.32×10^{-2}
CCl_3COOH	Trichloroacetic acid		56.3	198	Very sol.	2.00×10^{-1}
$CH_3CHClCOOH$	2-Chloropropanoic acid			186	Sol.	1.47×10^{-3}
CH_2ClCH_2COOH	3-Chloropropanoic acid		61	204	Sol.	1.04×10^{-4}
C_6H_5COOH	Benzoic acid		122	250	0.34	6.46×10^{-5}
p-$CH_3C_6H_4COOH$	*p*-Toluic acid		180	275	0.03	4.33×10^{-5}
p-ClC_6H_4COOH	*p*-Chlorobenzoic acid		242		0.009	1.04×10^{-4}
p-$NO_2C_6H_4COOH$	*p*-Nitrobenzoic acid		242		0.03	3.93×10^{-4}
(naphthalene ring with COOH at position 1)	1-Naphthoic acid		160		Insol.	2.00×10^{-4}
(naphthalene ring with COOH at position 2)	2-Naphthoic acid		185		Insol.	6.80×10^{-5}

acid is called formic acid (from the Latin, *formica,* or ant). Ethanoic acid is called acetic acid (from the Latin, *acetum,* or vinegar). Butanoic acid is one compound responsible for the odor of rancid butter, thus its common name is butyric acid (from the Latin, *butyrum,* or butter). Hexanoic acid is one compound associated with the odor of goats, hence its common name, caproic acid (from the Latin *caper,* or goat). Pentanoic acid, with odor in between that of rancid butter and goats, is appropriately named valeric acid (from the Latin, *valerum,* to be strong). Octadecanoic acid takes its common name, stearic acid, from the Greek word, *stear,* for tallow.

Most of these common names have been with us for a long time and some are likely to remain in common usage for even longer, so it is helpful to be familiar with them. In this text we will always refer to methanoic acid and ethanoic acid as formic acid and acetic acid. However, in almost all other instances we will use IUPAC names.

Carboxylic acids are polar substances. Their molecules can form strong hydrogen bonds with each other and with water. As a result, carboxylic acids generally have high boiling points, and low-molecular-weight carboxylic acids show appreciable solubility in water. The first four carboxylic acids (Table 17.2) are miscible with water in all proportions. As the length of the carbon chain increases, water solubility declines.

17.2B Carboxylate Salts

Salts of carboxylic acids are named as *-ates* in common nomenclature and as *-oates* in the IUPAC system. Thus, CH_3COONa is sodium acetate or sodium ethanoate. Other examples are:

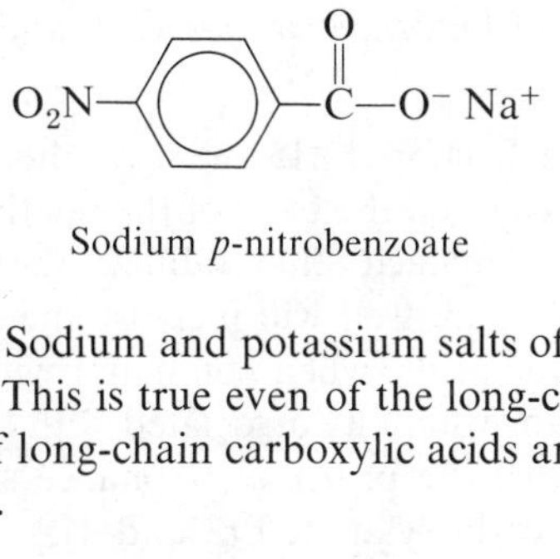

$$O_2N\text{—}C_6H_4\text{—}\overset{\overset{\displaystyle O}{\|}}{C}\text{—}O^-\ Na^+ \qquad CH_3CH_2\underset{\underset{\displaystyle CH_3}{|}}{C}HCH_2\overset{\overset{\displaystyle O}{\|}}{C}\text{—}O^-\ K^+$$

Sodium *p*-nitrobenzoate 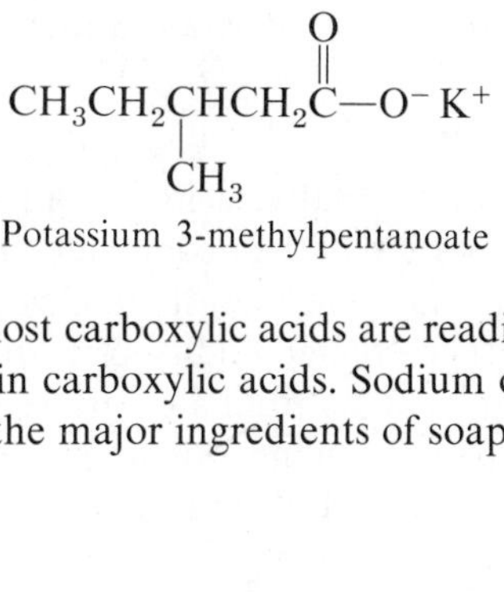 — Potassium 3-methylpentanoate

Sodium and potassium salts of most carboxylic acids are readily soluble in water. This is true even of the long-chain carboxylic acids. Sodium or potassium salts of long-chain carboxylic acids are the major ingredients of soap (cf., Section 21.2B).

17.2C Acidity of Carboxylic Acids

Most unsubstituted carboxylic acids have K_a's in the range of 10^{-4} to 10^{-5} as seen in Table 17.2. This means that carboxylic acids react readily with aqueous solutions of sodium hydroxide and sodium bicarbonate to form soluble sodium salts. We can use solubility tests, therefore, to distinguish water-insoluble carboxylic acids from water-insoluble phenols and alcohols. Water-insoluble carboxylic acids will dissolve in either aqueous sodium hydroxide or aqueous sodium bicarbonate:

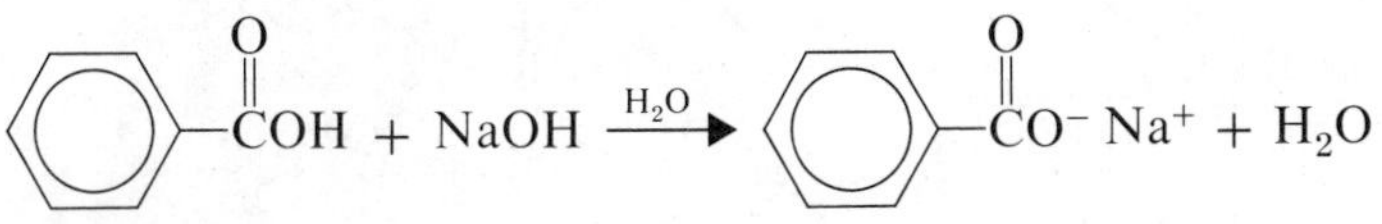

$$C_6H_5\text{—}\overset{\overset{\displaystyle O}{\|}}{C}OH + NaOH \xrightarrow{H_2O} C_6H_5\text{—}\overset{\overset{\displaystyle O}{\|}}{C}O^-\ Na^+ + H_2O$$

Benzoic acid (water insoluble) — Sodium benzoate (water soluble)

$$C_6H_5COOH + NaHCO_3 \xrightarrow{H_2O} C_6H_5CO_2^- Na^+ + CO_2\uparrow + H_2O$$

(water insoluble) (water soluble)

Water-insoluble phenols dissolve in aqueous sodium hydroxide but (except for the nitrophenols) do not dissolve in aqueous sodium bicarbonate:

$$CH_3C_6H_4OH + NaOH \xrightarrow{H_2O} CH_3C_6H_4O^- Na^+ + H_2O$$

p-Cresol (water insoluble) — Sodium *p*-cresoxide (water soluble)

$$CH_3C_6H_4OH + NaHCO_3 \xrightarrow{H_2O} \text{No reaction}$$

(water insoluble)

Water-insoluble alcohols do not dissolve in either aqueous sodium hydroxide or sodium bicarbonate.

Problem 17.1

(a) When excess carbon dioxide is passed into a solution of sodium benzoate and sodium *p*-cresoxide in aqueous sodium hydroxide, *p*-cresol separates from the solution (as an oil) but sodium benzoate remains in solution. Write equations that will provide an explanation for this. (b) Given the following reagents—aqueous sodium hydroxide, aqueous hydrochloric acid, ether, and carbon dioxide—explain how you would separate a mixture of benzoic acid, *p*-cresol, and cyclohexanol.

When we studied phenols in Section 15.15, we saw there that we can account for relative acidities of molecules on the basis of the relative stabilities of the acids and of their conjugate bases. Any factor that stabilizes the conjugate base of an acid more than it stabilizes the acid itself will increase the strength of the acid. The greater acidity of carboxylic acids, when compared with alcohols, for example, arises from an extra resonance stability associated with the carboxylate ion. To understand this let us examine the principal resonance structures for a carboxylic acid (**1** and **2**) and for a carboxylate ion (**3** and **4**).

$$R{-}C(=O){-}OH \;(\mathbf{1}) \longleftrightarrow R{-}C(-O^-)={O^+}{-}H \;(\mathbf{2}) \;=\; R{-}C(O^{\delta-})(O^{\delta+}{-}H)$$

1 **2**

This structure has separated charges

$$R{-}C(=O){-}O^- \;(\mathbf{3}) \longleftrightarrow R{-}C(-O^-)=O \;(\mathbf{4}) \;=\; R{-}C(O^{\delta-})(O^{\delta-})$$

3 **4**

These structures are equivalent

A carboxylic acid can be represented by two *nonequivalent* structures, one of which involves charge separation. The carboxylate anion, on the other hand, is a hybrid of two *equivalent* structures neither of which involves separated charges. Resonance theory (Section 10.11) tells us that because of this, we should expect much greater resonance stabilization for the carboxylate ion than for the acid. Since resonance stabilizes the carboxylate ion more than it does the acid, the ionization of a carboxylic acid is energetically more favorable than the corresponding ionization of an alcohol, where neither the alcohol nor the alkoxide ion is resonance stabilized.

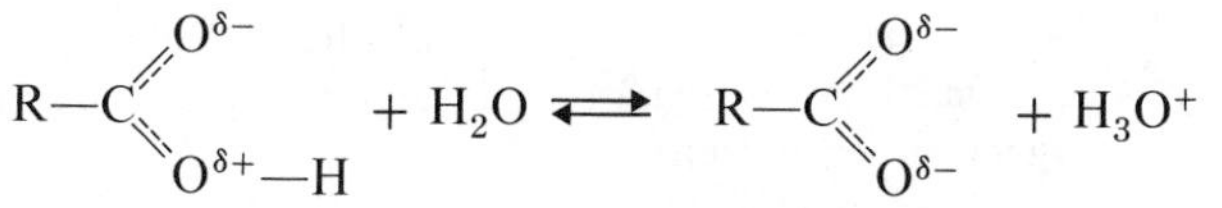

$$R{-}C(\overset{\delta-}{O})(\overset{\delta+}{O}{-}H) + H_2O \rightleftharpoons R{-}C(\overset{\delta-}{O})(\overset{\delta-}{O}) + H_3O^+$$

Small resonance stabilization — *Large resonance stabilization*

Carboxylate ion is stabilized more than the acid— K_a is larger

$$R{-}OH + H_2O \rightleftharpoons R{-}O^- + H_3O^+$$

No resonance stabilization — *No resonance stabilization*

Neither the acid nor the alkoxide ion is stabilized— K_a is smaller

Problem 17.2

We see in Table 17.2 that carboxylic acids having electron-withdrawing groups are stronger than unsubstituted acids. The chloroacetic acids, for example, show the following order of acidities:

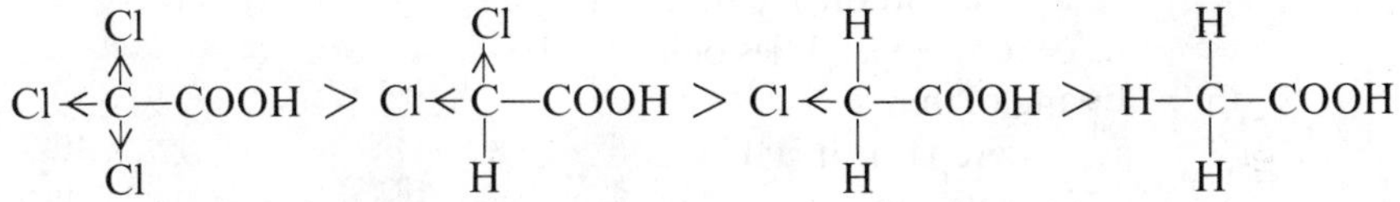

$$Cl_3C{-}COOH > Cl_2CH{-}COOH > ClCH_2{-}COOH > H_3C{-}COOH$$

	Cl_3CCOOH	$Cl_2CHCOOH$	$ClCH_2COOH$	CH_3COOH
K_a:	2.0×10^{-1}	3.32×10^{-2}	1.40×10^{-3}	1.76×10^{-5}

Explain this acid-strengthening effect of electron-withdrawing groups.

Problem 17.3

Which acid of each pair shown below would you expect to be stronger?

(a) CH_3COOH or CH_2FCOOH
(b) CH_2FCOOH or $CH_2ClCOOH$
(c) $CH_2ClCOOH$ or $CH_2BrCOOH$
(d) $CH_2ClCH_2CH_2COOH$ or $CH_3CHClCH_2COOH$
(e) $CH_3CH_2CHClCOOH$ or $CH_3CHClCH_2COOH$
(f) 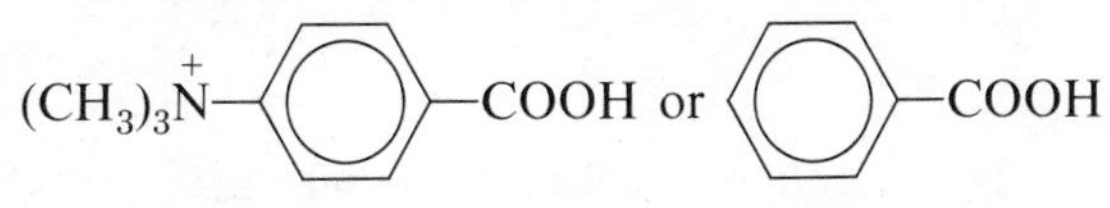 $(CH_3)_3\overset{+}{N}{-}C_6H_4{-}COOH$ or $C_6H_5{-}COOH$
(g) 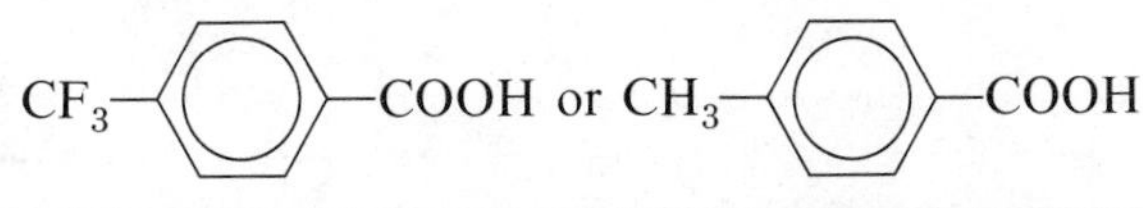 $CF_3{-}C_6H_4{-}COOH$ or $CH_3{-}C_6H_4{-}COOH$

Molecular orbital theory can also be used to describe the carboxylate ion. The molecular orbitals are similar to those of an allylic system (Section 10.3). The orbital that contains the extra electron (giving the carboxylate ion its negative charge) is an orbital that has a node at the central carbon:

Node → R—C(O)(O) [orbital lobes: (+) (−) on upper O; (−) (+) on lower O]

Molecular orbital theory then gives us the same picture of the carboxylate ion as resonance theory—one in which the charge is delocalized (giving it extra stability) and one in which the two oxygens are equivalent.

17.2D Dicarboxylic Acids

Dicarboxylic acids are named as *alkanedioic acids* in the IUPAC system. Most simple dicarboxylic acids have common names (Table 17.3), and these are the names that we will use.

Problem 17.4

Suggest explanations for the following facts.

(a) K_1 for all of the dicarboxylic acids in Table 17.3 are higher than the K_a for monocarboxylic acids with the same number of carbons.

TABLE 17.3 Dicarboxylic Acids

STRUCTURE	COMMON NAME	mp °C	K_a AT 25° K_1	K_2
HOOC—COOH	Oxalic acid	189 (dec.)	5.9×10^{-2}	6.4×10^{-5}
$HOOCCH_2COOH$	Malonic acid	136	1.4×10^{-3}	2.0×10^{-6}
$HOOC(CH_2)_2COOH$	Succinic acid	182	6.9×10^{-5}	2.5×10^{-6}
$HOOC(CH_2)_3COOH$	Glutaric acid	98	4.6×10^{-5}	3.9×10^{-6}
$HOOC(CH_2)_4COOH$	Adipic acid	153	3.7×10^{-5}	2.4×10^{-6}
cis-HOOC—CH=CH—COOH	Maleic acid	131	1.4×10^{-2}	8.6×10^{-7}
trans-HOOC—CH=CH—COOH	Fumaric acid	307	9.3×10^{-4}	3.6×10^{-5}
Benzene ring with COOH, COOH (ortho)	Phthalic acid	231	1.3×10^{-3}	3.9×10^{-6}
Benzene ring with COOH, COOH (meta)	Isophthalic acid	345	2.9×10^{-4}	2.5×10^{-5}
Benzene ring with COOH, COOH (para)	Terephthalic acid	Sublimes	3.1×10^{-4}	1.5×10^{-5}

(b) The difference between K_1 and K_2 for dicarboxylic acids of type $HOOC(CH_2)_nCOOH$ decreases as n increases.

17.2E Esters

The names of esters are derived from the names of the alcohol (with the ending *-yl*) and the acid (with the ending *-ate* or *-oate*). The portion of the name derived from the alcohol comes first.

$CH_3C(=O)—OCH_2CH_3$

Ethyl acetate or ethyl ethanoate

$CH_3CH_2C(=O)—OC(CH_3)_2—CH_3$

tert-Butyl propanoate

$Cl—C_6H_4—C(=O)OCH_3$

Methyl *p*-chlorobenzoate

$CH_3C(=O)OCH{=}CH_2$

Vinyl acetate or ethenyl ethanoate

$CH_3CH_2OC(=O)—C(=O)OCH_2CH_3$

Diethyl oxalate

$CH_3CH_2OC(=O)CH_2C(=O)OCH_2CH_3$

Diethyl malonate

Esters are polar compounds but their molecules cannot form strong hydrogen bonds to each other. As a result, esters have boiling points that are lower than those of acids and alcohols of comparable molecular weight. The boiling points (Table 17.4) of esters are about the same as those of comparable aldehydes and ketones. Esters have lower solubilities in water than acids and alcohols.

Unlike the low-molecular-weight acids, esters usually have pleasant odors,

TABLE 17.4 Carboxylic Esters

NAME	STRUCTURE	mp °C	bp °C	SOLUBILITY IN WATER (g/100 g at 20 °C)
Methyl formate	$HCOOCH_3$	−99	31.5	Very sol.
Ethyl formate	$HCOOCH_2CH_3$	−79	54	
Methyl acetate	CH_3COOCH_3	−99	57	24.4
Ethyl acetate	$CH_3COOCH_2CH_3$	−82	77	7.39 (25 °C)
n-Propyl acetate	$CH_3COOCH_2CH_2CH_3$	−93	102	1.89
n-Butyl acetate	$CH_3COOCH_2(CH_2)_2CH_3$	−78	125	1.0 (22 °C)
Ethyl propanoate	$CH_3CH_2COOCH_2CH_3$	−73	99	1.75
Ethyl butanoate	$CH_3(CH_2)_2COOCH_2CH_3$	−93	120	0.51
Ethyl pentanoate	$CH_3(CH_2)_3COOCH_2CH_3$	−91	145	0.22
Ethyl hexanoate	$CH_3(CH_2)_4COOCH_2CH_3$	−68	168	0.063
Methyl benzoate	$C_6H_5COOCH_3$	−12	199	Insol.
Ethyl benzoate	$C_6H_5COOCH_2CH_3$	−35	213	Insol.
Phenyl acetate	$CH_3COOC_6H_5$		196	Insol.
Methyl salicylate	o-$HOC_6H_4COOCH_3$	− 9	223	Insol.

some resembling those of fruits, and these are used in the manufacture of synthetic flavors:

$$CH_3\overset{O}{\overset{\|}{C}}OCH_2CH_2\underset{CH_3}{\underset{|}{C}}HCH_3$$

Isopentyl acetate
(used in synthetic banana flavor)

$$CH_3CH_2CH_2CH_2\overset{O}{\overset{\|}{C}}{-}OCH_2CH_2\underset{CH_3}{\underset{|}{C}}HCH_3$$

Isopentyl pentanoate
(used in synthetic apple flavor)

$$CH_3(CH_2)_2\overset{O}{\overset{\|}{C}}OCH_2CH_2CH_2CH_3$$

Butyl butanoate
(used in synthetic pineapple flavor)

$$CH_3CH_2\overset{O}{\overset{\|}{C}}OCH_2\underset{CH_3}{\underset{|}{C}}HCH_3$$

Isobutyl propanoate
(used in synthetic rum flavor)

Although it is not fragrant, glyceryl trimyristate is a major constituent of nutmeg and methyl salicylate (highly fragrant) is the chief component of oil of wintergreen.

$$\begin{array}{l} CH_2O\overset{O}{\overset{\|}{C}}(CH_2)_{12}CH_3 \\ | \\ CHO\overset{O}{\overset{\|}{C}}(CH_2)_{12}CH_3 \\ | \\ CH_2O\overset{O}{\overset{\|}{C}}(CH_2)_{12}CH_3 \end{array}$$

Glyceryl trimyristate
(in nutmeg)

$$o\text{-}HOC_6H_4\underset{O}{\underset{\|}{C}}{-}OCH_3$$

Methyl salicylate
(oil of wintergreen)

17.2F Carboxylic Anhydrides

Most anhydrides are named by dropping the word *acid* from the name of the carboxylic acid and then adding the word *anhydride*.

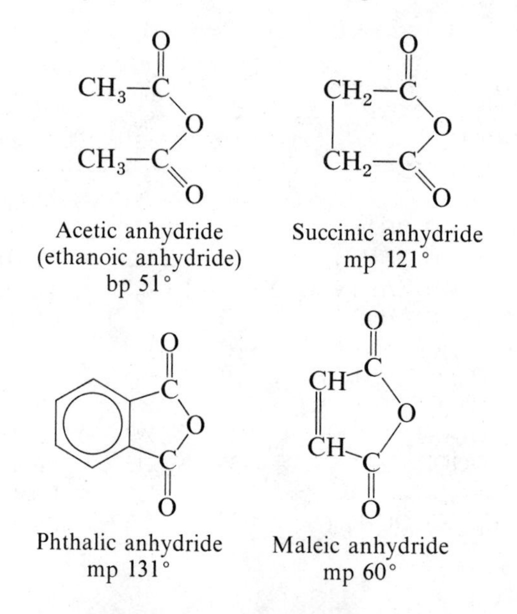

Acetic anhydride
(ethanoic anhydride)
bp 51°

Succinic anhydride
mp 121°

Phthalic anhydride
mp 131°

Maleic anhydride
mp 60°

17.2G Acyl Chlorides

Acyl chlorides are usually called *acid chlorides*. They are named by dropping *-ic* acid from the name of the acid and then adding *yl chloride*. Examples are:

$$CH_3\overset{\overset{\displaystyle O}{\|}}{C}{-}Cl \qquad CH_3CH_2\overset{\overset{\displaystyle O}{\|}}{C}{-}Cl \qquad C_6H_5\overset{\overset{\displaystyle O}{\|}}{C}{-}Cl$$

Acetyl chloride (ethanoyl chloride) mp −112°, bp 51°

Propanoyl chloride mp −94°, bp 80°

Benzoyl chloride mp −1°, bp 197°

Acyl chlorides and carboxylic anhydrides have boiling points in the same range as esters of comparable molecular weight.

17.2H Amides

Amides that have no substituent on nitrogen are named by dropping *-ic* acid from the common name of the acid (or *-oic* acid from the IUPAC name) and then adding *-amide*.

$$CH_3\overset{\overset{\displaystyle O}{\|}}{C}{-}NH_2 \qquad CH_3CH_2\overset{\overset{\displaystyle O}{\|}}{C}{-}NH_2 \qquad C_6H_5{-}\overset{\overset{\displaystyle O}{\|}}{C}{-}NH_2$$

Acetamide or ethanamide mp 82°, bp 221°

Propanamide mp 79°, bp 222°

Benzamide mp 130°, bp 290°

Substituents on the nitrogen atom of amides are named as alkyl groups and the named substituent is prefaced by *N*-, or *N,N*-. Examples are:

$$CH_3\overset{\overset{\displaystyle O}{\|}}{C}{-}N(CH_3)_2 \qquad CH_3\overset{\overset{\displaystyle O}{\|}}{C}{-}NHC_2H_5$$

N,N-Dimethylacetamide bp 185°

N-Ethylacetamide bp 205°

Molecules of amides with one (or no) substituent on nitrogen are able to form strong hydrogen bonds to each other and, consequently, such amides have high melting points and boiling points. Molecules of *N,N*-disubstituted amides cannot form strong hydrogen bonds to each other; they have lower melting points and boiling points.

$$R{-}C(=\ddot{O}^{\delta-}{:}\text{---}H^{\delta+}{-}N(R'){-}\overset{\overset{\displaystyle O}{\|}}{C}{-}R){-}NH{-}R'$$

Hydrogen bonding between molecules of an amide

17.21 Spectroscopic Properties of Acyl Compounds

Infrared spectroscopy is of considerable importance in identifying carboxylic acids and their derivatives. The C═O stretching band is one of the most prominent in their infrared spectra since it is always a strong band. The C═O stretching band occurs at different frequencies for acids, esters, and amides, and its precise location is often helpful in structure determination. Table 17.5 gives the location of this band for most acyl compounds.

The hydroxyl groups of carboxylic acids also give rise to a broad peak in the 2500 to 2700-cm^{-1} region arising from O—H stretching vibrations. The N—H stretching vibrations of amides absorb between 3140 and 3500 cm^{-1}.

The acidic protons of carboxylic acids usually absorb very far downfield (δ10–12) in their proton nmr spectra.

TABLE 17.5 Carbonyl Stretching Absorptions of Acyl Compounds

TYPE OF COMPOUND	FREQUENCY RANGE, cm^{-1}
Carboxylic Acids	
R—COOH	1700–1725
—C═C—COOH (with substituent bonds on C)	1690–1715
ArCOOH	1680–1700
Acid Anhydrides	
R—C(═O)—O—C(═O)—R	1800–1850 and 1740–1790
Ar—C(═O)—O—C(═O)—Ar	1780–1830 and 1730–1770
Acyl Chlorides	
R—C(═O)Cl and Ar—C(═O)Cl	1780–1850
Esters	
R—C(═O)—OR	1735–1750
Ar—C(═O)—OR	1715–1730
Amides	
RC(═O)—NH_2, RC(═O)NHR and RC(═O)NR_2	1630–1690
Carboxylate Ions	
$RCOO^-$	1550–1630

17.3 PREPARATION OF CARBOXYLIC ACIDS

Most of the methods for the preparation of carboxylic acids are familiar ones.

1. **By oxidation of alkenes** (discussed in Section 7.13).

$$R{-}CH{=}CHR' \xrightarrow[\text{heat}]{KMnO_4} RCOOH + R'COOH$$

2. **By oxidation of aldehydes and primary alcohols** (discussed in Sections 16.17 and 15.8).

$$R{-}CHO \xrightarrow[\text{or } Ag(NH_3)_2^+]{\text{dil. } KMnO_4} RCOOH$$

$$RCH_2OH \xrightarrow[\text{heat, (2) } H^+]{KMnO_4,\ OH^-} RCOOH$$

3. **By oxidation of alkylbenzenes** (discussed in Section 12.10).

$$C_6H_5{-}CH_3 \xrightarrow[\text{heat, (2) } H^+]{KMnO_4,\ OH^-} C_6H_5{-}COOH$$

4. **By oxidation of methyl ketones** (discussed in Section 16.16).

$$R{-}\overset{O}{\overset{\|}{C}}{-}CH_3 \xrightarrow[(2)\ H^+]{(1)\ X_2/NaOH} R{-}\overset{O}{\overset{\|}{C}}OH + CHX_3$$

5. **By hydrolysis of cyanohydrins and other nitriles.** We saw, in Section 16.9, that aldehydes and ketones can be converted to cyanohydrins, and that these can be hydrolyzed to α-hydroxy acids.

$$R_2C{=}O + HCN \rightleftharpoons R_2C(OH)CN \xrightarrow[H_2O]{H^+} R{-}\underset{R}{\underset{|}{\overset{OH}{\overset{|}{C}}}}{-}COOH$$

Nitriles can also be prepared by nucleophilic substitution reactions of alkyl halides with sodium cyanide. Hydrolysis of the nitrile yields a carboxylic acid:

General Reaction

$$R{-}CH_2X + CN^- \longrightarrow RCH_2CN \xrightarrow[H_2O,\ \text{heat}]{H^+} RCH_2COOH + NH_4^+$$

$$RCH_2CN \xrightarrow[H_2O,\ \text{heat}]{OH^-} RCH_2COO^- + NH_3$$

Specific Examples

$$HOCH_2CH_2Cl \xrightarrow[(80\%)]{NaCN} HOCH_2CH_2CN \xrightarrow[(2)\ H_3O^+\ (75\text{–}80\%)]{(1)\ OH^-,\ H_2O} HOCH_2CH_2COOH$$

2-Hydroxypropanenitrile → 2-Hydroxypropanoic acid

$$\mathrm{BrCH_2CH_2CH_2Br} \xrightarrow[(77\text{–}86\%)]{\mathrm{NaCN}} \underset{\text{Pentanedinitrile}}{\mathrm{NCCH_2CH_2CH_2CN}} \xrightarrow[(83\text{–}85\%)]{\mathrm{H_3O^+}} \underset{\text{Glutaric acid}}{\mathrm{HOOCCH_2CH_2CH_2COOH}}$$

This synthetic method is generally limited to the use of *primary alkyl* halides. The cyanide ion is a relatively strong base and the use of secondary and tertiary alkyl halides leads primarily to an alkene (through elimination) rather than to a nitrile (through substitution). Aryl halides (except for those with ortho and para nitro groups) do not react with sodium cyanide.

6. **By carbonation of Grignard reagents.** Grignard reagents react with carbon dioxide to yield magnesium carboxylates (Section 14.11). Acidification produces a carboxylic acid.

$$\mathrm{R{-}X + Mg} \xrightarrow[\text{Ether}]{} \mathrm{RMgX} \xrightarrow{\mathrm{CO_2}} \mathrm{RCOOMgX} \xrightarrow{\mathrm{H^+}} \mathrm{RCOOH}$$

or

$$\mathrm{Ar{-}Br + Mg} \xrightarrow[\text{Ether}]{} \mathrm{ArMgBr} \xrightarrow{\mathrm{CO_2}} \mathrm{ArCOOMgBr} \xrightarrow{\mathrm{H^+}} \mathrm{ArCOOH}$$

This synthesis of carboxylic acids is applicable to primary, secondary, tertiary, allyl, benzyl, and aryl halides, provided they have no groups incompatible with a Grignard reaction (cf. Section 14.12*B*).

$$\underset{\textit{tert}\text{-Butyl chloride}}{\mathrm{CH_3{-}C(CH_3)_2{-}Cl}} \xrightarrow[\text{ether}]{\mathrm{Mg}} \mathrm{CH_3C(CH_3)_2MgCl} \xrightarrow[\text{(2) } \mathrm{H_2O}]{\text{(1) } \mathrm{CO_2}} \underset{\substack{\text{2,2-Dimethylpropanoic acid} \\ (79\text{–}80\%\text{ overall})}}{\mathrm{CH_3C(CH_3)_2COOH}}$$

$$\underset{n\text{-Butyl chloride}}{\mathrm{CH_3CH_2CH_2CH_2Cl}} \xrightarrow[\text{ether}]{\mathrm{Mg}} \mathrm{CH_3CH_2CH_2CH_2MgCl} \xrightarrow[\text{(2) } \mathrm{H_2O}]{\text{(1) } \mathrm{CO_2}} \underset{\substack{\text{Pentanoic acid} \\ (80\%\text{ overall})}}{\mathrm{CH_3CH_2CH_2CH_2COOH}}$$

$$\mathrm{C_6H_5{-}Br} \xrightarrow[\text{ether}]{\mathrm{Mg}} \mathrm{C_6H_5{-}MgBr} \xrightarrow[\text{(2) } \mathrm{H_2O}]{\text{(1) } \mathrm{CO_2}} \underset{\substack{\text{Benzoic acid} \\ (85\%)}}{\mathrm{C_6H_5{-}COOH}}$$

Problem 17.5

Show how you would prepare each of the following carboxylic acids through a Grignard synthesis.

(a) Phenylacetic acid
(b) 2,2-Dimethylpentanoic acid
(c) 3-Butenoic acid
(d) *p*-Toluic acid
(e) Hexanoic acid

Problem 17.6

(a) Which of the carboxylic acids in Problem 17.5 could be prepared by a nitrile synthesis as well? (b) Which synthesis, Grignard or nitrile, would you choose to prepare $\mathrm{HOCH_2CH_2CH_2CH_2COOH}$ from $\mathrm{HOCH_2CH_2CH_2CH_2Br}$? Why?

17.4 NUCLEOPHILIC SUBSTITUTIONS AT ACYL CARBON

In our study of carbonyl compounds in the last chapter, we saw that a characteristic reaction of aldehydes and ketones is one of *nucleophilic addition* to the carbon-oxygen double bond.

$$\underset{\text{aldehyde or ketone}}{R'R\text{C}=\ddot{\text{O}}\text{:}} + \text{Nu:}^- \rightleftharpoons \text{Nu}-\text{C}(R')(R)-\ddot{\text{O}}\text{:}^- \overset{H^+}{\rightleftharpoons} \text{Nu}-\text{C}(R')(R)-\ddot{\text{O}}-\text{H} \quad \text{nucleophilic addition}$$

As we study carboxylic acids and their derivatives in this chapter we will find that their reactions are characterized by *nucleophilic substitution* reactions that take place at their acyl (carbonyl) carbons. We will encounter many reactions of the general type shown below:

$$\underset{\text{an acyl compound}}{R(L)\text{C}=\ddot{\text{O}}\text{:}} + \text{:Nu}-\text{H} \rightleftharpoons \text{H}-\text{Nu}^+-\text{C}(R)(L)-\ddot{\text{O}}\text{:}^- \rightleftharpoons \text{Nu}-\text{C}(R)(\text{HL}^+)-\ddot{\text{O}}\text{:}^- \longrightarrow \underset{\text{another acyl compound}}{\text{Nu}(R)\text{C}=\text{O}} \quad \text{nucleophilic substitution}$$

Although the final results obtained from the reactions of acyl compounds with nucleophiles (substitutions) differ from those obtained from aldehydes and ketones (additions), the two reactions have one characteristic in common. *The initial step in both reactions involves a nucleophilic attack on the carbonyl carbon.* With both groups of compounds this initial attack is facilitated by the same factors: the relative steric openness of the carbonyl carbon and the ability of the carbonyl oxygen to accommodate an electron pair of the carbon-oxygen double bond.

It is after the initial nucleophilic attack has taken place that the two reactions differ. The tetrahedral intermediate formed from an aldehyde or ketone usually accepts a proton to form an *addition product*. By contrast, the intermediate formed from an acyl compound usually ejects a leaving group; this leads to regeneration of the carbon-oxygen double bond and a *substitution product*.

Acyl compounds react as they do because they all have good leaving groups attached to the carbonyl carbon: an acyl chloride, for example, generally reacts by losing *a chloride ion*—a very weak base, and thus, a very good leaving group.

General Reaction

$$\text{R}-\text{C}(=\ddot{\text{O}}\text{:})-\ddot{\text{Cl}}\text{:} + \text{:Nu}-\text{H} \rightleftharpoons \text{R}-\text{C}(-\ddot{\text{O}}\text{:}^-)(-\ddot{\text{Cl}}\text{:})-\overset{+}{\text{Nu}}-\text{H} \longrightarrow \text{R}-\text{C}(=\ddot{\text{O}}\text{:})-\overset{+}{\text{Nu}}-\text{H} + \text{:}\ddot{\text{Cl}}\text{:}^- \longrightarrow \text{R}-\text{C}(=\text{O})-\text{Nu:} + \text{H}-\ddot{\text{Cl}}\text{:}$$

Ejection of a chloride ion

Example: The reaction of an acyl chloride with water.

R—C(=O:)—Cl: + :Ö—H (H) ⇌ R—C(—O:⁻)(—:Cl:)—O⁺—H (H) ⟶ R—C(=O:)—:O⁺—H (H) + :Cl:⁻

$H_2\ddot{O}$: ↓

R—C(=O:)—Ö—H + H_3O^+

An acid anhydride generally reacts by losing *a carboxylate ion* or a molecule of a *carboxylic acid*—both are weak bases and good leaving groups.

General Reaction

R—C(=O:)—:O:—C(=O)—R + :Nu—H ⇌ R—C(—O:⁻)(—Nu⁺—H)—:O:—C(=O:)—R ⇌ R—C(—O:⁻)(—Nu:)—:O:—C(=O⁺—H)—R

Leaving group is a carboxylate ion ↓

R—C(=O:)—Nu⁺—H + R—C(=O)—Ö:⁻ ⟶ R—C(=O)—Nu: + R—C(=O)—ÖH

Leaving group is a carboxylic acid ↓

R—C(=O:)—Nu: + R—C(=:O)—Ö—H

Example: The reaction of a carboxylic acid anhydride with an alcohol.

R—C(=O:)—:O:—C(=O:)—R + R′ÖH ⇌ R—C(—O:⁻)(—O⁺(H)—R′)—:O:—C(=O)—R ⇌ R—C(—O:⁻)(—Ö—R′)—:O:—C(=O⁺—H)—R

↓ ↓

$$R{-}C(=O){-}\overset{+}{O}(H){-}R' + RC(=O){-}O^{-} \rightleftharpoons R{-}C(=O){-}O{-}R' + R{-}C(=O){-}O{-}H$$

$$\downarrow$$

$$R{-}C(=O){-}O{-}H + R{-}C(=O){-}O{-}R'$$

As we will see later, esters generally undergo nucleophilic substitution by losing a molecule of an *alcohol,* acids react by losing a molecule of *water,* and amides react by losing a molecule of *ammonia* or *amine.* All of the molecules lost in these reactions are weak bases and are good leaving groups.

For an aldehyde or ketone to react by substitution, the tetrahedral intermediate would have to eject a hydride ion ($H:^-$) or an alkanide ion ($R:^-$). Both are *very powerful bases* and both are, therefore, *very poor leaving groups.*

$$R{-}C(O^-)(H){-}Nu \not\rightarrow R{-}C(=O){-}Nu + H:^-$$ (Hydride ion)

$$R{-}C(O^-)(R){-}Nu \not\rightarrow R{-}C(=O){-}Nu + R:^-$$ (Alkanide ion)

These reactions do not occur

17.4A Relative Reactivity of Acyl Compounds

Of the acid derivatives that we will study in this chapter, acyl chlorides are the most reactive toward nucleophilic substitution and amides are the least reactive. In general, the overall order of reactivity is:

$$R{-}C(=O){-}\mathbf{Cl} > R{-}C(=O){-}\mathbf{SR'} > R{-}C(=O){-}\mathbf{O}{-}C(=\mathbf{O}){-}R > R{-}C(=O){-}\mathbf{OR'} > R{-}C(=O){-}\mathbf{NH_2}$$

Acyl chloride > Thiol ester > Acid anhydride > Ester > Amide

We can account for this overall order of reactivity by taking into account three factors: (1) the basicities of the leaving groups, (2) resonance effects, and (3) inductive effects.

Let us look at the effect of the leaving groups first. The general order of reactivity of acid derivatives roughly parallels the basicities of the potential leaving groups. Chloride ions are the weakest bases and acyl chlorides are the most reactive acyl compounds. Amines are the strongest bases and amides are the least reactive acyl compounds.

All of the acid derivatives have an atom with an unshared electron pair adjacent to the carbonyl group. As a result of this, resonance contributions of the following kind can be important.

$$\mathrm{R{-}C({=}\ddot{O}{:}){-}\ddot{L}} \longleftrightarrow \mathrm{R{-}C({-}\ddot{O}{:}^{-}){=}L^{+}}$$

The effect of this type of resonance is to stabilize the molecule as a whole and to strengthen the bond between the carbonyl carbon and the leaving group. (Notice that the bond between carbon and the leaving group is a double bond in the second structure.)

Acyl chlorides and thiol esters are probably least affected by this kind of resonance because it requires overlap of a 2*p*-carbon orbital with a 3*p* orbital of chlorine or sulfur. Since these orbitals are of different sizes, overlap between them is not large.

$$\mathrm{R{-}C({=}\ddot{O}{:}){-}\ddot{Cl}{:}} \longleftrightarrow \mathrm{R{-}C({-}\ddot{O}{:}^{-}){=}\ddot{Cl}{:}^{+}} \qquad \mathrm{R{-}C({=}\ddot{O}){-}\ddot{S}{-}R'} \longleftrightarrow \mathrm{R{-}C({-}\ddot{O}{:}^{-}){=}\overset{+}{\ddot{S}}{-}R'}$$

These structures do not make a large contribution because of ineffective orbital overlap

On the basis of resonance, we would expect acyl chlorides and thiol esters to be the least stabilized and therefore to be the most reactive acyl compounds. The added inductive effect of the strongly electronegative chlorine explains why acyl chlorides are the most reactive of all.

With esters and amides, resonance effects (below) play a much larger part because overlap between a 2*p* orbital of carbon and a 2*p* orbital of oxygen or nitrogen is more favorable.

$$\mathrm{R{-}C({=}\ddot{O}{:}){-}\ddot{O}R} \longleftrightarrow \mathrm{R{-}C({-}\ddot{O}{:}^{-}){=}\overset{+}{\ddot{O}}R} \qquad \mathrm{R{-}C({=}\ddot{O}){-}\ddot{N}H_2} \longleftrightarrow \mathrm{R{-}C({-}\ddot{O}{:}^{-}){=}\overset{+}{N}H_2}$$

As a result, ester and amide groups are stabilized to a considerable extent, ($\sim$18 kcal/mole for an amide) and are, therefore, the least reactive. The greater inductive effect of oxygen versus nitrogen is in accord with the greater reactivity of an ester versus an amide.

With acid anhydrides, orbital overlap is favorable but the stabilization of a particular carbonyl group is less than that of an ester because the resonance effect is shared between two carbonyl groups.

$$\mathrm{R{-}C(=\ddot{O}:){-}\ddot{O}{-}C(=\ddot{O}:){-}R \longleftrightarrow R{-}C(\ddot{O}:^-){=}\overset{+}{O}{-}C(=\ddot{O}:){-}R \longleftrightarrow R{-}C(=\ddot{O}:){-}\overset{+}{O}{=}C(\ddot{O}:^-){-}R}$$

17.4B Synthesis of Acid Derivatives

As we begin now to explore the syntheses of carboxylic acid derivatives we will find that in many instances one acid derivative can be synthesized through a nucleophilic substitution reaction of another. The order of reactivities that we have presented gives us a clue as to which syntheses are practical and which are not. In general, *less reactive acyl compounds can be synthesized from more reactive ones, but the reverse is usually difficult* and, when possible, requires special conditions and/or a catalyst.

17.5 SYNTHESIS OF ACYL CHLORIDES

Since acyl chlorides are the most reactive of the acid derivatives, we must use highly reactive substances to prepare them. We use other acid chlorides, *the acid chlorides of inorganic acids:* We use $\mathbf{PCl_5}$ (an acid chloride of phosphoric acid), $\mathbf{PCl_3}$ (an acid chloride of phosphorous acid), and $\mathbf{SOCl_2}$ (an acid chloride of sulfurous acid).

All of these reagents react with carboxylic acids to give acyl chlorides in good yield.

General Reactions

$$\underset{}{\mathrm{R\overset{O}{\overset{\|}{C}}OH}} + \underset{\text{Thionyl chloride}}{\mathrm{SOCl_2}} \longrightarrow \mathrm{R{-}\overset{O}{\overset{\|}{C}}{-}Cl + SO_2 + HCl}$$

$$\mathrm{3R\overset{O}{\overset{\|}{C}}OH} + \underset{\text{Phosphorus trichloride'}}{\mathrm{PCl_3}} \longrightarrow \mathrm{3R\overset{O}{\overset{\|}{C}}Cl + H_3PO_3}$$

$$\mathrm{R\overset{O}{\overset{\|}{C}}OH} + \underset{\text{Phosphorus pentachloride}}{\mathrm{PCl_5}} \longrightarrow \mathrm{R\overset{O}{\overset{\|}{C}}Cl + POCl_3 + HCl}$$

Specific Examples

$$\underset{\text{Benzoic acid}}{\mathrm{C_6H_5\overset{O}{\overset{\|}{C}}OH}} + \mathrm{SOCl_2} \longrightarrow \underset{\substack{\text{Benzoyl chloride}\\(91\%)}}{\mathrm{C_6H_5\overset{O}{\overset{\|}{C}}Cl}} + \mathrm{SO_2 + HCl}$$

$$\underset{\text{Acetic acid}}{\mathrm{3CH_3\overset{O}{\overset{\|}{C}}OH}} + \mathrm{PCl_3} \longrightarrow \underset{\substack{\text{Acetyl chloride}\\(67\%)}}{\mathrm{3CH_3\overset{O}{\overset{\|}{C}}Cl}} + \mathrm{H_3PO_3}$$

$$CH_3(CH_2)_6\overset{O}{\overset{\|}{C}}OH + PCl_5 \longrightarrow CH_3(CH_2)_6\overset{O}{\overset{\|}{C}}Cl + POCl_3 + HCl$$

Octanoic acid — Octanoyl chloride (82%)

These reactions all involve nucleophilic substitutions by chloride ion on a highly reactive intermediate: an acyl chlorosulfite, an acyl chlorophosphite, or an acyl chlorophosphate. Thionyl chloride, for example, reacts with a carboxylic acid in the following way.

$$R-\overset{\ddot{O}:}{\overset{\|}{C}}-O-H + Cl-\overset{O}{\overset{\|}{S}}-Cl \longrightarrow R-\overset{\ddot{O}:\,H^+}{\overset{\|}{C}}-O-\overset{O}{\overset{\|}{S}}-Cl \longrightarrow$$

Acyl chlorosulfite

$$:\ddot{Cl}:^-$$

$$R-\underset{:\ddot{Cl}:}{\overset{:\ddot{O}-H}{\overset{|}{\underset{|}{C}}}}-O-\overset{O}{\overset{\|}{S}}-Cl \longrightarrow R-\overset{:\ddot{O}}{\overset{\|}{C}}-\ddot{Cl}: + SO_2 + HCl$$

17.6 SYNTHESIS OF CARBOXYLIC ACID ANHYDRIDES

Carboxylic acids react with acyl chlorides in the presence of pyridine to give carboxylic acid anhydrides.

$$R-\overset{O}{\overset{\|}{C}}-OH + R'-\overset{O}{\overset{\|}{C}}-Cl + \text{pyridine} \longrightarrow R-\overset{O}{\overset{\|}{C}}-O-\overset{O}{\overset{\|}{C}}-R' + \text{pyridinium (N}^+\text{—H)}\ Cl^-$$

This is the most frequently used laboratory method for the preparation of anhydrides. The method is quite general and can be used to prepare mixed anhydrides ($R \neq R'$) or simple anhydrides ($R = R'$).

Sodium salts of carboxylic acids also react with acyl chlorides to give anhydrides:

$$R-\overset{O}{\overset{\|}{C}}-O^-\,Na^+ + R'-\overset{O}{\overset{\|}{C}}-Cl \longrightarrow R-\overset{O}{\overset{\|}{C}}-O-\overset{O}{\overset{\|}{C}}-R' + Na^+\,Cl^-$$

In this reaction a carboxylate ion acts as a nucleophile and brings about a nucleophilic substitution reaction at the acyl carbon of the acyl chloride.

Acetic anhydride is an important industrial reagent. One industrial synthesis of acetic anhydride begins with dehydration of acetic acid at 700° to a highly reactive compound called *ketene*.

$$CH_3C(=O)OH \xrightarrow[700^\circ]{AlPO_4} CH_2{=}C{=}O + H_2O$$

Acetic acid — Ketene

Ketene is then allowed to react with acetic acid; this reaction gives acetic anhydride.

$$\underset{\text{Ketene}}{CH_2{=}C{=}O} + \underset{\text{Acetic acid}}{CH_3\overset{O}{\overset{\|}{C}}{-}OH} \longrightarrow \underset{\text{Acetic anhydride}}{CH_3\overset{O}{\overset{\|}{C}}{-}O{-}\overset{O}{\overset{\|}{C}}CH_3}$$

Acetic anhydride can be used as a dehydrating agent in the preparation of other anhydrides:

$$\underset{\text{Benzoic acid}}{2C_6H_5\overset{O}{\overset{\|}{C}}OH} + \underset{\text{Acetic anhydride}}{(CH_3\overset{O}{\overset{\|}{C}})_2O} \xrightleftharpoons{\text{heat}} \underset{\substack{\text{Benzoic anhydride}\\(74\%)}}{(C_6H_5\overset{O}{\overset{\|}{C}})_2O} + \underset{\text{Acetic acid}}{2CH_3\overset{O}{\overset{\|}{C}}OH}$$

This synthesis is successful because acetic acid can be distilled from the mixture as the reaction takes place. This shifts the equilibrium to the right.

Cyclic anhydrides can sometimes be prepared by simply heating the appropriate dicarboxylic acid. This is true, however, only when anhydride formation leads to a five- or six-membered ring.

$$HOOC{-}CH_2{-}CH_2{-}COOH \xrightarrow{300^\circ} \underset{\text{Succinic anhydride}}{\text{(cyclic } CH_2CH_2C(=O)OC(=O)\text{)}} + H_2O$$

$$\text{Phthalic acid } (C_6H_4(COOH)_2) \xrightarrow{230^\circ} \underset{(\sim 100\%)}{\text{Phthalic anhydride}} + H_2O$$

Problem 17.7

Give an equation showing a different method for the preparation of each of the following anhydrides.

(a) $CH_3\overset{O}{\overset{\|}{C}}O\overset{O}{\overset{\|}{C}}C_6H_5$

(b) $CH_3(CH_2)_4\overset{O}{\overset{\|}{C}}O\overset{O}{\overset{\|}{C}}(CH_2)_4CH_3$

(c) Glutaric anhydride

Problem 17.8

When maleic acid is heated to 200° it loses water and becomes maleic anhydride. Fumaric acid, a diastereomer of maleic acid, requires a much higher temperature before it dehydrates; when it does it also yields maleic anhydride. Explain.

17.7 ESTERS

17.7A Synthesis of Esters: Esterification

Carboxylic acids react with alcohols to form esters through a condensation reaction known as *esterification:*

General Reaction

$$R\text{—}\overset{\overset{\displaystyle O}{\|}}{C}\text{—}OH + R'\text{—}OH \underset{}{\overset{H^+}{\rightleftharpoons}} R\text{—}\overset{\overset{\displaystyle O}{\|}}{C}\text{—}OR' + H_2O$$

Specific Examples

$$\underset{\text{Acetic acid}}{CH_3\overset{\overset{\displaystyle O}{\|}}{C}OH} + \underset{\text{Ethyl alcohol}}{CH_3CH_2OH} \overset{H^+}{\rightleftharpoons} \underset{\text{Ethyl acetate}}{CH_3\overset{\overset{\displaystyle O}{\|}}{C}OCH_2CH_3} + H_2O$$

$$\underset{\text{Benzoic acid}}{C_6H_5\overset{\overset{\displaystyle O}{\|}}{C}OH} + \underset{\text{Methyl alcohol}}{CH_3OH} \overset{H^+}{\rightleftharpoons} \underset{\text{Methyl benzoate}}{C_6H_5\overset{\overset{\displaystyle O}{\|}}{C}OCH_3} + H_2O$$

Esterification reactions are acid catalyzed. They proceed very slowly in the absence of strong acids, but reach equilibrium within a matter of a few hours when an acid and an alcohol are refluxed with a small amount of concentrated sulfuric acid or hydrogen chloride. Since the position of equilibrium controls the amount of the ester formed, the use of an excess of either the carboxylic acid or the alcohol increases the yield based on the limiting reagent. Just which component we choose to use in excess will depend on its availability and cost. The yield of an esterification reaction can also be increased by removing water from the reaction mixture as it is formed.

When benzoic acid reacts with methanol that has been labeled with ^{18}O, the labeled oxygen appears in the ester: This result reveals just which bonds break in the esterification.

$$C_6H_5\overset{\overset{\displaystyle O}{\|}}{C}\,\vdots\,OH + CH_3O\,\vdots\,H \overset{H^+}{\rightleftharpoons} C_6H_5\overset{\overset{\displaystyle O}{\|}}{C}\text{—}OCH_3 + H_2O$$

The results of the labeling experiment and the fact that esterifications are acid catalyzed are both consistent with the mechanism that follows. This mechanism is typical of acid-catalyzed nucleophilic substitution reactions at acyl carbons.

$$R-C(=\ddot{O})-\ddot{O}-H \underset{-H^+}{\overset{+H^+}{\rightleftharpoons}} R-C(=\overset{+}{\ddot{O}}-H)-\ddot{O}-H \underset{-R'-\ddot{O}H}{\overset{+R'-\ddot{O}H}{\rightleftharpoons}} R-C(-\ddot{O}-H)(-\ddot{O}H)(-\overset{+}{\ddot{O}}(H)-R') \rightleftharpoons$$

$$R-C(-\ddot{O}-H)(-\overset{+}{\ddot{O}}H_2)(-\ddot{O}-R') \underset{+H_2O}{\overset{-H_2O}{\rightleftharpoons}} R-C(=\overset{+}{\ddot{O}}-H)-\ddot{O}-R' \underset{+H^+}{\overset{-H^+}{\rightleftharpoons}} R-C(=\ddot{O})-\ddot{O}-R'$$

If we follow the forward reactions in this mechanism, we have the mechanism for the *acid-catalyzed esterification of an acid.* If, however, we follow the reverse reactions, we have the mechanism for the *acid-catalyzed hydrolysis of an ester:*

$$R-\overset{O}{\overset{\|}{C}}-OR' + H_2O \overset{H_3O^+}{\rightleftharpoons} R-\overset{O}{\overset{\|}{C}}-OH + R'-OH$$

Which result we obtain will depend on the conditions we choose. If we want to esterify an acid, we use an excess of the alcohol and if possible remove the water as it is formed. If we want to hydrolyze an ester, we use a large excess of water; that is, we reflux the ester with dilute aqueous HCl or dilute aqueous H_2SO_4.

Problem 17.9

Where would you expect to find the labeled oxygen if you carried out an acid-catalyzed hydrolysis of methyl benzoate in ^{18}O-labeled water?

Steric factors strongly affect the rates of acid-catalyzed hydrolyses of esters. Large groups near the reaction site, whether in the alcohol component or the acid component, slow both reactions markedly. Tertiary alcohols, for example, react so slowly in esterifications that they usually undergo elimination instead. The dibromobenzoic acid shown below cannot be esterified by treatment with an alcohol and an acid catalyst, and methyl 2,6-dibromobenzoate cannot be hydrolyzed with aqueous acid. (Special techniques are required to bring about both reactions.)

[2,6-dibromobenzoic acid: C(=O)OH with Br, Br on benzene ring]

Bulky ortho substituents prevent acid-catalyzed esterification

[2,6-dibromobenzoate ester: C(=O)OR with Br, Br on benzene ring]

Bulky ortho substituents prevent acid-catalyzed hydrolysis

Esters from Acid Chlorides. Esters can also be synthesized by the reaction of acid chlorides with alcohols. Since acid chlorides are much more reactive toward nucleophilic substitution than carboxylic acids, the reaction of an acid chloride and an alcohol occurs rapidly and does not require an acid catalyst. Pyridine is usually added to the reaction mixture to react with the HCl that forms.

General Reaction

$$R{-}C(=\ddot{O}{:}){-}\ddot{Cl}{:} + R'{-}\ddot{O}{-}H \xrightarrow{-HCl} R{-}C(=\ddot{O}{:}){-}\ddot{O}{-}R'$$

Specific Examples

$$\underset{\text{Benzoyl chloride}}{C_6H_5\overset{O}{\overset{\|}{C}}{-}Cl} + CH_3CH_2OH + \underset{\text{(pyridine)}}{C_5H_5N} \longrightarrow \underset{\substack{\text{Ethyl benzoate}\\(80\%)}}{C_6H_5\overset{O}{\overset{\|}{C}}OCH_2CH_3} + C_5H_5\overset{+}{N}{-}H\ Cl^-$$

$$\underset{\text{Malonyl chloride}}{Cl{-}\overset{O}{\overset{\|}{C}}CH_2\overset{O}{\overset{\|}{C}}{-}Cl} + 2CH_3C(CH_3)_2{-}OH + 2\,C_5H_5N \longrightarrow$$

$$\underset{\substack{\text{Di-}\textit{tert}\text{-butyl malonate}\\(83\%)}}{CH_3C(CH_3)_2{-}O\overset{O}{\overset{\|}{C}}CH_2\overset{O}{\overset{\|}{C}}O{-}C(CH_3)_2CH_3} + 2\,C_5H_5\overset{+}{N}{-}H\ Cl^-$$

Esters from Acid Anhydrides. Acid anhydrides also react with alcohols to form esters in the absence of an acid catalyst.

General Reaction

$$(RC(=O))_2O + R'{-}OH \xrightarrow[-R\overset{O}{\overset{\|}{C}}OH]{} RC(=O){-}O{-}R'$$

Specific Example

$$\underset{\substack{\text{Acetic}\\\text{anhydride}}}{(CH_3\overset{O}{\overset{\|}{C}})_2O} + \underset{\substack{\text{Benzyl}\\\text{alcohol}}}{C_6H_5CH_2OH} \longrightarrow \underset{\text{Benzyl acetate}}{CH_3\overset{O}{\overset{\|}{C}}OCH_2C_6H_5} + CH_3COOH$$

Cyclic anhydrides react with one mole of an alcohol to form a compound that is both *an ester and an acid.*

$$\text{Phthalic anhydride} + CH_3CH(OH)CH_2CH_3 \xrightarrow{110^\circ} o\text{-}C_6H_4(COOCH(CH_3)CH_2CH_3)(COOH)$$

Phthalic anhydride — *sec*-Butyl alcohol — *sec*-Butyl hydrogen phthalate (97%)

Problem 17.10

Esters can also be synthesized by *transesterification:*

$$\underset{\text{High-boiling ester}}{R{-}\overset{O}{\overset{\|}{C}}{-}OR'} + \underset{\text{High-boiling alcohol}}{R''{-}OH} \underset{}{\overset{H^+}{\rightleftharpoons}} \underset{\text{Higher-boiling ester}}{R\overset{O}{\overset{\|}{C}}{-}OR''} + \underset{\text{Low-boiling alcohol}}{R'{-}OH}$$

In this procedure we shift the equilibrium to the right by allowing the low-boiling alcohol to distill from the reaction mixture. The mechanism for transesterification is similar to that for an acid-catalyzed esterification (or an acid-catalyzed ester hydrolysis). Write a mechanism for the transesterification shown below.

$$\underset{\text{Methyl acrylate}}{CH_2{=}CH\overset{O}{\overset{\|}{C}}OCH_3} + \underset{n\text{-Butyl alcohol}}{CH_3CH_2CH_2CH_2OH} \overset{H^+}{\rightleftharpoons}$$

$$\underset{n\text{-Butyl acrylate (94\%)}}{CH_2{=}CH\overset{O}{\overset{\|}{C}}OCH_2CH_2CH_2CH_3} + \underset{\text{Methyl alcohol}}{CH_3OH}$$

17.7B Base-Promoted Hydrolysis of Esters: Saponification

Esters not only undergo acid hydrolysis, they also undergo *base-promoted hydrolysis.* Base-promoted hydrolysis is sometimes called *saponification.* Refluxing an ester with aqueous sodium hydroxide, for example, produces an alcohol and the sodium salt of the acid:

$$\underset{\text{Ester}}{R\overset{O}{\overset{\|}{C}}{-}OR'} + NaOH \xrightarrow{H_2O} \underset{\text{Sodium carboxylate}}{R{-}\overset{O}{\overset{\|}{C}}{-}O^-Na^+} + \underset{\text{Alcohol}}{R'OH}$$

The carboxylate ion is very unreactive toward nucleophilic substitution because it is negatively charged. Base-promoted hydrolysis of an ester, as a result, is an essentially irreversible reaction.

The mechanism for the base-promoted hydrolysis of an ester also involves a nucleophilic substitution at the acyl carbon:

$$R-C(=O)-O-R' + {}^-:OH \rightleftharpoons R-C(O^-)(OH)-OR' \rightleftharpoons R-C(=O)-OH + R'-O^- \rightarrow R-C(=O)-O^- + R'-OH$$

Part of the evidence that nucleophilic attack occurs at the acyl carbon comes from studies in which esters of chiral alcohols were subjected to base-promoted hydrolysis. The reaction of an ester with hydroxide ion could conceivably occur in two ways: reaction could take place through a nucleophilic substitution at the acyl carbon of the acid component (path A) or through a nucleophilic substitution at the alkyl carbon of the alcohol portion (path B). Reaction by path A should lead to retention of configuration in the alcohol. Reaction by path B should lead to an inversion of configuration of the alcohol. *Inversion of configuration is almost never observed.* In almost every instance basic hydrolysis of a carboxylate ester of a chiral alcohol proceeds with *retention of configuration.*

Path A: Nucleophilic substitution at the acyl carbon

RC(=O)—O—C(R')(R'')H + $^-$:OH —(retention of configuration)→ R—C(=O)—O$^-$ + HO—C(R')(R'')H

Path B: Nucleophilic substitution at the alkyl carbon

R—C(=O)—O—C(R')(R'')H + $^-$:OH —(inversion of configuration)→ R—C(=O)—O$^-$ + C(R')(R'')(H)—OH

(*This reaction seldom occurs with esters of carboxylic acids.*)

Although nucleophilic attack at the alkyl carbon seldom occurs with esters of carboxylic acids, it is the preferred mode of attack with esters of sulfonic acids (Section 15.9).

R—S(=O)$_2$—O—C(R')(R'')H + $^-$:OH —(inversion of configuration)→ R—S(=O)$_2$—O$^-$ + C(R')(R'')(H)—OH

an alkyl sulfonate

Problem 17.11

(a) Write stereochemical formulas for compounds **A** to **F**.

1. *cis*-3-Methylcyclopentanol + $C_6H_5SO_2Cl$ ⟶ **A**

$\xrightarrow[\text{heat}]{^-OH}$ **B** + $C_6H_5SO_3^-$

2. *cis*-3-Methylcyclopentanol + $C_6H_5\overset{O}{\overset{\|}{C}}—Cl \longrightarrow$ **C**

$\xrightarrow[\text{reflux}]{^{-}OH}$ **D** + $C_6H_5CO_2^-$

3. *R*-2-Bromooctane + $CH_3COO^- Na^+ \longrightarrow$ **E** + NaBr

$\downarrow$ OH^-, H_2O (reflux)

F

4. *R*-2-Bromooctane + $OH^- \xrightarrow{\text{alcohol}}$ **F** + Br^-

(b) Which of the last two methods, (3) or (4), would you expect to give a higher yield of **F**? Why?

Problem 17.12

Base-promoted hydrolysis of methyl mesitoate occurs through an attack on the alcohol carbon instead of the acyl carbon.

$$CH_3-C_6H_2(CH_3)_2-\overset{O}{\overset{\|}{C}}-O-CH_3 + \overset{-}{O}H \longrightarrow CH_3-C_6H_2(CH_3)_2-CO_2^- + CH_3OH$$

Methyl mesitoate

(a) Can you suggest a reason that will account for this unusual behavior? (b) Suggest an experiment with labeled compounds that would confirm this mode of attack.

17.7C Lactones

Carboxylic acids whose molecules have a hydroxyl group on a γ- or δ-carbon undergo an intramolecular esterification to give cyclic esters known as γ- or δ-*lactones*.

$$\underset{\text{A }\gamma\text{-hydroxy acid}}{\overset{\gamma\ \ \beta\ \ \alpha}{R\underset{:OH}{C}HCH_2CH_2\overset{O}{\overset{\|}{C}}-OH}} \overset{+H^+}{\rightleftharpoons} \text{(cyclic intermediate: } CH_2-C(OH)_2,\ CH_2,\ \overset{+}{O}H,\ CH-R) \rightleftharpoons \text{(}\gamma\text{-lactone: } CH_2-C(=O),\ CH_2,\ O,\ CH-R) + H_3O^+$$

$$\underset{\text{A }\delta\text{-hydroxy acid}}{R-\overset{\delta\ \ \gamma\ \ \beta\ \ \alpha}{\underset{:OH}{C}HCH_2CH_2CH_2\overset{O}{\overset{\|}{C}}-OH}} \overset{H^+}{\rightleftharpoons} \underset{\text{A }\delta\text{-lactone}}{\text{(ring: } CH_2-C(=O),\ CH_2,\ O,\ CH_2-CH-R)} + H_2O$$

Lactones are hydrolyzed by aqueous base just as other esters are. Acidification of the sodium salt, however, may lead spontaneously back to the γ- or δ-lactone, particularly if excess acid is used.

$$\text{(γ-lactone, 5-phenyl)} \underset{H^+,\ \text{slight excess}}{\overset{^-OH/H_2O}{\rightleftharpoons}} C_6H_5CH(OH)CH_2CH_2C(=O){-}O^- \xrightarrow[0°]{H^+,\ \text{exactly one equivalent}} C_6H_5CH(OH)CH_2CH_2COOH$$

Many lactones occur in nature. Almost all are γ- or δ-lactones; that is, almost all contain five- or six-membered rings.

Beta-lactones (lactones with four-membered rings) have been detected as intermediates in some reactions. They are highly reactive, however. If one attempts to prepare a β-lactone from a β-hydroxy acid, β-elimination usually occurs instead:

$$RCH(OH)CH_2COOH \xrightarrow[\text{or acid}]{\text{heat}} RCH{=}CHCOOH + H_2O$$

β-Hydroxy acid → α,β-Unsaturated acid

β-Hydroxy acid —heat—✕→ β-Lactone (does not form)

When α-hydroxy acids are heated they form cyclic diesters called *lactides.*

$$2RCH(OH)COOH \xrightarrow{\text{heat}} \text{A lactide (cyclic diester of two } RCH(O)C{=}O \text{ units)}$$

α-Hydroxy acid → A lactide

Alpha-lactones also occur as intermediates in some reactions (cf. Special Topic O).

17.8 AMIDES

17.8A Synthesis of Amides

Amides can be prepared in a variety of ways starting with acyl chlorides, acid anhydrides, esters, carboxylic acids, and carboxylate salts. All of these methods involve nucleophilic substitution reactions by ammonia or an amine at an acyl carbon. As we might expect, acid chlorides are the most reactive and carboxylate ions are the least.

17.8B Amides from Acyl Chlorides

Primary amines, secondary amines, and ammonia all react rapidly with acid chlorides to form amides:

$$\text{R—C}(=\text{O})\text{Cl} + :\text{NH}_3(\text{excess}) \longrightarrow \text{R—C}(=\text{O})\ddot{\text{N}}\text{H}_2 + \text{NH}_4^+\,\text{Cl}^-$$

Ammonia — An amide

$$\text{R—C}(=\text{O})\text{Cl} + \text{R}'\ddot{\text{N}}\text{H}_2(\text{excess}) \longrightarrow \text{R—C}(=\text{O})\ddot{\text{N}}\text{HR}' + \text{R}'\text{NH}_3^+\,\text{Cl}^-$$

An *N*-substituted amide

$$\text{R—C}(=\text{O})\text{Cl} + \text{R}'\text{—}\ddot{\text{N}}\text{H}(\text{R}'')(\text{excess}) \longrightarrow \text{R—C}(=\text{O})\ddot{\text{N}}(\text{R}'')\text{—R}' + \text{R}'\text{NH}_2^+(\text{R}'')\,\text{Cl}^-$$

An *N,N*-disubstituted amide

Since acyl chlorides are easily prepared from carboxylic acids, this is one of the most widely used laboratory methods for the synthesis of amides. The reaction between the acyl chloride and the amine (or ammonia) usually takes place at room temperature (or below) and produces the amide in high yield.

$$\text{CH}_3\text{CH}_2\text{CH}(\text{CH}_2\text{CH}_3)\text{C}(=\text{O})\text{—Cl} + 2\text{NH}_3 \xrightarrow{\text{Benzene}} \text{CH}_3\text{CH}_2\text{CH}(\text{CH}_2\text{CH}_3)\text{C}(=\text{O})\text{—NH}_2 + \text{NH}_4\text{Cl}$$

(91%)

$$\text{CH}_2\text{=CHC}(=\text{O})\text{—Cl} + 2\text{NH}_3 \longrightarrow \text{CH}_2\text{=CHC}(=\text{O})\text{—NH}_2 + \text{NH}_4\text{Cl}$$

(80%)

Acyl chlorides also react with tertiary amines by a nucleophilic substitution reaction. The acylammonium ion that forms, however, is not stable in the presence of water or any hydroxylic solvent.

$$\text{R—C}(=\text{O})\text{Cl} + \text{R}_3\text{N}: \longrightarrow \text{R—C}(=\text{O})\text{—}\overset{+}{\text{N}}\text{R}_3\,\text{Cl}^-$$

Acyl chloride — 3° Amine — Acylammonium ion

$$\xrightarrow{\text{H}_2\text{O}} \text{R—C}(=\text{O})\text{OH} + \text{H}\overset{+}{\text{N}}\text{R}_3\,\text{Cl}^-$$

Acylpyridinium ions are probably involved as intermediates in those reactions of acyl chlorides that are carried out in the presence of pyridine.

17.8C Amides from Carboxylic Anhydrides

Acid anhydrides react with ammonia and with primary and secondary amines and form amides through reactions that are analogous to those of acyl chlorides.

$$(RC(=O))_2O + 2\ddot{N}H_3 \longrightarrow RC(=O)-\ddot{N}H_2 + RCOO^-NH_4^+$$

$$(RC(=O))_2O + 2R'-\ddot{N}H_2 \longrightarrow RC(=O)-\ddot{N}H-R' + RCOO^-R'NH_3^+$$

$$(RC(=O))_2O + 2R'-\ddot{N}H(R'') \longrightarrow RC(=O)-\ddot{N}(R'')-R' + RCOO^-R'R''NH_2^+$$

Cyclic anhydrides react with ammonia or an amine in the same general way as acyclic anhydrides; however, the reaction produces a product that is both an amide and an ammonium salt. Acidifying the ammonium salt gives a compound that is both an amide and an acid:

$$\text{Phthalic anhydride} + 2\ddot{N}H_3 \xrightarrow[\text{Warm}]{H_2O} C_6H_4(C(=O)-NH_2)(C(=O)-O^-\,\overset{+}{N}H_4) \xrightarrow[H_2O]{H^+} C_6H_4(C(=O)NH_2)(C(=O)OH)$$

Phthalic anhydride; Ammonium phthalamate (94%); Phthalamic acid (81%)

Heating the amide-acid causes dehydration to occur and gives an *imide*. Imides contain the linkage $-C(=O)-NH-C(=O)-$.

$$C_6H_4(C(=O)-\ddot{N}H_2)(C(=O)-OH) \xrightarrow{150\text{–}160^\circ} \text{Phthalimide (N–H)} + H_2O$$

Phthalamic acid; Phthalimide (~100%)

17.8D Amides from Esters

Esters undergo nucleophilic substitution at their acyl carbons when they are treated with ammonia or with primary and secondary amines. These reactions take place more slowly than those of acyl chlorides and anhydrides, but they are synthetically useful.

$$R{-}C({=}O){-}OR''' + H{-}\ddot{N}R'R'' \longrightarrow R{-}C({=}O){-}\ddot{N}R'R'' + R'''OH$$

R′ and/or R″ may be H

$$ClCH_2C({=}O)OC_2H_5 + NH_{3(aq)} \xrightarrow{0\text{–}5°} ClCH_2C({=}O)NH_2 + C_2H_5OH$$

Ethyl chloroacetate → Chloroacetamide (62–87%)

17.8E Amides from Carboxylic Acids and Ammonium Carboxylates

Carboxylic acids react with aqueous ammonia to form ammonium salts.

$$R{-}C({=}O){-}OH + \ddot{N}H_3 \longrightarrow R{-}C({=}O){-}O^- \; NH_4^+$$

An ammonium carboxylate

Because of the low reactivity of the carboxylate ion toward nucleophilic substitution, further reaction does not usually take place in aqueous solution. However, if we evaporate the water and subsequently heat the dry salt, dehydration produces an amide.

$$R{-}C({=}O)O^-NH_4^+{}_{(solid)} \xrightarrow{Heat} R{-}C({=}O){-}NH_2 + H_2O$$

An alternative method for synthesizing amides directly from acids involves adding solid ammonium carbonate to an excess of the acid and then slowly distilling the mixture.

$$2CH_3COOH + (NH_4)_2CO_3 \longrightarrow 2CH_3C({=}O)O^- NH_4^+ + CO_2 + H_2O$$

$$CH_3C({=}O)O^- NH_4^+ \xrightarrow{\text{Slow distillation}} CH_3C({=}O)NH_2 + H_2O$$

Acetamide (87–90%)

Amides are of great importance in biochemistry. The linkages that join individual amino acids together to form proteins are primarily amide linkages. As a consequence, much research has been done to find new and mild ways for amide synthesis. One especially useful reagent is the compound dicyclohexylcarbodiimide, C_6H_{11}—N=C=N—C_6H_{11}. Dicyclohexylcarbodiimide promotes amide formation by reacting with the carboxyl group of an acid and activating it toward nucleophilic substitution.

$$R-COOH + C_6H_{11}-N=C=N-C_6H_{11} \longrightarrow R-C(=O)-\overset{+}{O}(H)-C(=N-C_6H_{11})-\ddot{N}^{-}-C_6H_{11} \longrightarrow R-C(=O)-O-C(=N-C_6H_{11})NHC_6H_{11}$$

Dicyclohexylcarbodiimide (DCC)

Reactive intermediate

$$\xrightarrow{R'-\ddot{N}H_2} R-C(O^-)(NH_2^+R')-O-C(=N-C_6H_{11})NHC_6H_{11} \xrightarrow{(H^+ \text{ shift})} R-C(=O)NHR' + O=C(NHC_6H_{11})_2$$

An amide $\quad$ *N,N*-Dicyclohexylurea

The intermediate in this synthesis does not need to be isolated, and both steps take place at room temperature. Amides are produced in very high yield. In Chapter 22 we will see how dicyclohexylcarbodiimide can be used in an automated synthesis of proteins.

17.8F Hydrolysis of Amides

Amides undergo hydrolysis when they are heated with aqueous acid or aqueous base.

Acidic Hydrolysis

$$R-C(=O)\ddot{N}H_2 + H_3O^+ \xrightarrow[\text{heat}]{H_2O} R-C(=O)OH + NH_4^+$$

Basic Hydrolysis

$$R-C(=O)\ddot{N}H_2 + Na^+\,OH^- \xrightarrow[\text{heat}]{H_2O} R-C(=O)O^-\,Na^+ + \ddot{N}H_3$$

N-Substituted amides and *N,N*-disubstituted amides also undergo hydrolysis in aqueous acid or base. Amide hydrolysis by either method takes place more slowly than the corresponding hydrolysis of an ester. Thus, amide hydrolyses generally require more forcing conditions.

The mechanism for acid hydrolysis of an amide is similar to that given on p. 769 for the acid hydrolysis of an ester. Water acts as a nucleophile and attacks the protonated amide. The leaving group in the acidic hydrolysis of amide is ammonia (or an amine).

$$R{-}C(=O)NH_2 + H^+ \rightleftharpoons R{-}C(=\overset{+}{O}H)NH_2 \xrightleftharpoons{:\ddot{O}H_2} R{-}C(OH)(\overset{+}{O}H_2)(NH_2) \rightleftharpoons R{-}C(OH)(OH)(\overset{+}{N}H_3)$$

$$\rightleftharpoons R{-}C(=\overset{+}{O}H)(OH) + \ddot{N}H_3 \longrightarrow R{-}C(=O)OH + \overset{+}{N}H_4$$

There is evidence that in basic hydrolyses of amides, hydroxide ions act both as nucleophiles and as bases. In the first step (below) a hydroxide ion attacks the acyl carbon of the amide. In the second step, a hydroxide ion removes a proton to give a dianion. In the final step, the dianion loses a molecule of ammonia (or an amine); this step is synchronized with a proton transfer from water.

$$R{-}C(=O)NH_2 + :\ddot{O}H^- \rightleftharpoons R{-}C(O^-)(OH)(NH_2) \xrightleftharpoons{:\ddot{O}H^-} R{-}C(O^-)(O^-)(NH_2) \xrightarrow{H-\ddot{O}H} R{-}C(=O)O^- + :NH_3 + :\ddot{O}H^-$$

Problem 17.13

What products would you obtain from acidic and basic hydrolysis of each of the following amides?

(a) *N,N*-Diethylbenzamide

(b) Six-membered ring: $CH_2{-}CH_2{-}CH_2{-}CH_2{-}C(=O){-}NH$ (ring closed from N–H back to the first CH_2)

(c) 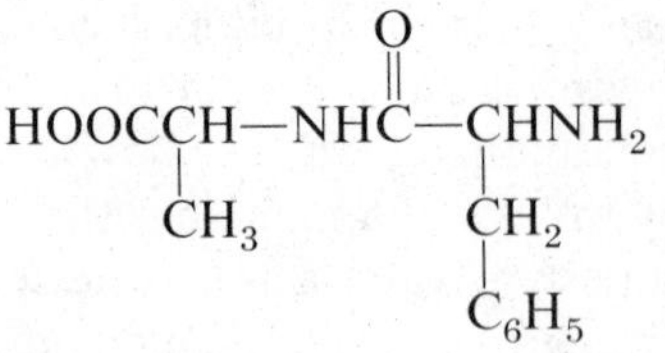 $HOOCCH(CH_3)-NHC(=O)-CHNH_2(CH_2C_6H_5)$ (a dipeptide)

17.8G Dehydration of Amides

Amides react with phosphorus pentoxide (P_2O_5) or with boiling acetic anhydride to form nitriles.

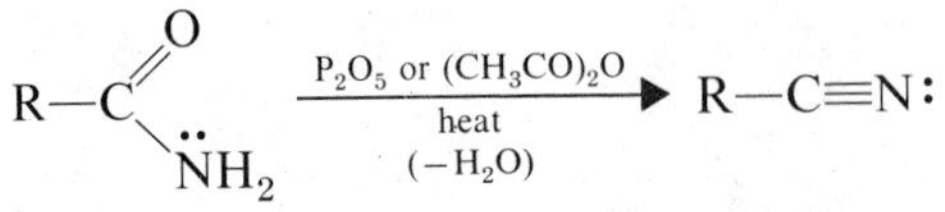

$$R-C(=O)\ddot{N}H_2 \xrightarrow[\text{heat}\ (-H_2O)]{P_2O_5 \text{ or } (CH_3CO)_2O} R-C\equiv N:$$

A nitrile

This is a useful synthetic method for preparing nitriles that are not available by nucleophilic substitution reactions between alkyl halides and cyanide ion.

Problem 17.14

(a) Show all steps in the synthesis of $(CH_3)_3CCN$ from $(CH_3)_3CCOOH$. (b) What product would you expect to obtain if you attempted to synthesize $(CH_3)_3CCN$ using the following reaction?

$$(CH_3)_3C-Br + CN^- \longrightarrow$$

17.8H Lactams

Cyclic amides are called lactams. The size of the lactam ring is designated by Greek letters in a way that is analogous to lactone nomenclature.

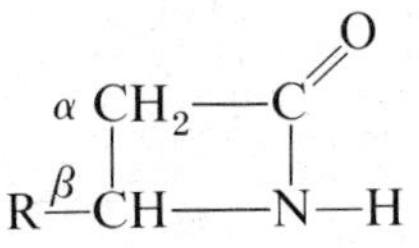

A β-lactam: α CH_2—C(=O) / R—βCH—N—H

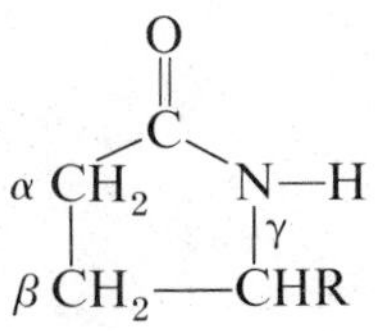

A γ-lactam: C(=O), α CH_2, N—H, γ, β CH_2—CHR

A δ-lactam: C(=O), α CH_2, N—H, δ, β CH_2 γ CH_2 CHR

Gamma-lactams and δ-lactams often form spontaneously from γ- and δ-amino acids. Beta-lactams, however, are highly reactive; their strained four-membered rings open easily in the presence of nucleophilic reagents. The penicillin antibiotics (below) contain a β-lactam ring.

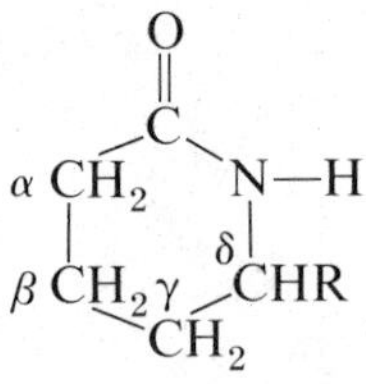

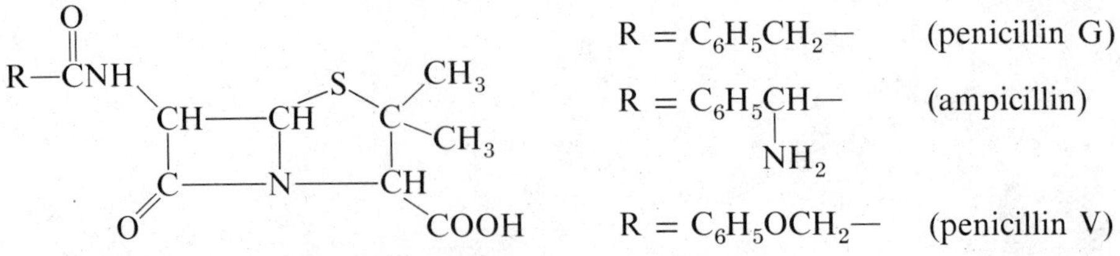

R—C(=O)NH—CH—CH—S—C(CH_3)(CH_3)—CH(COOH)—N—C(=O) (penicillin ring structure)

$R = C_6H_5CH_2-$ (penicillin G)

$R = C_6H_5CH(NH_2)-$ (ampicillin)

$R = C_6H_5OCH_2-$ (penicillin V)

The penicillins apparently act by interfering with the synthesis of the bacterial cell walls. It is thought that they do this by reacting with an amino group of an essential enzyme of the cell wall biosynthetic pathway. This reaction, which involves ring opening of the β-lactam and acylation of the amino group, inactivates the enzyme.

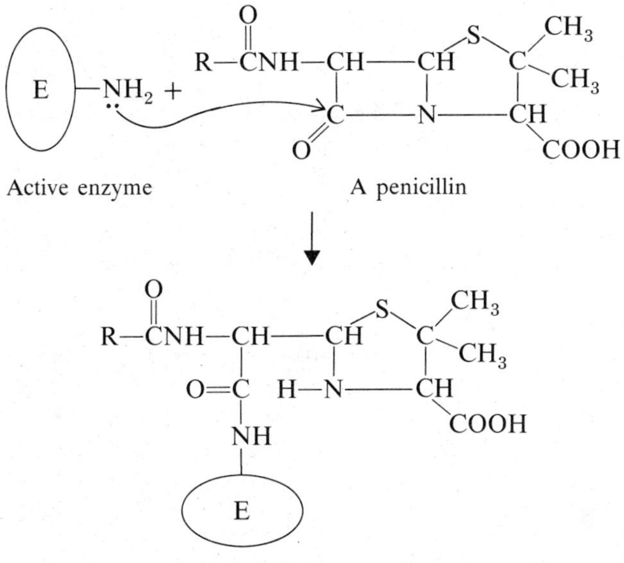

E—NH_2 + R—C(=O)NH—CH—CH(S—C(CH_3)$_2$—CH(COOH)—N)—C(=O)

Active enzyme

A penicillin

↓

R—C(=O)NH—CH—CH(S—C(CH_3)$_2$—CH(COOH)—N—H); O=C—NH—E

Inactive enzyme

17.9 α-HALO ACIDS: THE HELL-VOLHARD-ZELINSKI REACTION

Aliphatic carboxylic acids react with bromine or chlorine in the presence of phosphorus (or a phosphorus halide) to give α-halo acids through a reaction known as the Hell-Volhard-Zelinski reaction.

General Reaction

$$\text{R—CH}_2\text{COOH} \xrightarrow[\text{P}]{\text{X}_2} \text{RCHCOOH (X on the α-carbon)}$$

α-Halo acid

Specific Examples

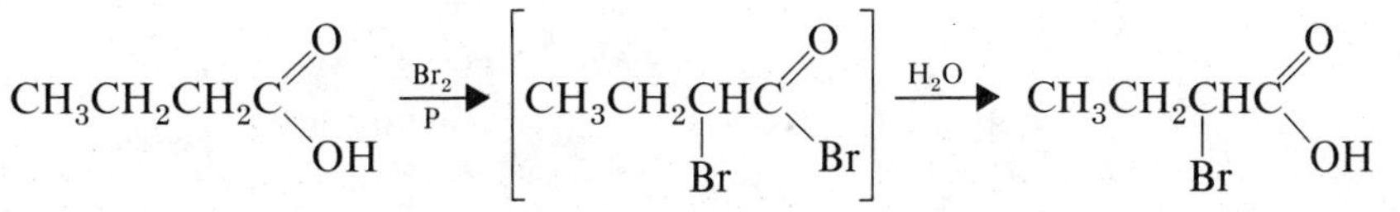

$$\text{CH}_3\text{CH}_2\text{CH}_2\text{C(=O)OH} \xrightarrow[\text{P}]{\text{Br}_2} \left[\text{CH}_3\text{CH}_2\text{CHBrC(=O)Br}\right] \xrightarrow{\text{H}_2\text{O}} \text{CH}_3\text{CH}_2\text{CHBrC(=O)OH}$$

Butanoic acid → 2-Bromobutanoic acid (77%)

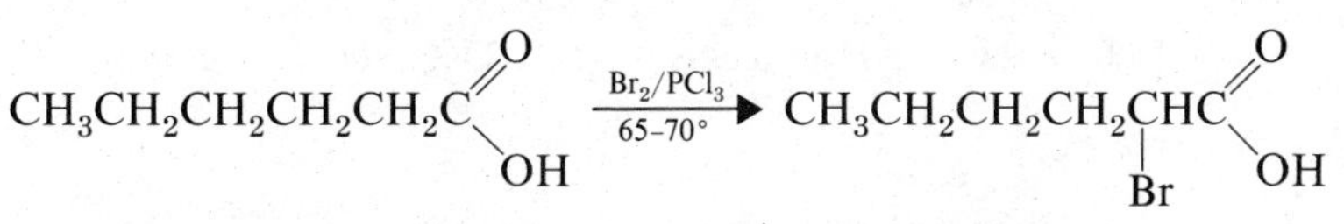

$$\text{CH}_3\text{CH}_2\text{CH}_2\text{CH}_2\text{CH}_2\text{C(=O)OH} \xrightarrow[\text{65–70°}]{\text{Br}_2/\text{PCl}_3} \text{CH}_3\text{CH}_2\text{CH}_2\text{CH}_2\text{CHBrC(=O)OH}$$

Hexanoic acid → 2-Bromohexanoic acid (83–89%)

Halogenation occurs specifically at the α-carbon. If more than one mole of bromine or chlorine is used in the reaction, the products obtained are α,α-dihalo acids or α,α,α-trihalo acids.

The mechanism for the Hell-Volhard-Zelinski reaction is outlined below. The key step involves the formation of an enol from an acyl halide. (Carboxylic acids do not form enols readily.) Enol formation accounts for specific halogenation at the α-position.

$$R{-}CH_2{-}C(=O)OH \xrightarrow[(PX_3)]{P + Br_2} \underset{\text{Acyl bromide}}{R{-}CH_2{-}C(=O)Br} \rightleftarrows \underset{\text{Enol form}}{R{-}CH{=}C(\ddot{O}{-}H)Br} \xrightarrow{Br{-}Br}$$

$$R{-}CH(Br){-}C(=O)OH \xleftarrow[(\text{or } H_2O)]{RCH_2COOH} R{-}CH(Br){-}C(=O)Br + HBr$$

Alpha-halo acids are important synthetic intermediates because they are capable of reacting with a variety of nucleophiles:

Conversion to α-Hydroxy Acids

$$\underset{\alpha\text{-Halo acid}}{R{-}CH(X)COOH} \xrightarrow[(2)\ H^+]{OH^-} \underset{\alpha\text{-Hydroxy acid}}{R{-}CH(OH)COOH} + X^-$$

Example:

$$CH_3CH_2CH(Br)COOH \xrightarrow[(2)\ H^+]{(1)\ K_2CO_3,\ H_2O,\ 100^\circ} \underset{\substack{\text{2-Hydroxybutanoic acid} \\ (69\%)}}{CH_3CH_2CH(OH)COOH}$$

Conversion to α-Amino Acids

$$\underset{\alpha\text{-Halo acid}}{R{-}CH(X)COOH} + 2NH_3 \longrightarrow \underset{\alpha\text{-Amino acid}}{RCH(NH_2)COOH} + NH_4X$$

Example:

$$CH_2(Br)COOH + NH_3(\text{excess}) \longrightarrow \underset{\substack{\alpha\text{-Aminoacetic acid} \\ \text{(glycine)} \\ (60\text{–}64\%)}}{CH_2(NH_2)COOH} + NH_4Br$$

17.10 DECARBOXYLATION OF CARBOXYLIC ACIDS

The reaction whereby a carboxylic acid loses CO_2 is called a *decarboxylation.*

$$\mathrm{R{-}\overset{\overset{\displaystyle O}{\|}}{C}{-}OH} \xrightarrow{\text{Decarboxylation}} \mathrm{R{-}H + CO_2}$$

Although the unusual stability of carbon dioxide means that decarboxylation of most acids is exothermic, in practice the reaction is not always easy to carry out. Special groups usually have to be present in the molecule for decarboxylation to be synthetically useful.

Acids whose molecules have a carbonyl group one carbon removed from the carboxylic acid group, **called β-keto acids,** decarboxylate readily when they are heated to 100–150°.

$$\underset{\text{A }\beta\text{-keto acid}}{\mathrm{R\overset{\overset{\displaystyle O}{\|}}{C}CH_2\overset{\overset{\displaystyle O}{\|}}{C}OH}} \xrightarrow{100\text{–}150^\circ} \mathrm{R\overset{\overset{\displaystyle O}{\|}}{C}CH_3 + CO_2}$$

There are two reasons for this:

1. When the carboxylate ion decarboxylates, it forms a resonance-stabilized anion:

$$\underset{\text{Acylacetate ion}}{\mathrm{R{-}\overset{\overset{\displaystyle \ddot{O}:}{\|}}{C}{-}CH_2{-}\overset{\overset{\displaystyle \ddot{O}:}{\|}}{C}{-}\ddot{\underset{..}{O}}:^-}} \xrightarrow[-CO_2]{} \mathrm{R{-}\overset{\overset{\displaystyle \ddot{O}:}{\|}}{C}{-}\ddot{C}H_2^-} \longleftrightarrow \mathrm{R{-}\overset{\overset{\displaystyle :\ddot{O}:^-}{|}}{C}{=}CH_2} \xrightarrow{H^+} \mathrm{R{-}\overset{\overset{\displaystyle \ddot{O}:}{\|}}{C}{-}CH_3}$$

Resonance-stabilized anion

This anion is much more stable than the anion, $RCH_2:^-$, that would be produced by decarboxylation of an ordinary carboxylic acid.

2. When the acid itself decarboxylates it can do so through a six-membered cyclic transition state:

$$\underset{\beta\text{-Keto acid}}{\text{(cyclic: R–C(=O)–CH}_2\text{–C(=O)–O–H)}} \xrightarrow{-CO_2} \underset{\text{Enol}}{\mathrm{R{-}\overset{\overset{\displaystyle OH}{|}}{C}{=}CH_2}} \longrightarrow \underset{\text{Ketone}}{\mathrm{R{-}\overset{\overset{\displaystyle O}{\|}}{C}{-}CH_3}}$$

This produces an enol directly and avoids an anionic intermediate. The enol then tautomerizes to a methyl ketone.

Malonic acids also decarboxylate readily and for similar reasons.

$$\underset{\text{A malonic acid}}{\mathrm{HO\overset{\overset{\displaystyle O}{\|}}{C}{-}\underset{\underset{\displaystyle R}{|}}{\overset{\overset{\displaystyle R}{|}}{C}}{-}\overset{\overset{\displaystyle O}{\|}}{C}OH}} \xrightarrow{100\text{–}150^\circ} \mathrm{H{-}\underset{\underset{\displaystyle R}{|}}{\overset{\overset{\displaystyle R}{|}}{C}}{-}\overset{\overset{\displaystyle O}{\|}}{C}OH + CO_2}$$

We will see in Chapter 20 how decarboxylations of β-keto acids and malonic acids are synthetically useful.

Aromatic carboxylic acids decarboxylate when their salts are fused with a mixture of NaOH and $Ca(OH)_2$:

$$ArCOO^- + {}^-OH \xrightarrow[\text{Heat}]{\text{NaOH—Ca(OH)}_2} ArH + CO_3^{2-}$$

These are extremely forcing conditions and the reaction is not appropriate for molecules containing base-sensitive groups.

17.10A Decarboxylation of Carboxylate Radicals

Although the carboxylate ions ($RCOO^-$) of simple aliphatic acids do not decarboxylate readily, carboxylate radicals ($RCOO\cdot$) do. They decarboxylate by losing CO_2 and producing an alkyl radical:

$$RCOO\cdot \longrightarrow R\cdot + CO_2$$

Carboxylate radicals can be generated by electrolysis, in a reaction known as the *Kolbe electrolysis,* or they can be generated chemically in a reaction known as the *Hunsdiecker reaction.*

In the **Kolbe electrolysis** an aqueous solution of the sodium or potassium salt of a carboxylic acid is subjected to electrolysis. At the anode the carboxylate ion loses an electron to become a carboxylate radical.

$$\textbf{1.}\quad R\text{—}\overset{\overset{\displaystyle :\ddot{O}}{\|}}{C}\text{—}\ddot{\underset{\cdot\cdot}{O}}:^- \xrightarrow[(-e^-)]{\text{Anode}} R\text{—}\overset{\overset{\displaystyle :\ddot{O}}{\|}}{C}\text{—}\ddot{\underset{\cdot\cdot}{O}}\cdot$$

Then the carboxylate radical decarboxylates and the alkyl radicals that are produced combine to form an alkane.

$$\textbf{2.}\quad R\text{—}\overset{\overset{\displaystyle :\ddot{O}}{\|}}{C}\text{—}\ddot{\underset{\cdot\cdot}{O}}\cdot \longrightarrow R\cdot + CO_2$$

$$\textbf{3.}\quad 2R\cdot \longrightarrow R\text{—}R$$

In the **Hunsdiecker reaction** the silver salt of a carboxylic acid is heated with bromine in CCl_4. A carboxylate radical is produced by steps 1 and 2 below.

$$\text{Step 1}\quad R\text{—}\overset{\overset{\displaystyle O}{\|}}{C}\text{—}OAg + Br_2 \xrightarrow{CCl_4} R\overset{\overset{\displaystyle O}{\|}}{C}\text{—}OBr + AgBr$$

$$\text{Step 2}\quad R\text{—}\overset{\overset{\displaystyle O}{\|}}{C}\text{—}OBr \longrightarrow R\text{—}\overset{\overset{\displaystyle O}{\|}}{C}\text{—}O\cdot + Br\cdot$$

Then the carboxylate radical decarboxylates. The resulting alkyl radical abstracts a bromine atom from $R\overset{\overset{O}{\|}}{C}OBr$ to produce an alkyl bromide and regenerate a carboxylate radical.

$$\text{Step 3}\quad R{-}\overset{\overset{\displaystyle O}{\|}}{C}{-}O\cdot \longrightarrow R\cdot + CO_2$$

$$\text{Step 4}\quad R\cdot + R{-}\overset{\overset{\displaystyle O}{\|}}{C}{-}OBr \longrightarrow R{-}Br + R{-}\overset{\overset{\displaystyle O}{\|}}{C}{-}O\cdot$$

Then steps 3 and 4 repeat and so on.

Overall the Hunsdiecker reaction amounts to the following:

$$R{-}COOAg + Br_2 \xrightarrow[\text{Heat}]{CCl_4} RBr + CO_2 + AgBr$$

Problem 17.15

Using decarboxylation reactions, outline syntheses of each of the following from appropriate starting materials.

(a) Decane
(b) 2-Hexanone
(c) 2-Methylbutanoic acid
(d) Benzyl bromide

Problem 17.16

Diacyl peroxides, $R\overset{\overset{O}{\|}}{C}{-}O{-}O{-}\overset{\overset{O}{\|}}{C}R$, decompose readily when heated. (a) What factor accounts for this? (b) The decomposition of a diacyl peroxide produces CO_2. How is it formed? (c) Diacyl peroxides are often used to initiate free-radical reactions, for example, the polymerization of an alkene:

$$nCH_2{=}CH_2 \xrightarrow[(-CO_2)]{R\overset{\overset{O}{\|}}{C}{-}O{-}O{-}\overset{\overset{O}{\|}}{C}R} R{-}(CH_2CH_2)_n{-}H$$

Show the steps involved.

17.11 THIOL ESTERS

Thiol esters can be prepared by reactions of a thiol with an acid chloride.

$$R{-}C(=O){-}Cl + R'{-}SH \longrightarrow \underset{\text{Thiol ester}}{R{-}C(=O){-}S{-}R'} + HCl$$

$$CH_3C(=O)Cl + CH_3SH \xrightarrow{\text{Pyridine}} CH_3C(=O)SCH_3 + C_5H_5N^{+}{-}H\ Cl^{-}$$

Although thiol esters are not often used in laboratory syntheses, they are of great importance in syntheses that occur within living cells. One of the important thiol esters in biochemistry is "acetylcoenzyme A."

$$\underbrace{CH_3\overset{O}{\overset{\|}{C}}SCH_2}_{\text{Thiol ester}}CH_2\overset{H}{\overset{|}{N}}-\overset{O}{\overset{\|}{C}}CH_2CH_2\overset{H}{\overset{|}{N}}-\overset{O}{\overset{\|}{C}}-CH(OH)-\underset{CH_3}{\underset{|}{\overset{CH_3}{\overset{|}{C}}}}CH_2O\underset{OH}{\underset{|}{\overset{O}{\overset{\|}{P}}}}O\underset{OH}{\underset{|}{\overset{O}{\overset{\|}{P}}}}OCH_2-\text{(ribose-3'-}HPO_3^-\text{, 2'-OH)}-\text{adenine}(\overset{+}{N}H_3)$$

Acetylcoenzyme A

The important part of this rather complicated structure is the thiol ester at the beginning of the chain; because of this, acetylcoenzyme A is usually abbreviated:

$$CoA-S\overset{O}{\overset{\|}{C}}CH_3$$

and coenzyme A, itself, is abbreviated:

$$CoA-SH$$

In certain biochemical reactions, an *acyl*coenzyme A operates as an *acylating agent;* it transfers an acyl group to another nucleophile in a reaction that involves a nucleophilic attack at the acyl carbon of the thiol ester. For example:

$$R-\overset{O}{\overset{\|}{C}}-SCoA + {}^-O-\underset{O^-}{\underset{|}{\overset{OH}{\overset{|}{P}}}}=O \xrightleftharpoons{\text{Enzyme}} \underset{\text{An acyl phosphate}}{R-\overset{O}{\overset{\|}{C}}-O-\underset{O^-}{\underset{|}{\overset{O^-}{\overset{|}{P}}}}=O} + CoA-SH$$

This reaction is catalyzed by the enzyme *phosphotransacetylase.*

The α-hydrogens of the acetyl group of acetylcoenzyme A are appreciably acidic. Acetylcoenzyme A, as a result, also functions as an *alkylating agent.* Acetylcoenzyme A, for example, reacts with oxaloacetate ion to form citrate ion in a reaction that resembles an aldol addition.

$$CoA-S\overset{O}{\overset{\|}{C}}CH_3 + \underset{\text{Oxaloacetate ion}}{O=\underset{\underset{COO^-}{|}}{\underset{CH_2}{\underset{|}{C}}}-COO^-} \rightleftharpoons \underset{\text{Citrate ion}}{HO\underset{\underset{COO^-}{|}}{\underset{CH_2}{\underset{|}{C}}}\overset{\overset{COO^-}{|}}{\overset{CH_2}{\overset{|}{}}}-COO^-} + CoA-SH$$

We will see other examples of the reactions of thiol esters in Special Topic N.

One might well ask, "Why has nature made such prominent use of thiol esters?" Or, "In contrast to ordinary esters, what advantages do thiol esters offer the cell?" In answering these questions we can consider three factors:

1. Resonance contributions of type (*b*) below stabilize an ordinary ester and make the carbonyl group less susceptible to nucleophilic attack.

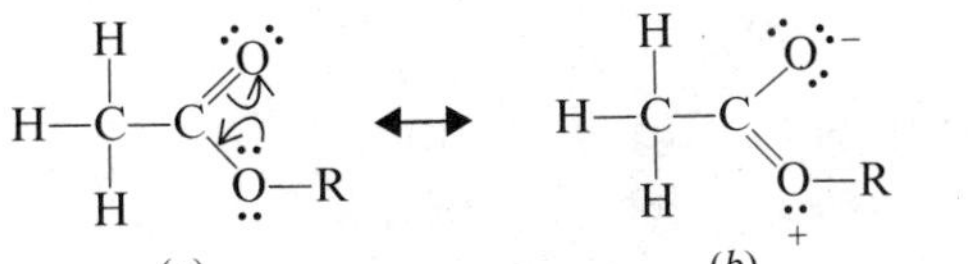

(*a*) ⟷ (*b*)

This structure makes an important contribution

By contrast, thiol esters are not as effectively stabilized by a similar resonance contribution because structure (*d*) below requires overlap between the 3*p* orbital of sulfur and a 2*p* orbital of carbon. Since this overlap is not large, resonance stabilization by (*d*) is not as effective. Structure (*e*) does, however, make an important contribution: one that makes the carbonyl group more susceptible to nucleophilic attack.

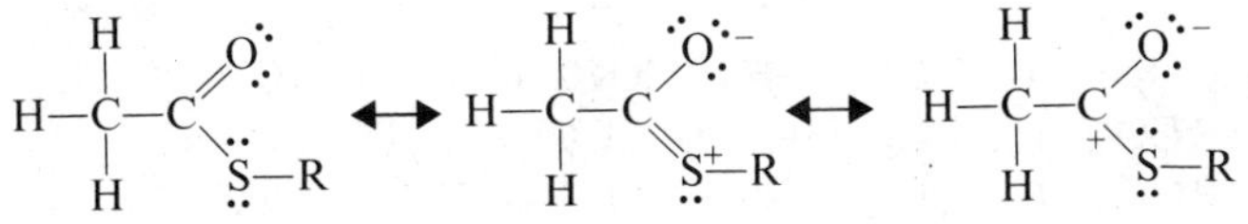

(*c*) ⟷ (*d*) ⟷ (*e*)

(*d*) *This structure is not an important contributor*

(*e*) *This structure makes the carbonyl carbon susceptible to nucleophilic attack*

2. A resonance contribution from the similar structure (*g*) below makes the α-hydrogens of thiol esters more acidic than those of ordinary esters.

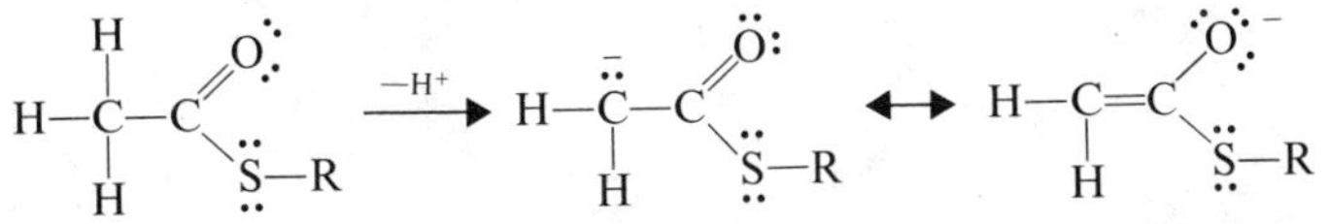

$\xrightarrow{-H^+}$ (*f*) ⟷ (*g*)

(*g*) *This structure effectively stabilizes the anion of a thiol ester*

3. The carbon-sulfur bond of a thiol ester is weaker than the carbon-oxygen bond of an ordinary ester; $^-$:S—R is a better leaving group than $^-$:OR.

Factors 1 and 3 make thiol esters effective *acylating agents;* factor 2 makes them effective *alkylating* agents. Thus, we should not be surprised when we encounter reactions like the one shown below:

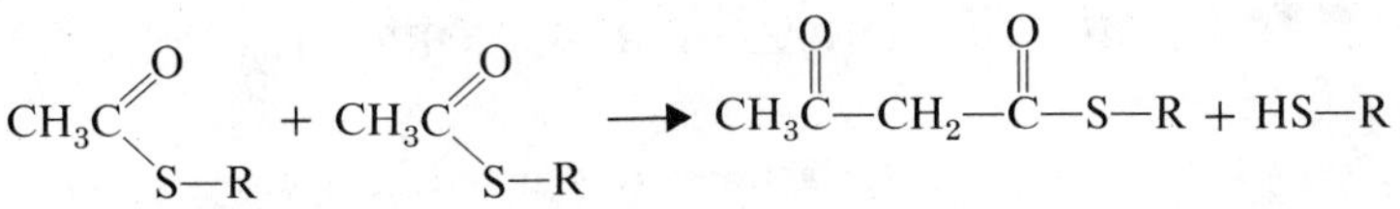

$$CH_3C(=O)S{-}R + CH_3C(=O)S{-}R \longrightarrow CH_3\overset{O}{\overset{\|}{C}}{-}CH_2{-}\overset{O}{\overset{\|}{C}}{-}S{-}R + HS{-}R$$

In this reaction one mole of a thiol ester acts as an acylating agent and the other acts as an alkylating agent (cf. p. 973).

17.12 SUMMARY OF THE REACTIONS OF CARBOXYLIC ACIDS AND THEIR DERIVATIVES

The reactions of carboxylic acids and their derivatives are summarized below.

17.12A Reactions of Carboxylic Acids

1. As acids (discussed in Section 17.2).

$$\mathrm{RCOOH + NaOH \longrightarrow RCOO^-Na^+ + H_2O}$$
$$\mathrm{RCOOH + NaHCO_3 \longrightarrow RCOO^-Na^+ + H_2O + CO_2}$$

2. Reduction (discussed in Section 15.7).

$$\mathrm{RCOOH + LiAlH_4 \xrightarrow[(2)\ H_2O]{} RCH_2OH}$$

3. Conversion to acid chlorides (discussed in Section 17.5).

$$\mathrm{RCOOH + SOCl_2 \longrightarrow RCOCl + SO_2 + HCl}$$
$$\mathrm{3RCOOH + PCl_3 \longrightarrow 3RCOCl + H_3PO_3}$$
$$\mathrm{RCOOH + PCl_5 \longrightarrow RCOCl + POCl_3 + HCl}$$

4. Conversion to acid anhydrides (discussed in Section 17.6).

$$\mathrm{R{-}C(=O)OH + R{-}C(=O)Cl \xrightarrow{pyridine} R{-}C(=O){-}O{-}C(=O){-}R}$$

or

$$\mathrm{HO{-}C(=O){-}CH_2{-}CH_2{-}C(=O){-}OH \xrightarrow{\Delta} \underbrace{CH_2{-}C(=O){-}O{-}C(=O){-}CH_2}_{ring} + H_2O}$$

5. Conversion to esters (discussed in Section 17.7).

$$\mathrm{R{-}\overset{O}{\overset{\|}{C}}OH + R'{-}OH \overset{H^+}{\rightleftharpoons} R{-}\overset{O}{\overset{\|}{C}}{-}OR' + H_2O}$$

6. Conversion to lactones (discussed in Section 17.7)

$$R{-}\underset{\displaystyle OH}{\underset{|}{C}H}{-}(CH_2)_n{-}\overset{\displaystyle O}{\overset{\|}{C}}OH \overset{H^+}{\rightleftharpoons} \text{(cyclic lactone: ring of } C{=}O,\ O,\ CH{-}R,\ (CH_2)_n)$$

$n = 2$, a γ-lactone
$n = 3$, a δ-lactone

7. Conversion to amides and imides (discussed in Section 17.8).

$$R{-}\overset{\displaystyle O}{\overset{\|}{C}}{-}OH + NH_3 \rightleftharpoons R{-}\overset{\displaystyle O}{\overset{\|}{C}}{-}O^-NH_4^+$$

$$\xrightarrow{\text{Isolate and heat}} R{-}\overset{\displaystyle O}{\overset{\|}{C}}{-}NH_2 + H_2O$$

An amide

$$\underset{CH_2{-}\overset{\displaystyle O}{\overset{\|}{C}}{-}OH}{CH_2{-}\overset{\displaystyle O}{\overset{\|}{C}}{-}NH_2} \xrightarrow{\text{Heat}} \text{(cyclic: } CH_2{-}CH_2 \text{ with two } C{=}O \text{ joined by } NH) + H_2O$$

A cyclic imide

8. Conversion to lactams (discussed in Section 17.8).

$$R{-}\overset{\displaystyle NH_2}{\overset{|}{C}}H(CH_2)_n\overset{\displaystyle O}{\overset{\|}{C}}OH \xrightarrow{\text{Heat}} \text{(cyclic: } RCH{-}N(H){-}C{=}O,\ (CH_2)_n) + H_2O$$

$n = 2$, a γ-lactam
$n = 3$, a δ-lactam

9. Alpha halogenation (discussed in Section 17.9).

$$R{-}CH_2COOH + X_2 \xrightarrow{P} R{-}\underset{\displaystyle X}{\underset{|}{C}}HCOOH$$

$X_2 = Cl_2$ or Br_2

10. Decarboxylation (discussed in Section 17.10)

$$R\overset{\displaystyle O}{\overset{\|}{C}}CH_2\overset{\displaystyle O}{\overset{\|}{C}}OH \xrightarrow{\text{Heat}} R\overset{\displaystyle O}{\overset{\|}{C}}CH_3 + CO_2$$

$$\mathrm{HO\overset{O}{\overset{\|}{C}}CH_2\overset{O}{\overset{\|}{C}}OH \xrightarrow{\text{Heat}} CH_3\overset{O}{\overset{\|}{C}}OH + CO_2}$$

$$\mathrm{R\overset{O}{\overset{\|}{C}}OAg + Br_2 \xrightarrow[\text{Heat}]{CCl_4} RBr + CO_2 + AgBr}$$

$$\mathrm{2R\overset{O}{\overset{\|}{C}}O^- \xrightarrow{\text{Electrolysis}} R{-}R + 2CO_2}$$

17.12B Reactions of Acyl Chlorides

1. Conversion to acids (discussed in Section 17.4).

$$\mathrm{R{-}\overset{O}{\overset{\|}{C}}{-}Cl + H_2O \longrightarrow R{-}\overset{O}{\overset{\|}{C}}{-}OH + HCl}$$

2. Conversion to anhydrides (discussed in Section 17.6).

$$\mathrm{R{-}\overset{O}{\overset{\|}{C}}{-}Cl + R{-}\overset{O}{\overset{\|}{C}}{-}O^- \longrightarrow R{-}\overset{O}{\overset{\|}{C}}{-}O{-}\overset{O}{\overset{\|}{C}}{-}R + Cl^-}$$

3. Conversion to esters (discussed in Section 17.7).

$$\mathrm{R{-}\overset{O}{\overset{\|}{C}}{-}Cl + R'{-}OH \xrightarrow{\text{Pyridine}} R{-}\overset{O}{\overset{\|}{C}}{-}OR'}$$

4. Conversion to amides (discussed in Section 17.8).

$$\mathrm{R{-}\overset{O}{\overset{\|}{C}}{-}Cl + NH_3(excess) \longrightarrow R{-}\overset{O}{\overset{\|}{C}}{-}NH_2 + NH_4Cl}$$

$$\mathrm{R{-}\overset{O}{\overset{\|}{C}}{-}Cl + NH_2R'(excess) \longrightarrow R{-}\overset{O}{\overset{\|}{C}}{-}NHR' + R'NH_3Cl}$$

$$\mathrm{R{-}\overset{O}{\overset{\|}{C}}{-}Cl + NHR'_2(excess) \longrightarrow R{-}\overset{O}{\overset{\|}{C}}{-}NR'_2 + R'_2NH_2Cl}$$

5. Conversion to thiol esters (discussed in Section 17.10).

$$\mathrm{R{-}\overset{O}{\overset{\|}{C}}{-}Cl + R'{-}SH \xrightarrow{\text{Base}} R{-}\overset{O}{\overset{\|}{C}}{-}SR'}$$

6. Conversion to ketones.

$$\mathrm{R{-}\overset{O}{\overset{\|}{C}}{-}Cl + C_6H_6 \xrightarrow{AlCl_3} C_6H_5{-}\overset{O}{\overset{\|}{C}}{-}R}$$

(discussed in Section 12.7)

$$2R-\overset{O}{\overset{\|}{C}}-Cl + R'_2Cd \longrightarrow 2R-\overset{O}{\overset{\|}{C}}-R'$$

$$R-\overset{O}{\overset{\|}{C}}-Cl + R'_2CuLi \longrightarrow R-\overset{O}{\overset{\|}{C}}-R'$$

(discussed in Section 16.5)

7. Conversion to aldehydes (discussed in Section 16.4).

$$R-\overset{O}{\overset{\|}{C}}-Cl + H_2 \xrightarrow{Pd(S)} R-\overset{O}{\overset{\|}{C}}-H$$

8. Reaction with carbanions (discussed in Section 20.3).

$$R-\overset{-}{C}H\overset{O}{\overset{\|}{C}}OR' + R''C(=O)Cl \longrightarrow R-CH(C(=O)R'')-\overset{O}{\overset{\|}{C}}OR'$$

17.12C Reactions of Acid Anhydrides

1. Conversion to acids.

$$(RCO)_2O + H_2O \longrightarrow 2R-\overset{O}{\overset{\|}{C}}-OH$$

2. Conversion to esters (discussed in Section 17.7).

$$(RCO)_2O + R'OH \longrightarrow R-\overset{O}{\overset{\|}{C}}-OR' + R-\overset{O}{\overset{\|}{C}}OH$$

or

$$\text{(cyclic) } CH_2C(=O)-O-C(=O)CH_2 + R'OH \longrightarrow CH_2(C(=O)-OR')-CH_2(C(=O)-OH)$$

3. Conversion to amides and imides (discussed in Section 17.8).

$$(RCO)_2O + H-N(R')(R'') \longrightarrow R-\overset{O}{\overset{\|}{C}}-N(R'')-R' + R-\overset{O}{\overset{\|}{C}}OH$$

R′ and/or R″ may be H

or

$$\text{(succinic anhydride: } CH_2C(=O)-O-C(=O)CH_2\text{ ring)} + R'-NH_2 \longrightarrow \begin{array}{l} CH_2-\overset{O}{\overset{\|}{C}}-NHR' \\ | \\ CH_2-\underset{O}{\underset{\|}{C}}-OH \end{array} \xrightarrow{\text{heat}} \text{(cyclic imide: } CH_2C(=O)-N(R')-C(=O)CH_2\text{ ring)}$$

R′ may be H

4. Conversion to ketones (discussed in Section 17.7).

$$(R-\overset{O}{\overset{\|}{C}})_2O + C_6H_6 \xrightarrow{AlCl_3} C_6H_5-\overset{O}{\overset{\|}{C}}-R + R-\overset{O}{\overset{\|}{C}}OH$$

17.12D Reactions of Esters

1. Hydrolysis (discussed in Section 17.7).

$$R-\overset{O}{\overset{\|}{C}}-O-R' + H_2O \xrightarrow{H^+} R-\overset{O}{\overset{\|}{C}}-OH + R'-OH$$

$$R-\overset{O}{\overset{\|}{C}}-O-R' + OH^- \xrightarrow[H_2O]{} R-\overset{O}{\overset{\|}{C}}-O^- + R'-OH$$

2. Conversion to other esters: transesterification (discussed in Section 17.7).

$$R-\overset{O}{\overset{\|}{C}}-O-R' + R''-OH \overset{H^+}{\rightleftharpoons} R\overset{O}{\overset{\|}{C}}-O-R'' + R'-OH$$

3. Conversion to amides (discussed in Section 17.8).

$$R-\overset{O}{\overset{\|}{C}}-OR' + HN\begin{smallmatrix} R'' \\ R''' \end{smallmatrix} \longrightarrow R-\overset{O}{\overset{\|}{C}}N\begin{smallmatrix} R'' \\ R''' \end{smallmatrix} + R'-OH$$

R″ and/or R‴ may be H

4. Reaction with Grignard reagents (discussed in Section 14.11).

$$R-\overset{O}{\overset{\|}{C}}-OR' + 2R''MgX \xrightarrow{\text{Ether}} R-\underset{R''}{\underset{|}{\overset{OMgX}{\overset{|}{C}}}}-R'' + R'OMgX$$

$$\downarrow H^+$$

$$R-\underset{R''}{\underset{|}{\overset{OH}{\overset{|}{C}}}}-R''$$

5. Reduction (discussed in Section 15.7).

$$\text{R}-\overset{\text{O}}{\overset{\|}{\text{C}}}-\text{O}-\text{R}' + \text{H}_2 \xrightarrow{\text{Ni}} \text{R}-\text{CH}_2\text{OH} + \text{R}'-\text{OH}$$

$$\text{R}-\overset{\text{O}}{\overset{\|}{\text{C}}}-\text{O}-\text{R}' + \text{LiAlH}_4 \xrightarrow[\text{(2) H}_2\text{O}]{} \text{R}-\text{CH}_2\text{OH} + \text{R}'-\text{OH}$$

$$\text{R}-\overset{\text{O}}{\overset{\|}{\text{C}}}-\text{OR}' + \text{Na} \xrightarrow{\text{C}_2\text{H}_5\text{OH}} \text{R}-\text{CH}_2\text{OH} + \text{R}'-\text{OH}$$

6. Reaction with carbanions: Claisen condensation (discussed in Section 20.2).

$$2\text{RCH}_2\overset{\text{O}}{\overset{\|}{\text{C}}}-\text{OR}' \xrightarrow[\text{(2) H}^+]{\text{(1) NaOC}_2\text{H}_5} \text{R}\underset{\underset{\text{RCH}_2}{|}}{\underset{\text{C=O}}{\underset{|}{\text{C}}}}\text{H}\overset{\text{O}}{\overset{\|}{\text{C}}}\text{OR}' + \text{R}'-\text{OH}$$

17.12E Reactions of Amides

1. Hydrolysis (discussed in Section 17.8).

$$\text{R}-\overset{\text{O}}{\overset{\|}{\text{C}}}-\underset{\underset{\text{R}''}{|}}{\text{N}}\text{R}' + \text{H}_3\text{O}^+ \xrightarrow{\text{H}_2\text{O}} \text{R}-\overset{\text{O}}{\overset{\|}{\text{C}}}-\text{OH} + \text{R}'-\underset{\underset{\text{R}''}{|}}{\text{N}}\text{H}_2{}^+$$

$$\text{R}-\overset{\text{O}}{\overset{\|}{\text{C}}}-\underset{\underset{\text{R}''}{|}}{\text{N}}\text{R}' + \text{OH}^- \xrightarrow{\text{H}_2\text{O}} \text{R}\overset{\text{O}}{\overset{\|}{\text{C}}}-\text{O}^- + \text{R}'-\underset{\underset{\text{R}''}{|}}{\text{N}}\text{H}$$

R, R′, and/or R″ may be H

2. Conversion to nitriles: dehydration (discussed in Section 17.8).

$$\text{R}-\overset{\text{O}}{\overset{\|}{\text{C}}}\text{NH}_2 \xrightarrow[\text{heat}\ (-\text{H}_2\text{O})]{\text{P}_2\text{O}_5} \text{R}-\text{C}\equiv\text{N}$$

3. Conversion to imides (discussed in Section 17.8).

$$\text{C}_6\text{H}_4(\overset{\text{O}}{\overset{\|}{\text{C}}}-\text{NHR})(\underset{\underset{\text{O}}{\|}}{\text{C}}\text{OH}) \xrightarrow[(-\text{H}_2\text{O})]{\text{heat}} \text{C}_6\text{H}_4(\overset{\text{O}}{\overset{\|}{\text{C}}})(\underset{\underset{\text{O}}{\|}}{\text{C}})\text{NR}$$

4. Conversion to amines (discussed in Section 18.5).

$$\mathrm{R\overset{O}{\overset{\|}{C}}{-}NH_2 + LiAlH_4 \longrightarrow RCH_2{-}NH_2}$$

$$\mathrm{R\overset{O}{\overset{\|}{C}}{-}NH_2 \xrightarrow{Br_2,\ OH^-} R{-}NH_2 + CO_3^{=}}$$

17.13 CHEMICAL TESTS FOR ACYL COMPOUNDS

Carboxylic acids are weak acids and their acidity helps us to detect them. Aqueous solutions of water-soluble carboxylic acids give an acid test with blue litmus paper. Water-insoluble carboxylic acids dissolve in aqueous sodium hydroxide and aqueous sodium bicarbonate. The latter reagent helps us distinguish carboxylic acids from most phenols. Except for the di- and trinitrophenols, phenols do not dissolve in aqueous sodium bicarbonate. Carboxylic acids not only dissolve in aqueous sodium bicarbonate, they also cause the evolution of carbon dioxide.

It is often helpful to determine the equivalent weight of a carboxylic acid by titrating a measured quantity of the acid with a standard solution of sodium hydroxide. For a monocarboxylic acid, the equivalent weight equals the molecular weight; for a dicarboxylic acid, the equivalent weight is one-half the molecular weight, and so on.

All acid derivatives can be hydrolyzed to carboxylic acids. The conditions required to bring about hydrolysis vary greatly, with acyl chlorides being the easiest to hydrolyze and amides being the most difficult.

Acyl chlorides hydrolyze in water and thus give a precipitate when treated with aqueous silver nitrate. Acid anhydrides dissolve when heated briefly with aqueous sodium hydroxide.

Esters and amides hydrolyze slowly when they are refluxed with sodium hydroxide: esters produce a carboxylate ion and an alcohol; amides produce a carboxylate ion and an amine or ammonia. The hydrolysis products, the acid and the alcohol or amine, can be isolated and identified. Since base-promoted hydrolysis of an unsubstituted amide produces ammonia, this can often be detected by holding moist red litmus in the vapors above the reaction mixture.

Base-promoted hydrolysis of an ester consumes one mole of hydroxide ion for each mole of the ester. It is often convenient, therefore, to carry out the hydrolysis quantitatively.

$$\underset{\text{1 mole}}{\mathrm{RCOOR'}} + \underset{\text{1 mole}}{\mathrm{OH^-}} \longrightarrow \mathrm{RCOO^- + R'OH}$$

This allows us to determine the *equivalent weight* of the ester. We can do this by hydrolyzing a known weight of the ester with an excess of a standard solution of sodium hydroxide. After the hydrolysis is complete, we can titrate the excess sodium hydroxide with a standard acid.

Amides can be distinguished from amines with dilute HCl. Most amines dissolve in dilute HCl whereas most amides do not (cf. Problem 17.35).

17.17

Write structural formulas for each of the following compounds.

(a) Hexanoic acid
(b) Hexanamide
(c) *N*-Ethylhexanamide
(d) *N,N*-Diethylhexanamide
(e) 3-Hexenoic acid
(f) 2-Methyl-4-hexenoic acid
(g) Hexanedioic acid
(h) Phthalic acid
(i) Isophthalic acid
(j) Terephthalic acid
(k) Diethyl oxalate
(l) Diethyl adipate
(m) Isobutyl propanoate
(n) 2-Naphthoic acid
(o) Maleic acid
(p) Malic acid
(q) Fumaric acid
(r) Succinic acid
(s) Succinimide
(t) Malonic acid
(u) Diethyl malonate

17.18

Give IUPAC or common names for each of the following compounds.

(a) C_6H_5COOH
(b) C_6H_5COCl
(c) $C_6H_5CONH_2$
(d) $(C_6H_5CO)_2O$
(e) $C_6H_5COOCH_2C_6H_5$
(f) $C_6H_5COOC_6H_5$
(g) $CH_3COOCH(CH_3)_2$
(h) $CH_3CON(CH_3)_2$
(i) CH_3CN

(j) O, O, O

(k) O, O, O

(l) O, NH, O

(m) $CH_2O\overset{O}{\overset{\|}{C}}(CH_2)_{14}CH_3$
$\quad|$
$CHO\overset{O}{\overset{\|}{C}}(CH_2)_{14}CH_3$
$\quad|$
$CH_2O\overset{O}{\overset{\|}{C}}(CH_2)_{14}CH_3$

(n) $HOOCCH_2\overset{O}{\overset{\|}{C}}COOH$

(o) OH, $\underset{\underset{O}{\|}}{C}OCH_3$

17.19

Show how benzoic acid can be synthesized from each of the following.

(a) Bromobenzene
(b) Toluene
(c) Benzonitrile, C_6H_5CN
(d) Acetophenone
(e) Benzaldehyde
(f) Styrene
(g) Benzyl alcohol

17.20

Show how phenylacetic acid can be prepared from

(a) Phenylacetaldehyde
(b) Benzyl bromide (two ways)

17.21

Show how pentanoic acid can be prepared from

(a) 1-Pentanol
(b) 1-Bromobutane (two ways)

(c) 2-Hexanone
(d) 5-Decene
(e) 1-Pentanal

17.22
What products would you expect to obtain when acetyl chloride reacts with each of the following?

(a) H_2O
(b) $AgNO_3/H_2O$
(c) $CH_3(CH_2)_2CH_2OH$ and pyridine
(d) NH_3 (excess)
(e) $C_6H_5CH_3$ and $AlCl_3$
(f) H_2 and Pd(S)
(g) $(CH_3)_2CuLi$
(h) $(CH_3CH_2)_2Cd$
(i) CH_3NH_2 (excess)
(j) $C_6H_5NH_2$ (excess)
(k) $(CH_3)_2NH$ (excess)
(l) CH_3CH_2SH + pyridine
(m) $CH_3COO^-Na^+$
(n) CH_3COOH and pyridine
(o) Phenol and pyridine

17.23
What products would you expect to obtain when acetic anhydride reacts with each of the following?

(a) NH_3 (excess)
(b) H_2O
(c) $CH_3CH_2CH_2OH$
(d) $C_6H_6 + AlCl_3$
(e) $CH_3CH_2NH_2$ (excess)
(f) $(CH_3CH_2)_2NH$ (excess)

17.24
What products would you expect to obtain when succinic anhydride reacts with each of the reagents given in Problem 17.23?

17.25
Show how you might carry out the following transformations.

(a) Succinic anhydride → (structure; O)

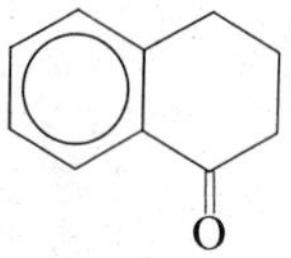

(b) 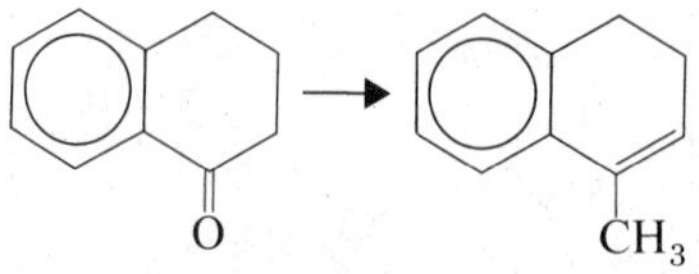 (structure; O) → (structure; CH_3)

(c) 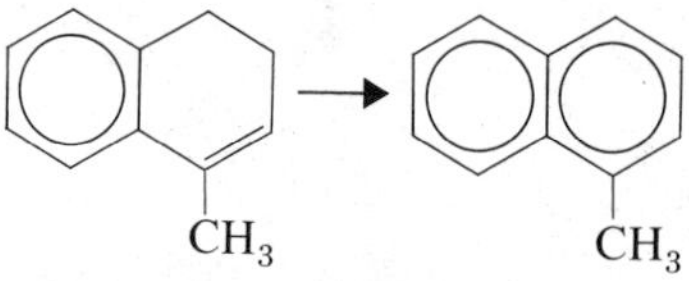 (structure; CH_3) → (structure; CH_3)

(d) Phthalic anhydride → (structure; O)

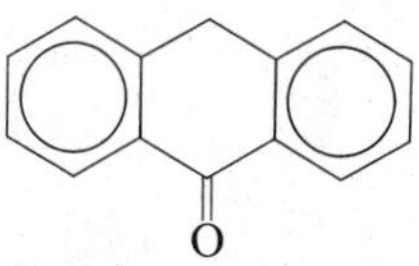

(e) Phthalic anhydride → *N*-methylphthalimide

(f) Maleic anhydride → (structure; O, O, O)

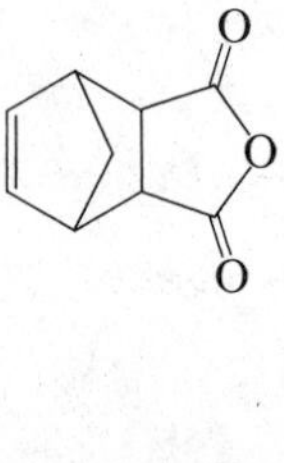

(g) Maleic anhydride →

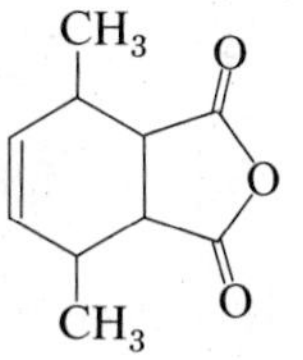

17.26

What products would you expect to obtain when ethyl propanoate reacts with each of the following?

(a) H_3O^+, H_2O
(b) OH^-, H_2O
(c) 1-octanol, HCl
(d) CH_3NH_2
(e) $LiAlH_4$, then H_2O
(f) C_6H_5MgBr, then H_2O

17.27

What products would you expect to obtain when propanamide reacts with each of the following?

(a) H_3O^+, H_2O
(b) OH^-, H_2O
(c) P_2O_5 and heat

17.28

Outline a simple chemical test that would serve to distinguish between

(a) Benzoic acid and methyl benzoate
(b) Benzoic acid and benzoyl chloride
(c) Benzoic acid and benzamide
(d) Benzoic acid and *p*-cresol
(e) Ethyl benzoate and benzamide
(f) Benzoic acid and cinnamic acid
(g) Ethyl benzoate and benzoyl chloride
(h) 2-Chlorobutanoic acid and butanoic acid

17.29

What products would you expect to obtain when each of the following compounds is heated?

(a) 4-Hydroxybutanoic acid
(b) 3-Hydroxybutanoic acid
(c) 2-Hydroxybutanoic acid
(d) Glutaric acid
(e) $CH_3CH(NH_2)CH_2CH_2CH_2C(=O)OH$
(f) 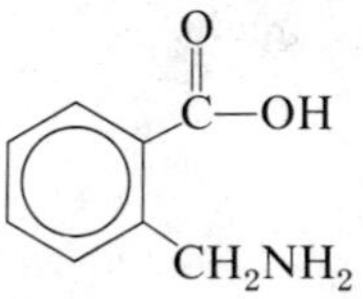 o-$(CH_2NH_2)C_6H_4C(=O)-OH$

17.30

Give stereochemical formulas for compounds **A** to **Q**.

(a) R-(−)-2-Butanol $\xrightarrow[\text{Pyridine}]{\text{TsCl}}$ **A** $\xrightarrow{\text{CN}^-}$ **B**(C_5H_9N) $\xrightarrow[H_2O]{H_2SO_4}$ (+)-**C**($C_5H_{10}O_2$) $\xrightarrow{LiAlH_4}$ (−)-**D**($C_5H_{12}O$)

(b) R-(−)-2-Butanol $\xrightarrow[\text{Pyridine}]{PBr_3}$ **E**(C_4H_9Br) $\xrightarrow{\text{CN}^-}$ **F**(C_5H_9N) $\xrightarrow[H_2O]{H_2SO_4}$ (−)-**C**($C_5H_{10}O_2$) $\xrightarrow{LiAlH_4}$ (+)-**D**($C_5H_{12}O$)

(c) **A** $\xrightarrow{CH_3COO^-}$ **G**($C_6H_{12}O_2$) $\xrightarrow{OH^-}$ (+)-**H**($C_4H_{10}O$) + CH_3COO^-

(d) (−)-**D** $\xrightarrow{PBr_3}$ **J**($C_5H_{11}Br$) $\xrightarrow[Ether]{Mg}$ **K**($C_5H_{11}MgBr$) $\xrightarrow[(2)\ H^+]{(1)\ CO_2}$ **L**($C_6H_{12}O_2$)

(e) *R*-(+)-Glyceraldehyde $\xrightarrow{HCN}$ **M**($C_4H_7NO_3$) + **N**($C_4H_7NO_3$)

Diastereomers, separated by fractional crystallization

(f) **M** $\xrightarrow[H_2O]{H_2SO_4}$ **P**($C_4H_8O_5$) $\xrightarrow[HNO_3]{(O)}$ *meso*-tartaric acid

(g) **N** $\xrightarrow[H_2O]{H_2SO_4}$ **Q**($C_4H_8O_5$) $\xrightarrow[HNO_3]{(O)}$ (−)-tartaric acid

17.31

(a) (±)-Pantotheine and (±)-pantothenic acid, important intermediates in the synthesis of coenzyme **A**, were prepared by the following route. Give structures for compounds **A** to **D**.

$CH_3CH(CH_3)CHO$ + $HCHO$ $\xrightarrow[H_2O]{K_2CO_3}$ (±)-**A**($C_5H_{10}O_2$) $\xrightarrow[(2)\ KCN]{(1)\ NaHSO_3}$

(±)-**B**($C_6H_{11}NO_2$) $\xrightarrow{H_3O^+}$ [(±)-**C**($C_6H_{12}O_4$)] $\xrightarrow{-H_2O}$

(±)-**D**($C_6H_{10}O_3$) $\xrightarrow{H_2NCH_2CH_2COOH}$ $(CH_3)_2C(CH_2OH)-CH(OH)-C(=O)-NHCH_2CH_2C(=O)OH$

A γ-lactone

(±)-Pantothenic acid

(±)-**D** $\xrightarrow{H_2NCH_2CH_2C(=O)NHCH_2CH_2SH}$ $(CH_3)_2C(CH_2OH)-CH(OH)-C(=O)-NHCH_2CH_2C(=O)NHCH_2CH_2SH$

(±)-Pantetheine

(b) The γ-lactone, (±) **D**, can be resolved. If the (−)-γ-lactone is used in the last step, the pantotheine that is obtained is identical with that obtained naturally. The (−)-γ-lactone has the *R* configuration. What is the stereochemistry of naturally occurring pantotheine?
(c) What products would you expect to obtain when (±)-pantotheine is refluxed with aqueous sodium hydroxide?

17.32

The infrared and proton nmr spectra of phenacetin ($C_{10}H_{13}NO_2$) are given in Fig. 17.1. Phenacetin is an analgesic and antipyretic compound, and is the P of A-P-C tablets (**A**spirin-**P**henacetin-**C**affeine). When phenacetin is refluxed with aqueous sodium hydroxide it yields phenetidine ($C_8H_{11}NO$) and sodium acetate. Propose structures for phenacetin and phenetidine.

17.33

Given below are the proton nmr spectra and carbonyl absorption peaks of five acyl compounds. Propose structures for each.

(a) $C_8H_{14}O_4$	proton nmr spectrum		IR spectrum
	Triplet	δ1.2 (6H)	1740 cm^{-1}

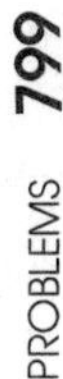

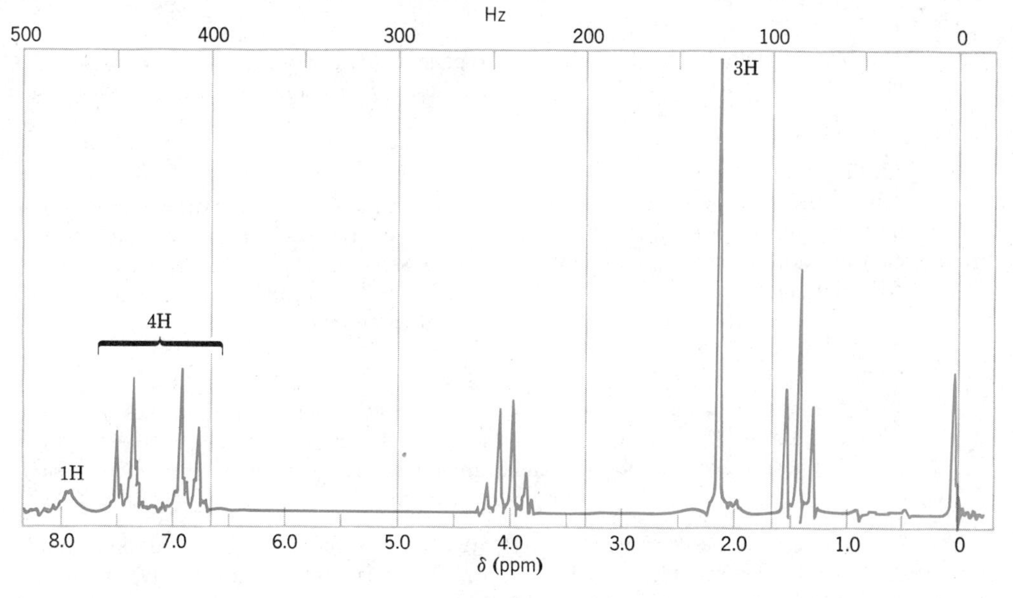

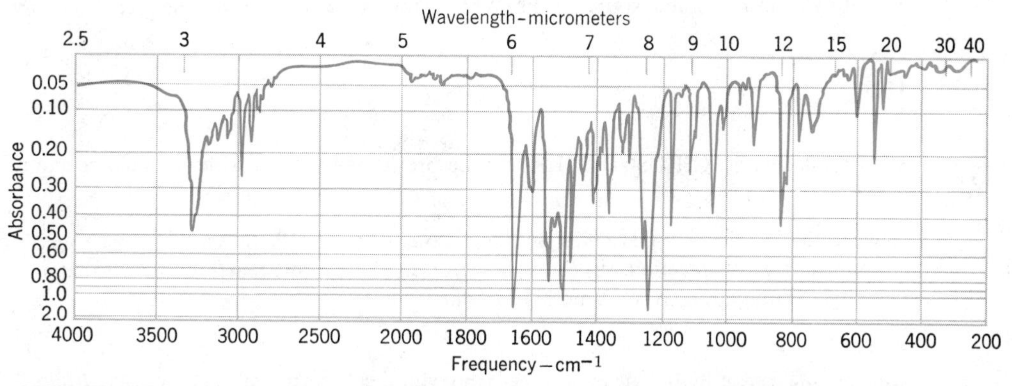

FIG. 17.1
The proton nmr and infrared spectra of phenacetin. (The proton nmr spectrum, courtesy Varian Associate, Palo Alto, Calif. Infrared spectrum, courtesy Sadtler Research Laboratories, Philadelphia, Pa.)

	Singlet	δ2.5 (4H)	
	Quartet	δ4.1 (4H)	
(b) $C_{11}H_{14}O_2$	proton nmr spectrum		IR spectrum
	Doublet	δ1.0 (6H)	1720 cm^{-1}
	Multiplet	δ2.1 (1H)	
	Doublet	δ4.1 (2H)	
	Multiplet	δ7.8 (5H)	
(c) $C_{10}H_{12}O_2$	proton nmr spectrum		IR spectrum
	Triplet	δ1.2 (3H)	1740 cm^{-1}
	Singlet	δ3.5 (2H)	
	Quartet	δ4.1 (2H)	
	Multiplet	δ7.3 (5H)	
(d) $C_2H_2Cl_2O_2$	proton nmr spectrum		IR spectrum
	Singlet	δ6.0	Broad peak 2500–2700 cm^{-1}
	Singlet	δ11.70	1705 cm^{-1}
(e) $C_4H_7ClO_2$	proton nmr spectrum		IR spectrum
	Triplet	δ1.3	1745 cm^{-1}
	Singlet	δ4.0	
	Quartet	δ4.2	

17.34

The active ingredient of the insect repellent "Off" is *N,N*-diethyl-*m*-toluamide, *m*-$CH_3C_6H_4CON(CH_2CH_3)_2$. Outline a synthesis of this compound starting with *meta*-toluic acid.

17.35

Amides are much weaker bases than corresponding amines. For example, most water-insoluble amines, RNH_2, will dissolve in dilute aqueous acids (e.g., aqueous HCl, H_2SO_4, etc.) by forming water-soluble alkylammonium salts, $RNH_3^+X^-$. Corresponding amides, $RCONH_2$, *do not dissolve in dilute aqueous acids,* however. Propose an explanation for the much lower basicity of amides when compared to amines.

17.36

While amides are much less basic than amines, they are much stronger acids. Amides have acidity constants (K_a's) in the range $10^{-14} - 10^{-16}$, while for amines, $K_a = 10^{-33} - 10^{-35}$. (a) What factor accounts for the much greater acidity of amides? (b) *Imides,* that is, compounds with the structure $(\overset{\text{O}}{\overset{\|}{\text{R}\text{C}}})_2\text{NH}$, are even stronger acids than amides. For imides $K_a = 10^{-9} - 10^{-10}$, and, as a consequence, water-insoluble imides dissolve in aqueous NaOH by forming soluble sodium salts. What extra factor accounts for the greater acidity of imides?

***17.37**

Alkyl thiolacetates, $CH_3\overset{\text{O}}{\overset{\|}{\text{C}}}SCH_2CH_2R$, can be prepared by a peroxide-initiated reaction between thiolacetic acid. $CH_3\overset{\text{O}}{\overset{\|}{\text{C}}}SH$, and an alkene, $CH_2{=}CHR$. (a) Outline a reasonable mechanism for this reaction. (b) Show how you might use this reaction in a synthesis of 3-methyl-2-butanethiol from 2-methyl-2-butene.

***17.38**

On heating, *cis*-4-hydroxycyclohexanecarboxylic acid forms a lactone but *trans*-4-hydroxycyclohexanecarboxylic acid does not. Explain.

***17.39**

R-(+)-Glyceraldehyde can be transformed into (+)-malic acid by the following synthetic route. Give stereochemical structures for all of the intermediates.

$$R\text{-}(+)\text{-Glyceraldehyde} \xrightarrow[\text{oxidation}]{Br_2,\ H_2O} (-)\text{-glyceric acid} \xrightarrow{PBr_3}$$

$$(-)\text{-3-bromo-2-hydroxypropanoic acid} \xrightarrow{NaCN} C_4H_5NO_3 \xrightarrow[\text{heat}]{H_3O^+} (+)\text{-malic acid}$$

***17.40**

R-(+)-Glyceraldehyde can also be transformed into (−)-malic acid. This synthesis begins with the conversion of *R*-(+)-glyceraldehyde into (−)-tartaric acid as shown in Problem 17.30 parts (e) and (g). Then (−)-tartaric acid is allowed to react with phosphorus tribromide in order to replace one alcoholic —OH group with —Br. This step takes place with inversion of configuration at the carbon that undergoes attack. Treating the product of this reaction with zinc and acid produces (−)-malic acid. (a) Outline all steps in this synthesis by writing stereochemical structures for each intermediate. (b) The step in which (−)-tartaric acid is treated with phosphorus tribromide produces only one stereoisomer even though there are two replaceable —OH groups. How is this possible? (c) Suppose that the step in which (−)-tartaric acid is treated with phosphorus tribromide had taken place with "mixed" stereochemistry—with both inversion and retention at the carbon under attack. How many stereoisomers would have been produced? (d) What difference would this have made to the overall outcome of the synthesis?

***17.41**

Cantharidin is a powerful vesicant that can be isolated from dried beetles (*Cantharis vesicatoria* or "Spanish fly"). Outlined below is the stereospecific synthesis of cantharidin reported by Professor Gilbert Stork of Columbia University in 1953. Supply the missing reagents (a) to (o).

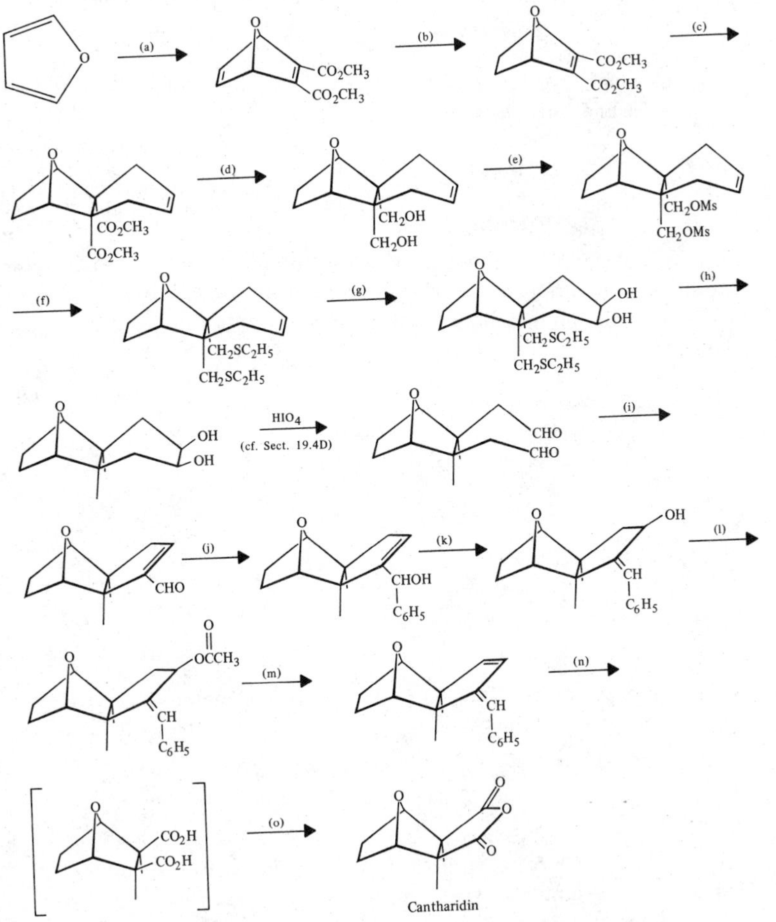

***17.42**

Examine the structure of cantharidin (Problem 17.41) carefully and (a) suggest a possible two-step synthesis of cantharidin starting with furan (p. 458). (b) F. von Bruchhausen and H. W. Bersch at the University of Münster attempted this two-step synthesis in 1928 only a few months after Diels and Alder published their first paper describing their new diene addition and found that the expected addition failed to take place. Von Bruchhausen and Bersch also found that although cantharidin is stable at relatively high temperatures, heating cantharidin with a palladium catalyst causes cantharidin to decompose. They identified furan and dimethylmaleic anhydride among the decomposition products. What has happened in the decomposition and what does this suggest about why the first step of their attempted synthesis failed?

SPECIAL TOPIC

CONDENSATION POLYMERS

We saw, in Special Topic C, that large molecules with many repeating subunits—called *polymers*—can be prepared by addition reactions of alkenes. These polymers, we noted, are called *addition polymers.*

Another broad group of polymers are those called *condensation polymers.* These polymers, as their name suggests, are prepared by condensation reactions—reactions in which monomeric subunits are joined through intermolecular eliminations of small molecules such as water or alcohols. Among the most important condensation polymers are *polyamides* and *polyesters.*

K.1 Polyamides

Silk and wool are two naturally occurring polymers that man has used for centuries to fabricate articles of clothing. They are examples of a family of compounds that are called *proteins*—a group of compounds that we will discuss in detail in Chapter 22. At this point we need only to notice (below) that the repeating subunits of proteins are derived from α-amino acids and that these subunits are joined by amide linkages. Proteins, therefore, are polyamides.

$$H_2N-\underset{\substack{|\\R}}{CH}-\overset{\substack{O\\ \|}}{C}-OH$$

An α-amino acid

Amide linkages

$$-NH-\underset{\substack{|\\R}}{CH}-\overset{\substack{O\\ \|}}{C}-NH-\underset{\substack{|\\R}}{CH}-\overset{\substack{O\\ \|}}{C}-NH-\underset{\substack{|\\R}}{CH}-\overset{\substack{O\\ \|}}{C}-NH-\underset{\substack{|\\R}}{CH}-$$

A portion of a polyamide chain as it might occur in a protein

The search for a synthetic material with properties similar to those of silk led to the discovery of a family of synthetic polyamides called nylons.

One of the most important nylons, called *nylon 6,6,* can be prepared from the six-carbon dicarboxylic acid, adipic acid, and the six-carbon diamine, hexamethylenediamine. In the commercial process these two compounds are allowed to react in equimolar proportions in order to produce a 1 : 1 salt,

$$n\underset{\text{Adipic acid}}{HO\overset{\substack{O\\ \|}}{C}-(CH_2)_4-\overset{\substack{O\\ \|}}{C}OH} + n\underset{\text{Hexamethylenediamine}}{H_2N-(CH_2)_6-NH_2} \longrightarrow$$

$$\underset{\text{1:1 salt (nylon salt)}}{n\left[{}^{-}O\overset{\substack{O\\ \|}}{C}-(CH_2)_4-\overset{\substack{O\\ \|}}{C}-O^{-}\quad H_3\overset{+}{N}-(CH_2)_6-\overset{+}{N}H_3\right]} \xrightarrow[\text{(polymerization)}]{\text{Heat}}$$

$${}^{-}O\overset{\substack{O\\ \|}}{C}-(CH_2)_4-\overset{\substack{O\\ \|}}{C}-\left[NH-(CH_2)_6-NH-\overset{\substack{O\\ \|}}{C}-(CH_2)_4-\overset{\substack{O\\ \|}}{C}\right]_{n-1}-NH-(CH_2)_6-NH_3^+ + (2n\text{-}1)H_2O$$

Nylon 6,6
(a polyamide)

Then, heating the 1 : 1 salt (nylon salt) to a temperature of 270 °C at a pressure of 250 pounds per square inch causes a polymerization to take place. Water molecules are lost as condensation reactions occur between $-\overset{O}{\overset{\|}{C}}-O^-$ and $-NH_3^+$ groups of the salt to give the polyamide.

The nylon 6,6 produced in this way has a molecular weight of ~10,000, has a melting point of ~250°, and when molten can be spun into fibers from a melt. The fibers are then stretched to about four times their original length. This orients the linear polyamide molecules so that they are parallel to the fiber axis and allows hydrogen bonds to form between —NH— and >C=O groups on adjacent chains. Called "cold drawing," stretching greatly increases the fibers' strength.

Another type of nylon, nylon 6, can be prepared by a ring-opening polymerization of ϵ-caprolactam:

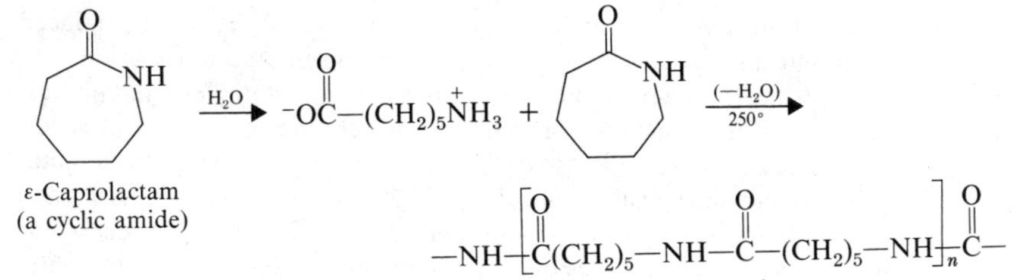

$$\text{ε-Caprolactam} \xrightarrow{H_2O} {}^{-}O\overset{O}{\overset{\|}{C}}-(CH_2)_5\overset{+}{N}H_3 + \text{ε-Caprolactam} \xrightarrow[250°]{(-H_2O)} -NH\left[\overset{O}{\overset{\|}{C}}(CH_2)_5-NH-\overset{O}{\overset{\|}{C}}-(CH_2)_5-NH\right]_n\overset{O}{\overset{\|}{C}}-$$

ε-Caprolactam
(a cyclic amide)

Nylon 6

In this process ϵ-caprolactam is allowed to react with water, converting some of it to ϵ-aminocaproic acid. Then heating this mixture at 250° drives off water as ϵ-caprolactam and ϵ-aminocaproic acid react to produce the polyamide. Nylon 6 can also be converted into fibers by melt spinning.

Problem K.1

The raw materials for the production of nylon 6,6 can be obtained in several ways as indicated on the next page. Give equations for each synthesis of adipic acid and of hexamethylenediamine.

(a) Cyclohexanone $\xrightarrow{(O)}$ adipic acid

(b) Adipic acid $\xrightarrow{2NH_3}$ a salt $\xrightarrow{\text{heat}}$ $C_6H_{12}N_2O_2$ $\xrightarrow[\text{catalyst}]{350°}$ $C_6H_8N_2$ $\xrightarrow[\text{catalyst}]{4H_2}$ hexamethylenediamine

(c) 1,3-Butadiene $\xrightarrow{Cl_2}$ $C_4H_6Cl_2$ $\xrightarrow{2NaCN}$ $C_6H_6N_2$ $\xrightarrow[Ni]{H_2}$ $C_6H_8N_2$ $\xrightarrow[\text{catalyst}]{4H_2}$ hexamethylenediamine

(d) Tetrahydrofuran $\xrightarrow{2HCl}$ $C_4H_8Cl_2$ $\xrightarrow{2NaCN}$ $C_6H_8N_2$ $\xrightarrow[\text{catalyst}]{4H_2}$ hexamethylenediamine

K.2 POLYESTERS

One of the most important polyesters is poly(ethylene terephthalate), a polymer that is marketed under the names *Dacron, Terylene,* and *Mylar.*

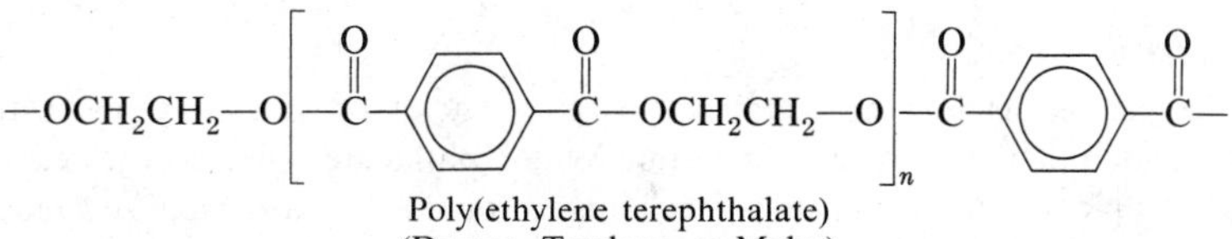

$$-OCH_2CH_2-O\left[-\overset{O}{\overset{\|}{C}}-C_6H_4-\overset{O}{\overset{\|}{C}}-OCH_2CH_2-O\right]_n-\overset{O}{\overset{\|}{C}}-C_6H_4-\overset{O}{\overset{\|}{C}}-$$

Poly(ethylene terephthalate)
(Dacron, Terylene, or Mylar)

Although one can obtain poly(ethylene terephthalate) by a direct acid-catalyzed esterification of ethylene glycol and terephthalic acid, the polymer produced in this way is of low molecular weight and is not generally useful.

$$\underset{\text{Ethylene glycol}}{HO\text{—}CH_2CH_2\text{—}OH} + \underset{\text{Terephthalic acid}}{HO\text{—}\overset{O}{\overset{\|}{C}}\text{—}C_6H_4\text{—}\overset{O}{\overset{\|}{C}}\text{—}OH} \xrightarrow[\text{heat}]{H^+} \underset{\text{of low molecular weight}}{\text{Poly(ethylene terephthalate)}} + H_2O$$

Moreover, high temperatures are required to drive off the water produced by the esterification and this causes the polymer to undergo some decomposition.

A much better method for synthesizing poly(ethylene terephthalate) is based on transesterification reactions—reactions in which one ester is converted into another. One commercial synthesis utilizes two transesterifications. In the first, dimethyl terephthalate and excess ethylene glycol are heated to 200° in the presence of a basic catalyst. Distillation of the mixture results in the loss of methanol (bp 78°) and the formation of a new ester, one formed from two moles of ethylene glycol and one mole of terephthalic acid. When this new ester is heated to a higher temperature (~280°), ethylene glycol distills and polymerization takes place.

$$\underset{\text{Dimethyl terephthalate}}{CH_3O\text{—}\overset{O}{\overset{\|}{C}}\text{—}C_6H_4\text{—}\overset{O}{\overset{\|}{C}}\text{—}OCH_3} + \underset{\text{Ethylene glycol}}{2HO\text{—}CH_2CH_2\text{—}OH} \xrightarrow[200^\circ]{\text{base}}$$

$$HO\text{—}CH_2CH_2\text{—}O\text{—}\overset{O}{\overset{\|}{C}}\text{—}C_6H_4\text{—}\overset{O}{\overset{\|}{C}}\text{—}O\text{—}CH_2CH_2\text{—}OH + 2CH_3OH$$

$$n HO\text{—}CH_2CH_2\text{—}O\text{—}C\text{—}C_6H_4\text{—}\overset{O}{\overset{\|}{C}}\text{—}O\text{—}CH_2CH_2\text{—}OH \xrightarrow{280^\circ}$$

$$\underset{\text{Poly(ethylene terephthalate)}}{\left[\overset{O}{\overset{\|}{C}}\text{—}C_6H_4\text{—}\overset{O}{\overset{\|}{C}}\text{—}O\text{—}CH_2CH_2\text{—}O\right]_n} + n HO\text{—}CH_2CH_2\text{—}OH$$

The poly(ethylene terephthalate) thus produced melts at ~270°. It can be melt spun into fibers to produce *Dacron* or *Terylene;* it can also be made into a film, in which form it is marketed as *Mylar.*

Problem K.2

Transesterifications are catalyzed by either acids or bases. Using the transesterification reaction that takes place when dimethyl terephthalate is heated with ethylene glycol as an example, outline reasonable mechanisms for (a) the base-catalyzed reaction and (b) the acid-catalyzed reaction.

Problem K.3

Kodel (below) is another polyester that enjoys wide commercial use.

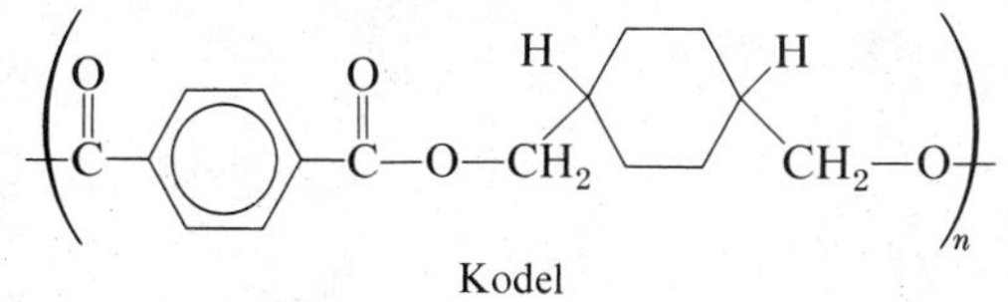

$$\left(\overset{O}{\overset{\|}{C}}-C_6H_4-\overset{O}{\overset{\|}{C}}-O-CH_2-C_6H_{10}(H)(H)-CH_2-O\right)_n$$

Kodel

Kodel is also produced by a transesterification. (a) What methyl ester and what alcohol are required for the synthesis of *Kodel?* (b) The alcohol can be prepared from dimethyl terephthalate. How might this be done?

Problem K.4

Heating phthalic anhydride and glycerol together yields a polyester called *Glyptal.* Glyptal is especially rigid because the polymer chains are "cross linked." Write a portion of the structure of *Glyptal* and show how cross linking occurs.

Problem K.5

Lexan, a high-molecular-weight "polycarbonate," is manufactured by mixing bisphenol A with phosgene in the presence of pyridine. Suggest a structure for Lexan.

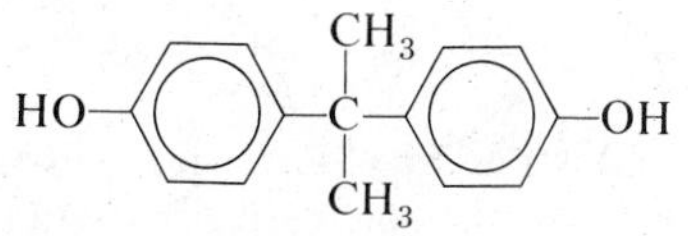

$$HO-C_6H_4-C(CH_3)_2-C_6H_4-OH \qquad Cl-\overset{O}{\overset{\|}{C}}-Cl$$

Bisphenol A 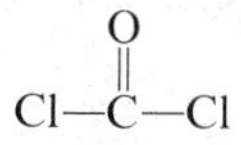 Phosgene

Problem K.6

The familiar "epoxy resins" or "epoxy glues" usually consist of two components that are sometimes labeled "resin" and "hardener." The resin is manufactured by allowing bisphenol A (Problem K.5) to react with an excess of epichlorohydrin, $\underset{\diagdown O \diagup}{CH_2-CH}CH_2Cl$, in the presence of a base until a low molecular weight polymer is obtained. (a) What is a likely structure for this polymer and (b) what is the purpose of using an excess of epichlorohydrin? The hardener is usually an amine such as $H_2NCH_2CH_2NHCH_2CH_2NH_2$. (c) What reaction takes place when the resin and hardener are mixed?

K.3 POLYURETHANES

A *urethane* is the product formed when an alcohol reacts with an isocyanate:

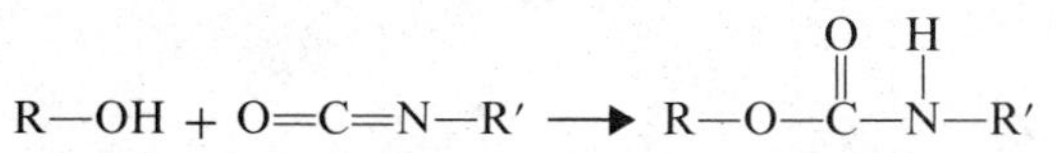

$$\underset{\text{Alcohol}}{R-OH} + \underset{\text{Isocyanate}}{O{=}C{=}N-R'} \longrightarrow \underset{\substack{\text{A urethane}\\ \text{(A carbamate)}}}{R-O-\overset{O}{\overset{\|}{C}}-\overset{H}{\overset{|}{N}}-R'}$$

The reaction probably takes place in the following way.

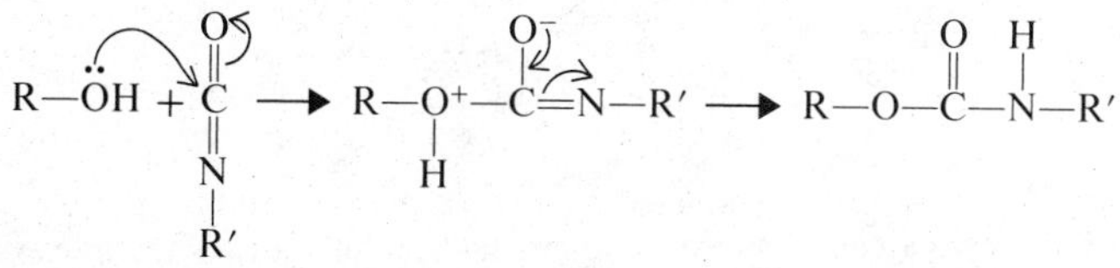

$$R-\ddot{O}H + \overset{O}{\overset{\|}{C}}{=}N-R' \longrightarrow R-\underset{H}{\underset{|}{O^+}}-\overset{O^-}{\overset{|}{C}}{=}N-R' \longrightarrow R-O-\overset{O}{\overset{\|}{C}}-\overset{H}{\overset{|}{N}}-R'$$

Urethanes are also called *carbamates* because formally they are esters of an alcohol (ROH) and a carbamic acid, R′NHCOOH.

Polyurethanes are usually made by allowing a *diol* to react with a *diisocyanate*. The diol is typically a polyester with $-CH_2OH$ end groups. The diisocyanate is usually toluene-2,4-diisocyanate.*

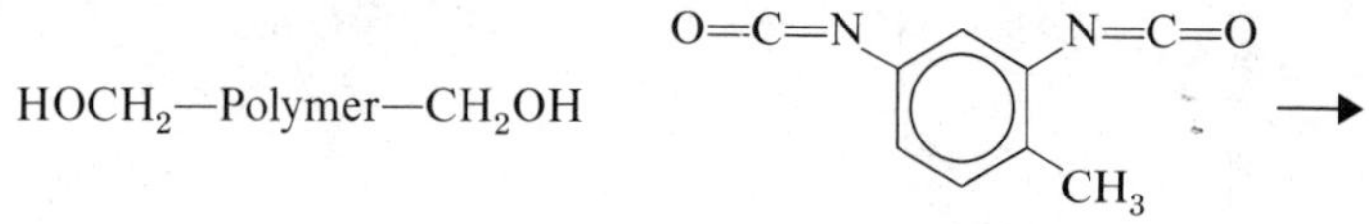

$$HOCH_2-Polymer-CH_2OH \quad + \quad O{=}C{=}N-C_6H_3(CH_3)-N{=}C{=}O \longrightarrow$$

Toluene-2,4-diisocyanate

$$\left[-C_6H_3(CH_3)-NH-\overset{\overset{\displaystyle O}{\|}}{C}-OCH_2-Polymer-CH_2O-\overset{\overset{\displaystyle O}{\|}}{C}-NH- \right]_n$$

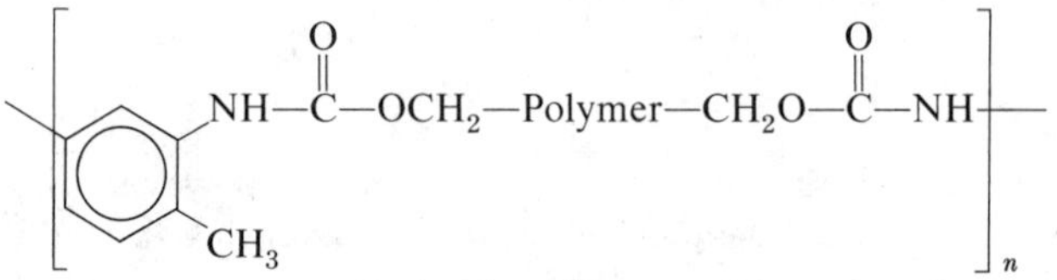

A polyurethane

Problem K.7

A typical polyurethane can be made in the following way. Adipic acid is polymerized with an excess of ethylene glycol. The resulting polyester is then treated with toluene-2,4-diisocyanate. (a) Write the structure of the polyurethane. (b) Why is an excess of ethylene glycol used in making the polyester?

Polyurethane foams, as used in pillows and paddings, are made by adding small amounts of water to the reaction mixture during the polymerization with the diisocyanate. Some of the isocyanate groups react with water to produce carbon dioxide and this gas acts as the foaming agent.

$$R-N{=}C{=}O + H_2O \longrightarrow R-NH_2 + CO_2\uparrow$$

K.4 PHENOL-FORMALDEHYDE POLYMERS

One of the first synthetic polymers to be produced was a polymer (or resin) known as *Bakelite*. Bakelite is made by a condensation reaction between phenol and formaldehyde; the reaction can be catalyzed by either acids or bases. The base-catalyzed reaction probably takes place in the general way shown below. Reaction can take place at the ortho and para positions of phenol.

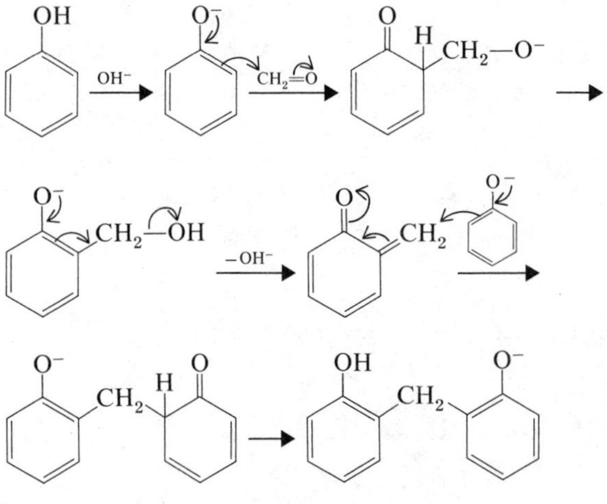

* Toluene diisocyanate is a hazardous chemical that has caused acute respiratory problems among workers synthesizing polyurethanes.

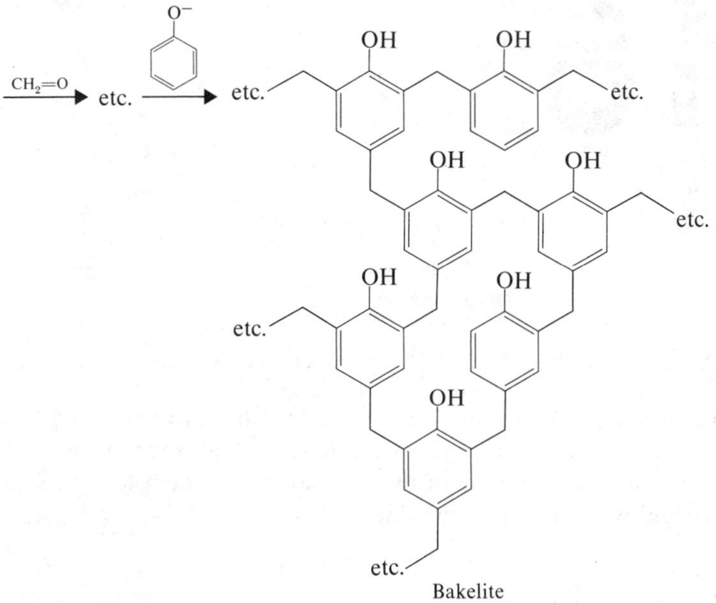

$\xrightarrow{CH_2=O}$ etc. $\xrightarrow{C_6H_5O^-}$ etc.

Bakelite

Generally, the polymerization is carried out in two stages. The first polymerization produces a low-molecular weight fusible (meltable) polymer called a *resole*. The resole can be molded to the desired shape, and then further polymerization produces very high-molecular-weight polymer, which, because it is highly cross-linked, is infusible.

Problem K.8

Using a para-substituted phenol such as *p*-cresol yields a phenol-formaldehyde polymer that is *thermoplastic* rather than *thermosetting*. That is, the polymer remains fusible, it does *not* become impossible to melt. What accounts for this?

Problem K.9

Outline a general mechanism for acid-catalyzed polymerization of phenol and formaldehyde.

18 AMINES

18.1 NOMENCLATURE

In common nomenclature most primary amines are named as *alkylamines*. In systematic nomenclature (in parentheses below) they are named by adding the suffix *amine* to the name of the chain or ring system to which the NH_2 group is attached with elision of the final *e*.

18.1A Primary Amines

$CH_3\ddot{N}H_2$
Methylamine
(Methanamine)

$CH_3CH_2\ddot{N}H_2$
Ethylamine
(Ethanamine)

$CH_3CH(CH_3)CH_2\ddot{N}H_2$
Isobutylamine
(2-Methyl-1-propanamine)

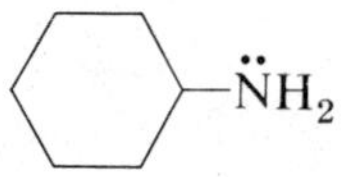

$C_6H_{11}\text{—}\ddot{N}H_2$
Cyclohexylamine
(Cyclohexanamine)

$(CH_3)_3C\text{—}\ddot{N}H_2$
tert-Butylamine
(2-Methyl-2-propanamine)

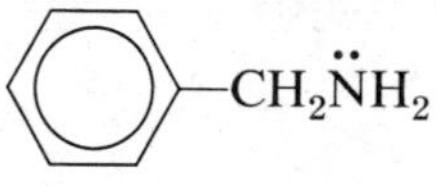

$C_6H_5\text{—}CH_2\ddot{N}H_2$
Benzylamine
(Phenylmethanamine)

$CH_2{=}CHCH_2\ddot{N}H_2$
Allylamine
(2-Propenamine)

Most secondary and tertiary amines are named in the same general way. In common nomenclature we either designate the organic groups individually if they are different, or use the prefixes *di-* or *tri-* if they are the same. In systematic nomenclature we use the locant *N* to designate substituents attached to nitrogen.

18.1B Secondary Amines

$CH_3\ddot{N}HCH_2CH_3$
Methylethylamine
(*N*-methylethanamine)

$(CH_3CH_2)_2\ddot{N}H$
Diethylamine
(*N*-ethylethanamine)

18.1C Tertiary Amines

$(CH_3CH_2)_3\ddot{N}$
Triethylamine
(*N*,*N*-Diethylethanamine)

$CH_3\ddot{N}(CH_2CH_3)CH_2CH_2CH_3$
Methylethylpropylamine
(*N*-Ethyl-*N*-methyl-1-propanamine)

18.1D Quaternary Ammonium Compounds and Amine Salts

Salts of amines and quaternary ammonium compounds are named by adding *-ammonium* and then the name of the anion.

$$(CH_3)_4N^+ \; I^-$$

Tetramethylammonium iodide
(a quaternary ammonium compound)

$$(CH_3)_3\overset{+}{N}{-}H \; Br^-$$

Trimethylammonium bromide
(a tertiary amine salt)

$$(CH_3CH_2\overset{+}{N}H_3)_2 \; SO_4^{=}$$

Ethylammonium sulfate
(a primary amine salt)

$$C_6H_5CH_2\overset{+}{N}H_2(CH_3) \; Cl^-$$

Methylbenzylammonium chloride
(a secondary amine salt)

In the IUPAC system, the —NH_2 group is named as the *amino* group. We often use this system for naming complicated amines.

$$H_2\ddot{N}CH_2CH_2OH$$

2-Aminoethanol

$$H_2\ddot{N}CH_2CH_2\overset{O}{\overset{\|}{C}}OH$$

3-Aminopropanoic acid

$$CH_3CH(\ddot{N}HCH_3)CH_2OH$$

2-(*N*-Methylamino)propanol

$$CH_3CH(\ddot{N}(CH_3)_2)CH_2\overset{O}{\overset{\|}{C}}OH$$

3-(*N,N*-Dimethylamino)-
butanoic acid

The more important *aromatic* amines have the following names:

$$C_6H_5NH_2$$

Aniline
(Benzenamine)

$$C_6H_5NHCH_3$$

N-Methylaniline
(*N*-Methylben-
zenamine)

$$p\text{-}CH_3C_6H_4NH_2$$

p-Toluidine
(4-Methylben-
zenamine)

The important *heterocyclic* amines all have common names. In systematic nomenclature the prefixes *aza, diaza,* and *triaza* are used to indicate that nitrogen has replaced carbon in the corresponding hydrocarbon.

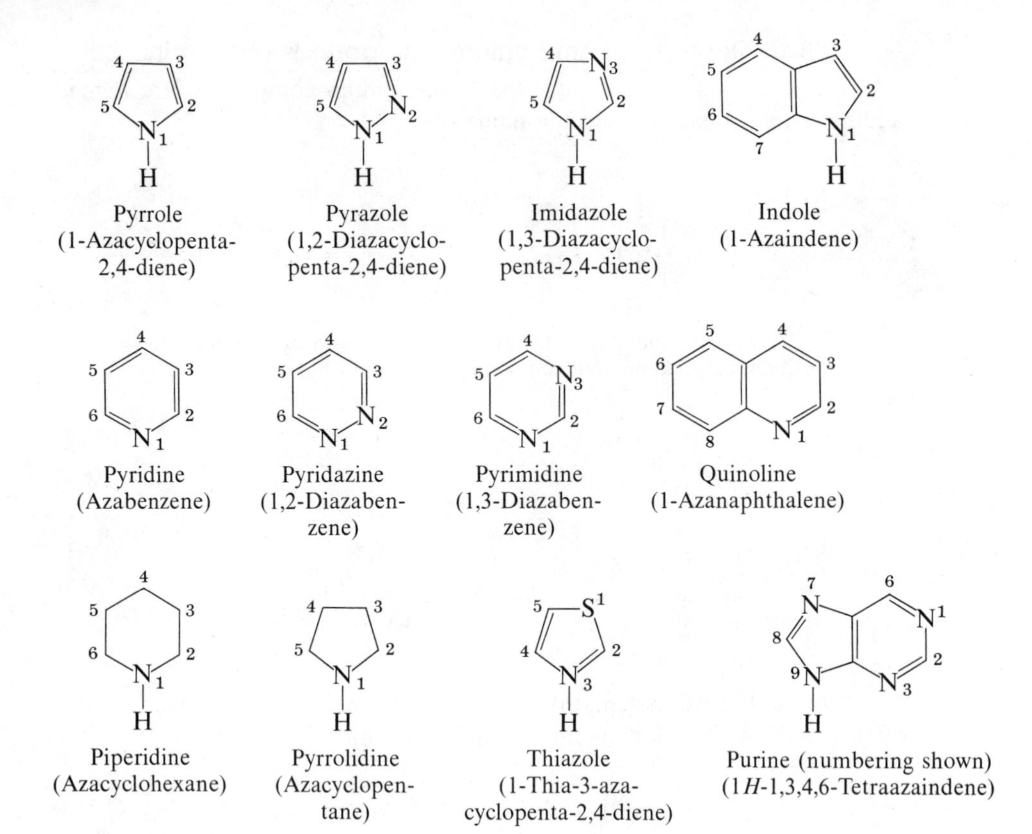

18.2 PHYSICAL PROPERTIES OF AMINES

Amines are moderately polar substances; they have boiling points that are higher than those of alkanes but generally lower than alcohols of comparable molecular weight. Molecules of primary and secondary amines can form strong hydrogen bonds to each other and to water. Molecules of tertiary amines cannot form hydrogen bonds to each other but they can form hydrogen bonds to molecules of water. As a result, tertiary amines generally boil at lower temperatures than primary and secondary amines of comparable molecular weight, but all low-molecular-weight amines are very water soluble.

Table 18.1 lists the physical properties of some common amines.

18.3 BASICITY OF AMINES: AMINE SALTS

Amines are relatively weak bases. They are stronger bases than water but are far weaker bases than hydroxide ions, alkoxide ions, and carbanions.

A convenient expression for relating basicities is a quantity called the *basicity constant, K_b*. When an amine dissolves in water, the following equilibrium is established.

$$R\ddot{N}H_2 + H_2O \rightleftharpoons RNH_3^+ + OH^-$$

and K_b is given by the expression:

$$K_b = \frac{[RNH_3^+][OH^-]}{[RNH_2]}$$

TABLE 18.1 Physical Properties of Amines

NAME	STRUCTURE	mp °C	bp °C	WATER SOLUBILITY (25°)	K_b (25°)
Primary Amines					
Methylamine	CH_3NH_2	−94	−6	Very sol.	4.4×10^{-4}
Ethylamine	$CH_3CH_2NH_2$	−84	17	Very sol.	5.6×10^{-4}
Propylamine	$CH_3CH_2CH_2NH_2$	−83	49	Very sol.	4.7×10^{-4}
Isopropylamine	$(CH_3)_2CHNH_2$	−101	33	Very sol.	5.3×10^{-4}
Butylamine	$CH_3(CH_2)_2CH_2NH_2$	−51	78	Very sol.	4.1×10^{-4}
Isobutylamine	$(CH_3)_2CHCH_2NH_2$	−86	68	Very sol.	3.1×10^{-4}
sec-Butylamine	$CH_3CH_2CH(CH_3)NH_2$	−104	63	Very sol.	3.6×10^{-4}
tert-Butylamine	$(CH_3)_3CNH_2$	−68	45	Very sol.	2.8×10^{-4}
Cyclohexylamine	Cyclo-$C_6H_{11}NH_2$	−18	134	Sl. sol.	4.4×10^{-4}
Benzylamine	$C_6H_5CH_2NH_2$		185	Sl. sol.	2.0×10^{-5}
Aniline	$C_6H_5NH_2$	−6	184	3.7 g/100 g	3.8×10^{-10}
p-Toluidine	*p*-$CH_3C_6H_4NH_2$	44	200	Sl. sol.	1.2×10^{-9}
p-Anisidine	*p*-$CH_3OC_6H_4NH_2$	57	244	V. sl. sol.	2.0×10^{-9}
p-Chloroaniline	*p*-$ClC_6H_4NH_2$	70	232	Insol.	1.0×10^{-10}
p-Nitroaniline	*p*-$NO_2C_6H_4NH_2$	148	232	Insol.	1.0×10^{-13}
Secondary Amines					
Dimethylamine	$(CH_3)_2NH$	−96	7	Very sol.	5.2×10^{-4}
Diethylamine	$(CH_3CH_2)_2NH$	−48	56	Very sol.	9.6×10^{-4}
Dipropylamine	$(CH_3CH_2CH_2)_2NH$	−40	110	Very sol.	9.5×10^{-4}
N-Methylaniline	$C_6H_5NHCH_3$	−57	196	Sl. sol.	5.0×10^{-10}
Diphenylamine	$(C_6H_5)_2NH$	53	302	Insol.	6.0×10^{-14}
Tertiary Amines					
Trimethylamine	$(CH_3)_3N$	−117	3.5	Very sol.	5.0×10^{-5}
Triethylamine	$(CH_3CH_2)_3N$	−115	90	14 g/100 g	5.7×10^{-4}
Tripropylamine	$(CH_3CH_2CH_2)_3N$	−90	156	Sl. sol.	4.4×10^{-4}
N,N-Dimethyl-aniline	$C_6H_5N(CH_3)_2$	3	194	Sl. sol.	11.5×10^{-10}

The larger the value of K_b, the greater is the tendency of the amine to accept a proton from water and, thus, the greater will be the concentration of RNH_3^+ and OH^- in the solution. Larger values of K_b, therefore, are associated with those amines that are stronger bases, and smaller values of K_b are associated with those amines that are weaker bases.

The basicity constant of ammonia at 25 °C is 1.8×10^{-5}.

$$\ddot{N}H_3 + H_2O \rightleftharpoons \overset{+}{N}H_4 + OH^-$$

$$K_b = 1.8 \times 10^{-5} = \frac{[NH_4^+][OH^-]}{[NH_3]}$$

When we examine the basicity constants of the amines given in Table 18.1, we see that most aliphatic primary amines (e.g., methylamine, ethylamine) are somewhat stronger bases than ammonia:

	$\ddot{N}H_3$	$CH_3\ddot{N}H_2$	$CH_3CH_2\ddot{N}H_2$	$CH_3CH_2CH_2\ddot{N}H_2$
K_b	1.8×10^{-5}	4.4×10^{-4}	5.6×10^{-4}	4.7×10^{-4}

We can account for this on the basis of the electron-releasing ability of an alkyl group. An alkyl group releases electrons, and it *stabilizes* the alkylammonium ion that results from the acid-base reaction *by dispersing its positive charge.* It stabilizes the alkylammonium ion to a greater extent than it stabilizes the amine.

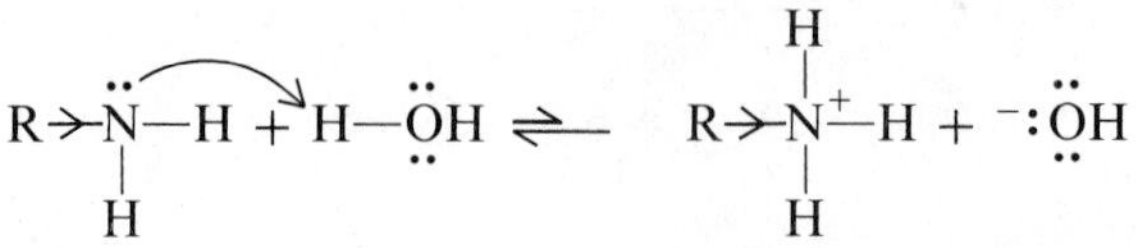

$$R\!\rightarrow\!\ddot{N}H_2 + H{-}\ddot{O}H \rightleftharpoons R\!\rightarrow\!\overset{+}{N}H_3 + {}^{-}{:}\ddot{O}H$$

By releasing electrons, R- stabilizes the alkyl-ammonium ion through dispersal of charge

This explanation is supported by measurements showing that in the *gas phase* the basicities of the amines (below) increase with increasing methyl substitution:

$$(CH_3)_3N > (CH_3)_2NH > CH_3NH_2 > NH_3$$

However, this is not the order of basicity of these amines in aqueous solution (cf. Table 18.1). Complicated solvent effects must be involved but as yet these are not fully understood.

When we examine the basicity constants of the aromatic amines (e.g., aniline, *p*-toluidine) in Table 18.1, we see that they are much weaker bases than the corresponding nonaromatic amine, cyclohexylamine.

	Cyclo-$C_6H_{11}NH_2$	$C_6H_5NH_2$	*p*-$CH_3C_6H_4NH_2$
K_b	4.4×10^{-4}	3.8×10^{-10}	1.2×10^{-9}

We can account for this effect on the basis of resonance contributions to the overall hybrid of an aromatic amine. For aniline, the following contributors are important.

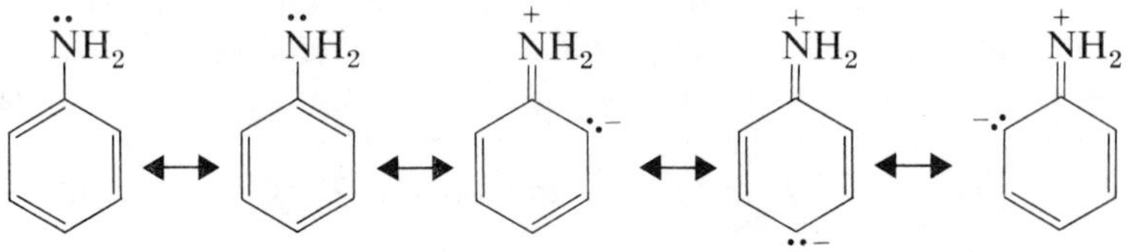

1 ↔ 2 ↔ 3 ↔ 4 ↔ 5

Structures **1** and **2** are the Kekulé structures that contribute to any benzene derivative. Structures **3-5,** however, *delocalize* the unshared electron pair of the nitrogen over the ortho and para positions of the ring. This delocalization of the electron pair makes it less available to a proton but, more importantly, *delocalization of the electron pair stabilizes aniline.*

When aniline accepts a proton it becomes an anilinium ion.

$$C_6H_5\ddot{N}H_2 + H_2O \rightleftharpoons C_6H_5\overset{+}{N}H_3 + \overset{-}{O}H$$

Anilinium ion

Since the electron pair of nitrogen accepts the proton, we are able to write only *two* resonance structures for the anilinium ion—the two Kekulé structures:

$^{+}NH_3$ $\longleftrightarrow$ $^{+}NH_3$

Structures corresponding to **3-5** are not possible for the anilinium ion and, consequently, resonance does not stabilize the anilinium ion to as great an extent as it does aniline itself. This greater stabilization of the reactant (aniline) when compared to that of the product (anilinium ion) means that $\Delta H°$ for the reaction,

$$\text{aniline} + H_2O \longrightarrow \text{anilinium ion} + OH^- \text{ (Fig. 18.1)},$$

will be a larger positive quantity than that for the reaction,

$$\text{cyclohexylamine} + H_2O \longrightarrow \text{cyclohexylammonium ion} + OH^-$$

Aniline, as a result, is the weaker base.

18.3A Amine Salts

When primary, secondary, and tertiary amines react with acids, they form amine salts:

$$CH_3CH_2\ddot{N}H_2 + HCl \xrightarrow[H_2O]{} CH_3CH_2\overset{+}{N}H_3\ \ Cl^-$$
Ethylammonium chloride
(an amine salt)

$$(CH_3CH_2)_2\ddot{N}H + HBr \xrightarrow[H_2O]{} (CH_3CH_2)_2\overset{+}{N}H_2\ \ Br^-$$
Diethylammonium bromide

$$(CH_3CH_2)_3\ddot{N} + HI \xrightarrow[H_2O]{} (CH_3CH_2)_3\overset{+}{N}H\ \ I^-$$
Triethylammonium iodide

FIG. 18.1
Potential energy diagram for the reaction of cyclohexylamine with H_2O (1), and for the reaction of an aniline with H_2O (2). The curves are aligned for comparison only.)

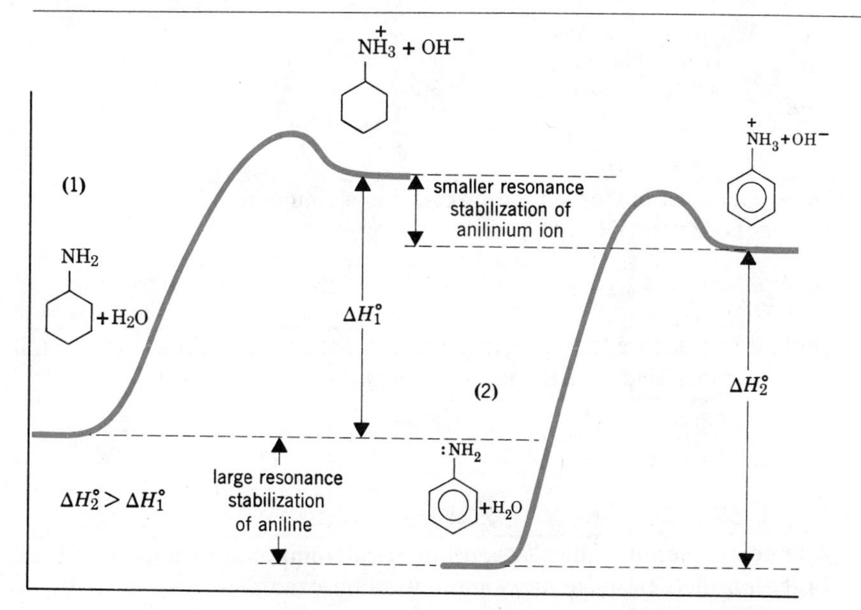

Quaternary ammonium halides do not undergo such a reaction because they are salts already and they do not have an unshared electron pair.

$$(CH_3CH_2)_4\overset{+}{N}\ Br^-$$

Tetraethylammonium bromide
(an amine salt—does not
undergo reaction with acid)

Quaternary ammonium hydroxides, however, are strong bases. As solids, or in solution, they consist *entirely* of quaternary ammonium cations (R_4N^+) and hydroxide ions, OH^-; they are, therefore, strong bases—as strong as sodium or potassium hydroxide. Quaternary ammonium hydroxides react with acids to form quaternary ammonium salts:

$$(CH_3)_4\overset{+}{N}\ OH^- + HCl \longrightarrow (CH_3)_4\overset{+}{N}\ Cl^- + H_2O$$

Almost all alkylammonium chlorides, bromides, iodides, and sulfates are soluble in water. Thus, primary, secondary, or tertiary amines that are not soluble in water will dissolve in aqueous HCl, HBr, HI, or H_2SO_4. Solubility in acid provides a convenient chemical method for distinguishing amines from nonbasic compounds that are insoluble in water. Solubility in acid also gives us a useful method for separating amines from nonbasic compounds that are insoluble in water.

$$-\ddot{N}: \quad + \quad HX\ (\text{or } H_2SO_4) \longrightarrow -\overset{+}{N}-H \quad X^-\ (\text{or } HSO_4^-)$$

Water-insoluble amine — Water-soluble salt

Although amides are superficially similar to amines, *amides are far less basic*. Water-insoluble amides *do not dissolve* in aqueous HCl, HBr, HI, or H_2SO_4.

$$R-\overset{O}{\overset{\|}{C}}-NH_2$$

Water-insoluble amide
(not soluble in
aqueous acids)

Problem 18.1

Outline a procedure for separating cyclohexylamine from cyclohexane using dilute HCl, aqueous NaOH, and ether.

Problem 18.2

Outline a procedure for separating a mixture of benzoic acid, *p*-cresol, aniline, and benzene using acids, bases, and organic solvents.

18.4 SOME BIOLOGICALLY IMPORTANT AMINES

A large number of medically and biologically important compounds are amines. Listed on the next page are some important examples.

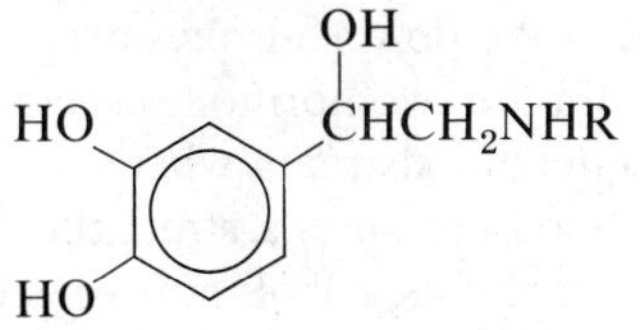

$R = CH_3$, Adrenalin (epinephrine)
$R = H$, Noradrenalin (norepinephrine)

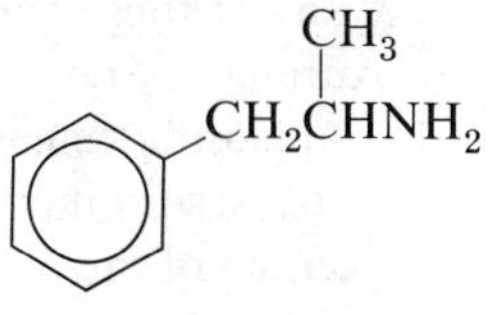

Amphetamine
(benzedrine)

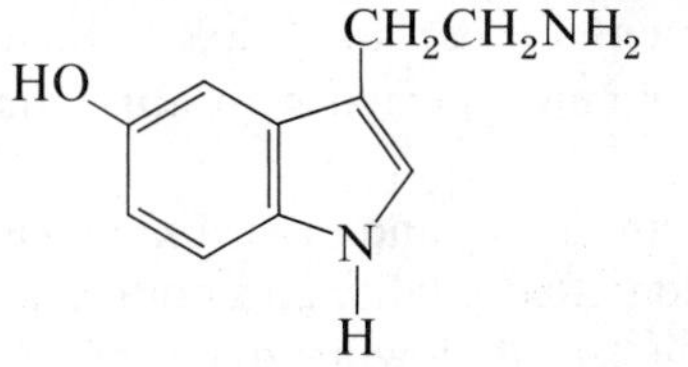

Serotonin

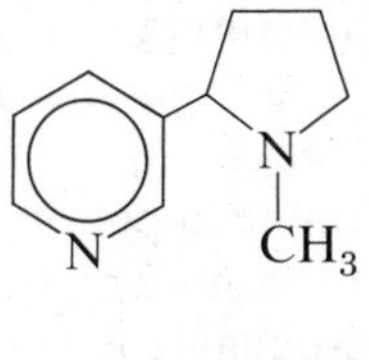

Nicotine

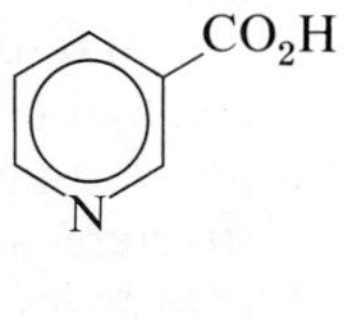

Nicotinic acid

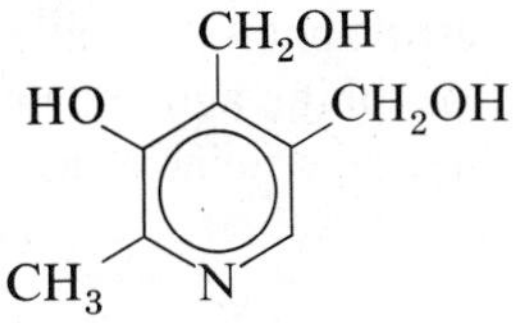

Pyridoxine
(vitamin B_6)

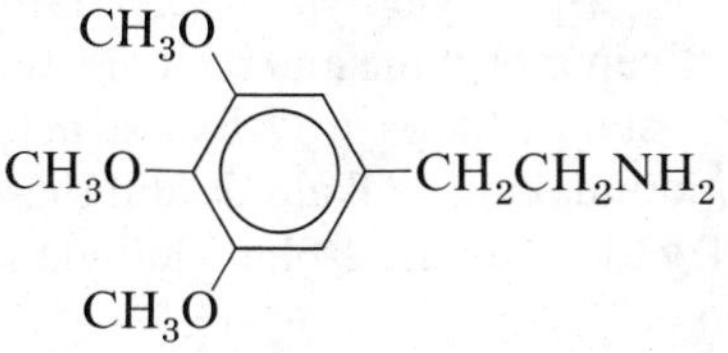

Mescaline

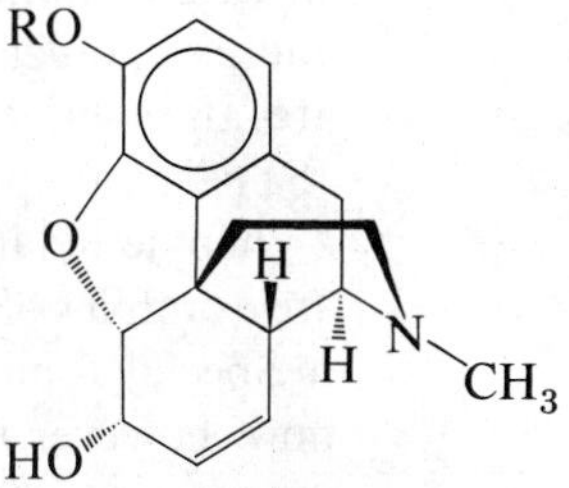

Morphine ($R = H$)
Codeine ($R = CH_3$)

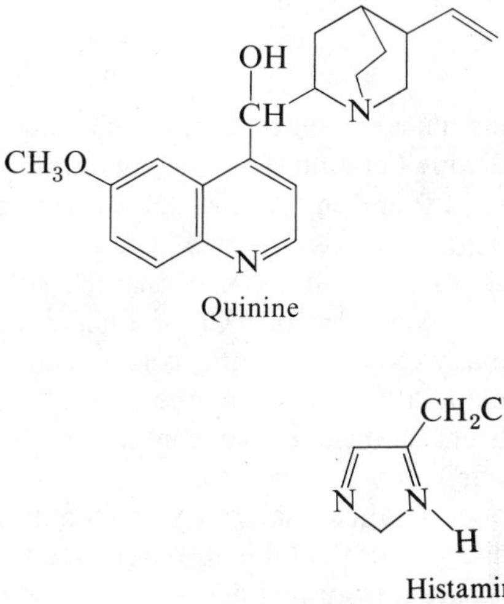

Quinine

Thiamine chloride
(Vitamin B_1)

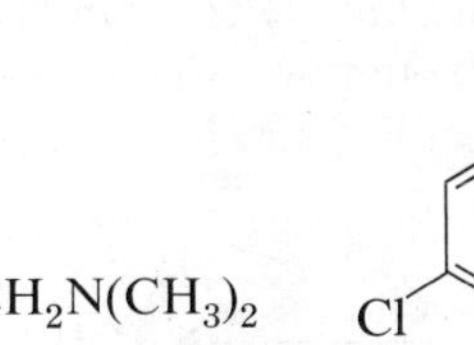

Histamine

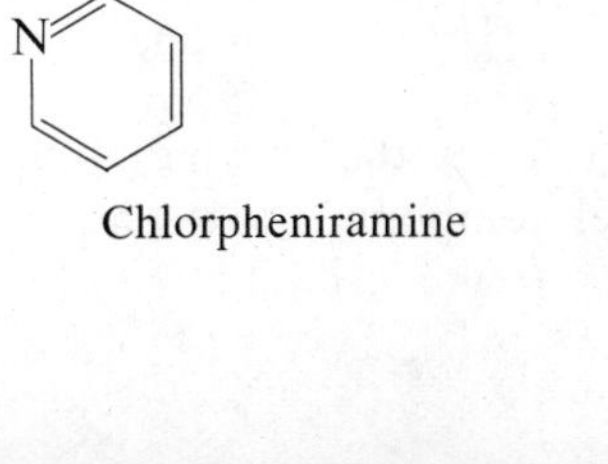

Chlorpheniramine

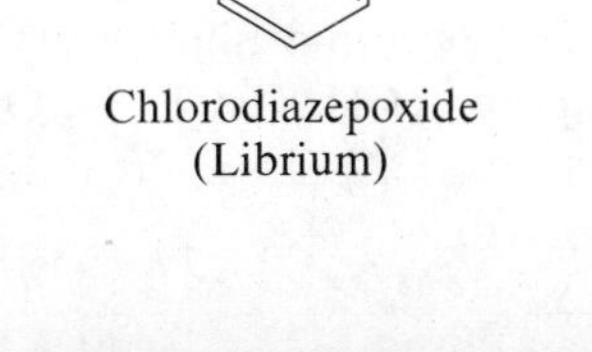

Chlorodiazepoxide
(Librium)

Many of these compounds have powerful physiological and psychological effects. Adrenalin and noradrenalin are two hormones secreted in the medulla of the adrenal gland. Released into the bloodstream when an animal senses danger, adrenalin causes an increase in blood pressure, a strengthening of the heart rate, and a widening of the passages of the lungs. All of these effects prepare the animal to fight or flee. Noradrenalin also causes an increase in blood pressure, and it is involved in the transmission of impulses from the end of one nerve fiber to the next. Serotonin is a compound of particular interest because it appears to be important in maintaining stable mental processes. It has been suggested that the mental disorder schizophrenia may be connected with abnormalities in the metabolism of serotonin.

Amphetamine (a powerful stimulant) and mescaline (a hallucinogen) have structures similar to those of serotonin, adrenalin, and noradrenalin. They are all derivatives of 2-phenylethylamine (see following diagram). (In serotonin the nitrogen is connected to the benzene ring to create a five-membered ring.) The structural similarities of these compounds must be related to their physiological and psychological effects because many other compounds with similar properties are also derivatives of 2-phenylethylamine. Examples (not shown) are LSD, "STP," and "speed." Even morphine and codeine, two powerful analgesics, have a 2-phenylethylamine system as a part of their structures. (Morphine and codeine are examples of compounds called alkaloids and these are discussed in more detail in Special Topic O. Try to locate the 2-phenylethylamine system in their structures now, however.)

$$C_6H_5CH_2CH_2NH_2$$

2-Phenylethylamine

A number of amines are vitamins. These include nicotinic acid and nicotinamide (the antipellagra factors), pyridoxine (Vitamin B_6), and thiamine chloride (Vitamin B_1). Nicotine is a toxic alkaloid found in tobacco that makes smoking habit forming. Histamine, another toxic amine, is found bound to proteins in nearly all tissues of the body. Release of free histamine causes the symptoms associated with allergic reactions and the common cold. Chloropheniramine, an "antihistamine," is an ingredient of many over-the-counter cold remedies.

Librium, an interesting compound with a seven-membered ring, is one of the most widely prescribed tranquilizers. (Librium also contains a positively charged nitrogen, present as an *N*-oxide.)

Acetylcholine and choline (below) contain a quaternary ammonium group. Being small and ionic, both compounds are highly soluble in water. Acetylcholine is vital in the process by which impulses are transmitted across junctions between nerves in muscles. After acetylcholine is released by the nerve and moves to a receptor site, contraction of the muscle is stimulated. For the muscle to contract again, the acetylcholine must be removed. This is done by an enzyme, cholinesterase, which hydrolyzes acetylcholine to choline and acetic acid (or acetate ion).

$$\underset{\text{Acetylcholine}}{(CH_3)_3\overset{+}{N}CH_2CH_2O\overset{O}{\overset{\|}{C}}CH_3} + H_2O \xrightarrow{\text{Cholinesterase}} \underset{\text{Choline}}{(CH_3)_3\overset{+}{N}CH_2CH_2OH} + CH_3COOH$$

The group that binds acetylcholine to the enzyme is the quaternary ammonium group $(CH_3)_3\overset{+}{N}CH_2$—. Other compounds that have this grouping can inhibit cholinesterase. Included among them are compounds used in surgery as muscle relaxants:

$$(CH_3)_3\overset{+}{N}CH_2CH_2CH_2CH_2CH_2CH_2CH_2CH_2CH_2CH_2\overset{+}{N}(CH_3)_3\ 2Br^-$$

Decamethonium bromide

$$(CH_3)_3\overset{+}{N}CH_2CH_2O\overset{O}{\overset{\|}{C}}CH_2CH_2\overset{O}{\overset{\|}{C}}OCH_2CH_2\overset{+}{N}(CH_3)_3\ 2Br^-$$

Succinylcholine bromide

Decamethonium bromide has a relatively long-lasting effect. Succinylcholine bromide, because it is an ester and can be hydrolyzed, has a muscle-relaxing effect of much shorter duration.

18.5 PREPARATION OF AMINES

1. Through Nucleophilic Substitution Reactions

Salts of primary amines can be prepared from ammonia and alkyl halides by nucleophilic substitution reactions. Subsequent treatment of the resulting ammonium salts with base gives primary amines.

$$\ddot{N}H_3 + R{-}X \longrightarrow R{-}NH_3^+\ X^- \xrightarrow{OH^-} RNH_2$$

This method is of very limited synthetic application because multiple alkylations occur. When ethyl bromide reacts with ammonia, for example, the ethylammonium bromide that is produced initially can react with ammonia to liberate ethylamine. Ethylamine can then compete with ammonia and react with ethyl bromide to give diethylammonium bromide. Repetitions of acid-base and alkylation reactions ultimately produce some tertiary amines and even some quaternary ammonium salts if the alkyl halide is present in excess.

$$C_2H_5Br + \ddot{N}H_3 \longrightarrow C_2H_5NH_3^+\,Br^-$$

$$C_2H_5NH_3^+\,Br^- + \ddot{N}H_3 \rightleftharpoons C_2H_5\ddot{N}H_2 + NH_4^+\,Br^-$$

$$C_2H_5\ddot{N}H_2 + C_2H_5Br \longrightarrow (C_2H_5)_2NH_2^+Br^-$$

$$(C_2H_5)_2NH_2^+\,Br^- + \ddot{N}H_3 \rightleftharpoons (C_2H_5)_2\ddot{N}H + NH_4^+\,Br^-$$

$$(C_2H_5)_2\ddot{N}H + C_2H_5Br \longrightarrow (C_2H_5)_3NH^+\,Br^-$$

etc.

Multiple alkylations can be minimized by using a large excess of ammonia. (Why?) An example of this technique can be seen in the synthesis of alanine from 2-bromopropanoic acid:

$$\underset{\text{(1 mole)}}{\underset{|\atop Br}{CH_3CHCOOH}} + \underset{\text{(70 moles)}}{NH_3} \longrightarrow \underset{\text{Alanine (65–70%)}}{\underset{|\atop NH_2}{CH_3CHCOO^-\,NH_4^+}}$$

Nucleophilic substitutions can also be carried out using primary, secondary, and tertiary amines:

$$C_6H_5\ddot{N}H_2 + \underset{C_6H_5}{CH_2}-Cl \xrightarrow[(2)\ OH^-]{} C_6H_5\ddot{N}H-CH_2C_6H_5$$

(4 moles) (1 mole) (96%)

$$(CH_3)_2\ddot{N}H + p\text{-}BrC_6H_4NO_2 \xrightarrow{NaHCO_3} p\text{-}(CH_3)_2\ddot{N}C_6H_4NO_2$$

(94–97%)

$$C_5H_5N: + CH_3-I \xrightarrow{CH_3CN} C_5H_5\overset{+}{N}-CH_3\ I^-$$

(95%)

A much better method for preparing a primary amine from an alkyl halide is first to convert the alkyl halide to an alkyl azide ($R-N_3$) by a nucleophilic substitution reaction:

$$R-X + \underset{\text{Azide ion}}{N_3^-} \xrightarrow[(-X^-)]{S_N2} \underset{\text{Alkyl azide}}{R-N=\overset{\oplus}{N}=\overset{\ominus}{N}} \xrightarrow[\text{or LiAlH}_4]{\text{Na/alcohol}} RNH_2$$

Then the alkyl azide can be reduced to a primary amine with sodium and alcohol or with lithium aluminum hydride. A word of *caution:* alkyl azides are explosive and low-molecular weight alkyl azides should not be isolated but should be kept in solution.

Potassium phthalimide (below) can also be used to prepare primary amines by a method known as the *Gabriel synthesis*. This synthesis also avoids the complications of multiple alkylations that occur when alkyl halides are treated with ammonia:

$$\text{Phthalimide} \xrightarrow[\text{KOH}]{\text{Step 1}} \text{Potassium phthalimide } (:\overset{-}{N}:K^+) \xrightarrow[(-KX)]{\text{Step 2: } R-X} N\text{-Alkylphthalimide} \xrightarrow[\text{reflux}]{\text{Step 3: } \ddot{N}H_2\ddot{N}H_2,\ \text{ethanol}}$$

$$C_6H_4(CONHNH_2)(CONHR) \longrightarrow \underset{\text{Phthalazine-1,4-dione}}{C_6H_4(CO)_2(NH)_2} + \underset{\text{Primary amine}}{R-\ddot{N}H_2}$$

Phthalimide is quite acidic ($K_a \simeq 10^{-9}$); it can be converted to potassium phthalimide by potassium hydroxide (step 1). The phthalimide anion is a strong nucleophile and (in step 2) it reacts with an alkyl halide to give an *N*-alkylphthalimide. Treating the *N*-alkylphthalimide with hydrazine (NH_2NH_2) in refluxing ethanol (step 3) gives a primary amine and phthalazine-1,4-dione.

Syntheses of amines using nucleophilic substitution reactions are, as we might expect, restricted to the use of methyl, primary, and secondary alkyl halides, and to aryl halides activated by strong *o*- or *p*-electron-withdrawing groups. The use of tertiary halides leads almost exclusively to eliminations.

Problem 18.3

(a) Write resonance structures for the phthalimide anion that will account for the acidity of phthalimide. (b) Would you expect phthalimide to be more or less acidic than benzamide? Why?

Problem 18.4

Outline a preparation of benzylamine using the Gabriel synthesis.

2. Preparation of Amines Through Reduction of Nitro Compounds

The most widely used method for preparing aromatic amines involves nitration of the ring and subsequent reduction of the nitro group to an amino group.

$$\text{Ar—H} \xrightarrow[\text{H}_2\text{SO}_4]{\text{HNO}_3} \text{Ar—NO}_2 \xrightarrow{\text{(H)}} \text{Ar—NH}_2$$

We studied ring nitration in Chapter 12 and saw there that it is applicable to a wide variety of aromatic compounds. Reduction of the nitro group can also be carried out in a number of ways. The most frequently used methods are based on catalytic hydrogenation, or treatment of the nitro compound with acid and iron, zinc, or tin, or a metal salt such as $SnCl_2$. Several specific examples are shown below.

$$p\text{-}O_2N\text{-}C_6H_4\text{-}CO_2C_2H_5 + 3H_2 \xrightarrow[\text{C}_2\text{H}_5\text{OH, }25°]{\text{Pt}} p\text{-}H_2N\text{-}C_6H_4\text{-}CO_2C_2H_5 \ (95\%) + 2H_2O$$

$$\text{2,4-dinitrotoluene } (CH_3, NO_2, NO_2) + \text{Fe} \xrightarrow[\text{C}_2\text{H}_5\text{OH, reflux}]{\text{HCl}} \text{2,4-diaminotoluene } (CH_3, NH_2, NH_2) \ (74\%)$$

OCH_3, NO_2, $C{=}O$, CH_3 + $SnCl_2$ $\xrightarrow[95^\circ]{\text{conc. HCl}}$ OCH_3, NH_2, $C{=}O$, CH_3

(76%)

Selective reduction of one nitro group of a dinitro compound can often be achieved through the use of hydrogen sulfide in aqueous (or alcoholic) ammonia:

NO_2, NO_2 $\xrightarrow[NH_3/C_2H_5OH]{H_2S}$ NO_2, NH_2

m-Dinitrobenzene *m*-Nitroaniline (70–80%)

When this method is used, the amount of the hydrogen sulfide must be carefully measured because the use of an excess may result in the reduction of more than one nitro group.

It is not always possible to predict just which nitro group will be reduced, however. Treating 2,4-dinitrotoluene with hydrogen sulfide and ammonia results in reduction of the 4-nitro group:

CH_3, NO_2, NO_2 $\xrightarrow{H_2S,NH_3}$ CH_3, NO_2, NH_2

On the other hand, monoreduction of 2,4-dinitroaniline causes reduction of the 2-nitro group:

NH_2, NO_2, NO_2 $\xrightarrow[50^\circ]{H_2S,NH_3}$ NH_2, NH_2, NO_2

(52–58%)

3. Preparation of Amines Through Reductive Amination

Aldehydes and ketones can be converted to primary amines through catalytic or chemical reduction in the presence of ammonia. This process, called *reductive amination,* proceeds through the formation of an imine.

General Reaction

$$\underset{\text{Aldehyde or ketone}}{R'(R)C{=}O} + NH_3 \xrightleftharpoons{(-H_2O)} \underset{\text{Imine}}{\left[R'(R)C{=}NH\right]} \xrightarrow{(H)} \underset{\text{A 1° amine}}{R{-}CH(R'){-}NH_2}$$

Specific Examples

$$\underset{\text{Benzaldehyde}}{C_6H_5CH{=}O} + NH_3 \xrightleftharpoons{(-H_2O)} \left[C_6H_5CH{=}NH\right] \xrightarrow[\text{90 atm. } 40\text{–}70°]{H_2,Ni} \underset{\text{Benzylamine (89\%)}}{C_6H_5CH_2NH_2}$$

$$\underset{\text{Cyclohexanone}}{C_6H_{10}{=}O} + NH_3 \xrightleftharpoons{(-H_2O)} \left[C_6H_{10}{=}NH\right] \xrightarrow{H_2,Ni} \underset{\text{Cyclohexylamine}}{C_6H_{11}{-}NH_2}$$

Secondary amines can be prepared through reductive amination of an aldehyde or ketone in the presence of a primary amine.

$$\underset{\text{Aldehyde or ketone}}{R(R')C{=}O} + \underset{\text{1° Amine}}{H_2NR''} \xrightleftharpoons{(-H_2O)} \underset{\text{An imine}}{R(R')C{=}NR''} \xrightarrow{(H)} \underset{\text{2° Amine}}{R'{-}CH(R)NHR''}$$

An example is the synthesis of ethylbenzylamine from benzaldehyde and ethylamine:

$$C_6H_5CH{=}O + H_2NCH_2CH_3 \xrightleftharpoons{(-H_2O)} C_6H_5CH{=}NCH_2CH_3 \xrightarrow{LiBH_3CN^{*}/CH_3OH} \underset{(72\%)}{C_6H_5CH_2NHCH_2CH_3}$$

Problem 18.5

Show how you might prepare each of the following amines through reductive amination.

(a) $CH_3(CH_2)_3CH_2NH_2$

(b) $C_6H_5CH(NH_2)CH_3$

(c) $CH_3(CH_2)_4CH_2NHC_6H_5$

* A reducing agent similar to $NaBH_4$. $LiBH_3CN$ and $NaBH_3CN$ reduce imine groups more rapidly than they reduce carbonyl groups.

Problem 18.6

Reductive amination of a ketone is almost always a better method for preparation of amines of the type $\overset{R'}{|}$ RCHNH$_2$, than treatment of an alkyl halide with ammonia. Why would this be true?

4. Preparation of Amines Through Reduction of Amides, Oximes, and Nitriles

Amides, oximes, and nitriles can be reduced to amines. Reduction of a nitrile or an oxime yields a primary amine; reduction of an amide can yield a primary, secondary or tertiary amine.

$$\underset{\text{Nitrile}}{R{-}C{\equiv}N} \xrightarrow{(H)} \underset{1^\circ \text{ Amine}}{RCH_2NH_2}$$

$$\underset{\text{Oxime}}{RCH{=}NOH} \xrightarrow{(H)} \underset{1^\circ \text{ Amine}}{RCH_2NH_2}$$

$$\underset{\text{Amide}}{R{-}\overset{O}{\overset{\|}{C}}N(R''){-}R'} \xrightarrow{(H)} \underset{3^\circ \text{ Amine}}{RCH_2N(R''){-}R'}$$

(If R′ and R″ = H, the product is a 1° amine,
if R′ = H, the product is a 2° amine.)

All of these reductions can be carried out with hydrogen and a catalyst or with $LiAlH_4$. Oximes are also conveniently reduced with sodium in alcohol—a safer method than the use of $LiAlH_4$.
Specific examples are the following:

$$C_6H_{10}{=}N{-}OH \xrightarrow{Na/C_2H_5OH} C_6H_{11}{-}NH_2 \quad (50\text{-}60\%)$$

$$\underset{\text{Benzyl cyanide}}{C_6H_5{-}CH_2C{\equiv}N} + 2H_2 \xrightarrow[140^\circ]{\text{Raney Ni}} \underset{\substack{\text{2-Phenylethylamine} \\ (71\%)}}{C_6H_5{-}CH_2CH_2NH_2}$$

$$\underset{\text{Benzonitrile}}{C_6H_5{-}C{\equiv}N} + LiAlH_4 \xrightarrow[(2)\ H_2O]{\text{Ether}} \underset{\substack{\text{Benzylamine} \\ (72\%)}}{C_6H_5{-}CH_2NH_2}$$

$$\underset{N\text{-Methylacetanilide}}{C_6H_5{-}N(CH_3){-}\overset{O}{\overset{\|}{C}}CH_3} + LiAlH_4 \xrightarrow[(2)\ H_2O]{\text{Ether}} \underset{N\text{-Ethyl-}N\text{-methylaniline}}{C_6H_5{-}\ddot{N}(CH_3)CH_2CH_3}$$

Problem 18.7

Show how you might utilize the reduction of an amide, oxime, or a nitrile to carry out each of the following transformations.

(a) Benzoic acid ⟶ *N*-ethyl-*N*-benzylamine
(b) 1-Bromopentane ⟶ hexylamine
(c) Propanoic acid ⟶ tripropylamine
(d) 2-Butanone ⟶ *sec*-butylamine

5. Preparation of Amines Through the Hofmann Degradation of Amides

Amides react with solutions of bromine or chlorine in sodium hydroxide to yield amines through a reaction known as the *Hofmann degradation:*

$$R-\overset{O}{\overset{\|}{C}}NH_2 + Br_2 + 4NaOH \xrightarrow{H_2O} RNH_2 + 2NaBr + Na_2CO_3 + 2H_2O$$

From the equation above we can see that the carbonyl carbon of the amide is lost (as CO_3^{2-}) and that the R group of the amide becomes attached to the nitrogen of the amine. Primary amines made this way are not contaminated by 2° or 3° amines.

The mechanism for this interesting reaction involves the following steps. In step 1, the amide undergoes a base-promoted *N-bromination.* In step 2, the *N*-bromo amide reacts with base to yield an anion which in step 3 simultaneously rearranges and loses a bromide ion to give an *isocyanate.* Finally, in step 4, the isocyanate undergoes hydrolysis to yield an amine and CO_3^{2-} ion.

Step 1 $$R-C(=O)-\ddot{N}H_2 + Br_2 + OH^- \longrightarrow R-C(=O)-\ddot{N}(H)-Br + H_2O + Br^-$$

N-Bromo amide

Step 2 $$R-C(=O)-\ddot{N}(H)-Br + :\ddot{O}H^- \longrightarrow R-C(=O)-\ddot{N}^{-}-Br + H_2O$$

Step 3 $$R-C(=O)-\ddot{N}^{-}-Br \longrightarrow R-\ddot{N}=C=O + Br^-$$

An isocyanate

Step 4 $$R-\ddot{N}=C=O + 2\overline{O}H \longrightarrow R-NH_2 + CO_3^{=}$$

Studies of Hofmann degradations of optically active amides in which the chiral carbon is directly attached to the carbonyl group have shown that they occur with *retention of configuration.* Thus, the R group migrates to nitrogen with its electrons, *but without inversion.*

With lower-molecular-weight amides, Hofmann degradations using bromine in sodium hydroxide give excellent yields as the first two examples shown below illustrate. With high-molecular-weight amides, the yields are not as good but can be improved through the use of bromine and sodium methoxide in methanol.

$$\underset{\text{Hexanamide}}{CH_3CH_2CH_2CH_2CH_2\overset{O}{\overset{\|}{C}}NH_2} \xrightarrow[H_2O]{Br_2, OH^-} \underset{\substack{\text{Pentylamine} \\ (88\%)}}{CH_3CH_2CH_2CH_2CH_2NH_2}$$

$$\text{1-}\overset{O}{\overset{\|}{C}}OH\text{-2-}\overset{O}{\overset{\|}{C}}\text{—}NH_2\text{-benzene} \xrightarrow[H_2O]{Br_2, OH^-} \underset{(85\%)}{\text{1-}\overset{O}{\overset{\|}{C}}O^-\text{-2-}NH_2\text{-benzene}}$$

$$CH_3(CH_2)_{14}\overset{O}{\overset{\|}{C}}NH_2 \xrightarrow[CH_3OH]{Br_2, CH_3O^-} \underset{(100\%)}{CH_3(CH_2)_{13}CH_2NH_2}$$

Problem 18.8

Using a different method for each part, but taking care in each case to select a *good* method, show how each of the following transformations might be accomplished.

(a) $CH_3O-C_6H_5 \longrightarrow CH_3O-C_6H_4-NH_2$ (para)

(b) $CH_3O-C_6H_5 \longrightarrow CH_3O-C_6H_4-\underset{NH_2}{\underset{|}{C}}HCH_3$ (para)

(c) $C_6H_5-CH_3 \longrightarrow C_6H_5-CH_2\overset{+}{N}(CH_3)_3\ Cl^-$

(d) $NO_2-C_6H_4-CH_3 \longrightarrow NO_2-C_6H_4-NH_2$ (para)

(e) $CH_3-C_6H_5 \longrightarrow C_6H_5-CH_2CH_2NH_2$

18.6 REACTIONS OF AMINES

We have encountered a number of important reactions of amines in earlier sections of this book. We saw reactions in which primary, secondary, and tertiary amines act *as bases* in Section 18.3. We saw their reactions as *nucleophiles* in *alkylation reactions* in Section 18.4, and as *nucleophiles* in *acylation reactions* in Chapter 17. In Chapter 12 we saw that an amino group on an aromatic ring acts as a powerful *activating group* and as an *ortho-para director.*

The structural feature of amines that underlies all of these reactions and that forms a basis for our understanding of most of the chemistry of amines is the ability of nitrogen to share an electron pair:

$$\text{>N:} + H^+ \rightleftharpoons \text{>}\overset{+}{N}\text{—H}$$

An amine acting as a base

$$\text{>N:} + R\text{—}CH_2\text{—}Br \longrightarrow \text{>}\overset{+}{N}\text{—}CH_2R + Br^-$$

An amine acting as a nucleophile in an alkylation reaction

$$\text{>N:}(H) + R\text{—}\overset{O}{\overset{\|}{C}}\text{—}Cl \xrightarrow[(-HCl)]{} \text{>}\ddot{N}\text{—}\overset{O}{\overset{\|}{C}}\text{—}R$$

An amine acting as a nucleophile in an acylation reaction

In the examples given above the amine acts as a nucleophile by donating its electron pair to an electrophilic reagent. In the following example resonance contributions involving the nitrogen electron pair make *carbon* atoms nucleophilic.

$$C_6H_5NH_2 + E^+ \longrightarrow [\text{ortho arenium ion: } H_2\overset{+}{N}=C_6H_5(H)(E)] \xrightarrow{-H^+} o\text{-}E\text{—}C_6H_4NH_2$$

$$C_6H_5NH_2 + E^+ \longrightarrow [\text{para arenium ion: } H_2\overset{+}{N}=C_6H_5(H)(E)] \xrightarrow{-H^+} p\text{-}E\text{—}C_6H_4NH_2$$

The amino group acting as an activating group and as an ortho-para director in electrophilic aromatic substitution

Problem 18.9

Review the chemistry of amines given in earlier sections and provide a specific example of each of the reactions illustrated above.

18.7 REACTIONS OF AMINES WITH NITROUS ACID

Nitrous acid, HONO, is a weak, unstable acid. It is usually prepared *in situ* by treating sodium nitrite ($NaNO_2$) with an aqueous solution of a strong acid:

$$HCl_{(aq)} + NaNO_{2(aq)} \longrightarrow HONO_{(aq)} + NaCl_{(aq)}$$
$$H_2SO_4 + 2NaNO_{2(aq)} \longrightarrow 2HONO_{(aq)} + Na_2SO_{4(aq)}$$

Nitrous acid reacts with all classes of amines. The products that we obtain from these reactions depend on whether the amine is primary, secondary, or tertiary and whether the amine is aliphatic or aromatic.

18.7A Reactions of Primary Aliphatic Amines with Nitrous Acid

Primary aliphatic amines react with nitrous acid through a reaction called *diazotization* to yield highly unstable aliphatic *diazonium salts.* Even at low temperatures, *aliphatic* diazonium salts decompose spontaneously by losing nitrogen to form carbocations. The carbocations go on to produce mixtures of alkenes, alcohols, and alkyl halides by elimination of H^+, reaction with H_2O, and reaction with X^-.

General Reaction

$$\underset{\text{1° Aliphatic amine}}{R{-}NH_2} + NaNO_2 + HX \xrightarrow[H_2O]{(HONO)} \underset{\substack{\text{Aliphatic diazonium salt} \\ (\textit{highly unstable})}}{\left[R{-}\overset{+}{N}{\equiv}N\!:\; X^-\right]}$$

$$\xrightarrow{-N_2\ (\text{i.e., } :N{\equiv}N:)} R^+ + X^- \longrightarrow \text{Alkenes, alcohols, alkyl halides}$$

Specific Examples

$$CH_3CH_2CH_2CH_2NH_2 + NaNO_2 + HCl \xrightarrow{H_2O} \left[CH_3CH_2CH_2CH_2{-}\overset{+}{N}{\equiv}N\!:\right]$$

$$\xrightarrow{-N_2} CH_3CH_2CH_2CH_2^{\ +} \longrightarrow$$

$$\underset{\substack{\text{1-Butene} \\ (26\%)}}{CH_2{=}CHCH_2CH_3} + \underset{\substack{\text{2-Butene} \\ (10\%)}}{CH_3CH{=}CHCH_3} + \underset{\substack{\text{1-Butanol} \\ (25\%)}}{CH_3CH_2CH_2CH_2OH} +$$

$$\underset{\substack{\text{2-Butanol} \\ (13\%)}}{CH_3CH_2\underset{OH}{\underset{|}{C}}HCH_3} + \underset{\substack{\text{1-Chlorobutane} \\ (5\%)}}{CH_3CH_2CH_2CH_2Cl} + \underset{\substack{\text{2-Chlorobutane} \\ (3\%)}}{CH_3CH_2\underset{Cl}{\underset{|}{C}}HCH_3}$$

Diazotizations of primary aliphatic amines are of little synthetic importance because they yield such a complex mixture of products. Diazotizations of primary aliphatic amines are used in some analytical procedures, however, because the evolution of nitrogen is quantitative. They can also be used to generate and thus study the behavior of carbocations in water, acetic acid, and other solvents.

Problem 18.10

Write mechanisms that account for the formation of each product in the reaction given above from the *n*-butyl cation.

18.7B Reactions of Primary Arylamines with Nitrous Acid

Primary aromatic amines react with nitrous acid to give arenediazonium salts. While arenediazonium salts are unstable, they are far more stable than aliphatic diazonium salts; they do not decompose at an appreciable rate when the temperature of the reaction mixture is kept below 5 °C.

$$\underset{\text{Primary arylamine}}{Ar{-}NH_2} + 2NaNO_2 + 2HX \longrightarrow \underset{\substack{\text{Arenediazonium salt} \\ (\textit{stable if kept below } 5\,°C)}}{Ar{-}\overset{+}{N}{\equiv}N{:}\ X^-} + NaX + 2H_2O$$

Diazotization reactions of primary aromatic amines are of considerable synthetic importance because the diazonium group, $-N{\equiv}N{:}$, can be replaced by a variety of other functional groups. We will examine these reactions in Section 18.8.

18.7C Reactions of Secondary Amines with Nitrous Acid

Aromatic and aliphatic secondary amines react with nitrous acid to yield *N*-nitrosoamines. *N*-Nitrosoamines usually separate from the reaction mixture as oily yellow liquids.

General Reaction

$$\underset{\text{Secondary aliphatic amine}}{R_2NH} + HONO \longrightarrow \underset{N\text{-Nitrosoamine}}{R_2N{-}N{=}O} + H_2O$$

$$\underset{\text{Secondary aromatic amine}}{ArNHR} + HONO \longrightarrow \underset{N\text{-Nitrosoamine}}{Ar{-}\underset{}{\overset{\overset{\displaystyle O}{\|}}{\overset{\displaystyle N}{|}}}{N}{-}R} + H_2O$$

Specific Examples

$$\underset{\text{Dimethylamine}}{(CH_3)_2\ddot{N}H} + HCl + NaNO_2 \xrightarrow[H_2O]{(HONO)} \underset{\substack{N\text{-Nitrosodimethylamine} \\ \text{(a yellow oil)}}}{(CH_3)_2\ddot{N}{-}\ddot{N}{=}O}$$

$$C_6H_5\ddot{N}(H)CH_3 + HCl + NaNO_2 \xrightarrow[H_2O]{(HONO)} C_6H_5\ddot{N}(N{=}O)CH_3$$

N-Methylaniline → *N*-Nitroso-*N*-methylaniline (87–93%, a yellow oil)

N-Nitrosoamines are very powerful carcinogens and many scientists fear that they may be present in many foods, especially in cooked meats that have been cured with sodium nitrite. Sodium nitrite is added to many meats (i.e., bacon, ham, frankfurter, sausages, and corned beef) to inhibit the growth of *Clostridium botulinum* (the bacterium that produces botulinus toxin) and to keep red meats from turning brown. (Food poisoning by botulinus toxin is often fatal.) In the presence of acid or under the influence of heat, sodium nitrite reacts with amines always present in the meat to produce *N*-nitrosoamines. Cooked bacon, for example, has been shown to contain *N*-nitrosodimethylamine and *N*-nitrosopyrrolidine. There is also concern that nitrites from food may produce nitrosoamines when they react with amines in the presence of the acid found in the stomach. In 1976, the FDA reduced the permissible amount of nitrite allowed in cured meats from 200 parts per million (ppm) to 50–125 ppm. Nitrites (and nitrates that can be converted to nitrites by bacteria) also occur naturally in many foods. Cigarette smoke is known to contain *N*-nitrosodimethylamine. Someone smoking a pack of cigarettes a day inhales about 0.8 microgram of *N*-nitrosodimethylamine and even more has been shown to be present in the side-stream smoke.

18.7D Reactions of Tertiary Amines with Nitrous Acid

When an aliphatic tertiary amine reacts with nitrous acid, an equilibrium is established between the tertiary amine, its salt, and an *N*-nitrosoammonium compound.

$$2R_3N\!: + HX + NaNO_2 \rightleftharpoons R_3\overset{+}{N}H\ X^- + R_3\overset{+}{N}{-}N{=}O\ X^-$$

Tertiary aliphatic amine; Amine salt; *N*-Nitrosoammonium compound

While *N*-nitrosoammonium compounds are stable at low temperatures, at higher temperatures and in aqueous acid they decompose to produce aldehydes or ketones. These reactions are of little synthetic importance, however.

Tertiary aromatic amines react with nitrous acid to form *C*-nitroso aromatic compounds. Nitrosation takes place almost exclusively at the para position if it is open; and, if not, at the ortho position.

General Reaction

$$R_2N{-}C_6H_5 + HONO \longrightarrow R_2N{-}C_6H_4{-}NO$$

Tertiary aromatic amine → *p*-Nitroso-*N*,*N*-dialkylaniline

Specific Example

$$(CH_3)_2\ddot{N}{-}C_6H_5 + HCl + NaNO_2 \xrightarrow[H_2O]{8^\circ} (CH_3)_2\ddot{N}{-}C_6H_4{-}\ddot{N}{=}O$$

p-Nitroso-*N*,*N*-dimethylaniline (80–90%)

Problem 18.11

Para nitrosation of *N,N*-dimethylaniline (*C*-nitrosation) is believed to take place through an electrophilic attack of NO^+ ions. (a) Show how NO^+ ions might be formed in an aqueous solution of $NaNO_2$ and HCl. (b) Write a mechanism for *p*-nitrosation of *N,N*-dimethylaniline. (c) Tertiary aromatic amines and phenols undergo *C*-nitrosation reaction, whereas most benzene derivatives do not. How can you account for this?

18.8 REPLACEMENT REACTIONS OF ARENEDIAZONIUM SALTS

Diazonium salts are highly useful intermediates in the synthesis of aromatic compounds because the diazonium group can be replaced by a number of other groups, including —F, —Cl, —Br, —I, —CN, —OH, and —H.

Diazonium salts are almost always prepared by diazotizing primary aromatic amines. Primary aromatic amines can be synthesized through reduction of nitro compounds that are readily available through direct nitration reactions.

18.8A Syntheses Using Diazonium Salts

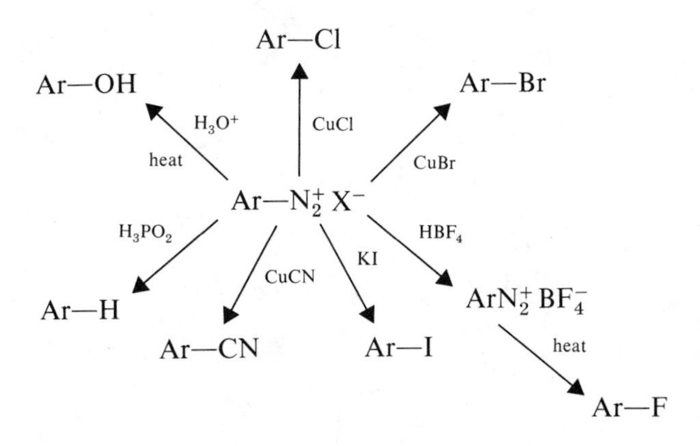

Most arenediazonium salts are unstable at temperatures above 5–10 °C, and many explode when dry. Fortunately, however, most of the replacement reactions of diazonium salts do not require their isolation. We simply add another reagent (CuCl, CuBr, KI, and so on) to the mixture, gently warm the solution, and the replacement (accompanied by the evolution of nitrogen) takes place.

Only in the replacement of the diazonium group by —F, need we isolate a diazonium salt. We do this by adding HBF_4 to the mixture; this causes the sparingly soluble and reasonably stable arenediazonium fluoroborate, $ArN_2^+BF_4^-$, to precipitate.

18.8B The Sandmeyer Reaction: Replacement of the Diazonium Group by —Cl, —Br, or —CN

Arenediazonium salts react with cuprous chloride, cuprous bromide, and cuprous cyanide to give products in which the diazonium group has been replaced by —Cl, —Br, and —CN, respectively. These reactions are known generally as *Sandmeyer reactions*. Several specific examples follow.

$$\text{o-Toluidine} \xrightarrow[H_2O\ (0\text{–}5°)]{HCl/NaNO_2} \text{o-}CH_3C_6H_4N_2^+\ Cl^- \xrightarrow[15\text{–}60°]{CuCl} \text{o-Chlorotoluene} + N_2$$

o-Toluidine → *o*-Chlorotoluene (74–79% overall)

$$\text{m-Benzenediamine} \xrightarrow[CH_3COOH,\ 30°]{H_2SO_4/NaNO_2} m\text{-}C_6H_4(N_2^+\ HSO_4^-)_2 \xrightarrow{CuCl\ (40\text{–}100°)} \text{m-Dichlorobenzene} + N_2$$

m-Benzenediamine → *m*-Dichlorobenzene (70% overall)

$$\text{m-Chloroaniline} \xrightarrow[H_2O\ (0\text{–}10°)]{HBr/NaNO_2} m\text{-}ClC_6H_4N_2^+\ Br^- \xrightarrow[100°]{CuBr} \text{m-Bromochlorobenzene} + N_2$$

m-Chloroaniline → *m*-Bromochlorobenzene (70% overall)

$$\text{o-Nitroaniline} \xrightarrow[H_2O\ (\text{room temp.})]{HCl/NaNO_2} o\text{-}NO_2C_6H_4N_2^+\ Cl^- \xrightarrow[90\text{–}100°]{CuCN} \text{o-Nitrobenzonitrile} + N_2$$

o-Nitroaniline → *o*-Nitrobenzo-nitrile (65% overall)

18.8C Replacement by —I

Arenediazonium salts react with potassium iodide to give products in which the diazonium group has been replaced by —I. An example is the synthesis of *p*-iodonitrobenzene:

$$p\text{-}NO_2C_6H_4NH_2 \xrightarrow[H_2O,\ 0\text{–}5°]{H_2SO_4/NaNO_2} p\text{-}NO_2C_6H_4N_2^+\ HSO_4^- \xrightarrow{KI} p\text{-}NO_2C_6H_4I + N_2$$

p-Iodonitrobenzene (81% overall)

18.8D Replacement by —F

The diazonium group can be replaced by fluorine by treating the diazonium salt with fluoroboric acid (HBF_4). The diazonium fluoroborate that precipitates is isolated, dried, and heated until decomposition occurs. An aryl fluoride is produced.

CH_3, NH_2 — *m*-Toluidine —(1) HONO, (2) HBF_4→ CH_3, N_2^+ BF_4^- — *m*-Toluenediazonium fluoroborate (79%) —Heat→ CH_3, F — *m*-Fluorotoluene (69%) + N_2

NH_2, Br — 4-Bromo-1-naphthylamine —(1) HONO, (2) HBF_4→ N_2^+ BF_4^-, Br — (isolate and dry) —150–155°→ F, Br — 1-Fluoro-4-bromonaphthalene (97% overall)

18.8E Replacement by —OH

The diazonium group can be replaced by a hydroxyl group simply by acidifying the mixture strongly and heating it:

NH_2, NO_2 — *m*-Nitroaniline —$H_2SO_4/NaNO_2$, H_2O, 0–5°→ N_2^+ HSO_4^-, NO_2 —H_2SO_4, H_2O, 100°→ OH, NO_2 — *m*-Nitrophenol (74–79%)

NH_2, Br — *m*-Bromoaniline —$H_2SO_4/NaNO_2$, H_2O→ N_2^+ HSO_4^-, Br —H_2SO_4, H_2O, 100°→ OH, Br — *m*-Bromophenol (78% overall)

Problem 18.12

In the preceding examples of diazonium reactions, we have illustrated syntheses beginning with the compounds (a)–(f) below. Show how you might prepare each of these compounds from benzene.

(a) *m*-Benzenediamine
(b) *m*-Nitroaniline
(c) *m*-Chloroaniline
(d) *m*-Bromoaniline
(e) *o*-Nitroaniline
(f) *p*-Nitroaniline

18.8F Replacement by —H. Deamination by Diazotization

Arenediazonium salts react with hypophosphorous acid (H_3PO_2) to yield products in which the diazonium group has been replaced by —H.

Since we usually begin a synthesis using diazonium salts by nitrating an aromatic compound, that is, replacing —H by $—NO_2$ and then by $—NH_2$, it may seem strange that we would ever want to replace a diazonium group by —H. However, replacement of the diazonium group by —H can be a useful reaction. We can introduce an amino group into an aromatic ring to influence the orientation of a subsequent reaction. Later we can remove the amino group (i.e., carry out a *deamination*) by diazotizing it and treating the diazonium salt with H_3PO_2.

Let us assume, for example, that we need to prepare 1,3,5-tribromobenzene:

1,3,5-Tribromobenzene

Direct bromination of benzene will not give us this product because bromine is an ortho-para director:

Br_2, $FeBr_3$ → Br_2, $FeBr_3$ → ... + ... → Br_2, $FeBr_3$ → ... + ... etc.

If, however, we begin our synthesis with aniline, we can take advantage of the activating and directing effect of the amino group:

Br_2, H_2O (Section 12.8A) → 2,4,6-Tribromobenzene (~100%) → H_2SO_4, $NaNO_2$, 0–5° → $N_2^+\ HSO_4^-$ → H_3PO_2 → + N_2

1,3,5-Tribromobenzene
(65–72% overall)

We can see another example of the usefulness of a deamination reaction in the synthesis of *m*-bromotoluene below. We cannot prepare *m*-bromotoluene by direct bromination of toluene or by a Friedel-Crafts alkylation of bromobenzene because both reactions give *o*- and *p*-bromotoluene. (Both CH_3— and Br— are ortho-para directors.) However, if we begin with *p*-toluidine (by nitrating toluene, separating the *p* isomer, and reducing the nitro group) we can carry out the following sequence of reactions and obtain *m*-bromotoluene in good yield. The first step, synthesis of the *N*-acetyl derivative of *p*-toluidine, is done to reduce the activating effect of the amino group. Later, the acetyl group is removed.)

CH_3, NH_2 — *p*-Toluidine $\xrightarrow{(CH_3CO)_2O}$ CH_3, $NHCOCH_3$ $\xrightarrow[(2)\ OH^-,\ H_2O]{(1)\ Br_2}$ CH_3, Br, NH_2 (65% from *p*-toluidine) $\xrightarrow[H_2O,\ 0\text{–}5°]{H_2SO_4, NaNO_2}$

CH_3, Br, N_2^+ $\xrightarrow[H_2O,\ 25°]{H_3PO_2}$ CH_3, Br + N_2

m-Bromotoluene (85% from 3-bromo-4-aminotoluene)

Problem 18.13

Suggest how you might modify the synthesis given above in order to prepare 3,5-dibromotoluene.

Problem 18.14

(a) On page 831 we showed a synthesis of *m*-fluorotoluene starting with *m*-toluidine. How would you prepare *m*-toluidine from toluene? (b) How would you prepare *m*-chlorotoluene? (c) *m*-Bromotoluene? (d) *m*-Iodotoluene? (e) *m*-Tolunitrile (*m*-$CH_3C_6H_4CN$)? (f) *m*-Toluic acid?

Problem 18.15

Starting with *p*-nitroaniline [Problem 18.12 (f)] show how you might synthesize 1,2,3-tribromobenzene.

18.9 COUPLING REACTIONS OF DIAZONIUM SALTS

Diazonium ions are weak electrophiles; they react with highly reactive aromatic compounds to yield *azo* compounds. This electrophilic aromatic substitution is often called a *diazo coupling reaction.*

General Reaction

$$\text{C}_6\text{H}_5\text{—}\overset{+}{\text{N}}\equiv\text{N:}\ \text{X}^- + \text{C}_6\text{H}_5\text{—G} \longrightarrow \left[\text{C}_6\text{H}_5\text{—N=N—}\overset{\text{H}}{}(\text{C}_6\text{H}_5^{+})\text{—G}\right]\ \text{X}^- \xrightarrow{-\text{HX}} \text{C}_6\text{H}_5\text{—N=N—C}_6\text{H}_4\text{—G}$$

$G = —NR_2$, or —OH

An *azo compound*

Specific Examples

$$\text{C}_6\text{H}_5\text{—N}_2^+\ \text{Cl}^- + \text{C}_6\text{H}_5\text{—OH} \xrightarrow[\text{NaOH, H}_2\text{O}]{0^\circ} \text{C}_6\text{H}_5\text{—N=N—C}_6\text{H}_4\text{—OH}$$

Benzenediazonium chloride; Phenol; *p*-(Phenylazo)phenol (orange solid)

$$\text{C}_6\text{H}_5\text{—N}_2^+\ \text{Cl}^- + \text{C}_6\text{H}_5\text{—N(CH}_3)_2 \xrightarrow[0^\circ,\ \text{H}_2\text{O}]{\text{CH}_3\overset{\text{O}}{\overset{\|}{\text{C}}}\text{O}^-\ \text{Na}^+} \text{C}_6\text{H}_5\text{—N=N—C}_6\text{H}_4\text{—N(CH}_3)_2$$

Benzenediazonium chloride; *N,N*-Dimethylaniline; *N,N*-Dimethyl-*p*-(phenylazo)aniline (yellow solid)

Couplings between diazonium cations and phenols take place most rapidly in *slightly* alkaline solution. Under these conditions an appreciable amount of the phenol is present as a phenoxide ion, ArO^-, and phenoxide ions are even more reactive toward electrophilic substitution than are phenols themselves. (Why?) If the solution is too alkaline, however, ($pH > 10$) the diazonium salt itself reacts with hydroxide ion to form an unreactive diazohydroxide or diazotate ion:

$$\text{C}_6\text{H}_5\text{—}\ddot{\text{O}}\text{H} \underset{\text{H}^+}{\overset{\text{OH}^-}{\rightleftharpoons}} \text{C}_6\text{H}_5\text{—}\ddot{\text{O}}\text{:}^-$$

Phenol (couples slowly); Phenoxide ion (couples rapidly)

$$\text{Ar—}\overset{+}{\text{N}}\equiv\text{N:} \underset{\text{H}^+}{\overset{\text{OH}^-}{\rightleftharpoons}} \text{Ar—}\ddot{\text{N}}\text{=}\ddot{\text{N}}\text{—OH} \underset{\text{H}^+}{\overset{\text{OH}^-}{\rightleftharpoons}} \text{Ar—}\ddot{\text{N}}\text{=}\ddot{\text{N}}\text{—}\ddot{\text{O}}\text{:}^-$$

Diazonium ion (couples); Diazohydroxide (does not couple); Diazotate ion (does not couple)

Couplings between diazonium cations and amines take place most rapidly in slightly acidic solutions (pH 5–7). Under these conditions the concentration of the diazonium salt is at a maximum; at the same time an excessive amount of the amine has not been converted to an unreactive amine salt:

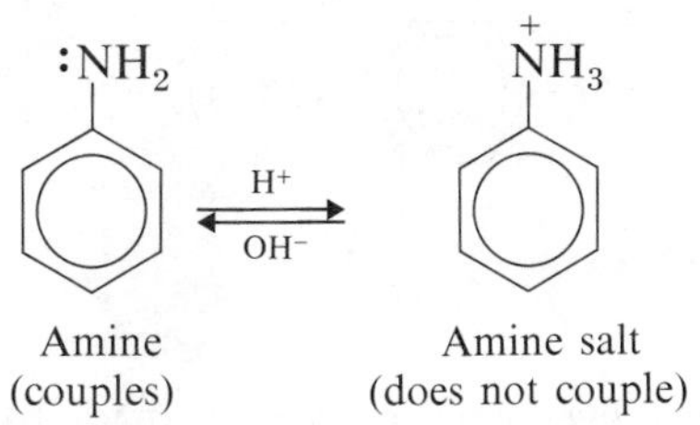

$$C_6H_5\ddot{N}H_2 \underset{OH^-}{\overset{H^+}{\rightleftharpoons}} C_6H_5\overset{+}{N}H_3$$

Amine (couples) — Amine salt (does not couple)

If the pH of the solution is lower than pH 5 the rate of amine coupling is low.

With phenols and aniline derivatives, coupling takes place almost exclusively at the para position if it is open. If it is not, coupling takes place at the ortho position.

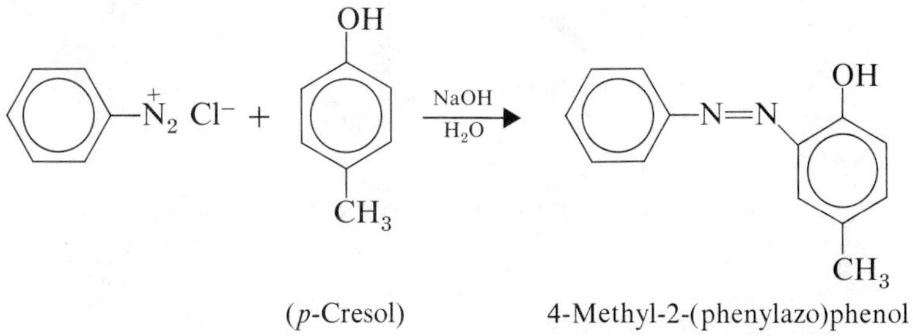

$$C_6H_5\overset{+}{N_2}\ Cl^- + p\text{-}CH_3C_6H_4OH \xrightarrow[H_2O]{NaOH} C_6H_5\text{—N=N—}C_6H_3(OH)(CH_3)$$

(*p*-Cresol) — 4-Methyl-2-(phenylazo)phenol

Azo compounds are usually intensely colored because the azo linkage —N=N— brings the two aromatic rings into conjugation. This gives an extended system of delocalized π electrons and allows absorption of light in visible regions. Azo compounds, because of their intense colors, and because they can be synthesized from relatively inexpensive compounds, are used extensively as *dyes*.

Azo dyes almost always contain one or more $—SO_3^-\ Na^+$ groups to confer water solubility on the dye and assist in binding the dye to the surfaces of polar fibers (wool, cotton, or nylon). Many dyes are made by coupling reactions of naphthylamines and naphthols. "H-acid" (below) is a particularly versatile component in dye manufacture; not only does it contain sulfonic acid groups, but it can also couple in two different ways depending on the pH of the medium.

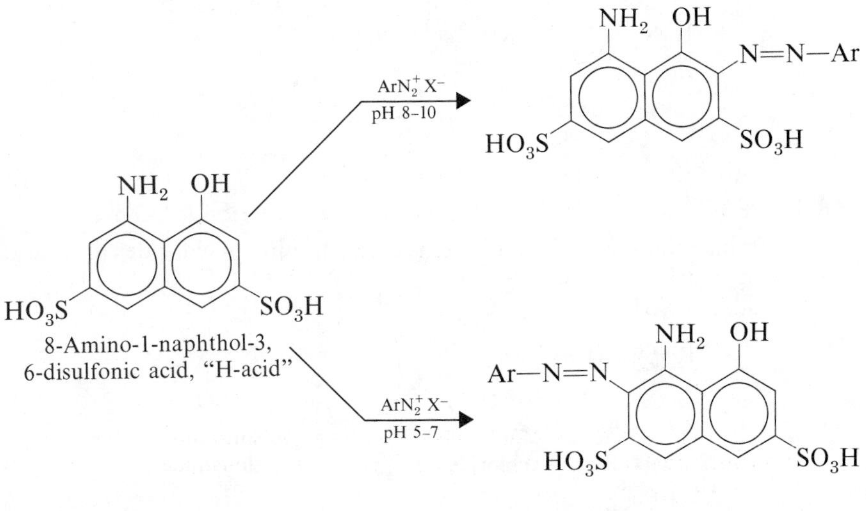

8-Amino-1-naphthol-3,6-disulfonic acid, "H-acid" (NH₂, OH, HO₃S, SO₃H)

$$\xrightarrow[\text{pH 8–10}]{ArN_2^+X^-}$$ coupling ortho to OH: N=N—Ar adjacent to OH (NH₂, OH, HO₃S, SO₃H)

$$\xrightarrow[\text{pH 5–7}]{ArN_2^+X^-}$$ coupling ortho to NH₂: Ar—N=N adjacent to NH₂ (NH₂, OH, HO₃S, SO₃H)

Diamine Green B, a dye introduced into commercial use in 1891, is made by a coupling reaction using H-acid.

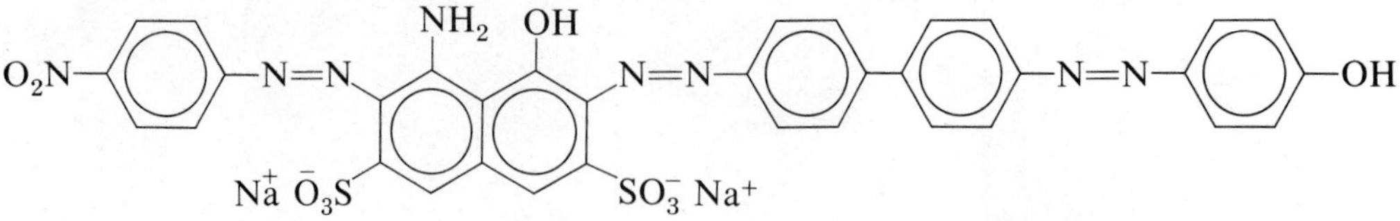

Diamine Green B

Orange II, a dye introduced even earlier (1876), is made from β-naphthol.

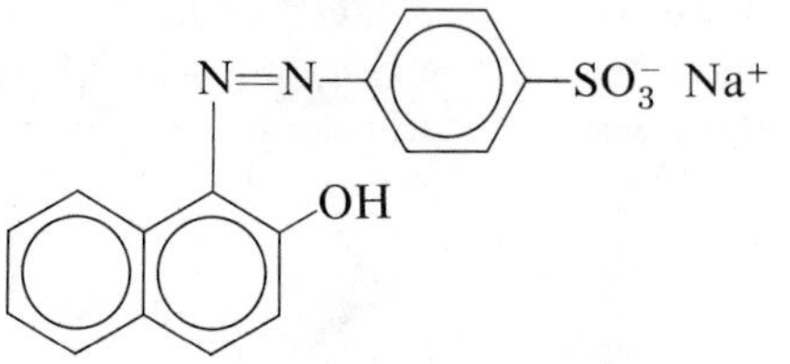

Orange II

Problem 18.16

Outline a synthesis of Orange II from β-naphthol and *p*-aminobenzenesulfonic acid.

Problem 18.17

Butter yellow is a dye once used to color margarine. It has since been shown to be carcinogenic and its use in food is no longer permitted. Outline a synthesis of butter yellow from benzene and *N,N*-dimethylaniline.

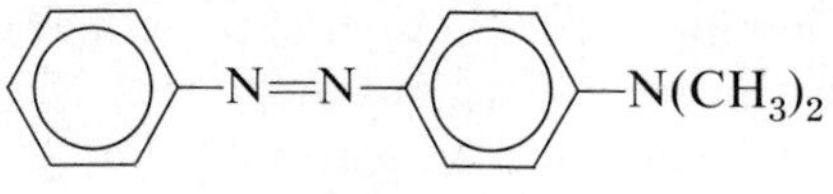

Butter Yellow

Problem 18.18

Outline a synthesis of Diamine Green B using, as starting materials, phenol, aniline, H-acid, and benzidine.

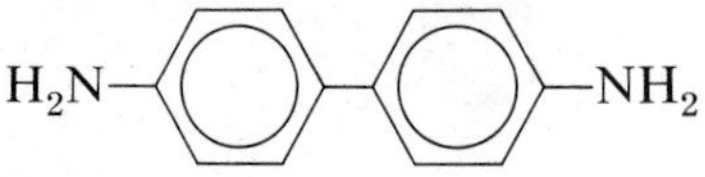

$H_2N-C_6H_4-C_6H_4-NH_2$

Benzidine

18.10 REACTIONS OF AMINES WITH SULFONYL CHLORIDES

Primary and secondary amines react with sulfonyl chlorides to form *sulfonamides.*

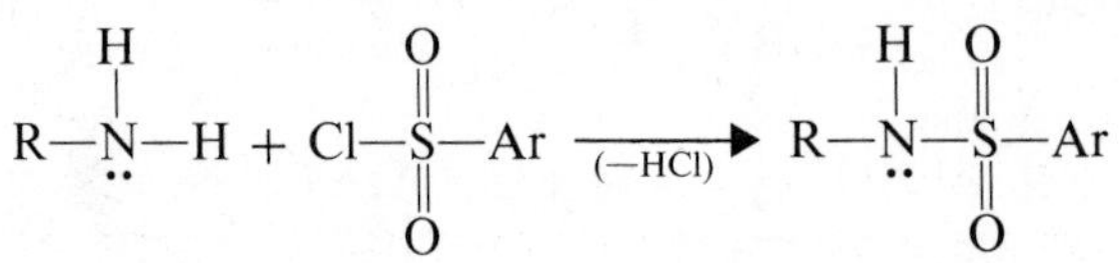

$$R-NH_2 + Cl-SO_2-Ar \xrightarrow[(-HCl)]{} R-NH-SO_2-Ar$$

1° Amine | Sulfonyl chloride | *N*-Substituted sulfonamide

$$\mathrm{R_2NH} + \mathrm{Cl{-}SO_2{-}Ar} \xrightarrow[(-\mathrm{HCl})]{} \mathrm{R_2N{-}SO_2{-}Ar}$$

2° Amine — *N,N*-Disubstituted sulfonamide

When heated with aqueous acid, sulfonamides are hydrolyzed to amines:

$$\mathrm{R_2N{-}SO_2{-}Ar} \xrightarrow[(2)\ \mathrm{OH^-}]{(1)\ \mathrm{H_3O^+,\ heat}} \mathrm{R_2N{-}H} + {}^{-}\mathrm{O{-}SO_2{-}Ar}$$

18.10A The Hinsberg Test

Sulfonamide formation is the basis for a chemical test, called the Hinsberg test, that we can use to demonstrate whether an amine is primary, secondary, or tertiary. We carry out a Hinsberg test in two steps. First, we shake a mixture containing a small amount of the amine and benzenesulfonyl chloride in *excess* potassium hydroxide. Next, we allow time for a reaction to take place, and then acidify the mixture. Each type of amine, 1°, 2°, or 3°, will give us a different *visible* result after each of these two stages of the test.

Primary amines react with benzenesulfonyl chloride to form *N*-substituted benzenesulfonamides. These, in turn, undergo acid-base reactions with the excess potassium hydroxide to form water-soluble potassium salts. (These reactions take place because the hydrogen attached to nitrogen is made acidic by the strongly electron-withdrawing $-SO_2-$ group.) At this stage our test tube will contain a clear solution. Acidification of this solution will, in the next stage, cause the water-insoluble *N*-substituted sulfonamide to precipitate.

$$\mathrm{RNH_2} + \mathrm{Cl{-}SO_2{-}C_6H_5} \xrightarrow[(-\mathrm{HCl})]{\mathrm{OH^-}} \left[\mathrm{R{-}NH{-}SO_2{-}C_6H_5}\right]$$

1° Amine — Acidic hydrogen (the H on N)

$$\left[\mathrm{R{-}NH{-}SO_2{-}C_6H_5}\right] \xrightarrow{\mathrm{KOH}} \mathrm{K^+\ R{-}\ddot{N}^{-}{-}SO_2{-}C_6H_5}$$

$$\mathrm{K^+\ R{-}\ddot{N}^{-}{-}SO_2{-}C_6H_5} \xrightarrow{\mathrm{HCl}} \mathrm{R{-}NH{-}SO_2{-}C_6H_5}$$

Water insoluble (precipitate) — Water-soluble salt (clear solution)

Some large primary aliphatic amines and cycloalkylamines with rings of seven carbons or more have low solubilities in aqueous KOH and a precipitate will form in the mixture even though the amine is primary. In these cases, the aqueous phase should be separated and then acidified. The appearance of a precipitate at this stage indicates that a primary amine was present.

Secondary amines react with benzenesulfonyl chloride in aqueous potassium hydroxide to form insoluble *N,N*-disubstituted sulfonamides that precipitate after the first stage. *N,N*-Disubstituted sulfonamides do not dissolve in aqueous potassium hydroxide because they do not have an acidic hydrogen. Acidification of the mixture obtained from a secondary amine produces no visible result—the *N,N*-disubstituted sulfonamide remains as a precipitate.

$$R_2N{-}H + Cl{-}SO_2{-}C_6H_5 \xrightarrow[(-HCl)]{OH^-} R_2\ddot{N}{-}SO_2{-}C_6H_5$$

Water insoluble
(precipitate)

↓ KOH

No reaction

↓ H^+

No reaction
(precipitate remains)

If the amine is a tertiary amine and if it is water-insoluble, no apparent change will take place in the mixture as we shake it with benzenesulfonyl chloride and aqueous KOH. When we acidify the mixture the tertiary amine will dissolve because it will form a water-soluble salt.

$$R_3N{:} \xrightarrow{HCl} R_3N^{+}{-}H \quad Cl^-$$

(insoluble) (soluble)

Tertiary amines do react with benzensulfonyl chloride to form *N*-benzenesulfonyl-*N,N,N*-trialkylammonium chlorides (see the following). In the presence of base (as in the Hinsberg test) these decompose to reform the tertiary amine:

$$R_3N + C_6H_5SO_2Cl \longrightarrow C_6H_5SO_2\overset{+}{N}R_3\ Cl^- \xrightarrow{OH^-} C_6H_5SO_2O^- + R_3N$$

Problem 18.19

An amine **A** has the molecular formula C_7H_9N. **A** reacts with benzenesulfonyl chloride in aqueous potassium hydroxide to give a clear solution; acidification of the solution gives a precipitate. When **A** is treated with $NaNO_2$ and HCl at 0–5 °C, and then with 2-naphthol; an intensely colored compound is formed. **A** gives a single strong absorption peak in the 680–840 cm^{-1} region at 815 cm^{-1}. What is the structure of **A**?

Problem 18.20

Sulfonamides of primary amines are often used to synthesize *pure* secondary amines. Suggest how this is done.

18.11 THE SULFA DRUGS: SULFANILAMIDE

18.11A Chemotherapy

Chemotherapy is defined as the use of chemical agents to destroy selectively infectious organisms without simultaneously destroying the host. Although it may be difficult to believe in this age of "wonder drugs," chemotherapy is a relatively modern phenomenon. Prior to 1900 only three specific chemical remedies were known: mercury (for syphilis—but often with disastrous results), cinchona bark (for malaria), and ipecacaunha (for dysentery).

Modern chemotherapy began with the work of Paul Ehrlich early in this century—particularly with his discovery in 1907 of the curative properties of a dye called Trypan Red I when used against experimental trypanosomiasis and with his discovery in 1909 of Salvarsan as a remedy for syphilis (Section I.5). Ehrlich invented the term "chemotherapy" and in his research sought what he called "magic bullets," that is, chemicals that would be toxic to infectious microorganisms but harmless to humans.*

As a medical student, Ehrlich had been impressed with the ability of certain dyes to stain tissues selectively. Working on the idea that "staining" was a result of a chemical reaction between the tissue and the dye, Ehrlich sought dyes with selective affinities for microorganisms. He hoped that in this way he might find a dye that could be modified so as to render it specifically lethal to microorganisms.

18.11B Sulfa Drugs

Between 1909 and 1935, tens of thousands of chemicals, including many dyes, were tested by Ehrlich and others in a search for such "magic bullets." Very few compounds, however, were found to have any promising effect. Then, in 1935, an amazing event happened. The daughter of Gerhard Domagk, a doctor employed by a German dye manufacturer, contracted a streptococcal infection from a pin prick. As his daughter neared death, Domagk decided to give her an oral dose of a dye called Prontosil. Prontosil had been developed at Domagk's firm (I. G. Farbenindustrie) and tests with mice had shown that Prontosil inhibited the growth of streptococci. Within a short time the little girl recovered. Domagk's gamble not only saved his daughter's life, but it also initiated a new and spectacularly productive phase in modern chemotherapy.*

A year later, in 1936, Ernest Fourneau of the Pasteur Institute in Paris demonstrated that Prontosil breaks down in the human body to produce sulfanilamide, and that sulfanilamide is the actual active agent against streptococci.

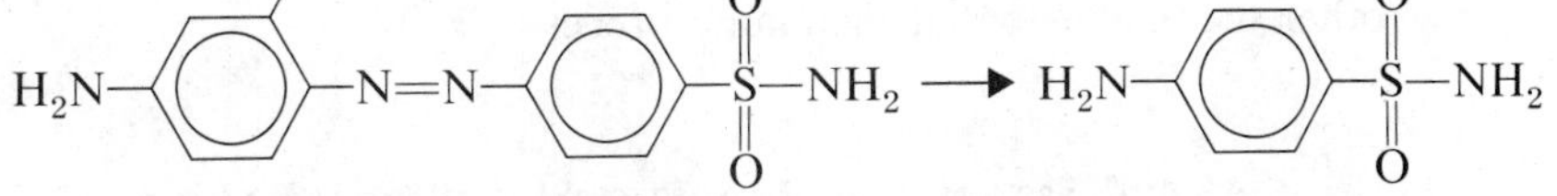

Prontosil → Sulfanilamide

Fourneau's announcement of this result set in motion a search for other chemicals (related to sulfanilamide) that might have even better chemotherapeutic

* Ehrlich was awarded the Nobel Prize for medicine in 1908.

* Domagk was awarded the Nobel Prize for medicine in 1939.

effects. Literally thousands of chemical variations were played on the sulfanilamide theme; the structure of sulfanilamide was varied in almost every imaginable way. The best therapeutic results were obtained from compounds in which one hydrogen of the $—SO_2NH_2$ group was replaced by some other group, usually a heterocyclic amine. Among the most successful variations were the compounds shown below.

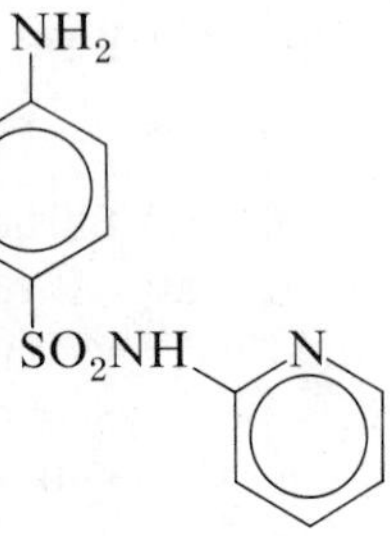

Sulfapyridine

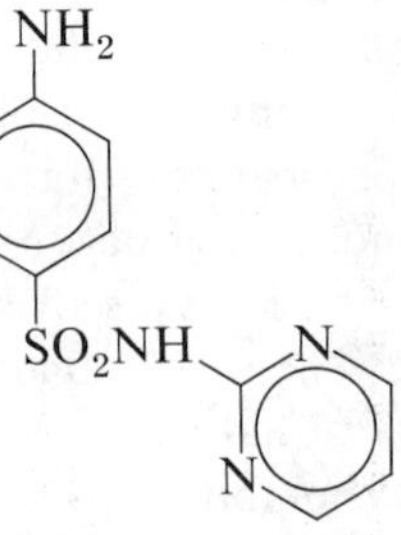

Sulfadiazine

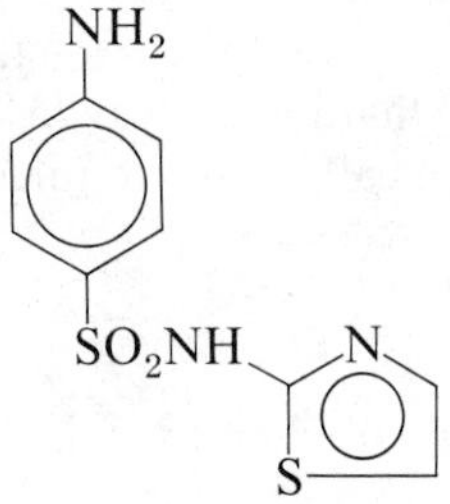

Sulfathiazole

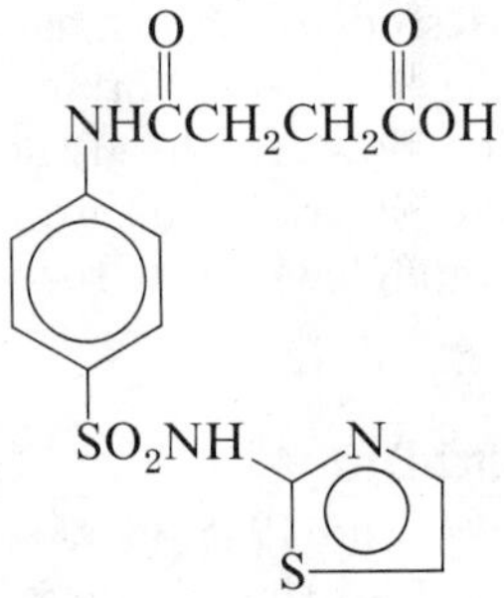

Succinoylsulfathiazole

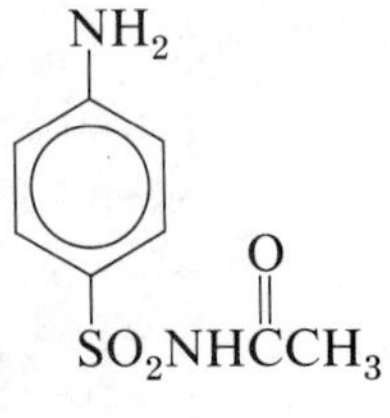

Sulfacetamide

Sulfapyridine was shown to be effective against pneumonia in 1938. (Prior to that time pneumonia epidemics had brought death to tens of thousands.) Sulfacetamide was used successfully in treating urinary tract infections in 1941. Succinoylsulfathiazole and the related compound phthalylsulfathiazole were used as chemotherapeutic agents against infections of the gastrointestinal tract beginning in 1942. (Both compounds are slowly hydrolyzed internally to sulfathiazole.) Sulfathiazole saved the lives of countless wounded soldiers during World War II.

In 1940 a discovery by D. D. Woods laid the groundwork for our understanding of how the sulfa drugs work. Woods observed that the inhibition of growth of certain microorganisms by sulfanilamide is competitively overcome by *p*-aminobenzoic acid. Woods noticed the structural similarity between the two compounds (Fig. 18.2) and reasoned that the two compounds compete with each other in some essential metabolic process.

18.11C Essential Nutrients and Antimetabolites

All higher animals and many microorganisms lack the biochemical ability to synthesize certain essential organic compounds. These essential nutrients include vitamins, certain amino acids, unsaturated carboxylic acids, purines, and pyrimidines. The aromatic amine *p*-aminobenzoic acid is an essential nutrient for those bacteria that are sensitive to sulfanilamide therapy. Enzymes within these bacteria use *p*-aminobenzoic acid to synthesize another essential compound called *folic acid.*

p-Aminobenzoic acid

Folic acid

Chemicals that inhibit the growth of microbes are called *antimetabolites.* The sulfanilamides are antimetabolites for those bacteria that require *p*-aminobenzoic acid. The sulfanilamides apparently inhibit those enzymic steps of the bacteria that are involved in the synthesis of folic acid. The bacterial enzymes are apparently unable to distinguish between a molecule of a sulfanilamide and a molecule of *p*-aminobenzoic acid; thus, sulfanilamide "inhibits" the bacterial enzyme. Because the microorganism is unable to synthesize enough folic acid when sulfanilamide is present, it dies. Humans are unaffected by sulfanilamide therapy because we derive our folic acid from dietary sources (folic acid is a vitamin) and do not synthesize it from *p*-aminobenzoic acid.

The discovery of the mode of action of the sulfanilamides has led to the discovery of many new and effective antimetabolites. One example is *Methotrexate,* a derivative of folic acid that has been used successfully in treating certain carcinomas:

Methotrexate

Methotrexate, by virtue of its resemblance to folic acid, can enter into some of the same reactions as folic acid, but it cannot serve the same function, particularly in important reactions involved in cell division. Although methotrexate is toxic to all dividing cells, those cells that divide most rapidly—*cancer cells*—are most vulnerable to its effect.

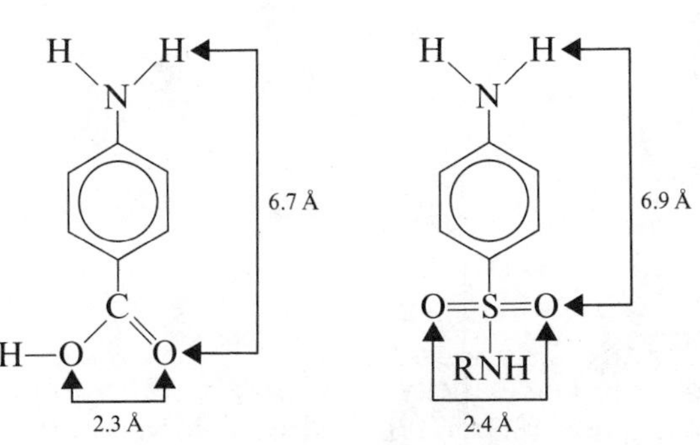

p-Aminobenzoic acid A sulfanilamide

FIG. 18.2
The structural similarity of p-amino-*benzoic acid and a sulfanilamide. (From A. Korolkovas,* Essentials of Molecular Pharmacology, *Wiley, New York, 1970, p. 105. Used with permission.)*

18.11D Synthesis of Sulfa Drugs

Sulfanilamides can be synthesized from aniline through the sequence of reactions shown below.

$$\underset{\substack{\text{Aniline}\\ \mathbf{1}}}{C_6H_5NH_2} \xrightarrow[(-CH_3COOH)]{(CH_3\overset{O}{\overset{\|}{C}})_2O} \underset{\substack{\text{Acetanilide}\\ \mathbf{2}}}{C_6H_5NH\overset{O}{\overset{\|}{C}}CH_3} \xrightarrow[\substack{80^\circ\\(-H_2O)}]{HOSO_2Cl} \underset{\substack{p\text{-Acetamidobenzene-}\\ \text{sulfonyl chloride}\\ \mathbf{3}}}{p\text{-}CH_3\overset{O}{\overset{\|}{C}}NHC_6H_4SO_2Cl} \xrightarrow[(-HCl)]{H_2N{-}R}$$

$$\underset{\mathbf{4}}{p\text{-}CH_3\overset{O}{\overset{\|}{C}}NHC_6H_4SO_2NHR} \xrightarrow[\text{heat}]{\text{dil. HCl}} \underset{\substack{\text{A sulfanilamide}\\ \mathbf{5}}}{p\text{-}H_2NC_6H_4SO_2NHR}$$

Acetylation of aniline produces acetanilide, **2**. Treatment of **2** with chlorosulfonic acid brings about an electrophilic aromatic substitution reaction and yields *p*-acetamidobenzenesulfonyl chloride, **3**. Addition of ammonia or a secondary amine gives the diamide, **4** (an amide of both a carboxylic acid and a sulfonic acid). Finally, refluxing **4** with dilute hydrochloric acid selectively hydrolyzes the carboxamide linkage and produces sulfanilamide.

Problem 18.21

(a) Starting with aniline and assuming that you have 2-aminothiazole available, show how you would synthesize sulfathiazole. (b) How would you convert sulfathiazole to succinoylsulfathiazole?

H_2N–(thiazole ring: N, S)

2-Aminothiazole

18.12 ANALYSIS OF AMINES

18.12A Chemical Analysis

Amines are characterized by their basicity, and thus, by their ability to dissolve in dilute aqueous acid (Section 18.3A). Primary, secondary, and tertiary amines can be distinguished from each other on the basis of the Hinsberg test (Section 18.10A). Primary aromatic amines are often detected through diazonium salt formation and subsequent coupling with β-naphthol to form a brightly colored azo dye (Section 18.9).

18.12B Spectroscopic Analysis

The *infrared spectra* of primary and secondary amines are characterized by absorption bands in the 3300–3500 cm^{-1} region that arise from N—H stretching vibration. Primary amines give two bands in this region; secondary amines generally give only one. Absorption bands arising from C—N stretching vibrations of aliphatic amines occur in the 1020–1220 cm^{-1} region but are usually weak and difficult to identify. Aromatic amines generally give a strong C—N stretching band in the 1250–1360 cm^{-1} region.

The *proton nmr spectra* of primary and secondary amines show N—H proton absorptions in region 1–5 δ. These peaks are sometimes difficult to identify and are best detected by proton counting.

18.13 THE HOFMANN ELIMINATION

All of the eliminations that we have described so far have involved electrically neutral substrates. However, a number of eliminations are known in which the substrate bears a positive charge. One of the most important of these is the elimination that takes place when a quaternary ammonium hydroxide is heated. The products are an alkene, water, and a tertiary amine.

$$HO:^- + H-\overset{|}{\underset{|}{C}}-\overset{|}{\underset{|}{C}}-\overset{+}{N}R_3 \longrightarrow \rangle C{=}C\langle + HOH + :NR_3$$

A quaternary ammonium hydroxide ⟶ An alkene + water + a tertiary amine

This reaction was discovered in 1851 by August W. von Hofmann and has since come to bear his name.

Quaternary ammonium hydroxides can be prepared from quaternary ammonium halides in aqueous solution through the use of silver oxide.

$$2RCH_2CH_2\overset{+}{N}(CH_3)_3\ X^- + Ag_2O + H_2O \longrightarrow 2RCH_2CH_2\overset{+}{N}(CH_3)_3\ OH^- + 2AgX\downarrow$$

A quaternary ammonium halide ⟶ quaternary ammonium hydroxide

Silver halide precipitates from the solution and can be removed by filtration. The quaternary ammonium hydroxide can then be obtained by evaporation of the water.

While most eliminations involving neutral substrates tend to follow the *Zaitzev rule* (Section 6.9), eliminations with charged substrates tend to follow the *Hofmann rule* (Section 6.9) and yield mainly the least substituted alkene. We can see an example of this behavior if we compare the following reactions.

$$C_2H_5O^-Na^+ + CH_3CH_2\underset{\substack{|\\Br}}{C}HCH_3 \xrightarrow[25^\circ]{C_2H_5OH}$$

$$\underset{(75\%)}{CH_3CH{=}CHCH_3} + \underset{(25\%)}{CH_3CH_2CH{=}CH_2} + NaBr + C_2H_5OH$$

$$\underset{\overset{|}{\underset{+}{N}(CH_3)_3}}{CH_3CH_2CHCH_3} \quad OH^- \xrightarrow{150°} \underset{(5\%)}{CH_3CH{=}CHCH_3} + \underset{(95\%)}{CH_3CH_2CH{=}CH_2} + (CH_3)_3N: + H_2O$$

$$\underset{\overset{|}{\underset{+}{S}(CH_3)_2}}{CH_3CH_2CHCH_3} \quad \bar{O}C_2H_5 \longrightarrow \underset{(26\%)}{CH_3CH{=}CHCH_3} + \underset{(74\%)}{CH_3CH_2CH{=}CH_2} + (CH_3)_2S + C_2H_5OH$$

The precise mechanistic reasons for these differences are complex and are not yet fully understood. One possible explanation is that the transition states of elimination reactions with charged substrates have considerable carbanion character. This means that these transition states show little resemblance to the final alkene product and, thus, are not stabilized appreciably by a developing double bond.

$$\overset{\delta-}{H\ddot{O}}\text{---}H \quad \overset{\delta-}{-C}-\underset{\underset{+}{N}(CH_3)_3}{C}-$$

Carbanionlike transition state
(gives Hofmann orientation)

$$\overset{\delta-}{H\ddot{O}}\text{---}H \quad -C{=}{=}{=}\underset{Br^{\delta-}}{C}-$$

Alkenelike transition state
(gives Zaitzev orientation)

With a charged substrate, the base attacks the most acidic hydrogen instead. Primary hydrogens are more acidic because their carbon bears only one electron-releasing group.

An elegant application of the Hofmann elimination can be seen in Richard Willstätter's synthesis of cyclooctatetraene (p. 845) from pseudopelletierine, a substance that can be obtained from the bark of the pomegranate tree. This synthesis was achieved in 1911, and in it Willstätter utilized *four* Hofmann eliminations. Even though most steps gave yields greater than 90%, the poor yield in the last step meant that the overall yield was only 3%. Still, Willstätter was able to obtain enough cyclooctatetraene to show that it was not aromatic. As we noted in Chapter 11, this was an important step in our understanding of aromatic compounds.

$$\text{Pseudopelletierine } (CH_3N:) \xrightarrow[\substack{(H)\\(84\%)}]{Na/C_2H_5OH} (CH_3N:, \text{-OH}) \xrightarrow[\substack{-H_2O\\(84\%)}]{H_2SO_4/CH_3COOH} (CH_3N:) \xrightarrow[(96\%)]{CH_3I}$$

$$CH_3\overset{+}{N}CH_3 \; I^- \xrightarrow[\substack{(2)\ Heat\\(90\%)}]{(1)\ Ag_2O/H_2O} \ddot{N}(CH_3)_2 \xrightarrow[(94.5\%)]{CH_3I} \overset{+}{N}(CH_3)_3\ I^- \xrightarrow[\substack{(2)\ Heat\\-(CH_3)_3\ddot{N},\ -H_2O\\(72\%)}]{(1)\ Ag_2O/H_2O}$$

$Br_2/CHCl_3$ (1,6-addition) → [Br, Br] (not isolated) → $(CH_3)_2\ddot{N}H$, —HBr (68% for both steps) → $(CH_3)_2\ddot{N}$, $\ddot{N}(CH_3)_2$ → CH_3I (97.5%) →

I^- $(CH_3)_3N^+$, $\overset{+}{N}(CH_3)_3$ I^- → (1) Ag_2O, H_2O (2) Heat (10–20%) → Cyclooctatetraene $+ 2(CH_3)_3\ddot{N} + 2H_2O$

Problem 18.22

What products would you expect to obtain from each of the following elimination reactions? If more than one product is possible, state which you think would be the major product:

(a) $CH_3\underset{CH_3}{CH}CH_2—\overset{CH_3}{\underset{CH_3}{N^+}}—CH_2CH_3 \quad OH^- \xrightarrow{heat}$

(b) $CH_3CH_2CH_2\overset{+}{\underset{CH_3}{S}}CH_2CH_3 \quad {}^-OC_2H_5 \xrightarrow{heat}$

(c) Pyrrolidinium ring with $\overset{+}{N}$ bearing CH_3 and CH_3 $\quad {}^-OH \xrightarrow{heat}$

(d) Product of (c) $\xrightarrow[(2)\ Ag_2O/H_2O\ (3)\ heat]{(1)\ CH_3I}$

Additional Problems

18.23

Write structural formulas for each of the following compounds.

(a) Methylbenzylamine
(b) Triisopropylamine
(c) *N*-Methyl-*N*-ethylaniline
(d) *m*-Toluidine
(e) 2-Methylpyrrole
(f) *N*-Ethylpiperidine
(g) *N*-Ethylpyridinium bromide
(h) 3-Pyridinecarboxylic acid
(i) Indole
(j) Acetanilide
(k) Dimethylammonium chloride
(l) 2-Methylimidazole
(m) Sulfapyridine
(n) Tetrapropylammonium chloride
(o) Pyrrolidine
(p) *N,N*-Dimethyl-*p*-toluidine
(q) *p*-Anisidine
(r) Benzidine
(s) *p*-Aminobenzoic acid
(t) Histidine

18.24

Give common or IUPAC names for each of the following compounds.

(a) $CH_3CH_2CH_2NH_2$ (b) $C_6H_5NHCH_3$

(c) $(CH_3)_2CH\overset{+}{N}(CH_3)_3$ I^-

(d) o-$CH_3C_6H_4NH_2$

(e) o-$CH_3OC_6H_4NH_2$

(f) 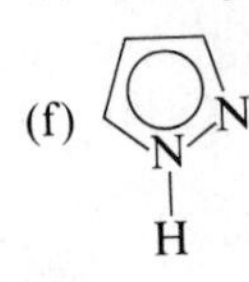

(g) 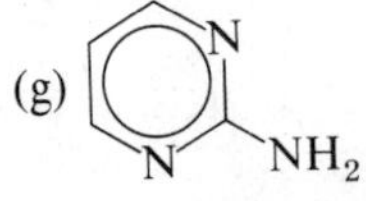

(h) $C_6H_5CH_2NH_3^+$ Cl^-

(i) $C_6H_5N(CH_2CH_2CH_3)_2$

(j) $C_6H_5SO_2NH_2$

(k) $CH_3NH_3^+$ CH_3COO^-

(l) $HOCH_2CH_2CH_2NH_2$

(m)

(n)

18.25

Show how you might prepare benzylamine from each of the following compounds.

(a) Benzonitrile

(b) Benzamide

(c) Benzyl bromide (two ways)

(d) Benzyl tosylate

(e) Benzaldehyde

(f) Phenylnitromethane

(g) Phenylacetamide

18.26

Show how you might prepare aniline from each of the following compounds.

(a) Benzene (b) Bromobenzene (c) Benzamide

18.27

Show how you might synthesize each of the following compounds from n-butyl alcohol.

(a) n-Butylamine (free of 2° and 3° amines)

(b) n-Pentylamine

(c) n-Propylamine

(d) Methyl-n-butylamine

18.28

Give structures for compounds **A**–**F**:

$$N\text{-Methylpiperidine} + CH_3I \longrightarrow \mathbf{A}\ (C_7H_{16}NI) \xrightarrow[H_2O]{Ag_2O} \mathbf{B}\ (C_7H_{17}NO) \xrightarrow[(-H_2O)]{\text{heat}}$$

$$\mathbf{C}\ (C_7H_{15}N) \xrightarrow{CH_3I} \mathbf{D}\ (C_8H_{18}NI) \xrightarrow[H_2O]{Ag_2O} \mathbf{E}\ (C_8H_{19}NO) \xrightarrow{\text{heat}}$$

$$\mathbf{F}\ (C_5H_8) + H_2O + (CH_3)_3N$$

18.29

Show how you might convert aniline into each of the following compounds. (You need not repeat steps carried out in earlier parts of this problem.)

(a) Acetanilide

(b) N-Phenylphthalimide

(c) p-Nitroaniline

(d) Sulfanilamide

(e) N,N-Dimethylaniline

(f) Fluorobenzene

(g) Chlorobenzene

(h) Bromobenzene

(i) Iodobenzene

(j) Benzonitrile

(k) Benzoic acid

(l) Phenol

(m) Benzene

(n) p-(Phenylazo)phenol

(o) p-N,N-Dimethyl-p-(phenylazo)aniline

18.30

What products would you expect to be formed when each of the following amines reacts with aqueous sodium nitrite and hydrochloric acid?

(a) Propylamine (b) Dipropylamine

(c) *N*-Propylaniline
(d) *N*-*N*-Dipropylaniline
(e) *p*-Propylaniline

18.31
(a) What products would you expect to be formed when each of the amines in the preceding problem reacts with benzenesulfonyl chloride and excess aqueous potassium hydroxide? (b) What would you observe in each reaction? (c) What would you observe when the resulting solution or mixture is acidified?

18.32
(a) What product would you expect to obtain from the reaction of piperidine with aqueous sodium nitrite and hydrochloric acid? (b) From the reaction of piperidine and benzenesulfonyl chloride in excess aqueous potassium hydroxide?

18.33
Give structures for the products of each of the following reactions.

(a) Ethylamine + benzoyl chloride $\longrightarrow$

(b) Methylamine + acetic anhydride $\longrightarrow$

(c) Methylamine + succinic anhydride $\longrightarrow$

(d) Product of (c) $\xrightarrow{\text{Heat}}$

(e) Pyrrolidine + phthalic anhydride $\longrightarrow$

(f) Pyrrole + acetic anhydride $\longrightarrow$

(g) Aniline + propanoyl chloride $\longrightarrow$

(h) Tetraethylammonium hydroxide $\xrightarrow{\text{Heat}}$

(i) *m*-Dinitrobenzene + H_2S $\xrightarrow[C_2H_5OH]{NH_3}$

(j) *p*-Toluidine + Br_2(excess) $\xrightarrow[H_2O]{}$

18.34
Starting with benzene or toluene outline syntheses of the following compounds using diazonium salts as intermediates. (You need not repeat syntheses carried out in earlier parts of this problem.)

(a) *o*-Cresol
(b) *m*-Cresol
(c) *p*-Cresol
(d) *m*-Dichlorobenzene
(e) *m*-$C_6H_4(CN)_2$
(f) *m*-Iodophenol
(g) *m*-Bromobenzonitrile
(h) 1,3-Dibromo-5-nitrobenzene
(i) 3,5-Dibromoaniline
(j) 3,4,5-Tribromophenol
(k) 3,4,5-Tribromobenzonitrile
(l) 2,6-Dibromobenzoic acid
(m) 1,3-Dibromo-2-iodobenzene
(n) 4-Bromo-2-nitrotoluene
(o) 3-Nitro-4-methylphenol
(p) 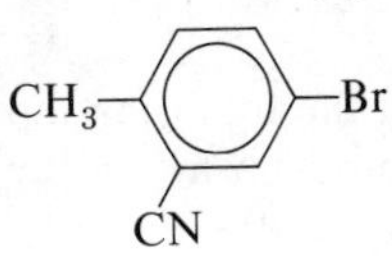 CH_3–C_6H_3(CN)–Br (CN ortho to CH_3, Br para to CH_3)
(q) 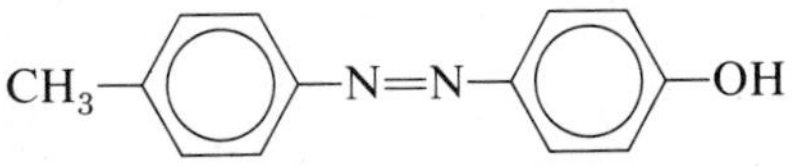 CH_3–C_6H_4–N=N–C_6H_4–OH
(r) 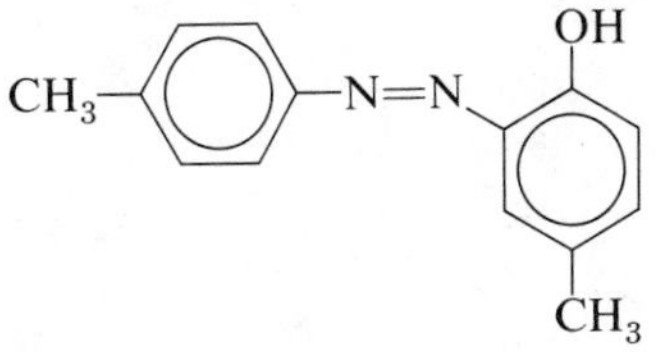 CH_3–C_6H_4–N=N–C_6H_3(OH)(CH_3) (OH ortho to N=N, CH_3 para to OH)

18.35

Write equations for simple chemical tests that would distinguish between

(a) Benzylamine and benzamide
(b) Allylamine and propylamine
(c) *p*-Toluidine and *N*-methylaniline
(d) Cyclohexylamine and piperidine
(e) Pyridine and benzene
(f) Cyclohexylamine and aniline
(g) Triethylamine and diethylamine
(h) Tripropylammonium chloride and tetrapropylammonium chloride
(i) Tetrapropylammonium chloride and tetrapropylammonium hydroxide

18.36

Describe with equations how you might separate a mixture of aniline, *p*-cresol, benzoic acid, and toluene using ordinary laboratory reagents.

18.37

Show how you might synthesize β-aminopropionic acid, $H_2NCH_2CH_2COOH$, from succinic anhydride (β-aminopropionic acid is used in the synthesis of pantothenic acid; cf. Problem 17.31).

18.38

Show how you might synthesize each of the following from the compounds indicated and any other needed reagents.

(a) Decamethonium bromide (p. 817) from 1,10-decanediol
(b) Succinylcholine bromide from succinic acid, 2-bromoethanol and trimethylamine
(c) Acetylcholine chloride from ethylene oxide

18.39

Basing your answer on the theory of antimetabolite activity presented on page 841, which of the following compounds would you expect to be *inactive* as inhibitors of folic acid biosynthesis?

(a) 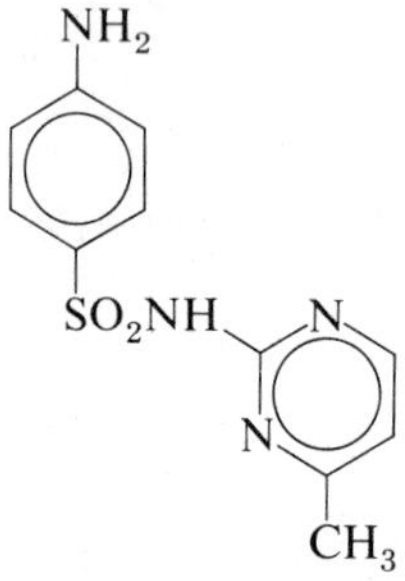 NH_2 / SO_2NH / N / N / CH_3

(c) 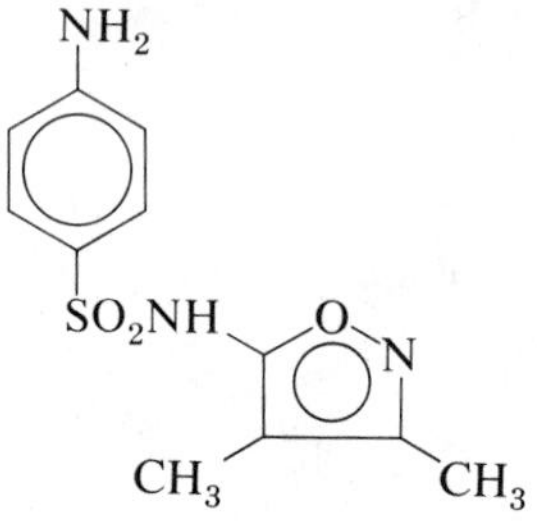 NH_2 / SO_2NH / O / N / CH_3 / CH_3

(b) 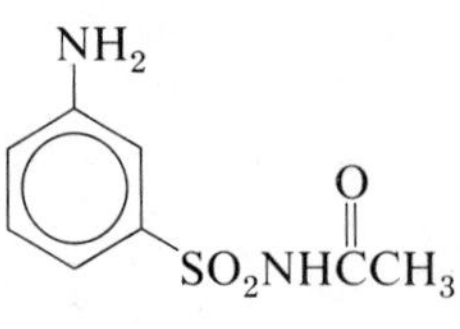 NH_2 / O / SO_2NHCCH_3

(d) 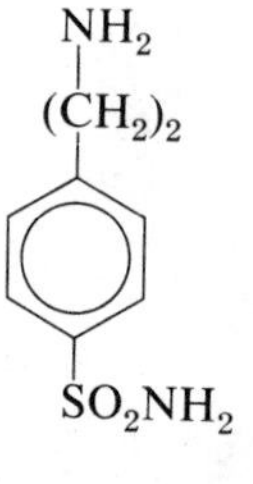 NH_2 / $(CH_2)_2$ / SO_2NH_2

(e) 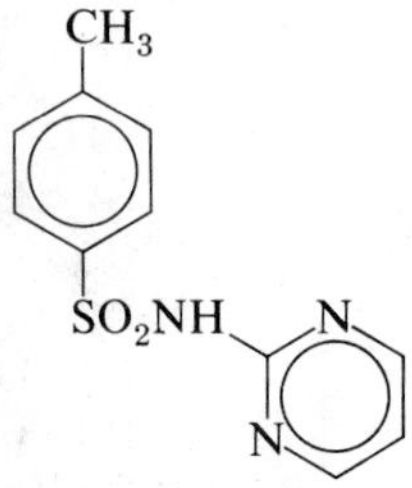 CH_3 / SO_2NH / N / N

18.40

A commercial synthesis of folic acid consists of heating the following three compounds with aqueous sodium bicarbonate. Propose reasonable mechanisms for the reactions that lead to folic acid.

$$\underset{\text{(pyrimidine: }H_2N,\ NH_2,\ NH_2,\ OH)}{} + CHBr_2\overset{O}{\overset{\|}{C}}CH_2Br + H_2N-C_6H_4-CONH\overset{COOH}{\underset{|}{C}}H\text{—}CH_2CH_2COOH \xrightarrow{HCO_3^-,\ H_2O} \underset{(\sim 10\%)}{\text{Folic acid}}$$

18.41
When compound W, $C_{15}H_{17}N$, is treated with benzenesulfonyl chloride and aqueous potassium hydroxide no apparent change occurs. Acidification of this mixture gives a clear solution. The proton nmr spectrum of W is shown in Fig. 18.3. Propose a structure for W.

18.42
Propose structures for compounds X, Y, and Z.

$$\mathbf{X}(C_7H_7Br) \xrightarrow{NaCN} \mathbf{Y}(C_8H_7N) \xrightarrow{LiAlH_4} \mathbf{Z}(C_8H_{11}N)$$

The proton nmr spectrum of X gives two signals, a multiplet at δ7.3 (5H) and a singlet at δ4.25 (2H); the 680–840 cm^{-1} region of the infrared spectrum of X shows a peak at 690 cm^{-1} and 770 cm^{-1}. The proton nmr spectrum of Y is similar to that of X: multiplet δ7.3 (5H), singlet δ3.7 (2H). The proton nmr and infrared spectra of Z are shown in Fig. 18.4 page 850.

FIG. 18.3
The proton nmr spectrum of W (Problem 18.41). (Courtesy Aldrich Chemical Company Inc., Milwaukee, Wis.)

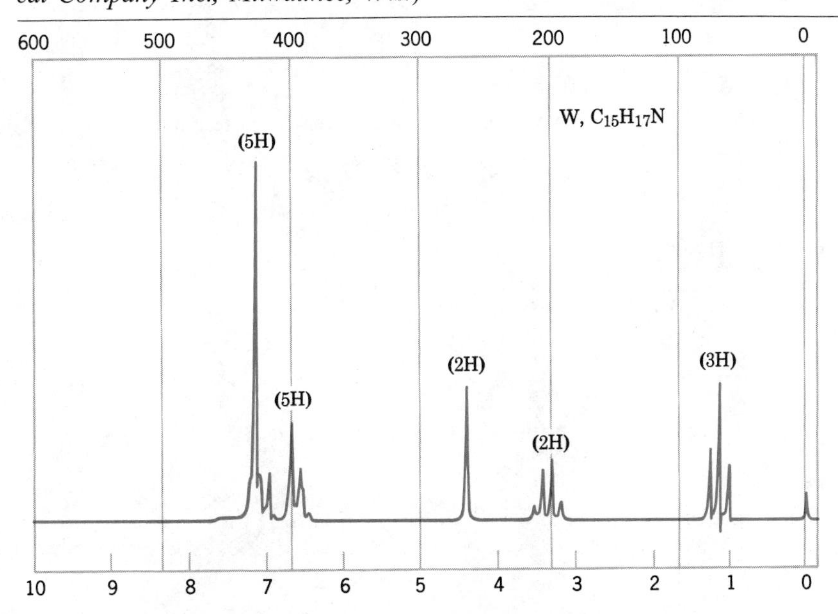

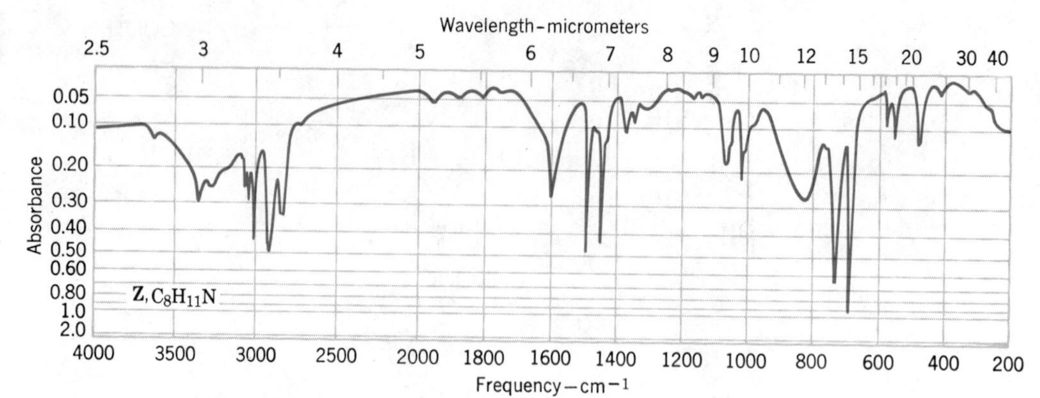

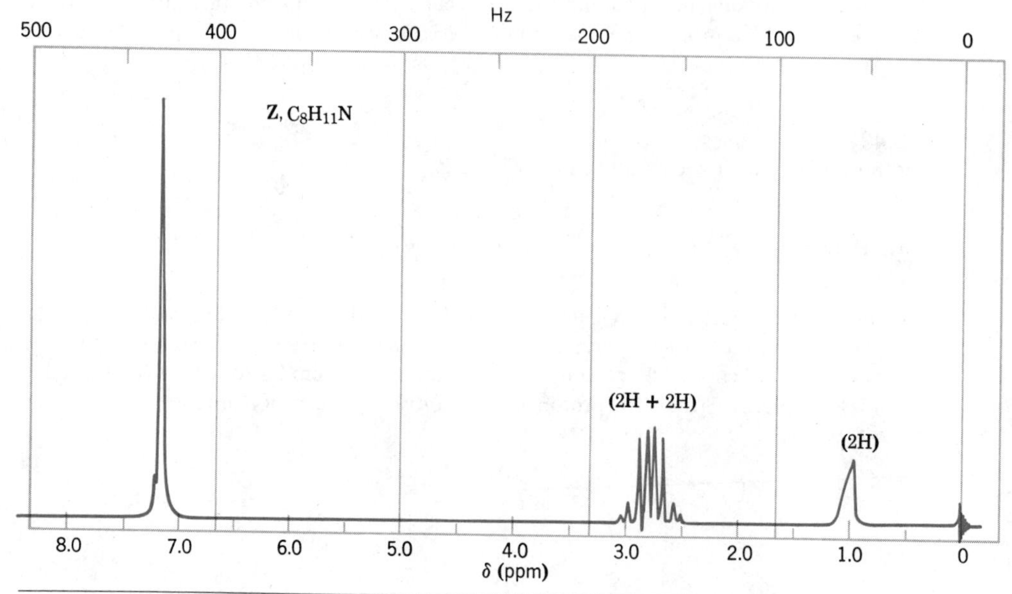

FIG. 18.4
Infrared and proton nmr for compound Z, Problem 18.42. (Courtesy Sadtler Research Laboratories, Inc., Philadelphia, Pa.)

*** 18.43**

Using reactions that we have studied in this chapter, propose a mechanism that accounts for the following reaction.

2-(CH_2CH_2CN)cyclohexanone $\xrightarrow{H_2,\ Pd}$ decahydroquinoline (N–H)

L

SPECIAL TOPIC

NUCLEOPHILIC SUBSTITUTION AND ELIMINATION REACTIONS—ANOTHER LOOK

L.1 S_N2 REACTIONS: THE ROLE OF ION PAIRS

For many years the Ingold mechanism (Section 5.5) was widely accepted as the only mechanism for bimolecular nucleophilic substitution reactions. In recent years, however, evidence has been advanced to support another possible mechanism: one involving *ion pairs.* The ion pair mechanism seems to be particularly important in those nucleophilic substitutions called *solvolyses;* however, it may operate in other S_N2 reactions. *A solvolysis is a reaction in which the nucleophile is a molecule of the solvent* (solvent + *lysis:* cleavage by solvent).

Solvolytic reactions are often described as being pseudo-first order since, by convention, the concentration of the solvent is not included in the rate expression. Even if it were included, we might not be able to detect a variation in rate with a change in concentration of the solvent since the solvent concentration is usually very large and is, therefore, essentially constant. As we will see, however, some solvolyses are actually bimolecular nucleophilic substitution reactions and occur with complete inversion of configuration.

The first clear demonstration of ion pair involvement in solvolytic reactions was given by Professor S. Winstein (of the University of California, Los Angeles).

Two examples of solvolytic reactions for which ion pair mechanisms have been proposed are the hydrolysis and acetolysis of methylheptyl sulfonates.

Hydrolysis of a 1-Methylheptyl Sulfonate

$$H_2O + \text{H}(C_6H_{13})(CH_3)C{-}OSO_2R \longrightarrow HO^{+}(H){-}C(C_6H_{13})(H)(CH_3) \xrightarrow{-H^+} HO{-}C(C_6H_{13})(H)(CH_3)$$

R–1–methylheptyl sulfonate — protonated alcohol + $^{-}OSO_2R$ — *S*–2–octanol

Acetolysis of a 1-Methylheptyl Sulfonate

$$CH_3\overset{O}{\overset{\|}{C}}OH + \text{H}(C_6H_{13})(CH_3)C{-}OSO_2R \longrightarrow CH_3\overset{O}{\overset{\|}{C}}{-}\overset{+}{O}(H){-}C(C_6H_{13})(H)(CH_3) \longrightarrow CH_3\overset{O}{\overset{\|}{C}}O{-}C(C_6H_{13})(H)(CH_3)$$

R–1–methylheptyl sulfonate — protonated acetate + $^{-}OSO_2R$ — *S*–2–octyl acetate

Both of these reactions have been shown to take place with *complete inversion of configuration.*

On the basis of kinetic evidence (beyond the scope of our treatment here) it has been proposed that these reactions take place with the rapid but reversible formation of an "*intimate*" (or tight) ion pair followed by a slow reaction with a solvent molecule. An intimate ion pair is an ionic intermediate in which the cation and anion are in close proximity and are not separated by solvent molecules.

$$\text{(C}_6\text{H}_{13}\text{, H, CH}_3\text{)C}-\text{OSO}_2\text{R} \underset{}{\overset{\text{fast}}{\rightleftharpoons}} \text{(C}_6\text{H}_{13}\text{, H, CH}_3\text{)C}^{+} \; {}^{-}\text{OSO}_2\text{R} \xrightarrow[\text{slow}]{\text{SolOH}} \text{Sol}\overset{+}{\text{O}}\text{(H)}-\text{C(C}_6\text{H}_{13}\text{, H, CH}_3\text{)} \xrightarrow{-\text{H}^+} \text{SolO}-\text{C(C}_6\text{H}_{13}\text{, H, CH}_3\text{)}$$

intimate ion pair
(retention of configuration)

(inversion of configuration)

Sol = H— or $CH_3\overset{O}{\overset{\|}{C}}$—

The intimate ion pair retains the configuration of the original sulfonate. However, when it reacts with the solvent, the displacement takes place from the back side and produces an inversion of configuration. In all likelihood, inversion occurs because the intimate ion pair still has partial bonding between the carbon and the leaving group, and attack of the nucleophile must occur from the back side in the same manner as for an S_N2 reaction of a neutral compound.

The difference between the Ingold mechanism and the ion pair mechanism is most apparent in their potential energy diagrams (Fig. L.1). In the one-step displacement mechanism there is a single transition state; in the ion pair mechanism there are two transition states and the ion pair occupies an energy minimum lying between them.

At the time of this writing not enough evidence is available to distinguish between these two mechanistic possibilities for most S_N2 reactions. However, with either mechanism the stereochemistry of S_N2 reactions is clear: S_N2 reactions, whether they take place in one step or through the formation of ion pairs, *occur with inversion of configuration.*

L.2 ION PAIRS AND S_N1 REACTIONS

When 1-phenylethyl chloride reacts with water in aqueous acetone, that is,

$$C_6H_5\underset{\displaystyle Cl}{\underset{|}{C}}HCH_3 + H_2O \xrightarrow[(-HCl)]{\text{Acetone}} C_6H_5\underset{\displaystyle OH}{\underset{|}{C}}HCH_3$$

the reaction rate is first order; it depends only on the concentration of 1-phenylethyl chloride and is essentially independent of the concentration of water.

$$\text{Rate} = k\,[C_6H_5\underset{\displaystyle Cl}{\underset{|}{C}}HCH_3]$$

FIG. L.1
Potential energy diagrams for the one-step mechanism (the Ingold mechanism) for an S_N2 reaction (a), *and for the intimate ion-pair mechanism* (b).

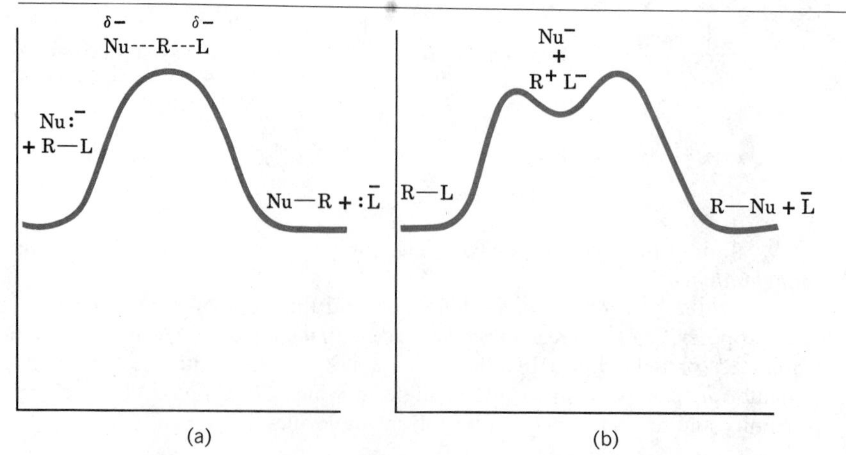

The stereochemistry of the reaction is shown below.

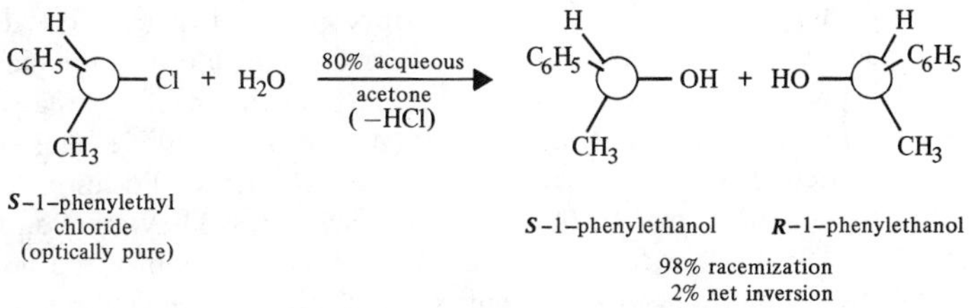

$$\text{S-1-phenylethyl chloride (optically pure)} + H_2O \xrightarrow[\text{acetone }(-HCl)]{\text{80\% acqueous}} \text{S-1-phenylethanol} + \text{R-1-phenylethanol}$$

98% racemization
2% net inversion

We see from the equation (above) that the reaction of 1-phenylethyl chloride of the *S* configuration gives a product, 1-phenylethanol, of which 51% has the opposite (*R*) configuration and 49% has the same (*S*) configuration. In other words, 51% of the 1-phenylethyl chloride molecules have had their configuration inverted by the reaction, while the remainder (49%) have retained their original configuration. We describe this situation by saying that the reaction has taken place with 98% *racemization* and 2% *net* inversion.

We can account for the fact that this reaction is first order if we assume that the rate-limiting step (or slow step) for the reaction involves the organic halide alone. A general mechanism is the following.

Step 1 $R—X \xrightarrow{\text{Slow}} R^+ + X^-$

Step 2 $R^+ + H_2O \xrightarrow{\text{Fast}} R\overset{+}{O}H_2$

Step 3 $R\overset{+}{O}H_2 \xrightarrow[-H^+]{\text{Fast}} ROH$

Step 1, the formation of a carbocation, is the slow step. Step 2 is a rapid reaction of the carbocation with water and step 3 is the rapid loss of a proton.

Since step 1 involves the organic halide alone (we are, for the moment, neglecting the involvement of solvent molecules) the overall rate of the reaction must correspond to the rate of this step,

$$\text{Rate} = k[RX]$$

and the reaction as a whole must show first-order kinetics.

A more detailed mechanism (shown below) illustrates one way in which we can account for the overall stereochemistry of the hydrolysis of 1-phenylethyl chloride.

$C_6H_5(H)(CH_3)C—Cl \underset{}{\overset{\text{slow}}{\rightleftharpoons}} [H_2\ddot{O}: \; C_6H_5(H)(CH_3)C^+ \; :\ddot{Cl}:^-] \overset{\text{fast}}{\rightleftharpoons} [(H_2\ddot{O}:)n \; C_6H_5(H)(CH_3)C^+ \; (:\ddot{O}H_2)n] \; :\ddot{Cl}:^-$

intimate ion pair (chiral)

dissociated carbocation (symmetrically solvated sp^2 cation)

Intimate ion pair → (moderately fast) → $Cl^- + H_2\overset{+}{O}—C(H)(C_6H_5)(CH_3)$

inversion (2%)

Dissociated carbocation → (fast) → $+Cl^-$ $H_2\overset{+}{O}—C(H)(C_6H_5)(CH_3)$ + $C_6H_5(H)(CH_3)C—\overset{+}{O}H_2$ $+Cl^-$

racemization (98%)

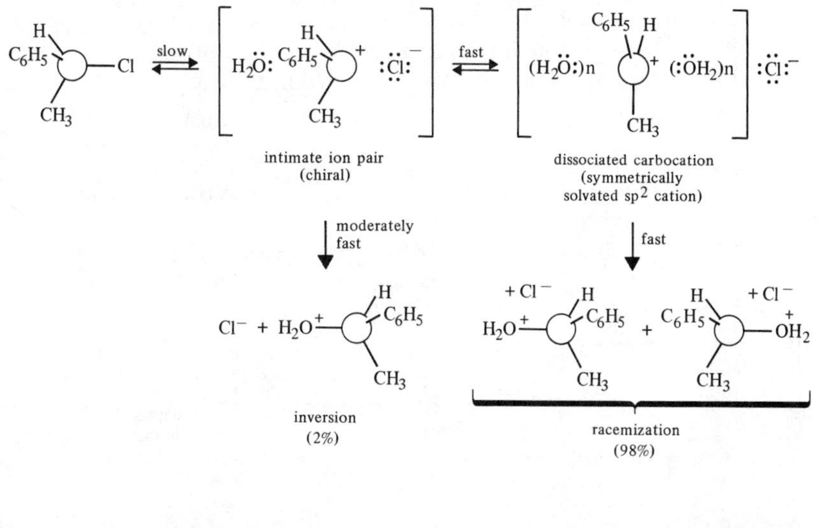

Here we see the important part played by solvent molecules and we also see the formation of two different cationic intermediates: an intimate ion pair and a more dissociated carbocation. The intimate ion pairs are formed first and their carbocations are solvated only on their back sides. A relatively small number of the intimate ion pairs collapse to give an inverted product. However, since the developing cation in this reaction is a relatively stable *benzyl* carbocation, most of the intimate ion pairs survive long enough to become more dissociated. The positive carbons of the dissociated carbocations are sp^2 hybridized and they are solvated on both the front and back sides. They react equally rapidly with water molecules at either face to give a racemic modification of the protonated alcohol.

Some evidence, however, suggests that a third type of cationic intermediate called a "solvent-separated" ion pair may intervene between the intimate ion pair and the dissociated carbocation in reactions of this type. A solvent-separated ion pair, as its name suggests, is one in which a molecule of solvent is situated between the carbocation and the anion. There is also evidence that solvent-separated ion pairs may preferentially react with

$$\left[\begin{array}{ccc} & \text{Sol} & \\ & | & \\ R^{+} & :\ddot{O}: & X^{-} \\ & | & \\ & \text{H} & \end{array}\right]$$

A solvent-separated
ion pair

the intervening solvent molecule by a mechanism that operates with *retention of configuration:*

$$\left[\begin{array}{ccc} & \text{Sol} & \\ & | & \\ R^{+} & :\ddot{O}: & X^{-} \\ & | & \\ & \text{H} & \end{array}\right] \xrightarrow[\text{of configuration of R}]{\text{Retention}} R\text{—}\overset{\text{Sol}}{\underset{\text{H}^{+}}{\ddot{O}:}} \quad + X^{-}$$

Therefore, it is possible that not all of the racemic product comes from dissociated carbocations and that more than 2% of the reaction may take place through intimate ion pairs. Part of the racemic product may result from a balanced collapse of intimate ion pairs (with inversion) and of solvent-separated ion pairs (with retention).

L.3 SUMMARY OF S_N REACTION MECHANISMS AT A SATURATED CARBON

Nucleophilic substitution reactions may very well take place through a spectrum of mechanisms ranging from a one-step displacement mechanism at one end and a mechanism involving fully dissociated carbocations at the other. Intervening between these limits may be mechanisms involving at least two kinds of ion pairs, intimate ion pairs and solvent-separated ion pairs. We can represent this spectrum as shown in Fig. L.2.

FIG. L.2
A spectrum of mechanisms for nucleophilic substitution reactions.

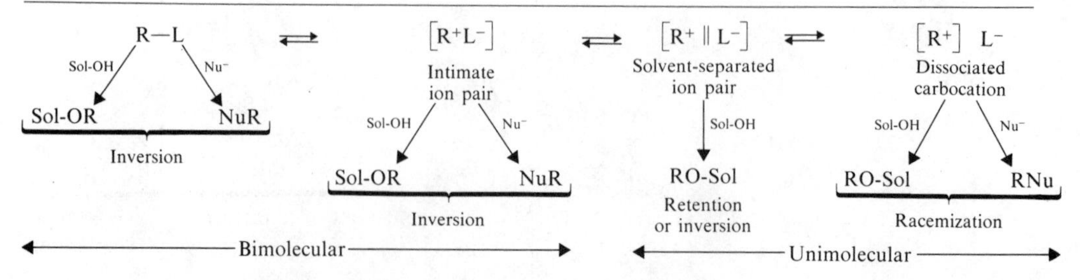

Whether a reaction gives a first-order or second-order rate equation will depend on just which of these mechanisms operates. We will obtain a second-order rate equation for those reactions that take place by a one-step displacement mechanism (the Ingold mechanism) or through an intimate ion pair since, in these instances, the rate-limiting step is bimolecular. We describe these reactions as being bimolecular or S_N2 reactions.

The mechanism involving an intimate ion pair is a true bimolecular reaction since the transition state of the rate-limiting step,

$$R^+ L^- + Nu\!:^- \longrightarrow NuR + L^-$$

or

$$R^+ L^- + \text{Sol-OH} \longrightarrow \text{Sol}\!-\!\underset{\displaystyle H}{\underset{|}{O^{+}}}\!-\!R + L^-$$

involves two species: the intimate ion pair and the nucleophile or solvent.

We will obtain a first-order rate equation for those reactions that involve dissociated carbocations or solvent-separated ion pairs, for in these reactions the rate-limiting step is unimolecular. We describe these reactions as being unimolecular or S_N1.

The only exception to these generalizations is a solvolysis. A solvolysis can be *bimolecular* because the transition state of its rate-limiting step can involve two species: the substrate and the solvent. Such a solvolysis, however, will show *pseudo first-order kinetics* because the solvent concentration is very large and is, consequently, essentially constant.

The stereochemical possibilities for nucleophilic substitution reactions are summarized in Table L.1.

L.4 APROTIC SOLVENTS, PHASE TRANSFER CATALYSIS, AND CROWN ETHERS

L.4A Polar Aprotic Solvents

In a solvent, such as *dimethylformamide,*

$$H\!-\!\overset{\displaystyle O}{\overset{\|}{C}}\!-\!\ddot{N}(CH_3)_2$$

Dimethylformamide
(DMF)

the relative order of reactivity of halide ions is:

$$Cl^- > Br^- > I^-$$

This is the opposite of their strength as nucleophiles in alcohol or water solutions:

$$I^- > Br^- > Cl^-$$

This effect seems to be related to the ability of the particular solvent to *solvate anions.* Alcohols and water are called *protic* solvents because their molecules have a hydrogen that is attached to a strongly electronegative atom (oxygen). Alcohols and water

TABLE L.1 The Stereochemistry of Nucleophilic Substitution Reactions

MOLECULARITY	SUBSTRATE	REPRESENTATION	STEREOCHEMISTRY
S_N2	(R—L) Alkyl halide, tosylate, etc.	R—L	Nucleophilic attack by the solvent or the nucleophile from the back side gives an inverted product by a one-step displacement mechanism.
S_N2	(R—L) Intimate ion pair	$[R^+L^-]$	Nucleophilic attack by the solvent or the nucleophile from the back side gives an inverted product.
S_N1	(R^+) S—O—H (L^-) Solvent-separated ion pair	$[R^+ \| L^-]$	Nucleophilic attack by the solvent from the front side occurs with retention of configuration. (Attack by another nucleophile may occur with inversion.)
S_N1	$(S—O—H)_n$ (R^+) $(S—O—H)_n$ (L^-) Dissociated carbocation	$[R^+]$ L^-	Nucleophilic attack by the solvent or the nucleophile from the front or back side gives a racemic product.

are very effective in solvating anions because these hydrogens can form strong bonds to the unshared electron pairs of anions. Smaller anions seem to be more effectively solvated (and thus, stabilized) by protic solvents than larger ones. This may be because with smaller anions the negative charge is more concentrated. Thus, in alcohol or water the smaller Cl^- ion has a more effective barrier of stabilizing solvent molecules surrounding it than the larger I^- ion. Therefore, in protic solvents the Cl^- ion is less reactive and is the weaker nucleophile.

Aprotic solvents are those solvents whose molecules do not have a hydrogen that is attached to a strongly electronegative element. Most aprotic solvents (benzene, the alkanes, and so on) are relatively nonpolar and they do not dissolve ionic compounds. In recent years, however, a number of *highly polar aprotic solvents* have been discovered. Dimethylformamide is one example of a highly polar aprotic solvent; dimethyl sulfoxide and dimethylacetamide are two others.

$$CH_3—\overset{O}{\overset{\|}{S}}—CH_3 \qquad CH_3\overset{O}{\overset{\|}{C}}N(CH_3)_2$$

Dimethyl sulfoxide (DMSO) — Dimethylacetamide (DMA)

All three solvents (DMF, DMSO, and DMA) can dissolve ionic compounds, and they solvate cations very well. However, they do not solvate anions to any appreciable extent. In these solvents anions are unencumbered by a layer of solvent molecules and, therefore, they are poorly stabilized by solvation. These "naked" anions are highly reactive both *as bases and as nucleophiles*. The unsolvated Cl^- ion is more basic than the unsolvated I^- ion. Therefore, in the aprotic solvent, dimethylformamide, chloride ions are stronger nucleophiles.

S_N2 reactions, in general, are strongly favored by the use of polar aprotic solvents.

L.4B Phase Transfer Catalysis

Until recently, nonpolar aprotic solvents such as benzene were seldom used for nucleophilic substitution reactions because of their inability to dissolve ionic compounds. This situation has changed with the development of a procedure called *phase transfer catalysis*.

This procedure uses two immiscible phases that are in contact—an aqueous phase containing an ionic reactant and an organic phase (benzene, $CHCl_3$, etc.) containing the organic substrate. Normally the reaction of two substances in separate phases like this is inhibited because of the inability of the reagents to come together. Adding a phase transfer catalyst solves this problem by transferring the ionic reactant into the organic phase. And again, because the reaction takes place in an aprotic medium, S_N2 reactions occur rapidly.

An example of phase transfer catalysis is outlined in Fig. L.3. The phase transfer catalyst, $Q^+ X^-$, is usually a quaternary ammonium halide such as tetrabutylammonium halide, $(CH_3CH_2CH_2CH_2)_4N^+ X^-$. The phase transfer catalyst causes the transfer of the nucleophile as an ion pair $[Q^+ Nu^-]$ into the organic phase. This transfer apparently takes place because the cation (Q^+) of the ion pair, with its four alkyl groups, resembles a hydrocarbon in spite of its positive charge. In the organic phase the nucleophile of the ion pair (Nu^-) reacts with the organic substrate RX. The cation (Q^+) then migrates back into the aqueous phase as $[Q^+ X^-]$, completing the cycle. This process continues until all of the nucleophile or the organic substrate has reacted.

Using phase transfer catalysis, for example, *n*-octyl methanesulfonate in a hydrocarbon solvent reacts with NaCN in just 20 minutes.

$$NaCN_{(H_2O)} + CH_3(CH_2)_6CH_2OSO_2CH_{3(RH)} \xrightarrow[\text{20 min.}]{\text{1\% } Q^+X^-} \underset{95\%}{CH_3(CH_2)_6CH_2CN_{(RH)}} + NaOSO_2CH_{3(H_2O)}$$

L.4C Crown Ethers*

Compounds called *crown* ethers are also able to transport ionic compounds into an organic phase. Crown ethers are cyclic polymers of ethylene glycol such as the 18-crown-6 that follows.

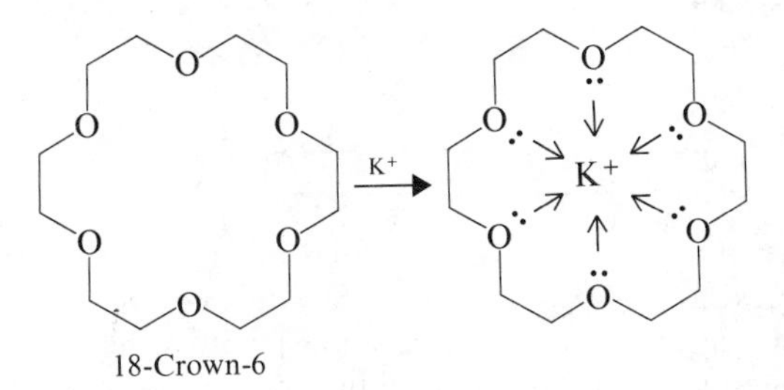

* Crown ethers are named as *x*-crown-*y* where *x* is the total number of atoms in the ring and *y* is the number of oxygens.

FIG. L.3
A mechanism for phase transfer catalysis.

Aqueous Phase: $Q^+ X^- + Nu^- \rightleftharpoons [Q^+ Nu^-] + X^-$

Organic phase: $[Q^+ X^-] + R\text{—}Nu \longleftarrow [Q^+ Nu^-] + R\text{—}X$

Crown ethers act by coordinating with a metal cation converting it into a species with a hydrocarbonlike exterior. The crown ether 18-crown-6, for example, coordinates very effectively with potassium ions because the six oxygens are ideally situated to donate their electron pairs to the central ion.

Crown ethers render many organic salts soluble in nonpolar solvents. Salts such as KF, KCN, and CH_3CO_2K, for example, can be transferred into aprotic solvents by using catalytic amounts of 18-crown-6. In the organic phase the anions of these salts can carry out a nucleophilic substitution reaction on an organic substate.

$$K^+CN^- + RCH_2X \xrightarrow[\text{benzene}]{\text{18-Crown-6}} RCH_2CN + K^+X^-$$

L.5 NEIGHBORING GROUP PARTICIPATION IN NUCLEOPHILIC SUBSTITUTION REACTIONS

Not all nucleophilic substitutions (Sections L.2 and L.3) take place with racemization or with inversion of configuration. Some take place with overall *retention of configuration.*

One factor that can lead to retention of configuration in a nucleophilic substitution is a phenomenon known as *neighboring group participation.* Let us see how this operates by examining the stereochemistry of two reactions in which 2-bromopropanoic acid is converted to lactic acid.

$$\underset{\text{2-Bromopropanoic acid}}{CH_3\underset{\underset{Br}{|}}{C}HCOOH} \longrightarrow \underset{\text{Lactic acid}}{CH_3\underset{\underset{OH}{|}}{C}HCOOH}$$

When *S*-2-bromopropanoic acid is treated with concentrated sodium hydroxide the reaction is *bimolecular* and it takes place with *inversion of configuration.* This, of course, is the normal stereochemical result for an S_N2 reaction.

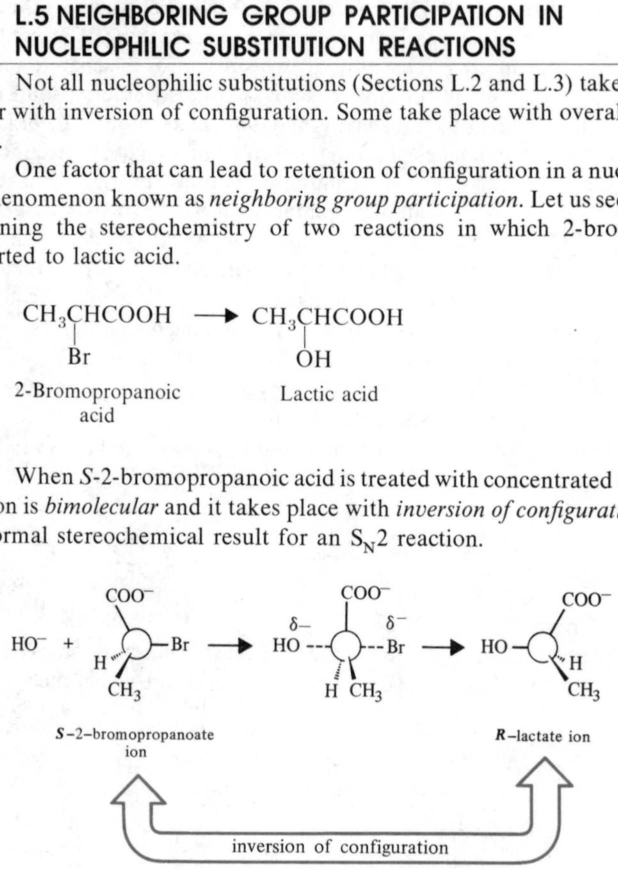

HO⁻ + [COO⁻, H, CH₃ carbon]—Br ⟶ [δ− HO---C(COO⁻)(H)(CH₃)---Br δ−] ⟶ HO—C(COO⁻)(H)(CH₃)

S-2-bromopropanoate ion — *R*-lactate ion

inversion of configuration

However, when the same reaction is carried out with a low concentration of hydroxide ion in the presence of Ag_2O, it takes place with an overall *retention of configuration.* In this case, the mechanism for the reaction involves the participation of the carboxylate group. In step 1 (below) an oxygen of the carboxylate group attacks the chiral carbon from the back side and displaces bromide ion. (Silver ion aids in this process in much the same way that protonation assists the ionization of an alcohol.) The configuration of the chiral carbon inverts in step 1, and a cyclic ester called an α-lactone forms.

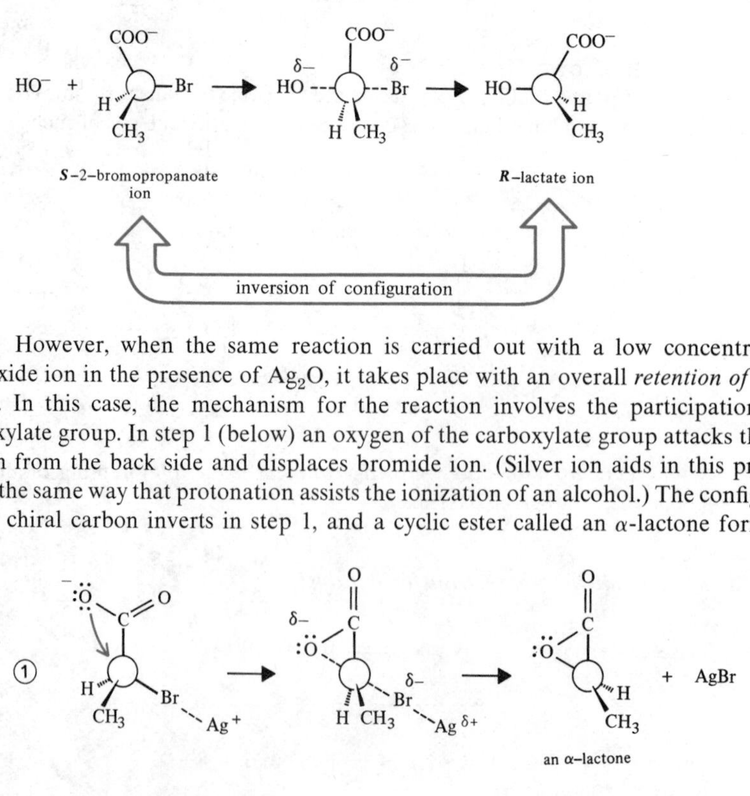

① [⁻:Ö:—C(=O) attached to chiral C with H, CH₃, Br ··· Ag⁺] ⟶ [δ− :Ö: bridging C=O and chiral C, H, CH₃, Br δ− ··· Ag δ+] ⟶ [:Ö: bridging C=O and chiral C, H, CH₃] + AgBr

an α-lactone

The highly strained three-membered ring of the α-lactone opens when it is attacked by a water molecule in step 2. *This step also takes place with an inversion of configuration.*

② α-lactone (H, CH_3) + $:\ddot{O}H_2$ ⟶ [$\delta-$ O ··· C ··· $\delta+$ OH_2; H CH_3] ⟶ $^{-}:\ddot{O}—C(=O)$ (H, OH, CH_3) + H^+

The net result of two inversions (in steps 1 and 2) is an overall *retention of configuration.*

Problem L.1

The phenomenon of configuration inversion in a chemical reaction was discovered in 1896 by Paul von Walden. (Configuration inversions are still called Walden inversions in his honor.) Walden's proof of configuration inversion was based on the following cycle.

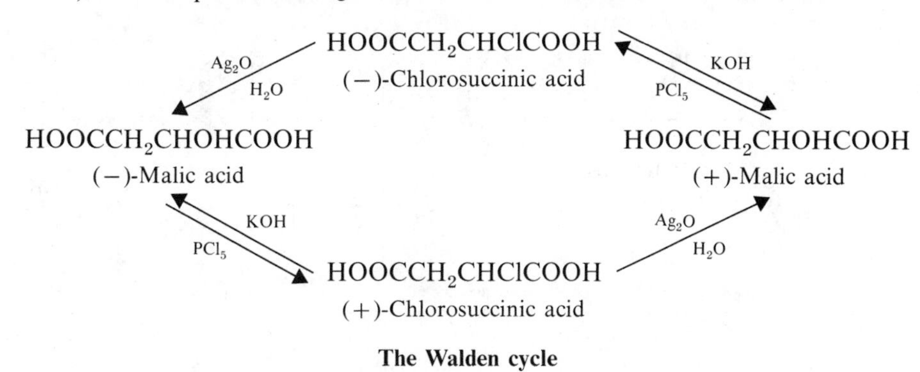

The Walden cycle

(a) Basing your answer on the preceding discussion, which reactions of the Walden cycle are likely to take place with overall inversion of configuration and which are likely to occur with overall retention of configuration? (b) Malic acid with a negative optical rotation is now known to have the *S* configuration. What are the configurations of the other compounds in the Walden cycle? (c) Walden also found that when (+)-malic acid is treated with thionyl chloride (rather than PCl_5) the product of the reaction is (+)-chlorosuccinic acid. How can you explain this result? (d) Assuming that the reaction of (−)-malic acid and thionyl chloride has the same stereochemistry, outline a Walden cycle based on the use of thionyl chloride instead of PCl_5.

Neighboring group participation can also lead to *cyclization reactions.* Epoxides, for example, can be prepared from 2-bromo alcohols by treating them with sodium hydroxide (see also Problem 15.20). This reaction involves the following steps.

$$\text{R—CH(OH)—CHR'(:\ddot{Br}:)} \xrightleftharpoons{OH^-} \text{R—CH(:\ddot{O}:^-)—CH(:\ddot{Br}:)—R'} \longrightarrow \text{RCH—CHR' (epoxide, } \ddot{O}\text{)} + :\ddot{Br}:^-$$

Problem L.2

(a) How would you synthesize a 2-halo alcohol (a halohydrin) from an alkene? (b) Show how you could use this method to synthesize propylene oxide from propylene.

When neighboring group participation occurs during the rate-determining step for a reaction, the rate is often markedly increased. This effect, called *anchimeric assistance* (Gr. *anchi* + *meros*, meaning neighboring parts) can be seen in the relative rates of S_N1 solvolysis reactions of isobutyl chloride and 2-phenyl-1-chloropropane. When 2-phenyl-1-chloropropane undergoes S_N1 solvolysis it reacts more rapidly than isobutyl chloride. The phenyl group is thought to assist in the ionization step by stabilizing the transition state

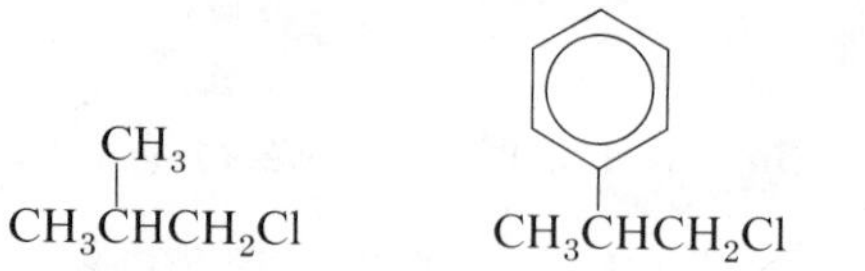

$CH_3CH(CH_3)CH_2Cl$
Isobutyl chloride

$CH_3CH(C_6H_5)CH_2Cl$
2-Phenyl-1-chlorobutane

leading to the phenonium ion intermediate below. The methyl group of isobutyl chloride is apparently unable to provide a similar kind of assistance when it undergoes solvolysis.

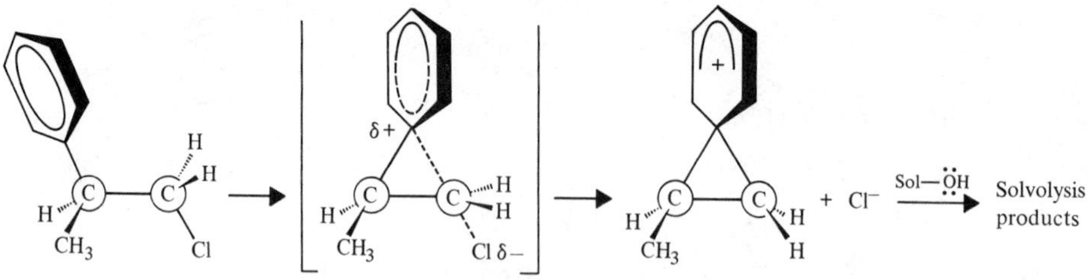

Transition state

A phenonium ion

+ Cl^- $\xrightarrow{Sol—\ddot{O}H}$ Solvolysis products

Problem L.3

The phenonium ion formed as an intermediate in this reaction (above) strongly resembles the arenium ions we saw in electrophilic substitution reactions. What relative order of reactivity would you expect the following compounds to show: 2-phenyl-1-chloropropane, 2-(*p*-nitrophenyl)-1-chloropropane; 2-(*p*-hydroxyphenyl)-1-chloropropane; 2-(*p*-tolyl)-1-chloropropane?

Neighboring group participation and anchimeric assistance are important in many reactions that are catalyzed by enzymes.

Problem L.4

In 1949, D. J. Cram published the first of a series of papers on the solvolysis of 1-methyl-2-phenylpropyl tosylates, **A** and **C.** These reactions displayed a remarkable stereospecificity: when the optically active tosylate **A** was heated in acetic acid, the reaction yielded almost exclusively the optically active acetate **B.** On the other hand, heating the optically active tosylate **C** in acetic acid gave the racemic acetate **D, E.** Provide an explanation for these results.

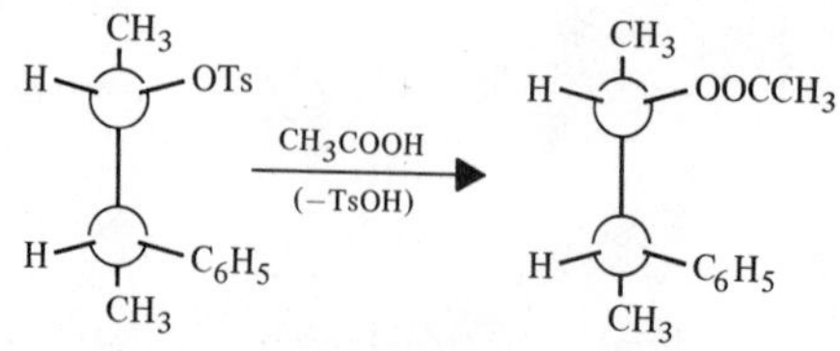

A $\xrightarrow[(-TsOH)]{CH_3COOH}$ **B**

Optically active

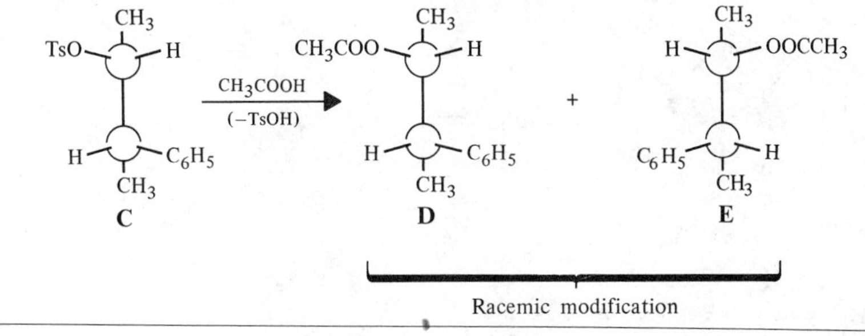

TsO, CH_3, H, H, C_6H_5, CH_3 **C** $\xrightarrow[(-TsOH)]{CH_3COOH}$ CH_3COO, CH_3, H, H, C_6H_5, CH_3 **D** + CH_3, H, $OOCCH_3$, C_6H_5, H, CH_3 **E**

Racemic modification

L.6 STEREOCHEMISTRY OF ELIMINATION REACTIONS

In our study of elimination reactions thus far we have not had much to say about their stereochemistry. We will begin to do this now with the familiar E2 reaction. After we have studied its stereochemistry we will move on to examine two new elimination reactions—acetate pyrolysis and the Cope elimination—where the base and the leaving group are part of the same molecule.

L.6A Stereochemistry of E2 Reactions

E2 reactions are usually stereospecific. With most open-chain compounds, E2 reactions take place from a transition state in which the proton being eliminated and the leaving group are *anti* to each other.

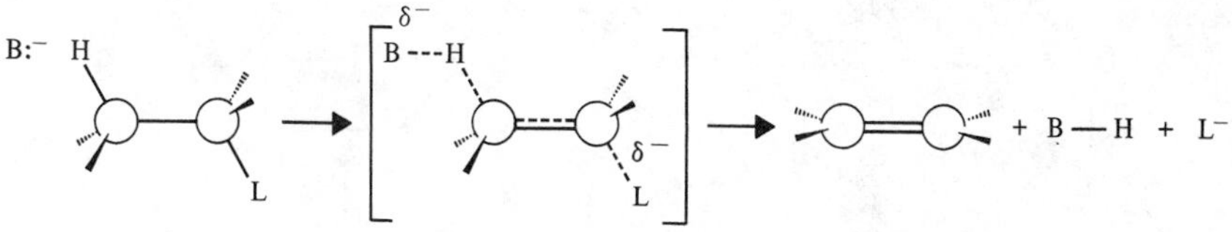

$B:^-$ H ... L → [δ^- B---H ... δ^- L] → + B—H + L^-

Anti relation between proton and leaving group

One explanation that has been advanced to account for a preferred anti orientation is that it allows the large lobe from the C—H bond to interact with the small lobe of the C—L bond (below). In this sense then, the E2 reaction resembles the S_N2 reaction in that an electron pair of the C—H bond displaces the leaving group from the back side.

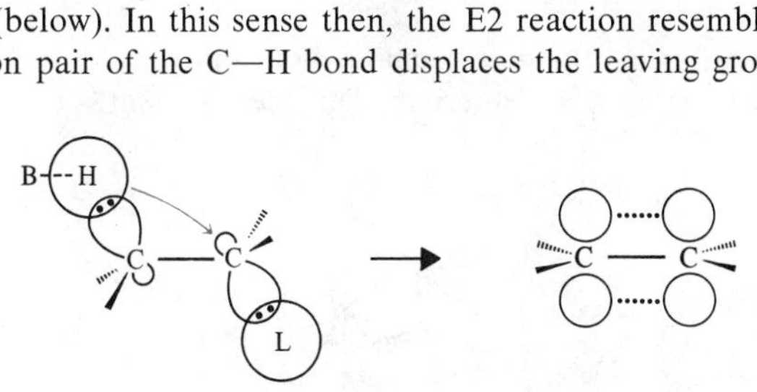

B--H, C, C, L → C — C

Problem L.5

When the deuterium-labeled isomer below (*erythro*-2-bromobutane-*3-d*) undergoes elimination, the reaction yields *trans*-2-butene and *cis*-2-butene-*2-d* (as well as some 1-butene-*3-d*). The reaction does not yield *cis*-2-butene or *trans*-2-butene-*2-d*.

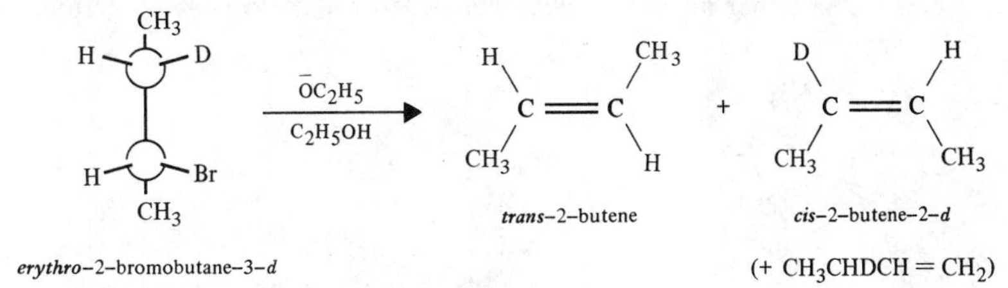

CH_3, H, D, H, Br, CH_3 $\xrightarrow[C_2H_5OH]{^-OC_2H_5}$ H, CH_3, C=C, CH_3, H + D, H, C=C, CH_3, CH_3

erythro-2-bromobutane-3-*d* — *trans*-2-butene — *cis*-2-butene-2-*d*

(+ $CH_3CHDCH=CH_2$)

but no

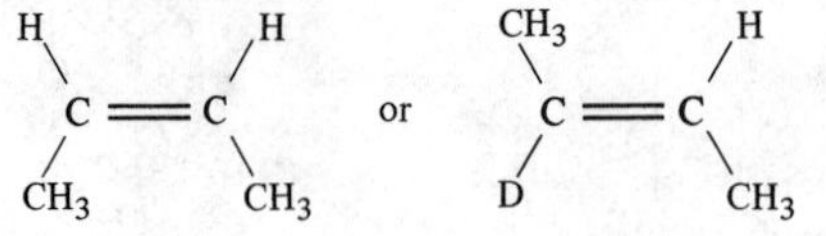

H, H / C=C / CH3, CH3 — *cis*-2-butene — or — CH3, H / C=C / D, CH3 — *trans*-2-butene-2-*d*

How can you explain these results?

With cyclic compounds the stereochemistry of elimination is more complex. *cis*-2-Phenylcyclohexyl tosylate, **1**, undergoes elimination to yield 1-phenylcyclohexene at a rate that has been estimated as being 10^4 times faster than that at which *trans*-2-phenylcyclohexyl tosylate, **2**, yields 1-phenylcyclohexene (**2** yields predominantly 3-phenylcyclohexene). This evidence indicates a preference for an *anti* relation between the proton at C-2 and the leaving group, because **1** can assume an anti conformation easily and **2** cannot.

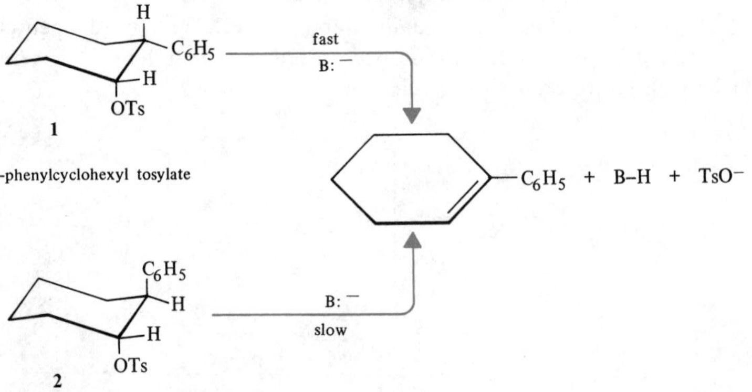

1 *cis*-2-phenylcyclohexyl tosylate $\xrightarrow[\text{B:}^-]{\text{fast}}$ 1-phenylcyclohexene ($-C_6H_5$) + B–H + TsO^-

2 *trans*-2-phenylcyclohexyl tosylate $\xrightarrow[\text{slow}]{\text{B:}^-}$ 1-phenylcyclohexene

On the other hand when the deuterium-labeled compound **7** undergoes elimination, 94% of the product contains no deuterium. This cyclic system is rigid and thus a *syn* elimination must take place.

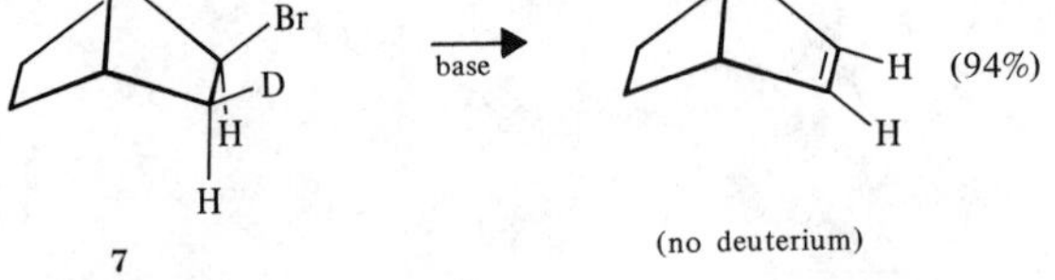

7 (Br, D, H, H) $\xrightarrow{\text{base}}$ H (94%), H (no deuterium)

Results such as this have led to the suggestion that elimination reactions *only require a transition state in which the proton and the leaving group are coplanar*. This can be achieved in either an *anti coplanar* relation or a *syn coplanar* relation.

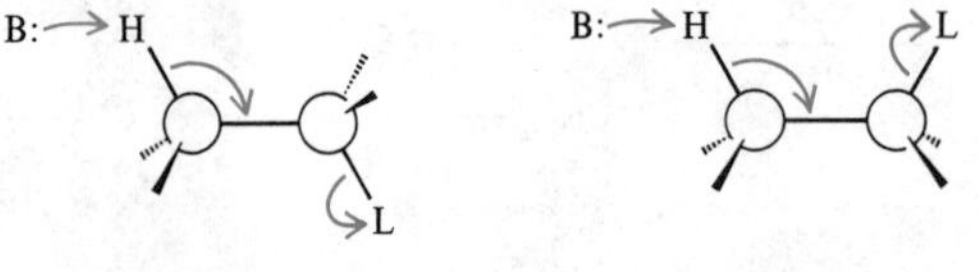

B:→H ... L

anti coplanar transition state

B:→H ... L

syn coplanar transition state

Therefore, open-chain compounds apparently proceed through an anti coplanar transition state because this avoids a high-energy eclipsed conformation. Cyclic molecules such as **7**, however, undergo elimination from a syn coplanar transition state because the molecular structure is such that it requires an eclipsed conformation.

L.6B Intramolecular Eliminations

There are several elimination reactions in which the leaving group and the base are a part of the same molecule. These *intramolecular* eliminations are *thermal eliminations;* they require only the application of heat. We will discuss two important examples: pyrolysis of acetates and the Cope elimination.

L.6C Pyrolysis of Acetates

When acetic esters of alcohols with a β-hydrogen are heated to a temperature of 200–500° acetic acid is eliminated and an alkene is formed.

$$\underset{\text{Pentyl acetate}}{CH_3CH_2CH_2CH_2CH_2O\overset{O}{\overset{\|}{C}}CH_3} \xrightarrow{470°} \underset{\text{1-Pentene}}{CH_3CH_2CH_2CH{=}CH_2} + \underset{\text{Acetic acid}}{HO\overset{O}{\overset{\|}{C}}CH_3}$$

The reaction appears to take place through a cyclic transition state in which the carbonyl oxygen of the acetate group acts as a base. Because the transition state is a six-membered ring, acetate pyrolyses are, of necessity, syn eliminations.

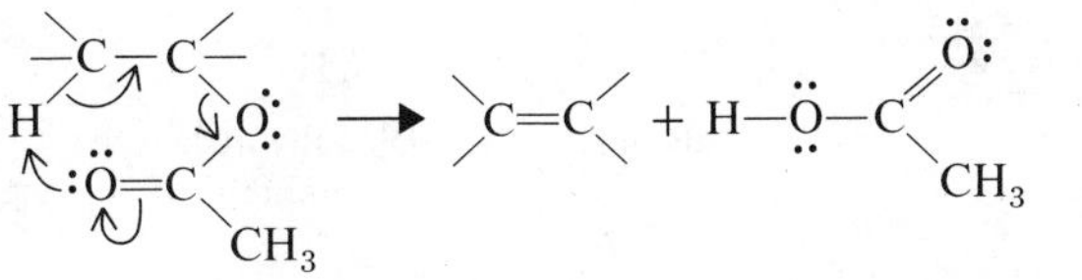

$$\text{(cyclic transition state: H–C–C–O–C(CH}_3\text{)=O:)} \longrightarrow \underset{\text{Alkene}}{>C{=}C<} + \underset{\text{Acetic acid}}{H{-}\ddot{O}{-}C(=\ddot{O}:)CH_3}$$

A cyclic anti elimination does not occur because the transition state would be highly strained and, thus, of very high energy. (A cyclic anti elimination would require a transition state that resembles a six-membered ring with a *trans* double bond.)

Problem L.6

When the diastereomeric acetates A and B are heated, they both yield *trans*-1,2-diphenylethene. The major product from A contains deuterium; however, that from B does not. Write transition states that explain these results.

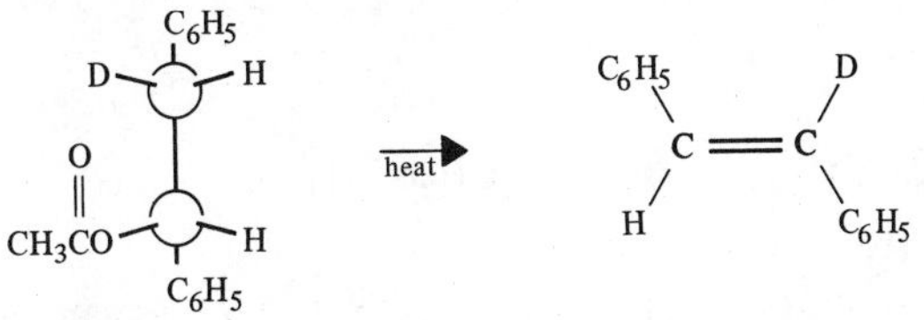

A: (Newman-type projection: front carbon with C_6H_5, D, H; rear carbon with $CH_3\overset{O}{\overset{\|}{C}}O$, H, C_6H_5) $\xrightarrow{\text{heat}}$ $(C_6H_5)(H)C{=}C(D)(C_6H_5)$

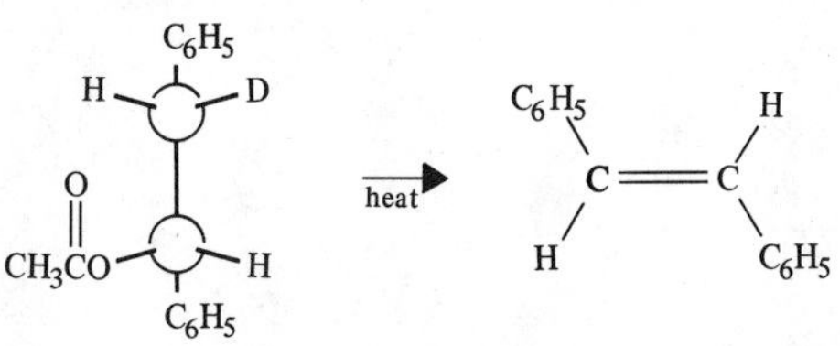

B: (front carbon with C_6H_5, H, D; rear carbon with $CH_3\overset{O}{\overset{\|}{C}}O$, H, C_6H_5) $\xrightarrow{\text{heat}}$ $(C_6H_5)(H)C{=}C(H)(C_6H_5)$

Problem L.7

Explain the different results obtained from the two reactions shown below.

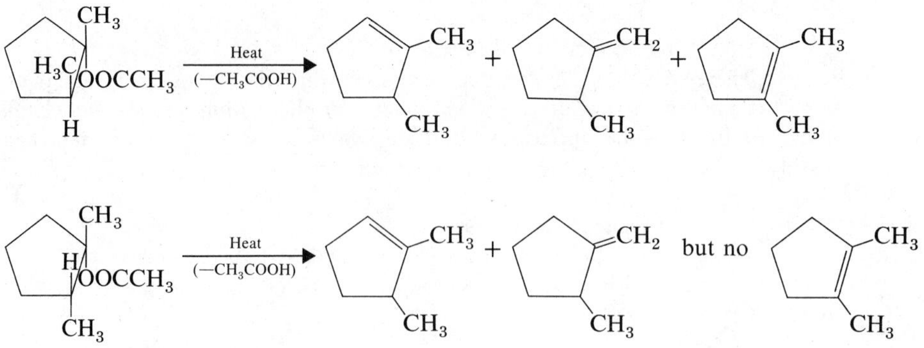

$$\text{(cis-1,2-dimethylcyclopentyl acetate: } H_3C, OOCCH_3, CH_3, H) \xrightarrow[(-CH_3COOH)]{\text{Heat}} \text{2,3-dimethylcyclopentene} + \text{1-methylene-2-methylcyclopentane} + \text{1,2-dimethylcyclopentene}$$

$$\text{(1,2-dimethylcyclopentyl acetate: } CH_3, H, OOCCH_3, CH_3) \xrightarrow[(-CH_3COOH)]{\text{Heat}} \text{2,3-dimethylcyclopentene} + \text{1-methylene-2-methylcyclopentane} \text{ but no } \text{1,2-dimethylcyclopentene}$$

L.6D The Cope Elimination

Tertiary amine *N*-oxides undergo the elimination of a dialkylhydroxylamine when they are heated. The temperatures required, however, are considerably lower than those required for acetate pyrolysis. This reaction is called the Cope elimination.

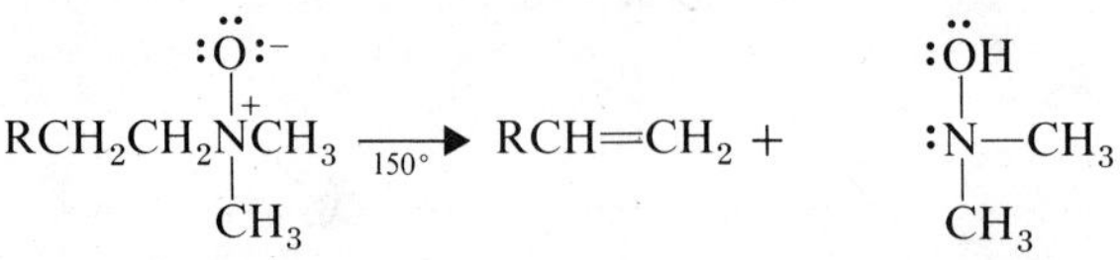

$$RCH_2CH_2\overset{+}{N}(O^-)(CH_3)CH_3 \xrightarrow{150°} RCH{=}CH_2 + (CH_3)_2N{-}OH$$

A tertiary amine *N*-oxide — An alkene — *N*,*N*-Dimethylhydroxylamine

The Cope elimination is also a syn elimination and proceeds through a cyclic transition state:

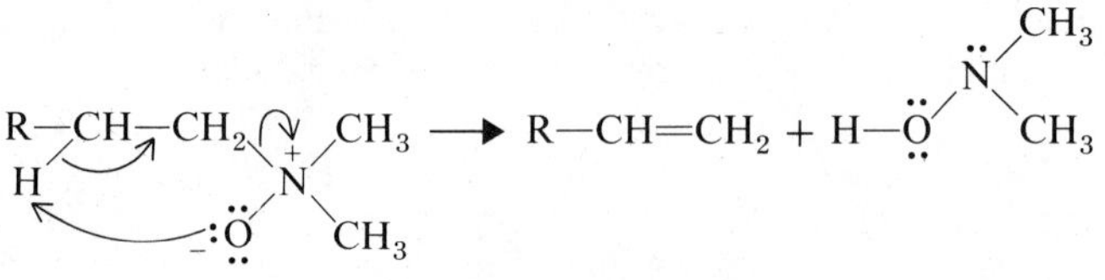

$$R{-}CH(H){-}CH_2{-}\overset{+}{N}(CH_3)_2{-}\ddot{O}:^- \longrightarrow R{-}CH{=}CH_2 + H{-}\ddot{O}{-}\ddot{N}(CH_3)_2$$

Tertiary amine *N*-oxides are easily prepared by treating tertiary amines with hydrogen peroxide:

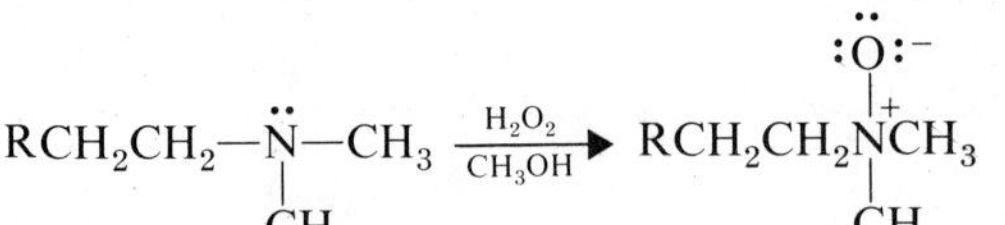

$$RCH_2CH_2{-}\ddot{N}(CH_3){-}CH_3 \xrightarrow[CH_3OH]{H_2O_2} RCH_2CH_2\overset{+}{N}(O^-)(CH_3)CH_3$$

SPECIAL TOPIC

REACTIONS OF HETEROCYCLIC AMINES

Heterocyclic amines undergo many reactions that are similar to those of the amines that we have studied in earlier chapters.

M.1 HETEROCYCLIC AMINES AS BASES

Nonaromatic heterocyclic amines have basicity constants that are approximately the same as those of acyclic amines.

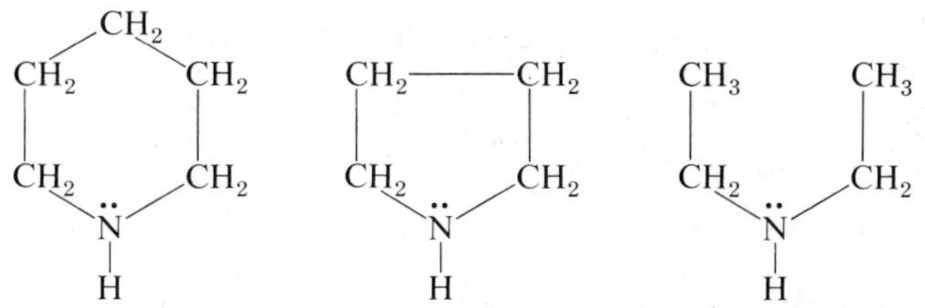

Piperidine	Pyrrolidine	Diethylamine
$K_b = 1.6 \times 10^{-3}$	$K_b = 1.3 \times 10^{-3}$	$K_b = 9.6 \times 10^{-4}$

In aqueous solution, aromatic heterocyclic amines such as pyridine, pyrimidine, and pyrrole are much weaker bases than nonaromatic amines or ammonia ($K_b = 1.8 \times 10^{-3}$). (In the gas phase, however, pyridine and pyrrole are more basic than ammonia, indicating that solvation has a very important effect on their relative basicities, cf. Section 9.4.)

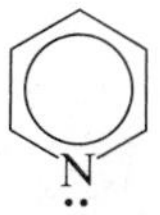

Pyridine	Pyrimidine	Pyrrole
$K_b = 1.7 \times 10^{-9}$	$K_b = 5 \times 10^{-12}$	$K_b = 2.5 \times 10^{-14}$

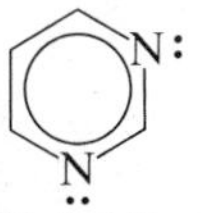

M.2 HETEROCYCLIC AMINES AS NUCLEOPHILES IN ALKYLATION ACYLATION REACTIONS

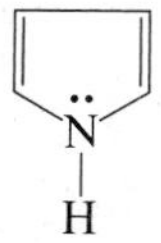

Most heterocyclic amines undergo alkylation and acylation reactions in much the same way as acyclic amines.

$$\text{Piperidine} + RCH_2\text{—Br} \xrightarrow{\text{Base}} N\text{-Alkylpiperidine} + N,N\text{-Dialkylpiperidinium bromide}$$

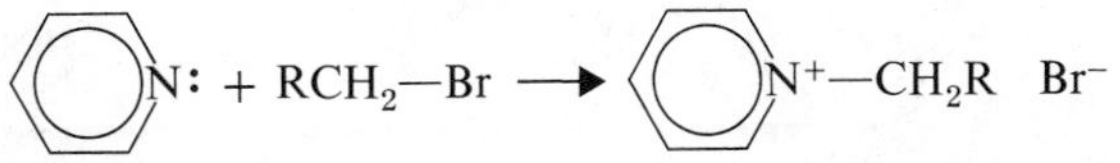

$$\text{Pyridine} + RCH_2\text{—Br} \longrightarrow N\text{-Alkylpyridinium bromide}$$

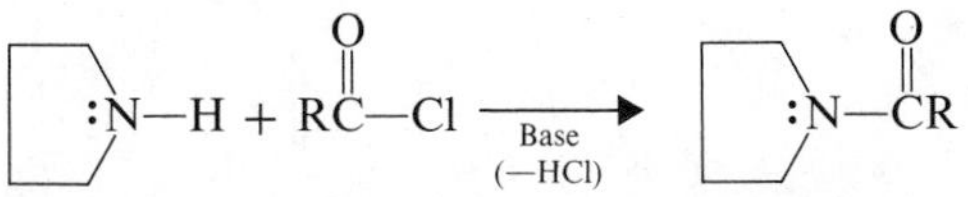

$$\text{Pyrrolidine} + RCOCl \xrightarrow[(\text{—HCl})]{\text{Base}} N\text{-Acylpyrrolidine (an amide)}$$

Problem M.1

What products would you expect to obtain from the following reactions?

(a) Piperidine + acetic anhydride ⟶

(b) Pyridine + methyl iodide ⟶

(c) Pyrrolidine + phthalic anhydride ⟶

(d) Pyrrolidine + (excess) methyl iodide $\xrightarrow[\text{(base)}]{}$

(e) Product of (d) + Ag_2O/H_2O, then heat ⟶

M.3 ELECTROPHILIC SUBSTITUTION REACTIONS OF AROMATIC HETEROCYCLIC AMINES

Pyrrole is highly reactive toward electrophilic substitution and substitution takes place primarily at position 2.

General Reaction

Pyrrole + E^+ (Electrophile) ⟶ 2-Substituted pyrrole + H^+

Specific Example

Pyrrole + $C_6H_5—\overset{+}{N}{\equiv}N\ X^-$ ⟶ 2-($N{=}N—C_6H_5$)pyrrole + HX

We can understand why electrophilic substitution at the 2 position is preferred if we examine the resonance structures that follow.

Substitution at the 2 position of pyrrole

(especially stable—every atom has an octet)

Positive charge is delocalized over three atoms

$-H^+$

Substitution at the 3 position of pyrrole

(especially stable—
every atom has an octet)

Positive charge is delocalized over only two atoms

$-H^+$

We see that while a relatively stable structure contributes to the hybrid for both intermediates, the intermediate arising from attack at the 2 position is stabilized by one additional resonance structure, and the positive charge is delocalized over three atoms rather than two. This means that this intermediate is more stable, and that attack at the 2 position has a lower activation energy.

Pyridine is much less reactive toward electrophilic substitution than benzene. Pyridine does not undergo Friedel-Crafts acylation or alkylation; it does not couple with diazonium compounds. Bromination of pyridine can be accomplished but only in the vapor phase at 200° where a free-radical mechanism may operate. Nitration and sulfonation also require forcing conditions. Electrophilic substitution, when it occurs, nearly always takes place at the 3 position.

Br_2, 200–220° → 3-Bromopyridine (37%) + 3,5-Dibromopyridine (26%)

KNO_3, H_2SO_4, 330° → 3-Nitropyridine (15%)

SO_3, H_2SO_4, $HgSO_4$, 220° → 3-Pyridinesulfonic acid

We can, in part, attribute the lower reactivity of pyridine (when compared to benzene) to the difference in electronegativity of nitrogen (when compared to carbon). Nitrogen, being more electronegative, is less able to accommodate the electron deficiency that characterizes the transition state leading to the positively charged ion (similar to an arenium ion) in electrophilic substitution.

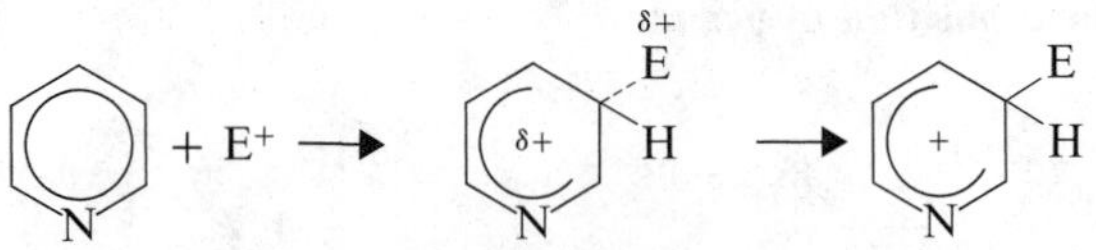

Pyridine + E^+ → [Transition state] → [Similar to an arenium ion]

Pyridine

Transition state is of higher energy because of greater electronegativity of nitrogen

Similar to an arenium ion

Benzene + E^+ → [Transition state] → [Arenium ion]

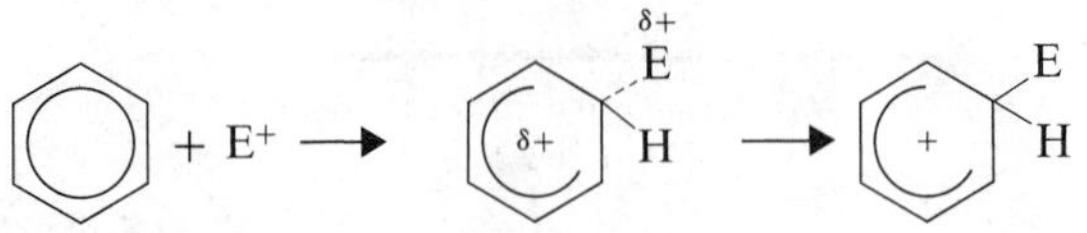

Benzene

Transition state is of lower energy because of lower electronegativity of carbon

Arenium ion

The low reactivity of pyridine toward electrophilic substitution may arise mainly from the fact that pyridine is converted initially to a pyridinium ion by an electrophile or proton.

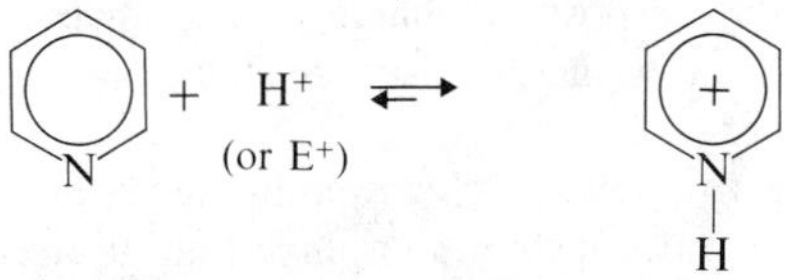

Pyridine + H^+ (or E^+) ⇌ Pyridinium ion

Pyridinium ion
(*highly unreactive because of positive charge*)

Electrophilic attack at the 4 position (or the 2 position) is unfavorable because an especially unstable resonance structure contributes to the intermediate hybrid.

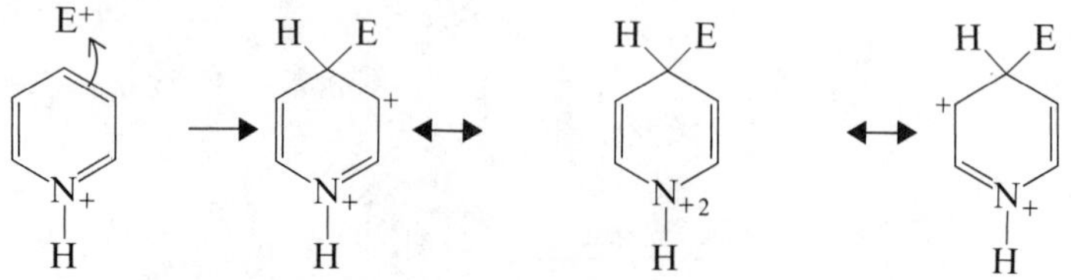

Especially unstable nitrogen has a sextet and two positive charges

Similar resonance structures can be written for attack at the 2 position.

No especially unstable *or stable* structure contributes to the hybrid arising from attack at the 3 position; as a result, attack at the 3 position is preferred but occurs slowly.

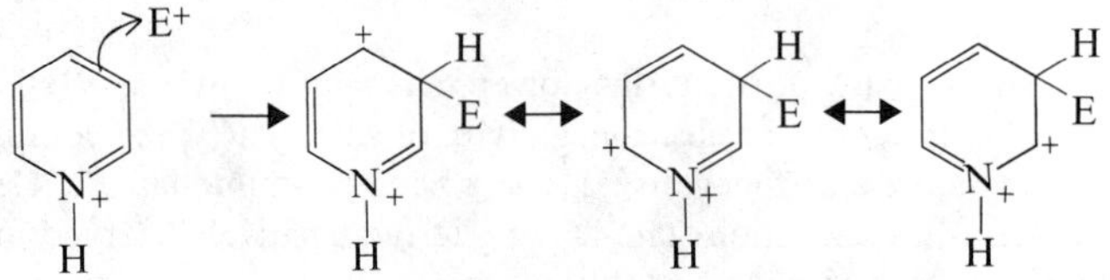

No especially unstable or stable structure contributes to the hybrid

Pyrimidine is even less reactive toward electrophilic substitution than pyridine. (Why?) When electrophilic substitution takes place, it occurs at the 5 position.

Electrophilic substitution takes place here

Pyrimidine

Imidazole is much more susceptible to electrophilic substitution than pyridine or pyrimidine, but is less reactive than pyrrole. Imidazoles with 1 substituents undergo electrophilic substitution at the 4 position.

$\xrightarrow{HNO_3}$

1-Methyl-4-nitroimidazole

Imidazole, itself, undergoes electrophilic substitution in a similar fashion. Tautomerism, however, makes the 4 and 5 positions equivalent.

$\xrightarrow{Br_2}$

4-(5)-Bromoimidazole

Problem M.2

Both pyrrole and imidazole are weak acids; they react with strong bases to form anions:

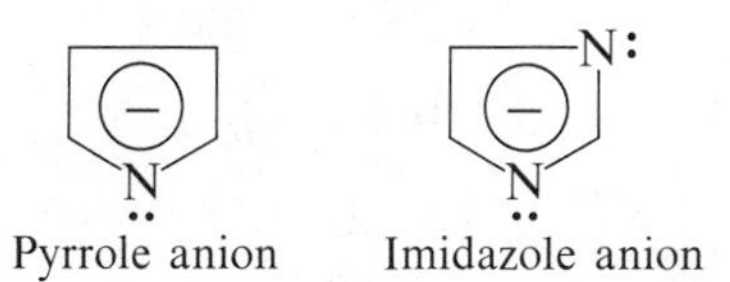

Pyrrole anion   Imidazole anion

(a) These anions resemble a carbocyclic anion that we have studied before. What is it?

(b) Write resonance structures that account for the stabilities of pyrrole and imidazole anions.

M.4 NUCLEOPHILIC SUBSTITUTIONS OF PYRIDINE

In its reactions, the pyridine ring resembles a benzene ring with a strong electron-withdrawing group; pyridine is unreactive toward electrophilic substitution but reactive toward nucleophilic substitution.

In the previous section we compared the reactivity of pyridine and benzene toward electrophilic substitution and there we attributed pyridine's lower reactivity to the greater electronegativity of its ring nitrogen. Because nitrogen is more electronegative than carbon, it is less able to accommodate the electron deficiency in the transition state of the rate-limiting step in electrophilic aromatic substitution. On the other hand, nitrogen's greater electronegativity makes it *more* able to accommodate the excess *negative* charge that an aromatic ring must accept in *nucleophilic substitution*.

Pyridine reacts with sodium amide, for example, to form 2-aminopyridine: In this remarkable reaction (called the Chichibabin reaction) amide ion, $:NH_2^-$, displaces a hydride ion, $H:^-$.

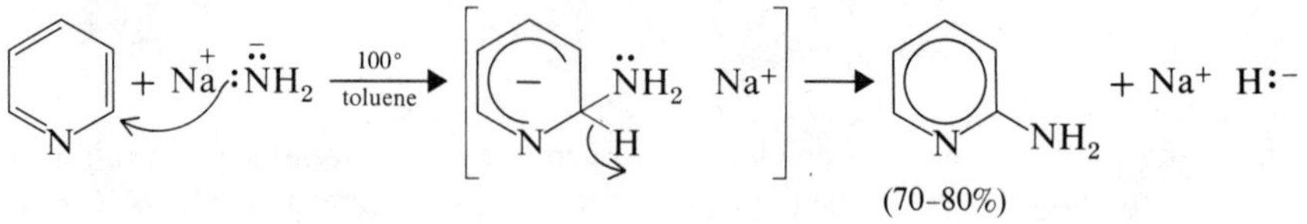

$+ Na^+ :\ddot{N}H_2^- \xrightarrow[\text{toluene}]{100°}$ [$\ddot{N}H_2$ Na^+] $\longrightarrow$ NH_2 $+ Na^+ \ H:^-$

(70–80%)

If we examine the resonance structures that contribute to the intermediate in this reaction we will be able to see how the ring nitrogen accommodates the negative charge:

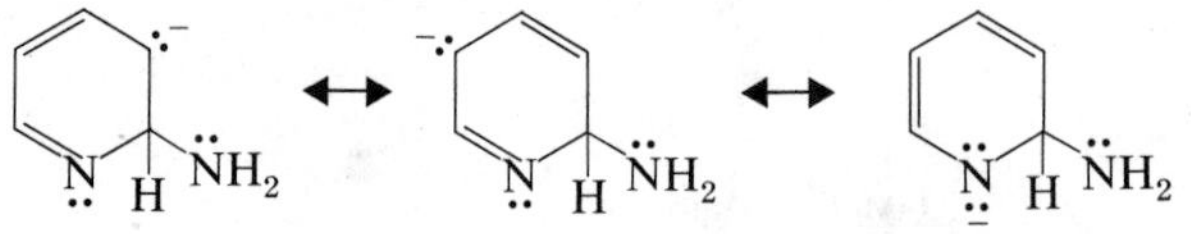

$\ddot{N}H_2$ H $\longleftrightarrow$ $\ddot{N}H_2$ H $\longleftrightarrow$ $\ddot{N}H_2$ H

Relatively stable because negative charge is on electro-negative nitrogen

In the next step the intermediate loses a hydride ion and becomes 2-aminopyridine.*

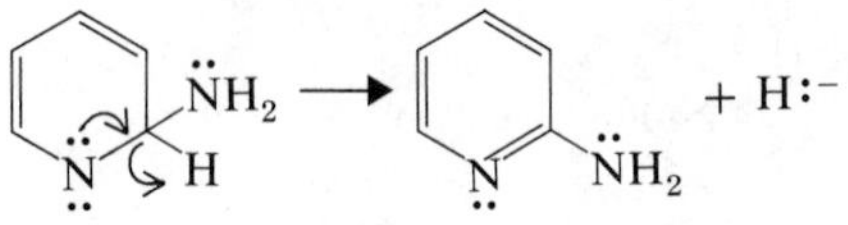

$\ddot{N}H_2$ H $\longrightarrow$ $\ddot{N}H_2$ $+ H:^-$

Pyridine undergoes similar nucleophilic substitution reactions with phenyllithium, butyllithium, and potassium hydroxide.

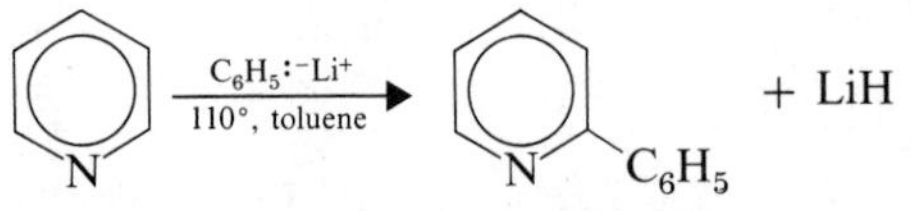

$\xrightarrow[\text{110°, toluene}]{C_6H_5:^-Li^+}$ C_6H_5 + LiH

2-Phenylpyridine

$\xrightarrow{CH_3(CH_2)_2CH_2:^-Li^+}$ $CH_2(CH_2)_2CH_3$ + LiH

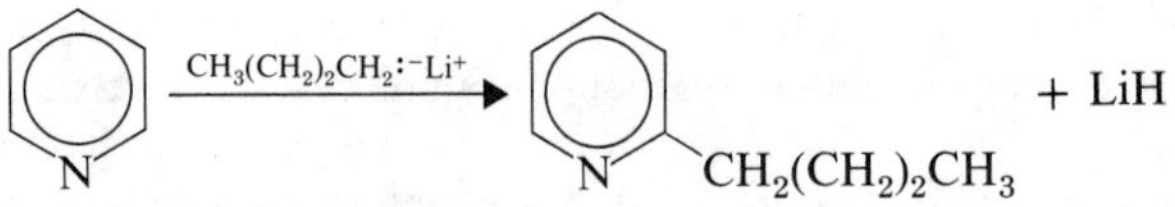

2-Butylpyridine

* In practice, a subsequent reaction occurs; 2-aminopyridine reacts with the sodium hydride to produce a sodio derivative:

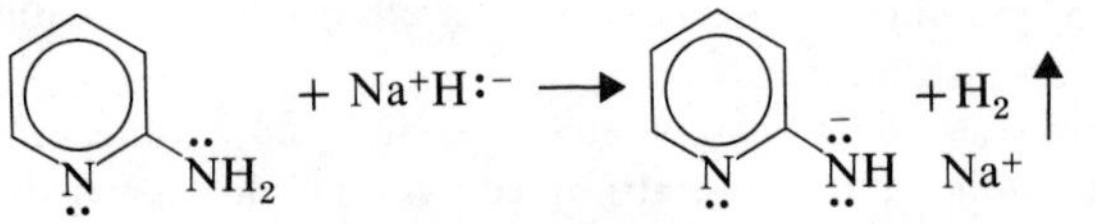

$\ddot{N}H_2$ $+ Na^+H:^- \longrightarrow$ $\ddot{N}H^-$ Na^+ $+ H_2\uparrow$

When the reaction is over, the addition of cold water to the reaction mixture converts the sodio derivative to 2-aminopyridine.

2-Chloropyridine reacts with sodium methoxide to yield 2-methoxypyridine:

$NaOCH_3$ (—NaCl)

KOH, O_2, 320°

2-Pyridinol (50%) ⇌ 2-Pyridone

M.5 NUCLEOPHILIC ADDITIONS TO PYRIDINIUM IONS

Pyridinium ions are especially susceptible to nucleophilic attack at the 2 or 4 position because of the contributions of the resonance forms shown below.

N-Alkylpyridinium halides, for example, react with hydroxide ions primarily at position 2; this causes the formation of an addition product called a *pseudo base:*

H_2O → Pseudo base; $K_3Fe(CN)_6$ → *N*-Methylpyridone (65–70%)

Oxidation of the pseudo base with potassium ferricyanide (above) produces an *N*-alkylpyridone.

Nucleophilic additions to pyridinium ions, especially the addition of *hydride ions,* have been of considerable interest to chemists because these reactions resemble the biological reduction of the important coenzyme, nicotinamide adenine dinucleotide, NAD^+ (Section 11.10).

A number of model reactions have been carried out in connection with these studies. Treating an *N*-alkylpyridinium ion with sodium borohydride, for example, brings about hydride addition, but addition occurs at position 2 and is usually accompanied by over-reduction:

N-Alkylpyridinium halide + $NaBH_4$ —(H_2O)→ 1,2-Dihydropyridine —($NaBH_4$, H_2O)→ 1,2,3,6-Tetrahydropyridine

Treating a pyridinium ion with basic sodium dithionite ($Na_2S_2O_4$), however, brings about specific addition to position 4:

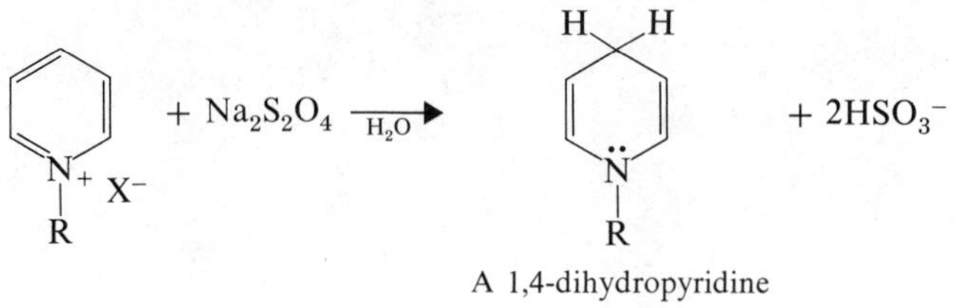

$$\text{pyridinium } (N^+\text{–R}, X^-) + Na_2S_2O_4 \xrightarrow{H_2O} \text{1,4-dihydropyridine} + 2HSO_3^-$$

A 1,4-dihydropyridine

Sodium dithionite in aqueous base also reduces NAD^+ to NADH. The NADH formed by

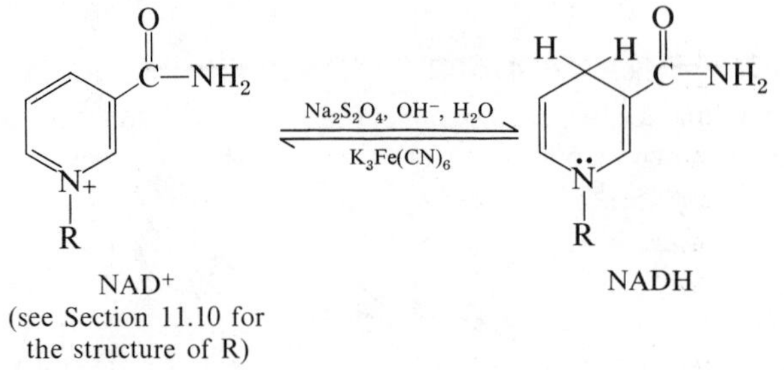

$$NAD^+ \underset{K_3Fe(CN)_6}{\overset{Na_2S_2O_4,\ OH^-,\ H_2O}{\rightleftharpoons}} NADH$$

NAD^+
(see Section 11.10 for the structure of R)

NADH

dithionite reduction has been shown to be biologically active and can be oxidized to NAD^+ with potassium ferricyanide.

Problem M.3

An alternative mechanism to the one given for the amination of pyridine on page 870, involves a "pyridyne" intermediate, that is,

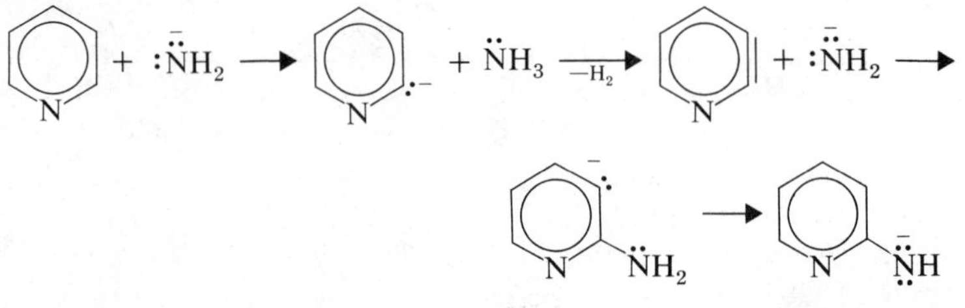

$$\text{pyridine} + :\ddot{N}H_2^- \longrightarrow \text{3-pyridyl anion} + \ddot{N}H_3 \xrightarrow{-H_2} \text{pyridyne} + :\ddot{N}H_2^- \longrightarrow \text{2-amino-3-pyridyl anion} \longrightarrow \text{2-pyridylamide anion } (\ddot{N}H^-)$$

This mechanism was disallowed on the basis of an experiment in which 3-deuteriopyridine was allowed to react with sodium amide. Consider the fate of deuterium in both mechanisms and explain.

Problem M.4

2-Halopyridines undergo nucleophilic substitution much more readily than pyridine itself. What factor accounts for this?

19 CARBOHYDRATES

19.1 INTRODUCTION

19.1A Classification of Carbohydrates

The group of compounds known as carbohydrates received their general name because of early observations that they often have the formula $C_x(H_2O)_y$—that is, they appear to be "hydrates of carbon." Simple carbohydrates are also known as sugars or saccharides (L. *saccharum,* sugar) and the ending of the names of most sugars is *ose*. Thus we have such names as *sucrose* for ordinary table sugar, *glucose* for the principal sugar in blood, and *maltose* for malt sugar.

Carbohydrates are usually defined as *polyhydroxy aldehydes and ketones or substances that hydrolyze to yield polyhydroxy aldehydes and ketones*. Although this definition draws attention to the important functional groups of carbohydrates, it is not entirely satisfactory. We will later find that because carbohydrates contain $\gt C{=}O$ groups and —OH groups, they exist, primarily, as *hemiacetals* and *acetals* or as *hemiketals* and *ketals*.

The simplest carbohydrates, those that cannot be hydrolyzed into smaller, simpler carbohydrates, are called *monosaccharides*. On a molar basis, carbohydrates that undergo hydrolysis to produce only two moles of a monosaccharide are called *disaccharides;* those that yield three moles of a monosaccharide are called *trisaccharides;* and so on. (Carbohydrates that hydrolyze to yield 2 to 10 moles of a monosaccharide are sometimes called *oligosaccharides*.) Carbohydrates that yield a larger number of moles of monosaccharide ($>$ 10) are known as *polysaccharides*.

Maltose and sucrose are examples of disaccharides. On hydrolysis, a mole of maltose yields two moles of the monosaccharide, glucose; sucrose undergoes hydrolysis to yield one mole of glucose and one mole of the monosaccharide, fructose. Starch and cellulose are examples of polysaccharides; both are glucose polymers. Hydrolysis of either yields a large number of glucose units.

$$\underset{\text{(a disaccharide)}}{\text{One mole of maltose}} \xrightarrow[H^+]{H_2O} \underset{\text{(a monosaccharide)}}{\text{two moles of glucose}}$$

$$\underset{\text{(a disaccharide)}}{\text{One mole of sucrose}} \xrightarrow[H^+]{H_2O} \text{one mole of glucose} + \underset{\text{(monosaccharides)}}{\text{one mole of fructose}}$$

$$\begin{array}{c}\text{One mole of starch}\\ \text{or}\\ \text{one mole of cellulose}\\ \text{(polysaccharides)}\end{array} \xrightarrow[H^+]{H_2O} \text{many moles of glucose}$$

Carbohydrates are the most abundant organic constituents of plants. They not only serve as an important source of chemical energy for living organisms (sugars and starches are important in this respect), but also in plants and in some animals they serve as important constituents of supporting tissues (this is the primary function of the cellulose found in wood, cotton, and flax, for example).

We encounter carbohydrates at almost every turn of our daily lives. The paper on which this book is written is largely cellulose; so too is the cotton of our clothes and the wood of our houses. The flour from which we make bread is mainly starch, and starch is also a major constituent of many other foodstuffs, such as potatoes, rice, beans, corn, and peas.

19.1B Photosynthesis and Carbohydrate Metabolism

Carbohydrates are synthesized in green plants by *photosynthesis*—a process that uses solar energy to reduce, or "fix," carbon dioxide. The overall equation for photosynthesis can be written as follows:

$$x\mathrm{CO_2} + y\mathrm{H_2O} + \text{solar energy} \longrightarrow \underset{\text{Carbohydrate}}{\mathrm{C}_x(\mathrm{H_2O})_y} + x\mathrm{O_2}$$

Many individual enzyme-catalyzed reactions take place in the general photosynthetic process and not all are fully understood. We know, however, that photosynthesis begins with the absorption of light by the important green pigment of plants, chlorophyll (Fig. 19.1). The green color of chlorophyll and, therefore, its ability to absorb sunlight in the visible region, is due primarily to its extended conjugated system. As photons of sunlight are trapped by chlorophyll, energy

FIG. 19.1
*Chlorophyll-*a. (*The structure of chlorophyll-*a *was established largely through the work of H. Fischer, R. Willstätter, and J. B. Conant. A synthesis of chlorophyll-*a *from simple organic compounds was achieved by R. B. Woodward in 1960, who won the Nobel Prize in 1965 for his outstanding contributions to synthetic organic chemistry.*)

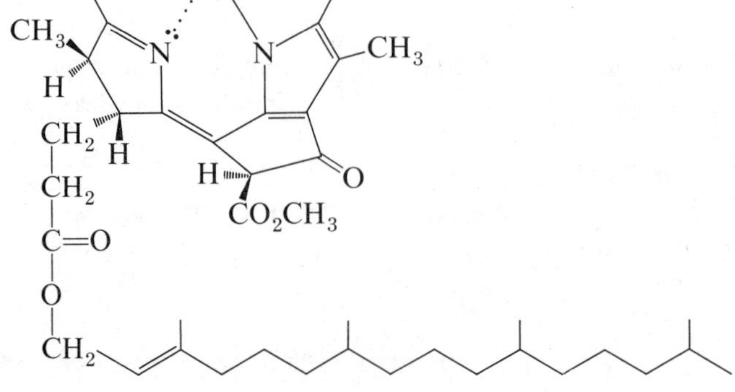

CH_2 CH CH_3 CH_2CH_3 CH_3 N N Mg CH_3 N N CH_3 H CH_2 H CH_2 H O CO_2CH_3 C=O O CH_2

becomes available to the plant in a chemical form that can be used to carry out the reactions that reduce carbon dioxide to carbohydrates and oxidize water to oxygen.

Carbohydrates act as a major chemical repository for solar energy. Their energy is released when animals or plants metabolize carbohydrates to carbon dioxide and water.

$$C_x(H_2O)_y + xO_2 \longrightarrow xCO_2 + yH_2O + \text{energy}$$

The metabolism of carbohydrates also takes place through a series of enzyme-catalyzed reactions in which each energy-yielding step is an oxidation (or the consequence of an oxidation).

Although some of the energy released in the oxidation of carbohydrates is inevitably converted to heat, much of it is conserved in a new chemical form through reactions that are coupled to the synthesis of adenosine triphosphate (ATP) from adenosine diphosphate (ADP) and inorganic phosphate (P_i) (Fig. 19.2). The phosphoric anhydride bond that forms between the terminal phosphate group of ADP and the phosphate ion becomes another repository of chemical

FIG. 19.2

The synthesis of adenosine triphosphate from adenosine diphosphate and (hydrogen) phosphate ion. This reaction takes place in all living organisms, and adenosine triphosphate is the major compound into which the chemical energy released by biological oxidations is transformed.

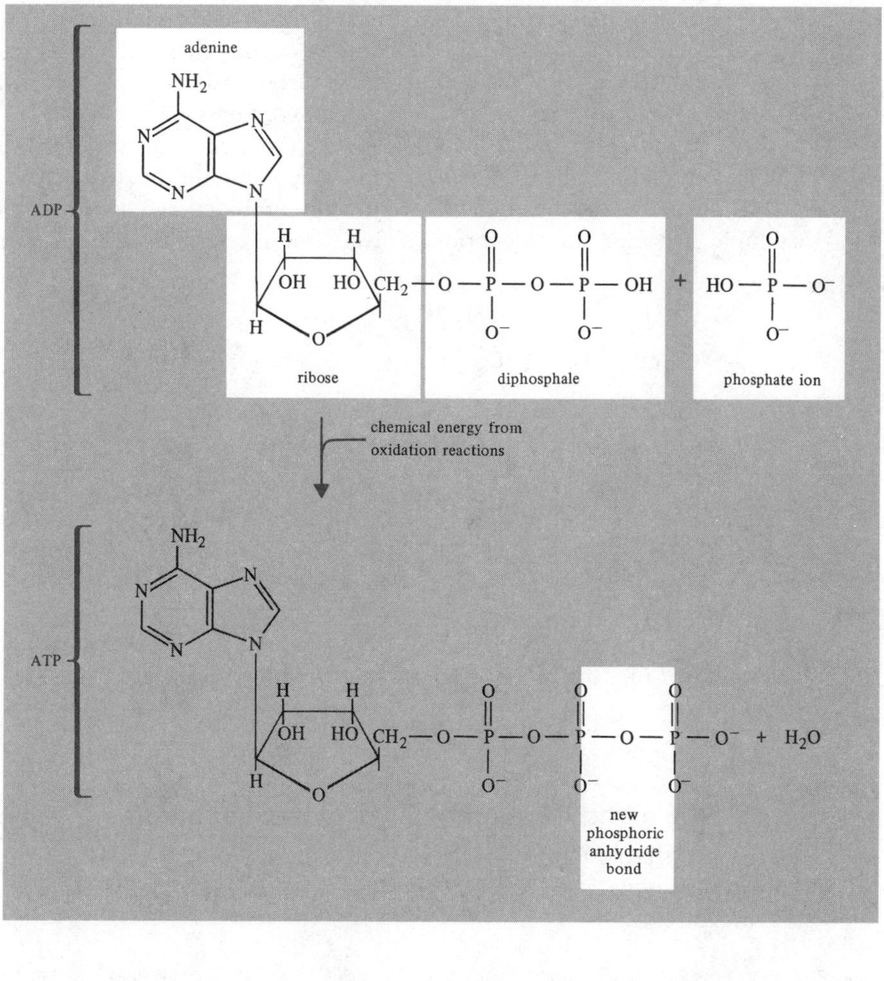

energy. Plants and animals can use the conserved energy of ATP (or very similar substances) to carry out all of their energy-requiring processes; the contraction of a muscle, the synthesis of a macromolecule, and so on. When the energy in ATP is used, a coupled reaction takes place in which ATP is hydrolyzed:

$$\text{ATP} + H_2O \xrightarrow[\text{energy}]{} \text{ADP} + P_i$$

or a new anhydride linkage is created:

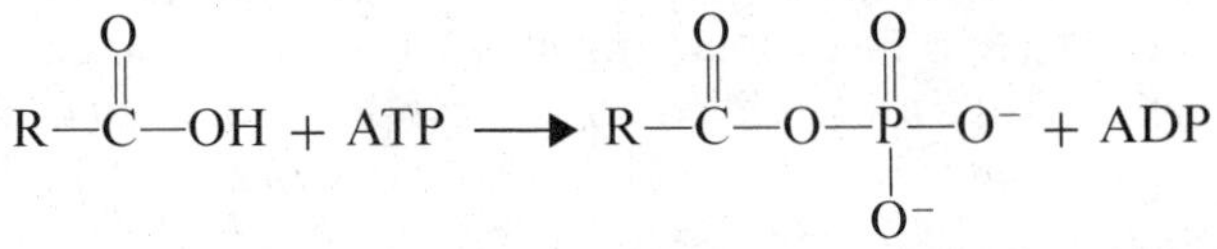

$$R{-}\overset{\overset{\displaystyle O}{\|}}{C}{-}OH + \text{ATP} \longrightarrow R{-}\overset{\overset{\displaystyle O}{\|}}{C}{-}O{-}\underset{\underset{\displaystyle O^-}{|}}{\overset{\overset{\displaystyle O}{\|}}{P}}{-}O^- + \text{ADP}$$

Acyl phosphate

19.2 MONOSACCHARIDES

19.2A Classification of Monosaccharides

Monosaccharides are classified according to (1) the number of carbon atoms present in the molecule and (2) whether they contain an aldehyde or keto group. Thus, a monosaccharide containing three carbon atoms is called a *triose;* one containing four carbons is called a *tetrose;* one containing five carbons is a *pentose;* and one containing six carbons is a *hexose.* A monosaccharide containing an aldehyde group is called an *aldose;* and one containing a keto group is called a *ketose.* These two classifications are frequently combined: a four-carbon aldose, for example, is called an *aldotetrose,* a five-carbon ketose is called a *ketopentose.*

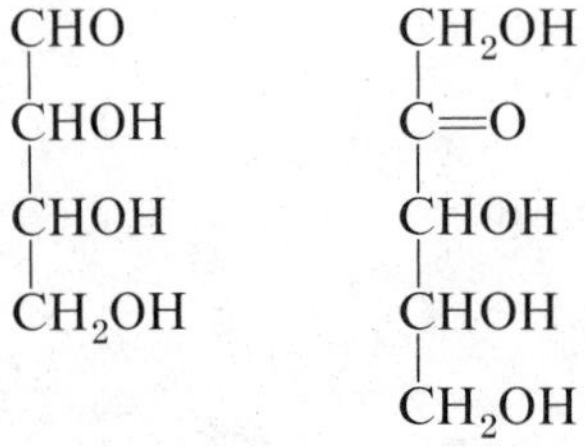

An aldotetrose	A ketopentose
CHO	CH_2OH
CHOH	C=O
CHOH	CHOH
CH_2OH	CHOH
	CH_2OH

An aldotetrose   A ketopentose

Problem 19.1

How many chiral carbons are contained by the (a) aldotetrose and (b) ketopentose given above? (c) How many stereoisomers would you expect from each general structure?

19.2B D and L Designations of Monosaccharides

The simplest monosaccharides are the compounds glyceraldehyde and dihydroxyacetone (p. 877). Of these two compounds, only glyceraldehyde contains a chiral carbon.

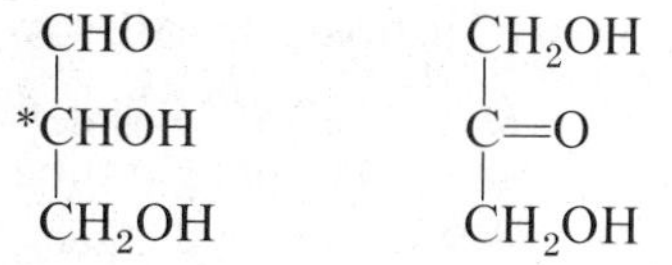

CHO
|
*CHOH
|
CH_2OH
Glyceraldehyde
(an aldotriose)

CH_2OH
|
C=O
|
CH_2OH
Dihydroxyacetone
(a ketotriose)

Glyceraldehyde exists, therefore, in two enantiomeric forms which are known to have the absolute configurations shown below.

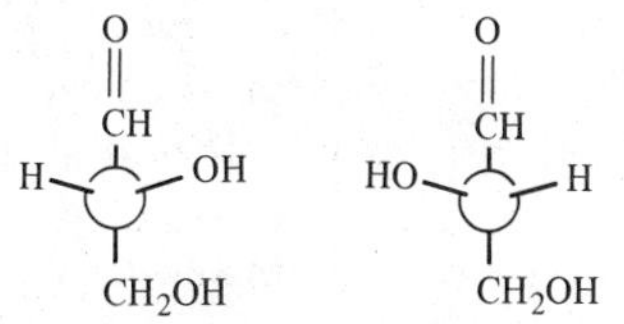

(+)-glyceraldehyde

(−)-glyceraldehyde

We saw in Section 8.5 that according to the Cahn-Ingold-Prelog convention, (+)-glyceraldehyde should be designated *R*-(+)-glyceraldehyde and (−)-glyceraldehyde should be designated *S*-(−)-glyceraldehyde.

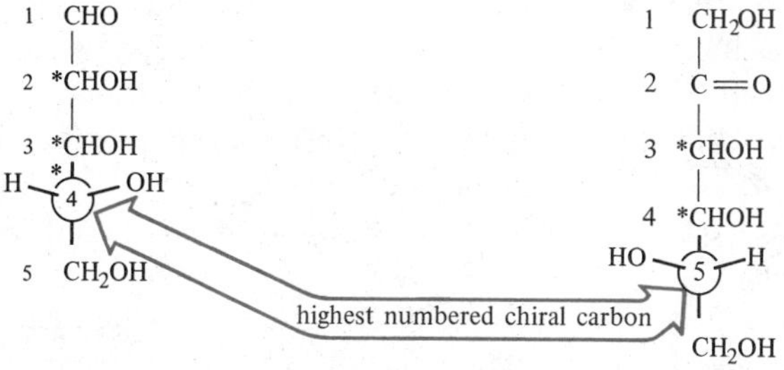

highest numbered chiral carbon

a D–aldopentose

an L-ketohexose

Early in this century, before the absolute configurations of any organic compounds were known, another system of stereochemical designations was introduced. According to this system (first suggested by M. A. Rosanoff in 1906), (+)-glyceraldehyde is designated D-(+)-glyceraldehyde and (−)-glyceraldehyde is designated L-(−)-glyceraldehyde. These two compounds, moreover, serve as configurational standards for all monosaccharides: a monosaccharide *whose highest-numbered chiral carbon* has the same configuration as D-(+)-glyceraldehyde is designated as a D sugar; one whose highest-numbered chiral carbon has the same configuration as L-glyceraldehyde is designated as an L sugar.

D and L designations are like *R* and *S* designations in that they are not necessarily related to the optical rotations of the sugars to which they are applied. Thus, one may encounter other sugars that are D-(+)- or D-(−)- and that are L-(+)- or L-(−)-.

The D-L system of stereochemical designations is thoroughly entrenched in the literature of carbohydrate chemistry, and even though it has the disadvantage of specifying the configuration of only one chiral center—that of the highest-numbered chiral center—we will employ the D-L system in our designations of carbohydrates.

Problem 19.2

Write three-dimensional formulas for each aldotetrose and ketopentose isomer in problem 19.1 and designate each as to whether it is a D or L sugar.

19.2C Structural Formulas for Monosaccharides

Later in this chapter we will see how the great carbohydrate chemist, Emil Fischer,* was able to establish the stereochemical configuration of the aldohexose D-(+)-glucose, the most abundant monosaccharide. In the meantime, however, we can use D-(+)-glucose as an example illustrating the various ways of presenting the structures of monosaccharides.

Fischer represented the structure of D-(+)-glucose with the cross formulation, **1**, in Fig. 19.3. This type of formulation is now called a Fischer projection formula and is still useful for carbohydrates. *When we use Fischer projection formulas, however, we must not* (in our mind's eye) *remove them from the plane of the page in order to test their superposability.* In terms of more familiar formulations, the Fischer projection formula translates into formulas **2** and **3**.†

In IUPAC nomenclature and using the Cahn-Ingold-Prelog system of stereochemical designations, the open-chain form of D-(+)-glucose is 2*R*, 3*S*, 4*R*, 5*R* -2,3,4,5,6-pentahydroxyhexanal.

Although many of the properties of D-(+)-glucose can be explained in terms of an open-chain structure (**1**, **2**, or **3**), a considerable body of evidence indicates that the open-chain structure exists, primarily, in equilibrium with two cyclic forms. These can be represented by structures **4** and **5** or **6** and **7**. The cyclic forms of D-(+)-glucose are *hemiacetals* formed by an intramolecular reaction of the —OH group at C-5 with the aldehyde group. Cyclization creates a new chiral carbon at C-1 and this explains how two cyclic forms are possible. These two cyclic forms are *diastereomers* that differ only in the configuration of C-1. In carbohydrate chemistry diastereomers of this type are called *anomers* and C-1 is called the *anomeric carbon.*

Structures **4** and **5** for the glucose anomers are called Haworth formulas‡ and although they do not give an accurate picture of the shape of the six-membered ring, they have many practical uses. Figure 19.4 demonstrates how the representation of each chiral carbon of the open-chain form can be correlated with its representation in the Haworth formula.

The two glucose anomers are designated as an α *anomer* or a β *anomer* depending on the location of the —OH group of C-1. When we draw the cyclic forms of a D sugar in the orientation shown in Figs. 19.3 or 19.4, the α anomer has the —OH *down* and the β anomer has the —OH *up*.

* Emil Fischer (1852–1919) was professor of organic chemistry at the University of Berlin. In addition to monumental work in the field of carbohydrate chemistry, where Fischer and his co-workers established the configuration of most of the monosaccharides, Fischer also made important contributions to studies of amino acids, proteins, purines, indoles, and to stereochemistry generally. As a graduate student Fischer discovered phenylhydrazine, a reagent that was highly important in his later work with carbohydrates. Fischer was the second recipient (in 1902) of the Nobel Prize for Chemistry.

† The meaning of formulas **1**, **2**, and **3** can be seen best through the use of molecular models: we first construct a chain of six carbon atoms with the —CHO group at the top and a —CH_2OH group at the bottom. We then bring the CH_2OH group up behind the chain until it almost touches the —CHO group. Holding this model so that the —CHO and —CH_2OH group are directed generally away from us, we then begin placing —H and —OH groups on each of the four remaining carbons. The —OH group of carbon-2 is placed on the right; that of carbon-3 on the left; and those of carbons-4 and -5 on the right.

‡ After the English chemist W. N. Haworth (1883–1950) who, in 1926, along with E. L. Hirst, demonstrated that the cyclic form of glucose acetals consists of a six-membered ring. Haworth received the Nobel Prize for his work in carbohydrate chemistry in 1937.

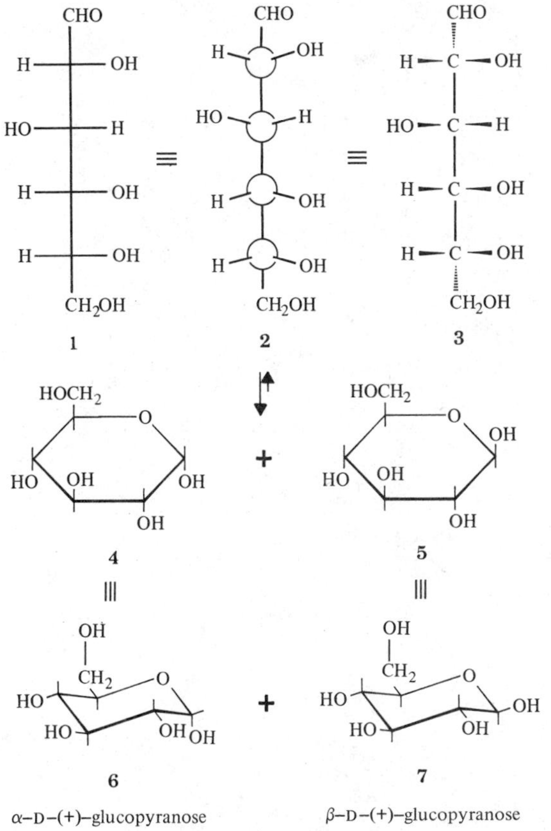

FIG. 19.3
(a) ***1–3*** *are formulas used for the open-chain structure of* D-(+)-*glucose.* (b) ***4–7*** *are formulas used for the two cyclic hemiacetal forms of* D-(+)-*glucose.*

α–D–(+)–glucopyranose

β–D–(+)–glucopyranose

A method for designating D-aldohexose anomers as α- or β- is the following. If the —OH group at the anomeric carbon and the —CH_2OH at C-5 are *trans,* the anomer is α-; if they are cis the anomer is β-.

Studies of the structures of the cyclic hemiacetal forms of D-(+)-glucose using X-ray analysis have demonstrated that the actual conformations of the rings are the chair forms represented by conformational formulas **6** and **7** in Fig. 19.3. This is exactly what we would expect from our studies of the conformations of cyclohexane (Chapter 3), and it is especially interesting to notice that in the β anomer of D-glucose all of the large substituents, —OH or —CH_2OH, are equatorial. In the α anomer the only bulky axial substituent is the —OH at C-1.

It is convenient at times to represent the cyclic structures of a monosaccharide without specifying whether the configuration of the anomeric carbon is α or β. When we do this we will use formulas such as the following.

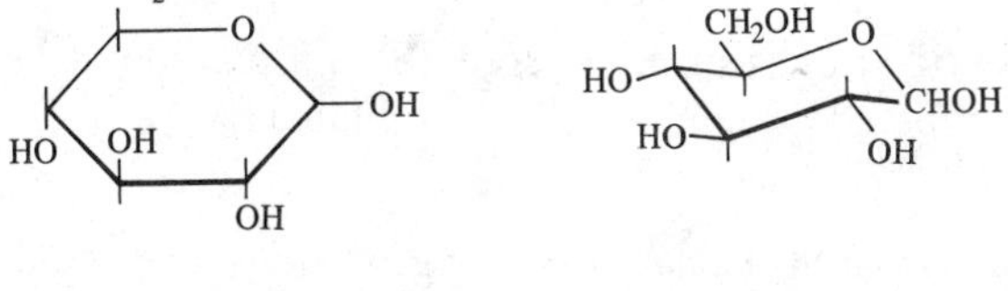

Not all carbohydrates exist in equilibrium with six-membered hemiacetal rings; in several instances the ring is five membered. (Even glucose exists, to a small extent, in equilibrium with five-membered hemiacetal rings.) Because of this, a system of nomenclature has been introduced to allow designation of the ring size.

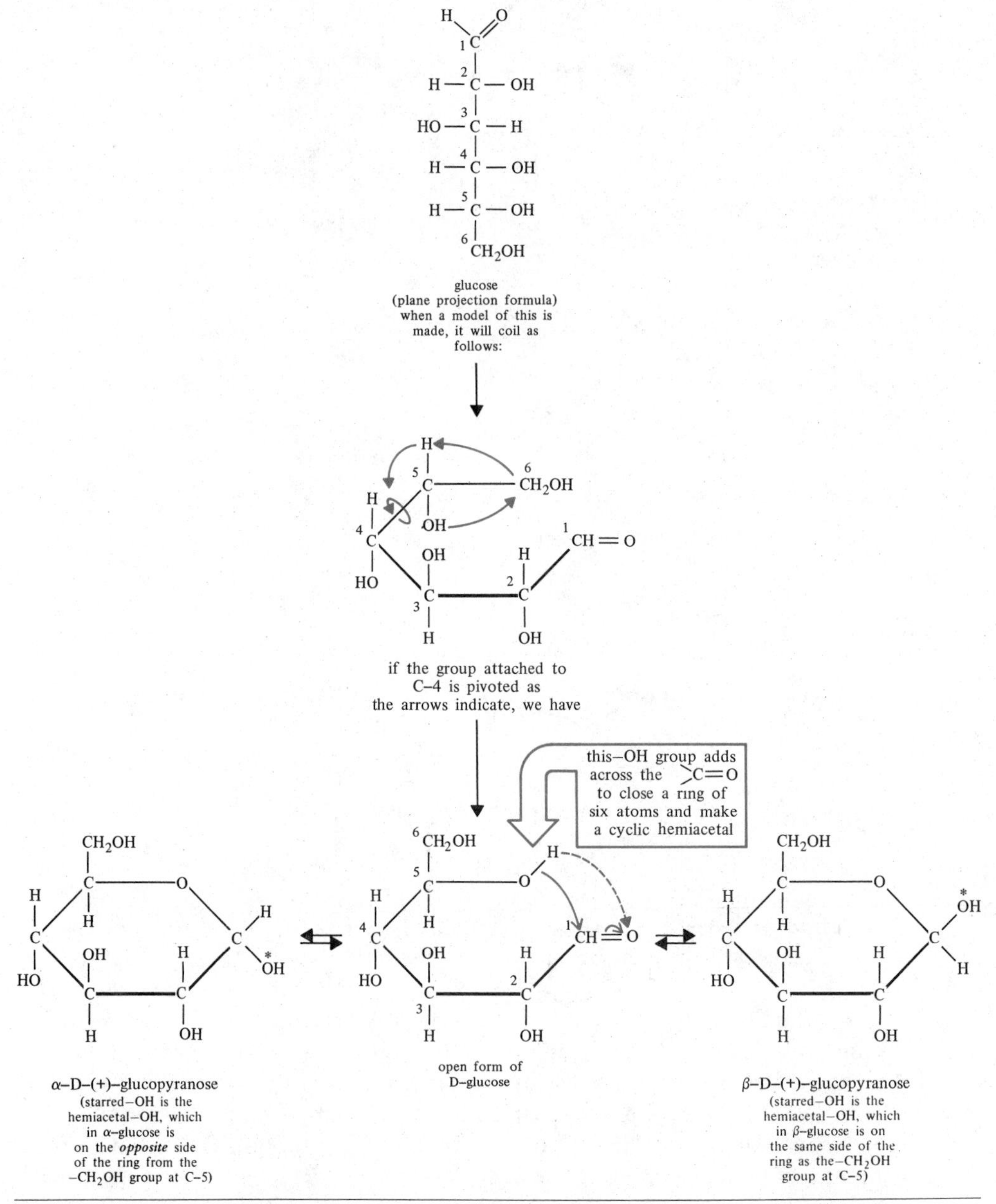

FIG. 19.4
The Haworth formulas for the cyclic hemiacetal forms of D-(+)-*glucose and their relation to the open-chain polyhydroxy aldehyde structure. (From John R. Holum,* Organic Chemistry: A Brief Course, *Wiley, New York, (1975), p. 332. Used by permission.)*

If the monosaccharide ring is six membered, the compound is called a *pyranose;* if the ring is five membered the compound is designated as a *furanose.** Thus, the full

* These names come from the names of the oxygen heterocycles *pyran* and *furan* + *ose.*

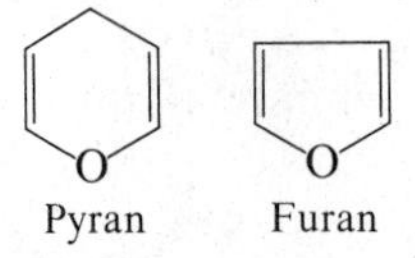

name of compound **4**, **5**, or **6** is α-D-(+)-glucopyranose, while that of **7**, **8**, or **9** is β-D-(+)-glucopyranose.

19.3 MUTAROTATION AND GLYCOSIDE FORMATION

Part of the evidence for the cyclic hemiacetal structure for D-(+)-glucose comes from experiments in which both α- and β-forms have been isolated. Ordinary D-(+)-glucose has a melting point of 146°. However, when D-(+)-glucose is crystallized by evaporating an aqueous solution kept above 98°, a second form of D-(+)-glucose with a melting point of 150° can be obtained. When the optical rotations of these two forms are measured they are found to be significantly different, but when an aqueous solution of either form is allowed to stand, its rotation changes—the specific rotation of one form decreases and the rotation of the other increases—*until both solutions show the same value.* A solution of ordinary D-(+)-glucose (mp 146°) has an initial specific rotation of +112° but ultimately the specific rotation of this solution falls to +52.7°. A solution of the second form of D-(+)-glucose (mp 150°) has an initial specific rotation of +19°; but on standing, the specific rotation of this solution rises to +52.7°. This phenomenon is called *mutarotation.*

The explanation for this mutarotation lies in the existence of an equilibrium between the open-chain form of D-(+)-glucose and the α and β forms of the cyclic hemiacetals.

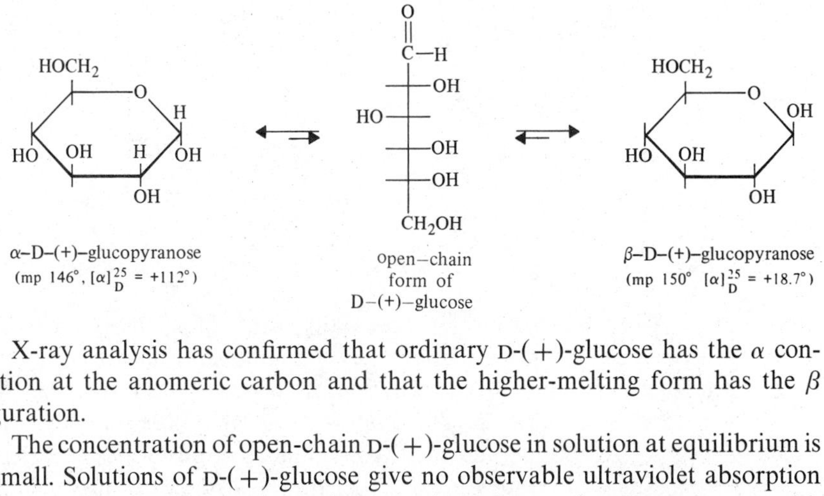

α-D-(+)-glucopyranose (mp 146°, $[\alpha]_D^{25}$ = +112°)

open-chain form of D-(+)-glucose

β-D-(+)-glucopyranose (mp 150° $[\alpha]_D^{25}$ = +18.7°)

X-ray analysis has confirmed that ordinary D-(+)-glucose has the α configuration at the anomeric carbon and that the higher-melting form has the β configuration.

The concentration of open-chain D-(+)-glucose in solution at equilibrium is very small. Solutions of D-(+)-glucose give no observable ultraviolet absorption band for a carbonyl group, and solutions of D-(+)-glucose give a negative test with Schiff's reagent—a special reagent that requires a relatively high concentration of a free aldehyde group (rather than a hemiacetal) in order to give a positive test.

Assuming that the concentration of the open-chain form is negligible, one can calculate the percentages of the α and β anomers present at equilibrium. These percentages, 36% α anomer and 64% β anomer, are in accord with a greater stability for β-D-(+)-glucopyranose. This is what we might expect on the basis of its having an equatorial anomeric hydroxyl:

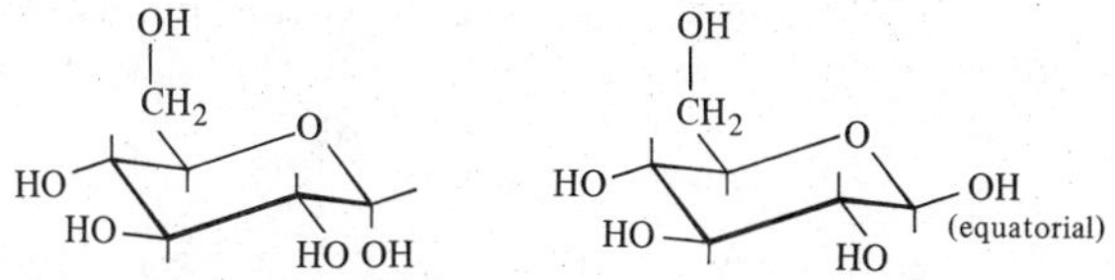

OH (axial)

α-D-(+)-glucopyranose (36% at equilibrium)

OH (equatorial)

β-D-(+)-glucopyranose (64% at equilibrium)

The β anomer of a pyranose is not always the more stable, however. With D-mannose the equilibrium favors the α anomer and this result—termed an *anomeric effect*—is at present not fully understood.

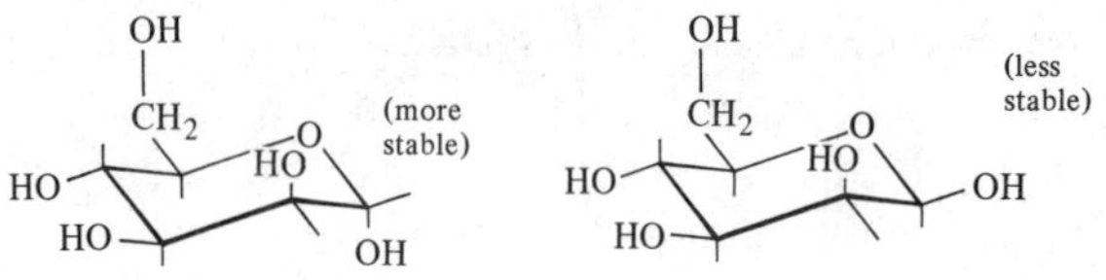

(more stable)

α–D–mannopyranose
(69% at equilibrium)

(less stable)

β–D–mannopyranose
(31% at equilibrium)

When a small amount of gaseous hydrogen chloride is passed into a solution of D-(+)-glucose in methanol, a reaction takes place that results in the formation of anomeric methyl *acetals:*

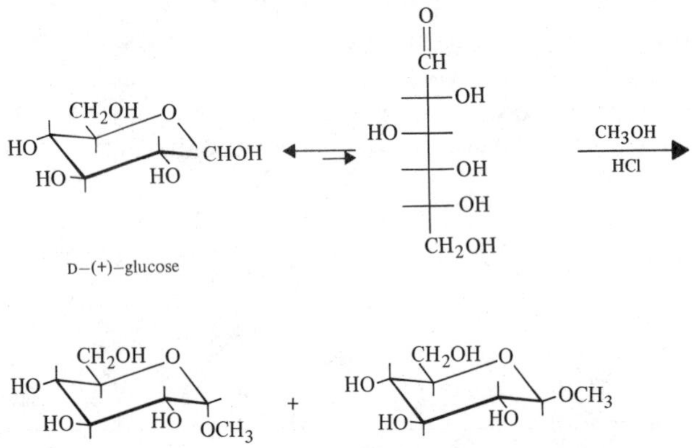

D–(+)–glucose

$\xrightarrow[HCl]{CH_3OH}$

methyl α–D–glucopyranoside
(mp 165°, $[\alpha]_D^{25}$ = +158°)

methyl β–D–glucopyranoside
(mp 107°, $[\alpha]_D^{25}$ = −33°)

Carbohydrate acetals, generally, are called *glycosides* and an acetal of glucose is called a *glucoside.* (Acetals of mannose are *mannosides,* ketals of fructose are *fructosides,* and so on.) The methyl D-glucosides have been shown to have six-membered rings (p. 889 and 898) so they are properly named methyl α-D-glucopyranoside and methyl β-D-glucopyranoside.

Since glycosides are acetals, they do not exist in equilibrium with an open-chain form in neutral or basic aqueous media. Under these conditions (that is, in neutral or basic solutions) glycosides do not show mutarotation.

Problem 19.3

Glycosides *do* exhibit mutarotation in aqueous acid. Explain.

Problem 19.4

(a) Write conformational formulas for methyl α-D-glucopyranoside and (b) methyl β-D-glucopyranoside.

Problem 19.5

Write Haworth and conformational formulas for methyl α-D-mannopyranoside.

A number of oxidizing agents are used to identify functional groups of carbohydrates and in elucidating their structures. The most important are (1) Benedict's or Tollens' reagent, (2) bromine water, (3) nitric acid, and (4) periodic acid. Each of these reagents produces a different and usually specific effect when it is allowed to react with a monosaccharide. We should now examine what these are.

19.4A Benedict's or Tollens' Reagent: Reducing Sugars

Benedict's reagent (an alkaline solution containing a cupric citrate complex ion) and Tollens' solution ($Ag(NH_3)_2^+OH^-$) oxidize and thus give positive tests with aldehydes *and with α-hydroxy ketones*. Both reagents, therefore, give positive tests with aldoses *and ketoses*. This is true even though aldoses and ketoses exist primarily as cyclic hemiacetals.

We studied the use of Tollens' silver mirror test in Section 16.7A. Benedict's solution (and the related Fehling's solution that contains a cupric tartrate complex ion) give brick-red precipitates of Cu_2O when they oxidize an aldehyde. (In alkaline solution α-hydroxy ketones are converted to aldehydes.) Since the solutions of cupric tartrates and citrates are blue, the appearance of a brick-red precipitate is a vivid and unmistakable indication of a positive test.

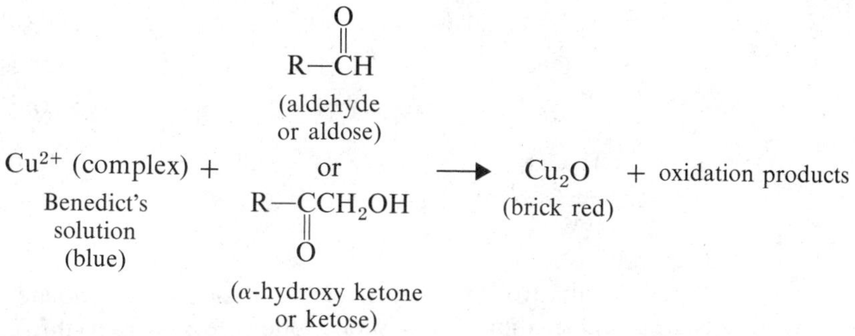

$$\underset{\substack{\text{Benedict's}\\\text{solution}\\\text{(blue)}}}{Cu^{2+}\text{ (complex)}} + \underset{\substack{\text{(aldehyde or aldose)}\\\text{or}\\R-C(=O)CH_2OH\\(\alpha\text{-hydroxy ketone or ketose)}}}{R-\overset{O}{\overset{\|}{C}}H} \longrightarrow \underset{\text{(brick red)}}{Cu_2O} + \text{oxidation products}$$

Sugars that give positive tests with Tollens' or Benedict's solutions are known as *reducing sugars,* and all carbohydrates that contain a *hemiacetal group* or a *hemiketal group* give positive tests. In aqueous solution these hemiacetals or hemiketals exist in equilibrium with relatively small, but not insignificant, concentrations of noncyclic aldehydes or α-hydroxy ketones. It is the latter that undergo the oxidation, perturbing the equilibrium to produce more aldehyde or α-hydroxy ketone, which undergoes oxidation and so forth, until one reactant is exhausted.

Carbohydrates that contain only acetal or ketal groups do not give positive tests with Benedict's or Tollens' solution and they are called *nonreducing sugars.* Acetals or ketals do not exist in equilibrium with aldehydes or α-hydroxy ketones in the basic aqueous media of the test reagents.

Reducing Sugar	Nonreducing Sugar
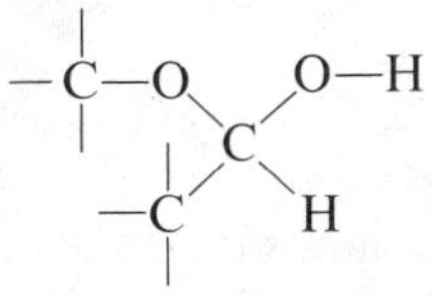 —C—O, O—H on C(H), —C—	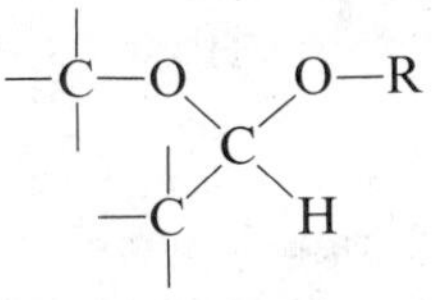 —C—O, O—R on C(H), —C—
hemiacetal (or hemiketal) (gives positive Tollens' or Benedict's test)	acetal (or ketal) (does not give a positive Tollens' or Benedict's test)

Problem 19.6

How might you distinguish between α-D-glucopyranose (i.e., D-glucose) and methyl α-D-glucopyranoside?

Although Benedict's and Tollens' reagent have some use as diagnostic tools (Benedict's solution can be used in quantitative determinations of glucose in blood or urine), none of these reagents is useful as a preparative reagent in carbohydrate oxidations. Oxidations with both reagents take place in alkaline solution, *and in alkaline solution sugars undergo a complex series of reactions that lead to isomerizations and fragmentations.*

Problem 19.7

If D-glucose is treated with aqueous calcium hydroxide and the solution is allowed to stand for several days, a mixture of products results, including D-mannose, D-fructose, glycolic aldehyde, and D-erythrose.

D-Mannose: CHO; HO—C—H; HO—C—H; H—C—OH; H—C—OH; CH_2OH

D-Fructose: CH_2OH; C=O; HO—C—H; H—C—OH; H—C—OH; CH_2OH

Glycolic aldehyde: CHO; CH_2OH

D-Erythrose: CHO; H—C—OH; H—C—OH; CH_2OH

D-Mannose results from the reversible formation of an enolate ion; D-fructose results from an enediol; and glycolic aldehyde and D-erythrose result from a *reverse* aldol addition. Using the open-chain forms of all the sugars involved, write mechanisms that explain the formation of each product.

19.4B Bromine Water: The Synthesis of Aldonic Acids

Monosaccharides do not undergo isomerization and fragmentation reactions in mildly acidic solution. Thus, a useful oxidizing reagent for preparative purposes is bromine in water (pH = 6.0). Bromine water is a general reagent that selectively oxidizes the —CHO group to a —COOH group. It converts an aldose to an *aldonic acid:*

$$\underset{\text{Aldose}}{\text{CHO}-(\text{CHOH})_n-\text{CH}_2\text{OH}} \xrightarrow[H_2O]{Br_2} \underset{\text{Aldonic acid}}{\text{COOH}-(\text{CHOH})_n-\text{CH}_2\text{OH}}$$

Experiments with aldopyranoses have shown that the actual course of the reaction is somewhat more complex than we have indicated above. Bromine water specifically oxidizes the β anomer, and the initial product that forms is a δ-*aldonolactone.* This compound may then hydrolyze to an aldonic acid, and the aldonic acid may undergo a subsequent ring closure to form a γ-*aldonolactone.*

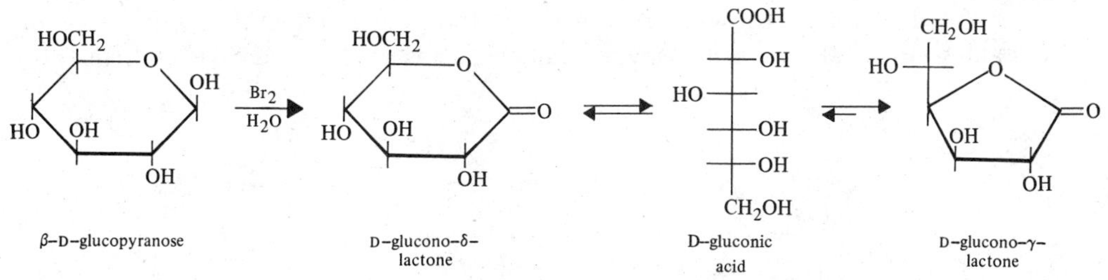

β-D-glucopyranose $\xrightarrow[H_2O]{Br_2}$ D-glucono-δ-lactone ⇌ D-gluconic acid ⇌ D-glucono-γ-lactone

Although α-aldopyranoses also undergo this oxidation, they react much more slowly. The rate of oxidation of β-D-glucopyranose, for example, is 250 times as fast as that of α-D-glucopyranose. The very slow rate of oxidation of α-D-glucopyranose reflects its conversion to the β anomer (followed by oxidation) rather than direct oxidation of the α anomer.

A mechanism that accounts for this behavior requires attack by Br^+ at the anomeric oxygen followed by loss of HBr from an intermediate in which the C—H and O—Br bonds are *antiparallel* to each other. Beta-anomers can achieve this stereochemistry easily.

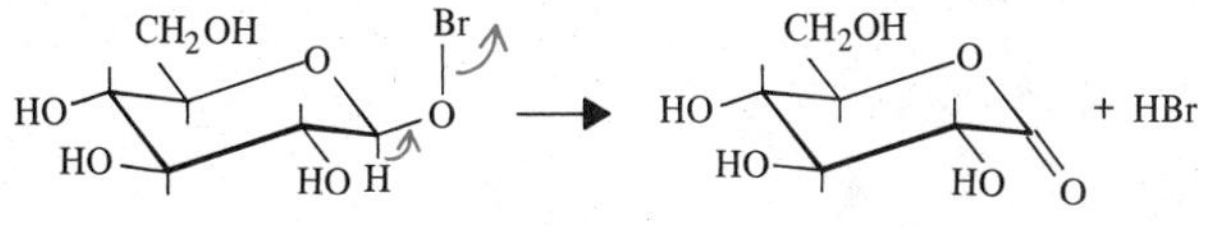

(from β anomer)

Antiparallel Arrangement

With α anomers the axial hydrogens at the 3 and 5 positions hinder an antiparallel arrangement:

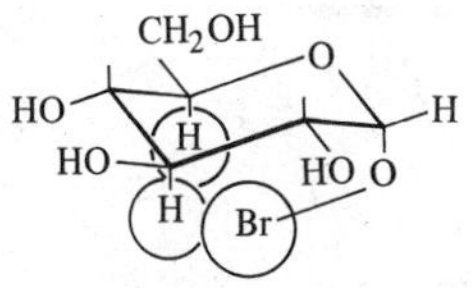

(from α anomer)

antiparallel arrangement is hindered

Problem 19.8

Write an equation to illustrate the oxidation of β-D-mannopyranose by bromine-water (cf. page 882). The name of the aldonic acid is D-mannonic acid and the names of the aldonolactones are δ- and γ-D-mannonolactone.

19.4C Nitric Acid Oxidation: Aldaric Acids

Dilute nitric acid—a stronger oxidizing agent than bromine water—oxidizes both the —CHO group and the terminal $—CH_2OH$ group of an aldose to —COOH groups. These dicarboxylic acids are known as *aldaric acids.*

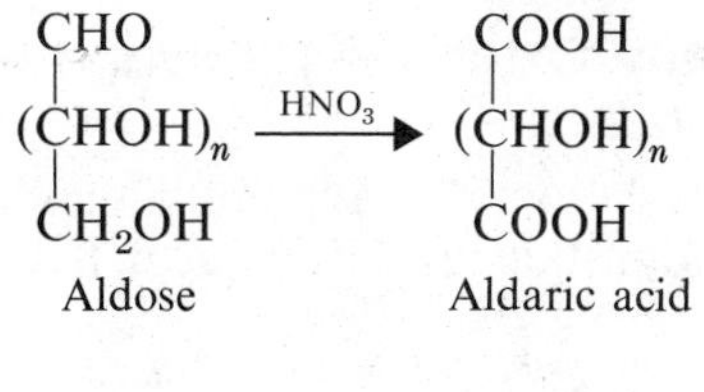

$$\begin{array}{c} CHO \\ | \\ (CHOH)_n \\ | \\ CH_2OH \\ \text{Aldose} \end{array} \xrightarrow{HNO_3} \begin{array}{c} COOH \\ | \\ (CHOH)_n \\ | \\ COOH \\ \text{Aldaric acid} \end{array}$$

It is not known whether a lactone is an intermediate in the oxidation of an aldose to an aldaric acid; however, aldaric acids form γ and δ lactones readily.

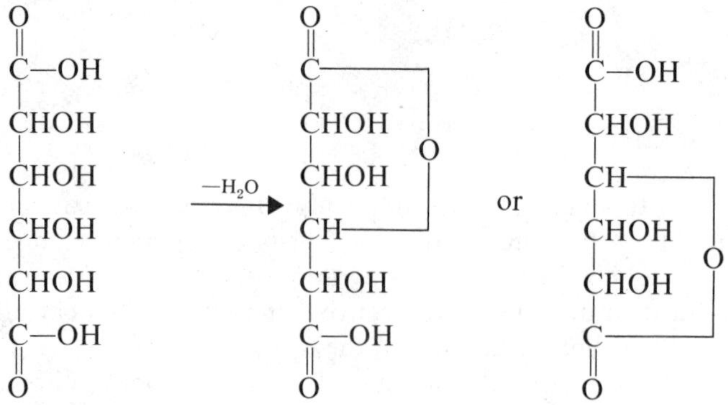

$-H_2O$ → or

Aldaric acid
(from an aldohexose)

γ-Lactones of an aldaric acid

The aldaric acid obtained from D-glucose is called D-glucaric acid.*

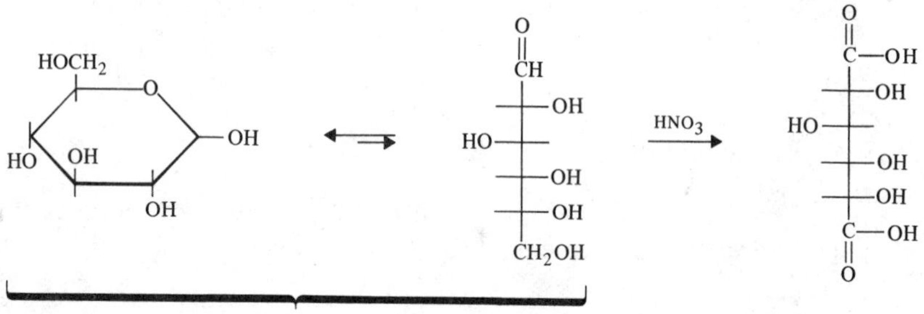

HNO_3 →

D-glucose

D-glucaric acid

Problem 19.9

(a) Would you expect D-glucaric acid to be optically active? (b) Write the open-chain structure for the aldaric acid (mannaric acid) that would be obtained by nitric acid oxidation of D-mannose. (c) Would you expect it to be optically active? (d) What aldaric acid would you expect to obtain from D-erythrose (cf. problem 19.7)? (e) Would it show optical activity? (f) D-Threose, a diastereomer of D-erythrose, yields an optically active aldaric acid when it is subjected to nitric acid oxidation. Write Fischer projection formulas for D-threose and its nitric acid oxidation product. (g) What are the names of the aldaric acids obtained from D-erythrose and D-threose? (See Sect. 8.18.)

Problem 19.10

D-Glucaric acid undergoes lactonization to yield two different γ lactones. What are their structures?

19.4D Periodate Oxidations: Oxidative Cleavage of Polyhydroxy Compounds

Compounds that have hydroxyl groups on adjacent atoms undergo oxidative cleavage when they are treated with aqueous periodic acid, HIO_4. The reaction

* Older terms for an aldaric acid are a *glycaric* acid or a *saccharic* acid.

breaks carbon-carbon bonds and produces carbonyl compounds (aldehydes, ketones, or acids). The stoichiometry of the reaction is:

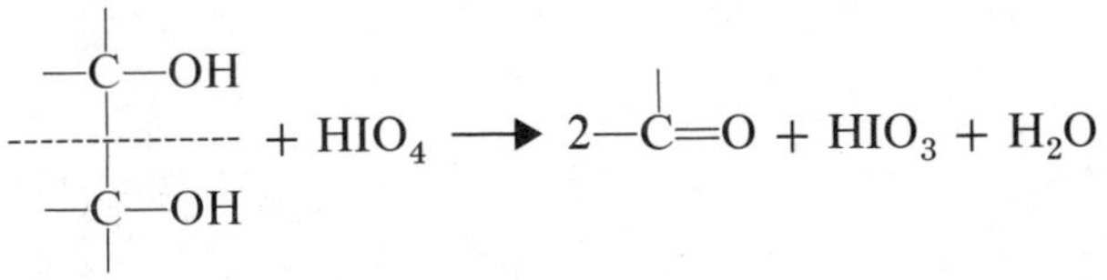

$$\begin{array}{c} -\overset{|}{C}-OH \\ \text{- - - - - -} \\ -\underset{|}{C}-OH \end{array} + HIO_4 \longrightarrow 2{-}\overset{|}{C}{=}O + HIO_3 + H_2O$$

Since the reaction usually takes place in quantitative yield, valuable information can often be gained by measuring the number of moles of periodic acid that are consumed in the reaction as well as by identifying the carbonyl products.*

Periodate oxidations are thought to take place through a cyclic intermediate:

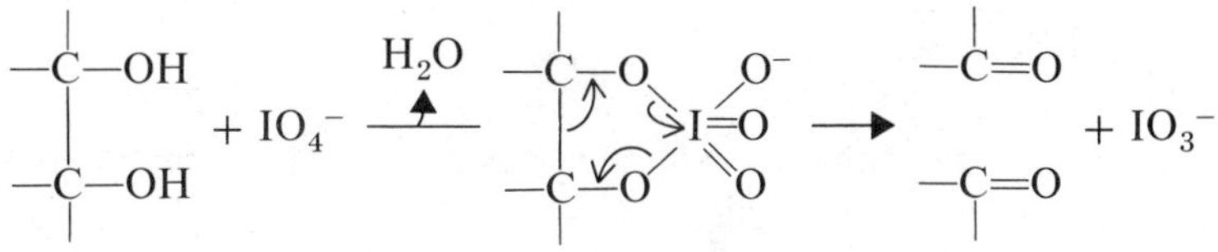

$$\begin{array}{c} -\overset{|}{C}-OH \\ | \\ -\underset{|}{C}-OH \end{array} + IO_4^- \overset{H_2O}{\rightleftharpoons} \begin{array}{c} -\overset{|}{C}-O \quad O^- \\ | \qquad \searrow I{=}O \\ -\underset{|}{C}-O \quad O \end{array} \longrightarrow \begin{array}{c} -\overset{|}{C}{=}O \\ \\ -\underset{|}{C}{=}O \end{array} + IO_3^-$$

Before we discuss the use of periodic acid in carbohydrate chemistry we should illustrate the course of the reaction with several simple examples. Notice in these periodate oxidations that *for every C—C bond broken, a C—O bond is formed at each carbon.*

1. When 1,2-propanediol is subjected to this oxidative cleavage the products are formaldehyde and acetaldehyde.

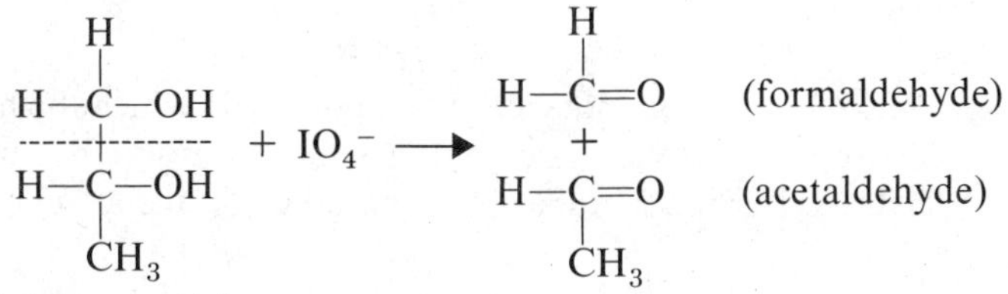

$$\begin{array}{c} H \\ | \\ H-C-OH \\ \text{- - - - - -} \\ H-C-OH \\ | \\ CH_3 \end{array} + IO_4^- \longrightarrow \begin{array}{cl} H & \\ | & \\ H-C{=}O & \text{(formaldehyde)} \\ + & \\ H-C{=}O & \text{(acetaldehyde)} \\ | & \\ CH_3 & \end{array}$$

1,2-Propanediol

2. When three or more —CHOH groups are contiguous, the internal ones are obtained as *formic acid.* Periodate oxidation of glycerol, for example, gives two moles of formaldehyde and one mole of formic acid.

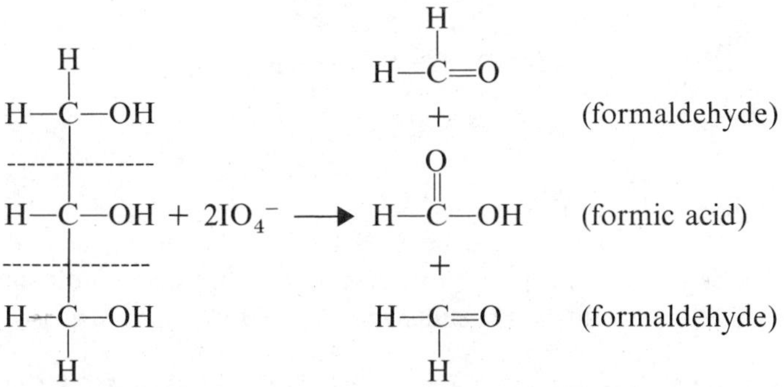

$$\begin{array}{c} H \\ | \\ H-C-OH \\ \text{- - - - - -} \\ H-C-OH \\ \text{- - - - - -} \\ H-C-OH \\ | \\ H \end{array} + 2IO_4^- \longrightarrow \begin{array}{cl} H & \\ | & \\ H-C{=}O & \text{(formaldehyde)} \\ + & \\ O & \\ \| & \\ H-C-OH & \text{(formic acid)} \\ + & \\ H-C{=}O & \text{(formaldehyde)} \\ | & \\ H & \end{array}$$

* The reagent lead tetraacetate, ($Pb(OOCCH_3)_4$), brings about cleavage reactions similar to those of periodic acid. The two reagents are complementary; periodic acid works well in aqueous solutions and lead tetraacetate gives good results in organic solvents.

3. Oxidative cleavage also takes place when an —OH group is adjacent to the carbonyl group of an aldehyde or ketone. Glyceraldehyde yields two moles of formic acid and one mole of formaldehyde,

```
       O                                 O
       ‖                                 ‖
       C—H                           H—C—OH        (formic acid)
       |                                 +
 ------|--------                         O
       |                                 ‖
   H—C—OH   + 2IO4⁻  ——▶           H—C—OH        (formic acid)
       |                                 +
 ------|--------
       |
   H—C—OH                            H—C=O         (formaldehyde)
       |                                 |
       H                                 H
Glyceraldehyde
```

while dihydroxyacetone gives two moles of formaldehyde and one mole of carbon dioxide.

```
       H                          H
       |                          |
   H—C—OH                     H—C=O        (formaldehyde)
 ------|--------                  +
       C=O   + 2IO4⁻  ——▶    O=C=O        (carbon dioxide)
 ------|--------                  +
   H—C—OH                     H—C=O        (formaldehyde)
       |                          |
       H                          H
```

4. Periodic acid does not cleave compounds in which the hydroxyl groups are separated by an intervening $—CH_2—$ group, nor those in which a hydroxyl group is adjacent to an ether or acetal function.

```
CH2OH
|
CH2      + IO4⁻  ——▶  No reaction
|
CH2OH
```

```
CH2OCH3
|
CHOH      + IO4⁻  ——▶  No reaction
|
CH2R
```

Problem 19.11

What products would you expect to be formed when each of the following compounds is treated with an appropriate amount of periodic acid? How many moles of HIO_4 would be consumed in each case?

(a) 2,3-Butanediol
(b) 1,2,3-Butanetriol
(c) $CH_2OHCHOHCH(OCH_3)_2$
(d) $CH_2OHCHOHCOCH_3$
(e) $CH_3COCHOHCOCH_3$
(f) *cis*-Cyclopentanediol

(g) 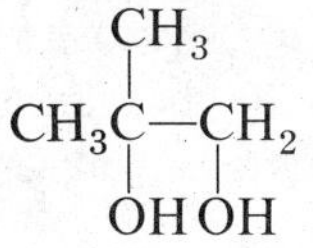 $CH_3C(CH_3)(OH)—CH_2OH$

(h) D-Erythrose

Problem 19.12

Show how periodic acid could be used to distinguish between an aldohexose and a ketohexose. What products would you obtain from each, and how many moles of HIO_4 would be consumed?

In 1936 E. L. Jackson and C. S. Hudson used periodate oxidations in an elegant procedure for determining the ring size of glycosides. Their method also permits us to demonstrate that different glycosides of the α-D-hexose series have the same configurations at C-1 and at C-5. This method is outlined in Fig. 19.5.

The fact that the methyl glycoside, **1**, consumes two moles of periodate ion and produces one mole of formic acid demonstrates that the glycoside contains a six-membered ring. The oxidation also eliminates three centers of chirality in **1**. (C-2 and C-4 are oxidized to aldehyde groups and C-3 is eliminated as formic

FIG. 19.5
*Jackson and Hudson's method. Placing the —OCH_3 group on the right in structure **1** indicates that the glycoside is an α anomer. The configuration of C-2, C-3, and C-4 is not specified, thus, **1** is any α-methyl D-hexopyranoside.*

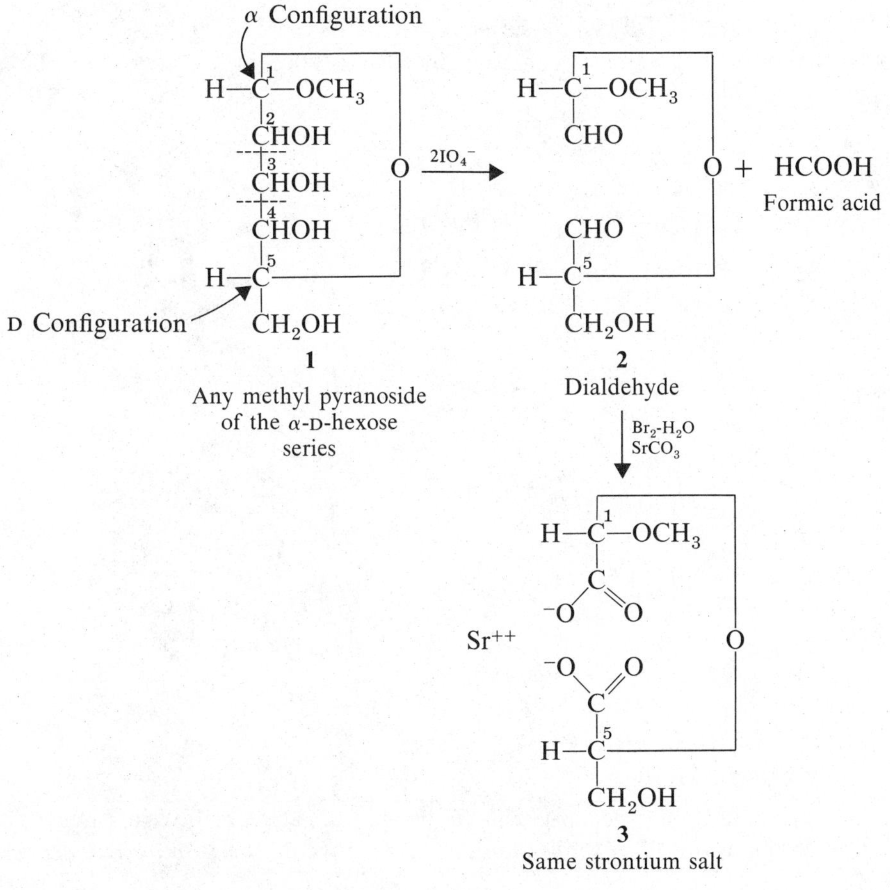

α Configuration

H—C1—OCH_3 / C2HOH / C3HOH / C4HOH / H—C5 (ring O) / CH_2OH

D Configuration

1
Any methyl pyranoside of the α-D-hexose series

$\xrightarrow{2IO_4^-}$

H—C1—OCH_3 / CHO / CHO / H—C5 (ring O) / CH_2OH + HCOOH (Formic acid)

2
Dialdehyde

$\downarrow$ Br_2-H_2O, $SrCO_3$

H—C1—OCH_3 / $C(=O)O^-$ / Sr^{++} / $^-O(O=)C$ / H—C5 (ring O) / CH_2OH

3
Same strontium salt

acid.) The fact that all methyl pyranosides of the α-D-hexose series yield the same dialdehyde, **2**, as can be shown by conversion to the same strontium salt, **3**, demonstrates that they all have the same configuration at C-1 and C-5.

Problem 19.13

When methyl furanosides of the α-D-pentose series are subjected to Jackson and Hudson's procedure they, too, yield the same dialdehyde, **2**, and strontium salt, **3**. (a) Outline the reactions involved. (b) In what respect would you obtain a different result from applying Jackson and Hudson's method to a methyl α-D-pentofuranoside and a methyl α-D-hexopyranoside?

Problem 19.14

(a) What are compounds **4** and **5** below?

$$\text{Strontium salt } \mathbf{3} \xrightarrow[H_2O]{H^+} \begin{cases} \mathbf{4}\ (C_2H_2O_3) \\ \mathbf{5}\ (C_3H_6O_4) \end{cases}$$

(b) Oxidation of (+)-glyceraldehyde by bromine water also produces compound **5**. What is the special significance of this fact?

19.5 REDUCTION OF MONOSACCHARIDES: ALDITOLS

Aldoses (and ketoses) can be reduced with sodium borohydride to compounds called *alditols*.

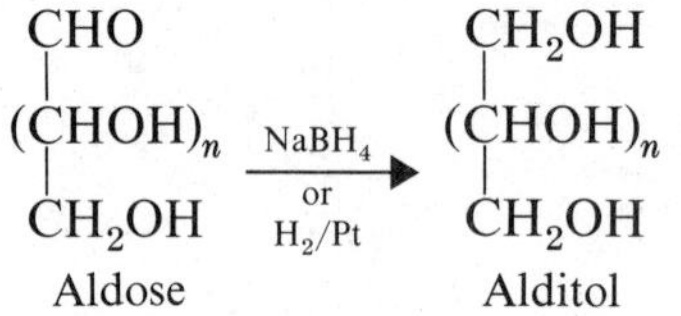

Aldose Alditol

Reduction of D-glucose, for example, yields D-glucitol.

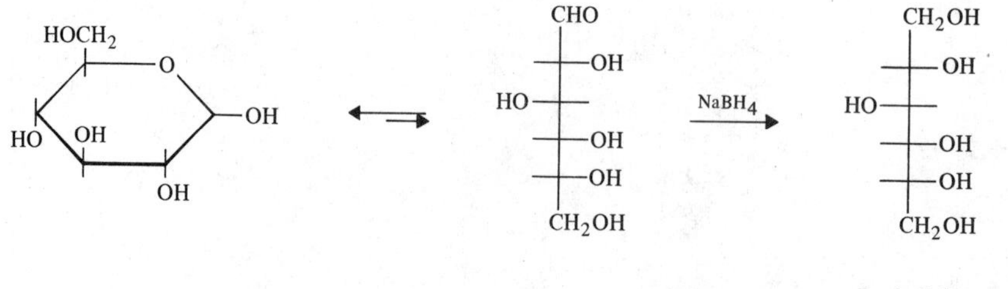

D–glucitol

Problem 19.15

(a) Would you expect D-glucitol to be optically active? (b) Write Fischer projection formulas for all of the D-aldohexoses that would yield *optically inactive alditols.*

19.6 REACTIONS OF MONOSACCHARIDES WITH PHENYLHYDRAZINE: OSAZONES

The aldehyde group of an aldose reacts with such carbonyl reagents as hydroxylamine and phenylhydrazine. With hydroxylamine, the product is the expected oxime. With enough phenylhydrazine, however, three moles of phenylhydrazine are consumed and a second phenylhydrazone group is introduced at C-2. The product is called an *phenylosazone.*

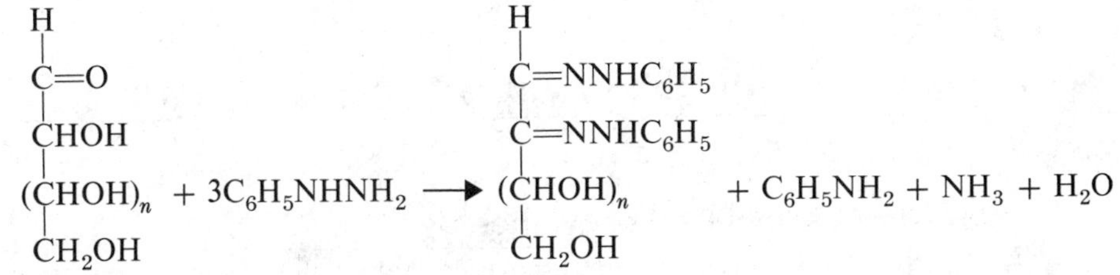

H
C=O
CHOH
$(CHOH)_n$ + $3C_6H_5NHNH_2$ ⟶
Aldose

H
$C{=}NNHC_6H_5$
$C{=}NNHC_6H_5$
$(CHOH)_n$ + $C_6H_5NH_2$ + NH_3 + H_2O
CH_2OH
Phenylosazone

Although the mechanism for osazone formation is not known with certainty, it probably depends on a series of reactions in which >C=N— behaves very much like >C=O.

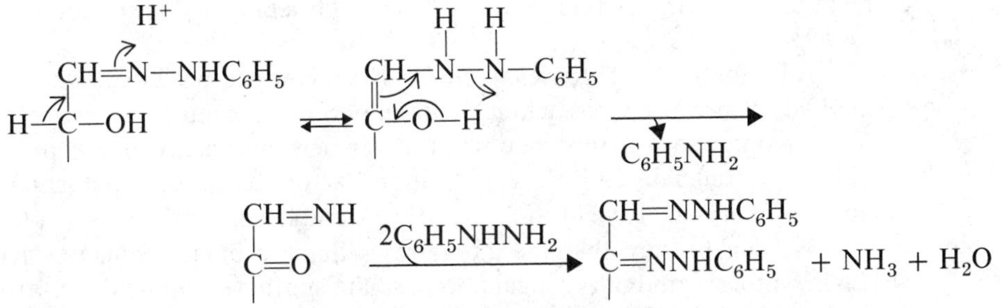

H^+
$CH{=}N{-}NHC_6H_5$
H—C—OH
⇌
$CH{-}N(H){-}N(H){-}C_6H_5$
C—O—H
⟶ $C_6H_5NH_2$

CH=NH
C=O
$\xrightarrow{2C_6H_5NHNH_2}$
$CH{=}NNHC_6H_5$
$C{=}NNHC_6H_5$ + NH_3 + H_2O

Osazone formation results in a loss of chirality at C-2 but does not affect other chiral centers; D-glucose and D-mannose, for example, yield the same phenylosazone:

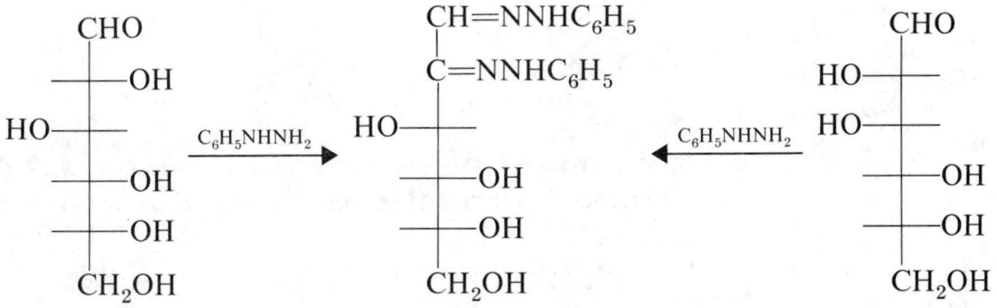

CHO
—OH
HO—
—OH
—OH
CH_2OH
D-Glucose

$\xrightarrow{C_6H_5NHNH_2}$

$CH{=}NNHC_6H_5$
$C{=}NNHC_6H_5$
HO—
—OH
—OH
CH_2OH
Same phenylosazone

$\xleftarrow{C_6H_5NHNH_2}$

CHO
HO—
HO—
—OH
—OH
CH_2OH
D-Mannose

This experiment, first done by Emil Fischer, establishes that D-glucose and D-mannose have the same configuration about C-3, C-4, and C-5. Diastereomeric aldoses (such as D-glucose and D-mannose) that differ only in configuration at C-2 are called *epimers.**

* The term epimer has taken on a broader meaning and is now often applied to any pair of diastereomers that differ only in the configuration of a single carbon.

Problem 19.16

Although D-fructose is not an epimer of D-glucose or D-mannose (D-fructose is a ketohexose), all three yield the same phenylosazone. (a) Using Fischer projection formulas, write an equation for the reaction of fructose with phenylhydrazine. (b) What information about the stereochemistry of D-fructose does this experiment yield?

19.7 SYNTHESIS AND DEGRADATION OF MONOSACCHARIDES

19.7A Kiliani-Fischer Synthesis

In 1886 Heinrich Kiliani discovered that an aldose can be converted to the epimeric aldonic acids of next higher carbon number through the addition of hydrogen cyanide and subsequent hydrolysis of the epimeric cyanohydrins. Fischer later extended this method by showing that aldonolactones obtained from the aldonic acids can be reduced to aldoses. Today this method for lengthening the carbon chain of an aldose is called the Kiliani-Fischer synthesis.

We can illustrate the Kiliani-Fischer synthesis with the synthesis of D-threose and D-erythrose (aldotetroses) from D-glyceraldehyde (an aldotriose) on page 893.

Addition of hydrogen cyanide to glyceraldehyde produces two epimeric cyanohydrins because the reaction creates a new chiral center. The cyanohydrins can be separated easily (since they are diastereomers) and each can be converted to an aldose through hydrolysis, acidification, lactonization, and reduction. One cyanohydrin ultimately yields D-(−)-erythrose and the other yields D-(−)-threose.

We can be sure that the aldotetroses that we obtain from this Kiliani-Fischer synthesis are both D sugars because the starting compound is D-glyceraldehyde and its chiral carbon is unaffected by the synthesis. On the basis of the Kiliani-Fischer synthesis we cannot know just which aldotetrose has both —OH groups on the right and which has the top —OH on the left. However, if we oxidize both aldotetroses to aldaric acids, one [D-(−)-erythrose] will yield an *optically inactive* product while the other [D-(−)-threose] will yield a product that is *optically active* (cf. Problem 19.9).

Problem 19.17

(a) What are the structures of L-(+)-threose and L-(+)-erythrose? (b) What aldotriose would you use to prepare them in a Kiliani-Fischer synthesis?

Problem 19.18

(a) Outline a Kiliani-Fischer synthesis of epimeric aldopentoses starting with D-(−)-erythrose (use Fischer projection formulas). (b) The two epimeric aldopentoses that one obtains are D-(−)-arabinose and D-(−)-ribose. Nitric acid oxidation of D-(−)-ribose yields an optically inactive aldaric acid, while similar oxidation of D-(−)-arabinose yields an optically active product. On the basis of this information alone, which Fischer projection formula represents D-(−)-arabinose and which represents D-(−)-ribose?

O
||
C—H
|
H—C—OH
|
CH_2OH

D-Glyceraldehyde

HCN

CN
|
H—C—OH
|
H—C—OH
|
CH_2OH

Epimeric cyanohydrins (separated)

CN
|
HO—C—H
|
H—C—OH
|
CH_2OH

(1) $Ba(OH)_2$
(2) H_3O^+

O
||
C—OH
|
H—C—OH
|
H—C—OH
|
CH_2OH

Epimeric aldonic acids

O
||
C—OH
|
HO—C—H
|
H—C—OH
|
CH_2OH

O, H, H, H, H, =O, OH, OH

Epimeric γ-aldonolactones

O, H, H, H, OH, =O, OH, H

Na-Hg,H_2O
pH 3–5

O
||
C—H
|
H—C—OH
|
H—C—OH
|
CH_2OH

D-(−)-Erythrose

O
||
C—H
|
HO—C—H
|
H—C—OH
|
CH_2OH

D-(−)-Threose

Problem 19.19

Subjecting D-(−)-threose to a Kiliani-Fischer synthesis yields two other epimeric aldopentoses, D-(+)-xylose and D-(−)-lyxose. D-(+)-Xylose can be oxidized (with nitric acid) to an optically inactive aldaric acid, while similar oxidation of D-(−)-lyxose gives an optically active product. What are the structures of D-(+)-xylose and D-(−)-lyxose?

Problem 19.20

There are eight aldopentoses. In Problems 19.18 and 19.19 you have arrived at the structures of four. What are the names and structures of the four that remain?

19.7B The Ruff Degradation

Just as the Kiliani-Fischer synthesis can be used to lengthen the chain of an aldose by one carbon, the Ruff degradation* can be used to shorten the chain by a similar unit. The Ruff degradation involves (1) oxidation of the aldose to an aldonic acid using bromine water and (2) oxidative decarboxylation of the aldonic acid to the next lower aldose using hydrogen peroxide and ferric sulfate. D-(−)-Ribose, for example, can be degraded to D-(−)-erythrose:

O=C—H, —OH, —OH, —OH, CH_2OH $\xrightarrow[H_2O]{Br_2}$ O=C—OH, —OH, —OH, —OH, CH_2OH $\xrightarrow[Fe_2(SO_4)_3]{H_2O_2}$ O=C—H, —OH, —OH, CH_2OH + CO_2

D-(−)-Ribose → D-Ribonic acid → D-(−)-Erythrose

Problem 19.21

The aldohexose D-(+)-galactose can be obtained by hydrolysis of *lactose,* a disaccharide found in milk. When D-(+)-galactose is treated with nitric acid it yields an optically inactive aldaric acid. When D-(+)-galactose is subjected to a Ruff degradation it yields D-(−)-lyxose (cf. problem 19.19). Using only these data, write the Fischer projection formula for D-(+)-galactose.

19.8 THE D-FAMILY OF ALDOSES

The Ruff degradation and the Kiliani-Fischer synthesis allow us to place all of the aldoses into families or "family trees" based on their relation to D- or L-glyceraldehyde. Such a tree is constructed in Fig. 19.6, and includes the structures of the D-aldohexoses, **1–8.**

Most, but not all, of the naturally occurring aldoses belong to the D family with D-(+)-glucose being by far the most common. D-(+)-Galactose can be obtained from milk sugar (lactose); but L-(−)-galactose occurs in a polysaccharide obtained from the vineyard snail, *Helix pomatia.* L-(+)-Arabinose is found widely, but D-(−)-arabinose is scarce, being found only in certain bacteria and sponges. Threose, lyxose, gulose, and allose do not occur naturally, but one or both forms (D or L) of each have been synthesized.

* Developed by Otto Ruff, a German chemist, 1871–1939.

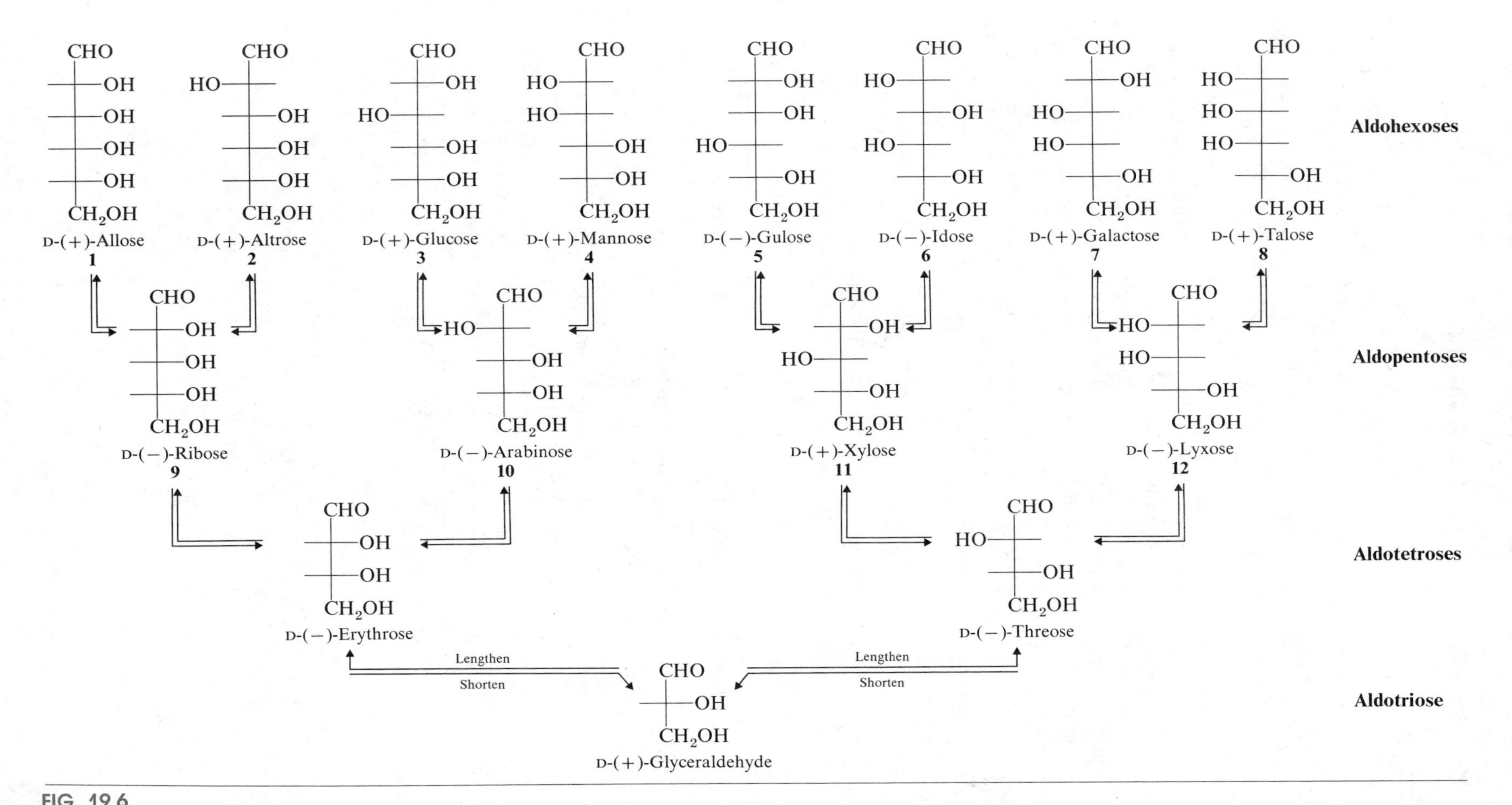

FIG. 19.6
*The D family of aldohexoses.**

* A useful mnemonic for the D-aldohexoses: all altruists gladly make gum in gallon tanks. Write the names in a line and above each write CH_2OH. Then, for C_5 write OH to the right all the way across. For C_4 write OH to the right four times, then four to the left; for C_3, write OH twice to the right, twice to the left, and repeat; C_2, alternate OH and H to the right. (from L. F. Fieser and Mary Fieser, *Organic Chemistry,* Reinhold Publishing Corp., New York, 1956, p. 359.)

19.9 FISCHER'S PROOF OF THE CONFIGURATION OF D-(+)-GLUCOSE

Emil Fischer began his work on the stereochemistry of (+)-glucose in 1888, only 12 years after van't Hoff and Le Bel had made their proposal concerning the tetrahedral structure of carbon. Only a small body of data was available to Fischer at the beginning: only a few monosaccharides were known, including (+)-glucose, (+)-arabinose, and (+)-mannose. [(+)-Mannose had just been synthesized by Fischer.] The sugars (+)-glucose and (+)-mannose were known to be aldohexoses; (+)-arabinose was known to be an aldopentose.

Since an aldohexose has four different chiral carbons, 2^4 (or 16) stereoisomers are possible—*one of which is (+)-glucose*. Fischer arbitrarily decided to limit his attention to the eight structures with the D configuration given in Fig. 19.6 (structures **1-8**).Fischer realized that we would be unable to differentiate between enantiomeric configurations because methods for determining the absolute configuration of organic compounds had not been developed. It was not until 1951, when Bijvoet (p. 317) determined the absolute configuration of D-(+)-tartaric acid (and, hence, D-(+)-glyceraldehyde) that Fischer's arbitrary assignment of (+)-glucose to the D family was known to be correct.

Fischer's assignment of structure **3** to (+)-glucose was based on the following reasoning.

1. Nitric acid oxidation of (+)-glucose gives an optically active aldaric acid. This eliminates structures **1** and **7** from consideration because both compounds would yield *meso*-aldaric acids.
2. *Degradation of (+)-glucose gives (−)-arabinose, and nitric acid oxidation of (−)-arabinose gives an optically active aldaric acid.* This means that (−)-arabinose cannot have configurations **9** or **11** and must have either structure **10** or **12.** It also establishes that (+)-glucose cannot have configuration **2, 5,** or **6.** This leaves structures **3, 4,** and **8** as possibilities for (+)-glucose.
3. A Kiliani-Fischer synthesis beginning with (−)-arabinose gives (+)-glucose and (+)-mannose; nitric acid oxidation of (+)-mannose gives an optically active aldaric acid. This together with the fact that (+)-glucose yields a different but also optically active aldaric acid establishes structure **10** as the structure of (−)-arabinose and eliminates structure **8** as a possible structure for (+)-glucose. Had (−)-arabinose been represented by structure **12** a Kiliani-Fischer synthesis would have given the two aldohexoses, **7** and **8,** one of which **(7)** would yield an optically inactive aldaric acid on nitric acid oxidation.

   Two structures now remain, **3** and **4;** one structure represents (+)-glucose and one represents (+)-mannose. Fischer realized that (+)-glucose and (+)-mannose were epimeric (at C-2) but a decision as to which compound was represented by which structure was most difficult.

5. Fischer had already developed a method for effectively *interchanging the two end groups* (CHO and CH_2OH) *of an aldose chain.* And, with brilliant logic, Fischer realized that if (+)-glucose has structure **4,** an interchange of end groups *will yield the same aldohexose:*

CHO / HO— / HO— / —OH / —OH / CH_2OH **4** —End-group interchange→ CH_2OH / HO— / HO— / —OH / —OH / CHO ≡ CHO / HO— / HO— / —OH / —OH / CH_2OH **4**

On the other hand, if (+)-glucose has structure **3,** *an end-group interchange will yield a different aldohexose,* **13:**

CHO / —OH / HO— / —OH / —OH / CH_2OH **3** —End-group interchange→ CH_2OH / —OH / HO— / —OH / —OH / CHO ≡ CHO / HO— / HO— / —OH / HO— / CH_2OH L-Gulose **13**

This new aldohexose, if it were formed, would be an L-sugar and it would be the mirror reflection of D-gulose. Thus its name would be L-gulose.

Fischer carried out the end-group interchange starting with (+)-glucose and *the product was the new aldohexose* **13.** This proved that (+)-glucose had structure **3.** It also established **4** as the structure for (+)-mannose and it proved the structure of L-(+)-gulose as **13.**

The procedure Fischer used for interchanging the ends of the (+)-glucose chain began with one of the γ-lactones of D-glucaric acid (cf. Problem 19.10) and was carried out as follows:

A γ-lactone of D-glucaric acid —Na-Hg→ L-Gulonic acid ⇌ γ-Aldonolactone —Na-Hg, pH 3–5→ L-(+)-Gulose **13**

Problem 19.22

Fischer actually had to subject both γ-lactones of D-glucaric acid (Problem 19.10) to the procedure just outlined. What product does the other γ-lactone yield?

19.10 METHYLATION OF MONOSACCHARIDES

Since methyl glycosides are acetals, they do not undergo isomerization or fragmentation in basic media. Thus, a methyl glucoside can be converted to a pentamethyl derivative by treating it with excess dimethyl sulfate in aqueous sodium hydroxide. Subsequent acid-catalyzed hydrolysis of the glycosidic methyl produces a tetramethyl-D-glucose.

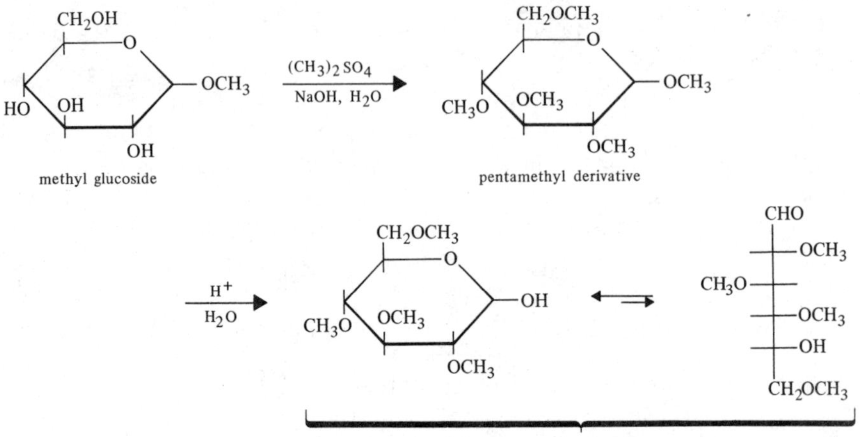

methyl glucoside $\xrightarrow[\text{NaOH, } H_2O]{(CH_3)_2SO_4}$ pentamethyl derivative $\xrightarrow[H_2O]{H^+}$ (2, 3, 4, 6-tetra-*O*-methyl-D-glucose)*

This method (developed by Haworth) allowed Hirst (in 1926) to demonstrate that the methyl glucosides have six-membered rings. Hirst oxidized the tetramethyl ether given by the reaction above and obtained a trimethoxyglutaric acid and a dimethoxysuccinic acid (below). These two products must come from cleavage of one of the two carbon-carbon bonds of the carbon bearing the hydroxyl group (in the open chain) and demonstrate that the 5-OH was involved in the oxide ring.

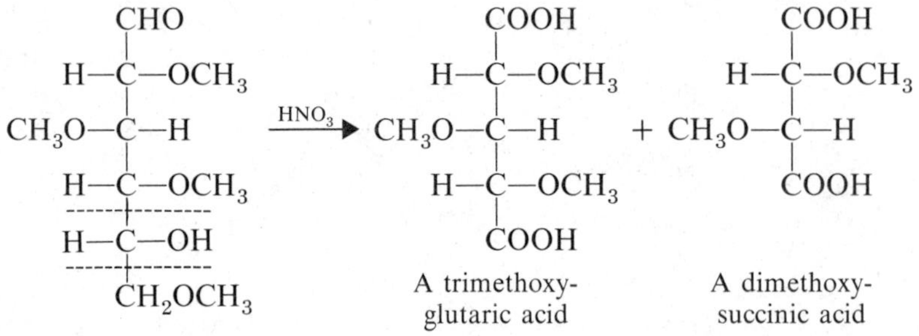

CHO, H—C—OCH_3, CH_3O—C—H, H—C—OCH_3, H—C—OH, CH_2OCH_3 $\xrightarrow{HNO_3}$ COOH, H—C—OCH_3, CH_3O—C—H, H—C—OCH_3, COOH + COOH, H—C—OCH_3, CH_3O—C—H, COOH

A trimethoxyglutaric acid; A dimethoxysuccinic acid

Problem 19.23

What products would have been obtained from the Haworth-Hirst procedure if the methyl glucoside had been a furanoside?

Methylation procedures, as we will see in the next section, are especially important in deciphering the structures of disaccharides.

* The designation ". . .—*O*—methyl. . ." in this name means that the methyl is attached to an oxygen.

19.11A Sucrose

Ordinary table sugar is a disaccharide called *sucrose*. Sucrose, the most widely occurring disaccharide, is found in all photosynthetic plants and is obtained commercially from sugar cane or sugar beets. Sucrose has the structure shown in Fig. 19.7.

The structure of sucrose is based on the following evidence.

1. Sucrose has the molecular formula $C_{12}H_{22}O_{11}$.
2. Acid-catalyzed hydrolysis of one mole sucrose yields one mole of D-glucose and one mole of D-fructose.
3. Sucrose is a nonreducing sugar; it gives negative tests with Benedict's and Tollens' solution. Sucrose does not form an osazone and does not undergo mutarotation. These facts mean that neither the glucose nor the fructose portions of sucrose has a hemiacetal or hemiketal group. Thus, the two hexoses must have a glycoside linkage that involves C-1 of glucose and C-2 of fructose, for only in this way will both carbonyl groups be present as full acetals (or glycosides).
4. The stereochemistry of the glycoside linkages can be inferred from experiments done with enzymes. Sucrose is hydrolyzed by an *α-glucosidase* obtained from yeast but not by *β*-glucosidases. This indicates *an α configuration at the glucoside portion*. Sucrose is also hydrolyzed by *sucrase,* an enzyme known to hydrolyze *β*-fructofuranosides but not *α*-fructofuranosides. This indicates *a β configuration at the fructoside portion.*
5. Methylation of sucrose gives an octamethyl derivative that, on hydrolysis, gives 2,3,4,6-tetra-*O*-methylglucose and 1,3,4,6-tetra-*O*-methylfructose. The identities of these two products demonstrate that the glucose portion is a *pyranoside* and that the fructose portion is a *furanoside.*

The structure of sucrose has been confirmed by X-ray analysis and by an unambiguous synthesis.

FIG. 19.7
Two representations of the formula for (+)-sucrose (α-D-glucopyranosyl β-D-fructofuranoside).

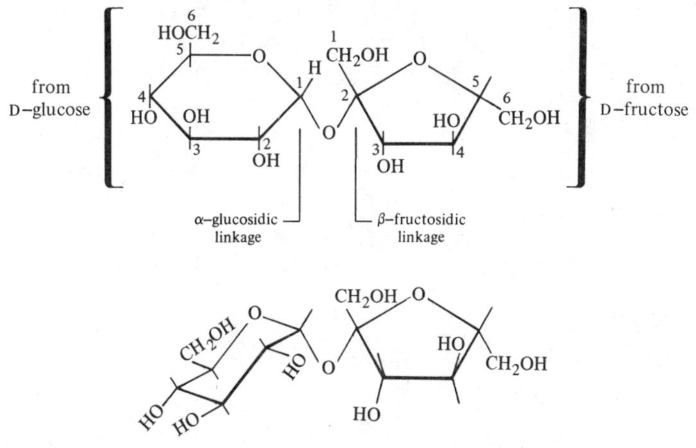

from D-glucose
from D-fructose
α-glucosidic linkage
β-fructosidic linkage

19.11B Maltose

When starch (p. 902) is hydrolyzed by the enzyme *diastase,* one product is a disaccharide known as *maltose* (Fig. 19.8).

1. When one mole maltose is subjected to acid-catalyzed hydrolysis it yields two moles of D-(+)-glucose.
2. Unlike sucrose, *maltose is a reducing sugar;* it gives positive tests with Fehling's, Benedict's, and Tollens' solutions. Maltose also reacts with phenylhydrazine to form a monophenylosazone (i.e., it incorporates two molecules of phenylhydrazine).
3. Maltose exists in two anomeric forms; α-maltose, $[\alpha]_D^{25} = +168°$, and β-maltose, $[\alpha]_D^{25} = +112°$. The maltose anomers undergo mutarotation to yield an equilibrium mixture, $[\alpha]_D^{25}$ $+136°$.

   Facts **2** and **3** demonstrate that one of the glucose residues of maltose is present in a hemiacetal form; the other, therefore, must be present as a glucoside. The configuration of this glucosidic linkage can be inferred as α, because maltose is hydrolyzed by α-glucosidases and not by β-glucosidases.
4. Maltose reacts with bromine water to form a monocarboxylic acid, maltonic acid (Fig. 19.9*a*). This fact, too, is consistent with the presence of only one hemiacetal group.
5. Methylation of maltonic acid followed by hydrolysis gives 2,3,4,6-tetra-*O*-methyl-D-glucose and 2,3,5,6-tetra-*O*-methyl-D-gluconic acid. That the first product has a free —OH at C-5 indicates that the nonreducing glucose portion is present as a pyranoside; that the second product, 2,3,5,6-tetra-*O*-methyl-D-gluconic acid, has a free —OH at C-4 indicates that this position was involved in a glycosidic linkage with the nonreducing glucose.

   Only the size of the reducing glucose ring needs to be determined.
6. Methylation of maltose itself, followed by hydrolysis (Fig. 19.9*b*), gives 2,3,4,6-tetra-*O*-methyl-D-glucose and 2,3,6-tri-*O*-methyl-D-glucose. The free —OH at C-5 in the latter product indicates that it must have been involved in the oxide ring and that the reducing glucose is present as a *pyranose*.

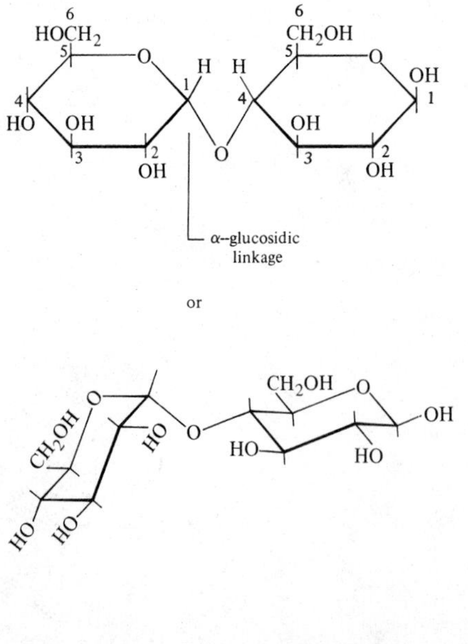

α–glucosidic linkage

or

FIG. 19.8
Two representations of the structure of the β anomer of (+)-maltose, 4-O-(α-D-glucopyranosyl) β-D-glucopyranose.

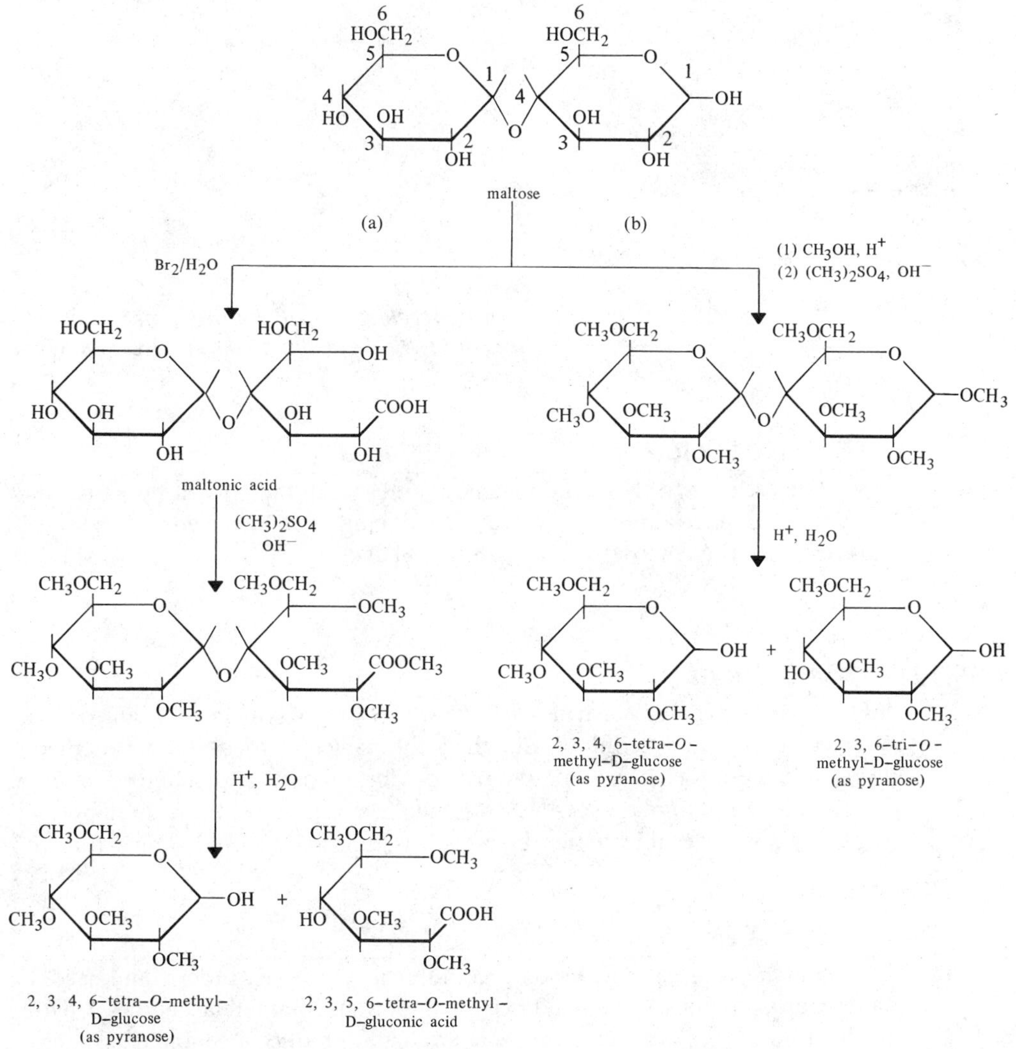

maltose

(a) (b)

Br_2/H_2O

(1) CH_3OH, H^+
(2) $(CH_3)_2SO_4$, OH^-

maltonic acid

$(CH_3)_2SO_4$
OH^-

H^+, H_2O

2, 3, 4, 6-tetra-*O*-methyl-D-glucose (as pyranose)

2, 3, 6-tri-*O*-methyl-D-glucose (as pyranose)

H^+, H_2O

2, 3, 4, 6-tetra-*O*-methyl-D-glucose (as pyranose)

2, 3, 5, 6-tetra-*O*-methyl-D-gluconic acid

FIG. 19.9
(a) *Oxidation of maltose to maltonic acid followed by methylation and hydrolysis.* (b) *Methylation and subsequent hydrolysis of maltose itself.*

19.11C Cellobiose

Partial hydrolysis of cellulose gives the disaccharide, cellobiose, $C_{12}H_{22}O_{11}$ (Fig. 19.10). Cellobiose resembles maltose in every respect except one: the configuration of its glycosidic linkage.

Cellobiose, like maltose, is a reducing sugar that, on acid-catalyzed hydrolysis, yields two moles of D-glucose. Cellobiose also undergoes mutarotation and forms a phenylosazone. Methylation studies show that C-1 of one glucose unit is connected in glycosidic linkage with C-4 of the other and that both rings are six membered. Unlike maltose, however, cellobiose is hydrolyzed by *β-glucosidases* and not by α-glucosidases: this indicates that the glycosidic linkage in cellobiose is β (Fig. 19.10).

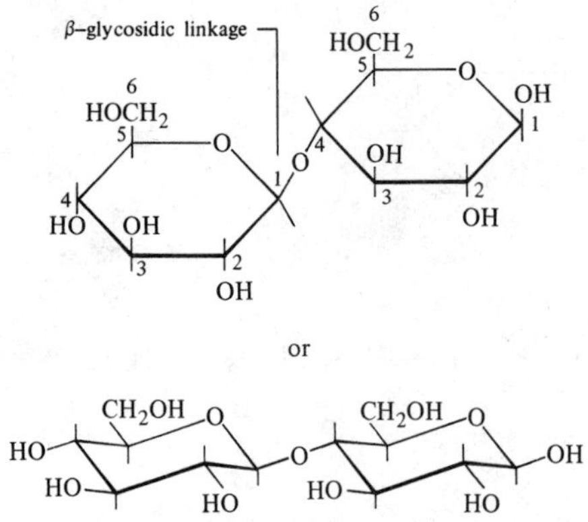

FIG. 19.10
Two representations of the β anomer of cellobiose, 4-O-(β-D-glucopyranosyl) β-D-glucopyranose.

19.11D Lactose

Lactose (Fig. 19.11) is a disaccharide present in the milk of humans, cows, and almost all other animals. Lactose is a reducing sugar that hydrolyzes to yield D-glucose and D-galactose; the glycosidic linkage is β.

19.12 POLYSACCHARIDES

Three important polysaccharides, all of which are polymers of D-glucose, are starch, glycogen, and cellulose. Starch is the principal food reserve of plants; glycogen functions as a carbohydrate reserve for animals; and cellulose serves as structural material in plants. As we examine the structures of these three polysaccharides we will be able to see how each is especially suited for its function.

19.12A Starch

Starch occurs as microscopic granules in the roots, tubers, and seeds of plants. Corn, potatoes, wheat, and rice are important commercial sources. Heating starch with water causes the granules to swell and produce a colloidal suspension from which two major components can be isolated. One fraction is called *amylose* and the other *amylopectin*. Most starches yield 10–20% amylose and 80–90% amylopectin.

FIG. 19.11
Two representations of the β anomer of lactose, 4-O-(β-D-galactopyranosyl) β-D-glucopyranose.

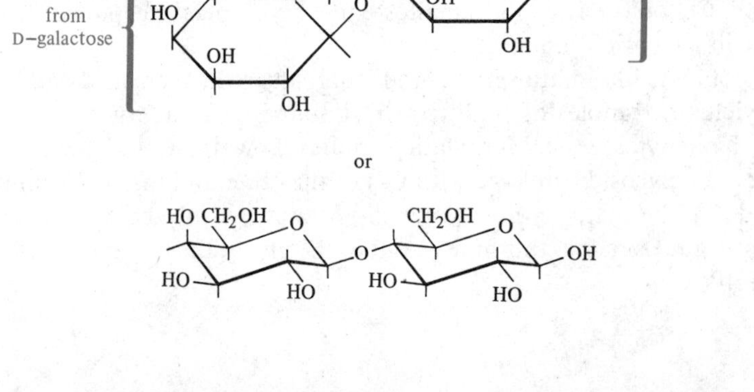

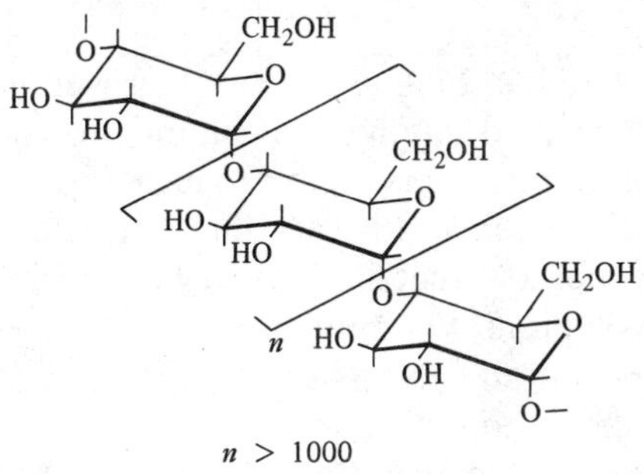

$n > 1000$

FIG. 19.12
Partial structure of amylose, an unbranched polymer of D*-glucose connected in* α*, 1:4-glycosidic linkages.*

Physical measurements show that amylose typically consists of more than 1000 D-glucopyranoside units *connected in* α *linkages* between C-1 of one unit and C-4 of the next (Fig. 19.12). Thus, in the ring size of its glucose units and in the configuration of the glycosidic linkages between them, amylose resembles maltose.

Chains of D-glucose units with α-glycosidic linkages such as those of amylose tend to assume a helical arrangement. This results in a compact shape for the amylose molecule even though its molecular weight is quite large (150,000–600,000).

Amylopectin has a structure similar to that of amylose (i.e., α,1 : 4 links), with the exception that in amylopectin the chains are branched. Branching takes place between C-6 of one glucose unit and C-1 of another and occurs at intervals of 20–25 glucose units (Fig. 19.13). Physical measurements indicate that amylopectin has a molecular weight of one to six million; thus amylopectin consists of hundreds of interconnecting chains of 20–25 glucose units each.

FIG. 19.13
Partial structure of amylopectin.

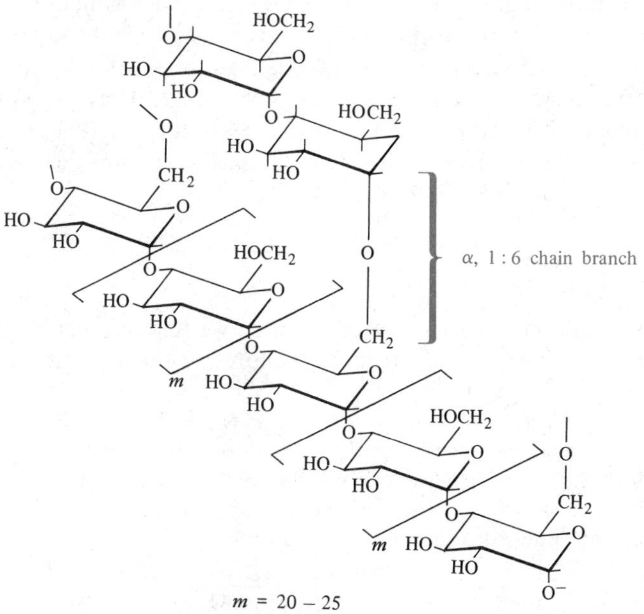

α, 1 : 6 chain branch

$m = 20 - 25$

19.12B Glycogen

Glycogen has a structure very much like that of amylopectin; however, in glycogen the chains are much more highly branched. Methylation and hydrolysis of glycogen indicates that there is one end group for every 10 to 12 glucose units; branches may occur as often as every 6. Glycogen has a very high molecular weight. Studies of glycogens isolated under conditions that minimize the likelihood of hydrolysis indicate molecular weights as high as 100 million.

The size and structure of glycogen beautifully suit its function as reserve carbohydrate for animals. First, its size makes it too large to diffuse across cell membranes; thus, glycogen remains inside the cell where it is needed as an energy source. Second, because glycogen incorporates tens of thousands of glucose units in a single molecule it solves an important osmotic problem for the cell. Were so many glucose units present in the cell as individual molecules, the osmotic pressure within the cell would be enormous—so large that the cell membrane would almost certainly break.* Finally, the localization of glucose units within a large, highly branched structure simplifies one of the cell's logistical problems: that of having a ready source of glucose when cellular glucose concentrations are low and of being able to store glucose rapidly when cellular glucose concentrations are high. There are enzymes within the cell that catalyze the reactions by which glucose units are detached from (or attached to) glycogen. These enzymes operate at end groups by hydrolyzing (or forming) α,1 : 4 glycosidic linkages. Because glycogen is so highly branched, a very large number of end groups are available at which these enzymes can operate. At the same time the overall concentration of glycogen (in moles per liter) is quite low because of its enormous molecular weight.

Amylopectin presumably serves a similar function in plants. The fact that amylopectin is less highly branched than glycogen is, however, not a serious disadvantage. Plants have a much lower metabolic rate than animals—and plants, of course, do not require sudden bursts of energy.

Animals store energy as fats (triacylglycerols) as well as in glycogen. Fats, because they are more highly reduced, are capable of furnishing much more energy. The metabolism of a typical fatty acid, for example, liberates more than twice as much energy per carbon as glucose or glycogen. Why then, we might ask, has Nature developed two different energy repositories? Glucose (from glycogen) is readily available and is highly water soluble.* Glucose, as a result, diffuses rapidly through the aqueous medium of the cell and serves as an ideal source of "ready energy." Long-chain fatty acids, by contrast, are almost insoluble in water and their concentration inside the cell could never be very high. They would be a poor source of energy if the cell were in an energy pinch. On the other hand, fatty acids (as triacylglycerols) because of their caloric richness are an excellent energy repository for long-term energy storage.

19.12C Cellulose

When we examine the structure of cellulose we find another example of a polysaccharide in which nature has arranged monomeric glucose units in a manner

* The phenomenon of osmotic pressure occurs whenever two solutions of different concentrations are separated by a membrane that will allow penetration (by osmosis) of the solvent but not of the solute. The osmotic pressure, π, on one side of the membrane is related to the number of moles of solute particles, n, the volume of the solution, V, and RT (the gas constant times the absolute temperature):

$$\pi V = n\,RT$$

* Glucose is actually liberated as glucose-6-phosphate, which is also water soluble.

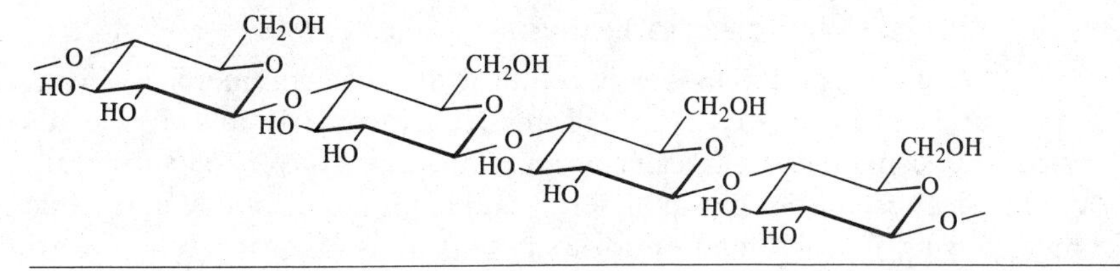

FIG. 19.14
A portion of a cellulose chain.

that suits its function. Cellulose contains D-glucopyranoside units linked in 1:4 fashion in very long unbranched chains. Unlike starch and glycogen, however, the linkages in cellulose are *β-glycosidic linkages* (Fig. 19.14). This configuration of the anomeric carbons of cellulose makes cellulose chains essentially linear; they do not tend to coil into helical structures as do glucose polymers when linked in an α,1:4 manner.

The linear arrangement of β-linked glucose units in cellulose presents a uniform distribution of —OH groups on the outside of each chain. When two or more cellulose chains make contact, the hydroxyl groups are ideally situated to "zip" the chains together by forming hydrogen bonds. Zipping many cellulose chains together in this way gives a highly insoluble, rigid, and fibrous polymer that is ideal as cell-wall material for plants.

This special property of cellulose chains, we should emphasize, is not just a result of β:1:4 glycosidic linkages; it is also a consequence of the precise stereochemistry of D-glucose at each chiral carbon. Were D-galactose or D-allose units linked in a similar fashion, they almost certainly would not give rise to a polymer with properties like cellulose. Thus, we get another glimpse of why D-glucose occupies such a special position in the chemistry of plants and animals. Not only is it the most stable aldohexose (because it can exist in a chair conformation that allows all of its bulky groups to occupy equatorial positions), but its special stereochemistry also allows it to form helical structures when α linked as in starches, and rigid linear structures when β linked as in cellulose.*

Perhaps we should ask ourselves one other question: Why has Nature "chosen" D-(+)-glucose for its special role rather than L-(−)-glucose, its mirror image? Here an answer cannot be given with any certainty. The selection of D-(+)-glucose may simply have been a random event early in the course of the evolution of enzyme catalysts. Once this selection was made, however, the chirality of the active sites of the enzymes involved would retain a bias toward D-(+)-glucose and against L-(−)-glucose (because of the improper fit of the latter). Once introduced, this bias would be perpetuated and extended to other catalysts.

Finally, when we speak of Nature selecting or choosing a particular molecule for a given function, we do not mean to imply that evolution operates on a molecular level. Evolution, of course, takes place at the level of organism populations, and molecules are selected only in the sense that their use gives the organism an increased likelihood of surviving and procreating.

* Another interesting and important fact about cellulose: the digestive enzymes of humans cannot attack its β, 1:4 linkages. Hence, cellulose cannot serve as a food source for humans, as can starch. Cows and termites, however, can use cellulose (of grass and wood) as a food source because symbiotic bacteria in their digestive systems furnish β-glucosidases.

19.12D Cellulose Derivatives

A number of derivatives of cellulose are used commercially. Most of these compounds in which two or three of the free hydroxyl groups of each glucose unit have been converted to an ester or an ether. This conversion substantially alters the physical properties of the material making it more soluble in organic solvents and allowing it to be made into fibers and films. Treating cellulose with acetic anhydride produces the triacetate known as "Arnel" used widely in the textile industry. Cellulose trinitrate, also called "gun cotton" or nitrocellulose, is used in explosives.

Rayon is made by treating cellulose (from cotton or wood pulp) with carbon disulfide in a basic solution. This converts cellulose to a soluble xanthate:

$$\text{Cellulose—OH} + CS_2 \xrightarrow{\text{NaOH}} \underset{\text{Cellulose xanthate}}{\text{Cellulose—O—}\overset{\overset{\displaystyle S}{\|}}{C}\text{—}S^- \, Na^+}$$

The solution of cellulose xanthate is then passed through a small orifice or slit into an acidic solution. This regenerates the —OH groups of cellulose causing it to precipitate as a fibers or a sheet.

$$\text{Cellulose—O—}\overset{\overset{\displaystyle S}{\|}}{C}\text{—}S^- \, Na^+ \xrightarrow{H^+} \underset{\text{Rayon or cellophane}}{\text{Cellulose—OH}}$$

The fibers are *rayon;* the sheets, after softening with glycerol, are *cellophane.*

19.13 SUGARS THAT CONTAIN NITROGEN

19.13A Glycosylamines

Sugars in which an amino group replaces the anomeric —OH are called glycosylamines. Examples are β-D-glucopyranosylamine and adenosine below.

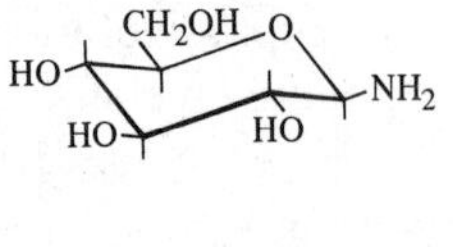

β-D-glucopyranosylamine

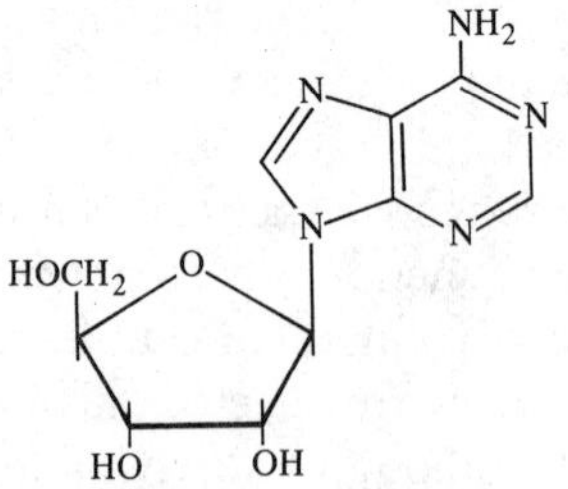

adenosine

Adenosine is an example of a glycosylamine that is also called a *nucleoside.* Nucleosides are glycosylamines in which the amino-component is a pyrimidine or a purine (Section 18.2) and in which the sugar component is either D-ribose or 2-deoxy-D-ribose (i.e., D-ribose minus the oxygen at the 2-position). Nucleosides are the important components of RNA (ribonucleic acid) and DNA (deoxyribonucleic acid). We will describe their properties in detail in Special Topic P.

19.13B Amino sugars

Sugars in which an amino group replaces an alcoholic —OH group are called amino sugars. Amino sugars occur widely in nature. Two frequently encountered are D-glucosamine and D-galactosamine. D-Glucosamine can be obtained by hydrolysis of chitin, a polysaccharide found in the shells of lobsters and crabs and the external skeletons of insects. D-Galactosamine can be prepared by hydrolysis of chondroitin, a polysaccharide found in cartilage and nasal mucus.

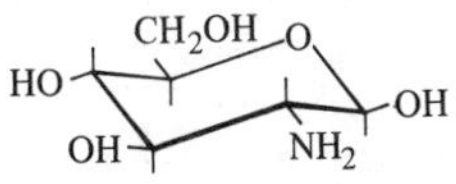

β–D–glucosamine (β–2–amino–2–deoxy–D–glucopyranose)

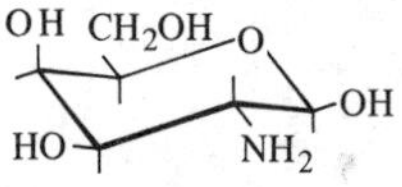

β–D–galactosamine (β–2–amino–2–deoxy–D–galatopyranose)

19.13C Carbohydrate Antibiotics

One of the important discoveries in carbohydrate chemistry was the isolation (in 1944) of the carbohydrate antibiotic called *streptomycin*. Streptomycin is made up of three subunits shown below:

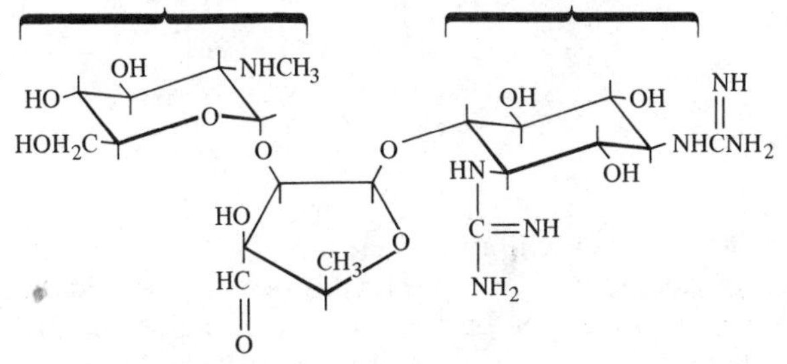

2–deoxy–2–methylamino–L–glucose

streptidine

streptose

All three components are unusual: the amino sugar is based on L-glucose; streptose is a branched-chain monosaccharide; and streptidine is not a sugar at all, it is a cyclohexane derivative called an amino cyclitol.

Other members of this family are antibiotics called kanamycins, neomycins, and gentamicins (not shown). All are based on an amino cyclitol linked to one or more amino sugars. The glycocidic linkage is nearly always α. These antibiotics are especially useful against bacteria that are resistant to penicillins.

Additional Problems

19.24

Give appropriate structural formulas to illustrate each of the following:

(a) An aldopentose
(b) A ketohexose
(c) An L-monosaccharide
(d) A glycoside
(e) An aldonic acid
(f) An aldaric acid
(g) An aldonolactone
(h) A pyranose
(i) A furanose
(j) A reducing sugar
(k) A pyranoside
(l) A furanoside
(m) Epimers
(n) Anomers
(o) A phenylosazone
(p) A disaccharide
(q) A polysaccharide
(r) A nonreducing sugar

19.25
Draw Haworth and conformational formulas for each of the following:
(a) α-D-Allopyranose
(b) Methyl β-D-allopyranoside
(c) Methyl 2,3,4,6-tetra-*O*-methyl-β-D-allopyranoside

19.26
Draw structures for furanose and pyranose forms of D-ribose. Show how you could use periodate oxidation to distinguish between a methyl ribofuranoside and a methyl ribopyranoside.

19.27
One reference book lists D-mannose as being dextrorotatory; another lists it as being levorotatory. Both references are correct. Explain.

19.28
The starting material for the commercial synthesis of vitamin C (Section 16.7B) is L-sorbose (below); it can be synthesized from D-glucose through the following reaction sequence:

$$\text{D-Glucose} \xrightarrow[\text{Ni}]{H_2} \text{D-Glucitol} \xrightarrow[\textit{Acetobacter suboxydans}]{O_2} \begin{array}{c} CH_2OH \\ | \\ C{=}O \\ | \\ HO{-}C{-}H \\ | \\ H{-}C{-}OH \\ | \\ HO{-}C{-}H \\ | \\ CH_2OH \end{array}$$

L-Sorbose

The second step of this sequence illustrates the use of a bacterial oxidation; the microorganism, *Acetobacter suboxydans,* accomplishes this step in 90% yield. The overall result of the synthesis is the transformation of a D-aldohexose, D-glucose, into an L-ketohexose, L-sorbose. What does this mean about the specificity of the bacterial oxidation?

19.29
What two aldoses would yield the same phenylosazone as L-sorbose (Problem 19.28)?

19.30
In addition to fructose (problem 19.16) and sorbose (Problem 19.28) there are two other 2-ketohexoses, *psicose* and *tagatose.* D-Psicose yields the same phenylosazone as D-allose (or D-altrose); D-tagatose yields the same osazone as D-galactose (or D-talose). What are the structures of D-psicose and D-tagatose?

19.31
A, B, and **C** are three aldohexoses. **A** and **B** yield the same optically active alditol when they are reduced with hydrogen and a catalyst; **A** and **B** yield different phenylosazones when treated with phenylhydrazine; **B** and **C** give the same phenylosazone but different alditols. Assuming that all are D-sugars, give names and structures for **A, B,** and **C.**

19.32
Although monosaccharides undergo complex isomerizations in base (cf. Problem 19.7), aldonic acids are epimerized specifically at C-2 when they are heated with pyridine. Show how you could make use of this reaction in a synthesis of D-mannose from D-glucose.

19.33
(a) The most stable conformation of most aldopyranoses is one in which the largest

group—the —CH_2OH group—is equatorial. However, D-idopyranose exists primarily in a conformation with an axial —CH_2OH group. Write formulas for the two chair conformations of α-D-idopyranose (one with the —CH_2OH group axial and one with the —CH_2OH group equatorial) and provide an explanation.

19.34

(a) Heating D-altrose with dilute acid produces a nonreducing *anhydro sugar,* $C_6H_{10}O_5$. Methylation of the anhydro sugar followed by acid hydrolysis yields 2,3,4-tri-*O*-methyl-D-altrose. The formation of the anhydro sugar takes place through a chair conformation of β-D-altropyranose in which the —CH_2OH group is axial. What is the structure of the anhydro sugar and how is it formed? (b) D-Glucose also forms an anhydro sugar but the conditions required are much more drastic than for the corresponding reaction of D-altrose. Explain.

19.35

Although mutarotation of the D-glucopyranoses is catalyzed by either acids or bases, one of the most effective catalysts is 2-hydroxypyridine:

HO N

2-Hydroxypyridine

Write a mechanism that accounts for this.

19.36

Show how the following experimental evidence can be used to deduce the structure of lactose (Sect. 19.11).

1. Acid hydrolysis of lactose, $C_{12}H_{22}O_{11}$, gives equimolar quantities of D-glucose and D-galactose. Lactose undergoes a similar hydrolysis in the presence of a *β-galactosidase.*
2. Lactose is a reducing sugar and forms a phenylosazone; it also undergoes mutarotation.
3. Oxidation of lactose with bromine water followed by hydrolysis with dilute acid gives D-galactose and D-gluconic acid.
4. Bromine water oxidation of lactose followed by methylation and hydrolysis gives 2,3,5,6-tetra-*O*-methyl-D-δ-gluconolactone and 2,3,4,6-tetra-*O*-methyl-D-galactose.
5. Methylation and hydrolysis of lactose gives 2,3,6-tri-*O*-methyl-D-glucose and 2,3,4,6-tetra-*O*-methyl-D-galactose.

19.37

Deduce the structure of the disaccharide, *melibiose,* from the following data.

1. Melibiose is a reducing sugar that undergoes mutarotation and forms a phenylosazone.
2. Hydrolysis of melibiose with acid or with an *α-galactosidase* gives D-galactose and D-glucose.
3. Bromine-water oxidation of melibiose gives *melibionic acid.* Hydrolysis of melibionic acid gives D-galactose and D-gluconic acid. Methylation of melibionic acid followed by hydrolysis gives 2,3,4,6-tetra-*O*-methyl-D-galactose and 2,3,4,5-tetra-*O*-methyl-D-gluconic acid.
4. Methylation and hydrolysis of melibiose gives 2,3,4,6-tetra-*O*-methyl-D-galactose and 2,3,4-tri-*O*-methyl-D-glucose.

19.38

Trehalose is a disaccharide that can be obtained from yeasts, fungi, sea urchins, algae, and insects. Deduce the structure of trehalose from the following information.

1. Acid hydrolysis of trehalose yields only D-glucose.
2. Trehalose is hydrolyzed by α-glucosidases but not by β-glucosidases.

3. Trehalose is a nonreducing sugar; it does not mutarotate, form a phenylosazone, or react with bromine water.
4. Methylation of trehalose followed by hydrolysis yields two moles of 2,3,4,6-tetra-*O*-methyl-D-glucose.

19.39

Outline chemical tests that will distinguish between each of the following:

(a) D-Glucose and D-glucitol
(b) D-Glucitol and D-glucaric acid
(c) D-Glucose and D-fructose
(d) D-Glucose and D-galactose
(e) Sucrose and maltose
(f) Maltose and maltonic acid
(g) Methyl β-D-glucopyranoside and 2,3,4,6-tetra-*O*-methyl-β-D-glucopyranose
(h) Methyl α-D-ribofuranoside (**I**) and methyl 2-deoxy-α-D-ribofuranoside (**II**)

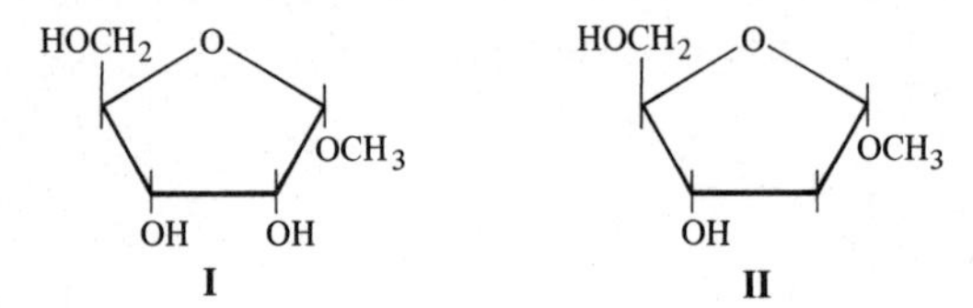

*** 19.40**

A group of oligosaccharides called *Schardinger dextrins* can be isolated from *Bacillus macerans* when the bacillus is grown on a medium rich in amylose. These oligosaccharides are all *nonreducing*. A typical Schardinger dextrin undergoes hydrolysis when treated with an acid or an α-glucosidase to yield six, seven, or eight molecules of D-glucose. Complete methylation of a Schardinger dextrin followed by acid hydrolysis yields only 2,3,6,tri-*O*-methyl-D-glucose. Propose a general structure for a Schardinger dextrin.

*** 19.41**

Isomaltose is a disaccharide that can be obtained by enzymic hydrolysis of amylopectin. Deduce the structure of isomaltose from the following data:

(1) Hydrolysis of one mole of isomaltose by acid or by an α-glucosidase gives two moles of D-glucose.
(2) Isomaltose is a reducing sugar.
(3) Isomaltose is oxidized by bromine water to isomaltonic acid. Methylation of isomaltonic acid and subsequent hydrolysis yields 2,3,4,6-tetra-*O*-methyl-D-glucose and 2,3,4,5-tetra-*O*-methyl-D-gluconic acid.
(4) Methylation of isomaltose itself followed by hydrolysis gives 2,3,4,6-tetra-*O*-methyl-D-glucose and 2,3,4-tri-*O*-methyl-D-glucose.

*** 19.42**

Stachyose occurs in the roots of several species of plants. Deduce the structure of stachyose from the following data:

(1) Acidic hydrolysis of one mole of stachyose yields two moles of D-galactose, one mole of D-glucose, and one mole of D-fructose.
(2) Stachyose is a nonreducing sugar.
(3) Treating stachyose with an α-galactosidase produces a mixture containing D-galactose, sucrose, and a nonreducing trisaccharide called *raffinose*.
(4) Acidic hydrolysis of raffinose gives D-glucose, D-fructose, and D-galactose. Treating raffinose with an α-galactosidase yields D-galactose and sucrose. Treating raffinose with invertase (an enzyme that hydrolyzes sucrose) yields fructose and *melibiose* (cf. problem 19.37).
(5) Methylation of stachyose followed by hydrolysis yields 2,3,4,6-tetra-*O*-methyl-D-galactose, 2,3,4-tri-*O*-methyl-D-galactose, 2,3,4-tri-*O*-methyl-D-glucose, and 1,3,4,6-tetra-*O*-methyl-D-fructose.

*** 19.43**

Arbutin, a compound that can be isolated from the leaves of bearberry, cranberry, and pear trees, has the molecular formula $C_{12}H_{16}O_7$. When arbutin is treated with aqueous acid or with a β-glucosidase, the reaction produces D-glucose and a compound **X** with the molecular formula $C_6H_6O_2$. The proton nmr spectrum of compound **X** consists of two singlets, one at δ6.8 (4H) and one at δ7.9 (2H). Methylation of arbutin followed by acidic hydrolysis yields 2,3,4,6-tetra-*O*-methyl-D-glucose and a compound **Y**, $C_7H_8O_2$. Compound **Y** is soluble in dilute aqueous NaOH but is insoluble in aqueous $NaHCO_3$. The proton nmr spectrum of **Y** shows a singlet at δ3.9 (3H), a singlet at δ4.8 (1H) and a multiplet (that resembles a singlet) at δ6.8 (4H). Treating compound **Y** with aqueous NaOH and $(CH_3)_2SO_4$ produces compound **Z**, $C_8H_{10}O_2$. The proton nmr spectrum of **Z** consists of two singlets, one at δ3.75 (6H) and one at δ6.8 (4H). Propose structures for arbutin and for compounds **X**, **Y**, and **Z**.

*** 19.44**

D-Glucose reacts with acetone in the presence of sulfuric acid to yield a compound with the molecular formula $C_{12}H_{19}O_6$ which has been given the common name "diacetone glucose." D-Galactose undergoes a similar reaction to yield an isomeric product. "Diacetone glucose" has three five-membered rings; the corresponding compound obtained from D-galactose has two five-membered rings and a six-membered ring. (a) Write structures for these two compounds and (b) explain their formation. (Hint: use models.)

20 SYNTHESIS AND REACTIONS OF β-DICARBONYL COMPOUNDS: MORE CHEMISTRY OF ENOLATE IONS

20.1 INTRODUCTION

Compounds having two carbonyl groups separated by an intervening carbon

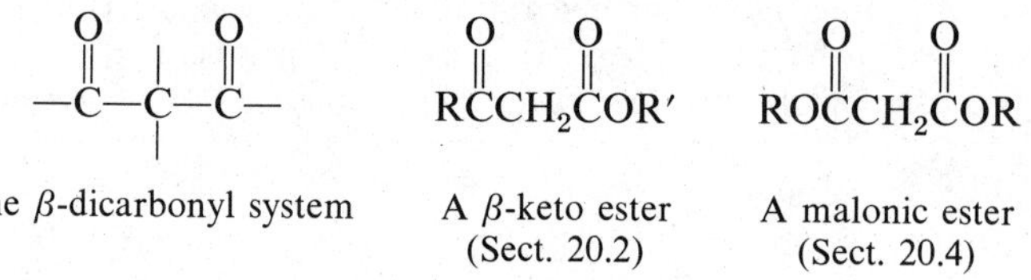

$$\underset{\text{The }\beta\text{-dicarbonyl system}}{-\overset{O}{\overset{\|}{C}}-\overset{|}{\underset{|}{C}}-\overset{O}{\overset{\|}{C}}-} \qquad \underset{\substack{\text{A }\beta\text{-keto ester}\\ \text{(Sect. 20.2)}}}{R\overset{O}{\overset{\|}{C}}CH_2\overset{O}{\overset{\|}{C}}OR'} \qquad \underset{\substack{\text{A malonic ester}\\ \text{(Sect. 20.4)}}}{RO\overset{O}{\overset{\|}{C}}CH_2\overset{O}{\overset{\|}{C}}OR}$$

are called β-dicarbonyl compounds and these compounds are highly versatile reagents for organic synthesis. In this chapter we will explore some of the methods for preparing β-dicarbonyl compounds and some of their important reactions.

20.2 THE CLAISEN CONDENSATION; THE SYNTHESIS OF β-KETO ESTERS

When ethyl acetate reacts with sodium ethoxide, it undergoes *a condensation reaction.* After acidification, the product is a β-keto ester, ethyl acetoacetate (commonly called *acetoacetic ester*).

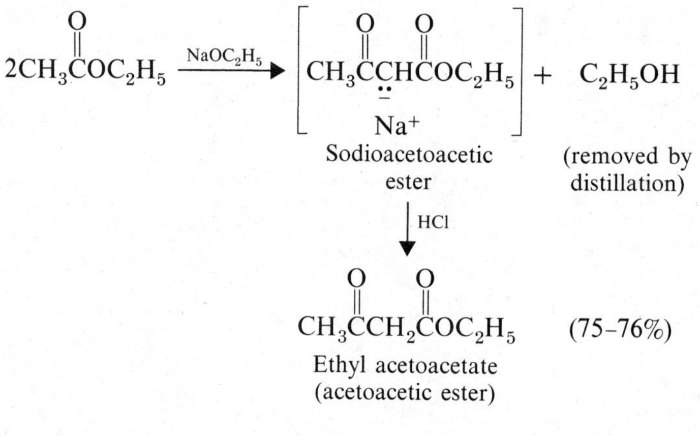

$$2CH_3\overset{O}{\overset{\|}{C}}OC_2H_5 \xrightarrow{NaOC_2H_5} \underset{\text{Sodioacetoacetic ester}}{\left[CH_3\overset{O}{\overset{\|}{C}}\underset{\ddot{-}}{C}H\overset{O}{\overset{\|}{C}}OC_2H_5 \; Na^+\right]} + \underset{\substack{\text{(removed by}\\ \text{distillation)}}}{C_2H_5OH}$$

$$\xrightarrow{HCl} \underset{\substack{\text{Ethyl acetoacetate}\\ \text{(acetoacetic ester)}}}{CH_3\overset{O}{\overset{\|}{C}}CH_2\overset{O}{\overset{\|}{C}}OC_2H_5} \quad (75\text{–}76\%)$$

Condensations of this type occur with many other esters and are known generally as *Claisen condensations.* Ethyl pentanoate, for example, reacts with sodium ethoxide to give a β-keto ester:

$$2CH_3CH_2CH_2CH_2\overset{O}{\overset{\|}{C}}OC_2H_5 \xrightarrow{NaOCH_2CH_3} \left[CH_3CH_2CH_2CH_2\overset{O}{\overset{\|}{C}}-\overset{Na^+}{\underset{\displaystyle CH_2CH_2CH_3}{\ddot{C}^-}}-\overset{O}{\overset{\|}{C}}OC_2H_5 \right] + C_2H_5OH$$

Ethyl pentanoate

$$\downarrow CH_3COOH$$

$$CH_3CH_2CH_2CH_2\overset{O}{\overset{\|}{C}}-\underset{\displaystyle CH_2CH_2CH_3}{CH}-\overset{O}{\overset{\|}{C}}OC_2H_5$$

(77%)

If we look closely at these two examples we can see that, overall, both reactions involve a condensation in which one ester loses an α-hydrogen and the other loses an ethoxide ion; that is,

$$R-CH_2\overset{O}{\overset{\|}{C}}\boxed{-OC_2H_5 + H}-\underset{\displaystyle R}{CH}\overset{O}{\overset{\|}{C}}-OC_2H_5 \xrightarrow[(2)\ H^+]{(1)\ NaOC_2H_5} R-CH_2\overset{O}{\overset{\|}{C}}-\underset{\displaystyle R}{CH}\overset{O}{\overset{\|}{C}}OC_2H_5 + C_2H_5OH$$

A β-keto ester

We can understand how this happens if we examine the reaction mechanism.

The first step of a Claisen condensation resembles that of an aldol addition (Section 16.13). Ethoxide ion abstracts an α-proton from the ester. Although the α-protons of an ester are not as acidic as those of aldehydes and ketones, the enolate anion that forms is stabilized by resonance in a similar way.

Step 1
$$\underset{\displaystyle H}{\overset{\alpha}{RCH}}-\overset{O}{\overset{\|}{C}}OC_2H_5 + {:}\ddot{\underset{..}{O}}^-C_2H_5 \rightleftharpoons R\underset{..}{\overset{-}{C}}H-\overset{\ddot{O}:}{\overset{\|}{C}}OC_2H_5 + C_2H_5OH$$

$$\updownarrow$$

$$R-CH{=}\overset{:\ddot{O}:^-}{\overset{|}{C}}OC_2H_5$$

In the second step the enolate anion attacks the carbonyl carbon of a second molecule of the ester. It is at this point that the Claisen condensation and the aldol addition *differ,* and they differ in an understandable way: in the aldol reaction nucleophilic attack leads to *addition;* in the Claisen condensation it leads to *substitution.*

Step 2 $RCH_2C(=O)OC_2H_5 + {}^{-}:CH(R)COC_2H_5 \rightleftharpoons RCH_2C(O^-)(OC_2H_5)-CH(R)-COC_2H_5 \rightleftharpoons RCH_2C(=O)-CH(R)-COC_2H_5 + {}^{-}:\ddot{O}C_2H_5$

Although the products of this second step are a β-keto ester and ethoxide ion, all of the equilibria up to this point have been unfavorable. Very little product would be formed if this were the last step in the reaction.

The final step of a Claisen condensation is an acid-base reaction that takes place between ethoxide ion and the β-keto ester. *The position of equilibrium for this step is favorable,* and we can make it even more favorable by distilling ethanol from the reaction mixture as it forms.

Step 3 $RCH_2C(=O)-CH(R)-COC_2H_5 + {}^{-}:\ddot{O}C_2H_5 \rightleftharpoons RCH_2C(=O)-\ddot{C}H(R)-COC_2H_5 + C_2H_5OH$

β-Keto ester (stronger acid); Ethoxide ion (stronger base); β-Keto ester anion (weaker base); Ethanol (weaker acid)

Beta-keto esters are stronger acids than ethanol. They react with ethoxide ion almost quantitatively to produce ethanol and anions of β-keto esters. (It is this that pulls the equilibrium to the right.) Beta-keto esters are much more acidic than ordinary esters because their enolate anions are more stabilized by resonance: their negative charge is delocalized into two carbonyl groups:

$$RCH_2-C(=O)-\ddot{C}^{-}(R)-COC_2H_5 \longleftrightarrow RCH_2-C(O^-)=C(R)-COC_2H_5 \longleftrightarrow RCH_2-C(=O)-C(R)=C(O^-)OC_2H_5$$

$$RCH_2-\overset{O^{\delta-}}{C}\!=\!\overset{\delta-}{C}(R)\!=\!\overset{O^{\delta-}}{C}OC_2H_5$$

Resonance hybrid

After steps 1–3 of a Claisen condensation have taken place, we add an acid to the reaction mixture. This brings about a rapid protonation of the anion and produces the β-keto ester as an equilibrium mixture of its keto and enol forms.

$$\text{Step 4}\quad \text{RCH}_2\text{—}\overset{\overset{\delta-}{\text{O}}}{\overset{\|}{\text{C}}}\text{==}\underset{\text{R}}{\underset{|}{\overset{\delta-}{\text{C}}}}\text{==}\overset{\overset{\delta-}{\text{O}}}{\overset{\|}{\text{C}}}\text{OC}_2\text{H}_5 \xrightarrow[\text{(rapid)}]{\text{H}^+} \text{RCH}_2\text{—}\overset{\text{O}}{\overset{\|}{\text{C}}}\text{—}\underset{\text{R}}{\underset{|}{\text{CH}}}\text{—}\overset{\text{O}}{\overset{\|}{\text{C}}}\text{OC}_2\text{H}_5$$

Keto form

$$\updownarrow$$

$$\text{RCH}_2\text{—}\overset{\text{OH}}{\overset{|}{\text{C}}}\text{=}\underset{\text{R}}{\underset{|}{\text{C}}}\text{—}\overset{\text{O}}{\overset{\|}{\text{C}}}\text{OC}_2\text{H}_5$$

Enol form

Problem 20.1

(a) Write a mechanism for all steps of the Claisen condensation that takes place when ethyl propanoate reacts with ethoxide ion. (b) What products form when the reaction mixture is acidified?

When diethyl hexanedioate is heated with sodium ethoxide, subsequent acidification of the reaction mixture gives 2-carbethoxycyclopentanone.

$$\text{C}_2\text{H}_5\text{O}\overset{\text{O}}{\overset{\|}{\text{C}}}(\text{CH}_2)_4\overset{\text{O}}{\overset{\|}{\text{C}}}\text{OC}_2\text{H}_5 \xrightarrow[\text{(2) H}^+]{\text{(1) NaOC}_2\text{H}_5} \text{2-carbethoxycyclopentanone (cyclopentanone ring with }-\overset{\text{O}}{\overset{\|}{\text{C}}}\text{OC}_2\text{H}_5\text{ at C-2)}$$

Diethyl hexanedioate (Diethyl adipate)

2-Carbethoxylcyclopentanone (74–81%)

The mechanism for this reaction, called the *Dieckmann condensation,* is the same as that of the Claisen condensation.

Problem 20.2

(a) Show all steps in the mechanism for the Dieckmann condensation. (b) What product would you expect from a Dieckmann condensation of diethyl heptanedioate (diethyl pimelate)? (c) Can you account for the fact that diethyl pentanedioate (diethyl glutarate) does not undergo a Dieckmann condensation?

20.2A Crossed Claisen Condensations

Crossed Claisen condensations (like crossed aldol condensations) are possible when one ester component has no α-hydrogens and is, therefore, unable to undergo self-condensation. Ethyl benzoate, for example, condenses with ethyl acetate to give ethyl benzoylacetate.

$$\text{C}_6\text{H}_5\text{—}\overset{\text{O}}{\overset{\|}{\text{C}}}\text{OC}_2\text{H}_5 + \text{CH}_3\overset{\text{O}}{\overset{\|}{\text{C}}}\text{OC}_2\text{H}_5 \xrightarrow[\text{(2) H}^+]{\text{(1) NaOC}_2\text{H}_5} \text{C}_6\text{H}_5\text{—}\overset{\text{O}}{\overset{\|}{\text{C}}}\text{CH}_2\overset{\text{O}}{\overset{\|}{\text{C}}}\text{OC}_2\text{H}_5$$

Ethyl benzoate

Ethyl benzoylacetate (60%)

Ethyl phenylacetate condenses with diethyl carbonate to give diethyl phenylmalonate.

$$C_6H_5CH_2\overset{O}{\overset{\|}{C}}OC_2H_5 + C_2H_5O\overset{O}{\overset{\|}{C}}OC_2H_5 \xrightarrow[(2)\ H^+]{(1)\ NaOC_2H_5} C_6H_5CH(\overset{O}{\overset{\|}{C}}-OC_2H_5)_2$$

Ethyl phenylacetate — Diethyl carbonate — Diethyl phenylmalonate (65%)

Problem 20.3

Write mechanisms that account for the products that are formed in the two crossed Claisen condensations illustrated above.

Problem 20.4

What products would you expect to obtain from each of the following crossed Claisen condensations?

(a) Ethyl propanoate + diethyl oxalate $\xrightarrow[(2)\ H^+]{(1)\ NaOCH_2CH_3}$

(b) Ethyl acetate + ethyl formate $\xrightarrow[(2)\ H^+]{(1)\ NaOCH_2CH_3}$

Esters that have only one α-hydrogen cannot be converted to β-keto esters by sodium ethoxide. (Why not?) However, they can be converted to β-keto esters by reactions that use very strong bases. The strong base converts the ester to its enolate anion in nearly quantitative yield. This allows us to *acylate* the enolate anion by treating it with an acyl chloride or an ester. An example of this technique that makes use of the very powerful base, sodium triphenylmethide, is shown next.

$$CH_3\underset{CH_3}{\overset{H}{C}}-\overset{O}{\overset{\|}{C}}-OC_2H_5 \xrightarrow[Ether]{Na^+:\bar{C}(C_6H_5)_3} CH_3\underset{CH_3}{\ddot{C}}-\overset{O}{\overset{\|}{C}}OC_2H_5 + H:C(C_6H_5)_3$$

$$\xrightarrow{C_6H_5\overset{O}{\overset{\|}{C}}-Cl} C_6H_5\overset{O}{\overset{\|}{C}}-\underset{CH_3}{\overset{CH_3}{C}}-\overset{O}{\overset{\|}{C}}OC_2H_5 + Cl^-$$

Ethyl 2,2-dimethyl-2-benzoylacetate (79%)

20.2B Acylation of Other Carbanions

Enolate anions derived from ketones also react with esters in nucleophilic substitution reactions that resemble Claisen condensations. In the first example below, although two anions are possible, the major product is derived from the primary carbanion.

$$CH_3\overset{O}{\overset{\|}{C}}(CH_2)_2CH_3 \xrightarrow[\text{Ether}]{NaNH_2} Na^+ \; ^-{:}CH_2\overset{O}{\overset{\|}{C}}(CH_2)_2CH_3$$

2-Pentanone

$$\downarrow CH_3(CH_2)_2C(=O)OC_2H_5$$

$$CH_3(CH_2)_2\overset{O}{\overset{\|}{C}}CH_2\overset{O}{\overset{\|}{C}}(CH_2)_2CH_3$$

4,6-Nonanedione (76%)

$$\text{cyclohexanone} \underset{}{\overset{NaOC_2H_5}{\rightleftharpoons}} \text{cyclohexanone enolate } Na^+ \xrightarrow[(2)\ H^+]{(1)\ C_2H_5O\overset{O}{\overset{\|}{C}}-\overset{O}{\overset{\|}{C}}OC_2H_5} \text{2-(}\overset{O}{\overset{\|}{C}}-\overset{O}{\overset{\|}{C}}OC_2H_5\text{)cyclohexanone}$$

(67%)

Problem 20.5

Show how you might synthesize each of the following compounds using, as your starting materials, esters, ketones, acyl halides, and so on.

(a) 2-oxocyclohexane with $\overset{O}{\overset{\|}{C}}H$ substituent

(b) 2-methylcyclohexane-1,3-dione (CH_3)

(c) cyclopentane-1,2-dione bearing $CO_2C_2H_5$, CH_3, CH_3, $CO_2C_2H_5$

Problem 20.6

Keto esters are capable of undergoing cyclization reactions similar to the Dieckmann condensation. Write mechanisms that account for the products formed in each of the following reactions. [Also account for the ring size of the product in part (b).]

$$\text{(a)}\ CH_3\overset{O}{\overset{\|}{C}}(CH_2)_4\overset{O}{\overset{\|}{C}}OC_2H_5 \xrightarrow[(2)\ H^+]{(1)\ NaOC_2H_5} \text{2-oxocyclopentyl-}\overset{O}{\overset{\|}{C}}CH_3$$

2-Acetylcyclopentanone

$$\text{(b)}\ CH_3\overset{O}{\overset{\|}{C}}(CH_2)_3\overset{O}{\overset{\|}{C}}OC_2H_5 \xrightarrow[(2)\ H^+]{(1)\ NaOC_2H_5}$$

Cyclohexane-1,3-dione

20.3 THE ACETOACETIC ESTER SYNTHESIS. SYNTHESIS OF SUBSTITUTED ACETONES

Acetoacetic esters are useful reagents for the preparation of methyl ketones of the types shown below:

$$\underset{\text{Monosubstituted acetone}}{CH_3-\overset{O}{\overset{\|}{C}}-CH_2-R} \quad \text{or} \quad \underset{\text{Disubstituted acetone}}{CH_3-\overset{O}{\overset{\|}{C}}-\underset{R}{\underset{|}{CH}}-R}$$

Two factors make such syntheses practical. (1) The methylene protons of β-keto ester are appreciably acidic and (2) β-keto acids decarboxylate readily (cf. Section 17.10).

As we have seen (Section 20.2) the methylene protons of acetoacetic ester are more acidic than the —OH proton of ethanol because they are located between two carbonyl groups and yield a highly stabilized enolate anion. This means that we can convert acetoacetic ester to an enolate anion using sodium ethoxide as a base. We can then carry out an alkylation reaction by treating the enolate anion with an alkyl halide.

Since the alkylation (below) is an S_N2 reaction, we are limited to the use of primary alkyl halides or methyl halides (cf. Problem 20.10).

$$\underset{\text{Acetoacetic ester}}{CH_3\overset{\ddot{O}:}{\overset{\|}{C}}-CH_2-\overset{\ddot{O}:}{\overset{\|}{C}}OC_2H_5} + \underset{\text{Sodium ethoxide}}{C_2H_5O^-Na^+} \rightleftharpoons \underset{\text{Sodioacetoacetic ester}}{CH_3\overset{\ddot{O}:}{\overset{\|}{C}}-\overset{Na^+}{\ddot{C}H^-}-\overset{\ddot{O}:}{\overset{\|}{C}}-OC_2H_5} + C_2H_5OH$$

$$\xrightarrow{R-X} \underset{\text{Monoalkylacetoacetic ester}}{CH_3\overset{\ddot{O}:}{\overset{\|}{C}}-\underset{R}{\underset{|}{CH}}-\overset{\ddot{O}:}{\overset{\|}{C}}-OC_2H_5} + NaX$$

The monoalkylacetoacetic ester still has one appreciably acidic hydrogen and, if we desire, we can carry out a second alkylation. Since the monoalkylacetoacetic ester is somewhat less acidic than acetoacetic ester itself (why?), it is usually helpful to use a stronger base.

$$\underset{\text{Monoalkylacetoacetic ester}}{CH_3\overset{O}{\overset{\|}{C}}-\underset{R}{\underset{|}{CH}}-\overset{O}{\overset{\|}{C}}-OC_2H_5} + \underset{\text{Potassium } tert\text{-butoxide}}{(CH_3)_3CO^-K^+} \rightleftharpoons CH_3\overset{O}{\overset{\|}{C}}-\underset{R}{\underset{|}{\overset{K^+}{\ddot{C}^-}}}-\overset{O}{\overset{\|}{C}}OC_2H_5 + (CH_3)_3COH$$

$$\xrightarrow{R'-X} \underset{\text{Dialkylacetoacetic ester}}{CH_3\overset{O}{\overset{\|}{C}}-\underset{R}{\underset{|}{\overset{R'}{\overset{|}{C}}}}-\overset{O}{\overset{\|}{C}}-OC_2H_5} + KX$$

If our goal is the preparation of a monosubstituted acetone we carry out only one alkylation reaction. We then hydrolyze the monoalkylacetoacetic ester using dilute sodium or potassium hydroxide. Subsequent acidification of the mixture gives an alkylacetoacetic acid and heating this β-keto acid to 100° brings about decarboxylation.

$$CH_3\overset{O}{\overset{\|}{C}}-\underset{R}{\underset{|}{C}H}-\overset{O}{\overset{\|}{C}}-O-C_2H_5 \xrightarrow{\text{Dil. KOH}} CH_3\overset{O}{\overset{\|}{C}}-\underset{R}{\underset{|}{C}H}-\overset{O}{\overset{\|}{C}}-O^-K^+$$

$$\downarrow H_3O^+$$

$$CH_3\overset{O}{\overset{\|}{C}}-\underset{R}{\underset{|}{C}H}-\overset{O}{\overset{\|}{C}}-OH$$

Alkylacetoacetic acid
(a β-keto acid)

$$\downarrow \text{Heat, } 100°$$

$$CH_3-\overset{O}{\overset{\|}{C}}-CH_2-R + CO_2$$

A specific example of an acetoacetic ester synthesis of a monosubstituted acetone is the following synthesis of 2-heptanone.

$$CH_3\overset{O}{\overset{\|}{C}}-CH_2-\overset{O}{\overset{\|}{C}}OC_2H_5 \xrightarrow[\text{(2) } CH_3CH_2CH_2CH_2Br]{\text{(1) } NaOC_2H_5/C_2H_5OH} CH_3\overset{O}{\overset{\|}{C}}-\underset{CH_2CH_2CH_2CH_3}{\underset{|}{C}H}-\overset{O}{\overset{\|}{C}}OC_2H_5 \xrightarrow[\text{(2) } H_3O^+]{\text{(1) dil. NaOH}}$$

Ethyl acetoacetate
(acetoacetic ester)

Ethyl *n*-butylacetoacetate
(69–72%)

$$CH_3\overset{O}{\overset{\|}{C}}-\underset{CH_2CH_2CH_2CH_3}{\underset{|}{C}H}-\overset{O}{\overset{\|}{C}}-OH \xrightarrow[-CO_2]{\text{Heat}} CH_3\overset{O}{\overset{\|}{C}}-CH_2CH_2CH_2CH_2CH_3$$

2-Heptanone
(52–61% overall from
ethyl acetoacetate)

If our goal is the preparation of a disubstituted acetone, we carry out two successive alkylations, we hydrolyze the dialkylacetoacetic ester that is produced, and then we decarboxylate the dialkylacetoacetic acid. An example of this procedure is the synthesis of 3-butyl-2-heptanone.

$$CH_3\overset{O}{\overset{\|}{C}}CH_2\overset{O}{\overset{\|}{C}}OC_2H_5 \xrightarrow[\text{[First alkylation]}]{\text{(1) } NaOC_2H_5/C_2H_5OH \quad \text{(2) } CH_3CH_2CH_2CH_2Br} CH_3\overset{O}{\overset{\|}{C}}\underset{\underset{CH_3}{|}}{\underset{(CH_2)_3}{\underset{|}{C}}}H\overset{O}{\overset{\|}{C}}OC_2H_5 \xrightarrow[\text{[Second alkylation]}]{\text{(1) } (CH_3)_3COK/(CH_3)_3COH \quad \text{(2) } CH_3CH_2CH_2CH_2Br}$$

Ethyl *n*-butylacetoacetate (69–72%)

$$CH_3\overset{O}{\overset{\|}{C}}-\overset{\overset{CH_3}{|}}{\underset{\underset{CH_3}{|}}{\underset{(CH_2)_3}{\overset{(CH_2)_3}{\underset{|}{\overset{|}{C}}}}}}-COOC_2H_5 \xrightarrow[\text{[Hydrolysis]}]{\text{(1) Dil. NaOH} \quad \text{(2) } H_3O^+} CH_3\overset{O}{\overset{\|}{C}}-\overset{\overset{CH_3}{|}}{\underset{\underset{CH_3}{|}}{\underset{(CH_2)_3}{\overset{(CH_2)_3}{\underset{|}{\overset{|}{C}}}}}}-COOH \xrightarrow[\text{[Decarboxylation]}]{\text{Heat} \quad -CO_2} CH_3\overset{O}{\overset{\|}{C}}-\underset{\underset{CH_3}{|}}{\underset{(CH_2)_3}{\underset{|}{C}}}H(CH_2)_3CH_3$$

Ethyl di-*n*-butylacetoacetate (77%) 3-Butyl-2-heptanone

Although both alkylations in the example just given were carried out with the same alkyl halide, we could have used different alkyl halides if our synthesis had required it.

Problem 20.7

Occasional side products of alkylations of sodioacetoacetic esters are compounds with the following general structure.

$$CH_3\overset{R\ddot{O}:}{\overset{|}{C}}=CH\overset{\ddot{O}:}{\overset{\|}{C}}OC_2H_5$$

Explain how these are formed.

Problem 20.8

Show how you would use the acetoacetic ester synthesis to prepare each of the following:

(a) 2-Pentanone
(b) 3-Propyl-2-hexanone
(c) 4-Phenyl-2-butanone

Problem 20.9

Attempts to bring about monoalkylation of an acetoacetic ester are sometimes complicated by competing dialkylation reactions even when only one mole of sodium ethoxide and one mole of alkyl halide are employed. (a) Write a mechanism showing how dialkylation might occur when ethyl acetoacetate is treated with one mole of sodium ethoxide and one mole of methyl iodide. (b) Does the greater acidity of ethyl acetoacetate versus that of ethyl methylacetoacetate favor monoalkylation or dialkylation? (c) Dialkylations are more common when methyl and ethyl halides are employed than when larger alkyl halides are used. What factor accounts for this? (d) Suppose you wanted to prepare 3-methyl-2-hexanone. Which alkylation would you carry out first to ensure the best yield?

Problem 20.10

The acetoacetic ester synthesis generally gives best yields when primary halides are used in the alkylation step. Secondary halides give low yields and tertiary halides give practically no alkylation product at all. (a) Explain. (b) What products would you expect from the reaction of sodioacetoacetic ester and *tert*-butyl bromide? (c) Bromobenzene cannot be used as an alkylating agent in an acetoacetic ester synthesis. Why not?

Problem 20.11

Since the products obtained from Claisen condensations are β-keto esters, subsequent hydrolysis and decarboxylation of these products gives a general method for the synthesis of ketones. Show how you would employ this technique in a synthesis of 4-heptanone.

The acetoacetic ester synthesis can also be carried out using halo esters and halo ketones. The use of a α-halo ester provides a convenient synthesis of γ-keto acids:

$$CH_3\overset{O}{\overset{\|}{C}}-CH_2-\overset{O}{\overset{\|}{C}}-OC_2H_5 \xrightarrow{C_2H_5ONa} CH_3\overset{O}{\overset{\|}{C}}-\ddot{C}H^{-}\ Na^{+}-\overset{O}{\overset{\|}{C}}-OC_2H_5 \xrightarrow{BrCH_2\overset{O}{\overset{\|}{C}}-OC_2H_5}$$

$$CH_3\overset{O}{\overset{\|}{C}}-\underset{CH_2\underset{O}{\underset{\|}{C}}-OC_2H_5}{\underset{|}{CH}}-\overset{O}{\overset{\|}{C}}-OC_2H_5 \xrightarrow[(2)\ H_3O^+]{(1)\ \text{Dil. NaOH}} CH_3\overset{O}{\overset{\|}{C}}-\underset{CH_2\underset{O}{\underset{\|}{C}}-OH}{\underset{|}{CH}}-\overset{O}{\overset{\|}{C}}-OH \xrightarrow[-CO_2]{\text{Heat}}$$

$$CH_3\overset{O}{\overset{\|}{C}}-CH_2CH_2-\overset{O}{\overset{\|}{C}}-OH$$

γ-Ketovaleric acid

The use of an α-halo ketone in an acetoacetic ester synthesis provides a general method for preparing γ-diketones:

$$CH_3\overset{O}{\overset{\|}{C}}-\ddot{C}H^{-}\ Na^{+}-\overset{O}{\overset{\|}{C}}-OC_2H_5 \xrightarrow[BrCH_2\underset{O}{\underset{\|}{C}}R]{} CH_3\overset{O}{\overset{\|}{C}}-\underset{\underset{\underset{R}{|}}{\underset{C=O}{|}}}{\underset{CH_2}{\underset{|}{CH}}}-\overset{O}{\overset{\|}{C}}-OC_2H_5 \xrightarrow[(2)\ H_3O^+]{(1)\ \text{Dil. NaOH}}$$

$$CH_3\overset{O}{\overset{\|}{C}}-\underset{\underset{\underset{R}{|}}{\underset{C=O}{|}}}{\underset{CH_2}{\underset{|}{CH}}}-\overset{O}{\overset{\|}{C}}-OH \xrightarrow[-CO_2]{\text{Heat}} CH_3\overset{O}{\overset{\|}{C}}-CH_2CH_2-\overset{O}{\overset{\|}{C}}-R$$

A γ-diketone

Problem 20.12

In the synthesis of the keto acid given above, the dicarboxylic acid decarboxylates in a specific way; it gives

$$CH_3\overset{O}{\overset{\|}{C}}CH_2CH_2\overset{O}{\overset{\|}{C}}OH \text{ rather than } CH_3\overset{O}{\overset{\|}{C}}\underset{CH_3}{\underset{|}{C}H}\overset{O}{\overset{\|}{C}}OH\text{. Explain.}$$

Problem 20.13

How would you use the acetoacetic ester synthesis to prepare the following?

$$C_6H_5\overset{O}{\overset{\|}{C}}CH_2CH_2\overset{O}{\overset{\|}{C}}CH_3$$

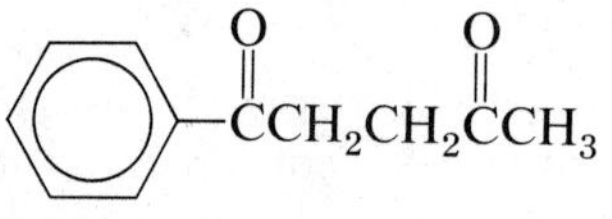

Anions obtained from acetoacetic esters undergo acylation when they are treated with acyl chlorides or acid anhydrides. Since both of these acylating agents react with alcohols, acylation reactions are carried out best in aprotic solvents. Sodium hydride can be used to generate the enolate anion.

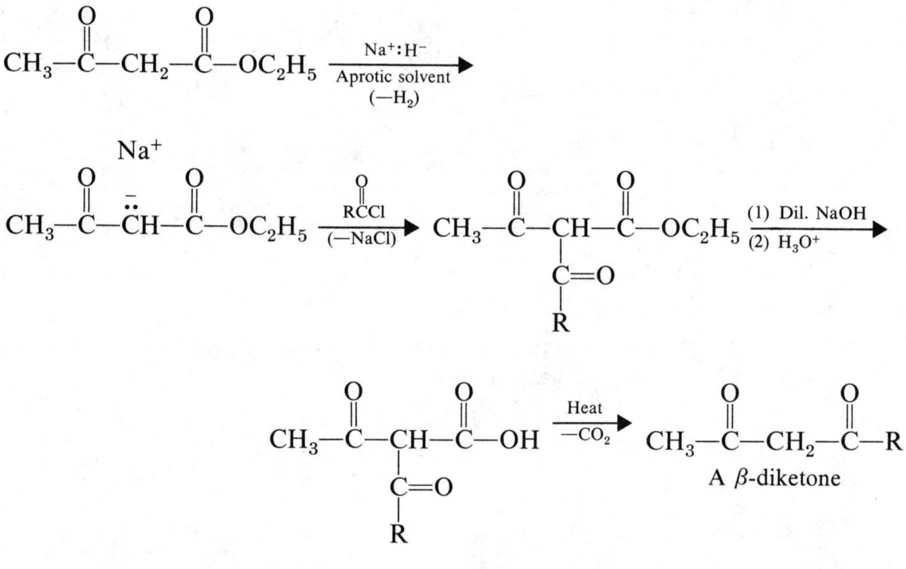

$$CH_3-\overset{O}{\overset{\|}{C}}-CH_2-\overset{O}{\overset{\|}{C}}-OC_2H_5 \xrightarrow[\text{Aprotic solvent}\ (-H_2)]{Na^+:H^-}$$

$$CH_3-\overset{O}{\overset{\|}{C}}-\overset{Na^+}{\ddot{C}H^-}-\overset{O}{\overset{\|}{C}}-OC_2H_5 \xrightarrow[(-NaCl)]{R\overset{O}{\overset{\|}{C}}Cl} CH_3-\overset{O}{\overset{\|}{C}}-\underset{\underset{R}{|}}{\underset{C=O}{\underset{|}{C}H}}-\overset{O}{\overset{\|}{C}}-OC_2H_5 \xrightarrow[(2)\ H_3O^+]{(1)\ \text{Dil. NaOH}}$$

$$CH_3-\overset{O}{\overset{\|}{C}}-\underset{\underset{R}{|}}{\underset{C=O}{\underset{|}{C}H}}-\overset{O}{\overset{\|}{C}}-OH \xrightarrow[-CO_2]{\text{Heat}} \underset{\text{A }\beta\text{-diketone}}{CH_3-\overset{O}{\overset{\|}{C}}-CH_2-\overset{O}{\overset{\|}{C}}-R}$$

Acylations of acetoacetic esters followed by hydrolysis and decarboxylation give us a method for preparing β-diketones.

Problem 20.14

How would you use the acetoacetic ester synthesis to prepare the following?

$$C_6H_5\overset{O}{\overset{\|}{C}}CH_2\overset{O}{\overset{\|}{C}}CH_3$$

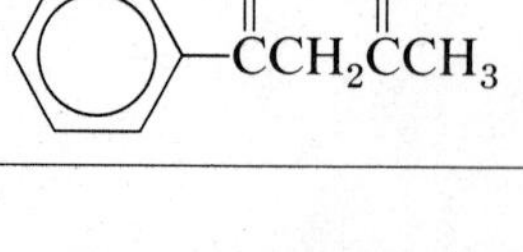

One further variation of the acetoacetic ester synthesis involves the conversion of an acetoacetic ester to a resonance-stabilized *dianion* by using a very strong base such as potassium amide in liquid ammonia.

$$CH_3-\overset{\overset{\large\ddot{O}:}{\|}}{C}-CH_2-\overset{\overset{\large\ddot{O}:}{\|}}{C}-OC_2H_5 \xrightarrow[\text{Liq. :NH}_3]{2K^+:\ddot{N}H_2^-} \left[^-:CH_2-\overset{\overset{\large\ddot{O}:}{\|}}{C}-\overset{\overline{\cdot\cdot}}{C}H-\overset{\overset{\large\ddot{O}:}{\|}}{C}-\ddot{O}C_2H_5 \right] 2K^+$$

$$\updownarrow$$

$$\left[CH_2=\overset{\overset{:\ddot{O}:^-}{|}}{C}-CH=\overset{\overset{:\ddot{O}:^-}{|}}{C}-\ddot{O}C_2H_5 \right] 2K^+$$

$$\updownarrow$$

$$\text{etc.}$$

When this dianion is treated with one mole of a primary (or methyl) halide it undergoes alkylation at its terminal carbon rather than at its interior one. This orientation of the alkylation reaction apparently results from the greater basicity (and thus nucleophilicity) of the terminal carbanion. After monoalkylation has taken place, the anion that remains can be protonated by adding ammonium chloride.

$$2K^+ \left[^-:CH_2-\overset{\overset{O}{\|}}{C}-\overset{\overline{\cdot\cdot}}{C}H-\overset{\overset{O}{\|}}{C}-OC_2H_5 \right] \xrightarrow[\text{liq. NH}_3]{R-X}$$

$$R-CH_2-\overset{\overset{O}{\|}}{C}-\overset{\overline{\cdot\cdot}}{C}H-\overset{\overset{O}{\|}}{C}-OC_2H_5 \; K^+ + KX \xrightarrow[\text{liq. NH}_3]{NH_4Cl}$$

$$RCH_2\overset{\overset{O}{\|}}{C}-CH_2-\overset{\overset{O}{\|}}{C}-OC_2H_5 + NH_3 + KCl$$

Problem 20.15

Show how you could use ethyl acetoacetate in a synthesis of

$$C_6H_5CH_2CH_2\overset{\overset{O}{\|}}{C}CH_2\overset{\overset{O}{\|}}{C}OC_2H_5$$

20.4 THE MALONIC ESTER SYNTHESIS. SYNTHESIS OF SUBSTITUTED ACETIC ACIDS

A useful counterpart of the acetoacetic ester synthesis—one that allows the synthesis of *mono- and disubstituted acetic acids*—is called the *malonic ester synthesis.*

The malonic ester synthesis resembles the acetoacetic ester synthesis in several respects.

1. Diethyl malonate, the starting compound, forms a relatively stable enolate ion:

$$\underset{\text{(:\ddot{O}:=)COC}_2\text{H}_5\text{—CH}_2\text{—COC}_2\text{H}_5\text{(=:\ddot{O}:)}}{} + C_2H_5O^- \rightleftharpoons {}^-\!:CH(COC_2H_5)_2 \longleftrightarrow CH(=C(\text{—}\ddot{O}:^-)OC_2H_5)COC_2H_5 \longleftrightarrow CH(COC_2H_5)(=C(\text{—}\ddot{O}:^-)OC_2H_5)$$

Resonance-stabilized anion

2. This enolate ion can be alkylated,

$$Na^+\ HC:^-(COC_2H_5)_2 + R\text{—}X \longrightarrow R\text{—}CH(COC_2H_5)_2 + NaX$$

Sodiomalonic ester — Monoalkylmalonic ester

and the product can be alkylated again if our synthesis requires it:

$$R\text{—}CH(COC_2H_5)_2 \underset{}{\overset{(CH_3)_3CO^-\ Na^+}{\rightleftharpoons}} R\text{—}C:^-(COC_2H_5)_2\ Na^+ \xrightarrow{R'\text{—}X} R\text{—}C(\text{—}R')(COC_2H_5)_2 + Na^+X^-$$

Dialkylmalonic ester

3. The mono- or dialkylmalonic ester can then be hydrolyzed to a mono- or dialkylmalonic acid, and substituted malonic acids decarboxylate readily. Decarboxylation gives a mono- or disubstituted acetic acid.

$$RCH(COC_2H_5)_2 \xrightarrow[(2)\ H_3O^+]{(1)\ OH^-/H_2O} RCH(COOH)_2 \xrightarrow[-CO_2]{\text{Heat}} RCH_2COOH$$

Monoalkylmalonic ester — Monoalkylacetic acid

$$\mathrm{R{-}C(COOC_2H_5)_2{-}R'} \xrightarrow[(2)\ H_3O^+]{(1)\ OH^-/H_2O} \mathrm{R{-}C(COOH)_2{-}R'} \xrightarrow[-CO_2]{\text{Heat}} \mathrm{R{-}CH(R')COOH}$$

Dialkylmalonic ester

Dialkylacetic acid

Two specific examples of the malonic ester synthesis are the syntheses of hexanoic acid and 2-ethylpentanoic acid given below.

$$\mathrm{CH_2(COOC_2H_5)_2} + \mathrm{NaOC_2H_5} + \mathrm{CH_3CH_2CH_2CH_2Br} \xrightarrow{\text{Reflux}}$$

$$\mathrm{CH_3CH_2CH_2CH_2CH(COOC_2H_5)_2} \xrightarrow[(2)\ \text{dil. } H_2SO_4,\ \text{reflux}]{(1)\ 50\%\ \text{KOH, reflux}} \mathrm{CH_3CH_2CH_2CH_2CH_2COOH}$$

Diethyl *n*-butylmalonate (80–90%)

Hexanoic acid (75%)

$$\mathrm{CH_2(COOC_2H_5)_2} \xrightarrow[CH_3CH_2I]{NaOC_2H_5} \mathrm{CH_3CH_2CH(COOC_2H_5)_2} \xrightarrow[CH_3CH_2CH_2Br]{NaOC(CH_3)_3}$$

Diethyl ethylmalonate

$$\mathrm{(CH_3CH_2)(CH_3CH_2CH_2)C(COOC_2H_5)_2} \xrightarrow[(2)\ H_3O^+]{(1)\ OH^-/H_2O} \mathrm{(CH_3CH_2)(CH_3CH_2CH_2)C(COOH)_2} \xrightarrow[180^\circ]{\text{heat}}$$

Ethylpropylmalonic acid

$$\mathrm{CH_3CH_2CH_2CH(CH_2CH_3)COOH}$$

2-Ethylpentanoic acid

Problem 20.16

Outline all steps in a malonic ester synthesis of each of the following:

(a) Pentanoic acid
(b) 2-Methylpentanoic acid
(c) 4-Methylpentanoic acid

Two variations of the malonic ester synthesis make use of dihaloalkanes. In the first of these, two moles of sodiomalonic ester are allowed to react with a dihaloalkane. Two consecutive alkylations occur giving a tetraester; hydrolysis and decarboxylation of the tetraester yields a dicarboxylic acid. An example is the synthesis of glutaric acid:

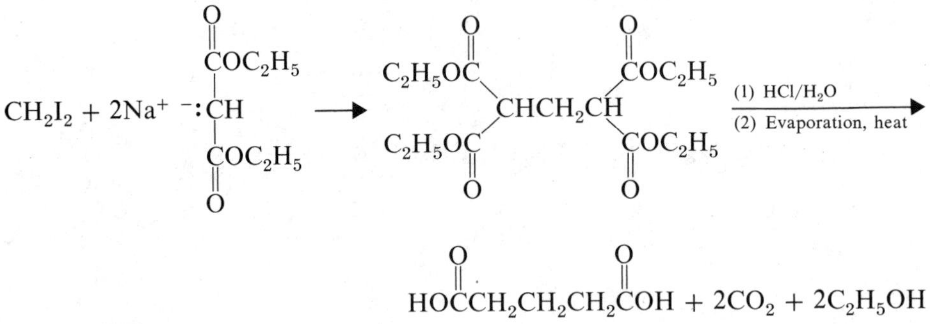

$$CH_2I_2 + 2Na^+\ {}^{-}\!:\!CH(COC_2H_5)_2 \longrightarrow (C_2H_5OC)_2CHCH_2CH(COC_2H_5)_2 \xrightarrow[\text{(2) Evaporation, heat}]{\text{(1) HCl/H}_2\text{O}}$$

(each COC_2H_5 group being $\overset{O}{\overset{\|}{C}}OC_2H_5$)

$$HO\overset{O}{\overset{\|}{C}}CH_2CH_2CH_2\overset{O}{\overset{\|}{C}}OH + 2CO_2 + 2C_2H_5OH$$

Glutaric acid
(80% from tetraester)

In a second variation one mole of sodiomalonic ester is allowed to react with one mole of a dihaloalkane. This gives a haloalkylmalonic ester, which when treated with sodium ethoxide undergoes an internal alkylation reaction. This method has been used to prepare three-, four-, five-, and six-membered rings. An example is the synthesis of cyclobutanecarboxylic acid.

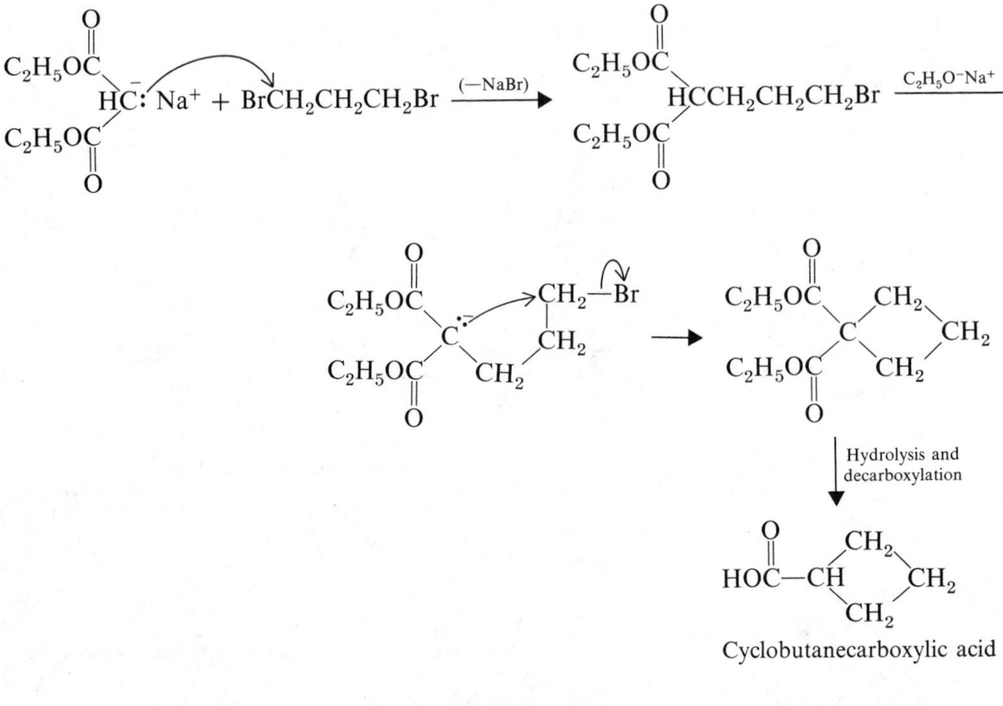

$$(C_2H_5O\overset{O}{\overset{\|}{C}})_2H\overset{-}{C}\!:\ Na^+ + BrCH_2CH_2CH_2Br \xrightarrow{(-NaBr)} (C_2H_5O\overset{O}{\overset{\|}{C}})_2HCCH_2CH_2CH_2Br \xrightarrow{C_2H_5O^-Na^+}$$

$$(C_2H_5O\overset{O}{\overset{\|}{C}})_2\overset{..}{\overset{-}{C}}\text{—}CH_2\text{—}CH_2\text{—}CH_2\text{—}Br \longrightarrow (C_2H_5O\overset{O}{\overset{\|}{C}})_2C\lt(CH_2)_3$$

Hydrolysis and decarboxylation

$$HO\overset{O}{\overset{\|}{C}}\text{—}CH\lt(CH_2)_3$$

Cyclobutanecarboxylic acid

20.5 FURTHER REACTIONS OF ACTIVE HYDROGEN COMPOUNDS

Because of the acidity of their methylene hydrogens, malonic esters, acetoacetic esters, and similar compounds are often called *active hydrogen compounds* or *active methylene compounds*. Generally speaking, active hydrogen compounds have two electron-withdrawing groups attached to the same carbon:

$$Z{-}CH_2{-}Z'$$

Active hydrogen compound
(Z and Z′ are electron-withdrawing groups)

The electron-withdrawing groups can be a variety of substituents including:

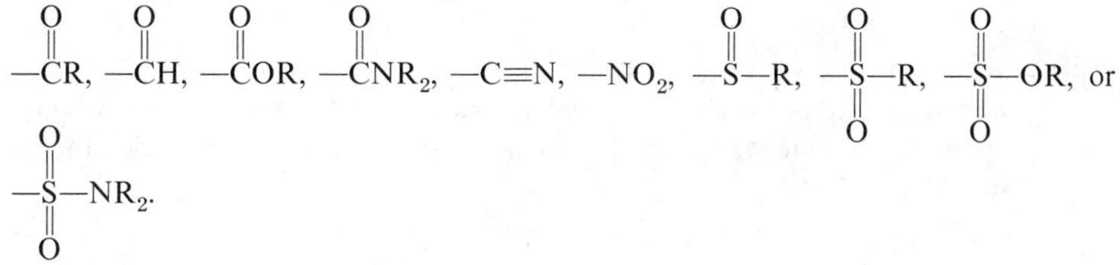

$$-\overset{O}{\overset{\|}{C}}R,\ -\overset{O}{\overset{\|}{C}}H,\ -\overset{O}{\overset{\|}{C}}OR,\ -\overset{O}{\overset{\|}{C}}NR_2,\ -C{\equiv}N,\ -NO_2,\ -\overset{O}{\overset{\|}{S}}{-}R,\ -\underset{O}{\underset{\|}{\overset{O}{\overset{\|}{S}}}}{-}R,\ -\underset{O}{\underset{\|}{\overset{O}{\overset{\|}{S}}}}{-}OR,\ \text{or}\ -\underset{O}{\underset{\|}{\overset{O}{\overset{\|}{S}}}}{-}NR_2.$$

For example, ethyl cyanoacetate reacts with base to yield a resonance-stabilized anion:

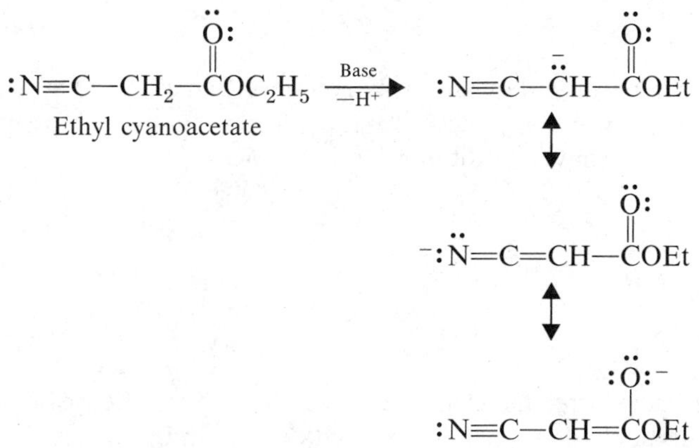

$$:N{\equiv}C{-}CH_2{-}\overset{\ddot{O}:}{\overset{\|}{C}}OC_2H_5 \xrightarrow[-H^+]{\text{Base}} :N{\equiv}C{-}\overset{-}{\ddot{C}}H{-}\overset{\ddot{O}:}{\overset{\|}{C}}OEt \longleftrightarrow {}^{-}:\ddot{N}{=}C{=}CH{-}\overset{\ddot{O}:}{\overset{\|}{C}}OEt \longleftrightarrow :N{\equiv}C{-}CH{=}\overset{:\ddot{O}:^{-}}{\overset{|}{C}}OEt$$

Ethyl cyanoacetate

Ethyl cyanoacetate anions also undergo alkylations. They can be dialkylated with isopropyl iodide, for example.

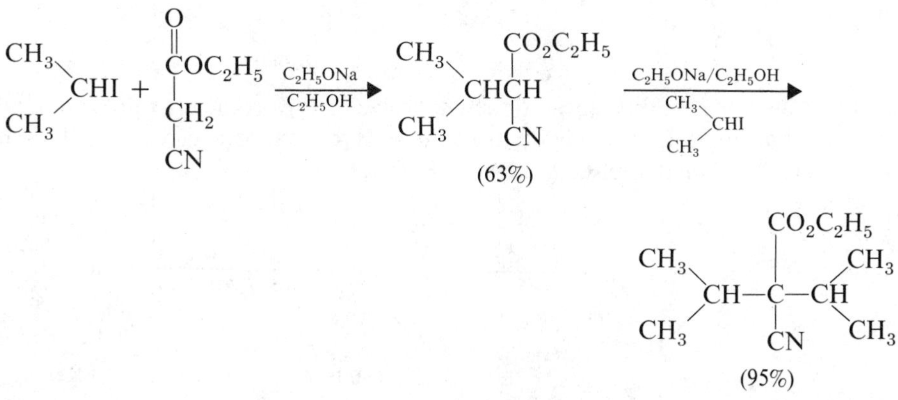

$$(CH_3)_2CHI + \overset{O}{\overset{\|}{C}}OC_2H_5{-}CH_2{-}CN \xrightarrow[C_2H_5OH]{C_2H_5ONa} (CH_3)_2CH{-}CH(CO_2C_2H_5)(CN) \xrightarrow[(CH_3)_2CHI]{C_2H_5ONa/C_2H_5OH} (CH_3)_2CH{-}C(CO_2C_2H_5)(CN){-}CH(CH_3)_2$$

(63%) (95%)

Another way of preparing ketones is to use a β-keto sulfoxide as an active hydrogen compound:

$$\underset{\text{A }\beta\text{-keto sulfoxide}}{\overset{\overset{O}{\|}}{RC}-CH_2-\overset{\overset{O}{\|}}{S}R'} \xrightarrow[(2)\ R''X]{(1)\ Base} \overset{\overset{O}{\|}}{RC}-\underset{\underset{R''}{|}}{CH}-\overset{\overset{O}{\|}}{S}R' \xrightarrow{Al-Hg} \overset{\overset{O}{\|}}{RC}-CH_2-R''$$

The β-keto sulfoxide is first converted to an anion and then the anion alkylated. Treating the product of these steps with aluminum amalgam (Al-Hg) causes cleavage of the carbon-sulfur bond and gives the ketone in high yield.

20.6 ALKYLATION OF 1,3-DITHIANES

Two sulfur atoms attached to the same carbon of 1,3-dithiane (below) cause the hydrogens of that carbon to be more acidic ($K_a \simeq 10^{-32}$) than those of most alkyl carbons.

Dithiane
($Ka \simeq 10^{-32}$)

Sulfur atoms, because they are easily polarized, can aid in stabilizing the negative charge of the anion (cf. Special Topic J). Strong bases such as butyllithium are usually used to convert a dithiane to its anion.

$$\text{dithiane (C}H_2\text{)} + C_4H_9{:}^-Li^+ \longrightarrow \text{dithiane anion (CH}{:}^-\ Li^+) + C_4H_{10}$$

Dithianes are thioacetals (cf. Section 16.7A); they can be prepared by treating an aldehyde with 1,3-propanedithiol in the presence of a trace of acid.

$$\overset{\overset{O}{\|}}{RC}H + HSCH_2CH_2CH_2SH \xrightarrow{H^+} \text{dithiane (C with R, H)}$$

A dithiane

Alkylating the dithiane and then hydrolyzing the product (a thioketal) is a method for converting an aldehyde to a ketone. Hydrolysis is usually carried out by using $HgCl_2$ in methanol or aqueous CH_3CN.

$$\text{dithiane (R, H)} \xrightarrow[(2)\ R'X(-LiX)]{(1)\ C_4H_9Li(-C_4H_{10})} \underset{\text{Thioketal}}{\text{dithiane (R, R')}} \xrightarrow[(-HSCH_2CH_2CH_2SH)]{HgCl_2, CH_3OH, H_2O} \underset{\text{Ketone}}{R-\overset{\overset{O}{\|}}{C}-R'}$$

The synthetic use of dithianes was developed by E. J. Corey (p. 116) and D. Seebach and is often called the *Corey-Seebach* method.

Problem 20.17

(a) Which aldehyde would you use to prepare dithiane itself? (b) How would you synthesize $C_6H_5CH_2CHO$ using dithiane as an intermediate? (c) How would you convert benzaldehyde to acetophenone?

Problem 20.18

The Corey-Seebach method can also be used to synthesize molecules with the structure RCH_2R'. (a) How might this be done?

Problem 20.19

(a) The Corey-Seebach method has been used to prepare the highly strained molecule called a metaparacyclophane below. What are the structures of the intermediates **A**–**D**?

$$\text{(isophthalaldehyde, } m\text{-}C_6H_4(CHO)_2) \xrightarrow[H^+]{2HSCH_2CH_2CH_2SH} \mathbf{A}\ (C_{14}H_{18}S_4) \xrightarrow[(2)\ BrCH_2\text{-}C_6H_4\text{-}CH_2Br]{(1)\ 2C_4H_9Li}$$

$$\mathbf{B}\ (C_{22}H_{20}S_4) \xrightarrow{\text{Hydrolysis}} \mathbf{C}\ (C_{16}H_{12}O_2) \xrightarrow{NaBH_4}$$

$$\mathbf{D}\ (C_{16}H_{16}O_2) \xrightarrow[(2)\ 2KOC(CH_3)_3]{(1)\ 2TsCl}$$

A metaparacyclophane

(b) What compound would be obtained by treating **B** with excess Raney Ni?

20.7 THE KNOEVENAGEL CONDENSATION

Active hydrogen compounds condense with aldehydes and ketones. Known as Knoevenagel condensations, these condensations are catalyzed by weak bases. Two examples are the following:

$$Cl\text{-}C_6H_4\text{-}CHO + CH_3\overset{O}{\overset{\|}{C}}CH_2\overset{O}{\overset{\|}{C}}OC_2H_5 \xrightarrow[C_2H_5OH]{(C_2H_5)_2NH}$$

$$\left[Cl-C_6H_4-CH(OH)-CH(COC_2H_5)(CCH_3) \right] \xrightarrow{-H_2O} Cl-C_6H_4-CH=C(\overset{O}{\overset{\|}{C}}OC_2H_5)(\overset{O}{\overset{\|}{C}}CH_3) \quad (86\%)$$

$$(CH_3CH_2)_2C=O + CH_2(CN)(CO_2H) \xrightarrow[\text{benzene}]{NH_4^+CH_3COO^-} (CH_3CH_2)_2C=C(CN)(CO_2H) + H_2O \quad (65\%)$$

$$(CH_3CH_2)_2C=C(CN)(CO_2H) \xrightarrow{150^\circ} (CH_3CH_2)_2C=CHCN + CO_2$$

When Knoevenagel condensations take place with malonic acid, decarboxylation usually occurs spontaneously. (These reactions are sometimes called the Doebner modification of the Knoevenagel condensation.)

$$HO-C_6H_3(OCH_3)-CHO + CH_2(\overset{O}{\overset{\|}{C}}OH)_2 \xrightarrow[\text{pyridine, } 100^\circ]{C_5H_5NH^+} HO-C_6H_3(OCH_3)-CH=CH\overset{O}{\overset{\|}{C}}OH + CO_2 \quad (80\%)$$

20.8 MICHAEL ADDITIONS

Active hydrogen compounds also undergo conjugate additions to α,β-unsaturated carbonyl compounds. These reactions are known as Michael additions. An example is the following:

$$CH_3\overset{CH_3}{\overset{|}{C}}=CH\overset{O}{\overset{\|}{C}}OC_2H_5 + CH_2(\overset{O}{\overset{\|}{C}}OC_2H_5)_2 \xrightarrow[\substack{C_2H_5OH \\ 25^\circ}]{C_2H_5O^-Na^+} CH_3\overset{CH_3}{\overset{|}{C}}(CH(CO_2C_2H_5)_2)-CH_2\overset{O}{\overset{\|}{C}}OC_2H_5 \quad (70\%)$$

The mechanism for this reaction involves (1) formation of an anion from the active hydrogen compound,

$$(1)\quad C_2H_5O^- + H-CH(\overset{O}{\overset{\|}{C}}OC_2H_5)_2 \rightleftarrows C_2H_5OH + {}^-{:}CH(\overset{O}{\overset{\|}{C}}OC_2H_5)_2$$

(2) conjugate addition of the anion to the α,β-unsaturated ester, followed by (3) the acceptance of a proton.

$$\text{(2)}\quad CH_3-\underset{\underset{\overset{\cdot\cdot}{C}H^-(COOC_2H_5)_2}{|}}{\overset{\overset{CH_3}{|}}{C}}=CH-\overset{\overset{\ddot{O}:}{\|}}{C}-OC_2H_5 \rightleftarrows CH_3-\underset{\underset{CH(COOC_2H_5)_2}{|}}{\overset{\overset{CH_3}{|}}{C}}-CH=\overset{\overset{:\ddot{O}:^-}{|}}{C}-OC_2H_5 \longleftrightarrow CH_3-\underset{\underset{CH(COOC_2H_5)_2}{|}}{\overset{\overset{CH_3}{|}}{C}}-\overset{\cdot\cdot}{\underset{-}{C}}H-\overset{\overset{\ddot{O}:}{\|}}{C}-OC_2H_5$$

$$\text{(3)}\ \overset{H^+}{\rightleftarrows}\ CH_3-\underset{\underset{CH(COOC_2H_5)_2}{|}}{\overset{\overset{CH_3}{|}}{C}}-CH_2-\overset{\overset{\ddot{O}:}{\|}}{C}-OC_2H_5$$

Problem 20.20

How would you prepare $HO\overset{\overset{O}{\|}}{C}CH_2\underset{\underset{CH_3}{|}}{\overset{\overset{CH_3}{|}}{C}}CH_2\overset{\overset{O}{\|}}{C}OH$ from the product of the Michael addition given above?

Michael additions take place with a variety of other reagents; these include acetylenic esters and α,β-unsaturated nitriles:

$$H-C\equiv C-\overset{\overset{O}{\|}}{C}-OC_2H_5 + CH_3\overset{\overset{O}{\|}}{C}-CH_2-\overset{\overset{O}{\|}}{C}-OC_2H_5 \xrightarrow{C_2H_5O^-} HC=CH-\overset{\overset{O}{\|}}{C}-OC_2H_5$$

$$\qquad\qquad\qquad\qquad\qquad |$$

$$CH_3-\underset{\underset{O}{\|}}{C}-CH-\underset{\underset{O}{\|}}{C}-OC_2H_5$$

$$CH_2{=}CH{-}C{\equiv}N + CH_2(COOC_2H_5)_2 \xrightarrow{C_2H_5O^-} N{\equiv}C{-}CH_2{-}CH_2{-}CH(COOC_2H_5)_2$$

20.9 THE MANNICH REACTION

Active hydrogen compounds react with formaldehyde and a primary or secondary amine to yield compounds called Mannich bases. An example is the reaction of acetone, formaldehyde, and diethylamine shown below.

$$CH_3{-}\overset{O}{\overset{\|}{C}}{-}CH_3 + H{-}\overset{O}{\overset{\|}{C}}{-}H + (C_2H_5)_2NH \xrightarrow{HCl} CH_3{-}\overset{O}{\overset{\|}{C}}{-}CH_2{-}CH_2{-}N(C_2H_5)_2 + H_2O$$

(a Mannich base)

The Mannich reaction apparently proceeds through a variety of mechanisms depending on the reactants and the conditions that are employed. One mechanism that appears to operate in neutral or acidic media involves (1) initial reaction of the secondary amine with formaldehyde to yield an iminium ion and (2) subsequent reaction of the iminium ion with the enol form of the active hydrogen compound.

$$1.\ R_2\ddot{N}H + H_2C{=}\ddot{O}{:} \rightleftharpoons R_2\ddot{N}{-}CH_2{-}\ddot{O}{-}H \overset{H^+}{\rightleftharpoons}$$

$$R_2\ddot{N}{-}CH_2{-}\overset{+}{O}H_2 \overset{-H_2O}{\rightleftharpoons} R_2\overset{+}{N}{=}CH_2$$

Iminium ion

$$2.\ CH_3{-}\overset{O}{\overset{\|}{C}}{-}CH_3 \overset{H^+}{\rightleftharpoons} CH_3{-}\overset{O-H}{\overset{|}{C}}{=}CH_2 \quad \text{Enol}$$

$$CH_2{=}\overset{+}{N}R_2 \quad \text{Iminium ion}$$

$$\downarrow$$

$$CH_3{-}\overset{O}{\overset{\|}{C}}{-}CH_2{-}CH_2{-}CH_2{-}\ddot{N}R_2$$

Mannich base

$+ H^+$

Problem 20.21

Outline reasonable mechanisms that account for the products of the following Mannich reactions:

(a) Cyclohexanone + CH_2O + $(CH_3)_2NH$ ⟶ 2-[$CH_2N(CH_3)_2$]cyclohexanone

(b) $C_6H_5\overset{O}{\overset{\|}{C}}CH_3$ + CH_2O + pyrrolidine (N—H) ⟶ $C_6H_5\overset{O}{\overset{\|}{C}}CH_2CH_2$—N(pyrrolidinyl)

(c) p-Cresol (OH, CH_3) + CH_2O + $(CH_3)_2NH$ ⟶ 2,6-bis[$CH_2N(CH_3)_2$]-4-methylphenol ($(CH_3)_2NCH_2$, OH, $CH_2N(CH_3)_2$, CH_3)

20.10 ACYLATION OF ENAMINES: SYNTHESIS OF β-DIKETONES

Aldehydes and ketones react with secondary amines to form compounds called *enamines* (Section 16.8B). The general reaction for enamine formation can be written as follows:

$$\underset{\text{Aldehyde or ketone}}{H-\overset{|}{\underset{|}{C}}-\overset{\ddot{O}:}{\overset{\|}{C}}-} + \underset{\text{2° Amine}}{H\ddot{N}(R)-R} \rightleftarrows -\overset{|}{\underset{H}{C}}-\overset{OH}{\overset{|}{C}}-\ddot{N}R_2 \rightleftarrows \underset{\text{Enamine}}{-C{=}C-\ddot{N}R_2} + H_2O$$

Since enamine formation requires the loss of a molecule of water, enamine preparations are usually carried out in a way that allows water to be removed as an azeotrope or by a drying agent. This drives the reversible reaction to completion. Enamine formation is also catalyzed by the presence of a trace of an acid. The secondary amines most commonly used to prepare enamines are cyclic amines such as pyrrolidine, piperidine, and morpholine.

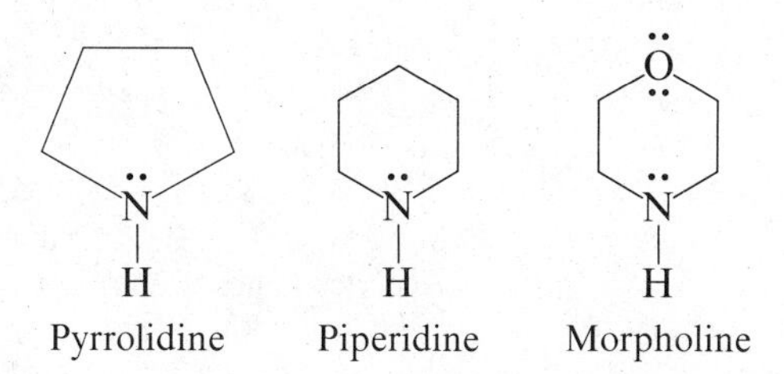

Cyclohexanone, for example, reacts with pyrrolidine in the following way.

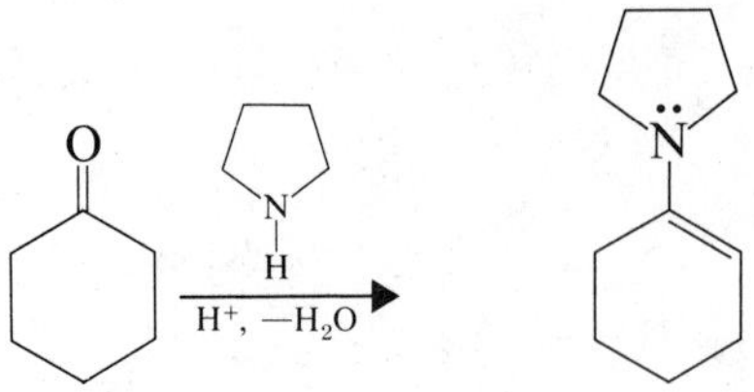

H^+, $-H_2O$

1-Pyrrolidinocyclohexene
(an enamine)

Enamines are good nucleophiles and an examination of the resonance structures below will show us that we should expect enamines to have both a nucleophilic nitrogen and a *nucleophilic carbon.*

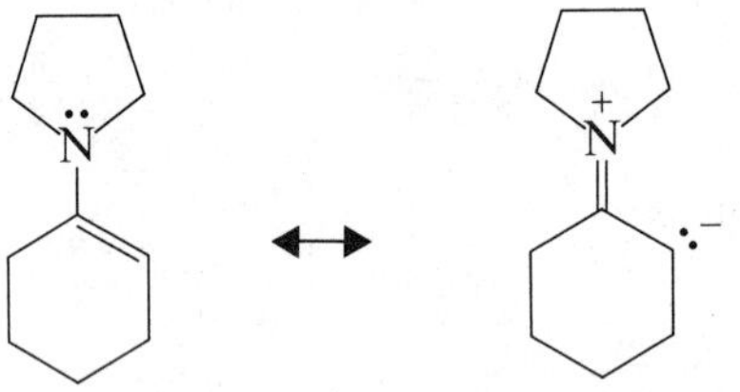

Contribution to the hybrid made by this structure confers nucleophilicity on nitrogen

Contribution to the hybrid made by this structure confers nucleophilicity on carbon and decreases nucleophilicity of nitrogen

The nucleophilicity of the carbon of enamines, because of the contribution made by the second structure above, makes them particularly useful reagents in organic synthesis.

When an enamine reacts with an acyl halide or an acid anhydride the product is the *C*-acylated compound. The iminium ion that forms, hydrolyzes when water is added and the overall reaction provides a synthesis of β-diketones.

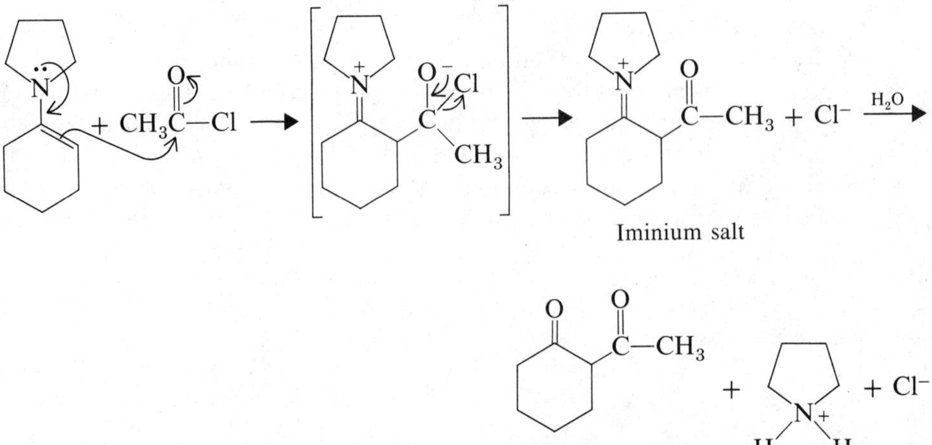

$+ CH_3\overset{O}{\overset{\|}{C}}-Cl \longrightarrow$ $\longrightarrow$ $\overset{O}{\overset{\|}{C}}-CH_3 + Cl^-$ $\xrightarrow{H_2O}$

Iminium salt

$\overset{O}{\overset{\|}{C}}-CH_3$ + + Cl^-

2-Acetylcyclohexanone
(a β-diketone)

Although *N*-acylation may occur in this synthesis, the *N*-acyl product is unstable and can act as an acylating agent itself.

Enamine + $CH_3C(=O)$–$\overset{+}{N}$ (*N*-Acylated enamine) Cl^- ⟶ *C*-Acylated iminium salt (CCH_3, Cl^-) + Enamine

As a consequence, the yields of *C*-acylated products are generally high.

Problem 20.22

One mole of a tertiary amine is generally added to the mixture during an enamine acylation reaction to neutralize any acid that might be formed. (a) Show how the iminium salt formed in the acylation reaction might react with the enamine to generate hydrogen chloride. (b) What effect would this acid-base reaction have on the yield if the tertiary amine were not present?

Enamines can be alkylated as well as acylated. While alkylation may lead to the formation of a considerable amount of *N*-alkylated product, heating the *N*-alkylated product often converts it to a *C*-alkyl compound. This is particularly true when the alkyl halide is an allyl or benzyl halide.

Enamine + $R{-}CH_2X$ ⟶ *N*-Alkylated product ($\overset{+}{N}$–CH_2R + X^-)

N-Alkylated product —Heat→ *C*-Alkylated product (CH_2R + X^-)

Enamine + $R{-}CH_2X$ ⟶ *C*-Alkylated product

$R = CH_2{=}CH{-}$ or $C_6H_5{-}$

C-Alkylated product —H_2O→ 2-(CH_2R)cyclohexanone + pyrrolidine (N–H)

Enamine alkylations are S_N2 reactions; thus, when we choose our alkylating agents, we are usually restricted to the use of methyl, primary, allyl, and benzyl halides. Alpha-halo esters can also be used as the alkylating agents, and this reaction provides a convenient synthesis of γ-keto esters:

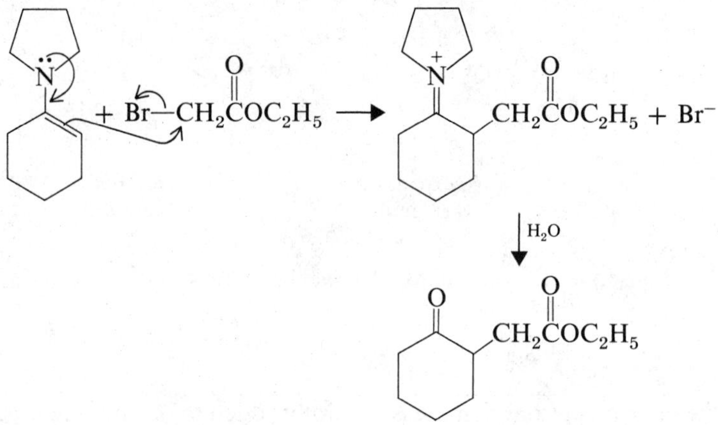

$$+ Br-CH_2\overset{O}{\overset{||}{C}}OC_2H_5 \longrightarrow CH_2\overset{O}{\overset{||}{C}}OC_2H_5 + Br^-$$

$$\xrightarrow{H_2O} CH_2\overset{O}{\overset{||}{C}}OC_2H_5$$

(75%)
(A γ-keto ester)

Problem 20.23

Show how you could employ enamines in syntheses of the following compounds.

(a) 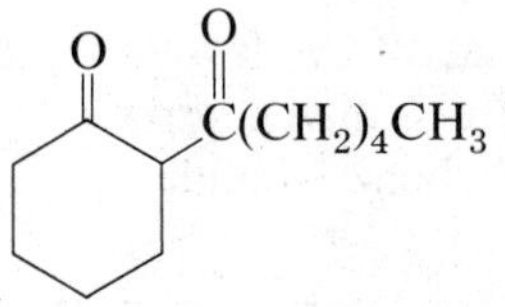 $\overset{O}{\overset{||}{C}}(CH_2)_4CH_3$

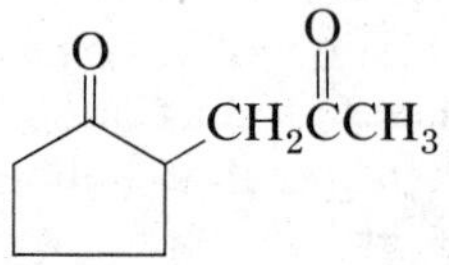

(b) 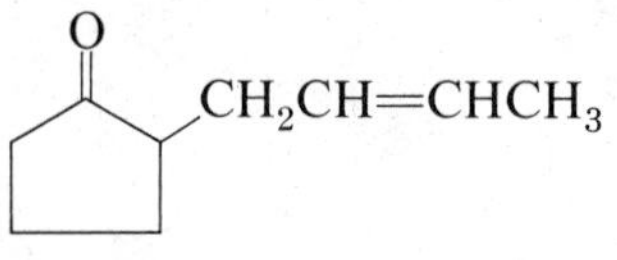 $CH_2CH{=}CHCH_3$

(c) $CH_2\overset{O}{\overset{||}{C}}CH_3$

(d) 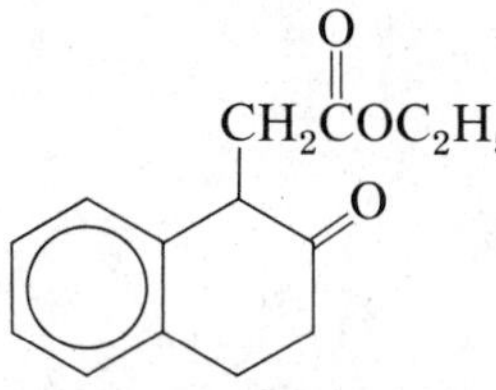 $CH_2\overset{O}{\overset{||}{C}}OC_2H_5$

An especially interesting set of enamine alkylations is shown below. The enamine is chiral. Alkylation from the bottom of the enamine is severely hindered by the methyl group. (Notice that this will be true even if rotation of the groups takes place about the bond connecting the two rings.) Consequently, alkylation takes place much more rapidly from the top side. This yields (after hydrolysis) 2-substituted cyclohexanones consisting almost entirely of a single enantiomer.

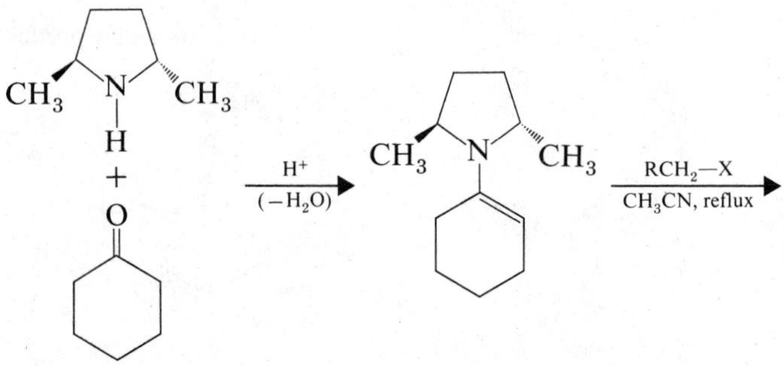

$$CH_3 \; N(H) \; CH_3 + O \xrightarrow[(-H_2O)]{H^+} CH_3 \; N \; CH_3 \xrightarrow[CH_3CN,\ reflux]{RCH_2-X}$$

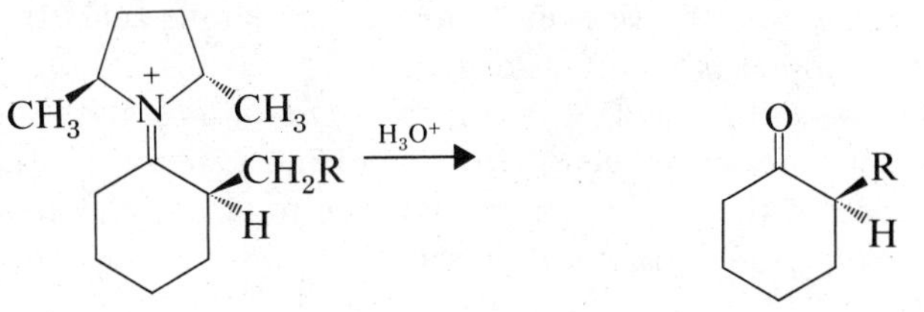

R group	Chemical yield	Optical purity
$CH_3—$	50%	83%
$CH_3CH_2CH_2—$	57%	93%
$CH_2{=}CHCH_2—$	80%	82%

Enamines can also be used in Michael additions. An example is the following.

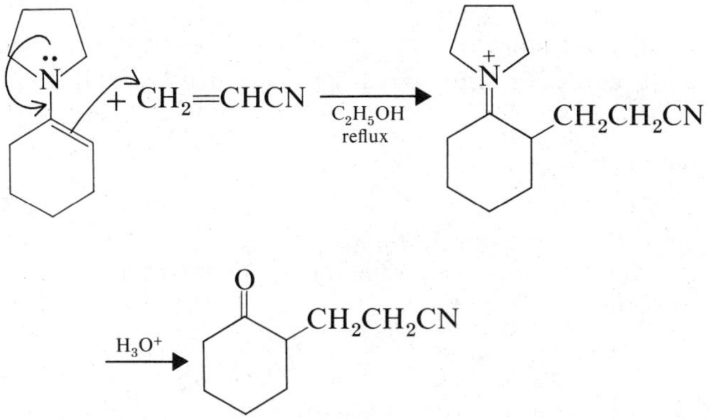

$$\xrightarrow[\text{reflux}]{C_2H_5OH}$$

$$\xrightarrow{H_3O^+}$$

20.11 BARBITURATES

In the presence of sodium ethoxide, diethyl malonate reacts with urea to yield a compound called barbituric acid.

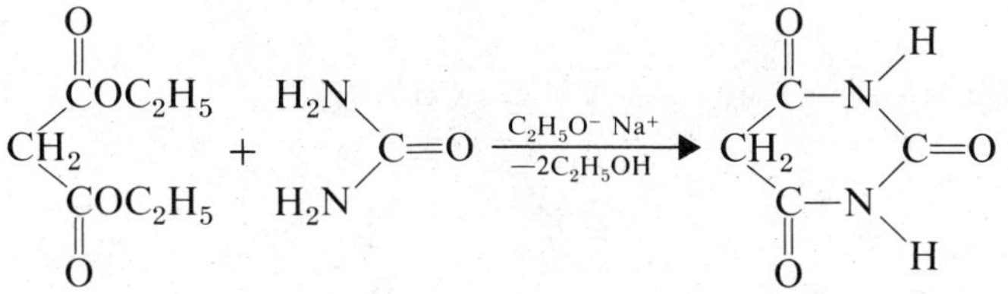

$$\xrightarrow[-2C_2H_5OH]{C_2H_5O^-\ Na^+}$$

Barbituric acid

Barbituric acid is a pyrimidine (cf. Section 18.1) and it exists in several tautomeric forms.

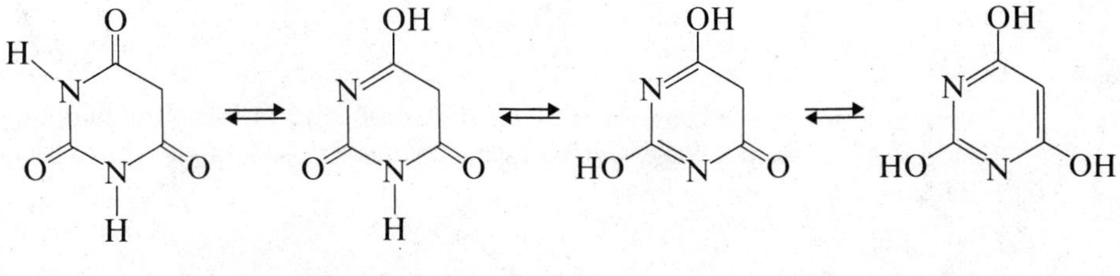

As its name suggests barbituric acid is a moderately strong acid, stronger even than acetic acid.* Its anion is highly resonance stabilized.

Salts of barbituric acid derivatives are *barbiturates*. Barbituric acid derivatives and barbiturates have been used in medicine as soporifics (sleep producers) since 1903. One of the earliest barbiturates introduced into medical use is the compound veronal (diethylbarbituric acid). Veronal is usually used as its sodium salt.

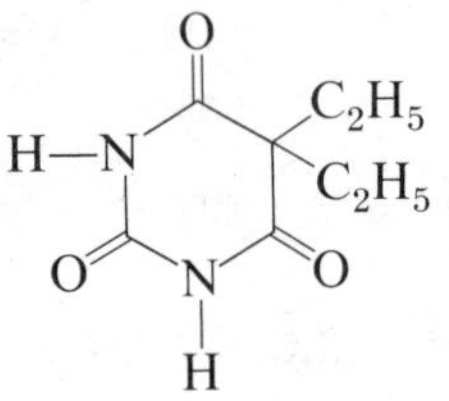

Veronal
(diethylbarbital)

Although barbiturates are very effective soporifics, their use is also hazardous. They are addictive, and overdosage, often with fatal results, is common.

Additional Problems

20.24

Show all steps in the following syntheses. You may use any other needed reagents but should begin with the compound given. You need not repeat steps carried out in earlier parts of this exercise.

(a) $CH_3CH_2CH_2\overset{O}{\overset{\|}{C}}OC_2H_5 \longrightarrow CH_3CH_2CH_2\overset{O}{\overset{\|}{C}}\underset{CH_2CH_3}{\underset{|}{C}H}\overset{O}{\overset{\|}{C}}OC_2H_5$

(b) $CH_3CH_2CH_2\overset{O}{\overset{\|}{C}}OC_2H_5 \longrightarrow CH_3CH_2CH_2\overset{O}{\overset{\|}{C}}CH_2CH_2CH_3$

(c) $C_6H_5CH_2\overset{O}{\overset{\|}{C}}OC_2H_5 \longrightarrow C_6H_5\overset{CH_3}{\overset{|}{C}H}COOH$

(d) $CH_3CH_2CH_2\overset{O}{\overset{\|}{C}}OC_2H_5 \longrightarrow CH_3CH_2\underset{\overset{\|}{O}\ \ \overset{\|}{O}}{\underset{C-COC_2H_5}{\underset{|}{C}H}}\overset{O}{\overset{\|}{C}}OC_2H_5$

(e) $CH_3CH_2CH_2\overset{O}{\overset{\|}{C}}OC_2H_5 \longrightarrow CH_3CH_2CH_2\overset{O}{\overset{\|}{C}}-\overset{O}{\overset{\|}{C}}OC_2H_5$

(f) $C_6H_5CH_2\overset{O}{\overset{\|}{C}}OC_2H_5 \longrightarrow C_6H_5\underset{\underset{O}{\underset{\|}{CH}}}{\underset{|}{C}H}\overset{O}{\overset{\|}{C}}OC_2H_5$

* Barbituric acid was given its name by Adolf von Baeyer in 1864. The barbituric part of the name is thought to have been motivated by Baeyer's gallantry toward a friend named Barbara.

(g) 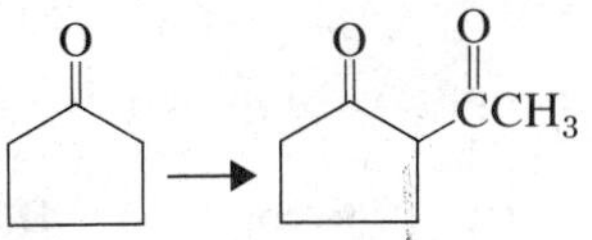 [cyclopentanone] → [2-acetylcyclopentanone: O, O, $\overset{\text{O}}{\overset{\|}{\text{C}}}\text{CH}_3$]

(h) 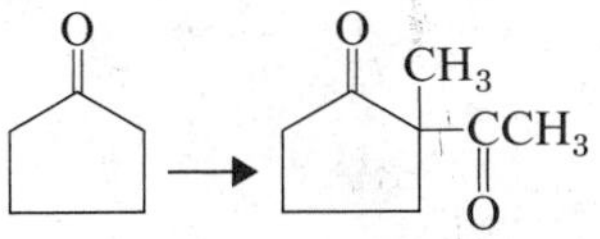 [cyclopentanone] → [cyclopentanone with CH_3, CCH_3, O substituent]

(i) 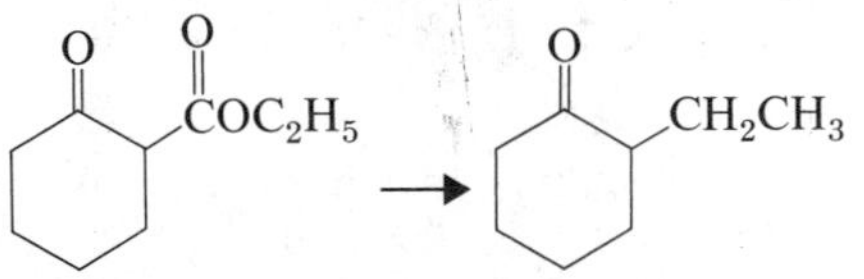 [ethyl 2-oxocyclohexanecarboxylate: O, O, $\overset{\text{O}}{\overset{\|}{\text{C}}}\text{OC}_2\text{H}_5$] → [2-ethylcyclohexanone: O, CH_2CH_3]

20.25

Outline syntheses of each of the following from acetoacetic ester and any other required reagents.

(a) Methyl *tert*-butyl ketone
(b) 2-Hexanone
(c) 2,5-Hexanedione
(d) 4-Hydroxypentanoic acid
(e) 2-Ethyl-1,3-butanediol
(f) 1-Phenyl-1,3-butanediol
(g) 1-Phenyl-1,3,5-pentanetriol

20.26

Outline syntheses of each of the following from diethyl malonate and any other required reagents..

(a) 2-Methylbutanoic acid
(b) 4-Methyl-1-pentanol
(c) $\text{CH}_3\text{CH}_2\underset{\underset{\text{CH}_2\text{OH}}{|}}{\text{C}}\text{HCH}_2\text{OH}$
(d) $\text{HOCH}_2\text{CH}_2\text{CH}_2\text{CH}_2\text{OH}$

20.27

The synthesis of cyclobutanecarboxylic acid given on p. 926 was first carried out by William Perkin, Jr. in 1883, and it represented the first synthesis of an organic compound with a ring smaller than six carbons. (There was a general feeling at the time that such compounds would be too unstable to exist.) Earlier in 1883, Perkin reported what he mistakenly believed to be a cyclobutane derivative obtained from the reaction of acetoacetic ester and 1,3-dibromopropane. The reaction that Perkin had expected to take place was the following:

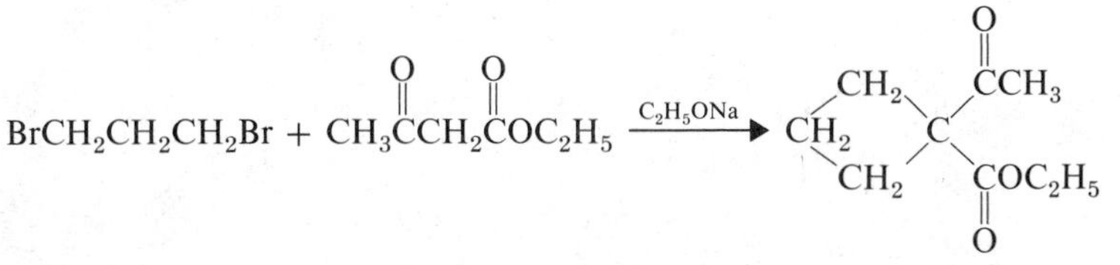

$$\text{BrCH}_2\text{CH}_2\text{CH}_2\text{Br} + \text{CH}_3\overset{\text{O}}{\overset{\|}{\text{C}}}\text{CH}_2\overset{\text{O}}{\overset{\|}{\text{C}}}\text{OC}_2\text{H}_5 \xrightarrow{\text{C}_2\text{H}_5\text{ONa}} \text{CH}_2\begin{matrix}\diagup \text{CH}_2 \diagdown \\ \diagdown \text{CH}_2 \diagup\end{matrix}\text{C}\begin{matrix}\diagup \overset{\text{O}}{\overset{\|}{\text{C}}}\text{CH}_3 \\ \diagdown \underset{\text{O}}{\underset{\|}{\text{C}}}\text{OC}_2\text{H}_5\end{matrix}$$

The molecular formula for his product agreed with the formulation given above and alkaline hydrolysis and acidification gave a nicely crystalline acid (also having the expected molecular formula). The acid, however, was quite stable to heat and resisted decarboxylation. Perkin later found that both the ester and the acid contained six-membered rings (five carbons and one oxygen). Recall the charge distribution in the enolate ion obtained from acetoacetic ester and propose structures for Perkin's ester and acid.

20.28
(a) In 1884 Perkin achieved a successful synthesis of cyclopropanecarboxylic acid from sodiomalonic ester and 1,2-dibromoethane. Outline the reactions involved in this synthesis. (b) In 1885 Perkin synthesized five-membered carbocyclic compounds **D** and **E** in the following way:

$$2Na^+ :\bar{C}H(CO_2C_2H_5)_2 + BrCH_2CH_2CH_2Br \longrightarrow \textbf{A}\ (C_{17}H_{28}O_8) \xrightarrow{2C_2H_5O^-Na^+} \xrightarrow{Br_2}$$

$$\textbf{B}\ (C_{17}H_{26}O_8) \xrightarrow[(2)\ H_3O^+]{(1)\ OH^-/H_2O} \textbf{C}\ (C_9H_{10}O_8) \xrightarrow{\text{Heat}} \textbf{D}\ (C_7H_{10}O_4) + \textbf{E}\ (C_7H_{10}O_4)$$

D and **E** are diastereomers; **D** can be resolved into enantiomeric forms while **E** cannot. What are the structures of **A, B, C, D,** and **E**? (c) Ten years later Perkin was able to synthesize 1,4-dibromobutane; he later used this and diethyl malonate to prepare cyclopentanecarboxylic acid. Show the reactions involved.

20.29
Thiol esters of the type $RCH_2\overset{O}{\overset{\|}{C}}SCH_2CH_3$ are more reactive in Claisen condensations than are ordinary esters. How can you account for this?

20.30
Write mechanisms that account for the products of the following reactions.

(a) $$C_6H_5CH{=}CH\overset{O}{\overset{\|}{C}}OC_2H_5 + CH_2(\overset{O}{\overset{\|}{C}}OC_2H_5)_2 \xrightarrow{NaOCH_2CH_3} C_6H_5CH{-}CH_2\overset{O}{\overset{\|}{C}}OC_2H_5 \text{ with } CH(\overset{O}{\overset{\|}{C}}{-}OC_2H_5)_2 \text{ on the benzylic carbon}$$

(b) $$CH_2{=}CH\overset{O}{\overset{\|}{C}}OCH_3 \xrightarrow{CH_3NH_2} CH_3N(CH_2CH_2\overset{O}{\overset{\|}{C}}OCH_3)_2 \xrightarrow{\text{base}} \text{1-methyl-4-oxopiperidine-3-}\overset{O}{\overset{\|}{C}}OCH_3$$

(c) $$CH_3{-}\underset{CH(CO_2C_2H_5)_2}{\overset{CH_3}{C}}{-}CH_2\overset{O}{\overset{\|}{C}}OC_2H_5 + C_2H_5O^- \longrightarrow CH_3\overset{CH_3}{C}{=}CH\overset{O}{\overset{\|}{C}}OC_2H_5 + :\bar{C}H(CO_2C_2H_5)_2 + C_2H_5OH$$

(d) $$\text{1-pyrrolidinylcyclohexene} + CH_2{=}CH\overset{O}{\overset{\|}{C}}OC_2H_5 \longrightarrow \text{1-pyrrolidinyl-6-}(CH_2CH_2\overset{O}{\overset{\|}{C}}OC_2H_5)\text{cyclohexene} \xrightarrow[H^+]{H_2O}$$

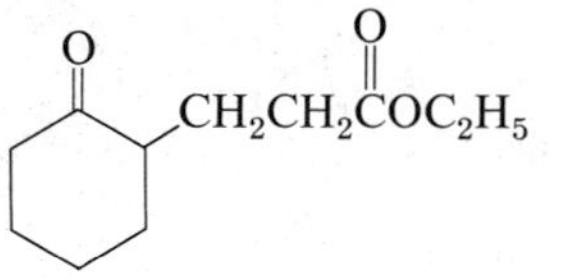

$$\text{(2-oxocyclohexyl)}-CH_2CH_2\overset{O}{\overset{\|}{C}}OC_2H_5$$

*(e) 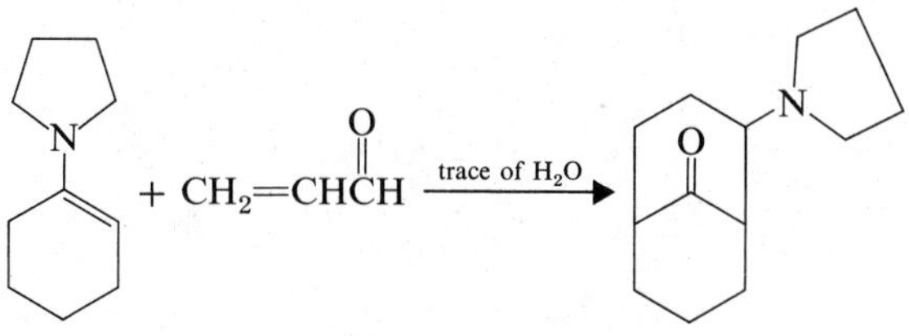 1-pyrrolidinocyclohexene $+ CH_2{=}CH\overset{O}{\overset{\|}{C}}H \xrightarrow{\text{trace of } H_2O}$ bicyclic keto product bearing a pyrrolidinyl (N) group

20.31

Knoevenagel condensations in which the active hydrogen compound is a β-keto ester or a β-diketone often yield products that result from one molecule of aldehyde or ketone and two molecules of the active methylene component. For example,

$$R{-}\underset{R'}{\underset{|}{C}}{=}O + CH_2(COCH_3)_2 \xrightarrow{\text{Base}} R{-}\underset{R'}{\underset{|}{C}}\begin{matrix}{\diagup}CH(COCH_3)_2 \\ {\diagdown}CH(COCH_3)_2\end{matrix}$$

Suggest a reasonable mechanism that will account for the formation of these products.

20.32

Thymine (below) is one of the heterocyclic bases found in DNA. Starting with ethyl propanoate and using any other needed reagents show how you might synthesize thymine.

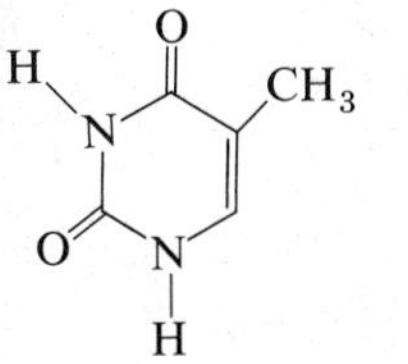

Thymine structure (ring with N–H, C=O, CH_3, N–H)

20.33

The mandibular glands of queen bees secrete a fluid that contains a remarkable compound known as "queen substance." When even an exceedingly small amount of the queen substance is transferred to worker bees it inhibits the development of their ovaries and prevents the workers from rearing new queens. Queen substance, a monocarboxylic acid with the molecular formula $C_{10}H_{16}O_3$, has been synthesized by the following route.

$$\text{Cycloheptanone} \xrightarrow[(2)\ H_3O^+]{(1)\ CH_3MgI} \textbf{A}\ (C_8H_{16}O) \xrightarrow{H^+,\ \text{heat}} \textbf{B}\ (C_8H_{14}) \xrightarrow[(2)\ Zn,\ H_2O]{(1)\ O_3}$$

$$\textbf{C}\ (C_8H_{14}O_2) \xrightarrow[\text{Pyridine}]{CH_2(CO_2H)_2} \text{Queen Substance}\ (C_{10}H_{16}O_3)$$

On catalytic hydrogenation queen substance yields compound **D**, which, on treatment with iodine in sodium hydroxide and subsequent acidification, yields a dicarboxylic acid **E**, that is,

$$\text{Queen substance} \xrightarrow[Pd]{H_2} \textbf{D}\ (C_{10}H_{18}O_3) \xrightarrow[(2)\ H_3O^+]{(1)\ I_2/NaOH} \textbf{E}\ (C_9H_{16}O_4)$$

Provide structures for the queen substance and compounds **A**–**E**.

20.34
Linalool, a fragrant compound that can be isolated from a variety of plants, is 3,7-dimethyl-1,6-octadien-3-ol. Linalool is used in making perfumes and it can be synthesized in the following way:

$$CH_2{=}\underset{\displaystyle CH_3}{\underset{|}{C}}{-}CH{=}CH_2 + HBr \longrightarrow \mathbf{F}(C_5H_9Br) \xrightarrow[\text{ester}]{\text{Sodioacetoacetic}}$$

$$\mathbf{G}(C_{11}H_{18}O_3) \xrightarrow[\text{(2) } H_3O^+\text{, (3) heat}]{\text{(1) dil. NaOH}} \mathbf{H}(C_8H_{14}O) \xrightarrow[\text{(2) } H_3O^+]{\text{(1) LiC}\equiv\text{CH}}$$

$$\mathbf{I}(C_{10}H_{16}O) \xrightarrow[\text{Lindlar's catalyst}]{H_2} \text{Linalool}$$

Outline the reactions involved. (Hint: compound **F** is the more stable isomer capable of being produced in the first step.)

20.35
Compound **J**, a compound with two four-membered rings, has been synthesized by the route shown below. Outline the steps that are involved.

$$NaCH(CO_2C_2H_5)_2 + BrCH_2CH_2CH_2Br \longrightarrow [C_{10}H_{17}BrO_4]$$

$$\xrightarrow{NaOC_2H_5} C_{10}H_{16}O_4 \xrightarrow{LiAlH_4} C_6H_{12}O_2 \xrightarrow{HBr}$$

$$C_6H_{10}Br_2 \xrightarrow[2\,NaOC_2H_5]{CH_2(CO_2C_2H_5)_2} C_{13}H_{20}O_4 \xrightarrow[\text{(2) } H_3O^+]{\text{(1) } OH^-,\ H_2O}$$

$$C_9H_{12}O_4 \xrightarrow{\text{Heat}} \mathbf{J}(C_8H_{12}O_2) + CO_2$$

* **20.36**
When an aldehyde or a ketone is condensed with ethyl α-chloroacetate in the presence of sodium ethoxide, the product is an α,β-epoxy ester called a *glycidic ester*. The synthesis is called the Darzens condensation.

$$R{-}\overset{\displaystyle R'}{\overset{|}{C}}{=}O + ClCH_2COOC_2H_5 \xrightarrow{C_2H_5ONa} R{-}\overset{\displaystyle R'}{\overset{|}{C}}\underset{O}{\diagdown\!\diagup}CHCOOC_2H_5 + NaCl + C_2H_5OH$$

A glycidic ester

(a) Outline a reasonable mechanism for the Darzens condensation. (b) Hydrolysis of the epoxy ester leads to an epoxy acid that, on heating with pyridine, furnishes an aldehyde.

$$R{-}\overset{\displaystyle R'}{\overset{|}{C}}\underset{O}{\diagdown\!\diagup}CHCOOH \xrightarrow[\text{Heat}]{C_5H_5N} R{-}\overset{\displaystyle R'}{\overset{|}{C}H}{-}\overset{\displaystyle O}{\overset{\|}{C}H} + CO_2$$

What is happening here? (c) Starting with β-ionone (p. 732) show how you might synthesize the aldehyde shown below. (This aldehyde is another intermediate in the industrial synthesis of Vitamin A that we began describing in Problem 15.49).

$$\text{CH}\underset{\displaystyle O}{\underset{\|}{}}$$

*** 20.37**

The *Perkin condensation* is an aldol-type condensation in which an aromatic aldehyde, ArCHO, reacts with a carboxylic acid anhydride, $(RCH_2CO)_2O$, to give an α,β-unsaturated acid, ArCH=CRCOOH. The catalyst that is usually employed is the potassium salt of the carboxylic acid, RCH_2COOK. (a) Outline the Perkin condensation that takes place when benzaldehyde reacts with propanoic anhydride in the presence of potassium propanoate. (b) How would you use a Perkin condensation to prepare *p*-chlorocinnamic acid, *p*-ClC_6H_4CH=CHCOOH?

*** 20.38**

(+)-Fenchone is a terpenoid that can be isolated from fennel oil. (±) Fenchone has been synthesized through the route shown below. Supply the missing intermediates and reagents.

CH_2=CH—CH=CH_2 + (a) ⟶ $COOCH_3$ (cyclohexene) $\xrightarrow[(2)\ (c)]{(1)\ (b)}$

$COOCH_3$, $COOCH_3$, $COOCH_3$ $\xrightarrow{(d)}$ mixture of (e) and (f) $\xrightarrow[(h)]{(g)}$

COOH (cyclopentanone, O) $\xrightarrow{(i)}$ $COOCH_3$ (cyclopentanone, O) $\xrightarrow[(2)\ H_3O^+]{(1)\ Zn,\ BrCH_2COOCH_3,\ ether}$ (cf. Problem 20.39) $COOCH_3$, CH_2COOCH_3, OH

$\xrightarrow[heat]{H^+}$ (j) $\xrightarrow{(k)}$ $COOCH_3$, CH_2COOCH_3 $\xrightarrow{(l)}$

O $\xrightarrow{(m)}$ O

(±)-Fenchone

*** 20.39**

One step in the preceding problem involves the following reaction.

$COOCH_3$ (cyclopentanone, O) $\xrightarrow[(2)\ H_3O^+]{(1)\ Zn,\ BrCH_2COOCH_3,\ ether}$ $COOCH_3$, CH_2COOCH_3, OH

This is an example of the *Reformatsky reaction*. In general terms, the Reformatsky reaction involves treating an α-bromo ester with zinc in the presence of an aldehyde or ketone; the product (after acidification) is a β-hydroxy ester. (a) What general synthesis does this resemble? (b) What kind of reagent is likely to be an intermediate in the Reformatsky reaction? (c) The success of the Reformatsky reaction depends on the selectivity of this reagent. Explain. (d) What factor accounts for this selectivity? (e) Show how you might use the Reformatsky reaction in a synthesis of $C_6H_5CH_2CH(CH_3)COOCH_3$ starting with benzaldehyde.

*** 20.40**

Using as your starting compounds benzaldehyde, acetone, diethyl malonate, ethanol, and any needed inorganic reagents show how you might synthesize the compound shown below.

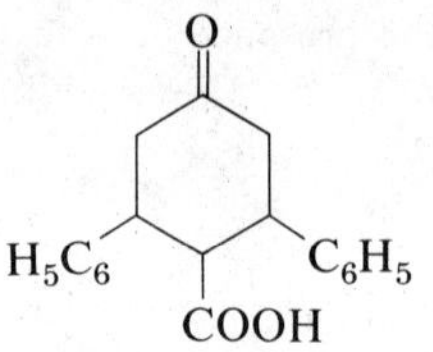

O

H_5C_6 C_6H_5

COOH

*** 20.41**

Give structures for compounds **A–H**.

(a) 2,5-Hexanedione + $(NH_4)_2CO_3$ $\xrightarrow{100°}$ **A** (C_6H_9N), a pyrrole

(b) $CH_3\overset{O}{\overset{\|}{C}}CH_2NH_2$ + acetone $\xrightarrow{Base}$ **B** (C_6H_9N), an isomer of **A**

(c) $CH_3NHNH_2 + (CH_3O)_2CHCH_2CH(OCH_3)_2 \xrightarrow[H_2O]{H^+}$ **C** ($C_4H_6N_2$), a pyrazole

(d) 2,5-Hexanedione + hydrazine $\xrightarrow{Heat}$

D ($C_6H_{10}N_2$), a dihydropyridazine $\xrightarrow{O_2}$ **E** ($C_6H_8N_2$), a pyridazine

(e) Aniline + $CH_2{=}CH\overset{O}{\overset{\|}{C}}CH_3 \xrightarrow[FeCl_3]{ZnCl_2}$ **F** ($C_{10}H_9N$), a quinoline

(f) 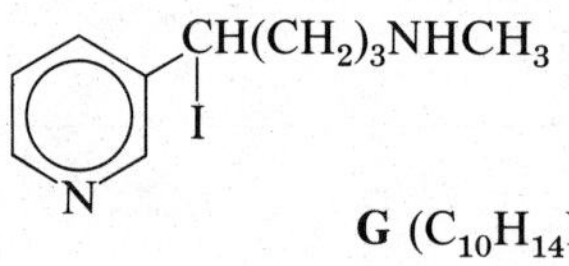 3-pyridyl–$\underset{I}{\underset{|}{C}H}(CH_2)_3NHCH_3 \xrightarrow{Heat} \xrightarrow{OH^-}$

G ($C_{10}H_{14}N_2$), nicotine $\xrightarrow[(2)\ H^+]{(1)\ KMnO_4,\ OH^-}$ **H** ($C_6H_5NO_2$), nicotinic acid

*** 20.42**

Antipyrine is a compound that has been used as an analgesic and to reduce fevers. It can be synthesized from ethyl acetoacetate and phenylhydrazine through the following steps.

$$CH_3\overset{O}{\overset{\|}{C}}CH_2COOC_2H_5 + C_6H_5NHNH_2 \xrightarrow{Heat} C_{12}H_{16}N_2O_2 \xrightarrow[heating]{Further}$$

$$C_{10}H_{10}N_2O \xrightarrow[NaOH]{CH_3OSO_2OCH_3} \text{Antipyrine, } C_{11}H_{12}N_2O$$

Outline the reactions that take place.

21 LIPIDS

21.1 INTRODUCTION

When plant or animal tissues are extracted with a nonpolar solvent, (e.g., ether, chloroform, benzene, or an alkane) a portion of the material usually dissolves. The components of this soluble fraction are called *lipids*.*

Lipids include a wide variety of structural types including the following:

Carboxylic acids (or "fatty" acids)
Triacylglycerols (or neutral fats)
Phospholipids
Glycolipids
Waxes
Terpenes
Steroids
Prostaglandins

We discussed the chemistry of terpenes earlier in Special Topic E. We can now examine the properties of other members of the lipid group.

21.2 FATTY ACIDS AND TRIACYLGLYCEROLS

Only a small portion of the total lipid fraction consists of free carboxylic acids. Most of the carboxylic acids in the lipid fraction are found as *esters of glycerol,* that is, as *triacylglycerols.*† The most common are *triacylglycerols of long-chain carboxylic acids* (Fig. 21.1).

Triacylglycerols are the oils and fats of plant or animal origin. They include such common substances as peanut oil, olive oil, soybean oil, corn oil, linseed oil,

* Lipids are defined in terms of the physical operation that we use to isolate them. In this respect, the definition of a lipid differs from that of proteins (Chapter 22) or carbohydrates (Chapter 19). The latter are defined on the basis of their structures.

† In older literature triacylglycerols were referred to as triglycerides or simply as glycerides.

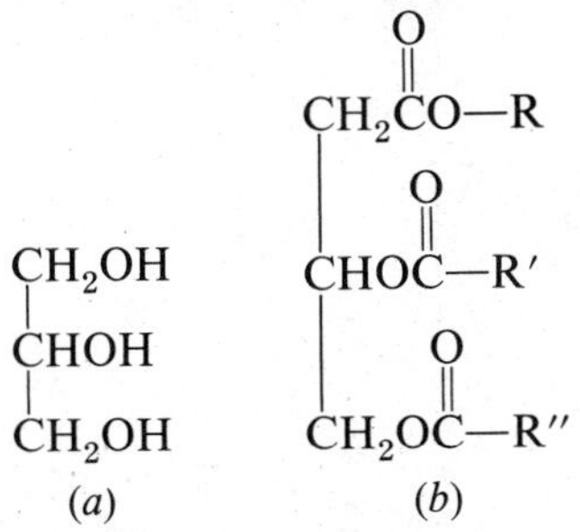

$$\begin{array}{l} CH_2OH \\ | \\ CHOH \\ | \\ CH_2OH \end{array} \qquad \begin{array}{l} CH_2OC(=O)\text{—}R \\ | \\ CHOC(=O)\text{—}R' \\ | \\ CH_2OC(=O)\text{—}R'' \end{array}$$

(*a*) (*b*)

FIG. 21.1
(a) *Glycerol.* (b) *A triacylglycerol. R, R′, and R″ are usually long-chain alkyl groups. R, R′, and R″ may also contain one or more carbon-carbon double bonds. In a typical triacylglycerol, R, R′, and R″ are all different.*

butter, lard, and tallow. Those triacylglycerols that are liquids at room temperature are generally known as *oils;* those that are solids are usually called *fats.*

The carboxylic acids that are obtained by hydrolysis of naturally occurring fats and oils usually have unbranched chains with an even number of carbon atoms. (We will see, later, that these facts give us important clues as to how they are synthesized in plants and animals.) The most common carboxylic acids obtained from fats and oils are the C_{14}-, C_{16}-, and C_{18}-acids shown in Table 21.1.

In addition to most of the common fatty acids given in Table 21.1, the hydrolysis of butter gives small amounts of saturated even-numbered carboxylic acids in the C_4-C_{12} range. These are butyric (butanoic), caproic (hexanoic), caprylic (octanoic), capric (decanoic), and lauric (dodecanoic) acids. The hydrolysis of coconut oil also gives short-chain carboxylic acids and a large amount of lauric acid.

Other less common fatty acids and their sources are given in Table 21.2. Table 21.3 gives the fatty acid composition of a number of common fats and oils.

TABLE 21.1 Common Fatty Acids

Saturated Carboxylic Acids	mp °C
$CH_3(CH_2)_{12}COOH$ Myristic acid (tetradecanoic acid)	54
$CH_3(CH_2)_{14}COOH$ Palmitic acid (hexadecanoic acid)	63
$CH_3(CH_2)_{16}COOH$ Stearic acid (octadecanoic acid)	70
Unsaturated Carboxylic Acids	
$CH_3(CH_2)_5(H)C{=}C(H)(CH_2)_7COOH$ Palmitoleic acid (*cis*-9-hexadecenoic acid)	32
$CH_3(CH_2)_7(H)C{=}C(H)(CH_2)_7COOH$ Oleic acid (*cis*-9-octadecenoic acid)	4
$CH_3(CH_2)_4(H)C{=}C(H)CH_2(H)C{=}C(H)(CH_2)_7COOH$ Linoleic acid (*cis, cis*-9, 12-octadecadienoic acid)	−5
$CH_3CH_2(H)C{=}C(H)CH_2(H)C{=}C(H)CH_2(H)C{=}C(H)(CH_2)_7COOH$ Linolenic acid (*cis, cis, cis*-9, 12, 15-octadecatrienoic acid)	−11

TABLE 21.2 Some Less Common Fatty Acids

NUMBER OF CARBONS	NAME	STRUCTURE	SOURCE
18	Eleostearic acid	$CH_3(CH_2)_3CH{=}CH{-}CH{=}CH{-}CH{=}CH(CH_2)_7COOH$	The main (80%) fatty acid obtained by hydrolysis of tung oil
18	Ricinoleic acid	$CH_3(CH_2)_5CH(OH)CH_2CH{=}CH(CH_2)_7COOH$	The main (80%) fatty acid obtained by hydrolysis of castor oil
18	Chaulmoogric acid	(cyclopent-2-enyl)$-(CH_2)_{12}COOH$	From oil of *Hydnocarpus kuazii* (used in the treatment of leprosy)
19	Sterculic acid	$CH_3(CH_2)_7C{=}C(CH_2)_7COOH$ (ring closed by CH_2)	From kernel oil of *Sterculia foetida*
19	Lactobacillic acid	$CH_3(CH_2)_5CH{-}CH(CH_2)_9COOH$ (ring closed by CH_2)	From phospholipids of bacteria
19	Tuberculostearic acid	$CH_3(CH_2)_7CH(CH_3)(CH_2)_8COOH$	From tubercule bacilli
20	Arachidonic acid	$CH_3(CH_2)_4(CH{=}CHCH_2)_4(CH_2)_2COOH$	From human fat (0.3–1.0%) liver, lecithins
22	Cetoleic acid	$CH_3(CH_2)_9CH{=}CH(CH_2)_9COOH$	From fish oils
22	Erucic acid	$CH_3(CH_2)_7CH{=}CH(CH_2)_{11}COOH$ (*cis*)	From seed oils of rape, wallflower-nasturtium, and mustard
24	Nervonic acid	$CH_3(CH_2)_7CH{=}CH(CH_2)_{13}COOH$ (*cis*)	From fish oils and brain tissues
27	Mycolipenic acid	$CH_3(CH_2)_{17}CH(CH_3)CH_2CH(CH_3)CH{=}C(CH_3){-}COOH$	From tubercule bacilli

21.2A Hydrogenation of Triacylglycerols

We see from the data given in Table 21.3 that most oils are made up of high percentages of unsaturated fatty acids. The fact that oils have lower melting points than fats is related to this factor; hydrogenation of an oil produces a solid fat. The *cis*-double bonds of unsaturated fatty acids causes their carbon chains to assume conformations that do not fit easily into an orderly crystal structure of a solid. The saturated chains produced by hydrogenating an oil fit much better.

TABLE 21.3 Fatty Acid Composition Obtained by Hydrolysis of Common Fats and Oils[a]

FAT OR OIL	AVERAGE COMPOSITION OF FATTY ACIDS (mole %)											
	SATURATED								UNSATURATED			
	C_4 BUTYRIC ACID	C_6 CAPROIC ACID	C_8 CAPRYLIC ACID	C_{10} CAPRIC ACID	C_{12} LAURIC ACID	C_{14} MYRISTIC ACID	C_{16} PALMITIC ACID	C_{18} STEARIC ACID	C_{16} PALMIT-OLEIC ACID	C_{18} OLEIC ACID	C_{18} LINOLEIC ACID	C_{18} LINOLENIC ACID
Animal Fats												
Butter	3–4	1–2	0–1	2–3	2–5	8–15	25–29	9–12	4–6	18–33	2–4	
Lard						1–2	25–30	12–18	4–6	48–60	6–12	0–1
Beef tallow						2–5	24–34	15–30		35–45	1–3	0–1
Vegetable Oils												
Olive						0–1	5–15	1–4		67–84	8–12	
Peanut							7–12	2–6		30–60	20–38	
Corn						1–2	7–11	3–4	1–2	25–35	50–60	
Cottonseed						1–2	18–25	1–2	1–3	17–38	45–55	
Soybean						1–2	6–10	2–4		20–30	50–58	5–10
Linseed							4–7	2–4		14–30	14–25	45–60
Coconut		0–1	5–7	7–9	40–50	15–20	9–12	2–4	0–1	6–9	0–1	
Marine Oils												
Cod liver						5–7	8–10	0–1	18–22	27–33	27–32	

[a] Data adapted from John R. Holum, *Organic and Biological Chemistry,* Wiley, New York, (1978), p. 220, and from *Biology Data Book,* Philip L. Altman and Dorothy S. Ditmer, eds., Federation of American Societies for Experimental Biology, Washington, D.C., (1964).

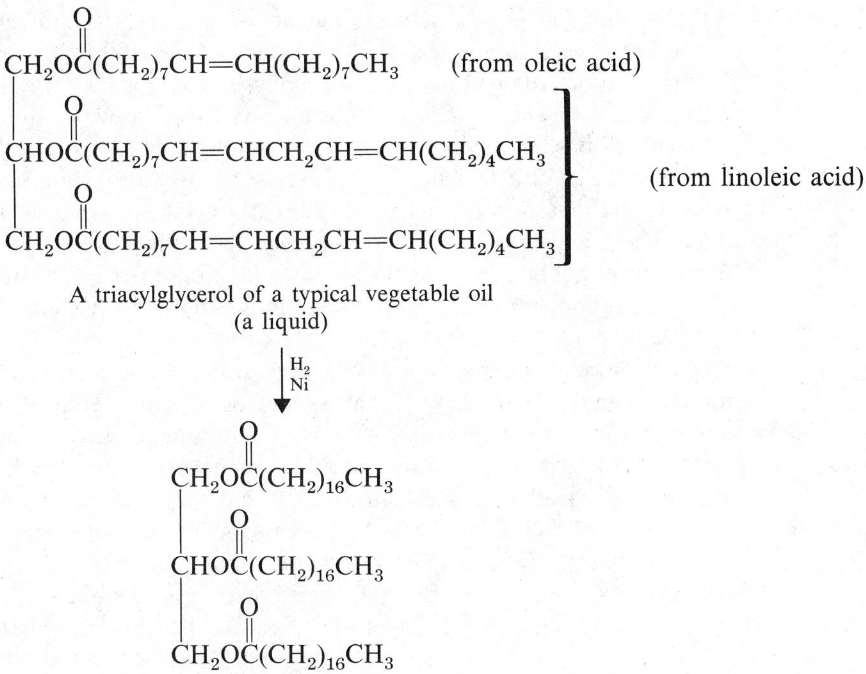

$$\begin{array}{ll}
\overset{\displaystyle O}{\overset{\|}{\text{CH}_2\text{OC}}}(\text{CH}_2)_7\text{CH}{=}\text{CH}(\text{CH}_2)_7\text{CH}_3 & \text{(from oleic acid)} \\
\overset{\displaystyle O}{\overset{\|}{\text{CHOC}}}(\text{CH}_2)_7\text{CH}{=}\text{CHCH}_2\text{CH}{=}\text{CH}(\text{CH}_2)_4\text{CH}_3 & \\
\overset{\displaystyle O}{\overset{\|}{\text{CH}_2\text{OC}}}(\text{CH}_2)_7\text{CH}{=}\text{CHCH}_2\text{CH}{=}\text{CH}(\text{CH}_2)_4\text{CH}_3 & \text{(from linoleic acid)}
\end{array}$$

A triacylglycerol of a typical vegetable oil
(a liquid)

$$\downarrow \; H_2, \; Ni$$

$$\begin{array}{l}
\overset{\displaystyle O}{\overset{\|}{\text{CH}_2\text{OC}}}(\text{CH}_2)_{16}\text{CH}_3 \\
\overset{\displaystyle O}{\overset{\|}{\text{CHOC}}}(\text{CH}_2)_{16}\text{CH}_3 \\
\overset{\displaystyle O}{\overset{\|}{\text{CH}_2\text{OC}}}(\text{CH}_2)_{16}\text{CH}_3
\end{array}$$

Glyceryl tristearate
(a solid fat)

Commercial cooking fats such as Crisco and Spry, for example, are manufactured in just this way. Vegetable oils are hydrogenated until a semisolid of an appealing consistency is obtained. Complete hydrogenation is avoided because a completely saturated triacylglycerol is very hard and brittle.

21.2B Saponification of Triacylglycerols: Soaps

Alkaline hydrolysis (i.e., "saponification") of triacylglycerols produces glycerol and a mixture of salts of long-chain carboxylic acids:

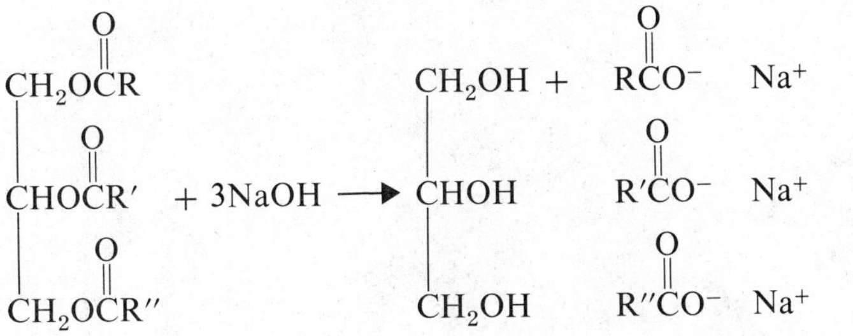

$$\begin{array}{lcll}
\overset{\displaystyle O}{\overset{\|}{\text{CH}_2\text{OCR}}} & & \text{CH}_2\text{OH} \;+ & \overset{\displaystyle O}{\overset{\|}{\text{RCO}^-}} \;\; \text{Na}^+ \\
\overset{\displaystyle O}{\overset{\|}{\text{CHOCR}'}} & + \; 3\text{NaOH} \longrightarrow & \text{CHOH} & \overset{\displaystyle O}{\overset{\|}{\text{R}'\text{CO}^-}} \;\; \text{Na}^+ \\
\overset{\displaystyle O}{\overset{\|}{\text{CH}_2\text{OCR}''}} & & \text{CH}_2\text{OH} & \overset{\displaystyle O}{\overset{\|}{\text{R}''\text{CO}^-}} \;\; \text{Na}^+ \\
 & & \text{Glycerol} & \text{Sodium carboxylates "soap"}
\end{array}$$

These salts of long-chain carboxylic acids are *soaps,* and this is the way most soaps are manufactured. Fats and oils are boiled in aqueous sodium hydroxide until hydrolysis is complete. Adding sodium chloride to the mixture then causes the soap to precipitate. (After the soap has been separated, glycerol can be isolated from the aqueous phase by distillation.) Crude soaps are usually purified by several reprecipitations. Perfumes can be added if a toilet soap is the desired product. Sand, sodium carbonate, and other fillers can be added to make a scouring soap and air can be blown into the molten soap if the manufacturer wants to market a soap that floats.

The sodium salts of long-chain carboxylic acids (soaps) are almost completely miscible with water. However, they do not dissolve as we might expect, that is, as individual ions. Except in very dilute solutions, soaps exist as *micelles* (Fig. 21.2). Soap micelles are usually spherical clusters of carboxylate ions that are dispersed throughout the aqueous phase. The carboxylate ions are packed together with their negatively charged (and thus, *polar*) carboxylate groups at the surface and with their nonpolar hydrocarbon chains on the interior. The sodium ions are scattered throughout the aqueous phase as individual solvated ions.

Micelle formation accounts for the fact that soaps dissolve in water. The nonpolar (and thus, *hydrophobic*) alkyl chains of the soap remain in a nonpolar environment—in the interior of the micelle. The polar (and therefore, *hydrophilic*) carboxylate groups are exposed to a polar environment—that of the aqueous phase. Because the surfaces of the micelles are negatively charged, individual micelles repel each other and remain dispersed throughout the aqueous phase.

Soaps serve their function as "dirt removers" in a similar way. Most dirt particles (on the skin, for example) become surrounded by a layer of an oil or fat. Water molecules alone are unable to disperse these greasy globules because they are unable to penetrate the oily layer and separate the individual particles from each other or from the surface to which they are stuck. Soap solutions, however, *are* able to separate the individual particles because their hydrocarbon chains can "dissolve" in the oily layer (Fig. 21.3). As this happens each individual particle develops an outer layer of carboxylate ions and presents the aqueous phase with a much more compatible exterior—a polar surface. The individual globules now repel each other and thus, become dispersed throughout the aqueous phase. Shortly thereafter they make their way down the drain.

Synthetic detergents (Fig. 21.4) function in the same way as soaps; they have long nonpolar alkane chains with polar groups at the end. The polar groups of most synthetic detergents are sodium sulfonates or sodium sulfates.

Synthetic detergents offer an advantage over soaps; they function well in

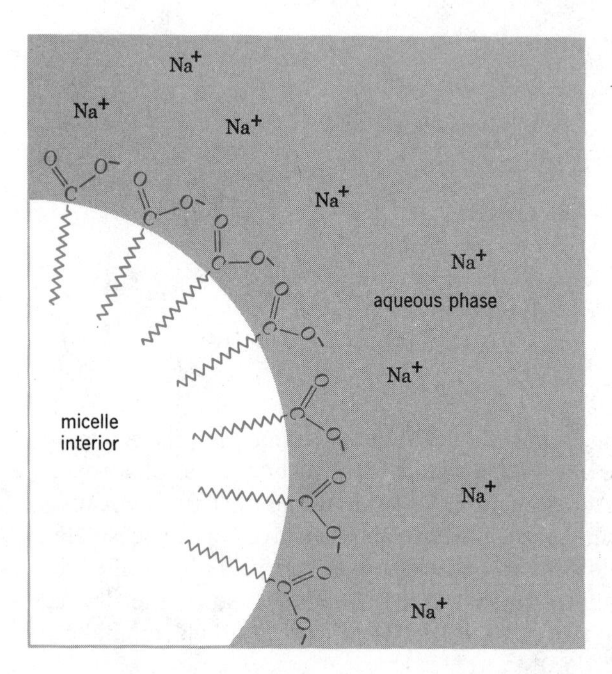

FIG. 21.2
A portion of a soap micelle.

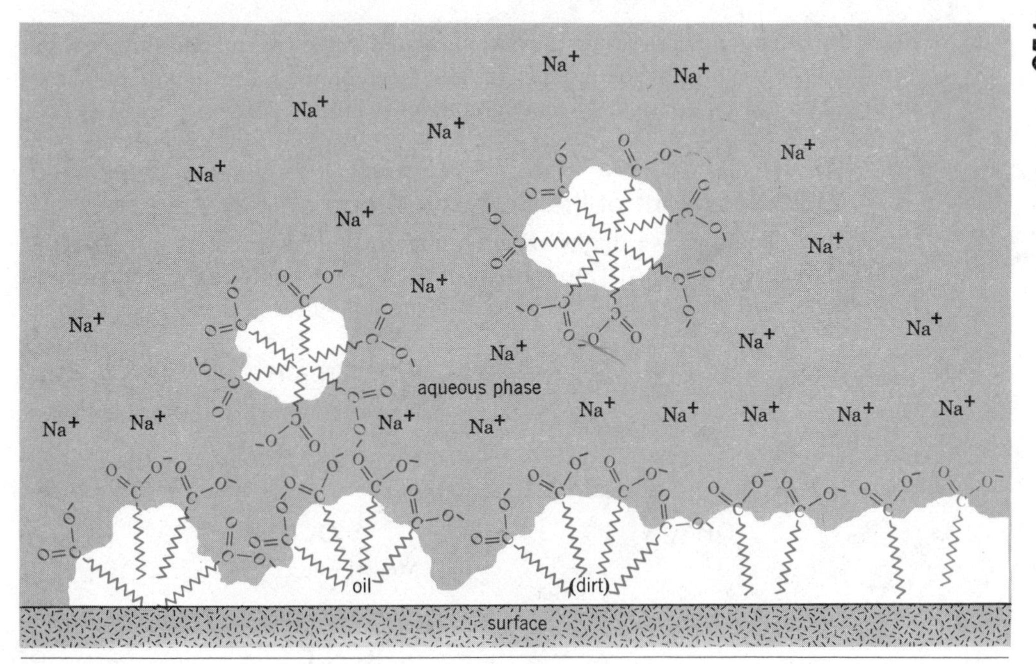

FIG. 21.3
Dispersal of oil-coated dirt particles by a soap.

"hard" water, that is, water containing Ca^{++}, Fe^{++}, Fe^{+++}, and Mg^{++} ions. Calcium, iron, and magnesium salts of alkane sulfonates and alkyl hydrogen sulfates are largely water soluble and, thus, synthetic detergents remain in solution. Soaps, by contrast, form precipitates—the ring around the bathtub—when they are used in hard water.

One serious disadvantage of some synthetic detergents is that they are not "biodegradable," that is, they are not broken down into harmless chemicals by microorganisms in the soil. For many years detergents were manufactured with the general structures shown below (cf. Problems 7.39 and 12.43).

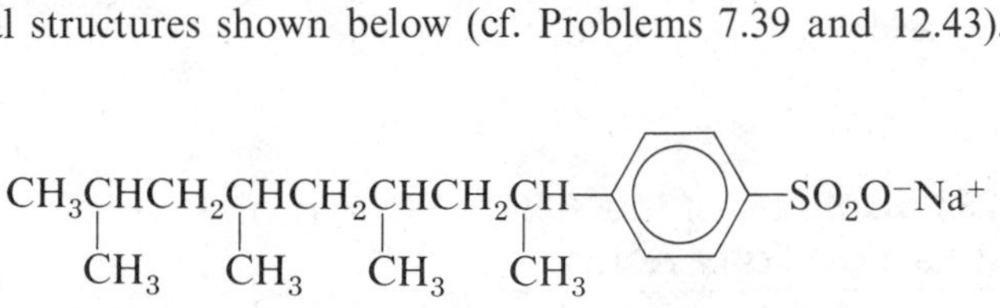

$$CH_3\underset{CH_3}{\underset{|}{C}}HCH_2\underset{CH_3}{\underset{|}{C}}HCH_2\underset{CH_3}{\underset{|}{C}}HCH_2\underset{CH_3}{\underset{|}{C}}H{-}C_6H_4{-}SO_2O^-Na^+$$

Tetrapropylene-based sodium alkylbenzenesulfonate

Soil bacteria are unable to degrade tetrapropylene-based detergents to any ap-

$$CH_3(CH_2)_nCH_2SO_2O^- \quad Na^+$$

Sodium alkanesulfonate

$$CH_3(CH_2)_nCH_2OSO_2O^- \quad Na^+$$

Sodium alkyl sulfate

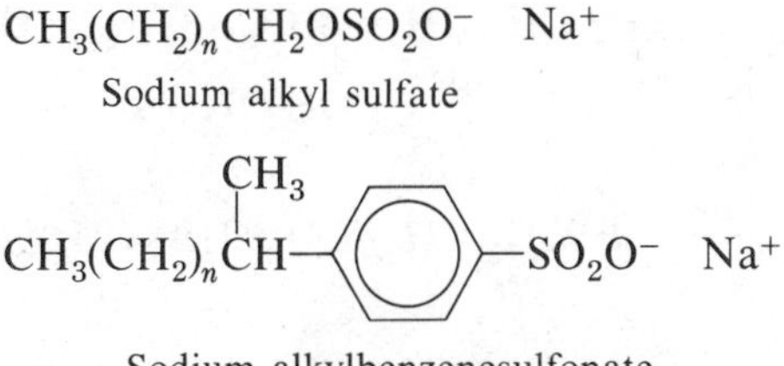

$$CH_3(CH_2)_n\overset{CH_3}{\overset{|}{C}}H{-}C_6H_4{-}SO_2O^- \quad Na^+$$

Sodium alkylbenzenesulfonate

FIG. 21.4
Typical synthetic detergents.

preciable extent and, as a result, they accumulated in the ground and surface water supplies. The use of these detergents has now been banned. The detergents shown in Fig. 21.4 are all completely biodegradable or largely so.

21.2C Reactions of the Carboxyl Group of Fatty Acids

Fatty acids, as we might expect, undergo typical reactions of carboxylic acids. They react with $LiAlH_4$ to form alcohols, with alcohols and mineral acid to form esters, and with thionyl chloride to form acyl chlorides:

$$RCH_2C(=O)OH \xrightarrow[(2)\ H_2O]{(1)\ LiAlH_4,\ ether} RCH_2CH_2OH$$

Fatty acid → Long-chain alcohol

$$RCH_2C(=O)OH \xrightarrow{CH_3OH,\ H^+} RCH_2C(=O)OCH_3$$

Methyl ester

$$RCH_2C(=O)OH \xrightarrow[Pyridine]{SOCl_2} RCH_2C(=O)Cl$$

Long-chain acyl chloride

21.2D Reactions of the Alkyl Chain of Saturated Fatty Acids

Fatty acids are like other carboxylic acids in that they undergo specific α-halogenation when they are treated with bromine or chlorine in the presence of phosphorus. This is the familiar Hell-Volhard-Zelinski reaction.

$$RCH_2\overset{O}{\overset{\|}{C}}OH + X_2 \xrightarrow{P_4} R\underset{X}{\underset{|}{C}H}\overset{O}{\overset{\|}{C}}OH + HX$$

Fatty acid

21.2E Reactions of the Alkenyl Chain of Unsaturated Fatty Acids

The double bonds of the carbon chains of fatty acids undergo characteristic alkene addition reactions:

$$CH_3(CH_2)_nCH{=}CH(CH_2)_mCOOH \xrightarrow[Ni]{H_2} CH_3(CH_2)_nCH_2CH_2(CH_2)_mCOOH$$

$$CH_3(CH_2)_nCH{=}CH(CH_2)_mCOOH \xrightarrow[CCl_4]{Br_2} CH_3(CH_2)_nCHBrCHBr(CH_2)_mCOOH$$

$$CH_3(CH_2)_nCH{=}CH(CH_2)_mCOOH \xrightarrow[KMnO_4]{dil.} CH_3(CH_2)_n\underset{OH}{\underset{|}{C}H}\,\underset{OH}{\underset{|}{C}H}(CH_2)_mCOOH$$

$$CH_3(CH_2)_nCH{=}CH(CH_2)_mCOOH \xrightarrow{HBr} CH_3(CH_2)_nCH_2CHBr(CH_2)_mCOOH + CH_3(CH_2)_nCHBrCH_2(CH_2)_mCOOH$$

Problem 21.1

(a) How many stereoisomers are possible for 9,10-dibromohexadecanoic acid? (b) The addition of bromine to palmitoleic acid yields primarily one set of enantiomers, (±)-*threo*-9,10-dibromohexadecanoic acid. The addition of bromine is an anti addition to the double bond (i.e., it apparently takes place through a bromonium ion intermediate). Taking into account the *cis* stereochemistry of the double bond of palmitoleic acid and the stereochemistry of the bromine addition, write three-dimensional structures for the (±)-*threo*-9,10-dibromohexadecanoic acids.

Oxidations of unsaturated fatty acids are often used to locate double bonds. Direct oxidative cleavage of a double bond with $KMnO_4$ in acetone sometimes leads to overoxidation; as a result, some of the carboxylic acids that one obtains may have had their chains shortened. A much better procedure, one that gives unambiguous results because overoxidation is avoided, involves (1) initial hydroxylation of the double bond using osmium tetroxide, dilute aqueous potassium permanganate, or a peroxy acid; (2) subsequent cleavage of the glycol using periodic acid. When this procedure is employed, the two carbons of the double bond are oxidized to the aldehyde stage. The following example is an illustration.

$$\underset{\text{Oleic acid}}{CH_3(CH_2)_7(H)C{=}C(H)(CH_2)_7COOH} \xrightarrow[\text{(2) } H_3O^+]{\overset{O}{\parallel}\text{HCOOH}} \underset{(\pm)\text{-}threo\text{-9,10-Dihydroxyoctadecanoic acid}}{CH_3(CH_2)_7{-}\underset{OH}{CH}{-}\underset{OH}{CH}{-}(CH_2)_7COOH}$$

$\downarrow$ HIO_4 (c.f. Section 19.4D.)

$$\underset{\text{Nonanal (pelargonic aldehyde) (89\%)}}{CH_3(CH_2)_7\overset{O}{\overset{\parallel}{C}}H} + \underset{\text{9-Oxononanoic acid (azelaic half aldehyde) (76\%)}}{H\overset{O}{\overset{\parallel}{C}}(CH_2)_7COOH}$$

Problem 21.2

When oleic acid is hydroxylated using dilute aqueous permanganate or osmium tetraoxide, the products obtained are (±)-*erythro*-9,10-dihydroxyoctadecanoic acids, diastereomers of the *threo*-9,10-dihydroxyoctadecanoic acids above. Recall the stereochemistry of hydroxylation reactions (Sections 7.13 and 8.9B) and give stereochemical formulas for *threo*- and *erythro*-9,10-dihydroxyoctadecanoic acids.

21.2F Biological Function of Triacylglycerols

The primary function of the triacylglycerols of mammals is as a source of chemical energy. When triacylglycerols are converted to carbon dioxide and water by biochemical reactions (i.e., when triacylglycerols are "metabolized") they yield more than twice as many kilocalories per gram as do carbohydrates or proteins. This is largely because of the high proportion of carbon-hydrogen bonds per molecule.

Triacylglycerols are distributed throughout nearly all types of body cells but they are stored primarily as "body fat" in certain depots of specialized connective tissue known as *adipose tissue*.

The saturated triacylglycerols of the body can be synthesized from all three major foodstuffs: proteins, carbohydrates, and fats or oils. Certain polyunsaturated fatty acids, however, are essential in the diets of higher animals.

21.3 STEROIDS

The lipid fractions obtained from plants and animals contain another important group of compounds known as *steroids*. Steroids are important "biological regulators" and steroids nearly always show dramatic physiological effects when they are administered to living organisms. Among these important compounds are male and female sex hormones, adrenocortical hormones, D vitamins, the bile acids, and certain cardiac poisons.

21.3A Structure and Systematic Nomenclature of Steroids

As we saw earlier (Section 3.14), steroids are derivatives of the perhydrocyclopentanophenanthrene ring system shown below.

The carbon atoms of this ring system are numbered as shown. The four rings are designated with letters.

In most steroids the **B,C** and **C,D** ring junctions are *trans*. The **A,B** ring junction, however, may be either *cis* or *trans* and this gives rise to two general groups of steroids having the three-dimensional structures shown in Fig. 21.5.

Problem 21.3

Draw the two basic ring systems given in Fig. 21.5 for the 5α and 5β series showing all hydrogens of the cyclohexane rings. Label each hydrogen as to whether it is axial or equatorial.

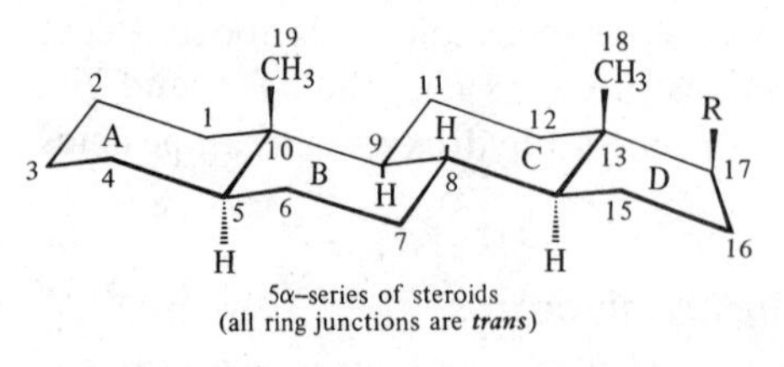
5α–series of steroids (all ring junctions are *trans*)

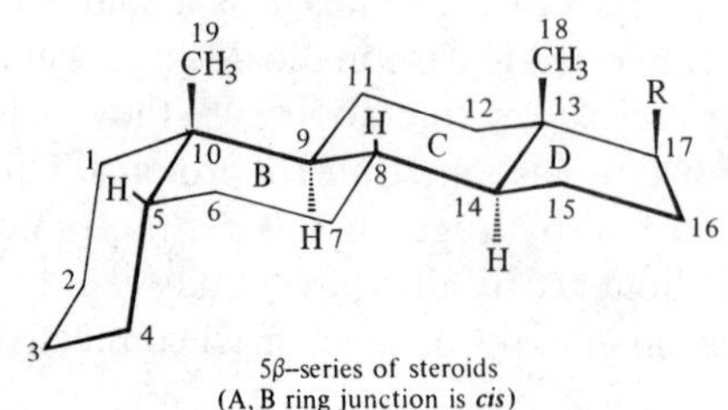
5β–series of steroids (A, B ring junction is *cis*)

FIG. 21.5
The basic ring systems of the 5α- and 5β- series of steroids.

The methyl groups that are attached at points of ring junction (i.e., those numbered 18 and 19) are called *angular methyl groups* and they serve as important reference points for stereochemical designations. The angular methyl groups protrude above the general plane of the ring system when it is written in the manner shown in Fig. 21.5. By convention, other groups that lie on the same general side of the molecule as the angular methyl groups (i.e., on the top side) are designated as *β substituents;* those that lie generally on the bottom (i.e., are *trans* to the angular methyl groups) are designated as *α substituents.* When these designations are applied to the hydrogen atom at position 5, the ring system in which the **A,B** ring junction is *trans* becomes the 5α series; and the ring system in which the **A,B** ring junction is *cis* becomes the 5β series.

In systematic nomenclature the nature of the R group at position 17 determines (primarily) the base name of an individual steroid. These names are derived from the steroid hydrocarbon names given in Table 21.4.

Two examples that illustrate the way these base names are used are shown below.

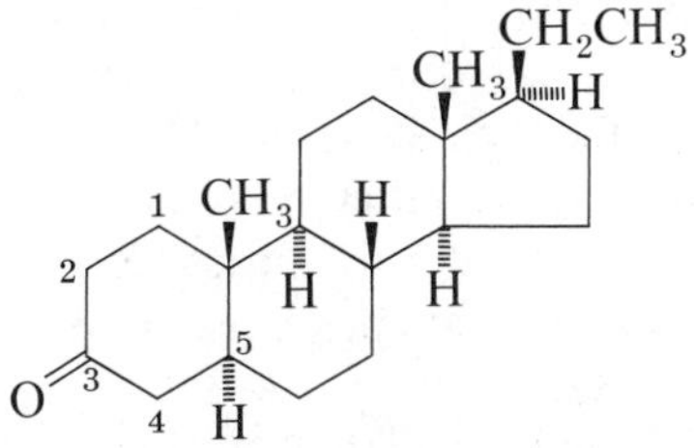

5α-Pregnan-3-one

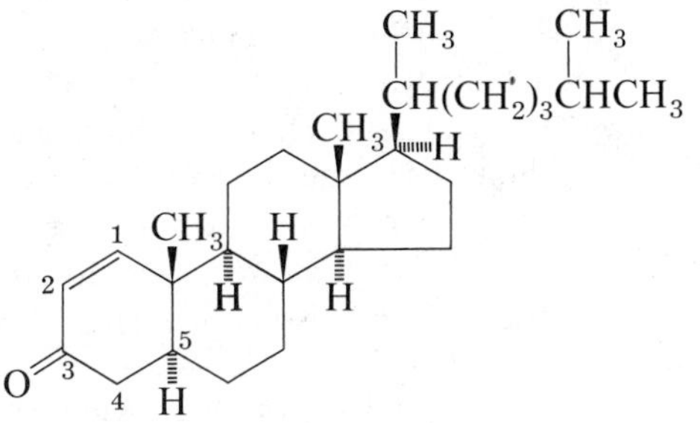

5α-Cholest-1-en-3-one

TABLE 21.4 Names of Steroid Hydrocarbons

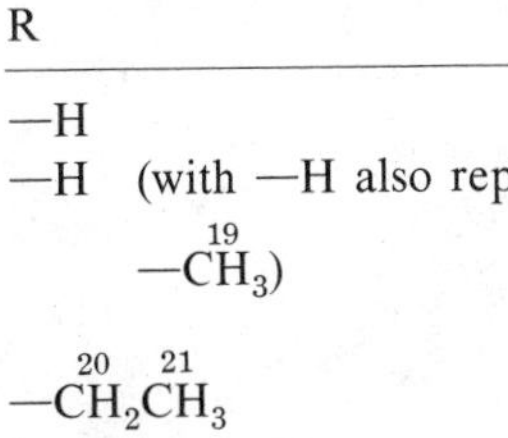

R	NAME
—H	Androstane
—H (with —H also replacing $-\overset{19}{C}H_3$)	Estrane
$-\overset{20}{C}H_2\overset{21}{C}H_3$	Pregnane
$-\overset{20}{C}H(\overset{21}{C}H_3)\overset{22}{C}H_2\overset{23}{C}H_2\overset{24}{C}H_3$	Cholane
$-\overset{20}{C}H(\overset{21}{C}H_3)\overset{22}{C}H_2\overset{23}{C}H_2\overset{24}{C}H_2\overset{25}{C}H(\overset{27}{C}H_3)\overset{26}{C}H_3$	Cholestane

We will see that many steroids also have common names and that the names of the steroid hydrocarbons given in Table 21.4 are derived from these common names.

Problem 21.4

(a) Androsterone, a secondary male sex hormone, has the systematic name 5α-androstan-3α-ol-17-one. Give a three-dimensional formula for androsterone. (b) Norethynodrel, a synthetic steroid that has been widely used in oral contraceptives, has the systematic name 17-α-ethynyl-17β-hydroxy-5(10)-estren-3-one. Give a three-dimensional formula for norethynodrel.

21.3B Cholesterol

Cholesterol, one of the most widely occurring steroids, can be isolated by extraction of nearly all animal tissues. Human gallstones are a particularly rich source.

Although cholesterol was isolated as early as 1770, the last stereochemical details of its structure were not known until 1955. Two German chemists, Adolf Windaus and Heinrich Wieland, were responsible for outlining the structure of cholesterol; they received the Nobel Prize for their work in 1928.*

Part of the difficulty in assigning an absolute structure to cholesterol is that cholesterol contains *eight* different chiral centers. This means that 2^8 or 256 possible stereoisomeric forms of the basic structure are possible, *only one of which is cholesterol.*

Cholesterol occurs widely in the body, but not all of the biological functions of cholesterol are yet known. Cholesterol is known to serve as an intermediate in the biosynthesis of steroid hormones and of bile acids but far more cholesterol is present in the body than is needed for these purposes. High levels of blood cholesterol have been implicated in the development of arteriosclerosis (hardening of the arteries) and in heart attacks that occur when cholesterol-containing plaques

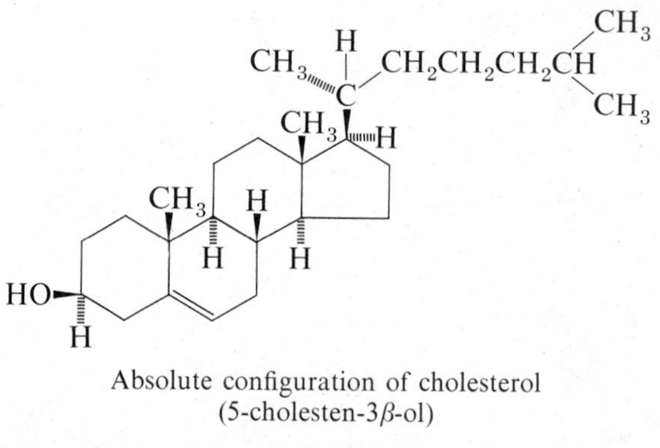

Absolute configuration of cholesterol
(5-cholesten-3β-ol)

block arteries of the heart. Considerable research is being carried out in the area of cholesterol metabolism with the hope of finding ways of minimizing cholesterol levels through the use of dietary adjustments or drugs.

* The original cholesterol structure proposed by Windaus and Wieland was incorrect. This became evident in 1932 as a result of x-ray diffraction studies done by the British physicist, J. D. Bernal. By the end of 1932, however, English scientists, and Wieland himself, using Bernal's results, were able to outline the correct structure of cholesterol.

21.3C Sex Hormones

The sex hormones can be classified into three major groups: (1) the female sex hormones, or *estrogens,* (2) the male sex hormones, or *androgens,* and (3) the pregnancy hormones, or *progestins.*

The first sex hormone to be isolated was the estrogen, *estrone.* Working independently, Adolf Butenandt (in Germany) and Edward Doisy (in the United States) isolated estrone from the urine of pregnant women. They published their discovery in 1929. Later, Doisy was able to isolate the much more potent estrogen, *estradiol* (the true female sex hormone). In this research Doisy had to extract *four tons* of sow ovaries in order to obtain just *12 milligrams* of estradiol.

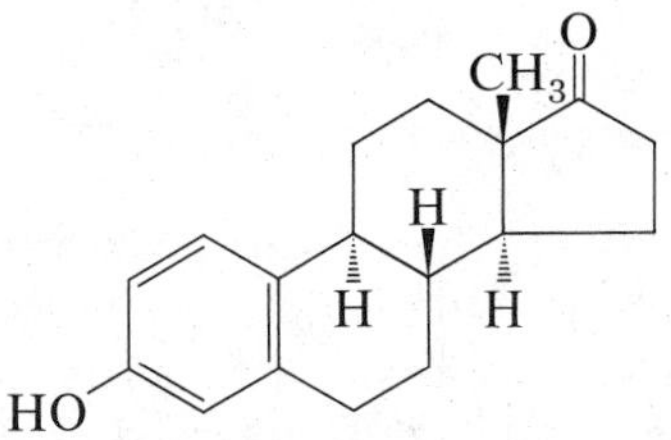

Estrone
(3-hydroxy-1,3,5(10)-estratrien-17-one)

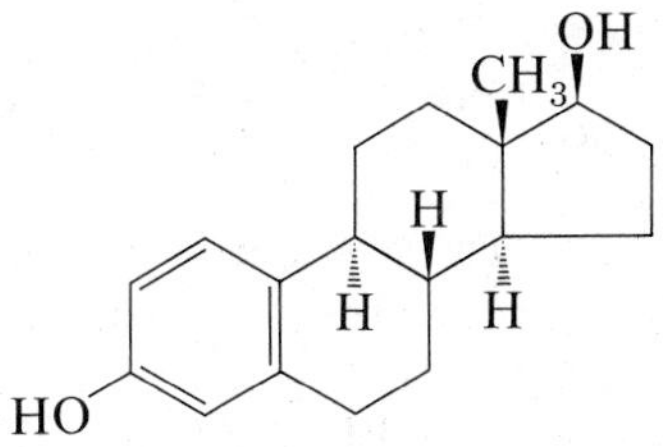

Estradiol
(3,17β-dihydroxyestra-1,3,5(10)-triene)

Estradiol is secreted by the ovaries and it promotes the development of the secondary female characteristics that appear at the onset of puberty. Estrogens also stimulate the development of the mammary glands during pregnancy and induce estrus (heat) in laboratory animals.

In 1931 Butenandt and Kurt Tscherning isolated the first androgen, *androsterone.* They were able to obtain 15 milligrams of this hormone by extracting approximately 15,000 liters of male urine. Soon afterwards (in 1935), Ernest Laqueur (in Holland) isolated another male sex hormone, *testosterone,* from steer testes. It soon became clear that testosterone is the true male sex hormone and that androsterone is a metabolized form of testosterone that is excreted in the urine.

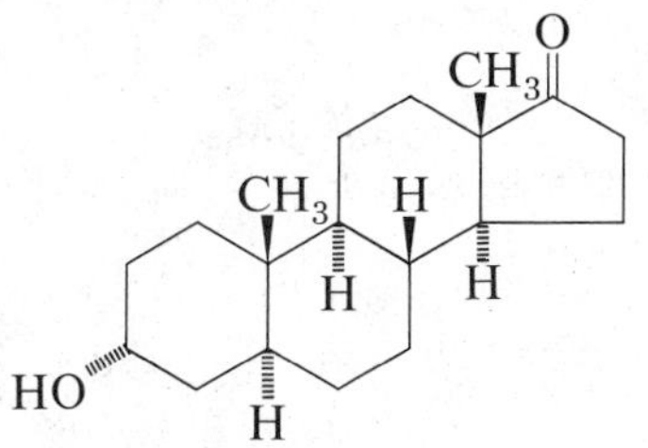

Androsterone
(5α-androstan-3α-ol-17-one)

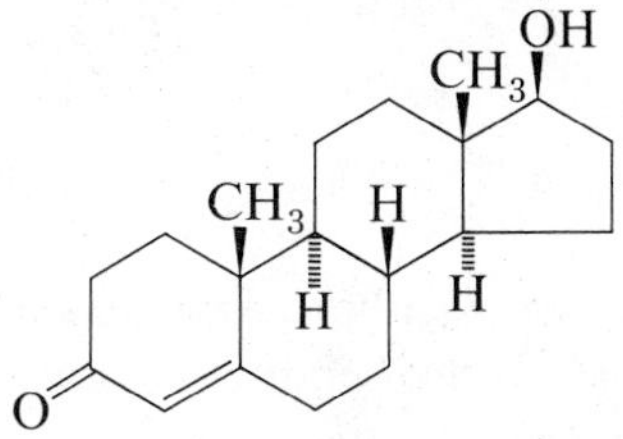

Testosterone
(17β-hydroxy-4-androsten-3-one)

Testosterone, secreted by the testes, is the hormone that promotes the development of secondary male characteristics: the growth of facial and body hair; the deepening of the voice; muscular development; and the maturation of the male sex organs.

Testosterone and estradiol, then, are the chemical compounds from which "maleness" and "femaleness" are derived. It is especially interesting to examine their structural formulas and see how very slightly these two compounds differ. Testosterone has an angular methyl group at the **A,B** ring junction that is missing in estradiol. Ring **A** of estradiol is a benzene ring and, as a result, estradiol is a phenol. Ring **A** of testosterone contains an α,β-unsaturated keto group.

Problem 21.5

The estrogens (estrone and estradiol) are easily separated from the androgens (androsterone and testosterone) on the basis of one of their chemical properties. What is the property and how could such a separation be accomplished?

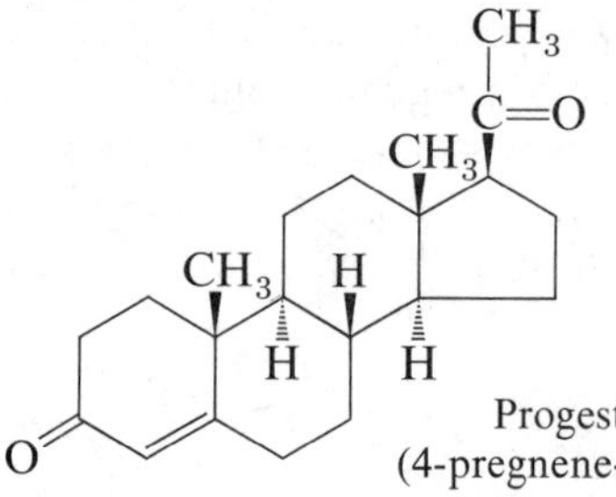

Progesterone
(4-pregnene-3,20-dione)

Progesterone is the most important *progestin* (pregnancy hormone). After ovulation occurs, the remnant of the ruptured ovarian follicle (called the *corpus luteum*) begins to secrete progesterone. This hormone prepares the lining of the uterus for implantation of the fertilized ovum, and continued progesterone secretion is necessary for the completion of pregnancy. (Progesterone is secreted by the placenta after secretion by the corpus luteum declines.)

Progesterone *also suppresses ovulation,* and it is the chemical agent that apparently accounts for the fact that pregnant women do not conceive again while pregnant. It was this observation that led to the search for synthetic progestins that could be used as oral contraceptives. (Progesterone, itself, requires very large doses to be effective in suppressing ovulation when taken orally.) A number of such compounds have been developed and are now widely used. In addition to norethynodrel (cf. Problem 21.4), another widely used synthetic progestin is its double bond isomer, *norethinodrone.*

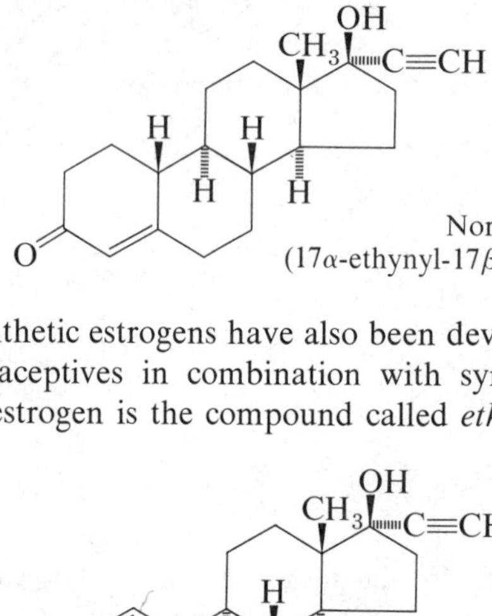

Norethinodrone
(17α-ethynyl-17β-hydroxy-4-estren-3-one)

Synthetic estrogens have also been developed and these are often used in oral contraceptives in combination with synthetic progestins. A very potent synthetic estrogen is the compound called *ethynylestradiol* or *novestrol.*

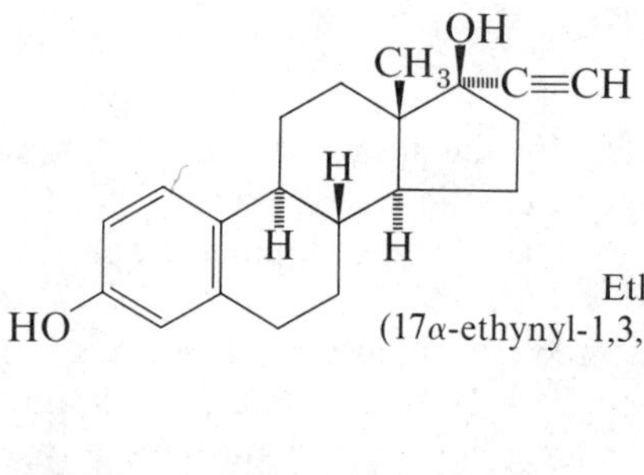

Ethynylestradiol
(17α-ethynyl-1,3,5(10)-estratriene-3,17β-diol)

21.3D Adrenocortical Hormones

At least 28 different hormones have been isolated from the adrenal cortex. Included in this group are the two steroids shown below.

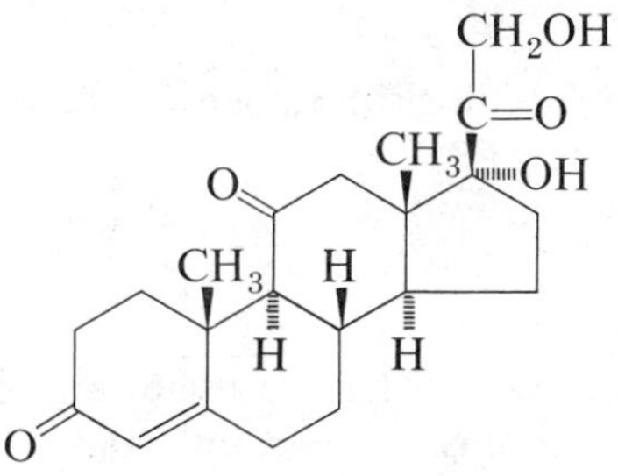

Cortisone
(17α,21-dihydroxy-4-pregnene-
3,11,20-trione)

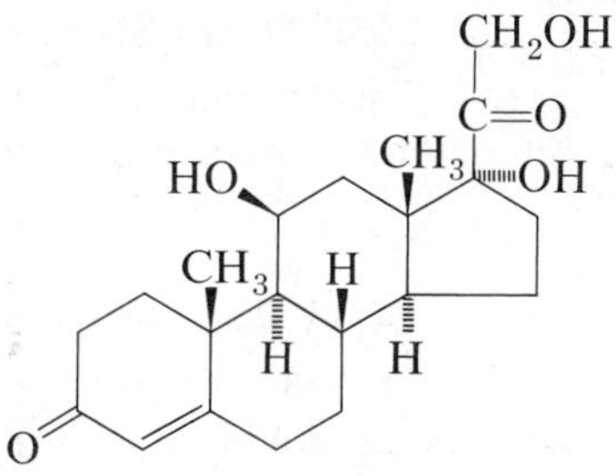

Cortisol
($11\beta,17\alpha$,21-trihydroxy-4-pregnene-
3,20-dione)

Most of the adrenocortical steroids have an oxygen function at position 11 (a keto group in cortisone, for example, and a β-hydroxyl in cortisol). Cortisol is the major hormone synthesized by the human adrenal cortex.

The adrenocortical steroids are apparently involved in the regulation of a large number of biological activities including carbohydrate, protein, and lipid metabolism, water and electrolyte balance, and reactions to allergic and inflammatory phenomena. Recognition of the antiinflammatory effect of cortisone and its usefulness in the treatment of rheumatoid arthritis, in 1949, has led to extensive research in this area. Many 11-oxygenated steroids are now used in the treatment of a variety of disorders ranging from Addison's disease, to asthma, and to skin inflammations.

21.3E D Vitamins

The demonstration, in 1919, that sunlight helped cure rickets—a childhood disease characterized by poor bone growth—began a long search for a chemical explanation. Soon, it was discovered that irradiation of certain foodstuffs increased their antirachitic properties and, in 1930, the search led to a steroid that can be isolated from yeast, called *ergosterol*. Irradiation of ergosterol was found to produce a highly active material. In 1932, Windaus and his co-workers in Germany demonstrated that this highly active substance was vitamin D_2. The photochemical reaction that takes place is one in which the dienoid ring **B** of ergosterol opens to produce a conjugated triene:

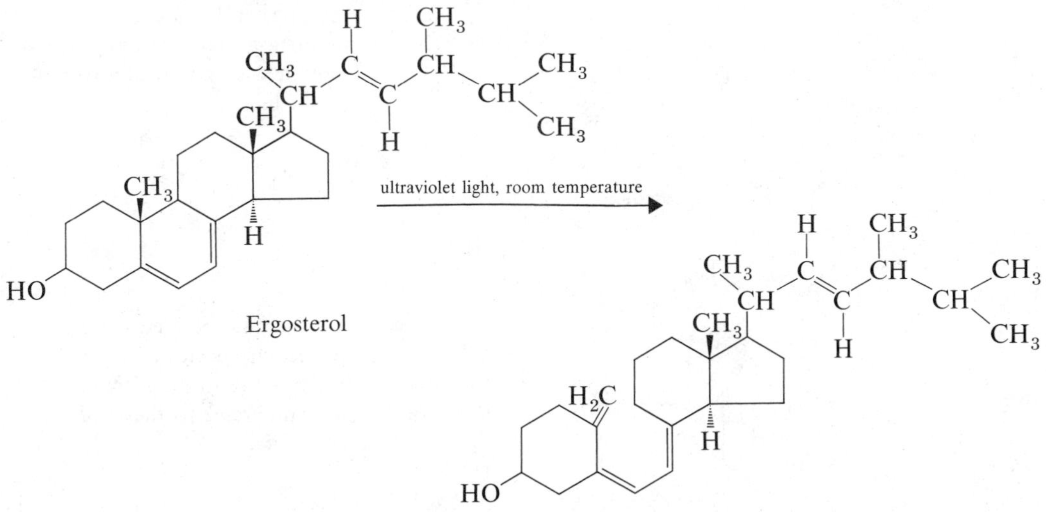

Ergosterol

ultraviolet light, room temperature

Vitamin D_2

21.3F Other Steroids

The structures, sources, and physiological properties of a number of other important steroids are given in Table 21.5.

21.3G Reactions of Steroids

Steroids undergo all of the reactions that we might expect of molecules containing double bonds, hydroxyl groups, keto groups, and so on. While the stereochemistry of steroid reactions is often quite complex, it is many times

TABLE 21.5 Other Important Steroids

Steroid	Description
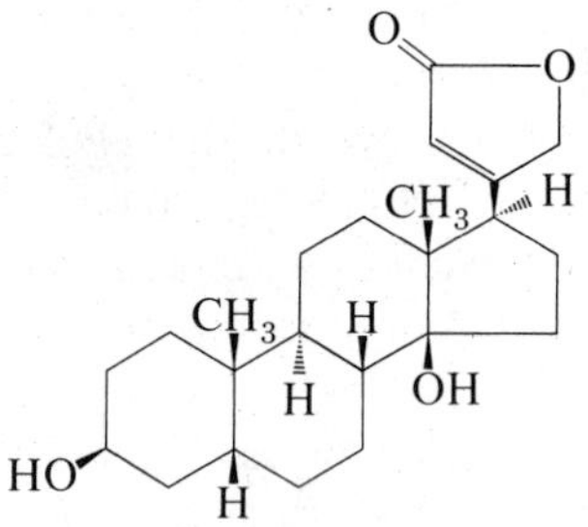 Digitoxigenin	Digitoxigenin is a cardiac aglycone that can be isolated by hydrolysis of digitalis, a pharmaceutical that has been used in treating heart disease since 1785. In digitalis, sugar molecules are joined in acetal linkages to the 3-OH of the steroid. In small doses digitalis strengthens the heart muscle, in larger doses it is a powerful heart poison.
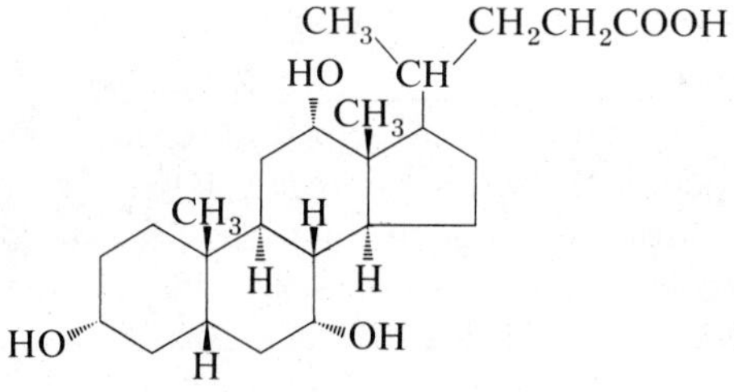 Cholic acid	Cholic acid is the most abundant acid obtained from the hydrolysis of human or ox bile. Bile is produced by the liver and stored in the gall bladder. When secreted into the small intestine, bile emulsifies lipids by acting as a soap. This aids in the digestive process.
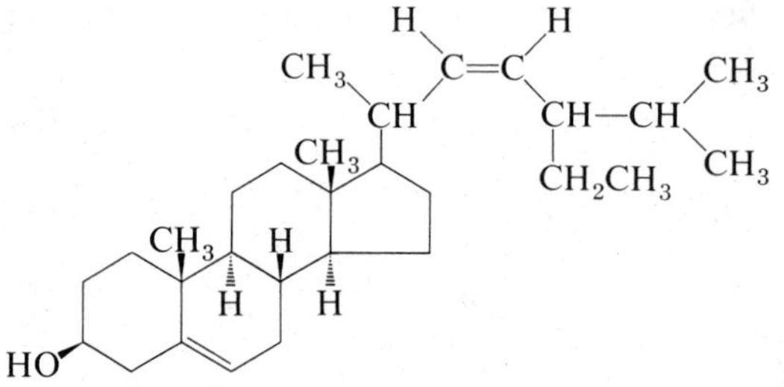 Stigmasterol	Stigmasterol is a widely occurring plant steroid that is obtained commercially from soybean oil.
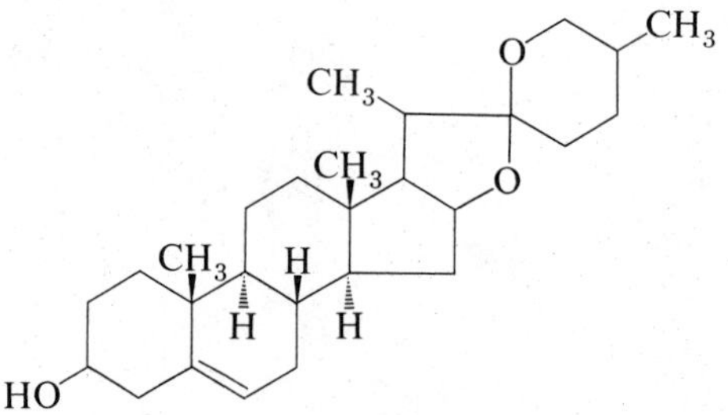 Diosgenin	Diosgenin, obtained from a particular species of yams, is used as the starting material for a commercial synthesis of cortisone.

strongly influenced by the steric hindrance presented at the β face of the molecule by the angular methyl groups. Many reagents react preferentially at the relatively unhindered α face especially when the reaction takes place at a functional group very near an angular methyl group and when the attacking reagent is bulky. Examples that illustrate this tendency are shown below.

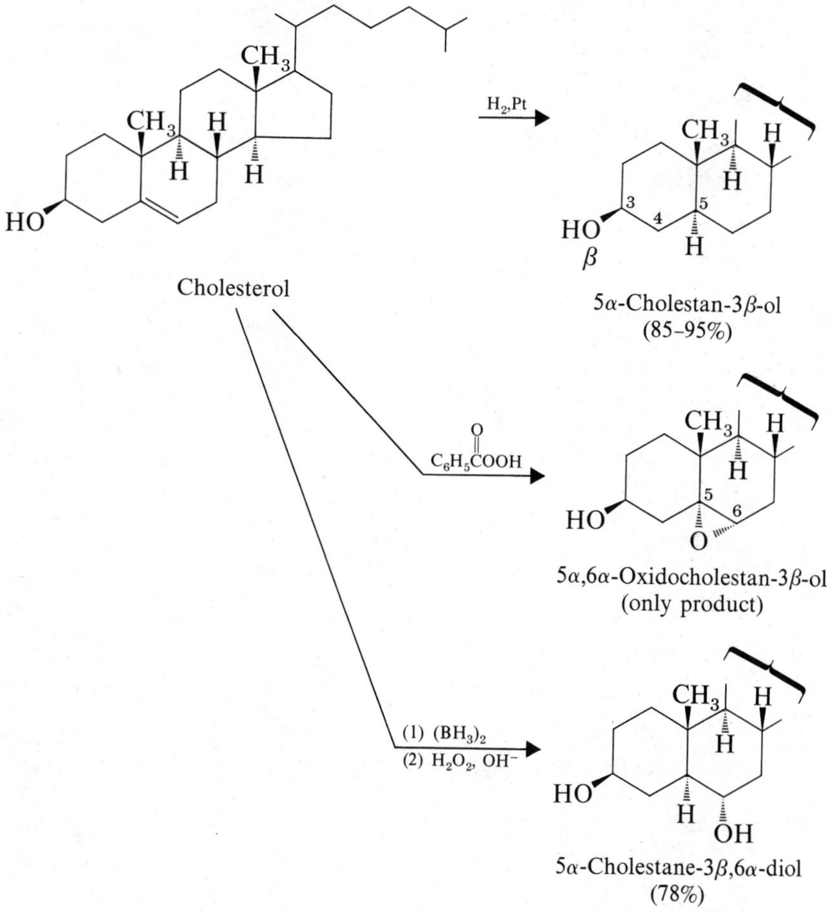

Cholesterol

H_2,Pt →

5α-Cholestan-3β-ol
(85–95%)

C_6H_5COOOH →

5α,6α-Oxidocholestan-3β-ol
(only product)

(1) $(BH_3)_2$
(2) H_2O_2, OH^- →

5α-Cholestane-3β,6α-diol
(78%)

When the epoxide ring of 5α,6α-epoxycholestan-3β-ol (above) is opened, attack by chloride ion must occur from the β face but it takes place at the more open 6 position.

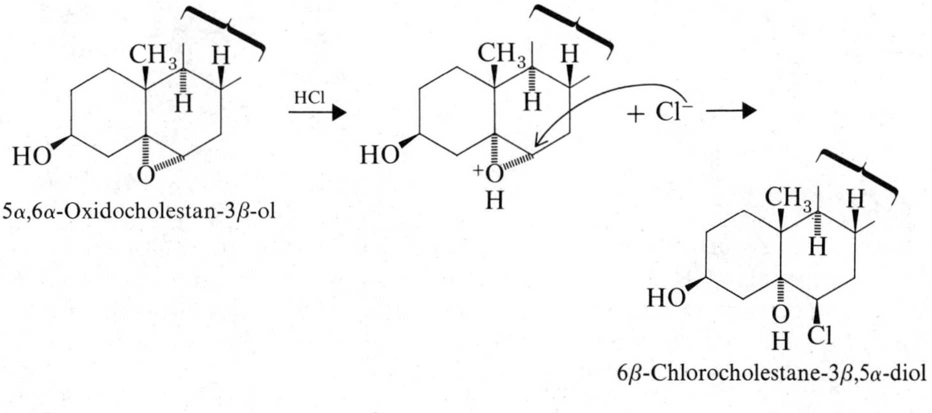

5α,6α-Oxidocholestan-3β-ol

HCl → + Cl^- →

6β-Chlorocholestane-3β,5α-diol

Problem 21.6

Show how you might convert cholesterol into each of the following compounds.

(a) 5α,6β-Dibromocholestan-3β-ol
(b) Cholestane-3β,5α,6β-triol
(c) 5α-Cholestan-3-one
(d) 6α-Deuterio-5α-cholestan-3β-ol
(e) 6β-Bromocholestane-3β-5α-diol

The relative openness of equatorial groups (when compared to axial groups) also influences the stereochemical course of steroid reactions. When 5α-cholestane-3β,7α-diol (below) is treated with excess ethyl chloroformate (C_2H_5OCOCl), only the equatorial 3β-hydroxyl becomes esterified. The axial 7α-hydroxyl is unaffected by the reaction.

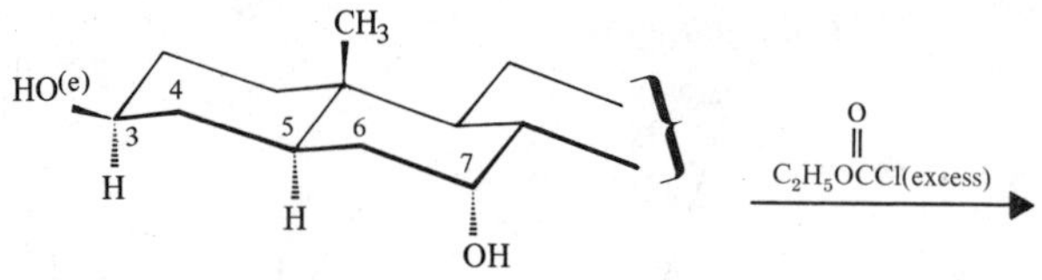

HO(e) 3 4 5 6 7 CH_3 H H OH

5α–cholestane–3β, 7α–diol

$\xrightarrow{C_2H_5OC(=O)Cl\text{(excess)}}$

$C_2H_5OC(=O)O$ CH_3 H H H OH

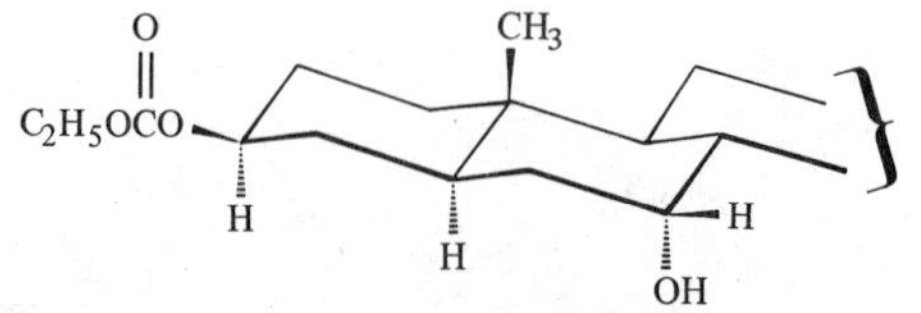

(only product)

By contrast, treating 5α-cholestane-3β,7β-diol with excess ethyl chloroformate esterifies both hydroxyl groups. In this instance both groups are equatorial.

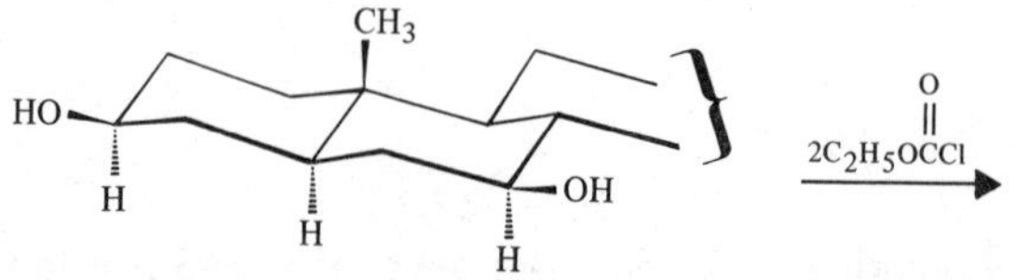

HO CH_3 OH H H H

5α–cholestane–3β, 7β–diol

$\xrightarrow{2C_2H_5OC(=O)Cl}$

$C_2H_5OC(=O)O$ CH_3 $OC(=O)OC_2H_5$ H H H

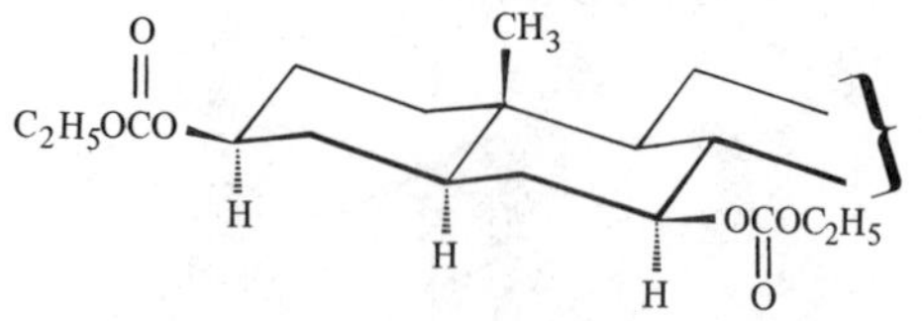

21.4 PROSTAGLANDINS

One very active area of current research is concerned with a group of lipids called prostaglandins. Prostaglandins are C_{20}-carboxylic acids that contain a five-membered ring, at least one double bond, and several oxygen-containing functional groups. Two of the most biologically active prostaglandins are prostaglandin E_2 and prostaglandin $F_{1\alpha}$.

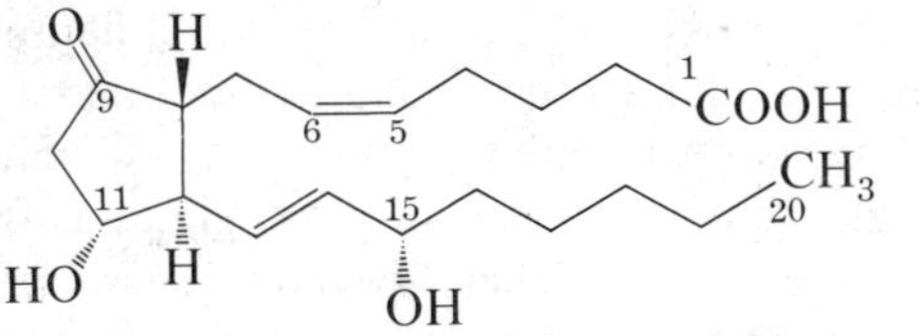

Prostaglandin E_2
(PGE_2)

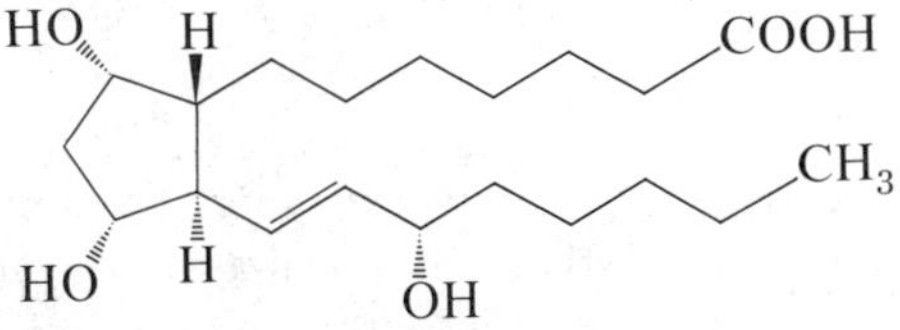

Prostaglandin $F_{1\alpha}$
($PGF_{1\alpha}$)

Prostaglandins of the E type have a carbonyl group at C-9 and a hydroxyl group at C-11; those of the F type have hydroxyl groups at both positions. Prostaglandins of the 2 series have a double bond between C-5 and C-6; in the 1 series this bond is a single bond.

First isolated from seminal fluid, prostaglandins have since been found in almost all animal tissues. The amounts vary from tissue to tissue but are almost always very small. Most prostaglandins have powerful physiological activity, however, and this activity covers a broad spectrum of effects. Prostaglandins are known to affect heart rate, blood pressure, blood clotting, conception, fertility, and allergic responses.

The recent finding that prostaglandins can prevent formation of blood clots has great clinical significance, because heart attacks and strokes often result from the formation of abnormal clots in blood vessels. An understanding of how prostaglandins affect the formation of clots may lead to the development of drugs to prevent heart attacks and strokes.

The biosynthesis of prostaglandins of the 2 series begins with the 20-carbon polyenoic acid, arachidonic acid. (Synthesis of prostaglandins of the 1 series begin with a fatty acid with one fewer double bond.) The first step requires two molecules of oxygen and is catalyzed by an enzyme called *cyclooxygenase*.

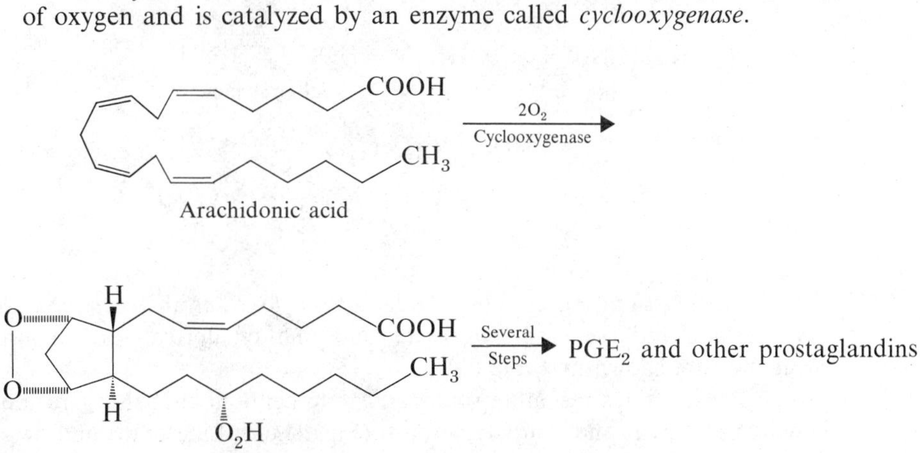

Arachidonic acid $\xrightarrow[\text{Cyclooxygenase}]{2O_2}$

PGG_2
(a cyclic endoperoxide)

$\xrightarrow[\text{Steps}]{\text{Several}}$ PGE_2 and other prostaglandins

The involvement of prostaglandins in allergic and inflammation responses has also been of special interest. Some prostaglandins induce inflammation; others relieve it. The most widely used antiinflammatory drug is ordinary aspirin (cf. Section 15.5D). Aspirin blocks the synthesis of prostaglandins from arachidonic acid, apparently by acetylating the enzyme cyclooxygenase, thus rendering it inactive. This may represent the origin of aspirin's antiinflammatory properties. Another prostaglandin (PGE_1) is a potent fever-inducing agent (pyrogen) and aspirin's ability to reduce fever may also arise from its inhibition of prostaglandin synthesis.

21.5 PHOSPHOLIPIDS

Another large class of lipids are those called *phospholipids*. Most phospholipids are structurally derived from a glycerol derivative known as a *phosphatidic acid*. In a phosphatidic acid, two hydroxyl groups of glycerol are ester linked to fatty acids and one terminal hydroxyl group is ester linked to *phosphoric acid*.

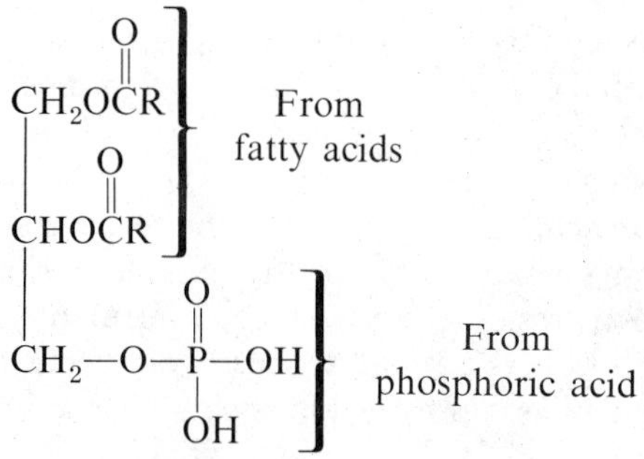

$$\begin{array}{l} CH_2OCOR \\ | \\ CHOCOR \\ | \\ CH_2\text{—O—}P(=O)(OH)\text{—OH} \end{array}$$

From fatty acids (the two $OCOR$ groups)

From phosphoric acid (the $P(=O)(OH)OH$ group)

A phosphatidic acid
(a diacylglycerophosphate)

Phosphatides

In *phosphatides*, the phosphate group of a phosphatidic acid is bound through another ester linkage to one of the nitrogen-containing compounds shown below.

$HOCH_2CH_2\overset{+}{N}(CH_3)_3\ \ OH^-$
Choline

$HOCH_2CH_2NH_2$
Ethanolamine

$HOCH_2CH(COO^-)NH_3^+$
L-Serine

The most important phosphatides are the *lecithins*, the *cephalins*, *phosphatidyl serines*, and the *plasmalogens* (a phosphatidyl derivative). Their general structures are shown in Table 21.6.

Phosphatides resemble soaps and detergents in that they are molecules having both polar and nonpolar groups (Fig. 21.6*a*). Like soaps and detergents, too, phosphatides "dissolve" in aqueous media by forming micelles. There is evidence that in biological systems the preferred micelles consist of three-dimensional arrays of "stacked" bimolecular micelles (Fig. 21.6).

TABLE 21.6 Phosphatides

Lecithins

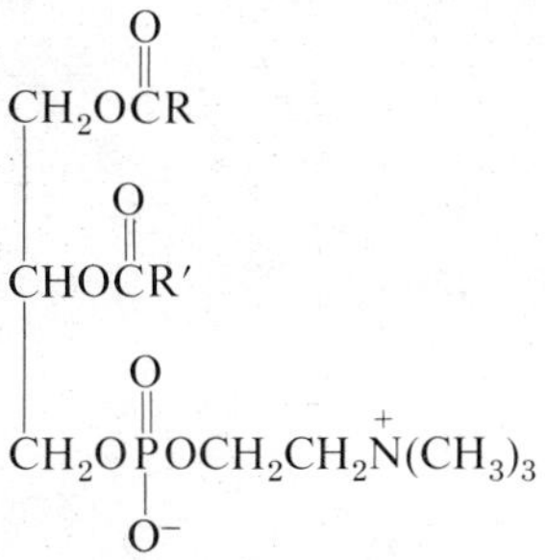

$$
\begin{array}{l}
\overset{\displaystyle O}{\|} \\
CH_2OCR \\
\;|\qquad\;\; \overset{\displaystyle O}{\|} \\
CHOCR' \\
\;|\qquad\;\; \overset{\displaystyle O}{\|}\qquad\qquad\qquad\;\; + \\
CH_2OPOCH_2CH_2N(CH_3)_3 \qquad \text{(from choline)} \\
\qquad\;\;\; | \\
\qquad\;\; O^-
\end{array}
$$

R is saturated and R′ is unsaturated

Cephalins

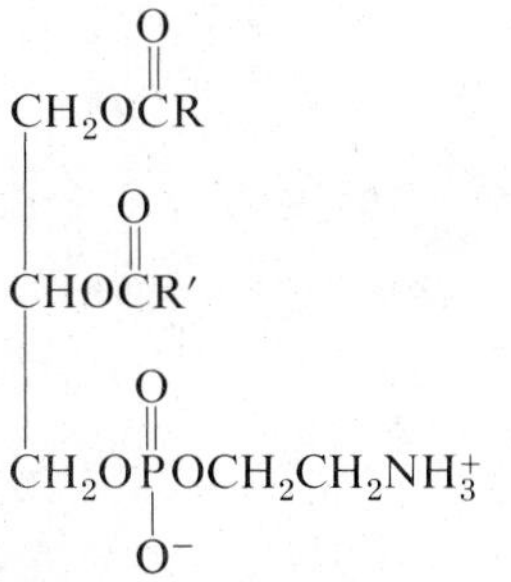

$$
\begin{array}{l}
\overset{\displaystyle O}{\|} \\
CH_2OCR \\
\;|\qquad\;\; \overset{\displaystyle O}{\|} \\
CHOCR' \\
\;|\qquad\;\; \overset{\displaystyle O}{\|} \\
CH_2OPOCH_2CH_2NH_3^+ \qquad \text{(from ethanolamine)} \\
\qquad\;\;\; | \\
\qquad\;\; O^-
\end{array}
$$

Phosphatidyl Serines

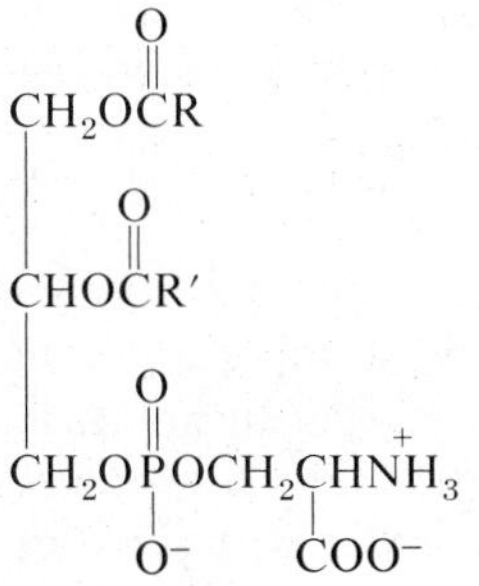

$$
\begin{array}{l}
\overset{\displaystyle O}{\|} \\
CH_2OCR \\
\;|\qquad\;\; \overset{\displaystyle O}{\|} \\
CHOCR' \\
\;|\qquad\;\; \overset{\displaystyle O}{\|}\qquad\qquad\qquad + \\
CH_2OPOCH_2CHNH_3 \qquad \text{(from L-serine)} \\
\qquad\;\;\; |\qquad\quad\;\; | \\
\qquad\;\; O^- \qquad COO^-
\end{array}
$$

R and R′ are like those of lecithins

Plasmalogens

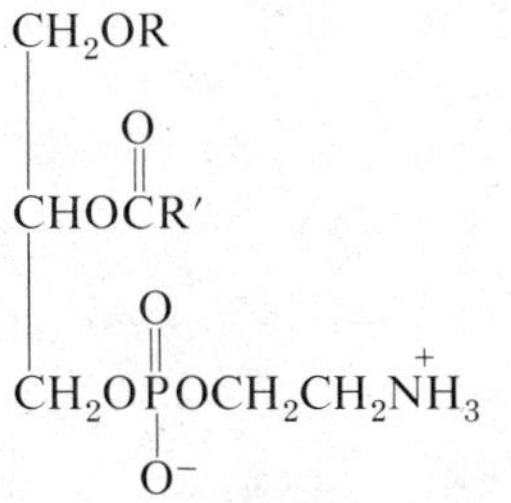

$$
\begin{array}{l}
CH_2OR \\
\;|\qquad\;\; \overset{\displaystyle O}{\|} \\
CHOCR' \\
\;|\qquad\;\; \overset{\displaystyle O}{\|}\qquad\qquad\qquad\;\; + \\
CH_2OPOCH_2CH_2NH_3 \qquad \text{(from ethanolamine)} \\
\qquad\;\;\; | \\
\qquad\;\; O^- \\
\qquad\qquad\qquad\qquad\qquad\quad + \\
\qquad\qquad\quad \text{or } OCH_2CH_2N(CH_3)_3 \qquad \text{(from choline)}
\end{array}
$$

R is $—CH{=}CH(CH_2)_nCH_3$ (this linkage is that of an α,β-unsaturated ether)
R′ is that of an unsaturated fatty acid

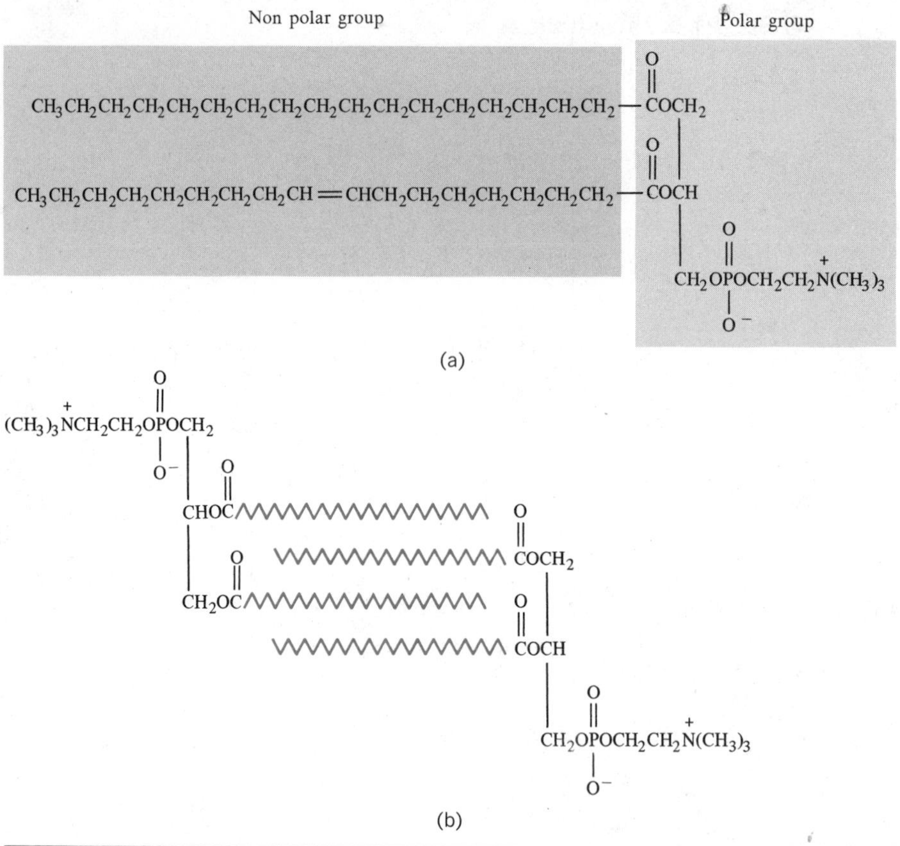

Non polar group

Polar group

$$CH_3CH_2CH_2CH_2CH_2CH_2CH_2CH_2CH_2CH_2CH_2CH_2CH_2CH_2CH_2CH_2CH_2{-}\overset{O}{\overset{\|}{C}}OCH_2$$

$$CH_3CH_2CH_2CH_2CH_2CH_2CH_2CH_2CH{=}CHCH_2CH_2CH_2CH_2CH_2CH_2CH_2{-}\overset{O}{\overset{\|}{C}}OCH$$

$$CH_2O\overset{O}{\overset{\|}{P}}OCH_2CH_2\overset{+}{N}(CH_3)_3 \quad (P{-}O^-)$$

(a)

$$(CH_3)_3\overset{+}{N}CH_2CH_2O\overset{O}{\overset{\|}{P}}OCH_2 \quad (P{-}O^-)$$

$$CHO\overset{O}{\overset{\|}{C}}\text{∧∧∧∧∧∧∧∧∧∧∧∧∧∧} \qquad \text{∧∧∧∧∧∧∧∧∧∧∧∧∧∧}\overset{O}{\overset{\|}{C}}OCH_2$$

$$CH_2O\overset{O}{\overset{\|}{C}}\text{∧∧∧∧∧∧∧∧∧∧∧∧∧∧} \qquad \text{∧∧∧∧∧∧∧∧∧∧∧∧∧∧}\overset{O}{\overset{\|}{C}}OCH$$

$$CH_2O\overset{O}{\overset{\|}{P}}OCH_2CH_2\overset{+}{N}(CH_3)_3 \quad (P{-}O^-)$$

(b)

FIG. 21.6
(a) *Polar and nonpolar sections of a phosphatide.* (b) *A bimolecular phosphatide micelle.*

The hydrophilic and hydrophobic portions of phosphatides make them perfectly suited for one of their most important biological functions: they form a portion of a structural unit that creates an interface between an organic and an aqueous environment. This is particularly true in cell walls and membranes where phosphatides are often found associated with proteins. Phosphatides also appear to be an essential factor in the formation of blood clots.

Problem 21.7

Under suitable conditions all of the ester (and ether) linkages of a phosphatide can be hydrolyzed. What organic compounds would you expect to obtain from the complete hydrolysis of (a) a lecithin, (b) an ethanolamine-based cephalin, (c) a choline-based plasmalogen. [Note: pay particular attention to the fate of the α,β-unsaturated ether in part (c).]

21.5A Derivatives of Sphingosine

Another important group of lipids are those that are derived from *sphingosine,* called *sphingolipids.* Two sphingolipids, a typical *sphingomyelin* and a typical *cerebroside,* are shown in Fig. 21.7.

Sphingomyelins are phosphatides. On hydrolysis they yield a sphingosine,

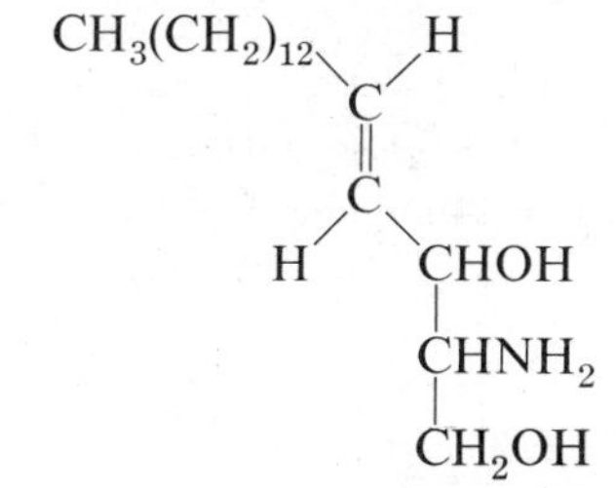

$CH_3(CH_2)_{12}CH{=}CHCHOHCHNH_2CH_2OH$

A sphingosine

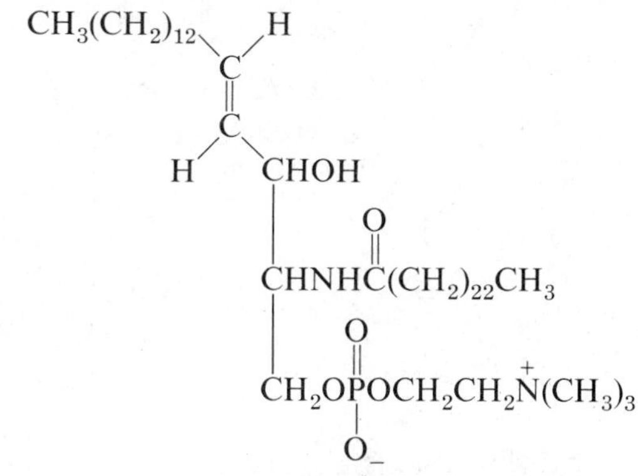

$CH_3(CH_2)_{12}CH{=}CHCHOHCH(NHC(=O)(CH_2)_{22}CH_3)CH_2OP(=O)(O^-)OCH_2CH_2\overset{+}{N}(CH_3)_3$

A sphingomyelin
(a sphingolipid)

$CH_3(CH_2)_{12}CH{=}CHCHOHCH(NHC(=O)(CH_2)_{22}CH_3)CH_2{-}O{-}$(galactose ring: CH_2OH, HO, H, HO, H, H, OH)

From a carbohydrate, D-galactose

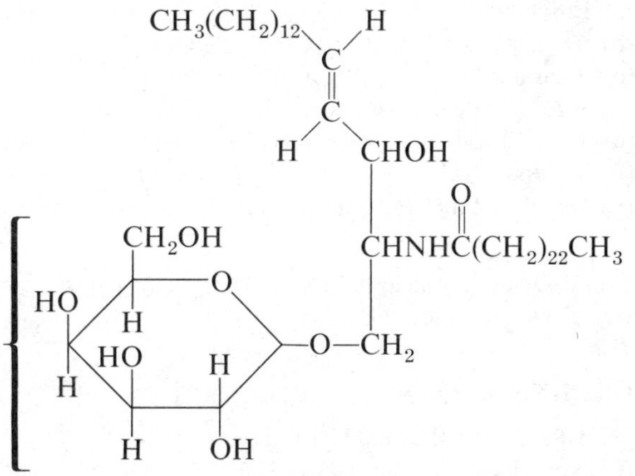

A cerebroside
(also a sphingolipid)

FIG. 21.7
A sphingosine and two sphingolipids.

choline, phosphoric acid, and a 24-carbon fatty acid called lignoceric acid. In a sphingomyelin this last component is bound to the $—NH_2$ group of a sphingosine. The sphingolipids do not yield glycerol when they are hydrolyzed.

The cerebroside shown above is an example of a *glycolipid.* Glycolipids have a polar group that is contributed by a *carbohydrate.* They do not yield phosphoric acid or choline when they are hydrolyzed.

The sphingolipids, together with proteins and polysaccharides, make up *myelin,* the protective coating that encloses nerve fibers or *axons.* The axons of nerve cells carry electrical nerve impulses. Myelin has been described as having a function relative to the axon similar to that of the insulation on an ordinary electric wire.

21.6 WAXES

Most waxes are esters of long-chain fatty acids and long-chain alcohols. Waxes are found as protective coatings on the skin, fur, or feathers of animals and on the leaves and fruits of plants. Several esters isolated from waxes are the following.

$$CH_3(CH_2)_{14}\overset{O}{\overset{\|}{C}}OCH_2(CH_2)_{14}CH_3$$

Cetyl palmitate
(from spermaceti)

$$CH_3(CH_2)_n\overset{O}{\overset{\|}{C}}OCH_2(CH_2)_mCH_3$$

$n = 24$ and 26; $m = 28$ and 30
(from beeswax)

$$HOCH_2(CH_2)_n\overset{O}{\overset{\|}{C}}{-}OCH_2(CH_2)_mCH_3$$

$n = 16$–28; $m = 30$ and 32
(from carnauba wax)

Additional Problems

21.8

How would you convert stearic acid, $CH_3(CH_2)_{16}COOH$, into each of the following?

(a) Ethyl stearate, $CH_3(CH_2)_{16}COOC_2H_5$ (two ways)
(b) *tert*-Butyl stearate, $CH_3(CH_2)_{16}COOC(CH_3)_3$
(c) Stearamide, $CH_3(CH_2)_{16}CONH_2$
(d) *N,N*-Dimethylstearamide, $CH_3(CH_2)_{16}CON(CH_3)_2$
(e) *n*-Octadecylamine, $CH_3(CH_2)_{16}CH_2NH_2$
(f) *n*-Heptadecylamine, $CH_3(CH_2)_{15}CH_2NH_2$
(g) Octadecanal, $CH_3(CH_2)_{16}CHO$
(h) 1-Octadecyl stearate, $CH_3(CH_2)_{16}\overset{O}{\overset{\|}{C}}OCH_2(CH_2)_{16}CH_3$
(i) 1-Octadecanol, $CH_3(CH_2)_{16}CH_2OH$ (two ways)
(j) 2-Nonadecanone, $CH_3(CH_2)_{16}\overset{O}{\overset{\|}{C}}CH_3$
(k) 1-Bromooctadecane, $CH_3(CH_2)_{16}CH_2Br$
(l) Nonadecanoic acid, $CH_3(CH_2)_{16}CH_2COOH$

21.9

How would you transform myristic acid into each of the following?

(a) $CH_3(CH_2)_{11}\underset{\underset{Br}{|}}{C}HCOOH$
(b) $CH_3(CH_2)_{11}\underset{\underset{OH}{|}}{C}HCOOH$
(c) $CH_3(CH_2)_{11}\underset{\underset{CN}{|}}{C}HCOOH$
(d) $CH_3(CH_2)_{11}\underset{\underset{NH_2}{|}}{C}HCOOH$

21.10

Using palmitoleic acid as an example and neglecting stereochemistry, illustrate each of the following reactions of the double bond.

(a) Addition of iodine
(b) Addition of hydrogen
(c) Hydroxylation
(d) Addition of HCl

21.11

When oleic acid is heated to 180–200° (in the presence of a small amount of selenium) an equilibrium is established between oleic acid (33%) and an isomeric compound called elaidic acid (67%). Suggest a possible structure for elaidic acid.

21.12

Gadoleic acid, $C_{20}H_{38}O_2$, a fatty acid that can be isolated from cod-liver oil, can be cleaved by hydroxylation and subsequent treatment with periodic acid to $CH_3(CH_2)_9CHO$ and $OHC(CH_2)_7COOH$. (a) What two stereoisomeric structures are possible for gadoleic acid? (b) What spectroscopic technique would make possible a decision as to the actual structure of gadoleic acid? (c) What peaks would you look for?

21.13
Vaccenic acid, a structural isomer of oleic acid, has been synthesized through the following reaction sequence:

$$\text{1-Octyne} + NaNH_2 \xrightarrow{NH_3} \mathbf{A}\ (C_8H_{13}Na) \xrightarrow{ICH_2(CH_2)_7CH_2Cl} \mathbf{B}\ (C_{17}H_{31}Cl) \xrightarrow{NaCN}$$

$$\mathbf{C}\ (C_{18}H_{31}N) \xrightarrow{KOH,H_2O} \mathbf{D}\ (C_{18}H_{31}O_2K) \xrightarrow{H_3O^+} \mathbf{E}\ (C_{18}H_{32}O_2) \xrightarrow{H_2,Pd}$$

$$\text{Vaccenic acid}\ (C_{18}H_{34}O_2).$$

Propose a structure for vaccenic acid and for the intermediates **A-E.**

21.14
ω-Fluorooleic acid can be isolated from a shrub, *Dechapetalum toxicarium,* that grows in Sierra Leone. The compound is highly toxic to warm-blooded animals; it has found use as an arrow poison in tribal warfare, in poisoning enemy water supplies, and by witch doctors "for terrorizing the native population." Powdered fruit of the plant has been used as a rat poison, hence ω-fluorooleic acid has the common name "ratsbane." A synthesis of ω-fluorooleic acid is outlined below. Give structures for compounds **F** and **I.**

$$\text{1-Bromo-8-fluorooctane} + \text{sodium acetylide} \longrightarrow \mathbf{F}\ (C_{10}H_{17}F) \xrightarrow[(2)\ I(CH_2)_7Cl]{(1)\ NaNH_2}$$

$$\mathbf{G}\ (C_{17}H_{30}FCl) \xrightarrow{NaCN} \mathbf{H}\ (C_{18}H_{30}NF) \xrightarrow[(2)\ H^+]{(1)\ KOH} \mathbf{I}\ (C_{18}H_{31}O_2F) \xrightarrow[Pd-BaSO_4]{H_2}$$

$$F{-}(CH_2)_8{-}CH{=}CH{-}(CH_2)_7COOH \quad (cis)$$

ω-Fluorooleic acid
(46% yield, overall)

21.15
Give structural formulas and names for compounds **A** and **B.**

$$5\alpha\text{-Cholest-2-ene} \xrightarrow{C_6H_5CO_3H} \mathbf{A}\ (\text{an epoxide}) \xrightarrow{HBr} \mathbf{B}$$

[Hint: **B** is not the most stable stereoisomer]

21.16
When cholesterol is treated with bromine in carbon tetrachloride the initial product that forms is the $5\alpha,6\beta$-dibromocholestan-3β-ol. If this product is allowed to stand in chloroform solution it slowly comes into equilibrium with the $5\beta,6\alpha$-dibromide. (The $5\beta,6\alpha$-dibromide predominates at equilibrium; the mixture contains 85% of the $5\beta,6\alpha$-dibromocholestan-3β-ol and only 15% of $5\alpha,6\beta$-dibromocholestan-3β-ol.) Consider conformational effects and account for this phenomenon.

*** 21.17**
One of the first laboratory syntheses of cholesterol was achieved by Professor R. B. Woodward and his students at Harvard University in 1951. Many of the steps of this synthesis are outlined below. Supply the missing reagents.

O CH$_3$ CH$_3$O O (a) → O CH$_3$ CH$_3$O O H (b) → O CH$_3$ CH$_3$O O H (c) →

HO CH$_3$ CH$_3$O HO H (d) → CH$_3$ O HO H Zn, CH$_3$COOH, Heat → CH$_3$ O H (e) →

CH$_3$ O H CHO Several steps → CH$_3$ H O H (f) → CH$_3$ OH H OH O H

(g) → CH$_3$ O O H O H (h) → CH$_3$ O O H O H

Several steps → CH$_3$ O CH$_3$ H O O H (i) → CH$_3$ OH CH$_3$ H OH O H

(j) → CH$_3$ CHO CH$_3$ H CHO O H (k) → CH$_3$ CHO CH$_3$ H O H

(l) (m) → CH$_3$ CO$_2$CH$_3$ CH$_3$ H O H (n) → CH$_3$ CO$_2$CH$_3$ H CH$_3$ H H O H H

(o), (p) (q) → CH$_3$ COOH CH$_3$ H H H CH$_3$CO$_2$ H (r) (s) →

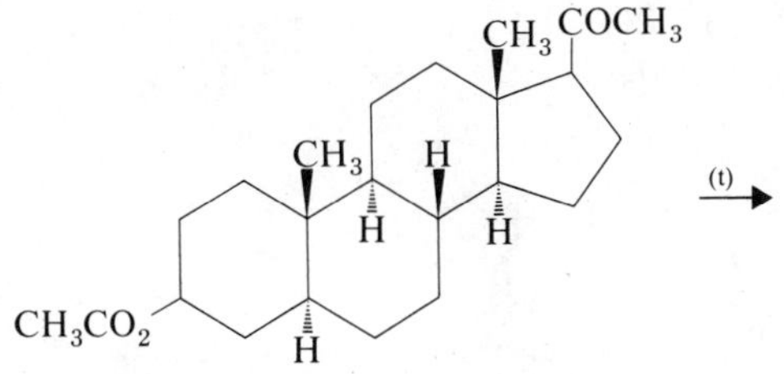

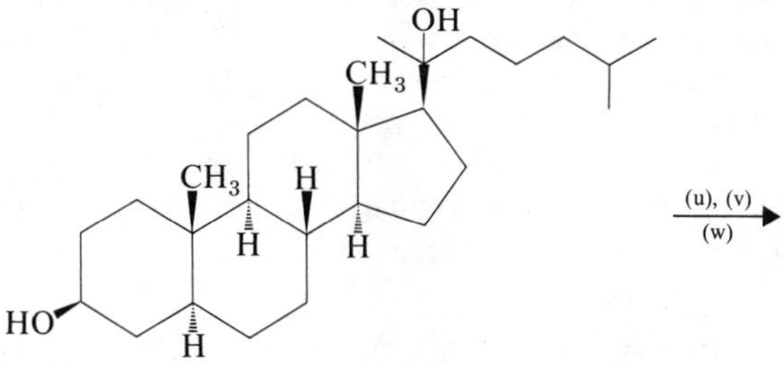

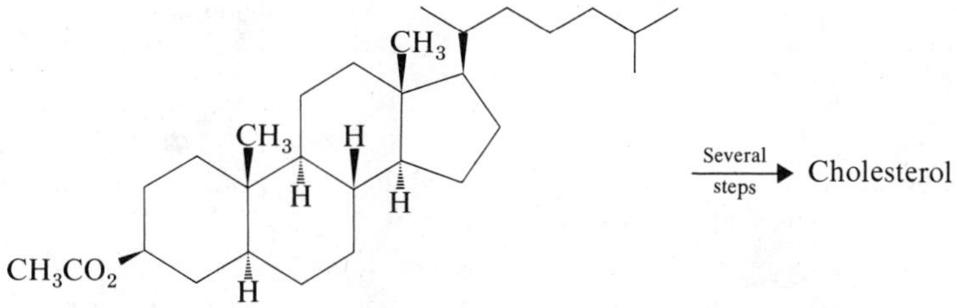

CH_3 $COCH_3$

CH_3 H

H H

CH_3CO_2 H

$\xrightarrow{(t)}$

OH

CH_3

CH_3 H

H H

HO H

$\xrightarrow[(w)]{(u),\ (v)}$

CH_3

CH_3 H

H H

CH_3CO_2 H

$\xrightarrow[\text{steps}]{\text{Several}}$ Cholesterol

*** 21.18**

The initial steps of a laboratory synthesis of several prostaglandins reported by E. J. Corey and his co-workers in 1968 are outlined below. Supply each of the missing reagents.

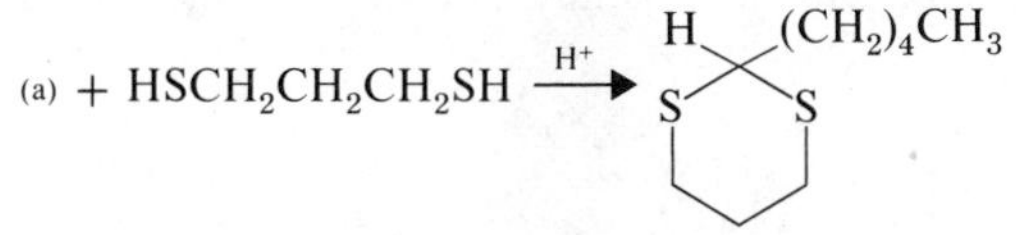

(a) + $HSCH_2CH_2CH_2SH \xrightarrow{H^+}$ H $(CH_2)_4CH_3$ S S

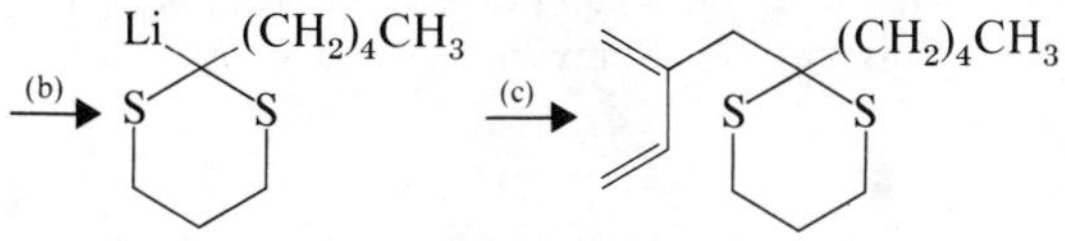

$\xrightarrow{(b)}$ Li $(CH_2)_4CH_3$ S S $\xrightarrow{(c)}$ $(CH_2)_4CH_3$ S S

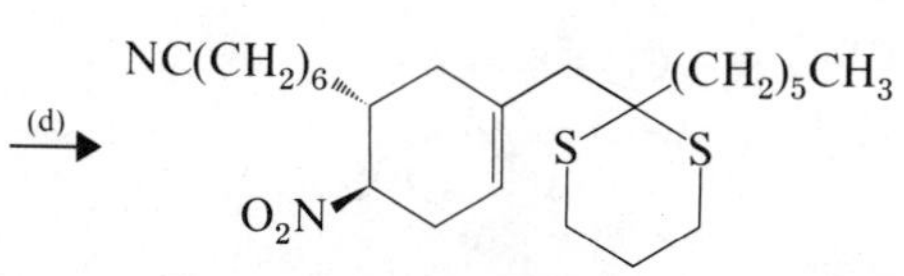

$\xrightarrow{(d)}$ $NC(CH_2)_6$ $(CH_2)_5CH_3$ S S O_2N

(e) The initial step in another prostaglandin synthesis is shown below. What kind of reaction—and catalyst—is needed here?

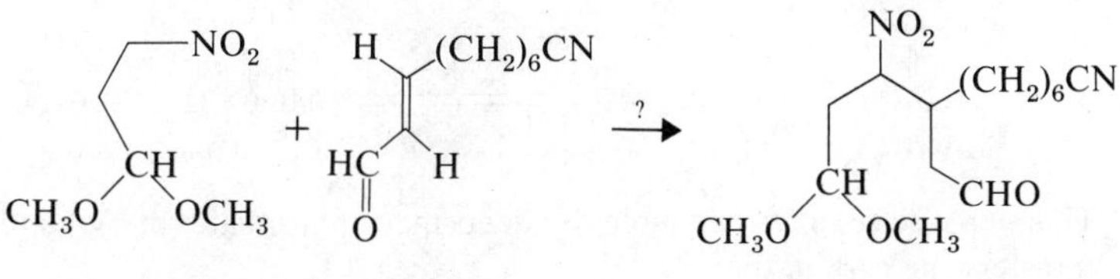

NO_2 CH CH_3O OCH_3 + H $(CH_2)_6CN$ HC H O $\xrightarrow{?}$ NO_2 $(CH_2)_6CN$ CH CHO CH_3O OCH_3

SPECIAL TOPIC

LIPID BIOSYNTHESIS

N.1 Biosynthesis of Fatty Acids

The fact that most naturally occurring fatty acids are made up of an even number of carbon atoms suggests that they are assembled from two-carbon units. The idea that these might be acetate (CH_3COO^-) units was put forth as early as 1893. Many years later, when radioactively labeled compounds became available, it became possible to test and confirm this hypothesis.

When an animal is fed acetic acid labeled with carbon-14 at the carboxyl group, the fatty acids that the animal synthesizes contain the label at alternate carbons beginning with the carboxyl carbon:

$$CH_3\overset{*}{C}OOH \qquad CH_3\overset{*}{C}H_2CH_3\overset{*}{C}H_2CH_2\overset{*}{C}H_2CH_2\overset{*}{C}H_2CH_2\overset{*}{C}H_2CH_2\overset{*}{C}H_2CH_2\overset{*}{C}H_2CH_2\overset{*}{C}OOH$$

Feeding carboxyl-labeled acetic acid ($C^* = {}^{14}C$)

Yields palmitic acid labeled at these positions

Conversely, feeding acetic acid labeled at the methyl carbon yields a fatty acid labeled at the other set of alternate carbons:

$$\overset{*}{C}H_3COOH \qquad \overset{*}{C}H_3CH_2\overset{*}{C}H_2CH_2\overset{*}{C}H_2CH_2\overset{*}{C}H_2CH_2\overset{*}{C}H_2CH_2\overset{*}{C}H_2CH_2\overset{*}{C}H_2CH_2\overset{*}{C}H_2COOH$$

Feeding methyl-labeled acetic acid

Yields palmitic acid labeled at these positions

The biosynthesis of fatty acids is now known to begin with acetyl coenzyme A:

$$CH_3\overset{\overset{O}{\|}}{C}-S-CoA$$

The acetyl portion of acetyl coenzyme A can be synthesized in the cell from acetic acid; it can also be synthesized from carbohydrates, proteins, and other fats.

$$\left.\begin{array}{l} CH_3\overset{\overset{O}{\|}}{C}OH \\ \text{Carbohydrates} \\ \text{Proteins} \\ \text{Fats} \end{array}\right\} \xrightarrow{\text{CoA—SH}} \underset{\text{Acetyl coenzyme A}}{CH_3\overset{\overset{O}{\|}}{C}S-CoA}$$

Although the methyl group of acetyl coenzyme A is already activated toward condensation reactions by virtue of its being a part of a thiol ester (Section 17.11), nature activates it again by converting it to *malonyl coenzyme A*.

$$\underset{\text{Acetyl CoA}}{CH_3\overset{\overset{O}{\|}}{C}S-CoA} + CO_2 \xrightleftharpoons{\text{Acetyl CoA carboxylase*}} \underset{\text{Malonyl CoA}}{HO\overset{\overset{O}{\|}}{C}CH_2\overset{\overset{O}{\|}}{C}S-CoA}$$

* This step also requires one mole of adenosine triphosphate (p. 875) and an enzyme that transfers the carbon dioxide.

The next steps in fatty acid synthesis involve the transfer of acyl groups of malonyl CoA and acetyl coenzyme A to the thiol group of a coenzyme called *acyl carrier protein* or ACP—SH.

$$\underset{\text{Malonyl CoA}}{HO\overset{O}{\overset{\|}{C}}CH_2\overset{O}{\overset{\|}{C}}S\text{—}CoA} + ACP\text{—}SH \rightleftharpoons \underset{\text{Malonyl—S—ACP}}{HO\overset{O}{\overset{\|}{C}}CH_2\overset{O}{\overset{\|}{C}}S\text{—}ACP} + CoA\text{—}SH$$

$$\underset{\text{Acetyl CoA}}{CH_3\overset{O}{\overset{\|}{C}}S\text{—}CoA} + ACP\text{—}SH \rightleftharpoons \underset{\text{Acetyl—S—ACP}}{CH_3\overset{O}{\overset{\|}{C}}S\text{—}ACP} + CoA\text{—}SH$$

Acetyl-S-ACP and malonyl-S-ACP then condense with each other to form acetoacetyl-S-ACP:

$$\underset{\text{Acetyl-S-ACP}}{CH_3\overset{O}{\overset{\|}{C}}S\text{—}ACP} + \underset{\text{Malonyl-S-ACP}}{HO\overset{O}{\overset{\|}{C}}CH_2\overset{O}{\overset{\|}{C}}S\text{—}ACP} \rightleftharpoons \underset{\text{Acetoacetyl-S-ACP}}{CH_3\overset{O}{\overset{\|}{C}}CH_2\overset{O}{\overset{\|}{C}}S\text{—}ACP} + CO_2 + ACP\text{—}SH$$

The molecule of CO_2 that is lost in this reaction is the same molecule that was incorporated into malonyl CoA in the acetyl CoA carboxylase reaction.

This remarkable reaction bears a strong resemblance to the malonic ester syntheses that we saw earlier (Section 20.4) and it deserves special comment. One can imagine, for example, a more economical synthesis of acetoacetyl-S-ACP, that is, a simple condensation between two moles of acetyl-S-ACP.

$$CH_3\overset{O}{\overset{\|}{C}}S\text{—}ACP + CH_3\overset{O}{\overset{\|}{C}}S\text{—}ACP \rightleftharpoons CH_3\overset{O}{\overset{\|}{C}}CH_2\overset{O}{\overset{\|}{C}}S\text{—}ACP + ACP\text{—}SH$$

Studies of this last reaction, however, have revealed that it is highly *endothermic* and that the position of equilibrium lies very far to the left. By contrast, the condensation of acetyl-S-ACP and malonyl-S-ACP is highly exothermic and the position of equilibrium lies far to the right. The favorable thermodynamics of the condensation utilizing malonyl-S-ACP comes about because *the reaction also produces a highly stable substance: carbon dioxide.* Thus, decarboxylation of the malonyl group provides the condensation with thermodynamic assistance.

The next three steps in fatty acid synthesis transform the acetoacetyl group of acetoacetyl-S-ACP into a butyryl (butanoyl) group. These steps involve (1) reduction of the keto group (utilizing NADPH* as the reducing agent), (2) dehydration of an alcohol, and (3) reduction of a double bond (again utilizing NADPH).

Reduction of the Keto Group

$$\underset{\text{Acetoacetyl-S-ACP}}{CH_3\overset{O}{\overset{\|}{C}}CH_2\overset{O}{\overset{\|}{C}}S\text{—}ACP} + NADPH + H^+ \rightleftharpoons \underset{\beta\text{-Hydroxybutyryl-S-ACP}}{CH_3\overset{OH}{\overset{|}{C}}HCH_2\overset{O}{\overset{\|}{C}}S\text{—}ACP} + NADP^+$$

* NADPH is *nicotinamide adenine dinucleotide phosphate,* a coenzyme that is very similar in structure and function to NADH, p. 462.

Dehydration of the Alcohol

$$\underset{\beta\text{-Hydroxybutyryl-S-ACP}}{CH_3\overset{OH}{\overset{|}{C}}HCH_2\overset{O}{\overset{\|}{C}}S{-}ACP} \rightleftharpoons \underset{\text{Crotonyl-S-ACP}}{CH_3CH{=}CH\overset{O}{\overset{\|}{C}}S{-}ACP} + H_2O$$

Reduction of the Double Bond

$$\underset{\text{Crotonyl-S-ACP}}{CH_3CH{=}CH\overset{O}{\overset{\|}{C}}S{-}ACP} + NADPH + H^+ \rightleftharpoons \underset{\text{Butyryl-S-ACP}}{CH_3CH_2CH_2\overset{O}{\overset{\|}{C}}S{-}ACP} + NADP^+$$

These steps complete one cycle of the overall fatty acid synthesis. Their net result is the conversion of two acetate units into the four-carbon butyrate unit of butyryl-S-ACP. (This conversion requires, of course, the crucial intervention of a molecule of carbon dioxide.) At this point, another cycle begins and the chain is lengthened by two more carbons:

Condensation

$$\underset{\text{(four carbons)}}{CH_3CH_2CH_2\overset{O}{\overset{\|}{C}}S{-}ACP} + HO\overset{O}{\overset{\|}{C}}CH_2\overset{O}{\overset{\|}{C}}S{-}ACP \longrightarrow CH_3CH_2CH_2\overset{O}{\overset{\|}{C}}CH_2\overset{O}{\overset{\|}{C}}S{-}ACP + CO_2 + ACP{-}SH$$

Reduction

$$CH_3CH_2CH_2\overset{O}{\overset{\|}{C}}CH_2\overset{O}{\overset{\|}{C}}S{-}ACP \xrightarrow{NADPH + H^+ \;\to\; NADP^+} CH_3CH_2CH_2\overset{OH}{\overset{|}{C}}HCH_2\overset{O}{\overset{\|}{C}}S{-}ACP$$

Dehydration

$$CH_3CH_2CH_2\overset{OH}{\overset{|}{C}}HCH_2\overset{O}{\overset{\|}{C}}S{-}ACP \xrightarrow{\;\to H_2O\;} CH_3CH_2CH_2CH{=}CH\overset{O}{\overset{\|}{C}}S{-}ACP$$

Reduction

$$CH_3CH_2CH_2CH{=}CH\overset{O}{\overset{\|}{C}}S{-}ACP \xrightarrow{NADPH + H^+ \;\to\; NADP^+} \underset{\text{(six carbons)}}{CH_3CH_2CH_2CH_2CH_2\overset{O}{\overset{\|}{C}}S{-}ACP}$$

Subsequent turns of the cycle continue to lengthen the chain by two-carbon units until a long-chain fatty acid is produced. The overall equation for the synthesis of palmitic acid, for example, can be written as follows:

$$CH_3\overset{O}{\overset{\|}{C}}S\text{—}CoA + 7HO\overset{O}{\overset{\|}{C}}CH_2\overset{O}{\overset{\|}{C}}S\text{—}CoA + 14NADH + 14H^+ \longrightarrow$$

$$CH_3(CH_2)_{14}COOH + 7CO_2 + 8CoA\text{—}SH + 14NAD^+ + 6H_2O$$

One of the most remarkable aspects of fatty acid synthesis is that the entire cycle appears to be carried out by a complex of enzymes that are clustered into a single unit. The molecular weight of this cluster of proteins, *called fatty acid synthetase,* has been estimated as 2,300,000.* The synthesis begins with a single molecule of acetyl-S-ACP serving as a primer. Then, in the synthesis of palmitic acid, for example, successive condensations of seven molecules of malonyl-S-ACP occur with each condensation followed by reduction, dehydration, and reduction. All of these steps, which result in the synthesis of a 16-carbon chain, take place before the fatty acid is released from the enzyme cluster.

The acyl carrier protein has been isolated and purified; its molecular weight is approximately 10,000. The protein contains a chain of groups called a *phosphopantetheine group* that is identical to that of coenzyme A (p. 786). In ACP this chain is attached to a protein (rather than to an adenosine phosphate as it is in coenzyme A):

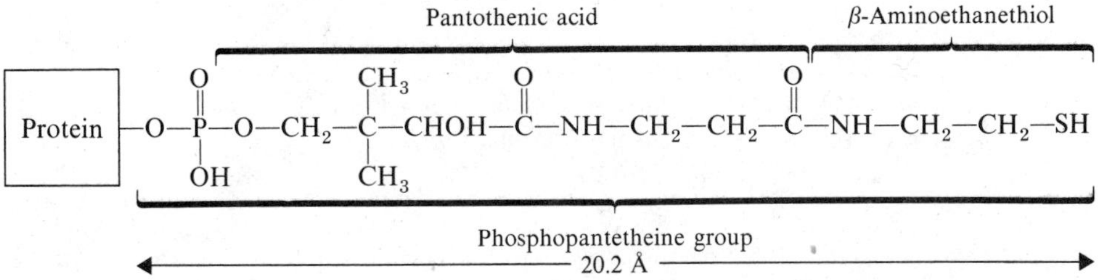

Pantothenic acid | β-Aminoethanethiol

$$\boxed{\text{Protein}}\text{—O—}\underset{OH}{\overset{O}{\overset{\|}{P}}}\text{—O—}CH_2\text{—}\underset{CH_3}{\overset{CH_3}{C}}\text{—CHOH—}\overset{O}{\overset{\|}{C}}\text{—NH—}CH_2\text{—}CH_2\text{—}\overset{O}{\overset{\|}{C}}\text{—NH—}CH_2\text{—}CH_2\text{—SH}$$

Phosphopantetheine group

← 20.2 Å →

The length of the phosphopantetheine group is 20.2 Å, and it has been postulated that it acts as a "swinging arm" in transferring the growing acyl chain from one enzyme of the cluster to the next (Fig. N.1).

N.2 Biosynthesis of Steroids

We saw in Special Topic E, that the five-carbon compound, 3-methyl-3-butenyl pyrophosphate, is the actual "isoprene unit" that nature uses in constructing terpenoids and carotenoids. We can now extend that biosynthetic pathway in two directions. We can show how 3-methyl-3-butenyl pyrophosphate (like the fatty acids) is ultimately derived from acetate units, and how cholesterol, the precursor of most of the important steroids, is synthesized from 3-methyl-3-butenyl pyrophosphate.

In the 1940's, Konrad Bloch of Harvard University used labeling experiments to demonstrate that all of the carbon atoms of cholesterol can be derived from acetic acid. Using *methyl-labeled* acetic acid, for example, Bloch found the following label distribution in the cholesterol that was synthesized.

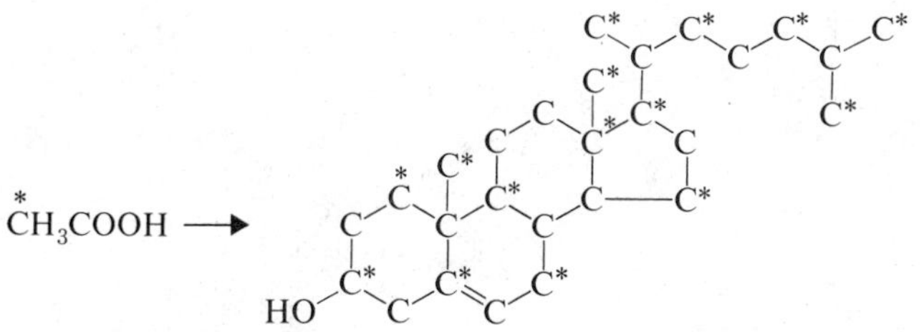

$$\overset{*}{C}H_3COOH \longrightarrow$$ (cholesterol skeleton with labeled carbons C* indicated)

* As isolated from yeast cells. Fatty acid synthetases from different sources have different molecular weights; that from pigeon liver, for example, has a molecular weight of 450,000.

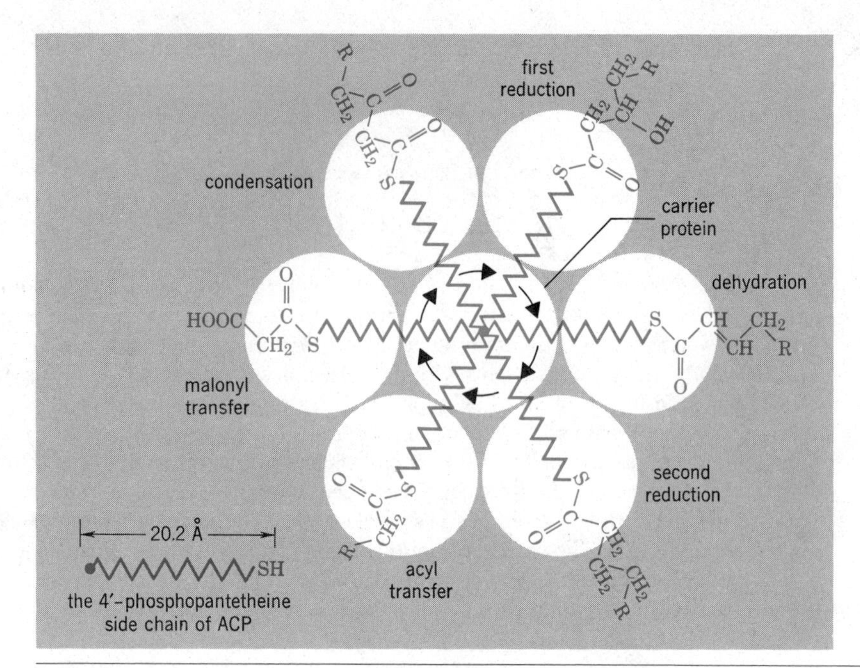

FIG. N.1
The phosphopantetheine group as a swinging arm in the fatty acid synthetase complex. (*From A.L. Lehninger, Biochemistry, Worth Publishers, Inc., New York, (1907), p. 519. Used with permission.*)

Bloch also found that feeding *carboxyl-labeled* acetic acid led to incorportion of the label into all of the other carbons of cholesterol (the unstarred carbons of the formula given above).

Subsequent research by a number of investigators has shown that 3-methyl-3-butenyl pyrophosphate is synthesized from acetate units through the following sequence of reactions:

$$\underset{\substack{(C_2)\\ \text{Acetyl CoA}}}{CH_3\overset{O}{\overset{\|}{C}}S{-}CoA} + \underset{\substack{(C_4)\\ \text{Acetoacetyl CoA}}}{CH_3\overset{O}{\overset{\|}{C}}CH_2\overset{O}{\overset{\|}{C}}S{-}CoA} \xrightarrow{\quad CoA{-}SH \quad} \underset{\substack{(C_6)\\ \beta\text{-Hydroxy-}\beta\text{-methylglutaryl CoA}}}{HO\overset{O}{\overset{\|}{C}}CH_2{-}\underset{\substack{/\quad\backslash\\ CH_3\quad OH}}{C}{-}CH_2\overset{O}{\overset{\|}{C}}S{-}CoA}$$

$$\xrightarrow[\text{2NADP}^+ + \text{CoA—SH}]{\text{2NADPH} + 2H^+}$$

$$\underset{\substack{(C_6)\\ \text{Mevalonic acid}}}{HO\overset{O}{\overset{\|}{C}}CH_2{-}\underset{\substack{/\quad\backslash\\ CH_3\quad OH}}{C}{-}CH_2CH_2OH}$$

$$\Big\downarrow \begin{matrix} \text{3ATP} \\ \text{3ADP} \end{matrix} \Bigg\} \text{ Three successive steps}$$

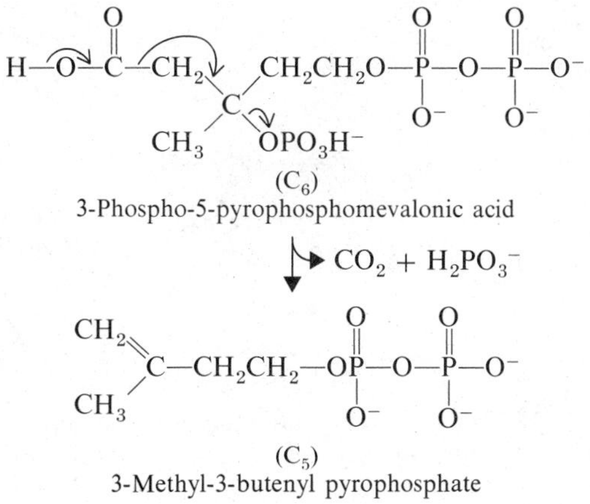

(C_6)

3-Phospho-5-pyrophosphomevalonic acid

$\rightarrow CO_2 + H_2PO_3^-$

(C_5)

3-Methyl-3-butenyl pyrophosphate

The first step of this synthetic pathway is straightforward. Acetyl CoA (from one mole of acetate) and acetoacetyl CoA (from two moles of acetate) condense to form the six-carbon compound, β-hydroxy-β-methylglutaryl CoA. This step is followed by an enzymatic reduction of the thiol ester group of β-hydroxy-β-methylglutaryl CoA to the primary alcohol of mevalonic acid. The key to finding this pathway was the discovery that mevalonic acid was an intermediate and that this six-carbon compound could be transformed into the five-carbon 3-methyl-3-butenyl pyrophosphate by successive phosphorylations and decarboxylation.

Then 3-methyl-3-butenyl pyrophosphate isomerizes to produce an equilibrium mixture that contains 3-methyl-2-butenyl pyrophosphate, and these two compounds condense to form geranylpyrophosphate, a 10-carbon compound. Geranylpyrophosphate subsequently condenses with another mole of 3-methyl-3-butenyl pyrophosphate to form farnesylpyrophosphate, a 15-carbon compound. (Geranylpyrophosphate and farnesylpyrophosphate are the precursors of the mono- and sesquiterpenes, cf. Special Topic E).

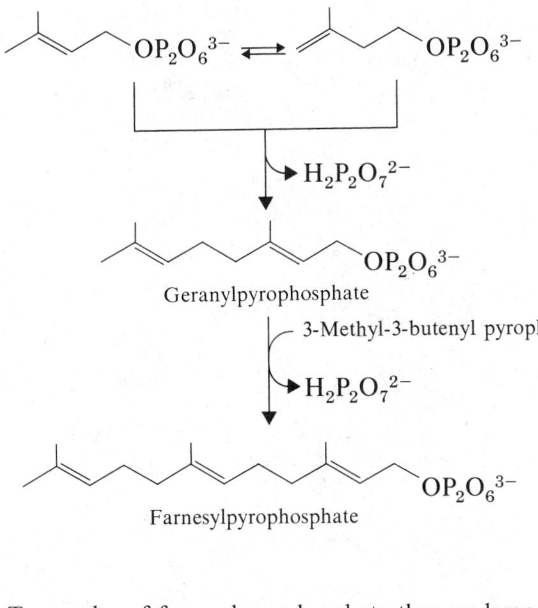

$OP_2O_6^{3-} \rightleftharpoons OP_2O_6^{3-}$

$\rightarrow H_2P_2O_7^{2-}$

$OP_2O_6^{3-}$

Geranylpyrophosphate

3-Methyl-3-butenyl pyrophosphate

$\rightarrow H_2P_2O_7^{2-}$

$OP_2O_6^{3-}$

Farnesylpyrophosphate

Two moles of farnesylpyrophosphate then undergo a reductive condensation to produce squalene.

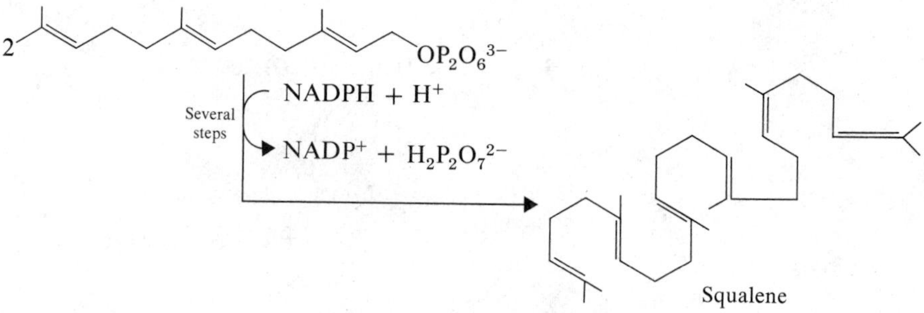

2 [farnesyl pyrophosphate] $OP_2O_6^{3-}$

Several steps: $NADPH + H^+$ → $NADP^+ + H_2P_2O_7^{2-}$

Squalene

Squalene is the direct precursor of cholesterol. Oxidation of squalene yields squalene 2,3-epoxide, which undergoes a remarkable series of ring closures accompanied by concerted methyl and hydride migrations to yield lanosterol (cf. Sect. 7.15). Lanosterol is then converted to cholesterol through a series of enzyme-catalyzed reactions.

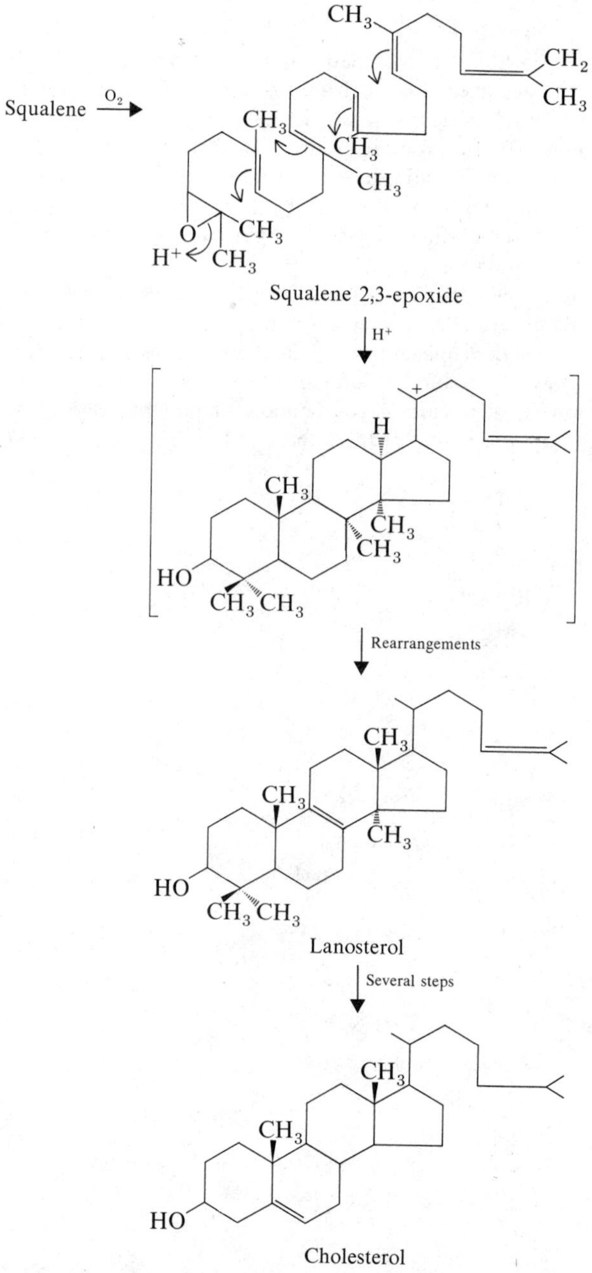

Squalene $\xrightarrow{O_2}$

CH_3 CH_2 CH_3 CH_3 CH_3 CH_3 CH_3 H^+ O CH_3

Squalene 2,3-epoxide

H^+

H CH_3 CH_3 CH_3 HO CH_3 CH_3

Rearrangements

CH_3 CH_3 CH_3 HO CH_3 CH_3

Lanosterol

Several steps

CH_3 CH_3 HO

Cholesterol

22 AMINO ACIDS AND PROTEINS

22.1 INTRODUCTION

The three major groups of biological polymers are polysaccharides, proteins, and nucleic acids. We studied polysaccharides in Chapter 19 and saw that they function primarily as energy reserves and, in plants, as structural materials. When we study nucleic acids in Special Topic P we will find that they serve two major purposes: storage and transmission of information. Of the three groups of biopolymers, proteins have the most diverse functions. As enzymes and hormones, proteins catalyze and regulate the reactions that occur in the body; as muscles and tendons they provide the body with the means for movement; as skin and hair they give it an outer covering; as hemoglobins they transfer all-important oxygen to its most remote corners; as antibodies they provide it with a means of protection against disease; and in combination with other substances in bone they provide it with structural support.

Given such diversity of functions, we should not be surprised to find that proteins come in all sizes and shapes. By the standard of most of the molecules we have studied, even small proteins have very high molecular weights. Lysozyme, an enzyme, is a relatively small protein and yet its molecular weight is 14,600. The molecular weights of most proteins are much larger. Their shapes cover a range from the globular proteins such as lysozyme and hemoglobin to the helical coils of α-keratin (hair, nails, wool) and the pleated sheets of silk fibroin.

And yet, in spite of such diversity of size, shape, and function, all proteins have common features that allow us to decipher their structures and understand their properties. Proteins are *polyamides* and their monomeric units are amino acids. Although hydrolysis of naturally occurring proteins may yield up to 22 different amino acids, all amino acids have two important structural features in common: They are all α-amino acids and, with the exception of glycine (whose molecules are achiral), almost all naturally occurring amino acids have the L configuration at the α-carbon.* That is, they have the same relative configuration as L-glyceraldehyde:

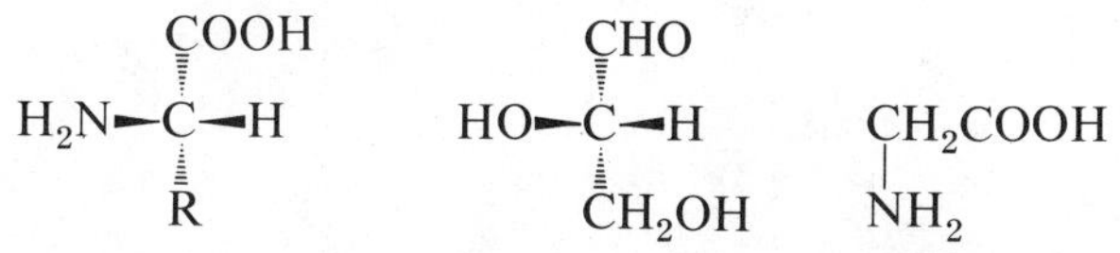

$$\begin{array}{ccc} \begin{array}{c} COOH \\ H_2N-C-H \\ R \end{array} & \begin{array}{c} CHO \\ HO-C-H \\ CH_2OH \end{array} & \begin{array}{l} CH_2COOH \\ | \\ NH_2 \end{array} \end{array}$$

An L-α-amino acid — L-Glyceraldehyde — Glycine

* Some D-amino acids have been obtained from the material comprising the cell walls of bacteria, and by hydrolysis of certain antibiotics.

22.2 AMINO ACIDS

22.2A Structures and Names

The 22 α-amino acids that can be obtained from proteins can be subdivided into five different groups on the basis of the structures of their side chains, *R*. These are given in Table 22.1.

TABLE 22.1 L-Amino Acids Found in Proteins

(Fischer projection: COOH at top, H_2N on left, H on right, **R** at bottom)

STRUCTURE OF **R**	NAME	ABBREVIATION	pK_{a_1} α-CO_2H	pK_{a_2} α-NH_3^+	pK_{a_3} R Group	p*I*
R group is neutral						
$-H$	Glycine	Gly	2.3	9.6		6.0
$-CH_3$	Alanine	Ala	2.3	9.7		6.0
$-CH(CH_3)_2$	Valine[e]	Val	2.3	9.6		6.0
$-CH_2CH(CH_3)_2$	Leucine[e]	Leu	2.4	9.6		6.0
$-CH(CH_3)CH_2CH_3$	Isoleucine[e]	Ile	2.4	9.7		6.1
$-CH_2-C_6H_5$ (phenyl)	Phenylalanine[e]	Phe	1.8	9.1		5.5
$-CH_2CONH_2$	Asparagine	Asn	2.0	8.8		5.4
$-CH_2CH_2CONH_2$	Glutamine	Gln	2.2	9.1		5.7
$-CH_2$-(3-indolyl, N–H)	Tryptophan[e]	Trp	2.4	9.4		5.9
$HO\overset{O}{\overset{\parallel}{C}}-CH-CH_2$, HN, CH_2, CH_2 (ring) (complete structure)	Proline	Pro	2.0	10.6		6.3
R contains an —OH group						
$-CH_2OH$	Serine	Ser	2.2	9.2		5.7
$-CH(OH)CH_3$	Threonine[e]	Thr	2.6	10.4		6.5
$-CH_2-C_6H_4-OH$	Tyrosine	Tyr	2.2	9.1	10.1	5.7
$HO\overset{O}{\overset{\parallel}{C}}-CH-CH_2$, HN, CH–OH, CH_2 (ring) (complete structure)	Hydroxyproline	Hyp	1.9	9.7		6.3

TABLE 25.1 L-Amino Acids Found in Proteins (Continued)

STRUCTURE OF **R**	NAME	ABBRE-VIATION	pK_{a_1} α-CO_2H	pK_{a_2} α-NH_3^+	pK_{a_3} R Group	p*I*
R contains sulfur						
$—CH_2SH$	Cysteine	Cys	1.7	10.8	8.3	5.0
$—CH_2—S$ / $—CH_2—S$ (S–S bonded)	Cystine	Cys-Cys	{1.6, 2.3}	{7.9, 9.9}		5.1
$—CH_2CH_2SCH_3$	Methionine[e]	Met	2.3	9.2		5.8
R contains a carboxyl group						
$—CH_2COOH$	Aspartic acid	Asp	2.1	9.8	3.9	3.0
$—CH_2CH_2COOH$	Glutamic acid	Glu	2.2	9.7	4.3	3.2
R contains a basic amino group						
$—CH_2CH_2CH_2CH_2NH_2$	Lysine[e]	Lys	2.2	9.0	10.5	9.8
$—CH_2CH_2CH_2NH—C(=NH)—NH_2$	Arginine	Arg	2.2	9.0	12.5	10.8
$—CH_2$–(imidazolyl, N–H)	Histidine	His	1.8	9.2	6.0	7.6

e = essential amino acids

22.2B Essential Amino Acids

Amino acids can be synthesized by all living organisms, plants and animals. Many higher animals, however, are deficient in their ability to synthesize all of the amino acids they need for their proteins. Thus these higher animals require certain amino acids as a part of their diet. For adult humans there are eight essential amino acids; these are designated with the superscript *e* in Table 22.1.

22.2C Amino Acids as Dipolar Ions

Since amino acids contain both a basic group ($—NH_2$) and an acidic group ($—COOH$), they are amphoteric. In the dry solid state, amino acids exist as *dipolar ions,* a form in which the carboxyl group is present as a carboxylate ion, $—COO^-$, and the amine group is present as an ammonium group, $—NH_3^+$. (Dipolar ions are also called *zwitterions.*) In aqueous solution, an equilibrium exists between the dipolar ion and the anionic and cationic forms of an amino acid.

$$H_3\overset{+}{N}CH(R)COOH \underset{+H^+}{\overset{-H^+}{\rightleftharpoons}} H_3\overset{+}{N}CH(R)COO^- \underset{+H^+}{\overset{-H^+}{\rightleftharpoons}} H_2NCH(R)COO^-$$

Cationic form (predominant in strongly acidic solutions, e.g., at pH 0) — **Dipolar Ion** — **Anionic form** (predominant in strongly basic solutions, e.g., at pH 14)

The predominant form of the amino acid present in a solution depends on the pH of the solution and on the nature of the amino acid. In strongly acidic solutions all amino acids are present primarily as cations; in strongly basic solutions they are present as anions. At some intermediate pH, called the *isoelectric point,* pI, the concentration of the dipolar ion is at its maximum and the concentrations of the anions and cations are equal.

Let us consider first an amino acid with a side chain that contains neither acidic nor basic groups—an amino acid, for example, such as alanine.

If alanine is present in a strongly acidic solution (e.g., at pH 0) it is present mainly in the cationic form shown below. The acidity constant, K_a, for the cationic form is 5×10^{-3}. This is considerably higher than the K_a of a corresponding carboxylic acid, (e.g., propanoic acid) and indicates that the cationic form of alanine is the stronger acid. But, we should expect it to be. After all it is a positively charged species and therefore should lose a proton more readily.

$$CH_3\underset{\displaystyle NH_3^+}{\underset{|}{C}}HCOOH \qquad\qquad CH_3CH_2COOH$$

Cationic form of alanine
$K_{a_1} = 5 \times 10^{-3}$
$pK_{a_1} = 2.3$

Propanoic acid
$K_a = 1.3 \times 10^{-5}$
$pK_a = 4.8$

The pK_a of an acid is the negative of the logarithm of the K_a:

$$pK_a = -\log K_a$$

For the cationic form of alanine pK_{a1} = 2.3; for propanoic acid, pK_a = 4.8.

The dipolar ion form of an amino acid is also a potential acid because the $—NH_3^+$ group can donate a proton. The pK_a of the dipolar ion form of alanine is 9.7 ($K_a = 2 \times 10^{-10}$).

$$CH_3\underset{\displaystyle NH_3^+}{\underset{|}{C}}HCOO^-$$

$K_{a_2} = 2 \times 10^{-10}$
$pK_{a_2} = 9.7$

The isoelectric point, pI, of an amino acid such as alanine is the average of pK_{a1} and pK_{a2}.

$$pI = \frac{2.3 + 9.7}{2} = 6.0 \quad \text{(isoelectric point of alanine)}$$

What does this mean about the behavior of alanine as the pH of a strongly acidic solution containing it is gradually raised by adding a base (i.e., OH^-)? At first, (pH 0) the predominant form will be the cationic form. But then as the acidity reaches pH 2.3 (the pK_a of the cationic form, pK_{a1}), half of the cationic form will be converted to the dipolar ion.* As the pH increases further—from pH

* It is easy to show that for an acid:

$$pK_a = pH + \log\frac{[\text{acid}]}{[\text{conjugate base}]}$$

When the acid is half neutralized, [acid] = [conjugate base] and $\log\dfrac{[\text{acid}]}{[\text{conjugate base}]} = 0$; thus pH = p$K_a$.

$$\underset{\substack{\text{Cationic form}\\(\text{p}K_{a_1} = 2.3)}}{\underset{\displaystyle NH_3^+}{CH_3\overset{|}{C}HCOOH}} \underset{H^+}{\overset{OH^-}{\rightleftharpoons}} \underset{\substack{\text{Dipolar ion}\\(\text{p}K_{a_2} = 9.7)}}{\underset{\displaystyle NH_3^+}{CH_3\overset{|}{C}HCOO^-}} \overset{OH}{\rightleftharpoons} \underset{\text{Anionic form}}{\underset{\displaystyle NH_2}{CH_3\overset{|}{C}HCOO^-}}$$

2.3 to pH 9.7—the predominant form will be the dipolar ion. At pH 6.0, the pH equals pI and the concentration of the dipolar ion is at its maximum. When the pH rises to pH 9.7 (the pK_a of the dipolar ion) the dipolar ion will be half-converted to the anionic form. Then, as the pH approaches pH 14, the anionic form becomes the predominant form present in the solution.

If the side chain of an amino acid contains an extra acidic or basic group, then the equilibria are more complex. Consider lysine, for example, an amino acid that has an extra —NH_2 group on its ϵ-carbon. In strongly acidic solution, lysine will be present as a di-cation because both amino groups will be protonated. The

$$\underset{\substack{\text{Dicationic form of}\\\text{lysine}\\(\text{p}K_{a_1} = 2.2)}}{\underset{\displaystyle NH_3^+}{H_3\overset{+}{N}(CH_2)_4\overset{|}{C}HCOOH}} \underset{H^+}{\overset{OH^-}{\rightleftharpoons}} \underset{\substack{\text{Monocationic}\\\text{form}\\(\text{p}K_{a_2} = 9.0)}}{\underset{\displaystyle NH_3^+}{H_3\overset{+}{N}(CH_2)_4\overset{|}{C}HCOO^-}} \underset{H^+}{\overset{OH^-}{\rightleftharpoons}}$$

$$\underset{\substack{\text{Dipolar ion}\\(\text{p}K_{a_3} = 10.5)}}{\underset{\displaystyle NH_2}{H_3\overset{+}{N}(CH_2)_4\overset{|}{C}HCOO^-}} \underset{H^+}{\overset{OH^-}{\rightleftharpoons}} \underset{\text{Anionic form}}{\underset{\displaystyle NH_2}{H_2N(CH_2)_4\overset{|}{C}HCOO^-}}$$

first proton to be lost as the pH is raised is a proton of the carboxyl group (pK_{a1} = 2.2), the next is from the α-ammonium group (pK_{a2} = 9.0) and the last is from the ϵ-ammonium group. The isoelectric point of lysine is the average of pK_{a2} (the monocation) and pK_{a3} (the dipolar ion).

$$\text{p}I = \frac{9.0 + 10.5}{2} = 9.8 \qquad \text{(isoelectric point of lysine)}$$

Problem 22.1

What form of glutamic acid would you expect to predominate in (a) strongly acid solution? (b) Strongly basic solution? (c) At its isoelectric point (pI 3.2)? (d) The isoelectric point of glutamine (pI 5.7) is considerably higher than that of glutamic acid. Explain.

Problem 22.2

The guanidino group, $—NH—\overset{\overset{\displaystyle NH}{\|}}{C}—NH_2$, of arginine is one of the most strongly basic of all organic groups. Explain.

22.3 LABORATORY SYNTHESIS OF α-AMINO ACIDS

A variety of methods have been developed for the laboratory synthesis of α-amino acids. We will describe here three general methods, all of which are based on reactions we have seen before.

22.3A Direct Ammonolysis of an α-Halo Acid

$$R{-}CH_2COOH \xrightarrow[X_2]{P} RCH(X)COOH \xrightarrow{NH_{3(excess)}} R{-}CH(\overset{+}{N}H_3)COO^-$$

This method is probably used least often because yields tend to be poor. We saw an example of this method in Section 17.9.

22.3B From Potassium Phthalimide

This method is a modification of the Gabriel synthesis of amines (Section 18.5). The yields are usually high and the products are easily purified.

$$\text{Potassium phthalimide (}C_6H_4(CO)_2NK\text{)} + ClCH_2COOC_2H_5 \longrightarrow$$

Potassium phthalimide — Ethyl chloroacetate

$$C_6H_4(CO)_2N{-}CH_2CO_2C_2H_5 \xrightarrow[(2)\ HCl]{(1)\ KOH/H_2O} \overset{+}{N}H_3CH_2COO^- + C_6H_4(COOH)_2$$

(97%) — Glycine (85%) — Phthalic acid

22.3C From Amidomalonic Esters and Imido Malonic Esters

Diethyl aminomalonate can be prepared from diethyl malonate by the following reactions.

$$CH_2(CO_2C_2H_5)_2 + CH_3(CH_2)_3ONO \xrightarrow[(2)\ H_2SO_4]{(1)\ 0\text{–}20^\circ} HON{=}C(CO_2C_2H_5)_2 + CH_3(CH_2)_3OH$$

Diethyl malonate — *n*-Butyl nitrite — Diethyl oximinomalonate (87%)

$$\xrightarrow{2H_2/Ni} H_2NCH(CO_2C_2H_5)_2$$

Diethyl aminomalonate (65%)

Converting diethyl aminomalonate to diethyl benzamidomalonate gives a very useful reagent for amino acid synthesis.

$$H_2NCH(CO_2C_2H_5)_2 + C_6H_5COCl \xrightarrow{\text{Pyridine}} \underset{\substack{\text{Diethyl benzamidomalonate} \\ (90\%)}}{C_6H_5CONHCH(CO_2C_2H_5)_2}$$

An example of its use is the following synthesis of DL-leucine.

$$\text{(1)}\ \underset{\text{Diethyl benzamidomalonate}}{C_6H_5CONHCH(CO_2CH_2CH_3)_2} \xrightarrow{NaOCH_2CH_3} C_6H_5CONH\ddot{\bar{C}}(CO_2C_2H_5)_2$$

$$\xrightarrow{(CH_3)_2CHCH_2I} C_6H_5CONHC(CH_2CH(CH_3)CH_3)(CO_2CH_2CH_3)_2$$

$$\text{(2)}\ C_6H_5CONHC(CH_2CH(CH_3)CH_3)(CO_2CH_2CH_3)_2 \xrightarrow[H_2O]{HBr} \left[H_3\overset{+}{N}{-}CH(CH_2CHCH_3CH_3)(COOH){-}COO^- \right] + C_6H_5COOH + 2CH_3CH_2OH$$

$$\xrightarrow{-CO_2} \underset{\substack{\text{DL-Leucine} \\ (78\%)}}{CH_3CH(CH_3)CH_2CH(^+NH_3)COO^-}$$

In the first step of the synthesis, diethyl benzamidomalonate reacts with sodium ethoxide to give an enolate ion; this then reacts with isobutyl iodide in an alkylation (S_N2) reaction. In the second step acidic hydrolysis cleaves both the amide and ester linkages, and the malonic acid that is produced decarboxylates spontaneously to give DL-leucine.

A variation of this procedure uses potassium phthalimide and bromomalonic ester to prepare an *imido* malonic ester. This method is illustrated with a synthesis of methionine.

$$\text{Potassium phthalimide } (C_6H_4(CO)_2N^-K^+) + \underset{\text{Ethyl }\alpha\text{-bromomalonate}}{BrCH(CO_2C_2H_5)_2} \xrightarrow{(82\text{–}85\%)}$$

$$\underset{\text{Phthalimidomalonic ester}}{C_6H_4(CO)_2NCH(CO_2C_2H_5)_2} \xrightarrow[\substack{ClCH_2CH_2SCH_3 \\ (96\text{–}98\%)}]{NaOCH_2CH_3} C_6H_4(CO)_2N\text{—}C(CO_2C_2H_5)_2CH_2CH_2SCH_3 \xrightarrow{NaOH}$$

$$C_6H_4(COO^-)\text{—}C(=O)\text{—}NHC(COO^-)_2CH_2CH_2SCH_3 \xrightarrow[(84\text{–}85\%)]{HCl} \underset{\text{DL-Methionine}}{CH_3SCH_2CH_2CH(\overset{+}{N}H_3)COO^-} + CO_2 + C_6H_4(COOH)_2$$

Problem 22.3

The following amino acids have been synthesized from diethyl benzamidomalonate and the appropriate alkyl halide. Outline each method. (The percentages in parentheses are the yields actually obtained.)

(a) DL-Phenylalanine (90%)
(b) DL-Aspartic acid (62%)
(c) DL-Valine (71%)

Problem 22.4

Starting with ethyl α-bromomalonate and potassium phthalimide and using any other necessary reagents show how you might synthesize:

(a) DL-Leucine
(b) DL-Alanine
(c) DL-Phenylalanine

22.3D Resolution of DL-Amino Acids

With the exception of glycine, which has no chiral carbon, the amino acids that are produced by the methods we have outlined are all produced as racemates. In order to obtain the naturally occurring L-amino acid we must, of course, resolve the racemate. This can be done in a variety of ways including the methods outlined in Section 8.18.

One especially interesting method for resolving amino acids is based on the use of enzymes called *deacylases*. These enzymes catalyze the hydrolysis of *N-acylamino acids* in living organisms. Since the active site of the enzyme is chiral it hydrolyzes only *N*-acylamino acids of the L configuration. When it is exposed to a racemic modification of *N*-acylamino acids, only the L-amino acid is affected and the products, as a result, are separated easily.

$$\text{DL-}\underset{\substack{|\\ ^{+}NH_3}}{RCHCOO^-} \xrightarrow{(CH_3CO)_2O} \text{DL-}\underset{\substack{|\\ CH_3CONH}}{RCHCOOH} \xrightarrow{\text{Deacylase}} CH_3COOH$$

(racemate)

$$\underbrace{\text{L-}\underset{\substack{|\\ \overset{+}{N}H_3}}{RCHCOO^-} + \text{D-}\underset{\substack{|\\ CH_3CONH}}{RCHCOOH}}_{\text{Easily separated}}$$

22.4 BIOSYNTHESIS OF AMINO ACIDS

Two of the most important methods used by living cells for the synthesis of amino acids are described below.

22.4A Reductive Amination

This biosynthesis bears a remarkable resemblance to one laboratory method that we have seen for the synthesis of amines (p. 820). In enzymatic reductive amination, α-ketoglutaric acid combines with ammonia in the presence of the reducing agent, NADH. The product is L-glutamic acid.

$$\underset{\alpha\text{-Ketoglutaric acid}}{HOOCCH_2CH_2\underset{\substack{\|\\ O}}{C}COOH} + H^+ + NH_3 \xrightleftharpoons[NAD^+]{NADH} \underset{\text{L-Glutamic acid}}{HOOCCH_2CH_2\underset{\substack{|\\ ^{+}NH_3}}{C}HCOO^-} + H_2O$$

The cell has α-ketoglutaric acid readily available because it is an intermediate in the metabolism of carbohydrates.

22.4B Transamination

The α-amino group of L-glutamic acid can be transferred to other α-keto acids whose carbon chains correspond in structure to those of other naturally occurring amino acids. These reactions are catalyzed by enzymes called *transaminases*. An example is the biosynthesis of L-aspartic acid from L-glutamic acid and oxaloacetic acid.

$$\underset{\text{L-Glutamic acid}}{HOOCCH_2CH_2\underset{\substack{|\\ ^{+}NH_3}}{C}HCOO^-} + \underset{\text{Oxaloacetic acid}}{HOOCCH_2\underset{\substack{\|\\ O}}{C}COOH} \xrightleftharpoons{\text{Transaminase}}$$

$$\underset{\alpha\text{-Ketoglutaric acid}}{HOOCCH_2CH_2\underset{\substack{\|\\ O}}{C}COOH} + \underset{\text{L-Aspartic acid}}{HOOCCH_2\underset{\substack{|\\ ^{+}NH_3}}{C}HCOO^-}$$

Oxaloacetic acid is also an intermediate in the metabolism of carbohydrates.

22.5 ANALYSIS OF AMINO ACID MIXTURES

The amide linkages that join α-amino acids in proteins are commonly called *peptide linkages,* and α-amino acid polymers with molecular weights less than 10,000 are usually called *polypeptides.* Those with molecular weights greater than 10,000 are called proteins. This division is quite arbitrary and the two terms are sometimes interchanged. In actuality, both proteins and polypeptides are *polyamides.*

We can represent the structures of polypeptides using the symbols for the amino acids. The dipeptide glycylvaline, for example, is represented as Gly · Val, and the dipeptide valylglycine as Val · Gly. In each case the amino acid whose carboxyl group is involved in an amide linkage is placed first.

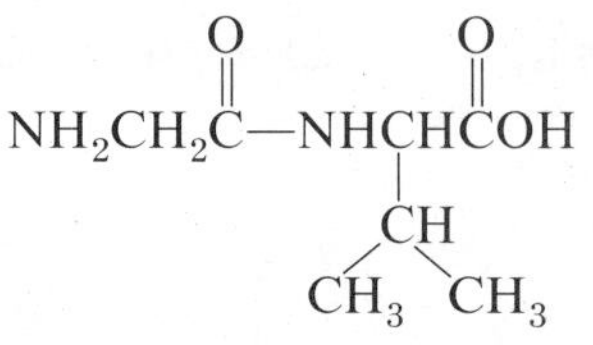

$$NH_2CH_2\overset{O}{\overset{\|}{C}}{-}NH\underset{CH(CH_3)_2}{CH}\overset{O}{\overset{\|}{C}}OH \qquad NH_2\underset{CH(CH_3)_2}{CH}\overset{O}{\overset{\|}{C}}{-}NHCH_2\overset{O}{\overset{\|}{C}}OH$$

Glycylvaline (Gly·Val)   Valylglycine (Val·Gly)

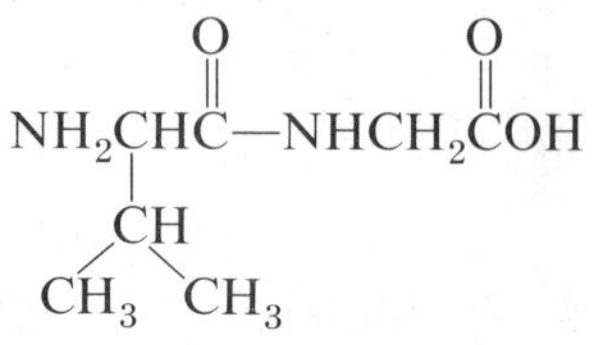

The tripeptide glycylvalylphenylalanine can be represented in the following way.

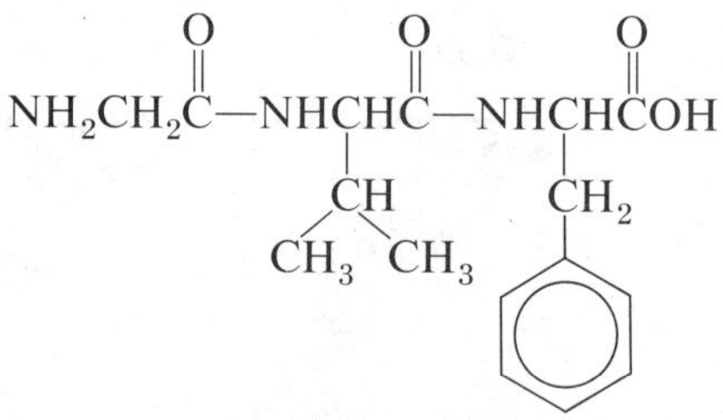

$$NH_2CH_2\overset{O}{\overset{\|}{C}}{-}NH\underset{CH(CH_3)_2}{CH}\overset{O}{\overset{\|}{C}}{-}NH\underset{CH_2C_6H_5}{CH}\overset{O}{\overset{\|}{C}}OH$$

Glycylvalylphenylalanine
(Gly·Val·Phe)

When a protein or polypeptide is refluxed with 6*N*-hydrochloric acid for 24 hours, hydrolysis of all of the amide linkages usually takes place and this produces a mixture of amino acids. One of the first tasks that we face when we attempt to determine the structure of a polypeptide or protein is the separation and identification of the individual amino acids in such a mixture. Since as many as 22 different amino acids may be present, this could be a formidable task if we are restricted to conventional methods.

Fortunately, techniques have been developed, based on the principle of elution chromatography, that simplify this problem immensely and even allow its solution to be automated. Automatic amino acid analyzers were developed at the Rockefeller Institute in 1950 and have since become commercially available. They are based on the use of insoluble polymers containing sulfonate groups, called *cation-exchange resins* (Fig. 22.1).

If an acidic solution containing a mixture of amino acids is passed through a column packed with a cation-exchange resin the amino acids will be adsorbed by the resin because of attractive forces between the negatively charged sulfonate groups and the positively charged amino acids. The strength of the adsorption will vary with the basicity of the individual amino acids; those that are most basic will be held most strongly. If the column is then washed with a buffered solution at a

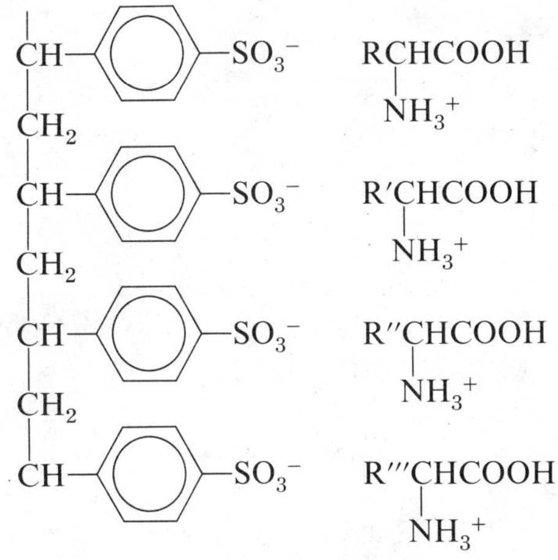

$$
\begin{array}{ll}
-CH-C_6H_4-SO_3^- & RCH(NH_3^+)COOH \\
CH_2 & \\
-CH-C_6H_4-SO_3^- & R'CH(NH_3^+)COOH \\
CH_2 & \\
-CH-C_6H_4-SO_3^- & R''CH(NH_3^+)COOH \\
CH_2 & \\
-CH-C_6H_4-SO_3^- & R'''CH(NH_3^+)COOH
\end{array}
$$

FIG. 22.1
A section of a cation-exchange resin with adsorbed amino acids.

given pH the individual amino acids will move down the column at different rates and ultimately become separated. At the end of the column the eluate is allowed to mix with *ninhydrin,* a reagent that reacts with the amino acids to give derivatives with an intense purple color (λ_{max} 570 nm). The amino acid analyzer is designed so that it can measure the absorbance of the eluate (at 570 nm) continuously and record this absorbance as a function of the volume of the effluent.

A typical graph obtained from an automatic amino acid analyzer is shown in Fig. 22.2. When the procedure is standardized, the positions of the peaks are characteristic of the individual amino acids and the areas under the peaks correspond to their relative amounts.

22.6 AMINO ACID SEQUENCE OF PROTEINS AND POLYPEPTIDES

Once we have determined the amino acid composition of a protein or a polypeptide we should then determine its molecular weight. A variety of methods are available for doing this, including chemical methods, ultracentrifugation, light scattering, osmotic pressure, and X-ray diffraction. With the molecular weight and amino acid composition we will now be able to calculate the "molecular formula" of the protein; that is, we will know how many of each type of amino acid are present as *amino acid residues* (i.e., $RCH(NH-)CO-$) in each protein molecule. Unfortunately, however, we have only begun our task of determining its structure. The next step is a formidable one, indeed. We must determine the order in which the amino acids are connected; that is, we must determine the *covalent structure of the polyamide.*

A simple tripeptide composed of 3 different amino acids can have 6 different amino acid sequences; a tetrapeptide composed of 4 different amino acids can have as many as 24. For a protein with a molecular weight of 10,000 or more, composed of 20 different amino acids, the number of possibilities approaches infinity.

In spite of this, a number of methods have been developed that allow the amino acid sequences to be determined and these, as we shall see, have been applied with amazing success. In our discussion here we will limit our attention to two methods that illustrate how sequence determinations can be done: *terminal residue analysis* and *partial hydrolysis.*

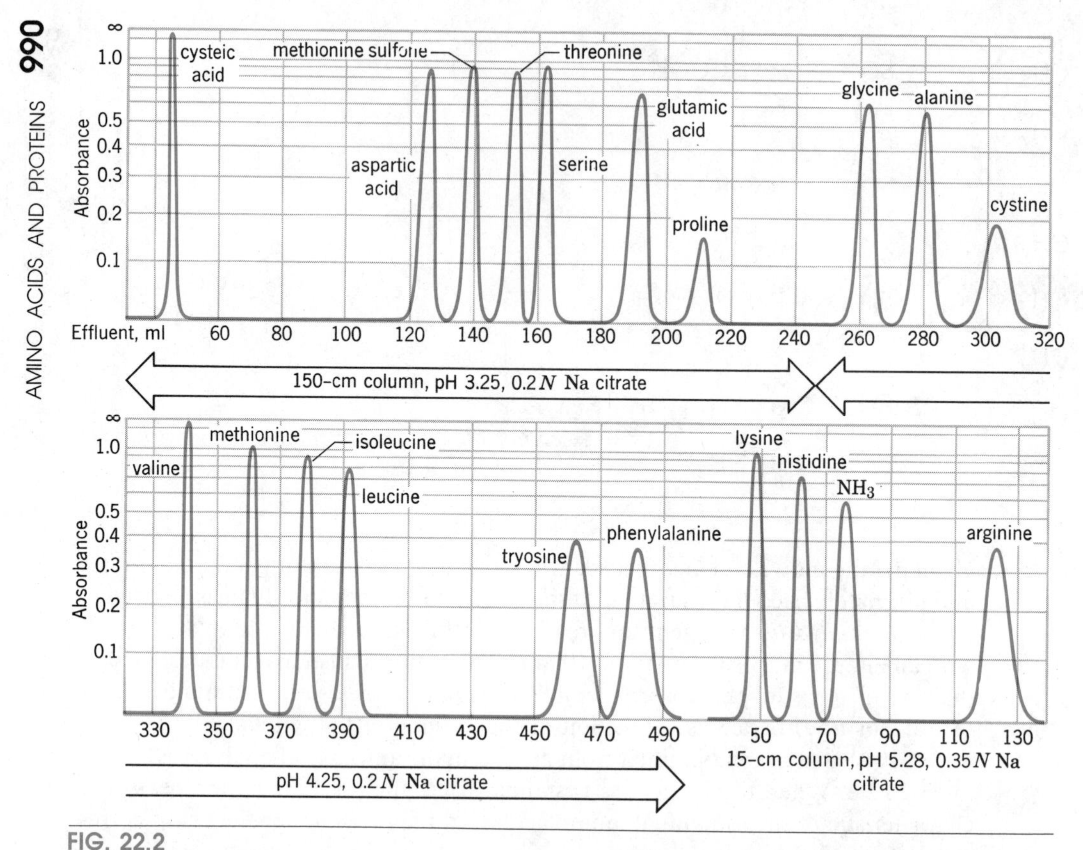

FIG. 22.2
Typical result given by an automatic amino acid analyzer. [*Reprinted with permission from D. H. Spackman, W. H. Stein, and S. Moore,* Anal. Chem. 30, *1190* (*1958*). *Copyright by the American Chemical Society.*]

22.6A Terminal Residue Analysis

One of the ends of a polypeptide chain terminates in an amino acid residue that has a free —NH_2 group; the other terminates in an amino acid residue with a free —COOH group. These two amino acids are called the *N-terminal* residue and the *C-terminal* residue, respectively.

$$\underbrace{H_2N\text{—}\underset{\displaystyle R}{\underset{|}{C}}HCO}_{N\text{-Terminal residue}} \left(NH\underset{\displaystyle R}{\underset{|}{C}}HCO \right) \underbrace{NH\underset{\displaystyle R}{\underset{|}{C}}HCOOH}_{C\text{-Terminal residue}}$$

One very useful method for determining the *N*-terminal amino acid residue, called the *Sanger method,* is based on the use of 2,4-dinitrofluorobenzene (DNFB).* When a polypeptide is treated with DNFB in mildly basic solution, a nucleophilic aromatic substitution reaction takes place involving the free amino group of the *N*-terminal residue. Subsequent hydrolysis of the polypeptide gives a mixture of amino acids in which the *N*-terminal amino acid bears a label, *the 2,4-dinitrophenyl group*. As a result, after separating this amino acid from the mixture, we can identify it.

* This method was introduced by Frederick Sanger of Cambridge University in 1945. Sanger made extensive use of this procedure in his determination of the amino acid sequence of insulin and won a Nobel Prize for the work in 1958.

$$O_2N\text{-}C_6H_3(NO_2)\text{-}F + H_2\ddot{N}CHCO\text{—}NHCHCO\sim \text{ etc.} \xrightarrow{HCO_3^-}$$
(R on the first CH, R′ on the second CH)

2,4-Dinitrofluorobenzene (DNFB) — Polypeptide

$$O_2N\text{-}C_6H_3(NO_2)\text{—}NHCHCO\text{—}NHCHCO\sim \xrightarrow{H_3O^+}$$
(R, R′)

Labeled polypeptide

$$O_2N\text{-}C_6H_3(NO_2)\text{—}NHCHCOOH + H_3\overset{+}{N}CHCOO^-$$
(R, R′)

Labeled *N*-terminal amino acid — Mixture of amino acids

Separate and identify

Problem 22.5

The electron-withdrawing property of the 2,4-dinitrophenyl group makes separation of the labeled amino acid very easy. Suggest how this is done.

Of course, 2,4-dinitrofluorobenzene will react with any free amino group present in a polypeptide including the ε-amino group of lysine. But, only the *N*-terminal amino acid residue will bear the label at the α-amino group.

A second method of *N*-terminal analysis is the *Edman degradation* (developed by Pehr Edman of the University of Lund, Sweden). This method offers an advantage over the Sanger method in that it removes the *N*-terminal residue and leaves the remainder of the peptide chain intact. The Edman degradation is based on a labeling reaction between the *N*-terminal amino group and phenyl isothiocyanate, $C_6H_5N{=}C{=}S$ (below). When the labeled polypeptide is treated with acid, the *N*-terminal amino acid residue splits off as a phenylthiohydantoin. The product can be identified by comparison with phenylthiohydantoins prepared from standard amino acids.

$$C_6H_5\text{—}N{=}C{=}S + H_2\ddot{N}CHCO\text{—}NHCHCO\sim \text{ etc.} \xrightarrow{OH^-,\ pH\ 9}$$
(R, R′)

$$C_6H_5\text{—}\ddot{N}H\text{—}\overset{S}{\overset{\|}{C}}\text{—}NHCHCO\text{—}NHCHCO\sim \text{ etc.} \xrightarrow{H^+}$$
(R, R′)

Labeled polypeptide

Phenylthiohydantoin (ring: C₆H₅–N, C=S, NH, CH–R, C=O) + $H_2NCHCO\sim$ (R′)

Phenylthiohydantoin — Polypeptide with one less amino acid residue

The polypeptide that remains after the first Edman degradation can be submitted to another degradation to identify the next amino acid in the sequence and this process has even been automated. Unfortunately, Edman degradations cannot be repeated indefinitely. As residues are successively removed, amino acids formed by hydrolysis during the acid treatment accumulate in the reaction mixture and interfere with the procedure. The Edman degradation, however, has been successfully applied to polypeptides with as many as 50 amino acid residues.

C-terminal residues can be identified through the use of digestive enzymes called *carboxypeptidases*. These enzymes specifically catalyze the hydrolysis of the amide bond of the amino acid residue containing a free —COOH group, liberating it as a free amino acid. A carboxypeptidase, however, will continue to attack the polypeptide chain that remains, successively lopping off *C*-terminal residues. As a consequence, it is necessary to follow the amino acids released as a function of time. The procedure can only be applied to a limited amino acid sequence, for, at best, after a time the situation becomes too confused to sort out.

Consider the polypeptide sequences shown in Fig. 22.3. Carboxypeptidase-catalyzed hydrolysis might give a result like that shown in Fig. 22.3*a* when the amino acids are released at the same rate. If the amino acids are released at different rates, however, we might obtain a result such as that shown in Fig. 22.3*b*; in this case we would even have difficulty in deciding which amino acid was the original *C*-terminal amino acid.

Problem 22.6

(a) Write a reaction showing how 2,4-dinitrofluorobenzene could be used to identify the *N*-terminal amino acid of Val · Ala · Gly. (b) What products would you expect (after hydrolysis) when Val · Lys · Gly is treated with 2,4-dinitrofluorobenzene?

Problem 22.7

Write the reactions involved in a sequential Edman degradation of Met · Ile · Arg.

22.6B Partial Hydrolysis

Sequential analysis using the Edman degradation or carboxypeptidase becomes impractical with proteins or polypeptides of appreciable size. Fortunately,

FIG. 22.3
The release by carboxypeptidase of amino acids from a peptide having the C-terminal sequence shown. (a) *All bonds cleaved at the same rate.* (b) *Ser removed slowly; Tyr cleaved off rapidly; Leu cleaved off at the same rate as Tyr.* (*From Robert Barker,* Organic Chemistry of Biological Compounds, *© 1971. Reprinted by permission of Prentice-Hall, Inc.*

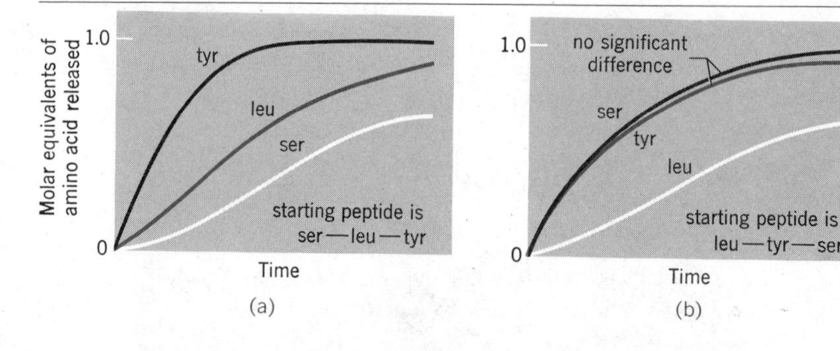

however, we can resort to another technique; that of *partial hydrolysis*. Using dilute acids or enzymes we attempt to break the polypeptide chain into small fragments, ones that we can identify using DNFB or the Edman degradation. Then we examine the structures of these smaller fragments looking for points of overlap and attempt to piece together the amino acid sequence of the original polypeptide.

Consider a simple example: we are given a pentapeptide known to contain valine (two residues), leucine (one residue), histidine (one residue) and phenylalanine (one residue). With this information, we can write the "molecular formula" of the protein in the following way, using commas to indicate that the sequence is unknown.

$$\text{Val}_2\text{, Leu, His, Phe}$$

Then let us assume that by using 2,4-dinitrofluorobenzene and carboxypeptidase we discover that valine and leucine are the *N*-terminal and *C*-terminal residues, respectively. This gives us:

Val(Val, His, Phe) Leu

But, the sequence of the three nonterminal amino acids is still unknown.

We then subject the pentapeptide to partial acid hydrolysis and obtain the following dipeptides.

Val · His + His · Val + Val · Phe + Phe · Leu

The points of overlap (i.e., His, Val, and Phe) tell us that the original pentapeptide must have been:

Val · His · Val · Phe · Leu

Two enzymes are also frequently used to cause partial hydrolyses. *Trypsin* preferentially catalyzes the hydrolysis of peptide bonds in which the carboxyl group is a part of a lysine or arginine residue. *Chymotrypsin* preferentially catalyzes the hydrolysis of peptide bonds at the carboxyl groups of phenylalanine, tyrosine, and tryptophan. It will also attack the peptide bonds at the carboxyl groups of leucine, methionine, asparagine, and glutamine.

Problem 22.8

Glutathione is a tripeptide found in most living cells. Partial acid-catalyzed hydrolysis of glutathione yields two dipeptides, Cys · Gly and one composed of Glu and Cys. When this second dipeptide was treated with DNFB, acid hydrolysis gave *N*-labeled Glu. (a) Based on this information alone, what structures are possible for glutathione? (b) Synthetic experiments have shown that the second dipeptide has the following structure.

$$\underset{\displaystyle \text{COO}^-}{\text{H}_3\overset{+}{\text{N}}\text{CH}}\text{CH}_2\text{CH}_2\text{CONH}\underset{\displaystyle \text{CH}_2\text{SH}}{\text{CH}}\text{COO}^-$$

What is the structure of glutathione?

Problem 22.9

Give the amino acid sequence of the following polypeptides using only the data given by partial acidic hydrolysis.

(a) Ser, Hyp, Pro, Thr $\xrightarrow[H_2O]{H^+}$ Ser · Thr · + Thr · Hyp + Pro · Ser

(b) Ala, Arg, Cys, Val, Leu $\xrightarrow[H_2O]{H^+}$

Ala · Cys · + Cys · Arg + Arg · Val + Leu · Ala

22.7 PRIMARY STRUCTURES OF POLYPEPTIDES AND PROTEINS

The covalent structure of a protein or polypeptide is called its *primary structure.* By using the techniques we described in the previous sections, chemists have had remarkable success in determining the primary structures of polypeptides and proteins. The compounds described in the following pages are important examples.

22.7A Oxytocin and Vasopressin

Oxytocin and vasopressin (Fig. 22.4) are two rather small polypeptides with strikingly similar structures (where oxytocin has leucine, vasopressin has arginine and where oxytocin has isoleucine, vasopressin has phenylalanine). In spite of the similarity of their amino acid sequences these two polypeptides have quite different physiological effects. Oxytocin occurs only in the female of a species and stimulates uterine contractions during childbirth. Vasopressin occurs in males and females; it causes contraction of peripheral blood vessels and an increase in blood pressure. Its major function, however, is as an *antidiuretic;* physiologists often refer to vasopressin as *antidiuretic hormone.*

The structures of oxytocin and vasopressin also illustrate the importance of the disulfide linkage between cysteine residues in the overall primary structure of a polypeptide. In these two molecules this disulfide linkage leads to a cyclic structure.*

Problem 22.10

Treating oxytocin with certain reducing agents (e.g., sodium in liquid ammonia) brings about a single chemical change that can be reversed by air oxidation. What chemical changes are involved?

22.7B Insulin

Insulin, a hormone secreted by the pancreas, regulates glucose metabolism. Insulin deficiency in humans is the major problem in diabetes mellitus.

The amino acid sequence of bovine insulin (Fig. 22.5) was determined by Sanger in 1953. Bovine insulin has a total of 51 amino acid residues in two polypeptide chains, called the A and B chains. These chains are joined by two

* Vincent du Vigneaud of Cornell Medical College synthesized oxytocin and vasopressin in 1953; he received the Nobel Prize in 1955.

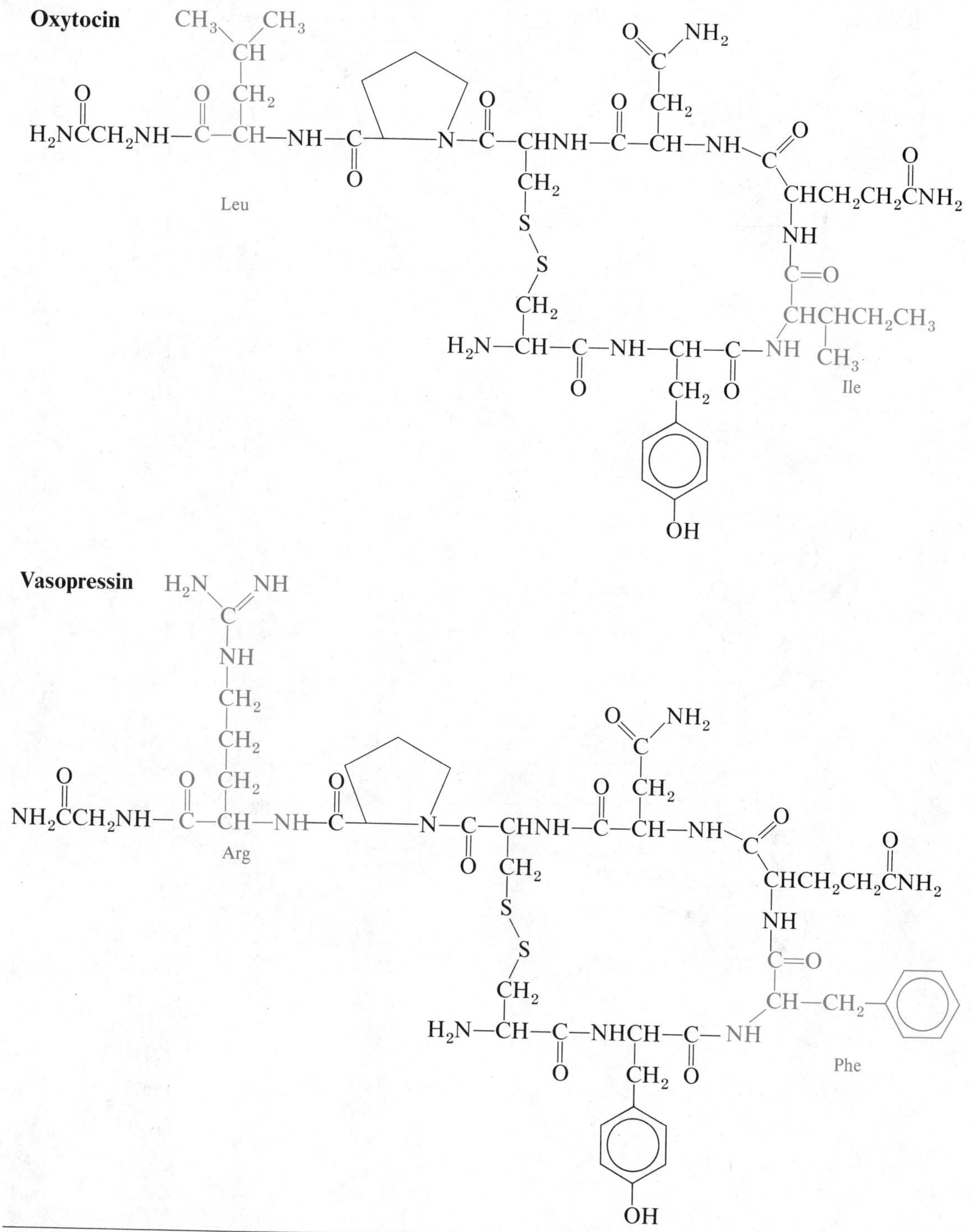

Oxytocin

Vasopressin

FIG. 22.4

The structures of oxytocin and vasopressin.

disulfide linkages. The A chain contains an additional disulfide linkage between cysteine residues at positions 6 and 11.

Since Sanger's determination of the amino acid sequence of bovine insulin, the amino acid sequences of insulin from a number of species have become known.

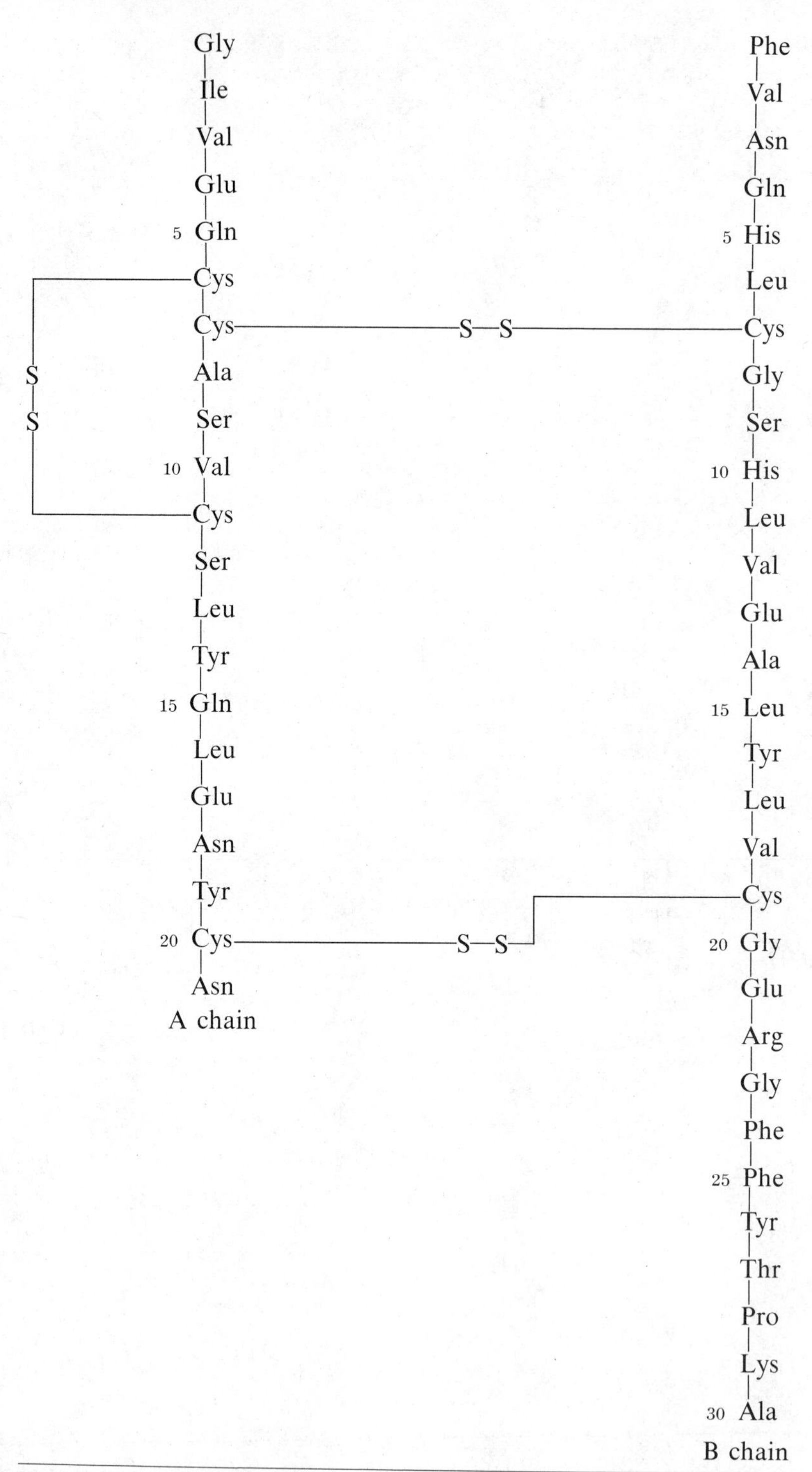

FIG. 22.5

The amino acid sequence of bovine insulin. (From A. L. Lehninger, Biochemistry, *Worth Publishers, Inc., New York, 1970, p. 92. Used with permission.)*

Those from vertebrates all have an A chain of 21 residues and a B chain of 30. The A chains of insulins from humans, pigs, dogs, rabbits, and sperm whales are the same. The B chains of cows, pigs, dogs, goats, sperm whales, and horses are identical. The B chains of insulins from humans and elephants are also identical.

The principal differences in the insulin A chains occur at positions 8, 9, and 10. In humans these are,

Thr · Ser · Ile

in cows,

Ala · Ser · Val

and in sheep,

Ala · Gly · Val

Similar studies have been carried out with other proteins. These have been very useful in constructing a phylogenetic tree that allows an estimation to be made of the probable times of divergence of the major genera and species of living organisms. As we might expect, the number of amino acid residue differences is proportional to their phylogenetic differences and to the time elapsed since divergence. One protein that has been studied extensively in this regard is *cytochrome c,* a protein important in biological oxidations. Cytochrome c isolated from monkeys shows only one different residue from that obtained from humans; that obtained from horses, however, differs by 12 residues. Cytochrome-c molecules isolated from vertebrates generally differ by as many as 43–48 residues from those isolated from a microorganism such as yeast.

22.7C Other Polypeptides and Proteins

Successful sequential analyses have been achieved with a number of other polypeptides and proteins including:

1. Bovine *ribonuclease.* This enzyme, which catalyzes the hydrolysis of ribonucleic acids (Special Topic P), has a single chain of 124 amino acid residues and four intrachain disulfide linkages.
2. Human *hemoglobin.* There are four peptide chains in this important oxygen-carrying protein. Two identical α chains have 141 residues each, and two identical β chains have 146 residues. The genetically based disease, sickle cell anemia, results from a single amino acid error in the β chain. In normal hemoglobin, position -6 has a glutamic acid residue, while in sickle cell hemoglobin position -6 is occupied by valine.

Red blood cells (erythrocytes) containing hemoglobin with this amino acid residue error tend to become crescent shaped ("sickle") when the partial pressure of oxygen is low, as it is in venous blood. These distorted cells are more difficult for the heart to pump through small capillaries. They may even block capillaries by clumping together; at other times the red cells may even split open.

Children who inherit this genetic trait from both parents suffer from a severe form of the disease and usually do not live past the age of two. Children who inherit the disease from only one parent generally have a much milder form.

Sickle cell anemia arose among the populations of central and western Africa where, ironically, it may have had a beneficial effect. People with a mild form of the disease are far less susceptible to malaria than those with normal hemoglobin. Malaria, a disease caused by an infectious microorganism, is especially prevalent in central and western Africa.

Mutational changes such as those that give rise to sickle cell anemia are very common. Approximately 150 different types of mutant hemoglobin have been detected in humans; fortunately, most are harmless.

3. Bovine *trypsinogen* and *chymotrypsinogen*. These two enzyme precursors have single chains of 229 and 245 residues, respectively.
4. *Gamma globulin*. This immunoprotein has a total of 1320 amino acid residues in four chains. Two chains have 214 residues each; the other two have 446.

22.8 PROTEIN AND POLYPEPTIDE SYNTHESIS

We saw in Chapter 17 that the synthesis of an amide linkage is a relatively simple one. We much first "activate" the carboxyl group of an acid by converting it to an anhydride or acid chloride and then allow it to react with an amine:

$$\underset{\text{Anhydride}}{R{-}\overset{O}{\overset{\|}{C}}{-}O{-}\overset{O}{\overset{\|}{C}}{-}R} + \underset{\text{Amine}}{R'{-}NH_2} \longrightarrow \underset{\text{Amide}}{R{-}\overset{O}{\overset{\|}{C}}{-}NHR'} + R{-}COOH$$

The problem becomes somewhat more complicated, however, when both the acid group and the amino group are present in the same molecule as they are in an amino acid and, especially, when our goal is the synthesis of a naturally occurring polyamide where the sequence of different amino acids is all-important. Let us consider, as an example, the synthesis of the simple dipeptide alanylglycine, Ala · Gly. We might first activate the carboxyl group of alanine by converting it to an acid chloride, and then we might allow it to react with glycine. Unfortunately, however, we cannot prevent alanyl chloride from reacting with itself. So our reaction would yield not only Ala · Gly but also Ala · Ala. It could also lead to Ala · Ala · Ala and Ala · Ala · Gly, and so on. The yield of our desired product would be low, and we would also have a difficult problem separating the dipeptides, tripeptides, and so on.

$$\underset{\text{Ala}}{CH_3\underset{^+NH_3}{\underset{|}{C}}H\overset{O}{\overset{\|}{C}}O^-} \xrightarrow[\text{(2) } H_3\overset{+}{N}CH_2COO^- \ (\text{Gly})]{\text{(1) } SOCl_2} \underset{\text{Ala}\cdot\text{Gly}}{CH_3\underset{\overset{+}{N}H_3}{\underset{|}{C}}H\overset{O}{\overset{\|}{C}}NHCH_2COO^-} + \underset{\text{Ala}\cdot\text{Ala}}{CH_3\underset{\overset{+}{N}H_3}{\underset{|}{C}}H\overset{O}{\overset{\|}{C}}NH\underset{CH_3}{\underset{|}{C}}HCOO^-} +$$

$$\underset{\text{Ala}\cdot\text{Ala}\cdot\text{Ala}}{CH_3\underset{\overset{+}{N}H_3}{\underset{|}{C}}H\overset{O}{\overset{\|}{C}}NH\underset{CH_3}{\underset{|}{C}}H\overset{O}{\overset{\|}{C}}NH\underset{CH_3}{\underset{|}{C}}HCOO^-} + \underset{\text{Ala}\cdot\text{Ala}\cdot\text{Gly}}{CH_3\underset{\overset{+}{N}H_3}{\underset{|}{C}}H\overset{O}{\overset{\|}{C}}NH\underset{CH_3}{\underset{|}{C}}H\overset{O}{\overset{\|}{C}}NHCH_2COO^-}$$

22.8A Protecting Groups

The solution to this problem is to "protect" the amino group of the first amino acid before we activate it and allow it to react with the second. By protecting the amino group we mean that we must convert it to some other group of low nucleophilicity—*one that will not react with a reactive acyl derivative*. The protecting group must be carefully chosen because after we have synthesized the amide linkage between the first amino acid and the second we will want to be able to remove the protecting group without disturbing the new amide bond.

A number of reagents have been developed to meet these requirements. Two that are often used are *benzyl chloroformate* and *tert-butyloxycarbonyl azide.*

$$C_6H_5CH_2{-}O{-}\overset{O}{\overset{\|}{C}}{-}Cl \qquad (CH_3)_3C{-}O{-}\overset{O}{\overset{\|}{C}}{-}N_3$$

Benzyl chloroformate — *tert*-Butyloxycarbonyl azide

Both reagents react with amino groups (below) to form derivatives that are unreactive toward further acylation. Both derivatives, however, are of a type that allow removal of the protecting group under conditions that do not affect peptide bonds. The benzyloxycarbonyl group (abbreviated Z-) can be removed by catalytic hydrogenation or by treating the derivative with cold HBr in acetic acid. The *tert*-butyloxycarbonyl group (abbreviated Boc-) can be removed through treatment with HCl or CF_3COOH in acetic acid.

Benzyloxycarbonyl Group

$$C_6H_5{-}CH_2O\overset{O}{\overset{\|}{C}}{-}Cl + H_2N{-}R \xrightarrow[25°]{OH^-} C_6H_5{-}CH_2O\overset{O}{\overset{\|}{C}}{-}NH{-}R + Cl^-$$

Benzyl chloroformate — Benzyloxycarbonyl or Z group

HBr, acetic acid (cold):

$$C_6H_5{-}CH_2Br + CO_2 + H_2N{-}R$$

H_2—Pd:

$$C_6H_5{-}CH_3 + CO_2 + H_2N{-}R$$

$$(CH_3)_3C{-}O\overset{O}{\overset{\|}{C}}{-}N_3 + H_2N{-}R \xrightarrow[25°]{base} (CH_3)_3C{-}O\overset{O}{\overset{\|}{C}}{-}NHR + N_3^-$$

tert-Butyloxycarbonyl azide — *tert*-Butyloxycarbonyl or Boc group

HCl or CF_3COOH in acetic acid, 25°:

$$(CH_3)_2C{=}CH_2 + CO_2 + H_2N{-}R$$

The easy removal of both groups (Z- and Boc-) in acidic media results from the exceptional stability of the carbocations that are formed initially. The benzyloxycarbonyl group gives a *benzyl cation;* the *tert*-butyloxycarbonyl group yields, initially, a *tert-butyl cation.*

Removal of the benzyloxycarbonyl group with hydrogen and a catalyst depends on the fact that benzyl-oxygen bonds are weak and are subject to hydrogenolysis at low temperatures.

$$C_6H_5CH_2{-}O\overset{O}{\overset{\|}{C}}R \xrightarrow[25°]{H_2,Pd} C_6H_5CH_3 + HO\overset{O}{\overset{\|}{C}}R$$

A benzyl ester

22.8B Activation of the Carboxyl Group

Perhaps the most obvious way to activate a carboxyl group is to convert it to an acyl chloride. This method was used in early peptide syntheses, but acyl chlorides are actually more reactive than necessary. As a result, their use leads to complicating side reactions. A much better method is to convert the carboxyl group of the "protected" amino acid to a mixed anhydride using ethyl chloroformate, $Cl-\overset{O}{\overset{\|}{C}}-OC_2H_5$

$$Z-NH\underset{R}{\underset{|}{C}}H\overset{O}{\overset{\|}{C}}-OH \xrightarrow[(2)\ ClCO_2C_2H_5]{(1)\ (C_2H_5)_3N} Z-NH\underset{R}{\underset{|}{C}}H-\overset{O}{\overset{\|}{C}}-O-\overset{O}{\overset{\|}{C}}-OC_2H_5$$

"Mixed anhydride"

The mixed anhydride can then be used to acylate another amino acid and form a peptide linkage.

$$Z-NH\underset{R}{\underset{|}{C}}H\overset{O}{\overset{\|}{C}}-O-\overset{O}{\overset{\|}{C}}OC_2H_5 \xrightarrow{H_3\overset{+}{N}-\underset{R'}{\underset{|}{C}}HCOO^-}$$

$$Z-NH\underset{R}{\underset{|}{C}}H\overset{O}{\overset{\|}{C}}-NH\underset{R'}{\underset{|}{C}}HCOOH + CO_2 + C_2H_5OH$$

22.8C Peptide Synthesis

Let us examine now how we might use these reagents in the preparation of the simple dipeptide, Ala · Leu. The principles involved here can, of course, be extended to the synthesis of much longer polypeptide chains.

$$\underset{\text{Ala}}{CH_3\underset{\overset{+}{N}H_3}{\underset{|}{C}}HCOO^-} + \underset{\text{Benzyl chloroformate}}{C_6H_5CH_2O\overset{O}{\overset{\|}{C}}-Cl} \xrightarrow[25^\circ]{OH^-} \underset{\text{Z-Ala}}{CH_3\underset{NH-C(=O)-OCH_2C_6H_5}{\underset{|}{C}}H-COOH} \xrightarrow[(2)\ ClCO_2C_2H_5]{(1)\ (C_2H_5)_3N}$$

$$\underset{\text{Mixed anhydride of Z-Ala}}{CH_3\underset{NH-C(=O)-OCH_2C_6H_5}{\underset{|}{C}}H-\overset{O}{\overset{\|}{C}}-O\overset{O}{\overset{\|}{C}}OC_2H_5} \xrightarrow{(CH_3)_2CHCH_2\underset{\text{Leu}}{\overset{\overset{+}{N}H_3}{\overset{|}{C}}}HCOO^-} \quad (\to CO_2 + C_2H_5OH)$$

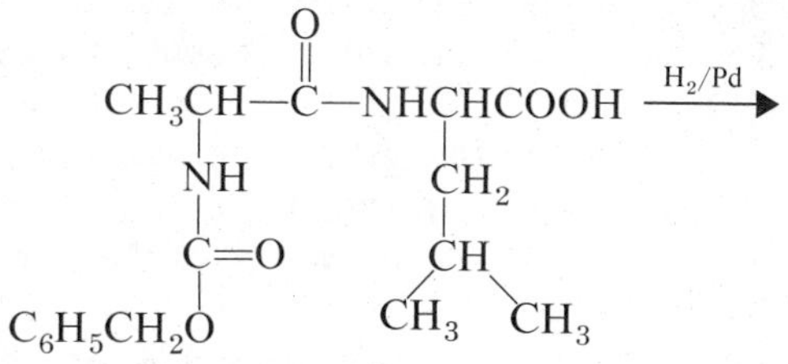

$$\underset{\text{Z-Ala} \cdot \text{Leu}}{\underset{C_6H_5CH_2O}{\overset{\displaystyle CH_3CH-\overset{\overset{O}{\|}}{C}-NHCHCOOH}{\underset{NH-C{=}O}{|}}}} \quad (\text{side chain } CH_2CH(CH_3)_2) \xrightarrow{H_2/Pd}$$

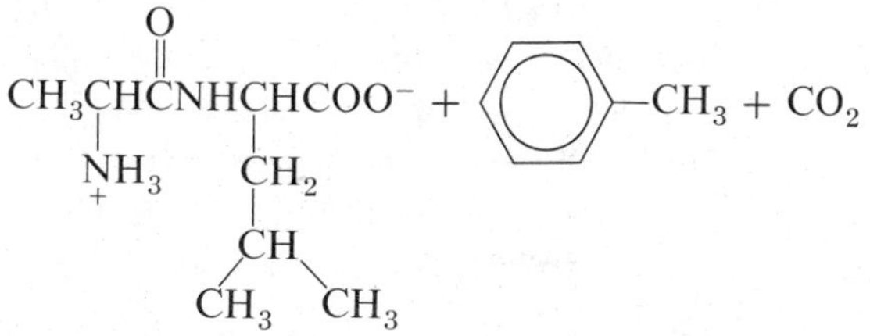

$$\underset{\text{Ala} \cdot \text{Leu}}{CH_3\underset{\overset{+}{N}H_3}{CH}\overset{\overset{O}{\|}}{C}NH\underset{CH_2CH(CH_3)_2}{CH}COO^-} + C_6H_5{-}CH_3 + CO_2$$

All of these reactions can be carried out at room temperature or below, and they avoid strongly acidic or basic conditions. This prevents racemization at the chiral carbons and also prevents hydrolysis.

Problem 22.11

Show all steps in the synthesis of Gly · Val · Ala using the *tert*-butyloxycarbonyl (Boc-) group as a protecting group.

Problem 22.12

The synthesis of a polypeptide containing lysine requires the protection of both amino groups. (a) Show how you might do this in a synthesis of Lys · Ile using the benzyloxycarbonyl group as a protecting group. (b) The benzyloxycarbonyl group can also be used to protect the guanidino group, $-NH\overset{\overset{NH}{\|}}{C}-NH_2$, of arginine. Show a synthesis of Arg · Ala.

Problem 22.13

The terminal carboxyl groups of glutamic acid and aspartic acid are often protected through their conversion to benzyl esters. What mild method could be used for removal of this protecting group?

22.8D Automated Peptide Synthesis

Although the methods that we have described thus far have been used to synthesize a number of polypeptides including ones as large as insulin, they are extremely time consuming. One must isolate and purify the product at almost every stage. Thus, a real advance in peptide synthesis came with the development by R. B. Merrifield (at Rockefeller University) of a procedure for automating peptide synthesis.

The Merrifield method is based on the use of a polystyrene resin similar to the one we saw on p. 989, *but one that contains* $-CH_2Cl$ *groups* instead of sulfonic acid groups. This resin is used in the form of small beads and is insoluble in most solvents.

The first step in automated peptide synthesis (Fig. 22.6) involves a reaction that attaches the first protected amino acid residue to the resin beads. After this step is complete, the protecting group is removed and the next amino acid (also protected) is condensed with the first using dicyclohexylcarbodiimide (p. 778) to activate its carboxyl group. Then removal of the protecting group of the second residue readies the resin-dipeptide for the next step.

The great advantage of this procedure is that purification of the resin with its attached polypeptide can be carried out at each stage by simply washing the resin with an appropriate solvent. Impurities, because they are not attached to the insoluble resin, are simply carried away by the solvent. In the automated procedure each cycle of the "protein-making machine" requires only four hours and attaches one new amino acid residue.*

The Merrifield technique has been applied successfully to the synthesis of ribonuclease, a protein with 124 amino acid residues. The synthesis involved 369

* Protein synthesis by enzymes as directed by DNA/RNA takes only one minute to add 150 amino acids in a specific sequence (cf. Section P.5).

FIG. 22.6
The Merrifield method for automated protein synthesis.

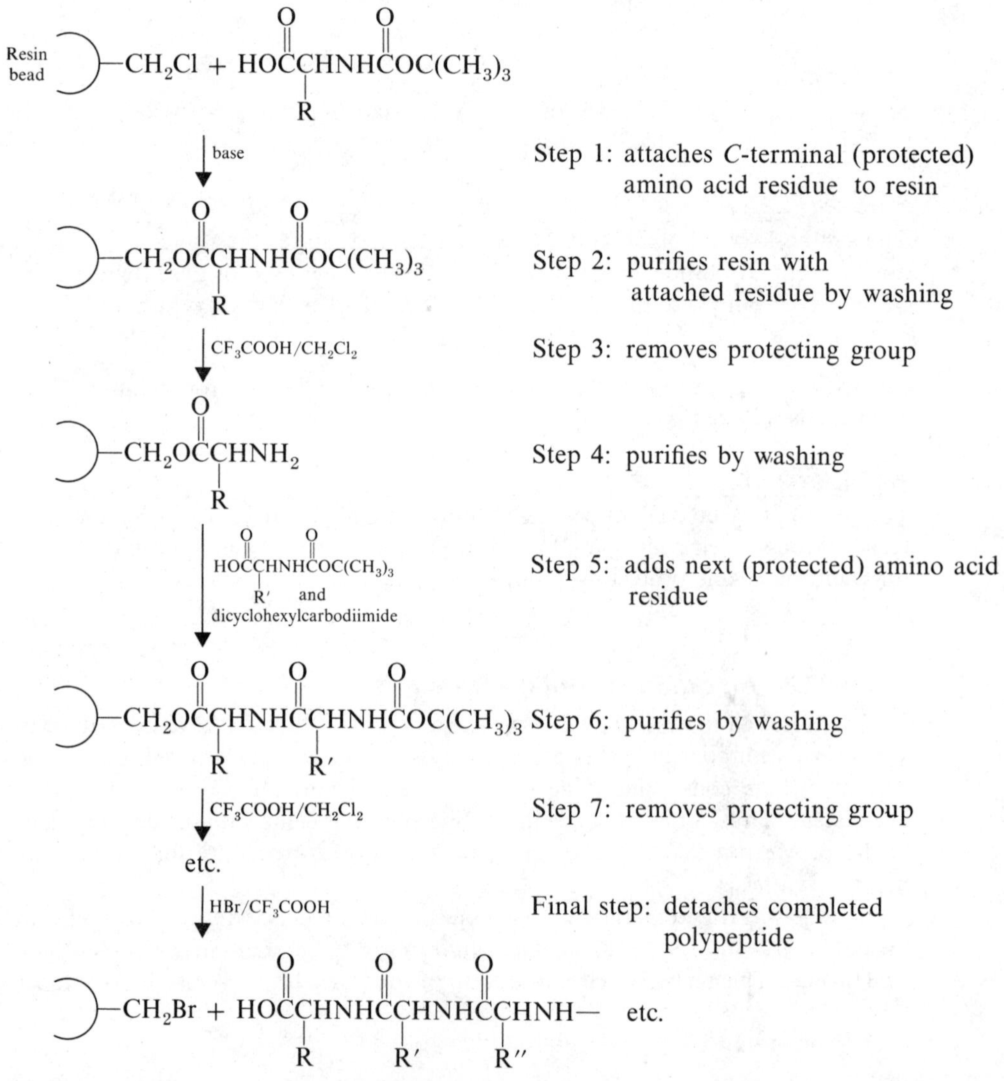

Resin bead ⟩–CH_2Cl + $HOC(=O)CH(R)NHC(=O)OC(CH_3)_3$

↓ base — Step 1: attaches *C*-terminal (protected) amino acid residue to resin

⟩–$CH_2OC(=O)CH(R)NHC(=O)OC(CH_3)_3$ — Step 2: purifies resin with attached residue by washing

↓ CF_3COOH/CH_2Cl_2 — Step 3: removes protecting group

⟩–$CH_2OC(=O)CH(R)NH_2$ — Step 4: purifies by washing

↓ $HOC(=O)CH(R')NHC(=O)OC(CH_3)_3$ and dicyclohexylcarbodiimide — Step 5: adds next (protected) amino acid residue

⟩–$CH_2OC(=O)CH(R)NHC(=O)CH(R')NHC(=O)OC(CH_3)_3$ — Step 6: purifies by washing

↓ CF_3COOH/CH_2Cl_2 — Step 7: removes protecting group

etc.

↓ HBr/CF_3COOH — Final step: detaches completed polypeptide

⟩–CH_2Br + $HOC(=O)CH(R)NHC(=O)CH(R')NHC(=O)CH(R'')NH$– etc.

chemical reactions and 11,931 automated steps—all were carried out without isolating an intermediate. The synthetic ribonuclease not only had the same physical characteristics of the natural enzyme; it possessed the biological activity as well.

Problem 22.14

The resin for the Merrifield procedure is prepared by treating polystyrene, $(CH_2CH)_n$ (with C_6H_5 attached to CH), with CH_3OCH_2Cl and a Lewis acid catalyst. (a) What reaction is involved? (b) After purification, the completed polypeptide or protein can be detached from the resin by treating it with HBr in trifluoroacetic acid under conditions mild enough not to affect the amide linkages. What structural feature of the resin makes this possible?

Problem 22.15

Outline the steps in the synthesis of Lys · Phe · Ala using the Merrifield procedure.

22.9 SECONDARY AND TERTIARY STRUCTURE OF PROTEINS

We have seen how amide and disulfide linkages constitute the covalent or *primary structure* of proteins. Of equal importance in understanding how proteins function is knowledge of the way in which the peptide chains are arranged in three dimensions. Involved here are the secondary and tertiary structures of proteins.

22.9A Secondary Structure

The major experimental technique that has been used in elucidating the secondary structures of proteins is X-ray analysis.

When X-rays pass through a crystalline substance they produce diffraction patterns. Analysis of these patterns indicates a regular repetition of particular structural units with certain specific distances between them, called *repeat distances.* The complete X-ray analysis of a molecule as complex as a protein often takes years of painstaking work. Nonetheless, many X-ray analyses have been done and they have revealed that the polypeptide chain of a natural protein can interact with itself in two major ways; through formation of a *pleated sheet* and an *α helix.**

To understand how these interactions occur let us look first at what X-ray analysis has revealed about the geometry at the peptide bond itself. Peptide bonds tend to assume a geometry such that six atoms of the amide linkage are coplanar (Fig. 22.7). The carbon-nitrogen bond of the amide linkage is unusually short, indicating that resonance contributions of the type shown below are important.

$$\ddot{N}-C(=O) \longleftrightarrow \overset{+}{N}=C(-O^-)$$

* Pioneers in the X-ray analysis of proteins were two American scientists, Linus Pauling (p. 667) and Robert B. Corey. Beginning in 1939, Pauling and Corey initiated a long series of studies of the conformations of peptide chains. At first they used crystals of single amino acids, then dipeptides and tripeptides, and so on. Moving on to larger and larger molecules and using the precisely constructed molecular models, they were able to understand the secondary structures of proteins for the first time.

The carbon-nitrogen bond, consequently, has considerable double bond character (~40%) and rotations of groups about this bond are severely hindered.

Rotations of groups attached to the amide nitrogen and the carbonyl carbon are relatively free, however, and these rotations allow peptide chains to form different conformations.

The transoid arrangement of groups around the relatively rigid amide bond would cause the R groups to alternate from side to side of a single fully extended peptide chain:

Calculations show that such a polypeptide chain would have a repeat distance (i.e., distance between alternating units) of 7.2 Å.

Fully extended polypeptide chains could conceivably form a flat-sheet structure with each alternating amino acid in each chain forming two hydrogen bonds with an amino acid in the adjacent chain:

hypothetical flat–sheet structure

This structure does not exist in naturally occurring proteins because of the crowding that would exist between R groups. If such a structure did exist, it would have the same repeat distance as the fully extended peptide chain, that is, 7.2 Å.

Slight rotations of bonds, however, can transform a flat-sheet structure into what is called the *pleated-sheet* or *β configuration*. The pleated-sheet structure gives small- and medium-sized R groups room enough to avoid van der Waals repulsions

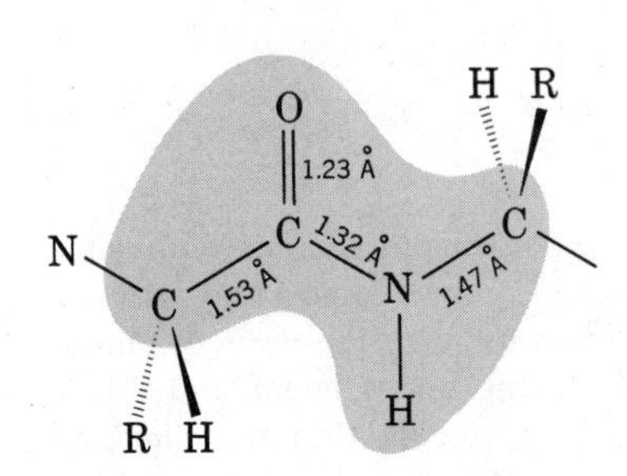

FIG. 22.7
The geometry and bond distance of the peptide linkage. The six enclosed atoms tend to be coplanar and assume a "transoid" arrangement.

and is the predominant structure of silk fibroin (48% glycine and 38% serine and alanine residues). The pleated-sheet structure has a slightly shorter repeat distance, 7.0 Å, than the flat sheet.

Of far more importance in naturally occurring proteins is the secondary structure called the α helix (Fig. 22.8). This structure is a right-handed helix with

FIG. 22.8
A representation of the α-helical structure of a polypeptide. Hydrogen bonds are denoted by dotted lines.

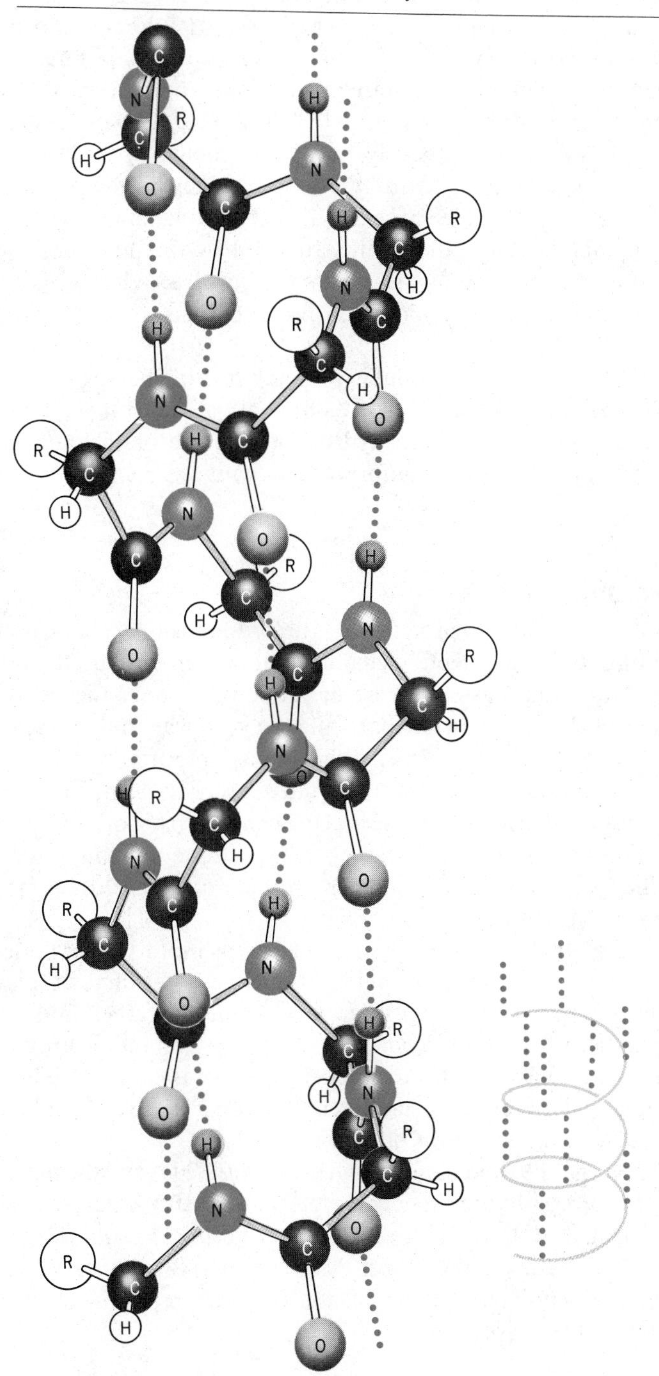

3.6 amino acid residues per turn. Each amide group in the chain has a hydrogen bond to an amide group at a distance of three amino acid residues in either direction and the R groups all extend away from the axis of the helix. The repeat distance of the α helix is 1.5 Å.

The α-helical structure is found in many proteins; it is the predominant structure of the polypeptide chains of fibrous proteins such as *myosin,* the protein of muscle, and of *α-keratin,* the protein of hair, unstretched wool, and nails.

Not all peptide chains can exist in an α-helical form. Certain peptide chains assume what is called a *random coil arrangement,* a structure that is flexible, changing, and statistically random. Synthetic polylysine, for example, exists as a random coil and does not normally form an α helix. At pH 7 the ϵ-amino groups of the lysine residues are positively charged and, as a result, repulsive forces between them are so large they overcome any stabilization that would be gained through hydrogen bond formation of an α helix. At pH 12, however, the ϵ-amino groups are uncharged and polylysine spontaneously forms an α helix.

The presence of proline or hydroxyproline residues in polypeptide chains produces another striking effect: because the nitrogen atoms of these amino acids are part of five-membered rings, the groups attached by the nitrogen—α-carbon bond cannot rotate enough to allow an α-helical structure. Wherever proline or hydroxyproline occur in a peptide chain their presence causes a kink or bend and interrupts the α helix.

The polypeptide chains of globular proteins such as hemoglobin, ribonuclease, α-chymotrypsin, and lysozyme (p. 1008) contain segments of α helix and segments of random coil. Proline and hydroxyproline are often found in those regions of the structure where the conformation changes.

22.9B Tertiary Structure

The tertiary structure of a protein is its three-dimensional shape that arises from further foldings of its polypeptide chains, foldings superimposed on the coils of the α-helices. These foldings do not occur randomly: under the proper environmental conditions they occur in one particular way—a way that is characteristic of a particular protein and one that is often highly important to its function.

A variety of forces are involved in stabilizing tertiary structures including the disulfide bonds of the primary structure. One characteristic of most proteins is that the folding takes place in such a way as to expose the maximum number of polar (hydrophilic) groups to the aqueous environment and enclose a maximum number of nonpolar (hydrophobic) groups within its interior.

The soluble globular proteins tend to be much more highly folded than fibrous proteins. However, fibrous proteins also have a tertiary structure; the α-helical strands of α-keratin, for example, are wound together into a "super helix." This super helix has a repeat distance of 5.1 Å units indicating that the super helix makes one complete turn for each 35 turns of the α helix. The tertiary structure does not end here, however. Even the super helixes can be wound together to give a ropelike structure of seven strands.

Myoglobin (Fig. 22.9) and hemoglobin (Section 22.11) were the first proteins (in 1957 and 1959) to be subjected to a completely successful X-ray analysis. This work was accomplished by J. C. Kendrew and Max Perutz at Cambridge University in England. (They received the Nobel Prize in 1962.) Since then a number of other proteins including lysozyme, ribonuclease, and α-chymotrypsin have yielded to complete structural analysis.

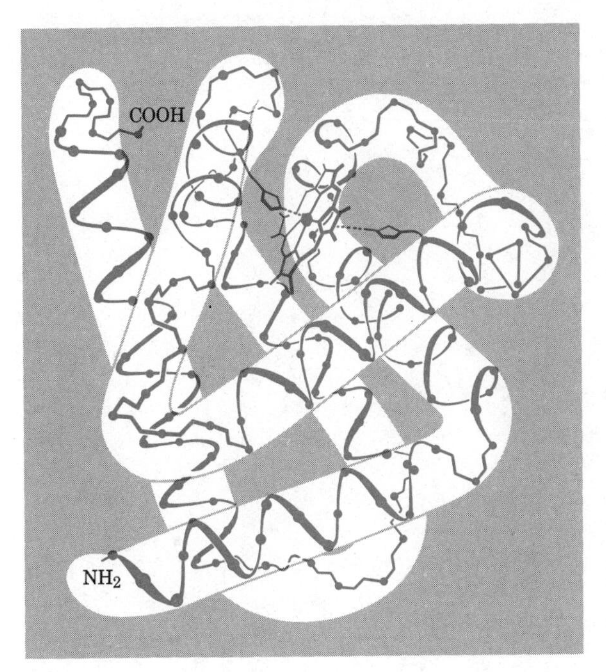

FIG. 22.9
The three-dimensional structure of myoglobin. (From R. E. Dickerson, The Proteins II, *H. Neurath, ed., Academic Press, New York, 1964, p. 634. Used with permission.)*

22.10 LYSOZYME: MODE OF ACTION OF AN ENZYME

Lysozyme is made up of 129 amino acid residues (Fig. 22.10). Three short segments of the chain between residues 5–15, 24–34, and 88–96 have the structure of an α helix; the residues between 41–45 and 50–54 form pleated sheets, and a hairpin turn occurs at residues 46–49. The remaining polypeptide segments of lysozyme have a random coil arrangement.

The discovery of lysozyme is an interesting story in itself:

"One day in 1922 Alexander Fleming was suffering from a cold. This is not unusual in London, but Fleming was a most unusual man and he took advantage of the cold in a characteristic way. He allowed a few drops of his nasal mucus to fall on a culture of bacteria he was working with and then put the plate to one side to see what would happen. Imagine his excitement when he discovered some time later that the bacteria near the mucus had dissolved away. For a while he thought his ambition of finding a universal antibiotic had been realized. In a burst of activity he quickly established that the antibacterial action of the mucus was due to the presence of an enzyme; he called this substance lysozyme because of its capacity to lyse, or dissolve the bacterial cells. Lysozyme was soon discovered in many tissues and secretions of the human body, in plants and most plentifully of all in the white of an egg. Unfortunately Fleming found that it is not effective against the most harmful bacteria. He had to wait seven years before a strangely similar experiment revealed the existence of a genuinely effective antibiotic: penicillin."

This story was related by Professor David C. Phillips of Oxford University who many years later used X-ray analysis to discover the three-dimensional structure of lysozyme.*

Phillips' X-ray diffraction studies of lysozyme are especially interesting

* Quotation from David C. Phillips, "The Three-Dimensional Structures of an Enzyme Molecule," Copyright © (1966) by Scientific American, Inc. All rights reserved.

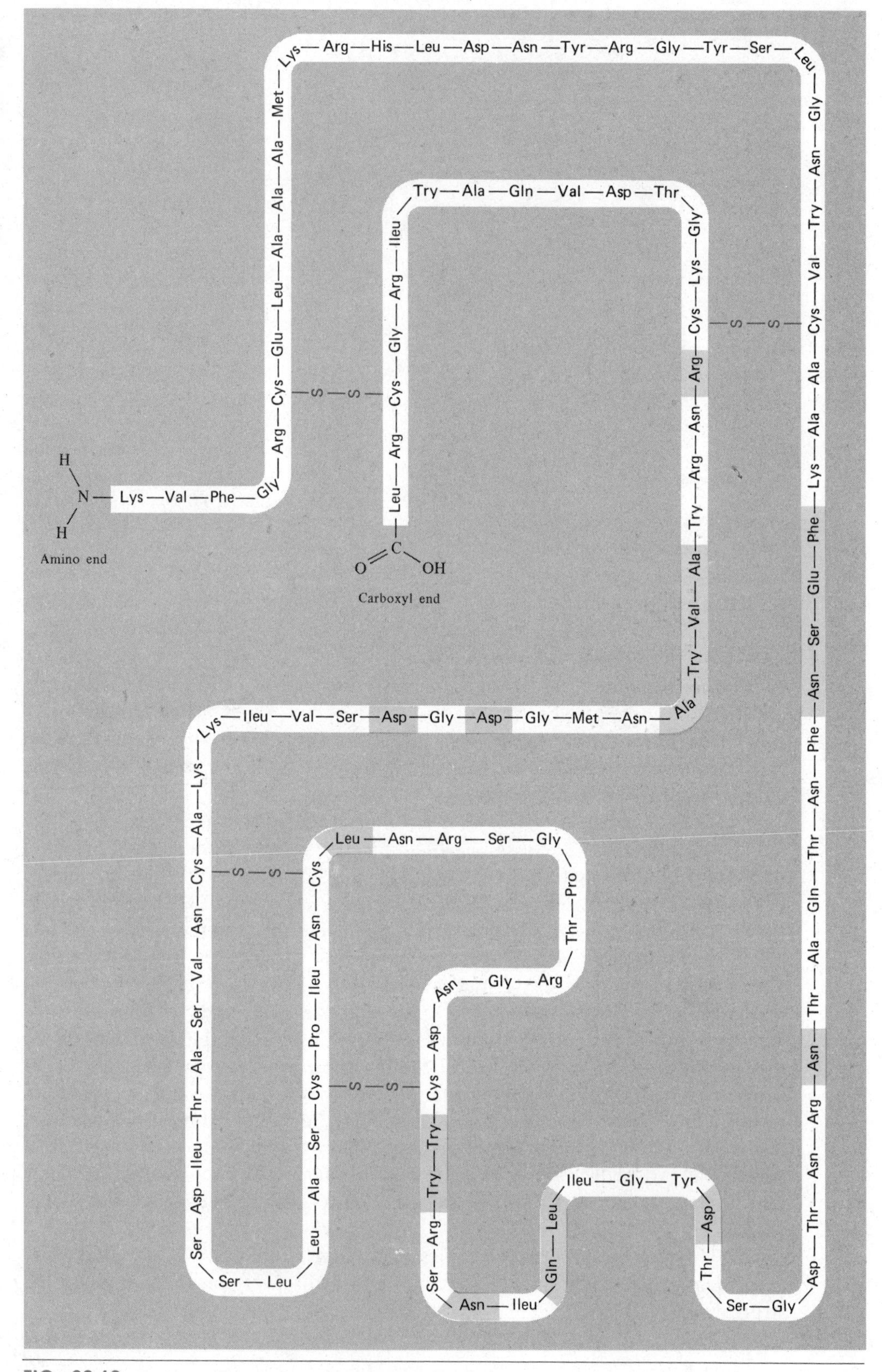

FIG. 22.10
The covalent structure of lysozyme. The amino acids that line the active site of lysozyme are shown in color.

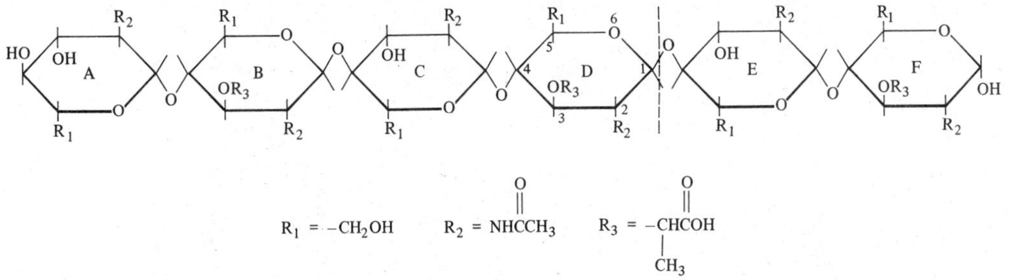

$R_1 = -CH_2OH \qquad R_2 = NHCCH_3 \text{ (with C=O)} \qquad R_3 = -CHCOH \text{ (with C=O and } CH_3\text{)}$

FIG. 22.11
A hexasaccharide that has the general structure as the cell wall polysaccharide on which lysozyme acts. Two different amino sugars are present: rings **A, C,** *and* **E** *are derived from a monosaccharide called* N-*acetylglucosamine; rings* **B, D,** *and* **F** *are derived from a monosaccharide called* N-*acetylmuramic acid. When lysozyme acts on this oligosaccharide, hydrolysis takes place and it results in cleavage of the glycosidic linkage between rings* **D** *and* **E.**

because they have also revealed important information about how this enzyme acts on its substrate. Lysozyme's substrate is a polysaccharide of amino sugars that makes up part of the bacterial cell wall. An oligosaccharide that has the same general structure as the cell wall polysaccharide is shown in Fig. 22.11.

By using oligosaccharides (made up of *N*-acetylglucosamine units only) on which lysozyme acts very slowly, Phillips and his co-workers were able to discover how the substrate fits into the enzyme's active site. This site is a deep cleft in the lysozyme structure (Fig. 22.12*a*). The oligosaccharide is held in this cleft by hydrogen bonds and as the enzyme binds the substrate two important changes take place: the cleft in the enzyme closes slightly and ring **D** of the oligosaccharide is "flattened" out of its stable chair conformation. This flattening causes atoms 1, 2, 5, and 6 of ring **D** to become coplanar; it also distorts ring **D** in such a way as to make the glycosidic linkage between it and ring E more susceptible to hydrolysis.†

Hydrolysis of the glycosidic linkage probably takes the course illustrated in Fig. 22.12*b*. The carboxyl group of glutamic acid (residue number 35) donates a proton to the oxygen between rings **D** and **E**. Protonation leads to cleavage of the glycosidic link and to the formation of a carbocation at C—1 of ring **D**. This carbocation is stabilized by the negatively charged carboxylate group of aspartic acid (residue number 52) which lies in close proximity. A water molecule diffuses in and supplies an OH^- ion to the carbocation and a proton to replace that lost by glutamic acid.

When the polysaccharide is a part of a bacterial cell wall, lysozyme probably first attaches itself to the cell wall by hydrogen bonds. After hydrolysis has taken place lysozyme falls away leaving behind a bacterium with a punctured cell wall.

22.11 HEMOGLOBIN: A CONJUGATED PROTEIN

Some proteins contain as a part of their structure a nonprotein group called a *prosthetic group*. An example is the oxygen-carrying protein, hemoglobin. Each of the four polypeptide chains of hemoglobin is bound to a prosthetic group called

† R. H. Lemieux and G. Huber of the National Research Council of Canada have shown that when an aldohexose is converted to a carbocation the ring of the carbocation assumes just this flattened conformation.

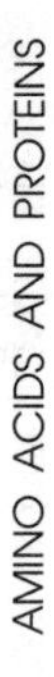

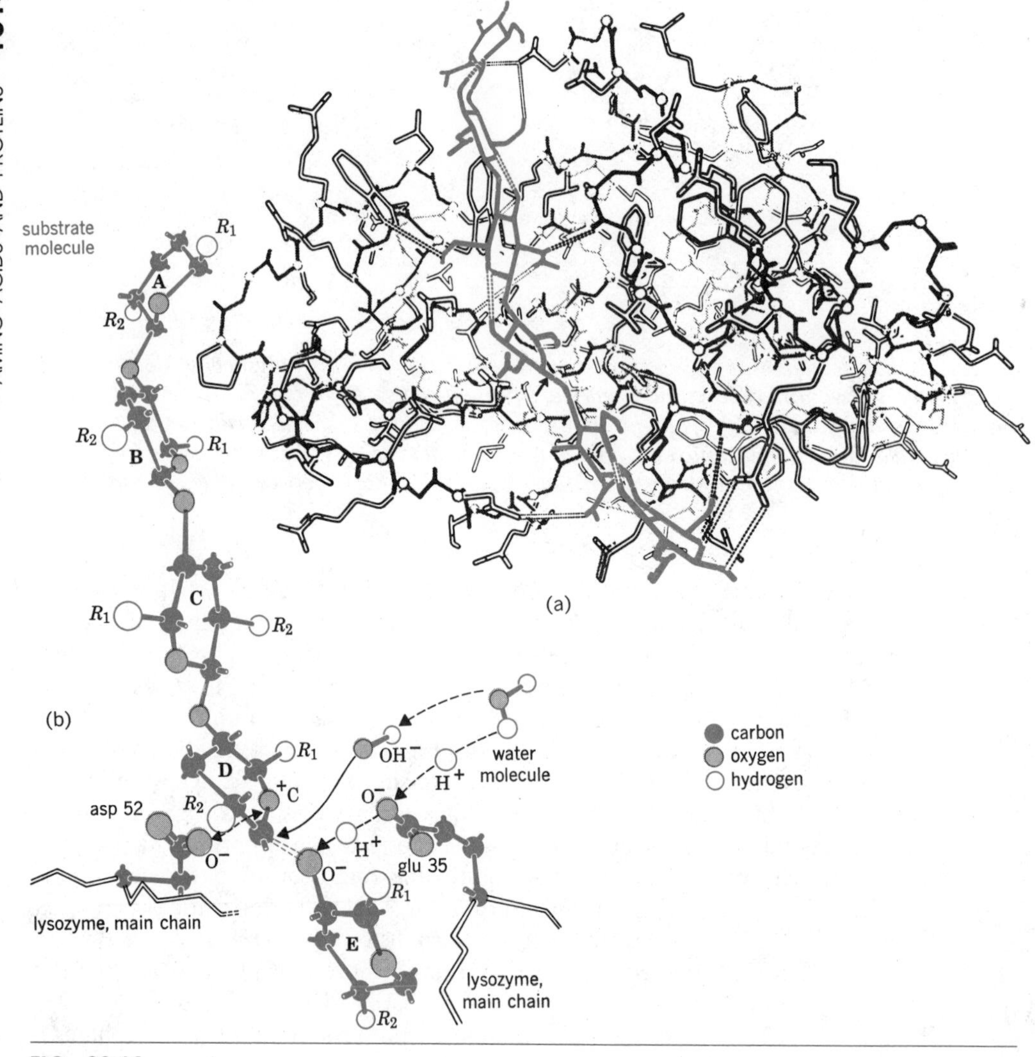

FIG. 22.12
(a) *This drawing shows the backbones of the lysozyme-substrate complex. The substrate (in this drawing a hexasaccharide) fits into a cleft in the lysozyme structure and is held in place by hydrogen bonds. As lysozyme binds the oligosaccharide, the cleft in its structure closes slightly. (Adapted with permission from* Atlas of Protein Sequence and Structure *1969, ed. Margaret O. Dayoff, National Biomedical Research Foundation, Washington, D.C. 1969. The drawing was made by Irving Geis based on his perspective painting of the molecule which appeared in* Scientific American, *November 1966. The painting was made of an actual model assembled at the Royal Institution, London, by D. C. Phillips and his colleagues, based on their X-ray crystallography results.)* (b) *A possible mechanism for lysozyme action. This drawing shows an expanded portion of the figure above and illustrates how hydrolysis of the acetal linkage between rings D and E of the substrate may occur. Glutamic acid (residue 35) donates a proton to the intervening oxygen. This causes the formation of a carbocation that is stabilized by the carboxylate ion aspartic acid (residue 52). A water molecule supplies an* OH^- *to the carbonium ion and* H^+ *to glutamic acid. (Adapted with permission from "The Three-Dimensional Structures of an Enzyme Molecule," by David C. Phillips. Copyright © Nov. 1966 by Scientific American, Inc., All rights reserved.)*

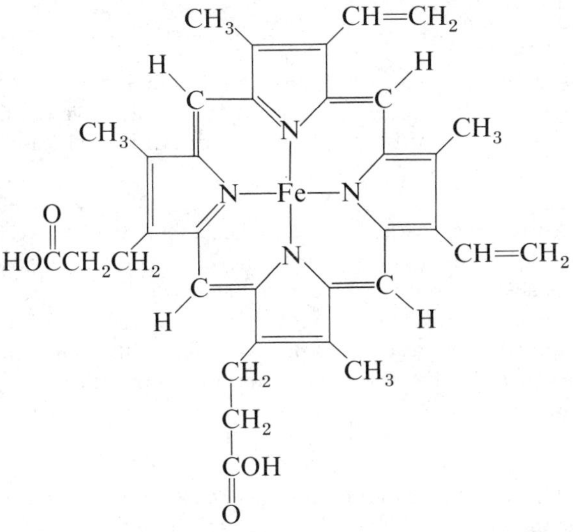

FIG. 22.13
The structure of heme, the prosthetic group of hemoglobin. Heme has a structure similar to that of chlorophyll (p. 874) in that each is derived from a heterocyclic, porphyrin, *ring. The iron of heme is in the ferrous (+2) oxidation state.*

heme (Fig. 22.13). The four polypeptide chains of hemoglobin are wound in such a way as to give hemoglobin a roughly spherical shape. Moreover, each heme group lies in a crevice with the hydrophobic vinyl groups of its porphyrin structure surrounded by side chains of hydrophobic amino residues. The two propanoate side chains of heme lie near positively charged amino groups of lysine and arginine residues.

The iron of the heme group is in the +2 (ferrous) oxidation state and it forms a coordinate bond to a nitrogen of the imidazole group of histidine of the polypeptide chain. This leaves one valence of the ferrous ion free to combine with oxygen as shown as follows.

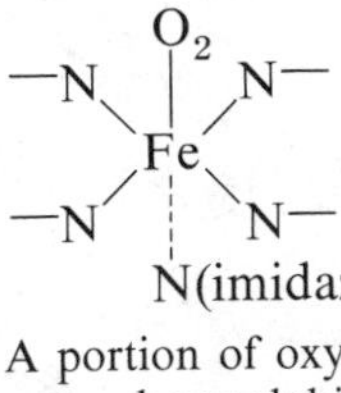

A portion of oxygenated hemoglobin

The fact that the ferrous ion of the heme group combines with oxygen is not particularly remarkable; many similar compounds do the same thing. What is remarkable about hemoglobin is that when the heme combines with oxygen the ferrous ion does not become readily oxidized to the ferric state. Studies with model heme compounds in water, for example, show that they undergo a rapid combination with oxygen but they also undergo a rapid oxidation of the iron from Fe^{+2} to Fe^{+3}. When these same compounds are embedded in the hydrophobic environment of a polystyrene resin, however, the iron is easily oxygenated and deoxygenated and this occurs *with no change in oxidation state of iron.* In this respect, it is especially interesting to note that X-ray studies of hemoglobin have revealed that the polypeptide chains provide each heme group with a similar hydrophobic environment.

Additional Problems

22.16

(a) Which amino acids in Table 22.1 have more than one chiral carbon? (b) Write projection formulas for the isomers of each of these amino acids that would have the L configuration at the α-carbon. (c) What kind of isomers have you drawn in each case?

22.17

(a) Which amino acid in Table 22.1 would react with nitrous acid (i.e., a solution of $NaNO_2$ and HCl) to yield lactic acid? (b) All of the amino acids in Table 22.1 liberate nitrogen when they are treated with nitrous acid except two; which are these? (c) What product would you expect to obtain from treating tyrosine with excess bromine water? (d) What product would you expect to be formed in the reaction of phenylalanine with ethanol in presence of hydrogen chloride? (e) What product would you expect from the reaction of alanine and benzoyl chloride in aqueous base?

22.18

(a) On the basis of the following sequence of reactions Emil Fischer was able to show that (−)serine and L-(+)-alanine have the same configuration. Write projection formulas for the intermediates **A**–**C**.

$$(-)\text{Serine} \xrightarrow[CH_3OH]{HCl} \mathbf{A}\ (C_4H_{10}ClNO_3) \xrightarrow{PCl_5} \mathbf{B}\ (C_4H_9Cl_2NO_2) \xrightarrow[(2)\ OH^-]{(1)\ H_3O^+,\ H_2O,\ heat}$$

$$\mathbf{C}\ (C_3H_6ClNO_2) \xrightarrow[dil.\ H_3O^+]{Na\text{-}Hg} \text{L-(+)-Alanine}$$

(b) The configuration of L-(−)-cysteine can be related to that of L-(−)-serine through the following reactions. Write projection formulas for **D** and **E**.

$$\mathbf{B}\ \text{[from part (a) above]} \xrightarrow{OH^-} \mathbf{D}\ (C_4H_8ClNO_2) \xrightarrow{NaSH}$$

$$\mathbf{E}\ (C_4H_9NO_2S) \xrightarrow[(2)\ OH^-]{(1)\ H_3O^+,\ H_2O,\ heat} \text{L-(+)-Cysteine}$$

(c) The configuration of L-(−)-asparagine can be related to that of L-(−)-serine in the following way. What is the structure of **F**?

$$\text{L-(−)-Asparagine} \xrightarrow[\text{Hofmann degradation}]{NaOBr/OH^-} \mathbf{F}\ (C_3H_7N_2O_2)$$

$$\mathbf{C}\ \text{[from part (a)]} \xrightarrow{NH_3} \mathbf{F}$$

22.19

(a) DL-Glutamic acid has been synthesized from diethyl acetamidomalonate in the following way: outline the reactions involved.

$$\underset{\text{Diethyl acetamidomalonate}}{CH_3\overset{O}{\overset{\|}{C}}NHCH(CO_2C_2H_5)_2} + CH_2{=}CH{-}C{\equiv}N \xrightarrow[C_2H_5OH\ (95\%\ yield)]{NaOC_2H_5}$$

$$\mathbf{G}\ (C_{12}H_{18}N_2O_5) \xrightarrow[\text{reflux 6 hr. (66\% yield)}]{conc.\ HCl} \text{DL-Glutamic acid}$$

(b) Compound **G** has also been used to prepare the amino acid DL-ornithine through the route shown below.

$$\mathbf{G}\ (C_{12}H_{18}N_2O_5) \xrightarrow[68°,\ 1000\ psi\ (90\%\ yield)]{H_2/Ni} \mathbf{H}\ (C_{10}H_{16}N_2O_4,\ \text{a}\ \delta\text{-lactam}) \xrightarrow[\text{reflux 4 hr. (97\% yield)}]{conc.\ HCl}$$

$$\text{DL-Ornithine hydrochloride, } C_5H_{13}ClN_2O_2$$

L-Ornithine is a naturally occurring amino acid but does not occur in proteins. In one metabolic pathway L-ornithine serves as a precursor for L-arginine.

22.20
The *Strecker synthesis* of DL-alanine is outlined below:

$$\underset{\text{Acetaldehyde}}{CH_3CHO} \xrightarrow[HCN]{NH_3} CH_3\underset{\displaystyle NH_2}{\underset{|}{C}}HC{\equiv}N \xrightarrow{H_3O^+} \underset{\text{DL-Alanine}}{CH_3\underset{\displaystyle {}^+NH_3}{\underset{|}{C}}HCOO^-}$$

(a) Outline a Strecker synthesis of DL-phenylalanine. (b) DL-Methionine can also be synthesized by a Strecker synthesis. The required starting aldehyde can be prepared from acrolein, $CH_2{=}CHCHO$, and methanethiol, CH_3SH. Outline all steps in this synthesis of DL-methionine.

22.21
Transamination reactions (p. 987) involving L-glutamic acid provide animals with pathways for redistributing amino groups. These pathways are important because in a given meal an animal may eat protein that, when hydrolyzed, provides it with a mixture of amino acids quite different from that which is optimal for its metabolism. It might, for example, ingest protein rich in phenylalanine but poor in aspartic acid. Assuming that α-ketoglutaric acid and oxaloacetic acid are available in the cell as a result of carbohydrate metabolism, show how two transamination reactions (the first synthesizing glutamic acid) would result in a net synthesis of aspartic acid from phenylalanine.

22.22
Bradykinin is a nonapeptide released by blood plasma globulins in response to a wasp sting. It is a very potent pain-causing agent. Its molecular formula is Arg_2, Gly, Phe_2, $Pro_3 \cdot$ Ser. The use of 2,4-dinitrofluorobenzene and carboxypeptidase show that both terminal residues are arginine. Partial acid hydrolysis of bradykinin gives the following di- and tripeptides:

Phe · Ser + Pro · Gly · Phe + Pro · Pro + Ser · Pro · Phe + Phe · Arg + Arg · Pro

What is the amino acid sequence of bradykinin?

22.23
Complete hydrolysis of a heptapeptide showed that it had the following molecular formula:

Ala_2, Glu, Leu, Lys, Phe, Val

Deduce the amino acid sequence of this heptapeptide from the following data.
1. Treatment of the heptapeptide with 2,4-dinitrofluorobenzene followed by incomplete hydrolysis gave, among other products: Val labeled at the α-amine group, lysine labeled at the ϵ-amino group, and a dipeptide, DNP—Val · Leu (DNP = 2,4-Dinitrophenyl-).
2. Hydrolysis of the heptapeptide with carboxypeptidase gives an initial high concentration of alanine, followed by a rising concentration of glutamic acid.
3. Partial enzymatic hydrolysis of the heptapeptide gave a dipeptide, **A**, and a tripeptide, **B**.

(a) Treatment of **A** with 2,4-dinitrofluorobenzene followed by hydrolysis gave DNP-labeled leucine and lysine labeled only at the ϵ-amino group.

(b) Complete hydrolysis of **B** gave phenylalanine, glutamic acid, and alanine. When **B** was allowed to react with carboxypeptidase the solution showed an initial high concentration of glutamic acid. Treatment of **B** with 2,4-dinitrofluorobenzene followed by hydrolysis gave labeled phenylalanine.

22.24
Synthetic polyglutamic acid exists as an α helix in solution at pH 2–3. When the pH of such

a solution is gradually raised through the addition of base, a dramatic change in optical rotation takes place at pH 5. This change has been associated with the unfolding of the α helix and the formation of a random coil. What structural feature of polyglutamic acid, and what chemical change, can you suggest as an explanation for this transformation?

*** 22.25**

Part of the evidence for restricted rotation about the carbon-nitrogen bond in a peptide linkage (see pp. 1003–1004) comes from proton nmr studies done with simple amides. For example, at room temperature and with the instrument operating at 60 MHz, the proton nmr spectrum of *N,N*-dimethylformamide, $(CH_3)_2NCHO$, shows a doublet at $\delta 2.80$ (3H), a doublet at $\delta 2.95$ (3H) and a multiplet at $\delta 8.05$ (1H). When the spectrum is determined at lower magnetic field strength (i.e., with the instrument operating at 30 MHz) the doublets are found to have shifted so that the distance (in Hertz) that separates one doublet from the other is smaller. When the temperature at which the spectrum is determined is raised, the doublets persist until a temperature of 111° is reached, then the doublets coalesce to become a single signal. Explain in detail how these observations are consistent with the existence of a relatively large barrier to rotation about the carbon-nitrogen bond of *N,N*-dimethylformamide.

O SPECIAL TOPIC

ALKALOIDS

Extracting the bark, roots, leaves, berries, and fruits of plants often yields nitrogen-containing bases called *alkaloids.* The name alkaloid comes from the fact that these substances are "alkalilike", that is, since alkaloids are amines they often react with acids to yield soluble salts. The nitrogen atoms of most alkaloids are present in heterocyclic rings. In a few instances, however, nitrogen may be present as a primary amine or as a quaternary ammonium group.

When administered to animals most alkaloids produce striking physiological effects and these effects *vary greatly* from alkaloid to alkaloid. Some alkaloids stimulate the central nervous system, others cause paralysis; some alkaloids elevate blood pressure, others lower it. Certain alkaloids act as pain relievers; others act as tranquilizers; still others act against infectious microorganisms. Most alkaloids are toxic when their dosage is large enough, and with some this dosage is very small. In spite of this, many alkaloids find use in medicine.

Systematic names are seldom used for alkaloids and their common names have a variety of origins. In many instances the common name reflects the botanical source of the compound. The alkaloid, strychnine, for example, comes from the seeds of the *Strychnos* plants. In other instances the names are more whimsical: the name of the opium alkaloid, morphine, comes from Morpheus, the ancient Greek god of dreams; the name of the tobacco alkaloid, nicotine, comes from Nicot, an early French ambassador who sent tobacco seeds to France. The one characteristic that alkaloid names have in common is the ending -ine, reflecting the fact that they are all amines.

Alkaloids have been of interest to chemists for nearly two centuries and in that time thousands of alkaloids have been isolated. Most of these have had their structures determined through the application of chemical and physical methods and in many instances these structures have been confirmed by independent synthesis. A complete account of the chemistry of the alkaloids would (and does) occupy volumes; here we have space to consider only a few representative examples.

O.1 Alkaloids Containing a Pyridine or Reduced Pyridine Ring

The predominant alkaloid of the tobacco plant is nicotine:

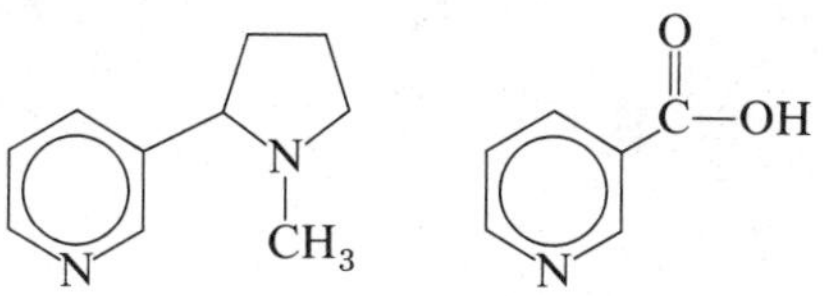

Nicotine — Nicotinic acid

In very small doses nicotine acts as a stimulant, but in larger doses it causes depression, nausea, and vomiting. In still larger doses it is a violent poison. Nicotine salts are used as insecticides.

Oxidation of nicotine by concentrated nitric acid produces pyridine-3-carboxylic acid—a compound that is called *nicotinic acid.* While the consumption of nicotine is of no benefit to humans, nicotinic acid is a vitamin; it is incorporated into the important coenzyme, nicotinamide adenine dinucleotide.

Problem O.1

Nicotine has been synthesized by the route shown on page 1016. All of the steps involve reactions that we have seen before. Suggest reagents that could be used for each.

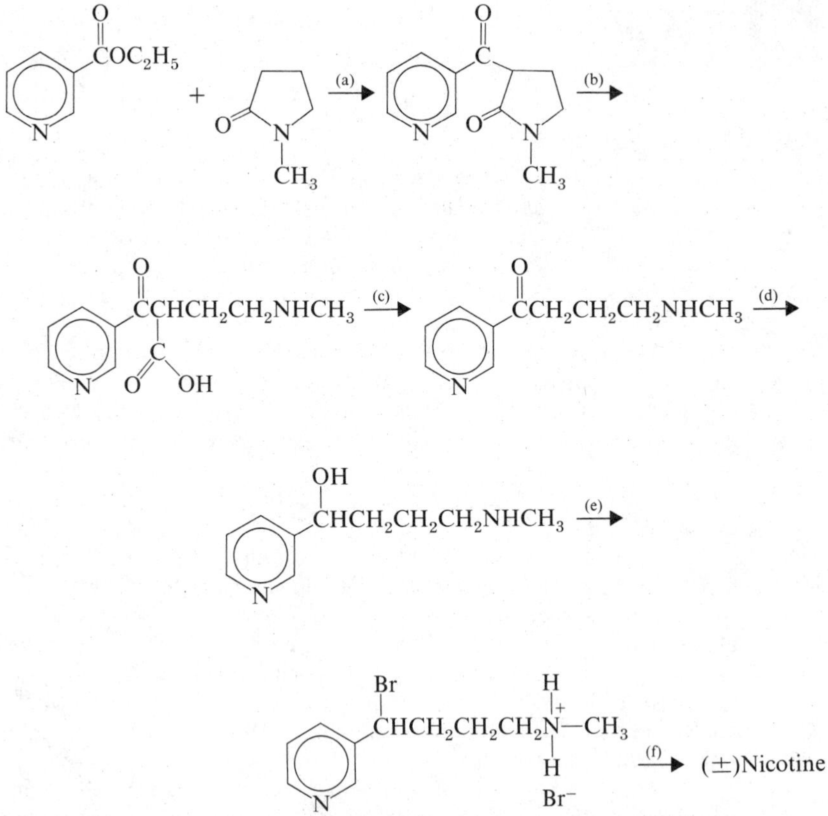

(Reaction scheme: ethyl nicotinate + N-methyl-2-pyrrolidone $\xrightarrow{(a)}$ 3-nicotinoyl-1-methyl-2-pyrrolidone $\xrightarrow{(b)}$ 3-pyridyl–$\overset{O}{\overset{\|}{C}}$CH($CO_2H$)$CH_2CH_2NHCH_3$ $\xrightarrow{(c)}$ 3-pyridyl–$\overset{O}{\overset{\|}{C}}CH_2CH_2CH_2NHCH_3$ $\xrightarrow{(d)}$ 3-pyridyl–$\overset{OH}{\overset{|}{C}}HCH_2CH_2CH_2NHCH_3$ $\xrightarrow{(e)}$ 3-pyridyl–$\overset{Br}{\overset{|}{C}}HCH_2CH_2CH_2\overset{+}{N}H_2$–$CH_3$ Br^- $\xrightarrow{(f)}$ (±)Nicotine)

A number of alkaloids contain a piperidine ring. These include coniine (from the water hemlock), atropine (from *Atropa belladonna* and other genera of the plant family Solanaceae), and cocaine (from *Erythroxylon coca*).

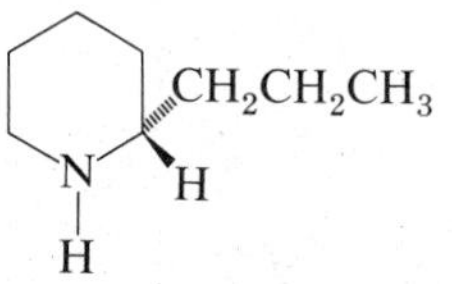

Coniine
((−)-2-propylpiperidine)

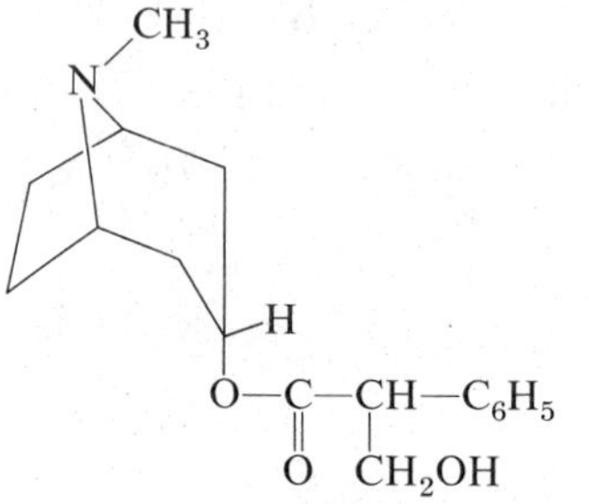

Atropine

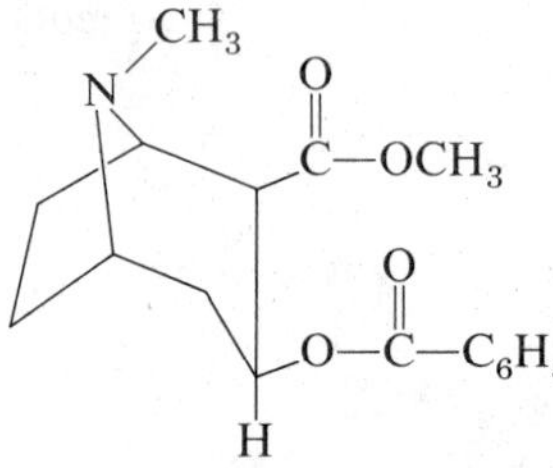

Cocaine

Coniine is highly toxic; its ingestion may cause weakness, drowsiness, nausea, labored respiration, paralysis, and death. Coniine was the toxic substance of the "hemlock" used in the execution of Socrates.

In small doses cocaine decreases fatigue, increases mental activity, and gives a general feeling of well-being. Prolonged use of cocaine, however, leads to psychical addiction and to periods of deep depression. Cocaine is also a local anesthetic and, for a time, it was used medically in that capacity. When its tendency to cause addiction was recognized, efforts were made to develop other local anesthetics. This led, in 1905, to the synthesis of novocaine, a compound that has some of the same structural features as cocaine (i.e., its benzoic ester and tertiary amine groups).

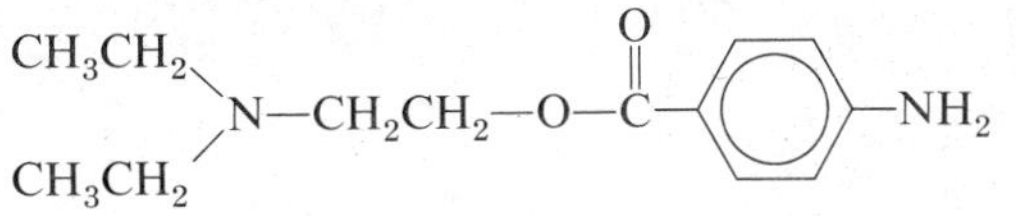

$$(CH_3CH_2)_2N{-}CH_2CH_2{-}O{-}\overset{O}{\overset{\|}{C}}{-}C_6H_4{-}NH_2$$

Novocaine

Atropine is an intense poison. In dilute solutions (0.5–1.0%) it is used to dilate the pupil of the eye in ophthalmic examinations. Alkaloids related to atropine are contained in the 12-hour continuous-release capsules used to relieve symptoms of the common cold.

Problem O.2

The principal alkaloid of *Atropa belladonna* is the optically active alkaloid *hyoscyamine.* During its isolation hyoscyamine is often racemized by bases to optically inactive atropine. (a) What chiral center is likely to be involved in the racemization? (b) In hyoscyamine this chiral center has the *S* configuration. Write a three-dimensional structure for hyoscyamine.

Problem O.3

Hydrolysis of atropine gives tropine and (±) tropic acid. (a) What are their structures? (b) Even though tropine has chiral carbons, it is optically inactive. Explain. (c) An isomeric form of tropine called ψ-tropine has also been prepared by heating tropine with base. ψ-Tropine is also optically inactive. What is its structure?

Problem O.4

In 1891, G. Merling transformed tropine (cf. Problem O.3) into 1,3,5-cycloheptatriene (tropylidene) through the following sequence of reactions.

$$\text{Tropine } (C_8H_{15}NO) \xrightarrow{-H_2O} C_8H_{13}N \xrightarrow{CH_3I} C_9H_{16}NI \xrightarrow[\text{(2) Heat}]{\text{(1) } Ag_2O/H_2O}$$

$$C_9H_{15}N \xrightarrow{CH_3I} C_{10}H_{18}NI \xrightarrow[\text{(2) Heat}]{\text{(1) } Ag_2O/H_2O} \text{1,3,5-cycloheptatriene} + (CH_3)_3N + H_2O$$

Write out all of the reactions that take place.

Problem O.5

Many alkaloids appear to be synthesized in plants by reactions that resemble the Mannich reaction (Section 20.9). Recognition of this (by R. Robinson in 1917) led to a synthesis of tropinone that takes place under "physiological conditions," that is, at room temperature and at pH values near neutrality. This synthesis is shown below. Propose reasonable mechanisms that account for the overall course of the reaction.

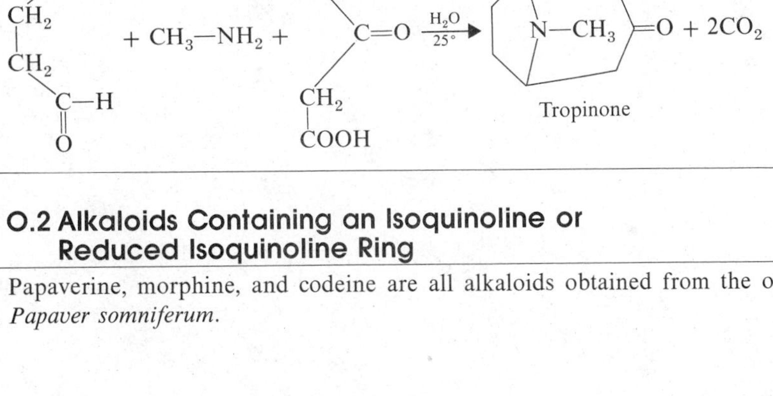

$$OHC{-}CH_2{-}CH_2{-}CHO + CH_3{-}NH_2 + (HOOC{-}CH_2)_2C{=}O \xrightarrow[25^\circ]{H_2O} \text{Tropinone (bicyclic, }N{-}CH_3\text{, }{=}O) + 2CO_2$$

Tropinone

O.2 Alkaloids Containing an Isoquinoline or Reduced Isoquinoline Ring

Papaverine, morphine, and codeine are all alkaloids obtained from the opium poppy, *Papaver somniferum.*

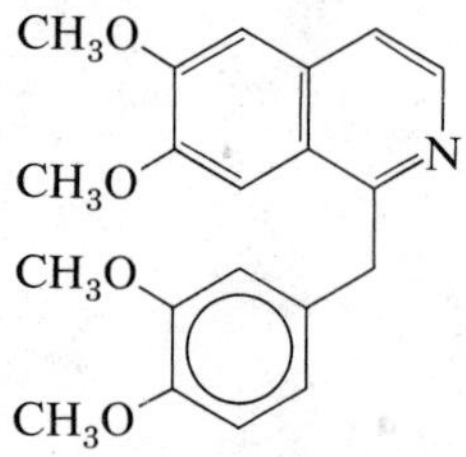

Papaverine

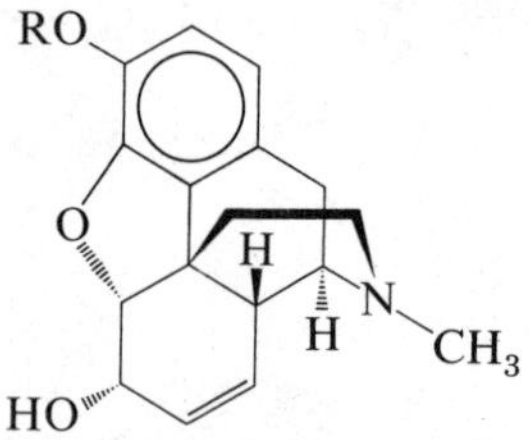

Morphine (R = H)
Codeine (R = CH_3)

Papaverine has an isoquinoline ring; in morphine and codeine the isoquinoline ring is partially hydrogenated (reduced).

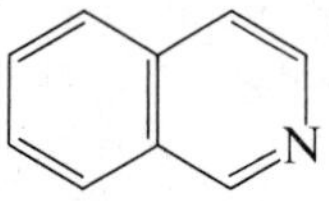

Isoquinoline

Opium has been used since earliest recorded history. Morphine was first isolated from opium in 1803 and its isolation represented one of the first instances of the purification of the active principle of a drug. One hundred and twenty years were to pass, however, before the complicated structure of morphine was deduced, and its final confirmation through independent synthesis (by Professor Marshall Gates of the University of Rochester) did not take place until 1952.

Morphine is one of the most potent analgesics known and it is still used extensively in medicine to relieve pain, especially "deep" pain. Its greatest drawbacks, however, are its tendencies to lead to addiction and to depress respiration. These disadvantages have brought about a search for morphinelike compounds that do not have these disadvantages. One of the newest candidates is the compound pentazocine. Pentazocine is a highly effective analgesic and it is nonaddictive; unfortunately however, like morphine, it depresses respiration.

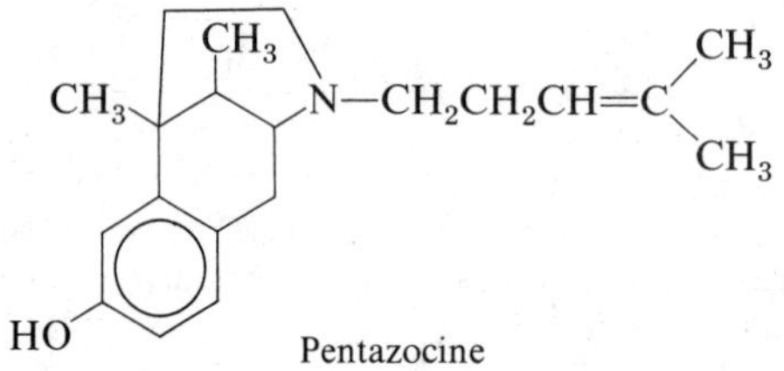

Pentazocine

Problem O.6

Papaverine has been synthesized by the following route:

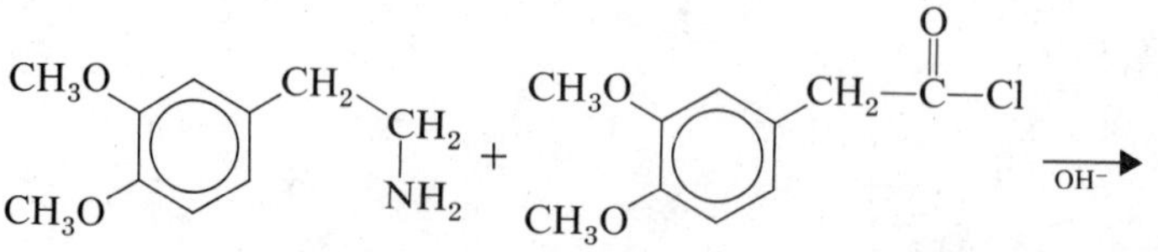

$$C_{20}H_{25}NO_5 \xrightarrow[\text{heat}\ (-H_2O)]{P_2O_5} \text{Dihydropapaverine} \xrightarrow[\text{heat}\ (-H_2)]{Pd} \text{Papaverine}$$

Outline the reactions involved.

Problem O.7

One of the important steps in the synthesis of morphine involved the following transformation:

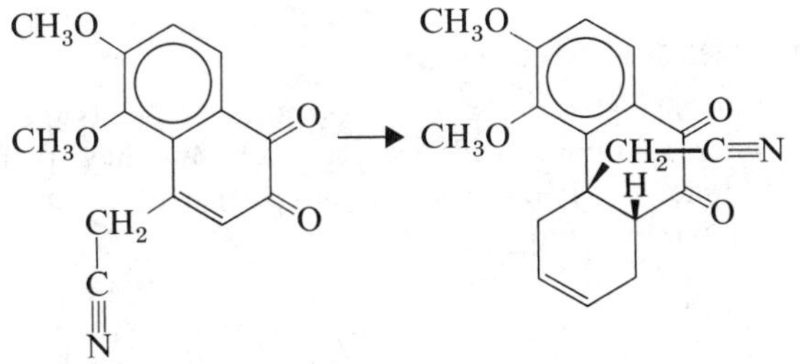

Suggest how this step was accomplished.

Problem O.8

When morphine reacts with two moles of acetic anhydride it is transformed into the highly addictive narcotic, heroin. What is the structure of heroin?

O.3 Alkaloids Containing Indole or Reduced Indole Rings

A large number of alkaloids are derivatives of an indole ring system. These range from the relatively simple *gramine* to the highly complicated structures of *strychnine* and *reserpine*.

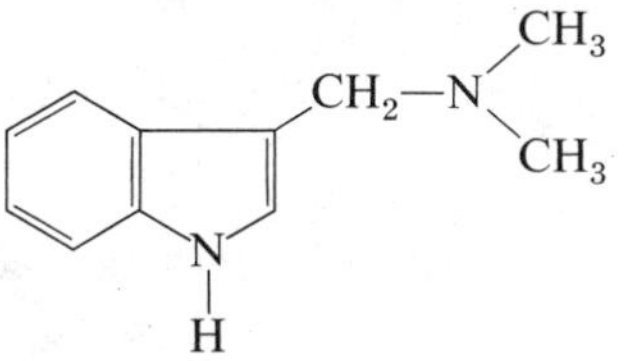

Gramine

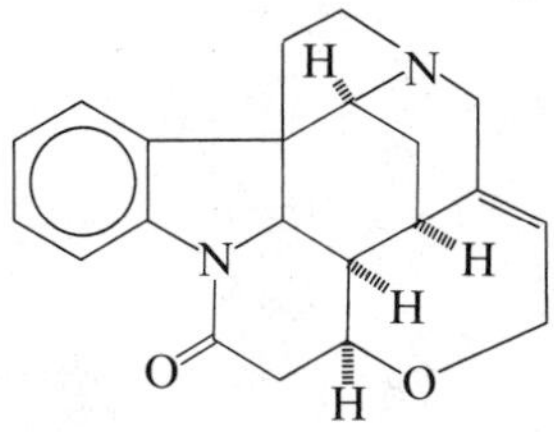

Strychnine

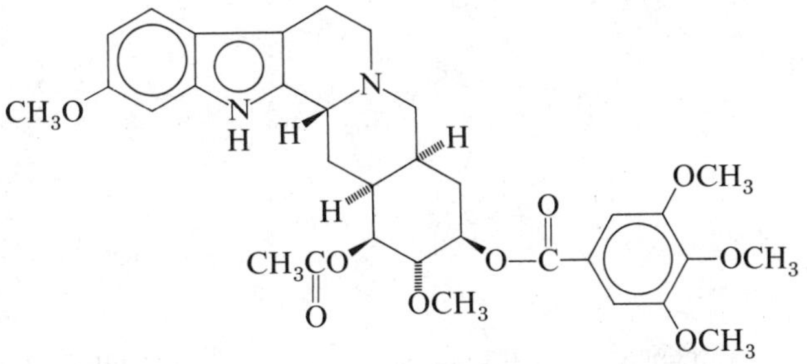

Reserpine

Gramine can be obtained from chlorophyll-deficient mutants of barley. Strychnine, a very bitter and highly poisonous compound, comes from the seeds of *Strychnos nux-vomica.* Strychnine is a central nervous system stimulant and has been used medically (in low dosage) to counteract poisoning by central nervous system depressants. Reserpine can be obtained from the Indian snake root *Rauwolfia serpentina,* a plant that has been used in native medicine for centuries. Reserpine is used in modern medicine as a tranquilizer and as an agent to lower blood pressure.

Problem O.9

Gramine has been synthesized by heating a mixture of indole, formaldehyde, and dimethylamine. (a) What general reaction is involved here? (b) Outline a reasonable mechanism for the gramine synthesis.

O.4 Biosynthesis of Alkaloids

The primary starting materials for alkaloid synthesis in plants appear to be α-amino acids. More than 20 different α-amino acids occur naturally and they are the main building blocks for proteins (Chapter 22). Two amino acids, in particular, are important in alkaloid biosynthesis. These are tyrosine and tryptophan:

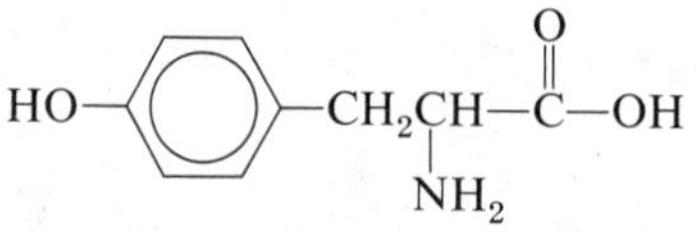

Tyrosine

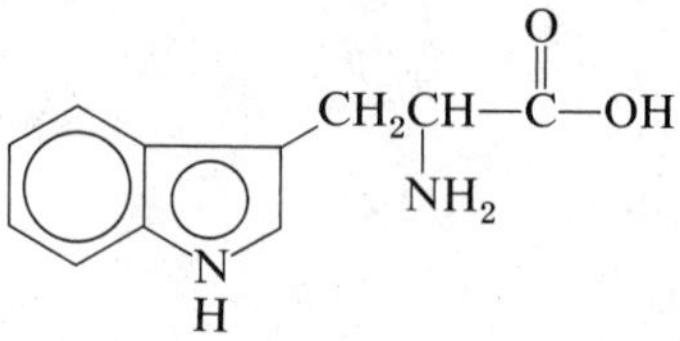

Tryptophan

Two general reactions appear to be of central importance in alkaloid biosynthesis—the Mannich reaction and the oxidative coupling of phenols. We studied the Mannich reaction earlier in Section 20.9 (cf. also Problem O.5). The oxidative coupling of phenols is a free-radical process that is catalyzed by enzymes in plants and that can also be carried out (usually less successfully) in the laboratory. A simple formulation of an oxidative phenol coupling is outlined below using phenol itself as the starting compound. Loss of an electron and a proton from phenol leads to a resonance-stabilized free radical. Two free radicals can then undergo coupling in a variety of ways:

Oxidation [O], $-e^-$, $-H^+$ ... etc.

Ortho-ortho coupling ... enolization

Ortho-para coupling

Para-para coupling

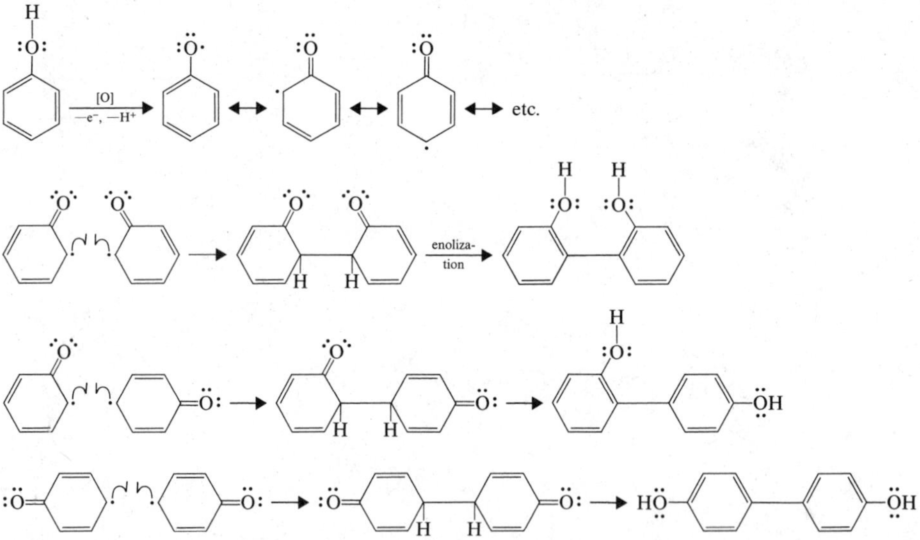

The biosynthetic route that the opium poppy uses to synthesize morphine is now known. Most of the morphine molecule, it turns out, is constructed from two molecules of tyrosine. In most cases, oxidative coupling occurs intramolecularly.

The synthesis begins with the oxidation of tyrosine to 3,4-dihydroxyphenylalanine:

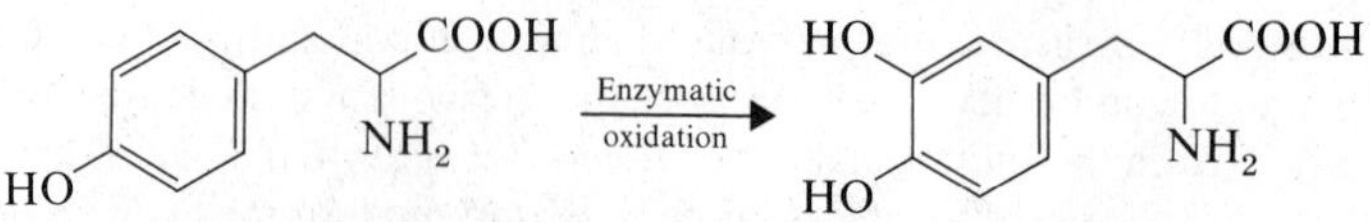

Tyrosine $\xrightarrow{\text{Enzymatic oxidation}}$ 3,4-Dihydroxyphenylalanine

Further enzyme-catalyzed reactions transform 3,4-dihydroxyphenylalanine into 3,4-dihydroxyphenylpyruvic acid, **I**, and into 3,4-dihydroxyphenylethylamine **II** (Fig. O.1). These two molecules then react in a Mannich-type condensation to yield norlaudanosoline, **III.**

Methylation of norlaudanosoline at two of its —OH groups and at its —N—H group yields reticuline, **IV*a*** (Fig. O.2). A reticuline molecule can be twisted into conformation **IV*b*,** one that allows an ortho-para oxidative phenolic coupling to take place yielding salutaridine. Reduction of salutaridine produces salutaridinol. Then salutaridinol is transformed into thebaine. (In this highly unusual step the oxygen bridge is installed through a reaction that is accompanied by *the displacement of a hydroxide ion.*) Finally, several additional enzymatic reactions transform thebaine into morphine.

FIG. O.1
The biosynthesis of norlaudanosoline from two moles of 3,4-dihydroxyphenylalanine.

3,4-Dihydroxyphenylalanine

3,4-Dihydroxyphenylalanine

Transamination
(cf. Section 22.4B)

I

Decarboxylation
($-CO_2$)

II

$-H_2O$
$-CO_2$

Mannich-type condensation

Enolization

Norlaudanosoline
III

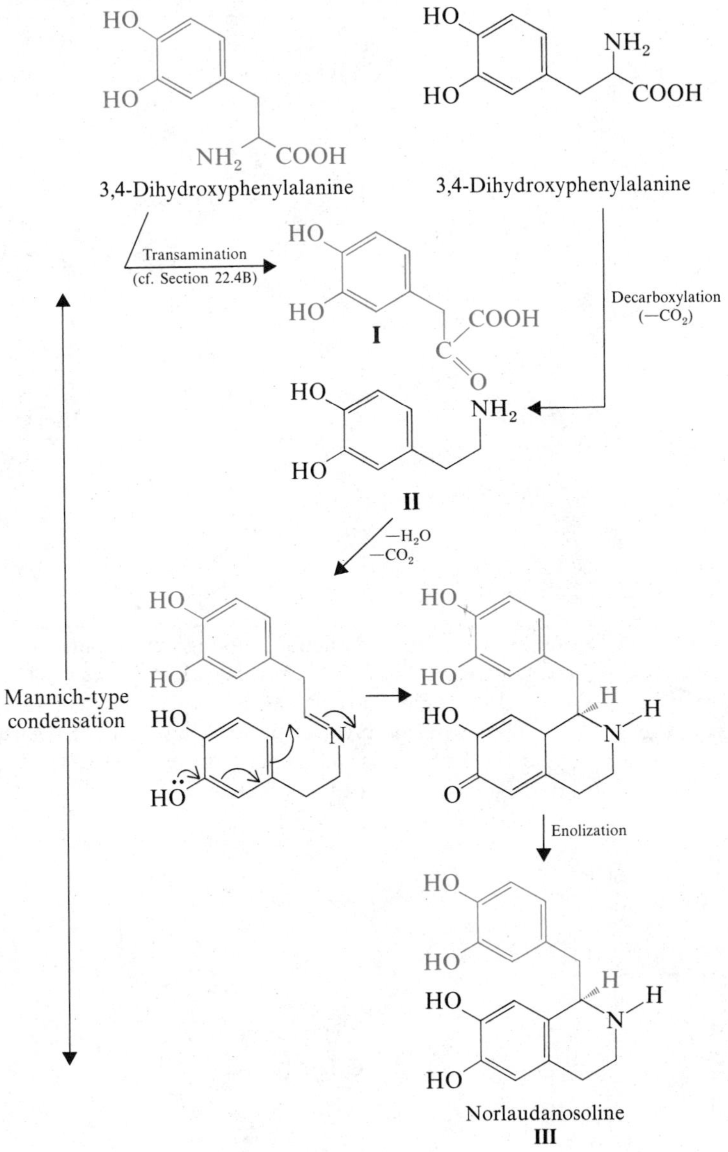

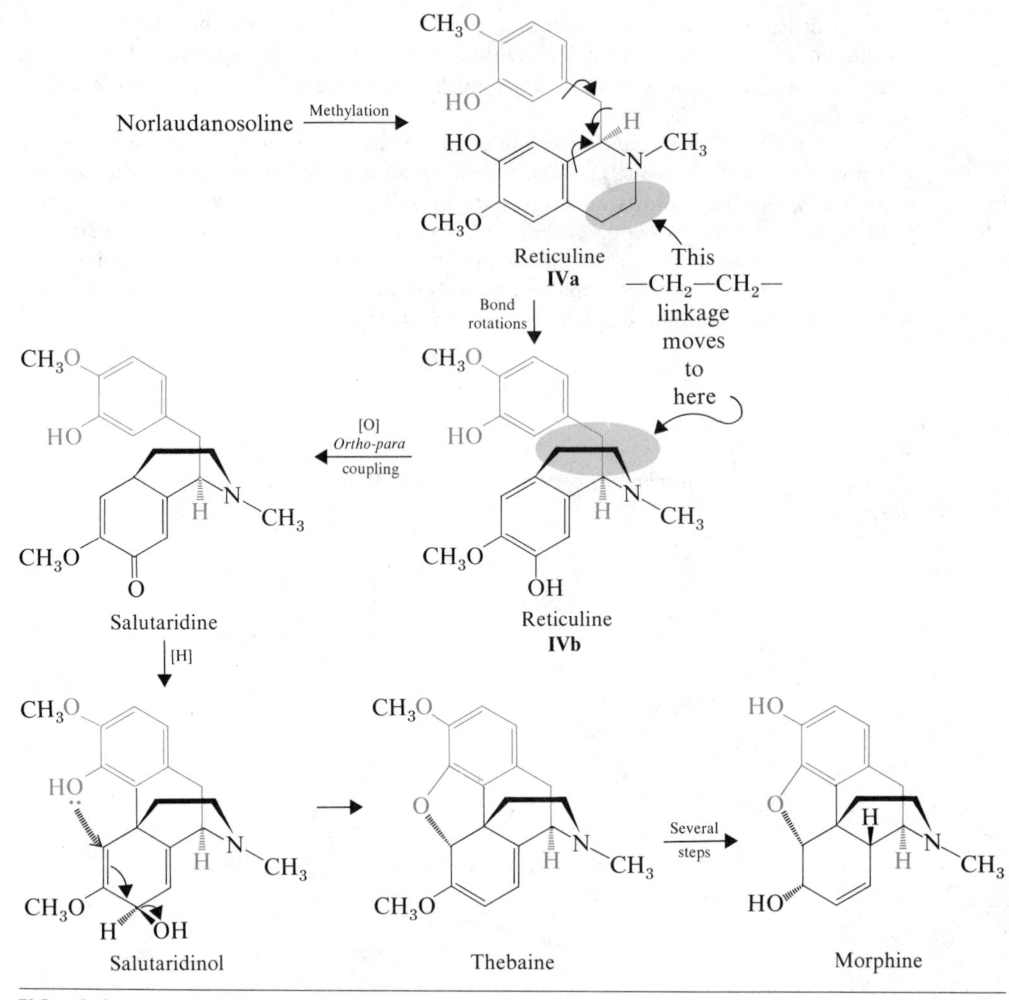

FIG. O.2
The biosynthesis of morphine from norlaudanosoline.

Problem O.10

There is considerable evidence that oxidative phenol couplings are important in the biosynthesis of bulbocapnine and galucine (p. 1020). (These two alkaloids have what is called an *apomorphine* ring system.) Both compounds appear to arise from reticuline. Show the type of oxidative phenol coupling that is involved in each biosynthesis. (Assume that methylation of —OH groups in both alkaloids and synthesis of the —O—CH_2—O— bridge in bulbocapnine occur after the oxidative phenol couplings.)

OCH_3, CH_3O, CH_3O, CH_3O, N—CH_3

Glaucine

CH_3O, HO, O, CH_2, O, N—CH_3

Bulbocapnine

Problem O.11

Much of the evidence for plant biosynthetic pathways comes from experiments in which labeled compounds are fed to the plant. Later, intermediates and alkaloids are isolated from the plant and the locations of the labeled atoms are determined. An example is an experiment in which tyrosine labeled at the α-carbon was administered to opium poppies (*Papaver somniferum*); the papaverine isolated from the poppies had the labeled atoms at positions 1 and 3:

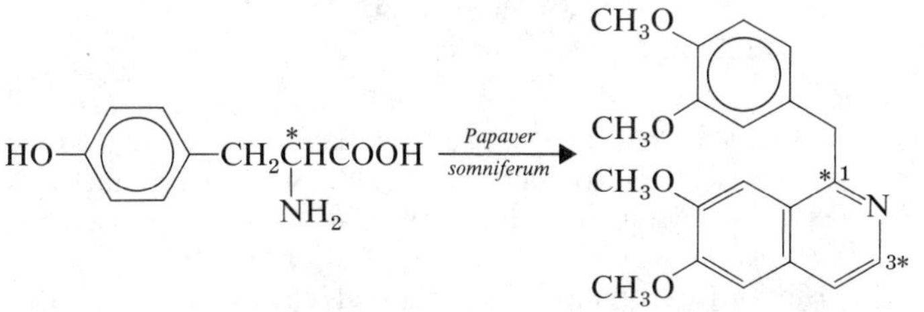

Labeled tyrosine

Labeled papaverine

(a) Is this evidence consistent with a biosynthesis of papaverine from tyrosine (i.e., with tyrosine leading to norlaudanosoline as in Fig. O.1) and then a conversion of norlaudanosoline to papaverine? (b) Experiments using labeled norlaudanosoline show that opium poppies efficiently convert it into papaverine. Does this support the pathway outlined in (a)? (c) What major steps must take place in the conversion of norlaudanosoline to papaverine?

Problem O.12

Harmine is an alkaloid isolated from *Peganum harmala L.* When tryptophan and pyruvic acid labeled in the positions shown below were fed to the plant, the harmine produced had the labeling pattern indicated.

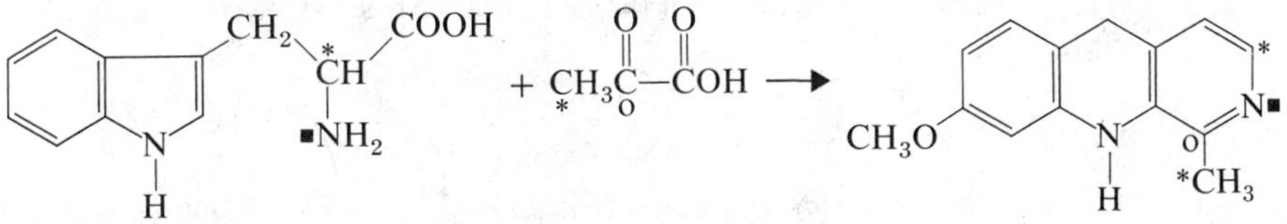

Tryptophan

* and o are ^{14}C labels

▪ is an ^{15}N label

Pyruvic acid

Harmine

Show how these results are consistent with the following pathway: (1) decarboxylation of tryptophan (to tryptamine), (2) a Mannich-type condensation of tryptamine and pyruvic acid, then (3) dehydrogenation, (4) hydroxylation, and (5) methylation.

P SPECIAL TOPIC

NUCLEIC ACIDS: PROTEIN SYNTHESIS

". . . I cannot help wondering whether some day an enthusiastic scientist will christen his newborn twins Adenine and Thymine."

F. H. C. Crick*

P.1 INTRODUCTION

The molecules that preserve hereditary information and that transcribe and translate that information in a way that allows the synthesis of all the varied enzymes of the cell are the nucleic acids, deoxyribonucleic acid (DNA) and ribonucleic acid (RNA). These biological polymers are sometimes found associated with proteins and in this form they are known as *nucleoproteins*.

It has been from studies of nucleic acids themselves that has come much of our knowledge of how genetic information is preserved, how it is passed on to succeeding generations of the organism, and how it is transformed into the working parts of the cell. For these reasons we will focus our attention on the structures and properties of nucleic acids and of their components, *nucleotides* and *nucleosides*.

P.2 NUCLEOTIDES AND NUCLEOSIDES

Mild degradations of nucleic acids yield their monomeric units, compounds that are called *nucleotides*. A general formula for a nucleotide and the specific structure of one, called adenylic acid, are shown in Fig. P.1.

Complete hydrolysis of a nucleotide furnishes:

1. A heterocyclic base, either a purine or pyrimidine.

2. A five-carbon monosaccharide, either D-ribose or 2-deoxy-D-ribose.

* Who along with J. D. Watson and Maurice Wilkins shared the Nobel Prize in 1962 for their proposal of (and evidence for) the double helix structure of DNA. (Taken from F. H. C. Crick, "The Structure of the Hereditary Material," *Scientific American,* October 1954.)

FIG. P.1

(a) *General structure of a nucleotide obtained from RNA. The heterocyclic base is a purine or pyrimidine. In nucleotides obtained from DNA, the sugar component is 2′-deoxyribose, that is, the —OH at position 2′ is replaced by —H. The phosphate group of the nucleotide shown below is attached to the 5′-carbon; it may also be attached to the 3′-carbon. The heterocyclic base is always attached through a β-glycosidic linkage at C-1′.* (b) *Adenylic acid, a typical nucleotide.*

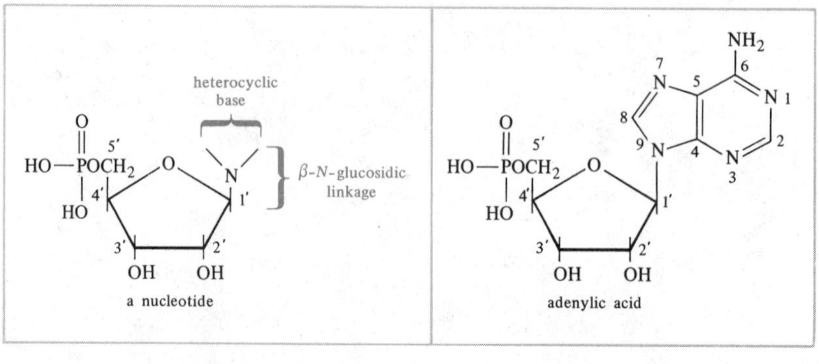

heterocyclic base

β-N-glucosidic linkage

a nucleotide

adenylic acid

3. A phosphate ion.

The central portion of the nucleotide is the monosaccharide and it is always present as a five-membered ring, that is, as a furanoside. The heterocyclic base of a nucleotide is attached through an *N*-glycosidic linkage to C-1′ of the ribose or deoxyribose unit and this linkage is always β. The phosphate group of a nucleotide is present as a phosphate ester and it may be attached at C-5′ or C-3′. (In nucleotides the carbons of the monosaccharide portion are designated with primed numbers, i.e., 1′, 2′, 3′, and so on.)

Removal of the phosphate group of a nucleotide converts it to a compound known as a *nucleoside*. The nucleosides that can be obtained from DNA all contain 2-deoxy-D-ribose as their sugar component and one of four heterocyclic bases, either adenine, guanine, cytosine, or thymine:

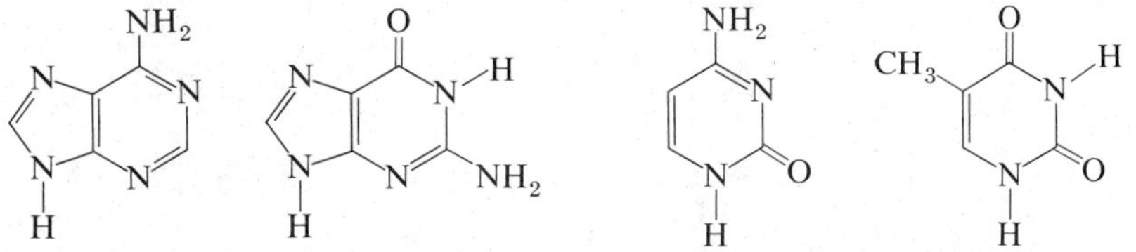

Adenine Guanine Cytosine Thymine

← Purines → ← Pyrimidines →

The nucleosides obtained from RNA contain D-ribose as their sugar component and either adenine, guanine, cytosine, or uracil as their heterocyclic base.*

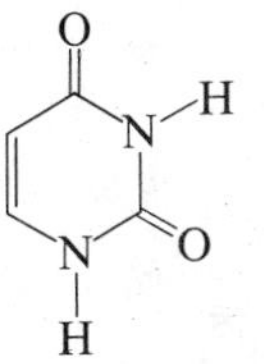

Uracil
(a pyrimidine)

The heterocyclic bases obtained from nucleosides are capable of existing in more than one tautomeric form. The forms that we have shown (above) are the predominant forms that the bases assume when they are present in nucleic acids.

Problem P.1

Write the structures of the other tautomeric forms of adenine, guanine, cytosine, thymine, and uracil.

The names and structures of the nucleosides found in DNA are shown in Fig. P.2; those found in RNA are given in Fig. P.3.

Problem P.2

The nucleosides shown in Figs. P.2 and P.3 are stable in dilute base. In dilute acid, however, they undergo rapid hydrolysis yielding a sugar (deoxyribose or ribose) and a heterocyclic base. (a) What structural feature of the nucleoside accounts for this behavior? (b) Propose a reasonable mechanism for the hydrolysis.

Nucleotides are named in several ways. Adenylic acid (Fig. P.1), for example, is

* Notice that in an RNA nucleoside (or nucleotide) uracil replaces thymine. (Some nucleosides obtained from specialized forms of RNA may also contain other, but similar, purines and pyrimidines.)

adenine
NH_2
$HOCH_2$
HO
deoxyadenosine

guanine
O
H
$HOCH_2$
NH_2
HO
deoxyguanosine

cytosine
NH_2
$HOCH_2$
HO
deoxycytidine

thymine
O
CH_3
H
$HOCH_2$
HO
deoxythymidine

FIG. P.2
Nucleosides that can be obtained from DNA. (Undesignated bonds are to —H.)

FIG. P.3
Nucleosides that can be obtained from RNA. (Undesignated bonds are to —H.)

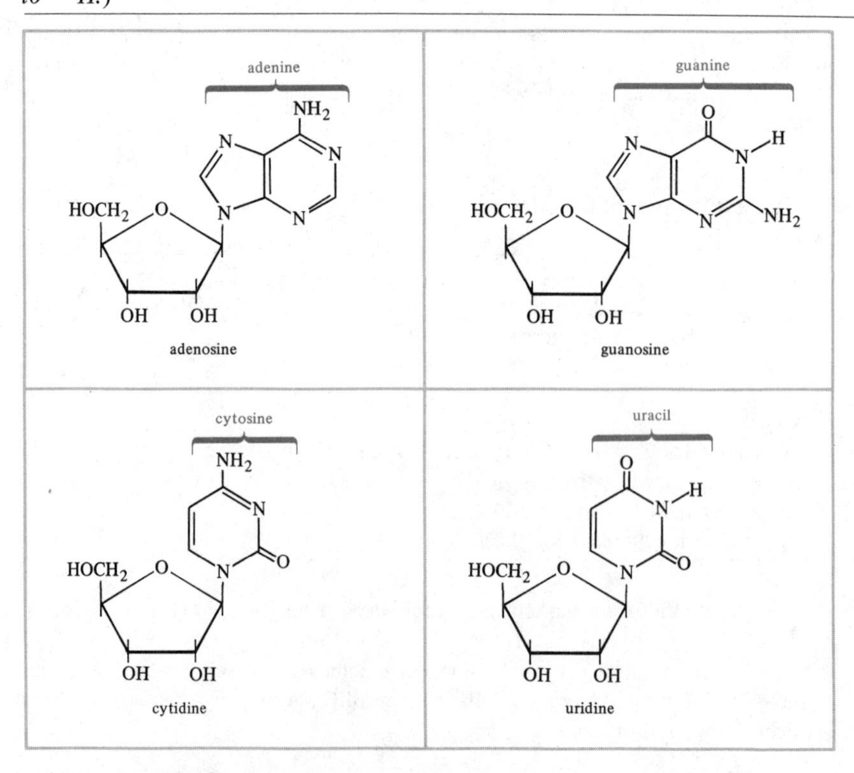

sometimes called 5′-adenylic acid in order to designate the position of the phosphate group; it is also called adenosine 5′-phosphate, or simply adenosine monophosphate (AMP). Uridylic acid is called 5′-uridylic acid, uridine 5′-phosphate, or uridine monophosphate (UMP), and so on.

Nucleosides and nucleotides are found in places other than as part of the structure of DNA and RNA. We have seen, for example, that adenosine units are part of the structures of the important coenzymes, NAD and coenzyme A (p. 786). The 5′-triphosphate of adenosine is, of course, the important energy source, ATP (p. 875). The compound called 3′,5′-cyclic adenylic acid (Fig. P.4) is an important regulator of hormone activity. Cells synthesize this compound from ATP through the action of an enzyme, *adenyl cyclase*. In the laboratory, 3′,5′-cyclic adenylic acid can be prepared through dehydration of 5′-adenylic acid with dicyclohexylcarbodimide.

Problem P.3

When 3′,5′-cyclic adenylic acid is treated with aqueous sodium hydroxide the major product that is obtained is 3′-adenylic (adenosine 3′-phosphate) rather than 5′-adenylic acid. Suggest an explanation that accounts for the course of this reaction.

P.3 LABORATORY SYNTHESIS OF NUCLEOSIDES AND NUCLEOTIDES

A variety of methods have been developed for the synthesis of nucleosides. One technique uses reactions that assemble the nucleoside from suitably activated and protected ribose derivatives and heterocyclic bases. An example is the following synthesis of adenosine.

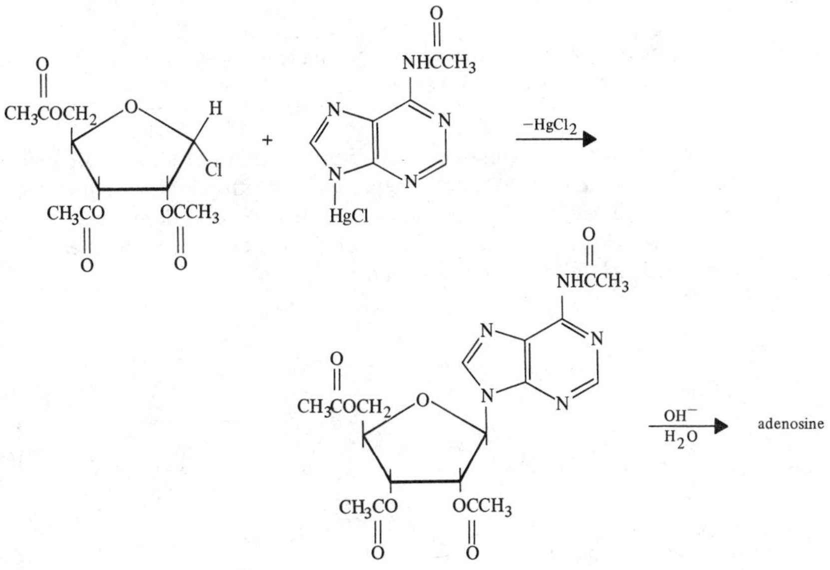

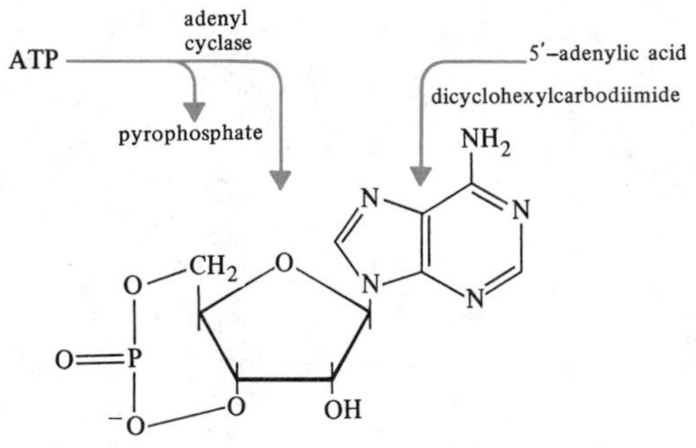

$\xrightarrow{-HgCl_2}$

$\xrightarrow[H_2O]{OH^-}$ adenosine

ATP → (adenyl cyclase) → pyrophosphate

5′-adenylic acid → (dicyclohexylcarbodiimide)

FIG. P.4
3′,5′-Cyclic adenylic acid and its laboratory and biosynthesis.

Another technique involves formation of the heterocyclic base on a protected *N*-aminoribose derivative:

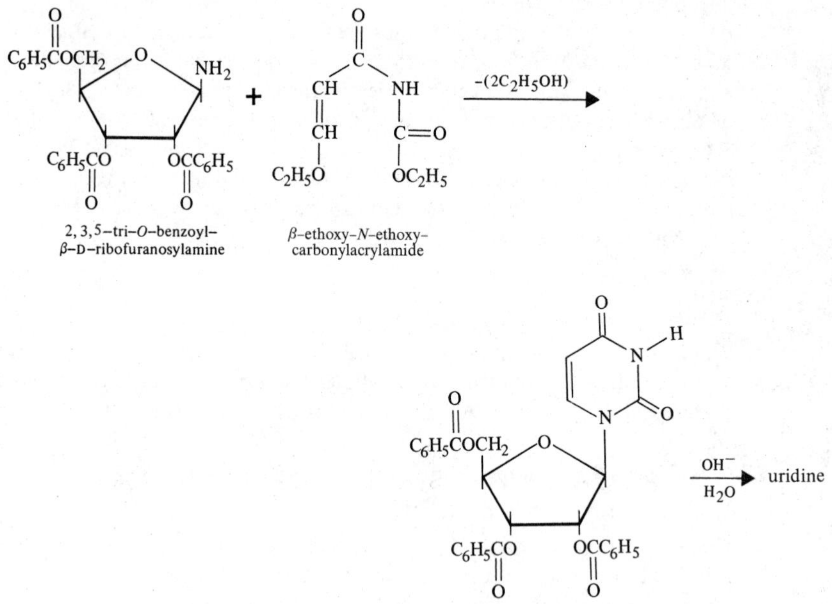

2,3,5-tri-*O*-benzoyl-β-D-ribofuranosylamine + β-ethoxy-*N*-ethoxycarbonylacrylamide $\xrightarrow{-(2C_2H_5OH)}$ $\xrightarrow[H_2O]{OH^-}$ uridine

Problem P.4

Basing your answer on reactions that you have seen before, propose a likely mechanism for the condensation reaction given above.

Still a third technique involves the synthesis of a nucleoside with a substituent in the heterocyclic ring that can be replaced with other groups. This method has been used extensively to synthesize unusual nucleosides that do not necessarily occur naturally. An example (below) makes use of a 6-chloropurine derivative obtained from the appropriate ribofuranosyl chloride and chloromercuripurine.

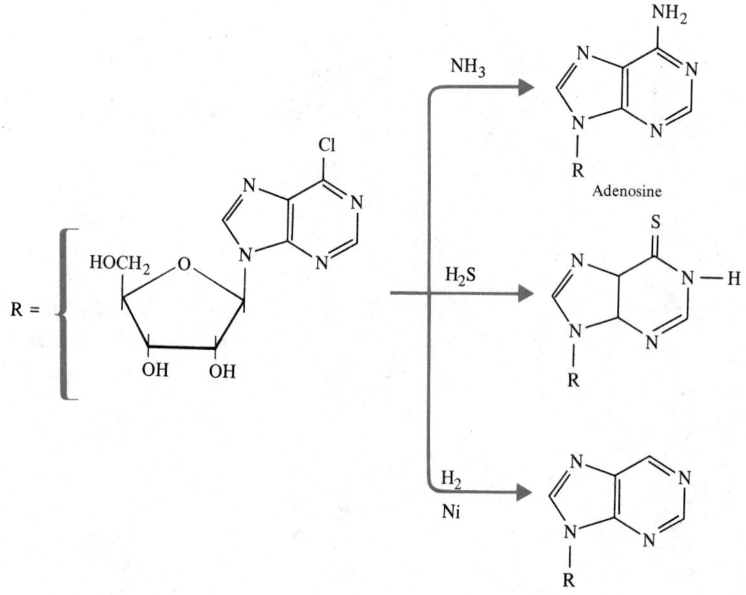

R = ; NH_3 → Adenosine; H_2S; H_2, Ni

Numerous phosphorylating agents have been used to convert nucleosides to nucleotides. One of the most useful is dibenzyl phosphochloridate.

$(C_6H_5CH_2O)_2P(=O)Cl$

Dibenzyl phosphochloridate

Specific phosphorylation of the 5′ —OH can be achieved if the 2′ and 3′ OH groups of the nucleoside are protected by an isopropylidene group (below).

$(C_6H_5CH_2O)_2P(=O)—Cl$ + $HOCH_2$ (ribose ring, Base; 2′,3′-O,O-$C(CH_3)_2$) —Pyridine→

Isopropylidene protecting group

$(C_6H_5CH_2O)_2P(=O)OCH_2$ (ribose ring, base; 2′,3′-O,O-$C(CH_3)_2$) —[1] H_3O^+, H_2O; [2] H_2/Pd→ HO—P(=O)OCH_2 (ribose ring, base; OH, OH)

nucleotide

Mild acid-catalyzed hydrolysis removes the isopropylidene group, and hydrogenolysis cleaves the benzyl phosphate bonds.

Problem P.5

(a) What kind of linkage is involved in the isopropylidene-protected nucleoside and why is it susceptible to mild acid-catalyzed hydrolysis? (b) How might such a protecting group be installed?

P.4 DEOXYRIBONUCLEIC ACID: DNA

P.4A Primary Structure

Nucleotides bear the same relation to a nucleic acid that amino acids do to a protein; they are its monomeric units. The connecting links in proteins are amide groups, in nucleic acids they are phosphate ester linkages. Phosphate esters link the 3′-hydroxyl of one ribose (or deoxyribose) with the 5′-hydroxyl of another. This makes the nucleic acid a long unbranched chain with a "backbone" of sugar and phosphate units with heterocyclic bases protruding from the chain at regular intervals (Fig. P.5).

It is, as we will see, the *base sequence* along the chain of DNA that contains the encoded genetic information. This sequence of bases can be determined through techniques based on selective enzymatic hydrolyses and the actual base sequences have been worked out for a number of smaller nucleic acids.

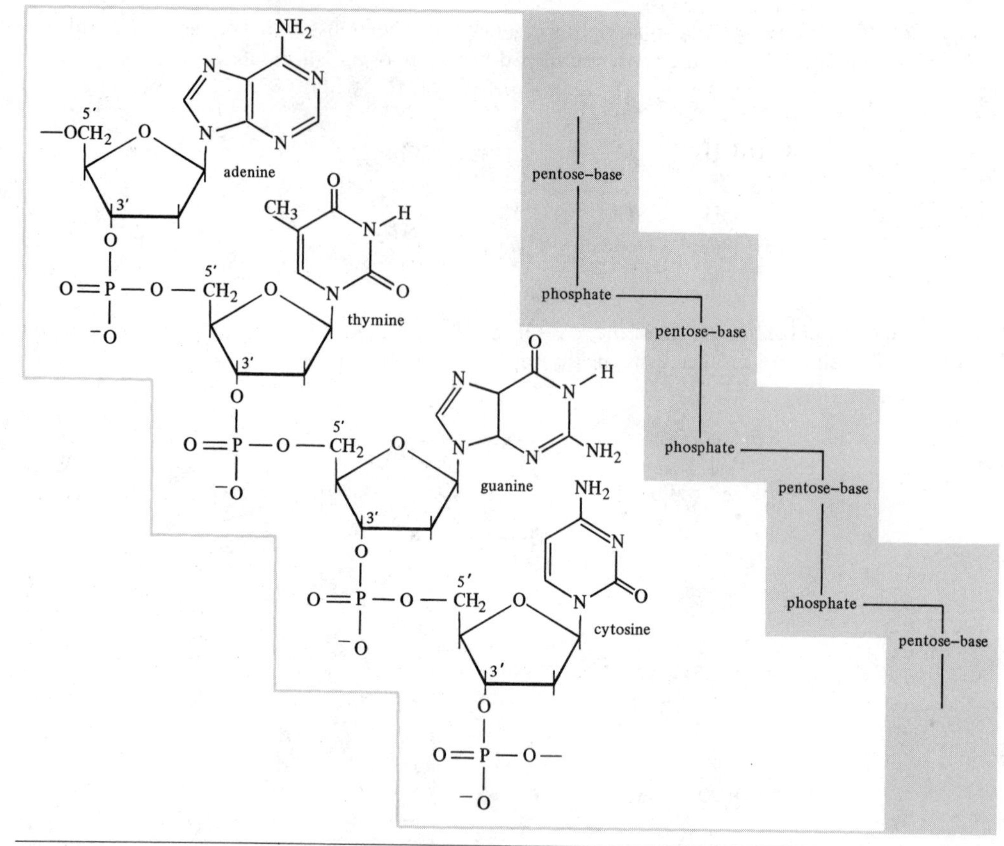

FIG. P.5

Hypothetical segment of a single DNA chain showing how phosphate ester groups link the 3′- and 5′-hydroxyls of deoxyribose units. RNA has a similar structure with two exceptions: a hydroxyl replaces hydrogen at the 2′-position of each ribose unit and uracil replaces thymine.

P.4B Secondary Structure

It was the now-classic proposal of Watson and Crick (made in 1953 and verified shortly thereafter by the X-ray analysis of Wilkins) that gave a model for the secondary structure of DNA. The secondary structure of DNA is especially important because it enables us to understand how the genetic information is preserved, how it can be passed on during the process of cell division, and how it can be transcribed to provide a template for protein synthesis.

Of prime importance to Watson and Crick's proposal was an earlier observation by E. Chargaff that certain regularities can be seen in the percentages of heterocyclic bases obtained from the DNA of a variety of species. Table P.1 gives results that are typical of those that can be obtained.

Chargaff pointed out that for all species examined:

1. The total mole percentage of purines is approximately equal to that of the pyrimidines, that is, $(\%G + \%A)(\%C + \%T) \simeq 1$.

2. The mole percentage of adenine is nearly equal to that of thymine (i.e., $\%A/\%T \simeq 1$) and that the mole percentage of guanine is nearly equal to that of cytosine (i.e., $\%G/\%C \simeq 1$).

Chargaff also noted that the ratio that varies from species to species is the ratio $(\%A + \%T)/(\%G + \%C)$. He noted, moreover, that while this ratio is characteristic of the DNA of a given species, it is the same for DNA obtained from different tissues of the same animal, and does not vary appreciably with the age or conditions of growth of individual organisms within the same species.

TABLE P.1 DNA Composition of Various Species

	BASE PROPORTIONS, MOLES %							
SPECIES	G	A	C	T	$\frac{G+A}{C+T}$	$\frac{A+T}{G+C}$	$\frac{A}{T}$	$\frac{G}{C}$
Sarcina lutea	37.1	13.4	37.1	12.4	1.02	0.35	1.08	1.00
Escherichia coli K12	24.9	26.0	25.2	23.9	1.08	1.00	1.09	0.99
Wheat germ	22.7	27.3	22.8[a]	27.1	1.00	1.19	1.01	1.00
Bovine thymus	21.5	28.2	22.5[a]	27.8	0.96	1.27	1.01	0.96
Staphylococcus aureus	21.0	30.8	19.0	29.2	1.11	1.50	1.05	1.11
Human thymus	19.9	30.9	19.8	29.4	1.01	1.52	1.05	1.01
Human liver	19.5	30.3	19.9	30.3	0.98	1.54	1.00	0.98

[a] Cytosine + methylcytosine.

Watson and Crick also had X-ray data that gave them the bond lengths and angles of the purines and pyrimidines of model compounds. In addition they had data from Wilkins that indicated an unusually long repeat distance, 34 Å, in natural DNA.

Reasoning from these data, Watson and Crick proposed a double helix as a model for the secondary structure of DNA. According to this model, two nucleic acid chains are held together by hydrogen bonds between base pairs on opposite strands. This double chain is wound into a helix with both chains sharing the same axis. The base pairs are on the inside of the helix and the sugar-phosphate backbone on the outside (Fig. P.6). The pitch of the helix is such that 10 successive nucleotide pairs give rise to one complete turn in 34 Å (the repeat distance). The exterior width of the spiral is ~20 Å and the internal distance between 1′-positions of ribose units on opposite chains is ~11 Å.

Using molecular scale models Watson and Crick observed that the internal distance of the double helix is such that it allows only a purine-pyrimidine type of hydrogen bonding between base pairs. Purine-purine base pairs do not occur because they would be too large to fit, and pyrimidine-pyrimidine base pairs do not occur because they would be too far apart to form effective hydrogen bonds.

Watson and Crick went one crucial step further in their proposal. Assuming that the oxygen-containing heterocyclic bases existed in keto forms they argued that base pairing through hydrogen bonds can only occur in a specific way:

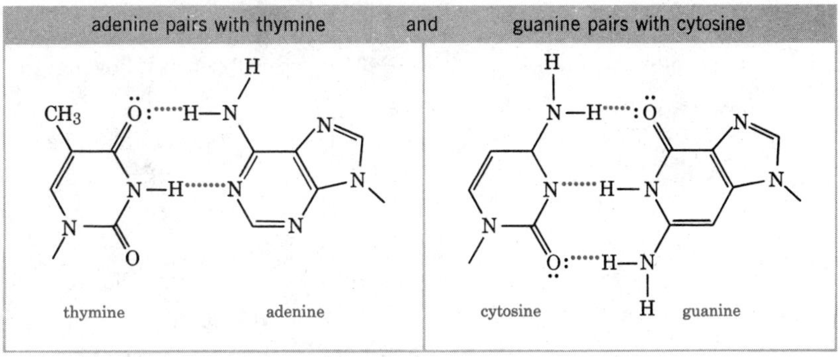

adenine pairs with thymine and guanine pairs with cytosine

thymine adenine cytosine guanine

The bond lengths of these base pairs are shown in Fig. P.7.

Specific base pairing of this kind is consistent with Chargaff's finding that %A/%T ≃ 1 and that %G/%C ≃ 1.

Specific base pairing also means that the two chains of DNA are complementary.

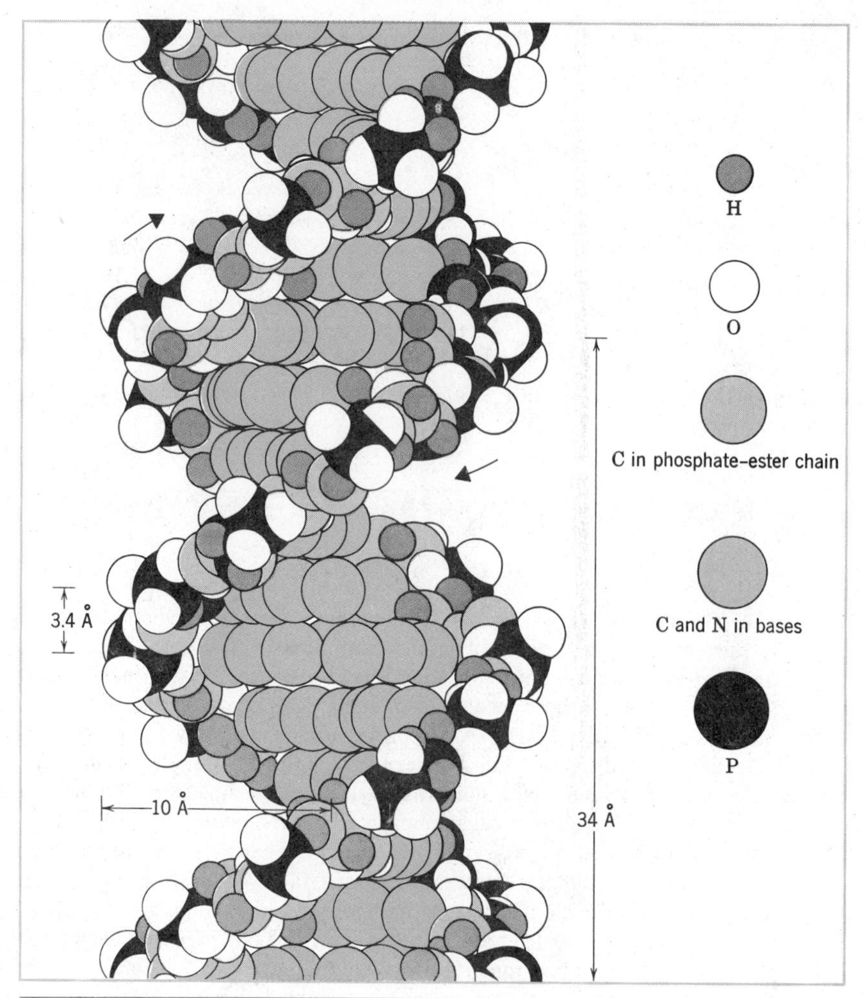

FIG. P.6
A molecular model of a portion of the DNA double helix. (From Chemistry and Biochemistry: A Comprehensive Introduction *by A. L. Neal. Copyright © 1971 by McGraw-Hill Inc. Used with permission of McGraw-Hill Book Company.**

Wherever adenine appears in one chain, thymine must appear opposite it in the other; wherever cytosine appears in one chain, guanine must appear in the other. Thus, a segment of a double chain might, if it were linear, resemble the following:

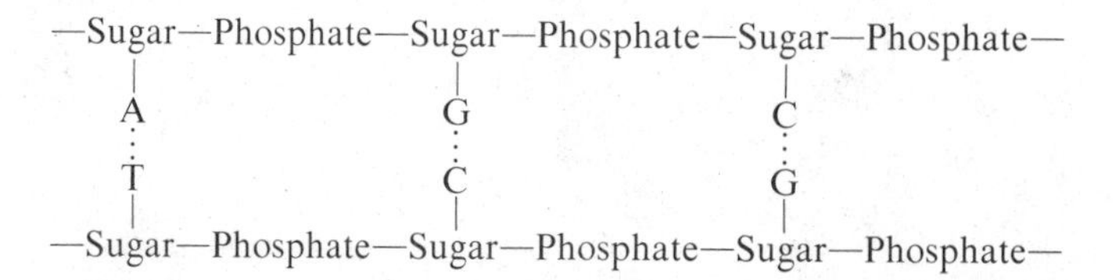

In Watson and Crick's own words, "The phosphate-sugar backbone of our model is completely regular, but any sequence of the pairs of bases can fit into the structure. It follows that, in a long molecule, many different permutations are possible, and it therefore seems likely that the precise sequence of the bases is the code which carries the genetic information. If the actual order of the bases on one of the pair of chains were given, one could write down the exact order of the bases on the other one, because of the specific

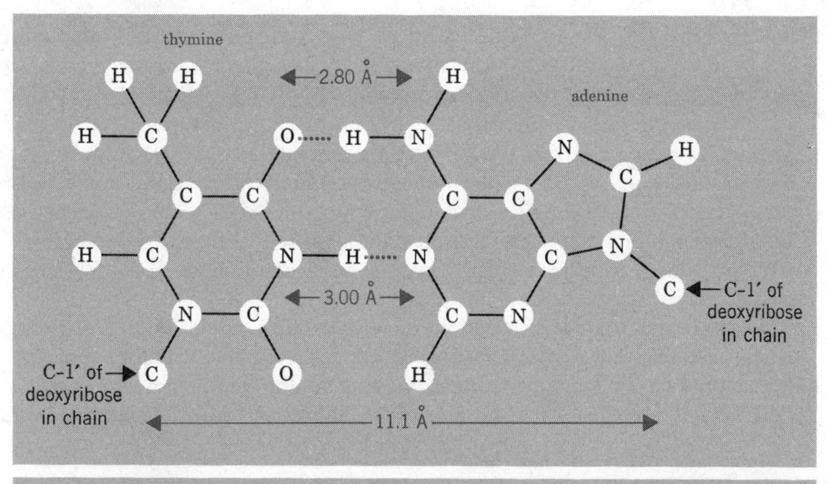

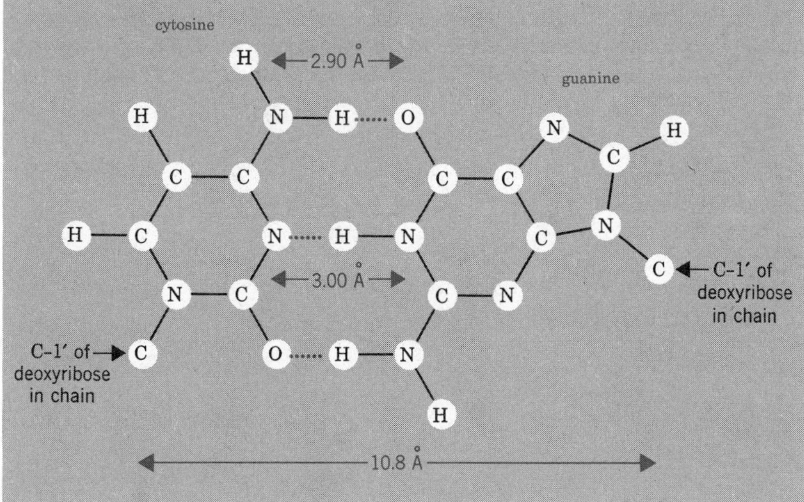

FIG. P.7
Dimensions of thymine-adenine and cytosine-guanine base pairs. The dimensions are such that they allow the formation of strong hydrogen bonds and also allow the base pairs to fit inside the two phosphate-ribose chains of the double helix. [Adapted from L. Pauling and R. B. Corey, Arch. Biochem. Biophys., 65, *164 (1956).*]

pairing. Thus, one chain is, as it were, the complement of the other, and it is this feature which suggests how the deoxyribonucleic acid molecule might replicate itself."

P.4C Replication of DNA

The Watson-Crick proposal gave, for the first time, a model that permitted an understanding of how the genetic information of a cell might be passed on to daughter cells at the time of cell division. Again, in their own words:

"Previous discussions of self duplication have usually involved the concept of a template or mould. Either the template or mould was supposed to copy itself directly or it was to produce a 'negative,' which in turn was to act as a template and produce the original 'positive' once again. In no case has it been explained in detail how it would do this in terms of atoms and molecules."

"Now our model for deoxyribonucleic acid is, in effect, a *pair* of templates, each of which is complementary to the other. We imagine that prior to duplication the hydrogen bonds are broken, and the two chains unwind and separate. Each chain then acts as a template for the formation on to itself of a new companion chain, so that eventually we shall have two pairs of chains, where we had only one before. Moreover, the sequence of pairs of bases will have been duplicated exactly." *

Thus, the genetic information is encoded in the pattern of bases (i.e., adenine, thymine, guanine, and cytosine), along the complementary DNA chains and all the directions necessary for the replication of a cell are written in the four-letters A,G,T,C. There are according to genetic calculations, approximately 1500 base pairs in a single gene. Four different bases give, therefore, a possibility of 4^{1500} different gene isomers. According to Watson, this number is so large that it is larger than all the different genes that have existed since life appeared on this planet.

Since the original publications of Watson and Crick, studies have been made that suggest that the chains of DNA do not unwind completely before replication occurs. Instead, the chains begin unwinding at one end and the complementary strands are formed as the unwinding takes place. An illustration of how this might occur is given in Fig. P.8.

Problem P.6

(a) There are approximately six billion base pairs in the DNA of a single human cell. Assuming that this DNA exists as a double helix calculate the length of all the DNA

FIG. P.8

Replication of DNA. The double strand unwinds from one end and complementary strands are formed along each chain.

```
Sugar—A···T—Sugar                              Sugar—A···T—Sugar
  |           |                                  |           |
Phosphate   Phosphate                          Phosphate   Phosphate
  |           |                                  |           |
Sugar—G···C—Sugar                              Sugar—G···C—Sugar
  |           |                                  |           |
Phosphate   Phosphate                          Phosphate   Phosphate
  |           |                                  |           |
Sugar—T···A—Sugar                              Sugar—T···A—Sugar
  |           |                                  |           |
Phosphate   Phosphate                          Phosphate   Phosphate
  |           |                                  |           |
Sugar—C···G—Sugar                              Sugar—C···G—Sugar
  \           |                                  |          /
  Phosphate                                             Phosphate
        Sugar—T                              A—Sugar
                   Phosphate      Phosphate
                        \            /
                     Sugar—C···G—Sugar
                       |           |
                     Phosphate   Phosphate
                       |           |
                     Sugar—A···T—Sugar
                       |           |
                     Phosphate   Phosphate
                       |           |
                     Sugar—G···C—Sugar
                       |           |
                     Phosphate   Phosphate
                       |           |
                     Sugar—C···G—Sugar
                       |           |
                     Phosphate   Phosphate
                       |           |
```

* This quotation and the previous ones were taken from J. D. Watson and F. H. C. Crick, "Genetical Implications of the Structure of Deoxyribonucleic Acid," *Nature,* 171 (1953), pp. 965, 966. Used with permission.

contained in a human cell. (b) The weight of DNA in a single human cell is 6×10^{-12} g. Assuming that the earth's population is about three billion we can conclude that all of the genetic information that gave rise to all human beings now alive was once contained in the DNA of a corresponding number of fertilized ova. What is the total weight of this DNA? (The volume that this DNA would occupy is approximately that of a raindrop, yet if the individual molecules were laid end to end they would stretch to the moon and back almost eight times.)

Problem P.7

(a) The most stable tautomeric form of guanine is the lactam form (below). This is the form normally present in DNA and, as we have seen, it pairs specifically with cytosine. If guanine tautomerizes to the abnormal lactim form, it pairs with thymine instead. Write structural formulas showing the hydrogen bonds in this abnormal base pair.

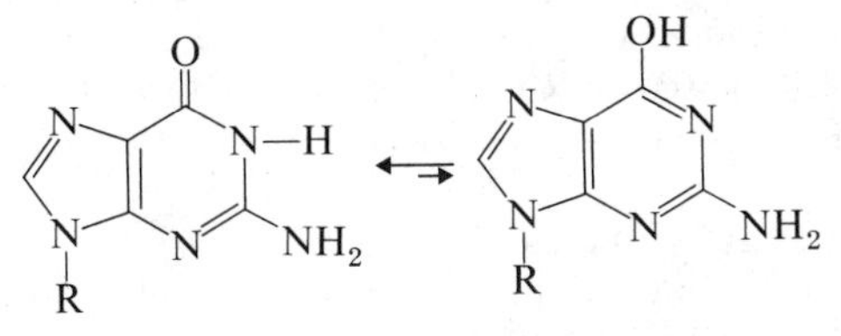

Lactam form of guanine ⇌ Lactim form of guanine

(b) Improper base pairings that result from tautomerizations occurring during the process of DNA replication have been suggested as a source of spontaneous mutations. We saw in part (a) that if a tautomerization of guanine occurred at the proper moment it could lead to the introduction of thymine (instead of cytosine) into its complementary DNA chain. What error would this new DNA chain introduce into *its* complementary strand during the next replication even if no further tautomerizations take place?

Problem P.8

Mutations can also be caused chemically, and nitrous acid is one of the most potent chemical *mutagens*. One explanation that has been suggested for the mutagenic effect of nitrous acid is the deamination reactions that it produces with purines and pyrimidines bearing amino groups. When, for example, an adenine-containing nucleotide is treated with nitrous acid, it is converted to a hypoxanthine derivative:

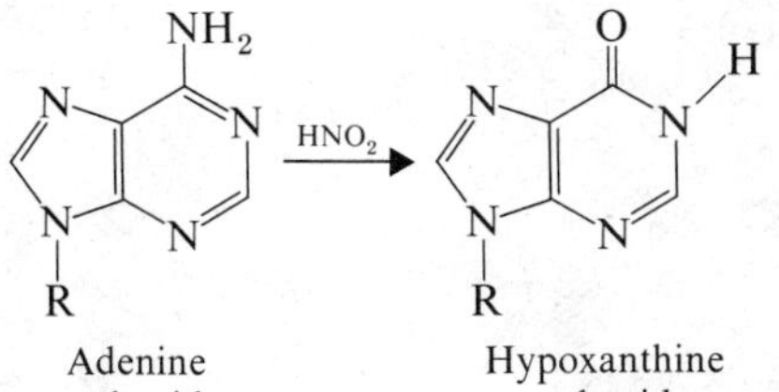

Adenine nucleotide $\xrightarrow{HNO_2}$ Hypoxanthine nucleotide

(a) Basing your answer on reactions you have seen before, what are likely intermediates in the adenine ⟶ hypoxanthine interconversion? (b) Adenine normally pairs with thymine in DNA, but hypoxanthine pairs with cytosine. Show the hydrogen bonds of a hypoxanthine-cytosine base pair. (c) Show what errors an adenine ⟶ hypoxanthine interconversion would generate in DNA through two replications.

P.5 RNA AND PROTEIN SYNTHESIS

Soon after the Watson-Crick hypothesis was published, scientists began to extend it to yield what Crick has called "the central dogma of molecular genetics." This dogma states that genetic information flows from:

$$\text{DNA} \longrightarrow \text{RNA} \longrightarrow \text{proteins}$$

The synthesis of proteins is, of course, all important to a cell's function because proteins (as enzymes) catalyze all its reactions. Even the very primitive cells of bacteria require as many as 3000 different enzymes. This means that the DNA molecules of these cells must contain a corresponding number of genes to direct the synthesis of these proteins. A gene is that segment of the DNA molecule that contains the information necessary to direct the synthesis of one protein (or one polypeptide).

DNA is found primarily in the nucleus of the cell. Protein synthesis takes place primarily in that part of the cell called the *cytoplasm*. Protein synthesis requires that two major processes take place; the first takes place in the cell nucleus, the second in the cytoplasm. The first is *transcription,* a process in which the genetic message is transcribed on to a form of RNA called messenger RNA (*m*RNA). The second process involves two other forms of RNA, called ribosomal RNA (*r*RNA) and transfer RNA (*t*RNA).

P.5A Messenger RNA Synthesis—Transcription

Protein synthesis begins in the cell nucleus with the synthesis of messenger RNA. Part of the DNA double helix unwinds sufficiently to expose on a single chain a portion corresponding to at least one gene. Ribonucleotides, present in the cell nucleus, assemble along the exposed DNA chain pairing with the bases of DNA. The pairing patterns are the same as those in DNA with the exception that in RNA uracil replaces thymine. The ribonucleotide units of messenger RNA are joined into a chain by an enzyme called RNA *polymerase*. This process is illustrated in Fig. P.9.

FIG. P.9
Transcription of the genetic code from DNA to messenger RNA.

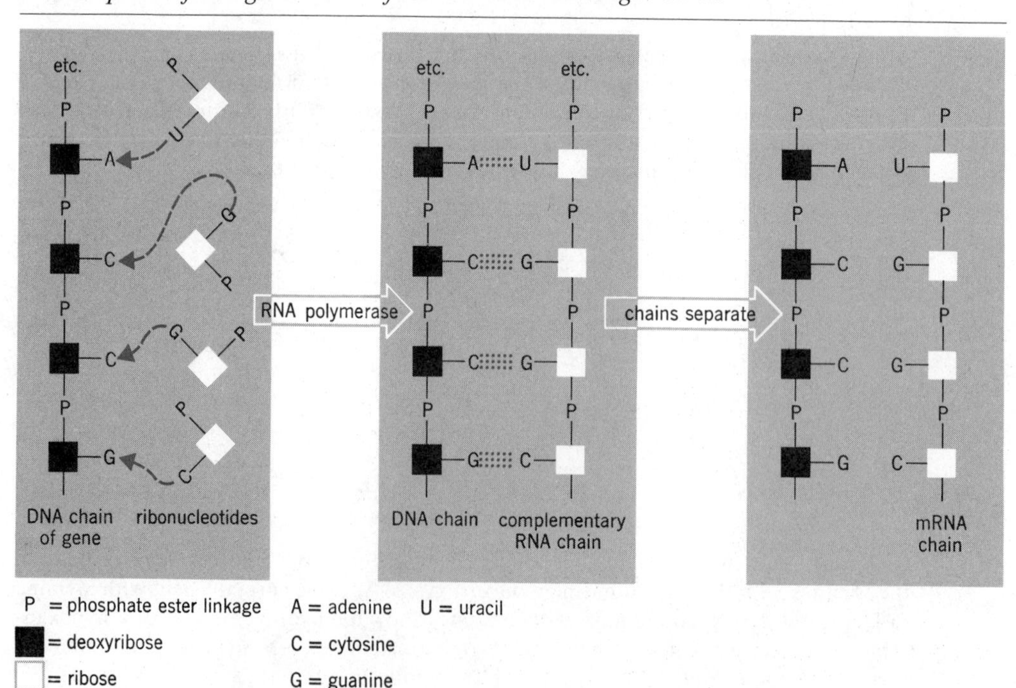

Problem P.9

Write structural formulas showing how the keto form of uracil in messenger RNA can pair with adenine in DNA through hydrogen bond formation.

After messenger RNA has been synthesized in the cell nucleus it migrates into the cytoplasm where, as we will see, it acts as a template for protein synthesis.

P.5B Ribosomes—rRNA

Scattered throughout the cytoplasm of most cells are small bodies called ribosomes. Ribosomes of *E. coli.* for example, are about 180 Å in diameter and are composed of approximately 60% RNA (ribosomal RNA) and 40% protein. They apparently exist as two associated subunits called the 50S and 30S subunits (Fig. P.10); together they form a 70S ribosome.* Although the ribosomes are at the site of protein synthesis, ribosomal RNA itself does not direct protein synthesis. Instead, a number of ribosomes become attached to a chain of messenger RNA and form what is called a *polysome*. It is along the polysome—with messenger RNA acting as the template—that protein synthesis takes place. One of the functions of ribosomal RNA is to bind the ribosome to the messenger RNA chain.

P.5C Transfer RNA

Transfer RNA has a very low molecular weight when compared to that of messenger RNA or ribosomal RNA. Transfer RNA, consequently, is much more soluble than messenger RNA or ribosomal RNA and is sometimes referred to as soluble RNA. The function of transfer RNA is to transport amino acids to specific areas of the messenger RNA of the polysome. There are, therefore, at least 20 different forms of transfer RNA, one for each of the 20 amino acids that are incorporated into proteins.†

The structures of a number of transfer RNA's have been determined. They are composed of a relatively small number of nucleotide units (70–90 units) folded into several loops or arms through base pairing along the chain (Fig. P.11). One arm always terminates in the sequence cytosine-cytosine-adenine. It is to this arm that a specific amino acid becomes attached *through an ester* linkage to the 3′—OH of the terminal adenine. This attachment reaction is catalyzed by an enzyme that is specific for the transfer RNA and for the amino acid. The specificity may grow out of the enzyme's ability to recognize base sequences along other arms of the transfer RNA.

At the loop of still another arm is a specific sequence of bases, called the *anticodon*. The anticodon is highly important because it allows the transfer RNA to bind with a specific site—called the *codon*—of messenger RNA. The order in which amino acids are brought by their *t*RNA units to the *m*RNA strand is therefore determined by the sequence of codons. This sequence, therefore, constitutes a genetic message. Individual units of that message (the individual words, each corresponding to an amino acid) are triplets of nucleotides.

* S stands for svedberg unit; it is used in describing the behavior of proteins in an ultracentrifuge.

† Although proteins are composed of 22 different amino acids, protein synthesis requires only 20. Proline is converted to hydroxyproline and cysteine is converted to cystine after synthesis of the polypeptide chain has taken place.

FIG. P.10
A 70 S ribosome showing the two subunits.

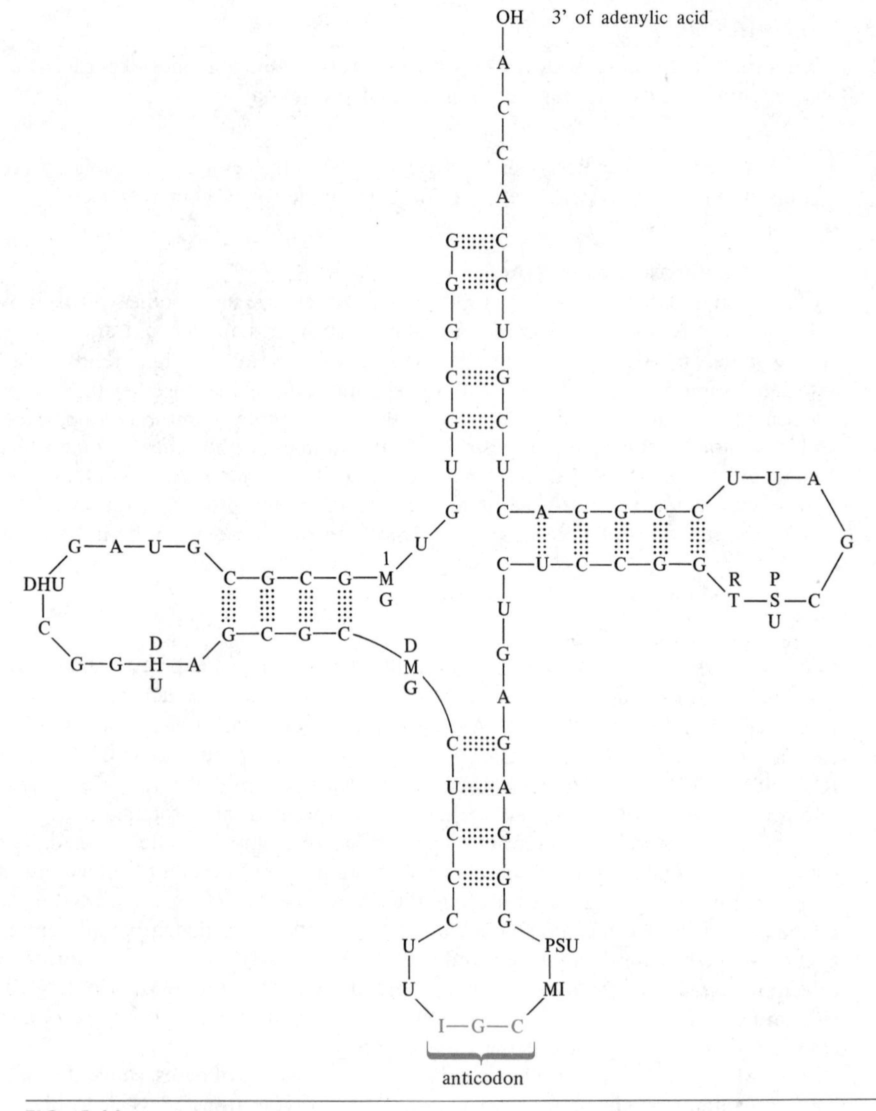

FIG. P.11
Structure of a transfer RNA isolated from yeast that has the specific function of transferring alanine residues. Transfer RNA's often contain unusual nucleosides. PSU = pseudouridine, RT = ribothymidine, MI = 1-methylinosine, I = inosine, DMG = N*-(2)-methylguanosine, DHU = 4,5-dihydrouridine, 1 MG = 1-methylguanosine.*

P.5D The Genetic Code

Which triplet on *m*RNA corresponds to which amino acid is called the genetic code (see Table P.2). The code must be in the form of three bases, not one or two because there are 20 different amino acids used in protein synthesis but there are only four different bases in messenger RNA. If only two bases were used, there would be only 4^2 or 16 possible combinations, a number too small to accommodate all of the possible amino acids. However, with a three-base code, 4^3 or 64 different sequences are possible. This is far more than are needed and it allows for multiple ways of specifying an amino acid. It also allows for sequences that punctuate protein synthesis, sequences that say, in effect, "start here" and "end here."

Both methionine (Met) and *N*-formylmethionine ($\text{Met}_{\text{formyl}}$) have the same mes-

TABLE P.2 The Messenger RNA Genetic Code

AMINO ACID	BASE SEQUENCE	AMINO ACID	BASE SEQUENCE	AMINO ACID	BASE SEQUENCE
Ala	GCA	His	CAC	Ser	AGC
	GCC		CAU		AGU
	GCG				UCA
	GCU	Ile	AUA		UCG
			AUC		UCC
Arg	AGA		AUU		UCU
	AGG				
	CGA	Leu	CUA	Thr	ACA
	CGC		CUC		ACC
	CGG		CUG		ACG
	CGU		CUU		ACU
			UUA		
Asn	AAC		UUG	Tyr	UGG
	AAU				UAC
		Lys	AAA		UAU
Asp	GAC		AAG		
	CAU			Val	GUA
					GUG
Cys	UGC	Met	AUG		GUC
	UGU				GUU
		Phe	UUU		
			UUC		
Gln	CAA			Chain initiation:	
	CAG	Pro	CCA		
			CCC	Met_{formyl}	AUG
Glu	GAA		CCG		
	GAG		CCU	Chain termination:	
Gly	GGA				
	GGC				UAA
	GGG				UAG
	GGU				

senger RNA code (AUG); however, *N*-formylmethionine (below) is carried by a different

$$\begin{array}{l} CH_3SCH_2CH_2CHCOOH \\ \qquad\qquad\quad | \\ \qquad\qquad\quad NH \\ \qquad\qquad\quad | \\ \qquad\qquad\quad C{=}O \\ \qquad\qquad\quad | \\ \qquad\qquad\quad H \end{array}$$

N-Formylmethionine

transfer RNA from that which carries methionine. *N*-Formylmethionine appears to be the first amino acid incorporated into the chain of proteins in bacteria and the transfer RNA that carries Met_{formyl} appears to be the punctuation mark that says "start here." Before the polypeptide synthesis is complete, *N*-formylmethionine is removed from the protein chain by an enzymatic hydrolysis.

We are now in a position to see how the synthesis of a hypothetical polypeptide might take place. Let us imagine that a long strand of messenger RNA is in the cytoplasm of a cell and that it is in contact with ribosomes. Also in the cytoplasm are the 20 different amino acids, each acylated to its own specific transfer RNA.

As shown in (Fig. P.12), a transfer RNA bearing Met_{formyl} uses its anticodon to associate with the proper codon (AUG) on that portion of messenger RNA that is in contact with a ribosome. The next triplet of bases on this particular messenger RNA chain is AAA; this is the codon that specifies lysine. A lysyl-transfer RNA with the matching anticodon UUU attaches itself to this site. The two amino acids, Met_{formyl} and Lys, are now in the proper position for an enzyme to join them in peptide linkage. After this happens, the ribosome moves down the chain so that it is in contact with the next codon. This one, GUA, specifies valine. A transfer RNA bearing valine (and with the proper anticodon) binds itself to this site. Another enzymatic reaction takes place attaching valine to the polypeptide chain. Then the whole process repeats itself again and again. The ribosome moves along the messenger RNA chain, other transfer RNA's move up with their amino acids, new peptide bonds are formed, and the polypeptide chain grows. At some point an enzymatic reaction removes Met_{formyl} from the beginning of the chain. Finally, when the chain is the proper length the ribosome reaches a punctuation mark, UAA, saying "stop here." The ribosome separates from the messenger RNA chain and so, too, does the protein.

Even before the polypeptide chain is fully grown, it begins to form its own specific secondary and tertiary structure (Fig. P.13). This happens because its primary structure is correct—its amino acids are ordered in just the right way. Hydrogen bonds form giving rise to specific segments of α helix, pleated sheet, and random coil. Then the whole thing folds and bends; enzymes install disulfide linkages, so that when the chain is fully grown, the whole protein has just the shape it needs to do its job.

If this protein happens to be lysozyme it has a deep cleft, or jaw, where a specific polysaccharide fits. And if it is lysozyme, and a certain bacterium wanders by, that jaw begins to work; it bites its first polysaccharide in half.

In the meantime other ribosomes nearer the beginning of the messenger RNA chain are already moving along, each one synthesizing another molecule of the polypeptide. The time required to synthesize a protein depends, of course, on the number of amino residues it contains, but indications are that each ribsome can cause 150 peptide bonds to be formed each minute. Thus, a protein, such as lysozyme, with 129 amino acid residues requires less than a minute for its synthesis. However, if four ribosomes are working their way along a single messenger RNA chain, the polysome can produce a lysozyme molecule every 13 seconds.

But why, we might ask, is all this protein synthesis necessary—particularly in a fully grown organism? The answer is that proteins are not permanent; they do not get themselves synthesized once and remain intact in the cell for the lifetime of the organism. They are synthesized when and where they are needed. Then they are taken apart, back to amino acids; enzymes disassemble enzymes. Some amino acids get metabolized for energy; others—new ones—come in from the food we have eaten and the whole process begins again.

Problem P.10

A segment of DNA has the following sequence of bases:

. . . A C C C C C A A A A T G T C G . . .

(a) What sequence of bases would appear in messenger RNA transcribed from this segment? (b) Assume that the first base in this messenger RNA is the beginning of a codon. What order of amino acids would be translated into a polypeptide synthesized along this segment? (c) Give anticodons for each transfer RNA associated with the translation in part (b).

Problem P.11

(a) Using the first codon given for each amino acid in Table P.2 write the base sequence of messenger RNA that would translate the synthesis of the following pentapeptide:

Arg · Ile · Cys · Tyr · Val

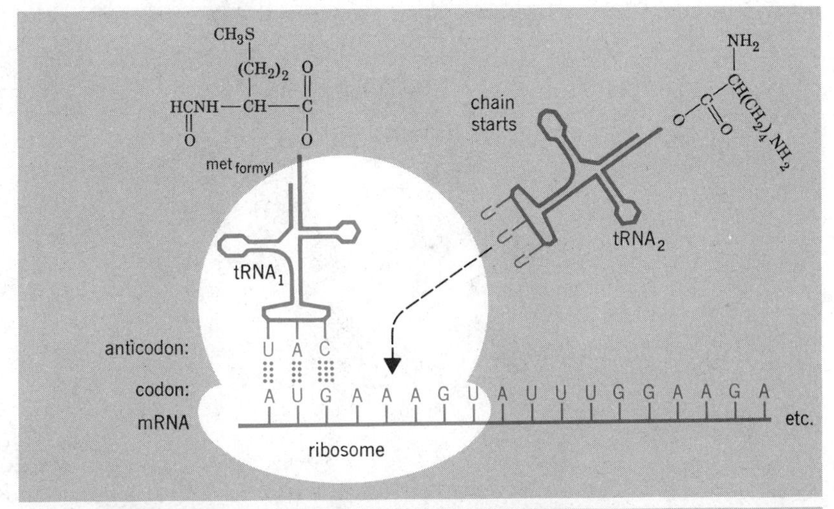

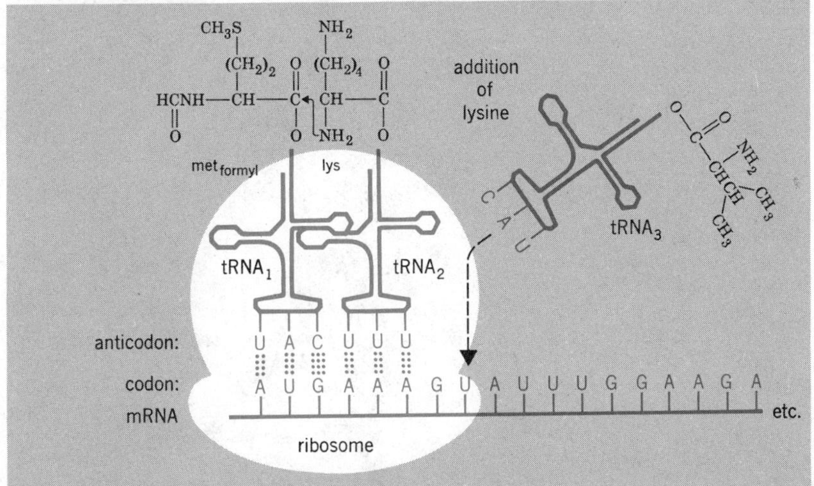

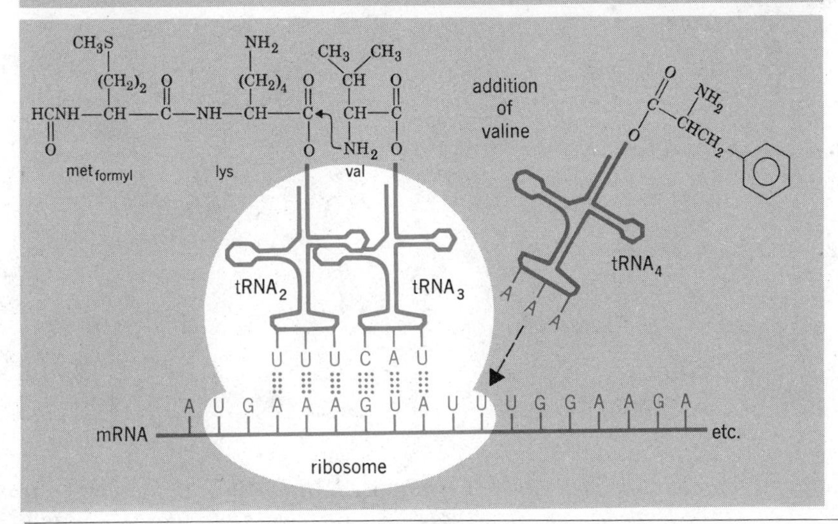

FIG. P.12

Step-by-step growth of a polypeptide chain with messenger RNA acting as a template. Transfer RNA's carry amino acid residue to the site of messenger RNA that is in contact with a ribosome. Codon-anticodon pairing occurs between messenger RNA and transfer RNA at the ribosomal surface. An enzymatic reaction joins the amino acid residues through an amide linkage. After the first amide bond is formed the ribosome moves to the next codon on messenger RNA. A new transfer RNA arrives, pairs, and transfers its amino acid residues to the growing peptide chain, and so on.

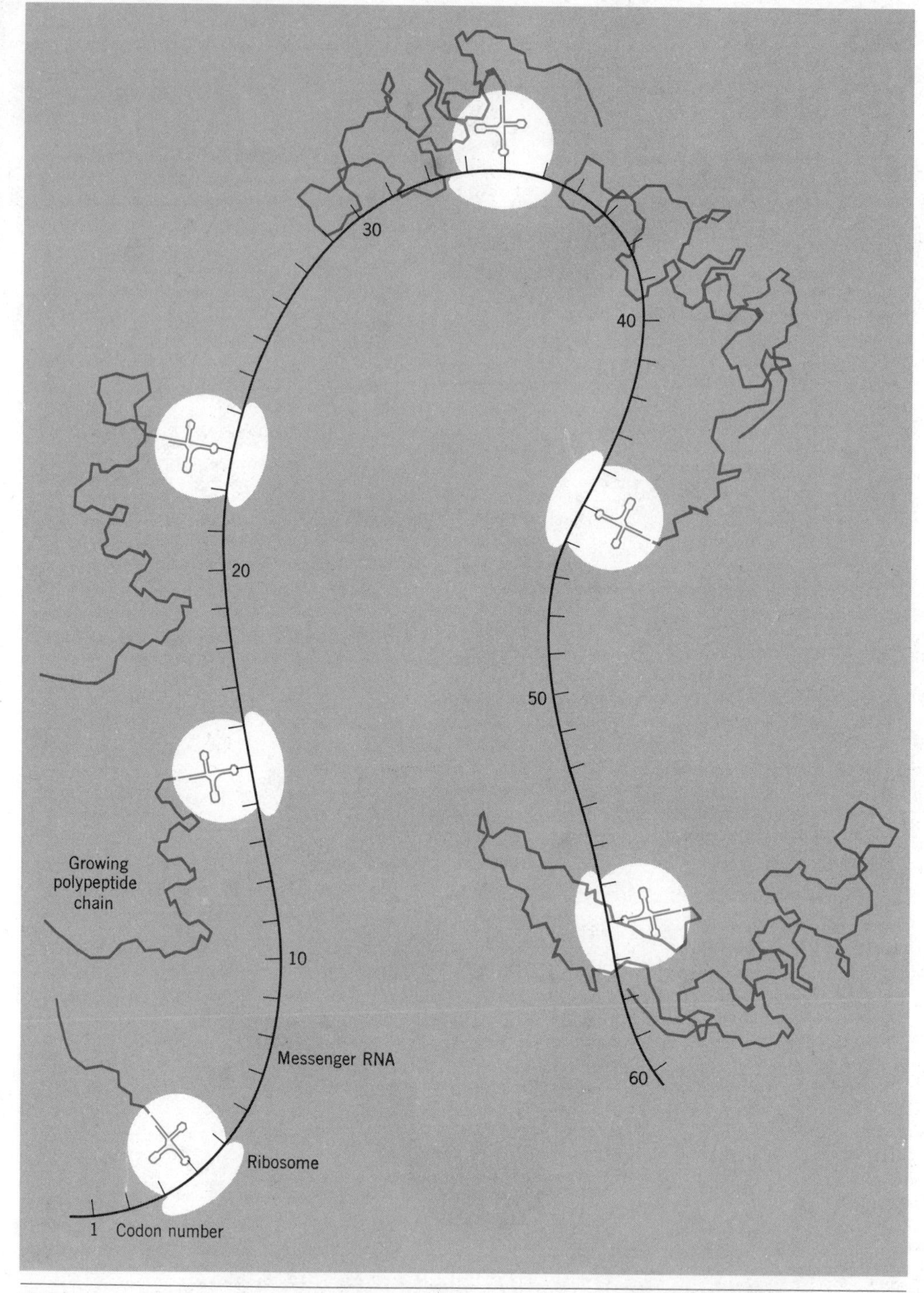

FIG. P.13

The folding of a protein molecule as it is synthesized. [*Adapted with permission from D. C. Phillips, "The Three-Dimensional Structure of an Enzyme Molecule,"* Copyright © *1966* Scientific American, Inc. *All rights reserved.*]

(b) What base sequence in DNA would transcribe a synthesis of the messenger RNA?
(c) What anticodons would appear in the transfer RNA's involved in the pentapeptide synthesis?

Problem P.12

Explain how an error of a single base in each strand of DNA could bring about the amino acid residue error that causes sickle cell anemia (p. 997).

SPECIAL TOPIC

REACTIONS CONTROLLED BY ORBITAL SYMMETRY

Q.1 INTRODUCTION

In recent years chemists have found that there are many reactions where certain symmetry characteristics of molecular orbitals control the overall course of the reaction. These reactions are often called *pericyclic reactions* because they take place through cyclic transition states. Now that we have a background knowledge of molecular orbital theory—especially as it applies to conjugated polyenes (e.g., dienes, trienes, etc.)—we are in a position to examine some of the intriguing aspects of these reactions. We will look in detail at two basic types: *electrocyclic reactions* and *cycloaddition reactions.*

Q.2 ELECTROCYCLIC REACTIONS

A number of reactions, like the one shown below, transform a conjugated polyene into a cyclic compound.

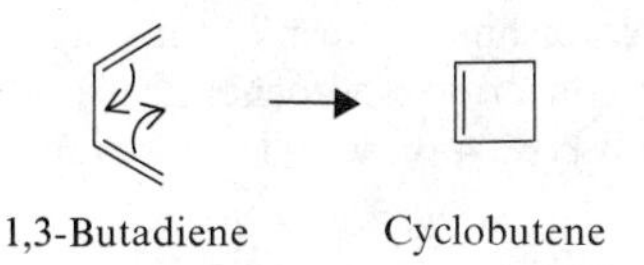

In many other reactions the ring of a cyclic compound opens and a conjugated polyene forms.*

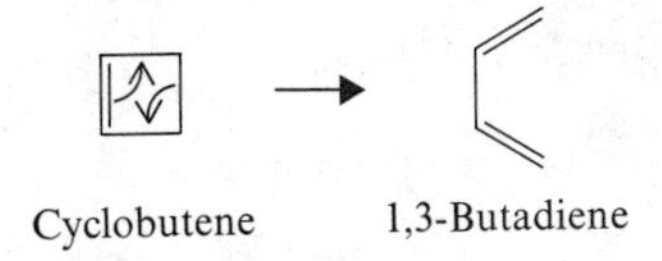

Reactions of either type are called *electrocyclic reactions.*

In electrocyclic reactions σ and π bonds are interconverted. In our first example, one π bond of 1,3-butadiene becomes a σ bond in cyclobutene. In our second example, the reverse is true: a σ bond of cyclobutene becomes a π bond in 1,3-butadiene.

Electrocyclic reactions have several characteristic features:

1. They require only heat or light for initiation.
2. Their mechanisms do not involve free-radical or ionic intermediates.

* The double-barbed arrows are included in these equations as an aid in keeping track of the electrons. They are not intended to indicate that this is the actual mechanism.

3. Bonds are made and broken in *a single concerted and cyclic transition state.*
4. The reactions are *highly stereospecific.*

The examples that follow demonstrate this last characteristic of electrocyclic reactions.

trans, trans-2,4-Hexadiene $\xrightarrow{h\nu}$ *cis*-3,4-Dimethylcyclobutene

trans, cis, trans-2,4,6-Octatriene $\xrightarrow{\text{Heat}}$ *cis*-5,6-Dimethyl-1,3-cyclohexadiene

cis-3,4-Dimethylcyclobutene $\xrightarrow{\text{Heat}}$ *cis, trans*-2,4-Hexadiene

In each of these three examples a single stereoisomeric form of the reactant yields a single stereoisomeric form of the product. The concerted photochemical cyclization of *trans, trans*-2,4-hexadiene, for example, yields only *cis*-3,4-dimethylcyclobutene; it does not yield *trans*-3,4-dimethylcyclobutene.

trans, trans-2,4-Hexadiene $\xrightarrow{h\nu}$ (blocked) *trans*-3,4-Dimethylcyclobutene (not formed)

The other two concerted reactions are characterized by the same stereospecificity.

The electrocyclic reactions that we will study here and the concerted cycloaddition reactions that we will study in the next section were poorly understood by chemists before 1960. Since then, several scientists, most notably K. Fukui in Japan, H. C. Longuet-Higgins in England, and R. B. Woodward and R. Hoffmann at Harvard University, have provided us with a basis for understanding how these reactions occur and why they take place with such remarkable stereospecificity.

All of these men worked from molecular orbital theory. In 1965, Woodward and Hoffmann formulated their theoretical insights into a set of rules that not only enabled chemists to understand reactions that were already known but that correctly predicted the outcome of many reactions that had not been attempted.

The Woodward-Hoffmann rules are formulated for concerted reactions only. Concerted reactions are reactions in which bonds are broken and formed simultaneously and thus no intermediates occur. The Woodward-Hoffmann rules are based on this hypothesis: *in concerted reactions molecular orbitals of the reactant are continuously converted into molecular orbitals of the product*. This conversion of molecular orbitals is not a random one, however. Molecular orbitals have symmetry characteristics. Because they do, restrictions exist on which molecular orbitals of the reactant may be transformed into particular molecular orbitals of the product.

According to Woodward and Hoffmann certain reaction paths are said to be *symmetry allowed* while others are said to be *symmetry forbidden*. To say that a particular path is symmetry forbidden does not necessarily mean, however, that the reaction will not occur. It simply means that if the reaction were to occur through a symmetry-forbidden path, the concerted reaction would have a much higher energy of activation. The reaction may occur, but it will probably do so in a different way: through another path that is symmetry allowed or through a nonconcerted path.

A complete analysis of electrocyclic reactions using the Woodward-Hoffmann rules requires a correlation of symmetry characteristics of *all* of the molecular orbitals of the reactants and products. Such analyses are beyond the scope of our discussion here. We will find, however, that a simplified approach can be undertaken; one that will be easy to visualize and, at the same time, will be accurate in most instances. In this simplified approach to electrocyclic reactions we focus our attention only on the *highest occupied molecular orbital* (HOMO) *of the conjugated polyene*. This approach is based on a method developed by Fukui called the *frontier orbital method*.

Q.2A Electrocyclic Reactions of 4*n*-π-Electron Systems

Let us begin with an analysis of the thermal interconversion of *cis*-3,4-dimethylcyclobutene and *cis,trans*-2,4-hexadiene shown below.

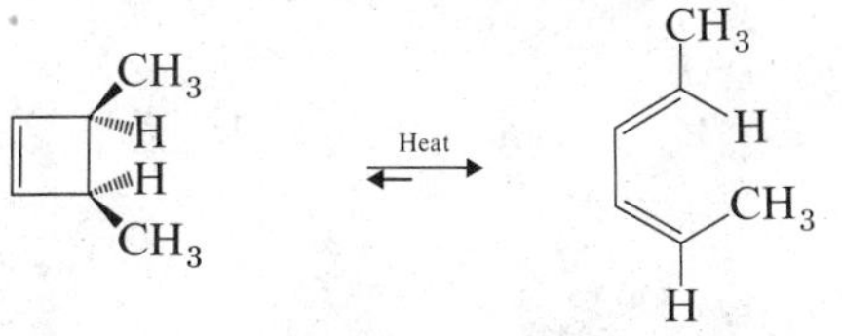

cis-3,4-Dimethylcyclobutene   *cis, trans*-2,4-Hexadiene

Electrocyclic reactions are reversible. According to the principle of microscopic reversibility, the path for the forward reaction is the same as that for the reverse reaction. In this example it is easier to see what happens to the orbitals, if we follow the *cyclization* reaction, *cis,trans*-2,4-hexadiene ⟶ *cis*-3,4-dimethylcyclobutene.

In this cyclization one π bond of the hexadiene is transformed into a σ bond of the cyclobutene. But which π bond? And, how does the conversion occur?

Let us begin by examining the π molecular orbitals of a hexadiene and, in particular, let us look at *the highest occupied molecular orbital of the ground state* (Fig. Q.1).

The cyclization that we are concerned with now, *cis,trans*-2,4-cyclohexadiene ⇌ *cis*-3,4-dimethylcyclobutene, requires heat alone. We conclude, therefore, that excited states of the hexadiene are not involved, for these would require the absorption of light. If we focus our attention on Ψ_2—the highest occupied molecular orbital of the ground state—we can see how the *p* orbitals at carbon-2 and carbon-5 can be transformed into a σ bond in the cyclobutene.

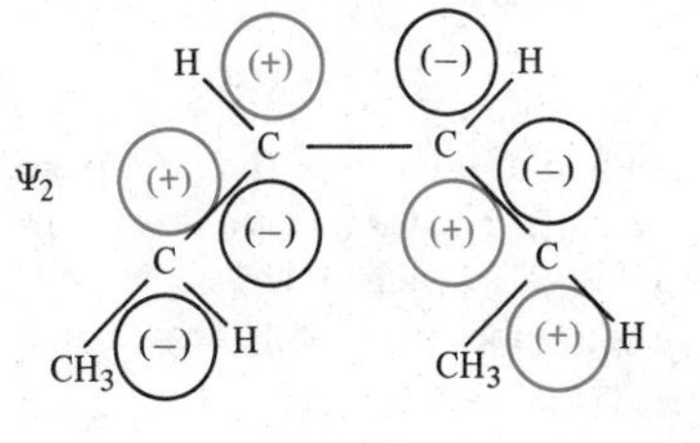

highest occupied molecular orbital (HOMO) of the ground state

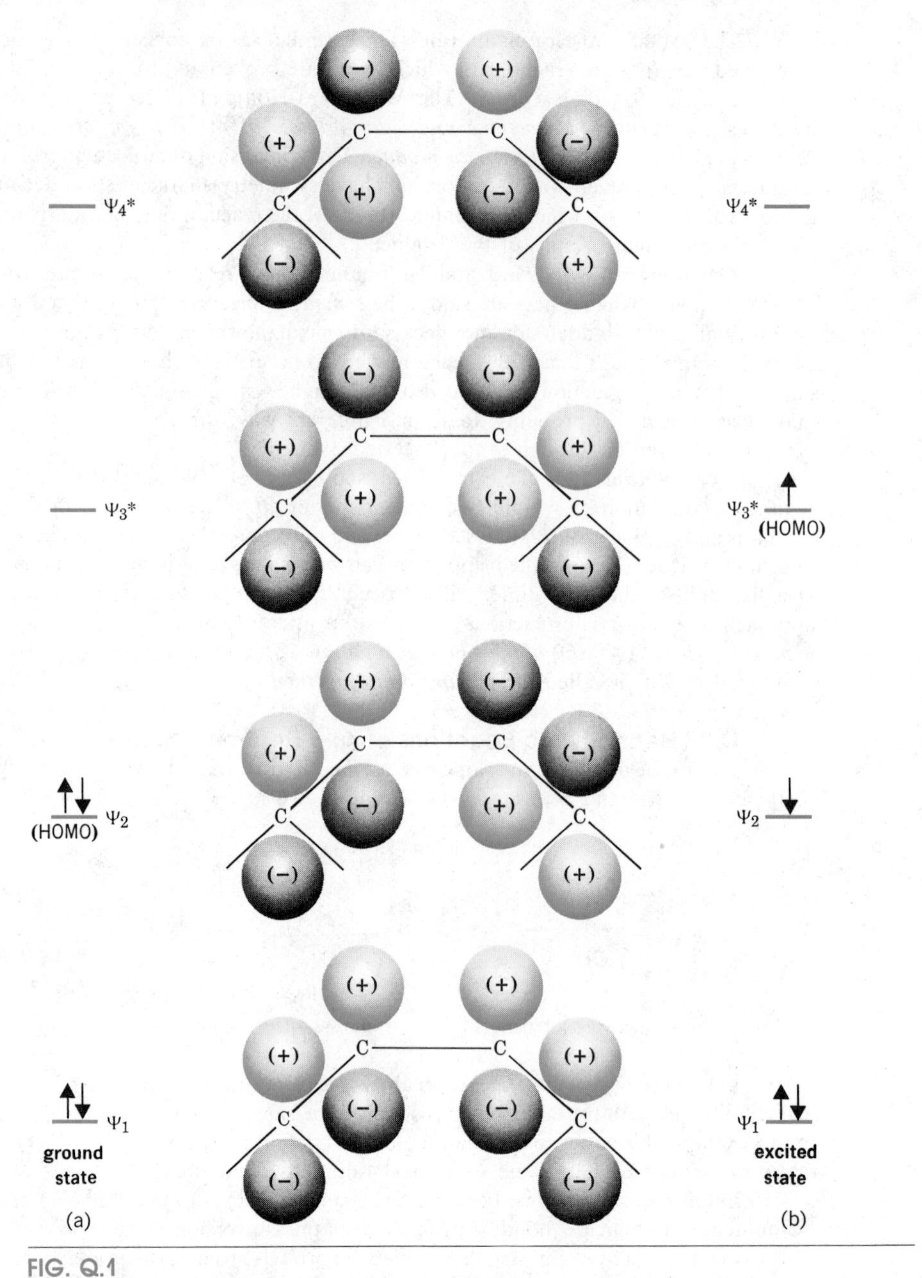

FIG. Q.1
The π molecular orbitals of a 2,4-hexadiene. (a) *The electron distribution of the ground state.* (b) *The electronic distribution of the first excited state. (The first excited state is formed when the molecule absorbs a photon of light of the proper wavelength.) Notice that the orbitals of a 2,4-hexadiene are like those of 1,3-butadiene shown in Figure 10.5.)*

A bonding σ-molecular orbital between C-2 and C-5 is formed when the *p* orbitals *rotate in the same direction* (both clockwise, as shown at the top of the next page, or both counterclockwise, which leads to an equivalent result). The term *conrotatory* is used to describe this type of motion of the two *p* orbitals relative to each other.

Conrotatory motion allows *p*-orbital lobes of the *same sign* to overlap. It also places

Ψ_2

conrotatory motion
(leads to bonding interaction between C-2 and C-5)

the two methyl groups on the same side of the molecule in the product, that is, in the *cis* configuration.*

The pathway with conrotatory motion of the methyl groups is consistent with what we know from experiments to be true: the *thermal reaction* results in the interconversion of *cis*-3,4-dimethylcyclobutene and *cis,trans*-2,4-hexadiene.

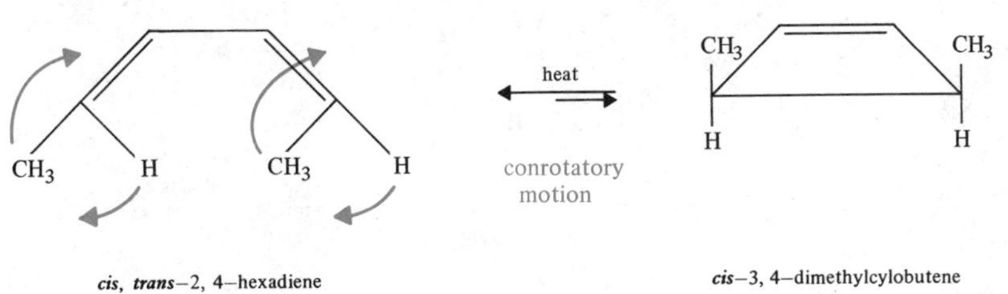

We can now examine another 2,4-hexadiene ⇄ dimethylcyclobutene interconversion: one that takes place under the influence of light. This reaction is shown below.

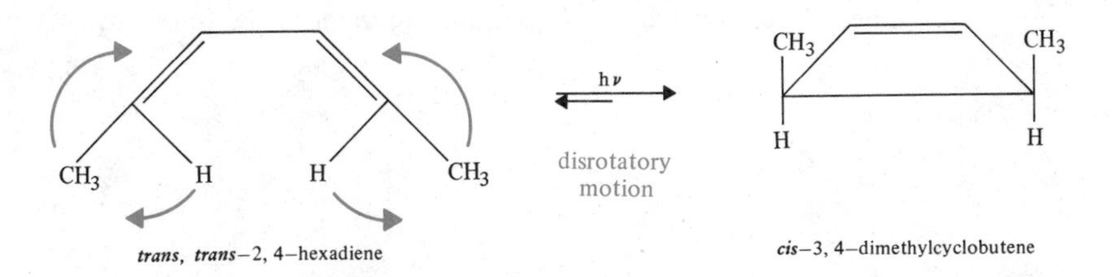

In the photochemical reaction *cis*-3,4-dimethylcyclobutene and *trans, trans*-2,4-hexadiene are interconverted. The photochemical interconversion occurs with the methyl groups rotating in *opposite directions,* that is, with the methyl groups undergoing *disrotatory motion*.

The photochemical reaction can also be understood by considering orbitals of the 2,4-hexadiene. In this reaction, however—since the absorption of light is involved—we

* Notice that if conrotatory motion occurs in the opposite (counterclockwise) direction, lobes of the same sign still overlap, and the methyl group are still *cis.*

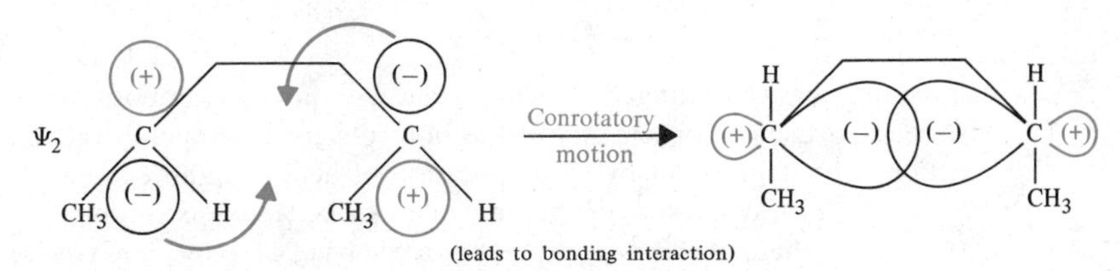

want to look at the first *excited state* of the hexadiene. We want to examine Ψ_3^*, because in the first excited state, Ψ_3^* *is the highest occupied molecular orbital.*

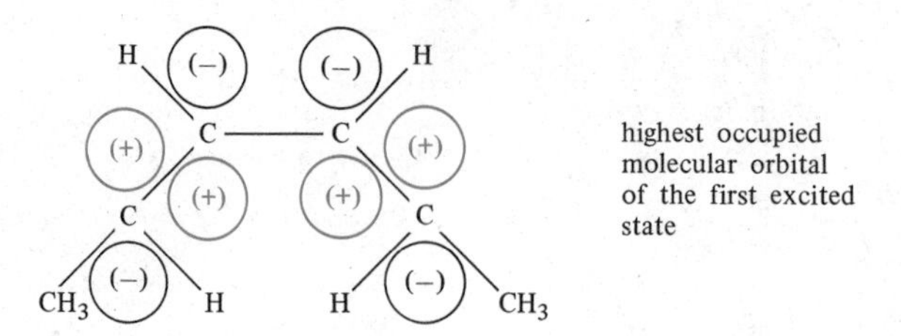

highest occupied molecular orbital of the first excited state

We find that disrotatory motion of the orbitals at carbons-2 and -5 of Ψ_3^* allows lobes of the same sign to overlap and form a bonding sigma molecular orbital between them.

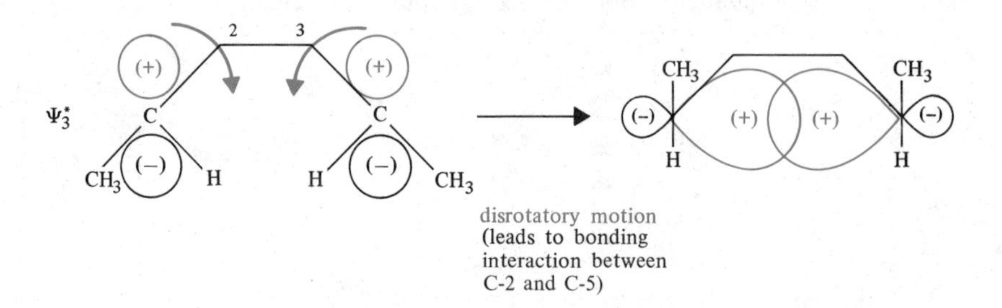

disrotatory motion (leads to bonding interaction between C-2 and C-5)

Disrotatory motion of the orbitals, of course, also requires disrotatory motion of the methyl groups and, once again, this is consistent with what we find experimentally. The *photochemical reaction* results in the interconversion of *cis*-3,4-dimethylcyclobutene and *trans, trans*-2,4-hexadiene.

Since both of the interconversions that we have presented, so far, involve *cis*-3,4-dimethylcyclobutene, we can summarize them in the following way.

Heat

Conrotatory

cis, trans-2,4-Hexadiene

Disrotatory

$h\nu$

trans, trans-2,4-Hexadiene

We see that these two interconversions occur with precisely opposite stereochemistry. We also see that the stereochemistry of the interconversions depends on whether or not the reaction is brought about by the application of heat or light.

The first Woodward-Hoffmann rule can be stated as follows:

1. A thermal electrocyclic reaction involving $4n$ π electrons (where $n = 1,2,3,\ldots$)

proceeds with conrotatory motion; the photochemical reaction proceeds with disrotatory motion.

Both of the interconversions that we have studied involve systems of 4 π electrons and both follow this rule. Many other $4n$-π-electron systems have been studied since Woodward and Hoffmann stated their rule. All were found to follow it.

Before we leave the subject of $4n$-π-electron systems let us illustrate an application of the rule with one other example.

When *trans*-3,4-dimethylcyclobutene is heated, ring opening occurs and *trans, trans*-2,4-hexadiene is formed.

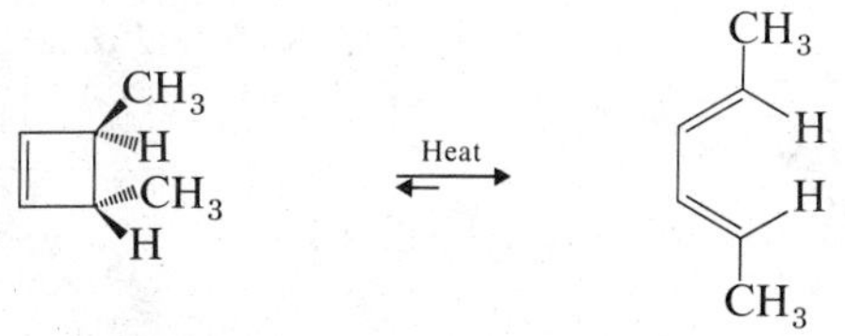

trans-3,4-Dimethylcyclobutene   *trans, trans*-2,4-Hexadiene

According to the Woodward-Hoffmann rule this thermal reaction of a 4-π electron system should occur with *conrotatory motion; and this is precisely what happens. trans*-3,4-Dimethylcyclobutene is transformed into *trans,trans*-2,4-hexadiene.

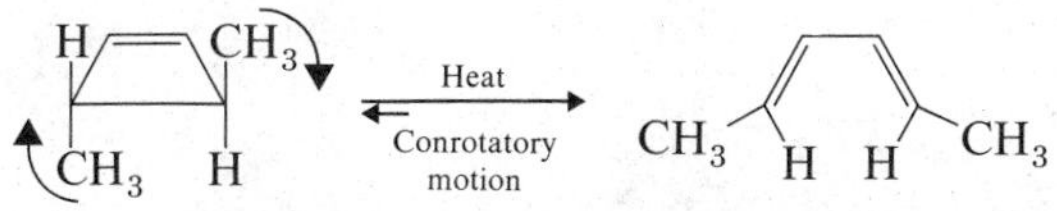

trans, trans-2,4-Hexadiene

Problem Q.1

In the example given above, another conrotatory path is available. This path would produce *cis,cis*-2,4-hexadiene. Can you suggest a reason that will account for the fact that this path is not followed to any appreciable extent?

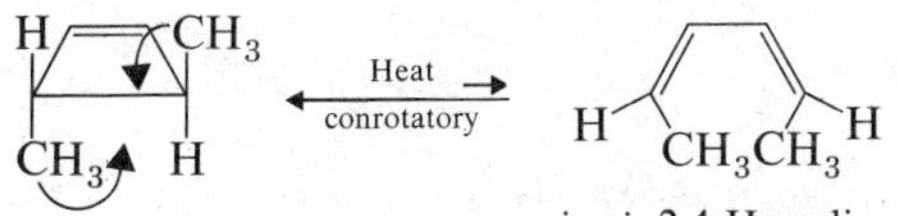

cis, cis-2,4-Hexadiene

Problem Q.2

What product would you expect from a concerted photochemical cyclization of *cis,trans*-2,4-hexadiene?

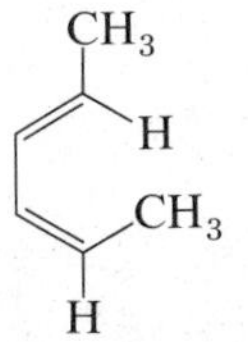

cis, trans-2,4-Hexadiene

Problem Q.3

(a) Show the orbitals involved in the following thermal electrocyclic reaction.

CH3 CH3 H H CH3 CH3 —200°→ CH3 H CH3 H CH3 CH3

(b) Do the groups rotate in a conrotatory or disrotatory manner?

Problem Q.4

Can you suggest a method for carrying out a stereospecific conversion of *trans,trans*-2,4-hexadiene into *cis,trans*-2,4-hexadiene?

Problem Q.5

The following 2,4,6,8-tetradecaenes undergo ring closure to dimethylcyclooctatrienes when heated. What product would you expect from each reaction?

(a) H CH3 H CH3 —*hv*→ ?

(b) H CH3 CH3 H —Heat→ ?

Problem Q.6

(a) For each of the following reactions, state whether conrotatory or disrotatory motion of the groups is involved and (b) state whether you would expect the reaction to occur under the influence of heat or light.

(a) CO_2CH_3 H H CO_2CH_3 → CO_2CH_3 H CO_2CH_3 H

(b) H CH3 H CH3 → CH3 H CH3 H

(c) → H H

Q.2B Electrocyclic Reactions of (4*n* + 2)-π-Electron Systems

The second Woodward-Hoffmann rule for electrocyclic reactions is stated as follows:

2. **A thermal electrocyclic reaction involving (4*n* + 2) π electrons (where *n* = 0, 1, 2, . . .) proceeds with disrotatory motion; the photochemical reaction proceeds with conrotatory motion.**

TABLE Q.1 Woodward-Hoffmann Rules for Electrocyclic Reactions

NUMBER OF ELECTRONS	MOTION	RULE
$4n$	Conrotatory	Thermally allowed, photochemically forbidden
$4n$	Disrotatory	Photochemically allowed, thermally forbidden
$4n + 2$	Disrotatory	Thermally allowed, photochemically forbidden
$4n + 2$	Conrotatory	Photochemically allowed, thermally forbidden

According to this rule the direction of rotation of the thermal and photochemical reactions of $(4n + 2)$-π-electron systems is the opposite of that for corresponding $4n$ systems. Thus, we can summarize both systems in the way shown in Table Q.1.

The interconversions of *trans*-5,6-dimethyl-1,3-cyclohexadiene and the two different 2,4,6-octatrienes shown below illustrate thermal and photochemical interconversions of six-π-electron systems ($4n + 2$ where $n = 1$).

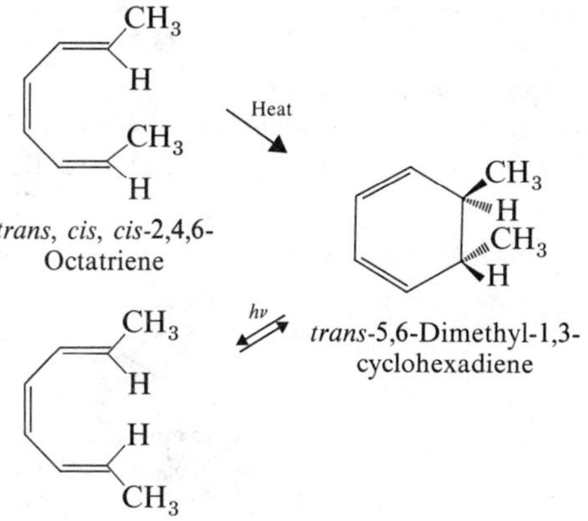

CH_3 H CH_3 H

trans, cis, cis-2,4,6-Octatriene

Heat

CH_3 H CH_3 H

trans-5,6-Dimethyl-1,3-cyclohexadiene

$h\nu$

CH_3 H H CH_3

trans, cis, trans-2,4,6-Octatriene

In the thermal reaction (below) the methyl groups rotate in a disrotatory fashion.

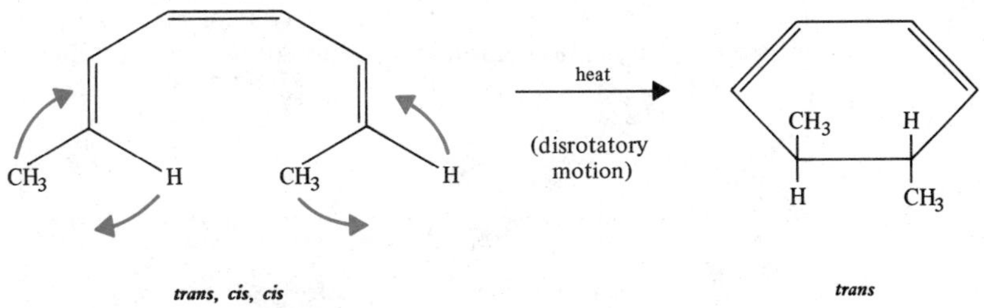

CH_3 H CH_3 H

heat

(disrotatory motion)

CH_3 H H CH_3

trans, cis, cis

trans

In the photochemical reaction the groups rotate in a conrotatory way.

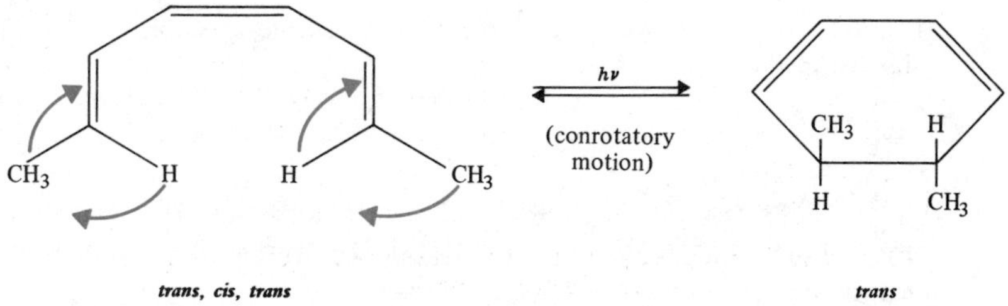

CH_3 H H CH_3

$h\nu$

(conrotatory motion)

CH_3 H H CH_3

trans, cis, trans

trans

We can understand how these reactions occur if we examine the π molecular orbitals shown in Fig. Q.2. Once again we want to pay attention to the highest occupied molecular orbitals. For the thermal reaction of a 2,4,6-octatriene, the highest occupied orbital will be Ψ_3 because the molecule reacts in its ground state.

We see below that disrotatory motion of orbitals at carbons-2 and -7 of Ψ_3 allows the formation of a bonding sigma molecular orbital between them. Disrotatory motion of

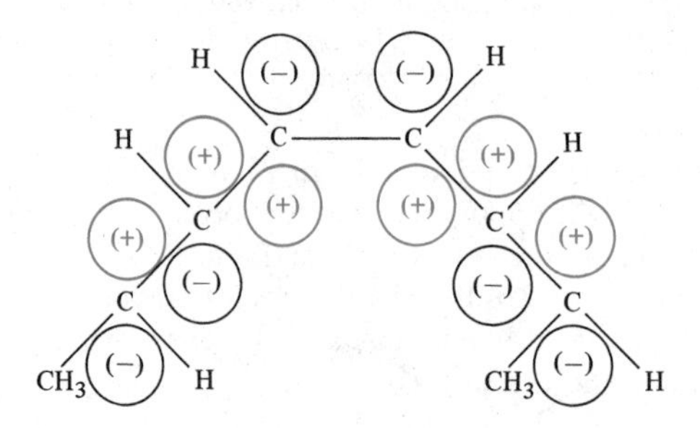

Ψ_3 of *trans, cis, cis*,–2, 4, 6–octatriene

the orbitals, of course, also requires disrotatory motion of the groups attached to carbons-2 and -7. And, disrotatory motion of the groups is what we observe in the thermal reaction: *trans,cis,cis*-2,4,6-octatriene ⟶ *trans*-5,6-dimethyl-1,3-cyclohexadiene.

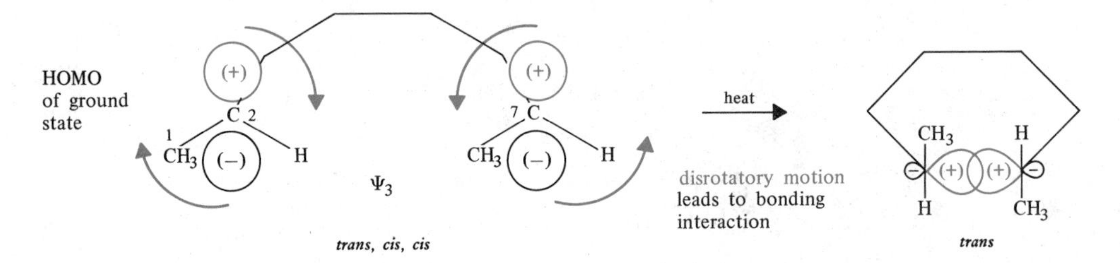

When we consider the photochemical reaction, *trans,cis, trans*-2,4-6-octatriene ⇄ *trans*-5,6-dimethyl-1,3-cyclohexadiene, we want to focus our attention on Ψ_4^*. In the photochemical reaction, light causes the promotion of an electron from Ψ_3 to Ψ_4^*, and thus Ψ_4^* becomes the highest occupied molecular orbital. We also want to look at the symmetry of the orbitals at C-2 and C-7 of Ψ_4*, for these are the orbitals that form a σ bond. In the interconversion shown below, conrotatory motion of the orbitals allows lobes of the same

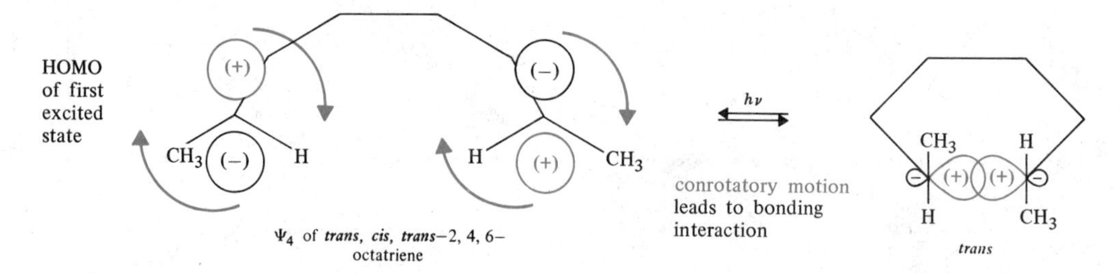

Ψ_4 of *trans, cis, trans*–2, 4, 6–octatriene

sign to overlap. Thus, we can understand why conrotatory motion of the groups is what we observe in the photochemical reaction.

Problem Q.7

Give the stereochemistry of the product that you would expect from each of the following electrocyclic reactions.

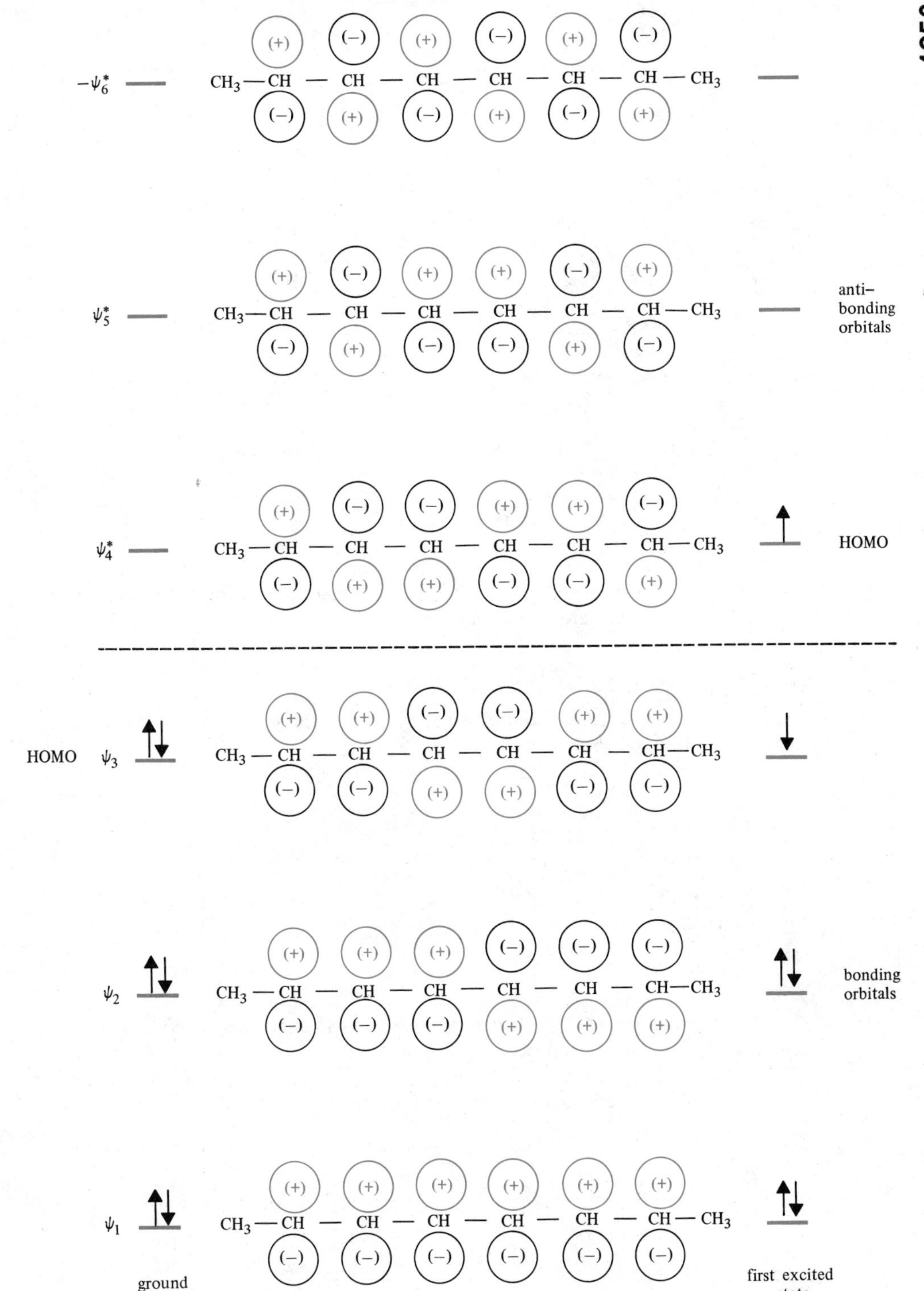

FIG. Q.2
The π molecular orbitals of 2,4,6-octatriene. The first excited state is formed when the molecule absorbs light of the proper wavelength. (These molecular orbitals are obtained from calculations that are beyond the scope of our discussions.)

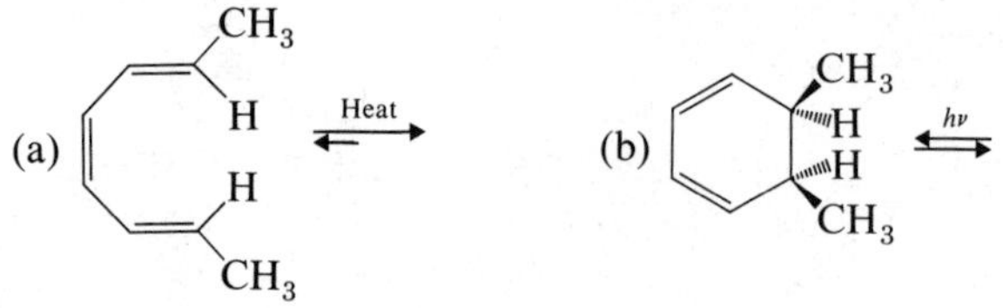

(a) CH_3, H, H, CH_3 —Heat⇄ (b) CH_3, H, H, CH_3 ⇄ $h\nu$

Problem Q.8

Can you suggest a stereospecific method for converting *trans*-5,6-dimethyl-1,3-cyclohexadiene into *cis*-5,6-dimethyl-1,3-cyclohexadiene?

Problem Q.9

When compound **A** is heated, compound **B** can be isolated from the reaction mixture. A sequence of two electrocyclic reactions occurs; the first involves a 4π-electron system, the

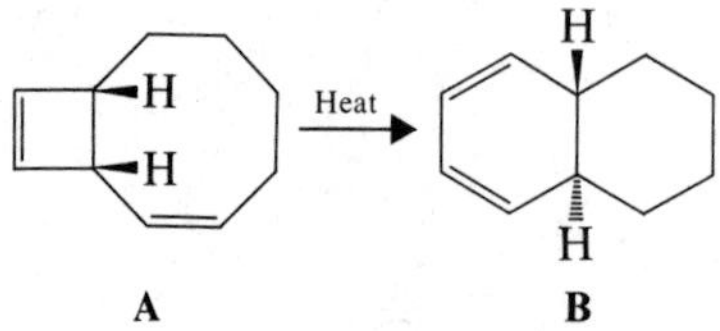

A —Heat→ **B**

second involves a 6π-electron system. Outline both electrocyclic reactions and give the structure of the intermediate that intervenes.

Problem Q.10

Cyclopropyl cations undergo rapid disrotatory ring-opening reactions when heated. A general example is shown below.

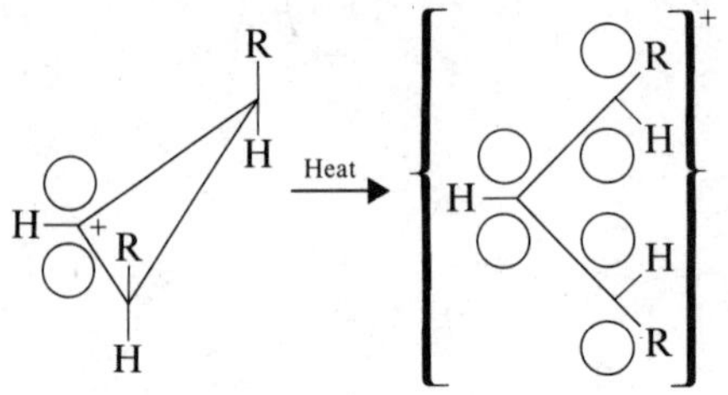

(a) What kind of π system is involved in this reaction, $4n$ or $4n + 2$?
(b) What kind of cation is formed?
(c) What kind of rotation would you expect the cyclopropyl *anion* to show in a thermal reaction, conrotatory or disrotatory?

Problem Q.11

Would you expect the following reactions to occur under the influence of heat or light?

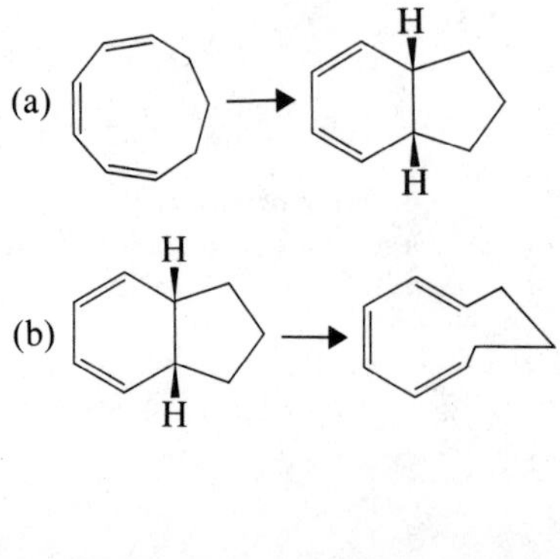

(a)

(b)

(c)

(d)

Problem Q.12

What are the structures of compounds **1** and **2** below?

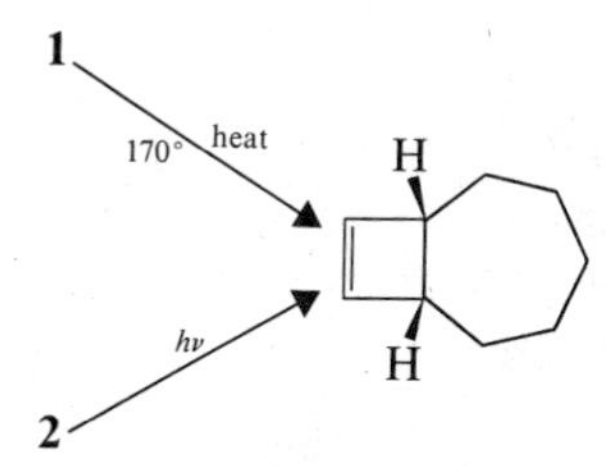

* Problem Q.13

(a) What is the product of the following reaction?

$\xrightarrow{h\nu}$ C_5H_6

(b) Although the product of this reaction is *highly strained,* it is remarkably stable to heat. (At room temperature it has a half-life of about 2 hours and it can be heated to 50° for short periods without extensive degradation.) What factor might account for this unusual thermal stability?

Q.3 CYCLOADDITION REACTIONS

There are a number of reactions of alkenes and polyenes in which two molecules react to form a cyclic product. These reactions, called *cycloaddition* reactions, are shown below.

Alkene + Alkene → Cyclobutane — A [2 + 2] cycloaddition

Diene + Alkene (dienophile) → Cyclohexene (adduct) — A [4 + 2] cycloaddition

Chemists classify cycloaddition reactions on the basis of the number of π electrons involved in each component. The reaction of two alkenes to form a cyclobutane is a

[2 + 2] cycloaddition; the reaction of a diene and an alkene to form a cyclohexene is called a [4 + 2] cycloaddition. We are already familiar with the [4 + 2] cycloaddition, because it is the Diels-Alder reaction that we studied in Section 10.8.

Cycloaddition reactions resemble electrocyclic reactions in the following important ways:

1. Sigma and pi bonds are interconverted.
2. Cycloaddition reactions require only heat or light for initiation.
3. Free radicals and ionic intermediates are not involved in the mechanisms for cycloadditions.
4. Bonds are made and broken in a single, concerted, and cyclic transition state.
5. Cycloaddition reactions are highly stereospecific.

As we might expect, concerted cycloaddition reactions resemble electrocyclic reactions in still another important way: the symmetry elements of the interacting molecular orbitals allow us to account for their stereochemistry. The symmetry elements of the interacting molecular orbitals also allow us to account for two other observations that have been made about cycloaddition reactions:

1. *Photochemical [2 + 2] cycloaddition reactions occur readily while thermal [2 + 2] cycloadditions only take place under extreme conditions.* When thermal [2 + 2] cycloadditions do take place they occur through free radical (or ionic) mechanisms, not through a concerted process.
2. *Thermal [4 + 2] cycloaddition reactions occur readily and photochemical [4 + 2] cycloadditions are difficult.*

Q.3A [2 + 2] Cycloadditions

Let us begin with an analysis of the [2 + 2] cycloaddition of two ethylene molecules to form a molecule of cyclobutane.

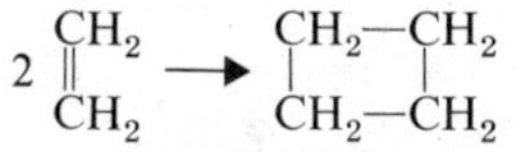

$$2\ \begin{matrix} CH_2 \\ \| \\ CH_2 \end{matrix} \longrightarrow \begin{matrix} CH_2-CH_2 \\ |\quad\quad | \\ CH_2-CH_2 \end{matrix}$$

In this reaction we see that two π bonds are converted into two σ bonds. But how does this conversion take place? In a concerted reaction we would expect that molecular orbitals of one reactant would overlap with molecular orbitals of the other. But which molecular orbitals overlap?

We can answer this last question if we consider two factors: the Pauli exclusion principle and effect of orbital symmetry.

According to the Pauli exclusion principle, a molecular orbital may not contain more than *two* (spin-paired) electrons. This means that if a molecular orbital of one reactant already contains two electrons it must overlap with an *unoccupied* molecular orbital of the other. (Otherwise, the Pauli principle would be violated, for the new σ molecular orbital would contain more than two electrons.)

We can see how orbital interactions come into play if we examine the possibility of a *concerted thermal* conversion of two ethene molecules into cyclobutane.

Thermal reactions involve molecules reacting in their ground states. The orbital diagram for ethene in its ground state is shown below.

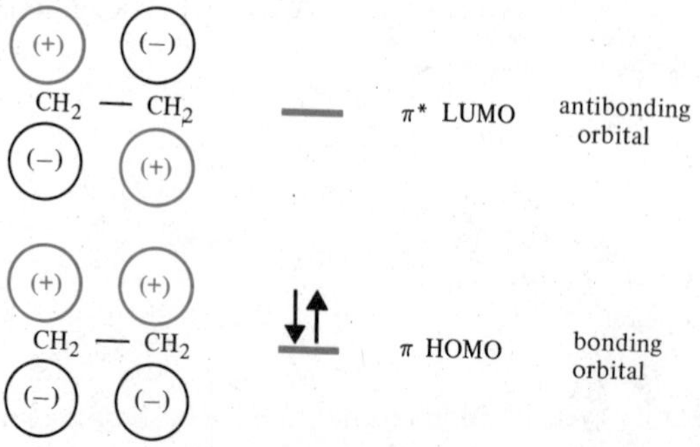

(+) (−)
CH_2 — CH_2 — π* LUMO antibonding orbital
(−) (+)

(+) (+)
CH_2 — CH_2 ↓↑ π HOMO bonding orbital
(−) (−)

the ground state of ethene

The highest occupied molecular orbital (HOMO) of ethene in its ground state is the π orbital. Since this orbital contains two electrons, it must interact with an *unoccupied* molecular orbital of another ethene molecule. The lowest unoccupied molecular orbital (LUMO) of the ground state of ethene is, of course, π^*.

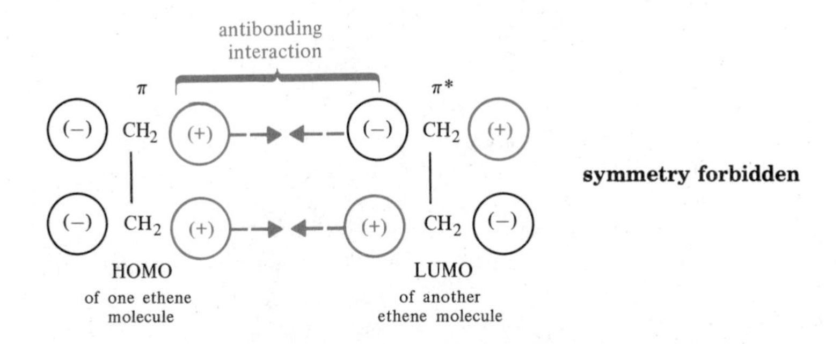

We see from the diagram above, however, that overlapping the π orbital of one ethene molecule with the π^* orbital of another does not lead to bonding between both sets of carbons. Moreover, complete correlation diagrams of all the molecular orbitals show that this reaction is *symmetry forbidden*. What does this mean? It means that a thermal (or ground state) cycloaddition of ethene would be unlikely to occur in a concerted process. This is exactly what we find experimentally; thermal cycloadditions of ethene, when they occur, take place through nonconcerted, free-radical mechanisms.

What, then, can we decide about the other possibility—a photochemical [2 + 2] cycloaddition? If an ethene molecule absorbs a photon of light of the proper wavelength, an electron is promoted from π to π^*. In this excited state the highest occupied molecular orbital of an ethene molecule is π^*. The diagram below shows how the highest occupied molecular orbital of an excited state ethene molecule interacts with the lowest unoccupied molecular orbital of a ground-state ethene molecule.

Here we find that bonding interactions occur between both CH_2 groups, that is, lobes of the same sign overlap between both sets of carbons. Complete correlation

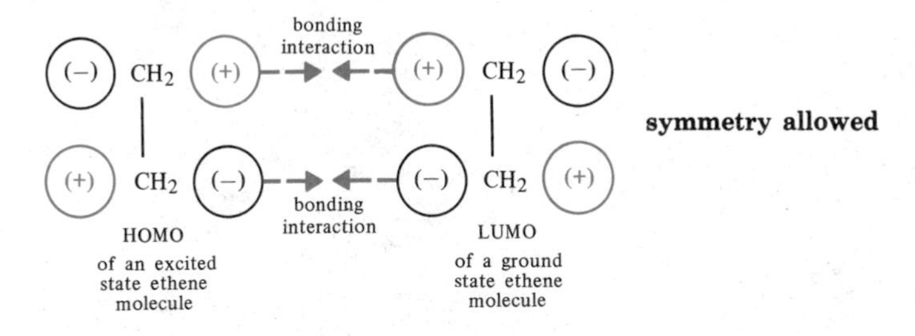

diagrams also show that the photochemical reaction is *symmetry allowed* and should occur readily through a concerted process. This, moreover, is what we observe experimentally: ethene reacts readily in a *photochemical* cycloaddition.

The analysis that we have given for the [2 + 2] ethene cycloaddition can be made for any alkene [2 + 2] cycloaddition because the symmetry elements of the π and π^* orbitals of all alkenes are the same.

Problem Q.14

What products would you expect from the following concerted cycloaddition reactions? (Give stereochemical formulas.)

(a) *cis*-2-Butene $\xrightarrow{h\nu}$

(b) *trans*-2-Butene $\xrightarrow{h\nu}$

Problem Q.15

Show what happens in the following reaction.

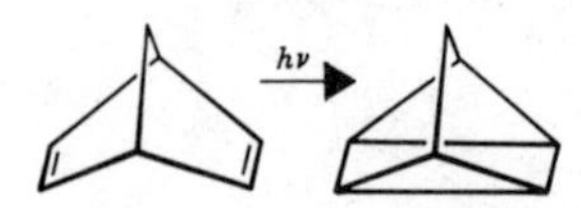

Q.3B [4 + 2] Cycloadditions

Concerted [4 + 2] cycloadditions—Diels-Alder reactions—are *thermal reactions.* Considerations of orbital interactions allow us to account for this fact as well. To see how, let us consider the diagrams shown in Figure Q.3.

Both modes of orbital overlap shown in Fig. Q.3 lead to bonding interactions and both involve *ground states* of the reactants. The ground state of a diene has two electrons in Ψ_2 (its highest occupied molecular orbital). The overlap shown in part (*a*) allows these two electrons to flow into the lowest unoccupied molecular orbital, π^*, of the dienophile. The overlap shown in part (*b*) allows two electrons to flow from the highest occupied molecular orbital of the dienophile, π, into the lowest unoccupied molecular orbital of the diene, Ψ_3^*. Complete correlation diagrams also show that the $4n + 2$ thermal reaction is symmetry allowed.

In Section 10.8 we saw that the Diels-Alder reaction proceeds with retention of configuration of the dienophile. Because the Diels-Alder reaction is usually concerted, it also proceeds with retention of configuration of the diene.

retention of configuration of the dienophile

retention of configuration of the diene

Problem Q.16

What products would you expect from the following reaction?

Problem Q.17

What are compounds **3**, **4**, and **5** in the following reaction sequence?

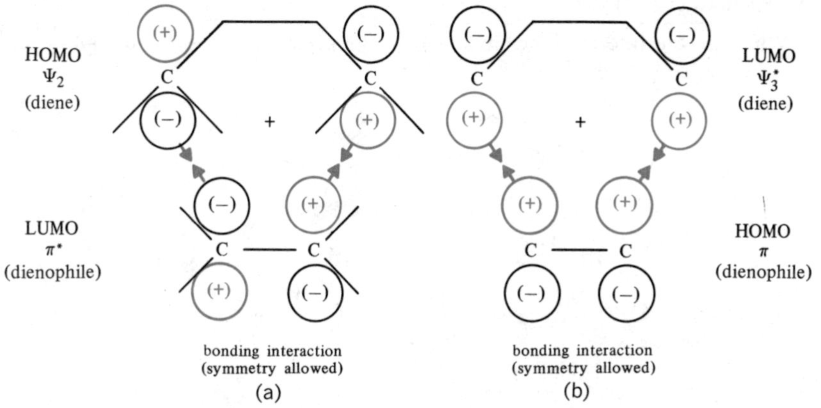

FIG. Q.3
Two symmetry-allowed interactions for a thermal [4 + 2] cycloaddition. (a) *Bonding interaction between the highest occupied molecular orbital of a diene and the lowest unoccupied orbital of a dienophile.* (b) *Bonding interaction between the lowest unoccupied molecular orbital of the diene and the highest occupied molecular orbital of the dienophile.*

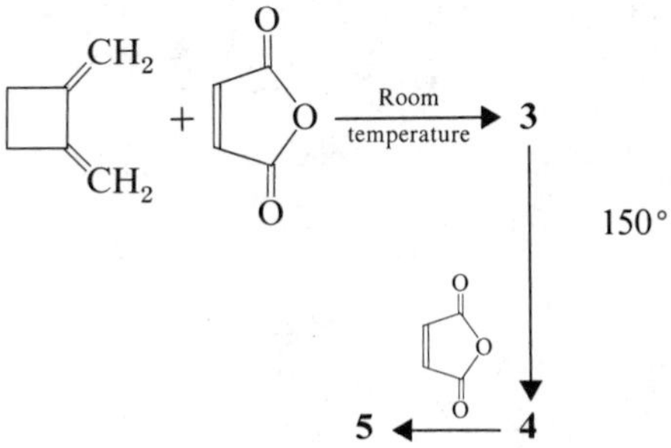

Problem Q.18

What is compound **7** in the following reaction sequence?

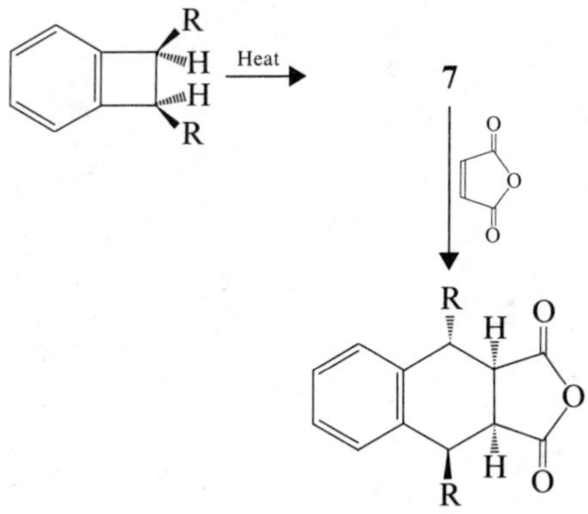

*Problem Q.19

Propose structures for compounds **A, B,** and **C.**

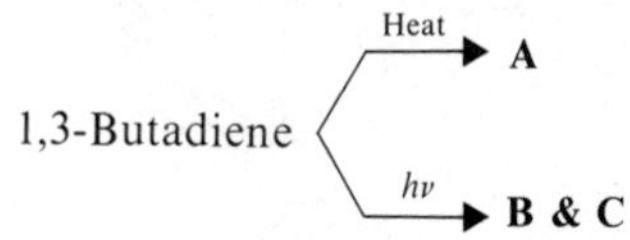

* Problem Q.20

What are the intermediates **A** and **B** in the synthesis of basketene given below?

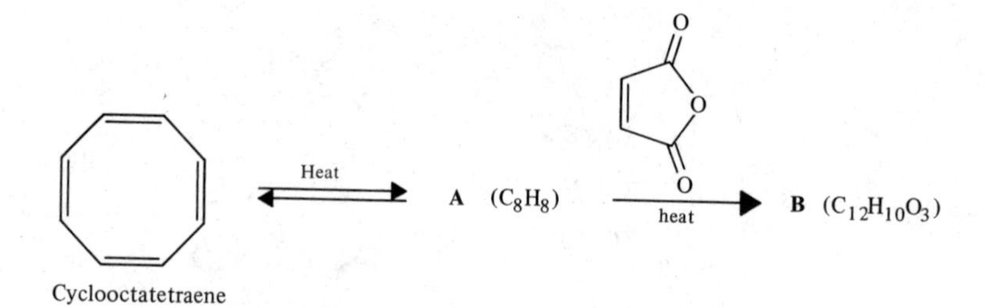

$h\nu$

O
O
O
H
H

Two steps

Basketene

ANSWERS TO SELECTED PROBLEMS

CHAPTER 1

1.6 (a) $\overset{H-Br;}{+\!\!\!\longrightarrow}$ (b) $\overset{I-Cl;}{+\!\!\!\longrightarrow}$ (c) H_2, μ=0;
(d) Cl_2, μ=0.
1.8 (a) (b) (d) (e) (g), tetrahedral;
(c) (i) trigonal planar; (f) (h) linear.
1.10 The bond moments cancel.
1.13 Trigonal planar structure causes bond moments to cancel.
1.16 (a) and (d), (e) and (f).
1.24 (a) an sp^3 orbital; (b) an sp^3 orbital.
1.33 The carbon atom of the methyl cation is sp^2 hybrized and uses sp^2 orbitals to form bonds to each hydrogen. The carbon also has a vacant *p* orbital.

CHAPTER 2

2.2 (a) $CH_3CH_2CH_2Cl$, $CH_3CHClCH_3$;
(b) $CH_3CH_2CHCl_2$, $CH_3CCl_2CH_3$;
$CH_3CHClCH_2Cl$, $CH_2ClCH_2CH_2Cl$.
2.7 RH.
2.8 (a) RCH_2OH; (b) R_2CHOH;
(c) R_3COH.
2.16 Molecules of trimethylamine cannot form hydrogen bonds to each other whereas molecules of *n*-propylamine can.
2.17 (a) alkyne; (d) aldehyde.
2.19 (c) tertiary; (e) secondary.
2.20 (a) secondary; (c) tertiary.
2.23 (b) ethylene glycol; (f) propionic acid.
2.24 (b) $CH_3CH_2CH_2OH$;
(e) $CH_3CH_2CH_2CH_2X$;
(l) $CH_3N(CH_3)CH_2CH_3$.
2.27 ester.

CHAPTER 3

3.14 (a) $CH_3CHClCHClCH_2CH_3$;
(k) $CH_3CH(CH_3)CH_2CH_2CH_2Cl$;
(m) $CH_3C(CH_3)_2CH_2Cl$;
(n) $CH_3CH(CH_3)CH_2CH_3$.
3.15 (a) 3,4-dimethylhexane;
(f) cyclopentylcyclopentane.
3.17 (a) neopentane (or 2,2-dimethylpropane);
(d) cyclopentane.
3.18 (a) 3-isopropyl-2,4-dimethylpentane ;
(f) 5,5-dibutylnonane.
3.23 (d) chloroethane.
3.26 (c) *trans*-1,4-dimethylcyclohexane.
3.29 (a) C, 40.0%; H, 6.66%; O, 53.3%.
(b) C, 32.0%; H, 6.66%; N, 18.7%, O, 42.7%.
(c) C, 12.8%; H, 1.78%; Br, 85.4%.
3.31 CH_2O.
3.32 C_2H_4.
3.34 $C_{11}H_{12}Cl_2N_2O_5$.

CHAPTER 4

4.1 (a) −25 kcal/mole;
(d) −24.5 kcal/mole.
4.11 (c) $\Delta H° = +88$ kcal/mole;
$E_{act} = +88$ kcal/mole.
4.13 Their hydrogens are all equivalent; replacing any one yields the same product.
4.14 (b) $\Delta H° = -8.5$ kcal/mole.
4.17 $CH_3\dot{C}(CH_3)CH_2CH_3 > CH_3CH(CH_3)\dot{C}HCH_3 >$
$\cdot CH_2CH(CH_3)CH_2CH_3 \simeq CH_3CH(CH_3)CH_2CH_2\cdot$
4.27 The mechanism involves protolysis of a carbon-carbon bond of isobutane.
4.29 (a) $CH_3{-}H \longrightarrow CH_3\cdot + H\cdot$,
$E_{act} = \Delta H° = +104$ kcal/mole
$CH_3CH_2{-}H \longrightarrow CH_3CH_2\cdot + H\cdot$,
$E_{act} = \Delta H° = +98$ kcal/mole
Since the E_{act} for the ethane C—H cracking reaction is lower, the ethane reaction occurs at a lower temperature;
(b) $CH_3CH_3 \longrightarrow 2CH_3\cdot$, $E_{act} = \Delta H° = +88$ kcal/mole. This reaction has a lower E_{act} than the ethane C—H cracking reaction given in part (a).
(c) $CH_3CH_2{-}CH_2CH_3 \longrightarrow 2CH_3CH_2\cdot$,
$E_{act} = \Delta H° = +82$ kcal/mole
$CH_3CH_2CH_2{-}CH_3 \longrightarrow CH_3CH_2CH_2\cdot + CH_3\cdot$,
$E_{act} = \Delta H° = +85$ kcal/mole.

CHAPTER 5

5.1 (a) $CH_3CH_2\ddot{\underset{..}{O}}H$, (c) $:NH_3$, (e) $-:C\equiv N:$.
5.2 cis-3-methylcyclopentanol.
5.5 (a) NH_2^-; (b) RS^-; (c) PH_3.
5.7 (b) decrease; (c) increase because the reactor is S_N1.
5.10 (a) $CH_3CH_2CH_2CH_2Br$ because it is a primary halide; (c) $CH_3CH_2CH_2—Br$ because bromide ion is a better leaving group.
5.12 (b) $(CH_3)_3CBr + H_2O \longrightarrow (CH_3)_3COH + HBr$ because water is a more polar medium than CH_3OH.

CHAPTER 6

6.6 (a) isobutane; (b) $4CO_2 + 4H_2O$; (c) yes; (d) cyclobutane and methylcyclopropane; (e) yes.
6.18 Add a drop of Br_2 in CCl_4 to a small amount of each isomer. Bromine will add to the double bond of 1-hexene producing a colorless solution. No reaction will take place with cyclohexane as long as the mixture is not heated and is protected from light; the red-brown bromine color, consequently, will persist.
6.19 (c), (g), (h), and (l) can exist as *cis-trans* isomers.
6.22 (a) *cis-trans* isomerization. This happens because at 300° the molecules have enough energy to surmount the rotational barrier of the carbon-carbon double bond. (b) *trans*-2-butene because it is more stable.
6.27 (a) *cis*-1,2-dimethylcyclopentane; (b) *cis*-1,2-dimethylcyclohexane; (c) *cis*-1,2-dideuteriocyclohexane.

CHAPTER 7

7.1 1-iodo-2-chloropropane.
7.6 (c) cyclohexene + $Hg(OAc)_2$ + CH_3OH, then $NaBH_4 + OH^-$.
7.8 (c) 1-methylcyclopentene + $(BH_3)_2$, then CH_3COOD.
7.11 (c) 3-methyl-1-pentene.
7.16 (a) 1,2-dibromobutane; (f) CH_3CH_2OH; (j) cyclopentene; (m) 2-chlorohexane; (q) $2CH_3CH_2COO^-$
7.22 2-methylpropene > propene > ethene.
7.25 4-methylcyclohexene.
7.32

$$CH_3\overset{\overset{CH_3}{|}}{C}{=}CHCH_2CH_2\overset{\overset{CH_3}{|}}{C}{=}CHCH_2CH_2\overset{\overset{CH_2}{\|}}{C}CH{=}CH_2.$$

CHAPTER 8

8.1 chiral: (a), (e), (f), (g), (h); achiral: (b), (c), (d).
8.6 (b), (c), (d).
8.13 75% (S)-(+)-2-butanol and 25% *R*-(−)-2-butanol.
8.15 (a) diastereomers; (b) diastereomers; (c) diastereomers; (e) yes; (f) no.
8.16 (a) and (b).
8.21 (a) *trans*-1,2-dibromocyclopentane; (b) as a racemic modification; (c) *cis*-1,2-dibromocyclopentane.
8.31 They are diastereomers.
8.34 (a) enantiomers; (d) diastereomers; (g) two molecules of the same compound; (j) enantiomers; (n) structural isomers.
8.39 (a) Retention because no bonds to the chiral carbon are broken.
(c) Inversion because an S_N2 reaction occurs at the chiral carbon.

CHAPTER 9

9.2 (a) $HC\equiv CNa + NH_3$; (d) $HC\equiv CH + CH_3CH_2ONa$.
9.4 (c) 2-butyne + $(BD_3)_2$, then CH_3COOD; (d) 1-butyne + Sia_2BH, then CH_3COOD.
9.8 (b) cyclooctyne.
9.10 (a) and (j).
9.13 (a) $CH_3CH_2CH_2CBr{=}CHBr$; (d) $CH_3CH_2CH_2CH{=}CHBr$; (j) $CH_3CH_2CH_2C\equiv CCH_3$; (k) $CH_3CH_2CH_2C\equiv C—C\equiv CCH_2CH_2CH_3$; (n) $CH_3CH_2CH_2COOH + CO_2$.
9.15 (a) 1-pentene + Br_2, then $2NaNH_2$ and heat; (e) 1-bromopropane + $HC\equiv CNa$.
9.16 (a) Br_2/CCl_4 or dil. $KMnO_4$; (e) sodium fusion, then HNO_3, then $AgNO_3$.

CHAPTER 10

10.1 (a) $^{14}CH_2{=}CH—CH_2—X$ + $X—^{14}CH_2—CH{=}CH_2$; (c) in equal amounts.
10.4 (a) $CH_3CH_2CHClCH{=}CHCH_3$ + $CH_3CHClCH{=}CHCH_2CH_3$; (b) $CH_3CHClCH{=}CHCH_3$ + $CH_3CH{=}CHCHClCH_3$, possibly, $CH_2ClCH{=}CHCH_2CH_3$ and $CH_2{=}CHCHClCH_2CH_3$.
10.13 (a) 1,4-dibromobutane + 2KOH, alcohol, and heat; (g) $HC\equiv CCH{=}CH_2$ + Sia_2BH, then CH_3COOH, or H_2 + P-2 catalyst.

10.16 (a) *1-butene* + *N*-bromosuccinimide, then KOH in alcohol and heat;
(e) cyclopentane + Br_2, $h\nu$, then KOH in alcohol and heat, then *N*-bromosuccinimide.

10.18 (a) $Ag(NH_3)_2OH$; (c) CrO_3 in H_2SO_4;
(e) $AgNO_3$ in alcohol.

10.24 **A** is 1,3-cyclohexadiene;
B is 1,4-cyclohexadiene.

10.27 This is another example of rate versus equilibrium control of a reaction. The *endo* adduct, **G**, is formed faster, and at the lower temperature it is the major product. The *exo* adduct, **H**, is more stable, and at the higher temperature it is the major product.

CHAPTER 11

11.1 (a) compound (d) only; (b) none.

11.5 Tropylium bromide is a largely ionic compound consisting of the cycloheptatrienyl (tropylium) cation and a bromide anion.

11.6 (a) $CH_2{=}CH{-}\overset{+}{C}H{-}CH{=}CH{-}CH{=}CH_2$.
(b) That the cycloheptatrienyl cation is aromatic.

11.10 The nitrogen atoms at positions 1-, 3- and 7- are of the pyridine type. The nitrogen at position 9- is of the pyrrole type.

11.17 Compound VII is the cyclononatetraenyl anion, a 10 π electron aromatic system.

CHAPTER 12

12.2 The following reaction takes place:
$2Fe + 3X_2 \longrightarrow 2FeX_3$.

12.3 $HO{-}NO_2 + HO{-}NO_2 \rightleftharpoons H_2O^+{-}NO_2 + {}^-O{-}NO_2$
$H_2O^+{-}NO_2 \rightleftharpoons H_2O + \overset{+}{N}O_2$.

12.8 (a) At the lower temperature the reaction is rate controlled; at the higher temperature it is equilibrium controlled. (b) *p*-toluenesulfonic acid.

12.10 (a) $CH_2ClCH_2\overset{+}{N}(CH_3)_3Cl^-$; (b) slower; (c) The positive nitrogen makes it strongly electron withdrawing; (d) the structures are similar to those for $C_6H_5CF_3$ given in Section 12.9C.

12.22 Introduce the chlorine into the benzene ring first, otherwise the double bond will undergo addition of chlorine when ring chlorination is attempted.

12.25 (a) benzene +

$$CH_3CH_2CH_2CH_2CH_2\overset{O}{\overset{\|}{C}}Cl \xrightarrow{AlCl_3}$$

$$C_6H_5\overset{O}{\overset{\|}{C}}CH_2CH_2CH_2CH_2CH_3 \xrightarrow[HCl]{Zn(Hg)}$$

$$C_6H_5CH_2CH_2CH_2CH_2CH_2CH_3.$$

12.28 (a) 2-methyl-5-acetylbenzenesulfonic acid;
(c) 2,4-dimethoxynitrobenzene;
(e) 3-nitro-4-hydroxybenzenesulfonic acid;
(g) *m*-chlorobenzotrichloride.

12.30 (a) toluene, $KMnO_4$, OH^-, heat; then H_3O^+; then Cl_2, $FeCl_3$.
(c) aniline, CH_3COCl; then conc. H_2SO_4; then Br_2, $FeBr_3$; then dilute H_2SO_4.
(f) toluene, CH_3COCl, $AlCl_3$; then isolate *ortho* isomer.
(g) product of (f), Br_2, OH^-; then H_3O^+.
(i) toluene, HNO_3, H_2SO_4; then isolate *para* isomer, then Br_2, $FeBr_3$.

12.32 $p\text{-}NO_2C_6H_4O{-}\overset{O}{\overset{\|}{C}}C_6H_5$ and

$o\text{-}NO_2C_6H_4O{-}\overset{O}{\overset{\|}{C}}C_6H_5$.

CHAPTER 13

13.2 (a) one; (b) two; (c) one; (d) three;
(e) two; (f) three.

13.7 a doublet (3H) downfield; a quartet (1H) upfield.

13.8 (a) CH_3CHICH_3; (b) CH_3CHBr_2;
(c) $CH_2ClCH_2CH_2Cl$.

13.10 (a) $C_6H_5CH(CH_3)_2$;
(b) $C_6H_5CH(NH_2)CH_3$;
(c) CH₂ CH₂ CH₂

13.16 **A**, *o*-bromotoluene; **B**, *p*-bromotoluene;
C, *m*-bromotoluene; **D**, benzyl bromide.

13.18 phenylacetylene.

13.21 $CH(CH_3)_2$

CH_3

F

13.24 **G**, $CH_3CH_2CHBrCH_3$; **H**, $CH_2{=}CBrCH_2Br$.

CHAPTER 14

14.6 (a) OCH_3 / NO_2 (b) NO_2 / $NHCH_3$

(c) NHC_6H_5 / NO_2 / NO_2

14.9 (a) $CH_3CH_2CH_2CH{=}CH_2 + HBr \xrightarrow{\text{Peroxides}}$.

(c) $(CH_3)_2CHCH_2CH_{3(excess)} + Br_2 \xrightarrow{\text{heat, } h\nu}$.

(f) $CH_3CH_2C{\equiv}CH + 2HBr$.

(h) 1-phenylheptane + *N*-bromosuccinimide $\xrightarrow[CCl_4]{h\nu}$.

(i) 1-phenylheptane + $Br_2 \xrightarrow[\text{heat}]{FeBr_3}$.

14.14 $(CH_3)_3CBr + Mg \xrightarrow{\text{ether}}$
$(CH_3)_3CMgBr \xrightarrow{D_2O} (CH_3)_3CD$.

14.16 Acetylation deactivates the first ring toward further electrophilic aromatic substitution.

14.30 (a) Br_2/CCl_4 or $KMnO_4/H_2O$;
(c) alcoholic $AgNO_3$;
(e) alcoholic $AgNO_3$.

CHAPTER 15

15.10 (a) $LiAlH_4$; (b) $NaBH_4$; (c) $LiAlH_4$.

15.21 In strong acid the leaving group is ROH; in a neutral or weakly acidic solution the leaving group would have to be RO^-.

15.27 (d), (e), and (f).

15.31. (a) $C_6H_5CH{=}CH_2$, H_2O, H^+, and heat; or $C_6H_5CH{=}CH_2$, $Hg(OOCCH_3)_2$, H_2O, then $NaBH_4, OH^-$; (e) $C_6H_5CH_2COOH$, $LiAlH_4$, ether; (h) C_6H_6, Br_2, $FeBr_3$; then Mg, Et_2O; then ethylene oxide; then H_2O.

15.33 (a) $CH_3CH_2CH_2ONa$;
(e) $CH_3CH_2CH_2OOCCH_3$;
(i) $(CH_3CH_2CH_2)_2O$;
(l) $C_6H_5CH(CH_3)_2 + C_6H_5CH_2CH_2CH_3$.

15.38 (a) aqueous NaOH; (c) Br_2/CCl_4, or $KMnO_4/H_2O$, or CrO_3/H_2SO_4;
(e) aqueous NaOH.

15.43 $CH_3CH_2COOH \xrightarrow{SOCl_2}$

$CH_3O{-}C_6H_4 + CH_3CH_2COCl \xrightarrow{Tl(OOCCF_3)_3}$

$CH_3O{-}C_6H_4{-}\overset{O}{\overset{\|}{C}}CH_2CH_3 \xrightarrow{NaBH_4}$

$CH_3O{-}C_6H_4{-}\overset{OH}{\overset{|}{C}}HCH_2CH_3 \xrightarrow[\text{heat}]{H^+}$

$CH_3O{-}C_6H_4{-}CH{=}CHCH_3$.

15.44 *p-tert*-butylphenol.

15.45 1-phenyl-1-propanol.

CHAPTER 16

16.2 (a) 1-pentanol; (c) pentanal;
(e) benzyl alcohol.

16.5 a hydride ion.

16.12 (a) $C_6H_5CH_2Br + (C_6H_5)_3P$, then strong base, then CH_3COCH_3.
(c) $CH_3CH_2Br + (C_6H_5)_3P$, then strong base, then $C_6H_5COCH_3$.
(e) $CH_3I + (C_6H_5)_3P$, then strong base, then cyclopentanone.
(g) $CH_2{=}CHCH_2Br + (C_6H_5)_3P$, then strong base, then C_6H_5CHO.

16.18 Because base is consumed as the reaction takes place.

16.30 (a), (b), (d), (f), (h), and (i).

16.32 (a) $CH_3CH_2CH_2OH$;
(c) $CH_3CH_2CH_2OH$;
(e) $CH_3CH_2CH{=}C(CH_3)CHO$;
(h) $CH_3CH_2CH{=}CHCH_3$;
(j) $CH_3CH_2COO^-$;
(l) $CH_3CH_2CH{=}NNHCONH_2$;
(n) CH_3CH_2COOH.

18.38 (a) Tollens' reagent; (c) $I_2/NaOH$;
(e) $I_2/NaOH$; (g) Br_2/CCl_4;
(i) Tollens' reagent; (k) Tollens' reagent.

CHAPTER 17

17.3 (a) CH_2FCOOH; (c) $CH_2ClCOOH$;
(e) $CH_3CH_2CHClCOOH$;
(g) $CF_3{-}C_6H_4{-}COOH$.

17.5 (a) $C_6H_5CH_2Br$ + Mg + ether, then CO_2, then H_3O^+.
(c) $CH_2{=}CHCH_2Br$ + Mg + ether, then CO_2, then H_3O^+.

17.6 (a), (c), and (e).

17.9 in the carboxyl group of benzoic acid.

17.14 (a) $(CH_3)_3CCOOH + SOCl_2$, then NH_3, then P_2O_5, heat.
(b) $CH_2{=}C(CH_3){-}CH_3$.

17.22 (a) CH_3COOH;
(c) $CH_3COOCH_2(CH_2)_2CH_3$;
(e) p-$CH_3COC_6H_4CH_3$ + o-$CH_3COC_6H_4CH_3$;
(g) CH_3COCH_3; (i) $CH_3CONHCH_3$;
(k) $CH_3CON(CH_3)_2$;
(m) $(CH_3CO)_2O$;
(o) $CH_3COOC_6H_5$.

17.28 (a) $NaHCO_3/H_2O$; (c) $NaHCO_3/H_2O$; (e) OH^-/H_2O, heat, detect NH_3 with litmus paper; (g) $AgNO_3$/alcohol.

17.33 (a) diethyl succinate; (c) ethyl phenylacetate; (e) ethyl chloroacetate.

CHAPTER 18

18.5 (a) $CH_3(CH_2)_3CHO + NH_3 \xrightarrow{H_2,\ Ni} CH_3(CH_2)_3CH_2NH_2$

(c) $CH_3(CH_2)_4CHO + C_6H_5NH_2 \xrightarrow{H_2,\ Ni} CH_3(CH_2)_4CH_2NHC_6H_5$

18.6 The reaction of a secondary halide with ammonia is almost always accompanied by some elimination.

18.8 (a) anisole + HNO_3 + H_2SO_4; then Fe + HCl (b) anisole + CH_3COCl + $AlCl_3$, then $NH_3 + H_2$ + Ni; (c) toluene + Cl_2 and light, then $(CH_3)_3N$; (d)p-nitrotoluene + $KMnO_4$ + OH^-; then H_3O^+, then $SOCl_2$ followed by NH_3; then NaOBr (Br_2 in NaOH); (e) toluene + N-bromosuccinimide in CCl_4, then KCN, then $LiAlH_4$.

18.15 p-nitroaniline + Br_2 + Fe, followed by $HCl/NaNO_2$ followed by CuBr, then Fe/HCl, then $HCl/NaNO_2$ followed by H_3PO_2.

18.41 **W** is N-benzyl-N-ethylaniline.

CHAPTER 19

19.1 (a) two; (b) two; (c) four

19.3 Acid catalyzes hydrolysis of the glycosidic (acetal) group.

19.11 (a) $2CH_3CHO$, one mole HIO_4;
(b) $HCHO + HCOOH + CH_3CHO$, two moles HIO_4;
(c) $HCHO + OHCCH(OCH_3)_2$, one mole HIO_4;
(d) $HCHO + HCOOH + CH_3COOH$, two moles HIO_4;
(e) $2CH_3COOH + HCOOH$, two moles, HIO_4.

19.22 D-(+)-glucose.

19.27 One anomeric form of D-mannose is dextrorotatory ($[\alpha]_D = +29.3°$), the other is levorotatory ($[\alpha]_D = -17.0°$).

19.28 The microorganism selectively oxidizes the—CHOH group of D-glucitol that corresponds to C-5 of D-glucose.

19.31 **A** is D-altrose; **B** is D-talose, **C** is D-galactose

CHAPTER 20

20.4 (a) $CH_3CH(CO_2C_2H_5)COCOOC_2H_5$;
(b) $HC(=O)CH_2CO_2C_2H_5$.

20.7 O-alkylation that results from the oxygen of the enolate ion acting as a nucleophile.

20.10 (a) Reactivity is the same as with any S_N2 reaction. With primary halides substitution is highly favored, with secondary halides elimination competes with substitution, and with tertiary halides elimination is the exclusive course of the reaction. (b) acetoacetic ester and 2-methyl propene; (c) Bromobenzene is unreactive toward nucleophilic substitution.

20.28 **D** is *trans*-1,2-cyclopentanedicarboxylic acid, **E** is *cis*-1,2-cyclopentanedicarboxylic acid.

CHAPTER 21

21.8 (a) C_2H_5OH, H^+, heat or $SOCl_2$, then C_2H_5OH;
(d) $SOCl_2$, then $(CH_3)_2NH$;
(g) $SOCl_2$, then $LiAlH[OCC(CH_3)_3]$;
(j) $SOCl_2$, then $(CH_3)_2CuLi$.

21.11 Elaidic acid is *trans*-9-octadecenoic acid.

21.13 **A** is $CH_3(CH_2)_5C{\equiv}CNa$, **B** is $CH_3(CH_2)_5C{\equiv}CCH_2(CH_2)_7CH_2Cl$, **C** is $CH_3(CH_2)_5C{\equiv}CCH_2(CH_2)_7CH_2CN$, **E** is $CH_3(CH_2)_5C{\equiv}CCH_2(CH_2)_7CH_2COOH$. Vaccenic acid is

$$\underset{H}{\overset{CH_3(CH_2)_5}{}}C{=}C\underset{H}{\overset{(CH_2)_9COOH}{}}$$

21.16 The 5α,6β-dibromocholestan-3β-ol that forms initially is unstable because both bromines are axial and because steric repulsions occur between the C-10 methyl and the bromine at C-6. Isomerization takes place to produce the more stable $5\beta,6\alpha$-dibromocholestan-3β-ol.

CHAPTER 22

22.5 The labeled amino acid no longer has a basic $-NH_2$ group; it is, therefore, insoluble in aqueous acid.

22.8

$$\overset{+}{H_3N}\underset{\displaystyle COO^-}{\underset{|}{C}}HCH_2CH_2CONH\underset{\displaystyle CH_2SH}{\underset{|}{C}}HCONHCH_2COOH$$

22.22 Arg·Pro·Pro·Gly·Phe·Ser·Pro·Phe·Arg.

22.23 Val·Leu·Lys·Phe·Ala·Glu·Ala.

BIBLIOGRAPHY OF SUGGESTED READINGS

CHAPTER 1

L. Salem, "A Faithful Couple: The Electron Pair," *J. Chem. Educ., 55,* 344 (1978).

M. B. Hall, "Valence Shell Electron Pair Repulsion and the Pauli Exclusion Principle," *J. Am. Chem. Soc., 100,* 6333 (1978).

D. Kolb, "The Chemical Formula, Part I: Development," *J. Chem. Educ., 55,* 44 (1978).

J. E. Fernandez and Robert D. Whitaker, *An Introduction to Chemical Principles.* Macmillan, New York, 1975.

J. E. Brady and G. E. Humiston, *General Chemistry: Principles and Structure,* 2nd ed., Wiley, New York, 1978.

R. E. Dickerson, H. P. Gray, and G. P. Haight, *Chemical Principles,* 2nd ed., W. A. Benjamin, Menlo Park, Calif., 1974.

W. L. Masterton and E. L. Slowinski, *Chemical Principles,* 4th ed., W. B. Saunders Company, Philadelphia, 1977.

O. T. Benfey, *From Vital Force to Structural Formulas,* Houghton Mifflin, Boston, 1964.

R. J. Gillespie, "The Electron-Pair Repulsion Model for Molecular Geometry," *J. Chem. Educ., 47,* 18 (1970).

P. E. Verkade, "August Kekulé," *Proc. Chem. Soc.,* 205 (1958).

L. Pauling, *The Nature of the Chemical Bond,* 3rd ed., Cornell University Press, Ithaca, N.Y., 1960.

G. W. Wheland, *Resonance in Organic Chemistry,* Wiley, New York, 1955.

J. B. Hendrickson, D. J. Cram, and G. S. Hammond, *Organic Chemistry,* McGraw-Hill, New York, 1970, Chapter 2.

SPECIAL TOPIC A

R. H. Maybury, "The Language of Quantum Mechanics," *J. Chem. Educ., 39,* 367 (1962).

L. C. Pauling, *The Chemical Bond; A Brief Introduction to Modern Structural Chemistry,* Cornell University Press, Ithaca, N.Y., 1967.

CHAPTER 2

D. Kolb, "Acids and Bases," *J. Chem. Educ. 55,* 459 (1978).

O. T. Benfey, *The Names and Structures of Organic Compounds,* Wiley, New York, 1966.

J. D. Roberts, R. Stewart, and M. C. Caserio, *Organic Chemistry: Methane to Macromolecules,* Benjamin, New York, 1971, Chapter 2.

CHAPTER 3

D. Kolb, "The Chemical Formula, Part II: Determination," *J. Chem. Educ., 55,* 109 (1978).

J. H. Fletcher, O. C. Dermer, and R. B. Fox, *Nomenclature of Organic Compounds,* American Chemical Society, Washington, D. C., 1973.

E. L. Eliel, *Conformational Analyses,* McGraw-Hill, New York, 1965.

G. W. Wheland, *Advanced Organic Chemistry,* 3rd ed., Wiley, New York, 1960.

Lloyd N. Ferguson, "Ring Strain and Reactivity of Alicycles," *J. Chem. Educ., 47,* 46 (1970).

C. A. Coulson, *Valence,* Oxford University Press, New York, 1952, Chapter VIII.

J. March, *Advanced Organic Chemistry,* 2nd ed., McGraw-Hill, New York, 1977, pp. 24–133.

G. H. Posner, "Substitution Reactions Using Organocopper Reagents," *Organic Reactions,* Vol. 22, Wiley, New York, 1975.

CHAPTER 4

J. March, *Advanced Organic Chemistry,* 2nd ed., McGraw-Hill, New York, 1977, Chapters 6 and 14.

S. W. Benson, "Bond Energies," *J. Chem. Educ., 42,* 502 (1965).

W. A. Pryor, *Free Radicals,* McGraw-Hill, New York, 1965.

E. S. Huyser, *Free-Radical Chain Reactions,* Wiley, New York, 1970.

C. Walling, *Free Radicals in Solution,* Wiley, New York, 1957.

W. A. Pryor, *Introduction to Free Radical Chemistry,* Prentice-Hall, Englewood Cliffs, N. J., 1965.

CHAPTER 5

W. H. Saunders, Jr., *Ionic Aliphatic Reactions,* Prentice-Hall, Englewood Cliffs, N. J., 1965, Chapters 3, 4, and 5.

Robert K. Boyd, "Some Common Oversimplifications in Teaching Chemical Kinetics," *J. Chem. Educ., 55,* 84 (1978).

C. K. Ingold, *Structure and Mechanism in Organic Chemistry,* 2nd ed., Cornell University Press, Ithaca, N.Y., 1969, Chapters 7 and 9.

CHAPTER 6

S. I. Miller, "Dissociation Energies of Pi Bonds," *J. Chem. Educ., 55,* 778 (1978).

G. Zweifel and H. C. Brown, "Hydration of Olefins, Dienes, and Acetylenes, via Hydroboration," *Organic Reactions, 13,* 1963.

N. Isenberg and M. Grdinic, "A Modern Look at Markovnikov's Rule and the Peroxide Effect," *J. Chem. Educ., 46,* 601 (1969).

L. F. Fieser and M. Fieser, *Advanced Organic Chemistry,* Reinhold, New York, 1961, Chapter 5.

G. A. Olah and P. v. R. Schleyer, eds., *Carbonium Ions,* Wiley, New York, 1968.

M. Orchin and H. H. Jaffé, *The Importance of Antibonding Orbitals,* Houghton Mifflin, Boston, 1967.

CHAPTER 7

F. C. Whitmore and J. M. Church, *J. Amer. Chem. Soc., 54,* 3710 (1932).

O. T. Benfey, *Introduction to Organic Reaction Mechanisms,* McGraw-Hill, New York, 1970, Chapter 5.

W. H. Saunders, *Ionic Aliphatic Reactions,* Prentice-Hall, Englewood Cliffs, N.J., 1965, Chapter 2.

H. C. Brown, *Hydroboration,* Benjamin, New York, 1962.

H. C. Brown and P. J. Geoghegan, Jr., "Solvomercuration-Demercuration. I.," *J. Org. Chem., 35,* 1844 (1970).

W. S. Johnson, "Non-enzymic Biogenetic-like Olefinic Cyclizations," *Accounts Chem. Res., 1,* 1 (1968).

J. G. MacConnell and Robert M. Silverstein, "Recent Results in Insect Pheromone Chemistry," *Angew. Chem. internat Edit., 12,* 644 (1973).

CHAPTER 8

M. Gielen, "From the Concept of Relative Configuration to the Definition of *Erythro* and *Threo,*" *J. Chem. Educ., 54,* 673 (1977).

J. March, *Advanced Organic Chemistry,* 2nd ed., McGraw-Hill, New York, 1977, Chapter 4.

E. L. Eliel, *Stereochemistry of Carbon Compounds,* McGraw-Hill, New York, 1962.

E. L. Eliel, "Recent Advances in Stereochemical Nomenclature," *J. Chem. Educ., 48,* 163 (1971).

"IUPAC Tentative Rules for the Nomenclature of Organic Chemistry, Section E. Fundamental Stereochemistry," *J. Org. Chem. 35,* 2849 (1970).

E. L. Eliel, *Elements of Stereochemistry,* Wiley, New York, 1969.

K. Mislow, *Introduction to Stereochemistry,* Benjamin, New York, 1965.

D. F. Mowery, Jr., "Criteria for Optical Activity in Organic Molecules," *J. Chem. Educ., 46,* 269 (1969).

D. Whittaker, *Stereochemistry and Mechanism,* Clarendon Press, Oxford, 1973, Chapters 1, 2, and 5.

SPECIAL TOPIC C

F. W. Bellmeyer, *Textbook of Polymer Science,* 2nd ed., Wiley, New York, 1973.

L. R. G. Treloar, *Introduction to Polymer Science,* Springer-Verlag, New York, 1970.

SPECIAL TOPIC D

M. Jones, Jr., "Carbenes," *Scientific American,* February 1976, p. 101.

J. Hine, *Divalent Carbon,* Ronald Press, New York, 1964.

G. L. Closs, "Structures of Carbenes and the Stereochemistry of Carbene Additions to Olefins," *Topics in Stereochemistry,* Vol. 3, Wiley, New York, 1968.

W. E. Parham and E. E. Schweizer, "Halocyclopropanes from Halocarbenes," *Organic Reactions,* Vol. 13, Wiley, New York, 1963.

H. E. Simmons, T. L. Cairns, S. A. Vladuchick, and C. M. Hoiness, "Cytopropanes from Unsaturated Compounds, Methylene Iodide, and Zinc-Copper Couple," *Organic Reactions,* Vol. 20, Wiley, New York, 1973.

CHAPTER 9

J. J. Lagowski, "The Chemistry of Liquid Ammonia," *J. Chem. Educ., 55,* 752 (1978).

L. F. Fieser and M. Fieser, *Advanced Organic Chemistry,* Reinhold, New York, 1961, Chapter 6.

T. F. Rutledge, *Acetylenic Compounds: Preparation and Substitution Reactions,* Reinhold, New York, 1968.

T. F. Rutledge, *Acetylenes and Allenes: Addition Cyclization and Polymerization Reactions,* Reinhold, New York, 1969.

R. L. Shriner, R. C. Fuson, and D. Y. Curtin, *Systematic Identification of Organic Compounds,* Wiley, New York, 1964.

T. L. Jacobs, "The Synthesis of Acetylenes," *Organic Reactions,* Vol. 5, Wiley, New York, 1949.

CHAPTER 10

J. March, *Advanced Organic Chemistry,* 2nd ed., McGraw-Hill, New York, 1977, pp. 29–41.

H. H. Jaffé and M. Orchin, *Theory and Applications of Ultraviolet Spectroscopy,* Wiley, New York, 1962.

A. Liberles, *Introduction to Molecular Orbital Theory,* Holt, Rinehart and Winston, New York, 1966.

M. Orchin and H. H. Jaffé, *The Importance of Antibonding Orbitals,* Houghton Mifflin, 1967.

J. Sauer, "Diels Alder Reactions, Part I." *Angew. Chem. internat Edit., 5,* 211 (1966); "Part II," *Ibid., 6,* 16 (1967).

SPECIAL TOPIC E

C. D. Poulter and H. C. Rilling, "The Prenyl Transfer Reaction. Enzymatic and Mechanistic Studies of 1′-4 Coupling Reaction in Terpene Biosynthetic Pathway," *Account. Chem. Res., 8,* 307 (1978).

J. W. Cornforth, "Terpenoid Biosynthesis," *Chemistry in Britain, 4,* 102 (1968).

J. B. Hendrickson, *The Molecules of Nature,* W. A. Benjamin, Menlo Park, Calif., 1965.

SPECIAL TOPIC F

R. Hubbard and A. Kropf, "Molecular Isomers in Vision," *Bio-organic Chemistry: Readings from Scientific American,* M. Calvin and M. Jorgenson, eds., W. H. Freeman, San Francisco, 1968.

R. H. Johnson and T. T. Williams, "Action of Light upon the Visual Pigment Rhodopsin," *J. Chem. Educ., 47,* 736 (1970).

E. L. Menger, ed., "Special Issue on the Chemistry of Vision," *Accounts Chem. Res., 8,* (3), 81–112, (1975).

CHAPTER 11

J. March, *Advanced Organic Chemistry,* 2nd ed., McGraw-Hill, New York, 1977, pp. 41–69.

G. M. Badger, *Aromatic Character and Aromaticity,* Cambridge University Press, 1969.

R. Breslow, "Antiaromaticity," *Accouts Chem. Res., 6,* 393 (1973).

F. Sondheimer, "The Annulenes," *Accounts Chem. Res., 5,* 81 (1972).

CHAPTER 12

L. M. Stock, *Aromatic Substitution Reactions,* Prentice-Hall, Englewood Cliffs, N.J., 1968.

J. March, *Advanced Organic Chemistry,* 2nd ed., McGraw-Hill, New York, 1977, Chapter 11.

G. A. Olah, *Friedel-Crafts and Related Reactions,* Vol. I, Wiley, New York, 1963.

W. R. Dolbier, Jr., "Electrophilic Additions to Alkenes," *J. Chem. Educ, 46,* 342 (1969).

SPECIAL TOPIC G

E. C. Taylor and Alexander McKellop, "Thallium in Organic Synthesis," *Accounts Chem. Res., 3,* 338 (1970).

CHAPTER 13

P. L. Fuchs and C. A. Bunnell, *Carbon-13 NMR Based Organic Spectral Problems,* Wiley, New York, 1979.

L. J. Bellamy, *The Infrared Spectra of Complex Molecules,* 3rd ed., Wiley, New York, 1975.

J. D. Roberts, *An Introduction to Spin-Spin Splitting in High Resolution Nuclear Magnetic Resonance Spectra,* Benjamin, Menlo Park, Calif., 1961.

F. A. Bovey, *Nuclear Magnetic Resonance Spectroscopy,* Academic Press, New York, 1969.

J. D. Roberts and M. C. Caserio, *Basic Principles of Organic Chemistry,* 2nd ed., Benjamin, Menlo Park, Calif., 1977, Chapters 9 and 27.

R. M. Silverstein and G. C. Bassler, *Spectrometric Identification of Organic Compounds,* Wiley, New York, 1967.

J. R. Dyer, *Applications of Absorption Spectroscopy of Organic Compounds,* Prentice-Hall, Englewood Cliffs, N.J., 1965.

J. D. Roberts, *Nuclear Magnetic Resonance,* McGraw-Hill, New York, 1959.

SPECIAL TOPIC H

W. F. MacLafferty, *Interpretation of Mass Spectroscopy,* 2nd ed., Benjamin, Reading, Mass., 1973.

CHAPTER 14

J. S. Thayer, "Teaching Bio-Organometal Chemistry, Part I," *J. Chem. Educ., 54,* 604 (1977); Part II, *ibid., 54,* 662 (1977).

SPECIAL TOPIC I

D. L. Rabenstein, "The Chemistry of Methylmercury Toxicology," *J. Chem. Educ., 55,* 292 (1978).

J. R. Holum, *Topics and Terms in Environmental Problems,* Wiley, New York, 1977.

CHAPTER 15

H. C. Brown, "Hydride Reductions: A 40-Year Revolution in Organic Chemistry," *Chem. and Eng. News,* March 5, 1979, p. 24.

S. Patai, ed., *Chemistry of the Hydroxyl Group,* Wiley, New York, 1971.

S. Patai, ed., *Chemistry of the Ether Linkage,* Wiley, New York, 1967.

L. B. Clapp, *The Chemistry of the OH Group,* Prentice-Hall, Englewood Cliffs, N.J., 1967.

CHAPTER 16

C. A. Buehler and D. E. Pearson, *Survey of Organic Synthesis,* Wiley, New York, 1970.

H. O. House, *Modern Synthetic Reactions,* 2nd ed., Benjamin, New York, 1972.

S. Patai, ed., *The Chemistry of the Carbonyl Group,* Vol. 1, Wiley, New York, 1966.

S. Patai and J. Zabicky, eds., *The Chemistry of the Carbonyl Group,* Vol. 2, Wiley, New York, 1970.

A. J. Nielson and W. J. Houlihan, "The Aldol Condensation," *Organic Reactions,* Vol. 16, Wiley, New York, 1968.

E. Vedejs, "Clemmensen Reduction of Ketones," *Organic Reactions,* Vol. 22, Wiley, New York, 1975.

M. W. Rathke, "The Reformatsky Reaction," *Organic Reactions,* Vol. 22, Wiley, New York, 1975.

CHAPTER 17

S. Patai, ed., *The Chemistry of Carboxylic Acids an Esters,* Wiley, New York, 1969.

L. F. Fieser and M. Fieser, *Advanced Organic Chemistry,* Reinhold, New York, 1961, Chapters 11, 23, and 24.

S. Patai, ed., *The Chemistry of Amides,* Wiley, New York, 1969.

C. D. Gutsche, *The Chemistry of Carbonyl Compounds,* Prentice-Hall, Englewood Cliffs, N.J., 1967.

SPECIAL TOPIC K

J. K. Stille, *Industrial Organic Chemistry,* Prentice-Hall, Englewood Cliffs, N.J., 1968.

CHAPTER 18

G. B. Kauffman, "Isoneazed-Destroyer of the White Plague," *J. Chem. Educ., 55,* 448–449 (1978).

S. Patai, ed., *The Chemistry of the Amino Group,* Wiley, New York, 1968.

L. F. Fieser and M. Fieser, *Advanced Organic Chemistry,* Reinhold, New York, 1961, Chapters 14 and 21.

H. K. Porter, "The Zinin Reduction of Nitroarenes," *Organic Reactions,* Vol. 20, Wiley, New York, 1973.

H. Zollinger, *Diazo and Azo Chemistry,* Wiley, New York, 1961.

SPECIAL TOPIC L

W. P. Weber and G. W. Gokel, "Phase Transfer Catalyses," *J. Chem. Educ., 55,* 350 (Part I) and 429 (Part II) (1970).

W. H. Saunders, Jr. and A. F. Cockerile, *Mechanisms of Elimination Reactions,* Wiley, New York, 1973.

W. H. Saunders, Jr., "Distinguishing between Concerted and Nonconcerted Eliminations," *Accounts Chem. Res., 9,* 19 (1976).

D. J. Raber and J. M. Harris, "Nucleophilic Substitution Reactions at Secondary Carbon Atoms," *J. Chem. Educ., 49,* 60 (1972).

R. A. Sneen, "Organic Ion Pairs as Intermediates in Nucleophilic Substitution and Elimination Reactions," *Accounts Chem. Res., 6,* 46 (1973).

F. G. Bordwell, "How Common are Base-Initiated, Concerted 1,2-Eliminations?" *Accounts Chem. Res., 5,* 374 (1972).

SPECIAL TOPIC M

L. A. Paguette, *Principles of Modern Heterocyclic Chemistry,* Benjamin, New York, 1968.

CHAPTER 19

R. J. Bergeron, "Cycloamyloses," *J. Chem. Educ., 54,* 204 (1977).

L. N. Ferguson et. al., "Sweet Organic Chemistry," *J. Chem. Educ., 55,* 281 (1978).

G. B. Kauffman and R. P. Ciula, "Emil Fischer's Discovery of Phenylhydrazine," *J. Chem. Educ., 54,* 295 (1977).

L. F. Fieser and M. Fieser, *Advanced Organic Chemistry,* Reinhold, New York, 1961, Chapter 29.

C. R. Noller, *Chemistry of Organic Compounds,* Saunders, New York, 1965, Chapter 18.

D. E. Green and R. F. Goldberger, *Molecular Insights into the Living Process,* Academic Press, 1967, Chapters 2 and 3.

C. S. Hudson, "Emil Fischer's Discovery of the Configuration of Glucose," *J. Chem. Educ., 18,* 353 (1941).

R. Barker, *Organic Chemistry of Biolgical Compounds,* Prentice-Hall, Englewood Cliffs, N. J., 1971, Chapter 5.

CHAPTER 20

C. R. Hauser and B. E. Hudson, "The Acetoacetic Ester Condensation and Certain Related Reactions," *Organic Reactions,* Vol. 1, Wiley, New York, 1942.

H. O. House, *Modern Synthetic Reactions,* Benjamin, New York, 1965, Chapters 7 and 9.

W. McCrae, *Basic Organic Reactions,* Heyden and Son, Ltd., London, 1973, Chapters 3 and 4.

J. P. Schaefer and J. J. Bloomfield, "The Dieckmann Condensation," *Organic Reactions,* Vol. 15, Wiley, New York, 1967.

G. Jones, "The Knovenagel Condensation," *Organic Reactions,* Vol. 15, Wiley, New York, 1967.

T. M. Harris and C. M. Harris, "The γ-Alkylation and γ-Arylation of Dianions of β-Dicarbonyl Compounds," *Organic Reactions,* Vol. 17, Wiley, New York, 1969.

A. G. Cook, *Enamines: Synthesis, Structure, and Reactions,* Dekker, New York, 1969.

CHAPTER 21

D. Kolb, "A Pill for Birth Control," *J. Chem. Educ., 55,* 591 (1978).

L. F. Fieser and M. Fieser, *Advanced Organic Chemistry,* Reinhold, New York, 1961, Chapter 30.

L. F. Fieser, "Steroids," *Bio-organic Chemistry: Readings from Scientific American,* Freeman, San Francisco, 1968.

A. White, P. Handler, E. L. Smith, *Principles of Biochemistry,* McGraw-Hill, New York, 1964, Chapters 5 and 6.

A. L. Lehninger, *Biochemistry,* Worth, New York, 1970, Chapter 23.

E. E. Conn and P. K. Stumpf, *Outlines of Biochemistry,* 3rd ed., Wiley, New York, 1972, Chapters 3 and 12.

J. R. Hanson, *Introduction to Steroid Chemistry,* Pergamon Press, New York, 1968.

SPECIAL TOPIC N

W. S. Johnson, "Nonenzymic Biogenetic-like Olefin Cyclizations," *Accounts Chem. Res., 1,* 1 (1968).

CHAPTER 22

C. R. Noller, *Chemistry of Organic Compounds,* Saunders, New York, 1965, Chapter 19.

L. F. Fieser and M. Fieser, *Advanced Organic Chemistry,* Reinhold, New York, 1961, Chapter 31.

J. R. Holum, *Organic Chemistry: A Brief Course,* Wiley, New York, 1975, Chapter 15.

The following articles from *Bio-organic Chemistry; Readings from Scientific American,* M. Calvin and M. J. Horgenson, eds., Freeman, San Francisco, 1968:

P. Doty, "Proteins," p. 15.

W. H. Stein and S. Moore, 'The Chemical Structure of Proteins," p. 23.

E. O. P. Thompson, "The Insulin Molecule," p. 34.

M. F. Perutz, "The Hemoglobin Molecule," p. 41.

E. Zucherkandl, "The Evolution of Hemoglobin," p. 53.

D. C. Phillips, "The Three-Dimensional Structure of an Enzyme Molecule," p. 67.

L. Pauling, R. B. Corey, and R. Hayward, "Structures of Protein Molecules," *Scientific American,* October 1954, p. 54.

H. D. Law, *The Organic Chemistry of Peptides,* Wiley, New York, 1970.

D. E. Green and R. F. Goldberger, *Molecular Insights into the Living Process,* Academic Press, 1967, Chapters 4 and 5.

E. E. Conn and P. K. Stumpf, *Outlines of Biochemistry,* 3rd ed., Wiley, New York, 1972, Chapter 4.

R. E. Dickerson and I. Geis, *The Structure and Action of Proteins,* Harper and Row, New York, 1969.

SPECIAL TOPIC O

G. A. Swan, *An Introduction to Alkaloids,* Wiley, New York, 1967.

T. A. Geissman and D. H. G. Crout, *Organic Chemistry of Secondary Plant Metabolism,* Freeman, Cooper and Co., San Francisco, 1969, Chapters 16 to 19.

H. Hart and J. L. Reilly, "Oxidative Coupling of Phenols," *J. Chem. Educ., 55,* 120 (1978).

SPECIAL TOPIC P

R. Barker, *Organic Chemistry of Biological Compounds,* Prentice-Hall, Englewood Cliffs, N.J., 1971, Chapter 8.

E. E. Conn and P. K. Stumpf, *Outlines of Biochemistry.* 2nd ed., Wiley, New York, 1972, Chapters 5, 18, and 19.

J. D. Watson, *Molecular Biology of the Gene,* 2nd ed., Benjamin, New York, 1970.

The following articles in *Bio-organic Chemistry: Readings from Scientific American,* M. Calvin and M. J. Jorgenson, eds., Freeman, San Francisco, 1968:

F. H. C. Crick, "The Structure of the Hereditary Material," p. 75.

R. W. Holley, "The Nucleotide Sequence of a Nucleic Acid," p. 82.

SPECIAL TOPIC Q

K. N. Houk, "The Frontier Molecular Orbital Theory of Cycloaddition Reactions," *Accounts Chem. Res., 8,* 361 (1975).

R. B. Woodward and R. Hoffman, *The Conservation of Orbital Symmetry,* Academic Press, New York, 1970.

INDEX

Page numbers printed in **boldface** refer to tables of physical properties.

The Modern Periodic Table of the Elements

Key: 1 (atomic number) H 1.0079 (atomic mass)

PERIODS	IA	IIA	IIIB	IVB	VB	VIB	VIIB	VIII	VIII	VIII	IB	IIB	IIIA	IVA	VA	VIA	VIIA	NOBLE GASES O
1	1 H 1.00797																	2 He 4.00260
2	3 Li 6.941	4 Be 9.01218											5 B 10.81	6 C 12.01115	7 N 14.0067	8 O 15.9994	9 F 18.99840	10 Ne 20.179
3	11 Na 22.98977	12 Mg 24.305											13 Al 26.98154	14 Si 28.086†	15 P 30.97376	16 S 32.06	17 Cl 35.453	18 Ar 39.948
4	19 K 39.098	20 Ca 40.08	21 Sc 44.9559	22 Ti 47.90	23 V 50.9414	24 Cr 51.996	25 Mn 54.9380	26 Fe 55.847	27 Co 58.9332	28 Ni 58.71	29 Cu 63.546	30 Zn 65.38	31 Ga 69.72	32 Ge 72.59	33 As 74.9216	34 Se 78.96	35 Br 79.904	36 Kr 83.80
5	37 Rb 85.4678	38 Sr 87.62	39 Y 88.9059	40 Zr 91.22	41 Nb 92.9064	42 Mo 95.94	43 Tc 98.9062	44 Ru 101.07	45 Rh 102.9055	46 Pd 106.4	47 Ag 107.868	48 Cd 112.40	49 In 114.82	50 Sn 118.69	51 Sb 121.75	52 Te 127.60	53 I 126.9045	54 Xe 131.30
6	55 Cs 132.9054	56 Ba 137.34	57 *La 138.9055	72 Hf 178.49	73 Ta 180.9479	74 W 183.85	75 Re 186.2	76 Os 190.2	77 Ir 192.22	78 Pt 195.09	79 Au 196.9665	80 Hg 200.59	81 Tl 204.37	82 Pb 207.19	83 Bi 208.9804	84 Po (210)	85 At (210)	86 Rn (222)
7	87 Fr (223)	88 Ra 226.0254	89 †Ac (227)	104 Ku (261)	105 Ha (260)													

*	58 Ce 140.12	59 Pr 140.9077	60 Nd 144.24	61 Pm (147)	62 Sm 150.4	63 Eu 151.96	64 Gd 157.25	65 Tb 158.9254	66 Dy 162.50	67 Ho 164.9304	68 Er 167.26	69 Tm 168.9342	70 Yb 173.04	71 Lu 174.97
†	90 Th 232.0381	91 Pa 231.0359	92 U 238.029	93 Np 237.0482	94 Pu (244)	95 Am (243)	96 Cm (247)	97 Bk (247)	98 Cf (251)	99 Es (254)	100 Fm (257)	101 Md (258)	102 No (255)	103 Lr (256)